Global warming of 1.5°C

An IPCC Special Report on the impacts of global warming of 1.5°C
above pre-industrial levels and related global greenhouse gas emission pathways,
in the context of strengthening the global response to the threat of climate change,
sustainable development, and efforts to eradicate poverty

Edited by

Valérie Masson-Delmotte
Co-Chair Working Group I

Panmao Zhai
Co-Chair Working Group I

Hans-Otto Pörtner
Co-Chair Working Group II

Debra Roberts
Co-Chair Working Group II

Jim Skea
Co-Chair Working Group III

Priyadarshi R. Shukla
Co-Chair Working Group III

Anna Pirani
Head of WGI TSU

Wilfran Moufouma-Okia
Head of Science

Clotilde Péan
Head of Operations

Roz Pidcock
Head of Communication

Sarah Connors
Science Officer

J. B. Robin Matthews
Science Officer

Yang Chen
Science Officer

Xiao Zhou
Science Assistant

Melissa I. Gomis
Graphics Officer

Elisabeth Lonnoy
Project Assistant

Tom Maycock
Science Editor

Melinda Tignor
Head of WGII TSU

Tim Waterfield
IT Officer

Working Group I Technical Support Unit

CAMBRIDGE
UNIVERSITY PRESS

University Printing House, Cambridge CB2 8BS, United Kingdom
One Liberty Plaza, 20th Floor, New York, NY 10006, USA
477 Williamstown Road, Port Melbourne, VIC 3207, Australia
314–321, 3rd Floor, Plot 3, Splendor Forum, Jasola District Centre, New Delhi – 110025, India
103 Penang Road, #05-06/07, Visioncrest Commercial, Singapore 238467

Cambridge University Press is part of the University of Cambridge.
It furthers the University's mission by disseminating knowledge in the pursuit of education,
learning, and research at the highest international levels of excellence.

www.cambridge.org
Information on this title: www.cambridge.org/9781009157957
DOI: 10.1017/9781009157940

First published 2022

Printed in the United Kingdom by TJ Books Limited, Padstow Cornwall

A catalogue record for this publication is available from the British Library.

ISBN 978-1-009-15795-7 Paperback

This report should be cited as:
IPCC, 2018: *Global Warming of 1.5°C. An IPCC Special Report on the impacts of global warming of 1.5°C above pre-industrial levels and related global greenhouse gas emission pathways, in the context of strengthening the global response to the threat of climate change, sustainable development, and efforts to eradicate poverty* [Masson-Delmotte, V., P. Zhai, H.-O. Pörtner, D. Roberts, J. Skea, P.R. Shukla, A. Pirani, W. Moufouma-Okia, C. Péan, R. Pidcock, S. Connors, J.B.R. Matthews, Y. Chen, X. Zhou, M.I. Gomis, E. Lonnoy, T. Maycock, M. Tignor, and T. Waterfield (eds.)]. Cambridge University Press, Cambridge, UK and New York, NY, USA, 616 pp. https://doi.org/ 10.1017/9781009157940.

The designations employed and the presentation of material on maps do not imply the expression of any opinion whatsoever on the part of the Intergovernmental Panel on Climate Change concerning the legal status of any country, territory, city or area or of its authorities, or concerning the delimitation of its frontiers or boundaries.

Front cover layout: Nigel Hawtin. Front cover artwork: Time to Choose by Alisa Singer - www.environmentalgraphiti.org - © Intergovernmental Panel on Climate Change. The artwork was inspired by a graphic from the SPM (Figure SPM.1).

Contents

Foreword
and Preface

Foreword

This IPCC Special Report on Global Warming of 1.5°C was formally approved by the world's governments in 2018 – the year of IPCC's 30th anniversary celebrations.

During its three decades of existence, the IPCC has shed light on climate change, contributing to the understanding of its causes and consequences and the options for risk management through adaptation and mitigation. In these three decades, global warming has continued unabated and we have witnessed an acceleration in sea-level rise. Emissions of greenhouse gases due to human activities, the root cause of global warming, continue to increase, year after year.

Five years ago, the IPCC's Fifth Assessment Report provided the scientific input into the Paris Agreement, which aims to strengthen the global response to the threat of climate change by holding the increase in the global average temperature to well below 2°C above pre-industrial levels and to pursue efforts to limit the temperature increase to 1.5°C above pre-industrial levels.

Many countries considered that a level of global warming close to 2°C would not be safe and, at that time, there was only limited knowledge about the implications of a level of 1.5°C of warming for climate-related risks and in terms of the scale of mitigation ambition and its feasibility. Parties to the Paris Agreement therefore invited the IPCC to assess the impacts of global warming of 1.5°C above pre-industrial levels and the related emissions pathways that would achieve this enhanced global ambition.

At the start of the Sixth Assessment cycle, governments, in a plenary IPCC session, decided to prepare three special reports, including this one, and expanded the scope of this special report by framing the assessment in the context of sustainable development and efforts to eradicate poverty.

Sustainable development goals provide a new framework to consider climate action within the multiple dimensions of sustainability. This report is innovative in multiple ways. It shows the importance of integration across the traditional IPCC working groups and across disciplines within each chapter. Transitions, integrating adaptation and mitigation for each sector, are explored within six dimensions of feasibility, showing both low hanging fruits and barriers to overcome. It also provides scientific guidance on strategies to embed climate action within development strategies, and how to optimize choices that maximize benefits for multiple sustainable development dimensions and implement ethical and just transitions.

In his address to the UN General Assembly in 2018, Secretary-General António Guterres quoted World Meteorological Organization (WMO) data showing that the past two decades have included eighteen of the twenty warmest years since record-keeping began in 1850.

"Climate change is moving faster than we are," said Secretary-General Guterres. *"We must listen to the Earth's best scientists,"* he added.

One month later the IPCC presented the Special Report on Global Warming of 1.5°C, based on the assessment of around 6,000 peer-review publications, most of them published in the last few years. This Special Report confirms that climate change is already affecting people, ecosystems and livelihoods all around the world. It shows that limiting warming to 1.5°C is possible within the laws of chemistry and physics but would require unprecedented transitions in all aspects of society. It finds that there are clear benefits to keeping warming to

1.5°C rather than 2°C or higher. Every bit of warming matters. And it shows that limiting warming to 1.5°C can go hand in hand with achieving other global goals such as the Sustainable Development Agenda. Every year matters and every choice matters.

This Special Report also shows that recent trends in emissions and the level of international ambition indicated by nationally determined contributions, within the Paris Agreement, deviate from a track consistent with limiting warming to well below 2°C. Without increased and urgent mitigation ambition in the coming years, leading to a sharp decline in greenhouse gas emissions by 2030, global warming will surpass 1.5°C in the following decades, leading to irreversible loss of the most fragile ecosystems, and crisis after crisis for the most vulnerable people and societies.

The Special Report on Global Warming of 1.5°C supports efforts by the WMO and United Nations Environment Programme for a comprehensive assessment of our understanding of climate change to help step up action to respond to climate change, achieve climate-resilient development and foster an integrated approach to the provision of climate services at all scales of governance.

The IPCC worked in record time to deliver this report for the 24th Conference of Parties (COP24) to the United Nations Framework Convention on Climate Change (UNFCCC) and the Talanoa Dialogue. We would like to thank Hoesung Lee, Chair of the IPCC, for his leadership and guidance in the preparation of this Special Report. We commend the work undertaken by the authors of this Special Report and the many contributing authors and reviewers within a timeline of unprecedented severity; the leadership of the Co-Chairs of Working Groups I, II and III: Valérie Masson-Delmotte, Panmao Zhai, Hans-Otto Pörtner, Debra Roberts, Jim Skea and Priyadarshi R. Shukla; the oversight by the Bureau members of Working Groups I, II and III; and the implementation by the Technical Support Unit of Working Group I, supported by the Technical Support Units of Working Groups II and III. We are also grateful for the responsiveness of the international research community, who produced the knowledge assessed in the report, and thank the reviewers of the report for the thousands of comments that helped the authors strengthen the assessment.

Every bit of warming matters, every year matters, every choice matters

Petteri Taalas
Secretary-General
World Meteorological Organization

Joyce Msuya
Acting Executive Director
United Nations Environment Programme

Preface

This Special Report on Global Warming of 1.5°C, an IPCC Special Report on the impacts of global warming of 1.5°C above pre-industrial levels and related global greenhouse gas emission pathways, in the context of strengthening the global response to the threat of climate change, sustainable development, and efforts to eradicate poverty, is the first publication in the Intergovernmental Panel on Climate Change (IPCC) Sixth Assessment Report (AR6). The Report was jointly prepared by Working Groups I, II and III. It is the first IPCC Report to be collectively produced by all three Working Groups, symbolizing the new level of integration sought between Working Groups during AR6. The Working Group I Technical Support Unit has been responsible for the logistical and technical support for the preparation of the Special Report. The Special Report builds upon the IPCC's Fifth Assessment Report (AR5) released in 2013–2014 and on relevant research subsequently published in the scientific, technical and socio-economic literature. It has been prepared following IPCC principles and procedures, following AR5 guidance on calibrated language for communicating the degree of certainty in key findings. This Special Report is the first of three cross-Working Group Special Reports to be published in AR6, accompanying the three main Working Group Reports, the Synthesis Report and a Refinement to the 2006 IPCC Guidelines for National Greenhouse Gas Inventories.

Scope of the Report

In its decision on the adoption of the Paris Agreement, the Conference of Parties (COP) to the United Nations Framework Convention on Climate Change (UNFCCC) at its 21st Session in Paris, France (30 November to 11 December 2015), invited the IPCC to provide a special report in 2018 on the impacts of global warming of 1.5°C above pre-industrial levels and related global greenhouse gas emission pathways. The Panel accepted the invitation and placed the Report in the context of strengthening the global response to the threat of climate change, sustainable development, and efforts to eradicate poverty.

The broad scientific community has also responded to the UNFCCC invitation. New knowledge and literature relevant to the topics of this report have been produced and published worldwide. The Special Report is an assessment of the relevant state of knowledge, based on the scientific and technical literature available and accepted for publication up to 15 May 2018. The Report draws on the findings of more than 6,000 published articles.

Structure of the Report

This report consists of a short Summary for Policymakers, a Technical Summary, five Chapters, and Annexes, as well as online chapter Supplementary Material.

Chapter 1 frames the context, knowledge base and assessment approaches used to understand the impacts of 1.5°C global warming above pre-industrial levels and related global greenhouse gas emission pathways, building on AR5, in the context of strengthening the global response to the threat of climate change, sustainable development, and efforts to eradicate poverty. The chapter provides an update on the current state of the climate system including the current level of warming.

Chapter 2 assesses the literature on mitigation pathways that limit or return global mean warming to 1.5°C (relative to the pre-industrial base period 1850–1900). Key questions addressed are: What types of mitigation pathways have been developed that could be consistent with 1.5°C? What changes in emissions, energy and land use do they entail? What do they imply for climate policy and implementation, and what impacts do they have on sustainable development? This chapter focuses on geophysical dimensions of feasibility and the technological and economic enabling conditions.

Chapter 3 builds on findings of AR5 and assesses new scientific evidence of changes in the climate system and the associated impacts on natural and human systems, with a specific focus on the magnitude and pattern of risks for global warming of 1.5°C above the pre-industrial period. It explores impacts and risks for a range of natural and human systems, including adaptation options, with a focus on how risk levels change between today and worlds where global mean temperature increases by 1.5°C and 2°C above pre-industrial levels. The chapter also revisits major categories of risk (Reasons for Concern) based on the assessment of the new knowledge available since AR5.

Chapter 4 discusses how the global economy and socio-technical and socio-ecological systems can transition to 1.5°C-consistent pathways and adapt to global warming of 1.5°C. In the context of systemic transitions across energy, land, urban and industrial systems, the chapter assesses adaptation and mitigation options, including carbon dioxide removal (CDR) measures, as well as the enabling conditions that would facilitate implementing the rapid and far-reaching global response.

Finally, Chapter 5 takes sustainable development, poverty eradication and reducing inequalities as the starting point and focus for analysis. It considers the complex interplay between sustainable development, including Sustainable Development Goals (SDGs) and climate actions related to a 1.5°C warmer world. The chapter also examines synergies and trade-offs of adaptation and mitigation options with sustainable development and the SDGs and offers insights into possible pathways, especially climate-resilient development pathways toward a 1.5°C warmer world.

The Process

The Special Report on 1.5°C of the IPCC AR6 has been prepared in accordance with the principles and procedures established by the IPCC and represents the combined efforts of leading experts in the field of climate change. A scoping meeting for the SR1.5°C was held in Geneva, Switzerland, in August 2016, and the final outline was approved by the Panel at its 44th Session in October 2016 in Bangkok, Thailand. Governments and IPCC observer organizations nominated 541 experts for the author team. The team of 74 Coordinating Lead Authors and Lead Authors plus 17 Review Editors were selected by the Working Group I, II and III Bureaux. In addition, 133 Contributing Authors were invited by chapter teams to provide technical information in the form of text, graphs or data for assessment. Report drafts prepared by the authors were subject to two rounds of formal review and revision followed by a final round of government comments on the Summary for Policymakers. The enthusiastic participation of the scientific community and governments to the review process resulted in 42,001 written review comments submitted by 796 individual expert reviewers and 65 governments.

The 17 Review Editors monitored the review process to ensure that all substantive review comments received appropriate consideration. The Summary for Policymakers was approved line-by-line at the joint meeting of Working Groups I, II and III; it and the underlying chapters were then accepted at the 48th Session of the IPCC from 01–06 October 2018 in Incheon, Republic of Korea.

Acknowledgements

We are very grateful for the expertise, rigour and dedication shown throughout by the volunteer Coordinating Lead Authors and Lead Authors, working across scientific disciplines in each chapter of the report, with essential help by the many Contributing Authors. The Review Editors have played a critical role in assisting the author teams and ensuring the integrity of the review process. We express our sincere appreciation to all the expert and government reviewers. A special thanks goes to the Chapter Scientists of this report who went above and beyond what was expected of them: Neville Ellis, Tania Guillén Bolaños, Daniel Huppmann, Kiane de Kleijne, Richard Millar and Chandni Singh.

We would also like to thank the three Intergovernmental Panel on Climate Change (IPCC) Vice-Chairs Ko Barrett, Thelma Krug, and Youba Sokona as well as the members of the WGI, WGII and WGIII Bureaux for their assistance, guidance, and wisdom throughout the preparation of the Report: Amjad Abdulla, Edvin Aldrian, Carlo Carraro, Diriba Korecha Dadi, Fatima Driouech, Andreas Fischlin, Gregory Flato, Jan Fuglestvedt, Mark Howden, Nagmeldin G. E. Mahmoud, Carlos Mendez, Joy Jacqueline Pereira, Ramón Pichs-Madruga, Andy Reisinger, Roberto Sánchez Rodríguez, Sergey Semenov, Muhammad I. Tariq, Diana Ürge-Vorsatz, Carolina Vera, Pius Yanda, Noureddine Yassaa, and Taha Zatari.

Our heartfelt thanks go to the hosts and organizers of the scoping meeting, the four Special Report on 1.5°C Lead Author Meetings and the 48th Session of the IPCC. We gratefully acknowledge the support from the host countries and institutions: World Meteorological Organization, Switzerland; Ministry of Foreign Affairs, and the National Institute for Space Research (INPE), Brazil; Met Office and the University of Exeter, the United Kingdom; Swedish Meteorological and Hydrological Institute (SMHI), Sweden; the Ministry of Environment Natural Resources Conservation and Tourism, the National Climate Change Committee in the Department of Meteorological Services and the Botswana Global Environmental Change Committee at the University of Botswana, Botswana; and Korea Meteorological Administration (KMA) and Incheon Metropolitan City, the Republic of Korea. The support provided by governments and institutions, as well as through contributions to the IPCC Trust Fund, is thankfully acknowledged as it enabled the participation of the author teams in the preparation of the Report. The efficient operation of the Working Group I Technical Support Unit was made possible by the generous financial support provided by the government of France and administrative and information technology support from the Université Paris Saclay (France), Institut Pierre Simon Laplace (IPSL) and the Laboratoire des Sciences du Climat et de l'Environnement (LSCE). We thank the Norwegian Environment Agency for supporting the preparation of the graphics for the Summary for Policymakers. We thank the UNEP Library, who supported authors throughout the drafting process by providing literature for the assessment.

We would also like to thank Abdalah Mokssit, Secretary of the IPCC, and the staff of the IPCC Secretariat: Kerstin Stendahl, Jonathan Lynn, Sophie Schlingemann, Judith Ewa, Mxolisi Shongwe, Jesbin Baidya, Werani Zabula, Nina Peeva, Joelle Fernandez, Annie Courtin, Laura Biagioni and Oksana Ekzarho. Thanks are due to Elhousseine Gouaini who served as the conference officer for the 48th Session of the IPCC.

Finally, our particular appreciation goes to the Working Group Technical Support Units whose tireless dedication, professionalism and enthusiasm led the production of this Special Report. This Report could not have been prepared without the commitment of members of the Working Group I Technical Support Unit, all new to the IPCC, who rose to the unprecedented Sixth Assessment Report challenge and were pivotal in all aspects of the preparation of the Report: Yang Chen, Sarah Connors, Melissa Gomis, Elisabeth Lonnoy, Robin Matthews, Wilfran Moufouma-Okia, Clotilde Péan, Roz Pidcock, Anna Pirani, Nicholas Reay, Tim Waterfield, and Xiao Zhou. Our warmest thanks go to the collegial and collaborative support provided by Marlies Craig, Andrew Okem, Jan Petzold, Melinda Tignor and Nora Weyer from the WGII Technical Support Unit and Bhushan Kankal, Suvadip Neogi and Joana Portugal Pereira from the WGIII Technical Support Unit. A special thanks goes to Kenny Coventry, Harmen Gudde, Irene Lorenzoni, and Stuart Jenkins for their support with the figures in the Summary for Policymakers, as well as Nigel Hawtin for graphical support of the Report. In addition, the following contributions are gratefully acknowledged: Jatinder Padda (copy edit), Melissa Dawes (copy edit), Marilyn Anderson (index), Vincent Grégoire (layout) and Sarah le Rouzic (intern).

The Special Report website has been developed by Habitat 7, led by Jamie Herring, and the report content has been prepared and managed for the website by Nicholas Reay and Tim Waterfield. We gratefully acknowledge the UN Foundation for supporting the website development.

Valérie Masson-Delmotte
IPCC Working Group I Co-Chair

Panmao Zhai
IPCC Working Group I Co-Chair

Hans-Otto Pörtner
IPCC Working Group II Co-Chair

Debra Roberts
IPCC Working Group II Co-Chair

Priyadarshi R. Shukla
IPCC Working Group III Co-Chair

Jim Skea
IPCC Working Group III Co-Chair

« Pour ce qui est de l'avenir, il ne s'agit pas de le prévoir, mais de le rendre possible. »

Antoine de Saint Exupéry, *Citadelle*, 1948

Summary for Policymakers

SPM

Summary for Policymakers

Drafting Authors:
Myles Allen (UK), Mustafa Babiker (Sudan), Yang Chen (China), Heleen de Coninck (Netherlands/EU), Sarah Connors (UK), Renée van Diemen (Netherlands), Opha Pauline Dube (Botswana), Kristie L. Ebi (USA), Francois Engelbrecht (South Africa), Marion Ferrat (UK/France), James Ford (UK/Canada), Piers Forster (UK), Sabine Fuss (Germany), Tania Guillén Bolaños (Germany/Nicaragua), Jordan Harold (UK), Ove Hoegh-Guldberg (Australia), Jean-Charles Hourcade (France), Daniel Huppmann (Austria), Daniela Jacob (Germany), Kejun Jiang (China), Tom Gabriel Johansen (Norway), Mikiko Kainuma (Japan), Kiane de Kleijne (Netherlands/EU), Elmar Kriegler (Germany), Debora Ley (Guatemala/Mexico), Diana Liverman (USA), Natalie Mahowald (USA), Valérie Masson-Delmotte (France), J. B. Robin Matthews (UK), Richard Millar (UK), Katja Mintenbeck (Germany), Angela Morelli (Norway/Italy), Wilfran Moufouma-Okia (France/Congo), Luis Mundaca (Sweden/Chile), Maike Nicolai (Germany), Chukwumerije Okereke (UK/Nigeria), Minal Pathak (India), Antony Payne (UK), Roz Pidcock (UK), Anna Pirani (Italy), Elvira Poloczanska (UK/Australia), Hans-Otto Pörtner (Germany), Aromar Revi (India), Keywan Riahi (Austria), Debra C. Roberts (South Africa), Joeri Rogelj (Austria/Belgium), Joyashree Roy (India), Sonia I. Seneviratne (Switzerland), Priyadarshi R. Shukla (India), James Skea (UK), Raphael Slade (UK), Drew Shindell (USA), Chandni Singh (India), William Solecki (USA), Linda Steg (Netherlands), Michael Taylor (Jamaica), Petra Tschakert (Australia/Austria), Henri Waisman (France), Rachel Warren (UK), Panmao Zhai (China), Kirsten Zickfeld (Canada).

This Summary for Policymakers should be cited as:
IPCC, 2018: Summary for Policymakers. In: *Global Warming of 1.5°C. An IPCC Special Report on the impacts of global warming of 1.5°C above pre-industrial levels and related global greenhouse gas emission pathways, in the context of strengthening the global response to the threat of climate change, sustainable development, and efforts to eradicate poverty* [Masson-Delmotte, V., P. Zhai, H.-O. Pörtner, D. Roberts, J. Skea, P.R. Shukla, A. Pirani, W. Moufouma-Okia, C. Péan, R. Pidcock, S. Connors, J.B.R. Matthews, Y. Chen, X. Zhou, M.I. Gomis, E. Lonnoy, T. Maycock, M. Tignor, and T. Waterfield (eds.)]. Cambridge University Press, Cambridge, UK and New York, NY, USA, pp. 3-24. https://doi.org/10.1017/9781009157940.001.

Introduction

This Report responds to the invitation for IPCC '... to provide a Special Report in 2018 on the impacts of global warming of 1.5°C above pre-industrial levels and related global greenhouse gas emission pathways' contained in the Decision of the 21st Conference of Parties of the United Nations Framework Convention on Climate Change to adopt the Paris Agreement.[1]

The IPCC accepted the invitation in April 2016, deciding to prepare this Special Report on the impacts of global warming of 1.5°C above pre-industrial levels and related global greenhouse gas emission pathways, in the context of strengthening the global response to the threat of climate change, sustainable development, and efforts to eradicate poverty.

This Summary for Policymakers (SPM) presents the key findings of the Special Report, based on the assessment of the available scientific, technical and socio-economic literature[2] relevant to global warming of 1.5°C and for the comparison between global warming of 1.5°C and 2°C above pre-industrial levels. The level of confidence associated with each key finding is reported using the IPCC calibrated language.[3] The underlying scientific basis of each key finding is indicated by references provided to chapter elements. In the SPM, knowledge gaps are identified associated with the underlying chapters of the Report.

A.   Understanding Global Warming of 1.5°C[4]

A.1   Human activities are estimated to have caused approximately 1.0°C of global warming[5] above pre-industrial levels, with a *likely* range of 0.8°C to 1.2°C. Global warming is *likely* to reach 1.5°C between 2030 and 2052 if it continues to increase at the current rate. (*high confidence*) (Figure SPM.1) {1.2}

A.1.1   Reflecting the long-term warming trend since pre-industrial times, observed global mean surface temperature (GMST) for the decade 2006–2015 was 0.87°C (*likely* between 0.75°C and 0.99°C)[6] higher than the average over the 1850–1900 period (*very high confidence*). Estimated anthropogenic global warming matches the level of observed warming to within ±20% (*likely range*). Estimated anthropogenic global warming is currently increasing at 0.2°C (*likely* between 0.1°C and 0.3°C) per decade due to past and ongoing emissions (*high confidence*). {1.2.1, Table 1.1, 1.2.4}

A.1.2   Warming greater than the global annual average is being experienced in many land regions and seasons, including two to three times higher in the Arctic. Warming is generally higher over land than over the ocean. (*high confidence*) {1.2.1, 1.2.2, Figure 1.1, Figure 1.3, 3.3.1, 3.3.2}

A.1.3   Trends in intensity and frequency of some climate and weather extremes have been detected over time spans during which about 0.5°C of global warming occurred (*medium confidence*). This assessment is based on several lines of evidence, including attribution studies for changes in extremes since 1950. {3.3.1, 3.3.2, 3.3.3}

---

1   Decision 1/CP.21, paragraph 21.

2   The assessment covers literature accepted for publication by 15 May 2018.

3   Each finding is grounded in an evaluation of underlying evidence and agreement. A level of confidence is expressed using five qualifiers: very low, low, medium, high and very high, and typeset in italics, for example, *medium confidence*. The following terms have been used to indicate the assessed likelihood of an outcome or a result: virtually certain 99–100% probability, very likely 90–100%, likely 66–100%, about as likely as not 33–66%, unlikely 0–33%, very unlikely 0–10%, exceptionally unlikely 0–1%. Additional terms (extremely likely 95–100%, more likely than not >50–100%, more unlikely than likely 0–<50%, extremely unlikely 0–5%) may also be used when appropriate. Assessed likelihood is typeset in italics, for example, *very likely*. This is consistent with AR5.

4   See also Box SPM.1: Core Concepts Central to this Special Report.

5   Present level of global warming is defined as the average of a 30-year period centred on 2017 assuming the recent rate of warming continues.

6   This range spans the four available peer-reviewed estimates of the observed GMST change and also accounts for additional uncertainty due to possible short-term natural variability. {1.2.1, Table 1.1}

A.2   Warming from anthropogenic emissions from the pre-industrial period to the present will persist for centuries to millennia and will continue to cause further long-term changes in the climate system, such as sea level rise, with associated impacts (*high confidence*), but these emissions alone are *unlikely* to cause global warming of 1.5°C (*medium confidence*). (Figure SPM.1) {1.2, 3.3, Figure 1.5}

A.2.1   Anthropogenic emissions (including greenhouse gases, aerosols and their precursors) up to the present are *unlikely* to cause further warming of more than 0.5°C over the next two to three decades (*high confidence*) or on a century time scale (*medium confidence*). {1.2.4, Figure 1.5}

A.2.2   Reaching and sustaining net zero global anthropogenic CO_2 emissions and declining net non-CO_2 radiative forcing would halt anthropogenic global warming on multi-decadal time scales (*high confidence*). The maximum temperature reached is then determined by cumulative net global anthropogenic CO_2 emissions up to the time of net zero CO_2 emissions (*high confidence*) and the level of non-CO_2 radiative forcing in the decades prior to the time that maximum temperatures are reached (*medium confidence*). On longer time scales, sustained net negative global anthropogenic CO_2 emissions and/ or further reductions in non-CO_2 radiative forcing may still be required to prevent further warming due to Earth system feedbacks and to reverse ocean acidification (*medium confidence*) and will be required to minimize sea level rise (*high confidence*). {Cross-Chapter Box 2 in Chapter 1, 1.2.3, 1.2.4, Figure 1.4, 2.2.1, 2.2.2, 3.4.4.8, 3.4.5.1, 3.6.3.2}

A.3   Climate-related risks for natural and human systems are higher for global warming of 1.5°C than at present, but lower than at 2°C (*high confidence*). These risks depend on the magnitude and rate of warming, geographic location, levels of development and vulnerability, and on the choices and implementation of adaptation and mitigation options (*high confidence*). (Figure SPM.2) {1.3, 3.3, 3.4, 5.6}

A.3.1   Impacts on natural and human systems from global warming have already been observed (*high confidence*). Many land and ocean ecosystems and some of the services they provide have already changed due to global warming (*high confidence*). (Figure SPM.2) {1.4, 3.4, 3.5}

A.3.2   Future climate-related risks depend on the rate, peak and duration of warming. In the aggregate, they are larger if global warming exceeds 1.5°C before returning to that level by 2100 than if global warming gradually stabilizes at 1.5°C, especially if the peak temperature is high (e.g., about 2°C) (*high confidence*). Some impacts may be long-lasting or irreversible, such as the loss of some ecosystems (*high confidence*). {3.2, 3.4.4, 3.6.3, Cross-Chapter Box 8 in Chapter 3}

A.3.3   Adaptation and mitigation are already occurring (*high confidence*). Future climate-related risks would be reduced by the upscaling and acceleration of far-reaching, multilevel and cross-sectoral climate mitigation and by both incremental and transformational adaptation (*high confidence*). {1.2, 1.3, Table 3.5, 4.2.2, Cross-Chapter Box 9 in Chapter 4, Box 4.2, Box 4.3, Box 4.6, 4.3.1, 4.3.2, 4.3.3, 4.3.4, 4.3.5, 4.4.1, 4.4.4, 4.4.5, 4.5.3}

Cumulative emissions of CO₂ and future non-CO₂ radiative forcing determine the probability of limiting warming to 1.5°C

a) Observed global temperature change and modeled responses to stylized anthropogenic emission and forcing pathways

Global warming relative to 1850-1900 (°C)

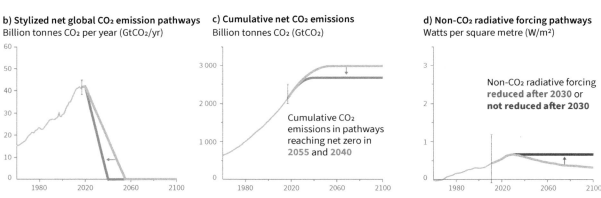

Observed monthly global mean surface temperature

Estimated anthropogenic warming to date and *likely* range

Likely range of modeled responses to stylized pathways

☐ Global CO₂ emissions reach **net zero in 2055** while net non-CO₂ radiative forcing is **reduced after 2030** (grey in **b, c & d**)

2017

☐ Faster CO₂ reductions (blue in **b & c**) result in a **higher probability** of limiting warming to 1.5°C

☐ **No reduction** of net non-CO₂ radiative forcing (purple in **d**) results in a **lower probability** of limiting warming to 1.5°C

b) Stylized net global CO₂ emission pathways
Billion tonnes CO₂ per year (GtCO₂/yr)

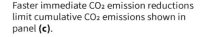

c) Cumulative net CO₂ emissions
Billion tonnes CO₂ (GtCO₂)

Cumulative CO₂ emissions in pathways reaching net zero in **2055** and **2040**

d) Non-CO₂ radiative forcing pathways
Watts per square metre (W/m²)

Non-CO₂ radiative forcing **reduced after 2030** or **not reduced after 2030**

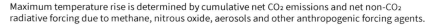

Faster immediate CO₂ emission reductions limit cumulative CO₂ emissions shown in panel **(c)**.

Maximum temperature rise is determined by cumulative net CO₂ emissions and net non-CO₂ radiative forcing due to methane, nitrous oxide, aerosols and other anthropogenic forcing agents.

Figure SPM.1 | Panel a: Observed monthly global mean surface temperature (GMST, grey line up to 2017, from the HadCRUT4, GISTEMP, Cowtan–Way, and NOAA datasets) change and estimated anthropogenic global warming (solid orange line up to 2017, with orange shading indicating assessed *likely* range). Orange dashed arrow and horizontal orange error bar show respectively the central estimate and *likely* range of the time at which 1.5°C is reached if the current rate of warming continues. The grey plume on the right of panel a shows the *likely* range of warming responses, computed with a simple climate model, to a stylized pathway (hypothetical future) in which net CO₂ emissions (grey line in panels b and c) decline in a straight line from 2020 to reach net zero in 2055 and net non-CO₂ radiative forcing (grey line in panel d) increases to 2030 and then declines. The blue plume in panel a) shows the response to faster CO₂ emissions reductions (blue line in panel b), reaching net zero in 2040, reducing cumulative CO₂ emissions (panel c). The purple plume shows the response to net CO₂ emissions declining to zero in 2055, with net non-CO₂ forcing remaining constant after 2030. The vertical error bars on right of panel a) show the *likely* ranges (thin lines) and central terciles (33rd – 66th percentiles, thick lines) of the estimated distribution of warming in 2100 under these three stylized pathways. Vertical dotted error bars in panels b, c and d show the *likely* range of historical annual and cumulative global net CO₂ emissions in 2017 (data from the Global Carbon Project) and of net non-CO₂ radiative forcing in 2011 from AR5, respectively. Vertical axes in panels c and d are scaled to represent approximately equal effects on GMST. {1.2.1, 1.2.3, 1.2.4, 2.3, Figure 1.2 and Chapter 1 Supplementary Material, Cross-Chapter Box 2 in Chapter 1}

B. Projected Climate Change, Potential Impacts and Associated Risks

B.1 **Climate models project robust[7] differences in regional climate characteristics between present-day and global warming of 1.5°C,[8] and between 1.5°C and 2°C.[8] These differences include increases in: mean temperature in most land and ocean regions (*high confidence*), hot extremes in most inhabited regions (*high confidence*), heavy precipitation in several regions (*medium confidence*), and the probability of drought and precipitation deficits in some regions (*medium confidence*). {3.3}**

B.1.1 Evidence from attributed changes in some climate and weather extremes for a global warming of about 0.5°C supports the assessment that an additional 0.5°C of warming compared to present is associated with further detectable changes in these extremes (*medium confidence*). Several regional changes in climate are assessed to occur with global warming up to 1.5°C compared to pre-industrial levels, including warming of extreme temperatures in many regions (*high confidence*), increases in frequency, intensity, and/or amount of heavy precipitation in several regions (*high confidence*), and an increase in intensity or frequency of droughts in some regions (*medium confidence*). {3.2, 3.3.1, 3.3.2, 3.3.3, 3.3.4, Table 3.2}

B.1.2 Temperature extremes on land are projected to warm more than GMST (*high confidence*): extreme hot days in mid-latitudes warm by up to about 3°C at global warming of 1.5°C and about 4°C at 2°C, and extreme cold nights in high latitudes warm by up to about 4.5°C at 1.5°C and about 6°C at 2°C (*high confidence*). The number of hot days is projected to increase in most land regions, with highest increases in the tropics (*high confidence*). {3.3.1, 3.3.2, Cross-Chapter Box 8 in Chapter 3}

B.1.3 Risks from droughts and precipitation deficits are projected to be higher at 2°C compared to 1.5°C of global warming in some regions (*medium confidence*). Risks from heavy precipitation events are projected to be higher at 2°C compared to 1.5°C of global warming in several northern hemisphere high-latitude and/or high-elevation regions, eastern Asia and eastern North America (*medium confidence*). Heavy precipitation associated with tropical cyclones is projected to be higher at 2°C compared to 1.5°C global warming (*medium confidence*). There is generally *low confidence* in projected changes in heavy precipitation at 2°C compared to 1.5°C in other regions. Heavy precipitation when aggregated at global scale is projected to be higher at 2°C than at 1.5°C of global warming (*medium confidence*). As a consequence of heavy precipitation, the fraction of the global land area affected by flood hazards is projected to be larger at 2°C compared to 1.5°C of global warming (*medium confidence*). {3.3.1, 3.3.3, 3.3.4, 3.3.5, 3.3.6}

B.2 **By 2100, global mean sea level rise is projected to be around 0.1 metre lower with global warming of 1.5°C compared to 2°C (*medium confidence*). Sea level will continue to rise well beyond 2100 (*high confidence*), and the magnitude and rate of this rise depend on future emission pathways. A slower rate of sea level rise enables greater opportunities for adaptation in the human and ecological systems of small islands, low-lying coastal areas and deltas (*medium confidence*). {3.3, 3.4, 3.6}**

B.2.1 Model-based projections of global mean sea level rise (relative to 1986–2005) suggest an indicative range of 0.26 to 0.77 m by 2100 for 1.5°C of global warming, 0.1 m (0.04–0.16 m) less than for a global warming of 2°C (*medium confidence*). A reduction of 0.1 m in global sea level rise implies that up to 10 million fewer people would be exposed to related risks, based on population in the year 2010 and assuming no adaptation (*medium confidence*). {3.4.4, 3.4.5, 4.3.2}

B.2.2 Sea level rise will continue beyond 2100 even if global warming is limited to 1.5°C in the 21st century (*high confidence*). Marine ice sheet instability in Antarctica and/or irreversible loss of the Greenland ice sheet could result in multi-metre rise in sea level over hundreds to thousands of years. These instabilities could be triggered at around 1.5°C to 2°C of global warming (*medium confidence*). (Figure SPM.2) {3.3.9, 3.4.5, 3.5.2, 3.6.3, Box 3.3}

---

7 Robust is here used to mean that at least two thirds of climate models show the same sign of changes at the grid point scale, and that differences in large regions are statistically significant.

8 Projected changes in impacts between different levels of global warming are determined with respect to changes in global mean surface air temperature.

B.2.3   Increasing warming amplifies the exposure of small islands, low-lying coastal areas and deltas to the risks associated with sea level rise for many human and ecological systems, including increased saltwater intrusion, flooding and damage to infrastructure (*high confidence*). Risks associated with sea level rise are higher at 2°C compared to 1.5°C. The slower rate of sea level rise at global warming of 1.5°C reduces these risks, enabling greater opportunities for adaptation including managing and restoring natural coastal ecosystems and infrastructure reinforcement (*medium confidence*). (Figure SPM.2) {3.4.5, Box 3.5}

B.3   On land, impacts on biodiversity and ecosystems, including species loss and extinction, are projected to be lower at 1.5°C of global warming compared to 2°C. Limiting global warming to 1.5°C compared to 2°C is projected to lower the impacts on terrestrial, freshwater and coastal ecosystems and to retain more of their services to humans (*high confidence*). (Figure SPM.2) {3.4, 3.5, Box 3.4, Box 4.2, Cross-Chapter Box 8 in Chapter 3}

B.3.1   Of 105,000 species studied,[9] 6% of insects, 8% of plants and 4% of vertebrates are projected to lose over half of their climatically determined geographic range for global warming of 1.5°C, compared with 18% of insects, 16% of plants and 8% of vertebrates for global warming of 2°C (*medium confidence*). Impacts associated with other biodiversity-related risks such as forest fires and the spread of invasive species are lower at 1.5°C compared to 2°C of global warming (*high confidence*). {3.4.3, 3.5.2}

B.3.2   Approximately 4% (interquartile range 2–7%) of the global terrestrial land area is projected to undergo a transformation of ecosystems from one type to another at 1°C of global warming, compared with 13% (interquartile range 8–20%) at 2°C (*medium confidence*). This indicates that the area at risk is projected to be approximately 50% lower at 1.5°C compared to 2°C (*medium confidence*). {3.4.3.1, 3.4.3.5}

B.3.3   High-latitude tundra and boreal forests are particularly at risk of climate change-induced degradation and loss, with woody shrubs already encroaching into the tundra (*high confidence*) and this will proceed with further warming. Limiting global warming to 1.5°C rather than 2°C is projected to prevent the thawing over centuries of a permafrost area in the range of 1.5 to 2.5 million km² (*medium confidence*). {3.3.2, 3.4.3, 3.5.5}

B.4   Limiting global warming to 1.5°C compared to 2°C is projected to reduce increases in ocean temperature as well as associated increases in ocean acidity and decreases in ocean oxygen levels (*high confidence*). Consequently, limiting global warming to 1.5°C is projected to reduce risks to marine biodiversity, fisheries, and ecosystems, and their functions and services to humans, as illustrated by recent changes to Arctic sea ice and warm-water coral reef ecosystems (*high confidence*). {3.3, 3.4, 3.5, Box 3.4, Box 3.5}

B.4.1   There is *high confidence* that the probability of a sea ice-free Arctic Ocean during summer is substantially lower at global warming of 1.5°C when compared to 2°C. With 1.5°C of global warming, one sea ice-free Arctic summer is projected per century. This likelihood is increased to at least one per decade with 2°C global warming. Effects of a temperature overshoot are reversible for Arctic sea ice cover on decadal time scales (*high confidence*). {3.3.8, 3.4.4.7}

B.4.2   Global warming of 1.5°C is projected to shift the ranges of many marine species to higher latitudes as well as increase the amount of damage to many ecosystems. It is also expected to drive the loss of coastal resources and reduce the productivity of fisheries and aquaculture (especially at low latitudes). The risks of climate-induced impacts are projected to be higher at 2°C than those at global warming of 1.5°C (*high confidence*). Coral reefs, for example, are projected to decline by a further 70–90% at 1.5°C (*high confidence*) with larger losses (>99%) at 2°C (*very high confidence*). The risk of irreversible loss of many marine and coastal ecosystems increases with global warming, especially at 2°C or more (*high confidence*). {3.4.4, Box 3.4}

---

9   Consistent with earlier studies, illustrative numbers were adopted from one recent meta-study.

B.4.3   The level of ocean acidification due to increasing CO_2 concentrations associated with global warming of 1.5°C is projected to amplify the adverse effects of warming, and even further at 2°C, impacting the growth, development, calcification, survival, and thus abundance of a broad range of species, for example, from algae to fish (*high confidence*). {3.3.10, 3.4.4}

B.4.4   Impacts of climate change in the ocean are increasing risks to fisheries and aquaculture via impacts on the physiology, survivorship, habitat, reproduction, disease incidence, and risk of invasive species (*medium confidence*) but are projected to be less at 1.5°C of global warming than at 2°C. One global fishery model, for example, projected a decrease in global annual catch for marine fisheries of about 1.5 million tonnes for 1.5°C of global warming compared to a loss of more than 3 million tonnes for 2°C of global warming (*medium confidence*). {3.4.4, Box 3.4}

B.5   Climate-related risks to health, livelihoods, food security, water supply, human security, and economic growth are projected to increase with global warming of 1.5°C and increase further with 2°C. (Figure SPM.2) {3.4, 3.5, 5.2, Box 3.2, Box 3.3, Box 3.5, Box 3.6, Cross-Chapter Box 6 in Chapter 3, Cross-Chapter Box 9 in Chapter 4, Cross-Chapter Box 12 in Chapter 5, 5.2}

B.5.1   Populations at disproportionately higher risk of adverse consequences with global warming of 1.5°C and beyond include disadvantaged and vulnerable populations, some indigenous peoples, and local communities dependent on agricultural or coastal livelihoods (*high confidence*). Regions at disproportionately higher risk include Arctic ecosystems, dryland regions, small island developing states, and Least Developed Countries (*high confidence*). Poverty and disadvantage are expected to increase in some populations as global warming increases; limiting global warming to 1.5°C, compared with 2°C, could reduce the number of people both exposed to climate-related risks and susceptible to poverty by up to several hundred million by 2050 (*medium confidence*). {3.4.10, 3.4.11, Box 3.5, Cross-Chapter Box 6 in Chapter 3, Cross-Chapter Box 9 in Chapter 4, Cross-Chapter Box 12 in Chapter 5, 4.2.2.2, 5.2.1, 5.2.2, 5.2.3, 5.6.3}

B.5.2   Any increase in global warming is projected to affect human health, with primarily negative consequences (*high confidence*). Lower risks are projected at 1.5°C than at 2°C for heat-related morbidity and mortality (*very high confidence*) and for ozone-related mortality if emissions needed for ozone formation remain high (*high confidence*). Urban heat islands often amplify the impacts of heatwaves in cities (*high confidence*). Risks from some vector-borne diseases, such as malaria and dengue fever, are projected to increase with warming from 1.5°C to 2°C, including potential shifts in their geographic range (*high confidence*). {3.4.7, 3.4.8, 3.5.5.8}

B.5.3   Limiting warming to 1.5°C compared with 2°C is projected to result in smaller net reductions in yields of maize, rice, wheat, and potentially other cereal crops, particularly in sub-Saharan Africa, Southeast Asia, and Central and South America, and in the CO_2-dependent nutritional quality of rice and wheat (*high confidence*). Reductions in projected food availability are larger at 2°C than at 1.5°C of global warming in the Sahel, southern Africa, the Mediterranean, central Europe, and the Amazon (*medium confidence*). Livestock are projected to be adversely affected with rising temperatures, depending on the extent of changes in feed quality, spread of diseases, and water resource availability (*high confidence*). {3.4.6, 3.5.4, 3.5.5, Box 3.1, Cross-Chapter Box 6 in Chapter 3, Cross-Chapter Box 9 in Chapter 4}

B.5.4   Depending on future socio-economic conditions, limiting global warming to 1.5°C compared to 2°C may reduce the proportion of the world population exposed to a climate change-induced increase in water stress by up to 50%, although there is considerable variability between regions (*medium confidence*). Many small island developing states could experience lower water stress as a result of projected changes in aridity when global warming is limited to 1.5°C, as compared to 2°C (*medium confidence*). {3.3.5, 3.4.2, 3.4.8, 3.5.5, Box 3.2, Box 3.5, Cross-Chapter Box 9 in Chapter 4}

B.5.5   Risks to global aggregated economic growth due to climate change impacts are projected to be lower at 1.5°C than at 2°C by the end of this century[10] (*medium confidence*). This excludes the costs of mitigation, adaptation investments and the benefits of adaptation. Countries in the tropics and Southern Hemisphere subtropics are projected to experience the largest impacts on economic growth due to climate change should global warming increase from 1.5°C to 2°C (*medium confidence*). {3.5.2, 3.5.3}

---

10 Here, impacts on economic growth refer to changes in gross domestic product (GDP). Many impacts, such as loss of human lives, cultural heritage and ecosystem services, are difficult to value and monetize.

B.5.6   Exposure to multiple and compound climate-related risks increases between 1.5°C and 2°C of global warming, with greater proportions of people both so exposed and susceptible to poverty in Africa and Asia (*high confidence*). For global warming from 1.5°C to 2°C, risks across energy, food, and water sectors could overlap spatially and temporally, creating new and exacerbating current hazards, exposures, and vulnerabilities that could affect increasing numbers of people and regions (*medium confidence*). {Box 3.5, 3.3.1, 3.4.5.3, 3.4.5.6, 3.4.11, 3.5.4.9}

B.5.7   There are multiple lines of evidence that since AR5 the assessed levels of risk increased for four of the five Reasons for Concern (RFCs) for global warming to 2°C (*high confidence*). The risk transitions by degrees of global warming are now: from high to very high risk between 1.5°C and 2°C for RFC1 (Unique and threatened systems) (*high confidence*); from moderate to high risk between 1°C and 1.5°C for RFC2 (Extreme weather events) (*medium confidence*); from moderate to high risk between 1.5°C and 2°C for RFC3 (Distribution of impacts) (*high confidence*); from moderate to high risk between 1.5°C and 2.5°C for RFC4 (Global aggregate impacts) (*medium confidence*); and from moderate to high risk between 1°C and 2.5°C for RFC5 (Large-scale singular events) (*medium confidence*). (Figure SPM.2) {3.4.13; 3.5, 3.5.2}

B.6   Most adaptation needs will be lower for global warming of 1.5°C compared to 2°C (*high confidence*). There are a wide range of adaptation options that can reduce the risks of climate change (*high confidence*). There are limits to adaptation and adaptive capacity for some human and natural systems at global warming of 1.5°C, with associated losses (*medium confidence*). The number and availability of adaptation options vary by sector (*medium confidence*). {Table 3.5, 4.3, 4.5, Cross-Chapter Box 9 in Chapter 4, Cross-Chapter Box 12 in Chapter 5}

B.6.1   A wide range of adaptation options are available to reduce the risks to natural and managed ecosystems (e.g., ecosystem-based adaptation, ecosystem restoration and avoided degradation and deforestation, biodiversity management, sustainable aquaculture, and local knowledge and indigenous knowledge), the risks of sea level rise (e.g., coastal defence and hardening), and the risks to health, livelihoods, food, water, and economic growth, especially in rural landscapes (e.g., efficient irrigation, social safety nets, disaster risk management, risk spreading and sharing, and community-based adaptation) and urban areas (e.g., green infrastructure, sustainable land use and planning, and sustainable water management) (*medium confidence*). {4.3.1, 4.3.2, 4.3.3, 4.3.5, 4.5.3, 4.5.4, 5.3.2, Box 4.2, Box 4.3, Box 4.6, Cross-Chapter Box 9 in Chapter 4}.

B.6.2   Adaptation is expected to be more challenging for ecosystems, food and health systems at 2°C of global warming than for 1.5°C (*medium confidence*). Some vulnerable regions, including small islands and Least Developed Countries, are projected to experience high multiple interrelated climate risks even at global warming of 1.5°C (*high confidence*). {3.3.1, 3.4.5, Box 3.5, Table 3.5, Cross-Chapter Box 9 in Chapter 4, 5.6, Cross-Chapter Box 12 in Chapter 5, Box 5.3}

B.6.3   Limits to adaptive capacity exist at 1.5°C of global warming, become more pronounced at higher levels of warming and vary by sector, with site-specific implications for vulnerable regions, ecosystems and human health (*medium confidence*). {Cross-Chapter Box 12 in Chapter 5, Box 3.5, Table 3.5}

How the level of global warming affects impacts and/or risks associated with the Reasons for Concern (RFCs) and selected natural, managed and human systems

Five Reasons For Concern (RFCs) illustrate the impacts and risks of different levels of global warming for people, economies and ecosystems across sectors and regions.

Impacts and risks associated with the Reasons for Concern (RFCs)

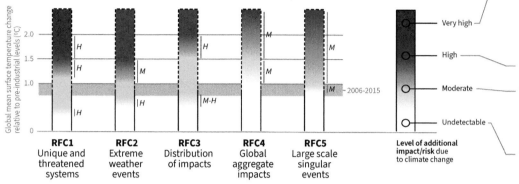

Purple indicates very high risks of severe impacts/risks and the presence of significant irreversibility or the persistence of climate-related hazards, combined with limited ability to adapt due to the nature of the hazard or impacts/risks.
Red indicates severe and widespread impacts/risks.
Yellow indicates that impacts/risks are detectable and attributable to climate change with at least medium confidence.
White indicates that no impacts are detectable and attributable to climate change.

Impacts and risks for selected natural, managed and human systems

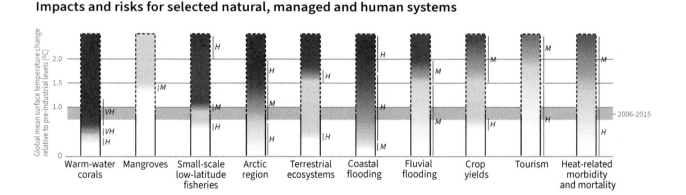

Confidence level for transition: *L*=Low, *M*=Medium, *H*=High and *VH*=Very high

Figure SPM.2 | Five integrative reasons for concern (RFCs) provide a framework for summarizing key impacts and risks across sectors and regions, and were introduced in the IPCC Third Assessment Report. RFCs illustrate the implications of global warming for people, economies and ecosystems. Impacts and/or risks for each RFC are based on assessment of the new literature that has appeared. As in AR5, this literature was used to make expert judgments to assess the levels of global warming at which levels of impact and/or risk are undetectable, moderate, high or very high. The selection of impacts and risks to natural, managed and human systems in the lower panel is illustrative and is not intended to be fully comprehensive. {3.4, 3.5, 3.5.2.1, 3.5.2.2, 3.5.2.3, 3.5.2.4, 3.5.2.5, 5.4.1, 5.5.3, 5.6.1, Box 3.4}

RFC1 Unique and threatened systems: ecological and human systems that have restricted geographic ranges constrained by climate-related conditions and have high endemism or other distinctive properties. Examples include coral reefs, the Arctic and its indigenous people, mountain glaciers and biodiversity hotspots.

RFC2 Extreme weather events: risks/impacts to human health, livelihoods, assets and ecosystems from extreme weather events such as heat waves, heavy rain, drought and associated wildfires, and coastal flooding.

RFC3 Distribution of impacts: risks/impacts that disproportionately affect particular groups due to uneven distribution of physical climate change hazards, exposure or vulnerability.

RFC4 Global aggregate impacts: global monetary damage, global-scale degradation and loss of ecosystems and biodiversity.

RFC5 Large-scale singular events: are relatively large, abrupt and sometimes irreversible changes in systems that are caused by global warming. Examples include disintegration of the Greenland and Antarctic ice sheets.

C. Emission Pathways and System Transitions Consistent with 1.5°C Global Warming

C.1 **In model pathways with no or limited overshoot of 1.5°C, global net anthropogenic CO$_2$ emissions decline by about 45% from 2010 levels by 2030 (40–60% interquartile range), reaching net zero around 2050 (2045–2055 interquartile range). For limiting global warming to below 2°C[11] CO$_2$ emissions are projected to decline by about 25% by 2030 in most pathways (10–30% interquartile range) and reach net zero around 2070 (2065–2080 interquartile range). Non-CO$_2$ emissions in pathways that limit global warming to 1.5°C show deep reductions that are similar to those in pathways limiting warming to 2°C. (*high confidence*) (Figure SPM.3a) {2.1, 2.3, Table 2.4}**

C.1.1 CO$_2$ emissions reductions that limit global warming to 1.5°C with no or limited overshoot can involve different portfolios of mitigation measures, striking different balances between lowering energy and resource intensity, rate of decarbonization, and the reliance on carbon dioxide removal. Different portfolios face different implementation challenges and potential synergies and trade-offs with sustainable development. (*high confidence*) (Figure SPM.3b) {2.3.2, 2.3.4, 2.4, 2.5.3}

C.1.2 Modelled pathways that limit global warming to 1.5°C with no or limited overshoot involve deep reductions in emissions of methane and black carbon (35% or more of both by 2050 relative to 2010). These pathways also reduce most of the cooling aerosols, which partially offsets mitigation effects for two to three decades. Non-CO$_2$ emissions[12] can be reduced as a result of broad mitigation measures in the energy sector. In addition, targeted non-CO$_2$ mitigation measures can reduce nitrous oxide and methane from agriculture, methane from the waste sector, some sources of black carbon, and hydrofluorocarbons. High bioenergy demand can increase emissions of nitrous oxide in some 1.5°C pathways, highlighting the importance of appropriate management approaches. Improved air quality resulting from projected reductions in many non-CO$_2$ emissions provide direct and immediate population health benefits in all 1.5°C model pathways. (*high confidence*) (Figure SPM.3a) {2.2.1, 2.3.3, 2.4.4, 2.5.3, 4.3.6, 5.4.2}

C.1.3 Limiting global warming requires limiting the total cumulative global anthropogenic emissions of CO$_2$ since the pre-industrial period, that is, staying within a total carbon budget (*high confidence*).[13] By the end of 2017, anthropogenic CO$_2$ emissions since the pre-industrial period are estimated to have reduced the total carbon budget for 1.5°C by approximately 2200 ± 320 GtCO$_2$ (*medium confidence*). The associated remaining budget is being depleted by current emissions of 42 ± 3 GtCO$_2$ per year (*high confidence*). The choice of the measure of global temperature affects the estimated remaining carbon budget. Using global mean surface air temperature, as in AR5, gives an estimate of the remaining carbon budget of 580 GtCO$_2$ for a 50% probability of limiting warming to 1.5°C, and 420 GtCO$_2$ for a 66% probability (*medium confidence*).[14] Alternatively, using GMST gives estimates of 770 and 570 GtCO$_2$, for 50% and 66% probabilities,[15] respectively (*medium confidence*). Uncertainties in the size of these estimated remaining carbon budgets are substantial and depend on several factors. Uncertainties in the climate response to CO$_2$ and non-CO$_2$ emissions contribute ±400 GtCO$_2$ and the level of historic warming contributes ±250 GtCO$_2$ (*medium confidence*). Potential additional carbon release from future permafrost thawing and methane release from wetlands would reduce budgets by up to 100 GtCO$_2$ over the course of this century and more thereafter (*medium confidence*). In addition, the level of non-CO$_2$ mitigation in the future could alter the remaining carbon budget by 250 GtCO$_2$ in either direction (*medium confidence*). {1.2.4, 2.2.2, 2.6.1, Table 2.2, Chapter 2 Supplementary Material}

C.1.4 Solar radiation modification (SRM) measures are not included in any of the available assessed pathways. Although some SRM measures may be theoretically effective in reducing an overshoot, they face large uncertainties and knowledge gaps

---

11 References to pathways limiting global warming to 2°C are based on a 66% probability of staying below 2°C.

12 Non-CO$_2$ emissions included in this Report are all anthropogenic emissions other than CO$_2$ that result in radiative forcing. These include short-lived climate forcers, such as methane, some fluorinated gases, ozone precursors, aerosols or aerosol precursors, such as black carbon and sulphur dioxide, respectively, as well as long-lived greenhouse gases, such as nitrous oxide or some fluorinated gases. The radiative forcing associated with non-CO$_2$ emissions and changes in surface albedo is referred to as non-CO$_2$ radiative forcing. {2.2.1}

13 There is a clear scientific basis for a total carbon budget consistent with limiting global warming to 1.5°C. However, neither this total carbon budget nor the fraction of this budget taken up by past emissions were assessed in this Report.

14 Irrespective of the measure of global temperature used, updated understanding and further advances in methods have led to an increase in the estimated remaining carbon budget of about 300 GtCO$_2$ compared to AR5. (*medium confidence*) {2.2.2}

15 These estimates use observed GMST to 2006–2015 and estimate future temperature changes using near surface air temperatures.

Global emissions pathway characteristics

General characteristics of the evolution of anthropogenic net emissions of CO_2, and total emissions of methane, black carbon, and nitrous oxide in model pathways that limit global warming to 1.5°C with no or limited overshoot. Net emissions are defined as anthropogenic emissions reduced by anthropogenic removals. Reductions in net emissions can be achieved through different portfolios of mitigation measures illustrated in Figure SPM.3b.

Global total net CO_2 emissions

Billion tonnes of CO_2/yr

*In pathways limiting global warming to 1.5°C with **no or limited overshoot** as well as in pathways with a **higher overshoot**, CO2 emissions are reduced to net zero globally around 2050.*

Four illustrative model pathways

P1
P2
P3
P4

Timing of net zero CO_2
Line widths depict the 5-95th percentile and the 25-75th percentile of scenarios

Pathways limiting global warming to 1.5°C with **no or limited** overshoot

Pathways with **higher overshoot**

Pathways limiting global warming below 2°C (Not shown above)

Non-CO_2 emissions relative to 2010

Emissions of non-CO_2 forcers are also reduced or limited in pathways limiting global warming to 1.5°C with **no or limited overshoot**, but they do not reach zero globally.

Methane emissions

Black carbon emissions

Nitrous oxide emissions

Figure SPM.3a | Global emissions pathway characteristics. The main panel shows global net anthropogenic CO_2 emissions in pathways limiting global warming to 1.5°C with no or limited (less than 0.1°C) overshoot and pathways with higher overshoot. The shaded area shows the full range for pathways analysed in this Report. The panels on the right show non-CO_2 emissions ranges for three compounds with large historical forcing and a substantial portion of emissions coming from sources distinct from those central to CO_2 mitigation. Shaded areas in these panels show the 5–95% (light shading) and interquartile (dark shading) ranges of pathways limiting global warming to 1.5°C with no or limited overshoot. Box and whiskers at the bottom of the figure show the timing of pathways reaching global net zero CO_2 emission levels, and a comparison with pathways limiting global warming to 2°C with at least 66% probability. Four illustrative model pathways are highlighted in the main panel and are labelled P1, P2, P3 and P4, corresponding to the LED, S1, S2, and S5 pathways assessed in Chapter 2. Descriptions and characteristics of these pathways are available in Figure SPM.3b. {2.1, 2.2, 2.3, Figure 2.5, Figure 2.10, Figure 2.11}

Characteristics of four illustrative model pathways

Different mitigation strategies can achieve the net emissions reductions that would be required to follow a pathway that limits global warming to 1.5°C with no or limited overshoot. All pathways use Carbon Dioxide Removal (CDR), but the amount varies across pathways, as do the relative contributions of Bioenergy with Carbon Capture and Storage (BECCS) and removals in the Agriculture, Forestry and Other Land Use (AFOLU) sector. This has implications for emissions and several other pathway characteristics.

Breakdown of contributions to global net CO_2 emissions in four illustrative model pathways

● Fossil fuel and industry ● AFOLU ○ BECCS

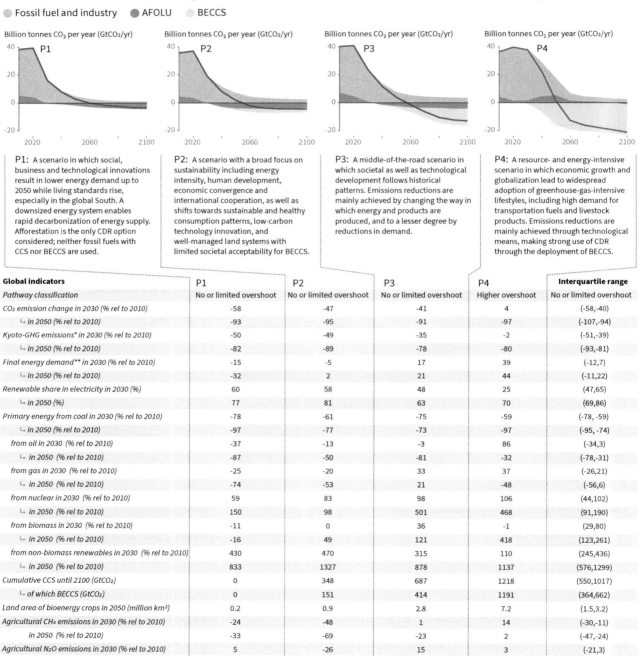

P1: A scenario in which social, business and technological innovations result in lower energy demand up to 2050 while living standards rise, especially in the global South. A downsized energy system enables rapid decarbonization of energy supply. Afforestation is the only CDR option considered; neither fossil fuels with CCS nor BECCS are used.

P2: A scenario with a broad focus on sustainability including energy intensity, human development, economic convergence and international cooperation, as well as shifts towards sustainable and healthy consumption patterns, low-carbon technology innovation, and well-managed land systems with limited societal acceptability for BECCS.

P3: A middle-of-the-road scenario in which societal as well as technological development follows historical patterns. Emissions reductions are mainly achieved by changing the way in which energy and products are produced, and to a lesser degree by reductions in demand.

P4: A resource- and energy-intensive scenario in which economic growth and globalization lead to widespread adoption of greenhouse-gas-intensive lifestyles, including high demand for transportation fuels and livestock products. Emissions reductions are mainly achieved through technological means, making strong use of CDR through the deployment of BECCS.

Global indicators	P1	P2	P3	P4	Interquartile range
Pathway classification	No or limited overshoot	No or limited overshoot	No or limited overshoot	Higher overshoot	No or limited overshoot
CO_2 emission change in 2030 (% rel to 2010)	-58	-47	-41	4	(-58,-40)
↳ in 2050 (% rel to 2010)	-93	-95	-91	-97	(-107,-94)
Kyoto-GHG emissions* in 2030 (% rel to 2010)	-50	-49	-35	-2	(-51,-39)
↳ in 2050 (% rel to 2010)	-82	-89	-78	-80	(-93,-81)
Final energy demand** in 2030 (% rel to 2010)	-15	-5	17	39	(-12,7)
↳ in 2050 (% rel to 2010)	-32	2	21	44	(-11,22)
Renewable share in electricity in 2030 (%)	60	58	48	25	(47,65)
↳ in 2050 (%)	77	81	63	70	(69,86)
Primary energy from coal in 2030 (% rel to 2010)	-78	-61	-75	-59	(-78, -59)
↳ in 2050 (% rel to 2010)	-97	-77	-73	-97	(-95, -74)
from oil in 2030 (% rel to 2010)	-37	-13	-3	86	(-34,3)
↳ in 2050 (% rel to 2010)	-87	-50	-81	-32	(-78,-31)
from gas in 2030 (% rel to 2010)	-25	-20	33	37	(-26,21)
↳ in 2050 (% rel to 2010)	-74	-53	21	-48	(-56,6)
from nuclear in 2030 (% rel to 2010)	59	83	98	106	(44,102)
↳ in 2050 (% rel to 2010)	150	98	501	468	(91,190)
from biomass in 2030 (% rel to 2010)	-11	0	36	-1	(29,80)
↳ in 2050 (% rel to 2010)	-16	49	121	418	(123,261)
from non-biomass renewables in 2030 (% rel to 2010)	430	470	315	110	(245,436)
↳ in 2050 (% rel to 2010)	833	1327	878	1137	(576,1299)
Cumulative CCS until 2100 (GtCO₂)	0	348	687	1218	(550,1017)
↳ of which BECCS (GtCO₂)	0	151	414	1191	(364,662)
Land area of bioenergy crops in 2050 (million km²)	0.2	0.9	2.8	7.2	(1.5,3.2)
Agricultural CH₄ emissions in 2030 (% rel to 2010)	-24	-48	1	14	(-30,-11)
in 2050 (% rel to 2010)	-33	-69	-23	2	(-47,-24)
Agricultural N₂O emissions in 2030 (% rel to 2010)	5	-26	15	3	(-21,3)
in 2050 (% rel to 2010)	6	-26	0	39	(-26,1)

NOTE: Indicators have been selected to show global trends identified by the Chapter 2 assessment. National and sectoral characteristics can differ substantially from the global trends shown above.

* Kyoto-gas emissions are based on IPCC Second Assessment Report GWP-100
** Changes in energy demand are associated with improvements in energy efficiency and behaviour change

Figure SPM.3b | Characteristics of four illustrative model pathways in relation to global warming of 1.5°C introduced in Figure SPM.3a. These pathways were selected to show a range of potential mitigation approaches and vary widely in their projected energy and land use, as well as their assumptions about future socio-economic developments, including economic and population growth, equity and sustainability. A breakdown of the global net anthropogenic CO_2 emissions into the contributions in terms of CO_2 emissions from fossil fuel and industry; agriculture, forestry and other land use (AFOLU); and bioenergy with carbon capture and storage (BECCS) is shown. AFOLU estimates reported here are not necessarily comparable with countries' estimates. Further characteristics for each of these pathways are listed below each pathway. These pathways illustrate relative global differences in mitigation strategies, but do not represent central estimates, national strategies, and do not indicate requirements. For comparison, the right-most column shows the interquartile ranges across pathways with no or limited overshoot of 1.5°C. Pathways P1, P2, P3 and P4 correspond to the LED, S1, S2 and S5 pathways assessed in Chapter 2 (Figure SPM.3a). {2.2.1, 2.3.1, 2.3.2, 2.3.3, 2.3.4, 2.4.1, 2.4.2, 2.4.4, 2.5.3, Figure 2.5, Figure 2.6, Figure 2.9, Figure 2.10, Figure 2.11, Figure 2.14, Figure 2.15, Figure 2.16, Figure 2.17, Figure 2.24, Figure 2.25, Table 2.4, Table 2.6, Table 2.7, Table 2.9, Table 4.1}

C.2 **Pathways limiting global warming to 1.5°C with no or limited overshoot would require rapid and far-reaching transitions in energy, land, urban and infrastructure (including transport and buildings), and industrial systems (*high confidence*). These systems transitions are unprecedented in terms of scale, but not necessarily in terms of speed, and imply deep emissions reductions in all sectors, a wide portfolio of mitigation options and a significant upscaling of investments in those options (*medium confidence*). {2.3, 2.4, 2.5, 4.2, 4.3, 4.4, 4.5}**

C.2.1   Pathways that limit global warming to 1.5°C with no or limited overshoot show system changes that are more rapid and pronounced over the next two decades than in 2°C pathways (*high confidence*). The rates of system changes associated with limiting global warming to 1.5°C with no or limited overshoot have occurred in the past within specific sectors, technologies and spatial contexts, but there is no documented historic precedent for their scale (*medium confidence*). {2.3.3, 2.3.4, 2.4, 2.5, 4.2.1, 4.2.2, Cross-Chapter Box 11 in Chapter 4}

C.2.2   In energy systems, modelled global pathways (considered in the literature) limiting global warming to 1.5°C with no or limited overshoot (for more details see Figure SPM.3b) generally meet energy service demand with lower energy use, including through enhanced energy efficiency, and show faster electrification of energy end use compared to 2°C (*high confidence*). In 1.5°C pathways with no or limited overshoot, low-emission energy sources are projected to have a higher share, compared with 2°C pathways, particularly before 2050 (*high confidence*). In 1.5°C pathways with no or limited overshoot, renewables are projected to supply 70–85% (interquartile range) of electricity in 2050 (*high confidence*). In electricity generation, shares of nuclear and fossil fuels with carbon dioxide capture and storage (CCS) are modelled to increase in most 1.5°C pathways with no or limited overshoot. In modelled 1.5°C pathways with limited or no overshoot, the use of CCS would allow the electricity generation share of gas to be approximately 8% (3–11% interquartile range) of global electricity in 2050, while the use of coal shows a steep reduction in all pathways and would be reduced to close to 0% (0–2% interquartile range) of electricity (*high confidence*). While acknowledging the challenges, and differences between the options and national circumstances, political, economic, social and technical feasibility of solar energy, wind energy and electricity storage technologies have substantially improved over the past few years (*high confidence*). These improvements signal a potential system transition in electricity generation. (Figure SPM.3b) {2.4.1, 2.4.2, Figure 2.1, Table 2.6, Table 2.7, Cross-Chapter Box 6 in Chapter 3, 4.2.1, 4.3.1, 4.3.3, 4.5.2}

C.2.3 CO_2 emissions from industry in pathways limiting global warming to 1.5°C with no or limited overshoot are projected to be about 65–90% (interquartile range) lower in 2050 relative to 2010, as compared to 50–80% for global warming of 2°C (*medium confidence*). Such reductions can be achieved through combinations of new and existing technologies and practices, including electrification, hydrogen, sustainable bio-based feedstocks, product substitution, and carbon capture, utilization and storage (CCUS). These options are technically proven at various scales but their large-scale deployment may be limited by economic, financial, human capacity and institutional constraints in specific contexts, and specific characteristics of large-scale industrial installations. In industry, emissions reductions by energy and process efficiency by themselves are insufficient for limiting warming to 1.5°C with no or limited overshoot (*high confidence*). {2.4.3, 4.2.1, Table 4.1, Table 4.3, 4.3.3, 4.3.4, 4.5.2}

C.2.4   The urban and infrastructure system transition consistent with limiting global warming to 1.5°C with no or limited overshoot would imply, for example, changes in land and urban planning practices, as well as deeper emissions reductions in transport and buildings compared to pathways that limit global warming below 2°C (*medium confidence*). Technical measures and practices enabling deep emissions reductions include various energy efficiency options. In pathways limiting global warming to 1.5°C with no or limited overshoot, the electricity share of energy demand in buildings would be about 55–75% in 2050 compared to 50–70% in 2050 for 2°C global warming (*medium confidence*). In the transport sector, the share of low-emission final energy would rise from less than 5% in 2020 to about 35–65% in 2050 compared to 25–45% for 2°C of global warming (*medium confidence*). Economic, institutional and socio-cultural barriers may inhibit these urban and infrastructure system transitions, depending on national, regional and local circumstances, capabilities and the availability of capital (*high confidence*). {2.3.4, 2.4.3, 4.2.1, Table 4.1, 4.3.3, 4.5.2}

C.2.5   Transitions in global and regional land use are found in all pathways limiting global warming to 1.5°C with no or limited overshoot, but their scale depends on the pursued mitigation portfolio. Model pathways that limit global warming to 1.5°C with no or limited overshoot project a 4 million km^2 reduction to a 2.5 million km^2 increase of non-pasture agricultural land for food and feed crops and a 0.5–11 million km^2 reduction of pasture land, to be converted into a 0–6 million km^2 increase of agricultural land for energy crops and a 2 million km^2 reduction to 9.5 million km^2 increase in forests by 2050 relative to 2010 (*medium confidence*).[16] Land-use transitions of similar magnitude can be observed in modelled 2°C pathways (*medium confidence*). Such large transitions pose profound challenges for sustainable management of the various demands on land for human settlements, food, livestock feed, fibre, bioenergy, carbon storage, biodiversity and other ecosystem services (*high confidence*). Mitigation options limiting the demand for land include sustainable intensification of land-use practices, ecosystem restoration and changes towards less resource-intensive diets (*high confidence*). The implementation of land-based mitigation options would require overcoming socio-economic, institutional, technological, financing and environmental barriers that differ across regions (*high confidence*). {2.4.4, Figure 2.24, 4.3.2, 4.3.7, 4.5.2, Cross-Chapter Box 7 in Chapter 3}

C.2.6   Additional annual average energy-related investments for the period 2016 to 2050 in pathways limiting warming to 1.5°C compared to pathways without new climate policies beyond those in place today are estimated to be around 830 billion USD2010 (range of 150 billion to 1700 billion USD2010 across six models[17]). This compares to total annual average energy supply investments in 1.5°C pathways of 1460 to 3510 billion USD2010 and total annual average energy demand investments of 640 to 910 billion USD2010 for the period 2016 to 2050. Total energy-related investments increase by about 12% (range of 3% to 24%) in 1.5°C pathways relative to 2°C pathways. Annual investments in low-carbon energy technologies and energy efficiency are upscaled by roughly a factor of six (range of factor of 4 to 10) by 2050 compared to 2015 (*medium confidence*). {2.5.2, Box 4.8, Figure 2.27}

C.2.7   Modelled pathways limiting global warming to 1.5°C with no or limited overshoot project a wide range of global average discounted marginal abatement costs over the 21st century. They are roughly 3-4 times higher than in pathways limiting global warming to below 2°C (*high confidence*). The economic literature distinguishes marginal abatement costs from total mitigation costs in the economy. The literature on total mitigation costs of 1.5°C mitigation pathways is limited and was not assessed in this Report. Knowledge gaps remain in the integrated assessment of the economy-wide costs and benefits of mitigation in line with pathways limiting warming to 1.5°C. {2.5.2; 2.6; Figure 2.26}

---

16 The projected land-use changes presented are not deployed to their upper limits simultaneously in a single pathway.

17 Including two pathways limiting warming to 1.5°C with no or limited overshoot and four pathways with higher overshoot.

C.3   **All pathways that limit global warming to 1.5°C with limited or no overshoot project the use of carbon dioxide removal (CDR) on the order of 100–1000 GtCO$_2$ over the 21st century. CDR would be used to compensate for residual emissions and, in most cases, achieve net negative emissions to return global warming to 1.5°C following a peak (*high confidence*). CDR deployment of several hundreds of GtCO$_2$ is subject to multiple feasibility and sustainability constraints (*high confidence*). Significant near-term emissions reductions and measures to lower energy and land demand can limit CDR deployment to a few hundred GtCO$_2$ without reliance on bioenergy with carbon capture and storage (BECCS) (*high confidence*). {2.3, 2.4, 3.6.2, 4.3, 5.4}**

C.3.1   Existing and potential CDR measures include afforestation and reforestation, land restoration and soil carbon sequestration, BECCS, direct air carbon capture and storage (DACCS), enhanced weathering and ocean alkalinization. These differ widely in terms of maturity, potentials, costs, risks, co-benefits and trade-offs (*high confidence*). To date, only a few published pathways include CDR measures other than afforestation and BECCS. {2.3.4, 3.6.2, 4.3.2, 4.3.7}

C.3.2   In pathways limiting global warming to 1.5°C with limited or no overshoot, BECCS deployment is projected to range from 0–1, 0–8, and 0–16 GtCO$_2$ yr^{-1} in 2030, 2050, and 2100, respectively, while agriculture, forestry and land-use (AFOLU) related CDR measures are projected to remove 0–5, 1–11, and 1–5 GtCO$_2$ yr^{-1} in these years (*medium confidence*). The upper end of these deployment ranges by mid-century exceeds the BECCS potential of up to 5 GtCO$_2$ yr^{-1} and afforestation potential of up to 3.6 GtCO$_2$ yr^{-1} assessed based on recent literature (*medium confidence*). Some pathways avoid BECCS deployment completely through demand-side measures and greater reliance on AFOLU-related CDR measures (*medium confidence*). The use of bioenergy can be as high or even higher when BECCS is excluded compared to when it is included due to its potential for replacing fossil fuels across sectors (*high confidence*). (Figure SPM.3b) {2.3.3, 2.3.4, 2.4.2, 3.6.2, 4.3.1, 4.2.3, 4.3.2, 4.3.7, 4.4.3, Table 2.4}

C.3.3   Pathways that overshoot 1.5°C of global warming rely on CDR exceeding residual CO$_2$ emissions later in the century to return to below 1.5°C by 2100, with larger overshoots requiring greater amounts of CDR (Figure SPM.3b) (*high confidence*). Limitations on the speed, scale, and societal acceptability of CDR deployment hence determine the ability to return global warming to below 1.5°C following an overshoot. Carbon cycle and climate system understanding is still limited about the effectiveness of net negative emissions to reduce temperatures after they peak (*high confidence*). {2.2, 2.3.4, 2.3.5, 2.6, 4.3.7, 4.5.2, Table 4.11}

C.3.4   Most current and potential CDR measures could have significant impacts on land, energy, water or nutrients if deployed at large scale (*high confidence*). Afforestation and bioenergy may compete with other land uses and may have significant impacts on agricultural and food systems, biodiversity, and other ecosystem functions and services (*high confidence*). Effective governance is needed to limit such trade-offs and ensure permanence of carbon removal in terrestrial, geological and ocean reservoirs (*high confidence*). Feasibility and sustainability of CDR use could be enhanced by a portfolio of options deployed at substantial, but lesser scales, rather than a single option at very large scale (*high confidence*). (Figure SPM.3b) {2.3.4, 2.4.4, 2.5.3, 2.6, 3.6.2, 4.3.2, 4.3.7, 4.5.2, 5.4.1, 5.4.2; Cross-Chapter Boxes 7 and 8 in Chapter 3, Table 4.11, Table 5.3, Figure 5.3}

C.3.5   Some AFOLU-related CDR measures such as restoration of natural ecosystems and soil carbon sequestration could provide co-benefits such as improved biodiversity, soil quality, and local food security. If deployed at large scale, they would require governance systems enabling sustainable land management to conserve and protect land carbon stocks and other ecosystem functions and services (*medium confidence*). (Figure SPM.4) {2.3.3, 2.3.4, 2.4.2, 2.4.4, 3.6.2, 5.4.1, Cross-Chapter Boxes 3 in Chapter 1 and 7 in Chapter 3, 4.3.2, 4.3.7, 4.4.1, 4.5.2, Table 2.4}

D. Strengthening the Global Response in the Context of Sustainable Development and Efforts to Eradicate Poverty

D.1 **Estimates of the global emissions outcome of current nationally stated mitigation ambitions as submitted under the Paris Agreement would lead to global greenhouse gas emissions[18] in 2030 of 52–58 GtCO$_2$eq yr^{-1} (*medium confidence*). Pathways reflecting these ambitions would not limit global warming to 1.5°C, even if supplemented by very challenging increases in the scale and ambition of emissions reductions after 2030 (*high confidence*). Avoiding overshoot and reliance on future large-scale deployment of carbon dioxide removal (CDR) can only be achieved if global CO$_2$ emissions start to decline well before 2030 (*high confidence*). {1.2, 2.3, 3.3, 3.4, 4.2, 4.4, Cross-Chapter Box 11 in Chapter 4}**

D.1.1 Pathways that limit global warming to 1.5°C with no or limited overshoot show clear emission reductions by 2030 (*high confidence*). All but one show a decline in global greenhouse gas emissions to below 35 GtCO$_2$eq yr^{-1} in 2030, and half of available pathways fall within the 25–30 GtCO$_2$eq yr^{-1} range (interquartile range), a 40–50% reduction from 2010 levels (*high confidence*). Pathways reflecting current nationally stated mitigation ambition until 2030 are broadly consistent with cost-effective pathways that result in a global warming of about 3°C by 2100, with warming continuing afterwards (*medium confidence*). {2.3.3, 2.3.5, Cross-Chapter Box 11 in Chapter 4, 5.5.3.2}

D.1.2 Overshoot trajectories result in higher impacts and associated challenges compared to pathways that limit global warming to 1.5°C with no or limited overshoot (*high confidence*). Reversing warming after an overshoot of 0.2°C or larger during this century would require upscaling and deployment of CDR at rates and volumes that might not be achievable given considerable implementation challenges (*medium confidence*). {1.3.3, 2.3.4, 2.3.5, 2.5.1, 3.3, 4.3.7, Cross-Chapter Box 8 in Chapter 3, Cross-Chapter Box 11 in Chapter 4}

D.1.3 The lower the emissions in 2030, the lower the challenge in limiting global warming to 1.5°C after 2030 with no or limited overshoot (*high confidence*). The challenges from delayed actions to reduce greenhouse gas emissions include the risk of cost escalation, lock-in in carbon-emitting infrastructure, stranded assets, and reduced flexibility in future response options in the medium to long term (*high confidence*). These may increase uneven distributional impacts between countries at different stages of development (*medium confidence*). {2.3.5, 4.4.5, 5.4.2}

D.2 **The avoided climate change impacts on sustainable development, eradication of poverty and reducing inequalities would be greater if global warming were limited to 1.5°C rather than 2°C, if mitigation and adaptation synergies are maximized while trade-offs are minimized (*high confidence*). {1.1, 1.4, 2.5, 3.3, 3.4, 5.2, Table 5.1}**

D.2.1 Climate change impacts and responses are closely linked to sustainable development which balances social well-being, economic prosperity and environmental protection. The United Nations Sustainable Development Goals (SDGs), adopted in 2015, provide an established framework for assessing the links between global warming of 1.5°C or 2°C and development goals that include poverty eradication, reducing inequalities, and climate action. (*high confidence*) {Cross-Chapter Box 4 in Chapter 1, 1.4, 5.1}

D.2.2 The consideration of ethics and equity can help address the uneven distribution of adverse impacts associated with 1.5°C and higher levels of global warming, as well as those from mitigation and adaptation, particularly for poor and disadvantaged populations, in all societies (*high confidence*). {1.1.1, 1.1.2, 1.4.3, 2.5.3, 3.4.10, 5.1, 5.2, 5.3. 5.4, Cross-Chapter Box 4 in Chapter 1, Cross-Chapter Boxes 6 and 8 in Chapter 3, and Cross-Chapter Box 12 in Chapter 5}

D.2.3 Mitigation and adaptation consistent with limiting global warming to 1.5°C are underpinned by enabling conditions, assessed in this Report across the geophysical, environmental-ecological, technological, economic, socio-cultural and institutional

---

18 GHG emissions have been aggregated with 100-year GWP values as introduced in the IPCC Second Assessment Report.

dimensions of feasibility. Strengthened multilevel governance, institutional capacity, policy instruments, technological innovation and transfer and mobilization of finance, and changes in human behaviour and lifestyles are enabling conditions that enhance the feasibility of mitigation and adaptation options for 1.5°C-consistent systems transitions. (*high confidence*) {1.4, Cross-Chapter Box 3 in Chapter 1, 2.5.1, 4.4, 4.5, 5.6}

D.3   Adaptation options specific to national contexts, if carefully selected together with enabling conditions, will have benefits for sustainable development and poverty reduction with global warming of 1.5°C, although trade-offs are possible (*high confidence*). {1.4, 4.3, 4.5}

D.3.1   Adaptation options that reduce the vulnerability of human and natural systems have many synergies with sustainable development, if well managed, such as ensuring food and water security, reducing disaster risks, improving health conditions, maintaining ecosystem services and reducing poverty and inequality (*high confidence*). Increasing investment in physical and social infrastructure is a key enabling condition to enhance the resilience and the adaptive capacities of societies. These benefits can occur in most regions with adaptation to 1.5°C of global warming (*high confidence*). {1.4.3, 4.2.2, 4.3.1, 4.3.2, 4.3.3, 4.3.5, 4.4.1, 4.4.3, 4.5.3, 5.3.1, 5.3.2}

D.3.2   Adaptation to 1.5°C global warming can also result in trade-offs or maladaptations with adverse impacts for sustainable development. For example, if poorly designed or implemented, adaptation projects in a range of sectors can increase greenhouse gas emissions and water use, increase gender and social inequality, undermine health conditions, and encroach on natural ecosystems (*high confidence*). These trade-offs can be reduced by adaptations that include attention to poverty and sustainable development (*high confidence*). {4.3.2, 4.3.3, 4.5.4, 5.3.2; Cross-Chapter Boxes 6 and 7 in Chapter 3}

D.3.3   A mix of adaptation and mitigation options to limit global warming to 1.5°C, implemented in a participatory and integrated manner, can enable rapid, systemic transitions in urban and rural areas (*high confidence*). These are most effective when aligned with economic and sustainable development, and when local and regional governments and decision makers are supported by national governments (*medium confidence*). {4.3.2, 4.3.3, 4.4.1, 4.4.2}

D.3.4   Adaptation options that also mitigate emissions can provide synergies and cost savings in most sectors and system transitions, such as when land management reduces emissions and disaster risk, or when low-carbon buildings are also designed for efficient cooling. Trade-offs between mitigation and adaptation, when limiting global warming to 1.5°C, such as when bioenergy crops, reforestation or afforestation encroach on land needed for agricultural adaptation, can undermine food security, livelihoods, ecosystem functions and services and other aspects of sustainable development. (*high confidence*) {3.4.3, 4.3.2, 4.3.4, 4.4.1, 4.5.2, 4.5.3, 4.5.4}

D.4   Mitigation options consistent with 1.5°C pathways are associated with multiple synergies and trade-offs across the Sustainable Development Goals (SDGs). While the total number of possible synergies exceeds the number of trade-offs, their net effect will depend on the pace and magnitude of changes, the composition of the mitigation portfolio and the management of the transition. (*high confidence*) (Figure SPM.4) {2.5, 4.5, 5.4}

D.4.1   1.5°C pathways have robust synergies particularly for the SDGs 3 (health), 7 (clean energy), 11 (cities and communities), 12 (responsible consumption and production) and 14 (oceans) (*very high confidence*). Some 1.5°C pathways show potential trade-offs with mitigation for SDGs 1 (poverty), 2 (hunger), 6 (water) and 7 (energy access), if not managed carefully (*high confidence*). (Figure SPM.4) {5.4.2; Figure 5.4, Cross-Chapter Boxes 7 and 8 in Chapter 3}

D.4.2   1.5°C pathways that include low energy demand (e.g., see P1 in Figure SPM.3a and SPM.3b), low material consumption, and low GHG-intensive food consumption have the most pronounced synergies and the lowest number of trade-offs with respect to sustainable development and the SDGs (*high confidence*). Such pathways would reduce dependence on CDR. In modelled pathways, sustainable development, eradicating poverty and reducing inequality can support limiting warming to 1.5°C (*high confidence*). (Figure SPM.3b, Figure SPM.4) {2.4.3, 2.5.1, 2.5.3, Figure 2.4, Figure 2.28, 5.4.1, 5.4.2, Figure 5.4}

Indicative linkages between mitigation options and sustainable development using SDGs (The linkages do not show costs and benefits)

Mitigation options deployed in each sector can be associated with potential positive effects (synergies) or negative effects (trade-offs) with the Sustainable Development Goals (SDGs). The degree to which this potential is realized will depend on the selected portfolio of mitigation options, mitigation policy design, and local circumstances and context. Particularly in the energy-demand sector, the potential for synergies is larger than for trade-offs. The bars group individually assessed options by level of confidence and take into account the relative strength of the assessed mitigation-SDG connections.

Length shows strength of connection

The overall **size of the coloured bars** depict **the relative potential** for synergies and trade-offs between the sectoral mitigation options and the SDGs.

Shades show level of confidence

The shades depict **the level of confidence** of the assessed potential for **Trade-offs/Synergies**.

Very High — Low

Figure SPM.4 | Potential synergies and trade-offs between the sectoral portfolio of climate change mitigation options and the Sustainable Development Goals (SDGs). The SDGs serve as an analytical framework for the assessment of the different sustainable development dimensions, which extend beyond the time frame of the 2030 SDG targets. The assessment is based on literature on mitigation options that are considered relevant for 1.5°C. The assessed strength of the SDG interactions is based on the qualitative and quantitative assessment of individual mitigation options listed in Table 5.2. For each mitigation option, the strength of the SDG-connection as well as the associated confidence of the underlying literature (shades of green and red) was assessed. The strength of positive connections (synergies) and negative connections (trade-offs) across all individual options within a sector (see Table 5.2) are aggregated into sectoral potentials for the whole mitigation portfolio. The (white) areas outside the bars, which indicate no interactions, have *low confidence* due to the uncertainty and limited number of studies exploring indirect effects. The strength of the connection considers only the effect of mitigation and does not include benefits of avoided impacts. SDG 13 (climate action) is not listed because mitigation is being considered in terms of interactions with SDGs and not vice versa. The bars denote the strength of the connection, and do not consider the strength of the impact on the SDGs. The energy demand sector comprises behavioural responses, fuel switching and efficiency options in the transport, industry and building sector as well as carbon capture options in the industry sector. Options assessed in the energy supply sector comprise biomass and non-biomass renewables, nuclear, carbon capture and storage (CCS) with bioenergy, and CCS with fossil fuels. Options in the land sector comprise agricultural and forest options, sustainable diets and reduced food waste, soil sequestration, livestock and manure management, reduced deforestation, afforestation and reforestation, and responsible sourcing. In addition to this figure, options in the ocean sector are discussed in the underlying report. {5.4, Table 5.2, Figure 5.2}

Information about the net impacts of mitigation on sustainable development in 1.5°C pathways is available only for a limited number of SDGs and mitigation options. Only a limited number of studies have assessed the benefits of avoided climate change impacts of 1.5°C pathways for the SDGs, and the co-effects of adaptation for mitigation and the SDGs. The assessment of the indicative mitigation potentials in Figure SPM.4 is a step further from AR5 towards a more comprehensive and integrated assessment in the future.

D.4.3   1.5°C and 2°C modelled pathways often rely on the deployment of large-scale land-related measures like afforestation and bioenergy supply, which, if poorly managed, can compete with food production and hence raise food security concerns (*high confidence*). The impacts of carbon dioxide removal (CDR) options on SDGs depend on the type of options and the scale of deployment (*high confidence*). If poorly implemented, CDR options such as BECCS and AFOLU options would lead to trade-offs. Context-relevant design and implementation requires considering people's needs, biodiversity, and other sustainable development dimensions (*very high confidence*). (Figure SPM.4) {5.4.1.3, Cross-Chapter Box 7 in Chapter 3}

D.4.4   Mitigation consistent with 1.5°C pathways creates risks for sustainable development in regions with high dependency on fossil fuels for revenue and employment generation (*high confidence*). Policies that promote diversification of the economy and the energy sector can address the associated challenges (*high confidence*). {5.4.1.2, Box 5.2}

D.4.5   Redistributive policies across sectors and populations that shield the poor and vulnerable can resolve trade-offs for a range of SDGs, particularly hunger, poverty and energy access. Investment needs for such complementary policies are only a small fraction of the overall mitigation investments in 1.5°C pathways. (*high confidence*) {2.4.3, 5.4.2, Figure 5.5}

D.5   Limiting the risks from global warming of 1.5°C in the context of sustainable development and poverty eradication implies system transitions that can be enabled by an increase of adaptation and mitigation investments, policy instruments, the acceleration of technological innovation and behaviour changes (*high confidence*). {2.3, 2.4, 2.5, 3.2, 4.2, 4.4, 4.5, 5.2, 5.5, 5.6}

D.5.1   Directing finance towards investment in infrastructure for mitigation and adaptation could provide additional resources. This could involve the mobilization of private funds by institutional investors, asset managers and development or investment banks, as well as the provision of public funds. Government policies that lower the risk of low-emission and adaptation investments can facilitate the mobilization of private funds and enhance the effectiveness of other public policies. Studies indicate a number of challenges, including access to finance and mobilization of funds. (*high confidence*) {2.5.1, 2.5.2, 4.4.5}

D.5.2   Adaptation finance consistent with global warming of 1.5°C is difficult to quantify and compare with 2°C. Knowledge gaps include insufficient data to calculate specific climate resilience-enhancing investments from the provision of currently underinvested basic infrastructure. Estimates of the costs of adaptation might be lower at global warming of 1.5°C than for 2°C. Adaptation needs have typically been supported by public sector sources such as national and subnational government budgets, and in developing countries together with support from development assistance, multilateral development banks, and United Nations Framework Convention on Climate Change channels (*medium confidence*). More recently there is a growing understanding of the scale and increase in non-governmental organizations and private funding in some regions (*medium confidence*). Barriers include the scale of adaptation financing, limited capacity and access to adaptation finance (*medium confidence*). {4.4.5, 4.6}

D.5.3   Global model pathways limiting global warming to 1.5°C are projected to involve the annual average investment needs in the energy system of around 2.4 trillion USD2010 between 2016 and 2035, representing about 2.5% of the world GDP (*medium confidence*). {4.4.5, Box 4.8}

D.5.4   Policy tools can help mobilize incremental resources, including through shifting global investments and savings and through market and non-market based instruments as well as accompanying measures to secure the equity of the transition, acknowledging the challenges related with implementation, including those of energy costs, depreciation of assets and impacts on international competition, and utilizing the opportunities to maximize co-benefits (*high confidence*). {1.3.3, 2.3.4, 2.3.5, 2.5.1, 2.5.2, Cross-Chapter Box 8 in Chapter 3, Cross-Chapter Box 11 in Chapter 4, 4.4.5, 5.5.2}

D.5.5   The systems transitions consistent with adapting to and limiting global warming to 1.5°C include the widespread adoption of new and possibly disruptive technologies and practices and enhanced climate-driven innovation. These imply enhanced technological innovation capabilities, including in industry and finance. Both national innovation policies and international cooperation can contribute to the development, commercialization and widespread adoption of mitigation and adaptation technologies. Innovation policies may be more effective when they combine public support for research and development with policy mixes that provide incentives for technology diffusion. (*high confidence*) {4.4.4, 4.4.5}.

D.5.6   Education, information, and community approaches, including those that are informed by indigenous knowledge and local knowledge, can accelerate the wide-scale behaviour changes consistent with adapting to and limiting global warming to 1.5°C. These approaches are more effective when combined with other policies and tailored to the motivations, capabilities and resources of specific actors and contexts (*high confidence*). Public acceptability can enable or inhibit the implementation of policies and measures to limit global warming to 1.5°C and to adapt to the consequences. Public acceptability depends on the individual's evaluation of expected policy consequences, the perceived fairness of the distribution of these consequences, and perceived fairness of decision procedures (*high confidence*). {1.1, 1.5, 4.3.5, 4.4.1, 4.4.3, Box 4.3, 5.5.3, 5.6.5}

D.6   Sustainable development supports, and often enables, the fundamental societal and systems transitions and transformations that help limit global warming to 1.5°C. Such changes facilitate the pursuit of climate-resilient development pathways that achieve ambitious mitigation and adaptation in conjunction with poverty eradication and efforts to reduce inequalities (*high confidence*). {Box 1.1, 1.4.3, Figure 5.1, 5.5.3, Box 5.3}

D.6.1   Social justice and equity are core aspects of climate-resilient development pathways that aim to limit global warming to 1.5°C as they address challenges and inevitable trade-offs, widen opportunities, and ensure that options, visions, and values are deliberated, between and within countries and communities, without making the poor and disadvantaged worse off (*high confidence*). {5.5.2, 5.5.3, Box 5.3, Figure 5.1, Figure 5.6, Cross-Chapter Boxes 12 and 13 in Chapter 5}

D.6.2   The potential for climate-resilient development pathways differs between and within regions and nations, due to different development contexts and systemic vulnerabilities (*very high confidence*). Efforts along such pathways to date have been limited (*medium confidence*) and enhanced efforts would involve strengthened and timely action from all countries and non-state actors (*high confidence*). {5.5.1, 5.5.3, Figure 5.1}

D.6.3   Pathways that are consistent with sustainable development show fewer mitigation and adaptation challenges and are associated with lower mitigation costs. The large majority of modelling studies could not construct pathways characterized by lack of international cooperation, inequality and poverty that were able to limit global warming to 1.5°C. (*high confidence*) {2.3.1, 2.5.1, 2.5.3, 5.5.2}

D.7 **Strengthening the capacities for climate action of national and sub-national authorities, civil society, the private sector, indigenous peoples and local communities can support the implementation of ambitious actions implied by limiting global warming to 1.5°C (*high confidence*). International cooperation can provide an enabling environment for this to be achieved in all countries and for all people, in the context of sustainable development. International cooperation is a critical enabler for developing countries and vulnerable regions (*high confidence*). {1.4, 2.3, 2.5, 4.2, 4.4, 4.5, 5.3, 5.4, 5.5, 5.6, 5, Box 4.1, Box 4.2, Box 4.7, Box 5.3, Cross-Chapter Box 9 in Chapter 4, Cross-Chapter Box 13 in Chapter 5}**

D.7.1   Partnerships involving non-state public and private actors, institutional investors, the banking system, civil society and scientific institutions would facilitate actions and responses consistent with limiting global warming to 1.5°C (*very high confidence*). {1.4, 4.4.1, 4.2.2, 4.4.3, 4.4.5, 4.5.3, 5.4.1, 5.6.2, Box 5.3}.

D.7.2   Cooperation on strengthened accountable multilevel governance that includes non-state actors such as industry, civil society and scientific institutions, coordinated sectoral and cross-sectoral policies at various governance levels, gender-sensitive policies, finance including innovative financing, and cooperation on technology development and transfer can ensure participation, transparency, capacity building and learning among different players (*high confidence*). {2.5.1, 2.5.2, 4.2.2, 4.4.1, 4.4.2, 4.4.3, 4.4.4, 4.4.5, 4.5.3, Cross-Chapter Box 9 in Chapter 4, 5.3.1, 5.5.3, Cross-Chapter Box 13 in Chapter 5, 5.6.1, 5.6.3}

D.7.3   International cooperation is a critical enabler for developing countries and vulnerable regions to strengthen their action for the implementation of 1.5°C-consistent climate responses, including through enhancing access to finance and technology and enhancing domestic capacities, taking into account national and local circumstances and needs (*high confidence*). {2.3.1, 2.5.1, 4.4.1, 4.4.2, 4.4.4, 4.4.5, 5.4.1 5.5.3, 5.6.1, Box 4.1, Box 4.2, Box 4.7}.

D.7.4   Collective efforts at all levels, in ways that reflect different circumstances and capabilities, in the pursuit of limiting global warming to 1.5°C, taking into account equity as well as effectiveness, can facilitate strengthening the global response to climate change, achieving sustainable development and eradicating poverty (*high confidence*). {1.4.2, 2.3.1, 2.5.1, 2.5.2, 2.5.3, 4.2.2, 4.4.1, 4.4.2, 4.4.3, 4.4.4, 4.4.5, 4.5.3, 5.3.1, 5.4.1, 5.5.3, 5.6.1, 5.6.2, 5.6.3}

Box SPM.1: Core Concepts Central to this Special Report

Global mean surface temperature (GMST): Estimated global average of near-surface air temperatures over land and sea ice, and sea surface temperatures over ice-free ocean regions, with changes normally expressed as departures from a value over a specified reference period. When estimating changes in GMST, near-surface air temperature over both land and oceans are also used.[19] {1.2.1.1}

Pre-industrial: The multi-century period prior to the onset of large-scale industrial activity around 1750. The reference period 1850–1900 is used to approximate pre-industrial GMST. {1.2.1.2}

Global warming: The estimated increase in GMST averaged over a 30-year period, or the 30-year period centred on a particular year or decade, expressed relative to pre-industrial levels unless otherwise specified. For 30-year periods that span past and future years, the current multi-decadal warming trend is assumed to continue. {1.2.1}

Net zero CO_2 emissions: Net zero carbon dioxide (CO_2) emissions are achieved when anthropogenic CO_2 emissions are balanced globally by anthropogenic CO_2 removals over a specified period.

Carbon dioxide removal (CDR): Anthropogenic activities removing CO_2 from the atmosphere and durably storing it in geological, terrestrial, or ocean reservoirs, or in products. It includes existing and potential anthropogenic enhancement of biological or geochemical sinks and direct air capture and storage, but excludes natural CO_2 uptake not directly caused by human activities.

Total carbon budget: Estimated cumulative net global anthropogenic CO_2 emissions from the pre-industrial period to the time that anthropogenic CO_2 emissions reach net zero that would result, at some probability, in limiting global warming to a given level, accounting for the impact of other anthropogenic emissions. {2.2.2}

Remaining carbon budget: Estimated cumulative net global anthropogenic CO_2 emissions from a given start date to the time that anthropogenic CO_2 emissions reach net zero that would result, at some probability, in limiting global warming to a given level, accounting for the impact of other anthropogenic emissions. {2.2.2}

Temperature overshoot: The temporary exceedance of a specified level of global warming.

Emission pathways: In this Summary for Policymakers, the modelled trajectories of global anthropogenic emissions over the 21st century are termed emission pathways. Emission pathways are classified by their temperature trajectory over the 21st century: pathways giving at least 50% probability based on current knowledge of limiting global warming to below 1.5°C are classified as 'no overshoot'; those limiting warming to below 1.6°C and returning to 1.5°C by 2100 are classified as '1.5°C limited-overshoot'; while those exceeding 1.6°C but still returning to 1.5°C by 2100 are classified as 'higher-overshoot'.

Impacts: Effects of climate change on human and natural systems. Impacts can have beneficial or adverse outcomes for livelihoods, health and well-being, ecosystems and species, services, infrastructure, and economic, social and cultural assets.

Risk: The potential for adverse consequences from a climate-related hazard for human and natural systems, resulting from the interactions between the hazard and the vulnerability and exposure of the affected system. Risk integrates the likelihood of exposure to a hazard and the magnitude of its impact. Risk also can describe the potential for adverse consequences of adaptation or mitigation responses to climate change.

Climate-resilient development pathways (CRDPs): Trajectories that strengthen sustainable development at multiple scales and efforts to eradicate poverty through equitable societal and systems transitions and transformations while reducing the threat of climate change through ambitious mitigation, adaptation and climate resilience.

---

19 Past IPCC reports, reflecting the literature, have used a variety of approximately equivalent metrics of GMST change.

Technical Summary

TS

Technical Summary

Coordinating Lead Authors:
Myles R. Allen (UK), Heleen de Coninck (Netherlands/EU), Opha Pauline Dube (Botswana), Ove Hoegh-Guldberg (Australia), Daniela Jacob (Germany), Kejun Jiang (China), Aromar Revi (India), Joeri Rogelj (Belgium/Austria), Joyashree Roy (India), Drew Shindell (USA), William Solecki (USA), Michael Taylor (Jamaica), Petra Tschakert (Australia/Austria), Henri Waisman (France)

Lead Authors:
Sharina Abdul Halim (Malaysia), Philip Antwi-Agyei (Ghana), Fernando Aragón–Durand (Mexico), Mustafa Babiker (Sudan), Paolo Bertoldi (Italy), Marco Bindi (Italy), Sally Brown (UK), Marcos Buckeridge (Brazil), Ines Camilloni (Argentina), Anton Cartwright (South Africa), Wolfgang Cramer (France/Germany), Purnamita Dasgupta (India), Arona Diedhiou (Ivory Coast/Senegal), Riyanti Djalante (Japan/Indonesia), Wenjie Dong (China), Kristie L. Ebi (USA), Francois Engelbrecht (South Africa), Solomone Fifita (Fiji), James Ford (UK/Canada), Piers Forster (UK), Sabine Fuss (Germany), Bronwyn Hayward (New Zealand), Jean-Charles Hourcade (France), Veronika Ginzburg (Russia), Joel Guiot (France), Collins Handa (Kenya), Yasuaki Hijioka (Japan), Stephen Humphreys (UK/Ireland), Mikiko Kainuma (Japan), Jatin Kala (Australia), Markku Kanninen (Finland), Haroon Kheshgi (USA), Shigeki Kobayashi (Japan), Elmar Kriegler (Germany), Debora Ley (Guatemala/Mexico), Diana Liverman (USA), Natalie Mahowald (USA), Reinhard Mechler (Germany), Shagun Mehrotra (USA/India), Yacob Mulugetta (UK/Ethiopia), Luis Mundaca (Sweden/Chile), Peter Newman (Australia), Chukwumerije Okereke (UK/Nigeria), Antony Payne (UK), Rosa Perez (Philippines), Patricia Fernanda Pinho (Brazil), Anastasia Revokatova (Russian Federation), Keywan Riahi (Austria), Seth Schultz (USA), Roland Séférian (France), Sonia I. Seneviratne (Switzerland), Linda Steg (Netherlands), Avelino G. Suarez Rodriguez (Cuba), Taishi Sugiyama (Japan), Adelle Thomas (Bahamas), Maria Virginia Vilariño (Argentina), Morgan Wairiu (Solomon Islands), Rachel Warren (UK), Guangsheng Zhou (China), Kirsten Zickfeld (Canada/Germany)

Contributing Authors:
Michelle Achlatis (Australia/Greece), Lisa V. Alexander (Australia), Malcolm Araos (Maldives/Canada), Stefan Bakker (Netherlands), Mook Bangalore (USA), Amir Bazaz (India), Ella Belfer (Canada), Tim Benton (UK), Peter Berry (Canada), Bishwa Bhaskar Choudhary (India), Christopher Boyer (USA), Lorenzo Brilli (Italy), Katherine Calvin (USA), William Cheung (Canada), Sarah Connors (France/UK), Joana Correia de Oliveira de Portugal Pereira (UK/Portugal), Marlies Craig (South Africa), Dipak Dasgupta (India), Kiane de Kleijne (Netherlands/EU), Maria del Mar Zamora Dominguez (Mexico), Michel den Elzen (Netherlands), Haile Eakin (USA), Oreane Edelenbosch (Netherlands/Italy), Neville Ellis (Australia), Johannes Emmerling (Italy/Germany), Jason Evans (Australia), Maria Figueroa (Denmark/Venezuela), Dominique Finon (France), Hubertus Fisher (Switzerland), Klaus Fraedrich (Germany), Jan Fuglestvedt (Norway), Anjani Ganase (Trinidad and Tobago), Thomas Gasser (Austria/France), Jean Pierre Gattuso (France), Frédéric Ghersi (France), Nathan Gillett (Canada), Adriana Grandis (Brazil), Peter Greve (Germany/Austria), Tania Guillén Bolaños (Germany/Nicaragua), Mukesh Gupta (India), Amaha Medhin Haileselassie (Ethiopia), Naota Hanasaki (Japan), Tomoko Hasegawa (Japan), Eamon Haughey (Ireland), Katie Hayes (Canada), Chenmin He (China), Edgar Hertwich (USA/Austria), Diana Hinge Salili (Vanuatu), Annette Hirsch (Australia/Switzerland), Lena Höglund-Isaksson (Austria/Sweden), Daniel Huppmann (Austria), Saleemul Huq (Bangladesh/UK), Rachel James (UK), Chris Jones (UK), Thomas Jung (Germany), Richard Klein (Netherlands/Germany), Gerhard Krinner (France), David Lawrence (USA), Tim Lenton (UK), Gunnar Luderer (Germany), Peter Marcotullio (USA), Anil Markandya (Spain/UK), Omar Massera (Mexico), David L. McCollum (Austria/USA), Kathleen McInnes (Australia), Malte Meinshausen (Australia/Germany), Katrin J. Meissner (Australia), Richard Millar (UK), Katja Mintenbeck (Germany), Dann Mitchell (UK), Alan C. Mix (USA), Dirk Notz (Germany), Leonard Nurse (Barbados), Andrew Okem (Nigeria), Lennart Olsson (Sweden), Carolyn Opio (Uganda), Michael Oppenheimer (USA), Karen Paiva Henrique (Brazil), Simon Parkinson (Canada), Shlomit Paz (Israel), Juliane Petersen (Germany), Jan Petzold (Germany), Maxime Plazzotta (France), Alexander Popp (Germany), Swantje Preuschmann (Germany), Pallav Purohit (Austria/India), Graciela Raga (Mexico/Argentina), Mohammad Feisal Rahman (Bangladesh), Andy Reisinger (New Zealand), Kevon Rhiney (Jamaica), Aurélien Ribes (France), Mark Richardson (USA/UK), Wilfried Rickels (Germany), Timmons Roberts (USA), Maisa Rojas (Chile), Harry Saunders (Canada/USA), Christina Schädel (USA/Switzerland), Hanna Scheuffele (Germany), Lisa Schipper (UK/Sweden), Carl-Friedrich Schleussner (Germany), Jörn Schmidt (Germany), Daniel Scott (Canada), Jana Sillmann (Germany/Norway), Chandni Singh (India), Raphael Slade (UK), Christopher Smith (UK), Pete Smith (UK), Shreya Some (India), Gerd Sparovek (Brazil), Will Steffen (Australia), Kimberly Stephenson (Jamaica), Tannecia Stephenson (Jamaica), Pablo Suarez (Argentina), Mouhamadou B. Sylla (Senegal), Nenenteiti Teariki-Ruatu (Kiribati), Mark Tebboth (UK), Peter Thorne (Ireland/UK), Evelina Trutnevyte (Switzerland/Lithuania), Penny Urquhart (South Africa), Arjan van Rooij (Netherlands), Anne M. van Valkengoed (Netherlands), Robert Vautard (France), Richard Wartenburger (Germany/Switzerland), Michael Wehner (USA), Margaretha Wewerinke-Singh (Netherlands), Nora M. Weyer (Germany), Felicia Whyte (Jamaica), Lini Wollenberg (USA), Yang Xiu (China), Gary Yohe (USA), Xuebin Zhang (Canada), Wenji Zhou (Austria/China), Robert B. Zougmoré (Burkina Faso/Mali)

Review Editors:

Amjad Abdulla (Maldives), Rizaldi Boer (Indonesia), Ismail Elgizouli Idris (Sudan), Andreas Fischlin (Switzerland), Greg Flato (Canada), Jan Fuglestvedt (Norway), Xuejie Gao (China), Mark Howden (Australia), Svitlana Krakovska (Ukraine), Ramon Pichs Madruga (Cuba), Jose Antonio Marengo (Brazil/Peru), Rachid Mrabet (Morocco), Joy Pereira (Malaysia), Roberto Sanchez (Mexico), Roberto Schaeffer (Brazil), Boris Sherstyukov (Russian Federation), Diana Ürge-Vorsatz (Hungary)

TS

Chapter Scientists:

Daniel Huppmann (Austria), Tania Guillén Bolaños (Germany/Nicaragua), Neville Ellis (Australia), Kiane de Kleijne (Netherlands/EU), Richard Millar (UK), Chandni Singh (India), Chris Smith (UK)

This Technical Summary should be cited as:

Allen, M.R., H. de Coninck, O.P. Dube, O. Hoegh-Guldberg, D. Jacob, K. Jiang, A. Revi, J. Rogelj, J. Roy, D. Shindell, W. Solecki, M. Taylor, P. Tschakert, H. Waisman, S. Abdul Halim, P. Antwi-Agyei, F. Aragón-Durand, M. Babiker, P. Bertoldi, M. Bindi, S. Brown, M. Buckeridge, I. Camilloni, A. Cartwright, W. Cramer, P. Dasgupta, A. Diedhiou, R. Djalante, W. Dong, K.L. Ebi, F. Engelbrecht, S. Fifita, J. Ford, P. Forster, S. Fuss, V. Ginzburg, J. Guiot, C. Handa, B. Hayward, Y. Hijioka, J.-C. Hourcade, S. Humphreys, M. Kainuma, J. Kala, M. Kanninen, H. Kheshgi, S. Kobayashi, E. Kriegler, D. Ley, D. Liverman, N. Mahowald, R. Mechler, S. Mehrotra, Y. Mulugetta, L. Mundaca, P. Newman, C. Okereke, A. Payne, R. Perez, P.F. Pinho, A. Revokatova, K. Riahi, S. Schultz, R. Séférian, S.I. Seneviratne, L. Steg, A.G. Suarez Rodriguez, T. Sugiyama, A. Thomas, M.V. Vilariño, M. Wairiu, R. Warren, K. Zickfeld, and G. Zhou, 2018: Technical Summary. In: *Global Warming of 1.5°C. An IPCC Special Report on the impacts of global warming of 1.5°C above pre-industrial levels and related global greenhouse gas emission pathways, in the context of strengthening the global response to the threat of climate change, sustainable development, and efforts to eradicate poverty* [Masson-Delmotte, V., P. Zhai, H.-O. Pörtner, D. Roberts, J. Skea, P.R. Shukla, A. Pirani, W. Moufouma-Okia, C. Péan, R. Pidcock, S. Connors, J.B.R. Matthews, Y. Chen, X. Zhou, M.I. Gomis, E. Lonnoy, T. Maycock, M. Tignor, and T. Waterfield (eds.)]. Cambridge University Press, Cambridge, UK and New York, NY, USA, pp. 27-46. https://doi.org/10.1017/9781009157940.002.

Table of Contents

TS

TS.1 Framing and Context

This chapter frames the context, knowledge-base and assessment approaches used to understand the impacts of 1.5°C global warming above pre-industrial levels and related global greenhouse gas emission pathways, building on the IPCC Fifth Assessment Report (AR5), in the context of strengthening the global response to the threat of climate change, sustainable development and efforts to eradicate poverty.

Human-induced warming reached approximately 1°C (*likely* between 0.8°C and 1.2°C) above pre-industrial levels in 2017, increasing at 0.2°C (*likely* between 0.1°C and 0.3°C) per decade (*high confidence*). Global warming is defined in this report as an increase in combined surface air and sea surface temperatures averaged over the globe and over a 30-year period. Unless otherwise specified, warming is expressed relative to the period 1850–1900, used as an approximation of pre-industrial temperatures in AR5. For periods shorter than 30 years, warming refers to the estimated average temperature over the 30 years centred on that shorter period, accounting for the impact of any temperature fluctuations or trend within those 30 years. Accordingly, warming from pre-industrial levels to the decade 2006–2015 is assessed to be 0.87°C (*likely* between 0.75°C and 0.99°C). Since 2000, the estimated level of human-induced warming has been equal to the level of observed warming with a likely range of ±20% accounting for uncertainty due to contributions from solar and volcanic activity over the historical period (*high confidence*). {1.2.1}

Warming greater than the global average has already been experienced in many regions and seasons, with higher average warming over land than over the ocean (*high confidence*). Most land regions are experiencing greater warming than the global average, while most ocean regions are warming at a slower rate. Depending on the temperature dataset considered, 20–40% of the global human population live in regions that, by the decade 2006–2015, had already experienced warming of more than 1.5°C above pre-industrial in at least one season (*medium confidence*). {1.2.1, 1.2.2}

Past emissions alone are unlikely to raise global-mean temperature to 1.5°C above pre-industrial levels (*medium confidence*), but past emissions do commit to other changes, such as further sea level rise (*high confidence*). If all anthropogenic emissions (including aerosol-related) were reduced to zero immediately, any further warming beyond the 1°C already experienced would *likely* be less than 0.5°C over the next two to three decades (*high confidence*), and *likely* less than 0.5°C on a century time scale (*medium confidence*), due to the opposing effects of different climate processes and drivers. A warming greater than 1.5°C is therefore not geophysically unavoidable: whether it will occur depends on future rates of emission reductions. {1.2.3, 1.2.4}

1.5°C emission pathways are defined as those that, given current knowledge of the climate response, provide a one-in-two to two-in-three chance of warming either remaining below 1.5°C or returning to 1.5°C by around 2100 following an overshoot. Overshoot pathways are characterized by the peak magnitude of the overshoot, which may have implications for impacts. All 1.5°C pathways involve limiting cumulative emissions of long-lived greenhouse gases, including carbon dioxide and nitrous oxide, and substantial reductions in other climate forcers (*high confidence*). Limiting cumulative emissions requires either reducing net global emissions of long-lived greenhouse gases to zero before the cumulative limit is reached, or net negative global emissions (anthropogenic removals) after the limit is exceeded. {1.2.3, 1.2.4, Cross-Chapter Boxes 1 and 2}

This report assesses projected impacts at a global average warming of 1.5°C and higher levels of warming. Global warming of 1.5°C is associated with global average surface temperatures fluctuating naturally on either side of 1.5°C, together with warming substantially greater than 1.5°C in many regions and seasons (*high confidence*), all of which must be considered in the assessment of impacts. Impacts at 1.5°C of warming also depend on the emission pathway to 1.5°C. Very different impacts result from pathways that remain below 1.5°C versus pathways that return to 1.5°C after a substantial overshoot, and when temperatures stabilize at 1.5°C versus a transient warming past 1.5°C (*medium confidence*). {1.2.3, 1.3}

Ethical considerations, and the principle of equity in particular, are central to this report, recognizing that many of the impacts of warming up to and beyond 1.5°C, and some potential impacts of mitigation actions required to limit warming to 1.5°C, fall disproportionately on the poor and vulnerable (*high confidence*). Equity has procedural and distributive dimensions and requires fairness in burden sharing both between generations and between and within nations. In framing the objective of holding the increase in the global average temperature rise to well below 2°C above pre-industrial levels, and to pursue efforts to limit warming to 1.5°C, the Paris Agreement associates the principle of equity with the broader goals of poverty eradication and sustainable development, recognising that effective responses to climate change require a global collective effort that may be guided by the 2015 United Nations Sustainable Development Goals. {1.1.1}

Climate adaptation refers to the actions taken to manage impacts of climate change by reducing vulnerability and exposure to its harmful effects and exploiting any potential benefits. Adaptation takes place at international, national and local levels. Subnational jurisdictions and entities, including urban and rural municipalities, are key to developing and reinforcing measures for reducing weather- and climate-related risks. Adaptation implementation faces several barriers including lack of up-to-date and locally relevant information, lack of finance and technology, social values and attitudes, and institutional constraints (*high confidence*). Adaptation is more likely to contribute to sustainable development when policies align with mitigation and poverty eradication goals (*medium confidence*). {1.1, 1.4}

Ambitious mitigation actions are indispensable to limit warming to 1.5°C while achieving sustainable development and poverty eradication (*high confidence*). Ill-designed responses, however, could pose challenges especially – but not exclusively – for countries and regions contending with poverty and those requiring significant transformation of their energy systems. This report focuses on 'climate-resilient development pathways', which aim to meet the goals of sustainable development, including climate adaptation and mitigation, poverty eradication and reducing inequalities. But any feasible pathway that remains within 1.5°C involves synergies and trade-offs (*high confidence*). Significant uncertainty remains as to which pathways are more consistent with the principle of equity. {1.1.1, 1.4}

Multiple forms of knowledge, including scientific evidence, narrative scenarios and prospective pathways, inform the understanding of 1.5°C. This report is informed by traditional evidence of the physical climate system and associated impacts and vulnerabilities of climate change, together with knowledge drawn from the perceptions of risk and the experiences of climate impacts and governance systems. Scenarios and pathways are used to explore conditions enabling goal-oriented futures while recognizing the significance of ethical considerations, the principle of equity, and the societal transformation needed. {1.2.3, 1.5.2}

There is no single answer to the question of whether it is feasible to limit warming to 1.5°C and adapt to the consequences. Feasibility is considered in this report as the capacity of a system as a whole to achieve a specific outcome. The global transformation that would be needed to limit warming to 1.5°C requires enabling conditions that reflect the links, synergies and trade-offs between mitigation, adaptation and sustainable development. These enabling conditions are assessed across many dimensions of feasibility – geophysical, environmental-ecological, technological, economic, socio-cultural and institutional – that may be considered through the unifying lens of the Anthropocene, acknowledging profound, differential but increasingly geologically significant human influences on the Earth system as a whole. This framing also emphasises the global interconnectivity of past, present and future human–environment relations, highlighting the need and opportunities for integrated responses to achieve the goals of the Paris Agreement. {1.1, Cross-Chapter Box 1}

TS.2 Mitigation Pathways Compatible with 1.5°C in the Context of Sustainable Development

This chapter assesses mitigation pathways consistent with limiting warming to 1.5°C above pre-industrial levels. In doing so, it explores the following key questions: What role do CO_2 and non-CO_2 emissions play? {2.2, 2.3, 2.4, 2.6} To what extent do 1.5°C pathways involve overshooting and returning below 1.5°C during the 21st century? {2.2, 2.3} What are the implications for transitions in energy, land use and sustainable development? {2.3, 2.4, 2.5} How do policy frameworks affect the ability to limit warming to 1.5°C? {2.3, 2.5} What are the associated knowledge gaps? {2.6}

The assessed pathways describe integrated, quantitative evolutions of all emissions over the 21st century associated with global energy and land use and the world economy. The assessment is contingent upon available integrated assessment literature and model assumptions, and is complemented by other studies with different scope, for example, those focusing on individual sectors. In recent years, integrated mitigation studies have improved the characterizations of mitigation pathways. However, limitations remain, as climate damages, avoided impacts, or societal co-benefits of the modelled transformations remain largely unaccounted for, while concurrent rapid technological changes, behavioural aspects, and uncertainties about input data present continuous challenges. (*high confidence*) {2.1.3, 2.3, 2.5.1, 2.6, Technical Annex 2}

The Chances of Limiting Warming to 1.5°C and the Requirements for Urgent Action

Pathways consistent with 1.5°C of warming above pre-industrial levels can be identified under a range of assumptions about economic growth, technology developments and lifestyles. However, lack of global cooperation, lack of governance of the required energy and land transformation, and increases in resource-intensive consumption are key impediments to achieving 1.5°C pathways. Governance challenges have been related to scenarios with high inequality and high population growth in the 1.5°C pathway literature. {2.3.1, 2.3.2, 2.5}

Under emissions in line with current pledges under the Paris Agreement (known as Nationally Determined Contributions, or NDCs), global warming is expected to surpass 1.5°C above pre-industrial levels, even if these pledges are supplemented with very challenging increases in the scale and ambition of mitigation after 2030 (*high confidence*). This increased action would need to achieve net zero CO_2 emissions in less than 15 years. Even if this is achieved, temperatures would only be expected to remain below the 1.5°C threshold if the actual geophysical response ends up being towards the low end of the currently estimated uncertainty range. Transition challenges as well as identified trade-offs can be reduced if global emissions peak before 2030 and marked emissions reductions compared to today are already achieved by 2030. {2.2, 2.3.5, Cross-Chapter Box 11 in Chapter 4}

Limiting warming to 1.5°C depends on greenhouse gas (GHG) emissions over the next decades, where lower GHG emissions in 2030 lead to a higher chance of keeping peak warming to 1.5°C (*high confidence*). Available pathways that aim for no or limited (less than 0.1°C) overshoot of 1.5°C keep GHG emissions in 2030 to 25–30 GtCO2e yr^{-1} in 2030 (interquartile range). This contrasts with median estimates for current unconditional NDCs of 52–58 GtCO$_2$e yr^{-1} in 2030. Pathways that aim for limiting warming to 1.5°C by 2100 after a temporary temperature overshoot rely on large-scale deployment of carbon dioxide removal (CDR) measures, which are uncertain and entail clear risks. In model pathways with no or limited overshoot of 1.5°C, global net anthropogenic CO$_2$ emissions decline by about 45% from 2010 levels by 2030 (40–60% interquartile range), reaching net zero around 2050 (2045–2055 interquartile range). For limiting global warming to below 2°C with at least 66% probability CO$_2$ emissions are projected to decline by about 25% by 2030 in most pathways (10–30% interquartile range) and reach net zero around 2070 (2065–2080 interquartile range).[1] {2.2, 2.3.3, 2.3.5, 2.5.3, Cross-Chapter Boxes 6 in Chapter 3 and 9 in Chapter 4, 4.3.7}

Limiting warming to 1.5°C implies reaching net zero CO$_2$ emissions globally around 2050 and concurrent deep reductions in emissions of non-CO$_2$ forcers, particularly methane (*high confidence*). Such mitigation pathways are characterized by energy-demand reductions, decarbonization of electricity and other fuels, electrification of energy end use, deep reductions in agricultural emissions, and some form of CDR with carbon storage on land or sequestration in geological reservoirs. Low energy demand and low demand for land- and GHG-intensive consumption goods facilitate limiting warming to as close as possible to 1.5°C. {2.2.2, 2.3.1, 2.3.5, 2.5.1, Cross-Chapter Box 9 in Chapter 4}.

In comparison to a 2°C limit, the transformations required to limit warming to 1.5°C are qualitatively similar but more pronounced and rapid over the next decades (*high confidence*). 1.5°C implies very ambitious, internationally cooperative policy environments that transform both supply and demand (*high confidence*). {2.3, 2.4, 2.5}

Policies reflecting a high price on emissions are necessary in models to achieve cost-effective 1.5°C pathways (*high confidence*). Other things being equal, modelling studies suggest the global average discounted marginal abatement costs for limiting warming to 1.5°C being about 3–4 times higher compared to 2°C over the 21st century, with large variations across models and socio-economic and policy assumptions. Carbon pricing can be imposed directly or implicitly by regulatory policies. Policy instruments, like technology policies or performance standards, can complement explicit carbon pricing in specific areas. {2.5.1, 2.5.2, 4.4.5}

Limiting warming to 1.5°C requires a marked shift in investment patterns (*medium confidence*). Additional annual average energy-related investments for the period 2016 to 2050 in pathways limiting warming to 1.5°C compared to pathways without new climate policies beyond those in place today (i.e., baseline) are estimated to be around 830 billion USD2010 (range of 150 billion to 1700 billion USD2010 across six models). Total energy-related investments increase by about 12% (range of 3% to 24%) in 1.5°C pathways relative to 2°C pathways. Average annual investment in low-carbon energy technologies and energy efficiency are upscaled by roughly a factor of six (range of factor of 4 to 10) by 2050 compared to 2015, overtaking fossil investments globally by around 2025 (*medium confidence*). Uncertainties and strategic mitigation portfolio choices affect the magnitude and focus of required investments. {2.5.2}

Future Emissions in 1.5°C Pathways

Mitigation requirements can be quantified using carbon budget approaches that relate cumulative CO$_2$ emissions to global mean temperature increase. Robust physical understanding underpins this relationship, but uncertainties become increasingly relevant as a specific temperature limit is approached. These uncertainties relate to the transient climate response to cumulative carbon emissions (TCRE), non-CO$_2$ emissions, radiative forcing and response, potential additional Earth system feedbacks (such as permafrost thawing), and historical emissions and temperature. {2.2.2, 2.6.1}

Cumulative CO$_2$ emissions are kept within a budget by reducing global annual CO$_2$ emissions to net zero. This assessment suggests a remaining budget of about 420 GtCO$_2$ for a two-thirds chance of limiting warming to 1.5°C, and of about 580 GtCO$_2$ for an even chance (*medium confidence*). The remaining carbon budget is defined here as cumulative CO$_2$ emissions from the start of 2018 until the time of net zero global emissions for global warming defined as a change in global near-surface air temperatures. Remaining budgets applicable to 2100 would be approximately 100 GtCO$_2$ lower than this to account for permafrost thawing and potential methane release from wetlands in the future, and more thereafter. These estimates come with an additional geophysical uncertainty of at least ±400 GtCO$_2$, related to non-CO$_2$ response and TCRE distribution. Uncertainties in the level of historic warming contribute ±250 GtCO$_2$. In addition, these estimates can vary by ±250 GtCO$_2$ depending on non-CO$_2$ mitigation strategies as found in available pathways. {2.2.2, 2.6.1}

Staying within a remaining carbon budget of 580 GtCO$_2$ implies that CO$_2$ emissions reach carbon neutrality in about 30 years, reduced to 20 years for a 420 GtCO$_2$ remaining carbon budget (*high confidence*). The ±400 GtCO$_2$ geophysical uncertainty range surrounding a carbon budget translates into a variation of this timing of carbon neutrality of roughly ±15–20 years. If emissions do not start declining in the next decade, the point of carbon neutrality would need to be reached at least two decades earlier to remain within the same carbon budget. {2.2.2, 2.3.5}

Non-CO$_2$ emissions contribute to peak warming and thus affect the remaining carbon budget. The evolution of methane and sulphur dioxide emissions strongly influences the chances of limiting warming to 1.5°C. In the near-term, a

---

[1] Kyoto-GHG emissions in this statement are aggregated with GWP-100 values of the IPCC Second Assessment Report.

weakening of aerosol cooling would add to future warming, but can be tempered by reductions in methane emissions (*high confidence*). Uncertainty in radiative forcing estimates (particularly aerosol) affects carbon budgets and the certainty of pathway categorizations. Some non-CO_2 forcers are emitted alongside CO_2, particularly in the energy and transport sectors, and can be largely addressed through CO_2 mitigation. Others require specific measures, for example, to target agricultural nitrous oxide (N_2O) and methane (CH_4), some sources of black carbon, or hydrofluorocarbons (*high confidence*). In many cases, non-CO2 emissions reductions are similar in 2°C pathways, indicating reductions near their assumed maximum potential by integrated assessment models. Emissions of N_2O and NH_3 increase in some pathways with strongly increased bioenergy demand. {2.2.2, 2.3.1, 2.4.2, 2.5.3}

The Role of Carbon Dioxide Removal (CDR)

All analysed pathways limiting warming to 1.5°C with no or limited overshoot use CDR to some extent to neutralize emissions from sources for which no mitigation measures have been identified and, in most cases, also to achieve net negative emissions to return global warming to 1.5°C following a peak (*high confidence*). The longer the delay in reducing CO_2 emissions towards zero, the larger the likelihood of exceeding 1.5°C, and the heavier the implied reliance on net negative emissions after mid-century to return warming to 1.5°C (*high confidence*). The faster reduction of net CO_2 emissions in 1.5°C compared to 2°C pathways is predominantly achieved by measures that result in less CO_2 being produced and emitted, and only to a smaller degree through additional CDR. Limitations on the speed, scale and societal acceptability of CDR deployment also limit the conceivable extent of temperature overshoot. Limits to our understanding of how the carbon cycle responds to net negative emissions increase the uncertainty about the effectiveness of CDR to decline temperatures after a peak. {2.2, 2.3, 2.6, 4.3.7}

CDR deployed at scale is unproven, and reliance on such technology is a major risk in the ability to limit warming to 1.5°C. CDR is needed less in pathways with particularly strong emphasis on energy efficiency and low demand. The scale and type of CDR deployment varies widely across 1.5°C pathways, with different consequences for achieving sustainable development objectives (*high confidence*). Some pathways rely more on bioenergy with carbon capture and storage (BECCS), while others rely more on afforestation, which are the two CDR methods most often included in integrated pathways. Trade-offs with other sustainability objectives occur predominantly through increased land, energy, water and investment demand. Bioenergy use is substantial in 1.5°C pathways with or without BECCS due to its multiple roles in decarbonizing energy use. {2.3.1, 2.5.3, 2.6.3, 4.3.7}

Properties of Energy and Land Transitions in 1.5°C Pathways

The share of primary energy from renewables increases while coal usage decreases across pathways limiting warming to 1.5°C with no or limited overshoot (*high confidence*). By 2050, renewables (including bioenergy, hydro, wind, and solar, with direct-equivalence method) supply a share of 52–67% (interquartile range) of primary energy in 1.5°C pathways with no or limited overshoot; while the share from coal decreases to 1–7% (interquartile range), with a large fraction of this coal use combined with carbon capture and storage (CCS). From 2020 to 2050 the primary energy supplied by oil declines in most pathways (−39 to −77% interquartile range). Natural gas changes by −13% to −62% (interquartile range), but some pathways show a marked increase albeit with widespread deployment of CCS. The overall deployment of CCS varies widely across 1.5°C pathways with no or limited overshoot, with cumulative CO_2 stored through 2050 ranging from zero up to 300 $GtCO_2$ (minimum–maximum range), of which zero up to 140 $GtCO_2$ is stored from biomass. Primary energy supplied by bioenergy ranges from 40–310 EJ yr^{-1} in 2050 (minimum-maximum range), and nuclear from 3–66 EJ yr^{-1} (minimum–maximum range). These ranges reflect both uncertainties in technological development and strategic mitigation portfolio choices. {2.4.2}

1.5°C pathways with no or limited overshoot include a rapid decline in the carbon intensity of electricity and an increase in electrification of energy end use (*high confidence*). By 2050, the carbon intensity of electricity decreases to −92 to +11 gCO_2 MJ^{-1} (minimum–maximum range) from about 140 gCO_2 MJ^{-1} in 2020, and electricity covers 34–71% (minimum–maximum range) of final energy across 1.5°C pathways with no or limited overshoot from about 20% in 2020. By 2050, the share of electricity supplied by renewables increases to 59–97% (minimum-maximum range) across 1.5°C pathways with no or limited overshoot. Pathways with higher chances of holding warming to below 1.5°C generally show a faster decline in the carbon intensity of electricity by 2030 than pathways that temporarily overshoot 1.5°C. {2.4.1, 2.4.2, 2.4.3}

Transitions in global and regional land use are found in all pathways limiting global warming to 1.5°C with no or limited overshoot, but their scale depends on the pursued mitigation portfolio (*high confidence*). Pathways that limit global warming to 1.5°C with no or limited overshoot project a 4 million km^2 reduction to a 2.5 million km2 increase of non-pasture agricultural land for food and feed crops and a 0.5–11 million km^2 reduction of pasture land, to be converted into 0-6 million km^2 of agricultural land for energy crops and a 2 million km^2 reduction to 9.5 million km^2 increase in forests by 2050 relative to 2010 (*medium confidence*). Land-use transitions of similar magnitude can be observed in modelled 2°C pathways (*medium confidence*). Such large transitions pose profound challenges for sustainable management of the various demands on land for human settlements, food, livestock feed, fibre, bioenergy, carbon storage, biodiversity and other ecosystem services (*high confidence*). {2.3.4, 2.4.4}

Demand-Side Mitigation and Behavioural Changes

Demand-side measures are key elements of 1.5°C pathways. Lifestyle choices lowering energy demand and the land- and GHG-intensity of food consumption can further support achievement of 1.5°C pathways (*high confidence*). By 2030 and 2050, all end-use sectors (including building, transport, and industry) show marked energy demand reductions in modelled 1.5°C pathways, comparable and beyond those projected in 2°C pathways. Sectoral models support the scale of these reductions. {2.3.4, 2.4.3, 2.5.1}

Links between 1.5°C Pathways and Sustainable Development

Choices about mitigation portfolios for limiting warming to 1.5°C can positively or negatively impact the achievement of other societal objectives, such as sustainable development (*high confidence*). In particular, demand-side and efficiency measures, and lifestyle choices that limit energy, resource, and GHG-intensive food demand support sustainable development (*medium confidence*). Limiting warming to 1.5°C can be achieved synergistically with poverty alleviation and improved energy security and can provide large public health benefits through improved air quality, preventing millions of premature deaths. However, specific mitigation measures, such as bioenergy, may result in trade-offs that require consideration. {2.5.1, 2.5.2, 2.5.3}

TS.3   Impacts of 1.5°C Global Warming on Natural and Human Systems

This chapter builds on findings of AR5 and assesses new scientific evidence of changes in the climate system and the associated impacts on natural and human systems, with a specific focus on the magnitude and pattern of risks linked for global warming of 1.5°C above temperatures in the pre-industrial period. Chapter 3 explores observed impacts and projected risks to a range of natural and human systems, with a focus on how risk levels change from 1.5°C to 2°C of global warming. The chapter also revisits major categories of risk (Reasons for Concern, RFC) based on the assessment of new knowledge that has become available since AR5.

1.5°C and 2°C Warmer Worlds

The global climate has changed relative to the pre-industrial period, and there are multiple lines of evidence that these changes have had impacts on organisms and ecosystems, as well as on human systems and well-being (*high confidence*). The increase in global mean surface temperature (GMST), which reached 0.87°C in 2006–2015 relative to 1850–1900, has increased the frequency and magnitude of impacts (*high confidence*), strengthening evidence of how an increase in GMST of 1.5°C or more could impact natural and human systems (1.5°C versus 2°C). {3.3, 3.4, 3.5, 3.6, Cross-Chapter Boxes 6, 7 and 8 in this chapter}

Human-induced global warming has already caused multiple observed changes in the climate system (*high confidence*). Changes include increases in both land and ocean temperatures, as well as more frequent heatwaves in most land regions (*high confidence*). There is also *high confidence* that global warming has resulted in an increase in the frequency and duration of marine heatwaves. Further, there is substantial evidence that human-induced global warming has led to an increase in the frequency, intensity and/or amount of heavy precipitation events at the global scale (*medium confidence*), as well as an increased risk of drought in the Mediterranean region (*medium confidence*). {3.3.1, 3.3.2, 3.3.3, 3.3.4, Box 3.4}

Trends in intensity and frequency of some climate and weather extremes have been detected over time spans during which about 0.5°C of global warming occurred (*medium confidence*). This assessment is based on several lines of evidence, including attribution studies for changes in extremes since 1950. {3.2, 3.3.1, 3.3.2, 3.3.3, 3.3.4}

Several regional changes in climate are assessed to occur with global warming up to 1.5°C as compared to pre-industrial levels, including warming of extreme temperatures in many regions (*high confidence*), increases in frequency, intensity and/or amount of heavy precipitation in several regions (*high confidence*), and an increase in intensity or frequency of droughts in some regions (*medium confidence*). {3.3.1, 3.3.2, 3.3.3, 3.3.4, Table 3.2}

There is no single '1.5°C warmer world' (*high confidence*). In addition to the overall increase in GMST, it is important to consider the

TS

size and duration of potential overshoots in temperature. Furthermore, there are questions on how the stabilization of an increase in GMST of 1.5°C can be achieved, and how policies might be able to influence the resilience of human and natural systems, and the nature of regional and subregional risks. Overshooting poses large risks for natural and human systems, especially if the temperature at peak warming is high, because some risks may be long-lasting and irreversible, such as the loss of some ecosystems (*high confidence*). The rate of change for several types of risks may also have relevance, with potentially large risks in the case of a rapid rise to overshooting temperatures, even if a decrease to 1.5°C can be achieved at the end of the 21st century or later (*medium confidence*). If overshoot is to be minimized, the remaining equivalent CO_2 budget available for emissions is very small, which implies that large, immediate and unprecedented global efforts to mitigate greenhouse gases are required (*high confidence*). {3.2, 3.6.2, Cross-Chapter Box 8 in this chapter}

Robust[1] global differences in temperature means and extremes are expected if global warming reaches 1.5°C versus 2°C above the pre-industrial levels (*high confidence*). For oceans, regional surface temperature means and extremes are projected to be higher at 2°C compared to 1.5°C of global warming (*high confidence*). Temperature means and extremes are also projected to be higher at 2°C compared to 1.5°C in most land regions, with increases being 2–3 times greater than the increase in GMST projected for some regions (*high confidence*). Robust increases in temperature means and extremes are also projected at 1.5°C compared to present-day values (*high confidence*) {3.3.1, 3.3.2}. There are decreases in the occurrence of cold extremes, but substantial increases in their temperature, in particular in regions with snow or ice cover (*high confidence*) {3.3.1}.

Climate models project robust[2] differences in regional climate between present-day and global warming up to 1.5°C[3], and between 1.5°C and 2°C[3] (*high confidence*), depending on the variable and region in question (*high confidence*). Large, robust and widespread differences are expected for temperature extremes (*high confidence*). Regarding hot extremes, the strongest warming is expected to occur at mid-latitudes in the warm season (with increases of up to 3°C for 1.5°C of global warming, i.e., a factor of two) and at high latitudes in the cold season (with increases of up to 4.5°C at 1.5°C of global warming, i.e., a factor of three) (*high confidence*). The strongest warming of hot extremes is projected to occur in central and eastern North America, central and southern Europe, the Mediterranean region (including southern Europe, northern Africa and the Near East), western and central Asia, and southern Africa (*medium confidence*). The number of exceptionally hot days are expected to increase the most in the tropics, where interannual temperature variability is lowest; extreme heatwaves are thus projected to emerge earliest in these regions, and they are expected to already become widespread there at 1.5°C global warming (*high confidence*). Limiting global warming to 1.5°C instead of 2°C could result in around 420 million fewer people being frequently exposed to extreme heatwaves,

and about 65 million fewer people being exposed to exceptional heatwaves, assuming constant vulnerability (*medium confidence*). {3.3.1, 3.3.2, Cross-Chapter Box 8 in this chapter}

Limiting global warming to 1.5°C would limit risks of increases in heavy precipitation events on a global scale and in several regions compared to conditions at 2°C global warming (*medium confidence*). The regions with the largest increases in heavy precipitation events for 1.5°C to 2°C global warming include: several high-latitude regions (e.g. Alaska/western Canada, eastern Canada/Greenland/Iceland, northern Europe and northern Asia); mountainous regions (e.g., Tibetan Plateau); eastern Asia (including China and Japan); and eastern North America (*medium confidence*). Tropical cyclones are projected to decrease in frequency but with an increase in the number of very intense cyclones (*limited evidence, low confidence*). Heavy precipitation associated with tropical cyclones is projected to be higher at 2°C compared to 1.5°C of global warming (*medium confidence*). Heavy precipitation, when aggregated at a global scale, is projected to be higher at 2°C than at 1.5°C of global warming (*medium confidence*) {3.3.3, 3.3.6}

Limiting global warming to 1.5°C is expected to substantially reduce the probability of extreme drought, precipitation deficits, and risks associated with water availability (i.e., water stress) in some regions (*medium confidence*). In particular, risks associated with increases in drought frequency and magnitude are projected to be substantially larger at 2°C than at 1.5°C in the Mediterranean region (including southern Europe, northern Africa and the Near East) and southern Africa (*medium confidence*). {3.3.3, 3.3.4, Box 3.1, Box 3.2}

Risks to natural and human systems are expected to be lower at 1.5°C than at 2°C of global warming (*high confidence*). This difference is due to the smaller rates and magnitudes of climate change associated with a 1.5°C temperature increase, including lower frequencies and intensities of temperature-related extremes. Lower rates of change enhance the ability of natural and human systems to adapt, with substantial benefits for a wide range of terrestrial, freshwater, wetland, coastal and ocean ecosystems (including coral reefs) (*high confidence*), as well as food production systems, human health, and tourism (*medium confidence*), together with energy systems and transportation (*low confidence*). {3.3.1, 3.4}

Exposure to multiple and compound climate-related risks is projected to increase between 1.5°C and 2°C of global warming with greater proportions of people both exposed and susceptible to poverty in Africa and Asia (*high confidence*). For global warming from 1.5°C to 2°C, risks across energy, food, and water sectors could overlap spatially and temporally, creating new – and exacerbating current – hazards, exposures, and vulnerabilities that could affect increasing numbers of people and regions (*medium confidence*). Small island states and economically disadvantaged populations are particularly at risk (*high confidence*). {3.3.1, 3.4.5.3, 3.4.5.6, 3.4.11, 3.5.4.9, Box 3.5}

---

[2]   Robust is used here to mean that at least two thirds of climate models show the same sign of changes at the grid point scale, and that differences in large regions are statistically significant.

[3]   Projected changes in impacts between different levels of global warming are determined with respect to changes in global mean near-surface air temperature.

Global warming of 2°C would lead to an expansion of areas with significant increases in runoff, as well as those affected by flood hazard, compared to conditions at 1.5°C (*medium confidence*). Global warming of 1.5°C would also lead to an expansion of the global land area with significant increases in runoff (*medium confidence*) and an increase in flood hazard in some regions (*medium confidence*) compared to present-day conditions. {3.3.5}

The probability of a sea-ice-free Arctic Ocean[4] during summer is substantially higher at 2°C compared to 1.5°C of global warming (*medium confidence*). Model simulations suggest that at least one sea-ice-free Arctic summer is expected every 10 years for global warming of 2°C, with the frequency decreasing to one sea-ice-free Arctic summer every 100 years under 1.5°C (*medium confidence*). An intermediate temperature overshoot will have no long-term consequences for Arctic sea ice coverage, and hysteresis is not expected (*high confidence*). {3.3.8, 3.4.4.7}

Global mean sea level rise (GMSLR) is projected to be around 0.1 m (0.04 – 0.16 m) less by the end of the 21st century in a 1.5°C warmer world compared to a 2°C warmer world (*medium confidence*). Projected GMSLR for 1.5°C of global warming has an indicative range of 0.26 – 0.77m, relative to 1986–2005, (*medium confidence*). A smaller sea level rise could mean that up to 10.4 million fewer people (based on the 2010 global population and assuming no adaptation) would be exposed to the impacts of sea level rise globally in 2100 at 1.5°C compared to at 2°C. A slower rate of sea level rise enables greater opportunities for adaptation (*medium confidence*). There is *high confidence* that sea level rise will continue beyond 2100. Instabilities exist for both the Greenland and Antarctic ice sheets, which could result in multi-meter rises in sea level on time scales of century to millennia. There is *medium confidence* that these instabilities could be triggered at around 1.5°C to 2°C of global warming. {3.3.9, 3.4.5, 3.6.3}

The ocean has absorbed about 30% of the anthropogenic carbon dioxide, resulting in ocean acidification and changes to carbonate chemistry that are unprecedented for at least the last 65 million years (*high confidence*). Risks have been identified for the survival, calcification, growth, development and abundance of a broad range of marine taxonomic groups, ranging from algae to fish, with substantial evidence of predictable trait-based sensitivities (*high confidence*). There are multiple lines of evidence that ocean warming and acidification corresponding to 1.5°C of global warming would impact a wide range of marine organisms and ecosystems, as well as sectors such as aquaculture and fisheries (*high confidence*). {3.3.10, 3.4.4}

Larger risks are expected for many regions and systems for global warming at 1.5°C, as compared to today, with adaptation required now and up to 1.5°C. However, risks would be larger at 2°C of warming and an even greater effort would be needed for adaptation to a temperature increase of that magnitude (*high confidence*). {3.4, Box 3.4, Box 3.5, Cross-Chapter Box 6 in this chapter}

Future risks at 1.5°C of global warming will depend on the mitigation pathway and on the possible occurrence of a transient overshoot (*high confidence*). The impacts on natural and human systems would be greater if mitigation pathways temporarily overshoot 1.5°C and return to 1.5°C later in the century, as compared to pathways that stabilize at 1.5°C without an overshoot (*high confidence*). The size and duration of an overshoot would also affect future impacts (e.g., irreversible loss of some ecosystems) (*high confidence*). Changes in land use resulting from mitigation choices could have impacts on food production and ecosystem diversity. {3.6.1, 3.6.2, Cross-Chapter Boxes 7 and 8 in this chapter}

Climate Change Risks for Natural and Human systems

Terrestrial and Wetland Ecosystems

Risks of local species losses and, consequently, risks of extinction are much less in a 1.5°C versus a 2°C warmer world (*high confidence*). The number of species projected to lose over half of their climatically determined geographic range at 2°C global warming (18% of insects, 16% of plants, 8% of vertebrates) is projected to be reduced to 6% of insects, 8% of plants and 4% of vertebrates at 1.5°C warming (*medium confidence*). Risks associated with other biodiversity-related factors, such as forest fires, extreme weather events, and the spread of invasive species, pests and diseases, would also be lower at 1.5°C than at 2°C of warming (*high confidence*), supporting a greater persistence of ecosystem services. {3.4.3, 3.5.2}

Constraining global warming to 1.5°C, rather than to 2°C and higher, is projected to have many benefits for terrestrial and wetland ecosystems and for the preservation of their services to humans (*high confidence*). Risks for natural and managed ecosystems are higher on drylands compared to humid lands. The global terrestrial land area projected to be affected by ecosystem transformations (13%, interquartile range 8–20%) at 2°C is approximately halved at 1.5°C global warming to 4% (interquartile range 2–7%) (*medium confidence*). Above 1.5°C, an expansion of desert terrain and vegetation would occur in the Mediterranean biome (*medium confidence*), causing changes unparalleled in the last 10,000 years (*medium confidence*). {3.3.2.2, 3.4.3.2, 3.4.3.5, 3.4.6.1, 3.5.5.10, Box 4.2}

Many impacts are projected to be larger at higher latitudes, owing to mean and cold-season warming rates above the global average (*medium confidence*). High-latitude tundra and boreal forest are particularly at risk, and woody shrubs are already encroaching into tundra (*high confidence*) and will proceed with further warming. Constraining warming to 1.5°C would prevent the thawing of an estimated permafrost area of 1.5 to 2.5 million km^2 over centuries compared to thawing under 2°C (*medium confidence*). {3.3.2, 3.4.3, 3.4.4}

---

4   Ice free is defined for the Special Report as when the sea ice extent is less than 106 km^2. Ice coverage less than this is considered to be equivalent to an ice-free Arctic Ocean for practical purposes in all recent studies.

Ocean Ecosystems

Ocean ecosystems are already experiencing large-scale changes, and critical thresholds are expected to be reached at 1.5°C and higher levels of global warming (*high confidence*). In the transition to 1.5°C of warming, changes to water temperatures are expected to drive some species (e.g., plankton, fish) to relocate to higher latitudes and cause novel ecosystems to assemble (*high confidence*). Other ecosystems (e.g., kelp forests, coral reefs) are relatively less able to move, however, and are projected to experience high rates of mortality and loss (*very high confidence*). For example, multiple lines of evidence indicate that the majority (70–90%) of warm water (tropical) coral reefs that exist today will disappear even if global warming is constrained to 1.5°C (*very high confidence*). {3.4.4, Box 3.4}

Current ecosystem services from the ocean are expected to be reduced at 1.5°C of global warming, with losses being even greater at 2°C of global warming (*high confidence*). The risks of declining ocean productivity, shifts of species to higher latitudes, damage to ecosystems (e.g., coral reefs, and mangroves, seagrass and other wetland ecosystems), loss of fisheries productivity (at low latitudes), and changes to ocean chemistry (e.g., acidification, hypoxia and dead zones) are projected to be substantially lower when global warming is limited to 1.5°C (*high confidence*). {3.4.4, Box 3.4}

Water Resources

The projected frequency and magnitude of floods and droughts in some regions are smaller under 1.5°C than under 2°C of warming (*medium confidence*). Human exposure to increased flooding is projected to be substantially lower at 1.5°C compared to 2°C of global warming, although projected changes create regionally differentiated risks (*medium confidence*). The differences in the risks among regions are strongly influenced by local socio-economic conditions (*medium confidence*). {3.3.4, 3.3.5, 3.4.2}

Risks of water scarcity are projected to be greater at 2°C than at 1.5°C of global warming in some regions (*medium confidence*). Depending on future socio-economic conditions, limiting global warming to 1.5°C, compared to 2°C, may reduce the proportion of the world population exposed to a climate change-induced increase in water stress by up to 50%, although there is considerable variability between regions (*medium confidence*). Regions with particularly large benefits could include the Mediterranean and the Caribbean (*medium confidence*). Socio-economic drivers, however, are expected to have a greater influence on these risks than the changes in climate (*medium confidence*). {3.3.5, 3.4.2, Box 3.5}

Land Use, Food Security and Food Production Systems

Limiting global warming to 1.5°C, compared with 2°C, is projected to result in smaller net reductions in yields of maize, rice, wheat, and potentially other cereal crops, particularly in sub-Saharan Africa, Southeast Asia, and Central and South America; and in the CO_2-dependent nutritional quality of rice and wheat

(*high confidence*). A loss of 7–10% of rangeland livestock globally is projected for approximately 2°C of warming, with considerable economic consequences for many communities and regions (*medium confidence*). {3.4.6, 3.6, Box 3.1, Cross-Chapter Box 6 in this chapter}

Reductions in projected food availability are larger at 2°C than at 1.5°C of global warming in the Sahel, southern Africa, the Mediterranean, central Europe and the Amazon (*medium confidence*). This suggests a transition from medium to high risk of regionally differentiated impacts on food security between 1.5°C and 2°C (*medium confidence*). Future economic and trade environments and their response to changing food availability (*medium confidence*) are important potential adaptation options for reducing hunger risk in low- and middle-income countries. {Cross-Chapter Box 6 in this chapter}

Fisheries and aquaculture are important to global food security but are already facing increasing risks from ocean warming and acidification (*medium confidence*). These risks are projected to increase at 1.5°C of global warming and impact key organisms such as fin fish and bivalves (e.g., oysters), especially at low latitudes (*medium confidence*). Small-scale fisheries in tropical regions, which are very dependent on habitat provided by coastal ecosystems such as coral reefs, mangroves, seagrass and kelp forests, are expected to face growing risks at 1.5°C of warming because of loss of habitat (*medium confidence*). Risks of impacts and decreasing food security are projected to become greater as global warming reaches beyond 1.5°C and both ocean warming and acidification increase, with substantial losses likely for coastal livelihoods and industries (e.g., fisheries and aquaculture) (*medium to high confidence*). {3.4.4, 3.4.5, 3.4.6, Box 3.1, Box 3.4, Box 3.5, Cross-Chapter Box 6 in this chapter}

Land use and land-use change emerge as critical features of virtually all mitigation pathways that seek to limit global warming to 1.5°C (*high confidence*). Most least-cost mitigation pathways to limit peak or end-of-century warming to 1.5°C make use of carbon dioxide removal (CDR), predominantly employing significant levels of bioenergy with carbon capture and storage (BECCS) and/or afforestation and reforestation (AR) in their portfolio of mitigation measures (*high confidence*). {Cross-Chapter Box 7 in this chapter}

Large-scale deployment of BECCS and/or AR would have a far-reaching land and water footprint (*high confidence*). Whether this footprint would result in adverse impacts, for example on biodiversity or food production, depends on the existence and effectiveness of measures to conserve land carbon stocks, measures to limit agricultural expansion in order to protect natural ecosystems, and the potential to increase agricultural productivity (*medium agreement*). In addition, BECCS and/or AR would have substantial direct effects on regional climate through biophysical feedbacks, which are generally not included in Integrated Assessments Models (*high confidence*). {3.6.2, Cross-Chapter Boxes 7 and 8 in this chapter}

The impacts of large-scale CDR deployment could be greatly reduced if a wider portfolio of CDR options were deployed, if a

holistic policy for sustainable land management were adopted, and if increased mitigation efforts were employed to strongly limit the demand for land, energy and material resources, including through lifestyle and dietary changes (*medium confidence*). In particular, reforestation could be associated with significant co-benefits if implemented in a manner than helps restore natural ecosystems (*high confidence*). {Cross-Chapter Box 7 in this chapter}

Human Health, Well-Being, Cities and Poverty

Any increase in global temperature (e.g., +0.5°C) is projected to affect human health, with primarily negative consequences (*high confidence*). Lower risks are projected at 1.5°C than at 2°C for heat-related morbidity and mortality (*very high confidence*), and for ozone-related mortality if emissions needed for ozone formation remain high (*high confidence*). Urban heat islands often amplify the impacts of heatwaves in cities (*high confidence*). Risks for some vector-borne diseases, such as malaria and dengue fever are projected to increase with warming from 1.5°C to 2°C, including potential shifts in their geographic range (*high confidence*). Overall for vector-borne diseases, whether projections are positive or negative depends on the disease, region and extent of change (*high confidence*). Lower risks of undernutrition are projected at 1.5°C than at 2°C (*medium confidence*). Incorporating estimates of adaptation into projections reduces the magnitude of risks (*high confidence*). {3.4.7, 3.4.7.1, 3.4.8, 3.5.5.8}

Global warming of 2°C is expected to pose greater risks to urban areas than global warming of 1.5°C (*medium confidence*). The extent of risk depends on human vulnerability and the effectiveness of adaptation for regions (coastal and non-coastal), informal settlements and infrastructure sectors (such as energy, water and transport) (*high confidence*). {3.4.5, 3.4.8}

Poverty and disadvantage have increased with recent warming (about 1°C) and are expected to increase for many populations as average global temperatures increase from 1°C to 1.5°C and higher (*medium confidence*). Outmigration in agricultural-dependent communities is positively and statistically significantly associated with global temperature (*medium confidence*). Our understanding of the links of 1.5°C and 2°C of global warming to human migration are limited and represent an important knowledge gap. {3.4.10, 3.4.11, 5.2.2, Table 3.5}

Key Economic Sectors and Services

Risks to global aggregated economic growth due to climate change impacts are projected to be lower at 1.5°C than at 2°C by the end of this century (*medium confidence*). {3.5.2, 3.5.3}

The largest reductions in economic growth at 2°C compared to 1.5°C of warming are projected for low- and middle-income countries and regions (the African continent, Southeast Asia, India, Brazil and Mexico) (*low to medium confidence*). Countries in the tropics and Southern Hemisphere subtropics are projected to experience the largest impacts on economic growth due to climate change should global warming increase from 1.5°C to 2°C (*medium confidence*). {3.5}

Global warming has already affected tourism, with increased risks projected under 1.5°C of warming in specific geographic regions and for seasonal tourism including sun, beach and snow sports destinations (*very high confidence*). Risks will be lower for tourism markets that are less climate sensitive, such as gaming and large hotel-based activities (*high confidence*). Risks for coastal tourism, particularly in subtropical and tropical regions, will increase with temperature-related degradation (e.g., heat extremes, storms) or loss of beach and coral reef assets (*high confidence*). {3.3.6, 3.4.4.12, 3.4.9.1, Box 3.4}

Small Islands, and Coastal and Low-lying areas

Small islands are projected to experience multiple inter-related risks at 1.5°C of global warming that will increase with warming of 2°C and higher levels (*high confidence*). Climate hazards at 1.5°C are projected to be lower compared to those at 2°C (*high confidence*). Long-term risks of coastal flooding and impacts on populations, infrastructures and assets (*high confidence*), freshwater stress (*medium confidence*), and risks across marine ecosystems (*high confidence*) and critical sectors (*medium confidence*) are projected to increase at 1.5°C compared to present-day levels and increase further at 2°C, limiting adaptation opportunities and increasing loss and damage (*medium confidence*). Migration in small islands (internally and internationally) occurs for multiple reasons and purposes, mostly for better livelihood opportunities (*high confidence*) and increasingly owing to sea level rise (*medium confidence*). {3.3.2.2, 3.3.6–9, 3.4.3.2, 3.4.4.2, 3.4.4.5, 3.4.4.12, 3.4.5.3, 3.4.7.1, 3.4.9.1, 3.5.4.9, Box 3.4, Box 3.5}

Impacts associated with sea level rise and changes to the salinity of coastal groundwater, increased flooding and damage to infrastructure, are projected to be critically important in vulnerable environments, such as small islands, low-lying coasts and deltas, at global warming of 1.5°C and 2°C (*high confidence*). Localized subsidence and changes to river discharge can potentially exacerbate these effects. Adaptation is already happening (*high confidence*) and will remain important over multi-centennial time scales. {3.4.5.3, 3.4.5.4, 3.4.5.7, 5.4.5.4, Box 3.5}

Existing and restored natural coastal ecosystems may be effective in reducing the adverse impacts of rising sea levels and intensifying storms by protecting coastal and deltaic regions (*medium confidence*). Natural sedimentation rates are expected to be able to offset the effect of rising sea levels, given the slower rates of sea level rise associated with 1.5°C of warming (*medium confidence*). Other feedbacks, such as landward migration of wetlands and the adaptation of infrastructure, remain important (*medium confidence*). {3.4.4.12, 3.4.5.4, 3.4.5.7}

Increased Reasons for Concern

There are multiple lines of evidence that since AR5 the assessed levels of risk increased for four of the five Reasons for Concern

(RFCs) for global warming levels of up to 2°C (*high confidence*). The risk transitions by degrees of global warming are now: from high to very high between 1.5°C and 2°C for RFC1 (Unique and threatened systems) (*high confidence*); from moderate to high risk between 1°C and 1.5°C for RFC2 (Extreme weather events) (*medium confidence*); from moderate to high risk between 1.5°C and 2°C for RFC3 (Distribution of impacts) (*high confidence*); from moderate to high risk between 1.5°C and 2.5°C for RFC4 (Global aggregate impacts) (*medium confidence*); and from moderate to high risk between 1°C and 2.5°C for RFC5 (Large-scale singular events) (*medium confidence*). {3.5.2}

1. **The category 'Unique and threatened systems' (RFC1) display a transition from high to very high risk which is now located between 1.5°C and 2°C of global warming** as opposed to at 2.6°C of global warming in AR5, owing to new and multiple lines of evidence for changing risks for coral reefs, the Arctic and biodiversity in general (*high confidence*). {3.5.2.1}

2. **In 'Extreme weather events' (RFC2), the transition from moderate to high risk is now located between 1.0°C and 1.5°C of global warming,** which is very similar to the AR5 assessment but is projected with greater confidence (*medium confidence*). The impact literature contains little information about the potential for human society to adapt to extreme weather events, and hence it has not been possible to locate the transition from 'high' to 'very high' risk within the context of assessing impacts at 1.5°C versus 2°C of global warming. There is thus *low confidence* in the level at which global warming could lead to very high risks associated with extreme weather events in the context of this report. {3.5}

3. **With respect to the 'Distribution of impacts' (RFC3) a transition from moderate to high risk is now located between 1.5°C and 2°C of global warming,** compared with between 1.6°C and 2.6°C global warming in AR5, owing to new evidence about regionally differentiated risks to food security, water resources, drought, heat exposure and coastal submergence (*high confidence*). {3.5}

4. **In 'global aggregate impacts' (RFC4) a transition from moderate to high levels of risk is now located between 1.5°C and 2.5°C of global warming,** as opposed to at 3.6°C of warming in AR5, owing to new evidence about global aggregate economic impacts and risks to Earth's biodiversity (*medium confidence*). {3.5}

5. **Finally, 'large-scale singular events' (RFC5), moderate risk is now located at 1°C of global warming and high risk is located at 2.5°C of global warming,** as opposed to at 1.6°C (moderate risk) and around 4°C (high risk) in AR5, because of new observations and models of the West Antarctic ice sheet (*medium confidence*). {3.3.9, 3.5.2, 3.6.3}

TS.4 Strengthening and Implementing the Global Response

Limiting warming to 1.5°C above pre-industrial levels would require transformative systemic change, integrated with sustainable development. Such change would require the upscaling and acceleration of the implementation of far-reaching, multilevel and cross-sectoral climate mitigation and addressing barriers. Such systemic change would need to be linked to complementary adaptation actions, including transformational adaptation, especially for pathways that temporarily overshoot 1.5°C (*medium evidence, high agreement***)** {Chapter 2, Chapter 3, 4.2.1, 4.4.5, 4.5}. Current national pledges on mitigation and adaptation are not enough to stay below the Paris Agreement temperature limits and achieve its adaptation goals. While transitions in energy efficiency, carbon intensity of fuels, electrification and land-use change are underway in various countries, limiting warming to 1.5°C will require a greater scale and pace of change to transform energy, land, urban and industrial systems globally. {4.3, 4.4, Cross-Chapter Box 9 in this Chapter}

Although multiple communities around the world are demonstrating the possibility of implementation consistent with 1.5°C pathways {Boxes 4.1-4.10}, very few countries, regions, cities, communities or businesses can currently make such a claim (*high confidence***). To strengthen the global response, almost all countries would need to significantly raise their level of ambition. Implementation of this raised ambition would require enhanced institutional capabilities in all countries, including building the capability to utilize indigenous and local knowledge (***medium evidence, high agreement***).** In developing countries and for poor and vulnerable people, implementing the response would require financial, technological and other forms of support to build capacity, for which additional local, national and international resources would need to be mobilized (*high confidence*). However, public, financial, institutional and innovation capabilities currently fall short of implementing far-reaching measures at scale in all countries (*high confidence*). Transnational networks that support multilevel climate action are growing, but challenges in their scale-up remain. {4.4.1, 4.4.2, 4.4.4, 4.4.5, Box 4.1, Box 4.2, Box 4.7}

Adaptation needs will be lower in a 1.5°C world compared to a 2°C world (*high confidence***) {Chapter 3; Cross-Chapter Box 11 in this chapter}.** Learning from current adaptation practices and strengthening them through adaptive governance {4.4.1}, lifestyle and behavioural change {4.4.3} and innovative financing mechanisms {4.4.5} can help their mainstreaming within sustainable development practices. Preventing maladaptation, drawing on bottom-up approaches {Box 4.6} and using indigenous knowledge {Box 4.3} would effectively engage and protect vulnerable people and communities. While adaptation finance has increased quantitatively, significant further expansion would be needed to adapt to 1.5°C. Qualitative gaps in the distribution of adaptation finance, readiness to absorb resources, and monitoring mechanisms undermine the potential of adaptation finance to reduce impacts. {Chapter 3, 4.4.2, 4.4.5, 4.6}

System Transitions

The energy system transition that would be required to limit global warming to 1.5°C above pre-industrial conditions is underway in many sectors and regions around the world (*medium evidence, high agreement*). The political, economic, social and technical feasibility of solar energy, wind energy and electricity storage technologies has improved dramatically over the past few years, while that of nuclear energy and carbon dioxide capture and storage (CCS) in the electricity sector have not shown similar improvements. {4.3.1}

Electrification, hydrogen, bio-based feedstocks and substitution, and, in several cases, carbon dioxide capture, utilization and storage (CCUS) would lead to the deep emissions reductions required in energy-intensive industries to limit warming to 1.5°C. However, those options are limited by institutional, economic and technical constraints, which increase financial risks to many incumbent firms (*medium evidence, high agreement*). Energy efficiency in industry is more economically feasible and helps enable industrial system transitions but would have to be complemented with greenhouse gas (GHG)-neutral processes or carbon dioxide removal (CDR) to make energy-intensive industries consistent with 1.5°C (*high confidence*). {4.3.1, 4.3.4}

Global and regional land-use and ecosystems transitions and associated changes in behaviour that would be required to limit warming to 1.5°C can enhance future adaptation and land-based agricultural and forestry mitigation potential. Such transitions could, however, carry consequences for livelihoods that depend on agriculture and natural resources {4.3.2, Cross-Chapter Box 6 in Chapter 3}. Alterations of agriculture and forest systems to achieve mitigation goals could affect current ecosystems and their services and potentially threaten food, water and livelihood security. While this could limit the social and environmental feasibility of land-based mitigation options, careful design and implementation could enhance their acceptability and support sustainable development objectives (*medium evidence, medium agreement*). {4.3.2, 4.5.3}

Changing agricultural practices can be an effective climate adaptation strategy. A diversity of adaptation options exists, including mixed crop-livestock production systems which can be a cost-effective adaptation strategy in many global agriculture systems (*robust evidence, medium agreement*). Improving irrigation efficiency could effectively deal with changing global water endowments, especially if achieved via farmers adopting new behaviours and water-efficient practices rather than through large-scale infrastructural interventions (*medium evidence, medium agreement*). Well-designed adaptation processes such as community-based adaptation can be effective depending upon context and levels of vulnerability. {4.3.2, 4.5.3}

Improving the efficiency of food production and closing yield gaps have the potential to reduce emissions from agriculture, reduce pressure on land, and enhance food security and future mitigation potential (*high confidence*). Improving productivity of existing agricultural systems generally reduces the emissions intensity of food production and offers strong synergies with rural development, poverty reduction and food security objectives, but options to reduce absolute emissions are limited unless paired with demand-side measures. Technological innovation including biotechnology, with adequate safeguards, could contribute to resolving current feasibility constraints and expand the future mitigation potential of agriculture. {4.3.2, 4.4.4}

Shifts in dietary choices towards foods with lower emissions and requirements for land, along with reduced food loss and waste, could reduce emissions and increase adaptation options (*high confidence*). Decreasing food loss and waste and changing dietary behaviour could result in mitigation and adaptation (*high confidence*) by reducing both emissions and pressure on land, with significant co-benefits for food security, human health and sustainable development {4.3.2, 4.4.5, 4.5.2, 4.5.3, 5.4.2}, but evidence of successful policies to modify dietary choices remains limited.

Mitigation and Adaptation Options and Other Measures

A mix of mitigation and adaptation options implemented in a participatory and integrated manner can enable rapid, systemic transitions – in urban and rural areas – that are necessary elements of an accelerated transition consistent with limiting warming to 1.5°C. Such options and changes are most effective when aligned with economic and sustainable development, and when local and regional governments are supported by national governments {4.3.3, 4.4.1, 4.4.3}. Various mitigation options are expanding rapidly across many geographies. Although many have development synergies, not all income groups have so far benefited from them. Electrification, end-use energy efficiency and increased share of renewables, amongst other options, are lowering energy use and decarbonizing energy supply in the built environment, especially in buildings. Other rapid changes needed in urban environments include demotorization and decarbonization of transport, including the expansion of electric vehicles, and greater use of energy-efficient appliances (*medium evidence, high agreement*). Technological and social innovations can contribute to limiting warming to 1.5°C, for example, by enabling the use of smart grids, energy storage technologies and general-purpose technologies, such as information and communication technology (ICT) that can be deployed to help reduce emissions. Feasible adaptation options include green infrastructure, resilient water and urban ecosystem services, urban and peri-urban agriculture, and adapting buildings and land use through regulation and planning (*medium evidence, medium to high agreement*). {4.3.3, 4.4.3, 4.4.4}

Synergies can be achieved across systemic transitions through several overarching adaptation options in rural and urban areas. Investments in health, social security and risk sharing and spreading are cost-effective adaptation measures with high potential for scaling up (*medium evidence, medium to high agreement*). Disaster risk management and education-based adaptation have lower prospects of scalability and cost-effectiveness (*medium evidence, high agreement*) but are critical for building adaptive capacity. {4.3.5, 4.5.3}

TS

Converging adaptation and mitigation options can lead to synergies and potentially increase cost-effectiveness, but multiple trade-offs can limit the speed of and potential for scaling up. Many examples of synergies and trade-offs exist in all sectors and system transitions. For instance, sustainable water management (*high evidence, medium agreement*) and investment in green infrastructure (*medium evidence, high agreement*) to deliver sustainable water and environmental services and to support urban agriculture are less cost-effective than other adaptation options but can help build climate resilience. Achieving the governance, finance and social support required to enable these synergies and to avoid trade-offs is often challenging, especially when addressing multiple objectives, and attempting appropriate sequencing and timing of interventions. {4.3.2, 4.3.4, 4.4.1, 4.5.2, 4.5.3, 4.5.4}

Though CO_2 dominates long-term warming, the reduction of warming short-lived climate forcers (SLCFs), such as methane and black carbon, can in the short term contribute significantly to limiting warming to 1.5°C above pre-industrial levels. Reductions of black carbon and methane would have substantial co-benefits (*high confidence*), including improved health due to reduced air pollution. This, in turn, enhances the institutional and socio-cultural feasibility of such actions. Reductions of several warming SLCFs are constrained by economic and social feasibility (*low evidence, high agreement*). As they are often co-emitted with CO_2, achieving the energy, land and urban transitions necessary to limit warming to 1.5°C would see emissions of warming SLCFs greatly reduced. {2.3.3.2, 4.3.6}

Most CDR options face multiple feasibility constraints, which differ between options, limiting the potential for any single option to sustainably achieve the large-scale deployment required in the 1.5°C-consistent pathways described in Chapter 2 (*high confidence*). Those 1.5°C pathways typically rely on bioenergy with carbon capture and storage (BECCS), afforestation and reforestation (AR), or both, to neutralize emissions that are expensive to avoid, or to draw down CO_2 emissions in excess of the carbon budget {Chapter 2}. Though BECCS and AR may be technically and geophysically feasible, they face partially overlapping yet different constraints related to land use. The land footprint per tonne of CO_2 removed is higher for AR than for BECCS, but given the low levels of current deployment, the speed and scales required for limiting warming to 1.5°C pose a considerable implementation challenge, even if the issues of public acceptance and absence of economic incentives were to be resolved (*high agreement, medium evidence*). The large potential of afforestation and the co-benefits if implemented appropriately (e.g., on biodiversity and soil quality) will diminish over time, as forests saturate (*high confidence*). The energy requirements and economic costs of direct air carbon capture and storage (DACCS) and enhanced weathering remain high (*medium evidence, medium agreement*). At the local scale, soil carbon sequestration has co-benefits with agriculture and is cost-effective even without climate policy (high confidence). Its potential feasibility and cost-effectiveness at the global scale appears to be more limited. {4.3.7}

Uncertainties surrounding solar radiation modification (SRM) measures constrain their potential deployment. These uncertainties include: technological immaturity; limited physical understanding about their effectiveness to limit global warming; and a weak capacity to govern, legitimize, and scale such measures. Some recent model-based analysis suggests SRM would be effective but that it is too early to evaluate its feasibility. Even in the uncertain case that the most adverse side-effects of SRM can be avoided, public resistance, ethical concerns and potential impacts on sustainable development could render SRM economically, socially and institutionally undesirable (*low agreement, medium evidence*). {4.3.8, Cross-Chapter Box 10 in this chapter}

Enabling Rapid and Far-Reaching Change

The speed of transitions and of technological change required to limit warming to 1.5°C above pre-industrial levels has been observed in the past within specific sectors and technologies {4.2.2.1}. But the geographical and economic scales at which the required rates of change in the energy, land, urban, infrastructure and industrial systems would need to take place are larger and have no documented historic precedent (*limited evidence, medium agreement*). To reduce inequality and alleviate poverty, such transformations would require more planning and stronger institutions (including inclusive markets) than observed in the past, as well as stronger coordination and disruptive innovation across actors and scales of governance. {4.3, 4.4}

Governance consistent with limiting warming to 1.5°C and the political economy of adaptation and mitigation can enable and accelerate systems transitions, behavioural change, innovation and technology deployment (*medium evidence, medium agreement*). For 1.5°C-consistent actions, an effective governance framework would include: accountable multilevel governance that includes non-state actors, such as industry, civil society and scientific institutions; coordinated sectoral and cross-sectoral policies that enable collaborative multi-stakeholder partnerships; strengthened global-to-local financial architecture that enables greater access to finance and technology; addressing climate-related trade barriers; improved climate education and greater public awareness; arrangements to enable accelerated behaviour change; strengthened climate monitoring and evaluation systems; and reciprocal international agreements that are sensitive to equity and the Sustainable Development Goals (SDGs). System transitions can be enabled by enhancing the capacities of public, private and financial institutions to accelerate climate change policy planning and implementation, along with accelerated technological innovation, deployment and upkeep. {4.4.1, 4.4.2, 4.4.3, 4.4.4}

Behaviour change and demand-side management can significantly reduce emissions, substantially limiting the reliance on CDR to limit warming to 1.5°C {Chapter 2, 4.4.3}. Political and financial stakeholders may find climate actions more cost-effective and socially acceptable if multiple factors affecting behaviour are considered, including aligning these actions with people's core values (*medium evidence, high agreement*). Behaviour- and lifestyle-related measures and demand-side management have already led to emission reductions around the world and can enable significant future reductions (*high confidence*). Social innovation through bottom-up initiatives can result in greater participation in the governance of systems transitions and increase support for technologies, practices

TS

and policies that are part of the global response to limit warming to 1.5°C . {Chapter 2, 4.4.1, 4.4.3, Figure 4.3}

This rapid and far-reaching response required to keep warming below 1.5°C and enhance the capacity to adapt to climate risks would require large increases of investments in low-emission infrastructure and buildings, along with a redirection of financial flows towards low-emission investments (*robust evidence, high agreement*). An estimated mean annual incremental investment of around 1.5% of global gross fixed capital formation (GFCF) for the energy sector is indicated between 2016 and 2035, as well as about 2.5% of global GFCF for other development infrastructure that could also address SDG implementation. Though quality policy design and effective implementation may enhance efficiency, they cannot fully substitute for these investments. {2.5.2, 4.2.1, 4.4.5}

Enabling this investment requires the mobilization and better integration of a range of policy instruments that include the reduction of socially inefficient fossil fuel subsidy regimes and innovative price and non-price national and international policy instruments. These would need to be complemented by de-risking financial instruments and the emergence of long-term low-emission assets. These instruments would aim to reduce the demand for carbon-intensive services and shift market preferences away from fossil fuel-based technology. Evidence and theory suggest that carbon pricing alone, in the absence of sufficient transfers to compensate their unintended distributional cross-sector, cross-nation effects, cannot reach the incentive levels needed to trigger system transitions (*robust evidence, medium agreement*). But, embedded in consistent policy packages, they can help mobilize incremental resources and provide flexible mechanisms that help reduce the social and economic costs of the triggering phase of the transition (*robust evidence, medium agreement*). {4.4.3, 4.4.4, 4.4.5}

Increasing evidence suggests that a climate-sensitive realignment of savings and expenditure towards low-emission, climate-resilient infrastructure and services requires an evolution of global and national financial systems. Estimates suggest that, in addition to climate-friendly allocation of public investments, a potential redirection of 5% to 10% of the annual capital revenues[5] is necessary for limiting warming to 1.5°C {4.4.5, Table 1 in Box 4.8}. This could be facilitated by a change of incentives for private day-to-day expenditure and the redirection of savings from speculative and precautionary investments towards long-term productive low-emission assets and services. This implies the mobilization of institutional investors and mainstreaming of climate finance within financial and banking system regulation. Access by developing countries to low-risk and low-interest finance through multilateral and national development banks would have to be facilitated (*medium evidence, high agreement*). New forms of public–private partnerships may be needed with multilateral, sovereign and sub-sovereign guarantees to de-risk climate-friendly investments, support new business models for small-scale enterprises and help households with limited access to capital. Ultimately, the aim is to promote a portfolio shift towards long-term low-emission assets that would help redirect capital away from potentially stranded assets (*medium evidence, medium agreement*). {4.4.5}

Knowledge Gaps

Knowledge gaps around implementing and strengthening the global response to climate change would need to be urgently resolved if the transition to a 1.5°C world is to become reality. Remaining questions include: how much can be realistically expected from innovation and behavioural and systemic political and economic changes in improving resilience, enhancing adaptation and reducing GHG emissions? How can rates of changes be accelerated and scaled up? What is the outcome of realistic assessments of mitigation and adaptation land transitions that are compliant with sustainable development, poverty eradication and addressing inequality? What are life-cycle emissions and prospects of early-stage CDR options? How can climate and sustainable development policies converge, and how can they be organised within a global governance framework and financial system, based on principles of justice and ethics (including 'common but differentiated responsibilities and respective capabilities' (CBDR-RC)), reciprocity and partnership? To what extent would limiting warming to 1.5°C require a harmonization of macro-financial and fiscal policies, which could include financial regulators such as central banks? How can different actors and processes in climate governance reinforce each other, and hedge against the fragmentation of initiatives? {4.1, 4.3.7, 4.4.1, 4.4.5, 4.6}

TS

---

[5] Annual capital revenues are the paid interests plus the increase of the asset value.

TS.5 Sustainable Development, Poverty Eradication and Reducing Inequalities

This chapter takes sustainable development as the starting point and focus for analysis. It considers the broad and multifaceted bi-directional interplay between sustainable development, including its focus on eradicating poverty and reducing inequality in their multidimensional aspects, and climate actions in a 1.5°C warmer world. These fundamental connections are embedded in the Sustainable Development Goals (SDGs). The chapter also examines synergies and trade-offs of adaptation and mitigation options with sustainable development and the SDGs and offers insights into possible pathways, especially climate-resilient development pathways towards a 1.5°C warmer world.

Sustainable Development, Poverty and Inequality in a 1.5°C Warmer World

Limiting global warming to 1.5°C rather than 2°C above pre-industrial levels would make it markedly easier to achieve many aspects of sustainable development, with greater potential to eradicate poverty and reduce inequalities (*medium evidence, high agreement*). Impacts avoided with the lower temperature limit could reduce the number of people exposed to climate risks and vulnerable to poverty by 62 to 457 million, and lessen the risks of poor people to experience food and water insecurity, adverse health impacts, and economic losses, particularly in regions that already face development challenges (*medium evidence, medium agreement*). {5.2.2, 5.2.3} Avoided impacts expected to occur between 1.5°C and 2°C warming would also make it easier to achieve certain SDGs, such as those that relate to poverty, hunger, health, water and sanitation, cities and ecosystems (SDGs 1, 2, 3, 6, 11, 14 and 15) (*medium evidence, high agreement*). {5.2.3, Table 5.2 available at the end of the chapter}

Compared to current conditions, 1.5°C of global warming would nonetheless pose heightened risks to eradicating poverty, reducing inequalities and ensuring human and ecosystem well-being (*medium evidence, high agreement*). Warming of 1.5°C is not considered 'safe' for most nations, communities, ecosystems and sectors and poses significant risks to natural and human systems as compared to the current warming of 1°C (*high confidence*). {Cross-Chapter Box 12 in Chapter 5} The impacts of 1.5°C of warming would disproportionately affect disadvantaged and vulnerable populations through food insecurity, higher food prices, income losses, lost livelihood opportunities, adverse health impacts and population displacements (*medium evidence, high agreement*). {5.2.1} Some of the worst impacts on sustainable development are expected to be felt among agricultural and coastal dependent livelihoods, indigenous people, children and the elderly, poor labourers, poor urban dwellers in African cities, and people and ecosystems in the Arctic and Small Island Developing States (SIDS) (*medium evidence, high agreement*). {5.2.1, Box 5.3, Chapter 3, Box 3.5, Cross-Chapter Box 9 in Chapter 4}

Climate Adaptation and Sustainable Development

Prioritization of sustainable development and meeting the SDGs is consistent with efforts to adapt to climate change (*high confidence*).** Many strategies for sustainable development enable transformational adaptation for a 1.5°C warmer world, provided attention is paid to reducing poverty in all its forms and to promoting equity and participation in decision-making (*medium evidence, high agreement*). As such, sustainable development has the potential to significantly reduce systemic vulnerability, enhance adaptive capacity, and promote livelihood security for poor and disadvantaged populations (*high confidence*). {5.3.1}

Synergies between adaptation strategies and the SDGs are expected to hold true in a 1.5°C warmer world, across sectors and contexts (*medium evidence, medium agreement*). Synergies between adaptation and sustainable development are significant for agriculture and health, advancing SDGs 1 (extreme poverty), 2 (hunger), 3 (healthy lives and well-being) and 6 (clean water) (*robust evidence, medium agreement*). {5.3.2} Ecosystem- and community-based adaptation, along with the incorporation of indigenous and local knowledge, advances synergies with SDGs 5 (gender equality), 10 (reducing inequalities) and 16 (inclusive societies), as exemplified in drylands and the Arctic (*high evidence, medium agreement*). {5.3.2, Box 5.1, Cross-Chapter Box 10 in Chapter 4}

Adaptation strategies can result in trade-offs with and among the SDGs (*medium evidence, high agreement*). Strategies that advance one SDG may create negative consequences for other SDGs, for instance SDGs 3 (health) versus 7 (energy consumption) and agricultural adaptation and SDG 2 (food security) versus SDGs 3 (health), 5 (gender equality), 6 (clean water), 10 (reducing inequalities), 14 (life below water) and 15 (life on the land) (*medium evidence, medium agreement*). {5.3.2}

Pursuing place-specific adaptation pathways towards a 1.5°C warmer world has the potential for significant positive outcomes for well-being in countries at all levels of development (*medium evidence, high agreement*). Positive outcomes emerge when adaptation pathways (i) ensure a diversity of adaptation options based on people's values and the trade-offs they consider acceptable, (ii) maximize synergies with sustainable development through inclusive, participatory and deliberative processes, and (iii) facilitate equitable transformation. Yet such pathways would be difficult to achieve without redistributive measures to overcome path dependencies, uneven power structures, and entrenched social inequalities (*medium evidence, high agreement*). {5.3.3}

Mitigation and Sustainable Development

The deployment of mitigation options consistent with 1.5°C pathways leads to multiple synergies across a range of sustainable development dimensions. At the same time, the rapid pace and magnitude of change that would be required to limit warming to 1.5°C, if not carefully managed, would lead to trade-offs with some sustainable development dimensions (*high confidence*). The number of synergies between mitigation response options and sustainable development exceeds the number of trade-offs in energy demand and supply sectors; agriculture, forestry and other land use (AFOLU); and for oceans (*very high confidence*). {Figure 5.2, Table 5.2 available at the end of the chapter} The 1.5°C pathways indicate robust synergies, particularly for the SDGs 3 (health), 7 (energy), 12 (responsible consumption and production) and 14 (oceans) (*very high confidence*). {5.4.2, Figure 5.3} For SDGs 1 (poverty), 2 (hunger), 6 (water) and 7 (energy), there is a risk of trade-offs or negative side effects from stringent mitigation actions compatible with 1.5°C of warming (*medium evidence, high agreement*). {5.4.2}

Appropriately designed mitigation actions to reduce energy demand can advance multiple SDGs simultaneously. Pathways compatible with 1.5°C that feature low energy demand show the most pronounced synergies and the lowest number of trade-offs with respect to sustainable development and the SDGs (*very high confidence*). Accelerating energy efficiency in all sectors has synergies with SDGs 7 (energy), 9 (industry, innovation and infrastructure), 11 (sustainable cities and communities), 12 (responsible consumption and production), 16 (peace, justice and strong institutions), and 17 (partnerships for the goals) (*robust evidence, high agreement*). {5.4.1, Figure 5.2, Table 5.2} Low-demand pathways, which would reduce or completely avoid the reliance on bioenergy with carbon capture and storage (BECCS) in 1.5°C pathways, would result in significantly reduced pressure on food security, lower food prices and fewer people at risk of hunger (*medium evidence, high agreement*). {5.4.2, Figure 5.3}

The impacts of carbon dioxide removal options on SDGs depend on the type of options and the scale of deployment (*high confidence*). If poorly implemented, carbon dioxide removal (CDR) options such as bioenergy, BECCS and AFOLU would lead to trade-offs. Appropriate design and implementation requires considering local people's needs, biodiversity and other sustainable development dimensions (*very high confidence*). {5.4.1.3, Cross-Chapter Box 7 in Chapter 3}

The design of the mitigation portfolios and policy instruments to limit warming to 1.5°C will largely determine the overall synergies and trade-offs between mitigation and sustainable development (*very high confidence*). Redistributive policies that shield the poor and vulnerable can resolve trade-offs for a range of SDGs (*medium evidence, high agreement*). Individual mitigation options are associated with both positive and negative interactions with the SDGs (*very high confidence*). {5.4.1} However, appropriate choices across the mitigation portfolio can help to maximize positive side effects while minimizing negative side effects (*high confidence*). {5.4.2, 5.5.2} Investment needs for complementary policies resolving trade-offs with a range of SDGs are only a small fraction of the overall mitigation investments in 1.5°C pathways (*medium evidence, high agreement*). {5.4.2, Figure 5.4} Integration of mitigation with adaptation and sustainable development compatible with 1.5°C warming requires a systems perspective (*high confidence*). {5.4.2, 5.5.2}

Mitigation consistent with 1.5°C of warming create high risks for sustainable development in countries with high dependency on fossil fuels for revenue and employment generation (*high confidence*). These risks are caused by the reduction of global demand affecting mining activity and export revenues and challenges to rapidly decrease high carbon intensity of the domestic economy (*robust evidence, high agreement*). {5.4.1.2, Box 5.2} Targeted policies that promote diversification of the economy and the energy sector could ease this transition (*medium evidence, high agreement*). {5.4.1.2, Box 5.2}

Sustainable Development Pathways to 1.5°C

Sustainable development broadly supports and often enables the fundamental societal and systems transformations that would be required for limiting warming to 1.5°C above pre-industrial levels (*high confidence*). Simulated pathways that feature the most sustainable worlds (e.g., Shared Socio-Economic Pathways (SSP) 1) are associated with relatively lower mitigation and adaptation challenges and limit warming to 1.5°C at comparatively lower mitigation costs. In contrast, development pathways with high fragmentation, inequality and poverty (e.g., SSP3) are associated with comparatively higher mitigation and adaptation challenges. In such pathways, it is not possible to limit warming to 1.5°C for the vast majority of the integrated assessment models (*medium evidence, high agreement*). {5.5.2} In all SSPs, mitigation costs substantially increase in 1.5°C pathways compared to 2°C pathways. No pathway in the literature integrates or achieves all 17 SDGs (*high confidence*). {5.5.2} Real-world experiences at the project level show that the actual integration between adaptation, mitigation and sustainable development is challenging as it requires reconciling trade-offs across sectors and spatial scales (*very high confidence*). {5.5.1}

Without societal transformation and rapid implementation of ambitious greenhouse gas reduction measures, pathways to limiting warming to 1.5°C and achieving sustainable development will be exceedingly difficult, if not impossible, to achieve (*high confidence*). The potential for pursuing such pathways differs between and within nations and regions, due to different development trajectories, opportunities and challenges (*very high confidence*). {5.5.3.2, Figure 5.1} Limiting warming to 1.5°C would require all countries and non-state actors to strengthen their contributions without delay. This could be achieved through sharing efforts based on bolder and more committed cooperation, with support for those with the least capacity to adapt, mitigate and transform (*medium evidence, high agreement*). {5.5.3.1, 5.5.3.2} Current efforts towards reconciling low-carbon trajectories and reducing inequalities, including those that avoid difficult trade-offs associated with transformation, are partially successful yet demonstrate notable obstacles (*medium evidence, medium agreement*). {5.5.3.3, Box 5.3, Cross-Chapter Box 13 in this chapter}

Social justice and equity are core aspects of climate-resilient development pathways for transformational social change. Addressing challenges and widening opportunities between and within countries and communities would be necessary to achieve sustainable development and limit warming to 1.5°C, without making the poor and disadvantaged worse off (*high confidence*). Identifying and navigating inclusive and socially acceptable pathways towards low-carbon, climate-resilient futures is a challenging yet important endeavour, fraught with moral, practical and political difficulties and inevitable trade-offs (*very high confidence*). {5.5.2, 5.5.3.3, Box 5.3} It entails deliberation and problem-solving processes to negotiate societal values, well-being, risks and resilience and to determine what is desirable and fair, and to whom (*medium evidence, high agreement*). Pathways that encompass joint, iterative planning and transformative visions, for instance in Pacific SIDS like Vanuatu and in urban contexts, show potential for liveable and sustainable futures (*high confidence*). {5.5.3.1, 5.5.3.3, Figure 5.5, Box 5.3, Cross-Chapter Box 13 in this chapter}

The fundamental societal and systemic changes to achieve sustainable development, eradicate poverty and reduce inequalities while limiting warming to 1.5°C would require meeting a set of institutional, social, cultural, economic and technological conditions (*high confidence*). The coordination and monitoring of policy actions across sectors and spatial scales is essential to support sustainable development in 1.5°C warmer conditions (*very high confidence*). {5.6.2, Box 5.3} External funding and technology transfer better support these efforts when they consider recipients' context-specific needs (*medium evidence, high agreement*). {5.6.1} Inclusive processes can facilitate transformations by ensuring participation, transparency, capacity building and iterative social learning (*high confidence*). {5.5.3.3, Cross-Chapter Box 13, 5.6.3} Attention to power asymmetries and unequal opportunities for development, among and within countries, is key to adopting 1.5°C-compatible development pathways that benefit all populations (*high confidence*). {5.5.3, 5.6.4, Box 5.3} Re-examining individual and collective values could help spur urgent, ambitious and cooperative change (*medium evidence, high agreement*). {5.5.3, 5.6.5}

Chapters

1 Framing and Context

Coordinating Lead Authors:
Myles R. Allen (UK), Opha Pauline Dube (Botswana), William Solecki (USA)

Lead Authors:
Fernando Aragón-Durand (Mexico), Wolfgang Cramer (France/Germany), Stephen Humphreys (UK/Ireland), Mikiko Kainuma (Japan), Jatin Kala (Australia), Natalie Mahowald (USA), Yacob Mulugetta (UK/Ethiopia), Rosa Perez (Philippines), Morgan Wairiu (Solomon Islands), Kirsten Zickfeld (Canada/Germany)

Contributing Authors:
Purnamita Dasgupta (India), Haile Eakin (USA), Bronwyn Hayward (New Zealand), Diana Liverman (USA), Richard Millar (UK), Graciela Raga (Mexico/Argentina), Aurélien Ribes (France), Mark Richardson (USA/UK), Maisa Rojas (Chile), Roland Séférian (France), Sonia I. Seneviratne (Switzerland), Christopher Smith (UK), Will Steffen (Australia), Peter Thorne (Ireland/UK)

Chapter Scientist:
Richard Millar (UK)

Review Editors:
Ismail Elgizouli Idris (Sudan), Andreas Fischlin (Switzerland), Xuejie Gao (China)

This chapter should be cited as:
Allen, M.R., O.P. Dube, W. Solecki, F. Aragón-Durand, W. Cramer, S. Humphreys, M. Kainuma, J. Kala, N. Mahowald, Y. Mulugetta, R. Perez, M. Wairiu, and K. Zickfeld, 2018: Framing and Context. In: *Global Warming of 1.5°C. An IPCC Special Report on the impacts of global warming of 1.5°C above pre-industrial levels and related global greenhouse gas emission pathways, in the context of strengthening the global response to the threat of climate change, sustainable development, and efforts to eradicate poverty* [Masson-Delmotte, V., P. Zhai, H.-O. Pörtner, D. Roberts, J. Skea, P.R. Shukla, A. Pirani, W. Moufouma-Okia, C. Péan, R. Pidcock, S. Connors, J.B.R. Matthews, Y. Chen, X. Zhou, M.I. Gomis, E. Lonnoy, T. Maycock, M. Tignor, and T. Waterfield (eds.)]. Cambridge University Press, Cambridge, UK and New York, NY, USA, pp. 49-92. https://doi.org/10.1017/9781009157940.003.

Table of Contents

Executive Summary

This chapter frames the context, knowledge-base and assessment approaches used to understand the impacts of 1.5°C global warming above pre-industrial levels and related global greenhouse gas emission pathways, building on the IPCC Fifth Assessment Report (AR5), in the context of strengthening the global response to the threat of climate change, sustainable development and efforts to eradicate poverty.

Human-induced warming reached approximately 1°C (*likely*** between 0.8°C and 1.2°C) above pre-industrial levels in 2017, increasing at 0.2°C (***likely*** between 0.1°C and 0.3°C) per decade (***high confidence***).** Global warming is defined in this report as an increase in combined surface air and sea surface temperatures averaged over the globe and over a 30-year period. Unless otherwise specified, warming is expressed relative to the period 1850–1900, used as an approximation of pre-industrial temperatures in AR5. For periods shorter than 30 years, warming refers to the estimated average temperature over the 30 years centred on that shorter period, accounting for the impact of any temperature fluctuations or trend within those 30 years. Accordingly, warming from pre-industrial levels to the decade 2006–2015 is assessed to be 0.87°C (*likely* between 0.75°C and 0.99°C). Since 2000, the estimated level of human-induced warming has been equal to the level of observed warming with a likely range of ±20% accounting for uncertainty due to contributions from solar and volcanic activity over the historical period (*high confidence*). {1.2.1}

Warming greater than the global average has already been experienced in many regions and seasons, with higher average warming over land than over the ocean (*high confidence***).** Most land regions are experiencing greater warming than the global average, while most ocean regions are warming at a slower rate. Depending on the temperature dataset considered, 20–40% of the global human population live in regions that, by the decade 2006–2015, had already experienced warming of more than 1.5°C above pre-industrial in at least one season (*medium confidence*). {1.2.1, 1.2.2}

Past emissions alone are *unlikely* **to raise global-mean temperature to 1.5°C above pre-industrial levels (***medium confidence***), but past emissions do commit to other changes, such as further sea level rise (***high confidence***).** If all anthropogenic emissions (including aerosol-related) were reduced to zero immediately, any further warming beyond the 1°C already experienced would *likely* be less than 0.5°C over the next two to three decades (*high confidence*), and *likely* less than 0.5°C on a century time scale (*medium confidence*), due to the opposing effects of different climate processes and drivers. A warming greater than 1.5°C is therefore not geophysically unavoidable: whether it will occur depends on future rates of emission reductions. {1.2.3, 1.2.4}

1.5°C emission pathways are defined as those that, given current knowledge of the climate response, provide a one-in-two to two-in-three chance of warming either remaining below 1.5°C or returning to 1.5°C by around 2100 following an overshoot. Overshoot pathways are characterized by the peak magnitude of the overshoot, which may have implications for impacts. All 1.5°C pathways involve limiting cumulative emissions of long-lived greenhouse gases, including carbon dioxide and nitrous oxide, and substantial reductions in other climate forcers (*high confidence*). Limiting cumulative emissions requires either reducing net global emissions of long-lived greenhouse gases to zero before the cumulative limit is reached, or net negative global emissions (anthropogenic removals) after the limit is exceeded. {1.2.3, 1.2.4, Cross-Chapter Boxes 1 and 2}

This report assesses projected impacts at a global average warming of 1.5°C and higher levels of warming. Global warming of 1.5°C is associated with global average surface temperatures fluctuating naturally on either side of 1.5°C, together with warming substantially greater than 1.5°C in many regions and seasons (*high confidence*), all of which must be considered in the assessment of impacts. Impacts at 1.5°C of warming also depend on the emission pathway to 1.5°C. Very different impacts result from pathways that remain below 1.5°C versus pathways that return to 1.5°C after a substantial overshoot, and when temperatures stabilize at 1.5°C versus a transient warming past 1.5°C (*medium confidence*). {1.2.3, 1.3}

Ethical considerations, and the principle of equity in particular, are central to this report, recognizing that many of the impacts of warming up to and beyond 1.5°C, and some potential impacts of mitigation actions required to limit warming to 1.5°C, fall disproportionately on the poor and vulnerable (*high confidence***).** Equity has procedural and distributive dimensions and requires fairness in burden sharing both between generations and between and within nations. In framing the objective of holding the increase in the global average temperature rise to well below 2°C above pre-industrial levels, and to pursue efforts to limit warming to 1.5°C, the Paris Agreement associates the principle of equity with the broader goals of poverty eradication and sustainable development, recognising that effective responses to climate change require a global collective effort that may be guided by the 2015 United Nations Sustainable Development Goals. {1.1.1}

Climate adaptation refers to the actions taken to manage impacts of climate change by reducing vulnerability and exposure to its harmful effects and exploiting any potential benefits. Adaptation takes place at international, national and local levels. Subnational jurisdictions and entities, including urban and rural municipalities, are key to developing and reinforcing measures for reducing weather- and climate-related risks. Adaptation implementation faces several barriers including lack of up-to-date and locally relevant information, lack of finance and technology, social values and attitudes, and institutional constraints (*high confidence*). Adaptation is more *likely* to contribute to sustainable development when policies align with mitigation and poverty eradication goals (*medium confidence*). {1.1, 1.4}

Ambitious mitigation actions are indispensable to limit warming to 1.5°C while achieving sustainable development and poverty eradication (*high confidence***).** Ill-designed responses, however, could pose challenges especially – but not exclusively – for countries and regions contending with poverty and those requiring significant transformation of their energy systems. This report focuses on 'climate-resilient development pathways', which aim to meet the goals of sustainable development, including climate adaptation and mitigation, poverty eradication and reducing inequalities. But any feasible pathway that remains within 1.5°C involves synergies and trade-offs (*high confidence*). Significant uncertainty remains as to which pathways are more consistent with the principle of equity. {1.1.1, 1.4}

Multiple forms of knowledge, including scientific evidence, narrative scenarios and prospective pathways, inform the understanding of 1.5°C. This report is informed by traditional evidence of the physical climate system and associated impacts and vulnerabilities of climate change, together with knowledge drawn from the perceptions of risk and the experiences of climate impacts and governance systems. Scenarios and pathways are used to explore conditions enabling goal-oriented futures while recognizing the significance of ethical considerations, the principle of equity, and the societal transformation needed. {1.2.3, 1.5.2}

There is no single answer to the question of whether it is feasible to limit warming to 1.5°C and adapt to the consequences. Feasibility is considered in this report as the capacity of a system as a whole to achieve a specific outcome. The global transformation that would be needed to limit warming to 1.5°C requires enabling conditions that reflect the links, synergies and trade-offs between mitigation, adaptation and sustainable development. These enabling conditions are assessed across many dimensions of feasibility – geophysical, environmental-ecological, technological, economic, socio-cultural and institutional – that may be considered through the unifying lens of the Anthropocene, acknowledging profound, differential but increasingly geologically significant human influences on the Earth system as a whole. This framing also emphasises the global interconnectivity of past, present and future human–environment relations, highlighting the need and opportunities for integrated responses to achieve the goals of the Paris Agreement. {1.1, Cross-Chapter Box 1}

1.1 Assessing the Knowledge Base for a 1.5°C Warmer World

Human influence on climate has been the dominant cause of observed warming since the mid-20th century, while global average surface temperature warmed by 0.85°C between 1880 and 2012, as reported in the IPCC Fifth Assessment Report, or AR5 (IPCC, 2013b). Many regions of the world have already greater regional-scale warming, with 20–40% of the global population (depending on the temperature dataset used) having experienced over 1.5°C of warming in at least one season (Figure 1.1; Chapter 3 Section 3.3.2.1). Temperature rise to date has already resulted in profound alterations to human and natural systems, including increases in droughts, floods, and some other types of extreme weather; sea level rise; and biodiversity loss – these changes are causing unprecedented risks to vulnerable persons and populations (IPCC, 2012a, 2014a; Mysiak et al., 2016; Chapter 3 Sections 3.4.5–3.4.13). The most affected people live in low and middle income countries, some of which have experienced a decline in food security, which in turn is partly linked to rising migration and poverty (IPCC, 2012a). Small islands, megacities, coastal regions, and high mountain ranges are likewise among the most affected (Albert et al., 2017). Worldwide, numerous ecosystems are at risk of severe impacts, particularly warm-water tropical reefs and Arctic ecosystems (IPCC, 2014a).

This report assesses current knowledge of the environmental, technical, economic, financial, socio-cultural, and institutional dimensions of a 1.5°C warmer world (meaning, unless otherwise specified, a world in which warming has been limited to 1.5°C relative to pre-industrial levels). Differences in vulnerability and exposure arise from numerous non-climatic factors (IPCC, 2014a). Global economic growth has been accompanied by increased life expectancy and income in much of the world; however, in addition to environmental degradation and pollution, many regions remain characterised by significant poverty and severe inequality in income distribution and access to resources, amplifying vulnerability to climate change (Dryzek, 2016; Pattberg and Zelli, 2016; Bäckstrand et al., 2017; Lövbrand et al., 2017). World population continues to rise, notably in hazard-prone small and medium-sized cities in low- and moderate-income countries (Birkmann et al., 2016). The spread of fossil-fuel-based material consumption and changing lifestyles is a major driver of global resource use, and the main contributor to rising greenhouse gas (GHG) emissions (Fleurbaey et al., 2014).

The overarching context of this report is this: human influence has become a principal agent of change on the planet, shifting the world out of the relatively stable Holocene period into a new geological era, often termed the Anthropocene (Box 1.1). Responding to climate change in the Anthropocene will require approaches that integrate multiple levels of interconnectivity across the global community.

This chapter is composed of seven sections linked to the remaining four chapters of the report. This introductory Section 1.1 situates the basic elements of the assessment within the context of sustainable development; considerations of ethics, equity and human rights; and the problem of poverty. Section 1.2 focuses on understanding 1.5°C, global versus regional warming, 1.5°C pathways, and associated emissions. Section 1.3 frames the impacts at 1.5°C and beyond on natural and human systems. The section on strengthening the global response (1.4) frames responses, governance and implementation, and trade-offs and synergies between mitigation, adaptation, and the Sustainable

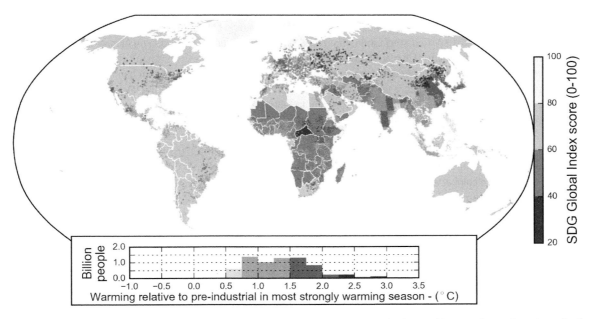

Figure 1.1 | Human experience of present-day warming. Different shades of pink to purple indicated by the inset histogram show estimated warming for the season that has warmed the most at a given location between the periods 1850–1900 and 2006–2015, during which global average temperatures rose by 0.91°C in this dataset (Cowtan and Way, 2014) and 0.87°C in the multi-dataset average (Table 1.1 and Figure 1.3). The density of dots indicates the population (in 2010) in any 1° × 1° grid box. The underlay shows national Sustainable Development Goal (SDG) Global Index Scores indicating performance across the 17 SDGs. Hatching indicates missing SDG index data (e.g., Greenland). The histogram shows the population (in 2010) living in regions experiencing different levels of warming (at 0.25°C increments). See Supplementary Material 1.SM for further details.

Development Goals (SDGs) under transformation, transformation pathways, and transition. Section 1.5 provides assessment frameworks and emerging methodologies that integrate climate change mitigation and adaptation with sustainable development. Section 1.6 defines approaches used to communicate confidence, uncertainty and risk, while 1.7 presents the storyline of the whole report.

Box 1.1 | The Anthropocene: Strengthening the Global Response to 1.5°C Global Warming

Introduction

The concept of the Anthropocene can be linked to the aspiration of the Paris Agreement. The abundant empirical evidence of the unprecedented rate and global scale of impact of human influence on the Earth System (Steffen et al., 2016; Waters et al., 2016) has led many scientists to call for an acknowledgement that the Earth has entered a new geological epoch: the Anthropocene (Crutzen and Stoermer, 2000; Crutzen, 2002; Gradstein et al., 2012). Although rates of change in the Anthropocene are necessarily assessed over much shorter periods than those used to calculate long-term baseline rates of change, and therefore present challenges for direct comparison, they are nevertheless striking. The rise in global CO_2 concentration since 2000 is about 20 ppm per decade, which is up to 10 times faster than any sustained rise in CO_2 during the past 800,000 years (Lüthi et al., 2008; Bereiter et al., 2015). AR5 found that the last geological epoch with similar atmospheric CO_2 concentration was the Pliocene, 3.3 to 3.0 Ma (Masson-Delmotte et al., 2013). Since 1970 the global average temperature has been rising at a rate of 1.7°C per century, compared to a long-term decline over the past 7,000 years at a baseline rate of 0.01°C per century (NOAA, 2016; Marcott et al., 2013). These global-level rates of human-driven change far exceed the rates of change driven by geophysical or biosphere forces that have altered the Earth System trajectory in the past (e.g., Summerhayes, 2015; Foster et al., 2017); even abrupt geophysical events do not approach current rates of human-driven change.

The Geological Dimension of the Anthropocene and 1.5°C Global Warming

The process of formalising the Anthropocene is on-going (Zalasiewicz et al., 2017), but a strong majority of the Anthropocene Working Group (AWG) established by the Subcommission on Quaternary Stratigraphy of the International Commission on Stratigraphy have agreed that: (i) the Anthropocene has a geological merit; (ii) it should follow the Holocene as a formal epoch in the Geological Time Scale; and, (iii) its onset should be defined as the mid-20th century. Potential markers in the stratigraphic record include an array of novel manufactured materials of human origin, and "these combined signals render the Anthropocene stratigraphically distinct from the Holocene and earlier epochs" (Waters et al., 2016). The Holocene period, which itself was formally adopted in 1885 by geological science community, began 11,700 years ago with a more stable warm climate providing for emergence of human civilisation and growing human-nature interactions that have expanded to give rise to the Anthropocene (Waters et al., 2016).

The Anthropocene and the Challenge of a 1.5° C Warmer World

The Anthropocene can be employed as a "boundary concept" (Brondizio et al., 2016) that frames critical insights into understanding the drivers, dynamics and specific challenges in responding to the ambition of keeping global temperature well below 2°C while pursuing efforts towards and adapting to a 1.5°C warmer world. The United Nations Framework Convention on Climate Change (UNFCCC) and its Paris Agreement recognize the ability of humans to influence geophysical planetary processes (Chapter 2, Cross-Chapter Box 1 in this chapter). The Anthropocene offers a structured understanding of the culmination of past and present human–environmental relations and provides an opportunity to better visualize the future to minimize pitfalls (Pattberg and Zelli, 2016; Delanty and Mota, 2017), while acknowledging the differentiated responsibility and opportunity to limit global warming and invest in prospects for climate-resilient sustainable development (Harrington, 2016) (Chapter 5). The Anthropocene also provides an opportunity to raise questions regarding the regional differences, social inequities, and uneven capacities and drivers of global social–environmental changes, which in turn inform the search for solutions as explored in Chapter 4 of this report (Biermann et al., 2016). It links uneven influences of human actions on planetary functions to an uneven distribution of impacts (assessed in Chapter 3) as well as the responsibility and response capacity to, for example, limit global warming to no more than a 1.5°C rise above pre-industrial levels. Efforts to curtail greenhouse gas emissions without incorporating the intrinsic interconnectivity and disparities associated with the Anthropocene world may themselves negatively affect the development ambitions of some regions more than others and negate sustainable development efforts (see Chapter 2 and Chapter 5).

1.1.1   Equity and a 1.5°C Warmer World

The AR5 suggested that equity, sustainable development, and poverty eradication are best understood as mutually supportive and co-achievable within the context of climate action and are underpinned by various other international hard and soft law instruments (Denton et al., 2014; Fleurbaey et al., 2014; Klein et al., 2014; Olsson et al., 2014; Porter et al., 2014; Stavins et al., 2014). The aim of the Paris Agreement under the UNFCCC to 'pursue efforts to limit' the rise in global temperatures to 1.5°C above pre-industrial levels raises ethical concerns that have long been central to climate debates (Fleurbaey et al., 2014; Kolstad et al., 2014). The Paris Agreement makes particular reference to the principle of equity, within the context of broader international goals of sustainable development and poverty eradication. Equity is a long-standing principle within international law and climate change law in particular (Shelton, 2008; Bodansky et al., 2017).

The AR5 describes equity as having three dimensions: intergenerational (fairness between generations), international (fairness between states), and national (fairness between individuals) (Fleurbaey et al., 2014). The principle is generally agreed to involve both procedural justice (i.e., participation in decision making) and distributive justice (i.e., how the costs and benefits of climate actions are distributed) (Kolstad et al., 2014; Savaresi, 2016; Reckien et al., 2017). Concerns regarding equity have frequently been central to debates around mitigation, adaptation and climate governance (Caney, 2005; Schroeder et al., 2012; Ajibade, 2016; Reckien et al., 2017; Shue, 2018). Hence, equity provides a framework for understanding the asymmetries between the distributions of benefits and costs relevant to climate action (Schleussner et al., 2016; Aaheim et al., 2017).

Four key framing asymmetries associated with the conditions of a 1.5°C warmer world have been noted (Okereke, 2010; Harlan et al., 2015; Ajibade, 2016; Savaresi, 2016; Reckien et al., 2017) and are reflected in the report's assessment. The first concerns differential contributions to the problem: the observation that the benefits from industrialization have been unevenly distributed and those who benefited most historically also have contributed most to the current climate problem and so bear greater responsibility (Shue, 2013; McKinnon, 2015; Otto et al., 2017; Skeie et al., 2017). The second asymmetry concerns differential impact: the worst impacts tend to fall on those least responsible for the problem, within states, between states, and between generations (Fleurbaey et al., 2014; Shue, 2014; Ionesco et al., 2016). The third is the asymmetry in capacity to shape solutions and response strategies, such that the worst-affected states, groups, and individuals are not always well represented (Robinson and Shine, 2018). Fourth, there is an asymmetry in future response capacity: some states, groups, and places are at risk of being left behind as the world progresses to a low-carbon economy (Fleurbaey et al., 2014; Shue, 2014; Humphreys, 2017).

A sizeable and growing literature exists on how best to operationalize climate equity considerations, drawing on other concepts mentioned in the Paris Agreement, notably its explicit reference to human rights (OHCHR, 2009; Caney, 2010; Adger et al., 2014; Fleurbaey et al., 2014; IBA, 2014; Knox, 2015; Duyck et al., 2018; Robinson and Shine, 2018). Human rights comprise internationally agreed norms that align with the Paris ambitions of poverty eradication, sustainable development, and the reduction of vulnerability (Caney, 2010; Fleurbaey et al., 2014; OHCHR, 2015). In addition to defining substantive rights (such as to life, health, and shelter) and procedural rights (such as to information and participation), human rights instruments prioritise the rights of marginalized groups, children, vulnerable and indigenous persons, and those discriminated against on grounds such as gender, race, age or disability (OHCHR, 2017). Several international human rights obligations are relevant to the implementation of climate actions and consonant with UNFCCC undertakings in the areas of mitigation, adaptation, finance, and technology transfer (Knox, 2015; OHCHR, 2015; Humphreys, 2017).

Much of this literature is still new and evolving (Holz et al., 2017; Dooley et al., 2018; Klinsky and Winkler, 2018), permitting the present report to examine some broader equity concerns raised both by possible failure to limit warming to 1.5°C and by the range of ambitious mitigation efforts that may be undertaken to achieve that limit. Any comparison between 1.5°C and higher levels of warming implies risk assessments and value judgements and cannot straightforwardly be reduced to a cost-benefit analysis (Kolstad et al., 2014). However, different levels of warming can nevertheless be understood in terms of their different implications for equity – that is, in the comparative distribution of benefits and burdens for specific states, persons, or generations, and in terms of their likely impacts on sustainable development and poverty (see especially Sections 2.3.4.2, 2.5, 3.4.5–3.4.13, 3.6, 5.4.1, 5.4.2, 5.6 and Cross-Chapter boxes 6 in Chapter 3 and 12 in Chapter 5).

1.1.2 Eradication of Poverty

This report assesses the role of poverty and its eradication in the context of strengthening the global response to the threat of climate change and sustainable development. A wide range of definitions for *poverty* exist. The AR5 discussed 'poverty' in terms of its multidimensionality, referring to 'material circumstances' (e.g., needs, patterns of deprivation, or limited resources), as well as to economic conditions (e.g., standard of living, inequality, or economic position), and/or social relationships (e.g., social class, dependency, lack of basic security, exclusion, or lack of entitlement; Olsson et al., 2014). The UNDP now uses a Multidimensional Poverty Index and estimates that about 1.5 billion people globally live in multidimensional poverty, especially in rural areas of South Asia and Sub-Saharan Africa, with an additional billion at risk of falling into poverty (UNDP, 2016).

A large and rapidly growing body of knowledge explores the connections between climate change and poverty. Climatic variability and climate change are widely recognized as factors that may exacerbate poverty, particularly in countries and regions where poverty levels are high (Leichenko and Silva, 2014). The AR5 noted that climate change-driven impacts often act as a threat multiplier in that the impacts of climate change compound other drivers of poverty (Olsson et al., 2014). Many vulnerable and poor people are dependent on activities such as agriculture that are highly susceptible to temperature increases and variability in precipitation patterns (Shiferaw et al., 2014; Miyan, 2015). Even modest changes in rainfall and temperature patterns can push marginalized people into poverty as they lack the means to recover from associated impacts. Extreme events, such as floods, droughts, and heat waves, especially when they occur in series, can significantly erode poor people's assets and further undermine their livelihoods in terms of labour productivity, housing, infrastructure and social networks (Olsson et al., 2014).

1.1.3 Sustainable Development and a 1.5°C Warmer World

AR5 (IPCC, 2014c) noted with *high confidence* that 'equity is an integral dimension of sustainable development' and that 'mitigation and adaptation measures can strongly affect broader sustainable development and equity objectives' (Fleurbaey et al., 2014). Limiting global warming to 1.5°C would require substantial societal and technological transformations, dependent in turn on global and regional sustainable development pathways. A range of pathways, both sustainable and not, are explored in this report, including implementation strategies to understand the enabling conditions and challenges required for such a transformation. These pathways and connected strategies are framed within the context of sustainable development, and in particular the United Nations 2030 Agenda for Sustainable Development (UN, 2015b) and Cross-Chapter Box 4 on SDGs (in this chapter). The feasibility of staying within 1.5°C depends upon a range of enabling conditions with geophysical, environmental–ecological, technological, economic, socio-cultural, and institutional dimensions. Limiting warming to 1.5°C also involves identifying technology and policy levers to accelerate the pace of transformation (see Chapter 4). Some pathways are more consistent than others with the requirements for sustainable development (see Chapter 5). Overall, the three-pronged emphasis on sustainable development, resilience, and transformation provides Chapter 5 an opportunity to assess the conditions of simultaneously reducing societal vulnerabilities, addressing entrenched inequalities, and breaking the circle of poverty.

The feasibility of any global commitment to a 1.5°C pathway depends, in part, on the cumulative influence of the nationally determined contributions (NDCs), committing nation states to specific GHG emission reductions. The current NDCs, extending only to 2030, do not limit warming to 1.5°C. Depending on mitigation decisions after 2030, they cumulatively track toward a warming of 3°-4°C above pre-industrial temperatures by 2100, with the potential for further warming thereafter (Rogelj et al., 2016a; UNFCCC, 2016). The analysis of pathways in this report reveals opportunities for greater decoupling of economic growth from GHG emissions. Progress towards limiting warming to 1.5°C requires a significant acceleration of this trend. AR5 concluded that climate change constrains possible development paths, that synergies and trade-offs exist between climate responses and socio-economic contexts, and that opportunities for effective climate responses overlap with opportunities for sustainable development, noting that many existing societal patterns of consumption are intrinsically unsustainable (Fleurbaey et al., 2014).

1.2 Understanding 1.5°C: Reference Levels, Probability, Transience, Overshoot, and Stabilization

1.2.1 Working Definitions of 1.5°C and 2°C Warming Relative to Pre-Industrial Levels

What is meant by 'the increase in global average temperature… above pre-industrial levels' referred to in the Paris Agreement depends on the choice of pre-industrial reference period, whether 1.5°C refers to total warming or the human-induced component of that warming, and which variables and geographical coverage are used to define global average temperature change. The cumulative impact of these definitional ambiguities (e.g., Hawkins et al., 2017; Pfleiderer et al., 2018) is comparable to natural multi-decadal temperature variability

on continental scales (Deser et al., 2012) and primarily affects the historical period, particularly that prior to the early 20th century when data is sparse and of less certain quality. Most practical mitigation and adaptation decisions do not depend on quantifying historical warming to this level of precision, but a consistent working definition is necessary to ensure consistency across chapters and figures. We adopt definitions that are as consistent as possible with key findings of AR5 with respect to historical warming.

This report defines 'warming', unless otherwise qualified, as an increase in multi-decade global mean surface temperature (GMST) above pre-industrial levels. Specifically, warming at a given point in time is defined as the global average of combined land surface air and sea surface temperatures for a 30-year period centred on that time, expressed relative to the reference period 1850–1900 (adopted for consistency with Box SPM.1 Figure 1 of IPCC (2014a)) 'as an approximation of pre-industrial levels', excluding the impact of natural climate fluctuations within that 30-year period and assuming any secular trend continues throughout that period, extrapolating into the future if necessary. There are multiple ways of accounting for natural fluctuations and trends (e.g., Foster and Rahmstorf, 2011; Haustein et al., 2017; Medhaug et al., 2017; Folland et al., 2018; Visser et al., 2018), but all give similar results. A major volcanic eruption might temporarily reduce observed global temperatures, but would not reduce warming as defined here (Bethke et al., 2017). Likewise, given that the level of warming is currently increasing at 0.3°C–0.7°C per 30 years (*likely* range quoted in Kirtman et al., 2013 and supported by Folland et al., 2018), the level of warming in 2017 was 0.15°C–0.35°C higher than average warming over the 30-year period 1988–2017.

In summary, this report adopts a working definition of '1.5°C relative to pre-industrial levels' that corresponds to global average combined land surface air and sea surface temperatures either 1.5°C warmer than the average of the 51-year period 1850–1900, 0.87°C warmer than the 20-year period 1986–2005, or 0.63°C warmer than the decade 2006–2015. These offsets are based on all available published global datasets, combined and updated, which show that 1986–2005 was 0.63°C warmer than 1850–1900 (with a 5–95% range of 0.57°C–0.69°C based on observational uncertainties alone), and 2006–2015 was 0.87°C warmer than 1850–1900 (with a *likely* range of 0.75°C–0.99°C, also accounting for the possible impact of natural fluctuations). Where possible, estimates of impacts and mitigation pathways are evaluated relative to these more recent periods. Note that the 5–95% intervals often quoted in square brackets in AR5 correspond to *very likely* ranges, while *likely* ranges correspond to 17–83%, or the central two-thirds, of the distribution of uncertainty.

1.2.1.1   Definition of global average temperature

The IPCC has traditionally defined changes in observed GMST as a weighted average of near-surface air temperature (SAT) changes over land and sea surface temperature (SST) changes over the oceans (Morice et al., 2012; Hartmann et al., 2013), while modelling studies have typically used a simple global average SAT. For ambitious mitigation goals, and under conditions of rapid warming or declining sea ice (Berger et al., 2017), the difference can be significant. Cowtan et al. (2015) and Richardson et al. (2016) show that the use of blended SAT/SST data and incomplete coverage together can give approximately 0.2°C less warming from the 19th century to the present relative to the use of complete global-average SAT (Stocker et al., 2013, Figure TFE8.1 and Figure 1.2). However, Richardson et al. (2018) show that this is primarily an issue for the interpretation of the historical record to date, with less absolute impact on projections of future changes, or estimated emissions budgets, under ambitious mitigation scenarios.

The three GMST reconstructions used in AR5 differ in their treatment of missing data. GISTEMP (Hansen et al., 2010) uses interpolation to infer trends in poorly observed regions like the Arctic (although even this product is spatially incomplete in the early record), while NOAAGlobalTemp (Vose et al., 2012) and HadCRUT (Morice et al., 2012) are progressively closer to a simple average of available observations. Since the AR5, considerable effort has been devoted to more sophisticated statistical modelling to account for the impact

of incomplete observation coverage (Rohde et al., 2013; Cowtan and Way, 2014; Jones, 2016). The main impact of statistical infilling is to increase estimated warming to date by about 0.1°C (Richardson et al., 2018 and Table 1.1).

We adopt a working definition of warming over the historical period based on an average of the four available global datasets that are supported by peer-reviewed publications: the three datasets used in the AR5, updated (Karl et al., 2015), together with the Cowtan-Way infilled dataset (Cowtan and Way, 2014). A further two datasets, Berkeley Earth (Rohde et al., 2013) and that of the Japan Meteorological Agency (JMA), are provided in Table 1.1. This working definition provides an updated estimate of 0.86°C for the warming over the period 1880–2012 based on a linear trend. This quantity was quoted as 0.85°C in the AR5. Hence the inclusion of the Cowtan-Way dataset does not introduce any inconsistency with the AR5, whereas redefining GMST to represent global SAT could increase this figure by up to 20% (Table 1.1, blue lines in Figure 1.2 and Richardson et al., 2016).

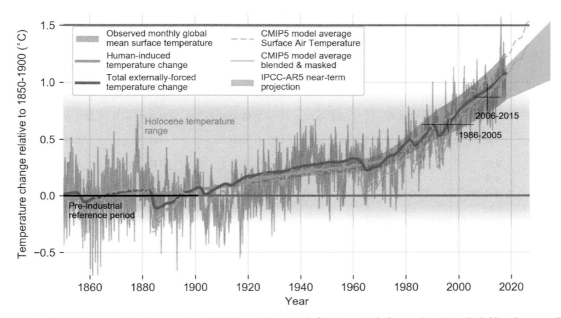

Figure 1.2 | Evolution of global mean surface temperature (GMST) over the period of instrumental observations. Grey shaded line shows monthly mean GMST in the HadCRUT4, NOAAGlobalTemp, GISTEMP and Cowtan-Way datasets, expressed as departures from 1850–1900, with varying grey line thickness indicating inter-dataset range. All observational datasets shown represent GMST as a weighted average of near surface air temperature over land and sea surface temperature over oceans. Human-induced (yellow) and total (human- and naturally-forced, orange) contributions to these GMST changes are shown calculated following Otto et al. (2015) and Haustein et al. (2017). Fractional uncertainty in the level of human-induced warming in 2017 is set equal to ±20% based on multiple lines of evidence. Thin blue lines show the modelled global mean surface air temperature (dashed) and blended surface air and sea surface temperature accounting for observational coverage (solid) from the CMIP5 historical ensemble average extended with RCP8.5 forcing (Cowtan et al., 2015; Richardson et al., 2018). The pink shading indicates a range for temperature fluctuations over the Holocene (Marcott et al., 2013). Light green plume shows the AR5 prediction for average GMST over 2016–2035 (Kirtman et al., 2013). See Supplementary Material 1.SM for further details.

1.2.1.2   Choice of reference period

Any choice of reference period used to approximate 'pre-industrial' conditions is a compromise between data coverage and representativeness of typical pre-industrial solar and volcanic forcing conditions. This report adopts the 51-year reference period, 1850–1900 inclusive, assessed as an approximation of pre-industrial levels in AR5 (Box TS.5, Figure 1 of Field et al., 2014). The years 1880–1900 are subject to strong but uncertain volcanic forcing, but

in the HadCRUT4 dataset, average temperatures over 1850–1879, prior to the largest eruptions, are less than 0.01°C from the average for 1850–1900. Temperatures rose by 0.0°C–0.2°C from 1720–1800 to 1850–1900 (Hawkins et al., 2017), but the anthropogenic contribution to this warming is uncertain (Abram et al., 2016; Schurer et al., 2017). The 18th century represents a relatively cool period in the context of temperatures since the mid-Holocene (Marcott et al., 2013; Lüning and Vahrenholt, 2017; Marsicek et al., 2018), which is indicated by the pink shaded region in Figure 1.2.

Projections of responses to emission scenarios, and associated impacts, may use a more recent reference period, offset by historical observations, to avoid conflating uncertainty in past and future changes (e.g., Hawkins et al., 2017; Millar et al., 2017b; Simmons et al., 2017). Two recent reference periods are used in this report: 1986–2005 and 2006–2015. In the latter case, when using a single decade to represent a 30-year average centred on that decade, it is important to consider the potential impact of internal climate variability. The years 2008–2013 were characterised by persistent cool conditions in the Eastern Pacific (Kosaka and Xie, 2013; Medhaug et al., 2017), related to both the El Niño-Southern Oscillation (ENSO) and, potentially, multi-decadal Pacific variability (e.g., England et al., 2014), but these were partially compensated for by El Niño conditions in 2006 and 2015. Likewise, volcanic activity depressed temperatures in 1986–2005, partly offset by the very strong El Niño event in 1998. Figure 1.2 indicates that natural variability (internally generated and externally driven) had little net impact on average temperatures over 2006–2015, in that the average temperature of the decade

is similar to the estimated externally driven warming. When solar, volcanic and ENSO-related variability is taken into account following the procedure of Foster and Rahmstorf (2011), there is no indication of average temperatures in either 1986–2005 or 2006–2015 being substantially biased by short-term variability (see Supplementary Material 1.SM.2). The temperature difference between these two reference periods (0.21°C–0.27°C over 15 years across available datasets) is also consistent with the AR5 assessment of the current warming rate of 0.3°C–0.7°C over 30 years (Kirtman et al., 2013).

On the definition of warming used here, warming to the decade 2006–2015 comprises an estimate of the 30-year average centred on this decade, or 1996–2025, assuming the current trend continues and that any volcanic eruptions that might occur over the final seven years are corrected for. Given this element of extrapolation, we use the AR5 near-term projection to provide a conservative uncertainty range. Combining the uncertainty in observed warming to 1986–2005 (±0.06°C) with the *likely* range in the current warming trend as

Table 1.1 | Observed increase in global average surface temperature in various datasets.
Numbers in square brackets correspond to 5–95% uncertainty ranges from individual datasets, encompassing known sources of observational uncertainty only.

Diagnostic / dataset	1850–1900 to (1) 2006–2015	1850–1900 to (2) 1986–2005	1986–2005 to (3) 2006–2015	1850–1900 to (4) 1981–2010	1850–1900 to (5) 1998–2017	Trend (6) 1880–2012	Trend (6) 1880–2015
HadCRUT4.6	0.84 [0.79–0.89]	0.60 [0.57–0.66]	0.22 [0.21–0.23]	0.62 [0.58–0.67]	0.83 [0.78–0.88]	0.83 [0.77–0.90]	0.88 [0.83–0.95]
NOAAGlobalTemp (7)	0.86	0.62	0.22	0.63	0.85	0.85	0.91
GISTEMP (7)	0.89	0.65	0.23	0.66	0.88	0.89	0.94
Cowtan-Way	0.91 [0.85–0.99]	0.65 [0.60–0.72]	0.26 [0.25–0.27]	0.65 [0.60–0.72]	0.88 [0.82–0.96]	0.88 [0.79–0.98]	0.93 [0.85–1.03]
Average (8)	**0.87**	0.63	0.23	0.64	0.86	0.86	0.92
Berkeley (9)	0.98	0.73	0.25	0.73	0.97	0.97	1.02
JMA (9)	0.82	0.59	0.17	0.60	0.81	0.82	0.87
ERA-Interim	N/A	N/A	0.26	N/A	N/A	N/A	N/A
JRA-55	N/A	N/A	0.23	N/A	N/A	N/A	N/A
CMIP5 global SAT (10)	0.99 [0.65–1.37]	0.62 [0.38–0.94]	0.38 [0.24–0.62]	0.62 [0.34–0.93]	0.89 [0.62–1.29]	0.81 [0.58–1.31]	0.86 [0.63–1.39]
CMIP5 SAT/SST blend-masked	0.86 [0.54–1.18]	0.50 [0.31–0.79]	0.34 [0.19–0.54]	0.48 [0.26–0.79]	0.75 [0.52–1.11]	0.68 [0.45–1.08]	0.74 [0.51–1.14]

Notes:

1) Most recent reference period used in this report.

2) Most recent reference period used in AR5.

3) Difference between recent reference periods.

4) Current WMO standard reference periods.

5) Most recent 20-year period.

6) Linear trends estimated by a straight-line fit, expressed in degrees yr−1 multiplied by 133 or 135 years respectively, with uncertainty ranges incorporating observational uncertainty only.

7) To estimate changes in the NOAAGlobalTemp and GISTEMP datasets relative to the 1850–1900 reference period, warming is computed relative to 1850–1900 using the HadCRUT4.6 dataset and scaled by the ratio of the linear trend 1880–2015 in the NOAAGlobalTemp or GISTEMP dataset with the corresponding linear trend computed from HadCRUT4.

8) Average of diagnostics derived – see (7) – from four peer-reviewed global datasets, HadCRUT4.6, NOAA, GISTEMP & Cowtan-Way. Note that differences between averages may not coincide with average differences because of rounding.

9) No peer-reviewed publication available for these global combined land–sea datasets.

10) CMIP5 changes estimated relative to 1861–80 plus 0.02°C for the offset in HadCRUT4.6 from 1850–1900. CMIP5 values are the mean of the RCP8.5 ensemble, with 5–95% ensemble range. They are included to illustrate the difference between a complete global surface air temperature record (SAT) and a blended surface air and sea surface temperature (SST) record accounting for incomplete coverage (masked), following Richardson et al. (2016). Note that 1986–2005 temperatures in CMIP5 appear to have been depressed more than observed temperatures by the eruption of Mount Pinatubo.

assessed by AR5 (±0.2°C/30 years), assuming these are uncorrelated, and using observed warming relative to 1850–1900 to provide the central estimate (no evidence of bias from short-term variability), gives an assessed warming to the decade 2006–2015 of 0.87°C with a ±0.12°C *likely* range. This estimate has the advantage of traceability to the AR5, but more formal methods of quantifying externally driven warming (e.g., Bindoff et al., 2013; Jones et al., 2016; Haustein et al., 2017; Ribes et al., 2017), which typically give smaller ranges of uncertainty, may be adopted in the future.

1.2.1.3   Total versus human-induced warming and warming rates

Total warming refers to the actual temperature change, irrespective of cause, while human-induced warming refers to the component of that warming that is attributable to human activities. Mitigation studies focus on human-induced warming (that is not subject to internal climate variability), while studies of climate change impacts typically refer to total warming (often with the impact of internal variability minimised through the use of multi-decade averages).

In the absence of strong natural forcing due to changes in solar or volcanic activity, the difference between total and human-induced warming is small: assessing empirical studies quantifying solar and volcanic contributions to GMST from 1890 to 2010, AR5 (Figure 10.6 of Bindoff et al., 2013) found their net impact on warming over the full period to be less than plus or minus 0.1°C. Figure 1.2 shows that the level of human-induced warming has been indistinguishable from total observed warming since 2000, including over the decade 2006–2015. Bindoff et al. (2013) assessed the magnitude of human-induced warming over the period 1951–2010 to be 0.7°C (*likely* between 0.6°C and 0.8°C), which is slightly greater than the 0.65°C observed warming over this period (Figures 10.4 and 10.5) with a *likely* range of ±14%. The key surface temperature attribution studies underlying this finding (Gillett et al., 2013; Jones et al., 2013; Ribes and Terray, 2013) used temperatures since the 19th century to constrain human-induced warming, and so their results are equally applicable to the attribution of causes of warming over longer periods. Jones et al. (2016) show (Figure 10) human-induced warming trends over the period 1905–2005 to be indistinguishable from the corresponding total observed warming trend accounting for natural variability using spatio-temporal detection patterns from 12 out of 15 CMIP5 models and from the multi-model average. Figures from Ribes and Terray (2013), show the anthropogenic contribution to the observed linear warming trend 1880–2012 in the HadCRUT4 dataset (0.83°C in Table 1.1) to be 0.86°C using a multi-model average global diagnostic, with a 5–95% confidence interval of 0.72°C–1.00°C (see figure 1.SM.6). In all cases, since 2000 the estimated combined contribution of solar and volcanic activity to warming relative to 1850–1900 is found to be less than ±0.1°C (Gillett et al., 2013), while anthropogenic warming is indistinguishable from, and if anything slightly greater than, the total observed warming, with 5–95% confidence intervals typically around ±20%.

Haustein et al. (2017) give a 5–95% confidence interval for human-induced warming in 2017 of 0.87°C–1.22°C, with a best estimate of 1.02°C, based on the HadCRUT4 dataset accounting

for observational and forcing uncertainty and internal variability. Applying their method to the average of the four datasets shown in Figure 1.2 gives an average level of human-induced warming in 2017 of 1.04°C. They also estimate a human-induced warming trend over the past 20 years of 0.17°C (0.13°C–0.33°C) per decade, consistent with estimates of the total observed trend of Foster and Rahmstorf (2011) (0.17° ± 0.03°C per decade, uncertainty in linear trend only), Folland et al. (2018) and Kirtman et al. (2013) (0.3°C–0.7°C over 30 years, or 0.1°C–0.23°C per decade, *likely* range), and a best-estimate warming rate over the past five years of 0.215°C/decade (Leach et al., 2018). Drawing on these multiple lines of evidence, human-induced warming is assessed to have reached 1.0°C in 2017, having increased by 0.13°C from the mid-point of 2006–2015, with a *likely* range of 0.8°C to 1.2°C (reduced from 5–95% to account for additional forcing and model uncertainty), increasing at 0.2°C per decade (with a *likely* range of 0.1°C to 0.3°C per decade: estimates of human-induced warming given to 0.1°C precision only).

Since warming is here defined in terms of a 30-year average, corrected for short-term natural fluctuations, when warming is considered to be at 1.5°C, global temperatures would fluctuate equally on either side of 1.5°C in the absence of a large cooling volcanic eruption (Bethke et al., 2017). Figure 1.2 indicates there is a substantial chance of GMST in a single month fluctuating over 1.5°C between now and 2020 (or, by 2030, for a longer period: Henley and King, 2017), but this would not constitute temperatures 'reaching 1.5°C' on our working definition. Rogelj et al. (2017) show limiting the probability of annual GMST exceeding 1.5°C to less than one-year-in-20 would require limiting warming, on the definition used here, to 1.31°C or lower.

1.2.2   Global versus Regional and Seasonal Warming

Warming is not observed or expected to be spatially or seasonally uniform (Collins et al., 2013). A 1.5°C increase in GMST will be associated with warming substantially greater than 1.5°C in many land regions, and less than 1.5°C in most ocean regions. This is illustrated by Figure 1.3, which shows an estimate of the observed change in annual and seasonal average temperatures between the 1850–1900 pre-industrial reference period and the decade 2006–2015 in the Cowtan-Way dataset. These regional changes are associated with an observed GMST increase of 0.91°C in the dataset shown here, or 0.87°C in the four-dataset average (Table 1.1). This observed pattern reflects an on-going transient warming: features such as enhanced warming over land may be less pronounced, but still present, in equilibrium (Collins et al., 2013). This figure illustrates the magnitude of spatial and seasonal differences, with many locations, particularly in Northern Hemisphere mid-latitude winter (December–February), already experiencing regional warming more than double the global average. Individual seasons may be substantially warmer, or cooler, than these expected changes in the long-term average.

1.2.3   Definition of 1.5°C Pathways: Probability, Transience, Stabilization and Overshoot

Pathways considered in this report, consistent with available literature on 1.5°C, primarily focus on the time scale up to 2100, recognising that the evolution of GMST after 2100 is also important. Two broad

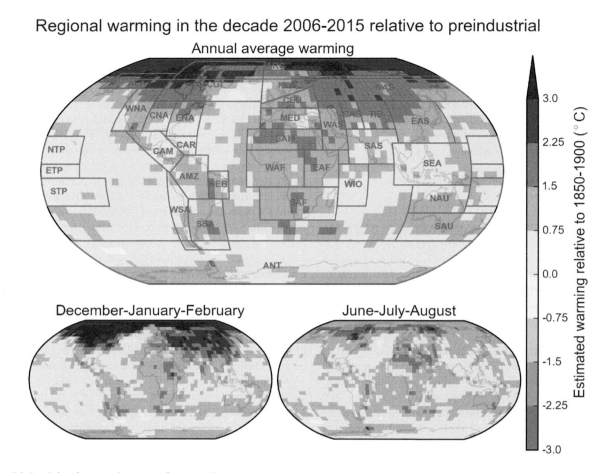

Figure 1.3 | Spatial and seasonal pattern of present-day warming: Regional warming for the 2006–2015 decade relative to 1850–1900 for the annual mean (top), the average of December, January, and February (bottom left) and for June, July, and August (bottom right). Warming is evaluated by regressing regional changes in the Cowtan and Way (2014) dataset onto the total (combined human and natural) externally forced warming (yellow line in Figure 1.2). See Supplementary Material 1.SM for further details and versions using alternative datasets. The definition of regions (green boxes and labels in top panel) is adopted from the AR5 (Christensen et al., 2013).

categories of 1.5°C pathways can be used to characterise mitigation options and impacts: pathways in which warming (defined as 30-year averaged GMST relative to pre-industrial levels, see Section 1.2.1) remains below 1.5°C throughout the 21st century, and pathways in which warming temporarily exceeds ('overshoots') 1.5°C and returns to 1.5°C either before or soon after 2100. Pathways in which warming exceeds 1.5°C before 2100, but might return to that level in some future century, are not considered 1.5°C pathways.

Because of uncertainty in the climate response, a 'prospective' mitigation pathway (see Cross-Chapter Box 1 in this chapter), in which emissions are prescribed, can only provide a level of probability of warming remaining below a temperature threshold. This probability cannot be quantified precisely since estimates depend on the method used (Rogelj et al., 2016b; Millar et al., 2017b; Goodwin et al., 2018; Tokarska and Gillett, 2018). This report defines a '1.5°C pathway' as a pathway of emissions and associated possible temperature responses in which the majority of approaches using presently available information assign a probability of approximately one-in-two to two-in-three to warming remaining below 1.5°C or, in the case of an overshoot pathway, to warming returning to 1.5°C by around 2100 or earlier. Recognizing the very different potential impacts and risks associated with high-overshoot pathways, this report singles

out 1.5°C pathways with no or limited (<0.1°C) overshoot in many instances and pursues efforts to ensure that when the term '1.5°C pathway' is used, the associated overshoot is made explicit where relevant. In Chapter 2, the classification of pathways is based on one modelling approach to avoid ambiguity, but probabilities of exceeding 1.5°C are checked against other approaches to verify that they lie within this approximate range. All these absolute probabilities are imprecise, depend on the information used to constrain them, and hence are expected to evolve in the future. Imprecise probabilities can nevertheless be useful for decision-making, provided the imprecision is acknowledged (Hall et al., 2007; Kriegler et al., 2009; Simpson et al., 2016). Relative and rank probabilities can be assessed much more consistently: approaches may differ on the absolute probability assigned to individual outcomes, but typically agree on which outcomes are more probable.

Importantly, 1.5°C pathways allow a substantial (up to one-in-two) chance of warming still exceeding 1.5°C. An 'adaptive' mitigation pathway in which emissions are continuously adjusted to achieve a specific temperature outcome (e.g., Millar et al., 2017b) reduces uncertainty in the temperature outcome while increasing uncertainty in the emissions required to achieve it. It has been argued (Otto et al., 2015; Xu and Ramanathan, 2017) that achieving very ambitious temperature goals will require such an adaptive approach to mitigation, but very few studies have been performed taking this approach (e.g., Jarvis et al., 2012).

Figure 1.4 illustrates categories of (a) 1.5°C pathways and associated (b) annual and (c) cumulative emissions of CO_2. It also shows (d) an example of a 'time-integrated impact' that continues to increase even after GMST has stabilised, such as sea level rise. This schematic assumes for the purposes of illustration that the fractional contribution of non-CO_2 climate forcers to total anthropogenic forcing (which is currently increasing, Myhre et al., 2017) is approximately constant from now on. Consequently, total human-induced warming is proportional to cumulative CO_2 emissions (solid line in c), and GMST stabilises when emissions reach zero. This is only the case in the most ambitious scenarios for non-CO_2 mitigation (Leach et al., 2018). A simple way of accounting for varying non-CO_2 forcing in Figure 1.4 would be to note that every 1 W m^{-2} increase in non-CO_2 forcing between now and the decade or two immediately prior to the time of peak warming reduces cumulative CO_2 emissions consistent with the same peak warming by approximately 1100 GtCO$_2$, with a range of 900-1500 GtCO$_2$ (using values from AR5: Myhre et al., 2013; Allen et al., 2018; Jenkins et al., 2018; Cross-Chapter Box 2 in this chapter).

1.2.3.1 Pathways remaining below 1.5°C

In this category of 1.5°C pathways, human-induced warming either rises monotonically to stabilise at 1.5°C (Figure 1.4, brown lines) or peaks at or below 1.5°C and then declines (yellow lines). Figure 1.4b demonstrates that pathways remaining below 1.5°C require net annual CO_2 emissions to peak and decline to near zero or below, depending on the long-term adjustment of the carbon cycle and non-CO_2 emissions (Bowerman et al., 2013; Wigley, 2018). Reducing emissions to zero corresponds to stabilizing cumulative CO_2 emissions (Figure 1.4c, solid lines) and falling concentrations of CO_2 in the atmosphere (panel c dashed lines) (Matthews and Caldeira, 2008; Solomon et al., 2009), which is required to stabilize GMST if non-CO_2 climate forcings are constant and positive. Stabilizing atmospheric greenhouse gas concentrations would result in continued warming (see Section 1.2.4).

If emission reductions do not begin until temperatures are close to the proposed limit, pathways remaining below 1.5°C necessarily involve much faster rates of net CO_2 emission reductions (Figure 1.4, green lines), combined with rapid reductions in non-CO_2 forcing and these pathways also reach 1.5°C earlier. Note that the emissions associated with these schematic temperature pathways may not correspond to feasible emission scenarios, but they do illustrate the fact that the timing of net zero emissions does not in itself determine peak warming: what matters is total cumulative emissions up to that

time. Hence every year's delay before initiating emission reductions decreases by approximately two years the remaining time available to reach zero emissions on a pathway still remaining below 1.5°C (Allen and Stocker, 2013; Leach et al., 2018).

1.2.3.2 Pathways temporarily exceeding 1.5°C

With the pathways in this category, also referred to as overshoot pathways, GMST rises above 1.5°C relative to pre-industrial before peaking and returning to 1.5°C around or before 2100 (Figure 1.4, blue lines), subsequently either stabilising or continuing to fall. This allows initially slower or delayed emission reductions, but lowering GMST requires net negative global CO_2 emissions (net anthropogenic removal of CO_2; Figure 1.4b). Cooling, or reduced warming, through sustained reductions of net non-CO_2 climate forcing (Cross-Chapter Box 2 in this chapter) is also required, but their role is limited because emissions of most non-CO_2 forcers cannot be reduced to below zero. Hence the feasibility and availability of large-scale CO_2 removal limits the possible rate and magnitude of temperature decline. In this report, overshoot pathways are referred to as 1.5°C pathways, but qualified by the amount of the temperature overshoot, which can have a substantial impact on irreversible climate change impacts (Mathesius et al., 2015; Tokarska and Zickfeld, 2015).

1.2.3.3 Impacts at 1.5°C warming associated with different pathways: transience versus stabilisation

Figure 1.4 also illustrates time scales associated with different impacts. While many impacts scale with the change in GMST itself, some (such as those associated with ocean acidification) scale with the change in atmospheric CO_2 concentration, indicated by the fraction of cumulative CO_2 emissions remaining in the atmosphere (dotted lines in Figure 1.4c). Others may depend on the rate of change of GMST, while 'time-integrated impacts', such as sea level rise, shown in Figure 1.4d continue to increase even after GMST has stabilised.

Hence impacts that occur when GMST reaches 1.5°C could be very different depending on the pathway to 1.5°C. CO_2 concentrations will be higher as GMST rises past 1.5°C (transient warming) than when GMST has stabilized at 1.5°C, while sea level and, potentially, global mean precipitation (Pendergrass et al., 2015) would both be lower (see Figure 1.4). These differences could lead to very different impacts on agriculture, on some forms of extreme weather (e.g., Baker et al., 2018), and on marine and terrestrial ecosystems (e.g., Mitchell et al., 2017 and Boxes 3.1 and 3.2). Sea level would be higher still if GMST returns to 1.5°C after an overshoot (Figure 1.4 d), with potentially significantly different impacts in vulnerable regions. Temperature overshoot could also cause irreversible impacts (see Chapter 3).

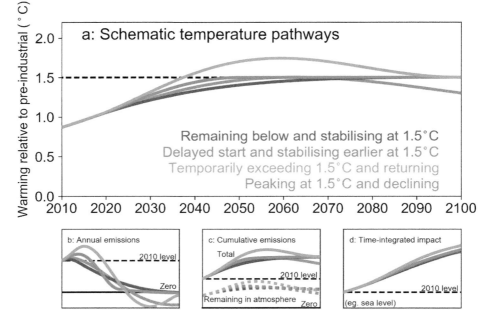

Figure 1.4 | Different 1.5°C pathways[1]: Schematic illustration of the relationship between (a) global mean surface temperature (GMST) change; (b) annual rates of CO_2 emissions, assuming constant fractional contribution of non-CO_2 forcing to total human-induced warming; (c) total cumulative CO_2 emissions (solid lines) and the fraction thereof remaining in the atmosphere (dashed lines; these also indicates changes in atmospheric CO_2 concentrations); and (d) a time-integrated impact, such as sea level rise, that continues to increase even after GMST has stabilized. Colours indicate different 1.5°C pathways. Brown: GMST remaining below and stabilizing at 1.5°C in 2100; Green: a delayed start but faster emission reductions pathway with GMST remaining below and reaching 1.5°C earlier; Blue: a pathway temporarily exceeding 1.5°C, with temperatures reduced to 1.5°C by net negative CO_2 emissions after temperatures peak; and Yellow: a pathway peaking at 1.5°C and subsequently declining. Temperatures are anchored to 1°C above pre-industrial in 2017; emissions–temperature relationships are computed using a simple climate model (Myhre et al., 2013; Millar et al., 2017a; Jenkins et al., 2018) with a lower value of the Transient Climate Response (TCR) than used in the quantitative pathway assessments in Chapter 2 to illustrate qualitative differences between pathways: this figure is not intended to provide quantitative information. The time-integrated impact is illustrated by the semi-empirical sea level rise model of Kopp et al. (2016).

Cross-Chapter Box 1 | Scenarios and Pathways

Contributing Authors:

Mikiko Kainuma (Japan), Kristie L. Ebi (USA), Sabine Fuss (Germany), Elmar Kriegler (Germany), Keywan Riahi (Austria), Joeri Rogelj (Austria/Belgium), Petra Tschakert (Australia/Austria), Rachel Warren (UK)

Climate change scenarios have been used in IPCC assessments since the First Assessment Report (Leggett et al., 1992). The **SRES scenarios** (named after the IPCC Special Report on Emissions Scenarios published in 2000; IPCC, 2000), consist of four scenarios that do not take into account any future measures to limit greenhouse gas (GHG) emissions. Subsequently, many policy scenarios have been developed based upon them (Morita et al., 2001). The SRES scenarios are superseded by a set of scenarios based on the Representative Concentration Pathways (RCPs) and Shared Socio-Economic Pathways (SSPs) (Riahi et al., 2017). The RCPs comprise a set of four GHG concentration trajectories that jointly span a large range of plausible human-caused climate forcing ranging from 2.6 W m⁻² (RCP2.6) to 8.5 W m⁻² (RCP8.5) by the end of the 21st century (van Vuuren et al., 2011). They were used to develop climate projections in the Coupled Model Intercomparison Project Phase 5 (CMIP5; Taylor et al., 2012) and were assessed in the IPCC Fifth Assessment Report (AR5). Based on the CMIP5 ensemble, RCP2.6, provides a better than two-in-three chance of staying below 2°C and a median warming of 1.6°C relative to 1850–1900 in 2100 (Collins et al., 2013).

The SSPs were developed to complement the RCPs with varying socio-economic challenges to adaptation and mitigation. SSP-based scenarios were developed for a range of climate forcing levels, including the end-of-century forcing levels of the RCPs (Riahi et al., 2017) and a level below RCP2.6 to explore pathways limiting warming to 1.5°C above pre-industrial levels (Rogelj et al., 2018). The SSP-based 1.5°C pathways are assessed in Chapter 2 of this report. These scenarios offer an integrated perspective on socio-economic, energy-system (Bauer et al., 2017), land use (Popp et al., 2017), air pollution (Rao et al., 2017) and, GHG emissions developments (Riahi et al.,

---

[1]   An animated version of Figure 1.4 will be embedded in the web-based version of this Special Report

Cross-Chapter Box 1 (continued)

2017). Because of their harmonised assumptions, scenarios developed with the SSPs facilitate the integrated analysis of future climate impacts, vulnerabilities, adaptation and mitigation.

Scenarios and Pathways in this Report

This report focuses on pathways that could limit the increase of global mean surface temperature (GMST) to 1.5°C above pre-industrial levels and pathways that align with the goals of sustainable development and poverty eradication. The pace and scale of mitigation and adaptation are assessed in the context of historical evidence to determine where unprecedented change is required (see Chapter 4). Other scenarios are also assessed, primarily as benchmarks for comparison of mitigation, impacts, and/or adaptation requirements. These include baseline scenarios that assume no climate policy; scenarios that assume some kind of continuation of current climate policy trends and plans, many of which are used to assess the implications of the nationally determined contributions (NDCs); and scenarios holding warming below 2°C above pre-industrial levels. This report assesses the spectrum from global mitigation scenarios to local adaptation choices – complemented by a bottom-up assessment of individual mitigation and adaptation options, and their implementation (policies, finance, institutions, and governance, see Chapter 4). Regional, national, and local scenarios, as well as decision-making processes involving values and difficult trade-offs are important for understanding the challenges of limiting GMST increase to 1.5°C and are thus indispensable when assessing implementation.

Different climate policies result in different temperature pathways, which result in different levels of climate risks and actual climate impacts with associated long-term implications. Temperature pathways are classified into continued warming pathways (in the cases of baseline and reference scenarios), pathways that keep the temperature increase below a specific limit (like 1.5°C or 2°C), and pathways that temporarily exceed and later fall to a specific limit (overshoot pathways). In the case of a temperature overshoot, net negative CO_2 emissions are required to remove excess CO_2 from the atmosphere (Section 1.2.3).

In a 'prospective' mitigation pathway, emissions (or sometimes concentrations) are prescribed, giving a range of GMST outcomes because of uncertainty in the climate response. Prospective pathways are considered '1.5°C pathways' in this report if, based on current knowledge, the majority of available approaches assign an approximate probability of one-in-two to two-in-three to temperatures either remaining below 1.5°C or returning to 1.5°C either before or around 2100. Most pathways assessed in Chapter 2 are prospective pathways, and therefore even '1.5°C pathways' are also associated with risks of warming higher than 1.5°C, noting that many risks increase non-linearly with increasing GMST. In contrast, the 'risks of warming of 1.5°C' assessed in Chapter 3 refer to risks in a world in which GMST is either passing through (transient) or stabilized at 1.5°C, without considering probabilities of different GMST levels (unless otherwise qualified). To stay below any desired temperature limit, mitigation measures and strategies would need to be adjusted as knowledge of the climate response is updated (Millar et al., 2017b; Emori et al., 2018). Such pathways can be called 'adaptive' mitigation pathways. Given there is always a possibility of a greater-than-expected climate response (Xu and Ramanathan, 2017), adaptive mitigation pathways are important to minimise climate risks, but need also to consider the risks and feasibility (see Cross-Chapter Box 3 in this chapter) of faster-than-expected emission reductions. Chapter 5 includes assessments of two related topics: aligning mitigation and adaptation pathways with sustainable development pathways, and transformative visions for the future that would support avoiding negative impacts on the poorest and most disadvantaged populations and vulnerable sectors.

Definitions of Scenarios and Pathways

Climate scenarios and pathways are terms that are sometimes used interchangeably, with a wide range of overlapping definitions (Rosenbloom, 2017).

A '**scenario**' is an internally consistent, plausible, and integrated description of a possible future of the human–environment system, including a narrative with qualitative trends and quantitative projections (IPCC, 2000). Climate change scenarios provide a framework for developing and integrating projections of emissions, climate change, and climate impacts, including an assessment of their inherent uncertainties. The long-term and multi-faceted nature of climate change requires climate scenarios to describe how socio-economic trends in the 21st century could influence future energy and land use, resulting emissions and the evolution of human vulnerability and exposure. Such driving forces include population, GDP, technological innovation, governance and lifestyles. Climate change scenarios are used for analysing and contrasting climate policy choices.

The notion of a '**pathway**' can have multiple meanings in the climate literature. It is often used to describe the temporal evolution of a set of scenario features, such as GHG emissions and socio-economic development. As such, it can describe individual scenario components or sometimes be used interchangeably with the word 'scenario'. For example, the RCPs describe GHG concentration trajectories (van Vuuren et al., 2011) and the SSPs are a set of narratives of societal futures augmented by quantitative projections of socio-economic determinants such as population, GDP and urbanization (Kriegler et al., 2012; O'Neill et al., 2014). Socio-economic

Cross-Chapter Box 1 (continued)

driving forces consistent with any of the SSPs can be combined with a set of climate policy assumptions (Kriegler et al., 2014) that together would lead to emissions and concentration outcomes consistent with the RCPs (Riahi et al., 2017). This is at the core of the scenario framework for climate change research that aims to facilitate creating scenarios integrating emissions and development pathways dimensions (Ebi et al., 2014; van Vuuren et al., 2014).

In other parts of the literature, 'pathway' implies a solution-oriented trajectory describing a pathway from today's world to achieving a set of future goals. **Sustainable Development Pathways** describe national and global pathways where climate policy becomes part of a larger sustainability transformation (Shukla and Chaturvedi, 2013; Fleurbaey et al., 2014; van Vuuren et al., 2015). The AR5 presented **climate-resilient pathways** as sustainable development pathways that combine the goals of adaptation and mitigation (Denton et al., 2014), more broadly defined as iterative processes for managing change within complex systems in order to reduce disruptions and enhance opportunities associated with climate change (IPCC, 2014a). The AR5 also introduced the notion of **climate-resilient development pathways**, with a more explicit focus on dynamic livelihoods, multi-dimensional poverty, structural inequalities, and equity among poor and non-poor people (Olsson et al., 2014). **Adaptation pathways** are understood as a series of adaptation choices involving trade-offs between short-term and long-term goals and values (Reisinger et al., 2014). They are decision-making processes sequenced over time with the purpose of deliberating and identifying socially salient solutions in specific places (Barnett et al., 2014; Wise et al., 2014; Fazey et al., 2016). There is a range of possible pathways for transformational change, often negotiated through iterative and inclusive processes (Harris et al., 2017; Fazey et al., 2018; Tàbara et al., 2018).

1.2.4   Geophysical Warming Commitment

It is frequently asked whether limiting warming to 1.5°C is 'feasible' (Cross-Chapter Box 3 in this chapter). There are many dimensions to this question, including the warming 'commitment' from past emissions of greenhouse gases and aerosol precursors. Quantifying commitment from past emissions is complicated by the very different behaviour of different climate forcers affected by human activity: emissions of long-lived greenhouse gases such as CO_2 and nitrous oxide (N_2O) have a very persistent impact on radiative forcing (Myhre et al., 2013), lasting from over a century (in the case of N_2O) to hundreds of thousands of years (for CO_2). The radiative forcing impact of short-lived climate forcers (SLCFs) such as methane (CH_4) and aerosols, in contrast, persists for at most about a decade (in the case of methane) down to only a few days. These different behaviours must be taken into account in assessing the implications of any approach to calculating aggregate emissions (Cross-Chapter Box 2 in this chapter).

Geophysical warming commitment is defined as the unavoidable future warming resulting from physical Earth system inertia. Different variants are discussed in the literature, including (i) the 'constant composition commitment' (CCC), defined by Meehl et al. (2007) as the further warming that would result if atmospheric concentrations of GHGs and other climate forcers were stabilised at the current level; and (ii) and the 'zero emissions commitment' (ZEC), defined as the further warming that would still occur if all future anthropogenic emissions of greenhouse gases and aerosol precursors were eliminated instantaneously (Meehl et al., 2007; Collins et al., 2013).

The CCC is primarily associated with thermal inertia of the ocean (Hansen et al., 2005), and has led to the misconception that substantial future warming is inevitable (Matthews and Solomon, 2013). The CCC takes into account the warming from past emissions, but also includes warming from future emissions (declining but still non-zero) that are required to maintain a constant atmospheric composition. It is therefore not relevant to the warming commitment from past emissions alone.

The ZEC, although based on equally idealised assumptions, allows for a clear separation of the response to past emissions from the effects of future emissions. The magnitude and sign of the ZEC depend on the mix of GHGs and aerosols considered. For CO_2, which takes hundreds of thousands of years to be fully removed from the atmosphere by natural processes following its emission (Eby et al., 2009; Ciais et al., 2013), the multi-century warming commitment from emissions to date in addition to warming already observed is estimated to range from slightly negative (i.e., a slight cooling relative to present-day) to slightly positive (Matthews and Caldeira, 2008; Lowe et al., 2009; Gillett et al., 2011; Collins et al., 2013). Some studies estimate a larger ZEC from CO_2, but for cumulative emissions much higher than those up to present day (Frölicher et al., 2014; Ehlert and Zickfeld, 2017). The ZEC from past CO_2 emissions is small because the continued warming effect from ocean thermal inertia is approximately balanced by declining radiative forcing due to CO_2 uptake by the ocean (Solomon et al., 2009; Goodwin et al., 2015; Williams et al., 2017). Thus, although present-day CO_2-induced warming is irreversible on millennial time scales (without human intervention such as active carbon dioxide removal or solar radiation modification; Section 1.4.1), past CO_2 emissions do not commit to substantial further warming (Matthews and Solomon, 2013).

Sustained net zero anthropogenic emissions of CO_2 and declining net anthropogenic non-CO_2 radiative forcing over a multi-decade period would halt anthropogenic global warming over that period, although it would not halt sea level rise or many other aspects of climate system adjustment. The rate of decline of non-CO_2 radiative forcing must be sufficient to compensate for the ongoing adjustment of the climate system to this forcing (assuming it remains positive) due to ocean thermal inertia. It therefore depends on deep ocean response time scales, which are uncertain but of order centuries, corresponding to decline rates of non-CO$_2$ radiative forcing of less than 1% per year. In the longer term, Earth system feedbacks such as the release of carbon from melting permafrost may require net negative CO$_2$ emissions to maintain stable temperatures (Lowe and Bernie, 2018).

For warming SLCFs, meaning those associated with positive radiative forcing such as methane, the ZEC is negative. Eliminating emissions of these substances results in an immediate cooling relative to the present (Figure 1.5, magenta lines) (Frölicher and Joos, 2010; Matthews and Zickfeld, 2012; Mauritsen and Pincus, 2017). Cooling SLCFs (those associated with negative radiative forcing) such as sulphate aerosols create a positive ZEC, as elimination of these forcers results in rapid increase in radiative forcing and warming (Figure 1.5, green lines) (Matthews and Zickfeld, 2012; Mauritsen and Pincus, 2017; Samset et al., 2018). Estimates of the warming commitment from eliminating aerosol emissions are affected by large uncertainties in net aerosol radiative forcing (Myhre et al., 2013, 2017) and the impact of other measures affecting aerosol loading (e.g., Fernández et al., 2017). If present-day emissions of all GHGs (short- and long-lived) and aerosols (including sulphate, nitrate and carbonaceous aerosols) are eliminated (Figure 1.5, yellow lines) GMST rises over the following decade, driven by the removal of negative aerosol radiative forcing. This initial warming is followed by a gradual cooling driven by the decline in radiative forcing of short-lived greenhouse gases (Matthews and Zickfeld, 2012; Collins et al., 2013). Peak warming following elimination of all emissions was assessed at a few tenths of a degree in AR5, and century-scale warming was assessed to change only slightly relative to the time emissions are reduced to zero (Collins et al., 2013). New evidence since AR5 suggests a larger methane forcing (Etminan et al., 2016) but no revision in the range of aerosol forcing (although this remains an active field of research, e.g., Myhre et al., 2017). This revised methane forcing estimate results in a smaller peak warming and a faster temperature decline than assessed in AR5 (Figure 1.5, yellow line).

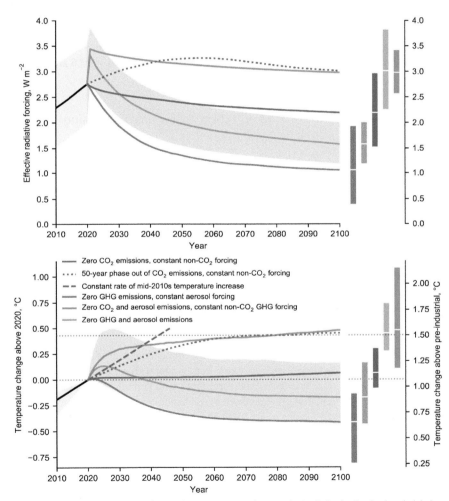

Figure 1.5 | Warming commitment from past emissions of greenhouse gases and aerosols: Radiative forcing (top) and global mean surface temperature change (bottom) for scenarios with different combinations of greenhouse gas and aerosol precursor emissions reduced to zero in 2020. Variables were calculated using a simple climate–carbon cycle model (Millar et al., 2017a) with a simple representation of atmospheric chemistry (Smith et al., 2018). The bars on the right-hand side indicate the median warming in 2100 and 5–95% uncertainty ranges (also indicated by the plume around the yellow line) taking into account one estimate of uncertainty in climate response, effective radiative forcing and carbon cycle sensitivity, and constraining simple model parameters with response ranges from AR5 combined with historical climate observations (Smith et al., 2018). Temperatures continue to increase slightly after elimination of CO$_2$ emissions (blue line) in response to constant non-CO$_2$ forcing. The dashed blue line extrapolates one estimate of the current rate of warming, while dotted blue lines show a case where CO$_2$ emissions are reduced linearly to zero assuming constant non-CO$_2$ forcing after 2020. Under these highly idealized assumptions, the time to stabilize temperatures at 1.5°C is approximately double the time remaining to reach 1.5°C at the current warming rate.

Expert judgement based on the available evidence (including model simulations, radiative forcing and climate sensitivity) suggests that if all anthropogenic emissions were reduced to zero immediately, any further warming beyond the 1°C already experienced would *likely* be less than 0.5°C over the next two to three decades, and also *likely* less than 0.5°C on a century time scale.

Since most sources of emissions cannot, in reality, be brought to zero instantaneously due to techno-economic inertia, the current rate of emissions also constitutes a conditional commitment to future emissions and consequent warming depending on achievable rates of emission reductions. The current level and rate of human-induced warming determines both the time left before a temperature threshold is exceeded if warming continues (dashed blue line in Figure 1.5) and the time over which the warming rate must be reduced to avoid exceeding that threshold (approximately indicated by the dotted blue line in Figure 1.5). Leach et al. (2018) use a central estimate of human-induced warming of 1.02°C in 2017, increasing at 0.215°C per decade (Haustein et al., 2017), to argue that it will take 13–32 years (one-standard-error range) to reach 1.5°C if the current warming rate continues, allowing 25–64 years to stabilise temperatures at 1.5°C if the warming rate is reduced at a constant

rate of deceleration starting immediately. Applying a similar approach to the multi-dataset average GMST used in this report gives an assessed *likely* range for the date at which warming reaches 1.5°C of 2030 to 2052. The lower bound on this range, 2030, is supported by multiple lines of evidence, including the AR5 assessment for the *likely* range of warming (0.3°C–0.7°C) for the period 2016–2035 relative to 1986–2005. The upper bound, 2052, is supported by fewer lines of evidence, so we have used the upper bound of the 5–95% confidence interval given by the Leach et al. (2018) method applied to the multi-dataset average GMST, expressed as the upper limit of the *likely* range, to reflect the reliance on a single approach. Results are sensitive both to the confidence level chosen and the number of years used to estimate the current rate of anthropogenic warming (5 years used here, to capture the recent acceleration due to rising non-CO_2 forcing). Since the rate of human-induced warming is proportional to the rate of CO_2 emissions (Matthews et al., 2009; Zickfeld et al., 2009) plus a term approximately proportional to the rate of increase in non-CO_2 radiative forcing (Gregory and Forster, 2008; Allen et al., 2018; Cross-Chapter Box 2 in this chapter), these time scales also provide an indication of minimum emission reduction rates required if a warming greater than 1.5°C is to be avoided (see Figure 1.5, Supplementary Material 1.SM.6 and FAQ 1.2).

Cross-Chapter Box 2 | Measuring Progress to Net Zero Emissions Combining Long-Lived and Short-Lived Climate Forcers

Contributing Authors:
Piers Forster (UK), Myles R. Allen (UK), Elmar Kriegler (Germany), Joeri Rogelj (Austria/Belgium), Seth Schultz (USA), Drew Shindell (USA), Kirsten Zickfeld (Canada/Germany)

Emissions of many different climate forcers will affect the rate and magnitude of climate change over the next few decades (Myhre et al., 2013). Since these decades will determine when 1.5°C is reached or whether a warming greater than 1.5°C is avoided, understanding the aggregate impact of different forcing agents is particularly important in the context of 1.5°C pathways. Paragraph 17 of Decision 1 of the 21st Conference of the Parties on the adoption of the Paris Agreement specifically states that this report is to identify aggregate greenhouse gas emission levels compatible with holding the increase in global average temperatures to 1.5°C above pre-industrial levels (see Chapter 2). This request highlights the need to consider the implications of different methods of aggregating emissions of different gases, both for future temperatures and for other aspects of the climate system (Levasseur et al., 2016; Ocko et al., 2017).

To date, reporting of GHG emissions under the UNFCCC has used Global Warming Potentials (GWPs) evaluated over a 100-year time horizon (GWP$_{100}$) to combine multiple climate forcers. IPCC Working Group 3 reports have also used GWP$_{100}$ to represent multi-gas pathways (Clarke et al., 2014). For reasons of comparability and consistency with current practice, Chapter 2 in this Special Report continues to use this aggregation method. Numerous other methods of combining different climate forcers have been proposed, such as the Global Temperature-change Potential (GTP; Shine et al., 2005) and the Global Damage Potential (Tol et al., 2012; Deuber et al., 2013).

Climate forcers fall into two broad categories in terms of their impact on global temperature (Smith et al., 2012): long-lived GHGs, such as CO_2 and nitrous oxide (N_2O), whose warming impact depends primarily on the total cumulative amount emitted over the past century or the entire industrial epoch; and short-lived climate forcers (SLCFs), such as methane and black carbon, whose warming impact depends primarily on current and recent annual emission rates (Reisinger et al., 2012; Myhre et al., 2013; Smith et al., 2013; Strefler et al., 2014). These different dependencies affect the emissions reductions required of individual forcers to limit warming to 1.5°C or any other level.

Natural processes that remove CO_2 permanently from the climate system are so slow that reducing the rate of CO_2-induced warming to zero requires net zero global anthropogenic CO_2 emissions (Archer and Brovkin, 2008; Matthews and Caldeira, 2008; Solomon et al.,

Cross-Chapter Box 2 (continued)

2009), meaning almost all remaining anthropogenic CO_2 emissions must be compensated for by an equal rate of anthropogenic carbon dioxide removal (CDR). Cumulative CO_2 emissions are therefore an accurate indicator of CO_2-induced warming, except in periods of high negative CO2 emissions (Zickfeld et al., 2016), and potentially in century-long periods of near-stable temperatures (Bowerman et al., 2011; Wigley, 2018). In contrast, sustained constant emissions of a SLCF such as methane, would (after a few decades) be consistent with constant methane concentrations and hence very little additional methane-induced warming (Allen et al., 2018; Fuglestvedt et al., 2018). Both GWP and GTP would equate sustained SLCF emissions with sustained constant CO_2 emissions, which would continue to accumulate in the climate system, warming global temperatures indefinitely. Hence nominally 'equivalent' emissions of CO_2 and SLCFs, if equated conventionally using GWP or GTP, have very different temperature impacts, and these differences are particularly evident under ambitious mitigation characterizing 1.5°C pathways.

Since the AR5, a revised usage of GWP has been proposed (Lauder et al., 2013; Allen et al., 2016), denoted GWP* (Allen et al., 2018), that addresses this issue by equating a permanently sustained change in the emission *rate* of an SLCF or SLCF-precursor (in tonnes-per-year), or other non-CO_2 forcing (in watts per square metre), with a one-off *pulse* emission (in tonnes) of a fixed amount of CO_2. Specifically, GWP* equates a 1 tonne-per-year increase in emission rate of an SLCF with a pulse emission of GWP_H x H tonnes of CO_2, where GWP_H is the conventional GWP of that SLCF evaluated over time GWP_H for SLCFs decreases with increasing time H, GWP_H x H for SLCFs is less dependent on the choice of time horizon. Similarly, a permanent 1 W m^{-2} increase in radiative forcing has a similar temperature impact as the cumulative emission of $H/AGWP_H$ tonnes of CO_2, where $AGWP_H$ is the Absolute Global Warming Potential of CO_2 (Shine et al., 2005; Myhre et al., 2013; Allen et al., 2018). This indicates approximately how future changes in non-CO_2 radiative forcing affect cumulative CO_2 emissions consistent with any given level of peak warming.

When combined using GWP*, cumulative aggregate GHG emissions are closely proportional to total GHG-induced warming, while the annual rate of GHG-induced warming is proportional to the annual rate of aggregate GHG emissions (see Cross-Chapter Box 2, Figure 1). This is not the case when emissions are aggregated using GWP or GTP, with discrepancies particularly pronounced when SLCF emissions are falling. Persistent net zero CO_2-equivalent emissions containing a residual positive forcing contribution from SLCFs and aggregated using GWP_{100} or GTP would result in a steady decline of GMST. Net zero global emissions aggregated using GWP* (which corresponds to zero net emissions of CO_2 and other long-lived GHGs like nitrous oxide, combined with near-constant SLCF forcing – see Figure 1.5) results in approximately stable GMST (Allen et al., 2018; Fuglestvedt et al., 2018 and Cross-Chapter Box 2, Figure 1, below).

Whatever method is used to relate emissions of different greenhouse gases, scenarios achieving stable GMST well below 2°C require both near-zero net emissions of long-lived greenhouse gases and deep reductions in warming SLCFs (Chapter 2), in part to compensate for the reductions in cooling SLCFs that are expected to accompany reductions in CO_2 emissions (Rogelj et al., 2016b; Hienola et al., 2018). Understanding the implications of different methods of combining emissions of different climate forcers is, however, helpful in tracking progress towards temperature stabilisation and 'balance between anthropogenic emissions by sources and removals by sinks of greenhouse gases' as stated in Article 4 of the Paris Agreement. Fuglestvedt et al. (2018) and Tanaka and O'Neill (2018) show that when, and even whether, aggregate GHG emissions need to reach net zero before 2100 to limit warming to 1.5°C depends on the scenario, aggregation method and mix of long-lived and short-lived climate forcers.

The comparison of the impacts of different climate forcers can also consider more than their effects on GMST (Johansson, 2012; Tol et al., 2012; Deuber et al., 2013; Myhre et al., 2013; Cherubini and Tanaka, 2016). Climate impacts arise from both magnitude and rate of climate change, and from other variables such as precipitation (Shine et al., 2015). Even if GMST is stabilised, sea level rise and associated impacts will continue to increase (Sterner et al., 2014), while impacts that depend on CO_2 concentrations such as ocean acidification may begin to reverse. From an economic perspective, comparison of different climate forcers ideally reflects the ratio of marginal economic damages if used to determine the exchange ratio of different GHGs under multi-gas regulation (Tol et al., 2012; Deuber et al., 2013; Kolstad et al., 2014).

Emission reductions can interact with other dimensions of sustainable development (see Chapter 5). In particular, early action on some SLCFs (including actions that may warm the climate, such as reducing sulphur dioxide emissions) may have considerable societal co-benefits, such as reduced air pollution and improved public health with associated economic benefits (OECD, 2016; Shindell et al., 2016). Valuation of broadly defined social costs attempts to account for many of these additional non-climate factors along with climate-related impacts (Shindell, 2015; Sarofim et al., 2017; Shindell et al., 2017). See Chapter 4, Section 4.3.6, for a discussions of mitigation options, noting that mitigation priorities for different climate forcers depend on multiple economic and social criteria that vary between sectors, regions and countries.

Cross-Chapter Box 2 (continued)

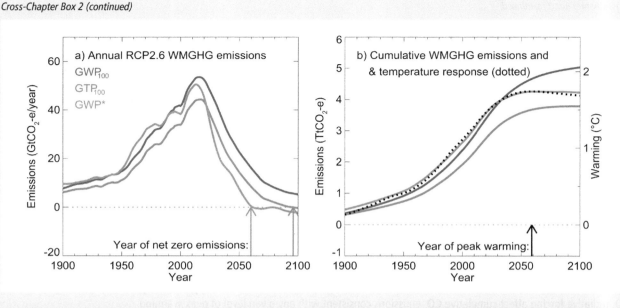

Cross-Chapter Box 2, Figure 1 | Implications of different approaches to calculating aggregate greenhouse gas emissions on a pathway to net zero. (a) Aggregate emissions of well-mixed greenhouse gases (WMGHGs) under the RCP2.6 mitigation scenario expressed as CO_2-equivalent using GWP_{100} (blue); GTP_{100} (green) and GWP* (yellow). Aggregate WMGHG emissions appear to fall more rapidly if calculated using GWP* than using either GWP or GTP, primarily because GWP* equates a falling methane emission rate with negative CO_2 emissions, as only active CO_2 removal would have the same impact on radiative forcing and GMST as a reduction in methane emission rate. (b) Cumulative emissions of WMGHGs combined as in panel (a) (blue, green and yellow lines & left hand axis) and warming response to combined emissions (black dotted line and right hand axis, Millar et al. (2017a). The temperature response under ambitious mitigation is closely correlated with cumulative WMGHG emissions aggregated using GWP*, but with neither emission rate nor cumulative emissions if aggregated using GWP or GTP.

1.3   Impacts at 1.5°C and Beyond

1.3.1   Definitions

Consistent with the AR5 (IPCC, 2014a), 'impact' in this report refers to the effects of climate change on human and natural systems. Impacts may include the effects of changing hazards, such as the frequency and intensity of heat waves. 'Risk' refers to potential negative impacts of climate change where something of value is at stake, recognizing the diversity of values. Risks depend on hazards, exposure, vulnerability (including sensitivity and capacity to respond) and likelihood. Climate change risks can be managed through efforts to mitigate climate change forcers, adaptation of impacted systems, and remedial measures (Section 1.4.1).

In the context of this report, *regional* impacts of *global* warming at 1.5°C and 2°C are assessed in Chapter 3. The '*warming experience at 1.5°C*' is that of regional climate change (temperature, rainfall, and other changes) at the time when global average temperatures, as defined in Section 1.2.1, reach 1.5°C above pre-industrial (the same principle applies to impacts at any other global mean temperature). Over the decade 2006–2015, many regions have experienced higher than average levels of warming and some are already now 1.5°C or more warmer with respect to the pre-industrial period (Figure 1.3).

At a global warming of 1.5°C, some seasons will be substantially warmer than 1.5°C above pre-industrial (Seneviratne et al., 2016). Therefore, most regional impacts of a global mean warming of 1.5°C will be different from those of a regional warming by 1.5°C.

The impacts of 1.5°C global warming will vary in both space and time (Ebi et al., 2016). For many regions, an increase in global mean temperature by 1.5°C or 2°C implies substantial increases in the occurrence and/or intensity of some extreme events (Fischer and Knutti, 2015; Karmalkar and Bradley, 2017; King et al., 2017; Chevuturi et al., 2018), resulting in different impacts (see Chapter 3). By comparing impacts at 1.5°C versus those at 2°C, this report discusses the 'avoided impacts' by maintaining global temperature increase at or below 1.5°C as compared to 2°C, noting that these also depend on the pathway taken to 1.5°C (see Section 1.2.3 and Cross-Chapter Box 8 in Chapter 3 on 1.5°C warmer worlds). Many impacts take time to observe, and because of the warming trend, impacts over the past 20 years were associated with a level of human-induced warming that was, on average, 0.1°C–0.23°C colder than its present level, based on the AR5 estimate of the warming trend over this period (Section 1.2.1 and Kirtman et al., 2013). Attribution studies (e.g., van Oldenborgh et al., 2017) can address this bias, but informal estimates of 'recent impact experience' in a rapidly warming world necessarily understate the temperature-related impacts of the current level of warming.

1.3.2 Drivers of Impacts

Impacts of climate change are due to multiple environmental drivers besides rising temperatures, such as rising atmospheric CO_2, shifting rainfall patterns (Lee et al., 2018), rising sea levels, increasing ocean acidification, and extreme events, such as floods, droughts, and heat waves (IPCC, 2014a). Changes in rainfall affect the hydrological cycle and water availability (Schewe et al., 2014; Döll et al., 2018; Saeed et al., 2018). Several impacts depend on atmospheric composition, increasing atmospheric carbon dioxide levels leading to changes in plant productivity (Forkel et al., 2016), but also to ocean acidification (Hoegh-Guldberg et al., 2007). Other impacts are driven by changes in ocean heat content such as the destabilization of coastal ice sheets and sea level rise (Bindoff et al., 2007; Chen et al., 2017), whereas impacts due to heat waves depend directly on ambient air or ocean temperature (Matthews et al., 2017). Impacts can be direct, such as coral bleaching due to ocean warming, and indirect, such as reduced tourism due to coral bleaching. Indirect impacts can also arise from mitigation efforts such as changed agricultural management (Section 3.6.2) or remedial measures such as solar radiation modification (Section 4.3.8, Cross-Chapter Box 10 in Chapter 4).

Impacts may also be triggered by combinations of factors, including 'impact cascades' (Cramer et al., 2014) through secondary consequences of changed systems. Changes in agricultural water availability caused by upstream changes in glacier volume are a typical example. Recent studies also identify compound events (e.g., droughts and heat waves), that is, when impacts are induced by the combination of several climate events (AghaKouchak et al., 2014; Leonard et al., 2014; Martius et al., 2016; Zscheischler and Seneviratne, 2017).

There are now techniques to attribute impacts formally to anthropogenic global warming and associated rainfall changes (Rosenzweig et al., 2008; Cramer et al., 2014; Hansen et al., 2016), taking into account other drivers such as land-use change (Oliver and Morecroft, 2014) and pollution (e.g., tropospheric ozone; Sitch et al., 2007). There are multiple lines of evidence that climate change has observable and often severely negative effects on people, especially where climate-sensitive biophysical conditions and socio-economic and political constraints on adaptive capacities combine to create high vulnerabilities (IPCC, 2012a, 2014a; World Bank, 2013). The character and severity of impacts depend not only on the hazards (e.g., changed climate averages and extremes) but also on the vulnerability (including sensitivities and adaptive capacities) of different communities and their exposure to climate threats. These impacts also affect a range of natural and human systems, such as terrestrial, coastal and marine ecosystems and their services; agricultural production; infrastructure; the built environment; human health; and other socio-economic systems (Rosenzweig et al., 2017).

Sensitivity to changing drivers varies markedly across systems and regions. Impacts of climate change on natural and managed ecosystems can imply loss or increase in growth, biomass or diversity at the level of species populations, interspecific relationships such as pollination, landscapes or entire biomes. Impacts occur in addition to the natural variation in growth, ecosystem dynamics, disturbance,

succession and other processes, rendering attribution of impacts at lower levels of warming difficult in certain situations. The same magnitude of warming can be lethal during one phase of the life of an organism and irrelevant during another. Many ecosystems (notably forests, coral reefs and others) undergo long-term successional processes characterised by varying levels of resilience to environmental change over time. Organisms and ecosystems may adapt to environmental change to a certain degree, through changes in physiology, ecosystem structure, species composition or evolution. Large-scale shifts in ecosystems may cause important feedbacks, in terms of changing water and carbon fluxes through impacted ecosystems – these can amplify or dampen atmospheric change at regional to continental scale. Of particular concern is the response of most of the world's forests and seagrass ecosystems, which play key roles as carbon sinks (Settele et al., 2014; Marbà et al., 2015).

Some ambitious efforts to constrain atmospheric greenhouse gas concentrations may themselves impact ecosystems. In particular, changes in land use, potentially required for massively enhanced production of biofuels (either as simple replacement of fossil fuels, or as part of bioenergy with carbon capture and storage, BECCS) impact all other land ecosystems through competition for land (e.g., Creutzig, 2016) (see Cross-Chapter Box 7 in Chapter 3, Section 3.6.2.1).

Human adaptive capacity to a 1.5°C warmer world varies markedly for individual sectors and across sectors such as water supply, public health, infrastructure, ecosystems and food supply. For example, density and risk exposure, infrastructure vulnerability and resilience, governance, and institutional capacity all drive different impacts across a range of human settlement types (Dasgupta et al., 2014; Revi et al., 2014; Rosenzweig et al., 2018). Additionally, the adaptive capacity of communities and human settlements in both rural and urban areas, especially in highly populated regions, raises equity, social justice and sustainable development issues. Vulnerabilities due to gender, age, level of education and culture act as compounding factors (Arora-Jonsson, 2011; Cardona et al., 2012; Resurrección, 2013; Olsson et al., 2014; Vincent et al., 2014).

1.3.3 Uncertainty and Non-Linearity of Impacts

Uncertainties in projections of future climate change and impacts come from a variety of different sources, including the assumptions made regarding future emission pathways (Moss et al., 2010), the inherent limitations and assumptions of the climate models used for the projections, including limitations in simulating regional climate variability (James et al., 2017), downscaling and bias-correction methods (Ekström et al., 2015), the assumption of a linear scaling of impacts with GMST used in many studies (Lewis et al., 2017; King et al., 2018b), and in impact models (e.g., Asseng et al., 2013). The evolution of climate change also affects uncertainty with respect to impacts. For example, the impacts of overshooting 1.5°C and stabilization at a later stage compared to stabilization at 1.5°C without overshoot may differ in magnitude (Schleussner et al., 2016).

AR5 (IPCC, 2013b) and World Bank (2013) underscored the non-linearity of risks and impacts as temperature rises from 2°C to 4°C of warming, particularly in relation to water availability, heat extremes, bleaching of coral reefs, and more. Recent studies (Schleussner et al., 2016; James et al., 2017; Barcikowska et al., 2018; King et al., 2018a) assess the impacts of 1.5°C versus 2°C warming, with the same message of non-linearity. The resilience of ecosystems, meaning their ability either to resist change or to recover after a disturbance, may change, and often decline, in a non-linear way. An example are reef ecosystems, with some studies suggesting that reefs will change, rather than disappear entirely, and with particular species showing greater tolerance to coral bleaching than others (Pörtner et al., 2014). A key issue is therefore whether ecosystems such as coral reefs survive an overshoot scenario, and to what extent they would be able to recover after stabilization at 1.5°C or higher levels of warming (see Box 3.4).

1.4 Strengthening the Global Response

This section frames the implementation options, enabling conditions (discussed further in Cross-Chapter Box 3 on feasibility in this chapter), capacities and types of knowledge and their availability (Blicharska et al., 2017) that can allow institutions, communities and societies to respond to the 1.5°C challenge in the context of sustainable development and the Sustainable Development Goals (SDGs). It also addresses other relevant international agreements such as the Sendai Framework for Disaster Risk Reduction. Equity and ethics are recognised as issues of importance in reducing vulnerability and eradicating poverty.

The connection between the enabling conditions for limiting global warming to 1.5°C and the ambitions of the SDGs are complex across scale and multi-faceted (Chapter 5). Climate mitigation–adaptation linkages, including synergies and trade-offs, are important when considering opportunities and threats for sustainable development. The IPCC AR5 acknowledged that 'adaptation and mitigation have the potential to both contribute to and impede sustainable development, and sustainable development strategies and choices have the potential to both contribute to and impede climate change responses' (Denton et al., 2014). Climate mitigation and adaptation measures and actions can reflect and enforce specific patterns of development and governance that differ amongst the world's regions (Gouldson et al., 2015; Termeer et al., 2017). The role of limited adaptation and mitigation capacity, limits to adaptation and mitigation, and conditions of mal-adaptation and mal-mitigation are assessed in this report (Chapters 4 and 5).

1.4.1 Classifying Response Options

Key broad categories of responses to the climate change problem are framed here. **Mitigation** refers to efforts to reduce or prevent the emission of greenhouse gases, or to enhance the absorption of gases already emitted, thus limiting the magnitude of future warming (IPCC, 2014b). Mitigation requires the use of new technologies, clean energy sources, reduced deforestation, improved sustainable agricultural methods, and changes in individual and collective behaviour. Many of these may provide substantial co-benefits for air quality, biodiversity and sustainable development. Mal-mitigation

includes changes that could reduce emissions in the short-term but could lock in technology choices or practices that include significant trade-offs for effectiveness of future adaptation and other forms of mitigation (Chapters 2 and 4).

Carbon dioxide removal (CDR) or 'negative emissions' activities are considered in this report as distinct from the above mitigation activities. While most mitigation activities focus on reducing the amount of carbon dioxide or other greenhouse gases emitted, CDR aims to reduce concentrations already in the atmosphere. Technologies for CDR are mostly in their infancy despite their importance to ambitious climate change mitigation pathways (Minx et al., 2017). Although some CDR activities such as reforestation and ecosystem restoration are well understood, the feasibility of massive-scale deployment of many CDR technologies remains an open question (IPCC, 2014b; Leung et al., 2014) (Chapters 2 and 4). Technologies for the active removal of other greenhouse gases, such as methane, are even less developed, and are briefly discussed in Chapter 4.

Climate change **adaptation** refers to the actions taken to manage the impacts of climate change (IPCC, 2014a). The aim is to reduce vulnerability and exposure to the harmful effects of climate change (e.g., sea level rise, more intense extreme weather events or food insecurity). It also includes exploring the potential beneficial opportunities associated with climate change (for example, longer growing seasons or increased yields in some regions). Different adaptation pathways can be undertaken. Adaptation can be incremental, or transformational, meaning fundamental attributes of the system are changed (Chapter 3 and 4). There can be limits to ecosystem-based adaptation or the ability of humans to adapt (Chapter 4). If there is no possibility for adaptive actions that can be applied to avoid an intolerable risk, these are referred to as hard adaptation limits, while soft adaptation limits are identified when there are currently no options to avoid intolerable risks, but they are theoretically possible (Chapter 3 and 4). While climate change is a global issue, impacts are experienced locally. Cities and municipalities are at the frontline of adaptation (Rosenzweig et al., 2018), focusing on reducing and managing disaster risks due to extreme and slow-onset weather and climate events, installing flood and drought early warning systems, and improving water storage and use (Chapters 3 and 4 and Cross-Chapter Box 12 in Chapter 5). Agricultural and rural areas, including often highly vulnerable remote and indigenous communities, also need to address climate-related risks by strengthening and making more resilient agricultural and other natural resource extraction systems.

Remedial measures are distinct from mitigation or adaptation, as the aim is to temporarily reduce or offset warming (IPCC, 2012b). One such measure is solar radiation modification (SRM), also referred to as solar radiation management in the literature, which involves deliberate changes to the albedo of the Earth system, with the net effect of increasing the amount of solar radiation reflected from the Earth to reduce the peak temperature from climate change (The Royal Society, 2009; Smith and Rasch, 2013; Schäfer et al., 2015). It should be noted that while some radiation modification measures, such as cirrus cloud thinning (Kristjánsson et al., 2016), aim at enhancing outgoing long-wave radiation, SRM is used in this report to refer to all direct interventions on the planetary radiation budget. This report does not use the term 'geo-engineering' because of inconsistencies in the literature, which uses this term to cover SRM, CDR or both, whereas this report explicitly differentiates between CDR and SRM. Large-scale SRM could potentially be used to supplement mitigation in overshoot scenarios to keep the global mean temperature below 1.5°C and temporarily reduce the severity of near-term impacts (e.g., MacMartin et al., 2018). The impacts of SRM (both biophysical and societal), costs, technical feasibility, governance and ethical issues associated need to be carefully considered (Schäfer et al., 2015; Section 4.3.8 and Cross-Chapter Box 10 in Chapter 4).

1.4.2 Governance, Implementation and Policies

A challenge in creating the enabling conditions of a 1.5°C warmer world is the governance capacity of institutions to develop, implement and evaluate the changes needed within diverse and highly interlinked global social-ecological systems (Busby, 2016) (Chapter 4). Policy arenas, governance structures and robust institutions are key enabling conditions for transformative climate action (Chapter 4). It is through governance that justice, ethics and equity within the adaptation–mitigation–sustainable development nexus can be addressed (von Stechow et al., 2016) (Chapter 5).

Governance capacity includes a wide range of activities and efforts needed by different actors to develop coordinated climate mitigation and adaptation strategies in the context of sustainable development, taking into account equity, justice and poverty eradication. Significant governance challenges include the ability to incorporate multiple stakeholder perspectives in the decision-making process to reach meaningful and equitable decisions, interactions and coordination between different levels of government, and the capacity to raise financing and support for both technological and human resource development. For example, Lövbrand et al. (2017), argue that the voluntary pledges submitted by states and non-state actors to meet the conditions of the Paris Agreement will need to be more firmly coordinated, evaluated and upscaled.

Barriers for transitioning from climate change mitigation and adaptation planning to practical policy implementation include finance, information, technology, public attitudes, social values and practices (Whitmarsh et al., 2011; Corner and Clarke, 2017), and human resource constraints. Institutional capacity to deploy available knowledge and resources is also needed (Mimura et al., 2014). Incorporating strong linkages across sectors, devolution of power and resources to sub-national and local governments with the support of national government, and facilitating partnerships among public, civic, private sectors and higher education institutions (Leal Filho et al., 2018) can help in the implementation of identified response options (Chapter 4). Implementation challenges of 1.5°C pathways are larger than for those that are consistent with limiting warming to well below 2°C, particularly concerning scale and speed of the transition and the distributional impacts on ecosystems and socio-economic actors. Uncertainties in climate change at different scales and capacities to respond combined with the complexities of coupled social and ecological systems point to a need for diverse and adaptive implementation options within and among different regions involving different actors. The large regional diversity between highly carbon-invested economies and emerging economies are important considerations for sustainable development and equity in pursuing efforts to limit warming to 1.5°C. Key sectors, including energy, food systems, health, and water supply, also are critical to understanding these connections.

Cross-Chapter Box 3 | Framing Feasibility: Key Concepts and Conditions for Limiting Global Temperature Increases to 1.5°C

Contributing Authors:
William Solecki (USA), Anton Cartwright (South Africa), Wolfgang Cramer (France/Germany), James Ford (UK/Canada), Kejun Jiang (China), Joana Portugal Pereira (UK/Portugal), Joeri Rogelj (Austria/Belgium), Linda Steg (Netherlands), Henri Waisman (France)

This Cross-Chapter Box describes the concept of feasibility in relation to efforts to limit global warming to 1.5°C in the context of sustainable development and efforts to eradicate poverty and draws from the understanding of feasibility emerging within the IPCC (IPCC, 2017). Feasibility can be assessed in different ways, and no single answer exists as to the question of whether it is feasible to limit warming to 1.5°C. This implies that an assessment of feasibility would go beyond a 'yes' or a 'no'. Rather, feasibility provides a frame to understand the different conditions and potential responses for implementing adaptation and mitigation pathways, and options compatible with a 1.5°C warmer world. This report assesses the overall feasibility of limiting warming to 1.5°C, and the feasibility of adaptation and mitigation options compatible with a 1.5°C warmer world, in six dimensions:

Geophysical: What global emission pathways could be consistent with conditions of a 1.5°C warmer world? What are the physical potentials for adaptation?

Environmental-ecological: What are the ecosystem services and resources, including geological storage capacity and related rate of needed land-use change, available to promote transformations, and to what extent are they compatible with enhanced resilience?

Technological: What technologies are available to support transformation?

Economic: What economic conditions could support transformation?

Cross-Chapter Box 3 (continued)

Socio-cultural: What conditions could support transformations in behaviour and lifestyles? To what extent are the transformations socially acceptable and consistent with equity?

Institutional: What institutional conditions are in place to support transformations, including multi-level governance, institutional capacity, and political support?

Assessment of feasibility in this report starts by evaluating the unavoidable warming from past emissions (Section 1.2.4) and identifying mitigation pathways that would lead to a 1.5°C world, which indicates that rapid and deep deviations from current emission pathways are necessary (Chapter 2). In the case of adaptation, an assessment of feasibility starts from an evaluation of the risks and impacts of climate change (Chapter 3). To mitigate and adapt to climate risks, system-wide technical, institutional and socio-economic transitions would be required, as well as the implementation of a range of specific mitigation and adaptation options. Chapter 4 applies various indicators categorised in these six dimensions to assess the feasibility of illustrative examples of relevant mitigation and adaptation options (Section 4.5.1). Such options and pathways have different effects on sustainable development, poverty eradication and adaptation capacity (Chapter 5).

The six feasibility dimensions interact in complex and place-specific ways. Synergies and trade-offs may occur between the feasibility dimensions, and between specific mitigation and adaptation options (Section 4.5.4). The presence or absence of enabling conditions would affect the options that comprise feasibility pathways (Section 4.4), and can reduce trade-offs and amplify synergies between options.

Sustainable development, eradicating poverty and reducing inequalities are not only preconditions for feasible transformations, but the interplay between climate action (both mitigation and adaptation options) and the development patterns to which they apply may actually enhance the feasibility of particular options (see Chapter 5).

The connections between the feasibility dimensions can be specified across three types of effects (discussed below). Each of these dimensions presents challenges and opportunities in realizing conditions consistent with a 1.5°C warmer world.

Systemic effects: Conditions that have embedded within them system-level functions that could include linear and non-linear connections and feedbacks. For example, the deployment of technology and large installations (e.g., renewable or low carbon energy mega-projects) depends upon economic conditions (costs, capacity to mobilize investments for R&D), social or cultural conditions (acceptability), and institutional conditions (political support; e.g., Sovacool et al., 2015). Case studies can demonstrate system-level interactions and positive or negative feedback effects between the different conditions (Jacobson et al., 2015; Loftus et al., 2015). This suggests that each set of conditions and their interactions need to be considered to understand synergies, inequities and unintended consequences.

Dynamic effects: Conditions that are highly dynamic and vary over time, especially under potential conditions of overshoot or no overshoot. Some dimensions might be more time sensitive or sequential than others (i.e., if conditions are such that it is no longer geophysically feasible to avoid overshooting 1.5°C, the social and institutional feasibility of avoiding overshoot will be no longer relevant). Path dependencies, risks of legacy lock-ins related to existing infrastructures, and possibilities of acceleration permitted by cumulative effects (e.g., dramatic cost decreases driven by learning-by-doing) are all key features to be captured. The effects can play out over various time scales and thus require understanding the connections between near-term (meaning within the next several years to two decades) and long-term implications (meaning over the next several decades) when assessing feasibility conditions.

Spatial effects: Conditions that are spatially variable and scale dependent, according to context-specific factors such as regional-scale environmental resource limits and endowment; economic wealth of local populations; social organisation, cultural beliefs, values and worldviews; spatial organisation, including conditions of urbanisation; and financial and institutional and governance capacity. This means that the conditions for achieving the global transformation required for a 1.5°C world will be heterogeneous and vary according to the specific context. On the other hand, the satisfaction of these conditions may depend upon global-scale drivers, such as international flows of finance, technologies or capacities. This points to the need for understanding feasibility to capture the interplay between the conditions at different scales.

With each effect, the interplay between different conditions influences the feasibility of both pathways (Chapter 2) and options (Chapter 4), which in turn affect the likelihood of limiting warming to 1.5°C. The complexity of these interplays triggers unavoidable uncertainties, requiring transformations that remain robust under a range of possible futures that limit warming to 1.5°C.

1.4.3 Transformation, Transformation Pathways, and Transition: Evaluating Trade-Offs and Synergies Between Mitigation, Adaptation and Sustainable Development Goals

Embedded in the goal of limiting warming to 1.5°C is the opportunity for intentional societal transformation (see Box 1.1 on the Anthropocene). The form and process of transformation are varied and multifaceted (Pelling, 2011; O'Brien et al., 2012; O'Brien and Selboe, 2015; Pelling et al., 2015). Fundamental elements of 1.5°C-related transformation include a decoupling of economic growth from energy demand and CO_2 emissions; leap-frogging development to new and emerging low-carbon, zero-carbon and carbon-negative technologies; and synergistically linking climate mitigation and adaptation to global scale trends (e.g., global trade and urbanization) that will enhance the prospects for effective climate action, as well as enhanced poverty reduction and greater equity (Tschakert et al., 2013; Rogelj et al., 2015; Patterson et al., 2017) (Chapters 4 and 5). The connection between transformative climate action and sustainable development illustrates a complex coupling of systems that have important spatial and time scale lag effects and implications for process and procedural equity, including intergenerational equity and for non-human species (Cross-Chapter Box 4 in this chapter, Chapter 5). Adaptation and mitigation transition pathways highlight the importance of cultural norms and values, sector-specific context, and proximate (i.e., occurrence of an extreme event) drivers that when acting together enhance the conditions for societal transformation (Solecki et al., 2017; Rosenzweig et al., 2018) (Chapters 4 and 5).

Diversity and flexibility in implementation choices exist for adaptation, mitigation (including carbon dioxide removal, CDR) and remedial measures (such as solar radiation modification, SRM), and a potential for trade-offs and synergies between these choices and sustainable development (IPCC, 2014d; Olsson et al., 2014). The responses chosen could act to synergistically enhance mitigation, adaptation and sustainable development, or they may result in trade-offs which positively impact some aspects and negatively impact others. Climate change is expected to decrease the likelihood of achieving the Sustainable Development Goals (SDGs). While some strategies limiting warming towards 1.5°C are expected to significantly increase the likelihood of meeting those goals while also providing synergies for climate adaptation and mitigation (Chapter 5).

Dramatic transformations required to achieve the enabling conditions for a 1.5°C warmer world could impose trade-offs on dimensions of development (IPCC, 2014d; Olsson et al., 2014). Some choices of adaptation methods also could adversely impact development (Olsson et al., 2014). This report recognizes the potential for adverse impacts and focuses on finding the synergies between limiting warming, sustainable development, and eradicating poverty, thus highlighting pathways that do not constrain other goals, such as sustainable development and eradicating poverty.

The report is framed to address these multiple goals simultaneously and assesses the conditions to achieve a cost-effective and socially acceptable solution, rather than addressing these goals piecemeal (von Stechow et al., 2016) (Section 4.5.4 and Chapter 5), although there may be different synergies and trade-offs between a 2°C (von Stechow et al., 2016) and 1.5°C warmer world (Kainuma et al., 2017). Climate-resilient development pathways (see Cross-Chapter Box 12 in Chapter 5 and Glossary) are trajectories that strengthen sustainable development, including mitigating and adapting to climate change and efforts to eradicate poverty while promoting fair and cross-scalar resilience in a changing climate. They take into account dynamic livelihoods; the multiple dimensions of poverty, structural inequalities; and equity between and among poor and non-poor people (Olsson et al., 2014). Climate-resilient development pathways can be considered at different scales, including cities, rural areas, regions or at global level (Denton et al., 2014; Chapter 5).

Cross-Chapter Box 4 | Sustainable Development and the Sustainable Development Goals

Contributing Authors:
Diana Liverman (USA), Mustafa Babiker (Sudan), Purnamita Dasgupta (India), Riyanti Djanlante (Japan/Indonesia), Stephen Humphreys (UK/Ireland), Natalie Mahowald (USA), Yacob Mulugetta (UK/Ethiopia), Virginia Villariño (Argentina), Henri Waisman (France)

Sustainable development is most often defined as 'development that meets the needs of the present without compromising the ability of future generations to meet their own needs' (WCED, 1987) and includes balancing social well-being, economic prosperity and environmental protection. The AR5 used this definition and linked it to climate change (Denton et al., 2014). The most significant step since AR5 is the adoption of the UN Sustainable Development Goals, and the emergence of literature that links them to climate (von Stechow et al., 2015; Wright et al., 2015; Epstein and Theuer, 2017; Hammill and Price-Kelly, 2017; Kelman, 2017; Lofts et al., 2017; Maupin, 2017; Gomez-Echeverri, 2018).

In September 2015, the UN endorsed a universal agenda – 'Transforming our World: the 2030 Agenda for Sustainable Development' – which aims 'to take the bold and transformative steps which are urgently needed to shift the world onto a sustainable and resilient path'. Based on a participatory process, the resolution in support of the 2030 agenda adopted 17 non-legally-binding Sustainable Development Goals (SDGs) and 169 targets to support people, prosperity, peace, partnerships and the planet (Kanie and Biermann, 2017).

1

Cross-Chapter Box 4 (continued)

The SDGs expanded efforts to reduce poverty and other deprivations under the UN Millennium Development Goals (MDGs). There were improvements under the MDGs between 1990 and 2015, including reducing overall poverty and hunger, reducing infant mortality, and improving access to drinking water (UN, 2015a). However, greenhouse gas emissions increased by more than 50% from 1990 to 2015, and 1.6 billion people were still living in multidimensional poverty with persistent inequalities in 2015 (Alkire et al., 2015).

The SDGs raise the ambition for eliminating poverty, hunger, inequality and other societal problems while protecting the environment. They have been criticised: as too many and too complex, needing more realistic targets, overly focused on 2030 at the expense of longer-term objectives, not embracing all aspects of sustainable development, and even contradicting each other (Horton, 2014; Death and Gabay, 2015; Biermann et al., 2017; Weber, 2017; Winkler and Satterthwaite, 2017).

Climate change is an integral influence on sustainable development, closely related to the economic, social and environmental dimensions of the SDGs. The IPCC has woven the concept of sustainable development into recent assessments, showing how climate change might undermine sustainable development, and the synergies between sustainable development and responses to climate change (Denton et al., 2014). Climate change is also explicit in the SDGs. SDG13 specifically requires 'urgent action to address climate change and its impacts'. The targets include strengthening resilience and adaptive capacity to climate-related hazards and natural disasters; integrating climate change measures into national policies, strategies and planning; and improving education, awareness-raising and human and institutional capacity.

Targets also include implementing the commitment undertaken by developed-country parties to the UNFCCC to the goal of mobilizing jointly 100 billion USD annually by 2020 and operationalizing the Green Climate Fund, as well as promoting mechanisms for raising capacity for effective climate change-related planning and management in least developed countries and Small Island Developing States, including focusing on women, youth and local and marginalised communities. SDG13 also acknowledges that the UNFCCC is the primary international, intergovernmental forum for negotiating the global response to climate change.

Climate change is also mentioned in SDGs beyond SDG13, for example in goal targets 1.5, 2.4, 11.B, 12.8.1 related to poverty, hunger, cities and education respectively. The UNFCCC addresses other SDGs in commitments to 'control, reduce or prevent anthropogenic emissions of greenhouse gases [...] in all relevant sectors, including the energy, transport, industry, agriculture, forestry and waste management sectors' (Art4, 1(c)) and to work towards 'the conservation and enhancement, as appropriate, of [...] biomass, forests and oceans as well as other terrestrial, coastal and marine ecosystems' (Art4, 1(d)). This corresponds to SDGs that seek clean energy for all (Goal 7), sustainable industry (Goal 9) and cities (Goal 11) and the protection of life on land and below water (14 and 15).

The SDGs and UNFCCC also differ in their time horizons. The SDGs focus primarily on 2030 whereas the Paris Agreement sets out that 'Parties aim [...] to achieve a balance between anthropogenic emissions by sources and removals by sinks of greenhouse gases in the second half of this century'.

The IPCC decision to prepare this report on the impacts of 1.5°C and associated emission pathways explicitly asked for the assessment to be in the context of sustainable development and efforts to eradicate poverty. Chapter 1 frames the interaction between sustainable development, poverty eradication and ethics and equity. Chapter 2 assesses how risks and synergies of individual mitigation measures interact with 1.5°C pathways within the context of the SDGs and how these vary according to the mix of measures in alternative mitigation portfolios (Section 2.5). Chapter 3 examines the impacts of 1.5°C global warming on natural and human systems with comparison to 2°C and provides the basis for considering the interactions of climate change with sustainable development in Chapter 5. Chapter 4 analyses strategies for strengthening the response to climate change, many of which interact with sustainable development. Chapter 5 takes sustainable development, eradicating poverty and reducing inequalities as its focal point for the analysis of pathways to 1.5°C and discusses explicitly the linkages between achieving SDGs while eradicating poverty and reducing inequality.

Cross-Chapter Box 4 (continued)

Cross-Chapter Box 4, Figure 1 | Climate action is number 13 of the UN Sustainable Development Goals.

1.5 Assessment Frameworks and Emerging Methodologies that Integrate Climate Change Mitigation and Adaptation with Sustainable Development

This report employs information and data that are global in scope and include region-scale analysis. It also includes syntheses of municipal, sub-national, and national case studies. Global level statistics including physical and social science data are used, as well as detailed and illustrative case study material of particular conditions and contexts. The assessment provides the state of knowledge, including an assessment of confidence and uncertainty. The main time scale of the assessment is the 21st century and the time is separated into the near-, medium-, and long-term. Near-term refers to the coming decade, medium-term to the period 2030–2050, while long-term refers to 2050–2100. Spatial and temporal contexts are illustrated throughout, including: assessment tools that include dynamic projections of emission trajectories and the underlying energy and land transformation (Chapter 2); methods for assessing observed impacts and projected risks in natural and managed ecosystems and at 1.5°C and higher levels of warming in natural and managed ecosystems and human systems (Chapter 3); assessments of the feasibility of mitigation and adaptation options (Chapter 4); and linkages of the Shared Socioeconomic Pathways (SSPs) and Sustainable Development Goals (SDGs) (Cross-Chapter Boxes 1 and 4 in this chapter, Chapter 2 and Chapter 5).

1.5.1 Knowledge Sources and Evidence Used in the Report

This report is based on a comprehensive assessment of documented evidence of the enabling conditions to pursuing efforts to limit the global average temperature rise to 1.5°C and adapting to this level of warming in the overarching context of the Anthropocene (Delanty and Mota, 2017). Two sources of evidence are used: peer-reviewed scientific literature and 'grey' literature in accordance with procedure on the use of literature in IPCC reports (IPCC, 2013a, Annex 2 to Appendix A), with the former being the dominant source. Grey literature is largely used on key issues not covered in peer-reviewed literature.

The peer-reviewed literature includes the following sources: 1) knowledge regarding the physical climate system and human-induced changes, associated impacts, vulnerabilities, and adaptation options, established from work based on empirical evidence, simulations, modelling, and scenarios, with emphasis on new information since the publication of the IPCC AR5 to the cut-off date for this report (15th of May 2018); 2) humanities and social science theory and knowledge from actual human experiences of climate change risks and vulnerability in the context of social-ecological systems, development, equity, justice, and governance, and from indigenous knowledge systems; and 3) mitigation pathways based on climate projections into the future.

The grey literature category extends to empirical observations, interviews, and reports from government, industry, research institutes, conference proceedings and international or other organisations. Incorporating knowledge from different sources, settings and information channels while building awareness at various levels will advance decision-making and motivate implementation of context-specific responses to 1.5°C warming (Somanathan et al., 2014). The assessment does not assess non-written evidence and does not use oral evidence, media reports or newspaper publications. With important exceptions, such as China, published knowledge from the most vulnerable parts of the world to climate change is limited (Czerniewicz et al., 2017).

1.5.2   Assessment Frameworks and Methodologies

Climate models and associated simulations

The multiple sources of climate model information used in this assessment are provided in Chapter 2 (Section 2.2) and Chapter 3 (Section 3.2). Results from global simulations, which have also been assessed in previous IPCC reports and that are conducted as part of the World Climate Research Programme (WCRP) Coupled Models Intercomparison Project (CMIP) are used. The IPCC AR4 and Managing the Risks of Extreme Events and Disasters to Advance Climate Change Adaptation (SREX) reports were mostly based on simulations from the CMIP3 experiment, while the AR5 was mostly based on simulations from the CMIP5 experiment. The simulations of the CMIP3 and CMIP5 experiments were found to be very similar (e.g., Knutti and Sedláček, 2012; Mueller and Seneviratne, 2014). In addition to the CMIP3 and CMIP5 experiments, results from coordinated regional climate model experiments (e.g., the Coordinated Regional Climate Downscaling Experiment, CORDEX) have been assessed and are available for different regions (Giorgi and Gutowski, 2015). For instance, assessments based on publications from an extension of the IMPACT2C project (Vautard et al., 2014; Jacob and Solman, 2017) are newly available for 1.5°C projections. Recently, simulations from the 'Half a degree Additional warming, Prognosis and Projected Impacts' (HAPPI) multimodel experiment have been performed to specifically assess climate changes at 1.5°C vs 2°C global warming (Mitchell et al., 2016). The HAPPI protocol consists of coupled land–atmosphere initial condition ensemble simulations with prescribed sea surface temperatures (SSTs); sea ice, GHG and aerosol concentrations; and solar and volcanic activity that coincide with three forced climate states: present-day (2006–2015) (see Section 1.2.1) and future (2091–2100) either with 1.5°C or 2°C global warming (prescribed by modified SSTs).

Detection and attribution of change in climate and impacted systems

Formalized scientific methods are available to detect and attribute impacts of greenhouse gas forcing on observed changes in climate (e.g., Hegerl et al., 2007; Seneviratne et al., 2012; Bindoff et al., 2013) and impacts of climate change on natural and human systems (e.g., Stone et al., 2013; Hansen and Cramer, 2015; Hansen et al., 2016). The reader is referred to these sources, as well as to the AR5 for more background on these methods.

Global climate warming has already reached approximately 1°C (see Section 1.2.1) relative to pre-industrial conditions, and thus 'climate at 1.5°C global warming' corresponds to approximately the addition of only half a degree of warming compared to the present day, comparable to the warming that has occurred since the 1970s (Bindoff et al., 2013). Methods used in the attribution of observed changes associate with this recent warming are therefore also applicable to assessments of future changes in climate at 1.5°C warming, especially in cases where no climate model simulations or analyses are available.

Impacts of 1.5°C global warming can be assessed in part from regional and global climate changes that have already been detected and attributed to human influence (e.g., Schleussner et al., 2017) and are components of the climate system that are most responsive to current and projected future forcing. For this reason, when specific projections are missing for 1.5°C global warming, some of the assessments of climate change provided in Chapter 3 (Section 3.3) build upon joint assessments of (i) changes that were observed and attributed to human influence up to the present, that is, for 1°C global warming and (ii) projections for higher levels of warming (e.g., 2°C, 3°C or 4°C) to assess the changes at 1.5°C. Such assessments are for transient changes only (see Chapter 3, Section 3.3).

Besides quantitative detection and attribution methods, assessments can also be based on indigenous and local knowledge (see Chapter 4, Box 4.3). While climate observations may not be available to assess impacts from a scientific perspective, local community knowledge can also indicate actual impacts (Brinkman et al., 2016; Kabir et al., 2016). The challenge is that a community's perception of loss due to the impacts of climate change is an area that requires further research (Tschakert et al., 2017).

Costs and benefits analysis

Cost–benefit analyses are common tools used for decision-making, whereby the costs of impacts are compared to the benefits from different response actions (IPCC, 2014a, b). However, for the case of climate change, recognising the complex inter-linkages of the Anthropocene, cost–benefit analysis tools can be difficult to use because of disparate impacts versus costs and complex interconnectivity within the global social-ecological system (see Box 1.1 and Cross-Chapter Box 5 in Chapter 2). Some costs are relatively easily quantifiable in monetary terms but not all. Climate change impacts human lives and livelihoods, culture and values, and whole ecosystems. It has unpredictable feedback loops and impacts on other regions (IPCC, 2014a), giving rise to indirect, secondary, tertiary and opportunity costs that are typically extremely difficult to quantify. Monetary quantification is further complicated by the fact that costs and benefits can occur in different regions at very different times, possibly spanning centuries, while it is extremely difficult if not impossible to meaningfully estimate discount rates for future costs and benefits. Thus standard cost–benefit analyses become difficult to justify (IPCC, 2014a; Dietz et al., 2016) and are not used as an assessment tool in this report.

1.6   Confidence, Uncertainty and Risk

This report relies on the IPCC's uncertainty guidance provided in Mastrandrea et al. (2011) and sources given therein. Two metrics for qualifying key findings are used:

Confidence: Five qualifiers are used to express levels of confidence in key findings, ranging from *very low*, through *low*, *medium*, *high*, to *very high*. The assessment of confidence involves at least two dimensions, one being the type, quality, amount or internal consistency of individual lines of evidence, and the second being the level of agreement between different lines of evidence. Very high confidence findings must either be supported by a high level of agreement across multiple lines of mutually independent and individually robust lines of evidence or, if only a single line of evidence is available, by a very high level of understanding underlying that evidence. Findings of low or very low confidence are presented only if they address a topic of major concern.

Likelihood: A calibrated language scale is used to communicate assessed probabilities of outcomes, ranging from *exceptionally unlikely* (<1%), *extremely unlikely* (<5%), *very unlikely* (<10%), *unlikely* (<33%), *about as likely as not* (33–66%), *likely* (>66%), *very likely* (>90%), *extremely likely* (>95%) to *virtually certain* (>99%). These terms are normally only applied to findings associated with high or very high confidence. Frequency of occurrence within a model ensemble does not correspond to actual assessed probability of outcome unless the ensemble is judged to capture and represent the full range of relevant uncertainties.

Three specific challenges arise in the treatment of uncertainty and risk in this report. First, the current state of the scientific literature on 1.5°C means that findings based on multiple lines of robust evidence for which quantitative probabilistic results can be expressed may be few in number, and those that do exist may not be the most policy-relevant. Hence many key findings are expressed using confidence qualifiers alone.

Second, many of the most important findings of this report are conditional because they refer to ambitious mitigation scenarios, potentially involving large-scale technological or societal transformation. Conditional probabilities often depend strongly on how conditions are specified, such as whether temperature goals are met through early emission reductions, reliance on negative emissions, or through a low climate response. Whether a certain risk is considered high at 1.5°C may therefore depend strongly on how 1.5°C is specified, whereas a statement that a certain risk may be substantially higher at 2°C relative to 1.5°C may be much more robust.

Third, achieving ambitious mitigation goals will require active, goal-directed efforts aiming explicitly for specific outcomes and incorporating new information as it becomes available (Otto et al., 2015). This shifts the focus of uncertainty from the climate outcome itself to the level of mitigation effort that may be required to achieve it. Probabilistic statements about human decisions are always problematic, but in the context of robust decision-making, many near-term policies that are needed to keep open the option of limiting warming to 1.5°C may be the same, regardless of the actual probability that the goal will be met (Knutti et al., 2015).

1.7   Storyline of the Report

The storyline of this report (Figure 1.6) includes a set of interconnected components. The report consists of five chapters (plus Supplementary Material for Chapters 1 through 4), a Technical Summary and a Summary for Policymakers. It also includes a set of boxes to elucidate specific or cross-cutting themes, as well as Frequently Asked Questions for each chapter, a Glossary, and several other Annexes.

At a time of unequivocal and rapid global warming, this report emerges from the long-term temperature goal of the Paris Agreement – strengthening the global response to the threat of climate change by pursuing efforts to limit warming to 1.5°C through reducing emissions to achieve a balance between anthropogenic emissions by sources and removals by sinks of greenhouse gases. The assessment focuses first, in Chapter 1, on how 1.5°C is defined and understood, what is the current level of warming to date, and the present trajectory of change. The framing presented in Chapter 1 provides the basis through which to understand the enabling conditions of a 1.5°C warmer world and connections to the SDGs, poverty eradication, and equity and ethics.

In Chapter 2, scenarios of a 1.5°C warmer world and the associated pathways are assessed. The pathways assessment builds upon the AR5 with a greater emphasis on sustainable development in mitigation pathways. All pathways begin now and involve rapid and unprecedented societal transformation. An important framing device for this report is the recognition that choices that determine emissions pathways, whether ambitious mitigation or 'no policy' scenarios, do not occur independently of these other changes and are, in fact, highly interdependent.

Projected impacts that emerge in a 1.5°C warmer world and beyond are dominant narrative threads of the report and are assessed in Chapter 3. The chapter focuses on observed and attributable global and regional climate changes and impacts and vulnerabilities. The projected impacts have diverse and uneven spatial, temporal, human, economic, and ecological system-level manifestations. Central to the assessment is the reporting of impacts at 1.5°C and 2°C, potential impacts avoided through limiting warming to 1.5°C, and, where possible, adaptation potential and limits to adaptive capacity.

Response options and associated enabling conditions emerge next, in Chapter 4. Attention is directed to exploring questions of adaptation and mitigation implementation, integration, and transformation in a highly interdependent world, with consideration of synergies and trade-offs. Emission pathways, in particular, are broken down into policy options and instruments. The role of technological choices, institutional capacity and global-scale trends like urbanization and changes in ecosystems are assessed.

Chapter 5 covers linkages between achieving the SDGs and a 1.5°C warmer world and turns toward identifying opportunities and challenges of transformation. This is assessed within a transition to climate-resilient development pathways and connection between the evolution towards 1.5°C, associated impacts, and emission pathways. Positive and negative effects of adaptation and mitigation response measures and pathways for a 1.5°C warmer world are examined.

Progress along these pathways involves inclusive processes, institutional integration, adequate finance and technology, and attention to issues of power, values, and inequalities to maximize the benefits of pursuing climate stabilisation at 1.5°C and the goals of sustainable development at multiple scales of human and natural systems from global, regional, national to local and community levels.

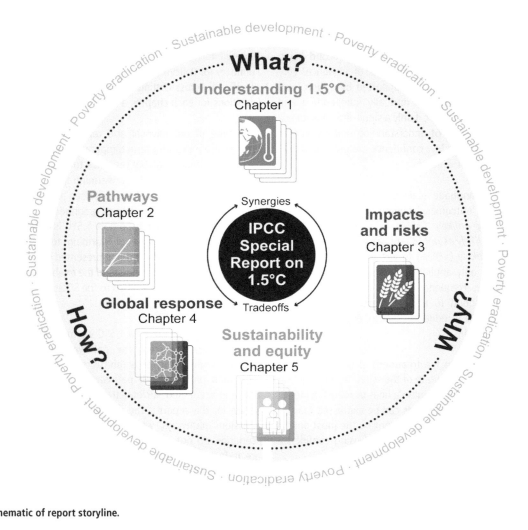

Figure 1.6 | Schematic of report storyline.

Frequently Asked Questions

FAQ 1.1 | Why are we Talking about 1.5°C?

Summary: Climate change represents an urgent and potentially irreversible threat to human societies and the planet. In recognition of this, the overwhelming majority of countries around the world adopted the Paris Agreement in December 2015, the central aim of which includes pursuing efforts to limit global temperature rise to 1.5°C. In doing so, these countries, through the United Nations Framework Convention on Climate Change (UNFCCC), also invited the IPCC to provide a Special Report on the impacts of global warming of 1.5°C above pre-industrial levels and related global greenhouse gas emissions pathways.

At the 21st Conference of the Parties (COP21) in December 2015, 195 nations adopted the Paris Agreement[2]. The first instrument of its kind, the landmark agreement includes the aim to strengthen the global response to the threat of climate change by 'holding the increase in the global average temperature to well below 2°C above pre-industrial levels and pursuing efforts to limit the temperature increase to 1.5°C above pre-industrial levels'.

The first UNFCCC document to mention a limit to global warming of 1.5°C was the Cancun Agreement, adopted at the sixteenth COP (COP16) in 2010. The Cancun Agreement established a process to periodically review the 'adequacy of the long-term global goal (LTGG) in the light of the ultimate objective of the Convention and the overall progress made towards achieving the LTGG, including a consideration of the implementation of the commitments under the Convention'. The definition of LTGG in the Cancun Agreement was 'to hold the increase in global average temperature below 2°C above pre-industrial levels'. The agreement also recognised the need to consider 'strengthening the long-term global goal on the basis of the best available scientific knowledge...to a global average temperature rise of 1.5°C'.

Beginning in 2013 and ending at the COP21 in Paris in 2015, the first review period of the long-term global goal largely consisted of the Structured Expert Dialogue (SED). This was a fact-finding, face-to-face exchange of views between invited experts and UNFCCC delegates. The final report of the SED[3] concluded that 'in some regions and vulnerable ecosystems, high risks are projected even for warming above 1.5°C'. The SED report also suggested that Parties would profit from restating the temperature limit of the long-term global goal as a 'defence line' or 'buffer zone', instead of a 'guardrail' up to which all would be safe, adding that this new understanding would 'probably also favour emission pathways that will limit warming to a range of temperatures below 2°C'. Specifically on strengthening the temperature limit of 2°C, the SED's key message was: 'While science on the 1.5°C warming limit is less robust, efforts should be made to push the defence line as low as possible'. The findings of the SED, in turn, fed into the draft decision adopted at COP21.

With the adoption of the Paris Agreement, the UNFCCC invited the IPCC to provide a Special Report in 2018 on 'the impacts of global warming of 1.5°C above pre-industrial levels and related global greenhouse gas emissions pathways'. The request was that the report, known as SR1.5, should not only assess what a 1.5°C warmer world would look like but also the different pathways by which global temperature rise could be limited to 1.5°C. In 2016, the IPCC accepted the invitation, adding that the Special Report would also look at these issues in the context of strengthening the global response to the threat of climate change, sustainable development and efforts to eradicate poverty.

The combination of rising exposure to climate change and the fact that there is a limited capacity to adapt to its impacts amplifies the risks posed by warming of 1.5°C and 2°C. This is particularly true for developing and island countries in the tropics and other vulnerable countries and areas. The risks posed by global warming of 1.5°C are greater than for present-day conditions but lower than at 2°C.

(continued on next page)

---

[2]   Paris Agreement FCCC/CP/2015/10/Add.1 https://unfccc.int/documents/9097

[3]   Structured Expert Dialogue (SED) final report FCCC/SB/2015/INF.1 https://unfccc.int/documents/8707

FAQ 1.1 (continued)

FAQ1.1: **Timeline of 1.5°C**

Milestones in the IPCC's preparation of the Special Report on Global Warming of 1.5°C and some relevant events in the history of international climate negotiations

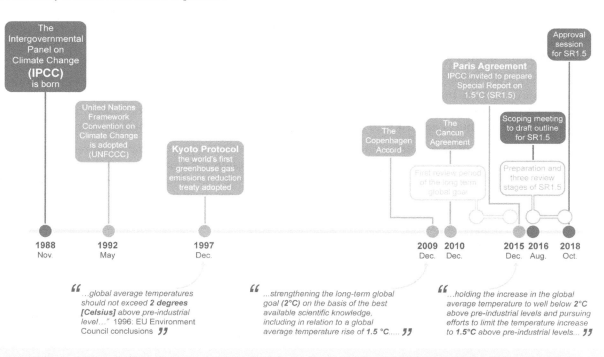

FAQ 1.1, Figure 1 | Timeline of notable dates in preparing the IPCC Special Report on Global Warming of 1.5°C (blue) embedded within processes and milestones of the United Nations Framework Convention on Climate Change (UNFCCC; grey), including events that may be relevant for discussion of temperature limits.

Frequently Asked Questions

FAQ 1.2 | How Close are we to 1.5°C?

Summary: Human-induced warming has already reached about 1°C above pre-industrial levels at the time of writing of this Special Report. By the decade 2006–2015, human activity had warmed the world by 0.87°C (±0.12°C) compared to pre-industrial times (1850–1900). If the current warming rate continues, the world would reach human-induced global warming of 1.5°C around 2040.

Under the 2015 Paris Agreement, countries agreed to cut greenhouse gas emissions with a view to 'holding the increase in the global average temperature to well below 2°C above pre-industrial levels and pursuing efforts to limit the temperature increase to 1.5°C above pre-industrial levels'. While the overall intention of strengthening the global response to climate change is clear, the Paris Agreement does not specify precisely what is meant by 'global average temperature', or what period in history should be considered 'pre-industrial'. To answer the question of how close are we to 1.5°C of warming, we need to first be clear about how both terms are defined in this Special Report.

The choice of pre-industrial reference period, along with the method used to calculate global average temperature, can alter scientists' estimates of historical warming by a couple of tenths of a degree Celsius. Such differences become important in the context of a global temperature limit just half a degree above where we are now. But provided consistent definitions are used, they do not affect our understanding of how human activity is influencing the climate.

In principle, 'pre-industrial levels' could refer to any period of time before the start of the industrial revolution. But the number of direct temperature measurements decreases as we go back in time. Defining a 'pre-industrial' reference period is, therefore, a compromise between the reliability of the temperature information and how representative it is of truly pre-industrial conditions. Some pre-industrial periods are cooler than others for purely natural reasons. This could be because of spontaneous climate variability or the response of the climate to natural perturbations, such as volcanic eruptions and variations in the sun's activity. This IPCC Special Report on Global Warming of 1.5°C uses the reference period 1850–1900 to represent pre-industrial temperature. This is the earliest period with near-global observations and is the reference period used as an approximation of pre-industrial temperatures in the IPCC Fifth Assessment Report.

Once scientists have defined 'pre-industrial', the next step is to calculate the amount of warming at any given time relative to that reference period. In this report, warming is defined as the increase in the 30-year global average of combined air temperature over land and water temperature at the ocean surface. The 30-year timespan accounts for the effect of natural variability, which can cause global temperatures to fluctuate from one year to the next. For example, 2015 and 2016 were both affected by a strong El Niño event, which amplified the underlying human-caused warming.

In the decade 2006–2015, warming reached 0.87°C (±0.12°C) relative to 1850–1900, predominantly due to human activity increasing the amount of greenhouse gases in the atmosphere. Given that global temperature is currently rising by 0.2°C (±0.1°C) per decade, human-induced warming reached 1°C above pre-industrial levels around 2017 and, if this pace of warming continues, would reach 1.5°C around 2040.

While the change in global average temperature tells researchers about how the planet as a whole is changing, looking more closely at specific regions, countries and seasons reveals important details. Since the 1970s, most land regions have been warming faster than the global average, for example. This means that warming in many regions has already exceeded 1.5°C above pre-industrial levels. Over a fifth of the global population live in regions that have already experienced warming in at least one season that is greater than 1.5°C above pre-industrial levels.

(continued on next page)

FAQ 1.2 (continued)

FAQ1.2:**How close are we to** 1.5°C**?**

Human-induced warming reached approximately 1°C above
pre-industrial levels in 2017

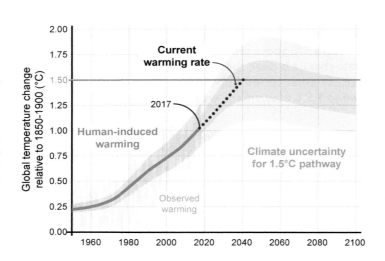

FAQ 1.2, Figure 1 | Human-induced warming reached approximately 1°C above pre-industrial levels in 2017. At the present rate, global temperatures would reach 1.5°C around 2040. Stylized 1.5°C pathway shown here involves emission reductions beginning immediately, and CO_2 emissions reaching zero by 2055.

References

Aaheim, A., T. Wei, and B. Romstad, 2017: Conflicts of economic interests by limiting global warming to +3°C. *Mitigation and Adaptation Strategies for Global Change*, **22(8)**, 1131–1148, doi:10.1007/s11027-016-9718-8.

Abram, N.J. et al., 2016: Early onset of industrial-era warming across the oceans and continents. *Nature*, **536**, 411–418, doi:10.1038/nature19082.

Adger, W.N. et al., 2014: Human Security. In: *Climate Change 2014: Impacts, Adaptation, and Vulnerability. Part A: Global and Sectoral Aspects. Contribution of Working Group II to the Fifth Assessment Report of the Intergovernmental Panel on Climate Change* [Field, C.B., V.R. Barros, D.J. Dokken, K.J. Mach, M.D. Mastrandrea, T.E. Bilir, M. Chatterjee, K.L. Ebi, Y.O. Estrada, R.C. Genova, B. Girma, E.S. Kissel, A.N. Levy, S. MacCracken, P.R. Mastrandrea, and L.L. White (eds.)]. Cambridge University Press, Cambridge, United Kingdom and New York, NY, USA, pp. 755–791.

AghaKouchak, A., L. Cheng, O. Mazdiyasni, and A. Farahmand, 2014: Global warming and changes in risk of concurrent climate extremes: Insights from the 2014 California drought. *Geophysical Research Letters*, **41(24)**, 8847–8852, doi:10.1002/2014gl062308.

Ajibade, I., 2016: Distributive justice and human rights in climate policy: the long road to Paris. *Journal of Sustainable Development Law and Policy (The)*, **7(2)**, 65–80, doi:10.4314/jsdlp.v7i2.4.

Albert, S. et al., 2017: Heading for the hills: climate-driven community relocations in the Solomon Islands and Alaska provide insight for a 1.5°C future. *Regional Environmental Change*, 1–12, doi:10.1007/s10113-017-1256-8.

Alkire, S., C. Jindra, G. Robles Aguilar, S. Seth, and A. Vaz, 2015: *Global Multidimensional Poverty Index 2015*. Briefing 31, Oxford Poverty & Human Development Initiative, University of Oxford, Oxford, UK, 8 pp.

Allen, M.R. and T.F. Stocker, 2013: Impact of delay in reducing carbon dioxide emissions. *Nature Climate Change*, **4(1)**, 23–26, doi:10.1038/nclimate2077.

Allen, M.R. et al., 2016: New use of global warming potentials to compare cumulative and short-lived climate pollutants. *Nature Climate Change*, **6**, 1–5, doi:10.1038/nclimate2998.

Allen, M.R. et al., 2018: A solution to the misrepresentations of CO_2-equivalent emissions of short-lived climate pollutants under ambitious mitigation. *npj Climate and Atmospheric Science*, **1(1)**, 16, doi:10.1038/s41612-018-0026-8.

Archer, D. and V. Brovkin, 2008: The millennial atmospheric lifetime of anthropogenic CO_2. *Climatic Change*, **90(3)**, 283–297, doi:10.1007/s10584-008-9413-1.

Arora-Jonsson, S., 2011: Virtue and vulnerability: Discourses on women, gender and climate change. *Global Environmental Change*, **21(2)**, 744–751, doi:10.1016/j.gloenvcha.2011.01.005.

Asseng, S. et al., 2013: Uncertainty in simulating wheat yields under climate change. *Nature Climate Change*, **3(9)**, 827–832, doi:10.1038/nclimate1916.

Bäckstrand, K., J.W. Kuyper, B.-O. Linnér, and E. Lövbrand, 2017: Non-state actors in global climate governance: from Copenhagen to Paris and beyond. *Environmental Politics*, **26(4)**, 561–579, doi:10.1080/09644016.2017.1327485.

Baker, H.S. et al., 2018: Higher CO_2 concentrations increase extreme event risk in a 1.5°C world. *Nature Climate Change*, **8(7)**, 604–608, doi:10.1038/s41558-018-0190-1.

Barcikowska, M.J. et al., 2018: Euro-Atlantic winter storminess and precipitation extremes under 1.5°C vs. 2°C warming scenarios. *Earth System Dynamics*, **9(2)**, 679–699, doi:10.5194/esd-9-679-2018.

Barnett, J. et al., 2014: A local coastal adaptation pathway. *Nature Climate Change*, **4(12)**, 1103–1108, doi:10.1038/nclimate2383.

Bauer, N. et al., 2017: Shared Socio-Economic Pathways of the Energy Sector – Quantifying the Narratives. *Global Environmental Change*, **42**, 316–330, doi:10.1016/j.gloenvcha.2016.07.006.

Bereiter, B. et al., 2015: Revision of the EPICA Dome C CO_2 record from 800 to 600-kyr before present. *Geophysical Research Letters*, **42(2)**, 542–549, doi:10.1002/2014gl061957.

Berger, A., Q. Yin, H. Nifenecker, and J. Poitou, 2017: Slowdown of global surface air temperature increase and acceleration of ice melting. *Earth's Future*, **5(7)**, 811–822, doi:10.1002/2017ef000554.

Bethke, I. et al., 2017: Potential volcanic impacts on future climate variability. *Nature Climate Change*, **7(11)**, 799–805, doi:10.1038/nclimate3394.

Biermann, F., N. Kanie, and R.E. Kim, 2017: Global governance by goal-setting: the novel approach of the UN Sustainable Development Goals. *Current Opinion in Environmental Sustainability*, **26–27**, 26–31, doi:10.1016/j.cosust.2017.01.010.

Biermann, F. et al., 2016: Down to Earth: Contextualizing the Anthropocene. *Global Environmental Change*, **39**, 341–350, doi:10.1016/j.gloenvcha.2015.11.004.

Bindoff, N.L. et al., 2007: Observations: Oceanic Climate Change and Sea Level. In: *Climate Change 2007: The Physical Science Basis. Contribution of Working Group I to the Fourth Assessment Report of the Intergovernmental Panel on Climate Change* [Solomon, S., D. Qin, M. Manning, Z. Chen, M. Marquis, K.B. Averyt, M. Tignor, and H.L. Miller (eds.)]. Cambridge University Press, Cambridge, United Kingdom and New York, NY, USA, pp. 385–432.

Bindoff, N.L. et al., 2013: Detection and Attribution of Climate Change: from Global to Regional. In: *Climate Change 2013: The Physical Science Basis. Contribution of Working Group I to the Fifth Assessment Report of the Intergovernmental Panel on Climate Change* [Stocker, T.F., D. Qin, G.-K. Plattner, M. Tignor, S.K. Allen, J. Boschung, A. Nauels, Y. Xia, V. Bex, and P.M. Midgley (eds.)]. Cambridge University Press, Cambridge, United Kingdom and New York, NY, USA, pp. 426–488.

Birkmann, J., T. Welle, W. Solecki, S. Lwasa, and M. Garschagen, 2016: Boost resilience of small and mid-sized cities. *Nature*, **537(7622)**, 605–608, doi:10.1038/537605a.

Blicharska, M. et al., 2017: Steps to overcome the North–South divide in research relevant to climate change policy and practice. *Nature Climate Change*, **7(1)**, 21–27, doi:10.1038/nclimate3163.

Bodansky, D., J. Brunnée, and L. Rajamani, 2017: *International Climate Change Law*. Oxford University Press, Oxford, UK, 416 pp.

Bowerman, N.H.A., D.J. Frame, C. Huntingford, J.A. Lowe, and M.R. Allen, 2011: Cumulative carbon emissions, emissions floors and short-term rates of warming: implications for policy. *Philosophical Transactions of the Royal Society A: Mathematical, Physical and Engineering Sciences*, **369(1934)**, 45–66, doi:10.1098/rsta.2010.0288.

Bowerman, N.H.A. et al., 2013: The role of short-lived climate pollutants in meeting temperature goals. *Nature Climate Change*, **3(12)**, 1021–1024, doi:10.1038/nclimate2034.

Brinkman, T.J. et al., 2016: Arctic communities perceive climate impacts on access as a critical challenge to availability of subsistence resources. *Climatic Change*, **139(3–4)**, 413–427, doi:10.1007/s10584-016-1819-6.

Brondizio, E.S. et al., 2016: Re-conceptualizing the Anthropocene: A call for collaboration. *Global Environmental Change*, **39**, 318–327, doi:10.1016/j.gloenvcha.2016.02.006.

Busby, J., 2016: After Paris: Good enough climate governance. *Current History*, **15(777)**, 3–9, www.currenthistory.com/busby_currenthistory.pdf.

Caney, S., 2005: Cosmopolitan Justice, Responsibility, and Global Climate Change. *Leiden Journal of International Law*, **18(04)**, 747–75, doi:10.1017/s0922156505002992.

Caney, S., 2010: Climate change and the duties of the advantaged. *Critical Review of International Social and Political Philosophy*, **13(1)**, 203–228, doi:10.1080/13698230903326331.

Cardona, O.D. et al., 2012: Determinants of Risk: Exposure and Vulnerablity. In: *Managing the Risks of Extreme Events and Disasters to Advance Climate Change Adaptation. A Special Report of Working Groups I and II of the Intergovernmental Panel on Climate Change (IPCC)* [Field, C.B., V.R. Barros, T.F. Stocker, D. Qin, D.J. Dokken, K.L. Ebi, M.D. Mastrandrea, K.J. Mach, G.-K. Plattner, S.K. Allen, M. Tignor, and P.M. Midgley (eds.)]. Cambridge University Press, Cambridge, United Kingdom and New York, NY, USA, pp. 65–108.

Chen, X. et al., 2017: The increasing rate of global mean sea-level rise during 1993–2014. *Nature Climate Change*, **7(7)**, 492–495, doi:10.1038/nclimate3325.

Cherubini, F. and K. Tanaka, 2016: Amending the Inadequacy of a Single Indicator for Climate Impact Analyses. *Environmental Science & Technology*, **50(23)**, 12530–12531, doi:10.1021/acs.est.6b05343.

Chevuturi, A., N.P. Klingaman, A.G. Turner, and S. Hannah, 2018: Projected Changes in the Asian-Australian Monsoon Region in 1.5°C and 2.0°C Global-Warming Scenarios. *Earth's Future*, **6(3)**, 339–358, doi:10.1002/2017ef000734.

Christensen, J.H. et al., 2013: Climate Phenomena and their Relevance for Future Regional Climate Change Supplementary Material. In: *Climate Change 2013: The Physical Science Basis. Contribution of Working Group I to the Fifth Assessment Report of the Intergovernmental Panel on Climate Change* [Stocker, T.F., D. Qin, G.K. Plattner, M. Tignor, S.K. Allen, J. Boschung, A. Nauels, Y. Xia, V. Bex, and P.M. Midgley (eds.)]. Cambridge University Press, Cambridge, United Kingdom and New York, NY, USA, pp. 1217–1308.

Ciais, P. et al., 2013: Carbon and Other Biogeochemical Cycles. In: *Climate Change 2013: The Physical Science Basis. Contribution of Working Group I to the Fifth Assessment Report of the Intergovernmental Panel on Climate Change* [Stocker, T.F., D. Qin, G.-K. Plattner, M. Tignor, S.K. Allen, J. Boschung, A. Nauels, Y. Xia, V. Bex, and P.M. Midgley (eds.)]. Cambridge University Press, Cambridge, United Kingdom and New York, NY, USA, pp. 465–570.

Clarke, L.E. et al., 2014: Assessing transformation pathways. In: *Climate Change 2014: Mitigation of Climate Change. Contribution of Working Group III to the Fifth Assessment Report of the Intergovernmental Panel on Climate Change* [Edenhofer, O., R. Pichs-Madruga, Y. Sokona, E. Farahani, S. Kadner, K. Seyboth, A. Adler, I. Baum, S. Brunner, P. Eickemeier, B. Kriemann, J. Savolainen, S. Schlömer, C. von Stechow, T. Zwickel, and J.C. Minx (eds.)]. Cambridge University Press, Cambridge, United Kingdom and New York, NY, USA, pp. 413–510.

Collins, M. et al., 2013: Long-term Climate Change: Projections, Commitments and Irreversibility. In: *Climate Change 2013: The Physical Science Basis. Contribution of Working Group I to the Fifth Assessment Report of the Intergovernmental Panel on Climate Change* [Stocker, T.F., D. Qin, G.-K. Plattner, M. Tignor, S.K. Allen, J. Boschung, A. Nauels, Y. Xia, V. Bex, and P.M. Midgley (eds.)]. Cambridge University Press, Cambridge, United Kingdom and New York, NY, USA, pp. 1029–1136.

Corner, A. and J. Clarke, 2017: *Talking Climate – From Research to Practice in Public Engagement*. Palgrave Macmillan, Oxford, UK, 146 pp., doi:10.1007/978-3-319-46744-3.

Cowtan, K. and R.G. Way, 2014: Coverage bias in the HadCRUT4 temperature series and its impact on recent temperature trends. *Quarterly Journal of the Royal Meteorological Society*, **140(683)**, 1935–1944, doi:10.1002/qj.2297.

Cowtan, K. et al., 2015: Robust comparison of climate models with observations using blended land air and ocean sea surface temperatures. *Geophysical Research Letters*, **42(15)**, 6526–6534, doi:10.1002/2015gl064888.

Cramer, W. et al., 2014: Detection and attribution of observed impacts. In: *Climate Change 2014: Impacts, Adaptation, and Vulnerability. Part A: Global and Sectoral Aspects. Contribution of Working Group II to the Fifth Assessment Report of the Intergovernmental Panel on Climate Change* [Field, C.B., V.R. Barros, D.J. Dokken, K.J. Mach, M.D. Mastrandrea, T.E. Bilir, M. Chatterjee, K.L. Ebi, Y.O. Estrada, R.C. Genova, B. Girma, E.S. Kissel, A.N. Levy, S. MacCracken, P.R. Mastrandrea, and L.L. White (eds.)]. Cambridge University Press, Cambridge, United Kingdom and New York, NY, USA, pp. 979–1037.

Creutzig, F., 2016: Economic and ecological views on climate change mitigation with bioenergy and negative emissions. *GCB Bioenergy*, **8(1)**, 4–10, doi:10.1111/gcbb.12235.

Crutzen, P.J., 2002: Geology of mankind. *Nature*, **415(6867)**, 23, doi:10.1038/415023a.

Crutzen, P.J. and E.F. Stoermer, 2000: The Anthropocene. *Global Change Newsletter*, **41**, 17–18, www.igbp.net/download/18.31 6f18321323470177580001401/1376383088452/nl41.pdf.

Czerniewicz, L., S. Goodier, and R. Morrell, 2017: Southern knowledge online? Climate change research discoverability and communication practices. *Information, Communication & Society*, **20(3)**, 386–405, doi:10.1080/1369118x.2016.1168473.

Dasgupta, P. et al., 2014: Rural areas. In: *Climate Change 2014: Impacts, Adaptation, and Vulnerability. Part A: Global and Sectoral Aspects. Contribution of Working Group II to the Fifth Assessment Report of the Intergovernmental Panel on Climate Change* [Field, C.B., V.R. Barros, D.J. Dokken, K.J. Mach, M.D. Mastrandrea, T.E. Bilir, M. Chatterjee, K.L. Ebi, Y.O. Estrada, R.C. Genova, B. Girma, E.S. Kissel, A.N. Levy, S. MacCracken, P.R. Mastrandrea, and L.L. White (eds.)]. Cambridge University Press, Cambridge, United Kingdom and New York, NY, USA, pp. 613–657.

Death, C. and C. Gabay, 2015: Doing biopolitics differently? Radical potential in the post-2015 MDG and SDG debates. *Globalizations*, **12(4)**, 597–612, doi:10.1080/14747731.2015.1033172.

Delanty, G. and A. Mota, 2017: Governing the Anthropocene. *European Journal of Social Theory*, **20(1)**, 9–38, doi:10.1177/1368431016668535.

Denton, F. et al., 2014: Climate-Resilient Pathways: Adaptation, Mitigation, and Sustainable Development. In: *Climate Change 2014: Impacts, Adaptation, and Vulnerability. Part A: Global and Sectoral Aspects. Contribution of Working Group II to the Fifth Assessment Report of the Intergovernmental Panel on Climate Change* [Field, C.B., V.R. Barros, D.J. Dokken, K.J. Mach, M.D. Mastrandrea, T.E. Bilir, M. Chatterjee, K.L. Ebi, Y.O. Estrada, R.C. Genova, B. Girma, E.S. Kissel, A.N. Levy, S. MacCracken, P.R. Mastrandrea, and L.L. White (eds.)]. Cambridge University Press, Cambridge, United Kingdom and New York, NY, USA, pp. 1101–1131.

Deser, C., R. Knutti, S. Solomon, and A.S. Phillips, 2012: Communication of the Role of Natural Variability in Future North American Climate. *Nature Climate Change*, **2(11)**, 775–779, doi:10.1038/nclimate1562.

Deuber, O., G. Luderer, and O. Edenhofer, 2013: Physico-economic evaluation of climate metrics: A conceptual framework. *Environmental Science & Policy*, **29(0)**, 37–45, doi:10.1016/j.envsci.2013.01.018.

Dietz, S., B. Groom, and W.A. Pizer, 2016: Weighing the Costs and Benefits of Climate Change to Our Children. *The Future of Children*, **26(1)**, 133–155, www.jstor.org/stable/43755234.

Döll, P. et al., 2018: Risks for the global freshwater system at 1.5°C and 2°C global warming. *Environmental Research Letters*, **13(4)**, 044038, doi:10.1088/1748-9326/aab792.

Dooley, K., J. Gupta, and A. Patwardhan, 2018: INEA editorial: Achieving 1.5°C and climate justice. *International Environmental Agreements: Politics, Law and Economics*, **18(1)**, 1–9, doi:10.1007/s10784-018-9389-x.

Dryzek, J.S., 2016: Institutions for the Anthropocene: Governance in a Changing Earth System. *British Journal of Political Science*, **46(04)**, 937–956, doi:10.1017/s0007123414000453.

Duyck, S., S. Jodoin, and A. Johl (eds.), 2018: *Routledge Handbook of Human Rights and Climate Governance*. Routledge, Abingdon, UK, 430 pp.

Ebi, K.L., L.H. Ziska, and G.W. Yohe, 2016: The shape of impacts to come: lessons and opportunities for adaptation from uneven increases in global and regional temperatures. *Climatic Change*, **139(3)**, 341–349, doi:10.1007/s10584-016-1816-9.

Ebi, K.L. et al., 2014: A new scenario framework for climate change research: Background, process, and future directions. *Climatic Change*, **122(3)**, 363–372, doi:10.1007/s10584-013-0912-3.

Eby, M. et al., 2009: Lifetime of anthropogenic climate change: Millennial time scales of potential CO_2 and surface temperature perturbations. *Journal of Climate*, **22(10)**, 2501–2511, doi:10.1175/2008jcli2554.1.

Ehlert, D. and K. Zickfeld, 2017: What determines the warming commitment after cessation of CO_2 emissions? *Environmental Research Letters*, **12(1)**, 015002, doi:10.1088/1748-9326/aa564a.

Ekström, M., M.R. Grose, and P.H. Whetton, 2015: An appraisal of downscaling methods used in climate change research. *Wiley Interdisciplinary Reviews: Climate Change*, **6(3)**, 301–319, doi:10.1002/wcc.339.

Emori, S. et al., 2018: Risk implications of long-term global climate goals: overall conclusions of the ICA-RUS project. *Sustainability Science*, **13(2)**, 279–289, doi:10.1007/s11625-018-0530-0.

England, M.H. et al., 2014: Recent intensification of wind-driven circulation in the Pacific and the ongoing warming hiatus. *Nature Climate Change*, **4(3)**, 222–227, doi:10.1038/nclimate2106.

Epstein, A.H. and S.L.H. Theuer, 2017: Sustainable development and climate action: thoughts on an integrated approach to SDG and climate policy implementation. In: *Papers from Interconnections 2017*. Interconnections 2017, pp. 50.

Etminan, M., G. Myhre, E.J. Highwood, and K.P. Shine, 2016: Radiative forcing of carbon dioxide, methane, and nitrous oxide: A significant revision of the methane radiative forcing. *Geophysical Research Letters*, **43(24)**, 12,614–12,623, doi:10.1002/2016gl071930.

Fazey, I. et al., 2016: Past and future adaptation pathways. *Climate and Development*, **8(1)**, 26–44, doi:10.1080/17565529.2014.989192.

Fazey, I. et al., 2018: Community resilience for a 1.5°C world. *Current Opinion in Environmental Sustainability*, **31**, 30–40, doi:10.1016/j.cosust.2017.12.006.

Fernández, A.J. et al., 2017: Aerosol optical, microphysical and radiative forcing properties during variable intensity African dust events in the Iberian Peninsula. *Atmospheric Research*, **196**, 129–141, doi:10.1016/j.atmosres.2017.06.019.

Field, C.B. et al., 2014: Technical Summary. In: *Climate Change 2014: Impacts, Adaptation, and Vulnerability. Part A: Global and Sectoral Aspects. Contribution of Working Group II to the Fifth Assessment Report of the Intergovernmental Panel on Climate Change* [Field, C.B., V.R. Barros, D.J. Dokken, K.J. Mach, M.D. Mastrandrea, T.E. Bilir, M. Chatterjee, K.L. Ebi, Y.O. Estrada, R.C. Genova, B. Girma, E.S. Kissel, A.N. Levy, S. MacCracken, P.R. Mastrandrea, and L.L. White (eds.)]. Cambridge University Press, Cambridge, United Kingdom and New York, NY, USA, pp. 35–94.

Fischer, E.M. and R. Knutti, 2015: Anthropogenic contribution to global occurrence of heavy-precipitation and high-temperature extremes. *Nature Climate Change*, **5(6)**, 560–564, doi:10.1038/nclimate2617.

Fleurbaey, M. et al., 2014: Sustainable Development and Equity. In: *Climate Change 2014: Mitigation of Climate Change. Contribution of Working Group III to the Fifth Assessment Report of the Intergovernmental Panel on Climate*

Change [Edenhofer, O., Pichs-Madruga, Y. Sokona, E. Farahani, S. Kadner, P.E. K. Seyboth, A. Adler, I. Baum, S. Brunner, and T.Z.J.C.M. B. Kriemann, J. Savolainen, S. Schlömer, C. von Stechow (eds.)]. Cambridge University Press, Cambridge, Cambridge, United Kingdom and New York, NY, USA, pp. 283–350.

Folland, C.K., O. Boucher, A. Colman, and D.E. Parker, 2018: Causes of irregularities in trends of global mean surface temperature since the late 19th century. *Science Advances*, **4(6)**, eaao5297, doi:10.1126/sciadv.aao5297.

Forkel, M. et al., 2016: Enhanced seasonal CO_2 exchange caused by amplified plant productivity in northern ecosystems. *Science*, **351(6274)**, 696–699, doi:10.1126/science.aac4971.

Foster, G. and S. Rahmstorf, 2011: Global temperature evolution 1979–2010. *Environmental Research Letters*, **6(4)**, 044022, doi:10.1088/1748-9326/6/4/044022.

Foster, G.L., D.L. Royer, and D.J. Lunt, 2017: Future climate forcing potentially without precedent in the last 420 million years. *Nature Communications*, **8**, 14845, doi: 10.1038/ncomms14845.

Frölicher, T.L. and F. Joos, 2010: Reversible and irreversible impacts of greenhouse gas emissions in multi-century projections with the NCAR global coupled carbon cycle-climate model. *Climate Dynamics*, **35(7)**, 1439–1459, doi:10.1007/s00382-009-0727-0.

Frölicher, T.L., M. Winton, and J.L. Sarmiento, 2014: Continued global warming after CO_2 emissions stoppage. *Nature Climate Change*, **4(1)**, 40–44, doi:10.1038/nclimate2060.

Fuglestvedt, J. et al., 2018: Implications of possible interpretations of 'greenhouse gas balance' in the Paris Agreement. *Philosophical Transactions of the Royal Society A: Mathematical, Physical and Engineering Sciences*, **376(2119)**, doi:10.1098/rsta.2016.0445.

Gillett, N.P., V.K. Arora, D. Matthews, and M.R. Allen, 2013: Constraining the ratio of global warming to cumulative CO_2 emissions using CMIP5 simulations. *Journal of Climate*, **26(18)**, 6844–6858, doi:10.1175/jcli-d-12-00476.1.

Gillett, N.P., V.K. Arora, K. Zickfeld, S.J. Marshall, and W.J. Merryfield, 2011: Ongoing climate change following a complete cessation of carbon dioxide emissions. *Nature Geoscience*, **4**, 83–87, doi:10.1038/ngeo1047.

Giorgi, F. and W.J. Gutowski, 2015: Regional Dynamical Downscaling and the CORDEX Initiative. *Annual Review of Environment and Resources*, **40(1)**, 467–490, doi:10.1146/annurev-environ-102014-021217.

Gomez-Echeverri, L., 2018: Climate and development: enhancing impact through stronger linkages in the implementation of the Paris Agreement and the Sustainable Development Goals (SDGs). *Philosophical Transactions of the Royal Society A: Mathematical, Physical and Engineering Sciences*, **376(2119)**, doi:10.1098/rsta.2016.0444.

Goodwin, P., R.G. Williams, and A. Ridgwell, 2015: Sensitivity of climate to cumulative carbon emissions due to compensation of ocean heat and carbon uptake. *Nature Geoscience*, **8(1)**, 29–34, doi:10.1038/ngeo2304.

Goodwin, P. et al., 2018: Pathways to 1.5°C and 2°C warming based on observational and geological constraints. *Nature Geoscience*, **11(2)**, 102–107, doi:10.1038/s41561-017-0054-8.

Gouldson, A. et al., 2015: Exploring the economic case for climate action in cities. *Global Environmental Change*, **35**, 93–105, doi:10.1016/j.gloenvcha.2015.07.009.

Gradstein, F.M., J.G. Ogg, M.D. Schmitz, and G.M. Ogg (eds.), 2012: *The Geologic Time Scale*. Elsevier BV, Boston, MA, USA, 1144 pp., doi:10.1016/b978-0-444-59425-9.01001-5.

Gregory, J.M. and P.M. Forster, 2008: Transient climate response estimated from radiative forcing and observed temperature change. *Journal of Geophysical Research: Atmospheres*, **113(D23)**, D23105, doi:10.1029/2008jd010405.

Hall, J., G. Fu, and J. Lawry, 2007: Imprecise probabilities of climate change: Aggregation of fuzzy scenarios and model uncertainties. *Climatic Change*, **81(3–4)**, 265–281, doi:10.1007/s10584-006-9175-6.

Hammill, A. and H. Price-Kelly, 2017: *Using NDCs , NAPs and the SDGs to Advance Climate-Resilient Development*. NDC Expert perspectives for the NDC Partnership, NDC Partnership, Washington, DC, USA and Bonn, Germany, 10 pp.

Hansen, G. and W. Cramer, 2015: Global distribution of observed climate change impacts. *Nature Climate Change*, **5(3)**, 182–185, doi:10.1038/nclimate2529.

Hansen, G., D. Stone, M. Auffhammer, C. Huggel, and W. Cramer, 2016: Linking local impacts to changes in climate: a guide to attribution. *Regional Environmental Change*, **16(2)**, 527–541, doi:10.1007/s10113-015-0760-y.

Hansen, J., R. Ruedy, M. Sato, and K. Lo, 2010: Global surface temperature change. *Reviews of Geophysics*, **48(4)**, RG4004, doi:10.1029/2010rg000345.

Hansen, J. et al., 2005: Earth's energy imbalance: confirmation and implications. *Science*, **308**, 1431–1435, doi:10.1126/science.1110252.

Harlan, S.L. et al., 2015: Climate Justice and Inequality: Insights from Sociology. In: *Climate Change and Society: Sociological Perspectives* [Dunlap, R.E. and R.J. Brulle (eds.)]. Oxford University Press, New York, NY, USA, pp. 127–163, doi:10.1093/acprof:oso/9780199356102.003.0005.

Harrington, C., 2016: The Ends of the World: International Relations and the Anthropocene. *Millennium: Journal of International Studies*, **44(3)**, 478–498, doi:10.1177/0305829816638745.

Harris, L.M., E.K. Chu, and G. Ziervogel, 2017: Negotiated resilience. *Resilience*, **3293**, 1–19, doi:10.1080/21693293.2017.1353196.

Hartmann, D.J. et al., 2013: Observations: Atmosphere and Surface. In: *Climate Change 2013: The Physical Science Basis. Contribution of Working Group I to the Fifth Assessment Report of the Intergovernmental Panel on Climate Change* [Stocker, T.F., D. Qin, G.-K. Plattner, M. Tignor, S.K. Allen, J. Boschung, A. Nauels, Y. Xia, V. Bex, and P.M. Midgley (eds.)]. Cambridge University Press, Cambridge, United Kingdom and New York, NY, USA, pp. 159–254.

Haustein, K. et al., 2017: A real-time Global Warming Index. *Scientific Reports*, **7(1)**, 15417, doi:10.1038/s41598-017-14828-5.

Hawkins, E. et al., 2017: Estimating changes in global temperature since the pre-industrial period. *Bulletin of the American Meteorological Society*, BAMS–D–16–0007.1, doi:10.1175/bams-d-16-0007.1.

Hegerl, G.C. et al., 2007: Understanding and Attributing Climate Change. In: *Climate Change 2007: The Physical Science Basis. Contribution of Working Group I to the Fourth Assessment Report of the Intergovernmental Panel on Climate Change* [Solomon, S., D. Qin, M. Manning, Z. Chen, M. Marquis, K.B. Averyt, M. Tignor, and H.L. Miller (eds.)]. Cambridge University Press, Cambridge, United Kingdom and New York, NY, USA, pp. 663–745.

Henley, B.J. and A.D. King, 2017: Trajectories toward the 1.5°C Paris target: Modulation by the Interdecadal Pacific Oscillation. *Geophysical Research Letters*, **44(9)**, 4256–4262, doi:10.1002/2017gl073480.

Hienola, A. et al., 2018: The impact of aerosol emissions on the 1.5°C pathways. *Environmental Research Letters*, **13(4)**, 044011.

Hoegh-Guldberg, O. et al., 2007: Coral Reefs Under Rapid Climate Change and Ocean Acidification. *Science*, **318(5857)**, 1737–1742, doi:10.1126/science.1152509.

Holz, C., S. Kartha, and T. Athanasiou, 2017: Fairly sharing 1.5: national fair shares of a 1.5°C-compliant global mitigation effort. *International Environmental Agreements: Politics, Law and Economics*, **18(1)**, 1–18, doi:10.1007/s10784-017-9371-z.

Horton, R., 2014: Why the sustainable development goals will fail. *The Lancet*, **383(9936)**, 2196, doi:10.1016/s0140-6736(14)61046-1.

Humphreys, S., 2017: Climate, Technology, 'Justice'. In: *Protecting the Environment for Future Generations – Principles and Actors in International Environmental Law* [Proelß, A. (ed.)]. Erich Schmidt Verlag, Berlin, Germany, pp. 171–190.

IBA, 2014: *Achieving Justice and Human Rights in an Era of Climate Disruption*. International Bar Association (IBA), London, UK, 240 pp.

Ionesco, D., D. Mokhnacheva, and F. Gemenne, 2016: *Atlas de Migrations Environnmentales* (in French). Presses de Sciences Po, Paris, France, 152 pp.

IPCC, 2000: Special Report on Emissions Scenarios: A Special Report of Working Group III of the Intergovernmental Panel on Climate Change. [Nakićenović, N. and R. Swart (eds.)]. Cambridge University Press, Cambridge, United Kingdom and New York, NY, USA, 570 pp.

IPCC, 2012a: Summary for Policymakers. In: *Managing the Risks of Extreme Events and Disasters to Advance Climate Change Adaptation* [Field, C.B., V.R. Barros, T.F. Stocker, D. Qin, D.J. Dokken, K.L. Ebi, M.D. Mastrandrea, K.J. Mach, G.-K. Plattner, S.K. Allen, M. Tignor, and P.M. Midgley (eds.)]. Cambridge University Press, Cambridge, United Kingdom and New York, NY, USA, pp. 3–21.

IPCC, 2012b: Meeting Report of the Intergovernmental Panel on Climate Change Expert Meeting on Geoengineering. [Edenhofer, O., R. Pichs-Madruga, Y. Sokona, C. Field, V. Barros, T.F. Stocker, Q. Dahe, J. Minx, K. Mach, G.-K. Plattner, S. Schlömer, G. Hansen, and M. Mastrandrea (eds.)]. IPCC Working Group III Technical Support Unit, Potsdam Institute for Climate Impact Research, Potsdam, Germany, 99 pp.

IPCC, 2013a: *Principles Governing IPCC Work*. Intergovernmental Panel on Climate Change (IPCC), Geneva, Switzerland, 2 pp.

IPCC, 2013b: Summary for Policymakers. In: *Climate Change 2013: The Physical Science Basis. Contribution of Working Group I to the Fifth Assessment Report of the Intergovernmental Panel on Climate Change* [Stocker, T.F., D. Qin, G.K. Plattner, M. Tignor, S.K. Allen, J. Boschung, A. Nauels, Y. Xia, V. Bex, and P.M. Midgley (eds.)]. Cambridge University Press, Cambridge, United Kingdom and New York, NY, USA, pp. 3–29.

IPCC, 2014a: Summary for Policymakers. In: *Climate Change 2014: Impacts, Adaptation, and Vulnerability. Part A: Global and Sectoral Aspects. Contribution of Working Group II to the Fifth Assessment Report of the Intergovernmental Panel on Climate Change* [Field, C.B., V.R. Barros, D.J. Dokken, K.J. Mach, M.D. Mastrandrea, T.E. Bilir, M. Chatterjee, K.L. Ebi, Y.O. Estrada, R.C. Genova, B. Girma, E.S. Kissel, A.N. Levy, S. MacCracken, P.R. Mastrandrea, and L.L. White (eds.)]. Cambridge University Press, Cambridge, United Kingdom and New York, NY, USA, pp. 1–32.

IPCC, 2014b: Summary for Policymakers. In: *Climate Change 2014: Mitigation of Climate Change. Contribution of Working Group III to the Fifth Assessment Report of the Intergovernmental Panel on Climate Change* [Edenhofer, O., R. Pichs-Madruga, Y. Sokona, E. Farahani, S. Kadner, K. Seyboth, A. Adler, I. Baum, S. Brunner, P. Eickemeier, B. Kriemann, J. Savolainen, S. Schlömer, C. von Stechow, T. Zwickel, and J.C. Minx (eds.)]. Cambridge University Press, Cambridge, United Kingdom and New York, NY, USA, pp. 1–30.

IPCC, 2014c: Climate Change 2014: Synthesis Report. Contribution of Working Groups I, II and III to the Fifth Assessment Report of the Intergovernmental Panel on Climate Change. [Core Writing Team, R.K. Pachauri, and L.A. Meyer (eds.)]. IPCC, Geneva, Switzerland, 151 pp.

IPCC, 2014d: Summary for Policymakers. In: *Climate Change 2014: Synthesis Report. Contribution of Working Groups I, II and III to the Fifth Assessment Report of the Intergovernmental Panel on Climate Change* [Core Writing Team, R.K. Pachauri, and L.A. Meyer (eds.)]. IPCC, Geneva, Switzerland, pp. 2–34.

IPCC, 2017: Meeting Report of the Intergovernmental Panel on Climate Change Expert Meeting on Mitigation, Sustainability and Climate Stabilization Scenarios. [Shukla, P.R., J. Skea, R. Diemen, E. Huntley, M. Pathak, J. Portugal-Pereira, J. Scull, and R. Slade (eds.)]. IPCC Working Group III Technical Support Unit, Imperial College London, London, UK, 44 pp.

Jacob, D. and S. Solman, 2017: IMPACT2C – An introduction. *Climate Services*, **7**, 1–2, doi:10.1016/j.cliser.2017.07.006.

Jacobson, M.Z. et al., 2015: 100% clean and renewable wind, water, and sunlight (WWS) all-sector energy roadmaps for the 50 United States. *Energy & Environmental Science*, **8(7)**, 2093–2117, doi:10.1039/c5ee01283j.

James, R., R. Washington, C.-F. Schleussner, J. Rogelj, and D. Conway, 2017: Characterizing half-a-degree difference: a review of methods for identifying regional climate responses to global warming targets. *Wiley Interdisciplinary Reviews: Climate Change*, **8(2)**, e457, doi:10.1002/wcc.457.

Jarvis, A.J., D.T. Leedal, and C.N. Hewitt, 2012: Climate-society feedbacks and the avoidance of dangerous climate change. *Nature Climate Change*, **2(9)**, 668–671, doi:10.1038/nclimate1586.

Jenkins, S., R.J. Millar, N. Leach, and M.R. Allen, 2018: Framing Climate Goals in Terms of Cumulative CO_2-Forcing-Equivalent Emissions. *Geophysical Research Letters*, **45(6)**, 2795–2804, doi:10.1002/2017gl076173.

Johansson, D.J.A., 2012: Economics- and physical-based metrics for comparing greenhouse gases. *Climatic Change*, **110(1–2)**, 123–141, doi:10.1007/s10584-011-0072-2.

Jones, G.S., P.A. Stott, and N. Christidis, 2013: Attribution of observed historical near-surface temperature variations to anthropogenic and natural causes using CMIP5 simulations. *Journal of Geophysical Research: Atmospheres*, **118(10)**, 4001–4024, doi:10.1002/jgrd.50239.

Jones, G.S., P.A. Stott, and J.F.B. Mitchell, 2016: Uncertainties in the attribution of greenhouse gas warming and implications for climate prediction. *Journal of Geophysical Research: Atmospheres*, **121(12)**, 6969–6992, doi:10.1002/2015jd024337.

Jones, P., 2016: The reliability of global and hemispheric surface temperature records. *Advances in Atmospheric Sciences*, **33(3)**, 269–282, doi:10.1007/s00376-015-5194-4.

Kabir, M.I. et al., 2016: Knowledge and perception about climate change and human health: findings from a baseline survey among vulnerable communities in Bangladesh. *BMC Public Health*, **16(1)**, 266, doi:10.1186/s12889-016-2930-3.

Kainuma, M., R. Pandey, T. Masui, and S. Nishioka, 2017: Methodologies for leapfrogging to low carbon and sustainable development in Asia. *Journal of Renewable and Sustainable Energy*, **9(2)**, 021406, doi:10.1063/1.4978469.

Kanie, N. and F. Biermann (eds.), 2017: *Governing through Goals: Sustainable Development Goals as Governance Innovation*. MIT Press, Cambridge, MA, USA, 352 pp.

Karl, T.R. et al., 2015: Possible artifacts of data biases in the recent global surface warming hiatus. *Science*, **348(6242)**, 1469–1472, doi:10.1126/science.aaa5632.

Karmalkar, A. and R.S. Bradley, 2017: Consequences of Global Warming of 1.5°C and 2°C for Regional Temperature and Precipitation Changes in the Contiguous United States. *PLOS ONE*, **12(1)**, e0168697, doi:10.1371/journal.pone.0168697.

Kelman, I., 2017: Linking disaster risk reduction, climate change, and the sustainable development goals. *Disaster Prevention and Management: An International Journal*, **26(3)**, 254–258, doi:10.1108/dpm-02-2017-0043.

King, A.D., D.J. Karoly, and B.J. Henley, 2017: Australian climate extremes at 1.5°C and 2°C of global warming. *Nature Climate Change*, **7(6)**, 412–416, doi:10.1038/nclimate3296.

King, A.D. et al., 2018a: Reduced heat exposure by limiting global warming to 1.5°C. *Nature Climate Change*, **8(7)**, 549–551, doi:10.1038/s41558-018-0191-0.

King, A.D. et al., 2018b: On the Linearity of Local and Regional Temperature Changes from 1.5°C to 2°C of Global Warming. *Journal of Climate*, **31(18)**, 7495–7514, doi:10.1175/jcli-d-17-0649.1.

Kirtman, B., et al., 2013: Near-term Climate Change: Projections and Predictability. In: *Climate Change 2013: The Physical Science Basis. Contribution of Working Group I to the Fifth Assessment Report of the Intergovernmental Panel on Climate Change* [Stocker, T.F., D. Qin, G.-K. Plattner, M. Tignor, S.K. Allen, J. Boschung, A. Nauels, Y. Xia, V. Bex, and P.M. Midgley (eds.)]. Cambridge University Press, Cambridge, United Kingdom and New York, NY, USA, pp. 953–1028.

Klein, R.J.T. et al., 2014: Adaptation opportunities, constraints, and limits. In: *Climate Change 2014: Impacts, Adaptation, and Vulnerability. Part A: Global and Sectoral Aspects. Contribution of Working Group II to the Fifth Assessment Report of the Intergovernmental Panel on Climate Change* [Field, C.B., V.R. Barros, D.J. Dokken, K.J. Mach, M.D. Mastrandrea, T.E. Bilir, M. Chatterjee, K.L. Ebi, Y.O. Estrada, R.C. Genova, B. Girma, E.S. Kissel, A.N. Levy, S. MacCracken, P.R. Mastrandrea, and L.L. White (eds.)]. Cambridge University Press, Cambridge, United Kingdom and New York, NY, USA, pp. 899–943.

Klinsky, S. and H. Winkler, 2018: Building equity in: strategies for integrating equity into modelling for a 1.5°C world. *Philosophical Transactions of the Royal Society A: Mathematical, Physical and Engineering Sciences*, **376(2119)**, doi:10.1098/rsta.2016.0461.

Knox, J.H., 2015: Human Rights Principles and Climate Change. In: *Oxford Handbook of International Climate Change Law* [Carlarne, C., K.R. Gray, and R. Tarasofsky (eds.)]. Oxford University Press, Oxford, UK, pp. 213–238, doi:10.1093/law/9780199684601.003.0011.

Knutti, R. and J. Sedláček, 2012: Robustness and uncertainties in the new CMIP5 climate model projections. *Nature Climate Change*, **3(4)**, 369–373, doi:10.1038/nclimate1716.

Knutti, R., J. Rogelj, J. Sedláček, and E.M. Fischer, 2015: A scientific critique of the two-degree climate change target. *Nature Geoscience*, **9(1)**, 13–18, doi:10.1038/ngeo2595.

Kolstad, C. et al., 2014: Social, Economic, and Ethical Concepts and Methods. In: *Climate Change 2014: Mitigation of Climate Change. Contribution of Working Group III to the Fifth Assessment Report of the Intergovernmental Panel on Climate Change* [Edenhofer, O., R. Pichs-Madruga, Y. Sokona, E. Farahani, S. Kadner, K. Seyboth, A. Adler, I. Baum, S. Brunner, P. Eickemeier, B. Kriemann, J. Savolainen, S. Schlömer, C. von Stechow, T. Zwickel, and J.C. Minx (eds.)]. Cambridge University Press, Cambridge, United Kingdom and New York, NY, USA, pp. 207–282.

Kopp, R.E. et al., 2016: Temperature-driven global sea-level variability in the Common Era. *Proceedings of the National Academy of Sciences*, **113(11)**, 1–8, doi:10.1073/pnas.1517056113.

Kosaka, Y. and S.P. Xie, 2013: Recent global-warming hiatus tied to equatorial Pacific surface cooling. *Nature*, **501(7467)**, 403–407, doi:10.1038/nature12534.

Kriegler, E., J.W. Hall, H. Held, R. Dawson, and H.J. Schellnhuber, 2009: Imprecise probability assessment of tipping points in the climate system. *Proceedings of the National Academy of Sciences*, **106(13)**, 5041–5046, doi:10.1073/pnas.0809117106.

Kriegler, E. et al., 2012: The need for and use of socio-economic scenarios for climate change analysis: A new approach based on shared socio-economic pathways. *Global Environmental Change*, **22(4)**, 807–822, doi:10.1016/j.gloenvcha.2012.05.005.

Kriegler, E. et al., 2014: A new scenario framework for climate change research: The concept of shared climate policy assumptions. *Climatic Change*, **122(3)**, 401–414, doi:10.1007/s10584-013-0971-5.

Kristjánsson, J.E., M. Helene, and S. Hauke, 2016: The hydrological cycle response to cirrus cloud thinning. *Geophysical Research Letters*, **42(24)**, 10,807–810,815, doi:10.1002/2015gl066795.

Lauder, A.R. et al., 2013: Offsetting methane emissions – An alternative to emission equivalence metrics. *International Journal of Greenhouse Gas Control*, **12**, 419–429, doi:10.1016/j.ijggc.2012.11.028.

Leach, N.J. et al., 2018: Current level and rate of warming determine emissions budgets under ambitious mitigation. *Nature Geoscience*, **11(8)**, 574–579, doi:10.1038/s41561-018-0156-y.

Leal Filho, W. et al., 2018: Implementing climate change research at universities: Barriers, potential and actions. *Journal of Cleaner Production*, **170**, 269–277, doi:10.1016/j.jclepro.2017.09.105.

Lee, D. et al., 2018: Impacts of half a degree additional warming on the Asian summer monsoon rainfall characteristics. *Environmental Research Letters*, **13(4)**, 044033, doi:10.1088/1748-9326/aab55d.

Leggett, J. et al., 1992: Emissions scenarios for the IPCC: an update. In: *Climate change 1992: The Supplementary Report to the IPCC Scientific Assessment* [Houghton, J.T., B.A. Callander, and S.K. Varney (eds.)]. Cambridge University Press, Cambridge, United Kingdom and New York, NY, USA, pp. 69–95.

Leichenko, R. and J.A. Silva, 2014: Climate change and poverty: Vulnerability, impacts, and alleviation strategies. *Wiley Interdisciplinary Reviews: Climate Change*, **5(4)**, 539–556, doi:10.1002/wcc.287.

Leonard, M. et al., 2014: A compound event framework for understanding extreme impacts. *Wiley Interdisciplinary Reviews: Climate Change*, **5(1)**, 113–128, doi:10.1002/wcc.252.

Leung, D.Y.C., G. Caramanna, and M.M. Maroto-Valer, 2014: An overview of current status of carbon dioxide capture and storage technologies. *Renewable and Sustainable Energy Reviews*, **39**, 426–443, doi:10.1016/j.rser.2014.07.093.

Levasseur, A. et al., 2016: Enhancing life cycle impact assessment from climate science: Review of recent findings and recommendations for application to LCA. *Ecological Indicators*, **71**, 163–174, doi:10.1016/j.ecolind.2016.06.049.

Lewis, S.C., A.D. King, and D.M. Mitchell, 2017: Australia's Unprecedented Future Temperature Extremes Under Paris Limits to Warming. *Geophysical Research Letters*, **44(19)**, 9947–9956, doi:10.1002/2017gl074612.

Lofts, K., S. Shamin, T.S. Zaman, and R. Kibugi, 2017: Brief on Sustainable Development Goal 13 on Taking Action on Climate Change and Its Impacts: Contributions of International Law, Policy and Governance,. *McGill Journal of Sustainable Development Law*, **11(1)**, 183–192, doi:10.3868/s050-004-015-0003-8.

Loftus, P.J., A.M. Cohen, J.C.S. Long, and J.D. Jenkins, 2015: A critical review of global decarbonization scenarios: What do they tell us about feasibility? *Wiley Interdisciplinary Reviews: Climate Change*, **6(1)**, 93–112, doi:10.1002/wcc.324.

Lövbrand, E., M. Hjerpe, and B.-O. Linnér, 2017: Making climate governance global: how UN climate summitry comes to matter in a complex climate regime. *Environmental Politics*, **26(4)**, 580–599, doi:10.1080/09644016.2017.1319019.

Lowe, J.A. and D. Bernie, 2018: The impact of Earth system feedbacks on carbon budgets and climate response. *Philosophical Transactions of the Royal Society A: Mathematical, Physical and Engineering Sciences*, **376(2119)**, doi:10.1098/rsta.2017.0263.

Lowe, J.A. et al., 2009: How difficult is it to recover from dangerous levels of global warming? *Environmental Research Letters*, **4(1)**, 014012, doi:10.1088/1748-9326/4/1/014012.

Lüning, S. and F. Vahrenholt, 2017: Paleoclimatological Context and Reference Level of the 2°C and 1.5°C Paris Agreement Long-Term Temperature Limits. *Frontiers in Earth Science*, **5**, 104, doi:10.3389/feart.2017.00104.

Lüthi, D. et al., 2008: High-resolution carbon dioxide concentration record 650,000–800,000 years before present. *Nature*, **453(7193)**, 379–382, doi:10.1038/nature06949.

MacMartin, D.G., K.L. Ricke, and D.W. Keith, 2018: Solar geoengineering as part of an overall strategy for meeting the 1.5°C Paris target. *Philosophical Transactions of the Royal Society A: Mathematical, Physical and Engineering Sciences*, **376(2119)**, doi:10.1098/rsta.2016.0454.

Marbà, N. et al., 2015: Impact of seagrass loss and subsequent revegetation on carbon sequestration and stocks. *Journal of Ecology*, **103(2)**, 296–302, doi:10.1111/1365-2745.12370.

Marcott, S.A., J.D. Shakun, P.U. Clark, and A.C. Mix, 2013: A reconstruction of regional and global temperature for the past 11,300 years. *Science*, **339(6124)**, 1198–201, doi:10.1126/science.1228026.

Marsicek, J., B.N. Shuman, P.J. Bartlein, S.L. Shafer, and S. Brewer, 2018: Reconciling divergent trends and millennial variations in Holocene temperatures. *Nature*,

554(7690), 92–96, doi:10.1038/nature25464.

Martius, O., S. Pfahl, and C. Chevalier, 2016: A global quantification of compound precipitation and wind extremes. *Geophysical Research Letters*, **43(14)**, 7709–7717, doi:10.1002/2016gl070017.

Masson-Delmotte, V. et al., 2013: Information from Paleoclimate Archives. In: *Climate Change 2013: The Physical Science Basis. Contribution of Working Group I to the Fifth Assessment Report of the Intergovernmental Panel on Climate Change* [Stocker, T.F., D. Qin, G.-K. Plattner, M. Tignor, S.K. Allen, J. Boschung, A. Nauels, Y. Xia, V. Bex, and P.M. Midgley (eds.)]. Cambridge University Press, Cambridge, United Kingdom and New York, NY, USA, pp. 383–464.

Mastrandrea, M.D. et al., 2011: The IPCC AR5 guidance note on consistent treatment of uncertainties: a common approach across the working groups. *Climatic Change*, **108(4)**, 675–691, doi:10.1007/s10584-011-0178-6.

Mathesius, S., M. Hofmann, K. Caldeira, and H.J. Schellnhuber, 2015: Long-term response of oceans to CO_2 removal from the atmosphere. *Nature Climate Change*, **5(12)**, 1107–1113, doi:10.1038/nclimate2729.

Matthews, H.D. and K. Caldeira, 2008: Stabilizing climate requires near-zero emissions. *Geophysical Research Letters*, **35(4)**, L04705, doi:10.1029/2007gl032388.

Matthews, H.D. and K. Zickfeld, 2012: Climate response to zeroed emissions of greenhouse gases and aerosols. *Nature Climate Change*, **2(5)**, 338–341, doi:10.1038/nclimate1424.

Matthews, H.D. and S. Solomon, 2013: Irreversible Does Not Mean Unavoidable. *Science*, **340(6131)**, 438–439, doi:10.1126/science.1236372.

Matthews, H.D., N.P. Gillett, P. Stott, and K. Zickfeld, 2009: The proportionality of global warming to cumulative carbon emissions. *Nature*, **459(7248)**, 829–32, doi:10.1038/nature08047.

Matthews, T.K.R., R.L. Wilby, and C. Murphy, 2017: Communicating the deadly consequences of global warming for human heat stress. *Proceedings of the National Academy of Sciences*, **114(15)**, 3861–3866, doi:10.1073/pnas.1617526114.

Maupin, A., 2017: The SDG13 to combat climate change: an opportunity for Africa to become a trailblazer? *African Geographical Review*, **36(2)**, 131–145, doi:10.1080/19376812.2016.1171156.

Mauritsen, T. and R. Pincus, 2017: Committed warming inferred from observations. *Nature Climate Change*, **2**, 1–5, doi:10.1038/nclimate3357.

McKinnon, C., 2015: Climate justice in a carbon budget. *Climatic Change*, **133(3)**, 375–384, doi:10.1007/s10584-015-1382-6.

Medhaug, I., M.B. Stolpe, E.M. Fischer, and R. Knutti, 2017: Reconciling controversies about the 'global warming hiatus'. *Nature*, **545(7652)**, 41–47, doi:10.1038/nature22315.

Meehl, G.A. et al., 2007: Global Climate Projections. In: *Climate Change 2007: The Physical Science Basis. Contribution of Working Group I to the Fourth Assessment Report of the Intergovernmental Panel on Climate Change* [Solomon, S., D. Qin, M. Manning, Z. Chen, M. Marquis, K.B. Averyt, M. Tignor, and H.L. Miller (eds.)]. Cambridge University Press, Cambridge, UK and New York, NY, USA, pp. 747–845.

Millar, R.J., Z.R. Nicholls, P. Friedlingstein, and M.R. Allen, 2017a: A modified impulse-response representation of the global near-surface air temperature and atmospheric concentration response to carbon dioxide emissions. *Atmospheric Chemistry and Physics*, **17(11)**, 7213–7228, doi:10.5194/acp-17-7213-2017.

Millar, R.J. et al., 2017b: Emission budgets and pathways consistent with limiting warming to 1.5°C. *Nature Geoscience*, **10(10)**, 741–747, doi:10.1038/ngeo3031.

Mimura, N. et al., 2014: Adaptation planning and implementation. In: *Climate Change 2014: Impacts, Adaptation, and Vulnerability. Part A: Global and Sectoral Aspects. Contribution of Working Group II to the Fifth Assessment Report of the Intergovernmental Panel on Climate Change* [Field, C.B., V.R. Barros, D.J. Dokken, K.J. Mach, M.D. Mastrandrea, T.E. Bilir, M. Chatterjee, K.L. Ebi, Y.O. Estrada, R.C. Genova, B. Girma, E.S. Kissel, A.N. Levy, S. MacCracken, P.R. Mastrandrea, and L.L. White (eds.)]. Cambridge University Press, Cambridge, UK and New York, NY, USA, pp. 869–898.

Minx, J.C., W.F. Lamb, M.W. Callaghan, L. Bornmann, and S. Fuss, 2017: Fast growing research on negative emissions. *Environmental Research Letters*, **12(3)**, 035007, doi:10.1088/1748-9326/aa5ee5.

Mitchell, D. et al., 2016: Realizing the impacts of a 1.5°C warmer world. *Nature Climate Change*, **6(8)**, 735–737, doi:10.1038/nclimate3055.

Mitchell, D. et al., 2017: Half a degree additional warming, prognosis and projected impacts (HAPPI): background and experimental design. *Geoscientific Model Development*, **10(2)**, 571–583, doi:10.5194/gmd-10-571-2017.

Miyan, M.A., 2015: Droughts in Asian Least Developed Countries: Vulnerability and sustainability. *Weather and Climate Extremes*, **7**, 8–23, doi:10.1016/j.wace.2014.06.003.

Morice, C.P., J.J. Kennedy, N.A. Rayner, and P.D. Jones, 2012: Quantifying uncertainties in global and regional temperature change using an ensemble of observational estimates: The HadCRUT4 data set. *Journal of Geophysical Research: Atmospheres*, **117(D8)**, D08101, doi:10.1029/2011jd017187.

Morita, T. et al., 2001: Greenhouse Gas Emission Mitigation Scenarios and Implications. In: *Climate Change 2001: Mitigation. Contribution of Working Group III to the Third Assessment Report of the Intergovernmental Panel on Climate Change* [B. Metz, O. Davidson, R. Swart, and J. Pan (eds.)]. Cambridge University Press, Cambridge, United Kingdom and New York, NY, USA, pp. 115–164.

Moss, R.H. et al., 2010: The next generation of scenarios for climate change research and assessment. *Nature*, **463(7282)**, 747–756, doi:10.1038/nature08823.

Mueller, B. and S.I. Seneviratne, 2014: Systematic land climate and evapotranspiration biases in CMIP5 simulations. *Geophysical Research Letters*, **41(1)**, 128–134, doi:10.1002/2013gl058055.

Myhre, G. et al., 2013: Anthropogenic and natural radiative forcing. In: *Climate Change 2013: The Physical Science Basis. Contribution of Working Group I to the Fifth Assessment Report of the Intergovernmental Panel on Climate Change* [Stocker, T.F., D. Qin, G.-K. Plattner, M. Tignor, S.K. Allen, J. Boschung, A. Nauels, Y. Xia, V. Bex, and P.M. Midgley (eds.)]. Cambridge University Press, Cambridge, United Kingdom and New York, NY, USA, pp. 658–740.

Myhre, G. et al., 2017: Multi-model simulations of aerosol and ozone radiative forcing due to anthropogenic emission changes during the period 1990–2015. *Atmospheric Chemistry and Physics*, **17(4)**, 2709–2720, doi:10.5194/acp-17-2709-2017.

Mysiak, J., S. Surminski, A. Thieken, R. Mechler, and J. Aerts, 2016: Brief communication: Sendai framework for disaster risk reduction – Success or warning sign for Paris? *Natural Hazards and Earth System Sciences*, **16(10)**, 2189–2193, doi:10.5194/nhess-16-2189-2016.

NOAA, 2016: State of the Climate: Global Climate Report for Annual 2015. National Oceanic and Atmospheric Administration (NOAA) National Centers for Environmental Information (NCEI). Retrieved from: www.ncdc.noaa.gov/sotc/global/201513.

O'Brien, K. and E. Selboe, 2015: Social transformation. In: *The Adaptive Challenge of Climate Change* [O'Brien, K. and E. Selboe (eds.)]. Cambridge University Press, Cambridge, United Kingdom and New York, NY , USA, pp. 311–324, doi:10.1017/cbo9781139149389.018.

O'Brien, K. et al., 2012: Toward a sustainable and resilient future. In: *Managing the Risks of Extreme Events and Disasters to Advance Climate Change Adaptation. A Special Report of Working Groups I and II of the Intergovernmental Panel on Climate Change (IPCC)* [Field, C.B., V. Barros, T.F. Stocker, D. Qin, D.J. Dokken, K.L. Ebi, M.D. Mastrandrea, K.J. Mach, G.-K. Plattner, S.K. Allen, M. Tignor, and P.M. Midgley (eds.)]. Cambridge University Press, Cambridge, United Kingdom and New York, NY, USA, pp. 437–486.

O'Neill, B.C. et al., 2014: A new scenario framework for climate change research: The concept of shared socioeconomic pathways. *Climatic Change*, **122(3)**, 387–400, doi:10.1007/s10584-013-0905-2.

Ocko, I.B. et al., 2017: Unmask temporal trade-offs in climate policy debates. *Science*, **356(6337)**, 492–493, doi:10.1126/science.aaj2350.

OECD, 2016: *The OECD supporting action on climate change*. Organisation for Economic Co-operation and Development (OECD), Paris, France, 18 pp.

OHCHR, 2009: *Report of the Office of the United Nations High Commissioner for Human Rights on the relationship between climate change and human rights*. A/HRC/10/61, Office of the United Nations High Commissioner for Human Rights (OHCHR), 32 pp.

OHCHR, 2015: *Understanding Human Rights and Climate Change*. Submission of the Office of the High Commissioner for Human Rights to the 21st Conference of the Parties to the United Nations Framework Convention on Climate Change, Office of the United Nations High Commissioner for Human Rights (OHCHR), 28 pp.

OHCHR, 2017: *Analytical study on the relationship between climate change and the full and effective enjoyment of the rights of the child*. A/HRC/35/13, Office of the United Nations High Commissioner for Human Rights (OHCHR), 18 pp.

Okereke, C., 2010: Climate justice and the international regime. *Wiley Interdisciplinary Reviews: Climate Change*, **1(3)**, 462–474, doi:10.1002/wcc.52.

Oliver, T.H. and M.D. Morecroft, 2014: Interactions between climate change and land use change on biodiversity: attribution problems, risks, and opportunities. *Wiley Interdisciplinary Reviews: Climate Change*, **5(3)**, 317–335, doi:10.1002/wcc.271.

Olsson, L. et al., 2014: Livelihoods and poverty. In: *Climate Change 2014: Impacts, Adaptation, and Vulnerability. Part A: Global and Sectoral Aspects. Contribution of Working Group II to the Fifth Assessment Report of the Intergovernmental Panel on Climate Change* [Field, C.B., V.R. Barros, D.J. Dokken, K.J. Mach, M.D. Mastrandrea, T.E. Bilir, M. Chatterjee, K.L. Ebi, Y.O. Estrada, R.C. Genova, B. Girma, E.S. Kissel, A.N. Levy, S. MacCracken, P.R. Mastrandrea, and L.L. White (eds.)]. Cambridge University Press, Cambridge, United Kingdom and New York, NY, USA, pp. 798–832.

Otto, F.E.L., D.J. Frame, A. Otto, and M.R. Allen, 2015: Embracing uncertainty in climate change policy. *Nature Climate Change*, **5**, 1–5, doi:10.1038/nclimate2716.

Otto, F.E.L., R.B. Skeie, J.S. Fuglestvedt, T. Berntsen, and M.R. Allen, 2017: Assigning historic responsibility for extreme weather events. *Nature Climate Change*, **7(11)**, 757–759, doi:10.1038/nclimate3419.

Pattberg, P. and F. Zelli (eds.), 2016: *Environmental politics and governance in the anthropocene: Institutions and legitimacy in a complex world*. Routledge, London, UK, 268 pp., doi:10.4324/9781315697468.

Patterson, J. et al., 2017: Exploring the governance and politics of transformations towards sustainability. *Environmental Innovation and Societal Transitions*, **24**, 1–16, doi:10.1016/j.eist.2016.09.001.

Pelling, M., 2011: *Adaptation to Climate Change: From Resilience to Transformation*. Routledge, Abingdon, Oxon, UK and New York, NY, USA, 224 pp.

Pelling, M., K. O'Brien, and D. Matyas, 2015: Adaptation and transformation. *Climatic Change*, **133(1)**, 113–127, doi:10.1007/s10584-014-1303-0.

Pendergrass, A.G., F. Lehner, B.M. Sanderson, and Y. Xu, 2015: Does extreme precipitation intensity depend on the emissions scenario? *Geophysical Research Letters*, **42(20)**, 8767–8774, doi:10.1002/2015gl065854.

Pfleiderer, P., C.-F. Schleussner, M. Mengel, and J. Rogelj, 2018: Global mean temperature indicators linked to warming levels avoiding climate risks. *Environmental Research Letters*, **13(6)**, 064015, doi:10.1088/1748-9326/aac319.

Popp, A. et al., 2017: Land-use futures in the shared socio-economic pathways. *Global Environmental Change*, **42**, 331–345, doi:10.1016/j.gloenvcha.2016.10.002.

Porter, J.R. et al., 2014: Food security and food production systems. In: *Climate Change 2014: Impacts, Adaptation, and Vulnerability. Part A: Global and Sectoral Aspects. Contribution of Working Group II to the Fifth Assessment Report of the Intergovernmental Panel on Climate Change* [Field, C.B., V.R. Barros, D.J. Dokken, K.J. Mach, M.D. Mastrandrea, T.E. Bilir, M. Chatterjee, K.L. Ebi, Y.O. Estrada, R.C. Genova, B. Girma, E.S. Kissel, A.N. Levy, S. MacCracken, P.R. Mastrandrea, and L.L. White (eds.)]. Cambridge University Press, Cambridge, United Kingdom and New York, NY, USA, pp. 485–533.

Pörtner, H.-O. et al., 2014: Ocean systems. In: *Climate Change 2014: Impacts, Adaptation, and Vulnerability. Part A: Global and Sectoral Aspects. Contribution of Working Group II to the Fifth Assessment Report of the Intergovernmental Panel on Climate Change* [Field, C.B., V.R. Barros, D.J. Dokken, K.J. Mach, M.D. Mastrandrea, T.E. Bilir, M. Chatterjee, K.L. Ebi, Y.O. Estrada, R.C. Genova, B. Girma, E.S. Kissel, A.N. Levy, S. MacCracken, P.R. Mastrandrea, and L.L. White (eds.)]. Cambridge University Press, Cambridge, United Kingdom and New York, NY, USA, pp. 411–484.

Rao, S. et al., 2017: Future Air Pollution in the Shared Socio-Economic Pathways. *Global Environmental Change*, **42**, 346–358, doi:10.1016/j.gloenvcha.2016.05.012.

Reckien, D. et al., 2017: Climate change, equity and the Sustainable Development Goals: an urban perspective. *Environment & Urbanization*, **29(1)**, 159–182, doi:10.1177/0956247816677778.

Reisinger, A. et al., 2012: Implications of alternative metrics for global mitigation costs and greenhouse gas emissions from agriculture. *Climatic Change*, 1–14, doi:10.1007/s10584-012-0593-3.

Reisinger, A. et al., 2014: Australasia. In: *Climate Change 2014: Impacts, Adaptation, and Vulnerability. Part B: Regional Aspects. Contribution of Working Group II to the Fifth Assessment Report of the Intergovernmental Panel on Climate Change* [Barros, V.R., C.B. Field, D.J. Dokken, M.D. Mastrandrea, K.J. Mach, T.E. Bilir, M. Chatterjee, K.L. Ebi, Y.O. Estrada, R.C. Genova, B. Girma, E.S. Kissel, A.N. Levy, S. MacCracken, P.R. Mastrandrea, and L.L. White (eds.)]. Cambridge University Press, Cambridge, United Kingdom and New York, NY, USA, pp. 1371–1438.

Resurrección, B.P., 2013: Persistent women and environment linkages in climate change and sustainable development agendas. *Women's Studies International Forum*, **40**, 33–43, doi:10.1016/j.wsif.2013.03.011.

Revi, A. et al., 2014: Urban areas. In: *Climate Change 2014: Impacts, Adaptation, and Vulnerability. Part A: Global and Sectoral Aspects. Contribution of Working Group II to the Fifth Assessment Report of the Intergovernmental Panel on Climate Change* [Field, C.B., V.R. Barros, D.J. Dokken, K.J. Mach, M.D. Mastrandrea, T.E. Bilir, M. Chatterjee, K.L. Ebi, Y.O. Estrada, R.C. Genova, B. Girma, E.S. Kissel, A.N. Levy, S. MacCracken, P.R. Mastrandrea, and L.L. White (eds.)]. Cambridge University Press, Cambridge, United Kingdom and New York, NY, USA, pp. 535–612.

Riahi, K. et al., 2017: The Shared Socioeconomic Pathways and their energy, land use, and greenhouse gas emissions implications: An overview. *Global Environmental Change*, **42**, 153–168, doi:10.1016/j.gloenvcha.2016.05.009.

Ribes, A. and L. Terray, 2013: Application of regularised optimal fingerprinting to attribution. Part II: application to global near-surface temperature. *Climate Dynamics*, **41(11)**, 2837–2853, doi:10.1007/s00382-013-1736-6.

Ribes, A., F.W. Zwiers, J.-M. Azaïs, and P. Naveau, 2017: A new statistical approach to climate change detection and attribution. *Climate Dynamics*, **48(1)**, 367–386, doi:10.1007/s00382-016-3079-6.

Richardson, M., K. Cowtan, and R.J. Millar, 2018: Global temperature definition affects achievement of long-term climate goals. *Environmental Research Letters*, **13(5)**, 054004, doi:10.1088/1748-9326/aab305.

Richardson, M., K. Cowtan, E. Hawkins, and M.B. Stolpe, 2016: Reconciled climate response estimates from climate models and the energy budget of Earth. *Nature Climate Change*, **6(10)**, 931–935, doi:10.1038/nclimate3066.

Robinson, M. and T. Shine, 2018: Achieving a climate justice pathway to 1.5°C. *Nature Climate Change*, **8(7)**, 564–569, doi:10.1038/s41558-018-0189-7.

Rogelj, J., C.-F. Schleussner, and W. Hare, 2017: Getting It Right Matters: Temperature Goal Interpretations in Geoscience Research. *Geophysical Research Letters*, **44(20)**, 10,662–10,665, doi:10.1002/2017gl075612.

Rogelj, J. et al., 2015: Energy system transformations for limiting end-of-century warming to below 1.5°C. *Nature Climate Change*, **5(6)**, 519–527, doi:10.1038/nclimate2572.

Rogelj, J. et al., 2016a: Paris Agreement climate proposals need boost to keep warming well below 2°C. *Nature Climate Change*, **534**, 631–639, doi:10.1038/nature18307.

Rogelj, J. et al., 2016b: Differences between carbon budget estimates unravelled. *Nature Climate Change*, **6(3)**, 245–252, doi:10.1038/nclimate2868.

Rogelj, J. et al., 2018: Scenarios towards limiting global mean temperature increase below 1.5°C. *Nature Climate Change*, **8(4)**, 325–332, doi:10.1038/s41558-018-0091-3.

Rohde, R. et al., 2013: Berkeley Earth Temperature Averaging Process. *Geoinformatics & Geostatistics: An Overview*, **1(2)**, 1–13, doi:10.4172/2327-4581.1000103.

Rosenbloom, D., 2017: Pathways: An emerging concept for the theory and governance of low-carbon transitions. *Global Environmental Change*, **43**, 37–50, doi:10.1016/j.gloenvcha.2016.12.011.

Rosenzweig, C. et al., 2008: Attributing physical and biological impacts to anthropogenic climate change. *Nature*, **453(7193)**, 353–357, doi:10.1038/nature06937.

Rosenzweig, C. et al., 2017: Assessing inter-sectoral climate change risks: the role of ISIMIP. *Environmental Research Letters*, **12(1)**, 010301.

Rosenzweig, C., W. Solecki, P. Romeo-Lankao, M. Shagun, S. Dhakal, and S. Ali Ibrahim (eds.), 2018: *Climate Change and Cities: Second Assessment Report of the Urban Climate Change Research Network*. Cambridge University Press, Cambridge, United Kingdom and New York, NY, USA, 811 pp.

Saeed, F. et al., 2018: Robust changes in tropical rainy season length at 1.5°C and 2°C. *Environmental Research Letters*, **13(6)**, 064024, doi:10.1088/1748-9326/aab797.

Samset, B.H. et al., 2018: Climate Impacts From a Removal of Anthropogenic Aerosol Emissions. *Geophysical Research Letters*, **45(2)**, 1020–1029, doi:10.1002/2017gl076079.

Sarofim, M.C., S.T. Waldhoff, and S.C. Anenberg, 2017: Valuing the Ozone-Related Health Benefits of Methane Emission Controls. *Environmental and Resource Economics*, **66(1)**, 45–63, doi:10.1007/s10640-015-9937-6.

Savaresi, A., 2016: The Paris Agreement: a new beginning? *Journal of Energy & Natural Resources Law*, **34(1)**, 16–26, doi:10.1080/02646811.2016.1133983.

Schäfer, S., M. Lawrence, H. Stelzer, W. Born, and S. Low (eds.), 2015: *The European Transdisciplinary Assessment of Climate Engineering (EuTRACE): Removing Greenhouse Gases from the Atmosphere and Reflecting Sunlight away from Earth*. The European Transdisciplinary Assessment of Climate Engineering (EuTRACE), 170 pp.

Schewe, J. et al., 2014: Multimodel assessment of water scarcity under climate change. *Proceedings of the National Academy of Sciences*, **111(9)**, 3245–3250, doi:10.1073/pnas.1222460110.

Schleussner, C.-F., P. Pfleiderer, and E.M. Fischer, 2017: In the observational record half a degree matters. *Nature Climate Change*, **7(7)**, 460–462, doi:10.1038/nclimate3320.

Schleussner, C.-F. et al., 2016: Differential climate impacts for policy relevant limits to global warming: the case of 1.5°C and 2°C. *Earth System Dynamics*, **7(2)**, 327–351, doi:10.5194/esd-7-327-2016.

Schroeder, H., M.T. Boykoff, and L. Spiers, 2012: Equity and state representations in climate negotiations. *Nature Climate Change*, **2**, 834–836, doi:10.1038/nclimate1742.

Schurer, A.P., M.E. Mann, E. Hawkins, S.F.B. Tett, and G.C. Hegerl, 2017: Importance of the pre-industrial baseline for likelihood of exceeding Paris goals. *Nature Climate Change*, **7(8)**, 563–567, doi:10.1038/nclimate3345.

Seneviratne, S.I., M.G. Donat, A.J. Pitman, R. Knutti, and R.L. Wilby, 2016: Allowable CO_2 emissions based on regional and impact-related climate targets. *Nature*, **529(7587)**, 477–483, doi:10.1038/nature16542.

Seneviratne, S.I. et al., 2012: Changes in climate extremes and their impacts on the natural physical environment. In: *Managing the Risks of Extreme Events and Disasters to Advance Climate Change Adaptation. A Special Report of Working Groups I and II of the Intergovernmental Panel on Climate Change (IPCC)* [Field, C.B., V. Barros, T.F. Stocker, D. Qin, D.J. Dokken, K.L. Ebi, M.D. Mastrandrea, K.J. Mach, G.-K. Plattner, S.K. Allen, M. Tignor, and P.M. Midgley (eds.)]. Cambridge University Press, Cambridge, United Kingdom and New York, NY, USA, pp. 109–230.

Settele, J. et al., 2014: Terrestrial and inland water systems. In: *Climate Change 2014: Impacts, Adaptation, and Vulnerability. Part A: Global and Sectoral Aspects. Contribution of Working Group II to the Fifth Assessment Report of the Intergovernmental Panel on Climate Change* [Field, C.B., V.R. Barros, D.J. Dokken, K.J. Mach, M.D. Mastrandrea, T.E. Bilir, M. Chatterjee, K.L. Ebi, Y.O. Estrada, R.C. Genova, B. Girma, E.S. Kissel, A.N. Levy, S. MacCracken, P.R. Mastrandrea, and L.L. White (eds.)]. Cambridge University Press, Cambridge, United Kingdom and New York, NY, USA, pp. 271–359.

Shelton, D., 2008: Equity. In: *The Oxford Handbook of International Environmental Law* [Bodansky, D., J. Brunnée, and E. Hey (eds.)]. Oxford University Press, Oxford, UK, pp. 639–662, doi:10.1093/oxfordhb/9780199552153.013.0027.

Shiferaw, B. et al., 2014: Managing vulnerability to drought and enhancing livelihood resilience in sub-Saharan Africa: Technological, institutional and policy options. *Weather and Climate Extremes*, **3**, 67–79, doi:10.1016/j.wace.2014.04.004.

Shindell, D.T., 2015: The social cost of atmospheric release. *Climatic Change*, **130(2)**, 313–326, doi:10.1007/s10584-015-1343-0.

Shindell, D.T., Y. Lee, and G. Faluvegi, 2016: Climate and health impacts of US emissions reductions consistent with 2°C. *Nature Climate Change*, **6**, 503–507, doi:10.1038/nclimate2935.

Shindell, D.T., J.S. Fuglestvedt, and W.J. Collins, 2017: The social cost of methane: theory and applications. *Faraday Discussions*, **200**, 429–451, doi:10.1039/c7fd00009j.

Shine, K.P., J.S. Fuglestvedt, K. Hailemariam, and N. Stuber, 2005: Alternatives to the Global Warming Potential for comparing climate impacts of emissions of greenhouse gases. *Climatic Change*, **68(3)**, 281–302, doi:10.1007/s10584-005-1146-9.

Shine, K.P., R.P. Allan, W.J. Collins, and J.S. Fuglestvedt, 2015: Metrics for linking emissions of gases and aerosols to global precipitation changes. *Earth System Dynamics*, **6(2)**, 525–540, doi:10.5194/esd-6-525-2015.

Shue, H., 2013: Climate Hope: Implementing the Exit Strategy. *Chicago Journal of International Law*, **13(2)**, 381–402, https://chicagounbound.uchicago.edu/cjil/vol13/iss2/6/.

Shue, H., 2014: *Climate Justice: Vulnerability and Protection*. Oxford University Press, Oxford, UK, 368 pp.

Shue, H., 2018: Mitigation gambles: uncertainty, urgency and the last gamble possible. *Philosophical Transactions of the Royal Society A: Mathematical, Physical and Engineering Sciences*, **376(2119)**, doi:10.1098/rsta.2017.0105.

1

Shukla, P.R. and V. Chaturvedi, 2013: Sustainable energy transformations in India under climate policy. *Sustainable Development*, **21(1)**, 48–59, doi:10.1002/sd.516.

Simmons, A.J. et al., 2017: A reassessment of temperature variations and trends from global reanalyses and monthly surface climatological datasets. *Quarterly Journal of the Royal Meteorological Society*, **143(702)**, 101–119, doi:10.1002/qj.2949.

Simpson, M. et al., 2016: Decision Analysis for Management of Natural Hazards. *Annual Review of Environment and Resources*, **41(1)**, 489–516, doi:10.1146/annurev-environ-110615-090011.

Sitch, S., P.M. Cox, W.J. Collins, and C. Huntingford, 2007: Indirect radiative forcing of climate change through ozone effects on the land-carbon sink. *Nature*, **448(7155)**, 791–794, doi:10.1038/nature06059.

Skeie, R.B. et al., 2017: Perspective has a strong effect on the calculation of historical contributions to global warming. *Environmental Research Letters*, **12(2)**, 024022, doi:10.1088/1748-9326/aa5b0a.

Smith, C.J. et al., 2018: FAIR v1.3: a simple emissions-based impulse response and carbon cycle model. *Geoscientific Model Development*, **11(6)**, 2273–2297, doi:10.5194/gmd-11-2273-2018.

Smith, S.J. and P.J. Rasch, 2013: The long-term policy context for solar radiation management. *Climatic Change*, **121(3)**, 487–497, doi:10.1007/s10584-012-0577-3.

Smith, S.J., J. Karas, J. Edmonds, J. Eom, and A. Mizrahi, 2013: Sensitivity of multi-gas climate policy to emission metrics. *Climatic Change*, **117(4)**, 663–675, doi:10.1007/s10584-012-0565-7.

Smith, S.M. et al., 2012: Equivalence of greenhouse-gas emissions for peak temperature limits. *Nature Climate Change*, **2(7)**, 535–538, doi:10.1038/nclimate1496.

Solecki, W., M. Pelling, and M. Garschagen, 2017: Transitions between risk management regimes in cities. *Ecology and Society*, **22(2)**, 38, doi:10.5751/es-09102-220238.

Solomon, S., G.-K.G. Plattner, R. Knutti, and P. Friedlingstein, 2009: Irreversible climate change due to carbon dioxide emissions. *Proceedings of the National Academy of Sciences*, **106(6)**, 1704–9, doi:10.1073/pnas.0812721106.

Somanathan, E. et al., 2014: National and Sub-national Policies and Institutions. In: *Climate Change 2014: Mitigation of Climate Change. Contribution of Working Group III to the Fifth Assessment Report of the Intergovernmental Panel on Climate Change* [Edenhofer, O., R. Pichs-Madruga, Y. Sokona, E. Farahani, S. Kadner, K. Seyboth, A. Adler, I. Baum, S. Brunner, P. Eickemeier, B. Kriemann, J. Savolainen, S. Schlömer, C. von Stechow, T. Zwickel, and J.C. Minx (eds.)]. Cambridge University Press, Cambridge, United Kingdom and New York, NY, USA, pp. 1141–1205.

Sovacool, B.K., B.-O. Linnér, and M.E. Goodsite, 2015: The political economy of climate adaptation. *Nature Climate Change*, **5(7)**, 616–618, doi:10.1038/nclimate2665.

Stavins, R. et al., 2014: International Cooperation: Agreements and Instruments. In: *Climate Change 2014: Mitigation of Climate Change. Contribution of Working Group III to the Fifth Assessment Report of the Intergovernmental Panel on Climate Change* [Edenhofer, O., R. Pichs-Madruga, Y. Sokona, E. Farahani, S. Kadner, K. Seyboth, A. Adler, I. Baum, S. Brunner, P. Eickemeier, B. Kriemann, J. Savolainen, S. Schlömer, C. von Stechow, T. Zwickel, and J.C. Minx (eds.)]. Cambridge University Press, Cambridge, United Kingdom and New York, NY, USA, pp. 1001–1082.

Steffen, W. et al., 2016: Stratigraphic and Earth System approaches to defining the Anthropocene. *Earth's Future*, **4(8)**, 324–345, doi:10.1002/2016ef000379.

Sterner, E., D.J.A. Johansson, and C. Azar, 2014: Emission metrics and sea level rise. *Climatic Change*, **127(2)**, 335–351, doi:10.1007/s10584-014-1258-1.

Stocker, T.F. et al., 2013: Technical Summary. In: *Climate Change 2013: The Physical Science Basis. Contribution of Working Group I to the Fifth Assessment Report of the Intergovernmental Panel on Climate Change* [Stocker, T.F., D. Qin, G.-K. Plattner, M. Tignor, S.K. Allen, J. Boschung, A. Nauels, Y. Xia, V. Bex, and P.M. Midgley (eds.)]. Cambridge University Press, Cambridge, United Kingdom and New York, NY, USA, pp. 33–115.

Stone, D. et al., 2013: The challenge to detect and attribute effects of climate change on human and natural systems. *Climatic Change*, **121(2)**, 381–395, doi:10.1007/s10584-013-0873-6.

Strefler, J., G. Luderer, T. Aboumahboub, and E. Kriegler, 2014: Economic impacts of alternative greenhouse gas emission metrics: a model-based assessment. *Climatic Change*, **125(3–4)**, 319–331, doi:10.1007/s10584-014-1188-y.

Summerhayes, C.P., 2015: *Earth's Climate Evolution*. John Wiley & Sons Ltd, Chichester, UK, 394 pp., doi:10.1002/9781118897362.

Tàbara, J.D. et al., 2018: Positive tipping points in a rapidly warming world. *Current Opinion in Environmental Sustainability*, **31**, 120–129, doi:10.1016/j.cosust.2018.01.012.

Tanaka, K. and B.C. O'Neill, 2018: The Paris Agreement zero-emissions goal is not always consistent with the 1.5°C and 2°C temperature targets. *Nature Climate Change*, **8(4)**, 319–324, doi:10.1038/s41558-018-0097-x.

Taylor, K.E., R.J. Stouffer, and G.A. Meehl, 2012: An overview of CMIP5 and the experiment design. *Bulletin of the American Meteorological Society*, **93(4)**, 485–498, doi:10.1175/bams-d-11-00094.1.

Termeer, C.J.A.M., A. Dewulf, and G.R. Biesbroek, 2017: Transformational change: governance interventions for climate change adaptation from a continuous change perspective. *Journal of Environmental Planning and Management*, **60(4)**, 558–576, doi:10.1080/09640568.2016.1168288.

The Royal Society, 2009: *Geoengineering the climate: science, governance and uncertainty*. RS Policy document 10/09, The Royal Society, London, UK, 82 pp.

Tokarska, K.B. and K. Zickfeld, 2015: The effectiveness of net negative carbon dioxide emissions in reversing anthropogenic climate change. *Environmental Research Letters*, **10(9)**, 094013, doi:10.1088/1748-9326/10/9/094013.

Tokarska, K.B. and N.P. Gillett, 2018: Cumulative carbon emissions budgets consistent with 1.5°C global warming. *Nature Climate Change*, **8(4)**, 296–299, doi:10.1038/s41558-018-0118-9.

Tol, R.S.J., T.K. Berntsen, B.C. O'Neill, Fuglestvedt, and P.S. Keith, 2012: A unifying framework for metrics for aggregating the climate effect of different emissions. *Environmental Research Letters*, **7(4)**, 044006.

Tschakert, P., B. van Oort, A.L. St. Clair, and A. LaMadrid, 2013: Inequality and transformation analyses: a complementary lens for addressing vulnerability to climate change. *Climate and Development*, **5(4)**, 340–350, doi:10.1080/17565529.2013.828583.

Tschakert, P. et al., 2017: Climate change and loss, as if people mattered: values, places, and experiences. *Wiley Interdisciplinary Reviews: Climate Change*, **8(5)**, e476, doi:10.1002/wcc.476.

UN, 2015a: *The Millennium Development Goals Report 2015*. United Nations (UN), New York, NY, USA, 75 pp.

UN, 2015b: *Transforming our world: The 2030 agenda for sustainable development*. A/RES/70/1, United Nations General Assembly (UNGA), 35 pp.

UNDP, 2016: *Human Development Report 2016: Human Development for Everyone*. United Nations Development Programme (UNDP), New York, NY, USA, 286 pp.

UNFCCC, 2016: *Aggregate effect of the intended nationally determined contributions: an update*. FCCC/CP/2016/2, United Nations Framework Convention on Climate Change (UNFCCC), 75 pp.

van Oldenborgh, G.J. et al., 2017: Attribution of extreme rainfall from Hurricane Harvey, August 2017. *Environmental Research Letters*, **12(12)**, 124009, doi:10.1088/1748-9326/aa9ef2.

van Vuuren, D.P. et al., 2011: The representative concentration pathways: An overview. *Climatic Change*, **109(1)**, 5–31, doi:10.1007/s10584-011-0148-z.

van Vuuren, D.P. et al., 2014: A new scenario framework for Climate Change Research: Scenario matrix architecture. *Climatic Change*, **122(3)**, 373–386, doi:10.1007/s10584-013-0906-1.

van Vuuren, D.P. et al., 2015: Pathways to achieve a set of ambitious global sustainability objectives by 2050: Explorations using the IMAGE integrated assessment model. *Technological Forecasting and Social Change*, **98**, 303–323, doi:10.1016/j.techfore.2015.03.005.

Vautard, R. et al., 2014: The European climate under a 2°C global warming. *Environmental Research Letters*, **9(3)**, 034006, doi:10.1088/1748-9326/9/3/034006.

Vincent, K.E., P. Tschakert, J. Barnett, M.G. Rivera-Ferre, and A. Woodward, 2014: Cross-chapter box on gender and climate change. In: *Climate Change 2014: Impacts, Adaptation, and Vulnerability. Part A: Global and Sectoral Aspects. Contribution of Working Group II to the Fifth Assessment Report of the Intergovernmental Panel on Climate Change* [Field, C.B., V.R. Barros, D.J. Dokken, K.J. Mach, M.D. Mastrandrea, T.E. Bilir, M. Chatterjee, K.L. Ebi, Y.O. Estrada, R.C. Genova, B. Girma, E.S. Kissel, A.N. Levy, S. MacCracken, P.R. Mastrandrea, and L.L. White (eds.)]. Cambridge University Press, Cambridge, United Kingdom and New York, NY, USA, pp. 105–107.

Visser, H., S. Dangendorf, D.P. van Vuuren, B. Bregman, and A.C. Petersen, 2018: Signal detection in global mean temperatures after "Paris": an uncertainty and sensitivity analysis. *Climate of the Past*, **14(2)**, 139–155, doi:10.5194/cp-14-139-2018.

von Stechow, C. et al., 2015: Integrating Global Climate Change Mitigation Goals with Other Sustainability Objectives: A Synthesis. *Annual Review of Environment and Resources*, **40(1)**, 363–394, doi:10.1146/annurev-environ-021113-095626.

von Stechow, C. et al., 2016: 2°C and the SDGs: United they stand, divided they fall? *Environmental Research Letters*, **11(3)**, 034022, doi:10.1088/1748-9326/11/3/034022.

Vose, R.S. et al., 2012: NOAA's merged land-ocean surface temperature analysis. *Bulletin of the American Meteorological Society*, **93(11)**, 1677–1685, doi:10.1175/bams-d-11-00241.1.

Waters, C.N. et al., 2016: The Anthropocene is functionally and stratigraphically distinct from the Holocene. *Science*, **351(6269)**, aad2622–aad2622, doi:10.1126/science.aad2622.

WCED, 1987: *Our Common Future*. World Commission on Environment and Development (WCED), Geneva, Switzerland, 383 pp., doi:10.2307/2621529.

Weber, H., 2017: Politics of 'Leaving No One Behind': Contesting the 2030 Sustainable Development Goals Agenda. *Globalizations*, **14(3)**, 399–414, doi:10.1080/14747731.2016.1275404.

Whitmarsh, L., S. O'Neill, and I. Lorenzoni (eds.), 2011: *Engaging the Public with Climate Change: Behaviour Change and Communication*. Earthscan, London, UK and Washington, DC, USA, 289 pp.

Wigley, T.M.L., 2018: The Paris warming targets: emissions requirements and sea level consequences. *Climatic Change*, **147(1–2)**, 31–45, doi:10.1007/s10584-017-2119-5.

Williams, R.G., V. Roussenov, T.L. Frölicher, and P. Goodwin, 2017: Drivers of Continued Surface Warming After Cessation of Carbon Emissions. *Geophysical Research Letters*, **44(20)**, 10,633–10,642, doi:10.1002/2017gl075080.

Winkler, I.T. and M.L. Satterthwaite, 2017: Leaving no one behind? Persistent inequalities in the SDGs. *The International Journal of Human Rights*, **21(8)**, 1073–1097, doi:10.1080/13642987.2017.1348702.

Wise, R.M. et al., 2014: Reconceptualising adaptation to climate change as part of pathways of change and response. *Global Environmental Change*, **28**, 325–336, doi:10.1016/j.gloenvcha.2013.12.002.

World Bank, 2013: *Turn Down the Heat: Climate Extremes, Regional Impacts, and the Case for Resilience*. The World Bank, Washington DC, USA, 254 pp.

Wright, H., S. Huq, and J. Reeves, 2015: *Impact of climate change on least developed countries: are the SDGs possible?* IIED Briefing May 2015, International Institute for Environment and Development (IIED), London, UK, 4 pp.

Xu, Y. and V. Ramanathan, 2017: Well below 2°C: Mitigation strategies for avoiding dangerous to catastrophic climate changes. *Proceedings of the National Academy of Sciences*, 1–9, doi:10.1073/pnas.1618481114.

Zalasiewicz, J. et al., 2017: Making the case for a formal Anthropocene Epoch: an analysis of ongoing critiques. *Newsletters on Stratigraphy*, **50(2)**, 205–226, doi:10.1127/nos/2017/0385.

Zickfeld, K., A.H. MacDougall, and H.D. Matthews, 2016: On the proportionality between global temperature change and cumulative CO_2 emissions during periods of net negative CO_2 emissions. *Environmental Research Letters*, **11(5)**, 055006, doi:10.1088/1748-9326/11/5/055006.

Zickfeld, K., M. Eby, H.D. Matthews, and A.J. Weaver, 2009: Setting cumulative emissions targets to reduce the risk of dangerous climate change. *Proceedings of the National Academy of Sciences*, **106(38)**, 16129–16134, doi:10.1073/pnas.0805800106.

Zscheischler, J. and S.I. Seneviratne, 2017: Dependence of drivers affects risks associated with compound events. *Science Advances*, **3(6)**, e1700263, doi:10.1126/sciadv.1700263.

Mitigation Pathways Compatible with 1.5°C in the Context of Sustainable Development

2

Coordinating Lead Authors:

Joeri Rogelj (Austria/Belgium), Drew Shindell (USA), Kejun Jiang (China)

Lead Authors:

Solomone Fifita (Fiji), Piers Forster (UK), Veronika Ginzburg (Russia), Collins Handa (Kenya), Haroon Kheshgi (USA), Shigeki Kobayashi (Japan), Elmar Kriegler (Germany), Luis Mundaca (Sweden/Chile), Roland Séférian (France), Maria Virginia Vilariño (Argentina)

Contributing Authors:

Katherine Calvin (USA), Joana Correia de Oliveira de Portugal Pereira (UK/Portugal), Oreane Edelenbosch (Netherlands/Italy), Johannes Emmerling (Italy/Germany), Sabine Fuss (Germany), Thomas Gasser (Austria/France), Nathan Gillett (Canada), Chenmin He (China), Edgar Hertwich (USA/Austria), Lena Höglund-Isaksson (Austria/Sweden), Daniel Huppmann (Austria), Gunnar Luderer (Germany), Anil Markandya (Spain/UK), David L. McCollum (USA/Austria), Malte Meinshausen (Australia/Germany), Richard Millar (UK), Alexander Popp (Germany), Pallav Purohit (Austria/India), Keywan Riahi (Austria), Aurélien Ribes (France), Harry Saunders (Canada/USA), Christina Schädel (USA/Switzerland), Chris Smith (UK), Pete Smith (UK), Evelina Trutnevyte (Switzerland/Lithuania), Yang Xiu (China), Wenji Zhou (Austria/China), Kirsten Zickfeld (Canada/Germany)

Chapter Scientists:

Daniel Huppmann (Austria), Chris Smith (UK)

Review Editors:

Greg Flato (Canada), Jan Fuglestvedt (Norway), Rachid Mrabet (Morocco), Roberto Schaeffer (Brazil)

This chapter should be cited as:

Rogelj, J., D. Shindell, K. Jiang, S. Fifita, P. Forster, V. Ginzburg, C. Handa, H. Kheshgi, S. Kobayashi, E. Kriegler, L. Mundaca, R. Séférian, and M.V. Vilariño, 2018: Mitigation Pathways Compatible with 1.5°C in the Context of Sustainable Development. In: *Global Warming of 1.5°C. An IPCC Special Report on the impacts of global warming of 1.5°C above pre-industrial levels and related global greenhouse gas emission pathways, in the context of strengthening the global response to the threat of climate change, sustainable development, and efforts to eradicate poverty* [Masson-Delmotte, V., P. Zhai, H.-O. Pörtner, D. Roberts, J. Skea, P.R. Shukla, A. Pirani, W. Moufouma-Okia, C. Péan, R. Pidcock, S. Connors, J.B.R. Matthews, Y. Chen, X. Zhou, M.I. Gomis, E. Lonnoy, T. Maycock, M. Tignor, and T. Waterfield (eds.)]. Cambridge University Press, Cambridge, UK and New York, NY, USA, pp. 93-174. https://doi.org/10.1017/9781009157940.004.

Table of Contents

Executive Summary

This chapter assesses mitigation pathways consistent with limiting warming to 1.5°C above pre-industrial levels. In doing so, it explores the following key questions: What role do CO_2 and non-CO_2 emissions play? {2.2, 2.3, 2.4, 2.6} To what extent do 1.5°C pathways involve overshooting and returning below 1.5°C during the 21st century? {2.2, 2.3} What are the implications for transitions in energy, land use and sustainable development? {2.3, 2.4, 2.5} How do policy frameworks affect the ability to limit warming to 1.5°C? {2.3, 2.5} What are the associated knowledge gaps? {2.6}

The assessed pathways describe integrated, quantitative evolutions of all emissions over the 21st century associated with global energy and land use and the world economy. The assessment is contingent upon available integrated assessment literature and model assumptions, and is complemented by other studies with different scope, for example, those focusing on individual sectors. In recent years, integrated mitigation studies have improved the characterizations of mitigation pathways. However, limitations remain, as climate damages, avoided impacts, or societal co-benefits of the modelled transformations remain largely unaccounted for, while concurrent rapid technological changes, behavioural aspects, and uncertainties about input data present continuous challenges. (*high confidence*) {2.1.3, 2.3, 2.5.1, 2.6, Technical Annex 2}

The Chances of Limiting Warming to 1.5°C and the Requirements for Urgent Action

Pathways consistent with 1.5°C of warming above pre-industrial levels can be identified under a range of assumptions about economic growth, technology developments and lifestyles. However, lack of global cooperation, lack of governance of the required energy and land transformation, and increases in resource-intensive consumption are key impediments to achieving 1.5°C pathways. Governance challenges have been related to scenarios with high inequality and high population growth in the 1.5°C pathway literature. {2.3.1, 2.3.2, 2.5}

Under emissions in line with current pledges under the Paris Agreement (known as Nationally Determined Contributions, or NDCs), global warming is expected to surpass 1.5°C above pre-industrial levels, even if these pledges are supplemented with very challenging increases in the scale and ambition of mitigation after 2030 (*high confidence*). This increased action would need to achieve net zero CO_2 emissions in less than 15 years. Even if this is achieved, temperatures would only be expected to remain below the 1.5°C threshold if the actual geophysical response ends up being towards the low end of the currently estimated uncertainty range. Transition challenges as well as identified trade-offs can be reduced if global emissions peak before 2030 and marked emissions reductions compared to today are already achieved by 2030. {2.2, 2.3.5, Cross-Chapter Box 11 in Chapter 4}

Limiting warming to 1.5°C depends on greenhouse gas (GHG) emissions over the next decades, where lower GHG emissions in 2030 lead to a higher chance of keeping peak warming to 1.5°C (*high confidence*). Available pathways that aim for no or limited (less than 0.1°C) overshoot of 1.5°C keep GHG emissions in 2030 to 25–30 GtCO2e yr^{-1} in 2030 (interquartile range). This contrasts with median estimates for current unconditional NDCs of 52–58 GtCO$_2$e yr^{-1} in 2030. Pathways that aim for limiting warming to 1.5°C by 2100 after a temporary temperature overshoot rely on large-scale deployment of carbon dioxide removal (CDR) measures, which are uncertain and entail clear risks. In model pathways with no or limited overshoot of 1.5°C, global net anthropogenic CO_2 emissions decline by about 45% from 2010 levels by 2030 (40–60% interquartile range), reaching net zero around 2050 (2045–2055 interquartile range). For limiting global warming to below 2°C with at least 66% probability CO_2 emissions are projected to decline by about 25% by 2030 in most pathways (10–30% interquartile range) and reach net zero around 2070 (2065–2080 interquartile range).[1] {2.2, 2.3.3, 2.3.5, 2.5.3, Cross-Chapter Boxes 6 in Chapter 3 and 9 in Chapter 4, 4.3.7}

Limiting warming to 1.5°C implies reaching net zero CO_2 emissions globally around 2050 and concurrent deep reductions in emissions of non-CO_2 forcers, particularly methane (*high confidence*). Such mitigation pathways are characterized by energy-demand reductions, decarbonization of electricity and other fuels, electrification of energy end use, deep reductions in agricultural emissions, and some form of CDR with carbon storage on land or sequestration in geological reservoirs. Low energy demand and low demand for land- and GHG-intensive consumption goods facilitate limiting warming to as close as possible to 1.5°C. {2.2.2, 2.3.1, 2.3.5, 2.5.1, Cross-Chapter Box 9 in Chapter 4}.

In comparison to a 2°C limit, the transformations required to limit warming to 1.5°C are qualitatively similar but more pronounced and rapid over the next decades (*high confidence*). 1.5°C implies very ambitious, internationally cooperative policy environments that transform both supply and demand *(high confidence).* {2.3, 2.4, 2.5}

Policies reflecting a high price on emissions are necessary in models to achieve cost-effective 1.5°C pathways (*high confidence*). Other things being equal, modelling studies suggest the global average discounted marginal abatement costs for limiting warming to 1.5°C being about 3–4 times higher compared to 2°C over the 21st century, with large variations across models and socio-economic and policy assumptions. Carbon pricing can be imposed directly or implicitly by regulatory policies. Policy instruments, like technology policies or performance standards, can complement explicit carbon pricing in specific areas. {2.5.1, 2.5.2, 4.4.5}

Limiting warming to 1.5°C requires a marked shift in investment patterns (*medium confidence*). Additional annual average energy-related investments for the period 2016 to 2050 in pathways limiting warming to 1.5°C compared to pathways without new climate policies beyond those in place today (i.e., baseline) are estimated to be around

---

[1] Kyoto-GHG emissions in this statement are aggregated with GWP-100 values of the IPCC Second Assessment Report.

830 billion USD2010 (range of 150 billion to 1700 billion USD2010 across six models). Total energy-related investments increase by about 12% (range of 3% to 24%) in 1.5°C pathways relative to 2°C pathways. Average annual investment in low-carbon energy technologies and energy efficiency are upscaled by roughly a factor of six (range of factor of 4 to 10) by 2050 compared to 2015, overtaking fossil investments globally by around 2025 (*medium confidence*). Uncertainties and strategic mitigation portfolio choices affect the magnitude and focus of required investments. {2.5.2}

Future Emissions in 1.5°C Pathways

Mitigation requirements can be quantified using carbon budget approaches that relate cumulative CO_2 emissions to global mean temperature increase. Robust physical understanding underpins this relationship, but uncertainties become increasingly relevant as a specific temperature limit is approached. These uncertainties relate to the transient climate response to cumulative carbon emissions (TCRE), non-CO_2 emissions, radiative forcing and response, potential additional Earth system feedbacks (such as permafrost thawing), and historical emissions and temperature. {2.2.2, 2.6.1}

Cumulative CO_2 emissions are kept within a budget by reducing global annual CO_2 emissions to net zero. This assessment suggests a remaining budget of about 420 GtCO$_2$ for a two-thirds chance of limiting warming to 1.5°C, and of about 580 GtCO$_2$ for an even chance (*medium confidence*). The remaining carbon budget is defined here as cumulative CO_2 emissions from the start of 2018 until the time of net zero global emissions for global warming defined as a change in global near-surface air temperatures. Remaining budgets applicable to 2100 would be approximately 100 GtCO$_2$ lower than this to account for permafrost thawing and potential methane release from wetlands in the future, and more thereafter. These estimates come with an additional geophysical uncertainty of at least ±400 GtCO$_2$, related to non-CO_2 response and TCRE distribution. Uncertainties in the level of historic warming contribute ±250 GtCO$_2$. In addition, these estimates can vary by ±250 GtCO$_2$ depending on non-CO_2 mitigation strategies as found in available pathways. {2.2.2, 2.6.1}

Staying within a remaining carbon budget of 580 GtCO$_2$ implies that CO_2 emissions reach carbon neutrality in about 30 years, reduced to 20 years for a 420 GtCO$_2$ remaining carbon budget (*high confidence*). The ±400 GtCO$_2$ geophysical uncertainty range surrounding a carbon budget translates into a variation of this timing of carbon neutrality of roughly ±15–20 years. If emissions do not start declining in the next decade, the point of carbon neutrality would need to be reached at least two decades earlier to remain within the same carbon budget. {2.2.2, 2.3.5}

Non-CO_2 emissions contribute to peak warming and thus affect the remaining carbon budget. The evolution of methane and sulphur dioxide emissions strongly influences the chances of limiting warming to 1.5°C. In the near-term, a weakening of aerosol cooling would add to future warming, but can be tempered by reductions in methane emissions (*high confidence*). Uncertainty in radiative forcing estimates (particularly

aerosol) affects carbon budgets and the certainty of pathway categorizations. Some non-CO_2 forcers are emitted alongside CO_2, particularly in the energy and transport sectors, and can be largely addressed through CO_2 mitigation. Others require specific measures, for example, to target agricultural nitrous oxide (N_2O) and methane (CH_4), some sources of black carbon, or hydrofluorocarbons (*high confidence*). In many cases, non-CO2 emissions reductions are similar in 2°C pathways, indicating reductions near their assumed maximum potential by integrated assessment models. Emissions of N_2O and NH_3 increase in some pathways with strongly increased bioenergy demand. {2.2.2, 2.3.1, 2.4.2, 2.5.3}

The Role of Carbon Dioxide Removal (CDR)

All analysed pathways limiting warming to 1.5°C with no or limited overshoot use CDR to some extent to neutralize emissions from sources for which no mitigation measures have been identified and, in most cases, also to achieve net negative emissions to return global warming to 1.5°C following a peak (*high confidence*). The longer the delay in reducing CO_2 emissions towards zero, the larger the likelihood of exceeding 1.5°C, and the heavier the implied reliance on net negative emissions after mid-century to return warming to 1.5°C (*high confidence*). The faster reduction of net CO_2 emissions in 1.5°C compared to 2°C pathways is predominantly achieved by measures that result in less CO_2 being produced and emitted, and only to a smaller degree through additional CDR. Limitations on the speed, scale and societal acceptability of CDR deployment also limit the conceivable extent of temperature overshoot. Limits to our understanding of how the carbon cycle responds to net negative emissions increase the uncertainty about the effectiveness of CDR to decline temperatures after a peak. {2.2, 2.3, 2.6, 4.3.7}

CDR deployed at scale is unproven, and reliance on such technology is a major risk in the ability to limit warming to 1.5°C. CDR is needed less in pathways with particularly strong emphasis on energy efficiency and low demand. The scale and type of CDR deployment varies widely across 1.5°C pathways, with different consequences for achieving sustainable development objectives (*high confidence*). Some pathways rely more on bioenergy with carbon capture and storage (BECCS), while others rely more on afforestation, which are the two CDR methods most often included in integrated pathways. Trade-offs with other sustainability objectives occur predominantly through increased land, energy, water and investment demand. Bioenergy use is substantial in 1.5°C pathways with or without BECCS due to its multiple roles in decarbonizing energy use. {2.3.1, 2.5.3, 2.6.3, 4.3.7}

Properties of Energy and Land Transitions in 1.5°C Pathways

The share of primary energy from renewables increases while coal usage decreases across pathways limiting warming to 1.5°C with no or limited overshoot (*high confidence*). By 2050, renewables (including bioenergy, hydro, wind, and solar, with direct-equivalence method) supply a share of 52–67% (interquartile range) of primary energy in 1.5°C pathways with no or limited overshoot; while the share from coal decreases to 1–7% (interquartile range), with a large fraction of this coal use combined with carbon capture and storage (CCS). From 2020 to 2050 the primary energy supplied by oil declines in most pathways (−39 to −77% interquartile range). Natural gas changes by −13% to −62% (interquartile range), but some pathways show a marked increase albeit with widespread deployment of CCS. The overall deployment of CCS varies widely across 1.5°C pathways with no or limited overshoot, with cumulative CO_2 stored through 2050 ranging from zero up to 300 $GtCO_2$ (minimum–maximum range), of which zero up to 140 $GtCO_2$ is stored from biomass. Primary energy supplied by bioenergy ranges from 40–310 EJ yr^{-1} in 2050 (minimum-maximum range), and nuclear from 3–66 EJ yr^{-1} (minimum–maximum range). These ranges reflect both uncertainties in technological development and strategic mitigation portfolio choices. {2.4.2}

1.5°C pathways with no or limited overshoot include a rapid decline in the carbon intensity of electricity and an increase in electrification of energy end use (*high confidence*). By 2050, the carbon intensity of electricity decreases to −92 to +11 gCO_2 MJ^{-1} (minimum–maximum range) from about 140 gCO_2 MJ^{-1} in 2020, and electricity covers 34–71% (minimum–maximum range) of final energy across 1.5°C pathways with no or limited overshoot from about 20% in 2020. By 2050, the share of electricity supplied by renewables increases to 59–97% (minimum-maximum range) across 1.5°C pathways with no or limited overshoot. Pathways with higher chances of holding warming to below 1.5°C generally show a faster decline in the carbon intensity of electricity by 2030 than pathways that temporarily overshoot 1.5°C. {2.4.1, 2.4.2, 2.4.3}

Transitions in global and regional land use are found in all pathways limiting global warming to 1.5°C with no or limited overshoot, but their scale depends on the pursued mitigation portfolio (*high confidence*). Pathways that limit global warming to 1.5°C with no or limited overshoot project a 4 million km² reduction to a 2.5 million km2 increase of non-pasture agricultural land for food and feed crops and a 0.5–11 million km² reduction of pasture land, to be converted into 0-6 million km² of agricultural land for energy crops and a 2 million km² reduction to 9.5 million km² increase in forests by 2050 relative to 2010 (*medium confidence*). Land-use transitions of similar magnitude can be observed in modelled 2°C pathways (*medium confidence*). Such large transitions pose profound challenges for sustainable management of the various demands on land for human settlements, food, livestock feed, fibre, bioenergy, carbon storage, biodiversity and other ecosystem services (*high confidence*). {2.3.4, 2.4.4}

Demand-Side Mitigation and Behavioural Changes

Demand-side measures are key elements of 1.5°C pathways. Lifestyle choices lowering energy demand and the land- and GHG-intensity of food consumption can further support achievement of 1.5°C pathways (*high confidence*). By 2030 and 2050, all end-use sectors (including building, transport, and industry) show marked energy demand reductions in modelled 1.5°C pathways, comparable and beyond those projected in 2°C pathways. Sectoral models support the scale of these reductions. {2.3.4, 2.4.3, 2.5.1}

Links between 1.5°C Pathways and Sustainable Development

Choices about mitigation portfolios for limiting warming to 1.5°C can positively or negatively impact the achievement of other societal objectives, such as sustainable development (*high confidence*). In particular, demand-side and efficiency measures, and lifestyle choices that limit energy, resource, and GHG-intensive food demand support sustainable development (*medium confidence*). Limiting warming to 1.5°C can be achieved synergistically with poverty alleviation and improved energy security and can provide large public health benefits through improved air quality, preventing millions of premature deaths. However, specific mitigation measures, such as bioenergy, may result in trade-offs that require consideration. {2.5.1, 2.5.2, 2.5.3}

2

2.1 Introduction to Mitigation Pathways and the Sustainable Development Context

This chapter assesses the literature on mitigation pathways to limit or return global mean warming to 1.5°C (relative to the pre-industrial base period 1850–1900). Key questions addressed are: What types of mitigation pathways have been developed that could be consistent with 1.5°C? What changes in emissions, energy and land use do they entail? What do they imply for climate policy and implementation, and what impacts do they have on sustainable development? In terms of feasibility (see Cross-Chapter Box 3 in Chapter 1), this chapter focuses on geophysical dimensions and technological and economic enabling factors. Social and institutional dimensions as well as additional aspects of technical feasibility are covered in Chapter 4.

Mitigation pathways are typically designed to reach a predefined climate target alone. Minimization of mitigation expenditures, but not climate-related damages or sustainable development impacts, is often the basis for these pathways to the desired climate target (see Cross-Chapter Box 5 in this chapter for additional discussion). However, there are interactions between mitigation and multiple other sustainable development goals (see Sections 1.1 and 5.4) that provide both challenges and opportunities for climate action. Hence there are substantial efforts to evaluate the effects of the various mitigation pathways on sustainable development, focusing in particular on aspects for which integrated assessment models (IAMs) provide relevant information (e.g., land-use changes and biodiversity, food security, and air quality). More broadly, there are efforts to incorporate climate change mitigation as one of multiple objectives that, in general, reflect societal concerns more completely and could potentially provide benefits at lower costs than simultaneous single-objective policies (e.g., Clarke et al., 2014). For example, with carefully selected policies, universal energy access can be achieved while simultaneously reducing air pollution and mitigating climate change (McCollum et al., 2011; Riahi et al., 2012; IEA, 2017d). This chapter thus presents both the pathways and an initial discussion of their context within sustainable development objectives (Section 2.5), with the latter, along with equity and ethical issues, discussed in more detail in Chapter 5.

As described in Cross-Chapter Box 1 in Chapter 1, scenarios are comprehensive, plausible, integrated descriptions of possible futures based on specified, internally consistent underlying assumptions, with pathways often used to describe the clear temporal evolution of specific scenario aspects or goal-oriented scenarios. We include both these usages of 'pathways' here.

2.1.1 Mitigation Pathways Consistent with 1.5°C

Emissions scenarios need to cover all sectors and regions over the 21st century to be associated with a climate change projection out to 2100. Assumptions regarding future trends in population, consumption of goods and services (including food), economic growth, behaviour, technology, policies and institutions are all required to generate scenarios (Section 2.3.1). These societal choices must then be linked to the drivers of climate change, including emissions of well-mixed greenhouse gases and aerosol and ozone precursors as well as land-use and land-cover changes. Deliberate solar radiation modification is not included in these scenarios (see Cross-Chapter Box 10 in Chapter 4).

Plausible developments need to be anticipated in many facets of the key sectors of energy and land use. Within energy, these scenarios consider energy resources like biofuels, energy supply and conversion technologies, energy consumption, and supply and end-use efficiency. Within land use, agricultural productivity, food demand, terrestrial carbon management, and biofuel production are all considered. Climate policies are also considered, including carbon pricing and technology policies such as research and development funding and subsidies. The scenarios incorporate regional differentiation in sectoral and policy development. The climate changes resulting from such scenarios are derived using models that typically incorporate physical understanding of the carbon cycle and climate response derived from complex geophysical models evaluated against observations (Sections 2.2 and 2.6).

The temperature response to a given emission pathway (see glossary) is uncertain and therefore quantified in terms of a probabilistic outcome. Chapter 1 assesses the climate objectives of the Paris Agreement in terms of human-induced warming, thus excluding potential impacts of natural forcing such as volcanic eruptions or solar output changes or unforced internal variability. Temperature responses in this chapter are assessed using simple geophysically based models that evaluate the anthropogenic component of future temperature change and do not incorporate internal natural variations and are thus fit for purpose in the context of this assessment (Section 2.2.1). Hence a scenario that is consistent with 1.5°C may in fact lead to either a higher or lower temperature change, but within quantified and generally well-understood bounds (see also Chapter 1, Section 1.2.3). Consistency with avoiding a human-induced temperature change limit must therefore also be defined probabilistically, with likelihood values selected based on risk-avoidance preferences. Responses beyond global mean temperature are not typically evaluated in such models and are assessed in Chapter 3.

2.1.2 The Use of Scenarios

Variations in scenario assumptions and design define to a large degree which questions can be addressed with a specific scenario set, for example, the exploration of implications of delayed climate mitigation action. In this assessment, the following classes of 1.5°C- and 2°C-consistent scenarios are of particular interest to the topics addressed in this chapter: (i) scenarios with the same climate target over the 21st century but varying socio-economic assumptions (Sections 2.3 and 2.4), (ii) pairs of scenarios with similar socio-economic assumptions but with forcing targets aimed at 1.5°C and 2°C (Section 2.3), and (iii) scenarios that follow the Nationally Determined Contributions or NDCs[2] until 2030 with much more stringent mitigation action thereafter (Section 2.3.5).

---

[2] Current pledges include those from the United States although they have stated their intention to withdraw in the future.

Characteristics of these pathways, such as emissions reduction rates, time of peaking, and low-carbon energy deployment rates, can be assessed as being consistent with 1.5°C. However, they cannot be assessed as 'requirements' for 1.5°C, unless a targeted analysis is available that specifically asked whether there could be other 1.5°C-consistent pathways without the characteristics in question. AR5 already assessed such targeted analyses, for example, asking which technologies are important in order to keep open the possibility of limiting warming to 2°C (Clarke et al., 2014). By now, several such targeted analyses are also available for questions related to 1.5°C (Luderer et al., 2013; Rogelj et al., 2013b; Bauer et al., 2018; Strefler et al., 2018b; van Vuuren et al., 2018). This assessment distinguishes between 'consistent' and the much stronger concept of required characteristics of 1.5°C pathways wherever possible.

Ultimately, society will adjust the choices it makes as new information becomes available and technical learning progresses, and these adjustments can be in either direction. Earlier scenario studies have shown, however, that deeper emissions reductions in the near term hedge against the uncertainty of both climate response and future technology availability (Luderer et al., 2013; Rogelj et al., 2013b; Clarke et al., 2014). Not knowing what adaptations might be put in place in the future, and due to limited studies, this chapter examines prospective rather than iteratively adaptive mitigation pathways (Cross-Chapter Box 1 in Chapter 1). Societal choices illustrated by scenarios may also influence what futures are envisioned as possible or desirable and hence whether those come into being (Beck and Mahony, 2017).

2.1.3   New Scenario Information since AR5

In this chapter, we extend the AR5 mitigation pathway assessment based on new scenario literature. Updates in understanding of climate sensitivity, transient climate response, radiative forcing, and the cumulative carbon budget consistent with 1.5°C are discussed in Sections 2.2.

Mitigation pathways developed with detailed process-based integrated assessment models (IAMs) covering all sectors and regions over the 21st century describe an internally consistent and calibrated (to historical trends) way to get from current developments to meeting long-term climate targets like 1.5°C (Clarke et al., 2014). The overwhelming majority of available 1.5°C pathways were generated by such IAMs, and these pathways can be directly linked to climate outcomes and their consistency with the 1.5°C goal evaluated. The AR5 similarly relied upon such studies, which were mainly discussed in Chapter 6 of Working Group III (WGIII) (Clarke et al., 2014).

Since the AR5, several new, integrated multimodel studies have appeared in the literature that explore specific characteristics of scenarios more stringent than the lowest scenario category assessed in AR5 that was assessed to limit warming below 2°C with greater than 66% likelihood (Rogelj et al., 2015b, 2018; Akimoto et al., 2017; Marcucci et al., 2017; Su et al., 2017; Bauer et al., 2018; Bertram et al., 2018; Grubler et al., 2018; Holz et al., 2018b; Kriegler et al., 2018a; Liu et al., 2018; Luderer et al., 2018; Strefler et al., 2018a; van Vuuren et al., 2018; Vrontisi et al., 2018; Zhang et al., 2018). Those scenarios explore 1.5°C-consistent pathways from multiple perspectives

(see Supplementary Material 2.SM.1.3), examining sensitivity to assumptions regarding:

- socio-economic drivers and developments including energy and food demand as, for example, characterized by the Shared Socio-Economic Pathways (SSPs; Cross-Chapter Box 1 in Chapter 1)
- near-term climate policies describing different levels of strengthening the NDCs
- the use of bioenergy and the availability and desirability of carbon dioxide removal (CDR) technologies

A large number of these scenarios were collected in a scenario database established for the assessment of this Special Report (Supplementary Material 2.SM.1.3). Mitigation pathways were classified by four factors: consistency with a temperature increase limit (as defined by Chapter 1), whether they temporarily overshoot that limit, the extent of this potential overshoot, and the likelihood of falling within these bounds.

Specifically, they were put into classes that either kept surface temperature increases below a given threshold throughout the 21st century or returned to a value below 1.5°C above pre-industrial levels at some point before 2100 after temporarily exceeding that level earlier – referred to as an overshoot (OS). Both groups were further separated based on the probability of being below the threshold and the degree of overshoot, respectively (Table 2.1). Pathways are uniquely classified, with 1.5°C-related classes given higher priority than 2°C classes in cases where a pathway would be applicable to either class.

The probability assessment used in the scenario classification is based on simulations using two reduced-complexity carbon cycle, atmospheric composition, and climate models: the 'Model for the Assessment of Greenhouse Gas-Induced Climate Change' (MAGICC) (Meinshausen et al., 2011a), and the 'Finite Amplitude Impulse Response' (FAIRv1.3) model (Smith et al., 2018). For the purpose of this report, and to facilitate comparison with AR5, the range of the key carbon cycle and climate parameters for MAGICC and its setup are identical to those used in AR5 WGIII (Clarke et al., 2014). For each mitigation pathway, MAGICC and FAIR simulations provide probabilistic estimates of atmospheric concentrations, radiative forcing and global temperature outcomes until 2100. However, the classification uses MAGICC probabilities directly for traceability with AR5 and because this model is more established in the literature. Nevertheless, the overall uncertainty assessment is based on results from both models, which are considered in the context of the latest radiative forcing estimates and observed temperatures (Etminan et al., 2016; Smith et al., 2018) (Section 2.2 and Supplementary Material 2.SM.1.1). The comparison of these lines of evidence shows *high agreement* in the relative temperature response of pathways, with *medium agreement* on the precise absolute magnitude of warming, introducing a level of imprecision in these attributes. Consideration of the combined evidence here leads to *medium confidence* in the overall geophysical characteristics of the pathways reported here.

In addition to the characteristics of the above-mentioned classes, four illustrative pathway archetypes have been selected and are used throughout this chapter to highlight specific features of and variations across 1.5°C pathways. These are chosen in particular to illustrate the spectrum of CO_2 emissions reduction patterns consistent with 1.5°C,

Table 2.1 | Classification of pathways that this chapter draws upon, along with the number of available pathways in each class. The definition of each class is based on probabilities derived from the MAGICC model in a setup identical to AR5 WGIII (Clarke et al., 2014), as detailed in Supplementary Material 2.SM.1.4.

Pathway group	Pathway Class	Pathway Selection Criteria and Description	Number of Scenarios	Number of Scenarios
1.5°C or 1.5°C-consistent**	Below-1.5°C	Pathways limiting peak warming to below 1.5°C during the entire 21st century with 50–66% likelihood*	9	90
	1.5°C-low-OS	Pathways limiting median warming to below 1.5°C in 2100 and with a 50–67% probability of temporarily overshooting that level earlier, generally implying less than 0.1°C higher peak warming than Below-1.5°C pathways	44	
	1.5°C-high-OS	Pathways limiting median warming to below 1.5°C in 2100 and with a greater than 67% probability of temporarily overshooting that level earlier, generally implying 0.1–0.4°C higher peak warming than Below-1.5°C pathways	37	
2°C or 2°C-consistent	Lower-2°C	Pathways limiting peak warming to below 2°C during the entire 21st century with greater than 66% likelihood	74	132
	Higher-2°C	Pathways assessed to keep peak warming to below 2°C during the entire 21st century with 50–66% likelihood	58	

*   No pathways were available that achieve a greater than 66% probability of limiting warming below 1.5°C during the entire 21st century based on the MAGICC model projections.

** This chapter uses the term 1.5°C-consistent pathways to refer to pathways with no overshoot, with limited (low) overshoot, and with high overshoot. However, the Summary for Policymakers focusses on pathways with no or limited (low) overshoot.

ranging from very rapid and deep near-term decreases, facilitated by efficiency and demand-side measures that lead to limited CDR requirements, to relatively slower but still rapid emissions reductions that lead to a temperature overshoot and necessitate large CDR deployment later in the century (Section 2.3).

2.1.4   Utility of Integrated Assessment Models (IAMs) in the Context of this Report

IAMs lie at the basis of the assessment of mitigation pathways in this chapter, as much of the quantitative global scenario literature is derived with such models. IAMs combine insights from various disciplines in a single framework, resulting in a dynamic description of the coupled energy–economy–land-climate system that cover the largest sources of anthropogenic greenhouse gas (GHG) emissions from different sectors. Many of the IAMs that contributed mitigation scenarios to this assessment include a process-based description of the land system in addition to the energy system (e.g., Popp et al., 2017), and several have been extended to cover air pollutants (Rao et al., 2017) and water use (Hejazi et al., 2014; Fricko et al., 2016; Mouratiadou et al., 2016). Such integrated pathways hence allow the exploration of the whole-system transformation, as well as the interactions, synergies, and trade-offs between sectors, and, increasingly, questions beyond climate mitigation (von Stechow et al., 2015). The models do not, however, fully account for all constraints that could affect realization of pathways (see Chapter 4).

Section 2.3 assesses the overall characteristics of 1.5°C pathways based on fully integrated pathways, while Sections 2.4 and 2.5 describe underlying sectoral transformations, including insights from sector-specific assessment models and pathways that are not derived from IAMs. Such models provide detail in their domain of application and make exogenous assumptions about cross-sectoral or global factors. They often focus on a specific sector, such as the energy (Bruckner et al., 2014; IEA, 2017a; Jacobson, 2017; OECD/IEA and IRENA, 2017), buildings (Lucon et al., 2014) or transport (Sims et al., 2014) sector, or

a specific country or region (Giannakidis et al., 2018). Sector-specific pathways are assessed in relation to integrated pathways because they cannot be directly linked to 1.5°C by themselves if they do not extend to 2100 or do not include all GHGs or aerosols from all sectors.

AR5 found sectoral 2°C decarbonization strategies from IAMs to be consistent with sector-specific studies (Clarke et al., 2014). A growing body of literature on 100%-renewable energy scenarios has emerged (e.g., see Creutzig et al., 2017; Jacobson et al., 2017), which goes beyond the wide range of IAM projections of renewable energy shares in 1.5°C and 2°C pathways. While the representation of renewable energy resource potentials, technology costs and system integration in IAMs has been updated since AR5, leading to higher renewable energy deployments in many cases (Luderer et al., 2017; Pietzcker et al., 2017), none of the IAM projections identify 100% renewable energy solutions for the global energy system as part of cost-effective mitigation pathways (Section 2.4.2). Bottom-up studies find higher mitigation potentials in the industry, buildings, and transport sectors in 2030 than realized in selected 2°C pathways from IAMs (UNEP 2017), indicating the possibility to strengthen sectoral decarbonization strategies until 2030 beyond the integrated 1.5°C pathways assessed in this chapter (Luderer et al., 2018).

Detailed, process-based IAMs are a diverse set of models ranging from partial equilibrium energy–land models to computable general equilibrium models of the global economy, from myopic to perfect foresight models, and from models with to models without endogenous technological change (Supplementary Material 2.SM.1.2). The IAMs used in this chapter have limited to no coverage of climate impacts. They typically use GHG pricing mechanisms to induce emissions reductions and associated changes in energy and land uses consistent with the imposed climate goal. The scenarios generated by these models are defined by the choice of climate goals and assumptions about near-term climate policy developments. They are also shaped by assumptions about mitigation potentials and technologies as well as baseline developments such as, for example, those represented by different Shared Socio-Economic Pathways (SSPs), especially those pertaining to energy and food demand (Riahi et al., 2017). See Section 2.3.1 for discussion of these assumptions. Since the AR5, the scenario literature has greatly expanded the exploration of these dimensions. This includes low-demand scenarios (Grubler et al., 2018; van Vuuren et al., 2018), scenarios taking into account a larger set of sustainable development goals (Bertram et al., 2018), scenarios with restricted availability of CDR technologies (Bauer et al., 2018; Grubler et al., 2018; Holz et al., 2018b; Kriegler et al., 2018a; Strefler et al., 2018b; van Vuuren et al., 2018), scenarios with near-term action dominated by regulatory policies (Kriegler et al., 2018a) and scenario variations across the SSPs (Riahi et al., 2017; Rogelj et al., 2018). IAM results depend upon multiple underlying assumptions, for example, the extent to which global markets and economies are assumed to operate frictionless and policies are cost-optimized, assumptions about technological progress and availability and costs of mitigation and CDR measures, assumptions about underlying socio-economic developments and future energy, food and materials demand, and assumptions about the geographic and temporal pattern of future regulatory and carbon pricing policies (see Supplementary Material 2.SM.1.2 for additional discussion on IAMs and their limitations).

2.2   Geophysical Relationships and Constraints

Emissions pathways can be characterized by various geophysical characteristics, such as radiative forcing (Masui et al., 2011; Riahi et al., 2011; Thomson et al., 2011; van Vuuren et al., 2011b), atmospheric concentrations (van Vuuren et al., 2007, 2011a; Clarke et al., 2014) or associated temperature outcomes (Meinshausen et al., 2009; Rogelj et al., 2011; Luderer et al., 2013). These attributes can be used to derive geophysical relationships for specific pathway classes, such as cumulative CO_2 emissions compatible with a specific level of warming, also known as 'carbon budgets' (Meinshausen et al., 2009; Rogelj et al., 2011; Stocker et al., 2013; Friedlingstein et al., 2014a), the consistent contributions of non-CO_2 GHGs and aerosols to the remaining carbon budget (Bowerman et al., 2011; Rogelj et al., 2015a, 2016b), or to temperature outcomes (Lamarque et al., 2011; Bowerman et al., 2013; Rogelj et al., 2014b). This section assesses geophysical relationships for both CO_2 and non-CO_2 emissions (see glossary).

2.2.1   Geophysical Characteristics of Mitigation Pathways

This section employs the pathway classification introduced in Section 2.1, with geophysical characteristics derived from simulations with the MAGICC reduced-complexity carbon cycle and climate model and supported by simulations with the FAIR reduced-complexity model (Section 2.1). Within a specific category and between models, there remains a large degree of variance. Most pathways exhibit a temperature overshoot which has been highlighted in several studies focusing on stringent mitigation pathways (Huntingford and Lowe, 2007; Wigley et al., 2007; Nohara et al., 2015; Rogelj et al., 2015d; Zickfeld and Herrington, 2015; Schleussner et al., 2016; Xu and Ramanathan, 2017). Only very few of the scenarios collected in the database for this report hold the average future warming projected by MAGICC below 1.5°C during the entire 21st century (Table 2.1, Figure 2.1). Most

1.5°C-consistent pathways available in the database overshoot 1.5°C around mid-century before peaking and then reducing temperatures so as to return below that level in 2100. However, because of numerous geophysical uncertainties and model dependencies (Section 2.2.1.1, Supplementary Material 2.SM.1.1), absolute temperature characteristics of the various pathway categories are more difficult to distinguish than relative features (Figure 2.1, Supplementary Material 2.SM.1.1), and actual probabilities of overshoot are imprecise. However, all lines of evidence available for temperature projections indicate a probability greater than 50% of overshooting 1.5°C by mid-century in all but the most stringent pathways currently available (Supplementary Material 2.SM.1.1, 2.SM.1.4).

Most 1.5°C-consistent pathways exhibit a peak in temperature by mid-century whereas 2°C-consistent pathways generally peak after 2050 (Supplementary Material 2.SM.1.4). The peak in median temperature in the various pathway categories occurs about ten years before reaching net zero CO_2 emissions due to strongly reduced annual CO_2 emissions and deep reductions in CH_4 emissions (Section 2.3.3). The two reduced-complexity climate models used in this assessment suggest that virtually all available 1.5°C-consistent pathways peak and then decline global mean temperature, but with varying rates of temperature decline after the peak (Figure 2.1). The estimated decadal rates of temperature change by the end of the century are smaller than the amplitude of the climate variability as assessed in AR5 (1 standard deviation of about ±0.1°C), which hence complicates the detection of a global peak and decline of warming in observations on time scales of one to two decades (Bindoff et al., 2013). In comparison, many pathways limiting warming to 2°C or higher by 2100 still have noticeable increasing trends at the end of the century, and thus imply continued warming.

By 2100, the difference between 1.5°C- and 2°C-consistent pathways becomes clearer compared to mid-century, not only for the temperature response (Figure 2.1) but also for atmospheric CO_2 concentrations. In 2100, the median CO_2 concentration in 1.5°C-consistent pathways is below 2016 levels (Le Quéré et al., 2018), whereas it remains higher by about 5–10% compared to 2016 in the 2°C-consistent pathways.

2.2.1.1   Geophysical uncertainties: non-CO_2 forcing agents

Impacts of non-CO_2 climate forcers on temperature outcomes are particularly important when evaluating stringent mitigation pathways (Weyant et al., 2006; Shindell et al., 2012; Rogelj et al., 2014b, 2015a; Samset et al., 2018). However, many uncertainties affect the role of non-CO_2 climate forcers in stringent mitigation pathways.

A first uncertainty arises from the magnitude of the radiative forcing attributed to non-CO_2 climate forcers. Figure 2.2 illustrates how, for one representative 1.5°C-consistent pathway (SSP2-1.9) (Fricko et al., 2017; Rogelj et al., 2018), the effective radiative forcings as estimated by MAGICC and FAIR can differ (see Supplementary Material 2.SM1.1 for further details). This large spread in non-CO_2 effective radiative forcings leads to considerable uncertainty in the predicted temperature response. This uncertainty ultimately affects the assessed temperature outcomes for pathway classes used in this chapter (Section 2.1) and also affects the carbon budget (Section 2.2.2). Figure 2.2 highlights

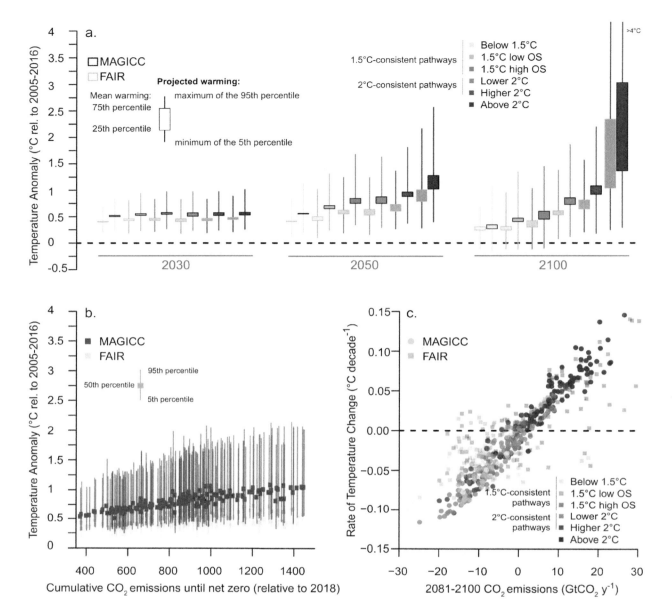

Figure 2.1 | Pathways classification overview. (a) Average global mean temperature increase relative to 2010 as projected by FAIR and MAGICC in 2030, 2050 and 2100; (b) response of peak warming to cumulative CO_2 emissions until net zero by MAGICC (red) and FAIR (blue); (c) decadal rate of average global mean temperature change from 2081 to 2100 as a function of the annual CO_2 emissions averaged over the same period as given by FAIR (transparent squares) and MAGICC (filled circles). In panel (a), horizontal lines at 0.63°C and 1.13°C are indicative of the 1.5°C and 2°C warming thresholds with the respect to 1850–1900, taking into account the assessed historical warming of 0.87°C ±0.12°C between the 1850–1900 and 2006–2015 periods (Chapter 1, Section 1.2.1). In panel (a), vertical lines illustrate both the physical and the scenario uncertainty as captured by MAGICC and FAIR and show the minimal warming of the 5th percentile of projected warming and the maximal warming of the 95th percentile of projected warming per scenario class. Boxes show the interquartile range of mean warming across scenarios, and thus represent scenario uncertainty only.

the important role of methane emissions reduction in this scenario, in agreement with the recent literature focussing on stringent mitigation pathways (Shindell et al., 2012; Rogelj et al., 2014b, 2015a; Stohl et al., 2015; Collins et al., 2018).

For mitigation pathways that aim at halting and reversing radiative forcing increase during this century, the aerosol radiative forcing is a considerable source of uncertainty (Figure 2.2) (Samset et al., 2018; Smith et al., 2018). Indeed, reductions in SO_2 (and NO_x) emissions largely associated with fossil-fuel burning are expected to reduce the cooling effects of both aerosol radiative interactions and aerosol cloud

interactions, leading to warming (Myhre et al., 2013; Samset et al., 2018). A multimodel analysis (Myhre et al., 2017) and a study based on observational constraints (Malavelle et al., 2017) largely support the AR5 best estimate and uncertainty range of aerosol forcing. The partitioning of total aerosol radiative forcing between aerosol precursor emissions is important (Ghan et al., 2013; Jones et al., 2018; Smith et al., 2018) as this affects the estimate of the mitigation potential from different sectors that have aerosol precursor emission sources. The total aerosol effective radiative forcing change in stringent mitigation pathways is expected to be dominated by the effects from the phase-out of SO_2, although the magnitude of this aerosol-warming effect depends on how much of the present-day aerosol cooling is attributable to SO_2, particularly the cooling associated with aerosol–cloud interaction (Figure 2.2). Regional differences in the linearity of aerosol–cloud interactions (Carslaw et al., 2013; Kretzschmar et al., 2017) make it difficult to separate the role of individual precursors. Precursors that are not fully mitigated will continue to affect the Earth system. If, for example, the role of nitrate aerosol cooling is at the strongest end of the assessed IPCC AR5 uncertainty range, future temperature increases may be more modest if ammonia emissions continue to rise (Hauglustaine et al., 2014).

Figure 2.2 shows that there are substantial differences in the evolution of estimated effective radiative forcing of non-CO_2 forcers between MAGICC and FAIR. These forcing differences result in MAGICC simulating a larger warming trend in the near term compared to both the FAIR model and the recent observed trends of 0.2°C per decade reported in Chapter 1 (Figure 2.1, Supplementary Material 2.SM.1.1, Chapter 1, Section 1.2.1.3). The aerosol effective forcing is stronger in MAGICC compared to either FAIR or the AR5 best estimate, though it is still well within the AR5 uncertainty range (Supplementary Material 2.SM.1.1.1). A recent revision (Etminan et al., 2016) increases the methane forcing by 25%. This revision is used in the FAIR but not in the AR5 setup of MAGICC that is applied here. Other structural differences exist in how the two models relate emissions to concentrations that contribute to differences in forcing (see Supplementary Material 2.SM.1.1.1).

Non-CO_2 climate forcers exhibit a greater geographical variation in radiative forcings than CO_2, which leads to important uncertainties in the temperature response (Myhre et al., 2013). This uncertainty increases the relative uncertainty of the temperature pathways associated with low emission scenarios compared to high emission scenarios (Clarke et al., 2014). It is also important to note that geographical patterns of temperature change and other climate responses, especially those related to precipitation, depend significantly on the forcing mechanism (Myhre et al., 2013; Shindell et al., 2015; Marvel et al., 2016; Samset et al., 2016) (see also Chapter 3, Section 3.6.2.2).

2.2.1.2   Geophysical uncertainties: climate and Earth system feedbacks

Climate sensitivity uncertainty impacts future projections as well as carbon-budget estimates (Schneider et al., 2017). AR5 assessed the equilibrium climate sensitivity (ECS) to be *likely* in the 1.5°–4.5°C range, *extremely unlikely* less than 1°C and *very unlikely* greater than 6°C. The lower bound of this estimate is lower than the range of CMIP5 models (Collins et al., 2013). The evidence for the 1.5°C lower bound on ECS in AR5 was based on analysis of energy-budget changes over the historical period. Work since AR5 has suggested that the climate sensitivity inferred from such changes has been lower than the $2 \times CO_2$ climate sensitivity for known reasons (Forster, 2016; Gregory and Andrews, 2016; Rugenstein et al., 2016; Armour, 2017; Ceppi and Gregory, 2017; Knutti et al., 2017; Proistosescu and Huybers, 2017). Both a revised interpretation of historical estimates and other lines of evidence based on analysis of climate models with the best representation of today's climate (Sherwood et al., 2014; Zhai et al., 2015; Tan et al., 2016; Brown and Caldeira, 2017; Knutti

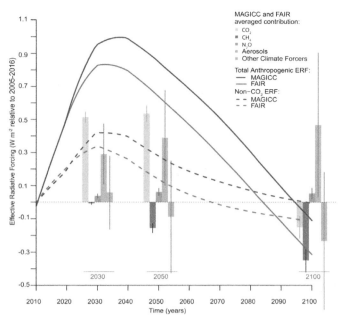

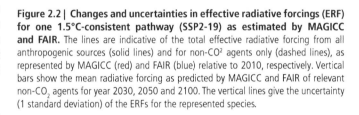

Figure 2.2 | Changes and uncertainties in effective radiative forcings (ERF) for one 1.5°C-consistent pathway (SSP2-19) as estimated by MAGICC and FAIR. The lines are indicative of the total effective radiative forcing from all anthropogenic sources (solid lines) and for non-CO^2 agents only (dashed lines), as represented by MAGICC (red) and FAIR (blue) relative to 2010, respectively. Vertical bars show the mean radiative forcing as predicted by MAGICC and FAIR of relevant non-CO_2 agents for year 2030, 2050 and 2100. The vertical lines give the uncertainty (1 standard deviation) of the ERFs for the represented species.

et al., 2017) suggest that the lower bound of ECS could be revised upwards, which would decrease the chances of limiting warming below 1.5°C in assessed pathways. However, such a reassessment has been challenged (Lewis and Curry, 2018), albeit from a single line of evidence. Nevertheless, it is premature to make a major revision to the lower bound. The evidence for a possible revision of the upper bound on ECS is less clear, with cases argued from different lines of evidence for both decreasing (Lewis and Curry, 2015, 2018; Cox et al., 2018) and increasing (Brown and Caldeira, 2017) the bound presented in the literature. The tools used in this chapter employ ECS ranges consistent with the AR5 assessment. The MAGICC ECS distribution has not been selected to explicitly reflect this but is nevertheless consistent (Rogelj et al., 2014a). The FAIR model used here to estimate carbon budgets explicitly constructs log-normal distributions of ECS and transient climate response based on a multi-parameter fit to the AR5 assessed ranges of climate sensitivity and individual historic effective radiative forcings (Smith et al., 2018) (Supplementary Material 2.SM.1.1.1).

Several feedbacks of the Earth system, involving the carbon cycle, non-CO_2 GHGs and/or aerosols, may also impact the future dynamics of the coupled carbon–climate system's response to anthropogenic emissions. These feedbacks are caused by the effects of nutrient limitation (Duce et al., 2008; Mahowald et al., 2017), ozone exposure (de Vries et al., 2017), fire emissions (Narayan et al., 2007) and changes associated with natural aerosols (Cadule et al., 2009; Scott et al., 2018). Among these Earth system feedbacks, the importance of the permafrost feedback's influence has been highlighted in recent studies. Combined evidence from both models (MacDougall et al., 2015; Burke et al., 2017; Lowe and Bernie, 2018) and field studies (like Schädel et al., 2014; Schuur et al., 2015) shows *high agreement* that permafrost thawing will release both CO_2 and CH_4 as the Earth warms, amplifying global warming. This thawing could also release N_2O (Voigt et al., 2017a, b). Field, laboratory and modelling studies estimate that the vulnerable fraction in permafrost is about 5–15% of the permafrost soil carbon (~5300–5600 $GtCO_2$ in Schuur et al., 2015) and that carbon emissions are expected to occur beyond 2100 because of system inertia and the large proportion of slowly decomposing carbon in permafrost (Schädel et al., 2014). Published model studies suggest that a large part of the carbon release to the atmosphere is in the form of CO_2 (Schädel et al., 2016), while the amount of CH4 released by permafrost thawing is estimated to be much smaller than that CO_2. Cumulative CH_4 release by 2100 under RCP2.6 ranges from 0.13 to 0.45 Gt of methane (Burke et al., 2012; Schneider von Deimling et al., 2012, 2015), with fluxes being the highest in the middle of the century because of maximum thermokarst lake extent by mid-century (Schneider von Deimling et al., 2015).

The reduced complexity climate models employed in this assessment do not take into account permafrost or non-CO_2 Earth system feedbacks, although the MAGICC model has a permafrost module that can be enabled. Taking the current climate and Earth system feedbacks understanding together, there is a possibility that these models would underestimate the longer-term future temperature response to stringent emission pathways (Section 2.2.2).

2.2.2   The Remaining 1.5°C Carbon Budget

2.2.2.1   Carbon budget estimates

Since the AR5, several approaches have been proposed to estimate carbon budgets compatible with 1.5°C or 2°C. Most of these approaches indirectly rely on the approximate linear relationship between peak global mean temperature and cumulative emissions of carbon (the transient climate response to cumulative emissions of carbon, TCRE) (Collins et al., 2013; Friedlingstein et al., 2014a; Rogelj et al., 2016b), whereas others base their estimates on equilibrium climate sensitivity (Schneider et al., 2017). The AR5 employed two approaches to determine carbon budgets. Working Group I (WGI) computed carbon budgets from 2011 onwards for various levels of warming relative to the 1861–1880 period using RCP8.5 (Meinshausen et al., 2011b; Stocker et al., 2013), whereas WGIII estimated their budgets from a set of available pathways that were assessed to have a >50% probability to exceed 1.5°C by mid-century, and return to 1.5°C or below in 2100 with greater than 66% probability (Clarke et al., 2014). These differences made AR5 WGI and WGIII carbon budgets difficult to compare as they are calculated over different time periods, are derived from a different sets of multi-gas and aerosol emission scenarios, and use different concepts of carbon budgets (exceedance for WGI, avoidance for WGIII) (Rogelj et al., 2016b; Matthews et al., 2017).

Carbon budgets can be derived from CO_2-only experiments as well as from multi-gas and aerosol scenarios. Some published estimates of carbon budgets compatible with 1.5°C or 2°C refer to budgets for CO_2-induced warming only, and hence do not take into account the contribution of non-CO_2 climate forcers (Allen et al., 2009;

Matthews et al., 2009; Zickfeld et al., 2009; IPCC, 2013a). However, because the projected changes in non-CO_2 climate forcers tend to amplify future warming, CO_2-only carbon budgets overestimate the total net cumulative carbon emissions compatible with 1.5°C or 2°C (Friedlingstein et al., 2014a; Rogelj et al., 2016b; Matthews et al., 2017; Mengis et al., 2018; Tokarska et al., 2018).

Since the AR5, many estimates of the remaining carbon budget for 1.5°C have been published (Friedlingstein et al., 2014a; MacDougall et al., 2015; Peters, 2016; Rogelj et al., 2016b, 2018; Matthews et al., 2017; Millar et al., 2017; Goodwin et al., 2018b; Kriegler et al., 2018b; Lowe and Bernie, 2018; Mengis et al., 2018; Millar and Friedlingstein, 2018; Schurer et al., 2018; Séférian et al., 2018; Tokarska and Gillett, 2018; Tokarska et al., 2018). These estimates cover a wide range as a result of differences in the models used, and of methodological choices, as well as physical uncertainties. Some estimates are exclusively model-based while others are based on observations or on a combination of both. Remaining carbon budgets limiting warming below 1.5°C or 2°C that are derived from Earth system models of intermediate complexity (MacDougall et al., 2015; Goodwin et al., 2018a), IAMs (Luderer et al., 2018; Rogelj et al., 2018), or are based on Earth-system model results (Lowe and Bernie, 2018; Séférian et al., 2018; Tokarska and Gillett, 2018) give remaining carbon budgets of the same order of magnitude as the IPCC AR5 Synthesis Report (SYR) estimates (IPCC, 2014a). This is unsurprising as similar sets of models were used for the AR5 (IPCC, 2013b). The range of variation across models stems mainly from either the inclusion or exclusion of specific Earth system feedbacks (MacDougall et al., 2015; Burke et al., 2017; Lowe and Bernie, 2018) or different budget definitions (Rogelj et al., 2018).

In contrast to the model-only estimates discussed above and employed in the AR5, this report additionally uses observations to inform its evaluation of the remaining carbon budget. Table 2.2 shows that the assessed range of remaining carbon budgets consistent with 1.5°C or 2°C is larger than the AR5 SYR estimate and is part way towards estimates constrained by recent observations (Millar et al., 2017; Goodwin et al., 2018a; Tokarska and Gillett, 2018). Figure 2.3 illustrates that the change since AR5 is, in very large part, due to the application of a more recent observed baseline to the historic temperature change and cumulative emissions; here adopting the baseline period of 2006–2015 (see Chapter 1, Section 1.2.1). AR5 SYR Figures SPM.10 and 2.3 already illustrated the discrepancy between models and observations, but did not apply this as a correction to the carbon budget because they were being used to illustrate the overall linear relationship between warming and cumulative carbon emissions in the CMIP5 models since 1870, and were not specifically designed to quantify residual carbon budgets relative to the present for ambitious temperature goals. The AR5 SYR estimate was also dependent on a subset of Earth system models illustrated in Figure 2.3 of this report. Although, as outlined below and in Table 2.2, considerably uncertainties remain, there is *high agreement* across various lines of evidence assessed in this report that the remaining carbon budget for 1.5°C or 2°C would be larger than the estimates at the time of the AR5. However, the overall remaining budget for 2100 is assessed to be smaller than that derived from the recent observational-informed estimates, as Earth system feedbacks such as permafrost thawing reduce the budget applicable to centennial scales (see Section 2.2.2.2).

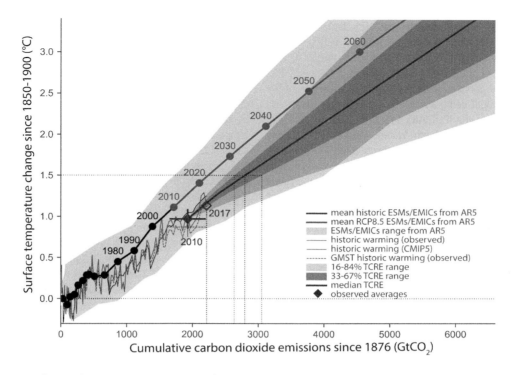

Figure 2.3 | Temperature changes from 1850–1900 versus cumulative CO₂ emissions since 1st January 1876. Solid lines with dots reproduce the globally averaged near-surface air temperature response to cumulative CO₂ emissions plus non-CO₂ forcers as assessed in Figure SPM10 of WGI AR5, except that points marked with years relate to a particular year, unlike in WGI AR5 Figure SPM.10, where each point relates to the mean over the previous decade. The AR5 data was derived from 15 Earth system models and 5 Earth system models of Intermediate Complexity for the historic observations (black) and RCP8.5 scenario (red), and the red shaded plume shows the range across the models as presented in the AR5. The purple shaded plume and the line are indicative of the temperature response to cumulative CO₂ emissions and non-CO₂ warming adopted in this report. The non-CO₂ warming contribution is averaged from the MAGICC and FAIR models, and the purple shaded range assumes the AR5 WGI TCRE distribution (Supplementary Material 2.SM.1.1.2). The 2010 observation of surface temperature change (0.97°C based on 2006–2015 mean compared to 1850–1900, Chapter 1, Section 1.2.1) and cumulative carbon dioxide emissions from 1876 to the end of 2010 of 1,930 GtCO₂ (Le Quéré et al., 2018) is shown as a filled purple diamond. The value for 2017 based on the latest cumulative carbon emissions up to the end of 2017 of 2,220 GtCO₂ (Version 1.3 accessed 22 May 2018) and a surface temperature anomaly of 1.1°C based on an assumed temperature increase of 0.2°C per decade is shown as a hollow purple diamond. The thin blue line shows annual observations, with CO₂ emissions from Le Quéré et al. (2018) and estimated globally averaged near-surface temperature from scaling the incomplete coverage and blended HadCRUT4 dataset in Chapter 1. The thin black line shows the CMIP5 multimodel mean estimate with CO₂ emissions also from (Le Quéré et al., 2018). The thin black line shows the GMST historic temperature trends from Chapter 1, which give lower temperature changes up to 2006–2015 of 0.87°C and would lead to a larger remaining carbon budget. The dotted black lines illustrate the remaining carbon budget estimates for 1.5°C given in Table 2.2. Note these remaining budgets exclude possible Earth system feedbacks that could reduce the budget, such as CO₂ and CH₄ release from permafrost thawing and tropical wetlands (see Section 2.2.2.2).

2.2.2.2 CO₂ and non-CO₂ contributions to the remaining carbon budget

A remaining carbon budget can be estimated from calculating the amount of CO₂ emissions consistent (given a certain value of TCRE) with an allowable additional amount of warming. Here, the allowable warming is the 1.5°C warming threshold minus the current warming taken as the 2006–2015 average, with a further amount removed to account for the estimated non-CO₂ temperature contribution to the remaining warming (Peters, 2016; Rogelj et al., 2016b). This assessment uses the TCRE range from AR5 WGI (Collins et al., 2013) supported by estimates of non-CO₂ contributions that are based on published methods and integrated pathways (Friedlingstein et al., 2014a; Allen et al., 2016, 2018; Peters, 2016; Smith et al., 2018). Table 2.2 and Figure 2.3 show the assessed remaining carbon budgets and key uncertainties for a set of additional warming levels relative to the 2006–2015 period (see Supplementary Material 2.SM.1.1.2 for details). With an assessed historical warming of 0.87°C ± 0.12°C from 1850–1900 to 2006–2015 (Chapter 1, Section 1.2.1), 0.63°C of additional warming would be

approximately consistent with a global mean temperature increase of 1.5°C relative to pre-industrial levels. For this level of additional warming, remaining carbon budgets have been estimated (Table 2.2, Supplementary Material 2.SM.1.1.2).

The remaining carbon budget calculation presented in the Table 2.2 and illustrated in Figure 2.3 does not consider additional Earth system feedbacks such as permafrost thawing. These are uncertain but estimated to reduce the remaining carbon budget by an order of magnitude of about 100 GtCO₂ and more thereafter. Accounting for such feedbacks would make the carbon budget more applicable for 2100 temperature targets, but would also increase uncertainty (Table 2.2 and see below). Excluding such feedbacks, the assessed range for the remaining carbon budget is estimated to be 840, 580, and 420 GtCO₂ for the 33rd, 50th and, 67th percentile of TCRE, respectively, with a median non-CO₂ warming contribution and starting from 1 January 2018 onward. Consistent with the approach used in the IPCC Fifth Assessment Report (IPCC, 2013b), the latter estimates use global near-surface air temperatures both over the ocean and over land to estimate global surface temperature change since pre-industrial. The global warming from the pre-industrial period until the 2006–2015 reference period is estimated to amount to 0.97°C with an uncertainty range of about ±0.1°C (see Chapter 1, Section 1.2.1). Three methodological improvements lead to these estimates of the remaining carbon budget being about 300 $GtCO_2$ larger than those reported in Table 2.2 of the IPCC AR5 SYR (IPCC, 2014a) (*medium confidence*). The AR5 used 15 Earth System Models (ESM) and 5 Earth-system Models of Intermediate Complexity (EMIC) to derive an estimate of the remaining carbon budget. Their approach hence made implicit assumptions about the level of warming to date, the future contribution of non-CO_2 emissions, and the temperature response to CO_2 (TCRE). In this report, each of these aspects are considered explicitly. When estimating global warming until the 2006–2015 reference period as a blend of near-surface air temperature over land and sea-ice regions, and sea-surface temperature over open ocean, by averaging the four global mean surface temperature time series listed in Chapter 1 Section 1.2.1, the global warming would amount to 0.87°C ±0.1°C. Using the latter estimate of historical warming and projecting global warming using global near-surface air temperatures from model projections leads to remaining carbon budgets for limiting global warming to 1.5°C of 1080, 770, and 570 $GtCO_2$ for the 33rd, 50th, and 67th percentile of TCRE, respectively. Note that future research and ongoing observations over the next years will provide a better indication as to how the 2006–2015 base period compares with the long-term trends and might affect the budget estimates. Similarly, improved understanding in Earth system feedbacks would result in a better quantification of their impacts on remaining carbon budgets for 1.5°C and 2°C.

After TCRE uncertainty, a major additional source of uncertainty is the magnitude of non-CO_2 forcing and its contribution to the temperature change between the present day and the time of peak warming. Integrated emissions pathways can be used to ensure consistency between CO_2 and non-CO_2 emissions (Bowerman et al., 2013; Collins et al., 2013; Clarke et al., 2014; Rogelj et al., 2014b, 2015a; Tokarska et al., 2018). Friedlingstein et al. (2014a) used pathways with limited to no climate mitigation to find a variation due to non-CO_2 contributions of about ±33% for a 2°C carbon budget. Rogelj et al. (2016b) showed no particular bias in non-CO_2 radiative forcing or warming at the time of exceedance of 2°C or at peak warming between scenarios with increasing emissions and strongly mitigated scenarios (consistent with Stocker et al., 2013). However, clear differences of the non-CO_2 warming contribution at the time of deriving a 2°C-consistent carbon budget were reported for the four RCPs. Although the spread in non-CO_2 forcing across scenarios can be smaller in absolute terms at lower levels of cumulative emissions, it can be larger in relative terms compared to the remaining carbon budget (Stocker et al., 2013; Friedlingstein et al., 2014a; Rogelj et al., 2016b). Tokarska and Gillett (2018) find no statistically significant differences in 1.5°C-consistent cumulative emissions budgets when calculated for different RCPs from consistent sets of CMIP5 simulations.

The mitigation pathways assessed in this report indicate that emissions of non-CO_2 forcers contribute an average additional warming of around 0.15°C relative to 2006–2015 at the time of net zero CO_2 emissions, reducing the remaining carbon budget by roughly 320 $GtCO_2$. This

arises from a weakening of aerosol cooling and continued emissions of non-CO_2 GHGs (Sections 2.2.1, 2.3.3). This non-CO_2 contribution at the time of net zero CO_2 emissions varies by about ±0.1°C across scenarios, resulting in a carbon budget uncertainty of about ±250 $GtCO_2$, and takes into account marked reductions in methane emissions (Section 2.3.3). If these reductions are not achieved, remaining carbon budgets are further reduced. Uncertainties in the non-CO_2 forcing and temperature response are asymmetric and can influence the remaining carbon budget by −400 to +200 $GtCO_2$, with the uncertainty in aerosol radiative forcing being the largest contributing factor (Table 2.2). The MAGICC and FAIR models in their respective parameter setups and model versions used to assess the non-CO_2 warming contribution give noticeable different non-CO_2 effective radiative forcing and warming for the same scenarios while both being within plausible ranges of future response (Figure 2.2 and Supplementary Material 2.SM.1.1, 2.SM.1.2). For this assessment, it is premature to assess the accuracy of their results, so it is assumed that both are equally representative of possible futures. Their non-CO_2 warming estimates are therefore averaged for the carbon budget assessment and their differences used to guide the uncertainty assessment of the role of non-CO_2 forcers. Nevertheless, the findings are robust enough to give *high confidence* that the changing emissions of non-CO_2 forcers (particularly the reduction in cooling aerosol precursors) cause additional near-term warming and reduce the remaining carbon budget compared to the CO_2-only budget.

TCRE uncertainty directly impacts carbon budget estimates (Peters, 2016; Matthews et al., 2017; Millar and Friedlingstein, 2018). Based on multiple lines of evidence, AR5 WGI assessed a *likely* range for TCRE of 0.2°–0.7°C per 1000 $GtCO_2$ (Collins et al., 2013). The TCRE of the CMIP5 Earth system models ranges from 0.23°C to 0.66°C per 1000 $GtCO_2$ (Gillett et al., 2013). At the same time, studies using observational constraints find best estimates of TCRE of 0.35°–0.41°C per 1000 $GtCO_2$ (Matthews et al., 2009; Gillett et al., 2013; Tachiiri et al., 2015; Millar and Friedlingstein, 2018). This assessment continues to use the assessed AR5 TCRE range under the working assumption that TCRE is normally distributed (Stocker et al., 2013). Observation-based estimates have reported log-normal distributions of TCRE (Millar and Friedlingstein, 2018). Assuming a log-normal instead of normal distribution of the assessed AR5 TCRE range would result in about a 200 $GtCO_2$ increase for the median budget estimates but only about half at the 67th percentile, while historical temperature uncertainty and uncertainty in recent emissions contribute ±150 and ±50 $GtCO_2$ to the uncertainty, respectively (Table 2.2).

Calculating carbon budgets from the TCRE requires the assumption that the instantaneous warming in response to cumulative CO_2 emissions equals the long-term warming or, equivalently, that the residual warming after CO_2 emissions cease is negligible. The magnitude of this residual warming, referred to as the zero-emission commitment, ranges from slightly negative (i.e., a slight cooling) to slightly positive for CO_2 emissions up to present-day (Chapter 1, Section 1.2.4) (Lowe et al., 2009; Frölicher and Joos, 2010; Gillett et al., 2011; Matthews and Zickfeld, 2012). The delayed temperature change from a pulse CO_2 emission introduces uncertainties in emission budgets, which have not been quantified in the literature for budgets consistent with limiting warming to 1.5°C. As a consequence, this uncertainty does not affect our carbon budget estimates directly but it is included as an additional factor in the assessed Earth system feedback uncertainty (as detailed below) of roughly 100 $GtCO_2$ on decadal time scales presented in Table 2.2.

Remaining carbon budgets are further influenced by Earth system feedbacks not accounted for in CMIP5 models, such as the permafrost carbon feedback (Friedlingstein et al., 2014b; MacDougall et al., 2015; Burke et al., 2017; Lowe and Bernie, 2018), and their influence on the TCRE. Lowe and Bernie (2018) used a simple climate sensitivity scaling approach to estimate that Earth system feedbacks (such as CO_2 released by permafrost thawing or methane released by wetlands) could reduce carbon budgets for 1.5°C and 2°C by roughly 100 $GtCO_2$ on centennial time scales. Their findings are based on an older understanding of Earth system feedbacks (Arneth et al., 2010). This estimate is broadly supported by more recent analysis of individual feedbacks. Schädel et al. (2014) suggest an upper bound of 24.4 PgC (90 $GtCO_2$) emitted from carbon release from permafrost over the next forty years for a RCP4.5 scenario. Burke et al. (2017) use a single model to estimate permafrost emissions between 0.3 and 0.6 $GtCO_2$ y^{-1} from the point of 1.5°C stabilization, which would reduce the budget by around 20 $GtCO_2$ by 2100. Comyn-Platt et al. (2018) include carbon and methane emissions from permafrost and wetlands and suggest the 1.5°C remaining carbon budget is reduced by 116 $GtCO_2$. Additionally, Mahowald et al. (2017) find there is possibility of 0.5–1.5 $GtCO_2$ y^{-1} being released from aerosol-biogeochemistry changes if aerosol emissions cease. In summary, these additional Earth system feedbacks taken together are assessed to reduce the remaining carbon budget applicable to 2100 by an order of magnitude of 100 $GtCO_2$, compared to the budgets based on the assumption of a constant TCRE presented in Table 2.2 (*limited evidence, medium agreement*), leading to overall *medium confidence* in their assessed impact. After 2100, the impact of additional Earth system feedbacks is expected to further reduce the remaining carbon budget (*medium confidence*).

The uncertainties presented in Table 2.2 cannot be formally combined, but current understanding of the assessed geophysical uncertainties suggests at least a ±50% possible variation for remaining carbon budgets for 1.5°C-consistent pathways. By the end of 2017, anthropogenic CO_2 emissions since the pre-industrial period are estimated to have amounted to approximately 2200 ±320 $GtCO_2$ (*medium confidence*) (Le Quéré et al., 2018). When put in the context of year-2017 CO_2 emissions (about 42 $GtCO_2$ yr^{-1}, ±3 $GtCO_2$ yr^{-1}, *high confidence*) (Le Quéré et al., 2018), a remaining carbon budget of 580 $GtCO_2$ (420 $GtCO_2$) suggests meeting net zero global CO_2 emissions in about 30 years (20 years) following a linear decline starting from 2018 (rounded to the nearest five years), with a variation of ±15–20 years due to the geophysical uncertainties mentioned above (*high confidence*).

The remaining carbon budgets assessed in this section are consistent with limiting peak warming to the indicated levels of additional warming. However, if these budgets are exceeded and the use of CDR (see Sections 2.3 and 2.4) is envisaged to return cumulative CO_2 emissions to within the carbon budget at a later point in time, additional uncertainties apply because the TCRE is different under increasing and decreasing atmospheric CO_2 concentrations due to

ocean thermal and carbon cycle inertia (Herrington and Zickfeld, 2014; Krasting et al., 2014; Zickfeld et al., 2016). This asymmetrical behaviour makes carbon budgets path-dependent in the case of a budget and/or temperature overshoot (MacDougall et al., 2015). Although potentially large for scenarios with large overshoot (MacDougall et al., 2015), this path-dependence of carbon budgets has not been well quantified for 1.5°C- and 2°C-consistent scenarios and as such remains an important knowledge gap. This assessment does not explicitly account for path dependence but takes it into consideration for its overall confidence assessment.

This assessment finds a larger remaining budget from the 2006–2015 base period than the 1.5°C and 2°C remaining budgets inferred from AR5 from the start of 2011, which were approximately 1000 $GtCO_2$ for the 2°C (66% of model simulations) and approximately 400 $GtCO_2$ for the 1.5°C budget (66% of model simulations). In contrast, this assessment finds approximately 1600 $GtCO_2$ for the 2°C (66th TCRE percentile) and approximately 860 $GtCO_2$ for the 1.5°C budget (66th TCRE percentile) from 2011. However, these budgets are not directly equivalent as AR5 reported budgets for fractions of CMIP5 simulations and other lines of evidence, while this report uses the assessed range of TCRE and an assessment of the non-CO_2 contribution at net zero CO_2 emissions to provide remaining carbon budget estimates at various percentiles of TCRE. Furthermore, AR5 did not specify remaining budgets to carbon neutrality as we do here, but budgets until the time the temperature limit of interest was reached, assuming negligible zero emission commitment and taking into account the non-CO_2 forcing at that point in time.

In summary, although robust physical understanding underpins the carbon budget concept, relative uncertainties become larger as a specific temperature limit is approached. For the budget, applicable to the mid-century, the main uncertainties relate to the TCRE, non-CO_2 emissions, radiative forcing and response. For 2100, uncertain Earth system feedbacks such as permafrost thawing would further reduce the available budget. The remaining budget is also conditional upon the choice of baseline, which is affected by uncertainties in both historical emissions, and in deriving the estimate of globally averaged human-induced warming. As a result, only *medium confidence* can be assigned to the assessed remaining budget values for 1.5°C and 2.0°C and their uncertainty.

Table 2.2 | **The assessed remaining carbon budget and its uncertainties.** Shaded blue horizontal bands illustrate the uncertainty in historical temperature increase from the 1850–1900 base period until the 2006–2015 period as estimated from global near-surface air temperatures, which impacts the additional warming until a specific temperature limit like 1.5°C or 2°C relative to the 1850–1900 period. Shaded grey cells indicate values for when historical temperature increase is estimated from a blend of near-surface air temperatures over land and sea ice regions and sea-surface temperatures over oceans.

Additional Warming since 2006–2015 [°C]*(1)	Approximate Warming since 1850–1900 [°C]*(1)	Remaining Carbon Budget (Excluding Additional Earth System Feedbacks*(5)) [GtCO₂ from 1.1.2018]*(2)			Key Uncertainties and Variations*(4)						
		Percentiles of TCRE *(3)			Earth System Feedbacks *(5)	Non-CO₂ scenario variation *(6)	Non-CO₂ forcing and response uncertainty	TCRE distribution uncertainty *(7)	Historical temperature uncertainty *(1)	Recent emissions uncertainty *(8)	
		33rd	50th	67th	[GtCO₂]	[GtCO₂]	[GtCO₂]	[GtCO₂]	[GtCO₂]	[GtCO₂]	
0.3		290	160	80	Budgets on the left are reduced by about –100 on centennial time scales	±250	–400 to +200	+100 to +200	±250	±20	
0.4		530	350	230							
0.5		770	530	380							
0.53	~1.5°C	840	580	420							
0.6		1010	710	530							
0.63		1080	770	570							
0.7		1240	900	680							
0.78		1440	1040	800							
0.8		1480	1080	830							
0.9		1720	1260	980							
1		1960	1450	1130							
1.03	~2°C	2030	1500	1170							
1.1		2200	1630	1280							
1.13		2270	1690	1320							
1.2		2440	1820	1430							

Notes:

*(1) Chapter 1 has assessed historical warming between the 1850–1900 and 2006–2015 periods to be 0.87°C with a ±0.12°C *likely* (1-standard deviation) range, and global near-surface air temperature to be 0.97°C. The temperature changes from the 2006–2015 period are expressed in changes of global near-surface air temperature.

*(2) Historical CO₂ emissions since the middle of the 1850–1900 historical base period (mid-1875) are estimated at 1940 GtCO₂ (1640–2240 GtCO2, one standard deviation range) until end 2010. Since 1 January 2011, an additional 290 GtCO₂ (270–310 GtCO₂, one sigma range) has been emitted until the end of 2017 (Le Quéré et al., 2018).

*(3) TCRE: transient climate response to cumulative emissions of carbon, assessed by AR5 to fall *likely* between 0.8–2.5°C/1000 PgC (Collins et al., 2013), considering a normal distribution consistent with AR5 (Stocker et al., 2013). Values are rounded to the nearest 10 GtCO₂.

*(4) Focussing on the impact of various key uncertainties on median budgets for 0.53°C of additional warming.

*(5) Earth system feedbacks include CO₂ released by permafrost thawing or methane released by wetlands, see main text.

*(6) Variations due to different scenario assumptions related to the future evolution of non-CO₂ emissions.

*(7) The distribution of TCRE is not precisely defined. Here the influence of assuming a lognormal instead of a normal distribution shown.

*(8) Historical emissions uncertainty reflects the uncertainty in historical emissions since 1 January 2011.

2.3   Overview of 1.5°C Mitigation Pathways

Limiting global mean temperature increase at any level requires global CO₂ emissions to become net zero at some point in the future (Zickfeld et al., 2009; Collins et al., 2013). At the same time, limiting the residual warming of short-lived non-CO₂ emissions can be achieved by reducing their annual emissions as much as possible (Section 2.2, Cross-Chapter Box 2 in Chapter 1). This would require large-scale transformations of the global energy–agriculture–land-economy system, affecting the way in which energy is produced, agricultural systems are organized, and food, energy and materials are consumed (Clarke et al., 2014). This section assesses key properties of pathways consistent with limiting global mean temperature to 1.5°C relative to pre-industrial levels, including their underlying assumptions and variations.

Since the AR5, an extensive body of literature has appeared on integrated pathways consistent with 1.5°C (Section 2.1) (Rogelj et al., 2015b, 2018; Akimoto et al., 2017; Löffler et al., 2017; Marcucci et al., 2017; Su et al., 2017; Bauer et al., 2018; Bertram et al., 2018; Grubler et al., 2018; Holz et al., 2018b; Kriegler et al., 2018a; Liu et al., 2018; Luderer et al., 2018; Strefler et al., 2018a; van Vuuren et al., 2018; Vrontisi et al., 2018; Zhang et al., 2018). These pathways have global coverage and represent all GHG-emitting sectors and their interactions. Such integrated pathways allow the exploration of the whole-system transformation, and hence provide the context in which the detailed sectoral transformations assessed in Section 2.4 of this chapter are taking place.

The overwhelming majority of published integrated pathways have been developed by global IAMs that represent key societal systems and their interactions, like the energy system, agriculture and land use, and the economy (see Section 6.2 in Clarke et al., 2014). Very often these models also include interactions with a representation of the geophysical system, for example, by including spatially explicit land models or carbon cycle and climate models. The complex features of these subsystems are approximated and simplified in these models. IAMs are briefly introduced in Section 2.1 and important knowledge gaps identified in Section 2.6. An overview to the use, scope and limitations of IAMs is provided in Supplementary Material 2.SM.1.2.

The pathway literature is assessed in two ways in this section. First, various insights on specific questions reported by studies can be assessed to identify robust or divergent findings. Second, the combined body of scenarios can be assessed to identify salient features of pathways in line with a specific climate goal across a wide range of models. The latter can be achieved by assessing pathways available in the database to this assessment (Section 2.1, Supplementary Material 2.SM.1.2–4). The ensemble of scenarios available to this assessment is an ensemble of opportunity: it is a collection of scenarios from a diverse set of studies that was not developed with a common set of questions and a statistical analysis of outcomes in mind. This means that ranges can be useful to identify robust and sensitive features across available scenarios and contributing modelling frameworks, but do not lend themselves to a statistical interpretation. To understand the reasons underlying the ranges, an assessment of the underlying scenarios and studies is required. To this end, this section highlights illustrative pathway archetypes that help to clarify the variation in assessed ranges for 1.5°C-consistent pathways.

2.3.1   Range of Assumptions Underlying 1.5°C Pathways

Earlier assessments have highlighted that there is no single pathway to achieve a specific climate objective (e.g., Clarke et al., 2014). Pathways depend on the underlying development processes, and societal choices, which affect the drivers of projected future baseline emissions. Furthermore, societal choices also affect climate change solutions in pathways, like the technologies that are deployed, the scale at which they are deployed, or whether solutions are globally coordinated. A key finding is that 1.5°C-consistent pathways could be identified under a considerable range of assumptions in model studies despite the tightness of the 1.5°C emissions budget (Figures 2.4, 2.5) (Rogelj et al., 2018).

The AR5 provided an overview of how differences in model structure and assumptions can influence the outcome of transformation pathways (Section 6.2 in Clarke et al., 2014, as well as Table A.II.14 in Krey et al., 2014b) and this was further explored by the modelling community in recent years with regard to, e.g., socio-economic drivers (Kriegler et al., 2016; Marangoni et al., 2017; Riahi et al., 2017), technology assumptions (Bosetti et al., 2015; Creutzig et al., 2017; Pietzcker et al., 2017), and behavioural factors (van Sluisveld et al., 2016; McCollum et al., 2017).

2.3.1.1   Socio-economic drivers and the demand for energy and land in 1.5°C pathways

There is deep uncertainty about the ways humankind will use energy and land in the 21st century. These ways are intricately linked to future population levels, secular trends in economic growth and income convergence, behavioural change and technological progress. These dimensions have been recently explored in the context of the SSPs (Kriegler et al., 2012; O'Neill et al., 2014), which provide narratives (O'Neill et al., 2017) and quantifications (Crespo Cuaresma, 2017; Dellink et al., 2017; KC and Lutz, 2017; Leimbach et al., 2017; Riahi et al., 2017) of different world futures across which scenario dimensions are varied to explore differential challenges to adaptation and mitigation (Cross-Chapter Box 1 in Chapter 1). This framework is increasingly adopted by IAMs to systematically explore the impact of socio-economic assumptions on mitigation pathways (Riahi et al., 2017), including 1.5°C-consistent pathways (Rogelj et al., 2018). The narratives describe five worlds (SSP1–5) with different socio-economic predispositions to mitigate and adapt to climate change (Table 2.3). As a result, population and economic growth projections can vary strongly across integrated scenarios, including available 1.5°C-consistent pathways (Figure 2.4). For example, based on alternative future fertility, mortality, migration and educational assumptions, population projections vary between 8.5 and 10.0 billion people by 2050 and between 6.9 and 12.6 billion people by 2100 across the SSPs. An important factor for these differences is future female educational attainment, with higher attainment leading to lower fertility rates and therefore decreased population growth up to a level of 1 billion people by 2050 (Lutz and KC, 2011; Snopkowski et al., 2016; KC and Lutz, 2017). Consistent with population development, GDP per capita also varies strongly in SSP baselines, ranging from about 20 to more than 50 thousand USD2010 per capita in 2050 (in purchasing power parity values, PPP), in part driven by assumptions on human development, technological progress and development convergence between and within regions (Crespo Cuaresma, 2017; Dellink et al., 2017; Leimbach et al., 2017). Importantly, none of the GDP projections in the mitigation pathway literature assessed in this chapter included the feedback of climate damages on economic growth (Hsiang et al., 2017).

Baseline projections for energy-related GHG emissions are sensitive to economic growth assumptions, while baseline projections for land-use emissions are more directly affected by population growth (assuming unchanged land productivity and per capita demand for agricultural products) (Kriegler et al., 2016). SSP-based modelling studies of mitigation pathways have identified high challenges to mitigation for worlds with a focus on domestic issues and regional security combined with high population growth (SSP3), and for worlds with rapidly growing resource and fossil-fuel intensive consumption (SSP5) (Riahi et al., 2017). No model could identify a 2°C-consistent pathway for SSP3, and high mitigation costs were found for SSP5. This picture translates to 1.5°C-consistent pathways that have to remain within even tighter emissions constraints (Rogelj et al., 2018). No model found a 1.5°C-consistent pathway for SSP3 and some models could not identify 1.5°C-consistent pathways for SSP5 (2 of 4 models, compared to 1 of 4 models for 2°C-consistent pathways). The modelling analysis also found that the effective control of land-use emissions becomes even more critical in 1.5°C-consistent pathways. Due to high inequality levels in SSP4, land use can be less well managed. This caused 2 of 3 models to no longer find an SSP4-based 1.5°C-consistent pathway even though they identified SSP4-based 2°C-consistent pathways at relatively moderate mitigation costs (Riahi et al., 2017). Rogelj et al. (2018) further reported that all six participating models identified

Table 2.3 | Key Characteristics of the Five Shared Socio-Economic Pathways (SSPs) (O'Neill et al., 2017).

Socio-Economic Challenges to Mitigation	Socio-Economic Challenges to Adaptation		
	Low	Medium	High
High	**SSP5: Fossil-fuelled development** • low population • very high economic growth per capita • high human development • high technological progress • ample fossil fuel resources • very resource intensive lifestyles • high energy and food demand per capita • economic convergence and global cooperation		**SSP3: Regional rivalry** • high population • low economic growth per capita • low human development • low technological progress • resource-intensive lifestyles • resource-constrained energy and food demand per capita • focus on regional food and energy security • regionalization and lack of global cooperation
Medium		**SSP2: Middle of the road** • medium population • medium and uneven economic growth • medium and uneven human development • medium and uneven technological progress • resource-intensive lifestyles • medium and uneven energy and food demand per capita • limited global cooperation and economic convergence	
Low	**SSP1: Sustainable development** • low population • high economic growth per capita • high human development • high technological progress • environmentally oriented technological and behavioural change • resource-efficient lifestyles • low energy and food demand per capita • economic convergence and global cooperation		**SSP4: Inequality** • Medium to high population • Unequal low to medium economic growth per capita • Unequal low to medium human development • unequal technological progress: high in globalized high-tech sectors, slow in domestic sectors • unequal lifestyles and energy /food consumption: resource intensity depending on income • Globally connected elite, disconnected domestic work forces

1.5°C-consistent pathways in a sustainability oriented world (SSP1) and four of six models found 1.5°C-consistent pathways for middle-of-the-road developments (SSP2). These results show that 1.5°C-consistent pathways can be identified under a broad range of assumptions, but that lack of global cooperation (SSP3), high inequality (SSP4) and/or high population growth (SSP3) that limit the ability to control land use emissions, and rapidly growing resource-intensive consumption (SSP5) are key impediments.

Figure 2.4 compares the range of underlying socio-economic developments as well as energy and food demand in available 1.5°C-consistent pathways with the full set of published scenarios that were submitted to this assessment. While 1.5°C-consistent pathways broadly cover the full range of population and economic growth developments (except for the high population development in SSP3-based scenarios), they tend to cluster on the lower end for energy and food demand. They still encompass, however, a wide range of developments from decreasing to increasing demand levels relative to today. For the purpose of this assessment, a set of four illustrative 1.5°C-consistent pathway archetypes were selected to show the variety of underlying assumptions and characteristics (Figure 2.4). They comprise three 1.5°C-consistent pathways based on the SSPs (Rogelj et al., 2018): a sustainability oriented scenario (S1 based on SSP1) developed with the AIM model (Fujimori, 2017), a fossil-fuel intensive

and high energy demand scenario (S5, based on SSP5) developed with the REMIND-MAgPIE model (Kriegler et al., 2017), and a middle-of-the-road scenario (S2, based on SSP2) developed with the MESSAGE-GLOBIOM model (Fricko et al., 2017). In addition, we include a scenario with low energy demand (LED) (Grubler et al., 2018), which reflects recent literature with a stronger focus on demand-side measures (Bertram et al., 2018; Grubler et al., 2018; Liu et al., 2018; van Vuuren et al., 2018). Pathways LED, S1, S2, and S5 are referred to as P1, P2, P3, and P4 in the Summary for Policymakers.

2.3.1.2   Mitigation options in 1.5°C pathways

In the context of 1.5°C pathways, the portfolio of mitigation options available to the model becomes an increasingly important factor. IAMs include a wide variety of mitigation options, as well as measures that achieve CDR from the atmosphere (Krey et al., 2014a, b) (see Chapter 4, Section 4.3 for a broad assessment of available mitigation measures). For the purpose of this assessment, we elicited technology availability in models that submitted scenarios to the database as summarized in Supplementary Material 2.SM.1.2, where a detailed picture of the technology variety underlying available 1.5°C-consistent pathways is provided. Modelling choices on whether a particular mitigation measure is included are influenced by an assessment of its global mitigation potential, the availability of data and literature describing

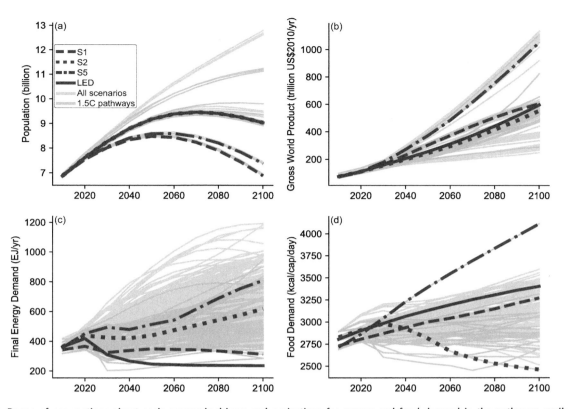

Figure 2.4 | Range of assumptions about socio-economic drivers and projections for energy and food demand in the pathways available to this assessment. 1.5°C-consistent pathways are blue, other pathways grey. Trajectories for the illustrative 1.5°C-consistent archetypes used in this Chapter (LED, S1, S2, S5; referred to as P1, P2, P3, and P4 in the Summary for Policymakers.) are highlighted. S1 is a sustainability oriented scenario, S2 is a middle-of-the-road scenario, and S5 is a fossil-fuel intensive and high energy demand scenario. LED is a scenario with particularly low energy demand. Population assumptions in S2 and LED are identical. Panels show (a) world population, (b) gross world product in purchasing power parity values, (c) final energy demand, and (d) food demand.

its techno-economic characteristics and future prospects, and the computational challenge of representing the measure, e.g., in terms of required spatio-temporal and process detail.

This elicitation (Supplementary Material 2.SM.1.2) confirms that IAMs cover most supply-side mitigation options on the process level, while many demand-side options are treated as part of underlying assumptions, which can be varied (Clarke et al., 2014). In recent years, there has been increasing attention on improving the modelling of integrating variable renewable energy into the power system (Creutzig et al., 2017; Luderer et al., 2017; Pietzcker et al., 2017) and of behavioural change and other factors influencing future demand for energy and food (van Sluisveld et al., 2016; McCollum et al., 2017; Weindl et al., 2017), including in the context of 1.5°C-consistent pathways (Grubler et al., 2018; van Vuuren et al., 2018). The literature on the many diverse CDR options only recently started to develop strongly (Minx et al., 2017) (see Chapter 4, Section 4.3.7 for a detailed assessment), and hence these options are only partially included in IAM analyses. IAMs mostly incorporate afforestation and bioenergy with carbon capture and storage (BECCS) and only in few cases also include direct air capture with CCS (DACCS) (Chen and Tavoni, 2013; Marcucci et al., 2017; Strefler et al., 2018b).

Several studies have either directly or indirectly explored the dependence of 1.5°C-consistent pathways on specific (sets of) mitigation and CDR technologies (Bauer et al., 2018; Grubler et al.,

2018; Holz et al., 2018b; Kriegler et al., 2018a; Liu et al., 2018; Rogelj et al., 2018; Strefler et al., 2018b; van Vuuren et al., 2018). However, there are a few potentially disruptive technologies that are typically not yet well covered in IAMs and that have the potential to alter the shape of mitigation pathways beyond the ranges in the IAM-based literature. Those are also included in Supplementary Material 2.SM.1.2. The configuration of carbon-neutral energy systems projected in mitigation pathways can vary widely, but they all share a substantial reliance on bioenergy under the assumption of effective land-use emissions control. There are other configurations with less reliance on bioenergy that are not yet comprehensively covered by global mitigation pathway modelling. One approach is to dramatically reduce and electrify energy demand for transportation and manufacturing to levels that make residual non-electric fuel use negligible or replaceable by limited amounts of electrolytic hydrogen. Such an approach is presented in a first-of-its kind low-energy-demand scenario (Grubler et al., 2018) which is part of this assessment. Other approaches rely less on energy demand reductions, but employ cheap renewable electricity to push the boundaries of electrification in the industry and transport sectors (Breyer et al., 2017; Jacobson, 2017). In addition, these approaches deploy renewable-based Power-2-X (read: Power to "x") technologies to substitute residual fossil-fuel use (Brynolf et al., 2018). An important element of carbon-neutral Power-2-X applications is the combination of hydrogen generated from renewable electricity and CO_2 captured from the atmosphere (Zeman and Keith, 2008). Alternatively, algae are considered as a bioenergy source with more limited implications for land use and agricultural systems than energy crops (Williams and Laurens, 2010; Walsh et al., 2016; Greene et al., 2017).

Furthermore, a range of measures could radically reduce agricultural and land-use emissions and are not yet well-covered in IAM modelling. This includes plant-based proteins (Joshi and Kumar, 2015) and cultured meat (Post, 2012) with the potential to substitute for livestock products at much lower GHG footprints (Tuomisto and Teixeira de Mattos, 2011). Large-scale use of synthetic or algae-based proteins for animal feed could free pasture land for other uses (Madeira et al., 2017; Pikaar et al., 2018). Novel technologies such as methanogen inhibitors and vaccines (Wedlock et al., 2013; Hristov et al., 2015; Herrero et al., 2016; Subharat et al., 2016) as well as synthetic and biological nitrification inhibitors (Subbarao et al., 2013; Di and Cameron, 2016) could substantially reduce future non-CO_2 emissions from agriculture if commercialized successfully. Enhancing carbon sequestration in soils (Paustian et al., 2016; Frank et al., 2017; Zomer et al., 2017) can provide the dual benefit of CDR and improved soil quality. A range of conservation, restoration and land management options can also increase terrestrial carbon uptake (Griscom et al., 2017). In addition, the literature discusses CDR measures to permanently sequester atmospheric carbon in rocks (mineralization and enhanced weathering, see Chapter 4, Section 4.3.7) as well as carbon capture and usage in long-lived products like plastics and carbon fibres (Mazzotti et al., 2005; Hartmann et al., 2013). Progress in the understanding of the technical viability, economics and sustainability of these ways to achieve and maintain carbon neutral energy and land use can affect the characteristics, costs and feasibility of 1.5°C-consistent pathways significantly.

2.3.1.3   Policy assumptions in 1.5°C pathways

Besides assumptions related to socio-economic drivers and mitigation technology, scenarios are also subject to assumptions about the mitigation policies that can be put in place. Mitigation policies can either be applied immediately in scenarios or follow staged or delayed approaches. Policies can span many sectors (e.g., economy-wide carbon pricing), or policies can be applicable to specific sectors only (like the energy sector) with other sectors (e.g., the agricultural or the land-use sector) treated differently. These variations can have an important impact on the ability of models to generate scenarios compatible with stringent climate targets like 1.5°C (Luderer et al., 2013; Rogelj et al., 2013b; Bertram et al., 2015b; Kriegler et al., 2018a; Michaelowa et al., 2018). In the scenario ensemble available to this assessment, several variations of near-term mitigation policy implementation can be found: immediate and cross-sectoral global cooperation from 2020 onward towards a global climate objective, a phase-in of globally coordinated mitigation policy from 2020 to 2040, and a more short-term oriented and regionally diverse global mitigation policy, following NDCs until 2030 (Kriegler et al., 2018a; Luderer et al., 2018; McCollum et al., 2018; Rogelj et al., 2018; Strefler et al., 2018b). For example, the above-mentioned SSP quantifications assume regionally scattered mitigation policies until 2020, and vary in global convergence thereafter (Kriegler et al., 2014a; Riahi et al., 2017). The impact of near-term policy choices on 1.5°C-consistent pathways is discussed in Section 2.3.5. The literature has also explored 1.5°C-consistent pathways that build on a portfolio of policy approaches until 2030, including the combination of regulatory policies and carbon pricing (Kriegler et al., 2018a),

and a variety of ancillary policies to safeguard other sustainable development goals (Bertram et al., 2018; van Vuuren et al., 2018). A further discussion of policy implications of 1.5°C-consistent pathways is provided in Section 2.5.1, while a general discussion of policies and options to strengthen action are subject of Chapter 4, Section 4.4.

2.3.2   Key Characteristics of 1.5°C Pathways

1.5°C-consistent pathways are characterized by a rapid phase out of CO_2 emissions and deep emissions reductions in other GHGs and climate forcers (Section 2.2.2 and 2.3.3). This is achieved by broad transformations in the energy; industry; transport; buildings; and agriculture, forestry and other land-use (AFOLU) sectors (Section 2.4) (Bauer et al., 2018; Grubler et al., 2018; Holz et al., 2018b; Kriegler et al., 2018b; Liu et al., 2018; Luderer et al., 2018; Rogelj et al., 2018; van Vuuren et al., 2018; Zhang et al., 2018). Here we assess 1.5°C-consistent pathways with and without overshoot during the 21st century. One study also explores pathways overshooting 1.5°C for longer than the 21st century (Akimoto et al., 2017), but these are not considered 1.5°C-consistent pathways in this report (Chapter 1, Section 1.1.3). This subsection summarizes robust and varying properties of 1.5°C-consistent pathways regarding system transformations, emission reductions and overshoot. It aims to provide an introduction to the detailed assessment of the emissions evolution (Section 2.3.3), CDR deployment (Section 2.3.4), energy (Section 2.4.1, 2.4.2), industry (2.4.3.1), buildings (2.4.3.2), transport (2.4.3.3) and land-use transformations (Section 2.4.4) in 1.5°C-consistent pathways. Throughout Sections 2.3 and 2.4, pathway properties are highlighted with four 1.5°C-consistent pathway archetypes (LED, S1, S2, S5; referred to as P1, P2, P3, and P4 in the Summary for Policymakers) covering a wide range of different socio-economic and technology assumptions (Figure 2.5, Section 2.3.1).

2.3.2.1   Variation in system transformations underlying 1.5°C pathways

Be it for the energy, transport, buildings, industry, or AFOLU sector, the literature shows that multiple options and choices are available in each of these sectors to pursue stringent emissions reductions (Section 2.3.1.2, Supplementary Material 2.SM.1.2, Chapter 4, Section 4.3). Because the overall emissions total under a pathway is limited by a geophysical carbon budget (Section 2.2.2), choices in one sector affect the efforts that are required from others (Clarke et al., 2014). A robust feature of 1.5°C-consistent pathways, as highlighted by the set of pathway archetypes in Figure 2.5, is a virtually full decarbonization of the power sector around mid-century, a feature shared with 2°C-consistent pathways. The additional emissions reductions in 1.5°C-consistent compared to 2°C-consistent pathways come predominantly from the transport and industry sectors (Luderer et al., 2018). Emissions can be apportioned differently across sectors, for example, by focussing on reducing the overall amount of CO_2 produced in the energy end-use sectors, and using limited contributions of CDR by the AFOLU sector (afforestation and reforestation, S1 and LED pathways in Figure 2.5) (Grubler et al., 2018; Holz et al., 2018b; van Vuuren et al., 2018), or by being more lenient about the amount of CO_2 that continues to be produced in the above-mentioned end-use sectors (both by 2030 and mid-century) and strongly relying on technological CDR options

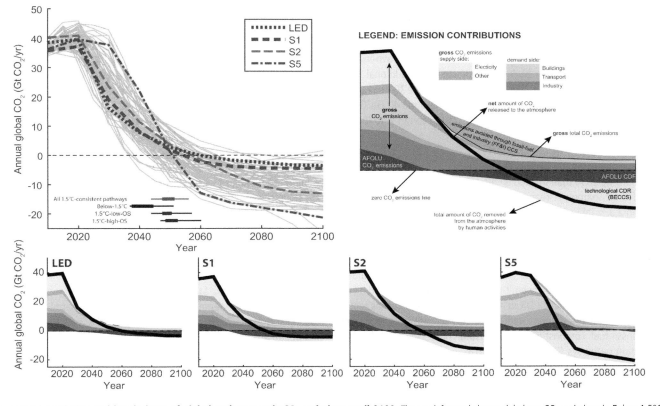

Figure 2.5 | Evolution and break down of global anthropogenic CO$_2$ emissions until 2100. The top-left panel shows global net CO$_2$ emissions in Below-1.5°C, 1.5°C-low-overshoot (OS), and 1.5°C-high-OS pathways, with the four illustrative 1.5°C-consistent pathway archetypes of this chapter highlighted. Ranges at the bottom of the top-left panel show the 10th–90th percentile range (thin line) and interquartile range (thick line) of the time that global CO$_2$ emissions reach net zero per pathway class, and for all pathways classes combined. The top-right panel provides a schematic legend explaining all CO$_2$ emissions contributions to global CO$_2$ emissions. The bottom row shows how various CO$_2$ contributions are deployed and used in the four illustrative pathway archetypes (LED, S1, S2, S5, referred to as P1, P2, P3, and P4 in the Summary for Policymakers) used in this chapter (see Section 2.3.1.1). Note that the S5 scenario reports the building and industry sector emissions jointly. Green-blue areas hence show emissions from the transport sector and the joint building and industry demand sector, respectively.

like BECCS (S2 and S5 pathways in Figure 2.5) (Luderer et al., 2018; Rogelj et al., 2018). Major drivers of these differences are assumptions about energy and food demand and the stringency of near-term climate policy (see the difference between early action in the scenarios S1, LED and more moderate action until 2030 in the scenarios S2, S5). Furthermore, the carbon budget in each of these pathways depends also on the non-CO$_2$ mitigation measures implemented in each of them, particularly for agricultural emissions (Sections 2.2.2, 2.3.3) (Gernaat et al., 2015). Those pathways differ not only in terms of their deployment of mitigation and CDR measures (Sections 2.3.4 and 2.4), but also in terms of the resulting temperature overshoot (Figure 2.1). Furthermore, they have very different implications for the achievement of sustainable development objectives, as further discussed in Section 2.5.3.

2.3.2.2 Pathways keeping warming below 1.5°C or temporarily overshooting it

This subsection explores the conditions that would need to be fulfilled to stay below 1.5°C warming without overshoot. As discussed in Section 2.2.2, to keep warming below 1.5°C with a two-in-three (one-in-two) chance, the cumulative amount of CO$_2$ emissions from 2018 onwards need to remain below a carbon budget of 420 (580) GtCO$_2$; accounting for the effects of additional Earth system feedbacks until 2100 reduces this estimate by 100 GtCO$_2$. Based on the current state of knowledge,

exceeding this remaining carbon budget at some point in time would give a one-in-three (one-in-two) chance that the 1.5°C limit is overshot (Table 2.2). For comparison, around 290 ± 20 (1 standard deviation range) GtCO$_2$ have been emitted in the years 2011–2017, with annual CO$_2$ emissions in 2017 around 42 ± 3 GtCO$_2$ yr^{-1} (Jackson et al., 2017; Le Quéré et al., 2018). Committed fossil-fuel emissions from existing fossil-fuel infrastructure as of 2010 have been estimated at around 500 ± 200 GtCO$_2$ (with about 200 GtCO$_2$ already emitted through 2017) (Davis and Caldeira, 2010). Coal-fired power plants contribute the largest part. Committed emissions from existing coal-fired power plants built through the end of 2016 are estimated to add up to roughly 200 GtCO$_2$, and a further 100–150 GtCO$_2$ from coal-fired power plants under construction or planned (González-Eguino et al., 2017; Edenhofer et al., 2018). However, there has been a marked slowdown of planned coal-power projects in recent years, and some estimates indicate that the committed emissions from coal plants that are under construction or planned have halved since 2015 (Shearer et al., 2018). Despite these uncertainties, the committed fossil-fuel emissions are assessed to already amount to more than two thirds (half) of the remaining carbon budget.

An important question is to what extent the nationally determined contributions (NDCs) under the Paris Agreement are aligned with the remaining carbon budget. It was estimated that the NDCs, if successfully implemented, imply a total of 400–560 $GtCO_2$ emissions over the 2018–2030 period (considering both conditional and unconditional NDCs) (Rogelj et al., 2016a). Thus, following an NDC trajectory would already exhaust 95–130% (70–95%) of the remaining two-in-three (one-in-two) 1.5°C carbon budget (unadjusted for additional Earth system feedbacks) by 2030. This would leave no time (0–9 years) to bring down global emissions from NDC levels of around 40 $GtCO_2$ yr^{-1} in 2030 (Fawcett et al., 2015; Rogelj et al., 2016a) to net zero (further discussion in Section 2.3.5).

Most 1.5°C-consistent pathways show more stringent emissions reductions by 2030 than implied by the NDCs (Section 2.3.5) The lower end of those pathways reach down to below 20 $GtCO_2$ yr^{-1} in 2030 (Section 2.3.3, Table 2.4), less than half of what is implied by the NDCs. Whether such pathways will be able to limit warming to 1.5°C without overshoot will depend on whether cumulative net CO_2 emissions over the 21st century can be kept below the remaining carbon budget at any time. Net global CO_2 emissions are derived from the gross amount of CO_2 that humans annually emit into the atmosphere reduced by the amount of anthropogenic CDR in each year. New research has looked more closely at the amount and the drivers of gross CO_2 emissions from fossil-fuel combustion and industrial processes (FFI) in deep mitigation pathways (Luderer et al., 2018), and found that the larger part of remaining CO_2 emissions come from direct fossil-fuel use in the transport and industry sectors, while residual energy supply sector emissions (mostly from the power sector) are limited by a rapid approach to net zero CO_2 emissions until mid-century. The 1.5°C pathways with no or limited (<0.1°C) overshoot that were reported in the scenario database project remaining FFI CO_2 emissions of 610–1260 $GtCO_2$ over the period 2018–2100 (5th–95th percentile range; median: 880 $GtCO_2$). Kriegler et al. (2018b) conducted a sensitivity analysis that explores the four central options for reducing fossil-fuel emissions: lowering energy demand, electrifying energy services, decarbonizing the power sector and decarbonizing non-electric fuel use in energy end-use sectors. By exploring these options to their extremes, they found a lowest value of 500 $GtCO_2$ (2018–2100) gross fossil-fuel CO_2 emissions for the hypothetical case of aligning the strongest assumptions for all four mitigation options. The two lines of evidence and the fact that available 1.5°C pathways cover a wide range of assumptions (Section 2.3.1) give a robust indication of a lower limit of about 500 $GtCO_2$ remaining fossil-fuel and industry CO_2 emissions in the 21st century.

To compare these numbers with the remaining carbon budget, CO_2 emissions from agriculture, forestry and other land use (AFOLU) need to be taken into account. In many of the 1.5°C-consistent pathways, AFOLU CO_2 emissions reach zero at or before mid-century and then turn to negative values (Table 2.4). This means human changes to the land lead to atmospheric carbon being stored in plants and soils. This needs to be distinguished from the natural CO_2 uptake by land, which is not accounted for in the anthropogenic AFOLU CO_2 emissions reported in the pathways. Given the difference in estimating the 'anthropogenic' sink between countries and the global integrated assessment and carbon modelling community (Grassi et al., 2017), the AFOLU CO_2 estimates included here are not necessarily directly comparable with countries' estimates at global level. The cumulated amount of AFOLU CO_2 emissions until the time they reach zero combine with the fossil-fuel and industry CO_2 emissions to give a total amount of gross emissions

of 650–1270 $GtCO_2$ for the period 2018–2100 (5th–95th percentile; median 950 $GtCO_2$) in 1.5°C pathways with no or limited overshoot. The lower end of the range is close to what emerges from a scenario of transformative change that halves CO_2 emissions every decade from 2020 to 2050 (Rockström et al., 2017). All these estimates are above the remaining carbon budget for a one-in-two chance of limiting warming below 1.5°C without overshoot, including the low end of the hypothetical sensitivity analysis of Kriegler et al. (2018b), who assumes 75 Gt AFOLU CO_2 emissions adding to a total of 575 $GtCO_2$ gross CO_2 emissions. As almost no cases have been identified that keep gross CO_2 emissions within the remaining carbon budget for a one-in-two chance of limiting warming to 1.5°C, and based on current understanding of the geophysical response and its uncertainties, the available evidence indicates that avoiding overshoot of 1.5°C will require some type of CDR in a broad sense, e.g., via net negative AFOLU CO_2 emissions (*medium confidence*). (Table 2.2).

Net CO_2 emissions can fall below gross CO_2 emissions, if CDR is brought into the mix. Studies have looked at mitigation and CDR in combination to identify strategies for limiting warming to 1.5°C (Sanderson et al., 2016; Ricke et al., 2017). CDR, which may include net negative AFOLU CO_2 emissions, is deployed by all 1.5°C-consistent pathways available to this assessment, but the scale of deployment and choice of CDR measures varies widely (Section 2.3.4). Furthermore, no CDR technology has been deployed at scale yet, and all come with concerns about their potential (Fuss et al., 2018), feasibility (Nemet et al., 2018) and/or sustainability (Smith et al., 2015; Fuss et al., 2018) (see Sections 2.3.4, 4.3.2 and 4.3.7 and Cross-Chapter Box 7 in Chapter 3 for further discussion). CDR can have two very different functions in 1.5°C-consistent pathways. If deployed in the first half of the century, before net zero CO_2 emissions are reached, it neutralizes some of the remaining CO_2 emissions year by year and thus slows the accumulation of CO_2 in the atmosphere. In this first function it can be used to remain within the carbon budget and avoid overshoot. If CDR is deployed in the second half of the century after carbon neutrality has been established, it can still be used to neutralize some residual emissions from other sectors, but also to create net negative emissions that actively draw down the cumulative amount of CO_2 emissions to return below a 1.5°C warming level. In the second function, CDR enables temporary overshoot. The literature points to strong limitations to upscaling CDR (limiting its first abovementioned function) and to sustainability constraints (limiting both abovementioned functions) (Fuss et al., 2018; Minx et al., 2018; Nemet et al., 2018). Large uncertainty hence exists about what amount of CDR could actually be available before mid-century. Kriegler et al. (2018b) explore a case limiting CDR to 100 $GtCO_2$ until 2050, and the 1.5°C pathways with no or limited overshoot available in the report's database project 40–260 $GtCO_2$ CDR until the point of carbon neutrality (5th to 95th percentile; median 110 $GtCO_2$). Because gross CO_2 emissions in most cases exceed the remaining carbon budget by several hundred $GtCO_2$ and given the limits to CDR deployment until 2050, most of the 1.5°C-consistent pathways available to this assessment are overshoot pathways. However, the scenario database also contains nine non-overshoot pathways that remain below 1.5°C throughout the 21st century (Table 2.1).

2.3.3   Emissions Evolution in 1.5°C Pathways

This section assesses the salient temporal evolutions of climate forcers over the 21st century. It uses the classification of 1.5°C pathways presented in Section 2.1, which includes a Below-1.5°C class, as well as other classes with varying levels of projected overshoot (1.5°C-low-OS and 1.5°C-high-OS). First, aggregate-GHG benchmarks for 2030 are assessed. Subsequent sections assess long-lived climate forcers (LLCF) and short-lived climate forcers (SLCF) separately because they contribute in different ways to near-term, peak and long-term warming (Section 2.2, Cross-Chapter Box 2 in Chapter 1).

Estimates of aggregated GHG emissions in line with specific policy choices are often compared to near-term benchmark values from mitigation pathways to explore their consistency with long-term climate goals (Clarke et al., 2014; UNEP, 2016, 2017; UNFCCC, 2016). Benchmark emissions or estimates of peak years derived from IAMs provide guidelines or milestones that are consistent with achieving a given temperature level. While they do not set mitigation requirements in a strict sense, exceeding these levels in a given year almost invariably increases the mitigation challenges afterwards by increasing the rates of change and increasing the reliance on speculative technologies, including the possibility that its implementation becomes unachievable (see Cross-Chapter Box 3 in Chapter 1 for a discussion of feasibility concepts) (Luderer et al., 2013; Rogelj et al., 2013b; Clarke et al., 2014; Fawcett et al., 2015; Riahi et al., 2015; Kriegler et al., 2018a). These trade-offs are particularly pronounced in 1.5°C pathways and are discussed in Section 2.3.5. This section assesses Kyoto-GHG emissions in 2030 expressed in CO_2 equivalent (CO_2e) emissions using 100-year global warming potentials.[3]

Appropriate benchmark values of aggregated GHG emissions depend on a variety of factors. First and foremost, they are determined by the desired likelihood to keep warming below 1.5°C and the extent to which projected temporary overshoot is to be avoided (Sections 2.2, 2.3.2, and 2.3.5). For instance, median aggregated 2030 GHG emissions are about 10 $GtCO_2$e yr^{-1} lower in 1.5°C-low-OS compared to 1.5°C-high-OS pathways, with respective interquartile ranges of 26–31 and 36–49 $GtCO_2$e yr^{-1} (Table 2.4). These ranges correspond to about 25–30 and 35–48 $GtCO_2$e yr^{-1} in 2030, respectively, when aggregated with 100-year Global Warming Potentials from the IPCC Second Assessment Report. The limited evidence available for pathways aiming to limit warming below 1.5°C without overshoot or with limited amounts of CDR (Grubler et al., 2018; Holz et al., 2018b; van Vuuren et al., 2018) indicates that under these conditions consistent emissions in 2030 would fall at the lower end and below the above mentioned ranges. Due to the small number of 1.5°C pathways with no overshoot in the report's database (Table 2.4) and the potential for a downward bias in the selection of underlying scenario assumptions, the headline range for 1.5°C pathways with no or limited overshoot is also assessed to be of the order of 25–30 $GtCO_2$e yr^{-1}. Ranges for the 1.5°C-low-OS and Lower-2°C classes only overlap outside their interquartile ranges,

highlighting the more accelerated reductions in 1.5°C-consistent compared to 2°C-consistent pathways.

Appropriate emissions benchmark values also depend on the acceptable or desired portfolio of mitigation measures, representing clearly identified trade-offs and choices (Sections 2.3.4, 2.4, and 2.5.3) (Luderer et al., 2013; Rogelj et al., 2013a; Clarke et al., 2014; Krey et al., 2014a; Strefler et al., 2018b). For example, lower 2030 GHG emissions correlate with a lower dependence on the future availability and desirability of CDR (Strefler et al., 2018b). On the other hand, pathways that assume or anticipate only limited deployment of CDR during the 21st century imply lower emissions benchmarks over the coming decades, which are achieved in models through further reducing CO_2 emissions in the coming decades. The pathway archetypes used in the chapter illustrate this further (Figure 2.6). Under middle-of-the-road assumptions of technological and socioeconomic development, pathway S2 suggests emission benchmarks of 34, 12 and –8 $GtCO_2$e yr^{-1} in the years 2030, 2050, and 2100, respectively. In contrast, a pathway that further limits overshoot and aims at eliminating the reliance on negative emissions technologies like BECCS as well as CCS (here labelled as the LED pathway) shows deeper emissions reductions in 2030 to limit the cumulative amount of CO_2 until net zero global CO_2 emissions (carbon neutrality). The LED pathway here suggests emission benchmarks of 25, 9 and 2 $GtCO_2$e yr^{-1} in the years 2030, 2050, and 2100, respectively. However, a pathway that allows and plans for the successful large-scale deployment of BECCS by and beyond 2050 (S5) shows a shift in the opposite direction. The variation within and between the abovementioned ranges of 2030 GHG benchmarks hence depends strongly on societal choices and preferences related to the acceptability and availability of certain technologies.

Overall these variations do not strongly affect estimates of the 1.5°C-consistent timing of global peaking of GHG emissions. Both Below-1.5°C and 1.5°C-low-OS pathways show minimum–maximum ranges in 2030 that do not overlap with 2020 ranges, indicating the global GHG emissions peaked before 2030 in these pathways. Also, 2020 and 2030 GHG emissions in 1.5°C-high-OS pathways only overlap outside their interquartile ranges.

Kyoto-GHG emission reductions are achieved by reductions in CO_2 and non-CO_2 GHGs. The AR5 identified two primary factors that influence the depth and timing of reductions in non-CO_2 Kyoto-GHG emissions: (i) the abatement potential and costs of reducing the emissions of these gases and (ii) the strategies that allow making trade-offs between them (Clarke et al., 2014). Many studies indicate low-cost, near-term mitigation options in some sectors for non-CO_2 gases compared to supply-side measures for CO_2 mitigation (Clarke et al., 2014). A large share of this potential is hence already exploited in mitigation pathways in line with 2°C. At the same time, by mid-century and beyond, estimates of further reductions of non-CO_2 Kyoto-GHGs – in particular CH_4 and N_2O – are hampered by the absence of mitigation

---

[3]   In this chapter GWP-100 values from the IPCC Fourth Assessement Report are used because emissions of fluorinated gases in the integrated pathways have been reported in this metric to the database. At a global scale, switching between GWP-100 values of the Second, Fourth or Fifth IPCC Assessment Reports could result in variations in aggregated Kyoto-GHG emissions of about ±5% in 2030 (UNFCCC, 2016).

options in the current generation of IAMs, which are hence not able to reduce residual emissions of sources linked to livestock production and fertilizer use (Clarke et al., 2014; Gernaat et al., 2015) (Sections 2.3.1.2, 2.4.4, Supplementary Material 2.SM.1.2). Therefore, while net CO2 emissions are projected to be markedly lower in 1.5°C-consistent compared to 2°C-consistent pathways, this is much less the case for methane (CH_4) and nitrous-oxide (N_2O) (Figures 2.6–2.7). This results in reductions of CO_2 being projected to take up the largest share of emissions reductions when moving between 1.5°C-consistent and 2°C-consistent pathways (Rogelj et al., 2015b, 2018; Luderer et al., 2018). If additional non-CO_2 mitigation measures are identified and adequately included in IAMs, they are expected to further contribute to mitigation efforts by lowering the floor of residual non-CO_2 emissions. However, the magnitude of these potential contributions has not been assessed as part of this report.

As a result of the interplay between residual CO_2 and non-CO_2 emissions and CDR, global GHG emissions reach net zero levels at different times in different 1.5°C-consistent pathways. Interquartile ranges of the years in which 1.5°C-low-OS and 1.5°C-high-OS reach net zero GHG emissions range from 2060 to 2080 (Table 2.4). A seesaw characteristic can be found between near-term emissions reductions and the timing of net zero GHG emissions. This is because pathways with limited emissions reductions in the next one to two decades require net negative CO_2 emissions later on (see earlier). Most 1.5°C-high-OS pathways lead to net zero GHG emissions in approximately the third quarter of this century, because all of them rely on significant amounts of annual net negative CO_2 emissions in the second half of the century to decline temperatures after overshoot (Table 2.4). However, in pathways that aim at limiting overshoot as much as possible or more slowly decline temperatures after their peak, emissions reach the point of net zero GHG emissions slightly later or at times never. Early emissions reductions in this case reduce the requirement for net negative CO_2 emissions. Estimates of 2030 GHG emissions in line with the current NDCs overlap with the highest quartile of 1.5°C-high-OS pathways (Cross-Chapter Box 9 in Chapter 4).

2.3.3.1   Emissions of long-lived climate forcers

Climate effects of long-lived climate forcers (LLCFs) are dominated by CO_2, with smaller contributions of N_2O and some fluorinated gases (Myhre et al., 2013; Blanco et al., 2014). Overall net CO_2 emissions in pathways are the result of a combination of various anthropogenic contributions (Figure 2.5) (Clarke et al., 2014): (i) CO_2 produced by fossil-fuel combustion and industrial processes, (ii) CO_2 emissions or removals from the agriculture, forestry and other land use (AFOLU) sector, (iii) CO_2 capture and sequestration (CCS) from fossil fuels or industrial activities before it is released to the atmosphere, (iv) CO_2 removal by technological means, which in current pathways is mainly achieved by BECCS and AFOLU-related CDR, although other options could be conceivable (see Chapter 4, Section 4.3.7). Pathways apply these four contributions in different configurations (Figure 2.5) depending on societal choices and preferences related to the acceptability and availability of certain technologies, the timing and stringency of near-term climate policy, and the ability to limit the demand that drives baseline emissions (Marangoni et al., 2017; Riahi et al., 2017; Grubler et al., 2018; Rogelj et al., 2018; van Vuuren et al., 2018), and come with

very different implication for sustainable development (Section 2.5.3).

All 1.5°C pathways see global CO_2 emissions embark on a steady decline to reach (near) net zero levels around 2050, with 1.5°C-low-OS pathways reaching net zero CO_2 emissions around 2045–2055 (Table 2.4; Figure 2.5). Near-term differences between the various pathway classes are apparent, however. For instance, Below-1.5°C and 1.5°C-low-OS pathways show a clear shift towards lower CO_2 emissions in 2030 relative to other 1.5°C and 2°C pathway classes, although in all 1.5°C classes reductions are clear (Figure 2.6). These lower near-term emissions levels are a direct consequence of the former two pathway classes limiting cumulative CO_2 emissions until carbon neutrality in order to aim for a higher probability of limiting peak warming to 1.5°C (Section 2.2.2 and 2.3.2.2). In some cases, 1.5°C-low-OS pathways achieve net zero CO_2 emissions one or two decades later, contingent on 2030 CO_2 emissions in the lower quartile of the literature range, that is, below about 18 $GtCO_2$ yr^{-1}. Median year-2030 global CO_2 emissions are of the order of 5–10 $GtCO_2$ yr^{-1} lower in Below-1.5°C compared to 1.5°C-low-OS pathways, which are in turn lower than 1.5°C-high-OS pathways (Table 2.4). Below-1.5°C and 1.5°C-low-OS pathways combined show a decline in global net anthropogenic CO_2 emissions of about 45% from 2010 levels by 2030 (40–60% interquartile range). Lower-2°C pathways show CO_2 emissions declining by about 25% by 2030 in most pathways (10–30% interquartile range). The 1.5°C-high-OS pathways show emissions levels that are broadly similar to the 2°C-consistent pathways in 2030.

The development of CO_2 emissions in the second half of the century in 1.5°C pathways is characterized by the need to stay or return within a carbon budget. Figure 2.6 shows net CO_2 and N_2O emissions from various sources in 2050 and 2100 in 1.5°C pathways in the literature. Virtually all 1.5°C pathways obtain net negative CO_2 emissions at some point during the 21st century, but the extent to which net negative emissions are relied upon varies substantially (Figure 2.6, Table 2.4). This net withdrawal of CO_2 from the atmosphere compensates for residual long-lived non-CO_2 GHG emissions that also accumulate in the atmosphere (like N2O) or cancels some of the build-up of CO_2 due to earlier emissions to achieve increasingly higher likelihoods that warming stays or returns below 1.5°C (see Section 2.3.4 for a discussion of various uses of CDR). Even non-overshoot pathways that aim at achieving temperature stabilization would hence deploy a certain amount of net negative CO_2 emissions to offset any accumulating long-lived non-CO_2 GHGs. The 1.5°C overshoot pathways display significantly larger amounts of annual net negative CO_2 emissions in the second half of the century. The larger the overshoot the more net negative CO_2 emissions are required to return temperatures to 1.5°C by the end of the century (Table 2.4, Figure 2.1).

N_2O emissions decline to a much lesser extent than CO_2 in currently available 1.5°C pathways (Figure 2.6). Current IAMs have limited emissions-reduction potentials (Gernaat et al., 2015) (Sections 2.3.1.2, 2.4.4, Supplementary Material 2.SM.1.2), reflecting the difficulty of eliminating N_2O emission from agriculture (Bodirsky et al., 2014). Moreover, the reliance of some pathways on significant amounts of bioenergy after mid-century (Section 2.4.2) coupled to a substantial use of nitrogen fertilizer (Popp et al., 2017) also makes reducing N_2O emissions harder (for example, see pathway S5 in Figure 2.6). As

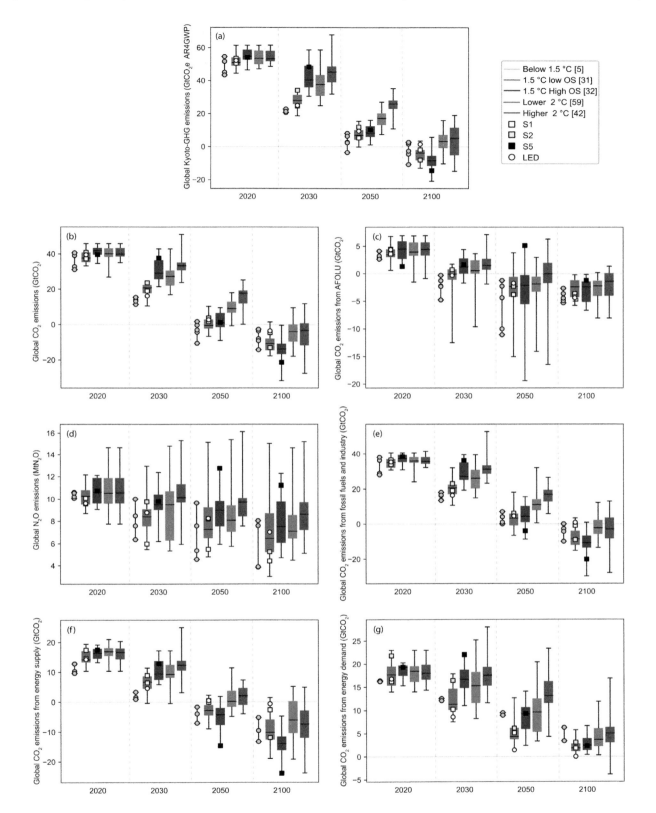

Figure 2.6 | Annual global emissions characteristics for 2020, 2030, 2050, 2100. Data are shown for (a) Kyoto-GHG emissions, and (b) global total CO_2 emissions, (c) CO_2 emissions from the agriculture, forestry and other land use (AFOLU) sector, (d) global N_2O emissions, and (e) CO_2 emissions from fossil fuel use and industrial processes. The latter is also split into (f) emissions from the energy supply sector (electricity sector and refineries) and (g) direct emissions from fossil-fuel use in energy demand sectors (industry, buildings, transport) (bottom row). Horizontal black lines show the median, boxes show the interquartile range, and whiskers the minimum–maximum range. Icons indicate the four pathway archetypes used in this chapter. In case less than seven data points are available in a class, the minimum–maximum range and single data points are shown. Kyoto-GHG, emissions in the top panel are aggregated with AR4 GWP-100 and contain CO_2, CH_4, N_2O, HFCs, PFCs, and SF_6. NF_3 is typically not reported by IAMs. Scenarios with year-2010 Kyoto-GHG emissions outside the range assessed by IPCC AR5 WGIII assessed are excluded (IPCC, 2014b).

a result, sizeable residual N_2O emissions are currently projected to continue throughout the century, and measures to effectively mitigate them will be of continued relevance for 1.5°C societies. Finally, the reduction of nitrogen use and N_2O emissions from agriculture is already a present-day concern due to unsustainable levels of nitrogen pollution (Bodirsky et al., 2012). Section 2.4.4 provides a further assessment of the agricultural non-CO_2 emissions reduction potential.

2.3.3.2   Emissions of short-lived climate forcers and fluorinated gases

SLCFs include shorter-lived GHGs like CH_4 and some fluorinated gases as well as particles (aerosols), their precursors and ozone precursors. SLCFs are strongly mitigated in 1.5°C pathways, as is the case for 2°C pathways (Figure 2.7). SLCF emissions ranges of 1.5°C and 2°C pathway classes strongly overlap, indicating that the main incremental mitigation contribution between 1.5°C and 2°C pathways comes from CO_2 (Luderer et al., 2018; Rogelj et al., 2018). CO_2 and SLCF emissions reductions are connected in situations where SLCF and CO_2 are co-emitted by the same process, for example, with coal-fired power plants (Shindell and Faluvegi, 2010) or within the transport sector (Fuglestvedt et al., 2010). Many CO_2-targeted mitigation measures in industry, transport and agriculture (Sections 2.4.3–4) hence also reduce non-CO_2 forcing (Rogelj et al., 2014b; Shindell et al., 2016).

Despite the fact that methane has a strong warming effect (Myhre et al., 2013; Etminan et al., 2016), current 1.5°C-consistent pathways still project significant emissions of CH_4 by 2050, indicating only a limited CH_4 mitigation potential in IAM analyses (Gernaat et al., 2015) (Sections 2.3.1.2, 2.4.4, Table 2.SM.2). The AFOLU sector contributes an important share of the residual CH_4 emissions until mid-century, with its relative share increasing from slightly below 50% in 2010 to around 55–70% in 2030, and 60–80% in 2050 in 1.5°C-consistent pathways (interquartile range across 1.5°C-consistent pathways for projections). Many of the proposed measures to target CH_4 (Shindell et al., 2012; Stohl et al., 2015) are included in 1.5°C pathways (Figure 2.7), though not all (Sections 2.3.1.2, 2.4.4, Table 2.SM.2). A detailed assessment of measures to further reduce AFOLU CH_4 emissions has not been conducted.

Overall reductions of SLCFs can have effects of either sign on temperature depending on the balance between cooling and warming agents. The reduction in SO_2 emissions is the dominant single effect as it weakens the negative total aerosol forcing. This means that reducing all SLCF emissions to zero would result in a short-term warming, although this warming is *unlikely* to be more than 0.5°C (Section 2.2 and Figure 1.5 (Samset et al., 2018)). Because of this effect, suggestions have been proposed that target the warming agents only (referred to as short-lived climate pollutants or SLCPs instead of the more general short-lived climate forcers; e.g., Shindell et al., 2012), though aerosols are often emitted in varying mixtures of warming and cooling species (Bond et al., 2013). Black carbon (BC) emissions reach similar levels across 1.5°C-consistent and 2°C-consistent pathways available in the literature, with interquartile ranges of emissions reductions across pathways of 16–34% and 48–58% in 2030 and 2050, respectively, relative to 2010 (Figure 2.7). Recent studies have identified further reduction potentials for the near term, with global reductions of about

80% being suggested (Stohl et al., 2015; Klimont et al., 2017). Because the dominant sources of certain aerosol mixtures are emitted during the combustion of fossil fuels, the rapid phase-out of unabated fossil fuels to avoid CO_2 emissions would also result in removal of these either warming or cooling SLCF air-pollutant species. Furthermore, SLCFs are also reduced by efforts to reduce particulate air pollution. For example, year-2050 SO_2 emissions (precursors of sulphate aerosol) in 1.5°C-consistent pathways are about 75–85% lower than their 2010 levels. Some caveats apply, for example, if residential biomass use would be encouraged in industrialised countries in stringent mitigation pathways without appropriate pollution control measures, aerosol concentrations could also increase (Sand et al., 2015; Stohl et al., 2015).

Emissions of fluorinated gases (IPCC/TEAP, 2005; US EPA, 2013; Velders et al., 2015; Purohit and Höglund-Isaksson, 2017) in 1.5°C-consistent pathways are reduced by roughly 75–80% relative to 2010 levels (interquartile range across 1.5°C-consistent pathways) in 2050, with no clear differences between the classes. Although unabated hydrofluorocarbon (HFC) emissions have been projected to increase (Velders et al., 2015), the Kigali Amendment recently added HFCs to the basket of gases controlled under the Montreal Protocol (Höglund-Isaksson et al., 2017). As part of the larger group of fluorinated gases, HFCs are also assumed to decline in 1.5°C-consistent pathways. Projected reductions by 2050 of fluorinated gases under 1.5°C-consistent pathways are deeper than published estimates of what a full implementation of the Montreal Protocol including its Kigali Amendment would achieve (Höglund-Isaksson et al., 2017), which project roughly a halving of fluorinated gas emissions in 2050 compared to 2010. Assuming the application of technologies that are currently commercially available and at least to a limited extent already tested and implemented, potential fluorinated gas emissions reductions of more than 90% have been estimated (Höglund-Isaksson et al., 2017).

There is a general agreement across 1.5°C-consistent pathways that until 2030 forcing from the warming SLCFs is reduced less strongly than the net cooling forcing from aerosol effects, compared to 2010. As a result, the net forcing contributions from all SLCFs combined are projected to increase slightly by about 0.2–0.3 W m^{-2}, compared to 2010. Also, by the end of the century, about 0.1–0.3 W m^{-2} of SLCF forcing is generally currently projected to remain in 1.5°C-consistent scenarios (Figure 2.8). This is similar to developments in 2°C-consistent pathways (Rose et al., 2014b; Riahi et al., 2017), which show median forcing contributions from these forcing agents that are generally no more than 0.1 W m^{-2} higher. Nevertheless, there can be additional gains from targeted deeper reductions of CH_4 emissions and tropospheric ozone precursors, with some scenarios projecting less than 0.1 W m^{-2} forcing from SLCFs by 2100.

2.3.4   CDR in 1.5°C Pathways

Deep mitigation pathways assessed in AR5 showed significant deployment of CDR, in particular through BECCS (Clarke et al., 2014). This has led to increased debate about the necessity, feasibility and desirability of large-scale CDR deployment, sometimes also called 'negative emissions technologies' in the literature (Fuss et al., 2014; Anderson and Peters, 2016; Williamson, 2016; van Vuuren et al.,

Table 2.4 | **Emissions in 2030, 2050 and 2100 in 1.5°C and 2°C scenario classes and absolute annual rates of change between 2010–2030, 2020–2030 and 2030–2050, respectively.**
Values show median and interquartile range across available scenarios (25th and 75th percentile given in brackets). If fewer than seven scenarios are available (*), the minimum–maximum range is given instead. Kyoto-GHG emissions are aggregated with GWP-100 values from IPCC AR4. Emissions in 2010 for total net CO_2, CO_2 from fossil-fuel use and industry, and AFOLU CO_2 are estimated at 38.5, 33.4, and 5 $GtCO_2$ yr^{-1}, respectively (Le Quéré et al., 2018). Percentage reduction numbers included in headline statement C.1 in the Summary for Policymakers are computed relative to 2010 emissions in each individual pathway, and hence differ slightly from a case where reductions are computed relative to the historical 2010 emissions reported above. A difference is reported in estimating the 'anthropogenic' sink by countries or the global carbon modelling community (Grassi et al., 2017), and AFOLU CO_2 estimates reported here are thus not necessarily comparable with countries' estimates. Scenarios with year-2010 Kyoto-GHG emissions outside the range assessed by IPCC AR5 WGIII are excluded (IPCC, 2014b), as are scenario duplicates that would bias ranges towards a single study.

Name	Category	#	Annual emissions/sequestration ($GtCO_2$ yr^{-1})			Absolute Annual Change ($GtCO_2$/yr^{-1})			Timing of Global Zero
			2030	2050	2100	2010–2030	2020–2030	2030–2050	Year
Total CO₂ (net)	Below-1.5°C	5*	13.4 (15.4, 11.4)	−3.0 (1.7, −10.6)	−8.0 (−2.6, −14.2)	−1.2 (−1.0, −1.3)	−2.5 (−1.8, −2.8)	−0.8 (−0.7, −1.2)	2044 (2037, 2054)
	1.5°C-low-OS	37	20.8 (22.2, 18.0)	−0.4 (2.7, −2.0)	−10.8 (−8.1, −14.3)	−0.8 (−0.7, −1.0)	−1.7 (−1.4, −2.3)	−1.0 (−0.8, −1.2)	2050 (2047, 2055)
	1.5°C with no or limited OS	42	20.3 (22.0, 15.9)	−0.5 (2.2, −2.8)	−10.2 (−7.6, −14.2)	−0.9 (−0.7, −1.1)	−1.8 (−1.5, −2.3)	−1.0 (−0.8, −1.2)	2050 (2046, 2055)
	1.5°C-high-OS	36	29.1 (36.4, 26.0)	1.0 (6.3, −1.2)	−13.8 (−11.1, −16.4)	−0.4 (0.0, −0.6)	−1.1 (−0.5, −1.5)	−1.3 (−1.1, −1.8)	2052 (2049, 2059)
	Lower-2°C	54	28.9 (33.7, 24.5)	9.9 (13.1, 6.5)	−5.1 (−2.6, −10.3)	−0.4 (−0.2, −0.6)	−1.1 (−0.8, −1.6)	−0.9 (−0.8, −1.2)	2070 (2063, 2079)
	Higher-2°C	54	33.5 (35.0, 31.0)	17.9 (19.1, 12.2)	−3.3 (0.6, −11.5)	−0.2 (−0.0, −0.4)	−0.7 (−0.5, −0.9)	−0.8 (−0.6, −1.0)	2085 (2070, post–2100)
CO₂ from fossil fuels and industry (gross)	Below-1.5°C	5*	18.0 (21.4, 13.8)	10.5 (20.9, 0.3)	8.3 (11.6, 0.1)	−0.7 (−0.6, −1)	−1.5 (−0.9, −2.2)	−0.4 (0, −0.7)	-
	1.5°C-low-OS	37	22.1 (24.4, 18.7)	10.3 (14.1, 7.8)	5.6 (8.1, 2.6)	−0.5 (−0.4, −0.6)	−1.3 (−0.9, −1.7)	−0.6 (−0.5, −0.7)	-
	1.5°C with no or limited OS	42	21.6 (24.2, 18.0)	10.3 (13.8, 7.7)	6.1 (8.4, 2.6)	−0.5 (−0.4, −0.7)	−1.3 (−0.9, −1.8)	−0.6 (−0.4, −0.7)	-
	1.5°C-high-OS	36	27.8 (37.1, 25.6)	13.1 (17.0, 11.6)	6.6 (8.8, 2.8)	−0.2 (0.2, −0.3)	−0.8 (−0.2, −1.1)	−0.7 (−0.6, −1.0)	-
	Lower-2°C	54	27.7 (31.5, 23.5)	15.4 (19.0, 11.1)	7.2 (10.4, 3.7)	−0.2 (−0.0, −0.4)	−0.8 (−0.5, −1.2)	−0.6 (−0.5, −0.8)	-
	Higher-2°C	54	31.3 (33.4, 28.7)	19.2 (22.6, 17.1)	8.1 (10.9, 5.0)	−0.1 (0.1, −0.2)	−0.5 (−0.2, −0.7)	−0.6 (−0.5, −0.7)	-
CO₂ from fossil fuels and industry (net)	Below-1.5°C	5*	16.4 (18.2, 13.5)	1.0 (7.0, 0)	−2.7 (0, −9.8)	−0.8 (−0.7, −1)	−1.8 (−1.2, −2.2)	−0.6 (−0.5, −0.9)	-
	1.5°C-low-OS	37	20.6 (22.2, 17.5)	3.2 (5.6, −0.6)	−8.5 (−4.1, −11.6)	−0.6 (−0.5, −0.7)	−1.4 (−1.1, −1.8)	−0.8 (−0.7, −1.1)	-
	1.5°C with no or limited OS	42	20.1 (22.1, 16.8)	3.0 (5.6, 0.0)	−8.3 (−3.5, −10.8)	−0.6 (−0.5, −0.8)	−1.4 (−1.1, −1.9)	−0.8 (−0.7, −1.1)	-
	1.5°C-high-OS	36	26.9 (34.7, 25.3)	4.2 (10.0, 1.2)	−10.7 (−6.9, −13.2)	−0.3 (0.1, −0.3)	−0.9 (−0.3, −1.2)	−1.2 (−0.9, −1.5)	-
	Lower-2°C	54	28.2 (31.0, 23.1)	11.8 (14.1, 6.2)	−3.1 (−0.7, −6.4)	−0.2 (−0.1, −0.4)	−0.8 (−0.5, −1.2)	−0.8 (−0.7, −1.0)	-
	Higher-2°C	54	31.0 (33.0, 28.7)	17.0 (19.3, 13.1)	−2.9 (3.3, −8.0)	−0.1 (0.1, −0.2)	−0.5 (−0.2, −0.7)	−0.7 (−0.5, −1.0)	-
CO₂ from AFOLU	Below-1.5°C	5*	−2.2 (−0.3, −4.8)	−4.4 (−1.2, −11.1)	−4.4 (−2.6, −5.3)	−0.3 (−0.2, −0.4)	−0.5 (−0.4, −0.8)	−0.1 (0, −0.4)	-
	1.5°C-low-OS	37	−0.1 (0.8, −1.0)	−2.3 (−0.6, −4.1)	−2.4 (−1.2, −4.2)	−0.2 (−0.2, −0.3)	−0.4 (−0.3, −0.5)	−0.1 (−0.1, −0.2)	-
	1.5°C with no or limited OS	42	−0.1 (0.7, −1.3)	−2.6 (−0.6, −4.5)	−2.6 (−1.3, −4.2)	−0.2 (−0.2, −0.3)	−0.4 (−0.3, −0.5)	−0.1 (−0.1, −0.2)	-
	1.5°C-high-OS	36	1.2 (2.7, 0.1)	−2.1 (−0.3, −5.4)	−2.4 (−1.5, −5.0)	−0.1 (−0.1, −0.3)	−0.2 (−0.1, −0.5)	−0.2 (−0.0, −0.3)	-
	Lower-2°C	54	1.4 (2.8, 0.3)	−1.4 (−0.5, −2.7)	−2.4 (−1.3, −4.2)	−0.2 (−0.1, −0.2)	−0.3 (−0.2, −0.4)	−0.1 (−0.1, −0.2)	-
	Higher-2°C	54	1.5 (2.7, 0.8)	−0.0 (1.9, −1.6)	−1.3 (0.1, −3.9)	−0.2 (−0.1, −0.2)	−0.2 (−0.1, −0.4)	−0.1 (−0.0, −0.1)	-
Bioenergy combined with carbon capture and storage (BECCS)	Below-1.5°C	5*	0.4 (1.1, 0)	3.4 (8.3, 0)	5.7 (13.4, 0)	0 (0.1, 0)	0 (0.1, 0)	0.2 (0.4, 0)	-
	1.5°C-low-OS	36	0.3 (1.1, 0.0)	4.6 (6.4, 3.8)	12.4 (15.6, 7.6)	0.0 (0.1, 0.0)	0.0 (0.1, 0.0)	0.2 (0.3, 0.2)	-
	1.5°C with no or limited OS	41	0.4 (1.0, 0.0)	4.5 (6.3, 3.4)	12.4 (15.0, 6.4)	0.0 (0.1, 0.0)	0.0 (0.1, 0.0)	0.2 (0.3, 0.2)	-
	1.5°C-high-OS	36	0.1 (0.4, 0.0)	6.8 (9.5, 3.7)	14.9 (16.3, 12.1)	0.0 (0.0, 0.0)	0.0 (0.0, 0.0)	0.3 (0.4, 0.2)	-
	Lower-2°C	54	0.1 (0.3, 0.0)	3.6 (4.6, 1.8)	9.5 (12.1, 6.9)	0.0 (0.0, 0.0)	0.0 (0.0, 0.0)	0.2 (0.2, 0.1)	-
	Higher-2°C	47	0.1 (0.2, 0.0)	3.0 (4.9, 1.6)	10.8 (15.3, 8.2) [46]	0.0 (0.0, 0.0)	0.0 (0.0, 0.0)	0.1 (0.2, 0.1)	-
Kyoto GHG (AR4) [GtCO₂e]	Below-1.5°C	5*	22.1 (22.8, 20.7)	2.7 (8.1, −3.5)	−2.6 (2.7, −10.7)	−1.4 (−1.3, −1.5)	−2.9 (−2.1, −3.3)	−0.9 (−0.7, −1.3)	2066 (2044, post–2100)
	1.5°C-low-OS	31	27.9 (31.1, 26.0)	7.0 (9.9, 4.5)	−3.8 (−2.1, −7.9)	−1.1 (−0.9, −1.2)	−2.3 (−1.8, −2.8)	−1.1 (−0.9, −1.2)	2068 (2061, 2080)
	1.5°C with no or limited OS	36	27.4 (30.9, 24.7)	6.5 (9.6, 4.2)	−3.7 (−1.8, −7.8)	−1.1 (−1.0, −1.3)	−2.4 (−1.9, −2.9)	−1.1 (−0.9, −1.2)	2067 (2061, 2084)
	1.5°C-high-OS	32	40.4 (48.9, 36.3)	8.4 (12.3, 6.2)	−8.5 (−5.7, −11.2)	−0.5 (−0.0, −0.7)	−1.3 (−0.6, −1.8)	−1.5 (−1.3, −2.1)	2063 (2058, 2067)
	Lower-2°C	46	39.6 (45.1, 35.7)	18.3 (20.4, 15.2)	2.1 (4.2, −2.4)	−0.5 (−0.1, −0.7)	−1.5 (−0.9, −2.2)	−1.1 (−0.9, −1.2)	post–2100 (2090 post–2100)
	Higher-2°C	42	45.3 (48.5, 39.3)	25.9 (27.9, 23.3)	5.2 (11.5, −4.8)	−0.2 (−0.0, −0.6)	−1.0 (−0.6, −1.2)	−1.0 (−0.7, −1.2)	post–2100 (2085 post–2100)

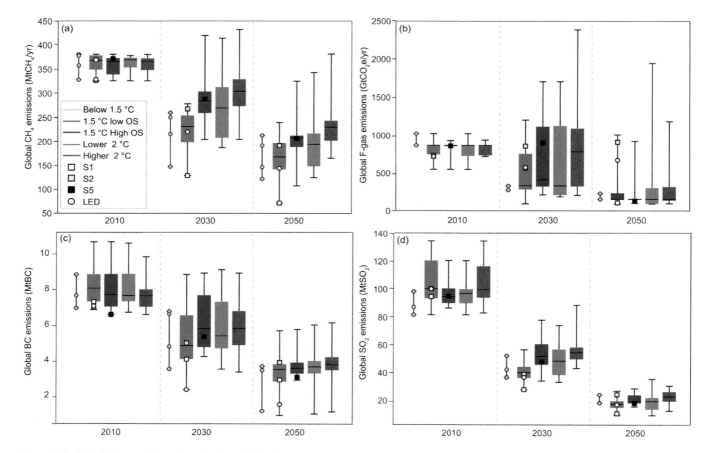

Figure 2.7 | Global characteristics of a selection of short-lived non-CO₂ emissions until mid-century for five pathway classes used in this chapter. Data are shown for (a) methane (CH₄), (b) fluorinated gases (F-gas), (c) black carbon (BC), and (d) sulphur dioxide (SO₂) emissions. Boxes with different colours refer to different scenario classes. Icons on top the ranges show four illustrative pathway archetypes that apply different mitigation strategies for limiting warming to 1.5°C. Boxes show the interquartile range, horizontal black lines the median, and whiskers the minimum–maximum range. F-gases are expressed in units of CO₂-equivalence computed with 100-year Global Warming Potentials reported in IPCC AR4.

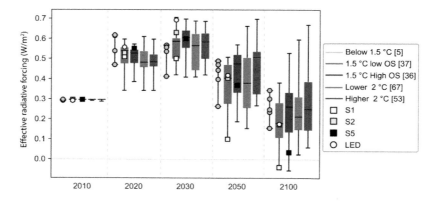

Figure 2.8 | Estimated aggregated effective radiative forcing of SLCFs for 1.5°C and 2°C pathway classes in 2010, 2020, 2030, 2050, and 2100, as estimated by the FAIR model (Smith et al., 2018). Aggregated short-lived climate forcer (SLCF) radiative forcing is estimated as the difference between total anthropogenic radiative forcing and the sum of CO₂ and N₂O radiative forcing over time, and is expressed relative to 1750. Symbols indicate the four pathways archetypes used in this chapter. Horizontal black lines indicate the median, boxes the interquartile range, and whiskers the minimum–maximum range per pathway class. Because very few pathways fall into the Below-1.5°C class, only the minimum–maximum is provided here.

2017a; Obersteiner et al., 2018). Most CDR technologies remain largely unproven to date and raise substantial concerns about adverse side-effects on environmental and social sustainability (Smith et al., 2015; Dooley and Kartha, 2018). A set of key questions emerge: how strongly do 1.5°C-consistent pathways rely on CDR deployment and what types of CDR measures are deployed at which scale? How does this vary across available 1.5°C-consistent pathways and on which factors does it depend? How does CDR deployment compare between 1.5°C- and 2°C-consistent pathways and how does it compare with the findings at the time of the AR5? How does CDR deployment in 1.5°C-consistent pathways relate to questions about availability, policy implementation and sustainable development implications that have been raised about CDR technologies? The first three questions are assessed in this section with the goal to provide an overview and assessment of CDR deployment in the 1.5°C pathway literature. The fourth question is only touched upon here and is addressed in greater depth in Chapter 4, Section 4.3.7, which assesses the rapidly growing literature on costs, potentials, availability and sustainability implications of individual CDR measures (Minx et al., 2017, 2018; Fuss et al., 2018; Nemet et al., 2018). In addition, Section 2.3.5 assesses the relationship between delayed mitigation action and increased CDR reliance. CDR deployment is intricately linked to the land-use transformation in 1.5°C-consistent pathways. This transformation is assessed in Section 2.4.4. Bioenergy and BECCS impacts on sustainable land management are further assessed in Chapter 3, Section 3.6.2 and Cross-Chapter Box 7 in Chapter 3. Ultimately, a comprehensive assessment of the land implication of land-based CDR measures will be provided in the IPCC AR6 Special Report on Climate Change and Land (SRCCL).

2.3.4.1 CDR technologies and deployment levels in 1.5°C pathways

A number of approaches to actively remove carbon-dioxide from the atmosphere are increasingly discussed in the literature (Minx et al., 2018) (see also Chapter 4, Section 4.3.7). Approaches under consideration include the enhancement of terrestrial and coastal carbon storage in plants and soils such as afforestation and reforestation (Canadell and Raupach, 2008), soil carbon enhancement (Paustian et al., 2016; Frank et al., 2017; Zomer et al., 2017), and other conservation, restoration, and management options for natural and managed land (Griscom et al., 2017) and coastal ecosystems (McLeod et al., 2011). Biochar sequestration (Woolf et al., 2010; Smith, 2016; Werner et al., 2018) provides an additional route for terrestrial carbon storage. Other approaches are concerned with storing atmospheric carbon dioxide in geological formations. They include the combination of biomass use for energy production with carbon capture and storage (BECCS) (Obersteiner et al., 2001; Keith and Rhodes, 2002; Gough and Upham, 2011) and direct air capture with storage (DACCS) using chemical solvents and sorbents (Zeman and Lackner, 2004; Keith et al., 2006; Socolow et al., 2011). Further approaches investigate the mineralization of atmospheric carbon dioxide (Mazzotti et al., 2005; Matter et al., 2016), including enhanced weathering of rocks (Schuiling and Krijgsman, 2006; Hartmann et al., 2013; Strefler et al., 2018a). A fourth group of approaches is concerned with the sequestration of carbon dioxide in the oceans, for example by means of ocean alkalinization (Kheshgi, 1995; Rau, 2011; Ilyina et al., 2013; Lenton et al., 2018). The costs, CDR potential and environmental side effects of

several of these measures are increasingly investigated and compared in the literature, but large uncertainties remain, in particular concerning the feasibility and impact of large-scale deployment of CDR measures (The Royal Society, 2009; Smith et al., 2015; Psarras et al., 2017; Fuss et al., 2018) (see Chapter 4.3.7). There are also proposals to remove methane, nitrous oxide and halocarbons via photocatalysis from the atmosphere (Boucher and Folberth, 2010; de Richter et al., 2017), but a broader assessment of their effectiveness, cost and sustainability impacts is lacking to date.

Only some of these approaches have so far been considered in IAMs (see Section 2.3.1.2). The mitigation scenario literature up to AR5 mostly included BECCS and, to a more limited extent, afforestation and reforestation (Clarke et al., 2014). Since then, some 2°C- and 1.5°C-consistent pathways including additional CDR measures such as DACCS (Chen and Tavoni, 2013; Marcucci et al., 2017; Lehtilä and Koljonen, 2018; Strefler et al., 2018b) and soil carbon sequestration (Frank et al., 2017) have become available. Other, more speculative approaches, in particular ocean-based CDR and removal of non-CO_2 gases, have not yet been taken up by the literature on mitigation pathways. See Supplementary Material 2.SM.1.2 for an overview on the coverage of CDR measures in models which contributed pathways to this assessment. Chapter 4.3.7 assesses the potential, costs, and sustainability implications of the full range of CDR measures.

Integrated assessment modelling has not yet explored land conservation, restoration and management options to remove carbon dioxide from the atmosphere in sufficient depth, despite land management having a potentially considerable impact on the terrestrial carbon stock (Erb et al., 2018). Moreover, associated CDR measures have low technological requirements, and come with potential environmental and social co-benefits (Griscom et al., 2017). Despite the evolving capabilities of IAMs in accounting for a wider range of CDR measures, 1.5°C-consistent pathways assessed here continue to predominantly rely on BECCS and afforestation/reforestation (see Supplementary Material 2.SM.1.2). However, IAMs with spatially explicit land-use modelling include a full accounting of land-use change emissions comprising carbon stored in the terrestrial biosphere and soils. Net CDR in the AFOLU sector, including but not restricted to afforestation and reforestation, can thus in principle be inferred by comparing AFOLU CO_2 emissions between a baseline scenario and a 1.5°C-consistent pathway from the same model and study. However, baseline AFOLU CO_2 emissions can not only be reduced by CDR in the AFOLU sector but also by measures to reduce deforestation and preserve land carbon stocks. The pathway literature and pathway data available to this assessment do not yet allow separating the two contributions. As a conservative approximation, the additional net negative AFOLU CO_2 emissions below the baseline are taken as a proxy for AFOLU CDR in this assessment. Because this does not include CDR that was deployed before reaching net zero AFOLU CO_2 emissions, this approximation is a lower-bound for terrestrial CDR in the AFOLU sector (including all mitigation-policy-related factors that lead to net negative AFOLU CO_2 emissions).

The scale and type of CDR deployment in 1.5°C-consistent pathways varies widely (Figure 2.9 and 2.10). Overall CDR deployment over the 21st century is substantial in most of the pathways, and deployment levels cover a wide range, on the order of 100–1000 Gt CO_2 in 1.5°C pathways with no or limited overshoot (730 [260–1030] GtCO$_2$, for median and 5th–95th percentile range). Both BECCS (480 [0–1000] GtCO$_2$ in 1.5°C pathways with no or limited overshoot) and AFOLU CDR measures including afforestation and reforestation (210 [10–540] GtCO$_2$ in 1.5°C pathways with no or limited overshoot) can play a major role,[4] but for both cases pathways exist where they play no role at all. This shows the flexibility in substituting between individual CDR measures, once a portfolio of options becomes available. The high end of the CDR deployment range is populated by high overshoot pathways, as illustrated by pathway archetype S5 based on SSP5 (fossil-fuelled development, see Section 2.3.1.1) and characterized by very large BECCS deployment to return warming to 1.5°C by 2100 (Kriegler et al., 2017). In contrast, the low end is populated by a few pathways with no or limited overshoot that limit CDR to on the order of 100–200 GtCO$_2$ over the 21st century, coming entirely from terrestrial CDR measures with no or small use of BECCS. These are pathways with very low energy demand facilitating the rapid phase-out of fossil fuels and process emissions that exclude BECCS and CCS use (Grubler et al., 2018) and/or pathways with rapid shifts to sustainable

food consumption freeing up sufficient land areas for afforestation and reforestation (Haberl et al., 2011; van Vuuren et al., 2018). Some pathways use neither BECCS nor afforestation but still rely on CDR through considerable net negative CO$_2$ emissions in the AFOLU sector around mid-century (Holz et al., 2018b). We conclude that the role of BECCS as a dominant CDR measure in deep mitigation pathways has been reduced since the time of the AR5. This is related to three factors: a larger variation of underlying assumptions about socio-economic drivers (Riahi et al., 2017; Rogelj et al., 2018) and associated energy (Grubler et al., 2018) and food demand (van Vuuren et al., 2018); the incorporation of a larger portfolio of mitigation and CDR options (Marcucci et al., 2017; Grubler et al., 2018; Lehtilä and Koljonen, 2018; Liu et al., 2018; van Vuuren et al., 2018); and targeted analysis of deployment limits for (specific) CDR measures (Holz et al., 2018b; Kriegler et al., 2018a; Strefler et al., 2018b), including the availability of bioenergy (Bauer et al., 2018), CCS (Krey et al., 2014a; Grubler et al., 2018) and afforestation (Popp et al., 2014b, 2017). As additional CDR measures are being built into IAMs, the prevalence of BECCS is expected to be further reduced.

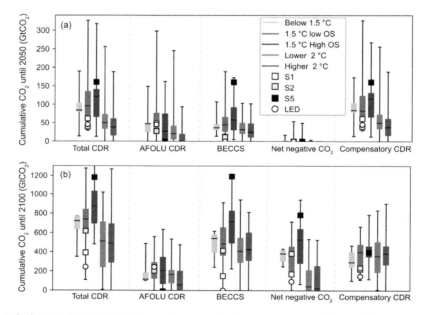

Figure 2.9 | Cumulative CDR deployment in 1.5°C-consistent pathways in the literature as reported in the database collected for this assessment until 2050 (panel a) and until 2100 (panel b). Total CDR comprises all forms of CDR, including AFOLU CDR and BECCS, and, in a few pathways, other CDR measures like DACCS. It does not include CCS combined with fossil fuels (which is not a CDR technology as it does not result in active removal of CO$_2$ from the atmosphere). AFOLU CDR has not been reported directly and is hence represented by means of a proxy: the additional amount of net negative CO$_2$ emissions in the AFOLU sector compared to a baseline scenario (see text for a discussion). 'Compensatory CO$_2$' depicts the cumulative amount of CDR that is used to neutralize concurrent residual CO$_2$ emissions. 'Net negative CO$_2$' describes the additional amount of CDR that is used to produce net negative CO$_2$ emissions, once residual CO$_2$ emissions are neutralized. The two quantities add up to total CDR for individual pathways (not for percentiles and medians, see Footnote 4).

As discussed in Section 2.3.2, CDR can be used in two ways in mitigation pathways: (i) to move more rapidly towards the point of carbon neutrality and maintain it afterwards in order to stabilize global mean temperature rise, and (ii) to produce net negative CO$_2$ emissions, drawing down anthropogenic CO$_2$ in the atmosphere in order to decline global mean temperature after an overshoot peak (Kriegler et al., 2018b; Obersteiner et al., 2018). Both uses are important in 1.5°C-consistent pathways (Figure 2.9 and 2.10). Because of the tighter remaining 1.5°C

carbon budget, and because many pathways in the literature do not restrict exceeding this budget prior to 2100, the relative weight of the net negative emissions component of CDR increases compared to 2°C-consistent pathways. The amount of compensatory CDR remains roughly the same over the century. This is the net effect of stronger deployment of compensatory CDR until mid-century to accelerate the approach to carbon neutrality and less compensatory CDR in the second half of the century due to deeper mitigation of end-use sectors

---

[4]   The median and percentiles of the sum of two quantities is in general not equal to the sum of the medians and percentiles, respectively, of the two quantitites.

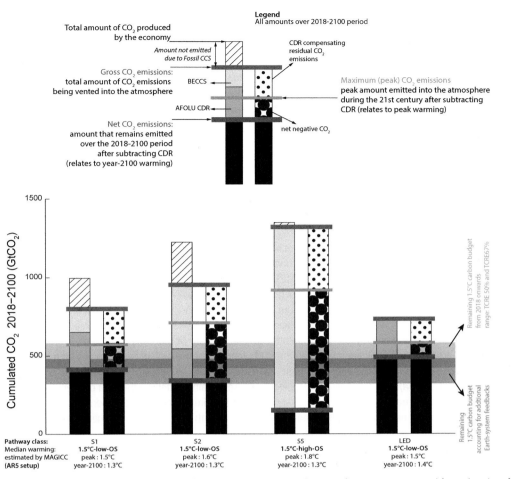

Figure 2.10 | Accounting of cumulative CO₂ emissions for the four 1.5°C-consistent pathway archetypes. See top panel for explanation of the bar plots. Total CDR is the difference between gross (red horizontal bar) and net (purple horizontal bar) cumulative CO₂ emissions over the period 2018–2100, and it is equal to the sum of the BECCS (grey) and AFOLU CDR (green) contributions. Cumulative net negative emissions are the difference between peak (orange horizontal bar) and net (purple) cumulative CO₂ emissions. The blue shaded area depicts the estimated range of the remaining carbon budget for a two-in-three to one-in-two chance of staying below1.5°C. The grey shaded area depicts the range when accounting for additional Earth system feedbacks.

in 1.5°C-consistent pathways (Luderer et al., 2018). Comparing median levels, end-of-century net cumulative CO₂ emissions are roughly 600 GtCO₂ smaller in 1.5°C compared to 2°C-consistent pathways, with approximately two thirds coming from further reductions of gross CO₂ emissions and the remaining third from increased CDR deployment. As a result, median levels of total CDR deployment in 1.5°C-consistent pathways are larger than in 2°C-consistent pathways (Figure 2.9), but with marked variations in each pathway class.

Ramp-up rates of individual CDR measures in 1.5°C-consistent pathways are provided in Table 2.4. BECCS deployment is still limited in 2030, but ramps up to median levels of 3 (Below-1.5°C), 5 (1.5°C-low-OS) and 7 GtCO₂ yr⁻¹ (1.5°C-high-OS) in 2050, and to 6 (Below-1.5°C), 12 (1.5°C-low-OS) and 15 GtCO₂ yr⁻¹ (1.5°C-high-OS) in 2100, respectively. In 1.5°C pathways with no or limited overshoot, this amounts to 0–1, 0–8, and 0–16 GtCO₂ yr⁻¹ in 2030, 2050, and 2100, respectively (ranges refer to the union of the min-max range of the Below-1.5°C and the interquartile range of the 1.5°C-low-OS class; see Table 2.4). Net CDR in the AFOLU sector reaches slightly lower levels in 2050, and stays more constant until 2100. In 1.5°C pathways with no or limited overshoot, AFOLU CDR amounts to 0–5,

1–11, and 1–5 GtCO₂ yr⁻¹ (see above for the definition of the ranges) in 2030, 2050, and 2100, respectively. In contrast to BECCS, AFOLU CDR is more strongly deployed in non-overshoot than overshoot pathways. This indicates differences in the timing of the two CDR approaches. Afforestation is scaled up until around mid-century, when the time of carbon neutrality is reached in 1.5°C-consistent pathways, while BECCS is projected to be used predominantly in the 2nd half of the century (Figure 2.5). This reflects the fact that afforestation is a readily available CDR technology, while BECCS is more costly and much less mature a technology. As a result, the two options contribute differently to compensating concurrent CO₂ emissions (until 2050) and to producing net negative CO₂ emissions (post-2050). BECCS deployment is particularly strong in pathways with high overshoots but can also feature in pathways with low overshoot (see Figure 2.5 and 2.10). Annual deployment levels until mid-century are not found to be significantly different between 2°C-consistent pathways and 1.5°C-consistent pathways with no or low overshoot. This suggests similar implementation challenges for ramping up BECCS deployment at the rates projected in the pathways (Honegger and Reiner, 2018; Nemet et al., 2018). The feasibility and sustainability of upscaling CDR at these rates is assessed in Chapter 4.3.7.

Concerns have been raised that building expectations about large-scale CDR deployment in the future can lead to an actual reduction of near-term mitigation efforts (Geden, 2015; Anderson and Peters, 2016; Dooley and Kartha, 2018). The pathway literature confirms that CDR availability influences the shape of mitigation pathways critically (Krey et al., 2014a; Holz et al., 2018b; Kriegler et al., 2018a; Strefler et al., 2018b). Deeper near-term emissions reductions are required to reach the 1.5°C–2°C target range if CDR availability is constrained. As a result, the least-cost benchmark pathways to derive GHG emissions gap estimates (UNEP, 2017) are dependent on assumptions about CDR availability. Using GHG benchmarks in climate policy makes implicit assumptions about CDR availability (Fuss et al., 2014; van Vuuren et al., 2017a). At the same time, the literature also shows that rapid and stringent mitigation as well as large-scale CDR deployment occur simultaneously in 1.5°C pathways due to the tight remaining carbon budget (Luderer et al., 2018). Thus, an emissions gap is identified even for high CDR availability (Strefler et al., 2018b), contradicting a wait-and-see approach. There are significant trade-offs between near-term action, overshoot and reliance on CDR deployment in the long-term which are assessed in Section 2.3.5.

Box 2.1 | Bioenergy and BECCS Deployment in Integrated Assessment Modelling

Bioenergy can be used in various parts of the energy sector of IAMs, including for electricity, liquid fuel, biogas, and hydrogen production. It is this flexibility that makes bioenergy and bioenergy technologies valuable for the decarbonization of energy use (Klein et al., 2014; Krey et al., 2014a; Rose et al., 2014a; Bauer et al., 2017, 2018). Most bioenergy technologies in IAMs are also available in combination with CCS (BECCS). Assumed capture rates differ between technologies, for example, about 90% for electricity and hydrogen production and about 40–50% for liquid fuel production. Decisions about bioenergy deployment in IAMs are based on economic considerations to stay within a carbon budget that is consistent with a long-term climate goal. IAMs consider both the value of bioenergy in the energy system and the value of BECCS in removing CO_2 from the atmosphere. Typically, if bioenergy is strongly limited, BECCS technologies with high capture rates are favoured. If bioenergy is plentiful IAMs tend to choose biofuel technologies with lower capture rates but high value for replacing fossil fuels in transport (Kriegler et al., 2013a; Bauer et al., 2018). Most bioenergy use in IAMs is combined with CCS if available (Rose et al., 2014a). If CCS is unavailable, bioenergy use remains largely unchanged or even increases due to the high value of bioenergy for the energy transformation (Bauer et al., 2018). As land impacts are tied to bioenergy use, the exclusion of BECCS from the mitigation portfolio will not automatically remove the trade-offs with food, water and other sustainability objectives due to the continued and potentially increased use of bioenergy.

IAMs assume bioenergy to be supplied mostly from second generation biomass feedstocks such as dedicated cellulosic crops (for example *Miscanthus* or poplar) as well as agricultural and forest residues. Detailed process IAMs include land-use models that capture competition for land for different uses (food, feed, fiber, bioenergy, carbon storage, biodiversity protection) under a range of dynamic factors including socio-economic drivers, productivity increases in crop and livestock systems, food demand, and land, environmental, biodiversity, and carbon policies. Assumptions about these factors can vary widely between different scenarios (Calvin et al., 2014; Popp et al., 2017; van Vuuren et al., 2018). IAMs capture a number of potential environmental impacts from bioenergy production, in particular indirect land-use change emissions from land conversion and nitrogen and water use for bioenergy production (Kraxner et al., 2013; Bodirsky et al., 2014; Bonsch et al., 2014; Obersteiner et al., 2016; Humpenöder et al., 2018). The impact of bioenergy production on soil degradation is an area of active IAM development and was not comprehensively accounted for in the mitigation pathways assessed in this report (but is, for example, in Frank et al., 2017). Whether bioenergy has large adverse impacts on environmental and societal goals depends in large parts on the governance of land use (Haberl et al., 2013; Erb et al., 2016b; Obersteiner et al., 2016; Humpenöder et al., 2018). Here IAMs often make idealized assumptions about effective land management, such as full protection of the land carbon stock by conservation measures and a global carbon price, respectively, but variations on these assumptions have also been explored (Calvin et al., 2014; Popp et al., 2014a).

2.3.4.2 Sustainability implications of CDR deployment in 1.5°C pathways

Strong concerns about the sustainability implications of large-scale CDR deployment in deep mitigation pathways have been raised in the literature (Williamson and Bodle, 2016; Boysen et al., 2017b; Dooley and Kartha, 2018; Heck et al., 2018), and a number of important knowledge gaps have been identified (Fuss et al., 2016). An assessment of the literature on implementation constraints and sustainable development implications of CDR measures is provided in Chapter 4, Section 4.3.7 and the Cross-chapter Box 7 in Chapter 3. An initial discussion of potential environmental side effects of CDR deployment in 1.5°C-consistent pathways is provided in this section. Chapter 4, Section 4.3.7 then contrasts CDR deployment in 1.5°C-consistent pathways with other branches of literature on limitations of CDR. Integrated modelling aims to explore a range of developments compatible with specific climate goals and often does not include the full set of broader environmental and societal concerns beyond climate change. This has given rise to the concept of sustainable development pathways (Cross-Chapter Box 1 in Chapter 1) (van Vuuren et al., 2015), and there is an increasing body of work to extend integrated modelling to cover a broader range of sustainable development goals (Section 2.6). However, only some of the available 1.5°C-consistent pathways were developed within a larger sustainable development context (Bertram et al., 2018; Grubler et al., 2018; Rogelj et al., 2018; van Vuuren et al., 2018). As discussed in Section 2.3.4.1, those pathways are characterized by low energy and/or food demand effectively limiting fossil-fuel substitution and alleviating land competition, respectively. They also include regulatory policies for deepening early action and ensuring environmental protection (Bertram et al., 2018). Overall sustainability implications of 1.5°C-consistent pathways are assessed in Section 2.5.3 and Chapter 5, Section 5.4.

Individual CDR measures have different characteristics and therefore would carry different risks for their sustainable deployment at scale (Smith et al., 2015). Terrestrial CDR measures, BECCS and enhanced weathering of rock powder distributed on agricultural lands require land. Those land-based measures could have substantial impacts on environmental services and ecosystems (Cross-Chapter Box 7 in Chapter 3) (Smith and Torn, 2013; Boysen et al., 2016; Heck et al., 2016; Krause et al., 2017). Measures like afforestation and bioenergy with and without CCS that directly compete with other land uses could have significant impacts on agricultural and food systems (Creutzig et al., 2012, 2015; Calvin et al., 2014; Popp et al., 2014b, 2017; Kreidenweis et al., 2016; Boysen et al., 2017a; Frank et al., 2017; Stevanović et al., 2017; Strapasson et al., 2017; Humpenöder et al., 2018). BECCS using dedicated bioenergy crops could substantially increase agricultural water demand (Bonsch et al., 2014; Séférian et al., 2018) and nitrogen fertilizer use (Bodirsky et al., 2014). DACCS and BECCS rely on CCS and would require safe storage space in geological formations, including management of leakage risks (Pawar et al., 2015) and induced seismicity (Nicol et al., 2013). Some approaches like DACCS have high energy demand (Socolow et al., 2011). Most of the CDR measures currently discussed could have significant impacts on either land, energy, water, or nutrients if deployed at scale (Smith et al., 2015). However, actual trade-offs depend on a multitude factors (Haberl et al., 2011; Erb et al., 2012; Humpenöder et al., 2018), including the modalities of CDR deployment (e.g., on marginal vs. productive land) (Bauer et al., 2018), socio-economic developments (Popp et al., 2017), dietary choices (Stehfest et al., 2009; Popp et al., 2010; van Sluisveld et al., 2016; Weindl et al., 2017; van Vuuren et al., 2018), yield increases, livestock productivity and other advances in agricultural technology (Havlik et al., 2013; Valin et al., 2013; Havlík et al., 2014; Weindl et al., 2015; Erb et al., 2016b), land policies (Schmitz et al., 2012; Calvin et al., 2014; Popp et al., 2014a), and governance of land use (Unruh, 2011; Buck, 2016; Honegger and Reiner, 2018).

Figure 2.11 shows the land requirements for BECCS and afforestation in the selected 1.5°C-consistent pathway archetypes, including the LED (Grubler et al., 2018) and S1 pathways (Fujimori, 2017; Rogelj et al., 2018) following a sustainable development paradigm. As discussed, these land-use patterns are heavily influenced by assumptions about, among other things, future population levels, crop yields, livestock production systems, and food and livestock demand, which all vary between the pathways (Popp et al., 2017) (Section 2.3.1.1). In pathways that allow for large-scale afforestation in addition to BECCS, land demand for afforestation can be larger than for BECCS (Humpenöder et al., 2014). This follows from the assumption in the modelled pathways that, unlike bioenergy crops, forests are not harvested to

allow unabated carbon storage on the same patch of land. If wood harvest and subsequent processing or burial are taken into account, this finding can change. There are also synergies between the various uses of land, which are not reflected in the depicted pathways. Trees can grow on agricultural land (Zomer et al., 2016), and harvested wood can be used with BECCS and pyrolysis systems (Werner et al., 2018). The pathways show a very substantial land demand for the two CDR measures combined, up to the magnitude of the current global cropland area. This is achieved in IAMs in particular by a conversion of pasture land freed by intensification of livestock production systems, pasture intensification and/or demand changes (Weindl et al., 2017), and to a more limited extent, cropland for food production, as well as expansion into natural land. However, pursuing such large-scale changes in land use would pose significant food supply, environmental and governance challenges, concerning both land management and tenure (Unruh, 2011; Erb et al., 2012, 2016b; Haberl et al., 2013; Haberl, 2015; Buck, 2016), particularly if synergies between land uses, the relevance of dietary changes for reducing land demand, and co-benefits with other sustainable development objectives are not fully recognized. A general discussion of the land-use transformation in 1.5°C-consistent pathways is provided in Section 2.4.4.

An important consideration for CDR which moves carbon from the atmosphere to the geological, oceanic or terrestrial carbon pools is the permanence of carbon stored in these different pools (Matthews and Caldeira, 2008; NRC, 2015; Fuss et al., 2016; Jones et al., 2016) (see also Chapter 4, Section 4.3.7 for a discussion). Terrestrial carbon can be returned to the atmosphere on decadal time scales by a variety of mechanisms, such as soil degradation, forest pest outbreaks and forest fires, and therefore requires careful consideration of policy frameworks to manage carbon storage, for example, in forests (Gren and Aklilu, 2016). There are similar concerns about outgassing of CO_2 from ocean storage (Herzog et al., 2003), unless it is transformed to a substance that does not easily exchange with the atmosphere, for example, ocean alkalinity or buried marine biomass (Rau, 2011). Understanding of the assessment and management of the potential risk of CO_2 release from geological storage of CO_2 has improved since the IPCC Special Report on Carbon Dioxide Capture and Storage (IPCC, 2005) with experience and the development of management practices in geological storage projects, including risk management to prevent sustentative leakage (Pawar et al., 2015). Estimates of leakage risk have been updated to include scenarios of unregulated drilling and limited wellbore integrity (Choi et al., 2013) and find that about 70% of stored CO_2 would still be retained after 10,000 years in these circumstances (Alcalde et al., 2018). The literature on the potential environmental impacts from the leakage of CO_2 – and approaches to minimize these impacts should a leak occur – has also grown and is reviewed by Jones et al. (2015). To the extent that non-permanence of terrestrial and geological carbon storage is driven by socio-economic and political factors, there are parallels to questions of fossil-fuel reservoirs remaining in the ground (Scott et al., 2015).

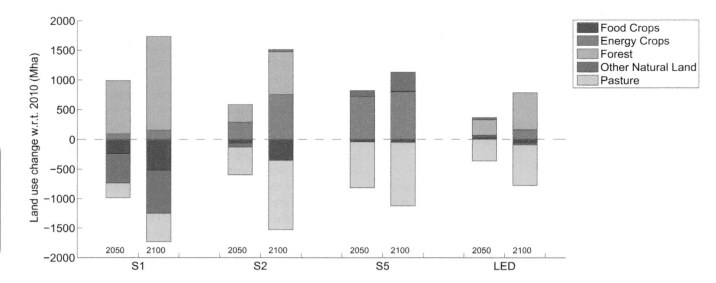

Figure 2.11 | Land-use changes in 2050 and 2100 in the illustrative 1.5°C-consistent pathway archetypes (Fricko et al., 2017; Fujimori, 2017; Kriegler et al., 2017; Grubler et al., 2018; Rogelj et al., 2018). Changes in land for food crops, energy crops, forest, pasture and other natural land are shown, compared to 2010.

2.3.5   Implications of Near-Term Action in 1.5°C Pathways

Less CO_2 emission reductions in the near term would require steeper and deeper reductions in the longer term in order to meet specific warming targets afterwards (Riahi et al., 2015; Luderer et al., 2016a). This is a direct consequence of the quasi-linear relationship between the total cumulative amount of CO_2 emitted into the atmosphere and global mean temperature rise (Matthews et al., 2009; Zickfeld et al., 2009; Collins et al., 2013; Knutti and Rogelj, 2015). Besides this clear geophysical trade-off over time, delaying GHG emissions reductions over the coming years also leads to economic and institutional lock-in into carbon-intensive infrastructure, that is, the continued investment in and use of carbon-intensive technologies that are difficult or costly to phase-out once deployed (Unruh and Carrillo-Hermosilla, 2006; Jakob et al., 2014; Erickson et al., 2015; Steckel et al., 2015; Seto et al., 2016; Michaelowa et al., 2018). Studies show that to meet stringent climate targets despite near-term delays in emissions reductions, models prematurely retire carbon-intensive infrastructure, in particular coal without CCS (Bertram et al., 2015a; Johnson et al., 2015). The AR5 reports that delaying mitigation action leads to substantially higher rates of emissions reductions afterwards, a larger reliance on CDR technologies in the long term, and higher transitional and long-term economic impacts (Clarke et al., 2014). The literature mainly focuses on delayed action until 2030 in the context of meeting a 2°C goal (den Elzen et al., 2010; van Vuuren and Riahi, 2011; Kriegler et al., 2013b; Luderer et al., 2013, 2016a; Rogelj et al., 2013b; Riahi et al., 2015; OECD/IEA and IRENA, 2017). However, because of the smaller carbon budget consistent with limiting warming to 1.5°C and the absence of a clearly declining long-term trend in global emissions to date, these general insights apply equally, or even more so, to the more stringent mitigation context of 1.5°C-consistent pathways. This

is further supported by estimates of committed emissions due to fossil fuel-based infrastructure (Seto et al., 2016; Edenhofer et al., 2018).

All available 1.5°C pathways that explore consistent mitigation action from 2020 onwards peak global Kyoto-GHG emissions in the next decade and already decline Kyoto-GHG emissions to below 2010 levels by 2030. The near-term emissions development in these pathways can be compared with estimated emissions in 2030 implied by the Nationally Determined Contributions (NDCs) submitted by Parties to the Paris Agreement (Figure 2.12). Altogether, the unconditional (conditional) NDCs are assessed to result in global Kyoto-GHG emissions on the order of 52–58 (50–54) GtCO2e yr–1 in 2030 (e.g., den Elzen et al., 2016; Fujimori et al., 2016; UNFCCC, 2016; Rogelj et al., 2017; Rose et al., 2017b; Benveniste et al., 2018; Vrontisi et al., 2018; see Cross-Chapter Box 11 in Chapter 4 for detailed assessment). In contrast, 1.5°C pathways with limited overshoot available to this assessment show an interquartile range of about 26–31 (median 28) $GtCO_2e$ yr^{-1} in 2030[5] (Table 2.4, Section 2.3.3). Based on these ranges, this report assesses the emissions gap for a two-in-three chance of limiting warming to 1.5°C to be 26 (19–29) and 28 (22–33) $GtCO_2e$ (median and interquartile ranges) for conditional and unconditional NDCs, respectively (Cross-Chapter Box 11, applying GWP-100 values from the IPCC Second Assessment Report).

The later emissions peak and decline, the more CO_2 will have accumulated in the atmosphere. Peak cumulated CO_2 emissions – and consequently peak temperatures – increase with higher 2030 emissions levels (Figure 2.12). Current NDCs (Cross-Chapter Box 11 in Chapter 4) are estimated to lead to CO_2 emissions of about 400–560 $GtCO_2$ from 2018 to 2030 (Rogelj et al., 2016a). Available 1.5°C- and 2°C-consistent pathways with 2030 emissions in the range estimated

---

[5]   Note that aggregated Kyoto-GHG emissions implied by the NDCs from Cross-Chapter Box 11 in Chapter 4 and Kyoto-GHG ranges from the pathway classes in Chapter 2 are only approximately comparable, because this chapter applies GWP-100 values from the IPCC Fourth Assessment Report while the NDC Cross-Chapter Box 11 applies GWP-100 values from the IPCC Second Assessment Report. At a global scale, switching between GWP-100 values of the Second to the Fourth IPCC Assessment Report would result in an increase in estimated aggregated Kyoto-GHG emissions of no more than about 3% in 2030 (UNFCCC, 2016).

for the NDCs rely on an assumed swift and widespread deployment of CDR after 2030, and show peak cumulative CO_2 emissions from 2018 of about 800–1000 $GtCO_2$, above the remaining carbon budget for a one-in-two chance of remaining below 1.5°C. These emissions reflect that no pathway is able to project a phase-out of CO_2 emissions starting from year-2030 NDC levels of about 40 $GtCO_2$ yr^{-1} (Fawcett et al., 2015; Rogelj et al., 2016a) to net zero in less than about 15 years. Based on the implied emissions until 2030, the high challenges of the assumed

post-2030 transition, and the assessment of carbon budgets in Section 2.2.2, global warming is assessed to exceed 1.5°C if emissions stay at the levels implied by the NDCs until 2030 (Figure 2.12). The chances of remaining below 1.5°C in these circumstances remain conditional upon geophysical properties that are uncertain, but these Earth system response uncertainties would have to serendipitously align beyond current median estimates in order for current NDCs to become consistent with limiting warming to 1.5°C.

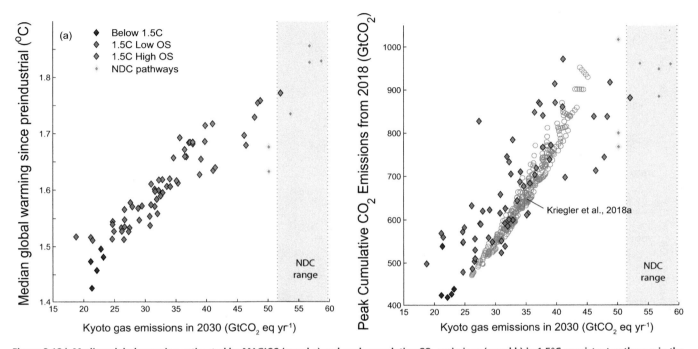

Figure 2.12 | Median global warming estimated by MAGICC (panel a) and peak cumulative CO_2 emissions (panel b) in 1.5°C-consistent pathways in the SR1.5 scenario database, as a function of CO_2-equivalent emissions (based on AR4 GWP-100) of Kyoto-GHGs in 2030. Pathways that were forced to go through the NDCs or a similarly high emissions point in 2030 by design are highlighted by yellow marker edges (see caption of Figure 2.13 and text for further details on the design of these pathways). The combined range of global Kyoto-GHG emissions in 2030 for the conditional and unconditional NDCs assessed in Cross-Chapter Box 11 is shown by the grey shaded area (adjusted to AR4 GWPs for comparison). As a second line of evidence, peak cumulative CO_2 emissions derived from a 1.5°C pathway sensitivity analysis (Kriegler et al., 2018b) are shown by grey circles in the right-hand panel. Circles show gross fossil-fuel and industry emissions of the sensitivity cases, increased by assumptions about the contributions from AFOLU (5 $GtCO_2$ yr^{-1} until 2020, followed by a linear phase out until 2040) and non-CO_2 Kyoto-GHGs (median non-CO_2 contribution from 1.5°C-consistent pathways available in the database: 10 $GtCO_2$e yr^{-1} in 2030), and reduced by assumptions about CDR deployment until the time of net zero CO_2 emissions (limiting case for CDR deployment assumed in (Kriegler et al., 2018b) (logistic growth to 1, 4, 10 $GtCO_2$ yr^{-1} in 2030, 2040, and 2050, respectively, leading to approximately 100 $GtCO_2$ of CDR by mid-century).

It is unclear whether following NDCs until 2030 would still allow global mean temperature to return to 1.5°C by 2100 after a temporary overshoot, due to the uncertainty associated with the Earth system response to net negative emissions after a peak (Section 2.2). Available IAM studies are working with reduced-form carbon cycle–climate models like MAGICC, which assume a largely symmetric Earth-system response to positive and net negative CO_2 emissions. The IAM findings on returning warming to 1.5°C from NDCs after a temporary temperature overshoot are hence all conditional on this assumption. Two types of pathways with 1.5°C-consistent action starting in 2030 have been considered in the literature (Luderer et al., 2018) (Figure 2.13): pathways aiming to obtain the same end-of-century carbon budget as 1.5°C-consistent pathways starting in 2020 despite higher emissions until 2030, and pathways assuming the same mitigation stringency after 2030 as in 1.5°C-consistent pathways starting in 2020 (approximated by using the same global price of emissions as

found in least-cost pathways starting from 2020). An IAM comparison study found increasing challenges to implementing pathways with the same end-of-century carbon budgets after following NDCs until 2030 (Luderer et al., 2018). The majority of model experiments (four out of seven) failed to produce NDC pathways that would return cumulative CO_2 emissions over the 2016–2100 period to 200 $GtCO_2$, indicating limitations to the availability and timing of CDR. The few such pathways that were identified show highly disruptive features in 2030 (including abrupt transitions from moderate to very large emissions reduction and low carbon energy deployment rates) indicating a high risk that the required post-2030 transformations are too steep and abrupt to be achieved by the mitigation measures in the models *(high confidence)*. NDC pathways aiming for a cumulative 2016–2100 CO_2 emissions budget of 800 $GtCO_2$ were more readily obtained (Luderer et al., 2018), and some were classified as 1.5°C-high-OS pathways in this assessment (Section 2.1).

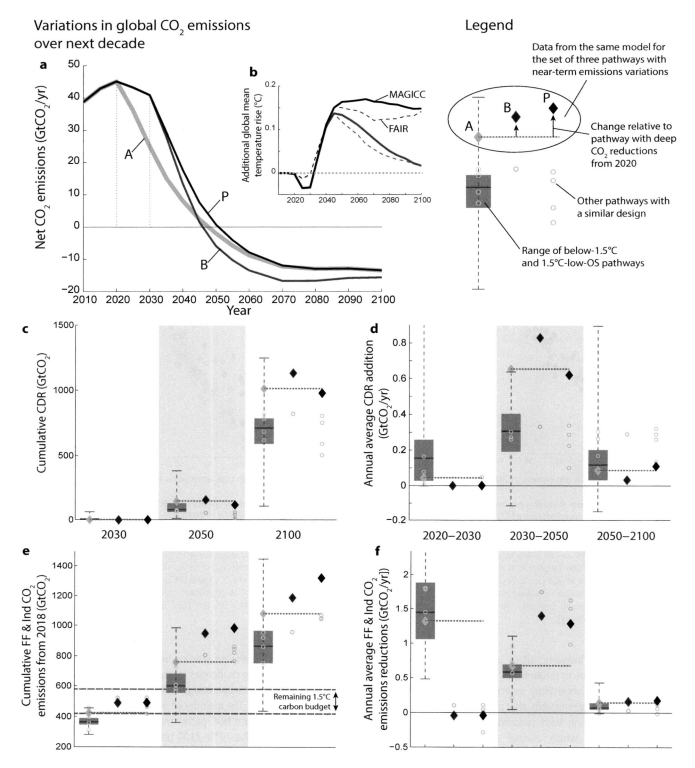

Figure 2.13 | Comparison of 1.5°C-consistent pathways starting action as of 2020 (A; light-blue diamonds) with pathways following the NDCs until 2030 and aiming to limit warming to 1.5°C thereafter. The 1.5°C pathways that follow the NDCs until 2030 either aim for the same cumulative CO₂ emissions by 2100 as the pathways that start action as of 2020 (B; red diamonds) or assume the same mitigation stringency as reflected by the price of emissions in associated least-cost 1.5°C-consistent pathways starting from 2020 (P; black diamonds). Panels show (a) the underlying emissions pathways, (b) additional warming in the delay scenarios compared to 2020 action case, (c) cumulated CDR, (d) CDR ramp-up rates, (e) cumulated gross CO₂ emissions from fossil-fuel combustion and industrial (FFI) processes over the 2018–2100 period, and (f) gross FFI CO₂ emissions reductions rates. Scenario pairs or triplets (circles and diamonds) with 2020 and 2030 action variants were calculated by six (out of seven) models in the ADVANCE study symbols (Luderer et al., 2018) and five of them (passing near-term plausibility checks) are shown by symbols. Only two of five models could identify pathways with post-2030 action leading to a 2016–2100 carbon budget of about 200 GtCO₂ (red). The range of all 1.5°C pathways with no and low overshoot is shown by the boxplots.

NDC pathways that apply a post-2030 price of emissions as found in least-cost pathways starting from 2020 show infrastructural carbon lock-in as a result of following NDCs instead of least-cost action until 2030. A key finding is that carbon lock-ins persist long after 2030, with the majority of additional CO_2 emissions occurring during the 2030–2050 period. Luderer et al. (2018) find 90 (80–120) $GtCO_2$ additional emissions until 2030, growing to 240 (190–260) $GtCO_2$ by 2050 and 290 (200–200) $GtCO_2$ by 2100. As a result, peak warming is about 0.2°C higher and not all of the modelled pathways return warming to 1.5°C by the end of the century. There is a four sided trade-off between (i) near-term ambition, (ii) degree of overshoot, (iii) transitional challenges during the 2030–2050 period, and (iv) the amount of CDR deployment required during the century (Figure 2.13) (Holz et al., 2018b; Strefler et al., 2018b). Transition challenges, overshoot, and CDR requirements can be significantly reduced if global emissions peak before 2030

and fall below levels in line with current NDCs by 2030. For example, Strefler et al. (2018b) find that CDR deployment levels in the second half of the century can be halved in 1.5°C-consistent pathways with similar CO_2 emissions reductions rates during the 2030–2050 period if CO_2 emissions by 2030 are reduced by an additional 30% compared to NDC levels. Kriegler et al. (2018a) investigate a global rollout of selected regulatory policies and moderate carbon pricing policies. They show that additional reductions of about 10 $GtCO_2$e yr^{-1} can be achieved in 2030 compared to the current NDCs. Such a 20% reduction of year-2030 emissions compared to current NDCs would effectively lower the disruptiveness of post-2030 action. The strengthening of short-term policies in deep mitigation pathways has hence been identified as a way of bridging options to keep the Paris climate goals within reach (Bertram et al., 2015b; IEA, 2015a; Spencer et al., 2015; Kriegler et al., 2018a).

2.4 Disentangling the Whole-System Transformation

Mitigation pathways map out prospective transformations of the energy, land and economic systems over this century (Clarke et al., 2014). There is a diversity of potential pathways consistent with 1.5°C, yet they share some key characteristics summarized in Table 2.5. To explore characteristics of 1.5°C pathways in greater detail, this section focuses on changes in energy supply and demand, and changes in the AFOLU sector.

2.4.1 Energy System Transformation

The energy system links energy supply (Section 2.4.2) with energy demand (Section 2.4.3) through final energy carriers, including electricity and liquid, solid or gaseous fuels, that are tailored to their end-uses. To chart energy-system transformations in mitigation pathways, four macro-level decarbonization indicators associated with final energy are useful: limits on the increase of final energy demand, reductions in the carbon intensity of electricity, increases in the share of final energy provided by electricity, and reductions in the carbon

Table 2.5 | Overview of Key Characteristics of 1.5°C Pathways.

1.5°C Pathway Characteristic	Supporting Information	Reference
Rapid and profound near-term decarbonisation of energy supply	Strong upscaling of renewables and sustainable biomass and reduction of unabated (no CCS) fossil fuels, along with the rapid deployment of CCS, lead to a zero-emission energy supply system by mid-century.	Section 2.4.1 Section 2.4.2
Greater mitigation efforts on the demand side	All end-use sectors show marked demand reductions beyond the reductions projected for 2°C pathways. Demand reductions from IAMs for 2030 and 2050 lie within the potential assessed by detailed sectoral bottom-up assessments.	Section 2.4.3
Switching from fossil fuels to electricity in end-use sectors	Both in the transport and the residential sector, electricity covers markedly larger shares of total demand by mid-century.	Section 2.4.3.2 Section 2.4.3.3
Comprehensive emission reductions are implemented in the coming decade	Virtually all 1.5°C-consistent pathways decline net annual CO_2 emissions between 2020 and 2030, reaching carbon neutrality around mid-century. In 2030, below-1.5°C and 1.5°C-low-OS pathways show maximum net CO_2 emissions of 18 and 28 $GtCO_2$ yr^{-1}, respectively. GHG emissions in these scenarios are not higher than 34 $GtCO_2$e yr^{-1} in 2030.	Section 2.3.4
Additional reductions, on top of reductions from both CO_2 and non-CO_2 required for 2°C, are mainly from CO_2	Both CO_2 and the non-CO_2 GHGs and aerosols are strongly reduced by 2030 and until 2050 in 1.5°C pathways. The greatest difference to 2°C pathways, however, lies in additional reductions of CO_2, as the non-CO_2 mitigation potential that is currently included in integrated pathways is mostly already fully deployed for reaching a 2°C pathway.	Section 2.3.1.2
Considerable shifts in investment patterns	Low-carbon investments in the energy supply side (energy production and refineries) are projected to average 1.6–3.8 trillion 2010USD yr^{-1} globally to 2050. Investments in fossil fuels decline, with investments in unabated coal halted by 2030 in most available 1.5°C-consistent projections, while the literature is less conclusive for investments in unabated gas and oil. Energy demand investments are a critical factor for which total estimates are uncertain.	Section 2.5.2
Options are available to align 1.5°C pathways with sustainable development	Synergies can be maximized, and risks of trade-offs limited or avoided through an informed choice of mitigation strategies. Particularly pathways that focus on a lowering of demand show many synergies and few trade-offs.	Section 2.5.3
CDR at scale before mid-century	By 2050, 1.5°C pathways project deployment of BECCS at a scale of 3–7 $GtCO_2$$yr^{-1}$ (range of medians across 1.5°C pathway classes), depending on the level of energy demand reductions and mitigation in other sectors. Some 1.5°C pathways are available that do not use BECCS, but only focus terrestrial CDR in the AFOLU sector.	Section 2.3.3, 2.3.4.1

intensity of final energy other than electricity (referred to in this section as the carbon intensity of the residual fuel mix). Figure 2.14 shows changes of these four indicators for the pathways in the scenario database (Section 2.1.3 and Supplementary Material 2.SM.1.3) for 1.5°C and 2°C pathways (Table 2.1).

Pathways in both the 1.5°C and 2°C classes (Figure 2.14) generally show rapid transitions until mid-century, with a sustained but slower evolution thereafter. Both show an increasing share of electricity accompanied by a rapid decline in the carbon intensity of electricity. Both also show a generally slower decline in the carbon intensity of the residual fuel mix, which arises from the decarbonization of liquids, gases and solids provided to industry, residential and commercial activities, and the transport sector.

The largest differences between 1.5°C and 2°C pathways are seen in the first half of the century (Figure 2.14), where 1.5°C pathways generally show lower energy demand, a faster electrification of energy end-use, and a faster decarbonization of the carbon intensity of electricity and the residual fuel mix. There are very few pathways in the Below-1.5°C class (Figure 2.14). Those scenarios that are available, however, show a faster decline in the carbon intensity of electricity generation and residual fuel mix by 2030 than most pathways that are projected to temporarily overshoot 1.5°C and return by 2100 (or 2°C pathways). The Below-1.5°C pathways also appear to differentiate themselves from the other pathways as early as 2030 through reductions in final energy demand and increases in electricity share (Figure 2.14).

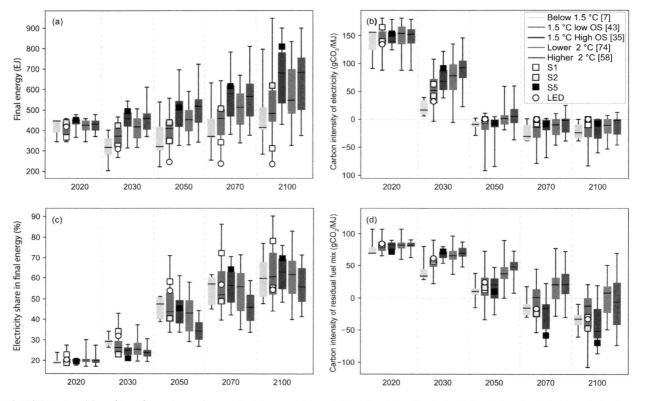

Figure 2.14 | Decomposition of transformation pathways into (a) energy demand, (b) carbon intensity of electricity, (c) the electricity share in final energy, and (d) the carbon intensity of the residual (non-electricity) fuel mix. Box plots show median, interquartile range and full range of pathways. Pathway temperature classes (Table 2.1) and illustrative pathway archetypes are indicated in the legend. Values following the class labels give the number of available pathways in each class.

2.4.2   Energy Supply

Several energy supply characteristics are evident in 1.5°C pathways assessed in this section: (i) growth in the share of energy derived from low-carbon-emitting sources (including renewables, nuclear and fossil fuel with CCS) and a decline in the overall share of fossil fuels without CCS (Section 2.4.2.1), (ii) rapid decline in the carbon intensity of electricity generation simultaneous with further electrification of energy end-use (Section 2.4.2.2), and (iii) the growth in the use of CCS applied to fossil and biomass carbon in most 1.5°C pathways (Section 2.4.2.3).

2.4.2.1   Evolution of primary energy contributions over time

By mid-century, the majority of primary energy comes from non-fossil-fuels (i.e., renewables and nuclear energy) in most 1.5°C pathways (Table 2.6). Figure 2.15 shows the evolution of primary energy supply over this century across 1.5°C pathways, and in detail for the four illustrative pathway archetypes highlighted in this chapter. Note that this section reports primary energy using the direct equivalent method on the basis of lower heating values (Bruckner et al., 2014).

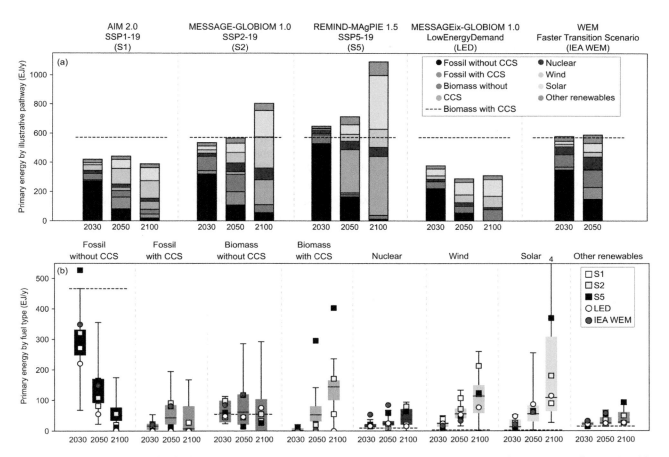

Figure 2.15 | Primary energy supply for the four illustrative pathway archetypes plus the IEA's Faster Transition Scenario (OECD/IEA and IRENA, 2017) (panel a), and their relative location in the ranges for pathways limiting warming to 1.5°C with no or limited overshoot (panel b). The category 'Other renewables' includes primary energy sources not covered by the other categories, for example, hydro and geothermal energy. The number of pathways that have higher primary energy than the scale in the bottom panel are indicated by the numbers above the whiskers. Black horizontal dashed lines indicates the level of primary energy supply in 2015 (IEA, 2017e). Box plots in the lower panel show the minimum–maximum range (whiskers), interquartile range (box), and median (vertical thin black line). Symbols in the lower panel show the four pathway archetypes S1 (white square), S2 (yellow square), S5 (black square), LED (white disc), as well as the IEA–(red disc). Pathways with no or limited overshoot included the Below-1.5°C and 1.5°C-low-OS classes.

The share of energy from renewable sources (including biomass, hydro, solar, wind and geothermal) increases in all 1.5°C pathways with no or limited overshoot, with the renewable energy share of primary energy reaching 38–88% in 2050 (Table 2.6), with an interquartile range of 52–67%. The magnitude and split between bioenergy, wind, solar, and hydro differ between pathways, as can be seen in the illustrative pathway archetypes in Figure 2.15. Bioenergy is a major supplier of primary energy, contributing to both electricity and other forms of final energy such as liquid fuels for transportation (Bauer et al., 2018). In 1.5°C pathways, there is a significant growth in bioenergy used in combination with CCS for pathways where it is included (Figure 2.15).

Nuclear power increases its share in most 1.5°C pathways with no or limited overshoot by 2050, but in some pathways both the absolute capacity and share of power from nuclear generators decrease (Table 2.15). There are large differences in nuclear power between models and across pathways (Kim et al., 2014; Rogelj et al., 2018). One of the reasons for this variation is that the future deployment of nuclear can be constrained by societal preferences assumed in narratives underlying the pathways (O'Neill et al., 2017; van Vuuren et al., 2017b). Some 1.5°C pathways with no or limited overshoot no longer see a role

for nuclear fission by the end of the century, while others project about 95 EJ yr⁻¹ of nuclear power in 2100 (Figure 2.15).

The share of primary energy provided by total fossil fuels decreases from 2020 to 2050 in all 1.5°C pathways, but trends for oil, gas and coal differ (Table 2.6). By 2050, the share of primary energy from coal decreases to 0–11% across 1.5°C pathways with no or limited overshoot, with an interquartile range of 1–7%. From 2020 to 2050 the primary energy supplied by oil changes by −93 to −9% (interquartile range −77 to −39%); natural gas changes by −88 to +85% (interquartile range −62 to −13%), with varying levels of CCS. Pathways with higher use of coal and gas tend to deploy CCS to control their carbon emissions (see Section 2.4.2.3). As the energy transition is accelerated by several decades in 1.5°C pathways compared to 2°C pathways, residual fossil-fuel use (i.e., fossil fuels not used for electricity generation) without CCS is generally lower in 2050 than in 2°C pathways, while combined hydro, solar, and wind power deployment is generally higher than in 2°C pathways (Figure 2.15).

In addition to the 1.5°C pathways included in the scenario database (Supplementary Material 2.SM.1.3), there are other analyses in the literature including, for example, sector-based analyses of energy demand and supply options. Even though they were not necessarily developed in the context of the 1.5°C target, they explore in greater detail some options for deep reductions in GHG emissions. For example, there are analyses of transitions to up to 100% renewable energy by 2050 (Creutzig et al., 2017; Jacobson et al., 2017), which describe what is entailed for a renewable energy share largely from solar and wind (and electrification) that is above the range of 1.5°C pathways available in the database, although there have been challenges to the assumptions used in high-renewable analyses (e.g., Clack et al., 2017). There are also analyses that result in a large role for nuclear energy in mitigation of GHGs (Hong et al., 2015; Berger et al., 2017a, b; Xiao and Jiang, 2018). BECCS could also contribute a larger share, but faces

challenges related to its land use and impact on food supply (Burns and Nicholson, 2017) (assessed in greater detail in Sections 2.3.4.2, 4.3.7 and 5.4). These analyses could, provided their assumptions prove plausible, expand the range of 1.5°C pathways.

In summary, the share of primary energy from renewables increases while that from coal decreases across 1.5°C pathways (*high confidence*). This statement is true for all 1.5°C pathways in the scenario database and associated literature (Supplementary Material 2.SM.1.3), and is consistent with the additional studies mentioned above, an increase in energy supply from lower-carbon-intensity energy supply, and a decrease in energy supply from higher-carbon-intensity energy supply.

Table 2.6 | Global primary energy supply of 1.5°C pathways from the scenario database (Supplementary Material 2.SM.1.3).
Values given for the median (maximum, minimum) across the full range of 85 available 1.5°C pathways. Growth Factor = [(primary energy supply in 2050)/(primary energy supply in 2020) − 1]

	Median (max, min)	Count	Primary Energy Supply (EJ)			Share in Primary Energy (%)			Growth (factor) 2020-2050
			2020	2030	2050	2020	2030	2050	
Below-1.5°C and 1.5°C-low-OS pathways	total primary	50	565.33 (619.70, 483.22)	464.50 (619.87, 237.37)	553.23 (725.40, 289.02)	NA	NA	NA	−0.05 (0.48, −0.51)
	renewables	50	87.14 (101.60, 60.16)	146.96 (203.90, 87.75)	291.33 (584.78, 176.77)	14.90 (20.39, 10.60)	29.08 (62.15, 18.24)	60.24 (87.89, 38.03)	2.37 (6.71, 0.91)
	biomass	50	60.41 (70.03, 40.54)	77.07 (113.02, 44.42)	152.30 (311.72, 40.36)	10.17 (13.66, 7.14)	17.22 (35.61, 9.08)	27.29 (54.10, 10.29)	1.71 (5.56, −0.42)
	non-biomass	50	26.35 (36.57, 17.78)	62.58 (114.41, 25.79)	146.23 (409.94, 53.79)	4.37 (7.19, 3.01)	13.67 (26.54, 5.78)	27.98 (61.61, 12.04)	4.28 (13.46, 1.45)
	wind & solar	44	10.93 (20.16, 2.61)	40.14 (82.66, 7.05)	121.82 (342.77, 27.95)	1.81 (3.66, 0.45)	9.73 (19.56, 1.54)	21.13 (51.52, 4.48)	10.00 (53.70, 3.71)
	nuclear	50	10.91 (18.55, 8.52)	16.26 (36.80, 6.80)	24.51 (66.30, 3.09)	2.10 (3.37, 1.45)	3.52 (9.61, 1.32)	4.49 (12.84, 0.44)	1.24 (5.01, −0.64)
	fossil	50	462.95 (520.41, 376.30)	310.36 (479.13, 70.14)	183.79 (394.71, 54.86)	82.53 (86.65, 77.73)	66.58 (77.30, 29.55)	32.79 (60.84, 8.58)	−0.59 (−0.21, −0.89)
	coal	50	136.89 (191.02, 83.23)	44.03 (127.98, 5.97)	24.15 (71.12, 0.92)	25.63 (30.82, 17.19)	9.62 (20.65, 1.31)	5.08 (11.43, 0.15)	−0.83 (−0.57, −0.99)
	gas	50	132.95 (152.80, 105.01)	112.51 (173.56, 17.30)	76.03 (199.18, 14.92)	23.10 (28.39, 18.09)	22.52 (35.05, 7.08)	13.23 (34.83, 3.68)	−0.40 (0.85, −0.88)
	oil	50	197.26 (245.15, 151.02)	156.16 (202.57, 38.94)	69.94 (167.52, 15.07)	34.81 (42.24, 29.00)	31.24 (39.84, 16.41)	12.89 (27.04, 2.89)	−0.66 (−0.09, −0.93)
1.5°C-high-OS	total primary	35	594.96 (636.98, 510.55)	559.04 (749.05, 419.28)	651.46 (1012.50, 415.31)	NA	NA	NA	0.13 (0.59, −0.27)
	renewables	35	89.84 (98.60, 66.57)	135.12 (159.84, 87.93)	323.21 (522.82, 177.66)	15.08 (18.58, 11.04)	23.65 (29.32, 13.78)	62.16 (86.26, 28.47)	2.68 (4.81, 1.17)
	biomass	35	62.59 (73.03, 48.42)	69.05 (98.27, 56.54)	160.16 (310.10, 71.17)	10.30 (14.23, 8.03)	13.64 (16.37, 9.03)	23.79 (45.79, 10.64)	1.71 (3.71, 0.19)
	non-biomass	35	28.46 (36.58, 17.60)	59.81 (92.12, 27.39)	164.91 (329.69, 55.72)	4.78 (6.64, 2.84)	10.23 (16.59, 4.49)	31.17 (45.86, 9.87)	6.10 (10.63, 1.38)
	wind & solar	26	11.32 (20.17, 1.91)	40.31 (65.50, 8.14)	139.20 (275.47, 30.92)	1.95 (3.66, 0.32)	7.31 (11.61, 1.83)	26.01 (38.79, 6.33)	16.06 (63.34, 3.13)
	nuclear	35	10.94 (14.27, 8.52)	16.12 (41.73, 6.80)	22.98 (115.80, 3.09)	1.86 (2.37, 1.45)	2.99 (5.57, 1.20)	4.17 (13.60, 0.43)	1.49 (7.22, −0.64)
	fossil	35	497.30 (543.29, 407.49)	397.76 (568.91, 300.63)	209.80 (608.39, 43.87)	83.17 (86.59, 79.39)	73.87 (82.94, 68.00)	33.58 (60.09, 7.70)	−0.56 (0.12, −0.91)
	coal	35	155.65 (193.55, 118.40)	70.99 (176.99, 19.15)	18.95 (134.69, 0.36)	25.94 (30.82, 19.10)	14.53 (26.35, 3.64)	4.14 (13.30, 0.05)	−0.87 (−0.30, −1.00)
	gas	35	138.01 (169.50, 107.07)	147.43 (208.55, 76.45)	97.71 (265.66, 15.96)	23.61 (27.35, 19.26)	25.79 (32.73, 14.69)	15.67 (33.80, 2.80)	−0.31 (0.99, −0.88)
	oil	35	195.02 (236.40, 154.66)	198.50 (319.80, 102.10)	126.20 (208.04, 24.68)	32.21 (38.87, 28.07)	33.27 (50.12, 24.35)	18.61 (27.30, 4.51)	−0.34 (0.06, −0.87)

Table 2.6 (continued)

Median (max, min)		Count	Primary Energy Supply (EJ)			Share in Primary Energy (%)			Growth (factor) 2020-2050
			2020	2030	2050	2020	2030	2050	
Two above classes combined	total primary	85	582.12 (636.98, 483.22)	502.81 (749.05, 237.37)	580.78 (1012.50, 289.02)	-	-	-	0.03 (0.59, −0.51)
	renewables	85	87.70 (101.60, 60.16)	139.48 (203.90, 87.75)	293.80 (584.78, 176.77)	15.03 (20.39, 10.60)	27.90 (62.15, 13.78)	60.80 (87.89, 28.47)	2.62 (6.71, 0.91)
	biomass	85	61.35 (73.03, 40.54)	75.28 (113.02, 44.42)	154.13 (311.72, 40.36)	10.27 (14.23, 7.14)	14.38 (35.61, 9.03)	26.38 (54.10, 10.29)	1.71 (5.56, −0.42)
	non-biomass	85	26.35 (36.58, 17.60)	61.60 (114.41, 25.79)	157.37 (409.94, 53.79)	4.40 (7.19, 2.84)	11.87 (26.54, 4.49)	28.60 (61.61, 9.87)	4.63 (13.46, 1.38)
	wind & solar	70	10.93 (20.17, 1.91)	40.17 (82.66, 7.05)	125.31 (342.77, 27.95)	1.81 (3.66, 0.32)	8.24 (19.56, 1.54)	22.10 (51.52, 4.48)	11.64 (63.34, 3.13)
	nuclear	85	10.93 (18.55, 8.52)	16.22 (41.73, 6.80)	24.48 (115.80, 3.09)	1.97 (3.37, 1.45)	3.27 (9.61, 1.20)	4.22 (13.60, 0.43)	1.34 (7.22, −0.64)
	fossil	85	489.52 (543.29, 376.30)	343.48 (568.91, 70.14)	198.58 (608.39, 43.87)	83.05 (86.65, 77.73)	69.19 (82.94, 29.55)	33.06 (60.84, 7.70)	−0.58 (0.12, −0.91)
	coal	85	147.09 (193.55, 83.23)	49.46 (176.99, 5.97)	23.84 (134.69, 0.36)	25.72 (30.82, 17.19)	10.76 (26.35, 1.31)	4.99 (13.30, 0.05)	−0.85 (−0.30, −1.00)
	gas	85	135.58 (169.50, 105.01)	127.99 (208.55, 17.30)	88.97 (265.66, 14.92)	23.28 (28.39, 18.09)	24.02 (35.05, 7.08)	13.46 (34.83, 2.80)	−0.37 (0.99, −0.88)
	oil	85	195.02 (245.15, 151.02)	175.69 (319.80, 38.94)	93.48 (208.04, 15.07)	33.79 (42.24, 28.07)	32.01 (50.12, 16.41)	16.22 (27.30, 2.89)	−0.54 (0.06, −0.93)

Table 2.7 | Global electricity generation of 1.5°C pathways from the scenarios database.
(Supplementary Material 2.SM.1.3). Values given for the median (maximum, minimum) values across the full range across 89 available 1.5°C pathways. Growth Factor = [(primary energy supply in 2050)/(primary energy supply in 2020) − 1].

Median (max, min)		Count	Electricity Generation (EJ)			Share in Electricity Generation (%)			Growth (factor) 2020–2050
			2020	2030	2050	2020	2030	2050	
TBelow -1.5°C and 1.5°C-low-OS pathways	total generation	50	98.45 (113.98, 83.53)	115.82 (152.40, 81.28)	215.58 (354.48, 126.96)	NA	NA	NA	1.15 (2.55, 0.28)
	renewables	50	26.28 (41.80, 18.50)	63.30 (111.70, 32.41)	145.50 (324.26, 90.66)	26.32 (41.84, 18.99)	53.68 (79.67, 37.30)	77.12 (96.65, 58.89)	4.48 (10.88, 2.65)
	biomass	50	2.02 (7.00, 0.76)	4.29 (11.96, 0.79)	20.35 (39.28, 0.24)	1.97 (6.87, 0.82)	3.69 (13.29, 0.73)	8.77 (30.28, 0.10)	6.42 (38.14, −0.93)
	non-biomass	50	24.21 (35.72, 17.70)	57.12 (101.90, 25.79)	135.04 (323.91, 53.79)	24.38 (40.43, 17.75)	49.88 (78.27, 29.30)	64.68 (96.46, 41.78)	4.64 (10.64, 1.45)
	wind & solar	50	1.66 (6.60, 0.38)	8.91 (48.04, 0.60)	39.04 (208.97, 2.68)	1.62 (7.90, 0.38)	8.36 (41.72, 0.53)	19.10 (60.11, 1.65)	26.31 (169.66, 5.23)
	nuclear	50	10.84 (18.55, 8.52)	15.46 (36.80, 6.80)	21.97 (64.72, 3.09)	12.09 (18.34, 8.62)	14.33 (31.63, 5.24)	8.10 (27.53, 1.02)	0.71 (4.97, −0.64)
	fossil	50	59.43 (68.75, 39.48)	36.51 (66.07, 2.25)	14.81 (57.76, 0.00)	61.32 (67.40, 47.26)	30.04 (52.86, 1.95)	8.61 (25.18, 0.00)	−0.74 (0.01, −1.00)
	coal	50	31.02 (42.00, 14.40)	8.83 (34.11, 0.00)	1.38 (17.39, 0.00)	32.32 (40.38, 17.23)	7.28 (27.29, 0.00)	0.82 (7.53, 0.00)	−0.96 (−0.56, −1.00)
	gas	50	24.70 (32.46, 13.44)	22.59 (42.08, 2.01)	12.79 (53.17, 0.00)	24.39 (35.08, 11.80)	20.18 (37.23, 1.75)	6.93 (24.87, 0.00)	−0.47 (1.27, −1.00)
	oil	50	2.48 (13.36, 1.12)	1.89 (7.56, 0.24)	0.10 (8.78, 0.00)	2.82 (11.73, 1.01)	1.95 (5.67, 0.21)	0.05 (3.80, 0.00)	−0.92 (0.36, −1.00)
1.5°C-high-OS	total generation	35	101.44 (113.96, 88.55)	125.26 (177.51, 89.60)	251.50 (363.10, 140.65)	NA	NA	NA	1.38 (2.19, 0.39)
	renewables	35	26.38 (31.83, 18.26)	53.32 (86.85, 30.06)	173.29 (273.92, 84.69)	28.37 (32.96, 17.38)	42.73 (65.73, 25.11)	82.39 (94.66, 35.58)	5.97 (8.68, 2.37)
	biomass	35	1.23 (6.47, 0.66)	2.14 (7.23, 0.86)	10.49 (40.32, 0.21)	1.22 (7.30, 0.63)	1.59 (6.73, 0.72)	3.75 (28.09, 0.08)	7.93 (33.32, −0.81)
	non-biomass	35	24.56 (30.70, 17.60)	47.96 (85.83, 27.39)	144.13 (271.17, 55.72)	26.77 (31.79, 16.75)	40.07 (64.96, 23.10)	69.72 (94.58, 27.51)	5.78 (8.70, 1.38)

Table 2.7 (continued next page)

Table 2.7 (continued)

	Median (max, min)	Count	Electricity Generation (EJ)			Share in Electricity Generation (%)			Growth (factor) 2020-2050
			2020	2030	2050	2020	2030	2050	
1.5°C-high-OS	wind & solar	35	2.24 (5.07, 0.42)	8.95 (36.52, 1.18)	65.08 (183.38, 13.79)	2.21 (5.25, 0.41)	7.48 (27.90, 0.99)	25.88 (61.24, 8.71)	30.70 (106.95, 4.87)
	nuclear	35	10.84 (14.08, 8.52)	16.12 (41.73, 6.80)	22.91 (115.80, 3.09)	10.91 (13.67, 8.62)	14.65 (23.51, 5.14)	11.19 (39.61, 1.12)	1.49 (7.22, −0.64)
	fossil	35	62.49 (76.76, 49.09)	48.08 (87.54, 30.99)	11.84 (118.12, 0.78)	61.58 (71.03, 54.01)	42.02 (59.48, 24.27)	6.33 (33.19, 0.27)	−0.80 (0.54, −0.99)
	coal	35	32.37 (46.20, 26.00)	16.22 (43.12, 1.32)	1.18 (46.72, 0.01)	32.39 (40.88, 24.41)	14.23 (29.93, 1.19)	0.55 (12.87, 0.00)	−0.96 (0.01, −1.00)
	gas	35	26.20 (41.20, 20.11)	26.45 (51.99, 16.45)	10.66 (67.94, 0.76)	26.97 (39.20, 19.58)	22.29 (43.43, 14.03)	5.29 (32.59, 0.26)	−0.57 (1.63, −0.97)
	oil	35	1.51 (6.28, 1.12)	0.61 (7.54, 0.36)	0.04 (7.47, 0.00)	1.51 (6.27, 1.01)	0.55 (6.20, 0.26)	0.02 (3.31, 0.00)	−0.99 (0.98, −1.00)
Two above classes combined	total generation	85	100.09 (113.98, 83.53)	120.01 (177.51, 81.28)	224.78 (363.10, 126.96)	NA	NA	NA	1.31 (2.55, 0.28)
	renewables	85	26.38 (41.80, 18.26)	59.50 (111.70, 30.06)	153.72 (324.26, 84.69)	27.95 (41.84, 17.38)	51.51 (79.67, 25.11)	77.52 (96.65, 35.58)	5.08 (10.88, 2.37)
	biomass	85	1.52 (7.00, 0.66)	3.55 (11.96, 0.79)	16.32 (40.32, 0.21)	1.55 (7.30, 0.63)	2.77 (13.29, 0.72)	8.02 (30.28, 0.08)	6.53 (38.14, −0.93)
	non-biomass	85	24.48 (35.72, 17.60)	55.68 (101.90, 25.79)	136.40 (323.91, 53.79)	25.00 (40.43, 16.75)	47.16 (78.27, 23.10)	66.75 (96.46, 27.51)	4.75 (10.64, 1.38)
	wind & solar	85	1.66 (6.60, 0.38)	8.95 (48.04, 0.60)	43.20 (208.97, 2.68)	1.67 (7.90, 0.38)	8.15 (41.72, 0.53)	19.70 (61.24, 1.65)	28.02 (169.66, 4.87)
	nuclear	85	10.84 (18.55, 8.52)	15.49 (41.73, 6.80)	22.64 (115.80, 3.09)	10.91 (18.34, 8.62)	14.34 (31.63, 5.14)	8.87 (39.61, 1.02)	1.21 (7.22, −0.64)
	fossil	85	61.35 (76.76, 39.48)	38.41 (87.54, 2.25)	14.10 (118.12, 0.00)	61.55 (71.03, 47.26)	33.96 (59.48, 1.95)	8.05 (33.19, 0.00)	−0.76 (0.54, −1.00)
	coal	85	32.37 (46.20, 14.40)	10.41 (43.12, 0.00)	1.29 (46.72, 0.00)	32.39 (40.88, 17.23)	8.95 (29.93, 0.00)	0.59 (12.87, 0.00)	−0.96 (0.01, −1.00)
	gas	85	24.70 (41.20, 13.44)	25.00 (51.99, 2.01)	11.92 (67.94, 0.00)	24.71 (39.20, 11.80)	21.03 (43.43, 1.75)	6.78 (32.59, 0.00)	−0.52 (1.63, −1.00)
	oil	85	1.82 (13.36, 1.12)	0.92 (7.56, 0.24)	0.08 (8.78, 0.00)	2.04 (11.73, 1.01)	0.71 (6.20, 0.21)	0.04 (3.80, 0.00)	−0.97 (0.98, −1.00)

2.4.2.2 Evolution of electricity supply over time

Electricity supplies an increasing share of final energy, reaching 34–71% in 2050, across 1.5°C pathways with no or limited overshoot (Figure 2.14), extending the historical increases in electricity share seen over the past decades (Bruckner et al., 2014). From 2020 to 2050, the quantity of electricity supplied in most 1.5°C pathways with no or limited overshoot more than doubles (Table 2.7). By 2050, the carbon intensity of electricity has fallen rapidly to −92 to +11 gCO_2 MJ^{-1} electricity across 1.5°C pathways with no or limited overshoot from a value of around 140 gCO_2 MJ^{-1} (range: 88–181 gCO_2 MJ^{-1}) in 2020 (Figure 2.14). A negative contribution to carbon intensity is provided by BECCS in most pathways (Figure 2.16).

By 2050, the share of electricity supplied by renewables increases from 23% in 2015 (IEA, 2017b) to 59–97% across 1.5°C pathways with no or limited overshoot. Wind, solar, and biomass together make a major contribution in 2050, although the share for each spans a wide range across 1.5°C pathways (Figure 2.16). Fossil fuels on the other hand have a decreasing role in electricity supply, with their share falling to 0–25% by 2050 (Table 2.7).

In summary, 1.5°C pathways include a rapid decline in the carbon intensity of electricity and an increase in electrification of energy end-use (*high confidence*). This is the case across all 1.5°C pathways and their associated literature (Supplementary Material 2.SM.1.3), with pathway trends that extend those seen in past decades, and results that are consistent with additional analyses (see Section 2.4.2.2).

2.4.2.3 Deployment of carbon capture and storage

Studies have shown the importance of CCS for deep mitigation pathways (Krey et al., 2014a; Kriegler et al., 2014b), based on its multiple roles to limit fossil-fuel emissions in electricity generation, liquids production, and industry applications along with the projected ability to remove CO_2 from the atmosphere when combined with bioenergy. This remains a valid finding for those 1.5°C and 2°C pathways that do not radically reduce energy demand or do not offer carbon-neutral alternatives to liquids and gases that do not rely on bioenergy.

There is a wide range of CCS that is deployed across 1.5°C pathways (Figure 2.17). A few 1.5°C pathways with very low energy demand do not include CCS at all (Grubler et al., 2018). For example, the LED pathway has no CCS, whereas other pathways, such as the S5 pathway,

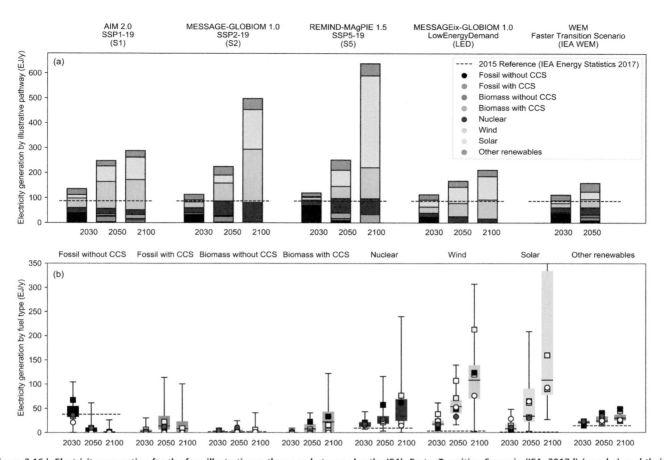

Figure 2.16 | Electricity generation for the four illustrative pathway archetypes plus the IEA's Faster Transition Scenario (IEA, 2017d) (panel a), and their relative location in the ranges for pathways limiting warming to 1.5°C with no or limited overshoot (panel b). The category 'Other renewables' includes electricity generation not covered by the other categories, for example, hydro and geothermal. The number of pathways that have higher primary energy than the scale in the bottom panel are indicated by the numbers above the whiskers. Black horizontal dashed lines indicate the level of primary energy supply in 2015 (IEA, 2017e). Box plots in the lower panel show the minimum–maximum range (whiskers), interquartile range (box), and median (vertical thin black line). Symbols in the lower panel show the four pathway archetypes – S1 (white square), S2 (yellow square), S5 (black square), LED (white disc) – as well as the IEA's Faster Transition Scenario (red disc). Pathways with no or limited overshoot included the Below-1.5°C and 1.5°C-low-OS classes.

rely on a large amount of BECCS to get to net-zero carbon emissions. The cumulative fossil and biomass CO_2 stored through 2050 ranges from zero to 300 $GtCO_2$ across 1.5°C pathways with no or limited overshoot, with zero up to 140 $GtCO_2$ from biomass captured and stored. Some pathways have very low fossil-fuel use overall, and consequently little CCS applied to fossil fuels. In 1.5°C pathways where the 2050 coal use remains above 20 EJ yr^{-1} in 2050, 33–100% is combined with CCS. While deployment of CCS for natural gas and coal vary widely across pathways, there is greater natural gas primary energy connected to CCS than coal primary energy connected to CCS in many pathways (Figure 2.17).

CCS combined with fossil-fuel use remains limited in some 1.5°C pathways (Rogelj et al., 2018), as the limited 1.5°C carbon budget penalizes CCS if it is assumed to have incomplete capture rates or if fossil fuels are assumed to continue to have significant lifecycle GHG emissions (Pehl et al., 2017). However, high capture rates are technically achievable now at higher cost, although efforts to date have focussed on reducing the costs of capture (IEAGHG, 2006; NETL, 2013).

The quantity of CO_2 stored via CCS over this century in 1.5°C pathways with no or limited overshoot ranges from zero to more than 1,200 $GtCO_2$, (Figure 2.17). The IPCC Special Report on Carbon Dioxide Capture and Storage (IPCC, 2005) found that that, worldwide, it is *likely* that there is a technical potential of at least about 2,000 $GtCO_2$ of storage capacity in geological formations. Furthermore, the IPCC (2005) recognized that there could be a much larger potential for geological storage in saline formations, but the upper limit estimates are uncertain due to lack of information and an agreed methodology. Since IPCC (2005), understanding has improved and there have been detailed regional surveys of storage capacity (Vangkilde-Pedersen et al., 2009; Ogawa et al., 2011; Wei et al., 2013; Bentham et al., 2014; Riis and Halland, 2014; Warwick et al., 2014; NETL, 2015) and improvement and standardization of methodologies (e.g., Bachu et al. 2007a, b). Dooley (2013) synthesized published literature on both the global geological storage resource as well as the potential demand for geologic storage in mitigation pathways, and found that the cumulative demand for CO_2 storage was small compared to a practical storage capacity estimate (as defined by Bachu et al., 2007a) of 3,900 $GtCO_2$ worldwide. Differences remain, however, in estimates of storage capacity due to, for example, the potential storage limitations of

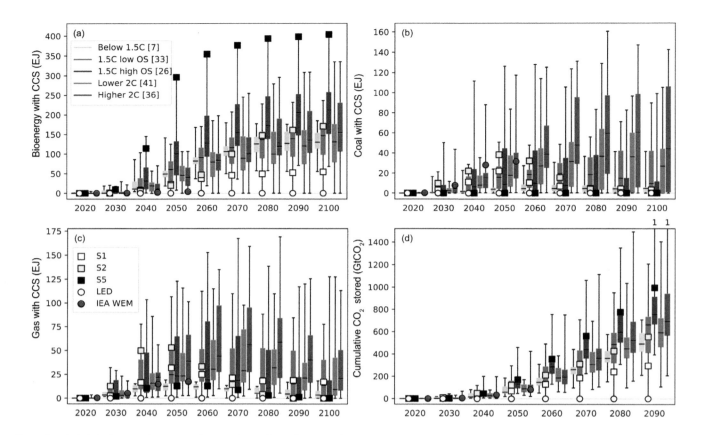

Figure 2.17 | CCS deployment in 1.5°C and 2°C pathways for (a) biomass, (b) coal and (c) natural gas (EJ of primary energy) and (d) the cumulative quantity of fossil (including from, e.g., cement production) and biomass CO_2 stored via CCS (in $GtCO_2$ stored). TBox plots show median, interquartile range and full range of pathways in each temperature class. Pathway temperature classes (Table 2.1), illustrative pathway archetypes, and the IEA's Faster Transition Scenario (IEA WEM) (OECD/IEA and IRENA, 2017) are indicated in the legend.

subsurface pressure build-up (Szulczewski et al., 2014) and assumptions on practices that could manage such issues (Bachu, 2015). Kearns et al. (2017) constructed estimates of global storage capacity of 8,000 to 55,000 $GtCO_2$ (accounting for differences in detailed regional and local estimates), which is sufficient at a global level for this century, but found that at a regional level, robust demand for CO_2 storage exceeds their lower estimate of regional storage available for some regions. However, storage capacity is not solely determined by the geological setting, and Bachu (2015) describes storage engineering practices that could further extend storage capacity estimates. In summary, the storage capacity of all of these global estimates is larger than the cumulative CO_2 stored via CCS in 1.5°C pathways over this century.

There is uncertainty in the future deployment of CCS given the limited pace of current deployment, the evolution of CCS technology that would be associated with deployment, and the current lack of incentives for large-scale implementation of CCS (Bruckner et al., 2014; Clarke et al., 2014; Riahi et al., 2017). Given the importance of CCS in most mitigation pathways and its current slow pace of improvement, the large-scale deployment of CCS as an option depends on the further development of the technology in the near term. Chapter 4 discusses how progress on CCS might be accelerated.

2.4.3 Energy End-Use Sectors

Since the power sector is almost decarbonized by mid-century in both 1.5°C and 2°C pathways, major differences come from CO_2 emission reductions in end-use sectors. Energy-demand reductions are key and common features in 1.5°C pathways, and they can be achieved by efficiency improvements and various specific demand-reduction measures. Another important feature is end-use decarbonization including by electrification, although the potential and challenges in each end-use sector vary significantly.

In the following sections, the potential and challenges of CO_2 emission reductions towards 1.5°C and 2°C- consistent pathways are discussed for each end-use energy sector (industry, buildings, and transport). For this purpose, two types of pathways are analysed and compared: IAM (integrated assessment modelling) studies and sectoral (detailed) studies. IAM data are extracted from the database that was compiled for this assessment (see Supplementary Material 2.SM.1.3), and the sectoral data are taken from a recent series of publications; 'Energy Technology Perspectives' (ETP) (IEA, 2014, 2015b, 2016a, 2017a), the IEA/IRENA report (OECD/IEA and IRENA, 2017), and the Shell Sky report (Shell International B.V., 2018). The IAM pathways are categorized according to their temperature rise in 2100 and the overshoot of temperature during the century (see Table 2.1 in Section 2.1). Since the number of Below-1.5°C pathways is small, the following analyses focus only on the features of the 1.5°C-low-OS and 1.5°C-high-OS pathways (hereafter denoted together as 1.5°C overshoot pathways or IAM-1.5DS-OS) and 2°C-consistent pathways (IAM-2DS). In order to show the diversity of IAM pathways, we again show specific data from the four illustrative pathways archetypes used throughout this chapter (see Sections 2.1 and 2.3).

IEA ETP-B2DS ('Beyond 2 Degrees') and ETP-2DS are pathways with a 50% chance of limiting temperature rise below 1.75°C and 2°C by 2100, respectively (IEA, 2017a). The IEA-66%2DS pathway keeps global mean temperature rise below 2°C, not just in 2100 but also over the course of the 21st century, with a 66% chance of being below 2°C by 2100 (OECD/IEA and IRENA, 2017). The comparison of CO_2 emission trajectories between ETP-B2DS and IAM-1.5DS-OS show that these are consistent up to 2060 (Figure 2.18). IEA scenarios assume that only a very low level of BECCS is deployed to help offset emissions in difficult-to-decarbonize sectors, and that global energy-related CO_2 emissions do not turn net negative at any time but stay at zero from 2060 to 2100 (IEA, 2017a). Therefore, although its temperature rise in 2100 is below 1.75°C rather than below 1.5°C, this scenario can give information related to a 1.5°C overshoot pathway up to 2050. The trajectory of IEA-66%2DS (also referred to in other publications as IEA's 'Faster Transition Scenario') lies between IAM-1.5DS-OS and IAM-2DS pathway ranges, and IEA-2DS stays in the range of 2°C-consistent IAM pathways. The Shell-Sky scenario aims to hold the temperature rise to well below 2°C, but it is a delayed action pathway relative to others, as can be seen in Figure 2.18.

Energy-demand reduction measures are key to reducing CO_2 emissions from end-use sectors for low-carbon pathways. The upstream energy reductions can be from several times to an order of magnitude larger than the initial end-use demand reduction. There are interdependencies among the end-use sectors and between energy-supply and end-use sectors, which elevate the importance of a wide, systematic approach. As shown in Figure 2.19, global final energy consumption grows by 30% and 10% from 2010 to 2050 for 2°C-consistent and 1.5°C overshoot pathways from IAMs, respectively, while much higher growth of 75% is projected for reference scenarios. The ranges within a specific pathway class are due to a variety of factors as introduced in Section 2.3.1, as well as differences between modelling frameworks. The important energy efficiency and conservation improvements that facilitate many of the 1.5°C pathways raise the issue of potential rebound effects (Saunders, 2015), which, while promoting development, can make the achievement of low-energy demand futures more difficult than modelling studies anticipate (see Sections 2.5 and 2.6).

Final energy demand is driven by demand in energy services for mobility, residential and commercial activities (buildings), and manufacturing. Projections of final energy demand depend heavily on assumptions about socio-economic futures as represented by the SSPs (Bauer et al., 2017) (see Sections 2.1, 2.3 and 2.5). The structure of this demand drives the composition of final energy use in terms of energy carriers (electricity, liquids, gases, solids, hydrogen etc.).

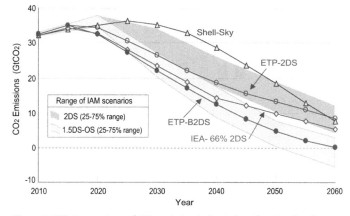

Figure 2.18 | Comparison of CO_2 emission trajectories of sectoral pathways (IEA ETP-B2DS, ETP-2DS, IEA-66%2DS, Shell-Sky) with the ranges of IAM pathway (2DS are 2°C-consistent pathways and 1.5DS-OS are 1.5°C overshoot pathways). The CO_2 emissions shown here are the energy-related emissions, including industrial process emissions.

Figure 2.19 shows the structure of global final energy demand in 2030 and 2050, indicating the trend toward electrification and fossil fuel usage reduction. This trend is more significant in 1.5°C pathways than 2°C pathways. Electrification continues throughout the second half of the century, leading to a 3.5- to 6-fold increase in electricity demand (interquartile range; median 4.5) by the end of the century relative to today (Grubler et al., 2018; Luderer et al., 2018). Since the electricity sector is completely decarbonized by mid-century in 1.5°C pathways (see Figure 2.20), electrification is the primary means to decarbonize energy end-use sectors.

The CO_2 emissions[6] of end-use sectors and carbon intensity are shown in Figure 2.20. The projections of IAMs and IEA studies show rather different trends, especially in the carbon intensity. These differences come from various factors, including the deployment of CCS, the level of fuel switching and efficiency improvements, and the effect of structural and behavioural changes. IAM projections are generally optimistic for the industry sectors, but not for buildings and transport sectors. Although GDP increases by a factor of 3.4 from 2010 to 2050, the total energy consumption of end-use sectors grows by only about 30% and 20% in 1.5°C overshoot and 2°C-consistent pathways, respectively. However, CO_2 emissions would need to be reduced further to achieve the stringent temperature limits. Figure 2.20 shows that the reduction in CO_2 emissions of end-use sectors is larger and more rapid in 1.5°C overshoot than 2°C-consistent pathways, while emissions from the power sector are already almost zero in 2050 in both sets of pathways, indicating that supply-side emissions reductions are almost fully exploited already in 2°C-consistent pathways (see Figure 2.20) (Rogelj et al., 2015b, 2018; Luderer et al., 2016b). The emission reductions in end-use sectors are largely made possible by efficiency improvements, demand reduction measures and electrification, but the level of emissions reductions varies across end-use sectors. While the carbon intensity of the industry and buildings sectors decreases

---

[6] This section reports 'direct' CO_2 emissions as reported for pathways in the database for the report. As shown below, the emissions from electricity are nearly zero around 2050, so the impact of indirect emissions on the whole emission contributions of each sector is very small in 2050.

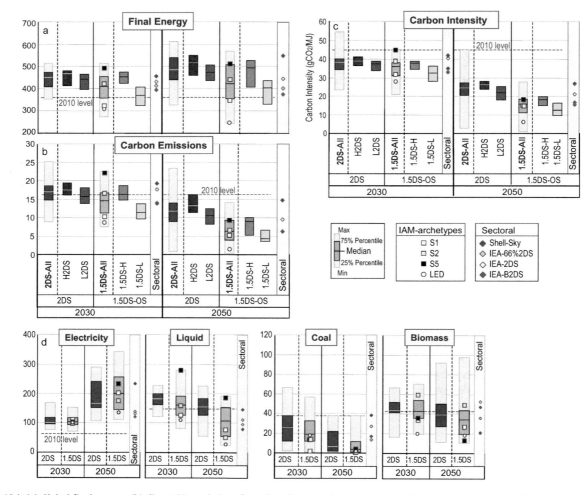

Figure 2.19 | (a) Global final energy, (b) direct CO_2 emissions from the all energy demand sectors, (c) carbon intensity, and (d) structure of final energy (electricity, liquid fuel, coal, and biomass). The squares and circles indicate the IAM archetype pathways and diamonds indicate the data of sectoral scenarios. The red dotted line indicates the 2010 level. H2DS = Higher-2°C, L2DS = Lower-2°C, 1.5DS-H = 1.5°C-high-OS, 1.5DS-L = 1.5°C-low-OS. The label 1.5DS combines both high and low overshoot 1.5°C-consistent pathway. See Section 2.1 for descriptions.

to a very low level of around 10 gCO_2 MJ^{-1}, the carbon intensity of transport becomes the highest of any sector by 2040 due to its higher reliance on oil-based fuels. In the following subsections, the potential and challenges of CO_2 emission reduction in each end-use sector are discussed in detail.

2.4.3.1   Industry

The industry sector is the largest end-use sector, both in terms of final energy demand and GHG emissions. Its direct CO_2 emissions currently account for about 25% of total energy-related and process CO_2 emissions, and emissions have increased at an average annual rate of 3.4% between 2000 and 2014, significantly faster than total CO_2 emissions (Hoesly et al., 2018). In addition to emissions from the combustion of fossil fuels, non-energy uses of fossil fuels in the petrochemical industry and metal smelting, as well as non-fossil fuel process emissions (e.g., from cement production) contribute a small amount (~5%) to the sector's CO_2 emissions inventory. Material industries are particularly energy and emissions intensive: together, the steel, non-ferrous metals, chemicals, non-metallic minerals, and

pulp and paper industries accounted for close to 66% of final energy demand and 72% of direct industry-sector emissions in 2014 (IEA, 2017a). In terms of end-uses, the bulk of energy in manufacturing industries is required for process heating and steam generation, while most electricity (but smaller shares of total final energy) is used for mechanical work (Banerjee et al., 2012; IEA, 2017a).

As shown in Figure 2.21, a major share of the additional emission reductions required for 1.5°C-overshoot pathways compared to those in 2°C-consistent pathways comes from industry. Final energy, CO_2 emissions, and carbon intensity are consistent in IAM and sectoral studies, but in IAM-1.5°C-overshoot pathways the share of electricity is higher than IEA-B2DS (40% vs. 25%) and hydrogen is also considered to have a share of about 5% versus 0%. In 2050, final energy is increased by 30% and 5% compared with the 2010 level (red dotted line) for 1.5°C-overshoot and 2°C-consistent pathways, respectively, but CO_2 emissions are decreased by 80% and 50% and carbon intensity by 80% and 60%, respectively. This additional decarbonization is brought by switching to low-carbon fuels and CCS deployment.

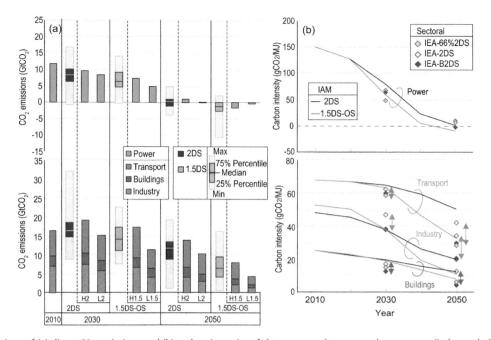

Figure 2.20 | Comparison of (a) direct CO₂ emissions and (b) carbon intensity of the power and energy end-use sectors (industry, buildings, and transport sectors) between IAMs and sectoral studies (IEA-ETP and IEA/IRENA). Diamond markers in panel (b) show data for IEA-ETP scenarios (2DS and B2DS), and IEA/IRENA scenario (66%2DS). Note: for the data from IAM studies, there is rather large variation of projections for each indicator. Please see the details in the following figures in each end-use sector section.

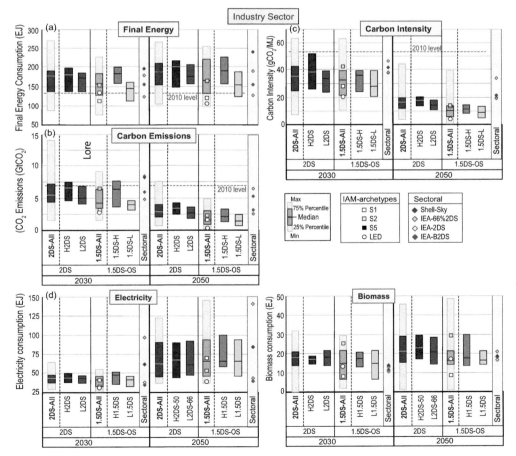

Figure 2.21 | Comparison of (a) final energy, (b) direct CO₂ emissions, (c) carbon intensity, (d) electricity and biomass consumption in the industry sector between IAM and sectoral studies. The squares and circles indicate the IAM archetype pathways and diamonds the data of sectoral scenarios. The red dotted line indicates the 2010 level. H2DS = Higher-2°C, L2DS = Lower-2°C, 1.5DS-H = 1.5°C-high-OS, 1.5DS-L = 1.5°C-low-OS. The label 1.5DS combines both high and low overshoot 1.5°C-consistent pathways. Section 2.1 for descriptions.

Broadly speaking, the industry sector's mitigation measures can be categorized in terms of the following five strategies: (i) reducing demand, (ii) energy efficiency, (iii) increasing electrification of energy demand, (iv) reducing the carbon content of non-electric fuels, and (v) deploying innovative processes and application of CCS. IEA ETP estimates the relative contribution of different measures for CO_2 emission reduction in their B2DS scenario compared with their reference scenario in 2050 as follows: energy efficiency 42%, innovative process and CCS 37%, switching to low-carbon fuels and feedstocks 13% and material efficiency (include efficient production and use to contribute to demand reduction) 8%. The remainder of this section delves more deeply into the potential mitigation contributions of these strategies as well as their limitations.

Reduction in the use of industrial materials, while delivering similar services, or improving the quality of products could help to reduce energy demand and overall system-level CO_2 emissions. Strategies include using materials more intensively, extending product lifetimes, increasing recycling, and increasing inter-industry material synergies, such as clinker substitution in cement production (Allwood et al., 2013; IEA, 2017a). Related to material efficiency, use of fossil-fuel feedstocks could shift to lower-carbon feedstocks, such as from oil to natural gas and biomass, and end-uses could shift to more sustainable materials, such as biomass-based materials, reducing the demand for energy-intensive materials (IEA, 2017a).

Reaping energy efficiency potentials hinges critically on advanced management practices, such as energy management systems, in industrial facilities as well as targeted policies to accelerate adoption of the best available technology (see Section 2.5). Although excess energy, usually as waste heat, is inevitable, recovering and reusing this waste heat under economically and technically viable conditions benefits the overall energy system. Furthermore, demand-side management strategies could modulate the level of industrial activity in line with the availability of resources in the power system. This could imply a shift away from peak demand and as power supply decarbonizes, this demand-shaping potential could shift some load to times with high portions of low-carbon electricity generation (IEA, 2017a).

In the industry sector, energy demand increases more than 40% between 2010 and 2050 in baseline scenarios. However, in the 1.5°C-overshoot and 2°C-consistent pathways from IAMs, the increase is only 30% and 5%, respectively (Figure 2.21). These energy-demand reductions encompass both efficiency improvements in production and reductions in material demand, as most IAMs do not discern these two factors.

CO_2 emissions from industry increase by 30% in 2050 compared to 2010 in baseline scenarios. By contrast, these emissions are reduced by 80% and 50% relative to 2010 levels in 1.5°C-overshoot and 2°C-consistent pathways from IAMs, respectively (Figure 2.21). By mid-

century, CO_2 emissions per unit of electricity are projected to decrease to near zero in both sets of pathways (see Figure 2.20). An accelerated electrification of the industry sector thus becomes an increasingly powerful mitigation option. In the IAM pathways, the share of electricity increases up to 30% by 2050 in 1.5°C-overshoot pathways (Figure 2.21) from 20% in 2010. Some industrial fuel uses are substantially more difficult to electrify than others, and electrification would have other effects on the process, including impacts on plant design, cost and available process integration options (IEA, 2017a).[7]

In 1.5°C-overshoot pathways, the carbon intensity of non-electric fuels consumed by industry decreases to 16 gCO_2 MJ^{-1} by 2050, compared to 25 gCO_2 MJ^{-1} in 2°C-consistent pathways. Considerable carbon intensity reductions are already achieved by 2030, largely via a rapid phase-out of coal. Biomass becomes an increasingly important energy carrier in the industry sector in deep-decarbonization pathways, but primarily in the longer term (in 2050, biomass accounts for only 10% of final energy consumption even in 1.5°C-overshoot pathways). In addition, hydrogen plays a considerable role as a substitute for fossil-based non-electric energy demands in some pathways.

Without major deployment of new sustainability-oriented low-carbon industrial processes, the 1.5°C-overshoot target is difficult to achieve. Bringing such technologies and processes to commercial deployment requires significant investment in research and development. Some examples of innovative low-carbon process routes include: new steelmaking processes such as upgraded smelt reduction and upgraded direct reduced iron, inert anodes for aluminium smelting, and full oxy-fuelling kilns for clinker production in cement manufacturing (IEA, 2017a).

CCS plays a major role in decarbonizing the industry sector in the context of 1.5°C and 2°C pathways, especially in industries with higher process emissions, such as cement, iron and steel industries. In 1.5°C-overshoot pathways, CCS in industry reaches 3 $GtCO_2$ yr^{-1} by 2050, albeit with strong variations across pathways. Given the projected long-lead times and need for technological innovation, early scale-up of industry-sector CCS is essential to achieving the stringent temperature target. Development and demonstration of such projects has been slow, however. Currently, only two large-scale industrial CCS projects outside of oil and gas processing are in operation (Global CCS Institute, 2016). The estimated current cost[8] of CO_2 avoided (in USD2015) ranges from $20–27 tCO_2^{-1} for gas processing and bio-ethanol production, and $60–138 tCO_2^{-1} for fossil fuel-fired power generation up to $104–188 tCO_2^{-1} for cement production (Irlam, 2017).

2.4.3.2   Buildings

In 2014, the buildings sector accounted for 31% of total global final energy use, 54% of final electricity demand, and 8% of energy-related CO_2 emissions (excluding indirect emissions due to electricity). When

---

[7]   Electrification can be linked with the heating and drying process by electric boilers and electro-thermal processes, and also with low-temperature heat demand by heat pumps. In the iron and steel industry, hydrogen produced by electrolysis can be used as a reduction agent of iron instead of coke. Excess resources, such as black liquor, will provide the opportunity to increase the systematic efficiency to use for electricity generation.

[8]   These are first-of-a-kind (FOAK) cost data.

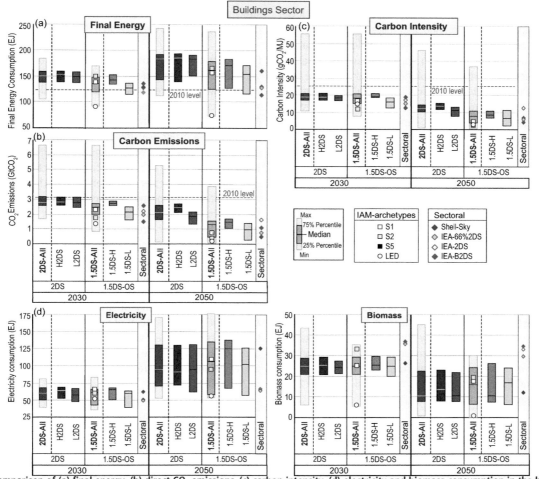

Figure 2.22 | Comparison of (a) final energy, (b) direct CO₂ emissions, (c) carbon intensity, (d) electricity and biomass consumption in the buildings sector between IAM and sectoral studies. The squares and circles indicate the IAM archetype pathways and diamonds the data of sectoral scenarios. The red dotted line indicates the 2010 level. H2DS = Higher-2°C, L2DS = Lower-2°C, 1.5DS-H = 1.5°C-high-OS, 1.5DS-L = 1.5°C-low-OS. The label 1.5DS combines both high and low overshoot 1.5°C-consistent pathways. Section 2.1 for descriptions.

upstream electricity generation is taken into account, buildings were responsible for 23% of global energy-related CO_2 emissions, with one-third of those from direct fossil fuel consumption (IEA, 2017a).

Past growth of energy consumption has been mainly driven by population and economic growth, with improved access to electricity, and higher use of electrical appliances and space cooling resulting from increasing living standards, especially in developing countries (Lucon et al., 2014). These trends will continue in the future and in 2050, energy consumption is projected to increase by 20% and 50% compared to 2010 in the IAM-1.5°C-overshoot and 2°C-consistent pathways, respectively (Figure 2.22). However, sectoral studies (IEA-ETP scenarios) show different trends. Energy consumption in 2050 decreases compared to 2010 in ETP-B2DS, and the reduction rate of CO_2 emissions is higher than in IAM pathways (Figure 2.22). Mitigation options are often more widely covered in sectoral studies (Lucon et al., 2014), leading to greater reductions in energy consumption and CO_2 emissions.

Emissions reductions are driven by a clear tempering of energy demand and a strong electrification of the buildings sector. The share of electricity in 2050 is 60% in 1.5°C-overshoot pathways, compared

with 50% in 2°C-consistent pathways (Figure 2.22). Electrification contributes to the reduction of direct CO_2 emissions by replacing carbon-intensive fuels, like oil and coal. Furthermore, when combined with a rapid decarbonization of the power system (see Section 2.4.1) it also enables further reduction of indirect CO_2 emissions from electricity. Sectoral bottom-up models generally estimate lower electrification potentials for the buildings sector in comparison to global IAMs (see Figure 2.22). Besides CO_2 emissions, increasing global demand for air conditioning in buildings may also lead to increased emissions of HFCs in this sector over the next few decades. Although these gases are currently a relatively small proportion of annual GHG emissions, their use in the air conditioning sector is expected to grow rapidly over the next few decades if alternatives are not adopted. However, their projected future impact can be significantly mitigated through better servicing and maintenance of equipment and switching of cooling gases (Shah et al., 2015; Purohit and Höglund-Isaksson, 2017).

IEA-ETP (IEA, 2017a) analysed the relative importance of various technology measures toward the reduction of energy and CO_2 emissions in the buildings sector. The largest energy savings potential is in heating and cooling demand, largely due to building envelope improvements and high efficiency and renewable equipment. In the

ETP-B2DS, energy demand for space heating and cooling is 33% lower in 2050 than in the reference scenario, and these reductions account for 54% of total reductions from the reference scenario. Energy savings from shifts to high-performance lighting, appliances, and water heating equipment account for a further 24% of the total reduction. The long-term, strategic shift away from fossil-fuel use in buildings, alongside the rapid uptake of energy efficient, integrated and renewable energy technologies (with clean power generation), leads to a drastic reduction of CO_2 emissions. In ETP-B2DS, the direct CO_2 emissions are 79% lower than the reference scenario in 2050, and the remaining emissions come mainly from the continued use of natural gas.

The buildings sector is characterized by very long-living infrastructure, and immediate steps are hence important to avoid lock-in of inefficient carbon and energy-intensive buildings. This applies both to new buildings in developing countries where substantial new construction is expected in the near future and to retrofits of existing building stock in developed regions. This represents both a significant risk and opportunity for mitigation.[9] A recent study highlights the benefits of deploying the most advanced renovation technologies, which would avoid lock-in into less efficient measures (Güneralp et al., 2017). Aside from the effect of building envelope measures, adoption of energy-efficient technologies such as heat pumps and, more recently, light-emitting diodes is also important for the reduction of energy and CO_2 emissions (IEA, 2017a). Consumer choices, behaviour and building operation can also significantly affect energy consumption (see Chapter 4, Section 4.3).

2.4.3.3   Transport

Transport accounted for 28% of global final energy demand and 23% of global energy-related CO_2 emissions in 2014. Emissions increased by 2.5% annually between 2010 and 2015, and over the past half century the sector has witnessed faster emissions growth than any other. The transport sector is the least diversified energy end-use sector; the sector consumed 65% of global oil final energy demand, with 92% of transport final energy demand consisting of oil products (IEA, 2017a), suggesting major challenges for deep decarbonization.

Final energy, CO_2 emissions, and carbon intensity for the transport sector are shown in Figure 2.23. The projections of IAMs are more pessimistic than IEA-ETP scenarios, though both clearly project deep cuts in energy consumption and CO_2 emissions by 2050. For example, 1.5°C-overshoot pathways from IAMs project a reduction of 15% in energy consumption between 2015 and 2050, while ETP-B2DS projects a reduction of 30% (Figure 2.23). Furthermore, IAM pathways are generally more pessimistic in the projections of CO_2 emissions and carbon intensity reductions. In AR5 (Clarke et al., 2014; Sims et al., 2014), similar comparisons between IAMs and sectoral studies were performed and these were in good agreement with each other. Since the AR5, two important changes can be identified: rapid growth of electric vehicle sales in passenger cars, and more attention towards

structural changes in this sector. The former contributes to reduction of CO_2 emissions and the latter to reduction of energy consumption.

Deep emissions reductions in the transport sector would be achieved by several means. Technology-focused measures such as energy efficiency and fuel-switching are two of these. Structural changes that avoid or shift transport activity are also important. While the former solutions (technologies) always tend to figure into deep decarbonization pathways in a major way, this is not always the case with the latter, especially in IAM pathways. Comparing different types of global transport models, Yeh et al. (2016) find that sectoral (intensive) studies generally envision greater mitigation potential from structural changes in transport activity and modal choice. Though, even there, it is primarily the switching of passengers and freight from less- to more-efficient travel modes (e.g., cars, trucks and airplanes to buses and trains) that is the main strategy; other actions, such as increasing vehicle load factors (occupancy rates) and outright reductions in travel demand (e.g., as a result of integrated transport, land-use and urban planning), figure much less prominently. Whether these dynamics accurately reflect the actual mitigation potential of structural changes in transport activity and modal choice is a point of investigation. According to the recent IEA-ETP scenarios, the share of avoid (reduction of mobility demand) and shift (shifting to more efficient modes) measures in the reduction of CO_2 emissions from the reference to B2DS scenarios in 2050 amounts to 20% (IEA, 2017a).

The potential and strategies to reduce energy consumption and CO_2 emissions differ significantly among transport modes. In ETP-B2DS, the shares of energy consumption and CO_2 emissions in 2050 for each mode are rather different (see Table 2.8), indicating the challenge of decarbonizing heavy-duty vehicles (HDV, trucks), aviation, and shipping. The reduction of CO_2 emissions in the whole sector from the reference scenario to ETP-B2DS is 60% in 2050, with varying contributions per mode (Table 2.8). Since there is no silver bullet for this deep decarbonization, every possible measure would be required to achieve this stringent emissions outcome. The contribution of various measures for the CO_2 emission reduction from the reference scenario to the IEA-B2DS in 2050 can be decomposed to efficiency improvement (29%), biofuels (36%), electrification (15%), and avoid/shift (20%) (IEA, 2017a). It is noted that the share of electrification becomes larger compared with older studies, reflected by the recent growth of electric vehicle sales worldwide. Another new trend is the allocation of biofuels to each mode of transport. In IEA-B2DS, the total amount of biofuels consumed in the transport sector is 24EJ[10] in 2060, and allocated to LDV (light-duty vehicles, 17%), HDV (35%), aviation (28%), and shipping (21%), that is, more biofuels is allocated to the difficult-to-decarbonize modes (see Table 2.8).

In road transport, incremental vehicle improvements (including engines) are relevant, especially in the short to medium term. Hybrid electric vehicles are also instrumental to enabling the transition from

---

9   In this section, we only discuss the direct emissions from the sector, but the selection of building materials has a significant impact on the reduction of energy and emissions during production, such as shift from the steel and concrete to wood-based materials.

10   This is estimated for the biofuels produced in a "sustainable manner" from non-food crop feedstocks, which are capable of delivering significant lifecycle GHG emissions savings compared with fossil fuel alternatives, and which do not directly compete with food and feed crops for agricultural land or cause adverse sustainability impacts.

Table 2.8 | Transport sector indicators by mode in 2050 (IEA, 2017a).
Share of energy consumption, biofuel consumption, CO_2 emissions, and reduction of energy consumption and CO_2 emissions from 2014. (CO_2 emissions are well-to-wheel emissions, including the emission during the fuel production.), LDV: light duty vehicle, HDV: heavy duty vehicle.

	Share of Each Mode (%)			Reduction from 2014 (%)	
	Energy	Biofuel	CO₂	Energy	CO₂
LDV	36	17	30	51	81
HDV	33	35	36	8	56
Rail	6	-	−1	−136	107
Aviation	12	28	14	14	56
Shipping	17	21	21	26	29

internal combustion engine vehicles to electric vehicles, especially plug-in hybrid electric vehicles. Electrification is a powerful measure to decarbonize short-distance vehicles (passenger cars and two and three wheelers) and the rail sector. In road freight transport (trucks), systemic improvements (e.g., in supply chains, logistics, and routing) would be effective measures in conjunction with efficiency improvement of vehicles. Shipping and aviation are more challenging to decarbonize, while their demand growth is projected to be higher than other

transport modes. Both modes would need to pursue highly ambitious efficiency improvements and use of low-carbon fuels. In the near and medium term, this would be advanced biofuels while in the long term it could be hydrogen as direct use for shipping or an intermediate product for synthetic fuels for both modes (IEA, 2017a).

The share of low-carbon fuels in the total transport fuel mix increases to 10% and 16% by 2030 and to 40% and 58% by 2050

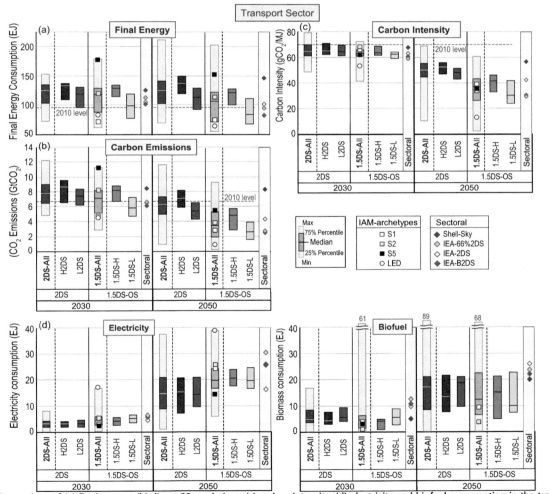

Figure 2.23 | Comparison of (a) final energy, (b) direct CO₂ emissions, (c) carbon intensity, (d) electricity and biofuel consumption in the transport sector between IAM and sectoral studies. The squares and circles indicate the IAM archetype pathways and diamonds the data of sectoral scenarios. The red dotted line indicates the 2010 level. H2DS = Higher-2°C, L2DS = Lower-2°C, 1.5DS-H = 1.5°C-high-OS, 1.5DS-L = 1.5°C-low-OS. The label 1.5DS combines both high and low overshoot 1.5°C-consistent pathways. Section 2.1 for descriptions.

in 1.5°C-overshoot pathways from IAMs and the IEA-B2DS pathway, respectively. The IEA-B2DS scenario is on the more ambitious side, especially in the share of electricity. Hence, there is wide variation among scenarios, including the IAM pathways, regarding changes in the transport fuel mix over the first half of the century. As seen in Figure 2.23, the projections of energy consumption, CO_2 emissions and carbon intensity are quite different between IAM and ETP scenarios. These differences can be explained by more weight on efficiency improvements and avoid/shift decreasing energy consumption, and the higher share of biofuels and electricity accelerating the speed of decarbonization in ETP scenarios. Although biofuel consumption and electric vehicle sales have increased significantly in recent years, the growth rates projected in these pathways would be unprecedented and far higher than has been experienced to date.

The 1.5°C pathways require an acceleration of the mitigation solutions already featured in 2°C-consistent pathways (e.g., more efficient vehicle technologies operating on lower-carbon fuels), as well as those having received lesser attention in most global transport decarbonization pathways up to now (e.g., mode-shifting and travel demand management). Current-generation, global pathways generally do not include these newer transport sector developments, whereby technological solutions are related to shifts in traveller's behaviour.

2.4.4 Land-Use Transitions and Changes in the Agricultural Sector

The agricultural and land system described together under the umbrella of the AFOLU (agriculture, forestry, and other land use) sector plays an important role in 1.5°C pathways (Clarke et al., 2014; Smith and Bustamante, 2014; Popp et al., 2017). On the one hand, its emissions need to be limited over the course of this century to be in line with pathways limiting warming to 1.5°C (see Sections 2.2-3). On the other hand, the AFOLU system is responsible for food and feed production; for wood production for pulp and construction; for the production of biomass that is used for energy, CDR or other uses; and for the supply of non-provisioning (ecosystem) services (Smith and Bustamante, 2014). Meeting all demands together requires changes in land use, as well as in agricultural and forestry practices, for which a multitude of potential options have been identified (Smith and Bustamante, 2014; Popp et al., 2017) (see also Supplementary Material 2.SM.1.2 and Chapter 4, Section 4.3.1, 4.3.2 and 4.3.7).

This section assesses the transformation of the AFOLU system, mainly making use of pathways from IAMs (see Section 2.1) that are based on quantifications of the SSPs and that report distinct land-use evolutions in line with limiting warming to 1.5°C (Calvin et al., 2017; Fricko et al., 2017; Fujimori, 2017; Kriegler et al., 2017; Popp et al., 2017; Riahi et al., 2017; van Vuuren et al., 2017b; Doelman et al., 2018; Rogelj et al., 2018). The SSPs were designed to vary mitigation challenges (O'Neill et al., 2014) (Cross-Chapter Box 1 in Chapter 1), including for the AFOLU sector (Popp et al., 2017; Riahi et al., 2017). The SSP pathway ensemble hence allows for a structured exploration of AFOLU transitions in the context of climate change mitigation in line with 1.5°C, taking into account technological and socio-economic aspects. Other considerations, like food security, livelihoods and biodiversity, are also of importance when identifying AFOLU strategies. These are

at present only tangentially explored by the SSPs. Further assessments of AFOLU mitigation options are provided in other parts of this report and in the IPCC Special Report on Climate Change and Land (SRCCL). Chapter 4 provides an assessment of bioenergy (including feedstocks, see Section 4.3.1), livestock management (Section 4.3.1), reducing rates of deforestation and other land-based mitigation options (as mitigation and adaptation option, see Section 4.3.2), and BECCS, afforestation and reforestation options (including the bottom-up literature of their sustainable potential, mitigation cost and side effects, Section 4.3.7). Chapter 3 discusses impacts land-based CDR (Cross-Chapter Box 7 in Chapter 3). Chapter 5 assesses the sustainable development implications of AFOLU mitigation, including impacts on biodiversity (Section 5.4). Finally, the SRCCL will undertake a more comprehensive assessment of land and climate change aspects. For the sake of complementarity, this section focusses on the magnitude and pace of land transitions in 1.5°C pathways, as well as on the implications of different AFOLU mitigation strategies for different land types. The interactions with other societal objectives and potential limitations of identified AFOLU measures link to these large-scale evolutions, but these are assessed elsewhere (see above).

Land-use changes until mid-century occur in the large majority of SSP pathways, both under stringent mitigation and in absence of mitigation (Figure 2.24). In the latter case, changes are mainly due to socio-economic drivers like growing demands for food, feed and wood products. General transition trends can be identified for many land types in 1.5°C pathways, which differ from those in baseline scenarios and depend on the interplay with mitigation in other sectors (Figure 2.24) (Popp et al., 2017; Riahi et al., 2017; Rogelj et al., 2018). Mitigation that demands land mainly occurs at the expense of agricultural land for food and feed production. Additionally, some biomass is projected to be grown on marginal land or supplied from residues and waste, but at lower shares. Land for second-generation energy crops (such as *Miscanthus* or poplar) expands by 2030 and 2050 in all available pathways that assume a cost-effective achievement of a 1.5°C temperature goal in 2100 (Figure 2.24), but the scale depends strongly on underlying socio-economic assumptions (see later discussion of land pathway archetypes). Reducing rates of deforestation restricts agricultural expansion, and forest cover can expand strongly in 1.5°C and 2°C pathways alike compared to its extent in no-climate-policy baselines due to reduced deforestation and afforestation and reforestation measures. However, the extent to which forest cover expands varies highly across models in the literature, with some models projecting forest cover to stay virtually constant or decline slightly. This is due to whether afforestation and reforestation is included as a mitigation technology in these pathways and interactions with other sectors.

As a consequence of other land-use changes, pasture land is generally projected to be reduced compared to both baselines in which no climate change mitigation action is undertaken and 2°C-consistent pathways. Furthermore, cropland for food and feed production decreases in most 1.5°C pathways, both compared to a no-climate baseline and relative to 2010. These reductions in agricultural land for food and feed production are facilitated by intensification on agricultural land and in livestock production systems (Popp et al., 2017), as well as changes in consumption patterns (Frank et al., 2017; Fujimori, 2017) (see

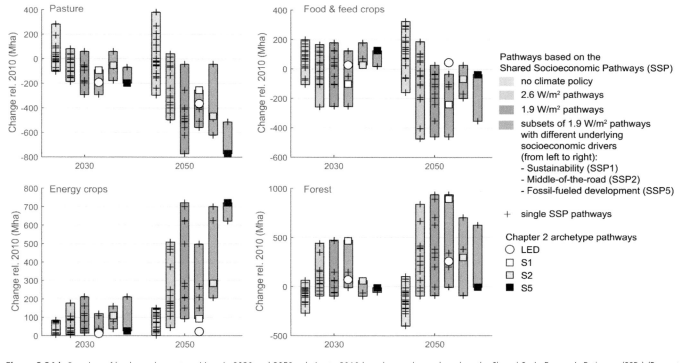

Figure 2.24 | Overview of land-use change transitions in 2030 and 2050, relative to 2010 based on pathways based on the Shared Socio-Economic Pathways (SSPs) (Popp et al., 2017; Riahi et al., 2017; Rogelj et al., 2018). Grey: no-climate-policy baseline; green: 2.6 W m⁻² pathways; blue: 1.9 W m⁻² pathways. Pink: 1.9 W m⁻² pathways grouped per underlying socio-economic assumption (from left to right: SSP1 sustainability, SSP2 middle-of-the-road, SSP5 fossil-fuelled development). Ranges show the minimum–maximum range across the SSPs. Single pathways are shown with plus signs. Illustrative archetype pathways are highlighted with distinct icons. Each panel shows the changes for a different land type. The 1.9 and 2.6 W m⁻² pathways are taken as proxies for 1.5°C and 2°C pathways, respectively. The 2.6 W m⁻² pathways are mostly consistent with the Lower-2°C and Higher-2°C pathway classes. The 1.9 W m⁻² pathways are consistent with the 1.5°C-low-OS (mostly SSP1 and SSP2) and 1.5°C-high-OS (SSP5) pathway classes. In 2010, pasture was estimated to cover about 3–3.5 10³ Mha, food and feed crops about 1.5–1.6 10³ Mha, energy crops about 0–14 Mha and forest about 3.7–4.2 10³ Mha, across the models that reported SSP pathways (Popp et al., 2017). When considering pathways limiting warming to 1.5°C with no or limited overshoot, the full set of scenarios shows a conversion of 50–1100 Mha of pasture into 0–600 Mha for energy crops, a 200 Mha reduction to 950 Mha increase forest, and a 400 Mha decrease to a 250 Mha increase in non-pasture agricultural land for food and feed crops by 2050 relative to 2010. The large range across the literature and the understanding of the variations across models and assumptions leads to *medium confidence* in the size of these ranges.

also Chapter 4, Section 4.3.2 for an assessment of these mitigation options). For example, in a scenario based on rapid technological progress (Kriegler et al., 2017), global average cereal crop yields in 2100 are assumed to be above 5 tDM ha⁻¹ yr⁻¹ in mitigation scenarios aiming at limiting end-of-century radiative forcing to 4.5 or 2.6 W m⁻², compared to 4 tDM ha⁻¹ yr⁻¹ in the SSP5 baseline to ensure the same food production. Similar improvements are present in 1.5°C variants of such scenarios. Historically, cereal crop yields are estimated at 1 tDM ha⁻¹ yr⁻¹ and about 3 tDM ha⁻¹ yr⁻¹ in 1965 and 2010, respectively (calculations based on FAOSTAT, 2018). For aggregate energy crops, models assume 4.2–8.9 tDM ha⁻¹ yr⁻¹ in 2010, increasing to about 6.9–17.4 tDM ha⁻¹ yr⁻¹ in 2050, which fall within the range found in the bottom-up literature yet depend on crop, climatic zone, land quality and plot size (Searle and Malins, 2014).

The pace of projected land transitions over the coming decades can differ strongly between 1.5°C and baseline scenarios without climate change mitigation and from historical trends (Table 2.9). However, there is uncertainty in the sign and magnitude of these future land-use changes (Prestele et al., 2016; Popp et al., 2017; Doelman et al., 2018). The pace of projected cropland changes overlaps with historical trends over the past four decades, but in several cases also goes well beyond this range. By the 2030–2050 period, the projected reductions

in pasture and potentially strong increases in forest cover imply a reversed dynamic compared to historical and baseline trends. This suggests that distinct policy and government measures would be needed to achieve forest increases, particularly in a context of projected increased bioenergy use.

Changes in the AFOLU sector are driven by three main factors: demand changes, efficiency of production, and policy assumptions (Smith et al., 2013; Popp et al., 2017). Demand for agricultural products and other land-based commodities is influenced by consumption patterns (including dietary preferences and food waste affecting demand for food and feed) (Smith et al., 2013; van Vuuren et al., 2018), demand for forest products for pulp and construction (including less wood waste), and demand for biomass for energy production (Lambin and Meyfroidt, 2011; Smith and Bustamante, 2014). Efficiency of agricultural and forestry production relates to improvements in agricultural and forestry practices (including product cascades, by-products and more waste- and residue-based biomass for energy production), agricultural and forestry yield increases, and intensification of livestock production systems leading to higher feed efficiency and changes in feed composition (Havlík et al., 2014; Weindl et al., 2015). Policy assumptions relate to the level of land protection, the treatment of food waste, policy choices about the timing of mitigation action (early vs late), the choice and

Table 2.9 | Annual pace of land-use change in baseline, 2°C and 1.5°C pathways.

All values in Mha yr⁻¹. The 2.6 W m⁻² pathways are mostly consistent with the Lower-2°C and Higher-2°C pathway classes. The 1.9 W m⁻² pathways are broadly consistent with the 1.5°C-low-OS (mostly SSP1 and SSP2) and 1.5°C-high-OS (SSP5) pathway classes. Baseline projections reflect land-use developments projected by integrated assessment models under the assumptions of the Shared Socio-Economic Pathways (SSPs) in absence of climate policies (Popp et al., 2017; Riahi et al., 2017; Rogelj et al., 2018). Values give the full range across SSP scenarios. According to the Food and Agriculture Organization of the United Nations (FAOSTAT, 2018), 4.9 billion hectares (approximately 40% of the land surface) was under agricultural use in 2005, either as cropland (1.5 billion hectares) or pasture (3.4 billion hectares). FAO data in the table are equally from FAOSTAT (2018).

Annual Pace of Land-Use Change [Mha yr⁻¹]					
Land Type	**Pathway**	**Time Window**		**Historical**	
		2010–2030	**2030–2050**	**1970–1990**	**1990–2010**
Pasture	1.9 W m⁻²	[–14.6/3.0]	[–28.7/–5.2]	8.7	0.9
	2.6 W m⁻²	[–9.3/4.1]	[–21.6/0.4]	Permanent meadows	Permanent meadows
	Baseline	[–5.1/14.1]	[–9.6/9.0]	and pastures (FAO)	and pastures (FAO)
Cropland for food, feed and material	1.9 W m⁻²	[–12.7/9.0]	[–18.5/0.1]		
	2.6 W m⁻²	[–12.9/8.3]	[–16.8/2.3]		
	Baseline	[–5.3/9.9]	[–2.7/6.7]		
Cropland for energy	1.9 W m⁻²	[0.7/10.5]	[3.9/34.8]		
	2.6 W m⁻²	[0.2/8.8]	[2.0/22.9]		
	Baseline	[0.2/4.2]	[–0.2/6.1]		
Total cropland (Sum of cropland for food and feed & energy)	1.9 W m⁻²	[–6.8/12.8]	[–5.8/26.7]	4.6	0.9
	2.6 W m⁻²	[–8.4/9.3]	[–7.1/17.8]	Arable land and	Arable land and
	Baseline	[–3.0/11.3]	[0.6/11.0]	Permanent crops	Permanent crops
Forest	1.9 W m⁻²	[–4.8/23.7]	[0.0/34.3]	N.A.	–5.6
	2.6 W m⁻²	[–4.7/22.2]	[–2.4/31.7]	Forest (FAO)	Forest (FAO)
	Baseline	[–13.6/3.3]	[–6.5/4.3]		

preference of land-based mitigation options (for example, the inclusion of afforestation and reforestation as mitigation options), interactions with other sectors (Popp et al., 2017), and trade (Schmitz et al., 2012; Wiebe et al., 2015).

A global study (Stevanović et al., 2017) reported similar GHG reduction potentials for both production-side (agricultural production measures in combination with reduced deforestation) and consumption-side (diet change in combination with lower shares of food waste) measures on the order of 40% in 2100[11] (compared to a baseline scenario without land-based mitigation). Lower consumption of livestock products by 2050 could also substantially reduce deforestation and cumulative carbon losses (Weindl et al., 2017). On the supply side, minor productivity growth in extensive livestock production systems is projected to lead to substantial CO_2 emission abatement, but the emission-saving potential of productivity gains in intensive systems is limited, mainly due to trade-offs with soil carbon stocks (Weindl et al., 2017). In addition, even within existing livestock production systems, a transition from extensive to more productive systems bears substantial GHG abatement potential, while improving food availability (Gerber et al., 2013; Havlík et al., 2014). Many studies highlight the capability of agricultural intensification for reducing GHG emissions in the AFOLU sector or even enhancing terrestrial carbon stocks (Valin et al., 2013; Popp et al., 2014a; Wise et al., 2014). Also the importance of immediate and global land-use regulations for a comprehensive reduction of

land-related GHG emissions (especially related to deforestation) has been shown by several studies (Calvin et al., 2017; Fricko et al., 2017; Fujimori, 2017). Ultimately, there are also interactions between these three factors and the wider society and economy, for example, if CDR technologies that are not land-based are deployed (like direct air capture – DACCS, see Chapter 4, Section 4.3.7) or if other sectors over- or underachieve their projected mitigation contributions (Clarke et al., 2014). Variations in these drivers can lead to drastically different land-use implications (Popp et al., 2014b) (Figure 2.24).

Stringent mitigation pathways inform general GHG dynamics in the AFOLU sector. First, CO_2 emissions from deforestation can be abated at relatively low carbon prices if displacement effects in other regions (Calvin et al., 2017) or other land-use types with high carbon density (Calvin et al., 2014; Popp et al., 2014a; Kriegler et al., 2017) can be avoided. However, efficiency and costs of reducing rates of deforestation strongly depend on governance performance, institutions and macroeconomic factors (Wang et al., 2016). Secondly, besides CO_2 reductions, the land system can play an important role for overall CDR efforts (Rogelj et al., 2018) via BECCS, afforestation and reforestation, or a combination of options. The AFOLU sector also provides further potential for active terrestrial carbon sequestration, for example, via land restoration, improved management of forest and agricultural land (Griscom et al., 2017), or biochar applications (Smith, 2016) (see also Chapter 4, Section 4.3.7). These options have so far

---

[11] Land-based mitigation options on the supply and the demand side are assessed in 4.3.2, and CDR options with a land component in 4.3.7. Chapter 5 (Section 5.4) assesses the implications of land-based mitigation for related SDGs, e.g., food security.

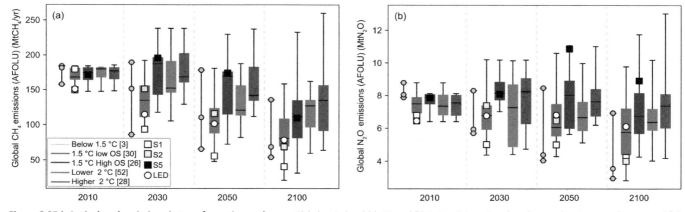

Figure 2.25 | Agricultural emissions in transformation pathways. Global agricultural (a) CH$_4$ and (b) N$_2$O emissions. Box plots show median, interquartile range and full range. Classes are defined in Section 2.1.

not been extensively integrated in the mitigation pathway literature (see Supplementary Material 2.SM.1.2), but in theory their availability would impact the deployment of other CDR technologies, like BECCS (Section 2.3.4) (Strefler et al., 2018a). These interactions will be discussed further in the SRCCL.

Residual agricultural non-CO$_2$ emissions of CH$_4$ and N$_2$O play an important role for temperature stabilization pathways, and their relative importance increases in stringent mitigation pathways in which CO$_2$ is reduced to net zero emissions globally (Gernaat et al., 2015; Popp et al., 2017; Stevanović et al., 2017; Rogelj et al., 2018), for example, through their impact on the remaining carbon budget (Section 2.2). Although agricultural non-CO$_2$ emissions show marked reduction potentials in 2°C-consistent pathways, complete elimination of these emission sources does not occur in IAMs based on the evolution of agricultural practice assumed in integrated models (Figure 2.25) (Gernaat et al., 2015). Methane emissions in 1.5°C pathways are reduced through improved agricultural management (e.g., improved management of water in rice production, manure and herds, and better livestock quality through breeding and improved feeding practices) as well as dietary shifts away from emissions-intensive livestock products. Similarly, N$_2$O emissions decrease due to improved N-efficiency and manure management (Frank et al., 2018). However, high levels of bioenergy production can also result in increased N$_2$O emissions (Kriegler et al., 2017), highlighting the importance of appropriate management approaches (Davis et al., 2013). Residual agricultural emissions can be further reduced by limiting demand for GHG-intensive foods through shifts to healthier and more sustainable diets (Tilman and Clark, 2014; Erb et al., 2016b; Springmann et al., 2016) and reductions in food waste (Bajželj et al., 2014; Muller et al., 2017; Popp et al., 2017) (see also Chapter 4 and SRCCL). Finally, several mitigation measures that could affect these agricultural non-CO$_2$ emissions are not, or only to a limited degree, considered in the current integrated pathway literature (see Supplementary Material 2.SM.1.2). Such measures (like plant-based and synthetic proteins, methane inhibitors and vaccines in livestock, alternate wetting and drying in paddy rice, or nitrification inhibitors) are very diverse and differ in their development or deployment stages. Their potentials have not been explicitly assessed here.

Pathways consistent with 1.5°C rely on one or more of the three strategies highlighted above (demand changes, efficiency gains, and

policy assumptions), and can apply these in different configurations. For example, among the four illustrative archetypes used in this chapter (Section 2.1), the LED and S1 pathways focus on generally low resource and energy consumption (including healthy diets with low animal-calorie shares and low food waste) as well as significant agricultural intensification in combination with high levels of nature protection. Under such assumptions, comparably small amounts of land are needed for land-demanding mitigation activities such as BECCS and afforestation and reforestation, leaving the land footprint for energy crops in 2050 virtually the same compared to 2010 levels for the LED pathway. In contrast, future land-use developments can look very different under the resource- and energy-intensive S5 pathway that includes less healthy diets with high animal shares and high shares of food waste (Tilman and Clark, 2014; Springmann et al., 2016) combined with a strong orientation towards technology solutions to compensate for high reliance on fossil-fuel resources and associated high levels of GHG emissions in the baseline. In such pathways, climate change mitigation strategies strongly depend on the availability of CDR through BECCS (Humpenöder et al., 2014). As a consequence, the S5 pathway sources significant amounts of biomass through bioenergy crop expansion in combination with agricultural intensification. Also, further policy assumptions can strongly affect land-use developments, highlighting the importance for land use of making appropriate policy choices. For example, within the SSP set, some pathways rely strongly on a policy to incentivize afforestation and reforestation for CDR together with BECCS, which results in an expansion of forest area and a corresponding increase in terrestrial carbon stock. Finally, the variety of pathways illustrates how policy choices in the AFOLU and other sectors strongly affect land-use developments and associated sustainable development interactions (Chapter 5, Section 5.4) in 1.5°C pathways.

The choice of strategy or mitigation portfolio impacts the GHG dynamics of the land system and other sectors (see Section 2.3), as well as the synergies and trade-offs with other environmental and societal objectives (see Section 2.5.3 and Chapter 5, Section 5.4). For example, AFOLU developments in 1.5°C pathways range from strategies that differ by almost an order of magnitude in their projected land requirements for bioenergy (Figure 2.24), and some strategies would allow an increase in forest cover over the 21st century compared to strategies under which forest cover remains approximately constant.

High agricultural yields and application of intensified animal husbandry, implementation of best-available technologies for reducing non-CO$_2$ emissions, or lifestyle changes including a less-meat-intensive diet and less CO$_2$-intensive transport modes, have been identified as allowing for such a forest expansion and reduced footprints from bioenergy without compromising food security (Frank et al., 2017; Doelman et al., 2018; van Vuuren et al., 2018).

The IAMs used in the pathways underlying this assessment (Popp et al., 2017; Riahi et al., 2017; Rogelj et al., 2018) do not include all potential land-based mitigation options and side-effects, and their results are hence subject to uncertainty. For example, recent research has highlighted the potential impact of forest management practices on land carbon content (Erb et al., 2016a; Naudts et al., 2016) and the uncertainty surrounding future crop yields (Haberl et al., 2013; Searle and Malins, 2014) and water availability (Liu et al., 2014). These aspects are included in IAMs in varying degrees but were not assessed in this report. Furthermore, land-use modules of some IAMs can depict spatially resolved climate damages to agriculture (Nelson et al., 2014), but this option was not used in the SSP quantifications (Riahi et al., 2017). Damages (e.g., due to ozone exposure or varying indirect fertilization due to atmospheric N and Fe deposition (e.g., Shindell et al., 2012; Mahowald et al., 2017) are also not included. Finally, this assessment did not look into the literature of agricultural sector models which could provide important additional detail and granularity to the discussion presented here.[12] This limits their ability to capture the full mitigation potentials and benefits between scenarios. An in-depth assessment of these aspects lies outside the scope of this Special Report. However, their existence affects the confidence assessment of the AFOLU transition in 1.5°C pathways.

Despite the limitations of current modelling approaches, there is *high agreement* and *robust evidence* across models and studies that the AFOLU sector plays an important role in stringent mitigation pathways. The findings from these multiple lines of evidence also result in *high confidence* that AFOLU mitigation strategies can vary significantly based on preferences and policy choices, facilitating the exploration of strategies that can achieve multiple societal objectives simultaneously (see also Section 2.5.3). At the same time, given the many uncertainties and limitations, only *low to medium confidence* can be attributed by this assessment to the more extreme AFOLU developments found in the pathway literature, and *low to medium confidence* to the level of residual non-CO$_2$ emissions.

2.5   Challenges, Opportunities and Co-Impacts of Transformative Mitigation Pathways

This section examines aspects other than climate outcomes of 1.5°C mitigation pathways. Focus is given to challenges and opportunities related to policy regimes, price of carbon and co-impacts, including sustainable development issues, which can be derived from the existing integrated pathway literature. Attention is also given to uncertainties and critical assumptions underpinning mitigation pathways. The challenges and opportunities identified in this section are further elaborated Chapter 4 (e.g., policy choice and implementation) and Chapter 5 (e.g., sustainable development). The assessment indicates unprecedented policy and geopolitical challenges.

2.5.1   Policy Frameworks and Enabling Conditions

Moving from a 2°C to a 1.5°C pathway implies bold integrated policies that enable higher socio-technical transition speeds, larger deployment scales, and the phase-out of existing systems that may lock in emissions for decades *(high confidence)* (Geels et al., 2017; Kuramochi et al., 2017; Rockström et al., 2017; Vogt-Schilb and Hallegatte, 2017; Kriegler et al., 2018a; Michaelowa et al., 2018). This requires higher levels of transformative policy regimes in the near term, which allow deep decarbonization pathways to emerge and a net zero carbon energy–economy system to emerge in the 2040–2060 period (Rogelj et al., 2015b; Bataille et al., 2016b). This enables accelerated levels of technological deployment and innovation (Geels et al., 2017; IEA, 2017a; Grubler et al., 2018) and assumes more profound behavioural, economic and political transformation (Sections 2.3, 2.4 and 4.4). Despite inherent levels of uncertainty attached to modelling studies (e.g., related to climate and carbon cycle response), studies stress the urgency for transformative policy efforts to reduce emissions in the short term (Riahi et al., 2015; Kuramochi et al., 2017; Rogelj et al., 2018).

The available literature indicates that mitigation pathways in line with 1.5°C pathways would require stringent and integrated policy interventions (*very high confidence*). Higher policy ambition often takes the form of stringent economy-wide emission targets (and resulting peak-and-decline of emissions), larger coverage of NDCs to more gases and sectors (e.g., land-use, international aviation), much lower energy and carbon intensity rates than historically seen, carbon prices much higher than the ones observed in real markets, increased climate finance, global coordinated policy action, and implementation of additional initiatives (e.g., by non-state actors) (Sections 2.3, 2.4 and 2.5.2). The diversity (beyond explicit carbon pricing) and effectiveness of policy portfolios are of prime importance, particularly in the short-term (Mundaca and Markandya, 2016; Kuramochi et al., 2017; OECD, 2017; Kriegler et al., 2018a; Michaelowa et al., 2018). For instance, deep decarbonization pathways in line with a 2°C target (covering 74% of global energy-system emissions) include a mix of stringent regulation (e.g., building codes, minimum performance standards), carbon pricing mechanisms and R&D (research and development) innovation policies (Bataille et al., 2016a). Explicit carbon pricing, direct regulation and public investment to enable innovation are critical for deep decarbonization pathways (Grubb et al., 2014). Effective planning (including compact city measures) and integrated regulatory frameworks are also key drivers in the IEA-ETP B2DS study for the transport sector (IEA, 2017a). Effective urban planning can reduce GHG emissions from urban transport between 20% and 50% (Creutzig, 2016). Comprehensive policy frameworks would be needed if the decarbonization of the power system is pursued while increasing end-use electrification (including transport) (IEA, 2017a). Technology policies (e.g., feed-in-tariffs), financing instruments, carbon pricing

---

[12]   For example, the GLEAM (http://www.fao.org/gleam/en/) model from the UN Food and Agricultural Organisation (FAO).

and system integration management driving the rapid adoption of renewable energy technologies are critical for the decarbonization of electricity generation (Bruckner et al., 2014; Luderer et al., 2014; Creutzig et al., 2017; Pietzcker et al., 2017). Likewise, low-carbon and resilient investments are facilitated by a mix of coherent policies, including fiscal and structural reforms (e.g., labour markets), public procurement, carbon pricing, stringent standards, information schemes, technology policies, fossil-fuel subsidy removal, climate risk disclosure, and land-use and transport planning (OECD, 2017). Pathways in which CDR options are restricted emphasize the strengthening of near-term policy mixes (Luderer et al., 2013; Kriegler et al., 2018a). Together with the decarbonization of the supply side, ambitious policies targeting fuel switching and energy efficiency improvements on the demand side play a major role across mitigation pathways (Clarke et al., 2014; Kriegler et al., 2014b; Riahi et al., 2015; Kuramochi et al., 2017; Brown and Li, 2018; Rogelj et al., 2018; Wachsmuth and Duscha, 2018).

The combined evidence suggests that aggressive policies addressing energy efficiency are central in keeping 1.5°C within reach and lowering energy system and mitigation costs (*high confidence*) (Luderer et al., 2013; Rogelj et al., 2013b, 2015b; Grubler et al., 2018). Demand-side policies that increase energy efficiency or limit energy demand at a higher rate than historically observed are critical enabling factors for reducing mitigation costs in stringent mitigation pathways across the board (Luderer et al., 2013; Rogelj et al., 2013b, 2015b; Clarke et al., 2014; Bertram et al., 2015a; Bataille et al., 2016b). Ambitious sector-specific mitigation policies in industry, transportation and residential sectors are needed in the short run for emissions to peak in 2030 (Méjean et al., 2018). Stringent demand-side policies (e.g., tightened efficiency standards for buildings and appliances) driving the expansion, efficiency and provision of high-quality energy services are essential to meet a 1.5°C mitigation target while reducing the reliance on CDR (Grubler et al., 2018). A 1.5°C pathway for the transport sector is possible using a mix of additional and stringent policy actions preventing (or reducing) the need for transport, encouraging shifts towards efficient modes of transport, and improving vehicle-fuel efficiency (Gota et al., 2018). Stringent demand-side policies also reduce the need for CCS (Wachsmuth and Duscha, 2018). Even in the presence of weak near term policy frameworks, increased energy efficiency lowers mitigation costs noticeably compared to pathways with reference energy intensity (Bertram et al., 2015a). Common issues in the literature relate to the rebound effect, the potential overestimation of the effectiveness of energy efficiency policy, and policies to counteract the rebound (Saunders, 2015; van den Bergh, 2017; Grubler et al., 2018) (Sections 2.4 and 4.4).

SSP-based modelling studies underline that socio-economic and climate policy assumptions strongly influence mitigation pathway characteristics and the economics of achieving a specific climate target (*very high confidence*) (Bauer et al., 2017; Guivarch and Rogelj, 2017; Riahi et al., 2017; Rogelj et al., 2018). SSP assumptions related to economic growth and energy intensity are critical determinants of projected CO_2 emissions (Marangoni et al., 2017). A multimodel inter-comparison study found that mitigation challenges in line with a 1.5°C target vary substantially across SSPs and policy assumptions (Rogelj et al., 2018). Under SSP1-SPA1 (sustainability) and SSP2-SPA2 (middle-of-the-road), the majority of IAMs were capable of producing

1.5°C pathways. On the contrary, none of the IAMs contained in the SR1.5 database could produce a 1.5°C pathway under SSP3-SPA3 assumptions. Preventing elements include, for instance, climate policy fragmentation, limited control of land-use emissions, heavy reliance on fossil fuels, unsustainable consumption and marked inequalities (Rogelj et al., 2018). Dietary aspects of the SSPs are also critical: climate-friendly diets were contained in 'sustainability' (SSP1) and meat-intensive diets in SSP3 and SSP5 (Popp et al., 2017). CDR requirements are reduced under 'sustainability' related assumptions (Strefler et al., 2018b). These are major policy-related reasons for why SSP1-SPA1 translates into relatively low mitigation challenges whereas SSP3-SPA3 and SSP5-SPA5 entail futures that pose the highest socio-technical and economic challenges. SSPs/SPAs assumptions indicate that policy-driven pathways that encompass accelerated change away from fossil fuels, large-scale deployment of low-carbon energy supplies, improved energy efficiency and sustainable consumption lifestyles reduce the risks of climate targets becoming unreachable (Clarke et al., 2014; Riahi et al., 2015, 2017; Marangoni et al., 2017; Rogelj et al., 2017, 2018; Strefler et al., 2018b).

Policy assumptions that lead to weak or delayed mitigation action from what would be possible in a fully cooperative world strongly influence the achievability of mitigation targets (*high confidence*) (Luderer et al., 2013; Rogelj et al., 2013b; OECD, 2017; Holz et al., 2018a; Strefler et al., 2018b). Such regimes also include current NDCs (Fawcett et al., 2015; Aldy et al., 2016; Rogelj et al., 2016a, 2017; Hof et al., 2017; van Soest et al., 2017), which have been reported to make achieving a 2°C pathway unattainable without CDR (Strefler et al., 2018b). Not strengthening NDCs would make it very challenging to keep 1.5°C within reach (see Section 2.3 and Cross-Chapter Box 11 in Chapter 4). One multimodel inter-comparison study (Luderer et al., 2016b, 2018) explored the effects on 1.5°C pathways assuming the implementation of current NDCs until 2030 and stringent reductions thereafter. It finds that delays in globally coordinated actions lead to various models reaching no 1.5°C pathways during the 21st century. Transnational emission reduction initiatives (TERIs) outside the UNFCCC have also been assessed and found to overlap (70–80%) with NDCs and be inadequate to bridge the gap between NDCs and a 2°C pathway (Roelfsema et al., 2018). Weak and fragmented short-term policy efforts use up a large share of the long-term carbon budget before 2030–2050 (Bertram et al., 2015a; van Vuuren et al., 2016) and increase the need for the full portfolio of mitigation measures, including CDR (Clarke et al., 2014; Riahi et al., 2015; Xu and Ramanathan, 2017). Furthermore, fragmented policy scenarios also exhibit 'carbon leakage' via energy and capital markets (Arroyo-Currás et al., 2015; Kriegler et al., 2015b). A lack of integrated policy portfolios can increase the risks of trade-offs between mitigation approaches and sustainable development objectives (see Sections 2.5.3 and 5.4). However, more detailed analysis is needed about realistic (less disruptive) policy trajectories until 2030 that can strengthen near-term mitigation action and meaningfully decrease post-2030 challenges (see Chapter 4, Section 4.4).

Whereas the policy frameworks and enabling conditions identified above pertain to the 'idealized' dimension of mitigation pathways, aspects related to 1.5°C mitigation pathways in practice are of prime importance. For example, issues related to second-best stringency levels, international cooperation, public acceptance, distributional consequences, multilevel governance, non-state actions, compliance levels, capacity building, rebound effects, linkages across highly heterogeneous policies, sustained behavioural change, finance and intra- and inter-generational issues need to be considered (see Chapter 4, Section 4.4) (Bataille et al., 2016a; Mundaca and Markandya, 2016; Baranzini et al., 2017; MacDougall et al., 2017; van den Bergh, 2017; Vogt-Schilb and Hallegatte, 2017; Chan et al., 2018; Holz et al., 2018a; Klinsky and Winkler, 2018; Michaelowa et al., 2018; Patterson et al., 2018). Furthermore, policies interact with a wide portfolio of pre-existing policy instruments that address multiple areas (e.g., technology markets, economic growth, poverty alleviation, climate adaptation) and deal with various market failures (e.g., information asymmetries) and behavioural aspects (e.g., heuristics) that prevent or hinder mitigation actions (Kolstad et al., 2014; Mehling and Tvinnereim, 2018). The socio-technical transition literature points to multiple complexities in real-world settings that prevent reaching 'idealized' policy conditions but at the same time can still accelerate transformative change through other co-evolutionary processes of technology and society (Geels et

al., 2017; Rockström et al., 2017). Such co-processes are complex and go beyond the role of policy (including carbon pricing) and comprise the role of citizens, businesses, stakeholder groups or governments, as well as the interplay of institutional and socio-political dimensions (Michaelowa et al., 2018; Veland et al., 2018). It is argued that large system transformations, similar to those in 1.5°C pathways, require prioritizing an evolutionary and behavioural framework in economic theory rather than an optimization or equilibrium framework as is common in current IAMs (Grubb et al., 2014; Patt, 2017). Accumulated know-how, accelerated innovation and public investment play a key role in (rapid) transitions (see Sections 4.2 and 4.4) (Geels et al., 2017; Michaelowa et al., 2018).

In summary, the emerging literature supports the AR5 on the need for integrated, robust and stringent policy frameworks targeting both the supply and demand-side of energy-economy systems (*high confidence*). Continuous ex-ante policy assessments provide learning opportunities for both policy makers and stakeholders.

Cross-Chapter Box 5 | Economics of 1.5°C Pathways and the Social Cost of Carbon

Contributing Authors:
Luis Mundaca (Sweden/Chile), Mustafa Babiker (Sudan), Johannes Emmerling (Italy/Germany), Sabine Fuss (Germany), Jean-Charles Hourcade (France), Elmar Kriegler (Germany), Anil Markandya (Spain/UK), Joyashree Roy (India), Drew Shindell (USA)

Two approaches have been commonly used to assess alternative emissions pathways: **cost-effectiveness analysis (CEA)** and **cost–benefit analysis (CBA)**. **CEA** aims at identifying emissions pathways minimising the total mitigation costs of achieving a given warming or GHG limit (Clarke et al., 2014). **CBA** has the goal to identify the optimal emissions trajectory minimising the discounted flows of abatement expenditures and monetized climate change damages (Boardman et al., 2006; Stern, 2007). A third concept, the **Social Cost of Carbon (SCC)** measures the total net damages of an extra metric ton of CO_2 emissions due to the associated climate change (Nordhaus, 2014; Pizer et al., 2014; Rose et al., 2017a). Negative and positive impacts are monetized, discounted and the net value is expressed as an equivalent loss of consumption today. The SCC can be evaluated for any emissions pathway under policy consideration (Rose, 2012; NASEM, 2016, 2017).

Along the optimal trajectory determined by CBA, the SCC equals the discounted value of the marginal abatement cost of a metric ton of CO_2 emissions. Equating the present value of future damages and marginal abatement costs includes a number of critical value judgements in the formulation of the social welfare function (SWF), particularly in how non-market damages and the distribution of damages across countries and individuals and between current and future generations are valued (Kolstad et al., 2014). For example, since climate damages accrue to a larger extent farther in the future and can persist for many years, assumptions and approaches to determine the social discount rate (normative 'prescriptive' vs. positive 'descriptive') and social welfare function (e.g., discounted utilitarian SWF vs. undiscounted prioritarian SWF) can heavily influence CBA outcomes and associated estimates of SCC (Kolstad et al., 2014; Pizer et al., 2014; Adler and Treich, 2015; Adler et al., 2017; NASEM, 2017; Nordhaus, 2017; Rose et al., 2017a).

In CEA, the marginal abatement cost of carbon is determined by the climate goal under consideration. It equals the shadow price of carbon associated with the goal which in turn can be interpreted as the willingness to pay for imposing the goal as a political constraint. Emissions prices are usually expressed in carbon (equivalent) prices using the GWP-100 metric as the exchange rate for pricing emissions of non-CO_2 GHGs controlled under internationally climate agreements (like CH_4, N_2O and fluorinated gases, see Cross-Chapter Box 2 in Chapter 1).[13] Since policy goals like the goals of limiting warming to 1.5°C or well below 2°C do not directly result from a money metric trade-off between mitigation and damages, associated shadow prices can differ from the SCC in a CBA. In CEA, value judgments are to a large extent concentrated in the choice of climate goal and related implications, while more explicit assumptions about social values are required to perform CBA. For example, in CEA assumptions about the social discount rate no longer affect the overall abatement levels now set by the climate goal, but the choice and timing of investments in individual measures to reach these levels.

---

[13] Also other metrics to compare emissions have been suggested and adopted by governments nationally (Kandlikar, 1995; Marten et al., 2015; Shindell, 2015; IWG, 2016).

Cross Chapter Box 5 (continued)

Although CBA-based and CEA-based assessment are both subject to large uncertainty about socio-techno-economic trends, policy developments and climate response, the range of estimates for the SCC along an optimal trajectory determined by CBA is far wider than for estimates of the shadow price of carbon in CEA-based approaches. In CBA, the value judgments about inter- and intra-generational equity combined with uncertainties in the climate damage functions assumed, including their empirical basis, are important (Pindyck, 2013; Stern, 2013; Revesz et al., 2014). In a CEA-based approach, the value judgments about the aggregate welfare function matter less, and uncertainty about climate response and impacts can be tied into various climate targets and related emissions budgets (Clarke et al., 2014).

The CEA- and CBA-based carbon cost estimates are derived with a different set of tools. They are all summarised as integrated assessment models (IAMs) but in fact are of very different nature (Weyant, 2017). Detailed process IAMs such as AIM (Fujimori, 2017), GCAM (Thomson et al., 2011; Calvin et al., 2017), IMAGE (van Vuuren et al., 2011b, 2017b), MESSAGE-GLOBIOM (Riahi et al., 2011; Havlík et al., 2014; Fricko et al., 2017), REMIND-MAgPIE (Popp et al., 2010; Luderer et al., 2013; Kriegler et al., 2017) and WITCH (Bosetti et al., 2006, 2008, 2009) include a process-based representation of energy and land systems, but in most cases lack a comprehensive representation of climate damages, and are typically used for CEA. Diagnostic analyses across CBA-IAMs indicate important dissimilarities in modelling assembly, implementation issues and behaviour (e.g., parametric uncertainty, damage responses, income sensitivity) that need to be recognized to better understand SCC estimates (Rose et al., 2017a).

CBA-IAMs such as DICE (Nordhaus and Boyer, 2000; Nordhaus, 2013, 2017), PAGE (Hope, 2006) and FUND (Tol, 1999; Anthoff and Tol, 2009) attempt to capture the full feedback from climate response to socio-economic damages in an aggregated manner, but are usually much more stylised than detailed process IAMs. In a nutshell, the methodological framework for estimating SCC involves projections of population growth, economic activity and resulting emissions; computations of atmospheric composition and global mean temperatures as a result of emissions; estimations of physical impacts of climate changes; monetization of impacts (positive and negative) on human welfare; and the discounting of the future monetary value of impacts to year of emission (Kolstad et al., 2014; Revesz et al., 2014; NASEM, 2017; Rose et al., 2017a). There has been a discussion in the literature to what extent CBA-IAMs underestimate the SCC due to, for example, a limited treatment or difficulties in addressing damages to human well-being, labour productivity, value of capital stock, ecosystem services and the risks of catastrophic climate change for future generations (Ackerman and Stanton, 2012; Revesz et al., 2014; Moore and Diaz, 2015; Stern, 2016). However, there has been progress in 'bottom-up' empirical analyses of climate damages (Hsiang et al., 2017), the insights of which could be integrated into these models (Dell et al., 2014). Most of the models used in Chapter 2 on 1.5°C mitigation pathways are detailed process IAMs and thus deal with CEA.

An important question is how results from CEA- and CBA-type approaches can be compared and synthesized. Such synthesis needs to be done with care, since estimates of the shadow price of carbon under the climate goal and SCC estimates from CBA might not be directly comparable due to different tools, approaches and assumptions used to derive them. Acknowledging this caveat, the SCC literature has identified a range of factors, assumptions and value judgements that support SCC values above $100 tCO_2^{-1} that are also found as net present values of the shadow price of carbon in 1.5°C pathways. These factors include accounting for tipping points in the climate system (Lemoine and Traeger, 2014; Cai et al., 2015; Lontzek et al., 2015), a low social discount rate (Nordhaus, 2007a; Stern, 2007) and inequality aversion (Schmidt et al., 2013; Dennig et al., 2015; Adler et al., 2017).

The SCC and the shadow price of carbon are not merely theoretical concepts but used in regulation (Pizer et al., 2014; Revesz et al., 2014; Stiglitz et al., 2017). As stated by the report of the High-Level Commission on Carbon Pricing (Stiglitz et al., 2017), in the real world there is a distinction to be made between the implementable and efficient explicit carbon prices and the implicit (notional) carbon prices to be retained for policy appraisal and the evaluation of public investments, as is already done in some jurisdictions such as the USA, UK and France. Since 2008, the U.S. government has used SCC estimates to assess the benefits and costs related to CO_2 emissions resulting from federal policymaking (NASEM, 2017; Rose et al., 2017a).

The use of the SCC for policy appraisals is, however, not straightforward in an SDG context. There are suggestions that a broader range of polluting activities than only CO_2 emissions, for example emissions of air pollutants, and a broader range of impacts than only climate change, such as impacts on air quality, health and sustainable development in general (see Chapter 5 for a detailed discussion), would need to be included in social costs (Sarofim et al., 2017; Shindell et al., 2017a). Most importantly, a consistent valuation of the SCC in a sustainable development framework would require accounting for the SDGs in the social welfare formulation (see Chapter 5).

2.5.2 Economic and Investment Implications of 1.5°C Pathways

2.5.2.1 Price of carbon emissions

The price of carbon assessed here is fundamentally different from the concepts of optimal carbon price in a cost–benefit analysis, or the social cost of carbon (see Cross-Chapter Box 5 in this chapter and Chapter 3, Section 3.5.2). Under a cost-effectiveness analysis (CEA) modelling framework, prices for carbon (mitigation costs) reflect the stringency of mitigation requirements at the margin (i.e., cost of mitigating one extra unit of emission). Explicit carbon pricing is briefly addressed here to the extent it pertains to the scope of Chapter 2. For detailed policy issues about carbon pricing see Section 4.4.5.

Based on data available for this special report, the price of carbon varies substantially across models and scenarios, and their values increase with mitigation efforts (see Figure 2.26) (high confidence). For instance, undiscounted values under a Higher-2°C pathway range from 15–220 USD2010 $tCO_{2\text{-eq}}^{-1}$ in 2030, 45–1050 USD2010 $tCO_{2\text{-eq}}^{-1}$ in 2050, 120–1100 USD2010 $tCO_{2\text{-eq}}^{-1}$ in 2070 and 175–2340 USD2010 $tCO_{2\text{-eq}}^{-1}$ in 2100. On the contrary, estimates for a Below-1.5°C pathway range from 135–6050 USD2010 $tCO_{2\text{-eq}}^{-1}$ in 2030, 245–14300 USD2010 $tCO_{2\text{-eq}}^{-1}$ in 2050, 420–19300 USD2010 $tCO_{2\text{-eq}}^{-1}$ in 2070 and 690–30100 USD2010 $tCO_{2\text{-eq}}^{-1}$ in 2100. Values for 1.5°C-low-OS pathway are relatively higher than 1.5°C-high-OS pathway in 2030, but the difference decreases over time, particularly between 2050 and 2070. This is because in 1.5°C-high-OS pathways there is relatively less mitigation activity in the first half of the century, but more in the second half. The low energy demand (LED, P1 in the Summary for Policymakers) scenario exhibits the lowest values across the illustrative pathway archetypes. As a whole, the global average discounted price of emissions across 1.5°C- and 2°C pathways differs by a factor of four across models (assuming a 5% annual discount rate, comparing to Below-1.5°C and 1.5°C-low-OS pathways). If 1.5°C-high-OS pathways (with peak warming 0.1–0.4°C higher than 1.5°C) or pathways with very large land-use sinks are also considered, the differential value is reduced to a limited degree, from a factor 4 to a factor 3. The increase in mitigation costs between 1.5°C and 2°C pathways is based on a direct comparison of pathway pairs from the same model and the same study in which the 1.5°C pathway assumes a significantly smaller carbon budget compared to the 2°C pathway (e.g., 600 $GtCO_2$ smaller in the CD-LINKS and ADVANCE studies). This assumption is the main driver behind the increase in the price of carbon (Luderer et al., 2018; McCollum et al., 2018).[14]

The wide range of values depends on numerous aspects, including methodologies, projected energy service demands, mitigation targets, fuel prices and technology availability (high confidence) (Clarke et al., 2014; Kriegler et al., 2015b; Rogelj et al., 2015c; Riahi et al., 2017; Stiglitz et al., 2017). The characteristics of the technology portfolio, particularly in terms of investment costs and deployment rates, play a key role (Luderer et al., 2013, 2016a; Clarke et al., 2014; Bertram et al., 2015a; Riahi et al., 2015; Rogelj et al., 2015c). Models that encompass

a higher degree of technology granularity and that entail more flexibility regarding mitigation response often produce relatively lower mitigation costs than those that show less flexibility from a technology perspective (Bertram et al., 2015a; Kriegler et al., 2015a). Pathways providing high estimates often have limited flexibility of substituting fossil fuels with low-carbon technologies and the associated need to compensate fossil-fuel emissions with CDR. The price of carbon is also sensitive to the non-availability of BECCS (Bauer et al., 2018). Furthermore, and due to the treatment of future price anticipation, recursive-dynamic modelling approaches (with 'myopic anticipation') exhibit higher prices in the short term but modest increases in the long term compared to optimization modelling frameworks with 'perfect foresight' that show exponential pricing trajectories (Guivarch and Rogelj, 2017). The chosen social discount rate in CEA studies (range of 2–8% per year in the reported data, varying over time and sectors) can also affect the choice and timing of investments in mitigation measures (Clarke et al., 2014; Kriegler et al., 2015b; Weyant, 2017). However, the impacts of varying discount rates on 1.5°C (and 2°C) mitigation strategies can only be assessed to a limited degree. The above highlights the importance of sampling bias in pathway analysis ensembles towards outcomes derived from models which are more flexible, have more mitigation options and cheaper cost assumptions and thus can provide feasible pathways in contrast to other who are unable to do so (Tavoni and Tol, 2010; Clarke et al., 2014; Bertram et al., 2015a; Kriegler et al., 2015a; Guivarch and Rogelj, 2017). All CEA-based IAM studies reveal no unique path for the price of emissions (Bertram et al., 2015a; Kriegler et al., 2015b; Akimoto et al., 2017; Riahi et al., 2017).

Socio-economic conditions and policy assumptions also influence the price of carbon (very high confidence) (Bauer et al., 2017; Guivarch and Rogelj, 2017; Hof et al., 2017; Riahi et al., 2017; Rogelj et al., 2018). A multimodel study (Riahi et al., 2017) estimated the average discounted price of carbon (2010–2100, 5% discount rate) for a 2°C target to be nearly three times higher in the SSP5 marker than in the SSP1 marker. Another multimodel study (Rogelj et al., 2018) estimated the average discounted price of carbon (2020–2100, 5%) to be 35–65% lower in SSP1 compared to SSP2 in 1.5°C pathways. Delayed near-term mitigation policies and measures, including the limited extent of international global cooperation, result in increases in total economic mitigation costs and corresponding prices of carbon (Luderer et al., 2013; Clarke et al., 2014). This is because stronger efforts are required in the period after the delay to counterbalance the higher emissions in the near term. Staged accession scenarios also produce higher mitigation costs than immediate action mitigation scenarios under the same stringency level of emissions (Kriegler et al., 2015b).

It has been long argued that an explicit carbon pricing mechanism (whether via a tax or cap-and-trade scheme) can theoretically achieve cost-effective emission reductions (Nordhaus, 2007b; Stern, 2007; Aldy and Stavins, 2012; Goulder and Schein, 2013; Somanthan et al., 2014; Weitzman, 2014; Tol, 2017). Whereas the integrated assessment literature is mostly focused on the role of carbon pricing to reduce emissions (Clarke et al., 2014; Riahi et al., 2017; Weyant, 2017), there

---

[14] Unlike AR5, which only included cost-effective scenarios for estimating discounted average carbon prices for 2015–2100 (also using a 5% discount rate) (see Clarke et al., 2014, p.450), please note that values shown in Figure 2.26b include delays or technology constraint cases (see Sections 2.1 and 2.3).

is an emerging body of studies (including bottom-up approaches) that focuses on the interaction and performance of various policy mixes (e.g., regulation, subsidies, standards). Assuming global implementation of a mix of regionally existing best-practice policies (mostly regulatory policies in the electricity, industry, buildings, transport and agricultural sectors) and moderate carbon pricing (between 5–20 USD2010 tCO$_2^{-1}$ in 2025 in most world regions and average prices around 25 USD2010 tCO$_2^{-1}$ in 2030), early action mitigation pathways are generated that reduce global CO2 emissions by an additional 10 GtCO$_2$e in 2030 compared to the NDCs (Kriegler et al., 2018a) (see Section 2.3.5). Furthermore, a mix of stringent energy efficiency policies (e.g., minimum performance standards, building codes) combined with a carbon tax (rising from 10 USD2010 tCO$_2^{-1}$ in 2020 to 27 USD2010 tCO$_2^{-1}$ in 2040) is more cost-effective than a carbon tax alone (from 20 to 53 USD2010 tCO$_2^{-1}$) to generate a 1.5°C pathway for the U.S. electric sector (Brown and Li, 2018). Likewise, a policy mix encompassing a moderate carbon price (7 USD2010 tCO$_2^{-1}$ in 2015) combined with a ban on new coal-based power plants and dedicated policies addressing renewable electricity generation capacity and electric vehicles reduces efficiency losses compared with an optimal carbon pricing in 2030 (Bertram et al., 2015b). One study estimates the carbon prices in high energy-intensive pathways to be 25–50% higher than in low energy-intensive pathways that assume ambitious regulatory instruments, economic incentives (in addition to a carbon price) and voluntary initiatives (Méjean et al., 2018). A bottom-up approach shows that stringent minimum performance standards (MEPS) for appliances (e.g., refrigerators) can effectively complement explicit carbon pricing, as tightened MEPS can achieve ambitious efficiency improvements that cannot be assured by carbon prices of 100 USD2010 tCO$_2^{-1}$ or higher (Sonnenschein et al., 2018). In addition, the revenue recycling effect of carbon pricing can reduce mitigation costs by displacing distortionary taxes (Baranzini et al., 2017; OECD, 2017; McFarland et al., 2018; Sands, 2018; Siegmeier et al., 2018), and the reduction of capital tax (compared to a labour tax) can yield greater savings in welfare costs (Sands, 2018). The effect on public budgets is particularly important in the near term; however, it can decline in the long term as carbon neutrality is achieved (Sands, 2018). The literature indicates that explicit carbon pricing is relevant but needs to be complemented with other policies to drive the required changes in line with 1.5°C cost-effective pathways (*low to medium evidence, high agreement*) (see Chapter 4, Section 4.4.5) (Stiglitz et al., 2017; Mehling and Tvinnereim, 2018; Méjean et al., 2018; Michaelowa et al., 2018).

In summary, new analyses are consistent with AR5 and show that the price of carbon increases significantly if a higher level of stringency is pursued (*high confidence*). Values vary substantially across models, scenarios and socio-economic, technology and policy assumptions. While an explicit carbon pricing mechanism is central to prompt mitigation scenarios compatible with 1.5°C pathways, a complementary mix of stringent policies is required.

2.5.2.2 Investments

Realizing the transformations towards a 1.5°C world would require a major shift in investment patterns (McCollum et al., 2018). Literature on global climate change mitigation investments is relatively sparse, with most detailed literature having focused on 2°C pathways (McCollum

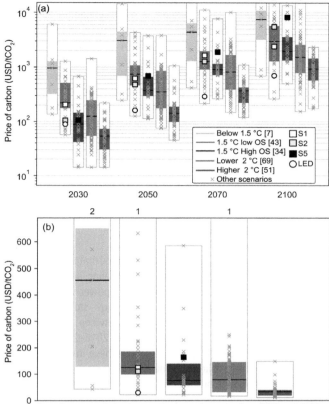

Figure 2.26 | Global price of carbon emissions consistent with mitigation pathways. Panels show (a) undiscounted price of carbon (2030–2100) and (b) average price of carbon (2030–2100) discounted at a 5% discount rate to 2020 in USD2010. AC: Annually compounded. NPV: Net present value. Median values in floating black line. The number of pathways included in box plots is indicated in the legend. Number of pathways outside the figure range is noted at the top.

et al., 2013; Bowen et al., 2014; Gupta and Harnisch, 2014; Marangoni and Tavoni, 2014; OECD/IEA and IRENA, 2017).

Global energy-system investments in the year 2016 are estimated at approximately 1.7 trillion USD2010 (approximately 2.2% of global GDP and 10% of gross capital formation), of which 0.23 trillion USD2010 was for incremental end-use energy efficiency and the remainder for supply-side capacity installations (IEA, 2017c). There is some uncertainty surrounding this number because not all entities making investments report them publicly, and model-based estimates show an uncertainty range of about ±15% (McCollum et al., 2018). Notwithstanding, the trend for global energy investments has been generally upward over the last two decades: increasing about threefold between 2000 and 2012, then levelling off for three years before declining in both 2015 and 2016 as a result of the oil price collapse and simultaneous capital cost reductions for renewables (IEA, 2017c).

Estimates of demand-side investments, either in total or for incremental efficiency efforts, are more uncertain, mainly due to a lack of reliable statistics and definitional issues about what exactly is counted towards a demand-side investment and what the reference should be for estimating incremental efficiency (McCollum et al., 2013). Grubler and

Wilson (2014) use two working definitions (a broader and a narrower one) to provide a first-order estimate of historical end-use technology investments in total. The broad definition defines end-use technologies as the technological systems purchasable by final consumers in order to provide a useful service, for example, heating and air conditioning systems, cars, freezers, or aircraft. The narrow definition sets the boundary at the specific energy-using components or subsystems of the larger end-use technologies (e.g., compressor, car engine, heating element). Based on these two definitions, demand-side energy investments for the year 2005 were estimated about 1–3.5 trillion USD2010 (central estimate 1.7 trillion USD2010) using the broad definition and 0.1–0.6 trillion USD2010 (central estimate 0.3 trillion USD2010) using the narrower definition. Due to these definitional issues, demand-side investment projections are uncertain, often underreported, and difficult to compare. Global IAMs often do not fully and explicitly represent all the various measures that could improve end-use efficiency.

Research carried out by six global IAM teams found that 1.5°C-consistent climate policies would require a marked upscaling of energy system supply-side investments (resource extraction, power generation, fuel conversion, pipelines/transmission, and energy storage) between now and mid-century, reaching levels of between 1.6–3.8 trillion USD2010 yr^{-1} globally on average over the 2016–2050 timeframe (McCollum et al., 2018) (Figure 2.27). How these investment needs compare to those in a policy baseline scenario is uncertain: they could be higher, much higher, or lower. Investments in the policy baselines from these same models are 1.6–2.7 trillion USD2010 yr^{-1}. Much hinges on the reductions in energy demand growth embodied in the 1.5°C pathways, which require investing in energy efficiency. Studies suggest that annual supply-side investments by mid-century could be lowered by around 10% (McCollum et al., 2018) and in some cases up to 50% (Grubler et al., 2018) if strong policies to limit energy demand growth are successfully implemented. However, the degree to which these supply-side reductions would be partially offset by an increase in demand-side investments is unclear.

Some trends are robust across scenarios (Figure 2.27). First, pursuing 1.5°C mitigation efforts requires a major reallocation of the investment portfolio, implying a financial system aligned to mitigation challenges. The path laid out by countries' current NDCs until 2030 will not drive these structural changes; and despite increasing low-carbon investments in recent years (IEA, 2016b; Frankfurt School-UNEP Centre/BNEF, 2017), these are not yet aligned with 1.5°C. Second, additional annual average energy-related investments for the period 2016 to 2050 in pathways limiting warming to 1.5°C compared to the baseline (i.e., pathways without new climate policies beyond those in place today) are estimated by the models employed in McCollum et al. (2018) to be around 830 billion USD2010 (range of 150 billion to 1700 billion USD2010 across six models). This compares to total annual average energy *supply* investments in 1.5°C pathways of 1460 to 3510 billion USD2010 and total annual average energy *demand* investments of 640 to 910 billion USD2010 for the period 2016 to 2050. Total energy-related investments increase by about 12% (range of 3% to 24%) in 1.5°C pathways relative to 2°C pathways. Average annual investment in low-carbon energy technologies and energy efficiency are upscaled by roughly a factor of six (range of factor of 4 to 10) by 2050 compared to 2015. Specifically, annual investments in low-carbon energy are

projected to average 0.8–2.9 trillion USD2010 yr^{-1} globally to 2050 in 1.5°C pathways, overtaking fossil investments globally already by around 2025 (McCollum et al., 2018). The bulk of these investments are projected to be for clean electricity generation, particularly solar and wind power (0.09–1.0 trillion USD2010 yr^{-1} and 0.1–0.35 trillion USD2010 yr^{-1}, respectively) as well as nuclear power (0.1–0.25 trillion USD2010 yr^{-1}). Third, the precise apportioning of these investments depends on model assumptions and societal preferences related to mitigation strategies and policy choices (see Sections 2.1 and 2.3). Investments for electricity transmission and distribution and storage are also scaled up in 1.5°C pathways (0.3–1.3 trillion USD2010 yr^{-1}), given their widespread electrification of the end-use sectors (see Section 2.4). Meanwhile, 1.5°C pathways see a reduction in annual investments for fossil-fuel extraction and unabated fossil electricity generation (to 0.3–0.85 trillion USD2010 yr^{-1} on average over the 2016–2050 period). Investments in unabated coal are halted by 2030 in most 1.5°C projections, while the literature is less conclusive for investments in unabated gas (McCollum et al., 2018). This illustrates how mitigation strategies vary between models, but in the real world should be considered in terms of their societal desirability (see Section 2.5.3). Furthermore, some fossil investments made over the next few years – or those made in the last few – will *likely* need to be retired prior to fully recovering their capital investment or before the end of their operational lifetime (Bertram et al., 2015a; Johnson et al., 2015; OECD/IEA and IRENA, 2017). How the pace of the energy transition will be affected by such dynamics, namely with respect to politics and society, is not well captured by global IAMs at present. Modelling studies have, however, shown how the reliability of institutions influences investment risks and hence climate mitigation investment decisions (Iyer et al., 2015), finding that a lack of regulatory credibility or policy commitment fails to stimulate low-carbon investments (Bosetti and Victor, 2011; Faehn and Isaksen, 2016).

Low-carbon supply-side investment needs are projected to be largest in OECD countries and those of developing Asia. The regional distribution of investments in 1.5°C pathways estimated by the multiple models in (McCollum et al., 2018) are the following (average over 2016–2050 timeframe): 0.30–1.3 trillion USD2010 yr^{-1}(ASIA), 0.35–0.85 trillion USD2010 yr^{-1} (OECD), 0.08–0.55 trillion USD2010 yr^{-1} (MAF), 0.07–0.25 trillion USD2010 yr^{-1} (LAM), and 0.05–0.15 trillion USD2010 yr^{-1} (REF) (regions are defined consistent with their use in AR5 WGIII, see Table A.II.8 in Krey et al., 2014b).

Until now, IAM investment analyses of 1.5°C pathways have focused on middle-of-the-road socio-economic and technological development futures (SSP2) (Fricko et al., 2017). Consideration of a broader range of development futures would yield different outcomes in terms of the magnitudes of the projected investment levels. Sensitivity analyses indicate that the magnitude of supply-side investments as well as the investment portfolio do not change strongly across the SSPs for a given level of climate policy stringency (McCollum et al., 2018). With only one dedicated multimodel comparison study published, there is *limited to medium evidence* available. For some features, there is *high agreement* across modelling frameworks leading, for example, to *medium to high confidence* that limiting global temperature increase to 1.5°C would require a major reallocation of the investment portfolio. Given the limited amount of sensitivity cases available compared to the default SSP2 assumptions, *medium confidence* can be assigned to the specific energy and climate mitigation investment estimates reported here.

Assumptions in modelling studies indicate a number of challenges. For instance, access to finance and mobilization of funds are critical (Fankhauser et al., 2016; OECD, 2017). In turn, policy efforts need to be effective in redirecting financial resources (UNEP, 2015; OECD, 2017) and reducing transaction costs for bankable mitigation projects (i.e. projects that have adequate future cash flow, collateral, etc. so lenders are willing to finance it), particularly on the demand side (Mundaca et al., 2013; Brunner and Enting, 2014; Grubler et al., 2018). Assumptions also imply that policy certainty, regulatory oversight mechanisms and fiduciary duty need to be robust and effective to safeguard credible and stable financial

markets and de-risk mitigation investments in the long term (Clarke et al., 2014; Mundaca et al., 2016; EC, 2017; OECD, 2017). Importantly, the different time horizons that actors have in the competitive finance industry are typically not explicitly captured by modelling assumptions (Harmes, 2011). See Chapter 4, Section 4.4.5 for details of climate finance in practice.

In summary and despite inherent uncertainties, the emerging literature indicates a gap between current investment patterns and those compatible with 1.5°C (or 2°C) pathways (*limited to medium evidence, high agreement*). Estimates and assumptions from modelling frameworks suggest a major shift in investment patterns and entail a financial system effectively aligned with mitigation challenges (*high confidence*).

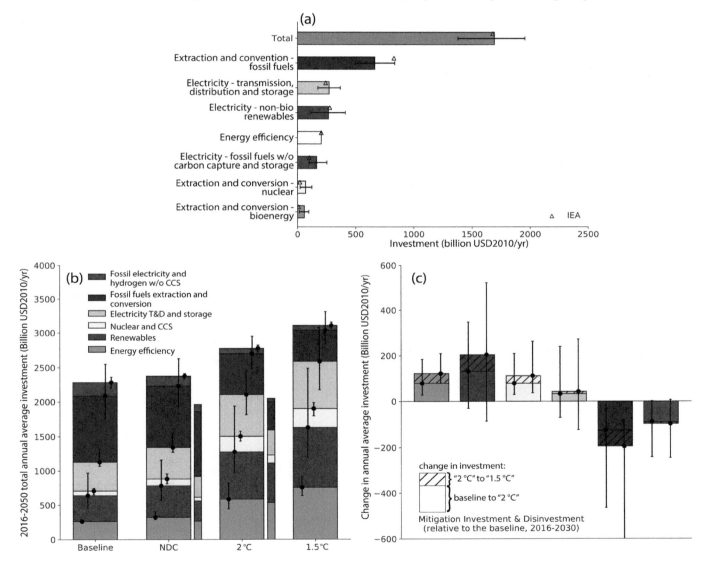

Figure 2.27 | Historical and projected global energy investments. (a) Historical investment estimates across six global models from (McCollum et al., 2018) (bars = model means, whiskers full model range) compared to historical estimates from IEA (International Energy Agency (IEA) 2016) (triangles). (b) Average annual investments over the 2016–2050 period in the "baselines" (i.e., pathways without new climate policies beyond those in place today), scenarios which implement the NDCs ('NDC', including conditional NDCs), scenarios consistent with the Lower-2°C pathway class ('2°C'), and scenarios in line with the 1.5°C-low-OS pathway class ('1.5°C'). Whiskers show the range of models; wide bars show the multimodel means; narrow bars represent analogous values from individual IEA scenarios (OECD/IEA and IRENA, 2017). (c) Average annual mitigation investments and disinvestments for the 2016–2030 periods relative to the baseline. The solid bars show the values for '2°C' pathways, while the hatched areas show the additional investments for the pathways labelled with '1.5°C'. Whiskers show the full range around the multimodel means. T&D stands for transmission and distribution, and CCS stands for carbon capture and storage. Global cumulative carbon dioxide emissions, from fossil fuels and industrial processes (FF&I) but excluding land use, over the 2016-2100 timeframe range from 880 to 1074 GtCO$_2$ (multimodel mean: 952 GtCO$_2$) in the '2°C' pathway and from 206 to 525 GtCO$_2$ (mean: 390 GtCO$_2$) in the '1.5°C' pathway.

2.5.3   Sustainable Development Features of 1.5°C Pathways

Potential synergies and trade-offs between 1.5°C mitigation pathways and different sustainable development (SD) dimensions (see Cross-Chapter Box 4 in Chapter 1) are an emerging field of research. Chapter 5, Section 5.4 assesses interactions between individual mitigation measures with other societal objectives, as well as the Sustainable

Development Goals (SDGs) (Table 5.1). This section synthesized the Chapter 5 insights to assess how these interactions play out in integrated 1.5°C pathways, and the four illustrative pathway archetypes of this chapter in particular (see Section 2.1). Information from integrated pathways is combined with the interactions assessed in Chapter 5 and aggregated for each SDG, with a level of confidence attributed to each interaction based on the amount and agreement of the scientific evidence (see Chapter 5).

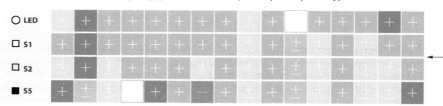

Figure 2.28 | Interactions of individual mitigation measures and alternative mitigations portfolios for 1.5°C with Sustainable Development Goals (SDGs).
The assessment of interactions between mitigation measures and individual SDGs is based on the assessment of Chapter 5, Section 5.4. Proxy indicators and synthesis method are described in Supplementary Material 2.SM.1.5.

Figure 2.28 shows how the scale and combination of individual mitigation measures (i.e., their mitigation portfolios) influence the extent of synergies and trade-offs with other societal objectives. All pathways generate multiple synergies with sustainable development dimensions and can advance several other SDGs simultaneously. Some, however, show higher risks for trade-offs. An example is increased biomass production and its potential to increase pressure on land and water resources, food production, and biodiversity and to reduce air quality when combusted inefficiently. At the same time, mitigation actions in energy-demand sectors and behavioural response options with appropriate management of rebound effects can advance multiple SDGs simultaneously, more so than energy supply-side mitigation actions (see Chapter 5, Section 5.4, Table 5.1 and Figure 5.3 for more examples). Of the four pathway archetypes used in this chapter (LED, S1, S2, and S5, referred to as P1, P2, P3, and P4 in the Summary for Policymakers), the S1 and LED pathways show the largest number of synergies and least number of potential trade-offs, while for the S5 pathway more potential trade-offs are identified. In general, pathways with emphasis on demand reductions and policies that incentivize behavioural change, sustainable consumption patterns, healthy diets and relatively low use of CDR (or only afforestation) show relatively more synergies with individual SDGs than other pathways.

There is *robust evidence* and *high agreement* in the pathway literature that multiple strategies can be considered to limit warming to 1.5°C (see Sections 2.1.3, 2.3 and 2.4). Together with the extensive evidence on the existence of interactions of mitigation measures with other societal objectives (Chapter 5, Section 5.4), this results in *high confidence* that the choice of mitigation portfolio or strategy can markedly affect the achievement of other societal objectives. For instance, action on SLCFs has been suggested to facilitate the achievement of SDGs (Shindell et al., 2017b) and to reduce regional impacts, for example, from black carbon sources on snow and ice loss in the Arctic and alpine regions (Painter et al., 2013), with particular focus on the warming sub-set of SLCFs. Reductions in both surface aerosols and ozone through methane reductions provide health and ecosystem co-benefits (Jacobson, 2002, 2010; Anenberg et al., 2012; Shindell et al., 2012; Stohl et al., 2015; Collins et al., 2018). Public health benefits of stringent mitigation pathways in line with 1.5°C pathways can be sizeable. For instance, a study examining a more rapid reduction of fossil-fuel usage to achieve 1.5°C relative to 2°C, similar to that of other recent studies (Grubler et al., 2018; van Vuuren et al., 2018), found that improved air quality would lead to more than 100 million avoided premature deaths over the 21st century (Shindell et al., 2018). These benefits are assumed to be in addition to those occurring under 2°C pathways (e.g., Silva et al., 2016), and could in monetary terms offset either a large portion or all of the initial mitigation costs (West et al., 2013; Shindell et al., 2018). However, some sources of SLCFs with important impacts for public health (e.g., traditional biomass burning) are only mildly affected by climate policy in the available integrated pathways and are more strongly impacted by baseline assumptions about future societal development and preferences, and technologies instead (Rao et al., 2016, 2017).

At the same time, the literature on climate–SDG interactions is still an emergent field of research and hence there is *low to medium confidence* in the precise magnitude of the majority of these interactions. Very limited literature suggests that achieving co-benefits is not automatically assured but results from conscious and carefully coordinated policies and implementation strategies (Shukla and Chaturvedi, 2012; Clarke et al., 2014; McCollum et al., 2018). Understanding these mitigation–SDG interactions is key for selecting mitigation options that maximize synergies and minimize trade-offs towards the 1.5°C and sustainable development objectives (van Vuuren et al., 2015; Hildingsson and Johansson, 2016; Jakob and Steckel, 2016; von Stechow et al., 2016; Delponte et al., 2017).

In summary, the combined evidence indicates that the chosen mitigation portfolio can have a distinct impact on the achievement of other societal policy objectives (*high confidence*); however, there is uncertainty regarding the specific extent of climate–SDG interactions.

2.6   Knowledge Gaps

This section summarizes the knowledge gaps articulated in earlier sections of the chapter.

2.6.1   Geophysical Understanding

Knowledge gaps are associated with the carbon cycle response, the role of non-CO_2 emissions and the evaluation of an appropriate historic baseline.

Quantifying how the carbon cycle responds to negative emissions is an important knowledge gap for strong mitigation pathways (Section 2.2). Earth system feedback uncertainties are important to consider for the longer-term response, particularly in how permafrost melting might affect the carbon budget (Section 2.2). Future research and ongoing observations over the next years will provide a better indication as to how the 2006-2015 base period compares with the long-term trends and might at present bias the carbon budget estimates.

The future emissions of short-lived climate forcers and their temperature response are a large source of uncertainty in 1.5°C pathways, having a greater relative uncertainty than in higher CO_2 emission pathways. Their global emissions, their sectoral and regional disaggregation, and their climate response are generally less well quantified than for CO_2 (Sections 2.2 and 2.3). Emissions from the agricultural sector, including land-use based mitigation options, in 1.5°C pathways constitute the main source of uncertainty here and are an important gap in understanding the potential achievement of stringent mitigation scenarios (Sections 2.3 and 2.4). This also includes uncertainties surrounding the mitigation potential of the long-lived GHG nitrous oxide (Sections 2.3 and 2.4).

There is considerable uncertainty in how future emissions of aerosol precursors will affect the effective radiative forcing from aerosol–cloud interaction. The potential future warming from mitigation of these emissions reduces remaining carbon budgets and increases peak temperatures (Section 2.2). The potential co-benefits of mitigating air pollutants and how the reduction in air pollution may affect the carbon sink are also important sources of uncertainty (Sections 2.2 and 2.5).

2

The pathway classification employed in this chapter employs results from the MAGICC model with its AR5 parameter sets. The alternative representation of the relationship between emissions and effective radiative forcing and response in the FAIR model would lead to a different classification that would make 1.5°C targets more achievable (Section 2.2 and Supplementary Material 2.SM.1.1). Such a revision would significantly alter the temperature outcomes for the pathways and, if the result is found to be robust, future research and assessments would need to adjust their classifications accordingly. Any possible high bias in the MAGICC response may be partly or entirely offset by missing Earth system feedbacks that are not represented in either climate emulator and that would act to increase the temperature response (Section 2.2). For this assessment report, any possible bias in the MAGICC setup applied in this and earlier reports is not established enough in the literature to change the classification approach. However, we only place *medium confidence* in the classification adopted by the chapter.

2.6.2   Integrated Assessment Approaches

IAMs attempt to be as broad as possible in order to explore interactions between various societal subsystems, like the economy, land, and energy system. They hence include stylized and simplified representations of these subsystems. Climate damages, avoided impacts and societal co-benefits of the modelled transformations remain largely unaccounted for and are important knowledge gaps. Furthermore, rapid technological changes and uncertainties about input data present continuous challenges.

The IAMs used in this report do not account for climate impacts (Section 2.1), and similarly, none of the Gross Domestic Product (GDP) projections in the mitigation pathway literature assessed in this chapter included the feedback of climate damages on economic growth (Section 2.3). Although some IAMs do allow for climate impact feedbacks in their modelling frameworks, particularly in their land components, such feedbacks were by design excluded in pathways developed in the context of the SSP framework. The SSP framework aims at providing an integrative framework for the assessment of climate change adaptation and mitigation. IAMs are typically developed to inform the mitigation component of this question, while the assessment of impacts is carried out by specialized impact models. However, the use of a consistent set of socio-economic drivers embodied by the SSPs allows for an integrated assessment of climate change impacts and mitigation challenges at a later stage. Further integration of these two strands of research will allow a better understanding of climate impacts on mitigation studies.

Many of the IAMs that contributed mitigation pathways to this assessment include a process-based description of the land system in addition to the energy system, and several have been extended to cover air pollutants and water use. These features make them increasingly fit to explore questions beyond those that touch upon climate mitigation only. The models do not, however, fully account for all constraints that could affect realization of pathways (Section 2.1).

While the representation of renewable energy resource potentials, technology costs and system integration in IAMs has been updated since AR5, bottom-up studies find higher mitigation potentials in the industry, buildings, and transport sector in that realized by selected pathways from IAMs, indicating the possibility to strengthen sectoral decarbonization strategies compared to the IAM 1.5°C pathways assessed in this chapter (Section 2.1).

Studies indicate that a major shift in investment patterns is required to limit global warming to 1.5°C. This assessment would benefit from a more explicit representation and understanding of the financial sector within the modelling approaches. Assumptions in modelling studies imply low-to-zero transaction costs for market agents and that regulatory oversight mechanisms and fiduciary duty need to be highly robust to guarantee stable and credible financial markets in the long term. This area can be subject to high uncertainty, however. The heterogeneity of actors (e.g., banks, insurance companies, asset managers, or credit rating agencies) and financial products also needs to be taken into account, as does the mobilization of capital and financial flows between countries and regions (Section 2.5).

The literature on interactions between 1.5°C mitigation pathways and SDGs is an emergent field of research (Section 2.3.5, 2.5 and Chapter 5). Whereas the choice of mitigation strategies can noticeably affect the attainment of various societal objectives, there is uncertainty regarding the extent of the majority of identified interactions. Understanding climate–SDG interactions helps inform the choice of mitigation options that minimize trade-offs and risks and maximize synergies towards sustainable development objectives and the 1.5°C goal (Section 2.5).

2.6.3   Carbon Dioxide Removal (CDR)

Most 1.5°C and 2°C pathways are heavily reliant on CDR at a speculatively large scale before mid-century. There are a number of knowledge gaps associated which such technologies. Chapter 4 performs a detailed assessment of CDR technologies.

There is uncertainty in the future deployment of CCS given the limited pace of current deployment, the evolution of CCS technology that would be associated with deployment, and the current lack of incentives for large-scale implementation of CCS (Chapter 4, Section 4.2.7). Technologies other than BECCS and afforestation have yet to be comprehensively assessed in integrated assessment approaches. No proposed technology is close to deployment at scale, and regulatory frameworks are not established. This limits how they can be realistically implemented within IAMs. (Section 2.3)

Evaluating the potential from BECCS is problematic due to large uncertainties in future land projections due to differences in modelling approaches in current land-use models, and these differences are at least as great as the differences attributed to climate scenario variations. (Section 2.3)

There is substantial uncertainty about the adverse effects of large-scale CDR deployment on the environment and societal sustainable development goals. It is not fully understood how land-use and land-management choices for large-scale BECCS will affect various ecosystem services and sustainable development, and how they further translate into indirect impacts on climate, including GHG emissions other than CO_2. (Section 2.3, Section 2.5.3)

Frequently Asked Questions

FAQ 2.1 | What Kind of Pathways Limit Warming to 1.5°C and are we on Track?

Summary: *There is no definitive way to limit global temperature rise to 1.5°C above pre-industrial levels. This Special Report identifies two main conceptual pathways to illustrate different interpretations. One stabilizes global temperature at, or just below, 1.5°C. Another sees global temperature temporarily exceed 1.5°C before coming back down. Countries' pledges to reduce their emissions are currently not in line with limiting global warming to 1.5°C.*

Scientists use computer models to simulate the emissions of greenhouse gases that would be consistent with different levels of warming. The different possibilities are often referred to as 'greenhouse gas emission pathways'. There is no single, definitive pathway to limiting warming to 1.5°C.

This IPCC special report identifies two main pathways that explore global warming of 1.5°C. The first involves global temperature stabilizing at or below before 1.5°C above pre-industrial levels. The second pathway sees warming exceed 1.5°C around mid-century, remain above 1.5°C for a maximum duration of a few decades, and return to below 1.5°C before 2100. The latter is often referred to as an 'overshoot' pathway. Any alternative situation in which global temperature continues to rise, exceeding 1.5°C permanently until the end of the 21st century, is not considered to be a 1.5°C pathway.

The two types of pathway have different implications for greenhouse gas emissions, as well as for climate change impacts and for achieving sustainable development. For example, the larger and longer an 'overshoot', the greater the reliance on practices or technologies that remove CO_2 from the atmosphere, on top of reducing the sources of emissions (mitigation). Such ideas for CO_2 removal have not been proven to work at scale and, therefore, run the risk of being less practical, effective or economical than assumed. There is also the risk that the use of CO_2 removal techniques ends up competing for land and water, and if these trade-offs are not appropriately managed, they can adversely affect sustainable development. Additionally, a larger and longer overshoot increases the risk for irreversible climate impacts, such as the onset of the collapse of polar ice shelves and accelerated sea level rise.

Countries that formally accept or 'ratify' the Paris Agreement submit pledges for how they intend to address climate change. Unique to each country, these pledges are known as Nationally Determined Contributions (NDCs). Different groups of researchers around the world have analysed the combined effect of adding up all the NDCs. Such analyses show that current pledges are not on track to limit global warming to 1.5°C above pre-industrial levels. If current pledges for 2030 are achieved but no more, researchers find very few (if any) ways to reduce emissions after 2030 sufficiently quickly to limit warming to 1.5°C. This, in turn, suggests that with the national pledges as they stand, warming would exceed 1.5°C, at least for a period of time, and practices and technologies that remove CO_2 from the atmosphere at a global scale would be required to return warming to 1.5°C at a later date.

A world that is consistent with holding warming to 1.5°C would see greenhouse gas emissions rapidly decline in the coming decade, with strong international cooperation and a scaling up of countries' combined ambition beyond current NDCs. In contrast, delayed action, limited international cooperation, and weak or fragmented policies that lead to stagnating or increasing greenhouse gas emissions would put the possibility of limiting global temperature rise to 1.5°C above pre-industrial levels out of reach.

(continued on next page)

FAQ 2.1 (continued)

FAQ2.1:Conceptual pathways that limit global warming to 1.5°C

Two main pathways illustrate different interpretations for limiting global warming to 1.5°C.
The consequences will be different depending on the pathway

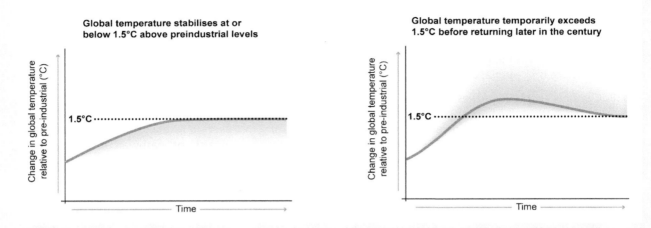

FAQ 2.1, Figure 1 | Two main pathways for limiting global temperature rise to 1.5°C above pre-industrial levels are discussed in this Special Report. These are: stabilizing global temperature at, or just below, 1.5°C (left) and global temperature temporarily exceeding 1.5°C before coming back down later in the century (right). Temperatures shown are relative to pre-industrial but pathways are illustrative only, demonstrating conceptual not quantitative characteristics.

Frequently Asked Questions

FAQ 2.2 | What do Energy Supply and Demand have to do with Limiting Warming to 1.5°C?

Summary: Limiting global warming to 1.5°C above pre-industrial levels would require major reductions in greenhouse gas emissions in all sectors. But different sectors are not independent of each other, and making changes in one can have implications for another. For example, if we as a society use a lot of energy, then this could mean we have less flexibility in the choice of mitigation options available to limit warming to 1.5°C. If we use less energy, the choice of possible actions is greater – for example, we could be less reliant on technologies that remove carbon dioxide (CO_2) from the atmosphere.

To stabilize global temperature at any level, 'net' CO_2 emissions would need to be reduced to zero. This means the amount of CO_2 entering the atmosphere must equal the amount that is removed. Achieving a balance between CO_2 'sources' and 'sinks' is often referred to as 'net zero' emissions or 'carbon neutrality'. The implication of net zero emissions is that the concentration of CO_2 in the atmosphere would slowly decline over time until a new equilibrium is reached, as CO_2 emissions from human activity are redistributed and taken up by the oceans and the land biosphere. This would lead to a near-constant global temperature over many centuries.

Warming will not be limited to 1.5°C or 2°C unless transformations in a number of areas achieve the required greenhouse gas emissions reductions. Emissions would need to decline rapidly across all of society's main sectors, including buildings, industry, transport, energy, and agriculture, forestry and other land use (AFOLU). Actions that can reduce emissions include, for example, phasing out coal in the energy sector, increasing the amount of energy produced from renewable sources, electrifying transport, and reducing the 'carbon footprint' of the food we consume.

The above are examples of 'supply-side' actions. Broadly speaking, these are actions that can reduce greenhouse gas emissions through the use of low-carbon solutions. A different type of action can reduce how much energy human society uses, while still ensuring increasing levels of development and well-being. Known as 'demand-side' actions, this category includes improving energy efficiency in buildings and reducing consumption of energy- and greenhouse-gas intensive products through behavioural and lifestyle changes, for example. Demand- and supply-side measures are not an either-or question, they work in parallel with each other. But emphasis can be given to one or the other.

Making changes in one sector can have consequences for another, as they are not independent of each other. In other words, the choices that we make now as a society in one sector can either restrict or expand our options later on. For example, a high demand for energy could mean we would need to deploy almost all known options to reduce emissions in order to limit global temperature rise to 1.5°C above pre-industrial levels, with the potential for adverse side-effects. In particular, a pathway with high energy demand would increase our reliance on practices and technologies that remove CO_2 from the atmosphere. As of yet, such techniques have not been proven to work on a large scale and, depending on how they are implemented, could compete for land and water. By leading to lower overall energy demand, effective demand-side measures could allow for greater flexibility in how we structure our energy system. However, demand-side measures are not easy to implement and barriers have prevented the most efficient practices being used in the past.

(continued on next page)

FAQ 2.2 (continued)

FAQ2.2: **Energy demand and supply in 1.5°C world**

Lower energy demand could allow for greater flexibility in how we structure our energy system.

Low energy demand allows more choice about which low-carbon energy supply options to use to limit warming to 1.5°C.

With high energy demand, there is less flexibility as virtually all available options would need to be considered.

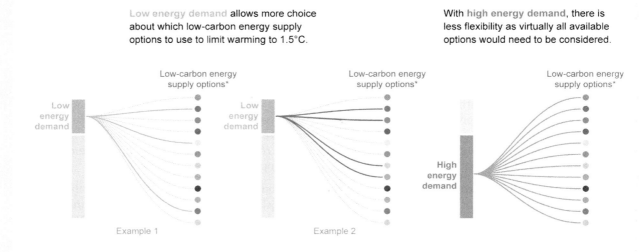

* Options include renewable energy (such as bioenergy, hydro, wind and solar), nuclear and the use of carbon dioxide removal techniques

FAQ 2.2, Figure 1 | Having a lower energy demand increases the flexibility in choosing options for supplying energy. A larger energy demand means many more low carbon energy supply options would need to be used.

References

Ackerman, F. and E.A. Stanton, 2012: Climate Risks and Carbon Prices: Revising the Social Cost of Carbon. *Economics: The Open-Access, Open-Assessment E-Journal*, **6(2012-10)**, 1–25, doi:10.5018/economics-ejournal.ja.2012-10.

Adler, M.D. and N. Treich, 2015: Prioritarianism and Climate Change. *Environmental and Resource Economics*, **62(2)**, 279–308, doi:10.1007/s10640-015-9960-7.

Adler, M.D. et al., 2017: Priority for the worse-off and the social cost of carbon. *Nature Climate Change*, **7(6)**, 443–449, doi:10.1038/nclimate3298.

Akimoto, K., F. Sano, and T. Tomoda, 2017: GHG emission pathways until 2300 for the 1.5°C temperature rise target and the mitigation costs achieving the pathways. *Mitigation and Adaptation Strategies for Global Change*, 1–14, doi:10.1007/s11027-017-9762-z.

Alcalde, J. et al., 2018: Estimating geological CO_2 storage security to deliver on climate mitigation. *Nature Communications*, **9(1)**, 2201, doi:10.1038/s41467-018-04423-1.

Aldy, J.E. and R.N. Stavins, 2012: The Promise and Problems of Pricing Carbon. *The Journal of Environment & Development*, **21(2)**, 152–180, doi:10.1177/1070496512442508.

Aldy, J.E. et al., 2016: Economic tools to promote transparency and comparability in the Paris Agreement. *Nature Climate Change*, **6(11)**, 1000–1004, doi:10.1038/nclimate3106.

Allen, M.R. et al., 2009: Warming caused by cumulative carbon emissions towards the trillionth tonne. *Nature*, **458(7242)**, 1163–1166, doi:10.1038/nature08019.

Allen, M.R. et al., 2016: New use of global warming potentials to compare cumulative and short-lived climate pollutants. *Nature Climate Change*, **6(5)**, 1–5, doi:10.1038/nclimate2998.

Allen, M.R. et al., 2018: A solution to the misrepresentations of CO_2-equivalent emissions of short climate pollutants under ambitious mitigation. *npj Climate and Atmospheric Science*, **1(1)**, 16, doi:10.1038/s41612-018-0026-8.

Allwood, J.M., M.F. Ashby, T.G. Gutowski, and E. Worrell, 2013: Material efficiency: providing material services with less material production. *Philosophical Transactions of the Royal Society A: Mathematical, Physical and Engineering Sciences*, **371(1986)**, 20120496, doi:10.1098/rsta.2012.0496.

Anderson, K. and G. Peters, 2016: The trouble with negative emissions. *Science*, **354(6309)**, 182–183, doi:10.1126/science.aah4567.

Anenberg, S.C. et al., 2012: Global Air Quality and Health Co-benefits of Mitigating Near-Term Climate Change through Methane and Black Carbon Emission Controls. *Environmental Health Perspectives*, **120(6)**, 831–839, doi:10.1289/ehp.1104301.

Anthoff, D. and R.S.J. Tol, 2009: The impact of climate change on the balanced growth equivalent: An application of FUND. *Environmental and Resource Economics*, **43(3)**, 351–367, doi:10.1007/s10640-009-9269-5.

Armour, K.C., 2017: Energy budget constraints on climate sensitivity in light of inconstant climate feedbacks. *Nature Climate Change*, **7(5)**, 331–335, doi:10.1038/nclimate3278.

Arneth, A. et al., 2010: Terrestrial biogeochemical feedbacks in the climate system. *Nature Geoscience*, **3(8)**, 525–532, doi:10.1038/ngeo905.

Arroyo-Currás, T. et al., 2015: Carbon leakage in a fragmented climate regime: The dynamic response of global energy markets. *Technological Forecasting and Social Change*, **90**, 192–203, doi:10.1016/j.techfore.2013.10.002.

Bachu, S., 2015: Review of CO_2 storage efficiency in deep saline aquifers. *International Journal of Greenhouse Gas Control*, **40**, 188–202, doi:10.1016/j.ijggc.2015.01.007.

Bachu, S. et al., 2007a: *Phase II Final Report from the Task Force for Review and Identification of Standards for CO_2 Storage Capacity Estimation*. CSLF-T-2007-04, Task Force on CO_2 Storage Capacity Estimation for the Technical Group (TG) of the Carbon Sequestration Leadership Forum (CSLF), 43 pp.

Bachu, S. et al., 2007b: CO_2 storage capacity estimation: Methodology and gaps. *International Journal of Greenhouse Gas Control*, **1(4)**, 430–443, doi:10.1016/s1750-5836(07)00086-2.

Bajželj, B. et al., 2014: Importance of food-demand management for climate mitigation. *Nature Climate Change*, **4(10)**, 924–929, doi:10.1038/nclimate2353.

Banerjee, R. et al., 2012: Energy End-Use: Industry. In: *Global Energy Assessment – Toward a Sustainable Future* [Johansson, T.B., N. Nakicenovic, A. Patwardhan, and L. Gomez-Echeverri (eds.)]. Cambridge University Press, Cambridge, United Kingdom and New York, NY, USA and the International Institute for Applied Systems Analysis, Laxenburg, Austria, pp. 513–574, doi:10.1017/cbo9780511793677.014.

Baranzini, A. et al., 2017: Carbon pricing in climate policy: seven reasons, complementary instruments, and political economy considerations. *Wiley Interdisciplinary Reviews: Climate Change*, **8(4)**, e462, doi:10.1002/wcc.462.

Bataille, C., H. Waisman, M. Colombier, L. Segafredo, and J. Williams, 2016a: The Deep Decarbonization Pathways Project (DDPP): insights and emerging issues. *Climate Policy*, **16(sup1)**, S1–S6, doi:10.1080/14693062.2016.1179620.

Bataille, C. et al., 2016b: The need for national deep decarbonization pathways for effective climate policy. *Climate Policy*, **16(sup1)**, S7–S26, doi:10.1080/14693062.2016.1173005.

Bauer, N. et al., 2017: Shared Socio-Economic Pathways of the Energy Sector – Quantifying the Narratives. *Global Environmental Change*, **42**, 316–330, doi:10.1016/j.gloenvcha.2016.07.006.

Bauer, N. et al., 2018: Global energy sector emission reductions and bioenergy use: overview of the bioenergy demand phase of the EMF-33 model comparison. *Climatic Change*, 1–16, doi:10.1007/s10584-018-2226-y.

Beck, S. and M. Mahony, 2017: The IPCC and the politics of anticipation. *Nature Climate Change*, **7(5)**, 311–313, doi:10.1038/nclimate3264.

Bentham, M., T. Mallows, J. Lowndes, and A. Green, 2014: CO_2 STORage evaluation database (CO_2 Stored). The UK's online storage atlas. *Energy Procedia*, **63**, 5103–5113, doi:10.1016/j.egypro.2014.11.540.

Benveniste, H. et al., 2018: Impacts of nationally determined contributions on 2030 global greenhouse gas emissions: Uncertainty analysis and distribution of emissions. *Environmental Research Letters*, **13(1)**, 014022, doi:10.1088/1748-9326/aaa0b9.

Berger, A. et al., 2017a: Nuclear energy and bio energy carbon capture and storage, keys for obtaining 1.5°C mean surface temperature limit. *International Journal of Global Energy Issues*, **40(3/4)**, 240–254, doi:10.1504/ijgei.2017.086622.

Berger, A. et al., 2017b: How much can nuclear energy do about global warming? *International Journal of Global Energy Issues*, **40(1/2)**, 43–78, doi:10.1504/ijgei.2017.080766.

Bertram, C. et al., 2015a: Carbon lock-in through capital stock inertia associated with weak near-term climate policies. *Technological Forecasting and Social Change*, **90(Part A)**, 62–72, doi:10.1016/j.techfore.2013.10.001.

Bertram, C. et al., 2015b: Complementing carbon prices with technology policies to keep climate targets within reach. *Nature Climate Change*, **5(3)**, 235–239, doi:10.1038/nclimate2514.

Bertram, C. et al., 2018: Targeted policies can compensate most of the increased sustainability risks in 1.5°C mitigation scenarios. *Environmental Research Letters*, **13(6)**, 064038, doi:10.1088/1748-9326/aac3ec.

Bindoff, N. et al., 2013: Detection and Attribution of Climate Change: from Global to Regional. In: *Climate Change 2013: The Physical Science Basis. Contribution of Working Group I to the Fifth Assessment Report of the Intergovernmental Panel on Climate Change* [Stocker, T.F., D. Qin, G.-K. Plattner, M. Tignor, S.K. Allen, J. Boschung, A. Nauels, Y. Xia, V. Bex, and P.M. Midgley (eds.)]. Cambridge University Press, Cambridge, United Kingdom and New York, NY, USA, pp. 867–952.

Blanco, G. et al., 2014: Drivers, Trends and Mitigation. In: *Climate Change 2014: Mitigation of Climate Change. Contribution of Working Group III to the Fifth Assessment Report of the Intergovernmental Panel on Climate Change* [Edenhofer, O., R. Pichs-Madruga, Y. Sokona, E. Farahani, S. Kadner, K. Seyboth, A. Adler, I. Baum, S. Brunner, P. Eickemeier, B. Kriemann, J. Savolainen, S. Schlömer, C. von Stechow, T. Zwickel, and J.C. Minx (eds.)]. Cambridge University Press, Cambridge, United Kingdom and New York, NY, USA, pp. 351–412.

Boardman, A.E., D.H. Greenberg, A.R. Vining, and D.L. Weimer, 2006: *Cost-Benefit Analysis: Concepts and Practice (3rd edition)*. Pearson/Prentice Hall, Upper Saddle River, NJ, USA, 560 pp.

Bodirsky, B.L. et al., 2012: N_2O emissions from the global agricultural nitrogen cycle-current state and future scenarios. *Biogeosciences*, **9(10)**, 4169–4197, doi:10.5194/bg-9-4169-2012.

Bodirsky, B.L. et al., 2014: Reactive nitrogen requirements to feed the world in 2050 and potential to mitigate nitrogen pollution. *Nature Communications*, **5**, 3858, doi:10.1038/ncomms4858.

Bond, T.C. et al., 2013: Bounding the role of black carbon in the climate system: A scientific assessment. *Journal of Geophysical Research: Atmospheres*, **118(11)**, 5380–5552, doi:10.1002/jgrd.50171.

Bonsch, M. et al., 2014: Trade-offs between land and water requirements for large-scale bioenergy production. *GCB Bioenergy*, **8(1)**, 11–24, doi:10.1111/gcbb.12226.

Bosetti, V. and D.G. Victor, 2011: Politics and Economics of Second-Best Regulation of Greenhouse Gases: The Importance of Regulatory Credibility. *The Energy Journal*, **32(1)**, 1–24, www.jstor.org/stable/41323391.

Bosetti, V., C. Carraro, E. Massetti, and M. Tavoni, 2008: International energy R&D spillovers and the economics of greenhouse gas atmospheric stabilization. *Energy Economics*, **30(6)**, 2912–2929, doi:10.1016/j.eneco.2008.04.008.

Bosetti, V., C. Carraro, A. Sgobbi, and M. Tavoni, 2009: Delayed action and uncertain stabilisation targets. How much will the delay cost? *Climatic Change*, **96(3)**, 299–312, doi:10.1007/s10584-009-9630-2.

Bosetti, V., C. Carraro, M. Galeotti, E. Massetti, and M. Tavoni, 2006: WITCH – A World Induced Technical Change Hybrid Model. *The Energy Journal*, **27**, 13–37, www.jstor.org/stable/23297044.

Bosetti, V. et al., 2015: Sensitivity to energy technology costs: A multi-model comparison analysis. *Energy Policy*, **80**, 244–263, doi:10.1016/j.enpol.2014.12.012.

Boucher, O. and G.A. Folberth, 2010: New Directions: Atmospheric methane removal as a way to mitigate climate change? *Atmospheric Environment*, **44(27)**, 3343–3345, doi:10.1016/j.atmosenv.2010.04.032.

Bowen, A., E. Campiglio, and M. Tavoni, 2014: A macroeconomic perspective on climate change mitigation: Meeting the financing challenge. *Climate Change Economics*, **5(1)**, 1440005, doi:10.1142/s2010007814400053.

Bowerman, N.H.A., D.J. Frame, C. Huntingford, J.A. Lowe, and M.R. Allen, 2011: Cumulative carbon emissions, emissions floors and short-term rates of warming: implications for policy. *Philosophical Transactions of the Royal Society A: Mathematical, Physical and Engineering Sciences*, **369(1934)**, 45–66, doi:10.1098/rsta.2010.0288.

Bowerman, N.H.A. et al., 2013: The role of short-lived climate pollutants in meeting temperature goals. *Nature Climate Change*, **3(12)**, 1021–1024, doi:10.1038/nclimate2034.

Boysen, L.R., W. Lucht, and D. Gerten, 2017a: Trade-offs for food production, nature conservation and climate limit the terrestrial carbon dioxide removal potential. *Global Change Biology*, **23(10)**, 4303–4317, doi:10.1111/gcb.13745.

Boysen, L.R., W. Lucht, D. Gerten, and V. Heck, 2016: Impacts devalue the potential of large-scale terrestrial CO_2 removal through biomass plantations. *Environmental Research Letters*, **11(9)**, 1–10, doi:10.1088/1748-9326/11/9/095010.

Boysen, L.R. et al., 2017b: The limits to global-warming mitigation by terrestrial carbon removal. *Earth's Future*, **5(5)**, 463–474, doi:10.1002/2016ef000469.

Breyer, C. et al., 2017: On the role of solar photovoltaics in global energy transition scenarios. *Progress in Photovoltaics: Research and Applications*, **25(8)**, 727–745, doi:10.1002/pip.2885.

Brown, M.A. and Y. Li, 2018: Carbon pricing and energy efficiency: pathways to deep decarbonization of the US electric sector. *Energy Efficiency*, 1–19, doi:10.1007/s12053-018-9686-9.

Brown, P.T. and K. Caldeira, 2017: Greater future global warming inferred from Earth's recent energy budget. *Nature*, **552(7683)**, 45–50, doi:10.1038/nature24672.

Bruckner, T., I.A. Bashmakov, and Y. Mulugetta, 2014: Energy Systems. In: *Climate Change 2014: Mitigation of Climate Change. Contribution of Working Group III to the Fifth Assessment Report of the Intergovernmental Panel on Climate Change* [Edenhofer, O., R. Pichs-Madruga, Y. Sokona, E. Farahani, S. Kadner, K. Seyboth, A. Adler, I. Baum, S. Brunner, P. Eickemeier, B. Kriemann, J. Savolainen, S. Schlömer, C. Stechow, T. Zwickel, and J.C. Minx (eds.)]. Cambridge University Press, Cambridge, United Kingdom and New York, NY, USA, pp. 511–598.

Brunner, S. and K. Enting, 2014: Climate finance: A transaction cost perspective on the structure of state-to-state transfers. *Global Environmental Change*, **27**, 138–143, doi:10.1016/j.gloenvcha.2014.05.005.

Brynolf, S., M. Taljegard, M. Grahn, and J. Hansson, 2018: Electrofuels for the transport sector: A review of production costs. *Renewable and Sustainable Energy Reviews*, **81**, 1887–1905, doi:10.1016/j.rser.2017.05.288.

Buck, H.J., 2016: Rapid scale-up of negative emissions technologies: social barriers and social implications. *Climatic Change*, 1–13, doi:10.1007/s10584-016-1770-6.

Burke, E.J., I.P. Hartley, and C.D. Jones, 2012: Uncertainties in the global temperature change caused by carbon release from permafrost thawing. *Cryosphere*, **6(5)**, 1063–1076, doi:10.5194/tc-6-1063-2012.

Burke, E.J. et al., 2017: Quantifying uncertainties of permafrost carbon-climate feedbacks. *Biogeosciences*, **14(12)**, 3051–3066, doi:10.5194/bg-14-3051-2017.

Burns, W. and S. Nicholson, 2017: Bioenergy and carbon capture with storage (BECCS): the prospects and challenges of an emerging climate policy response. *Journal of Environmental Studies and Sciences*, **7(4)**, 527–534, doi:10.1007/s13412-017-0445-6.

Cadule, P., L. Bopp, and P. Friedlingstein, 2009: A revised estimate of the processes contributing to global warming due to climate-carbon feedback. *Geophysical Research Letters*, **36(14)**, L14705, doi:10.1029/2009gl038681.

Cai, Y., K.L. Judd, T.M. Lenton, T.S. Lontzek, and D. Narita, 2015: Environmental tipping points significantly affect the cost-benefit assessment of climate policies. *Proceedings of the National Academy of Sciences*, **112(15)**, 4606–4611, doi:10.1073/pnas.1503890112.

Calvin, K. et al., 2014: Trade-offs of different land and bioenergy policies on the path to achieving climate targets. *Climatic Change*, **123(3–4)**, 691–704, doi:10.1007/s10584-013-0897-y.

Calvin, K. et al., 2017: The SSP4: A world of deepening inequality. *Global Environmental Change*, **42**, 284–296, doi:10.1016/j.gloenvcha.2016.06.010.

Canadell, J.G. and M.R. Raupach, 2008: Managing Forests for Climate Change Mitigation. *Science*, **320(5882)**, 1456–1457, doi:10.1126/science.1155458.

Carslaw, K.S. et al., 2013: Large contribution of natural aerosols to uncertainty in indirect forcing. *Nature*, **503(7474)**, 67–71, doi:10.1038/nature12674.

Ceppi, P. and J.M. Gregory, 2017: Relationship of tropospheric stability to climate sensitivity and Earth's observed radiation budget. *Proceedings of the National Academy of Sciences*, **114(50)**, 13126–13131, doi:10.1073/pnas.1714308114.

Chan, S., P. Ellinger, and O. Widerberg, 2018: Exploring national and regional orchestration of non-state action for a <1.5°C world. *International Environmental Agreements: Politics, Law and Economics*, **18(1)**, 135–152, doi:10.1007/s10784-018-9384-2.

Chen, C. and M. Tavoni, 2013: Direct air capture of CO_2 and climate stabilization: A model based assessment. *Climatic Change*, **118(1)**, 59–72, doi:10.1007/s10584-013-0714-7.

Choi, Y.-S., D. Young, S. Nešić, and L.G.S. Gray, 2013: Wellbore integrity and corrosion of carbon steel in CO_2 geologic storage environments: A literature review. *International Journal of Greenhouse Gas Control*, **16**, S70–S77, doi:10.1016/j.ijggc.2012.12.028.

Clack, C.T.M. et al., 2017: Evaluation of a proposal for reliable low-cost grid power with 100% wind, water, and solar. *Proceedings of the National Academy of Sciences*, **114(26)**, 6722–6727, doi:10.1073/pnas.1610381114.

Clarke, L. et al., 2014: Assessing transformation pathways. In: *Climate Change 2014: Mitigation of Climate Change. Contribution of Working Group III to the Fifth Assessment Report of the Intergovernmental Panel on Climate Change* [Edenhofer, O., R. Pichs-Madruga, Y. Sokona, E. Farahani, S. Kadner, K. Seyboth, A. Adler, I. Baum, S. Brunner, P. Eickemeier, B. Kriemann, J. Savolainen, S. Schlömer, C. von Stechow, T. Zwickel, and J.C. Minx (eds.)]. Cambridge University Press, Cambridge, United Kingdom and New York, NY, USA, pp. 413–510.

Collins, M. et al., 2013: Long-term Climate Change: Projections, Commitments and Irreversibility. In: *Climate Change 2013: The Physical Science Basis. Contribution of Working Group I to the Fifth Assessment Report of the Intergovernmental Panel on Climate Change* [Stocker, T.F., D. Qin, G.-K. Plattner, M. Tignor, S.K. Allen, J. Boschung, A. Nauels, Y. Xia, V. Bex, and P.M. Midgley (eds.)]. Cambridge University Press, Cambridge, United Kingdom and New York, NY, USA, pp. 1029–1136.

Collins, W.J. et al., 2018: Increased importance of methane reduction for a 1.5 degree target. *Environmental Research Letters*, **13(5)**, 054003, doi:10.1088/1748-9326/aab89c.

Comyn-Platt, E. et al., 2018: Carbon budgets for 1.5 and 2°C targets lowered by natural wetland and permafrost feedbacks. *Nature Geoscience*, **11(8)**, 568–573, doi:10.1038/s41561-018-0174-9.

Cox, P.M., C. Huntingford, and M.S. Williamson, 2018: Emergent constraint on equilibrium climate sensitivity from global temperature variability. *Nature*, **553(7688)**, 319–322, doi:10.1038/nature25450.

Crespo Cuaresma, J., 2017: Income projections for climate change research: A framework based on human capital dynamics. *Global Environmental Change*, **42**, 226–236, doi:10.1016/j.gloenvcha.2015.02.012.

Creutzig, F., 2016: Evolving Narratives of Low-Carbon Futures in Transportation. *Transport Reviews*, **36(3)**, 341–360, doi:10.1080/01441647.2015.1079277.

Creutzig, F. et al., 2012: Reconciling top-down and bottom-up modelling on future bioenergy deployment. *Nature Climate Change*, **2(5)**, 320–327, doi:10.1038/nclimate1416.

Creutzig, F. et al., 2015: Bioenergy and climate change mitigation: an assessment. *GCB Bioenergy*, **7(5)**, 916–944, doi:10.1111/gcbb.12205.

Creutzig, F. et al., 2017: The underestimated potential of solar energy to mitigate climate change. *Nature Energy*, **2(9)**, 17140, doi:10.1038/nenergy.2017.140.

Davis, S.C. et al., 2013: Management swing potential for bioenergy crops. *GCB Bioenergy*, **5(6)**, 623–638, doi:10.1111/gcbb.12042.

Davis, S.J. and K. Caldeira, 2010: Consumption-based accounting of CO_2 emissions. *Proceedings of the National Academy of Sciences*, **107(12)**, 5687–92, doi:10.1073/pnas.0906974107.

de Richter, R., T. Ming, P. Davies, W. Liu, and S. Caillol, 2017: Removal of non-CO₂ greenhouse gases by large-scale atmospheric solar photocatalysis. *Progress in Energy and Combustion Science*, **60**, 68–96, doi:10.1016/j.pecs.2017.01.001.

de Vries, W., M. Posch, D. Simpson, and G.J. Reinds, 2017: Modelling long-term impacts of changes in climate, nitrogen deposition and ozone exposure on carbon sequestration of European forest ecosystems. *Science of The Total Environment*, **605–606**, 1097–1116, doi:10.1016/j.scitotenv.2017.06.132.

Dell, M., B.F. Jones, and B.A. Olken, 2014: What Do We Learn from the Weather? The New Climate–Economy Literature. *Journal of Economic Literature*, **52(3)**, 740–798, www.jstor.org/stable/24434109.

Dellink, R., J. Chateau, E. Lanzi, and B. Magné, 2017: Long-term economic growth projections in the Shared Socioeconomic Pathways. *Global Environmental Change*, **42**, 1–15, doi:10.1016/j.gloenvcha.2015.06.004.

Delponte, I., I. Pittaluga, and C. Schenone, 2017: Monitoring and evaluation of Sustainable Energy Action Plan: Practice and perspective. *Energy Policy*, **100**, 9–17, doi:10.1016/j.enpol.2016.10.003.

den Elzen, M.G.J., D.P. van Vuuren, and J. van Vliet, 2010: Postponing emission reductions from 2020 to 2030 increases climate risks and long-term costs. *Climatic Change*, **99(1)**, 313–320, doi:10.1007/s10584-010-9798-5.

den Elzen, M.G.J. et al., 2016: Contribution of the G20 economies to the global impact of the Paris agreement climate proposals. *Climatic Change*, **137(3–4)**, 655–665, doi:10.1007/s10584-016-1700-7.

Dennig, F., M.B. Budolfson, M. Fleurbaey, A. Siebert, and R.H. Socolow, 2015: Inequality, climate impacts on the future poor, and carbon prices. *Proceedings of the National Academy of Sciences*, **112(52)**, 15827–15832, doi:10.1073/pnas.1513967112.

Di, H.J. and K. Cameron, 2016: Inhibition of nitrification to mitigate nitrate leaching and nitrous oxide emissions in grazed grassland: A review. *Journal of Soils and Sediments*, **16**, 1401–1420, doi:10.1007/s11368-016-1403-8.

Doelman, J.C. et al., 2018: Exploring SSP land-use dynamics using the IMAGE model: Regional and gridded scenarios of land-use change and land-based climate change mitigation. *Global Environmental Change*, **48**, 119–135, doi:10.1016/j.gloenvcha.2017.11.014.

Dooley, J.J., 2013: Estimating the supply and demand for deep geologic CO₂ storage capacity over the course of the 21st century: A meta-analysis of the literature. *Energy Procedia*, **37**, 5141–5150, doi:10.1016/j.egypro.2013.06.429.

Dooley, K. and S. Kartha, 2018: Land-based negative emissions: risks for climate mitigation and impacts on sustainable development. *International Environmental Agreements: Politics, Law and Economics*, **18(1)**, 79–98, doi:10.1007/s10784-017-9382-9.

Duce, R.A. et al., 2008: Impacts of Atmospheric Anthropogenic Nitrogen on the Open Ocean. *Science*, **320(5878)**, 893–897, doi:10.1126/science.1150369.

EC, 2017: *High-Level Expert Group on Sustainable Finance interim report – Financing a sustainable European economy*. European Commission, Brussels, Belgium, 72 pp.

Edenhofer, O., J.C. Steckel, M. Jakob, and C. Bertram, 2018: Reports of coal's terminal decline may be exaggerated. *Environmental Research Letters*, **13(2)**, 024019, doi:10.1088/1748-9326/aaa3a2.

Erb, K.-H., H. Haberl, and C. Plutzar, 2012: Dependency of global primary bioenergy crop potentials in 2050 on food systems, yields, biodiversity conservation and political stability. *Energy Policy*, **47**, 260–269, doi:10.1016/j.enpol.2012.04.066.

Erb, K.-H. et al., 2016a: Biomass turnover time in terrestrial ecosystems halved by land use. *Nature Geoscience*, **9(9)**, 674–678, doi:10.1038/ngeo2782.

Erb, K.-H. et al., 2016b: Exploring the biophysical option space for feeding the world without deforestation. *Nature Communications*, **7**, 11382, doi:10.1038/ncomms11382.

Erb, K.-H. et al., 2018: Unexpectedly large impact of forest management and grazing on global vegetation biomass. *Nature*, **553**, 73–76, doi:10.1038/nature25138.

Erickson, P., S. Kartha, M. Lazarus, and K. Tempest, 2015: Assessing carbon lock-in. *Environmental Research Letters*, **10(8)**, 084023, doi:10.1088/1748-9326/10/8/084023.

Etminan, M., G. Myhre, E.J. Highwood, and K.P. Shine, 2016: Radiative forcing of carbon dioxide, methane, and nitrous oxide: A significant revision of the methane radiative forcing. *Geophysical Research Letters*, **43(24)**, 12,614–12,623, doi:10.1002/2016gl071930.

Faehn, T. and E. Isaksen, 2016: Diffusion of Climate Technologies in the Presence of Commitment Problems. *The Energy Journal*, **37(2)**, 155–180, doi:10.5547/01956574.37.2.tfae.

Fankhauser, S., A. Sahni, A. Savvas, and J. Ward, 2016: Where are the gaps in climate finance? *Climate and Development*, **8(3)**, 203–206, doi:10.1080/17565529.2015.1064811.

FAOSTAT, 2018: Database Collection of the Food and Agriculture Organization of the United Nations. FAO. Retrieved from: www.fao.org/faostat.

Fawcett, A.A. et al., 2015: Can Paris pledges avert severe climate change? *Science*, **350(6265)**, 1168–1169, doi:10.1126/science.aad5761.

Forster, P.M., 2016: Inference of Climate Sensitivity from Analysis of Earth's Energy Budget. *Annual Review of Earth and Planetary Sciences*, **44(1)**, 85–106, doi:10.1146/annurev-earth-060614-105156.

Frank, S. et al., 2017: Reducing greenhouse gas emissions in agriculture without compromising food security? *Environmental Research Letters*, **12(10)**, 105004, doi:10.1088/1748-9326/aa8c83.

Frank, S. et al., 2018: Structural change as a key component for agricultural non-CO₂ mitigation efforts. *Nature Communications*, **9(1)**, 1060, doi:10.1038/s41467-018-03489-1.

Frankfurt School-UNEP Centre/BNEF, 2017: *Global Trends in Renewable Energy Investment 2017*. Frankfurt School of Finance & Management gGmbH, Frankfurt, Germany, 90 pp.

Fricko, O. et al., 2016: Energy sector water use implications of a 2°C climate policy. *Environmental Research Letters*, **11(3)**, 034011, doi:10.1088/1748-9326/11/3/034011.

Fricko, O. et al., 2017: The marker quantification of the Shared Socioeconomic Pathway 2: A middle-of-the-road scenario for the 21st century. *Global Environmental Change*, **42**, 251–267, doi:10.1016/j.gloenvcha.2016.06.004.

Friedlingstein, P. et al., 2014a: Persistent growth of CO₂ emissions and implications for reaching climate targets. *Nature Geoscience*, **7(10)**, 709–715, doi:10.1038/ngeo2248.

Friedlingstein, P. et al., 2014b: Uncertainties in CMIP5 climate projections due to carbon cycle feedbacks. *Journal of Climate*, **27(2)**, 511–526, doi:10.1175/jcli-d-12-00579.1.

Frölicher, T.L. and F. Joos, 2010: Reversible and irreversible impacts of greenhouse gas emissions in multi-century projections with the NCAR global coupled carbon cycle-climate model. *Climate Dynamics*, **35(7)**, 1439–1459, doi:10.1007/s00382-009-0727-0.

Fuglestvedt, J.S. et al., 2010: Transport impacts on atmosphere and climate: Metrics. *Atmospheric Environment*, **44(37)**, 4648–4677, doi:10.1016/j.atmosenv.2009.04.044.

Fujimori, S., 2017: SSP3: AIM Implementation of Shared Socioeconomic Pathways. *Global Environmental Change*, **42**, 268–283, doi:10.1016/j.gloenvcha.2016.06.009.

Fujimori, S. et al., 2016: Implication of Paris Agreement in the context of long-term climate mitigation goals. *SpringerPlus*, **5(1)**, 1620, doi:10.1186/s40064-016-3235-9.

Fuss, S. et al., 2014: Betting on negative emissions. *Nature Climate Change*, **4(10)**, 850–853, doi:10.1038/nclimate2392.

Fuss, S. et al., 2016: Research priorities for negative emissions. *Environmental Research Letters*, **11(11)**, 115007, doi:10.1088/1748-9326/11/11/115007.

Fuss, S. et al., 2018: Negative emissions – Part 2: Costs, potentials and side effects. *Environmental Research Letters*, **13(6)**, 063002, doi:10.1088/1748-9326/aabf9f.

Geden, O., 2015: Policy: Climate advisers must maintain integrity. *Nature*, **521(7550)**, 27–28, doi:10.1038/521027a.

Geels, F.W., B.K. Sovacool, T. Schwanen, and S. Sorrell, 2017: Sociotechnical transitions for deep decarbonization. *Science*, **357(6357)**, 1242–1244, doi:10.1126/science.aao3760.

Gerber, P.J. et al., 2013: *Tackling climate change through livestock – A global assessment of emissions and mitigation opportunities*. Food and Agriculture Organization of the United Nations (FAO), Rome, Italy, 115 pp.

Gernaat, D.E.H.J. et al., 2015: Understanding the contribution of non-carbon dioxide gases in deep mitigation scenarios. *Global Environmental Change*, **33**, 142–153, doi:10.1016/j.gloenvcha.2015.04.010.

Ghan, S.J. et al., 2013: A simple model of global aerosol indirect effects. *Journal of Geophysical Research: Atmospheres*, **118(12)**, 6688–6707, doi:10.1002/jgrd.50567.

Giannakidis, G., K. Karlsson, M. Labriet, and B. Ó Gallachóir (eds.), 2018: *Limiting Global Warming to Well Below 2°C: Energy System Modelling and Policy Development*. Springer International Publishing, Cham, Switzerland, 423 pp., doi:10.1007/978-3-319-74424-7.

Gillett, N.P., V.K. Arora, D. Matthews, and M.R. Allen, 2013: Constraining the Ratio of Global Warming to Cumulative CO₂ Emissions Using CMIP5 Simulations. *Journal of Climate*, **26(18)**, 6844–6858, doi:10.1175/jcli-d-12-00476.1.

Gillett, N.P., V.K. Arora, K. Zickfeld, S.J. Marshall, and W.J. Merryfield, 2011: Ongoing climate change following a complete cessation of carbon dioxide emissions. *Nature Geoscience*, **4(2)**, 83–87, doi:10.1038/ngeo1047.

Global CCS Institute, 2016: *The Global Status of CCS: 2016 Summary Report*. Global CCS Institute, Melbourne, Australia, 28 pp.

González-Eguino, M., A. Olabe, and T. Ribera, 2017: New Coal-Fired Plants Jeopardise Paris Agreement. *Sustainability*, **9(2)**, 168, doi:10.3390/su9020168.

Goodwin, P., S. Brown, I.D. Haigh, R.J. Nicholls, and J.M. Matter, 2018a: Adjusting Mitigation Pathways to Stabilize Climate at 1.5°C and 2.0°C Rise in Global Temperatures to Year 2300. *Earth's Future*, 0–3, doi:10.1002/2017ef000732.

Goodwin, P. et al., 2018b: Pathways to 1.5 and 2°C warming based on observational and geological constraints. *Nature Geoscience*, **11(1)**, 1–22, doi:10.1038/s41561-017-0054-8.

Gota, S., C. Huizenga, K. Peet, N. Medimorec, and S. Bakker, 2018: Decarbonising transport to achieve Paris Agreement targets. *Energy Efficiency*, 1–24, doi:10.1007/s12053-018-9671-3.

Gough, C. and P. Upham, 2011: Biomass energy with carbon capture and storage (BECCS or Bio-CCS). *Greenhouse Gases: Science and Technology*, **1(4)**, 324–334, doi:10.1002/ghg.34.

Goulder, L. and A. Schein, 2013: Carbon Taxes vs. Cap and Trade: A Critical Review. *Climate Change Economics*, **04(03)**, 1350010, doi:10.1142/s2010007813500103.

Grassi, G. et al., 2017: The key role of forests in meeting climate targets requires science for credible mitigation. *Nature Climate Change*, **7(3)**, 220–226, doi:10.1038/nclimate3227.

Greene, C.H. et al., 2017: Geoengineering, marine microalgae, and climate stabilization in the 21st century. *Earth's Future*, **5(3)**, 278–284, doi:10.1002/2016ef000486.

Gregory, J.M. and T. Andrews, 2016: Variation in climate sensitivity and feedback parameters during the historical period. *Geophysical Research Letters*, **43(8)**, 3911–3920, doi:10.1002/2016gl068406.

Gren, I.-M. and A.Z. Aklilu, 2016: Policy design for forest carbon sequestration: A review of the literature. *Forest Policy and Economics*, **70**, 128–136, doi:10.1016/j.forpol.2016.06.008.

Griscom, B.W. et al., 2017: Natural climate solutions. *Proceedings of the National Academy of Sciences*, **114(44)**, 11645–11650, doi:10.1073/pnas.1710465114.

Grubb, M., J.C. Hourcade, and K. Neuhoff, 2014: *Planetary economics: Energy, climate change and the three domains of sustainable development*. Routledge Earthscan, Abingdon, UK and New York, NY, USA, 520 pp.

Grubler, A. and C. Wilson (eds.), 2014: *Energy Technology Innovation: Learning from Historical Successes and Failures*. Cambridge University Press, Cambridge, United Kingdom and New York, NY, USA, 400 pp., doi:10.1017/cbo9781139150880.

Grubler, A. et al., 2018: A low energy demand scenario for meeting the 1.5°C target and sustainable development goals without negative emission technologies. *Nature Energy*, **3(6)**, 515–527, doi:10.1038/s41560-018-0172-6.

Guivarch, C. and J. Rogelj, 2017: *Carbon price variations in 2°C scenarios explored*. Carbon Pricing Leadership Coalition (CPLC), 15 pp.

Güneralp, B. et al., 2017: Global scenarios of urban density and its impacts on building energy use through 2050. *Proceedings of the National Academy of Sciences*, **114(34)**, 8945–8950, doi:10.1073/pnas.1606035114.

Gupta, S. and J. Harnisch, 2014: Cross-cutting Investment and Finance Issues. In: *Climate Change 2014: Mitigation of Climate Change. Contribution of Working Group III to the Fifth Assessment Report of the Intergovernmental Panel on Climate Change* [Edenhofer, O., R. Pichs-Madruga, Y. Sokona, E. Farahani, S. Kadner, K. Seyboth, A. Adler, I. Baum, S. Brunner, P. Eickemeier, B. Kriemann, J. Savolainen, S. Schlömer, C. von Stechow, T. Zwickel, and J.C. Minx (eds.)]. Cambridge University Press, Cambridge, United Kingdom and New York, NY, USA, pp. 1207–1246.

Haberl, H., 2015: Competition for land: A sociometabolic perspective. *Ecological Economics*, **119**, 424–431, doi:10.1016/j.ecolecon.2014.10.002.

Haberl, H. et al., 2011: Global bioenergy potentials from agricultural land in 2050: Sensitivity to climate change, diets and yields. *Biomass and Bioenergy*, **35(12)**, 4753–4769, doi:10.1016/j.biombioe.2011.04.035.

Haberl, H. et al., 2013: Bioenergy: how much can we expect for 2050? *Environmental Research Letters*, **8(3)**, 031004, doi:10.1088/1748-9326/8/3/031004.

Harmes, A., 2011: The Limits of Carbon Disclosure: Theorizing the Business Case for Investor Environmentalism. *Global Environmental Politics*, **11(2)**, 98–119, doi:10.1162/glep_a_00057.

Hartmann, J. et al., 2013: Enhanced chemical weathering as a geoengineering strategy to reduce atmospheric carbon dioxide, supply nutrients, and mitigate ocean acidification. *Reviews of Geophysics*, **51(2)**, 113–149, doi:10.1002/rog.20004.

Hauglustaine, D.A., Y. Balkanski, and M. Schulz, 2014: A global model simulation of present and future nitrate aerosols and their direct radiative forcing of climate. *Atmospheric Chemistry and Physics*, **14**, 11031–11063, doi:10.5194/acp-14-11031-2014.

Havlik, P. et al., 2013: Crop Productivity and the Global Livestock Sector: Implications for Land Use Change and Greenhouse Gas Emissions. *American Journal of Agricultural Economics*, **95(2)**, 442–448, doi:10.1093/ajae/aas085.

Havlík, P. et al., 2014: Climate change mitigation through livestock system transitions. *Proceedings of the National Academy of Sciences*, **111(10)**, 3709–3714, doi:10.1073/pnas.1308044111.

Heck, V., D. Gerten, W. Lucht, and L.R. Boysen, 2016: Is extensive terrestrial carbon dioxide removal a 'green' form of geoengineering? A global modelling study. *Global and Planetary Change*, **137**, 123–130, doi:10.1016/j.gloplacha.2015.12.008.

Heck, V., D. Gerten, W. Lucht, and A. Popp, 2018: Biomass-based negative emissions difficult to reconcile with planetary boundaries. *Nature Climate Change*, **8(2)**, 151–155, doi:10.1038/s41558-017-0064-y.

Hejazi, M. et al., 2014: Long-term global water projections using six socioeconomic scenarios in an integrated assessment modeling framework. *Technological Forecasting and Social Change*, **81**, 205–226, doi:10.1016/j.techfore.2013.05.006.

Herrero, M. et al., 2016: Greenhouse gas mitigation potentials in the livestock sector. *Nature Climate Change*, **6(5)**, 452–461, doi:10.1038/nclimate2925.

Herrington, T. and K. Zickfeld, 2014: Path independence of climate and carbon cycle response over a broad range of cumulative carbon emissions. *Earth System Dynamics*, **5**, 409–422, doi:10.5194/esd-5-409-2014.

Herzog, H., K. Caldeira, and J. Reilly, 2003: An Issue of Permanence: Assessing the Effectiveness of Temporary Carbon Storage. *Climatic Change*, **59(3)**, 293–310, doi:10.1023/a:1024801618900.

Hildingsson, R. and B. Johansson, 2016: Governing low-carbon energy transitions in sustainable ways: Potential synergies and conflicts between climate and environmental policy objectives. *Energy Policy*, **88**, 245–252, doi:10.1016/j.enpol.2015.10.029.

Hoesly, R.M. et al., 2018: Historical (1750–2014) anthropogenic emissions of reactive gases and aerosols from the Community Emissions Data System (CEDS). *Geoscientific Model Development*, **11(1)**, 369–408, doi:10.5194/gmd-11-369-2018.

Hof, A.F. et al., 2017: Global and regional abatement costs of Nationally Determined Contributions (NDCs) and of enhanced action to levels well below 2°C and 1.5°C. *Environmental Science & Policy*, **71**, 30–40, doi:10.1016/j.envsci.2017.02.008.

Höglund-Isaksson, L. et al., 2017: Cost estimates of the Kigali Amendment to phase-down hydrofluorocarbons. *Environmental Science & Policy*, **75**, 138–147, doi:10.1016/j.envsci.2017.05.006.

Holz, C., S. Kartha, and T. Athanasiou, 2018a: Fairly sharing 1.5: national fair shares of a 1.5°C-compliant global mitigation effort. *International Environmental Agreements: Politics, Law and Economics*, **18(1)**, 117–134, doi:10.1007/s10784-017-9371-z.

Holz, C., L.S. Siegel, E. Johnston, A.P. Jones, and J. Sterman, 2018b: Ratcheting ambition to limit warming to 1.5°C – trade-offs between emission reductions and carbon dioxide removal. *Environmental Research Letters*, **13(6)**, 064028, doi:10.1088/1748-9326/aac0c1.

Honegger, M. and D. Reiner, 2018: The political economy of negative emissions technologies: consequences for international policy design. *Climate Policy*, **18(3)**, 306–321, doi:10.1080/14693062.2017.1413322.

Hong, S., C.J.A. Bradshaw, and B.W. Brook, 2015: Global zero-carbon energy pathways using viable mixes of nuclear and renewables. *Applied Energy*, **143**, 451–459, doi:10.1016/j.apenergy.2015.01.006.

Hope, C., 2006: The Marginal Impact of CO_2 from PAGE2002: An Integrated Assessment Model Incorporating the IPCC's Five Reasons for Concern. *The Integrated Assessment Journal*, **6(1)**, 19–56, http://journals.sfu.ca/int_assess/index.php/iaj/article/viewarticle/227.

Hristov, A.N. et al., 2015: An inhibitor persistently decreased enteric methane emission from dairy cows with no negative effect on milk production. *Proceedings of the National Academy of Sciences*, **112(34)**, 10663–8, doi:10.1073/pnas.1504124112.

Hsiang, S. et al., 2017: Estimating economic damage from climate change in the United States. *Science*, **356(6345)**, 1362–1369, doi:10.1126/science.aal4369.

Humpenöder, F. et al., 2014: Investigating afforestation and bioenergy CCS as climate change mitigation strategies. *Environmental Research Letters*, **9(6)**, 064029, doi:10.1088/1748-9326/9/6/064029.

Humpenöder, F. et al., 2018: Large-scale bioenergy production: how to resolve sustainability trade-offs? *Environmental Research Letters*, **13(2)**, 024011, doi:10.1088/1748-9326/aa9e3b.

Huntingford, C. and J. Lowe, 2007: "Overshoot" Scenarios and Climate Change. *Science*, **316(5826)**, 829, doi:10.1126/science.316.5826.829b.

IEA, 2014: *Energy Technology Perspectives 2014: Harnessing Electricity's Potential*. International Energy Agency (IEA), Paris, France, 382 pp.

IEA, 2015a: *Energy and Climate Change. World Energy Outlook Special Report*. International Energy Agency (IEA), Paris, France, 200 pp.

IEA, 2015b: *Energy Technology Perspectives 2015: Mobilising Innovation to Accelerate Climate Action*. International Energy Agency (IEA), Paris, France, 418 pp.

IEA, 2016a: *Energy Technology Perspectives 2016: Towards Sustainable Urban Energy Systems*. International Energy Agency (IEA), Paris, France, 418 pp.

IEA, 2016b: *World Energy Investment 2016*. International Energy Agency (IEA), Paris, France, 177 pp.

IEA, 2017a: *Energy Technology Perspectives 2017: Catalyzing Energy Technology Transformations*. International Energy Agency (IEA), Paris, France, 443 pp.

IEA, 2017b: *Renewables Information – Overview (2017 edition)*. International Energy Agency (IEA), Paris, 11 pp.

IEA, 2017c: *World Energy Investment 2017*. International Energy Agency (IEA), Paris, France, 191 pp.

IEA, 2017d: *World Energy Outlook 2017*. International Energy Agency (IEA), Paris, France, 782 pp.

IEA, 2017e: *World Energy Statistics 2017*. International Energy Agency (IEA). OECD Publishing, Paris, France, 847 pp., doi:10.1787/world_energy_stats-2017-en.

IEAGHG, 2006: *Near zero emission technology for CO_2 capture from power plant*. IEAGHG 2006/13, IEA Greenhouse Gas R&D Programme, Cheltenham, UK, 114 pp.

Ilyina, T., D. Wolf-Gladrow, G. Munhoven, and C. Heinze, 2013: Assessing the potential of calcium-based artificial ocean alkalinization to mitigate rising atmospheric CO_2 and ocean acidification. *Geophysical Research Letters*, **40(22)**, 5909–5914, doi:10.1002/2013gl057981.

IPCC, 2005: IPCC Special Report on Carbon Dioxide Capture and Storage. [Metz, B., O. Davidson, H.C. de Coninck, M. Loos, and L.A. Meyer (eds.)]. Prepared by Working Group III of the Intergovernmental Panel on Climate Change. Cambridge University Press, Cambridge, United Kingdom and New York, NY, USA, 442 pp.

IPCC, 2013a: Climate Change 2013: The Physical Science Basis. Contribution of Working Group I to the Fifth Assessment Report of the Intergovernmental Panel on Climate Change. [Stocker, T.F., D. Qin, G.-K. Plattner, M. Tignor, S.K. Allen, J. Boschung, A. Nauels, Y. Xia, V. Bex, and P.M. Midgley (eds.)]. Cambridge University Press, Cambridge, United Kingdom and New York, NY, USA, 1535 pp.

IPCC, 2013b: Climate Change 2013: The Physical Science Basis. Working Group I Contribution to the Fifth Assessment Report of the Intergovernmental Panel on Climate Change. [Stocker, T.F., D. Qin, G.-K. Plattner, M. Tignor, S.K. Allen, J. Boschung, A. Nauels, Y. Xia, V. Bex, and P.M. Midgley (eds.)]. Cambridge University Press, Cambridge, United Kingdom and New York, NY, USA, 1535 pp.

IPCC, 2014a: Climate Change 2014: Synthesis Report. Contribution of Working Groups I, II and III to the Fifth Assessment Report of the Intergovernmental Panel on Climate Change. [Core Writing Team, R.K. Pachauri, and L.A. Meyer (eds.)]. Cambridge University Press, Cambridge, United Kingdom and New York, NY, USA, 151 pp.

IPCC, 2014b: Summary for Policymakers. In: *Climate Change 2014: Mitigation of Climate Change. Contribution of Working Group III to the Fifth Assessment Report of the Intergovernmental Panel on Climate Change* [Edenhofer, O., R. Pichs-Madruga, Y. Sokona, E. Farahani, S. Kadne, K. Seyboth, A. Adler, I. Baum, S. Brunner, P. Eickemeier, B. Kriemann, J. Savolainen, S. Schlömer, C. Stechow, T. Zwickel, and J.C. Minx (eds.)]. Cambridge University Press, Cambridge, United Kingdom and New York, NY, USA, pp. 1–30.

IPCC/TEAP, 2005: Safeguarding the Ozone Layer and the Global Climate System: Issues Related to Hydrofluorocarbons and Perfluorocarbons. [Metz, B., L. Kuijpers, S. Solomon, S.O. Andersen, O. Davidson, J. Pons, D. Jager, T. Kestin, M. Manning, and L. Meyer (eds.)]. A Special Report of the Intergovernmental Panel on Climate Change (IPCC) and Technology and Economic Assessment Panel (TEAP). Cambridge University Press, Cambridge, United Kingdom and New York, NY, USA, 485 pp.

Irlam, L., 2017: *Global Costs of Carbon Capture and Storage*. Global CCS Institute, Melbourne, Australia, 14 pp.

IWG, 2016: *Technical Support Document: Technical Update of the Social Cost of Carbon for Regulatory Impact Analysis - Under Executive Order 12866*. Interagency Working Group on Social Cost of Greenhouse Gases, United States, 35 pp.

Iyer, G.C. et al., 2015: Improved representation of investment decisions in assessments of CO_2 mitigation. *Nature Climate Change*, **5(5)**, 436–440, doi:10.1038/nclimate2553.

Jackson, R.B. et al., 2017: Warning signs for stabilizing global CO_2 emissions. *Environmental Research Letters*, **12(11)**, 110202, doi:10.1088/1748-9326/aa9662.

Jacobson, M.Z., 2002: Control of fossil-fuel particulate black carbon and organic matter, possibly the most effective method of slowing global warming. *Journal of Geophysical Research: Atmospheres*, **107(D19)**, 4410, doi:10.1029/2001jd001376.

Jacobson, M.Z., 2010: Short-term effects of controlling fossil-fuel soot, biofuel soot and gases, and methane on climate, Arctic ice, and air pollution health. *Journal of Geophysical Research: Atmospheres*, **115(D14)**, D14209, doi:10.1029/2009jd013795.

Jacobson, M.Z., 2017: Roadmaps to Transition Countries to 100% Clean, Renewable Energy for All Purposes to Curtail Global Warming, Air Pollution, and Energy Risk. *Earth's Future*, **5(10)**, 948–952, doi:10.1002/2017ef000672.

Jacobson, M.Z. et al., 2017: 100% Clean and Renewable Wind, Water, and Sunlight All-Sector Energy Roadmaps for 139 Countries of the World. *Joule*, **1(1)**, 108–121, doi:10.1016/j.joule.2017.07.005.

Jakob, M. and J.C. Steckel, 2016: Implications of climate change mitigation for sustainable development. *Environmental Research Letters*, **11(10)**, 104010, doi:10.1088/1748-9326/11/10/104010.

Jakob, M. et al., 2014: Feasible mitigation actions in developing countries. *Nature Climate Change*, **4(11)**, 961–968, doi:10.1038/nclimate2370.

Johnson, N. et al., 2015: Stranded on a low-carbon planet: Implications of climate policy for the phase-out of coal-based power plants. *Technological Forecasting and Social Change*, **90**, 89–102, doi:10.1016/j.techfore.2014.02.028.

Jones, A., J.M. Haywood, and C.D. Jones, 2018: Can reducing black carbon and methane below RCP2.6 levels keep global warming below 1.5°C? *Atmospheric Science Letters*, **19(6)**, e821, doi:10.1002/asl.821.

Jones, C.D. et al., 2016: Simulating the Earth system response to negative emissions. *Environmental Research Letters*, **11(9)**, 095012, doi:10.1088/1748-9326/11/9/095012.

Jones, D.G. et al., 2015: Developments since 2005 in understanding potential environmental impacts of CO_2 leakage from geological storage. *International Journal of Greenhouse Gas Control*, **40**, 350–377, doi:10.1016/j.ijggc.2015.05.032.

Joshi, V. and S. Kumar, 2015: Meat Analogues: Plant based alternatives to meat products – A review. *International Journal of Food Fermentation and Technology*, **5(2)**, 107–119, doi:10.5958/2277-9396.2016.00001.5.

Kandlikar, M., 1995: The relative role of trace gas emissions in greenhouse abatement policies. *Energy Policy*, **23(10)**, 879–883, doi:10.1016/0301-4215(95)00108-u.

KC, S. and W. Lutz, 2017: The human core of the shared socioeconomic pathways: Population scenarios by age, sex and level of education for all countries to 2100. *Global Environmental Change*, **42**, 181–192, doi:10.1016/j.gloenvcha.2014.06.004.

Kearns, J. et al., 2017: Developing a consistent database for regional geologic CO_2 storage capacity worldwide. *Energy Procedia*, **114**, 4697–4709, doi:10.1016/j.egypro.2017.03.1603.

Keith, D.W. and J.S. Rhodes, 2002: Bury, Burn or Both: A Two-for-One Deal on Biomass Carbon and Energy. *Climatic Change*, **54(3)**, 375–377, doi:10.1023/a:1016187420442.

Keith, D.W., M. Ha-Duong, and J.K. Stolaroff, 2006: Climate Strategy with CO_2 Capture from the Air. *Climatic Change*, **74(1–3)**, 17–45, doi:10.1007/s10584-005-9026-x.

Kheshgi, H.S., 1995: Sequestering atmospheric carbon dioxide by increasing ocean alkalinity. *Energy*, **20(9)**, 915–922, doi:10.1016/0360-5442(95)00035-f.

Kim, S.H., K. Wada, A. Kurosawa, and M. Roberts, 2014: Nuclear energy response in the EMF27 study. *Climatic Change*, **123(3–4)**, 443–460, doi:10.1007/s10584-014-1098-z.

Klein, D. et al., 2014: The value of bioenergy in low stabilization scenarios: An assessment using REMIND-MAgPIE. *Climatic Change*, **123(3–4)**, 705–718, doi:10.1007/s10584-013-0940-z.

Klimont, Z. et al., 2017: Global anthropogenic emissions of particulate matter including black carbon. *Atmospheric Chemistry and Physics*, **17(14)**, 8681–8723, doi:10.5194/acp-17-8681-2017.

Klinsky, S. and H. Winkler, 2018: Building equity in: strategies for integrating equity into modelling for a 1.5°C world. *Philosophical Transactions of the Royal Society A: Mathematical, Physical and Engineering Sciences*, **376(2119)**, 20160461, doi:10.1098/rsta.2016.0461.

Knutti, R. and J. Rogelj, 2015: The legacy of our CO_2 emissions: a clash of scientific facts, politics and ethics. *Climatic Change*, **133(3)**, 361–373, doi:10.1007/s10584-015-1340-3.

Knutti, R., M.A.A. Rugenstein, and G.C. Hegerl, 2017: Beyond equilibrium climate sensitivity. *Nature Geoscience*, **10(10)**, 727–736, doi:10.1038/ngeo3017.

Kolstad, C. et al., 2014: Social, Economic and Ethical Concepts and Methods. In: *Climate Change 2014: Mitigation of Climate Change. Contribution of Working Group III to the Fifth Assessment Report of the Intergovernmental Panel on Climate Change* [Edenhofer, O., R. Pichs-Madruga, Y. Sokona, E. Farahani, S. Kadne, K. Seyboth, A. Adler, I. Baum, S. Brunner, P. Eickemeier, B. Kriemann, J. Savolainen, S. Schlömer, C. von Stechow, T. Zwickel, and J.C. Minx (eds.)]. Cambridge University Press, Cambridge, United Kingdom and New York, NY, USA, pp. 207–282.

Krasting, J.P., J.P. Dunne, E. Shevliakova, and R.J. Stouffer, 2014: Trajectory sensitivity of the transient climate response to cumulative carbon emissions. *Geophysical Research Letters*, **41(7)**, 2520–2527, doi:10.1002/(issn)1944-8007.

Krause, A. et al., 2017: Global consequences of afforestation and bioenergy cultivation on ecosystem service indicators. *Biogeosciences*, **14(21)**, 4829–4850, doi:10.5194/bg-14-4829-2017.

Kraxner, F. et al., 2013: Global bioenergy scenarios – Future forest development, land-use implications, and trade-offs. *Biomass and Bioenergy*, **57**, 86–96, doi:10.1016/j.biombioe.2013.02.003.

Kreidenweis, U. et al., 2016: Afforestation to mitigate climate change: impacts on food prices under consideration of albedo effects. *Environmental Research Letters*, **11(8)**, 085001, doi:10.1088/1748-9326/11/8/085001.

Kretzschmar, J. et al., 2017: Comment on "Rethinking the Lower Bound on Aerosol Radiative Forcing". *Journal of Climate*, **30(16)**, 6579–6584, doi:10.1175/jcli-d-16-0668.1.

Krey, V., G. Luderer, L. Clarke, and E. Kriegler, 2014a: Getting from here to there – energy technology transformation pathways in the EMF27 scenarios. *Climatic Change*, **123**, 369–382, doi:10.1007/s10584-013-0947-5.

Krey, V. et al., 2014b: Annex II: Metrics & Methodology. In: *Climate Change 2014: Mitigation of Climate Change. Contribution of Working Group III to the Fifth Assessment Report of the Intergovernmental Panel on Climate Change* [Edenhofer, O., R. Pichs-Madruga, Y. Sokona, E. Farahani, S. Kadner, K. Seyboth, A. Adler, I. Baum, S. Brunner, P. Eickemeier, B. Kriemann, J. Savolainen, S. Schlömer, C. von Stechow, T. Zwickel, and J.C. Minx (eds.)]. Cambridge University Press, Cambridge, United Kingdom and New York, NY, USA, pp. 1281–1328.

Kriegler, E., O. Edenhofer, L. Reuster, G. Luderer, and D. Klein, 2013a: Is atmospheric carbon dioxide removal a game changer for climate change mitigation? *Climatic Change*, **118(1)**, 45–57, doi:10.1007/s10584-012-0681-4.

Kriegler, E. et al., 2012: The need for and use of socio-economic scenarios for climate change analysis: A new approach based on shared socio-economic pathways. *Global Environmental Change*, **22(4)**, 807–822, doi:10.1016/j.gloenvcha.2012.05.005.

Kriegler, E. et al., 2013b: What Does the 2°C Target Imply for a Global Climate Agreement in 2020? The Limits Study on Durban Platform Scenarios. *Climate Change Economics*, **4(4)**, 1340008, doi:10.1142/s2010007813400083.

Kriegler, E. et al., 2014a: A new scenario framework for climate change research: the concept of shared climate policy assumptions. *Climatic Change*, **122(3)**, 401–414, doi:10.1007/s10584-013-0971-5.

Kriegler, E. et al., 2014b: The role of technology for achieving climate policy objectives: Overview of the EMF 27 study on global technology and climate policy strategies. *Climatic Change*, **123(3–4)**, 353–367, doi:10.1007/s10584-013-0953-7.

Kriegler, E. et al., 2015a: Diagnostic indicators for integrated assessment models of climate policy. *Technological Forecasting and Social Change*, **90(Part A)**, 45–61, doi:10.1016/j.techfore.2013.09.020.

Kriegler, E. et al., 2015b: Making or breaking climate targets: The AMPERE study on staged accession scenarios for climate policy. *Technological Forecasting and Social Change*, **90(Part A)**, 24–44, doi:10.1016/j.techfore.2013.09.021.

Kriegler, E. et al., 2016: Will economic growth and fossil fuel scarcity help or hinder climate stabilization?: Overview of the RoSE multi-model study. *Climatic Change*, **136(1)**, 7–22, doi:10.1007/s10584-016-1668-3.

Kriegler, E. et al., 2017: Fossil-fueled development (SSP5): An energy and resource intensive scenario for the 21st century. *Global Environmental Change*, **42**, 297–315, doi:10.1016/j.gloenvcha.2016.05.015.

Kriegler, E. et al., 2018a: Short term policies to keep the door open for Paris climate goals. *Environmental Research Letters*, **13(7)**, 074022, doi:10.1088/1748-9326/aac4f1.

Kriegler, E. et al., 2018b: Pathways limiting warming to 1.5°C: a tale of turning around in no time? *Philosophical Transactions of the Royal Society A: Mathematical, Physical and Engineering Sciences*, **376(2119)**, 20160457, doi:10.1098/rsta.2016.0457.

Kuramochi, T. et al., 2017: Ten key short-term sectoral benchmarks to limit warming to 1.5°C. *Climate Policy*, **18(3)**, 1–19, doi:10.1080/14693062.2017.1397495.

Lamarque, J.-F. et al., 2011: Global and regional evolution of short-lived radiatively-active gases and aerosols in the Representative Concentration Pathways. *Climatic Change*, **109(1–2)**, 191–212, doi:10.1007/s10584-011-0155-0.

Lambin, E.F. and P. Meyfroidt, 2011: Global land use change, economic globalization, and the looming land scarcity. *Proceedings of the National Academy of Sciences*, **108(9)**, 3465–72, doi:10.1073/pnas.1100480108.

Le Quéré, C. et al., 2018: Global Carbon Budget 2017. *Earth System Science Data*, **10(1)**, 405–448, doi:10.5194/essd-10-405-2018.

Lehtilä, A. and T. Koljonen, 2018: Pathways to Post-fossil Economy in a Well Below 2°C World. In: *Limiting Global Warming to Well Below 2°C: Energy System Modelling and Policy Development* [Giannakidis, G., K. Karlsson, M. Labriet, and B. Gallachóir (eds.)]. Springer International Publishing, Cham, Switzerland, pp. 33–49, doi:10.1007/978-3-319-74424-7_3.

Leimbach, M., E. Kriegler, N. Roming, and J. Schwanitz, 2017: Future growth patterns of world regions – A GDP scenario approach. *Global Environmental Change*, **42**, 215–225, doi:10.1016/j.gloenvcha.2015.02.005.

Lemoine, D. and C. Traeger, 2014: Watch Your Step: Optimal Policy in a Tipping Climate. *American Economic Journal: Economic Policy*, **6(1)**, 137–166, doi:10.1257/pol.6.1.137.

Lenton, A., R.J. Matear, D.P. Keller, V. Scott, and N.E. Vaughan, 2018: Assessing carbon dioxide removal through global and regional ocean alkalinization under high and low emission pathways. *Earth System Dynamics*, **9(2)**, 339–357, doi:10.5194/esd-9-339-2018.

Lewis, N. and J.A. Curry, 2015: The implications for climate sensitivity of AR5 forcing and heat uptake estimates. *Climate Dynamics*, **45(3–4)**, 1009–1023, doi:10.1007/s00382-014-2342-y.

Lewis, N. and J. Curry, 2018: The impact of recent forcing and ocean heat uptake data on estimates of climate sensitivity. *Journal of Climate*, JCLI–D–17–0667.1, doi:10.1175/jcli-d-17-0667.1.

Liu, J., T.W. Hertel, F. Taheripour, T. Zhu, and C. Ringler, 2014: International trade buffers the impact of future irrigation shortfalls. *Global Environmental Change*, **29**, 22–31, doi:10.1016/j.gloenvcha.2014.07.010.

Liu, J.-Y. et al., 2018: Socioeconomic factors and future challenges of the goal of limiting the increase in global average temperature to 1.5°C. *Carbon Management*, 1–11, doi:10.1080/17583004.2018.1477374.

Löffler, K. et al., 2017: Designing a Model for the Global Energy System– GENeSYS-MOD: An Application of the Open-Source Energy Modeling System (OSeMOSYS). *Energies*, **10(10)**, 1468, doi:10.3390/en10101468.

Lontzek, T.S., Y. Cai, K.L. Judd, and T.M. Lenton, 2015: Stochastic integrated assessment of climate tipping points indicates the need for strict climate policy. *Nature Climate Change*, **5(5)**, 441–444, doi:10.1038/nclimate2570.

Lowe, J.A. and D. Bernie, 2018: The impact of Earth system feedbacks on carbon budgets and climate response. *Philosophical Transactions of the Royal Society A: Mathematical, Physical and Engineering Sciences*, **376(2119)**, 20170263, doi:10.1098/rsta.2017.0263.

Lowe, J.A. et al., 2009: How difficult is it to recover from dangerous levels of global warming? *Environmental Research Letters*, **4(1)**, 014012, doi:10.1088/1748-9326/4/1/014012.

Lucon, O. et al., 2014: Buildings. In: *Climate Change 2014: Mitigation of Climate Change. Contribution of Working Group III to the Fifth Assessment Report of the Intergovernmental Panel on Climate Change* [Edenhofer, O., R. Pichs-Madruga, Y. Sokona, E. Farahani, S. Kadner, K. Seyboth, A. Adler, I. Baum, S. Brunner, P. Eickemeier, B. Kriemann, J. Savolainen, S. Schlömer, C. von Stechow, T. Zwickel, and J.C. Minx (eds.)]. Cambridge University Press, Cambridge, United Kingdom and New York, NY, USA, pp. 671–738.

Luderer, G., C. Bertram, K. Calvin, E. De Cian, and E. Kriegler, 2016a: Implications of weak near-term climate policies on long-term mitigation pathways. *Climatic Change*, **136(1)**, 127–140, doi:10.1007/s10584-013-0899-9.

Luderer, G. et al., 2013: Economic mitigation challenges: how further delay closes the door for achieving climate targets. *Environmental Research Letters*, **8(3)**, 034033, doi:10.1088/1748-9326/8/3/034033.

Luderer, G. et al., 2014: The role of renewable energy in climate stabilization: results from the {EMF}27 scenarios. *Climatic Change*, **123(3–4)**, 427–441, doi:10.1007/s10584-013-0924-z.

2

Luderer, G. et al., 2016b: *Deep Decarbonisation towards 1.5°C – 2°C stabilization: Policy findings from the ADVANCE project*. The ADVANCE Consortium, 42 pp.

Luderer, G. et al., 2017: Assessment of wind and solar power in global low-carbon energy scenarios: An introduction. *Energy Economics*, **64**, 542–551, doi:10.1016/j.eneco.2017.03.027.

Luderer, G. et al., 2018: Residual fossil CO_2 emissions in 1.5–2°C pathways. *Nature Climate Change*, **8(7)**, 626–633, doi:10.1038/s41558-018-0198-6.

Lutz, W. and S. KC, 2011: Global Human Capital: Integrating Education and Population. *Science*, **333(6042)**, 587–592, doi:10.1126/science.1206964.

MacDougall, A.H., N.C. Swart, and R. Knutti, 2017: The Uncertainty in the Transient Climate Response to Cumulative CO 2 Emissions Arising from the Uncertainty in Physical Climate Parameters. *Journal of Climate*, **30(2)**, 813–827, doi:10.1175/jcli-d-16-0205.1.

MacDougall, A.H., K. Zickfeld, R. Knutti, and H.D. Matthews, 2015: Sensitivity of carbon budgets to permafrost carbon feedbacks and non-CO_2 forcings. *Environmental Research Letters*, **10(12)**, 125003, doi:10.1088/1748-9326/10/12/125003.

Madeira, M.S. et al., 2017: Microalgae as feed ingredients for livestock production and meat quality: A review. *Livestock Science*, **205**, 111–121, doi:10.1016/j.livsci.2017.09.020.

Mahowald, N.M. et al., 2017: Aerosol Deposition Impacts on Land and Ocean Carbon Cycles. *Current Climate Change Reports*, **3(1)**, 16–31, doi:10.1007/s40641-017-0056-z.

Malavelle, F.F. et al., 2017: Strong constraints on aerosol–cloud interactions from volcanic eruptions. *Nature*, **546(7659)**, 485–491, doi:10.1038/nature22974.

Marangoni, G. and M. Tavoni, 2014: The Clean Energy R&D Strategy for 2°C. *Climate Change Economics*, **5(1)**, 1440003, doi:10.1142/s201000781440003x.

Marangoni, G. et al., 2017: Sensitivity of projected long-term CO_2 emissions across the Shared Socioeconomic Pathways. *Nature Climate Change*, **7(1)**, 113–119, doi:10.1038/nclimate3199.

Marcucci, A., S. Kypreos, and E. Panos, 2017: The road to achieving the long-term Paris targets: Energy transition and the role of direct air capture. *Climatic Change*, **144(2)**, 181–193, doi:10.1007/s10584-017-2051-8.

Marten, A.L., E.A. Kopits, C.W. Griffiths, S.C. Newbold, and A. Wolverton, 2015: Incremental CH_4 and N_2O mitigation benefits consistent with the US Government's SC-CO_2 estimates. *Climate Policy*, **15(2)**, 272–298, doi:10.1080/14693062.2014.912981.

Marvel, K., G.A. Schmidt, R.L. Miller, and L.S. Nazarenko, 2016: Implications for climate sensitivity from the response to individual forcings. *Nature Climate Change*, **6(4)**, 386–389, doi:10.1038/nclimate2888.

Masui, T. et al., 2011: An emission pathway for stabilization at 6 W m^{-2} radiative forcing. *Climatic Change*, **109(1)**, 59–76, doi:10.1007/s10584-011-0150-5.

Matter, J.M. et al., 2016: Rapid carbon mineralization for permanent disposal of anthropogenic carbon dioxide emissions. *Science*, **352(6291)**, 1312–1314, doi:10.1126/science.aad8132.

Matthews, H.D. and K. Caldeira, 2008: Stabilizing climate requires near-zero emissions. *Geophysical Research Letters*, **35(4)**, 1–5, doi:10.1029/2007gl032388.

Matthews, H.D. and K. Zickfeld, 2012: Climate response to zeroed emissions of greenhouse gases and aerosols. *Nature Climate Change*, **2(5)**, 338–341, doi:10.1038/nclimate1424.

Matthews, H.D., N.P. Gillett, P.A. Stott, and K. Zickfeld, 2009: The proportionality of global warming to cumulative carbon emissions. *Nature*, **459(7248)**, 829–832, doi:10.1038/nature08047.

Matthews, H.D. et al., 2017: Estimating Carbon Budgets for Ambitious Climate Targets. *Current Climate Change Reports*, **3**, 69–77, doi:10.1007/s40641-017-0055-0.

Mazzotti, M. et al., 2005: Mineral carbonation and industrial uses of carbon dioxide. In: *IPCC Special Report on Carbon Dioxide Capture and Storage* [Metz, B., O. Davidson, H.C. de Coninck, M. Loos, and L.A. Meyer (eds.)]. Prepared by Working Group III of the Intergovernmental Panel on Climate Change. Cambridge University Press, Cambridge, United Kingdom and New York, NY, USA, pp. 319–338.

McCollum, D.L., V. Krey, and K. Riahi, 2011: An integrated approach to energy sustainability. *Nature Climate Change*, **1(9)**, 428–429, doi:10.1038/nclimate1297.

McCollum, D.L. et al., 2013: Energy Investments under Climate Policy: A Comparison of Global Models. *Climate Change Economics*, **4(4)**, 1340010, doi:10.1142/s2010007813400101.

McCollum, D.L. et al., 2017: Improving the behavioral realism of global integrated assessment models: An application to consumers' vehicle choices. *Transportation Research Part D: Transport and Environment*, **55**, 322–342, doi:10.1016/j.trd.2016.04.003.

McCollum, D.L. et al., 2018: Energy investment needs for fulfilling the Paris Agreement and achieving the Sustainable Development Goals. *Nature Energy*, **3(7)**, 589–599, doi:10.1038/s41560-018-0179-z.

McFarland, J.R., A.A. Fawcett, A.C. Morris, J.M. Reilly, and P.J. Wilcoxen, 2018: Overview of the EMF 32 Study on U.S. Carbon Tax Scenarios. *Climate Change Economics*, **9(1)**, 1840002, doi:10.1142/s201000781840002x.

McLeod, E. et al., 2011: A blueprint for blue carbon: Toward an improved understanding of the role of vegetated coastal habitats in sequestering CO_2. *Frontiers in Ecology and the Environment*, **9(10)**, 552–560, doi:10.1890/110004.

Mehling, M. and E. Tvinnereim, 2018: Carbon Pricing and the 1.5°C Target: Near-Term Decarbonisation and the Importance of an Instrument Mix. *Carbon & Climate Law Review*, **12(1)**, 50–61, doi:10.21552/cclr/2018/1/9.

Meinshausen, M., S.C.B. Raper, and T.M.L. Wigley, 2011a: Emulating coupled atmosphere-ocean and carbon cycle models with a simpler model, MAGICC6 – Part 1: Model description and calibration. *Atmospheric Chemistry and Physics*, **11(4)**, 1417–1456, doi:10.5194/acp-11-1417-2011.

Meinshausen, M. et al., 2009: Greenhouse-gas emission targets for limiting global warming to 2°C. *Nature*, **458(7242)**, 1158–1162, doi:10.1038/nature08017.

Meinshausen, M. et al., 2011b: The RCP greenhouse gas concentrations and their extensions from 1765 to 2300. *Climatic Change*, **109(1–2)**, 213–241, doi:10.1007/s10584-011-0156-z.

Méjean, A., C. Guivarch, J. Lefèvre, and M. Hamdi-Cherif, 2018: The transition in energy demand sectors to limit global warming to 1.5°C. *Energy Efficiency*, 1–22, doi:10.1007/s12053-018-9682-0.

Mengis, N., A.- Partanen, J. Jalbert, and H.D. Matthews, 2018: 1.5°C carbon budget dependent on carbon cycle uncertainty and future non-CO_2 forcing. *Scientific Reports*, **8(1)**, 5831, doi:10.1038/s41598-018-24241-1.

Michaelowa, A., M. Allen, and F. Sha, 2018: Policy instruments for limiting global temperature rise to 1.5°C – can humanity rise to the challenge? *Climate Policy*, **18(3)**, 275–286, doi:10.1080/14693062.2018.1426977.

Millar, R.J. and P. Friedlingstein, 2018: The utility of the historical record for assessing the transient climate response to cumulative emissions. *Philosophical Transactions of the Royal Society A: Mathematical, Physical and Engineering Sciences*, **376(2119)**, 20160449, doi:10.1098/rsta.2016.0449.

Millar, R.J. et al., 2017: Emission budgets and pathways consistent with limiting warming to 1.5°C. *Nature Geoscience*, **10(10)**, 741–747, doi:10.1038/ngeo3031.

Minx, J.C., W.F. Lamb, M.W. Callaghan, L. Bornmann, and S. Fuss, 2017: Fast growing research on negative emissions. *Environmental Research Letters*, **12(3)**, 035007, doi:10.1088/1748-9326/aa5ee5.

Minx, J.C. et al., 2018: Negative emissions-Part 1: Research landscape and synthesis. *Environmental Research Letters*, **13(6)**, 063001, doi:10.1088/1748-9326/aabf9b.

Moore, F.C. and D.B. Diaz, 2015: Temperature impacts on economic growth warrant stringent mitigation policy. *Nature Climate Change*, **5(1)**, 127–132, doi:10.1038/nclimate2481.

Mouratiadou, I. et al., 2016: The impact of climate change mitigation on water demand for energy and food: An integrated analysis based on the Shared Socioeconomic Pathways. *Environmental Science & Policy*, **64**, 48–58, doi:10.1016/j.envsci.2016.06.007.

Muller, A. et al., 2017: Strategies for feeding the world more sustainably with organic agriculture. *Nature Communications*, **8(1)**, 1290, doi:10.1038/s41467-017-01410-w.

Mundaca, L. and A. Markandya, 2016: Assessing regional progress towards a Green Energy Economy. *Applied Energy*, **179**, 1372–1394, doi:10.1016/j.apenergy.2015.10.098.

Mundaca, L., M. Mansoz, L. Neij, and G. Timilsina, 2013: Transaction costs analysis of low-carbon technologies. *Climate Policy*, **13(4)**, 490–513, doi:10.1080/14693062.2013.781452.

Mundaca, L., L. Neij, A. Markandya, P. Hennicke, and J. Yan, 2016: Towards a Green Energy Economy? Assessing policy choices, strategies and transitional pathways. *Applied Energy*, **179**, 1283–1292, doi:10.1016/j.apenergy.2016.08.086.

Myhre, G. et al., 2013: Anthropogenic and Natural Radiative Forcing. In: *Climate Change 2013: The Physical Science Basis. Contribution of Working Group I to the Fifth Assessment Report of the Intergovernmental Panel on Climate Change* [Stocker, T.F., D. Qin, G.-K. Plattner, M. Tignor, S.K. Allen, J. Boschung, A. Nauels, Y. Xia, V. Bex, and P.M. Midgley (eds.)]. Cambridge University Press, Cambridge, United Kingdom and New York, NY, USA, pp. 659–740.

Myhre, G. et al., 2017: Multi-model simulations of aerosol and ozone radiative forcing due to anthropogenic emission changes during the period 1990-2015. *Atmospheric Chemistry and Physics*, **17(4)**, 2709–2720, doi:10.5194/acp-17-2709-2017.

2

Narayan, C., P.M. Fernandes, J. van Brusselen, and A. Schuck, 2007: Potential for CO_2 emissions mitigation in Europe through prescribed burning in the context of the Kyoto Protocol. *Forest Ecology and Management*, **251(3)**, 164–173, doi:10.1016/j.foreco.2007.06.042.

NASEM, 2016: *Assessment of Approaches to Updating the Social Cost of Carbon: Phase 1 Report on a Near-Term Update*. National Academies of Sciences, Engineering, and Medicine (NASEM). The National Academies Press, Washington DC, USA, 72 pp., doi:10.17226/21898.

NASEM, 2017: *Valuing Climate Damages: Updating estimation of the social costs of carbon dioxide*. National Academies of Sciences, Engineering, and Medicine (NASEM), The National Academies Press, Washington DC, USA, 262 pp., doi:10.17226/24651.

Naudts, K. et al., 2016: Europe's forest management did not mitigate climate warming. *Science*, **351(6273)**, 597–600, doi:10.1126/science.aad7270.

Nelson, G.C. et al., 2014: Climate change effects on agriculture: economic responses to biophysical shocks. *Proceedings of the National Academy of Sciences*, **111(9)**, 3274–9, doi:10.1073/pnas.1222465110.

Nemet, G.F. et al., 2018: Negative emissions – Part 3: Innovation and upscaling. *Environmental Research Letters*, **13(6)**, 063003, doi:10.1088/1748-9326/aabff4.

NETL, 2013: *Cost and performance of PC and IGCC plants for a range of carbon dioxide capture: Revision 1*. DOE/NETL-2011/1498, U.S. Department of Energy (DOE) National Energy Technology Laboratory (NETL), 500 pp.

NETL, 2015: *Carbon Storage Atlas – Fifth Edition (Atlas V)*. U.S. Department of Energy (DOE) National Energy Technology Laboratory (NETL), 114 pp.

Nicol, A. et al., 2013: Induced seismicity; observations, risks and mitigation measures at CO_2 storage sites. *Energy Procedia*, **37**, 4749–4756, doi:10.1016/j.egypro.2013.06.384.

Nohara, D. et al., 2015: Examination of a climate stabilization pathway via zero-emissions using Earth system models. *Environmental Research Letters*, **10(9)**, 095005, doi:10.1088/1748-9326/10/9/095005.

Nordhaus, W.D., 2007a: A Review of The Stern Review on the Economics of Climate Change. *Journal of Economic Literature*, **45(3)**, 686–702, www.jstor.org/stable/27646843.

Nordhaus, W.D., 2007b: To Tax or Not to Tax: Alternative Approaches to Slowing Global Warming. *Review of Environmental Economics and Policy*, **1(1)**, 26–44, doi:10.1093/reep/rem008.

Nordhaus, W.D., 2013: *The Climate Casino: Risk, Uncertainty, and Economics for a Warming World*. Yale University Press, New Haven, CT, USA, 392 pp.

Nordhaus, W.D., 2014: Estimates of the Social Cost of Carbon: Concepts and Results from the DICE-2013R Model and Alternative Approaches. *Journal of the Association of Environmental and Resource Economists*, **1(1–2)**, 273–312, doi:10.1086/676035.

Nordhaus, W.D., 2017: Revisiting the social cost of carbon. *Proceedings of the National Academy of Sciences*, **114(7)**, 1518–1523, doi:10.1073/pnas.1609244114.

Nordhaus, W.D. and J. Boyer, 2000: *Warming the World: Economic Models of Global Warming*. MIT Press, Cambridge, MA, USA and London, UK, 244 pp.

NRC, 2015: *Climate Intervention: Carbon Dioxide Removal and Reliable Sequestration*. National Research Council (NRC). The National Academies Press, Washington DC, USA, 140 pp., doi:10.17226/18805.

O'Neill, B.C. et al., 2014: A new scenario framework for climate change research: The concept of shared socioeconomic pathways. *Climatic Change*, **122(3)**, 387–400, doi:10.1007/s10584-013-0905-2.

O'Neill, B.C. et al., 2017: The roads ahead: Narratives for shared socioeconomic pathways describing world futures in the 21st century. *Global Environmental Change*, **42**, 169–180, doi:10.1016/j.gloenvcha.2015.01.004.

Obersteiner, M. et al., 2001: Managing Climate Risk. *Science*, **294(5543)**, 786–787, doi:10.1126/science.294.5543.786b.

Obersteiner, M. et al., 2016: Assessing the land resource–food price nexus of the Sustainable Development Goals. *Science Advances*, **2(9)**, e1501499, doi:10.1126/sciadv.1501499.

Obersteiner, M. et al., 2018: How to spend a dwindling greenhouse gas budget. *Nature Climate Change*, **8(1)**, 7–10, doi:10.1038/s41558-017-0045-1.

OECD, 2017: *Investing in Climate, Investing in Growth*. OECD Publishing, Paris, France, 314 pp., doi:10.1787/9789264273528-en.

OECD/IEA and IRENA, 2017: *Perspectives for the Energy Transition: Investment Needs for a Low-Carbon Energy System*. OECD/IEA and IRENA, 204 pp.

Ogawa, T., S. Nakanishi, T. Shidahara, T. Okumura, and E. Hayashi, 2011: Saline-aquifer CO_2 sequestration in Japan-methodology of storage capacity assessment. *International Journal of Greenhouse Gas Control*, **5(2)**, 318–326, doi:10.1016/j.ijggc.2010.09.009.

Painter, T.H. et al., 2013: End of the Little Ice Age in the Alps forced by industrial black carbon. *Proceedings of the National Academy of Sciences*, **110(38)**, 15216–21, doi:10.1073/pnas.1302570110.

Patt, A., 2017: Beyond the tragedy of the commons: Reframing effective climate change governance. *Energy Research & Social Science*, **34**, 1–3, doi:10.1016/j.erss.2017.05.023.

Patterson, J.J. et al., 2018: Political feasibility of 1.5°C societal transformations: the role of social justice. *Current Opinion in Environmental Sustainability*, **31**, 1–9, doi:10.1016/j.cosust.2017.11.002.

Paustian, K. et al., 2016: Climate-smart soils. *Nature*, **532(7597)**, 49–57, doi:10.1038/nature17174.

Pawar, R.J. et al., 2015: Recent advances in risk assessment and risk management of geologic CO_2 storage. *International Journal of Greenhouse Gas Control*, **40**, 292–311, doi:10.1016/j.ijggc.2015.06.014.

Pehl, M. et al., 2017: Understanding future emissions from low-carbon power systems by integration of life-cycle assessment and integrated energy modelling. *Nature Energy*, **2(12)**, 939–945, doi:10.1038/s41560-017-0032-9.

Peters, G.P., 2016: The 'best available science' to inform 1.5°C policy choices. *Nature Climate Change*, **6(7)**, 646–649, doi:10.1038/nclimate3000.

Pietzcker, R.C. et al., 2017: System integration of wind and solar power in integrated assessment models: A cross-model evaluation of new approaches. *Energy Economics*, **64**, 583–599, doi:10.1016/j.eneco.2016.11.018.

Pikaar, I. et al., 2018: Decoupling Livestock from Land Use through Industrial Feed Production Pathways. *Environmental Science & Technology*, **52(13)**, 7351–7359, doi:10.1021/acs.est.8b00216.

Pindyck, R.S., 2013: Climate Change Policy: What Do the Models Tell Us? *Journal of Economic Literature*, **51(3)**, 1–23, doi:10.1257/jel.51.3.860.

Pizer, W. et al., 2014: Using and improving the social cost of carbon. *Science*, **346(6214)**, 1189–1190, doi:10.1126/science.1259774.

Popp, A., H. Lotze-Campen, and B. Bodirsky, 2010: Food consumption, diet shifts and associated non-CO_2 greenhouse gases from agricultural production. *Global Environmental Change*, **20(3)**, 451–462, doi:10.1016/j.gloenvcha.2010.02.001.

Popp, A. et al., 2014a: Land-use protection for climate change mitigation. *Nature Climate Change*, **4(12)**, 1095–1098, doi:10.1038/nclimate2444.

Popp, A. et al., 2014b: Land-use transition for bioenergy and climate stabilization: Model comparison of drivers, impacts and interactions with other land use based mitigation options. *Climatic Change*, **123(3–4)**, 495–509, doi:10.1007/s10584-013-0926-x.

Popp, A. et al., 2017: Land-use futures in the shared socio-economic pathways. *Global Environmental Change*, **42**, 331–345, doi:10.1016/j.gloenvcha.2016.10.002.

Post, M.J., 2012: Cultured meat from stem cells: Challenges and prospects. *Meat Science*, **92(3)**, 297–301, doi:10.1016/j.meatsci.2012.04.008.

Prestele, R. et al., 2016: Hotspots of uncertainty in land-use and land-cover change projections: a global-scale model comparison. *Global Change Biology*, **22(12)**, 3967–3983, doi:10.1111/gcb.13337.

Proistosescu, C. and P.J. Huybers, 2017: Slow climate mode reconciles historical and model-based estimates of climate sensitivity. *Science Advances*, **3(7)**, e1602821, doi:10.1126/sciadv.1602821.

Psarras, P. et al., 2017: Slicing the pie: how big could carbon dioxide removal be? *Wiley Interdisciplinary Reviews: Energy and Environment*, **6(5)**, e253, doi:10.1002/wene.253.

Purohit, P. and L. Höglund-Isaksson, 2017: Global emissions of fluorinated greenhouse gases 2005–2050 with abatement potentials and costs. *Atmospheric Chemistry and Physics*, **17(4)**, 2795–2816, doi:10.5194/acp-17-2795-2017.

Rao, S. et al., 2016: A multi-model assessment of the co-benefits of climate mitigation for global air quality. *Environmental Research Letters*, **11(12)**, 124013, doi:10.1088/1748-9326/11/12/124013.

Rao, S. et al., 2017: Future air pollution in the Shared Socio-economic Pathways. *Global Environmental Change*, **42**, 346–358, doi:10.1016/j.gloenvcha.2016.05.012.

Rau, G.H., 2011: CO_2 Mitigation via Capture and Chemical Conversion in Seawater. *Environmental Science & Technology*, **45(3)**, 1088–1092, doi:10.1021/es102671x.

Revesz, R. et al., 2014: Global warming: Improve economic models of climate change. *Nature*, **508(7495)**, 173–175, doi:10.1038/508173a.

Riahi, K. et al., 2011: RCP 8.5 – A scenario of comparatively high greenhouse gas emissions. *Climatic Change*, **109(1)**, 33, doi:10.1007/s10584-011-0149-y.

Riahi, K. et al., 2012: Energy Pathways for Sustainable Development. In: *Global Energy Assessment – Toward a Sustainable Future*. Cambridge, United Kingdom and New York, NY, USA and the International Institute for Applied Systems Analysis, Laxenburg, Austria, pp. 1203–1306, doi:10.1017/cbo9780511793677.023.

Riahi, K. et al., 2015: Locked into Copenhagen pledges – Implications of short-term emission targets for the cost and feasibility of long-term climate goals. *Technological Forecasting and Social Change*, **90(Part A)**, 8–23, doi:10.1016/j.techfore.2013.09.016.

Riahi, K. et al., 2017: The Shared Socioeconomic Pathways and their energy, land use, and greenhouse gas emissions implications: An overview. *Global Environmental Change*, **42**, 153–168, doi:10.1016/j.gloenvcha.2016.05.009.

Ricke, K.L., R.J. Millar, and D.G. MacMartin, 2017: Constraints on global temperature target overshoot. *Scientific Reports*, **7(1)**, 1–7, doi:10.1038/s41598-017-14503-9.

Riis, F. and E. Halland, 2014: CO_2 storage atlas of the Norwegian Continental shelf: Methods used to evaluate capacity and maturity of the CO_2 storage potential. *Energy Procedia*, **63**, 5258–5265, doi:10.1016/j.egypro.2014.11.557.

Rockström, J. et al., 2017: A roadmap for rapid decarbonization. *Science*, **355(6331)**, 1269–1271, doi:10.1126/science.aah3443.

Roelfsema, M., M. Harmsen, J.J.G. Olivier, A.F. Hof, and D.P. van Vuuren, 2018: Integrated assessment of international climate mitigation commitments outside the UNFCCC. *Global Environmental Change*, **48**, 67–75, doi:10.1016/j.gloenvcha.2017.11.001.

Rogelj, J., D.L. McCollum, B.C. O'Neill, and K. Riahi, 2013a: 2020 emissions levels required to limit warming to below 2 C. *Nature Climate Change*, **3(4)**, 405–412, doi:10.1038/nclimate1758.

Rogelj, J., M. Meinshausen, J. Sedláček, and R. Knutti, 2014a: Implications of potentially lower climate sensitivity on climate projections and policy. *Environmental Research Letters*, **9(3)**, 031003, doi:10.1088/1748-9326/9/3/031003.

Rogelj, J., D.L. McCollum, A. Reisinger, M. Meinshausen, and K. Riahi, 2013b: Probabilistic cost estimates for climate change mitigation. *Nature*, **493(7430)**, 79–83, doi:10.1038/nature11787.

Rogelj, J., M. Meinshausen, M. Schaeffer, R. Knutti, and K. Riahi, 2015a: Impact of short-lived non-CO_2 mitigation on carbon budgets for stabilizing global warming. *Environmental Research Letters*, **10(7)**, 075001, doi:10.1088/1748-9326/10/7/075001.

Rogelj, J. et al., 2011: Emission pathways consistent with a 2°C global temperature limit. *Nature Climate Change*, **1(8)**, 413–418, doi:10.1038/nclimate1258.

Rogelj, J. et al., 2014b: Disentangling the effects of CO_2 and short-lived climate forcer mitigation. *Proceedings of the National Academy of Sciences*, **111(46)**, 16325–16330, doi:10.1073/pnas.1415631111.

Rogelj, J. et al., 2015b: Energy system transformations for limiting end-of-century warming to below 1.5°C. *Nature Climate Change*, **5(6)**, 519–527, doi:10.1038/nclimate2572.

Rogelj, J. et al., 2015c: Mitigation choices impact carbon budget size compatible with low temperature goals. *Environmental Research Letters*, **10(7)**, 075003, doi:10.1088/1748-9326/10/7/075003.

Rogelj, J. et al., 2015d: Zero emission targets as long-term global goals for climate protection. *Environmental Research Letters*, **10(10)**, 105007, doi:10.1088/1748-9326/10/10/105007.

Rogelj, J. et al., 2016a: Paris Agreement climate proposals need a boost to keep warming well below 2°C. *Nature*, **534(7609)**, 631–639, doi:10.1038/nature18307.

Rogelj, J. et al., 2016b: Differences between carbon budget estimates unravelled. *Nature Climate Change*, **6(3)**, 245–252, doi:10.1038/nclimate2868.

Rogelj, J. et al., 2017: Understanding the origin of Paris Agreement emission uncertainties. *Nature Communications*, **8**, 15748, doi:10.1038/ncomms15748.

Rogelj, J. et al., 2018: Scenarios towards limiting global mean temperature increase below 1.5°C. *Nature Climate Change*, **8(4)**, 325–332, doi:10.1038/s41558-018-0091-3.

Rose, S.K., 2012: The role of the social cost of carbon in policy. *Wiley Interdisciplinary Reviews: Climate Change*, **3(2)**, 195–212, doi:10.1002/wcc.163.

Rose, S.K., D.B. Diza, and G.J. Blanford, 2017a: Understanding the Social Cost of Carbon: A Model diagnostic and Inter-comparison Study. *Climate Change Economics*, **8(2)**, 1750009, doi:10.1142/s2010007817500099.

Rose, S.K., R. Richels, G. Blanford, and T. Rutherford, 2017b: The Paris Agreement and next steps in limiting global warming. *Climatic Change*, **142(1–2)**, 1–16, doi:10.1007/s10584-017-1935-y.

Rose, S.K. et al., 2014a: Bioenergy in energy transformation and climate management. *Climatic Change*, **123(3–4)**, 477–493, doi:10.1007/s10584-013-0965-3.

Rose, S.K. et al., 2014b: Non-Kyoto radiative forcing in long-run greenhouse gas emissions and climate change scenarios. *Climatic Change*, **123(3–4)**, 511–525, doi:10.1007/s10584-013-0955-5.

Rugenstein, M.A.A. et al., 2016: Multiannual Ocean-Atmosphere Adjustments to Radiative Forcing. *Journal of Climate*, **29(15)**, 5643–5659, doi:10.1175/jcli-d-16-0312.1.

Samset, B.H. et al., 2016: Fast and slow precipitation responses to individual climate forcers: A PDRMIP multimodel study. *Geophysical Research Letters*, **43(6)**, 2782–2791, doi:10.1002/2016gl068064.

Samset, B.H. et al., 2018: Climate impacts from a removal of anthropogenic aerosol emissions. *Geophysical Research Letters*, **45(2)**, 1020–1029, doi:10.1002/2017gl076079.

Sand, M. et al., 2015: Response of Arctic temperature to changes in emissions of short-lived climate forcers. *Nature Climate Change*, **6(3)**, 286–289, doi:10.1038/nclimate2880.

Sanderson, B.M., B.C. O'Neill, and C. Tebaldi, 2016: What would it take to achieve the Paris temperature targets? *Geophysical Research Letters*, **43(13)**, 7133–7142, doi:10.1002/2016gl069563.

Sands, R.D., 2018: U.S. Carbon Tax Scenarios and Bioenergy. *Climate Change Economics*, **9(1)**, 1840010, doi:10.1142/s2010007818400109.

Sarofim, M.C., S.T. Waldhoff, and S.C. Anenberg, 2017: Valuing the Ozone-Related Health Benefits of Methane Emission Controls. *Environmental and Resource Economics*, **66(1)**, 45–63, doi:10.1007/s10640-015-9937-6.

Saunders, H.D., 2015: Recent Evidence for Large Rebound: Elucidating the Drivers and their Implications for Climate Change Models. *The Energy Journal*, **36(1)**, 23–48, doi:10.5547/01956574.36.1.2.

Schädel, C. et al., 2014: Circumpolar assessment of permafrost C quality and its vulnerability over time using long-term incubation data. *Global Change Biology*, **20(2)**, 641–652, doi:10.1111/gcb.12417.

Schädel, C. et al., 2016: Potential carbon emissions dominated by carbon dioxide from thawed permafrost soils. *Nature Climate Change*, **6(10)**, 950–953, doi:10.1038/nclimate3054.

Schleussner, C.-F. et al., 2016: Science and policy characteristics of the Paris Agreement temperature goal. *Nature Climate Change*, **6(7)**, 827–835, doi:10.1038/nclimate3096.

Schmidt, M.G.W., H. Held, E. Kriegler, and A. Lorenz, 2013: Climate Policy Under Uncertain and Heterogeneous Climate Damages. *Environmental and Resource Economics*, **54(1)**, 79–99, doi:10.1007/s10640-012-9582-2.

Schmitz, C. et al., 2012: Trading more food: Implications for land use, greenhouse gas emissions, and the food system. *Global Environmental Change*, **22(1)**, 189–209, doi:10.1016/j.gloenvcha.2011.09.013.

Schneider, T. et al., 2017: Climate goals and computing the future of clouds. *Nature Climate Change*, **7(1)**, 3–5, doi:10.1038/nclimate3190.

Schneider von Deimling, T. et al., 2012: Estimating the near-surface permafrost-carbon feedback on global warming. *Biogeosciences*, **9(2)**, 649–665, doi:10.5194/bg-9-649-2012.

Schneider von Deimling, T. et al., 2015: Observation-based modelling of permafrost carbon fluxes with accounting for deep carbon deposits and thermokarst activity. *Biogeosciences*, **12(11)**, 3469–3488, doi:10.5194/bg-12-3469-2015.

Schuiling, R.D. and P. Krijgsman, 2006: Enhanced Weathering: An Effective and Cheap Tool to Sequester CO_2. *Climatic Change*, **74(1–3)**, 349–354, doi:10.1007/s10584-005-3485-y.

Schurer, A.P. et al., 2018: Interpretations of the Paris climate target. *Nature Geoscience*, **11(4)**, 220–221, doi:10.1038/s41561-018-0086-8.

Schuur, E.A.G. et al., 2015: Climate change and the permafrost carbon feedback. *Nature*, **520(7546)**, 171–179, doi:10.1038/nature14338.

Scott, C.E. et al., 2018: Substantial large-scale feedbacks between natural aerosols and climate. *Nature Geoscience*, **11(1)**, 44–48, doi:10.1038/s41561-017-0020-5.

Scott, V., R.S. Haszeldine, S.F.B. Tett, and A. Oschlies, 2015: Fossil fuels in a trillion tonne world. *Nature Climate Change*, **5(5)**, 419–423, doi:10.1038/nclimate2578.

Searle, S.Y. and C.J. Malins, 2014: Will energy crop yields meet expectations? *Biomass and Bioenergy*, **65**, 3–12, doi:10.1016/j.biombioe.2014.01.001.

Séférian, R., M. Rocher, C. Guivarch, and J. Colin, 2018: Constraints on biomass energy deployment in mitigation pathways: the case of water limitation. *Environmental Research Letters*, 1–32, doi:10.1088/1748-9326/aabcd7.

Seto, K.C. et al., 2016: Carbon Lock-In: Types, Causes, and Policy Implications. *Annual Review of Environment and Resources*, **41(1)**, 425–452, doi:10.1146/annurev-environ-110615-085934.

Shah, N., M. Wei, V. Letschert, and A. Phadke, 2015: *Benefits of Leapfrogging to Superefficiency and Low Global Warming Potential Refrigerants in Room Air Conditioning*. LBNL-1003671, Ernest Orlando Lawrence Berkeley National Laboratory, Berkeley, CA, USA, 58 pp.

Shearer, C., N. Mathew-Shah, L. Myllyvirta, A. Yu, and T. Nace, 2018: *Boom and Bust 2018: Tracking the Global Coal Plant Pipeline*. CoalSwarm, Greenpeace USA, and Sierra Club, 16 pp.

Shell International B.V., 2018: *Shell Scenarios: Sky – Meeting the Goals of the Paris Agreement*. Shell International B.V. 36 pp.

Sherwood, S.C., S. Bony, and J.-L. Dufresne, 2014: Spread in model climate sensitivity traced to atmospheric convective mixing. *Nature*, **505(7481)**, 37–42, doi:10.1038/nature12829.

Shindell, D.T., 2015: The social cost of atmospheric release. *Climatic Change*, **130(2)**, 313–326, doi:10.1007/s10584-015-1343-0.

Shindell, D.T. and G. Faluvegi, 2010: The net climate impact of coal-fired power plant emissions. *Atmospheric Chemistry and Physics*, **10(7)**, 3247–3260, doi:10.5194/acp-10-3247-2010.

Shindell, D.T., Y. Lee, and G. Faluvegi, 2016: Climate and health impacts of US emissions reductions consistent with 2°C. *Nature Climate Change*, **6(5)**, 503–507, doi:10.1038/nclimate2935.

Shindell, D.T., J.S. Fuglestvedt, and W.J. Collins, 2017a: The social cost of methane: theory and applications. *Faraday Discussions*, **200**, 429–451, doi:10.1039/c7fd00009j.

Shindell, D.T., G. Faluvegi, L. Rotstayn, and G. Milly, 2015: Spatial patterns of radiative forcing and surface temperature response. *Journal of Geophysical Research: Atmospheres*, **120(11)**, 5385–5403, doi:10.1002/2014jd022752.

Shindell, D.T., G. Faluvegi, K. Seltzer, and C. Shindell, 2018: Quantified, localized health benefits of accelerated carbon dioxide emissions reductions. *Nature Climate Change*, **8(4)**, 291–295, doi:10.1038/s41558-018-0108-y.

Shindell, D.T. et al., 2012: Simultaneously Mitigating Near-Term Climate Change and Improving Human Health and Food Security. *Science*, **335(6065)**, 183–189, doi:10.1126/science.1210026.

Shindell, D.T. et al., 2017b: A climate policy pathway for near- and long-term benefits. *Science*, **356(6337)**, 493–494, doi:10.1126/science.aak9521.

Shukla, P.R. and V. Chaturvedi, 2012: Low carbon and clean energy scenarios for India: Analysis of targets approach. *Energy Economics*, **34(sup3)**, S487–S495, doi:10.1016/j.eneco.2012.05.002.

Siegmeier, J. et al., 2018: The fiscal benefits of stringent climate change mitigation: an overview. *Climate Policy*, **18(3)**, 352–367, doi:10.1080/14693062.2017.1400943.

Silva, R.A. et al., 2016: The effect of future ambient air pollution on human premature mortality to 2100 using output from the ACCMIP model ensemble. *Atmospheric Chemistry and Physics*, **16(15)**, 9847–9862, doi:10.5194/acp-16-9847-2016.

Sims, R. et al., 2014: Transport. In: *Climate Change 2014: Mitigation of Climate Change. Contribution of Working Group III to the Fifth Assessment Report of the Intergovernmental Panel on Climate Change* [Edenhofer, O., R. Pichs-Madruga, Y. Sokona, E. Farahani, S. Kadne, K. Seyboth, A. Adler, I. Baum, S. Brunner, P. Eickemeier, B. Kriemann, J. Savolainen, S. Schlömer, C. Stechow, T. Zwickel, and J.C. Minx (eds.)]. Cambridge University Press, Cambridge, United Kingdom and New York, NY, USA, pp. 599–670.

Smith, C.J. et al., 2018: FAIR v1.3: a simple emissions-based impulse response and carbon cycle model. *Geoscientific Model Development*, **11(6)**, 2273–2297, doi:10.5194/gmd-11-2273-2018.

Smith, L.J. and M.S. Torn, 2013: Ecological limits to terrestrial biological carbon dioxide removal. *Climatic Change*, **118(1)**, 89–103, doi:10.1007/s10584-012-0682-3.

Smith, P., 2016: Soil carbon sequestration and biochar as negative emission technologies. *Global Change Biology*, **22(3)**, 1315–1324, doi:10.1111/gcb.13178.

Smith, P. and M. Bustamante, 2014: Agriculture, Forestry and Other Land Use (AFOLU). In: *Climate Change 2014: Mitigation of Climate Change. Contribution of Working Group III to the Fifth Assessment Report of the Intergovernmental Panel on Climate Change* [Edenhofer, O., R. Pichs-Madruga, Y. Sokona, E. Farahani, S. Kadner, K. Seyboth, A. Adler, I. Baum, S. Brunner, P. Eickemeier, B. Kriemann, J. Savolainen, S. Schlömer, C. von Stechow, T. Zwickel, and J.C. Minx (eds.)]. Cambridge University Press, Cambridge, United Kingdom and New York, NY, USA, pp. 811–922.

Smith, P. et al., 2013: How much land-based greenhouse gas mitigation can be achieved without compromising food security and environmental goals? *Global Change Biology*, **19(8)**, 2285–2302, doi:10.1111/gcb.12160.

Smith, P. et al., 2015: Biophysical and economic limits to negative CO_2 emissions. *Nature Climate Change*, **6(1)**, 42–50, doi:10.1038/nclimate2870.

Snopkowski, K., M.C. Towner, M.K. Shenk, and H. Colleran, 2016: Pathways from education to fertility decline: a multi-site comparative study. *Philosophical Transactions of the Royal Society B: Biological Sciences*, **371(1692)**, 20150156, doi:10.1098/rstb.2015.0156.

Socolow, R. et al., 2011: *Direct Air Capture of CO_2 with Chemicals: A Technology Assessment for the APS Panel on Public Affairs*. American Physical Society (APS), College Park, MD, USA, 100 pp.

Somanthan, E. et al., 2014: National and Sub-national Policies and Institutions. In: *Climate Change 2014: Mitigation of Climate Change. Contribution of Working Group III to the Fifth Assessment Report of the Intergovernmental Panel on Climate Change* [Edenhofer, O., R. Pichs-Madruga, Y. Sokona, E. Farahani, S. Kadner, K. Seyboth, A. Adler, I. Baum, S. Brunner, P. Eickemeier, B. Kriemann, J. Savolainen, S. Schlömer, C. von Stechow, T. Zwickel, and J.C. Minx (eds.)]. Cambridge University Press, Cambridge, United Kingdom and New York, NY, USA, pp. 1141–1206.

Sonnenschein, J., R. Van Buskirk, J.L. Richter, and C. Dalhammar, 2018: Minimum energy performance standards for the 1.5°C target: an effective complement to carbon pricing. *Energy Efficiency*, 1–16, doi:10.1007/s12053-018-9669-x.

Spencer, T., R. Pierfederici, H. Waisman, and M. Colombier, 2015: *Beyond the Numbers. Understanding the Transformation Induced by INDCs*. Study N°05/15, MILES Project Consortium, Paris, France, 80 pp.

Springmann, M., H.C.J. Godfray, M. Rayner, and P. Scarborough, 2016: Analysis and valuation of the health and climate change cobenefits of dietary change. *Proceedings of the National Academy of Sciences*, **113(15)**, 4146–4151, doi:10.1073/pnas.1523119113.

Steckel, J.C., O. Edenhofer, and M. Jakob, 2015: Drivers for the renaissance of coal. *Proceedings of the National Academy of Sciences*, **112(29)**, E3775–E3781, doi:10.1073/pnas.1422722112.

Stehfest, E. et al., 2009: Climate benefits of changing diet. *Climatic Change*, **95(1–2)**, 83–102, doi:10.1007/s10584-008-9534-6.

Stern, N., 2007: *The Economics of Climate Change: The Stern Review*. Cambridge University Press, Cambridge, United Kingdom and New York, NY, USA, 692 pp., doi:10.1017/cbo9780511817434.

Stern, N., 2013: The Structure of Economic Modeling of the Potential Impacts of Climate Change: Grafting Gross Underestimation of Risk onto Already Narrow Science Models. *Journal of Economic Literature*, **51(3)**, 838–859, doi:10.1257/jel.51.3.838.

Stern, N., 2016: Current climate models are grossly misleading. *Nature*, **530**, 407–409, doi:10.1038/530407a.

Stevanović, M. et al., 2017: Mitigation Strategies for Greenhouse Gas Emissions from Agriculture and Land-Use Change: Consequences for Food Prices. *Environmental Science & Technology*, **51(1)**, 365–374, doi:10.1021/acs.est.6b04291.

Stiglitz, J.E. et al., 2017: *Report of the High-Level Commission on Carbon Prices*. Carbon Pricing Leadership Coalition (CPLC), 68 pp.

Stocker, T.F. et al., 2013: Technical Summary. In: *Climate Change 2013: The Physical Science Basis. Contribution of Working Group I to the Fifth Assessment Report of the Intergovernmental Panel on Climate Change* [Stocker, T.F., D. Qin, G.-K. Plattner, M. Tignor, S.K. Allen, J. Boschung, A. Nauels, Y. Xia, V. Bex, and P.M. Midgley (eds.)]. Cambridge University Press, Cambridge, United Kingdom and New York, NY, USA, pp. 33–115.

Stohl, A. et al., 2015: Evaluating the climate and air quality impacts of short-lived pollutants. *Atmospheric Chemistry and Physics*, **15(18)**, 10529–10566, doi:10.5194/acp-15-10529-2015.

Strapasson, A. et al., 2017: On the global limits of bioenergy and land use for climate change mitigation. *GCB Bioenergy*, **9(12)**, 1721–1735, doi:10.1111/gcbb.12456.

Strefler, J., T. Amann, N. Bauer, E. Kriegler, and J. Hartmann, 2018a: Potential and costs of carbon dioxide removal by enhanced weathering of rocks. *Environmental Research Letters*, **13(3)**, 034010, doi:10.1088/1748-9326/aaa9c4.

Strefler, J. et al., 2018b: Between Scylla and Charybdis: Delayed mitigation narrows the passage between large-scale CDR and high costs. *Environmental Research Letters*, **13(4)**, 044015, doi:10.1088/1748-9326/aab2ba.

Su, X. et al., 2017: Emission pathways to achieve 2.0°C and 1.5°C climate targets. *Earth's Future*, **5(6)**, 592–604, doi:10.1002/2016ef000492.

Subbarao, G. et al., 2013: Potential for biological nitrification inhibition to reduce nitrification and N_2O emissions in pasture crop–livestock systems. *Animal*, **7(s2)**, 322–332, doi:10.1017/s1751731113000761.

Subharat, S. et al., 2016: Vaccination of Sheep with a Methanogen Protein Provides Insight into Levels of Antibody in Saliva Needed to Target Ruminal Methanogens. *PLOS ONE*, **11(7)**, e0159861, doi:10.1371/journal.pone.0159861.

Szulczewski, M.L., C.W. MacMinn, and R. Juanes, 2014: Theoretical analysis of how pressure buildup and CO_2 migration can both constrain storage capacity in deep saline aquifers. *International Journal of Greenhouse Gas Control*, **23**, 113–118, doi:10.1016/j.ijggc.2014.02.006.

Tachiiri, K., T. Hajima, and M. Kawamiya, 2015: Increase of uncertainty in transient climate response to cumulative carbon emissions after stabilization of atmospheric CO_2 concentration. *Environmental Research Letters*, **10(12)**, 125018, doi:10.1088/1748-9326/10/12/125018.

Tan, I., T. Storelvmo, and M.D. Zelinka, 2016: Observational constraints on mixed-phase clouds imply higher climate sensitivity. *Science*, **352(6282)**, 224–227, doi:10.1126/science.aad5300.

Tavoni, M. and R.S.J. Tol, 2010: Counting only the hits? The risk of underestimating the costs of stringent climate policy. *Climatic Change*, **100(3–4)**, 769–778, doi:10.1007/s10584-010-9867-9.

The Royal Society, 2009: *Geoengineering the climate: science, governance and uncertainty*. RS Policy document 10/09, The Royal Society, London, UK, 82 pp.

Thomson, A.M. et al., 2011: RCP4.5: a pathway for stabilization of radiative forcing by 2100. *Climatic Change*, **109(1–2)**, 77–94, doi:10.1007/s10584-011-0151-4.

Tilman, D. and M. Clark, 2014: Global diets link environmental sustainability and human health. *Nature*, **515(7528)**, 518–522, doi:10.1038/nature13959.

Tokarska, K.B. and N.P. Gillett, 2018: Cumulative carbon emissions budgets consistent with 1.5°C global warming. *Nature Climate Change*, **8(4)**, 296–299, doi:10.1038/s41558-018-0118-9.

Tokarska, K.B., N.P. Gillett, V.K. Arora, W.G. Lee, and K. Zickfeld, 2018: The influence of non-CO_2 forcings on cumulative carbon emissions budgets. *Environmental Research Letters*, **13(3)**, 034039, doi:10.1088/1748-9326/aaafdd.

Tol, R.S.J., 1999: Spatial and Temporal Efficiency in Climate Policy: Applications of FUND. *Environmental and Resource Economics*, **14(1)**, 33–49, doi:10.1023/a:1008314205375.

Tol, R.S.J., 2017: The structure of the climate debate. *Energy Policy*, **104**, 431–438, doi:10.1016/j.enpol.2017.01.005.

Tuomisto, H.L. and M.J. Teixeira de Mattos, 2011: Environmental Impacts of Cultured Meat Production. *Environmental Science & Technology*, **45(14)**, 6117–6123, doi:10.1021/es200130u.

UNEP, 2015: *Aligning the Financial System with Sustainable Development: Pathways to Scale*. Inquiry: Design of a Sustainable Financial System and United Nations Environment Programme (UNEP), Geneva, Switzerland, 25 pp.

UNEP, 2016: *The Emissions Gap Report 2016: A UNEP Synthesis Report*. United Nations Environment Programme (UNEP), Nairobi, Kenya, 85 pp.

UNEP, 2017: *The Emissions Gap Report 2017: A UN Environment Synthesis Report*. United Nations Environment Programme (UNEP), Nairobi, Kenya, 116 pp.

UNFCCC, 2016: *Aggregate effect of the intended nationally determined contributions: an update*. FCCC/CP/2016/2, The Secretariat of the United Nations Framework Convention on Climate Change (UNFCCC), Bonn, Germany, 75 pp.

Unruh, G.C. and J. Carrillo-Hermosilla, 2006: Globalizing carbon lock-in. *Energy Policy*, **34(10)**, 1185–1197, doi:10.1016/j.enpol.2004.10.013.

Unruh, J.D., 2011: Tree-Based Carbon Storage in Developing Countries: Neglect of the Social Sciences. *Society & Natural Resources*, **24(2)**, 185–192, doi:10.1080/08941920903410136.

US EPA, 2013: *Global Mitigation of Non-CO_2 Greenhouse Gases: 2010–2030*. Report No. EPA-430-R-13-011. United States Environmental Protection Agency (US EPA), Washington DC, USA, 410 pp.

Valin, H. et al., 2013: Agricultural productivity and greenhouse gas emissions: trade-offs or synergies between mitigation and food security? *Environmental Research Letters*, **8(3)**, 035019, doi:10.1088/1748-9326/8/3/035019.

van den Bergh, J.C.J.M., 2017: Rebound policy in the Paris Agreement: instrument comparison and climate-club revenue offsets. *Climate Policy*, **17(6)**, 801–813, doi:10.1080/14693062.2016.1169499.

van Sluisveld, M.A.E., S.H. Martínez, V. Daioglou, and D.P. van Vuuren, 2016: Exploring the implications of lifestyle change in 2°C mitigation scenarios using the IMAGE integrated assessment model. *Technological Forecasting and Social Change*, **102**, 309–319, doi:10.1016/j.techfore.2015.08.013.

van Soest, H.L. et al., 2017: Low-emission pathways in 11 major economies: comparison of cost-optimal pathways and Paris climate proposals. *Climatic Change*, **142(3–4)**, 491–504, doi:10.1007/s10584-017-1964-6.

van Vuuren, D.P. and K. Riahi, 2011: The relationship between short-term emissions and long-term concentration targets. *Climatic Change*, **104(3)**, 793–801, doi:10.1007/s10584-010-0004-6.

van Vuuren, D.P., A.F. Hof, M.A.E. van Sluisveld, and K. Riahi, 2017a: Open discussion of negative emissions is urgently needed. *Nature Energy*, **2(12)**, 902–904, doi:10.1038/s41560-017-0055-2.

van Vuuren, D.P. et al., 2007: Stabilizing greenhouse gas concentrations at low levels: An assessment of reduction strategies and costs. *Climatic Change*, **81(2)**, 119–159, doi:10.1007/s10584-006-9172-9.

van Vuuren, D.P. et al., 2011a: The representative concentration pathways: An overview. *Climatic Change*, **109(1)**, 5–31, doi:10.1007/s10584-011-0148-z.

van Vuuren, D.P. et al., 2011b: RCP2.6: Exploring the possibility to keep global mean temperature increase below 2°C. *Climatic Change*, **109(1)**, 95–116, doi:10.1007/s10584-011-0152-3.

van Vuuren, D.P. et al., 2015: Pathways to achieve a set of ambitious global sustainability objectives by 2050: Explorations using the IMAGE integrated assessment model. *Technological Forecasting and Social Change*, **98**, 303–323, doi:10.1016/j.techfore.2015.03.005.

van Vuuren, D.P. et al., 2016: Carbon budgets and energy transition pathways. *Environmental Research Letters*, **11(7)**, 075002, doi:10.1088/1748-9326/11/7/075002.

van Vuuren, D.P. et al., 2017b: Energy, land-use and greenhouse gas emissions trajectories under a green growth paradigm. *Global Environmental Change*, **42**, 237–250, doi:10.1016/j.gloenvcha.2016.05.008.

van Vuuren, D.P. et al., 2018: Alternative pathways to the 1.5°C target reduce the need for negative emission technologies. *Nature Climate Change*, **8(5)**, 391–397, doi:10.1038/s41558-018-0119-8.

Vangkilde-Pedersen, T. et al., 2009: Assessing European capacity for geological storage of carbon dioxide-the EU GeoCapacity project. *Energy Procedia*, **1(1)**, 2663–2670, doi:10.1016/j.egypro.2009.02.034.

Veland, S. et al., 2018: Narrative matters for sustainability: the transformative role of storytelling in realizing 1.5°C futures. *Current Opinion in Environmental Sustainability*, **31**, 41–47, doi:10.1016/j.cosust.2017.12.005.

Velders, G.J.M., D.W. Fahey, J.S. Daniel, S.O. Andersen, and M. McFarland, 2015: Future atmospheric abundances and climate forcings from scenarios of global and regional hydrofluorocarbon (HFC) emissions. *Atmospheric Environment*, **123**, 200–209, doi:10.1016/j.atmosenv.2015.10.071.

Vogt-Schilb, A. and S. Hallegatte, 2017: Climate policies and nationally determined contributions: reconciling the needed ambition with the political economy. *Wiley Interdisciplinary Reviews: Energy and Environment*, **6(6)**, 1–23, doi:10.1002/wene.256.

Voigt, C. et al., 2017a: Warming of subarctic tundra increases emissions of all three important greenhouse gases – carbon dioxide, methane, and nitrous oxide. *Global Change Biology*, **23(8)**, 3121–3138, doi:10.1111/gcb.13563.

Voigt, C. et al., 2017b: Increased nitrous oxide emissions from Arctic peatlands after permafrost thaw. *Proceedings of the National Academy of Sciences*, **114(24)**, 6238–6243, doi:10.1073/pnas.1702902114.

von Stechow, C. et al., 2015: Integrating Global Climate Change Mitigation Goals with Other Sustainability Objectives: A Synthesis. *Annual Review of Environment and Resources*, **40(1)**, 363–394, doi:10.1146/annurev-environ-021113-095626.

von Stechow, C. et al., 2016: 2°C and SDGs: united they stand, divided they fall? *Environmental Research Letters*, **11(3)**, 034022, doi:10.1088/1748-9326/11/3/034022.

Vrontisi, Z. et al., 2018: Enhancing global climate policy ambition towards a 1.5°C stabilization: a short-term multi-model assessment. *Environmental Research Letters*, **13(4)**, 044039, doi:10.1088/1748-9326/aab53e.

Wachsmuth, J. and V. Duscha, 2018: Achievability of the Paris targets in the EU-the role of demand-side-driven mitigation in different types of scenarios. *Energy Efficiency*, 1–19, doi:10.1007/s12053-018-9670-4.

Walsh, M.J. et al., 2016: Algal food and fuel coproduction can mitigate greenhouse gas emissions while improving land and water-use efficiency. *Environmental Research Letters*, **11(11)**, 114006, doi:10.1088/1748-9326/11/11/114006.

Wang, X. et al., 2016: Taking account of governance: Implications for land-use dynamics, food prices, and trade patterns. *Ecological Economics*, **122**, 12–24, doi:10.1016/j.ecolecon.2015.11.018.

Warwick, P.D., M.K. Verma, P.A. Freeman, M.D. Corum, and S.H. Hickman, 2014: U.S. geological survey carbon sequestration – geologic research and assessments. *Energy Procedia*, **63**, 5305–5309, doi:10.1016/j.egypro.2014.11.561.

Wedlock, D.N., P.H. Janssen, S.C. Leahy, D. Shu, and B.M. Buddle, 2013: Progress in the development of vaccines against rumen methanogens. *Animal*, **7(s2)**, 244–252, doi:10.1017/s1751731113000682.

Wei, N. et al., 2013: A preliminary sub-basin scale evaluation framework of site suitability for onshore aquifer-based CO_2 storage in China. *International Journal of Greenhouse Gas Control*, **12**, 231–246, doi:10.1016/j.ijggc.2012.10.012.

Weindl, I. et al., 2015: Livestock in a changing climate: production system transitions as an adaptation strategy for agriculture. *Environmental Research Letters*, **10(9)**, 094021, doi:10.1088/1748-9326/10/9/094021.

Weindl, I. et al., 2017: Livestock and human use of land: Productivity trends and dietary choices as drivers of future land and carbon dynamics. *Global and Planetary Change*, **159**, 1–10, doi:10.1016/j.gloplacha.2017.10.002.

Weitzman, M.L., 2014: Can Negotiating a Uniform Carbon Price Help to Internalize the Global Warming Externality? *Journal of the Association of Environmental and Resource Economists*, **1(1/2)**, 29–49, doi:10.1086/676039.

Werner, C., H.-P. Schmidt, D. Gerten, W. Lucht, and C. Kammann, 2018: Biogeochemical potential of biomass pyrolysis systems for limiting global warming to 1.5°C. *Environmental Research Letters*, **13(4)**, 044036, doi:10.1088/1748-9326/aabb0e.

West, J.J. et al., 2013: Co-benefits of mitigating global greenhouse gas emissions for future air quality and human health. *Nature Climate Change*, **3(10)**, 885–889, doi:10.1038/nclimate2009.

Weyant, J., 2017: Some Contributions of Integrated Assessment Models of Global Climate Change. *Review of Environmental Economics and Policy*, **11(1)**, 115–137, doi:10.1093/reep/rew018.

Weyant, J.P., F.C. Chesnaye, and G.J. Blanford, 2006: Overview of EMF-21: Multigas Mitigation and Climate Policy. *The Energy Journal*, **27**, 1–32, doi:10.5547/issn0195-6574-ej-volsi2006-nosi3-1.

Wiebe, K. et al., 2015: Climate change impacts on agriculture in 2050 under a range of plausible socioeconomic and emissions scenarios. *Environmental Research Letters*, **10(8)**, 085010, doi:10.1088/1748-9326/10/8/085010.

Wigley, T.M.L., R. Richels, and J. Edmonds, 2007: Overshoot Pathways to CO_2 stabilization in a multi-gas context. In: *Human Induced Climate Change: An Interdisciplinary Perspective* [Schlesinger, M.E., H.S. Kheshgi, J. Smith, F.C. de la Chesnaye, J.M. Reilly, T. Wilson, and C. Kolstad (eds.)]. Cambridge University Press, Cambridge, pp. 84–92, doi:10.1017/cbo9780511619472.009.

Williams, P.J. and L.M.L. Laurens, 2010: Microalgae as biodiesel & biomass feedstocks: Review & analysis of the biochemistry, energetics & economics. *Energy & Environmental Science*, **3(5)**, 554–590, doi:10.1039/b924978h.

Williamson, P., 2016: Emissions reduction: Scrutinize CO_2 removal methods. *Nature*, **530(153)**, 153–155, doi:10.1038/530153a.

Williamson, P. and R. Bodle, 2016: *Update on Climate Geoengineering in Relation to the Convention on Biological Diversity: Potential Impacts and Regulatory Framework*. CBD Technical Series No. 84, Secretariat of the Convention on Biological Diversity, Montreal, QC, Canada, 158 pp.

Wise, M., K. Calvin, P. Kyle, P. Luckow, and J. Edmonds, 2014: Economic and physical modeling of land use in GCAM 3.0 and an application to agricultural productivity, land, and terrestrial carbon. *Climate Change Economics*, **5(2)**, 1450003, doi:10.1142/s2010007814500031.

Woolf, D., J.E. Amonette, F.A. Street-Perrott, J. Lehmann, and S. Joseph, 2010: Sustainable biochar to mitigate global climate change. *Nature Communications*, **1(5)**, 1–9, doi:10.1038/ncomms1053.

Xiao, X.-J. and K. Jiang, 2018: China's nuclear power under the global 1.5°C target: Preliminary feasibility study and prospects. *Advances in Climate Change Research*, **9(2)**, 138–143, doi:10.1016/j.accre.2018.05.002.

Xu, Y. and V. Ramanathan, 2017: Well below 2°C: Mitigation strategies for avoiding dangerous to catastrophic climate changes. *Proceedings of the National Academy of Sciences*, **114(39)**, 10315–10323, doi:10.1073/pnas.1618481114.

Yeh, S. et al., 2016: Detailed assessment of global transport-energy models' structures and projections. *Transportation Research Part D: Transport and Environment*, **55**, 294–309, doi:10.1016/j.trd.2016.11.001.

Zeman, F.S. and K. Lackner, 2004: Capturing Carbon Dioxide directly from the Atmosphere. *World Resource Review*, **16(2)**, 157–172.

Zeman, F.S. and D.W. Keith, 2008: Carbon Neutral Hydrocarbons. *Philosophical Transactions of the Royal Society A: Mathematical, Physical and Engineering Sciences*, **366(1882)**, 3901–3918, doi:10.1098/rsta.2008.0143.

Zhai, C., J.H. Jiang, and H. Su, 2015: Long-term cloud change imprinted in seasonal cloud variation: More evidence of high climate sensitivity. *Geophysical Research Letters*, **42(20)**, 8729–8737, doi:10.1002/2015gl065911.

Zhang, R., S. Fujimori, and T. Hanaoka, 2018: The contribution of transport policies to the mitigation potential and cost of 2°C and 1.5°C goals. *Environmental Research Letters*, **13(5)**, 054008, doi:10.1088/1748-9326/aabb0d.

Zickfeld, K. and T. Herrington, 2015: The time lag between a carbon dioxide emission and maximum warming increases with the size of the emission. *Environmental Research Letters*, **10(3)**, 031001, doi:10.1088/1748-9326/10/3/031001.

Zickfeld, K., A.H. MacDougall, and H.D. Matthews, 2016: On the proportionality between global temperature change and cumulative CO_2 emissions during periods of net negative CO_2 emissions. *Environmental Research Letters*, **11(5)**, 055006, doi:10.1088/1748-9326/11/5/055006.

Zickfeld, K., M. Eby, H.D. Matthews, and A.J. Weaver, 2009: Setting cumulative emissions targets to reduce the risk of dangerous climate change. *Proceedings of the National Academy of Sciences*, **106(38)**, 16129–16134, doi:10.1073/pnas.0805800106.

Zomer, R.J., D.A. Bossio, R. Sommer, and L.V. Verchot, 2017: Global Sequestration Potential of Increased Organic Carbon in Cropland Soils. *Scientific Reports*, **7(1)**, 15554, doi:10.1038/s41598-017-15794-8.

Zomer, R.J. et al., 2016: Global Tree Cover and Biomass Carbon on Agricultural Land: The contribution of agroforestry to global and national carbon budgets. *Scientific Reports*, **6**, 29987, doi:10.1038/srep29987.

Impacts of 1.5°C of Global Warming on Natural and Human Systems

3

Coordinating Lead Authors:
Ove Hoegh-Guldberg (Australia), Daniela Jacob (Germany), Michael Taylor (Jamaica)

Lead Authors:
Marco Bindi (Italy), Sally Brown (UK), Ines Camilloni (Argentina), Arona Diedhiou (Ivory Coast/Senegal), Riyanti Djalante (Japan/Indonesia), Kristie L. Ebi (USA), Francois Engelbrecht (South Africa), Joel Guiot (France), Yasuaki Hijioka (Japan), Shagun Mehrotra (USA/India), Antony Payne (UK), Sonia I. Seneviratne (Switzerland), Adelle Thomas (Bahamas), Rachel Warren (UK), Guangsheng Zhou (China)

Contributing Authors:
Sharina Abdul Halim (Malaysia), Michelle Achlatis (Australia/Greece), Lisa V. Alexander (Australia), Myles R. Allen (UK), Peter Berry (Canada), Christopher Boyer (USA), Lorenzo Brilli (Italy), Marcos Buckeridge (Brazil), Edward Byers (Austria/Brazil), William Cheung (Canada), Marlies Craig (South Africa), Neville Ellis (Australia), Jason Evans (Australia), Hubertus Fischer (Switzerland), Klaus Fraedrich (Germany), Sabine Fuss (Germany), Anjani Ganase (Australia/Trinidad and Tobago), Jean-Pierre Gattuso (France), Peter Greve (Austria/Germany), Tania Guillén Bolaños (Germany/Nicaragua), Naota Hanasaki (Japan), Tomoko Hasegawa (Japan), Katie Hayes (Canada), Annette Hirsch (Switzerland/Australia), Chris Jones (UK), Thomas Jung (Germany), Markku Kanninen (Finland), Gerhard Krinner (France), David Lawrence (USA), Tim Lenton (UK), Debora Ley (Guatemala/Mexico), Diana Liverman (USA), Natalie Mahowald (USA), Kathleen McInnes (Australia), Katrin J. Meissner (Australia), Richard Millar (UK), Katja Mintenbeck (Germany), Dann Mitchell (UK), Alan C. Mix (US), Dirk Notz (Germany), Leonard Nurse (Barbados), Andrew Okem (Nigeria), Lennart Olsson (Sweden), Michael Oppenheimer (USA), Shlomit Paz (Israel), Juliane Petersen (Germany), Jan Petzold (Germany), Swantje Preuschmann (Germany), Mohammad Feisal Rahman (Bangladesh), Joeri Rogelj (Austria/Belgium), Hanna Scheuffele (Germany), Carl-Friedrich Schleussner (Germany), Daniel Scott (Canada), Roland Séférian (France), Jana Sillmann (Germany/Norway), Chandni Singh (India), Raphael Slade (UK), Kimberly Stephenson (Jamaica), Tannecia Stephenson (Jamaica), Mouhamadou B. Sylla (Senegal), Mark Tebboth (UK), Petra Tschakert (Australia/Austria), Robert Vautard (France), Richard Wartenburger (Switzerland/Germany), Michael Wehner (USA), Nora M. Weyer (Germany), Felicia Whyte (Jamaica), Gary Yohe (USA), Xuebin Zhang (Canada), Robert B. Zougmoré (Burkina Faso/Mali)

Review Editors:
Jose Antonio Marengo (Brazil/Peru), Joy Pereira (Malaysia), Boris Sherstyukov (Russian Federation)

Chapter Scientist:
Tania Guillén Bolaños (Germany/Nicaragua)

This chapter should be cited as:
Hoegh-Guldberg, O., D. Jacob, M. Taylor, M. Bindi, S. Brown, I. Camilloni, A. Diedhiou, R. Djalante, K.L. Ebi, F. Engelbrecht, J. Guiot, Y. Hijioka, S. Mehrotra, A. Payne, S.I. Seneviratne, A. Thomas, R. Warren, and G. Zhou, 2018: Impacts of 1.5°C Global Warming on Natural and Human Systems. In: *Global Warming of 1.5°C. An IPCC Special Report on the impacts of global warming of 1.5°C above pre-industrial levels and related global greenhouse gas emission pathways, in the context of strengthening the global response to the threat of climate change, sustainable development, and efforts to eradicate poverty* [Masson-Delmotte, V., P. Zhai, H.-O. Pörtner, D. Roberts, J. Skea, P.R. Shukla, A. Pirani, W. Moufouma-Okia, C. Péan, R. Pidcock, S. Connors, J.B.R. Matthews, Y. Chen, X. Zhou, M.I. Gomis, E. Lonnoy, T. Maycock, M. Tignor, and T. Waterfield (eds.)]. Cambridge University Press, Cambridge, UK and New York, NY, USA, pp. 175-312. https://doi.org/10.1017/9781009157940.005.

Table of Contents

Executive Summary

This chapter builds on findings of AR5 and assesses new scientific evidence of changes in the climate system and the associated impacts on natural and human systems, with a specific focus on the magnitude and pattern of risks linked for global warming of 1.5°C above temperatures in the pre-industrial period. Chapter 3 explores observed impacts and projected risks to a range of natural and human systems, with a focus on how risk levels change from 1.5°C to 2°C of global warming. The chapter also revisits major categories of risk (Reasons for Concern, RFC) based on the assessment of new knowledge that has become available since AR5.

1.5°C and 2°C Warmer Worlds

The global climate has changed relative to the pre-industrial period, and there are multiple lines of evidence that these changes have had impacts on organisms and ecosystems, as well as on human systems and well-being (*high confidence*). The increase in global mean surface temperature (GMST), which reached 0.87°C in 2006–2015 relative to 1850–1900, has increased the frequency and magnitude of impacts (*high confidence*), strengthening evidence of how an increase in GMST of 1.5°C or more could impact natural and human systems (1.5°C versus 2°C). {3.3, 3.4, 3.5, 3.6, Cross-Chapter Boxes 6, 7 and 8 in this chapter}

Human-induced global warming has already caused multiple observed changes in the climate system (*high confidence*). Changes include increases in both land and ocean temperatures, as well as more frequent heatwaves in most land regions (*high confidence*). There is also *high confidence* that global warming has resulted in an increase in the frequency and duration of marine heatwaves. Further, there is substantial evidence that human-induced global warming has led to an increase in the frequency, intensity and/or amount of heavy precipitation events at the global scale (*medium confidence*), as well as an increased risk of drought in the Mediterranean region (*medium confidence*). {3.3.1, 3.3.2, 3.3.3, 3.3.4, Box 3.4}

Trends in intensity and frequency of some climate and weather extremes have been detected over time spans during which about 0.5°C of global warming occurred (*medium confidence*). This assessment is based on several lines of evidence, including attribution studies for changes in extremes since 1950. {3.2, 3.3.1, 3.3.2, 3.3.3, 3.3.4}

Several regional changes in climate are assessed to occur with global warming up to 1.5°C as compared to pre-industrial levels, including warming of extreme temperatures in many regions (*high confidence*), increases in frequency, intensity and/or amount of heavy precipitation in several regions (*high confidence*), and an increase in intensity or frequency of droughts in some regions (*medium confidence*). {3.3.1, 3.3.2, 3.3.3, 3.3.4, Table 3.2}

There is no single '1.5°C warmer world' (*high confidence*). In addition to the overall increase in GMST, it is important to consider the size and duration of potential overshoots in temperature. Furthermore, there are questions on how the stabilization of an increase in GMST of 1.5°C can be achieved, and how policies might be able to influence the resilience of human and natural systems, and the nature of regional and subregional risks. Overshooting poses large risks for natural and human systems, especially if the temperature at peak warming is high, because some risks may be long-lasting and irreversible, such as the loss of some ecosystems (*high confidence*). The rate of change for several types of risks may also have relevance, with potentially large risks in the case of a rapid rise to overshooting temperatures, even if a decrease to 1.5°C can be achieved at the end of the 21st century or later (*medium confidence*). If overshoot is to be minimized, the remaining equivalent CO_2 budget available for emissions is very small, which implies that large, immediate and unprecedented global efforts to mitigate greenhouse gases are required (*high confidence*). {3.2, 3.6.2, Cross-Chapter Box 8 in this chapter}

Robust[1] global differences in temperature means and extremes are expected if global warming reaches 1.5°C versus 2°C above the pre-industrial levels (*high confidence*). For oceans, regional surface temperature means and extremes are projected to be higher at 2°C compared to 1.5°C of global warming (*high confidence*). Temperature means and extremes are also projected to be higher at 2°C compared to 1.5°C in most land regions, with increases being 2–3 times greater than the increase in GMST projected for some regions (*high confidence*). Robust increases in temperature means and extremes are also projected at 1.5°C compared to present-day values (*high confidence*) {3.3.1, 3.3.2}. There are decreases in the occurrence of cold extremes, but substantial increases in their temperature, in particular in regions with snow or ice cover (*high confidence*) {3.3.1}.

Climate models project robust[1] differences in regional climate between present-day and global warming up to 1.5°C[2], and between 1.5°C and 2°C[2] (*high confidence*), depending on the variable and region in question (*high confidence*). Large, robust and widespread differences are expected for temperature extremes (*high confidence*). Regarding hot extremes, the strongest warming is expected to occur at mid-latitudes in the warm season (with increases of up to 3°C for 1.5°C of global warming, i.e., a factor of two) and at high latitudes in the cold season (with increases of up to 4.5°C at 1.5°C of global warming, i.e., a factor of three) (*high confidence*). The strongest warming of hot extremes is projected to occur in central and eastern North America, central and southern Europe, the Mediterranean region (including southern Europe, northern Africa and the Near East), western and central Asia, and southern Africa (*medium confidence*). The number of exceptionally hot days are expected to increase the most in the tropics, where interannual temperature variability is lowest; extreme heatwaves are thus projected to emerge earliest in these regions, and they are expected to already become widespread there at 1.5°C global warming (*high confidence*). Limiting global warming to 1.5°C instead of 2°C could result in around 420

---

[1] Robust is used here to mean that at least two thirds of climate models show the same sign of changes at the grid point scale, and that differences in large regions are statistically significant.

[2] Projected changes in impacts between different levels of global warming are determined with respect to changes in global mean near-surface air temperature.

million fewer people being frequently exposed to extreme heatwaves, and about 65 million fewer people being exposed to exceptional heatwaves, assuming constant vulnerability (*medium confidence*). {3.3.1, 3.3.2, Cross-Chapter Box 8 in this chapter}

Limiting global warming to 1.5°C would limit risks of increases in heavy precipitation events on a global scale and in several regions compared to conditions at 2°C global warming (*medium confidence*). The regions with the largest increases in heavy precipitation events for 1.5°C to 2°C global warming include: several high-latitude regions (e.g. Alaska/western Canada, eastern Canada/Greenland/Iceland, northern Europe and northern Asia); mountainous regions (e.g., Tibetan Plateau); eastern Asia (including China and Japan); and eastern North America (*medium confidence*). Tropical cyclones are projected to decrease in frequency but with an increase in the number of very intense cyclones (*limited evidence, low confidence*). Heavy precipitation associated with tropical cyclones is projected to be higher at 2°C compared to 1.5°C of global warming (*medium confidence*). Heavy precipitation, when aggregated at a global scale, is projected to be higher at 2°C than at 1.5°C of global warming (*medium confidence*) {3.3.3, 3.3.6}

Limiting global warming to 1.5°C is expected to substantially reduce the probability of extreme drought, precipitation deficits, and risks associated with water availability (i.e., water stress) in some regions (*medium confidence*). In particular, risks associated with increases in drought frequency and magnitude are projected to be substantially larger at 2°C than at 1.5°C in the Mediterranean region (including southern Europe, northern Africa and the Near East) and southern Africa (*medium confidence*). {3.3.3, 3.3.4, Box 3.1, Box 3.2}

Risks to natural and human systems are expected to be lower at 1.5°C than at 2°C of global warming (*high confidence*). This difference is due to the smaller rates and magnitudes of climate change associated with a 1.5°C temperature increase, including lower frequencies and intensities of temperature-related extremes. Lower rates of change enhance the ability of natural and human systems to adapt, with substantial benefits for a wide range of terrestrial, freshwater, wetland, coastal and ocean ecosystems (including coral reefs) (*high confidence*), as well as food production systems, human health, and tourism (*medium confidence*), together with energy systems and transportation (*low confidence*). {3.3.1, 3.4}

Exposure to multiple and compound climate-related risks is projected to increase between 1.5°C and 2°C of global warming with greater proportions of people both exposed and susceptible to poverty in Africa and Asia (*high confidence*). For global warming from 1.5°C to 2°C, risks across energy, food, and water sectors could overlap spatially and temporally, creating new – and exacerbating current – hazards, exposures, and vulnerabilities that could affect increasing numbers of people and regions (*medium confidence*). Small island states and economically disadvantaged populations are particularly at risk (*high confidence*). {3.3.1, 3.4.5.3, 3.4.5.6, 3.4.11, 3.5.4.9, Box 3.5}

Global warming of 2°C would lead to an expansion of areas with significant increases in runoff, as well as those affected by flood hazard, compared to conditions at 1.5°C (*medium confidence*). Global warming of 1.5°C would also lead to an expansion of the global land area with significant increases in runoff (*medium confidence*) and an increase in flood hazard in some regions (*medium confidence*) compared to present-day conditions. {3.3.5}

The probability of a sea-ice-free Arctic Ocean[3] during summer is substantially higher at 2°C compared to 1.5°C of global warming (*medium confidence*). Model simulations suggest that at least one sea-ice-free Arctic summer is expected every 10 years for global warming of 2°C, with the frequency decreasing to one sea-ice-free Arctic summer every 100 years under 1.5°C (*medium confidence*). An intermediate temperature overshoot will have no long-term consequences for Arctic sea ice coverage, and hysteresis is not expected (*high confidence*). {3.3.8, 3.4.4.7}

Global mean sea level rise (GMSLR) is projected to be around 0.1 m (0.04 – 0.16 m) less by the end of the 21st century in a 1.5°C warmer world compared to a 2°C warmer world (*medium confidence*). Projected GMSLR for 1.5°C of global warming has an indicative range of 0.26 – 0.77m, relative to 1986–2005, (*medium confidence*). A smaller sea level rise could mean that up to 10.4 million fewer people (based on the 2010 global population and assuming no adaptation) would be exposed to the impacts of sea level rise globally in 2100 at 1.5°C compared to at 2°C. A slower rate of sea level rise enables greater opportunities for adaptation (*medium confidence*). There is *high confidence* that sea level rise will continue beyond 2100. Instabilities exist for both the Greenland and Antarctic ice sheets, which could result in multi-meter rises in sea level on time scales of century to millennia. There is *medium confidence* that these instabilities could be triggered at around 1.5°C to 2°C of global warming. {3.3.9, 3.4.5, 3.6.3}

The ocean has absorbed about 30% of the anthropogenic carbon dioxide, resulting in ocean acidification and changes to carbonate chemistry that are unprecedented for at least the last 65 million years (*high confidence*). Risks have been identified for the survival, calcification, growth, development and abundance of a broad range of marine taxonomic groups, ranging from algae to fish, with substantial evidence of predictable trait-based sensitivities (*high confidence*). There are multiple lines of evidence that ocean warming and acidification corresponding to 1.5°C of global warming would impact a wide range of marine organisms and ecosystems, as well as sectors such as aquaculture and fisheries (*high confidence*). {3.3.10, 3.4.4}

Larger risks are expected for many regions and systems for global warming at 1.5°C, as compared to today, with adaptation required now and up to 1.5°C. However, risks would be larger at 2°C of warming and an even greater effort would be needed for adaptation to a temperature increase of that magnitude (*high confidence*). {3.4, Box 3.4, Box 3.5, Cross-Chapter Box 6 in this chapter}

---

[3]   Ice free is defined for the Special Report as when the sea ice extent is less than 106 km². Ice coverage less than this is considered to be equivalent to an ice-free Arctic Ocean for practical purposes in all recent studies.

Future risks at 1.5°C of global warming will depend on the mitigation pathway and on the possible occurrence of a transient overshoot (*high confidence*). The impacts on natural and human systems would be greater if mitigation pathways temporarily overshoot 1.5°C and return to 1.5°C later in the century, as compared to pathways that stabilize at 1.5°C without an overshoot (*high confidence*). The size and duration of an overshoot would also affect future impacts (e.g., irreversible loss of some ecosystems) (*high confidence*). Changes in land use resulting from mitigation choices could have impacts on food production and ecosystem diversity. {3.6.1, 3.6.2, Cross-Chapter Boxes 7 and 8 in this chapter}

Climate Change Risks for Natural and Human systems

Terrestrial and Wetland Ecosystems

Risks of local species losses and, consequently, risks of extinction are much less in a 1.5°C versus a 2°C warmer world (*high confidence*). The number of species projected to lose over half of their climatically determined geographic range at 2°C global warming (18% of insects, 16% of plants, 8% of vertebrates) is projected to be reduced to 6% of insects, 8% of plants and 4% of vertebrates at 1.5°C warming (*medium confidence*). Risks associated with other biodiversity-related factors, such as forest fires, extreme weather events, and the spread of invasive species, pests and diseases, would also be lower at 1.5°C than at 2°C of warming (*high confidence*), supporting a greater persistence of ecosystem services. {3.4.3, 3.5.2}

Constraining global warming to 1.5°C, rather than to 2°C and higher, is projected to have many benefits for terrestrial and wetland ecosystems and for the preservation of their services to humans (*high confidence*). Risks for natural and managed ecosystems are higher on drylands compared to humid lands. The global terrestrial land area projected to be affected by ecosystem transformations (13%, interquartile range 8–20%) at 2°C is approximately halved at 1.5°C global warming to 4% (interquartile range 2–7%) (*medium confidence*). Above 1.5°C, an expansion of desert terrain and vegetation would occur in the Mediterranean biome (*medium confidence*), causing changes unparalleled in the last 10,000 years (*medium confidence*). {3.3.2.2, 3.4.3.2, 3.4.3.5, 3.4.6.1, 3.5.5.10, Box 4.2}

Many impacts are projected to be larger at higher latitudes, owing to mean and cold-season warming rates above the global average (*medium confidence*). High-latitude tundra and boreal forest are particularly at risk, and woody shrubs are already encroaching into tundra (*high confidence*) and will proceed with further warming. Constraining warming to 1.5°C would prevent the thawing of an estimated permafrost area of 1.5 to 2.5 million km^2 over centuries compared to thawing under 2°C (*medium confidence*). {3.3.2, 3.4.3, 3.4.4}

Ocean Ecosystems

Ocean ecosystems are already experiencing large-scale changes, and critical thresholds are expected to be reached at 1.5°C and higher levels of global warming (*high confidence*). In the transition to 1.5°C of warming, changes to water temperatures are expected to drive some species (e.g., plankton, fish) to relocate to higher latitudes and cause novel ecosystems to assemble (*high confidence*). Other ecosystems (e.g., kelp forests, coral reefs) are relatively less able to move, however, and are projected to experience high rates of mortality and loss (*very high confidence*). For example, multiple lines of evidence indicate that the majority (70–90%) of warm water (tropical) coral reefs that exist today will disappear even if global warming is constrained to 1.5°C (*very high confidence*). {3.4.4, Box 3.4}

Current ecosystem services from the ocean are expected to be reduced at 1.5°C of global warming, with losses being even greater at 2°C of global warming (*high confidence*). The risks of declining ocean productivity, shifts of species to higher latitudes, damage to ecosystems (e.g., coral reefs, and mangroves, seagrass and other wetland ecosystems), loss of fisheries productivity (at low latitudes), and changes to ocean chemistry (e.g., acidification, hypoxia and dead zones) are projected to be substantially lower when global warming is limited to 1.5°C (*high confidence*). {3.4.4, Box 3.4}

Water Resources

The projected frequency and magnitude of floods and droughts in some regions are smaller under 1.5°C than under 2°C of warming (*medium confidence*). Human exposure to increased flooding is projected to be substantially lower at 1.5°C compared to 2°C of global warming, although projected changes create regionally differentiated risks (*medium confidence*). The differences in the risks among regions are strongly influenced by local socio-economic conditions (*medium confidence*). {3.3.4, 3.3.5, 3.4.2}

Risks of water scarcity are projected to be greater at 2°C than at 1.5°C of global warming in some regions (*medium confidence*). Depending on future socio-economic conditions, limiting global warming to 1.5°C, compared to 2°C, may reduce the proportion of the world population exposed to a climate change-induced increase in water stress by up to 50%, although there is considerable variability between regions (*medium confidence*). Regions with particularly large benefits could include the Mediterranean and the Caribbean (*medium confidence*). Socio-economic drivers, however, are expected to have a greater influence on these risks than the changes in climate (*medium confidence*). {3.3.5, 3.4.2, Box 3.5}

Land Use, Food Security and Food Production Systems

Limiting global warming to 1.5°C, compared with 2°C, is projected to result in smaller net reductions in yields of maize, rice, wheat, and potentially other cereal crops, particularly in sub-Saharan Africa, Southeast Asia, and Central and South America; and in the CO_2-dependent nutritional quality of rice and wheat (*high confidence*). A loss of 7–10% of rangeland livestock globally is projected for approximately 2°C of warming, with considerable economic consequences for many communities and regions (*medium confidence*). {3.4.6, 3.6, Box 3.1, Cross-Chapter Box 6 in this chapter}

Reductions in projected food availability are larger at 2°C than at 1.5°C of global warming in the Sahel, southern Africa, the Mediterranean, central Europe and the Amazon (*medium confidence*). This suggests a transition from medium to high risk of regionally differentiated impacts on food security between 1.5°C and 2°C (*medium confidence*). Future economic and trade environments and their response to changing food availability (*medium confidence*) are important potential adaptation options for reducing hunger risk in low- and middle-income countries. {Cross-Chapter Box 6 in this chapter}

Fisheries and aquaculture are important to global food security but are already facing increasing risks from ocean warming and acidification *(medium confidence)*. These risks are projected to increase at 1.5°C of global warming and impact key organisms such as fin fish and bivalves (e.g., oysters), especially at low latitudes (*medium confidence*). Small-scale fisheries in tropical regions, which are very dependent on habitat provided by coastal ecosystems such as coral reefs, mangroves, seagrass and kelp forests, are expected to face growing risks at 1.5°C of warming because of loss of habitat (*medium confidence*). Risks of impacts and decreasing food security are projected to become greater as global warming reaches beyond 1.5°C and both ocean warming and acidification increase, with substantial losses likely for coastal livelihoods and industries (e.g., fisheries and aquaculture) (*medium to high confidence*). {3.4.4, 3.4.5, 3.4.6, Box 3.1, Box 3.4, Box 3.5, Cross-Chapter Box 6 in this chapter}

Land use and land-use change emerge as critical features of virtually all mitigation pathways that seek to limit global warming to 1.5°C (*high confidence*). Most least-cost mitigation pathways to limit peak or end-of-century warming to 1.5°C make use of carbon dioxide removal (CDR), predominantly employing significant levels of bioenergy with carbon capture and storage (BECCS) and/or afforestation and reforestation (AR) in their portfolio of mitigation measures (*high confidence*). {Cross-Chapter Box 7 in this chapter}

Large-scale deployment of BECCS and/or AR would have a far-reaching land and water footprint (*high confidence*). Whether this footprint would result in adverse impacts, for example on biodiversity or food production, depends on the existence and effectiveness of measures to conserve land carbon stocks, measures to limit agricultural expansion in order to protect natural ecosystems, and the potential to increase agricultural productivity (*medium agreement*). In addition, BECCS and/or AR would have substantial direct effects on regional climate through biophysical feedbacks, which are generally not included in Integrated Assessments Models (*high confidence*). {3.6.2, Cross-Chapter Boxes 7 and 8 in this chapter}

The impacts of large-scale CDR deployment could be greatly reduced if a wider portfolio of CDR options were deployed, if a holistic policy for sustainable land management were adopted, and if increased mitigation efforts were employed to strongly limit the demand for land, energy and material resources, including through lifestyle and dietary changes (*medium confidence*). In particular, reforestation could be associated with significant co-benefits if implemented in a manner than helps restore natural ecosystems (*high confidence*). {Cross-Chapter Box 7 in this chapter}

Human Health, Well-Being, Cities and Poverty

Any increase in global temperature (e.g., +0.5°C) is projected to affect human health, with primarily negative consequences (*high confidence*). Lower risks are projected at 1.5°C than at 2°C for heat-related morbidity and mortality (*very high confidence*), and for ozone-related mortality if emissions needed for ozone formation remain high (*high confidence*). Urban heat islands often amplify the impacts of heatwaves in cities (*high confidence*). Risks for some vector-borne diseases, such as malaria and dengue fever are projected to increase with warming from 1.5°C to 2°C, including potential shifts in their geographic range (*high confidence*). Overall for vector-borne diseases, whether projections are positive or negative depends on the disease, region and extent of change (*high confidence*). Lower risks of undernutrition are projected at 1.5°C than at 2°C (*medium confidence*). Incorporating estimates of adaptation into projections reduces the magnitude of risks (*high confidence*). {3.4.7, 3.4.7.1, 3.4.8, 3.5.5.8}

Global warming of 2°C is expected to pose greater risks to urban areas than global warming of 1.5°C (*medium confidence*). The extent of risk depends on human vulnerability and the effectiveness of adaptation for regions (coastal and non-coastal), informal settlements and infrastructure sectors (such as energy, water and transport) (*high confidence*). {3.4.5, 3.4.8}

Poverty and disadvantage have increased with recent warming (about 1°C) and are expected to increase for many populations as average global temperatures increase from 1°C to 1.5°C and higher (*medium confidence*). Outmigration in agricultural-dependent communities is positively and statistically significantly associated with global temperature (*medium confidence*). Our understanding of the links of 1.5°C and 2°C of global warming to human migration are limited and represent an important knowledge gap. {3.4.10, 3.4.11, 5.2.2, Table 3.5}

Key Economic Sectors and Services

Risks to global aggregated economic growth due to climate change impacts are projected to be lower at 1.5°C than at 2°C by the end of this century (*medium confidence*). {3.5.2, 3.5.3}

The largest reductions in economic growth at 2°C compared to 1.5°C of warming are projected for low- and middle-income countries and regions (the African continent, Southeast Asia, India, Brazil and Mexico) (*low to medium confidence*). Countries in the tropics and Southern Hemisphere subtropics are projected to experience the largest impacts on economic growth due to climate change should global warming increase from 1.5°C to 2°C (*medium confidence*). {3.5}

Global warming has already affected tourism, with increased risks projected under 1.5°C of warming in specific geographic regions and for seasonal tourism including sun, beach and snow sports destinations (*very high confidence*). Risks will be lower for tourism markets that are less climate sensitive, such as gaming and large hotel-based activities (*high confidence*). Risks for coastal tourism, particularly in subtropical and tropical regions, will increase with temperature-related degradation (e.g., heat extremes, storms) or loss of beach and coral reef assets (*high confidence*). {3.3.6, 3.4.4.12, 3.4.9.1, Box 3.4}

Small Islands, and Coastal and Low-lying areas

Small islands are projected to experience multiple inter-related risks at 1.5°C of global warming that will increase with warming of 2°C and higher levels (*high confidence*). Climate hazards at 1.5°C are projected to be lower compared to those at 2°C (*high confidence*). Long-term risks of coastal flooding and impacts on populations, infrastructures and assets (*high confidence*), freshwater stress (*medium confidence*), and risks across marine ecosystems (*high confidence*) and critical sectors (*medium confidence*) are projected to increase at 1.5°C compared to present-day levels and increase further at 2°C, limiting adaptation opportunities and increasing loss and damage (*medium confidence*). Migration in small islands (internally and internationally) occurs for multiple reasons and purposes, mostly for better livelihood opportunities (*high confidence*) and increasingly owing to sea level rise (*medium confidence*). {3.3.2.2, 3.3.6–9, 3.4.3.2, 3.4.4.2, 3.4.4.5, 3.4.4.12, 3.4.5.3, 3.4.7.1, 3.4.9.1, 3.5.4.9, Box 3.4, Box 3.5}

Impacts associated with sea level rise and changes to the salinity of coastal groundwater, increased flooding and damage to infrastructure, are projected to be critically important in vulnerable environments, such as small islands, low-lying coasts and deltas, at global warming of 1.5°C and 2°C (*high confidence*). Localized subsidence and changes to river discharge can potentially exacerbate these effects. Adaptation is already happening (*high confidence*) and will remain important over multi-centennial time scales. {3.4.5.3, 3.4.5.4, 3.4.5.7, 5.4.5.4, Box 3.5}

Existing and restored natural coastal ecosystems may be effective in reducing the adverse impacts of rising sea levels and intensifying storms by protecting coastal and deltaic regions (*medium confidence*). Natural sedimentation rates are expected to be able to offset the effect of rising sea levels, given the slower rates of sea level rise associated with 1.5°C of warming (*medium confidence*). Other feedbacks, such as landward migration of wetlands and the adaptation of infrastructure, remain important (*medium confidence*). {3.4.4.12, 3.4.5.4, 3.4.5.7}

Increased Reasons for Concern

There are multiple lines of evidence that since AR5 the assessed levels of risk increased for four of the five Reasons for Concern (RFCs) for global warming levels of up to 2°C (*high confidence*). The risk transitions by degrees of global warming are now: from high to very high between 1.5°C and 2°C for RFC1 (Unique and threatened systems) (*high confidence*); from moderate to high risk between 1°C and 1.5°C for RFC2 (Extreme weather events) (*medium confidence*); from moderate to high risk between 1.5°C and 2°C for RFC3 (Distribution of impacts) (*high confidence*); from moderate to high risk between 1.5°C and 2.5°C for RFC4 (Global aggregate impacts) (*medium confidence*); and from moderate to high risk between 1°C and 2.5°C for RFC5 (Large-scale singular events) (*medium confidence*). {3.5.2}

1. **The category 'Unique and threatened systems' (RFC1) display a transition from high to very high risk which is now located between 1.5°C and 2°C of global warming** as opposed to at 2.6°C of global warming in AR5, owing to new and multiple lines of evidence for changing risks for coral reefs, the Arctic and biodiversity in general (*high confidence*). {3.5.2.1}

2. **In 'Extreme weather events' (RFC2), the transition from moderate to high risk is now located between 1.0°C and 1.5°C of global warming,** which is very similar to the AR5 assessment but is projected with greater confidence (*medium confidence*). The impact literature contains little information about the potential for human society to adapt to extreme weather events, and hence it has not been possible to locate the transition from 'high' to 'very high' risk within the context of assessing impacts at 1.5°C versus 2°C of global warming. There is thus *low confidence* in the level at which global warming could lead to very high risks associated with extreme weather events in the context of this report. {3.5}

3. **With respect to the 'Distribution of impacts' (RFC3) a transition from moderate to high risk is now located between 1.5°C and 2°C of global warming,** compared with between 1.6°C and 2.6°C global warming in AR5, owing to new evidence about regionally differentiated risks to food security, water resources, drought, heat exposure and coastal submergence (*high confidence*). {3.5}

4. **In 'global aggregate impacts' (RFC4) a transition from moderate to high levels of risk is now located between 1.5°C and 2.5°C of global warming,** as opposed to at 3.6°C of warming in AR5, owing to new evidence about global aggregate economic impacts and risks to Earth's biodiversity (*medium confidence*). {3.5}

5. **Finally, 'large-scale singular events' (RFC5), moderate risk is now located at 1°C of global warming and high risk is located at 2.5°C of global warming,** as opposed to at 1.6°C (moderate risk) and around 4°C (high risk) in AR5, because of new observations and models of the West Antarctic ice sheet (*medium confidence*). {3.3.9, 3.5.2, 3.6.3}

3

3.1 About the Chapter

Chapter 3 uses relevant definitions of a potential 1.5°C warmer world from Chapters 1 and 2 and builds directly on their assessment of gradual versus overshoot scenarios. It interacts with information presented in Chapter 2 via the provision of specific details relating to the mitigation pathways (e.g., land-use changes) and their implications for impacts. Chapter 3 also includes information needed for the assessment and implementation of adaptation options (presented in Chapter 4), as well as the context for considering the interactions of climate change with sustainable development and for the assessment of impacts on sustainability, poverty and inequalities at the household to subregional level (presented in Chapter 5).

This chapter is necessarily transdisciplinary in its coverage of the climate system, natural and managed ecosystems, and human systems and responses, owing to the integrated nature of the natural and human experience. While climate change is acknowledged as a centrally important driver, it is not the only driver of risks to human and natural systems, and in many cases, it is the interaction between these two broad categories of risk that is important (Chapter 1).

The flow of the chapter, linkages between sections, a list of chapter- and cross-chapter boxes, and a content guide for reading according to focus or interest are given in Figure 3.1. Key definitions used in the chapter are collected in the Glossary. Confidence language is used throughout this chapter and likelihood statements (e.g., *likely*, *very likely*) are provided when there is *high confidence* in the assessment.

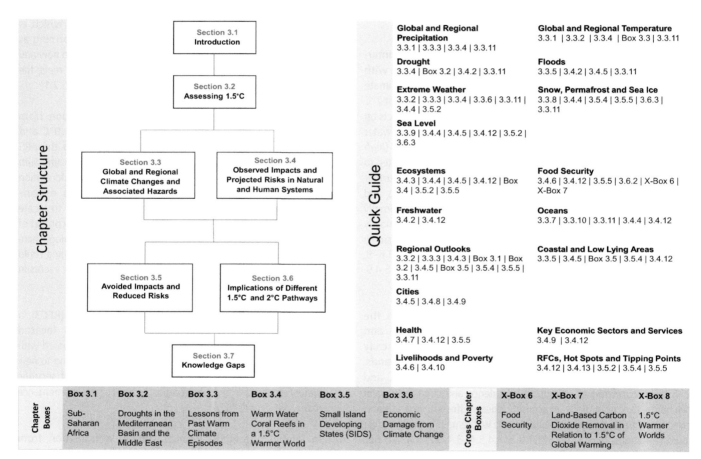

Figure 3.1 | Chapter 3 structure and quick guide.

The underlying literature assessed in Chapter 3 is broad and includes a large number of recent publications specific to assessments for 1.5°C of warming. The chapter also utilizes information covered in prior IPCC special reports, for example the Special Report on Managing the Risks of Extreme Events and Disasters to Advance Climate Change Adaptation (SREX; IPCC, 2012), and many chapters from the IPCC WGII Fifth Assessment Report (AR5) that assess impacts on natural and managed ecosystems and humans, as well as adaptation options (IPCC, 2014b). For this reason, the chapter provides information based

on a broad range of assessment methods. Details about the approaches used are presented in Section 3.2.

Section 3.3 gives a general overview of recent literature on observed climate change impacts as the context for projected future risks. With a few exceptions, the focus here is the analysis of transient responses at 1.5°C and 2°C of global warming, with simulations *of short-term stabilization scenarios* (Section 3.2) also assessed in some cases. In general, *long-term equilibrium stabilization responses* could not be assessed owing to a lack of data and analysis. A detailed analysis of detection and attribution is not provided but will be the focus of the next IPCC assessment report (AR6). Furthermore, possible interventions in the climate system through radiation modification measures, which are not tied to reductions of greenhouse gas emissions or concentrations, are not assessed in this chapter.

Understanding the observed impacts and projected risks of climate change is crucial to comprehending how the world is likely to change under global warming of 1.5°C above temperatures in the pre-industrial period (with reference to 2°C). Section 3.4 explores the new literature and updates the assessment of impacts and projected risks for a large number of natural and human systems. By also exploring adaptation opportunities, where the literature allows, the section prepares the reader for discussions in subsequent chapters about opportunities to tackle both mitigation and adaptation. The section is mostly globally focused because of limited research on regional risks and adaptation options at 1.5°C and 2°C. For example, the risks of 1.5°C and 2°C of warming in urban areas, as well as the risks of health outcomes under these two warming scenarios (e.g. climate-related diseases, air quality impacts and mental health problems), were not considered because of a lack of projections of how these risks might change in a 1.5°C or 2°C warmer world. In addition, the complexity of many interactions of climate change with drivers of poverty, along with a paucity of relevant studies, meant it was not possible to detect and attribute many dimensions of poverty and disadvantage to climate change. Even though there is increasing documentation of climate-related impacts on places where indigenous people live and where subsistence-oriented communities are found, relevant projections of the risks associated with warming of 1.5°C and 2°C are necessarily limited.

To explore avoided impacts and reduced risks at 1.5°C compared with at 2°C of global warming, the chapter adopts the AR5 'Reasons for Concern' aggregated projected risk framework (Section 3.5). Updates in terms of the aggregation of risks are informed by the most recent literature and the assessments offered in Sections 3.3 and 3.4, with a focus on the impacts at 2°C of warming that could potentially be avoided if warming were constrained to 1.5°C. Economic benefits that would be obtained (Section 3.5.3), climate change 'hotspots' that could be avoided or reduced (Section 3.5.4 as guided by the assessments of Sections 3.3, 3.4 and 3.5), and tipping points that could be circumvented (Section 3.5.5) at 1.5°C compared to higher degrees of global warming are all examined. The latter assessments are, however, constrained to regional analyses, and hence this particular section does not include an assessment of specific losses and damages.

Section 3.6 provides an overview on specific aspects of the mitigation pathways considered compatible with 1.5°C of global warming, including some scenarios involving temperature overshoot above 1.5°C global warming during the 21st century. Non-CO$_2$ implications and projected risks of mitigation pathways, such as changes to land use and atmospheric compounds, are presented and explored. Finally, implications for sea ice, sea level and permafrost beyond the end of the century are assessed.

The exhaustive assessment of literature specific to global warming of 1.5°C above the pre-industrial period, presented across all the

sections in Chapter 3, highlights knowledge gaps resulting from the heterogeneous information available across systems, regions and sectors. Some of these gaps are described in Section 3.7.

3.2   How are Risks at 1.5°C and Higher Levels of Global Warming Assessed in this Chapter?

The methods that are applied for assessing observed and projected changes in climate and weather are presented in Section 3.2.1, while those used for assessing the observed impacts on and projected risks to natural and managed systems, and to human settlements, are described in Section 3.2.2. Given that changes in climate associated with 1.5°C of global warming were not the focus of past IPCC reports, dedicated approaches based on recent literature that are specific to the present report are also described. Background on specific methodological aspects (climate model simulations available for assessments at 1.5°C global warming, attribution of observed changes in climate and their relevance for assessing projected changes at 1.5°C and 2°C global warming, and the propagation of uncertainties from climate forcing to impacts on ecosystems) are provided in the Supplementary Material 3.SM.

3.2.1   How are Changes in Climate and Weather at 1.5°C versus Higher Levels of Warming Assessed?

Evidence for the assessment of changes to climate at 1.5°C versus 2°C can be drawn both from observations and model projections. Global mean surface temperature (GMST) anomalies were about +0.87°C (±0.10°C likely range) above pre-industrial industrial (1850–1900) values in the 2006–2015 decade, with a recent warming of about 0.2°C (±0.10°C) per decade (Chapter 1). Human-induced global warming reached approximately 1°C (±0.2°C *likely* range) in 2017 (Chapter 1). While some of the observed trends may be due to internal climate variability, methods of detection and attribution can be applied to assess which part of the observed changes may be attributed to anthropogenic forcing (Bindoff et al., 2013b). Hence, evidence from attribution studies can be used to assess changes in the climate system that are already detectable at lower levels of global warming and would thus continue to change with a further 0.5°C or 1°C of global warming (see Supplementary Material 3.SM.1 and Sections 3.3.1, 3.3.2, 3.3.3, 3.3.4 and 3.3.11). A recent study identified significant changes in extremes for a 0.5°C difference in global warming based on the historical record (Schleussner et al., 2017). It should also be noted that attributed changes in extremes since 1950 that were reported in the IPCC AR5 report (IPCC, 2013) generally correspond to changes in global warming of about 0.5°C (see 3.SM.1)

Climate model simulations are necessary for the investigation of the response of the climate system to various forcings, in particular to forcings associated with higher levels of greenhouse gas concentrations. Model simulations include experiments with global and regional climate models, as well as impact models – driven with output from climate models – to evaluate the risk related to climate change for natural and human systems (Supplementary Material 3.SM.1). Climate model simulations were generally used in the context of particular 'climate scenarios' from previous IPCC reports (e.g., IPCC, 2007, 2013). This means that emissions scenarios (IPCC, 2000) were used to drive climate models, providing different projections for given emissions pathways. The results were consequently used in a 'storyline' framework, which presents the development of climate in the course of the 21st century and beyond for a given emissions pathway. Results were assessed for different time slices within the model projections such as 2016–2035 ('near term', which is slightly below a global warming of 1.5°C according to most scenarios, Kirtman et al., 2013), 2046–2065 (mid-21st century, Collins et al., 2013), and 2081–2100 (end of 21st century, Collins et al., 2013). Given that this report focuses on climate change for a given mean global temperature response (1.5°C or 2°C), methods of analysis had to be developed and/or adapted from previous studies in order to provide assessments for the specific purposes here.

A major challenge in assessing climate change under 1.5°C, or 2°C (and higher levels), of global warming pertains to the **definition of a '1.5°C or 2°C climate projection'** (see also Cross-Chapter Box 8 in this chapter). Resolving this challenge includes the following considerations:

A. The need to distinguish between (i) **transient climate responses** (i.e., those that 'pass through' 1.5°C or 2°C of global warming), (ii) **short-term stabilization responses** (i.e., scenarios for the late 21st century that result in stabilization at a mean global warming of 1.5°C or 2°C by 2100), and (iii) **long-term equilibrium stabilization responses** (i.e., those occurring after several millennia once climate (temperature) equilibrium at 1.5°C or 2°C is reached). These responses can be very different in terms of climate variables and the inertia associated with a given climate forcing. A striking example is sea level rise (SLR). In this case, projected increases within the 21st century are minimally dependent on the scenario considered, yet they stabilize at very different levels for a long-term warming of 1.5°C versus 2°C (Section 3.3.9).

B. The '1.5°C or 2°C emissions scenarios' presented in Chapter 2 are targeted to hold warming below 1.5°C or 2°C with a certain probability (generally two-thirds) over the course, or at the end, of the 21st century. These scenarios should be seen as the operationalization of 1.5°C or 2°C warmer worlds. However, when these emission scenarios are used to drive climate models, some of the resulting simulations lead to warming above these respective thresholds (typically with a probability of one-third, see Chapter 2 and Cross-Chapter Box 8 in this chapter). This is due both to discrepancies between models and to internal climate variability. For this reason, the climate outcome for any of these scenarios, even those excluding an overshoot (see next point, C.), include some probability of reaching a global climate warming of more than 1.5°C or 2°C. Hence, a comprehensive assessment of climate risks associated with '1.5°C or 2°C climate scenarios' needs to include consideration of higher levels of warming (e.g., up to 2.5°C to 3°C, see Chapter 2 and Cross-Chapter Box 8 in this chapter).

C. Most of the '1.5°C scenarios', and some of the '2°C emissions scenarios' presented in Chapter 2 include a temperature overshoot during the course of the 21st century. This means that median temperature projections under these scenarios exceed the target warming levels over the course of the century (typically 0.5°C–1°C higher than the respective target levels at most), before warming returns to below 1.5°C or 2°C by 2100. During the overshoot phase, impacts would therefore correspond to higher transient temperature increases than 1.5°C or 2°C. For this reason, impacts of transient responses at these higher warming levels are also partly addressed in Cross-Chapter Box 8 in this chapter (on a 1.5°C warmer world), and some analyses for changes in extremes are also presented for higher levels of warming in Section 3.3 (Figures 3.5, 3.6, 3.9, 3.10, 3.12 and 3.13). Most importantly, different overshoot scenarios may have very distinct impacts depending on (i) the peak temperature of the overshoot, (ii) the length of the overshoot period, and (iii) the associated rate of change in global temperature over the time period of the overshoot. While some of these issues are briefly addressed in Sections 3.3 and 3.6, and in the Cross-Chapter Box 8, the definition of overshoot and related questions will need to be more comprehensively addressed in the IPCC AR6 report.

D. The levels of global warming that are the focus of this report (1.5°C and 2°C) are measured relative to the pre-industrial period. This definition requires an agreement on the exact reference time period (for 0°C of warming) and the time frame over which the global warming is assessed, typically 20 to 30 years in length. As discussed in Chapter 1, a climate with 1.5°C global warming is one in which temperatures averaged over a multi-decade time scale are 1.5°C above those in the pre-industrial reference period. Greater detail is provided in Cross-Chapter Box 8 in this chapter. Inherent to this is the observation that the mean temperature of a '1.5°C warmer world' can be regionally and temporally much higher (e.g., with regional annual temperature extremes involving warming of more than 6°C; see Section 3.3 and Cross-Chapter Box 8 in this chapter).

E. The interference of factors unrelated to greenhouse gases with mitigation pathways can strongly affect regional climate. For example, biophysical feedbacks from changes in land use and irrigation (e.g., Hirsch et al., 2017; Thiery et al., 2017), or projected changes in short-lived pollutants (e.g., Z. Wang et al., 2017), can have large influences on local temperatures and climate conditions. While these effects are not explicitly integrated into the scenarios developed in Chapter 2, they may affect projected changes in climate under 1.5°C of global warming. These issues are addressed in more detail in Section 3.6.2.2.

The assessment presented in the current chapter largely focuses on the analysis of **transient responses in climate at 1.5°C versus 2°C** and higher levels of global warming (see point A. above and Section 3.3). It generally uses the empirical scaling relationship (ESR) approach (Seneviratne et al., 2018c), also termed the 'time sampling' approach (James et al., 2017), which consists of sampling the response at 1.5°C and other levels of global warming from all available global climate model scenarios for the 21st century (e.g., Schleussner et al., 2016b;

Seneviratne et al., 2016; Wartenburger et al., 2017). The ESR approach focuses more on the derivation of a continuous relationship, while the term 'time sampling' is more commonly used when comparing a limited number of warming levels (e.g., 1.5°C versus 2°C). A similar approach in the case of regional climate model (RCM) simulations consists of sampling the RCM model output corresponding to the time frame at which the driving general circulation model (GCM) reaches the considered temperature level, for example, as done within IMPACT2C (Jacob and Solman, 2017), see description in Vautard et al. (2014). As an alternative to the ESR or time sampling approach, pattern scaling may be used. Pattern scaling is a statistical approach that describes relationships of specific climate responses as a function of global temperature change. Some assessments presented in this chapter are based on this method. The disadvantage of pattern scaling, however, is that the relationship may not perfectly emulate the models' responses at each location and for each global temperature level (James et al., 2017). Expert judgement is a third methodology that can be used to assess probable changes at 1.5°C or 2°C of global warming by combining changes that have been attributed to the observed time period (corresponding to warming of 1°C or less if assessed over a shorter period) with known projected changes at 3°C or 4°C above pre-industrial temperatures (Supplementary Material 3.SM.1). In order to assess effects induced by a 0.5°C difference in global warming, the historical record can be used at first approximation as a proxy, meaning that conditions are compared for two periods that have a 0.5°C difference in GMST warming (such as 1991–2010 and 1960–1979, e.g., Schleussner et al., 2017). This in particular also applies to attributed changes in extremes since 1950 that were reported in the IPCC AR5 report (IPCC, 2013; see also 3.SM.1). Using observations, however, it is not possible to account for potential non-linear changes that could occur above 1°C of global warming or as 1.5°C of warming is reached.

In some cases, assessments of **short-term stabilization responses** are also presented, derived using a subset of model simulations that reach a given temperature limit by 2100, or driven by sea surface temperature (SST) values consistent with such scenarios. This includes new results from the 'Half a degree additional warming, prognosis and projected impacts' (HAPPI) project (Section 1.5.2; Mitchell et al., 2017). Notably, there is evidence that for some variables (e.g., temperature and precipitation extremes), responses after short-term stabilization (i.e., approximately equivalent to the RCP2.6 scenario) are very similar to the transient response of higher-emissions scenarios (Seneviratne et al., 2016, 2018c; Wartenburger et al., 2017; Tebaldi and Knutti, 2018). This is, however, less the case for mean precipitation (e.g., Pendergrass et al., 2015), for which other aspects of the emissions scenarios appear relevant.

For the assessment of **long-term equilibrium stabilization responses**, this chapter uses results from existing simulations where available (e.g., for sea level rise), although the available data for this type of projection is limited for many variables and scenarios and will need to be addressed in more depth in the IPCC AR6 report.

Supplementary Material 3.SM.1 of this chapter includes further details of the climate models and associated simulations that were used to support the present assessment, as well as a background on detection

and attribution approaches of relevance to assessing changes in climate at 1.5°C of global warming.

3.2.2   How are Potential Impacts on Ecosystems Assessed at 1.5°C versus Higher Levels of Warming?

Considering that the impacts observed so far are for a global warming lower than 1.5°C (generally up to the 2006–2015 decade, i.e., for a global warming of 0.87°C or less; see above), direct information on the impacts of a global warming of 1.5°C is not yet available. The global distribution of observed impacts shown in AR5 (Cramer et al., 2014), however, demonstrates that methodologies now exist which are capable of detecting impacts on systems strongly influenced by factors (e.g., urbanization and human pressure in general) or where climate may play only a secondary role in driving impacts. Attribution of observed impacts to greenhouse gas forcing is more rarely performed, but a recent study (Hansen and Stone, 2016) shows that most of the detected temperature-related impacts that were reported in AR5 (Cramer et al., 2014) can be attributed to anthropogenic climate change, while the signals for precipitation-induced responses are more ambiguous.

One simple approach for assessing possible impacts on natural and managed systems at 1.5°C versus 2°C consists of identifying impacts of a global 0.5°C of warming in the observational record (e.g., Schleussner et al., 2017) assuming that the impacts would scale linearly for higher levels of warming (although this may not be appropriate). Another approach is to use conclusions from analyses of past climates combined with modelling of the relationships between climate drivers and natural systems (Box 3.3). A more complex approach relies on laboratory or field experiments (Dove et al., 2013; Bonal et al., 2016), which provide useful information on the causal effect of a few factors, which can be as diverse as climate, greenhouse gases (GHG), management practices, and biological and ecological variables, on specific natural systems that may have unusual physical and chemical characteristics (e.g., Fabricius et al., 2011; Allen et al., 2017). This last approach can be important in helping to develop and calibrate impact mechanisms and models through empirical experimentation and observation.

Risks for natural and human systems are often assessed with impact models where climate inputs are provided by representative concentration pathway (RCP)-based climate projections. The number of studies projecting impacts at 1.5°C or 2°C of global warming has increased in recent times (see Section 3.4), even if the four RCP scenarios used in AR5 are not strictly associated with these levels of global warming. Several approaches have been used to extract the required climate scenarios, as described in Section 3.2.1. As an example, Schleussner et al. (2016b) applied a time sampling (or ESR) approach, described in Section 3.2.1, to estimate the differential effect of 1.5°C and 2°C of global warming on water availability and impacts on agriculture using an ensemble of simulations under the RCP8.5 scenario. As a further example using a different approach, Iizumi et al. (2017) derived a 1.5°C scenario from simulations with a crop model using an interpolation between the no-change (approximately 2010) conditions and the RCP2.6 scenario (with a global warming of 1.8°C in 2100), and they derived the corresponding 2°C scenario from RCP2.6 and RCP4.5 simulations in 2100. The Inter-Sectoral Impact Model

Integration and Intercomparison Project Phase 2 (ISIMIP2; Frieler et al., 2017) extended this approach to investigate a number of sectoral impacts on terrestrial and marine ecosystems. In most cases, risks are assessed by impact models coupled offline to climate models after bias correction, which may modify long-term trends (Grillakis et al., 2017).

Assessment of local impacts of climate change necessarily involves a change in scale, such as from the global scale to that of natural or human systems (Frieler et al., 2017; Reyer et al., 2017d; Jacob et al., 2018). An appropriate method of downscaling (Supplementary Material 3.SM.1) is crucial for translating perspectives on 1.5°C and 2°C of global warming to scales and impacts relevant to humans and ecosystems. A major challenge associated with this requirement is the correct reproduction of the variance of local to regional changes, as well as the frequency and amplitude of extreme events (Vautard et al., 2014). In addition, maintaining physical consistency between downscaled variables is important but challenging (Frost et al., 2011).

Another major challenge relates to the propagation of the uncertainties at each step of the methodology, from the global forcings to the global climate and from regional climate to impacts at the ecosystem level, considering local disturbances and local policy effects. The risks for natural and human systems are the result of complex combinations of global and local drivers, which makes quantitative uncertainty analysis difficult. Such analyses are partly done using multimodel approaches, such as multi-climate and multi-impact models (Warszawski et al., 2013, 2014; Frieler et al., 2017). In the case of crop projections, for example, the majority of the uncertainty is caused by variation among crop models rather than by downscaling outputs of the climate models used (Asseng et al., 2013). Error propagation is an important issue for coupled models. Dealing correctly with uncertainties in a robust probabilistic model is particularly important when considering the potential for relatively small changes to affect the already small signal associated with 0.5°C of global warming (Supplementary Material 3.SM.1). The computation of an impact per unit of climatic change, based either on models or on data, is a simple way to present the probabilistic ecosystem response while taking into account the various sources of uncertainties (Fronzek et al., 2011).

In summary, in order to assess risks at 1.5°C and higher levels of global warming, several things need to be considered. Projected climates under 1.5°C of global warming differ depending on temporal aspects and emission pathways. Considerations include whether global temperature is (i) temporarily at this level (i.e., is a transient phase on its way to higher levels of warming), (ii) arrives at 1.5°C, with or without overshoot, after stabilization of greenhouse gas concentrations, or (iii) is at this level as part of long-term climate equilibrium (complete only after several millennia). Assessments of impacts of 1.5°C of warming are generally based on climate simulations for these different possible pathways. Most existing data and analyses focus on transient impacts (i). Fewer data are available for dedicated climate model simulations that are able to assess pathways consistent with (ii), and very few data are available for the assessment of changes at climate equilibrium (iii). In some cases, inferences regarding the impacts of further warming of 0.5°C above present-day temperatures (i.e., 1.5°C of global warming) can also be drawn from observations of similar sized changes (0.5°C) that have occurred in the past, such as during the last 50 years.

However, impacts can only be partly inferred from these types of observations, given the strong possibility of non-linear changes, as well as lag effects for some climate variables (e.g., sea level rise, snow and ice melt). For the impact models, three challenges are noted about the coupling procedure: (i) the bias correction of the climate model, which may modify the simulated response of the ecosystem, (ii) the necessity to downscale the climate model outputs to reach a pertinent scale for the ecosystem without losing physical consistency of the downscaled climate fields, and (iii) the necessity to develop an integrated study of the uncertainties.

3.3   Global and Regional Climate Changes and Associated Hazards

This section provides the assessment of changes in climate at 1.5°C of global warming relative to changes at higher global mean temperatures. Section 3.3.1 provides a brief overview of changes to global climate. Sections 3.3.2–3.3.11 provide assessments for specific aspects of the climate system, including regional assessments for temperature (Section 3.3.2) and precipitation (Section 3.3.3) means and extremes. Analyses of regional changes are based on the set of regions displayed in Figure 3.2. A synthesis of the main conclusions of this section is provided in Section 3.3.11. The section builds upon assessments from the IPCC AR5 WGI report (Bindoff et al., 2013a; Christensen et al., 2013; Collins et al., 2013; Hartmann et al., 2013; IPCC, 2013) and Chapter 3 of the IPCC Special Report on Managing the Risks of Extreme Events and Disasters to Advance Climate Change Adaptation (SREX; Seneviratne et al., 2012), as well as a substantial body of new literature related to projections of climate at 1.5°C and 2°C of warming above the pre-industrial period (e.g., Vautard et al., 2014; Fischer and Knutti, 2015; Schleussner et al., 2016b, 2017; Seneviratne et al., 2016, 2018c; Déqué et al., 2017; Maule et al., 2017; Mitchell et al., 2017, 2018a; Wartenburger et al., 2017; Zaman et al., 2017; Betts et al., 2018; Jacob et al., 2018; Kharin et al., 2018; Wehner et al., 2018b). The main assessment methods are as already detailed in Section 3.2.

3.3.1   Global Changes in Climate

There is *high confidence* that the increase in global mean surface temperature (GMST) has reached 0.87°C (±0.10°C *likely* range) above pre-industrial values in the 2006–2015 decade (Chapter 1). AR5 assessed that the globally averaged temperature (combined over land and ocean) displayed a warming of about 0.85°C [0.65°C to 1.06°C] during the period 1880–2012, with a large fraction of the detected global warming being attributed to anthropogenic forcing (Bindoff et al., 2013a; Hartmann et al., 2013; Stocker et al., 2013). While new evidence has highlighted that sampling biases and the choice of approaches used to estimate GMST (e.g., using water versus air temperature over oceans and using model simulations versus observations-based estimates) can affect estimates of GMST increase (Richardson et al., 2016; see also Supplementary Material 3.SM.2), the present assessment is consistent with that of AR5 regarding a detectable and dominant effect of anthropogenic forcing on observed trends in global temperature (also confirmed in Ribes et al., 2017). As highlighted in Chapter 1, human-induced warming

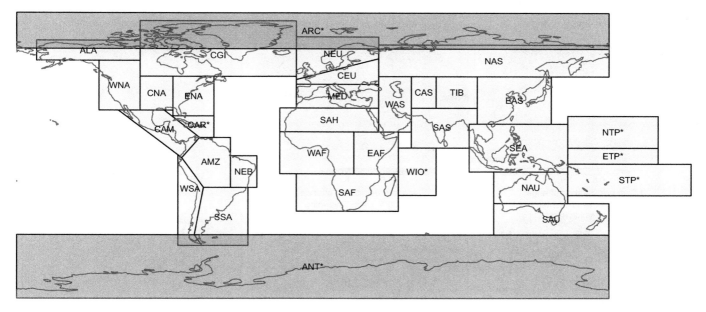

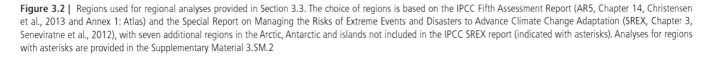

Abbreviation	Name	Abbreviation	Name	Abbreviation	Name	Abbreviation	Name
ALA	Alaska/N.W. Canada	CNA	Central North America	NEU	North Europe	TIB	Tibetan Plateau
AMZ	Amazon	EAF	East Africa	NTP*	Pacific Islands region[2]	WAF	West Africa
ANT*	Antarctica	EAS	East Asia	SAF	Southern Africa	WAS	West Asia
ARC*	Arctic	ENA	East North America	SAH	Sahara	WIO*	West Indian Ocean
CAM	Central America/Mexico	ETP*	Pacific Islands region[3]	SAS	South Asia	WNA	West North America
CAR*	small islands regions Caribbean	MED	South Europe/Mediterranean	SAU	South Australia/New Zealand	WSA	West Coast South America
CAS	Central Asia	NAS	North Asia	SEA	Southeast Asia		
CEU	Central Europe	NAU	North Australia	SSA	Southeastern South America		
CGI	Canada/Greenland/Iceland	NEB	North−East Brazil	STP*	Southern Topical Pacific		

Figure 3.2 | Regions used for regional analyses provided in Section 3.3. The choice of regions is based on the IPCC Fifth Assessment Report (AR5, Chapter 14, Christensen et al., 2013 and Annex 1: Atlas) and the Special Report on Managing the Risks of Extreme Events and Disasters to Advance Climate Change Adaptation (SREX, Chapter 3, Seneviratne et al., 2012), with seven additional regions in the Arctic, Antarctic and islands not included in the IPCC SREX report (indicated with asterisks). Analyses for regions with asterisks are provided in the Supplementary Material 3.SM.2

reached approximately 1°C (±0.2°C *likely* range) in 2017. More background on recent observed trends in global climate is provided in the Supplementary Material 3.SM.2.

A global warming of 1.5°C implies higher mean temperatures compared to during pre-industrial times in almost all locations, both on land and in oceans (*high confidence*) (Figure 3.3). In addition, a global warming of 2°C versus 1.5°C results in robust differences in the mean temperatures in almost all locations, both on land and in the ocean (*high confidence*). The land–sea contrast in warming is important and implies particularly large changes in temperature over land, with mean warming of more than 1.5°C in most land regions (*high confidence;* see Section 3.3.2 for more details). The largest increase in mean temperature is found in the high latitudes of the Northern Hemisphere (*high confidence;* Figure 3.3, see Section 3.3.2 for more details). Projections for precipitation are more uncertain, but they highlight robust increases in mean precipitation in the Northern Hemisphere high latitudes at 1.5°C global warming

versus pre-industrial conditions, as well as at 2°C global warming versus pre-industrial conditions (*high confidence*) (Figure 3.3). There are consistent but less robust signals when comparing changes in mean precipitation at 2°C versus 1.5°C of global warming. Hence, it is assessed that there is *medium confidence* in an increase of mean precipitation in high-latitudes at 2°C versus 1.5°C of global warming (Figure 3.3). For droughts, changes in evapotranspiration and precipitation timing are also relevant (see Section 3.3.4). Figure 3.4 displays changes in temperature extremes (the hottest daytime temperature of the year, TXx, and the coldest night-time temperature of the year, TNn) and heavy precipitation (the annual maximum 5-day precipitation, Rx5day). These analyses reveal distinct patterns of changes, with the largest changes in TXx occurring on mid-latitude land and the largest changes in TNn occurring at high latitudes (both on land and in oceans). Differences in TXx and TNn compared to pre-industrial climate are robust at both global warming levels. Differences in TXx and TNn at 2°C versus 1.5°C of global warming are robust across most of the globe. Changes in heavy precipitation

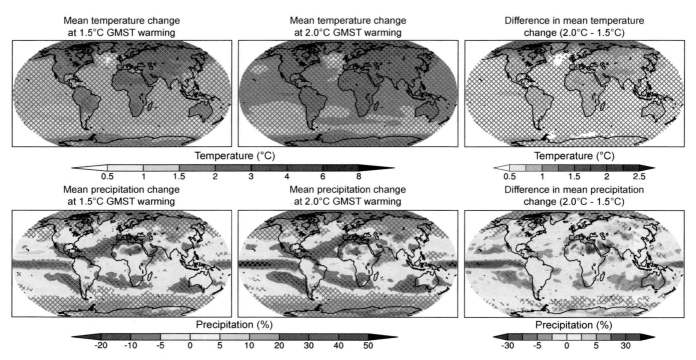

Figure 3.3 | Projected changes in mean temperature (top) and mean precipitation (bottom) at 1.5°C (left) and 2°C (middle) of global warming compared to the pre-industrial period (1861–1880), and the difference between 1.5°C and 2°C of global warming (right). Cross-hatching highlights areas where at least two-thirds of the models agree on the sign of change as a measure of robustness (18 or more out of 26). Values were assessed from the transient response over a 10-year period at a given warming level, based on Representative Concentration Pathway (RCP)8.5 Coupled Model Intercomparison Project Phase 5 (CMIP5) model simulations (adapted from Seneviratne et al., 2016 and Wartenburger et al., 2017, see Supplementary Material 3.SM.2 for more details). Note that the responses at 1.5°C of global warming are similar for RCP2.6 simulations (see Supplementary Material 3.SM.2). Differences compared to 1°C of global warming are provided in the Supplementary Material 3.SM.2.

are less robust, but particularly strong increases are apparent at high latitudes as well as in the tropics at both 1.5°C and 2°C of global warming compared to pre-industrial conditions. The differences in heavy precipitation at 2°C versus 1.5°C global warming are generally not robust at grid-cell scale, but they display consistent increases in most locations (Figure 3.4). However, as addressed in Section 3.3.3, statistically significant differences are found in several large regions and when aggregated over the global land area. We thus assess that there is *high confidence* regarding global-scale differences in temperature means and extremes at 2°C versus 1.5°C global warming, and *medium confidence* regarding global-scale differences in precipitation means and extremes. Further analyses, including differences at 1.5°C and 2°C global warming versus 1°C (i.e., present-day) conditions are provided in the Supplementary Material 3.SM.2.

These projected changes at 1.5°C and 2°C of global warming are consistent with the attribution of observed historical global trends in temperature and precipitation means and extremes (Bindoff et al., 2013a), as well as with some observed changes under the recent global warming of 0.5°C (Schleussner et al., 2017). These comparisons are addressed in more detail in Sections 3.3.2 and 3.3.3. Attribution studies have shown that there is *high confidence* that anthropogenic forcing has had a detectable influence on trends in global warming (*virtually certain* since the mid-20th century), in land warming on all continents except Antarctica (*likely* since the mid-20th century), in ocean warming since 1970 (*very likely*), and in increases in hot extremes and decreases in cold extremes since the mid-20th century

(*very likely*) (Bindoff et al., 2013a). In addition, there is *medium confidence* that anthropogenic forcing has contributed to increases in mean precipitation at high latitudes in the Northern Hemisphere since the mid-20th century and to global-scale increases in heavy precipitation in land regions with sufficient observations over the same period (Bindoff et al., 2013a). Schleussner et al. (2017) showed, through analyses of recent observed tendencies, that changes in temperature extremes and heavy precipitation indices are detectable in observations for the 1991–2010 period compared with those for 1960–1979, with a global warming of approximately 0.5°C occurring between these two periods (*high confidence*). The observed tendencies over that time frame are thus consistent with attributed changes since the mid-20th century (*high confidence*).

The next sections assess changes in several different types of climate-related hazards. It should be noted that the different types of hazards are considered in isolation but some regions are projected to be affected by collocated and/or concomitant changes in several types of hazards (*high confidence*). Two examples are sea level rise and heavy precipitation in some regions, possibly leading together to more flooding, and droughts and heatwaves, which can together increase the risk of fire occurrence. Such events, also called compound events, may substantially increase risks in some regions (e.g., AghaKouchak et al., 2014; Van Den Hurk et al., 2015; Martius et al., 2016; Zscheischler et al., 2018). A detailed assessment of physically-defined compound events was not possible as part of this report, but aspects related to overlapping multi-sector risks are highlighted in Sections 3.4 and 3.5.

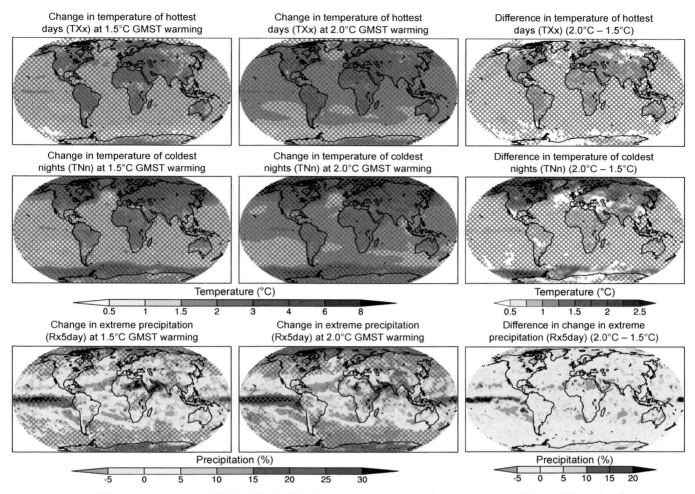

Figure 3.4 | Projected changes in extremes at 1.5°C (left) and 2°C (middle) of global warming compared to the pre-industrial period (1861–1880), and the difference between 1.5°C and 2°C of global warming (right). Cross-hatching highlights areas where at least two-thirds of the models agree on the sign of change as a measure of robustness (18 or more out of 26): temperature of annual hottest day (maximum temperature), TXx (top), and temperature of annual coldest night (minimum temperature), TNn (middle), and annual maximum 5-day precipitation, Rx5day (bottom). The underlying methodology and data basis are the same as for Figure 3.3 (see Supplementary Material 3.SM.2 for more details). Note that the responses at 1.5°C of global warming are similar for Representative Concentration Pathway (RCP)2.6 simulations (see Supplementary Material 3.SM.2). Differences compared to 1°C of global warming are provided in the Supplementary Material 3.SM.2.

3.3.2 Regional Temperatures on Land, Including Extremes

3.3.2.1 Observed and attributed changes in regional temperature means and extremes

While the quality of temperature measurements obtained through ground observational networks tends to be high compared to that of measurements for other climate variables (Seneviratne et al., 2012), it should be noted that some regions are undersampled. Cowtan and Way (2014) highlighted issues regarding undersampling, which is most problematic at the poles and over Africa, and which may lead to biases in estimated changes in GMST (see also Supplementary Material 3.SM.2 and Chapter 1). This undersampling also affects the confidence of assessments regarding regional observed and projected changes in both mean and extreme temperature. Despite this partly limited coverage, the attribution chapter of AR5 (Bindoff et al., 2013a) and recent papers (e.g., Sun et al., 2016; Wan et al., 2018) assessed that, over every continental region and in many sub-continental

regions, anthropogenic influence has made a substantial contribution to surface temperature increases since the mid-20th century.

Based on the AR5 and SREX, as well as recent literature (see Supplementary Material 3.SM), there is *high confidence* (*very likely*) that there has been an overall decrease in the number of cold days and nights and an overall increase in the number of warm days and nights at the global scale on land. There is also *high confidence* (*likely*) that consistent changes are detectable on the continental scale in North America, Europe and Australia. There is *high confidence* that these observed changes in temperature extremes can be attributed to anthropogenic forcing (Bindoff et al., 2013a). As highlighted in Section 3.2, the observational record can be used to assess past changes associated with a global warming of 0.5°C. Schleussner et al. (2017) used this approach to assess observed changes in extreme indices for the 1991–2010 versus the 1960–1979 period, which corresponds to just about a 0.5°C GMST difference in the observed record (based on the Goddard Institute for Space Studies Surface Temperature Analysis

(GISTEMP) dataset, Hansen et al., 2010). They found that substantial changes due to 0.5°C of warming are apparent for indices related to hot and cold extremes, as well as for the Warm Spell Duration Indicator (WSDI). In particular, they identified that one-quarter of the land has experienced an intensification of hot extremes (maximum temperature on the hottest day of the year, TXx) by more than 1°C and a reduction in the intensity of cold extremes by at least 2.5°C (minimum temperature on the coldest night of the year, TNn). In addition, the same study showed that half of the global land mass has experienced changes in WSDI of more than six days, as well as an emergence of extremes outside the range of natural variability (Schleussner et al., 2017). Analyses from Schleussner et al. (2017) for temperature extremes are provided in the Supplementary Material 3.SM, Figure 3.SM.6. It should be noted that assessments of attributed changes in the IPCC SREX and AR5 reports were generally provided since 1950, for time frames also approximately corresponding to a 0.5°C global warming (3.SM).

3.3.2.2   Projected changes in regional temperature means and extremes at 1.5°C versus 2°C of global warming

There are several lines of evidence available for providing a regional assessment of projected changes in temperature means and extremes at 1.5°C versus 2°C of global warming (see Section 3.2). These include: analyses of changes in extremes as a function of global warming based on existing climate simulations using the empirical scaling relationship (ESR) and variations thereof (e.g., Schleussner et al., 2017; Dosio and Fischer, 2018; Seneviratne et al., 2018c; see Section 3.2 for details about the methodology); dedicated simulations of 1.5°C versus 2°C of global warming, for instance based on the Half a degree additional warming, prognosis and projected impacts (HAPPI) experiment (Mitchell et al., 2017) or other model simulations (e.g., Dosio et al., 2018; Kjellström et al., 2018); and analyses based on statistical pattern scaling approaches (e.g., Kharin et al., 2018). These different lines of evidence lead to qualitatively consistent results regarding changes in temperature means and extremes at 1.5°C of global warming compared to the pre-industrial climate and 2°C of global warming.

There are statistically significant differences in temperature means and extremes at 1.5°C versus 2°C of global warming, both in the global average (Schleussner et al., 2016b; Dosio et al., 2018; Kharin et al., 2018), as well as in most land regions (*high confidence*) (Wartenburger et al., 2017; Seneviratne et al., 2018c; Wehner et al., 2018b). Projected temperatures over oceans display significant increases in means and extremes between 1.5°C and 2°C of global warming (Figures 3.3 and 3.4). A general background on the available evidence on regional changes in temperature means and extremes at 1.5°C versus 2°C of global warming is provided in the Supplementary Material 3.SM.2. As an example, Figure 3.5 shows regionally-based analyses for the IPCC SREX regions (see Figure 3.2) of changes in the temperature of hot extremes as a function of global warming (corresponding analyses for changes in the temperature of cold extremes are provided in the Supplementary Material 3.SM.2). As demonstrated in these analyses, the mean response of the intensity of temperature extremes in climate models to changes in the global mean temperature is approximately linear and independent of the considered emissions scenario (Seneviratne et al., 2016; Wartenburger et al., 2017). Nonetheless, in the case of changes in the number of days exceeding a given threshold,

changes are approximately exponential, with higher increases for rare events (Fischer and Knutti, 2015; Kharin et al., 2018); see also Figure 3.6. This behaviour is consistent with a linear increase in absolute temperature for extreme threshold exceedances (Whan et al., 2015).

As mentioned in Section 3.3.1, there is an important land–sea warming contrast, with stronger warming on land (see also Christensen et al., 2013; Collins et al., 2013; Seneviratne et al., 2016), which implies that regional warming on land is generally more than 1.5°C even when mean global warming is at 1.5°C. As highlighted in Seneviratne et al. (2016), this feature is generally stronger for temperature extremes (Figures 3.4 and 3.5; Supplementary Material 3.SM.2). For differences in regional temperature extremes at a mean global warming of 1.5°C versus 2°C, that is, a difference of 0.5°C in global warming, this implies differences of as much as 1°C–1.5°C in some locations, which are two to three times larger than the differences in global mean temperature. For hot extremes, the strongest warming is found in central and eastern North America, central and southern Europe, the Mediterranean, western and central Asia, and southern Africa (Figures 3.4 and 3.5) (*medium confidence*). These regions are all characterized by a strong soil-moisture–temperature coupling and projected increased dryness (Vogel et al., 2017), which leads to a reduction in evaporative cooling in the projections. Some of these regions also show a wide range of responses to temperature extremes, in particular central Europe and central North America, owing to discrepancies in the representation of the underlying processes in current climate models (Vogel et al., 2017). For mean temperature and cold extremes, the strongest warming is found in the northern high-latitude regions (*high confidence*). This is due to substantial ice-snow-albedo-temperature feedbacks (Figure 3.3 and Figure 3.4, middle) related to the known 'polar amplification' mechanism (e.g., IPCC, 2013; Masson-Delmotte et al., 2013).

Figure 3.7 displays maps of changes in the number of hot days (NHD) at 1.5°C and 2°C of GMST increase. Maps of changes in the number of frost days (FD) can be found in Supplementary Material 3.SM.2. These analyses reveal clear patterns of changes between the two warming levels, which are consistent with analysed changes in heatwave occurrence (e.g., Dosio et al., 2018). For the NHD, the largest differences are found in the tropics (*high confidence*), owing to the low interannual temperature variability there (Mahlstein et al., 2011), although absolute changes in hot temperature extremes tended to be largest at mid-latitudes (*high confidence*) (Figures 3.4 and 3.5). Extreme heatwaves are thus projected to emerge earliest in the tropics and to become widespread in these regions already at 1.5°C of global warming (*high confidence*). These results are consistent with other recent assessments. Coumou and Robinson (2013) found that 20% of the global land area, centred in low-latitude regions, is projected to experience highly unusual monthly temperatures during Northern Hemisphere summers at 1.5°C of global warming, with this number nearly doubling at 2°C of global warming.

Figure 3.8 features an objective identification of 'hotspots' / key risks in temperature indices subdivided by region, based on the ESR approach applied to Coupled Model Intercomparison Project Phase 5 (CMIP5) simulations (Wartenburger et al., 2017). Note that results based on the HAPPI multimodel experiment (Mitchell et al., 2017) are similar (Seneviratne et al., 2018c). The considered regions follow the classification used in Figure 3.2 and also include the global land areas. Based on these analyses, the following can be stated: significant changes in responses are found in all regions for most temperature indices, with the exception of i) the diurnal temperature range (DTR) in most regions, ii) ice days (ID), frost days (FD) and growing season length (GSL) (mostly in regions where differences are zero, because, e.g., there are no ice or frost days), iii) the minimum yearly value of the maximum daily temperature (TXn) in very few regions. In terms of the sign of the changes, warm extremes display an increase in intensity, frequency and duration (e.g., an increase in the temperature of the hottest day of the year (TXx) in all regions, an increase in the proportion of days with a maximum temperature above the 90th percentile of Tmax (TX90p) in all regions, and an increase in the length of the WSDI in all regions), while cold extremes display a decrease in intensity, frequency and duration (e.g., an increase in the temperature of the coldest night of the year (TNn) in all regions, a decrease in the proportion of days with a minimum temperature below the 10th percentile of Tmin (TN10p), and a decrease in the cold spell duration index (CSDI) in all regions). Hence, while warm extremes are intensified, cold extremes become less intense in affected regions.

Overall, large increases in hot extremes occur in many densely inhabited regions (Figure 3.5), for both warming scenarios compared to pre-industrial and present-day climate, as well as for 2°C versus 1.5°C GMST warming. For instance, Dosio et al. (2018) concluded, based on a modelling study, that 13.8% of the world population would be exposed to 'severe heatwaves' at least once every 5 years under 1.5°C of global warming, with a threefold increase (36.9%) under 2°C of GMST warming, corresponding to a difference of about 1.7 billion people between the two global warming levels. They also concluded that limiting global warming to 1.5°C would result in about 420 million fewer people being frequently exposed to extreme heatwaves, and about 65 million fewer people being exposed to 'exceptional heatwaves' compared to conditions at 2°C GMST warming. However, changes in vulnerability were not considered in their study. For this reason, we assess that there is *medium confidence* in their conclusions.

In summary, there is *high confidence* that there are robust and statistically significant differences in the projected temperature means and extremes at 1.5°C versus 2°C of global warming, both in the global average and in nearly all land regions[4] (*likely*). Further, the observational record reveals that substantial changes due to a 0.5°C GMST warming are apparent for indices related to hot and cold extremes, as well as for the WSDI (*likely*). A global warming of 2°C versus 1.5°C would lead to more frequent and more intense hot extremes in all land regions[4], as well as longer warm spells, affecting many densely inhabited regions (*very likely*). The strongest increases in the frequency of hot extremes are projected for the rarest events (*very likely*). On the other hand, cold extremes would become less intense and less frequent, and cold spells would be shorter (*very likely*). Temperature extremes on land would generally increase more than the global average temperature (*very likely*). Temperature increases of extreme hot days in mid-latitudes are projected to be up to two times the increase in GMST, that is, 3°C at 1.5°C GMST warming (*high confidence*). The highest levels of warming for extreme hot days are expected to occur in central and eastern North

America, central and southern Europe, the Mediterranean, western and central Asia, and southern Africa (*medium confidence*). These regions have a strong soil-moisture-temperature coupling in common as well as increased dryness and, consequently, a reduction in evaporative cooling. However, there is a substantial range in the representation of these processes in models, in particular in central Europe and central North America (*medium confidence*). The coldest nights in high latitudes warm by as much as 1.5°C for a 0.5°C increase in GMST, corresponding to a threefold stronger warming (*high confidence*). NHD shows the largest differences between 1.5°C and 2°C in the tropics, because of the low interannual temperature variability there (*high confidence*); extreme heatwaves are thus projected to emerge earliest in these regions, and they are expected to become widespread already at 1.5°C of global warming (*high confidence*). Limiting global warming to 1.5°C instead of 2°C could result in around 420 million fewer people being frequently exposed to extreme heatwaves, and about 65 million fewer people being exposed to exceptional heatwaves, assuming constant vulnerability (*medium confidence*).

3.3.3   Regional Precipitation, Including Heavy Precipitation and Monsoons

This section addresses regional changes in precipitation on land, with a focus on heavy precipitation and consideration of changes to the key features of monsoons.

3.3.3.1   Observed and attributed changes in regional precipitation

Observed global changes in the water cycle, including precipitation, are more uncertain than observed changes in temperature (Hartmann et al., 2013; Stocker et al., 2013). There is *high confidence* that mean precipitation over the mid-latitude land areas of the Northern Hemisphere has increased since 1951 (Hartmann et al., 2013). For other latitudinal zones, area-averaged long-term positive or negative trends have *low confidence* because of poor data quality, incomplete data or disagreement amongst available estimates (Hartmann et al., 2013). There is, in particular, *low confidence* regarding observed trends in precipitation in monsoon regions, according to the SREX report (Seneviratne et al., 2012) and AR5 (Hartmann et al., 2013), as well as more recent publications (Singh et al., 2014; Taylor et al., 2017; Bichet and Diedhiou, 2018; see Supplementary Material 3.SM.2).

For heavy precipitation, AR5 (Hartmann et al., 2013) assessed that observed trends displayed more areas with increases than decreases in the frequency, intensity and/or amount of heavy precipitation (*likely*). In addition, for land regions where observational coverage is sufficient for evaluation, it was assessed that there is *medium confidence* that anthropogenic forcing has contributed to a global-scale intensification of heavy precipitation over the second half of the 20th century (Bindoff et al., 2013a).

Regarding changes in precipitation associated with global warming of 0.5°C, the observed record suggests that increases in precipitation extremes can be identified for annual maximum 1-day precipitation

---

[4]   Using the SREX definition of regions (Figure 3.2)   *Continued page 194 >*

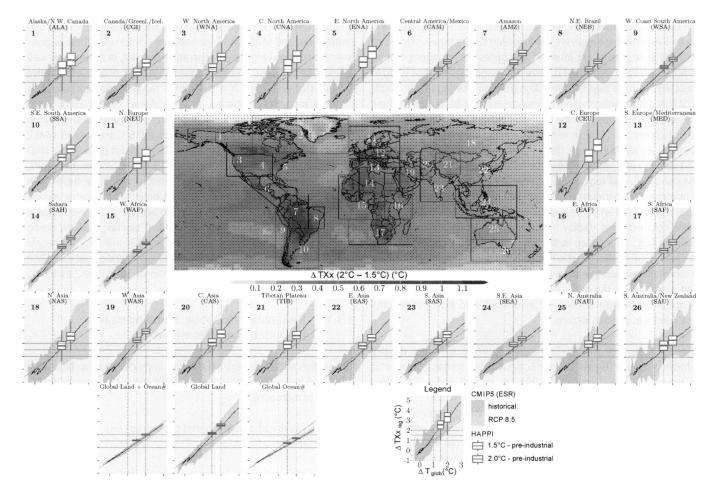

Figure 3.5 | Projected changes in annual maximum daytime temperature (TXx) as a function of global warming for IPCC Special Report on Managing the Risk of Extreme Events and Disasters to Advance Climate Change Adaptation (SREX) regions (see Figure 3.2), based on an empirical scaling relationship applied to Coupled Model Intercomparison Project Phase 5 (CMIP5) data (adapted from Seneviratne et al., 2016 and Wartenburger et al., 2017) together with projected changes from the Half a degree additional warming, prognosis and projected impacts (HAPPI) multimodel experiment (Mitchell et al., 2017; based on analyses in Seneviratne et al., 2018c) (bar plots on regional analyses and central plot, respectively). For analyses for other regions from Figure 3.2 (with asterisks), see Supplementary Material 3.SM.2. (The stippling indicates significance of the differences in changes between 1.5°C and 2°C of global warming based on all model simulations, using a two-sided paired Wilcoxon test (P = 0.01, after controlling the false discovery rate according to Benjamini and Hochberg, 1995). See Supplementary Material 3.SM.2 for details.

Probability ratio of temperature extremes as function of global warming and event probability

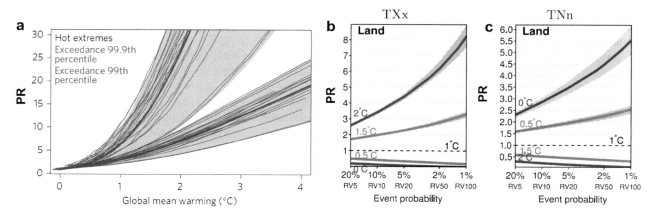

Figure 3.6 | Probability ratio (PR) of exceeding extreme temperature thresholds. (a) PR of exceeding the 99th (blue) and 99.9th (red) percentile of pre-industrial daily temperatures at a given warming level, averaged across land (from Fischer and Knutti, 2015). (b) PR for the hottest daytime temperature of the year (TXx). (c) PR for the coldest night of the year (TNn) for different event probabilities (with RV indicating return values) in the current climate (1°C of global warming). Shading shows the interquartile (25–75%) range (from Kharin et al., 2018).

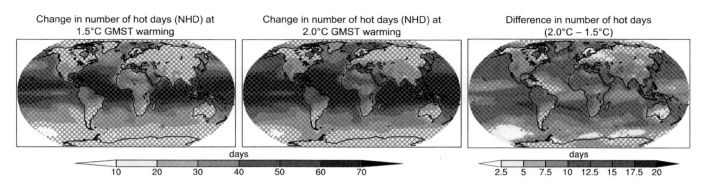

Change in number of hot days (NHD) at 1.5°C GMST warming

Change in number of hot days (NHD) at 2.0°C GMST warming

Difference in number of hot days (2.0°C – 1.5°C)

Figure 3.7 | Projected changes in the number of hot days (NHD; 10% warmest days) at 1.5°C (left) and at 2°C (middle) of global warming compared to the pre-industrial period (1861–1880), and the difference between 1.5°C and 2°C of warming (right). Cross-hatching highlights areas where at least two-thirds of the models agree on the sign of change as a measure of robustness (18 or more out of 26). The underlying methodology and the data basis are the same as for Figure 3.2 (see Supplementary Material 3.SM.2 for more details). Differences compared to 1°C global warming are provided in the Supplementary Material 3.SM.2.

	Global Land	ALA	AMZ	CAM	CAS	CEU	CGI	CNA	EAF	EAS	ENA	MED	NAS	NAU	NEB	NEU	SAF	SAH	SAS	SAU	SEA	SSA	TIB	WAF	WAS	WNA	WSA
T	+	+	+	+	+	+	+	+	+	+	+	+	+	+	+	+	+	+	+	+	+	+	+	+	+	+	+
CSDI	−	−	−	−	−	−	−	−	−	−	−	−	−	−	−	−	−	−	−	−	−	−	−	−	−	−	−
DTR	−	−	+	+	+	+	−	+	+	+	−	+	−	+	+	−	+	−	−	+	−	−	−	−	−	−	+
FD	−	−	−	−	−	−	−	−		−	−	−	−	−	+	−	−		−	−	−	−		−	−	−	−
GSL	+	+	+	+	+	+	+	+			+	+	+	+		+	+		+	+	+	−	+	+		+	+
ID	−	−		−	−	−	−	−			−	−	−			−			−			−	−		−		−
SU	+	+	+	+	+	+	+	+	+	+	+	+	+	+	+	+	+	+	+	+	+	+	+	+	+	+	+
TN10p	−	−	−	−	−	−	−	−	−	−	−	−	−	−	−	−	−	−	−	−	−	−	−	−	−	−	−
TN90p	+	+	+	+	+	+	+	+	+	+	+	+	+	+	+	+	+	+	+	+	+	+	+	+	+	+	+
TNn	+	+	+	+	+	+	+	+	+	+	+	+	+	+	+	+	+	+	+	+	+	+	+	+	+	+	+
TNx	+	+	+	+	+	+	+	+	+	+	+	+	+	+	+	+	+	+	+	+	+	+	+	+	+	+	+
TR	+	+	+	+	+	+	+	+	+	+	+	+	+	+	+	+	+	+	+	+	+	+	+	+	+	+	+
TX10p	−	−	−	−	−	−	−	−	−	−	−	−	−	−	−	−	−	−	−	−	−	−	−	−	−	−	−
TX90p	+	+	+	+	+	+	+	+	+	+	+	+	+	+	+	+	+	+	+	+	+	+	+	+	+	+	+
TXn	+	+	+	+	+	+	+	+	+	+	+	+	+	+	+	+	+	+	+	+	+	+	+	+	+	+	+
TXx	+	+	+	+	+	+	+	+	+	+	+	+	+	+	+	+	+	+	+	+	+	+	+	+	+	+	+
WSDI	+	+	+	+	+	+	+	+	+	+	+	+	+	+	+	+	+	+	+	+	+	+	+	+	+	+	+

Figure 3.8 | Significance of differences in regional mean temperature and range of temperature indices between the 1.5°C and 2°C global mean temperature targets (rows). Definitions of indices: T: mean temperature; CSDI: cold spell duration index; DTR: diurnal temperature range; FD: frost days; GSL: growing season length; ID: ice days; SU: summer days; TN10p: proportion of days with a minimum temperature (TN) lower than the 10th percentile of TN; TN90p: proportion of days with TN higher than the 90th percentile of TN; TNn: minimum yearly value of TN; TNx: maximum yearly value of TN; TR: tropical nights; TX10p: proportion of days with a maximum temperature (TX) lower than the 10th percentile of TX; TX90p: proportion of days with TX higher than the 90th percentile of TX; TXn: minimum yearly value of TX; TXx: maximum yearly value of TX; WSDI: warm spell duration index. Columns indicate analysed regions and global land (see Figure 3.2 for definitions). Significant differences are shown in red shading, with increases indicated with + and decreases indicated with −, while non-significant differences are shown in grey shading. White shading indicates when an index is the same at the two global warming levels (i.e., zero changes). Note that decreases in CSDI, FD, ID, TN10p and TX10p are linked to increased temperatures on cold days or nights. Significance was tested using a two-sided paired Wilcoxon test (P=0.01, after controlling the false discovery rate according to Benjamini and Hochberg, 1995) (adapted from Wartenburger et al., 2017).

3.3.3.1 (continued)

(RX1day) and consecutive 5-day precipitation (RX5day) for GMST changes of this magnitude (Supplementary Material 3.SM.2, Figure 3.SM.7; Schleussner et al., 2017). It should be noted that assessments of attributed changes in the IPCC SREX and AR5 reports were generally provided since 1950, for time frames also approximately corresponding to a 0.5°C global warming (3.SM).

3.3.3.2 Projected changes in regional precipitation at 1.5°C versus 2°C of global warming

Figure 3.3 in Section 3.3.1 summarizes the projected changes in mean precipitation at 1.5°C and 2°C of global warming. Both warming levels display robust differences in mean precipitation compared to the pre-industrial period. Regarding differences at 2°C vs 1.5°C global warming, some regions are projected to display changes in mean precipitation at 2°C compared with that at 1.5°C of global warming in the CMIP5 multimodel average, such as decreases in the Mediterranean area, including southern Europe, the Arabian Peninsula and Egypt, or increases in high latitudes. The results, however, are less robust across models than for mean temperature. For instance, Déqué et al. (2017) investigated the impact of 2°C of global warming on precipitation over tropical Africa and found that average precipitation does not show a significant response, owing to two phenomena: (i) the number of days with rain decreases whereas the precipitation intensity increases, and (ii) the rainy season occurs later during the year, with less precipitation in early summer and more precipitation in late summer. The results from Déqué et al. (2017) regarding insignificant differences between 1.5°C and 2°C scenarios for tropical Africa are consistent with the results presented in Figure 3.3. For Europe, recent studies (Vautard et al., 2014; Jacob et al., 2018; Kjellström et al., 2018) have shown that 2°C of global warming was associated with a robust increase in mean precipitation over central and northern Europe in winter but only over northern Europe in summer, and with decreases in mean precipitation in central/southern Europe in summer. Precipitation changes reaching 20% have been projected for the 2°C scenario (Vautard et al., 2014) and are overall more pronounced than with 1.5°C of global warming (Jacob et al., 2018; Kjellström et al., 2018).

Regarding changes in heavy precipitation, Figure 3.9 displays projected changes in the 5-day maximum precipitation (Rx5day) as a function of global temperature increase, using a similar approach as in Figure 3.5. Further analyses are available in Supplementary Material 3.SM.2. These analyses show that projected changes in heavy precipitation are more uncertain than those for temperature extremes. However, the mean response of model simulations is generally robust and linear (see also Fischer et al., 2014; Seneviratne et al., 2016). As observed for temperature extremes, this response is also mostly independent of the considered emissions scenario (e.g., RCP2.6 versus RCP8.5; see also Section 3.2). This feature appears to be specific to heavy precipitation, possibly due to a stronger coupling with temperature, as the scaling of projections of mean precipitation changes with global warming shows some scenario dependency (Pendergrass et al., 2015).

Robust changes in heavy precipitation compared to pre-industrial conditions are found at both 1.5°C and 2°C global warming (Figure 3.4). This is also consistent with results for, for example, the European

continent, although different indices for heavy precipitation changes have been analysed. Based on regional climate simulations, Vautard et al. (2014) found a robust increase in heavy precipitation everywhere in Europe and in all seasons, except southern Europe in summer at 2°C versus 1971–2000. Their findings are consistent with those of Jacob et al. (2014), who used more recent downscaled climate scenarios (EURO-CORDEX) and a higher resolution (12 km), but the change is not so pronounced in Teichmann et al. (2018). There is consistent agreement in the direction of change in heavy precipitation at 1.5°C of global warming over much of Europe, compared to 1971–2000 (Jacob et al., 2018).

Differences in heavy precipitation are generally projected to be small between 1.5°C and 2°C GMST warming (Figure 3.4 and 3.9 and Supplementary Material 3.SM.2, Figure 3.SM.10). Some regions display substantial increases, for instance southern Asia, but generally in less than two-thirds of the CMIP5 models (Figure 3.4, Supplementary Material 3.SM.2, Figure 3.SM.10). Wartenburger et al. (2017) suggested that there are substantial differences in heavy precipitation in eastern Asia at 1.5°C versus 2°C. Overall, while there is variation among regions, the global tendency is for heavy precipitation to increase at 2°C compared with at 1.5°C (see e.g., Fischer and Knutti, 2015 and Kharin et al., 2018, as illustrated in Figure 3.10 from this chapter; see also Betts et al., 2018).

AR5 assessed that the global monsoon, aggregated over all monsoon systems, is *likely* to strengthen, with increases in its area and intensity, while the monsoon circulation weakens (Christensen et al., 2013). A few publications provide more recent evaluations of projections of changes in monsoons for high-emission scenarios (e.g., Jiang and Tian, 2013; Jones and Carvalho, 2013; Sylla et al., 2015, 2016; Supplementary Material 3.SM.2). However, scenarios at 1.5°C or 2°C global warming would involve a substantially smaller radiative forcing than those assessed in AR5 and these more recent studies, and there appears to be no specific assessment of changes in monsoon precipitation at 1.5°C versus 2°C of global warming in the literature. Consequently, the current assessment is that there is *low confidence* regarding changes in monsoons at these lower global warming levels, as well as regarding differences in monsoon responses at 1.5°C versus 2°C.

Similar to Figure 3.8, Figure 3.11 features an objective identification of 'hotspots' / key risks outlined in heavy precipitation indices subdivided by region, based on the approach by Wartenburger et al. (2017). The considered regions follow the classification used in Figure 3.2 and also include global land areas. Hotspots displaying statistically significant changes in heavy precipitation at 1.5°C versus 2°C global warming are located in high-latitude (Alaska/western Canada, eastern Canada/Greenland/Iceland, northern Europe, northern Asia) and high-elevation (e.g., Tibetan Plateau) regions, as well as in eastern Asia (including China and Japan) and in eastern North America. Results are less consistent for other regions. Note that analyses for meteorological drought (lack of precipitation) are provided in Section 3.3.4.

In summary, observations and projections for mean and heavy precipitation are less robust than for temperature means and extremes (*high confidence*). Observations show that there are more areas with increases than decreases in the frequency, intensity and/or amount of

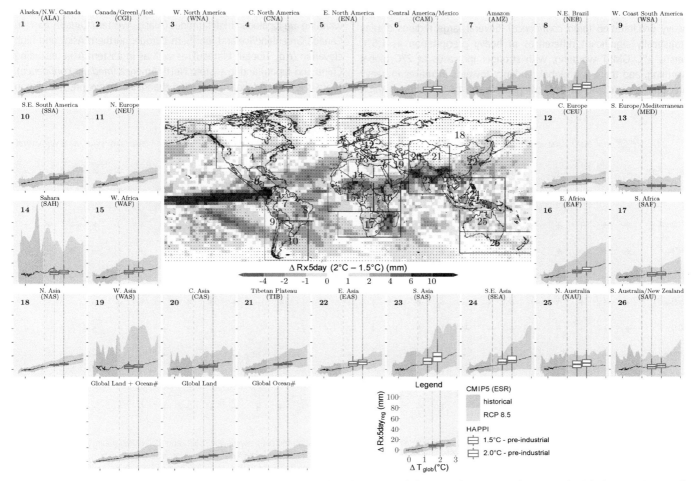

Figure 3.9 | Projected changes in annual 5-day maximum precipitation (Rx5day) as a function of global warming for IPCC Special Report on the Risk of Extreme Events and Disasters to Advance Climate Change Adaptation (SREX) regions (see Figure 3.2), based on an empirical scaling relationship applied to Coupled Model Intercomparison Project Phase 5 (CMIP5) data together with projected changes from the HAPPI multimodel experiment (bar plots on regional analyses and central plot). The underlying methodology and data basis are the same as for Figure 3.5 (see Supplementary Material 3.SM.2 for more details).

Probability ratio of heavy precipitation as function of global warming and event probability

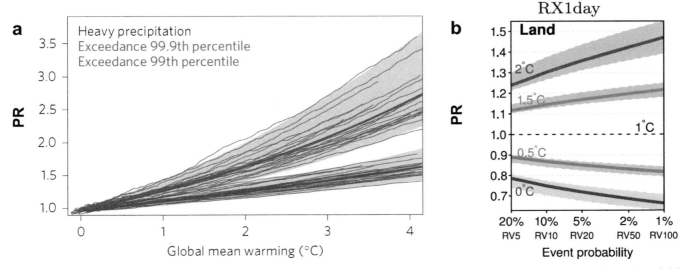

Figure 3.10 | Probability ratio (PR) of exceeding (heavy precipitation) thresholds. (a) PR of exceeding the 99th (blue) and 99.9th (red) percentile of pre-industrial daily precipitation at a given warming level, averaged across land (from Fischer and Knutti, 2015). (b) PR for precipitation extremes (RX1day) for different event probabilities (with RV indicating return values) in the current climate (1°C of global warming). Shading shows the interquartile (25–75%) range (from Kharin et al., 2018).

3.3.3.2 (continued)

heavy precipitation (*high confidence*). Several large regions display statistically significant differences in heavy precipitation at 1.5°C versus 2°C GMST warming, with stronger increases at 2°C global warming, and there is a global tendency towards increases in heavy precipitation on land at 2°C compared with 1.5°C warming (*high confidence*). Overall, regions that display statistically significant

changes in heavy precipitation between 1.5°C and 2°C of global warming are located in high latitudes (Alaska/western Canada, eastern Canada/Greenland/Iceland, northern Europe, northern Asia) and high elevation (e.g., Tibetan Plateau), as well as in eastern Asia (including China and Japan) and in eastern North America (*medium confidence*). There is *low confidence* in projected changes in heavy precipitation in other regions.

	Global Land	ALA	AMZ	CAM	CAS	CEU	CGI	CNA	EAF	EAS	ENA	MED	NAS	NAU	NEB	NEU	SAF	SAH	SAS	SAU	SEA	SSA	TIB	WAF	WAS	WNA	WSA
PRCPTOT	+	+	-	+	+	+	+	+	-	+	+	-	+	-	+	-	+	-	+	-	+	-	+	+	-	+	-
CWD	-	+	-	-	-	-	+	+	-	-	+	-	+	-	-	-	-	-	-	-	-	-	+	-	-	-	-
R95ptot	+	+	+	+	+	+	+	+	+	+	+	+	+	+	+	+	-	+	-	+	+	+	+	+	+	+	+
R99ptot	+	+	+	+	+	+	+	+	+	+	+	+	+	+	+	+	-	+	+	+	+	+	+	+	+	+	+
Rx1day	+	+	+	+	+	+	+	+	+	+	+	-	+	-	+	+	-	+	+	+	+	+	+	+	+	+	+
Rx5day	+	+	+	+	+	+	+	+	+	+	+	+	-	+	+	-	+	+	+	-	+	+	+	-	+	+	+
SDII	+	+	-	+	+	+	+	+	+	+	+	+	+	+	+	+	-	+	+	+	+	+	+	+	-	+	+
R1mm	+	+	-	-	-	+	-	+	-	+	-	+	-	-	-	-	-	-	-	-	+	-	+	-	-	+	-
R10mm	+	+	-	-	+	+	+	+	-	+	+	-	+	-	+	-	+	-	+	-	+	-	+	+	-	+	-
R20mm	+	+	+	+	+	+	+	+	+	+	+	+	+	-	-	+	+	-	+	-	+	+	+	+	+	+	+

Figure 3.11 | Significance of differences in regional mean precipitation and range of precipitation indices between the 1.5°C and 2°C global mean temperature targets (rows). Definition of indices: PRCPTOT: mean precipitation; CWD: consecutive wet days; R10mm: number of days with precipitation >10 mm; R1mm: number of days with precipitation >1 mm; R20mm: number of days with precipitation >20 mm; R95ptot: proportion of rain falling as 95th percentile or higher; R99ptot: proportion of rain falling as 99th percentile or higher; RX1day: intensity of maximum yearly 1-day precipitation; RX5day: intensity of maximum yearly 5-day precipitation; SDII: Simple Daily Intensity Index. Columns indicate analysed regions and global land (see Figure 3.2 for definitions). Significant differences are shown in light blue (wetting tendency) or brown (drying tendency) shading, with increases indicated with '+' and decreases indicated with '–', while non-significant differences are shown in grey shading. The underlying methodology and the data basis are the same as in Figure 3.8 (see Supplementary Material 3.SM.2 for more details).

3.3.4 Drought and Dryness

3.3.4.1 Observed and attributed changes

The IPCC AR5 assessed that there was *low confidence* in the sign of drought trends since 1950 at the global scale, but that there was *high confidence* in observed trends in some regions of the world, including drought increases in the Mediterranean and West Africa and drought decreases in central North America and northwest Australia (Hartmann et al., 2013; Stocker et al., 2013). AR5 assessed that there was *low confidence* in the attribution of global changes in droughts and did not provide assessments for the attribution of regional changes in droughts (Bindoff et al., 2013a).

The recent literature does not suggest that the SREX and AR5 assessment of drought trends should be revised, except in the Mediterranean region. Recent publications based on observational and modelling evidence suggest that human emissions have substantially increased the probability of drought years in the Mediterranean region (Gudmundsson and Seneviratne, 2016; Gudmundsson et al., 2017). Based on this evidence, there is *medium confidence* that enhanced

greenhouse forcing has contributed to increased drying in the Mediterranean region (including southern Europe, northern Africa and the Near East) and that this tendency will continue to increase under higher levels of global warming.

3.3.4.2 Projected changes in drought and dryness at 1.5°C versus 2°C

There is *medium confidence* in projections of changes in drought and dryness. This is partly consistent with AR5, which assessed these projections as being '*likely (medium confidence)*' (Collins et al., 2013; Stocker et al., 2013). However, given this *medium confidence*, the current assessment does not include a likelihood statement, thereby maintaining consistency with the IPCC uncertainty guidance document (Mastrandrea et al., 2010) and the assessment of the IPCC SREX report (Seneviratne et al., 2012). The technical summary of AR5 (Stocker et al., 2013) assessed that soil moisture drying in the Mediterranean, southwestern USA and southern African regions was consistent with projected changes in the Hadley circulation and increased surface temperatures, and it concluded that there was *high confidence* in *likely* surface drying in these regions by the end of this century

Box 3.1 | Sub-Saharan Africa: Changes in Temperature and Precipitation Extremes

Sub-Saharan Africa has experienced the dramatic consequences of climate extremes becoming more frequent and more intense over the past decades (Paeth et al., 2010; Taylor et al., 2017). In order to join international efforts to reduce climate change, all African countries signed the Paris Agreement. In particular, through their nationally determined contributions (NDCs), they committed to contribute to the global effort to mitigate greenhouse gas (GHG) emissions with the aim to constrain global temperature increases to 'well below 2°C' and to pursue efforts to limit warming to '1.5°C above pre-industrial levels'. The target of limiting global warming to 1.5°C above pre-industrial levels is useful for conveying the urgency of the situation. However, it focuses the climate change debate on a temperature threshold (Section 3.3.2), while the potential impacts of these global warming levels on key sectors at local to regional scales, such as agriculture, energy and health, remain uncertain in most regions and countries of Africa (Sections 3.3.3, 3.3.4, 3.3.5 and 3.3.6).

Weber et al. (2018) found that at regional scales, temperature increases in sub-Saharan Africa are projected to be higher than the global mean temperature increase (at global warming of 1.5°C and at 2°C; see Section 3.3.2 for further background and analyses of climate model projections). Even if the mean global temperature anomaly is kept below 1.5°C, regions between 15°S and 15°N are projected to experience an increase in hot nights, as well as longer and more frequent heatwaves (e.g., Kharin et al., 2018). Increases would be even larger if the global mean temperature were to reach 2°C of global warming, with significant changes in the occurrence and intensity of temperature extremes in all sub-Saharan regions (Sections 3.3.1 and 3.3.2; Figures 3.4, 3.5 and 3.8).

West and Central Africa are projected to display particularly large increases in the number of hot days, both at 1.5°C and 2°C of global warming (Section 3.3.2). This is due to the relatively small interannual present-day variability in this region, which implies that climate-change signals can be detected earlier there (Section 3.3.2; Mahlstein et al., 2011). Projected changes in total precipitation exhibit uncertainties, mainly in the Sahel (Section 3.3.3 and Figure 3.8; Diedhiou et al., 2018). In the Guinea Coast and Central Africa, only a small change in total precipitation is projected, although most models (70%) indicate a decrease in the length of wet periods and a slight increase in heavy rainfall. Western Sahel is projected by most models (80%) to experience the strongest drying, with a significant increase in the maximum length of dry spells (Diedhiou et al., 2018). Above 2°C, this region could become more vulnerable to drought and could face serious food security issues (Cross-Chapter Box 6 and Section 3.4.6 in this chapter; Salem et al., 2017; Parkes et al., 2018). West Africa has thus been identified as a climate-change hotspot with negative impacts from climate change on crop yields and production (Cross-Chapter Box 6 and Section 3.4.6; Sultan and Gaetani, 2016; Palazzo et al., 2017). Despite uncertainty in projections for precipitation in West Africa, which is essential for rain-fed agriculture, robust evidence of yield loss might emerge. This yield loss is expected to be mainly driven by increased mean temperature, while potential wetter or drier conditions – as well as elevated CO_2 concentrations – could modulate this effect (Roudier et al., 2011; see also Cross-Chapter Box 6 and Section 3.4.6). Using Representative Concentration Pathway (RCP)8.5 Coordinated Regional Climate Downscaling Experiment (CORDEX) scenarios from 25 regional climate models (RCMs) forced with different general circulation models (GCMs), Klutse et al. (2018) noted a decrease in mean rainfall over West Africa in models with stronger warming for this region at 1.5°C of global warming (Section 3.3.4). Mba et al. (2018) used a similar approach and found a lack of consensus in the changes in precipitation over Central Africa (Figure 3.8 and Section 3.3.4), although there was a tendency towards a decrease in the maximum number of consecutive wet days (CWD) and a significant increase in the maximum number of consecutive dry days (CDD).

Over southern Africa, models agree on a positive sign of change for temperature, with temperature rising faster at 2°C (1.5°C–2.5°C) as compared to 1.5°C (0.5°C–1.5°C) of global warming. Areas in the south-western region, especially in South Africa and parts of Namibia and Botswana, are expected to experience the largest increases in temperature (Section 3.3.2; Engelbrecht et al., 2015; Maúre et al., 2018). The western part of southern Africa is projected to become drier with increasing drought frequency and number of heatwaves towards the end of the 21st century (Section 3.3.4; Engelbrecht et al., 2015; Dosio, 2017; Maúre et al., 2018). At 1.5°C, a robust signal of precipitation reduction is found over the Limpopo basin and smaller areas of the Zambezi basin in Zambia, as well as over parts of Western Cape in South Africa, while an increase is projected over central and western South Africa, as well as in southern Namibia (Section 3.3.4). At 2°C, the region is projected to face robust precipitation decreases of about 10–20% and increases in the number of CDD, with longer dry spells projected over Namibia, Botswana, northern Zimbabwe and southern Zambia. Conversely, the number of CWD is projected to decrease, with robust signals over Western Cape (Maúre et al., 2018). Projected reductions in stream flow of 5–10% in the Zambezi River basin have been associated with increased evaporation and transpiration rates resulting from a rise in temperature (Section 3.3.5; Kling et al., 2014), with issues for hydroelectric power across the region of southern Africa.

For Eastern Africa, Osima et al. (2018) found that annual rainfall projections show a robust increase in precipitation over Somalia and a less robust decrease over central and northern Ethiopia (Section 3.3.3). The number of CDD and CWD are projected to increase and decrease, respectively (Section 3.3.4). These projected changes could impact the agricultural and water sectors in the region (Cross-Chapter Box 6 in this chapter and Section 3.4.6).

under the RCP8.5 scenario. However, more recent assessments have highlighted uncertainties in dryness projections due to a range of factors, including variations between the drought and dryness indices considered, and the effects of enhanced CO_2 concentrations on plant water-use efficiency (Orlowsky and Seneviratne, 2013; Roderick et al., 2015). Overall, projections of changes in drought and dryness for high-emissions scenarios (e.g., RCP8.5, corresponding to about 4°C of global warming) are uncertain in many regions, although a few regions display consistent drying in most assessments (e.g., Seneviratne et al., 2012; Orlowsky and Seneviratne, 2013). Uncertainty is expected to be even larger for conditions with a smaller signal-to-noise ratio, such as for global warming levels of 1.5°C and 2°C.

Some published literature is now available on the evaluation of differences in drought and dryness occurrence at 1.5°C and 2°C of global warming for (i) precipitation minus evapotranspiration (P–E, a general measure of water availability; Wartenburger et al., 2017; Greve et al., 2018), (ii) soil moisture anomalies (Lehner et al., 2017; Wartenburger et al., 2017), (iii) consecutive dry days (CDD) (Schleussner et al., 2016b; Wartenburger et al., 2017), (iv) the 12-month standardized precipitation index (Wartenburger et al., 2017), (v) the Palmer drought severity index (Lehner et al., 2017), and (vi) annual mean runoff (Schleussner et al., 2016b, see also next section). These analyses have produced consistent findings overall, despite the known sensitivity of drought assessments to chosen drought indices (see above paragraph). These analyses suggest that increases in drought, dryness or precipitation deficits are projected at 1.5°C or 2°C global warming in some regions compared to the pre-

industrial or present-day conditions, as well as between these two global warming levels, although there is substantial variability in signals depending on the considered indices or climate models (Lehner et al., 2017; Schleussner et al., 2017; Greve et al., 2018) (*medium confidence*). Generally, the clearest signals are found for the Mediterranean region (*medium confidence*).

Greve et al. (2018, Figure 3.12) derives the sensitivity of regional changes in precipitation minus evapotranspiration to global temperature changes. The simulations analysed span the full range of available emission scenarios, and the sensitivities are derived using a modified pattern scaling approach. The applied approach assumes linear dependencies on global temperature changes while thoroughly addressing associated uncertainties via resampling methods. Northern high-latitude regions display robust responses tending towards increased wetness, while subtropical regions display a tendency towards drying but with a large range of responses. While the internal variability and the scenario choice play an important role in the overall spread of the simulations, the uncertainty stemming from the climate model choice usually dominates, accounting for about half of the total uncertainty in most regions (Wartenburger et al., 2017; Greve et al., 2018). The sign of projections, that is, whether there might be increases or decreases in water availability under higher global warming levels, is particularly uncertain in tropical and mid-latitude regions. An assessment of the implications of limiting the global mean temperature increase to values below (i) 1.5°C or (ii) 2°C shows that constraining global warming to the 1.5°C target might slightly influence the mean

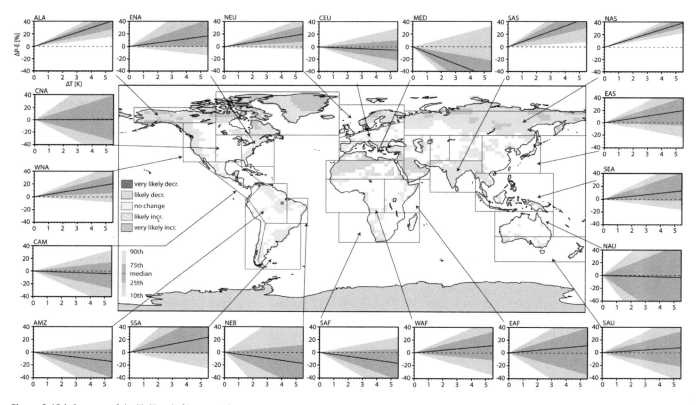

Figure 3.12 | Summary of the likelihood of increases/decreases in precipitation minus evapotranspiration (P–E) in Coupled Model Intercomparison Project Phase 5 (CMIP5) simulations considering all scenarios and a representative subset of 14 climate models (one from each modelling centre). Panel plots show the uncertainty distribution of the sensitivity of P–E to global temperature change, averaged for most IPCC Special Report on Managing the Risk of Extreme Events and Disasters to Advance Climate Change Adaptation (SREX) regions (see Figure 3.2) outlined in the map (from Greve et al., 2018).

response but could substantially reduce the risk of experiencing extreme changes in regional water availability (Greve et al., 2018).

The findings from the analysis for the mean response by Greve et al. (2018) are qualitatively consistent with results from Wartenburger et al. (2017), who used an ESR (Section 3.2) rather than a pattern scaling approach for a range of drought and dryness indices. They are also consistent with a study by Lehner et al. (2017), who assessed changes in droughts based on soil moisture changes and the Palmer-Drought Severity Index. Notably, these two publications do not provide a

specific assessment of changes in the tails of the drought and dryness distribution. The conclusions of Lehner et al. (2017) are that (i) 'risks of consecutive drought years show little change in the US Southwest and Central Plains, but robust increases in Europe and the Mediterranean', and that (ii) 'limiting warming to 1.5°C may have benefits for future drought risk, but such benefits are regional, and in some cases highly uncertain'.

Figure 3.13 features projected changes in CDD as a function of global temperature increase, using a similar approach as for Figures 3.5 (based

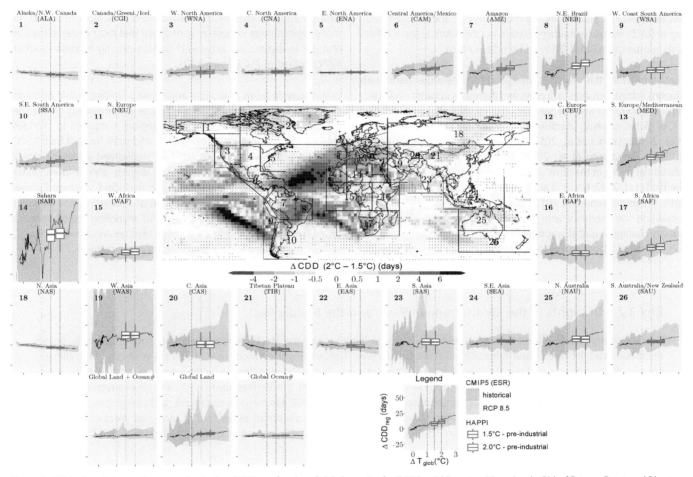

Figure 3.13 | Projected changes in consecutive dry days (CDD) as a function of global warming for IPCC Special Report on Managing the Risk of Extreme Events and Disasters to Advance Climate Change Adaptation (SREX) regions, based on an empirical scaling relationship applied to Coupled Model Intercomparison Project Phase 5 (CMIP5) data together with projected changes from the HAPPI multimodel experiment (bar plots on regional analyses and central plot, respectively). The underlying methodology and the data basis are the same as for Figure 3.5 (see Supplementary Material 3.SM.2 for more details).

	Global Land	ALA	AMZ	CAM	CAS	CEU	CGI	CNA	EAF	EAS	ENA	MED	NAS	NAU	NEB	NEU	SAF	SAH	SAS	SAU	SEA	SSA	TIB	WAF	WAS	WNA	WSA
CDD	+	-		+	+	+	+	-	+	+	-	-	+	-	+	+	+	+	+	+	-	+	-	+	+	-	+
P–E	+	+	+	+	-	+	+	+	+	+	+	-	-	+	-	-	+	-	-	+	-	+	+	-	+	-	+
SMA	-	+	+	-	-		-	+	+		-	-		-	+	-	-		-	-	+	+	-	-	+	-	-
SPI12	+	+	-	+	+	+	+	+	-	+	+	-	-	-	+	-	-	-	+	-	+	+	-	+	-	+	+

Figure 3.14 | Significance of differences in regional drought and dryness indices between the 1.5°C and 2°C global mean temperature targets (rows). Definition of indices: CDD: consecutive dry days; P–E: precipitation minus evapotranspiration; SMA: soil moisture anomalies; SPI12: 12-month Standardized Precipitation Index. Columns indicate analysed regions and global land (see Figure 3.2 for definitions). Significant differences are shown in light blue/brown shading (increases indicated with +, decreases indicated with –; light blue shading indicates decreases in dryness (decreases in CDD, or increases in P–E, SMA or SPI12) and light brown shading indicates increases in dryness (increases in CDD, or decreases in P–E, SMA or SPI12). Non-significant differences are shown in grey shading. The underlying methodology and the data basis are the same as for Figure 3.7 (see Supplementary Material 3.SM.2 for more details).

on Wartenburger et al., 2017). The figure also include results from the HAPPI experiment (Mitchell et al., 2017). Again, the CMIP5-based ESR estimates and the results of the HAPPI experiment agree well. Note that the responses vary widely among the considered regions.

Similar to Figures 3.8 and 3.11, Figure 3.14 features an objective identification of 'hotspots' / key risks in dryness indices subdivided by region, based on the approach by Wartenburger et al. (2017). This analysis reveals the following hotspots of drying (i.e. increases in CDD and/or decreases in P–E, soil moisture anomalies (SMA) and 12-month Standardized Precipitation Index (SPI12), with at least one of the indices displaying statistically significant drying): the Mediterranean region (MED; including southern Europe, northern Africa, and the Near East), northeastern Brazil (NEB) and southern Africa.

Consistent with this analysis, the available literature particularly supports robust increases in dryness and decreases in water availability in southern Europe and the Mediterranean with a shift from 1.5°C to 2°C of global warming (*medium confidence*) (Figure 3.13; Schleussner et al., 2016b; Lehner et al., 2017; Wartenburger et al., 2017; Greve et al., 2018; Samaniego et al., 2018). This region is already displaying substantial drying in the observational record (Seneviratne et al., 2012; Sheffield et al., 2012; Greve et al., 2014; Gudmundsson and Seneviratne, 2016; Gudmundsson et al., 2017), which provides additional evidence supporting this tendency and suggests that it will be a hotspot of dryness change at global warming levels beyond 1.5°C (see also Box 3.2). The other identified hotspots, southern Africa and northeastern

Brazil, also consistently display drying trends under higher levels of forcing in other publications (e.g., Orlowsky and Seneviratne, 2013), although no published studies could be found reporting observed drying trends in these regions. There are substantial increases in the risk of increased dryness (*medium confidence*) in both the Mediterranean region and Southern Africa at 2°C versus 1.5°C of global warming because these regions display significant changes in two dryness indicators (CDD and SMA) between these two global warming levels (Figure 3.14); the strongest effects are expected for extreme droughts (*medium confidence*) (Figure 3.12). There is *low confidence* elsewhere, owing to a lack of consistency in analyses with different models or different dryness indicators. However, in many regions there is *medium confidence* that most extreme risks of changes in dryness are avoided if global warming is constrained at 1.5°C instead of 2°C (Figure 3.12).

In summary, in terms of drought and dryness, limiting global warming to 1.5°C is expected to substantially reduce the probability of extreme changes in water availability in some regions compared to changes under 2°C of global warming (*medium confidence*). For shift from 1.5°C to 2°C of GMST warming, the available studies and analyses suggest strong increases in the probability of dryness and reduced water availability in the Mediterranean region (including southern Europe, northern Africa and the Near East) and in southern Africa (*medium confidence*). Based on observations and modelling experiments, a drying trend is already detectable in the Mediterranean region, that is, at global warming of less than 1°C (*medium confidence*).

Box 3.2 | Droughts in the Mediterranean Basin and the Middle East

Human society has developed in tandem with the natural environment of the Mediterranean basin over several millennia, laying the groundwork for diverse and culturally rich communities. Even if advances in technology may offer some protection from climatic hazards, the consequences of climatic change for inhabitants of this region continue to depend on the long-term interplay between an array of societal and environmental factors (Holmgren et al., 2016). As a result, the Mediterranean is an example of a region with high vulnerability where various adaptation responses have emerged. Previous IPCC assessments and recent publications project regional changes in climate under increased temperatures, including consistent climate model projections of increased precipitation deficit amplified by strong regional warming (Section 3.3.3; Seneviratne et al., 2012; Christensen et al., 2013; Collins et al., 2013; Greve and Seneviratne, 2015).

The long history of resilience to climatic change is especially apparent in the eastern Mediterranean region, which has experienced a strong negative trend in precipitation since 1960 (Mathbout et al., 2017) and an intense and prolonged drought episode between 2007 and 2010 (Kelley et al., 2015). This drought was the longest and most intense in the last 900 years (Cook et al., 2016). Some authors (e.g., Trigo et al., 2010; Kelley et al., 2015) assert that very low precipitation levels have driven a steep decline in agricultural productivity in the Euphrates and Tigris catchment basins, and displaced hundreds of thousands of people, mainly in Syria. Impacts on the water resources (Yazdanpanah et al., 2016) and crop performance in Iran have also been reported (Saeidi et al., 2017). Many historical periods of turmoil have coincided with severe droughts, for example the drought which occurred at the end of the Bronze Age approximately 3200 years ago (Kaniewski et al., 2015). In this instance, a number of flourishing eastern Mediterranean civilizations collapsed, and rural settlements re-emerged with agro-pastoral activities and limited long-distance trade. This illustrates how some vulnerable regions are forced to pursue drastic adaptive responses, including migration and societal structure changes.

The potential evolution of drought conditions under 1.5°C or 2°C of global warming (Section 3.3.4) can be analysed by comparing the 2008 drought (high temperature, low precipitation) with the 1960 drought (low temperature, low precipitation) (Kelley et al., 2015). Though the precipitation deficits were comparable, the 2008 drought was amplified by increased evapotranspiration induced by much higher temperatures (a mean increase of 1°C compared with the 1931–2008 period in Syria) and a large population increase (from

Box 3.2 (continued)

5 million in 1960 to 22 million in 2008). Koutroulis et al. (2016) reported that only 6% out of the total 18% decrease in water availability projected for Crete under 2°C of global warming at the end of the 21st century would be due to decreased precipitation, with the remaining 12% due to an increase in evapotranspiration. This study and others like it confirm an important risk of extreme drought conditions for the Middle East under 1.5°C of global warming (Jacob et al., 2018), with risks being even higher in continental locations than on islands; these projections are consistent with current observed changes (Section 3.3.4; Greve et al., 2014). Risks of drying in the Mediterranean region could be substantially reduced if global warming is limited to 1.5°C compared to 2°C or higher levels of warming (Section 3.4.3; Guiot and Cramer, 2016). Higher warming levels may induce high levels of vulnerability exacerbated by large changes in demography.

3.3.5 Runoff and Fluvial Flooding

3.3.5.1 Observed and attributed changes in runoff and river flooding

There has been progress since AR5 in identifying historical changes in streamflow and continental runoff. Using the available streamflow data, Dai (2016) showed that long-term (1948–2012) flow trends are statistically significant only for 27.5% of the world's 200 major rivers, with negative trends outnumbering the positive ones. Although streamflow trends are mostly not statistically significant, they are consistent with observed regional precipitation changes. From 1950 to 2012, precipitation and runoff have increased over southeastern South America, central and northern Australia, the central and northeastern United States, central and northern Europe, and most of Russia, and they have decreased over most of Africa, East and South Asia, eastern coastal Australia, the southeastern and northwestern United States, western and eastern Canada, the Mediterranean region and some regions of Brazil (Dai, 2016).

A large part of the observed regional trends in streamflow and runoff might have resulted from internal multi-decadal and multi-year climate variations, especially the Pacific decadal variability (PDV), the Atlantic Multi-Decadal Oscillation (AMO) and the El Niño–Southern Oscillation (ENSO), although the effect of anthropogenic greenhouse gases and aerosols could also be important (Hidalgo et al., 2009; Gu and Adler, 2013, 2015; Chiew et al., 2014; Luo et al., 2016; Gudmundsson et al., 2017). Additionally, other human activities can influence the hydrological cycle, such as land-use/land-cover change, modifications in river morphology and water table depth, construction and operation of hydropower plants, dikes and weirs, wetland drainage, and agricultural practices such as water withdrawal for irrigation. All of these activities can also have a large impact on runoff at the river basin scale, although there is less agreement over their influence on global mean runoff (Gerten et al., 2008; Sterling et al., 2012; Hall et al., 2014; Betts et al., 2015; Arheimer et al., 2017). Some studies suggest that increases in global runoff resulting from changes in land cover or land use (predominantly deforestation) are counterbalanced by decreases resulting from irrigation (Gerten et al., 2008; Sterling et al., 2012). Likewise, forest and grassland fires can modify the hydrological response at the watershed scale when the burned area is significant (Versini et al., 2013; Springer et al., 2015; Wine and Cadol, 2016).

Few studies have explored observed changes in extreme streamflow and river flooding since the IPCC AR5. Mallakpour and Villarini (2015)

analysed changes of flood magnitude and frequency in the central United States by considering stream gauge daily records with at least 50 years of data ending no earlier than 2011. They showed that flood frequency has increased, whereas there was limited evidence of a decrease in flood magnitude in this region. Stevens et al. (2016) found a rise in the number of reported floods in the United Kingdom during the period 1884–2013, with flood events appearing more frequently towards the end of the 20th century. A peak was identified in 2012, when annual rainfall was the second highest in over 100 years. Do et al. (2017) computed the trends in annual maximum daily streamflow data across the globe over the 1966–2005 period. They found decreasing trends for a large number of stations in western North America and Australia, and increasing trends in parts of Europe, eastern North America, parts of South America, and southern Africa.

In summary, streamflow trends since 1950 are not statistically significant in most of the world's largest rivers (*high confidence*), while flood frequency and extreme streamflow have increased in some regions (*high confidence*).

3.3.5.2 Projected changes in runoff and river flooding at 1.5°C versus 2°C of global warming

Global-scale assessments of projected changes in freshwater systems generally suggest that areas with either positive or negative changes in mean annual streamflow are smaller for 1.5°C than for 2°C of global warming (Betts et al., 2018; Döll et al., 2018). Döll et al. (2018) found that only 11% of the global land area (excluding Greenland and Antarctica) shows a statistically significantly larger hazard at 2°C than at 1.5°C. Significant decreases are found for 13% of the global land area for both global warming levels, while significant increases are projected to occur for 21% of the global land area at 1.5°C, and rise to between 26% (Döll et al., 2018) and approximately 50% (Betts et al., 2018) at 2°C.

At the regional scale, projected runoff changes generally follow the spatial extent of projected changes in precipitation (see Section 3.3.3). Emerging literature includes runoff projections for different warming levels. For 2°C of global warming, an increase in runoff is projected for much of the high northern latitudes, Southeast Asia, East Africa, northeastern Europe, India, and parts of, Austria, China, Hungary, Norway, Sweden, the northwest Balkans and Sahel (Schleussner et al., 2016b; Donnelly et al., 2017; Döll et al., 2018; Zhai et al., 2018). Additionally, decreases are projected in the Mediterranean region, southern Australia, Central America, and central and southern South

America (Schleussner et al., 2016b; Donnelly et al., 2017; Döll et al., 2018). Differences between 1.5°C and 2°C would be most prominent in the Mediterranean, where the median reduction in annual runoff is expected to be about 9% (likely range 4.5–15.5%) at 1.5°C, while at 2°C of warming runoff could decrease by 17% (likely range 8–25%) (Schleussner et al., 2016b). Consistent with these projections, Döll et al. (2018) found that statistically insignificant changes in the mean annual streamflow around the Mediterranean region became significant when the global warming scenario was changed from 1.5°C to 2°C, with decreases of 10–30% between these two warming levels. Donnelly et al. (2017) found an intense decrease in runoff along both the Iberian and Balkan coasts with an increase in warming level.

Basin-scale projections of river runoff at different warming levels are available for many regions. Betts et al. (2018) assessed runoff changes in 21 of the world's major river basins at 1.5°C and 2°C of global warming (Figure 3.15). They found a general tendency towards increased runoff, except in the Amazon, Orange, Danube and Guadiana basins where the range of projections indicate decreased mean flows (Figure 3.13). In the case of the Amazon, mean flows are projected to decline by up to 25% at 2°C global warming (Betts et al., 2018).

Gosling et al. (2017) analysed the impact of global warming of 1°C, 2°C and 3°C above pre-industrial levels on river runoff at the catchment scale, focusing on eight major rivers in different continents: Upper Amazon, Darling, Ganges, Lena, Upper Mississippi, Upper Niger, Rhine and Tagus. Their results show that the sign and magnitude of change with global warming for the Upper Amazon, Darling, Ganges, Upper Niger and Upper Mississippi is unclear, while the Rhine and Tagus may experience decreases in projected runoff and the Lena may experience increases. Donnelly et al. (2017) analysed the mean flow response to different warming levels for six major European rivers: Glomma, Wisla, Lule, Ebro, Rhine and Danube. Consistent with the increases in mean runoff projected for large parts of northern Europe, the Glomma, Wisla and Lule rivers could experience increased discharges with global warming while discharges from the Ebro could decrease, in part due to a decrease in runoff in southern Europe. In the case of the Rhine and Danube rivers, Donnelly et al. (2017) did not find clear results. Mean annual runoff of the Yiluo River catchment in northern China is projected to decrease by 22% at 1.5°C and by 21% at 2°C, while the mean annual runoff for the Beijiang River catchment in southern China is projected to increase by less than 1% at 1.5°C and 3% at 2°C in comparison to the studied baseline period (L. Liu et al., 2017).

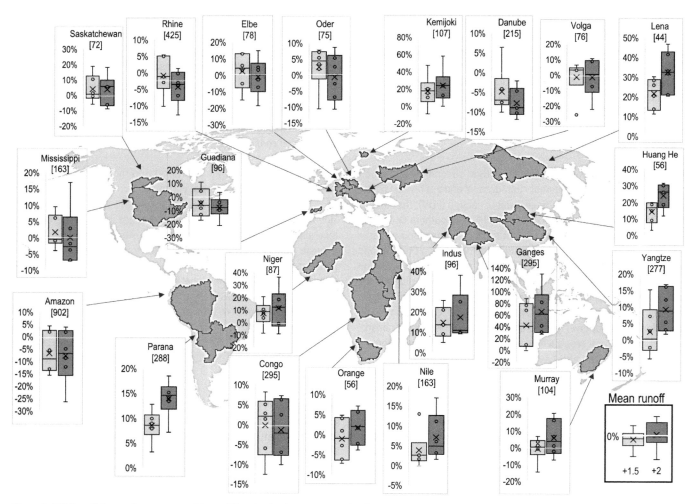

Figure 3.15 | Runoff changes in twenty-one of the world's major river basins at 1.5°C (blue) and 2°C (orange) of global warming, simulated by the Joint UK Land Environment Simulator (JULES) ecosystem–hydrology model under the ensemble of six climate projections. Boxes show the 25th and 75th percentile changes, whiskers show the range, circles show the four projections that do not define the ends of the range, and crosses show the ensemble means. Numbers in square brackets show the ensemble-mean flow in the baseline (millimetres of rain equivalent) (Source: Betts et al., 2018).

Chen et al. (2017) assessed the future changes in water resources in the Upper Yangtze River basin for the same warming levels and found a slight decrease in the annual discharge at 1.5°C but a slight increase at 2°C. Montroull et al. (2018) studied the hydrological impacts of the main rivers (Paraguay, Paraná, Iguazú and Uruguay) in La Plata basin in South America under 1.5°C and 2°C of global warming and for two emissions scenarios. The Uruguay basin shows increases in streamflow for all scenarios/warming targets except for the combination of RCP8.5/1.5°C of warming. The increase is approximately 15% above the 1981–2000 reference period for 2°C of global warming and the RCP4.5 scenario. For the other three rivers the sign of the change in mean streamflow depends strongly on the RCP and GCM used.

Marx et al. (2018) analysed how hydrological low flows in Europe are affected under different global warming levels (1.5°C, 2°C and 3°C). The Alpine region showed the strongest low flow increase, from 22% at 1.5°C to 30% at 2°C, because of the relatively large snow melt contribution, while in the Mediterranean low flows are expected to decrease because of the decreases in annual precipitation projected for that region. Döll et al. (2018) found that extreme low flows in the tropical Amazon, Congo and Indonesian basins could decrease by 10% at 1.5°C, whereas they could increase by 30% in the southwestern part of Russia under the same warming level. At 2°C, projected increases in extreme low flows are exacerbated in the higher northern latitudes and in eastern Africa, India and Southeast Asia, while projected decreases intensify in the Amazon basin, western United States, central Canada, and southern and western Europe, although not in the Congo basin or Indonesia, where models show less agreement.

Recent analyses of projections in river flooding and extreme runoff and flows are available for different global warming levels. At the global scale, Alfieri et al. (2017) assessed the frequency and magnitude of river floods and their impacts under 1.5°C, 2°C and 4°C global warming scenarios. They found that flood events with an occurrence interval longer than the return period of present-day flood protections are projected to increase in all continents under all considered warming levels, leading to a widespread increment in the flood hazard. Döll et al. (2018) found that high flows are projected to increase significantly on 11% and 21% of the global land area at 1.5°C and 2°C, respectively. Significantly increased high flows are expected to occur in South and Southeast Asia and Central Africa at 1.5°C, with this effect intensifying and including parts of South America at 2°C.

Regarding the continental scale, Donnelly et al. (2017) and Thober et al. (2018) explored climate change impacts on European high flows and/or floods under 1.5°C, 2°C and 3°C of global warming. Thober et al. (2018) identified the Mediterranean region as a hotspot of change, with significant decreases in high flows of −11% and −13% at 1.5°C and 2°C, respectively, mainly resulting from reduced precipitation (Box 3.2). In northern regions, high flows are projected to rise by 1% and 5% at 1.5°C and 2°C, respectively, owing to increasing precipitation, although floods could decrease by 6% in both scenarios because of less snowmelt. Donnelly et al. (2017) found that high runoff levels could rise in intensity, robustness and spatial extent over large parts of continental Europe with an increasing warming level. At 2°C, flood magnitudes are expected to increase significantly in Europe south of 60°N, except for some regions (Bulgaria, Poland and southern Spain);

in contrast, they are projected to decrease at higher latitudes (e.g., in most of Finland, northwestern Russia and northern Sweden), with the exception of southern Sweden and some coastal areas in Norway where flood magnitudes may increase (Roudier et al., 2016). At the basin scale, Mohammed et al. (2017) found that floods are projected to be more frequent and flood magnitudes greater at 2°C than at 1.5°C in the Brahmaputra River in Bangladesh. In coastal regions, increases in heavy precipitation associated with tropical cyclones (Section 3.3.6) combined with increased sea levels (Section 3.3.9) may lead to increased flooding (Section 3.4.5).

In summary, there is *medium confidence* that global warming of 2°C above the pre-industrial period would lead to an expansion of the area with significant increases in runoff, as well as the area affected by flood hazard, compared to conditions at 1.5°C of global warming. A global warming of 1.5°C would also lead to an expansion of the global land area with significant increases in runoff (*medium confidence*) and to an increase in flood hazard in some regions (*medium confidence*) compared to present-day conditions.

3.3.6 Tropical Cyclones and Extratropical Storms

Most recent studies on observed trends in the attributes of tropical cyclones have focused on the satellite era starting in 1979 (Rienecker et al., 2011), but the study of observed trends is complicated by the heterogeneity of constantly advancing remote sensing techniques and instrumentation during this period (e.g., Landsea, 2006; Walsh et al., 2016). Numerous studies leading up to and after AR5 have reported a decreasing trend in the global number of tropical cyclones and/or the globally accumulated cyclonic energy (Emanuel, 2005; Elsner et al., 2008; Knutson et al., 2010; Holland and Bruyère, 2014; Klotzbach and Landsea, 2015; Walsh et al., 2016). A theoretical physical basis for such a decrease to occur under global warming was recently provided by Kang and Elsner (2015). However, using a relatively short (20 year) and relatively homogeneous remotely sensed record, Klotzbach (2006) reported no significant trends in global cyclonic activity, consistent with more recent findings of Holland and Bruyère (2014). Such contradictions, in combination with the fact that the almost four-decade-long period of remotely sensed observations remains relatively short to distinguish anthropogenically induced trends from decadal and multi-decadal variability, implies that there is only *low confidence* regarding changes in global tropical cyclone numbers under global warming over the last four decades.

Studies in the detection of trends in the occurrence of very intense tropical cyclones (category 4 and 5 hurricanes on the Saffir-Simpson scale) over recent decades have yielded contradicting results. Most studies have reported increases in these systems (Emanuel, 2005; Webster et al., 2005; Klotzbach, 2006; Elsner et al., 2008; Knutson et al., 2010; Holland and Bruyère, 2014; Walsh et al., 2016), in particular for the North Atlantic, North Indian and South Indian Ocean basins (e.g., Singh et al., 2000; Singh, 2010; Kossin et al., 2013; Holland and Bruyère, 2014; Walsh et al., 2016). In the North Indian Ocean over the Arabian Sea, an increase in the frequency of extremely severe cyclonic storms has been reported and attributed to anthropogenic warming (Murakami et al., 2017). However, to the east over the Bay of Bengal, tropical cyclones and severe tropical cyclones have exhibited decreasing trends over the period 1961–2010, although the ratio between severe tropical cyclones and all tropical cyclones is increasing (Mohapatra et al., 2017). Moreover, studies that have used more homogeneous records, but were consequently limited to rather short periods of 20 to 25 years, have reported no statistically significant trends or decreases in the global number of these systems (Kamahori et al., 2006; Klotzbach and Landsea, 2015). Likewise, CMIP5 model simulations of the historical period have not produced anthropogenically induced trends in very intense tropical cyclones (Bender et al., 2010; Knutson et al., 2010, 2013; Camargo, 2013; Christensen et al., 2013), consistent with the findings of Klotzbach and Landsea (2015). There is consequently *low confidence* in the conclusion that the number of very intense cyclones is increasing globally.

General circulation model (GCM) projections of the changing attributes of tropical cyclones under high levels of greenhouse gas forcing (3°C to 4°C of global warming) consistently indicate decreases in the global number of tropical cyclones (Knutson et al., 2010, 2015; Sugi and Yoshimura, 2012; Christensen et al., 2013; Yoshida et al., 2017). A smaller number of studies based on statistical downscaling methodologies contradict these findings, however, and indicate increases in the global number of tropical cyclones under climate change (Emanuel, 2017). Most studies also indicate increases in the global number of very intense tropical cyclones under high levels of global warming (Knutson et al., 2015; Sugi et al., 2017), consistent with dynamic theory (Kang and Elsner, 2015), although a few studies contradict this finding (e.g., Yoshida et al., 2017). Hence, it is assessed that under 3°C to 4°C of warming that the global number of tropical cyclones would decrease whilst the number of very intense cyclones would increase (*medium confidence*).

To date, only two studies have directly explored the changing tropical cyclone attributes under 1.5°C versus 2°C of global warming. Using a high resolution global atmospheric model, Wehner et al. (2018a) concluded that the differences in tropical cyclone statistics under 1.5°C versus 2°C stabilization scenarios, as defined by the HAPPI protocols (Mitchell et al., 2017) are small. Consistent with the majority of studies performed for higher degrees of global warming, the total number of tropical cyclones is projected to decrease under global warming, whilst the most intense (categories 4 and 5) cyclones are projected to occur more frequently. These very intense storms are projected to be associated with higher peak wind speeds and lower central pressures under 2°C versus 1.5°C of global warming. The accumulated cyclonic energy is projected to decrease globally from 1.5°C to 2°C, in association with a decrease in the global number of tropical cyclones under progressively higher levels of global warming. It is also noted that heavy rainfall associated with tropical cyclones was assessed in the IPCC SREX as *likely* to increase under increasing global warming (Seneviratne et al., 2012). Two recent articles suggest that there is *high confidence* that the current level of global warming (i.e., about 1°C, see Section 3.3.1) increased the heavy precipitation associated with the 2017 Hurricane Harvey by about 15% or more (Risser and Wehner, 2017; van Oldenborgh et al., 2017). Hence, it can be inferred, under the assumption of linear dynamics, that further increases in heavy precipitation would occur under 1.5°C, 2°C and higher levels of global warming (*medium confidence*). Using a high resolution regional climate model, Muthige et al. (2018) explored the effects of different

degrees of global warming on tropical cyclones over the southwest Indian Ocean, using transient simulations that downscaled a number of RCP8.5 GCM projections. Decreases in tropical cyclone frequencies are projected under both 1.5°C and 2°C of global warming. The decreases in cyclone frequencies under 2°C of global warming are somewhat larger than under 1.5°C, but no further decreases are projected under 3°C. This suggests that 2°C of warming, at least in these downscaling simulations, represents a type of stabilization level in terms of tropical cyclone formation over the southwest Indian Ocean and landfall over southern Africa (Muthige et al., 2018). There is thus *limited evidence* that the global number of tropical cyclones will be lower under 2°C compared to 1.5°C of global warming, but with an increase in the number of very intense cyclones (*low confidence*).

The global response of the mid-latitude atmospheric circulation to 1.5°C and 2°C of warming was investigated using the HAPPI ensemble with a focus on the winter season (Li et al., 2018). Under 1.5°C of global warming a weakening of storm activity over North America, an equatorward shift of the North Pacific jet exit and an equatorward intensification of the South Pacific jet are projected. Under an additional 0.5°C of warming a poleward shift of the North Atlantic jet exit and an intensification on the flanks of the Southern Hemisphere storm track are projected to become more pronounced. The weakening of the Mediterranean storm track that is projected under low mitigation emerges in the 2°C warmer world (Li et al., 2018). AR5 assessed that under high greenhouse gas forcing (3°C or 4°C of global warming) there is *low confidence* in projections of poleward shifts of the Northern Hemisphere storm tracks, while there is *high confidence* that there would be a small poleward shift of the Southern Hemisphere storm tracks (Stocker et al., 2013). In the context of this report, the assessment is that there is *limited evidence* and *low confidence* in whether any projected signal for higher levels of warming would be clearly manifested under 2°C of global warming.

3.3.7   Ocean Circulation and Temperature

It is *virtually certain* that the temperature of the upper layers of the ocean (0–700 m in depth) has been increasing, and that the global mean for sea surface temperature (SST) has been changing at a rate just behind that of GMST. The surfaces of three ocean basins has warmed over the period 1950–2016 (by 0.11°C, 0.07°C and 0.05°C per decade for the Indian, Atlantic and Pacific Oceans, respectively; Hoegh-Guldberg et al., 2014), with the greatest changes occurring at the highest latitudes. Isotherms (i.e., lines of equal temperature) of sea surface temperature (SST) are shifting to higher latitudes at rates of up to 40 km per year (Burrows et al., 2014; García Molinos et al., 2015). Long-term patterns of variability make detecting signals due to climate change complex, although the recent acceleration of changes to the temperature of the surface layers of the ocean has made the climate signal more distinct (Hoegh-Guldberg et al., 2014). There is also evidence of significant increases in the frequency of marine heatwaves in the observational record (Oliver et al., 2018), consistent with changes in mean ocean temperatures (*high confidence*). Increasing climate extremes in the ocean are associated with the general rise in global average surface temperature, as well as more intense patterns of climate variability (e.g., climate change intensification of ENSO) (Section 3.5.2.5). Increased heat in the upper layers of the ocean is also driving more intense storms and greater rates of inundation in some regions, which, together with sea level rise, are already driving significant impacts to sensitive coastal and low-lying areas (Section 3.3.6).

Increasing land–sea temperature gradients have the potential to strengthen upwelling systems associated with the eastern boundary currents (Benguela, Canary, Humboldt and Californian Currents; Bakun, 1990). Observed trends support the conclusion that a general strengthening of longshore winds has occurred (Sydeman et al., 2014), but the implications of trends detected in upwelling currents themselves are unclear (Lluch-Cota et al., 2014). Projections of the scale of changes between 1°C and 1.5°C of global warming and between 1.5°C and 2°C are only informed by the changes during the past increase in GMST of 0.5°C (*low confidence*). However, evidence from GCM projections of future climate change indicates that a general strengthening of the Benguela, Canary and Humboldt upwelling systems under enhanced anthropogenic forcing (D. Wang et al., 2015) is projected to occur (*medium confidence*). This strengthening is projected to be stronger at higher latitudes. In fact, evidence from regional climate modelling is supportive of an increase in long-shore winds at higher latitudes, whereas long-shore winds may decrease at lower latitudes as a consequence of the poleward displacement of the subtropical highs under climate change (Christensen et al., 2007; Engelbrecht et al., 2009).

It is more likely than not that the Atlantic Meridional Overturning Circulation (AMOC) has been weakening in recent decades, given the detection of the cooling of surface waters in the North Atlantic and evidence that the Gulf Stream has slowed since the late 1950s (Rahmstorf et al., 2015b; Srokosz and Bryden, 2015; Caesar et al., 2018). There is only *limited evidence* linking the current anomalously weak state of AMOC to anthropogenic warming (Caesar et al., 2018). It is *very likely* that the AMOC will weaken over the 21st century. The best estimates and ranges for the reduction based on CMIP5 simulations are 11% (1– 24%) in RCP2.6 and 34% (12– 54%) in RCP8.5 (AR5). There is *no evidence* indicating significantly different amplitudes of AMOC weakening for 1.5°C versus 2°C of global warming.

3.3.8 Sea Ice

Summer sea ice in the Arctic has been retreating rapidly in recent decades. During the period 1997 to 2014, for example, the monthly mean sea ice extent during September (summer) decreased on average by 130,000 km² per year (Serreze and Stroeve, 2015). This is about four times as fast as the September sea ice loss during the period 1979 to 1996. Sea ice thickness has also decreased substantially, with an estimated decrease in ice thickness of more than 50% in the central Arctic (Lindsay and Schweiger, 2015). Sea ice coverage and thickness also decrease in CMIP5 simulations of the recent past, and are projected to decrease in the future (Collins et al., 2013). However, the modelled sea ice loss in most CMIP5 models is much smaller than observed losses. Compared to observations, the simulations are less sensitive to both global mean temperature rise (Rosenblum and Eisenman, 2017) and anthropogenic CO_2 emissions (Notz and Stroeve, 2016). This mismatch between the observed and modelled sensitivity of Arctic sea ice implies that the multi-model-mean responses of future sea ice evolution probably underestimates the sea ice loss for a given amount of global warming. To address this issue, studies estimating the future evolution of Arctic sea ice tend to bias correct the model simulations based on the observed evolution of Arctic sea ice in response to global warming. Based on such bias correction, pre-AR5 and post-AR5 studies generally agree that for 1.5°C of global warming relative to pre-industrial levels, the Arctic Ocean will maintain a sea ice cover throughout summer in most years (Collins et al., 2013; Notz and Stroeve, 2016; Screen and Williamson, 2017; Jahn, 2018; Niederdrenk and Notz, 2018; Sigmond et al., 2018). For 2°C of global warming, chances of a sea ice-free Arctic during summer are substantially higher (Screen and Williamson, 2017; Jahn, 2018; Niederdrenk and Notz, 2018; Screen et al., 2018; Sigmond et al., 2018). Model simulations suggest that there will be at least one sea ice-free Arctic[5] summer after approximately 10 years of stabilized warming at 2°C, as compared to one sea ice-free summer after 100 years of stabilized warming at 1.5°C above pre-industrial temperatures (Jahn, 2018; Screen et al., 2018; Sigmond et al., 2018). For a specific given year under stabilized warming of 2°C, studies based on large ensembles of simulations with a single model estimate the likelihood of ice-free conditions as 35% without a bias correction of the underlying model (Sanderson et al., 2017; Jahn, 2018); as between 10% and >99% depending on the observational record used to correct the sensitivity of sea ice decline to global warming in the underlying model (Niederdrenk and Notz, 2018); and as 19% based on a procedure to correct for biases in the climatological sea ice coverage in the underlying model (Sigmond et al., 2018). The uncertainty of the first year of the occurrence of an ice-free Arctic Ocean arising from internal variability is estimated to be about 20 years (Notz, 2015; Jahn et al., 2016).

The more recent estimates of the warming necessary to produce an ice-free Arctic Ocean during summer are lower than the ones given in AR5 (about 2.6°C–3.1°C of global warming relative to pre-industrial levels or 1.6°C–2.1°C relative to present-day conditions), which were similar to the estimate of 3°C of global warming relative to pre-industrial levels (or 2°C relative to present-day conditions) by Mahlstein and Knutti (2012) based on bias-corrected CMIP3 models. Rosenblum and Eisenman (2016) explained why the sensitivity estimated by Mahlstein and Knutti (2012) might be too low, estimating instead that September sea ice in the Arctic would disappear at 2°C of global warming relative to pre-industrial levels (or about 1°C relative to present-day conditions), in line with the other recent estimates. Notz and Stroeve (2016) used the observed correlation between September sea ice extent and cumulative CO_2 emissions to estimate that the Arctic Ocean would become nearly free of sea ice during September with a further 1000 Gt of emissions, which also implies a sea ice loss at about 2°C of global warming. Some of the uncertainty in these numbers stems from the possible impact of aerosols (Gagne et al., 2017) and of volcanic forcing (Rosenblum and Eisenman, 2016). During winter, little Arctic sea ice is projected to be lost for either 1.5°C or 2°C of global warming (Niederdrenk and Notz, 2018).

---

[5] Ice free is defined for the Special Report as when the sea ice extent is less than 106 km². Ice coverage less than this is considered to be equivalent to an ice-free Arctic Ocean for practical purposes in all recent studies.

A substantial number of pre-AR5 studies found that there is no indication of hysteresis behaviour of Arctic sea ice under decreasing temperatures following a possible overshoot of a long-term temperature target (Holland et al., 2006; Schröder and Connolley, 2007; Armour et al., 2011; Sedláček et al., 2011; Tietsche et al., 2011; Boucher et al., 2012; Ridley et al., 2012). In particular, the relationship between Arctic sea ice coverage and GMST was found to be indistinguishable between a warming scenario and a cooling scenario. These results have been confirmed by post-AR5 studies (Li et al., 2013; Jahn, 2018), which implies *high confidence* that an intermediate temperature overshoot has no long-term consequences for Arctic sea ice coverage.

In the Antarctic, sea ice shows regionally contrasting trends, such as a strong decrease in sea ice coverage near the Antarctic peninsula but increased sea ice coverage in the Amundsen Sea (Hobbs et al., 2016). Averaged over these contrasting regional trends, there has been a slow long-term increase in overall sea ice coverage in the Southern Ocean, although with comparably low ice coverage from September 2016 onwards. Collins et al. (2013) assessed *low confidence* in Antarctic sea ice projections because of the wide range of model projections and an inability of almost all models to reproduce observations such as the seasonal cycle, interannual variability and the long-term slow increase. No existing studies have robustly assessed the possible future evolution of Antarctic sea ice under low-warming scenarios.

In summary, the probability of a sea-ice-free Arctic Ocean during summer is substantially higher at 2°C compared to 1.5°C of global warming relative to pre-industrial levels, and there is *medium confidence* that there will be at least one sea ice-free Arctic summer after about 10 years of stabilized warming at 2°C, while about 100 years are required at 1.5°C. There is *high confidence* that an intermediate temperature overshoot has no long-term consequences for Arctic sea ice coverage with regrowth on decadal time scales.

3.3.9   Sea Level

Sea level varies over a wide range of temporal and spatial scales, which can be divided into three broad categories. These are global mean sea level (GMSL), regional variation about this mean, and the occurrence of sea-level extremes associated with storm surges and tides. GMSL has been rising since the late 19th century from the low rates of change that characterized the previous two millennia (Church et al., 2013). Slowing in the reported rate over the last two decades (Cazenave et al., 2014) may be attributable to instrumental drift in the observing satellite system (Watson et al., 2015) and increased volcanic activity (Fasullo et al., 2016). Accounting for the former results in rates (1993 to mid-2014) between 2.6 and 2.9 mm yr^{-1} (Watson et al., 2015). The relative contributions from thermal expansion, glacier and ice-sheet mass loss, and freshwater storage on land are relatively well understood (Church et al., 2013; Watson et al., 2015) and their attribution is dominated by anthropogenic forcing since 1970 (15 ± 55% before 1950, 69 ± 31% after 1970) (Slangen et al., 2016).

There has been a significant advance in the literature since AR5, which has included the development of semi-empirical models (SEMs) into a broader emulation-based approach (Kopp et al., 2014; Mengel et al., 2016; Nauels et al., 2017) that is partially based on the results from

more detailed, process-based modelling Church et al. (2013) assigned *low confidence* to SEMs because these models assume that the relation between climate forcing and GMSL is the same in the past (calibration) and future (projection). Probable future changes in the relative contributions of thermal expansion, glaciers and (in particular) ice sheets invalidate this assumption. However, recent emulation-based studies overcame this shortcoming by considering individual GMSL contributors separately, and they are therefore employed in this assessment. In this subsection, the process-based literature of individual contributors to GMSL is considered for scenarios close to 1.5°C and 2°C of global warming before emulation-based approaches are assessed.

A limited number of processes-based studies are relevant to GMSL in 1.5°C and 2°C worlds. Marzeion et al. (2018) used a global glacier model with temperature-scaled scenarios based on RCP2.6 to investigate the difference between 1.5°C and 2°C of global warming and found little difference between scenarios in the glacier contribution to GMSL for the year 2100 (54–97 mm relative to present-day levels for 1.5°C and 63–112 mm for 2°C, using a 90% confidence interval). This arises because glacier melt during the remainder of the century is dominated by the response to warming from pre-industrial to present-day levels, which is in turn a reflection of the slow response times of glaciers. Fürst et al. (2015) made projections of the Greenland ice sheet's contribution to GMSL using an ice-flow model forced by the regional climate model Modèle Atmosphérique Régional (MAR; considered by Church et al. (2013) to be the 'most realistic' such model). They projected an RCP2.6 range of 24–60 mm (1 standard deviation) by the end of the century (relative to the year 2000 and consistent with the assessment of Church et al. (2013); however, their projections do not allow the difference between 1.5°C and 2°C worlds to be evaluated.

The Antarctic ice sheet can contribute both positively, through increases in outflow (solid ice lost directly to the ocean), and negatively, through increases in snowfall (owing to the increased moisture-bearing capacity of a warmer atmosphere), to future GMSL rise. Frieler et al. (2015) suggested a range of 3.5–8.7% °C^{-1} for this effect, which is consistent with AR5. Observations from the Amundsen Sea sector of Antarctica suggest an increase in outflow (Mouginot et al., 2014) over recent decades associated with grounding line retreat (Rignot et al., 2014) and the influx of relatively warm Circumpolar Deepwater (Jacobs et al., 2011). Literature on the attribution of these changes to anthropogenic forcing is still in its infancy (Goddard et al., 2017; Turner et al., 2017a). RCP2.6-based projections of Antarctic outflow (Levermann et al., 2014; Golledge et al., 2015; DeConto and Pollard, 2016, who include snowfall changes) are consistent with the AR5 assessment of Church et al. (2013) for end-of-century GMSL for RCP2.6, and do not support substantial additional GMSL rise by Marine Ice Sheet Instability or associated instabilities (see Section 3.6). While agreement is relatively good, concerns about the numerical fidelity of these models still exist, and this may affect the quality of their projections (Drouet et al., 2013; Durand and Pattyn, 2015). An assessment of Antarctic contributions beyond the end of the century, in particular related to the Marine Ice Sheet Instability, can be found in Section 3.6.

While some literature on process-based projections of GMSL for the period up to 2100 is available, it is insufficient for distinguishing between emissions scenarios associated with 1.5°C and 2°C warmer worlds. This literature is, however, consistent with the assessment by Church et al. (2013) of a *likely* range of 0.28–0.61 m in 2100 (relative to 1986–2005), suggesting that the AR5 assessment is still appropriate.

Recent emulation-based studies show convergence towards this AR5 assessment (Table 3.1) and offer the advantage of allowing a comparison between 1.5°C and 2°C warmer worlds. Table 3.1 features a compilation of recent emulation-based and SEM studies.

Table 3.1 | Compilation of recent projections for sea level at 2100 (in cm) for Representative Concentration Pathway (RCP)2.6, and 1.5°C and 2°C scenarios. Upper and lower limits are shown for the 17-84% and 5-95% confidence intervals quoted in the original papers.

Study	Baseline	RCP2.6		1.5°C		2°C	
		67%	90%	67%	90%	67%	90%
AR5	1986–2005	28–61					
Kopp et al. (2014)	2000	37–65	29–82				
Jevrejeva et al. (2016)	1986–2005		29–58				
Kopp et al. (2016)	2000	28–51	24–61				
Mengel et al. (2016)	1986–2005	28–56					
Nauels et al. (2017)	1986–2005	35–56					
Goodwin et al. (2017)	1986–2005		31–59 45–70 45–72				
Schaeffer et al. (2012)	2000		52–96		54–99		56–105
Schleussner et al. (2016b)	2000			26–53		36–65	
Bittermann et al. (2017)	2000				29–46		39–61
Jackson et al. (2018)	1986–2005			30–58 40–77	20–67 28–93	35–64 47–93	24–74 32–117
Sanderson et al. (2017)					50–80		60–90
Nicholls et al. (2018)	1986–2005				24–54		31–65
Rasmussen et al. (2018)	2000			35–64	28–82	39–76	28–96
Goodwin et al. (2018)	1986–2005				26–62		30–69

There is little consensus between the reported ranges of GMSL rise (Table 3.1). Projections vary in the range 0.26–0.77 m and 0.35–0.93 m for 1.5°C and 2°C respectively for the 17–84% confidence interval (0.20–0.99 m and 0.24–1.17 m for the 5–95% confidence interval). There is, however, *medium agreement* that GMSL in 2100 would be 0.04–0.16 m higher in a 2°C warmer world compared to a 1.5°C warmer world based on the 17–84% confidence interval (0.00–0.24 m based on 5–95% confidence interval) with a value of around 0.1 m. There is *medium confidence* in this assessment because of issues associated with projections of the Antarctic contribution to GMSL that are employed in emulation-based studies (see above) and the issues previously identified with SEMs (Church et al., 2013).

Translating projections of GMSL to the scale of coastlines and islands requires two further steps. The first step accounts for regional changes associated with changing water and ice loads (such as Earth's gravitational field and rotation, and vertical land movement), as well as spatial differences in ocean heat uptake and circulation. The second step maps regional sea level to changes in the return periods of particular flood events to account for effects not included in global climate models, such as tides, storm surges, and wave setup and runup. Kopp et al. (2014) presented a framework to do this and gave an example application for nine sites located in the US, Japan, northern Europe and Chile. Of these sites, seven (all except those in northern Europe) were found to experience at least a quadrupling in the number of years in the 21st century with 1-in-100-year floods under RCP2.6 compared to under no future sea level rise. Rasmussen

et al. (2018) used this approach to investigate the difference between 1.5°C and 2°C warmer worlds up to 2200. They found that the reduction in the frequency of 1-in-100-year floods in a 1.5°C compared to a 2°C warmer world would be greatest in the eastern USA and Europe, with ESL event frequency amplification being reduced by about a half and with smaller reductions for small island developing states (SIDS). This last result contrasts with the finding of Vitousek et al. (2017) that regions with low variability in extreme water levels (such as SIDS in the tropics) are particularly sensitive to GMSL rise, such that a doubling of frequency may be expected for even small (0.1–0.2 m) rises. Schleussner et al. (2011) emulated the AMOC based on a subset of CMIP-class climate models. When forced using global temperatures appropriate for the CP3-PD scenario (1°C of warming in 2100 relative to 2000 or about 2°C of warming relative to pre-industrial) the emulation suggests an 11% median reduction in AMOC strength at 2100 (relative to 2000) with an associated 0.04 m dynamic sea level rise along the New York City coastline.

In summary, there is *medium confidence* that GMSL rise will be about 0.1 m (within a 0.00–0.20 m range based on 17–84% confidence-interval projections) less by the end of the 21st century in a 1.5°C compared to a 2°C warmer world. Projections for 1.5°C and 2°C global warming cover the ranges 0.2–0.8 m and 0.3–1.00 m relative to 1986–2005, respectively (*medium confidence*). Sea level rise beyond 2100 is discussed in Section 3.6; however, recent literature strongly supports the assessment by Church et al. (2013) that sea level rise will continue well beyond 2100 (*high confidence*).

Box 3.3 | Lessons from Past Warm Climate Episodes

Climate projections and associated risk assessments for a future warmer world are based on climate model simulations. However, Coupled Model Intercomparison Project Phase 5 (CMIP5) climate models do not include all existing Earth system feedbacks and may therefore underestimate both rates and extents of changes (Knutti and Sedláček, 2012). Evidence from natural archives of three moderately warmer (1.5°C–2°C) climate episodes in Earth's past help to assess such long-term feedbacks (Fischer et al., 2018).

While evidence over the last 2000 years and during the Last Glacial Maximum (LGM) was discussed in detail in the IPCC Fifth Assessment Report (Masson-Delmotte et al., 2013), the climate system response during past warm intervals was the focus of a recent review paper (Fischer et al., 2018) summarized in this Box. Examples of past warmer conditions with essentially modern physical geography include the Holocene Thermal Maximum (HTM; broadly defined as about 10–5 kyr before present (BP), where present is defined as 1950), the Last Interglacial (LIG; about 129–116 kyr BP) and the Mid Pliocene Warm Period (MPWP; 3.3–3.0 Myr BP).

Changes in insolation forcing during the HTM (Marcott et al., 2013) and the LIG (Hoffman et al., 2017) led to a global temperature up to 1°C higher than that in the pre-industrial period (1850–1900); high-latitude warming was 2°C–4°C (Capron et al., 2017), while temperature in the tropics changed little (Marcott et al., 2013). Both HTM and LIG experienced atmospheric CO_2 levels similar to pre-industrial conditions (Masson-Delmotte et al. 2013). During the MPWP, the most recent time period when CO_2 concentrations were similar to present-day levels, the global temperature was >1°C and Arctic temperatures about 8°C warmer than pre-industrial (Brigham-Grette et al., 2013).

Although imperfect as analogues for the future, these regional changes can inform risk assessments such as the potential for crossing irreversible thresholds or amplifying anthropogenic changes (Box 3.3, Figure 1). For example, HTM and LIG greenhouse gas (GHG) concentrations show no evidence of runaway greenhouse gas releases under limited global warming. Transient releases of CO_2 and CH_4 may follow permafrost melting, but these occurrences may be compensated by peat growth over longer time scales (Yu et al., 2010). Warming may release CO_2 by enhancing soil respiration, counteracting CO_2 fertilization of plant growth (Frank et al., 2010). Evidence of a collapse of the Atlantic Meridional Overturning Circulation (AMOC) during these past events of limited global warming could not be found (Galaasen et al., 2014).

The distribution of ecosystems and biomes (major ecosystem types) changed significantly during past warming events, both in the ocean and on land. For example, some tropical and temperate forests retreated because of increased aridity, while savannas expanded (Dowsett et al., 2016). Further, poleward shifts of marine and terrestrial ecosystems, upward shifts in alpine regions, and reorganizations of marine productivity during past warming events are recorded in natural archives (Williams et al., 2009; Haywood et al., 2016). Finally, past warming events are associated with partial sea ice loss in the Arctic. The limited amount of data collected so far on Antarctic sea ice precludes firm conclusions about Southern Hemisphere sea ice losses (de Vernal et al., 2013).

Reconstructed global sea level rise of 6–9 m during the LIG and possibly >6 m during the MPWP requires a retreat of either the Greenland or Antarctic ice sheets or both (Dutton et al., 2015). While ice sheet and climate models suggest a substantial retreat of the West Antarctic ice sheet (WAIS) and parts of the East Antarctic ice sheet (DeConto and Pollard, 2016) during these periods, direct observational evidence is still lacking. Evidence for ice retreat in Greenland is stronger, although a complete collapse of the Greenland ice sheet during the LIG can be excluded (Dutton et al., 2015). Rates of past sea level rises under modest warming were similar to or up to two times larger than rises observed over the past two decades (Kopp et al., 2013). Given the long time scales required to reach equilibrium in a warmer world, sea level rise will likely continue for millennia even if warming is limited to 2°C.

Finally, temperature reconstructions from these past warm intervals suggest that current climate models underestimate regional warming at high latitudes (polar amplification) and long-term (multi-millennial) global warming. None of these past warm climate episodes involved the high rate of change in atmospheric CO_2 and temperatures that we are experiencing today (Fischer et al., 2018).

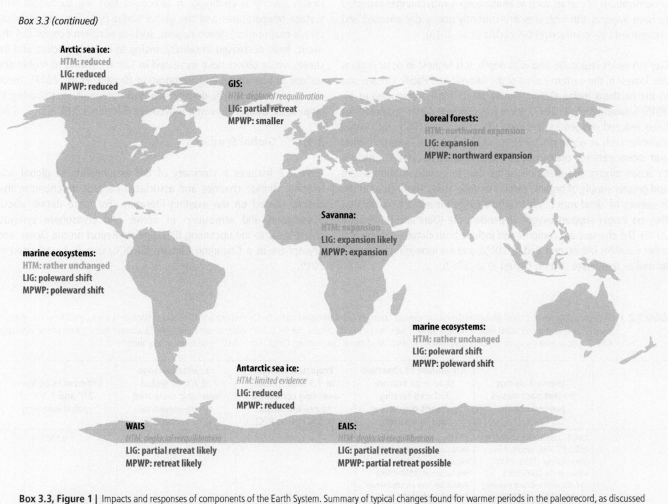

Box 3.3 (continued)

Arctic sea ice:
HTM: reduced
LIG: reduced
MPWP: reduced

GIS:
HTM: deglacial reequilibration
LIG: partial retreat
MPWP: smaller

boreal forests:
HTM: northward expansion
LIG: expansion
MPWP: northward expansion

Savanna:
HTM: expansion
LIG: expansion likely
MPWP: expansion

marine ecosystems:
HTM: rather unchanged
LIG: poleward shift
MPWP: poleward shift

marine ecosystems:
HTM: rather unchanged
LIG: poleward shift
MPWP: poleward shift

Antarctic sea ice:
HTM: limited evidence
LIG: reduced
MPWP: reduced

WAIS
HTM: deglacial reequilibration
LIG: partial retreat likely
MPWP: retreat likely

EAIS:
HTM: deglacial reequilibration
LIG: partial retreat possible
MPWP: partial retreat possible

Box 3.3, Figure 1 | Impacts and responses of components of the Earth System. Summary of typical changes found for warmer periods in the paleorecord, as discussed by Fischer et al. (2018). All statements are relative to pre-industrial conditions. Statements in italics indicate that no conclusions can be drawn for the future. Note that significant spatial variability and uncertainty exists in the assessment of each component, and this figure therefore should not be referred to without reading the publication in detail. HTM: Holocene Thermal Maximum, LIG: Last Interglacial, MPWP: Mid Pliocene Warm Period. (Adapted from Fischer et al., 2018).

3.3.10   Ocean Chemistry

Ocean chemistry includes pH, salinity, oxygen, CO_2, and a range of other ions and gases, which are in turn affected by precipitation, evaporation, storms, river runoff, coastal erosion, up-welling, ice formation, and the activities of organisms and ecosystems (Stocker et al., 2013). Ocean chemistry is changing alongside increasing global temperature, with impacts projected at 1.5°C and, more so, at 2°C of global warming (Doney et al., 2014) (*medium to high confidence*). Projected changes in the upper layers of the ocean include altered pH, oxygen content and sea level. Despite its many component processes, ocean chemistry has been relatively stable for long periods of time prior to the industrial period (Hönisch et al., 2012). Ocean chemistry is changing under the influence of human activities and rising greenhouse gases (*virtually certain*; Rhein et al., 2013; Stocker et al., 2013). About 30% of CO_2 emitted by human activities, for example, has been absorbed by the upper layers of the ocean, where it has combined with water to produce a dilute acid that dissociates and drives ocean acidification

(*high confidence*) (Cao et al., 2007; Stocker et al., 2013). Ocean pH has decreased by 0.1 pH units since the pre-industrial period, a shift that is unprecedented in the last 65 Ma (*high confidence*) (Ridgwell and Schmidt, 2010) or even 300 Ma of Earth's history (*medium confidence*) (Hönisch et al., 2012).

Ocean acidification is a result of increasing CO_2 in the atmosphere (*very high confidence*) and is most pronounced where temperatures are lowest (e.g., polar regions) or where CO_2-rich water is brought to the ocean surface by upwelling (Feely et al., 2008). Acidification can also be influenced by effluents from natural or disturbed coastal land use (Salisbury et al., 2008), plankton blooms (Cai et al., 2011), and the atmospheric deposition of acidic materials (Omstedt et al., 2015). These sources may not be directly attributable to climate change, but they may amplify the impacts of ocean acidification (Bates and Peters, 2007; Duarte et al., 2013). Ocean acidification also influences the ionic composition of seawater by changing the organic and inorganic speciation of trace metals (e.g., 20-fold increases in free ion concentrations of metals such as aluminium) – with changes expected to have impacts although they are currently poorly documented and understood (*low confidence*) (Stockdale et al., 2016).

Oxygen varies regionally and with depth; it is highest in polar regions and lowest in the eastern basins of the Atlantic and Pacific Oceans and in the northern Indian Ocean (Doney et al., 2014; Karstensen et al., 2015; Schmidtko et al., 2017). Increasing surface water temperatures have reduced oxygen in the ocean by 2% since 1960, with other variables such as ocean acidification, sea level rise, precipitation, wind and storm patterns playing roles (Schmidtko et al., 2017). Changes to ocean mixing and metabolic rates, due to increased temperature and greater supply of organic carbon to deep areas, has increased the frequency of 'dead zones', areas where oxygen levels are so low that they no longer support oxygen dependent life (Diaz and Rosenberg, 2008). The changes are complex and include both climate change and other variables (Altieri and Gedan, 2015), and are increasing in tropical as well as temperate regions (Altieri et al., 2017).

Ocean salinity is changing in directions that are consistent with surface temperatures and the global water cycle (i.e., precipitation versus evaporation). Some regions, such as northern oceans and the Arctic, have decreased in salinity, owing to melting glaciers and ice sheets, while others have increased in salinity, owing to higher sea surface temperatures and evaporation (Durack et al., 2012). These changes in salinity (i.e., density) are also potentially contributing to large-scale changes in water movement (Section 3.3.8).

3.3.11   Global Synthesis

Table 3.2 features a summary of the assessments of global and regional climate changes and associated hazards described in this chapter, based on the existing literature. For more details about observation and attribution in ocean and cryosphere systems, please refer to the upcoming IPCC Special Report on the Ocean and Cryosphere in a Changing Climate (SROCC) due to be released in 2019.

Table 3.2 | Summary of assessments of global and regional climate changes and associated hazards. Confidence and likelihood statements are quoted from the relevant chapter text and are omitted where no assessment was made, in which case the IPCC Fifth Assessment Report (AR5) assessment is given where available. GMST: global mean surface temperature, AMOC: Atlantic Meridional Overturning Circulation, GMSL: global mean sea level.

	Observed change (recent past versus pre-industrial)	Attribution of observed change to human-induced forcing (present-day versus pre-industrial)	Projected change at 1.5°C of global warming compared to pre-industrial (1.5°C versus 0°C)	Projected change at 2°C of global warming compared to pre-industrial (2°C versus 0°C)	Differences between 2°C and 1.5°C of global warming
GMST anomaly	GMST anomalies were 0.87°C (±0.10°C *likely* range) above pre-industrial (1850–1900) values in the 2006–2015 decade, with a recent warming of about 0.2°C (±0.10°C) per decade (*high confidence*) [Chapter 1]	The observed 0.87°C GMST increase in the 2006–2015 decade compared to pre-industrial (1850–1900) conditions was mostly human-induced (*high confidence*) Human-induced warming reached about 1°C (±0.2°C *likely* range) above pre-industrial levels in 2017 [Chapter 1]	1.5°C	2°C	0.5°C
Temperature extremes	Overall decrease in the number of cold days and nights and overall increase in the number of warm days and nights at the global scale on land (*very likely*) Continental-scale increase in intensity and frequency of hot days and nights, and decrease in intensity and frequency of cold days and nights, in North America, Europe and Australia (*very likely*) Increases in frequency or duration of warm spell lengths in large parts of Europe, Asia and Australia (*high confidence* (*likely*)), as well as at the global scale *(medium confidence)* [Section 3.3.2]	Anthropogenic forcing has contributed to the observed changes in frequency and intensity of daily temperature extremes on the global scale since the mid-20th century (*very likely*) [Section 3.3.2]	Global-scale increased intensity and frequency of hot days and nights, and decreased intensity and frequency of cold days and nights (*very likely*) Warming of temperature extremes highest over land, including many inhabited regions (*high confidence*), with increases of up to 3°C in the mid-latitude warm season and up to 4.5°C in the high-latitude cold season (*high confidence*) Largest increase in frequency of unusually hot extremes in tropical regions (*high confidence*) [Section 3.3.2]	Global-scale increased intensity and frequency of hot days and nights, and decreased intensity and frequency of cold days and nights (*very likely*) Warming of temperature extremes highest over land, including many inhabited regions (*high confidence*), with increases of up to 4°C in the mid-latitude warm season and up to 6°C in the high-latitude cold season (*high confidence*) Largest increase in frequency of unusually hot extremes in tropical regions (*high confidence*) [Section 3.3.2]	Global-scale increased intensity and frequency of hot days and nights, and decreased intensity and frequency of cold days and nights (*high confidence*) Global-scale increase in length of warm spells and decrease in length of cold spells (*high confidence*) Strongest increase in frequency for the rarest and most extreme events (*high confidence*) Particularly large increases in hot extremes in inhabited regions (*high confidence*) [Section 3.3.2]

Table 3.2 (continued)

	Observed change (recent past versus pre-industrial)	Attribution of observed change to human-induced forcing (present-day versus pre-industrial)	Projected change at 1.5°C of global warming compared to pre-industrial (1.5°C versus 0°C)	Projected change at 2°C of global warming compared to pre-industrial (2°C versus 0°C)	Differences between 2°C and 1.5°C of global warming
Heavy precipitation	More areas with increases than decreases in the frequency, intensity and/or amount of heavy precipitation (*likely*) [Section 3.3.3]	Human influence contributed to the global-scale tendency towards increases in the frequency, intensity and/or amount of heavy precipitation events (*medium confidence*) [Section 3.3.3; AR5 Chapter 10 (Bindoff et al., 2013a)]	Increases in frequency, intensity and/or amount heavy precipitation when averaged over global land, with positive trends in several regions (*high confidence*) [Section 3.3.3]	Increases in frequency, intensity and/or amount heavy precipitation when averaged over global land, with positive trends in several regions (*high confidence*) [Section 3.3.3]	Higher frequency, intensity and/or amount of heavy precipitation when averaged over global land, with positive trends in several regions (*medium confidence*) Several regions are projected to experience increases in heavy precipitation at 2°C versus 1.5°C (*medium confidence*), in particular in high-latitude and mountainous regions, as well as in eastern Asia and eastern North America (*medium confidence*) [Section 3.3.3]
Drought and dryness	*High confidence* in dryness trends in some regions, especially drying in the Mediterranean region (including southern Europe, northern Africa and the Near East) *Low confidence* in drought and dryness trends at the global scale [Section 3.3.4]	*Medium confidence* in attribution of drying trends in southern Europe (Mediterranean region) *Low confidence* elsewhere, in part due to large interannual variability and longer duration (and thus lower frequency) of drought events, as well as dependency on the dryness index definition applied [Section 3.3.4]	*Medium confidence* in drying trends in the Mediterranean region *Low confidence* elsewhere, in part due to large interannual variability and longer duration (and thus lower frequency) of drought events, as well as to dependency on the dryness index definition applied Increases in drought, dryness or precipitation deficits projected in some regions compared to the pre-industrial or present-day conditions, but substantial variability in signals depending on considered indices or climate model (*medium confidence*) [Section 3.3.4]	*Medium confidence* in drying trends in the Mediterranean region and Southern Africa *Low confidence* elsewhere, in part due to large interannual variability and longer duration (and thus lower frequency) of drought events, as well as to dependency on the dryness index definition applied Increases in drought, dryness or precipitation deficits projected in some regions compared to the pre-industrial or present-day conditions, but substantial variability in signals depending on considered indices or climate model (*medium confidence*). [Section 3.3.4]	*Medium confidence* in stronger drying trends in the Mediterranean region and Southern Africa *Low confidence* elsewhere, in part due to large interannual variability and longer duration (and thus lower frequency) of drought events, as well as to dependency on the dryness index definition applied [Section 3.3.4]
Runoff and river flooding	Streamflow trends mostly not statistically significant (*high confidence*) Increase in flood frequency and extreme streamflow in some regions (*high confidence*) [Section 3.3.5]	Not assessed in this report	Expansion of the global land area with a significant increase in runoff (*medium confidence*) Increase in flood hazard in some regions (*medium confidence*) [Section 3.3.5]	Expansion of the global land area with a significant increase in runoff (*medium confidence*) Increase in flood hazard in some regions (*medium confidence*) [Section 3.3.5]	Expansion of the global land area with significant increase in runoff (*medium confidence*) Expansion in the area affected by flood hazard (*medium confidence*) [Section 3.3.5]
Tropical and extra-tropical cyclones	*Low confidence* in the robustness of observed changes [Section 3.3.6]	Not meaningful to assess given *low confidence* in changes, due to large interannual variability, heterogeneity of the observational record and contradictory findings regarding trends in the observational record	Increases in heavy precipitation associated with tropical cyclones (*medium confidence*)	Further increases in heavy precipitation associated with tropical cyclones (*medium confidence*)	Heavy precipitation associated with tropical cyclones is projected to be higher at 2°C compared to 1.5°C global warming (*medium confidence*). Limited evidence that the global number of tropical cyclones will be lower under 2°C of global warming compared to under 1.5°C of warming, but an increase in the number of very intense cyclones (*low confidence*)

3

Table 3.2 (continued)

	Observed change (recent past versus pre-industrial)	Attribution of observed change to human-induced forcing (present-day versus pre-industrial)	Projected change at 1.5°C of global warming compared to pre-industrial (1.5°C versus 0°C)	Projected change at 2°C of global warming compared to pre-industrial (2°C versus 0°C)	Differences between 2°C and 1.5°C of global warming
Ocean circulation and temperature	Observed warming of the upper ocean, with slightly lower rates than global warming (*virtually certain*) Increased occurrence of marine heatwaves (*high confidence*) AMOC has been weakening over recent decades (*more likely than not*) [Section 3.3.7]	*Limited evidence* attributing the weakening of AMOC in recent decades to anthropogenic forcing [Section 3.3.7]	Further increases in ocean temperatures, including more frequent marine heatwaves (*high confidence*) AMOC will weaken over the 21st century and substantially so under high levels (more than 2°C) of global warming (*very likely*) [Section 3.3.7]		
Sea ice	Continuing the trends reported in AR5, the annual Arctic sea ice extent decreased over the period 1979–2012. The rate of this decrease was *very likely* between 3.5 and 4.1% per decade (0.45 to 0.51 million km² per decade) [AR5 Chapter 4 (Vaughan et al., 2013)]	Anthropogenic forcings are *very likely* to have contributed to Arctic sea ice loss since 1979 [AR5 Chapter 10 (Bindoff et al., 2013a)]	At least one sea-ice-free Arctic summer after about 100 years of stabilized warming (*medium confidence*) [Section 3.3.8]	At least one sea-ice-free Arctic summer after about 10 years of stabilized warming (*medium confidence*) [Section 3.3.8]	Probability of sea-ice-free Arctic summer greatly reduced at 1.5°C versus 2°C of global warming (*medium confidence*) [Section 3.3.8]
			Intermediate temperature overshoot has no long-term consequences for Arctic sea ice cover (*high confidence*) [3.3.8]		
Sea level	It is *likely* that the rate of GMSL rise has continued to increase since the early 20th century, with estimates that range from 0.000 [–0.002 to 0.002] mm yr⁻² to 0.013 [0.007 to 0.019] mm yr⁻² [AR5 Chapter 13 (Church et al., 2013)]	It is very *likely* that there is a substantial contribution from anthropogenic forcings to the global mean sea level rise since the 1970s [AR5 Chapter 10 (Bindoff et al., 2013a)]	Not assessed in this report	Not assessed in this report	GMSL rise will be about 0.1 m (0.00–0.20 m) less at 1.5°C versus 2°C global warming (*medium confidence*) [Section 3.3.9]
Ocean chemistry	Ocean acidification due to increased CO_2 has resulted in a 0.1 pH unit decrease since the pre-industrial period, which is unprecedented in the last 65 Ma (*high confidence*) [Section 3.3.10]	The oceanic uptake of anthropogenic CO_2 has resulted in acidification of surface waters (*very high confidence*). [Section 3.3.10]	Ocean chemistry is changing with global temperature increases, with impacts projected at 1.5°C and, more so, at 2°C of warming (*high confidence*) [Section 3.3.10]		

Note: "It is *likely* that the rate of GMSL rise has continued to increase since the early 20th century, with estimates that range from 0.000 [–0.002 to 0.002] mm yr⁻² to 0.013 [0.007 to 0.019] mm yr⁻²"

3.4　Observed Impacts and Projected Risks in Natural and Human Systems

3.4.1　Introduction

In Section 3.4, new literature is explored and the assessment of impacts and projected risks is updated for a large number of natural and human systems. This section also includes an exploration of adaptation opportunities that could be important steps towards reducing climate change, thereby laying the ground for later discussions on opportunities to tackle both mitigation and adaptation while at the same time recognising the importance of sustainable development and reducing the inequities among people and societies facing climate change.

Working Group II (WGII) of the IPCC Fifth Assessment Report (AR5) provided an assessment of the literature on the climate risk for natural and human systems across a wide range of environments, sectors and greenhouse gas scenarios, as well as for particular geographic

regions (IPCC, 2014a, b). The comprehensive assessment undertaken by AR5 evaluated the evidence of changes to natural systems, and the impact on human communities and industry. While impacts varied substantially among systems, sectors and regions, many changes over the past 50 years could be attributed to human driven climate change and its impacts. In particular, AR5 attributed observed impacts in natural ecosystems to anthropogenic climate change, including changes in phenology, geographic and altitudinal range shifts in flora and fauna, regime shifts and increased tree mortality, all of which can reduce ecosystem functioning and services thereby impacting people. AR5 also reported increasing evidence of changing patterns of disease and invasive species, as well as growing risks for communities and industry, which are especially important with respect to sea level rise and human vulnerability.

One of the important themes that emerged from AR5 is that previous assessments may have under-estimated the sensitivity of natural and human systems to climate change. A more recent analysis of attribution to greenhouse gas forcing at the global scale (Hansen and Stone, 2016) confirmed that many impacts related to changes in regional atmospheric and ocean temperature can be confidently attributed to anthropogenic forcing, while attribution to anthropogenic forcing of changes related to precipitation are by comparison less clear. Moreover, there is no strong direct relationship between the robustness of climate attribution and that of impact attribution (Hansen and Stone, 2016). The observed changes in human systems are amplified by the loss of ecosystem services (e.g., reduced access to safe water) that are supported by biodiversity (Oppenheimer et al., 2014). Limited research on the risks of warming of 1.5°C and 2°C was conducted following AR5 for most key economic sectors and services, for livelihoods and poverty, and for rural areas. For these systems, climate is one of many drivers that result in adverse outcomes. Other factors include patterns of demographic change, socio-economic development, trade and tourism. Further, consequences of climate change for infrastructure, tourism, migration, crop yields and other impacts interact with underlying vulnerabilities, such as for individuals and communities engaged in pastoralism, mountain farming and artisanal fisheries, to affect livelihoods and poverty (Dasgupta et al., 2014).

Incomplete data and understanding of these lower-end climate scenarios have increased the need for more data and an improved understanding of the projected risks of warming of 1.5°C and 2°C for reference. In this section, the available literature on the projected risks, impacts and adaptation options is explored, supported by additional information and background provided in Supplementary Material 3.SM.3.1, 3.SM.3.2, 3.SM.3.4, and 3.SM.3.5. A description of the main assessment methods of this chapter is given in Section 3.2.2.

3.4.2   Freshwater Resources (Quantity and Quality)

3.4.2.1   Water availability

Working Group II of AR5 concluded that about 80% of the world's population already suffers from serious threats to its water security, as measured by indicators including water availability, water demand and pollution (Jiménez Cisneros et al., 2014). UNESCO (2011) concluded that climate change can alter the availability of water and threaten water security.

Although physical changes in streamflow and continental runoff that are consistent with climate change have been identified (Section 3.3.5), water scarcity in the past is still less well understood because the scarcity assessment needs to take into account various factors, such as the operations of water supply infrastructure and human water use behaviour (Mehran et al., 2017), as well as green water, water quality and environmental flow requirements (J. Liu et al., 2017). Over the past century, substantial growth in populations, industrial and agricultural activities, and living standards have exacerbated water stress in many parts of the world, especially in semi-arid and arid regions such as California in the USA (AghaKouchak et al., 2015; Mehran et al., 2015). Owing to changes in climate and water consumption behaviour, and particularly effects of the spatial distribution of population growth relative to water resources, the population under water scarcity increased from 0.24 billion (14% of the global population) in the 1900s to 3.8 billion (58%) in the 2000s. In that last period (2000s), 1.1

billion people (17% of the global population) who mostly live in South and East Asia, North Africa and the Middle East faced serious water shortage and high water stress (Kummu et al., 2016).

Over the next few decades, and for increases in global mean temperature less than about 2°C, AR5 concluded that changes in population will generally have a greater effect on water resource availability than changes in climate. Climate change, however, will regionally exacerbate or offset the effects of population pressure (Jiménez Cisneros et al., 2014).

The differences in projected changes to levels of runoff under 1.5°C and 2°C of global warming, particularly those that are regional, are described in Section 3.3.5. Constraining warming to 1.5°C instead of 2°C might mitigate the risks for water availability, although socio-economic drivers could affect water availability more than the risks posed by variation in warming levels, while the risks are not homogeneous among regions (*medium confidence*) (Gerten et al., 2013; Hanasaki et al., 2013; Arnell and Lloyd-Hughes, 2014; Schewe et al., 2014; Karnauskas et al., 2018). Assuming a constant population in the models used in his study, Gerten et al. (2013) determined that an additional 8% of the world population in 2000 would be exposed to new or aggravated water scarcity at 2°C of global warming. This value was almost halved – with 50% greater reliability – when warming was constrained to 1.5°C. People inhabiting river basins, particularly in the Middle East and Near East, are projected to become newly exposed to chronic water scarcity even if global warming is constrained to less than 2°C. Many regions, especially those in Europe, Australia and southern Africa, appear to be affected at 1.5°C if the reduction in water availability is computed for non-water-scarce basins as well as for water-scarce regions. Out of a contemporary population of approximately 1.3 billion exposed to water scarcity, about 3% (North America) to 9% (Europe) are expected to be prone to aggravated scarcity at 2°C of global warming (Gerten et al., 2013). Under the Shared Socio-Economic Pathway (SSP)2 population scenario, about 8% of the global population is projected to experience a severe reduction in water resources under warming of 1.7°C in 2021–2040, increasing to 14% of the population under 2.7°C in 2043–2071, based on the criteria of discharge reduction of either >20% or >1 standard deviation (Schewe et al., 2014). Depending on the scenarios of SSP1–5, exposure to the increase in water scarcity in 2050 will be globally reduced by 184–270 million people at about 1.5°C of warming compared to the impacts at about 2°C. However, the variation between socio-economic levels is larger than the variation between warming levels (Arnell and Lloyd-Hughes, 2014).

On many small islands (e.g., those constituting SIDS), freshwater stress is expected to occur as a result of projected aridity change. Constraining warming to 1.5°C, however, could avoid a substantial fraction of water stress compared to 2°C, especially across the Caribbean region, particularly on the island of Hispaniola (Dominican Republic and Haiti) (Karnauskas et al., 2018). Hanasaki et al. (2013) concluded that the projected range of changes in global irrigation water withdrawal (relative to the baseline of 1971–2000), using human configuration fixing non-meteorological variables for the period around 2000, are 1.1–2.3% and 0.6–2.0% lower at 1.5°C and 2°C, respectively. In the same study, Hanasaki et al. (2013) highlighted the importance of water use scenarios in water scarcity assessments, but neither quantitative nor qualitative information regarding water use is available.

When the impacts on hydropower production at 1.5°C and 2°C are compared, it is found that mean gross potential increases in northern, eastern and western Europe, and decreases in southern Europe (Jacob et al., 2018; Tobin et al., 2018). The Baltic and Scandinavian countries are projected to experience the most positive impacts on hydropower production. Greece, Spain and Portugal are expected to be the most negatively impacted countries, although the impacts could be reduced by limiting warming to 1.5°C (Tobin et al., 2018). In Greece, Spain and Portugal, warming of 2°C is projected to decrease hydropower potential below 10%, while limiting global warming to 1.5°C would keep the reduction to 5% or less. There is, however, substantial uncertainty associated with these results due to a large spread between the climate models (Tobin et al., 2018).

Due to a combination of higher water temperatures and reduced summer river flows, the usable capacity of thermoelectric power plants using river water for cooling is expected to reduce in all European countries (Jacob et al., 2018; Tobin et al., 2018), with the magnitude of decreases being about 5% for 1.5°C and 10% for 2°C of global warming for most European countries (Tobin et al., 2018). Greece, Spain and Bulgaria are projected to have the largest reduction at 2°C of warming (Tobin et al., 2018).

Fricko et al. (2016) assessed the direct water use of the global energy sector across a broad range of energy system transformation pathways in order to identify the water impacts of a 2°C climate policy. This study revealed that there would be substantial divergence in water withdrawal for thermal power plant cooling under conditions in which the distribution of future cooling technology for energy generation is fixed, whereas adopting alternative cooling technologies and water resources would make the divergence considerably smaller.

3.4.2.2   Extreme hydrological events (floods and droughts)

Working Group II of AR5 concluded that socio-economic losses from flooding since the mid-20th century have increased mainly because of greater exposure and vulnerability (*high confidence*) (Jiménez Cisneros et al., 2014). There was *low confidence* due to limited evidence, however, that anthropogenic climate change has affected the frequency and magnitude of floods. WGII AR5 also concluded that there is no evidence that surface water and groundwater drought frequency has changed over the last few decades, although impacts of drought have increased mostly owing to increased water demand (Jiménez Cisneros et al., 2014).

Since AR5, the number of studies related to fluvial flooding and meteorological drought based on long-term observed data has been gradually increasing. There has also been progress since AR5 in identifying historical changes in streamflow and continental runoff (Section 3.3.5). As a result of population and economic growth, increased exposure of people and assets has caused more damage due to flooding. However, differences in flood risks among regions reflect the balance among the magnitude of the flood, the populations, their vulnerabilities, the value of assets affected by flooding, and the

capacity to cope with flood risks, all of which depend on socio-economic development conditions, as well as topography and hydro-climatic conditions (Tanoue et al., 2016). AR5 concluded that there was *low confidence* in the attribution of global changes in droughts (Bindoff et al., 2013b). However, recent publications based on observational and modelling evidence assessed that human emissions have substantially increased the probability of drought years in the Mediterranean region (Section 3.3.4).

WGII AR5 assessed that global flood risk will increase in the future, partly owing to climate change (*low to medium confidence*), with projected changes in the frequency of droughts longer than 12 months being more uncertain because of their dependence on accumulated precipitation over long periods (Jiménez Cisneros et al., 2014).

Increases in the risks associated with runoff at the global scale (*medium confidence*), and in flood hazard in some regions (*medium confidence*), can be expected at global warming of 1.5°C, with an overall increase in the area affected by flood hazard at 2°C (*medium confidence*) (Section 3.3.5). There are studies, however, that indicate that socio-economic conditions will exacerbate flood impacts more than global climate change, and that the magnitude of these impacts could be larger in some regions (Arnell and Lloyd-Hughes, 2014; Winsemius et al., 2016; Alfieri et al., 2017; Arnell et al., 2018; Kinoshita et al., 2018). Assuming constant population sizes, countries representing 73% of the world population will experience increasing flood risk, with an average increase of 580% at 4°C compared to the impact simulated over the baseline period 1976–2005. This impact is projected to be reduced to a 100% increase at 1.5°C and a 170% increase at 2°C (Alfieri et al., 2017). Alfieri et al. (2017) additionally concluded that the largest increases in flood risks would be found in the US, Asia, and Europe in general, while decreases would be found in only a few countries in eastern Europe and Africa. Overall, Alfieri et al. (2017) reported that the projected changes are not homogeneously distributed across the world land surface. Alfieri et al. (2018) studied the population affected by flood events using three case studies in European states, specifically central and western Europe, and found that the population affected could be limited to 86% at 1.5°C of warming compared to 93% at 2°C. Under the SSP2 population scenario, Arnell et al. (2018) found that 39% (range 36–46%) of impacts on populations exposed to river flooding globally could be avoided at 1.5°C compared to 2°C of warming.

Under scenarios SSP1–5, Arnell and Lloyd-Hughes (2014) found that the number of people exposed to increased flooding in 2050 under warming of about 1.5°C could be reduced by 26–34 million compared to the number exposed to increased flooding associated with 2°C of warming. Variation between socio-economic levels, however, is projected to be larger than variation between the two levels of global warming. Kinoshita et al. (2018) found that a serious increase in potential flood fatality (5.7%) is projected without any adaptation if global warming increases from 1.5°C to 2°C, whereas the projected increase in potential economic loss (0.9%) is relatively small. Nevertheless, their study indicates that socio-economic changes make a larger contribution to the potentially increased consequences of future floods, and about half of the increase in potential economic losses could be mitigated by autonomous adaptation.

There is limited information about the global and regional projected risks posed by droughts at 1.5°C and 2°C of global warming. However, hazards by droughts at 1.5°C could be reduced compared to the hazards at 2°C in some regions, in particular in the Mediterranean region and southern Africa (Section 3.3.4). Under constant socio-economic conditions, the population exposed to drought at 2°C of warming is projected to be larger than at 1.5°C (*low to medium confidence*) (Smirnov et al., 2016; Sun et al., 2017; Arnell et al., 2018; Liu et al., 2018). Under the same scenario, the global mean monthly number of people expected to be exposed to extreme drought at 1.5°C in 2021–2040 is projected to be 114.3 million, compared to 190.4 million at 2°C in 2041–2060 (Smirnov et al., 2016). Under the SSP2 population scenario, Arnell et al. (2018) projected that 39% (range 36–51%) of impacts on populations exposed to drought could be globally avoided at 1.5°C compared to 2°C warming.

Liu et al. (2018) studied the changes in population exposure to severe droughts in 27 regions around the globe for 1.5°C and 2°C of warming using the SSP1 population scenario compared to the baseline period of 1986–2005 based on the Palmer Drought Severity Index (PDSI). They concluded that the drought exposure of urban populations in most regions would be decreased at 1.5°C (350.2 ± 158.8 million people) compared to 2°C (410.7 ± 213.5 million people). Liu et al. (2018) also suggested that more urban populations would be exposed to severe droughts at 1.5°C in central Europe, southern Europe, the Mediterranean, West Africa, East and West Asia, and Southeast Asia, and that number of affected people would increase further in these regions at 2°C. However, it should be noted that the PDSI is known to have limitations (IPCC SREX, Seneviratne et al., 2012), and drought projections strongly depend on considered indices (Section 3.3.4); thus only *medium confidence* is assigned to these projections. In the Haihe River basin in China, a study has suggested that the proportion of the population exposed to droughts is projected to be reduced by 30.4% at 1.5°C but increased by 74.8% at 2°C relative to the baseline value of 339.65 million people in the 1986–2005 period, when assessing changes in droughts using the Standardized Precipitation-Evaporation Index, using a Penman–Monteith estimate of potential evaporation (Sun et al., 2017) .

Alfieri et al. (2018) estimated damage from flooding in Europe for the baseline period (1976–2005) at 5 billion euro of losses annually, with projections of relative changes in flood impacts that will rise with warming levels, from 116% at 1.5°C to 137% at 2°C.

Kinoshita et al. (2018) studied the increase of potential economic loss under SSP3 and projected that the smaller loss at 1.5°C compared to 2°C (0.9%) is marginal, regardless of whether the vulnerability is fixed at the current level or not. By analysing the differences in results with and without flood protection standards, Winsemius et al. (2016) showed that adaptation measures have the potential to greatly reduce present-day and future flood damage. They concluded that increases in flood-induced economic impacts (% gross domestic product, GDP) in African countries are mainly driven by climate change and that Africa's growing assets would become increasingly exposed to floods. Hence, there is an increasing need for long-term and sustainable investments in adaptation in Africa.

3.4.2.3 Groundwater

Working Group II of AR5 concluded that the detection of changes in groundwater systems, and attribution of those changes to climatic changes, are rare, owing to a lack of appropriate observation wells and an overall small number of studies (Jiménez Cisneros et al., 2014).

Since AR5, the number of studies based on long-term observed data continues to be limited. The groundwater-fed lakes in northeastern central Europe have been affected by climate and land-use changes, and they showed a predominantly negative lake-level trend in 1999–2008 (Kaiser et al., 2014).

WGII AR5 concluded that climate change is projected to reduce groundwater resources significantly in most dry subtropical regions (*high confidence*) (Jiménez Cisneros et al., 2014).

In some regions, groundwater is often intensively used to supplement the excess demand, often leading to groundwater depletion. Climate change adds further pressure on water resources and exaggerates human water demands by increasing temperatures over agricultural lands (Wada et al., 2017). Very few studies have projected the risks of groundwater depletion under 1.5°C and 2°C of global warming. Under 2°C of warming, impacts posed on groundwater are projected to be greater than at 1.5°C (*low confidence*) (Portmann et al., 2013; Salem et al., 2017).

Portmann et al. (2013) indicated that 2% (range 1.1–2.6%) of the global land area is projected to suffer from an extreme decrease in renewable groundwater resources of more than 70% at 2°C, with a clear mitigation at 1.5°C. These authors also projected that 20% of the global land surface would be affected by a groundwater reduction of more than 10% at 1.5°C of warming, with the percentage of land impacted increasing at 2°C. In a groundwater-dependent irrigated region in northwest Bangladesh, the average groundwater level during the major irrigation period (January–April) is projected to decrease in accordance with temperature rise (Salem et al., 2017).

3.4.2.4 Water quality

Working Group II of AR5 concluded that most observed changes to water quality from climate change are from isolated studies, mostly of rivers or lakes in high-income countries, using a small number of variables (Jiménez Cisneros et al., 2014). AR5 assessed that climate change is projected to reduce raw water quality, posing risks to drinking water quality with conventional treatment (*medium to high confidence*) (Jiménez Cisneros et al., 2014).

Since AR5, studies have detected climate change impacts on several indices of water quality in lakes, watersheds and regions (e.g., Patiño et al., 2014; Aguilera et al., 2015; Watts et al., 2015; Marszelewski and Pius, 2016; Capo et al., 2017). The number of studies utilising RCP scenarios at the regional or watershed scale have gradually increased since AR5 (e.g., Boehlert et al., 2015; Teshager et al., 2016; Marcinkowski et al., 2017). Few studies, have explored projected impacts on water quality under 1.5°C versus 2°C of warming, however, the differences are unclear (*low confidence*) (Bonte and

Zwolsman, 2010; Hosseini et al., 2017). The daily probability of exceeding the chloride standard for drinking water taken from Lake IJsselmeer (Andijk, the Netherlands) is projected to increase by a factor of about five at 2°C relative to the present-day warming level of 1°C since 1990 (Bonte and Zwolsman, 2010). Mean monthly dissolved oxygen concentrations and nutrient concentrations in the upper Qu'Appelle River (Canada) in 2050–2055 are projected to decrease less at about 1.5°C of warming (RCP2.6) compared to concentrations at about 2°C (RCP4.5) (Hosseini et al., 2017). In three river basins in Southeast Asia (Sekong, Sesan and Srepok), about 2°C of warming (corresponding to a 1.05°C increase in the 2030s relative to the baseline period 1981–2008, RCP8.5), impacts posed by land-use change on water quality are projected to be greater than at 1.5°C (corresponding to a 0.89°C increase in the 2030s relative to the baseline period 1981–2008, RCP4.5) (Trang et al., 2017). Under the same warming scenarios, Trang et al. (2017) projected changes in the annual nitrogen (N) and phosphorus (P) yields in the 2030s, as well as with combinations of two land-use change scenarios: (i) conversion of forest to grassland, and (ii) conversion of forest to agricultural land. The projected changes in N (P) yield are +7.3% (+5.1%) under a 1.5°C scenario and −6.6% (−3.6%) under 2°C, whereas changes under the combination of land-use scenarios are (i) +5.2% (+12.6%) at 1.5°C and +8.8% (+11.7%) at 2°C, and (ii) +7.5% (+14.9%) at 1.5°C and +3.7% (+8.8%) at 2°C (Trang et al., 2017).

3.4.2.5   Soil erosion and sediment load

Working Group II of AR5 concluded that there is little or no observational evidence that soil erosion and sediment load have been altered significantly by climate change (*low to medium confidence*) (Jiménez Cisneros et al., 2014). As the number of studies on climate change impacts on soil erosion has increased where rainfall is an important driver (Lu et al., 2013), studies have increasingly considered other factors, such as rainfall intensity (e.g., Shi and Wang, 2015; Li and Fang, 2016), snow melt, and change in vegetation cover resulting from temperature rise (Potemkina and Potemkin, 2015), as well as crop management practices (Mullan et al., 2012). WGII AR5 concluded that increases in heavy rainfall and temperature are projected to change soil erosion and sediment yield, although the extent of these changes is highly uncertain and depends on rainfall seasonality, land cover, and soil management practices (Jiménez Cisneros et al., 2014).

While the number of published studies of climate change impacts on soil erosion have increased globally since 2000 (Li and Fang, 2016), few articles have addressed impacts at 1.5°C and 2°C of global warming. The existing studies have found few differences in projected risks posed on sediment load under 1.5°C and 2°C (*low confidence*) (Cousino et al., 2015; Shrestha et al., 2016). The differences between average annual sediment load under 1.5°C and 2°C of warming are not clear, owing to complex interactions among climate change, land cover/surface and soil management (Cousino et al., 2015; Shrestha et al., 2016). Averages of annual sediment loads are projected to be similar under 1.5°C and 2°C of warming, in particular in the Great Lakes region in the USA and in the Lower Mekong region in Southeast Asia (Cross-Chapter Box 6 in this chapter, Cousino et al., 2015; Shrestha et al., 2016).

3.4.3   Terrestrial and Wetland Ecosystems

3.4.3.1   Biome shifts

Latitudinal and elevational shifts of biomes (major ecosystem types) in boreal, temperate and tropical regions have been detected (Settele et al., 2014) and new studies confirm these changes (e.g., shrub encroachment on tundra; Larsen et al., 2014). Attribution studies indicate that anthropogenic climate change has made a greater contribution to these changes than any other factor (*medium confidence*) (Settele et al., 2014).

An ensemble of seven Dynamic Vegetation Models driven by projected climates from 19 alternative general circulation models (GCMs) (Warszawski et al., 2013) shows 13% (range 8–20%) of biomes transforming at 2°C of global warming, but only 4% (range 2–7%) doing so at 1°C, suggesting that about 6.5% may be transformed at 1.5°C; these estimates indicate a doubling of the areal extent of biome shifts between 1.5°C and 2°C of warming (*medium confidence*) (Figure 3.16a). A study using the single ecosystem model LPJmL (Gerten et al., 2013) illustrated that biome shifts in the Arctic, Tibet, Himalayas, southern Africa and Australia would be avoided by constraining warming to 1.5°C compared with 2°C (Figure 3.16b). Seddon et al. (2016) quantitatively identified ecologically sensitive regions to climate change in most of the continents from tundra to tropical rainforest. Biome transformation may in some cases be associated with novel climates and ecological communities (Prober et al., 2012).

3.4.3.2   Changes in phenology

Advancement in spring phenology of 2.8 ± 0.35 days per decade has been observed in plants and animals in recent decades in most Northern Hemisphere ecosystems (between 30°N and 72°N), and these shifts have been attributed to changes in climate (*high confidence*) (Settele et al., 2014). The rates of change are particularly high in the Arctic zone owing to the stronger local warming (Oberbauer et al., 2013), whereas phenology in tropical forests appears to be more responsive to moisture stress (Zhou et al., 2014). While a full review cannot be included here, trends consistent with this earlier finding continue to be detected, including in the flowering times of plants (Parmesan and Hanley, 2015), in the dates of egg laying and migration in birds (newly reported in China; Wu and Shi, 2016), in the emergence dates of butterflies (Roy et al., 2015), and in the seasonal greening-up of vegetation as detected by satellites (i.e., in the normalized difference vegetation index, NDVI; Piao et al., 2015).

The potential for decoupling species–species interactions owing to differing phenological responses to climate change is well established (Settele et al., 2014), for example for plants and their insect pollinators (Willmer, 2012; Scaven and Rafferty, 2013). Mid-century projections of plant and animal phenophases in the UK clearly indicate that the timing of phenological events could change more for primary consumers (6.2 days earlier on average) than for higher trophic levels (2.5–2.9 days earlier on average) (Thackeray et al., 2016). This indicates the potential for phenological mismatch and associated risks for ecosystem functionality in the future under global warming of 2.1°C–2.7°C above pre-industrial levels. Further, differing responses

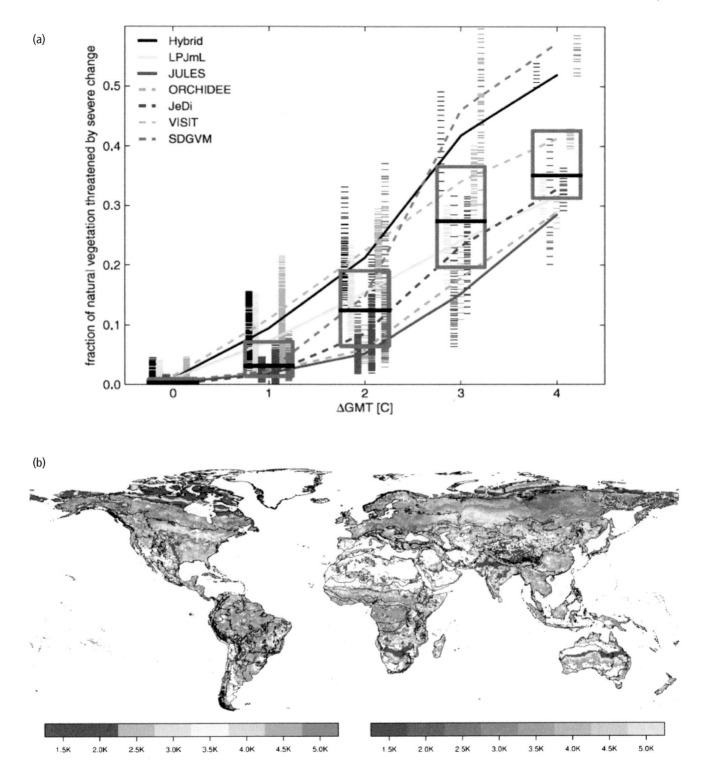

Figure 3.16 | (a) Fraction of global natural vegetation (including managed forests) at risk of severe ecosystem change as a function of global mean temperature change for all ecosystems, models, global climate change models and Representative Concentration Pathways (RCPs). The colours represent the different ecosystem models, which are also horizontally separated for clarity. Results are collated in unit-degree bins, where the temperature for a given year is the average over a 30-year window centred on that year. The boxes span the 25th and 75th percentiles across the entire ensemble. The short, horizontal stripes represent individual (annual) data points, the curves connect the mean value per ecosystem model in each bin. The solid (dashed) curves are for models with (without) dynamic vegetation composition changes. Source: (Warszawski et al., 2013) (b) Threshold level of global temperature anomaly above pre-industrial levels that leads to significant local changes in terrestrial ecosystems. Regions with severe (coloured) or moderate (greyish) ecosystem transformation; delineation refers to the 90 biogeographic regions. All values denote changes found in >50% of the simulations. Source: (Gerten et al., 2013). Regions coloured in dark red are projected to undergo severe transformation under a global warming of 1.5°C while those coloured in light red do so at 2°C; other colours are used when there is no severe transformation unless global warming exceeds 2°C.

could alter community structure in temperate forests (Roberts et al., 2015). Specifically, temperate forest phenology is projected to advance by 14.3 days in the near term (2010–2039) and 24.6 days in the medium term (2040–2069), so as a first approximation the difference between 2°C and 1.5°C of global warming is about 10 days (Roberts et al., 2015). This phenological plasticity is not always adaptive and must be interpreted cautiously (Duputié et al., 2015), and considered in the context of accompanying changes in climate variability (e.g., increased risk of frost damage for plants or earlier emergence of insects resulting in mortality during cold spells). Another adaptive response of some plants is range expansion with increased vigour and altered herbivore resistance in their new range, analogous to invasive plants (Macel et al., 2017).

In summary, limiting warming to 1.5°C compared with 2°C may avoid advance in spring phenology (*high confidence*) by perhaps a few days (*medium confidence*) and hence decrease the risks of loss of ecosystem functionality due to phenological mismatch between trophic levels, and also of maladaptation coming from the sensitivity of many species to increased climate variability. Nevertheless, this difference between 1.5°C and 2°C of warming might be limited for plants that are able to expand their range.

3.4.3.3   Changes in species range, abundance and extinction

AR5 (Settele et al., 2014) concluded that the geographical ranges of many terrestrial and freshwater plant and animal species have moved over the last several decades in response to warming: approximately 17 km poleward and 11 m up in altitude per decade. Recent trends confirm this finding; for example, the spatial and interspecific variance in bird populations in Europe and North America since 1980 were found to be well predicted by trends in climate suitability (Stephens et al., 2016). Further, a recent meta-analysis of 27 studies concerning a total of 976 species (Wiens, 2016) found that 47% of local extinctions (extirpations) reported across the globe during the 20th century could be attributed to climate change, with significantly more extinctions occurring in tropical regions, in freshwater habitats and for animals. IUCN (2018) lists 305 terrestrial animal and plant species from Pacific Island developing nations as being threatened by climate change and severe weather. Owing to lags in the responses of some species to climate change, shifts in insect pollinator ranges may result in novel assemblages with unknown implications for biodiversity and ecosystem function (Rafferty, 2017).

Warren et al. (2013) simulated climatically determined geographic range loss under 2°C and 4°C of global warming for 50,000 plant and animal species, accounting for uncertainty in climate projections and for the potential ability of species to disperse naturally in an attempt to track their geographically shifting climate envelope. This earlier study has now been updated and expanded to incorporate 105,501 species, including 19,848 insects, and new findings indicate that warming of 2°C by 2100 would lead to projected bioclimatic range losses of >50% in 18% (6–35%) of the 19,848 insects species, 8% (4–16%) of the 12,429 vertebrate species, and 16% (9–28%) of the 73,224 plant species studied (Warren et al., 2018a). At 1.5°C of warming, these values fall to 6% (1–18%) of the insects, 4% (2–9%) of the vertebrates and 8% (4–15%) of the plants studied. Hence, the number of insect species projected to lose over half of their geographic range is reduced by two-thirds when warming is limited to 1.5°C compared with 2°C, while the number of vertebrate

and plant species projected to lose over half of their geographic range is halved (Warren et al., 2018a) (*medium confidence*). These findings are consistent with estimates made from an earlier study suggesting that range losses at 1.5°C were significantly lower for plants than those at 2°C of warming (Smith et al., 2018). It should be noted that at 1.5°C of warming, and if species' ability to disperse naturally to track their preferred climate geographically is inhibited by natural or anthropogenic obstacles, there would still remain 10% of the amphibians, 8% of the reptiles, 6% of the mammals, 5% of the birds, 10% of the insects and 8% of the plants which are projected to lose over half their range, while species on average lose 20–27% of their range (Warren et al., 2018a). Given that bird and mammal species can disperse more easily than amphibians and reptiles, a small proportion can expand their range as climate changes, but even at 1.5°C of warming the total range loss integrated over all birds and mammals greatly exceeds the integrated range gain (Warren et al., 2018a).

A number of caveats are noted for studies projecting changes to climatic range. This approach, for example, does not incorporate the effects of extreme weather events and the role of interactions between species. As well, trophic interactions may locally counteract the range expansion of species towards higher altitudes (Bråthen et al., 2018). There is also the potential for highly invasive species to become established in new areas as the climate changes (Murphy and Romanuk, 2014), but there is no literature that quantifies this possibility for 1.5°C of global warming.

Pecl et al. (2017) summarized at the global level the consequences of climate-change-induced species redistribution for economic development, livelihoods, food security, human health and culture. These authors concluded that even if anthropogenic greenhouse gas emissions stopped today, the effort for human systems to adapt to the most crucial effects of climate-driven species redistribution will be far-reaching and extensive. For example, key insect crop pollinator families (Apidae, Syrphidae and Calliphoridae; i.e., bees, hoverflies and blowflies) are projected to retain significantly greater geographic ranges under 1.5°C of global warming compared with 2°C (Warren et al., 2018a). In some cases, when species (such as pest and disease species) move into areas which have become climatically suitable they may become invasive or harmful to human or natural systems (Settele et al., 2014). Some studies are beginning to locate 'refugial' areas where the climate remains suitable in the future for most of the species currently present. For example, Smith et al. (2018) estimated that 5.5–14% more of the globe's terrestrial land area could act as climatic refugia for plants under 1.5°C of warming compared to 2°C.

There is no literature that directly estimates the proportion of species at increased risk of global (as opposed to local) commitment to extinction as a result of climate change, as this is inherently difficult to quantify. However, it is possible to compare the proportions of species at risk of very high range loss; for example, a discernibly smaller number of terrestrial species are projected to lose over 90% of their range at 1.5°C of global warming compared with 2°C (Figure 2 in Warren et al., 2018a). A link between very high levels of range loss and greatly increased extinction risk may be inferred (Urban, 2015). Hence, limiting global warming to 1.5°C compared with 2°C would be expected to reduce both range losses and associated extinction risks in terrestrial species (*high confidence*).

3.4.3.4   Changes in ecosystem function, biomass and carbon stocks

Working Group II of AR5 (Settele et al., 2014) concluded that there is *high confidence* that net terrestrial ecosystem productivity at the global scale has increased relative to the pre-industrial era and that rising CO_2 concentrations are contributing to this trend through stimulation of photosynthesis. There is, however, no clear and consistent signal of a climate change contribution. In northern latitudes, the change in productivity has a lower velocity than the warming, possibly because of a lack of resource and vegetation acclimation mechanisms (M. Huang et al., 2017). Biomass and soil carbon stocks in terrestrial ecosystems are currently increasing (*high confidence*), but they are vulnerable to loss of carbon to the atmosphere as a result of projected increases in the intensity of storms, wildfires, land degradation and pest outbreaks (Settele et al., 2014; Seidl et al., 2017). These losses are expected to contribute to a decrease in the terrestrial carbon sink. Anderegg et al. (2015) demonstrated that total ecosystem respiration at the global scale has increased in response to increases in night-time temperature (1 PgC yr^{-1} °C^{-1}, P=0.02).

The increase in total ecosystem respiration in spring and autumn, associated with higher temperatures, may convert boreal forests from carbon sinks to carbon sources (Hadden and Grelle, 2016). In boreal peatlands, for example, increased temperature may diminish carbon storage and compromise the stability of the peat (Dieleman et al., 2016). In addition, J. Yang et al. (2015) showed that fires reduce the carbon sink of global terrestrial ecosystems by 0.57 PgC yr^{-1} in ecosystems with large carbon stores, such as peatlands and tropical forests. Consequently, for adaptation purposes, it is necessary to enhance carbon sinks, especially in forests which are prime regulators within the water, energy and carbon cycles (Ellison et al., 2017). Soil can also be a key compartment for substantial carbon sequestration (Lal, 2014; Minasny et al., 2017), depending on the net biome productivity and the soil quality (Bispo et al., 2017).

AR5 assessed that large uncertainty remains regarding the land carbon cycle behaviour of the future (Ciais et al., 2013), with most, but not all, CMIP5 models simulating continued terrestrial carbon uptake under all four RCP scenarios (Jones et al., 2013). Disagreement between models outweighs differences between scenarios even up to the year 2100 (Hewitt et al., 2016; Lovenduski and Bonan, 2017). Increased atmospheric CO_2 concentrations are expected to drive further increases in the land carbon sink (Ciais et al., 2013; Schimel et al., 2015), which could persist for centuries (Pugh et al., 2016). Nitrogen, phosphorus and other nutrients will limit the terrestrial carbon cycle response to both elevated CO_2 and altered climate (Goll et al., 2012; Yang et al., 2014; Wieder et al., 2015; Zaehle et al., 2015; Ellsworth et al., 2017). Climate change may accelerate plant uptake of carbon (Gang et al., 2015) but also increase the rate of decomposition (Todd-Brown et al., 2014; Koven et al., 2015; Crowther et al., 2016). Ahlström et al. (2012) found a net loss of carbon in extra-tropical regions and the largest spread across model results in the tropics. The projected net effect of climate change is to reduce the carbon sink expected under CO_2 increase alone (Settele et al., 2014). Friend et al. (2014) found substantial uptake of carbon by vegetation under future scenarios when considering the effects of both climate change and elevated CO_2.

There is limited published literature examining modelled land carbon changes specifically under 1.5°C of warming, but existing CMIP5 models and published data are used in this report to draw some conclusions. For systems with significant inertia, such as vegetation or soil carbon stores, changes in carbon storage will depend on the rate of change of forcing and thus depend on the choice of scenario (Jones et al., 2009; Ciais et al., 2013; Sihi et al., 2017). To avoid legacy effects of the choice of scenario, this report focuses on the response of gross primary productivity (GPP) – the rate of photosynthetic carbon uptake – by the models, rather than by changes in their carbon store.

Figure 3.17 shows different responses of the terrestrial carbon cycle to climate change in different regions. The models show a consistent response of increased GPP in temperate latitudes of approximately 2 GtC yr^{-1} °C^{-1}. Similarly, Gang et al. (2015) projected a robust increase in the net primary productivity (NPP) of temperate forests. However, Ahlström et al. (2012) showed that this effect could be offset or reversed by increases in decomposition. Globally, most models project that GPP will increase or remain approximately unchanged (Hashimoto et al., 2013). This projection is supported by findings by Sakalli et al. (2017) for Europe using Euro-CORDEX regional models under a 2°C global warming for the period 2034–2063, which indicated that storage will increase by 5% in soil and by 20% in vegetation. However, using the same models Jacob et al. (2018) showed that limiting warming to 1.5°C instead of 2°C avoids an increase in ecosystem vulnerability (compared to a no-climate change scenario) of 40–50%.

At the global level, linear scaling is acceptable for net primary production, biomass burning and surface runoff, and impacts on terrestrial carbon storage are projected to be greater at 2°C than at 1.5°C (Tanaka et al., 2017). If global CO_2 concentrations and temperatures stabilize, or peak and decline, then both land and ocean carbon sinks – which are primarily driven by the continued increase in atmospheric CO_2 – will also decline and may even become carbon sources (Jones et al., 2016). Consequently, if a given amount of anthropogenic CO_2 is removed from the atmosphere, an equivalent amount of land and ocean anthropogenic CO_2 will be released to the atmosphere (Cao and Caldeira, 2010).

In conclusion, ecosystem respiration is expected to increase with increasing temperature, thus reducing soil carbon storage. Soil carbon storage is expected to be larger if global warming is restricted to 1.5°C, although some of the associated changes will be countered by enhanced gross primary production due to elevated CO_2 concentrations (i.e., the 'fertilization effect') and higher temperatures, especially at mid- and high latitudes (*medium confidence*).

3.4.3.5   Regional and ecosystem-specific risks

A large number of threatened systems, including mountain ecosystems, highly biodiverse tropical wet and dry forests, deserts, freshwater systems and dune systems, were assessed in AR5. These include Mediterranean areas in Europe, Siberian, tropical and desert ecosystems in Asia, Australian rainforests, the Fynbos and succulent Karoo areas of South Africa, and wetlands in Ethiopia, Malawi, Zambia and Zimbabwe. In all these systems, it has been shown that impacts accrue with greater warming, and thus impacts at 2°C are expected to be greater than those at 1.5°C (*medium confidence*).

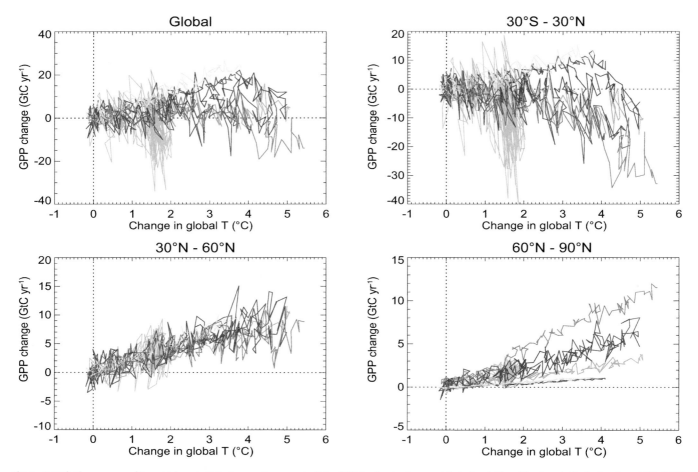

Figure 3.17 | The response of terrestrial productivity (gross primary productivity, GPP) to climate change, globally (top left) and for three latitudinal regions: 30°S–30°N; 30–60°N and 60–90°N. Data come from the Coupled Model Intercomparison Project Phase 5 (CMIP5) archive (http://cmip-pcmdi.llnl.gov/cmip5/). Seven Earth System Models were used: Norwegian Earth System Model (NorESM-ME, yellow); Community Earth System Model (CESM, red); Institute Pierre Simon Laplace (IPLS)-CM5-LR (dark blue); Geophysical Fluid Dynamics Laboratory (GFDL, pale blue); Max Plank Institute-Earth System Model (MPI-ESM, pink); Hadley Centre New Global Environmental Model 2-Earth System (HadGEM2-ES, orange); and Canadian Earth System Model 2 (CanESM2, green). Differences in GPP between model simulations with ('1pctCO$_2$') and without ('esmfixclim1') the effects of climate change are shown. Data are plotted against the global mean temperature increase above pre-industrial levels from simulations with a 1% per year increase in CO$_2$ ('1pctCO$_2$').

The High Arctic region, with tundra-dominated landscapes, has warmed more than the global average over the last century (Section 3.3; Settele et al., 2014). The Arctic tundra biome is experiencing increasing fire disturbance and permafrost degradation (Bring et al., 2016; DeBeer et al., 2016; Jiang et al., 2016; Yang et al., 2016). Both of these processes facilitate the establishment of woody species in tundra areas. Arctic terrestrial ecosystems are being disrupted by delays in winter onset and mild winters associated with global warming (*high confidence*) (Cooper, 2014). Observational constraints suggest that stabilization at 1.5°C of warming would avoid the thawing of approximately 1.5 to 2.5 million km² of permafrost (*medium confidence*) compared with stabilization at 2°C (Chadburn et al., 2017), but the time scale for release of thawed carbon as CO$_2$ or CH$_4$ should be many centuries (Burke et al., 2017). In northern Eurasia, the growing season length is projected to increase by about 3–12 days at 1.5°C and 6–16 days at 2°C of warming (*medium confidence*) (Zhou et al., 2018). Aalto et al. (2017) predicted a 72% reduction in cryogenic land surface processes in northern Europe for RCP2.6 in 2040–2069 (corresponding to a global warming of approximately 1.6°C), with only slightly larger losses for RCP4.5 (2°C of global warming).

Projected impacts on forests as climate change occurs include increases in the intensity of storms, wildfires and pest outbreaks (Settele et al., 2014), potentially leading to forest dieback (*medium confidence*). Warmer and drier conditions in particular facilitate fire, drought and insect disturbances, while warmer and wetter conditions increase disturbances from wind and pathogens (Seidl et al., 2017). Particularly vulnerable regions are Central and South America, Mediterranean Basin, South Africa, South Australia where the drought risk will increase (see Figure 3.12). Including disturbances in simulations may influence productivity changes in European forests in response to climate change (Reyer et al., 2017b). There is additional evidence for the attribution of increased forest fire frequency in North America to anthropogenic climate change during 1984–2015, via the mechanism of increasing fuel aridity almost doubling the western USA forest fire area compared to what would have been expected in the absence of climate change (Abatzoglou and Williams, 2016). This projection is in line with expected fire risks, which indicate that fire frequency could increase over 37.8% of the global land area during 2010–2039 (Moritz et al., 2012), corresponding to a global warming level of approximately 1.2°C, compared with over 61.9% of the global land area in 2070–2099, corresponding to a warming of approximately 3.5°C.[6] The values in Table 26-1 in a recent paper by Romero-Lankao et al. (2014) also indicate significantly lower wildfire risks in North America for near-term warming (2030–2040, considered a proxy for 1.5°C of warming) than at 2°C (*high confidence*).

The Amazon tropical forest has been shown to be close to its climatic limits (Hutyra et al., 2005), but this threshold may move under elevated CO_2 (Good et al., 2011). Future changes in rainfall, especially dry season length, will determine responses of the Amazon forest (Good et al., 2013). The forest may be especially vulnerable to combined pressure from multiple stressors, namely changes in climate and continued anthropogenic disturbance (Borma et al., 2013; Nobre et al., 2016). Modelling (Huntingford et al., 2013) and observational constraints (Cox et al., 2013) suggest that large-scale forest dieback is less likely than suggested under early coupled modelling studies (Cox et al., 2000; Jones et al., 2009). Nobre et al. (2016) estimated a climatic threshold of 4°C of warming and a deforestation threshold of 40%.

In many places around the world, the savanna boundary is moving into former grasslands. Woody encroachment, including increased tree cover and biomass, has increased over the past century, owing to changes in land management, rising CO_2 levels, and climate variability and change (often in combination) (Settele et al., 2014). For plant species in the Mediterranean region, shifts in phenology, range contraction and health decline have been observed with precipitation decreases and temperature increases (*medium confidence*) (Settele et al., 2014). Recent studies using independent complementary approaches have shown that there is a regional-scale threshold in the Mediterranean region between 1.5°C and 2°C of warming (Guiot and Cramer, 2016; Schleussner et al., 2016b). Further, Guiot and Cramer (2016) concluded that biome shifts unprecedented in the last 10,000 years can only be avoided if global warming is constrained to 1.5°C (*medium confidence*) – whilst 2°C of warming will result in a decrease of 12–15% of the Mediterranean biome area. The Fynbos biome in southwestern South Africa is vulnerable to the increasing impact of fires under increasing temperatures and drier winters. It is projected to lose about 20%, 45% and 80% of its current suitable climate area under 1°C, 2°C and 3°C of global warming, respectively, compared to 1961–1990 (*high confidence*) (Engelbrecht and Engelbrecht, 2016). In Australia, an increase in the density of trees and shrubs at the expense of grassland species is occurring across all major ecosystems and is projected to be amplified (NCCARF, 2013). Regarding Central America, Lyra et al. (2017) showed that the tropical rainforest biomass would be reduced by about 40% under global warming of 3°C, with considerable replacement by savanna and grassland. With a global warming of close to 1.5°C in 2050, a biomass decrease of 20% is projected for tropical rainforests of Central America (Lyra et al., 2017). If a linear response is assumed, this decrease may reach 30% (*medium confidence*).

Freshwater ecosystems are considered to be among the most threatened on the planet (Settele et al., 2014). Although peatlands cover only about 3% of the land surface, they hold one-third of the world's soil carbon stock (400 to 600 Pg) (Settele et al., 2014). When drained, this carbon is released to the atmosphere. At least 15% of peatlands have drained,

mostly in Europe and Southeast Asia, and are responsible for 5% of human derived CO_2 emissions (Green and Page, 2017). Moreover, in the Congo basin (Dargie et al., 2017) and in the Amazonian basin (Draper et al., 2014), the peatlands store the equivalent carbon as that of a tropical forest. However, stored carbon is vulnerable to land-use change and future risk of drought, for example in northeast Brazil (*high confidence*) (Figure 3.12, Section 3.3.4.2). At the global scale, these peatlands are undergoing rapid major transformations through drainage and burning in preparation for oil palm and other crops or through unintentional burning (Magrin et al., 2014). Wetland salinization, a widespread threat to the structure and ecological functioning of inland and coastal wetlands, is occurring at a high rate and large geographic scale (Section 3.3.6; Herbert et al., 2015). Settele et al. (2014) found that rising water temperatures are projected to lead to shifts in freshwater species distributions and worsen water quality. Some of these ecosystems respond non-linearly to changes in temperature. For example, Johnson and Poiani (2016) found that the wetland function of the Prairie Pothole region in North America is projected to decline at temperatures beyond a local warming of 2°C–3°C above present-day values (1°C local warming, corresponding to 0.6°C of global warming). If the ratio of local to global warming remains similar for these small levels of warming, this would indicate a global temperature threshold of 1.2°C–1.8°C of warming. Hence, constraining global warming to approximately 1.5°C would maintain the functioning of prairie pothole ecosystems in terms of their productivity and biodiversity, although a 20% increase of precipitation could offset 2°C of global warming (*high confidence*) (Johnson and Poiani, 2016).

3.4.3.6   Summary of implications for ecosystem services

In summary, constraining global warming to 1.5°C rather than 2°C has strong benefits for terrestrial and wetland ecosystems and their services (*high confidence*). These benefits include avoidance or reduction of changes such as biome transformations, species range losses, increased extinction risks (all *high confidence*) and changes in phenology (*high confidence*), together with projected increases in extreme weather events which are not yet factored into these analyses (Section 3.3). All of these changes contribute to disruption of ecosystem functioning and loss of cultural, provisioning and regulating services provided by these ecosystems to humans. Examples of such services include soil conservation (avoidance of desertification), flood control, water and air purification, pollination, nutrient cycling, sources of food, and recreation.

3.4.4   Ocean Ecosystems

The ocean plays a central role in regulating atmospheric gas concentrations, global temperature and climate. It also provides habitat to a large number of organisms and ecosystems that provide goods and services worth trillions of USD per year (e.g., Costanza et al., 2014; Hoegh-Guldberg et al., 2015). Together with local stresses (Halpern et al., 2015), climate change poses a major threat to an increasing number of ocean ecosystems (e.g., warm water or tropical coral reefs: *virtually certain*, WGII AR5) and consequently to many

---

[6]   The approximate temperatures are derived from Figure 10.5a in Meehl et al. (2007), which indicates an ensemble average projection of 0.7°C or 3°C above 1980–1999 temperatures, which were already 0.5°C above pre-industrial values.

coastal communities that depend on marine resources for food, livelihoods and a safe place to live. Previous sections of this report have described changes in the ocean, including rapid increases in ocean temperature down to a depth of at least 700 m (Section 3.3.7). In addition, anthropogenic carbon dioxide has decreased ocean pH and affected the concentration of ions in seawater such as carbonate (Sections 3.3.10 and 3.4.4.5), both over a similar depth range. Increased ocean temperatures have intensified storms in some regions (Section 3.3.6), expanded the ocean volume and increased sea levels globally (Section 3.3.9), reduced the extent of polar summer sea ice (Section 3.3.8), and decreased the overall solubility of the ocean for oxygen (Section 3.3.10). Importantly, changes in the response to climate change rarely operate in isolation. Consequently, the effect of global warming of 1.5°C versus 2°C must be considered in the light of multiple factors that may accumulate and interact over time to produce complex risks, hazards and impacts on human and natural systems.

3.4.4.1 Observed impacts

Physical and chemical changes to the ocean resulting from increasing atmospheric CO_2 and other GHGs are already driving significant changes to ocean systems (*very high confidence*) and will continue to do so at 1.5°C, and more so at 2°C, of global warming above pre-industrial temperatures (Section 3.3.11). These changes have been accompanied by other changes such as ocean acidification, intensifying storms and deoxygenation (Levin and Le Bris, 2015). Risks are already significant at current greenhouse gas concentrations and temperatures, and they vary significantly among depths, locations and ecosystems, with impacts being singular, interactive and/or cumulative (Boyd et al., 2015).

3.4.4.2 Warming and stratification of the surface ocean

As atmospheric greenhouse gases have increased, the global mean surface temperature (GMST) has reached about 1°C above the pre-industrial period, and oceans have rapidly warmed from the ocean surface to the deep sea (*high confidence*) (Sections 3.3.7; Hughes and Narayanaswamy, 2013; Levin and Le Bris, 2015; Yasuhara and Danovaro, 2016; Sweetman et al., 2017). Marine organisms are already responding to these changes by shifting their biogeographical ranges to higher latitudes at rates that range from approximately 0 to 40 km yr^{-1} (Burrows et al., 2014; Chust, 2014; Bruge et al., 2016; Poloczanska et al., 2016), which has consequently affected the structure and function of the ocean, along with its biodiversity and foodwebs (*high confidence*). Movements of organisms does not necessarily equate to the movement of entire ecosystems. For example, species of reef-building corals have been observed to shift their geographic ranges, yet this has not resulted in the shift of entire coral ecosystems (*high confidence*) (Woodroffe et al., 2010; Yamano et al., 2011). In the case of 'less mobile' ecosystems (e.g., coral reefs, kelp forests and intertidal communities), shifts in biogeographical ranges may be limited, with mass mortalities and disease outbreaks increasing in frequency as the exposure to extreme temperatures increases (*very high confidence*) (Hoegh-Guldberg, 1999; Garrabou et al., 2009; Rivetti et al., 2014; Maynard et al., 2015; Krumhansl et al., 2016; Hughes et al., 2017b; see also Box 3.4). These trends are projected to become more pronounced at warming of 1.5°C, and

more so at 2°C, above the pre-industrial period (Hoegh-Guldberg et al., 2007; Donner, 2009; Frieler et al., 2013; Horta E Costa et al., 2014; Vergés et al., 2014, 2016; Zarco-Perello et al., 2017) and are *likely* to result in decreases in marine biodiversity at the equator but increases in biodiversity at higher latitudes (Cheung et al., 2009; Burrows et al., 2014).

While the impacts of species shifting their ranges are mostly negative for human communities and industry, there are instances of short-term gains. Fisheries, for example, may expand temporarily at high latitudes in the Northern Hemisphere as the extent of summer sea ice recedes and NPP increases (*medium confidence*) (Cheung et al., 2010; Lam et al., 2016; Weatherdon et al., 2016). High-latitude fisheries are not only influenced by the effect of temperature on NPP but are also strongly influenced by the direct effects of changing temperatures on fish and fisheries (Section 3.4.4.9; Barange et al., 2014; Pörtner et al., 2014; Cheung et al., 2016b; Weatherdon et al., 2016). Temporary gains in the productivity of high-latitude fisheries are offset by a growing number of examples from low and mid-latitudes where increases in sea temperature are driving decreases in NPP, owing to the direct effects of elevated temperatures and/or reduced ocean mixing from reduced ocean upwelling, that is, increased stratification (*low-medium confidence*) (Cheung et al., 2010; Ainsworth et al., 2011; Lam et al., 2012, 2014, 2016; Bopp et al., 2013; Boyd et al., 2014; Chust et al., 2014; Hoegh-Guldberg et al., 2014; Poloczanska et al., 2014; Pörtner et al., 2014; Signorini et al., 2015). Reduced ocean upwelling has implications for millions of people and industries that depend on fisheries for food and livelihoods (Bakun et al., 2015; FAO, 2016; Kämpf and Chapman, 2016), although there is *low confidence* in the projection of the size of the consequences at 1.5°C. It is also important to appreciate these changes in the context of large-scale ocean processes such as the ocean carbon pump. The export of organic carbon to deeper layers of the ocean increases as NPP changes in the surface ocean, for example, with implications for foodwebs and oxygen levels (Boyd et al., 2014; Sydeman et al., 2014; Altieri and Gedan, 2015; Bakun et al., 2015; Boyd, 2015).

3.4.4.3 Storms and coastal runoff

Storms, wind, waves and inundation can have highly destructive impacts on ocean and coastal ecosystems, as well as the human communities that depend on them (IPCC, 2012; Seneviratne et al., 2012). The intensity of tropical cyclones across the world's oceans has increased, although the overall number of tropical cyclones has remained the same or decreased (*medium confidence*) (Section 3.3.6; Elsner et al., 2008; Holland and Bruyère, 2014). The direct force of wind and waves associated with larger storms, along with changes in storm direction, increases the risks of physical damage to coastal communities and to ecosystems such as mangroves (*low to medium confidence*) (Long et al., 2016; Primavera et al., 2016; Villamayor et al., 2016; Cheal et al., 2017) and tropical coral reefs (De'ath et al., 2012; Bozec et al., 2015; Cheal et al., 2017). These changes are associated with increases in maximum wind speed, wave height and the inundation, although trends in these variables vary from region to region (Section 3.3.5). In some cases, this can lead to increased exposure to related impacts, such as flooding, reduced water quality and increased sediment runoff (*medium-high confidence*) (Brodie et al., 2012; Wong et al., 2014; Anthony, 2016; AR5, Table 5.1).

3

Sea level rise also amplifies the impacts of storms and wave action (Section 3.3.9), with robust evidence that storm surges and damage are already penetrating farther inland than a few decades ago, changing conditions for coastal ecosystems and human communities. This is especially true for small islands (Box 3.5) and low-lying coastal communities, where issues such as storm surges can transform coastal areas (Section 3.4.5; Brown et al., 2018a). Changes in the frequency of extreme events, such as an increase in the frequency of intense storms, have the potential (along with other factors, such as disease, food web changes, invasive organisms and heat stress-related mortality; Burge et al., 2014; Maynard et al., 2015; Weatherdon et al., 2016; Clements et al., 2017) to overwhelm the capacity for natural and human systems to recover following disturbances. This has recently been seen for key ecosystems such as tropical coral reefs (Box 3.4), which have changed from coral-dominated ecosystems to assemblages dominated by other organisms such as seaweeds, with changes in associated organisms and ecosystem services (*high confidence*) (De'ath et al., 2012; Bozec et al., 2015; Cheal et al., 2017; Hoegh-Guldberg et al., 2017; Hughes et al., 2017a, b). The impacts of storms are amplified by sea level rise (Section 3.4.5), leading to substantial challenges today and in the future for cities, deltas and small island states in particular (Sections 3.4.5.2 to 3.4.5.4), as well as for coastlines and their associated ecosystems (Sections 3.4.5.5 to 3.4.5.7).

3.4.4.4 Ocean circulation

The movement of water within the ocean is essential to its biology and ecology, as well to the circulation of heat, water and nutrients around the planet (Section 3.3.7). The movement of these factors drives local and regional climates, as well as primary productivity and food production. Firmly attributing recent changes in the strength and direction of ocean currents to climate change, however, is complicated by long-term patterns and variability (e.g., Pacific decadal oscillation, PDO; Signorini et al., 2015) and a lack of records that match the long-term nature of these changes in many cases (Lluch-Cota et al., 2014). An assessment of the literature since AR5 (Sydeman et al., 2014), however, concluded that (overall) upwelling-favourable winds have intensified in the California, Benguela and Humboldt upwelling systems, but have weakened in the Iberian system and have remained neutral in the Canary upwelling system in over 60 years of records (1946–2012) (*medium confidence*). These conclusions are consistent with a growing consensus that wind-driven upwelling systems are likely to intensify under climate change in many upwelling systems (Sydeman et al., 2014; Bakun et al., 2015; Di Lorenzo, 2015), with potentially positive and negative consequences (Bakun et al., 2015).

Changes in ocean circulation can have profound impacts on marine ecosystems by connecting regions and facilitating the entry and establishment of species in areas where they were unknown before (e.g., 'tropicalization' of temperate ecosystems; Wernberg et al., 2012; Vergés et al., 2014, 2016; Zarco-Perello et al., 2017), as well as the arrival of novel disease agents (*low-medium confidence*) (Burge et al., 2014; Maynard et al., 2015; Weatherdon et al., 2016). For example, the herbivorous sea urchin *Centrostephanus rodgersii* has been reached Tasmania from the Australian mainland, where it was previously unknown, owing to a strengthening of the East Australian Current (EAC) that connects the two regions (*high confidence*) (Ling et al., 2009). As a consequence, the

distribution and abundance of kelp forests has rapidly decreased, with implications for fisheries and other ecosystem services (Ling et al., 2009). These risks to marine ecosystems are projected to become greater at 1.5°C, and more so at 2°C (*medium confidence*) (Cheung et al., 2009; Pereira et al., 2010; Pinsky et al., 2013; Burrows et al., 2014).

Changes to ocean circulation can have even larger influence in terms of scale and impacts. Weakening of the Atlantic Meridional Overturning Circulation (AMOC), for example, is projected to be highly disruptive to natural and human systems as the delivery of heat to higher latitudes via this current system is reduced (Collins et al., 2013). Evidence of a slowdown of AMOC has increased since AR5 (Smeed et al., 2014; Rahmstorf et al., 2015a, b; Kelly et al., 2016), yet a strong causal connection to climate change is missing (*low confidence*) (Section 3.3.7).

3.4.4.5 Ocean acidification

Ocean chemistry encompasses a wide range of phenomena and chemical species, many of which are integral to the biology and ecology of the ocean (Section 3.3.10; Gattuso et al., 2014, 2015; Hoegh-Guldberg et al., 2014; Pörtner et al., 2014). While changes to ocean chemistry are likely to be of central importance, the literature on how climate change might influence ocean chemistry over the short and long term is limited (*medium confidence*). By contrast, numerous risks from the specific changes associated with ocean acidification have been identified (Dove et al., 2013; Kroeker et al., 2013; Pörtner et al., 2014; Gattuso et al., 2015; Albright et al., 2016), with the consensus that resulting changes to the carbonate chemistry of seawater are having, and are likely to continue to have, fundamental and substantial impacts on a wide variety of organisms (*high confidence*). Organisms with shells and skeletons made out of calcium carbonate are particularly at risk, as are the early life history stages of a large number of organisms and processes such as de-calcification, although there are some taxa that have not shown high-sensitivity to changes in CO_2, pH and carbonate concentrations (Dove et al., 2013; Fang et al., 2013; Kroeker et al., 2013; Pörtner et al., 2014; Gattuso et al., 2015). Risks of these impacts also vary with latitude and depth, with the greatest changes occurring at high latitudes as well as deeper regions. The aragonite saturation horizon (i.e., where concentrations of calcium and carbonate fall below the saturation point for aragonite, a key crystalline form of calcium carbonate) is decreasing with depth as anthropogenic CO_2 penetrates deeper into the ocean over time. Under many models and scenarios, the aragonite saturation is projected to reach the surface by 2030 onwards, with a growing list of impacts and consequences for ocean organisms, ecosystems and people (Orr et al., 2005; Hauri et al., 2016).

Further, it is difficult to reliably separate the impacts of ocean warming and acidification. As ocean waters have increased in sea surface temperature (SST) by approximately 0.9°C they have also decreased by 0.2 pH units since 1870–1899 ('pre-industrial'; Table 1 in Gattuso et al., 2015; Bopp et al., 2013). As CO_2 concentrations continue to increase along with other GHGs, pH will decrease while sea temperature will increase, reaching 1.7°C and a decrease of 0.2 pH units (by 2100 under RCP4.5) relative to the pre-industrial period. These changes are likely to continue given the negative correlation of temperature and pH. Experimental manipulation of CO_2, temperature and consequently acidification indicate that these impacts will continue to increase in size and scale as CO_2 and SST continue to increase in tandem (Dove et al., 2013; Fang et al., 2013; Kroeker et al., 2013).

While many risks have been defined through laboratory and mesocosm experiments, there is a growing list of impacts from the field (*medium confidence*) that include community-scale impacts on bacterial assemblages and processes (Endres et al., 2014), coccolithophores (K.J.S. Meier et al., 2014), pteropods and polar foodwebs (Bednaršek et al., 2012, 2014), phytoplankton (Moy et al., 2009; Riebesell et al., 2013; Richier et al., 2014), benthic ecosystems (Hall-Spencer et al., 2008; Linares et al., 2015), seagrass (Garrard et al., 2014), and macroalgae (Webster et al., 2013; Ordonez et al., 2014), as well as excavating sponges, endolithic microalgae and reef-building corals (Dove et al., 2013; Reyes-Nivia et al., 2013; Fang et al., 2014), and coral reefs (Box 3.4; Fabricius et al., 2011; Allen et al., 2017). Some ecosystems, such as those from bathyal areas (i.e., 200–3000 m below the surface), are likely to undergo very large reductions in pH by the year 2100 (0.29 to 0.37 pH units), yet evidence of how deep-water ecosystems will respond is currently limited despite the potential planetary importance of these areas (*low to medium confidence*) (Hughes and Narayanaswamy, 2013; Sweetman et al., 2017).

3.4.4.6 Deoxygenation

Oxygen levels in the ocean are maintained by a series of processes including ocean mixing, photosynthesis, respiration and solubility (Boyd et al., 2014, 2015; Pörtner et al., 2014; Breitburg et al., 2018). Concentrations of oxygen in the ocean are declining (*high confidence*) owing to three main factors related to climate change: (i) heat-related stratification of the water column (less ventilation and mixing), (ii) reduced oxygen solubility as ocean temperature increases, and (iii) impacts of warming on biological processes that produce or consume oxygen such as photosynthesis and respiration (*high confidence*) (Bopp et al., 2013; Pörtner et al., 2014; Altieri and Gedan, 2015; Deutsch et al., 2015; Schmidtko et al., 2017; Shepherd et al., 2017; Breitburg et al., 2018). Further, a range of processes (Section 3.4.11) are acting synergistically, including factors not related to climate change, such as runoff and coastal eutrophication (e.g., from coastal farming and intensive aquaculture). These changes can lead to increased phytoplankton productivity as a result of the increased concentration of dissolved nutrients. Increased supply of organic carbon molecules from coastal run-off can also increase the metabolic activity of coastal microbial communities (Altieri and Gedan, 2015; Bakun et al., 2015; Boyd, 2015). Deep sea areas are likely to experience some of the greatest challenges, as abyssal seafloor habitats in areas of deep-water formation are projected to experience decreased water column oxygen concentrations by as much as 0.03 mL L^{-1} by 2100 (Levin and Le Bris, 2015; Sweetman et al., 2017).

The number of 'dead zones' (areas where oxygenated waters have been replaced by hypoxic conditions) has been growing strongly since the 1990s (Diaz and Rosenberg, 2008; Altieri and Gedan, 2015; Schmidtko et al., 2017). While attribution can be difficult because of the complexity of the processes involved, both related and unrelated to climate change, some impacts associated to deoxygenation (*low-medium confidence*) include the expansion of oxygen minimum

zones (OMZ) (Turner et al., 2008; Carstensen et al., 2014; Acharya and Panigrahi, 2016; Lachkar et al., 2018), physiological impacts (Pörtner et al., 2014), and mortality and/or displacement of oxygen dependent organisms such as fish (Hamukuaya et al., 1998; Thronson and Quigg, 2008; Jacinto, 2011) and invertebrates (Hobbs and Mcdonald, 2010; Bednaršek et al., 2016; Seibel, 2016; Altieri et al., 2017). In addition, deoxygenation interacts with ocean acidification to present substantial separate and combined challenges for fisheries and aquaculture (*medium confidence*) (Hamukuaya et al., 1998; Bakun et al., 2015; Rodrigues et al., 2015; Feely et al., 2016; S. Li et al., 2016; Asiedu et al., 2017a; Clements and Chopin, 2017; Clements et al., 2017; Breitburg et al., 2018). Deoxygenation is expected to have greater impacts as ocean warming and acidification increase (*high confidence*), with impacts being larger and more numerous than today (e.g., greater challenges for aquaculture and fisheries from hypoxia), and as the number of hypoxic areas continues to increase. Risks from deoxygenation are *virtually certain* to increase as warming continues, although our understanding of risks at 1.5°C versus 2°C is incomplete (*medium confidence*). Reducing coastal pollution, and consequently the penetration of organic carbon into deep benthic habitats, is expected to reduce the loss of oxygen in coastal waters and hypoxic areas in general (*high confidence*) (Breitburg et al., 2018).

3.4.4.7 Loss of sea ice

Sea ice is a persistent feature of the planet's polar regions (Polyak et al., 2010) and is central to marine ecosystems, people (e.g., food, culture and livelihoods) and industries (e.g., fishing, tourism, oil and gas, and shipping). Summer sea ice in the Arctic, however, has been retreating rapidly in recent decades (Section 3.3.8), with an assessment of the literature revealing that a fundamental transformation is occurring in polar organisms and ecosystems, driven by climate change (*high confidence*) (Larsen et al., 2014). These changes are strongly affecting people in the Arctic who have close relationships with sea ice and associated ecosystems, and these people are facing major adaptation challenges as a result of sea level rise, coastal erosion, the accelerated thawing of permafrost, changing ecosystems and resources, and many other issues (Ford, 2012; Ford et al., 2015).

There is considerable and compelling evidence that a further increase of 0.5°C beyond the present-day average global surface temperature will lead to multiple levels of impact on a variety of organisms, from phytoplankton to marine mammals, with some of the most dramatic changes occurring in the Arctic Ocean and western Antarctic Peninsula (Turner et al., 2014, 2017b; Steinberg et al., 2015; Piñones and Fedorov, 2016).

The impacts of climate change on sea ice are part of the focus of the IPCC Special Report on the Ocean and Cryosphere in a Changing Climate (SROCC), due to be released in 2019, and hence are not covered comprehensively here. However, there is a range of responses to the loss of sea ice that are occurring and which increase at 1.5°C and further so with 2°C of global warming. Some of these changes are described briefly here. Photosynthetic communities, such as macroalgae, phytoplankton and microalgae dwelling on the underside of floating sea ice are changing, owing to increased temperatures, light and nutrient levels. As sea ice retreats, mixing of the water column increases, and phototrophs have increased access to seasonally high levels of solar radiation (*medium confidence*) (Dalpadado et al., 2014; W.N. Meier et al., 2014). These changes are expected to stimulate fisheries productivity in high-latitude regions by mid-century (*high confidence*) (Cheung et al., 2009, 2010, 2016b; Lam et al., 2014), with evidence that this is already happening for several high-latitude fisheries in the Northern Hemisphere, such as the Bering Sea, although these 'positive' impacts may be relatively short-lived (Hollowed and Sundby, 2014; Sundby et al., 2016). In addition to the impact of climate change on fisheries via impacts on net primary productivity (NPP), there are also direct effects of temperature on fish, which may in turn have a range of impacts (Pörtner et al., 2014). Sea ice in Antarctica is undergoing changes that exceed those seen in the Arctic (Maksym et al., 2011; Reid et al., 2015), with increases in sea ice coverage in the western Ross Sea being accompanied by strong decreases in the Bellingshausen and Amundsen Seas (Hobbs et al., 2016). While Antarctica is not permanently populated, the ramifications of changes to the productivity of vast regions, such as the Southern Ocean, have substantial implications for ocean foodwebs and fisheries globally.

3.4.4.8   Sea level rise

Mean sea level is increasing (Section 3.3.9), with substantial impacts already being felt by coastal ecosystems and communities (Wong et al., 2014) (*high confidence*). These changes are interacting with other factors, such as strengthening storms, which together are driving larger storm surges, infrastructure damage, erosion and habitat loss (Church et al., 2013; Stocker et al., 2013; Blankespoor et al., 2014). Coastal wetland ecosystems such as mangroves, sea grasses and salt marshes are under pressure from rising sea level (*medium confidence*) (Section 3.4.5; Di Nitto et al., 2014; Ellison, 2014; Lovelock et al., 2015; Mills et al., 2016; Nicholls et al., 2018), as well as from a wide range of other risks and impacts unrelated to climate change, with the ongoing loss of wetlands recently estimated at approximately 1% per annum across a large number of countries (Blankespoor et al., 2014; Alongi, 2015). While some ecosystems (e.g., mangroves) may be able to shift shoreward as sea levels increase, coastal development (e.g., buildings, seawalls and agriculture) often interrupts shoreward shifts, as well as reducing sediment supplies down some rivers (e.g., dams) due to coastal development (Di Nitto et al., 2014; Lovelock et al., 2015; Mills et al., 2016).

Responses to sea level rise challenges for ocean and coastal systems include reducing the impact of other stresses, such as those arising from tourism, fishing, coastal development, reduced sediment supply and unsustainable aquaculture/agriculture, in order to build ecological resilience (Hossain et al., 2015; Sutton-Grier and Moore, 2016; Asiedu et al., 2017a). The available literature largely concludes that these impacts will intensify under a 1.5°C warmer world but will be even higher at 2°C, especially when considered in the context of changes occurring beyond the end of the current century. In some cases, restoration of coastal habitats and ecosystems may be a cost-effective way of responding to changes arising from increasing levels of exposure to rising sea levels, intensifying storms, coastal inundation and salinization (Section 3.4.5 and Box 3.5; Arkema et al., 2013), although limitations of these strategies have been identified (e.g., Lovelock et al., 2015; Weatherdon et al., 2016).

3.4.4.9   Projected risks and adaptation options for oceans under global warming of 1.5°C or 2°C above pre-industrial levels

A comprehensive discussion of risk and adaptation options for all natural and human systems is not possible in the context and length of this report, and hence the intention here is to illustrate key risks and adaptation options for ocean ecosystems and sectors. This assessment builds on the recent expert consensus of Gattuso et al. (2015) by assessing new literature from 2015–2017 and adjusting the levels of risk from climate change in the light of literature since 2014. The original expert group's assessment (Supplementary Material 3.SM.3.2) was used as input for this new assessment, which focuses on the implications of global warming of 1.5°C as compared to 2°C. A discussion of potential adaptation options is also provided, the details of which will be further explored in later chapters of this special report. The section draws on the extensive analysis and literature presented in the Supplementary Material of this report (3.SM.3.2, 3.SM.3.3) and has a summary in Figures 3.18 and 3.20 which outline the added relative risks of climate change.

3.4.4.10   Framework organisms (tropical corals, mangroves and seagrass)

Marine organisms ('ecosystem engineers'), such as seagrass, kelp, oysters, salt marsh species, mangroves and corals, build physical structures or frameworks (i.e., sea grass meadows, kelp forests, oyster reefs, salt marshes, mangrove forests and coral reefs) which form the habitat for a large number of species (Gutiérrez et al., 2012). These organisms in turn provide food, livelihoods, cultural significance, and services such as coastal protection to human communities (Bell et al., 2011, 2018; Cinner et al., 2012; Arkema et al., 2013; Nurse et al., 2014; Wong et al., 2014; Barbier, 2015; Bell and Taylor, 2015; Hoegh-Guldberg et al., 2015; Mycoo, 2017; Pecl et al., 2017).

Risks of climate change impacts for seagrass and mangrove ecosystems were recently assessed by an expert group led by Short et al. (2016). Impacts of climate change were assessed to be similar across a range of submerged and emerged plants. Submerged plants such as seagrass were affected mostly by temperature extremes (Arias-Ortiz et al., 2018), and indirectly by turbidity, while emergent communities such as mangroves and salt marshes were most susceptible to sea level variability and temperature extremes, which is consistent with other evidence (Di Nitto et al., 2014; Sierra-Correa and Cantera Kintz, 2015; Osorio et al., 2016; Sasmito et al., 2016), especially in the context of human activities that reduce sediment supply (Lovelock et al., 2015) or interrupt the shoreward movement of mangroves though the construction of coastal infrastructure. This in turn leads to 'coastal squeeze' where coastal ecosystems are trapped between changing ocean conditions and coastal infrastructure (Mills et al., 2016). Projections of the future distribution of seagrasses suggest a poleward shift, which raises concerns that low-latitude seagrass communities may contract as a result of increasing stress levels (Valle et al., 2014).

Climate change (e.g., sea level rise, heat stress, storms) presents risk for coastal ecosystems such as seagrass (*high confidence*) and reef-building corals (*very high confidence*) (Figure 3.18, Supplementary Material 3.SM.3.2), with evidence of increasing concern since AR5 and the conclusion that tropical corals may be even more vulnerable to climate change than indicated in assessments made in 2014 (Hoegh-Guldberg et al., 2014; Gattuso et al., 2015). The current assessment also considered the heatwave-related loss of 50% of shallow-water corals across hundreds of kilometres of the world's largest continuous coral reef system, the Great Barrier Reef. These large-scale impacts, plus the observation of back-to-back bleaching events on the Great Barrier Reef (predicted two decades ago, Hoegh-Guldberg, 1999) and arriving sooner than predicted (Hughes et al., 2017b, 2018), suggest that the research community may have underestimated climate risks for coral reefs (Figure 3.18). The general assessment of climate risks for mangroves prior to this special report was that they face greater risks from deforestation and unsustainable coastal development than from climate change (Alongi, 2008; Hoegh-Guldberg et al., 2014; Gattuso et al., 2015). Recent large-scale die-offs (Duke et al., 2017; Lovelock et al., 2017), however, suggest that risks from climate change may have been underestimated for mangroves as well. With the events of the last past three years in mind, risks are now considered to be undetectable to moderate (i.e., moderate risks now start at 1.3°C as opposed to 1.8°C; *medium confidence*). Consequently, when average global warming reaches 1.3°C above pre-industrial levels, the risk of climate change to mangroves are projected to be moderate (Figure 3.18) while tropical coral reefs will have reached a high level of risk as examplified by increasing damage from heat stress since the early 1980s. At global warming of 1.8°C above pre-industrial levels, seagrasses are projected to reach moderate to high levels of risk (e.g., damage resulting from sea level rise, erosion, extreme temperatures, and storms), while risks to mangroves from climate change are projected to remain moderate (e.g., not keeping up with sea level rise, and more frequent heat stress mortality) although there is *low certainty* as to when or if this important ecosystem is likely to transition to higher levels of additional risk from climate change (Figure 3.18).

Warm water (tropical) coral reefs are projected to reach a very high risk of impact at 1.2°C (Figure 3.18), with most available evidence suggesting that coral-dominated ecosystems will be non-existent at this temperature or higher (*high confidence*). At this point, coral abundance will be near zero at many locations and storms will contribute to 'flattening' the three-dimensional structure of reefs without recovery, as already observed for some coral reefs (Alvarez-Filip et al., 2009). The impacts of warming, coupled with ocean acidification, are expected to undermine the ability of tropical coral reefs to provide habitat for thousand of species, which together provide a range of ecosystem services (e.g., food, livelihoods, coastal protection, cultural services) that are important for millions of people (*high confidence*) (Burke et al., 2011).

Strategies for reducing the impact of climate change on framework organisms include reducing stresses not directly related to climate change (e.g., coastal pollution, overfishing and destructive coastal development) in order to increase their ecological resilience in the face of accelerating climate change impacts (World Bank, 2013; Ellison, 2014; Anthony et al., 2015; Sierra-Correa and Cantera Kintz, 2015; Kroon et al., 2016; O'Leary et al., 2017), as well as protecting locations where organisms may be more robust (Palumbi et al., 2014) or less exposed to climate change (Bongaerts et al., 2010; van Hooidonk et al., 2013; Beyer et al., 2018). This might involve cooler areas due to

upwelling, or involve deep-water locations that experience less extreme conditions and impacts. Given the potential value of such locations for promoting the survival of coral communities under climate change, efforts to prevent their loss resulting from other stresses are important (Bongaerts et al., 2010, 2017; Chollett et al., 2010, 2014; Chollett and Mumby, 2013; Fine et al., 2013; van Hooidonk et al., 2013; Cacciapaglia and van Woesik, 2015; Beyer et al., 2018). A full understanding of the role of refugia in reducing the loss of ecosystems has yet to be developed (*low to medium confidence*). There is also interest in *ex situ* conservation approaches involving the restoration of corals via aquaculture (Shafir et al., 2006; Rinkevich, 2014) or the use of 'assisted evolution' to help corals adapt to changing sea temperatures (van Oppen et al., 2015, 2017), although there are numerous challenges that must be surpassed if these approaches are to be cost-effective responses to preserving coral reefs under rapid climate change (*low confidence*) (Hoegh-Guldberg, 2012, 2014a; Bayraktarov et al., 2016).

High levels of adaptation are expected to be required to prevent impacts on food security and livelihoods in coastal populations (*medium confidence*). Integrating coastal infrastructure with changing ecosystems such as mangroves, seagrasses and salt marsh, may offer adaptation strategies as they shift shoreward as sea levels rise (*high confidence*). Maintaining the sediment supply to coastal areas would also assist mangroves in keeping pace with sea level rise (Shearman et al., 2013; Lovelock et al., 2015; Sasmito et al., 2016). For this reason, habitat for mangroves can be strongly affected by human actions such as building dams which reduce the sediment supply and hence the ability of mangroves to escape 'drowning' as sea level rises (Lovelock et al., 2015). In addition, integrated coastal zone management should recognize the importance and economic expediency of using natural ecosystems such as mangroves and tropical coral reefs to protect coastal human communities (Arkema et al., 2013; Temmerman et al., 2013; Ferrario et al., 2014; Hinkel et al., 2014; Elliff and Silva, 2017). Adaptation options include developing alternative livelihoods and food sources, ecosystem-based management/adaptation such as ecosystem restoration, and constructing coastal infrastructure that reduces the impacts of rising seas and intensifying storms (Rinkevich, 2015; Weatherdon et al., 2016; Asiedu et al., 2017a; Feller et al., 2017). Clearly, these options need to be carefully assessed in terms of feasibility, cost and scalability, as well as in the light of the coastal ecosystems involved (Bayraktarov et al., 2016).

3.4.4.11   Ocean foodwebs (pteropods, bivalves, krill and fin fish)

Ocean foodwebs are vast interconnected systems that transfer solar energy and nutrients from phytoplankton to higher trophic levels, including apex predators and commercially important species such as tuna. Here, we consider four representative groups of marine organisms which are important within foodwebs across the ocean, and which illustrate the impacts and ramifications of 1.5°C or higher levels of warming.

The first group of organisms, pteropods, are small pelagic molluscs that suspension feed and produce a calcium carbonate shell. They are highly abundant in temperate and polar waters where they are an important link in the foodweb between phytoplankton and a range of other organisms including fish, whales and birds. The second group, bivalve molluscs (e.g., clams, oysters and mussels), are filter-feeding invertebrates. These invertebrate organisms underpin important fisheries and aquaculture industries, from polar to tropical regions, and are important food sources for a range of organisms including humans. The third group of organisms considered here is a globally significant group of invertebrates known as *euphausiid crustaceans* (krill), which are a key food source for many marine organisms and hence a major link between primary producers and higher trophic levels (e.g., fish, mammals and sea birds). Antarctic krill, *Euphausia superba*, are among the most abundant species in terms of mass and are consequently an essential component of polar foodwebs (Atkinson et al., 2009). The last group, fin fishes, is vitally important components of ocean foodwebs, contribute to the income of coastal communities, industries and nations, and are important to the foodsecurity and livelihood of hundreds of millions of people globally (FAO, 2016). Further background for this section is provided in Supplementary Material 3.SM.3.2.

There is a moderate risk to ocean foodwebs under present-day conditions (*medium to high confidence*) (Figure 3.18). Changing water chemistry and temperature are already affecting the ability of pteropods to produce their shells, swim and survive (Bednaršek et al., 2016). Shell dissolution, for example, has increased by 19–26% in both nearshore and offshore populations since the pre-industrial period (Feely et al., 2016). There is considerable concern as to whether these organisms are declining further, especially given the central importance in ocean foodwebs (David et al., 2017). Reviewing the literature reveals that pteropods are projected to face high risks of impact at average global temperatures 1.5°C above pre-industrial levels and increasing risks of impacts at 2°C (*medium confidence*).

As GMST increases by 1.5°C and more, the risk of impacts from ocean warming and acidification are expected to be moderate to high, except in the case of bivalves (mid-latitudes) where the risks of impacts are projected to be high to very high (Figure 3.18). Ocean warming and acidification are already affecting the life history stages of bivalve molluscs (e.g., Asplund et al., 2014; Mackenzie et al., 2014; Waldbusser et al., 2014; Zittier et al., 2015; Shi et al., 2016; Velez et al., 2016; Q. Wang et al., 2016; Castillo et al., 2017; Lemasson et al., 2017; Ong et al., 2017; X. Zhao et al., 2017). Impacts on adult bivalves include decreased growth, increased respiration and reduced calcification, whereas larval stages tend to show greater developmental abnormalities and increased mortality after exposure to these conditions (*medium to high confidence*) (Q. Wang et al., 2016; Lemasson et al., 2017; Ong et al., 2017; X. Zhao et al., 2017). Risks are expected to accumulate at higher temperatures for bivalve molluscs, with very high risks expected at 1.8°C of warming or more. This general pattern applies to low-latitude fin fish, which are expected to experience moderate to high risks of impact at 1.3°C of global warming (*medium confidence*), and very high risks at 1.8°C at low latitudes (*medium confidence*) (Figure 3.18).

Large-scale changes to foodweb structure are occurring in all oceans. For example, record levels of sea ice loss in the Antarctic (Notz and Stroeve, 2016; Turner et al., 2017b) translate into a loss of habitat and hence reduced abundance of krill (Piñones and Fedorov, 2016), with negative ramifications for the seabirds and whales which feed on krill (Croxall, 1992; Trathan and Hill, 2016) (*low-medium confidence*). Other influences,

such as high rates of ocean acidification coupled with shoaling of the aragonite saturation horizon, are likely to also play key roles (Kawaguchi et al., 2013; Piñones and Fedorov, 2016). As with many risks associated with impacts at the ecosystem scale, most adaptation options focus on the management of stresses unrelated to climate change but resulting from human activities, such as pollution and habitat destruction. Reducing these stresses will be important in efforts to maintain important foodweb components. Fisheries management at local to regional scales will be important in reducing stress on foodweb organisms, such as those discussed here, and in helping communities and industries adapt to changing foodweb structures and resources (see further discussion of fisheries *per se* below; Section 3.4.6.3). One strategy is to maintain larger population levels of fished species in order to provide more resilient stocks in the face of challenges that are increasingly driven by climate change (Green et al., 2014; Bell and Taylor, 2015).

3.4.4.12 Key ecosystem services (e.g., carbon uptake, coastal protection, and tropical coral reef recreation)

The ocean provides important services, including the regulation of atmospheric composition via gas exchange across the boundary between ocean and atmosphere, and the storage of carbon in vegetation and soils associated with ecosystems such as mangroves, salt marshes and coastal peatlands. These services involve a series of physicochemical processes which are influenced by ocean chemistry, circulation, biology, temperature and biogeochemical components, as well as by factors other than climate (Boyd, 2015). The ocean is also a net sink for CO_2 (another important service), absorbing approximately 30% of human emissions from the burning of fossil fuels and modification of land use (IPCC, 2013). Carbon uptake by the ocean is decreasing (Iida et al., 2015), and there is increasing concern from observations and models regarding associated changes to ocean circulation (Sections 3.3.7 and 3.4.4., Rahmstorf et al., 2015b);. Biological components of carbon uptake by the ocean are also changing, with observations of changing net primary productivity (NPP) in equatorial and coastal upwelling systems (*medium confidence*) (Lluch-Cota et al., 2014; Sydeman et al., 2014; Bakun et al., 2015), as well as subtropical gyre systems (*low confidence*) (Signorini et al., 2015). There is general agreement that NPP will decline as ocean warming and acidification increase (*medium confidence*) (Bopp et al., 2013; Boyd et al., 2014; Pörtner et al., 2014; Boyd, 2015).

Projected risks of impacts from reductions in carbon uptake, coastal protection and services contributing to coral reef recreation suggest a transition from moderate to high risks at 1.5°C and higher (*low confidence*). At 2°C, risks of impacts associated with changes to carbon uptake are high (*high confidence*), while the risks associated with reduced coastal protection and recreation on tropical coral reefs are high, especially given the vulnerability of this ecosystem type, and others (e.g., seagrass and mangroves), to climate change (*medium confidence*) (Figure 3.18). Coastal protection is a service provided by natural barriers such as mangroves, seagrass meadows, coral reefs, and other coastal ecosystems, and it is important for protecting human communities and infrastructure against the impacts associated with rising sea levels, larger waves and intensifying storms (*high confidence*) (Gutiérrez et al., 2012; Kennedy et al., 2013; Ferrario et al., 2014; Barbier, 2015; Cooper et al., 2016; Hauer et al., 2016; Narayan et al., 2016). Both natural and human coastal

3

protection have the potential to reduce these impacts (Fu and Song, 2017). Tropical coral reefs, for example, provide effective protection by dissipating about 97% of wave energy, with 86% of the energy being dissipated by reef crests alone (Ferrario et al., 2014; Narayan et al., 2016). Mangroves similarly play an important role in coastal protection, as well as providing resources for coastal communities, but they are already under moderate risk of not keeping up with sea level rise due to climate change and to contributing factors, such as reduced sediment supply or obstacles to shoreward shifts (Saunders et al., 2014; Lovelock et al., 2015). This implies that coastal areas currently protected by mangroves may experience growing risks over time.

Risks for specific marine and coastal organisms, ecosystems and sectors

The key elements are presented here as a function of the risk level assessed between 1.5 and 2°C (Average global sea surface temperature).

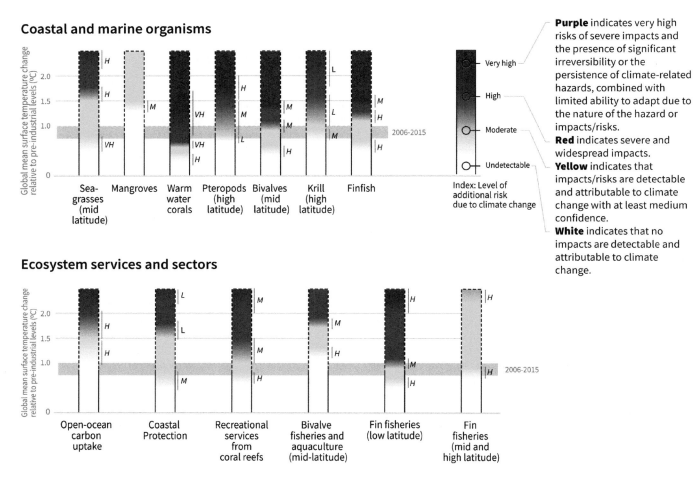

Confidence level for transition: L=Low, M=Medium, H=High and VH=Very high

Figure 3.18 | Summary of additional risks of impacts from ocean warming (and associated climate change factors such ocean acidification) for a range of ocean organisms, ecosystems and sectors at 1.0°C, 1.5°C and 2.0°C of warming of the average sea surface temperature (SST) relative to the pre-industrial period. The grey bar represents the range of GMST for the most recent decade: 2006–2015. The assessment of changing risk levels and associated confidence were primarily derived from the expert judgement of Gattuso et al. (2015) and the lead authors and relevant contributing authors of Chapter 3 (SR1.5), while additional input was received from the many reviewers of the ocean systems section of SR1.5. Notes: (i) The analysis shown here is not intended to be comprehensive. The examples of organisms, ecosystems and sectors included here are intended to illustrate the scale, types and projection of risks for representative natural and human ocean systems. (ii) The evaluation of risks by experts did not consider genetic adaptation, acclimatization or human risk reduction strategies (mitigation and societal adaptation). (iii) As discussed elsewhere (Sections 3.3.10 and 3.4.4.5, Box 3.4; Gattuso et al., 2015), ocean acidification is also having impacts on organisms and ecosystems as carbon dioxide increases in the atmosphere. These changes are part of the responses reported here, although partitioning the effects of the two drivers is difficult at this point in time and hence was not attempted. (iv) Confidence levels for location of transition points between levels of risk (L = low, M = moderate, H = high and VH = very high) are assessed and presented here as in the accompanying study by Gattuso et al. (2015). Three transitions in risk were possible: W–Y (white to yellow), Y–R (yellow to red), and R–P (red to purple), with the colours corresponding to the level of additional risk posed by climate change. The confidence levels for these transitions were assessed, based on level of agreement and extent of evidence, and appear as letters associated with each transition (see key in diagram).

Tourism is one of the largest industries globally (Rosselló-Nadal, 2014; Markham et al., 2016; Spalding et al., 2017). A substantial part of the global tourist industry is associated with tropical coastal regions and islands, where tropical coral reefs and related ecosystems play important roles (Section 3.4.9.1) (*medium confidence*). Coastal tourism can be a dominant money earner in terms of foreign exchange for many countries, particularly small island developing states (SIDS) (Section 3.4.9.1, Box 3.5; Weatherdon et al., 2016; Spalding et al., 2017). The direct relationship between increasing global temperatures, intensifying storms, elevated thermal stress, and the loss of tropical coral reefs has raised concern about the risks of climate change for local economies and industries based on tropical coral reefs. Risks to coral reef recreational services from climate change are considered here, as well as in Box 3.5, Section 3.4.9 and Supplementary Material 3.SM.3.2.

Adaptations to the broad global changes in carbon uptake by the ocean are limited and are discussed later in this report with respect to changes in NPP and implications for fishing industries. These adaptation options are broad and indirect, and the only other solution at large scale is to reduce the entry of CO_2 into the ocean. Strategies for adapting to reduced coastal protection involve (a) avoidance of vulnerable areas and hazards, (b) managed retreat from threatened locations, and/or (c) accommodation of impacts and loss of services (Bell, 2012; André et al., 2016; Cooper et al., 2016; Mills et al., 2016; Raabe and Stumpf, 2016; Fu and Song, 2017). Within these broad options, there are some strategies that involve direct human intervention, such as coastal hardening and the construction of seawalls and artificial reefs (Rinkevich, 2014, 2015; André et al., 2016; Cooper et al., 2016; Narayan et al., 2016), while others exploit opportunities for increasing coastal protection by involving naturally occurring oyster banks, coral reefs, mangroves, seagrass and other ecosystems (UNEP-WCMC, 2006; Scyphers et al., 2011; Zhang et

al., 2012; Ferrario et al., 2014; Cooper et al., 2016). Natural ecosystems, when healthy, also have the ability to repair themselves after being damaged, which sets them apart from coastal hardening and other human structures that require constant maintenance (Barbier, 2015; Elliff and Silva, 2017). In general, recognizing and restoring coastal ecosystems may be more cost-effective than installing human structures, in that creating and maintaining structures is typically expensive (Temmerman et al., 2013; Mycoo, 2017).

Recent studies have increasingly stressed the need for coastal protection to be considered within the context of coastal land management, including protecting and ensuring that coastal ecosystems are able to undergo shifts in their distribution and abundance as climate change occurs (Clausen and Clausen, 2014; Martínez et al., 2014; Cui et al., 2015; André et al., 2016; Mills et al., 2016). Facilitating these changes will require new tools in terms of legal and financial instruments, as well as integrated planning that involves not only human communities and infrastructure, but also associated ecosystem responses and values (Bell, 2012; Mills et al., 2016). In this regard, the interactions between climate change, sea level rise and coastal disasters are increasingly being informed by models (Bosello and De Cian, 2014) with a widening appreciation of the role of natural ecosystems as an alternative to hardened coastal structures (Cooper et al., 2016). Adaptation options for tropical coral reef recreation include: (i) protecting and improving biodiversity and ecological function by minimizing the impact of stresses unrelated to climate change (e.g., pollution and overfishing), (ii) ensuring adequate levels of coastal protection by supporting and repairing ecosystems that protect coastal regions, (iii) ensuring fair and equitable access to the economic opportunities associated with recreational activities, and (iv) seeking and protecting supplies of water for tourism, industry and agriculture alongside community needs.

Box 3.4 | Warm-Water (Tropical) Coral Reefs in a 1.5°C Warmer World

Warm-water coral reefs face very high risks (Figure 3.18) from climate change. A world in which global warming is restricted to 1.5°C above pre-industrial levels would be a better place for coral reefs than that of a 2°C warmer world, in which coral reefs would mostly disappear (Donner et al., 2005; Hoegh-Guldberg et al., 2014; Schleussner et al., 2016b; van Hooidonk et al., 2016; Frieler et al., 2017; Hughes et al., 2017a). Even with warming up until today (GMST for decade 2006–2015: 0.87°C; Chapter 1), a substantial proportion of coral reefs have experienced large-scale mortalities that have lead to much reduced coral populations (Hoegh-Guldberg et al., 2014). In the last three years alone (2016–2018), large coral reef systems such as the Great Barrier Reef (Australia) have lost as much as 50% of their shallow water corals (Hughes et al., 2017b).

Coral-dominated reefs are found along coastlines between latitudes 30°S and 30°N, where they provide habitat for over a million species (Reaka-Kudla, 1997) and food, income, coastal protection, cultural context and many other services for millions of people in tropical coastal areas (Burke et al., 2011; Cinner et al., 2012; Kennedy et al., 2013; Pendleton et al., 2016). Ultimately, coral reefs are underpinned by a mutualistic symbiosis between reef-building corals and dinoflagellates from the genus *Symbiodinium* (Hoegh-Guldberg et al., 2017). Warm-water coral reefs are found down to depths of 150 m and are dependent on light, making them distinct from the cold deep-water reef systems that extend down to depths of 2000 m or more. The difficulty in accessing deep-water reefs also means that the literature on the impacts of climate change on these systems is very limited by comparison to those on warm-water coral reefs (Hoegh-Guldberg et al., 2017). Consequently, this Box focuses on the impacts of climate change on warm-water (tropical) coral reefs, particularly with respect to their prospects under average global surface temperatures of 1.5°C and 2°C above the pre-industrial period.

Box 3.4 (continued)

The distribution and abundance of coral reefs has decreased by approximately 50% over the past 30 years (Gardner et al., 2005; Bruno and Selig, 2007; De'ath et al., 2012) as a result of pollution, storms, overfishing and unsustainable coastal development (Burke et al., 2011; Halpern et al., 2015; Cheal et al., 2017). More recently, climate change (i.e., heat stress; Hoegh-Guldberg, 1999; Baker et al., 2008; Spalding and Brown, 2015; Hughes et al., 2017b) has emerged as the greatest threat to coral reefs, with temperatures of just 1°C above the long-term summer maximum for an area (reference period 1985–1993) over 4–6 weeks being enough to cause mass coral bleaching (loss of the symbionts) and mortality (*very high confidence*) (WGII AR5, Box 18-2; Cramer et al., 2014). Ocean warming and acidification can also slow growth and calcification, making corals less competitive compared to other benthic organisms such as macroalgae or seaweeds (Dove et al., 2013; Reyes-Nivia et al., 2013, 2014). As corals disappear, so do fish and many other reef-dependent species, which directly impacts industries such as tourism and fisheries, as well as the livelihoods for many, often disadvantaged, coastal people (Wilson et al., 2006; Graham, 2014; Graham et al., 2015; Cinner et al., 2016; Pendleton et al., 2016). These impacts are exacerbated by increasingly intense storms (Section 3.3.6), which physically destroy coral communities and hence reefs (Cheal et al., 2017), and by ocean acidification (Sections 3.3.10 and 3.4.4.5), which can weaken coral skeletons, contribute to disease, and slow the recovery of coral communities after mortality events (*low to medium confidence*) (Gardner et al., 2005; Dove et al., 2013; Kennedy et al., 2013; Webster et al., 2013; Hoegh-Guldberg, 2014b; Anthony, 2016). Ocean acidification also leads to enhanced activity by decalcifying organisms such as excavating sponges (Kline et al., 2012; Dove et al., 2013; Fang et al., 2013, 2014; Reyes-Nivia et al., 2013, 2014).

The predictions of back-to-back bleaching events (Hoegh-Guldberg, 1999) have become the reality in the summers of 2016–2017 (e.g., Hughes et al., 2017b), as have projections of declining coral abundance (*high confidence*). Models have also become increasingly capable and are currently predicting the large-scale loss of coral reefs by mid-century under even low-emissions scenarios (Hoegh-Guldberg, 1999; Donner et al., 2005; Donner, 2009; van Hooidonk and Huber, 2012; Frieler et al., 2013; Hoegh-Guldberg et al., 2014; van Hooidonk et al., 2016). Even achieving emissions reduction targets consistent with the ambitious goal of 1.5°C of global warming under the Paris Agreement will result in the further loss of 70–90% of reef-building corals compared to today, with 99% of corals being lost under warming of 2°C or more above the pre-industrial period (Frieler et al., 2013; Hoegh-Guldberg, 2014b; Hoegh-Guldberg et al., 2014; Schleussner et al., 2016b; Hughes et al., 2017a).

The assumptions underpinning these assessments are considered to be highly conservative. In some cases, 'optimistic' assumptions in models include rapid thermal adaptation by corals of 0.2°C–1°C per decade (Donner et al., 2005) or 0.4°C per decade (Schleussner et al., 2016b), as well as very rapid recovery rates from impacts (e.g., five years in the case of Schleussner et al., 2016b). Adaptation to climate change at these high rates, has not been documented, and recovery from mass mortality tends to take much longer (>15 years; Baker et al., 2008). Probability analysis also indicates that the underlying increases in sea temperatures that drive coral bleaching and mortality are 25% less likely under 1.5°C when compared to 2°C (King et al., 2017). Spatial differences between the rates of heating suggest the possibility of temporary climate refugia (Caldeira, 2013; van Hooidonk et al., 2013; Cacciapaglia and van Woesik, 2015; Keppel and Kavousi, 2015), which may play an important role in terms of the regeneration of coral reefs, especially if these refuges are protected from risks unrelated to climate change. Locations at higher latitudes are reporting the arrival of reef-building corals, which may be valuable in terms of the role of limited refugia and coral reef structures but will have low biodiversity (*high confidence*) when compared to present-day tropical reefs (Kersting et al., 2017). Similarly, deep-water (30–150 m) or mesophotic coral reefs (Bongaerts et al., 2010; Holstein et al., 2016) may play an important role because they avoid shallow water extremes (i.e., heat and storms) to some extent, although the ability of these ecosystems to assist in repopulating damaged shallow water areas may be limited (Bongaerts et al., 2017).

Given the sensitivity of corals to heat stress, even short periods of overshoot (i.e., decades) are expected to be extremely damaging to coral reefs. Losing 70–90% of today's coral reefs, however, will remove resources and increase poverty levels across the world's tropical coastlines, highlighting the key issue of equity for the millions of people that depend on these valuable ecosystems (Cross-Chapter Box 6; Spalding et al., 2014; Halpern et al., 2015). Anticipating these challenges to food and livelihoods for coastal communities will become increasingly important, as will adaptation options, such as the diversification of livelihoods and the development of new sustainable industries, to reduce the dependency of coastal communities on threatened ecosystems such as coral reefs (Cinner et al., 2012, 2016; Pendleton et al., 2016). At the same time, coastal communities will need to pre-empt changes to other services provided by coral reefs such as coastal protection (Kennedy et al., 2013; Hoegh-Guldberg et al., 2014; Pörtner et al., 2014; Gattuso et al., 2015). Other threats and challenges to coastal living, such as sea level rise, will amplify challenges from declining coral reefs, specially for SIDS and low-lying tropical nations. Given the scale and cost of these interventions, implementing them earlier rather than later would be expedient.

3.4.5 Coastal and Low-Lying Areas, and Sea Level Rise

Sea level rise (SLR) is accelerating in response to climate change (Section 3.3.9; Church et al., 2013) and will produce significant impacts (*high confidence*). In this section, impacts and projections of SLR are reported at global and city scales (Sections 3.4.5.1 and 3.4.5.2) and for coastal systems (Sections 3.4.5.3 to 3.4.5.6). For some sectors, there is a lack of precise evidence of change at 1.5°C and 2°C of global warming. Adaptation to SLR is discussed in Section 3.4.5.7.

3.4.5.1 Global / sub-global scale

Sea level rise (SLR) and other oceanic climate changes are already resulting in salinization, flooding, and erosion and in the future are projected to affect human and ecological systems, including health, heritage, freshwater availability, biodiversity, agriculture, fisheries and other services, with different impacts seen worldwide (*high confidence*). Owing to the commitment to SLR, there is an overlapping uncertainty in projections at 1.5°C and 2°C (Schleussner et al., 2016b; Sanderson et al., 2017; Goodwin et al., 2018; Mengel et al., 2018; Nicholls et al., 2018; Rasmussen et al., 2018) and about 0.1 m difference in global mean sea level (GMSL) rise between 1.5°C and 2°C worlds in the year 2100 (Section 3.3.9, Table 3.3). Exposure and impacts at 1.5°C and 2°C differ at different time horizons (Schleussner et al., 2016b; Brown et al., 2018a, b; Nicholls et al., 2018; Rasmussen et al., 2018). However, these are distinct from impacts associated with higher increases in temperature (e.g., 4°C or more, as discussed in Brown et al., 2018a) over centennial scales. The benefits of climate change mitigation reinforce findings of earlier IPCC reports (e.g., Wong et al., 2014).

Table 3.3 shows the land and people exposed to SLR (assuming there is no adaptation or protection at all) using the Dynamic Interactive Vulnerability Assessment (DIVA) model (extracted from Brown et al., 2018a and Goodwin et al., 2018; see also Supplementary Material 3.SM, Table 3.SM.4). Thus, exposure increases even with temperature stabilization. The exposed land area is projected to at least double by 2300 using a RCP8.5 scenario compared with a mitigation scenario (Brown et al., 2018a). In the 21st century, land area exposed to sea level rise (assuming there is no adaptation or protection at all) is projected to be at least an order of magnitude larger than the cumulative land loss due to submergence (which takes into account defences) (Brown et al., 2016, 2018a) regardless of the SLR scenario applied. Slower rates of rise due to climate change mitigation may provide a greater opportunity for adaptation (*medium confidence*), which could substantially reduce impacts.

In agreement with the assessment in WGII AR5 Section 5.4.3.1 (Wong et al., 2014), climate change mitigation may reduce or delay coastal exposure and impacts (*very high confidence*). Adaptation has the potential to substantially reduce risk through a portfolio of available options (Sections 5.4.3.1 and 5.5 of Wong et al., 2014; Sections 6.4.2.3 and 6.6 of Nicholls et al., 2007). At 1.5°C in 2100, 31–69 million people (2010 population values) worldwide are projected to be exposed to flooding, assuming no adaptation or protection at all, compared with 32–79 million people (2010 population values) at 2°C in 2100 (Supplementary Material 3.SM, Table 3.SM.4; Rasmussen et al., 2018). As a result, up to 10.4 million more people would be exposed to sea level rise at 2°C compared with 1.5°C in 2100 (*medium confidence*). With a 1.5°C stabilization scenario in 2100, 62.7 million people per year are at risk from flooding, with this value increasing to 137.6 million people per year in 2300 (50th percentile, average across SSP1–5, no socio-economic change after 2100). These projections assume that no upgrade to current protection levels occurs (Nicholls et al., 2018). The number of people at risk increases by approximately 18% in 2030 if a 2°C scenario is used and by 266% in 2300 if an RCP8.5 scenario is considered (Nicholls et al., 2018). Through prescribed IPCC Special Report on Emissions Scenarios (SRES) SLR scenarios, Arnell et al. (2016) also found that the number of people exposed to flooding increased substantially at warming levels higher than 2°C, assuming no adaptation beyond current protection levels. Additionally, impacts increased in the second half of the 21st century.

Coastal flooding is projected to cost thousands of billions of USD annually, with damage costs under constant protection estimated at 0.3–5.0% of global gross domestic product (GDP) in 2100 under an RCP2.6 scenario (Hinkel et al., 2014). Risks are projected to be highest in South and Southeast Asia, assuming there is no upgrade to current protection levels, for all levels of climate warming (Arnell et al., 2016; Brown et al., 2016). Countries with at least 50 million people exposed to SLR (assuming no adaptation or protection at all) based on a 1,280 Pg C emissions scenario (approximately a 1.5°C temperature rise above today's level) include China, Bangladesh, Egypt, India, Indonesia, Japan, Philippines, United States and Vietnam (Clark et al., 2016). Rasmussen et al. (2018) and Brown et al. (2018a) project that similar countries would have high exposure to SLR in the 21st century using 1.5°C and 2°C scenarios. Thus, there is *high confidence* that SLR will have significant impacts worldwide in this century and beyond.

3.4.5.2 Cities

Observations of the impacts of SLR in cities are difficult to record because multiple drivers of change are involved. There are observations of ongoing and planned adaptation to SLR and extreme water levels in some cities (Araos et al., 2016; Nicholls et al., 2018), whilst other cities have yet to prepare for these impacts (*high confidence*) (see Section 3.4.8 and Cross-Chapter Box 9 in Chapter 4). There are limited observations and analyses of how cities will cope with higher and/or multi-centennial SLR, with the exception of Amsterdam, New York and London (Nicholls et al., 2018).

Coastal urban areas are projected to see more extreme water levels due to rising sea levels, which may lead to increased flooding and damage of infrastructure from extreme events (unless adaptation is undertaken), plus salinization of groundwater. These impacts may be enhanced through localized subsidence (Wong et al., 2014), which causes greater relative SLR. At least 136 megacities (port cities with a population greater than 1 million in 2005) are at risk from flooding due to SLR (with magnitudes of rise possible under 1.5°C or 2°C in the 21st century, as indicated in Section 3.3.9) unless further adaptation is undertaken (Hanson et al., 2011; Hallegatte et al., 2013). Many of these cities are located in South and Southeast Asia (Hallegatte et al., 2013; Cazenave and Cozannet, 2014; Clark et al., 2016; Jevrejeva et al., 2016). Jevrejeva et al. (2016) projected that more than 90% of global coastlines could experience SLR greater than 0.2 m with 2°C of warming by 2040 (RCP8.5). However, for scenarios where 2°C is stabilized or occurs later in time, this figure is likely to differ because of the commitment to SLR. Raising existing dikes helps protect against SLR, substantially reducing risks, although other forms of adaptation exist. By 2300, dike heights under a non-mitigation scenario (RCP8.5) could be more than 2 m higher (on average for 136 megacities) than under climate change mitigation scenarios at 1.5°C or 2°C (Nicholls et al., 2018). Thus, rising sea levels commit coastal cities to long-term adaptation (*high confidence*).

3.4.5.3   Small islands

Qualitative physical observations of SLR (and other stresses) include inundation of parts of low-lying islands, land degradation due to saltwater intrusion in Kiribati and Tuvalu (Wairiu, 2017), and shoreline change in French Polynesia (Yates et al., 2013), Tuvalu (Kench et al., 2015, 2018) and Hawaii (Romine et al., 2013). Observations, models and other evidence indicate that unconstrained Pacific atolls have kept pace with SLR, with little reduction in size or net gain in land (Kench et al., 2015, 2018; McLean and Kench, 2015; Beetham et al., 2017). Whilst islands are highly vulnerable to SLR (*high confidence*), they are also reactive to change. Small islands are impacted by multiple climatic stressors, with SLR being a more important stressor to some islands than others (Sections 3.4.10, 4.3.5.6, 5.2.1, 5.5.3.3, Boxes 3.5, 4.3 and 5.3).

Observed adaptation to multiple drivers of coastal change, including SLR, includes retreat (migration), accommodation and defence. Migration (internal and international) has always been important on small islands (Farbotko and Lazrus, 2012; Weir et al., 2017), with changing environmental and weather conditions being just one factor in the choice to migrate (Sections 3.4.10, 4.3.5.6 and 5.3.2; Campbell and Warrick, 2014). Whilst flooding may result in migration or relocation, for example in Vunidogoloa, Fiji (McNamara and Des Combes, 2015; Gharbaoui and Blocher, 2016) and the Solomon Islands (Albert et al., 2017), in situ adaptation may be tried or preferred, for example stilted housing or raised floors in Tubigon, Bohol, Philippines (Jamero et al., 2017), raised roads and floors in Batasan and Ubay, Philippines (Jamero et al., 2018), and raised platforms for faluw in Leang, Federated States of Micronesia (Nunn et al., 2017). Protective features, such as seawalls or beach nourishment, are observed to locally reduce erosion and flood risk but can have other adverse implications (Sovacool, 2012; Mycoo, 2014, 2017; Nurse et al., 2014; AR5 Section 29.6.22).

There is a lack of precise, quantitative studies of projected impacts of SLR at 1.5°C and 2°C. Small islands are projected to be at risk and very sensitive to coastal climate change and other stressors (*high confidence*) (Nurse et al., 2014; Benjamin and Thomas, 2016; Ourbak and Magnan, 2017; Brown et al., 2018a; Nicholls et al., 2018; Rasmussen et al., 2018; AR5 Sections 29.3 and 29.4), such as oceanic warming, SLR (resulting in salinization, flooding and erosion), cyclones and mass coral bleaching and mortality (Section 3.4.4, Boxes 3.4 and 3.5). These impacts can have significant socio-economic and ecological implications, such as on health, agriculture and water resources, which in turn have impacts on livelihoods (Sovacool, 2012; Mycoo, 2014, 2017; Nurse et al., 2014). Combinations of drivers causing adverse impacts are important. For example, Storlazzi et al. (2018) found that

the impacts of SLR and wave-induced flooding (within a temperature horizon equivalent of 1.5°C), could affect freshwater availability on Roi-Namur, Marshall Islands, but is also dependent on other extreme weather events. Freshwater resources may also be affected by a 0.40 m rise in sea level (which may be experienced with a 1.5°C warming) in other Pacific atolls (Terry and Chui, 2012). Whilst SLR is a major hazard for atolls, islands reaching higher elevations are also threatened given that there is often a lot of infrastructure located near the coast (*high confidence*) (Kumar and Taylor, 2015; Nicholls et al., 2018). Tens of thousands of people on small islands are exposed to SLR (Rasmussen et al., 2018). Giardino et al. (2018) found that hard defence structures on the island of Ebeye in the Marshall Islands were effective in reducing damage due to SLR at 1.5°C and 2°C. Additionally, damage was also reduced under mitigation scenarios compared with non-mitigation scenarios. In Jamaica and St Lucia, SLR and extreme sea levels are projected to threaten transport system infrastructure at 1.5°C unless further adaptation is undertaken (Monioudi et al., 2018). Slower rates of SLR will provide a greater opportunity for adaptation to be successful (*medium confidence*), but this may not be substantial enough on islands with a very low mean elevation. Migration and/or relocation may be an adaptation option (Section 3.4.10). Thomas and Benjamin (2017) highlight three areas of concern in the context of loss and damage at 1.5°C: a lack of data, gaps in financial assessments, and a lack of targeted policies or mechanisms to address these issues (Cross-Chapter Box 12 in Chapter 5). Small islands are projected to remain vulnerable to SLR (*high confidence*).

3.4.5.4   Deltas and estuaries

Observations of SLR and human influence are felt through salinization, which leads to mixing in deltas and estuaries, aquifers, leading to flooding (also enhanced by precipitation and river discharge), land degradation and erosion. Salinization is projected to impact freshwater sources and pose risks to ecosystems and human systems (Section 5.4; Wong et al., 2014). For instance, in the Delaware River estuary on the east coast of the USA, upward trends of salinity (measured since the 1900s), accounting for the effects of streamflow and seasonal variations, have been detected and SLR is a potential cause (Ross et al., 2015).

Z. Yang et al. (2015) found that future climate scenarios for the USA (A1B 1.6°C and B1 2°C in the 2040s) had a greater effect on salinity intrusion than future land-use/land-cover change in the Snohomish River estuary in Washington state (USA). This resulted in a shift in the salinity both upstream and downstream in low flow conditions. Projecting impacts in deltas needs an understanding of both fluvial discharge and SLR, making projections complex because the drivers operate on different temporal and spatial scales (Zaman et al., 2017; Brown et al., 2018b). The mean annual flood depth when 1.5°C is first projected to be reached in the Ganges-Brahmaputra delta may be less than the most extreme annual flood depth seen today, taking into account SLR, surges, tides, bathymetry and local river flows (Brown et al., 2018b). Further, increased river salinity and saline intrusion in the Ganges-Brahmaputra-Meghna is likely with 2°C of warming (Zaman et al., 2017). Salinization could impact agriculture and food security (Cross-Chapter Box 6 in this chapter). For 1.5°C or 2°C stabilization conditions in 2200 or 2300 plus surges, a minimum of 44% of the

Bangladeshi Ganges-Brahmaputra, Indian Bengal, Indian Mahanadi and Ghanese Volta delta land area (without defences) would be exposed unless sedimentation occurs (Brown et al., 2018b). Other deltas are similarly vulnerable. SLR is only one factor affecting deltas, and assessment of numerous geophysical and anthropogenic drivers of geomorphic change is important (Tessler et al., 2018). For example, dike building to reduce flooding and dam building (Gupta et al., 2012) restricts sediment movement and deposition, leading to enhanced subsidence, which can occur at a greater rate than SLR (Auerbach et al., 2015; Takagi et al., 2016). Although dikes remain essential for reducing flood risk today, promoting sedimentation is an advisable strategy (Brown et al., 2018b) which may involve nature-based solutions. Transformative decisions regarding the extent of sediment restrictive infrastructure may need to be considered over centennial scales (Brown et al., 2018b). Thus, in a 1.5°C or 2°C warmer world, deltas, which are home to millions of people, are expected to be highly threatened from SLR and localized subsidence (*high confidence*).

3.4.5.5   Wetlands

Observations indicate that wetlands, such as saltmarshes and mangrove forests, are disrupted by changing conditions (Sections 3.4.4.8; Wong et al., 2014; Lovelock et al., 2015), such as total water levels and sediment availability. For example, saltmarshes in Connecticut and New York, USA, measured from 1900 to 2012, have accreted with SLR but have lost marsh surface relative to tidal datums, leading to increased marsh flooding and further accretion (Hill and Anisfeld, 2015). This change stimulated marsh carbon storage and aided climate change mitigation.

Salinization may lead to shifts in wetland communities and their ecosystem functions (Herbert et al., 2015). Some projections of wetland change, with magnitudes (but not necessarily rates or timing) of SLR analogous to 1.5°C and 2°C of global warming, indicate a net loss of wetlands in the 21st century (e.g., Blankespoor et al., 2014; Cui et al., 2015; Arnell et al., 2016; Crosby et al., 2016), whilst others report a net gain with wetland transgression (e.g., Raabe and Stumpf, 2016 in the Gulf of Mexico). However, the feedback between wetlands and sea level is complex, with parameters such as a lack of accommodation space restricting inland migration, or sediment supply and feedbacks between plant growth and geomorphology (Kirwan and Megonigal, 2013; Ellison, 2014; Martínez et al., 2014; Spencer et al., 2016) still being explored. Reducing global warming from 2°C to 1.5°C will deliver long-term benefits, with natural sedimentation rates more likely keep up with SLR. It remains unclear how wetlands will respond and under what conditions (including other climate parameters) to a global temperature rise of 1.5°C and 2°C. However, they have great potential to aid and benefit climate change mitigation and adaptation (*medium confidence*) (Sections 4.3.2.2 and 4.3.2.3).

3.4.5.6   Other coastal settings

Numerous impacts have not been quantified at 1.5°C or 2°C but remain important. This includes systems identified in WGII AR5 (AR5 – Section 5.4 of Wong et al., 2014), such as beaches, barriers, sand dunes, rocky coasts, aquifers, lagoons and coastal ecosystems (for the last system, see Section 3.4.4.12). For example, SLR potentially affects erosion and accretion, and therefore sediment movement, instigating shoreline

change (Section 5.4.2.1 of Wong et al., 2014), which could affect land-based ecosystems. Global observations indicate no overall clear effect of SLR on shoreline change (Le Cozannet et al., 2014), as it is highly site specific (e.g., Romine et al., 2013). Infrastructure and geological constraints reduce shoreline movement, causing coastal squeeze. In Japan, for example, SLR is projected to cause beach losses under an RCP2.6 scenario, which will worsen under RCP8.5 (Udo and Takeda, 2017). Further, compound flooding (the combined risk of flooding from multiple sources) has increased significantly over the past century in major coastal cities (Wahl et al., 2015) and is likely to increase with further development and SLR at 1.5°C and 2°C unless adaptation is undertaken. Thus, overall SLR will have a wide range of adverse effects on coastal zones (*medium confidence*).

3.4.5.7   Adapting to coastal change

Adaptation to coastal change from SLR and other drivers is occurring today (*high confidence*) (see Cross-Chapter Box 9 in Chapter 4), including migration, ecosystem-based adaptation, raising infrastructure and defences, salt-tolerant food production, early warning systems, insurance and education (Section 5.4.2.1 of Wong et al., 2014). Climate change mitigation will reduce the rate of SLR this century, decreasing the need for extensive and, in places, immediate adaptation. Adaptation will reduce impacts in human settings (*high confidence*) (Hinkel et al., 2014; Wong et al., 2014), although there is less certainty for natural ecosystems (Sections 4.3.2 and 4.3.3.3). While some ecosystems (e.g., mangroves) may be able to move shoreward as sea levels increase, coastal development (e.g., coastal building, seawalls and agriculture) often interrupt these transitions (Saunders et al., 2014). Options for responding to these challenges include reducing the impact of other stresses such as those arising from tourism, fishing, coastal development and unsustainable aquaculture/agriculture. In some cases, restoration of coastal habitats and ecosystems can be a cost-effective way of responding to changes arising from increasing levels of exposure from rising sea levels, changes in storm conditions, coastal inundation and salinization (Arkema et al., 2013; Temmerman et al., 2013; Ferrario et al., 2014; Hinkel et al., 2014; Spalding et al., 2014; Elliff and Silva, 2017).

Since AR5, planned and autonomous adaptation and forward planning have become more widespread (Araos et al., 2016; Nicholls et al., 2018), but continued efforts are required as many localities are in the early stages of adapting or are not adapting at all (Cross-Chapter Box 9 in Chapter 4; Araos et al., 2016). This is region and sub-sector specific, and also linked to non-climatic factors (Ford et al., 2015; Araos et al., 2016; Lesnikowski et al., 2016). Adaptation pathways (e.g., Ranger et al., 2013; Barnett et al., 2014; Rosenzweig and Solecki, 2014; Buurman and Babovic, 2016) assist long-term planning but are not widespread practices despite knowledge of long-term risks (Section 4.2.2). Furthermore, human retreat and migration are increasingly being considered as an adaptation response (Hauer et al., 2016; Geisler and Currens, 2017), with a growing emphasis on green adaptation. There are few studies on the adaptation limits to SLR where transformation change may be required (AR5-Section 5.5 of Wong et al., 2014; Nicholls et al., 2015). Sea level rise poses a long-term threat (Section 3.3.9), and adaptation will remain essential at the centennial scale under 1.5°C and 2°C of warming (*high confidence*).

Table 3.3 | Land and people exposed to sea level rise (SLR), assuming no protection at all. Extracted from Brown et al. (2018a) and Goodwin et al. (2018). SSP: Shared Socio-Economic Pathway; wrt: with respect to; *:Population held constant at 2100 level.

Climate scenario	Impact factor, assuming there is no adaptation or protection at all (50th, [5th-95th percentiles])	Year			
		2050	2100	2200	2300
1.5°C	Temperature rise wrt 1850–1900 (°C)	1.71 (1.44–2.16)	1.60 (1.26–2.33)	1.41 (1.15–2.10)	1.32 (1.12–1.81)
	SLR (m) wrt 1986–2005	0.20 (0.14–0.29)	0.40 (0.26–0.62)	0.73 (0.47–1.25)	1.00 (0.59–1.55)
	Land exposed (x10³ km²)	574 [558–597]	620 [575–669]	666 [595–772]	702 [666–853]
	People exposed, SSP1–5 (millions)	127.9–139.0 [123.4–134.0, 134.5–146.4]	102.7–153.5 [94.8–140.7, 102.7–153.5]	--	133.8–207.1 [112.3–169.6, 165.2–263.4]*
2°C	Temperature rise wrt 1850–1900 (° C)	1.76 (1.51–2.16)	2.03 (1.72–2.64)	1.90 (1.66–2.57)	1.80 (1.60–2.20)
	SLR (m) wrt 1986-2005	0.20 (0.14–0.29)	0.46 (0.30–0.69)	0.90 (0.58–1.50)	1.26 (0.74–1.90)
	Land exposed (x10³ km²)	575 [558–598]	637 [585–686]	705 [618–827]	767 [642–937]
	People exposed, SSP1–5 (millions)	128.1–139.2 [123.6–134.2, 134.7–146.6]	105.5–158.1 [97.0–144.1, 118.1–179.0]	--	148.3–233.0 [120.3–183.4, 186.4–301.8]*

Box 3.5 | Small Island Developing States (SIDS)

Global warming of 1.5°C is expected to prove challenging for small island developing states (SIDS) that are already experiencing impacts associated with climate change (*high confidence*). At 1.5°C, compounding impacts from interactions between climate drivers may contribute to the loss of, or change in, critical natural and human systems (*medium to high confidence*). There are a number of reduced risks at 1.5°C versus 2°C, particularly when coupled with adaptation efforts (*medium to high confidence*).

Changing climate hazards for SIDS at 1.5°C

Mean surface temperature is projected to increase in SIDS at 1.5°C of global warming (*high confidence*). The Caribbean region will experience 0.5°C–1.5°C of warming compared to a 1971–2000 baseline, with the strongest warming occurring over larger land masses (Taylor et al., 2018). Under the Representative Concentration Pathway (RCP)2.6 scenario, the western tropical Pacific is projected to experience warming of 0.5°C–1.7°C relative to 1961–1990. Extreme temperatures will also increase, with potential for elevated impacts as a result of comparably small natural variability (Reyer et al., 2017a). Compared to the 1971–2000 baseline, up to 50% of the year is projected to be under warm spell conditions in the Caribbean at 1.5°C, with a further increase of up to 70 days at 2°C (Taylor et al., 2018).

Changes in precipitation patterns, freshwater availability and drought sensitivity differ among small island regions (*medium to high confidence*). Some western Pacific islands and those in the northern Indian Ocean may see increased freshwater availability, while islands in most other regions are projected to see a substantial decline (Holding et al., 2016; Karnauskas et al., 2016). For several SIDS, approximately 25% of the overall freshwater stress projected under 2°C at 2030 could be avoided by limiting global warming to 1.5°C (Karnauskas et al., 2018). In accordance with an overall drying trend, an increasing drought risk is projected for Caribbean SIDS (Lehner et al., 2017), and moderate to extreme drought conditions are projected to be about 9% longer on average at 2°C versus 1.5°C for islands in this region (Taylor et al., 2018).

Projected changes in the ocean system at higher warming targets (Section 3.4.4), including potential changes in circulation (Section 3.3.7) and increases in both surface temperatures (Section 3.3.7) and ocean acidification (Section 3.3.10), suggest increasing risks for SIDS associated with warming levels close to and exceeding 1.5°C.

Differences in global sea level between 1.5°C and 2°C depend on the time scale considered and are projected to fully materialize only after 2100 (Section 3.3.9). Projected changes in regional sea level are similarly time dependent, but generally found to be above the global average for tropical regions including small islands (Kopp et al., 2014; Jevrejeva et al., 2016). Threats related to sea level rise (SLR) for SIDS, for example from salinization, flooding, permanent inundation, erosion and pressure on ecosystems, will therefore persist well beyond the 21st century even under 1.5°C of warming (Section 3.4.5.3; Nicholls et al., 2018). Prolonged interannual sea level inundations may increase throughout the tropical Pacific with ongoing warming and in the advent of an

Box 3.5 (continued)

increased frequency of extreme La Niña events, exacerbating coastal impacts of projected global mean SLR (Widlansky et al., 2015). Changes to the frequency of extreme El Niño and La Niña events may also increase the frequency of droughts and floods in South Pacific islands (Box 4.2, Section 3.5.2; Cai et al., 2012).

Extreme precipitation in small island regions is often linked to tropical storms and contributes to the climate hazard (Khouakhi et al., 2017). Similarly, extreme sea levels for small islands, particularly in the Caribbean, are linked to tropical cyclone occurrence (Khouakhi and Villarini, 2017). Under a 1.5°C stabilization scenario, there is a projected decrease in the frequency of weaker tropical storms and an increase in the number of intense cyclones (Section 3.3.6; Wehner et al., 2018a). There are not enough studies to assess differences in tropical cyclone statistics for 1.5°C versus 2°C (Section 3.3.6). There are considerable differences in the adaptation responses to tropical cyclones across SIDS (Cross-Chapter Box 11 in Chapter 4).

Impacts on key natural and human systems

Projected increases in aridity and decreases in freshwater availability at 1.5°C of warming, along with additional risks from SLR and increased wave-induced run-up, might leave several atoll islands uninhabitable (Storlazzi et al., 2015; Gosling and Arnell, 2016). Changes in the availability and quality of freshwater, linked to a combination of changes to climate drivers, may adversely impact SIDS' economies (White and Falkland, 2010; Terry and Chui, 2012; Holding and Allen, 2015; Donk et al., 2018). Growth-rate projections based on temperature impacts alone indicate robust negative impacts on gross domestic product (GDP) per capita growth for SIDS (Sections 3.4.7.1, 3.4.9.1 and 3.5.4.9; Pretis et al., 2018). These impacts would be reduced considerably under 1.5°C but may be increased by escalating risks from climate-related extreme weather events and SLR (Sections 3.4.5.3, 3.4.9.4 and 3.5.3)

Marine systems and associated livelihoods in SIDS face higher risks at 2°C compared to 1.5°C (*medium to high confidence*). Mass coral bleaching and mortality are projected to increase because of interactions between rising ocean temperatures, ocean acidification, and destructive waves from intensifying storms (Section 3.4.4 and 5.2.3, Box 3.4). At 1.5°C, approximately 70–90% of global coral reefs are projected to be at risk of long-term degradation due to coral bleaching, with these values increasing to 99% at 2°C (Frieler et al., 2013; Schleussner et al., 2016b). Higher temperatures are also related to an increase in coral disease development, leading to coral degradation (Maynard et al., 2015). For marine fisheries, limiting warming to 1.5°C decreases the risk of species extinction and declines in maximum catch potential, particularly for small islands in tropical oceans (Cheung et al., 2016a).

Long-term risks of coastal flooding and impacts on populations, infrastructure and assets are projected to increase with higher levels of warming (*high confidence*). Tropical regions including small islands are expected to experience the largest increases in coastal flooding frequency, with the frequency of extreme water-level events in small islands projected to double by 2050 (Vitousek et al., 2017). Wave-driven coastal flooding risks for reef-lined islands may increase as a result of coral reef degradation and SLR (Quataert et al., 2015). Exposure to coastal hazards is particularly high for SIDS, with a significant share of population, infrastructure and assets at risk (Sections 3.4.5.3 and 3.4.9; Scott et al., 2012; Kumar and Taylor, 2015; Rhiney, 2015; Byers et al., 2018). Limiting warming to 1.5°C instead of 2°C would spare the inundation of lands currently home to 60,000 individuals in SIDS by 2150 (Rasmussen et al., 2018). However, such estimates do not consider shoreline response (Section 3.4.5) or adaptation.

Risks of impacts across sectors are projected to be higher at 1.5°C compared to the present, and will further increase at 2°C (*medium to high confidence*). Projections indicate that at 1.5°C there will be increased incidents of internal migration and displacement (Sections 3.5.5, 4.3.6 and 5.2.2; Albert et al., 2017), limited capacity to assess loss and damage (Thomas and Benjamin, 2017) and substantial increases in the risk to critical transportation infrastructure from marine inundation (Monioudi et al., 2018). The difference between 1.5°C and 2°C might exceed limits for normal thermoregulation of livestock animals and result in persistent heat stress for livestock animals in SIDS (Lallo et al., 2018).

At 1.5°C, limits to adaptation will be reached for several key impacts in SIDS, resulting in residual impacts, as well as loss and damage (Section 1.1.1, Cross-Chapter Box 12 in Chapter 5). Limiting temperature increase to 1.5°C versus 2°C is expected to reduce a number of risks, particularly when coupled with adaptation efforts that take into account sustainable development (Section 3.4.2 and 5.6.3.1, Box 4.3 and 5.3, Mycoo, 2017; Thomas and Benjamin, 2017). Region-specific pathways for SIDS exist to address climate change (Section 5.6.3.1, Boxes 4.6 and 5.3, Cross-Chapter Box 11 in Chapter 4).

3.4.6 Food, Nutrition Security and Food Production Systems (Including Fisheries and Aquaculture)

3.4.6.1 Crop production

Quantifying the observed impacts of climate change on food security and food production systems requires assumptions about the many non-climate variables that interact with climate change variables. Implementing specific strategies can partly or greatly alleviate the climate change impacts on these systems (Wei et al., 2017), whilst the degree of compensation is mainly dependent on the geographical area and crop type (Rose et al., 2016). Despite these uncertainties, recent studies confirm that observed climate change has already affected crop suitability in many areas, resulting in changes in the production levels of the main agricultural crops. These impacts are evident in many areas of the world, ranging from Asia (C. Chen et al., 2014; Sun et al., 2015; He and Zhou, 2016) to America (Cho and McCarl, 2017) and Europe (Ramirez-Cabral et al., 2016), and they particularly affect the typical local crops cultivated in specific climate conditions (e.g., Mediterranean crops like olive and grapevine, Moriondo et al., 2013a, b).

Temperature and precipitation trends have reduced crop production and yields, with the most negative impacts being on wheat and maize (Lobell et al., 2011), whilst the effects on rice and soybean yields are less clear and may be positive or negative (Kim et al., 2013; van Oort and Zwart, 2018). Warming has resulted in positive effects on crop yield in some high-latitude areas (Jaggard et al., 2007; Supit et al., 2010; Gregory and Marshall, 2012; C. Chen et al., 2014; Sun et al., 2015; He and Zhou, 2016; Daliakopoulos et al., 2017), and may make it possible to have more than one harvest per year (B. Chen et al., 2014; Sun et al., 2015). Climate variability has been found to explain more than 60% of the of maize, rice, wheat and soybean yield variations in the main global breadbaskets areas (Ray et al., 2015), with the percentage varying according to crop type and scale (Moore and Lobell, 2015; Kent et al., 2017). Climate trends also explain changes in the length of the growing season, with greater modifications found in the northern high-latitude areas (Qian et al., 2010; Mueller et al., 2015).

The rise in tropospheric ozone has already reduced yields of wheat, rice, maize and soybean by 3–16% globally (Van Dingenen et al., 2009). In some studies, increases in atmospheric CO_2 concentrations were found to increase yields by enhancing radiation and water use efficiencies (Elliott et al., 2014; Durand et al., 2018). In open-top chamber experiments with a combination of elevated CO_2 and 1.5°C of warming, maize and potato yields were observed to increase by 45.7% and 11%, respectively (Singh et al., 2013; Abebe et al., 2016). However, observations of trends in actual crop yields indicate that reductions as a result of climate change remain more common than crop yield increases, despite increased atmospheric CO_2 concentrations (Porter et al., 2014). For instance, McGrath and Lobell (2013) indicated that production stimulation at increased atmospheric CO_2 concentrations was mostly driven by differences in climate and crop species, whilst yield variability due to elevated CO_2 was only about 50–70% of the variability due to climate. Importantly, the faster growth rates induced by elevated CO_2 have been found to coincide with lower protein content in several important C3 cereal grains (Myers et al., 2014), although this may not always be the case for C4 grains, such as sorghum, under

drought conditions (De Souza et al., 2015). Elevated CO_2 concentrations of 568–590 ppm (a range that corresponds approximately to RCP6 in the 2080s and hence a warming of 2.3°C–3.3°C (van Vuuren et al., 2011a, AR5 WGI Table 12.2) alone reduced the protein, micronutrient and B vitamin content of the 18 rice cultivars grown most widely in Southeast Asia, where it is a staple food source, by an amount sufficient to create nutrition-related health risks for 600 million people (Zhu et al., 2018). Overall, the effects of increased CO_2 concentrations alone during the 21st century are therefore expected to have a negative impact on global food security (*medium confidence*).

Crop yields in the future will also be affected by projected changes in temperature and precipitation. Studies of major cereals showed that maize and wheat yields begin to decline with 1°C–2°C of local warming and under nitrogen stress conditions at low latitudes (*high confidence*) (Porter et al., 2014; Rosenzweig et al., 2014). A few studies since AR5 have focused on the impacts on cropping systems for scenarios where the global mean temperature increase is within 1.5°C. Schleussner et al. (2016b) projected that constraining warming to 1.5°C rather than 2°C would avoid significant risks of declining tropical crop yield in West Africa, Southeast Asia, and Central and South America. Ricke et al. (2016) highlighted that cropland stability declines rapidly between 1°C and 3°C of warming, whilst Bassu et al. (2014) found that an increase in air temperature negatively influences the modelled maize yield response by –0.5 t ha^{-1} °C^{-1} and Challinor et al. (2014) reported similar effect for tropical regions. Niang et al. (2014) projected significantly lower risks to crop productivity in Africa at 1.5°C compared to 2°C of warming. Lana et al. (2017) indicated that the impact of temperature increases on crop failure of maize hybrids would be much greater as temperatures increase by 2°C compared to 1.5°C (*high confidence*). J. Huang et al. (2017) found that limiting warming to 1.5°C compared to 2°C would reduce maize yield losses over drylands. Although Rosenzweig et al. (2017, 2018) did not find a clear distinction between yield declines or increases in some breadbasket regions between the two temperature levels, they generally did find projections of decreasing yields in breadbasket regions when the effects of CO_2 fertilization were excluded. Iizumi et al. (2017) found smaller reductions in maize and soybean yields at 1.5°C than at 2°C of projected warming, higher rice production at 2°C than at 1.5°C, and no clear differences for wheat on a global mean basis. These results are largely consistent with those of other studies (Faye et al., 2018; Ruane et al., 2018). In the western Sahel and southern Africa, moving from 1.5°C to 2°C of warming has been projected to result in a further reduction of the suitability of maize, sorghum and cocoa cropping areas and yield losses, especially for C3 crops, with rainfall change only partially compensating these impacts (Läderach et al., 2013; World Bank, 2013; Sultan and Gaetani, 2016).

A significant reduction has been projected for the global production of wheat (by 6.0 ± 2.9%), rice (by 3.2 ± 3.7%), maize (by 7.4 ± 4.5%), and soybean, (by 3.1%) for each degree Celsius increase in global mean temperature (Asseng et al., 2015; C. Zhao et al., 2017). Similarly, Li et al. (2017) indicated a significant reduction in rice yields for each degree Celsius increase, by about 10.3%, in the greater Mekong subregion (*medium confidence*; Cross-Chapter Box 6: Food Security in this chapter). Large rice and maize yield losses are to be expected in China, owing to climate extremes (*medium confidence*) (Wei et al., 2017; Zhang et al., 2017).

While not often considered, crop production is also negatively affected by the increase in both direct and indirect climate extremes. Direct extremes include changes in rainfall extremes (Rosenzweig et al., 2014), increases in hot nights (Welch et al., 2010; Okada et al., 2011), extremely high daytime temperatures (Schlenker and Roberts, 2009; Jiao et al., 2016; Lesk et al., 2016), drought (Jiao et al., 2016; Lesk et al., 2016), heat stress (Deryng et al., 2014, Betts et al., 2018), flooding (Betts et al., 2018; Byers et al., 2018), and chilling damage (Jiao et al., 2016), while indirect effects include the spread of pests and diseases (Jiao et al., 2014; van Bruggen et al., 2015), which can also have detrimental effects on cropping systems.

Taken together, the findings of studies on the effects of changes in temperature, precipitation, CO_2 concentration and extreme weather events indicate that a global warming of 2°C is projected to result in a greater reduction in global crop yields and global nutrition than global warming of 1.5°C (*high confidence*; Section 3.6).

3.4.6.2   Livestock production

Studies of climate change impacts on livestock production are few in number. Climate change is expected to directly affect yield quantity and quality (Notenbaert et al., 2017), as well as indirectly impacting the livestock sector through feed quality changes and spread of pests and diseases (Kipling et al., 2016) (*high confidence*). Increased warming and its extremes are expected to cause changes in physiological processes in livestock (i.e., thermal distress, sweating and high respiratory rates) (Mortola and Frappell, 2000) and to have detrimental effects on animal feeding, growth rates (André et al., 2011; Renaudeau et al., 2011; Collier and Gebremedhin, 2015) and reproduction (De Rensis et al., 2015). Wall et al. (2010) observed reduced milk yields and increased cow mortality as the result of heat stress on dairy cow production over some UK regions.

Further, a reduction in water supply might increase cattle water demand (Masike and Urich, 2008). Generally, heat stress can be responsible for domestic animal mortality increase and economic losses (Vitali et al., 2009), affecting a wide range of reproductive parameters (e.g., embryonic development and reproductive efficiency in pigs, Barati et al., 2008; ovarian follicle development and ovulation in horses, Mortensen et al., 2009). Much attention has also been dedicated to ruminant diseases (e.g., liver fluke, Fox et al., 2011; blue-tongue virus, Guis et al., 2012; foot-and-mouth disease (FMD), Brito et al. (2017); and zoonotic diseases, Njeru et al., 2016; Simulundu et al., 2017).

Climate change impacts on livestock are expected to increase. In temperate climates, warming is expected to lengthen the forage growing season but decrease forage quality, with important variations due to rainfall changes (Craine et al., 2010; Hatfield et al., 2011; Izaurralde et al., 2011). Similarly, a decrease in forage quality is expected for both natural grassland in France (Graux et al., 2013) and sown pastures in Australia (Perring et al., 2010). Water resource availability for livestock is expected to decrease owing to increased runoff and reduced groundwater resources. Increased temperature will likely induce changes in river discharge and the amount of water in basins, leading human and livestock populations to experience water stress, especially in the driest areas (i.e., sub-Saharan Africa and South Asia)

(*medium confidence*) (Palmer et al., 2008). Elevated temperatures are also expected to increase methane production (Knapp et al., 2014; M.A. Lee et al., 2017). Globally, a decline in livestock of 7–10% is expected at about 2°C of warming, with associated economic losses between $9.7 and $12.6 billion (Boone et al., 2018).

3.4.6.3   Fisheries and aquaculture production

Global fisheries and aquaculture contribute a total of 88.6 and 59.8 million tonnes of fish and other products annually (FAO, 2016), and play important roles in the food security of a large number of countries (McClanahan et al., 2015; Pauly and Charles, 2015) as well as being essential for meeting the protein demand of a growing global population (Cinner et al., 2012, 2016; FAO, 2016; Pendleton et al., 2016). A steady increase in the risks associated with bivalve fisheries and aquaculture at mid-latitudes is coincident with increases in temperature, ocean acidification, introduced species, disease and other drivers (Lacoue-Labarthe et al., 2016; Clements and Chopin, 2017; Clements et al., 2017; Parker et al., 2017). Sea level rise and storm intensification pose a risk to hatcheries and other infrastructure (Callaway et al., 2012; Weatherdon et al., 2016), whilst others risks are associated with the invasion of parasites and pathogens (Asplund et al., 2014; Castillo et al., 2017). Specific human strategies have reduced these risks, which are expected to be moderate under RCP2.6 and very high under RCP8.5 (Gattuso et al., 2015). The risks related to climate change for fin fish (Section 3.4.4) are producing a number of challenges for small-scale fisheries (e.g., Kittinger, 2013; Pauly and Charles, 2015; Bell et al., 2018). Recent literature from 2015 to 2017 has described growing threats from rapid shifts in the biogeography of key species (Poloczanska et al., 2013, 2016; Burrows et al., 2014; García Molinos et al., 2015) and the ongoing rapid degradation of key ecosystems such as coral reefs, seagrass and mangroves (Section 3.4.4, Box 3.4). The acceleration of these changes, coupled with non-climate stresses (e.g., pollution, overfishing and unsustainable coastal development), are driving many small-scale fisheries well below the sustainable harvesting levels required to maintain these resources as a source of food (McClanahan et al., 2009, 2015; Cheung et al., 2010; Pendleton et al., 2016). As a result, future scenarios surrounding climate change and global population growth increasingly project shortages of fish protein for many regions, such as the Pacific Ocean (Bell et al., 2013, 2018) and Indian Ocean (McClanahan et al., 2015). Mitigation of these risks involves marine spatial planning, fisheries repair, sustainable aquaculture, and the development of alternative livelihoods (Kittinger, 2013; McClanahan et al., 2015; Song and Chuenpagdee, 2015; Weatherdon et al., 2016). Other threats concern the increasing incidence of alien species and diseases (Kittinger et al., 2013; Weatherdon et al., 2016).

Risks of impacts related to climate change on low-latitude small-scale fin fisheries are moderate today but are expected to reach very high levels by 1.1°C of global warming. Projections for mid- to high-latitude fisheries include increases in fishery productivity in some cases (Cheung et al., 2013; Hollowed et al., 2013; Lam et al., 2014; FAO, 2016). These projections are associated with the biogeographical shift of species towards higher latitudes (Fossheim et al., 2015), which brings benefits as well as challenges (e.g., increased production yet a greater risk of disease and invasive species; *low confidence*). Factors underpinning the expansion of fisheries production to high-latitude locations include warming, increased light levels and mixing due to retreating sea ice (Cheung et al., 2009), which result in substantial increases in primary productivity and fish harvesting in the North Pacific and North Atlantic (Hollowed and Sundby, 2014).

Present-day risks for mid-latitude bivalve fisheries and aquaculture become undetectible up to 1.1°C of global warming, moderate at 1.3°C, and moderate to high up to 1.9°C (Figure 3.18). For instance, Cheung et al. (2016a), simulating the loss in fishery productivity at 1.5°C, 2°C and 3.5°C above the pre-industrial period, found that the potential global catch for marine fisheries will *likely* decrease by more than three million metric tonnes for each degree of warming. Low-latitude fin-fish fisheries have higher risks of impacts, with risks being moderate under present-day conditions and becoming high above 0.9°C and very high at 2°C of global warming. High-latitude

fisheries are undergoing major transformations, and while production is increasing, present-day risk is moderate and is projected to remain moderate at 1.5°C and 2°C (Figure 3.18).

Adaptation measures can be applied to shellfish, large pelagic fish resources and biodiversity, and they include options such as protecting reproductive stages and brood stocks from periods of high ocean acidification (OA), stock selection for high tolerance to OA (*high confidence*) (Ekstrom et al., 2015; Rodrigues et al., 2015; Handisyde et al., 2016; Lee, 2016; Weatherdon et al., 2016; Clements and Chopin, 2017), redistribution of highly migratory resources (e.g., Pacific tuna) (*high confidence*), governance instruments such as international fisheries agreements (Lehodey et al., 2015; Matear et al., 2015), protection and regeneration of reef habitats, reduction of coral reef stresses, and development of alternative livelihoods (e.g., aquaculture; Bell et al., 2013, 2018).

Cross-Chapter Box 6 | Food Security

Lead Authors:
Ove Hoegh-Guldberg (Australia), Sharina Abdul Halim (Malaysia), Marco Bindi (Italy), Marcos Buckeridge (Brazil), Arona Diedhiou (Ivory Coast/Senegal), Kristie L. Ebi (USA), Deborah Ley (Guatemala/Mexico), Diana Liverman (USA), Chandni Singh (India), Rachel Warren (UK), Guangsheng Zhou (China).

Contributing Author:
Lorenzo Brilli (Italy)

Climate change influences food and nutritional security through its effects on food availability, quality, access and distribution (Paterson and Lima, 2010; Thornton et al., 2014; FAO, 2016). More than 815 million people were undernourished in 2016, and 11% of the world's population has experienced recent decreases in food security, with higher percentages in Africa (20%), southern Asia (14.4%) and the Caribbean (17.7%) (FAO et al., 2017). Overall, food security is expected to be reduced at 2°C of global warming compared to 1.5°C, owing to projected impacts of climate change and extreme weather on yields, crop nutrient content, livestock, fisheries and aquaculture and land use (cover type and management) (Sections 3.4.3.6, 3.4.4.12 and 3.4.6), (*high confidence*). The effects of climate change on crop yield, cultivation area, presence of pests, food price and supplies are projected to have major implications for sustainable development, poverty eradication, inequality and the ability of the international community to meet the United Nations sustainable development goals (SDGs; Cross-Chapter Box 4 in Chapter 1).

Goal 2 of the SDGs is to end hunger, achieve food security, improve nutrition and promote sustainable agriculture by 2030. This goal builds on the first millennium development goal (MDG-1) which focused on eradicating extreme poverty and hunger, through efforts that reduced the proportion of undernourished people in low- and middle-income countries from 23.3% in 1990 to 12.9% in 2015. Climate change threatens the capacity to achieve SDG 2 and could reverse the progress made already. Food security and agriculture are also critical to other aspects of sustainable development, including poverty eradication (SDG 1), health and well-being (SDG 3), clean water (SDG 6), decent work (SDG 8), and the protection of ecosystems on land (SDG 14) and in water (SDG 15) (UN, 2015, 2017; Pérez-Escamilla, 2017).

Increasing global temperature poses large risks to food security globally and regionally, especially in low-latitude areas (*medium confidence*) (Cheung et al., 2010; Rosenzweig et al., 2013; Porter et al., 2014; Rosenzweig and Hillel, 2015; Lam et al., 2016), with warming of 2°C projected to result in a greater reduction in global crop yields and global nutrition than warming of 1.5°C (*high confidence*) (Section 3.4.6), owing to the combined effects of changes in temperature, precipitation and extreme weather events, as well as increasing CO_2 concentrations. Climate change can exacerbate malnutrition by reducing nutrient availability and the quality of food products (*medium confidence*) (Cramer et al., 2014; Zhu et al., 2018). Generally, vulnerability to decreases in water and food availability is projected to be reduced at 1.5°C versus 2°C (Cheung et al., 2016a; Betts et al., 2018), especially in regions such as the African Sahel, the Mediterranean, central Europe, the Amazon, and western and southern Africa (*medium confidence*) (Sultan and Gaetani, 2016; Lehner et al., 2017; Betts et al., 2018; Byers et al., 2018; Rosenzweig et al., 2018).

Cross-Chapter Box 6 (continued)

Rosenzweig et al. (2018) and Ruane et al. (2018) reported that the higher CO_2 concentrations associated with 2°C as compared to those at 1.5°C of global warming are projected to drive positive effects in some regions. Production can also benefit from warming in higher latitudes, with more fertile soils, favouring crops, and grassland production, in contrast to the situation at low latitudes (Section 3.4.6), and similar benefits could arise for high-latitude fisheries production (*high confidence*) (Section 3.4.6.3). Studies exploring regional climate change risks on crop production are strongly influenced by the use of different regional climate change projections and by the assumed strength of CO_2 fertilization effects (Section 3.6), which are uncertain. For C3 crops, theoretically advantageous CO_2 fertilization effects may not be realized in the field; further, they are often accompanied by losses in protein and nutrient content of crops (Section 3.6), and hence these projected benefits may not be realized. In addition, some micronutrients such as iron and zinc will accumulate less and be less available in food (Myers et al., 2014). Together, the impacts on protein availability may bring as many as 150 million people into protein deficiency by 2050 (Medek et al., 2017). However, short-term benefits could arise for high-latitude fisheries production as waters warm, sea ice contracts and primary productivity increases under climate change (*high confidence*) (Section 3.4.6.3; Cheung et al., 2010; Hollowed and Sundby, 2014; Lam et al., 2016; Sundby et al., 2016; Weatherdon et al., 2016).

Factors affecting the projections of food security include variability in regional climate projections, climate change mitigation (where land use is involved; see Section 3.6 and Cross-Chapter Box 7 in this chapter) and biological responses (*medium confidence*) (Section 3.4.6.1; McGrath and Lobell, 2013; Elliott et al., 2014; Pörtner et al., 2014; Durand et al., 2018), extreme events such as droughts and floods (*high confidence*) (Sections 3.4.6.1, 3.4.6.2; Rosenzweig et al., 2014; Wei et al., 2017), financial volatility (Kannan et al., 2000; Ghosh, 2010; Naylor and Falcon, 2010; HLPE, 2011), and the distributions of pests and disease (Jiao et al., 2014; van Bruggen et al., 2015). Changes in temperature and precipitation are projected to increase global food prices by 3–84% by 2050 (IPCC, 2013). Differences in price impacts of climate change are accompanied by differences in land-use change (Nelson et al., 2014b), energy policies and food trade (Mueller et al., 2011; Wright, 2011; Roberts and Schlenker, 2013). Fisheries and aquatic production systems (aquaculture) face similar challenges to those of crop and livestock sectors (Section 3.4.6.3; Asiedu et al., 2017a, b; Utete et al., 2018). Human influences on food security include demography, patterns of food waste, diet shifts, incomes and prices, storage, health status, trade patterns, conflict, and access to land and governmental or other assistance (Chapters 4 and 5). Across all these systems, the efficiency of adaptation strategies is uncertain because it is strongly linked with future economic and trade environments and their response to changing food availability (*medium confidence*) (Lobell et al., 2011; von Lampe et al., 2014; d'Amour et al., 2016; Wei et al., 2017).

Climate change impacts on food security can be reduced through adaptation (Hasegawa et al., 2014). While climate change is projected to decrease agricultural yield, the consequences could be reduced substantially at 1.5°C versus 2°C with appropriate investment (*high confidence*) (Neumann et al., 2010; Muller, 2011; Roudier et al., 2011), awareness-raising to help inform farmers of new technologies for maintaining yield, and strong adaptation strategies and policies that develop sustainable agricultural choices (Sections 4.3.2 and 4.5.3). In this regard, initiatives such as 'climate-smart' food production and distribution systems may assist via technologies and adaptation strategies for food systems (Lipper et al., 2014; Martinez-Baron et al., 2018; Whitfield et al., 2018), as well as helping meet mitigation goals (Harvey et al., 2014).

K.R. Smith et al. (2014) concluded that climate change will exacerbate current levels of childhood undernutrition and stunting through reduced food availability. As well, climate change can drive undernutrition-related childhood mortality, and increase disability-adjusted life years lost, with the largest risks in Asia and Africa (Supplementary Material 3.SM, Table 3.SM.12; Ishida et al., 2014; Hasegawa et al., 2016; Springmann et al., 2016). Studies comparing the health risks associated with reduced food security at 1.5°C and 2°C concluded that risks would be higher and the globally undernourished population larger at 2°C (Hales et al., 2014; Ishida et al., 2014; Hasegawa et al., 2016). Climate change impacts on dietary and weight-related risk factors are projected to increase mortality, owing to global reductions in food availability and consumption of fruit, vegetables and red meat (Springmann et al., 2016). Further, temperature increases are projected to reduce the protein and micronutrient content of major cereal crops, which is expected to further affect food and nutritional security (Myers et al., 2017; Zhu et al., 2018).

Strategies for improving food security often do so in complex settings such as the Mekong River basin in Southeast Asia. The Mekong is a major food bowl (Smajgl et al., 2015) but is also a climate change hotspot (de Sherbinin, 2014; Lebel et al., 2014). This area is also a useful illustration of the complexity of adaptation choices and actions in a 1.5°C warmer world. Climate projections include increased annual average temperatures and precipitation in the Mekong (Zhang et al., 2017), as well as increased flooding and related disaster risks (T.F. Smith et al., 2013; Ling et al., 2015; Zhang et al., 2016). Sea level rise and saline intrusion are ongoing risks to agricultural systems in this area by reducing soil fertility and limiting the crop productivity (Renaud et al., 2015). The main climate impacts in the Mekong are expected to be on ecosystem health, through salinity intrusion, biomass reduction and biodiversity losses (Le Dang et al., 2013; Smajgl et al., 2015); agricultural productivity and food security (Smajgl et al., 2015); livelihoods such as fishing and farming (D. Wu et al., 2013); and disaster risk (D. Wu et al., 2013; Hoang et al., 2016), with implications for human mortality and economic and infrastructure losses.

Adaptation imperatives and costs in the Mekong will be higher under higher temperatures and associated impacts on agriculture and aquaculture, hazard exposure, and infrastructure. Adaptation measures to meet food security include greater investment in crop diversification and integrated agriculture–aquaculture practices (Renaud et al., 2015), improvement of water-use technologies (e.g., irrigation, pond capacity improvement and rainwater harvesting), soil management, crop diversification, and strengthening allied sectors such as livestock rearing and aquaculture (ICEM, 2013). Ecosystem-based approaches, such as integrated water resources management, demonstrate successes in mainstreaming adaptation into existing strategies (Sebesvari et al., 2017). However, some of these adaptive strategies can have negative impacts that deepen the divide between land-wealthy and land-poor farmers (Chapman et al., 2016). Construction of high dikes, for example, has enabled triple-cropping, which benefits land-wealthy farmers but leads to increasing debt for land-poor farmers (Chapman and Darby, 2016).

Institutional innovation has happened through the Mekong River Commission (MRC), which is an intergovernmental body between Cambodia, Lao PDR, Thailand and Viet Nam that was established in 1995. The MRC has facilitated impact assessment studies, regional capacity building and local project implementation (Schipper et al., 2010), although the mainstreaming of adaptation into development policies has lagged behind needs (Gass et al., 2011). Existing adaptation interventions can be strengthened through greater flexibility of institutions dealing with land-use planning and agricultural production, improved monitoring of saline intrusion, and the installation of early warning systems that can be accessed by the local authorities or farmers (Renaud et al., 2015; Hoang et al., 2016; Tran et al., 2018). It is critical to identify and invest in synergistic strategies from an ensemble of infrastructural options (e.g., building dikes); soft adaptation measures (e.g., land-use change) (Smajgl et al., 2015; Hoang et al., 2018); combinations of top-down government-led (e.g., relocation) and bottom-up household strategies (e.g., increasing house height) (Ling et al., 2015); and community-based adaptation initiatives that merge scientific knowledge with local solutions (Gustafson et al., 2016, 2018; Tran et al., 2018). Special attention needs to be given to strengthening social safety nets and livelihood assets whilst ensuring that adaptation plans are mainstreamed into broader development goals (Sok and Yu, 2015; Kim et al., 2017). The combination of environmental, social and economic pressures on people in the Mekong River basin highlights the complexity of climate change impacts and adaptation in this region, as well as the fact that costs are projected to be much lower at 1.5°C than 2°C of global warming.

3.4.7   Human Health

Climate change adversely affects human health by increasing exposure and vulnerability to climate-related stresses, and decreasing the capacity of health systems to manage changes in the magnitude and pattern of climate-sensitive health outcomes (Cramer et al., 2014; Hales et al., 2014). Changing weather patterns are associated with shifts in the geographic range, seasonality and transmission intensity of selected climate-sensitive infectious diseases (e.g., Semenza and Menne, 2009), and increasing morbidity and mortality are associated with extreme weather and climate events (e.g., K.R. Smith et al., 2014). Health detection and attribution studies conducted since AR5 have provided evidence, using multistep attribution, that climate change is negatively affecting adverse health outcomes associated with heatwaves, Lyme disease in Canada, and *Vibrio* emergence in northern Europe (Mitchell, 2016; Mitchell et al., 2016; Ebi et al., 2017). The IPCC AR5 concluded there is *high* to *very high confidence* that climate change will lead to greater risks of injuries, disease and death, owing to more intense heatwaves and fires, increased risks of undernutrition, and consequences of reduced labour productivity in vulnerable populations (K.R. Smith et al., 2014).

3.4.7.1   Projected risk at 1.5°C and 2°C of global warming

The projected risks to human health of warming of 1.5°C and 2°C, based on studies of temperature-related morbidity and mortality, air quality and vector borne diseases assessed in and since AR5, are summarized in Supplementary Material 3.SM, Tables 3.SM.8, 3.SM.9

and 3.SM.10 (based on Ebi et al., 2018). Other climate-sensitive health outcomes, such as diarrheal diseases, mental health issues and the full range of sources of poor air quality, were not considered because of the lack of projections of how risks could change at 1.5°C and 2°C. Few projections were available for specific temperatures above pre-industrial levels; Supplementary Material 3.SM, Table 3.SM.7 provides the conversions used to translate risks projected for particular time slices to those for specific temperature changes (Ebi et al., 2018).

Temperature-related morbidity and mortality: The magnitude of projected heat-related morbidity and mortality is greater at 2°C than at 1.5°C of global warming (*very high confidence*)(Doyon et al., 2008; Jackson et al., 2010; Hanna et al., 2011; Huang et al., 2012; Petkova et al., 2013; Hajat et al., 2014; Hales et al., 2014; Honda et al., 2014; Vardoulakis et al., 2014; Garland et al., 2015; Huynen and Martens, 2015; Li et al., 2015; Schwartz et al., 2015; L. Wang et al., 2015; Guo et al., 2016; T. Li et al., 2016; Chung et al., 2017; Kendrovski et al., 2017; Mishra et al., 2017; Arnell et al., 2018; Mitchell et al., 2018b). The number of people exposed to heat events is projected to be greater at 2°C than at 1.5°C (Russo et al., 2016; Mora et al., 2017; Byers et al., 2018; Harrington and Otto, 2018; King et al., 2018). The extent to which morbidity and mortality are projected to increase varies by region, presumably because of differences in acclimatization, population vulnerability, the built environment, access to air conditioning and other factors (Russo et al., 2016; Mora et al., 2017; Byers et al., 2018; Harrington and Otto, 2018; King et al., 2018). Populations at highest risk include older adults, children, women, those with chronic diseases, and people taking certain medications (*very high confidence*). Assuming adaptation takes place reduces the projected magnitude of risks (Hales et al., 2014; Huynen and Martens, 2015; T. Li et al., 2016).

In some regions, cold-related mortality is projected to decrease with increasing temperatures, although increases in heat-related mortality generally are projected to outweigh any reductions in cold-related mortality with warmer winters, with the heat-related risks increasing with greater degrees of warming (Huang et al., 2012; Hajat et al., 2014; Vardoulakis et al., 2014; Gasparrini et al., 2015; Huynen and Martens, 2015; Schwartz et al., 2015).

Occupational health: Higher ambient temperatures and humidity levels place additional stress on individuals engaging in physical activity. Safe work activity and worker productivity during the hottest months of the year would be increasingly compromised with additional climate change (*medium confidence*) (Dunne et al., 2013; Kjellstrom et al., 2013, 2018; Sheffield et al., 2013; Habibi Mohraz et al., 2016). Patterns of change may be complex; for example, at 1.5°C, there could be about a 20% reduction in areas experiencing severe heat stress in East Asia, compared to significant increases in low latitudes at 2°C (Lee and Min, 2018). The costs of preventing workplace heat-related illnesses through worker breaks suggest that the difference in economic loss between 1.5°C and 2°C could be approximately 0.3% of global gross domestic product (GDP) in 2100 (Takakura et al., 2017). In China, taking into account population growth and employment structure, high temperature subsidies for employees working on extremely hot days are projected to increase from 38.6 billion yuan yr^{-1} in 1979–2005 to 250 billion yuan yr^{-1} in the 2030s (about 1.5°C) (Zhao et al., 2016).

Air quality: Because ozone formation is temperature dependent, projections focusing only on temperature increase generally conclude that ozone-related mortality will increase with additional warming, with the risks higher at 2°C than at 1.5°C (*high confidence*) (Supplementary Material 3.SM, Table 3.SM.9; Heal et al., 2013; Tainio et al., 2013; Likhvar et al., 2015; Silva et al., 2016; Dionisio et al., 2017; J.Y. Lee et al., 2017). Reductions in precursor emissions would reduce future ozone concentrations and associated mortality. Mortality associated with exposure to particulate matter could increase or decrease in the future, depending on climate projections and emissions assumptions (Supplementary Material 3.SM, Table 3.SM.8; Tainio et al., 2013; Likhvar et al., 2015; Silva et al., 2016).

Malaria: Recent projections of the potential impacts of climate change on malaria globally and for Asia, Africa, and South America (Supplementary Material 3.SM, Table 3.SM.10) confirm that weather and climate are among the drivers of the geographic range, intensity of transmission, and seasonality of malaria, and that the relationships are not necessarily linear, resulting in complex patterns of changes in risk with additional warming (*very high confidence*) (Ren et al., 2016; Song et al., 2016; Semakula et al., 2017). Projections suggest that the burden of malaria could increase with climate change because of a greater geographic range of the Anopheles vector, longer season, and/or increase in the number of people at risk, with larger burdens at higher levels of warming, but with regionally variable patterns (*medium to high confidence*). Vector populations are projected to shift with climate

change, with expansions and reductions depending on the degree of local warming, the ecology of the mosquito vector, and other factors (Ren et al., 2016).

Aedes **(mosquito vector for dengue fever, chikungunya, yellow fever and Zika virus):** Projections of the geographic distribution of *Aedes aegypti* and Ae. *albopictus* (principal vectors) or of the prevalence of dengue fever generally conclude that there will be an increase in the number of mosquitos and a larger geographic range at 2°C than at 1.5°C, and they suggest that more individuals will be at risk of dengue fever, with regional differences (*high confidence*) (Fischer et al., 2011, 2013; Colón-González et al., 2013, 2018; Bouzid et al., 2014; Ogden et al., 2014a; Mweya et al., 2016). The risks increase with greater warming. Projections suggest that climate change is projected to expand the geographic range of chikungunya, with greater expansions occurring at higher degrees of warming (Tjaden et al., 2017).

Other vector-borne diseases: Increased warming in North America and Europe could result in geographic expansions of regions (latitudinally and altitudinally) climatically suitable for West Nile virus transmission, particularly along the current edges of its transmission areas, and extension of the transmission season, with the magnitude and pattern of changes varying by location and level of warming (Semenza et al., 2016). Most projections conclude that climate change could expand the geographic range and seasonality of Lyme and other tick-borne diseases in parts of North America and Europe (Ogden et al., 2014b; Levi et al., 2015). The projected changes are larger with greater warming and under higher greenhouse gas emissions pathways. Projections of the impacts of climate change on leishmaniasis and Chagas disease indicate that climate change could increase or decrease future health burdens, with greater impacts occurring at higher degrees of warming (González et al., 2014; Ceccarelli and Rabinovich, 2015).

In summary, warming of 2°C poses greater risks to human health than warming of 1.5°C, often with the risks varying regionally, with a few exceptions (*high confidence*). There is *very high confidence* that each additional unit of warming could increase heat-related morbidity and mortality, and that adaptation would reduce the magnitude of impacts. There is *high confidence* that ozone-related mortality could increase if precursor emissions remain the same, and that higher temperatures could affect the transmission of some infectious diseases, with increases and decreases projected depending on the disease (e.g., malaria, dengue fever, West Nile virus and Lyme disease), region and degree of temperature change.

3.4.8   Urban Areas

There is new literature on urban climate change and its differential impacts on and risks for infrastructure sectors – energy, water, transport and buildings – and vulnerable populations, including those living in informal settlements (UCCRN, 2018). However, there is limited literature on the risks of warming of 1.5°C and 2°C in urban areas. Heat-related extreme events (Matthews et al., 2017), variability in precipitation (Yu et al., 2018) and sea level rise can directly affect urban areas (Section 3.4.5, Bader et al., 2018; Dawson et al., 2018). Indirect risks may arise from interactions between urban and natural systems.

Future warming and urban expansion could lead to more extreme heat stress (Argüeso et al., 2015; Suzuki-Parker et al., 2015). At 1.5°C of warming, twice as many megacities (such as Lagos, Nigeria and Shanghai, China) could become heat stressed, exposing more than 350 million more people to deadly heat by 2050 under midrange population growth. Without considering adaptation options, such as cooling from more reflective roofs, and overall characteristics of urban agglomerations in terms of land use, zoning and building codes (UCCRN, 2018), Karachi (Pakistan) and Kolkata (India) could experience conditions equivalent to the deadly 2015 heatwaves on an annual basis under 2°C of warming (Akbari et al., 2009; Oleson et al., 2010; Matthews et al., 2017). Warming of 2°C is expected to increase the risks of heatwaves in China's urban agglomerations (Yu et al., 2018). Stabilizing at 1.5°C of warming instead of 2°C could decrease mortality related to extreme temperatures in key European cities, assuming no adaptation and constant vulnerability (Jacob et al., 2018; Mitchell et al., 2018a). Holding temperature change to below 2°C but taking urban heat islands (UHI) into consideration, projections indicate that there could be a substantial increase in the occurrence of deadly heatwaves in cities. The urban impacts of these heatwaves are expected to be similar at 1.5°C and 2°C and substantially larger than under the present climate (Matthews et al., 2017; Yu et al., 2018). Increases in the intensity of UHI could exacerbate warming of urban areas, with projections ranging from a 6% decrease to a 30% increase for a doubling of CO_2 (McCarthy et al., 2010). Increases in population and city size, in the context of a warmer climate, are projected to increase UHI (Georgescu et al., 2012; Argüeso et al., 2014; Conlon et al., 2016; Kusaka et al., 2016; Grossman-Clarke et al., 2017).

For extreme heat events, an additional 0.5°C of warming implies a shift from the upper bounds of observed natural variability to a new global climate regime (Schleussner et al., 2016b), with distinct implications for the urban poor (Revi et al., 2014; Jean-Baptiste et al., 2018; UCCRN, 2018). Adverse impacts of extreme events could arise in tropical coastal areas of Africa, South America and Southeast Asia (Schleussner et al., 2016b). These urban coastal areas in the tropics are particularly at risk given their large informal settlements and other vulnerable urban populations, as well as vulnerable assets, including businesses and critical urban infrastructure (energy, water, transport and buildings) (McGranahan et al., 2007; Hallegatte et al., 2013; Revi et al., 2014; UCCRN, 2018). Mediterranean water stress is projected to increase from 9% at 1.5°C to 17% at 2°C compared to values in 1986–2005 period. Regional dry spells are projected to expand from 7% at 1.5°C to 11% at 2°C for the same reference period. Sea level rise is expected to be lower at 1.5°C than 2°C, lowering risks for coastal metropolitan agglomerations (Schleussner et al., 2016b).

Climate models are better at projecting implications of greenhouse gas forcing on physical systems than at assessing differential risks associated with achieving a specific temperature target (James et al., 2017). These challenges in managing risks are amplified when combined with the scale of urban areas and assumptions about socio-economic pathways (Krey et al., 2012; Kamei et al., 2016; Yu et al., 2016; Jiang and Neill, 2017).

In summary, in the absence of adaptation, in most cases, warming of 2°C poses greater risks to urban areas than warming of 1.5°C,

depending on the vulnerability of the location (coastal or non-coastal) (*high confidence*), businesses, infrastructure sectors (energy, water and transport), levels of poverty, and the mix of formal and informal settlements.

3.4.9   Key Economic Sectors and Services

Climate change could affect tourism, energy systems and transportation through direct impacts on operations (e.g., sea level rise) and through impacts on supply and demand, with the risks varying significantly with geographic region, season and time. Projected risks also depend on assumptions with respect to population growth, the rate and pattern of urbanization, and investments in infrastructure. Table 3.SM.11 in Supplementary Material 3.SM summarizes the cited publications.

3.4.9.1   Tourism

The implications of climate change for the global tourism sector are far-reaching and are impacting sector investments, destination assets (environment and cultural), operational and transportation costs, and tourist demand patterns (Scott et al., 2016a; Scott and Gössling, 2018). Since AR5, observed impacts on tourism markets and destination communities continue to be not well analysed, despite the many analogue conditions (e.g., heatwaves, major hurricanes, wild fires, reduced snow pack, coastal erosion and coral reef bleaching) that are anticipated to occur more frequently with climate change. There is some evidence that observed impacts on tourism assets, such as environmental and cultural heritage, are leading to the development of 'last chance to see' tourism markets, where travellers visit destinations before they are substantially degraded by climate change impacts or to view the impacts of climate change on landscapes (Lemelin et al., 2012; Stewart et al., 2016; Piggott-McKellar and McNamara, 2017).

There is limited research on the differential risks of a 1.5° versus 2°C temperature increase and resultant environmental and socio-economic impacts in the tourism sector. The translation of these changes in climate resources for tourism into projections of tourism demand remains geographically limited to Europe. Based on analyses of tourist comfort, summer and spring/autumn tourism in much of western Europe may be favoured by 1.5°C of warming, but with negative effects projected for Spain and Cyprus (decreases of 8% and 2%, respectively, in overnight stays) and most coastal regions of the Mediterranean (Jacob et al., 2018). Similar geographic patterns of potential tourism gains (central and northern Europe) and reduced summer favourability (Mediterranean countries) are projected under 2°C (Grillakis et al., 2016). Considering potential changes in natural snow only, winter overnight stays at 1.5°C are projected to decline by 1–2% in Austria, Italy and Slovakia, with an additional 1.9 million overnight stays lost under 2°C of warming (Jacob et al., 2018). Using an econometric analysis of the relationship between regional tourism demand and climate conditions, Ciscar et al. (2014) projected that a 2°C warmer world would reduce European tourism by 5% (€15 billion yr[-1]), with losses of up to 11% (€6 billion yr[-1]) for southern Europe and a potential gain of €0.5 billion yr[-1] in the UK.

There is growing evidence that the magnitude of projected impacts is temperature dependent and that sector risks could be much greater with higher temperature increases and resultant environmental and socio-economic impacts (Markham et al., 2016; Scott et al., 2016a; Jones, 2017; Steiger et al., 2017). Studies from 27 countries consistently project substantially decreased reliability of ski areas that are dependent on natural snow, increased snowmaking requirements and investment in snowmaking systems, shortened and more variable ski seasons, a contraction in the number of operating ski areas, altered competitiveness among and within regional ski markets, and subsequent impacts on employment and the value of vacation properties (Steiger et al., 2017). Studies that omit snowmaking do not reflect the operating realities of most ski areas and overestimate impacts at 1.5°C–2°C. In all regional markets, the extent and timing of these impacts depend on the magnitude of climate change and the types of adaptive responses by the ski industry, skiers and destination communities. The decline in the number of former Olympic Winter Games host locations that could remain climatically reliable for future Olympic and Paralympic Winter Games has been projected to be much greater under scenarios warmer than 2°C (Scott et al., 2015; Jacob et al., 2018).

The tourism sector is also affected by climate-induced changes in environmental assets critical for tourism, including biodiversity, beaches, glaciers and other features important for environmental and cultural heritage. Limited analyses of projected risks associated with 1.5°C versus 2°C are available (Section 3.4.4.12). A global analysis of sea level rise (SLR) risk to 720 UNESCO Cultural World Heritage sites projected that about 47 sites might be affected under 1°C of warming, with this number increasing to 110 and 136 sites under 2°C and 3°C, respectively (Marzeion and Levermann, 2014). Similar risks to vast worldwide coastal tourism infrastructure and beach assets remain unquantified for most major tourism destinations and small island developing states (SIDS) that economically depend on coastal tourism. One exception is the projection that an eventual 1 m SLR could partially or fully inundate 29% of 900 coastal resorts in 19 Caribbean countries, with a substantially higher proportion (49–60%) vulnerable to associated coastal erosion (Scott and Verkoeyen, 2017).

A major barrier to understanding the risks of climate change for tourism, from the destination community scale to the global scale, has been the lack of integrated sectoral assessments that analyse the full range of potential compounding impacts and their interactions with other major drivers of tourism (Rosselló-Nadal, 2014; Scott et al., 2016b). When applied to 181 countries, a global vulnerability index including 27 indicators found that countries with the lowest risk are located in western and northern Europe, central Asia, Canada and New Zealand, while the highest sector risks are projected for Africa, the Middle East, South Asia and SIDS in the Caribbean, Indian and Pacific Oceans (Scott and Gössling, 2018). Countries with the highest risks and where tourism represents a significant proportion of the national economy (i.e., more than 15% of GDP) include many SIDS and least developed countries. Sectoral climate change risk also aligns strongly with regions where tourism growth is projected to be the strongest over the coming decades, including sub-Saharan Africa and South Asia, pointing to an important potential barrier to tourism development. The transnational implications of these impacts on the highly interconnected global tourism sector and the contribution of tourism to achieving the 2030 sustainable development goals (SDGs) remain important uncertainties.

In summary, climate is an important factor influencing the geography and seasonality of tourism demand and spending globally (*very high confidence*). Increasing temperatures are projected to directly impact climate-dependent tourism markets, including sun, beach and snow sports tourism, with lesser risks for other tourism markets that are less climate sensitive (*high confidence*). The degradation or loss of beach and coral reef assets is expected to increase risks for coastal tourism, particularly in subtropical and tropical regions (*high confidence*).

3.4.9.2   Energy systems

Climate change is projected to lead to an increased demand for air conditioning in most tropical and subtropical regions (Arent et al., 2014; Hong and Kim, 2015) (*high confidence*). Increasing temperatures will decrease the thermal efficiency of fossil, nuclear, biomass and solar power generation technologies, as well as buildings and other infrastructure (Arent et al., 2014). For example, in Ethiopia, capital expenditures through 2050 might either decrease by approximately 3% under extreme wet scenarios or increase by up to 4% under a severe dry scenario (Block and Strzepek, 2012).

Impacts on energy systems can affect gross domestic product (GDP). The economic damage in the United States from climate change is estimated to be, on average, roughly 1.2% cost of GDP per year per 1°C increase under RCP8.5 (Hsiang et al., 2017). Projections of GDP indicate that negative impacts of energy demand associated with space heating and cooling in 2100 will be greatest (median: –0.94% change in GDP) under 4°C (RCP8.5) compared with under 1.5°C (median: –0.05%), depending on the socio-economic conditions (Park et al., 2018). Additionally, projected total energy demands for heating and cooling at the global scale do not change much with increases in global mean surface temperature (GMST) of up to 2°C. A high degree of variability is projected between regions (Arnell et al., 2018).

Evidence for the impact of climate change on energy systems since AR5 is limited. Globally, gross hydropower potential is projected to increase (by 2.4% under RCP2.6 and by 6.3% under RCP8.5 for the 2080s), with the most growth expected in Central Africa, Asia, India and northern high latitudes (van Vliet et al., 2016). Byers et al. (2018) found that energy impacts at 2°C increase, including more cooling degree days, especially in tropical regions, as well as increased hydro-climatic risk to thermal and hydropower plants predominantly in Europe, North America, South and Southeast Asia and southeast Brazil. Donk et al. (2018) assessed future climate impacts on hydropower in Suriname and projected a decrease of approximately 40% in power capacity for a global temperature increase in the range of 1.5°C. At minimum and maximum increases in global mean temperature of 1.35°C and 2°C, the overall stream flow in Florida, USA is projected to increase by an average of 21%, with pronounced seasonal variations, resulting in increases in power generation in winter (+72%) and autumn (+15%) and decreases in summer (–14%; Chilkoti et al., 2017). Greater changes are projected at higher temperature increases. In a reference scenario with global mean temperatures rising by 1.7°C from 2005 to 2050, U.S. electricity demand in 2050 was 1.6–6.5% higher than in a control scenario with constant temperatures (McFarland et al., 2015). Decreased electricity generation of –15% is projected for Brazil starting in 2040, with values expected to decline to –28% later in the century (de Queiroz et al., 2016). In large parts of Europe, electricity demand is projected to decrease, mainly owing to reduced heating demand (Jacob et al., 2018).

In Europe, no major differences in large-scale wind energy resources or in inter- or intra-annual variability are projected for 2016–2035 under RCP8.5 and RCP4.5 (Carvalho et al., 2017). However, in 2046–2100, wind energy density is projected to decrease in eastern Europe (−30%) and increase in Baltic regions (+30%). Intra-annual variability is expected to increase in northern Europe and decrease in southern Europe. Under RCP4.5 and RCP8.5, the annual energy yield of European wind farms as a whole, as projected to be installed by 2050, will remain stable (±5 yield for all climate models). However, wind farm yields are projected to undergo changes of up to 15% in magnitude at country and local scales and of 5% at the regional scale (Tobin et al., 2015, 2016). Hosking et al. (2018) assessed wind power generation over Europe for 1.5°C of warming and found the potential for wind energy to be greater than previously assumed in northern Europe. Additionally, Tobin et al. (2018) assessed impacts under 1.5°C and 2°C of warming on wind, solar photovoltaic and thermoelectric power generation across Europe. These authors found that photovoltaic and wind power might be reduced by up to 10%, and hydropower and thermoelectric generation might decrease by up to 20%, with impacts being limited at 1.5°C of warming but increasing as temperature increases (Tobin et al., 2018).

3.4.9.3   Transportation

Road, air, rail, shipping and pipeline transportation can be impacted directly or indirectly by weather and climate, including increases in precipitation and temperature; extreme weather events (flooding and storms); SLR; and incidence of freeze–thaw cycles (Arent et al., 2014). Much of the published research on the risks of climate change for the transportation sector has been qualitative.

The limited new research since AR5 supports the notion that increases in global temperatures will impact the transportation sector. Warming is projected to result in increased numbers of days of ice-free navigation and a longer shipping season in cold regions, thus affecting shipping and reducing transportation costs (Arent et al., 2014). In the North Sea Route, large-scale commercial shipping might not be possible until 2030 for bulk shipping and until 2050 for container shipping under RCP8.5. A 0.05% increase in mean temperature is projected from an increase in short-lived pollutants, as well as elevated CO_2 and non-CO_2 emissions, associated with additional economic growth enabled by the North Sea Route. (Yumashev et al., 2017). Open water vessel transit has the potential to double by mid-century, with a two to four month longer season (Melia et al., 2016).

3.4.10   Livelihoods and Poverty, and the Changing Structure of Communities

Multiple drivers and embedded social processes influence the magnitude and pattern of livelihoods and poverty, as well as the changing structure of communities related to migration, displacement and conflict (Adger et al., 2014). In AR5, evidence of a climate change

signal was limited, with more evidence of impacts of climate change on the places where indigenous people live and use traditional ecological knowledge (Olsson et al., 2014).

3.4.10.1   Livelihoods and poverty

At approximately 1.5°C of global warming (2030), climate change is expected to be a poverty multiplier that makes poor people poorer and increases the poverty head count (Hallegatte et al., 2016; Hallegatte and Rozenberg, 2017). Poor people might be heavily affected by climate change even when impacts on the rest of population are limited. Climate change alone could force more than 3 million to 16 million people into extreme poverty, mostly through impacts on agriculture and food prices (Hallegatte et al., 2016; Hallegatte and Rozenberg, 2017). Unmitigated warming could reshape the global economy later in the century by reducing average global incomes and widening global income inequality (Burke et al., 2015b). The most severe impacts are projected for urban areas and some rural regions in sub-Saharan Africa and Southeast Asia.

3.4.10.2   The changing structure of communities: migration, displacement and conflict

Migration: In AR5, the potential impacts of climate change on migration and displacement were identified as an emerging risk (Oppenheimer et al., 2014). The social, economic and environmental factors underlying migration are complex and varied; therefore, detecting the effect of observed climate change or assessing its possible magnitude with any degree of confidence is challenging (Cramer et al., 2014).

No studies have specifically explored the difference in risks between 1.5°C and 2°C of warming on human migration. The literature consistently highlights the complexity of migration decisions and the difficulties in attributing causation (e.g., Nicholson, 2014; Baldwin and Fornalé, 2017; Bettini, 2017; Constable, 2017; Islam and Shamsuddoha, 2017; Suckall et al., 2017). The studies on migration that have most closely explored the probable impacts of 1.5°C and 2°C have mainly focused on the direct effects of temperature and precipitation anomalies on migration or the indirect effects of these climatic changes through changing agriculture yield and livelihood sources (Mueller et al., 2014; Piguet and Laczko, 2014; Mastrorillo et al., 2016; Sudmeier-Rieux et al., 2017).

Temperature has had a positive and statistically significant effect on outmigration over recent decades in 163 countries, but only for agriculture-dependent countries (R. Cai et al., 2016). A 1°C increase in average temperature in the International Migration Database of the Organisation for Economic Co-operation and Development (OECD) was associated with a 1.9% increase in bilateral migration flows from 142 sending countries and 19 receiving countries, and an additional millimetre of average annual precipitation was associated with an increase in migration by 0.5% (Backhaus et al., 2015). In another study, an increase in precipitation anomalies from the long-term mean, was strongly associated with an increase in outmigration, whereas no significant effects of temperature anomalies were reported (Coniglio and Pesce, 2015).

Internal and international migration have always been important for small islands (Farbotko and Lazrus, 2012; Weir et al., 2017). There is rarely a single cause for migration (Constable, 2017). Numerous factors are important, including work, education, quality of life, family ties, access to resources, and development (Bedarff and Jakobeit, 2017; Speelman et al., 2017; Nicholls et al., 2018). Depending on the situation, changing weather, climate or environmental conditions might each be a factor in the choice to migrate (Campbell and Warrick, 2014).

Displacement: At 2°C of warming, there is a potential for significant population displacement concentrated in the tropics (Hsiang and Sobel, 2016). Tropical populations may have to move distances greater than 1000 km if global mean temperature rises by 2°C from 2011–2030 to the end of the century. A disproportionately rapid evacuation from the tropics could lead to a concentration of population in tropical margins and the subtropics, where population densities could increase by 300% or more (Hsiang and Sobel, 2016).

Conflict: A recent study has called for caution in relating conflict to climate change, owing to sampling bias (Adams et al., 2018). Insufficient consideration of the multiple drivers of conflict often leads to inconsistent associations being reported between climate change and conflict (e.g., Hsiang et al., 2013; Hsiang and Burke, 2014; Buhaug, 2015, 2016; Carleton and Hsiang, 2016; Carleton et al., 2016). There also are inconsistent relationships between climate change, migration and conflict (e.g., Theisen et al., 2013; Buhaug et al., 2014; Selby, 2014; Christiansen, 2016; Brzoska and Fröhlich, 2016; Burrows and Kinney, 2016; Reyer et al., 2017c; Waha et al., 2017). Across world regions and from the international to micro level, the relationship between drought and conflict is weak under most circumstances (Buhaug, 2016; von Uexkull et al., 2016). However, drought significantly increases the likelihood of sustained conflict for particularly vulnerable nations or groups, owing to the dependence of their livelihood on agriculture. This is particularly relevant for groups in the least developed countries (von Uexkull et al., 2016), in sub-Saharan Africa (Serdeczny et al., 2016; Almer et al., 2017) and in the Middle East (Waha et al., 2017). Hsiang et al. (2013) reported causal evidence and convergence across studies that climate change is linked to human conflicts across all major regions of the world, and across a range of spatial and temporal scales. A 1°C increase in temperature or more extreme rainfall increases the frequency of intergroup conflicts by 14% (Hsiang et al., 2013). If the world warms by 2°C–4°C by 2050, rates of human conflict could increase. Some causal associations between violent conflict and socio-political instability were reported from local to global scales and from hour to millennium time frames (Hsiang and Burke, 2014). A temperature increase of one standard deviation increased the risk of interpersonal conflict by 2.4% and intergroup conflict by 11.3% (Burke et al., 2015a). Armed-conflict risks and climate-related disasters are both relatively common in ethnically fractionalized countries, indicating that there is no clear signal that environmental disasters directly trigger armed conflicts (Schleussner et al., 2016a).

In summary, average global temperatures that extend beyond 1.5°C are projected to increase poverty and disadvantage in many populations globally (*medium confidence*). By the mid- to late 21st century, climate change is projected to be a poverty multiplier that makes poor people

poorer and increases poverty head count, and the association between temperature and economic productivity is not linear (*high confidence*). Temperature has a positive and statistically significant effect on outmigration for agriculture-dependent communities (*medium confidence*).

3.4.11   Interacting and Cascading Risks

The literature on compound as well as interacting and cascading risks at warming of 1.5°C and 2°C is limited. Spatially compound risks, often referred to as hotspots, involve multiple hazards from different sectors overlapping in location (Piontek et al., 2014). Global exposures were assessed for 14 impact indicators, covering water, energy and land sectors, from changes including drought intensity and water stress index, cooling demand change and heatwave exposure, habitat degradation, and crop yields using an ensemble of climate and impact models (Byers et al., 2018). Exposures are projected to approximately double between 1.5°C and 2°C, and the land area affected by climate risks is expected to increase as warming progresses. For populations vulnerable to poverty, the exposure to climate risks in multiple sectors could be an order of magnitude greater (8–32 fold) in the high poverty and inequality scenarios (SSP3; 765–1,220 million) compared to under sustainable socio-economic development (SSP1; 23–85 million). Asian and African regions are projected to experience 85–95% of global exposure, with 91–98% of the exposed and vulnerable population (depending on SSP/GMT combination), approximately half of which are in South Asia. Figure 3.19 shows that moderate and large multi-sector impacts are prevalent at 1.5°C where vulnerable people live, predominantly in South Asia (mostly Pakistan, India and China), but that impacts spread to sub-Saharan Africa, the Middle East and East Asia at higher levels of warming. Beyond 2°C and at higher risk thresholds, the world's poorest populations are expected to be disproportionately impacted, particularly in cases (SSP3) of great inequality in Africa and southern Asia. Table 3.4 shows the number of exposed and vulnerable people at 1.5°C and 2°C of warming, with 3°C shown for context, for selected multi-sector risks.

3.4.12   Summary of Projected Risks at 1.5°C and 2°C of Global Warming

The information presented in Section 3.4 is summarized below in Table 3.5, which illustrates the growing evidence of increasing risks across a broad range of natural and human systems at 1.5°C and 2°C of global warming.

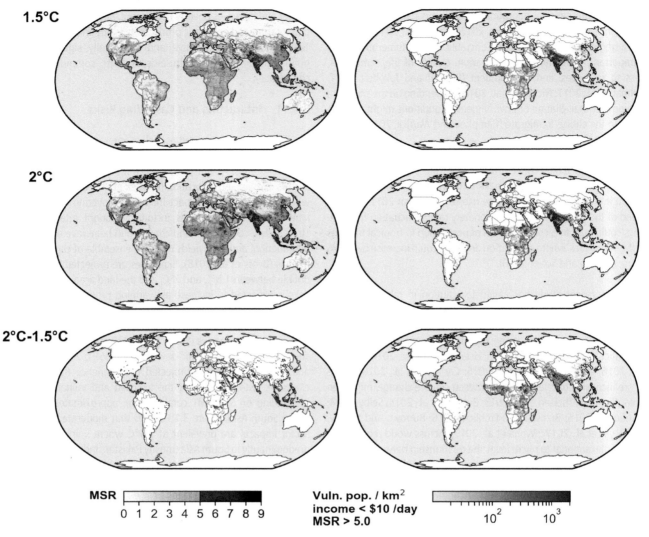

Figure 3.19 | Multi-sector risk maps for 1.5°C (top), 2°C (middle), and locations where 2°C brings impacts not experienced at 1.5°C (2°C–1.5°C; bottom). The maps in the left column show the full range of the multi-sector risk (MSR) score (0–9), with scores ≤5.0 shown with a transparency gradient and scores >5.0 shown with a colour gradient. Score must be >4.0 to be considered 'multi-sector'. The maps in the right column overlay the 2050 vulnerable populations (low income) under Shared Socio-Economic Pathway (SSP)2 (greyscale) with the multi-sector risk score >5.0 (colour gradient), thus indicating the concentrations of exposed and vulnerable populations to risks in multiple sectors. Source: Byers et al. (2018).

Table 3.4 | Number of exposed and vulnerable people at 1.5°C, 2°C, and 3°C for selected multi-sector risks under shared socioeconomic pathways (SSPs).
Source: Byers et al., 2018

SSP2 (SSP1 to SSP3 range), millions	1.5°C		2°C		3°C	
Indicator	Exposed	Exposed and vulnerable	Exposed	Exposed and vulnerable	Exposed	Exposed and vulnerable
Water stress index	3340 (3032–3584)	496 (103–1159)	3658 (3080–3969)	586 (115–1347)	3920 (3202–4271)	662 (146–1480)
Heatwave event exposure	3960 (3546–4508)	1187 (410–2372)	5986 (5417–6710)	1581 (506–3218)	7909 (7286–8640)	1707 (537–3575)
Hydroclimate risk to power production	334 (326–337)	30 (6–76)	385 (374–389)	38 (9–94)	742 (725–739)	72 (16–177)
Crop yield change	35 (32–36)	8 (2–20)	362 (330–396)	81 (24–178)	1817 (1666–1992)	406 (118–854)
Habitat degradation	91 (92–112)	10 (4–31)	680 (314–706)	102 (23–234)	1357 (809–1501)	248 (75–572)
Multi-sector exposure						
Two indicators	1129 (1019–1250)	203 (42–487)	2726 (2132–2945)	562 (117–1220)	3500 (3212–3864)	707 (212–1545)
Three indicators	66 (66–68)	7 (0.9–19)	422 (297–447)	54 (8–138)	1472 (1177–1574)	237 (48–538)
Four indicators	5 (0.3–5.7)	0.3 (0–1.2)	11 (5–14)	0.5 (0–2)	258 (104–280)	33 (4–86)

Table 3.5 | Summary of projected risks to natural and human systems at 1.5°C and 2°C of global warming, and of the potential to adapt to these risks. Table summarizes the chapter text and with references supporting table entries found in the main chapter text. Risk magnitude is provided either as assessed levels of risk (very high: vh, high: h, medium: m, or low: l) or as quantitative examples of risk levels taken from the literature. Further compilations of quantified levels of risk taken from the literature may be found Tables 3.SM1-5 in the Supplementary Material. Similarly, potential to adapt is assessed from the literature by expert judgement as either high (h), medium (m), or low (l). Confidence in each assessed level/quantification of risk, or in each assessed adaptation potential, is indicated as very high (VH), high (H), medium (M), or low (L). Note that the use of l, m, h and vh here is distinct from the use of L, M, H and VH in Figures 3.18, 3.20 and 3.21.

Sector	Physical climate change drivers	Nature of risk	Global risks at 1.5°C of global warming above pre-industrial	Global risks at 2°C of global warming above pre-industrial	Change in risk when moving from 1.5°C to 2°C of warming	Confidence in risk statements	Regions where risks are particularly high with 2°C of global warming	Regions where the change in risk when moving from 1.5°C to 2°C are particularly high	Regions with little or no information	RFC*	Adaptation potential at 1.5°C	Adaptation potential at 2°C	Confidence in assigning adaptation potential
Freshwater	Precipitation, temperature, snowmelt	Water Stress	Around half compared to the risks at 2°C[1]	Additional 8% of the world population in 2000 exposed to new or aggravated water scarcity[1]	Up to 100% increase	M		Europe, Australia, southern Africa		3	l	l	M
		Fluvial flood	100% increase in the population affected compared to the impact simulated over the baseline period 1976–2005[2]	170% increase in the population affected compared to the impact simulated over the baseline period 1976–2005[2]	70% increase	M	USA, Asia, Europe		Africa, Oceania	2	l/m	l/m	M
		Drought	350.2 ± 158.8 million, changes in urban population exposure to severe drought at the globe scale[3]	410.7 ± 213.5 million, changes in urban population exposure to severe drought at the globe scale[3]	60.5 ± 84.1 million (±84.1 based on the SSP1 scenario) (based on PDSI estimate)	M	Central Europe, southern Europe, Mediterranean, West Africa, East and West Asia, Southeast Asia (based on PDSI estimate[a])			2	l/m	l/m	L
Terrestrial ecosystems	Temperature, precipitation	Species range loss	6% insects, 4% vertebrates, 8% plants, lose >50% range[4]	18% insects, 8% vertebrates, 16% plants lose >50% range[4]	Double or triple	M		Amazon, Europe, southern Africa		1,4	m	l	H
		Loss of ecosystem functioning and services	m	h		M				4			
		Shifts of biomes (major ecosystem types)	About 7% transformed[5]	13% (range 8–20%) transformed[5]	About double	M		Arctic, Tibet, Himalayas, South Africa, Australia		4	l		
	Heat and cold stress, warming, precipitation drought	Wildfire	h	h	Increased risk	M	Canada, USA and Mediterranean	Mediterranean	Central and South America, Australia, Russia, China, Africa	1, 2, 4, 5	l		M

3

3

Table 3.5 (continued)

Sector	Physical climate change drivers	Nature of risk	Global risks at 1.5°C of global warming above pre-industrial	Global risks at 2°C of global warming above pre-industrial	Change in risk when moving from 1.5°C to 2°C of warming	Confidence in risk statements	Regions where risks are particularly high with 2°C of global warming	Regions where the change in risk when moving from 1.5°C to 2°C are particularly high	Regions with little or no information	RFC*	Adaptation potential at 1.5°C	Adaptation potential at 2°C	Confidence in assigning adaptation potential
Ocean	Warming and stratification of the surface ocean	Loss of framework species (coral reefs)	vh	vh	Greater rate of loss: from 70–90% loss at 1.5°C to 99% loss at 2°C and above	H/very H	Tropical/subtropical countries	Tropical/subtropical countries	Southern Red Sea, Somalia, Yemen, deep water coral reefs	1,2	h	l	H
		Loss of framework species (seagrass)	m	h	Increase in risk	M	Tropical/subtropical countries	Tropical/subtropical countries	Southern Red Sea, Somalia, Yemen, Myanmar	1,2	m	l	M/H
		Loss of framework species (mangroves)	m	m	Uncertain and depends on other human activities	M/H	Tropical/subtropical countries	Tropical/subtropical countries	Southern Red Sea, Somalia, Yemen, Myanmar	1,3	m	l	L/M
		Disruption of marine foodwebs	h	vh	Large increase in risk	M	Global	Global	Deep sea	4	m	l	M/H
		Range migration of marine species and ecosystems	m	h	Large increase in risk	H	Global	Global	Deep sea	1	m	l	H
		Loss of fin fish and fisheries	h	h/vh	Large increase in risk	H	Global	Global	Deep sea, up-welling systems	4	m	m/l	M/H
		Loss of coastal ecosystems and protection	m	h	Increase in risk	M	Low-latitude tropical/subtropical countries	Low-latitude tropical/subtropical countries	Most regions – risks not well defined	1	m	m/l	M
	Ocean acidification and elevated sea temperatures	Loss of bivalves and bivalve fisheries	m/h	h/vh	Large increase in risk	H	Temperate countries with upwelling	Temperate countries with upwelling	Most regions – risks not well defined	4	m/h	l/m	M/H
		Changes to physiology and ecology of marine species	l/m	m	Increase in risk	L/M	Global	Global	Most regions – risks not well defined	4	l	l	M/H
	Reduced bulk ocean circulation and de-oxygenation	Increased hypoxic dead zones	l	l/m	Large increase in risk	L/M	Temperate countries with upwelling	Temperate countries with upwelling	Deep sea	4	m	l	M
		Changes to upwelling productivity	l	m	Increase in risk	L/M	Most upwelling regions	Most upwelling regions	Some upwelling systems	4	l	l	M

Table 3.5 (continued)

Sector	Physical climate change drivers	Nature of risk	Global risks at 1.5°C of global warming above pre-industrial	Global risks at 2°C of global warming above pre-industrial	Change in risk when moving from 1.5°C to 2°C of warming	Confidence in risk statements	Regions where risks are particularly high with 2°C of global warming	Regions where the change in risk when moving from 1.5°C to 2°C are particularly high	Regions with little or no information	RFC*	Adaptation potential at 1.5°C	Adaptation potential at 2°C	Confidence in assigning adaptation potential
Ocean	Intensified storms, precipitation plus sea level rise	Loss of coastal ecosystems	h	h/vh	Large increase in risk	H	Tropical/subtropical countries	Tropical/subtropical countries		1, 4	m	l	M
		Inundation and destruction of human/coastal infrastructure and livelihoods	h	h/vh	Large increase in risk	H	Global	Global		1, 5	m/h	m	M/L
	Loss of sea ice	Loss of habitat	h	vh	Large increase in risk	H	Polar regions	Polar regions		1	l	very l	H
		Increased productivity but changing fisheries	l/m	m/h	Large increase in risk	very H	Polar regions	Polar regions		1, 4	l	m/l	H
Coastal	Sea level rise, increased storminess	Area exposed (assuming no defences)	562–575th km² when 1.5°C first reached[6,7,8]	590–613th km² when 2°C first reached[6,7,8]	Increasing; 25–38th km² when temperatures are first reached, 10–17th km² in 2100 increasing to 16–230th km² in 2300[6,7,8]	M/H (dependent on population datasets)	Asia, small islands	Asia, small islands	Small islands	2, 3	m	m	M
		Population exposed (assuming no defences)	128–143 million when 1.5°C first reached	141–151 million when 2°C first reached	Increasing; 13–8 million when temperatures are first reached, 0–6 million people in 2100, increasing to 35–95 million people in 2300[6]	M/H (dependent on population datasets)	Asia, small islands	Asia, small islands	Small islands	2, 3	m	m	M
		People at risk accounting for defences (modelled in 1995)	2–28 million people yr⁻¹ if defences are not upgraded from the modelled 1995 baseline[9]	15–53 million people yr⁻¹ if defences are not upgraded from the modelled 1995 baseline[9]	Increasing with time, but highly dependent on adaptation[9]	M/H (dependent on adaptation)	Asia, small islands, potentially African nations	Asia, small islands	Small islands	2, 3, 4	m	m	M

3

Table 3.5 (continued)

Sector	Physical climate change drivers	Nature of risk	Global risks at 1.5°C of global warming above pre-industrial	Global risks at 2°C of global warming above pre-industrial	Change in risk when moving from 1.5°C to 2°C of warming	Confidence in risk statements	Regions where risks are particularly high with 2°C of global warming	Regions where the change in risk when moving from 1.5°C to 2°C are particularly high	Regions with little or no information	RFC*	Adaptation potential at 1.5°C	Adaptation potential at 2°C	Confidence in assigning adaptation potential
Food security and food production systems	Heat and cold stress, warming, precipitation, drought	Changes in ecosystem production	m/h	h	Large increase	M/H	Global	North America, Central and South America, Mediterranean basin, South Africa, Australia, Asia		2, 4, 5	h	m/h	M/H
	Heat and cold stress, warming, precipitation drought	Shift and composition change of biomes (major ecosystem types)	m/h	h	Moderate increase	L/M	Global	Global, tropical areas, Mediterranean	Africa, Asia	1, 2, 3, 4	l/m	l	L/M
Human health	Temperature	Heat-related morbidity and mortality	m	m/h	Risk increased	VH	All regions at risk	All regions	Africa	2, 3, 4	h	h	H
	Temperature	Occupational heat stress	m	m/h	Risk increased	M	Tropical regions	Tropical regions	Africa	2, 3, 4	h	m	M
	Air quality	Ozone-related mortality	m (if precursor emissions remain the same)	m/h (if precursor emissions remain the same)	Risk increased	H	High income and emerging economies	High income and emerging economies	Africa, parts of Asia	2, 3, 4	l	m	M
	Temperature, precipitation	Undernutrition	m	m/h	Risk increased	H	Low-income countries in Africa and Asia	Low-income countries in Africa and Asia	Small islands	2, 3, 4	m	l	M
Key economic sectors	Temperature	Tourism (sun, beach, and snow sports)	m/h	h	Risk increased	VH	Coastal tourism, particularly in subtropical and tropical regions	Coastal tourism, particularly in subtropical and tropical regions	Africa	1, 2, 3	m	l	H

*RFC: 1 = unique and threatened systems, 2 = extreme events, 3 = unequal distribution of impacts, 4 = global aggregate impacts (economic + biodiversity), 5 = large-scale singular events.

PDSI-based drought estimates tend to overestimate drought impacts (see Section 3.4); hence projections with other drought indices may differ. Further quantifications may be found in Table 3.SM.1

[1] Gerten et al., 2013; [2] Alfieri et al., 2017; [3] Liu et al., 2018; [4] Warren et al., 2018a; [5] Warzawski et al., 2013; [6] Brown et al., 2018a; [7] Rasmussen et al., 2018; [8] Yokoki et al., 2018; [9] Nicholls et al., 2018.

3.4.13   Synthesis of Key Elements of Risk

Some elements of the assessment in Section 3.4 were synthesized into Figure 3.18 and 3.20, indicating the overall risk for a representative set of natural and human systems from increases in global mean surface temperature (GMST) and anthropogenic climate change. The elements included are supported by a substantive enough body of literature providing at least *medium confidence* in the assessment. The format for Figures 3.18 and 3.20 match that of Figure 19.4 of WGII AR5 Chapter 19 (Oppenheimer et al., 2014) indicating the levels of additional risk as colours: undetectable (white) to moderate (detected and attributed; yellow), from moderate to high (severe and widespread; red), and from high to very high (purple), the last of which indicates significant irreversibility or persistence of climate-related hazards combined with a much reduced capacity to adapt. Regarding the transition from undetectable to moderate, the impact literature assessed in AR5 focused on describing and quantifying linkages between weather and climate patterns and impact outcomes, with limited detection and attribution to anthropogenic climate change (Cramer et al., 2014). A more recent analysis of attribution to greenhouse gas forcing at the global scale (Hansen and Stone, 2016) confirmed that the impacts related to changes in regional atmospheric and ocean temperature can be confidently attributed to anthropogenic forcing, while attribution to anthropogenic forcing of those impacts related to precipitation is only weakly evident or absent. Moreover, there is no strong direct relationship between the robustness of climate attribution and that of impact attribution (Hansen and Stone, 2016).

The current synthesis is complementary to the synthesis in Section 3.5.2 that categorizes risks into 'Reasons for Concern' (RFCs), as described in Oppenheimer et al. (2014). Each element, or burning ember, presented here (Figures 3.18, 3.20) maps to one or more RFCs (Figure 3.21). It should be emphasized that risks to the elements assessed here are only a subset of the full range of risks that contribute to the RFCs. Figures 3.18 and 3.20 are not intended to replace the RFCs but rather to indicate how risks to particular elements of the Earth system accrue with global warming, through the visual burning embers format, with a focus on levels of warming of 1.5°C and 2°C. Key evidence assessed in earlier parts of this chapter is summarized to indicate the transition points between the levels of risk. In this regard, the assessed confidence in assigning the transitions between risk levels are as follows: L=Low, M=Medium, H=High, and VH=Very high levels of confidence. A detailed account of the procedures involved is provided in the Supplementary Material (3.SM.3.2 and 3.SM.3.3).

In terrestrial ecosystems (feeding into RFC1 and RFC4), detection and attribution studies show that impacts of climate change on terrestrial ecosystems began to take place over the past few decades, indicating a transition from no risk (white areas in Figure 3.20) to moderate risk below recent temperatures (*high confidence*) (Section 3.4.3). Risks to unique and threatened terrestrial ecosystems are generally projected to be higher under warming of 2°C compared to 1.5°C (Section 3.5.2.1), while at the global scale severe and widespread risks are projected to occur by 2°C of warming. These risks are associated with biome shifts and species range losses (Sections 3.4.3 and 3.5.2.4); however, because many systems and species are projected to be unable to adapt to levels of warming below 2°C, the transition to high risk (red areas

in Figure 3.20) is located below 2°C (*high confidence*). With 3°C of warming, however, biome shifts and species range losses are expected to escalate to very high levels, and the systems are projected to have very little capacity to adapt (Figure 3.20) (*high confidence*) (Section 3.4.3).

In the Arctic (related to RFC1), the increased rate of summer sea ice melt was detected and attributed to climate change by the year 2000 (corresponding to warming of 0.7°C), indicating moderate risk. At 1.5°C of warming an ice-free Arctic Ocean is considered *unlikely*, whilst by 2°C of warming it is considered *likely* and this unique ecosystem is projected to be unable to adapt. Hence, a transition from high to very high risk is expected between 1.5°C and 2°C of warming.

For warm-water coral reefs, there is *high confidence* in the transitions between risk levels, especially in the growing impacts in the transition of warming from non-detectable (0.2°C to 0.4°C), and then successively higher levels risk until high and very high levels of risks by 1.2°C (Section 3.4.4 and Box 3.4). This assessment considered the heatwave-related loss of 50% of shallow water corals across hundreds of kilometres of the world's largest continuous coral reef system, the Great Barrier Reef, as well as losses at other sites globally. The major increase in the size and loss of coral reefs over the past three years, plus sequential mass coral bleaching and mortality events on the Great Barrier Reef, (Hoegh-Guldberg, 1999; Hughes et al., 2017b, 2018), have reinforced the scale of climate-change related risks to coral reefs. General assessments of climate-related risks for mangroves prior to this special report concluded that they face greater risks from deforestation and unsustainable coastal development than from climate change (Alongi, 2008; Hoegh-Guldberg et al., 2014; Gattuso et al., 2015). Recent climate-related die-offs (Duke et al., 2017; Lovelock et al., 2017), however, suggest that climate change risks may have been underestimated for mangroves as well, and risks have thus been assessed as undetectable to moderate, with the transition now starting at 1.3°C as opposed to 1.8°C as assessed in 2015 (Gattuso et al., 2015). Risks of impacts related to climate change on small-scale fisheries at low latitudes, many of which are dependent on ecosystems such as coral reefs and mangroves, are moderate today but are expected to reach high levels of risk around 0.9°C–1.1°C (*high confidence*) (Section 3.4.4.10).

The transition from undetectable to moderate risk (related to RFCs 3 and 4), shown as white to yellow in Figure 3.20, is based on AR5 WGII Chapter 7, which indicated with *high confidence* that climate change impacts on crop yields have been detected and attributed to climate change, and the current assessment has provided further evidence to confirm this (Section 3.4.6). Impacts have been detected in the tropics (AR5 WGII Chapters 7 and 18), and regional risks are projected to become high in some regions by 1.5°C of warming, and in many regions by 2.5°C, indicating a transition from moderate to high risk between 1.5°C and 2.5°C of warming (*medium confidence*).

Impacts from fluvial flooding (related to RFCs 2, 3 and 4) depend on the frequency and intensity of the events, as well as the extent of exposure and vulnerability of society (i.e., socio-economic conditions and the effect of non-climate stressors). Moderate risks posed by 1.5°C of warming are expected to continue to increase with higher levels of warming (Sections 3.3.5 and 3.4.2), with projected risks being threefold the current risk in economic damages due to flooding in 19 countries for warming of 2°C, indicating a transition to high risk at this level (*medium confidence*). Because few studies have assessed the potential to adapt to these risks, there was insufficient evidence to locate a transition to very high risk (purple).

Climate-change induced sea level rise (SLR) and associated coastal flooding (related to RFCs 2, 3 and 4) have been detectable and attributable since approximately 1970 (Slangen et al., 2016), during which time temperatures have risen by 0.3°C (*medium confidence*) (Section 3.3.9). Analysis suggests that impacts could be more widespread in sensitive systems such as small islands (*high confidence*) (Section 3.4.5.3) and increasingly widespread by the 2070s (Brown et al., 2018a) as temperatures rise from 1.5°C to 2°C, even when adaptation measures are considered, suggesting a transition to high

risk (Section 3.4.5). With 2.5°C of warming, adaptation limits are expected to be exceeded in sensitive areas, and hence a transition to very high risk is projected. Additionally, at this temperature, sea level rise could have adverse effects for centuries, posing significant risk to low-lying areas (*high confidence*) (Sections 3.4.5.7 and 3.5.2.5).

For heat-related morbidity and mortality (related to RFCs 2, 3 and 4), detection and attribution studies show heat-related mortality in some locations increasing with climate change (*high confidence*) (Section 3.4.7; Ebi et al., 2017). The projected risks of heat-related morbidity and mortality are generally higher under warming of 2°C than 1.5°C (*high confidence*), with projections of greater exposure to high ambient temperatures and increased morbidity and mortality (Section 3.4.7). Risk levels will depend on the rate of warming and the (related) level of adaptation, so a transition in risk from moderate (yellow) to high (red) is located between 1°C and 3°C (*medium confidence*).

Risks and/or impacts for specific natural, managed and human systems

The key elements are presented here as a function of the risk level assessed between 1.5°C and 2°C.

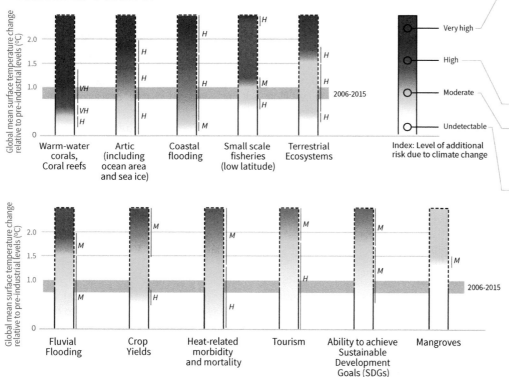

Confidence level for transition: *L*=Low, *M*=Medium, *H*=High and *VH*=Very high

Figure 3.20 | The dependence of risks and/or impacts associated with selected elements of human and natural systems on the level of climate change, adapted from Figure 3.21 and from AR5 WGII Chapter 19, Figure 19.4, and highlighting the nature of this dependence between 0°C and 2°C warming above pre-industrial levels. The selection of impacts and risks to natural, managed and human systems is illustrative and is not intended to be fully comprehensive. Following the approach used in AR5, literature was used to make expert judgements to assess the levels of global warming at which levels of impact and/or risk are undetectable (white), moderate (yellow), high (red) or very high (purple). The colour scheme thus indicates the additional risks due to climate change. The transition from red to purple, introduced for the first time in AR4, is defined by a very high risk of severe impacts and the presence of significant irreversibility or persistence of climate-related hazards combined with limited ability to adapt due to the nature of the hazard or impact. Comparison of the increase of risk across RFCs indicates the relative sensitivity of RFCs to increases in GMST. As was done previously, this assessment takes autonomous adaptation into account, as well as limits to adaptation independently of development pathway. The levels of risk illustrated reflect the judgements of the authors of Chapter 3 and Gattuso et al. (2015; for three marine elements). The grey bar represents the range of GMST for the most recent decade: 2006–2015.

For tourism (related to RFCs 3 and 4), changing weather patterns, extreme weather and climate events, and sea level rise are affecting many – but not all – global tourism investments, as well as environmental and cultural destination assets (Section 3.4.4.12), with 'last chance to see' tourism markets developing based on observed impacts on environmental and cultural heritage (Section 3.4.9.1), indicating a transition from undetectable to moderate risk between 0°C and 1.5°C of warming (*high confidence*). Based on limited analyses, risks to the tourism sector are projected to be larger at 2°C than at 1.5°C, with impacts on climate-sensitive sun, beach and snow sports tourism markets being greatest. The degradation or loss of coral reef systems is expected to increase the risks to coastal tourism in subtropical and tropical regions. A transition in risk from moderate to high levels of added risk from climate change is projcted to occur between 1.5°C and 3°C (*medium confidence*).

Climate change is already having large scale impacts on ecosystems, human health and agriculture, which is making it much more difficult to reach goals to eradicate poverty and hunger, and to protect health and life on land (Sections 5.1 and 5.2.1 in Chapter 5), suggesting a transition from undetectable to moderate risk for recent temperatures at 0.5°C of warming (*medium confidence*). Based on the limited analyses available, there is evidence and agreement that the risks to sustainable development are considerably less at 1.5°C than 2°C (Section 5.2.2), including impacts on poverty and food security. It is easier to achieve many of the sustainable development goals (SDGs) at 1.5°C, suggesting that a transition to higher risk will not begin yet at this level. At 2°C and higher levels of warming (e.g., RCP8.5), however, there are high risks of failure to meet SDGs such as eradicating poverty and hunger, providing safe water, reducing inequality and protecting ecosystems, and these risks are projected to become severe and widespread if warming increases further to about 3°C (*medium confidence*) (Section 5.2.3).

Disclosure statement: The selection of elements depicted in Figures 3.18 and 3.20 is not intended to be fully comprehensive and does not necessarily include all elements for which there is a substantive body of literature, nor does it necessarily include all elements which are of particular interest to decision-makers.

3.5 Avoided Impacts and Reduced Risks at 1.5°C Compared with 2°C of Global Warming

3.5.1 Introduction

Oppenheimer et al. (2014, AR5 WGII Chapter 19) provided a framework that aggregates projected risks from global mean temperature change into five categories identified as 'Reasons for Concern'. Risks are classified as moderate, high or very high and coloured yellow, red or purple, respectively, in Figure 19.4 of that chapter (AR5 WGII Chapter 19 for details and findings). The framework's conceptual basis and the risk judgements made by Oppenheimer et al. (2014) were recently reviewed, and most judgements were confirmed in the light of more recent literature (O'Neill et al., 2017). The approach

of Oppenheimer et al. (2014) was adopted, with updates to the aggregation of risk informed by the most recent literature, for the analysis of avoided impacts at 1.5°C compared to 2°C of global warming presented in this section.

The regional economic benefits that could be obtained by limiting the global temperature increase to 1.5°C of warming, rather than 2°C or higher levels, are discussed in Section 3.5.3 in the light of the five RFCs explored in Section 3.5.2. Climate change hotspots that could be avoided or reduced by achieving the 1.5°C target are summarized in Section 3.5.4. The section concludes with a discussion of regional tipping points that could be avoided at 1.5°C compared to higher degrees of global warming (Section 3.5.5).

3.5.2 Aggregated Avoided Impacts and Reduced Risks at 1.5°C versus 2°C of Global Warming

A brief summary of the accrual of RFCs with global warming, as assessed in WGII AR5, is provided in the following sections, which leads into an update of relevant literature published since AR5. The new literature is used to confirm the levels of global warming at which risks are considered to increase from undetectable to moderate, from moderate to high, and from high to very high. Figure 3.21 modifies Figure 19.4 from AR5 WGII, and the following text in this subsection provides justification for the modifications. O'Neill et al. (2017) presented a very similar assessment to that of WGII AR5, but with further discussion of the potential to create 'embers' specific to socio-economic scenarios in the future. There is insufficient literature to do this at present, so the original, simple approach has been used here. As the focus of the present assessment is on the consequences of global warming of 1.5°C–2°C above the pre-industrial period, no assessment for global warming of 3°C or more is included in the figure (i.e., analysis is discontinued at 2.5°C).

3.5.2.1 RFC 1 – Unique and threatened systems

WGII AR5 Chapter 19 found that some unique and threatened systems are at risk from climate change at current temperatures, with increasing numbers of systems at potential risk of severe consequences at global warming of 1.6°C above pre-industrial levels. It was also observed that many species and ecosystems have a limited ability to adapt to the very large risks associated with warming of 2.6°C or more, particularly Arctic sea ice and coral reef systems (*high confidence*). In the AR5 analysis, a transition from white to yellow indicated that the onset of moderate risk was located below present-day global temperatures (*medium confidence*); a transition from yellow to red indicated that the onset of high risk was located at 1.6°C, and a transition from red to purple indicated that the onset of very high risk was located at about 2.6°C. This WGII AR5 analysis already implied that there would be a significant reduction in risks to unique and threatened systems if warming were limited to 1.5°C compared with 2°C. Since AR5, evidence of present-day impacts in these systems has continued to grow (Sections 3.4.2, 3.4.4 and 3.4.5), whilst new evidence has also accumulated for reduced risks at 1.5°C compared to 2°C of warming in Arctic ecosystems (Section 3.3.9), coral reefs (Section 3.4.4) and some other unique ecosystems (Section 3.4.3), as well as for biodiversity.

Risks and/or impacts associated with Reasons for Concern

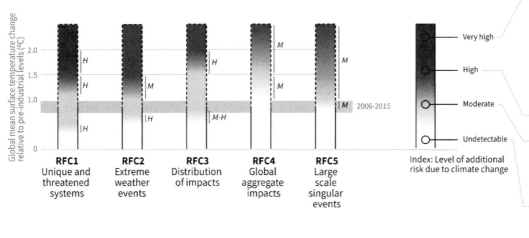

Confidence level for transition: *L*=Low, *M*=Medium, *H*=High and *VH*=Very high

Figure 3.21 | The dependence of risks and/or impacts associated with the Reasons for Concern (RFCs) on the level of climate change, updated and adapted from WGII AR5 Ch 19, Figure 19.4 and highlighting the nature of this dependence between 0°C and 2°C warming above pre-industrial levels. As in the AR5, literature was used to make expert judgements to assess the levels of global warming at which levels of impact and/or risk are undetectable (white), moderate (yellow), high (red) or very high (purple). The colour scheme thus indicates the additional risks due to climate change. The transition from red to purple, introduced for the first time in AR4, is defined by very high risk of severe impacts and the presence of significant irreversibility, or persistence of climate-related hazards combined with a limited ability to adapt due to the nature of the hazard or impact. Comparison of the increase of risk across RFCs indicates the relative sensitivity of RFCs to increases in GMST. As was done previously, this assessment takes autonomous adaptation into account, as well as limits to adaptation (RFC 1, 3, 5) independently of development pathway. The rate and timing of impacts were taken into account in assessing RFC 1 and 5. The levels of risk illustrated reflect the judgements of the Ch 3 authors. **RFC1 Unique and threatened systems:** ecological and human systems that have restricted geographic ranges constrained by climate related conditions and have high endemism or other distinctive properties. Examples include coral reefs, the Arctic and its indigenous people, mountain glaciers and biodiversity hotspots. **RFC2 Extreme weather events:** risks/impacts to human health, livelihoods, assets and ecosystems from extreme weather events such as heatwaves, heavy rain, drought and associated wildfires, and coastal flooding. **RFC3 Distribution of impacts:** risks/impacts that disproportionately affect particular groups due to uneven distribution of physical climate change hazards, exposure or vulnerability. **RFC4 Global aggregate impacts:** global monetary damage, global scale degradation and loss of ecosystems and biodiversity. **RFC5 Large-scale singular events:** are relatively large, abrupt and sometimes irreversible changes in systems that are caused by global warming. Examples include disintegration of the Greenland and Antarctic ice sheets. The grey bar represents the range of GMST for the most recent decade: 2006–2015.

New literature since AR5 has provided a closer focus on the comparative levels of risk to coral reefs at 1.5°C versus 2°C of global warming. As assessed in Section 3.4.4 and Box 3.4, reaching 2°C will increase the frequency of mass coral bleaching and mortality to a point at which it will result in the total loss of coral reefs from the world's tropical and subtropical regions. Restricting overall warming to 1.5°C will still see a downward trend in average coral cover (70–90% decline by mid-century) but will prevent the total loss of coral reefs projected with warming of 2°C (Frieler et al., 2013). The remaining reefs at 1.5°C will also benefit from increasingly stable ocean conditions by the mid-to-late 21st century. Limiting global warming to 1.5°C during the course of the century may, therefore, open the window for many ecosystems to adapt or reassort geographically. This indicates a transition in risk in this system from high to very high (*high confidence*) at 1.5°C of warming and contributes to a lowering of the transition from high to very high (Figure 3.21) in this RFC1 compared to in AR5. Further details of risk transitions for ocean systems are described in Figure 3.18.

Substantial losses of Arctic Ocean summer ice were projected in WGI AR5 for global warming of 1.6°C, with a nearly ice-free Arctic Ocean being projected for global warming of more than 2.6°C. Since AR5, the importance of a threshold between 1°C and 2°C has been further emphasized in the literature, with sea ice projected to persist throughout the year for a global warming of less than 1.5°C,

yet chances of an ice-free Arctic during summer being high at 2°C of warming (Section 3.3.8). Less of the permafrost in the Arctic is projected to thaw under 1.5°C of warming (17–44%) compared with under 2°C (28–53%) (Section 3.3.5.2; Chadburn et al., 2017), which is expected to reduce risks to both social and ecological systems in the Arctic. This indicates a transition in the risk in this system from high to very high between 1.5°C and 2°C of warming and contributes to a lowering of the transition from high to very high in this RFC1 compared to in AR5.

AR5 identified a large number of threatened systems, including mountain ecosystems, highly biodiverse tropical wet and dry forests, deserts, freshwater systems and dune systems. These include Mediterranean areas in Europe, Siberian, tropical and desert ecosystems in Asia, Australian rainforests, the Fynbos and succulent Karoo areas of South Africa, and wetlands in Ethiopia, Malawi, Zambia and Zimbabwe. In all these systems, impacts accrue with greater warming and impacts at 2°C are expected to be greater than those at 1.5°C (*medium confidence*). One study since AR5 has shown that constraining global warming to 1.5°C would maintain the functioning of prairie pothole ecosystems in North America in terms of their productivity and biodiversity, whilst warming of 2°C would not do so (Johnson et al., 2016). The large proportion of insects projected to lose over half their range at 2°C of warming (25%) compared to at 1.5°C (9%) also suggests a significant loss of functionality in these threatened systems at 2°C of warming, owing to the critical role of insects in nutrient cycling, pollination, detritivory and other important ecosystem processes (Section 3.4.3).

Unique and threatened systems in small island states and in systems fed by glacier meltwater were also considered to contribute to this RFC in AR5, but there is little new information about these systems that pertains to 1.5°C or 2°C of global warming. Taken together, the evidence suggests that the transition from high to very high risk in unique and threatened systems occurs at a lower level of warming, between 1.5°C and 2°C (*high confidence*), than in AR5, where this transition was located at 2.6°C. The transition from moderate to high risk relocates very slightly from 1.6°C to 1.5°C (*high confidence*). There is also *high confidence* in the location of the transition from low to moderate risk below present-day global temperatures.

3.5.2.2   RFC 2 – Extreme weather events

Reduced risks in terms of the likelihood of occurrence of extreme weather events are discussed in this sub-subsection for 1.5°C as compared to 2°C of global warming, for those extreme events where evidence is currently available based on the assessments of Section 3.3. AR5 assigned a moderate level of risk from extreme weather events at recent temperatures (1986–2005) owing to the attribution of heat and precipitation extremes to climate change, and a transition to high risk beginning below 1.6°C of global warming based on the magnitude, likelihood and timing of projected changes in risk associated with extreme events, indicating more severe and widespread impacts. The AR5 analysis already suggested a significant benefit of limiting warming to 1.5°C, as doing so might keep risks closer to the moderate level. New literature since AR5 has provided greater confidence in a reduced level of risks due to extreme weather events at 1.5°C versus 2°C of warming for some types of extremes (Section 3.3 and below; Figure 3.21).

Temperature: It is expected that further increases in the number of warm days/nights and decreases in the number of cold days/nights, and an increase in the overall temperature of hot and cold extremes would occur under 1.5°C of global warming relative to pre-industrial levels (*high confidence*) compared to under the present-day climate (1°C of warming), with further changes occurring towards 2°C of global warming (Section 3.3). As assessed in Sections 3.3.1 and 3.3.2, impacts of 0.5°C of global warming can be identified for temperature extremes at global scales, based on observations and the analysis of climate models. At 2°C of global warming, it is *likely* that temperature increases of more than 2°C would occur over most land regions in terms of extreme temperatures (up to 4°C–6°C depending on region and considered extreme index) (Section 3.3.2, Table 3.2). Regional increases in temperature extremes can be robustly limited if global warming is constrained to 1.5°C, with regional warmings of up to 3°C–4.5°C (Section 3.3.2, Table 3.2). Benefits obtained from this general reduction in extremes depend to a large extent on whether the lower range of increases in extremes at 1.5°C is sufficient for critical thresholds to be exceeded, within the context of wide-ranging aspects such as crop yields, human health and the sustainability of ecosystems.

Heavy precipitation: AR5 assessed trends in heavy precipitation for land regions where observational coverage was sufficient for assessment. It concluded with *medium confidence* that anthropogenic forcing has contributed to a global-scale intensification of heavy precipitation over the second half of the 20th century, for a global warming of approximately 0.5°C (Section 3.3.3). A recent observation-based study likewise showed that a 0.5°C increase in global mean temperature has had a detectable effect on changes in precipitation extremes at the global scale (Schleussner et al., 2017), thus suggesting that there would be detectable differences in heavy precipitation at 1.5°C and 2°C of global warming. These results are consistent with analyses of climate projections, although they also highlight a large amount of regional variation in the sensitivity of changes in heavy precipitation (Section 3.3.3).

Droughts: When considering the difference between precipitation and evaporation (P–E) as a function of global temperature changes, the subtropics generally display an overall trend towards drying, whilst the northern high latitudes display a robust response towards increased wetting (Section 3.3.4, Figure 3.12). Limiting global mean temperature increase to 1.5°C as opposed to 2°C could substantially reduce the risk of reduced regional water availability in some regions (Section 3.3.4). Regions that are projected to benefit most robustly from restricted warming include the Mediterranean and southern Africa (Section 3.3.4).

Fire: Increasing evidence that anthropogenic climate change has already caused significant increases in fire area globally (Section 3.4.3) is in line with projected fire risks. These risks are projected to increase further under 1.5°C of global warming relative to the present day (Section 3.4.3). Under 1.2°C of global warming, fire frequency has been estimated to increase by over 37.8% of global land areas, compared to 61.9% of global land areas under 3.5°C of warming. For in-depth discussion and uncertainty estimates, see Meehl et al. (2007), Moritz et al. (2012) and Romero-Lankao et al. (2014).

Regarding extreme weather events (RFC2), the transition from moderate to high risk is located between 1°C and 1.5°C of global warming (Figure 3.21), which is very similar to the AR5 assessment but is assessed with greater confidence (*medium confidence*). The impact literature contains little information about the potential for human society to adapt to extreme weather events, and hence it has not been possible to locate the transition from high to very high risk within the context of assessing impacts at 1.5°C and 2°C of global warming. There is thus *low confidence* in the level at which global warming could lead to very high risks associated with extreme weather events in the context of this report.

3.5.2.3   RFC 3 – Distribution of impacts

Risks due to climatic change are unevenly distributed and are generally greater at lower latitudes and for disadvantaged people and communities in countries at all levels of development. AR5 located the transition from undetectable to moderate risk below recent temperatures, owing to the detection and attribution of regionally differentiated changes in crop yields (*medium to high confidence*; Figure 3.20), and new literature has continued to confirm this finding. Based on the assessment of risks to regional crop production and water resources, AR5 located the transition from moderate to high risk between 1.6°C and 2.6°C above pre-industrial levels. Cross-Chapter Box 6 in this chapter highlights that at 2°C of warming, new literature shows that risks of food shortage are projected to emerge in the African Sahel, the Mediterranean, central Europe, the Amazon, and western and southern Africa, and that these are much larger than the corresponding risks at 1.5°C. This suggests a transition from moderate to high risk of regionally differentiated impacts between 1.5°C and 2°C above pre-industrial levels for food security (*medium confidence*) (Figure 3.20). Reduction in the availability of water resources at 2°C is projected to be greater than 1.5°C of global warming, although changes in socio-economics could have a greater influence (Section 3.4.2), with larger risks in the Mediterranean (Box 3.2); estimates of the magnitude of the risks remain similar to those cited in AR5. Globally, millions of people may be at risk from sea level rise (SLR) during the 21st century (Hinkel et al., 2014; Hauer et al., 2016), particularly if adaptation is limited. At 2°C of warming, more than 90% of global coastlines are projected to experience SLR greater than 0.2 m, suggesting regional differences in the risks of coastal flooding. Regionally differentiated multi-sector risks are already apparent at 1.5°C of warming, being more prevalent where vulnerable people live, predominantly in South Asia (mostly Pakistan, India and China), but these risks are projected to spread to sub-Saharan Africa, the Middle East and East Asia as temperature rises, with the world's poorest people disproportionately impacted at 2°C of warming (Byers et al., 2018). The hydrological impacts of climate change in Europe are projected to increase in spatial extent and intensity across increasing global warming levels of 1.5°C, 2°C and 3°C (Donnelly et al., 2017). Taken together, a transition from moderate to high risk is now located between 1.5°C and 2°C above pre-industrial levels, based on the assessment of risks to food security, water resources, drought, heat exposure and coastal submergence (*high confidence*; Figure 3.21).

3.5.2.4   RFC 4 – Global aggregate impacts

Oppenheimer et al. (2014) explained the inclusion of non-economic metrics related to impacts on ecosystems and species at the global level, in addition to economic metrics in global aggregate impacts. The degradation of ecosystem services by climate change and ocean acidification have generally been excluded from previous global aggregate economic analyses.

Global economic impacts: WGII AR5 found that overall global aggregate impacts become moderate at 1°C–2°C of warming, and the transition to moderate risk levels was therefore located at 1.6°C above pre-industrial levels. This was based on the assessment of literature using model simulations which indicated that the global aggregate economic impact will become significantly negative between 1°C and 2°C of warming (*medium confidence*), whilst there will be a further increase in the magnitude and likelihood of aggregate economic risks at 3°C of warming (*low confidence*).

Since AR5, three studies have emerged using two entirely different approaches which indicate that economic damages are projected to be higher by 2100 if warming reaches 2°C than if it is constrained to 1.5°C. The study by Warren et al. (2018c) used the integrated assessment model PAGE09 to estimate that avoided global economic damages of 22% (10–26%) accrue from constraining warming to 1.5°C rather than 2°C, 90% (77–93%) from 1.5°C rather than 3.66°C,

and 87% (74–91%) from 2°C rather than 3.66°C. In the second study, Pretis et al. (2018) identified several regions where economic damages are projected to be greater at 2°C compared to 1.5°C of warming, further estimating that projected damages at 1.5°C remain similar to today's levels of economic damage. The third study, by M. Burke et al. (2018) used an empirical, statistical approach and found that limiting warming to 1.5°C instead of 2°C would save 1.5–2.0% of the gross world product (GWP) by mid-century and 3.5% of the GWP by end-of-century (see Figure 2A in M. Burke et al., 2018). Based on a 3% discount rate, this corresponds to 8.1–11.6 trillion USD and 38.5 trillion USD in avoided damages by mid- and end-of-century, respectively, agreeing closely with the estimate by Warren et al. (2018c) of 15 trillion USD. Under the no-policy baseline scenario, temperature rises by 3.66°C by 2100, resulting in a global gross domestic product (GDP) loss of 2.6% (5–95% percentile range 0.5–8.2%), compared with 0.3% (0.1–0.5%) by 2100 under the 1.5°C scenario and 0.5% (0.1–1.0%) in the 2°C scenario. Limiting warming to 1.5°C rather than 2°C by 2060 has also been estimated to result in co-benefits of 0.5–0.6% of the world GDP, owing to reductions in air pollution (Shindell et al., 2018), which is similar to the avoided damages identified for the USA (Box 3.6).

Two studies focusing only on the USA found that economic damages are projected to be higher by 2100 if warming reaches 2°C than if it is constrained to 1.5°C. Hsiang et al. (2017) found a mean difference of 0.35% GDP (range 0.2–0.65%), while Yohe (2017) identified a GDP loss of 1.2% per degree of warming, hence approximately 0.6% for half a degree. Further, the avoided risks compared to a no-policy baseline are greater in the 1.5°C case (4%, range 2–7%) compared to the 2°C case (3.5%, range 1.8–6.5%). These analyses suggest that the point at which global aggregates of economic impacts become negative is below 2°C (*medium confidence*), and that there is a possibility that it is below 1.5°C of warming.

Oppenheimer et al. (2014) noted that the global aggregated damages associated with large-scale singular events has not been explored, and reviews of integrated modelling exercises have indicated a potential underestimation of global aggregate damages due to the lack of consideration of the potential for these events in many studies. Since AR5, further analyses of the potential economic consequences of triggering these large-scale singular events have indicated a two to eight fold larger economic impact associated with warming of 3°C than estimated in most previous analyses, with the extent of increase depending on the number of events incorporated. Lemoine and Traeger (2016) included only three known singular events whereas Y. Cai et al. (2016) included five.

Biome shifts, species range loss, increased risks of species extinction and risks of loss of ecosystem functioning and services: 13% (range 8–20%) of Earth's land area is projected to undergo biome shifts at 2°C of warming compared to approximately 7% at 1.5°C (*medium confidence*) (Section 3.4.3; Warszawski et al., 2013), implying a halving of biome transformations. Overall levels of species loss at 2°C of warming are similar to values found in previous studies for plants and vertebrates (Warren et al., 2013, 2018a), but insects have been found to be more sensitive to climate change, with 18% (6–35%) projected to lose over half their range at 2°C of warming compared to 6% (1–18%) under 1.5°C of warming, corresponding to a difference of 66% (Section 3.4.3). The critical role of insects in ecosystem functioning therefore suggests that there will be impacts on global ecosystem functioning already at 2°C of warming, whilst species that lose large proportions of their range are considered to be at increased risk of extinction (Section 3.4.3.3). Since AR5, new literature has indicated that impacts on marine fish stocks and fisheries are lower under 1.5°C–2°C of global warming relative to pre-industrial levels compared to under higher warming scenarios (Section 3.4.6), especially in tropical and polar systems.

In AR5, the transition from undetectable to moderate impacts was considered to occur between 1.6°C and 2.6°C of global warming reflecting impacts on the economy and on biodiversity globally, whereas high risks were associated with 3.6°C of warming to reflect the high risks to biodiversity and accelerated effects on the global economy. New evidence suggests moderate impacts on the global aggregate economy and global biodiversity by 1.5°C of warming, suggesting a lowering of the temperature level for the transition to moderate risk to 1.5°C (Figure 3.21). Further, recent literature points to higher risks than previously assessed for the global aggregate economy and global biodiversity by 2°C of global warming, suggesting that the transition to a high risk level is located between 1.5°C and 2.5°C of warming (Figure 3.21), as opposed to at 3.6°C as previously assessed (*medium confidence*).

3.5.2.5   RFC 5 – Large-scale singular events

Large-scale singular events are components of the global Earth system that are thought to hold the risk of reaching critical tipping points under climate change, and that can result in or be associated with major shifts in the climate system. These components include:

- the cryosphere: West Antarctic ice sheet, Greenland ice sheet
- the thermohaline circulation: slowdown of the Atlantic Meridional Overturning Circulation (AMOC)
- the El Niño–Southern Oscillation (ENSO) as a global mode of climate variability
- role of the Southern Ocean in the global carbon cycle

AR5 assessed that the risks associated with these events become moderate between 0.6°C and 1.6°C above pre-industrial levels, based on early warning signs, and that risk was expected to become high between 1.6°C and 4.6°C based on the potential for commitment to large irreversible sea level rise from the melting of land-based ice sheets (*low to medium confidence*). The increase in risk between 1.6°C and 2.6°C above pre-industrial levels was assessed to be disproportionately large. New findings since AR5 are described in detail below.

Greenland and West Antarctic ice sheets and marine ice sheet instability (MISI): Various feedbacks between the Greenland ice sheet and the wider climate system, most notably those related to the dependence of ice melt on albedo and surface elevation, make irreversible loss of the ice sheet a possibility. Church et al. (2013) assessed this threshold to be at 2°C of warming or higher levels relative to pre-industrial temperature. Robinson et al. (2012) found a range for this threshold of 0.8°C–3.2°C (95% confidence). The threshold of global

temperature increase that may initiate irreversible loss of the West Antarctic ice sheet and marine ice sheet instability (MISI) is estimated to lie between 1.5°C and 2°C. The time scale for eventual loss of the ice sheets varies between millennia and tens of millennia and assumes constant surface temperature forcing during this period. If temperature were to decline subsequently the ice sheets might regrow, although the amount of cooling required is likely to be highly dependent on the duration and rate of the previous retreat. The magnitude of global sea level rise that could occur over the next two centuries under 1.5°C–2°C of global warming is estimated to be in the order of several tenths of a metre according to most studies (*low confidence*) (Schewe et al., 2011; Church et al., 2013; Levermann et al., 2014; Marzeion and Levermann, 2014; Fürst et al., 2015; Golledge et al., 2015), although a smaller number of investigations (Joughin et al., 2014; Golledge et al., 2015; DeConto and Pollard, 2016) project increases of 1–2 m. This body of evidence suggests that the temperature range of 1.5°C–2°C may be regarded as representing moderate risk, in that it may trigger MISI in Antarctica or irreversible loss of the Greenland ice sheet and it may be associated with sea level rise by as much as 1–2 m over a period of two centuries.

Thermohaline circulation (slowdown of AMOC): It is *more likely than not* that the AMOC has been weakening in recent decades, given the detection of cooling of surface waters in the North Atlantic and evidence that the Gulf Stream has slowed since the late 1950s (Rahmstorf et al., 2015b; Srokosz and Bryden, 2015; Caesar et al., 2018). There is limited evidence linking the recent weakening of the AMOC to anthropogenic warming (Caesar et al., 2018). It is *very likely* that the AMOC will weaken over the 21st century. Best estimates and ranges for the reduction based on CMIP5 simulations are 11% (1–24%) in RCP2.6 and 34% (12–54%) in RCP8.5 (AR5). There is no evidence indicating significantly different amplitudes of AMOC weakening for 1.5°C versus 2°C of global warming, or of a shutdown of the AMOC at these global temperature thresholds. Associated risks are classified as low to moderate.

El Niño–Southern Oscillation (ENSO): Extreme El Niño events are associated with significant warming of the usually cold eastern Pacific Ocean, and they occur about once every 20 years (Cai et al., 2015). Such events reorganize the distribution of regions of organized convection and affect weather patterns across the globe. Recent research indicates that the frequency of extreme El Niño events increases linearly with the global mean temperature, and that the number of such events might double (one event every ten years) under 1.5°C of global warming (G. Wang et al., 2017). This pattern is projected to persist for a century after stabilization at 1.5°C, thereby challenging the limits to adaptation, and thus indicates high risk even at the 1.5°C threshold. La Niña event (the opposite or balancing event to El Niño) frequency is projected to remain similar to that of the present day under 1.5°C–2°C of global warming.

Role of the Southern Ocean in the global carbon cycle: The critical role of the Southern Ocean as a net sink of carbon might decline under global warming, and assessing this effect under 1.5°C compared to 2°C of global warming is a priority. Changes in ocean chemistry (e.g., oxygen content and ocean acidification), especially those associated with the deep sea, are associated concerns (Section 3.3.10).

For large-scale singular events (RFC5), moderate risk is now located at 1°C of warming and high risk is located at 2.5°C (Figure 3.21), as opposed to at 1.6°C (moderate risk) and around 4°C (high risk) in AR5, because of new observations and models of the West Antarctic ice sheet (*medium confidence*), which suggests that the ice sheet may be in the early stages of marine ice sheet instability (MISI). Very high risk is assessed as lying above 5°C because the growing literature on process-based projections of the West Antarctic ice sheet predominantly supports the AR5 assessment of an MISI contribution of several additional tenths of a metre by 2100.

3.5.3   Regional Economic Benefit Analysis for the 1.5°C versus 2°C Global Goals

This section reviews recent literature that has estimated the economic benefits of constraining global warming to 1.5°C compared to 2°C. The focus here is on evidence pertaining to specific regions, rather than on global aggregated benefits (Section 3.5.2.4). At 2°C of global warming, lower economic growth is projected for many countries than at 1.5°C of global warming, with low-income countries projected to experience the greatest losses (*low to medium confidence*) (M. Burke et al., 2018; Pretis et al., 2018). A critical issue for developing countries in particular is that advantages in some sectors are projected to be offset by increasing mitigation costs (Rogelj et al., 2013; M. Burke et al., 2018), with food production being a key factor. That is, although restraining the global temperature increase to 2°C is projected to reduce crop losses under climate change relative to higher levels of warming, the associated mitigation costs may increase the risk of hunger in low-income countries (*low confidence*) (Hasegawa et al., 2016). It is *likely* that the even more stringent mitigation measures required to restrict global warming to 1.5°C (Rogelj et al., 2013) will further increase these mitigation costs and impacts. International trade in food might be a key response measure for alleviating hunger in developing countries under 1.5°C and 2°C stabilization scenarios (IFPRI, 2018).

Although warming is projected to be the highest in the Northern Hemisphere under 1.5°C or 2°C of global warming, regions in the tropics and Southern Hemisphere subtropics are projected to experience the largest impacts on economic growth (*low to medium confidence*) (Gallup et al., 1999; M. Burke et al., 2018; Pretis et al., 2018). Despite the uncertainties associated with climate change projections and econometrics (e.g., M. Burke et al., 2018), it is *more likely than not* that there will be large differences in economic growth under 1.5°C and 2°C of global warming for developing versus developed countries (M. Burke et al., 2018; Pretis et al., 2018). Statistically significant reductions in gross domestic product (GDP) per capita growth are projected across much of the African continent, Southeast Asia, India, Brazil and Mexico (*low to medium confidence*). Countries in the western parts of tropical Africa are projected to benefit most from restricting global warming to 1.5°C, as opposed to 2°C, in terms of future economic growth (Pretis et al., 2018). An important reason why developed countries in the tropics and subtropics are projected to benefit substantially from restricting global warming to 1.5°C is that present-day temperatures in these regions are above the threshold thought to be optimal for economic production (M. Burke et al., 2015b, 2018).

The world's largest economies are also projected to benefit from restricting warming to 1.5°C as opposed to 2°C (*medium confidence*), with the likelihood of such benefits being realized estimated at 76%, 85% and 81% for the USA, China and Japan, respectively (M. Burke et al., 2018). Two studies focusing only on the USA found that economic damages are projected to be higher by 2100 if warming reaches 2°C than if it is constrained to 1.5°C. Yohe (2017) found a mean difference of 0.35% GDP (range 0.2–0.65%), while Hsiang et al. (2017) identified a GDP loss of 1.2% per degree of warming, hence approximately 0.6% for half a degree. Overall, no statistically significant changes in GDP are projected to occur over most of the developed world under 1.5°C of global warming in comparison to present-day conditions, but under 2°C of global warming impacts on GDP are projected to be generally negative (*low confidence*) (Pretis et al., 2018).

A caveat to the analyses of Pretis et al. (2018) and M. Burke et al. (2018) is that the effects of sea level rise were not included in the estimations of damages or future economic growth, implying a potential underestimation of the benefits of limiting warming to 1.5°C for the case where significant sea level rise is avoided at 1.5°C but not at 2°C.

3.5.4   Reducing Hotspots of Change for 1.5°C and 2°C of Global Warming

This subsection integrates Sections 3.3 and 3.4 in terms of climate-change-induced hotspots that occur through interactions across the physical climate system, ecosystems and socio-economic human systems, with a focus on the extent to which risks can be avoided or reduced by achieving the 1.5°C global warming goal (as opposed to the 2°C goal). Findings are summarized in Table 3.6.

3.5.4.1   Arctic sea ice

Ice-free Arctic Ocean summers are *very likely* at levels of global warming higher than 2°C (Notz and Stroeve, 2016; Rosenblum and Eisenman, 2016; Screen and Williamson, 2017; Niederdrenk and Notz, 2018). Some studies even indicate that the entire Arctic Ocean summer period will become ice free under 2°C of global warming, whilst others more conservatively estimate this probability to be in the order of 50% (Section 3.3.8; Sanderson et al., 2017). The probability of an ice-free Arctic in September at 1.5°C of global warming is low and substantially lower than for the case of 2°C of global warming (*high confidence*) (Section 3.3.8; Screen and Williamson, 2017; Jahn, 2018; Niederdrenk and Notz, 2018). There is, however, a single study that questions the validity of the 1.5°C threshold in terms of maintaining summer Arctic Ocean sea ice (Niederdrenk and Notz, 2018). In contrast to summer, little ice is projected to be lost during winter for either 1.5°C or 2°C of global warming (*medium confidence*) (Niederdrenk and Notz, 2018). The losses in sea ice at 1.5°C and 2°C of warming will result in habitat losses for organisms such as seals, polar bears, whales and sea birds (e.g., Larsen et al., 2014). There is *high agreement and robust evidence* that photosynthetic species will change because of sea ice retreat and related changes in temperature and radiation (Section 3.4.4.7), and this is *very likely* to benefit fisheries productivity in the Northern Hemisphere spring bloom system (Section 3.4.4.7).

3.5.4.2 Arctic land regions

In some Arctic land regions, the warming of cold extremes and the increase in annual minimum temperature at 1.5°C are stronger than the global mean temperature increase by a factor of two to three, meaning 3°C–4.5°C of regional warming at 1.5°C of global warming (e.g., northern Europe in Supplementary Material 3.SM, Figure 3.SM.5 see also Section 3.3.2.2 and Seneviratne et al., 2016). Moreover, over much of the Arctic, a further increase of 0.5°C in the global surface temperature, from 1.5°C to 2°C, may lead to further temperature increases of 2°C–2.5°C (Figure 3.3). As a consequence, biome (major ecosystem type) shifts are *likely* in the Arctic, with increases in fire frequency, degradation of permafrost, and tree cover *likely* to occur at 1.5°C of warming and further amplification of these changes expected under 2°C of global warming (e.g., Gerten et al., 2013; Bring et al., 2016). Rising temperatures, thawing permafrost and changing weather patterns are projected to increasingly impact people, infrastructure and industries in the Arctic (W.N. Meier et al., 2014) with these impacts larger at 2°C than at 1.5°C of warming (*medium confidence*).

3.5.4.3 Alpine regions

Alpine regions are generally regarded as climate change hotspots given that rich biodiversity has evolved in their cold and harsh climate, but with many species consequently being vulnerable to increases in temperature. Under regional warming, alpine species have been found to migrate upwards on mountain slopes (Reasoner and Tinner, 2009), an adaptation response that is obviously limited by mountain height and habitability. Moreover, many of the world's alpine regions are important from a water security perspective through associated glacier melt, snow melt and river flow (see Section 3.5.5.2 for a discussion of these aspects). Projected biome shifts are *likely* to be severe in alpine regions already at 1.5°C of warming and to increase further at 2°C (Gerten et al., 2013, Figure 1b; B. Chen et al., 2014).

3.5.4.4 Southeast Asia

Southeast Asia is a region highly vulnerable to increased flooding in the context of sea level rise (Arnell et al., 2016; Brown et al., 2016, 2018a). Risks from increased flooding are projected to rise from 1.5°C to 2°C of warming (*medium confidence*), with substantial increases projected beyond 2°C (Arnell et al., 2016). Southeast Asia displays statistically significant differences in projected changes in heavy precipitation, runoff and high flows at 1.5°C versus 2°C of warming, with stronger increases occurring at 2°C (Section 3.3.3; Wartenburger et al., 2017; Döll et al., 2018; Seneviratne et al., 2018c); thus, this region is considered a hotspot in terms of increases in heavy precipitation between these two global temperature levels (*medium confidence*) (Schleussner et al., 2016b; Seneviratne et al., 2016). For Southeast Asia, 2°C of warming by 2040 could lead to a decline by one-third in per capita crop production associated with general decreases in crop yields (Nelson et al., 2010). However, under 1.5°C of warming, significant risks for crop yield reduction in the region are avoided (Schleussner et al., 2016b). These changes pose significant risks for poor people in both rural regions and urban areas of Southeast Asia (Section 3.4.10.1), with these risks being larger at 2°C of global warming compared to 1.5°C (*medium confidence*).

3.5.4.5 Southern Europe and the Mediterranean

The Mediterranean is regarded as a climate change hotspot, both in terms of projected stronger warming of the regional land-based hot extremes compared to the mean global temperature increase (e.g., Seneviratne et al., 2016) and in terms of of robust increases in the probability of occurrence of extreme droughts at 2°C vs 1.5°C global warming (Section 3.3.4). Low river flows are projected to decrease in the Mediterranean under 1.5°C of global warming (Marx et al., 2018), with associated significant decreases in high flows and floods (Thober et al., 2018), largely in response to reduced precipitation. The median reduction in annual runoff is projected to almost double from about 9% (*likely* range 4.5–15.5%) at 1.5°C to 17% (*likely* range 8–25%) at 2°C (Schleussner et al., 2016b). Similar results were found by Döll et al. (2018). Overall, there is *high confidence* that strong increases in dryness and decreases in water availability in the Mediterranean and southern Europe would occur from 1.5°C to 2°C of global warming. Sea level rise is expected to be lower for 1.5°C versus 2°C, lowering risks for coastal metropolitan agglomerations. The risks (assuming current adaptation) related to water deficit in the Mediterranean are high for global warming of 2°C but could be substantially reduced if global warming were limited to 1.5°C (Section 3.3.4; Guiot and Cramer, 2016; Schleussner et al., 2016b; Donnelly et al., 2017).

3.5.4.6 West Africa and the Sahel

West Africa and the Sahel are *likely* to experience increases in the number of hot nights and longer and more frequent heatwaves even if the global temperature increase is constrained to 1.5°C, with further increases expected at 2°C of global warming and beyond (e.g., Weber et al., 2018). Moreover, daily rainfall intensity and runoff is expected to increase (*low confidence*) towards 2°C and higher levels of global warming (Schleussner et al., 2016b; Weber et al., 2018), with these changes also being relatively large compared to the projected changes at 1.5°C of warming. Moreover, increased risks are projected in terms of drought, particularly for the pre-monsoon season (Sylla et al., 2015), with both rural and urban populations affected, and more so at 2°C of global warming as opposed to 1.5°C (Liu et al., 2018). Based on a World Bank (2013) study for sub-Saharan Africa, a 1.5°C warming by 2030 might reduce the present maize cropping areas by 40%, rendering these areas no longer suitable for current cultivars. Substantial negative impacts are also projected for sorghum suitability in the western Sahel (Läderach et al., 2013; Sultan and Gaetani, 2016). An increase in warming to 2°C by 2040 would result in further yield losses and damages to crops (i.e., maize, sorghum, wheat, millet, groundnut and cassava). Schleussner et al. (2016b) found consistently reduced impacts on crop yield for West Africa under 2°C compared to 1.5°C of global warming. There is *medium confidence* that vulnerabilities to water and food security in the African Sahel will be higher at 2°C compared to 1.5°C of global warming (Cheung et al., 2016a; Betts et al., 2018), and at 2°C these vulnerabilities are expected to be worse (*high evidence*) (Sultan and Gaetani, 2016; Lehner et al., 2017; Betts et al., 2018; Byers et al., 2018; Rosenzweig et al., 2018). Under global warming of more than 2°C, the western Sahel might experience the strongest drying and experience serious food security issues (Ahmed et al., 2015; Parkes et al., 2018).

3.5.4.7   Southern Africa

The southern African region is projected to be a climate change hotspot in terms of both hot extremes (Figures 3.5 and 3.6) and drying (Figure 3.12). Indeed, temperatures have been rising in the subtropical regions of southern Africa at approximately twice the global rate over the last five decades (Engelbrecht et al., 2015). Associated elevated warming of the regional land-based hot extremes has occurred (Section 3.3; Seneviratne et al., 2016). Increases in the number of hot nights, as well as longer and more frequent heatwaves, are projected even if the global temperature increase is constrained to 1.5°C (*high confidence*), with further increases expected at 2°C of global warming and beyond (*high confidence*) (Weber et al., 2018).

Moreover, southern Africa is *likely* to generally become drier with reduced water availability under low mitigation (Niang et al., 2014; Engelbrecht et al., 2015; Karl et al., 2015; James et al., 2017), with this particular risk being prominent under 2°C of global warming and even under 1.5°C (Gerten et al., 2013). Risks are significantly reduced, however, under 1.5°C of global warming compared to under higher levels (Schleussner et al., 2016b). There are consistent and statistically significant increases in projected risks of increased meteorological drought in southern Africa at 2°C versus 1.5°C of warming (*medium confidence*). Despite the general rainfall reductions projected for southern Africa, daily rainfall intensities are expected to increase over much of the region (*medium confidence*), and increasingly so with higher levels of global warming. There is *medium confidence* that livestock in southern Africa will experience increased water stress under both 1.5°C and 2°C of global warming, with negative economic consequences (e.g., Boone et al., 2018). The region is also projected to experience reduced maize, sorghum and cocoa cropping area suitability, as well as yield losses under 1.5°C of warming, with further decreases occurring towards 2°C of warming (World Bank, 2013). Generally, there is *high confidence* that vulnerability to decreases in water and food availability is reduced at 1.5°C versus 2°C for southern Africa (Betts et al., 2018), whilst at 2°C these are expected to be higher (*high confidence*) (Lehner et al., 2017; Betts et al., 2018; Byers et al., 2018; Rosenzweig et al., 2018).

3.5.4.8   Tropics

Worldwide, the largest increases in the number of hot days are projected to occur in the tropics (Figure 3.7). Moreover, the largest differences in the number of hot days for 1.5°C versus 2°C of global warming are projected to occur in the tropics (Mahlstein et al., 2011). In tropical Africa, increases in the number of hot nights, as well as longer and more frequent heatwaves, are projected under 1.5°C of global warming, with further increases expected under 2°C of global warming (Weber et al., 2018). Impact studies for major tropical cereals reveal that yields of maize and wheat begin to decline with 1°C to 2°C of local warming in the tropics. Schleussner et al. (2016b) project that constraining warming to 1.5°C rather than 2°C would avoid significant risks of tropical crop yield declines in West Africa, Southeast Asia, and Central and South America. There is *limited evidence* and thus *low confidence* that these changes may result in significant population displacement from the tropics to the subtropics (e.g., Hsiang and Sobel, 2016).

3.5.4.9   Small islands

It is widely recognized that small islands are very sensitive to climate change impacts such as sea level rise, oceanic warming, heavy precipitation, cyclones and coral bleaching *(high confidence)* (Nurse et al., 2014; Ourbak and Magnan, 2017). Even at 1.5°C of global warming, the compounding impacts of changes in rainfall, temperature, tropical cyclones and sea level are likely to be significant across multiple natural and human systems. There are potential benefits to small island developing states (SIDS) from avoided risks at 1.5°C versus 2°C, especially when coupled with adaptation efforts. In terms of sea level rise, by 2150, roughly 60,000 fewer people living in SIDS will be exposed in a 1.5°C world than in a 2°C world (Rasmussen et al., 2018). Constraining global warming to 1.5°C may significantly reduce water stress (by about 25%) compared to the projected water stress at 2°C, for example in the Caribbean region (Karnauskas et al., 2018), and may enhance the ability of SIDS to adapt (Benjamin and Thomas, 2016). Up to 50% of the year is projected to be very warm in the Caribbean at 1.5°C, with a further increase by up to 70 days at 2°C versus 1.5°C (Taylor et al., 2018). By limiting warming to 1.5°C instead of 2°C in 2050, risks of coastal flooding (measured as the flood amplification factors for 100-year flood events) are reduced by 20–80% for SIDS (Rasmussen et al., 2018). A case study of Jamaica with lessons for other Caribbean SIDS demonstrated that the difference between 1.5°C and 2°C is *likely* to challenge livestock thermoregulation, resulting in persistent heat stress for livestock (Lallo et al., 2018).

3.5.4.10   Fynbos and shrub biomes

The Fynbos and succulent Karoo biomes of South Africa are threatened systems that were assessed in AR5. Similar shrublands exist in the semi-arid regions of other continents, with the Sonora-Mojave creosotebush-white bursage desert scrub ecosystem in the USA being a prime example. Impacts accrue across these systems with greater warming, with impacts at 2°C likely to be greater than those at 1.5°C (*medium confidence*). Under 2°C of global warming, regional warming in drylands is projected to be 3.2°C–4°C, and under 1.5°C of global warming, mean warming in drylands is projected to still be about 3°C. The Fynbos biome in southwestern South Africa is vulnerable to the increasing impact of fires under increasing temperatures and drier winters (*high confidence*). The Fynbos biome is projected to lose about 20%, 45% and 80% of its current suitable climate area relative to its present-day area under 1°C, 2°C and 3°C of warming, respectively (Engelbrecht and Engelbrecht, 2016), demonstrating the value of climate change mitigation in protecting this rich centre of biodiversity.

Table 3.6 | Emergence and intensity of climate change hotspots under different degrees of global warming.

Region and/or Phenomenon	Warming of 1.5°C or less	Warming of 1.5°C–2°C	Warming of 2°C–3°C
Arctic sea ice	Arctic summer sea ice is *likely* to be maintained	The risk of an ice-free Arctic in summer is about 50% or higher	The Arctic is *very likely* to be ice free in summer
	Habitat losses for organisms such as polar bears, whales, seals and sea birds	Habitat losses for organisms such as polar bears, whales, seals and sea birds may be critical if summers are ice free	Critical habitat losses for organisms such as polar bears, whales, seals and sea birds
	Benefits for Arctic fisheries	Benefits for Arctic fisheries	Benefits for Arctic fisheries
Arctic land regions	Cold extremes warm by a factor of 2–3, reaching up to 4.5°C (*high confidence*)	Cold extremes warm by as much as 8°C (*high confidence*)	Drastic regional warming is *very likely*
	Biome shifts in the tundra and permafrost deterioration are *likely*	Larger intrusions of trees and shrubs in the tundra than under 1.5°C of warming are *likely*; larger but constrained losses in permafrost are *likely*	A collapse in permafrost may occur (*low confidence*); a drastic biome shift from tundra to boreal forest is possible (*low confidence*)
Alpine regions	Severe shifts in biomes are *likely*	Even more severe shifts are *likely*	Critical losses in alpine habitats are *likely*
Southeast Asia	Risks for increased flooding related to sea level rise	Higher risks of increased flooding related to sea level rise (*medium confidence*)	Substantial increases in risks related to flooding from sea level rise
	Increases in heavy precipitation events	Stronger increases in heavy precipitation events (*medium confidence*)	Substantial increase in heavy precipitation and high-flow events
	Significant risks of crop yield reductions are avoided	One-third decline in per capita crop production (*medium confidence*)	Substantial reductions in crop yield
Mediterranean	Increase in probability of extreme drought (*medium confidence*)	Robust increase in probability of extreme drought (*medium confidence*)	Robust and large increases in extreme drought. Substantial reductions in precipitation and in runoff (*medium confidence*)
	Medium confidence in reduction in runoff of about 9% (*likely* range 4.5–15.5%)	*Medium confidence* in further reductions (about 17%) in runoff (*likely* range 8–28%)	
	Risk of water deficit (*medium confidence*)	Higher risks of water deficit (*medium confidence*)	Very high risks of water deficit (*medium confidence*)
West Africa and the Sahel	Increases in the number of hot nights and longer and more frequent heatwaves are *likely*	Further increases in number of hot nights and longer and more frequent heatwaves are *likely*	Substantial increases in the number of hot nights and heatwave duration and frequency (*very likely*)
	Reduced maize and sorghum production is *likely*, with area suitable for maize production reduced by as much as 40%	Negative impacts on maize and sorghum production *likely* larger than at 1.5°C; *medium confidence* that vulnerabilities to food security in the African Sahel will be higher at 2°C compared to 1.5°C	Negative impacts on crop yield may result in major regional food insecurities (*medium confidence*)
	Increased risks of undernutrition	Higher risks of undernutrition	High risks of undernutrition
Southern Africa	Reductions in water availability (*medium confidence*)	Larger reductions in rainfall and water availability (*medium confidence*)	Large reductions in rainfall and water availability (*medium confidence*)
	Increases in number of hot nights and longer and more frequent heatwaves (*high confidence*)	Further increases in number of hot nights and longer and more frequent heatwaves (*high confidence*), associated increases in risks of increased mortality from heatwaves compared to 1.5°C warming (*high confidence*)	Drastic increases in the number of hot nights, hot days and heatwave duration and frequency to impact substantially on agriculture, livestock and human health and mortality (*high confidence*)
	High risks of increased mortality from heatwaves		
	High risk of undernutrition in communities dependent on dryland agriculture and livestock	Higher risks of undernutrition in communities dependent on dryland agriculture and livestock	Very high risks of undernutrition in communities dependent on dryland agriculture and livestock
Tropics	Increases in the number of hot days and hot nights as well as longer and more frequent heatwaves (*high confidence*)	The largest increase in hot days under 2°C compared to 1.5°C is projected for the tropics.	Oppressive temperatures and accumulated heatwave duration *very likely* to directly impact human health, mortality and productivity
	Risks to tropical crop yields in West Africa, Southeast Asia and Central and South America are significantly less than under 2°C of warming	Risks to tropical crop yields in West Africa, Southeast Asia and Central and South America could be extensive	Substantial reductions in crop yield *very likely*
Small islands	Land of 60,000 less people exposed by 2150 on SIDS compared to impacts under 2°C of global warming	Tens of thousands of people displaced owing to inundation of SIDS	Substantial and widespread impacts through inundation of SIDS, coastal flooding, freshwater stress, persistent heat stress and loss of most coral reefs (*very likely*)
	Risks for coastal flooding reduced by 20–80% for SIDS compared to 2°C of global warming	High risks for coastal flooding	
	Freshwater stress reduced by 25%	Freshwater stress reduced by 25% compared to 2°C of global warming	
		Freshwater stress from projected aridity	
	Increase in the number of warm days for SIDS in the tropics	Further increase of about 70 warm days per year	
	Persistent heat stress in cattle avoided	Persistent heat stress in cattle in SIDS	
	Loss of 70–90% of coral reefs	Loss of most coral reefs and weaker remaining structures owing to ocean acidification	
Fynbos biome	About 30% of suitable climate area lost (*medium confidence*)	Increased losses (about 45%) of suitable climate area (*medium confidence*)	Up to 80% of suitable climate area lost (*medium confidence*)

3

3.5.5 Avoiding Regional Tipping Points by Achieving More Ambitious Global Temperature Goals

Tipping points refer to critical thresholds in a system that, when exceeded, can lead to a significant change in the state of the system, often with an understanding that the change is irreversible. An understanding of the sensitivities of tipping points in the physical climate system, as well as in ecosystems and human systems, is essential for understanding the risks associated with different degrees of global warming. This subsection reviews tipping points across these three areas within the context of the different sensitivities to 1.5°C versus 2°C of global warming. Sensitivities to less ambitious global temperature goals are also briefly reviewed. Moreover, an analysis is provided of how integrated risks across physical, natural and human systems may accumulate to lead to the exceedance of thresholds for particular systems. The emphasis in this section is on the identification of regional tipping points and their sensitivity to 1.5°C and 2°C of global warming, whereas tipping points in the global climate system, referred to as large-scale singular events, were already discussed in Section 3.5.2. A summary of regional tipping points is provided in Table 3.7.

3.5.5.1 Arctic sea ice

Collins et al. (2013) discussed the loss of Arctic sea ice in the context of potential tipping points. Climate models have been used to assess whether a bifurcation exists that would lead to the irreversible loss of Arctic sea ice (Armour et al., 2011; Boucher et al., 2012; Ridley et al., 2012) and to test whether the summer sea ice extent can recover after it has been lost (Schröder and Connolley, 2007; Sedláček et al., 2011; Tietsche et al., 2011). These studies did not find evidence of bifurcation or indicate that sea ice returns within a few years of its loss, leading Collins et al. (2013) to conclude that there is little evidence for a tipping point in the transition from perennial to seasonal ice cover. No evidence has been found for irreversibility or tipping points, suggesting that year-round sea ice will return given a suitable climate (*medium confidence*) (Schröder and Connolley, 2007; Sedláček et al., 2011; Tietsche et al., 2011).

3.5.5.2 Tundra

Tree growth in tundra-dominated landscapes is strongly constrained by the number of days with mean air temperature above 0°C. A potential tipping point exists where the number of days below 0°C decreases to the extent that the tree fraction increases significantly. Tundra-dominated landscapes have warmed more than the global average over the last century (Settele et al., 2014), with associated increases in fires and permafrost degradation (Bring et al., 2016; DeBeer et al., 2016; Jiang et al., 2016; Yang et al., 2016). These processes facilitate conditions for woody species establishment in tundra areas, and for the eventual transition of the tundra to boreal forest. The number of investigations into how the tree fraction may respond in the Arctic to different degrees of global warming is limited, and studies generally indicate that substantial increases will *likely* occur gradually (e.g., Lenton et al., 2008). Abrupt changes are only plausible at levels of warming significantly higher than 2°C (*low confidence*) and would occur in conjunction with a collapse in permafrost (Drijfhout et al., 2015).

3.5.5.3 Permafrost

Widespread thawing of permafrost potentially makes a large carbon store (estimated to be twice the size of the atmospheric store; Dolman et al., 2010) vulnerable to decomposition, which could lead to further increases in atmospheric carbon dioxide and methane and hence to further global warming. This feedback loop between warming and the release of greenhouse gas from thawing tundra represents a potential tipping point. However, the carbon released to the atmosphere from thawing permafrost is projected to be restricted to 0.09–0.19 Gt C yr^{-1} at 2°C of global warming and to 0.08–0.16 Gt C yr^{-1} at 1.5°C (E.J. Burke et al., 2018), which does not indicate a tipping point (*medium confidence*). At higher degrees of global warming, in the order of 3°C, a different type of tipping point in permafrost may be reached. A single model projection (Drijfhout et al., 2015) suggested that higher temperatures may induce a smaller ice fraction in soils in the tundra, leading to more rapidly warming soils and a positive feedback mechanism that results in permafrost collapse (*low confidence*). The disparity between the multi-millennial time scales of soil carbon accumulation and potentially rapid decomposition in a warming climate implies that the loss of this carbon to the atmosphere would be essentially irreversible (Collins et al., 2013).

3.5.5.4 Asian monsoon

At a fundamental level, the pressure gradient between the Indian Ocean and Asian continent determines the strength of the Asian monsoon. As land masses warm faster than the oceans, a general strengthening of this gradient, and hence of monsoons, may be expected under global warming (e.g., Lenton et al., 2008). Additional factors such as changes in albedo induced by aerosols and snow-cover change may also affect temperature gradients and consequently pressure gradients and the strength of the monsoon. In fact, it has been estimated that an increase of the regional land mass albedo to 0.5 over India would represent a tipping point resulting in the collapse of the monsoon system (Lenton et al., 2008). The overall impacts of the various types of radiative forcing under different emissions scenarios are more subtle, with a weakening of the monsoon north of about 25°N in East Asia but a strengthening south of this latitude projected by Jiang and Tian (2013) under high and modest emissions scenarios. Increases in the intensity of monsoon precipitation are *likely* under low mitigation (AR5). Given that scenarios of 1.5°C or 2°C of global warming would include a substantially smaller radiative forcing than those assessed in the study by Jiang and Tian (2013), there is *low confidence* regarding changes in monsoons at these low global warming levels, as well as regarding the differences between responses at 1.5°C versus 2°C of warming.

3.5.5.5 West African monsoon and the Sahel

Earlier work has identified 3°C of global warming as the tipping point leading to a significant strengthening of the West African monsoon and subsequent wettening (and greening) of the Sahel and Sahara (Lenton et al., 2008). AR5 (Niang et al., 2014), as well as more recent research through the Coordinated Regional Downscaling Experiment for Africa (CORDEX–AFRICA), provides a more uncertain view, however, in terms of the rainfall futures of the Sahel under low mitigation futures. Even if a wetter Sahel should materialize under 3°C of global warming (*low*

3

confidence), it should be noted that there would be significant offsets in the form of strong regional warming and related adverse impacts on crop yield, livestock mortality and human health under such low mitigation futures (Engelbrecht et al., 2015; Sylla et al., 2016; Weber et al., 2018).

3.5.5.6   Rainforests

A large portion of rainfall over the world's largest rainforests is recirculated (e.g., Lenton et al., 2008), which raises the concern that deforestation may trigger a threshold in reduced forest cover, leading to pronounced forest dieback. For the Amazon, this deforestation threshold has been estimated to be 40% (Nobre et al., 2016). Global warming of 3°C–4°C may also, independent of deforestation, represent a tipping point that results in a significant dieback of the Amazon forest, with a key forcing mechanism being stronger El Niño events bringing more frequent droughts to the region (Nobre et al., 2016). Increased fire frequencies under global warming may interact with and accelerate deforestation, particularly during periods of El Niño-induced droughts (Lenton et al., 2008; Nobre et al., 2016). Global warming of 3°C is projected to reduce the extent of tropical rainforest in Central America, with biomass being reduced by about 40%, which can lead to a large replacement of rainforest by savanna and grassland (Lyra et al., 2017). Overall, modelling studies (Huntingford et al., 2013; Nobre et al., 2016) and observational constraints (Cox et al., 2013) suggest that pronounced rainforest dieback may only be triggered at 3°C–4°C (*medium confidence*), although pronounced biomass losses may occur at 1.5°C– 2°C of global warming.

3.5.5.7   Boreal forests

Boreal forests are likely to experience stronger local warming than the global average (WGII AR5; Collins et al., 2013). Increased disturbance from fire, pests and heat-related mortality may affect, in particular, the southern boundary of boreal forests (*medium confidence*) (Gauthier et al., 2015), with these impacts accruing with greater warming and thus impacts at 2°C would be expected to be greater than those at 1.5°C (*medium confidence*). A tipping point for significant dieback of the boreal forests is thought to exist, where increased tree mortality would result in the creation of large regions of open woodlands and grasslands, which would favour further regional warming and increased fire frequencies, thus inducing a powerful positive feedback mechanism (Lenton et al., 2008; Lenton, 2012). This tipping point has been estimated to exist between 3°C and 4°C of global warming (*low confidence*) (Lucht et al., 2006; Kriegler et al., 2009), but given the complexities of the various forcing mechanisms and feedback processes involved, this is thought to be an uncertain estimate.

3.5.5.8   Heatwaves, unprecedented heat and human health

Increases in ambient temperature are linearly related to hospitalizations and deaths once specific thresholds are exceeded (so there is not a tipping point per se). It is plausible that coping strategies will not be in place for many regions, with potentially significant impacts on communities with low adaptive capacity, effectively representing the occurrence of a local/regional tipping point. In fact, even if global warming is restricted to below 2°C, there could be a substantial increase in the occurrence of deadly heatwaves in cities if urban heat island effects are considered, with impacts being similar at 1.5°C and 2°C but substantially larger than under the present climate (Matthews et al., 2017). At 1.5°C of warming, twice as many megacities (such as Lagos, Nigeria, and Shanghai, China) than at present are *likely* to become heat stressed, potentially exposing more than 350 million more people to deadly heat stress by 2050. At 2°C of warming, Karachi (Pakistan) and Kolkata (India) could experience conditions equivalent to their deadly 2015 heatwaves on an annual basis (*medium confidence*). These statistics imply a tipping point in the extent and scale of heatwave impacts. However, these projections do not integrate adaptation to projected warming, for instance cooling that could be achieved with more reflective roofs and urban surfaces in general (Akbari et al., 2009; Oleson et al., 2010).

3.5.5.9   Agricultural systems: key staple crops

A large number of studies have consistently indicated that maize crop yield will be negatively affected under increased global warming, with negative impacts being higher at 2°C of warming than at 1.5°C (e.g., Niang et al., 2014; Schleussner et al., 2016b; J. Huang et al., 2017; Iizumi et al., 2017). Under 2°C of global warming, losses of 8–14% are projected in global maize production (Bassu et al., 2014). Under global warming of more than 2°C, regional losses are projected to be about 20% if they co-occur with reductions in rainfall (Lana et al., 2017). These changes may be classified as incremental rather than representing a tipping point. Large-scale reductions in maize crop yield, including the potential collapse of this crop in some regions, may exist under 3°C or more of global warming (*low confidence*) (e.g., Thornton et al., 2011).

3.5.5.10   Agricultural systems: livestock in the tropics and subtropics

The potential impacts of climate change on livestock (Section 3.4.6), in particular the direct impacts through increased heat stress, have been less well studied than impacts on crop yield, especially from the perspective of critical thresholds being exceeded. A case study from Jamaica revealed that the difference in heat stress for livestock between 1.5°C and 2°C of warming is likely to exceed the limits for normal thermoregulation and result in persistent heat stress for these animals (Lallo et al., 2018). It is plausible that this finding holds for livestock production in both tropical and subtropical regions more generally (*medium confidence*) (Section 3.4.6). Under 3°C of global warming, significant reductions in the areas suitable for livestock production could occur (*low confidence*), owing to strong increases in regional temperatures in the tropics and subtropics (*high confidence*). Thus, regional tipping points in the viability of livestock production may well exist, but little evidence quantifying such changes exists.

Table 3.7 | Summary of enhanced risks in the exceedance of regional tipping points under different global temperature goals.

Tipping point	Warming of 1.5°C or less	Warming of 1.5°C–2°C	Warming of up to 3°C
Arctic sea ice	Arctic summer sea ice is *likely* to be maintained Sea ice changes reversible under suitable climate restoration	The risk of an ice-free Arctic in summer is about 50% or higher Sea ice changes reversible under suitable climate restoration	Arctic is *very likely* to be ice free in summer Sea ice changes reversible under suitable climate restoration
Tundra	Decrease in number of growing degree days below 0°C Abrupt increases in tree cover are *unlikely*	Further decreases in number of growing degree days below 0°C Abrupt increased in tree cover are *unlikely*	Potential for an abrupt increase in tree fraction (*low confidence*)
Permafrost	17–44% reduction in permafrost Approximately 2 million km² more permafrost maintained than under 2°C of global warming (*medium confidence*) Irreversible loss of stored carbon	28–53% reduction in permafrost Irreversible loss of stored carbon	Potential for permafrost collapse (*low confidence*)
Asian monsoon	*Low confidence* in projected changes	*Low confidence* in projected changes	Increases in the intensity of monsoon precipitation *likely*
West African monsoon and the Sahel	Uncertain changes; *unlikely* that a tipping point is reached	Uncertain changes; *unlikely* that tipping point is reached	Strengthening of monsoon with wettening and greening of the Sahel and Sahara (*low confidence*) Negative associated impacts through increases in extreme temperature events
Rainforests	Reduced biomass, deforestation and fire increases pose uncertain risks to forest dieback	Larger biomass reductions than under 1.5°C of warming; deforestation and fire increases pose uncertain risk to forest dieback	Reduced extent of tropical rainforest in Central America and large replacement of rainforest by savanna and grassland Potential tipping point leading to pronounced forest dieback (*medium confidence*)
Boreal forests	Increased tree mortality at southern boundary of boreal forest (*medium confidence*)	Further increases in tree mortality at southern boundary of boreal forest (*medium confidence*)	Potential tipping point at 3°C–4°C for significant dieback of boreal forest (*low confidence*)
Heatwaves, unprecedented heat and human health	Substantial increase in occurrence of potentially deadly heatwaves (*likely*) More than 350 million more people exposed to deadly heat by 2050 under a midrange population growth scenario (*likely*)	Substantial increase in potentially deadly heatwaves (*likely*) Annual occurrence of heatwaves similar to the deadly 2015 heatwaves in India and Pakistan (*medium confidence*)	Substantial increase in potentially deadly heatwaves *very likely*
Agricultural systems: key staple crops	Global maize crop reductions of about 10%	Larger reductions in maize crop production than under 1.5°C of about 15%	Drastic reductions in maize crop globally and in Africa (*high confidence*) Potential tipping point for collapse of maize crop in some regions (*low confidence*)
Livestock in the tropics and subtropics	Increased heat stress	Onset of persistent heat stress (*medium confidence*)	Persistent heat stress *likely*

Box 3.6 | Economic Damages from Climate Change

Balancing the costs and benefits of mitigation is challenging because estimating the value of climate change damages depends on multiple parameters whose appropriate values have been debated for decades (for example, the appropriate value of the discount rate) or that are very difficult to quantify (for example, the value of non-market impacts; the economic effects of losses in ecosystem services; and the potential for adaptation, which is dependent on the rate and timing of climate change and on the socio-economic content). See Cross-Chapter Box 5 in Chapter 2 for the definition of the social cost of carbon and for a discussion of the economics of 1.5°C-consistent pathways and the social cost of carbon, including the impacts of inequality on the social cost of carbon.

Global economic damages of climate change are projected to be smaller under warming of 1.5°C than 2°C in 2100 (Warren et al., 2018c). The mean net present value of the costs of damages from warming in 2100 for 1.5°C and. 2°C (including costs associated with climate change-induced market and non-market impacts, impacts due to sea level rise, and impacts associated with large-scale discontinuities) are $54 and $69 trillion, respectively, relative to 1961–1990.

Box 3.6 (continued)

Values of the social cost of carbon vary when tipping points are included. The social cost of carbon in the default setting of the Dynamic Integrated Climate-Economy (DICE) model increases from $15 tCO_2^{-1} to $116 (range 50–166) tCO_2^{-1} when large-scale singularities or 'tipping elements' are incorporated (Y. Cai et al., 2016; Lemoine and Traeger, 2016). Lemoine and Traeger (2016) included optimization calculations that minimize welfare impacts resulting from the combination of climate change risks and climate change mitigation costs, showing that welfare is minimized if warming is limited to 1.5°C. These calculations excluded the large health co-benefits that accrue when greenhouse gas emissions are reduced (Section 3.4.7.1; Shindell et al., 2018).

The economic damages of climate change in the USA are projected to be large (Hsiang et al., 2017; Yohe, 2017). Hsiang et al. (2017) shows that the USA stand to lose -0.1 to 1.7% of the Gross Domestic Product (GDP) at 1.5°C warming. Yohe (2017) calculated transient temperature trajectories from a linear relationship with contemporaneous cumulative emissions under a median no-policy baseline trajectory that brings global emissions to roughly 93 $GtCO_2$ yr^{-1} by the end of the century (Fawcett et al., 2015), with 1.75°C per 1000 $GtCO_2$ as the median estimate. Associated aggregate economic damages in decadal increments through the year 2100 are estimated in terms of the percentage loss of GDP at the median, 5th percentile and 95th percentile transient temperature (Hsiang et al., 2017). The results for the baseline no-policy case indicate that economic damages along median temperature change and median damages (median-median) reach 4.5% of GDP by 2100, with an uncertainty range of 2.5% and 8.5% resulting from different combinations of temperature change and damages. Avoided damages from achieving a 1.5°C temperature limit along the median-median case are nearly 4% (range 2–7%) by 2100. Avoided damages from achieving a 2°C temperature limit are only 3.5% (range 1.8–6.5%). Avoided damages from achieving 1.5°C versus 2°C are modest at about 0.35% (range 0.20–0.65%) by 2100. The values of achieving the two temperature limits do not diverge significantly until 2040, when their difference tracks between 0.05 and 0.13%; the differences between the two temperature targets begin to diverge substantially in the second half of the century.

3.6   Implications of Different 1.5°C and 2°C Pathways

This section provides an overview on specific aspects of the mitigation pathways considered compatible with 1.5°C of global warming. Some of these aspects are also addressed in more detail in Cross-Chapter Boxes 7 and 8 in this chapter.

3.6.1   Gradual versus Overshoot in 1.5°C Scenarios

All 1.5°C scenarios from Chapter 2 include some overshoot above 1.5°C of global warming during the 21st century (Chapter 2 and Cross-Chapter Box 8 in this chapter). The level of overshoot may also depend on natural climate variability. An overview of possible outcomes of 1.5°C-consistent mitigation scenarios for changes in the physical climate at the time of overshoot and by 2100 is provided in Cross-Chapter Box 8 on '1.5°C warmer worlds'. Cross-Chapter Box 8 also highlights the implications of overshoots.

3.6.2   Non-CO$_2$ Implications and Projected Risks of Mitigation Pathways

3.6.2.1   Risks arising from land-use changes in mitigation pathways

In mitigation pathways, land-use change is affected by many different mitigation options. First, mitigation of non-CO$_2$ emissions from agricultural production can shift agricultural production between regions via trade of agricultural commodities. Second, protection of carbon-rich ecosystems such as tropical forests constrains the area for agricultural expansion. Third, demand-side mitigation measures,

such as less consumption of resource-intensive commodities (animal products) or reductions in food waste, reduce pressure on land (Popp et al., 2017; Rogelj et al., 2018). Finally, carbon dioxide removal (CDR) is a key component of most, but not all, mitigation pathways presented in the literature to date which constrain warming to 1.5°C or 2°C. Carbon dioxide removal measures that require land include bioenergy with carbon capture and storage (BECCS), afforestation and reforestation (AR), soil carbon sequestration, direct air capture, biochar and enhanced weathering (see Cross-Chapter Box 7 in this chapter). These potential methods are assessed in Section 4.3.7.

In cost-effective integrated assessment modelling (IAM) pathways recently developed to be consistent with limiting warming to 1.5°C, use of CDR in the form of BECCS and AR are fundamental elements (Chapter 2; Popp et al., 2017; Hirsch et al., 2018; Rogelj et al., 2018; Seneviratne et al., 2018c). The land-use footprint of CDR deployment in 1.5°C-consistent pathways can be substantial (Section 2.3.4, Figure 2.11), even though IAMs predominantly rely on second-generation biomass and assume future productivity increases in agriculture.

A body of literature has explored potential consequences of large-scale use of CDR. In this case, the corresponding land footprint by the end of the century could be extremely large, with estimates including: up to 18% of the land surface being used (Wiltshire and Davies-Barnard, 2015); vast acceleration of the loss of primary forest and natural grassland (Williamson, 2016) leading to increased greenhouse gas emissions (P. Smith et al., 2013, 2015); and potential loss of up to 10% of the current forested lands to biofuels (Yamagata et al., 2018). Other estimates reach 380–700 Mha or 21–64% of current arable cropland (Section 4.3.7). Boysen et al. (2017) found that in a scenario in which emissions reductions were sufficient only to limit warming to 2.5°C, use of CDR to further limit warming to 1.7°C would result in the conversion of 1.1–1.5 Gha of land – implying enormous losses of both cropland and natural ecosystems. Newbold et al. (2015) found that biodiversity loss in the Representative Concentration Pathway (RCP)2.6 scenario could be greater than that in RCP4.5 and RCP6, in which there is more climate change but less land-use change. Risks to biodiversity conservation and agricultural production are therefore projected to result from large-scale bioenergy deployment pathways (P. Smith et al., 2013; Tavoni and Socolow, 2013). One study explored an extreme mitigation strategy encouraging biofuel expansion sufficient to limit warming to 1.5°C and found that this would be more disruptive to land use and crop prices than the impacts of a 2°C warmer world which has a larger climate signal and lower mitigation requirement (Ruane et al., 2018). However, it should again be emphasized that many of the pathways explored in Chapter 2 of this report follow strategies that explore how to reduce these issues. Chapter 4 provides an assessment of the land footprint of various CDR technologies (Section 4.3.7).

The degree to which BECCS has these large land-use footprints depends on the source of the bioenergy used and the scale at which BECCS is deployed. Whether there is competition with food production and biodiversity depends on the governance of land use, agricultural intensification, trade, demand for food (in particular meat), feed and timber, and the context of the whole supply chain (Section 4.3.7, Fajardy and Mac Dowell, 2017; Booth, 2018; Sterman et al., 2018).

The more recent literature reviewed in Chapter 2 explores pathways which limit warming to 2°C or below and achieve a balance between sources and sinks of CO_2 by using BECCS that relies on second-generation (or even third-generation) biofuels, changes in diet or more generally, management of food demand, or CDR options such as forest restoration (Chapter 2; Bajželj et al., 2014). Overall, this literature explores how to reduce the issues of competition for land with food production and with natural ecosystems (in particular forests) (Cross-Chapter Box 1 in Chapter 1; van Vuuren et al., 2009; Haberl et al., 2010, 2013; Bajželj et al., 2014; Daioglou et al., 2016; Fajardy and Mac Dowell, 2017).

Some IAMs manage this transition by effectively protecting carbon stored on land and focusing on the conversion of pasture area into both forest area and bioenergy cropland. Some IAMs explore 1.5°C-consistent pathways with demand-side measures such as dietary changes and efficiency gains such as agricultural changes (Sections 2.3.4 and 2.4.4), which lead to a greatly reduced CDR deployment and consequently land-use impacts (van Vuuren et al., 2018). In reality, however, whether this CDR (and bioenergy in general) has large adverse impacts on environmental and societal goals depends in large part on the governance of land use (Section 2.3.4; Obersteiner et al., 2016; Bertram et al., 2018; Humpenöder et al., 2018).

Rates of sequestration of 3.3 GtC ha^{-1} require 970 Mha of afforestation and reforestation (Smith et al., 2015). Humpenöder et al. (2014) estimated that in least-cost pathways afforestation would cover 2800 Mha by the end of the century to constrain warming to 2°C. Hence, the amount of land considered if least-cost mitigation is implemented by afforestation and reforestation could be up to three to five times greater than that required by BECCS, depending on the forest

management used. However, not all of the land footprint of CDR is necessarily to be in competition with biodiversity protection. Where reforestation is the restoration of natural ecosystems, it benefits both carbon sequestration and conservation of biodiversity and ecosystem services (Section 4.3.7) and can contribute to the achievement of the Aichi targets under the Convention on Biological Diversity (CBD) (Leadley et al., 2016). However, reforestation is often not defined in this way (Section 4.3.8; Stanturf et al., 2014) and the ability to deliver biodiversity benefits is strongly dependent on the precise nature of the reforestation, which has different interpretations in different contexts and can often include agroforestry rather than restoration of pristine ecosystems (Pistorious and Kiff, 2017). However, 'natural climate solutions', defined as conservation, restoration, and improved land management actions that increase carbon storage and/or avoid greenhouse gas emissions across global forests, wetlands, grasslands and agricultural lands, are estimated to have the potential to provide 37% of the cost-effective CO_2 mitigation needed by southern Europe and the Mediterranean by 2030 – in order to have a >66% chance of holding warming to below 2°C (Griscom et al., 2017).

Any reductions in agricultural production driven by climate change and/or land management decisions related to CDR may (e.g., Nelson et al., 2014a; Dalin and Rodríguez-Iturbe, 2016) or may not (Muratori et al., 2016) affect food prices. However, these studies did not consider the deployment of second-generation (instead of first-generation) bioenergy crops, for which the land footprint can be much smaller.

Irrespective of any mitigation-related issues, in order for ecosystems to adapt to climate change, land use would also need to be carefully managed to allow biodiversity to disperse to areas that become newly climatically suitable for it (Section 3.4.1) and to protect the areas where the future climate will still remain suitable. This implies a need for considerable expansion of the protected area network (Warren et al., 2018b), either to protect existing natural habitat or to restore it (perhaps through reforestation, see above). At the same time, adaptation to climate change in the agricultural sector (Rippke et al., 2016) can require transformational as well as new approaches to land-use management; in order to meet the rising food demand of a growing human population, it is projected that additional land will need to be brought into production unless there are large increases in agricultural productivity (Tilman et al., 2011). However, future rates of deforestation may be underestimated in the existing literature (Mahowald et al., 2017a), and reforestation may therefore be associated with significant co-benefits if implemented to restore natural ecosystems (*high confidence*).

3.6.2.2   Biophysical feedbacks on regional climate associated with land-use changes

Changes in the biophysical characteristics of the land surface are known to have an impact on local and regional climates through changes in albedo, roughness, evapotranspiration and phenology, which can lead to a change in temperature and precipitation. This includes changes in land use through agricultural expansion/intensification (e.g., Mueller et al., 2016), reforestation/revegetation endeavours (e.g., Feng et al., 2016; Sonntag et al., 2016; Bright et al., 2017) and changes in land management (e.g., Luyssaert et al., 2014; Hirsch et al., 2017) that can

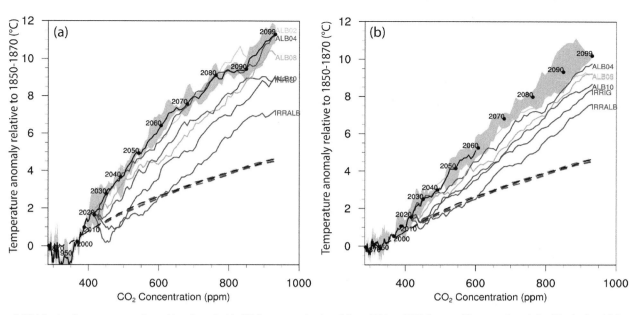

Figure 3.22 | Regional temperature scaling with carbon dioxide (CO₂) concentration (ppm) from 1850 to 2099 for two different regions defined in the Special Report on Managing the Risks of Extreme Events and Disasters to Advance Climate Change Adaptation (SREX) for central Europe (CEU) (a) and central North America (CNA) (b). Solid lines correspond to the regional average annual maximum daytime temperature (TXx) anomaly, and dashed lines correspond to the global mean temperature anomaly, where all temperature anomalies are relative to 1850–1870 and units are degrees Celsius. The black line in all panels denotes the three-member control ensemble mean, with the grey shaded regions corresponding to the ensemble range. The coloured lines represent the three-member ensemble means of the experiments corresponding to albedo +0.02 (cyan), albedo +0.04 (purple), albedo + 0.08 (orange), albedo +0.10 (red), irrigation (blue), and irrigation with albedo +0.10 (green). Adapted from Hirsch et al. (2017).

involve double cropping (e.g., Jeong et al., 2014; Mueller et al., 2015; Seifert and Lobell, 2015), irrigation (e.g., Lobell et al., 2009; Sacks et al., 2009; Cook et al., 2011; Qian et al., 2013; de Vrese et al., 2016; Pryor et al., 2016; Thiery et al., 2017), no-till farming and conservation agriculture (e.g., Lobell et al., 2006; Davin et al., 2014), and wood harvesting (e.g., Lawrence et al., 2012). Hence, the biophysical impacts of land-use changes are an important topic to assess in the context of low-emissions scenarios (e.g., van Vuuren et al., 2011b), in particular for 1.5°C warming levels (see also Cross-Chapter Box 7 in this chapter).

The magnitude of the biophysical impacts is potentially large for temperature extremes. Indeed, changes induced both by modifications in moisture availability and irrigation and by changes in surface albedo tend to be larger (i.e., stronger cooling) for hot extremes than for mean temperatures (e.g., Seneviratne et al., 2013; Davin et al., 2014; Wilhelm et al., 2015; Hirsch et al., 2017; Thiery et al., 2017). The reasons for reduced moisture availability are related to a strong contribution of moisture deficits to the occurrence of hot extremes in mid-latitude regions (Mueller and Seneviratne, 2012; Seneviratne et al., 2013). In the case of surface albedo, cooling associated with higher albedo (e.g., in the case of no-till farming) is more effective at cooling hot days because of the higher incoming solar radiation for these days (Davin et al., 2014). The overall effect of either irrigation or albedo has been found to be at the most in the order of about 1°C–2°C regionally for temperature extremes. This can be particularly important in the context of low-emissions scenarios because the overall effect is in this case of similar magnitude to the response to the greenhouse gas forcing (Figure 3.22; Hirsch et al., 2017; Seneviratne et al., 2018a,c).

In addition to the biophysical feedbacks from land-use change and land management on climate, there are potential consequences for particular

ecosystem services. This includes climate change-induced changes in crop yield (e.g., Schlenker and Roberts, 2009; van der Velde et al., 2012; Asseng et al., 2013, 2015; Butler and Huybers, 2013; Lobell et al., 2014) which may be further exacerbated by competing demands for arable land between reforestation mitigation activities, crop growth for BECCS (Chapter 2), increasing food production to support larger populations, and urban expansion (see review by Smith et al., 2010). In particular, some land management practices may have further implications for food security, for instance throughincreases or decreases in yield when tillage is ceased in some regions (Pittelkow et al., 2014).

We note that the biophysical impacts of land use in the context of mitigation pathways constitute an emerging research topic. This topic, as well as the overall role of land-use change in climate change projections and socio-economic pathways, will be addressed in depth in the upcoming IPCC Special Report on Climate Change and Land Use due in 2019.

3.6.2.3   Atmospheric compounds (aerosols and methane)

There are multiple pathways that could be used to limit anthropogenic climate change, and the details of the pathways will influence the impacts of climate change on humans and ecosystems. Anthropogenic-driven changes in aerosols cause important modifications to the global climate (Bindoff et al., 2013a; Boucher et al., 2013b; P. Wu et al., 2013; Sarojini et al., 2016; H. Wang et al., 2016). Enforcement of strict air quality policies may lead to a large decrease in cooling aerosol emissions in the next few decades. These aerosol emission reductions may cause a warming comparable to that resulting from the increase in greenhouse gases by mid-21st century under low CO₂ pathways (Kloster et al., 2009; Acosta Navarro et al., 2017). Further background is provided in Sections 2.2.2 and 2.3.1; Cross Chapter Box 1 in Chapter 1). Because aerosol effects on the energy budget are regional, strong regional changes in precipitation from aerosols may occur if aerosol emissions are reduced for air quality reasons or as a co-benefit from switches to sustainable energy sources (H. Wang et al., 2016). Thus, regional impacts, especially on precipitation, are very sensitive to 1.5°C-consistent pathways (Z. Wang et al., 2017).

Pathways which rely heavily on reductions in methane (CH_4) instead of CO_2 will reduce warming in the short term because CH_4 is such a stronger and shorter-lived greenhouse gas than CO_2, but will lead to stronger warming in the long term because of the much longer residence time of CO_2 (Myhre et al., 2013; Pierrehumbert, 2014). In addition, the dominant loss mechanism for CH_4 is atmospheric photo-oxidation. This conversion modifies ozone formation and destruction in the troposphere and stratosphere, therefore modifying the contribution of ozone to radiative forcing, as well as feedbacks on the oxidation rate of methane itself (Myhre et al., 2013). Focusing on pathways and policies which both improve air quality and reduce impacts of climate change can provide multiple co-benefits (Shindell et al., 2017). These pathways are discussed in detail in Sections 4.3.7 and 5.4.1 and in Cross-Chapter Box 12 in Chapter 5.

Atmospheric aerosols and gases can also modify the land and ocean uptake of anthropogenic CO_2; some compounds enhance uptake while others reduce it (Section 2.6.2; Ciais et al., 2013). While CO_2 emissions tend to encourage greater uptake of carbon by the land and the ocean (Ciais et al., 2013), CH_4 emissions can enhance ozone pollution, depending on nitrogen oxides, volatile organic compounds and other organic species concentrations, and ozone pollution tends to reduce land productivity (Myhre et al., 2013; B. Wang et al., 2017). Aside from inhibiting land vegetation productivity, ozone may also alter the CO_2, CH_4 and nitrogen (N_2O) exchange at the land–atmosphere interface and transform the global soil system from a sink to a source of carbon (B. Wang et al., 2017). Aerosols and associated nitrogen-based compounds tend to enhance the uptake of CO_2 in land and ocean systems through deposition of nutrients and modification of climate (Ciais et al., 2013; Mahowald et al., 2017b).

Cross-Chapter Box 7 | Land-Based Carbon Dioxide Removal in Relation to 1.5°C of Global Warming

Lead Authors:
Rachel Warren (United Kingdom), Marcos Buckeridge (Brazil), Sabine Fuss (Germany), Markku Kanninen (Finland), Joeri Rogelj (Austria/Belgium), Sonia I. Seneviratne (Switzerland), Raphael Slade (United Kingdom)

Climate and land form a complex system characterized by multiple feedback processes and the potential for non-linear responses to perturbation. Climate determines land cover and the distribution of vegetation, affecting above- and below-ground carbon stocks. At the same time, land cover influences global climate through altered biogeochemical processes (e.g., atmospheric composition and nutrient flow into oceans), and regional climate through changing biogeophysical processes including albedo, hydrology, transpiration and vegetation structure (Forseth, 2010).

Greenhouse gas (GHG) fluxes related to land use are reported in the 'agriculture, forestry and other land use' sector (AFOLU) and comprise about 25% (about 10–12 $GtCO_2eq$ yr^{-1}) of anthropogenic GHG emissions (P. Smith et al., 2014). Reducing emissions from land use, as well as land-use change, are thus an important component of low-emissions mitigation pathways (Clarke et al., 2014), particularly as land-use emissions can be influenced by human actions such as deforestation, afforestation, fertilization, irrigation, harvesting, and other aspects of cropland, grazing land and livestock management (Paustian et al., 2006; Griscom et al., 2017; Houghton and Nassikas, 2018).

In the IPCC Fifth Assessment Report, the vast majority of scenarios assessed with a 66% or better chance of limiting global warming to 2°C by 2100 included carbon dioxide removal (CDR) – typically about 10 $GtCO_2$ yr^{-1} in 2100 or about 200–400 $GtCO_2$ over the course of the century (Smith et al., 2015; van Vuuren et al., 2016). These integrated assessment model (IAM) results were predominately achieved by using bioenergy with carbon capture and storage (BECCS) and/or afforestation and reforestation (AR). Virtually all scenarios that limit either peak or end-of-century warming to 1.5°C also use land-intensive CDR technologies (Rogelj et al., 2015; Holz et al., 2017; Kriegler et al., 2017; Fuss et al., 2018; van Vuuren et al., 2018). Again, AR (Sections 2.3 and 4.3.7) and BECCS (Sections 4.3.2. and 4.3.7) predominate. Other CDR options, such as the application of biochar to soil, soil carbon sequestration, and enhanced weathering (Section 4.3.7) are not yet widely incorporated into IAMs, but their deployment would also necessitate the use of land and/or changes in land management.

Integrated assessment models provide a simplified representation of land use and, with only a few exceptions, do not include biophysical feedback processes (e.g., albedo and evapotranspiration effects) (Kreidenweis et al., 2016) despite the importance of these processes for regional climate, in particular hot extremes (Section 3.6.2.2; Seneviratne et al., 2018c). The extent, location and impacts of large-scale land-use change described by existing IAMs can also be widely divergent, depending on model structure, scenario parameters, modelling objectives and assumptions (including regarding land availability and productivity) (Prestele et

Cross-Chapter Box 7 (continued)

al., 2016; Alexander et al., 2017; Popp et al., 2017; Seneviratne et al., 2018c). Despite these limitations, IAM scenarios effectively highlight the extent and nature of potential land-use transitions implicit in limiting warming to 1.5°C.

Cross-Chapter Box 7 Table 1 presents a comparison of the five CDR options assessed in this report. This illustrates that if BECCS and AR were to be deployed at a scale of 12 GtCO$_2$ yr^{-1} in 2100, for example, they would have a substantial land and water footprint. Whether this footprint would result in adverse impacts, for example on biodiversity or food production, depends on the existence and effectiveness of measures to conserve land carbon stocks, limit the expansion of agriculture at the expense of natural ecosystems, and increase agriculture productivity (Bonsch et al., 2016; Obersteiner et al., 2016; Bertram et al., 2018; Humpenöder et al., 2018). In comparison, the land and water footprints of enhanced weathering, soil carbon sequestration and biochar application are expected to be far less per GtCO$_2$ sequestered. These options may offer potential co-benefits by providing an additional source of nutrients or by reducing N$_2$O emissions, but they are also associated with potential side effects. Enhanced weathering would require massive mining activity, and providing feedstock for biochar would require additional land, even though a proportion of the required biomass is expected to come from residues (Woolf et al., 2010; Smith, 2016). For the terrestrial CDR options, permanence and saturation are important considerations, making their viability and long-term contributions to carbon reduction targets uncertain.

The technical, political and social feasibility of scaling up and implementing land-intensive CDR technologies (Cross-Chapter Box 3 in Chapter 1) is recognized to present considerable potential barriers to future deployment (Boucher et al., 2013a; Fuss et al., 2014, 2018; Anderson and Peters, 2016; Vaughan and Gough, 2016; Williamson, 2016; Minx et al., 2017, 2018; Nemet et al., 2018; Strefler et al., 2018; Vaughan et al., 2018). To investigate the implications of restricting CDR options should these barriers prove difficult to overcome, IAM studies (Section 2.3.4) have developed scenarios that limit – either implicitly or explicitly – the use of BECCS and bioenergy (Krey et al., 2014; Bauer et al., 2018; Rogelj et al., 2018) or the use of BECCS and afforestation (Strefler et al., 2018). Alternative strategies to limit future reliance on CDR have also been examined, including increased electrification, agricultural intensification, behavioural change, and dramatic improvements in energy and material efficiency (Bauer et al., 2018; Grubler et al., 2018; van Vuuren et al., 2018). Somewhat counterintuitively, scenarios that seek to limit the deployment of BECCs may result in increased land use, through greater deployment of bioenergy, and afforestation (Chapter 2, Box 2.1; Krey et al., 2014; Krause et al., 2017; Bauer et al., 2018; Rogelj et al., 2018). Scenarios aiming to minimize the total human land footprint (including land for food, energy and climate mitigation) also result in land-use change, for example by increasing agricultural efficiency and dietary change (Grubler et al., 2018).

The impacts of changing land use are highly context, location and scale dependent (Robledo-Abad et al., 2017). The supply of biomass for CDR (e.g., energy crops) has received particular attention. The literature identifies regional examples of where the use of land to produce biofuels might be sustainably increased (Jaiswal et al., 2017), where biomass markets could contribute to the provision of ecosystem services (Dale et al., 2017), and where bioenergy could increase the resilience of production systems and contribute to rural development (Kline et al., 2017). However, studies of global biomass potential provide only limited insight into the local feasibility of supplying large quantities of biomass on a global scale (Slade et al., 2014). Concerns about large-scale use of biomass for CDR include a range of potential consequences including greatly increased demand for freshwater use, increased competition for land, loss of biodiversity and/or impacts on food security (Section 3.6.2.1; Heck et al., 2018). The short- versus long-term carbon impacts of substituting biomass for fossil fuels, which are largely determined by feedstock choice, also remain a source of contention (Schulze et al., 2012; Jonker et al., 2014; Booth, 2018; Sterman et al., 2018).

Afforestation and reforestation can also present trade-offs between biodiversity, carbon sequestration and water use, and these strategies have a higher land footprint per tonne of CO$_2$ removed (Cunningham, 2015; Naudts et al., 2016; Smith et al., 2018). For example, changing forest management to strategies favouring faster growing species, greater residue extraction and shorter rotations may have a negative impact on biodiversity (de Jong et al., 2014). In contrast, reforestation of degraded land with native trees can have substantial benefits for biodiversity (Section 3.6). Despite these constraints, the potential for increased carbon sequestration through improved land stewardship measures is considered to be substantial (Griscom et al., 2017).

Evaluating the synergies and trade-offs between mitigation and adaptation actions, resulting land and climate impacts, and the myriad issues related to land-use governance will be essential to better understand the future role of CDR technologies. This topic will be addressed further in the IPCC Special Report on Climate Change and Land (SRCCL) due to be published in 2019.

Cross-Chapter Box 7 (continued next page)

Cross-Chapter Box 7 (continued)

Key messages:

Cost-effective strategies to limit peak or end-of-century warming to 1.5°C all include enhanced GHG removals in the AFOLU sector as part of their portfolio of measures (*high confidence*).

Large-scale deployment of land-based CDR would have far-reaching implications for land and water availability (*high confidence*). This may impact food production, biodiversity and the provision of other ecosystem services (*high confidence*).

The impacts of deploying land-based CDR at large scales can be reduced if a wider portfolio of CDR options is deployed, and if increased mitigation effort focuses on strongly limiting demand for land, energy and material resources, including through lifestyle and dietary changes (*medium confidence*).

Afforestation and reforestation may be associated with significant co-benefits if implemented appropriately, but they feature large land and water footprints if deployed at large scales (*medium confidence*).

Cross-Chapter Box 7, Table 1 | Comparison of land-based carbon removal options.
Sources: ª assessed ranges by Fuss et al. (2018), see Figures in Section 4.3.7 for full literature range; ᵇ based on the 2100 estimate for mean potentials by Smith et al. (2015). Note that biophysical impacts of land-based CDR options besides albedo changes (e.g., through changes in evapotranspiration related to irrigation or land cover/use type) are not displayed.

Option	Potentials [a]	Cost [a]	Required land [b]	Required water [b]	Impact on nutrients [b]	Impact on albedo [b]	Saturation and permanence [a]
	$GtCO_2\, y^{-1}$	$\$\, tCO_2^{-1}$	$Mha\, GtCO_2^{-1}$	$km^3\, GtCO_2^{-1}$	$Mt\, N, P, K\, y^{-1}$	*No units*	*No units*
BECCS	0.5–5	100–200	31–58	60	Variable	Variable; depends on source of biofuel (higher albedo for crops than for forests) and on land management (e.g., no-till farming for crops)	Long-term governance of storage; limits on rates of bioenergy production and carbon sequestration
Afforestation & reforestation	0.5–3.6	5–50	80	92	0.5	Negative, or reduced GHG benefit where not negative	Saturation of forests; vulnerable to disturbance; post-AR forest management essential
Enhanced weathering	2–4	50–200	3	0.4	0	0	Saturation of soil; residence time from months to geological timescale
Biochar	0.3–2	30–120	16–100	0	N: 8.2, P: 2.7, K: 19.1	0.08–0.12	Mean residence times between decades to centuries, depending on soil type, management and environmental conditions
Soil carbon sequestration	2.3–5	0–100	0	0	N: 21.8, P: 5.5, K: 4.1	0	Soil sinks saturate and can reverse if poor management practices resume

3.6.3   Implications Beyond the End of the Century

3.6.3.1   Sea ice

Sea ice is often cited as a tipping point in the climate system (Lenton, 2012). Detailed modelling of sea ice (Schröder and Connolley, 2007; Sedláček et al., 2011; Tietsche et al., 2011), however, suggests that summer sea ice can return within a few years after its artificial removal for climates in the late 20th and early 21st centuries. Further studies (Armour et al., 2011; Boucher et al., 2012; Ridley et al., 2012) modelled the removal of sea ice by raising CO_2 concentrations and studied subsequent regrowth by lowering CO_2. These studies suggest that changes in Arctic sea ice are neither irreversible nor exhibit bifurcation behaviour. It is therefore plausible that the extent of Arctic sea ice may quickly re-equilibrate to the end-of-century climate under an overshoot scenario.

3.6.3.2 Sea level

Policy decisions related to anthropogenic climate change will have a profound impact on sea level, not only for the remainder of this century but for many millennia to come (Clark et al., 2016). On these long time scales, 50 m of sea level rise (SLR) is possible (Clark et al., 2016). While it is *virtually certain* that sea level will continue to rise well beyond 2100, the amount of rise depends on future cumulative emissions (Church et al., 2013) as well as their profile over time (Bouttes et al., 2013; Mengel et al., 2018). Marzeion et al. (2018) found that 28–44% of present-day glacier volume is unsustainable in the present-day climate and that it would eventually melt over the course of a few centuries, even if there were no further climate change. Some components of SLR, such as thermal expansion, are only considered reversible on centennial time scales (Bouttes et al., 2013; Zickfeld et al., 2013), while the contribution from ice sheets may not be reversible under any plausible future scenario (see below).

Based on the sensitivities summarized by Levermann et al. (2013), the contributions of thermal expansion (0.20–0.63 m °C^{-1}) and glaciers (0.21 m °C^{-1} but falling at higher degrees of warming mostly because of the depletion of glacier mass, with a possible total loss of about 0.6 m) amount to 0.5–1.2 m and 0.6–1.7 m in 1.5°C and 2°C warmer worlds, respectively. The bulk of SLR on greater than centennial time scales will therefore be caused by contributions from the continental ice sheets of Greenland and Antarctica, whose existence is threatened on multi-millennial time scales.

For Greenland, where melting from the ice sheet's surface is important, a well-documented instability exists where the surface of a thinning ice sheet encounters progressively warmer air temperatures that further promote melting and thinning. A useful indicator associated with this instability is the threshold at which annual mass loss from the ice sheet by surface melt exceeds mass gain by snowfall. Previous estimates put this threshold at about 1.9°C to 5.1°C above pre-industrial temperatures (Gregory and Huybrechts, 2006). More recent analyses, however, suggest that this threshold sits between 0.8°C and 3.2°C, with a best estimate at 1.6°C (Robinson et al., 2012). The continued decline of the ice sheet after this threshold has been passed is highly dependent on the future climate and varies between about 80% loss after 10,000 years to complete loss after as little as 2000 years (contributing about 6 m to SLR). Church et al. (2013) were unable to quantify a *likely* range for this threshold. They assigned *medium confidence* to a range greater than 2°C but less than 4°C, and had *low confidence* in a threshold of about 1°C. There is insufficient new literature to change this assessment.

The Antarctic ice sheet, in contrast, loses the mass gained by snowfall as outflow and subsequent melt to the ocean, either directly from the underside of floating ice shelves or indirectly by the melting of calved icebergs. The long-term existence of this ice sheet will also be affected by a potential instability (the marine ice sheet instability, MISI), which links outflow (or mass loss) from the ice sheet to water depth at the grounding line (i.e., the point at which grounded ice starts to float and becomes an ice shelf) so that retreat into deeper water (the bedrock underlying much of Antarctica slopes downwards towards the centre of the ice sheet) leads to further increases in outflow and promotes

yet further retreat (Schoof, 2007). More recently, a variant on this mechanism was postulated in which an ice cliff forms at the grounding line and retreats rapidly though fracture and iceberg calving (DeConto and Pollard, 2016). There is a growing body of evidence (Golledge et al., 2015; DeConto and Pollard, 2016) that large-scale retreat may be avoided in emissions scenarios such as Representative Concentration Pathway (RCP)2.6 but that higher-emissions RCP scenarios could lead to the loss of the West Antarctic ice sheet and sectors in East Antarctica, although the duration (centuries or millennia) and amount of mass loss during such a collapse is highly dependent on model details and no consensus exists yet. Schoof (2007) suggested that retreat may be irreversible, although a rigorous test has yet to be made. In this context, overshoot scenarios, especially of higher magnitude or longer duration, could increase the risk of such irreversible retreat.

Church et al. (2013) noted that the collapse of marine sectors of the Antarctic ice sheet could lead to a global mean sea level (GMSL) rise above the likely range, and that there was *medium confidence* that this additional contribution 'would not exceed several tenths of a metre during the 21st century'.

The multi-centennial evolution of the Antarctic ice sheet has been considered in papers by DeConto and Pollard (2016) and Golledge et al. (2015). Both suggest that RCP2.6 is the only RCP scenario leading to long-term contributions to GMSL of less than 1.0 m. The long-term committed future of Antarctica and the GMSL contribution at 2100 are complex and require further detailed process-based modelling; however, a threshold in this contribution may be located close to 1.5°C to 2°C of global warming.

In summary, there is *medium confidence* that a threshold in the long-term GMSL contribution of both the Greenland and Antarctic ice sheets lies around 1.5°C to 2°C of global warming relative to pre-industrial; however, the GMSL associated with these two levels of global warming cannot be differentiated on the basis of the existing literature.

3.6.3.3 Permafrost

The slow rate of permafrost thaw introduces a lag between the transient degradation of near-surface permafrost and contemporary climate, so that the equilibrium response is expected to be 25–38% greater than the transient response simulated in climate models (Slater and Lawrence, 2013). The long-term, equilibrium Arctic permafrost loss to global warming was analysed by Chadburn et al. (2017). They used an empirical relation between recent mean annual air temperatures and the area underlain by permafrost coupled to Coupled Model Intercomparison Project Phase 5 (CMIP5) stabilization projections to 2300 for RCP2.6 and RCP4.5. Their estimate of the sensitivity of permafrost to warming is 2.9–5.0 million km^2 °C^{-1} (1 standard deviation confidence interval), which suggests that stabilizing climate at 1.5°C as opposed to 2°C would reduce the area of eventual permafrost loss by 1.5 to 2.5 million km^2 (stabilizing at 56–83% as opposed to 43–72% of 1960–1990 levels). This work, combined with the assessment of Collins et al. (2013) on the link between global warming and permafrost loss, leads to the assessment that permafrost extent would be appreciably greater in a 1.5°C warmer world compared to in a 2°C warmer world (*low to medium confidence*).

3.7   Knowledge Gaps

Most scientific literature specific to global warming of 1.5°C is only just emerging. This has led to differences in the amount of information available and gaps across the various sections of this chapter. In general, the number of impact studies that specifically focused on 1.5°C lags behind climate-change projections in general, due in part to the dependence of the former on the latter. There are also insufficient studies focusing on regional changes, impacts and consequences at 1.5°C and 2°C of global warming.

The following gaps have been identified with respect to tools, methodologies and understanding in the current scientific literature specific to Chapter 3. The gaps identified here are not comprehensive but highlight general areas for improved understanding, especially regarding global warming at 1.5°C compared to 2°C and higher levels.

3.7.1   Gaps in Methods and Tools

- Regional and global climate model simulations for low-emissions scenarios such as a 1.5°C warmer world.

- Robust probabilistic models which separate the relatively small signal between 1.5°C versus 2°C from background noise, and which handle the many uncertainties associated with non-linearities, innovations, overshoot, local scales, and latent or lagging responses in climate.

- Projections of risks under a range of climate and development pathways required to understand how development choices affect the magnitude and pattern of risks, and to provide better estimates of the range of uncertainties.

- More complex and integrated socio-ecological models for predicting the response of terrestrial as well as coastal and oceanic ecosystems to climate and models which are more capable of separating climate effects from those associated with human activities.

- Tools for informing local and regional decision-making, especially when the signal is ambiguous at 1.5°C and/or reverses sign at higher levels of global warming.

3.7.2   Gaps in Understanding

3.7.2.1   Earth systems and 1.5°C of global warming

- The cumulative effects of multiple stresses and risks (e.g., increased storm intensity interacting with sea level rise and the effect on coastal people; feedbacks on wetlands due to climate change and human activities).

- Feedbacks associated with changes in land use/cover for low-emissions scenarios, for example feedback from changes in forest cover, food production, biofuel production, bio-energy with carbon capture and storage (BECCS), and associated unquantified biophysical impacts.

- The distinct impacts of different overshoot scenarios, depending on (i) the peak temperature of the overshoot, (ii) the length of the overshoot period, and (iii) the associated rate of change in global temperature over the time period of the overshoot.

3.7.2.2   Physical and chemical characteristics of a 1.5°C warmer world

- Critical thresholds for extreme events (e.g., drought and inundation) between 1.5°C and 2°C of warming for different climate models and projections. All aspects of storm intensity and frequency as a function of climate change, especially for 1.5°C and 2°C warmer worlds, and the impact of changing storminess on storm surges, damage, and coastal flooding at regional and local scales.

- The timing and implications of the release of stored carbon in Arctic permafrost in a 1.5°C warmer world and for climate stabilization by the end of the century.

- Antarctic ice sheet dynamics, global sea level, and links between seasonal and year-long sea ice in both polar regions.

3.7.2.3   Terrestrial and freshwater systems

- The dynamics between climate change, freshwater resources and socio-economic impacts for lower levels of warming.

- How the health of vegetation is likely to change, carbon storage in plant communities and landscapes, and phenomena such as the fertilization effect.

- The risks associated with species' maladaptation in response to climatic changes (e.g., effects of late frosts). Questions associated with issues such as the consequences of species advancing their spring phenology in response to warming, as well as the interaction between climate change, range shifts and local adaptation in a 1.5°C warmer world.

- The biophysical impacts of land use in the context of mitigation pathways.

3.7.2.4   Ocean Systems

- Deep sea processes and risks to deep sea habitats and ecosystems.

- How changes in ocean chemistry in a 1.5°C warmer world, including decreasing ocean oxygen content, ocean acidification and changes in the activity of multiple ion species, will affect natural and human systems.

- How ocean circulation is changing towards 1.5°C and 2°C warmer worlds, including vertical mixing, deep ocean processes, currents, and their impacts on weather patterns at regional to local scales.

- The impacts of changing ocean conditions at 1.5°C and 2°C of warming on foodwebs, disease, invading species, coastal protection, fisheries and human well-being, especially as organisms modify their biogeographical ranges within a changing ocean.

- Specific linkages between food security and changing coastal and ocean resources.

3.7.2.5   Human systems

- The impacts of global and regional climate change at 1.5°C on food distribution, nutrition, poverty, tourism, coastal infrastructure and public health, particularly for developing nations.

- Health and well-being risks in the context of socio-economic and climate change at 1.5°C, especially in key areas such as occupational health, air quality and infectious disease.

- Micro-climates at urban/city scales and their associated risks

for natural and human systems, within cities and in interaction with surrounding areas. For example, current projections do not integrate adaptation to projected warming by considering cooling that could be achieved through a combination of revised building codes, zoning and land use to build more reflective roofs and urban surfaces that reduce urban heat island effects.

- Implications of climate change at 1.5°C on livelihoods and poverty, as well as on rural communities, indigenous groups and marginalized people.

- The changing levels of risk in terms of extreme events, including storms and heatwaves, especially with respect to people being displaced or having to migrate away from sensitive and exposed systems such as small islands, low-lying coasts and deltas.

3

Cross-Chapter Box 8 | 1.5°C Warmer Worlds

Lead Authors:

Sonia I. Seneviratne (Switzerland), Joeri Rogelj (Austria/Belgium), Roland Séférian (France), Myles R. Allen (United Kingdom), Marcos Buckeridge (Brazil), Kristie L. Ebi (United States of America), Ove Hoegh-Guldberg (Australia), Richard J. Millar (United Kingdom), Antony J. Payne (United Kingdom), Petra Tschakert (Australia), Rachel Warren (United Kingdom)

Contributing Authors:

Neville Ellis (Australia), Richard Wartenburger (Germany/Switzerland)

Introduction

The Paris Agreement includes goals of stabilizing global mean surface temperature (GMST) well below 2°C and 1.5°C above pre-industrial levels in the longer term. There are several aspects, however, that remain open regarding what a '1.5°C warmer world' could be like, in terms of mitigation (Chapter 2) and adaptation (Chapter 4), as well as in terms of projected warming and associated regional climate change (Chapter 3), which are overlaid on anticipated and differential vulnerabilities (Chapter 5). **Alternative '1.5°C warmer worlds' resulting from mitigation and adaptation choices, as well as from climate variability (climate 'noise'), can be vastly different,** as highlighted in this Cross-Chapter Box. In addition, the range of models underlying 1.5°C projections can be substantial and needs to be considered.

Key questions[7]:

- **What is a 1.5°C global mean warming, how is it measured, and what temperature increase does it imply for single locations and at specific times?** Global mean surface temperature (GMST) corresponds to the globally averaged temperature of Earth derived from point-scale ground observations or computed in climate models (Chapters 1 and 3). Global mean surface temperature is additionally defined over a given time frame, for example averaged over a month, a year, or multiple decades. Because of climate variability, a climate-based GMST typically needs to be defined over several decades (typically 20 or 30 years; Chapter 3, Section 3.2). Hence, whether or when global warming reaches 1.5°C depends to some extent on the choice of pre-industrial reference period, whether 1.5°C refers to total or human-induced warming, and which variables and coverage are used to define GMST change (Chapter 1). By definition, because GMST is an average in time and space, there will be locations and time periods in which 1.5°C of warming is exceeded, even if the global mean warming is at 1.5°C. In some locations, these differences can be particularly large (Cross-Chapter Box 8, Figure 1).

- **What is the impact of different climate models for projected changes in climate at 1.5°C of global warming?** The range between single model simulations of projected regional changes at 1.5°C GMST increase can be substantial for regional responses (Chapter 3, Section 3.3). For instance, for the warming of cold extremes in a 1.5°C warmer world, some model simulations project a 3°C warming while others project more than 6°C of warming in the Arctic land areas (Cross-Chapter Box 8, Figure 2). For hot temperature extremes in the contiguous United States, the range of model simulations includes temperatures lower than pre-industrial values (−0.3°C) and a warming of 3.5°C (Cross-Chapter Box 8, Figure 2). Some regions display an even larger range (e.g., 1°C–6°C regional warming in hot extremes in central Europe at 1.5°C of warming; Chapter 3, Sections 3.3.1 and 3.3.2). This large spread is due to both modelling uncertainty and internal climate variability. While the range is large, it also highlights risks that can be avoided with near certainty in a 1.5°C warmer world compared to worlds at higher levels of warming (e.g., an 8°C warming of cold extremes in the Arctic is not reached at 1.5°C of global warming in the multimodel ensemble but could happen at 2°C of global warming; Cross-Chapter Box 8, Figure 2). Inferred projected ranges of regional responses (mean value, minimum and maximum) for different mitigation scenarios from Chapter 2 are displayed in Cross-Chapter Box 8, Table 1.

- **What is the impact of emissions pathways with, versus without, an overshoot?** All mitigation pathways projecting less than 1.5°C of global warming over or at the end of the 21st century include some probability of overshooting 1.5°C. These pathways include some periods with warming stronger than 1.5°C in the course of the coming decades and/or some probability of not reaching 1.5°C (Chapter 2, Section 2.2). This is inherent to the difficulty of limiting global warming to 1.5°C, given that we are already very close to this warming level. The implications of overshooting are large for risks to natural and human

---

[7] Part of this discussion is based on Seneviratne et al. (2018b).

Temperatures with 25% chance of occurring in any 10-year period with ΔT = 1.5°C (CMIP5 ensemble)

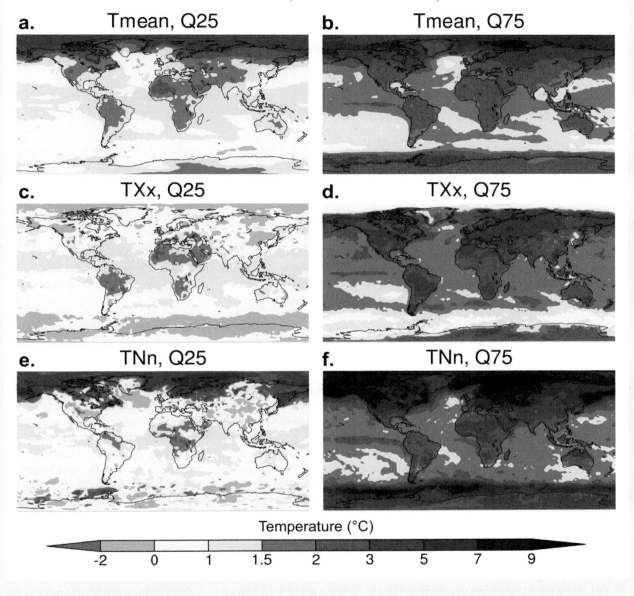

Cross-Chapter Box 8, Figure 1 | Range of projected realized temperatures at 1.5°C of global warming (due to stochastic noise and model-based spread). Temperatures with a 25% chance of occurrence at any location within a 10-year time frame are shown, corresponding to GMST anomalies of 1.5°C (Coupled Model Intercomparison Project Phase 5 (CMIP5) multimodel ensemble). The plots display the 25th percentile (Q25, left) and 75th percentile (Q75, right) values of mean temperature (Tmean), yearly maximum daytime temperature (TXx) and yearly minimum night-time temperature (TNn), sampled from all time frames with GMST anomalies of 1.5°C in Representative Concentration Pathway (RCP)8.5 model simulations of the CMIP5 ensemble. From Seneviratne et al. (2018b).

Cross-Chapter Box 8 (continued next page)

Cross-Chapter Box 8 (continued)

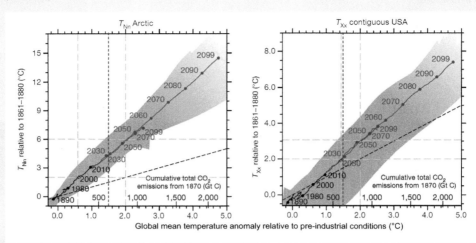

Cross-Chapter Box 8, Figure 2 | Spread of projected multimodel changes in minimum annual night-time temperature (TNn) in Arctic land (left) and in maximum annual daytime temperature (TXx) in the contiguous United States as a function of mean global warming in climate simulations. The multimodel range (due to model spread and internal climate variability) is indicated in red shading (minimum and maximum value based on climate model simulations). The multimodel mean value is displayed with solid red and blue lines for two emissions pathways (blue: Representative Concentration Pathway (RCP)4.5; red: RCP8.5). The dashed red line indicates projections for a 1.5°C warmer world. The dashed black line displays the 1:1 line. The figure is based on Figure 3 of Seneviratne et al. (2016).

systems, especially if the temperature at peak warming is high, because some risks may be long lasting and irreversible, such as the loss of some ecosystems (Chapter 3, Box 3.4). The chronology of emissions pathways and their implied warming is also important for the more slowly evolving parts of the Earth system, such as those associated with sea level rise. In addition, for several types of risks the rate of change may be most relevant (Loarie et al., 2009; LoPresti et al., 2015), with potentially large risks occurring in the case of a rapid rise to overshooting temperatures, even if a decrease to 1.5°C may be achieved at the end of the 21st century or later. On the other hand, if overshoot is to be minimized, the remaining equivalent CO_2 budget available for emissions has to be very small, which implies that large, immediate and unprecedented global efforts to mitigate GHGs are required (Cross-Chapter Box 8, Table 1; Chapter 4).

• **What is the probability of reaching 1.5°C of global warming if emissions compatible with 1.5°C pathways are followed?** Emissions pathways in a 'prospective scenario' (see Chapter 1, Section 1.2.3, and Cross-Chapter Box 1 in Chapter 1 on 'Scenarios and pathways') compatible with 1.5°C of global warming are determined based on their probability of reaching 1.5°C by 2100 (Chapter 2, Section 2.1), given current knowledge of the climate system response. These probabilities cannot be quantified precisely but are typically 50–66% in 1.5°C-consistent pathways (Section 1.2.3). This implies a one-in-two to one-in-three probability that global warming would exceed 1.5°C even under a 1.5°C-consistent pathway, including some possibility that global warming would be substantially over this value (generally about 5–10% probability; see Cross-Chapter Box 8, Table 1 and Seneviratne et al., 2018b). These alternative outcomes need to be factored into the decision-making process. To address this issue, 'adaptive' mitigation scenarios have been proposed in which emissions are continually adjusted to achieve a temperature goal (Millar et al., 2017). The set of dimensions involved in mitigation options (Chapter 4) is complex and need system-wide approaches to be successful. Adaptive scenarios could be facilitated by the global stocktake mechanism established in the Paris Agreement, and thereby transfer the risk of higher-than-expected warming to a risk of faster-than-expected mitigation efforts. However, there are some limits to the feasibility of such approaches because some investments, for example in infrastructure, are long term and also because the actual departure from an aimed pathway will need to be detected against the backdrop of internal climate variability, typically over several decades (Haustein et al., 2017; Seneviratne et al., 2018b). Avoiding impacts that depend on atmospheric composition as well as GMST (Baker et al., 2018) would also require limits on atmospheric CO_2 concentrations in the event of a lower-than-expected GMST response.

• **How can the transformation towards a 1.5°C warmer world be implemented?** This can be achieved in a variety of ways, such as decarbonizing the economy with an emphasis on demand reductions and sustainable lifestyles, or, alternatively, with an emphasis on large-scale technological solutions, amongst many other options (Chapter 2, Sections 2.3 and 2.4; Chapter 4, Sections 4.1 and 4.4.4). Different portfolios of mitigation measures come with distinct synergies and trade-offs with respect to other societal objectives. Integrated solutions and approaches are required to achieve multiple societal objectives simultaneously (see Chapter 4, Section 4.5.4 for a set of synergies and trade-offs).

Cross-Chapter Box 8 (continued)

- **What determines risks and opportunities in a 1.5°C warmer world?** The risks to natural, managed and human systems in a 1.5°C warmer world will depend not only on uncertainties in the regional climate that results from this level of warming, but also very strongly on the methods that humanity uses to limit global warming to 1.5°C. This is particularly the case for natural ecosystems and agriculture (see Cross-Chapter Box 7 in this chapter and Chapter 4, Section 4.3.2). The risks to human systems will also depend on the magnitude and effectiveness of policies and measures implemented to increase resilience to the risks of climate change and on development choices over coming decades, which will influence the underlying vulnerabilities and capacities of communities and institutions for responding and adapting.

- **Which aspects are not considered, or only partly considered, in the mitigation scenarios from Chapter 2?** These include biophysical impacts of land use, water constraints on energy infrastructure, and regional implications of choices of specific scenarios for tropospheric aerosol concentrations or the modulation of concentrations of short-lived climate forcers, that is, greenhouse gases (Chapter 3, Section 3.6.3). Such aspects of development pathways need to be factored into comprehensive assessments of the regional implications of mitigation and adaptation measures. On the other hand, some of these aspects are assessed in Chapter 4 as possible options for mitigation and adaptation to a 1.5°C warmer world.

- **Are there commonalities to all alternative 1.5°C warmer worlds?** Human-driven warming linked to CO_2 emissions is nearly irreversible over time frames of 1000 years or more (Matthews and Caldeira, 2008; Solomon et al., 2009). The GSMT of the Earth responds to the cumulative amount of CO_2 emissions. Hence, all 1.5°C stabilization scenarios require both net CO_2 emissions and multi-gas CO_2-forcing-equivalent emissions to be zero at some point (Chapter 2, Section 2.2). This is also the case for stabilization scenarios at higher levels of warming (e.g., at 2°C); the only difference is the projected time at which the net CO_2 budget is zero.

  Hence, a transition to decarbonization of energy use is necessary in all scenarios. It should be noted that **all scenarios of Chapter 2 include approaches for carbon dioxide removal (CDR)** in order to achieve the net zero CO_2 emissions budget. **Most of these use carbon capture and storage (CCS)** in addition to reforestation, although to varying degrees (Chapter 4, Section 4.3.7). Some potential pathways to 1.5°C of warming in 2100 would minimize the need for CDR (Obersteiner et al., 2018; van Vuuren et al., 2018). Taking into account the implementation of CDR, the CO_2-induced warming by 2100 is determined by the difference between the total amount of CO_2 generated (that can be reduced by early decarbonization) and the total amount permanently stored out of the atmosphere, for example by geological sequestration (Chapter 4, Section 4.3.7).

- **What are possible storylines of 'warmer worlds' at 1.5°C versus higher levels of global warming?** Cross-Chapter Box 8, Table 2 features possible storylines based on the scenarios of Chapter 2, the impacts of Chapters 3 and 5, and the options of Chapter 4. These storylines are not intended to be comprehensive of all possible future outcomes. Rather, they are intended as plausible scenarios of alternative warmer worlds, with two storylines that include stabilization at 1.5°C (Scenario 1) or close to 1.5°C (Scenario 2), and one storyline missing this goal and consequently only including reductions of CO_2 emissions and efforts towards stabilization at higher temperatures (Scenario 3).

Summary:

There is no single '1.5°C warmer world'. Impacts can vary strongly for different worlds characterized by a 1.5°C global warming. Important aspects to consider (besides the changes in global temperature) are the possible occurrence of an overshoot and its associated peak warming and duration, how stabilization of the increase in global surface temperature at 1.5°C could be achieved, how policies might be able to influence the resilience of human and natural systems, and the nature of regional and subregional risks.

The implications of overshooting are large for risks to natural and human systems, especially if the temperature at peak warming is high, because some risks may be long lasting and irreversible, such as the loss of some ecosystems. In addition, for several types of risks, the rate of change may be most relevant, with potentially large risks occurring in the case of a rapid rise to overshooting temperatures, even if a decrease to 1.5°C may be achieved at the end of the 21st century or later. If overshoot is to be minimized, the remaining equivalent CO_2 budget available for emissions has to be very small, which implies that large, immediate and unprecedented global efforts to mitigate GHGs are required.

The time frame for initiating major mitigation measures is essential in order to reach a 1.5°C (or even a 2°C) global stabilization of climate warming (see consistent cumulative CO_2 emissions up to peak warming in Cross-Chapter Box 8, Table 1). If mitigation pathways are not rapidly activated, much more expensive and complex adaptation measures will have to be taken to avoid the impacts of higher levels of global warming on the Earth system.   *Cross-Chapter Box 8 (continued next page)*

Cross-Chapter Box 8 (continued)

Cross-Chapter Box 8, Table 1 | Different worlds resulting from 1.5°C and 2°C mitigation (prospective) pathways, including 66% (probable) best-case outcome, and 5% worst-case outcome, based on Chapter 2 scenarios and Chapter 3 assessments of changes in regional climate. Note that the pathway characteristics estimates are based on computations with the MAGICC model (Meinshausen et al., 2011) consistent with the set-up used in AR5 WGIII (Clarke et al., 2014), but are uncertain and will be subject to updates and adjust-ments (see Chapter 2 for details). Updated from Seneviratne et al. (2018b).

		B1.5_LOS (below 1.5°C with low overshoot) with 2/3 ´probable best-case outcome´ᵃ	B1.5_LOS (below 1.5°C with low overshoot) with 1/20 ´worst-case outcome´ᵇ	L20 (lower than 2°C) with 2/3 ´probable best-case outcome´ᵃ	L20 (lower than 2°C) with 1/20 ´worst-case outcome´ᵇ
General characteristics of pathway	Overshoot > 1.5°C in 21st centuryᶜ	Yes (51/51)	Yes (51/51)	Yes (72/72)	Yes (72/72)
	Overshoot > 2°C in 21st century	No (0/51)	Yes (37/51)	No (72/72)	Yes (72/72)
	Cumulative CO_2 emissions up to peak warming (relative to 2016)ᵈ [$GtCO_2$]	610–760	590–750	1150–1460	1130–1470
	Cumulative CO_2 emissions up to 2100 (relative to 2016)ᵈ [$GtCO_2$]	170–560		1030–1440	
	Global GHG emissions in 2030ᵈ [$GtCO_2$ y-1]	19–23		31–38	
	Years of global net zero CO_2 emissionsᵈ	2055–2066		2082–2090	
Possible climate range at peak warming (regional+global)	**Global mean temperature anomaly at peak warming**	**1.7°C (1.66°C–1.72°C)**	**2.05°C (2.00°C–2.09°C)**	**2.11°C (2.05°C–2.17°C)**	**2.67°C (2.59°C–2.76°C)**
	Warming in the Arcticᵉ (TNnᶠ)	4.93°C (4.36, 5.52)	6.02°C (5.12, 6.89)	6.24°C (5.39, 7.21)	7.69°C (6.69, 8.93)
	Warming in Central North Americaᵉ (TXxᵍ)	2.65°C (1.92, 3.15)	3.11°C (2.37, 3.63)	3.18°C (2.50, 3.71)	4.06°C (3.35, 4.63)
	Warming in Amazon regionᵉ (TXx)	2.55°C (2.23, 2.83)	3.07°C (2.74, 3.46)	3.16°C (2.84, 3.57)	4.05°C (3.62, 4.46)
	Drying in the Mediterranean regionᵉ,ʰ	−1.11 (−2.24, −0.41)	−1.28 (−2.44, −0.51)	−1.38 (−2.58, −0.53)	−1.56 (−3.19, −0.67)
	Increase in heavy precipitation eventsᵉ in Southern Asiaⁱ	9.94% (6.76, 14.00)	11.94% (7.52, 18.86)	12.68% (7.71, 22.39)	19.67% (11.56, 27.24)
Possible climate range in 2100 (regional+global)	**Global mean temperature warming in 2100**	**1.46°C (1.41°C–1.51°C)**	**1.87°C (1.81°C–1.94°C)**	**2.06°C (1.99°C–2.15°C)**	**2.66°C (2.56°C–2.76°C)**
	Warming in the Arcticʲ (TNn)	4.28°C (3.71, 4.77)	5.50°C (4.74, 6.21)	6.08°C (5.20, 6.94)	7.63°C (6.66, 8.90)
	Warming in Central North Americaʲ (TXx)	2.31°C (1.56, 2.66)	2.83°C (2.03, 3.49)	3.12°C (2.38, 3.67)	4.06°C (3.33, 4.59)
	Warming in Amazon regionʲ (TXx)	2.22°C (2.00, 2.45)	2.76°C (2.50, 3.07)	3.10°C (2.75, 3.49)	4.03°C (3.62, 4.45)
	Drying in the Mediterranean regionʲ	−0.95 (−1.98, −0.30)	−1.10 (−2.17, −0.51)	−1.26 (−2.43, −0.52)	−1.55 (−3.17, −0.67)
	Increase in heavy precipitation events in Southern Asiaʲ	8.38% (4.63, 12.68)	10.34% (6.64, 16.07)	12.02% (7.41, 19.62)	19.72% (11.34, 26.95)

Notes:

a) 66th percentile for global temperature (that is, 66% likelihood of being at or below values)

b) 95th percentile for global temperature (that is, 5% likelihood of being at or above values)

c) All 1.5°C scenarios include a substantial probability of overshooting above 1.5°C global warming before returning to 1.5°C.

d) Interquartile range (25th percentile, q25, and 75th percentile, q75)

e) The regional projections in these rows provide the median and the range [q25, q75] associated with the median global temperature outcomes of the considered mitigation scenarios at peak warming.

f) TNn: Annual minimum night-time temperature

g) TXx: Annual maximum day-time temperature

h) Indicates drying of soil moisture expressed in units of standard deviations of pre-industrial climate (1861–1880) variability (where −1 is dry; −2 is severely dry; and −3 is very severely dry);

i) Rx5day: the annual maximum consecutive 5-day precipitation.

j) As for footnote e, but for the regional responses associated with the median global temperature outcomes of the considered mitigation scenarios in 2100

Cross-Chapter Box 8 (continued)

Cross-Chapter Box 8, Table 2 | Storylines of possible worlds resulting from different mitigation options. The storylines build upon Cross-Chapter Box 8, Table 1 and the assessments of Chapters 1–5. Only a few of the many possible storylines were chosen and they are presented for illustrative purposes.

Scenario 1 [one possible storyline among best-case scenarios]: **Mitigation:** early move to decarbonization, decarbonization designed to minimize land footprint, coordination and rapid action of the world's nations towards 1.5°C goal by 2100 **Internal climate variability:** probable (66%) best-case outcome for global and regional climate responses	**In 2020, strong participation and support for the Paris Agreement and its ambitious goals for reducing CO₂ emissions by an almost unanimous international community led to a time frame for net zero emissions that is compatible with halting global warming at 1.5°C by 2100.** There is strong participation in all major world regions at the national, state and/or city levels. Transport is strongly decarbonized through a shift to electric vehicles, with more cars with electric than combustion engines being sold by 2025 (Chapter 2, Section 2.4.3; Chapter 4, Section 4.3.3). Several industry-sized plants for carbon capture and storage are installed and tested in the 2020s (Chapter 2, Section 2.4.2; Chapter 4, Sections 4.3.4 and 4.3.7). Competition for land between bioenergy cropping, food production, and biodiversity conservation is minimized by sourcing bioenergy for carbon capture and storage from agricultural wastes, algae and kelp farms (Cross-Chapter Box 7 in Chapter 3; Chapter 4, Section 4.3.2). Agriculture is intensified in countries with coordinated planning associated with a drastic decrease in food waste (Chapter 2, Section 2.4.4; Chapter 4, Section 4.3.2). This leaves many natural ecosystems relatively intact, supporting continued provision of most ecosystem services, although relocation of species towards higher latitudes and elevations still results in changes in local biodiversity in many regions, particularly in mountain, tropical, coastal and Arctic ecosystems (Chapter 3, Section 3.4.3). Adaptive measures such as the establishment of corridors for the movement of species and parts of ecosystems become a central practice within conservation management (Chapter 3, Section 3.4.3; Chapter 4, Section 4.3.2). The movement of species presents new challenges for resource management as novel ecosystems, as well as pests and disease, increase (Cross-Chapter Box 6 in Chapter 3). Crops are grown on marginal land, no-till agriculture is deployed, and large areas are reforested with native trees (Chapter 2, Section 2.4.4; Chapter 3, Section 3.6.2; Cross-Chapter Box 7 in Chapter 3; Chapter 4, Section 4.3.2). Societal preference for healthy diets reduces meat consumption and associated GHG emissions (Chapter 2, Section 2.4.4; Chapter 4, Section 4.3.2; Cross-Chapter Box 6 in Chapter 3). By 2100, global mean temperature is on average 0.5°C warmer than it was in 2018 (Chapter 1, Section 1.2.1). Only a minor temperature overshoot occurs during the century (Chapter 2, Section 2.2). In mid-latitudes, frequent hot summers and precipitation events tend to be more intense (Chapter 3, Section 3.3). Coastal communities struggle with increased inundation associated with rising sea levels and more frequent and intense heavy rainfall (Chapter 3, Sections 3.3.2 and 3.3.9; Chapter 4, Section 4.3.2; Chapter 5, Box 5.3 and Section 5.3.2; Cross-Chapter Box 12 in Chapter 5), and some respond by moving, in many cases with consequences for urban areas. In the tropics, in particular in megacities, there are frequent deadly heatwaves whose risks are reduced by proactive adaptation (Chapter 3, Sections 3.3.1 and 3.4.8; Chapter 4, Section 4.3.8), overlaid on a suite of development challenges and limits in disaster risk management (Chapter 4, Section 4.3.3; Chapter 5, Sections 5.2.1 and 5.2.2; Cross-Chapter Box 12 in Chapter 5). Glaciers extent decreases in most mountainous areas (Chapter 3, Sections 3.3.5 and 3.5.4). Reduced Arctic sea ice opens up new shipping lanes and commercial corridors (Chapter 3, Section 3.3.8; Chapter 4, Box 4.3). Small island developing states (SIDS), as well as coastal and low-lying areas, have faced significant changes but have largely persisted in most regions (Chapter 3, Sections 3.3.9 and 3.5.4, Box 3.5). The Mediterranean area becomes drier (Chapter 3, Section 3.3.4 and Box 3.2) and irrigation of crops expands, drawing the water table down in many areas (Chapter 3, Section 3.4.6). The Amazon is reasonably well preserved, through avoided risk of droughts (Chapter 3, Sections 3.3.4 and 3.4.3; Chapter 4, Box 4.3) and reduced deforestation (Chapter 2, Section 2.4.4; Cross-Chapter Box 7 in Chapter 3; Chapter 4, Section 4.3.2), and the forest services are working with the pattern observed at the beginning of the 21st century (Chapter 4, Box 4.3). While some climate hazards become more frequent (Chapter 3, Section 3.3), timely adaptation measures help reduce the associated risks for most, although poor and disadvantaged groups continue to experience high climate risks to their livelihoods and well-being (Chapter 5, Section 5.3.1; Cross-Chapter Box 12 in Chapter 5; Chapter 3, Boxes 3.4 and 3.5; Cross-Chapter Box 6 in Chapter 3). Summer sea ice has not completely disappeared from the Arctic (Chapter 3, Section 3.4.4.7) and coral reefs, having been driven to a low level (10–30% of levels in 2018), have partially recovered by 2100 after extensive dieback (Chapter 3, Section 3.4.4.10 and Box 3.4). The Earth system, while warmer, is still recognizable compared to the 2000s, and no major tipping points are reached (Chapter 3, Section 3.5.2.5). Crop yields remain relatively stable (Chapter 3, Section 3.4). Aggregate economic damage of climate change impacts is relatively small, although there are some local losses associated with extreme weather events (Chapter 3, Section 3.5; Chapter 4). Human well-being remains overall similar to that in 2020 (Chapter 5, Section 5.2.2).
Scenario 2 [one possible storyline among mid-case scenarios]: **Mitigation:** delayed action (ambitious targets reached only after warmer decade in the 2020s due to internal climate variability), overshoot at 2°C, decrease towards 1.5°C afterward, no efforts to minimize the land and water footprints of bioenergy **Internal climate variability:** 10% worst-case outcome (2020s) followed by normal internal climate variability	**The international community continues to largely support the Paris Agreement and agrees in 2020 on reduction targets for CO₂ emissions and time frames for net zero emissions. However, these targets are not ambitious enough to reach stabilization at 2°C of warming, let alone 1.5°C.** In the 2020s, internal climate variability leads to higher warming than projected, in a reverse development to what happened in the so-called 'hiatus' period of the 2000s. Temperatures are regularly above 1.5°C of warming, although radiative forcing is consistent with a warming of 1.2°C or 1.3°C. Deadly heatwaves in major cities (Chicago, Kolkata, Beijing, Karachi, São Paulo), droughts in southern Europe, southern Africa and the Amazon region, and major flooding in Asia, all intensified by the global and regional warming (Chapter 3, Sections 3.3.1, 3.3.2, 3.3.3, 3.3.4 and 3.4.8; Cross-Chapter Box 11 in Chapter 4), lead to increasing levels of public unrest and political destabilization (Chapter 5, Section 5.2.1). An emergency global summit in 2025 moves to much more ambitious climate targets. Costs for rapidly phasing out fossil fuel use and infrastructure, while rapidly expanding renewables to reduce emissions, are much higher than in Scenario 1, owing to a failure to support economic measures to drive the transition (Chapter 4). Disruptive technologies become crucial to face up to the adaptation measures needed (Chapter 4, Section 4.4.4).

3

Cross-Chapter Box 8, Table 2 *(continued)*

Scenario 2 [one possible storyline among mid-case scenarios]: **Mitigation:** delayed action (ambitious targets reached only after warmer decade in the 2020s due to internal climate variability), overshoot at 2°C, decrease towards 1.5°C afterward, no efforts to minimize the land and water footprints of bioenergy **Internal climate variability:** 10% worst-case outcome (2020s) followed by normal internal climate variability	Temperature peaks at 2°C of warming by the middle of the century before decreasing again owing to intensive implementation of bioenergy plants with carbon capture and storage (Chapter 2), without efforts to minimize the land and water footprint of bioenergy production (Cross-Chapter Box 7 in Chapter 3). Reaching 2°C of warming for several decades eliminates or severely damages key ecosystems such as coral reefs and tropical forests (Chapter 3, Section 3.4). The elimination of coral reef ecosystems and the deterioration of their calcified frameworks, as well as serious losses of coastal ecosystems such as mangrove forests and seagrass beds (Chapter 3, Boxes 3.4 and 3.5, Sections 3.4.4.10 and 3.4.5), leads to much reduced levels of coastal defence from storms, winds and waves. These changes increase the vulnerability and risks facing communities in tropical and subtropical regions, with consequences for many coastal communities (Cross-Chapter Box 12 in Chapter 5). These impacts are being amplified by steadily rising sea levels (Chapter 3, Section 3.3.9) and intensifying storms (Chapter 3, Section 3.4.4.3). The intensive area required for the production of bioenergy, combined with increasing water stress, puts pressure on food prices (Cross-Chapter Box 6 in Chapter 3), driving elevated rates of food insecurity, hunger and poverty (Chapter 4, Section 4.3.2; Cross-Chapter Box 6 in Chapter 3; Cross-Chapter Box 11 in Chapter 4). Crop yields decline significantly in the tropics, leading to prolonged famines in some African countries (Chapter 3, Section 3.4; Chapter 4, Section 4.3.2). Food trumps environment in terms of importance in most countries, with the result that natural ecosystems decrease in abundance, owing to climate change and land-use change (Cross-Chapter Box 7 in Chapter 3). The ability to implement adaptive action to prevent the loss of ecosystems is hindered under the circumstances and is consequently minimal (Chapter 3, Sections 3.3.6 and 3.4.4.10). Many natural ecosystems, in particular in the Mediterranean, are lost because of the combined effects of climate change and land-use change, and extinction rates increase greatly (Chapter 3, Section 3.4 and Box 3.2). By 2100, warming has decreased but is still stronger than 1.5°C, and the yields of some tropical crops are recovering (Chapter 3, Section 3.4.3). Several of the remaining natural ecosystems experience irreversible climate change-related damages whilst others have been lost to land-use change, with very rapid increases in the rate of species extinctions (Chapter 3, Section 3.4; Cross-Chapter Box 7 in Chapter 3; Cross-Chapter Box 11 in Chapter 4). Migration, forced displacement, and loss of identity are extensive in some countries, reversing some achievements in sustainable development and human security (Chapter 5, Section 5.3.2). Aggregate economic impacts of climate change damage are small, but the loss in ecosystem services creates large economic losses (Chapter 4, Sections 4.3.2 and 4.3.3). The health and well-being of people generally decrease from 2020, while the levels of poverty and disadvantage increase considerably (Chapter 5, Section 5.2.1).
Scenario 3 [one possible storyline among worst-case scenarios]: **Mitigation:** uncoordinated action, major actions late in the 21st century, 3°C of warming in 2100 **Internal climate variability:** unusual (ca. 10%) best-case scenario for one decade, followed by normal internal climate variability	**In 2020, despite past pledges, the international support for the Paris Agreement starts to wane. In the years that follow, CO$_2$ emissions are reduced at the local and national level but efforts are limited and not always successful.** Radiative forcing increases and, due to chance, the most extreme events tend to happen in less populated regions and thus do not increase global concerns. Nonetheless, there are more frequent heatwaves in several cities and less snow in mountain resorts in the Alps, Rockies and Andes (Chapter 3, Section 3.3). Global warming of 1.5°C is reached by 2030 but no major changes in policies occur. Starting with an intense El Niño–La Niña phase in the 2030s, several catastrophic years occur while global warming starts to approach 2°C. There are major heatwaves on all continents, with deadly consequences in tropical regions and Asian megacities, especially for those ill-equipped for protecting themselves and their communities from the effects of extreme temperatures (Chapter 3, Sections 3.3.1, 3.3.2 and 3.4.8). Droughts occur in regions bordering the Mediterranean Sea, central North America, the Amazon region and southern Australia, some of which are due to natural variability and others to enhanced greenhouse gas forcing (Chapter 3, Section 3.3.4; Chapter 4, Section 4.3.2; Cross-Chapter Box 11 in Chapter 4). Intense flooding occurs in high-latitude and tropical regions, in particular in Asia, following increases in heavy precipitation events (Chapter 3, Section 3.3.3). Major ecosystems (coral reefs, wetlands, forests) are destroyed over that period (Chapter 3, Section 3.4), with massive disruption to local livelihoods (Chapter 5, Section 5.2.2 and Box 5.3; Cross-Chapter Box 12 in Chapter 5). An unprecedented drought leads to large impacts on the Amazon rainforest (Chapter 3, Sections 3.3.4 and 3.4), which is also affected by deforestation (Chapter 2). A hurricane with intense rainfall and associated with high storm surges (Chapter 3, Section 3.3.6) destroys a large part of Miami. A two-year drought in the Great Plains in the USA and a concomitant drought in eastern Europe and Russia decrease global crop production (Chapter 3, Section 3.3.4), resulting in major increases in food prices and eroding food security. Poverty levels increase to a very large scale, and the risk and incidence of starvation increase considerably as food stores dwindle in most countries; human health suffers (Chapter 3, Section 3.4.6.1; Chapter 4, Sections 4.3.2 and 4.4.3; Chapter 5, Section 5.2.1). There are high levels of public unrest and political destabilization due to the increasing climatic pressures, resulting in some countries becoming dysfunctional (Chapter 4, Sections 4.4.1 and 4.4.2). The main countries responsible for the CO$_2$ emissions design rapidly conceived mitigation plans and try to install plants for carbon capture and storage, in some cases without sufficient prior testing (Chapter 4, Section 4.3.6). Massive investments in renewable energy often happen too late and are uncoordinated; energy prices soar as a result of the high demand and lack of infrastructure. In some cases, demand cannot be met, leading to further delays. Some countries propose to consider sulphate-aerosol based Solar Radiation Modification (SRM) (Chapter 4, Section 4.3.8); however, intensive international negotiations on the topic take substantial time and are inconclusive because of overwhelming concerns about potential impacts on monsoon rainfall and risks in case of termination (Cross-Chapter Box 10 in Chapter 5). Global and regional temperatures continue to increase strongly while mitigation solutions are being developed and implemented.

3

Cross-Chapter Box 8 (continued)

Cross-Chapter Box 8, Table 2 *(continued)*

Scenario 3 [one possible storyline among worst-case scenarios]: **Mitigation:** uncoordinated action, major actions late in the 21st century, 3°C of warming in 2100 **Internal climate variability:** unusual (ca. 10%) best-case scenario for one decade, followed by normal internal climate variability	Global mean warming reaches 3°C by 2100 but is not yet stabilized despite major decreases in yearly CO_2 emissions, as a net zero CO_2 emissions budget could not yet be achieved and because of the long lifetime of CO_2 concentrations (Chapters 1, 2 and 3). The world as it was in 2020 is no longer recognizable, with decreasing life expectancy, reduced outdoor labour productivity, and lower quality of life in many regions because of too frequent heatwaves and other climate extremes (Chapter 4, Section 4.3.3). Droughts and stress on water resources renders agriculture economically unviable in some regions (Chapter 3, Section 3.4; Chapter 4, Section 4.3.2) and contributes to increases in poverty (Chapter 5, Section 5.2.1; Cross-Chapter Box 12 in Chapter 5). Progress on the sustainable development goals is largely undone and poverty rates reach new highs (Chapter 5, Section 5.2.3). Major conflicts take place (Chapter 3, Section 3.4.9.6; Chapter 5, Section 5.2.1). Almost all ecosystems experience irreversible impacts, species extinction rates are high in all regions, forest fires escalate, and biodiversity strongly decreases, resulting in extensive losses to ecosystem services. These losses exacerbate poverty and reduce quality of life (Chapter 3, Section 3.4; Chapter 4, Section 4.3.2). Life for many indigenous and rural groups becomes untenable in their ancestral lands (Chapter 4, Box 4.3; Cross-Chapter Box 12 in Chapter 5). The retreat of the West Antarctic ice sheet accelerates (Chapter 3, Sections 3.3 and 3.6), leading to more rapid sea level rise (Chapter 3, Section 3.3.9; Chapter 4, Section 4.3.2). Several small island states give up hope of survival in their locations and look to an increasingly fragmented global community for refuge (Chapter 3, Box 3.5; Cross-Chapter Box 12 in Chapter 5). Aggregate economic damages are substantial, owing to the combined effects of climate changes, political instability, and losses of ecosystem services (Chapter 4, Sections 4.4.1 and 4.4.2; Chapter 3, Box 3.6 and Section 3.5.2.4). The general health and well-being of people is substantially reduced compared to the conditions in 2020 and continues to worsen over the following decades (Chapter 5, Section 5.2.3).

3

Frequently Asked Questions

FAQ 3.1 | What are the Impacts of 1.5°C and 2°C of Warming?

Summary: The impacts of climate change are being felt in every inhabited continent and in the oceans. However, they are not spread uniformly across the globe, and different parts of the world experience impacts differently. An average warming of 1.5°C across the whole globe raises the risk of heatwaves and heavy rainfall events, amongst many other potential impacts. Limiting warming to 1.5°C rather than 2°C can help reduce these risks, but the impacts the world experiences will depend on the specific greenhouse gas emissions 'pathway' taken. The consequences of temporarily overshooting 1.5°C of warming and returning to this level later in the century, for example, could be larger than if temperature stabilizes below 1.5°C. The size and duration of an overshoot will also affect future impacts.

Human activity has warmed the world by about 1°C since pre-industrial times, and the impacts of this warming have already been felt in many parts of the world. This estimate of the increase in global temperature is the average of many thousands of temperature measurements taken over the world's land and oceans. Temperatures are not changing at the same speed everywhere, however: warming is strongest on continents and is particularly strong in the Arctic in the cold season and in mid-latitude regions in the warm season. This is due to self-amplifying mechanisms, for instance due to snow and ice melt reducing the reflectivity of solar radiation at the surface, or soil drying leading to less evaporative cooling in the interior of continents. This means that some parts of the world have already experienced temperatures greater than 1.5°C above pre-industrial levels.

Extra warming on top of the approximately 1°C we have seen so far would amplify the risks and associated impacts, with implications for the world and its inhabitants. This would be the case even if the global warming is held at 1.5°C, just half a degree above where we are now, and would be further amplified at 2°C of global warming. Reaching 2°C instead of 1.5°C of global warming would lead to substantial warming of extreme hot days in all land regions. It would also lead to an increase in heavy rainfall events in some regions, particularly in the high latitudes of the Northern Hemisphere, potentially raising the risk of flooding. In addition, some regions, such as the Mediterranean, are projected to become drier at 2°C versus 1.5°C of global warming. The impacts of any additional warming would also include stronger melting of ice sheets and glaciers, as well as increased sea level rise, which would continue long after the stabilization of atmospheric CO_2 concentrations.

Change in climate means and extremes have knock-on effects for the societies and ecosystems living on the planet. Climate change is projected to be a poverty multiplier, which means that its impacts are expected to make the poor poorer and the total number of people living in poverty greater. The 0.5°C rise in global temperatures that we have experienced in the past 50 years has contributed to shifts in the distribution of plant and animal species, decreases in crop yields and more frequent wildfires. Similar changes can be expected with further rises in global temperature.

Essentially, the lower the rise in global temperature above pre-industrial levels, the lower the risks to human societies and natural ecosystems. Put another way, limiting warming to 1.5°C can be understood in terms of 'avoided impacts' compared to higher levels of warming. Many of the impacts of climate change assessed in this report have lower associated risks at 1.5°C compared to 2°C.

Thermal expansion of the ocean means sea level will continue to rise even if the increase in global temperature is limited to 1.5°C, but this rise would be lower than in a 2°C warmer world. Ocean acidification, the process by which excess CO_2 is dissolving into the ocean and increasing its acidity, is expected to be less damaging in a world where CO_2 emissions are reduced and warming is stabilized at 1.5°C compared to 2°C. The persistence of coral reefs is greater in a 1.5°C world than that of a 2°C world, too.

The impacts of climate change that we experience in future will be affected by factors other than the change in temperature. The consequences of 1.5°C of warming will additionally depend on the specific greenhouse gas emissions 'pathway' that is followed and the extent to which adaptation can reduce vulnerability. This IPCC Special Report uses a number of 'pathways' to explore different possibilities for limiting global warming to 1.5°C above pre-industrial levels. One type of pathway sees global temperature stabilize at, or just below, 1.5°C. Another sees global temperature temporarily exceed 1.5°C before declining later in the century (known as an 'overshoot' pathway).

(continued on next page)

Such pathways would have different associated impacts, so it is important to distinguish between them for planning adaptation and mitigation strategies. For example, impacts from an overshoot pathway could be larger than impacts from a stabilization pathway. The size and duration of an overshoot would also have consequences for the impacts the world experiences. For instance, pathways that overshoot 1.5°C run a greater risk of passing through 'tipping points', thresholds beyond which certain impacts can no longer be avoided even if temperatures are brought back down later on. The collapse of the Greenland and Antarctic ice sheets on the time scale of centuries and millennia is one example of a tipping point.

FAQ3.1:Impact of 1.5°C and 2.0°C global warming

Temperature rise is not uniform across the world. Some regions will experience greater increases in the temperature of hot days and cold nights than others.

+ 1.5°C: Change in average temperature of hottest days

+ 2.0°C: Change in average temperature of hottest days

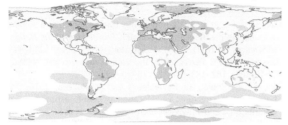
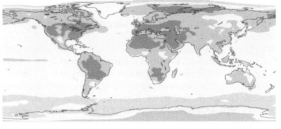

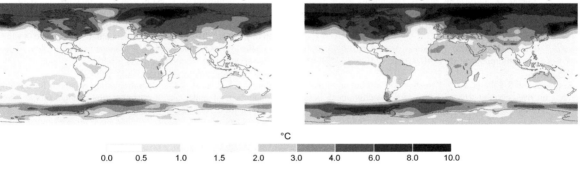

FAQ 3.1, Figure 1 | Temperature change is not uniform across the globe. Projected changes are shown for the average temperature of the annual hottest day (top) and the annual coldest night (bottom) with 1.5°C of global warming (left) and 2°C of global warming (right) compared to pre-industrial levels.

References

Aalto, J., S. Harrison, and M. Luoto, 2017: Statistical modelling predicts almost complete loss of major periglacial processes in Northern Europe by 2100. *Nature Communications*, 1–8, doi:10.1038/s41467-017-00669-3.

Abatzoglou, J.T. and A.P. Williams, 2016: Impact of anthropogenic climate change on wildfire across western US forests. *Proceedings of the National Academy of Sciences*, **113(42)**, 11770–11775, doi:10.1073/pnas.1607171113.

Abebe, A. et al., 2016: Growth, yield and quality of maize with elevated atmospheric carbon dioxide and temperature in north-west India. *Agriculture, Ecosystems & Environment*, **218**, 66–72, doi:10.1016/j.agee.2015.11.014.

Acharya, S.S. and M.K. Panigrahi, 2016: Eastward shift and maintenance of Arabian Sea oxygen minimum zone: Understanding the paradox. *Deep-Sea Research Part I: Oceanographic Research Papers*, **115**, 240–252, doi:10.1016/j.dsr.2016.07.004.

Acosta Navarro, J. et al., 2017: Future Response of Temperature and Precipitation to Reduced Aerosol Emissions as Compared with Increased Greenhouse Gas Concentrations. *Journal of Climate*, **30**, 939–954, doi:10.1175/jcli-d-16-0466.1.

Adams, C., T. Ide, J. Barnett, and A. Detges, 2018: Sampling bias in climate-conflict research. *Nature Climate Change*, **8(3)**, 200–203, doi:10.1038/s41558-018-0068-2.

Adger, W.N. et al., 2014: Human Security. In: *Climate Change 2014: Impacts, Adaptation, and Vulnerability. Part A: Global and Sectoral Aspects. Contribution of Working Group II to the Fifth Assessment Report of the Intergovernmental Panel on Climate Change* [Field, C.B., V.R. Barros, D.J. Dokken, K.J. Mach, M.D. Mastrandrea, T.E. Bilir, M. Chatterjee, K.L. Ebi, Y.O. Estrada, R.C. Genova, B. Girma, E.S. Kissel, A.N. Levy, S. MacCracken, P.R. Mastrandrea, and L.L. White (eds.)]. Cambridge University Press, Cambridge, United Kingdom and New York, NY, USA, pp. 755–791.

AghaKouchak, A., L. Cheng, O. Mazdiyasni, and A. Farahmand, 2014: Global warming and changes in risk of concurrent climate extremes: Insights from the 2014 California drought. *Geophysical Research Letters*, **41(24)**, 8847–8852, doi:10.1002/2014gl062308.

AghaKouchak, A., D. Feldman, M. Hoerling, T. Huxman, and J. Lund, 2015: Water and climate: Recognize anthropogenic drought. *Nature*, **524(7566)**, 409–411, doi:10.1038/524409a.

Aguilera, R., R. Marcé, and S. Sabater, 2015: Detection and attribution of global change effects on river nutrient dynamics in a large Mediterranean basin. *Biogeosciences*, **12**, 4085–4098, doi:10.5194/bg-12-4085-2015.

Ahlström, A., G. Schurgers, A. Arneth, and B. Smith, 2012: Robustness and uncertainty in terrestrial ecosystem carbon response to CMIP5 climate change projections. *Environmental Research Letters*, **7(4)**, 044008, doi:10.1088/1748-9326/7/4/044008.

Ahmed, K.F., G. Wang, M. Yu, J. Koo, and L. You, 2015: Potential impact of climate change on cereal crop yield in West Africa. *Climatic Change*, **133(2)**, 321–334, doi:10.1007/s10584-015-1462-7.

Ainsworth, C.H. et al., 2011: Potential impacts of climate change on Northeast Pacific marine foodwebs and fisheries. *ICES Journal of Marine Science*, **68(6)**, 1217–1229, doi:10.1093/icesjms/fsr043.

Akbari, H., S. Menon, and A. Rosenfeld, 2009: Global cooling: Increasing world-wide urban albedos to offset CO_2. *Climatic Change*, **94(3–4)**, 275–286, doi:10.1007/s10584-008-9515-9.

Albert, S. et al., 2017: Heading for the hills: climate-driven community relocations in the Solomon Islands and Alaska provide insight for a 1.5°C future. *Regional Environmental Change*, 1–12, doi:10.1007/s10113-017-1256-8.

Albright, R. et al., 2016: Ocean acidification: Linking science to management solutions using the Great Barrier Reef as a case study. *Journal of Environmental Management*, **182**, 641–650, doi:10.1016/j.jenvman.2016.07.038.

Alexander, P. et al., 2017: Assessing uncertainties in land cover projections. *Global Change Biology*, **23(2)**, 767–781, doi:10.1111/gcb.13447.

Alfieri, L., F. Dottori, R. Betts, P. Salamon, and L. Feyen, 2018: Multi-Model Projections of River Flood Risk in Europe under Global Warming. *Climate*, **6(1)**, 6, doi:10.3390/cli6010006.

Alfieri, L. et al., 2017: Global projections of river flood risk in a warmer world. *Earth's Future*, **5(2)**, 171–182, doi:10.1002/2016ef000485.

Allen, R., A. Foggo, K. Fabricius, A. Balistreri, and J.M. Hall-Spencer, 2017: Tropical CO_2 seeps reveal the impact of ocean acidification on coral reef invertebrate recruitment. *Marine Pollution Bulletin*, **124(2)**, 607–613, doi:10.1016/j.marpolbul.2016.12.031.

Almer, C., J. Laurent-Lucchetti, and M. Oechslin, 2017: Water scarcity and rioting: Disaggregated evidence from Sub-Saharan Africa. *Journal of Environmental Economics and Management*, **86**, 193–209, doi:10.1016/j.jeem.2017.06.002.

Alongi, D.M., 2008: Mangrove forests: Resilience, protection from tsunamis, and responses to global climate change. *Estuarine, Coastal and Shelf Science*, **76(1)**, 1–13, doi:10.1016/j.ecss.2007.08.024.

Alongi, D.M., 2015: The Impact of Climate Change on Mangrove Forests. *Current Climate Change Reports*, **1(1)**, 30–39, doi:10.1007/s40641-015-0002-x.

Altieri, A.H. and K.B. Gedan, 2015: Climate change and dead zones. *Global Change Biology*, **21(4)**, 1395–1406, doi:10.1111/gcb.12754.

Altieri, A.H. et al., 2017: Tropical dead zones and mass mortalities on coral reefs. *Proceedings of the National Academy of Sciences*, **114(14)**, 3660–3665, doi:10.1073/pnas.1621517114.

Alvarez-Filip, L., N.K. Dulvy, J.A. Gill, I.M. Cote, and A.R. Watkinson, 2009: Flattening of Caribbean coral reefs: region-wide declines in architectural complexity. *Proceedings of the Royal Society B: Biological Sciences*, **276(1669)**, 3019–3025, doi:10.1098/rspb.2009.0339.

Anderegg, W.R.L. et al., 2015: Tropical nighttime warming as a dominant driver of variability in the terrestrial carbon sink. *Proceedings of the National Academy of Sciences*, **112(51)**, 15591–15596, doi:10.1073/pnas.1521479112.

Anderson, K. and G. Peters, 2016: The trouble with negative emissions. *Science*, **354(6309)**, 182–183, doi:10.1126/science.aah4567.

André, C., D. Boulet, H. Rey-Valette, and B. Rulleau, 2016: Protection by hard defence structures or relocation of assets exposed to coastal risks: Contributions and drawbacks of cost-benefit analysis for long-term adaptation choices to climate change. *Ocean and Coastal Management*, **134**, 173–182, doi:10.1016/j.ocecoaman.2016.10.003.

André, G., B. Engel, P.B.M. Berentsen, T.V. Vellinga, and A.G.J.M. Oude Lansink, 2011: Quantifying the effect of heat stress on daily milk yield and monitoring dynamic changes using an adaptive dynamic model. *Journal of Dairy Science*, **94(9)**, 4502–4513, doi:10.3168/jds.2010-4139.

Anthony, K.R.N., 2016: Coral Reefs Under Climate Change and Ocean Acidification: Challenges and Opportunities for Management and Policy. *Annual Review of Environment and Resources*, **41(1)**, 59–81, doi:10.1146/annurev-environ-110615-085610.

Anthony, K.R.N. et al., 2015: Operationalizing resilience for adaptive coral reef management under global environmental change. *Global Change Biology*, **21(1)**, 48–61, doi:10.1111/gcb.12700.

Araos, M. et al., 2016: Climate change adaptation planning in large cities: A systematic global assessment. *Environmental Science & Policy*, **66**, 375–382, doi:10.1016/j.envsci.2016.06.009.

Arent, D.J. et al., 2014: Key economic sectors and services. In: *Climate Change 2014: Impacts, Adaptation, and Vulnerability. Part A: Global and Sectoral Aspects. Contribution of Working Group II to the Fifth Assessment Report of the Intergovernmental Panel on Climate Change* [Field, C.B., V.R. Barros, D.J. Dokken, K.J. Mach, M.D. Mastrandrea, T.E. Bilir, M. Chatterjee, K.L. Ebi, Y.O. Estrada, R.C. Genova, B. Girma, E.S. Kissel, A.N. Levy, S. MacCracken, P.R. Mastrandrea, and L.L. White (eds.)]. Cambridge University Press, Cambridge, United Kingdom and New York, NY, USA, pp. 659–708.

Argüeso, D., J.P. Evans, L. Fita, and K.J. Bormann, 2014: Temperature response to future urbanization and climate change. *Climate Dynamics*, **42(7–8)**, 2183–2199, doi:10.1007/s00382-013-1789-6.

Argüeso, D., J.P. Evans, A.J. Pitman, A. Di Luca, and A. Luca, 2015: Effects of city expansion on heat stress under climate change conditions. *PLOS ONE*, **10(2)**, e0117066, doi:10.1371/journal.pone.0117066.

Arheimer, B., C. Donnelly, and G. Lindström, 2017: Regulation of snow-fed rivers affects flow regimes more than climate change. *Nature Communications*, **8(1)**, 62, doi:10.1038/s41467-017-00092-8.

Arias-Ortiz, A. et al., 2018: A marine heatwave drives massive losses from the world's largest seagrass carbon stocks. *Nature Climate Change*, **8(4)**, 338–344, doi:10.1038/s41558-018-0096-v.

Arkema, K.K. et al., 2013: Coastal habitats shield people and property from sea-level rise and storms. *Nature Climate Change*, **3(10)**, 913–918, doi:10.1038/nclimate1944.

Armour, K.C., I. Eisenman, E. Blanchard-Wrigglesworth, K.E. McCusker, and C.M. Bitz, 2011: The reversibility of sea ice loss in a state-of-the-art climate model. *Geophysical Research Letters*, **38(16)**, L16705, doi:10.1029/2011gl048739.

Arnell, N.W. and B. Lloyd-Hughes, 2014: The global-scale impacts of climate change on water resources and flooding under new climate and socio-economic scenarios. *Climatic Change*, **122(1–2)**, 127–140, doi:10.1007/s10584-013-0948-4.

Arnell, N.W., J.A. Lowe, B. Lloyd-Hughes, and T.J. Osborn, 2018: The impacts avoided with a 1.5°C climate target: a global and regional assessment. *Climatic Change*, **147(1–2)**, 61–76, doi:10.1007/s10584-017-2115-9.

Arnell, N.W. et al., 2016: Global-scale climate impact functions: the relationship between climate forcing and impact. *Climatic Change*, **134(3)**, 475–487, doi:10.1007/s10584-013-1034-7.

Asiedu, B., J.-O. Adetola, and I. Odame Kissi, 2017a: Aquaculture in troubled climate: Farmers' perception of climate change and their adaptation. *Cogent Food & Agriculture*, **3(1)**, 1296400, doi:10.1080/23311932.2017.1296400.

Asiedu, B., F.K.E. Nunoo, and S. Iddrisu, 2017b: Prospects and sustainability of aquaculture development in Ghana, West Africa. *Cogent Food & Agriculture*, **3(1)**, 1349531, doi:10.1080/23311932.2017.1349531.

Asplund, M.E. et al., 2014: Ocean acidification and host-pathogen interactions: Blue mussels, *Mytilus edulis*, encountering *Vibrio tubiashii*. *Environmental Microbiology*, **16(4)**, 1029–1039, doi:10.1111/1462-2920.12307.

Asseng, S. et al., 2013: Uncertainty in simulating wheat yields under climate change. *Nature Climate Change*, **3(9)**, 827–832, doi:10.1038/nclimate1916.

Asseng, S. et al., 2015: Rising temperatures reduce global wheat production. *Nature Climate Change*, **5(2)**, 143–147, doi:10.1038/nclimate2470.

Atkinson, A., V. Siegel, E.A. Pakhomov, M.J. Jessopp, and V. Loeb, 2009: A re-appraisal of the total biomass and annual production of Antarctic krill. *Deep-Sea Research Part I: Oceanographic Research Papers*, **56(5)**, 727–740, doi:10.1016/j.dsr.2008.12.007.

Auerbach, L.W. et al., 2015: Flood risk of natural and embanked landscapes on the Ganges-Brahmaputra tidal delta plain. *Nature Climate Change*, **5**, 153–157, doi:10.1038/nclimate2472.

Backhaus, A., I. Martinez-Zarzoso, and C. Muris, 2015: Do climate variations explain bilateral migration? A gravity model analysis. *IZA Journal of Migration*, **4(1)**, 3, doi:10.1186/s40176-014-0026-3.

Bader, D.A. et al., 2018: Urban Climate Science. In: *Climate Change and Cities: Second Assessment Report of the Urban Climate Change Research Network* [Rosenzweig, C., W. Solecki, P. Romero-Lankao, S. Mehrotra, S. Dhakal, and S.A. Ibrahim (eds.)]. Cambridge University Press, Cambridge, United Kingdom and New York, NY, USA, pp. 27–60.

Bajželj, B. et al., 2014: Importance of food-demand management for climate mitigation. *Nature Climate Change*, **4(10)**, 924–929, doi:10.1038/nclimate2353.

Baker, A.C., P.W. Glynn, and B. Riegl, 2008: Climate change and coral reef bleaching: An ecological assessment of long-term impacts, recovery trends and future outlook. *Estuarine, Coastal and Shelf Science*, **80(4)**, 435–471, doi:10.1016/j.ecss.2008.09.003.

Baker, H.S. et al., 2018: Higher CO_2 concentrations increase extreme event risk in a 1.5°C world. *Nature Climate Change*, **8(7)**, 604–608, doi:10.1038/s41558-018-0190-1.

Bakun, A., 1990: Global climate change and intensification of coastal ocean upwelling. *Science*, **247(4939)**, 198–201, doi:10.1126/science.247.4939.198.

Bakun, A. et al., 2015: Anticipated Effects of Climate Change on Coastal Upwelling Ecosystems. *Current Climate Change Reports*, **1(2)**, 85–93, doi:10.1007/s40641-015-0008-4.

Baldwin, A. and E. Fornalé, 2017: Adaptive migration: pluralising the debate on climate change and migration. *The Geographical Journal*, **183(4)**, 322–328, doi:10.1111/geoj.12242.

Barange, M. et al., 2014: Impacts of climate change on marine ecosystem production in societies dependent on fisheries. *Nature Climate Change*, **4(3)**, 211–216, doi:10.1038/nclimate2119.

Barati, F. et al., 2008: Meiotic competence and DNA damage of porcine oocytes exposed to an elevated temperature. *Theriogenology*, **69(6)**, 767–772, doi:10.1016/j.theriogenology.2007.08.038.

Barbier, E.B., 2015: Valuing the storm protection service of estuarine and coastal ecosystems. *Ecosystem Services*, **11**, 32–38, doi:10.1016/j.ecoser.2014.06.010.

Barnett, J. et al., 2014: A local coastal adaptation pathway. *Nature Climate Change*, **4(12)**, 1103–1108, doi:10.1038/nclimate2383.

Bassu, S. et al., 2014: How do various maize crop models vary in their responses to climate change factors? *Global Change Biology*, **20(7)**, 2301–2320, doi:10.1111/gcb.12520.

Bates, N.R. and A.J. Peters, 2007: The contribution of atmospheric acid deposition to ocean acidification in the subtropical North Atlantic Ocean. *Marine Chemistry*, **107(4)**, 547–558, doi:10.1016/j.marchem.2007.08.002.

Bauer, N. et al., 2018: Global energy sector emission reductions and bioenergy use: overview of the bioenergy demand phase of the EMF-33 model comparison. *Climatic Change*, 1–16, doi:10.1007/s10584-018-2226-y.

Bayraktarov, E. et al., 2016: The cost and feasibility of marine coastal restoration. *Ecological Applications*, **26(4)**, 1055–1074, doi:10.1890/15-1077.

Bedarff, H. and C. Jakobeit, 2017: *Climate Change, Migration, and Displacement. The Underestimated Disaster*. 38 pp.

Bednaršek, N., C.J. Harvey, I.C. Kaplan, R.A. Feely, and J. Možina, 2016: Pteropods on the edge: Cumulative effects of ocean acidification, warming, and deoxygenation. *Progress in Oceanography*, **145**, 1–24, doi:10.1016/j.pocean.2016.04.002.

Bednaršek, N. et al., 2012: Extensive dissolution of live pteropods in the Southern Ocean. *Nature Geoscience*, **5(12)**, 881–885, doi:10.1038/ngeo1635.

Bednaršek, N. et al., 2014: *Limacina helicina* shell dissolution as an indicator of declining habitat suitability owing to ocean acidification in the California Current Ecosystem. *Proceedings of the Royal Society B: Biological Sciences*, **281(1785)**, 20140123, doi:10.1098/rspb.2014.0123.

Beetham, E., P.S. Kench, and S. Popinet, 2017: Future Reef Growth Can Mitigate Physical Impacts of Sea-Level Rise on Atoll Islands. *Earth's Future*, **5(10)**, 1002–1014, doi:10.1002/2017ef000589.

Bell, J., 2012: Planning for Climate Change and Sea Level Rise: Queensland's New Coastal Plan. *Environmental and Planning Law Journal*, **29(1)**, 61–74, http://ssrn.com/abstract=2611478.

Bell, J. and M. Taylor, 2015: *Building Climate-Resilient Food Systems for Pacific Islands*. Program Report: 2015–15, WorldFish, Penang, Malaysia, 72 pp.

Bell, J.D., J.E. Johnson, and A.J. Hobday, 2011: *Vulnerability of tropical pacific fisheries and aquaculture to climate change*. Secretariat of the Pacific Community (SPC), Noumea, New Caledonia, 925 pp.

Bell, J.D. et al., 2013: Mixed responses of tropical Pacific fisheries and aquaculture to climate change. *Nature Climate Change*, **3(6)**, 591–599, doi:10.1038/nclimate1838.

Bell, J.D. et al., 2018: Adaptations to maintain the contributions of small-scale fisheries to food security in the Pacific Islands. *Marine Policy*, **88**, 303–314, doi:10.1016/j.marpol.2017.05.019.

Bender, M.A. et al., 2010: Modeled Impact of Anthropogenic Warming on the Frequency of Intense Atlantic Hurricanes. *Science*, **327(5964)**, 454–458, doi:10.1126/science.1180568.

Benjamin, L. and A. Thomas, 2016: 1.5°C To Stay Alive?: AOSIS and the Long Term Temperature Goal in the Paris Agreement. *IUCNAEL eJournal*, **7**, 122–129, www.iucnael.org/en/documents/1324-iucn-ejournal-issue-7.

Benjamini, Y. and Y. Hochberg, 1995: Controlling the False Discovery Rate: A Practical and Powerful Approach to Multiple Testing. *Journal of the Royal Statistical Society. Series B (Methodological)*, **57**, 289–300, www.jstor.org/stable/2346101.

Bertram, C. et al., 2018: Targeted policies can compensate most of the increased sustainability risks in 1.5°C mitigation scenarios. *Environmental Research Letters*, **13(6)**, 064038, doi:10.1088/1748-9326/aac3ec.

Bettini, G., 2017: Where Next? Climate Change, Migration, and the (Bio)politics of Adaptation. *Global Policy*, **8(S1)**, 33–39, doi:10.1111/1758-5899.12404.

Betts, R.A. et al., 2015: Climate and land use change impacts on global terrestrial ecosystems and river flows in the HadGEM2-ES Earth system model using the representative concentration pathways. *Biogeosciences*, **12(5)**, 1317–1338, doi:10.5194/bg-12-1317-2015.

Betts, R.A. et al., 2018: Changes in climate extremes, fresh water availability and vulnerability to food insecurity projected at 1.5°C and 2°C global warming with a higher-resolution global climate model. *Philosophical Transactions of the Royal Society A: Mathematical, Physical and Engineering Sciences*, **376(2119)**, 20160452, doi:10.1098/rsta.2016.0452.

Beyer, H.L. et al., 2018: Risk-sensitive planning for conserving coral reefs under rapid climate change. *Conservation Letters*, e12587, doi:10.1111/conl.12587.

Bichet, A. and A. Diedhiou, 2018: West African Sahel has become wetter during the last 30 years, but dry spells are shorter and more frequent. *Climate Research*, **75(2)**, 155–162, doi:10.3354/cr01515.

Bindoff, N.L. et al., 2013a: Detection and Attribution of Climate Change: from Global to Regional. In: *Climate Change 2013: The Physical Science Basis. Contribution of Working Group I to the Fifth Assessment Report of the Intergovernmental Panel on Climate Change* [Stocker, T.F., D. Qin, G.-K. Plattner, M. Tignor, S.K. Allen, J. Boschung, A. Nauels, Y. Xia, V. Bex, and P.M. Midgley (eds.)]. Cambridge University Press, Cambridge, United Kingdom and New York, NY, USA, pp. 867–952.

3

Bindoff, N.L. et al., 2013b: Detection and Attribution of Climate Change: from Global to Regional – Supplementary Material. In: *Climate Change 2013: The Physical Science Basis. Contribution of Working Group I to the Fifth Assessment Report of the Intergovernmental Panel on Climate Change* [Stocker, T.F., D. Qin, G.-K. Plattner, M. Tignor, S.K. Allen, J. Boschung, A. Nauels, Y. Xia, V. Bex, and P.M. Midgley (eds.)]. Cambridge University Press, Cambridge, United Kingdom and New York, NY, USA, pp. 1–25.

Bispo, A. et al., 2017: Accounting for Carbon Stocks in Soils and Measuring GHGs Emission Fluxes from Soils: Do We Have the Necessary Standards? *Frontiers in Environmental Science*, **5**, 1–12, doi:10.3389/fenvs.2017.00041.

Bittermann, K., S. Rahmstorf, R.E. Kopp, and A.C. Kemp, 2017: Global mean sea-level rise in a world agreed upon in Paris. *Environmental Research Letters*, **12(12)**, 124010, doi:10.1088/1748-9326/aa9def.

Blankespoor, B., S. Dasgupta, and B. Laplante, 2014: Sea-Level Rise and Coastal Wetlands. *Ambio*, **43(8)**, 996–1005, doi:10.1007/s13280-014-0500-4.

Block, P. and K. Strzepek, 2012: Power Ahead: Meeting Ethiopia's Energy Needs Under a Changing Climate. *Review of Development Economics*, **16(3)**, 476–488, doi:10.1111/j.1467-9361.2012.00675.x.

Boehlert, B. et al., 2015: Climate change impacts and greenhouse gas mitigation effects on US water quality. *Journal of Advances in Modeling Earth Systems*, **7(3)**, 1326–1338, doi:10.1002/2014ms000400.

Bonal, D., B. Burban, C. Stahl, F. Wagner, and B. Hérault, 2016: The response of tropical rainforests to drought – lessons from recent research and future prospects. *Annals of Forest Science*, **73(1)**, 27–44, doi:10.1007/s13595-015-0522-5.

Bongaerts, P., T. Ridgway, E.M. Sampayo, and O. Hoegh-Guldberg, 2010: Assessing the 'deep reef refugia' hypothesis: focus on Caribbean reefs. *Coral Reefs*, **29(2)**, 309–327, doi:10.1007/s00338-009-0581-x.

Bongaerts, P., C. Riginos, R. Brunner, N. Englebert, and S.R. Smith, 2017: Deep reefs are not universal refuges: reseeding potential varies among coral species. *Science Advances*, **3(2)**, e1602373, doi:10.1126/sciadv.1602373.

Bonsch, M. et al., 2016: Trade-offs between land and water requirements for large-scale bioenergy production. *GCB Bioenergy*, **8(1)**, 11–24, doi:10.1111/gcbb.12226.

Bonte, M. and J.J.G. Zwolsman, 2010: Climate change induced salinisation of artificial lakes in the Netherlands and consequences for drinking water production. *Water Research*, **44(15)**, 4411–4424, doi:10.1016/j.watres.2010.06.004.

Boone, R.B., R.T. Conant, J. Sircely, P.K. Thornton, and M. Herrero, 2018: Climate change impacts on selected global rangeland ecosystem services. *Global Change Biology*, **24(3)**, 1382–1393, doi:10.1111/gcb.13995.

Booth, M.S., 2018: Not carbon neutral: Assessing the net emissions impact of residues burned for bioenergy. *Environmental Research Letters*, **13(3)**, 035001, doi:10.1088/1748-9326/aaac88.

Bopp, L. et al., 2013: Multiple stressors of ocean ecosystems in the 21st century: Projections with CMIP5 models. *Biogeosciences*, **10(10)**, 6225–6245, doi:10.5194/bg-10-6225-2013.

Borma, L.S., C.A. Nobre, and M.F. Cardoso, 2013: 2.15 – Response of the Amazon Tropical Forests to Deforestation, Climate, and Extremes, and the Occurrence of Drought and Fire. In: *Climate Vulnerability* [Pielke, R.A. (ed.)]. Academic Press, Oxford, UK, pp. 153–163, doi:10.1016/b978-0-12-384703-4.00228-8.

Bosello, F. and E. De Cian, 2014: Climate change, sea level rise, and coastal disasters. A review of modeling practices. *Energy Economics*, **46**, 593–605, doi:10.1016/j.eneco.2013.09.002.

Boucher, O. et al., 2012: Reversibility in an Earth System model in response to CO_2 concentration changes. *Environmental Research Letters*, **7(2)**, 024013, doi:10.1088/1748-9326/7/2/024013.

Boucher, O. et al., 2013a: Rethinking climate engineering categorization in the context of climate change mitigation and adaptation. *Wiley Interdisciplinary Reviews: Climate Change*, **5(1)**, 23–35, doi:10.1002/wcc.261.

Boucher, O. et al., 2013b: Clouds and Aerosols. In: *Climate Change 2013: The Physical Science Basis. Contribution of Working Group I to the Fifth Assessment Report of the Intergovernmental Panel on Climate Change* [Stocker, T.F., D. Qin, G.-K. Plattner, M. Tignor, S.K. Allen, J. Boschung, A. Nauels, Y. Xia, V. Bex, and P.M. Midgley (eds.)]. Cambridge University Press, Cambridge, United Kingdom and New York, NY, USA, pp. 573–657.

Bouttes, N., J.M. Gregory, and J.A. Lowe, 2013: The Reversibility of Sea Level Rise. *Journal of Climate*, **26(8)**, 2502–2513, doi:10.1175/jcli-d-12-00285.1.

Bouzid, M., F.J. Colón-González, T. Lung, I.R. Lake, and P.R. Hunter, 2014: Climate change and the emergence of vector-borne diseases in Europe: case study of dengue fever. *BMC Public Health*, **14(1)**, 781, doi:10.1186/1471-2458-14-781.

Boyd, P.W., 2015: Toward quantifying the response of the oceans' biological pump to climate change. *Frontiers in Marine Science*, **2**, 77, doi:10.3389/fmars.2015.00077.

Boyd, P.W., S. Sundby, and H.-O. Pörtner, 2014: Cross-chapter box on net primary production in the ocean. In: *Climate Change 2014: Impacts, Adaptation, and Vulnerability. Part A: Global and Sectoral Aspects. Contribution of Working Group II to the Fifth Assessment Report of the Intergovernmental Panel on Climate Change* [Field, C.B., V.R. Barros, D.J. Dokken, K.J. Mach, M.D. Mastrandrea, T.E. Bilir, M. Chatterjee, K.L. Ebi, Y.O. Estrada, R.C. Genova, B. Girma, E.S. Kissel, A.N. Levy, S. MacCracken, P.R. Mastrandrea, and L.L. White (eds.)]. Cambridge University Press, Cambridge, United Kingdom and New York, NY, USA, pp. 133–136.

Boyd, P.W., S.T. Lennartz, D.M. Glover, and S.C. Doney, 2015: Biological ramifications of climate-change-mediated oceanic multi-stressors. *Nature Climate Change*, **5(1)**, 71–79, doi:10.1038/nclimate2441.

Boysen, L. et al., 2017: The limits to global-warming mitigation by terrestrial carbon removal. *Earth's Future*, **5(5)**, 463–474, doi:10.1002/2016ef000469.

Bozec, Y.-M.M., L. Alvarez-Filip, and P.J. Mumby, 2015: The dynamics of architectural complexity on coral reefs under climate change. *Global Change Biology*, **21(1)**, 223–235, doi:10.1111/gcb.12698.

Bråthen, K., V. González, and N. Yoccoz, 2018: Gatekeepers to the effects of climate warming? Niche construction restricts plant community changes along a temperature gradient. *Perspectives in Plant Ecology, Evolution and Systematics*, **30**, 71–81, doi:10.1016/j.ppees.2017.06.005.

Breitburg, D. et al., 2018: Declining oxygen in the global ocean and coastal waters. *Science*, **359(6371)**, eaam7240, doi:10.1126/science.aam7240.

Brigham-Grette, J. et al., 2013: Pliocene Warmth, Polar Amplification, and Stepped Pleistocene Cooling Recorded in NE Arctic Russia. *Science*, **340(6139)**, 1421–1427, doi:10.1126/science.1233137.

Bright, R.M. et al., 2017: Local temperature response to land cover and management change driven by non-radiative processes. *Nature Climate Change*, **7(4)**, 296–302, doi:10.1038/nclimate3250.

Bring, A. et al., 2016: Arctic terrestrial hydrology: A synthesis of processes, regional effects, and research challenges. *Journal of Geophysical Research: Biogeosciences*, **121(3)**, 621–649, doi:10.1002/2015jg003131.

Brito, B.P., L.L. Rodriguez, J.M. Hammond, J. Pinto, and A.M. Perez, 2017: Review of the Global Distribution of Foot-and-Mouth Disease Virus from 2007 to 2014. *Transboundary and Emerging Diseases*, **64(2)**, 316–332, doi:10.1111/tbed.12373.

Brodie, J.E. et al., 2012: Terrestrial pollutant runoff to the Great Barrier Reef: An update of issues, priorities and management responses. *Marine Pollution Bulletin*, **65(4–9)**, 81–100, doi:10.1016/j.marpolbul.2011.12.012.

Brown, S., R.J. Nicholls, J.A. Lowe, and J. Hinkel, 2016: Spatial variations of sea-level rise and impacts: An application of DIVA. *Climatic Change*, **134(3)**, 403–416, doi:10.1007/s10584-013-0925-y.

Brown, S. et al., 2018a: Quantifying Land and People Exposed to Sea-Level Rise with No Mitigation and 1.5°C and 2.0°C Rise in Global Temperatures to Year 2300. *Earth's Future*, **6(3)**, 583–600, doi:10.1002/2017ef000738.

Brown, S. et al., 2018b: What are the implications of sea-level rise for a 1.5, 2 and 3°C rise in global mean temperatures in the Ganges-Brahmaputra-Meghna and other vulnerable deltas? *Regional Environmental Change*, 1–14, doi:10.1007/s10113-018-1311-0.

Bruge, A., P. Alvarez, A. Fontán, U. Cotano, and G. Chust, 2016: Thermal Niche Tracking and Future Distribution of Atlantic Mackerel Spawning in Response to Ocean Warming. *Frontiers in Marine Science*, **3**, 86, doi:10.3389/fmars.2016.00086.

Bruno, J.F. and E.R. Selig, 2007: Regional decline of coral cover in the Indo-Pacific: Timing, extent, and subregional comparisons. *PLOS ONE*, **2(8)**, e711, doi:10.1371/journal.pone.0000711.

Brzoska, M. and C. Fröhlich, 2016: Climate change, migration and violent conflict: vulnerabilities, pathways and adaptation strategies. *Migration and Development*, **5(2)**, 190–210, doi:10.1080/21632324.2015.1022973.

Buhaug, H., 2015: Climate-conflict research: some reflections on the way forward. *Wiley Interdisciplinary Reviews: Climate Change*, **6(3)**, 269–275, doi:10.1002/wcc.336.

Buhaug, H., 2016: Climate Change and Conflict: Taking Stock. *Peace Economics, Peace Science and Public Policy*, **22(4)**, 331–338, doi:10.1515/peps-2016-0034.

Buhaug, H. et al., 2014: One effect to rule them all? A comment on climate and conflict. *Climatic Change*, **127(3–4)**, 391–397, doi:10.1007/s10584-014-1266-1.

Burge, C.A. et al., 2014: Climate Change Influences on Marine Infectious Diseases: Implications for Management and Society. *Annual Review of Marine Science*, **6**(1), 249–277, doi:10.1146/annurev-marine-010213-135029.

Burke, E.J., S.E. Chadburn, C. Huntingford, and C.D. Jones, 2018: CO_2 loss by permafrost thawing implies additional emissions reductions to limit warming to 1.5 or 2°C. *Environmental Research Letters*, **13**(2), 024024, doi:10.1088/1748-9326/aaa138.

Burke, E.J. et al., 2017: Quantifying uncertainties of permafrost carbon-climate feedbacks. *Biogeosciences*, **14**(12), 3051–3066, doi:10.5194/bg-14-3051-2017.

Burke, L., K. Reytar, M. Spalding, and A. Perry, 2011: *Reefs at risk: Revisited*. World Resources Institute, Washington DC, USA, 115 pp.

Burke, M., S.M. Hsiang, and E. Miguel, 2015a: Climate and Conflict. *Annual Review of Economics*, **7**(1), 577–617, doi:10.1146/annurev-economics-080614-115430.

Burke, M., S.M. Hsiang, and E. Miguel, 2015b: Global non-linear effect of temperature on economic production. *Nature*, **527**(7577), 235–239, doi:10.1038/nature15725.

Burke, M., W.M. Davis, and N.S. Diffenbaugh, 2018: Large potential reduction in economic damages under UN mitigation targets. *Nature*, **557**(7706), 549–553, doi:10.1038/s41586-018-0071-9.

Burrows, K. and P. Kinney, 2016: Exploring the Climate Change, Migration and Conflict Nexus. *International Journal of Environmental Research and Public Health*, **13**(4), 443, doi:10.3390/ijerph13040443.

Burrows, M.T. et al., 2014: Geographical limits to species-range shifts are suggested by climate velocity. *Nature*, **507**(7493), 492–495, doi:10.1038/nature12976.

Butler, E.E. and P. Huybers, 2013: Adaptation of US maize to temperature variations. *Nature Climate Change*, **3**(1), 68–72, doi:10.1038/nclimate1585.

Buurman, J. and V. Babovic, 2016: Adaptation Pathways and Real Options Analysis: An approach to deep uncertainty in climate change adaptation policies. *Policy and Society*, **35**(2), 137–150, doi:10.1016/j.polsoc.2016.05.002.

Byers, E. et al., 2018: Global exposure and vulnerability to multi-sector development and climate change hotspots. *Environmental Research Letters*, **13**(5), 055012, doi:10.1088/1748-9326/aabf45.

Cacciapaglia, C. and R. van Woesik, 2015: Reef-coral refugia in a rapidly changing ocean. *Global Change Biology*, **21**(6), 2272–2282, doi:10.1111/gcb.12851.

Caesar, L., S. Rahmstorf, A. Robinson, G. Feulner, and V. Saba, 2018: Observed fingerprint of a weakening Atlantic Ocean overturning circulation. *Nature*, **556**(7700), 191–196, doi:10.1038/s41586-018-0006-5.

Cai, R., S. Feng, M. Oppenheimer, and M. Pytlikova, 2016: Climate variability and international migration: The importance of the agricultural linkage. *Journal of Environmental Economics and Management*, **79**, 135–151, doi:10.1016/j.jeem.2016.06.005.

Cai, W. et al., 2012: More extreme swings of the South Pacific convergence zone due to greenhouse warming. *Nature*, **488**(7411), 365–369, doi:10.1038/nature11358.

Cai, W. et al., 2015: Increased frequency of extreme La Niña events under greenhouse warming. *Nature Climate Change*, **5**, 132–137, doi:10.1038/nclimate2492.

Cai, W.-J. et al., 2011: Acidification of subsurface coastal waters enhanced by eutrophication. *Nature Geoscience*, **4**(11), 766–770, doi:10.1038/ngeo1297.

Cai, Y., T.M. Lenton, and T.S. Lontzek, 2016: Risk of multiple interacting tipping points should encourage rapid CO_2 emission reduction. *Nature Climate Change*, **6**(5), 520–525, doi:10.1038/nclimate2964.

Caldeira, K., 2013: Coral Bleaching: Coral 'refugia' amid heating seas. *Nature Climate Change*, **3**(5), 444–445, doi:10.1038/nclimate1888.

Callaway, R. et al., 2012: Review of climate change impacts on marine aquaculture in the UK and Ireland. *Aquatic Conservation: Marine and Freshwater Ecosystems*, **22**(3), 389–421, doi:10.1002/aqc.2247.

Camargo, S.J., 2013: Global and Regional Aspects of Tropical Cyclone Activity in the CMIP5 Models. *Journal of Climate*, **26**(24), 9880–9902, doi:10.1175/jcli-d-12-00549.1.

Campbell, J. and O. Warrick, 2014: *Climate Change and Migration Issues in the Pacific*. United Nations Economic and Social Commission for Asia and the Pacific (UNESCAP) Pacific Office, Suva, Fiji, 34 pp.

Cao, L. and K. Caldeira, 2010: Atmospheric carbon dioxide removal: long-term consequences and commitment. *Environmental Research Letters*, **5**(2), 024011, doi:10.1088/1748-9326/5/2/024011.

Cao, L., K. Caldeira, and A.K. Jain, 2007: Effects of carbon dioxide and climate change on ocean acidification and carbonate mineral saturation. *Geophysical Research Letters*, **34**(5), L05607, doi:10.1029/2006gl028605.

Capo, E. et al., 2017: Tracking a century of changes in microbial eukaryotic diversity in lakes driven by nutrient enrichment and climate warming. *Environmental Microbiology*, **19**(7), 2873–2892, doi:10.1111/1462-2920.13815.

Capron, E., A. Govin, R. Feng, B.L. Otto-Bliesner, and E.W. Wolff, 2017: Critical evaluation of climate syntheses to benchmark CMIP6/PMIP4 127 ka Last Interglacial simulations in the high-latitude regions. *Quaternary Science Reviews*, **168**, 137–150, doi:10.1016/j.quascirev.2017.04.019.

Carleton, T.A. and S.M. Hsiang, 2016: Social and economic impacts of climate. *Science*, **353**(6304), aad9837, doi:10.1126/science.aad9837.

Carleton, T.A., S.M. Hsiang, and M. Burke, 2016: Conflict in a changing climate. *The European Physical Journal Special Topics*, **225**(3), 489–511, doi:10.1140/epjst/e2015-50100-5.

Carstensen, J., J.H. Andersen, B.G. Gustafsson, and D.J. Conley, 2014: Deoxygenation of the Baltic Sea during the last century. *Proceedings of the National Academy of Sciences*, **111**(15), 5628–33, doi:10.1073/pnas.1323156111.

Carvalho, D., A. Rocha, M. Gómez-Gesteira, and C. Silva Santos, 2017: Potential impacts of climate change on European wind energy resource under the CMIP5 future climate projections. *Renewable Energy*, **101**, 29–40, doi:10.1016/j.renene.2016.08.036.

Castillo, N., L.M. Saavedra, C.A. Vargas, C. Gallardo-Escárate, and C. Détrée, 2017: Ocean acidification and pathogen exposure modulate the immune response of the edible mussel *Mytilus chilensis*. *Fish and Shellfish Immunology*, **70**, 149–155, doi:10.1016/j.fsi.2017.08.047.

Cazenave, A. and G. Cozannet, 2014: Sea level rise and its coastal impacts. *Earth's Future*, **2**(2), 15–34, doi:10.1002/2013ef000188.

Cazenave, A. et al., 2014: The rate of sea-level rise. *Nature Climate Change*, **4**(5), 358–361, doi:10.1038/nclimate2159.

Ceccarelli, S. and J.E. Rabinovich, 2015: Global Climate Change Effects on Venezuela's Vulnerability to Chagas Disease is Linked to the Geographic Distribution of Five Triatomine Species. *Journal of Medical Entomology*, **52**(6), 1333–1343, doi:10.1093/jme/tjv119.

Chadburn, S.E. et al., 2017: An observation-based constraint on permafrost loss as a function of global warming. *Nature Climate Change*, 1–6, doi:10.1038/nclimate3262.

Challinor, A.J. et al., 2014: A meta-analysis of crop yield under climate change and adaptation. *Nature Climate Change*, **4**(4), 287–291, doi:10.1038/nclimate2153.

Chapman, A. and S. Darby, 2016: Evaluating sustainable adaptation strategies for vulnerable mega-deltas using system dynamics modelling: Rice agriculture in the Mekong Delta's An Giang Province, Vietnam. *Science of The Total Environment*, **559**, 326–338, doi:10.1016/j.scitotenv.2016.02.162.

Chapman, A.D., S.E. Darby, H.M. Hông, E.L. Tompkins, and T.P.D. Van, 2016: Adaptation and development trade-offs: fluvial sediment deposition and the sustainability of rice-cropping in An Giang Province, Mekong Delta. *Climatic Change*, **137**(3–4), 1–16, doi:10.1007/s10584-016-1684-3.

Cheal, A.J., M.A. MacNeil, M.J. Emslie, and H. Sweatman, 2017: The threat to coral reefs from more intense cyclones under climate change. *Global Change Biology*, **23**(4), 1511–1524, doi:10.1111/gcb.13593.

Chen, B. et al., 2014: The impact of climate change and anthropogenic activities on alpine grassland over the Qinghai-Tibet Plateau. *Agricultural and Forest Meteorology*, **189**, 11–18, doi:10.1016/j.agrformet.2014.01.002.

Chen, C., G.S. Zhou, and L. Zhou, 2014: Impacts of climate change on rice yield in china from 1961 to 2010 based on provincial data. *Journal of Integrative Agriculture*, **13**(7), 1555–1564, doi:10.1016/s2095-3119(14)60816-9.

Chen, J. et al., 2017: Assessing changes of river discharge under global warming of 1.5°C and 2°C in the upper reaches of the Yangtze River Basin: Approach by using multiple- GCMs and hydrological models. *Quaternary International*, **453**, 1–11, doi:10.1016/j.quaint.2017.01.017.

Cheung, W.W.L., R. Watson, and D. Pauly, 2013: Signature of ocean warming in global fisheries catch. *Nature*, **497**(7449), 365–368, doi:10.1038/nature12156.

Cheung, W.W.L., G. Reygondeau, and T.L. Frölicher, 2016a: Large benefits to marine fisheries of meeting the 1.5°C global warming target. *Science*, **354**(6319), 1591–1594, doi:10.1126/science.aag2331.

Cheung, W.W.L. et al., 2009: Projecting global marine biodiversity impacts under climate change scenarios. *Fish and Fisheries*, **10**(3), 235–251, doi:10.1111/j.1467-2979.2008.00315.x.

Cheung, W.W.L. et al., 2010: Large-scale redistribution of maximum fisheries catch potential in the global ocean under climate change. *Global Change Biology*, **16**(1), 24–35, doi:10.1111/j.1365-2486.2009.01995.x.

Cheung, W.W.L. et al., 2016b: Structural uncertainty in projecting global fisheries catches under climate change. *Ecological Modelling*, **325**, 57–66, doi:10.1016/j.ecolmodel.2015.12.018.

3

Chiew, F.H.S. et al., 2014: Observed hydrologic non-stationarity in far south-eastern Australia: Implications for modelling and prediction. *Stochastic Environmental Research and Risk Assessment*, **28(1)**, 3–15, doi:10.1007/s00477-013-0755-5.

Chilkoti, V., T. Bolisetti, and R. Balachandar, 2017: Climate change impact assessment on hydropower generation using multi-model climate ensemble. *Renewable Energy*, **109**, 510–517, doi:10.1016/j.renene.2017.02.041.

Cho, S.J. and B.A. McCarl, 2017: Climate change influences on crop mix shifts in the United States. *Scientific Reports*, **7**, 40845, doi:10.1038/srep40845.

Chollett, I. and P.J. Mumby, 2013: Reefs of last resort: Locating and assessing thermal refugia in the wider Caribbean. *Biological Conservation*, **167(2013)**, 179–186, doi:10.1016/j.biocon.2013.08.010.

Chollett, I., P.J. Mumby, and J. Cortés, 2010: Upwelling areas do not guarantee refuge for coral reefs in a warming ocean. *Marine Ecology Progress Series*, **416**, 47–56, doi:10.2307/24875251.

Chollett, I., S. Enríquez, and P.J. Mumby, 2014: Redefining Thermal Regimes to Design Reserves for Coral Reefs in the Face of Climate Change. *PLOS ONE*, **9(10)**, e110634, doi:10.1371/journal.pone.0110634.

Christensen, J.H. et al., 2007: Regional Climate Projections. In: *Climate Change 2007: The Physical Science Basis. Contribution of Working Group I to the Fourth Assessment Report of the Intergovernmental Panel on Climate Change* [Solomon, S., D. Qin, M. Manning, Z. Chen, M. Marquis, K.B. Averyt, M. Tignor, and H.L. Miller (eds.)]. Cambridge University Press, Cambridge, United Kingdom and New York, NY, USA, pp. 847–940.

Christensen, J.H. et al., 2013: Climate Phenomena and their Relevance for Future Regional Climate Change. In: *Climate Change 2013: The Physical Science Basis. Contribution of Working Group I to the Fifth Assessment Report of the Intergovernmental Panel on Climate Change* [Stocker, T.F., D. Qin, G.-K. Plattner, M. Tignor, S.K. Allen, J. Boschung, A. Nauels, Y. Xia, V. Bex, and P.M. Midgley (eds.)]. Cambridge University Press, Cambridge, United Kingdom and New York, NY, USA, pp. 1217–1308.

Christiansen, S.M., 2016: Introduction. In: *Climate Conflicts – A Case of International Environmental and Humanitarian Law*. Springer International Publishing, Cham, Switzerland, pp. 1–17, doi:10.1007/978-3-319-27945-9_1.

Chung, E.S., H.-K. Cheong, J.-H. Park, J.-H. Kim, and H. Han, 2017: Current and Projected Burden of Disease From High Ambient Temperature in Korea. *Epidemiology*, **28**, S98–S105, doi:10.1097/ede.0000000000000731.

Church, J. et al., 2013: Sea level change. In: *Climate Change 2013: The Physical Science Basis. Contribution of Working Group I to the Fifth Assessment Report of the Intergovernmental Panel on Climate Change* [Stocker, T.F., D. Qin, G.-K. Plattner, M. Tignor, S.K. Allen, J. Boschung, A. Nauels, Y. Xia, V. Bex, and P.M. Midgley (eds.)]. Cambridge University Press, Cambridge, United Kingdom and New York, NY, USA, pp. 1137–1216.

Chust, G., 2014: Are *Calanus* spp. shifting poleward in the North Atlantic? A habitat modelling approach. *ICES Journal of Marine Science*, **71(2)**, 241–253, doi:10.1093/icesjms/fst147.

Chust, G. et al., 2014: Biomass changes and trophic amplification of plankton in a warmer ocean. *Global Change Biology*, **20(7)**, 2124–2139, doi:10.1111/gcb.12562.

Ciais, P. et al., 2013: Carbon and Other Biogeochemical Cycles. In: *Climate Change 2013: The Physical Science Basis. Contribution of Working Group I to the Fifth Assessment Report of the Intergovernmental Panel on Climate Change* [Stocker, T.F., D. Qin, G.K. Plattner, M. Tignor, S.K. Allen, J. Boschung, A. Nauels, Y. Xia, V. Bex, and P.M. Midgley (eds.)]. Cambridge University Press, Cambridge, United Kingdom and New York, NY, USA, pp. 465–570.

Cinner, J.E. et al., 2012: Vulnerability of coastal communities to key impacts of climate change on coral reef fisheries. *Global Environmental Change*, **22(1)**, 12–20, doi:10.1016/j.gloenvcha.2011.09.018.

Cinner, J.E. et al., 2016: A framework for understanding climate change impacts on coral reef social-ecological systems. *Regional Environmental Change*, **16(4)**, 1133–1146, doi:10.1007/s10113-015-0832-z.

Ciscar, J.C. (ed.), 2014: *Climate impacts in Europe – The JRC PESETA II project*. EUR – Scientific and Technical Research, Publications Office of the European Union, doi:10.2791/7409.

Clark, P.U. et al., 2016: Consequences of twenty-first-century policy for multi-millennial climate and sea-level change. *Nature Climate Change*, **6(4)**, 360–369, doi:10.1038/nclimate2923.

Clarke, L.E. et al., 2014: Assessing Transformation Pathways. In: *Climate Change 2014: Mitigation of Climate Change. Contribution of Working Group III to the Fifth Assessment Report of the Intergovernmental Panel on Climate Change* [Edenhofer, O., R. Pichs-Madruga, Y. Sokona, E. Farahani, S. Kadner, K. Seyboth, A. Adler, I. Baum, S. Brunner, P. Eickemeier, B. Kriemann, J. Savolainen, S. Schlömer, C. von Stechow, T. Zwickel, and J.C. Minx (eds.)]. Cambridge University Press, Cambridge, United Kingdom and New York, NY, USA, pp. 413–510.

Clausen, K.K. and P. Clausen, 2014: Forecasting future drowning of coastal waterbird habitats reveals a major conservation concern. *Biological Conservation*, **171**, 177–185, doi:10.1016/j.biocon.2014.01.033.

Clements, J.C. and T. Chopin, 2017: Ocean acidification and marine aquaculture in North America: Potential impacts and mitigation strategies. *Reviews in Aquaculture*, **9(4)**, 326341, doi:10.1111/raq.12140.

Clements, J.C., D. Bourque, J. McLaughlin, M. Stephenson, and L.A. Comeau, 2017: Extreme ocean acidification reduces the susceptibility of eastern oyster shells to a polydorid parasite. *Journal of Fish Diseases*, **40(11)**, 1573–1585, doi:10.1111/jfd.12626.

Collier, R.J. and K.G. Gebremedhin, 2015: Thermal Biology of Domestic Animals. *Annual Review of Animal Biosciences*, **3(1)**, 513–532, doi:10.1146/annurev-animal-022114-110659.

Collins, M. et al., 2013: Long-term Climate Change: Projections, Commitments and Irreversibility. In: *Climate Change 2013: The Physical Science Basis. Contribution of Working Group I to the Fifth Assessment Report of the Intergovernmental Panel on Climate Change* [Stocker, T.F., D. Qin, G.-K. Plattner, M. Tignor, S.K. Allen, J. Boschung, A. Nauels, Y. Xia, V. Bex, and P.M. Midgley (eds.)]. Cambridge University Press, Cambridge, United Kingdom and New York, NY, USA, pp. 1029–1136.

Colón-González, F.J., C. Fezzi, I.R. Lake, P.R. Hunter, and Y. Sukthana, 2013: The Effects of Weather and Climate Change on Dengue. *PLOS Neglected Tropical Diseases*, **7(11)**, e2503, doi:10.1371/journal.pntd.0002503.

Colón-González, F.J. et al., 2018: Limiting global-mean temperature increase to 1.5–2°C could reduce the incidence and spatial spread of dengue fever in Latin America. *Proceedings of the National Academy of Sciences*, 115(24), 6243–6248, doi:10.1073/pnas.1718945115.

Coniglio, N.D. and G. Pesce, 2015: Climate variability and international migration: an empirical analysis. *Environment and Development Economics*, **20(04)**, 434–468, doi:10.1017/s1355770x14000722.

Conlon, K., A. Monaghan, M. Hayden, and O. Wilhelmi, 2016: Potential impacts of future warming and land use changes on intra-urban heat exposure in Houston, Texas. *PLOS ONE*, **11(2)**, e0148890, doi:10.1371/journal.pone.0148890.

Constable, A.L., 2017: Climate change and migration in the Pacific: options for Tuvalu and the Marshall Islands. *Regional Environmental Change*, **17(4)**, 1029–1038, doi:10.1007/s10113-016-1004-5.

Cook, B.I., M.J. Puma, and N.Y. Krakauer, 2011: Irrigation induced surface cooling in the context of modern and increased greenhouse gas forcing. *Climate Dynamics*, **37(7–8)**, 1587–1600, doi:10.1007/s00382-010-0932-x.

Cook, B.I., K.J. Anchukaitis, R. Touchan, D.M. Meko, and E.R. Cook, 2016: Spatiotemporal drought variability in the Mediterranean over the last 900 years. *Journal of Geophysical Research: Atmospheres*, **121(5)**, 2060–2074, doi:10.1002/2015jd023929.

Cooper, E.J., 2014: Warmer Shorter Winters Disrupt Arctic Terrestrial Ecosystems. *Annual Review of Ecology, Evolution, and Systematics*, **45**, 271–295, doi:10.1146/annurev-ecolsys-120213-091620.

Cooper, J.A.G., M.C. O'Connor, and S. McIvor, 2016: Coastal defences versus coastal ecosystems: A regional appraisal. *Marine Policy*, doi:10.1016/j.marpol.2016.02.021.

Costanza, R. et al., 2014: Changes in the global value of ecosystem services. *Global Environmental Change*, **26(1)**, 152–158, doi:10.1016/j.gloenvcha.2014.04.002.

Coumou, D. and A. Robinson, 2013: Historic and future increase in the global land area affected by monthly heat extremes. *Environmental Research Letters*, **8(3)**, 034018, doi:10.1088/1748-9326/8/3/034018.

Cousino, L.K., R.H. Becker, and K.A. Zmijewski, 2015: Modeling the effects of climate change on water, sediment, and nutrient yields from the Maumee River watershed. *Journal of Hydrology: Regional Studies*, **4**, 762–775, doi:10.1016/j.ejrh.2015.06.017.

Cowtan, K. and R.G. Way, 2014: Coverage bias in the HadCRUT4 temperature series and its impact on recent temperature trends. *Quarterly Journal of the Royal Meteorological Society*, **140(683)**, 1935–1944, doi:10.1002/qj.2297.

Cox, P.M., R.A. Betts, C.D. Jones, S.A. Spall, and I.J. Totterdell, 2000: Acceleration of global warming due to carbon-cycle feedbacks in a coupled climate model. *Nature*, **408**, 184–187, doi:10.1038/35041539.

Cox, P.M. et al., 2013: Sensitivity of tropical carbon to climate change constrained by carbon dioxide variability. *Nature*, **494**, 341–344, doi:10.1038/nature11882.

Craine, J.M., A.J. Elmore, K.C. Olson, and D. Tolleson, 2010: Climate change and cattle nutritional stress. *Global Change Biology*, **16(10)**, 2901–2911, doi:10.1111/j.1365-2486.2009.02060.x.

3

Cramer, W. et al., 2014: Detection and Attribution of Observed Impacts. In: *Climate Change 2014: Impacts, Adaptation, and Vulnerability. Part A: Global and Sectoral Aspects. Contribution of Working Group II to the Fifth Assessment Report of the Intergovernmental Panel on Climate Change* [Field, C.B., V.R. Barros, D.J. Dokken, K.J. Mach, and M.D. Mastrandrea (eds.)]. Cambridge University Press, Cambridge, United Kingdom and New York, NY, USA, pp. 979–1037.

Crosby, S.C. et al., 2016: Salt marsh persistence is threatened by predicted sea-level rise. *Estuarine, Coastal and Shelf Science*, **181**, 93–99, doi:10.1016/j.ecss.2016.08.018.

Crowther, T.W. et al., 2016: Quantifying global soil carbon losses in response to warming. *Nature*, **540(7631)**, 104–108, doi:10.1038/nature20150.

Croxall, J.P., 1992: Southern-Ocean Environmental-Changes – Effects on Seabird, Seal and Whale Populations. *Philosophical Transactions of the Royal Society B: Biological Sciences*, **338(1285)**, 319–328, doi:10.1098/rstb.1992.0152.

Cui, L., Z. Ge, L. Yuan, and L. Zhang, 2015: Vulnerability assessment of the coastal wetlands in the Yangtze Estuary, China to sea-level rise. *Estuarine, Coastal and Shelf Science*, **156**, 42–51, doi:10.1016/j.ecss.2014.06.015.

Cunningham, S.C., 2015: Balancing the environmental benefits of reforestation in agricultural regions. *Perspectives in Plant Ecology, Evolution and Systematics*, **17(4)**, 301–317, doi:10.1016/j.ppees.2015.06.001.

d'Amour, C.B., L. Wenz, M. Kalkuhl, J.C. Steckel, and F. Creutzig, 2016: Teleconnected food supply shocks. *Environmental Research Letters*, **11(3)**, 035007, doi:10.1088/1748-9326/11/3/035007.

Dai, A., 2016: Historical and Future Changes in Streamflow and Continental Runoff. In: *Terrestrial Water Cycle and Climate Change: Natural and Human-Induced Impacts* [Tang, Q. and T. Oki (eds.)]. American Geophysical Union (AGU), Washington DC, USA, pp. 17–37, doi:10.1002/9781118971772.ch2.

Daioglou, V. et al., 2016: Projections of the availability and cost of residues from agriculture and forestry. *GCB Bioenergy*, **8(2)**, 456–470, doi:10.1111/gcbb.12285.

Dale, V.H. et al., 2017: Status and prospects for renewable energy using wood pellets from the southeastern United States. *GCB Bioenergy*, **9(8)**, 1296–1305, doi:10.1111/gcbb.12445.

Daliakopoulos, I.N. et al., 2017: Yield Response of Mediterranean Rangelands under a Changing Climate. *Land Degradation & Development*, **28(7)**, 1962–1972, doi:10.1002/ldr.2717.

Dalin, C. and I. Rodríguez-Iturbe, 2016: Environmental impacts of food trade via resource use and greenhouse gas emissions. *Environmental Research Letters*, **11(3)**, 035012, doi:10.1088/1748-9326/11/3/035012.

Dalpadado, P. et al., 2014: Productivity in the Barents Sea – Response to recent climate variability. *PLOS ONE*, **9(5)**, e95273, doi:10.1371/journal.pone.0095273.

Dargie, G.C. et al., 2017: Age, extent and carbon storage of the central Congo Basin peatland complex. *Nature*, **542(7639)**, 86–90, doi:10.1038/nature21048.

Dasgupta, P. et al., 2014: Rural areas. In: *Climate Change 2014: Impacts, Adaptation, and Vulnerability. Part A: Global and Sectoral Aspects. Contribution of Working Group II to the Fifth Assessment Report of the Intergovernmental Panel on Climate Change* [Field, C.B., V.R. Barros, D.J. Dokken, K.J. Mach, M.D. Mastrandrea, T.E. Bilir, M. Chatterjee, K.L. Ebi, Y.O. Estrada, R.C. Genova, B. Girma, E.S. Kissel, A.N. Levy, S. MacCracken, P.R. Mastrandrea, and L.L. White (eds.)]. Cambridge University Press, Cambridge, United Kingdom and New York, NY, USA, pp. 613–657.

David, C. et al., 2017: Community structure of under-ice fauna in relation to winter sea-ice habitat properties from the Weddell Sea. *Polar Biology*, **40(2)**, 247–261, doi:10.1007/s00300-016-1948-4.

Davin, E.L., S.I. Seneviratne, P. Ciais, A. Olioso, and T. Wang, 2014: Preferential cooling of hot extremes from cropland albedo management. *Proceedings of the National Academy of Sciences*, **111(27)**, 9757–9761, doi:10.1073/pnas.1317323111.

Dawson, R.J. et al., 2018: Urban areas in coastal zones. In: *Climate Change and Cities: Second Assessment Report of the Urban Climate Change Research Network* [Rosenzweig, C., W.D. Solecki, P. Romero-Lankao, S. Mehrotra, S. Dhakal, and S.A. Ibrahim (eds.)]. Cambridge Univeristy Press, Cambridge, United Kingdom and New York, NY, USA, pp. 319–362.

de Jong, J. et al., 2014: *Consequences of an increased extraction of forest biofuel in Sweden – A synthesis from the biofuel research programme 2007–2011, supported by Swedish Energy Agency. Summary of the synthesis report.* ER 2014:09, Swedish Energy Agency, Eskilstuna, Sweden, 37 pp.

de Queiroz, A.R., L.M. Marangon Lima, J.W. Marangon Lima, B.C. da Silva, and L.A. Scianni, 2016: Climate change impacts in the energy supply of the Brazilian hydro-dominant power system. *Renewable Energy*, **99**, 379–389, doi:10.1016/j.renene.2016.07.022.

De Rensis, F., I. Garcia-Ispierto, and F. López-Gatius, 2015: Seasonal heat stress: Clinical implications and hormone treatments for the fertility of dairy cows. *Theriogenology*, **84(5)**, 659–666, doi:10.1016/j.theriogenology.2015.04.021.

de Sherbinin, A., 2014: Climate change hotspots mapping: what have we learned? *Climatic Change*, **123(1)**, 23–37, doi:10.1007/s10584-013-0900-7.

De Souza, A.P., J.-C. Cocuron, A.C. Garcia, A.P. Alonso, and M.S. Buckeridge, 2015: Changes in Whole-Plant Metabolism during the Grain-Filling Stage in Sorghum Grown under Elevated CO_2 and Drought. *Plant Physiology*, **169(3)**, 1755–1765, doi:10.1104/pp.15.01054.

de Vernal, A., R. Gersonde, H. Goosse, M.-S. Seidenkrantz, and E.W. Wolff, 2013: Sea ice in the paleoclimate system: the challenge of reconstructing sea ice from proxies – an introduction. *Quaternary Science Reviews*, **79**, 1–8, doi:10.1016/j.quascirev.2013.08.009.

de Vrese, P., S. Hagemann, and M. Claussen, 2016: Asian irrigation, African rain: Remote impacts of irrigation. *Geophysical Research Letters*, **43(8)**, 3737–3745, doi:10.1002/2016gl068146.

De'ath, G., K.E. Fabricius, H. Sweatman, and M. Puotinen, 2012: The 27-year decline of coral cover on the Great Barrier Reef and its causes. *Proceedings of the National Academy of Sciences*, **109(44)**, 17995–9, doi:10.1073/pnas.1208909109.

DeBeer, C.M., H.S. Wheater, S.K. Carey, and K.P. Chun, 2016: Recent climatic, cryospheric, and hydrological changes over the interior of western Canada: a review and synthesis. *Hydrology and Earth System Sciences*, **20(4)**, 1573–1598, doi:10.5194/hess-20-1573-2016.

DeConto, R.M. and D. Pollard, 2016: Contribution of Antarctica to past and future sea-level rise. *Nature*, **531(7596)**, 591–7, doi:10.1038/nature17145.

Déqué, M. et al., 2017: A multi-model climate response over tropical Africa at +2°C. *Climate Services*, **7**, 87–95, doi:10.1016/j.cliser.2016.06.002.

Deryng, D., D. Conway, N. Ramankutty, J. Price, and R. Warren, 2014: Global crop yield response to extreme heat stress under multiple climate change futures. *Environmental Research Letters*, **9(3)**, 034011, doi:10.1088/1748-9326/9/3/034011.

Deutsch, C., A. Ferrel, B. Seibel, H.-O. Pörtner, and R.B. Huey, 2015: Climate change tightens a metabolic constraint on marine habitats. *Science*, **348(6239)**, 1132–1135, doi:10.1126/science.aaa1605.

Di Lorenzo, E., 2015: Climate science: The future of coastal ocean upwelling. *Nature*, **518(7539)**, 310–311, doi:10.1038/518310a.

Di Nitto, D. et al., 2014: Mangroves facing climate change: Landward migration potential in response to projected scenarios of sea level rise. *Biogeosciences*, **11(3)**, 857–871, doi:10.5194/bg-11-857-2014.

Diaz, R.J. and R. Rosenberg, 2008: Spreading Dead Zones and Consequences for Marine Ecosystems. *Science*, **321(5891)**, 926–929, doi:10.1126/science.1156401.

Diedhiou, A. et al., 2018: Changes in climate extremes over West and Central Africa at 1.5°C and 2°C global warming. *Environmental Research Letters*, **13(6)**, 065020, doi:10.1088/1748-9326/aac3e5.

Dieleman, C.M., Z. Lindo, J.W. McLaughlin, A.E. Craig, and B.A. Branfireun, 2016: Climate change effects on peatland decomposition and porewater dissolved organic carbon biogeochemistry. *Biogeochemistry*, **128(3)**, 385–396, doi:10.1007/s10533-016-0214-8.

Dionisio, K.L. et al., 2017: Characterizing the impact of projected changes in climate and air quality on human exposures to ozone. *Journal of Exposure Science and Environmental Epidemiology*, **27**, 260–270, doi:10.1038/jes.2016.81.

Do, H.X., S. Westra, and M. Leonard, 2017: A global-scale investigation of trends in annual maximum streamflow. *Journal of Hydrology*, **552**, 28–43, doi:10.1016/j.jhydrol.2017.06.015.

Döll, P. et al., 2018: Risks for the global freshwater system at 1.5°C and 2°C global warming. *Environmental Research Letters*, **13(4)**, 044038, doi:10.1088/1748-9326/aab792.

Dolman, A.J. et al., 2010: A Carbon Cycle Science Update Since IPCC AR4. *Ambio*, **39(5–6)**, 402–412, doi:10.1007/s13280-010-0083-7.

Doney, S., L. Bopp, and M. Long, 2014: Historical and Future Trends in Ocean Climate and Biogeochemistry. *Oceanography*, **27(1)**, 108–119, doi:10.5670/oceanog.2014.14.

Donk, P., E. Van Uytven, P. Willems, and M.A. Taylor, 2018: Assessment of the potential implications of a 1.5°C versus higher global temperature rise for the Afobaka hydropower scheme in Suriname. *Regional Environmental Change*, 1–13, doi:10.1007/s10113-018-1339-1.

Donnelly, C. et al., 2017: Impacts of climate change on European hydrology at 1.5, 2 and 3 degrees mean global warming above preindustrial level. *Climatic Change*, **143(1–2)**, 13–26, doi:10.1007/s10584-017-1971-7.

Donner, S.D., 2009: Coping with commitment: Projected thermal stress on coral reefs under different future scenarios. *PLOS ONE*, **4**(6), e5712, doi:10.1371/journal.pone.0005712.

Donner, S.D., W.J. Skirving, C.M. Little, M. Oppenheimer, and O. Hoegh-Guldberg, 2005: Global assessment of coral bleaching and required rates of adaptation under climate change. *Global Change Biology*, **11**(12), 2251–2265, doi:10.1111/j.1365-2486.2005.01073.x.

Dosio, A., 2017: Projection of temperature and heat waves for Africa with an ensemble of CORDEX Regional Climate Models. *Climate Dynamics*, **49**(1), 493–519, doi:10.1007/s00382-016-3355-5.

Dosio, A. and E.M. Fischer, 2018: Will half a degree make a difference? Robust projections of indices of mean and extreme climate in Europe under 1.5°C, 2°C, and 3°C global warming. *Geophysical Research Letters*, **45**, 935–944, doi:10.1002/2017gl076222.

Dosio, A., L. Mentaschi, E.M. Fischer, and K. Wyser, 2018: Extreme heat waves under 1.5°C and 2°C global warming. *Environmental Research Letters*, **13**(5), 054006, doi:10.1088/1748-9326/aab827.

Dove, S.G. et al., 2013: Future reef decalcification under a business-as-usual CO_2 emission scenario. *Proceedings of the National Academy of Sciences*, **110**(38), 15342–15347, doi:10.1073/pnas.1302701110.

Dowsett, H. et al., 2016: The PRISM4 (mid-Piacenzian) paleoenvironmental reconstruction. *Climate of the Past*, **12**(7), 1519–1538, doi:10.5194/cp-12-1519-2016.

Doyon, B., D. Belanger, and P. Gosselin, 2008: The potential impact of climate change on annual and seasonal mortality for three cities in Québec, Canada. *International Journal of Health Geographics*, **7**, 23, doi:10.1186/1476-072x-7-23.

Draper, F.C. et al., 2014: The distribution and amount of carbon in the largest peatland complex in Amazonia. *Environmental Research Letters*, **9**(12), 124017, doi:10.1088/1748-9326/9/12/124017.

Drijfhout, S. et al., 2015: Catalogue of abrupt shifts in Intergovernmental Panel on Climate Change climate models. *Proceedings of the National Academy of Sciences*, **112**(43), E5777–E5786, doi:10.1073/pnas.1511451112.

Drouet, A.S. et al., 2013: Grounding line transient response in marine ice sheet models. *The Cryosphere*, **7**(2), 395–406, doi:10.5194/tc-7-395-2013.

Duarte, C.M. et al., 2013: Is Ocean Acidification an Open-Ocean Syndrome? Understanding Anthropogenic Impacts on Seawater pH. *Estuaries and Coasts*, **36**(2), 221–236, doi:10.1007/s12237-013-9594-3.

Duke, N.C. et al., 2017: Large-scale dieback of mangroves in Australia's Gulf of Carpentaria: A severe ecosystem response, coincidental with an unusually extreme weather event. *Marine and Freshwater Research*, **68**(10), 1816–1829, doi:10.1071/mf16322.

Dunne, J.P., R.J. Stouffer, and J.G. John, 2013: Reductions in labour capacity from heat stress under climate warming. *Nature Climate Change*, **3**(4), 1–4, doi:10.1038/nclimate1827.

Duputié, A., A. Rutschmann, O. Ronce, and I. Chuine, 2015: Phenological plasticity will not help all species adapt to climate change. *Global Change Biology*, **21**(8), 3062–3073, doi:10.1111/gcb.12914.

Durack, P.J., S.E. Wijffels, and R.J. Matear, 2012: Ocean Salinities Reveal Strong Global Water Cycle Intensification During 1950 to 2000. *Science*, **336**(6080), 455–458, doi:10.1126/science.1212222.

Durand, G. and F. Pattyn, 2015: Reducing uncertainties in projections of Antarctic ice mass loss. *The Cryosphere*, **9**(6), 2043–2055, doi:10.5194/tc-9-2043-2015.

Durand, J.-L. et al., 2018: How accurately do maize crop models simulate the interactions of atmospheric CO_2 concentration levels with limited water supply on water use and yield? *European Journal of Agronomy*, **100**, 67–75, doi:10.1016/j.eja.2017.01.002.

Dutton, A. et al., 2015: Sea-level rise due to polar ice-sheet mass loss during past warm periods. *Science*, **349**(6244), doi:10.1126/science.aaa4019.

Ebi, K.L., N.H. Ogden, J.C. Semenza, and A. Woodward, 2017: Detecting and Attributing Health Burdens to Climate Change. *Environmental Health Perspectives*, **125**(8), 085004, doi:10.1289/ehp1509.

Ebi, K.L. et al., 2018: Health risks of warming of 1.5°C, 2°C, and higher, above pre-industrial temperatures. *Environmental Research Letters*, **13**(6), 063007, doi:10.1088/1748-9326/aac4bd.

Ekstrom, J.A. et al., 2015: Vulnerability and adaptation of US shellfisheries to ocean acidification. *Nature Climate Change*, **5**(3), 207–214, doi:10.1038/nclimate2508.

Elliff, C.I. and I.R. Silva, 2017: Coral reefs as the first line of defense: Shoreline protection in face of climate change. *Marine Environmental Research*, **127**, 148–154, doi:10.1016/j.marenvres.2017.03.007.

Elliott, J. et al., 2014: Constraints and potentials of future irrigation water availability on agricultural production under climate change. *Proceedings of the National Academy of Sciences*, **111**(9), 3239–3244, doi:10.1073/pnas.1222474110.

Ellison, D. et al., 2017: Trees, forests and water: Cool insights for a hot world. *Global Environmental Change*, **43**, 51–61, doi:10.1016/j.gloenvcha.2017.01.002.

Ellison, J.C., 2014: Climate Change Adaptation: Management Options for Mangrove Areas. In: *Mangrove Ecosystems of Asia: Status, Challenges and Management Strategies* [Faridah-Hanum, I., A. Latiff, K.R. Hakeem, and M. Ozturk (eds.)]. Springer New York, New York, NY, USA, pp. 391–413, doi:10.1007/978-1-4614-8582-7_18.

Ellsworth, D.S. et al., 2017: Elevated CO_2 does not increase eucalypt forest productivity on a low-phosphorus soil. *Nature Climate Change*, **7**(4), 279–282, doi:10.1038/nclimate3235.

Elsner, J.B., J.P. Kossin, and T.H. Jagger, 2008: The increasing intensity of the strongest tropical cyclones. *Nature*, **455**(7209), 92–95, doi:10.1038/nature07234.

Emanuel, K., 2005: Increasing destructiveness of tropical cyclones over the past 30 years. *Nature*, **436**(7051), 686–688, doi:10.1038/nature03906.

Emanuel, K., 2017: A fast intensity simulator for tropical cyclone risk analysis. *Natural Hazards*, **88**(2), 779–796, doi:10.1007/s11069-017-2890-7.

Endres, S., L. Galgani, U. Riebesell, K.G. Schulz, and A. Engel, 2014: Stimulated bacterial growth under elevated pCO_2: Results from an off-shore mesocosm study. *PLOS ONE*, **9**(6), e99228, doi:10.1371/journal.pone.0099228.

Engelbrecht, C.J. and F.A. Engelbrecht, 2016: Shifts in Köppen-Geiger climate zones over southern Africa in relation to key global temperature goals. *Theoretical and Applied Climatology*, **123**(1–2), 247–261, doi:10.1007/s00704-014-1354-1.

Engelbrecht, F.A., J.L. McGregor, and C.J. Engelbrecht, 2009: Dynamics of the Conformal-Cubic Atmospheric Model projected climate-change signal over southern Africa. *International Journal of Climatology*, **29**(7), 1013–1033, doi:10.1002/joc.1742.

Engelbrecht, F.A. et al., 2015: Projections of rapidly rising surface temperatures over Africa under low mitigation. *Environmental Research Letters*, **10**(8), 085004, doi:10.1088/1748-9326/10/8/085004.

Fabricius, K.E. et al., 2011: Losers and winners in coral reefs acclimatized to elevated carbon dioxide concentrations. *Nature Climate Change*, **1**(3), 165–169, doi:10.1038/nclimate1122.

Fajardy, M. and N. Mac Dowell, 2017: Can BECCS deliver sustainable and resource efficient negative emissions? *Energy & Environmental Science*, **10**(6), 1389–1426, doi:10.1039/c7ee00465f.

Fang, J.K.H., C.H.L. Schönberg, M.A. Mello-Athayde, O. Hoegh-Guldberg, and S. Dove, 2014: Effects of ocean warming and acidification on the energy budget of an excavating sponge. *Global Change Biology*, **20**(4), 1043–1054, doi:10.1111/gcb.12369.

Fang, J.K.H. et al., 2013: Sponge biomass and bioerosion rates increase under ocean warming and acidification. *Global Change Biology*, **19**(12), 3581–3591, doi:10.1111/gcb.12334.

FAO, 2016: *The State of World Fisheries and Aquaculture 2016. Contributing to food security and nutrition for all*. Food and Agriculture Organization of the United Nations (FAO), Rome, Italy, 200 pp.

FAO, IFAD, UNICEF, WFP, and WHO, 2017: *The State of Food Security and Nutrition in the World 2017. Building resilience for peace and food security*. Food and Agricultural Organization of the United Nations (FAO), Rome, Italy, 117 pp.

Farbotko, C. and H. Lazrus, 2012: The first climate refugees? Contesting global narratives of climate change in Tuvalu. *Global Environmental Change*, **22**(2), 382–390, doi:10.1016/j.gloenvcha.2011.11.014.

Fasullo, J.T., R.S. Nerem, and B. Hamlington, 2016: Is the detection of accelerated sea level rise imminent? *Scientific Reports*, **6**, 1–7, doi:10.1038/srep31245.

Fawcett, A.A. et al., 2015: Can Paris pledges avert severe climate change? *Science*, **350**(6265), 1168–1169, doi:10.1126/science.aad5761.

Faye, B. et al., 2018: Impacts of 1.5 versus 2.0°C on cereal yields in the West African Sudan Savanna. *Environmental Research Letters*, **13**(3), 034014, doi:10.1088/1748-9326/aaab40.

Feely, R.A., C.L. Sabine, J.M. Hernandez-Ayon, D. Ianson, and B. Hales, 2008: Evidence for Upwelling of Corrosive "Acidified" Water onto the Continental Shelf. *Science*, **320**(5882), 1490–1492, doi:10.1126/science.1155676.

Feely, R.A. et al., 2016: Chemical and biological impacts of ocean acidification along the west coast of North America. *Estuarine, Coastal and Shelf Science*, **183**, 260–270, doi:10.1016/j.ecss.2016.08.043.

3

Feller, I.C., D.A. Friess, K.W. Krauss, and R.R. Lewis, 2017: The state of the world's mangroves in the 21st century under climate change. *Hydrobiologia*, **803**(1), 1–12, doi:10.1007/s10750-017-3331-z.

Feng, X. et al., 2016: Revegetation in China's Loess Plateau is approaching sustainable water resource limits. *Nature Climate Change*, **6**(11), 1019–1022, doi:10.1038/nclimate3092.

Ferrario, F. et al., 2014: The effectiveness of coral reefs for coastal hazard risk reduction and adaptation. *Nature Communications*, **5**, 3794, doi:10.1038/ncomms4794.

Fine, M., H. Gildor, and A. Genin, 2013: A coral reef refuge in the Red Sea. *Global Change Biology*, **19**(12), 3640–3647, doi:10.1111/gcb.12356.

Fischer, D., S.M. Thomas, F. Niemitz, B. Reineking, and C. Beierkuhnlein, 2011: Projection of climatic suitability for *Aedes albopictus* Skuse (Culicidae) in Europe under climate change conditions. *Global and Planetary Change*, **78**(1–2), 54–64, doi:10.1016/j.gloplacha.2011.05.008.

Fischer, D. et al., 2013: Climate change effects on Chikungunya transmission in Europe: geospatial analysis of vector's climatic suitability and virus' temperature requirements. *International Journal of Health Geographics*, **12**(1), 51, doi:10.1186/1476-072x-12-51.

Fischer, E.M. and R. Knutti, 2015: Anthropogenic contribution to global occurrence of heavy-precipitation and high-temperature extremes. *Nature Climate Change*, **5**(6), 560–564, doi:10.1038/nclimate2617.

Fischer, E.M., J. Sedláček, E. Hawkins, and R. Knutti, 2014: Models agree on forced response pattern of precipitation and temperature extremes. *Geophysical Research Letters*, **41**(23), 8554–8562, doi:10.1002/2014gl062018.

Fischer, H. et al., 2018: Palaeoclimate constraints on the impact of 2°C anthropogenic warming and beyond. *Nature Geoscience*, **11**, 1–12, doi:10.1038/s41561-018-0146-0.

Ford, J.D., 2012: Indigenous health and climate change. *American Journal of Public Health*, **102**(7), 1260–1266, doi:10.2105/ajph.2012.300752.

Ford, J.D., G. McDowell, and T. Pearce, 2015: The adaptation challenge in the Arctic. *Nature Climate Change*, **5**(12), 1046–1053, doi:10.1038/nclimate2723.

Forseth, I., 2010: Terrestrial Biomes. *Nature Education Knowledge*, **3**(10), 11, www.nature.com/scitable/knowledge/library/terrestrial-biomes-13236757.

Fossheim, M. et al., 2015: Recent warming leads to a rapid borealization of fish communities in the Arctic. *Nature Climate Change*, **5**(7), 673–677, doi:10.1038/nclimate2647.

Fox, N.J. et al., 2011: Predicting Impacts of Climate Change on *Fasciola hepatica* Risk. *PLOS ONE*, **6**(1), e16126, doi:10.1371/journal.pone.0016126.

Frank, D.C. et al., 2010: Ensemble reconstruction constraints on the global carbon cycle sensitivity to climate. *Nature*, **463**, 527, doi:10.1038/nature08769.

Fricko, O. et al., 2016: Energy sector water use implications of a 2°C climate policy. *Environmental Research Letters*, **11**(3), 034011, doi:10.1088/1748-9326/11/3/034011.

Frieler, K. et al., 2013: Limiting global warming to 2° C is unlikely to save most coral reefs. *Nature Climate Change*, **3**(2), 165–170, doi:10.1038/nclimate1674.

Frieler, K. et al., 2015: Consistent evidence of increasing Antarctic accumulation with warming. *Nature Climate Change*, **5**(4), 348–352, doi:10.1038/nclimate2574.

Frieler, K. et al., 2017: Assessing the impacts of 1.5°C global warming – simulation protocol of the Inter-Sectoral Impact Model Intercomparison Project (ISIMIP2b). *Geoscientific Model Development*, **10**, 4321–4345, doi:10.5194/gmd-10-4321-2017.

Friend, A.D. et al., 2014: Carbon residence time dominates uncertainty in terrestrial vegetation responses to future climate and atmospheric CO_2. *Proceedings of the National Academy of Sciences*, **111**(9), 3280–3285, doi:10.1073/pnas.1222477110.

Fronzek, S., T.R. Carter, and M. Luoto, 2011: Evaluating sources of uncertainty in modelling the impact of probabilistic climate change on sub-arctic palsa mires. *Natural Hazards and Earth System Science*, **11**(11), 2981–2995, doi:10.5194/nhess-11-2981-2011.

Frost, A.J. et al., 2011: A comparison of multi-site daily rainfall downscaling techniques under Australian conditions. *Journal of Hydrology*, **408**(1), 1–18, doi:10.1016/j.jhydrol.2011.06.021.

Fu, X. and J. Song, 2017: Assessing the economic costs of sea level rise and benefits of coastal protection: A spatiotemporal approach. *Sustainability*, **9**(8), 1495, doi:10.3390/su9081495.

Fürst, J.J., H. Goelzer, and P. Huybrechts, 2015: Ice-dynamic projections of the Greenland ice sheet in response to atmospheric and oceanic warming. *The Cryosphere*, **9**(3), 1039–1062, doi:10.5194/tc-9-1039-2015.

Fuss, S. et al., 2014: Betting on negative emissions. *Nature Climate Change*, **4**(10), 850–853, doi:10.1038/nclimate2392.

Fuss, S. et al., 2018: Negative emissions-Part 2: Costs, potentials and side effects. *Environmental Research Letters*, **13**(6), 063002, doi:10.1088/1748-9326/aabf9f.

Gagne, M.-E., J.C. Fyfe, N.P. Gillett, I. Polyakov, and G.M. Flato, 2017: Aerosol-driven increase in Arctic sea ice over the middle of the twentieth century. *Geophysical Research Letters*, **44**(14), 7338–7346, doi:10.1002/2016gl071941.

Galaasen, E.V. et al., 2014: Rapid Reductions in North Atlantic Deep Water During the Peak of the Last Interglacial Period. *Science*, **343**(6175), 1129–1132, doi:10.1126/science.1248667.

Gallup, J.L., J.D. Sachs, and A.D. Mellinger, 1999: Geography and Economic Development. *International Regional Science Review*, **22**(2), 179–232, doi:10.1177/016001799761012334.

Gang, C. et al., 2015: Projecting the dynamics of terrestrial net primary productivity in response to future climate change under the RCP2.6 scenario. *Environmental Earth Sciences*, **74**(7), 5949–5959, doi:10.1007/s12665-015-4618-x.

García Molinos, J. et al., 2015: Climate velocity and the future global redistribution of marine biodiversity. *Nature Climate Change*, **6**(1), 83–88, doi:10.1038/nclimate2769.

Gardner, T.A., I.M. Côté, J.A. Gill, A. Grant, and A.R. Watkinson, 2005: Hurricanes and caribbean coral reefs: Impacts, recovery patterns, and role in long-term decline. *Ecology*, **86**(1), 174–184, doi:10.1890/04-0141.

Garland, R.M. et al., 2015: Regional Projections of Extreme Apparent Temperature Days in Africa and the Related Potential Risk to Human Health. *International Journal of Environmental Research and Public Health*, **12**(10), 12577–12604, doi:10.3390/ijerph121012577.

Garrabou, J. et al., 2009: Mass mortality in Northwestern Mediterranean rocky benthic communities: Effects of the 2003 heat wave. *Global Change Biology*, **15**(5), 1090–1103, doi:10.1111/j.1365-2486.2008.01823.x.

Garrard, S.L. et al., 2014: Indirect effects may buffer negative responses of seagrass invertebrate communities to ocean acidification. *Journal of Experimental Marine Biology and Ecology*, **461**, 31–38, doi:10.1016/j.jembe.2014.07.011.

Gasparrini, A. et al., 2015: Mortality risk attributable to high and low ambient temperature: A multicountry observational study. *The Lancet*, **386**(9991), 369–375, doi:10.1016/s0140-6736(14)62114-0.

Gass, P., H. Hove, and P. Jo-Ellen, 2011: *Review of Current and Planned Adaptation Action: East and Southeast Asia*. Adaptation Partnership, 217 pp.

Gattuso, J.-P. et al., 2014: Cross-chapter box on ocean acidification. In: *Climate Change 2014: Impacts, Adaptation, and Vulnerability. Part A: Global and Sectoral Aspects. Contribution of Working Group II to the Fifth Assessment Report of the Intergovernmental Panel on Climate Change* [Field, C.B., V.R. Barros, D.J. Dokken, K.J. Mach, M.D. Mastrandrea, T.E. Bilir, M. Chatterjee, K.L. Ebi, Y.O. Estrada, R.C. Genova, B. Girma, E.S. Kissel, A.N. Levy, S. MacCracken, P.R. Mastrandrea, and L.L. White (eds.)]. Cambridge University Press, Cambridge, United Kingdom and New York, NY, USA, pp. 129–131.

Gattuso, J.-P. et al., 2015: Contrasting futures for ocean and society from different anthropogenic CO_2 emissions scenarios. *Science*, **349**(6243), aac4722, doi:10.1126/science.aac4722.

Gauthier, S., P. Bernier, T. Kuuluvainen, A.Z. Shvidenko, and D.G. Schepaschenko, 2015: Boreal forest health and global change. *Science*, **349**(6250), 819–822, doi:10.1126/science.aaa9092.

Geisler, C. and B. Currens, 2017: Impediments to inland resettlement under conditions of accelerated sea level rise. *Land Use Policy*, **66**, 322–330, doi:10.1016/j.landusepol.2017.03.029.

Georgescu, M., M. Moustaoui, A. Mahalov, and J. Dudhia, 2012: Summer-time climate impacts of projected megapolitan expansion in Arizona. *Nature Climate Change*, **3**(1), 37–41, doi:10.1038/nclimate1656.

Gerten, D., S. Rost, W. von Bloh, and W. Lucht, 2008: Causes of change in 20th century global river discharge. *Geophysical Research Letters*, **35**(20), L20405, doi:10.1029/2008gl035258.

Gerten, D. et al., 2013: Asynchronous exposure to global warming: freshwater resources and terrestrial ecosystems. *Environmental Research Letters*, **8**(3), 034032, doi:10.1088/1748-9326/8/3/034032.

Gharbaoui, D. and J. Blocher, 2016: The Reason Land Matters: Relocation as Adaptation to Climate Change in Fiji Islands. In: *Migration, Risk Management and Climate Change: Evidence and Policy Responses* [Milan, A., B. Schraven, K. Warner, and N. Cascone (eds.)]. Springer, Cham, Switzerland, pp. 149–173, doi:10.1007/978-3-319-42922-9_8.

Ghosh, J., 2010: The unnatural coupling: Food and global finance. *Journal of Agrarian Change*, **10**(1), 72–86, doi:10.1111/j.1471-0366.2009.00249.x.

3

Giardino, A., K. Nederhoff, and M. Vousdoukas, 2018: Coastal hazard risk assessment for small islands: assessing the impact of climate change and disaster reduction measures on Ebeye (Marshall Islands). *Regional Environmental Change*, 1–12, doi:10.1007/s10113-018-1353-3.

Goddard, P.B., C.O. Dufour, J. Yin, S.M. Griffies, and M. Winton, 2017: CO$_2$-Induced Ocean Warming of the Antarctic Continental Shelf in an Eddying Global Climate Model. *Journal of Geophysical Research: Oceans*, **122(10)**, 8079–8101, doi:10.1002/2017jc012849.

Goll, D.S. et al., 2012: Nutrient limitation reduces land carbon uptake in simulations with a model of combined carbon, nitrogen and phosphorus cycling. *Biogeosciences*, **9(9)**, 3547–3569, doi:10.5194/bg-9-3547-2012.

Golledge, N.R. et al., 2015: The multi-millennial Antarctic commitment to future sea-level rise. *Nature*, **526(7573)**, 421–425, doi:10.1038/nature15706.

González, C., A. Paz, and C. Ferro, 2014: Predicted altitudinal shifts and reduced spatial distribution of *Leishmania infantum* vector species under climate change scenarios in Colombia. *Acta Tropica*, **129**, 83–90, doi:10.1016/j.actatropica.2013.08.014.

Good, P., C. Jones, J. Lowe, R. Betts, and N. Gedney, 2013: Comparing Tropical Forest Projections from Two Generations of Hadley Centre Earth System Models, HadGEM2-ES and HadCM3LC. *Journal of Climate*, **26(2)**, 495–511, doi:10.1175/jcli-d-11-00366.1.

Good, P. et al., 2011: Quantifying Environmental Drivers of Future Tropical Forest Extent. *Journal of Climate*, **24(5)**, 1337–1349, doi:10.1175/2010jcli3865.1.

Goodwin, P., I.D. Haigh, E.J. Rohling, and A. Slangen, 2017: A new approach to projecting 21st century sea-level changes and extremes. *Earth's Future*, **5(2)**, 240–253, doi:10.1002/2016ef000508.

Goodwin, P., S. Brown, I.D. Haigh, R.J. Nicholls, and J.M. Matter, 2018: Adjusting Mitigation Pathways to Stabilize Climate at 1.5°C and 2.0°C Rise in Global Temperatures to Year 2300. *Earth's Future*, **6(3)**, 601–615, doi:10.1002/2017ef000732.

Gosling, S.N. and N.W. Arnell, 2016: A global assessment of the impact of climate change on water scarcity. *Climatic Change*, **134(3)**, 371–385, doi:10.1007/s10584-013-0853-x.

Gosling, S.N. et al., 2017: A comparison of changes in river runoff from multiple global and catchment-scale hydrological models under global warming scenarios of 1°C, 2°C and 3°C. *Climatic Change*, **141(3)**, 577–595, doi:10.1007/s10584-016-1773-3.

Graham, N.A.J., 2014: Habitat complexity: Coral structural loss leads to fisheries declines. *Current Biology*, **24(9)**, R359–R361, doi:10.1016/j.cub.2014.03.069.

Graham, N.A.J., S. Jennings, M.A. MacNeil, D. Mouillot, and S.K. Wilson, 2015: Predicting climate-driven regime shifts versus rebound potential in coral reefs. *Nature*, **518(7537)**, 1–17, doi:10.1038/nature14140.

Graux, A.-I., G. Bellocchi, R. Lardy, and J.-F. Soussana, 2013: Ensemble modelling of climate change risks and opportunities for managed grasslands in France. *Agricultural and Forest Meteorology*, **170**, 114–131, doi:10.1016/j.agrformet.2012.06.010.

Green, A.L. et al., 2014: Designing Marine Reserves for Fisheries Management, Biodiversity Conservation, and Climate Change Adaptation. *Coastal Management*, **42(2)**, 143–159, doi:10.1080/08920753.2014.877763.

Green, S.M. and S. Page, 2017: Tropical peatlands: current plight and the need for responsible management. *Geology Today*, **33(5)**, 174–179, doi:10.1111/gto.12197.

Gregory, J.M. and P. Huybrechts, 2006: Ice-sheet contributions to future sea-level change. *Philosophical Transactions of the Royal Society A: Mathematical, Physical and Engineering Sciences*, **364(1844)**, 1709–1731, doi:10.1098/rsta.2006.1796.

Gregory, P.J. and B. Marshall, 2012: Attribution of climate change: a methodology to estimate the potential contribution to increases in potato yield in Scotland since 1960. *Global Change Biology*, **18(4)**, 1372–1388, doi:10.1111/j.1365-2486.2011.02601.x.

Greve, P. and S.I. Seneviratne, 2015: Assessment of future changes in water availability and aridity. *Geophysical Research Letters*, **42(13)**, 5493–5499, doi:10.1002/2015gl064127.

Greve, P., L. Gudmundsson, and S.I. Seneviratne, 2018: Regional scaling of annual mean precipitation and water availability with global temperature change. *Earth System Dynamics*, **9(1)**, 227–240, doi:10.5194/esd-9-227-2018.

Greve, P. et al., 2014: Global assessment of trends in wetting and drying over land. *Nature Geoscience*, **7(10)**, 716–721, doi:10.1038/ngeo2247.

Grillakis, M.G., A.G. Koutroulis, K.D. Seiradakis, and I.K. Tsanis, 2016: Implications of 2°C global warming in European summer tourism. *Climate Services*, **1**, 30–38, doi:10.1016/j.cliser.2016.01.002.

Grillakis, M.G., A.G. Koutroulis, I.N. Daliakopoulos, and I.K. Tsanis, 2017: A method to preserve trends in quantile mapping bias correction of climate modeled temperature. *Earth System Dynamics*, **8**, 889–900, doi:10.5194/esd-8-889-2017.

Griscom, B.W. et al., 2017: Natural climate solutions. *Proceedings of the National Academy of Sciences*, **114(44)**, 11645–11650, doi:10.1073/pnas.1710465114.

Grossman-Clarke, S., S. Schubert, and D. Fenner, 2017: Urban effects on summertime air temperature in Germany under climate change. *International Journal of Climatology*, **37(2)**, 905–917, doi:10.1002/joc.4748.

Grubler, A. et al., 2018: A low energy demand scenario for meeting the 1.5°C target and sustainable development goals without negative emission technologies. *Nature Energy*, **3(6)**, 515–527, doi:10.1038/s41560-018-0172-6.

Gu, G. and R.F. Adler, 2013: Interdecadal variability/long-term changes in global precipitation patterns during the past three decades: global warming and/or pacific decadal variability? *Climate Dynamics*, **40(11–12)**, 3009–3022, doi:10.1007/s00382-012-1443-8.

Gu, G. and R.F. Adler, 2015: Spatial patterns of global precipitation change and variability during 1901–2010. *Journal of Climate*, **28(11)**, 4431–4453, doi:10.1175/jcli-d-14-00201.1.

Gudmundsson, L. and S.I. Seneviratne, 2016: Anthropogenic climate change affects meteorological drought risk in Europe. *Environmental Research Letters*, **11(4)**, 044005, doi:10.1088/1748-9326/11/4/044005.

Gudmundsson, L., S.I. Seneviratne, and X. Zhang, 2017: Anthropogenic climate change detected in European renewable freshwater resources. *Nature Climate Change*, **7(11)**, 813–816, doi:10.1038/nclimate3416.

Guiot, J. and W. Cramer, 2016: Climate change: The 2015 Paris Agreement thresholds and Mediterranean basin ecosystems. *Science*, **354(6311)**, 4528–4532, doi:10.1126/science.aah5015.

Guis, H. et al., 2012: Modelling the effects of past and future climate on the risk of bluetongue emergence in Europe. *Journal of The Royal Society Interface*, **9(67)**, 339–350, doi:10.1098/rsif.2011.0255.

Guo, Y. et al., 2016: Projecting future temperature-related mortality in three largest Australian cities. *Environmental Pollution*, **208**, 66–73, doi:10.1016/j.envpol.2015.09.041.

Gupta, H., S.-J. Kao, and M. Dai, 2012: The role of mega dams in reducing sediment fluxes: A case study of large Asian rivers. *Journal of Hydrology*, **464–465**, 447–458, doi:10.1016/j.jhydrol.2012.07.038.

Gustafson, S., A. Joehl Cadena, and P. Hartman, 2016: Adaptation planning in the Lower Mekong Basin: merging scientific data with local perspective to improve community resilience to climate change. *Climate and Development*, 1–15, doi:10.1080/17565529.2016.1223593.

Gustafson, S. et al., 2018: Merging science into community adaptation planning processes: a cross-site comparison of four distinct areas of the Lower Mekong Basin. *Climatic Change*, **149(1)**, 91–106, doi:10.1007/s10584-016-1887-7.

Gutiérrez, J.L. et al., 2012: Physical Ecosystem Engineers and the Functioning of Estuaries and Coasts. In: *Treatise on Estuarine and Coastal Science Vol. 7*. Academic Press, Waltham, MA, USA, pp. 53–81, doi:10.1016/b978-0-12-374711-2.00705-1.

Haberl, H., T. Beringer, S.C. Bhattacharya, K.-H. Erb, and M. Hoogwijk, 2010: The global technical potential of bio-energy in 2050 considering sustainability constraints. *Current Opinion in Environmental Sustainability*, **2(5)**, 394–403, doi:10.1016/j.cosust.2010.10.007.

Haberl, H. et al., 2013: Bioenergy: how much can we expect for 2050? *Environmental Research Letters*, **8(3)**, 031004, doi:10.1088/1748-9326/8/3/031004.

Habibi Mohraz, M., A. Ghahri, M. Karimi, and F. Golbabaei, 2016: The Past and Future Trends of Heat Stress Based On Wet Bulb Globe Temperature Index in Outdoor Environment of Tehran City, Iran. *Iranian Journal of Public Health*, **45(6)**, 787–794, http://ijph.tums.ac.ir/index.php/ijph/article/view/7085.

Hadden, D. and A. Grelle, 2016: Changing temperature response of respiration turns boreal forest from carbon sink into carbon source. *Agricultural and Forest Meteorology*, **223**, 30–38, doi:10.1016/j.agrformet.2016.03.020.

Hajat, S., S. Vardoulakis, C. Heaviside, and B. Eggen, 2014: Climate change effects on human health: Projections of temperature-related mortality for the UK during the 2020s, 2050s and 2080s. *Journal of Epidemiology and Community Health*, **68(7)**, 641–648, doi:10.1136/jech-2013-202449.

Hales, S., S. Kovats, S. Lloyd, and D. Campbell-Lendrum (eds.), 2014: *Quantitative risk assessment of the effects of climate change on selected causes of death, 2030s and 2050s*. World Health Organization (WHO), Geneva, Switzerland, 115 pp.

Hall, J. et al., 2014: Understanding flood regime changes in Europe: a state-of-the-art assessment. *Hydrology and Earth System Sciences*, **18(7)**, 2735–2772, doi:10.5194/hess-18-2735-2014.

Hallegatte, S. and J. Rozenberg, 2017: Climate change through a poverty lens. *Nature Climate Change*, **7(4)**, 250–256, doi:10.1038/nclimate3253.

Hallegatte, S., C. Green, R.J. Nicholls, and J. Corfee-Morlot, 2013: Future flood losses in major coastal cities. *Nature Climate Change*, **3(9)**, 802–806, doi:10.1038/nclimate1979.

Hallegatte, S. et al., 2016: *Shock Waves: Managing the Impacts of Climate Change on Poverty*. Climate Change and Development Series, World Bank, Washington DC, USA, 227 pp., doi:10.1596/978-1-4648-0673-5.

Hall-Spencer, J.M. et al., 2008: Volcanic carbon dioxide vents show ecosystem effects of ocean acidification. *Nature*, **454(7200)**, 96–99, doi:10.1038/nature07051.

Halpern, B.S. et al., 2015: Spatial and temporal changes in cumulative human impacts on the world's ocean. *Nature Communications*, **6(1)**, 7615, doi:10.1038/ncomms8615.

Hamukuaya, H., M. O'Toole, and P. Woodhead, 1998: Observations of severe hypoxia and offshore displacement of Cape hake over the Namibian shelf in 1994. *South African Journal of Marine Science*, **19**, 57–59, doi:10.2989/025776198784126809.

Hanasaki, N. et al., 2013: A global water scarcity assessment under Shared Socio-economic Pathways – Part 2: Water availability and scarcity. *Hydrology and Earth System Sciences*, **17(7)**, 2393–2413, doi:10.5194/hess-17-2393-2013.

Handisyde, N., T.C. Telfer, and L.G. Ross, 2016: Vulnerability of aquaculture-related livelihoods to changing climate at the global scale. *Fish and Fisheries*, **18(3)**, 466–488, doi:10.1111/faf.12186.

Hanna, E.G., T. Kjellstrom, C. Bennett, and K. Dear, 2011: Climate Change and Rising Heat: Population Health Implications for Working People in Australia. *Asia-Pacific Journal of Public Health*, **23(2)**, 14s–26s, doi:10.1177/1010539510391457.

Hansen, G. and D. Stone, 2016: Assessing the observed impact of anthropogenic climate change. *Nature Climate Change*, **6(5)**, 532–537, doi:10.1038/nclimate2896.

Hansen, J., R. Ruedy, M. Sato, and K. Lo, 2010: Global surface temperature change. *Reviews of Geophysics*, **48(4)**, RG4004, doi:10.1029/2010rg000345.

Hanson, S. et al., 2011: A global ranking of port cities with high exposure to climate extremes. *Climatic Change*, **104(1)**, 89–111, doi:10.1007/s10584-010-9977-4.

Harrington, L.J. and F.E.L. Otto, 2018: Changing population dynamics and uneven temperature emergence combine to exacerbate regional exposure to heat extremes under 1.5°C and 2°C of warming. *Environmental Research Letters*, **13(3)**, 034011, doi:10.1088/1748-9326/aaaa99.

Hartmann, D.L. et al., 2013: Observations: Atmosphere and Surface. In: *Climate Change 2013: The Physical Science Basis. Contribution of Working Group I to the Fifth Assessment Report of the Intergovernmental Panel on Climate Change* [Stocker, T.F., D. Qin, G.-K. Plattner, M. Tignor, S.K. Allen, J. Boschung, A. Nauels, Y. Xia, V. Bex, and P.M. Midgley (eds.)]. Cambridge University Press, Cambridge, United Kingdom and New York, NY, USA, pp. 159–254.

Harvey, C.A. et al., 2014: Climate-Smart Landscapes: Opportunities and Challenges for Integrating Adaptation and Mitigation in Tropical Agriculture. *Conservation Letters*, **7(2)**, 77–90, doi:10.1111/conl.12066.

Hasegawa, T., S. Fujimori, K. Takahashi, T. Yokohata, and T. Masui, 2016: Economic implications of climate change impacts on human health through undernourishment. *Climatic Change*, **136(2)**, 189–202, doi:10.1007/s10584-016-1606-4.

Hasegawa, T. et al., 2014: Climate Change Impact and Adaptation Assessment on Food Consumption Utilizing a New Scenario Framework. *Environmental Science & Technology*, **48(1)**, 438–445, doi:10.1021/es4034149.

Hashimoto, H. et al., 2013: Structural Uncertainty in Model-Simulated Trends of Global Gross Primary Production. *Remote Sensing*, **5(3)**, 1258–1273, doi:10.3390/rs5031258.

Hatfield, J.L. et al., 2011: Climate Impacts on Agriculture: Implications for Crop Production. *Agronomy Journal*, **103(2)**, 351–370, doi:10.2134/agronj2010.0303.

Hauer, M.E., J.M. Evans, and D.R. Mishra, 2016: Millions projected to be at risk from sea-level rise in the continental United States. *Nature Climate Change*, **6(7)**, 691–695, doi:10.1038/nclimate2961.

Hauri, C., T. Friedrich, and A. Timmermann, 2016: Abrupt onset and prolongation of aragonite undersaturation events in the Southern Ocean. *Nature Climate Change*, **6(2)**, 172–176, doi:10.1038/nclimate2844.

Haustein, K. et al., 2017: A real-time Global Warming Index. *Scientific Reports*, **7(1)**, 15417, doi:10.1038/s41598-017-14828-5.

Haywood, A.M., H.J. Dowsett, and A.M. Dolan, 2016: Integrating geological archives and climate models for the mid-Pliocene warm period. *Nature Communications*, **7**, 10646, doi:10.1038/ncomms10646.

He, Q. and G. Zhou, 2016: Climate-associated distribution of summer maize in China from 1961 to 2010. *Agriculture, Ecosystems & Environment*, **232**, 326–335, doi:10.1016/j.agee.2016.08.020.

Heal, M.R. et al., 2013: Health burdens of surface ozone in the UK for a range of future scenarios. *Environment International*, **61**, 36–44, doi:10.1016/j.envint.2013.09.010.

Heck, V., D. Gerten, W. Lucht, and A. Popp, 2018: Biomass-based negative emissions difficult to reconcile with planetary boundaries. *Nature Climate Change*, **8(2)**, 151–155, doi:10.1038/s41558-017-0064-y.

Herbert, E.R. et al., 2015: A global perspective on wetland salinization: ecological consequences of a growing threat to freshwater wetlands. *Ecosphere*, **6(10)**, 1–43, doi:10.1890/es14-00534.1.

Hewitt, A.J. et al., 2016: Sources of Uncertainty in Future Projections of the Carbon Cycle. *Journal of Climate*, **29(20)**, 7203–7213, doi:10.1175/jcli-d-16-0161.1.

Hidalgo, H.G. et al., 2009: Detection and Attribution of Streamflow Timing Changes to Climate Change in the Western United States. *Journal of Climate*, **22(13)**, 3838–3855, doi:10.1175/2009jcli2470.1.

Hill, T.D. and S.C. Anisfeld, 2015: Coastal wetland response to sea level rise in Connecticut and New York. *Estuarine, Coastal and Shelf Science*, **163**, 185–193, doi:10.1016/j.ecss.2015.06.004.

Hinkel, J. et al., 2014: Coastal flood damage and adaptation costs under 21st century sea-level rise. *Proceedings of the National Academy of Sciences*, **111(9)**, 3292–7, doi:10.1073/pnas.1222469111.

Hirsch, A.L., M. Wilhelm, E.L. Davin, W. Thiery, and S.I. Seneviratne, 2017: Can climate-effective land management reduce regional warming? *Journal of Geophysical Research: Atmospheres*, **122(4)**, 2269–2288, doi:10.1002/2016jd026125.

Hirsch, A.L. et al., 2018: Biogeophysical Impacts of Land-Use Change on Climate Extremes in Low-Emission Scenarios: Results From HAPPI-Land. *Earth's Future*, **6(3)**, 396–409, doi:10.1002/2017ef000744.

HLPE, 2011: *Price volatility and food security. A report by the High Level Panel of Experts on Food Security and Nutrition of the Committee on World Food Security*. The High Level Panel of Experts on Food Security and Nutrition (HLPE), 79 pp.

Hoang, L.P. et al., 2016: Mekong River flow and hydrological extremes under climate change. *Hydrology and Earth System Sciences*, **20(7)**, 3027–3041, doi:10.5194/hess-20-3027-2016.

Hoang, L.P. et al., 2018: Managing flood risks in the Mekong Delta: How to address emerging challenges under climate change and socioeconomic developments. *Ambio*, **47(6)**, 635–649, doi:10.1007/s13280-017-1009-4.

Hobbs, J.P.A. and C.A. Mcdonald, 2010: Increased seawater temperature and decreased dissolved oxygen triggers fish kill at the Cocos (Keeling) Islands, Indian Ocean. *Journal of Fish Biology*, **77(6)**, 1219–1229, doi:10.1111/j.1095-8649.2010.02726.x.

Hobbs, W.R. et al., 2016: A review of recent changes in Southern Ocean sea ice, their drivers and forcings. *Global and Planetary Change*, **143**, 228–250, doi:10.1016/j.gloplacha.2016.06.008.

Hoegh-Guldberg, O., 1999: Climate change, coral bleaching and the future of the world's coral reefs. *Marine and Freshwater Research*, **50(8)**, 839–866, doi:10.1071/mf99078.

Hoegh-Guldberg, O., 2012: The adaptation of coral reefs to climate change: Is the Red Queen being outpaced? *Scientia Marina*, **76(2)**, 403–408, doi:10.3989/scimar.03660.29a.

Hoegh-Guldberg, O., 2014a: Coral reef sustainability through adaptation: Glimmer of hope or persistent mirage? *Current Opinion in Environmental Sustainability*, **7**, 127–133, doi:10.1016/j.cosust.2014.01.005.

Hoegh-Guldberg, O., 2014b: Coral reefs in the Anthropocene: persistence or the end of the line? *Geological Society, London, Special Publications*, **395(1)**, 167–183, doi:10.1144/sp395.17.

Hoegh-Guldberg, O., E.S. Poloczanska, W. Skirving, and S. Dove, 2017: Coral Reef Ecosystems under Climate Change and Ocean Acidification. *Frontiers in Marine Science*, **4**, 158, doi:10.3389/fmars.2017.00158.

Hoegh-Guldberg, O. et al., 2007: Coral Reefs Under Rapid Climate Change and Ocean Acidification. *Science*, **318(5857)**, 1737–1742, doi:10.1126/science.1152509.

Hoegh-Guldberg, O. et al., 2014: The Ocean. In: *Climate Change 2014: Impacts, Adaptation, and Vulnerability. Part B: Regional Aspects. Contribution of Working Group II to the Fifth Assessment Report of the Intergovernmental Panel on Climate Change* [Barros, V.R., C.B. Field, D.J. Dokken, M.D. Mastrandrea, K.J.

3

Mach, T.E. Bilir, M. Chatterjee, K.L. Ebi, Y.O. Estrada, R.C. Genova, B. Girma, E.S. Kissel, A.N. Levy, S. MacCracken, P.R. Mastrandrea, and L.L. White (eds.)]. Cambridge University Press, Cambridge, United Kingdom and New York, NY, USA, pp. 1655–1731.

Hoegh-Guldberg, O. et al., 2015: *Reviving the Ocean Economy: the case for action – 2015*. WWF International, Gland, Switzerland, 60 pp.

Hoffman, J.S., P.U. Clark, A.C. Parnell, and F. He, 2017: Regional and global sea-surface temperatures during the last interglaciation. *Science*, **355(6322)**, 276–279, doi:10.1126/science.aai8464.

Holding, S. and D.M. Allen, 2015: Wave overwash impact on small islands: Generalised observations of freshwater lens response and recovery for multiple hydrogeological settings. *Journal of Hydrology*, **529(Part 3)**, 1324–1335, doi:10.1016/j.jhydrol.2015.08.052.

Holding, S. et al., 2016: Groundwater vulnerability on small islands. *Nature Climate Change*, **6**, 1100–1103, doi:10.1038/nclimate3128.

Holland, G. and C.L. Bruyère, 2014: Recent intense hurricane response to global climate change. *Climate Dynamics*, **42(3–4)**, 617–627, doi:10.1007/s00382-013-1713-0.

Holland, M.M., C.M. Bitz, and B. Tremblay, 2006: Future abrupt reductions in the summer Arctic sea ice. *Geophysical Research Letters*, **33(23)**, L23503, doi:10.1029/2006gl028024.

Hollowed, A.B. and S. Sundby, 2014: Change is coming to the northern oceans. *Science*, **344(6188)**, 1084–1085, doi:10.1126/science.1251166.

Hollowed, A.B. et al., 2013: Projected impacts of climate change on marine fish and fisheries. *ICES Journal of Marine Science*, **70(510)**, 1023–1037, doi:10.1093/icesjms/fst081.

Holmgren, K. et al., 2016: Mediterranean Holocene climate, environment and human societies. *Quaternary Science Reviews*, **136**, 1–4, doi:10.1016/j.quascirev.2015.12.014.

Holstein, D.M., C.B. Paris, A.C. Vaz, and T.B. Smith, 2016: Modeling vertical coral connectivity and mesophotic refugia. *Coral Reefs*, **35(1)**, 23–37, doi:10.1007/s00338-015-1339-2.

Holz, C., L. Siegel, E.B. Johnston, A.D. Jones, and J. Sterman, 2017: Ratcheting Ambition to Limit Warming to 1.5°C – Trade-offs Between Emission Reductions and Carbon Dioxide Removal. *Environmental Research Letters*, doi:10.1088/1748-9326/aac0c1.

Honda, Y. et al., 2014: Heat-related mortality risk model for climate change impact projection. *Environmental Health and Preventive Medicine*, **19(1)**, 56–63, doi:10.1007/s12199-013-0354-6.

Hong, J. and W.S. Kim, 2015: Weather impacts on electric power load: partial phase synchronization analysis. *Meteorological Applications*, **22(4)**, 811–816, doi:10.1002/met.1535.

Hönisch, B. et al., 2012: The geological record of ocean acidification. *Science*, **335(6072)**, 1058–1063, doi:10.1126/science.1208277.

Horta E Costa, B. et al., 2014: Tropicalization of fish assemblages in temperate biogeographic transition zones. *Marine Ecology Progress Series*, **504**, 241–252, doi:10.3354/meps10749.

Hosking, J.S. et al., 2018: Changes in European wind energy generation potential within a 1.5°C warmer world. *Environmental Research Letters*, **13(5)**, 054032, doi:10.1088/1748-9326/aabf78.

Hossain, M.S., L. Hein, F.I. Rip, and J.A. Dearing, 2015: Integrating ecosystem services and climate change responses in coastal wetlands development plans for Bangladesh. *Mitigation and Adaptation Strategies for Global Change*, **20(2)**, 241–261, doi:10.1007/s11027-013-9489-4.

Hosseini, N., J. Johnston, and K.-E. Lindenschmidt, 2017: Impacts of Climate Change on the Water Quality of a Regulated Prairie River. *Water*, **9(3)**, 199, doi:10.3390/w9030199.

Houghton, R.A. and A.A. Nassikas, 2018: Negative emissions from stopping deforestation and forest degradation, globally. *Global Change Biology*, **24(1)**, 350–359, doi:10.1111/gcb.13876.

Hsiang, S. et al., 2017: Estimating economic damage from climate change in the United States. *Science*, **356(6345)**, 1362–1369, doi:10.1126/science.aal4369.

Hsiang, S.M. and M. Burke, 2014: Climate, conflict, and social stability: what does the evidence say? *Climatic Change*, **123(1)**, 39–55, doi:10.1007/s10584-013-0868-3.

Hsiang, S.M. and A.H. Sobel, 2016: Potentially Extreme Population Displacement and Concentration in the Tropics Under Non-Extreme Warming. *Scientific Reports*, **6**, 25697, doi:10.1038/srep25697.

Hsiang, S.M., M. Burke, and E. Miguel, 2013: Quantifying the influence of climate on human conflict. *Science*, **341(6151)**, 1235367, doi:10.1126/science.1235367.

Huang, C.R., A.G. Barnett, X.M. Wang, and S.L. Tong, 2012: The impact of temperature on years of life lost in Brisbane, Australia. *Nature Climate Change*, **2(4)**, 265–270, doi:10.1038/nclimate1369.

Huang, J., H. Yu, A. Dai, Y. Wei, and L. Kang, 2017: Drylands face potential threat under 2°C global warming target. *Nature Climate Change*, **7(6)**, 417–422, doi:10.1038/nclimate3275.

Huang, M. et al., 2017: Velocity of change in vegetation productivity over northern high latitudes. *Nature Ecology & Evolution*, **1(11)**, 1649–1654, doi:10.1038/s41559-017-0328-y.

Hughes, D.J. and B.E. Narayanaswamy, 2013: Impacts of climate change on deep-sea habitats. In: *MCCIP Science Review*. Marine Climate Change Impacts Partnership (MCCIP), pp. 204–210, doi:10.14465/2013.arc17.155-166.

Hughes, T.P., J.T. Kerry, and T. Simpson, 2018: Large-scale bleaching of corals on the Great Barrier Reef. *Ecology*, **99(2)**, 501, doi:10.1002/ecy.2092.

Hughes, T.P. et al., 2017a: Coral reefs in the Anthropocene. *Nature*, **546(7656)**, 82–90, doi:10.1038/nature22901.

Hughes, T.P. et al., 2017b: Global warming and recurrent mass bleaching of corals. *Nature*, **543(7645)**, 373–377, doi:10.1038/nature21707.

Humpenöder, F. et al., 2014: Investigating afforestation and bioenergy CCS as climate change mitigation strategies. *Environmental Research Letters*, **9(6)**, 064029, doi:10.1088/1748-9326/9/6/064029.

Humpenöder, F. et al., 2018: Large-scale bioenergy production: how to resolve sustainability trade-offs? *Environmental Research Letters*, **13(2)**, 024011, doi:10.1088/1748-9326/aa9e3b.

Huntingford, C. et al., 2013: Simulated resilience of tropical rainforests to CO_2-induced climate change. *Nature Geoscience*, **6**, 268–273, doi:10.1038/ngeo1741.

Hutyra, L.R. et al., 2005: Climatic variability and vegetation vulnerability in Amazônia. *Geophysical Research Letters*, **32(24)**, L24712, doi:10.1029/2005gl024981.

Huynen, M.M.T.E. and P. Martens, 2015: Climate Change Effects on Heat- and Cold-Related Mortality in the Netherlands: A Scenario-Based Integrated Environmental Health Impact Assessment. *International Journal of Environmental Research and Public Health*, **12(10)**, 13295–13320, doi:10.3390/ijerph121013295.

ICEM, 2013: *USAID Mekong ARCC Climate Change Impact and Adaptation: Summary*. Prepared for the United States Agency for International Development by ICEM – International Centre for Environmental Management, 61 pp.

IFPRI, 2018: *2018 Global Food Policy Report*. International Food Policy Research Institute (IFPRI), Washington DC, USA, 150 pp., doi:10.2499/9780896292970.

Iida, Y. et al., 2015: Trends in pCO_2 and sea-air CO_2 flux over the global open oceans for the last two decades. *Journal of Oceanography*, **71(6)**, 637–661, doi:10.1007/s10872-015-0306-4.

Iizumi, T. et al., 2017: Responses of crop yield growth to global temperature and socioeconomic changes. *Scientific Reports*, **7(1)**, 7800, doi:10.1038/s41598-017-08214-4.

IPCC, 2000: Special Report on Emissions Scenarios. [Nakicenovic, N. and R. Swart (eds.)]. A Special Report of Working Group III of the Intergovernmental Panel on Climate Change. Cambridge University Press, Cambridge, United Kingdom and New York, NY, USA, 599 pp.

IPCC, 2007: Climate Change 2007: Synthesis Report. Contribution of Working Groups I, II, III to the Fourth Assessment Report of the International Panel on Climate Change. [Core Writing Team, R.K. Pachauri, and A. Reisinger (eds.)]. IPCC, Geneva, Switzerland, 104 pp.

IPCC, 2012: Managing the Risks of Extreme Events and Disasters to Advance Climate Change Adaptation. [Field, C.B., V. Barros, T.F. Stocker, D. Qin, D.J. Dokken, K.L. Ebi, M.D. Mastrandrea, K.J. Mach, G.-K. Plattner, S.K. Allen, M. Tignor, and P.M. Midgley (eds.)]. A Special Report of Working Groups I and II of IPCC Intergovernmental Panel on Climate Change. Cambridge University Press, Cambridge, United Kingdom and New York, USA, 594 pp.

IPCC, 2013: Climate Change 2013: The Physical Science Basis. Working Group I Contribution to the Fifth Assessment Report of the Intergovernmental Panel on Climate Change. [Stocker, T.F., D. Qin, G.-K. Plattner, M. Tignor, S.K. Allen, J. Boschung, A. Nauels, Y. Xia, V. Bex, and P.M. Midgley (eds.)]. Cambridge University Press, Cambridge, United Kingdom and New York, NY, USA, 1535 pp.

IPCC, 2014a: Climate Change 2014: Impacts, Adaptation, and Vulnerability. Part A: Global and Sectoral Aspects. Contribution of Working Group II to the Fifth Assessment Report of the Intergovernmental Panel on Climate Change. [Field, C.B., V.R. Barros, D.J. Dokken, K.J. Mach, M.D. Mastrandrea, T.E. Bilir, M. Chatterjee, K.L. Ebi, Y.O. Estrada, R.C. Genova, B. Girma, E.S. Kissel, A.N. Levy,

S. MacCracken, P.R. Mastrandrea, and L.L. White (eds.)]. Cambridge University Press, Cambridge, United Kingdom and New York, NY, USA, 1132 pp.

IPCC, 2014b: Climate Change 2014: Impacts, Adaptation, and Vulnerability. Part B: Regional Aspects. Contribution of Working Group II to the Fifth Assessment Report of the Intergovernmental Panel on Climate Change. [Field, C.B., V.R. Barros, D.J. Dokken, K.J. Mach, M.D. Mastrandrea, T.E. Bilir, M. Chatterjee, K.L. Ebi, Y.O. Estrada, R.C. Genova, B. Girma, E.S. Kissel, A.N. Levy, S. MacCracken, P.R. Mastrandrea, and L.L. White (eds.)]. Cambridge University Press, Cambridge, United Kingdom and New York, NY, USA, 688 pp.

Ishida, H. et al., 2014: Global-scale projection and its sensitivity analysis of the health burden attributable to childhood undernutrition under the latest scenario framework for climate change research. Environmental Research Letters, 9(6), 064014, doi:10.1088/1748-9326/9/6/064014.

Islam, M.R. and M. Shamsuddoha, 2017: Socioeconomic consequences of climate induced human displacement and migration in Bangladesh. International Sociology, 32(3), 277–298, doi:10.1177/0268580917693173.

IUCN, 2018: The IUCN Red List of Threatened Species. International Union for Conservation of Nature (IUCN). Retrieved from: www.iucnredlist.org.

Izaurralde, R.C. et al., 2011: Climate Impacts on Agriculture: Implications for Forage and Rangeland Production. Agronomy Journal, 103(2), 371–381, doi:10.2134/agronj2010.0304.

Jacinto, G.S., 2011: Fish Kill in the Philippines – Déjà Vu. Science Diliman, 23, 1–3, http://journals.upd.edu.ph/index.php/sciencediliman/article/view/2835.

Jackson, J.E. et al., 2010: Public health impacts of climate change in Washington State: projected mortality risks due to heat events and air pollution. Climatic Change, 102(1–2), 159–186, doi:10.1007/s10584-010-9852-3.

Jackson, L.P., A. Grinsted, and S. Jevrejeva, 2018: 21st Century Sea-Level Rise in Line with the Paris Accord. Earth's Future, 6(2), 213–229, doi:10.1002/2017ef000688.

Jacob, D. and S. Solman, 2017: IMPACT2C – An introduction. Climate Services, 7, 1–2, doi:10.1016/j.cliser.2017.07.006.

Jacob, D. et al., 2014: EURO-CORDEX: new high-resolution climate change projections for European impact research. Regional Environmental Change, 14(2), 563–578, doi:10.1007/s10113-013-0499-2.

Jacob, D. et al., 2018: Climate Impacts in Europe Under +1.5°C Global Warming. Earth's Future, 6(2), 264–285, doi:10.1002/2017ef000710.

Jacobs, S.S., A. Jenkins, C.F. Giulivi, and P. Dutrieux, 2011: Stronger ocean circulation and increased melting under Pine Island Glacier ice shelf. Nature Geoscience, 4(8), 519–523, doi:10.1038/ngeo1188.

Jaggard, K.W., A. Qi, and M.A. Semenov, 2007: The impact of climate change on sugarbeet yield in the UK: 1976–2004. The Journal of Agricultural Science, 145(4), 367–375, doi:10.1017/s0021859607006922.

Jahn, A., 2018: Reduced probability of ice-free summers for 1.5°C compared to 2°C warming. Nature Climate Change, 8(5), 409–413, doi:10.1038/s41558-018-0127-8.

Jahn, A., J.E. Kay, M.M. Holland, and D.M. Hall, 2016: How predictable is the timing of a summer ice-free Arctic? Geophysical Research Letters, 43(17), 9113–9120, doi:10.1002/2016gl070067.

Jaiswal, D. et al., 2017: Brazilian sugarcane ethanol as an expandable green alternative to crude oil use. Nature Climate Change, 7(11), 788–792, doi:10.1038/nclimate3410.

Jamero, M.L., M. Onuki, M. Esteban, and N. Tan, 2018: Community-based adaptation in low-lying islands in the Philippines: challenges and lessons learned. Regional Environmental Change, 1–12, doi:10.1007/s10113-018-1332-8.

Jamero, M.L. et al., 2017: Small-island communities in the Philippines prefer local measures to relocation in response to sea-level rise. Nature Climate Change, 7, 581–586, doi:10.1038/nclimate3344.

James, R., R. Washington, C.-F. Schleussner, J. Rogelj, and D. Conway, 2017: Characterizing half-a-degree difference: a review of methods for identifying regional climate responses to global warming targets. Wiley Interdisciplinary Reviews: Climate Change, 8(2), e457, doi:10.1002/wcc.457.

Jean-Baptiste, N. et al., 2018: Housing and Informal Settlements. In: Climate Change and Cities: Second Assessment Report of the Urban Climate Change Research Network [Rosenzweig, C., W.D. Solecki, P. Romero-Lankao, S. Mehrotra, S. Dhakal, and S.E. Ali Ibrahim (eds.)]. Cambridge University Press, Cambridge, United Kingdom and New York, USA, pp. 399–431, doi:10.1017/9781316563878.018.

Jeong, S.-J. et al., 2014: Effects of double cropping on summer climate of the North China Plain and neighbouring regions. Nature Climate Change, 4(7), 615–619, doi:10.1038/nclimate2266.

Jevrejeva, S., L.P. Jackson, R.E.M. Riva, A. Grinsted, and J.C. Moore, 2016: Coastal sea level rise with warming above 2°C. Proceedings of the National Academy of Sciences, 113(47), 13342–13347, doi:10.1073/pnas.1605312113.

Jiang, D.B. and Z.P. Tian, 2013: East Asian monsoon change for the 21st century: Results of CMIP3 and CMIP5 models. Chinese Science Bulletin, 58(12), 1427–1435, doi:10.1007/s11434-012-5533-0.

Jiang, L. and B.C.O. Neill, 2017: Global urbanization projections for the Shared Socioeconomic Pathways. Global Environmental Change, 42, 193–199, doi:10.1016/j.gloenvcha.2015.03.008.

Jiang, Y. et al., 2016: Importance of soil thermal regime in terrestrial ecosystem carbon dynamics in the circumpolar north. Global and Planetary Change, 142, 28–40, doi:10.1016/j.gloplacha.2016.04.011.

Jiao, M., G. Zhou, and Z. Chen (eds.), 2014: Blue book of agriculture for addressing climate change: Assessment report of climatic change impacts on agriculture in China (No.1) (in Chinese). Social Sciences Academic Press, Beijing, China.

Jiao, M., G. Zhou, and Z. Zhang (eds.), 2016: Blue book of agriculture for addressing climate change: Assessment report of agro-meteorological disasters and yield losses in China (No.2) (in Chinese). Social Sciences Academic Press, Beijing, China.

Jiménez Cisneros, B.E. et al., 2014: Freshwater Resources. In: Climate Change 2014: Impacts, Adaptation, and Vulnerability. Part A: Global and Sectoral Aspects. Contribution of Working Group II to the Fifth Assessment Report of the Intergovernmental Panel on Climate Change [Field, C.B., V.R. Barros, D.J. Dokken, K.J. Mach, M.D. Mastrandrea, T.E. Bilir, M. Chatterjee, K.L. Ebi, Y.O. Estrada, R.C. Genova, B. Girma, E.S. Kissel, A.N. Levy, S. MacCracken, P.R. Mastrandrea, and L.L. White (eds.)]. Cambridge University Press, Cambridge, United Kingdom and New York, NY, USA, pp. 229–269.

Johnson, W.C. and K.A. Poiani, 2016: Climate Change Effects on Prairie Pothole Wetlands: Findings from a Twenty-five Year Numerical Modeling Project. Wetlands, 36(2), 273–285, doi:10.1007/s13157-016-0790-3.

Johnson, W.C., B. Werner, and G.R. Guntenspergen, 2016: Non-linear responses of glaciated prairie wetlands to climate warming. Climatic Change, 134(1–2), 209–223, doi:10.1007/s10584-015-1534-8.

Jones, A. and M. Phillips (eds.), 2017: Global Climate Change and Coastal Tourism: Recognizing Problems, Managing Solutions and Future Expectations. Centre for Agriculture and Biosciences International (CABI), 360 pp.

Jones, C. and L.M. Carvalho, 2013: Climate change in the South American monsoon system: Present climate and CMIP5 projections. Journal of Climate, 26(17), 6660–6678, doi:10.1175/jcli-d-12-00412.1.

Jones, C.D., J. Lowe, S. Liddicoat, and R. Betts, 2009: Committed terrestrial ecosystem changes due to climate change. Nature Geoscience, 2(7), 484–487, doi:10.1038/ngeo555.

Jones, C.D. et al., 2013: Twenty-First-Century Compatible CO2 Emissions and Airborne Fraction Simulated by CMIP5 Earth System Models under Four Representative Concentration Pathways. Journal of Climate, 26(13), 4398–4413, doi:10.1175/jcli-d-12-00554.1.

Jones, C.D. et al., 2016: Simulating the Earth system response to negative emissions. Environmental Research Letters, 11(9), 095012, doi:10.1088/1748-9326/11/9/095012.

Jonker, J.G.G., M. Junginger, and A. Faaij, 2014: Carbon payback period and carbon offset parity point of wood pellet production in the South-eastern United States. GCB Bioenergy, 6(4), 371–389, doi:10.1111/gcbb.12056.

Joughin, I., B.E. Smith, and B. Medley, 2014: Marine Ice Sheet Collapse Potentially Under Way for the Thwaites Glacier Basin, West Antarctica. Science, 344(6185), 735–738, doi:10.1126/science.1249055.

Kaiser, K. et al., 2014: Detection and attribution of lake-level dynamics in north-eastern central Europe in recent decades. Regional Environmental Change, 14(4), 1587–1600, doi:10.1007/s10113-014-0600-5.

Kamahori, H., N. Yamazaki, N. Mannoji, and K. Takahashi, 2006: Variability in Intense Tropical Cyclone Days in the Western North Pacific. SOLA, 2, 104–107, doi:10.2151/sola.2006-027.

Kamei, M., K. Hanaki, and K. Kurisu, 2016: Tokyo's long-term socioeconomic pathways: Towards a sustainable future. Sustainable Cities and Society, 27, 73–82, doi:10.1016/j.scs.2016.07.002.

Kämpf, J. and P. Chapman, 2016: The Functioning of Coastal Upwelling Systems. In: Upwelling Systems of the World. Springer International Publishing, Cham, Switzerland, pp. 31–65, doi:10.1007/978-3-319-42524-5_2.

Kang, N.-Y. and J.B. Elsner, 2015: Trade-off between intensity and frequency of global tropical cyclones. Nature Climate Change, 5(7), 661–664, doi:10.1038/nclimate2646.

Kaniewski, D., J.J.J. Guiot, and E. Van Campo, 2015: Drought and societal collapse 3200 years ago in the Eastern Mediterranean: A review. *Wiley Interdisciplinary Reviews: Climate Change*, **6**(4), 369–382, doi:10.1002/wcc.345.

Kannan, K.P., S.M. Dev, and A.N. Sharma, 2000: Concerns on Food Security. *Economic And Political Weekly*, **35**(45), 3919–3922, www.jstor.org/stable/4409916.

Karl, T.R. et al., 2015: Possible artifacts of data biases in the recent global surface warming hiatus. *Science*, **348**(6242), 1469–1472, doi:10.1126/science.aaa5632.

Karnauskas, K.B., J.P. Donnelly, and K.J. Anchukaitis, 2016: Future freshwater stress for island populations. *Nature Climate Change*, **6**(7), 720–725, doi:10.1038/nclimate2987.

Karnauskas, K.B. et al., 2018: Freshwater Stress on Small Island Developing States: Population Projections and Aridity Changes at 1.5°C and 2°C. *Regional Environmental Change*, 1–10, doi:10.1007/s10113-018-1331-9.

Karstensen, J. et al., 2015: Open ocean dead zones in the tropical North Atlantic Ocean. *Biogeosciences*, **12**(8), 2597–2605, doi:10.5194/bg-12-2597-2015.

Kawaguchi, S. et al., 2013: Risk maps for Antarctic krill under projected Southern Ocean acidification. *Nature Climate Change*, **3**(9), 843–847, doi:10.1038/nclimate1937.

Kelley, C.P., S. Mohtadi, M.A. Cane, R. Seager, and Y. Kushnir, 2015: Climate change in the Fertile Crescent and implications of the recent Syrian drought. *Proceedings of the National Academy of Sciences*, **112**(11), 3241–6, doi:10.1073/pnas.1421533112.

Kelly, K.A., K. Drushka, L.A. Thompson, D. Le Bars, and E.L. McDonagh, 2016: Impact of slowdown of Atlantic overturning circulation on heat and freshwater transports. *Geophysical Research Letters*, **43**(14), 7625–7631, doi:10.1002/2016gl069789.

Kench, P., D. Thompson, M. Ford, H. Ogawa, and R. Mclean, 2015: Coral islands defy sea-level rise over the past century: Records from a central Pacific atoll. *Geology*, **43**(6), 515–518, doi:10.1130/g36555.1.

Kench, P.S., M.R. Ford, and S.D. Owen, 2018: Patterns of island change and persistence offer alternate adaptation pathways for atoll nations. *Nature Communications*, **9**(1), 605, doi:10.1038/s41467-018-02954-1.

Kendrovski, V. et al., 2017: Quantifying Projected Heat Mortality Impacts under 21st-Century Warming Conditions for Selected European Countries. *International Journal of Environmental Research and Public Health*, **14**(7), 729, doi:10.3390/ijerph14070729.

Kennedy, E. et al., 2013: Avoiding Coral Reef Functional Collapse Requires Local and Global Action. *Current Biology*, **23**(10), 912–918, doi:10.1016/j.cub.2013.04.020.

Kent, C. et al., 2017: Using climate model simulations to assess the current climate risk to maize production. *Environmental Research Letters*, **12**(5), 054012, doi:10.1088/1748-9326/aa6cb9.

Keppel, G. and J. Kavousi, 2015: Effective climate change refugia for coral reefs. *Global Change Biology*, **21**(8), 2829–2830, doi:10.1111/gcb.12936.

Kersting, D.K., E. Cebrian, J. Verdura, and E. Ballesteros, 2017: A new cladocora caespitosa population with unique ecological traits. *Mediterranean Marine Science*, **18**(1), 38–42, doi:10.12681/mms.1955.

Kharin, V. et al., 2018: Risks from Climate Extremes Change Differently from 1.5°C to 2.0°C Depending on Rarity. *Earth's Future*, **6**(5), 704–715, doi:10.1002/2018ef000813.

Khouakhi, A. and G. Villarini, 2017: Attribution of annual maximum sea levels to tropical cyclones at the global scale. *International Journal of Climatology*, **37**(1), 540–547, doi:10.1002/joc.4704.

Khouakhi, A., G. Villarini, and G.A. Vecchi, 2017: Contribution of Tropical Cyclones to Rainfall at the Global Scale. *Journal of Climate*, **30**(1), 359–372, doi:10.1175/jcli-d-16-0298.1.

Kim, H.-Y., J. Ko, S. Kang, and J. Tenhunen, 2013: Impacts of climate change on paddy rice yield in a temperate climate. *Global Change Biology*, **19**(2), 548–562, doi:10.1111/gcb.12047.

Kim, Y. et al., 2017: A perspective on climate-resilient development and national adaptation planning based on USAID's experience. *Climate and Development*, **9**(2), 141–151, doi:10.1080/17565529.2015.1124037.

King, A.D., D.J. Karoly, and B.J. Henley, 2017: Australian climate extremes at 1.5°C and 2°C of global warming. *Nature Climate Change*, **7**(6), 412–416, doi:10.1038/nclimate3296.

King, A.D. et al., 2018: Reduced heat exposure by limiting global warming to 1.5°C. *Nature Climate Change*, **8**(7), 549–551, doi:10.1038/s41558-018-0191-0.

Kinoshita, Y., M. Tanoue, S. Watanabe, and Y. Hirabayashi, 2018: Quantifying the effect of autonomous adaptation to global river flood projections: Application to future flood risk assessments. *Environmental Research Letters*, **13**(1), 014006, doi:10.1088/1748-9326/aa9401.

Kipling, R.P. et al., 2016: Modeling European ruminant production systems: Facing the challenges of climate change. *Agricultural Systems*, **147**, 24–37, doi:10.1016/j.agsy.2016.05.007.

Kirtman, B. et al., 2013: Near-term Climate Change: Projections and Predictability. In: *Climate Change 2013: The Physical Science Basis. Contribution of Working Group I to the Fifth Assessment Report of the Intergovernmental Panel on Climate Change* [Stocker, T.F., D. Qin, G.-K. Plattner, M. Tignor, S.K. Allen, J. Boschung, A. Nauels, Y. Xia, V. Bex, and P.M. Midgley (eds.)]. Cambridge University Press, Cambridge, United Kingdom and New York, NY, USA, pp. 953–1028.

Kirwan, M. and P. Megonigal, 2013: Tidal wetland stability in the face of human impacts and sea-level rise. *Nature*, **504**, 53–60, doi:10.1038/nature12856.

Kittinger, J.N., 2013: Human Dimensions of Small-Scale and Traditional Fisheries in the Asia-Pacific Region. *Pacific Science*, **67**(3), 315–325, doi:10.2984/67.3.1.

Kittinger, J.N. et al., 2013: Emerging frontiers in social-ecological systems research for sustainability of small-scale fisheries. *Current Opinion in Environmental Sustainability*, **5**(3–4), 352–357, doi:10.1016/j.cosust.2013.06.008.

Kjellstrom, T., B. Lemke, and M. Otto, 2013: Mapping Occupational Heat Exposure and Effects in South-East Asia: Ongoing Time Trends 1980–2011 and Future Estimates to 2050. *Industrial Health*, **51**(1), 56–67, doi:10.2486/indhealth.2012-0174.

Kjellstrom, T., C. Freyberg, B. Lemke, M. Otto, and D. Briggs, 2018: Estimating population heat exposure and impacts on working people in conjunction with climate change. *International Journal of Biometeorology*, **62**(3), 291–306, doi:10.1007/s00484-017-1407-0.

Kjellström, E. et al., 2018: European climate change at global mean temperature increases of 1.5 and 2°C above pre-industrial conditions as simulated by the EURO-CORDEX regional climate models. *Earth System Dynamics*, **9**(2), 459–478, doi:10.5194/esd-9-459-2018.

Kline, D.I. et al., 2012: A short-term in situ CO_2 enrichment experiment on Heron Island (GBR). *Scientific Reports*, **2**, 413, doi:10.1038/srep00413.

Kline, K.L. et al., 2017: Reconciling food security and bioenergy: priorities for action. *GCB Bioenergy*, **9**(3), 557–576, doi:10.1111/gcbb.12366.

Kling, H., P. Stanzel, and M. Preishuber, 2014: Impact modelling of water resources development and climate scenarios on Zambezi River discharge. *Journal of Hydrology: Regional Studies*, **1**, 17–43, doi:10.1016/j.ejrh.2014.05.002.

Kloster, S. et al., 2009: A GCM study of future climate response to aerosol pollution reductions. *Climate Dynamics*, **34**(7–8), 1177–1194, doi:10.1007/s00382-009-0573-0.

Klotzbach, P.J., 2006: Trends in global tropical cyclone activity over the past twenty years (1986–2005). *Geophysical Research Letters*, **33**(10), L10805, doi:10.1029/2006gl025881.

Klotzbach, P.J. and C.W. Landsea, 2015: Extremely Intense Hurricanes: Revisiting Webster et al. (2005) after 10 Years. *Journal of Climate*, **28**(19), 7621–7629, doi:10.1175/jcli-d-15-0188.1.

Klutse, N.A.B. et al., 2018: Potential Impact of 1.5°C and 2°C global warming on extreme rainfall over West Africa. *Environmental Research Letters*, **13**(5), 055013, doi:10.1088/1748-9326/aab37b.

Knapp, J.R., G.L. Laur, P.A. Vadas, W.P. Weiss, and J.M. Tricarico, 2014: Enteric methane in dairy cattle production: Quantifying the opportunities and impact of reducing emissions. *Journal of Dairy Science*, **97**(6), 3231–3261, doi:10.3168/jds.2013-7234.

Knutson, T.R. et al., 2010: Tropical cyclones and climate change. *Nature Geoscience*, **3**(3), 157–163, doi:10.1038/ngeo779.

Knutson, T.R. et al., 2013: Dynamical Downscaling Projections of Twenty-First-Century Atlantic Hurricane Activity: CMIP3 and CMIP5 Model-Based Scenarios. *Journal of Climate*, **26**(17), 6591–6617, doi:10.1175/jcli-d-12-00539.1.

Knutson, T.R. et al., 2015: Global Projections of Intense Tropical Cyclone Activity for the Late Twenty-First Century from Dynamical Downscaling of CMIP5/RCP4.5 Scenarios. *Journal of Climate*, **28**(18), 7203–7224, doi:10.1175/jcli-d-15-0129.1.

Knutti, R. and J. Sedláček, 2012: Robustness and uncertainties in the new CMIP5 climate model projections. *Nature Climate Change*, **3**(4), 369–373, doi:10.1038/nclimate1716.

Kopp, R.E., F.J. Simons, J.X. Mitrovica, A.C. Maloof, and M. Oppenheimer, 2013: A probabilistic assessment of sea level variations within the last interglacial stage. *Geophysical Journal International*, **193(2)**, 711–716, doi:10.1093/gji/ggt029.

Kopp, R.E. et al., 2014: Probabilistic 21st and 22nd century sea-level projections at a global network of tide-gauge sites. *Earth's Future*, **2(8)**, 383–406, doi:10.1002/2014ef000239.

Kopp, R.E. et al., 2016: Temperature-driven global sea-level variability in the Common Era. *Proceedings of the National Academy of Sciences*, **113(11)**, E1434–E1441, doi:10.1073/pnas.1517056113.

Kossin, J.P. et al., 2013: Trend Analysis with a New Global Record of Tropical Cyclone Intensity. *Journal of Climate*, **26(24)**, 9960–9976, doi:10.1175/jcli-d-13-00262.1.

Koutroulis, A.G., M.G. Grillakis, I.N. Daliakopoulos, I.K. Tsanis, and D. Jacob, 2016: Cross sectoral impacts on water availability at +2°C and +3°C for east Mediterranean island states: The case of Crete. *Journal of Hydrology*, **532**, 16–28, doi:10.1016/j.jhydrol.2015.11.015.

Koven, C.D. et al., 2015: Controls on terrestrial carbon feedbacks by productivity versus turnover in the CMIP5 Earth System Models. *Biogeosciences*, **12(17)**, 5211–5228, doi:10.5194/bg-12-5211-2015.

Krause, A. et al., 2017: Global consequences of afforestation and bioenergy cultivation on ecosystem service indicators. *Biogeosciences*, **14(21)**, 4829–4850, doi:10.5194/bg-14-4829-2017.

Kreidenweis, U. et al., 2016: Afforestation to mitigate climate change: impacts on food prices under consideration of albedo effects. *Environmental Research Letters*, **11(8)**, 085001, doi:10.1088/1748-9326/11/8/085001.

Krey, V., G. Luderer, L. Clarke, and E. Kriegler, 2014: Getting from here to there – energy technology transformation pathways in the EMF27 scenarios. *Climatic Change*, **123(3)**, 369–382, doi:10.1007/s10584-013-0947-5.

Krey, V. et al., 2012: Urban and rural energy use and carbon dioxide emissions in Asia. *Energy Economics*, **34**, S272–S283, doi:10.1016/j.eneco.2012.04.013.

Kriegler, E., J.W. Hall, H. Held, R. Dawson, and H.J. Schellnhuber, 2009: Imprecise probability assessment of tipping points in the climate system. *Proceedings of the National Academy of Sciences*, **106(13)**, 5041–5046, doi:10.1073/pnas.0809117106.

Kriegler, E. et al., 2017: Fossil-fueled development (SSP5): An energy and resource intensive scenario for the 21st century. *Global Environmental Change*, **42**, 297–315, doi:10.1016/j.gloenvcha.2016.05.015.

Kroeker, K.J. et al., 2013: Impacts of ocean acidification on marine organisms: Quantifying sensitivities and interaction with warming. *Global Change Biology*, **19(6)**, 1884–1896, doi:10.1111/gcb.12179.

Kroon, F.J., P. Thorburn, B. Schaffelke, and S. Whitten, 2016: Towards protecting the Great Barrier Reef from land-based pollution. *Global Change Biology*, **22(6)**, 1985–2002, doi:10.1111/gcb.13262.

Krumhansl, K.A. et al., 2016: Global patterns of kelp forest change over the past half-century. *Proceedings of the National Academy of Sciences*, **113(48)**, 13785–13790, doi:10.1073/pnas.1606102113.

Kumar, L. and S. Taylor, 2015: Exposure of coastal built assets in the South Pacific to climate risks. *Nature Climate Change*, **5(11)**, 992–996, doi:10.1038/nclimate2702.

Kummu, M. et al., 2016: The world's road to water scarcity: shortage and stress in the 20th century and pathways towards sustainability. *Scientific Reports*, **6(1)**, 38495, doi:10.1038/srep38495.

Kusaka, H., A. Suzuki-Parker, T. Aoyagi, S.A. Adachi, and Y. Yamagata, 2016: Assessment of RCM and urban scenarios uncertainties in the climate projections for August in the 2050s in Tokyo. *Climatic Change*, **137(3–4)**, 427–438, doi:10.1007/s10584-016-1693-2.

Lachkar, Z., M. Lévy, and S. Smith, 2018: Intensification and deepening of the Arabian Sea oxygen minimum zone in response to increase in Indian monsoon wind intensity. *Biogeosciences*, **15(1)**, 159–186, doi:10.5194/bg-15-159-2018.

Lacoue-Labarthe, T. et al., 2016: Impacts of ocean acidification in a warming Mediterranean Sea: An overview. *Regional Studies in Marine Science*, **5**, 1–11, doi:10.1016/j.rsma.2015.12.005.

Läderach, P., A. Martinez-Valle, G. Schroth, and N. Castro, 2013: Predicting the future climatic suitability for cocoa farming of the world's leading producer countries, Ghana and Côte d'Ivoire. *Climatic Change*, **119(3)**, 841–854, doi:10.1007/s10584-013-0774-8.

Lal, R., 2014: Soil Carbon Management and Climate Change. In: *Soil Carbon* [Hartemink, A.E. and K. McSweeney (eds.)]. Progress in Soil Science, Springer International Publishing, Cham, Switzerland, pp. 339–361, doi:10.1007/978-3-319-04084-4_35.

Lallo, C.H.O. et al., 2018: Characterizing heat stress on livestock using the temperature humidity index (THI) – prospects for a warmer Caribbean. *Regional Environmental Change*, 1–12, doi:10.1007/s10113-018-1359-x.

Lam, V.W.Y., W.W.L. Cheung, and U.R. Sumaila, 2014: Marine capture fisheries in the Arctic: Winners or losers under climate change and ocean acidification? *Fish and Fisheries*, **17**, 335–357, doi:10.1111/faf.12106.

Lam, V.W.Y., W.W.L. Cheung, W. Swartz, and U.R. Sumaila, 2012: Climate change impacts on fisheries in West Africa: implications for economic, food and nutritional security. *African Journal of Marine Science*, **34(1)**, 103–117, doi:10.2989/1814232x.2012.673294.

Lam, V.W.Y., W.W.L. Cheung, G. Reygondeau, and U.R. Sumaila, 2016: Projected change in global fisheries revenues under climate change. *Scientific Reports*, **6(1)**, 32607, doi:10.1038/srep32607.

Lana, M.A. et al., 2017: Yield stability and lower susceptibility to abiotic stresses of improved open-pollinated and hybrid maize cultivars. *Agronomy for Sustainable Development*, **37(4)**, 30, doi:10.1007/s13593-017-0442-x.

Landsea, C.W., 2006: Can We Detect Trends in Extreme Tropical Cyclones? *Science*, **313(5786)**, 452–454, doi:10.1126/science.1128448.

Larsen, J.N. et al., 2014: Polar regions. In: *Climate Change 2014: Impacts, Adaptation, and Vulnerability. Part B: Regional Aspects. Contribution of Working Group II to the Fifth Assessment Report of the Intergovernmental Panel on Climate Change* [Barros, V.R., C.B. Field, D.J. Dokken, M.D. Mastrandrea, K.J. Mach, T.E. Bilir, M. Chatterjee, K.L. Ebi, Y.O. Estrada, R.C. Genova, B. Girma, E.S. Kissel, A.N. Levy, S. MacCracken, P.R. Mastrandrea, and L.L. White (eds.)]. Cambridge University Press, Cambridge, United Kingdom and New York, NY, USA, pp. 1567–1612.

Lawrence, P.J. et al., 2012: Simulating the biogeochemical and biogeophysical impacts of transient land cover change and wood harvest in the Community Climate System Model (CCSM4) from 1850 to 2100. *Journal of Climate*, **25(9)**, 3071–3095, doi:10.1175/jcli-d-11-00256.1.

Le Cozannet, G., M. Garcin, M. Yates, D. Idier, and B. Meyssignac, 2014: Approaches to evaluate the recent impacts of sea-level rise on shoreline changes. *Earth-Science Reviews*, **138**, 47–60, doi:10.1016/j.earscirev.2014.08.005.

Le Dang, H., E. Li, J. Bruwer, and I. Nuberg, 2013: Farmers' perceptions of climate variability and barriers to adaptation: lessons learned from an exploratory study in Vietnam. *Mitigation and Adaptation Strategies for Global Change*, **19(5)**, 531–548, doi:10.1007/s11027-012-9447-6.

Leadley, P. et al., 2016: *Relationships between the Aichi Targets and land-based climate mitigation*. Convention on Biological Diversity (CBD), 26 pp.

Lebel, L., C.T. Hoanh, C. Krittasudthacheewa, and R. Daniel (eds.), 2014: *Climate risks, regional integration and sustainability in the Mekong region*. Strategic Information and Research Development Centre (SIRD), Selangor, Malaysia, 417 pp.

Lee, J., 2016: *Valuation of Ocean Acidification Effects on Shellfish Fisheries and Aquaculture*. DP 132, Centre for Financial and Management Studies (CeFiMS), School of Oriental and African Studies (SOAS), University of London, London, UK, 14 pp.

Lee, J.Y., S. Hyun Lee, S.-C. Hong, and H. Kim, 2017: Projecting future summer mortality due to ambient ozone concentration and temperature changes. *Atmospheric Environment*, **156**, 88–94, doi:10.1016/j.atmosenv.2017.02.034.

Lee, M.A., A.P. Davis, M.G.G. Chagunda, and P. Manning, 2017: Forage quality declines with rising temperatures, with implications for livestock production and methane emissions. *Biogeosciences*, **14(6)**, 1403–1417, doi:10.5194/bg-14-1403-2017.

Lee, S.-M. and S.-K. Min, 2018: Heat Stress Changes over East Asia under 1.5° and 2.0°C Global Warming Targets. *Journal of Climate*, **31(7)**, 2819–2831, doi:10.1175/jcli-d-17-0449.1.

Lehner, F. et al., 2017: Projected drought risk in 1.5°C and 2°C warmer climates. *Geophysical Research Letters*, **44(14)**, 7419–7428, doi:10.1002/2017gl074117.

Lehodey, P., I. Senina, S. Nicol, and J. Hampton, 2015: Modelling the impact of climate change on South Pacific albacore tuna. *Deep Sea Research Part II: Topical Studies in Oceanography*, **113**, 246–259, doi:10.1016/j.dsr2.2014.10.028.

Lemasson, A.J., S. Fletcher, J.M. Hall-Spencer, and A.M. Knights, 2017: Linking the biological impacts of ocean acidification on oysters to changes in ecosystem services: A review. *Journal of Experimental Marine Biology and Ecology*, **492**, 49–62, doi:10.1016/j.jembe.2017.01.019.

Lemelin, H., J. Dawson, and E.J. Stewart (eds.), 2012: *Last Chance Tourism: Adapting Tourism Opportunities in a Changing World*. Routledge, Abingdon, UK, 238 pp., doi:10.1080/14927713.2012.747359.

Lemoine, D. and C.P. Traeger, 2016: Economics of tipping the climate dominoes. *Nature Climate Change*, **6(5)**, 514–519, doi:10.1038/nclimate2902.

3

Lenton, T.M., 2012: Arctic Climate Tipping Points. *Ambio*, **41(1, SI)**, 10–22, doi:10.1007/s13280-011-0221-x.

Lenton, T.M. et al., 2008: Tipping elements in the Earth's climate system. *Proceedings of the National Academy of Sciences*, **105(6)**, 1786–93, doi:10.1073/pnas.0705414105.

Lesk, C., P. Rowhani, and N. Ramankutty, 2016: Influence of extreme weather disasters on global crop production. *Nature*, **529(7584)**, 84–87, doi:10.1038/nature16467.

Lesnikowski, A., J. Ford, R. Biesbroek, L. Berrang-Ford, and S.J. Heymann, 2016: National-level progress on adaptation. *Nature Climate Change*, **6(3)**, 261–264, doi:10.1038/nclimate2863.

Levermann, A. et al., 2013: The multimillennial sea-level commitment of global warming. *Proceedings of the National Academy of Sciences*, **110(34)**, 13745–13750, doi:10.1073/pnas.1219414110.

Levermann, A. et al., 2014: Projecting Antarctic ice discharge using response functions from SeaRISE ice-sheet models. *Earth System Dynamics*, **5(2)**, 271–293, doi:10.5194/esd-5-271-2014.

Levi, T., F. Keesing, K. Oggenfuss, and R.S. Ostfeld, 2015: Accelerated phenology of blacklegged ticks under climate warming. *Philosophical Transactions of the Royal Society B: Biological Sciences*, **370(1665)**, 20130556, doi:10.1098/rstb.2013.0556.

Levin, L.A. and N. Le Bris, 2015: The deep ocean under climate change. *Science*, **350(6262)**, 766–768, doi:10.1126/science.aad0126.

Li, C., D. Notz, S. Tietsche, and J. Marotzke, 2013: The Transient versus the Equilibrium Response of Sea Ice to Global Warming. *Journal of Climate*, **26(15)**, 5624–5636, doi:10.1175/jcli-d-12-00492.1.

Li, C. et al., 2018: Midlatitude atmospheric circulation responses under 1.5 and 2.0°C warming and implications for regional impacts. *Earth System Dynamics*, **9(2)**, 359–382, doi:10.5194/esd-9-359-2018.

Li, S., Q. Wang, and J.A. Chun, 2017: Impact assessment of climate change on rice productivity in the Indochinese Peninsula using a regional-scale crop model. *International Journal of Climatology*, **37**, 1147–1160, doi:10.1002/joc.5072.

Li, S. et al., 2016: Interactive Effects of Seawater Acidification and Elevated Temperature on the Transcriptome and Biomineralization in the Pearl Oyster *Pinctada fucata*. *Environmental Science & Technology*, **50(3)**, 1157–1165, doi:10.1021/acs.est.5b05107.

Li, T. et al., 2015: Heat-related mortality projections for cardiovascular and respiratory disease under the changing climate in Beijing, China. *Scientific Reports*, **5(1)**, 11441, doi:10.1038/srep11441.

Li, T. et al., 2016: Aging Will Amplify the Heat-related Mortality Risk under a Changing Climate: Projection for the Elderly in Beijing, China. *Scientific Reports*, **6(1)**, 28161, doi:10.1038/srep28161.

Li, Z. and H. Fang, 2016: Impacts of climate change on water erosion: A review. *Earth-Science Reviews*, **163**, 94–117, doi:10.1016/j.earscirev.2016.10.004.

Likhvar, V. et al., 2015: A multi-scale health impact assessment of air pollution over the 21st century. *Science of The Total Environment*, **514**, 439–449, doi:10.1016/j.scitotenv.2015.02.002.

Linares, C. et al., 2015: Persistent natural acidification drives major distribution shifts in marine benthic ecosystems. *Proceedings of the Royal Society B: Biological Sciences*, **282(1818)**, 20150587, doi:10.1098/rspb.2015.0587.

Lindsay, R. and A. Schweiger, 2015: Arctic sea ice thickness loss determined using subsurface, aircraft, and satellite observations. *The Cryosphere*, **9(1)**, 269–283, doi:10.5194/tc-9-269-2015.

Ling, F.H., M. Tamura, K. Yasuhara, K. Ajima, and C. Van Trinh, 2015: Reducing flood risks in rural households: survey of perception and adaptation in the Mekong delta. *Climatic change*, **132(2)**, 209–222, doi:10.1007/s10584-015-1416-0.

Ling, S.D., C.R. Johnson, K. Ridgway, A.J. Hobday, and M. Haddon, 2009: Climate-driven range extension of a sea urchin: Inferring future trends by analysis of recent population dynamics. *Global Change Biology*, **15(3)**, 719–731, doi:10.1111/j.1365-2486.2008.01734.x.

Lipper, L. et al., 2014: Climate-smart agriculture for food security. *Nature Climate Change*, **4(12)**, 1068–1072, doi:10.1038/nclimate2437.

Liu, J. et al., 2017: Water scarcity assessments in the past, present, and future. *Earth's Future*, **5(6)**, 545–559, doi:10.1002/2016ef000518.

Liu, L., H. Xu, Y. Wang, and T. Jiang, 2017: Impacts of 1.5 and 2°C global warming on water availability and extreme hydrological events in Yiluo and Beijiang River catchments in China. *Climatic Change*, **145**, 145–158, doi:10.1007/s10584-017-2072-3.

Liu, W. et al., 2018: Global drought and severe drought-affected populations in 1.5 and 2°C warmer worlds. *Earth System Dynamics*, **9(1)**, 267–283, doi:10.5194/esd-9-267-2018.

Lluch-Cota, S.E. et al., 2014: Cross-chapter box on uncertain trends in major upwelling ecosystems. In: *Climate Change 2014: Impacts, Adaptation, and Vulnerability. Part A: Global and Sectoral Aspects. Contribution of Working Group II to the Fifth Assessment Report of the Intergovernmental Panel on Climate Change* [Field, C.B., V.R. Barros, D.J. Dokken, K.J. Mach, M.D. Mastrandrea, T.E. Bilir, M. Chatterjee, K.L. Ebi, Y.O. Estrada, R.C. Genova, B. Girma, E. Kissel, A. Levy, S. MacCracken, P.R. Mastrandrea, and L.L. White (eds.)]. Cambridge University Press, Cambridge, UK and New York, NY, USA, pp. 149–151.

Loarie, S.R. et al., 2009: The velocity of climate change. *Nature*, **462(7276)**, 1052–1055, doi:10.1038/nature08649.

Lobell, D. et al., 2009: Regional differences in the influence of irrigation on climate. *Journal of Climate*, **22(8)**, 2248–2255, doi:10.1175/2008jcli2703.1.

Lobell, D.B., G. Bala, and P.B. Duffy, 2006: Biogeophysical impacts of cropland management changes on climate. *Geophysical Research Letters*, **33(6)**, L06708, doi:10.1029/2005gl025492.

Lobell, D.B., W. Schlenker, and J. Costa-Roberts, 2011: Climate Trends and Global Crop Production Since 1980. *Science*, **333(6042)**, 616–620, doi:10.1126/science.1204531.

Lobell, D.B. et al., 2014: Greater Sensitivity to Drought Accompanies Maize Yield Increase in the U.S. Midwest. *Science*, **344(6183)**, 516–519, doi:10.1126/science.1251423.

Long, J., C. Giri, J. Primavera, and M. Trivedi, 2016: Damage and recovery assessment of the Philippines' mangroves following Super Typhoon *Haiyan*. *Marine Pollution Bulletin*, **109(2)**, 734–743, doi:10.1016/j.marpolbul.2016.06.080.

LoPresti, A. et al., 2015: Rate and velocity of climate change caused by cumulative carbon emissions. *Environmental Research Letters*, **10(9)**, 095001, doi:10.1088/1748-9326/10/9/095001.

Lovelock, C.E., I.C. Feller, R. Reef, S. Hickey, and M.C. Ball, 2017: Mangrove dieback during fluctuating sea levels. *Scientific Reports*, **7(1)**, 1680, doi:10.1038/s41598-017-01927-6.

Lovelock, C.E. et al., 2015: The vulnerability of Indo-Pacific mangrove forests to sea-level rise. *Nature*, **526(7574)**, 559–563, doi:10.1038/nature15538.

Lovenduski, N.S. and G.B. Bonan, 2017: Reducing uncertainty in projections of terrestrial carbon uptake. *Environmental Research Letters*, **12(4)**, 044020, doi:10.1088/1748-9326/aa66b8.

Lu, X.X. et al., 2013: Sediment loads response to climate change: A preliminary study of eight large Chinese rivers. *International Journal of Sediment Research*, **28(1)**, 1–14, doi:10.1016/s1001-6279(13)60013-x.

Lucht, W., S. Schaphoff, T. Erbrecht, U. Heyder, and W. Cramer, 2006: Terrestrial vegetation redistribution and carbon balance under climate change. *Carbon Balance and Management*, **1(1)**, 6, doi:10.1186/1750-0680-1-6.

Luo, K., F. Tao, J.P. Moiwo, D. Xiao, and J. Zhang, 2016: Attribution of hydrological change in Heihe River Basin to climate and land use change in the past three decades. *Scientific Reports*, **6(1)**, 33704, doi:10.1038/srep33704.

Luyssaert, S. et al., 2014: Land management and land-cover change have impacts of similar magnitude on surface temperature. *Nature Climate Change*, **4(5)**, 1–5, doi:10.1038/nclimate2196.

Lyra, A. et al., 2017: Projections of climate change impacts on central America tropical rainforest. *Climatic Change*, **141(1)**, 93–105, doi:10.1007/s10584-016-1790-2.

Macel, M., T. Dostálek, S. Esch, and A. Bucharová, 2017: Evolutionary responses to climate change in a range expanding plant. *Oecologia*, **184(2)**, 543–554, doi:10.1007/s00442-017-3864-x.

Mackenzie, C.L. et al., 2014: Ocean Warming, More than Acidification, Reduces Shell Strength in a Commercial Shellfish Species during Food Limitation. *PLOS ONE*, **9(1)**, e86764, doi:10.1371/journal.pone.0086764.

Magrin, G.O. et al., 2014: Central and South America. In: *Climate Change 2014: Impacts, Adaptation, and Vulnerability. Part B: Regional Aspects. Contribution of Working Group II to the Fifth Assessment Report of the Intergovernmental Panel on Climate Change* [Barros, V.R., C.B. Field, D.J. Dokken, M.D. Mastrandrea, K.J. Mach, T.E. Bilir, M. Chatterjee, K.L. Ebi, Y.O. Estrada, R.C. Genova, B. Girma, E.S. Kissel, A.N. Levy, S. MacCracken, P.R. Mastrandrea, and L.L. White (eds.)]. Cambridge University Press, Cambridge, United Kingdom and New York, NY, USA, pp. 1499–1566.

Mahlstein, I. and R. Knutti, 2012: September Arctic sea ice predicted to disappear near 2°C global warming above present. *Journal of Geophysical Research: Atmospheres*, **117(D6)**, D06104, doi:10.1029/2011jd016709.

Mahlstein, I., R. Knutti, S. Solomon, and R.W. Portmann, 2011: Early onset of significant local warming in low latitude countries. *Environmental Research Letters*, **6(3)**, 034009, doi:10.1088/1748-9326/6/3/034009.

3

Mahowald, N.M., D.S. Ward, S.C. Doney, P.G. Hess, and J.T. Randerson, 2017a: Are the impacts of land use on warming underestimated in climate policy? *Environmental Research Letters*, **12**(9), 94016.

Mahowald, N.M. et al., 2017b: Aerosol Deposition Impacts on Land and Ocean Carbon Cycles. *Current Climate Change Reports*, **3**(1), 16–31, doi:10.1007/s40641-017-0056-z.

Maksym, T., S.E. Stammerjohn, S. Ackley, and R. Massom, 2011: Antarctic Sea ice – A polar opposite? *Oceanography*, **24**(3), 162–173, doi:10.5670/oceanog.2012.88.

Mallakpour, I. and G. Villarini, 2015: The changing nature of flooding across the central United States. *Nature Climate Change*, **5**(3), 250–254, doi:10.1038/nclimate2516.

Marcinkowski, P. et al., 2017: Effect of Climate Change on Hydrology, Sediment and Nutrient Losses in Two Lowland Catchments in Poland. *Water*, **9**(3), 156, doi:10.3390/w9030156.

Marcott, S.A., J.D. Shakun, P.U. Clark, and A.C. Mix, 2013: A Reconstruction of Regional and Global Temperature for the Past 11,300 Years. *Science*, **339**(6124), 1198–1201, doi:10.1126/science.1228026.

Markham, A., E. Osipova, K. Lafrenz Samuels, and A. Caldas, 2016: *World Heritage and Tourism in a Changing Climate*. United Nations Environment Programme (UNEP), United Nations Educational, Scientific and Cultural Organization (UNESCO) and Union of Concerned Scientists, 108 pp.

Marszelewski, W. and B. Pius, 2016: Long-term changes in temperature of river waters in the transitional zone of the temperate climate: a case study of Polish rivers. *Hydrological Sciences Journal*, **61**(8), 1430–1442, doi:10.1080/02626667.2015.1040800.

Martínez, M.L., G. Mendoza-González, R. Silva-Casarín, and E. Mendoza-Baldwin, 2014: Land use changes and sea level rise may induce a "coastal squeeze" on the coasts of Veracruz, Mexico. *Global Environmental Change*, **29**, 180–188, doi:10.1016/j.gloenvcha.2014.09.009.

Martinez-Baron, D., G. Orjuela, G. Renzoni, A.M. Loboguerrero Rodríguez, and S.D. Prager, 2018: Small-scale farmers in a 1.5°C future: The importance of local social dynamics as an enabling factor for implementation and scaling of climate-smart agriculture. *Current Opinion in Environmental Sustainability*, **31**, 112–119, doi:10.1016/j.cosust.2018.02.013.

Martius, O., S. Pfahl, and C. Chevalier, 2016: A global quantification of compound precipitation and wind extremes. *Geophysical Research Letters*, **43**(14), 7709–7717, doi:10.1002/2016gl070017.

Marx, A. et al., 2018: Climate change alters low flows in Europe under global warming of 1.5, 2, and 3°C. *Hydrology and Earth System Sciences*, **22**(2), 1017–1032, doi:10.5194/hess-22-1017-2018.

Marzeion, B. and A. Levermann, 2014: Loss of cultural world heritage and currently inhabited places to sea-level rise. *Environmental Research Letters*, **9**(3), 034001, doi:10.1088/1748-9326/9/3/034001.

Marzeion, B., G. Kaser, F. Maussion, and N. Champollion, 2018: Limited influence of climate change mitigation on short-term glacier mass loss. *Nature Climate Change*, **8**, 305–308, doi:10.1038/s41558-018-0093-1.

Masike, S. and P. Urich, 2008: Vulnerability of traditional beef sector to drought and the challenges of climate change: The case of Kgatleng District, Botswana. *Journal of Geography and Regional Planning*, **1**(1), 12–18, https://academicjournals.org/journal/jgrp/article-abstract/93cd326741.

Masson-Delmotte, V. et al., 2013: Information from Paleoclimate Archives. In: *Climate Change 2013: The Physical Science Basis. Contribution of Working Group I to the Fifth Assessment Report of the Intergovernmental Panel on Climate Change* [Stocker, T.F., D. Qin, G.-K. Plattner, M. Tignor, S.K. Allen, J. Boschung, A. Nauels, Y. Xia, V. Bex, and P.M. Midgley (eds.)]. Cambridge University Press, Cambridge, United Kingdom and New York, NY, USA, pp. 383–464.

Mastrandrea, M.D. et al., 2010: *Guidance Note for Lead Authors of the IPCC Fifth Assessment Report on Consistent Treatment of Uncertainties*. Intergovernmental Panel on Climate Change (IPCC), Geneva, Switzerland, 7 pp.

Mastrorillo, M. et al., 2016: The influence of climate variability on internal migration flows in South Africa. *Global Environmental Change*, **39**, 155–169, doi:10.1016/j.gloenvcha.2016.04.014.

Matear, R.J., M.A. Chamberlain, C. Sun, and M. Feng, 2015: Climate change projection for the western tropical Pacific Ocean using a high-resolution ocean model: Implications for tuna fisheries. *Deep Sea Research Part II: Topical Studies in Oceanography*, **113**, 22–46, doi:10.1016/j.dsr2.2014.07.003.

Mathbout, S., J.A. Lopez-bustins, J. Martin-vide, and F.S. Rodrigo, 2017: Spatial and temporal analysis of drought variability at several time scales in Syria during 1961–2012. *Atmospheric Research*, 1–39, doi:10.1016/j.atmosres.2017.09.016.

Matthews, D. and K. Caldeira, 2008: Stabilizing climate requires near-zero emissions. *Geophysical Research Letters*, **35**(4), L04705, doi:10.1029/2007gl032388.

Matthews, T.K.R., R.L. Wilby, and C. Murphy, 2017: Communicating the deadly consequences of global warming for human heat stress. *Proceedings of the National Academy of Sciences*, **114**(15), 3861–3866, doi:10.1073/pnas.1617526114.

Maule, C.F., T. Mendlik, and O.B. Christensen, 2017: The effect of the pathway to a two degrees warmer world on the regional temperature change of Europe. *Climate Services*, **7**, 3–11, doi:10.1016/j.cliser.2016.07.002.

Maúre, G. et al., 2018: The southern African climate under 1.5°C and 2°C of global warming as simulated by CORDEX regional climate models. *Environmental Research Letters*, **13**(6), 065002, doi:10.1088/1748-9326/aab190.

Maynard, J. et al., 2015: Projections of climate conditions that increase coral disease susceptibility and pathogen abundance and virulence. *Nature Climate Change*, **5**(7), 688–694, doi:10.1038/nclimate2625.

Mba, W.P. et al., 2018: Consequences of 1.5°C and 2°C global warming levels for temperature and precipitation changes over Central Africa. *Environmental Research Letters*, **13**(5), 055011, doi:10.1088/1748-9326/aab048.

McCarthy, M.P., M.J. Best, and R.A. Betts, 2010: Climate change in cities due to global warming and urban effects. *Geophysical Research Letters*, **37**(9), L09705, doi:10.1029/2010gl042845.

McClanahan, T.R., E.H. Allison, and J.E. Cinner, 2015: Managing fisheries for human and food security. *Fish and Fisheries*, **16**(1), 78–103, doi:10.1111/faf.12045.

McClanahan, T.R., J.C. Castilla, A.T. White, and O. Defeo, 2009: Healing small-scale fisheries by facilitating complex socio-ecological systems. *Reviews in Fish Biology and Fisheries*, **19**(1), 33–47, doi:10.1007/s11160-008-9088-8.

McFarland, J. et al., 2015: Impacts of rising air temperatures and emissions mitigation on electricity demand and supply in the United States: a multi-model comparison. *Climatic Change*, **131**(1), 111–125, doi:10.1007/s10584-015-1380-8.

McGranahan, G., D. Balk, and B. Anderson, 2007: The rising tide: Assessing the risks of climate change and human settlements in low elevation coastal zones. *Environment and Urbanization*, **19**(1), 17–37, doi:10.1177/0956247807076960.

McGrath, J.M. and D.B. Lobell, 2013: Regional disparities in the CO_2 fertilization effect and implications for crop yields. *Environmental Research Letters*, **8**(1), 014054, doi:10.1088/1748-9326/8/1/014054.

McLean, R. and P. Kench, 2015: Destruction or persistence of coral atoll islands in the face of 20th and 21st century sea-level rise? *Wiley Interdisciplinary Reviews: Climate Change*, **6**(5), 445–463, doi:10.1002/wcc.350.

McNamara, K.E. and H.J. Des Combes, 2015: Planning for Community Relocations Due to Climate Change in Fiji. *International Journal of Disaster Risk Science*, **6**(3), 315–319, doi:10.1007/s13753-015-0065-2.

Medek, D.E., J. Schwartz, and S.S. Myers, 2017: Estimated Effects of Future Atmospheric CO_2 Concentrations on Protein Intake and the Risk of Protein Deficiency by Country and Region. *Environmental Health Perspectives*, **125**(8), 087002, doi:10.1289/ehp41.

Meehl, G.A. et al., 2007: Global Climate Projections. In: *Climate Change 2007: The Physical Science Basis. Contribution of Working Group I to the Fourth Assessment Report of the Intergovernmental Panel on Climate Change* [Solomon, S., D. Qin, M. Manning, Z. Chen, M. Marquis, K.B. Averyt, M. Tignor, and H.L. Miller (eds.)]. Cambridge University Press, Cambridge, United Kingdom and New York, NY, USA, pp. 747–846.

Mehran, A., O. Mazdiyasni, and A. Aghakouchak, 2015: A hybrid framework for assessing socioeconomic drought: Linking climate variability, local resilience, and demand. *Journal of Geophysical Research: Atmospheres*, **120**(15), 7520–7533, doi:10.1002/2015jd023147.

Mehran, A. et al., 2017: Compounding Impacts of Human-Induced Water Stress and Climate Change on Water Availability. *Scientific Reports*, **7**(1), 1–9, doi:10.1038/s41598-017-06765-0.

Meier, K.J.S., L. Beaufort, S. Heussner, and P. Ziveri, 2014: The role of ocean acidification in Emiliania huxleyi coccolith thinning in the Mediterranean Sea. *Biogeosciences*, **11**(10), 2857–2869, doi:10.5194/bg-11-2857-2014.

Meier, W.N. et al., 2014: Arctic sea ice in transformation: A review of recent observed changes and impacts on biology and human activity. *Reviews of Geophysics*, **52**(3), 185–217, doi:10.1002/2013rg000431.

Meinshausen, M., S.C.B. Raper, and T.M.L. Wigley, 2011: Emulating coupled atmosphere-ocean and carbon cycle models with a simpler model, MAGICC6 – Part 1: Model description and calibration. *Atmospheric Chemistry and Physics*, **11**(4), 1417–1456, doi:10.5194/acp-11-1417-2011.

3

Melia, N., K. Haines, and E. Hawkins, 2016: Sea ice decline and 21st century trans-Arctic shipping routes. *Geophysical Research Letters*, **43(18)**, 9720–9728, doi:10.1002/2016gl069315.

Mengel, M., A. Nauels, J. Rogelj, and C.F. Schleussner, 2018: Committed sea-level rise under the Paris Agreement and the legacy of delayed mitigation action. *Nature Communications*, **9(1)**, 1–10, doi:10.1038/s41467-018-02985-8.

Mengel, M. et al., 2016: Future sea level rise constrained by observations and long-term commitment. *Proceedings of the National Academy of Sciences*, **113(10)**, 2597–2602, doi:10.1073/pnas.1500515113.

Millar, R.J. et al., 2017: Emission budgets and pathways consistent with limiting warming to 1.5°C. *Nature Geoscience*, **10(10)**, 741–747, doi:10.1038/ngeo3031.

Mills, M. et al., 2016: Reconciling Development and Conservation under Coastal Squeeze from Rising Sea Level. *Conservation Letters*, **9(5)**, 361–368, doi:10.1111/conl.12213.

Minasny, B. et al., 2017: Soil carbon 4 per mille. *Geoderma*, **292**, 59–86, doi:10.1016/j.geoderma.2017.01.002.

Minx, J.C., W.F. Lamb, M.W. Callaghan, L. Bornmann, and S. Fuss, 2017: Fast growing research on negative emissions. *Environmental Research Letters*, **12(3)**, 035007, doi:10.1088/1748-9326/aa5ee5.

Minx, J.C. et al., 2018: Negative emissions – Part 1: Research landscape and synthesis. *Environmental Research Letters*, **13(6)**, 063001, doi:10.1088/1748-9326/aabf9b.

Mishra, V., M. Sourav, R. Kumar, and D.A. Stone, 2017: Heat wave exposure in India in current, 1.5°C, and 2.0°C worlds. *Environmental Research Letters*, **12(12)**, 124012, doi:10.1088/1748-9326/aa9388.

Mitchell, D., 2016: Human influences on heat-related health indicators during the 2015 Egyptian heat wave. *Bulletin of the American Meteorological Society*, **97(12)**, S70–S74, doi:10.1175/bams-d-16-0132.1.

Mitchell, D. et al., 2016: Attributing human mortality during extreme heat waves to anthropogenic climate change. *Environmental Research Letters*, **11(7)**, 074006, doi:10.1088/1748-9326/11/7/074006.

Mitchell, D. et al., 2017: Half a degree additional warming, prognosis and projected impacts (HAPPI): background and experimental design. *Geoscientific Model Development*, **10**, 571–583, doi:10.5194/gmd-10-571-2017.

Mitchell, D. et al., 2018a: The myriad challenges of the Paris Agreement. *Philosophical Transactions of the Royal Society A: Mathematical, Physical and Engineering Sciences*, **376(2119)**, 20180066, doi:10.1098/rsta.2018.0066.

Mitchell, D. et al., 2018b: Extreme heat-related mortality avoided under Paris Agreement goals. *Nature Climate Change*, **8(7)**, 551–553, doi:10.1038/s41558-018-0210-1.

Mohammed, K. et al., 2017: Extreme flows and water availability of the Brahmaputra River under 1.5 and 2°C global warming scenarios. *Climatic Change*, **145(1–2)**, 159–175, doi:10.1007/s10584-017-2073-2.

Mohapatra, M., A.K. Srivastava, S. Balachandran, and B. Geetha, 2017: Inter-annual Variation and Trends in Tropical Cyclones and Monsoon Depressions Over the North Indian Ocean. In: *Observed Climate Variability and Change over the Indian Region* [Rajeevan, M.N. and S. Nayak (eds.)]. Springer Singapore, Singapore, pp. 89–106, doi:10.1007/978-981-10-2531-0_6.

Monioudi, I. et al., 2018: Climate change impacts on critical international transportation assets of Caribbean small island developing states: The case of Jamaica and Saint Lucia. *Regional Environmental Change*, 1–15, doi:10.1007/s10113-018-1360-4.

Montroull, N.B., R.I. Saurral, and I.A. Camilloni, 2018: Hydrological impacts in La Plata basin under 1.5, 2 and 3°C global warming above the pre-industrial level. *International Journal of Climatology*, **38(8)**, 3355–3368, doi:10.1002/joc.5505.

Moore, F.C. and D.B. Lobell, 2015: The fingerprint of climate trends on European crop yields. *Proceedings of the National Academy of Sciences*, **112(9)**, 2670–2675, doi:10.1073/pnas.1409606112.

Mora, C. et al., 2017: Global risk of deadly heat. *Nature Climate Change*, **7(7)**, 501–506, doi:10.1038/nclimate3322.

Moriondo, M. et al., 2013a: Projected shifts of wine regions in response to climate change. *Climatic Change*, **119(3–4)**, 825–839, doi:10.1007/s10584-013-0739-y.

Moriondo, M. et al., 2013b: Olive trees as bio-indicators of climate evolution in the Mediterranean Basin. *Global Ecology and Biogeography*, **22(7)**, 818–833, doi:10.1111/geb.12061.

Moritz, M.A. et al., 2012: Climate change and disruptions to global fire activity. *Ecosphere*, **3(6)**, art49, doi:10.1890/es11-00345.1.

Mortensen, C.J. et al., 2009: Embryo recovery from exercised mares. *Animal Reproduction Science*, **110(3)**, 237–244, doi:10.1016/j.anireprosci.2008.01.015.

Mortola, J.P. and P.B. Frappell, 2000: Ventilatory Responses to Changes in Temperature in Mammals and Other Vertebrates. *Annual Review of Physiology*, **62(1)**, 847–874, doi:10.1146/annurev.physiol.62.1.847.

Mouginot, J., E. Rignot, and B. Scheuchl, 2014: Sustained increase in ice discharge from the Amundsen Sea Embayment, West Antarctica, from 1973 to 2013. *Geophysical Research Letters*, **41(5)**, 1576–1584, doi:10.1002/2013gl059069.

Moy, A.D., W.R. Howard, S.G. Bray, and T.W. Trull, 2009: Reduced calcification in modern Southern Ocean planktonic foraminifera. *Nature Geoscience*, **2(4)**, 276–280, doi:10.1038/ngeo460.

Mueller, B. and S.I. Seneviratne, 2012: Hot days induced by precipitation deficits at the global scale. *Proceedings of the National Academy of Sciences*, **109(31)**, 12398–12403, doi:10.1073/pnas.1204330109.

Mueller, B. et al., 2015: Lengthening of the growing season in wheat and maize producing regions. *Weather and Climate Extremes*, **9**, 47–56, doi:10.1016/j.wace.2015.04.001.

Mueller, N.D. et al., 2016: Cooling of US Midwest summer temperature extremes from cropland intensification. *Nature Climate Change*, **6(3)**, 317–322, doi:10.1038/nclimate2825.

Mueller, S.A., J.E. Anderson, and T.J. Wallington, 2011: Impact of biofuel production and other supply and demand factors on food price increases in 2008. *Biomass and Bioenergy*, **35(5)**, 1623–1632, doi:10.1016/j.biombioe.2011.01.030.

Mueller, V., C. Gray, and K. Kosec, 2014: Heat stress increases long-term human migration in rural Pakistan. *Nature Climate Change*, **4(3)**, 182–185, doi:10.1038/nclimate2103.

Mullan, D., D. Favis-Mortlock, and R. Fealy, 2012: Addressing key limitations associated with modelling soil erosion under the impacts of future climate change. *Agricultural and Forest Meteorology*, **156**, 18–30, doi:10.1016/j.agrformet.2011.12.004.

Muller, C., 2011: Agriculture: Harvesting from uncertainties. *Nature Climate Change*, **1(5)**, 253–254, doi:10.1038/nclimate1179.

Murakami, H., G.A. Vecchi, and S. Underwood, 2017: Increasing frequency of extremely severe cyclonic storms over the Arabian Sea. *Nature Climate Change*, **7(12)**, 885–889, doi:10.1038/s41558-017-0008-6.

Muratori, M., K. Calvin, M. Wise, P. Kyle, and J. Edmonds, 2016: Global economic consequences of deploying bioenergy with carbon capture and storage (BECCS). *Environmental Research Letters*, **11(9)**, 095004, doi:10.1088/1748-9326/11/9/095004.

Murphy, G.E.P. and T.N. Romanuk, 2014: A meta-analysis of declines in local species richness from human disturbances. *Ecology and Evolution*, **4(1)**, 91–103, doi:10.1002/ece3.909.

Muthige, M.S. et al., 2018: Projected changes in tropical cyclones over the South West Indian Ocean under different extents of global warming. *Environmental Research Letters*, **13(6)**, 065019, doi:10.1088/1748-9326/aabc60.

Mweya, C.N. et al., 2016: Climate Change Influences Potential Distribution of Infected *Aedes aegypti* Co-Occurrence with Dengue Epidemics Risk Areas in Tanzania. *PLOS ONE*, **11(9)**, e0162649, doi:10.1371/journal.pone.0162649.

Mycoo, M., 2014: Sustainable tourism, climate change and sea level rise adaptation policies in Barbados. *Natural Resources Forum*, **38(1)**, 47–57, doi:10.1111/1477-8947.12033.

Mycoo, M.A., 2017: Beyond 1.5°C: vulnerabilities and adaptation strategies for Caribbean Small Island Developing States. *Regional Environmental Change*, 1–13, doi:10.1007/s10113-017-1248-8.

Myers, S.S. et al., 2014: Increasing CO_2 threatens human nutrition. *Nature*, **510(7503)**, 139–142, doi:10.1038/nature13179.

Myers, S.S. et al., 2017: Climate Change and Global Food Systems: Potential Impacts on Food Security and Undernutrition. *Annual Review of Public Health*, **38(1)**, 259–277, doi:10.1146/annurev-publhealth-031816-044356.

Myhre, G. et al., 2013: Anthropogenic and natural radiative forcing. In: *Climate Change 2013: The Physical Science Basis. Contribution of Working Group I to the Fifth Assessment Report of the Intergovernmental Panel on Climate Change* [Stocker, T.F., D. Qin, G.-K. Plattner, M. Tignor, S.K. Allen, J. Boschung, A. Nauels, Y. Xia, V. Bex, and P.M. Midgley (eds.)]. Cambridge University Press, Cambridge, United Kingdom and New York, NY, USA, pp. 658–740.

Narayan, S. et al., 2016: The effectiveness, costs and coastal protection benefits of natural and nature-based defences. *PLOS ONE*, **11(5)**, e0154735, doi:10.1371/journal.pone.0154735.

Naudts, K. et al., 2016: Europe's forest management did not mitigate climate warming. *Science*, **351(6273)**, 597–600, doi:10.1126/science.aad7270.

3

Nauels, A., M. Meinshausen, M. Mengel, K. Lorbacher, and T.M.L. Wigley, 2017: Synthesizing long-term sea level rise projections – the MAGICC sea level model v2.0. *Geoscientific Model Development*, **10(6)**, 2495–2524, doi:10.5194/gmd-10-2495-2017.

Naylor, R.L. and W.P. Falcon, 2010: Food security in an era of economic volatility. *Population and Development Review*, **36(4)**, 693–723, doi:10.1111/j.1728-4457.2010.00354.x.

NCCARF, 2013: *Terrestrial Report Card 2013: Climate change impacts and adaptation on Australian biodiversity*. National Climate Change Adaptation Research Facility (NCCARF), Gold Coast, Australia, 8 pp.

Nelson, G.C. et al., 2010: *Food Security, Farming, and Climate Change to 2050: Scenarios, Results, Policy Options*. IFPRI Research Monograph, International Food Policy Research Institute (IFPRI), Washington DC, USA, 140 pp., doi:10.2499/9780896291867.

Nelson, G.C. et al., 2014a: Climate change effects on agriculture: Economic responses to biophysical shocks. *Proceedings of the National Academy of Sciences*, **111(9)**, 3274–3279, doi:10.1073/pnas.1222465110.

Nelson, G.C. et al., 2014b: Agriculture and climate change in global scenarios: why don't the models agree. *Agricultural Economics*, **45(1)**, 85–101, doi:10.1111/agec.12091.

Nemet, G.F. et al., 2018: Negative emissions-Part 3: Innovation and upscaling. *Environmental Research Letters*, **13(6)**, 063003, doi:10.1088/1748-9326/aabff4.

Neumann, K., P.H. Verburg, E. Stehfest, and C. Müller, 2010: The yield gap of global grain production: A spatial analysis. *Agricultural Systems*, **103(5)**, 316–326, doi:10.1016/j.agsy.2010.02.004.

Newbold, T. et al., 2015: Global effects of land use on local terrestrial biodiversity. *Nature*, **520(7545)**, 45–50, doi:10.1038/nature14324.

Niang, I. et al., 2014: Africa. In: *Climate Change 2014: Impacts, Adaptation, and Vulnerability. Part B: Regional Aspects. Contribution of Working Group II to the Fifth Assessment Report of the Intergovernmental Panel on Climate Change* [Barros, V.R., C.B. Field, D.J. Dokken, M.D. Mastrandrea, K.J. Mach, T.E. Bilir, M. Chatterjee, K.L. Ebi, Y.O. Estrada, R.C. Genova, B. Girma, E.S. Kissel, A.N. Levy, S. MacCracken, P.R. Mastrandrea, and L.L. White (eds.)]. Cambridge University Press, Cambridge, United Kingdom and New York, NY, USA, pp. 1199–1265.

Nicholls, R.J., T. Reeder, S. Brown, and I.D. Haigh, 2015: The risks of sea-level rise in coastal cities. In: *Climate Change: A risk assessment* [King, D., D. Schrag, Z. Dadi, Q. Ye, and A. Ghosh (eds.)]. Centre for Science and Policy (CSaP), University of Cambridge, Cambridge, UK, pp. 94–98.

Nicholls, R.J. et al., 2007: Coastal systems and low-lying areas. In: *Climate Change 2007: Impacts, Adaptation, and Vulnerability. Contribution of Working Group II to the Fourth Assessment Report of the Intergovernmental Panel on Climate Change* [Parry, M.L., O.F. Canziani, J.P. Palutikof, P.J. Linden, and C.E. Hanson (eds.)]. Cambridge University Press, Cambridge, UK, pp. 315–356.

Nicholls, R.J. et al., 2018: Stabilization of global temperature at 1.5°C and 2.0°C: implications for coastal areas. *Philosophical Transactions of the Royal Society A: Mathematical, Physical and Engineering Sciences*, **376(2119)**, 20160448, doi:10.1098/rsta.2016.0448.

Nicholson, C.T.M., 2014: Climate change and the politics of causal reasoning: the case of climate change and migration. *The Geographical Journal*, **180(2)**, 151–160, doi:10.1111/geoj.12062.

Niederdrenk, A.L. and D. Notz, 2018: Arctic Sea Ice in a 1.5°C Warmer World. *Geophysical Research Letters*, **45(4)**, 1963–1971, doi:10.1002/2017gl076159.

Njeru, J., K. Henning, M.W. Pletz, R. Heller, and H. Neubauer, 2016: Q fever is an old and neglected zoonotic disease in Kenya: a systematic review. *BMC Public Health*, **16**, 297, doi:10.1186/s12889-016-2929-9.

Nobre, C.A. et al., 2016: Land-use and climate change risks in the Amazon and the need of a novel sustainable development paradigm. *Proceedings of the National Academy of Sciences*, **113(39)**, 10759–68, doi:10.1073/pnas.1605516113.

Notenbaert, A.M.O., J.A. Cardoso, N. Chirinda, M. Peters, and A. Mottet, 2017: *Climate change impacts on livestock and implications for adaptation*. International Center for Tropical Agriculture (CIAT), Rome, Italy, 30 pp.

Notz, D., 2015: How well must climate models agree with observations? *Philosophical Transactions of the Royal Society A: Mathematical, Physical and Engineering Sciences*, **373(2052)**, 20140164, doi:10.1098/rsta.2014.0164.

Notz, D. and J. Stroeve, 2016: Observed Arctic sea-ice loss directly follows anthropogenic CO_2 emission. *Science*, **354(6313)**, 747–750, doi:10.1126/science.aag2345.

Nunn, P.D., J. Runman, M. Falanruw, and R. Kumar, 2017: Culturally grounded responses to coastal change on islands in the Federated States of Micronesia, northwest Pacific Ocean. *Regional Environmental Change*, **17(4)**, 959–971, doi:10.1007/s10113-016-0950-2.

Nurse, L.A. et al., 2014: Small islands. In: *Climate Change 2014: Impacts, Adaptation, and Vulnerability. Part B: Regional Aspects. Contribution of Working Group II to the Fifth Assessment Report of the Intergovernmental Panel on Climate Change* [Barros, V.R., C.B. Field, D.J. Dokken, M.D. Mastrandrea, K.J. Mach, T.E. Bilir, M. Chatterjee, K.L. Ebi, Y.O. Estrada, R.C. Genova, B. Girma, E.S. Kissel, A.N. Levy, S. MacCracken, P.R. Mastrandrea, and L.L. White (eds.)]. Cambridge University Press, Cambridge, United Kingdom and New York, NY, USA, pp. 1613–1654.

O'Leary, J.K. et al., 2017: The Resilience of Marine Ecosystems to Climatic Disturbances. *BioScience*, **67(3)**, 208–220, doi:10.1093/biosci/biw161.

O'Neill, B.C. et al., 2017: IPCC Reasons for Concern regarding climate change risks. *Nature Climate Change*, **7**, 28–37, doi:10.1038/nclimate3179.

Oberbauer, S.F. et al., 2013: Phenological response of tundra plants to background climate variation tested using the International Tundra Experiment. *Philosophical Transactions of the Royal Society B: Biological Sciences*, **368(1624)**, 20120481, doi:10.1098/rstb.2012.0481.

Obersteiner, M. et al., 2016: Assessing the land resource-food price nexus of the Sustainable Development Goals. *Science Advances*, **2(9)**, e1501499, doi:10.1126/sciadv.1501499.

Obersteiner, M. et al., 2018: How to spend a dwindling greenhouse gas budget. *Nature Climate Change*, **8(1)**, 7–10, doi:10.1038/s41558-017-0045-1.

Ogden, N.H., R. Milka, C. Caminade, and P. Gachon, 2014a: Recent and projected future climatic suitability of North America for the Asian tiger mosquito *Aedes albopictus*. *Parasites & Vectors*, **7(1)**, 532, doi:10.1186/s13071-014-0532-4.

Ogden, N.H. et al., 2014b: Estimated effects of projected climate change on the basic reproductive number of the Lyme disease vector *Ixodes scapularis*. *Environmental Health Perspectives*, **122(6)**, 631–638, doi:10.1289/ehp.1307799.

Okada, M., T. Iizumi, Y. Hayashi, and M. Yokozawa, 2011: Modeling the multiple effects of temperature and precipitation on rice quality. *Environmental Research Letters*, **6(3)**, 034031, doi:10.1088/1748-9326/6/3/034031.

Oleson, K.W., G.B. Bonan, and J. Feddema, 2010: Effects of white roofs on urban temperature in a global climate model. *Geophysical Research Letters*, **37(3)**, L03701, doi:10.1029/2009gl042194.

Oliver, E.C.J. et al., 2018: Longer and more frequent marine heatwaves over the past century. *Nature Communications*, **9(1)**, 1324, doi:10.1038/s41467-018-03732-9.

Olsson, L. et al., 2014: Livelihoods and Poverty. In: *Climate Change 2014: Impacts, Adaptation, and Vulnerability. Part A: Global and Sectoral Aspects. Contribution of working Group II to the Fifth Assessment Report of the Intergovernmental Panel on Climate Change* [Field, C.B., V.R. Barros, D.J. Dokken, K.J. Mach, M.D. Mastrandrea, T.E. Bilir, M. Chatterjee, K.L. Ebi, Y.O. Estrada, R.C. Genova, B. Girma, E.S. Kissel, A.N. Levy, S. MacCracken, P.R. Mastrandrea, and L.L. White (eds.)]. Cambridge University Press, Cambridge, United Kingdom and New York, NY, USA, pp. 793–832.

Omstedt, A., M. Edman, B. Claremar, and A. Rutgersson, 2015: Modelling the contributions to marine acidification from deposited SOx, NOx, and NHx in the Baltic Sea: Past and present situations. *Continental Shelf Research*, **111**, 234–249, doi:10.1016/j.csr.2015.08.024.

Ong, E.Z., M. Briffa, T. Moens, and C. Van Colen, 2017: Physiological responses to ocean acidification and warming synergistically reduce condition of the common cockle *Cerastoderma edule*. *Marine Environmental Research*, **130**, 38–47, doi:10.1016/j.marenvres.2017.07.001.

Oppenheimer, M. et al., 2014: Emergent risks and key vulnerabilities. In: *Climate Change 2014: Impacts, Adaptation, and Vulnerability. Part A: Global and Sectoral Aspects. Contribution of Working Group II to the Fifth Assessment Report of the Intergovernmental Panel on Climate Change* [Field, C.B., V.R. Barros, D.J. Dokken, K.J. Mach, M.D. Mastrandrea, T.E. Bilir, M. Chatterjee, K.L. Ebi, Y.O. Estrada, R.C. Genova, B. Girma, E.S. Kissel, A.N. Levy, S. MacCracken, P.R. Mastrandrea, and L.L.White (eds.)]. Cambridge University Press, Cambridge, United Kingdom and New York, NY, USA, pp. 1039–1099.

Ordonez, A. et al., 2014: Effects of Ocean Acidification on Population Dynamics and Community Structure of Crustose Coralline Algae. *The Biological Bulletin*, **226**, 255–268, doi:10.1086/bblv226n3p255.

Orlowsky, B. and S.I. Seneviratne, 2013: Elusive drought: uncertainty in observed trends and short- and long-term CMIP5 projections. *Hydrology and Earth System Sciences*, **17(5)**, 1765–1781, doi:10.5194/hess-17-1765-2013.

Orr, J.C. et al., 2005: Anthropogenic ocean acidification over the twenty-first century and its impact on calcifying organisms. *Nature*, **437(7059)**, 681–686, doi:10.1038/nature04095.

Osima, S. et al., 2018: Projected climate over the Greater Horn of Africa under 1.5°C and 2°C global warming. *Environmental Research Letters*, **13(6)**, 065004, doi:10.1088/1748-9326/aaba1b.

3

Osorio, J.A., M.J. Wingfield, and J. Roux, 2016: A review of factors associated with decline and death of mangroves, with particular reference to fungal pathogens. *South African Journal of Botany*, **103**, 295–301, doi:10.1016/j.sajb.2014.08.010.

Ourbak, T. and A.K. Magnan, 2017: The Paris Agreement and climate change negotiations: Small Islands, big players. *Regional Environmental Change*, 1–7, doi:10.1007/s10113-017-1247-9.

Paeth, H. et al., 2010: Meteorological characteristics and potential causes of the 2007 flood in sub-Saharan Africa. *International Journal of Climatology*, **31(13)**, 1908–1926, doi:10.1002/joc.2199.

Palazzo, A. et al., 2017: Linking regional stakeholder scenarios and shared socioeconomic pathways: Quantified West African food and climate futures in a global context. *Global Environmental Change*, **45**, 227–242, doi:10.1016/j.gloenvcha.2016.12.002.

Palmer, M.A. et al., 2008: Climate change and the world's river basins: Anticipating management options. *Frontiers in Ecology and the Environment*, **6(2)**, 81–89, doi:10.1890/060148.

Palumbi, S.R., D.J. Barshis, N. Traylor-Knowles, and R.A. Bay, 2014: Mechanisms of reef coral resistance to future climate change. *Science*, **344(6186)**, 895–8, doi:10.1126/science.1251336.

Park, C. et al., 2018: Avoided economic impacts of energy demand changes by 1.5 and 2°C climate stabilization. *Environmental Research Letters*, **13(4)**, 045010, doi:10.1088/1748-9326/aab724.

Parker, L.M. et al., 2017: Ocean acidification narrows the acute thermal and salinity tolerance of the Sydney rock oyster *Saccostrea glomerata*. *Marine Pollution Bulletin*, **122(1–2)**, 263–271, doi:10.1016/j.marpolbul.2017.06.052.

Parkes, B., D. Defrance, B. Sultan, P. Ciais, and X. Wang, 2018: Projected changes in crop yield mean and variability over West Africa in a world 1.5 K warmer than the pre-industrial era. *Earth System Dynamics*, **9(1)**, 119–134, doi:10.5194/esd-9-119-2018.

Parmesan, C. and M.E. Hanley, 2015: Plants and climate change: complexities and surprises. *Annals of Botany*, **116(6, SI)**, 849–864, doi:10.1093/aob/mcv169.

Paterson, R.R.M. and N. Lima, 2010: How will climate change affect mycotoxins in food? *Food Research International*, **43(7)**, 1902–1914, doi:10.1016/j.foodres.2009.07.010.

Patiño, R., D. Dawson, and M.M. VanLandeghem, 2014: Retrospective analysis of associations between water quality and toxic blooms of golden alga (*Prymnesium parvum*) in Texas reservoirs: Implications for understanding dispersal mechanisms and impacts of climate change. *Harmful Algae*, **33**, 1–11, doi:10.1016/j.hal.2013.12.006.

Pauly, D. and A. Charles, 2015: Counting on small-scale fisheries. *Science*, **347(6219)**, 242–243, doi:10.1126/science.347.6219.242-b.

Paustian, K. et al., 2006: Introduction. In: *2006 IPCC Guidelines for National Greenhouse Gas Inventories, Prepared by the National Greenhouse Gas Inventories Programme: Vol. 4* [Eggleston, H.S., L. Buendia, K. Miwa, T. Ngara, and K. Tanabe (eds.)]. IGES, Japan, pp. 1–21.

Pecl, G.T. et al., 2017: Biodiversity redistribution under climate change: Impacts on ecosystems and human well-being. *Science*, **355(6332)**, eaai9214, doi:10.1126/science.aai9214.

Pendergrass, A.G., F. Lehner, B.M. Sanderson, and Y. Xu, 2015: Does extreme precipitation intensity depend on the emissions scenario? *Geophysical Research Letters*, **42(20)**, 8767–8774, doi:10.1002/2015gl065854.

Pendleton, L. et al., 2016: Coral reefs and people in a high-CO_2 world: Where can science make a difference to people? *PLOS ONE*, **11(11)**, 1–21, doi:10.1371/journal.pone.0164699.

Pereira, H.M. et al., 2010: Scenarios for global biodiversity in the 21st century. *Science*, **330(6010)**, 1496–1501, doi:10.1126/science.1196624.

Pérez-Escamilla, R., 2017: Food Security and the 2015–2030 Sustainable Development Goals: From Human to Planetary Health. *Current Developments in Nutrition*, **1(7)**, e000513, doi:10.3945/cdn.117.000513.

Perring, M.P., B.R. Cullen, I.R. Johnson, and M.J. Hovenden, 2010: Modelled effects of rising CO_2 concentration and climate change on native perennial grass and sown grass-legume pastures. *Climate Research*, **42(1)**, 65–78, doi:10.3354/cr00863.

Petkova, E.P., R.M. Horton, D.A. Bader, and P.L. Kinney, 2013: Projected Heat-Related Mortality in the U.S. Urban Northeast. *International Journal of Environmental Research and Public Health*, **10(12)**, 6734–6747, doi:10.3390/ijerph10126734.

Piao, S. et al., 2015: Detection and attribution of vegetation greening trend in China over the last 30 years. *Global Change Biology*, **21(4)**, 1601–1609, doi:10.1111/gcb.12795.

Pierrehumbert, R.T., 2014: Short-Lived Climate Pollution. *Annual Review of Earth and Planetary Sciences*, **42(1)**, 341–379, doi:10.1146/annurev-earth-060313-054843.

Piggott-McKellar, A.E. and K.E. McNamara, 2017: Last chance tourism and the Great Barrier Reef. *Journal of Sustainable Tourism*, **25(3)**, 397–415, doi:10.1080/09669582.2016.1213849.

Piguet, E. and F. Laczko (eds.), 2014: *People on the Move in a Changing Climate: The Regional Impact of Environmental Change on Migration*. Global Migration Issues, Springer Netherlands, Dordrecht, The Netherlands, 253 pp., doi:10.1007/978-94-007-6985-4.

Piñones, A. and A. Fedorov, 2016: Projected changes of Antarctic krill habitat by the end of the 21st century. *Geophysical Research Letters*, **43(16)**, 8580–8589, doi:10.1002/2016gl069656.

Pinsky, M.L., B. Worm, M.J. Fogarty, J.L. Sarmiento, and S.A. Levin, 2013: Marine Taxa Track Local Climate Velocities. *Science*, **341(6151)**, 1239–1242, doi:10.1126/science.1239352.

Piontek, F. et al., 2014: Multisectoral climate impact hotspots in a warming world. *Proceedings of the National Academy of Sciences*, **111(9)**, 3233–3238, doi:10.1073/pnas.1222471110.

Pistorious, T. and L. Kiff, 2017: *From a biodiversity perspective: risks, trade- offs, and international guidance for Forest Landscape Restoration*. UNIQUE forestry and land use GmbH, Freiburg, Germany, 66 pp.

Pittelkow, C.M. et al., 2014: Productivity limits and potentials of the principles of conservation agriculture. *Nature*, **517(7534)**, 365–367, doi:10.1038/nature13809.

Poloczanska, E.S., O. Hoegh-Guldberg, W. Cheung, H.-O. Pörtner, and M. Burrows, 2014: Cross-chapter box on observed global responses of marine biogeography, abundance, and phenology to climate change. In: *Climate Change 2014: Impacts, Adaptation, and Vulnerability. Part A: Global and Sectoral Aspects. Contribution of Working Group II to the Fifth Assessment Report of the Intergovernmental Panel on Climate Change* [Field, C.B., V.R. Barros, D.J. Dokken, K.J. Mach, M.D. Mastrandrea, T.E. Bilir, M. Chatterjee, K.L. Ebi, Y.O. Estrada, R.C. Genova, B. Girma, E.S. Kissel, A.N. Levy, S. MacCracken, P.R. Mastrandrea, and L.L. White (eds.)]. Cambridge University Press, Cambridge, United Kingdom and New York, NY, USA, pp. 123–127.

Poloczanska, E.S. et al., 2013: Global imprint of climate change on marine life. *Nature Climate Change*, **3(10)**, 919–925, doi:10.1038/nclimate1958.

Poloczanska, E.S. et al., 2016: Responses of Marine Organisms to Climate Change across Oceans. *Frontiers in Marine Science*, **3**, 62, doi:10.3389/fmars.2016.00062.

Polyak, L. et al., 2010: History of sea ice in the Arctic. *Quaternary Science Reviews*, **29(15–16)**, 1757–1778, doi:10.1016/j.quascirev.2010.02.010.

Popp, A. et al., 2017: Land-use futures in the shared socio-economic pathways. *Global Environmental Change*, **42**, 331–345, doi:10.1016/j.gloenvcha.2016.10.002.

Porter, J.R. et al., 2014: Food security and food production systems. In: *Climate Change 2014: Impacts, Adaptation, and Vulnerability. Part A: Global and Sectoral Aspects. Contribution of Working Group II to the Fifth Assessment Report of the Intergovernmental Panel on Climate Change* [Field, C.B., V.R. Barros,, D.J. Dokken, K.J. March, M.D. Mastrandrea, T.E. Bilir, M. Chatterjee, K.L. Ebi, Y.O. Estrada, R.C. Genova, B. Girma, E.S. Kissel, A. Levy, S. MacCracken, P.R. Mastrandrea, and L.L. White Field (eds.)]. Cambridge University Press, Cambridge, United Kingdom and New York, NY, USA, pp. 485–533.

Portmann, F.T., P. Döll, S. Eisner, and M. Flörke, 2013: Impact of climate change on renewable groundwater resources: assessing the benefits of avoided greenhouse gas emissions using selected CMIP5 climate projections. *Environmental Research Letters*, **8(2)**, 024023, doi:10.1088/1748-9326/8/2/024023.

Pörtner, H.O. et al., 2014: Ocean Systems. In: *Climate Change 2014: Impacts, Adaptation, and Vulnerability. Part A: Global and Sectoral Aspects. Contribution of Working Group II to the Fifth Assessment Report of the Intergovernmental Panel on Climate Change* [Field, C.B., V.R. Barros, D.J. Dokken, K.J. Mach, M.D. Mastrandrea, T.E. Bilir, M. Chatterjee, K.L. Ebi, Y.O. Estrada, R.C. Genova, B. Girma, E.S. Kissel, A.N. Levy, S. MacCracken, P.R. Mastrandrea, and L.L. White (eds.)]. Cambridge University Press, Cambridge, United Kingdom and New York, NY, USA, pp. 411–484.

Potemkina, T.G. and V.L. Potemkin, 2015: Sediment load of the main rivers of Lake Baikal in a changing environment (east Siberia, Russia). *Quaternary International*, **380–381**, 342–349, doi:10.1016/j.quaint.2014.08.029.

3

Prestele, R. et al., 2016: Hotspots of uncertainty in land-use and land-cover change projections: a global-scale model comparison. *Global Change Biology*, **22**(12), 3967–3983, doi:10.1111/gcb.13337.

Pretis, F., M. Schwarz, K. Tang, K. Haustein, and M.R. Allen, 2018: Uncertain impacts on economic growth when stabilizing global temperatures at 1.5°C or 2°C warming. *Philosophical Transactions of the Royal Society A: Mathematical, Physical and Engineering Sciences*, **376**(2119), 20160460, doi:10.1098/rsta.2016.0460.

Primavera, J.H. et al., 2016: Preliminary assessment of post-*Haiyan* mangrove damage and short-term recovery in Eastern Samar, central Philippines. *Marine Pollution Bulletin*, **109**(2), 744–750, doi:10.1016/j.marpolbul.2016.05.050.

Prober, S.M., D.W. Hilbert, S. Ferrier, M. Dunlop, and D. Gobbett, 2012: Combining community-level spatial modelling and expert knowledge to inform climate adaptation in temperate grassy eucalypt woodlands and related grasslands. *Biodiversity and Conservation*, **21**(7), 1627–1650, doi:10.1007/s10531-012-0268-4.

Pryor, S.C., R.C. Sullivan, and T. Wright, 2016: Quantifying the Roles of Changing Albedo, Emissivity, and Energy Partitioning in the Impact of Irrigation on Atmospheric Heat Content. *Journal of Applied Meteorology and Climatology*, **55**(8), 1699–1706, doi:10.1175/jamc-d-15-0291.1.

Pugh, T.A.M., C. Müller, A. Arneth, V. Haverd, and B. Smith, 2016: Key knowledge and data gaps in modelling the influence of CO_2 concentration on the terrestrial carbon sink. *Journal of Plant Physiology*, **203**, 3–15, doi:10.1016/j.jplph.2016.05.001.

Qian, B., X. Zhang, K. Chen, Y. Feng, and T. O'Brien, 2010: Observed Long-Term Trends for Agroclimatic Conditions in Canada. *Journal of Applied Meteorology and Climatology*, **49**(4), 604–618, doi:10.1175/2009jamc2275.1.

Qian, Y. et al., 2013: A Modeling Study of Irrigation Effects on Surface Fluxes and Land-Air-Cloud Interactions in the Southern Great Plains. *Journal of Hydrometeorology*, **14**(3), 700–721, doi:10.1175/jhm-d-12-0134.1.

Quataert, E., C. Storlazzi, A. van Rooijen, O. Cheriton, and A. van Dongeren, 2015: The influence of coral reefs and climate change on wave-driven flooding of tropical coastlines. *Geophysical Research Letters*, **42**(15), 6407–6415, doi:10.1002/2015gl064861.

Raabe, E.A. and R.P. Stumpf, 2016: Expansion of Tidal Marsh in Response to Sea-Level Rise: Gulf Coast of Florida, USA. *Estuaries and Coasts*, **39**(1), 145–157, doi:10.1007/s12237-015-9974-y.

Rafferty, N.E., 2017: Effects of global change on insect pollinators: multiple drivers lead to novel communities. *Current Opinion in Insect Science*, **23**, 1–6, doi:10.1016/j.cois.2017.06.009.

Rahmstorf, S. et al., 2015a: Corrigendum: Evidence for an exceptional twentieth-century slowdown in Atlantic Ocean overturning. *Nature Climate Change*, **5**(10), 956–956, doi:10.1038/nclimate2781.

Rahmstorf, S. et al., 2015b: Exceptional twentieth-century slowdown in Atlantic Ocean overturning circulation. *Nature Climate Change*, **5**(5), 475–480, doi:10.1038/nclimate2554.

Ramirez-Cabral, N.Y.Z., L. Kumar, and S. Taylor, 2016: Crop niche modeling projects major shifts in common bean growing areas. *Agricultural and Forest Meteorology*, **218–219**, 102–113, doi:10.1016/j.agrformet.2015.12.002.

Ranger, N., T. Reeder, and J. Lowe, 2013: Addressing 'deep' uncertainty over long-term climate in major infrastructure projects: four innovations of the Thames Estuary 2100 Project. *EURO Journal on Decision Processes*, **1**(3–4), 233–262, doi:10.1007/s40070-013-0014-5.

Rasmussen, D.J. et al., 2018: Extreme sea level implications of 1.5°C, 2.0°C, and 2.5°C temperature stabilization targets in the 21st and 22nd centuries. *Environmental Research Letters*, **13**(3), 034040, doi:10.1088/1748-9326/aaac87.

Ray, D.K., J.S. Gerber, G.K. MacDonald, and P.C. West, 2015: Climate variation explains a third of global crop yield variability. *Nature Communications*, **6**(1), 5989, doi:10.1038/ncomms6989.

Reaka-Kudla, M.L., 1997: The Global Biodiversity of Coral Reefs: A comparison with Rain Forests. In: *Biodiversity II: Understanding and Protecting Our Biological Resources* [Reaka-Kudla, M., D.E. Wilson, and E.O. Wilson (eds.)]. Joseph Henry Press, Washington DC, USA, pp. 83–108, doi:10.2307/1791071.

Reasoner, M.A. and W. Tinner, 2009: Holocene Treeline Fluctuations. In: *Encyclopedia of Paleoclimatology and Ancient Environments*. Springer, Dordrecht, The Netherlands, pp. 442–446, doi:10.1007/978-1-4020-4411-3_107.

Reid, P., S. Stammerjohn, R. Massom, T. Scambos, and J. Lieser, 2015: The record 2013 Southern Hemisphere sea-ice extent maximum. *Annals of Glaciology*, **56**(69), 99–106, doi:10.3189/2015aog69a892.

Ren, Z. et al., 2016: Predicting malaria vector distribution under climate change scenarios in China: Challenges for malaria elimination. *Scientific Reports*, **6**, 20604, doi:10.1038/srep20604.

Renaud, F.G., T.T.H. Le, C. Lindener, V.T. Guong, and Z. Sebesvari, 2015: Resilience and shifts in agro-ecosystems facing increasing sea-level rise and salinity intrusion in Ben Tre Province, Mekong Delta. *Climatic Change*, **133**(1), 69–84, doi:10.1007/s10584-014-1113-4.

Renaudeau, D., J.L. Gourdine, and N.R. St-Pierre, 2011: A meta-analysis of the effects of high ambient temperature on growth performance of growing-finishing pigs. *Journal of Animal Science*, **89**(7), 2220–2230, doi:10.2527/jas.2010-3329.

Revi, A. et al., 2014: Urban areas. In: *Climate Change 2014: Impacts, Adaptation, and Vulnerability. Part A: Global and Sectoral Aspects. Contribution of Working Group II to the Fifth Assessment Report of the Intergovernmental Panel on Climate Change* [Field, C.B., V.R. Barros, D.J. Dokken, K.J. Mach, M.D. Mastrandrea, T.E. Bilir, M. Chatterjee, K.L. Ebi, Y.O. Estrada, R.C. Genova, B. Girma, E.S. Kissel, A.N. Levy, S. MacCracken, P.R. Mastrandrea, and L.L. White (eds.)]. Cambridge University Press, Cambridge, United Kingdom and New York, NY, USA, pp. 535–612.

Reyer, C.P.O. et al., 2017a: Climate change impacts in Latin America and the Caribbean and their implications for development. *Regional Environmental Change*, **17**(6), 1601–1621, doi:10.1007/s10113-015-0854-6.

Reyer, C.P.O. et al., 2017b: Are forest disturbances amplifying or canceling out climate change-induced productivity changes in European forests? *Environmental Research Letters*, **12**(3), 034027, doi:10.1088/1748-9326/aa5ef1.

Reyer, C.P.O. et al., 2017c: Climate change impacts in Central Asia and their implications for development. *Regional Environmental Change*, **17**(6), 1639–1650, doi:10.1007/s10113-015-0893-z.

Reyer, C.P.O. et al., 2017d: Turn down the heat: regional climate change impacts on development. *Regional Environmental Change*, **17**(6), 1563–1568, doi:10.1007/s10113-017-1187-4.

Reyes-Nivia, C., G. Diaz-Pulido, and S. Dove, 2014: Relative roles of endolithic algae and carbonate chemistry variability in the skeletal dissolution of crustose coralline algae. *Biogeosciences*, **11**(17), 4615–4626, doi:10.5194/bg-11-4615-2014.

Reyes-Nivia, C., G. Diaz-Pulido, D. Kline, O.H. Guldberg, and S. Dove, 2013: Ocean acidification and warming scenarios increase microbioerosion of coral skeletons. *Global Change Biology*, **19**(6), 1919–1929, doi:10.1111/gcb.12158.

Rhein, M., S.R. Rintoul, S. Aoki, E. Campos, and D. Chambers, 2013: Observations: Ocean. In: *Climate Change 2013: The Physical Science Basis. Contribution of Working Group I to the Fifth Assessment Report of the Intergovernmental Panel on Climate Change* [Stocker, T.F., D. Qin, G.-K. Plattner, M. Tignor, S.K. Allen, J. Boschung, A. Nauels, Y. Xia, V. Bex, and P.M. Midgley (eds.)]. Cambridge University Press, Cambridge, United Kingdom and New York, NY, USA, pp. 255–316.

Rhiney, K., 2015: Geographies of Caribbean Vulnerability in a Changing Climate: Issues and Trends. *Geography Compass*, **9**(3), 97–114, doi:10.1111/gec3.12199.

Ribes, A., F.W. Zwiers, J.-M. Azaïs, and P. Naveau, 2017: A new statistical approach to climate change detection and attribution. *Climate Dynamics*, **48**(1), 367–386, doi:10.1007/s00382-016-3079-6.

Richardson, M., K. Cowtan, E. Hawkins, and M.B. Stolpe, 2016: Reconciled climate response estimates from climate models and the energy budget of Earth. *Nature Climate Change*, **6**(10), 931–935, doi:10.1038/nclimate3066.

Richier, S. et al., 2014: Phytoplankton responses and associated carbon cycling during shipboard carbonate chemistry manipulation experiments conducted around Northwest European shelf seas. *Biogeosciences*, **11**(17), 4733–4752, doi:10.5194/bg-11-4733-2014.

Ricke, K.L., J.B. Moreno-cruz, J. Schewe, A. Levermann, and K. Caldeira, 2016: Policy thresholds in mitigation. *Nature Geoscience*, **9**(1), 5–6, doi:10.1038/ngeo2607.

Ridgwell, A. and D.N. Schmidt, 2010: Past constraints on the vulnerability of marine calcifiers to massive carbon dioxide release. *Nature Geoscience*, **3**(3), 196–200, doi:10.1038/ngeo755.

Ridley, J.K., J.A. Lowe, and H.T. Hewitt, 2012: How reversible is sea ice loss? *The Cryosphere*, **6**(1), 193–198, doi:10.5194/tc-6-193-2012.

Riebesell, U., J.-P. Gattuso, T.F. Thingstad, and J.J. Middelburg, 2013: Arctic ocean acidification: pelagic ecosystem and biogeochemical responses during a mesocosm study. *Biogeosciences*, **10**(8), 5619–5626, doi:10.5194/bg-10-5619-2013.

3

Rienecker, M.M. et al., 2011: MERRA: NASA's Modern-Era Retrospective Analysis for Research and Applications. *Journal of Climate*, **24**(14), 3624–3648, doi:10.1175/jcli-d-11-00015.1.

Rignot, E., J. Mouginot, M. Morlighem, H. Seroussi, and B. Scheuchl, 2014: Widespread, rapid grounding line retreat of Pine Island, Thwaites, Smith, and Kohler glaciers, West Antarctica, from 1992 to 2011. *Geophysical Research Letters*, **41**(10), 3502–3509, doi:10.1002/2014gl060140.

Rinkevich, B., 2014: Rebuilding coral reefs: Does active reef restoration lead to sustainable reefs? *Current Opinion in Environmental Sustainability*, **7**, 28–36, doi:10.1016/j.cosust.2013.11.018.

Rinkevich, B., 2015: Climate Change and Active Reef Restoration-Ways of Constructing the "Reefs of Tomorrow". *Journal of Marine Science and Engineering*, **3**(1), 111–127, doi:10.3390/jmse3010111.

Rippke, U. et al., 2016: Timescales of transformational climate change adaptation in sub-Saharan African agriculture. *Nature Climate Change*, **6**(6), 605–609, doi:10.1038/nclimate2947.

Risser, M.D. and M.F. Wehner, 2017: Attributable Human-Induced Changes in the Likelihood and Magnitude of the Observed Extreme Precipitation during Hurricane Harvey. *Geophysical Research Letters*, **44**(24), 12457–12464, doi:10.1002/2017gl075888.

Rivetti, I., S. Fraschetti, P. Lionello, E. Zambianchi, and F. Boero, 2014: Global warming and mass mortalities of benthic invertebrates in the Mediterranean Sea. *PLOS ONE*, **9**(12), e115655, doi:10.1371/journal.pone.0115655.

Roberts, A.M.I., C. Tansey, R.J. Smithers, and A.B. Phillimore, 2015: Predicting a change in the order of spring phenology in temperate forests. *Global Change Biology*, **21**(7), 2603–2611, doi:10.1111/gcb.12896.

Roberts, M.J. and W. Schlenker, 2013: Identifying Supply and Demand Elasticities of Agricultural Commodities: Implications for the US Ethanol Mandate. *American Economic Review*, **103**(6), 2265–2295, doi:10.1257/aer.103.6.2265.

Robinson, A., R. Calov, and A. Ganopolski, 2012: Multistability and critical thresholds of the Greenland ice sheet. *Nature Climate Change*, **2**(6), 429–432, doi:10.1038/nclimate1449.

Robledo-Abad, C. et al., 2017: Bioenergy production and sustainable development: science base for policymaking remains limited. *GCB Bioenergy*, **9**(3), 541–556, doi:10.1111/gcbb.12338.

Roderick, M., G. Peter, and F.G. D., 2015: On the assessment of aridity with changes in atmospheric CO_2. *Water Resources Research*, **51**(7), 5450–5463, doi:10.1002/2015wr017031.

Rodrigues, L.C. et al., 2015: Sensitivity of Mediterranean Bivalve Mollusc Aquaculture to Climate Change, Ocean Acidification, and Other Environmental Pressures: Findings from a Producer Survey. *Journal of Shellfish Research*, **34**(3), 1161–1176, doi:10.2983/035.034.0341.

Rogelj, J., D.L. McCollum, A. Reisinger, M. Meinshausen, and K. Riahi, 2013: Probabilistic cost estimates for climate change mitigation. *Nature*, **493**(7430), 79–83, doi:10.1038/nature11787.

Rogelj, J. et al., 2015: Energy system transformations for limiting end-of-century warming to below 1.5°C. *Nature Climate Change*, **5**(6), 519–527, doi:10.1038/nclimate2572.

Rogelj, J. et al., 2018: Scenarios towards limiting global mean temperature increase below 1.5°C. *Nature Climate Change*, **8**(4), 325–332, doi:10.1038/s41558-018-0091-3.

Romero-Lankao, P. et al., 2014: North America. In: *Climate Change 2014: Impacts, Adaptation, and Vulnerability. Part B: Regional Aspects. Contribution of Working Group II to the Fifth Assessment Report of the Intergovernmental Panel on Climate Change* [Barros, V.R., C.B. Field, D.J. Dokken, M.D. Mastrandrea, K.J. Mach, T.E. Bilir, M. Chatterjee, K.L. Ebi, Y.O. Estrada, R.C. Genova, B. Girma, E.S. Kissel, A.N. Levy, S. MacCracken, P.R. Mastrandrea, and L.L. White (eds.)]. Cambridge University Press, Cambridge, United Kingdom and New York, NY, USA, pp. 1439–1498.

Romine, B.M., C.H. Fletcher, M.M. Barbee, T.R. Anderson, and L.N. Frazer, 2013: Are beach erosion rates and sea-level rise related in Hawaii? *Global and Planetary Change*, **108**, 149–157, doi:10.1016/j.gloplacha.2013.06.009.

Rose, G., T. Osborne, H. Greatrex, and T. Wheeler, 2016: Impact of progressive global warming on the global-scale yield of maize and soybean. *Climatic Change*, **134**(3), 417–428, doi:10.1007/s10584-016-1601-9.

Rosenblum, E. and I. Eisenman, 2016: Faster Arctic Sea Ice Retreat in CMIP5 than in CMIP3 due to Volcanoes. *Journal of Climate*, **29**(24), 9179–9188, doi:10.1175/jcli-d-16-0391.1.

Rosenblum, E. and I. Eisenman, 2017: Sea Ice Trends in Climate Models Only Accurate in Runs with Biased Global Warming. *Journal of Climate*, **30**(16), 6265–6278, doi:10.1175/jcli-d-16-0455.1.

Rosenzweig, C. and W. Solecki, 2014: Hurricane Sandy and adaptation pathways in New York: Lessons from a first-responder city. *Global Environmental Change*, **28**, 395–408, doi:10.1016/j.gloenvcha.2014.05.003.

Rosenzweig, C. and D. Hillel (eds.), 2015: *Handbook of Climate Change and Agroecosystems*. Imperial College Press, London, UK, 1160 pp., doi:10.1142/p970.

Rosenzweig, C. et al., 2013: The Agricultural Model Intercomparison and Improvement Project (AgMIP): Protocols and pilot studies. *Agricultural and Forest Meteorology*, **170**, 166–182, doi:10.1016/j.agrformet.2012.09.011.

Rosenzweig, C. et al., 2014: Assessing agricultural risks of climate change in the 21st century in a global gridded crop model intercomparison. *Proceedings of the National Academy of Sciences*, **111**(9), 3268–73, doi:10.1073/pnas.1222463110.

Rosenzweig, C. et al., 2017: Assessing inter-sectoral climate change risks: the role of ISIMIP. *Environmental Research Letters*, **12**(1), 10301, doi:10.1088/1748-9326/12/1/010301.

Rosenzweig, C. et al., 2018: Coordinating AgMIP data and models across global and regional scales for 1.5°C and 2°C assessments. *Philosophical Transactions of the Royal Society A: Mathematical, Physical and Engineering Sciences*, **376**(2119), 20160455, doi:10.1098/rsta.2016.0455.

Ross, A.C. et al., 2015: Sea-level rise and other influences on decadal-scale salinity variability in a coastal plain estuary. *Estuarine, Coastal and Shelf Science*, **157**, 79–92, doi:10.1016/j.ecss.2015.01.022.

Rosselló-Nadal, J., 2014: How to evaluate the effects of climate change on tourism. *Tourism Management*, **42**, 334–340, doi:10.1016/j.tourman.2013.11.006.

Roudier, P., B. Sultan, P. Quirion, and A. Berg, 2011: The impact of future climate change on West African crop yields: What does the recent literature say? *Global Environmental Change*, **21**(3), 1073–1083, doi:10.1016/j.gloenvcha.2011.04.007.

Roudier, P. et al., 2016: Projections of future floods and hydrological droughts in Europe under a +2°C global warming. *Climatic Change*, **135**(2), 341–355, doi:10.1007/s10584-015-1570-4.

Roy, D.B. et al., 2015: Similarities in butterfly emergence dates among populations suggest local adaptation to climate. *Global Change Biology*, **21**(9), 3313–3322, doi:10.1111/gcb.12920.

Ruane, A.C. et al., 2018: Biophysical and economic implications for agriculture of +1.5° and +2.0°C global warming using AgMIP Coordinated Global and Regional Assessments. *Climate Research*, **76**(1), 17–39, doi:10.3354/cr01520.

Russo, S., A.F. Marchese, J. Sillmann, and G. Immé, 2016: When will unusual heat waves become normal in a warming Africa? *Environmental Research Letters*, **11**(5), 1–22, doi:10.1088/1748-9326/11/5/054016.

Sacks, W.J., B.I. Cook, N. Buenning, S. Levis, and J.H. Helkowski, 2009: Effects of global irrigation on the near-surface climate. *Climate Dynamics*, **33**(2–3), 159–175, doi:10.1007/s00382-008-0445-z.

Saeidi, M., F. Moradi, and M. Abdoli, 2017: Impact of drought stress on yield, photosynthesis rate, and sugar alcohols contents in wheat after anthesis in semiarid region of Iran. *Arid Land Research and Management*, **31**(2), 1–15, doi:10.1080/15324982.2016.1260073.

Sakalli, A., A. Cescatti, A. Dosio, and M.U. Gücel, 2017: Impacts of 2°C global warming on primary production and soil carbon storage capacity at pan-European level. *Climate Services*, **7**, 64–77, doi:10.1016/j.cliser.2017.03.006.

Salem, G.S.A., S. Kazama, S. Shahid, and N.C. Dey, 2017: Impact of temperature changes on groundwater levels and irrigation costs in a groundwater-dependent agricultural region in Northwest Bangladesh. *Hydrological Research Letters*, **11**(1), 85–91, doi:10.3178/hrl.11.85.

Salisbury, J., M. Green, C. Hunt, and J. Campbell, 2008: Coastal acidification by rivers: A threat to shellfish? *Eos*, **89**(50), 513, doi:10.1029/2008eo500001.

Samaniego, L. et al., 2018: Anthropogenic warming exacerbates European soil moisture droughts. *Nature Climate Change*, **8**(5), 421–426, doi:10.1038/s41558-018-0138-5.

Sanderson, B.M. et al., 2017: Community climate simulations to assess avoided impacts in 1.5°C and 2°C futures. *Earth System Dynamics*, **8**(3), 827–847, doi:10.5194/esd-8-827-2017.

Sarojini, B.B., P.A. Stott, and E. Black, 2016: Detection and attribution of human influence on regional precipitation. *Nature Climate Change*, **6**(7), 669–675, doi:10.1038/nclimate2976.

Sasmito, S.D., D. Murdiyarso, D.A. Friess, and S. Kurnianto, 2016: Can mangroves keep pace with contemporary sea level rise? A global data review. *Wetlands Ecology and Management*, **24**(2), 263–278, doi:10.1007/s11273-015-9466-7.

Saunders, M.I. et al., 2014: Interdependency of tropical marine ecosystems in response to climate change. *Nature Climate Change*, **4**(8), 724–729, doi:10.1038/nclimate2274.

Scaven, V.L. and N.E. Rafferty, 2013: Physiological effects of climate warming on flowering plants and insect pollinators and potential consequences for their interactions. *Current Zoology*, **59**(3), 418–426, doi:10.1093/czoolo/59.3.418.

Schaeffer, M., W. Hare, S. Rahmstorf, and M. Vermeer, 2012: Long-term sea-level rise implied by 1.5°C and 2°C warming levels. *Nature Climate Change*, **2**(12), 867–870, doi:10.1038/nclimate1584.

Schewe, J., A. Levermann, and M. Meinshausen, 2011: Climate change under a scenario near 1.5 degrees C of global warming: monsoon intensification, ocean warming and steric sea level rise. *Earth System Dynamics*, **2**(1), 25–35, doi:10.5194/esd-2-25-2011.

Schewe, J. et al., 2014: Multimodel assessment of water scarcity under climate change. *Proceedings of the National Academy of Sciences*, **111**(9), 3245–3250, doi:10.1073/pnas.1222460110.

Schimel, D., B.B. Stephens, and J.B. Fisher, 2015: Effect of increasing CO_2 on the terrestrial carbon cycle. *Proceedings of the National Academy of Sciences*, **112**(2), 436–441, doi:10.1073/pnas.1407302112.

Schipper, L., W. Liu, D. Krawanchid, and S. Chanthy, 2010: *Review of climate change adaptation methods and tools*. MRC Technical Paper No. 34, Mekong River Commission (MRC), Vientiane, Lao PDR, 76 pp.

Schlenker, W. and M.J. Roberts, 2009: Nonlinear temperature effects indicate severe damages to U.S. crop yields under climate change. *Proceedings of the National Academy of Sciences*, **106**(37), 15594–15598, doi:10.1073/pnas.0906865106.

Schleussner, C.-F., P. Pfleiderer, and E.M. Fischer, 2017: In the observational record half a degree matters. *Nature Climate Change*, **7**, 460–462, doi:10.1038/nclimate3320.

Schleussner, C.-F., J.F. Donges, R. Donner, and H.J. Schellnhuber, 2016a: Armed-conflict risks enhanced by climate-related disasters in ethnically fractionalized countries. *Proceedings of the National Academy of Sciences*, **113**(33), 9216–21, doi:10.1073/pnas.1601611113.

Schleussner, C.-F., K. Frieler, M. Meinshausen, J. Yin, and A. Levermann, 2011: Emulating Atlantic overturning strength for low emission scenarios: Consequences for sea-level rise along the North American east coast. *Earth System Dynamics*, **2**(2), 191–200, doi:10.5194/esd-2-191-2011.

Schleussner, C.-F. et al., 2016b: Differential climate impacts for policy-relevant limits to global warming: The case of 1.5°C and 2°C. *Earth System Dynamics*, **7**(2), 327–351, doi:10.5194/esd-7-327-2016.

Schmidtko, S., L. Stramma, and M. Visbeck, 2017: Decline in global oceanic oxygen content during the past five decades. *Nature*, **542**(7641), 335–339, doi:10.1038/nature21399.

Schoof, C., 2007: Ice sheet grounding line dynamics: Steady states, stability, and hysteresis. *Journal of Geophysical Research*, **112**(F3), F03S28, doi:10.1029/2006jf000664.

Schröder, D. and W.M. Connolley, 2007: Impact of instantaneous sea ice removal in a coupled general circulation model. *Geophysical Research Letters*, **34**(14), L14502, doi:10.1029/2007gl030253.

Schulze, E.-D., C. Körner, B.E. Law, H. Haberl, and S. Luyssaert, 2012: Large-scale bioenergy from additional harvest of forest biomass is neither sustainable nor greenhouse gas neutral. *GCB Bioenergy*, **4**(6), 611–616, doi:10.1111/j.1757-1707.2012.01169.x.

Schwartz, J.D. et al., 2015: Projections of temperature-attributable premature deaths in 209 US cities using a cluster-based Poisson approach. *Environmental Health*, **14**(1), 85, doi:10.1186/s12940-015-0071-2.

Scott, D. and S. Verkoeyen, 2017: Assessing the Climate Change Risk of a Coastal-Island Destination. In: *Global climate change and coastal tourism: recognizing problems, managing solutions and future expectations* [Jones, A. and M. Phillips (eds.)]. Centre for Agriculture and Biosciences International (CABI), Wallingford, UK, pp. 62–73, doi:10.1079/9781780648439.0062.

Scott, D. and S. Gössling, 2018: *Tourism and Climate Change Mitigation. Embracing the Paris Agreement: Pathways to Decarbonisation*. European Travel Commission (ETC), Brussels, Belgium, 39 pp.

Scott, D., M.C. Simpson, and R. Sim, 2012: The vulnerability of Caribbean coastal tourism to scenarios of climate change related sea level rise. *Journal of Sustainable Tourism*, **20**(6), 883–898, doi:10.1080/09669582.2012.699063.

Scott, D., C.M. Hall, and S. Gössling, 2016a: A review of the IPCC Fifth Assessment and implications for tourism sector climate resilience and decarbonization. *Journal of Sustainable Tourism*, **24**(1), 8–30, doi:10.1080/09669582.2015.1062021.

Scott, D., R. Steiger, M. Rutty, and P. Johnson, 2015: The future of the Olympic Winter Games in an era of climate change. *Current Issues in Tourism*, **18**(10), 913–930, doi:10.1080/13683500.2014.887664.

Scott, D., M. Rutty, B. Amelung, and M. Tang, 2016b: An Inter-Comparison of the Holiday Climate Index (HCI) and the Tourism Climate Index (TCI) in Europe. *Atmosphere*, **7**(6), 80, doi:10.3390/atmos7060080.

Screen, J.A. and D. Williamson, 2017: Ice-free Arctic at 1.5°C? *Nature Climate Change*, **7**(4), 230–231, doi:10.1038/nclimate3248.

Screen, J.A. et al., 2018: Consistency and discrepancy in the atmospheric response to Arctic sea-ice loss across climate models. *Nature Geoscience*, **11**(3), 155–163, doi:10.1038/s41561-018-0059-y.

Scyphers, S.B., S.P. Powers, K.L. Heck, and D. Byron, 2011: Oyster reefs as natural breakwaters mitigate shoreline loss and facilitate fisheries. *PLOS ONE*, **6**(8), e22396, doi:10.1371/journal.pone.0022396.

Sebesvari, Z., S. Rodrigues, and F. Renaud, 2017: Mainstreaming ecosystem-based climate change adaptation into integrated water resources management in the Mekong region. *Regional Environmental Change*, 1–14, doi:10.1007/s10113-017-1161-1.

Seddon, A.W.R., M. Macias-Fauria, P.R. Long, D. Benz, and K.J. Willis, 2016: Sensitivity of global terrestrial ecosystems to climate variability. *Nature*, **531**(7593), 229–232, doi:10.1038/nature16986.

Sedláček, J., O. Martius, and R. Knutti, 2011: Influence of subtropical and polar sea-surface temperature anomalies on temperatures in Eurasia. *Geophysical Research Letters*, **38**(12), L12803, doi:10.1029/2011gl047764.

Seibel, B.A., 2016: Cephalopod Susceptibility to Asphyxiation via Ocean Incalescence, Deoxygenation, and Acidification. *Physiology*, **31**(6), 418–429, doi:10.1152/physiol.00061.2015.

Seidl, R. et al., 2017: Forest disturbances under climate change. *Nature Climate Change*, **7**, 395–402, doi:10.1038/nclimate3303.

Seifert, C.A. and D.B. Lobell, 2015: Response of double cropping suitability to climate change in the United States. *Environmental Research Letters*, **10**(2), 024002, doi:10.1088/1748-9326/10/2/024002.

Selby, J., 2014: Positivist Climate Conflict Research: A Critique. *Geopolitics*, **19**(4), 829–856, doi:10.1080/14650045.2014.964865.

Semakula, H.M. et al., 2017: Prediction of future malaria hotspots under climate change in sub-Saharan Africa. *Climatic Change*, **143**(3), 415–428, doi:10.1007/s10584-017-1996-y.

Semenza, J.C. and B. Menne, 2009: Climate change and infectious diseases in Europe. *The Lancet Infectious Diseases*, **9**(6), 365–375, doi:10.1016/s1473-3099(09)70104-5.

Semenza, J.C. et al., 2016: Climate change projections of West Nile virus infections in Europe: implications for blood safety practices. *Environmental Health*, **15**(S1), S28, doi:10.1186/s12940-016-0105-4.

Seneviratne, S.I., M.G. Donat, A.J. Pitman, R. Knutti, and R.L. Wilby, 2016: Allowable CO_2 emissions based on regional and impact-related climate targets. *Nature*, **529**(7587), 477–83, doi:10.1038/nature16542.

Seneviratne, S.I. et al., 2012: Changes in Climate Extremes and their Impacts on the Natural Physical Environment. In: *Managing the Risks of Extreme Events and Disasters to Advance Climate Change Adaptation* [Field, C.B., V. Barros, T.F. Stocker, D. Qin, D.J. Dokken, K.L. Ebi, M.D. Mastrandrea, K.J. Mach, G.-K. Plattner, S.K. Allen, M. Tignor, and P.M. Midgley (eds.)]. A Special Report of Working Groups I and II of the Intergovernmental Panel on Climate Change (IPCC). Cambridge University Press, Cambridge, United Kingdom and New York, NY, USA, pp. 109–230.

Seneviratne, S.I. et al., 2013: Impact of soil moisture-climate feedbacks on CMIP5 projections: First results from the GLACE-CMIP5 experiment. *Geophysical Research Letters*, **40**(19), 5212–5217, doi:10.1002/grl.50956.

Seneviratne, S.I. et al., 2018a: Land radiative management as contributor to regional-scale climate adaptation and mitigation. *Nature Geoscience*, **11**, 88–96, doi:10.1038/s41561-017-0057-5.

Seneviratne, S.I. et al., 2018b: The many possible climates from the Paris Agreement's aim of 1.5°C warming. *Nature*, **558**(7708), 41–49, doi:10.1038/s41586-018-0181-4.

Seneviratne, S.I. et al., 2018c: Climate extremes, land-climate feedbacks and land-use forcing at 1.5°C. *Philosophical Transactions of the Royal Society A: Mathematical, Physical and Engineering Sciences*, **376**(2119), 20160450, doi:10.1098/rsta.2016.0450.

Serdeczny, O. et al., 2016: Climate change impacts in Sub-Saharan Africa: from physical changes to their social repercussions. *Regional Environmental Change*, 1–16, doi:10.1007/s10113-015-0910-2.

Serreze, M.C. and J. Stroeve, 2015: Arctic sea ice trends, variability and implications for seasonal ice forecasting. *Philosophical Transactions of the Royal Society A: Mathematical, Physical and Engineering Sciences*, **373(2045)**, 20140159, doi:10.1098/rsta.2014.0159.

Settele, J. et al., 2014: Terrestrial and Inland Water Systems. In: *Climate Change 2014: Impacts, Adaptation, and Vulnerability. Part A: Global and Sectoral Aspects. Contribution of Working Group II to the Fifth Assessment Report of the Intergovernmental Panel on Climate Change* [Field, C.B., V.R. Barros, D.J. Dokken, K.J. Mach, M.D. Mastrandrea, T.E. Bilir, M. Chatterjee, K.L. Ebi, Y.O. Estrada, R.C. Genova, B. Girma, E.S. Kissel, A.N. Levy, S. MacCracken, P.R. Mastrandrea, and L.L. White (eds.)]. Cambridge University Press, Cambridge, United Kingdom and New York, NY, USA, pp. 271–359.

Shafir, S., J. Van Rijn, and B. Rinkevich, 2006: Steps in the construction of underwater coral nursery, an essential component in reef restoration acts. *Marine Biology*, **149(3)**, 679–687, doi:10.1007/s00227-005-0236-6.

Shearman, P., J. Bryan, and J.P. Walsh, 2013: Trends in Deltaic Change over Three Decades in the Asia-Pacific Region. *Journal of Coastal Research*, **290**, 1169–1183, doi:10.2112/jcoastres-d-12-00120.1.

Sheffield, J., E.F. Wood, and M.L. Roderick, 2012: Little change in global drought over the past 60 years. *Nature*, **491(7424)**, 435–438, doi:10.1038/nature11575.

Sheffield, P.E. et al., 2013: Current and future heat stress in Nicaraguan work places under a changing climate. *Industrial Health*, **51**, 123–127, doi:10.2486/indhealth.2012-0156.

Shepherd, J.G., P.G. Brewer, A. Oschlies, and A.J. Watson, 2017: Ocean ventilation and deoxygenation in a warming world: introduction and overview. *Philosophical Transactions of the Royal Society A: Mathematical, Physical and Engineering Sciences*, **375(2102)**, 20170240, doi:10.1098/rsta.2017.0240.

Shi, H. and G. Wang, 2015: Impacts of climate change and hydraulic structures on runoff and sediment discharge in the middle Yellow River. *Hydrological Processes*, **29(14)**, 3236–3246, doi:10.1002/hyp.10439.

Shi, W. et al., 2016: Ocean acidification increases cadmium accumulation in marine bivalves: a potential threat to seafood safety. *Scientific Reports*, **6(1)**, 20197, doi:10.1038/srep20197.

Shindell, D.T., G. Faluvegi, K. Seltzer, and C. Shindell, 2018: Quantified, localized health benefits of accelerated carbon dioxide emissions reductions. *Nature Climate Change*, **8**, 1–5, doi:10.1038/s41558-018-0108-y.

Shindell, D.T. et al., 2017: A climate policy pathway for near- and long-term benefits. *Science*, **356(6337)**, 493–494, doi:10.1126/science.aak9521.

Short, F.T., S. Kosten, P.A. Morgan, S. Malone, and G.E. Moore, 2016: Impacts of climate change on submerged and emergent wetland plants. *Aquatic Botany*, **135**, 3–17, doi:10.1016/j.aquabot.2016.06.006.

Shrestha, B., T.A. Cochrane, B.S. Caruso, M.E. Arias, and T. Piman, 2016: Uncertainty in flow and sediment projections due to future climate scenarios for the 3S Rivers in the Mekong Basin. *Journal of Hydrology*, **540**, 1088–1104, doi:10.1016/j.jhydrol.2016.07.019.

Sierra-Correa, P.C. and J.R. Cantera Kintz, 2015: Ecosystem-based adaptation for improving coastal planning for sea-level rise: A systematic review for mangrove coasts. *Marine Policy*, **51**, 385–393, doi:10.1016/j.marpol.2014.09.013.

Sigmond, M., J.C. Fyfe, and N.C. Swart, 2018: Ice-free Arctic projections under the Paris Agreement. *Nature Climate Change*, **8**, 404–408, doi:10.1038/s41558-018-0124-y.

Signorini, S.R., B.A. Franz, and C.R. McClain, 2015: Chlorophyll variability in the oligotrophic gyres: mechanisms, seasonality and trends. *Frontiers in Marine Science*, **2**, 1–11, doi:10.3389/fmars.2015.00001.

Sihi, D., P.W. Inglett, S. Gerber, and K.S. Inglett, 2017: Rate of warming affects temperature sensitivity of anaerobic peat decomposition and greenhouse gas production. *Global Change Biology*, 24(1), e259–e274, doi:10.1111/gcb.13839.

Silva, R.A. et al., 2016: The effect of future ambient air pollution on human premature mortality to 2100 using output from the ACCMIP model ensemble. *Atmospheric Chemistry and Physics*, **16(15)**, 9847–9862, doi:10.5194/acp-16-9847-2016.

Simulundu, E. et al., 2017: Genetic characterization of orf virus associated with an outbreak of severe orf in goats at a farm in Lusaka, Zambia (2015). *Archives of Virology*, **162(8)**, 2363–2367, doi:10.1007/s00705-017-3352-y.

Singh, B.P., V.K. Dua, P.M. Govindakrishnan, and S. Sharma, 2013: Impact of Climate Change on Potato. In: *Climate-Resilient Horticulture: Adaptation and Mitigation Strategies* [Singh, H.C.P., N.K.S. Rao, and K.S. Shivashankar (eds.)]. Springer India, India, pp. 125–135, doi:10.1007/978-81-322-0974-4_12.

Singh, D., M. Tsiang, B. Rajaratnam, and N.S. Diffenbaugh, 2014: Observed changes in extreme wet and dry spells during the South Asian summer monsoon season. *Nature Climate Change*, **4(6)**, 456–461, doi:10.1038/nclimate2208.

Singh, O.P., 2010: Recent Trends in Tropical Cyclone Activity in the North Indian Ocean. In: *Indian Ocean Tropical Cyclones and Climate Change* [Charabi, Y. (ed.)]. Springer Netherlands, Dordrecht, pp. 51–54, doi:10.1007/978-90-481-3109-9_8.

Singh, O.P., T.M. Ali Khan, and M.S. Rahman, 2000: Changes in the frequency of tropical cyclones over the North Indian Ocean. *Meteorology and Atmospheric Physics*, **75(1–2)**, 11–20, doi:10.1007/s007030070011.

Slade, R., A. Bauen, and R. Gross, 2014: Global bioenergy resources. *Nature Climate Change*, **4(2)**, 99–105, doi:10.1038/nclimate2097.

Slangen, A.B.A. et al., 2016: Anthropogenic forcing dominates global mean sea-level rise since 1970. *Nature Climate Change*, **6(7)**, 701–705, doi:10.1038/nclimate2991.

Slater, A.G. and D.M. Lawrence, 2013: Diagnosing present and future permafrost from climate models. *Journal of Climate*, **26(15)**, 5608–5623, doi:10.1175/jcli-d-12-00341.1.

Smajgl, A. et al., 2015: Responding to rising sea levels in the Mekong Delta. *Nature Climate Change*, **5(2)**, 167–174, doi:10.1038/nclimate2469.

Smeed, D.A. et al., 2014: Observed decline of the Atlantic meridional overturning circulation 2004–2012. *Ocean Science*, **10(1)**, 29–38, doi:10.5194/os-10-29-2014.

Smirnov, O. et al., 2016: The relative importance of climate change and population growth for exposure to future extreme droughts. *Climatic Change*, **138(1–2)**, 41–53, doi:10.1007/s10584-016-1716-z.

Smith, K.R. et al., 2014: Human Health: Impacts, Adaptation, and Co-Benefits. In: *Climate Change 2014: Impacts, Adaptation, and Vulnerability. Part A: Global and Sectoral Aspects. Contribution of Working Group II to the Fifth Assessment Report of the Intergovernmental Panel on Climate Change* [Field, C.B., V.R. Barros, D.J. Dokken, K.J. Mach, M.D. Mastrandrea, T.E. Bilir, M. Chatterjee, K.L. Ebi, Y.O. Estrada, R.C. Genova, B. Girma, E.S. Kissel, A.N. Levy, S. MacCracken, P.R. Mastrandrea, and L.L. White (eds.)]. Cambridge University Press, Cambridge, United Kingdom and New York, NY, USA, pp. 709–754.

Smith, P., 2016: Soil carbon sequestration and biochar as negative emission technologies. *Global Change Biology*, **22(3)**, 1315–1324, doi:10.1111/gcb.13178.

Smith, P., J. Price, A. Molotoks, R. Warren, and Y. Malhi, 2018: Impacts on terrestrial biodiversity of moving from a 2°C to a 1.5°C target. *Philosophical Transactions of the Royal Society A: Mathematical, Physical and Engineering Sciences*, **376(2119)**, 20160456, doi:10.1098/rsta.2016.0456.

Smith, P. et al., 2010: Competition for land. *Philosophical Transactions of the Royal Society B: Biological Sciences*, **365(1554)**, 2941–2957, doi:10.1098/rstb.2010.0127.

Smith, P. et al., 2013: How much land-based greenhouse gas mitigation can be achieved without compromising food security and environmental goals? *Global Change Biology*, **19(8)**, 2285–2302, doi:10.1111/gcb.12160.

Smith, P. et al., 2014: Agriculture, Forestry and Other Land Use (AFOLU). In: *Climate Change 2014: Mitigation of Climate Change. Contribution of Working Group III to the Fifth Assessment Report of the Intergovernmental Panel on Climate Change* [Edenhofer, O., R. Pichs-Madruga, Y. Sokona, E. Farahani, S. Kadner, K. Seyboth, A. Adler, I. Baum, S. Brunner, P. Eickemeier, B. Kriemann, J. Savolainen, S. Schlömer, C. von Stechow, T. Zwickel, and J.C. Minx (eds.)]. Cambridge University Press, Cambridge, UK and New York, NY, USA, pp. 811–922.

Smith, P. et al., 2015: Biophysical and economic limits to negative CO_2 emissions. *Nature Climate Change*, **6(1)**, 42–50, doi:10.1038/nclimate2870.

Smith, T.F., D.C. Thomsen, S. Gould, K. Schmitt, and B. Schlegel, 2013: Cumulative Pressures on Sustainable Livelihoods: Coastal Adaptation in the Mekong Delta. *Sustainability*, **5(1)**, 228–241, doi:10.3390/su5010228.

Sok, S. and X. Yu, 2015: Adaptation, resilience and sustainable livelihoods in the communities of the Lower Mekong Basin, Cambodia. *International Journal of Water Resources Development*, **31(4)**, 575–588, doi:10.1080/07900627.2015.1012659.

Solomon, S., G.-K. Plattner, R. Knutti, and P. Friedlingstein, 2009: Irreversible climate change due to carbon dioxide emissions. *Proceedings of the National Academy of Sciences*, **106(6)**, 1704–1709, doi:10.1073/pnas.0812721106.

Song, A.M. and R. Chuenpagdee, 2015: Interactive Governance for Fisheries. *Interactive Governance for Small-Scale Fisheries*, **5**, 435–456, doi:10.1007/978-3-319-17034-3.

Song, Y. et al., 2016: Spatial distribution estimation of malaria in northern China and its scenarios in 2020, 2030, 2040 and 2050. *Malaria journal*, **15(1)**, 345, doi:10.1186/s12936-016-1395-2.

3

Sonntag, S., J. Pongratz, C.H. Reick, and H. Schmidt, 2016: Reforestation in a high-CO_2 world – Higher mitigation potential than expected, lower adaptation potential than hoped for. *Geophysical Research Letters*, **43(12)**, 6546–6553, doi:10.1002/2016gl068824.

Sovacool, B.K., 2012: Perceptions of climate change risks and resilient island planning in the Maldives. *Mitigation and Adaptation Strategies for Global Change*, **17(7)**, 731–752, doi:10.1007/s11027-011-9341-7.

Spalding, M.D. and B.E. Brown, 2015: Warm-water coral reefs and climate change. *Science*, **350(6262)**, 769–771, doi:10.1126/science.aad0349.

Spalding, M.D. et al., 2014: The role of ecosystems in coastal protection: Adapting to climate change and coastal hazards. *Ocean and Coastal Management*, **90**, 50–57, doi:10.1016/j.ocecoaman.2013.09.007.

Spalding, M.D. et al., 2017: Mapping the global value and distribution of coral reef tourism. *Marine Policy*, **82**, 104–113, doi:10.1016/j.marpol.2017.05.014.

Speelman, L.H., R.J. Nicholls, and J. Dyke, 2017: Contemporary migration intentions in the Maldives: the role of environmental and other factors. *Sustainability Science*, **12(3)**, 433–451, doi:10.1007/s11625-016-0410-4.

Spencer, T. et al., 2016: Global coastal wetland change under sea-level rise and related stresses: The DIVA Wetland Change Model. *Global and Planetary Change*, **139**, 15–30, doi:10.1016/j.gloplacha.2015.12.018.

Springer, J., R. Ludwig, and S. Kienzle, 2015: Impacts of Forest Fires and Climate Variability on the Hydrology of an Alpine Medium Sized Catchment in the Canadian Rocky Mountains. *Hydrology*, **2(1)**, 23–47, doi:10.3390/hydrology2010023.

Springmann, M. et al., 2016: Global and regional health effects of future food production under climate change: a modelling study. *The Lancet*, **387(10031)**, 1937–1946, doi:10.1016/s0140-6736(15)01156-3.

Srokosz, M.A. and H.L. Bryden, 2015: Observing the Atlantic Meridional Overturning Circulation yields a decade of inevitable surprises. *Science*, **348(6241)**, 1255575, doi:10.1126/science.1255575.

Stanturf, J.A., B.J. Palik, and R.K. Dumroese, 2014: Contemporary forest restoration: A review emphasizing function. *Forest Ecology and Management*, **331**, 292–323, doi:10.1016/j.foreco.2014.07.029.

Steiger, R., D. Scott, B. Abegg, M. Pons, and C. Aall, 2017: A critical review of climate change risk for ski tourism. *Current Issues in Tourism*, 1–37, doi:10.1080/13683500.2017.1410110.

Steinberg, D.K. et al., 2015: Long-term (1993–2013) changes in macrozooplankton off the Western Antarctic Peninsula. *Deep Sea Research Part I: Oceanographic Research Papers*, **101**, 54–70, doi:10.1016/j.dsr.2015.02.009.

Stephens, P.A. et al., 2016: Consistent response of bird populations to climate change on two continents. *Science*, **352(6281)**, 84–87, doi:10.1126/science.aac4858.

Sterling, S.M., A. Ducharne, and J. Polcher, 2012: The impact of global land-cover change on the terrestrial water cycle. *Nature Climate Change*, **3(4)**, 385–390, doi:10.1038/nclimate1690.

Sterman, J.D., L. Siegel, and J.N. Rooney-Varga, 2018: Does replacing coal with wood lower CO_2 emissions? Dynamic lifecycle analysis of wood bioenergy. *Environmental Research Letters*, **13(1)**, 015007, doi:10.1088/1748-9326/aaa512.

Stevens, A.J., D. Clarke, and R.J. Nicholls, 2016: Trends in reported flooding in the UK: 1884–2013. *Hydrological Sciences Journal*, **61(1)**, 50–63, doi:10.1080/02626667.2014.950581.

Stewart, E.J. et al., 2016: Implications of climate change for glacier tourism. *Tourism Geographies*, **18(4)**, 377–398, doi:10.1080/14616688.2016.1198416.

Stockdale, A., E. Tipping, S. Lofts, and R.J.G. Mortimer, 2016: Effect of Ocean Acidification on Organic and Inorganic Speciation of Trace Metals. *Environmental Science & Technology*, **50(4)**, 1906–1913, doi:10.1021/acs.est.5b05624.

Stocker, T.F. et al., 2013: Technical Summary. In: *Climate Change 2013: The Physical Science Basis. Contribution of Working Group I to the Fifth Assessment Report of the Intergovernmental Panel on Climate Change* [Stocker, T.F., D. Qin, G.-K. Plattner, M. Tignor, S.K. Allen, J. Boschung, A. Nauels, Y. Xia, V. Bex, and P.M. Midgley (eds.)]. Cambridge University Press, Cambridge, United Kingdom and New York, NY, USA, pp. 33–115.

Storlazzi, C.D., E.P.L. Elias, and P. Berkowitz, 2015: Many Atolls May be Uninhabitable Within Decades Due to Climate Change. *Scientific Reports*, **5(1)**, 14546, doi:10.1038/srep14546.

Storlazzi, C.D. et al., 2018: Most atolls will be uninhabitable by the mid-21st century because of sea-level rise exacerbating wave-driven flooding. *Science Advances*, **4(4)**, eaap9741, doi:10.1126/sciadv.aap9741.

Strefler, J. et al., 2018: Between Scylla and Charybdis: Delayed mitigation narrows the passage between large-scale CDR and high costs. *Environmental Research Letters*, **13(044015)**, 1–6, doi:10.1088/1748-9326/aab2ba.

Suckall, N., E. Fraser, and P. Forster, 2017: Reduced migration under climate change: evidence from Malawi using an aspirations and capabilities framework. *Climate and Development*, **9(4)**, 298–312, doi:10.1080/17565529.2016.1149441.

Sudmeier-Rieux, K., M. Fernández, J.C. Gaillard, L. Guadagno, and M. Jaboyedoff, 2017: Introduction: Exploring Linkages Between Disaster Risk Reduction, Climate Change Adaptation, Migration and Sustainable Development. In: *Identifying Emerging Issues in Disaster Risk Reduction, Migration, Climate Change and Sustainable Development* [Sudmeier-Rieux, K., M. Fernández, I.M. Penna, M. Jaboyedoff, and J.C. Gaillard (eds.)]. Springer International Publishing, Cham, pp. 1–11, doi:10.1007/978-3-319-33880-4_1.

Sugi, M. and J. Yoshimura, 2012: Decreasing trend of tropical cyclone frequency in 228-year high-resolution AGCM simulations. *Geophysical Research Letters*, **39(19)**, L19805, doi:10.1029/2012gl053360.

Sugi, M., H. Murakami, and K. Yoshida, 2017: Projection of future changes in the frequency of intense tropical cyclones. *Climate Dynamics*, **49(1)**, 619–632, doi:10.1007/s00382-016-3361-7.

Sultan, B. and M. Gaetani, 2016: Agriculture in West Africa in the Twenty-First Century: Climate Change and Impacts Scenarios, and Potential for Adaptation. *Frontiers in Plant Science*, **7**, 1262, doi:10.3389/fpls.2016.01262.

Sun, H. et al., 2017: Exposure of population to droughts in the Haihe River Basin under global warming of 1.5 and 2.0°C scenarios. *Quaternary International*, **453**, 74–84, doi:10.1016/j.quaint.2017.05.005.

Sun, S., X.-G. Yang, J. Zhao, and F. Chen, 2015: The possible effects of global warming on cropping systems in China XI The variation of potential light-temperature suitable cultivation zone of winter wheat in China under climate change. *Scientia Agricultura Sinica*, **48(10)**, 1926–1941.

Sun, Y., X. Zhang, G. Ren, F.W. Zwiers, and T. Hu, 2016: Contribution of urbanization to warming in China. *Nature Climate Change*, **6(7)**, 706–709, doi:10.1038/nclimate2956.

Sundby, S., K.F. Drinkwater, and O.S. Kjesbu, 2016: The North Atlantic Spring-Bloom System-Where the Changing Climate Meets the Winter Dark. *Frontiers in Marine Science*, **3**, 28, doi:10.3389/fmars.2016.00028.

Supit, I. et al., 2010: Recent changes in the climatic yield potential of various crops in Europe. *Agricultural Systems*, **103(9)**, 683–694, doi:10.1016/j.agsy.2010.08.009.

Sutton-Grier, A.E. and A. Moore, 2016: Leveraging Carbon Services of Coastal Ecosystems for Habitat Protection and Restoration. *Coastal Management*, **44(3)**, 259–277, doi:10.1080/08920753.2016.1160206.

Suzuki-Parker, A., H. Kusaka, and Y. Yamagata, 2015: Assessment of the Impact of Metropolitan-Scale Urban Planning Scenarios on the Moist Thermal Environment under Global Warming: A Study of the Tokyo Metropolitan Area Using Regional Climate Modeling. *Advances in Meteorology*, **2015**, 1–11, doi:10.1155/2015/693754.

Sweetman, A.K. et al., 2017: Major impacts of climate change on deep-sea benthic ecosystems. *Elementa: Science of the Anthropocene*, **5**, 4, doi:10.1525/elementa.203.

Sydeman, W.J. et al., 2014: Climate change and wind intensification in coastal upwelling ecosystems. *Science*, **345(6192)**, 77–80, doi:10.1126/science.1251635.

Sylla, M.B., N. Elguindi, F. Giorgi, and D. Wisser, 2016: Projected robust shift of climate zones over West Africa in response to anthropogenic climate change for the late 21st century. *Climatic Change*, **134(1)**, 241–253, doi:10.1007/s10584-015-1522-z.

Sylla, M.B. et al., 2015: Projected Changes in the Annual Cycle of High-Intensity Precipitation Events over West Africa for the Late Twenty-First Century. *Journal of Climate*, **28(16)**, 6475–6488, doi:10.1175/jcli-d-14-00854.1.

Tainio, M. et al., 2013: Future climate and adverse health effects caused by fine particulate matter air pollution: case study for Poland. *Regional Environmental Change*, **13(3)**, 705–715, doi:10.1007/s10113-012-0366-6.

Takagi, H., N. Thao, and L. Anh, 2016: Sea-Level Rise and Land Subsidence: Impacts on Flood Projections for the Mekong Delta's Largest City. *Sustainability*, **8(9)**, 959, doi:10.3390/su8090959.

Takakura, J. et al., 2017: Cost of preventing workplace heat-related illness through worker breaks and the benefit of climate-change mitigation. *Environmental Research Letters*, **12(6)**, 064010, doi:10.1088/1748-9326/aa72cc.

Tanaka, A. et al., 2017: On the scaling of climate impact indicators with global mean temperature increase: a case study of terrestrial ecosystems and water resources. *Climatic Change*, **141(4)**, 775–782, doi:10.1007/s10584-017-1911-6.

3

Tanoue, M., Y. Hirabayashi, H. Ikeuchi, E. Gakidou, and T. Oki, 2016: Global-scale river flood vulnerability in the last 50 years. *Scientific Reports*, **6**(1), 36021, doi:10.1038/srep36021.

Tavoni, M. and R. Socolow, 2013: Modeling meets science and technology: an introduction to a special issue on negative emissions. *Climatic Change*, **118**(1), 1–14, doi:10.1007/s10584-013-0757-9.

Taylor, C.M. et al., 2017: Frequency of extreme Sahelian storms tripled since 1982 in satellite observations. *Nature*, **544**(7651), 475–478, doi:10.1038/nature22069.

Taylor, M.A. et al., 2018: Future Caribbean Climates in a World of Rising Temperatures: The 1.5 vs 2.0 Dilemma. *Journal of Climate*, **31**(7), 2907–2926, doi:10.1175/jcli-d-17-0074.1.

Tebaldi, C. and R. Knutti, 2018: Evaluating the accuracy of climate change pattern emulation for low warming targets. *Environmental Research Letters*, **13**(5), 055006, doi:10.1088/1748-9326/aabef2.

Teichmann, C. et al., 2018: Avoiding extremes: Benefits of staying below +1.5°C compared to +2.0°C and +3.0°C global warming. *Atmosphere*, **9**(4), 1–19, doi:10.3390/atmos9040115.

Temmerman, S. et al., 2013: Ecosystem-based coastal defence in the face of global change. *Nature*, **504**(7478), 79–83, doi:10.1038/nature12859.

Terry, J.P. and T.F.M. Chui, 2012: Evaluating the fate of freshwater lenses on atoll islands after eustatic sea-level rise and cyclone-driven inundation: A modelling approach. *Global and Planetary Change*, **88–89**, 76–84, doi:10.1016/j.gloplacha.2012.03.008.

Teshager, A.D., P.W. Gassman, J.T. Schoof, and S. Secchi, 2016: Assessment of impacts of agricultural and climate change scenarios on watershed water quantity and quality, and crop production. *Hydrology and Earth System Sciences*, **20**(8), 3325–3342, doi:10.5194/hess-20-3325-2016.

Tessler, Z.D., C.J. Vörösmarty, I. Overeem, and J.P.M. Syvitski, 2018: A model of water and sediment balance as determinants of relative sea level rise in contemporary and future deltas. *Geomorphology*, **305**, 209–220, doi:10.1016/j.geomorph.2017.09.040.

Thackeray, S.J. et al., 2016: Phenological sensitivity to climate across taxa and trophic levels. *Nature*, **535**(7611), 241–245, doi:10.1038/nature18608.

Theisen, O.M., N.P. Gleditsch, and H. Buhaug, 2013: Is climate change a driver of armed conflict? *Climatic Change*, **117**(3), 613–625, doi:10.1007/s10584-012-0649-4.

Thiery, W. et al., 2017: Present-day irrigation mitigates heat extremes. *Journal of Geophysical Research: Atmospheres*, **122**(3), 1403–1422, doi:10.1002/2016jd025740.

Thober, T. et al., 2018: Multi-model ensemble projections of European river floods and high flows at 1.5, 2, and 3 degrees global warming. *Environmental Research Letters*, **13**(1), 014003, doi:10.1088/1748-9326/aa9e35.

Thomas, A. and L. Benjamin, 2017: Management of loss and damage in small island developing states: implications for a 1.5°C or warmer world. *Regional Environmental Change*, 1–10, doi:10.1007/s10113-017-1184-7.

Thornton, P.K., P.G. Jones, P.J. Ericksen, and A.J. Challinor, 2011: Agriculture and food systems in sub-Saharan Africa in a 4°C+ world. *Philosophical Transactions of the Royal Society A: Mathematical, Physical and Engineering Sciences*, **369**(1934), 117–136, doi:10.1098/rsta.2010.0246.

Thornton, P.K., P.J. Ericksen, M. Herrero, and A.J. Challinor, 2014: Climate variability and vulnerability to climate change: a review. *Global Change Biology*, **20**(11), 3313–3328, doi:10.1111/gcb.12581.

Thronson, A. and A. Quigg, 2008: Fifty-five years of fish kills in coastal Texas. *Estuaries and Coasts*, **31**(4), 802–813, doi:10.1007/s12237-008-9056-5.

Tietsche, S., D. Notz, J.H. Jungclaus, and J. Marotzke, 2011: Recovery mechanisms of Arctic summer sea ice. *Geophysical Research Letters*, **38**(2), L02707, doi:10.1029/2010gl045698.

Tilman, D., C. Balzer, J. Hill, and B.L. Befort, 2011: Global food demand and the sustainable intensification of agriculture. *Proceedings of the National Academy of Sciences*, **108**(50), 20260–20264, doi:10.1073/pnas.1116437108.

Tjaden, N.B. et al., 2017: Modelling the effects of global climate change on Chikungunya transmission in the 21st century. *Scientific Reports*, **7**(1), 3813, doi:10.1038/s41598-017-03566-3.

Tobin, I. et al., 2015: Assessing climate change impacts on European wind energy from ENSEMBLES high-resolution climate projections. *Climatic Change*, **128**(1), 99–112, doi:10.1007/s10584-014-1291-0.

Tobin, I. et al., 2016: Climate change impacts on the power generation potential of a European mid-century wind farms scenario. *Environmental Research Letters*, **11**(3), 034013, doi:10.1088/1748-9326/11/3/034013.

Tobin, I. et al., 2018: Vulnerabilities and resilience of European power generation to 1.5°C, 2°C and 3°C warming. *Environmental Research Letters*, **13**(4), 044024, doi:10.1088/1748-9326/aab211.

Todd-Brown, K.E.O. et al., 2014: Changes in soil organic carbon storage predicted by Earth system models during the 21st century. *Biogeosciences*, **11**(8), 2341–2356, doi:10.5194/bg-11-2341-2014.

Tran, T.A., J. Pittock, and L.A. Tuan, 2018: Adaptive co-management in the Vietnamese Mekong Delta: examining the interface between flood management and adaptation. *International Journal of Water Resources Development*, 1–17, doi:10.1080/07900627.2018.1437713.

Trang, N.T.T., S. Shrestha, M. Shrestha, A. Datta, and A. Kawasaki, 2017: Evaluating the impacts of climate and land-use change on the hydrology and nutrient yield in a transboundary river basin: A case study in the 3S River Basin (Sekong, Sesan, and Srepok). *Science of The Total Environment*, **576**, 586–598, doi:10.1016/j.scitotenv.2016.10.138.

Trathan, P.N. and S.L. Hill, 2016: The Importance of Krill Predation in the Southern Ocean. In: *Biology and Ecology of Antarctic Krill* [Siegel, V. (ed.)]. Springer, Cham, Switzerland, pp. 321–350, doi:10.1007/978-3-319-29279-3_9.

Trigo, R.M., C.M. Gouveia, and D. Barriopedro, 2010: The intense 2007–2009 drought in the Fertile Crescent: Impacts and associated atmospheric circulation. *Agricultural and Forest Meteorology*, **150**(9), 1245–1257, doi:10.1016/j.agrformet.2010.05.006.

Turner, J. et al., 2014: Antarctic climate change and the environment: An update. *Polar Record*, **50**(3), 237–259, doi:10.1017/s0032247413000296.

Turner, J. et al., 2017a: Atmosphere-ocean-ice interactions in the Amundsen Sea Embayment, West Antarctica. *Reviews of Geophysics*, **55**(1), 235–276, doi:10.1002/2016rg000532.

Turner, J. et al., 2017b: Unprecedented springtime retreat of Antarctic sea ice in 2016. *Geophysical Research Letters*, **44**(13), 6868–6875, doi:10.1002/2017gl073656.

Turner, R.E., N.N. Rabalais, and D. Justic, 2008: Gulf of Mexico hypoxia: Alternate states and a legacy. *Environmental Science & Technology*, **42**(7), 2323–2327, doi:10.1021/es071617k.

Rosenzweig, C., W. Solecki, P. Romero-Lankao, S. Mehrotra, S. Dhakal, and S. Ali Ibrahim (eds.), 2018: *Climate Change and Cities: Second Assessment Report of the Urban Climate Change Research Network*. Urban Climate Change Research Network (UCCRN). Cambridge University Press, Cambridge, United Kingdom and New York, NY, USA, 811 pp., doi:10.1017/9781316563878.

Udo, K. and Y. Takeda, 2017: Projections of Future Beach Loss in Japan Due to Sea-Level Rise and Uncertainties in Projected Beach Loss. *Coastal Engineering Journal*, **59**(2), 1740006, doi:10.1142/s057856341740006x.

UN, 2015: *Transforming our world: The 2030 agenda for sustainable development*. A/RES/70/1, United Nations General Assembly (UNGA), 35 pp., doi:10.1007/s13398-014-0173-7.2.

UN, 2017: *The Sustainable Development Goals Report 2017*. United Nations (UN), New York, NY, USA, 64 pp.

UNEP-WCMC, 2006: *In the front line: shoreline protection and other ecosystem services from mangroves and coral reefs*. UNEP-WCMC Biodiversity Series 24, UNEP World Conservation Monitoring Centre (UNEP-WCMC), Cambridge, UK, 33 pp.

UNESCO, 2011: The Impact of Global Change on Water Resources: The Response of UNESCO'S International Hydrology Programme. United Nations Educational Scientific and Cultural Organization (UNESCO) International Hydrological Programme (IHP), Paris, France, 20 pp.

Urban, M.C., 2015: Accelerating extinction risk from climate change. *Science*, **348**(6234), 571–573, doi:10.1126/science.aaa4984.

Utete, B., C. Phiri, S.S. Mlambo, N. Muboko, and B.T. Fregene, 2018: Vulnerability of fisherfolks and their perceptions towards climate change and its impacts on their livelihoods in a peri-urban lake system in Zimbabwe. *Environment, Development and Sustainability*, 1–18, doi:10.1007/s10668-017-0067-x.

Valle, M. et al., 2014: Projecting future distribution of the seagrass *Zostera noltii* under global warming and sea level rise. *Biological Conservation*, **170**, 74–85, doi:10.1016/j.biocon.2013.12.017.

van Bruggen, A.H.C., J.W. Jones, J.M.C. Fernandes, K. Garrett, and K.J. Boote, 2015: Crop Diseases and Climate Change in the AgMIP Framework. In: *Handbook of Climate Change and Agroecosystems* [Rosenzweig, C. and D. Hillel (eds.)]. ICP Series on Climate Change Impacts, Adaptation, and Mitigation Volume 3, Imperial College Press, London, UK, pp. 297–330, doi:10.1142/9781783265640_0012.

3

Van Den Hurk, B., E. Van Meijgaard, P. De Valk, K.J. Van Heeringen, and J. Gooijer, 2015: Analysis of a compounding surge and precipitation event in the Netherlands. *Environmental Research Letters*, **10**(3), 035001, doi:10.1088/1748-9326/10/3/035001.

van der Velde, M., F.N. Tubiello, A. Vrieling, and F. Bouraoui, 2012: Impacts of extreme weather on wheat and maize in France: evaluating regional crop simulations against observed data. *Climatic Change*, **113**(3–4), 751–765, doi:10.1007/s10584-011-0368-2.

Van Dingenen, R. et al., 2009: The global impact of ozone on agricultural crop yields under current and future air quality legislation. *Atmospheric Environment*, **43**(3), 604–618, doi:10.1016/j.atmosenv.2008.10.033.

van Hooidonk, R. and M. Huber, 2012: Effects of modeled tropical sea surface temperature variability on coral reef bleaching predictions. *Coral Reefs*, **31**(1), 121–131, doi:10.1007/s00338-011-0825-4.

van Hooidonk, R., J.A. Maynard, and S. Planes, 2013: Temporary refugia for coral reefs in a warming world. *Nature Climate Change*, **3**(5), 508–511, doi:10.1038/nclimate1829.

van Hooidonk, R. et al., 2016: Local-scale projections of coral reef futures and implications of the Paris Agreement. *Scientific Reports*, **6**(1), 39666, doi:10.1038/srep39666.

van Oldenborgh, G.J. et al., 2017: Attribution of extreme rainfall from Hurricane Harvey, August 2017. *Environmental Research Letters*, **12**(12), 124009, doi:10.1088/1748-9326/aa9ef2.

van Oort, P.A.J. and S.J. Zwart, 2018: Impacts of climate change on rice production in Africa and causes of simulated yield changes. *Global Change Biology*, **24**(3), 1029–1045, doi:10.1111/gcb.13967.

van Oppen, M.J.H., J.K. Oliver, H.M. Putnam, and R.D. Gates, 2015: Building coral reef resilience through assisted evolution. *Proceedings of the National Academy of Sciences*, **112**(8), 2307–2313, doi:10.1073/pnas.1422301112.

van Oppen, M.J.H. et al., 2017: Shifting paradigms in restoration of the world's coral reefs. *Global Change Biology*, **23**(9), 3437–3448, doi:10.1111/gcb.13647.

van Vliet, M.T.H. et al., 2016: Multi-model assessment of global hydropower and cooling water discharge potential under climate change. *Global Environmental Change*, **40**, 156–170, doi:10.1016/j.gloenvcha.2016.07.007.

van Vuuren, D.P. et al., 2009: Comparison of top-down and bottom-up estimates of sectoral and regional greenhouse gas emission reduction potentials. *Energy Policy*, **37**(12), 5125–5139, doi:10.1016/j.enpol.2009.07.024.

van Vuuren, D.P. et al., 2011a: The representative concentration pathways: an overview. *Climatic Change*, **109**(1), 5, doi:10.1007/s10584-011-0148-z.

van Vuuren, D.P. et al., 2011b: RCP2.6: exploring the possibility to keep global mean temperature increase below 2°C. *Climatic Change*, **109**(1), 95, doi:10.1007/s10584-011-0152-3.

van Vuuren, D.P. et al., 2016: Carbon budgets and energy transition pathways. *Environmental Research Letters*, **11**(7), 075002, doi:10.1088/1748-9326/11/7/075002.

van Vuuren, D.P. et al., 2018: Alternative pathways to the 1.5°C target reduce the need for negative emission technologies. *Nature Climate Change*, **8**(5), 391–397, doi:10.1038/s41558-018-0119-8.

Vardoulakis, S. et al., 2014: Comparative Assessment of the Effects of Climate Change on Heat-and Cold-Related Mortality in the United Kingdom and Australia. *Environmental Health Perspectives*, **122**(12), 1285–1292, doi:10.1289/ehp.1307524.

Vaughan, D.G. et al., 2013: Observations: Cryosphere. In: *Climate Change 2013: The Physical Science Basis. Contribution of Working Group I to the Fifth Assessment Report of the Intergovernmental Panel on Climate Change* [Stocker, T.F., D. Qin, G.-K. Plattner, M. Tignor, S.K. Allen, J. Boschung, A. Nauels, Y. Xia, V.B. And, P.M. Midgley, and Midgley (eds.)]. Cambridge University Press, Cambridge, United Kingdom and New York, NY, USA, pp. 317–382.

Vaughan, N.E. and C. Gough, 2016: Expert assessment concludes negative emissions scenarios may not deliver. *Environmental Research Letters*, **11**(9), 095003, doi:10.1088/1748-9326/11/9/095003.

Vaughan, N.E. et al., 2018: Evaluating the use of biomass energy with carbon capture and storage in low emission scenarios. *Environmental Research Letters*, **13**(4), 044014, doi:10.1088/1748-9326/aaaa02.

Vautard, R. et al., 2014: The European climate under a 2°C global warming. *Environmental Research Letters*, **9**(3), 034006, doi:10.1088/1748-9326/9/3/034006.

Velez, C., E. Figueira, A.M.V.M. Soares, and R. Freitas, 2016: Combined effects of seawater acidification and salinity changes in *Ruditapes philippinarum*. *Aquatic Toxicology*, **176**, 141–150, doi:10.1016/j.aquatox.2016.04.016.

Vergés, A. et al., 2014: The tropicalization of temperate marine ecosystems: climate-mediated changes in herbivory and community phase shifts. *Proceedings of the Royal Society B: Biological Sciences*, **281**(1789), 20140846, doi:10.1098/rspb.2014.0846.

Vergés, A. et al., 2016: Long-term empirical evidence of ocean warming leading to tropicalization of fish communities, increased herbivory, and loss of kelp. *Proceedings of the National Academy of Sciences*, **113**(48), 13791–13796, doi:10.1073/pnas.1610725113.

Versini, P.-A., M. Velasco, A. Cabello, and D. Sempere-Torres, 2013: Hydrological impact of forest fires and climate change in a Mediterranean basin. *Natural Hazards*, **66**(2), 609–628, doi:10.1007/s11069-012-0503-z.

Villamayor, B.M.R., R.N. Rollon, M.S. Samson, G.M.G. Albano, and J.H. Primavera, 2016: Impact of *Haiyan* on Philippine mangroves: Implications to the fate of the widespread monospecific *Rhizophora* plantations against strong typhoons. *Ocean and Coastal Management*, **132**, 1–14, doi:10.1016/j.ocecoaman.2016.07.011.

Vitali, A. et al., 2009: Seasonal pattern of mortality and relationships between mortality and temperature-humidity index in dairy cows. *Journal of Dairy Science*, **92**(8), 3781–3790, doi:10.3168/jds.2009-2127.

Vitousek, S. et al., 2017: Doubling of coastal flooding frequency within decades due to sea-level rise. *Scientific Reports*, **7**(1), 1399, doi:10.1038/s41598-017-01362-7.

Vogel, M.M. et al., 2017: Regional amplification of projected changes in extreme temperatures strongly controlled by soil moisture-temperature feedbacks. *Geophysical Research Letters*, **44**(3), 1511–1519, doi:10.1002/2016gl071235.

von Lampe, M. et al., 2014: Why do global long-term scenarios for agriculture differ? An overview of the AgMIP Global Economic Model Intercomparison. *Agricultural Economics*, **45**(1), 3–20, doi:10.1111/agec.12086.

von Uexkull, N., M. Croicu, H. Fjelde, and H. Buhaug, 2016: Civil conflict sensitivity to growing-season drought. *Proceedings of the National Academy of Sciences*, **113**(44), 12391–12396, doi:10.1073/pnas.1607542113.

Wada, Y. et al., 2017: Human-water interface in hydrological modelling: current status and future directions. *Hydrology and Earth System Sciences*, **215194**, 4169–4193, doi:10.5194/hess-21-4169-2017.

Waha, K. et al., 2017: Climate change impacts in the Middle East and Northern Africa (MENA) region and their implications for vulnerable population groups. *Regional Environmental Change*, **17**(6), 1623–1638, doi:10.1007/s10113-017-1144-2.

Wahl, T., S. Jain, J. Bender, S.D. Meyers, and M.E. Luther, 2015: Increasing risk of compound flooding from storm surge and rainfall for major US cities. *Nature Climate Change*, **5**(12), 1093–1097, doi:10.1038/nclimate2736.

Wairiu, M., 2017: Land degradation and sustainable land management practices in Pacific Island Countries. *Regional Environmental Change*, **17**(4), 1053–1064, doi:10.1007/s10113-016-1041-0.

Waldbusser, G.G. et al., 2014: Saturation-state sensitivity of marine bivalve larvae to ocean acidification. *Nature Climate Change*, **5**(3), 273–280, doi:10.1038/nclimate2479.

Wall, E., A. Wreford, K. Topp, and D. Moran, 2010: Biological and economic consequences heat stress due to a changing climate on UK livestock. *Advances in Animal Biosciences*, **1**(01), 53, doi:10.1017/s2040470010001962.

Walsh, K.J.E. et al., 2016: Tropical cyclones and climate change. *Wiley Interdisciplinary Reviews: Climate Change*, **7**(1), 65–89, doi:10.1002/wcc.371.

Wan, H., X. Zhang, and F. Zwiers, 2018: Human influence on Canadian temperatures. *Climate Dynamics*, 1–16, doi:10.1007/s00382-018-4145-z.

Wang, B., H.H. Shugart, and M.T. Lerdau, 2017: Sensitivity of global greenhouse gas budgets to tropospheric ozone pollution mediated by the biosphere. *Environmental Research Letters*, **12**(8), 084001, doi:10.1088/1748-9326/aa7885.

Wang, D., T.C. Gouhier, B.A. Menge, and A.R. Ganguly, 2015: Intensification and spatial homogenization of coastal upwelling under climate change. *Nature*, **518**(7539), 390–394, doi:10.1038/nature14235.

Wang, G. et al., 2017: Continued increase of extreme El Niño frequency long after 1.5°C warming stabilization. *Nature Climate Change*, **7**(8), 568–572, doi:10.1038/nclimate3351.

Wang, H., S.P. Xie, and Q. Liu, 2016: Comparison of climate response to anthropogenic aerosol versus greenhouse gas forcing: Distinct patterns. *Journal of Climate*, **29**(14), 5175–5188, doi:10.1175/jcli-d-16-0106.1.

Wang, L., J.B. Huang, Y. Luo, Y. Yao, and Z.C. Zhao, 2015: Changes in Extremely Hot Summers over the Global Land Area under Various Warming Targets. *PLOS ONE*, **10**(6), e0130660, doi:10.1371/journal.pone.0130660.

Wang, Q. et al., 2016: Effects of ocean acidification on immune responses of the Pacific oyster *Crassostrea gigas*. *Fish & shellfish immunology*, **49**, 24–33, doi:10.1016/j.fsi.2015.12.025.

Wang, Z. et al., 2017: Scenario dependence of future changes in climate extremes under 1.5°C and 2°C global warming.. *Scientific Reports*, **7**, 46432, doi:10.1038/srep46432.

Warren, R., J. Price, E. Graham, N. Forstenhaeusler, and J. VanDerWal, 2018a: The projected effect on insects, vertebrates, and plants of limiting global warming to 1.5°C rather than 2°C. *Science*, **360(6390)**, 791–795, doi:10.1126/science.aar3646.

Warren, R., J. Price, J. VanDerWal, S. Cornelius, and H. Sohl, 2018b: The implications of the United Nations Paris Agreement on Climate Change for Key Biodiversity Areas. *Climatic change*, **147(3–4)**, 395–409, doi:10.1007/s10584-018-2158-6.

Warren, R. et al., 2013: Quantifying the benefit of early climate change mitigation in avoiding biodiversity loss. *Nature Climate Change*, **3(7)**, 678–682, doi:10.1038/nclimate1887.

Warren, R. et al., 2018c: *Risks associated with global warming of 1.5°C or 2°C*. Briefing Note, Tyndall Centre for Climate Change Research, UK, 4 pp.

Warszawski, L. et al., 2013: A multi-model analysis of risk of ecosystem shifts under climate change. *Environmental Research Letters*, **8(4)**, 044018, doi:10.1088/1748-9326/8/4/044018.

Warszawski, L. et al., 2014: The Inter-Sectoral Impact Model Intercomparison Project (ISI-MIP): project framework. *Proceedings of the National Academy of Sciences*, **111(9)**, 3228–32, doi:10.1073/pnas.1312330110.

Wartenburger, R. et al., 2017: Changes in regional climate extremes as a function of global mean temperature: an interactive plotting framework. *Geoscientific Model Development*, **10**, 3609–3634, doi:10.5194/gmd-2017-33.

Watson, C.S. et al., 2015: Unabated global mean sea-level rise over the satellite altimeter era. *Nature Climate Change*, **5(6)**, 565–568, doi:10.1038/nclimate2635.

Watts, G. et al., 2015: Climate change and water in the UK – past changes and future prospects. *Progress in Physical Geography*, **39(1)**, 6–28, doi:10.1177/0309133314542957.

Weatherdon, L., A.K. Magnan, A.D. Rogers, U.R. Sumaila, and W.W.L. Cheung, 2016: Observed and Projected Impacts of Climate Change on Marine Fisheries, Aquaculture, Coastal Tourism, and Human Health: An Update. *Frontiers in Marine Science*, **3**, 48, doi:10.3389/fmars.2016.00048.

Weber, T. et al., 2018: Analysing regional climate change in Africa in a 1.5°C, 2°C and 3°C global warming world. *Earth's Future*, **6**, 1–13, doi:10.1002/2017ef000714.

Webster, N.S., S. Uthicke, E.S. Botté, F. Flores, and A.P. Negri, 2013: Ocean acidification reduces induction of coral settlement by crustose coralline algae. *Global Change Biology*, **19(1)**, 303–315, doi:10.1111/gcb.12008.

Webster, P.J., G.J. Holland, J.A. Curry, and H.-R. Chang, 2005: Changes in Tropical Cyclone Number, Duration, and Intensity in a Warming Environment. *Science*, **309(5742)**, 1844–1846, doi:10.1126/science.1116448.

Wehner, M.F., K.A. Reed, B. Loring, D. Stone, and H. Krishnan, 2018a: Changes in tropical cyclones under stabilized 1.5 and 2.0°C global warming scenarios as simulated by the Community Atmospheric Model under the HAPPI protocols. *Earth System Dynamics*, **9(1)**, 187–195, doi:10.5194/esd-9-187-2018.

Wehner, M.F. et al., 2018b: Changes in extremely hot days under stabilized 1.5 and 2.0°C global warming scenarios as simulated by the HAPPI multi-model ensemble. *Earth System Dynamics*, **9(1)**, 299–311, doi:10.5194/esd-9-299-2018.

Wei, T., S. Glomsrød, and T. Zhang, 2017: Extreme weather, food security and the capacity to adapt – the case of crops in China. *Food Security*, **9(3)**, 523–535, doi:10.1007/s12571-015-0420-6.

Weir, T., L. Dovey, and D. Orcherton, 2017: Social and cultural issues raised by climate change in Pacific Island countries: an overview. *Regional Environmental Change*, **17(4)**, 1017–1028, doi:10.1007/s10113-016-1012-5.

Welch, J.R. et al., 2010: Rice yields in tropical/subtropical Asia exhibit large but opposing sensitivities to minimum and maximum temperatures. *Proceedings of the National Academy of Sciences*, **107(33)**, 14562–7, doi:10.1073/pnas.1001222107.

Wernberg, T. et al., 2012: An extreme climatic event alters marine ecosystem structure in a global biodiversity hotspot. *Nature Climate Change*, **3(1)**, 78–82, doi:10.1038/nclimate1627.

Whan, K. et al., 2015: Impact of soil moisture on extreme maximum temperatures in Europe. *Weather and Climate Extremes*, **9**, 57–67, doi:10.1016/j.wace.2015.05.001.

White, I. and T. Falkland, 2010: Management of freshwater lenses on small Pacific islands. *Hydrogeology Journal*, **18(1)**, 227–246, doi:10.1007/s10040-009-0525-0.

Whitfield, S., A.J. Challinor, and R.M. Rees, 2018: Frontiers in Climate Smart Food Systems: Outlining the Research Space. *Frontiers in Sustainable Food Systems*, **2**, 2, doi:10.3389/fsufs.2018.00002.

Widlansky, M.J., A. Timmermann, and W. Cai, 2015: Future extreme sea level seesaws in the tropical Pacific. *Science Advances*, **1(8)**, e1500560, doi:10.1126/sciadv.1500560.

Wieder, W.R., C.C. Cleveland, W.K. Smith, and K. Todd-Brown, 2015: Future productivity and carbon storage limited by terrestrial nutrient availability. *Nature Geoscience*, **8**, 441–444, doi:10.1038/ngeo2413.

Wiens, J.J., 2016: Climate-Related Local Extinctions Are Already Widespread among Plant and Animal Species. *PLOS Biology*, **14(12)**, e2001104, doi:10.1371/journal.pbio.2001104.

Wilhelm, M., E.L. Davin, and S.I. Seneviratne, 2015: Climate engineering of vegetated land for hot extremes mitigation: An Earth system model sensitivity study. *Journal of Geophysical Research: Atmospheres*, **120(7)**, 2612–2623, doi:10.1002/2014jd022293.

Williams, J.W., B. Shuman, and P.J. Bartlein, 2009: Rapid responses of the prairie-forest ecotone to early Holocene aridity in mid-continental North America. *Global and Planetary Change*, **66(3)**, 195–207, doi:10.1016/j.gloplacha.2008.10.012.

Williamson, P., 2016: Emissions reduction: Scrutinize CO_2 removal methods. *Nature*, **530(7589)**, 153–155, doi:10.1038/530153a.

Willmer, P., 2012: Ecology: Pollinator-Plant Synchrony Tested by Climate Change. *Current Biology*, **22(4)**, R131–R132, doi:10.1016/j.cub.2012.01.009.

Wilson, S.K., N.A.J. Graham, M.S. Pratchett, G.P. Jones, and N.V.C. Polunin, 2006: Multiple disturbances and the global degradation of coral reefs: are reef fishes at risk or resilient? *Global Change Biology*, **12(11)**, 2220–2234, doi:10.1111/j.1365-2486.2006.01252.x.

Wiltshire, A. and T. Davies-Barnard, 2015: *Planetary limits to BECCS negative emissions*. AVOID 2, UK, 24 pp.

Wine, M.L. and D. Cadol, 2016: Hydrologic effects of large southwestern USA wildfires significantly increase regional water supply: fact or fiction? *Environmental Research Letters*, **11(8)**, 085006, doi:10.1088/1748-9326/11/8/085006.

Winsemius, H.C. et al., 2016: Global drivers of future river flood risk. *Nature Climate Change*, **6(4)**, 381–385, doi:10.1038/nclimate2893.

Wong, P.P. et al., 2014: Coastal Systems and Low-Lying Areas. In: *Climate Change 2014: Impacts, Adaptation, and Vulnerability. Part A: Global and Sectoral Aspects. Contribution of Working Group II to the Fifth Assessment Report of the Intergovernmental Panel on Climate Change* [Field, C.B., V.R. Barros, D.J. Dokken, K.J. Mach, M.D. Mastrandrea, T.E. Bilir, M. Chatterjee, K.L. Ebi, Y.O. Estrada, R.C. Genova, B. Girma, E.S. Kissel, A.N. Levy, S. MacCracken, P.R. Mastrandrea, and L.L. White (eds.)]. Cambridge University Press, Cambridge, United Kingdom and New York, NY, USA, pp. 361–409.

Woodroffe, C.D. et al., 2010: Response of coral reefs to climate change: Expansion and demise of the southernmost pacific coral reef. *Geophysical Research Letters*, **37(15)**, L15602, doi:10.1029/2010gl044067.

Woolf, D., J.E. Amonette, F.A. Street-Perrott, J. Lehmann, and S. Joseph, 2010: Sustainable biochar to mitigate global climate change. *Nature Communications*, **1(5)**, 56, doi:10.1038/ncomms1053.

World Bank, 2013: *Turn Down The Heat: Climate Extremes, Regional Impacts and the Case for Resilience*. World Bank, Washington DC, USA, 255 pp., doi:10.1017/cbo9781107415324.004.

Wright, B.D., 2011: The Economics of Grain Price Volatility. *Applied Economic Perspectives and Policy*, **33(1)**, 32–58, doi:10.1093/aepp/ppq033.

Wu, D., Y. Zhao, Y.-S. Pei, and Y.-J Bi, 2013: Climate Change and its Effects on Runoff in Upper and Middle Reaches of Lancang-Mekong river (in Chinese). *Journal of Natural Resources*, **28(9)**, 1569–1582, doi:10.11849/zrzyxb.2013.09.012.

Wu, J. and Y.-J. Shi, 2016: Attribution index for changes in migratory bird distributions: The role of climate change over the past 50 years in China. *Ecological Informatics*, **31**, 147–155, doi:10.1016/j.ecoinf.2015.11.013.

Wu, P., N. Christidis, and P. Stott, 2013: Anthropogenic impact on Earth's hydrological cycle. *Nature Climate Change*, **3(9)**, 807–810, doi:10.1038/nclimate1932.

Yamagata, Y. et al., 2018: Estimating water-food-ecosystem trade-offs for the global negative emission scenario (IPCC-RCP2.6). *Sustainability Science*, **13(2)**, 301–313, doi:10.1007/s11625-017-0522-5.

Yamano, H., K. Sugihara, and K. Nomura, 2011: Rapid poleward range expansion of tropical reef corals in response to rising sea surface temperatures. *Geophysical Research Letters*, **38(4)**, L04601, doi:10.1029/2010gl046474.

3

Yang, J. et al., 2015: Century-scale patterns and trends of global pyrogenic carbon emissions and fire influences on terrestrial carbon balance. *Global Biogeochemical Cycles*, **29(9)**, 1549–1566, doi:10.1002/2015gb005160.

Yang, X., P.E. Thornton, D.M. Ricciuto, and W.M. Post, 2014: The role of phosphorus dynamics in tropical forests – a modeling study using CLM-CNP. *Biogeosciences*, **11(6)**, 1667–1681, doi:10.5194/bg-11-1667-2014.

Yang, Z., T. Wang, N. Voisin, and A. Copping, 2015: Estuarine response to river flow and sea-level rise under future climate change and human development. *Estuarine, Coastal and Shelf Science*, **156**, 19–30, doi:10.1016/j.ecss.2014.08.015.

Yang, Z. et al., 2016: Warming increases methylmercury production in an Arctic soil. *Environmental Pollution*, **214**, 504–509, doi:10.1016/j.envpol.2016.04.069.

Yasuhara, M. and R. Danovaro, 2016: Temperature impacts on deep-sea biodiversity. *Biological Reviews*, **91(2)**, 275–287, doi:10.1111/brv.12169.

Yates, M., G. Le Cozannet, M. Garcin, E. Salai, and P. Walker, 2013: Multidecadal Atoll Shoreline Change on Manihi and Manuae, French Polynesia. *Journal of Coastal Research*, **29**, 870–882, doi:10.2112/jcoastres-d-12-00129.1.

Yazdanpanah, M., M. Thompson, and J. Linnerooth-Bayer, 2016: Do Iranian Policy Makers Truly Understand And Dealing with the Risk of Climate Change Regarding Water Resource Management? *IDRiM*, 367–368.

Yohe, G.W., 2017: Characterizing transient temperature trajectories for assessing the value of achieving alternative temperature targets. *Climatic Change*, **145(3–4)**, 469–479, doi:10.1007/s10584-017-2100-3.

Yokoki, H., M.Tamura, M.Yotsukuri, N.Kumano, and Y.Kuwahara, 2018: Global distribution of projected sea level changes using multiple climate models and economic assessment of sea level rise. CLIVAR Exchanges: A joint special edition on Sea Level Rise, 36–39,doi: 10.5065/d6445k82.

Yoshida, K., M. Sugi, M. Ryo, H. Murakami, and I. Masayoshi, 2017: Future Changes in Tropical Cyclone Activity in High-Resolution Large-Ensemble Simulations. *Geophysical Research Letters*, **44(19)**, 9910–9917, doi:10.1002/2017gl075058.

Yu, R., Z. Jiang, and P. Zhai, 2016: Impact of urban land-use change in eastern China on the East Asian subtropical monsoon: A numerical study. *Journal of Meteorological Research*, **30(2)**, 203–216, doi:10.1007/s13351-016-5157-4.

Yu, R., P. Zhai, and Y. Lu, 2018: Implications of differential effects between 1.5 and 2°C global warming on temperature and precipitation extremes in China's urban agglomerations. *International Journal of Climatology*, **38(5)**, 2374–2385, doi:10.1002/joc.5340.

Yu, Z., J. Loisel, D. Brosseau, D. Beilman, and S. Hunt, 2010: Global peatland dynamics since the Last Glacial Maximum. *Geophysical Research Letters*, **37(13)**, L13402, doi:10.1029/2010gl043584.

Yumashev, D., K. van Hussen, J. Gille, and G. Whiteman, 2017: Towards a balanced view of Arctic shipping: estimating economic impacts of emissions from increased traffic on the Northern Sea Route. *Climatic Change*, **143(1–2)**, 143–155, doi:10.1007/s10584-017-1980-6.

Zaehle, S., C.D. Jones, B. Houlton, J.-F. Lamarque, and E. Robertson, 2015: Nitrogen Availability Reduces CMIP5 Projections of Twenty-First-Century Land Carbon Uptake. *Journal of Climate*, **28(6)**, 2494–2511, doi:10.1175/jcli-d-13-00776.1.

Zaman, A.M. et al., 2017: Impacts on river systems under 2°C warming: Bangladesh Case Study. *Climate Services*, **7**, 96–114, doi:10.1016/j.cliser.2016.10.002.

Zarco-Perello, S., T. Wernberg, T.J. Langlois, and M.A. Vanderklift, 2017: Tropicalization strengthens consumer pressure on habitat-forming seaweeds. *Scientific Reports*, **7(1)**, 820, doi:10.1038/s41598-017-00991-2.

Zhai, R., F. Tao, and Z. Xu, 2018: Spatial-temporal changes in runoff and terrestrial ecosystem water retention under 1.5 and 2°C warming scenarios across China. *Earth System Dynamics*, **9(2)**, 717–738, doi:10.5194/esd-9-717-2018.

Zhang, F., J. Tong, B. Su, J. Huang, and X. Zhu, 2016: Simulation and projection of climate change in the south Asian River basin by CMIP5 multi-model ensembles (in Chinese). *Journal of Tropical Meteorology*, **32(5)**, 734–742.

Zhang, K. et al., 2012: The role of mangroves in attenuating storm surges. *Estuarine, Coastal and Shelf Science*, **102–103**, 11–23, doi:10.1016/j.ecss.2012.02.021.

Zhang, Z., Y. Chen, C. Wang, P. Wang, and F. Tao, 2017: Future extreme temperature and its impact on rice yield in China. *International Journal of Climatology*, **37(14)**, 4814–4827, doi:10.1002/joc.5125.

Zhao, C. et al., 2017: Temperature increase reduces global yields of major crops in four independent estimates. *Proceedings of the National Academy of Sciences*, **114(35)**, 9326–9331, doi:10.1073/pnas.1701762114.

Zhao, X. et al., 2017: Ocean acidification adversely influences metabolism, extracellular pH and calcification of an economically important marine bivalve, *Tegillarca granosa*. *Marine Environmental Research*, **125**, 82–89, doi:10.1016/j.marenvres.2017.01.007.

Zhao, Y. et al., 2016: Potential escalation of heat-related working costs with climate and socioeconomic changes in China. *Proceedings of the National Academy of Sciences*, **113(17)**, 4640–4645, doi:10.1073/pnas.1521828113.

Zhou, B., P. Zhai, Y. Chen, and R. Yu, 2018: Projected changes of thermal growing season over Northern Eurasia in a 1.5°C and 2°C warming world. *Environmental Research Letters*, **13(3)**, 035004, doi:10.1088/1748-9326/aaa6dc.

Zhou, L. et al., 2014: Widespread decline of Congo rainforest greenness in the past decade. *Nature*, **508(7498)**, 86–90, doi:10.1038/nature13265.

Zhu, C. et al., 2018: Carbon dioxide (CO_2) levels this century will alter the protein, micronutrients, and vitamin content of rice grains with potential health consequences for the poorest rice-dependent countries. *Science Advances*, **4(5)**, eaaq1012, doi:10.1126/sciadv.aaq1012.

Zickfeld, K. et al., 2013: Long-Term climate change commitment and reversibility: An EMIC intercomparison. *Journal of Climate*, **26(16)**, 5782–5809, doi:10.1175/jcli-d-12-00584.1.

Zittier, Z.M.C., C. Bock, G. Lannig, and H.O. Pörtner, 2015: Impact of ocean acidification on thermal tolerance and acid-base regulation of *Mytilus edulis* (L.) from the North Sea. *Journal of Experimental Marine Biology and Ecology*, **473**, 16–25, doi:10.1016/j.jembe.2015.08.001.

Zscheischler, J. et al., 2018: Future climate risk from compound events. *Nature Climate Change*, **8(6)**, 469–477, doi:10.1038/s41558-018-0156-3.

3

Strengthening and Implementing the Global Response

4

Coordinating Lead Authors:
Heleen de Coninck (Netherlands/EU), Aromar Revi (India)

Lead Authors:
Mustafa Babiker (Sudan), Paolo Bertoldi (Italy), Marcos Buckeridge (Brazil), Anton Cartwright (South Africa), Wenjie Dong (China), James Ford (UK/Canada), Sabine Fuss (Germany), Jean-Charles Hourcade (France), Debora Ley (Guatemala/Mexico), Reinhard Mechler (Germany), Peter Newman (Australia), Anastasia Revokatova (Russian Federation), Seth Schultz (USA), Linda Steg (Netherlands), Taishi Sugiyama (Japan)

Contributing Authors:
Malcolm Araos (Canada), Stefan Bakker (Netherlands), Amir Bazaz (India), Ella Belfer (Canada), Tim Benton (UK), Sarah Connors (France/UK), Joana Correia de Oliveira de Portugal Pereira (UK/Portugal), Dipak Dasgupta (India), Kiane de Kleijne (Netherlands/EU), Maria del Mar Zamora Dominguez (Mexico), Michel den Elzen (Netherlands), Kristie L. Ebi (USA), Dominique Finon (France), Piers Forster (UK), Jan Fuglestvedt (Norway), Frédéric Ghersi (France), Adriana Grandis (Brazil), Eamon Haughey (Ireland), Bronwyn Hayward (New Zealand), Ove Hoegh-Guldberg (Australia), Daniel Huppmann (Austria), Kejun Jiang (China), Richard Klein (Netherlands/Germany), Shagun Mehrotra (USA/India), Luis Mundaca (Sweden/Chile), Carolyn Opio (Uganda), Maxime Plazzotta (France), Andy Reisinger (New Zealand), Kevon Rhiney (Jamaica), Timmons Roberts (USA), Joeri Rogelj (Austria/Belgium), Arjan van Rooij (Netherlands), Roland Séférian (France), Drew Shindell (USA), Jana Sillmann (Germany/Norway), Chandni Singh (India), Raphael Slade (UK), Gerd Sparovek (Brazil), Pablo Suarez (Argentina), Adelle Thomas (Bahamas), Evelina Trutnevyte (Switzerland/Lithuania), Anne van Valkengoed (Netherlands), Maria Virginia Vilariño (Argentina), Eva Wollenberg (USA)

Review Editors:
Amjad Abdulla (Maldives), Rizaldi Boer (Indonesia), Mark Howden (Australia), Diana Ürge-Vorsatz (Hungary)

Chapter Scientists:
Kiane de Kleijne (Netherlands/EU), Chandni Singh (India)

This chapter should be cited as:
de Coninck, H., A. Revi, M. Babiker, P. Bertoldi, M. Buckeridge, A. Cartwright, W. Dong, J. Ford, S. Fuss, J.-C. Hourcade, D. Ley, R. Mechler, P. Newman, A. Revokatova, S. Schultz, L. Steg, and T. Sugiyama, 2018: Strengthening and Implementing the Global Response. In: *Global Warming of 1.5°C. An IPCC Special Report on the impacts of global warming of 1.5°C above pre-industrial levels and related global greenhouse gas emission pathways, in the context of strengthening the global response to the threat of climate change, sustainable development, and efforts to eradicate poverty* [Masson-Delmotte, V., P. Zhai, H.-O. Pörtner, D. Roberts, J. Skea, P.R. Shukla, A. Pirani, W. Moufouma-Okia, C. Péan, R. Pidcock, S. Connors, J.B.R. Matthews, Y. Chen, X. Zhou, M.I. Gomis, E. Lonnoy, T. Maycock, M. Tignor, and T. Waterfield (eds.)]. Cambridge University Press, Cambridge, UK and New York, NY, USA, pp. 313-444. https://doi.org/10.1017/9781009157940.006.

Table of Contents

4

Executive Summary

Limiting warming to 1.5°C above pre-industrial levels would require transformative systemic change, integrated with sustainable development. Such change would require the upscaling and acceleration of the implementation of far-reaching, multilevel and cross-sectoral climate mitigation and addressing barriers. Such systemic change would need to be linked to complementary adaptation actions, including transformational adaptation, especially for pathways that temporarily overshoot 1.5°C (*medium evidence, high agreement*) {Chapter 2, Chapter 3, 4.2.1, 4.4.5, 4.5}. Current national pledges on mitigation and adaptation are not enough to stay below the Paris Agreement temperature limits and achieve its adaptation goals. While transitions in energy efficiency, carbon intensity of fuels, electrification and land-use change are underway in various countries, limiting warming to 1.5°C will require a greater scale and pace of change to transform energy, land, urban and industrial systems globally. {4.3, 4.4, Cross-Chapter Box 9 in this Chapter}

Although multiple communities around the world are demonstrating the possibility of implementation consistent with 1.5°C pathways {Boxes 4.1-4.10}, very few countries, regions, cities, communities or businesses can currently make such a claim (*high confidence*). To strengthen the global response, almost all countries would need to significantly raise their level of ambition. Implementation of this raised ambition would require enhanced institutional capabilities in all countries, including building the capability to utilize indigenous and local knowledge (*medium evidence, high agreement*). In developing countries and for poor and vulnerable people, implementing the response would require financial, technological and other forms of support to build capacity, for which additional local, national and international resources would need to be mobilized (*high confidence*). However, public, financial, institutional and innovation capabilities currently fall short of implementing far-reaching measures at scale in all countries (*high confidence*). Transnational networks that support multilevel climate action are growing, but challenges in their scale-up remain. {4.4.1, 4.4.2, 4.4.4, 4.4.5, Box 4.1, Box 4.2, Box 4.7}

Adaptation needs will be lower in a 1.5°C world compared to a 2°C world (*high confidence*) {Chapter 3; Cross-Chapter Box 11 in this chapter}. Learning from current adaptation practices and strengthening them through adaptive governance {4.4.1}, lifestyle and behavioural change {4.4.3} and innovative financing mechanisms {4.4.5} can help their mainstreaming within sustainable development practices. Preventing maladaptation, drawing on bottom-up approaches {Box 4.6} and using indigenous knowledge {Box 4.3} would effectively engage and protect vulnerable people and communities. While adaptation finance has increased quantitatively, significant further expansion would be needed to adapt to 1.5°C. Qualitative gaps in the distribution of adaptation finance, readiness to absorb resources, and monitoring mechanisms undermine the potential of adaptation finance to reduce impacts. {Chapter 3, 4.4.2, 4.4.5, 4.6}

System Transitions

The energy system transition that would be required to limit global warming to 1.5°C above pre-industrial conditions is underway in many sectors and regions around the world (*medium evidence, high agreement*). The political, economic, social and technical feasibility of solar energy, wind energy and electricity storage technologies has improved dramatically over the past few years, while that of nuclear energy and carbon dioxide capture and storage (CCS) in the electricity sector have not shown similar improvements. {4.3.1}

Electrification, hydrogen, bio-based feedstocks and substitution, and, in several cases, carbon dioxide capture, utilization and storage (CCUS) would lead to the deep emissions reductions required in energy-intensive industries to limit warming to 1.5°C. However, those options are limited by institutional, economic and technical constraints, which increase financial risks to many incumbent firms (*medium evidence, high agreement*). Energy efficiency in industry is more economically feasible and helps enable industrial system transitions but would have to be complemented with greenhouse gas (GHG)-neutral processes or carbon dioxide removal (CDR) to make energy-intensive industries consistent with 1.5°C (*high confidence*). {4.3.1, 4.3.4}

Global and regional land-use and ecosystems transitions and associated changes in behaviour that would be required to limit warming to 1.5°C can enhance future adaptation and land-based agricultural and forestry mitigation potential. Such transitions could, however, carry consequences for livelihoods that depend on agriculture and natural resources {4.3.2, Cross-Chapter Box 6 in Chapter 3}. Alterations of agriculture and forest systems to achieve mitigation goals could affect current ecosystems and their services and potentially threaten food, water and livelihood security. While this could limit the social and environmental feasibility of land-based mitigation options, careful design and implementation could enhance their acceptability and support sustainable development objectives (*medium evidence, medium agreement*). {4.3.2, 4.5.3}

Changing agricultural practices can be an effective climate adaptation strategy. A diversity of adaptation options exists, including mixed crop-livestock production systems which can be a cost-effective adaptation strategy in many global agriculture systems (*robust evidence, medium agreement*). Improving irrigation efficiency could effectively deal with changing global water endowments, especially if achieved via farmers adopting new behaviours and water-efficient practices rather than through large-scale infrastructural interventions (*medium evidence, medium agreement*). Well-designed adaptation processes such as community-based adaptation can be effective depending upon context and levels of vulnerability. {4.3.2, 4.5.3}

Improving the efficiency of food production and closing yield gaps have the potential to reduce emissions from agriculture, reduce pressure on land, and enhance food security and future mitigation potential (*high confidence*). Improving productivity of existing agricultural systems generally reduces the emissions intensity of food production and offers strong synergies with rural development, poverty reduction and food security objectives, but options to reduce absolute emissions are limited unless paired with demand-side measures. Technological innovation including biotechnology, with adequate safeguards, could contribute to resolving current feasibility constraints and expand the future mitigation potential of agriculture. {4.3.2, 4.4.4}

Shifts in dietary choices towards foods with lower emissions and requirements for land, along with reduced food loss and waste, could reduce emissions and increase adaptation options (*high confidence*). Decreasing food loss and waste and changing dietary behaviour could result in mitigation and adaptation (*high confidence*) by reducing both emissions and pressure on land, with significant co-benefits for food security, human health and sustainable development {4.3.2, 4.4.5, 4.5.2, 4.5.3, 5.4.2}, but evidence of successful policies to modify dietary choices remains limited.

Mitigation and Adaptation Options and Other Measures

A mix of mitigation and adaptation options implemented in a participatory and integrated manner can enable rapid, systemic transitions – in urban and rural areas – that are necessary elements of an accelerated transition consistent with limiting warming to 1.5°C. Such options and changes are most effective when aligned with economic and sustainable development, and when local and regional governments are supported by national governments {4.3.3, 4.4.1, 4.4.3}. Various mitigation options are expanding rapidly across many geographies. Although many have development synergies, not all income groups have so far benefited from them. Electrification, end-use energy efficiency and increased share of renewables, amongst other options, are lowering energy use and decarbonizing energy supply in the built environment, especially in buildings. Other rapid changes needed in urban environments include demotorization and decarbonization of transport, including the expansion of electric vehicles, and greater use of energy-efficient appliances (*medium evidence, high agreement*). Technological and social innovations can contribute to limiting warming to 1.5°C, for example, by enabling the use of smart grids, energy storage technologies and general-purpose technologies, such as information and communication technology (ICT) that can be deployed to help reduce emissions. Feasible adaptation options include green infrastructure, resilient water and urban ecosystem services, urban and peri-urban agriculture, and adapting buildings and land use through regulation and planning (*medium evidence, medium to high agreement*). {4.3.3, 4.4.3, 4.4.4}

Synergies can be achieved across systemic transitions through several overarching adaptation options in rural and urban areas. Investments in health, social security and risk sharing and spreading are cost-effective adaptation measures with high potential for scaling up (*medium evidence, medium to high agreement*). Disaster risk management and education-based adaptation have lower prospects of scalability and cost-effectiveness (*medium evidence, high agreement*) but are critical for building adaptive capacity. {4.3.5, 4.5.3}

Converging adaptation and mitigation options can lead to synergies and potentially increase cost-effectiveness, but multiple trade-offs can limit the speed of and potential for scaling up. Many examples of synergies and trade-offs exist in all sectors and system transitions. For instance, sustainable water management (*high evidence, medium agreement*) and investment in green infrastructure (*medium evidence, high agreement*) to deliver sustainable water and environmental services and to support urban agriculture are less cost-effective than other adaptation options but can help build climate resilience. Achieving the governance, finance and social support required to enable these synergies and to avoid trade-offs is often challenging, especially when addressing multiple objectives, and attempting appropriate sequencing and timing of interventions. {4.3.2, 4.3.4, 4.4.1, 4.5.2, 4.5.3, 4.5.4}

Though CO_2 dominates long-term warming, the reduction of warming short-lived climate forcers (SLCFs), such as methane and black carbon, can in the short term contribute significantly to limiting warming to 1.5°C above pre-industrial levels. Reductions of black carbon and methane would have substantial co-benefits (*high confidence*), including improved health due to reduced air pollution. This, in turn, enhances the institutional and socio-cultural feasibility of such actions. Reductions of several warming SLCFs are constrained by economic and social feasibility (*low evidence, high agreement*). As they are often co-emitted with CO_2, achieving the energy, land and urban transitions necessary to limit warming to 1.5°C would see emissions of warming SLCFs greatly reduced. {2.3.3.2, 4.3.6}

Most CDR options face multiple feasibility constraints, which differ between options, limiting the potential for any single option to sustainably achieve the large-scale deployment required in the 1.5°C-consistent pathways described in Chapter 2 (*high confidence*). Those 1.5°C pathways typically rely on bioenergy with carbon capture and storage (BECCS), afforestation and reforestation (AR), or both, to neutralize emissions that are expensive to avoid, or to draw down CO_2 emissions in excess of the carbon budget {Chapter 2}. Though BECCS and AR may be technically and geophysically feasible, they face partially overlapping yet different constraints related to land use. The land footprint per tonne of CO_2 removed is higher for AR than for BECCS, but given the low levels of current deployment, the speed and scales required for limiting warming to 1.5°C pose a considerable implementation challenge, even if the issues of public acceptance and absence of economic incentives were to be resolved (*high agreement, medium evidence*). The large potential of afforestation and the co-benefits if implemented appropriately (e.g., on biodiversity and soil quality) will diminish over time, as forests saturate (*high confidence*). The energy requirements and economic costs of direct air carbon capture and storage (DACCS) and enhanced weathering remain high (*medium evidence, medium agreement*). At the local scale, soil carbon sequestration has co-benefits with agriculture and is cost-effective even without climate policy (*high confidence*). Its potential feasibility and cost-effectiveness at the global scale appears to be more limited. {4.3.7}

Uncertainties surrounding solar radiation modification (SRM) measures constrain their potential deployment. These uncertainties include: technological immaturity; limited physical understanding about their effectiveness to limit global warming; and a weak capacity to govern, legitimize, and scale such measures. Some recent model-based analysis suggests SRM would be effective but that it is too early to evaluate its feasibility. Even in the uncertain case that the most adverse side-effects of SRM can be avoided, public resistance, ethical concerns and potential impacts on sustainable development could render SRM economically, socially and institutionally undesirable (*low agreement, medium evidence*). {4.3.8, Cross-Chapter Box 10 in this chapter}

Enabling Rapid and Far-Reaching Change

The speed of transitions and of technological change required to limit warming to 1.5°C above pre-industrial levels has been observed in the past within specific sectors and technologies {4.2.2.1}. But the geographical and economic scales at which the required rates of change in the energy, land, urban, infrastructure and industrial systems would need to take place are larger and have no documented historic precedent (*limited evidence, medium agreement*). To reduce inequality and alleviate poverty, such transformations would require more planning and stronger institutions (including inclusive markets) than observed in the past, as well as stronger coordination and disruptive innovation across actors and scales of governance. {4.3, 4.4}

Governance consistent with limiting warming to 1.5°C and the political economy of adaptation and mitigation can enable and accelerate systems transitions, behavioural change, innovation and technology deployment (*medium evidence, medium agreement*). For 1.5°C-consistent actions, an effective governance framework would include: accountable multilevel governance that includes non-state actors, such as industry, civil society and scientific institutions; coordinated sectoral and cross-sectoral policies that enable collaborative multi-stakeholder partnerships; strengthened global-to-local financial architecture that enables greater access to finance and technology; addressing climate-related trade barriers; improved climate education and greater public awareness; arrangements to enable accelerated behaviour change; strengthened climate monitoring and evaluation systems; and reciprocal international agreements that are sensitive to equity and the Sustainable Development Goals (SDGs). System transitions can be enabled by enhancing the capacities of public, private and financial institutions to accelerate climate change policy planning and implementation, along with accelerated technological innovation, deployment and upkeep. {4.4.1, 4.4.2, 4.4.3, 4.4.4}

Behaviour change and demand-side management can significantly reduce emissions, substantially limiting the reliance on CDR to limit warming to 1.5°C {Chapter 2, 4.4.3}. Political and financial stakeholders may find climate actions more cost-effective and socially acceptable if multiple factors affecting behaviour are considered, including aligning these actions with people's core values (*medium evidence, high agreement*). Behaviour- and lifestyle-related measures and demand-side management have already led to emission reductions around the world and can enable significant future reductions (*high confidence*). Social innovation through bottom-up initiatives can result in greater participation in the governance of systems transitions and increase support for technologies, practices and policies that are part of the global response to limit warming to 1.5°C . {Chapter 2, 4.4.1, 4.4.3, Figure 4.3}

This rapid and far-reaching response required to keep warming below 1.5°C and enhance the capacity to adapt to climate risks would require large increases of investments in low-emission infrastructure and buildings, along with a redirection of financial flows towards low-emission investments (*robust evidence, high agreement*). An estimated mean annual incremental investment of around 1.5% of global gross fixed capital formation (GFCF) for the energy sector is indicated between 2016 and 2035, as well as about 2.5% of global GFCF for other development infrastructure that could also address SDG implementation. Though quality policy design and effective implementation may enhance efficiency, they cannot fully substitute for these investments. {2.5.2, 4.2.1, 4.4.5}

Enabling this investment requires the mobilization and better integration of a range of policy instruments that include the reduction of socially inefficient fossil fuel subsidy regimes and innovative price and non-price national and international policy instruments. These would need to be complemented by de-risking financial instruments and the emergence of long-term low-emission assets. These instruments would aim to reduce the demand for carbon-intensive services and shift market preferences away from fossil fuel-based technology. Evidence and theory suggest that carbon pricing alone, in the absence of sufficient transfers to compensate their unintended distributional cross-sector, cross-nation effects, cannot reach the incentive levels needed to trigger system transitions (*robust evidence, medium agreement*). But, embedded in consistent policy packages, they can help mobilize incremental resources and provide flexible mechanisms that help reduce the social and economic costs of the triggering phase of the transition (*robust evidence, medium agreement*). {4.4.3, 4.4.4, 4.4.5}

Increasing evidence suggests that a climate-sensitive realignment of savings and expenditure towards low-emission, climate-resilient infrastructure and services requires an evolution of global and national financial systems. Estimates suggest that, in addition to climate-friendly allocation of public investments, a potential redirection of 5% to 10% of the annual capital revenues[1] is necessary for limiting warming to 1.5°C {4.4.5, Table 1 in Box 4.8}. This could be facilitated by a change of incentives for private day-to-day expenditure and the redirection of savings from speculative and precautionary investments towards long-term productive low-emission assets and services. This implies the mobilization of institutional investors and mainstreaming of climate finance within financial and banking system regulation. Access by developing countries to low-risk and low-interest finance through multilateral and national development banks would have to be facilitated (*medium evidence, high agreement*). New forms of public–private partnerships may be needed with multilateral, sovereign and sub-sovereign guarantees to de-risk climate-friendly investments, support new business models for small-scale enterprises and help households with limited access to capital. Ultimately, the aim is to

---

[1]　Annual capital revenues are paid interests plus an increase of asset value.

promote a portfolio shift towards long-term low-emission assets that would help redirect capital away from potentially stranded assets (*medium evidence, medium agreement*). {4.4.5}

Knowledge Gaps

Knowledge gaps around implementing and strengthening the global response to climate change would need to be urgently resolved if the transition to a 1.5°C world is to become reality. Remaining questions include: how much can be realistically expected from innovation and behavioural and systemic political and economic changes in improving resilience, enhancing adaptation and reducing GHG emissions? How can rates of changes be accelerated and scaled up? What is the outcome of realistic assessments of mitigation and adaptation land transitions that are compliant with sustainable development, poverty eradication and addressing inequality? What are life-cycle emissions and prospects of early-stage CDR options? How can climate and sustainable development policies converge, and how can they be organised within a global governance framework and financial system, based on principles of justice and ethics (including 'common but differentiated responsibilities and respective capabilities' (CBDR-RC)), reciprocity and partnership? To what extent would limiting warming to 1.5°C require a harmonization of macro-financial and fiscal policies, which could include financial regulators such as central banks? How can different actors and processes in climate governance reinforce each other, and hedge against the fragmentation of initiatives? {4.1, 4.3.7, 4.4.1, 4.4.5, 4.6}

4

4.1 Accelerating the Global Response to Climate Change

This chapter discusses how the global economy and socio-technical and socio-ecological systems can transition to 1.5°C-consistent pathways and adapt to warming of 1.5°C above pre-industrial levels. In the context of systemic transitions, the chapter assesses adaptation and mitigation options, including carbon dioxide removal (CDR), and potential solar radiation modification (SRM) remediative measures (Section 4.3), as well as the enabling conditions that would be required for implementing the rapid and far-reaching global response of limiting warming to 1.5°C (Section 4.4), and render the options more or less feasible (Section 4.5).

The impacts of a 1.5°C-warmer world, while less than in a 2°C-warmer world, would require complementary adaptation and development action, typically at local and national scale. From a mitigation perspective, 1.5°C-consistent pathways require immediate action on a greater and global scale so as to achieve net zero emissions by mid-century, or earlier (Chapter 2). This chapter and Chapter 5 highlight the potential that combined mitigation, development and poverty reduction offer for accelerated decarbonization.

The global context is an increasingly interconnected world, with the human population growing from the current 7.6 billion to over 9 billion by mid-century (UN DESA, 2017). There has been a consistent growth of global economic output, wealth and trade with a significant reduction in extreme poverty. These trends could continue for the next few decades (Burt et al., 2014), potentially supported by new and disruptive information and communication, and nano- and bio-technologies. However, these trends co-exist with rising inequality (Piketty, 2014), exclusion and social stratification, and regions locked in poverty traps (Deaton, 2013) that could fuel social and political tensions.

The aftermath of the 2008 financial crisis generated a challenging environment in which leading economists have issued repeated alerts about the 'discontents of globalisation' (Stiglitz, 2002), 'depression economics' (Krugman, 2009), an excessive reliance of export-led development strategies (Rajan, 2011), and risks of 'secular stagnation' due to the 'saving glut' that slows down the flow of global savings towards productive 1.5°C-consistent investments (Summers, 2016). Each of these affects the implementation of both 1.5°C-consistent pathways and sustainable development (Chapter 5).

The range of mitigation and adaptation actions that can be deployed in the short run are well-known: for example, low-emission technologies, new infrastructure, and energy efficiency measures in buildings, industry and transport; transformation of fiscal structures; reallocation of investments and human resources towards low-emission assets; sustainable land and water management; ecosystem restoration; enhancement of adaptive capacities to climate risks and impacts; disaster risk management; research and development; and mobilization of new, traditional and indigenous knowledge.

The convergence of short-term development co-benefits from mitigation and adaptation to address 'everyday development failures' (e.g., institutions, market structures and political processes) (Hallegatte et al., 2016; Pelling et al., 2018) could enhance the adaptive capacity of key systems at risk (e.g., water, energy, food, biodiversity, urban, regional and coastal systems) to 1.5°C climate impacts (Chapter 3). The issue is whether aligning 1.5°C-consistent pathways with the Sustainable Development Goals (SDGs) will secure support for accelerated change and a new growth cycle (Stern, 2013, 2015). It is difficult to imagine how a 1.5°C world would be attained unless the SDG on cities and sustainable urbanization is achieved in developing countries (Revi, 2016), or without reforms in the global financial intermediation system.

Unless affordable and environmentally and socially acceptable CDR becomes feasible and available at scale well before 2050, 1.5°C-consistent pathways will be difficult to realize, especially in overshoot scenarios. The social costs and benefits of 1.5°C-consistent pathways depend on the depth and timing of policy responses and their alignment with short term and long-term development objectives, through policy packages that bring together a diversity of policy instruments, including public investment (Grubb et al., 2014; Winkler and Dubash, 2015; Campiglio, 2016).

Whatever its potential long-term benefits, a transition to a 1.5°C world may suffer from a lack of broad political and public support, if it exacerbates existing short-term economic and social tensions, including unemployment, poverty, inequality, financial tensions, competitiveness issues and the loss of economic value of carbon-intensive assets (Mercure et al., 2018). The challenge is therefore how to strengthen climate policies without inducing economic collapse or hardship, and to make them contribute to reducing some of the 'fault lines' of the world economy (Rajan, 2011).

This chapter reviews literature addressing the alignment of climate with other public policies (e.g., fiscal, trade, industrial, monetary, urban planning, infrastructure, and innovation) and with a greater access to basic needs and services, defined by the SDGs. It also reviews how de-risking low-emission investments and the evolution of the financial intermediation system can help reduce the 'savings glut' (Arezki et al., 2016) and the gap between cash balances and long-term assets (Aglietta et al., 2015b) to support more sustainable and inclusive growth.

As the transitions associated with 1.5°C-consistent pathways require accelerated and coordinated action, in multiple systems across all world regions, they are inherently exposed to risks of freeriding and moral hazards. A key governance challenge is how the convergence of voluntary domestic policies can be organized via aligned global, national and sub-national governance, based on reciprocity (Ostrom and Walker, 2005) and partnership (UN, 2016), and how different actors and processes in climate governance can reinforce each other to enable this (Gupta, 2014; Andonova et al., 2017). The emergence of polycentric sources of climate action and transnational and subnational networks that link these efforts (Abbott, 2012) offer the opportunity to experiment and learn from different approaches, thereby accelerating approaches led by national governments (Cole, 2015; Jordan et al., 2015).

4

Section 4.2 of this chapter outlines existing rates of change and attributes of accelerated change. Section 4.3 identifies global systems, and their components, that offer options for this change. Section 4.4 documents the enabling conditions that influence the feasibility of those options, including economic, financial and policy instruments that could trigger the transition to 1.5°C-consistent pathways. Section 4.5 assesses mitigation and adaptation options for feasibility, strategies for implementation and synergies and trade-offs between mitigation and adaptation.

4.2　Pathways Compatible with 1.5°C: Starting Points for Strengthening Implementation

4.2.1　Implications for Implementation of 1.5°C-Consistent Pathways

The 1.5°C-consistent pathways assessed in Chapter 2 form the basis for the feasibility assessment in section 4.5. A wide range of 1.5°C-consistent pathways from integrated assessment modelling (IAM), supplemented by other literature, are assessed in Chapter 2 (Sections 2.1, 2.3, 2.4, and 2.5). The most common feature shared by these pathways is their requirement for faster and more radical changes compared to 2°C and higher warming pathways.

A variety of 1.5°C-consistent technological options and policy targets is identified in the assessed modelling literature (Sections 2.3, 2.4, 2.5). These technology and policy options include energy demand reduction, greater penetration of low-emission and carbon-free technologies as well as electrification of transport and industry, and reduction of land-use change. Both the detailed integrated modelling pathway literature and a number of broader sectoral and bottom-up studies provide examples of how these sectoral technological and policy characteristics can be broken down sectorally for 1.5°C-consistent pathways (see Table 4.1).

Both the integrated pathway literature and the sectoral studies agree on the need for rapid transitions in the production and use of energy across various sectors, to be consistent with limiting global warming to 1.5°C. The pace of these transitions is particularly significant for the supply mix and electrification (Table 4.1). Individual, sectoral studies may show higher rates of change compared to IAMs (Figueres et al., 2017; Rockström et al., 2017; WBCSD, 2017; Kuramochi et al., 2018). These trends and transformation patterns create opportunities and challenges for both mitigation and adaptation (Sections 4.2.1.1 and 4.2.1.2) and have significant implications for the assessment of feasibility and enablers, including governance, institutions, and policy instruments addressed in Sections 4.3 and 4.4.

Table 4.1 | Sectoral indicators of the pace of transformation in 1.5°C-consistent pathways, based on selected integrated pathways assessed in Chapter 2 (from the scenario database) and several other studies reviewed in Chapter 2 that assess mitigation transitions consistent with limiting warming to 1.5°C. Values for '1.5°C-no or -low-OS' and '1.5C-high-OS' indicate the median and the interquartile ranges for 1.5°C scenarios. If a number in square brackets is indicated, this is the number of scenarios for this indicator. S1, S2, S5 and LED represent the four illustrative pathway archetypes selected for this assessment (see Chapter 2, Section 2.1 and Supplementary Material 4.SM.1 for detailed description).

Pathways		Number of scenarios	Energy		Buildings	Transport		Industry
			Share of renewables in primary energy [%]	Share of renewables in electricity [%]	Change in energy demand for buildings (2010 baseline) [%]	Share of low-carbon fuels (electricity, hydrogen and biofuel) in transport [%]	Share of electricity in transport [%]	Industrial emissions reductions (2010 baseline) [%]
IAM Pathways 2030	1.5°C-no or low-OS	50	29 (37; 26)	54 (65; 47)	0 (7; −7) [42]	12 (18; 9) [29]	5 (7; 3) [49]	42 (55; 34) [42]
	1.5°C-high-OS	35	24 (27; 20)	43 (54; 37)	−17 (−12; −20) [29]	7 (8; 6) [23]	3 (5; 3)	18 (28; −13) [29]
	S1		29	58	−8		4	49
	S2		29	48	−14	5	4	19
	S5		14	25		3	1	
	LED		37	60	30		21	42
Other Studies 2030	Löffler et al. (2017)		46	79				
	IEA (2017c) (ETP)		31	47	2	14	5	22
	IEA (2017g) (WEM)		27	50	−6	17	6	15
IAM Pathways 2050	1.5°C-no or low-OS	50	60 (67; 52)	77 (86; 69)	−17 (3; −36) [42]	55 (66; 35) [29]	23 (29; 17) [49]	79 (91; 67) [42]
	1.5°C-high-OS	35	62 (68; 47)	82 (88; 64)	−37 (−13; −51) [29]	38 (44; 27) [23]	18 (23; 14)	68 (81; 54) [29]
	S1		58	81	−21		34	74
	S2		53	63	−25	26	23	73
	S5		67	70		53	10	
	LED		73	77	45		59	91
Other Studies 2050	Löffler et al. (2017)		100	100				
	IEA (2017c) (ETP)		58	74	5	55	30	57
	IEA (2017g) (WEM)		47	69	−5	58	32	55

4.2.1.1 Challenges and Opportunities for Mitigation Along the Reviewed Pathways

Greater scale, speed and change in investment patterns. There is agreement in the literature reviewed by Chapter 2 that staying below 1.5°C would entail significantly greater transformation in terms of energy systems, lifestyles and investments patterns compared to 2°C-consistent pathways. Yet there is *limited evidence* and *low agreement* regarding the magnitudes and costs of the investments (Sections 2.5.1, 2.5.2 and 4.4.5). Based on the IAM literature reviewed in Chapter 2, climate policies in line with limiting warming to 1.5°C would require a marked upscaling of supply-side energy system investments between now and mid-century, reaching levels of between 1.6–3.8 trillion USD yr^{-1} globally with an average of about 3.5 trillion USD yr^{-1} over 2016–2050 (see Figure 2.27). This can be compared to an average of about 3.0 trillion USD yr^{-1} over the same period for 2°C-consistent pathways (also in Figure 2.27).

Not only the level of investment but also the type and speed of sectoral transformation would be impacted by the transitions associated with 1.5°C-consistent pathways. IAM literature projects that investments in low-emission energy would overtake fossil fuel investments globally by 2025 in 1.5°C-consistent pathways (Chapter 2, Section 2.5.2). The projected low-emission investments in electricity generation allocations over the period 2016–2050 are: solar (0.09–1.0 trillion USD yr^{-1}), wind (0.1–0.35 trillion USD yr^{-1}), nuclear (0.1–0.25 trillion USD yr^{-1}), and transmission, distribution, and storage (0.3–1.3 trillion USD yr^{-1}). In contrast, investments in fossil fuel extraction and unabated fossil electricity generation along a 1.5°C-consistent pathway are projected to drop by 0.3–0.85 trillion USD yr^{-1} over the period 2016–2050, with investments in unabated coal generation projected to halt by 2030 in most 1.5°C-consistent pathways (Chapter 2, Section 2.5.2). Estimates of investments in other infrastructure are currently unavailable, but they could be considerably larger in volume than solely those in the energy sector (Section 4.4.5).

Greater policy design and decision-making implications. The 1.5°C-consistent pathways raise multiple challenges for effective policy design and responses to address the scale, speed, and pace of mitigation technology, finance and capacity building needs. These policies and responses would also need to deal with their distributional implications while addressing adaptation to residual climate impacts (see Chapter 5). The available literature indicates that 1.5°C-consistent pathways would require robust, stringent and urgent transformative policy interventions targeting the decarbonization of energy supply, electrification, fuel switching, energy efficiency, land-use change, and lifestyles (Chapter 2, Section 2.5, 4.4.2, 4.4.3). Examples of effective approaches to integrate mitigation with adaptation in the context of sustainable development and to deal with distributional implications proposed in the literature include the utilization of dynamic adaptive policy pathways (Haasnoot et al., 2013; Mathy et al., 2016) and transdisciplinary knowledge systems (Bendito and Barrios, 2016). Yet, even with good policy design and effective implementation, 1.5°C-consistent pathways would incur higher costs. Projections of the magnitudes of global economic costs associated with 1.5°C-consistent pathways and their sectoral and regional distributions from the

currently assessed literature are scant, yet suggestive. For example, IAM simulations assessed in Chapter 2 project (with a probability greater than 50%) that marginal abatement costs, typically represented in IAMs through a carbon price, would increase by about 3–4 times by 2050 under a 1.5°C-consistent pathway compared to a 2°C-consistent pathway (Chapter 2, Section 2.5.2, Figure 2.26). Managing these costs and distributional effects would require an approach that takes account of unintended cross-sector, cross-nation, and cross-policy trade-offs during the transition (Droste et al., 2016; Stiglitz et al., 2017; Pollitt, 2018; Sands, 2018; Siegmeier et al., 2018).

Greater sustainable development implications. Few studies address the relations between the Shared Socio-Economic Pathways (SSPs) and the Sustainable Developments Goals (SDGs) (O'Neill et al., 2015; Riahi et al., 2017). Nonetheless, literature on potential synergies and trade-offs between 1.5°C-consistent mitigation pathways and sustainable development dimensions is emerging (Chapter 2, Section 2.5.3, Chapter 5, Section 5.4). Areas of potential trade-offs include reduction in final energy demand in relation to SDG 7 (the universal clean energy access goal) and increase of biomass production in relation to land use, water resources, food production, biodiversity and air quality (Chapter 2, Sections 2.4.3, 2.5.3). Strengthening the institutional and policy responses to deal with these challenges is discussed in Section 4.4 together with the linkage between disruptive changes in the energy sector and structural changes in other infrastructure (transport, building, water and telecommunication) sectors. A more in-depth assessment of the complexity and interfaces between 1.5°C-consistent pathways and sustainable development is presented in Chapter 5.

4.2.1.2 Implications for Adaptation Along the Reviewed Pathways

Climate variability and uncertainties in the underlying assumptions in Chapter 2's IAMs as well as in model comparisons complicate discerning the implications for climate impacts, adaptation options and avoided adaptation investments at the global level of 2°C compared to 1.5°C warming (James et al., 2017; Mitchell et al., 2017).

Incremental warming from 1.5°C to 2°C would lead to significant increases in temperature and precipitation extremes in many regions (Chapter 3, Section 3.3.2, 3.3.3). Those projected changes in climate extremes under both warming levels, however, depend on the emissions pathways, as they have different greenhouse gas (GHG)/ aerosol forcing ratios. Impacts are sector-, system- and region-specific, as described in Chapter 3. For example, precipitation-related impacts reveal distinct regional differences (Chapter 3, Sections 3.3.3, 3.3.4, 3.3.5, 3.4.2). Similarly, regional reduction in water availability and the lengthening of regional dry spells have negative implications for agricultural yields depending on crop types and world regions (see for example Chapter 3, Sections 3.3.4, 3.4.2, 3.4.6).

Adaptation helps reduce impacts and risks. However, adaptation has limits. Not all systems can adapt, and not all impacts can be reversed (Cross-Chapter Box 12 in Chapter 5). For example, tropical coral reefs are projected to be at risk of severe degradation due to temperature-induced bleaching (Chapter 3, Box 3.4).

4.2.2   System Transitions and Rates of Change

Society-wide transformation involves socio-technical transitions and social-ecological resilience (Gillard et al., 2016). Transitional adaptation pathways would need to respond to low-emission energy and economic systems, and the socio-technical transitions for mitigation involve removing barriers in social and institutional processes that could also benefit adaptation (Pant et al., 2015; Geels et al., 2017; Ickowitz et al., 2017). In this chapter, transformative change is framed in mitigation around socio-technical transitions, and in adaptation around socio-ecological transitions. In both instances, emphasis is placed on the enabling role of institutions (including markets, and formal and informal regulation). 1.5°C-consistent pathways and adaptation needs associated with warming of 1.5°C imply both incremental and rapid, disruptive and transformative changes.

4.2.2.1   Mitigation: historical rates of change and state of decoupling

Realizing 1.5°C-consistent pathways would require rapid and systemic changes on unprecedented scales (see Chapter 2 and Section 4.2.1). This section examines whether the needed rates of change have historical precedents and are underway.

Some studies conduct a de-facto validation of IAM projections. For CO_2 emission intensity over 1990–2010, this resulted in the IAMs projecting declining emission intensities while actual observations showed an increase. For individual technologies (in particular solar energy), IAM projections have been conservative regarding deployment rates and cost reductions (Creutzig et al., 2017), suggesting that IAMs do not always impute actual rates of technological change resulting from influence of shocks, broader changes and mutually reinforcing factors in society and politics (Geels and Schot, 2007; Daron et al., 2015; Sovacool, 2016; Battiston et al., 2017).

Other studies extrapolate historical trends into the future (Höök et al., 2011; Fouquet, 2016), or contrast the rates of change associated with specific temperature limits in IAMs (such as those in Chapter 2) with historical trends to investigate plausibility of emission pathways and associated temperature limits (Wilson et al., 2013; Gambhir et al., 2017; Napp et al., 2017). When metrics are normalized to gross domestic product (GDP; as opposed to other normalization metrics such as primary energy), low-emission technology deployment rates used by IAMs over the course of the coming century are shown to be broadly consistent with past trends, but rates of change in emission intensity are typically overestimated (Wilson et al., 2013; Loftus et al., 2014; van Sluisveld et al., 2015). This bias is consistent with the findings from the 'validation' studies cited above, suggesting that IAMs may under-report the potential for supply-side technological change assumed in 1.5°-consistent pathways, but may be more optimistic about the systemic ability to realize incremental changes in reduction of emission intensity as a consequence of favourable energy efficiency payback times (Wilson et al., 2013). This finding suggests that barriers and enablers other than costs and climate limits play a role in technological change, as also found in the innovation literature (Hekkert et al., 2007; Bergek et al., 2008; Geels et al., 2016b).

One barrier to a greater rate of change in energy systems is that economic growth in the past has been coupled to the use of fossil fuels. Disruptive innovation and socio-technical changes could enable the decoupling of economic growth from a range of environmental drivers, including the consumption of fossil fuels, as represented by 1.5°C-consistent pathways (UNEP, 2014; Newman, 2017). This may be relative decoupling due to rebound effects that see financial savings generated by renewable energy used in the consumption of new products and services (Jackson and Senker, 2011; Gillingham et al., 2013), but in 2015 and 2016 total global GHG emissions have decoupled absolutely from economic growth (IEA, 2017g; Peters et al., 2017). A longer data trend would be needed before stable decoupling can be established. The observed decoupling in 2015 and 2016 was driven by absolute declines in both coal and oil use since the early 2000s in Europe, in the past seven years in the United States and Australia, and more recently in China (Newman, 2017). In 2017, decoupling in China reversed by 2% due to a drought and subsequent replacement of hydropower with coal-fired power (Tollefson, 2017), but this reversal is expected to be temporary (IEA, 2017c). Oil consumption in China is still rising slowly, but absolute decoupling is ongoing in megacities like Beijing (Gao and Newman, 2018) (see Box 4.9).

4.2.2.2   Transformational adaptation

In some regions and places, incremental adaptation would not be sufficient to mitigate the impacts of climate change on social-ecological systems (see Chapter 3). Transformational adaptation would then be required (Bahadur and Tanner, 2014; Pant et al., 2015; Gillard, 2016; Gillard et al., 2016; Colloff et al., 2017; Termeer et al., 2017). Transformational adaptation refers to actions aiming at adapting to climate change resulting in significant changes in structure or function that go beyond adjusting existing practices (Dowd et al., 2014; IPCC, 2014a; Few et al., 2017), including approaches that enable new ways of decision-making on adaptation (Colloff et al., 2017). Few studies have assessed the potentially transformative character of adaptation options (Pelling et al., 2015; Rippke et al., 2016; Solecki et al., 2017), especially in the context of warming of 1.5°C.

Transformational adaptation can be adopted at a large scale, can lead to new strategies in a region or resource system, transform places and potentially shift locations (Kates et al., 2012). Some systems might require transformational adaptation at 1.5°C. Implementing adaptation policies in anticipation of 1.5°C would require transformation and flexible planning of adaptation (sometimes called adaptation pathways) (Rothman et al., 2014; Smucker et al., 2015; Holland, 2017; Gajjar et al., 2018), an understanding of the varied stakeholders involved and their motives, and knowledge of less visible aspects of vulnerability based on social, cultural, political, and economic factors (Holland, 2017). Transformational adaptation would seek deep and long-term societal changes that influence sustainable development (Chung Tiam Fook, 2017; Few et al., 2017).

Adaptation requires multidisciplinary approaches integrating scientific, technological and social dimensions. For example, a framework for transformational adaptation and the integration of mitigation and adaptation pathways can transform rural indigenous communities to address risks of climate change and other stressors (Thornton and Comberti, 2017). In villages in rural Nepal, transformational adaptation has taken place, with villagers changing their agricultural and pastoralist livelihood strategies after years of lost crops due to changing rain patterns and degradation of natural resources (Thornton and Comberti, 2017). Instead, they are now opening stores, hotels, and tea shops. In another case, the arrival of an oil pipeline altered traditional Alaskan communities' livelihoods. With growth of oil production, investments were made for rural development. A later drop in oil production decreased these investments. Alaskan indigenous populations are also dealing with impacts of climate change, such as sea level rise, which is altering their livelihood sources. Transformational adaptation is taking place by changing the energy matrix to renewable energy, in which indigenous people apply their knowledge to achieve environmental, economic, and social benefits (Thornton and Comberti, 2017).

4.2.2.3 Disruptive innovation

Demand-driven disruptive innovations that emerge as the product of political and social changes across multiple scales can be transformative (Seba, 2014; Christensen et al., 2015; Green and Newman, 2017a). Such innovations would lead to simultaneous, profound changes in behaviour, economies and societies (Seba, 2014; Christensen et al. 2015), but are difficult to predict in supply-focused economic models (Geels et al., 2016a; Pindyck, 2017). Rapid socio-technical change has been observed in the solar industry (Creutzig et al. (2017). Similar changes to socio-ecological systems can stimulate adaptation and mitigation options that lead to more climate-resilient systems (Adger et al., 2005; Ostrom, 2009; Gillard et al., 2016) (see the Alaska and Nepal examples in Section 4.2.2.2). The increase in roof-top solar and energy storage technology as well as the increase in passive housing and net zero-emissions buildings are further examples of such disruptions (Green and Newman, 2017b). Both roof-top solar and energy storage have benefitted from countries' economic growth strategies and associated price declines in photovoltaic technologies, particularly in China (Shrivastava and Persson, 2018), as well as from new information and communication technologies (Koomey et al., 2013), rising demand for electricity in urban areas, and global concern regarding greenhouse gas emissions (Azeiteiro and Leal Filho, 2017; Lutz and Muttarak, 2017; Wamsler, 2017).

System co-benefits can create the potential for mutually enforcing and demand-driven climate responses (Jordan et al., 2015; Hallegatte and Mach, 2016; Pelling et al., 2018), and for rapid and transformational change (Cole, 2015; Geels et al., 2016b; Hallegatte and Mach, 2016). Examples of co-benefits include gender equality, agricultural productivity (Nyantakyi-Frimpong and Bezner-Kerr, 2015), reduced indoor air pollution (Satterthwaite and Bartlett, 2017), flood buffering (Colenbrander et al., 2017), livelihood support (Shaw et al., 2014; Ürge-Vorsatz et al., 2014), economic growth (GCEC, 2014; Stiglitz et al., 2017), social progress (Steg et al., 2015; Hallegatte and Mach, 2016) and social justice (Ziervogel et al., 2017; Patterson et al., 2018).

Innovations that disrupt entire systems may leave firms and utilities with stranded assets, as the transition can happen very quickly (IPCC, 2014b; Kossoy et al., 2015). This may have consequences for fossil fuels that are rendered 'unburnable' (McGlade and Ekins, 2015) and fossil fuel-fired power and industry assets that would become obsolete (Caldecott, 2017; Farfan and Breyer, 2017). The presence of multiple barriers and enablers operating in a system implies that rapid change, whether the product of many small changes (Termeer et al., 2017) or large-scale disruptions, is seldom an insular or discrete process (Sterling et al., 2017). This finding informs the multidimensional nature of feasibility in Cross-Chapter Box 3 in Chapter 1 which is applied in Section 4.5. Climate responses that are aligned with multiple feasibility dimensions and combine adaptation and mitigation interventions with non-climate benefits can accelerate change and reduce risks and costs (Fazey et al., 2018). Also political, social and technological influences on energy transitions, for example, can accelerate them faster than narrow techno-economic analysis suggests is possible (Kern and Rogge, 2016), but could also introduce new constraints and risks (Geels et al., 2016b; Sovacool, 2016; Eyre et al., 2018).

Disruptive innovation and technological change may play a role in mitigation and in adaptation. The next section assesses mitigation and adaptation options in energy, land and ecosystem, urban and infrastructure and industrial systems.

4.3 Systemic Changes for 1.5°C-Consistent Pathways

Section 4.2 emphasizes the importance of systemic change for 1.5°C-consistent pathways. This section translates this into four main system transitions: energy, land and ecosystem, urban and infrastructure, and industrial system transitions. This section assesses the mitigation, adaptation and carbon dioxide removal options that offer the potential for such change within those systems, based on options identified by Chapter 2 and risks and impacts in Chapter 3.

The section puts more emphasis on those adaptation options (Sections 4.3.1–4.3.5) and mitigation options (Sections 4.3.1–4.3.4, 4.3.6 and 4.3.7) that are 1.5°C-relevant and have developed considerably since AR5. They also form the basis for the mitigation and adaptation feasibility assessments in Section 4.5. Section 4.3.8 discusses solar radiation modification methods.

This section emphasizes that no single solution or option can enable a global transition to 1.5°C-consistent pathways or adapting to projected impacts. Rather, accelerating change, much of which is already starting or underway, in multiple global systems, simultaneously and at different scales, could provide the impetus for these system transitions. The feasibility of individual options as well as the potential for synergies and reducing trade-offs will vary according to context and the local enabling conditions. These are explored at a high level in Section 4.5. Policy packages that bring together multiple enabling conditions can provide building blocks for a strategy to scale up implementation and intervention impacts.

4

4.3.1   Energy System Transitions

This section discusses the feasibility of mitigation and adaptation options related to the energy system transition. Only options relevant to 1.5°C and with significant changes since AR5 are discussed, which means that for options like hydropower and geothermal energy, the chapter refers to AR5 and does not provide a discussion. Socio-technical inertia of energy options for 1.5°C-consistent pathways are increasingly being surmounted as fossil fuels start to be phased out. Supply-side mitigation and adaptation options and energy demand-side options, including energy efficiency in buildings and transportation, are discussed in Section 4.3.3; options around energy use in industry are discussed in Section 4.3.4.

Section 4.5 assesses the feasibility in a systematic manner based on the approach outlined in Cross-Chapter Box 3 in Chapter 1.

4.3.1.1   Renewable electricity: solar and wind

All renewable energy options have seen considerable advances over the years since AR5, but solar energy and both onshore and offshore wind energy have had dramatic growth trajectories. They appear well underway to contribute to 1.5°C-consistent pathways (IEA, 2017c; IRENA, 2017b; REN21, 2017).

The largest growth driver for renewable energy since AR5 has been the dramatic reduction in the cost of solar photovoltaics (PV) (REN21, 2017). This has made rooftop solar competitive in sunny areas between 45° north and south latitude (Green and Newman, 2017b), though IRENA (2018) suggests it is cost effective in many other places too. Solar PV with batteries has been cost effective in many rural and developing areas (Pueyo and Hanna, 2015; Szabó et al., 2016; Jimenez, 2017), for example 19 million people in Bangladesh now have solar-battery electricity in remote villages and are reporting positive experiences on safety and ease of use (Kabir et al., 2017). Small-scale distributed energy projects are being implemented in developed and developing cities where residential and commercial rooftops offer potential for consumers becoming producers (called prosumers) (ACOLA, 2017; Kotilainen and Saari, 2018). Such prosumers could contribute significantly to electricity generation in sun-rich areas like California (Kurdgelashvili et al., 2016) or sub-Saharan Africa in combination with micro-grids and mini-grids (Bertheau et al., 2017). It could also contribute to universal energy access (SDG 7) as shown by (IEA, 2017c).

The feasibility of renewable energy options depends to a large extent on geophysical characteristics of the area where the option is implemented. However, technological advances and policy instruments make renewable energy options increasingly attractive in other areas. For example, solar PV is deployed commercially in areas with low solar insolation, like northwest Europe (Nyholm et al., 2017). Feasibility also depends on grid adaptations (e.g., storage, see below) as renewables grow (IEA, 2017c). For regions with high energy needs, such as industrial areas (see Section 4.3.4), high-voltage DC transmission across long distances would be needed (MacDonald et al., 2016).

Another important factor affecting feasibility is public acceptance, in particular for wind energy and other large-scale renewable facilities

(Yenneti and Day, 2016; Rand and Hoen, 2017; Gorayeb et al., 2018) that raise landscape management (Nadaï and Labussière, 2017) and distributional justice (Yenneti and Day, 2016) challenges. Research indicates that financial participation and community engagement can be effective in mitigating resistance (Brunes and Ohlhorst, 2011; Rand and Hoen, 2017) (see Section 4.4.3).

Bottom-up studies estimating the use of renewable energy in the future, either at the global or at the national level, are plentiful, especially in the grey literature. It is hotly debated whether a fully renewable energy or electricity system, with or without biomass, is possible (Jacobson et al., 2015, 2017) or not (Clack et al., 2017; Heard et al., 2017), and by what year. Scale-up estimates vary with assumptions about costs and technological maturity, as well as local geographical circumstances and the extent of storage used (Ghorbani et al., 2017; REN21, 2017). Several countries have adopted targets of 100% renewable electricity (IEA, 2017c) as this meets multiple social, economic and environmental goals and contributes to mitigation of climate change (REN21, 2017).

4.3.1.2   Bioenergy and biofuels

Bioenergy is renewable energy from biomass. Biofuel is biomass-based energy used in transport. Chapter 2 suggests that pathways limiting warming to 1.5°C would enable supply of 67–310 (median 150) EJ yr^{-1} (see Table 2.8) from biomass. Most scenarios find that bioenergy is combined with carbon dioxide capture and storage (CCS, BECCS) if it is available but also find robust deployment of bioenergy independent of the availability of CCS (see Chapter 2, Section 2.3.4.2 and Section 4.3.7 for a discussion of BECCS). Detailed assessments indicate that deployment is similar for pathways limiting global warming to below 2°C (Chum et al., 2011; P. Smith et al., 2014; Creutzig et al., 2015b). There is however *high agreement* that the sustainable bioenergy potential in 2050 would be restricted to around 100 EJ yr^{-1} (Slade et al., 2014; Creutzig et al., 2015b). Sustainable deployment at such or higher levels envisioned by 1.5°C-consistent pathways may put significant pressure on available land, food production and prices (Popp et al., 2014b; Persson, 2015; Kline et al., 2017; Searchinger et al., 2017), preservation of ecosystems and biodiversity (Creutzig et al., 2015b; Holland et al., 2015; Santangeli et al., 2016), and potential water and nutrient constraints (Gerbens-Leenes et al., 2009; Gheewala et al., 2011; Bows and Smith, 2012; Smith and Torn, 2013; Bonsch et al., 2016; Lampert et al., 2016; Mouratiadou et al., 2016; Smith et al., 2016b; Wei et al., 2016; Mathioudakis et al., 2017); but there is still *low agreement* on these interactions (Robledo-Abad et al., 2017). Some of the disagreement on the sustainable capacity for bioenergy stems from global versus local assessments. Global assessments may mask local dynamics that exacerbate negative impacts and shortages while at the same time niche contexts for deployment may avoid trade-offs and exploit co-benefits more effectively. In some regions of the world (e.g., the case of Brazilian ethanol, see Box 4.7, where land may be less of a constraint, the use of bioenergy is mature and the industry is well developed), land transitions could be balanced with food production and biodiversity to enable a global impact on CO_2 emissions (Jaiswal et al., 2017).

The carbon intensity of bioenergy, key for both bioenergy as an emission-neutral energy option and BECCS as a CDR measure, is still a matter of debate (Buchholz et al., 2016; Liu et al., 2018) and depends on management (Pyörälä et al., 2014; Torssonen et al., 2016; Baul et al., 2017; Kilpeläinen et al., 2017); direct and indirect land-use change emissions (Plevin et al., 2010; Schulze et al., 2012; Harris et al., 2015; Repo et al., 2015; DeCicco et al., 2016; Qin et al., 2016)[2]; the feedstock considered; and time frame (Zanchi et al., 2012; Daioglou et al., 2017; Booth, 2018; Sterman et al., 2018), as well as the availability of coordinated policies and management to minimize negative side effects and trade-offs, particularly those around food security (Stevanović et al., 2017) and livelihood and equity considerations (Creutzig et al., 2013; Calvin et al., 2014) .

Biofuels are a part of the transport sector in some cities and countries, and may be deployed as a mitigation option for aviation, shipping and freight transport (see Section 4.3.3.5) as well as industrial decarbonization (IEA, 2017g) (Section 4.3.4), though only Brazil has mainstreamed ethanol as a substantial, commercial option. Lower emissions and reduced urban air pollution have been achieved there by use of ethanol and biodiesel as fuels (Hill et al., 2006; Salvo et al., 2017) (see Box 4.7).

4.3.1.3 Nuclear energy

Many scenarios in Chapter 2 and in AR5 (Bruckner et al., 2014) project an increase in the use of nuclear power, while others project a decrease. The increase can be realized through existing mature nuclear technologies or new options (generation III/IV reactors, breeder reactors, new uranium and thorium fuel cycles, small reactors or nuclear cogeneration).

Even though scalability and speed of scaling of nuclear plants have historically been high in many nations, such rates are currently not achieved anymore. In the 1960s and 1970s, France implemented a programme to rapidly get 80% of its power from nuclear in about 25 years (IAEA, 2018), but the current time lag between the decision date and the commissioning of plants is observed to be 10-19 years (Lovins et al., 2018). The current deployment pace of nuclear energy is constrained by social acceptability in many countries due to concerns over risks of accidents and radioactive waste management (Bruckner et al., 2014). Though comparative risk assessment shows health risks are low per unit of electricity production (Hirschberg et al., 2016), and land requirement is lower than that of other power sources (Cheng and Hammond, 2017), the political processes triggered by societal concerns depend on the country-specific means of managing the political debates around technological choices and their environmental impacts (Gregory et al., 1993). Such differences in perception explain why the 2011 Fukushima incident resulted in a confirmation or acceleration of phasing out nuclear energy in five countries (Roh, 2017) while 30 other countries have continued using nuclear energy, amongst which 13 are building new nuclear capacity, including China, India and the United Kingdom (IAEA, 2017; Yuan et al., 2017).

Costs of nuclear power have increased over time in some developed nations, principally due to market conditions where increased investment risks of high-capital expenditure technologies have become significant. 'Learning by doing' processes often failed to compensate for this trend because they were slowed down by the absence of standardization and series effects (Grubler, 2010). What the costs of nuclear power are and have been is debated in the literature (Lovering et al., 2016; Koomey et al., 2017). Countries with liberalized markets that continue to develop nuclear employ de-risking instruments through long-term contracts with guaranteed sale prices (Finon and Roques, 2013). For instance, the United Kingdom works with public guarantees covering part of the upfront investment costs of newly planned nuclear capacity. This dynamic differs in countries such as China and South Korea, where monopolistic conditions in the electric system allow for reducing investment risks, deploying series effects and enhancing the engineering capacities of users due to stable relations between the security authorities and builders (Schneider et al., 2017).

The safety of nuclear plants depends upon the public authorities of each country. However, because accidents affect worldwide public acceptance of this industry, questions have been raised about the risk of economic and political pressures weakening the safety of the plants (Finon, 2013; Budnitz, 2016). This raises the issue of international governance of civil nuclear risks and reinforced international cooperation involving governments, companies and engineering (Walker and Lönnroth, 1983; Thomas, 1988; Finon, 2013), based on the experience of the International Atomic Energy Agency.

4.3.1.4 Energy storage

The growth in electricity storage for renewables has been around grid flexibility resources (GFR) that would enable several places to source more than half their power from non-hydro renewables (Komarnicki, 2016). Ten types of GFRs within smart grids have been developed (largely since AR5)(Blaabjerg et al., 2004; IRENA, 2013; IEA, 2017d; Majzoobi and Khodaei, 2017), though how variable renewables can be balanced without hydro or natural gas-based power back-up at a larger scale would still need demonstration. Pumped hydro comprised 150 GW of storage capacity in 2016, and grid-connected battery storage just 1.7 GW, but the latter grew between 2015 to 2016 by 50% (REN21, 2017). Battery storage has been the main growth feature in energy storage since AR5 (Breyer et al., 2017). This appears to the result of significant cost reductions due to mass production for electric vehicles (EVs) (Nykvist and Nilsson, 2015; Dhar et al., 2017). Although costs and technical maturity look increasingly positive, the feasibility of battery storage is challenged by concerns over the availability of resources and the environmental impacts of its production (Peters et al., 2017). Lithium, a common element in the earth's crust, does not appear to be restricted and large increases in production have happened in recent years with eight new mines in Western Australia where most lithium is produced (GWA, 2016). Emerging battery technologies may provide greater efficiency and recharge rates (Belmonte et al., 2016) but remain significantly more expensive due to speed and scale issues compared to lithium ion batteries (Dhar et al., 2017; IRENA, 2017a).

---

[2]   While there is *high agreement* that indirect land use change (iLUC) could occur, there is *low agreement* about the actual extent of iLUC (P. Smith et al., 2014; Verstegen et al., 2015; Zilberman, 2017)

Research and demonstration of energy storage in the form of thermal and chemical systems continues, but large-scale commercial systems are rare (Pardo et al., 2014). Renewably derived synthetic liquid (like methanol and ammonia) and gas (like methane and hydrogen) are increasingly being seen as a feasible storage options for renewable energy (producing fuel for use in industry during times when solar and wind are abundant) (Bruce et al., 2010; Jiang et al., 2010; Ezeji, 2017) but, in the case of carbonaceous storage media, would need a renewable source of carbon to make a positive contribution to GHG reduction (von der Assen et al., 2013; Abanades et al., 2017) (see also Section 4.3.4.5). The use of electric vehicles as a form of storage has been modelled and evaluated as an opportunity, and demonstrations are emerging (Dhar et al., 2017; Green and Newman, 2017a), but challenges to upscaling remain.

4.3.1.5   Options for adapting electricity systems to 1.5°C

Climate change has started to disrupt electricity generation and, if climate change adaptation options are not considered, it is predicted that these disruptions will be lengthier and more frequent (Jahandideh-Tehrani et al., 2014; Bartos and Chester, 2015; Kraucunas et al., 2015; van Vliet et al., 2016). Adaptation would both secure vulnerable infrastructure and ensure the necessary generation capacity (Minville et al., 2009; Eisenack and Stecker, 2012; Schaeffer et al., 2012; Cortekar and Groth, 2015; Murrant et al., 2015; Panteli and Mancarella, 2015; Goytia et al., 2016). The literature shows *high agreement* that climate change impacts need to be planned for in the design of any kind of infrastructure, especially in the energy sector (Nierop, 2014), including interdependencies with other sectors that require electricity to function, including water, data, telecommunications and transport (Fryer, 2017).

Recent research has developed new frameworks and models that aim to assess and identify vulnerabilities in energy infrastructure and create more proactive responses (Francis and Bekera, 2014; Ouyang and Dueñas-Osorio, 2014; Arab et al., 2015; Bekera and Francis, 2015; Knight et al., 2015; Jeong and An, 2016; Panteli et al., 2016; Perrier, 2016; Erker et al., 2017; Fu et al., 2017). Assessments of energy infrastructure adaptation, while limited, emphasize the need for redundancy (Liu et al., 2017). The implementation of controllable and islandable microgrids, including the use of residential batteries, can increase resiliency, especially after extreme weather events (Qazi and Young Jr., 2014; Liu et al., 2017). Hybrid renewables-based power systems with non-hydro capacity, such as with high-penetration wind generation, could provide the required system flexibility (Canales et al., 2015). Overall, there is *high agreement* that hybrid systems, taking advantage of an array of sources and time of use strategies, can help make electricity generation more resilient (Parkinson and Djilali, 2015), given that energy security standards are in place (Almeida Prado et al., 2016).

Interactions between water and energy are complex (IEA, 2017g). Water scarcity patterns and electricity disruptions will differ across regions. There is *high agreement* that mitigation and adaptation options for thermal electricity generation (if that remains fitted with CCS) need to consider increasing water shortages, taking into account other factors such as ambient water resources and demand changes in irrigation water (Hayashi et al., 2018). Increasing the efficiency of power

plants can reduce emissions and water needs (Eisenack and Stecker, 2012; van Vliet et al., 2016), but applying CCS would increase water consumption (Koornneef et al., 2012). The technological, economic, social and institutional feasibility of efficiency improvements is high, but insufficient to limit temperature rise to 1.5°C (van Vliet et al., 2016).

In addition, a number of options for water cooling management systems have been proposed, such as hydraulic measures (Eisenack and Stecker, 2012) and alternative cooling technologies (Chandel et al., 2011; Eisenack and Stecker, 2012; Bartos and Chester, 2015; Murrant et al., 2015; Bustamante et al., 2016; van Vliet et al., 2016; Huang et al., 2017b). There is *high agreement* on the technological and economic feasibility of these technologies, as their absence can severely impact the functioning of the power plant as well as safety and security standards.

4.3.1.6   Carbon dioxide capture and storage in the power sector

The AR5 (IPCC, 2014b) as well as Chapter 2, Section 2.4.2, assign significant emission reductions over the course of this century to CO_2 capture and storage (CCS) in the power sector. This section focuses on CCS in the fossil-fuelled power sector; Section 4.3.4 discusses CCS in non-power industry, and Section 4.3.7 discusses bioenergy with CCS (BECCS). Section 2.4.2 puts the cumulative CO_2 stored from fossil-fuelled power at 410 (199–470 interquartile range) $GtCO_2$ over this century. Such modelling suggests that CCS in the power sector can contribute to cost-effective achievement of emission reduction requirements for limiting warming to 1.5°C. CCS may also offer employment and political advantages for fossil fuel-dependent economies (Kern et al., 2016), but may entail more limited co-benefits than other mitigation options (that, e.g., generate power) and therefore relies on climate policy incentives for its business case and economic feasibility. Since 2017, two CCS projects in the power sector capture 2.4 $MtCO_2$ annually, while 30 $MtCO_2$ is captured annually in all CCS projects (Global CCS Institute, 2017).

The technological maturity of CO_2 capture options in the power sectors has improved considerably (Abanades et al., 2015; Bui et al., 2018), but costs have not come down between 2005 and 2015 due to limited learning in commercial settings and increased energy and resources costs (Rubin et al., 2015). Storage capacity estimates vary greatly, but Section 2.4.2 as well as literature (V. Scott et al., 2015) indicate that perhaps 10,000 $GtCO_2$ could be stored in underground reservoirs. Regional availability of this may not be sufficient, and it requires efforts to have this storage and the corresponding infrastructure available at the necessary rates and times (de Coninck and Benson, 2014). CO_2 retention in the storage reservoir was recently assessed as 98% over 10,000 years for well-managed reservoirs, and 78% for poorly regulated ones (Alcalde et al., 2018). A paper reviewing 42 studies on public perception of CCS (Seigo et al., 2014) found that social acceptance of CCS is predicted by trust, perceived risks and benefits. The technology itself mattered less than the social context of the project. Though insights on communication of CCS projects to the general public and inhabitants of the area around the CO_2 storage sites have been documented over the years, project stakeholders are not consistently implementing these lessons, although some projects have observed good practices (Ashworth et al., 2015).

CCS in the power sector is hardly being realized at scale, mainly because the incremental costs of capture, and the development of transport and storage infrastructures are not sufficiently compensated by market or government incentives (IEA, 2017c). In the two full-scale projects in the power sector mentioned above, part of the capture costs are compensated for by revenues from enhanced oil recovery (EOR) (Global CCS Institute, 2017), demonstrating that EOR helps developing CCS further. EOR is a technique that uses CO_2 to mobilize more oil out of depleting oil fields, leading to additional CO_2 emissions by combusting the additionally recovered oil (Cooney et al., 2015).

4.3.2 Land and Ecosystem Transitions

This section assesses the feasibility of mitigation and adaptation options related to land use and ecosystems. Land transitions are grouped around agriculture and food, ecosystems and forests, and coastal systems.

4.3.2.1 Agriculture and food

In a 1.5°C world, local yields are projected to decrease in tropical regions that are major food producing areas of the world (West Africa, Southeast Asia, South Asia, and Central and northern South America) (Schleussner et al., 2016). Some high-latitude regions may benefit from the combined effects of elevated CO_2 and temperature because their average temperatures are below optimal temperature for crops. In both cases there are consequences for food production and quality (Cross-Chapter Box 6 in Chapter 3 on Food Security), conservation agriculture, irrigation, food wastage, bioenergy and the use of novel technologies.

Food production and quality. Increased temperatures, including 1.5°C warming, would affect the production of cereals such as wheat and rice, impacting food security (Schleussner et al., 2016). There is *medium agreement* that elevated CO_2 concentrations can change food composition, with implications for nutritional security (Taub et al., 2008; Högy et al., 2009; DaMatta et al., 2010; Loladze, 2014; De Souza et al., 2015), with the effects being different depending on the region (Medek et al., 2017).

Meta-analyses of the effects of drought, elevated CO_2, and temperature conclude that at 2°C local warming and above, aggregate production of wheat, maize, and rice are expected to decrease in both temperate and tropical areas (Challinor et al., 2014). These production losses could be lowered if adaptation measures are taken (Challinor et al., 2014), such as developing varieties better adapted to changing climate conditions.

Adaptation options can help ensure access to sufficient, quality food. Such options include conservation agriculture, improved livestock management, increasing irrigation efficiency, agroforestry and management of food loss and waste. Complementary adaptation and mitigation options, for example, the use of climate services (Section 4.3.5), bioenergy (Section 4.3.1) and biotechnology (Section 4.4.4) can also serve to reduce emissions intensity and the carbon footprint of food production.

Conservation agriculture (CA) is a soil management approach that reduces the disruption of soil structure and biotic processes by minimising tillage. A recent meta-analysis showed that no-till practices

work well in water-limited agroecosystems when implemented jointly with residue retention and crop rotation, but when used independently, may decrease yields in other situations (Pittelkow et al., 2014). Additional climate adaptations include adjusting planting times and crop varietal selection and improving irrigation efficiency. Adaptations such as these may increase wheat and maize yields by 7–12% under climate change (Challinor et al., 2014). CA can also help build adaptive capacity (*medium evidence, medium agreement*) (H. Smith et al., 2017; Pradhan et al., 2018) and have mitigation co-benefits through improved fertiliser use or efficient use of machinery and fossil fuels (Harvey et al., 2014; Cui et al., 2018; Pradhan et al., 2018). CA practices can also raise soil carbon and therefore remove CO_2 from the atmosphere (Aguilera et al., 2013; Poeplau and Don, 2015; Vicente-Vicente et al., 2016). However, CA adoption can be constrained by inadequate institutional arrangements and funding mechanisms (Harvey et al., 2014; Baudron et al., 2015; Li et al., 2016; Dougill et al., 2017; H. Smith et al., 2017).

Sustainable intensification of agriculture consists of agricultural systems with increased production per unit area but with management of the range of potentially adverse impacts on the environment (Pretty and Bharucha, 2014). Sustainable intensification can increase the efficiency of inputs and enhance health and food security (Ramankutty et al., 2018).

Livestock management. Livestock are responsible for more GHG emissions than all other food sources. Emissions are caused by feed production, enteric fermentation, animal waste, land-use change and livestock transport and processing. Some estimates indicate that livestock supply chains could account for 7.1 GtCO_2 per year, equivalent to 14.5% of global anthropogenic greenhouse gas emissions (Gerber et al., 2013). Cattle (beef, milk) are responsible for about two-thirds of that total, largely due to methane emissions resulting from rumen fermentation (Gerber et al., 2013; Opio et al., 2013).

Despite ongoing gains in livestock productivity and volumes, the increase of animal products in global diets is restricting overall agricultural efficiency gains because of inefficiencies in the conversion of agricultural primary production (e.g., crops) in the feed-animal products pathway (Alexander et al., 2017), offsetting the benefits of improvements in livestock production systems (Clark and Tilman, 2017).

There is increasing agreement that overall emissions from food systems could be reduced by targeting the demand for meat and other livestock products, particularly where consumption is higher than suggested by human health guidelines. Adjusting diets to meet nutritional targets could bring large co-benefits, through GHG mitigation and improvements in the overall efficiency of food systems (Erb et al., 2009; Tukker et al., 2011; Tilman and Clark, 2014; van Dooren et al., 2014; Ranganathan et al., 2016). Dietary shifts could contribute one-fifth of the mitigation needed to hold warming below 2°C, with one-quarter of low-cost options (Griscom et al., 2017). There, however, remains *limited evidence* of effective policy interventions to achieve such large-scale shifts in dietary choices, and prevailing trends are for increasing rather than decreasing demand for livestock products at the global scale (Alexandratos and Bruinsma, 2012; OECD/FAO, 2017). How the role of dietary shift could change in 1.5°C-consistent pathways is also not clear (see Chapter 2).

4

Adaptation of livestock systems can include a suite of strategies such as using different breeds and their wild relatives to develop a genetic pool resilient to climatic shocks and longer-term temperature shifts (Thornton and Herrero, 2014), improving fodder and feed management (Bell et al., 2014; Havet et al., 2014) and disease prevention and control (Skuce et al., 2013; Nguyen et al., 2016). Most interventions that improve the productivity of livestock systems and enhance adaptation to climate changes would also reduce the emissions intensity of food production, with significant co-benefits for rural livelihoods and the security of food supplies (Gerber et al., 2013; FAO and NZAGRC, 2017a, b, c). Whether such reductions in emission intensity result in lower or higher absolute GHG emissions depends on overall demand for livestock products, indicating the relevance of integrating supply-side with demand-side measures within food security objectives (Gerber et al., 2013; Bajželj et al., 2014). Transitions in livestock production systems (e.g., from extensive to intensive) can also result in significant emission reductions as part of broader land-based mitigation strategies (Havlik et al., 2014).

Overall, there is *high agreement* that farm strategies that integrate mixed crop–livestock systems can improve farm productivity and have positive sustainability outcomes (Havet et al., 2014; Thornton and Herrero, 2014; Herrero et al., 2015; Weindl et al., 2015). Shifting towards mixed crop–livestock systems is estimated to reduce agricultural adaptation costs to 0.3% of total production costs while abating deforestation by 76 Mha globally, making it a highly cost-effective adaptation option with mitigation co-benefits (Weindl et al., 2015). Evidence from various regions supports this (Thornton and Herrero, 2015), although the feasible scale varies between regions and systems, as well as being moderated by overall demand in specific food products. In Australia, some farmers have successfully shifted to crop–livestock systems where, each year, they allocate land and forage resources in response to climate and price trends (Bell et al., 2014) . However, there can be some unintended negative impacts of such integration, including increased burdens on women, higher requirements of capital, competing uses of crop residues (e.g., feed vs. mulching vs. carbon sequestration) and higher requirements of management skills, which can be a challenge across several low income countries (Thornton and Herrero, 2015; Thornton et al., 2018). Finally, the feasibility of improving livestock efficiency is dependent on socio-cultural context and acceptability: there remain significant issues around widespread adoption of crossbred animals, especially by smallholders (Thornton et al., 2018).

Irrigation efficiency. Irrigation efficiency is especially critical since water endowments are expected to change, with 20–60 Mha of global cropland being projected to revert from irrigated to rain-fed land, while other areas will receive higher precipitation in shorter time spans, thus affecting irrigation demand (Elliott et al., 2014). While increasing irrigation system efficiency is necessary, there is mixed evidence on how to enact efficiency improvements (Fader et al., 2016; Herwehe and Scott, 2018). Physical and technical strategies include building large-scale reservoirs or dams, renovating or deepening irrigation channels, building on-farm rainwater harvesting structures, lining ponds, channels and tanks to reduce losses through percolation and evaporation, and investing in small infrastructure such as sprinkler or drip irrigation sets (Varela-Ortega et al., 2016; Sikka et al., 2018).

Each strategy has differing costs and benefits relating to unique biophysical, social, and economic contexts. Also, increasing irrigation efficiency may foster higher dependency on irrigation, resulting in a heightened sensitivity to climate that may be maladaptive in the long term (Lindoso et al., 2014).

Improvements in irrigation efficiency would need to be supplemented with ancillary activities, such as shifting to crops that require less water and improving soil and moisture conservation (Fader et al., 2016; Hong and Yabe, 2017; Sikka et al., 2018). Currently, the feasibility of improving irrigation efficiency is constrained by issues of replicability across scale and sustainability over time (Burney and Naylor, 2012), institutional barriers and inadequate market linkages (Pittock et al., 2017).

Growing evidence suggests that investing in behavioural shifts towards using irrigation technology such as micro-sprinklers or drip irrigation, is an effective and quick adaptation strategy (Varela-Ortega et al., 2016; Herwehe and Scott, 2018; Sikka et al., 2018) as opposed to large dams which have high financial, ecological and social costs (Varela-Ortega et al., 2016). While improving irrigation efficiency is technically feasible (R. Fishman et al., 2015) and has clear benefits for environmental values (Pfeiffer and Lin, 2014; R. Fishman et al., 2015), feasibility is regionally differentiated as shown by examples as diverse as Kansas (Jägermeyr et al., 2015), India (R. Fishman et al., 2015) and Africa (Pittock et al., 2017).

Agroforestry. The integration of trees and shrubs into crop and livestock systems, when properly managed, can potentially restrict soil erosion, facilitate water infiltration, improve soil physical properties and buffer against extreme events (Lasco et al., 2014; Mbow et al., 2014; Quandt et al., 2017; Sida et al., 2018). There is *medium evidence* and *high agreement* on the feasibility of agroforestry practices that enhance productivity, livelihoods and carbon storage (Lusiana et al., 2012; Murthy, 2013; Coulibaly et al., 2017; Sida et al., 2018), including from indigenous production systems (Coq-Huelva et al., 2017), with variation by region, agroforestry type, and climatic conditions (Place et al., 2012; Coe et al., 2014; Mbow et al., 2014; Iiyama et al., 2017; Abdulai et al., 2018). Long-term studies examining the success of agroforestry, however, are rare (Coe et al., 2014; Meijer et al., 2015; Brockington et al., 2016; Zomer et al., 2016).

The extent to which agroforestry practices employed at the farm level could be scaled up globally while satisfying growing food demand is relatively unknown. Agroforestry adoption has been relatively low and uneven (Jacobi et al., 2017; Hernández-Morcillo et al., 2018), with constraints including the expense of establishment and lack of reliable financial support, insecure land tenure, landowner's lack of experience with trees, complexity of management practices, fluctuating market demand and prices for different food and fibre products, the time and knowledge required for management, low intermediate benefits to offset revenue lags, and inadequate market access (Pattanayak et al., 2003; Mercer, 2004; Sendzimir et al., 2011; Valdivia et al., 2012; Coe et al., 2014; Meijer et al., 2015; Coulibaly et al., 2017; Jacobi et al., 2017).

Managing food loss and waste. The way food is produced, processed and transported strongly influences GHG emissions. Around

4

one-third of the food produced on the planet is not consumed (FAO, 2013), affecting food security and livelihoods (See Cross-Chapter Box 6 on Food Security in Chapter 3). Food wastage is a combination of food loss – the decrease in mass and nutritional value of food due to poor infrastructure, logistics, and lack of storage technologies and management – and food waste that derives from inappropriate human consumption that leads to food spoilage associated with inferior quality or overproduction. Food wastage could lead to an increase in emissions estimated to 1.9–2.5 $GtCO_2$-eq yr^{-1} (Hiç et al., 2016).

Decreasing food wastage has high mitigation and adaptation potential and could play an important role in land transitions towards 1.5°C, provided that reduced food waste results in lower production-side emissions rather than increased consumption (Foley et al., 2011). There is *medium agreement* that a combination of individual–institutional behaviour (Refsgaard and Magnussen, 2009; Thornton and Herrero, 2014), and improved technologies and management (Lin et al., 2013; Papargyropoulou et al., 2014) can transform food waste into products with marketable value. Institutional behaviour depends on investment and policies, which if adequately addressed could enable mitigation and adaptation co-benefits in a relatively short time.

Novel technologies. New molecular biology tools have been developed that can lead to fast and precise genome modification (De Souza et al., 2016; Scheben et al., 2016) (e.g., CRISPR Cas9; Ran et al., 2013; Schaeffer and Nakata, 2015). Such genome editing tools may moderately assist in mitigation and adaptation of agriculture in relation to climate changes, elevated CO_2, drought and flooding (DaMatta et al., 2010; De Souza et al., 2015, 2016). These tools could contribute to developing new plant varieties that can adapt to warming of 1.5°C and overshoot, potentially avoiding some of the costs of crop shifting (Schlenker and Roberts, 2009; De Souza et al., 2016). However, biosafety concerns and government regulatory systems can be a major barrier to the use of these tools as this increases the time and cost of turning scientific discoveries into ready applicable technologies (Andow and Zwahlen, 2006; Maghari and Ardekani, 2011).

The strategy of reducing enteric methane emissions by ruminants through the development of inhibitors or vaccines has already been attempted with some successes, although the potential for application at scale and in different situations remains uncertain. A methane inhibitor has been demonstrated to reduce methane from feedlot systems by 30% over a 12-week period (Hristov et al., 2015) with some productivity benefits, but the ability to apply it in grazing systems will depend on further technological developments as well as costs and incentives. A vaccine could potentially modify the microbiota of the rumen and be applicable even in extensive grazing systems by reducing the presence of methanogenic micro-organisms (Wedlock et al., 2013) but has not yet been successfully demonstrated to reduce emissions in live animals. Selective breeding for lower-emitting ruminants is becoming rapidly feasible, offering small but cumulative emissions reductions without requiring substantial changes in farm systems (Pickering et al., 2015).

Technological innovation in culturing marine and freshwater micro and macro flora has significant potential to expand food, fuel and fibre resources, and could reduce impacts on land and conventional agriculture (Greene et al., 2017).

Technological innovation could assist in increased agricultural efficiency (e.g., via precision agriculture), decrease food wastage and genetics that enhance plant adaptation traits (Section 4.4.4). Technological and associated management improvements may be ways to increase the efficiency of contemporary agriculture to help produce enough food to cope with population increases in a 1.5°C warmer world, and help reduce the pressure on natural ecosystems and biodiversity.

4.3.2.2   Forests and other ecosystems

Ecosystem restoration. Biomass stocks in tropical, subtropical, temperate and boreal biomes currently hold 1085, 194, 176, 190 Gt CO_2, respectively. Conservation and restoration can enhance these natural carbon sinks (Erb et al., 2017).

Recent studies explore options for conservation, restoration and improved land management estimating up to 23 $GtCO_2$ (Griscom et al., 2017). Mitigation potentials are dominated by reduced rates of deforestation, reforestation and forest management, and concentrated in tropical regions (Houghton, 2013; Canadell and Schulze, 2014; Grace et al., 2014; Houghton et al., 2015; Griscom et al., 2017). Much of the literature focuses on REDD+ (reducing emissions from deforestation and forest degradation) as an institutional mechanism. However, restoration and management activities need not be limited to REDD+, and locally adapted implementation may keep costs low, capitalize on co-benefits and ensure consideration of competing for socio-economic goals (Jantke et al., 2016; Ellison et al., 2017; Perugini et al., 2017; Spencer et al., 2017).

Half of the estimated potential can be achieved at <100 USD/tCO_2; and a third of the cost-effective potential at <10 USD/tCO_2 (Griscom et al., 2017). Variation of costs in projects aiming to reduce emissions from deforestation is high when considering opportunity and transaction costs (Dang Phan et al., 2014; Overmars et al., 2014; Ickowitz et al., 2017; Rakatama et al., 2017).

However, the focus on forests raises concerns of cross-biome leakage (*medium evidence, low agreement*) (Popp et al., 2014a; Strassburg et al., 2014; Jayachandran et al., 2017) and encroachment on other ecosystems (Veldman et al., 2015). Reducing rates of deforestation constrains the land available for agriculture and grazing, with trade-offs between diets, higher yields and food prices (Erb et al., 2016a; Kreidenweis et al., 2016). Forest restoration and conservation are compatible with biodiversity (Rey Benayas et al., 2009; Jantke et al., 2016) and available water resources; in the tropics, reducing rates of deforestation maintains cooler surface temperatures (Perugini et al., 2017) and rainfall (Ellison et al., 2017).

Its multiple potential co-benefits have made REDD+ important for local communities, biodiversity and sustainable landscapes (Ngendakumana et al., 2017; Turnhout et al., 2017). There is *low agreement* on whether climate impacts will reverse mitigation benefits of restoration (Le Page et al., 2013) by increasing the likelihood of disturbance (Anderegg et al., 2015), or reinforce them through carbon fertilization (P. Smith et al., 2014).

Emerging regional assessments offer new perspectives for upscaling. Strengthening coordination, additional funding sources, and access and disbursement points increase the potential of REDD+ in working towards 2°C and 1.5°C limits (Well and Carrapatoso, 2017). While there are indications that land tenure has a positive impact (Sunderlin et al., 2014), a meta-analysis by Wehkamp et al. (2018a) shows that there is *medium evidence* and *low agreement* on which aspects of governance improvements are supportive of conservation. Local benefits, especially for indigenous communities, will only be accrued if land tenure is respected and legally protected, which is not often the case (Sunderlin et al., 2014; Brugnach et al., 2017). Although payments for reduced rates of deforestation may benefit the poor, the most vulnerable populations could have limited, uneven access (Atela et al., 2014) and face lower opportunity costs from deforestation (Ickowitz et al., 2017).

Community-based adaptation (CbA). There is *medium evidence* and *high agreement* for the use of CbA. The specific actions to take will depend upon the location, context, and vulnerability of the specific community. CbA is defined as 'a community-led process, based on communities' priorities, needs, knowledge, and capacities, which aim to empower people to plan for and cope with the impacts of climate change' (Reid et al., 2009). The integration of CbA with ecosystems-based adaptation (EbA) has been increasingly promoted, especially in efforts to alleviate poverty (Mannke, 2011; Reid, 2016).

Despite the potential and advantages of both CbA and EbA, including knowledge exchange, information access and increased social capital and equity; institutional and governance barriers still constitute a challenge for local adaptation efforts (Wright et al., 2014; Fernández-Giménez et al., 2015).

Wetland management. In wetland ecosystems, temperature rise has direct and irreversible impacts on species functioning and distribution, ecosystem equilibrium and services, and second-order impacts on local livelihoods (see Chapter 3, Section 3.4.3). The structure and function of wetland systems are changing due to climate change. Wetland management strategies, including adjustments in infrastructural, behavioural, and institutional practices have clear implications for adaptation (Colloff et al., 2016b; Finlayson et al., 2017; Wigand et al., 2017)

Despite international initiatives on wetland restoration and management through the Ramsar Convention on Wetlands, policies have not been effective (Finlayson, 2012; Finlayson et al., 2017). Institutional reform, such as flexible, locally relevant governance, drawing on principles of adaptive co-management, and multi-stakeholder participation becomes increasingly necessary for effective wetland management (Capon et al., 2013; Finlayson et al., 2017).

4.3.2.3   Coastal systems

Managing coastal stress. Particularly to allow for the landward relocation of coastal ecosystems under a transition to a 1.5°C warmer world, planning for climate change would need to be integrated with the use of coastlines by humans (Saunders et al., 2014; Kelleway et al., 2017). Adaptation options for managing coastal stress include coastal

hardening through the building of seawalls and the re-establishment of coastal ecosystems such as mangroves (André et al., 2016; Cooper et al., 2016). While the feasibility of the solutions is high, they are expensive to scale (*robust evidence, medium agreement*).

There is *low evidence* and *high agreement* that reducing the impact of local stresses (Halpern et al., 2015) will improve the resilience of marine ecosystems as they transition to a 1.5°C world (O'Leary et al., 2017). Approaches to reducing local stresses are considered feasible, cost-effective and highly scalable. Ecosystem resilience may be increased through alternative livelihoods (e.g., sustainable aquaculture), which are among a suite of options for building resilience in coastal ecosystems. These options enjoy high levels of feasibility yet are expensive, which stands in the way of scalability (*robust evidence, medium agreement*) (Hiwasaki et al., 2015; Brugnach et al., 2017).

Working with coastal communities has the potential for improving the resilience of coastal ecosystems. Combined with the advantages of using indigenous knowledge to guide transitions, solutions can be more effective when undertaken in partnership with local communities, cultures, and knowledge (See Box 4.3).

Restoration of coastal ecosystems and fisheries. Marine restoration is expensive compared to terrestrial restoration, and the survival of projects is currently low, with success depending on the ecosystem and site, rather than the size of the financial investment (Bayraktarov et al., 2016). Mangrove replanting shows evidence of success globally, with numerous examples of projects that have established forests (Kimball et al., 2015; Bayraktarov et al., 2016).

Efforts with reef-building corals have been attempted with a low level of success (Bayraktarov et al., 2016). Technologies to help re-establish coral communities are limited (Rinkevich, 2014), as are largely untested disruptive technologies (e.g., genetic manipulation, assisted evolution) (van Oppen et al., 2015). Current technologies also have trouble scaling given the substantial costs and investment required (Bayraktarov et al., 2016).

Johannessen and Macdonald (2016) report the 'blue carbon' sink to be 0.4–0.8% of global anthropogenic emissions. However, this does not adequately account for post-depositional processes and could overestimate removal potentials, subject to a risk of reversal. Seagrass beds will thus not contribute significantly to enabling 1.5°C-consistent pathways.

4.3.3   Urban and Infrastructure System Transitions

There will be approximately 70 million additional urban residents every year through to the middle part of this century (UN DESA, 2014). The majority of these new urban citizens will reside in small and medium-sized cities in low- and middle-income countries (Cross-Chapter Box 13 in Chapter 5). The combination of urbanization and economic and infrastructure development could account for an additional 226 $GtCO_2$ by 2050 (Bai et al. 2018). However, urban systems can harness the mega-trends of urbanization, digitalization, financialization and growing sub-national commitment to smart cities, green cities, resilient cities, sustainable cities and adaptive cities, for the type of transformative change required by 1.5°C-consistent pathways (SDSN, 2013; Parag and Sovacool, 2016; Roberts, 2016; Wachsmuth et al., 2016; Revi, 2017; Solecki et al., 2018). There is a growing number of urban climate responses driven by cost-effectiveness, development, work creation and inclusivity considerations (Solecki et al., 2013; Ahern et al., 2014; Floater et al., 2014; Revi et al., 2014a; Villarroel Walker et al., 2014; Kennedy et al., 2015; Rodríguez, 2015; McGranahan et al., 2016; Dodman et al., 2017a; Newman et al., 2017; UN-Habitat, 2017; Westphal et al., 2017).

In addition, low-carbon cities could reduce the need to deploy carbon dioxide removal (CDR) and solar radiation modification (SRM) (Fink, 2013; Thomson and Newman, 2016).

Cities are also places in which the risks associated with warming of 1.5°C, such as heat stress, terrestrial and coastal flooding, new disease vectors, air pollution and water scarcity, will coalesce (see Chapter 3, Section 3.3) (Dodman et al., 2017a; Satterthwaite and Bartlett, 2017). Unless adaptation and mitigation efforts are designed around the need to decarbonize urban societies in the developed world and provide low-carbon solutions to the needs of growing urban populations in developing countries, they will struggle to deliver the pace or scale of change required by 1.5°C-consistent pathways (Hallegatte et al., 2013; Villarroel Walker et al., 2014; Roberts, 2016; Solecki et al., 2018). The pace and scale of urban climate responses can be enhanced by attention to social equity (including gender equity), urban ecology (Brown and McGranahan, 2016; Wachsmuth et al., 2016; Ziervogel et al., 2016a) and participation in sub-national networks for climate action (Cole, 2015; Jordan et al., 2015).

The long-lived urban transport, water and energy systems that will be constructed in the next three decades to support urban populations in developing countries and to retrofit cities in developed countries will have to be different to those built in Europe and North America in the 20th century, if they are to support the required transitions (Freire et al., 2014; Cartwright, 2015; McPhearson et al., 2016; Roberts, 2016; Lwasa, 2017). Recent literature identifies energy, infrastructure, appliances, urban planning, transport and adaptation options as capable of facilitating systemic change. It is these aspects of the urban system that are discussed below and from which options in Section 4.5 are selected.

4.3.3.1 Urban energy systems

Urban economies tend to be more energy intensive than national economies due to higher levels of per capita income, mobility and consumption (Kennedy et al., 2015; Broto, 2017; Gota et al., 2018). However, some urban systems have begun decoupling development from the consumption of fossil fuel-powered energy through energy efficiency, renewable energy and locally managed smart grids (Dodman, 2009; Freire et al., 2014; Eyre et al., 2018; Glazebrook and Newman, 2018).

The rapidly expanding cities of Africa and Asia, where energy poverty currently undermines adaptive capacity (Westphal et al., 2017; Satterthwaite et al., 2018), have the opportunity to benefit from recent

price changes in renewable energy technologies to enable clean energy access to citizens (SDG 7) (Cartwright, 2015; Watkins, 2015; Lwasa, 2017; Kennedy et al., 2018; Teferi and Newman, 2018). This will require strengthened energy governance in these countries (Eberhard et al., 2017). Where renewable energy displaces paraffin, wood fuel or charcoal feedstocks in informal urban settlements, it provides the co-benefits of improved indoor air quality, reduced fire risk and reduced deforestation, all of which can enhance adaptive capacity and strengthen demand for this energy (Newham and Conradie, 2013; Winkler, 2017; Kennedy et al., 2018; Teferi and Newman, 2018).

4.3.3.2 Urban infrastructure, buildings and appliances

Buildings are responsible for 32% of global energy consumption (IEA, 2016c) and have a large energy saving potential with available and demonstrated technologies such as energy efficiency improvements in technical installations and in thermal insulation (Toleikyte et al., 2018) and energy sufficiency (Thomas et al., 2017). Kuramochi et al. (2018) show that 1.5°C-consistent pathways require building emissions to be reduced by 80–90% by 2050, new construction to be fossil-free and near-zero energy by 2020, and an increased rate of energy refurbishment of existing buildings to 5% per annum in OECD (Organisation for Economic Co-operation and Development) countries (see also Section 4.2.1).

Based on the IEA-ETP (IEA, 2017g), Chapter 2 identifies large saving potential in heating and cooling through improved building design, efficient equipment, lighting and appliances. Several examples of net zero energy in buildings are now available (Wells et al., 2018). In existing buildings, refurbishment enables energy saving (Semprini et al., 2017; Brambilla et al., 2018; D'Agostino and Parker, 2018; Sun et al., 2018) and cost savings (Toleikyte et al., 2018; Zangheri et al., 2018).

Reducing the energy embodied in building materials provides further energy and GHG savings (Cabeza et al., 2013; Oliver and Morecroft, 2014; Koezjakov et al., 2018), in particular through increased use of bio-based materials (Lupíšek et al., 2015) and wood construction (Ramage et al., 2017). The United Nations Environment Programme (UNEP[3]) estimates that improving embodied energy, thermal performance, and direct energy use of buildings can reduce emissions by 1.9 GtCO$_2$e yr^{-1} (UNEP, 2017b), with an additional reduction of 3 GtCO$_2$e yr^{-1} through energy efficient appliances and lighting (UNEP, 2017b). Further increasing the energy efficiency of appliances and lighting, heating and cooling offers the potential for further savings (Parikh and Parikh, 2016; Garg et al., 2017).

Smart technology, drawing on the internet of things (IoT) and building information modelling, offers opportunities to accelerate energy efficiency in buildings and cities (Moreno-Cruz and Keith, 2013; Hoy, 2016) (see also Section 4.4.4). Some cities in developing countries are drawing on these technologies to adopt 'leapfrog' infrastructure, buildings and appliances to pursue low-carbon development (Newman et al., 2017; Teferi and Newman, 2017) (Cross-Chapter Box 13 in Chapter 5).

---

[3] Currently called UN Environment.

4.3.3.3   Urban transport and urban planning

Urban form impacts demand for energy (Sims et al., 2014) and other welfare related factors: a meta-analysis of 300 papers reported energy savings of 26 USD per person per year attributable to a 10% increase in urban population density (Ahlfeldt and Pietrostefani, 2017). Significant reductions in car use are associated with dense, pedestrianized cities and towns and medium-density transit corridors (Newman and Kenworthy, 2015; Newman et al., 2017) relative to low-density cities in which car dependency is high (Schiller and Kenworthy, 2018). Combined dense urban forms and new mass transit systems in Shanghai and Beijing have yielded less car use (Gao and Newman, 2018) (see Box 4.9). Compact cities also create the passenger density required to make public transport more financially viable (Rode et al., 2017; Ahlfeldt and Pietrostefani, 2017) and enable combinations of cleaner fuel feedstocks and urban smart grids, in which vehicles form part of the storage capacity (Oldenbroek et al., 2017). Similarly, the spatial organization of urban energy influenced the trajectories of urban development in cities as diverse as Hong Kong, Bengaluru and Maputo (Broto, 2017).

The informal settlements of middle- and low-income cities, where urban density is more typically associated with a range of water- and vector-borne health risks, may provide a notable exception to the adaptive advantages of urban density (Mitlin and Satterthwaite, 2013; Lilford et al., 2017) unless new approaches and technologies are harnessed to accelerate slum upgrading (Teferi and Newman, 2017).

Scenarios consistent with 1.5°C depend on a roughly 15% reduction in final energy use by the transport sector by 2050 relative to 2015 (Chapter 2, Figure 2.12). In one analysis the phasing out of fossil fuel passenger vehicle sales by 2035–2050 was identified as a benchmark for aligning with 1.5°C-consistent pathways (Kuramochi et al., 2018). Reducing emissions from transport has lagged the power sector (Sims et al., 2014; Creutzig et al., 2015a), but evidence since AR5 suggests that cities are urbanizing and re-urbanizing in ways that coordinate transport sector adaptation and mitigation (Colenbrander et al., 2017; Newman et al., 2017; Salvo et al., 2017; Gota et al., 2018). The global transport sector could reduce 4.7 GtCO$_2$e yr^{-1} (4.1–5.3) by 2030. This is significantly more than is predicted by integrated assessment models (UNEP, 2017b). Such a transition depends on cities that enable modal shifts and avoided journeys and that provide incentives for uptake of improved fuel efficiency and changes in urban design that encourage walkable cities, non-motorized transport and shorter commuter distances (IEA, 2016a; Mittal et al., 2016; Zhang et al., 2016; Li and Loo, 2017). In at least 4 African cities, 43 Asian cities and 54 Latin American cities, transit-oriented development (TOD), has emerged as an organizing principle for urban growth and spatial planning (Colenbrander et al., 2017; Lwasa, 2017; BRTData, 2018). This trend is important to counter the rising demand for private cars in developing-country cities (AfDB/OECD/UNDP, 2016). In India, TOD has been combined with localized solar PV installations and new ways of financing rail expansion (Sharma, 2018).

Cities pursuing sustainable transport benefit from reduced air pollution, congestion and road fatalities and are able to harness the relationship between transport systems, urban form, urban energy intensity and social cohesion (Goodwin and Van Dender, 2013; Newman and Kenworthy, 2015; Wee, 2015).

Technology and electrification trends since AR5 make carbon-efficient urban transport easier (Newman et al., 2016), but realizing urban transport's contribution to a 1.5°C-consistent pathways will require the type of governance that can overcome the financial, institutional, behavioural and legal barriers to change (Geels, 2014; Bakker et al., 2017).

Adaptation to a 1.5°C world is enabled by urban design and spatial planning policies that consider extreme weather conditions and reduce displacement by climate related disasters (UNISDR, 2009; UN-Habitat, 2011; Mitlin and Satterthwaite, 2013).

Building codes and technology standards for public lighting, including traffic lights (Beccali et al., 2015), play a critical role in reducing carbon emissions, enhancing urban climate resilience and managing climate risk (Steenhof and Sparling, 2011; Parnell, 2015; Shapiro, 2016; Evans et al., 2017). Building codes can support the convergence to zero emissions from buildings (Wells et al., 2018) and can be used retrofit the existing building stock for energy efficiency (Ruparathna et al., 2016).

The application of building codes and standards for 1.5°C-consistent pathways will require improved enforcement, which can be a challenge in developing countries where inspection resources are often limited and codes are poorly tailored to local conditions (Ford et al., 2015c; Chandel et al., 2016; Eisenberg, 2016; Shapiro, 2016; Hess and Kelman, 2017; Mavhura et al., 2017). In all countries, building codes can be undermined by industry interests and can be maladaptive if they prevent buildings or land use from evolving to reduce climate impacts (Eisenberg, 2016; Shapiro, 2016).

The deficit in building codes and standards in middle-income and developing-country cities need not be a constraint to more energy-efficient and resilient buildings (Tait and Euston-Brown, 2017). For example, the relatively high price that poor households pay for unreliable and at times dangerous household energy in African cities has driven the uptake of renewable energy and energy efficiency technologies in the absence of regulations or fiscal incentives (Eberhard et al., 2011, 2016; Cartwright, 2015; Watkins, 2015). The Kuyasa Housing Project in Khayelitsha, one of Cape Town's poorest suburbs, created significant mitigation and adaptation benefits by installing ceilings, solar water heaters and energy-efficient lightbulbs in houses independent of the formal housing or electrification programme (Winkler, 2017).

4.3.3.4   Electrification of cities and transport

The electrification of urban systems, including transport, has shown global progress since AR5 (IEA, 2016a; Kennedy et al., 2018; Schiller and Kenworthy, 2018). High growth rates are now appearing in electric vehicles (Figure 4.1), electric bikes and electric transit (IEA, 2018), which would need to displace fossil fuel-powered passenger vehicles by 2035–2050 to remain in line with 1.5°C-consistent pathways. China's 2017 Road Map calls for 20% of new vehicle

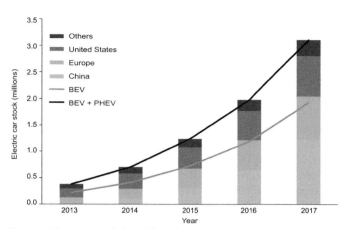

Figure 4.1 | Increase of the global electric car stock by country (2013–2017). The grey line is battery electric vehicles (BEV) only while the black line includes both BEV and plug-in hybrid vehicles (PHEV). Source: (IEA, 2018). Based on IEA data from Global EV Outlook 2018 © OECD/IEA 2018, IEA Publishing.

sales to be electric. India is aiming for exclusively electric vehicles (EVs) by 2032 (NITI Aayog and RMI, 2017). Globally, EV sales were up 42% in 2016 relative to 2015, and in the United States EV sales were up 36% over the same period (Johnson and Walker, 2016).

The extent of electric railways in and between cities has expanded since AR5 (IEA, 2016a; Mittal et al., 2016; Zhang et al., 2016; Li and Loo, 2017). In high-income cities there is *medium evidence* for the decoupling of car use and wealth since AR5 (Newman, 2017). In cities where private vehicle ownership is expected to increase, less carbon-intensive fuel sources and reduced car journeys will be necessary as well as electrification of all modes of transport (Mittal et al., 2016; van Vuuren et al., 2017). Some recent urban data show a decoupling of urban growth and GHG emissions (Newman and Kenworthy, 2015) and that 'peak car' has been reached in Shanghai and Beijing (Gao and Kenworthy, 2017) and beyond (Manville et al., 2017) (also see Box 4.9).

An estimated 800 cities globally have operational bike-share schemes (E. Fishman et al., 2015), and China had 250 million electric bicycles in 2017 (Newman et al., 2017). Advances in information and communication technologies (ICT) offer cities the chance to reduce urban transport congestion and fuel consumption by making better use of the urban vehicle fleet through car sharing, driverless cars and coordinated public transport, especially when electrified (Wee, 2015; Glazebrook and Newman, 2018). Advances in 'big-data' can assist in creating a better understanding of the connections between cities, green infrastructure, environmental services and health (Jennings et al., 2016) and improve decision-making in urban development (Lin et al., 2017).

4.3.3.5 Shipping, freight and aviation

International transport hubs, including airports and ports, and the associated mobility of people are major economic contributors to most large cities even while under the governance of national authorities and international legislation. Shipping, freight and aviation systems have grown rapidly, and little progress has been made since AR5 on replacing fossil fuels, though some trials are continuing (Zhang, 2016; Bouman et al., 2017; EEA, 2017). Aviation emissions do not yet feature

in IAMs (Bows-Larkin, 2015), but could be reduced by between a third and two-thirds through energy efficiency measures and operational changes (Dahlmann et al., 2016). On shorter intercity trips, aviation could be replaced by high-speed electric trains drawing on renewable energy (Åkerman, 2011). Some progress has been made on the use of electricity in planes and shipping (Grewe et al., 2017) though no commercial applications have arisen. Studies indicate that biofuels are the most viable means of decarbonizing intercontinental travel, given their technical characteristics, energy content and affordability (Wise et al., 2017). The lifecycle emissions of bio-based jet fuels and marine fuels can be considerable (Cox et al., 2014; IEA, 2017g) depending on their location (Elshout et al., 2014), but can be reduced by feedstock and conversion technology choices (de Jong et al., 2017).

In recent years the potential for transport to use synfuels, such as ethanol, methanol, methane, ammonia and hydrogen, created from renewable electricity and CO_2, has gained momentum but has not yet demonstrated benefits on a scale consistent with 1.5°C pathways (Ezeji, 2017; Fasihi et al., 2017). Decarbonizing the fuel used by the world's 60,000 large ocean vessels faces governance barriers and the need for a global policy (Bows and Smith, 2012; IRENA, 2015; Rehmatulla and Smith, 2015). Low-emission marine fuels could simultaneously address sulphur and black carbon issues in ports and around waterways and accelerate the electrification of all large ports (Bouman et al., 2017; IEA, 2017g).

4.3.3.6 Climate-resilient land use

Urban land use influences energy intensity, risk exposure and adaptive capacity (Carter et al., 2015; Araos et al., 2016a; Ewing et al., 2016; Newman et al., 2016; Broto, 2017). Accordingly, urban land-use planning can contribute to climate mitigation and adaptation (Parnell, 2015; Francesch-Huidobro et al., 2017) and the growing number of urban climate adaptation plans provide instruments to do this (Carter et al., 2015; Dhar and Khirfan, 2017; Siders, 2017; Stults and Woodruff, 2017). Adaptation plans can reduce exposure to urban flood risk (which, in a 1.5°C world, could double relative to 1976–2005; Alfieri et al., 2017), heat stress (Chapter 3, Section 3.5.5.8), fire risk (Chapter 3, Section 3.4.3.4) and sea level rise (Chapter 3, Section 3.4.5.1) (Schleussner et al., 2016).

Cities can reduce their risk exposure by considering investment in infrastructure and buildings that are more resilient to warming of 1.5°C or beyond. Where adaptation planning and urban planning generate the type of local participation that enhances capacity to cope with risks, they can be mutually supportive processes (Archer et al., 2014; Kettle et al., 2014; Campos et al., 2016; Chu et al., 2017; Siders, 2017; Underwood et al., 2017). Not all adaptation plans are reported as effective (Measham et al., 2011; Hetz, 2016; Woodruff and Stults, 2016; Mahlkow and Donner, 2017), especially in developing-country cities (Kiunsi, 2013). In cases where adaptation planning may further marginalize poor citizens, either through limited local control over adaptation priorities or by displacing impacts onto poorer communities, successful urban risk management would need to consider factors such as justice, equity, and inclusive participation, as well as recognize the political economy of adaptation (Archer, 2016; Shi et al., 2016; Ziervogel et al., 2016a, 2017; Chu et al., 2017).

4

4.3.3.7   Green urban infrastructure and ecosystem services

Integrating and promoting green urban infrastructure (including street trees, parks, green roofs and facades, and water features) into city planning can be difficult (Leck et al., 2015) but increases urban resilience to impacts of 1.5°C warming (Table 4.2) in ways that can be more cost-effective than conventional infrastructure (Cartwright et al., 2013; Culwick and Bobbins, 2016).

Realizing climate benefits from urban green infrastructure sometimes requires a city-region perspective (Wachsmuth et al., 2016). Where the urban impact on ecological systems in and beyond the city is appreciated, the potential for transformative change exists (Soderlund and Newman, 2015; Ziervogel et al., 2016a), and a locally appropriate combination of green space, ecosystem goods and services and the built environment can increase the set of urban adaptation options (Puppim de Oliveira et al., 2013).

Milan, Italy, a city with deliberate urban greening policies, planted 10,000 hectares of new forest and green areas over the last two decades (Sanesi et al., 2017). The accelerated growth of urban trees, relative to rural trees, in several regions of the world is expected to decrease tree longevity (Pretzsch et al., 2017), requiring monitoring and additional management of urban trees if their contribution to urban ecosystem-based adaptation and mitigation is to be maintained in a 1.5°C world (Buckeridge, 2015; Pretzsch et al., 2017).

4.3.3.8   Sustainable urban water and environmental services

Urban water supply and wastewater treatment is energy intensive and currently accounts for significant GHG emissions (Nair et al., 2014). Cities can integrate sustainable water resource management and the supply of water services in ways that support mitigation, adaptation and development through waste water recycling and storm water diversion (Xue et al., 2015; Poff et al., 2016). Governance and finance challenges complicate balancing sustainable water supply and rising urban demand, particularly in low-income cities (Bettini et al., 2015; Deng and Zhao, 2015; Hill Clarvis and Engle, 2015; Lemos, 2015; Margerum and Robinson, 2015).

Urban surface-sealing with impervious materials affects the volume and velocity of runoff and flooding during intense rainfall (Skougaard Kaspersen et al., 2015), but urban design in many cities now seeks to mediate runoff, encourage groundwater recharge and enhance water quality (Liu et al., 2014; Lamond et al., 2015; Voskamp and Van de Ven, 2015; Costa et al., 2016; Mguni et al., 2016; Xie et al., 2017). Challenges remain for managing intense rainfall events that are reported to be increasing in frequency and intensity in some locations (Ziervogel et al., 2016b), and urban flooding is expected to increase at 1.5°C of warming (Alfieri et al., 2017). This risk falls disproportionately on women and poor people in cities (Mitlin, 2005; Chu et al., 2016; Ziervogel et al., 2016b; Chant et al., 2017; Dodman et al., 2017a, b).

Nexus approaches that highlight urban areas as socio-ecological systems can support policy coherence (Rasul and Sharma, 2016) and sustainable urban livelihoods (Biggs et al., 2015). The water–energy–food (WEF) nexus is especially important to growing urban populations (Tacoli et al., 2013; Lwasa et al., 2014; Villarroel Walker et al., 2014).

4.3.4   Industrial Systems Transitions

Industry consumes about one-third of global final energy and contributes, directly and indirectly, about one-third of global GHG emissions (IPCC, 2014b). If the increase in global mean temperature is to remain under 1.5°C, modelling indicates that industry cannot emit more than 2 $GtCO_2$ in 2050, corresponding to a reduction of between 67 and 91% (interquartile range) in GHG emissions compared to 2010 (see Chapter 2, Figures 2.20 and 2.21, and Table 4.1). Moreover, the consequences of warming of 1.5°C or more pose substantial challenges for industrial diversity. This section will first briefly discuss the limited literature on adaptation options for industry. Subsequently, new literature since AR5 on the feasibility of industrial mitigation options will be discussed.

Research assessing adaptation actions by industry indicates that only a small fraction of corporations has developed adaptation measures. Studies of adaptation in the private sector remain limited (Agrawala et al., 2011; Linnenluecke et al., 2015; Averchenkova et al., 2016; Bremer and Linnenluecke, 2016; Pauw et al., 2016a) and for 1.5°C are largely

Table 4.2 | Green urban infrastructure and benefits

Green Infrastructure	Adaptation Benefits	Mitigation Benefits	References
Urban tree planting, urban parks	Reduced heat island effect, psychological benefits	Less cement, reduced air-conditioning use	Demuzere et al., 2014; Mullaney et al., 2015; Soderlund and Newman, 2015; Beaudoin and Gosselin, 2016; Green et al., 2016; Lin et al., 2017
Permeable surfaces	Water recharge	Less cement in city, some bio-sequestration, less water pumping	Liu et al., 2014; Lamond et al., 2015; Skougaard Kaspersen et al., 2015; Voskamp and Van de Ven, 2015; Costa et al., 2016; Mguni et al., 2016; Xie et al., 2017
Forest retention, urban agricultural land	Flood mediation, healthy lifestyles	Reduced air pollution	Nowak et al., 2006; Tallis et al., 2011; Elmqvist et al., 2013; Buckeridge, 2015; Culwick and Bobbins, 2016; Panagopoulos et al., 2016; Stevenson et al., 2016; R. White et al., 2017
Wetland restoration, riparian buffer zones	Reduced urban flooding, low-skilled local work, sense of place	Some bio-sequestration, less energy spent on water treatment	Cartwright et al., 2013; Elmqvist et al., 2015; Brown and McGranahan, 2016; Camps-Calvet et al., 2016; Culwick and Bobbins, 2016; McPhearson et al., 2016; Ziervogel et al., 2016b; Collas et al., 2017; F. Li et al., 2017
Biodiverse urban habitat	Psychological benefits, inner-city recreation	Carbon sequestration	Beatley, 2011; Elmqvist et al., 2015; Brown and McGranahan, 2016; Camps-Calvet et al., 2016; McPhearson et al., 2016; Collas et al., 2017; F. Li et al., 2017

absent. This knowledge gap is particularly evident for medium-sized enterprises and in low- and middle-income nations (Surminski, 2013).

Depending on the industrial sector, mitigation consistent with 1.5°C would mean, across industries, a reduction of final energy demand by one-third, an increase of the rate of recycling of materials and the development of a circular economy in industry (Lewandowski, 2016; Linder and Williander, 2017), the substitution of materials in high-carbon products with those made up of renewable materials (e.g., wood instead of steel or cement in the construction sector, natural textile fibres instead of plastics), and a range of deep emission reduction options, including use of bio-based feedstocks, low-emission heat sources, electrification of production processes, and/or capture and storage of all CO_2 emissions by 2050 (Åhman et al., 2016). Some of the choices for mitigation options and routes for GHG-intensive industry are discrete and potentially subject to path dependency: if an industry goes one way (e.g., in keeping existing processes), it will be harder to transition to process change (e.g., electrification) (Bataille et al., 2018). In the context of rising demand for construction, an increasing share of industrial production may be based in developing countries (N. Li et al., 2017), where current efficiencies may be lower than in developed countries, and technical and institutional feasibility may differ (Ma et al., 2015).

Except for energy efficiency, costs of disruptive change associated with hydrogen- or electricity-based production, bio-based feedstocks and carbon dioxide capture, (utilization) and storage (CC(U)S) for trade-sensitive industrial sectors (in particular the iron and steel, petrochemical and refining industries) make policy action by individual countries challenging because of competitiveness concerns (Åhman et al., 2016; Nabernegg et al., 2017).

Table 4.3 provides an overview of applicable mitigation options for key industrial sectors.

4.3.4.1   Energy efficiency

Isolated efficiency implementation in energy-intensive industries is a necessary but insufficient condition for deep emission reductions (Napp et al., 2014; Aden, 2018). Various options specific to different industries

are available. In general, their feasibility depends on lowering capital costs and raising awareness and expertise (Wesseling et al., 2017). General-purpose technologies, such as ICT, and energy management tools can improve the prospects of energy efficiency in industry (see Section 4.4.4).

Cross-sector technologies and practices, which play a role in all industrial sectors including small- and medium-sized enterprises (SMEs) and non-energy intensive industry, also offer potential for considerable energy efficiency improvements. They include: (i) motor systems (for example electric motors, variable speed drives, pumps, compressors and fans), responsible for about 10% of worldwide industrial energy consumption, with a global energy efficiency improvement potential of around 20–25% (Napp et al., 2014); and (ii) steam systems, responsible for about 30% of industrial energy consumption and energy saving potentials of about 10% (Hasanbeigi et al., 2014; Napp et al., 2014). Waste heat recovery from industry has substantial potential for energy efficiency and emission reduction (Forman et al., 2016). Low awareness and competition from other investments limit the feasibility of such options (Napp et al., 2014).

4.3.4.2   Substitution and circularity

Recycling materials and developing a circular economy can be institutionally challenging, as it requires advanced capabilities (Henry et al., 2006) and organizational changes (Cooper-Searle et al., 2018), but has advantages in terms of cost, health, governance and environment (Ali et al., 2017). An assessment of the impacts on energy use and environmental issues is not available, but substitution could play a large role in reducing emissions (Åhman et al., 2016) although its potential depends on the demand for material and the turnover rate of, for example, buildings (Haas et al., 2015). Material substitution and CO_2 storage options are under development, for example, the use of algae and renewable energy for carbon fibre production, which could become a net sink of CO_2 (Arnold et al., 2018).

4.3.4.3   Bio-based feedstocks

Bio-based feedstock processes could be seen as part of the circular materials economy (see section above). In several sectors, bio-based

Table 4.3 | Overview of different mitigation options potentially consistent with limiting warming to 1.5°C and applicable to main industrial sectors, including examples of application (Napp et al., 2014; Boulamanti and Moya, 2017; Wesseling et al., 2017).

Industrial mitigation option	Iron/Steel	Cement	Refineries and Petrochemicals	Chemicals
Process and Energy Efficiency	Can make a difference of between 10% and 50%, depending on the plant. Relevant but not enough for 1.5°C			
Bio-based	Coke can be made from biomass instead of coal	Partial (only energy-related emissions)	Biomass can replace fossil feedstocks	
Circularity & Substitution	More recycling and replacement by low-emission materials, including alternative chemistries for cement		Limited potential	
Electrification & Hydrogen	Direct reduction with hydrogen Heat generation through electricity	Partial (only electrified heat generation)	Electrified heat and hydrogen generation	
Carbon dioxide capture, utilization and storage	Possible for process emissions and energy. Reduces emissions by 80–95%, and net emissions can become negative when combined with biofuel		Can be applied to energy emissions and different stacks but not on emissions of products in the use phase (e.g., gasoline)	

feedstocks would leave the production process of materials relatively untouched, and a switch would not affect the product quality, making the option more attractive. However, energy requirements for processing bio-based feedstocks are often high, costs are also still higher, and the emissions over the full life cycle, both upstream and downstream, could be significant (Wesseling et al., 2017). Bio-based feedstocks may put pressure on natural resources by increasing land demand by biodiversity impacts beyond bioenergy demand for electricity, transport and buildings (Slade et al., 2014), and, partly as a result, face barriers in public acceptance (Sleenhoff et al., 2015).

4.3.4.4 Electrification and hydrogen

Electrification of manufacturing processes would constitute a significant technological challenge and would entail a more disruptive innovation in industry than bio-based or CCS options to get to very low or zero emissions, except potentially in steel-making (Philibert, 2017). The disruptive characteristics could potentially lead to stranded assets, and could reduce political feasibility and industry support (Åhman et al., 2016). Electrification of manufacturing would require further technological development in industry, as well as an ample supply of cost-effective low-emission electricity (Philibert, 2017).

Low-emission hydrogen can be produced by natural gas with CCS, by electrolysis of water powered by zero-emission electricity, or potentially in the future by generation IV nuclear reactors. Feasibility of electrification and use of hydrogen in production processes or fuel cells is affected by technical development (in terms of efficient hydrogen production and electrification of processes), by geophysical factors related to the availability of low-emission electricity (MacKay, 2013), by associated public perception and by economic feasibility, except in areas with ample solar and/or wind resources (Philibert, 2017; Wesseling et al., 2017).

4.3.4.5 CO_2 capture, utilization and storage in industry

CO_2 capture in industry is generally considered more feasible than CCS in the power sector (Section 4.3.1) or from bioenergy sources (Section 4.3.7), although CCS in industry faces similar barriers. Almost all of the current full-scale ($>1MtCO_2$ yr^{-1}) CCS projects capture CO_2 from industrial sources, including the Sleipner project in Norway, which has been injecting CO_2 from a gas facility in an offshore saline formation since 1996 (Global CCS Institute, 2017). Compared to the power sector, retrofitting CCS on existing industrial plants would leave the production process of materials relatively untouched (Åhman et al., 2016), though significant investments and modifications still have to be made. Some industries, in particular cement, emit CO_2 as inherent process emissions and can therefore not reduce emissions to zero without CC(U)S. CO_2 stacks in some industries have a high economic and technical feasibility for CO_2 capture as the CO_2 concentration in the exhaust gases is relatively high (IPCC, 2005b; Leeson et al., 2017), but others require strong modifications in the production process, limiting technical and economic feasibility, though costs remain lower than other deep GHG reduction options (Rubin et al., 2015). There are indications that the energy use in CO_2 capture through amine solvents (for solvent regeneration) can decrease by around 60%, from 5 GJ tCO_2^{-1} in 2005 to 2 GJ tCO_2^{-1} in the best-performing

current pilot plants (Idem et al., 2015), increasing both technical and economic potential for this option. The heterogeneity of industrial production processes might point to the need for specific institutional arrangements to incentivize industrial CCS (Mikunda et al., 2014), and may decrease institutional feasibility.

Whether carbon dioxide utilization (CCU) can contribute to limiting warming to 1.5°C depends on the origin of the CO_2 (fossil, biogenic or atmospheric), the source of electricity for converting the CO_2 or regenerating catalysts, and the lifetime of the product. Review studies indicate that CO_2 utilization in industry has a small role to play in limiting warming to 1.5°C because of the limited potential of reusing CO_2 with currently available technologies and the re-emission of CO_2 when used as a fuel (IPCC, 2005b; Mac Dowell et al., 2017). However, new developments could make CCU more feasible, in particular in CO_2 use as a feedstock for carbon-based materials that would isolate CO_2 from the atmosphere for a long time, and in low-cost, low-emission electricity that would make the energy use of CO_2 capture more sustainable. The conversion of CO_2 to fuels using zero-emission electricity has a lower technical, economic and environmental feasibility than direct CO_2 capture and storage from industry (Abanades et al., 2017), although the economic prospects have improved recently (Philibert, 2017).

4.3.5 Overarching Adaptation Options Supporting Adaptation Transitions

This section assesses overarching adaptation options –specific solutions from which actors can choose and make decisions to reduce climate vulnerability and build resilience. We examine their feasibility in the context of transitions of energy, land and ecosystem, urban and infrastructure, and industrial systems here, and further in Section 4.5. These options can contribute to creating an enabling environment for adaptation (see Table 4.4 and Section 4.4).

4.3.5.1 Disaster risk management (DRM)

DRM is a process for designing, implementing and evaluating strategies, policies and measures to improve the understanding of disaster risk, and promoting improvement in disaster preparedness, response and recovery (IPCC, 2012). There is increased demand to integrate DRM and adaptation (Howes et al., 2015; Kelman et al., 2015; Serrao-Neumann et al., 2015; Archer, 2016; Rose, 2016; van der Keur et al., 2016; Kelman, 2017; Wallace, 2017) to reduce vulnerability, but institutional, technical and financial capacity challenges in frontline agencies constitute constraints (*medium evidence, high agreement*) (Eakin et al., 2015; Kita, 2017; Wallace, 2017).

4.3.5.2 Risk sharing and spreading

Risks associated with 1.5°C warming (Chapter 3, Section 3.4) may increase the demand for options that share and spread financial burdens. Formal, market-based (re)insurance spreads risk and provides a financial buffer against the impacts of climate hazards (Linnerooth-Bayer and Hochrainer-Stigler, 2015; Wolfrom and Yokoi-Arai, 2015; O'Hare et al., 2016; Glaas et al., 2017; Patel et al., 2017). As an alternative to traditional indemnity-based insurance, index-based micro-crop and livestock insurance programmes have been rolled out in regions with less developed insurance markets (Akter et al., 2016, 2017; Jensen and Barrett, 2017). There is *medium evidence* and *medium agreement* on the feasibility of insurance for adaptation, with financial, social, and institutional barriers to implementation and uptake, especially in low-income nations (García Romero and Molina, 2015; Joyette et al., 2015; Lashley and Warner, 2015; Jin et al., 2016). Social protection programmes include cash and in-kind transfers to protect poor and vulnerable households from the impact of economic shocks, natural disasters and other crises (World Bank, 2017b), and can build generic adaptive capacity and reduce vulnerability when combined with a comprehensive climate risk management approach (*medium evidence, medium agreement*) (Devereux, 2016; Lemos et al., 2016).

4.3.5.3 Education and learning

Educational adaptation options motivate adaptation through building awareness (Butler et al., 2016; Myers et al., 2017), leveraging multiple knowledge systems (Pearce et al., 2015; Janif et al., 2016), developing participatory action research and social learning processes (Butler and Adamowski, 2015; Ensor and Harvey, 2015; Butler et al., 2016; Thi Hong Phuong et al., 2017; Ford et al., 2018), strengthening extension services, and building mechanisms for learning and knowledge sharing through community-based platforms, international conferences and knowledge networks (Vinke-de Kruijf and Pahl-Wostl, 2016) (*medium evidence, high agreement*).

4.3.5.4 Population health and health system adaptation options

Climate change will exacerbate existing health challenges (Chapter 3, Section 3.4.7). Options for enhancing current health services include providing access to safe water and improved sanitation, enhancing access to essential services such as vaccination, and developing or strengthening integrated surveillance systems (WHO, 2015). Combining these with iterative management can facilitate effective adaptation (*medium evidence, high agreement*).

4.3.5.5 Indigenous knowledge

There is *medium evidence* and *high agreement* that indigenous knowledge is critical for adaptation, underpinning adaptive capacity through the diversity of indigenous agro-ecological and forest management systems, collective social memory, repository of accumulated experience and social networks (Hiwasaki et al., 2015; Pearce et al., 2015; Mapfumo et al., 2016; Sherman et al., 2016; Ingty, 2017) (Box 4.3). Indigenous knowledge is threatened by acculturation, dispossession of land rights and land grabbing, rapid environmental changes, colonization and social change, resulting in increasing vulnerability to climate change – which climate policy can exacerbate if based on limited understanding of indigenous worldviews (Thornton and Manasfi, 2010; Ford, 2012; Nakashima et al., 2012; McNamara and Prasad, 2014). Many scholars argue that recognition of indigenous rights, governance systems and laws is central to adaptation, mitigation and sustainable development (Magni, 2017; Thornton and Comberti, 2017; Pearce, 2018).

4.3.5.6 Human migration

Human migration, whether planned, forced or voluntary, is increasingly gaining attention as a response, particularly where climatic risks are becoming severe (Chapter 3, Section 3.4.10.2). There is *medium evidence* and *low agreement* as to whether migration is adaptive, in relation to cost effectiveness concerns (Grecequet et al., 2017) and scalability (Brzoska and Fröhlich, 2016; Gemenne and Blocher, 2017; Grecequet et al., 2017). Migrating can have mixed outcomes on reducing socio-economic vulnerability (Birk and Rasmussen, 2014; Kothari, 2014; Adger et al., 2015; Betzold, 2015; Kelman, 2015; Grecequet et al., 2017; Melde et al., 2017; World Bank, 2017a; Kumari Rigaud et al., 2018) and its feasibility is constrained by low political and legal acceptability and inadequate institutional capacity (Betzold, 2015; Methmann and Oels, 2015; Brzoska and Fröhlich, 2016; Gemenne and Blocher, 2017; Grecequet et al., 2017; Yamamoto et al., 2017).

4.3.5.7 Climate services

There is *medium evidence* and *high agreement* that climate services can play a critical role in aiding adaptation decision-making (Vaughan and Dessai, 2014; Wood et al., 2014; Lourenço et al., 2016; Trenberth et al., 2016; Singh et al., 2017; Vaughan et al., 2018). The higher uptake of short-term climate information such as weather advisories and daily forecasts contrast with lesser use of longer-term information such as seasonal forecasts and multi-decadal projections (Singh et al., 2017; Vaughan et al., 2018). Climate service interventions have met challenges with scaling up due to low capacity, inadequate institutions, and difficulties in maintaining systems beyond pilot project stage (Sivakumar et al., 2014; Tall et al., 2014; Gebru et al., 2015; Singh et al., 2016b), and technical, institutional, design, financial and capacity barriers to the application of climate information for better decision-making remain (Briley et al., 2015; WMO, 2015; L. Jones et al., 2016; Lourenço et al., 2016; Snow et al., 2016; Harjanne, 2017; Singh et al., 2017; C.J. White et al., 2017).

4

Table 4.4 | Assessment of overarching adaptation options in relation to enabling conditions. For more details, see Supplementary Material 4.SM.2.

Option	Enabling Conditions	Examples
Disaster risk management (DRM)	Governance and institutional capacity: supports post-disaster recovery and reconstruction (Kelman et al., 2015; Kull et al., 2016).	Early warning systems (Anacona et al., 2015), and monitoring of dangerous lakes and surrounding slopes (including using remote sensing) offer DRM opportunities (Emmer et al., 2016; Milner et al., 2017).
Risk sharing and spreading: insurance	Institutional capacity and finance: buffers climate risk (Wolfrom and Yokoi-Arai, 2015; O'Hare et al., 2016; Glaas et al., 2017; Jenkins et al., 2017; Patel et al., 2017).	In 2007, the Caribbean Catastrophe Risk Insurance Facility was formed to pool risk from tropical cyclones, earthquakes, and excess rainfalls (Murphy et al., 2012; CCRIF, 2017).
Social safety nets	Institutional capacity and finance: builds generic adaptive capacity and reduces social vulnerability (Weldegebriel and Prowse, 2013; Eakin et al., 2014; Lemos et al., 2016; Schwan and Yu, 2017).	In sub-Saharan Africa, cash transfer programmes targeting poor communities have proven successful in smoothing household welfare and food security during droughts, strengthening community ties, and reducing debt levels (del Ninno et al., 2016; Asfaw et al., 2017; Asfaw and Davis, 2018).
Education and learning	Behavioural change and institutional capacity: social learning strengthens adaptation and affects longer-term change (Clemens et al., 2015; Ensor and Harvey, 2015; Henly-Shepard et al., 2015).	Participatory scenario planning is a process by which multiple stakeholders work together to envision future scenarios under a range of climatic conditions (Oteros-Rozas et al., 2015; Butler et al., 2016; Flynn et al., 2018).
Population health and health system	Institutional capacity: 1.5°C warming will primarily exacerbate existing health challenges (K.R. Smith et al., 2014), which can be targeted by enhancing health services.	Heatwave early warning and response systems coordinate the implementation of multiple measures in response to predicted extreme temperatures (e.g., public announcements, opening public cooling shelters, distributing information on heat stress symptoms) (Knowlton et al., 2014; Takahashi et al., 2015; Nitschke et al., 2016, 2017).
Indigenous knowledge	Institutional capacity and behavioural change: knowledge of environmental conditions helps communities detect and monitor change (Johnson et al., 2015; Mistry and Berardi, 2016; Williams et al., 2017).	Options such as integration of indigenous knowledge into resource management systems and school curricula, are identified as potential adaptations (Cunsolo Willox et al., 2013; McNamara and Prasad, 2014; MacDonald et al., 2015; Pearce et al., 2015; Chambers et al., 2017; Inamara and Thomas, 2017).
Human migration	Governance: revising and adopting migration issues in national disaster risk management policies, National Adaptation Plans and NDCs (Kuruppu and Willie, 2015; Yamamoto et al., 2017).	In dryland India, populations in rural regions already experiencing 1.5°C warming are migrating to cities (Gajjar et al., 2018) but are inadequately covered by existing policies (Bhagat, 2017).
Climate services	Technological innovation: rapid technical development (due to increased financial inputs and growing demand) is improving quality of climate information provided (Rogers and Tsirkunov, 2010; Clements et al., 2013; Perrels et al., 2013; Gasc et al., 2014; WMO, 2015; Roudier et al., 2016).	Climate services are seeing wide application in sectors such as agriculture, health, disaster management and insurance (Lourenço et al., 2016; Vaughan et al., 2018), with implications for adaptation decision-making (Singh et al., 2017).

4

Cross-Chapter Box 9 | Risks, Adaptation Interventions, and Implications for Sustainable Development and Equity Across Four Social-Ecological Systems: Arctic, Caribbean, Amazon, and Urban

Authors:

Debora Ley (Guatemala/Mexico), Malcolm E. Araos (Canada), Amir Bazaz (India), Marcos Buckeridge (Brazil), Ines Camilloni (Argentina), James Ford (UK/Canada), Bronwyn Hayward (New Zealand), Shagun Mehrotra (USA/India), Antony Payne (UK), Patricia Pinho (Brazil), Aromar Revi (India), Kevon Rhiney (Jamaica), Chandni Singh (India), William Solecki (USA), Avelino Suarez (Cuba), Michael Taylor (Jamaica), Adelle Thomas (Bahamas).

This box presents four case studies from different social-ecological systems as examples of risks of 1.5°C warming and higher (Chapter 3); adaptation options that respond to these risks (Chapter 4); and their implications for poverty, livelihoods and sustainability (Chapter 5). It is not yet possible to generalize adaptation effectiveness across regions due to a lack of empirical studies and monitoring and evaluation of current efforts.

Arctic

The Arctic is undergoing the most rapid climate change globally (Larsen et al., 2014), warming by 1.9°C over the last 30 years (Walsh, 2014; Grosse et al., 2016). For 2°C of global warming relative to pre-industrial levels, chances of an ice-free Arctic during summer are substantially higher than at 1.5°C (see Chapter 3, Sections 3.3.5 and 3.3.8), with permafrost melt, increased instances of storm surge, and extreme weather events anticipated along with later ice freeze up, earlier break up, and a longer ice-free open water season (Bring et al., 2016; DeBeer et al., 2016; Jiang et al., 2016; Chadburn et al., 2017; Melvin et al., 2017). Negative impacts on health, infrastructure, and economic sectors (AMAP, 2017a, b, 2018) are projected, although the extension of the summer ocean-shipping season has potential economic opportunities (Ford et al., 2015b; Dawson et al., 2016; Ng et al., 2018).

Cross Chapter Box 9 (continued)

Communities, many with indigenous roots, have adapted to environmental change, developing or shifting harvesting activities and patterns of travel and transitioning economic systems (Forbes et al., 2009; Wenzel, 2009; Ford et al., 2015b; Pearce et al., 2015), although emotional and psychological effects have been documented (Cunsolo Willox et al., 2012; Cunsolo and Ellis, 2018). Besides climate change (Keskitalo et al., 2011; Loring et al., 2016), economic and social conditions can constrain the capacity to adapt unless resources and cooperation are available from public and private sector actors (AMAP, 2017a, 2018) (see Chapter 5, Box 5.3). In Alaska, the cumulative economic impacts of climate change on public infrastructure are projected at 4.2 billion USD to 5.5 billion USD from 2015 to 2099, with adaptation efforts halving these estimates (Melvin et al., 2017). Marginalization, colonization, and land dispossession provide broader underlying challenges facing many communities across the circumpolar north in adapting to change (Ford et al., 2015a; Sejersen, 2015) (see Section 4.3.5).

Adaptation opportunities include alterations to building codes and infrastructure design, disaster risk management, and surveillance (Ford et al., 2014a; AMAP, 2017a, b; Labbé et al., 2017). Most adaptation initiatives are currently occurring at local levels in response to both observed and projected environmental changes as well as social and economic stresses (Ford et al., 2015a). In a recent study of Canada, most adaptations were found to be in the planning stages (Labbé et al., 2017). Studies have suggested that a number of the adaptation actions are not sustainable, lack evaluation frameworks, and hold potential for maladaptation (Loboda, 2014; Ford et al., 2015a; Larsson et al., 2016). Utilizing indigenous and local knowledge and stakeholder engagement can aid the development of adaptation policies and broader sustainable development, along with more proactive and regionally coherent adaptation plans and actions, and regional cooperation (e.g., through the Arctic Council) (Larsson et al., 2016; AMAP, 2017a; Melvin et al., 2017; Forbis Jr and Hayhoe, 2018) (see Section 4.3.5).

Caribbean Small Island Developing States (SIDS) and Territories

Extreme weather, linked to tropical storms and hurricanes, represent one of the largest risks facing Caribbean island nations (Chapter 3, Section 3.4.5.3). Non-economic damages include detrimental health impacts, forced displacement and destruction of cultural heritages. Projections of increased frequency of the most intense storms at 1.5°C and higher warming levels (Wehner et al., 2018; Chapter 3, Section 3.3.6; Box 3.5) are a significant cause for concern, making adaptation a matter of survival (Mycoo and Donovan, 2017).

Despite a shared vulnerability arising from commonalities in location, circumstance and size (Bishop and Payne, 2012; Nurse et al., 2014), adaptation approaches are nuanced by differences in climate governance, affecting vulnerability and adaptive capacity (see Section 4.4.1). Three cases exemplify differences in disaster risk management.

Cuba: Together with a robust physical infrastructure and human-resource base (Kirk, 2017), Cuba has implemented an effective civil defence system for emergency preparedness and disaster response, centred around community mobilization and preparedness (Kirk, 2017). Legislation to manage disasters, an efficient and robust early warning system, emergency stockpiles, adequate shelter system and continuous training and education of the population help create a 'culture of risk' (Isayama and Ono, 2015; Lizarralde et al., 2015) which reduces vulnerability to extreme events (Pichler and Striessnig, 2013). Cuba's infrastructure is still susceptible to devastation, as seen in the aftermath of the 2017 hurricane season.

United Kingdom Overseas Territories (UKOT): All UKOT have developed National Disaster Preparedness Plans (PAHO/WHO, 2016) and are part of the Caribbean Disaster Risk Management Program which aims to improve disaster risk management within the health sector. Different vulnerability levels across the UKOT (Lam et al., 2015) indicate the benefits of greater regional cooperation and capacity-building, not only within UKOT, but throughout the Caribbean (Forster et al., 2011). While sovereign states in the region can directly access climate funds and international support, Dependent Territories are reliant on their controlling states (Bishop and Payne, 2012). There tends to be low-scale management for environmental issues in UKOT, which increases UKOT's vulnerability. Institutional limitations, lack of human and financial resources, and limited long-term planning are identified as barriers to adaptation (Forster et al., 2011).

Jamaica: Disaster management is coordinated through a hierarchy of national, parish and community disaster committees under the leadership of the Office of Disaster Preparedness and Emergency Management (ODPEM). ODPEM coordinates disaster preparedness and risk-reduction efforts among key state and non-state agencies (Grove, 2013). A National Disaster Committee provides technical and policy oversight to the ODPEM and is composed of representatives from multiple stakeholders (Osei, 2007). Most initiatives are primarily funded through a mix of multilateral and bilateral loan and grant funding focusing on strengthening technical and institutional capacities of state- and research-based institutions and supporting integration of climate change considerations into national and sectoral development plans (Robinson, 2017).

4

Cross Chapter Box 9 (continued)

To improve climate change governance in the region, Pittman et al. (2015) suggest incorporating holistic and integrated management systems, improving flexibility in collaborative processes, implementing monitoring programs, and increasing the capacity of local authorities. Implementation of the 2030 Sustainable Development Agenda and the Sustainable Development Goals (SDGs) can contribute to addressing the risks related with extreme events (Chapter 5, Box 5.3).

The Amazon
Terrestrial forests, such as the Amazon, are sensitive to changes in the climate, particularly drought (Laurance and Williamson, 2001) which might intensify through the 21st century (Marengo and Espinoza, 2016) (Chapter 3, Section 3.5.5.6).

The poorest communities in the region face substantial risks with climate change, and barriers and limits to adaptive capacity (Maru et al., 2014; Pinho et al., 2014, 2015; Brondízio et al., 2016). The Amazon is considered a hotspot, with interconnections between increasing temperature, decreased precipitation and hydrological flow (Betts et al., 2018) (Sections 3.3.2.2, 3.3.3.2 and 3.3.5); low levels of socio-economic development (Pinho et al., 2014); and high levels of climate vulnerability (Darela et al., 2016). Limiting global warming to 1.5°C could increase food and water security in the region compared to 2°C (Betts et al., 2018), reduce the impact on poor people and sustainable development, and make adaptation easier (O'Neill et al., 2017), particularly in the Amazon (Bathiany et al., 2018) (Chapter 5, Section 5.2.2).

Climate policy in many Amazonian nations has focused on forests as carbon sinks (Soares-Filho et al., 2010). In 2009, the Brazilian National Policy on Climate Change acknowledged adaptation as a concern, and the government sought to mainstream adaptation into public administration. Brazil's National Adaptation Plan sets guidelines for sectoral adaptation measures, primarily by developing capacity building, plans, assessments and tools to support adaptive decision-making. Adaptation is increasingly being presented as having mitigation co-benefits in the Brazilian Amazon (Gregorio et al., 2016), especially within ecosystem-based adaptation (Locatelli et al., 2011). In Peru's Framework Law for Climate Change, every governmental sector will consider climatic conditions as potential risks and/or opportunities to promote economic development and to plan adaptation.

Drought and flood policies have had limited effectiveness in reducing vulnerability (Marengo et al., 2013). In the absence of effective adaptation, achieving the SDGs will be challenging, mainly in poverty, health, water and sanitation, inequality and gender equality (Chapter 5, Section 5.2.3).

Urban systems
Around 360 million people reside in urban coastal areas where precipitation variability is exposing inadequacies of urban infrastructure and governance, with the poor being especially vulnerable (Reckien et al., 2017) (Cross-Chapter Box 13 in Chapter 5). Urban systems have seen growing adaptation action (Revi et al., 2014b; Araos et al., 2016b; Amundsen et al., 2018). Developing cities spend more on health and agriculture-related adaptation options while developed cities spend more on energy and water (Georgeson et al., 2016). Current adaptation activities are lagging in emerging economies, which are major centres of population growth facing complex interrelated pressures on investment in health, housing and education (Georgeson et al., 2016; Reckien et al., 2017).

New York, United States: Adaptation plans are undertaken across government levels, sectors and departments (NYC Parks, 2010; Vision 2020 Project Team, 2011; PlaNYC, 2013), and have been advanced by an expert science panel that is obligated by local city law to provide regular updates on policy-relevant climate science (NPCC, 2015). Federal initiatives include 2013's Rebuild By Design competition to promote resilience through infrastructural projects (HSRTF, 2013). In 2013 the Mayor's office, in response to Hurricane Sandy, published the city's adaptation strategy (PlaNYC, 2013). In 2015, the OneNYC Plan for a Strong and Just City (OneNYC Team, 2015) laid out a strategy for urban planning through a justice and equity lens. In 2017, new climate resiliency guidelines proposed that new construction must include sea level rise projections into planning and development (ORR, 2018). Although this attention to climate-resilient development may help reduce income inequality, its full effect could be constrained if a policy focus on resilience obscures analysis of income redistribution for the poor (Fainstein, 2018).

Kampala, Uganda: Kampala Capital City Authority (KCCA) has the statutory responsibility for managing the city. The Kampala Climate Change Action Strategy (KCCAS) is responding to climatic impacts of elevated temperature and more intense, erratic rain. KCCAS has considered multi-scale and temporal aspects of response (Chelleri et al., 2015; Douglas, 2017; Fraser et al., 2017), strengthened community adaptation (Lwasa, 2010; Dobson, 2017), responded to differential adaptive capacities (Waters and Adger, 2017) and believes in participatory processes and bridging of citywide linkages (KCCA, 2016). Analysis of the implications of uniquely adapted local solutions (e.g., motorcycle taxis) suggests sustainability can be enhanced when planning recognizes the need to adapt to uniquely local solutions (Evans et al., 2018).

Cross Chapter Box 9 (continued)

Rotterdam, The Netherlands: The Rotterdam Climate Initiative (RCI) was launched to reduce greenhouse gas emissions and climate-proof Rotterdam (RCI, 2017). Rotterdam has an integrated adaptation strategy, built on flood management, accessibility, adaptive building, urban water systems and urban climate, defined through the Rotterdam Climate Proof programme and the Rotterdam Climate Change Adaptation Strategy (RCI, 2008, 2013). Governance mechanisms that enabled integration of flood risk management plans with other policies, citizen participation, institutional eco-innovation, and focusing on green infrastructure (Albers et al., 2015; Dircke and Molenaar, 2015; de Boer et al., 2016a; Huang-Lachmann and Lovett, 2016) have contributed to effective adaptation (Ward et al., 2013). Entrenched institutional characteristics constrain the response framework (Francesch-Huidobro et al., 2017), but emerging evidence suggests that new governance arrangements and structures can potentially overcome these barriers in Rotterdam (Hölscher et al., 2018).

4.3.6   Short-Lived Climate Forcers

The main short-lived climate forcer (SLCF) emissions that cause warming are methane (CH_4), other precursors of tropospheric ozone (i.e., carbon monoxide (CO), non-methane volatile organic compounds (NMVOC), black carbon (BC) and hydrofluorocarbons (HFCs); Myhre et al., 2013). SLCFs also include emissions that lead to cooling, such as sulphur dioxide (SO_2) and organic carbon (OC). Nitrogen oxides (NOx) can have both warming and cooling effects, by affecting ozone (O_3) and CH_4, depending on time scale and location (Myhre et al., 2013).

Cross-Chapter Box 2 in Chapter 1 provides a discussion of role of SLCFs in comparison to long-lived GHGs. Chapter 2 shows that 1.5°C-consistent pathways require stringent reductions in CO_2 and CH_4, and that non-CO_2 climate forcers reduce carbon budgets by about 2200 $GtCO_2$ per degree of warming attributed to them (see the Supplementary Material to Chapter 2).

Reducing non-CO_2 emissions is part of most mitigation pathways (IPCC, 2014c). All current GHG emissions and other forcing agents affect the rate and magnitude of climate change over the next few decades, while long-term warming is mainly driven by CO_2 emissions. CO_2 emissions result in a virtually permanent warming, while temperature change from SLCFs disappears within decades after emissions of SLCFs are ceased. Any scenario that fails to reduce CO_2 emissions to net zero would not limit global warming, even if SLCFs are reduced, due to accumulating CO_2-induced warming that overwhelms SLCFs' mitigation benefits in a couple of decades (Shindell et al., 2012; Schmale et al., 2014) (and see Chapter 2, Section 2.3.3.2).

Mitigation options for warming SLCFs often overlap with other mitigation options, especially since many warming SLCFs are co-emitted with CO_2. SLCFs are generally mitigated in 1.5°C- or 2°C-consistent pathways as an integral part of an overall mitigation strategy (Chapter 2). For example, Section 2.3 indicates that most very-low-emissions pathways include a transition away from the use of coal and natural gas in the energy sector and oil in transportation, which coincides with emission-reduction strategies related to methane from the fossil fuel sector and BC from the transportation sector. Much SLCF emission reduction aims at BC-rich sectors and considers the impacts of several co-emitted SLCFs (Bond et al., 2013; Sand et al., 2015; Stohl et al., 2015). The benefits of such strategies depend greatly upon the assumed level of progression of access to modern energy for the

poorest populations who still rely on biomass fuels, as this affects the reference level of BC emissions (Rogelj et al., 2014).

Some studies have evaluated the focus on SLCFs in mitigation strategies and point towards trade-offs between short-term SLCF benefits and lock-in of long-term CO_2 warming (Smith and Mizrahi, 2013; Pierrehumbert, 2014). Reducing fossil fuel combustion will reduce aerosols levels, and thereby cause warming from removal of aerosol cooling effects (Myhre et al., 2013; Xu and Ramanathan, 2017; Samset et al., 2018). While some studies have found a lower temperature effect from BC mitigation, thus questioning the effectiveness of targeted BC mitigation for climate change mitigation (Myhre et al., 2013; Baker et al., 2015; Stjern et al., 2017; Samset et al., 2018), other models and observationally constrained estimates suggest that these widely-used models do not fully capture observed effects of BC and co-emissions on climate (e.g., Bond et al., 2013; Cui et al., 2016; Peng et al., 2016).

Table 4.5 provides an overview of three warming SLCFs and their emission sources, with examples of options for emission reductions and associated co-benefits.

A wide range of options to reduce SLCF emissions was extensively discussed in AR5 (IPCC, 2014b). Fossil fuel and waste sector methane mitigation options have high cost-effectiveness, producing a net profit over a few years, considering market costs only. Moreover, reducing roughly one-third to one-half of all human-caused emissions has societal benefits greater than mitigation costs when considering environmental impacts only (UNEP, 2011; Höglund-Isaksson, 2012; IEA, 2017b; Shindell et al., 2017a). Since AR5, new options for methane, such as those related to shale gas, have been included in mitigation portfolios (e.g., Shindell et al., 2017a).

Reducing BC emissions and co-emissions has sustainable development co-benefits, especially around human health (Stohl et al., 2015; Haines et al., 2017; Aakre et al., 2018), avoiding premature deaths and increasing crop yields (Scovronick et al., 2015; Peng et al., 2016). Additional benefits include lower likelihood of non-linear climate changes and feedbacks (Shindell et al., 2017b) and temporarily slowing down the rate of sea level rise (Hu et al., 2013). Interventions to reduce BC offer tangible local air quality benefits, increasing the likelihood of local public support (Eliasson, 2014; Venkataraman et al., 2016) (see Chapter 5, Section 5.4.2.1). Limited interagency co-ordination, poor science-policy interactions (Zusman et al., 2015), and weak policy and

Table 4.5 | Overview of main characteristics of three warming short-lived climate forcers (SLCFs) (core information based on Pierrehumbert, 2014 and Schmale et al., 2014; rest of the details as referenced).

SLCF Compound	Atmospheric Lifetime	Annual Global Emission	Main Anthropogenic Emission Sources	Examples of Options to Reduce Emissions Consistent with 1.5°C	Examples of Co-Benefits Based on Haines et al. (2017) Unless Specified Otherwise
Methane	On the order of 10 years	0.3 GtCH$_4$ (2010) (Pierrehumbert, 2014)	Fossil fuel extraction and transportation; Land-use change; Livestock and rice cultivation; Waste and wastewater	Managing manure from livestock; Intermittent irrigation of rice; Capture and usage of fugitive methane; Dietary change; For more: see Section 4.3.2	Reduction of tropospheric ozone (Shindell et al., 2017a); Health benefits of dietary changes; Increased crop yields; Improved access to drinking water
HFCs	Months to decades, depending on the gas	0.35 GtCO$_2$-eq (2010) (Velders et al., 2015)	Air conditioning; Refrigeration; Construction material	Alternatives to HFCs in air-conditioning and refrigeration applications	Greater energy efficiency (Mota-Babiloni et al., 2017)
Black Carbon	Days	~7 Mt (2010) (Klimont et al., 2017)	Incomplete combustion of fossil fuels or biomass in vehicles (esp. diesel), cook stoves or kerosene lamps; Field and biomass burning	Fewer and cleaner vehicles; Reducing agricultural biomass burning; Cleaner cook stoves, gas-based or electric cooking; Replacing brick and coke ovens; Solar lamps; For more see Section 4.3.3	Health benefits of better air quality; Increased education opportunities; Reduced coal consumption for modern brick kilns; Reduced deforestation

absence of inspections and enforcement (Kholod and Evans, 2016) are among barriers that reduce the institutional feasibility of options to reduce vehicle-induced BC emissions. A case study for India shows that switching from biomass cook stoves to cleaner gas stoves (based on liquefied petroleum gas or natural gas) or to electric cooking stoves is technically and economically feasible in most areas, but faces barriers in user preferences, costs and the organization of supply chains (Jeuland et al., 2015). Similar feasibility considerations emerge in switching from kerosene wick lamps for lighting to solar lanterns, from current low-efficiency brick kilns and coke ovens to cleaner production technologies; and from field burning of crop residues to agricultural practices using deep-sowing and mulching technologies (Williams et al., 2011; Wong, 2012).

The radiative forcing from HFCs are currently small but have been growing rapidly (Myhre et al., 2013). The Kigali Amendment (from 2016) to the Montreal Protocol set out a global accord for phasing out these compounds (Höglund-Isaksson et al., 2017). HFC mitigation options include alternatives with reduced warming effects, ideally combined with improved energy efficiency so as to simultaneously reduce CO$_2$ and co-emissions (Shah et al., 2015). Costs for most of HFC's mitigation potential are estimated to be below USD$_{2010}$ 60 tCO$_2$-eq^{-1}, and the remainder below roughly double that number (Höglund-Isaksson et al., 2017).

Reductions in SLCFs can provide large benefits towards sustainable development, beneficial for social, institutional and economic feasibility. Strategies that reduce SLCFs can provide benefits that include improved air quality (e.g., Anenberg et al., 2012) and crop yields (e.g., Shindell et al., 2012), energy access, gender equality and poverty eradication (e.g.,Shindell et al., 2012; Haines et al., 2017). Institutional feasibility can be negatively affected by an information deficit, with

the absence of international frameworks for integrating SLCFs into emissions accounting and reporting mechanisms being a barrier to developing policies for addressing SLCF emissions (Venkataraman et al., 2016). The incentives for reducing SLCFs are particularly strong for small groups of countries, and such collaborations could increase the feasibility and effectiveness of SLCF mitigation options (Aakre et al., 2018).

4.3.7 Carbon Dioxide Removal (CDR)

CDR methods refer to a set of techniques for removing CO$_2$ from the atmosphere. In the context of 1.5°C-consistent pathways (Chapter 2), they serve to offset residual emissions and, in most cases, achieve net negative emissions to return to 1.5°C from an overshoot. See Cross-Chapter Box 7 in Chapter 3 for a synthesis of land-based CDR options. Cross-cutting issues and uncertainties are summarized in Table 4.6.

4.3.7.1   Bioenergy with carbon capture and storage (BECCS)

BECCS has been assessed in previous IPCC reports (IPCC, 2005b, 2014b; P. Smith et al., 2014; Minx et al., 2017) and has been incorporated into integrated assessment models (Clarke et al., 2014), but also 1.5°C-consistent pathways without BECCS have emerged (Bauer et al., 2018; Grubler et al., 2018; Mousavi and Blesl, 2018; van Vuuren et al., 2018). Still, the overall set of pathways limiting global warming to 1.5°C with limited or no overshoot indicates that 0–1, 0–8, and 0–16 GtCO$_2$ yr^{-1} would be removed by BECCS by 2030, 2050 and 2100, respectively (Chapter 2, Section 2.3.4). BECCS is constrained by sustainable bioenergy potentials (Section 4.3.1.2, Chapter 5, Section 5.4.1.3 and Cross-Chapter Box 6 in Chapter 3), and availability of safe storage for CO$_2$ (Section 4.3.1.6). Literature estimates for BECCS mitigation potentials in 2050 range from 1–85 GtCO$_2$[4]. Fuss et al.

---

[4]   As more bottom-up literature exists on bioenergy potentials, this exercise explored the bioenergy literature and converted those estimates to BECCS potential with 1EJ of bioenergy yielding 0.02–0.05 GtCO$_2$ emission reduction. For the bottom-up literature references for the potentials range, please refer to Supplementary Material 4.SM.3 Table 1.

(2018) narrow this range to 0.5–5 GtCO$_2$ yr^{-1} (*medium agreement, high evidence*) (Figure 4.3), meaning that BECCS mitigation potentials are not necessarily sufficient for 1.5°C-consistent pathways. This is, among other things, related to sustainability concerns (Boysen et al., 2017; Heck et al., 2018; Henry et al., 2018).

Assessing BECCS deployment in 2°C pathways (of about 12 GtCO$_2$-eq yr^{-1} by 2100, considered as a conservative deployment estimate for BECCS-accepting pathways consistent with 1.5°C), Smith et al. (2016b) estimate a land-use intensity of 0.3–0.5 ha tCO$_2$-eq^{-1} yr^{-1} using forest residues, 0.16 ha CO$_2$-eq^{-1} yr^{-1} for agricultural residues, and 0.03–0.1 ha tCO$_2$-eq^{-1} yr^{-1} for purpose-grown energy crops. The average amount of BECCS in these pathways requires 25–46% of arable and permanent crop area in 2100. Land area estimates differ in scale and are not necessarily a good indicator of competition with, for example, food production, because requiring a smaller land area for the same potential could indicate that high-productivity agricultural land is used. In general, the literature shows *low agreement* on the availability of land (Fritz et al., 2011; see Erb et al., 2016b for recent advances). Productivity, food production and competition with other ecosystem services and land use by local communities are important factors for designing regulation. These potentials and trade-offs are not homogenously distributed across regions. However, Robledo-Abad et al. (2017) find that regions with higher potentials are understudied, given their potential contribution. Researchers have expressed the need to complement global assessments with regional, geographically explicit bottom-up studies of biomass potentials and socio-economic impacts (e.g., de Wit and Faaij, 2010; Kraxner et al., 2014; Baik et al., 2018).

Energy production and land and water footprints show wide ranges in bottom-up assessments due to differences in technology, feedstock and other parameters (−1–150 EJ yr^{-1} of energy, 109–990 Mha, 6–79 MtN, 218–4758 km^3 yr^{-1} of water per GtCO$_2$ yr^{-1}; Smith and Torn, 2013; Smith et al., 2016b; Fajardy and Mac Dowell, 2017) and are not comparable to IAM pathways which consider system effects (Bauer et al., 2018). Global impacts on nutrients and albedo are difficult to quantify (Smith et al., 2016b). BECCS competes with other land-based CDR and mitigation measures for resources (Chapter 2).

There is uncertainty about the feasibility of timely upscaling (Nemet et al., 2018). CCS (see Section 4.3.1) is largely absent from the Nationally Determined Contributions (Spencer et al., 2015) and lowly ranked in investment priorities (Fridahl, 2017). Although there are dozens of small-scale BECCS demonstrations (Kemper, 2015) and a full-scale project capturing 1 MtCO$_2$ exists (Finley, 2014), this is well below the numbers associated with 1.5°C or 2°C-compatible pathways (IEA, 2016a; Peters et al., 2017). Although the majority of BECCS cost estimates are below 200 USD tCO$_2^{-1}$ (Figure 4.2), estimates vary widely. Economic incentives for ramping up large CCS or BECCS infrastructure are weak (Bhave et al., 2017). The 2050 average investment costs for such a BECCS infrastructure for bio-electricity and biofuels are estimated at 138 and 123 billion USD yr^{-1}, respectively (Smith et al., 2016b).

BECCS deployment is further constrained by bioenergy's carbon accounting, land, water and nutrient requirements (Section 4.3.1), its compatibility with other policy goals and limited public acceptance of

both bioenergy and CCS (Section 4.3.1). Current pathways are believed to have inadequate assumptions on the development of societal support and governance structures (Vaughan and Gough, 2016). However, removing BECCS and CCS from the portfolio of available options significantly raises modelled mitigation costs (Kriegler et al., 2013; Bauer et al., 2018).

4.3.7.2 Afforestation and reforestation (AR)

Afforestation implies planting trees on land not forested for a long time (e.g., over the last 50 years in the context of the Kyoto Protocol), while reforestation implies re-establishment of forest formations after a temporary condition with less than 10% canopy cover due to human-induced or natural perturbations. Houghton et al. (2015) estimate about 500 Mha could be available for the re-establishment of forests on lands previously forested, but not currently used productively. This could sequester at least 3.7 GtCO$_2$ yr^{-1} for decades. The full literature range gives 2050 potentials of 1–7 GtCO$_2$ yr^{-1} (*low evidence, medium agreement*), narrowed down to 0.5–3.6 GtCO$_2$ yr^{-1} based on a number of constraints (Fuss et al., 2018). Abatement costs are estimated to be low compared to other CDR options, 5–50 USD tCO$_2$-eq^{-1} (*robust evidence, high agreement*). Yet, realizing such large potentials comes at higher land and water footprints than BECCS, although there would be a positive impact on nutrients and the energy requirement would be negligible (Smith et al., 2016b; Cross-Chapter Box 7 in Chapter 3). The 2030 estimate by Griscom et al. (2017) is up to 17.9 GtCO$_2$ yr^{-1} for reforestation with significant co-benefits (Cross-Chapter Box 7 in Chapter 3).

Biogenic storage is not as permanent as emission reductions by geological storage. In addition, forest sinks saturate, a process which typically occurs in decades to centuries compared to the thousands of years of residence time of CO$_2$ stored geologically (Smith et al., 2016a) and is subject to disturbances that can be exacerbated by climate change (e.g., drought, forest fires and pests) (Seidl et al., 2017). Handling these challenges requires careful forest management. There is much practical experience with AR, facilitating upscaling but with two caveats: AR potentials are heterogeneously distributed (Bala et al., 2007), partly because the planting of less reflective forests results in higher net absorbed radiation and localised surface warming in higher latitudes (Bright et al., 2015; Jones et al., 2015), and forest governance structures and monitoring capacities can be bottlenecks and are usually not considered in models (Wang et al., 2016; Wehkamp et al., 2018b). There is *medium agreement* on the positive impacts of AR on ecosystems and biodiversity due to different forms of afforestation discussed in the literature: afforestation of grassland ecosystems or diversified agricultural landscapes with monocultures or invasive alien species can have significant negative impacts on biodiversity, water resources, etc. (P. Smith et al., 2014), while forest ecosystem restoration (forestry and agroforestry) with native species can have positive social and environmental impacts (Cunningham et al., 2015; Locatelli et al., 2015; Paul et al., 2016; See Section 4.3.2).

Synergies with other policy goals are possible (see also Section 4.5.4); for example, land spared by diet shifts could be afforested (Röös et al., 2017) or used for energy crops (Grubler et al., 2018). Such land-sparing strategies could also benefit other land-based CDR options.

4

Panel A - Estimated costs and 2050 potentials

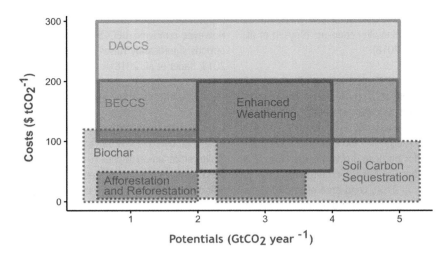

Panel B - Literature estimates on costs, potentials (2050) and side effects

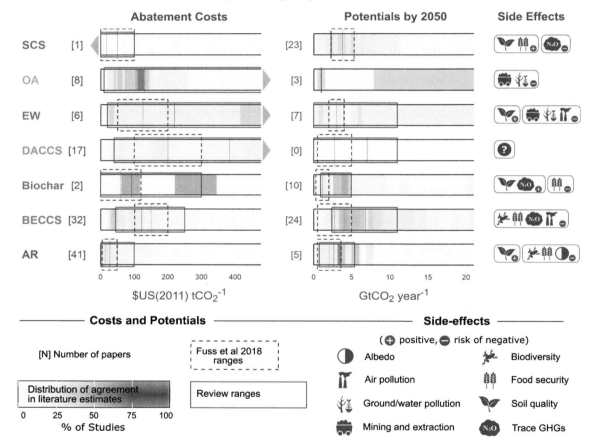

Figure 4.2 | Evidence on carbon dioxide removal (CDR) abatement costs, 2050 deployment potentials, and key side effects. Panel A presents estimates based on a systematic review of the bottom up literature (Fuss et al., 2018), corresponding to dashed blue boxes in Panel B. Dashed lines represent saturation limits for the corresponding technology. Panel B shows the percentage of papers at a given cost or potential estimate. Reference year for all potential estimates is 2050, while all cost estimates preceding 2050 have been included (as early as 2030, older estimates are excluded if they lack a base year and thus cannot be made comparable). Ranges have been trimmed to show detail (see Fuss et al., 2018 for the full range). Costs refer only to abatement costs. Icons for side-effects are allocated only if a critical mass of papers corroborates their occurrence

Notes: For references please see Supplementary Material Table 4.SM.3. Direct air carbon dioxide capture and storage (DACCS) is theoretically only constrained by geological storage capacity, estimates presented are considering upscaling and cost challenges (Nemet et al., 2018). BECCS potential estimates are based on bioenergy estimates in the literature (EJ yr⁻¹), converted to GtCO₂ following footnote 4. Potentials cannot be added up, as CDR options would compete for resources (e.g., land). SCS - soil carbon sequestration; OA - ocean alkalinization; EW- enhanced weathering; DACCS - direct air carbon dioxide capture and storage; BECCS - bioenergy with carbon capture and storage; AR - afforestation.

4.3.7.3 Soil carbon sequestration and biochar

At local scales there is *robust evidence* that soil carbon sequestration (SCS, e.g., agroforestry, De Stefano and Jacobson, 2018), restoration of degraded land (Griscom et al., 2017), or conservation agriculture management practices (Aguilera et al., 2013; Poeplau and Don, 2015; Vicente-Vicente et al., 2016) have co-benefits in agriculture and that many measures are cost-effective even without supportive climate policy. Evidence at global scale for potentials and especially costs is much lower. The literature spans cost ranges of −45–100 USD tCO_2^{-1} (negative costs relating to the multiple co-benefits of SCS, such as increased productivity and resilience of soils; P. Smith et al., 2014), and 2050 potentials are estimated at between 0.5 and 11 $GtCO_2$ yr^{-1}, narrowed down to 2.3–5.3 $GtCO_2$ yr^{-1} considering that studies above 5 $GtCO_2$ yr^{-1} often do not apply constraints, while estimates lower than 2 $GtCO_2$ yr^{-1} mostly focus on single practices (Fuss et al., 2018).

SCS has negligible water and energy requirements (Smith, 2016), affects nutrients and food security favourably (*high agreement, robust evidence*) and can be applied without changing current land use, thus making it socially more acceptable than CDR options with a high land footprint. However, soil sinks saturate after 10–100 years, depending on the SCS option, soil type and climate zone (Smith, 2016).

Biochar is formed by recalcitrant (i.e., very stable) organic carbon obtained from pyrolysis, which, applied to soil, can increase soil carbon sequestration leading to improved soil fertility properties.[5] Looking at the full literature range, the global potential in 2050 lies between 1 and 35 Gt CO_2 yr^{-1} (*low agreement, low evidence*), but considering limitations in biomass availability and uncertainties due to a lack of large-scale trials of biochar application to agricultural soils under field conditions, Fuss et al. (2018) lower the 2050 range to 0.3–2 $GtCO_2$ yr^{-1}. This potential is below previous estimates (e.g., Woolf et al., 2010), which additionally consider the displacement of fossil fuels through biochar. Permanence depends on soil type and biochar production temperatures, varying between a few decades and several centuries (Fang et al., 2014). Costs are 30–120 USD tCO_2^{-1} (*medium agreement, medium evidence*) (McCarl et al., 2009; McGlashan et al., 2012; McLaren, 2012; Smith, 2016).

Water requirements are low and at full theoretical deployment, up to 65 EJ yr^{-1} of energy could be generated as a side product (Smith, 2016). Positive side effects include a favourable effect on nutrients and reduced N_2O emissions (Cayuela et al., 2014; Kammann et al., 2017). However, 40–260 Mha are needed to grow the biomass for biochar for implementation at 0.3 $GtCO_2$-eq yr^{-1} (Smith, 2016), even though it is also possible to use residues (e.g., Windeatt et al., 2014). Biochar is further constrained by the maximum safe holding capacity of soils (Lenton, 2010) and the labile nature of carbon sequestrated in plants and soil at higher temperatures (Wang et al., 2013).

4.3.7.4 Enhanced weathering (EW) and ocean alkalinization

Weathering is the natural process of rock decomposition via chemical and physical processes in which CO_2 is spontaneously consumed and converted to solid or dissolved alkaline bicarbonates and/or carbonates (IPCC, 2005a). The process is controlled by temperature, reactive surface area, interactions with biota and, in particular, water solution composition. CDR can be achieved by accelerating mineral weathering through the distribution of ground-up rock material over land (Hartmann and Kempe, 2008; Wilson et al., 2009; Köhler et al., 2010; Renforth, 2012; ten Berge et al., 2012; Manning and Renforth, 2013; Taylor et al., 2016), shorelines (Hangx and Spiers, 2009; Montserrat et al., 2017) or the open ocean (House et al., 2007; Harvey, 2008; Köhler et al., 2013; Hauck et al., 2016). Ocean alkalinization adds alkalinity to marine areas to locally increase the CO_2 buffering capacity of the ocean (González and Ilyina, 2016; Renforth and Henderson, 2017).

In the case of land application of ground minerals, the estimated CDR potential range is 0.72–95 $GtCO_2$ yr^{-1} (*low evidence, low agreement*) (Hartmann and Kempe, 2008; Köhler et al., 2010; Hartmann et al., 2013; Taylor et al., 2016; Strefler et al., 2018a). Marine application of ground minerals is limited by feasible rates of mineral extraction, grinding and delivery, with estimates of 1–6 $GtCO_2$ yr^{-1} (*low evidence, low agreement*) (Köhler et al., 2013; Hauck et al., 2016; Renforth and Henderson, 2017). Agreement is low due to a variety of assumptions and unknown parameter ranges in the applied modelling procedures that would need to be verified by field experiments (Fuss et al., 2018). As with other CDR options, scaling and maturity are challenges, with deployment at scale potentially requiring decades (NRC, 2015a), considerable costs in transport and disposal (Hangx and Spiers, 2009; Strefler et al., 2018a) and mining (NRC, 2015a; Strefler et al., 2018a).[6]

Site-specific cost estimates vary depending on the chosen technology for rock grinding (an energy-intensive process; Köhler et al., 2013; Hauck et al., 2016), material transport, and rock source (Renforth, 2012; Hartmann et al., 2013), and range from 15–40 USD tCO_2^{-1} to 3,460 USD tCO_2^{-1} (*limited evidence, low agreement*; Figure 4.2) (Schuiling and Krijgsman, 2006; Köhler et al., 2010; Taylor et al., 2016). The evidence base for costs of ocean alkalinization and marine enhanced weathering is sparser than the land applications. The ocean alkalinization potential is assessed to be 0.1–10 $GtCO_2$ yr^{-1} with costs of 14–>500 USD tCO_2^{-1} (Renforth and Henderson, 2017).

The main side effects of terrestrial EW are an increase in water pH (Taylor et al., 2016), the release of heavy metals like Ni and Cr and plant nutrients like K, Ca, Mg, P and Si (Hartmann et al., 2013), and changes in hydrological soil properties. Respirable particle sizes, though resulting in higher potentials, can have impacts on health (Schuiling and Krijgsman, 2006; Taylor et al., 2016); utilization of wave-assisted decomposition through deployment on coasts could avert the need for fine grinding (Hangx and Spiers, 2009; Schuiling and de Boer, 2010). Side effects

---

[5] Other pyrolysis products that can achieve net CO_2 removals are bio-oil (pumped into geological storages) and permanent-pyrogas (capture and storage of CO_2 from gas combustion) (Werner et al., 2018)

[6] It has also been suggested that ocean alkalinity can be increased through accelerated weathering of limestone (Rau and Caldeira, 1999; Rau, 2011; Chou et al., 2015) or electrochemical processes (House et al., 2007; Rau, 2008; Rau et al., 2013; Lu et al., 2015). However, these techniques have not been proven at large scale either (Renforth and Henderson, 2017).

of marine EW and ocean alkalinization are the potential release of heavy metals like Ni and Cr (Montserrat et al., 2017). Increasing ocean alkalinity helps counter ocean acidification (Albright et al., 2016; Feng et al., 2016). Ocean alkalinization could affect ocean biogeochemical functioning (González and Ilyina, 2016). A further caveat of relates to saturation state and the potential to trigger spontaneous carbonate precipitation.[7] While the geochemical potential to remove and store CO_2 is quite large, *limited evidence* on the preceding topics makes it difficult to assess the true capacity, net benefits and desirability of EW and ocean alkalinity addition in the context of CDR.

4.3.7.5   Direct air carbon dioxide capture and storage (DACCS)

Capturing CO_2 from ambient air through chemical processes with subsequent storage of the CO_2 in geological formations is independent of source and timing of emissions and can avoid competition for land. Yet, this is also the main challenge: while the theoretical potential for DACCS is mainly limited by the availability of safe and accessible geological storage, the CO_2 concentration in ambient air is 100–300 times lower than at gas- or coal-fired power plants (Sanz-Pérez et al., 2016) thus requiring more energy than flue gas CO_2 capture (Pritchard et al., 2015). This appears to be the main challenge to DACCS (Sanz-Pérez et al., 2016; Barkakaty et al., 2017).

Studies explore alternative techniques to reduce the energy penalty of DACCS (van der Giesen et al., 2017). Energy consumption could be up to 12.9 GJ tCO_2-eq^{-1}; translating into an average of 156 EJ yr^{-1} by 2100 (current annual global primary energy supply is 600 EJ); water requirements are estimated to average 0.8–24.8 km^3 GtCO_2-eq^{-1} yr^{-1} (Smith et al., 2016b, based on Socolow et al., 2011).

However, the literature shows *low agreement* and is fragmented (Broehm et al., 2015). This fragmentation is reflected in a large range of cost estimates: from 20–1,000 USD tCO_2^{-1} (Keith et al., 2006; Pielke, 2009; House et al., 2011; Ranjan and Herzog, 2011; Simon et al., 2011; Goeppert et al., 2012; Holmes and Keith, 2012; Zeman, 2014; Sanz-Pérez et al., 2016; Sinha et al., 2017). There is lower agreement and a smaller evidence base at the lower end of the cost range. Fuss et al. (2018) narrow this range to 100–300 USD tCO_2^{-1}.

Research and efforts by small-scale commercialization projects focus on utilization of captured CO_2 (Wilcox et al., 2017). Given that only a few IAM scenarios incorporate DACCS (e.g., Chen and Tavoni, 2013; Strefler et al., 2018b) its possible role in cost-optimized 1.5°C scenarios is not yet fully explored. Given the technology's early stage of development (McLaren, 2012; NRC, 2015a; Nemet et al., 2018) and few demonstrations (Holmes et al., 2013; Rau et al., 2013; Agee

et al., 2016), deploying the technology at scale is still a considerable challenge, though both optimistic (Lackner et al., 2012) and pessimistic outlooks exist (Pritchard et al., 2015).

4.3.7.6   Ocean fertilization

Nutrients can be added to the ocean resulting in increased biologic production, leading to carbon fixation in the sunlit ocean and subsequent sequestration in the deep ocean or sea floor sediments. The added nutrients can be either micronutrients (such as iron) or macronutrients (such as nitrogen and/or phosphorous) (Harrison, 2017). There is *limited evidence* and *low agreement* on the readiness of this technology to contribute to rapid decarbonization (Williamson et al., 2012). Only small-scale field experiments and theoretical modelling have been conducted (e.g., McLaren, 2012). The full range of CDR potential estimates is from 15.2 ktCO_2 yr^{-1} (Bakker et al., 2001) for a spatially constrained field experiment up to 44 GtCO_2 yr^{-1} (Sarmiento and Orr, 1991) following a modelling approach, but Fuss et al. (2018) consider the potential to be extremely limited given the evidence and existing barriers. Due to scavenging of iron, the iron addition only leads to inefficient use of the nitrogen in exporting carbon (Zeebe, 2005; Aumont and Bopp, 2006; Zahariev et al., 2008).

Cost estimates range from 2 USD tCO_2^{-1} (for iron fertilization) (Boyd and Denman, 2008) to 457 USD tCO_2^{-1} (Harrison, 2013). Jones (2014) proposed values greater than 20 USD tCO_2^{-1} for nitrogen fertilization. Fertilization is expected to impact food webs by stimulating its base organisms (Matear, 2004), and extensive algal blooms may cause anoxia (Sarmiento and Orr, 1991; Matear, 2004; Russell et al., 2012) and deep water oxygen decline (Matear, 2004), with negative impacts on biodiversity. Nutrient inputs can shift ecosystem production from an iron-limited system to a P, N-, or Si-limited system depending on the location (Matear, 2004; Bertram, 2010) and non-CO_2 GHGs may increase (Sarmiento and Orr, 1991; Matear, 2004; Bertram, 2010). The greatest theoretical potential for this practice is the Southern Ocean, posing challenges for monitoring and governance (Robinson et al., 2014). The London Protocol of the International Maritime Organization has asserted authority for regulation of ocean fertilization (Strong et al., 2009), which is widely viewed as a de facto moratorium on commercial ocean fertilization activities.

There is *low agreement* in the technical literature on the permanence of CO_2 in the ocean, with estimated residence times of 1,600 years to millennia, especially if injected or buried in or below the sea floor (Williams and Druffel, 1987; Jones, 2014). Storage at the surface would mean that the carbon would be rapidly released after cessation (Zeebe, 2005; Aumont and Bopp, 2006).

---

[7]   This analysis relies on the assessment in Fuss et al. (2018), which provides more detail on saturation and permanence.

Table 4.6 | Cross-cutting issues and uncertainties across carbon dioxide removal (CDR) options, aspects and uncertainties

Area of Uncertainty	Cross-Cutting Issues and Uncertainties
Technology upscaling	• CDR options are at different stages of technological readiness (McLaren, 2012) and differ with respect to scalability. • Nemet et al. (2018) find >50% of the CDR innovation literature concerned with the earliest stages of the innovation process (R&D), identifying a dissonance between the large CO_2 removals needed in 1.5°C pathways and the long -time periods involved in scaling up novel technologies. • Lack of post-R&D literature, including incentives for early deployment, niche markets, scale up, demand, and public acceptance.
Emerging and niche technologies	• For BECCS, there are niche opportunities with high efficiencies and fewer trade-offs, for example, sugar and paper processing facilities (Möllersten et al., 2003), district heating (Kärki et al., 2013; Ericsson and Werner, 2016), and industrial and municipal waste (Sanna et al., 2012). Turner et al. (2018) constrain potential using sustainability considerations and overlap with storage basins to avoid the CO_2 transportation challenge, providing a possible, though limited entry point for BECCS. • The impacts on land use, water, nutrients and albedo of BECCS could be alleviated using marine sources of biomass that could include aquacultured micro and macro flora (Hughes et al., 2012; Lenton, 2014). • Regarding captured CO_2 as a resource is discussed as an entry point for CDR. However, this does not necessarily lead to carbon removals, particularly if the CO_2 is sourced from fossil fuels and/or if the products do not store the CO_2 for climate-relevant horizons (von der Assen et al., 2013) (see also Section 4.3.4.5). • Methane[8] is a much more potent GHG than CO_2 (Montzka et al., 2011), associated with difficult-to-abate emissions in industry and agriculture and with outgassing from lakes, wetlands, and oceans (Lockley, 2012; Stolaroff et al., 2012). Enhancing processes that naturally remove methane, either by chemical or biological decomposition (Sundqvist et al., 2012), has been proposed to remove CH_4. There is *low confidence* that existing technologies for CH_4 removal are economically or energetically suitable for large-scale air capture (Boucher and Folberth, 2010). Methane removal potentials are limited due to its low atmospheric concentration and its low chemical reactivity at ambient conditions.
Ethical aspects	• Preston (2013) identifies distributive and procedural justice, permissibility, moral hazard (Shue, 2018), and hubris as ethical aspects that could apply to large-scale CDR deployment. • There is a lack of reflection on the climate futures produced by recent modelling and implying very different ethical costs/risks and benefits (Minx et al., 2018).
Governance	• Existing governance mechanisms are scarce and either targeted at particular CDR options (e.g., ocean-based) or aspects (e.g., concerning indirect land-use change (iLUC)) associated with bioenergy upscaling, and often the mechanisms are at national or regional scale (e.g., EU). Regulation accounting for iLUC by formulating sustainability criteria (e.g., the EU Renewable Energy Directive) has been assessed as insufficient in avoiding leakage (e.g., Frank et al., 2013). • An international governance mechanism is only in place for R&D of ocean fertilization within the Convention on Biological Diversity (IMO, 1972, 1996; CBD, 2008, 2010). • Burns and Nicholson (2017) propose a human rights-based approach to protect those potentially adversely impacted by CDR options.
Policy	• The CDR potentials that can be realized are constrained by the lack of policy portfolios incentivising large-scale CDR (Peters and Geden, 2017). • Near-term opportunities could be supported through modifying existing policy mechanisms (Lomax et al., 2015). • Scott and Geden (2018) sketch three possible routes for limited progress, (i) at EU-level, (ii) at EU Member State level, and (iii) at private sector level, noting the implied paradigm shift this would entail. • EU may struggle to adopt policies for CDR deployment on the scale or time-frame envisioned by IAMs (Geden et al., 2018). • Social impacts of large-scale CDR deployment (Buck, 2016) require policies taking these into account.
Carbon cycle	• On long time scales, natural sinks could reverse (C.D. Jones et al., 2016) • No robust assessments yet of the effectiveness of CDR in reverting climate change (Tokarska and Zickfeld, 2015; Wu et al., 2015; Keller et al., 2018), see also Chapter 2, Section 2.2.2.2.

4.3.8   Solar Radiation Modification (SRM)

This report refrains from using the term 'geoengineering' and separates SRM from CDR and other mitigation options (see Chapter 1, Section 1.4.1 and Glossary).

Table 4.7 gives an overview of SRM methods and characteristics. For a more comprehensive discussion of currently proposed SRM methods, and their implications for geophysical quantities and sustainable development, see also Cross-Chapter Box 10 in this Chapter. This section assesses the feasibility, from an institutional, technological, economic and social-cultural viewpoint, focusing on stratospheric aerosol injection (SAI) unless otherwise indicated, as most available literature is about SAI.

Some of the literature on SRM appears in the forms of commentaries, policy briefs, viewpoints and opinions (e.g., (Horton et al., 2016; Keith et al., 2017; Parson, 2017). This assessment covers original research rather than viewpoints, even if the latter appear in peer-reviewed journals.

SRM could reduce some of the global risks of climate change related to temperature rise (Izrael et al., 2014; MacMartin et al., 2014), rate of sea level rise (Moore et al., 2010), sea-ice loss (Berdahl et al., 2014) and frequency of extreme storms in the North Atlantic and heatwaves in Europe (Jones et al., 2018). SRM also holds risks of changing precipitation and ozone concentrations and potentially reductions in biodiversity (Pitari et al., 2014; Visioni et al., 2017a; Trisos et al., 2018). Literature only supports SRM as a supplement to deep mitigation, for example in overshoot scenarios (Smith and Rasch, 2013; MacMartin et al., 2018).

4.3.8.1   Governance and institutional feasibility

There is *robust evidence* but *medium agreement* for unilateral action potentially becoming a serious SRM governance issue (Weitzman, 2015; Rabitz, 2016), as some argue that enhanced collaboration might emerge around SRM (Horton, 2011). An equitable institutional or governance arrangement around SRM would have to reflect views of different countries (Heyen et al., 2015) and be multilateral because of the risk of termination, and risks that implementation or unilateral action by one country or organization will produce negative precipitation or extreme weather effects across borders (Lempert and

---

8   Current work (e.g., de Richter et al., 2017) examines other technologies considering non-CO_2 GHGs like N_2O.

Table 4.7 | Overview of the main characteristics of the most-studied SRM methods.

SRM indicator	Stratospheric Aerosol injection (SAI)	Marine Cloud Brightening (MCB)	Cirrus Cloud Thinning (CCT)	Ground-Based Albedo Modification (GBAM)
Description of SRM method	Injection of a gas in the stratosphere, which then converts to aerosols. Injection of other particles also considered.	Spraying sea salt or other particles into marine clouds, making them more reflective.	Seeding to promote nucleation, reducing optical thickness and cloud lifetime, to allow more outgoing longwave radiation to escape into space.	Whitening roofs, changes in land use management (e.g., no-till farming), change of albedo at a larger scale (covering glaciers or deserts with reflective sheeting and changes in ocean albedo).
Radiative forcing efficiencies	1–4 TgS W^{-1} m^2 yr^{-1}	100–295 Tg dry sea salt W^{-1} m^2 yr^{-1}	Not known	Small on global scale, up to 1°C–3°C on regional scale
Amount needed for 1°C overshoot	2–8 TgS yr^{-1}	70 Tg dry sea salt yr^{-1}	Not known	0.04–0.1 albedo change in agricultural and urban areas
SRM specific impacts on climate variables	Changes in precipitation patterns and circulation regimes; in case of SO_2 injection, disruption to stratospheric chemistry (for instance NOx depletion and changes in methane lifetime); increase in stratospheric water vapour and tropospheric-stratospheric ice formation affecting cloud microphysics	Regional rainfall responses; reduction in hurricane intensity	Low-level cloud changes; tropospheric drying; intensification of the hydrological cycle	Impacts on precipitation in monsoon areas; could target hot extremes
SRM specific impacts on human/ natural systems	In case of SO_2 injection, stratospheric ozone loss (which could also have a positive effect – a net reduction in global mortality due to competing health impact pathways) and significant increase of surface UV	Reduction in the number of mild crop failures	Not known	Not known
Maturity of science	Volcanic analogues; *high agreement* amongst simulations; *robust evidence* on ethical, governance and sustainable development limitations	Observed in ships tracks; several simulations confirm mechanism; regionally limited	No clear physical mechanism; *limited evidence* and *low agreement*; several simulations	Natural and land-use analogues; several simulations confirm mechanism; *high agreement* to influence on regional temperature; land use costly
Key references	Robock et al., 2008; Heckendorn et al., 2009; Tilmes et al., 2012, 2016; Pitari et al., 2014; Crook et al., 2015; C.J. Smith et al., 2017; Visioni et al., 2017a, b; Eastham et al., 2018; Plazzotta et al., 2018	Salter et al., 2008; Alterskjær et al., 2012; Jones and Haywood, 2012; Latham et al., 2012, 2013; Kravitz et al., 2013; Crook et al., 2015; Parkes et al., 2015; Ahlm et al., 2017	Storelvmo et al., 2014; Kristjánsson et al., 2015; Jackson et al., 2016; Kärcher, 2017; Lohmann and Gasparini, 2017	Irvine et al., 2011; Akbari et al., 2012; Jacobson and Ten Hoeve, 2012; Davin et al., 2014; Crook et al., 2015, 2016; Seneviratne et al., 2018

Prosnitz, 2011; Dilling and Hauser, 2013; NRC, 2015b). Some have suggested that the governance of research and field experimentation can help clarify uncertainties surrounding deployment of SRM (Long and Shepherd, 2014; Parker, 2014; NRC, 2015c; Caldeira and Bala, 2017; Lawrence and Crutzen, 2017), and that SRM is compatible with democratic processes (Horton et al., 2018) or not (Szerszynski et al., 2013; Owen, 2014).

Several possible institutional arrangements have been considered for SRM governance: under the UNFCCC (in particular under the Subsidiary Body on Scientific and Technological Advice (SBSTA)) or the United Nations Convention on Biological Diversity (UNCBD) (Honegger et al., 2013; Nicholson et al., 2018), or through a consortium of states (Bodansky, 2013; Sandler, 2017). Reasons for states to join an international governance framework for SRM include having a voice in SRM diplomacy, prevention of unilateral action by others and benefits from research collaboration (Lloyd and Oppenheimer, 2014).

Alongside SBSTA, the WMO, UNESCO and UN Environment could play a role in governance of SRM (Nicholson et al., 2018). Each of these organizations has relevance with respect to the regulatory framework (Bodle et al., 2012; Williamson and Bodle, 2016). The UNCBD gives guidance that 'that no climate-related geo-engineering activities that may affect biodiversity take place' (CBD, 2010).

4.3.8.2 Economic and technological feasibility

The literature on the engineering costs of SRM is limited and may be unreliable in the absence of testing or deployment. There is *high agreement* that costs of SAI (not taking into account indirect and social costs, research and development costs and monitoring expenses) may be in the range of 1–10 billion USD yr^{-1} for injection of 1–5 MtS to achieve cooling of 1–2 W m^{-2} (Robock et al., 2009; McClellan et al., 2012; Ryaboshapko and Revokatova, 2015; Moriyama et al., 2016), suggesting that cost-effectiveness may be high if side-effects are low or neglected (McClellan et al., 2012). The overall economic feasibility of SRM also depends on externalities and social costs (Moreno-Cruz and Keith, 2013; Mackerron, 2014), climate sensitivity (Kosugi, 2013), option value (Arino et al., 2016), presence of climate tipping points (Eric Bickel, 2013) and damage costs as a function of the level of SRM (Bahn et al., 2015; Heutel et al., 2018). Modelling of game-theoretic, strategic interactions of states under heterogeneous climatic impacts shows *low agreement* on the outcome and viability of a cost-benefit analysis for SRM (Ricke et al., 2015; Weitzman, 2015).

For SAI, there is *high agreement* that aircrafts could, after some modifications, inject millions of tons of SO_2 in the lower stratosphere (at approximately 20 km; (Davidson et al., 2012; McClellan et al., 2012; Irvine et al., 2016).

4.3.8.3 Social acceptability and ethics

Ethical questions around SRM include those of international responsibilities for implementation, financing, compensation for negative effects, the procedural justice questions of who is involved in decisions, privatization and patenting, welfare, informed consent by affected publics, intergenerational ethics (because SRM requires sustained action in order to avoid termination hazards), and the so-called 'moral hazard' (Burns, 2011; Whyte, 2012; Gardiner, 2013; Lin, 2013; Buck et al., 2014; Klepper and Rickels, 2014; Morrow, 2014; Wong, 2014; Reynolds, 2015; Lockley and Coffman, 2016; McLaren, 2016; Suarez and van Aalst, 2017; Reynolds et al., 2018). The literature

shows *low agreement* on whether SRM research and deployment may lead policy-makers to reduce mitigation efforts and thus imply a moral hazard (Linnér and Wibeck, 2015). SRM might motivate individuals (as opposed to policymakers) to reduce their GHG emissions, but even a subtle difference in the articulation of information about SRM can influence subsequent judgements of favourability (Merk et al., 2016). The argument that SRM research increases the likelihood of deployment (the 'slippery slope' argument), is also made (Quaas et al., 2017), but some also found an opposite effect (Bellamy and Healey, 2018).

Unequal representation and deliberate exclusion are plausible in decision-making on SRM, given diverging regional interests and the anticipated low resource requirements to deploy SRM (Ricke et al., 2013). Whyte (2012) argues that the concerns, sovereignties, and experiences of indigenous peoples may particularly be at risk.

The general public can be characterized as oblivious to and worried about SRM (Carr et al., 2013; Parkhill et al., 2013; Wibeck et al., 2017). An emerging literature discusses public perception of SRM, showing a lack of knowledge and unstable opinions (Scheer and Renn, 2014). The perception of controllability affects legitimacy and public acceptability of SRM experiments (Bellamy et al., 2017). In Germany, laboratory work on SRM is generally approved of, field research much less so, and immediate deployment is largely rejected (Merk et al., 2015; Braun et al., 2017). Various factors could explain variations in the degree of rejection of SRM between Canada, China, Germany, Switzerland, the United Kingdom, and the United States (Visschers et al., 2017).

Cross-Chapter Box 10 | Solar Radiation Modification in the Context of 1.5°C Mitigation Pathways

Contributing Authors:
Anastasia Revokatova (Russian Federation), Heleen de Coninck (Netherlands/EU), Piers Forster (UK), Veronika Ginzburg (Russian Federation), Jatin Kala (Australia), Diana Liverman (USA), Maxime Plazzotta (France), Roland Séférian (France), Sonia I. Seneviratne (Switzerland), Jana Sillmann (Norway).

Solar radiation modification (SRM) refers to a range of radiation modification measures not related to greenhouse gas (GHG) mitigation that seek to limit global warming (see Chapter 1, Section 1.4.1). Most methods involve reducing the amount of incoming solar radiation reaching the surface, but others also act on the longwave radiation budget by reducing optical thickness and cloud lifetime (see Table 4.7). In the context of this report, SRM is assessed in terms of its potential to limit warming below 1.5°C in temporary overshoot scenarios as a way to reduce elevated temperatures and associated impacts (Irvine et al., 2016; Keith and Irvine, 2016; Chen and Xin, 2017; Sugiyama et al., 2017a; Visioni et al., 2017a; MacMartin et al., 2018). The inherent variability of the climate system would make it difficult to detect the efficacy or side-effects of SRM intervention when deployed in such a temporary scenario (Jackson et al., 2015).

A. Potential SRM timing and magnitude
Published SRM approaches are summarized in Table 4.7. The timing and magnitude of potential SRM deployment depends on the temperature overshoot associated with mitigation pathways. All overshooting pathways make use of carbon dioxide removal. Therefore, if considered, SRM would only be deployed as a supplemental measure to large-scale carbon dioxide removal (Chapter 2, Section 2.3).

Cross-Chapter Box 10, Figure 1 below illustrates an example of how a hypothetical SRM deployment based on stratospheric aerosols injection (SAI) could be used to limit warming below 1.5°C using an 'adaptive SRM' approach (e.g., Kravitz et al., 2011; Tilmes et al., 2016), where global mean temperature rise exceeds 1.5°C compared to pre-industrial level by mid-century and returns below 1.5°C before 2100 with a 66% likelihood (see Chapter 2). In all such limited adaptive deployment scenarios, deployment of SRM only

Cross Chapter Box 10 (continued)

commences under conditions in which CO_2 emissions have already fallen substantially below their peak level and are continuing to fall. In order to hold warming to 1.5°C, a hypothetical SRM deployment could span from one to several decades, with the earliest possible threshold exceedance occurring before mid-century. Over this duration, SRM has to compensate for warming that exceeds 1.5°C (displayed with hatching on panel a) with a decrease in radiative forcing (panel b) which could be achieved with a rate of SAI varying between 0–5.9 $MtSO_2$ yr^{-1} (panel c) (Robock et al., 2008; Heckendorn et al., 2009).

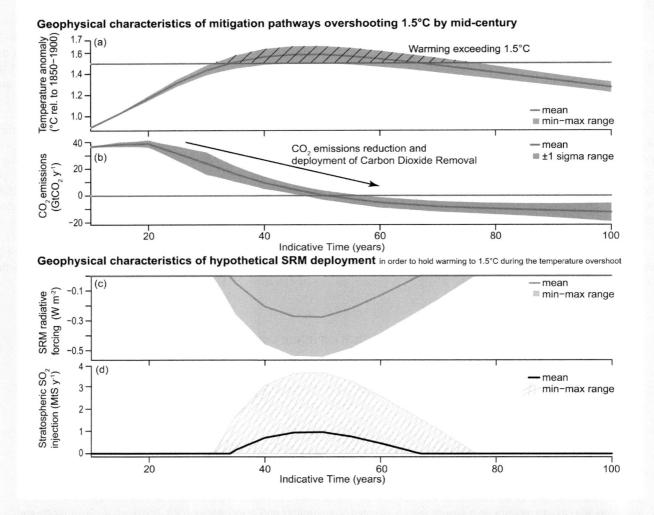

Cross-Chapter Box 10, Figure 1 | Evolution of hypothetical SRM deployment (based on stratospheric aerosols injection, or SAI) in the context of 1.5°C-consistent pathways. (a) Range of median temperature outcomes as simulated by MAGICC (see in Chapter 2, Section 2.2) given the range of CO_2 emissions and (b) other climate forcers for mitigation pathways exceeding 1.5°C at mid-century and returning below by 2100 with a 66% likelihood. Geophysical characteristics are represented by (c) the magnitude of radiative forcing and (d) the amount of stratospheric SO_2 injection that are required to keep the global median temperature below 1.5°C during the temperature overshoot (given by the blue hatching on panel a). SRM surface radiative forcing has been diagnosed using a mean cooling efficiency of 0.3°C (W– m²) of Plazzotta et al. (2018). Magnitude and timing of SO_2 injection have been derived from published estimates of Heckendorn et al. (2009) and Robock et al. (2008).

SAI is the most-researched SRM method, with *high agreement* that it could limit warming to below 1.5°C (Tilmes et al., 2016; Jones et al., 2018). The response of global temperature to SO_2 injection, however, is uncertain and varies depending on the model parametrization and emission scenarios (Jones et al., 2011; Kravitz et al., 2011; Izrael et al., 2014; Crook et al., 2015; Niemeier and Timmreck, 2015; Tilmes et al., 2016; Kashimura et al., 2017). Uncertainty also arises due to the nature and the optical properties of injected aerosols.

Cross Chapter Box 10 (continued)

Other approaches are less well researched, but the literature suggests that ground-based albedo modification (GBAM), marine cloud brightening (MCB) or cirrus cloud thinning (CCT) are not assessed to be able to substantially reduce overall global temperature (Irvine et al., 2011; Seneviratne et al., 2018). However, these SRM approaches are known to create spatially heterogeneous forcing and potentially more spatially heterogeneous climate effects, which may be used to mitigate regional climate impacts. This may be of most relevance in the case of GBAM when applied to crop and urban areas (Seneviratne et al., 2018). Most of the literature on regional mitigation has focused on GBAM in relationship with land-use and land-cover change scenarios. Both models and observations suggest that there is a *high agreement* that GBAM would result in cooling over the region of changed albedo, and in particular would reduce hot extremes (Irvine et al., 2011; Akbari et al., 2012; Jacobson and Ten Hoeve, 2012; Davin et al., 2014; Crook et al., 2015, 2016; Alkama and Cescatti, 2016; Seneviratne et al., 2018). In comparison, there is a *limited evidence* on the ability of MCB or CCT to mitigate regional climate impacts of 1.5°C warming because the magnitude of the climate response to MCB or CCT remains uncertain and the processes are not fully understood (Lohmann and Gasparini, 2017).

B. General consequences and impacts of solar radiation modification

It has been proposed that deploying SRM as a supplement to mitigation may reduce increases in global temperature-related extremes and rainfall intensity, and lessen the loss of coral reefs from increasing sea-surface temperatures (Keith and Irvine, 2016), but it would not address, or could even worsen (Tjiputra et al., 2016), negative effects from continued ocean acidification.

Another concern with SRM is the risk of a 'termination shock' or 'termination effect' when suddenly stopping SRM, which might cause rapid temperature rise and associated impacts (Jones et al., 2013; Izrael et al., 2014; McCusker et al., 2014), most noticeably biodiversity loss (Trisos et al., 2018). The severity of the termination effect has recently been debated (Parker and Irvine, 2018) and depends on the degree of SRM cooling. This report only considers limited SRM in the context of mitigation pathways to 1.5°C. Other risks of SRM deployment could be associated with the lack of testing of the proposed deployment schemes (e.g., Schäfer et al., 2013). Ethical aspects and issues related to the governance and economics are discussed in Section 4.3.8.

C. Consequences and impacts of SRM on the carbon budget

Because of its effects on surface temperature, precipitation and surface shortwave radiation, SRM would also alter the carbon budget pathways to 1.5°C or 2°C (Eliseev, 2012; Keller et al., 2014; Keith et al., 2017; Lauvset et al., 2017).

Despite the large uncertainties in the simulated climate response to SRM, current model simulations suggest that SRM would lead to altered carbon budgets compatible with 1.5°C or 2°C. The 6 CMIP5 models investigated simulated an increase of natural carbon uptake by land biosphere and, to a smaller extent, by the oceans (*high agreement*). The multimodel mean of this response suggests an increase of the RCP4.5 carbon budget of about 150 $GtCO_2$ after 50 years of SO_2 injection with a rate of 4 TgS yr^{-1}, which represents about 4 years of CO_2 emissions at the current rate (36 $GtCO_2$ yr^{-1}). However, there is uncertainty around quantitative determination of the effects that SRM or its cessation has on the carbon budget due to a lack of understanding of the radiative processes driving the global carbon cycle response to SRM (Ramachandran et al., 2000; Mercado et al., 2009; Eliseev, 2012; Xia et al., 2016), uncertainties about how the carbon cycle will respond to termination effects of SRM, and uncertainties in climate–carbon cycle feedbacks (Friedlingstein et al., 2014).

D. Sustainable development and SRM

There are few studies investigating potential implications of SRM for sustainable development. These are based on a limited number of scenarios and hypothetical considerations, mainly referring to benefits from lower temperatures (Irvine et al., 2011; Nicholson, 2013; Anshelm and Hansson, 2014; Harding and Moreno-Cruz, 2016). Other studies suggest negative impacts from SRM implementation concerning issues related to regional disparities (Heyen et al., 2015), equity (Buck, 2012), fisheries, ecosystems, agriculture, and termination effects (Robock, 2012; Morrow, 2014; Wong, 2014). If SRM is initiated by the richer nations, there might be issues with local agency, and possibly worsening conditions for those suffering most under climate change (Buck et al., 2014). In addition, ethical issues related to testing SRM have been raised (e.g., Lenferna et al., 2017). Overall, there is *high agreement* that SRM would affect many development issues but *limited evidence* on the degree of influence, and how it manifests itself across regions and different levels of society.

E. Overall feasibility of SRM

If mitigation efforts do not keep global mean temperature below 1.5°C, SRM can potentially reduce the climate impacts of a temporary temperature overshoot, in particular extreme temperatures, rate of sea level rise and intensity of tropical cyclones, alongside intense mitigation and adaptation efforts. While theoretical developments show that SRM is technically feasible (see Section 4.3.8.2), global field experiments have not been conducted and most of the knowledge about SRM is based on imperfect model simulations and some natural analogues. There are also considerable challenges to the implementation of SRM associated with disagreements over the governance, ethics, public perception, and distributional development impacts (see Section 4.3.8) (Boyd, 2016; Preston, 2016; Asayama et al., 2017; Sugiyama et al., 2017b; Svoboda, 2017; McKinnon, 2018; Talberg et al., 2018). Overall, the combined uncertainties surrounding the various SRM approaches, including technological maturity, physical understanding, potential impacts, and challenges of governance, constrain the ability to implement SRM in the near future.

4.4 Implementing Far-Reaching and Rapid Change

The feasibility of 1.5°C-compatible pathways is contingent upon enabling conditions for systemic change (see Cross Chapter Box 3 in Chapter 1). Section 4.3 identifies the major systems, and options within those systems, that offer the potential for change to align with 1.5°C pathways.

AR5 identifies enabling conditions as influencing the feasibility of climate responses (Kolstad et al., 2014). This section draws on 1.5°C-specific and related literature on rapid and scaled up change to identify the enabling conditions that influence the feasibility of adaptation and mitigation options assessed in Section 4.5. Examples from diverse regions and sectors are provided in Boxes 4.1 to 4.10 to illustrate how these conditions could enable or constrain the implementation of incremental, rapid, disruptive and transformative mitigation and adaptation consistent with 1.5°C pathways.

Coherence between the enabling conditions holds potential to enhance the feasibility of 1.5°C-consistent pathways and adapting to the consequences. This includes better alignment across governance scales (OECD, 2015a; Geels et al., 2017), enabling multilevel governance (Cheshmehzangi, 2016; Revi, 2017; Tait and Euston-Brown, 2017) and nested institutions (Abbott, 2012). It also includes interdisciplinary actions, combined adaptation and mitigation action (Göpfert et al., 2018), and science–policy partnerships (Vogel et al., 2007; Hering et al., 2014; Roberts, 2016; Figueres et al., 2017; Leal Filho et al., 2018). These partnerships are difficult to establish and sustain, but can generate trust (Cole, 2015; Jordan et al., 2015) and inclusivity that ultimately can provide durability and the realization of co-benefits for sustained rapid change (Blanchet, 2015; Ziervogel et al., 2016a).

4.4.1 Enhancing Multilevel Governance

Addressing climate change and implementing responses to 1.5°C-consistent pathways would require engagement between various levels and types of governance (Betsill and Bulkeley, 2006; Kern and Alber, 2009; Christoforidis et al., 2013; Romero-Lankao et al., 2018). AR5 highlighted the significance of governance as a means of strengthening adaptation and mitigation and advancing sustainable development (Fleurbaey et al., 2014). Governance is defined in the broadest sense as the 'processes of interaction and decision-making among actors involved in a common problem' (Kooiman, 2003; Hufty, 2011; Fleurbaey et al., 2014). This definition goes beyond notions of formal government or political authority and integrates other actors, networks, informal institutions and communities.

4.4.1.1 Institutions and their capacity to invoke far-reaching and rapid change

Institutions – the rules and norms that guide human interactions (Section 4.4.2) – enable or impede the structures, mechanisms and measures that guide mitigation and adaptation. Institutions, understood as the 'rules of the game' (North, 1990), exert direct and indirect influence over the viability of 1.5°C-consistent pathways (Munck et al., 2014; Willis, 2017). Governance would be needed to support wide-scale and effective adoption of mitigation and adaptation options. Institutions and governance structures are strengthened when the principle of the 'commons' is explored as a way of sharing management and responsibilities (Ostrom et al., 1999; Chaffin et al., 2014; Young, 2016). Institutions would need to be strengthened to interact amongst themselves, and to share responsibilities for the development and implementation of rules, regulations and policies (Ostrom et al., 1999; Wejs et al., 2014; Craig et al., 2017), with the goal of ensuring that these embrace equity, justice, poverty alleviation and sustainable development, enabling a 1.5°C world (Reckien et al., 2017; Wood et al., 2017).

Several authors have identified different modes of cross-stakeholder interaction in climate policy, including the role played by large multinational corporations, small enterprises, civil society and non-state actors. Ciplet et al. (2015) argue that civil society is to a great extent the only reliable motor for driving institutions to change at the pace required. Kern and Alber (2009) recognize different forms of collaboration relevant to successful climate policies beyond the local level. Horizontal collaboration (e.g., transnational city networks) and vertical collaboration within nation-states can play an enabling role (Ringel, 2017). Vertical and horizontal collaboration requires synergistic relationships between stakeholders (Ingold and Fischer, 2014; Hsu et al., 2017). The importance of community participation is emphasized in literature, and in particular the need to take into account equity and gender considerations (Chapter 5) (Graham et al., 2015; Bryan et al., 2017; Wangui and Smucker, 2017). Participation often faces implementation challenges and may not always result in better policy outcomes. Stakeholders, for example, may not view climate change as a priority and may not share the same preferences, potentially creating a policy deadlock (Preston et al., 2013, 2015; Ford et al., 2016).

4.4.1.2 International governance

International treaties help strengthen policy implementation, providing a medium- and long-term vision (Obergassel et al., 2016). International climate governance is organized via many mechanisms, including international organizations, treaties and conventions, for example,

UNFCCC, the Paris Agreement and the Montreal Protocol. Other multilateral and bilateral agreements, such as trade agreements, also have a bearing on climate change.

There are significant differences between global mitigation and adaptation governance frames. Mitigation tends to be global by its nature and based on the principle of the climate system as a global commons (Ostrom et al., 1999). Adaptation has traditionally been viewed as a local process, involving local authorities, communities, and stakeholders (Khan, 2013; Preston et al., 2015), although it is now recognized to be a multi-scaled, multi-actor process that transcends scales from local and sub-national to national and international (Mimura et al., 2014; UNEP, 2017a). National governments provide a central pivot for coordination, planning, determining policy priorities and distributing resources. National governments are accountable to the international community through international agreements. Yet, many of the impacts of climate change are transboundary, so that bilateral and multilateral cooperation are needed (Nalau et al., 2015; Donner et al., 2016; Magnan and Ribera, 2016; Tilleard and Ford, 2016; Lesnikowski et al., 2017). The Kigali Amendment to the Montreal Protocol demonstrates that a global environmental agreement facilitating common but differentiated responsibilities is possible (Sharadin, 2018). This was operationalized by developed countries acting first, with developing countries following and benefiting from leap-frogging the trial-and-error stages of innovative technology development.

Work on international climate governance has focused on the nature of 'climate regimes' and coordinating the action of nation-states (Aykut, 2016) organized around a diverse set of instruments: (i) binding limits allocated by principles of historical responsibility and equity, (ii) carbon prices, emissions quotas, (iii) pledges and review of policies and measures or (iv) a combination of these options (Stavins, 1988; Grubb, 1990; Pizer, 2002; Newell and Pizer, 2003).

Literature on the Kyoto Protocol provides two important insights for the 1.5°C transition: the challenge of agreeing on rules to allocate emissions quotas (Shukla, 2005; Caney, 2012; Winkler et al., 2013; Gupta, 2014; Méjean et al., 2015) and a climate-centric vision (Shukla, 2005; BASIC experts, 2011), separated from development issues which drove resistance from many developing nations (Roberts and Parks, 2006). For the former, a burden-sharing approach led to an adversarial process among nations to decide who should be allocated 'how much' of the remainder of the emissions budget (Caney, 2014; Ohndorf et al., 2015; Roser et al., 2015; Giménez-Gómez et al., 2016). Industry group lobbying further contributed to reducing space for manoeuvre of some major emitting nations (Newell and Paterson, 1998; Levy and Egan, 2003; Dunlap and McCright, 2011; Michaelowa, 2013; Geels, 2014).

Given the political unwillingness to continue with the Kyoto Protocol approach a new approach was introduced in the Copenhagen Accord, the Cancun Agreements, and finally in the Paris Agreement. The transition to 1.5°C requires carbon neutrality and thus going beyond the traditional framing of climate as a 'tragedy of the commons' to be addressed via cost-optimal allocation rules, which demonstrated a low probability of enabling a transition to 1.5°C-consistent pathways (Patt, 2017). The Paris Agreement, built on a 'pledge and review' system,

is thought be more effective in securing trust (Dagnet et al., 2016) and enables effective monitoring and timely reporting on national actions (including adaptation), allowing for international scrutiny and persistent efforts of civil society and non-state actors to encourage action in both national and international contexts (Allan and Hadden, 2017; Bäckstrand and Kuyper, 2017; Höhne et al., 2017; Lesnikowski et al., 2017; Maor et al., 2017; UNEP, 2017a), with some limitations (Nieto et al., 2018).

The paradigm shift enabled at Cancun succeeded by focusing on the objective of 'equitable access to sustainable development' (Hourcade et al., 2015). The use of 'pledge and review' now underpins the Paris Agreement. This consolidates multiple attempts to define a governance approach that relies on Nationally Determined Contributions (NDCs) and on means for a 'facilitative model' (Bodansky and Diringer, 2014) to reinforce them. This enables a regular, iterative, review of NDCs allowing countries to set their own ambitions after a global stocktake and more flexible, experimental forms of climate governance, which may provide room for higher ambition and be consistent with the needs of governing for a rapid transition to close the emission gap (Clémençon, 2016; Falkner, 2016) (Cross-Chapter Box 11 in this chapter). Beyond a general consensus on the necessity of measurement, reporting and verification (MRV) mechanisms as a key element of a climate regime (Ford et al., 2015b; van Asselt et al., 2015), some authors emphasize different governance approaches to implement the Paris Agreement. Through the new proposed sustainable development mechanism in Article 6, the Paris Agreement allows the space to harness the lowest cost mitigation options worldwide. This may incentivize policymakers to enhance mitigation ambition by speeding up climate action as part of a 'climate regime complex' (Keohane and Victor, 2011) of loosely interrelated global governance institutions. In the Paris Agreement, the 'common but differentiated responsibilities and respective capabilities' (CBDR-RC) principle could be expanded and revisited under a 'sharing the pie' paradigm (Ji and Sha, 2015) as a tool to open innovation processes towards alternative development pathways (Chapter 5).

COP 16 in Cancun was also the first time in the UNFCCC that adaptation was recognized to have similar priority as mitigation. The Paris Agreement recognizes the importance of adaptation action and cooperation to enhance such action. Chung Tiam Fook (2017) and Lesnikowski et al. (2017) suggest that the Paris Agreement is explicit about multilevel adaptation governance, outlines stronger transparency mechanisms, links adaptation to development and climate justice, and is therefore suggestive of greater inclusiveness of non-state voices and the broader contexts of social change.

1.5°C-consistent pathways require further exploration of conditions of trust and reciprocity amongst nation states (Schelling, 1991; Ostrom and Walker, 2005). Some authors (Colman et al., 2011; Courtois et al., 2015) suggest a departure from the vision of actors acting individually in the pursuit of self-interest to that of iterated games with actors interacting over time showing that reciprocity, with occasional forgiveness and initial good faith, can lead to win-win outcomes and to cooperation as a stable strategy (Axelrod and Hamilton, 1981).

Regional cooperation plays an important role in the context of global governance. Literature on climate regimes has only started exploring innovative governance arrangements, including coalitions of transnational actors including state, market and non-state actors (Bulkeley et al., 2012; Hovi et al., 2016; Hagen et al., 2017; Hermwille et al., 2017; Roelfsema et al., 2018) and groupings of countries, as a complement to the UNFCCC (Abbott and Snidal, 2009; Biermann, 2010; Zelli, 2011; Nordhaus, 2015). Climate action requires multilevel governance from the local and community level to national, regional and international levels. Box 4.1 shows the role of sub-national authorities (e.g., regions and provinces) in facilitating urban climate action, while Box 4.2 shows that climate governance can be organized across hydrological as well as political units.

4.4.1.3   Sub-national governance

Local governments can play a key role (Melica et al., 2018; Romero-Lankao et al., 2018) in influencing mitigation and adaptation strategies. It is important to understand how rural and urban areas, small islands, informal settlements and communities might intervene to reduce climate impacts (Bulkeley et al., 2011), either by implementing climate objectives defined at higher government levels or by taking initiative autonomously or collectively (Aall et al., 2007; Reckien et al., 2014; Araos et al., 2016a; Heidrich et al., 2016). Local governance faces the challenge of reconciling local concerns with global objectives. Local governments could coordinate and develop effective local responses, and could pursue procedural justice in ensuring community engagement and more effective policies around energy and vulnerability reduction (Moss et al., 2013; Fudge et al., 2016). They can enable more participative decision-making (Barrett, 2015; Hesse, 2016). Fudge et al. (2016) argue that local authorities are well-positioned to involve the wider community in: designing and implementing climate policies, engaging with sustainable energy generation (e.g., by supporting energy communities) (Slee, 2015), and the delivery of demand-side measures and adaptation implementation.

By 2050, it is estimated three billion people will be living in slums and informal settlements: neighbourhoods without formal governance, on un-zoned land developments and in places that are exposed to climate-related hazards (Bai et al., 2018). Emerging research is examining how citizens can contribute informally to governance with rapid urbanization and weaker government regulation (Sarmiento and Tilly, 2018). It remains to be seen how the possibilities and consequences of alternative urban governance models will be managed for large, complex problems and for addressing inequality and urban adaptation (Amin and Cirolia, 2018; Bai et al., 2018; Sarmiento and Tilly, 2018).

Expanding networks of cities are sharing experiences on coping with climate change and drawing economic and development benefits from climate change responses – a recent institutional innovation. This could be complemented by efforts of national governments to enhance local climate action through national urban policies (Broekhoff et al., 2018). Over the years, non-state actors have set up several transnational climate governance initiatives to accelerate the climate response, for example, ICLEI (1990), C–40 (2005), the Global Island Partnership (2006) and the Covenant of Mayors (2008) (Gordon and Johnson, 2017; Hsu et al., 2017; Ringel, 2017; Kona et al., 2018; Melica et al., 2018) and to exert influence on national governments and the UNFCCC

(Bulkeley, 2005). However, Michaelowa and Michaelowa (2017) find low effectiveness for over 100 of such mitigation initiatives.

4.4.1.4   Interactions and processes for multilevel governance

Literature has proposed multilevel governance in climate change as an enabler for systemic transformation and effective governance, as the concept is thought to allow for combining decisions across levels and sectors and across institutional types at the same level (Romero-Lankao et al., 2018), with multilevel reinforcement and the mobilization of economic interests at different levels of governance (Jänicke and Quitzow, 2017). These governance mechanisms are based on accountability and transparency rules and participation and coordination across and within these levels.

A study of 29 European countries showed that the rapid adoption and diffusion of adaptation policymaking is largely driven by internal factors, at the national and sub-national levels (Massey et al., 2014). An assessment of national-level adaptation in 117 countries (Berrang-Ford et al., 2014) found good governance to be the one of the strongest predictors of national adaptation policy. An analysis of the climate responses of 200 large and medium-sized cities across eleven European countries found that factors such as membership of climate networks, population size, gross domestic product (GDP) per capita and adaptive capacity act as drivers of mitigation and adaptation plans (Reckien et al., 2015).

Adaptation policy has seen growth in some areas (Massey et al., 2014; Lesnikowski et al., 2016), although efforts to track adaptation progress are constrained by an absence of data sources on adaptation (Berrang-Ford et al., 2011; Ford and Berrang-Ford, 2016; Magnan, 2016; Magnan and Ribera, 2016). Many developing countries have made progress in formulating national policies, plans and strategies on responding to climate change. The NDCs have been identified as one such institutional mechanism (Cross-Chapter Box 11 in this Chapter) (Magnan et al., 2015; Kato and Ellis, 2016; Peters et al., 2017).

To overcome barriers to policy implementation, local conflicts of interest or vested interests, strong leadership and agency is needed by political leaders. As shown by the Covenant of Mayors initiative (Box 4.1), political leaders with a vision for the future of the local community can succeed in reducing GHG emissions, when they are supported by civil society (Rivas et al., 2015; Croci et al., 2017; Kona et al., 2018). Any political vision would need to be translated into an action plan, which could include elements describing policies and measures needed to achieve transition, the human and financial resources needed, milestones, and appropriate measurement and verification processes (Azevedo and Leal, 2017). Discussing the plan with stakeholders and civil society, including citizens and allowing for participation for minorities, and having them provide input and endorse it, has been found to increase the likelihood of success (Rivas et al., 2015; Wamsler, 2017). However, as described by Nightingale (2017) and Green (2016), struggles over natural resources and adaptation governance both at the national and community levels would also need to be addressed 'in politically unstable contexts, where power and politics shape adaptation outcomes'.

Multilevel governance includes adaptation across local, regional, and national scales (Adger et al., 2005). The whole-of-government approach to understanding and influencing climate change policy design and implementation puts analytical emphasis on how different levels of government and different types of actors (e.g., public and private) can constrain or support local adaptive capacity (Corfee-Morlot et al., 2011), including the role of the civil society. National governments, for example, have been associated with enhancing adaptive capacity through building awareness of climate impacts, encouraging economic growth, providing incentives, establishing legislative frameworks conducive to adaptation, and communicating climate change information (Berrang-Ford et al., 2014; Massey et al., 2014; Austin et al., 2015; Henstra, 2016; Massey and Huitema, 2016). Local governments, on the other hand, are responsible for delivering basic services and utilities to the urban population, and protecting their integrity from the impacts of extreme weather (Austin et al., 2015; Cloutier et al., 2015; Nalau et al., 2015; Araos et al., 2016b). National policies and transnational governance could be seen as complementary, rather than competitors, and strong national policies favour transnational engagement of sub- and non-state actors (Andonova et al., 2017). Local initiatives are complementary with higher level policies and can be integrated in the multilevel governance system (Fuhr et al., 2018).

A multilevel approach considers that adaptation planning is affected by scale mismatches between the local manifestation of climate impacts and the diverse scales at which the problem is driven (Shi et al., 2016). Multilevel approaches may be relevant in low-income countries where limited financial resources and human capabilities within local governments often lead to greater dependency on national governments and other (donor) organizations, to strengthen adaptation responses (Donner et al., 2016; Adenle et al., 2017). National governments or international organizations may motivate urban adaptation externally through broad policy directives or projects by international donors. Municipal governments on the other hand work within the city to spur progress on adaptation. Individual political leadership in municipal government, for example, has been cited as a factor driving the adaptation policies of early adapters in Quito, Ecuador, and Durban, and South Africa (Anguelovski et al., 2014), and for adaptation more generally (Smith et al., 2009). Adaptation pathways can help identify maladaptive actions (Juhola et al., 2016; Magnan et al., 2016; Gajjar et al., 2018) and encourage social learning approaches across multiple levels of stakeholders in sectors such as marine biodiversity and water supply (Bosomworth et al., 2015; Butler et al., 2015; van der Brugge and Roosjen, 2015).

Box 4.1 | Multilevel Governance in the EU Covenant of Mayors: Example of the Provincia di Foggia

Since 2005, cities have emerged as a locus of institutional and governance climate innovation (Melica et al., 2018) and are driving responses to climate change (Roberts, 2016). Many cities have adopted more ambitious greenhouse gas (GHG) emission reduction targets than countries (Kona et al., 2018), with an overall commitment of GHG emission reduction targets by 2020 of 27%, almost 7 percentage points higher than the minimum target for 2020 (Kona et al., 2018). The Covenant of Mayors (CoM) is an initiative in which municipalities voluntarily commit to CO_2 emission reduction. The participation of small municipalities has been facilitated by the development and testing of a new multilevel governance model involving Covenant Territorial Coordinators (CTCs), i.e., provinces and regions, which commit to providing strategic guidance and financial and technical support to municipalities in their territories. Results from the 315 monitoring inventories submitted show an achievement of 23% reduction in emissions (compared to an average year 2005) for more than half of the cities under a CTC schema (Kona et al., 2018).

The Province of Foggia, acting as a CTC, gave support to 36 municipalities to participate in the CoM and to prepare Sustainable Energy Action Plans (SEAPs). The Province developed a common approach to prepare SEAPs, provided data to compile municipal emission inventories (Bertoldi et al., 2018) and guided the signatory to identify an appropriate combination of measures to curb GHG emissions. The local Chamber of Commerce also had a key role in the implementation of these projects by the municipalities (Lombardi et al., 2016). The joint action by the province and the municipalities in collaboration with the local business community could be seen as an example of multilevel governance (Lombardi et al., 2016).

Researchers have investigated local forms of collaboration within local government, with the active involvement of citizens and stakeholders, and acknowledge that public acceptance is key to the successful implementation of policies (Larsen and Gunnarsson-Östling, 2009; Musall and Kuik, 2011; Pollak et al., 2011; Christoforidis et al., 2013; Pasimeni et al., 2014; Lee and Painter, 2015). Achieving ambitious targets would need leadership, enhanced multilevel governance, vision and widespread participation in transformative change (Castán Broto and Bulkeley, 2013; Rosenzweig et al., 2015; Castán Broto, 2017; Fazey et al., 2017; Wamsler, 2017; Romero-Lankao et al., 2018). The Chapter 5, Section 5.6.4 case studies of climate-resilient development pathways, at state and community scales, show that participation, social learning and iterative decision-making are governance features of strategies that deliver mitigation, adaptation, and sustainable development in a fair and equitable manner. Another insight is the finding that incremental voluntary changes are amplified through community networking, polycentric governance (Dorsch and Flachsland, 2017), partnerships, and long-term change to governance systems at multiple levels (Stevenson and Dryzek, 2014; Lövbrand et al., 2017; Pichler et al., 2017; Termeer et al., 2017).

4

Box 4.2 | Watershed Management in a 1.5°C World

Water management is necessary in order for the global community to adapt to 1.5°C-consistent pathways. Cohesive planning that includes numerous stakeholders would be required to improve access, utilization and efficiency of water use and to ensure hydrologic viability.

Response to drought and El Niño–Southern Oscillation (ENSO) in southern Guatemala

Hydro-meteorological events, including ENSO, have impacted Central America (Steinhoff et al., 2014; Chang et al., 2015; Maggioni et al., 2016) and are projected to increase in frequency during a 1.5°C transition (Wang et al., 2017). The 2014–2016 ENSO damaged agriculture, seriously impacting rural communities.

In 2016, the Climate Change Institute, in conjunction with local governments, the private sector, communities and human rights organizations, established dialogue tables for different watersheds to discuss water usage amongst stakeholders and plans to mitigate the effects of drought, alleviate social tension, and map water use of watersheds at risk. The goal was to encourage better water resource management and to enhance ecological flow through improved communication, transparency, and coordination amongst users. These goals were achieved in 2017 when each previously affected river reached the Pacific Ocean with at least its minimum ecological flow (Guerra, 2017).

Drought management through the Limpopo Watercourse Commission

The governments sharing the Limpopo river basin (Botswana, Mozambique, South Africa and Zimbabwe) formed the Limpopo Watercourse Commission in 2003 (Nyagwambo et al., 2008; Mitchell, 2013). It has an advisory body composed of working groups that assess water use and sustainability, decide national level distribution of water access, and support disaster and emergency planning. The Limpopo basin delta is highly vulnerable (Tessler et al., 2015), and is associated with a lack of infrastructure and investment capacity, requiring increased economic development together with plans for vulnerability reduction (Tessler et al., 2015) and water rights (Swatuk, 2015). The high vulnerability is influenced by gender inequality, limited stakeholder participation and limited institutional capacity to address unequal water access (Mehta et al., 2014). The implementation of integrated water resources management (IWRM) would need to consider pre-existing social, economic, historical and cultural contexts (Merrey, 2009; Mehta et al., 2014). The Commission therefore could play a role in improving participation and in providing an adaptable and equitable strategy for cross-border water sharing (Ekblom et al., 2017).

Flood management in the Danube

The Danube River Protection Convention is the official instrument for cooperation on transboundary water governance between the countries that share the Danube Basin. The International Commission for the Protection of the Danube River (ICPDR) provides a strong science–policy link through expert working groups dealing with issues including governance, monitoring and assessment, and flood protection (Schmeier, 2014). The Trans-National Monitoring Network (TNMN) was developed to undertake comprehensive monitoring of water quality (Schmeier, 2014). Monitoring of water quality constitutes almost 50% of ICPDR's scientific publications, although ICPDR also works on governance, basin planning, monitoring, and IWRM, indicating its importance. The ICPDR is an example of IWRM 'coordinating groundwater, surface water abstractions, flood management, energy production, navigation, and water quality' (Hering et al., 2014).

Box 4.2 exemplifies how multilevel governance has been used for watershed management in different basins, given the impacts on water sources (Chapter 3, Section 3.4.2).

Cross-Chapter Box 11 | Consistency Between Nationally Determined Contributions and 1.5°C Scenarios

Contributing Authors:

Paolo Bertoldi (Italy), Michel den Elzen (Netherlands), James Ford (Canada/UK), Richard Klein (Netherlands/Germany), Debora Ley (Guatemala/Mexico), Timmons Roberts (USA), Joeri Rogelj (Austria/Belgium).

Mitigation

1. Introduction

There is *high agreement* that Nationally Determined Contributions (NDCs) are important for the global response to climate change and represent an innovative bottom-up instrument in climate change governance (Section 4.4.1), with contributions from all signatory countries (den Elzen et al., 2016; Rogelj et al., 2016; Vandyck et al., 2016; Luderer et al., 2018; Vrontisi et al., 2018). The global emission projections resulting from full implementation of the NDCs represent an improvement compared to business as usual (Rogelj et al., 2016) and current policies scenarios to 2030 (den Elzen et al., 2016; Vrontisi et al., 2018). Most G20 economies would require new policies and actions to achieve their NDC targets (den Elzen et al., 2016; Vandyck et al., 2016; UNEP, 2017b; Kuramochi et al., 2018).

2. The effect of NDCs on global greenhouse gas (GHG) emissions

Several studies estimate global emission levels that would be achieved under the NDCs (e.g., den Elzen et al., 2016; Luderer et al., 2016; Rogelj et al., 2016, 2017; Vandyck et al., 2016; Rose et al., 2017; Vrontisi et al., 2018). Rogelj et al. (2016) and UNEP (2017b) concluded that the full implementation of the unconditional and conditional NDCs are expected to result in global GHG emissions of about 55 (52–58) and 53 (50–54) GtCO$_2$-eq yr^{-1}, respectively (Cross-Chapter Box 11, Figure 1 below).

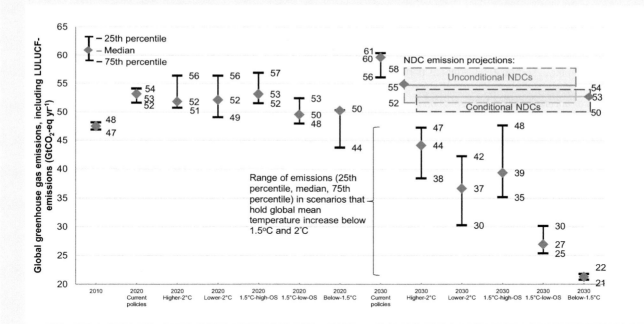

Cross-Chapter Box 11, Figure 1 | GHG emissions are all expressed in units of CO$_2$-equivalence computed with 100-year global warming potentials (GWPs) reported in IPCC SAR, while the emissions for the 1.5°C and 2°C scenarios in Table 2.4 are reported using the 100-year GWPs reported in IPCC AR4, and are hence about 3% higher. Using IPCC AR4 instead of SAR GWP values is estimated to result in a 2–3% increase in estimated 1.5°C and 2°C emissions levels in 2030. Source: based on Rogelj et al. (2016) and UNEP (2017b).

3. The effect of NDCs on temperature increase and carbon budget

Estimates of global average temperature increase are 2.9°C–3.4°C above preindustrial levels with a greater than 66% probability by 2100 (Rogelj et al., 2016; UNEP, 2017b), under a full implementation of unconditional NDCs and a continuation of climate action similar to that of the NDCs. Full implementation of the conditional NDCs would lower the estimates by about 0.2°C by 2100. As an indication of the carbon budget implications of NDC scenarios, Rogelj et al. (2016) estimated cumulative emissions in the range

Cross Chapter Box 11 (continued)

of 690 to 850 $GtCO_2$ for the period 2011–2030 if the NDCs are successfully implemented. The carbon budget for post-2010 till 2100 compatible with staying below 1.5°C with a 50–66% probability was estimated at 550–600 $GtCO_2$ (Clarke et al., 2014; Rogelj et al., 2016), which will be well exceeded by 2030 at full implementation of the NDCs (Chapter 2, Section 2.2 and Section 2.3.1).

4. The 2030 emissions gap with 1.5°C and urgency of action
As the 1.5°C pathways require reaching carbon neutrality by mid-century, the NDCs alone are not sufficient, as they have a time horizon until 2030. Rogelj et al. (2016) and Hof et al. (2017) have used results or compared NDC pathways with emissions pathways produced by integrated assessment models (IAMs) assessing the contribution of NDCs to achieve the 1.5°C targets. There is *high agreement* that current NDC emissions levels are not in line with pathways that limit warming to 1.5°C by the end of the century (Rogelj et al., 2016, 2017; Hof et al., 2017; UNEP, 2017b; Vrontisi et al., 2018). The median 1.5°C emissions gap (>66% chance) for the full implementation of both the conditional and unconditional NDCs for 2030 is 26 (19–29) to 28 (22–33) $GtCO_2$-eq (Cross-Chapter Box 11, Figure 1 above).

Studies indicate important trade-offs of delaying global emissions reductions (Chapter 2, Sections 2.3.5 and 2.5.1). AR5 identified flexibility in 2030 emission levels when pursuing a 2°C objective (Clarke et al., 2014) indicating that strongest trade-offs for 2°C pathways could be avoided if emissions are limited to below 50 $GtCO_2$-eq yr^{-1} in 2030 (here computed with the GWP–100 metric of the IPCC SAR). New scenario studies show that full implementation of the NDCs by 2030 would imply the need for deeper and faster emission reductions beyond 2030 in order to meet 2°C, and also higher costs and efforts of negative emissions (Fujimori et al., 2016; Sanderson et al., 2016; Rose et al., 2017; van Soest et al., 2017; Luderer et al., 2018). However, no flexibility has been found for 1.5°C-consistent pathways (Luderer et al., 2016; Rogelj et al., 2017), indicating that if emissions through 2030 are at NDC levels, the resulting post-2030 reductions required to remain within a 1.5°C-consistent carbon budget during the 21st century (Chapter 2, Section 2.2) are not within the feasible operating space of IAMs. This indicates that the chances of failing to reach a 1.5°C pathway are significantly increased (Riahi et al., 2015), if near-term ambition is not strengthened beyond the level implied by current NDCs.

Accelerated and stronger short-term action and enhanced longer-term national ambition going beyond the NDCs would be needed for 1.5°C-consistent pathways. Implementing deeper emissions reductions than current NDCs would imply action towards levels identified in Chapter 2, Section 2.3.3, either as part of or over-delivering on NDCs.

5. The impact of uncertainties on NDC emission levels
The measures proposed in NDCs are not legally binding (Nemet et al., 2017), further impacting estimates of anticipated 2030 emission levels. The aggregation of targets results in high uncertainty (Rogelj et al., 2017), which could be reduced with clearer guidelines for compiling future NDCs focused more on energy accounting (Rogelj et al., 2017) and increased transparency and comparability (Pauw et al., 2018).

Many factors would influence NDCs global aggregated effects, including: (1) variations in socio-economic conditions (GDP and population growth), (2) uncertainties in historical emission inventories, (3) conditionality of certain NDCs, (4) definition of NDC targets as ranges instead of single values, (5) the way in which renewable energy targets are expressed, and (6) the way in which traditional biomass use is accounted for. Additionally, there are land-use mitigation uncertainties (Forsell et al., 2016; Grassi et al., 2017). Land-use options play a key role in many country NDCs; however, many analyses on NDCs do not use country estimates on land-use emissions, but use model estimates, mainly because of the large difference in estimating the 'anthropogenic' forest sink between countries and models (Grassi et al., 2017).

6. Comparing countries' NDC ambition (equity, cost optimal allocation and other indicators)
Various assessment frameworks have been proposed to analyse, benchmark and compare NDCs, and indicate possible strengthening, based on equity and other indicators (Aldy et al., 2016; den Elzen et al., 2016; Höhne et al., 2017; Jiang et al., 2017; Holz et al., 2018).There is large variation in conformity/fulfilment with equity principles across NDCs and countries. Studies use assessment frameworks based on six effort sharing categories in the AR5 (Clarke et al., 2014) with the principles of 'responsibility', 'capability' and 'equity' (Höhne et al., 2017; Pan et al., 2017; Robiou du Pont et al., 2017). There is an important methodological gap in the assessment of the NDCs' fairness and equity implications, partly due to lack of information on countries' own assessments (Winkler et al., 2017). Implementation of Article 2.2 of the Paris Agreement could reflect equity and the principle of 'common but differentiated responsibilities and respective capabilities', due to different national circumstances and different interpretations of equity principles (Lahn and Sundqvist, 2017; Lahn, 2018).

Cross Chapter Box 11 (continued)

Adaptation

The Paris Agreement recognizes adaptation by establishing a global goal for adaptation (Kato and Ellis, 2016; Rajamani, 2016; Kinley, 2017; Lesnikowski et al., 2017; UNEP, 2017a). This is assessed qualitatively, as achieving a temperature goal would determine the level of adaptation ambition required to deal with the consequent risks and impacts (Rajamani, 2016). Countries can include domestic adaptation goals in their NDCs, which together with national adaptation plans (NAPs) give countries flexibility to design and adjust their adaptation trajectories as their needs evolve and as progress is evaluated over time. A challenge for assessing progress on adaptation globally is the aggregation of many national adaptation actions and approaches. Knowledge gaps still remain about how to design measurement frameworks that generate and integrate national adaptation data without placing undue burdens on countries (UNEP, 2017a).

The Paris Agreement stipulates that adaptation communications shall be submitted as a component of or in conjunction with other communications, such as an NDC, a NAP, or a national communication. Of the 197 Parties to the UNFCCC, 140 NDCs have an adaptation component, almost exclusively from developing countries. NDC adaptation components could be an opportunity for enhancing adaptation planning and implementation by highlighting priorities and goals (Kato and Ellis, 2016). At the national level they provide momentum for the development of NAPs and raise the profile of adaptation (Pauw et al., 2016b, 2018). The Paris Agreement's transparency framework includes adaptation, through which 'adaptation communication' and accelerated adaptation actions are submitted and reviewed every five years (Hermwille, 2016; Kato and Ellis, 2016). This framework, unlike others used in the past, is applicable to all countries taking into account differing capacities amongst Parties (Rajamani, 2016).

Adaptation measures presented in qualitative terms include sectors, risks and vulnerabilities that are seen as priorities by the Parties. Sectoral coverage of adaptation actions identified in NDCs is uneven, with adaptation primarily reported to focus on the water sector (71% of NDCs with adaptation component), agriculture (63%), health (54%), and biodiversity/ecosystems (50%) (Pauw et al., 2016b, 2018).

4.4.2   Enhancing Institutional Capacities

The implementation of sound responses and strategies to enable a transition to 1.5°C world would require strengthening governance and scaling up institutional capacities, particularly in developing countries (Adenle et al., 2017; Rosenbloom, 2017). Building on the characterization of governance in Section 4.4.1, this section examines the necessary institutional capacity to implement actions to limit warming to 1.5°C and adapt to the consequences. This takes into account a plurality of regional and local responses, as institutional capacity is highly context-dependent (North, 1990; Lustick et al., 2011).

Institutions would need to interact with one another and align across scales to ensure that rules and regulations are followed (Chaffin and Gunderson, 2016; Young, 2016). The institutional architecture required for a 1.5°C world would include the growing proportion of the world's population that live in peri-urban and informal settlements and engage in informal economic activity (Simone and Pieterse, 2017). This population, amongst the most exposed to perturbed climates in the world (Hallegatte et al., 2017), is also beyond the direct reach of some policy instruments (Jaglin, 2014; Thieme, 2018). Strategies that accommodate the informal rules of the game adopted by these populations have large chances of success (McGranahan et al., 2016; Kaika, 2017).

The goal for strengthening implementation is to ensure that these rules and regulations embrace equity, equality and poverty alleviation along

1.5°C-consistent pathways (mitigation) and enables the building of adaptive capacity that together, will enable sustainable development and poverty reduction.

Rising to the challenge of a transition to a 1.5°C world would require enhancing institutional climate change capacities along multiple dimensions presented below.

4.4.2.1   Capacity for policy design and implementation

The enhancement of institutional capacity for integrated policy design and implementation has long been among the top items on the UN agenda of addressing global environmental problems and sustainable development (see Chapter 5, Section 5.5) (UNEP, 2005).

Political stability, an effective regulatory and enforcement framework (e.g., institutions to impose sanctions, collect taxes and to verify building codes), access to a knowledge base and the availability of resources, would be needed at various governance levels to address a wide range of stakeholders and their concerns. The strengthening of the global response would need to support these with different interventions, in the context of sustainable development (Chapter 5, Section 5.5.1) (Pasquini et al., 2015).

Given the scale of change needed to limit warming to 1.5°C, strengthening the response capacity of relevant institutions is best addressed in ways that take advantage of existing decision-making processes in local and regional governments and within cities and communities (Romero-Lankao et al., 2013), and draws upon diverse knowledge sources including indigenous and local knowledge (Nakashima et al., 2012; Smith and Sharp, 2012; Mistry and Berardi, 2016; Tschakert et al., 2017). Examples of successful local institutional processes and the integration of local knowledge in climate-related decision-making are provided in Box 4.3 and Box 4.4.

Implementing 1.5°C-consistent strategies would require well-functioning legal frameworks to be in place, in conjunction with clearly defined mandates, rights and responsibilities to enable the institutional capacity to deliver (Romero-Lankao et al., 2013). As an example, current rates of urbanization occurring in cities with a lack of institutional capacity for effective land-use planning, zoning and infrastructure development result in unplanned, informal urban settlements which are vulnerable to climate impacts. It is common for 30–50% of urban populations in low-income nations to live in informal settlements with no regulatory infrastructure (Revi et al., 2014b). For example, in Huambo (Angola), a classified 'urban' area extends 20 km west of the city and is predominantly made up of 'unplanned' urban settlements (Smith and Jenkins, 2015).

Internationally, the Paris Agreement process has aimed at enhancing the capacity of decision-making institutions in developing countries to support effective implementation. These efforts are particularly reflected in Article 11 of the Paris Agreement on capacity building (the creation of the Paris Committee on Capacity Building), Article 13 (the creation of the Capacity Building Initiative on Transparency), and Article 15 on compliance (UNFCCC, 2016).

Box 4.3 | Indigenous Knowledge and Community Adaptation

Indigenous knowledge refers to the understandings, skills and philosophies developed by societies with long histories of interaction with their natural surroundings (UNESCO, 2017). This knowledge can underpin the development of adaptation and mitigation strategies (Ford et al., 2014b; Green and Minchin, 2014; Pearce et al., 2015; Savo et al., 2016).

Climate change is an important concern for the Maya, who depend on climate knowledge for their livelihood. In Guatemala, the collaboration between the Mayan K'iché population of the Nahualate river basin and the Climate Change Institute has resulted in a catalogue of indigenous knowledge, used to identify indicators for watershed meteorological forecasts (López and Álvarez, 2016). These indicators are relevant but would need continuous assessment if their continued reliability is to be confirmed (Nyong et al., 2007; Alexander et al., 2011; Mistry and Berardi, 2016). For more than ten years, Guatemala has maintained an 'Indigenous Table for Climate Change', to enable the consideration of indigenous knowledge in disaster management and adaptation development.

In Tanzania, increased variability of rainfall is challenging indigenous and local communities (Mahoo et al., 2015; Sewando et al., 2016). The majority of agro-pastoralists use indigenous knowledge to forecast seasonal rainfall, relying on observations of plant phenology, bird, animal, and insect behaviour, the sun and moon, and wind (Chang'a et al., 2010; Elia et al., 2014; Shaffer, 2014). Increased climate variability has raised concerns about the reliability of these indicators (Shaffer, 2014); therefore, initiatives have focused on the co-production of knowledge by involving local communities in monitoring and discussing the implications of indigenous knowledge and meteorological forecasts (Shaffer, 2014), and creating local forecasts by utilizing the two sources of knowledge (Mahoo et al., 2013). This has resulted in increased documentation of indigenous knowledge, understanding of relevant climate information amongst stakeholders, and adaptive capacity at the community level (Mahoo et al., 2013, 2015; Shaffer, 2014).

The Pacific Islands and small island developing states (SIDS) are vulnerable to the effects of climate change, but the cultural resilience of Pacific Island inhabitants is also recognized (Nunn et al., 2017). In Fiji and Vanuatu, strategies used to prepare for cyclones include building reserve emergency supplies and utilizing farming techniques to ensure adequate crop yield to combat potential losses from a cyclone or drought (McNamara and Prasad, 2014; Granderson, 2017; Pearce et al., 2017). Social cohesion and kinship are important in responding and preparing for climate-related hazards, including the role of resource sharing, communal labour, and remittances (McMillen et al., 2014; Gawith et al., 2016; Granderson, 2017). There is a concern that indigenous knowledge will weaken, a process driven by westernization and disruptions in established bioclimatic indicators and traditional planning calendars (Granderson, 2017). In some urban settlements, it has been noted that cultural practices (e.g., prioritizing the quantity of food over the quality of food) can lower food security through dispersing limited resources and by encouraging the consumption of cheap but nutrient-poor foods (Mccubbin et al., 2017) (See Cross-Chapter Box 6 on Food Security in Chapter 3). Indigenous practices also encounter limitations, particularly in relation to sea level rise (Nunn et al., 2017).

Box 4.4 | Manizales, Colombia: Supportive National Government and Localized Planning and Integration as an Enabling Condition for Managing Climate and Development Risks

Institutional reform in the city of Manizales, Colombia, helps identify three important features of an enabling environment: integrating climate change adaptation, mitigation and disaster risk management at the city-scale; the importance of decentralized planning and policy formulation within a supportive national policy environment; and the role of a multi-sectoral framework in mainstreaming climate action in development activities.

Manizales is exposed to risks caused by rapid development and expansion in a mountainous terrain exposed to seismic activity and periodic wet and dry spells. Local assessments expect climate change to amplify the risk of disasters (Carreño et al., 2017). The city is widely recognized for its longstanding urban environmental policy (Biomanizales) and local environmental action plan (Bioplan), and has been integrating environmental planning in its development agenda for nearly two decades (Velásquez Barrero, 1998; Hardoy and Velásquez Barrero, 2014). When the city's environmental agenda was updated in 2014 to reflect climate change risks, assessments were conducted in a participatory manner at the street and neighbourhood level (Hardoy and Velásquez Barrero, 2016).

The creation of a new Environmental Secretariat assisted in coordination and integration of environmental policies, disaster risk management, development and climate change (Leck and Roberts, 2015). Planning in Manizales remains mindful of steep gradients through its longstanding Slope Guardian programme that trains women and keeps records of vulnerable households. Planning also looks to include mitigation opportunities and enhance local capacity through participatory engagement (Hardoy and Velásquez Barrero, 2016).

Manizales' mayors were identified as important champions for much of these early integration and innovation efforts. Their role may have been enabled by Colombia's history of decentralized approaches to planning and policy formulation, including establishing environmental observatories (for continuous environmental assessment) and participatory tracking of environmental indicators. Multi-stakeholder involvement has both enabled and driven progress, and has enabled the integration of climate risks in development planning (Hardoy and Velásquez Barrero, 2016).

4.4.2.2 Monitoring, reporting, and review institutions

One of the novel features of the new climate governance architecture emerging from the 2015 Paris Agreement is the transparency framework in Article 13 committing countries, based on capacity, to provide regular progress reports on national pledges to address climate change (UNFCCC, 2016). Many countries will rely on public policies and existing national reporting channels to deliver on their NDCs under the Paris Agreement. Scaling up the mitigation and adaptation efforts in these countries to be consistent with 1.5°C would put significant pressure on the need to develop, enhance and streamline local, national and international climate change reporting and monitoring methodologies and institutional capacity in relation to mitigation, adaptation, finance, and GHG inventories (Ford et al., 2015b; Lesnikowski et al., 2015; Schoenefeld et al., 2016). Consistent with this direction, the provision of the information to the stocktake under Article 14 of the Paris Agreement would contribute to enhancing reporting and transparency (UNFCCC, 2016). Nonetheless, approaches, reporting procedures, reference points, and data sources to assess progress on implementation across and within nations are still largely underdeveloped (Ford et al., 2015b; Araos et al., 2016b; Magnan and Ribera, 2016; Lesnikowski et al., 2017). The availability of independent private and public reporting and statistical institutions are integral to oversight, effective monitoring, reporting and review. The creation and enhancement of these institutions would be an important contribution to an effective transition to a low-emission world.

4.4.2.3 Financial institutions

IPCC AR5 assessed that in order to enable a transition to a 2°C pathway, the volume of climate investments would need to be transformed along with changes in the pattern of general investment behaviour towards low emissions. The report argued that, compared to 2012, annually up to a trillion dollars in additional investment in low-emission energy and energy efficiency measures may be required until 2050 (Blanco et al., 2014; IEA, 2014a). Financing of 1.5°C would present an even greater challenge, addressing financing of both existing and new assets, which would require significant transitions to the type and structure of financial institutions as well as to the method of financing (Cochrani et al., 2014; Ma, 2014). Both public and private financial institutions would be needed to contribute to the large resource mobilization needed for 1.5°C, yet, in the ordinary course of business, these transitions may not be expected. On the one hand, private financial institutions could face scale-up risk, for example, the risks associated with commercialization and scaling up of renewable technologies to accelerate mitigation (Wilson, 2012; Hartley and Medlock, 2013) and/or price risk, such as carbon price volatility that carbon markets could face. In contrast, traditional public financial institutions are limited by both structure and instruments, while concessional financing would require taxpayer support for subsidization. Special efforts and innovative approaches would be needed to address these challenges, for example the creation of special institutions that underwrite the value of emission reductions using auctioned price floors (Bodnar et al., 2018) to deal with price volatility.

Financial institutions are equally important for adaptation. Linnerooth-Bayer and Hochrainer-Stigler (2015) discussed the benefits of financial instruments in adaptation, including the provision of post-disaster finances for recovery and pre-disaster security necessary for climate adaptation and poverty reduction. Pre-disaster financial instruments and options include insurance, such as index-based weather insurance schemes, catastrophe bonds, and laws to encourage insurance purchasing. The development and enhancement of microfinance institutions to ensure social resilience and smooth transitions in the adaptation to climate change impacts could be an important local institutional innovation (Hammill et al., 2008).

4.4.2.4 Co-operative institutions and social safety nets

Effective cooperative institutions and social safety nets may help address energy access and adaptation, as well as distributional impacts during the transition to 1.5°C-consistent pathways and enabling sustainable development. Not all countries have the institutional capabilities to design and manage these. Social capital for adaptation in the form of bonding, bridging, and linking social institutions has proved to be effective in dealing with climate crises at the local, regional and national levels (Aldrich et al., 2016).

The shift towards sustainable energy systems in transitioning economies could impact the livelihoods of large populations in traditional and legacy employment sectors. The transition of selected EU Member States to biofuels, for example, caused anxiety among farmers, who lacked confidence in the biofuel crop market. Enabling contracts between farmers and energy companies, involving local governments, helped create an atmosphere of confidence during the transition (McCormick and Kåberger, 2007).

How do broader socio-economic processes influence urban vulnerabilities and thereby underpin climate change adaptation? This is a systemic challenge originating from a lack of collective societal ownership of the responsibility for climate risk management. Explanations for this situation include competing time-horizons due

to self-interest of stakeholders to a more 'rational' conception of risk assessment, measured across a risk-tolerance spectrum (Moffatt, 2014).

Self-governing and self-organ¬ised institutional settings, where equipment and resource systems are commonly owned and managed, can poten¬tially generate a much higher diversity of administration solutions, than other institutional arrangements, where energy technology and resource systems are either owned and administered individually in market settings or via a central authority (e.g., the state). They can also increase the adaptability of technological systems while reducing their burden on the environment (Labanca, 2017). Educational, learning and awareness-building institutions can help strengthen the societal response to climate change (Butler et al., 2016; Thi Hong Phuong et al., 2017).

4.4.3 Enabling Lifestyle and Behavioural Change

Humans are at the centre of global climate change: their actions cause anthropogenic climate change, and social change is key to effectively responding to climate change (Vlek and Steg, 2007; Dietz et al., 2013; ISSC and UNESCO, 2013; Hackmann et al., 2014). Chapter 2 shows that 1.5°C-consistent pathways assume substantial changes in behaviour. This section assesses the potential of behaviour change, as the integrated assessment models (IAMs) applied in Chapter 2 do not comprehensively asses this potential.

Table 4.8 shows examples of mitigation and adaption actions relevant for 1.5°C-consistent pathways. Reductions in population growth can reduce overall carbon demand and mitigate climate change (Bridgeman, 2017), particularly when population growth is accompanied by increases in affluence and carbon-intensive consumption (Rosa and Dietz, 2012; Clayton et al., 2017). Mitigation actions with a substantial carbon emission reduction potential (see Figure 4.3) that individuals may readily adopt would have the most climate impact (Dietz et al., 2009).

Various policy approaches and strategies can encourage and enable climate actions by individuals and organizations. Policy approaches would be more effective when they address key contextual and psycho-

Table 4.8 | Examples of mitigation and adaptation behaviours relevant for 1.5°C (Dietz et al., 2009; Jabeen, 2014; Taylor et al., 2014; Araos et al., 2016b; Steg, 2016; Stern et al., 2016b; Creutzig et al., 2018)

Climate action	Type of action	Examples
Mitigation	Implementing resource efficiency in buildings	Insulation Low-carbon building materials
	Adopting low-emission innovations	Electric vehicles Heat pumps, district heating and cooling
	Adopting energy efficient appliances	Energy-efficient heating or cooling Energy-efficient appliances
	Energy-saving behaviour	Walking or cycling rather than drive short distances Using mass transit rather than flying Lower temperature for space heating Line drying of laundry Reducing food waste
	Buying products and materials with low GHG emissions during production and transport	Reducing meat and dairy consumption Buying local, seasonal food Replacing aluminium products by low-GHG alternatives
	Organisational behaviour	Designing low-emission products and procedures Replacing business travel by videoconferencing

Table 4.8 (continued)

Climate action	Type of action	Examples
Adaptation	Growing different crops and raising different animal varieties	Using crops with higher tolerance for higher temperatures or CO_2 elevation
	Flood protective behaviour	Elevating barriers between rooms Building elevated storage spaces Building drainage channels outside the home
	Heat protective behaviour	Staying hydrated Moving to cooler places Installing green roofs
	Efficient water use during water shortage crisis	Rationing water Constructing wells or rainwater tanks
Mitigation & adaptation	Adoption of renewable energy sources	Solar PV Solar water heaters
	Citizenship behaviour	Engage through civic channels to encourage or support planning for low-carbon climate-resilient development

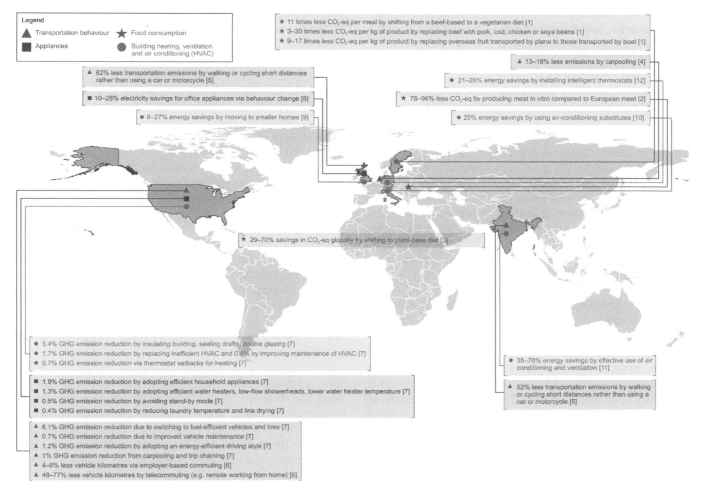

Figure 4.3 | Examples of mitigation behaviour and their GHG emission reduction potential. Mitigation potential assessments are printed in different units. Based on [1] Carlsson-Kanyama and González (2009); [2] Tuomisto and Teixeira de Mattos (2011); [3] Springmann et al. (2016); [4] Nijland and Meerkerk (2017); [5] Woodcock et al. (2009); [6] Salon et al. (2012); [7] Dietz et al. (2009); [8] Mulville et al. (2017); [9] Huebner and Shipworth (2017); [10] Jaboyedoff et al. (2004); [11] Pellegrino et al. (2016); [12] Nägele et al. (2017).

social factors influencing climate actions, which differ across contexts and individuals (Steg and Vlek, 2009; Stern, 2011). This suggests that diverse policy approaches would be needed in 1.5°C-consistent pathways in different contexts and regions. Combinations of policies that target multiple barriers and enabling factors simultaneously can be more effective (Nissinen et al., 2015).

In the United States and Europe, GHG emissions are lower when legislators have strong environmental records (Jensen and Spoon, 2011; Dietz et al., 2015). Political elites affect public concern about climate change: pro-climate action statements increased concern, while anti-climate action statements and anti-environment voting reduced public concern about climate change (Brulle et al., 2012). In the European

Union (EU), individuals worry more about climate change and engage more in climate actions in countries where political party elites are united rather than divided in their support for environmental issues (Sohlberg, 2017).

This section discusses how to enable and encourage behaviour and lifestyle changes that strengthen implementation of 1.5°C-consistent pathways by assessing psycho-social factors related to climate action, as well as the effects and acceptability of policy approaches targeting climate actions that are consistent with 1.5°C. Box 4.5 and Box 4.6 illustrate how these have worked in practice.

4.4.3.1   Factors related to climate actions

Mitigation and adaptation behaviour is affected by many factors that shape which options are feasible and considered by individuals. Besides contextual factors (see other sub-sections in Section 4.4), these include abilities and different types of motivation to engage in behaviour.

Ability to engage in climate action. Individuals more often engage in adaptation (Gebrehiwot and van der Veen, 2015; Koerth et al., 2017) and mitigation behaviour (Pisano and Lubell, 2017) when they are or feel more capable to do so. Hence, it is important to enhance ability to act on climate change, which depends on income and knowledge, among other things. A higher income is related to higher CO_2 emissions; higher income groups can afford more carbon-intensive lifestyles (Lamb et al., 2014; Dietz et al., 2015; Wang et al., 2015). Yet low-income groups may lack resources to invest in energy-efficient technology and refurbishments (Andrews-Speed and Ma, 2016) and adaptation options (Wamsler, 2007; Fleming et al., 2015b; Takahashi et al., 2016). Adaptive capacity further depends on gender roles (Jabeen, 2014; Bunce and Ford, 2015), technical capacities and knowledge (Feola et al., 2015; Eakin et al., 2016; Singh et al., 2016b).

Knowledge of the causes and consequences of climate change and of ways to reduce GHG emissions is not always accurate (Bord et al., 2000; Whitmarsh et al., 2011; Tobler et al., 2012), which can inhibit climate actions, even when people would be motivated to act. For example, people overestimate savings from low-energy activities, and underestimate savings from high-energy activities (Attari et al., 2010). They know little about 'embodied' energy (i.e., energy needed to produce products; Tobler et al., 2011), including meat (de Boer et al., 2016b). Some people mistake weather for climate (Reynolds et al., 2010), or conflate climate risks with other hazards, which can inhibit adequate adaptation (Taylor et al., 2014).

More knowledge on adaptation is related to higher engagement in adaptation actions in some circumstances (Bates et al., 2009; van Kasteren, 2014; Hagen et al., 2016). How adaptation is framed in the media can influence the types of options viewed as important in different contexts (Boykoff et al., 2013; Moser, 2014; Ford and King, 2015).

Knowledge is important, but is often not sufficient to motivate action (Trenberth et al., 2016). Climate change knowledge and perceptions are not strongly related to mitigation actions (Hornsey et al., 2016). Direct experience of events related to climate change influences

climate concerns and actions (Blennow et al., 2012; Taylor et al., 2014), more so than second-hand information (Spence et al., 2011; Myers et al., 2012; Demski et al., 2017); high impact events with low frequency are remembered more than low impact regular events (Meze-Hausken, 2004; Singh et al., 2016b; Sullivan-Wiley and Short Gianotti, 2017). Personal experience with climate hazards strengthens motivation to protect oneself (Jabeen, 2014) and enhances adaptation actions (Bryan et al., 2009; Berrang-Ford et al., 2011; Demski et al., 2017), although this does not always translate into proactive adaptation (Taylor et al., 2014). Collectively constructed notions of risk and expectations of future climate variability shape risk perception and adaptation behaviour (Singh et al., 2016b). People with particular political views and those who emphasize individual autonomy may reject climate science knowledge and believe that there is widespread scientific disagreement about climate change (Kahan, 2010; O'Neill et al., 2013), inhibiting support for climate policy (Ding et al., 2011; McCright et al., 2013). This may explain why extreme weather experiences enhances preparedness to reduce energy use among left- but not right-leaning voters (Ogunbode et al., 2017).

Motivation to engage in climate action. Climate actions are more strongly related to motivational factors than to knowledge, reflecting individuals' reasons for actions, such as values, ideology and worldviews (Hornsey et al., 2016). People consider various types of costs and benefits of actions (Gölz and Hahnel, 2016) and focus on consequences that have implications for the values they find most important (Dietz et al., 2013; Hahnel et al., 2015; Steg, 2016). This implies that different individuals consider different consequences when making choices. People who strongly value protecting the environment and other people generally more strongly consider climate impacts and act more on climate change than those who strongly endorse hedonic and egoistic values (Taylor et al., 2014; Steg, 2016). People are more prone to adopt sustainable innovations when they are more open to new ideas (Jansson, 2011; Wolske et al., 2017). Further, a free-market ideology is associated with weaker climate change beliefs (McCright and Dunlap, 2011; Hornsey et al., 2016), and a capital-oriented culture tends to promote activity associated with GHG emissions (Kasser et al., 2007).

Some indigenous populations believe it is arrogant to predict the future, and some cultures have belief systems that interpret natural phenomena as sentient, where thoughts and words are believed to influence the future, with people reluctant to talk about negative future possibilities (Natcher et al., 2007; Flynn et al., 2018). Integrating these considerations into the design of adaptation and mitigation policy is important (Cochran et al., 2013; Chapin et al., 2016; Brugnach et al., 2017; Flynn et al., 2018).

People are more prone to act on climate change when individual benefits of actions exceed costs (Steg and Vlek, 2009; Kardooni et al., 2016; Wolske et al., 2017). For this reason, people generally prefer adoption of energy-efficient appliances above energy-consumption reductions; the latter is perceived as more costly (Poortinga et al., 2003; Steg et al., 2006), although transaction costs can inhibit the uptake of mitigation technology (Mundaca, 2007). Decentralized renewable energy systems are evaluated most favourably when they guarantee independence, autonomy, control and supply security (Ecker et al., 2017).

4

Besides, social costs and benefits affect climate action (Farrow et al., 2017). People engage more in climate actions when they think others expect them to do so and when others act as well (Nolan et al., 2008; Le Dang et al., 2014; Truelove et al., 2015; Rai et al., 2016), and when they experience social support (Singh et al., 2016a; Burnham and Ma, 2017; Wolske et al., 2017). Discussing effective actions with peers also encourages climate action (Esham and Garforth, 2013), particularly when individuals strongly identify with their peers (Biddau et al., 2012; Fielding and Hornsey, 2016). Further, individuals may engage in mitigation actions when they think doing so would enhance their reputation (Milinski et al., 2006; Noppers et al., 2014; Kastner and Stern, 2015). Such social costs and benefits can be addressed in climate policy (see Section 4.4.3.2).

Feelings affect climate action (Brosch et al., 2014). Negative feelings related to climate change can encourage adaptation action (Kerstholt et al., 2017; Zhang et al., 2017), while positive feelings associated with climate risks may inhibit protective behaviour (Lefevre et al., 2015). Individuals are more prone to engage in mitigation actions when they worry about climate change (Verplanken and Roy, 2013) and when they expect to derive positive feelings from such actions (Pelletier et al., 1998; Taufik et al., 2016).

Furthermore, collective consequences affect climate actions (Balcombe et al., 2013; Dóci and Vasileiadou, 2015; Kastner and Stern, 2015). People are motivated to see themselves as morally right, which encourages mitigation actions (Steg et al., 2015), particularly when long-term goals are salient (Zaval et al., 2015) and behavioural costs are not too high (Diekmann and Preisendörfer, 2003). Individuals are more prone to engage in climate actions when they believe climate change is occurring, when they are aware of threats caused by climate change and by their inaction, and when they think they can engage in actions that will reduce these threats (Esham and Garforth, 2013; Arunrat et al., 2017; Chatrchyan et al., 2017). The more individuals are concerned about climate change and aware of the negative climate impact of their behaviour, the more they feel responsible for their actions and think that their actions can help reduce such negative impacts, which can strengthen their moral norms to act accordingly (Steg and de Groot, 2010; Jakovcevic and Steg, 2013; Chen, 2015; Ray et al., 2017; Wolske et al., 2017; Woods et al., 2017). Individuals may engage in mitigation actions when they see themselves as supportive of the environment (i.e., strong environmental self-identity) (Fielding et al., 2008; van der Werff et al., 2013b; Kashima et al., 2014; Barbarossa et al., 2017); a strong environmental identity strengthens intrinsic motivation to engage in mitigation actions both at home (van der Werff et al., 2013a) and at work (Ruepert et al., 2016). Environmental self-identity is strengthened when people realize they have engaged in mitigation actions, which can in turn promote further mitigation actions (van der Werff et al., 2014b).

Individuals are less prone to engage in adaptation behaviour themselves when they rely on external measures such as government interventions (Grothmann and Reusswig, 2006; Wamsler and Brink, 2014a; Armah et al., 2015; Burnham and Ma, 2017) or perceive themselves as protected by god (Gandure et al., 2013; Dang et al., 2014; Cannon, 2015).

Habits, heuristics and biases. Decisions are often not based on weighing costs and benefits, but on habit or automaticity, both of individuals (Aarts and Dijksterhuis, 2000; Kloeckner et al., 2003) and within organizations (Dooley, 2017) and institutions (Munck et al., 2014). When habits are strong, individuals are less perceptive of information (Verplanken et al., 1997; Aarts et al., 1998) and may not consider alternatives as long as outcomes are good enough (Maréchal, 2010). Habits are mostly only reconsidered when the situation changed significantly (Fujii and Kitamura, 2003; Maréchal, 2010; Verplanken and Roy, 2016). Hence, strategies that create the opportunity for reflection and encourage active decisions can break habits (Steg et al., 2018).

Individuals can follow heuristics, or 'rules of thumb', in making inferences, which demand less cognitive resources, knowledge and time than thinking through all implications of actions (Preston et al., 2013; Frederiks et al., 2015; Gillingham and Palmer, 2017). For example, people tend to think that larger and more visible appliances use more energy, which is not always accurate (Cowen and Gatersleben, 2017). They underestimate energy used for water heating and overestimate energy used for lighting (Stern, 2014). When facing choice overload, people may choose the easiest or first available option, which can inhibit energy-saving behaviour (Stern and Gardner, 1981; Frederiks et al., 2015). As a result, individuals and firms often strive for satisficing ('good enough') outcomes with regard to energy decisions (Wilson and Dowlatabadi, 2007; Klotz, 2011), which can inhibit investments in energy efficiency (Decanio, 1993; Frederiks et al., 2015).

Biases also play a role. In Mozambique, farmers displayed omission biases (unwillingness to take adaptation actions with potentially negative consequences to avoid personal responsibility for losses), while policymakers displayed action biases (wanting to demonstrate positive action despite potential negative consequences; Patt and Schröter, 2008). People tend to place greater value on relative losses than gains (Kahneman, 2003). Perceived gains and losses depend on the reference point or status-quo (Kahneman, 2003). Loss aversion and the status-quo bias prevent consumers from switching electricity suppliers (Ek and Söderholm, 2008), to time-of-use electricity tariffs (Nicolson et al., 2017), and to accept new energy systems (Leijten et al., 2014).

Owned inefficient appliances and fossil fuel-based electricity can act as endowments, increasing their value compared to alternatives (Pichert and Katsikopoulos, 2008; Dinner et al., 2011). Uncertainty and loss aversion lead consumers to undervalue future energy savings (Greene, 2011) and savings from energy efficient technologies (Kolstad et al., 2014). Uncertainties about the performance of products and illiquidity of investments can drive consumers to postpone (profitable) energy-efficient investments (Sutherland, 1991; van Soest and Bulte, 2001). People with a higher tendency to delay decisions may engage less in energy saving actions (Lillemo, 2014). Training energy auditors in loss-aversion increased their clients' investments in energy efficiency improvements (Gonzales et al., 1988). Engagement in energy saving and renewable energy programmes can be enhanced if participation is set as a default option (Pichert and Katsikopoulos, 2008; Ölander and Thøgersen, 2014; Ebeling and Lotz, 2015).

4

4.4.3.2 Strategies and policies to promote actions on climate change

Policy can enable and strengthen motivation to act on climate change via top-down or bottom-up approaches, through informational campaigns, regulatory measures, financial (dis)incentives, and infrastructural and technological changes (Adger et al., 2003; Steg and Vlek, 2009; Henstra, 2016).

Adaptation efforts tend to focus on infrastructural and technological solutions (Ford and King, 2015) with lower emphasis on socio-cognitive and finance aspects of adaptation. For example, flooding policies in cities focus on infrastructure projects and regulation such as building codes, and hardly target individual or household behaviour (Araos et al., 2016b; Georgeson et al., 2016).

Current mitigation policies emphasize infrastructural and technology development, regulation, financial incentives and information provision (Mundaca and Markandya, 2016) that can create conditions enabling climate action, but target only some of the many factors influencing climate actions (see Section 4.4.5.1). They fall short of their true potential if their social and psychological implications are overlooked (Stern et al., 2016a). For example, promising energy-saving or low-carbon technology may not be adopted or not be used as intended (Pritoni et al., 2015) when people lack resources and trustworthy information (Stern, 2011; Balcombe et al., 2013).

Financial incentives or feedback on financial savings can encourage climate action (Santos, 2008; Bolderdijk et al., 2011; Maki et al., 2016) (see Box 4.5), but are not always effective (Delmas et al., 2013) and can be less effective than social rewards (Handgraaf et al., 2013) or emphasising benefits for people and the environment (Bolderdijk et al., 2013b; Asensio and Delmas, 2015; Schwartz et al., 2015). The latter can happen when financial incentives reduce a focus on environmental considerations and weaken intrinsic motivation to engage in climate action (Evans et al., 2012; Agrawal et al., 2015; Schwartz et al., 2015). In addition, pursuing small financial gains is perceived to be less worth the effort than pursuing equivalent CO_2 emission reductions (Bolderdijk et al., 2013b; Dogan et al., 2014). Also, people may not respond to financial incentives (e.g., to improve energy efficiency) because they do not trust the organization sponsoring incentive programmes (Mundaca, 2007) or when it takes too much effort to receive the incentive (Stern et al., 2016a).

Box 4.5 | How Pricing Policy has Reduced Car Use in Singapore, Stockholm and London

In Singapore, Stockholm and London, car ownership, car use, and GHG emissions have reduced because of pricing and regulatory policies and policies facilitating behaviour change. Notably, acceptability of these policies has increased as people experienced their positive effects.

Singapore implemented electronic road pricing in the central business district and at major expressways, a vehicle quota and registration fee system, and investments in mass transit. In the vehicle quota system introduced in 1990, registration of new vehicles is conditional upon a successful bid (via auctioning) (Chu, 2015), costing about 50,000 USD in 2014 (LTA, 2015). The registration tax incentivizes purchases of low-emission vehicles via a feebate system. As a result, per capita transport emissions (approximately 1.25 tCO_2yr^{-1}) and car ownership (107 vehicles per 1000 capita) (LTA, 2017) are substantially lower than in cities with comparable income levels. Modal share of public transport was 63% during peak hours in 2013 (LTA, 2013).

The Stockholm congestion charge implemented in 2007 (after a trial in 2006) reduced kilometres driven in the inner city by 16%, and outside the city by 5%; traffic volumes reduced by 20% and remained constant over time despite economic and population growth (Eliasson, 2014). CO_2 emissions from traffic reduced by 2–3% in Stockholm county. Vehicles entering or leaving the city centre were charged during weekdays (except for holidays). Charges were 1–2€ (maximum 6€ per day), being higher during peak hours; taxis, emergency vehicles and buses were exempted. Before introducing the charge, public transport and parking places near mass transit stations were extended. The aim and effects of the charge were extensively communicated to the public. Acceptability of the congestion charge was initially low, but the scheme gained support of about two-thirds of the population and all political parties after it was implemented (Eliasson, 2014), which may be related to the fact that the revenues were earmarked for constructing a motorway tunnel. After the trial, people believed that the charge had more positive effects on environmental, congestion and parking problems while costs increased less than they anticipated beforehand (Schuitema et al., 2010a). The initially hostile media eventually declared the scheme to be a success.

In 2003, a congestion charge was implemented in the Greater London area, with an enforcement and compliance scheme and an information campaign on the functioning of the scheme. Vehicles entering, leaving, driving or parking on a public road in the zone at weekdays at daytime pay a congestion charge of 8£ (until 2005, 5£), with some exemptions. Revenues were invested in London's bus network (80%), cycling facilities, and road safety measures (Leape, 2006). The number of cars entering the zone decreased by 18% in 2003 and 2004. In the charging zone, vehicle kilometres driven decreased by 15% in the first year and a further 6% a year later, while CO_2 emissions from road traffic reduced by 20% (Santos, 2008).

While providing information on the causes and consequences of climate change or on effective climate actions generally increases knowledge, it often does not encourage engagement in climate actions by individuals (Abrahamse et al., 2005; Ünal et al., 2017) or organizations (Anderson and Newell, 2004). Similarly, media coverage on the UN Climate Summit slightly increased knowledge about the conference but did not enhance motivation to engage personally in climate protection (Brüggemann et al., 2017). Fear-inducing representations of climate change may inhibit action when they make people feel helpless and overwhelmed (O'Neill and Nicholson-Cole, 2009). Energy-related recommendations and feedback (e.g., via performance contracts, energy audits, smart metering) are more effective for promoting energy conservation, load shifting in electricity use and sustainable travel choices when framed in terms of losses rather than gains (Gonzales et al., 1988; Wolak, 2011; Bradley et al., 2016; Bager and Mundaca, 2017).

Credible and targeted information at the point of decision can promote climate action (Stern et al., 2016a). For example, communicating the impacts of climate change is more effective when provided right before adaptation decisions are taken (e.g., before the agricultural season) and when bundled with information on potential actions to ameliorate impacts, rather than just providing information on climate projections with little meaning to end users (e.g., weather forecasts, seasonal forecasts, decadal climate trends) (Dorward et al., 2015; Singh et al., 2017). Similarly, heat action plans that provide early alerts and advisories combined with emergency public health measures can reduce heat-related morbidity and mortality (Benmarhnia et al., 2016).

Information provision is more effective when tailored to the personal situation of individuals, demonstrating clear impacts, and resonating with individuals' core values (Daamen et al., 2001; Abrahamse et al., 2007; Bolderdijk et al., 2013a; Dorward et al., 2015; Singh et al., 2017). Tailored information prevents information overload, and people are more motivated to consider and act upon information that aligns with their core values and beliefs (Campbell and Kay, 2014; Hornsey et al., 2016). Also, tailored information can remove barriers to receive and interpret information faced by vulnerable groups, such as the elderly during heatwaves (Vandentorren et al., 2006; Keim, 2008). Further, prompts can be effective when they serve as reminders to perform a planned action (Osbaldiston and Schott, 2012).

Feedback provision is generally effective in promoting mitigation behaviour within households (Abrahamse et al., 2005; Delmas et al., 2013; Karlin et al., 2015) and at work (Young et al., 2015), particularly when provided in real-time or immediately after the action (Abrahamse et al., 2005), which makes the implications of one's behaviour more salient (Tiefenbeck et al., 2016). Simple information is more effective than detailed and technical data (Wilson and Dowlatabadi, 2007; Ek and Söderholm, 2010; Frederiks et al., 2015). Energy labels (Banerjee and Solomon, 2003; Stadelmann, 2017), visualization techniques (Pahl et al., 2016), and ambient persuasive technology (Midden and Ham, 2012) can encourage mitigation actions by providing information and feedback in a format that immediately makes sense and hardly requires users' conscious attention.

Social influence approaches that emphasize what other people do or think can encourage climate action (Clayton et al., 2015), particularly

when they involve face-to-face interaction (Abrahamse and Steg, 2013). For example, community approaches, where change is initiated from the bottom-up, can promote adaptation (see Box 4.6) and mitigation actions (Middlemiss, 2011; Seyfang and Haxeltine, 2012; Abrahamse and Steg, 2013), especially when community ties are strong (Weenig and Midden, 1991). Furthermore, providing social models of desired actions can encourage mitigation action (Osbaldiston and Schott, 2012; Abrahamse and Steg, 2013). Social influence approaches that do not involve social interaction, such as social norm, social comparison and group feedback, are less effective, but can be easily administered on a large scale at low costs (Allcott, 2011; Abrahamse and Steg, 2013).

Goal setting can promote mitigation action when goals are not set too low or too high (Loock et al., 2013). Commitment strategies where people make a pledge to engage in climate actions can encourage mitigation behaviour (Abrahamse and Steg, 2013; Lokhorst et al., 2013), particularly when individuals also indicate how and when they will perform the relevant action and anticipate how to cope with possible barriers (i.e., implementation intentions) (Bamberg, 2000, 2002). Such strategies take advantage of individuals' desire to be consistent (Steg, 2016). Similarly, hypocrisy-related strategies that make people aware of inconsistencies between their attitudes and behaviour can encourage mitigation actions (Osbaldiston and Schott, 2012).

Actions that reduce climate risks can be rewarded and facilitated, while actions that increase climate risks can be punished and inhibited, and behaviour change can be voluntary (e.g., information provision) or imposed (e.g., by law); voluntary changes that involve rewards are more acceptable than imposed changes that restrict choices (Eriksson et al., 2006, 2008; Steg et al., 2006; Dietz et al., 2007). Policies punishing maladaptive behaviour can increase vulnerability when they reinforce socio-economic inequalities that typically produce the maladaptive behaviour in the first place (Adger et al., 2003). Change can be initiated by governments at various levels, but also by individuals, communities, profit-making organizations, trade organizations, and other non-governmental actors (Lindenberg and Steg, 2013; Robertson and Barling, 2015; Stern et al., 2016b).

Strategies can target intrinsic versus extrinsic motivation. It may be particularly important to enhance intrinsic motivation so that people voluntarily engage in climate action over and again (Steg, 2016). Endorsement of mitigation and adaptation actions are positively related (Brügger et al., 2015; Carrico et al., 2015); both are positively related to concern about climate change (Brügger et al., 2015). Strategies that target general antecedents that affect a wide range of actions, such as values, identities, worldviews, climate change beliefs, awareness of the climate impacts of one's actions, and feelings of responsibility to act on climate change, can encourage consistent actions on climate change (van Der Werff and Steg, 2015; Hornsey et al., 2016; Steg, 2016). Initial climate actions can lead to further commitment to climate action (Juhl et al., 2017), when people learn that such actions are easy and effective (Lauren et al., 2016), when they engaged in the initial behaviour for environmental reasons (Peters et al., 2018), hold strong pro-environmental values and norms (Thøgersen and Ölander, 2003), and when initial actions make them realise they are an environmentally sensitive person, motivating them to act on climate change in subsequent situations so as to be consistent (van der

4

Box 4.6 | Bottom-up Initiatives: Adaptation Responses Initiated by Individuals and Communities

To effectively adapt to climate change, bottom-up initiatives by individuals and communities are essential, in addition to efforts of governments, organizations, and institutions (Wamsler and Brink, 2014a). This box presents examples of bottom-up adaptation responses and behavioural change.

Fiji increasingly faces a lack of freshwater due to decreasing rainfall and rising temperatures (Deo, 2011; IPCC, 2014a). While some villages have access to boreholes, these are not sufficient to supply the population with freshwater. Villagers are adapting by rationing water, changing diets, and setting up inter-village sharing networks (Pearce et al., 2017). Some villagers take up wage employment to buy food instead of growing it themselves (Pearce et al., 2017). In Kiribati, residents adapt to drought by purchasing rainwater tanks and constructing additional wells (Kuruppu and Liverman, 2011). An important factor that motivated residents of Kiribati to adapt to drought was the perception that they could effectively adapt to the negative consequences of climate change (Kuruppu and Liverman, 2011).

In the Philippines, seismic activity has caused some islands to flood during high tide. While the municipal government offered affected island communities the possibility to relocate to the mainland, residents preferred to stay and implement measures themselves in their local community to reduce flood damage (Laurice Jamero et al., 2017). Migration is perceived as undesirable because island communities have strong place-based identities (Mortreux and Barnett, 2009). Instead, these island communities have adapted to flooding by constructing stilted houses and raising floors, furniture, and roads to prevent water damage (Laurice Jamero et al., 2017). While inundation was in this case caused by seismic activity, this example indicates how island-based communities may respond to rising sea levels caused by climate change.

Adaptation initiatives by individuals may temporarily reduce the impacts of climate change and enable residents to cope with changing environmental circumstances. However, they may not be sufficient to sustain communities' way of life in the long term. For instance, in Fiji and Kiribati, freshwater and food are projected to become even scarcer in the future, rendering individual adaptations ineffective. Moreover, individuals can sometimes engage in behaviour that may be maladaptive over larger spatio-temporal scales. For example, in the Philippines, many islanders adapt to flooding by elevating their floors using coral stone (Laurice Jamero et al., 2017). Over time, this can harm the survivability of their community, as coral reefs are critical for reducing flood vulnerability (Ferrario et al., 2014). In Maharashtra, India, on-farm ponds are promoted as rainwater harvesting structures to adapt to dry spells during the monsoon season. However, some individuals fill these ponds with groundwater, leading to depletion of water tables and potentially maladaptive outcomes in the long run (Kale, 2015).

Integration of individuals' adaptation initiatives with top-down adaptation policy is critical (Butler et al., 2015), as failing to do so may lead individual actors to mistrust authority and can discourage them from undertaking adequate adaptive actions (Wamsler and Brink, 2014a).

Werff et al., 2014a; Lacasse, 2015, 2016). Yet some studies suggest that people may feel licensed not to engage in further mitigation actions when they believe they have already done their part (Truelove et al., 2014).

4.4.3.3   Acceptability of policy and system changes

Public acceptability can shape, enable or prevent policy and system changes. Acceptability reflects the extent to which policy or system changes are evaluated (un)favourably. Acceptability is higher when people expect more positive and less negative effects of policy and system changes (Perlaviciute and Steg, 2014; Demski et al., 2015; Drews and Van den Bergh, 2016), including climate impacts (Schuitema et al., 2010b). Because of this, policy 'rewarding' climate actions is more acceptable than policy 'punishing' actions that increase climate risks (Steg et al., 2006; Eriksson et al., 2008). Pricing policy is more acceptable when revenues are earmarked for environmental purposes (Steg et al., 2006; Sælen and Kallbekken,

2011) or redistributed towards those affected (Schuitema and Steg, 2008). Acceptability can increase when people experience positive effects after a policy has been implemented (Schuitema et al., 2010a; Eliasson, 2014; Weber, 2015); effective policy trials can thus build public support for climate policy (see Box 4.8).

Climate policy and renewable energy systems are more acceptable when people strongly value other people and the environment, or support egalitarian worldviews, left-wing or green political ideologies (Drews and Van den Bergh, 2016), and less acceptable when people strongly endorse self-enhancement values, or support individualistic and hierarchical worldviews (Dietz et al., 2007; Perlaviciute and Steg, 2014; Drews and Van den Bergh, 2016). Solar radiation modification is more acceptable when people strongly endorse self-enhancement values, and less acceptable when they strongly value other people and the environment (Visschers et al., 2017). Climate policy is more acceptable when people believe climate change is real, when they are concerned about climate change (Hornsey et al., 2016), when they think their actions may reduce climate risks, and when they feel responsible to act on climate change (Steg et al., 2005; Eriksson et al., 2006; Jakovcevic and Steg, 2013; Drews and Van den Bergh, 2016; Kim and Shin, 2017). Stronger environmental awareness is associated with a preference for governmental regulation and behaviour change rather than free-market and technological solutions (Poortinga et al., 2002).

Climate policy is more acceptable when costs and benefits are distributed equally, when nature and future generations are protected (Sjöberg and Drottz-Sjöberg, 2001; Schuitema et al., 2011; Drews and Van den Bergh, 2016), and when fair procedures have been followed, including participation by the public (Dietz, 2013; Bernauer et al., 2016a; Bidwell, 2016) or public society organizations (Bernauer and Gampfer, 2013). Providing benefits to compensate affected communities for losses due to policy or systems changes enhanced public acceptability in some cases (Perlaviciute and Steg, 2014), although people may disagree on what would be a worthwhile compensation (Aitken, 2010; Cass et al., 2010), or feel they are being bribed (Cass et al., 2010; Perlaviciute and Steg, 2014).

Public support is higher when individuals trust responsible parties (Perlaviciute and Steg, 2014; Drews and Van den Bergh, 2016). Yet, public support for multilateral climate policy is not higher than for unilateral policy (Bernauer and Gampfer, 2015); public support for unilateral, non-reciprocal climate policy is rather strong and robust (Bernauer et al., 2016b). Public opposition may result from a culturally valued landscape being affected by adaptation or mitigation options, such as renewable energy development (Warren et al., 2005; Devine-wright and Howes, 2010) or coastal protection measures (Kimura, 2016), particularly when people have formed strong emotional bonds with the place (Devine-Wright, 2009, 2013).

Climate actions may reduce human well-being when such actions involve more costs, effort or discomfort. Yet some climate actions enhance well-being, such as technology that improves daily comfort and nature-based solutions for climate adaptation (Wamsler and Brink, 2014b). Further, climate action may enhance well-being (Kasser and Sheldon, 2002; Xiao et al., 2011; Schmitt et al., 2018) because pursuing meaning by acting on climate change can make people feel good (Venhoeven et al., 2013, 2016; Taufik et al., 2015), more so than merely pursuing pleasure.

4.4.4   Enabling Technological Innovation

This section focuses on the role of technological innovation in limiting warming to 1.5°C, and how innovation can contribute to strengthening implementation to move towards or to adapt to 1.5°C worlds. This assessment builds on information of technological innovation and related policy debates in and after AR5 (Somanathan et al., 2014).

4.4.4.1   The nature of technological innovations

Technological systems have their own dynamics. New technologies have been described as emerging as part of a 'socio-technical system' that is integrated with social structures and that itself evolves over time (Geels and Schot, 2007). This progress is cumulative and accelerating

(Kauffman, 2002; Arthur, 2009). To illustrate such a process of co-evolution: the progress of computer simulation enables us to better understand climate, agriculture, and material sciences, contributing to upgrading food production and quality, microscale manufacturing techniques, and leading to much faster computing technologies, resulting, for instance, in better performing photovoltaic (PV) cells.

A variety of technological developments have and will contribute to 1.5°C-consistent climate action or the lack of it. They can do this, for example, in the form of applications such as smart lighting systems, more efficient drilling techniques that make fossil fuels cheaper, or precision agriculture. As discussed in Section 4.3.1, costs of PV (IEA, 2017f) and batteries (Nykvist and Nilsson, 2015) have sharply dropped. In addition, costs of fuel cells (Iguma and Kidoshi, 2015; Wei et al., 2017) and shale gas and oil (Wang et al., 2014; Mills, 2015) have come down as a consequence of innovation.

4.4.4.2   Technologies as enablers of climate action

Since AR5, literature has emerged as to how much future GHG emission reductions can be enabled by the rapid progress of general purpose technologies (GPTs), consisting of information and communication technologies (ICT), including artificial intelligence (AI) and the internet of things (IoT), nanotechnologies, biotechnologies, robotics, and so forth (WEF, 2015; OECD, 2017c). Although these may contribute to limiting warming to 1.5°C, the potential environmental, social and economic impacts of new technologies are uncertain.

Rapid improvement of performance and cost reduction is observed for many GPTs. They include AI, sensors, internet, memory storage and microelectromechanical systems. The latter GPTs are not usually categorized as climate technologies, but they can impact GHG emissions.

Progress of GPT could help reduce GHG emissions more cost-effectively. Examples are shown in Table 4.9. It may however, result in more emissions by increasing the volume of economic activities, with unintended negative consequence on sustainable development. While ICT increases electricity consumption (Aebischer and Hilty, 2015), the energy consumption of ICT is usually dwarfed by the energy saving by ICT (Koomey et al., 2013; Malmodin et al., 2014), but rebound effects and other sustainable development impacts may be significant. An appropriate policy framework that accommodates such impacts and their uncertainties could address the potential negative impacts by GPT (Jasanoff, 2007).

GHG emission reduction potentials in relation to GPTs were estimated for passenger cars using a combination of three emerging technologies: electric vehicles, car sharing, and self-driving. GHG emission reduction potential is reported, assuming generation of electricity with low GHG emissions (Greenblatt and Saxena, 2015; ITF, 2015; Viegas et al., 2016; Fulton et al., 2017). It is also possible that GHG emissions increase due to an incentive to car use. Appropriate policies such as urban planning and efficiency regulations could contain such rebound effects (Wadud et al., 2016).

Estimating emission reductions by GPT is difficult due to substantial uncertainties, including projections of future technological performance,

4

Table 4.9 | Examples of technological innovations relevant to 1.5°C enabled by general purpose technologies (GPT). Note: lists of enabling GPT or adaptation/mitigation options are not exhaustive, and the GPTs by themselves do not reduce emissions or increase climate change resilience.

Sector	Examples of Mitigation/Adaptation Technological Innovation	Enabling GPT
Buildings	Energy and CO_2 efficiency of logistics, warehouse and shops (GeSI, 2015; IEA, 2017a)	IoT, AI
	Smart lighting and air conditioning (IEA, 2016b, 2017a)	IoT, AI
Industry	Energy efficiency improvement by industrial process optimization (IEA, 2017a)	Robots, IoT
	Bio-based plastic production by biorefinery (OECD, 2017c)	Biotechnology
	New materials from biorefineries (Fornell et al., 2013; McKay et al., 2016)	ICT, biotechnology
Transport	Electric vehicles, car sharing, automation (Greenblatt and Saxena, 2015; Fulton et al., 2017)	Biotechnology
	Bio-based diesel fuel by biorefinery (OECD, 2017c)	ICT, biotechnology
	Second generation bioethanol potentially coupled to carbon capture systems (De Souza et al., 2014; Rochedo et al., 2016)	Biotechnology
	Logistical optimization, and electrification of trucks by overhead line (IEA, 2017e)	ICT, biotechnology
	Reduction of transport needs by remote education, health and other services (GeSI, 2015; IEA, 2017a)	Biotechnology
	Energy saving by lightweight aircraft components (Beyer, 2014; Faludi et al., 2015; Verhoef et al., 2018)	Additive manufacturing (3D printing)
Electricity	Solar PV manufacturing (Nemet, 2014)	Nanotechnology
	Smart grids and grid flexibility to accommodate intermittent renewables (Heard et al., 2017)	IoT, AI
	Plasma confinement for nuclear fusion (Baltz et al., 2017)	AI
Agriculture	Precision agriculture (improvement of energy and resource efficiency including reduction of fertilizer use and N_2O emissions) (Pierpaoli et al., 2013; Brown et al., 2016; Schimmelpfennig and Ebel, 2016)	Biotechnology ICT, AI
	Methane inhibitors (and methane-suppressing vaccines) that reduce livestock emissions from enteric fermentation (Wedlock et al. 2013; Hristov et al. 2015; Wollenberg et al. 2016)	Biotechnology
	Engineering C3 into C4 photosynthesis to improve agricultural production and productivity (Schuler et al., 2016)	Biotechnology
	Genome editing using CRISPR to improve/adapt crops to a changing climate (Gao, 2018)	Biotechnology
Disaster Reduction and Adaptation	Weather forecasting and early warning systems, in combination with user knowledge (Hewitt et al., 2012; Lourenço et al., 2016)	ICT
	Climate risk reduction (Upadhyay and Bijalwan, 2015)	ICT
	Rapid assessment of disaster damage (Kryvasheyeu et al., 2016)	ICT

costs, penetration rates, and induced human activity. Even if a technology is available, the establishment of business models might not be feasible (Linder and Williander, 2017). Indeed, studies show a wide range of estimates, ranging from deep emission reductions to possible increases in emissions due to the rebound effect (Larson and Zhao, 2017).

GPT could also enable climate adaptation, in particular through more effective climate disaster risk management and improved weather forecasting.

Government policy usually plays a role in promoting or limiting GPTs, or science and technology in general. It has impacts on climate action, because the performance of further climate technologies will partly depend on the progress of GPTs. Governments have established institutions for achieving many social, and sometimes conflicting goals, including economic growth and addressing climate change (OECD, 2017c), which include investment in basic research and development (R&D) that can help develop game-changing technologies (Shayegh et al., 2017). Governments are also needed to create an enabling environment for the growth of scientific and technological ecosystems necessary for GPT development (Tassey, 2014).

4.4.4.3   The role of government in 1.5°C-consistent climate technology policy

While literature on 1.5°C-specific innovation policy is absent, a growing body of literature indicates that governments aim to achieve social, economic and environmental goals by promoting science and a broad range of technologies through 'mission-driven' innovation policies, based on differentiated national priorities (Edler and Fagerberg, 2017). Governments can play a role in advancing climate technology via a 'technology push' policy on the technology supply side (e.g., R&D subsidies), and by 'demand pull' policy on the demand side (e.g., energy-efficiency regulation), and these policies can be complemented by enabling environments (Somanathan et al., 2014). Governments may also play a role in removing existent support for incumbents (Kivimaa and Kern, 2016). A growing literature indicates that policy mixes, rather than single policy instruments, are more effective in addressing climate innovation challenges ranging from technologies in the R&D phase to those ready for diffusion (Veugelers, 2012; Quitzow, 2015; Rogge et al., 2017; Rosenow et al., 2017). Such innovation policies can help address two kinds of externalities: environmental externalities and proprietary problems (GEA, 2012; IPCC, 2014b; Mazzucato and Semieniuk, 2017). To avoid 'picking winners', governments often maintain a broad portfolio of technological options (Kverndokk and Rosendahl, 2007) and work in close collaboration with the industrial sector and society in general. Some governments have achieved relative success in supporting innovation policies (Grubler et al., 2012; Mazzucato, 2013) that addressed climate-related R&D (see Box 4.7 on bioethanol in Brazil).

Funding for R&D could come from various sources, including the general budget, energy or resource taxation, or emission trading schemes (see Section 4.4.5). Investing in climate-related R&D has as an additional benefit of building capabilities to implement climate mitigation and adaptation technologies (Ockwell et al., 2015). Countries regard innovation in general and climate technology specifically as a national interests issue and addressing climate change primarily as being in the global interest. Reframing part of climate policy as technology or industrial policy might therefore contribute to resolving the difficulties that continue to plague emission target negotiations (Faehn and Isaksen, 2016; Fischer et al., 2017; Lachapelle et al., 2017).

Climate technology transfer to emerging economies has happened regardless of international treaties, as these countries have been keen to acquire them, and companies have an incentive to access emerging markets to remain competitive (Glachant and Dechezleprêtre, 2016).

However, the complexity of these transfer processes is high, and they have to be conducted carefully by governments and institutions (Favretto et al., 2017). It is noticeable that the impact of the EU emission trading scheme (EU ETS) on innovation is contested; recent work (based on lower carbon prices than anticipated for 1.5°C-consistent pathways) indicates that it is limited (Calel and Dechezleprêtre, 2016), but earlier assessments (Blanco et al., 2014) indicate otherwise.

4.4.4.4 Technology transfer in the Paris Agreement

Technology development and transfer is recognized as an enabler of both mitigation and adaptation in Article 10 in the Paris Agreement (UNFCCC, 2016) as well as in Article 4.5 of the original text of the UNFCCC (UNFCCC, 1992). As previous sections have focused on technology development and diffusion, this section focuses on technology transfer. Technology transfer can adapt technologies to local circumstances, reduce financing costs, develop indigenous technology, and build capabilities to operate, maintain, adapt and innovate on technology globally (Ockwell et al., 2015; de Coninck and Sagar, 2017). Technology cooperation could decrease global mitigation cost, and enhance developing countries' mitigation contributions (Huang et al., 2017a).

Box 4.7 | Bioethanol in Brazil: Innovation and Lessons for Technology Transfer

The use of sugarcane as a bioenergy source started in Brazil in the 1970s. Government and multinational car factories modified car engines nationwide so that vehicles running only on ethanol could be produced. As demand grew, production and distribution systems matured and costs came down (Soccol et al., 2010). After a transition period in which both ethanol-only and gasoline-only cars were used, the flex-fuel era started in 2003, when all gasoline was blended with 25% ethanol (de Freitas and Kaneko, 2011). By 2010, around 80% of the car fleet in Brazil had been converted to use flex-fuel (Goldemberg, 2011; Su et al., 2015).

More than forty years of combining technology push and market pull measures led to the deployment of ethanol production, transportation and distribution systems across Brazil, leading to a significant decrease in CO_2 emissions (Macedo et al., 2008). Examples of innovations include: (i) the development of environmentally well-adapted varieties of sugarcane; (ii) the development and scaling up of sugar fermentation in a non-sterile environment, and (iii) the development of adaptations of car engines to use ethanol as a fuel in isolation or in combination with gasoline (Amorim et al., 2011; de Freitas and Kaneko, 2011; De Souza et al., 2014). Public procurement, public investment in R&D and mandated fuel blends accompanying these innovations were also crucial (Hogarth, 2017). In the future, innovation could lead to viable partial CO_2 removal through deployment of BECCS associated with the bioethanol refineries (Fuss et al., 2014; Rochedo et al., 2016) (see Section 4.3.7).

Ethanol appears to reduce urban car emission of health-affecting ultrafine particles by 30% compared to gasoline-based cars, but increases ozone (Salvo et al., 2017). During the 1990s, when sugarcane burning was still prevalent, particulate pollution had negative consequences for human health and the environment (Ribeiro, 2008; Paraiso and Gouveia, 2015). While Jaiswal et al. (2017) report bioethanol's limited impact on food production and forests in Brazil, despite the large scale, and attribute this to specific agro-ecological zoning legislation, various studies report adverse effects of bioenergy production through forest substitution by croplands (Searchinger et al., 2008), as well as impacts on biodiversity, water resources and food security (Rathore et al., 2016). For new generation biofuels, feasibility and life cycle assessment studies can provide information on their impacts on environmental, economic and social factors (Rathore et al., 2016).

Brazil and the European Union have tried to replicate Brazil's bioethanol experience in climatically suitable African countries. Although such technology transfer achieved relative success in Angola and Sudan, the attempts to set up bioethanol value chains did not pass the phase of political deliberations and feasibility studies elsewhere in Africa. Lessons learned include the need for political and economic stability of the donor country (Brazil) and the necessity for market creation to attract investments in first-generation biofuels alongside a safe legal and policy environment for improved technologies (Afionis et al., 2014; Favretto et al., 2017).

4

The international institutional landscape around technology development and transfer includes the UNFCCC (via its technology framework and Technology Mechanism including the Climate Technology Centre and Network (CTCN)), the United Nations (a technology facilitation mechanism for the SDGs) and a variety of non-UN multilateral and bilateral cooperation initiatives such as the Consultative Group on International Agricultural Research (CGIAR, founded in the 1970s), and numerous initiatives of companies, foundations, governments and non-governmental and academic organizations. Moreover, in 2015, twenty countries launched an initiative called 'Mission Innovation', seeking to double their energy R&D funding. At this point it is difficult to evaluate whether Mission Innovation achieved its objective (Sanchez and Sivaram, 2017). At the same time, the private sector started an innovation initiative called the 'Breakthrough Energy Coalition'.

Most technology transfer is driven by through markets by the interests of technology seekers and technology holders, particularly in regions with well-developed institutional and technological capabilities such as developed and emerging nations (Glachant and Dechezleprêtre, 2016). However, the current international technology transfer landscape has gaps, in particular in reaching out to least-developed countries, where institutional and technology capabilities are limited (de Coninck and Puig, 2015; Ockwell and Byrne, 2016). On the one hand, literature suggests that the management or even monitoring of all these UN, bilateral, private and public initiatives may fail to lead to better results. On the other hand, it is probably more cost-effective to adopt a strategy of 'letting a thousand flowers bloom', by challenging and enticing researchers in the public and the private sector to direct innovation towards low-emission and adaptation options (Haselip et al., 2015). This can be done at the same time as mission-oriented research is adopted in parallel by the scientific community (Mazzucato, 2018).

At COP 21, the UNFCCC requested the Subsidiary Body for Scientific and Technological Advice (SBSTA) to initiate the elaboration of the technology framework established under the Paris Agreement (UNFCCC, 2016). Among other things, the technology framework would 'provide overarching guidance for the work of the Technology Mechanism in promoting and facilitating enhanced action on technology development and transfer in order to support the implementation of this Agreement' (this Agreement being the Paris Agreement). An enhanced guidance issued by the Technology Executive Committee (TEC) for preparing a technology action plan (TAP) supports the new technology framework as well as the Parties' long-term vision on technology development and transfer, reflected in the Paris Agreement (TEC, 2016).

4.4.5   Strengthening Policy Instruments and Enabling Climate Finance

Triggering rapid and far-reaching change in technical choices and institutional arrangements, consumption and lifestyles, infrastructure, land use, and spatial patterns implies the ability to scale up policy signals to enable the decoupling of GHGs emission, and economic growth and development (Section 4.2.2.3). Such a scale-up would also imply that potential short-term negative responses by populations and interest groups, which could block these changes from the outset, would need to be prevented or overcome. This section describes the size and nature of investment needs and the financial challenge over the coming two decades in the context of 1.5°C warmer worlds, assesses the potential and constraints of three categories of policy instruments that respond to the challenge, and explains the conditions for using them synergistically. The policy and finance instruments discussed in this section relate to Section 4.4.1 (on governance) and other Sections in 4.4.

4.4.5.1   The core challenge: cost-efficiency, coordination of expectations and distributive effects

Box 4.8 shows that the average estimate by seven models of annual investment needs in the energy system is around 2.38 trillion USD2010 (1.38 to 3.25) between 2016 and 2035. This represents between 2.53% (1.6–4%) of the world GDP in market exchange rates (MER) and 1.7% of the world GDP in purchasing power parity (PPP). OECD investment assessments for a 2°C-consistent transition suggest that including investments in transportation and in other infrastructure would increase the investment needs by a factor of three. Other studies not included in Box 4.8, in particular by the World Economic Forum (WEF, 2013) and the Global Commission on the Economy and Climate (GCEC, 2014) confirm these orders of magnitude of investment.

The average increase of investment in the energy sector resulting from Box 4.8 represents a mean value of 1.5% of the total world investment compared with the baselines scenario in MER and a little over 1% in PPP. Including infrastructure investments would raise this to 2.5% and 1.7% respectively.[9]

These incremental investments could be funded through a drain on consumption (Bowen et al., 2017), which would necessitate between 0.68% and 0.45% lower global consumption than in the baseline. But, consumption at a constant savings/consumption ratio can alternatively be funded by shifting savings towards productive adaptation and mitigation investments, instead of real-estate sector and liquid financial products. This response depends upon whether it is possible to close the global investment funding gap for infrastructure that potentially inhibits growth, through structural changes in the global economy. In this case, investing more in infrastructure would not be an incremental cost in terms of development and welfare (IMF, 2014; Gurara et al., 2017)

---

[9]   A calculation in MER tends indeed to underestimate the world GDP and its growth by giving a lower weight to fast-growing developing countries, whereas a calculation in PPP tends to overestimate it. The difference between the value of two currencies in PPP and MER should vanish as the gap of the income levels of the two concerned countries decreases. Accounting for this trend in modelling is challenging.

Box 4.8 | Investment Needs and the Financial Challenge of Limiting Warming to 1.5°C

Peer-reviewed literature that estimates the investment needs over the next two decades to scale up the response to limit warming to 1.5°C is very limited (see Section 4.6). This box attempts to bring together available estimates of the order of magnitude of these investments, after consultation with the makers of those estimates, to provide the context for global and national financial mobilization policy and related institutional arrangements.

Table 1 in this box presents mean annual investments up to 2035, based on three studies (after clarifying their scope and harmonizing their metrics): an ensemble of four integrated assessment models (here denoted IAM, see Chapter 2), an Organization for Economic Co-operation and Development (OECD) scenario for a 2°C limit (OECD, 2017a) and scenarios from the International Energy Agency (IEA, 2016c). All three sources provide estimates for the energy sector for various mitigation scenarios. They give a mean value of 2.38 trillion USD of yearly investments in the energy sector over the period, with minimum and maximum values of 1.38 and 3.25 respectively. We also report the OECD estimate for 2°C because it also covers transportation and other infrastructure (water, sanitation, and telecommunication), which are essential to deliver the Sustainable Development Goals (SDGs), including SDG 7 on clean energy access, and enhance the adaptive capacity to climate change.

Box 4.8, Table 1 | Estimated annualized world mitigation investment needed to limit global warming to 2°C or 1.5°C (2015–2035 in trillions of USD at market exchange rates) from different sources. The top four lines indicate the results of Integrated Assessment Models (IAMs) as reported in Chapter 2 for their Baseline, Nationally Determined Contributions (NDC), 2°C- and 1.5°C-consistent pathways. These numbers only cover the energy sector and the second row includes energy efficiency in all sectors. The final two rows indicate the mitigation investment needs for the energy, transport and other infrastructure according to the Organization for Economic Co-operation and Development (OECD) for a Baseline pathway and a 2°C-consistent pathway. Sources: IEA, 2016c; OECD, 2017a.

	Energy Investments	Of which Demand Side	Transport	Other Infra-structures	Total	Ratio to MER GDP
IAM Baseline (mean)	1.96	0.24			1.96	1.8%
IAM NDC (mean)	2.04	0.28			2.04	1.9%
IAM 2°C (mean)	2.19	0.38			2.19	2.1%
IAM 1.5°C (mean)	2.32	0.45			2.32	2.2%
IEA NDC	2.40	0.72			2.40	2.3%
IEA 1.5°C	2.76	1.13			2.76	2.7%
Mean IAM-IEA, 1.5°C	**2.38**	**0.54**			**2.38**	**2.53%**
Min IAM-IEA, 1.5°C	1.38	0.38			1.38	1.6%
Max IAM-IEA, 1.5°C	3.25	1.13			3.25	4.0%
OECD Baseline					5.74	5.4%
OECD 2°C	2.13	0.40	2.73	1.52	6.38	6.0%

The mean incremental share of annual energy investments to stay well below 2°C is 0.36% (between 0.2–1%) of global GDP between 2016 and 2035. Since total world investment (also called gross fixed capital formation (GFCF)) is about 24% of global GDP, the estimated incremental energy investments between a baseline and a 1.5°C transition would be approximately 1.5% (between 0.8–4.2%) of projected total world investments. As the higher ends of these ranges reflect pessimistic assumptions in 1.5°C-consistent pathways on technological change, the implementation of policies to accelerate technical change (see the remainder of Section 4.4.5) could lower the probability of higher incremental investment.

If we assume the amounts of investments given by the OECD for transportation and other infrastructure for warming of 2°C to be a lower limit for an 1.5°C pathway, then total incremental investments for all sectors for a 1.5°C-consistent pathway would be estimated at 2.4% of total world investments. This total incremental investment reaches 2.53% if the investments in transportation are scaled up proportionally with the investments in the energy sector and if all other investments are kept constant. Comparing this 2.4% or 2.53% number for all sectors to the 1.5% number for energy only (see previous paragraph) suggests that the investments in sectors other than energy contribute significantly to incremental world investments, even though a comprehensive study or estimate of these investments for a 1.5°C limit is not available.

The issue, from a macroeconomic perspective, is whether these investments would be funded by higher savings at the costs of lower consumption. This would mean a 0.5% reduction in consumption for the energy sector for 1.5°C. Note that for a 2°C scenario, this

Box 4.8 (continued)

reduction would be 0.8% if we account for the investment needs of all infrastructure sectors. Assuming conversely a constant savings ratio, this would necessitate reallocating existing capital flows towards infrastructure. In addition to these incremental investments, the amount of redirected investments is relevant from a financial perspective. In the reported IAM energy sector scenarios, about three times the incremental investments is redirected. There is no such assessment for the other sectors. The OECD report suggests that these ratios might be higher.

These orders of magnitude of investment can be compared to the available statistics of the global stock of 386 trillion USD of financial capital, which consists of 100 trillion USD in bonds (SIFMA, 2017), around 60 trillion USD in equity (World Bank, 2018b), and 226 trillion USD of loans managed by the banking system (IIF, 2017; World Bank, 2018a). The long-term rate of return (interest plus increase of shareholder value) is about 3% on bonds, 5% on bank lending and 7% on equity, leading to a weighted mean return on capital of 3.4% in real terms (5.4% in nominal terms). Using 3.4% as a lower bound and 5% as a higher bound (following Piketty, 2014) and taking a conservative assumption that global financial capital grows at the same rate as global GDP, the estimated yearly financial capital revenues would be between 16.8 and 25.4 trillion USD.

Assuming that a quarter of these investments comes from public funds (as estimated by the World Bank; World Bank, 2018a), the amount of private resources needed to enable an energy sector transition is between 3.3% and 5.3% of annual capital income and between 5.6% and 8.3% of these revenues for all infrastructure to meet the 2°C limit and the SDGs.

Since the financial system has limited fungibility across budget lines, changing the partitioning of investments is not a zero-sum game. An effective policy regime could encourage investment managers to change their asset allocation. Part of the challenge may lie in increasing the pace of financing of low-emission assets to compensate for a possible 38% decrease, by 2035, in the value of fossil fuel assets (energy sector and indirect holdings in downstream uses like automobiles) (Mercure et al., 2018).

Investments in other (non-energy system) infrastructure to meet development and poverty-reduction goals can strengthen the adaptive capacity to address climate change, and are difficult to separate from overall sustainable development and poverty-alleviation investments (Hallegatte and Rozenberg, 2017). The magnitude of potential climate change damages is related to pre-existing fragility of impacted societies (Hallegatte et al., 2007). Enhancing infrastructure and service provision would lower this fragility, for example, through the provision of universal (water, sanitation, telecommunication) service access (Arezki et al., 2016).

The main challenge is thus not just a lack of mobilization of aggregate resources but of redirection of savings towards infrastructure, and the further redirection of these infrastructure investments towards low-emission options. If emission-free assets emerge fast enough to compensate for the devaluation of high-emission assets, the sum of the required incremental and redirected investments in the energy sector would (up to 2035) be equivalent to between 3.3% and 5.3% of the average annual revenues of the private capital stock (see Box 4.8) and to between 5.6% and 8.3%, including all infrastructure investments.

The interplay between mechanisms of financial intermediation and the private risk-return calculus is a major barrier to realizing these investments (Sirkis et al., 2015). This obstacle is not specific to climate mitigation investments but also affects infrastructure and has been characterised as the gap between the 'propensity to save' and the 'propensity to invest' (Summers, 2016). The issue is whether new financial instruments could close this gap and inject liquidity into the low-emission transition, thereby unlocking new economic opportunities (GCEC, 2014; NCE, 2016). By offsetting the crowding-out of other private and public investments (Pollitt and Mercure, 2017), the ensuing ripple effect could reinforce growth and the sustainability of development (King, 2011; Teulings and Baldwin, 2014) and potentially trigger a new growth cycle (Stern, 2013, 2015). In this case, a massive mobilization of low-emission investments would require a significant effort but may be complementary to sustainable development investments.

This uncertain but potentially positive outcome might be constrained by the higher energy costs of low-emission options in the energy and transportation sectors. The envelope of worldwide marginal abatement costs for 1.5°C-consistent pathways reported in Chapter 2 is 135–5500 USD2010 tCO_2^{-1} in 2030 and 245–13000 USD2010 tCO_2^{-1} in 2050, which is between three to four times higher than for a 2°C limit.

These figures are consistent with the dramatic reduction in the unit costs of some low-emission technical options (for example solar PV, LED lighting) over the past decade (see Section 4.3.1) (OECD, 2017c). Yet there are multiple constraints to a system-wide energy transition. Lower costs of some supply- and demand-side options do not always result in a proportional decrease in energy system costs. The adoption of alternative options can be slowed down by increasing costs of decommissioning existing infrastructure, the inertia of market structures, cultural habits and risk-adverse user behaviour (see Sections 4.4.1 to 4.4.3). Learning-by-doing processes and R&D can accelerate the cost-efficiency of low-emission technology but often imply higher early-phase costs. The German energy transition resulted in high consumer prices for electricity in Germany (Kreuz and Müsgens, 2017) and needed strong accompanying measures to succeed.

One key issue is that energy costs can propagate across sectors and amplify overall production costs. During the early stage of a low-emission transition, an increase in the prices of non-energy goods could reduce consumer purchasing power and final demand. A rise in energy prices has a proportionally greater impact in developing countries that are in a catch-up phase, as they have a stronger dependence on energy-intensive sectors (Crassous et al., 2006; Luderer et al., 2012) and a higher ratio of energy to labour cost (Waisman et al., 2012). This explains why with lower carbon prices, similar emission reductions are reached in South Africa (Altieri et al., 2016) and Brazil (La Rovere et al., 2017a) compared to developed countries. However, three distributional issues emerge.

First, in the absence of countervailing policies, higher energy costs have an adverse effect on the distribution of welfare (see also Chapter 5). The negative impact is inversely correlated with the level of income (Harberger, 1984; Fleurbaey and Hammond, 2004) and positively correlated with the share of energy in the households budget, which is high for low- and middle-income households (Proost and Van Regemorter, 1995; Barker and Kohler, 1998; West and Williams, 2004; Chiroleu-Assouline and Fodha, 2011). Moreover, climatic conditions and the geographical conditions of human settlements matter for heating and mobility needs (see Chapter 5). Medium-income populations in the suburbs, in remote areas, and in low-density regions can be as vulnerable as residents of low-income urban areas. Poor households with low levels of energy consumption are also impacted by price increases of non-energy goods caused by the propagation of energy costs (Combet et al., 2010; Dubois, 2012). These impacts are generally not offset by non-market co-benefits of climate policies for the poor (Baumgärtner et al., 2017).

A second matter of concern is the distortion of international competition and employment implications in the case of uneven carbon constraints, especially for energy-intensive industries (Demailly and Quirion, 2008). Some of these industries are not highly exposed to international competition because of their very high transportation costs per unit value added (Sartor, 2013; Branger et al., 2016), but other industries could suffer severe shocks, generate 'carbon leakage' through cheaper imports from countries with lower carbon constraints (Branger and Quirion, 2014), and weaken the surrounding regional industrial fabric with economy-wide and employment implications.

A third challenge is the depreciation of assets whose value is based on the valuation of fossil energy resources, of which future revenues may decline precipitously with higher carbon prices (Waisman et al., 2013; Jakob and Hilaire, 2015; McGlade and Ekins, 2015), and on emission-intensive capital stocks (Guivarch and Hallegatte, 2011; OECD, 2015a; Pfeiffer et al., 2016). This raises issues of changes in industrial structure, adaptation of worker skills, and of stability of financial, insurance and social security systems. These systems are in part based on current holdings of carbon-based assets whose value might decrease by about 38% by the mid-2030s (Mercure et al., 2018). This stranded asset challenge may be exacerbated by a decline of export revenues of fossil fuel producing countries and regions (Waisman et al., 2013; Jakob and Hilaire, 2015; McGlade and Ekins, 2015).

These distributional issues, if addressed carefully and expeditiously, could affect popular sensitivity towards climate policies. Addressing them could mitigate adverse macroeconomic effects on economic growth and employment that could undermine the potential benefits of a redirection of savings and investments towards 1.5°C-consistent pathways.

Strengthening policy instruments for a low-emission transition would thus need to reconcile three objectives: (i) handling the short-term frictions inherent to this transition in an equitable way, (ii) minimizing these frictions by lowering the cost of avoided GHGs emissions, and (iii) coordinating expectations of multiple stakeholders at various decision-making levels to accelerate the decline in costs of emission reduction, efficiency and decoupling options and maximizing their co-benefits (see the practical example of lowering car use in cities in Box 4.9).

Three categories of policy tools would be available to meet the distributional challenges: carbon pricing, regulatory instruments and information and financial tools. Each of them has its own strengths and weaknesses, from a 1.5°C perspective, policy tools would have to be both scaled up and better coordinated in packages in a synergistic manner.

4.4.5.2 Carbon pricing: necessity and constraints

Economic literature has long argued that climate and energy policy grounded only in regulation, standards and public funding of R&D is at risk of being influenced by political and administrative arbitrariness, which could raise the costs of implementation. This literature has argued that it may be more efficient to make these costs explicit through carbon taxes and carbon trading, securing the abatement of emissions in places and sectors where it is cheapest (IPCC, 1995, 2001; Gupta et al., 2007; Somanathan et al., 2014).

In a frictionless world, a uniform world carbon price could minimize the social costs of the low-carbon transition by equating the marginal costs of abatement across all sources of emissions. This implies that investors will be able to make the right choices under perfect foresight and that domestic and international compensatory transfers offset the adverse distributional impacts of higher energy prices and their consequences on economic activity. In the absence of such transfers, carbon prices would have to be differentiated by jurisdiction (Chichilnisky and Heal, 2000; Sheeran, 2006; Böhringer et al., 2009; Böhringer and Alexeeva-Talebi, 2013). This differentiation could in turn raise concerns of distortions in international competition (Hourcade et al., 2001; Stavins et al., 2014).

Obstacles to enforcing a uniform world carbon price in the short run would not necessarily crowd out explicit national carbon pricing, for three reasons. First, a uniform carbon price would limit an emissions rebound resulting from a higher consumption of energy services enabled by efficiency gains, if energy prices do not change (Greening et al., 2000; Fleurbaey and Hammond, 2004; Sorrell et al., 2009; Guivarch and Hallegatte, 2011; Chitnis and Sorrell, 2015; Freire-González, 2017). Second, it could hedge against the arbitrariness of regulatory policies. Third, 'revenue neutral' recycling, at a constant share of taxes on GDP, into lowering some existing taxes would compensate for at least part of the propagation effect of higher energy costs (Stiglitz et al., 2017). The substitution by carbon taxes of taxes that cause distortions on the

4

Box 4.9 | Emerging Cities and 'Peak Car Use': Evidence of Decoupling in Beijing

The phenomenon of 'peak car use', or reductions in per capita car use, provides hope for continuing reductions in greenhouse gases from oil consumption (Millard-Ball and Schipper, 2011; Newman and Kenworthy, 2011; Goodwin and Van Dender, 2013). The phenomenon has been mostly associated with developed cities apart from some early signs in Eastern Europe, Latin America and China (Newman and Kenworthy, 2015). New research indicates that peak car is now also underway in China (Gao and Newman, 2018).

China's rapid urban motorization was a result of strong economic growth, fast urban development and the prosperity of the Chinese automobile industry (Gao et al., 2015). However, recent data (Gao and Newman, 2018) (expressed as a percentage of daily trips) suggest the first signs of a break in the growth of car use along with the growth in mass transit, primarily the expansion of Metro systems (see Box 4.9, Figure 1).

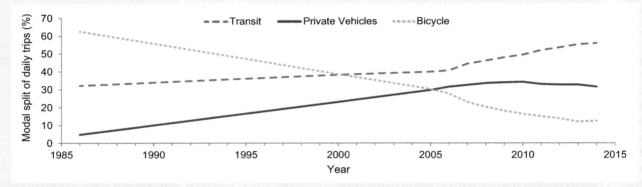

Box 4.9, Figure 1 | The modal split data in Beijing between 1986 and 2014. Source: (Gao and Newman, 2018).

Chinese urban fabrics, featuring traditional dense linear forms and mixed land use, favour mass transit systems over automobiles (Gao and Newman, 2018). The data show that the decline in car use did not impede economic development, but the growth in vehicle kilometres of travel (VKT) has decoupled absolutely from GDP as shown in Box 4.9, Figure 2 below.

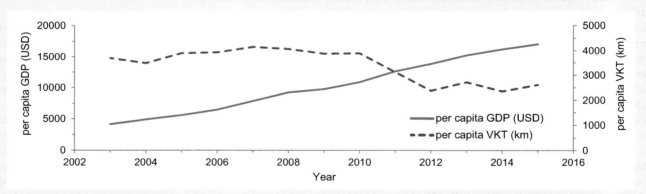

Box 4.9, Figure 2 | Peak car in Beijing: relationships between economic performance and private automobile use in Beijing from 1986 to 2014. VKT is vehicle kilometres of travel. Source: (Gao and Newman, 2018).

economy can counteract the regressive effect of higher energy prices. For example, offsetting increased carbon prices with lower labour taxes can potentially decrease labour costs (without affecting salaries), enhance employment and reduce the attractiveness of informal economic activity (Goulder, 2013).

The conditions under which an economic gain along with climate benefit (a 'double dividend') can be expected are well documented

(Goulder, 1995; Bovenberg, 1999; Mooij, 2000). In the context of OECD countries, the literature examines how carbon taxation could substitute for other taxes to fund the social security system (Combet, 2013). The same general principles apply for countries that are building their social welfare system, such as China (Li and Wang, 2012) or Brazil (La Rovere et al., 2017a), but an optimal recycling scheme could differ based on the structure of the economy (Lefèvre et al., 2018).

In every country the design of carbon pricing policy implies a balance between incentivizing low-carbon behaviour and mitigating the adverse distributional consequences of higher energy prices (Combet et al., 2010). Carbon taxes can offset these effects if their revenues are redistributed through rebates to poor households. Other options include the reduction of value-added taxes for basic products or direct benefit transfers to enable poverty reduction (see Winkler et al. (2017) for South Africa and Grottera et al. (2016) for Brazil). This is possible because higher-income households pay more in absolute terms, even though their carbon tax burden is a relatively smaller share of their income (Arze del Granado et al., 2012).

Ultimately, the pace of increase of carbon prices would depend on the pace at which they can be embedded in a consistent set of fiscal and social policies. This is specifically critical in the context of the 1.5°C limit (Michaelowa et al., 2018). This is why, after a quarter century of academic debate and experimentation (see IPCC WGIII reports since the SAR), a gap persists with respect to 'switching carbon prices' needed to trigger rapid changes. In 2016, only 15% of global emissions are covered by carbon pricing, three-quarters of which with prices below 10 USD tCO_2^{-1} (World Bank, 2016). This is too low to outweigh the 'noise' from the volatility of oil markets (in the range of 100 USD tCO_2^{-1} over the past decade), of other price dynamics (interest rates, currency exchange rates and real estate prices) and of regulatory policies in energy, transportation and industry. For example, the dynamics of mobility depend upon a trade-off between housing prices and transportation costs in which the price of real estate and the inert endowments in public transport play as important a role as liquid fuel prices (Lampin et al., 2013).

These considerations apply to attempts to secure a minimum price in carbon trading systems (Wood and Jotzo, 2011; Fell et al., 2012; Fuss et al., 2018) and to the reduction of fossil fuel subsidies. Estimated at 650 billion USD in 2015 (Coady et al., 2017), these subsidies represent 25–30% of government expenditures in forty (mostly developing) countries (IEA, 2014b). Reducing these subsidies would contribute to reaching 1.5°C-consistent pathways, but raises similar issues as carbon pricing around long-term benefits and short-term costs (Jakob et al., 2015; Zeng and Chen, 2016), as well as social impacts.

Explicit carbon prices remain a necessary condition of ambitious climate policies, and some authors highlight the potential benefit brought by coordination among groups of countries (Weischer et al., 2012; Hermwille et al., 2017; Keohane et al., 2017). They could take the form of carbon pricing corridors (Bhattacharya et al., 2015). They are a necessary 'lubricant' through fiscal reforms or direct compensating transfers to accommodate the general equilibrium effects of higher energy prices but may not suffice to trigger the low-carbon transition because of a persistent 'implementation gap' between the aspirational carbon prices and those that can practically be enforced. When systemic changes, such as those needed for 1.5°C-consistent pathways, are at play on many dimensions of development, price levels 'depend on the path and the path depends on political decisions' (Drèze and Stern, 1990).

4.4.5.3 Regulatory measures and information flows

Regulatory instruments are a common tool for improving energy efficiency and enhancing renewable energy in OECD countries (e.g., the USA, Japan, Korea, Australia, the EU) and, more recently, in developing countries (M.J. Scott et al., 2015; Brown et al., 2017). Such instruments include constraints on the import of products banned in other countries (Knoop and Lechtenböhmer, 2017).

For energy efficiency, these instruments include end-use standards and labelling for domestic appliances, lighting, electric motors, water heaters and air-conditioners. They are often complemented by mandatory efficiency labels to attract consumers' attention and stimulate the manufacture of more efficient products (Girod et al., 2017). Experience shows that these policy instruments are effective only if they are regularly reviewed to follow technological developments, as in the 'Top Runner' programme for domestic appliances in Japan (Sunikka-Blank and Iwafune, 2011).

In four countries, efficiency standards (e.g. miles per gallon or level of CO_2 emission per kilometre) have been used in the transport sector, for light- and heavy-duty vehicles, which have spillovers for the global car industry. In the EU (Ajanovic and Haas, 2017) and the USA (Sen et al., 2017), vehicle manufacturers need to meet an annual CO_2 emission target for their entire new vehicle fleet. This allows them to compensate through the introduction of low-emission vehicles for the high-emission ones in the fleet. This leads to increasingly efficient fleets of vehicles over time but does not necessarily limit the driven distance.

Building codes that prescribe efficiency requirements for new and existing buildings have been adopted in many OECD countries (Evans et al., 2017) and are regularly revised to increase their efficiency per unit of floor space. Building codes can avoid locking rapidly urbanizing countries into poorly performing buildings that remain in use for the next 50–100 years (Ürge-Vorsatz et al., 2014). In OECD countries, however, their main role is to incentivize the retrofit of existing buildings. In addition of the convergence of these codes to net zero energy buildings (D'Agostino, 2015), a new focus should be placed, in the context of 1.5°C-consistent pathways, on public and private coordination to achieve better integration of building policies with the promotion of low-emission transportation modes (Bertoldi, 2017).

The efficacy of regulatory instruments can be reinforced by economic incentives, such as feed-in tariffs based on the quantity of renewable energy produced, subsidies or tax exemptions for energy savings (Bertoldi et al., 2013; Ritzenhofen and Spinler, 2016; García-Álvarez et al., 2017; Pablo-Romero et al., 2017), fee-bates, and 'bonus-malus' that foster the penetration of low-emission options (Butler and Neuhoff, 2008). Economic incentives can also be combined with direct-use market-based instruments, for example combining, in the United States and, in some EU countries, carbon trading schemes with energy savings obligations for energy retailers (Haoqi et al., 2017), or with green certificates for renewable energy portfolio standards (Upton and Snyder, 2017). Scholars have investigated caps on utilities' energy sales (Thomas et al., 2017) and emission caps implemented at a personal level (Fawcett et al., 2010).

In combination with the funding of public research institutes, grants or subsidies also support R&D, where risk and the uncertainty about long-term perspectives can reduce the private sector's willingness to invest in low-emission innovation (see also Section 4.4.4). Subsidies can take the form of rebates on value-added tax (VAT), of direct support to investments (e.g., renewable energy or refurbishment of buildings) or feed-in tariffs (Mir-Artigues and del Río, 2014). They can be provided by the public budget, via consumption levies, or via the revenues of carbon taxes or pricing. Fee-bates, introduced in some countries (e.g., for cars), have had a neutral impact on public budgets by incentivizing low-emission products and penalizing high-emission ones (de Haan et al., 2009).

All policy instruments can benefit from information campaigns (e.g., TV ads) tailored to specific end-users. A vast majority of public campaigns on energy and climate have been delivered through mass-media channels and advertising-based approaches (Corner and Randall, 2011; Doyle, 2011). Although some authors report large savings obtained by such campaigns, most agree that the effects are short-lived and decrease over time (Bertoldi et al., 2016). Recently, focus has been placed on the use of social norms to motivate behavioural changes (Allcott, 2011; Alló and Loureiro, 2014). More on strategies to change behaviour can be found in Section 4.4.3.

4.4.5.4 Scaling up climate finance and de-risking low-emission investments

The redirection of savings towards low-emission investments may be constrained by enforceable carbon prices, implementation of technical standards and the short-term bias of financial systems (Miles, 1993; Bushee, 2001; Black and Fraser, 2002). The many causes of this bias are extensively analysed in economic literature (Tehranian and Waegelein, 1985; Shleifer and Vishny, 1990; Bikhchandani and Sharma, 2000), including their link with prevailing patterns of economic globalization (Krugman, 2009; Rajan, 2011) and the chronic underinvestment in long-term infrastructure (IMF, 2014). Emerging literature explores how to overcome this through reforms targeted to bridge the gap between short-term cash balances and long-term low-emission assets and to reduce the risk-weighted capital costs of climate-resilient investments. This gap, which was qualified by the Governor of the Bank of England as a 'tragedy of the horizon' (Carney, 2016) that constitutes a threat to the stability of the financial system, is confirmed by the literature (Arezki et al., 2016; Christophers, 2017). This potential threat would encompass the impact of climate events on the value of assets (Battiston et al., 2017), liability risks (Heede, 2014) and the transition risk due to devaluation of certain classes of assets (Platinga and Scholtens, 2016).

The financial community's attention to climate change grew after COP 15 (ESRB ASC, 2016). This led to the introduction of climate-related risk disclosure in financial portfolios (UNEP, 2015), placing it on the agenda of G20 Green Finance Study Group and of the Financial Stability Board. This led to the creation of low-carbon financial indices that investors could consider as a 'free option on carbon' to hedge against risks of stranded carbon-intensive assets (Andersson et al., 2016). This could also accelerate the emergence of climate-friendly financial products such as

green or climate bonds. The estimated value of the green bonds market in 2017 is 155 billion USD (BNEF, 2018). The bulk of these investments are in renewable energy, energy efficiency and low-emission transport (Lazurko and Venema, 2017), with only 4% for adaptation (OECD, 2017b). One major question is whether individual strategies based on improved climate-related information alone will enable the financial system to allocate capital in an optimal way (Christophers, 2017) since climate change is a systemic risk (CISL, 2015; Schoenmaker and van Tilburg, 2016).

The readiness of financial actors to reduce investments in fossil fuels is a real trend (Platinga and Scholtens, 2016; Ayling and Gunningham, 2017), but they may not resist the attractiveness of carbon-intensive investments in many regions. Hence, decarbonizing an investment portfolio is not synonymous with investing massively in low-emission infrastructure. Scaling up climate-friendly financial products may depend upon a business context conducive to the reduction of the risk-weighted capital costs of low-emission projects. The typical leverage of public funding mechanisms for low-emission investment is low (2 to 4) compared with other sectors (10 to 15) (Maclean et al., 2008; Ward et al., 2009; MDB, 2016). This is due to the interplay of the uncertainty of emerging low-emission technologies in the midst of their learning-by-doing cycle with uncertain future revenues due to volatility of fossil fuel prices (Roques et al., 2008; Gross et al., 2010) as well as uncertainty around regulatory policies. This inhibits low-emission investments by corporations functioning under a 'shareholder value business regime' (Berle and Means, 1932; Roe, 1996; Froud et al., 2000) and actors with restricted access to capital (e.g. cities, local authorities, SMEs and households).

De-risking policy instruments to enable low-emission investment encompasses interest rate subsidies, fee-bates, tax breaks, concessional loans from development banks, and public investment funds, including revolving funds. Given the constraints on public budgets, public guarantees can be used to increase the leverage effect of public financing on private financing. Such de-risking instruments imply indeed a full direct burden on public budgets only in case of default of the project. They could back for example various forms of green infrastructure funds (de Gouvello and Zelenko, 2010; Emin et al., 2014; Studart and Gallagher, 2015).[10]

The risk of defaulting can be mitigated by strong measurement, reporting and verifying (MRV) systems (Bellassen et al., 2015) and by the use of notional prices recommended in public economics (and currently in use in France and the UK) to calibrate public support to the provision of public goods in case of persisting distortions in pricing (Stiglitz et al., 2017). Some suggest linking these notional prices to 'social, economic and environmental value of voluntary mitigation actions' recognized by the COP 21 Decision accompanying the Paris Agreement (paragraph 108) (Hourcade et al., 2015; La Rovere et al., 2017b; Shukla et al., 2017), in order to incorporate the co-benefits of mitigation.

Such public guarantees ultimately amount to money issuance backed by low-emission projects as collateral. This explains the potentially strong link between global climate finance and the evolution of the financial

---

[10] One prototype is the World Bank's Pilot Auction Facility on Methane and Climate Change

and monetary system. Amongst suggested mechanisms for this evolution are the use of International Monetary Fund's (IMF's) Special Drawing Rights to fund the paid-in capital of the Green Climate Fund (Bredenkamp and Pattillo, 2010) and the creation of carbon remediation assets at a predetermined face value per avoided tonne of emissions (Aglietta et al., 2015a, b). Such a predetermined value could hedge against the fragmentation of climate finance initiatives and support the emergence of financial products backed by a new class of long-term assets.

Combining public guarantees at a predetermined value of avoided emissions, in addition to improving the consistency of non-price measures, could support the emergence of financial products backed by a new class of certified assets to attract savers in search of safe and ethical investments (Aglietta et al., 2015b). It could hedge against the fragmentation of climate finance initiatives and provide a mechanism to compensate for the 'stranded' assets caused by divestment in carbon-based activities and in lowering the systemic risk of stranded assets (Safarzyńska and van den Bergh, 2017). These new assets could also facilitate a low-carbon transition for fossil fuel producers and help them to overcome the 'resource curse' (Ross, 2015; Venables, 2016).

Blended injection of liquidity has monetary implications. Some argue that this questions the premise that money should remain neutral (Annicchiarico and Di Dio, 2015, 2016; Nikiforos and Zezza, 2017). Central banks or financial regulators could act as a facilitator of last resort for low-emission financing instruments, which could in turn lower the systemic risk of stranded assets (Safarzyńska and van den Bergh, 2017). This may, in time, lead to the use of carbon-based monetary instruments to diversify reserve currencies (Jaeger et al., 2013) and differentiate reserve requirements (Rozenberg et al., 2013) in the context of a climate-friendly Bretton Woods (Sirkis et al., 2015; Stua, 2017).

4.4.5.5 Financial challenge for basic needs and adaptation finance

Adaptation finance is difficult to quantify for two reasons. The first is that it is very difficult to isolate specific investment needs to enhance climate resilience from the provision of basic infrastructure that are currently underinvested (IMF, 2014; Gurara et al., 2017). The UNEP (2016) estimate of investment needs on adaptation in developing countries between 140–300 billion USD yr^{-1} in 2030, a major part being investment expenditures that are complementary with SDG-related investments focused on universal access to infrastructure and services and meeting basic needs. Many climate-adaptation-centric financial incentives are relevant to non-market services, offering fewer opportunities for market revenues while they contribute to creating resilience to climate impacts.

Hence, adaptation investments and the provision of basic needs would typically have to be supported by national and sub-national government budgets together with support from overseas development assistance and multilateral development banks (Fankhauser and Schmidt-Traub, 2011; Adenle et al., 2017; Robinson and Dornan, 2017), and a slow increase of dedicated NGO and private climate funds (Nakhooda and Watson, 2016). Even though the UNEP estimates of the costs of

adaptation might be lower in a 1.5°C world (UNEP/Climate Analytics, 2015) they would be higher than the UNEP estimate of 22.5 billion USD of bilateral and multilateral funding for climate change adaptation in 2014. Currently, 18–25% of climate finance flows to adaptation in developing countries (OECD, 2015b, 2016; Shine and Campillo, 2016). It remains fragmented, with small proportions flowing through UNFCCC channels (AdaptationWatch, 2015; Roberts and Weikmans, 2017).

Means of raising resources for adaptation, achieving the SDGs and meeting basic needs (Durand et al., 2016; Roberts et al., 2017) include the reduction of fossil fuel subsidies (Jakob et al., 2016), increasing revenues from carbon taxes (Jakob et al., 2016), levies on international aviation and maritime transport, and sharing of the proceeds of financial arrangements supporting mitigation activities (Keen et al., 2013). Each have different redistribution implications. Challenges, however, include the efficient use of resources, the emergence of long-term assets using infrastructure as collateral and the capacity to implement small-scale adaptation and the mainstreaming of adaptation in overall development policies. There is thus a need for greater policy coordination (Fankhauser and McDermott, 2014; Morita and Matsumoto, 2015; Sovacool et al., 2015, 2017; Lemos et al., 2016; Adenle et al., 2017; Peake and Ekins, 2017) that includes robust mechanisms for tracking, reporting and ensuring transparency of adaptation finance (Donner et al., 2016; Pauw et al., 2016a; Roberts and Weikmans, 2017; Trabacchi and Buchner, 2017) and its consistency with the provision of basic needs (Hallegatte et al., 2016).

4.4.5.6 Towards integrated policy packages and innovative forms of financial cooperation

Carbon prices, regulation and standards, improved information and appropriate financial instruments can work synergistically to meet the challenge of 'making finance flows consistent with a pathway towards low greenhouse gas emissions and climate-resilient development', as in Article 2 in the Paris Agreement.

There is growing attention to the combination of policy instruments that address three domains of action: behavioural changes, economic optimization and long-term strategies (Grubb et al., 2014). For example, de-risking low-emission investments would result in higher volumes of low-emission investments, and would in turn lead to a lower switching price for the same climate ambition (Hirth and Steckel, 2016). In the reverse direction, higher explicit carbon prices may generate more low-emission projects for a given quantum of de-risking. For example, efficiency standards for housing can increase the efficacy of carbon prices and overcome the barriers coming from the high discount rates used by households (Parry et al., 2014), while explicit and notional carbon prices can lower the risk of arbitrary standards. The calibration of innovative financial instruments to notional carbon prices could encourage large multinational companies to increase their level of internal carbon prices (UNEP, 2016). These notional prices could be higher than explicit carbon prices because they redirect new hardware investments without an immediate impact on existing capital stocks and associated interests.

Literature, however, shows that conflicts between poorly articulated policy instruments can undermine their efficiency (Lecuyer and Quirion, 2013; Bhattacharya et al., 2017; García-Álvarez et al., 2017).

As has been illustrated in Europe, commitment uncertainty and lack of credibility of regulation have consistently led to low carbon prices in the case of the EU Emission Trading System (Koch et al., 2014, 2016). A comparative study shows how these conflicts can be avoided by policy packages that integrate many dimensions of public policies and are designed to match institutional and social context of each country and region (Bataille et al., 2015).

Even though policy packages depend upon domestic political processes, they might not reinforce the NDCs at a level consistent with the 1.5°C transition without a conducive international setting where international development finance plays a critical role. Section 4.4.1 explores the means of mainstreaming climate finance in the current evolution of the lending practices of national and multilateral banks (Badré, 2018). This could facilitate the access of developing countries to loans via bond markets at low interest rates, encouragement of the emergence of new business models for infrastructure, and encouragement of financial markets to support small-scale investments (Déau and Touati, 2017).

These financial innovations may involve non-state public actors like cities and regional public authorities that govern infrastructure investment, enable energy and food systems transitions and manage urban dynamics (Cartwright, 2015). They would help, for example, in raising the 4.5–5.4 trillion USD yr^{-1} from 2015 to 2030 announced by the Cities Climate Finance Leadership Alliance (CCFLA, 2016) to achieve the commitments by the Covenant of Mayors of many cities to long-term climate targets (Kona et al., 2018).

The evolution of global climate financial cooperation may involve central banks, financial regulatory authorities, and multilateral and commercial banks. There are still knowledge gaps about the form, structure and potential of these arrangements. They could be viewed as a form of a burden-sharing between high-, medium- and low-income countries to enhance the deployment of ambitious Nationally Determined Contributions (NDCs) and new forms of 'common but differentiated responsibility and respective capabilities' (Edenhofer et al., 2015; Hourcade et al., 2015; Ji and Sha, 2015).

4.5   Integration and Enabling Transformation

4.5.1   Assessing Feasibility of Options for Accelerated Transitions

Chapter 2 shows that 1.5°C-consistent pathways involve rapid, global climate responses to reach net zero emissions by mid-century or earlier. Chapter 3 identifies climate change risks and impacts to which the world would need to adapt during these transitions and additional risks and impacts during potential 1.5°C overshoot pathways. The feasibility of these pathways is contingent upon systemic change (Section 4.3) and enabling conditions (Section 4.4), including policy packages. This section assesses the feasibility of options (technologies, actions and measures) that form part of global systems under transition that make up 1.5°C-consistent pathways.

Following the assessment framework developed in Chapter 1, economic and technological, institutional and socio-cultural, and environmental and geophysical feasibility are considered and applied to system transitions (Sections 4.3.1–4.3.4), overarching adaptation options (Section 4.3.5) and carbon dioxide removal (CDR) options (Section 4.3.7). This is done to assess the multidimensional feasibility of mitigation and adaptation options that have seen considerable development and change since AR5. In the case of adaptation, the assessed AR5 options are typically clustered. For example, all options related to energy infrastructure resilience, independently of the generation source, are categorized as 'resilience of power infrastructure'.

Table 4.10 presents sets of indicators against which the multidimensional feasibility of individual adaptation options relevant to warming of 1.5°C, and mitigation options along 1.5°C-consistent pathways, is assessed.

The feasibility assessment takes the following steps. First, each of the mitigation and adaptation options is assessed along the relevant indicators grouped around six feasibility dimensions: economic, technological, institutional, socio-cultural, environmental/ecological and geophysical. Three types of feasibility groupings were assessed from the underlying literature: first, if the indicator could block the feasibility of this option; second, if the indicator has neither a positive nor a negative effect on the feasibility of the option or the evidence is mixed; and third, if the indicator does not pose any barrier to the feasibility of this option. The full assessment of each option under each indicator, including the literature references on which the assessment is based, can be found in supplementary materials 4.SM.4.2 and 4.SM.4.3. When appropriate, it is indicated that there is no evidence (NE), limited evidence (LE) or that the indicator is not applicable to the option (NA).

Next, for each feasibility dimension and option, the overall feasibility for a given dimension is assessed as the mean of combined scores of the relevant underlying indicators and classified into 'insignificant barriers' (2.5 to 3), 'mixed or moderate but still existent barriers' (1.5 to 2.5) or 'significant barriers' (below 1.5) to feasibility. Indicators assessed as NA, LE or NE are not included in this overall assessment (see supplementary material 4.SM.4.1 for the averaging and weighing guidance).

The results are summarized in Table 4.11 (for mitigation options) and Table 4.12 (for adaptation options) for each of the six feasibility dimensions: where dark shading indicates few feasibility barriers; moderate shading indicates that there are mixed or moderate but still existent barriers, and light shading indicates that multiple barriers, in this dimension, may block implementation.

A three-step process of independent validation and discussion by authors was undertaken to make this assessment as robust as possible within the scope of this Special Report. It must, however, be recognized that this is an indicative assessment at global scale, and both policy and implementation at regional, national and local level would need to adapt and build on this knowledge, within the particular local context and constraints. Some contextual factors are indicated in the rightmost column in Tables 4.11 and 4.12.

Table 4.10 | Sets of indicators against which the feasibility of adaptation and mitigation options are assessed for each feasibility dimension. The options are discussed in Sections 4.3.1-4.3.5 and 4.3.7.

Feasibility Dimensions	Adaptation Indicators	Mitigation Indicators
Economic	Microeconomic viability Macroeconomic viability Socio-economic vulnerability reduction potential Employment & productivity enhancement potential	Cost-effectiveness Absence of distributional effects Employment & productivity enhancement potential
Technological	Technical resource availability Risks mitigation potential	Technical scalability Maturity Simplicity Absence of risk
Institutional	Political acceptability Legal & regulatory feasibility Institutional capacity & administrative feasibility Transparency & accountability potential	Political acceptability Legal & administrative feasibility Institutional capacity Transparency & accountability potential
Socio-cultural	Social co-benefits (health, education) Socio-cultural acceptability Social & regional inclusiveness Intergenerational equity	Social co-benefits (health, education) Public acceptance Social & regional inclusiveness Intergenerational equity Human capabilities
Environmental/Ecological	Ecological capacity Adaptive capacity/ resilience building potential	Reduction of air pollution Reduction of toxic waste Reduction of water use Improved biodiversity
Geophysical	Physical feasibility Land use change enhancement potential Hazard risk reduction potential	Physical feasibility (physical potentials) Limited use of land Limited use of scarce (geo)physical resources Global spread

4.5.2   Implementing Mitigation

This section builds on the insights on mitigation options in Section 4.3, applies the assessment methodology along feasibility dimensions and indicators explained in Section 4.5.1, and synthesizes the assessment of the enabling conditions in Section 4.4.

4.5.2.1   Assessing mitigation options for limiting warming to 1.5°C against feasibility dimensions

An assessment of the degree to which examples of 1.5°C-relevant mitigation options face barriers to implementation, and on which contexts this depends, is summarized in Table 4.11. An explanation of the approach is given in Section 4.5.1 and in supplementary material 4.SM.4.1. Selected options were mapped onto system transitions and clustered through an iterative process of literature review, expert feedback, and responses to reviewer comments. The detailed assessment and the literature underpinning the assessment can be found in supplementary material 4.SM.4.2.

The feasibility framework in Cross-Chapter Box 3 in Chapter 1 highlights that the feasibility of mitigation and adaptation options depends on many factors. Many of those are captured in the indicators in Table 4.10, but many depend on the specific context in which an option features. This Special Report did not have the mandate, space or the literature base to undertake a regionally specific assessment. Hence the assessment is caveated as providing a broad indication of the likely global barriers, ignoring significant regional diversity. Regional and context-specific literature is also just emerging as is noted in the knowledge gaps

section (Section 4.6). Nevertheless, in Table 4.11, an indicative attempt has been made to capture relevant contextual information. The 'context' column indicates which contextual factors may affect the feasibility of an option, including regional differences. For instance, solar irradiation in an area impacts the cost-effectiveness of solar photovoltaic energy, so solar irradiation is mentioned in this column.

4.5.2.2   Enabling conditions for implementation of mitigation options towards 1.5°C

The feasibility assessment highlights six dimensions that could help inform an agenda that could be addressed by the areas discussed in Section 4.4: governance, behaviour and lifestyles, innovation, enhancing institutional capacities, policy and finance. For instance, Section 4.4.3 on behaviour offers strategies for addressing public acceptance problems, and how changes can be more effective when communication and actions relate to people's values. This section synthesizes the findings in Section 4.4 in an attempt to link them to the assessment in Table 4.11. The literature on which the discussion is based is found in Section 4.4.

From Section 4.4, including the case studies presented in the Boxes 4.1 to 4.10, several main messages can be constructed. For instance, governance would have to be multilevel and engaging different actors, while being efficient, and choosing the form of cooperation based on the specific systemic challenge or option at hand. If institutional capacity for financing and governing the various transitions is not urgently built, many countries would lack the ability to change pathways from a high-emission scenario to a low- or zero-emission scenario. In terms of innovation, governments, both national and multilateral, can contribute

Table 4.11 | Feasibility assessment of examples of 1.5°C-relevant mitigation options, with dark shading signifying the absence of barriers in the feasibility dimension, moderate shading indicating that, on average, the dimension does not have a positive or negative effect on the feasibility of the option, or the evidence is mixed, and faint shading the presence of potentially blocking barriers. No shading means that the literature found was not sufficient to make an assessment. Evidence and agreement assessment is undertaken at the option level. The context column on the far right indicates how the assessment might change if contextual factors were different. For the methodology and literature basis, see supplementary material 4.SM.4.1 and 4.SM.4.2.
Abbreviations used: Ec: Economic - Tec: Technological - Inst: Institutional - Soc: Socio-cultural - Env: Environmental/Ecological - Geo: Geophysical

System	Mitigation Option	Evidence	Agreement	Ec	Tec	Inst	Soc	Env	Geo	Context
Energy System Transitions	Wind energy (on-shore & off-shore)	Robust	Medium							Wind regime, economic status, space for wind farms, and the existence of a legal framework for independent power producers affect uptake; cost-effectiveness affected by incentive regime
	Solar PV	Robust	High							Cost-effectiveness affected by solar irradiation and incentive regime. Also enhanced by legal framework for independent power producers, which affects uptake
	Bioenergy	Robust	Medium							Depends on availability of biomass and land and the capability to manage sustainable land use. Distributional effects depend on the agrarian (or other) system used to produce feedstock
	Electricity storage	Robust	High							Batteries universal, but grid-flexible resources vary with area's level of development
	Power sector carbon dioxide capture and storage	Robust	High							Varies with local CO$_2$ storage capacity, presence of legal framework, level of development and quality of public engagement
	Nuclear energy	Robust	High							Electricity market organization, legal framework, standardization & know-how, country's 'democratic fabric', institutional and technical capacity, and safety culture of public and private institutions
Land & Ecosystem Transitions	Reduced food wastage & efficient food production	Robust	High							Will depend on the combination of individual and institutional behaviour
	Dietary shifts	Medium	High							Depends on individual behaviour, education, cultural factors and institutional support
	Sustainable intensification of agriculture	Medium	High							Depends on development and deployment of new technologies
	Ecosystems restoration	Medium	High							Depends on location and institutional factors
Urban & Infrastructure System Transitions	Land-use & urban planning	Robust	Medium							Varies with urban fabric, not geography or economy; requires capacitated local government and legitimate tenure system
	Electric cars and buses	Medium	High							Varies with degree of government intervention; requires capacity to retrofit "fuelling" stations
	Sharing schemes	Limited	Medium							Historic schemes universal, but new ones depend on ICT status; undermined by high crime and low levels of law enforcement
	Public transport	Robust	Medium							Depends on presence of existing 'informal' taxi systems, which may be more cost-effective and affordable than capital-intensive new build schemes, as well as (local) government capabilities
	Non-motorized transport	Robust	High							Viability rests on linkages with public transport, cultural factors, climate and geography
	Aviation & shipping	Medium	Medium							Varies with technology, governance and accountability
	Smart grids	Medium	Medium							Varies with economic status and presence or quality of existing grid
	Efficient appliances	Medium	High							Adoption varies with economic status and policy framework
	Low/zero-energy buildings	Medium	High							Depends on size of existing building stock and growth of building stock

4

Table 4.11 (continued)

System	Mitigation Option	Evidence	Agreement	Ec	Tec	Inst	Soc	Env	Geo	Context
Industrial System Transitions	Energy efficiency	Robust	High							Potential and adoption depend on existing efficiency, energy prices and interest rates, as well as government incentives
	Bio-based & circularity	Medium	Medium							Faces barriers in terms of pressure on natural resources and biodiversity. Product substitution depends on market organization and government incentivization
	Electrification & hydrogen	Medium	High							Depends on availability of large-scale, cheap, emission-free electricity (electrification, hydrogen) or CO_2 storage nearby (hydrogen). Manufacturers' appetite to embrace disruptive innovations
	Industrial carbon dioxide capture, utilization and storage	Robust	High							High concentration of CO_2 in exhaust gas improve economic and technical feasibility of CCUS in industry. CO_2 storage or reuse possibilities
Carbon Dioxide Removal	Bioenergy and carbon dioxide capture and storage	Robust	Medium							Depends on biomass availability, CO_2 storage capacity, legal framework, economic status and social acceptance
	Direct air carbon dioxide capture and storage	Medium	Medium							Depends on CO_2-free energy, CO_2 storage capacity, legal framework, economic status and social acceptance
	Afforestation & reforestation	Robust	High							Depends on location, mode of implementation, and economic and institutional factors
	Soil carbon sequestration & biochar	Robust	High							Depends on location, soil properties, time span
	Enhanced weathering	Medium	Low							Depends on CO_2-free energy, economic status and social acceptance

to applying general-purpose technologies to mitigation purposes. If this is not managed, some reduction in emissions could happen autonomously, but it may not lead to a 1.5°C-consistent pathway. International cooperation on technology, including technology transfer where this does not happen autonomously, is needed and can help create innovation capabilities in all countries that allow them to operate, maintain, adapt and regulate a portfolio of mitigation technologies. Case studies in the various subsections highlight the opportunities and challenges of doing this in practice. They indicate that it can be done in specific circumstances, which can be created.

A combination of behaviour-oriented pricing policies and financing options can help change technologies and social behaviour as it would challenge the existing, high-emission socio-technical regime on multiple levels across feasibility characteristics. For instance, for dietary change, combining supply-side measures with value-driven communication and economic instruments may help make a lasting transition, while an economic instrument, such as enhanced prices or taxation, on its own may not be as robust.

Governments could benefit from enhanced carbon prices, as a price and innovation incentive and also a source of additional revenue to correct distributional effects and subsidize the development of new, cost-effective negative-emission technology and infrastructure. However, there is *high evidence* and *medium agreement* that pricing alone is insufficient. Even if prices rise significantly, they typically incentivize incremental change, but typically fail to provide the impetus for private actors to take the risk of engaging in the transformational changes that would be needed to limit warming to 1.5°C. Apart from the

incentives to change behaviour and technology, financial systems are an indispensable element of a systemic transition. If financial markets do not acknowledge climate risk and the risk of transitions, regulatory financial institutions, such as central banks, could intervene.

Strengthening implementation revolves around more than addressing barriers to feasibility. A system transition, be it in energy, industry, land or a city, requires changing the core parameters of a system. These relate, as introduced in Section 4.2 and further elaborated in Section 4.4, to how actors cooperate, how technologies are embedded, how resources are linked, how cultures relate and what values people associate with the transition and the current regime.

4.5.3   Implementing Adaptation

Article 7 of the Paris Agreement provides an aspirational global goal for adaptation, of 'enhancing adaptive capacity, strengthening resilience, and reducing vulnerability' (UNFCCC, 2016). Adaptation implementation is gathering momentum in many regions, guided by national NDC's and national adaptation plans (see Cross-Chapter Box 11 in this Chapter).

Operationalizing adaptation in a set of regional environments on pathways to a 1.5°C world requires strengthened global and differentiated regional and local capacities. It also needs rapid and decisive adaptation actions to reduce the costs and magnitude of potential climate impacts (Vergara et al., 2015).

This could be facilitated by: (i) enabling conditions, especially improved governance, economic measures and financing (Section 4.4); (ii) enhanced clarity on adaptation options to help identify strategic priorities, sequencing and timing of implementation (Section 4.3); (iii) robust monitoring and evaluation frameworks; and (iv) political leadership (Magnan et al., 2015; Magnan and Ribera, 2016; Lesnikowski et al., 2017; UNEP, 2017a).

4.5.3.1   Feasible adaptation options

This section summarizes the feasibility (defined in Cross-Chapter Box 3, Table 1 in Chapter 1 and Table 4.4) of select adaptation options using evidence presented across this chapter and in supplementary material 4.SM.4.3 and the expert-judgement of its authors (Table 4.12). The options assessed respond to risks and impacts identified in Chapter 3. They were selected based on options identified in AR5 (Noble et al., 2014), focusing on those relevant to 1.5°C-compatible pathways, where sufficient literature exists. Selected options were mapped onto system transitions and clustered through an iterative process of literature review, expert feedback, and responses to reviewer comments.

Besides gaps in the literature around crucial adaptation questions on the transition to a 1.5°C world (Section 4.6), there is inadequate current literature to undertake a spatially differentiated assessment (Cross-Chapter Box 3 in Chapter 1). There are also limited baselines for exposure, vulnerability and risk to help policy and implementation prioritization. Hence, the compiled results can at best provide a broad framework to inform policymaking. Given the bottom-up nature of most adaptation implementation evidence, care needs to be taken in generalizing these findings.

Options are considered as part of a systemic approach, recognizing that no single solution exists to limit warming to 1.5°C and adapting to its impacts. To respond to the local and regional context – and to synergies and trade-offs between adaptation, mitigation and sustainable development – packages of options suited to local enabling conditions can be implemented.

Table 4.12 summarizes the feasibility assessment through its six dimensions with levels of evidence and agreement and indicates how the feasibility of an adaptation option may be differentiated by certain contextual factors (last column).

When considered jointly, the description of adaptation options (Section 4.3), the feasibility assessment (summarized in Table 4.12), and discussion of enabling conditions (Section 4.4) show us how options can be implemented and lead towards transformational adaptation if and when needed.

The adaptation options for energy system transitions focus on existing power infrastructure resilience and water management, when required, for any type of generation source. These options are not sufficient for the far-reaching transformations required in the energy sector, which have tended to focus on technologies to shift from a fossil-based to a renewable energy system (Erlinghagen and Markard, 2012; Muench et al., 2014; Brand and von Gleich, 2015; Monstadt and Wolff, 2015; Child and Breyer, 2017; Hermwille et al., 2017). There is also need for integration of such energy system transitions with social-ecological systems transformations to increase the resilience of the energy sector,

for which appropriate enabling conditions, such as for technological innovations, are fundamentally important. Institutional capacities can be enhanced by expanding the role of actors as transformation catalysts (Erlinghagen and Markard, 2012). The integration of ethics and justice within these transformations can help attain SDG7 on clean energy access (Jenkins et al., 2018), while inclusion of the cultural dimension and cultural legitimacy (Amars et al., 2017) can provide a more substantial base for societal transformation. Strengthening policy instruments and regulatory frameworks and enhancing multilevel governance that focuses on resilience components can help secure these transitions (Exner et al., 2016).

For land and ecosystem transitions, the options of conservation agriculture, efficient irrigation, agroforestry, ecosystem restoration and avoided deforestation, and coastal defence and hardening have between *medium and robust evidence* with *medium to high agreement*. The other options assessed have limited or no evidence across one or more of the feasibility dimensions. Community-based adaptation is assessed as having *medium evidence* and *high agreement* to face scaling barriers. Scaling community-based adaptation may require structural changes, implying the need for transformational adaptation in some regions. This would involve enhanced multilevel governance and institutional capacities by enabling anticipatory and flexible decision-making systems that access and develop collaborative networks (Dowd et al., 2014), tackling root causes of vulnerability (Chung Tiam Fook, 2017), and developing synergies between development and climate change (Burch et al., 2017). Case studies show the use of transformational adaptation approaches for fire management (Colloff et al., 2016a), floodplain and wetland management (Colloff et al., 2016b), and forest management (Chung Tiam Fook, 2017), in which the strengthening of policy instruments and climate finance are also required.

There is growing recognition of the need for transformational adaptation within the agricultural sector but limited evidence on how to facilitate processes of deep, systemic change (Dowd et al., 2014). Case studies demonstrate that transformational adaptation in agriculture requires a sequencing and overlap between incremental and transformational adaptation actions (Hadarits et al., 2017; Termeer et al., 2017), e.g., incremental improvements to crop management while new crop varieties are being researched and field-tested (Rippke et al., 2016). Broader considerations include addressing stakeholder values and attitudes (Fleming et al., 2015a), understanding and leveraging the role of social capital, collaborative networks, and information (Dowd et al., 2014), and being inclusive with rural and urban communities, and the social, political, and cultural environment (Rickards and Howden, 2012). Transformational adaptation in agriculture systems could have significant economic and institutional costs (Mushtaq, 2016), along with potential unintended negative consequences (Davidson, 2016; Rippke et al., 2016; Gajjar et al., 2018; Mushtaq, 2018), and a need to focus on the transitional space between incremental and transformational adaptation (Hadarits et al., 2017), as well as the timing of the shift from one to the other (Läderach et al., 2017).

Within urban and infrastructure transitions, green infrastructure and sustainable water management are assessed as the most feasible options, followed by sustainable land-use and urban planning. The

Table 4.12 | Feasibility assessment of examples of 1.5°C-relevant adaptation options, with dark shading signifying the absence of barriers in the feasibility dimension, moderate shading indicating that, on average, the dimension does not have a positive or negative effect on the feasibility of the option, or the evidence is mixed, and light shading indicating the presence of potentially blocking barriers. No shading means that sufficient literature could not be found to make the assessment. NA signifies that the dimension is not applicable to that adaptation option. For methodology and literature basis, see supplementary material 4.SM.4.

Abbreviations used: Ec: Economic - Tec: Technological - Inst: Institutional - Soc: Socio-cultural - Env: Environmental/Ecological - Geo: Geophysical

System	Adaptation Option	Evidence	Agreement	Ec	Tec	Inst	Soc	Env	Geo	Context
Energy System Transitions	Power infrastructure, including water	Medium	High							Depends on existing power infrastructure, all generation sources and those with intensive water requirements
Land & Ecosystem Transitions	Conservation agriculture	Medium	Medium							Depends on irrigated/rainfed system, ecosystem characteristics, crop type, other farming practices
	Efficient irrigation	Medium	Medium							Depends on agricultural system, technology used, regional institutional and biophysical context
	Efficient livestock systems	Limited	High							Dependent on livestock breeds, feed practices, and biophysical context (e.g., carrying capacity)
	Agroforestry	Medium	High							Depends on knowledge, financial support, and market conditions
	Community-based adaptation	Medium	High							Focus on rural areas and combined with ecosystems-based adaptation, does not include urban settings
	Ecosystem restoration & avoided deforestation	Robust	Medium							Mostly focused on existing and evaluated REDD+ projects
	Biodiversity management	Medium	Medium							Focus on hotspots of biodiversity vulnerability and high connectivity
	Coastal defence & hardening	Robust	Medium							Depends on locations that require it as a first adaptation option
	Sustainable aquaculture	Limited	Medium							Depends on locations at risk and socio-cultural context
Urban & Infrastructure System Transitions	Sustainable land-use & urban planning	Medium	Medium							Depends on nature of planning systems and enforcement mechanisms
	Sustainable water management	Robust	Medium							Balancing sustainable water supply and rising demand, especially in low-income countries
	Green infrastructure & ecosystem services	Medium	High							Depends on reconciliation of urban development with green infrastructure
	Building codes & standards	Limited	Medium							Adoption requires legal, educational, and enforcement mechanisms to regulate buildings
Industrial System Transitions	Intensive industry infrastructure resilience and water management	Limited	High							Depends on intensive industry, existing infrastructure and using or requiring high demand of water
Overarching Adaptation Options	Disaster risk management	Medium	High							Requires institutional, technical, and financial capacity in frontline agencies and government
	Risk spreading and sharing: insurance	Medium	Medium							Requires well-developed financial structures and public understanding
	Social safety nets	Medium	Medium							Type and mechanism of safety net, political priorities, institutional transparency
	Climate services	Medium	High							Depends on climate information availability and usability, local infrastructure and institutions, national priorities
	Indigenous knowledge	Medium	High							Dependent on recognition of indigenous rights, laws, and governance systems
	Education and learning	Medium	High							Existing education system, funding
	Population health and health system	Medium	High						NA	Requires basic health services and infrastructure
	Human migration	Medium	Low							Hazard exposure, political and socio-cultural acceptability (in destination), migrant skills and social networks

4

need for transformational adaptation in urban settings arises from the root causes of poverty, failures in sustainable development, and a lack of focus on social justice (Revi et al., 2014a; Parnell, 2015; Simon and Leck, 2015; Shi et al., 2016; Ziervogel et al., 2016a; Burch et al., 2017), and necessitates a focus on governance structures and the inclusion of equity and justice concerns (Bos et al., 2015; Shi et al., 2016; Hölscher et al., 2018).

Current implementation of urban ecosystems-based adaptation (EbA) lacks a systems perspective of transformations and consideration of the normative and ethical aspects of EbA (Brink et al., 2016). Flexibility within urban planning could help deal with the multiple uncertainties of implementing adaptation (Rosenzweig and Solecki, 2014; Radhakrishnan et al., 2018), for example, urban adaptation pathways were implemented in the aftermath of Superstorm Sandy in New York, which is considered as tipping point that led to the implementation of transformational adaptation practices.

Adaptation options for industry focus on infrastructure resilience and water management. Like with energy system transitions, technological innovation would be required, but also the enhancement of institutional capacities. Recent research illustrates transformational adaptation within industrial transitions focusing on the role of different actors and tools driving innovation, and points to the role of nationally appropriate mitigation actions in avoiding lock-ins and promoting system innovation (Boodoo and Olsen, 2017), the role of private sector in sustainability governance in the socio-political context (Burch et al., 2016), and of green entrepreneurs driving transformative change in the green economy (Gibbs and O'Neill, 2014). Lim-Camacho et al. (2015) suggest an analysis of the complete lifecycle of supply chains as a means of identifying additional adaptation strategies, as opposed to the current focus on a part of the supply chain. Chain-wide strategies can modify the rest of the chain and present a win-win with commercial objectives.

The assessed adaptation options also have mitigation synergies and trade-offs (assessed in Section 4.5.4) that need to be carefully considered, while planning climate action.

4.5.3.2   Monitoring and evaluation

Monitoring and evaluation (M&E) in adaptation implementation can promote accountability and transparency of adaptation financing, facilitate policy learning and sharing good practices, pressure laggards, and guide adaptation planning. The majority of research on M&E focuses on specific policies or programmes, and has typically been driven by the needs of development organizations, donors, and governments to measure the impact and attribution of adaptation initiatives (Ford and Berrang-Ford, 2016). There is growing research examining adaptation progress across nations, sectors, and scales (Reckien et al., 2014; Araos et al., 2016a, b; Austin et al., 2016; Heidrich et al., 2016; Lesnikowski et al., 2016; Robinson, 2017). In response to a need for global, regional and local adaptation, the development of indicators and standardized approaches to evaluate and compare adaptation over time and across regions, countries, and sectors would enhance comparability and learning. A number of constraints continue to hamper progress on adaptation M&E, including a debate on what actually constitutes

adaptation for the purposes of assessing progress (Dupuis and Biesbroek, 2013; Biesbroek et al., 2015), an absence of comprehensive and systematically collected data on adaptation to support longitudinal assessment and comparison (Ford et al., 2015b; Lesnikowski et al., 2016), a lack of agreement on indicators to measure (Brooks et al., 2013; Bours et al., 2015; Lesnikowski et al., 2015), and challenges of attributing altered vulnerability to adaptation actions (Ford et al., 2013; Bours et al., 2015; UNEP, 2017a).

4.5.4   Synergies and Trade-Offs between Adaptation and Mitigation

Implementing a particular mitigation or adaptation option may affect the feasibility and effectiveness of other mitigation and adaptation options. Supplementary Material 4.SM.5.1 provides examples of possible positive impacts (synergies) and negative impacts (trade-offs) of mitigation options for adaptation. For example, renewable energy sources such as wind energy and solar PV combined with electricity storage can increase resilience due to distributed grids, thereby enhancing both mitigation and adaptation. Yet, as another example, urban densification may reduce GHG emissions, enhancing mitigation, but can also intensify heat island effects and inhibit restoration of local ecosystems if not accounted for, thereby increasing adaptation challenges.

The table in Supplementary Material 4.SM.5.2 provides examples of synergies and trade-offs of adaptation options for mitigation. It shows, for example, that conservation agriculture can reduce some GHG emissions and thus enhance mitigation, but at the same time can increase other GHG emissions, thereby reducing mitigation potential. As another example, agroforestry can reduce GHG emissions through reduced deforestation and fossil fuel consumption but has a lower carbon sequestration potential compared with natural and secondary forest.

Maladaptive actions could increase the risk of adverse climate-related outcomes. For example, biofuel targets could lead to indirect land use change and influence local food security, through a shift in land use abroad in response to increased domestic biofuel demand, increasing global GHG emissions rather than decreasing them.

Various options enhance both climate change mitigation and adaptation, and would hence serve two 1.5°C-related goals: reducing emissions while adapting to the associated climate change. Examples of such options are reforestation, urban and spatial planning, and land and water management.

Synergies between mitigation and adaptation may be enhanced, and trade-offs reduced, by considering enabling conditions (Section 4.4), while trade-offs can be amplified when enabling conditions are not considered (C.A. Scott et al., 2015). For example, information that is tailored to the personal situation of individuals and communities, including climate services that are credible and targeted at the point of decision-making, can enable and promote both mitigation and adaptation actions (Section 4.4.3). Similarly, multilevel governance and community participation, respectively, can enable and promote both adaptation and mitigation actions (Section 4.4.1). Governance, policies and institutions can facilitate the implementation of the water–

4

energy–food (WEF) nexus (Rasul and Sharma, 2016). The WEF nexus can enhance food, water and energy security, particularly in cities with agricultural production areas (Biggs et al., 2015), electricity generation with intensive water requirements (Conway et al 2015), and in agriculture (El Gafy et al., 2017) and livelihoods (Biggs et al., 2015). Such a nexus approach can reduce the transport energy that is embedded in food value chains (Villarroel Walker et al., 2014), providing diverse sources of food in the face of changing climates (Tacoli et al., 2013). Urban agriculture, where integrated, can mitigate climate change and support urban flood management (Angotti, 2015; Bell et al., 2015; Biggs

et al., 2015; Gwedla and Shackleton, 2015; Lwasa et al., 2015; Yang et al., 2016; Sanesi et al., 2017). In the case of electricity generation, enabling conditions through a combination of carefully selected policy instruments can maximize the synergic benefits between low GHG energy production and water for energy (Shang et al., 2018). Despite the multiple benefits of maximizing synergies between mitigation and adaptations options through the WEF nexus approach (Chen and Chen, 2016), there are implementation challenges given institutional complexity, political economy, and interdependencies between actors (Leck et al., 2015).

Box 4.10 | Bhutan: Synergies and Trade-Offs in Economic Growth, Carbon Neutrality and Happiness

Bhutan has three national goals: improving its gross national happiness index (GNHI), improving its economic growth (gross domestic product, GDP) and maintaining its carbon neutrality. These goals increasingly interact and raise questions about whether they can be sustainably maintained into the future. Interventions in this enabling environment are required to comply with all three goals.

Bhutan is well known for its GNHI, which is based on a variety of indicators covering psychological well-being, health, education, cultural and community vitality, living standards, ecological issues and good governance (RGoB, 2012; Schroeder and Schroeder, 2014; Ura, 2015). The GNHI is a precursor to the Sustainable Development Goals (SDGs) (Allison, 2012; Brooks, 2013) and reflects local enabling environments. The GNHI has been measured twice, in 2010 and 2015, and this showed an increase of 1.8% (CBS & GNH, 2016). Like most emerging countries, Bhutan wants to increase its wealth and become a middle-income country (RGoB, 2013, 2016), while remaining carbon-neutral – a goal which has been in place since 2009 at COP15 and was reiterated in its Intended Nationally Determined Contribution (NEC, 2015). Bhutan achieves its current carbon-neutral status through hydropower and forest cover (Yangka and Diesendorf, 2016), which are part of its resilience and adaptation strategy.

Nevertheless, Bhutan faces rising GHG emissions. Transport and industry are the largest growth areas (NEC, 2011). Bhutan's carbon-neutral status would be threatened by 2044 with business-as-usual approaches to economic growth (Yangka and Newman, 2018). Increases in hydropower are being planned based on climate change scenarios that suggest sufficient water supply will be available (NEC, 2011). Forest cover is expected to remain sufficient to maintain co-benefits. The biggest challenge is to electrify both freight and passenger transport (ADB, 2013). Bhutan wants to be a model for achieving economic growth consistent with limiting climate change to 1.5°C and improving its GNHI (Michaelowa et al., 2018) through synthesizing all three goals and improving its adaptive capacity.

4.6   Knowledge Gaps and Key Uncertainties

The global response to limiting warming to 1.5°C is a new knowledge area, which has emerged after the Paris Agreement. This section presents a number of knowledge gaps that have emerged from the assessment of mitigation, adaptation and carbon dioxide removal (CDR) options and solar radiation modification (SRM) measures; enabling conditions; and synergies and trade-offs. Illustrative questions that emerge synthesizing the more comprehensive Table 4.13 below include: how much can be realistically expected from innovation, behaviour and systemic political and economic change in improving resilience, enhancing adaptation and reducing GHG emissions? How can rates of changes be accelerated and scaled up? What is the outcome of realistic assessments of mitigation and adaptation

land transitions that are compliant with sustainable development, poverty eradication and addressing inequality? What are life-cycle emissions and prospects of early-stage CDR options? How can climate and sustainable development policies converge, and how can they be organized within a global governance framework and financial system, based on principles of justice and ethics (CBDR-RC), reciprocity and partnership? To what extent would limiting warming to 1.5°C require a harmonization of macro-financial and fiscal policies, which could include central banks? How can different actors and processes in climate governance reinforce each other, and hedge against the fragmentation of initiatives?

These knowledge gaps are highlighted in Table 4.13 along with a cross-reference to the respective sections in the last column.

Table 4.13 | Knowledge gaps and uncertainties

Knowledge Area		Mitigation	Adaptation	Reference
1.5°C Pathways and Ensuing Change		• Lack of literature specific to 1.5°C on investment costs with detailed breakdown by technology • Lack of literature specific to 1.5°C on mitigation costs in terms of GDP and welfare • Lack of literature on distributional implications of 1.5°C compared to 2°C or business-as-usual at sectoral and regional levels • Limited 1.5°C-specific case studies for mitigation • Limited knowledge on the systemic and dynamic aspects of transitions to 1.5°C, including how vicious or virtuous circles might work, how self-reinforcing aspects can be actively introduced and managed	• Lack of literature specific to 1.5°C on adaptation costs and need • Lack of literature on what overshoot means for adaptation • Lack of knowledge on avoided adaptation investments associated with limiting warming to 1.5°C, 2°C or business-as-usual • Limited 1.5°C-specific case studies for adaptation • Scant literature examining current or future adaptation options, or examining what different climate pathways mean for adaptation success • Need for transformational adaptation at 1.5°C and beyond remains largely unexplored	4.2
Options to Achieve and Adapt to 1.5°C	Energy Systems	• The shift to variable renewables that many countries are implementing is just reaching a level where large-scale storage systems or other grid flexibility options, e.g., demand response, are required to enable resilient grid systems. Thus, new knowledge on the opportunities and issues associated with scaling up zero-carbon grids would be needed, including knowledge about how zero-carbon electric grids can integrate with the full-scale electrification of transport systems • CCS suffers mostly from uncertainty about the feasibility of timely upscaling, both due to lack of regulatory capacity and concerns about storage safety and cost • There is not much literature on the distributional implications of large-scale bioenergy deployment, the assessment of environmental feasibility is hampered by a diversity of contexts of individual studies (type of feedstock, technology, land availability), which could be improved through emerging meta-studies	• Relatively little literature on individual adaptation options since AR5 • No evidence on socio-cultural acceptability of adaptation options • Lack of regional research on the implementation of adaptation options	4.3.1
	Land & ecosystems	• More knowledge would be needed on how land-based mitigation can be reconciled with land demands for adaptation and development • While there is now more literature on the underlying mechanisms of land transitions, data is often insufficient to draw robust conclusions, and there is uncertainty about land availability • The lack of data on social and institutional information (largest knowledge gap indicated for ecosystems restoration in Table 4.11), which are therefore not widely integrated in land use modelling • Examples of successful policy implementation and institutions related to land-based mitigation leading to co-benefits for adaptation and development are missing from the literature • There is relatively little scientific literature on the effects of dietary shifts and reduction of food wastage on mitigation, especially regarding the institutional, technical and environmental concerns	• Regional information on some options does not exist, especially in the case of land-use transitions • Limited research examining socio-cultural perspectives and impacts of adaptation options, especially for efficient irrigation, coastal defence and hardening, agroforestry and biodiversity management • Lack of longitudinal, regional studies assessing the impacts of certain adaptation options, such as conservation agriculture and shifting to efficient livestock systems • More knowledge is needed on the cost-effectiveness and scalability of various adaptation options. For example, there is no evidence for the macro-economic viability of community-based adaptation (CbA) and biodiversity management, or on employment and productivity enhancement potential for biodiversity management and coastal defence and hardening. • More knowledge is needed on risk mitigation and the potential of biodiversity management • Lack of evidence of the political acceptability of efficient livestock systems • *Limited evidence* on legal and regulatory feasibility of conservation agriculture and no evidence on coastal defence and hardening • For transparency and accountability potential, there is *limited evidence* for conservation agriculture and no evidence for biodiversity management, coastal defence and hardening and sustainable aquaculture • No evidence on hazard risk reduction potential of conservation agriculture and biodiversity management	4.3.2
	Urban & infrastructure systems	• *Limited evidence* of effective land-use planning in low-income cities where tenure and land zoning are contested, and the risks of trying to implement land-use planning under communal tenure • *Limited evidence* on the governance of public transport from an accountability and transparency perspective	• Regional and sectoral adaptation cost assessments are missing, particularly in the context of welfare losses of households, across time and space • More knowledge is needed on the political economy of adaptation, particularly on how to impute different types of cost and benefit in a consistent manner, on adaptation performance indicators that could stimulate investment, and the impact of adaptation interventions on socio-economic and other types of inequality	4.3.3

Table 4.13 (continued)

Knowledge Area		Mitigation	Adaptation	Reference
Options to Achieve and Adapt to 1.5°C	Urban & infrastructure systems	• *Limited evidence* on relationship between toxic waste and public transport • *Limited evidence* on the impacts of electric vehicles and non-motorized urban transport, as most schemes are too new • As changes in shipping and aviation have been limited to date, limited evidence of social impacts • Knowledge about how to facilitate disruptive, demand-based innovations that may be transformative in urban systems, is needed • Understanding of the urban form implications of combined changes from electric, autonomous and shared/public mobility systems, is needed • Considering distributional consequences of climate responses is an on-going need • Knowledge gaps in the application and scale up of combinations of new smart technologies, sustainable design, advanced construction techniques and new insulation materials, renewable energy and behaviour change in urban settlements • The potential for leapfrog technologies to be applied to slums and new urban developments in developing countries is weak.	• More evidence would be needed on hot-spots, for example the growth of peri-urban areas populated by large informal settlements • Major uncertainties emanate from the lack of knowledge on the integration of climate adaptation and mitigation, disaster risk management, and urban poverty alleviation • There is *limited evidence* on the institutional, technological and economic feasibility of green infrastructure and environmental services and for socio-cultural and environmental feasibility of codes and standards • In general, there is no evidence for the employment and productivity enhancement potential of most adaptation options. • There is *limited evidence* on the economic feasibility of sustainable water management	4.3.3
	Industrial systems	• Lack of knowledge on potential for scaling up and global diffusion of zero- and low-emission technologies in industry • Questions remain on the socio-cultural feasibility of industry options, including human capacity and private sector acceptance of new, radically different technologies from current well-developed practices, as well as distributional effects of potential new business models • As the industrial transition unfolds, lack of knowledge on its dynamic interactions with other sectors, in particular with the power sector (and infrastructure) for electrification of industry, with food production and other users of biomass in case of bio-based industry developments, and with CDR technologies in the case of CC(U)S • Life-cycle assessment-based comparative analyses of CCUS options are missing, as well as life-cycle information on electrification and hydrogen • Impacts of industrial system transitions are not well understood, especially on employment, identity and well-being, in particular in the case of substitution of conventional, high-carbon industrial products with lower-carbon alternatives, as well as electrification and use of hydrogen	• *Very limited evidence* on how industry would adapt to the consequences of 1.5°C or 2°C temperature increases, in particular large and immobile industrial clusters in low-lying areas as well as availability of transportation and (cooling) water resources and infrastructure • There is *limited evidence* on the economic, institutional and socio-cultural feasibility of adaptation options available to industry	4.3.4
	Overarching adaptation options	• There is no evidence on technical and institutional feasibility of educational options • There is *limited evidence* on employment and productivity enforcement potential of climate services • There is *limited evidence* on socio-cultural acceptability of social safety nets • There is a small but growing literature on human migration as an adaptation strategy. Scant literature on the cost-effectiveness of migration		4.3.5
	Short-lived climate forcers	• *Limited evidence* of co-benefits and trade-offs of SLCF reduction (e.g., better health outcomes, agricultural productivity improvements) • Integration of SLCFs into emissions accounting and international reporting mechanisms enabling a better understanding of the links between black carbon, air pollution, climate change and agricultural productivity		4.3.6

4

Table 4.13 (continued)

Knowledge Area		Mitigation	Adaptation	Reference
Options to Achieve and Adapt to 1.5°C	Carbon dioxide removal	• A bottom-up analysis of CDR options indicates that there are still key uncertainties around the individual technologies. Ocean-based options will be assessed in depth in the IPCC Special Report on the Ocean and Cryosphere in a Changing Climate (SROCC) • Assessments of environmental aspects are missing, especially for 'newer' options like enhanced weathering or direct air carbon capture • In order to obtain more information on realistically available and sustainable removal potentials, more bottom-up, regional studies, also taking into account also social issues, would be needed. These can better inform the modelling of 1.5°C pathways • Knowledge gaps on issues of governance and public acceptance, the impacts of large-scale removals on the carbon cycle, the potential to accelerate deployment and upscaling, and means of incentivization • Knowledge gaps on integrated systems of renewable energy and CDR technologies such as enhanced weathering and DACCS • Knowledge gaps on under which conditions the use of captured CO_2 is generating negative emissions and would qualify as a mitigation option		4.3.7
Solar radiation modification (SRM)		• In spite of increasing attention to the different SRM measures and their potential to keep global temperature below 1.5°C, knowledge gaps remain, not only with respect to the physical understanding of SRM measures but also concerning ethical issues • We do not know how to govern SRM in order to avoid unilateral action and how to prevent possible reductions in mitigation ('moral hazard')		4.3.8
Enabling Conditions	Governance	• As technological changes have begun to accelerate, there is a lack of knowledge on new mechanisms that can enable private enterprise to mainstream this activity, and reasons for success and failure need to be researched • Research is thin on effective multilevel governance, in particular in developing countries, including participation by civil society, women and minorities • Gaps in knowledge remain pertaining to partnerships within local governance arrangements that may act as mediators and drivers for achieving global ambition and local action • Methods for assessing contribution and aggregation of non-state actors in limiting warming to 1.5°C • Knowledge gap on an enhanced framework for assessment of the ambition of NDCs	• The ability to identify explanatory factors affecting the progress of climate policy is constrained by a lack of data on adaptation actions across nations, regions, and sectors, compounded by an absence of frameworks for assessing progress. Most hypotheses on what drives adaptation remain untested • Limited empirical assessment of how governance affects adaptation across cases • Focus on 'success' stories and leading adaptors overlooks lessons from situations where no or unsuccessful adaptation is taking place	4.4.1
	Institutions	• Lack of 1.5°C-specific literature • Role of regulatory financial institutions and their capacity to guarantee financial stability of economies when investments potentially face risks, both because of climate impacts and because of the systems transitions if lower temperature scenarios are pursued • Knowledge gaps on how to build capabilities across all countries and regions globally to implement, maintain, manage, govern and further develop mitigation options for 1.5°C. • While importance of indigenous and local knowledge is recognized, the ability to scale up beyond the local remains challenging and little examined • There is a lack of monitoring and evaluation (M&E) of adaptation measures, with most studies enumerating M&E challenges and emphasising the importance of context and social learning. Very few studies evaluate whether and why an adaptation initiative has been effective. One of the challenges of M&E for both mitigation and adaptation is a lack of high quality information for modelling. Adaptation M&E is additionally challenged by limited understanding on what indicators to measure and how to attribute altered vulnerability to adaptation actions		4.4.2
	Lifestyle and behavioural change	• Whereas mitigation pathways studies address (implicitly or explicitly) the reduction or elimination of market failures (e.g., external costs, information asymmetries) via climate or energy policies, no study addresses behavioural change strategies in the relationship with mitigation and adaptation actions in the 1.5°C context • Limited knowledge on GHG emissions reduction potential of diverse mitigation behaviour across the world • Most studies on factors enabling lifestyle changes have been conducted in high-income countries, more knowledge needed from low- and middle-income countries, and the focus is typically on enabling individual behaviour change, far less on enabling change in organizations and political systems	• Knowledge gaps on factors enabling adaptation behaviour, except for behaviour in agriculture. • Little is known about cognitive and motivational factors promoting adaptive behaviour. • Little is known about how potential adaptation actions might affect behaviour to influence vulnerability outcomes	4.4.3

Table 4.13 (continued)

Knowledge Area		Mitigation	Adaptation	Reference
Enabling Conditions	Lifestyle and behavioural change	• Limited understanding and treatment of behavioural change and the potential effects of related policies in ambitious mitigation pathways, e.g., in Integrated Assessment Models		4.4.3
		Lack of insight on what can enable changes in adaptation and mitigation behaviour in organizations and political systems		
	Technological innovation	• Quantitative estimates for mitigation and adaptation potentials at economy or sector scale as a result of the combination of general purpose technologies and mitigation technologies have been scarce, except for some evidence in the transport sector • Evidence on the role of international organizations, including the UNFCCC, in building capabilities and enhancing technological innovation for 1.5°C, except for some parts of the transport sector • Technology transfer trials to enable leapfrog applications in developing countries have *limited evidence*		4.4.4
	Policy	• More empirical research would be needed to derive robust conclusions on effectiveness of policies for enabling transitions to 1.5°C and on which factors aid decision-makers seeking to ratchet up their NDCs	• Understanding of what policies work (and do not work) is limited for adaptation in general and for 1.5°C in particular, beyond specific case studies	4.4.5
	Finance	Knowledge gaps persist with respect to the instruments to match finance to its most effective use in mitigation and adaptation		4.4.5
Synergies and Trade-Offs Between Adaptation and Mitigation		• Strong claims are made with respect to synergies and trade-offs, but there is little knowledge to underpin these, especially of co-benefits by region • Water–energy conservation relationships of individual conservation measures in industries other than the water and energy sectors have not been investigated in detail • There is no evidence on synergies with adaptation of CCS in the power sector and of enhanced weathering under carbon dioxide removal • There is no evidence on trade-offs with adaptation of low- and zero-energy buildings, and circularity and substitution and bio-based industrial system transitions • There is no evidence of synergies or trade-offs with mitigation of CbA • There is no evidence of trade-offs with mitigation of the built environment, on adaptation options for industrial energy, and climate services		4.5.4

4

Frequently Asked Questions

FAQ 4.1 | What Transitions could Enable Limiting Global Warming to 1.5°C?

Summary: In order to limit warming to 1.5°C above pre-industrial levels, the world would need to transform in a number of complex and connected ways. While transitions towards lower greenhouse gas emissions are underway in some cities, regions, countries, businesses and communities, there are few that are currently consistent with limiting warming to 1.5°C. Meeting this challenge would require a rapid escalation in the current scale and pace of change, particularly in the coming decades. There are many factors that affect the feasibility of different adaptation and mitigation options that could help limit warming to 1.5°C and with adapting to the consequences.

There are actions across all sectors that can substantially reduce greenhouse gas emissions. This Special Report assesses energy, land and ecosystems, urban and infrastructure, and industry in developed and developing nations to see how they would need to be transformed to limit warming to 1.5°C. Examples of actions include shifting to low- or zero-emission power generation, such as renewables; changing food systems, such as diet changes away from land-intensive animal products; electrifying transport and developing 'green infrastructure', such as building green roofs, or improving energy efficiency by smart urban planning, which will change the layout of many cities.

Because these different actions are connected, a 'whole systems' approach would be needed for the type of transformations that could limit warming to 1.5°C. This means that all relevant companies, industries and stakeholders would need to be involved to increase the support and chance of successful implementation. As an illustration, the deployment of low-emission technology (e.g., renewable energy projects or a bio-based chemical plants) would depend upon economic conditions (e.g., employment generation or capacity to mobilize investment), but also on social/cultural conditions (e.g., awareness and acceptability) and institutional conditions (e.g., political support and understanding).

To limit warming to1.5°C, mitigation would have to be large-scale and rapid. Transitions can be transformative or incremental, and they often, but not always, go hand in hand. Transformative change can arise from growth in demand for a new product or market, such that it displaces an existing one. This is sometimes called 'disruptive innovation'. For example, high demand for LED lighting is now making more energy-intensive, incandescent lighting near-obsolete, with the support of policy action that spurred rapid industry innovation. Similarly, smart phones have become global in use within ten years. But electric cars, which were released around the same time, have not been adopted so quickly because the bigger, more connected transport and energy systems are harder to change. Renewable energy, especially solar and wind, is considered to be disruptive by some as it is rapidly being adopted and is transitioning faster than predicted. But its demand is not yet uniform. Urban systems that are moving towards transformation are coupling solar and wind with battery storage and electric vehicles in a more incremental transition, though this would still require changes in regulations, tax incentives, new standards, demonstration projects and education programmes to enable markets for this system to work.

Transitional changes are already underway in many systems, but limiting warming to 1.5°C would require a rapid escalation in the scale and pace of transition, particularly in the next 10–20 years. While limiting warming to 1.5°C would involve many of the same types of transitions as limiting warming to 2°C, the pace of change would need to be much faster. While the pace of change that would be required to limit warming to 1.5°C can be found in the past, there is no historical precedent for the scale of the necessary transitions, in particular in a socially and economically sustainable way. Resolving such speed and scale issues would require people's support, public-sector interventions and private-sector cooperation.

Different types of transitions carry with them different associated costs and requirements for institutional or governmental support. Some are also easier to scale up than others, and some need more government support than others. Transitions between, and within, these systems are connected and none would be sufficient on its own to limit warming to 1.5°C.

The 'feasibility' of adaptation and mitigation options or actions within each system that together can limit warming to 1.5°C within the context of sustainable development and efforts to eradicate poverty requires careful consideration of multiple different factors. These factors include: (i) whether sufficient natural systems and resources are available to support the various options for transitioning (known as *environmental feasibility*); (ii) the degree to which the required technologies are developed and available (known as *technological feasibility*);

FAQ 4.1 (continued)

(iii) the economic conditions and implications (known as *economic feasibility*); (iv) what are the implications for human behaviour and health (known as *social/cultural feasibility*); and (v) what type of institutional support would be needed, such as governance, institutional capacity and political support (known as *institutional feasibility*). An additional factor (vi – known as the *geophysical feasibility*) addresses the capacity of physical systems to carry the option, for example, whether it is geophysically possible to implement large-scale afforestation consistent with 1.5°C.

Promoting enabling conditions, such as finance, innovation and behaviour change, would reduce barriers to the options, make the required speed and scale of the system transitions more likely, and therefore would increase the overall feasibility limiting warming to 1.5°C.

FAQ4.1: **The different feasibility dimensions towards limiting warming to 1.5°C**
Assessing the feasibility of different adaptation and mitigation options/actions requires consideration across six dimensions.

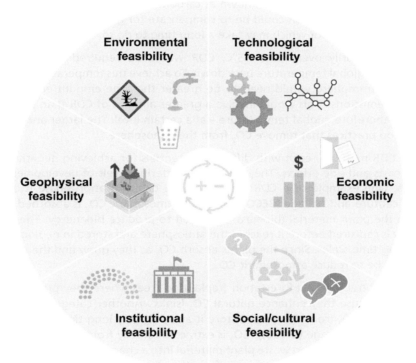

FAQ 4.1, Figure 1 | The different dimensions to consider when assessing the 'feasibility' of adaptation and mitigation options or actions within each system that can help to limit warming to 1.5°C. These are: (i) the environmental feasibility; (ii) the technological feasibility; (iii) the economic feasibility; (iv) the social/cultural feasibility; (v) the institutional feasibility; and (vi) the geophysical feasibility.

Frequently Asked Questions

FAQ 4.2 | What are Carbon Dioxide Removal and Negative Emissions?

Summary: *Carbon dioxide removal (CDR) refers to the process of removing CO_2 from the atmosphere. Since this is the opposite of emissions, practices or technologies that remove CO_2 are often described as achieving 'negative emissions'. The process is sometimes referred to more broadly as greenhouse gas removal if it involves removing gases other than CO_2. There are two main types of CDR: either enhancing existing natural processes that remove carbon from the atmosphere (e.g., by increasing its uptake by trees, soil, or other 'carbon sinks') or using chemical processes to, for example, capture CO_2 directly from the ambient air and store it elsewhere (e.g., underground). All CDR methods are at different stages of development and some are more conceptual than others, as they have not been tested at scale.*

Limiting warming to 1.5°C above pre-industrial levels would require unprecedented rates of transformation in many areas, including in the energy and industrial sectors, for example. Conceptually, it is possible that techniques to draw CO_2 out of the atmosphere (known as carbon dioxide removal, or CDR) could contribute to limiting warming to 1.5°C. One use of CDR could be to compensate for greenhouse gas emissions from sectors that cannot completely decarbonize, or which may take a long time to do so.

If global temperature temporarily overshoots 1.5°C, CDR would be required to reduce the atmospheric concentration of CO_2 to bring global temperature back down. To achieve this temperature reduction, the amount of CO_2 drawn out of the atmosphere would need to be greater than the amount entering the atmosphere, resulting in 'net negative emissions'. This would involve a greater amount of CDR than stabilizing atmospheric CO_2 concentration – and, therefore, global temperature – at a certain level. The larger and longer an overshoot, the greater the reliance on practices that remove CO_2 from the atmosphere.

There are a number of CDR methods, each with different potentials for achieving negative emissions, as well as different associated costs and side effects. They are also at differing levels of development, with some more conceptual than others. One example of a CDR method in the demonstration phase is a process known as bioenergy with carbon capture and storage (BECCS), in which atmospheric CO_2 is absorbed by plants and trees as they grow, and then the plant material (biomass) is burned to produce bioenergy. The CO_2 released in the production of bioenergy is captured before it reaches the atmosphere and stored in geological formations deep underground on very long time scales. Since the plants absorb CO_2 as they grow and the process does not emit CO_2, the overall effect can be to reduce atmospheric CO_2.

Afforestation (planting new trees) and reforestation (replanting trees where they previously existed) are also considered forms of CDR because they enhance natural CO_2 'sinks'. Another category of CDR techniques uses chemical processes to capture CO_2 from the air and store it away on very long time scales. In a process known as direct air carbon capture and storage (DACCS), CO_2 is extracted directly from the air and stored in geological formations deep underground. Converting waste plant material into a charcoal-like substance called biochar and burying it in soil can also be used to store carbon away from the atmosphere for decades to centuries.

There can be beneficial side effects of some types of CDR, other than removing CO_2 from the atmosphere. For example, restoring forests or mangroves can enhance biodiversity and protect against flooding and storms. But there could also be risks involved with some CDR methods. For example, deploying BECCS at large scale would require a large amount of land to cultivate the biomass required for bioenergy. This could have consequences for sustainable development if the use of land competes with producing food to support a growing population, biodiversity conservation or land rights. There are also other considerations. For example, there are uncertainties about how much it would cost to deploy DACCS as a CDR technique, given that removing CO_2 from the air requires considerable energy.

FAQ 4.2 (continued)

FAQ4.2: **Carbon dioxide removal and negative emissions**

Examples of some CDR / negative emissions techniques and practices

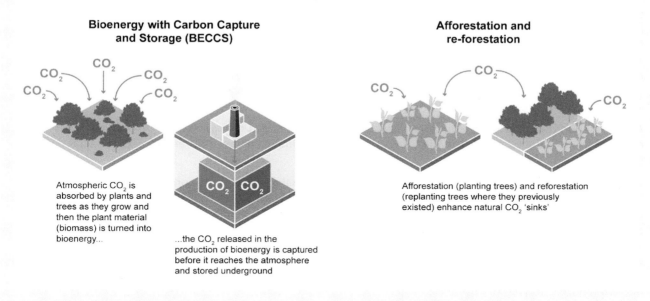

Bioenergy with Carbon Capture and Storage (BECCS)

Atmospheric CO_2 is absorbed by plants and trees as they grow and then the plant material (biomass) is turned into bioenergy...

...the CO_2 released in the production of bioenergy is captured before it reaches the atmosphere and stored underground

Afforestation and re-forestation

Afforestation (planting trees) and reforestation (replanting trees where they previously existed) enhance natural CO_2 'sinks'

FAQ 4.2, Figure 1 | Carbon dioxide removal (CDR) refers to the process of removing CO_2 from the atmosphere. There are a number of CDR techniques, each with different potential for achieving 'negative emissions', as well as different associated costs and side effects.

4

Frequently Asked Questions

FAQ 4.3 | Why is Adaptation Important in a 1.5°C-Warmer World?

Summary: Adaptation is the process of adjusting to current or expected changes in climate and its effects. Even though climate change is a global problem, its impacts are experienced differently across the world. This means that responses are often specific to the local context, and so people in different regions are adapting in different ways. A rise in global temperature from the current 1°C above pre-industrial levels to 1.5°C, and beyond, increases the need for adaptation. Therefore, stabilizing global temperatures at 1.5°C above pre-industrial levels would require a smaller adaptation effort than at 2°C. Despite many successful examples around the world, progress in adaptation is, in many regions, in its infancy and unevenly distributed globally.

Adaptation refers to the process of adjustment to actual or expected changes in climate and its effects. Since different parts of the world are experiencing the impacts of climate change differently, there is similar diversity in how people in a given region are adapting to those impacts.

The world is already experiencing the impacts from 1°C of global warming above pre-industrial levels, and there are many examples of adaptation to impacts associated with this warming. Examples of adaptation efforts taking place around the world include investing in flood defences such as building sea walls or restoring mangroves, efforts to guide development away from high risk areas, modifying crops to avoid yield reductions, and using social learning (social interactions that change understanding on the community level) to modify agricultural practices, amongst many others. Adaptation also involves building capacity to respond better to climate change impacts, including making governance more flexible and strengthening financing mechanisms, such as by providing different types of insurance.

In general, an increase in global temperature from present day to 1.5°C or 2°C (or higher) above pre-industrial temperatures would increase the need for adaptation. Stabilizing global temperature increase at 1.5°C would require a smaller adaptation effort than for 2°C.

Since adaptation is still in early stages in many regions, there are questions about the capacity of vulnerable communities to cope with any amount of further warming. Successful adaptation can be supported at the national and sub-national levels, with national governments playing an important role in coordination, planning, determining policy priorities, and distributing resources and support. However, given that the need for adaptation can be very different from one community to the next, the kinds of measures that can successfully reduce climate risks will also depend heavily on the local context.

When done successfully, adaptation can allow individuals to adjust to the impacts of climate change in ways that minimize negative consequences and to maintain their livelihoods. This could involve, for example, a farmer switching to drought-tolerant crops to deal with increasing occurrences of heatwaves. In some cases, however, the impacts of climate change could result in entire systems changing significantly, such as moving to an entirely new agricultural system in areas where the climate is no longer suitable for current practices. Constructing sea walls to stop flooding due to sea level rise from climate change is another example of adaptation, but developing city planning to change how flood water is managed throughout the city would be an example of transformational adaptation. These actions require significantly more institutional, structural, and financial support. While this kind of transformational adaptation would not be needed everywhere in a 1.5°C world, the scale of change needed would be challenging to implement, as it requires additional support, such as through financial assistance and behavioural change. Few empirical examples exist to date.

Examples from around the world show that adaptation is an iterative process. Adaptation pathways describe how communities can make decisions about adaptation in an ongoing and flexible way. Such pathways allow for pausing, evaluating the outcomes of specific adaptation actions, and modifying the strategy as appropriate. Due to their flexible nature, adaptation pathways can help to identify the most effective ways to minimise the impacts of present and future climate change for a given local context. This is important since adaptation can sometimes exacerbate vulnerabilities and existing inequalities if poorly designed. The unintended negative consequences of adaptation that can sometimes occur are known as 'maladaptation'. Maladaptation can be seen if a particular adaptation option has negative consequences for some (e.g., rainwater harvesting upstream might reduce water availability downstream) or if an adaptation intervention in the present has trade-offs in the future (e.g., desalination plants may improve water availability in the present but have large energy demands over time).

FAQ 4.3 (continued)

While adaptation is important to reduce the negative impacts from climate change, adaptation measures on their own are not enough to prevent climate change impacts entirely. The more global temperature rises, the more frequent, severe, and erratic the impacts will be, and adaptation may not protect against all risks. Examples of where limits may be reached include substantial loss of coral reefs, massive range losses for terrestrial species, more human deaths from extreme heat, and losses of coastal-dependent livelihoods in low lying islands and coasts.

FAQ4.3: **Adaptation in a warming world**
Adapting to further warming requires action at national & sub-national levels and can mean different things to different people in different contexts

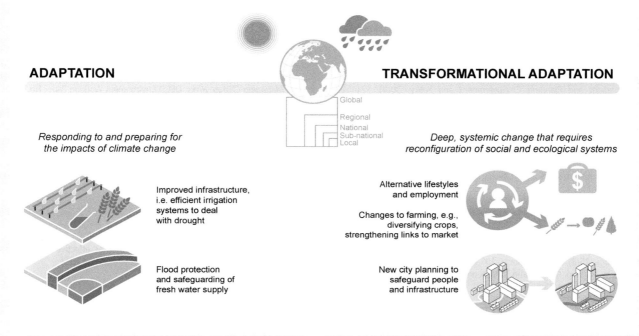

FAQ 4.3, Figure 1 | Why is adaptation important in a world with global warming of 1.5°C? Examples of adaptation and transformational adaptation. Adapting to further warming requires action at national and sub-national levels and can mean different things to different people in different contexts. While transformational adaptation would not be needed everywhere in a world limited to 1.5°C warming, the scale of change needed would be challenging to implement.

References

Aakre, S., S. Kallbekken, R. Van Dingenen, and D.G. Victor, 2018: Incentives for small clubs of Arctic countries to limit black carbon and methane emissions. *Nature Climate Change*, **8(1)**, 85–90, doi:10.1038/s41558-017-0030-8.

Aall, C., K. Groven, and G. Lindseth, 2007: The Scope of Action for Local Climate Policy: The Case of Norway. *Global Environmental Politics*, **7(2)**, 83–101, doi:10.1162/glep.2007.7.2.83.

Aarts, H. and A.P. Dijksterhuis, 2000: The Automatic Activation of Goal-directed Behaviour: The case of travel habit. *Journal of Environmental Psychology*, **20(1)**, 75–82, doi:10.1006/jevp.1999.0156.

Aarts, H., B. Verplanken, and A. Knippenberg, 1998: Predicting Behavior From Actions in the Past: Repeated Decision Making or a Matter of Habit? *Journal of Applied Social Psychology*, **28(15)**, 1355–1374, doi:10.1111/j.1559-1816.1998.tb01681.x.

Abanades, J.C., E.S. Rubin, M. Mazzotti, and H.J. Herzog, 2017: On the climate change mitigation potential of CO_2 conversion to fuels. *Energy & Environmental Science*, **10(12)**, 2491–2499, doi:10.1039/c7ee02819a.

Abanades, J.C. et al., 2015: Emerging CO_2 capture systems. *International Journal of Greenhouse Gas Control*, **40**, 126–166, doi:10.1016/j.ijggc.2015.04.018.

Abbott, K.W., 2012: The transnational regime complex for climate change. *Environment and Planning C: Government and Policy*, **30(4)**, 571–590, doi:10.1068/c11127.

Abbott, K.W. and D. Snidal, 2009: Strengthening International Regulation Through Transnational New Governance: Overcoming the Orchestration Deficit. *Vanderbilt Journal of Transnational Law*, **42**, 501–578, https://wp0.vanderbilt.edu/wp-content/uploads/sites/78/abbott-cr_final.pdf.

Abdulai, I. et al., 2018: Cocoa agroforestry is less resilient to sub-optimal and extreme climate than cocoa in full sun. *Global Change Biology*, **24(1)**, 273–286, doi:10.1111/gcb.13885.

Abrahamse, W. and L. Steg, 2013: Social influence approaches to encourage resource conservation: A meta-analysis. *Global Environmental Change*, **23**, 1773–1785, doi:10.1016/j.gloenvcha.2013.07.029.

Abrahamse, W., L. Steg, C. Vlek, and T. Rothengatter, 2005: A review of intervention studies aimed at household energy conservation. *Journal of Environmental Psychology*, **25(3)**, 273–291, doi:10.1016/j.jenvp.2005.08.002.

Abrahamse, W., L. Steg, C. Vlek, and T. Rothengatter, 2007: The effect of tailored information, goal setting, and tailored feedback on household energy use, energy-related behaviors, and behavioral antecedents. *Journal of Environmental Psychology*, **27**, 265–276, doi:10.1016/j.jenvp.2007.08.002.

ACOLA, 2017: *The Role of Energy Storage in Australia's Future Energy Supply Mix*. Australian Council of Learned Academies (ACOLA), Melbourne, Australia, 158 pp.

AdaptationWatch, 2015: *Toward Mutual Accountability: The 2015 Adaptation Finance Transparency Gap Report*. AdaptationWatch, 97 pp.

ADB, 2013: *Bhutan Transport 2040: Integrated Strategic Vision*. Asian Development Bank (ADB), Manila, Philippines, 24 pp.

Aden, N., 2018: Necessary but not sufficient: the role of energy efficiency in industrial sector low-carbon transformation. *Energy Efficiency*, **11(5)**, 1083–1101, doi:10.1007/s12053-017-9570-z.

Adenle, A.A. et al., 2017: Managing Climate Change Risks in Africa – A Global Perspective. *Ecological Economics*, **141**, 190–201, doi:10.1016/j.ecolecon.2017.06.004.

Adger, W.N., N.W. Arnell, and E.L. Tompkins, 2005: Successful adaptation to climate change across scales. *Global Environmental Change*, **15(2)**, 77–86, doi:10.1016/j.gloenvcha.2004.12.005.

Adger, W.N., S. Huq, K. Brown, D. Conway, and M. Hulme, 2003: Adaptation to climate change in the developing world. *Progress in Development Studies*, **3(3)**, 179–195, doi:10.1191/1464993403ps060oa.

Adger, W.N. et al., 2015: Focus on environmental risks and migration: causes and consequences. *Environmental Research Letters*, **10(6)**, 060201, doi:10.1088/1748-9326/10/6/060201.

Aebischer, B. and L.M. Hilty, 2015: The Energy Demand of ICT: A Historical Perspective and Current Methodological Challenges. In: *ICT Innovations for Sustainability* [Hilty, L.M. and B. Aebischer (eds.)]. Springer International Publishing, Cham, Switzerland, pp. 71–103, doi:10.1007/978-3-319-09228-7_4.

AfDB/OECD/UNDP, 2016: *African Economic Outlook 2016: Sustainable Cities and Structural Transformation*. African Economic Outlook, OECD Publishing, Paris, France, 400 pp., doi:10.1787/aeo-2016-en.

Afionis, S., L.C. Stringer, N. Favretto, J. Tomei, and M.S. Buckeridge, 2014: Unpacking Brazil's Leadership in the Global Biofuels Arena: Brazilian Ethanol Diplomacy in Africa. *Global Environmental Politics*, **14(2)**, 82–101, doi:10.1162/glep.

Agee, E.M., A. Orton, E.M. Agee, and A. Orton, 2016: An Initial Laboratory Prototype Experiment for Sequestration of Atmospheric CO_2. *Journal of Applied Meteorology and Climatology*, **55(8)**, 1763–1770, doi:10.1175/jamc-d-16-0135.1.

Aglietta, M., E. Espagne, and B. Perrissin Fabert, 2015a: *A proposal to finance low carbon investment in Europe*. La Note D'Analyse Feb. 2015 No. 4, France Stratégie, Paris, France, 8 pp.

Aglietta, M., J.-C. Hourcade, C. Jaeger, and B.P. Fabert, 2015b: Financing transition in an adverse context: climate finance beyond carbon finance. *International Environmental Agreements: Politics, Law and Economics*, **15(4)**, 403–420, doi:10.1007/s10784-015-9298-1.

Agrawal, A., C. E.R., and E.R. Gerber, 2015: Motivational Crowding in Sustainable Development Interventions. *American Political Science Review*, **109(3)**, 470–487, doi:10.1017/s0003055415000209.

Agrawala, S., M. Carraro, N. Kingsmill, E. Lanzi, and G. Prudent-richard, 2011: *Private Sector Engagement in Adaptation to Climate Change: Approaches to Managing Climate Risks*. OECD Environment Working Paper No 39, OECD Publishing, Paris, France, 56 pp., doi:10.1787/5kg221jkf1g7-en.

Aguilera, E., L. Lassaletta, A. Gattinger, and B.S. Gimeno, 2013: Managing soil carbon for climate change mitigation and adaptation in Mediterranean cropping systems: A meta-analysis. *Agriculture, Ecosystems & Environment*, **168**, 25–36, doi:10.1016/j.agee.2013.02.003.

Ahern, J., S. Cilliers, and J. Niemelä, 2014: The concept of ecosystem services in adaptive urban planning and design: A framework for supporting innovation. *Landscape and Urban Planning*, **125**, 254–259, doi:10.1016/j.landurbplan.2014.01.020.

Ahlfeldt, G. and E. Pietrostefani, 2017: *Demystifying Compact Urban Growth: Evidence From 300 Studies From Across the World*. Coalition for Urban Transitions, London, UK and Washington DC, USA, 84 pp.

Ahlm, L. et al., 2017: Marine cloud brightening – as effective without clouds. *Atmospheric Chemistry and Physics*, **17(21)**, 13071–13087, doi:10.5194/acp-17-13071-2017.

Åhman, M., L.J. Nilsson, and B. Johansson, 2016: Global climate policy and deep decarbonization of energy-intensive industries. *Climate Policy*, **17(5)**, 634–649, doi:10.1080/14693062.2016.1167009.

Aitken, M., 2010: Wind power and community benefits: Challenges and opportunities. *Energy Policy*, **38(10)**, 6066–6075, doi:10.1016/j.enpol.2010.05.062.

Ajanovic, A. and R. Haas, 2017: The impact of energy policies in scenarios on GHG emission reduction in passenger car mobility in the EU-15. *Renewable and Sustainable Energy Reviews*, **68**, 1088–1096, doi:10.1016/j.rser.2016.02.013.

Akbari, H., H.D. Matthews, and D. Seto, 2012: The long-term effect of increasing the albedo of urban areas. *Environmental Research Letters*, **7(2)**, 024004, doi:10.1088/1748-9326/7/2/024004.

Åkerman, J., 2011: The role of high-speed rail in mitigating climate change – The Swedish case Europabanan from a life cycle perspective. *Transportation Research Part D: Transport and Environment*, **16(3)**, 208–217, doi:10.1016/j.trd.2010.12.004.

Akter, S., T.J. Krupnik, and F. Khanam, 2017: Climate change skepticism and index versus standard crop insurance demand in coastal Bangladesh. *Regional Environmental Change*, **17(8)**, 2455–2466, doi:10.1007/s10113-017-1174-9.

Akter, S., T.J. Krupnik, F. Rossi, and F. Khanam, 2016: The influence of gender and product design on farmers' preferences for weather-indexed crop insurance. *Global Environmental Change*, **38**, 217–229, doi:10.1016/j.gloenvcha.2016.03.010.

Albers, R.A.W. et al., 2015: Overview of challenges and achievements in the climate adaptation of cities and in the Climate Proof Cities program. *Building and Environment*, **83**, 1–10, doi:10.1016/j.buildenv.2014.09.006.

Albright, R. et al., 2016: Reversal of ocean acidification enhances net coral reef calcification. *Nature*, **531(7594)**, 362–365, doi:10.1038/nature17155.

Alcalde, J. et al., 2018: Estimating geological CO_2 storage security to deliver on climate mitigation. *Nature Communications*, **9(1)**, 2201, doi:10.1038/s41467-018-04423-1.

4

Aldrich, D.P., C. Page, and C.J. Paul, 2016: Social Capital and Climate Change Adaptation. In: *Oxford Research Encyclopedia of Climate Science*. Oxford University Press, Oxford, UK, doi:10.1093/acrefore/9780190228620.013.342.

Aldy, J. et al., 2016: Economic tools to promote transparency and comparability in the Paris Agreement. *Nature Climate Change*, **6**, 1000–1004, doi:10.1038/nclimate3106.

Alexander, C. et al., 2011: Linking indigenous and scientific knowledge of climate change. *BioScience*, **61(6)**, 477–484, doi:10.1525/bio.2011.61.6.10.

Alexander, P. et al., 2017: Losses, inefficiencies and waste in the global food system. *Agricultural Systems*, **153**, 190–200, doi:10.1016/j.agsy.2017.01.014.

Alexandratos, N. and J. Bruinsma, 2012: *World agriculture towards 2030/2050: the 2012 revision*. ESA Working paper No. 12-03, Agricultural Development Economics Division, Food and Agriculture Organization of the United Nations, Rome, Italy, 147 pp.

Alfieri, L. et al., 2017: Global projections of river flood risk in a warmer world. *Earth's Future*, **5(2)**, 171–182, doi:10.1002/2016ef000485.

Ali, S.H. et al., 2017: Mineral supply for sustainable development requires resource governance. *Nature*, **543**, 367–372, doi:10.1038/nature21359.

Alkama, R. and A. Cescatti, 2016: Biophysical climate impacts of recent changes in global forest cover. *Science*, **351(6273)**, 600–604, doi:10.1126/science.aac8083.

Allan, J.I. and J. Hadden, 2017: Exploring the framing power of NGOs in global climate politics. *Environmental Politics*, **26(4)**, 600–620, doi:10.1080/09644016.2017.1319017.

Allcott, H., 2011: Social norms and energy conservation. *Journal of Public Economics*, **95(9–10)**, 1082–1095, doi:10.1016/j.jpubeco.2011.03.003.

Allison, E., 2012: Gross National Happiness. In: *The Berkshire Encyclopedia of Sustainability: Measurements, Indicators, and Research Methods for Sustainability* [Spellerberg, I., D.S. Fogel, L.M. Butler Harrington, and S.E. Fredericks (eds.)]. Berkshire Publishing Group, Great Barrington, MA, USA, pp. 180–184.

Alló, M. and M.L. Loureiro, 2014: The role of social norms on preferences towards climate change policies: A meta-analysis. *Energy Policy*, **73**, 563–574, doi:10.1016/j.enpol.2014.04.042.

Almeida Prado, F. et al., 2016: How much is enough? An integrated examination of energy security, economic growth and climate change related to hydropower expansion in Brazil. *Renewable and Sustainable Energy Reviews*, **53**, 1132–1136, doi:10.1016/j.rser.2015.09.050.

Alterskjær, K., J.E. Kristjánsson, and O. Seland, 2012: Sensitivity to deliberate sea salt seeding of marine clouds – Observations and model simulations. *Atmospheric Chemistry and Physics*, **12(5)**, 2795–2807, doi:10.5194/acp-12-2795-2012.

Altieri, K.E. et al., 2016: Achieving development and mitigation objectives through a decarbonization development pathway in South Africa. *Climate Policy*, **16(Sup 1)**, s78–s91, doi:10.1080/14693062.2016.1150250.

AMAP, 2017a: *Adaptation Actions for a Changing Arctic: Barents Area Overview Report*. Arctic Monitoring and Assessment Programme (AMAP), Oslo, Norway, 24 pp.

AMAP, 2017b: *Adaptation Actions for a Changing Arctic: Perspectives from the Bering-Chukchi-Beaufort Region*. Arctic Monitoring and Assessment Programme (AMAP), Oslo, Norway, 269 pp.

AMAP, 2018: *Adaptation Actions for a Changing Arctic: Perspectives from the Baffin Bay/Davis Strait Region*. Arctic Monitoring and Assessment Programme (AMAP), Oslo, Norway, 370 pp.

Amars, L., M. Fridahl, M. Hagemann, F. Röser, and B.-O. Linnér, 2017: The transformational potential of Nationally Appropriate Mitigation Actions in Tanzania: assessing the concept's cultural legitimacy among stakeholders in the solar energy sector. *Local Environment*, **22(1)**, 86–105, doi:10.1080/13549839.2016.1161607.

Amin, A. and L.R. Cirolia, 2018: Politics/matter: Governing Cape Town's informal settlements. *Urban Studies*, **55(2)**, 274–295, doi:10.1177/0042098017694133.

Amorim, H., M.L. Lopes, J.V. De Castro Oliveira, M.S. Buckeridge, and G.H. Goldman, 2011: Scientific challenges of bioethanol production in Brazil. *Applied Microbiology and Biotechnology*, **91(5)**, 1267–1275, doi:10.1007/s00253-011-3437-6.

Amundsen, H., G.K. Hovelsrud, C. Aall, M. Karlsson, and H. Westskog, 2018: Local governments as drivers for societal transformation: towards the 1.5°C ambition. *Current Opinion in Environmental Sustainability*, **31**, 23–29, doi:10.1016/j.cosust.2017.12.004.

Anacona, P.I., A. Mackintosh, and K. Norton, 2015: Reconstruction of a glacial lake outburst flood (GLOF) in the Engaño Valley, Chilean Patagonia: Lessons for GLOF risk management. *Science of The Total Environment*, **527–528**, 1–11, doi:10.1016/j.scitotenv.2015.04.096.

Anderegg, W.R.L. et al., 2015: Tropical nighttime warming as a dominant driver of variability in the terrestrial carbon sink. *Proceedings of the National Academy of Sciences*, **112(51)**, 15591–15596, doi:10.1073/pnas.1521479112.

Anderson, S.T. and R.G. Newell, 2004: Information programs for technology adoption: the case of energy-efficiency audits. *Resource and Energy Economics*, **26(1)**, 27–50, doi:10.1016/j.reseneeco.2003.07.001.

Andersson, M., P. Bolton, and F. Samama, 2016: Hedging Climate Risk. *Financial Analysts Journal*, **72(3)**, 13–32, doi:10.2469/faj.v72.n3.4.

Andonova, L., T.N. Hale, and C. Roger, 2017: National Policies and Transnational Governance of Climate Change: Substitutes or Complements? *International Studies Quarterly*, 1–16, doi:10.1093/isq/sqx014.

Andow, D.A. and C. Zwahlen, 2006: Assessing environmental risks of transgenic plants. *Ecology Letters*, **9(2)**, 196–214, doi:10.1111/j.1461-0248.2005.00846.x.

André, C., D. Boulet, H. Rey-Valette, and B. Rulleau, 2016: Protection by hard defence structures or relocation of assets exposed to coastal risks: Contributions and drawbacks of cost-benefit analysis for long-term adaptation choices to climate change. *Ocean and Coastal Management*, **134**, 173–182, doi:10.1016/j.ocecoaman.2016.10.003.

Andrews-Speed, P. and G. Ma, 2016: Household Energy Saving in China: The Challenge of Changing Behaviour. In: *China's Energy Efficiency and Conservation: Household Behaviour, Leglisation, Regional Analysis and Impacts* [Su, B. and E. Thomson (eds.)]. SpringerBriefs in Environment, Security, Development and Peace, pp. 23–39, doi:10.1007/978-981-10-0928-0_3.

Anenberg, S.C. et al., 2012: Global Air Quality and Health Co-benefits of Mitigating Near-Term Climate Change through Methane and Black Carbon Emission Controls. *Environmental Health Perspectives*, **120(6)**, 831–839, doi:10.1289/ehp.1104301.

Angotti, T., 2015: Urban agriculture: long-term strategy or impossible dream? *Public Health*, **129(4)**, 336–341, doi:10.1016/j.puhe.2014.12.008.

Anguelovski, I., E. Chu, and J.A. Carmin, 2014: Variations in approaches to urban climate adaptation: Experiences and experimentation from the global South. *Global Environmental Change*, **27(1)**, 156–167, doi:10.1016/j.gloenvcha.2014.05.010.

Annicchiarico, B. and F. Di Dio, 2015: Environmental policy and macroeconomic dynamics in a new Keynesian model. *Journal of Environmental Economics and Management*, **69(1)**, 1–21, doi:10.1016/j.jeem.2014.10.002.

Annicchiarico, B. and F. Di Dio, 2016: GHG Emissions Control and Monetary Policy. *Environmental and Resource Economics*, 1–29, doi:10.1007/s10640-016-0007-5.

Anshelm, J. and A. Hansson, 2014: Battling Promethean dreams and Trojan horses: Revealing the critical discourses of geoengineering. *Energy Research & Social Science*, **2**, 135–144, doi:10.1016/j.erss.2014.04.001.

Arab, A., A. Khodaei, Z. Han, and S.K. Khator, 2015: Proactive Recovery of Electric Power Assets for Resiliency Enhancement. *IEEE Access*, **3**, 99–109, doi:10.1109/access.2015.2404215.

Araos, M., J. Ford, L. Berrang-Ford, R. Biesbroek, and S. Moser, 2016a: Climate change adaptation planning for Global South megacities: the case of Dhaka. *Journal of Environmental Policy & Planning*, **19(6)**, 682–696, doi:10.1080/1523908x.2016.1264873.

Araos, M. et al., 2016b: Climate change adaptation planning in large cities: A systematic global assessment. *Environmental Science & Policy*, **66**, 375–382, doi:10.1016/j.envsci.2016.06.009.

Archer, D., 2016: Building urban climate resilience through community-driven approaches to development. *International Journal of Climate Change Strategies and Management*, **8(5)**, 654–669, doi:10.1108/ijccsm-03-2014-0035.

Archer, D. et al., 2014: Moving towards inclusive urban adaptation: approaches to integrating community-based adaptation to climate change at city and national scale. *Climate and Development*, **6(4)**, 345–356, doi:10.1080/17565529.2014.918868.

Arezki, R., P. Bolton, S. Peters, F. Samama, and J. Stiglitz, 2016: *From Global Savings Glut to Financing Infrastructure: The Advent of Investment Platforms*. IMF Working Paper WP/16/18, International Monetary Fund (IMF), Washington DC, USA, 46 pp.

4

Arino, Y. et al., 2016: Estimating option values of solar radiation management assuming that climate sensitivity is uncertain. *Proceedings of the National Academy of Sciences*, **113(21)**, 5886–91, doi:10.1073/pnas.1520795113.

Armah, F.A., I. Luginaah, H. Hambati, R. Chuenpagdee, and G. Campbell, 2015: Assessing barriers to adaptation to climate change in coastal Tanzania: Does where you live matter? *Population and Environment*, **37(2)**, 231–263, doi:10.1007/s11111-015-0232-9.

Arnold, U., A. De Palmenaer, T. Brück, and K. Kuse, 2018: Energy-Efficient Carbon Fiber Production with Concentrated Solar Power: Process Design and Techno-economic Analysis. *Industrial & Engineering Chemistry Research*, **57(23)**, 7934–7945, doi:10.1021/acs.iecr.7b04841.

Arthur, W.B., 2009: *The Nature of Technology: What It Is and How It Evolves*. Free Press, New York, NY, USA, 256 pp.

Arunrat, N., C. Wang, N. Pumijumnong, S. Sereenonchai, and W. Cai, 2017: Farmers' intention and decision to adapt to climate change: A case study in the Yom and Nan basins, Phichit province of Thailand. *Journal of Cleaner Production*, **143**, 672–685, doi:10.1016/j.jclepro.2016.12.058.

Arze del Granado, F.J., D. Coady, and R. Gillingham, 2012: The Unequal Benefits of Fuel Subsidies: A Review of Evidence for Developing Countries. *World Development*, **40(11)**, 2234–2248, doi:10.1016/j.worlddev.2012.05.005.

Asayama, S., M. Sugiyama, and A. Ishii, 2017: Ambivalent climate of opinions: Tensions and dilemmas in understanding geoengineering experimentation. *Geoforum*, **80**, 82–92, doi:10.1016/j.geoforum.2017.01.012.

Asensio, O.I. and M.A. Delmas, 2015: Nonprice incentives and energy conservation. *Proceedings of the National Academy of Sciences*, **112(6)**, 1–6, doi:10.1073/pnas.1401880112.

Asfaw, S. and B. Davis, 2018: Can Cash Transfer Programmes Promote Household Resilience? Cross-Country Evidence from Sub-Saharan Africa. In: *Climate Smart Agriculture : Building Resilience to Climate Change* [Lipper, L., N. McCarthy, D. Zilberman, S. Asfaw, and G. Branca (eds.)]. Springer International Publishing, Cham, Switzerland, pp. 227–250, doi:10.1007/978-3-319-61194-5_11.

Asfaw, S., A. Carraro, B. Davis, S. Handa, and D. Seidenfeld, 2017: Cash transfer programmes, weather shocks and household welfare: evidence from a randomised experiment in Zambia. *Journal of Development Effectiveness*, **9(4)**, 419–442, doi:10.1080/19439342.2017.1377751.

Ashworth, P., S. Wade, D. Reiner, and X. Liang, 2015: Developments in public communications on CCS. *International Journal of Greenhouse Gas Control*, **40**, 449–458, doi:10.1016/j.ijggc.2015.06.002.

Atela, J.O., C.H. Quinn, and P.A. Minang, 2014: Are REDD projects pro-poor in their spatial targeting? Evidence from Kenya. *Applied Geography*, **52**, 14–24, doi:10.1016/j.apgeog.2014.04.009.

Attari, S.Z., M.L. DeKay, C.I. Davidson, and W. Bruine de Bruin, 2010: Public perceptions of energy consumption and savings. *Proceedings of the National Academy of Sciences*, **107(37)**, 16054–16059, doi:10.1073/pnas.1001509107.

Aumont, O. and L. Bopp, 2006: Globalizing results from ocean in situ iron fertilization studies. *Global Biogeochemical Cycles*, **20(2)**, GB2017, doi:10.1029/2005gb002591.

Austin, S.E. et al., 2015: Public Health Adaptation to Climate Change in Canadian Jurisdictions. *International Journal of Environmental Research and Public Health*, **12(1)**, 623–651, doi:10.3390/ijerph120100623.

Austin, S.E. et al., 2016: Public health adaptation to climate change in OECD countries. *International Journal of Environmental Research and Public Health*, **13(9)**, 889, doi:10.3390/ijerph13090889.

Averchenkova, A., F. Crick, A. Kocornik-Mina, H. Leck, and S. Surminski, 2016: Multinational and large national corporations and climate adaptation: are we asking the right questions? A review of current knowledge and a new research perspective. *Wiley Interdisciplinary Reviews: Climate Change*, **7(4)**, 517–536, doi:10.1002/wcc.402.

Axelrod, R. and W.D. Hamilton, 1981: The Evolution of Cooperation. *Science*, **211(4489)**, 1390–1396, doi:10.1126/science.7466396.

Aykut, S.C., 2016: Taking a wider view on climate governance: moving beyond the 'iceberg', the 'elephant', and the 'forest'. *Wiley Interdisciplinary Reviews: Climate Change*, **7(3)**, 318–328, doi:10.1002/wcc.391.

Ayling, J. and N. Gunningham, 2017: Non-state governance and climate policy: the fossil fuel divestment movement. *Climate Policy*, **17(2)**, 131–149, doi:10.1080/14693062.2015.1094729.

Azeiteiro, U.M. and W. Leal Filho, 2017: E*ditorial. International Journal of Global Warming*, **12(3/4)**, 297–298.

Azevedo, I. and V.M.S. Leal, 2017: Methodologies for the evaluation of local climate change mitigation actions: A review. *Renewable and Sustainable Energy Reviews*, **79**, 681–690, doi:10.1016/j.rser.2017.05.100.

Bäckstrand, K. and J.W. Kuyper, 2017: The democratic legitimacy of orchestration: the UNFCCC, non-state actors, and transnational climate governance. *Environmental Politics*, **26(4)**, 764–788, doi:10.1080/09644016.2017.1323579.

Badré, B., 2018: *Can Finance Save the World?: Regaining Power over Money to Serve the Common Good*. Berrett-Koehler Publishers, Oakland, CA, USA, 288 pp.

Bager, S. and L. Mundaca, 2017: Making 'Smart Meters' smarter? Insights from a behavioural economics pilot field experiment in Copenhagen, Denmark. *Energy Research & Social Science*, **28**, 68–76, doi:10.1016/j.erss.2017.04.008.

Bahadur, A. and T. Tanner, 2014: Transformational resilience thinking: putting people, power and politics at the heart of urban climate resilience. *Environment and Urbanization*, **26(1)**, 200–214, doi:10.1177/0956247814522154.

Bahn, O., M. Chesney, J. Gheyssens, R. Knutti, and A.C. Pana, 2015: Is there room for geoengineering in the optimal climate policy mix? *Environmental Science & Policy*, **48**, 67–76, doi:10.1016/j.envsci.2014.12.014.

Bai, X. et al., 2018: Six research priorities for cities and climate change. *Nature*, **555(7694)**, 23–25, doi:10.1038/d41586-018-02409-z.

Baik, E. et al., 2018: Geospatial analysis of near-term potential for carbon-negative bioenergy in the United States. *Proceedings of the National Academy of Sciences*, **115(13)**, 3290–3295, doi:10.1073/pnas.1720338115.

Bajželj, B. et al., 2014: Importance of food-demand management for climate mitigation. *Nature Climate Change*, **4(10)**, 924–929, doi:10.1038/nclimate2353.

Baker, L.H. et al., 2015: Climate responses to anthropogenic emissions of short-lived climate pollutants. *Atmospheric Chemistry and Physics*, **15(14)**, 8201–8216, doi:10.5194/acp-15-8201-2015.

Bakker, D.C.E., A.J. Watson, and C.S. Law, 2001: Southern Ocean iron enrichment promotes inorganic carbon drawdown. *Deep Sea Research Part II: Topical Studies in Oceanography*, **48(11)**, 2483–2507, doi:10.1016/s0967-0645(01)00005-4.

Bakker, S. et al., 2017: Low-Carbon Transport Policy in Four ASEAN Countries: Developments in Indonesia, the Philippines, Thailand and Vietnam. *Sustainability*, **9(7)**, 1217, doi:10.3390/su9071217.

Bala, G. et al., 2007: Combined climate and carbon-cycle effects of large-scale deforestation. *Proceedings of the National Academy of Sciences*, **104(16)**, 6550–6555, doi:10.1073/pnas.0608998104.

Balcombe, P., D. Rigby, and A. Azapagic, 2013: Motivations and barriers associated with adopting microgeneration energy technologies in the UK. *Renewable and Sustainable Energy Reviews*, **22**, 655–666, doi:10.1016/j.rser.2013.02.012.

Baltz, E.A. et al., 2017: Achievement of Sustained Net Plasma Heating in a Fusion Experiment with the Optometrist Algorithm. *Scientific Reports*, **7(1)**, 6425, doi:10.1038/s41598-017-06645-7.

Bamberg, S., 2000: The Promotion of New Behavior by Forming an Implementation Intention: Results of a Field Experiment in the Domain of Travel Mode Choice. *Journal of Applied Social Psychology*, **30**, 1903–1922, doi:10.1111/j.1559-1816.2000.tb02474.x.

Bamberg, S., 2002: Implementation intention versus monetary incentive comparing the effects of interventions to promote the purchase of organically produced food. *Journal of Economic Psychology*, **23**, 573–587, doi:10.1016/s0167-4870(02)00118-6.

Banerjee, A. and B.D. Solomon, 2003: Eco-labeling for energy efficiency and sustainability: a meta-evaluation of US programs. *Energy Policy*, **31**, 109–123, doi:10.1016/s0301-4215(02)00012-5.

Barbarossa, C., P. De Pelsmacker, and I. Moons, 2017: Personal Values, Green Self-identity and Electric Car Adoption. *Ecological Economics*, **140**, 190–200, doi:10.1016/j.ecolecon.2017.05.015.

Barkakaty, B. et al., 2017: Emerging materials for lowering atmospheric carbon. *Environmental Technology & Innovation*, **7**, 30–43, doi:10.1016/j.eti.2016.12.001.

Barker, T. and J. Kohler, 1998: Equity and Ecotax Reform in the EU: Achieving a 10 per cent Reduction in CO_2 Emissions Using Excise Duties. *Fiscal Studies*, **19(4)**, 375–402, doi:10.1111/j.1475-5890.1998.tb00292.x.

4

Barrett, S., 2015: Subnational Adaptation Finance Allocation: Comparing Decentralized and Devolved Political Institutions in Kenya. *Global Environmental Politics*, **15(3)**, 118–139, doi:10.1162/glep_a_00314.

Bartos, M.D. and M. Chester, 2015: Impacts of climate change on electric power supply in the Western United States. *Nature Climate Change*, **5(8)**, 748–752, doi:10.1038/nclimate2648.

BASIC experts, 2011: *Equitable access to sustainable development: Contribution to the body of scientific knowledge*. BASIC Expert group, Beijing, Brasilia, Cape Town, and Mumbai, 91 pp.

Bataille, C., D. Sawyer, and N. Melton, 2015: *Pathways to deep decarbonization in Canada*. Sustainable Development Solutions Network (SDSN) and Institute for Sustainable Development and International Relations (IDDRI), 54 pp.

Bataille, C. et al., 2018: A review of technology and policy deep decarbonization pathway options for making energy-intensive industry production consistent with the Paris Agreement. *Journal of Cleaner Production*, **187**, 960–973, doi:10.1016/j.jclepro.2018.03.107.

Bates, B.R., B.L. Quick, and A.A. Kloss, 2009: Antecedents of intention to help mitigate wildfire: Implications for campaigns promoting wildfire mitigation to the general public in the wildland-urban interface. *Safety Science*, **47(3)**, 374–381, doi:10.1016/j.ssci.2008.06.002.

Bathiany, S., V. Dakos, M. Scheffer, and T.M. Lenton, 2018: Climate models predict increasing temperature variability in poor countries. *Science Advances*, **4(5)**, eaar5809, doi:10.1126/sciadv.aar5809.

Battiston, S. et al., 2017: A climate stress-test of the financial system. *Nature Climate Change*, **7(4)**, 283–288, doi:10.1038/nclimate3255.

Baudron, F., C. Thierfelder, I. Nyagumbo, and B. Gérard, 2015: Where to Target Conservation Agriculture for African Smallholders? How to Overcome Challenges Associated with its Implementation? Experience from Eastern and Southern Africa. *Environments*, **2(3)**, 338–357, doi:10.3390/environments2030338.

Bauer, N. et al., 2018: Global energy sector emission reductions and bioenergy use: overview of the bioenergy demand phase of the EMF-33 model comparison. *Climatic Change*, 1–16, doi:10.1007/s10584-018-2226-y.

Baul, T.K., A. Alam, H. Strandman, and A. Kilpeläinen, 2017: Net climate impacts and economic profitability of forest biomass production and utilization in fossil fuel and fossil-based material substitution under alternative forest management. *Biomass and Bioenergy*, **98**, 291–305, doi:10.1016/j.biombioe.2017.02.007.

Baumgärtner, S., M.A. Drupp, J.N. Meya, J.M. Munz, and M.F. Quaas, 2017: Income inequality and willingness to pay for environmental public goods. *Journal of Environmental Economics and Management*, **85**, 35–61, doi:10.1016/j.jeem.2017.04.005.

Bayraktarov, E. et al., 2016: The cost and feasibility of marine coastal restoration. *Ecological Applications*, **26(4)**, 1055–1074, doi:10.1890/15-1077.

Beatley, T., 2011: *Biophilic Cities: Integrating Nature into Urban Design and Planning*. Island Press, Washington DC, USA, 208 pp.

Beaudoin, M. and P. Gosselin, 2016: An effective public health program to reduce urban heat islands in Québec, Canada. *Revista Panamericana de Salud Pública*, **40(3)**, 160–166.

Beccali, M., M. Bonomolo, G. Ciulla, A. Galatioto, and V. Brano, 2015: Improvement of energy efficiency and quality of street lighting in South Italy as an action of Sustainable Energy Action Plans. The case study of Comiso (RG). *Energy*, **92(3)**, 394–408, doi:10.1016/j.energy.2015.05.003.

Bekera, B. and R.A. Francis, 2015: A Bayesian method for thermo-electric power generation drought risk assessment. In: *Safety and Reliability of Complex Engineered Systems – Proceedings of the 25th European Safety and Reliability Conference, ESREL 2015, Zurich, Switzerland, 7–10 September 2015* [Podofillini, L., B. Sudret, B. Stojadinovic, E. Zio, and W. Kröger (eds.)]. CRC Press/Balkema, Leiden, The Netherlands, 1921–1927 pp.

Bell, L.W., A.D. Moore, and J.A. Kirkegaard, 2014: Evolution in crop-livestock integration systems that improve farm productivity and environmental performance in Australia. *European Journal of Agronomy*, **57**, 10–20, doi:10.1016/j.eja.2013.04.007.

Bell, T., R. Briggs, R. Bachmayer, and S. Li, 2015: Augmenting Inuit knowledge for safe sea-ice travel – The SmartICE information system. In: *2014 Oceans - St. John's*. IEEE, pp. 1–9, doi:10.1109/oceans.2014.7003290.

Bellamy, R. and P. Healey, 2018: 'Slippery slope' or 'uphill struggle'? Broadening out expert scenarios of climate engineering research and development. *Environmental Science & Policy*, **83**, 1–10, doi:10.1016/j.envsci.2018.01.021.

Bellamy, R., J. Lezaun, and J. Palmer, 2017: Public perceptions of geoengineering research governance: An experimental deliberative approach. *Global Environmental Change*, **45**, 194–202, doi:10.1016/j.gloenvcha.2017.06.004.

Bellassen, V. et al., 2015: Monitoring, reporting and verifying emissions in the climate economy. *Nature Climate Change*, **5**, 319–328, doi:10.1007/s10584-017-2089-7.

Belmonte, N. et al., 2016: A comparison of energy storage from renewable sources through batteries and fuel cells: A case study in Turin, Italy. *International Journal of Hydrogen Energy*, **41(46)**, 21427–21438, doi:10.1016/j.ijhydene.2016.07.260.

Bendito, A. and E. Barrios, 2016: Convergent Agency: Encouraging Transdisciplinary Approaches for Effective Climate Change Adaptation and Disaster Risk Reduction. *International Journal of Disaster Risk Science*, **7(4)**, 430–435, doi:10.1007/s13753-016-0102-9.

Benmarhnia, T. et al., 2016: A Difference-in-Differences Approach to Assess the Effect of a Heat Action Plan on Heat-Related Mortality, and Differences in Effectiveness According to Sex, Age, and Socioeconomic Status (Montreal, Quebec). *Environmental Health Perspectives*, **124(11)**, 1694–1699, doi:10.1289/ehp203.

Berdahl, M. et al., 2014: Arctic cryosphere response in the Geoengineering Model Intercomparison Project G3 and G4 scenarios. *Journal of Geophysical Research: Atmospheres*, **119**, 1308–1321, doi:10.1002/2013jd021264.received.

Bergek, A., S. Jacobsson, B. Carlsson, S. Lindmark, and A. Rickne, 2008: Analyzing the functional dynamics of technological innovation systems: A scheme of analysis. *Research Policy*, **37(3)**, 407–429, doi:10.1016/j.respol.2007.12.003.

Berle, A.A. and G.C. Means, 1932: *The Modern Corporation and Private Property*. Harcourt, Brace and World, New York, NY, USA, 380 pp.

Bernauer, T. and R. Gampfer, 2013: Effects of civil society involvement on popular legitimacy of global environmental governance. *Global Environmental Change*, **23(2)**, 439–449, doi:10.1016/j.gloenvcha.2013.01.001.

Bernauer, T. and R. Gampfer, 2015: How robust is public support for unilateral climate policy? *Environmental Science & Policy*, **54**, 316–330, doi:10.1016/j.envsci.2015.07.010.

Bernauer, T., R. Gampfer, T. Meng, and Y.-S. Su, 2016a: Could more civil society involvement increase public support for climate policy-making ? Evidence from a survey experiment in China. *Global Environmental Change*, **40**, 1–12, doi:10.1016/j.gloenvcha.2016.06.001.

Bernauer, T., L. Dong, L.F. McGrath, I. Shaymerdenova, and H. Zhang, 2016b: Unilateral or Reciprocal Climate Policy? Experimental Evidence from China. *Politics and Governance*, **4(3)**, 152–171, doi:10.17645/pag.v4i3.650.

Berrang-Ford, L., J.D. Ford, and J. Paterson, 2011: Are we adapting to climate change? *Global Environmental Change*, **21(1)**, 25–33, doi:10.1016/j.gloenvcha.2010.09.012.

Berrang-Ford, L. et al., 2014: What drives national adaptation? A global assessment. *Climatic Change*, **124(1–2)**, 441–450, doi:10.1007/s10584-014-1078-3.

Bertheau, P., A. Oyewo, C. Cader, C. Breyer, and P. Blechinger, 2017: Visualizing National Electrification Scenarios for Sub-Saharan African Countries. *Energies*, **10(11)**, 1899, doi:10.3390/en10111899.

Bertoldi, P., 2017: Are current policies promoting a change in behaviour, conservation and sufficiency? An analysis of existing policies and recommendations for new and effective policies. In: *Proceedings of the ECEEE 2017 Summer Study on Consumption, Efficiency & Limits*. European Council for an Energy Efficient Economy (ECEEE) Secretariat, Stockholm, Sweden, pp. 201–211.

Bertoldi, P., S. Rezessy, and V. Oikonomou, 2013: Rewarding energy savings rather than energy efficiency: Exploring the concept of a feed-in tariff for energy savings. *Energy Policy*, **56**, 526–535, doi:10.1016/j.enpol.2013.01.019.

Bertoldi, P., T. Ribeiro Serrenho, and P. Zangheri, 2016: Consumer Feedback Systems: How Much Energy Saving Will They Deliver and for How Long? In: *Proceedings of the 2016 ACEEE Summer Study on Energy Efficiency in Buildings*. American Council for an Energy-Efficient Economy (ACEEE), Washington DC, USA, pp. 1–13.

Bertoldi, P., A. Kona, S. Rivas, and J.F. Dallemand, 2018: Towards a global comprehensive and transparent framework for cities and local governments enabling an effective contribution to the Paris climate agreement. *Current Opinion in Environmental Sustainability*, **30**, 67–74, doi:10.1016/j.cosust.2018.03.009.

Bertram, C., 2010: Ocean iron fertilization in the context of the Kyoto protocol and the post-Kyoto process. *Energy Policy*, **38(2)**, 1130–1139, doi:10.1016/j.enpol.2009.10.065.

Betsill, M.M. and H. Bulkeley, 2006: Cities and the Multilevel Governance of Global Climate Change. *Global Governance*, **12(2)**, 141–159, doi:10.2307/27800607.

4

Bettini, Y., R.R. Brown, and F.J. de Haan, 2015: Exploring institutional adaptive capacity in practice: examining water governance adaptation in Australia. *Ecology and Society*, **20(1)**, art47, doi:10.5751/es-07291-200147.

Betts, R.A. et al., 2018: Changes in climate extremes, fresh water availability and vulnerability to food insecurity projected at 1.5°C and 2°C global warming with a higher-resolution global climate model. *Philosophical Transactions of the Royal Society A: Mathematical, Physical and Engineering Sciences*, **376(2119)**, 20160452, doi:10.1098/rsta.2016.0452.

Betzold, C., 2015: Adapting to climate change in small island developing states. *Climatic Change*, **133(3)**, 481–489, doi:10.1007/s10584-015-1408-0.

Beyer, C., 2014: Strategic Implications of Current Trends in Additive Manufacturing. *Journal of Manufacturing Science and Engineering*, **136(6)**, 064701, doi:10.1115/1.4028599.

Bhagat, R., 2017: Migration, Gender and Right to the City. *Economic & Political Weekly*, **52(32)**, 35–40, www.epw.in/journal/2017/32/perspectives/migration-gender-and-right-city.html.

Bhattacharya, A., J. Oppenheim, and N. Stern, 2015: *Driving sustainable development through better infrastructure: key elements of a transformation program*. Global Economy & Development Working Paper 91, Brookings, Washington DC, USA, 38 pp.

Bhattacharya, S., K. Giannakas, and K. Schoengold, 2017: Market and welfare effects of renewable portfolio standards in United States electricity markets. *Energy Economics*, **64**, 384–401, doi:10.1016/j.eneco.2017.03.011.

Bhave, A. et al., 2017: Screening and techno-economic assessment of biomass-based power generation with CCS technologies to meet 2050 CO_2 targets. *Applied Energy*, **190**, 481–489, doi:10.1016/j.apenergy.2016.12.120.

Biddau, F., A. Armenti, and P. Cottone, 2012: Socio-Psychological Aspects of Grassroots Participation in the Transition Movement: An Italian Case Study. *Journal of Social and Political Psychology*, **4(1)**, 142–165, doi:10.5964/jspp.v4i1.518.

Bidwell, D., 2016: Thinking through participation in renewable energy decisions. *Nature Energy*, **1(5)**, 16051, doi:10.1038/nenergy.2016.51.

Biermann, F., 2010: Beyond the intergovernmental regime: Recent trends in global carbon governance. *Current Opinion in Environmental Sustainability*, **2(4)**, 284–288, doi:10.1016/j.cosust.2010.05.002.

Biesbroek, R. et al., 2015: Opening up the black box of adaptation decision-making. *Nature Climate Change*, **5(6)**, 493–494, doi:10.1038/nclimate2615.

Biggs, E.M. et al., 2015: Sustainable development and the water–energy–food nexus: A perspective on livelihoods. *Environmental Science & Policy*, **54**, 389–397, doi:10.1016/j.envsci.2015.08.002.

Bikhchandani, S. and S. Sharma, 2000: *Herd Behavior in Financial Markets*. IMF Working Paper WP/00/48, International Monetary Fund (IMF), 32 pp.

Birk, T. and K. Rasmussen, 2014: Migration from atolls as climate change adaptation: Current practices, barriers and options in Solomon Islands. *Natural Resources Forum*, **38(1)**, 1–13, doi:10.1111/1477-8947.12038.

Bishop, M.L. and A. Payne, 2012: Climate Change and the Future of Caribbean Development. *The Journal of Development Studies*, **48(10)**, 1536–1553, doi:10.1080/00220388.2012.693166.

Blaabjerg, F., Z. Chen, and S.B. Kjaer, 2004: Power Electronics as Efficient Interface in Dispersed Power Generation Systems. *IEEE Transactions on Power Electronics*, **19(5)**, 1184–1194, doi:10.1109/tpel.2004.833453.

Black, A. and P. Fraser, 2002: Stock market short-termism-an international perspective. *Journal of Multinational Financial Management*, **12(2)**, 135–158, doi:10.1016/s1042-444x(01)00044-5.

Blanchet, T., 2015: Struggle over energy transition in Berlin: How do grassroots initiatives affect local energy policy-making? *Energy Policy*, **78**, 246–254, doi:10.1016/j.enpol.2014.11.001.

Blanco, G. et al., 2014: Drivers, Trends and Mitigation. In: *Climate Change 2014: Mitigation of Climate Change. Contribution of Working Group III to the Fifth Assessment Report of the Intergovernmental Panel on Climate Change* [Edenhofer, O., R. Pichs-Madruga, Y. Sokona, E. Farahani, S. Kadner, K. Seyboth, A. Adler, I. Baum, S. Brunner, P. Eickemeier, B. Kriemann, J. Savolainen, S. Schlömer, C. von Stechow, T. Zwickel, and J.C. Minx (eds.)]. Cambridge University Press, Cambridge, United Kingdom and New York, NY, USA, pp. 351–411.

Blennow, K., J. Persson, M. Tomé, and M. Hanewinkel, 2012: Climate Change: Believing and Seeing Implies Adapting. *PLOS ONE*, **7(11)**, 1435–1439, doi:10.1371/journal.pone.0050182.

BNEF, 2018: New Energy Outlook 2018. Bloomberg New Energy Finance (BNEF), New York, NY, USA. Retrieved from: https://about.bnef.com/new-energy-outlook.

Bodansky, D., 2013: The who, what, and wherefore of geoengineering governance. *Climatic Change*, **121(3)**, 539–551, doi:10.1007/s10584-013-0759-7.

Bodansky, D. and E. Diringer, 2014: *Alternative Models for the 2015 Climate Change Agreement*. FNI Climate Policy Perspectives 13, Fridtjof Nansen Institute (FNI), Lysaker, Norway, 1-8 pp.

Bodle, R., G. Homan, S. Schiele, and E. Tedsen, 2012: Part II: The Regulatory Framework for Climate-Related Geoengineering Relevant to the Convention on Biological Diversity. In: *Geoengineering in Relation to the Convention on Biological Diversity: Technical and Regulatory Matters*. CBD Technical Series No. 66, Secretariat of the Convention on Biological Diversity, Montreal, QC, Canada, pp. 99-145.

Bodnar, P. et al., 2018: Underwriting 1.5°C: competitive approaches to financing accelerated climate change mitigation. *Climate Policy*, **18(3)**, 368–382, doi:10.1080/14693062.2017.1389687.

Böhringer, C. and V. Alexeeva-Talebi, 2013: Unilateral Climate Policy and Competitiveness: Economic Implications of Differential Emission Pricing. *The World Economy*, **36(2)**, 121–154, doi:10.1111/j.1467-9701.2012.01470.x.

Böhringer, C., A. Löschel, U. Moslener, and T.F. Rutherford, 2009: EU climate policy up to 2020: An economic impact assessment. *Energy Economics*, **31**, S295 – S305, doi:10.1016/j.eneco.2009.09.009.

Bolderdijk, J.W., J. Knockaert, E.M. Steg, and E.T. Verhoef, 2011: Effects of Pay-As-You-Drive vehicle insurance on young drivers' speed choice: Results of a Dutch field experiment. *Accident Analysis & Prevention*, **43(3)**, 1181–1186, doi:10.1016/j.aap.2010.12.032.

Bolderdijk, J.W., M. Gorsira, K. Keizer, and L. Steg, 2013a: Values determine the (in)effectiveness of informational interventions in promoting pro-environmental behavior. *PLOS ONE*, **8(12)**, e83911, doi:10.1371/journal.pone.0083911.

Bolderdijk, J.W., L. Steg, E.S. Geller, P.K. Lehman, and T. Postmes, 2013b: Comparing the effectiveness of monetary versus moral motives in environmental campaigning. *Nature Climate Change*, **3(1)**, 1–4, doi:10.1038/nclimate1767.

Bond, T.C. et al., 2013: Bounding the role of black carbon in the climate system: A scientific assessment. *Journal of Geophysical Research: Atmospheres*, **118(11)**, 5380–5552, doi:10.1002/jgrd.50171.

Bonsch, M. et al., 2016: Trade-offs between land and water requirements for large-scale bioenergy production. *GCB Bioenergy*, **8(1)**, 11–24, doi:10.1111/gcbb.12226.

Boodoo, Z. and K.H. Olsen, 2017: Assessing transformational change potential: the case of the Tunisian cement Nationally Appropriate Mitigation Action (NAMA). *Climate Policy*, 1–19, doi:10.1080/14693062.2017.1386081.

Booth, M.S., 2018: Not carbon neutral: Assessing the net emissions impact of residues burned for bioenergy. *Environmental Research Letters*, **13(3)**, 035001, doi:10.1088/1748-9326/aaac88.

Bord, R.J., R.E. O'Connor, and A. Fisher, 2000: In what sense does the public need to understand global climate change? *Public Understanding of Science*, **9(3)**, 205–218, doi:10.1088/0963-6625/9/3/301.

Bos, J.J., R.R. Brown, and M.A. Farrelly, 2015: Building networks and coalitions to promote transformational change: Insights from an Australian urban water planning case study. *Environmental Innovation and Societal Transitions*, **15**, 11–25, doi:10.1016/j.eist.2014.10.002.

Bosomworth, K., A. Harwood, P. Leith, and P. Wallis, 2015: *Adaptation Pathways: a playbook for developing options for climate change adaptation in NRM*. Southern Slopes Climate Adaptation Research Partnership (SCARP): RMIT University, University of Tasmania, and Monash University, Australia, 26 pp., doi:978-1-86295-792-3.

Boucher, O. and G.A. Folberth, 2010: New Directions: Atmospheric methane removal as a way to mitigate climate change? *Atmospheric Environment*, **44(27)**, 3343–3345, doi:10.1016/j.atmosenv.2010.04.032.

Boulamanti, A. and J.A. Moya, 2017: *Energy efficiency and GHG emissions: Prospective scenarios for the Chemical and Petrochemical Industry*. 237 pp., doi:10.2760/20486.

Bouman, E.A., E. Lindstad, A.I. Rialland, and A.H. Strømman, 2017: State-of-the-art technologies, measures, and potential for reducing GHG emissions from shipping – A review. *Transportation Research Part D: Transport and Environment*, **52(Part A)**, 408–421, doi:10.1016/j.trd.2017.03.022.

Bours, D., C. McGinn, and P. Pringle (eds.), 2015: *Monitoring and Evaluation of Climate Change Adaptation: A Review of the Landscape: New Directions for Evaluation, Number 147*. John Wiley & Sons, Hoboken, NJ, USA, 160 pp.

Bovenberg, A.L., 1999: Green Tax Reforms and the Double Dividend: an Updated Reader's Guide. *International Tax and Public Finance*, **6(3)**, 421–443, doi:10.1023/a:1008715920337.

Bowen, K.J. et al., 2017: Implementing the "Sustainable Development Goals": towards addressing three key governance challenges – collective action, trade-offs, and accountability. *Current Opinion in Environmental Sustainability*, **26–27**, 90–96, doi:10.1016/j.cosust.2017.05.002.

Bows, A. and T. Smith, 2012: The (low-carbon) shipping forecast: opportunities on the high seas. *Carbon Management*, **3(6)**, 525–528, doi:10.4155/cmt.12.68.

Bows-Larkin, A., 2015: All adrift: aviation, shipping, and climate change policy. *Climate Policy*, **15(6)**, 681–702, doi:10.1080/14693062.2014.965125.

Boyd, P.W., 2016: Development of geopolitically relevant ranking criteria for geoengineering methods. *Earth's Future*, **4(11)**, 523–531, doi:10.1002/2016ef000447.

Boyd, P.W. and K.L. Denman, 2008: Implications of large-scale iron fertilization of the oceans. *Marine Ecology Progress Series*, **364**, 213–218, doi:10.3354/meps07541.

Boykoff, M.T., A. Ghoshi, and K. Venkateswaran, 2013: Media Coverage of Discourse on Adaptation: Competing Visions of "Success" in the Indian Context. In: *Successful Adaptation to Climate Change: Linking Science and Policy in a Rapidly Changing World* [Moser, S.C. and M.T. Boykoff (eds.)]. Routledge, Abingdon, UK and New York, NY, USA, pp. 237–252.

Boysen, L.R. et al., 2017: The limits to global-warming mitigation by terrestrial carbon removal. *Earth's Future*, **5(5)**, 463–474, doi:10.1002/2016ef000469.

Bradley, P., A. Coke, and M. Leach, 2016: Financial incentive approaches for reducing peak electricity demand, experience from pilot trials with a UK energy provider. *Energy Policy*, **98**, 108–120, doi:10.1016/j.enpol.2016.07.022.

Brambilla, A., G. Salvalai, M. Imperadori, and M.M. Sesana, 2018: Nearly zero energy building renovation: From energy efficiency to environmental efficiency, a pilot case study. *Energy and Buildings*, **166**, 271–283, doi:10.1016/j.enbuild.2018.02.002.

Brand, U. and A. von Gleich, 2015: Transformation toward a Secure and Precaution-Oriented Energy System with the Guiding Concept of Resilience – Implementation of Low-Exergy Solutions in Northwestern Germany. *Energies*, **8(7)**, 6995–7019, doi:10.3390/en8076995.

Branger, F. and P. Quirion, 2014: Climate policy and the 'carbon haven' effect. *Wiley Interdisciplinary Reviews: Climate Change*, **5(1)**, 53–71, doi:10.1002/wcc.245.

Branger, F., P. Quirion, and J. Chevallier, 2016: Carbon leakage and competitiveness of cement and steel industries under the EU ETS: Much ado about nothing. *Energy Journal*, **37(3)**, 109–135, doi:10.5547/01956574.37.3.fbra.

Braun, C., C. Merk, G. Pönitzsch, K. Rehdanz, and U. Schmidt, 2017: Public perception of climate engineering and carbon capture and storage in Germany: survey evidence. *Climate Policy*, **3062**, 1–14, doi:10.1080/14693062.2017.1304888.

Bredenkamp, H. and C. Pattillo, 2010: *Financing the Response to Climate Change*. IMF Station Position Note SPN10/06, International Monetary Fund (IMF), Washington DC, USA, 14 pp.

Bremer, J. and M.K. Linnenluecke, 2016: Determinants of the perceived importance of organisational adaptation to climate change in the Australian energy industry. *Australian Journal of Management*, **42(3)**, 502–521, doi:10.1177/0312896216672273.

Breyer, C. et al., 2017: On the role of solar photovoltaics in global energy transition scenarios. *Progress in Photovoltaics: Research and Applications*, **25(8)**, 727–745, doi:10.1002/pip.2885.

Bridgeman, B., 2017: Population growth underlies most other environmental problems: Comment on Clayton et al. (2016). *American Psychologist*, **72(4)**, 386–387, doi:10.1037/amp0000137.

Bright, R.M., K. Zhao, R.B. Jackson, and F. Cherubini, 2015: Quantifying surface albedo and other direct biogeophysical climate forcings of forestry activities. *Global Change Biology*, **21(9)**, 3246–3266, doi:10.1111/gcb.12951.

Briley, L., D. Brown, and S.E. Kalafatis, 2015: Overcoming barriers during the co-production of climate information for decision-making. *Climate Risk Management*, **9**, 41–49, doi:10.1016/j.crm.2015.04.004.

Bring, A. et al., 2016: Arctic terrestrial hydrology: A synthesis of processes, regional effects, and research challenges. *Journal of Geophysical Research: Biogeosciences*, **121(3)**, 621–649, doi:10.1002/2015jg003131.

Brink, E. et al., 2016: Cascades of green: A review of ecosystem-based adaptation in urban areas. *Global Environmental Change*, **36**, 111–123, doi:10.1016/j.gloenvcha.2015.11.003.

Brockington, J.D., I.M. Harris, and R.M. Brook, 2016: Beyond the project cycle: a medium-term evaluation of agroforestry adoption and diffusion in a south Indian village. *Agroforestry Systems*, **90(3)**, 489–508, doi:10.1007/s10457-015-9872-0.

Broehm, M., J. Strefler, and N. Bauer, 2015: *Techno-Economic Review of Direct Air Capture Systems for Large Scale Mitigation of Atmospheric CO₂*. Potsdam Institute for Climate Impact Research, Potsdam, Germany, 28 pp., doi:10.2139/ssrn.2665702.

Broekhoff, D., G. Piggot, and P. Erickson, 2018: *Building Thriving, Low-Carbon Cities: An Overview of Policy Options for National Governments*. Coalition for Urban Transitions, London, UK and Washington DC, USA, 124 pp.

Brondízio, E.S., A.C.B. de Lima, S. Schramski, and C. Adams, 2016: Social and health dimensions of climate change in the Amazon. *Annals of Human Biology*, **43(4)**, 405–414, doi:10.1080/03014460.2016.1193222.

Brooks, N. et al., 2013: *An operational framework for Tracking Adaptation and Measuring Development (TAMD)*. IIED Climate Change Working Paper Series No. 5, International Institute for Environment and Development (IIED), London and Edinburgh, UK, 39 pp.

Brooks, S.J., 2013: Avoiding the *Limits to Growth*: Gross National Happiness in Bhutan as a Model for Sustainable Development. *Sustainability*, **5(9)**, 3640–3664, doi:10.3390/su5093640.

Brosch, T., M.K. Patel, and D. Sander, 2014: Affective influences on energy-related decisions and behaviors. *Frontiers in Energy Research*, **2**, 1–12, doi:10.3389/fenrg.2014.00011.

Broto, V.C., 2017: Energy landscapes and urban trajectories towards sustainability. *Energy Policy*, **108**, 755–764, doi:10.1016/j.enpol.2017.01.009.

Brown, D. and G. McGranahan, 2016: The urban informal economy, local inclusion and achieving a global green transformation. *Habitat International*, **53**, 97–105, doi:10.1016/j.habitatint.2015.11.002.

Brown, M.A., G. Kim, A.M. Smith, and K. Southworth, 2017: Exploring the impact of energy efficiency as a carbon mitigation strategy in the U.S.. *Energy Policy*, **109**, 249–259, doi:10.1016/j.enpol.2017.06.044.

Brown, R.M., C.R. Dillon, J. Schieffer, and J.M. Shockley, 2016: The carbon footprint and economic impact of precision agriculture technology on a corn and soybean farm. *Journal of Environmental Economics and Policy*, **5(3)**, 335–348, doi:10.1080/21606544.2015.1090932.

BRTData, 2018: Key Indicators Per Region. World Resources Institute (WRI) Brasil Ross Center for Sustainable Cities. Retrieved from: https://brtdata.org.

Bruce, P., R. Catlow, and P. Edwards, 2010: Energy materials to combat climate change. *Philosophical Transactions of the Royal Society A: Mathematical, Physical and Engineering Sciences*, **368(1923)**, 3225, doi:10.1098/rsta.2010.0105.

Bruckner, T. et al., 2014: Energy Systems. In: *Climate Change 2014: Mitigation of Climate Change. Contribution of Working Group III to the Fifth Assessment Report of the Intergovernmental Panel on Climate Change* [Edenhofer, O., R. Pichs-Madruga, Y. Sokona, E. Farahani, S. Kadner, K. Seyboth, A. Adler, I. Baum, S. Brunner, P. Eickemeier, B. Kriemann, J. Savolainen, S. Schlömer, C. von Stechow, T. Zwickel, and J.C. Minx (eds.)]. Cambridge University Press, Cambridge, United Kingdom and New York, NY, USA, pp. 511–597.

Brüggemann, M., F. De Silva-Schmidt, I. Hoppe, D. Arlt, and J.B. Schmitt, 2017: The appeasement effect of a United Nations climate summit on the German public. *Nature Climate Change*, 1–7, doi:10.1038/nclimate3409.

Brügger, A., T.A. Morton, and S. Dessai, 2015: Hand in hand: Public endorsement of climate change mitigation and adaptation. *PLOS ONE*, **10(4)**, 1–18, doi:10.1371/journal.pone.0124843.

Brugnach, M., M. Craps, and A. Dewulf, 2017: Including indigenous peoples in climate change mitigation: addressing issues of scale, knowledge and power. *Climatic Change*, **140(1)**, 19–32, doi:10.1007/s10584-014-1280-3.

Brulle, R.J., J. Carmichael, and J.C. Jenkins, 2012: Shifting public opinion on climate change: an empirical assessment of factors influencing concern over climate change in the U.S., 2002–2010. *Climatic Change*, **114(2)**, 169–188, doi:10.1007/s10584-012-0403-y.

Brunes, E. and D. Ohlhorst, 2011: Wind Power Generation in Germany – a transdisciplinary view on the innovation biography. *The Journal of Transdisciplinary Environmental Studies*, **10(1)**, 45–67, http://journal-tes.dk/vol_10_no_1_page_17/no_5_elke_bruns.html.

Bryan, E., T.T. Deressa, G.A. Gbetibouo, and C. Ringler, 2009: Adaptation to climate change in Ethiopia and South Africa: options and constraints. *Environmental*

Science & Policy, **12**(4), 413–426, doi:10.1016/j.envsci.2008.11.002.

Bryan, E., Q. Bernier, M. Espinal, and C. Ringler, 2017: Making climate change adaptation programmes in sub-Saharan Africa more gender responsive: insights from implementing organizations on the barriers and opportunities. *Climate and Development*, 1–15, doi:10.1080/17565529.2017.1301870.

Brzoska, M. and C. Fröhlich, 2016: Climate change, migration and violent conflict: vulnerabilities, pathways and adaptation strategies. *Migration and Development*, **5**(2), 190–210, doi:10.1080/21632324.2015.1022973.

Buchholz, T., M.D. Hurteau, J. Gunn, and D. Saah, 2016: A global meta-analysis of forest bioenergy greenhouse gas emission accounting studies. *GCB Bioenergy*, **8**(2), 281–289, doi:10.1111/gcbb.12245.

Buck, H.J., 2012: Geoengineering: Re-making Climate for Profit or Humanitarian Intervention? *Development and Change*, **43**(1), 253–270, doi:10.1111/j.1467-7660.2011.01744.x.

Buck, H.J., 2016: Rapid scale-up of negative emissions technologies: social barriers and social implications. *Climatic Change*, **139**(2), 155–167, doi:10.1007/s10584-016-1770-6.

Buck, H.J., A.R. Gammon, and C.J. Preston, 2014: Gender and geoengineering. *Hypatia*, **29**(3), 651–669, doi:10.1111/hypa.12083.

Buckeridge, M.S., 2015: Árvores urbanas em São Paulo: planejamento, economia e água (in Portugese). *Estudos Avançados*, **29**(84), 85–101, doi:10.1590/s0103-40142015000200006.

Budnitz, R.J., 2016: Nuclear power: Status report and future prospects. *Energy Policy*, **96**, 735–739, doi:10.1016/j.enpol.2016.03.011.

Bui, M. et al., 2018: Carbon capture and storage (CCS): the way forward. *Energy & Environmental Science*, **11**(5), 1062–1176, doi:10.1039/c7ee02342a.

Bulkeley, H., 2005: Reconfiguring environmental governance: Towards a politics of scales and networks. *Political Geography*, **24**(8), 875–902, doi:10.1016/j.polgeo.2005.07.002.

Bulkeley, H. et al., 2011: The Role of Institutions, Governance, and Urban Planning for Mitigation and Adaptation. In: *Cities and Climate Change*. The World Bank, pp. 125–159, doi:10.1596/9780821384930_ch05.

Bulkeley, H. et al., 2012: Governing climate change transnationally: Assessing the evidence from a database of sixty initiatives. *Environment and Planning C: Government and Policy*, **30**(4), 591–612, doi:10.1068/c11126.

Bunce, A. and J. Ford, 2015: How is adaptation, resilience, and vulnerability research engaging with gender? *Environmental Research Letters*, **10**(10), 123003, doi:10.1088/1748-9326/10/12/123003.

Burch, S., C. Mitchell, M. Berbes-Blazquez, and J. Wandel, 2017: Tipping Toward Transformation: Progress, Patterns and Potential for Climate Change Adaptation in the Global South. *Journal of Extreme Events*, **04**(01), 1750003, doi:10.1142/s2345737617500038.

Burch, S. et al., 2016: Governing and accelerating transformative entrepreneurship: exploring the potential for small business innovation on urban sustainability transitions. *Current Opinion in Environmental Sustainability*, **22**, 26–32, doi:10.1016/j.cosust.2017.04.002.

Burney, J.A. and R.L. Naylor, 2012: Smallholder Irrigation as a Poverty Alleviation Tool in Sub-Saharan Africa. *World Development*, **40**(1), 110–123, doi:10.1016/j.worlddev.2011.05.007.

Burnham, M. and Z. Ma, 2017: Climate change adaptation: factors influencing Chinese smallholder farmers' perceived self-efficacy and adaptation intent. *Regional Environmental Change*, **17**(1), 171–186, doi:10.1007/s10113-016-0975-6.

Burns, W. and S. Nicholson, 2017: Bioenergy and carbon capture with storage (BECCS): the prospects and challenges of an emerging climate policy response. *Journal of Environmental Studies and Sciences*, **15**(2), 527–534, doi:10.1007/s13412-017-0445-6.

Burns, W.C.G., 2011: Climate Geoengineering: Solar Radiation Management and its Implications for Intergenerational Equity. *Stanford Journal of Law, Science & Policy*, **4**(1), 39–55, https://law.stanford.edu/publications/climate-geoengineering-solar-radiation-management-and-its-implications-for-intergenerational-equity.

Burt, A., B. Hughes, and G. Milante, 2014: *Eradicating Poverty in Fragile States: Prospects of Reaching The 'High-Hanging' Fruit by 2030*. WPS7002, World Bank, Washington DC, USA, 35 pp.

Bushee, B.J., 2001: Do Institutional Investors Prefer Near-Term Earnings over Long-Run Value? *Contemporary Accounting Research*, **18**(2), 207–246, doi:10.1506/j4gu-bwwh-8hme-le0x.

Bustamante, J.G., A.S. Rattner, and S. Garimella, 2016: Achieving near-water-cooled power plant performance with air-cooled condensers. *Applied Thermal Engineering*, **105**, 362–371, doi:10.1016/j.applthermaleng.2015.05.065.

Butler, C. and J. Adamowski, 2015: Empowering marginalized communities in water resources management: Addressing inequitable practices in Participatory Model Building. *Journal of Environmental Management*, **153**, 153–162, doi:10.1016/j.jenvman.2015.02.010.

Butler, J.R.A. et al., 2015: Integrating Top-Down and Bottom-Up Adaptation Planning to Build Adaptive Capacity: A Structured Learning Approach. *Coastal Management*, **43**(4), 346–364, doi:10.1080/08920753.2015.1046802.

Butler, J.R.A. et al., 2016: Scenario planning to leap-frog the Sustainable Development Goals: An adaptation pathways approach. *Climate Risk Management*, **12**, 83–99, doi:10.1016/j.crm.2015.11.003.

Butler, L. and K. Neuhoff, 2008: Comparison of feed-in tariff, quota and auction mechanisms to support wind power development. *Renewable Energy*, **33**(8), 1854–1867, doi:10.1016/j.renene.2007.10.008.

Cabeza, L.F. et al., 2013: Affordable construction towards sustainable buildings: review on embodied energy in building materials. *Current Opinion in Environmental Sustainability*, **5**(2), 229–236, doi:10.1016/j.cosust.2013.05.005.

Caldecott, B., 2017: Introduction to special issue: stranded assets and the environment. *Journal of Sustainable Finance & Investment*, **7**(1), 1–13, doi:10.1080/20430795.2016.1266748.

Caldeira, K. and G. Bala, 2017: Reflecting on 50 years of geoengineering research. *Earth's Future*, **5**(1), 10–17, doi:10.1002/2016ef000454.

Calel, R. and A. Dechezleprêtre, 2016: Environmental Policy and Directed Technological Change: Evidence from the European Carbon Market. *Review of Economics and Statistics*, **98**(1), 173–191, doi:10.1162/rest_a_00470.

Calvin, K. et al., 2014: Trade-offs of different land and bioenergy policies on the path to achieving climate targets. *Climatic Change*, **123**(3–4), 691–704, doi:10.1007/s10584-013-0897-y.

Campbell, T.H. and A.C. Kay, 2014: Solution aversion: On the relation between ideology and motivated disbelief. *Journal of Personality and Social Psychology*, **107**(5), 809–824, doi:10.1037/a0037963.

Campiglio, E., 2016: Beyond carbon pricing: The role of banking and monetary policy in financing the transition to a low-carbon economy. *Ecological Economics*, **121**, 220–230, doi:10.1016/j.ecolecon.2015.03.020.

Campos, I.S. et al., 2016: Climate adaptation, transitions, and socially innovative action-research approaches. *Ecology and Society*, **21**(1), art13, doi:10.5751/es-08059-210113.

Camps-Calvet, M., J. Langemeyer, L. Calvet-Mir, and E. Gómez-Baggethun, 2016: Ecosystem services provided by urban gardens in Barcelona, Spain: Insights for policy and planning. *Environmental Science & Policy*, **62**, 14–23, doi:10.1016/j.envsci.2016.01.007.

Canadell, J.G. and E.D. Schulze, 2014: Global potential of biospheric carbon management for climate mitigation. *Nature Communications*, **5**, 1–12, doi:10.1038/ncomms6282.

Canales, F.A., A. Beluco, and C.A.B. Mendes, 2015: A comparative study of a wind hydro hybrid system with water storage capacity: Conventional reservoir or pumped storage plant? *Journal of Energy Storage*, **4**, 96–105, doi:10.1016/j.est.2015.09.007.

Caney, S., 2012: Just Emissions. *Philosophy & Public Affairs*, **40**(4), 255–300, doi:10.1111/papa.12005.

Caney, S., 2014: Two Kinds of Climate Justice: Avoiding Harm and Sharing Burdens. *Journal of Political Philosophy*, **22**(2), 125–149, doi:10.1111/jopp.12030.

Cannon, T., 2015: Disasters, climate change and the significance of 'culture'. In: *Cultures and Disasters: Understanding Cultural Framings in Disaster Risk Reduction* [Krüger, F., G. Bankoff, T. Cannon, B. Orlowski, and E.L.F. Schipper (eds.)]. Routledge, Abingdon, UK and New York, NY, USA, pp. 88–106.

Capon, S.J. et al., 2013: Riparian Ecosystems in the 21st Century: Hotspots for Climate Change Adaptation? *Ecosystems*, **16**(3), 359–381, doi:10.1007/s10021-013-9656-1.

Carlsson-Kanyama, A. and A.D. González, 2009: Potential contributions of food consumption patterns to climate change. *The American Journal of Clinical Nutrition*, **89**(5), 1704S–1709S, doi:10.3945/ajcn.2009.26736aa.

Carney, M., 2016: Breaking the tragedy of the horizon climate change and financial stability. Speech given at Lloyd's of London, London, UK. Retrieved from: www.bankofengland.co.uk/publications/pages/speeches/2015/844.aspx.

Carr, W.A. et al., 2013: Public engagement on solar radiation management and why it needs to happen now. *Climatic Change*, **121**(3), 567–577,

4

doi:10.1007/s10584-013-0763-y.

Carreño, M.L. et al., 2017: Holistic Disaster Risk Evaluation for the Urban Risk Management Plan of Manizales, Colombia. *International Journal of Disaster Risk Science*, **8(3)**, 258–269, doi:10.1007/s13753-017-0136-7.

Carrico, A.R., H.B. Truelove, M.P. Vandenbergh, and D. Dana, 2015: Does learning about climate change adaptation change support for mitigation? *Journal of Environmental Psychology*, **41**, 19–29, doi:10.1016/j.jenvp.2014.10.009.

Carter, J.G. et al., 2015: Climate change and the city: Building capacity for urban adaptation. *Progress in Planning*, **95**, 1–66, doi:10.1016/j.progress.2013.08.001.

Cartwright, A., 2015: *Better Growth, Better Cities: Rethinking and Redirecting Urbanisation in Africa*. New Climate Economy, Washington DC, USA and London, UK, 44 pp.

Cartwright, A. et al., 2013: Economics of climate change adaptation at the local scale under conditions of uncertainty and resource constraints: the case of Durban, South Africa. *Environment and Urbanization*, **25(1)**, 139–156, doi:10.1177/0956247813477814.

Cass, N., G. Walker, and P. Devine-Wright, 2010: Good neighbours, public relations and bribes: The politics and perceptions of community benefit provision in renewable energy development in the UK. *Journal of Environmental Policy & Planning*, **12(3)**, 255–275, doi:10.1080/1523908x.2010.509558.

Castán Broto, V., 2017: Urban Governance and the Politics of Climate change. *World Development*, **93**, 1–15, doi:10.1016/j.worlddev.2016.12.031.

Castán Broto, V. and H. Bulkeley, 2013: A survey of urban climate change experiments in 100 cities. *Global Environmental Change*, **23(1)**, 92–102, doi:10.1016/j.gloenvcha.2012.07.005.

Cayuela, M.L. et al., 2014: Biochar's role in mitigating soil nitrous oxide emissions: A review and meta-analysis. *Agriculture, Ecosystems & Environment*, **191**, 5–16, doi:10.1016/j.agee.2013.10.009.

CBD, 2008: *Decision IX/16: Biodiversity and Climate Change. Decision Adopted by the Conference of the Partis to the Convention on Biological Diversity at its Ninth Meeting*. UNEP/CBD/COP/DEC/IX/16, Convention on Biological Diversity (CBD), 12 pp.

CBD, 2010: *Decision X/33: Biodiversity and climate change. Decision adopted by the Conference of the Parties to the Convention on Biological Diversity at its Tenth Meeting*. UNEP/CBD/COP/DEC/X/33, Convention on Biological Diversity (CBD), 9 pp.

CBS & GNH, 2016: *A Compass Towards a Just and Harmonious Society: 2015 GNH Survey Report*. Centre for Bhutan Studies & Gross National Happiness Research (CBS & GNH), Thimphu, Bhutan, 342 pp.

CCFLA, 2016: *Localizing Climate Finance: Mapping Gaps and Opportunities, Designing solutions*. Cities Climate Finance Leadership Alliance (CCFLA), 29 pp.

CCRIF, 2017: *Annual Report 2016–2017*. The Caribbean Catastrophe Risk Insurance Facility Segregated Portfolio Company (CCRIF SPC), Grand Cayman, Cayman Islands, 107 pp.

Chadburn, S.E. et al., 2017: An observation-based constraint on permafrost loss as a function of global warming. *Nature Climate Change*, **7(5)**, 340–344, doi:10.1038/nclimate3262.

Chaffin, B.C. and L.H. Gunderson, 2016: Emergence, institutionalization and renewal: Rhythms of adaptive governance in complex social-ecological systems. *Journal of Environmental Management*, **165**, 81–87, doi:10.1016/j.jenvman.2015.09.003.

Chaffin, B.C., H. Gosnell, and B.A. Cosens, 2014: A decade of adaptive governance scholarship: synthesis and future directions. *Ecology and Society*, **19(3)**, 56, doi:10.5751/es-06824-190356.

Challinor, A.J. et al., 2014: A meta-analysis of crop yield under climate change and adaptation. *Nature Climate Change*, **4(4)**, 287–291, doi:10.1038/nclimate2153.

Chambers, L.E. et al., 2017: A database for traditional knowledge of weather and climate in the Pacific. *Meteorological Applications*, **24(3)**, 491–502, doi:10.1002/met.1648.

Chandel, M.K., L.F. Pratson, and R.B. Jackson, 2011: The potential impacts of climate-change policy on freshwater use in thermoelectric power generation. *Energy Policy*, **39(10)**, 6234–6242, doi:10.1016/j.enpol.2011.07.022.

Chandel, S.S., A. Sharma, and B.M. Marwaha, 2016: Review of energy efficiency initiatives and regulations for residential buildings in India. *Renewable and Sustainable Energy Reviews*, **54**, 1443–1458, doi:10.1016/j.rser.2015.10.060.

Chang, N., M.V. Vasquez, C.F. Chen, S. Imen, and L. Mullon, 2015: Global nonlinear and nonstationary climate change effects on regional precipitation and forest phenology in Panama, Central America. *Hydrological Processes*, **29(3)**, 339–355, doi:10.1002/hyp.10151.

Chang'a, L.B., P.Z. Yanda, and J. Ngana, 2010: Indigenous knowledge in seasonal rainfall prediction in Tanzania: A case of the South-western Highland of Tanzania. *Journal of Geography and Regional Planning*, **3**, 66–72.

Chant, S., M. Klett-davies, and J. Ramalho, 2017: *Challenges and potential solutions for adolescent girls in urban settings: a rapid evidence review*. Gender & Adolescence: Global Evidence (GAGE), 47 pp.

Chapin, F.S., C.N. Knapp, T.J. Brinkman, R. Bronen, and P. Cochran, 2016: Community-empowered adaptation for self-reliance. *Current Opinion in Environmental Sustainability*, **19**, 67–75, doi:10.1016/j.cosust.2015.12.008.

Chatrchyan, A.M. et al., 2017: United States agricultural stakeholder views and decisions on climate change. *Wiley Interdisciplinary Reviews: Climate Change*, **8(5)**, e469, doi:10.1002/wcc.469.

Chelleri, L., J.J. Waters, M. Olazabal, and G. Minucci, 2015: Resilience trade-offs: addressing multiple scales and temporal aspects of urban resilience. *Environment and Urbanization*, **27(1)**, 181–198, doi:10.1177/0956247814550780.

Chen, C. and M. Tavoni, 2013: Direct air capture of CO_2 and climate stabilization: A model based assessment. *Climatic Change*, **118(1)**, 59–72, doi:10.1007/s10584-013-0714-7.

Chen, M.-F., 2015: An examination of the value-belief-norm theory model in predicting pro-environmental behaviour in Taiwan. *Asian Journal of Social Psychology*, **18**, 145–151, doi:10.1111/ajsp.12096.

Chen, S. and B. Chen, 2016: Urban energy–water nexus: A network perspective. *Applied Energy*, **184**, 905–914, doi:10.1016/j.apenergy.2016.03.042.

Chen, Y. and Y. Xin, 2017: Implications of geoengineering under the 1.5°C target: Analysis and policy suggestions. *Advances in Climate Change Research*, **7**, 1–7, doi:10.1016/j.accre.2017.05.003.

Cheng, V.K.M. and G.P. Hammond, 2017: Life-cycle energy densities and land-take requirements of various power generators: A UK perspective. *Journal of the Energy Institute*, **90(2)**, 201–213, doi:10.1016/j.joei.2016.02.003.

Cheshmehzangi, A., 2016: China's New-type Urbanisation Plan (NUP) and the Foreseeing Challenges for Decarbonization of Cities: A Review. *Energy Procedia*, **104(5)**, 146–152, doi:10.1016/j.egypro.2016.12.026.

Chichilnisky, G. and G. Heal (eds.), 2000: *Equity and Efficiency in Environmental Markets: Global Trade in Carbon Dioxide Emissions*. Columbia University Press, New York, NY, USA, 280 pp.

Child, M. and C. Breyer, 2017: Transition and transformation: A review of the concept of change in the progress towards future sustainable energy systems. *Energy Policy*, **107**, 11–26, doi:10.1016/j.enpol.2017.04.022.

Chiroleu-Assouline, M. and M. Fodha, 2011: Environmental Tax and the Distribution of Income among Heterogeneous Workers. *Annals of Economics and Statistics*, **103/104(103)**, 71–92, doi:10.2307/41615494.

Chitnis, M. and S. Sorrell, 2015: Living up to expectations: Estimating direct and indirect rebound effects for UK households. *Energy Economics*, **52**, S100–S116, doi:10.1016/j.eneco.2015.08.026.

Chou, W.-C. et al., 2015: Potential impacts of effluent from accelerated weathering of limestone on seawater carbon chemistry: A case study for the Hoping power plant in northeastern Taiwan. *Marine Chemistry*, **168**, 27–36, doi:10.1016/j.marchem.2014.10.008.

Christensen, C., M. Raynor, and R. McDonald, 2015: What is Disruptive Innovation? *Harvard Business Review*, **December**, 44–53.

Christoforidis, G.C., K.C. Chatzisavvas, S. Lazarou, and C. Parisses, 2013: Covenant of Mayors initiative – Public perception issues and barriers in Greece. *Energy Policy*, **60**, 643–655, doi:10.1016/j.enpol.2013.05.079.

Christophers, B., 2017: Climate Change and Financial Instability: Risk Disclosure and the Problematics of Neoliberal Governance. *Annals of the American Association of Geographers*, **107(5)**, 1108–1127, doi:10.1080/24694452.2017.1293502.

Chu, E., I. Anguelovski, and J.A. Carmin, 2016: Inclusive approaches to urban climate adaptation planning and implementation in the Global South. *Climate Policy*, **16(3)**, 372–392, doi:10.1080/14693062.2015.1019822.

Chu, E., I. Anguelovski, and D. Roberts, 2017: Climate adaptation as strategic urbanism: assessing opportunities and uncertainties for equity and inclusive development in cities. *Cities*, **60**, 378–387, doi:10.1016/j.cities.2016.10.016.

Chu, S., 2015: Car restraint policies and mileage in Singapore. *Transportation Research Part A: Policy and Practice*, **77**, 404–412, doi:10.1016/j.tra.2015.04.028.

Chum, H. et al., 2011: Bioenergy. In: *IPCC Special Report on Renewable Energy Sources and Climate Change Mitigation* [Edenhofer, O., R. Pichs-Madruga, Y. Sokona, K. Seyboth, P. Matschoss, S. Kadner, T. Zwickel, P. Eickemeier, G. Hansen, S. Schlömer, and C. von Stechow (eds.)]. Cambridge University Press, Cambridge, United Kingdom and New York, NY, USA, pp. 209–332.

4

Chung Tiam Fook, T., 2017: Transformational processes for community-focused adaptation and social change: a synthesis. *Climate and Development*, **9(1)**, 5–21, doi:10.1080/17565529.2015.1086294.

Ciplet, D., J.T. Roberts, and M.R. Khan, 2015: *Power in a Warming World: The New Global Politics of Climate Change and the Remaking of Environmental Inequality*. MIT Press, Cambridge, MA, USA, 342 pp.

CISL, 2015: *Unhedgeable Risk: How climate change sentiment impacts investment*. University of Cambridge Institute for Sustainability Leadership (CISL), Cambridge, UK, 64 pp.

Clack, C.T.M. et al., 2017: Evaluation of a proposal for reliable low-cost grid power with 100% wind, water, and solar. *Proceedings of the National Academy of Sciences*, **114(26)**, 6722–6727, doi:10.1073/pnas.1610381114.

Clark, M. and D. Tilman, 2017: Comparative analysis of environmental impacts of agricultural production systems, agricultural input efficiency, and food choice. *Environmental Research Letters*, **12(6)**, 064016, doi:10.1088/1748-9326/aa6cd5.

Clarke, L. et al., 2014: Assessing transformation pathways. In: *Climate Change 2014: Mitigation of Climate Change. Contribution of Working Group III to the Fifth Assessment Report of the Intergovernmental Panel on Climate Change* [Edenhofer, O., R. Pichs-Madruga, Y. Sokona, E. Farahani, S. Kadner, K. Seyboth, A. Adler, I. Baum, S. Brunner, P. Eickemeier, B. Kriemann, J. Savolainen, S. Schlömer, C. von Stechow, T. Zwickel, and J.C. Minx (eds.)]. Cambridge University Press, Cambridge, United Kingdom and New York, NY, USA, pp. 413–510.

Clayton, S. et al., 2015: Psychological research and global climate change. *Nature Climate Change*, **5(7)**, 640–646, doi:10.1038/nclimate2622.

Clayton, S. et al., 2017: Psychologists and the Problem of Population Growth: Reply to Bridgeman (2017). *American Psychologist*, **72(4)**, 388–389, doi:10.1037/amp0000152.

Clémençon, R., 2016: The two sides of the Paris climate agreement: Dismal failure or historic breakthrough? *Journal of Environment & Development*, **25(1)**, 3–24, doi:10.1177/1070496516631362.

Clemens, M., J. Rijke, A. Pathirana, J. Evers, and N. Hong Quan, 2015: Social learning for adaptation to climate change in developing countries: insights from Vietnam. *Journal of Water and Climate Change*, **8(4)**, 365–378, doi:10.2166/wcc.2015.004.

Clements, J., A. Ray, and G. Anderson, 2013: *The Value of Climate Services Across Economic and Public Sectors: A Review of Relevant Literature*. United States Agency for International Development (USAID), Washington DC, USA, 43 pp.

Cloutier, G. et al., 2015: Planning adaptation based on local actors' knowledge and participation: a climate governance experiment. *Climate Policy*, **15(4)**, 458–474, doi:10.1080/14693062.2014.937388.

Coady, D., I. Parry, L. Sears, and B. Shang, 2017: How Large Are Global Fossil Fuel Subsidies? *World Development*, **91**, 11–27, doi:10.1016/j.worlddev.2016.10.004.

Cochran, P. et al., 2013: Indigenous frameworks for observing and responding to climate change in Alaska. *Climatic Change*, **120(3)**, 557–567, doi:10.1007/s10584-013-0735-2.

Cochrani, I., R. Hubert, V. Marchal, and R. Youngman, 2014: *Public Financial Institutions and the Low-carbon Transition: Five Case Studies on Low-Carbon Infrastructure and Project Investment*. OECD Environment Working Papers No. 72, OECD Publishing, Paris, France, 93 pp., doi:10.1787/5jxt3rhpgn9t-en.

Coe, R., F. Sinclair, and E. Barrios, 2014: Scaling up agroforestry requires research 'in' rather than 'for' development. *Current Opinion in Environmental Sustainability*, **6(1)**, 73–77, doi:10.1016/j.cosust.2013.10.013.

Cole, D.H., 2015: Advantages of a polycentric approach to climate change policy. *Nature Climate Change*, **5(2)**, 114–118, doi:10.1038/nclimate2490.

Colenbrander, S. et al., 2017: Can low-carbon urban development be pro-poor? The case of Kolkata, India. *Environment and Urbanization*, **29(1)**, 139–158, doi:10.1177/0956247816677775.

Collas, L., R.E. Green, A. Ross, J.H. Wastell, and A. Balmford, 2017: Urban development, land sharing and land sparing: the importance of considering restoration. *Journal of Applied Ecology*, **54(6)**, 1865–1873, doi:10.1111/1365-2664.12908.

Colloff, M.J. et al., 2016a: Adaptation services and pathways for the management of temperate montane forests under transformational climate change. *Climatic Change*, **138(1–2)**, 267–282, doi:10.1007/s10584-016-1724-z.

Colloff, M.J. et al., 2016b: Adaptation services of floodplains and wetlands under transformational climate change. *Ecological Applications*, **26(4)**, 1003–1017, doi:10.1890/15-0848.

Colloff, M.J. et al., 2017: An integrative research framework for enabling transformative adaptation. *Environmental Science & Policy*, **68**, 87–96, doi:10.1016/j.envsci.2016.11.007.

Colman, A.M., T.W. Körner, O. Musy, and T. Tazdaït, 2011: Mutual support in games: Some properties of Berge equilibria. *Journal of Mathematical Psychology*, **55(2)**, 166–175, doi:10.1016/j.jmp.2011.02.001.

Combet, E., 2013: Fiscalité carbone et progrès social. Application au cas français., École des Hautes Études en Sciences Sociales (EHESS), Paris, France, 412 pp.

Combet, E., F. Ghersi, J.C. Hourcade, and D. Théry, 2010: Carbon Tax and Equity: The Importance of Policy Design. In: *Critical Issues In Environmental Taxation Volume VIII* [Dias Soares, C., J. Milne, H. Ashiabor, K. Deketelaere, and L. Kreiser (eds.)]. Oxford University Press, Oxford, UK, pp. 277–295.

Cooney, G., J. Littlefield, J. Marriott, and T.J. Skone, 2015: Evaluating the Climate Benefits of CO_2-Enhanced Oil Recovery Using Life Cycle Analysis. *Environmental Science & Technology*, **49(12)**, 7491–7500, doi:10.1021/acs.est.5b00700.

Cooper, J.A.G., M.C. O'Connor, and S. McIvor, 2016: Coastal defences versus coastal ecosystems: A regional appraisal. *Marine Policy*, doi:10.1016/j.marpol.2016.02.021.

Cooper-Searle, S., F. Livesey, and J.M. Allwood, 2018: Why are Material Efficiency Solutions a Limited Part of the Climate Policy Agenda? An application of the Multiple Streams Framework to UK policy on CO_2 emissions from cars. *Environmental Policy and Governance*, **28(1)**, 51–64, doi:10.1002/eet.1782.

Coq-Huelva, D., A. Higuchi, R. Alfalla-Luque, R. Burgos-Morán, and R. Arias-Gutiérrez, 2017: Co-Evolution and Bio-Social Construction: The Kichwa Agroforestry Systems (Chakras) in the Ecuadorian Amazonia. *Sustainability*, **9(11)**, 1920, doi:10.3390/su9101920.

Corfee-Morlot, J., I. Cochran, S. Hallegatte, and P.J. Teasdale, 2011: Multilevel risk governance and urban adaptation policy. *Climatic Change*, **104(1)**, 169–197, doi:10.1007/s10584-010-9980-9.

Corner, A. and A. Randall, 2011: Selling climate change? The limitations of social marketing as a strategy for climate change public engagement. *Global Environmental Change*, **21(3)**, 1005–1014, doi:10.1016/j.gloenvcha.2011.05.002.

Cortekar, J. and M. Groth, 2015: Adapting energy infrastructure to climate change – Is there a need for government interventions and legal obligations within the German "energiewende"? *Energy Procedia*, **73**, 12–17, doi:10.1016/j.egypro.2015.07.552.

Costa, D., P. Burlando, and C. Priadi, 2016: The importance of integrated solutions to flooding and water quality problems in the tropical megacity of Jakarta. *Sustainable Cities and Society*, **20**, 199–209, doi:10.1016/j.scs.2015.09.009.

Coulibaly, J.Y., B. Chiputwa, T. Nakelse, and G. Kundhlande, 2017: Adoption of agroforestry and the impact on household food security among farmers in Malawi. *Agricultural Systems*, **155**, 52–69, doi:10.1016/j.agsy.2017.03.017.

Courtois, P., R. Nessah, and T. Tazdaït, 2015: How to play games? Nash versus Berge Behaviour Rules. *Economics and Philosophy*, **31(1)**, 123–139, doi:10.1017/s026626711400042x.

Cowen, L. and B. Gatersleben, 2017: Testing for the size heuristic in householders' perceptions of energy consumption. *Journal of Environmental Psychology*, **54**, 103–115, doi:10.1016/j.jenvp.2017.10.002.

Cox, K., M. Renouf, A. Dargan, C. Turner, and D. Klein-Marcuschamer, 2014: Environmental life cycle assessment (LCA) of aviation biofuel from microalgae, *Pongamia pinnata*, and sugarcane molasses. *Biofuels, Bioproducts and Biorefining*, **8(4)**, 579–593, doi:10.1002/bbb.1488.

Craig, R.K. et al., 2017: Balancing stability and flexibility in adaptive governance: an analysis of tools available in U.S. environmental law. *Ecology and Society*, **22(2)**, 3, doi:10.5751/es-08983-220203.

Crassous, R., J.C. Hourcade, and O. Sassi, 2006: Endogenous Structural Change and Climate Targets Modeling Experiments with Imaclim-R. *The Energy Journal*, **27**, 259–276, www.jstor.org/stable/23297067.

Creutzig, F., E. Corbera, S. Bolwig, and C. Hunsberger, 2013: Integrating place-specific livelihood and equity outcomes into global assessments of bioenergy deployment. *Environmental Research Letters*, **8(3)**, 035047, doi:10.1088/1748-9326/8/3/035047.

Creutzig, F., G. Baiocchi, R. Bierkandt, P.-P. Pichler, and K.C. Seto, 2015a: Global typology of urban energy use and potentials for an urbanization mitigation

4

wedge. *Proceedings of the National Academy of Sciences*, **112(20)**, 6283–6288, doi:10.1073/pnas.1315545112.

Creutzig, F. et al., 2015b: Bioenergy and climate change mitigation: an assessment. *GCB Bioenergy*, **7(5)**, 916–944, doi:10.1111/gcbb.12205.

Creutzig, F. et al., 2017: The underestimated potential of solar energy to mitigate climate change. *Nature Energy*, **2**, 17140, doi:10.1038/nenergy.2017.140.

Creutzig, F. et al., 2018: Towards demand-side solutions for mitigating climate change. *Nature Climate Change*, **8(4)**, 268–271, doi:10.1038/s41558-018-0121-1.

Croci, E., B. Lucchitta, G. Janssens-Maenhout, S. Martelli, and T. Molteni, 2017: Urban CO_2 mitigation strategies under the Covenant of Mayors: An assessment of 124 European cities. *Journal of Cleaner Production*, **169**, 161–177, doi:10.1016/j.jclepro.2017.05.165.

Crook, J.A., L.S. Jackson, and P.M. Forster, 2016: Can increasing albedo of existing ship wakes reduce climate change? *Journal of Geophysical Research: Atmospheres*, **121(4)**, 1549–1558, doi:10.1002/2015jd024201.

Crook, J.A., L.S. Jackson, S.M. Osprey, and P.M. Forster, 2015: A comparison of temperature and precipitation responses to different Earth radiation management geoengineering schemes. *Journal of Geophysical Research: Atmospheres*, **120(18)**, 9352–9373, doi:10.1002/2015jd023269.

Cui, X. et al., 2016: Radiative absorption enhancement from coatings on black carbon aerosols. *Science of The Total Environment*, **551–552**, 51–56, doi:10.1016/j.scitotenv.2016.02.026.

Cui, Z. et al., 2018: Pursuing sustainable productivity with millions of smallholder farmers. *Nature*, **555(7696)**, 363–366, doi:10.1038/nature25785.

Culwick, C. and K. Bobbins, 2016: *A Framework for a Green Infrastructure Planning Approach in the Gauteng City-Region*. GCRO Research Report No. 04, Gauteng City-Region Observatory (GCRO), Johannesburg, South Africa, 132 pp.

Cunningham, S.C. et al., 2015: Balancing the environmental benefits of reforestation in agricultural regions. *Perspectives in Plant Ecology, Evolution and Systematics*, **17(4)**, 301–317, doi:10.1016/j.ppees.2015.06.001.

Cunsolo, A. and N.R. Ellis, 2018: Ecological grief as a mental health response to climate change-related loss. *Nature Climate Change*, **8(4)**, 275–281, doi:10.1038/s41558-018-0092-2.

Cunsolo Willox, A., S.L. Harper, and V.L. Edge, 2013: Storytelling in a digital age: digital storytelling as an emerging narrative method for preserving and promoting indigenous oral wisdom. *Qualitative Research*, **13(2)**, 127–147, doi:10.1177/1468794112446105.

Cunsolo Willox, A. et al., 2012: "From this place and of this place:" Climate change, sense of place, and health in Nunatsiavut, Canada. *Social Science & Medicine*, **75(3)**, 538–547, doi:10.1016/j.socscimed.2012.03.043.

D'Agostino, D., 2015: Assessment of the progress towards the establishment of definitions of Nearly Zero Energy Buildings (nZEBs) in European Member States. *Journal of Building Engineering*, **1**, 20–32, doi:10.1016/j.jobe.2015.01.002.

D'Agostino, D. and D. Parker, 2018: A framework for the cost-optimal design of nearly zero energy buildings (NZEBs) in representative climates across Europe. *Energy*, **149**, 814–829, doi:10.1016/j.energy.2018.02.020.

Daamen, D.D.L., H. Staats, H.A.M. Wilke, and M. Engelen, 2001: Improving Environmental Behavior in Companies. *Environment and Behavior*, **33(2)**, 229–248, doi:10.1177/00139160121972963.

Dagnet, Y. et al., 2016: *Staying on Track from Paris: Advancing the Key Elements of the Paris Agreement*. World Resources Institute, Washington DC, USA.

Dahlmann, K. et al., 2016: Climate-Compatible Air Transport System – Climate Impact Mitigation Potential for Actual and Future Aircraft. *Aerospace*, **3(4)**, 38, doi:10.3390/aerospace3040038.

Daioglou, V. et al., 2017: Greenhouse gas emission curves for advanced biofuel supply chains. *Nature Climate Change*, **7(12)**, 920–924, doi:10.1038/s41558-017-0006-8.

DaMatta, F.M., A. Grandis, B.C. Arenque, and M.S. Buckeridge, 2010: Impacts of climate changes on crop physiology and food quality. *Food Research International*, **43(7)**, 1814–1823, doi:10.1016/j.foodres.2009.11.001.

Dang, H., E. Li, I. Nuberg, and J. Bruwer, 2014: Understanding farmers' adaptation intention to climate change: A structural equation modelling study in the Mekong Delta, Vietnam. *Environmental Science & Policy*, **41**, 11–22, doi:10.1016/j.envsci.2014.04.002.

Dang Phan, T.-H., R. Brouwer, and M. Davidson, 2014: The economic costs of avoided deforestation in the developing world: A meta-analysis. *Journal of Forest Economics*, **20(1)**, 1–16, doi:10.1016/j.jfe.2013.06.004.

Darela, J.P., D.M. Lapola, R.R. Torres, and M.C. Lemos, 2016: Socio-climatic hotspots in Brazil: how do changes driven by the new set of IPCC climatic projections affect their relevance for policy? *Climatic Change*, **136(3–4)**, 413–425, doi:10.1007/s10584-016-1635-z.

Daron, J.D., K. Sutherland, C. Jack, and B.C. Hewitson, 2015: The role of regional climate projections in managing complex socio-ecological systems. *Regional Environmental Change*, **15(1)**, 1–12, doi:10.1007/s10113-014-0631-y.

Davidson, D., 2016: Gaps in agricultural climate adaptation research. *Nature Climate Change*, **6(5)**, 433–435, doi:10.1038/nclimate3007.

Davidson, P., C. Burgoyne, H. Hunt, and M. Causier, 2012: Lifting options for stratospheric aerosol geoengineering: advantages of tethered balloon systems. *Philosophical Transactions of the Royal Society A: Mathematical, Physical and Engineering Sciences*, 4263–4300, doi:10.1098/rsta.2011.0639.

Davin, E.L., S.I. Seneviratne, P. Ciais, A. Olioso, and T. Wang, 2014: Preferential cooling of hot extremes from cropland albedo management. *Proceedings of the National Academy of Sciences*, **111(27)**, 9757–9761, doi:10.1073/pnas.1317323111.

Dawson, J., E.J. Stewart, M.E. Johnston, and C.J. Lemieux, 2016: Identifying and evaluating adaptation strategies for cruise tourism in Arctic Canada. *Journal of Sustainable Tourism*, **24(10)**, 1425–1441, doi:10.1080/09669582.2015.1125358.

de Boer, J., W.J.W. Botzen, and T. Terpstra, 2016a: Flood risk and climate change in the Rotterdam area, The Netherlands: enhancing citizen's climate risk perceptions and prevention responses despite skepticism. *Regional Environmental Change*, **16(6)**, 1613–1622, doi:10.1007/s10113-015-0900-4.

de Boer, J., A. de Witt, and H. Aiking, 2016b: Help the climate, change your diet: A cross-sectional study on how to involve consumers in a transition to a low-carbon society. *Appetite*, **98**, 19–27, doi:10.1016/j.appet.2015.12.001.

de Coninck, H.C. and S.M. Benson, 2014: Carbon Dioxide Capture and Storage: Issues and Prospects. *Annual Review of Environment and Resources*, **39**, 243–70, doi:10.1146/annurev-environ-032112-095222.

de Coninck, H.C. and D. Puig, 2015: Assessing climate change mitigation technology interventions by international institutions. *Climatic Change*, **131(3)**, 417–433, doi:10.1007/s10584-015-1344-z.

de Coninck, H.C. and A. Sagar, 2017: Technology Development and Transfer (Article 10). In: *The Paris Agreement on Climate Change* [Klein, D., M.P. Carazo, M. Doelle, J. Bulmer, and A. Higham (eds.)]. Oxford University Press, Oxford, UK, pp. 258–276.

de Freitas, L.C. and S. Kaneko, 2011: Ethanol demand under the flex-fuel technology regime in Brazil. *Energy Economics*, **33(6)**, 1146–1154, doi:10.1016/j.eneco.2011.03.011.

de Gouvello, C. and I. Zelenko, 2010: *A Financing Facility for Low-Carbon Development in Developing Countries*. World Bank Working Papers 203, The World Bank, Washington DC, USA, doi:10.1596/978-0-8213-8521-0.

de Haan, P., M.G. Mueller, and R.W. Scholz, 2009: How much do incentives affect car purchase? Agent-based microsimulation of consumer choice of new cars – Part II: Forecasting effects of feebates based on energy-efficiency. *Energy Policy*, **37(3)**, 1083–1094, doi:10.1016/j.enpol.2008.11.003.

de Jong, S. et al., 2017: Life-cycle analysis of greenhouse gas emissions from renewable jet fuel production. *Biotechnology for Biofuels*, **10(1)**, 64, doi:10.1186/s13068-017-0739-7.

de Richter, R., T. Ming, P. Davies, W. Liu, and S. Caillol, 2017: Removal of non-CO_2 greenhouse gases by large-scale atmospheric solar photocatalysis. *Progress in Energy and Combustion Science*, **60**, 68–96, doi:10.1016/j.pecs.2017.01.001.

De Souza, A.P., A. Grandis, D.C.C. Leite, and M.S. Buckeridge, 2014: Sugarcane as a Bioenergy Source: History, Performance, and Perspectives for Second-Generation Bioethanol. *Bioenergy Research*, **7(1)**, 24–35, doi:10.1007/s12155-013-9366-8.

De Souza, A.P., B.C. Arenque, E.Q.P. Tavares, and M.S. Buckeridge, 2016: Transcriptomics and Genetics Associated with Plant Responses to Elevated CO_2 Atmospheric Concentrations. In: *Plant Genomics and Climate Change* [Edwards, D. and J. Batley (eds.)]. Springer New York, New York, NY, USA, pp. 67–83, doi:10.1007/978-1-4939-3536-9_4.

De Souza, A.P., J.-C. Cocuron, A.C. Garcia, A.P. Alonso, and M.S. Buckeridge, 2015: Changes in Whole-Plant Metabolism during the Grain-Filling Stage in Sorghum Grown under Elevated CO_2 and Drought. *Plant physiology*, **169(3)**, 1755–65, doi:10.1104/pp.15.01054.

4

De Stefano, A. and M.G. Jacobson, 2018: Soil carbon sequestration in agroforestry systems: a meta-analysis. *Agroforestry Systems*, **92(2)**, 285–299, doi:10.1007/s10457-017-0147-9.

de Wit, M. and A. Faaij, 2010: European biomass resource potential and costs. *Biomass and Bioenergy*, **34(2)**, 188–202, doi:10.1016/j.biombioe.2009.07.011.

Deaton, A., 2013: *The Great Escape Health, Wealth, and the Origins of Inequality*. Princeton University Press, Princeton, NJ, USA, 376 pp.

Déau, T. and J. Touati, 2017: Financing Sustainable Infrastructure. In: *Coping with the Climate Crisis Mitigation Policies and Global Coordination* [Arezki, R., P. Bolton, K. Aynaoui, and M. Obstfeld (eds.)]. Columbia University Press, New York, NY, USA, pp. 167–178.

DeBeer, C.M., H.S. Wheater, S.K. Carey, and K.P. Chun, 2016: Recent climatic, cryospheric, and hydrological changes over the interior of western Canada: A review and synthesis. *Hydrology and Earth System Sciences*, **20(4)**, 1573–1598, doi:10.5194/hess-20-1573-2016.

Decanio, S.J., 1993: Barriers within firms to energy- efficient investments. *Energy Policy*, **21(9)**, 906–914, doi:10.1016/0301-4215(93)90178-j.

DeCicco, J.M. et al., 2016: Carbon balance effects of U.S. biofuel production and use. *Climatic Change*, **138(3–4)**, 667–680, doi:10.1007/s10584-016-1764-4.

del Ninno, C., S. Coll-Black, and P. Fallavier, 2016: Social Protection: Building Resilience Among the Poor and Protecting the Most Vulnerable. In: *Confronting Drought in Africa's Drylands: Opportunities for Enhancing Resilience*. The World Bank, Washington DC, USA, pp. 165–184, doi:10.1596/978-1-4648-0817-3_ch10.

Delmas, M.A., M. Fischlein, and O.I. Asensio, 2013: Information strategies and energy conservation behavior: A meta-analysis of experimental studies from 1975 to 2012. *Energy Policy*, **61**, 729–739, doi:10.1016/j.enpol.2013.05.109.

Demailly, D. and P. Quirion, 2008: European Emission Trading Scheme and competitiveness: A case study on the iron and steel industry. *Energy Economics*, **30(4)**, 2009–2027, doi:10.1016/j.eneco.2007.01.020.

Demski, C., C. Butler, K.A. Parkhill, A. Spence, and N.F. Pidgeon, 2015: Public values for energy system change. *Global Environmental Change*, **34**, 59–69, doi:10.1016/j.gloenvcha.2015.06.014.

Demski, C., S. Capstick, N. Pidgeon, N. Frank, and A. Spence, 2017: Experience of extreme weather affects climate change mitigation and adaptation responses. *Climatic Change*, **140(2)**, 149–1164, doi:10.1007/s10584-016-1837-4.

Demuzere, M. et al., 2014: Mitigating and adapting to climate change: Multi-functional and multi-scale assessment of green urban infrastructure. *Journal of Environmental Management*, **146**, 107–115, doi:10.1016/j.jenvman.2014.07.025.

den Elzen, M. et al., 2016: Contribution of the G20 economies to the global impact of the Paris agreement climate proposals. *Climatic Change*, **137(3–4)**, 655–665, doi:10.1007/s10584-016-1700-7.

Deng, X. and C. Zhao, 2015: Identification of Water Scarcity and Providing Solutions for Adapting to Climate Changes in the Heihe River Basin of China. *Advances in Meteorology*, **2015**, 1–13, doi:10.1155/2015/279173.

Deo, R.C., 2011: On meteorological droughts in tropical Pacific Islands: Time-series analysis of observed rainfall using Fiji as a case study. *Meteorological Applications*, **18(2)**, 171–180, doi:10.1002/met.216.

Devereux, S., 2016: Social protection for enhanced food security in sub-Saharan Africa. *Food Policy*, **60**, 52–62, doi:10.1016/j.foodpol.2015.03.009.

Devine-Wright, P., 2009: Rethinking NIMBYism: The role of place attachment and place identity in explaining and supporting place-protective action. *Journal of Community & Applied Social Psychology*, **19(6)**, 426–441, doi:10.1002/casp.1004.

Devine-Wright, P., 2013: Think global, act local? The relevance of place attachments and place identities in a climate changed world. *Global Environmental Change*, **23(1)**, 61–69, doi:10.1016/j.gloenvcha.2012.08.003.

Devine-wright, P. and Y. Howes, 2010: Disruption to place attachment and the protection of restorative environments: A wind energy case study. *Journal of Environmental Psychology*, **30(3)**, 271–280, doi:10.1016/j.jenvp.2010.01.008.

Dhar, S., M. Pathak, and P.R. Shukla, 2017: Electric vehicles and India's low carbon passenger transport: a long-term co-benefits assessment. *Journal of Cleaner Production*, **146**, 139–148, doi:10.1016/j.jclepro.2016.05.111.

Dhar, T.K. and L. Khirfan, 2017: Climate change adaptation in the urban planning and design research: missing links and research agenda. *Journal of Environmental Planning and Management*, **60(4)**, 602–627, doi:10.1080/09640568.2016.1178107.

Diekmann, A. and P. Preisendörfer, 2003: Rationality and Society. *Rational*, **15**, 441–472, doi:10.1177/1043463103154002.

Dietz, T., 2013: Bringing values and deliberation to science communication. *Proceedings of the National Academy of Sciences*, **110**, 14081–14087, doi:10.1073/pnas.1212740110.

Dietz, T., A. Dan, and R. Shwom, 2007: Support for Climate Change Policy: Social Psychological and Social Structural Influences. *Rural Sociology*, **72(2)**, 185–214, doi:10.1526/003601107781170026.

Dietz, T., P.C. Stern, and E.U. Weber, 2013: Reducing Carbon-Based Energy Consumption through Changes in Household Behavior. *Daedalus*, **142(1)**, 78–89, doi:10.1162/daed_a_00186.

Dietz, T., G.T. Gardner, J. Gilligan, P.C. Stern, and M.P. Vandenbergh, 2009: Household actions can provide a behavioral wedge to rapidly reduce US carbon emissions. *Proceedings of the National Academy of Sciences*, **106(44)**, 18452–18456, doi:10.1073/pnas.0908738106.

Dietz, T., K.A. Frank, C.T. Whitley, J. Kelly, and R. Kelly, 2015: Political influences on greenhouse gas emissions from US states. *Proceedings of the National Academy of Sciences*, **112(27)**, 8254–8259, doi:10.1073/pnas.1417806112.

Dilling, L. and R. Hauser, 2013: Governing geoengineering research: Why, when and how? *Climatic Change*, **121(3)**, 553–565, doi:10.1007/s10584-013-0835-z.

Ding, D., E.W. Maibach, X. Zhao, C. Roser-Renouf, and A. Leiserowitz, 2011: Support for climate policy and societal action are linked to perceptions about scientific agreement. *Nature Climate Change*, **1(9)**, 462–466, doi:10.1038/nclimate1295.

Dinner, I., E.J. Johnson, D.G. Goldstein, and K. Liu, 2011: Partitioning default effects: Why people choose not to choose. *Journal of Experimental Psychology: Applied*, **17(4)**, 432–432, doi:10.1037/a0026470.

Dircke, P. and A. Molenaar, 2015: Climate change adaptation; innovative tools and strategies in Delta City Rotterdam. *Water Practice and Technology*, **10(4)**, 674–680, doi:10.2166/wpt.2015.080.

Dobson, S., 2017: Community-driven pathways for implementation of global urban resilience goals in Africa. *International Journal of Disaster Risk Reduction*, **26**, 78–84, doi:10.1016/j.ijdrr.2017.09.028.

Dóci, G. and E. Vasileiadou, 2015: "Let's do it ourselves" Individual motivations for investing in renewables at community level. *Renewable and Sustainable Energy Reviews*, **49**, 41–50, doi:10.1016/j.rser.2015.04.051.

Dodman, D., 2009: Blaming cities for climate change? An analysis of urban greenhouse gas emissions inventories. *Environment and Urbanization*, **21(1)**, 185–201, doi:10.1177/0956247809103016.

Dodman, D., S. Colenbrander, and D. Archer, 2017a: Conclusion: towards adaptive urban governance. In: *Responding to climate change in Asian cities: Governance for a more resilient urban future* [Archer, D., S. Colenbrander, and D. Dodman (eds.)]. Routledge Earthscan, Abingdon, UK and New York, NY, USA, pp. 200–217.

Dodman, D., H. Leck, M. Rusca, and S. Colenbrander, 2017b: African Urbanisation and Urbanism: Implications for risk accumulation and reduction. *International Journal of Disaster Risk Reduction*, **26**, 7–15, doi:10.1016/j.ijdrr.2017.06.029.

Dogan, E., J.W. Bolderdijk, and L. Steg, 2014: Making Small Numbers Count: Environmental and Financial Feedback in Promoting Eco-driving Behaviours. *Journal of Consumer Policy*, **37(3)**, 413–422, doi:10.1007/s10603-014-9259-z.

Donner, S.D., M. Kandlikar, and S. Webber, 2016: Measuring and tracking the flow of climate change adaptation aid to the developing world. *Environmental Research Letters*, **11(5)**, 054006, doi:10.1088/1748-9326/11/5/054006.

Dooley, K., 2017: Routines, Rigidity and Real Estate: Organisational Innovations in the Workplace. *Sustainability*, **9(6)**, 998, doi:10.3390/su9060998.

Dorsch, M.J. and C. Flachsland, 2017: A Polycentric Approach to Global Climate Governance. *Global Environmental Politics*, **17(2)**, 45–64, doi:10.1162/glep_a_00400.

Dorward, P., G. Clarkson, and R. Stern, 2015: *Participatory integrated climate services for agriculture (PICSA): Field manual*. Walker Institute, University of Reading, Reading, UK, 65 pp.

Dougill, A.J. et al., 2017: Mainstreaming conservation agriculture in Malawi: Knowledge gaps and institutional barriers. *Journal of Environmental Management*, **195**, 25–34, doi:10.1016/j.jenvman.2016.09.076.

Douglas, I., 2017: Flooding in African cities, scales of causes, teleconnections, risks, vulnerability and impacts. *International Journal of Disaster Risk Reduction*, **26**, 34–42, doi:10.1016/j.ijdrr.2017.09.024.

Dowd, A.-M. et al., 2014: The role of networks in transforming Australian agriculture. *Nature Climate Change*, **4(7)**, 558–563, doi:10.1038/nclimate2275.

Doyle, J., 2011: Acclimatizing nuclear? Climate change, nuclear power and the reframing of risk in the UK news media. *International Communication Gazette*, **73(1–2)**, 107–125, doi:10.1177/1748048510386744.

Drews, S. and J.C.J.M. Van den Bergh, 2016: What explains public support for climate policies ? A review of empirical and experimental studies review of empirical and experimental studies. *Climate Policy*, **16**(7), 855–876, doi:10.1080/14693062.2015.1058240.

Drèze, J. and N. Stern, 1990: Policy Reform, Shadow Prices, and Market Prices. *Journal of Public Economics*, **42**, 1–45, doi:10.1016/0047-2727(90)90042-g.

Droste, N. et al., 2016: Steering innovations towards a green economy: Understanding government intervention. *Journal of Cleaner Production*, **135**, 426–434, doi:10.1016/j.jclepro.2016.06.123.

Dubois, U., 2012: From targeting to implementation: The role of identification of fuel poor households. *Energy Policy*, **49**, 107–115, doi:10.1016/j.enpol.2011.11.087.

Dunlap, R.E. and A.M. McCright, 2011: Organized Climate Change Denial. In: *The Oxford Handbook of Climate Change and Society* [Dryzek, J.S., R.B. Norgaard, and D. Schlosberg (eds.)]. Oxford University Press, Oxford, UK, pp. 144–160, doi:10.1093/oxfordhb/9780199566600.001.0001.

Dupuis, J. and R. Biesbroek, 2013: Comparing apples and oranges: The dependent variable problem in comparing and evaluating climate change adaptation policies. *Global Environmental Change*, **23**(6), 1476–1487, doi:10.1016/j.gloenvcha.2013.07.022.

Durand, A. et al., 2016: *Financing Options for Loss and Damage: a Review and Roadmap*. Deutsches Institut für Entwicklungspolitik gGmbH, Bonn, Germany, 41 pp.

Eakin, H. et al., 2015: Information and communication technologies and climate change adaptation in Latin America and the Caribbean: a framework for action. *Climate and Development*, **7**(3), 208–222, doi:10.1080/17565529.2014.951021.

Eakin, H. et al., 2016: Cognitive and institutional influences on farmers' adaptive capacity: insights into barriers and opportunities for transformative change in central Arizona. *Regional Environmental Change*, **16**(3), 801–814, doi:10.1007/s10113-015-0789-y.

Eakin, H.C.C., M.C.C. Lemos, and D.R.R. Nelson, 2014: Differentiating capacities as a means to sustainable climate change adaptation. *Global Environmental Change*, **27**(1), 1–8, doi:10.1016/j.gloenvcha.2014.04.013.

Eastham, S.D., D.W. Keith, and S.R.H. Barrett, 2018: Mortality tradeoff between air quality and skin cancer from changes in stratospheric ozone. *Environmental Research Letters*, **13**(3), 034035, doi:10.1088/1748-9326/aaad2e.

Ebeling, F. and S. Lotz, 2015: Domestic uptake of green energy promoted by opt-out tariffs. *Nature Climate Change*, **5**, 868–871, doi:10.1038/nclimate2681.

Eberhard, A., O. Rosnes, M. Shkaratan, and H. Vennemo, 2011: *Africa's Power Infrastructure: Investment, Integration, Efficiency*. The World Bank, Washington DC, USA, 352 pp., doi:10.1596/978-0-8213-8455-8.

Eberhard, A., K. Gratwick, E. Morella, and P. Antmann, 2016: *Independent Power Projects in Sub-Saharan Africa: Lessons from Five Key Countries*. The World Bank, Washington DC, USA, 382 pp., doi:10.1596/978-1-4648-0800-5.

Eberhard, A., K. Gratwick, E. Morella, and P. Antmann, 2017: Accelerating investments in power in sub-Saharan Africa. *Nature Energy*, **2**(2), 17005, doi:10.1038/nenergy.2017.5.

Ecker, F., U.J.J. Hahnel, and H. Spada, 2017: Promoting Decentralized Sustainable Energy Systems in Different Supply Scenarios: The Role of Autarky Aspiration. *Frontiers in Energy Research*, **5**, 14, doi:10.3389/fenrg.2017.00014.

Edenhofer, O. et al., 2015: Closing the emission price gap. *Global Environmental Change*, **31**, 132–143, doi:10.1016/j.gloenvcha.2015.01.003.

Edler, J. and J. Fagerberg, 2017: Innovation policy: what, why, and how. *Oxford Review of Economic Policy*, **33**(1), 2–23, doi:10.1093/oxrep/grx001.

EEA, 2017: *Aviation and shipping – impacts on Europe's environment: TERM 2017: Transport and Environment Reporting Mechanism (TERM) report*. European Environment Agency (EEA), Copenhagen, 70 pp.

Eisenack, K. and R. Stecker, 2012: A framework for analyzing climate change adaptations as actions. *Mitigation and Adaptation Strategies for Global Change*, **17**(3), 243–260, doi:10.1007/s11027-011-9323-9.

Eisenberg, D.A., 2016: Transforming building regulatory systems to address climate change. *Building Research & Information*, **44**(5–6), 468–473, doi:10.1080/09613218.2016.1126943.

Ek, K. and P. Söderholm, 2008: Households' switching behavior between electricity suppliers in Sweden. *Utilities Policy*, **16**(4), 254–261, doi:10.1016/j.jup.2008.04.005.

Ek, K. and P. Söderholm, 2010: The devil is in the details: Household electricity saving behavior and the role of information. *Energy Policy*, **38**(3), 1578–1587, doi:10.1016/j.enpol.2009.11.041.

Ekblom, A., L. Gillson, and M. Notelid, 2017: Water flow, ecological dynamics, and management in the lower Limpopo Valley: a long-term view. *Wiley Interdisciplinary Reviews: Water*, **4**(5), e1228, doi:10.1002/wat2.1228.

El Gafy, I., N. Grigg, and W. Reagan, 2017: Dynamic Behaviour of the Water-Food-Energy Nexus: Focus on Crop Production and Consumption. *Irrigation and Drainage*, **66**(1), 19–33, doi:10.1002/ird.2060.

Elia, E.F., S. Mutula, and C. Stilwell, 2014: Indigenous Knowledge use in seasonal weather forecasting in Tanzania : the case of semi-arid central Tanzania. *South African Journal of Libraries and Information Science*, **80**(1), 18–27, doi:10.7553/80-1-180.

Eliasson, J., 2014: The role of attitude structures, direct experience and reframing for the success of congestion pricing. *Transportation Research Part A: Policy and Practice*, **67**, 81–95, doi:10.1016/j.tra.2014.06.007.

Eliseev, A., 2012: Climate change mitigation via sulfate injection to the stratosphere: impact on the global carbon cycle and terrestrial biosphere. *Atmospheric and Oceanic Optics*, **25**(6), 405–413, doi:10.1134/s1024856012060024.

Elliott, J. et al., 2014: Constraints and potentials of future irrigation water availability on agricultural production under climate change. *Proceedings of the National Academy of Sciences*, **111**(9), 3239–3244, doi:10.1073/pnas.1222474110.

Ellison, D. et al., 2017: Trees, forests and water: Cool insights for a hot world. *Global Environmental Change*, **43**, 51–61, doi:10.1016/j.gloenvcha.2017.01.002.

Elmqvist, T. et al., 2013: *Urbanization, Biodiversity and Ecosystem Services: Challenges and Opportunities: A Global Assessment*. Springer Netherlands, Dordrecht, The Netherlands, 755 pp.

Elmqvist, T. et al., 2015: Benefits of restoring ecosystem services in urban areas. *Current Opinion in Environmental Sustainability*, **14**, 101–108, doi:10.1016/j.cosust.2015.05.001.

Elshout, P.M.F., R. van Zelm, R. Karuppiah, I.J. Laurenzi, and M.A.J. Huijbregts, 2014: A spatially explicit data-driven approach to assess the effect of agricultural land occupation on species groups. *The International Journal of Life Cycle Assessment*, **19**(4), 758–769, doi:10.1007/s11367-014-0701-x.

Emin, G., M. Lepetit, A. Grandjean, and O. Ortega, 2014: *Massive financing of the energy transition. SFTE feasibility study: synthesis report. Energy renovation of public buildings*. Association for the Financing of the EneRgy Transition (A.F.T.E.R), 37 pp.

Emmer, A., J. Klimeš, M. Mergili, V. Vilímek, and A. Cochachin, 2016: 882 lakes of the Cordillera Blanca: An inventory, classification, evolution and assessment of susceptibility to outburst floods. *CATENA*, **147**, 269–279, doi:10.1016/j.catena.2016.07.032.

Ensor, J. and B. Harvey, 2015: Social learning and climate change adaptation: evidence for international development practice. *Wiley Interdisciplinary Reviews: Climate Change*, **6**(5), 509–522, doi:10.1002/wcc.348.

Erb, K.-H. et al., 2009: *Eating the Planet: Feeding and fuelling the world sustainably, fairly and humanely – a scoping study*. Social Ecology Working Paper No. 116, Institute of Social Ecology, Alpen-Adria Universität, Vienna, Austria, 132 pp.

Erb, K.-H. et al., 2016a: Exploring the biophysical option space for feeding the world without deforestation. *Nature Communications*, **7**, 11382, doi:10.1038/ncomms11382.

Erb, K.-H. et al., 2016b: Land management: data availability and process understanding for global change studies. *Global Change Biology*, **23**(2), 512–533, doi:10.1111/gcb.13443.

Erb, K.-H. et al., 2017: Unexpectedly large impact of forest management and grazing on global vegetation biomass. *Nature*, **553**(7686), 73–76, doi:10.1038/nature25138.

Eric Bickel, J., 2013: Climate engineering and climate tipping-point scenarios. *Environment Systems & Decisions*, **33**(1), 152–167, doi:10.1007/s10669-013-9435-8.

Ericsson, K. and S. Werner, 2016: The introduction and expansion of biomass use in Swedish district heating systems. *Biomass and Bioenergy*, **94**, 57–65, doi:10.1016/j.biombioe.2016.08.011.

Eriksson, L., J. Garvill, and A.M. Nordlund, 2006: Acceptability of travel demand management measures: The importance of problem awareness, personal norm, freedom, and fairness. *Journal of Environmental Psychology*, **26**, 15–26, doi:10.1016/j.jenvp.2006.05.003.

Eriksson, L., J. Garvill, and A.M. Nordlund, 2008: Acceptability of single and combined transport policy measures: The importance of environmental and policy specific beliefs. *Transportation Research Part A: Policy and Practice*, **42**(8), 1117–1128, doi:10.1016/j.tra.2008.03.006.

4

Erker, S., R. Stangl, and G. Stoeglehner, 2017: Resilience in the light of energy crises – Part II: Application of the regional energy resilience assessment. *Journal of Cleaner Production*, **164**, 495–507, doi:10.1016/j.jclepro.2017.06.162.

Erlinghagen, S. and J. Markard, 2012: Smart grids and the transformation of the electricity sector: ICT firms as potential catalysts for sectoral change. *Energy Policy*, **51**, 895–906, doi:10.1016/j.enpol.2012.09.045.

Esham, M. and C. Garforth, 2013: Agricultural adaptation to climate change: insights from a farming community in Sri Lanka. *Mitigation and Adaptation Strategies for Global Change*, **18(5)**, 535–549, doi:10.1007/s11027-012-9374-6.

ESRB ASC, 2016: *Too late, too sudden: Transition to a low-carbon economy and systemic risk*. ESRB ASC Report No 6, European Systemic Risk Board, Frankfurt, Germany, 22 pp.

Evans, J., J. O'Brien, and B. Ch Ng, 2018: Towards a geography of informal transport: Mobility, infrastructure and urban sustainability from the back of a motorbike. *Transactions of the Institute of British Geographers*, **43(4)**, 674–688, doi:10.1111/tran.12239.

Evans, L. et al., 2012: Self-interest and pro-environmental behaviour. *Nature Climate Change*, **3(2)**, 122–125, doi:10.1038/nclimate1662.

Evans, M., V. Roshchanka, and P. Graham, 2017: An international survey of building energy codes and their implementation. *Journal of Cleaner Production*, **158**, 382–389, doi:10.1016/j.jclepro.2017.01.007.

Ewing, R., S. Hamidi, and J.B. Grace, 2016: Compact development and VMT-Environmental determinism, self-selection, or some of both? *Environment and Planning B: Planning and Design*, **43(4)**, 737–755, doi:10.1177/0265813515594811.

Exner, A. et al., 2016: Measuring regional resilience towards fossil fuel supply constraints. Adaptability and vulnerability in socio-ecological Transformations- the case of Austria. *Energy Policy*, **91**, 128–137, doi:10.1016/j.enpol.2015.12.031.

Eyre, N., S.J. Darby, P. Grünewald, E. McKenna, and R. Ford, 2018: Reaching a 1.5°C target: socio-technical challenges for a rapid transition to low-carbon electricity systems. *Philosophical Transactions of the Royal Society A: Mathematical, Physical and Engineering Sciences*, **376(2119)**, 20160462, doi:10.1098/rsta.2016.0462.

Ezeji, T., 2017: Production of Bio-Derived Fuels and Chemicals. *Fermentation*, **3(3)**, 42, doi:10.3390/fermentation3030042.

Fader, M., S. Shi, W. von Bloh, A. Bondeau, and W. Cramer, 2016: Mediterranean irrigation under climate change: more efficient irrigation needed to compensate for increases in irrigation water requirements. *Hydrology and Earth System Sciences*, **20(2)**, 953–973, doi:10.5194/hess-20-953-2016.

Faehn, T. and E.T. Isaksen, 2016: Diffusion of Climate Technologies in the Presence of Commitment Problems. *The Energy Journal*, **37(2)**, 155–180, doi:10.5547/01956574.37.2.tfae.

Fainstein, S.S., 2018: Resilience and justice: planning for New York City. *Urban Geography*, 1–8, doi:10.1080/02723638.2018.1448571.

Fajardy, M. and N. Mac Dowell, 2017: Can BECCS deliver sustainable and resource efficient negative emissions? *Energy & Environmental Science*, **10(6)**, 1389–1426, doi:10.1039/c7ee00465f.

Falkner, R., 2016: The Paris Agreement and the new logic of international climate politics. *International Affairs*, **92(5)**, 1107–1125, doi:10.1111/1468-2346.12708.

Faludi, J., C. Bayley, S. Bhogal, and M. Iribarne, 2015: Comparing environmental impacts of additive manufacturing vs traditional machining via life-cycle assessment. *Rapid Prototyping Journal*, **21(1)**, 14–33, doi:10.1108/rpj-07-2013-0067.

Fang, Y., B. Singh, B.P. Singh, and E. Krull, 2014: Biochar carbon stability in four contrasting soils. *European Journal of Soil Science*, **65(1)**, 60–71, doi:10.1111/ejss.12094.

Fankhauser, S. and G. Schmidt-Traub, 2011: From adaptation to climate-resilient development: The costs of climate-proofing the Millennium Development Goals in Africa. *Climate and Development*, **3(2)**, 94–113, doi:10.1080/17565529.2011.582267.

Fankhauser, S. and T.K.J. McDermott, 2014: Understanding the adaptation deficit: Why are poor countries more vulnerable to climate events than rich countries? *Global Environmental Change*, **27(1)**, 9–18, doi:10.1016/j.gloenvcha.2014.04.014.

FAO, 2013: *Food wastage footprint. Impacts on natural resources. Summary Report*. Food and Agriculture Organisation of the United Nations (FAO), Rome, Italy, 63 pp.

FAO and NZAGRC, 2017a: *Low emissions development of the beef cattle sector in Uruguay – reducing enteric methane for food security and livelihoods*. Food and Agriculture Organization of the United Nations (FAO), Rome, Italy and New Zealand Agricultural Greenhouse Gas Research Centre (NZAGRC), 34 pp.

FAO and NZAGRC, 2017b: *Options for low emission development in the Kenya dairy sector – reducing enteric methane for food security and livelihoods*. Food and Agriculture Organization of the United Nations (FAO), Rome, Italy and New Zealand Agricultural Greenhouse Gas Research Centre (NZAGRC), 43 pp.

FAO and NZAGRC, 2017c: *Supporting low emissions development in the Ethiopian dairy cattle sector – reducing enteric methane for food security and livelihoods*. Food and Agriculture Organization of the United Nations (FAO), Rome, Italy and New Zealand Agricultural Greenhouse Gas Research Centre (NZAGRC), 34 pp.

Farfan, J. and C. Breyer, 2017: Structural changes of global power generation capacity towards sustainability and the risk of stranded investments supported by a sustainability indicator. *Journal of Cleaner Production*, **141**, 370–384, doi:10.1016/j.jclepro.2016.09.068.

Farrow, K., G. Grolleau, and L. Ibanez, 2017: Social Norms and Pro-environmental Behavior: A Review of the Evidence. *Ecological Economics*, **140**, 1–13, doi:10.1016/j.ecolecon.2017.04.017.

Fasihi, M., D. Bogdanov, and C. Breyer, 2017: Long-Term Hydrocarbon Trade Options for the Maghreb Region and Europe-Renewable Energy Based Synthetic Fuels for a Net Zero Emissions World. *Sustainability*, **9(2)**, 306, doi:10.3390/su9020306.

Favretto, N., L.C. Stringer, M.S. Buckeridge, and S. Afionis, 2017: Policy and Diplomacy in the Production of Second Generation Ethanol in Brazil: International Relations with the EU, the USA and Africa. In: *Advances of Basic Science for Second Generation from Sugarcane* [Buckeridge, M.S. and A.P. De Souza (eds.)]. Springer International Publishing, New York, pp. 197–212, doi:10.1007/978-3-319-49826-3_11.

Fawcett, T., F. Hvelplund, and N.I. Meyer, 2010: Making It Personal: Per Capita Carbon Allowances. In: *Generating Electricity in a Carbon-Constrained World* [Sioshansi, F.P. (ed.)]. Academic Press, Boston, MA, USA, pp. 87–107, doi:10.1016/b978-1-85617-655-2.00004-3.

Fazey, I. et al., 2017: Transformation in a changing climate: a research agenda. *Climate and Development*, 1–21, doi:10.1080/17565529.2017.1301864.

Fazey, I. et al., 2018: Energy Research & Social Science Ten essentials for action-oriented and second order energy transitions, transformations and climate change research. *Energy Research & Social Science*, **40**, 54–70, doi:10.1016/j.erss.2017.11.026.

Fell, H., D. Burtraw, R.D. Morgenstern, and K.L. Palmer, 2012: Soft and hard price collars in a cap-and-trade system: A comparative analysis. *Journal of Environmental Economics and Management*, **64(2)**, 183–198, doi:10.1016/j.jeem.2011.11.004.

Feng, E.Y., D.P. Keller, W. Koeve, and A. Oschlies, 2016: Could artificial ocean alkalinization protect tropical coral ecosystems from ocean acidification? *Environmental Research Letters*, **11(7)**, 074008, doi:10.1088/1748-9326/11/7/074008.

Feola, G., A.M. Lerner, M. Jain, M.J.F. Montefrio, and K. Nicholas, 2015: Researching farmer behaviour in climate change adaptation and sustainable agriculture: Lessons learned from five case studies. *Journal of Rural Studies*, **39**, 74–84, doi:10.1016/j.jrurstud.2015.03.009.

Fernández-Giménez, M.E., B. Batkhishig, B. Batbuyan, and T. Ulambayar, 2015: Lessons from the Dzud: Community-Based Rangeland Management Increases the Adaptive Capacity of Mongolian Herders to Winter Disasters. *World Development*, **68**, 48–65, doi:10.1016/j.worlddev.2014.11.015.

Ferrario, F. et al., 2014: The effectiveness of coral reefs for coastal hazard risk reduction and adaptation. *Nature Communications*, **5**, 1–9, doi:10.1038/ncomms4794.

Few, R., D. Morchain, D. Spear, A. Mensah, and R. Bendapudi, 2017: Transformation, adaptation and development: relating concepts to practice. *Palgrave Communications*, **3**, 17092, doi:10.1057/palcomms.2017.92.

Fielding, K.S. and M.J. Hornsey, 2016: A Social Identity Analysis of Climate Change and Environmental Attitudes and Behaviors: Insights and Opportunities. *Frontiers in Psychology*, **7(FEB)**, 1–12, doi:10.3389/fpsyg.2016.00121.

Fielding, K.S., R. Mcdonald, and W.R. Louis, 2008: Theory of planned behaviour,

4

identity and intentions to engage in environmental activism. *Journal of Environmental Psychology*, **28**, 318–326, doi:10.1016/j.jenvp.2008.03.003.

Figueres, C. et al., 2017: Three years to safeguard our climate. *Nature*, **546(7660)**, 593–595, doi:10.1038/546593a.

Fink, J.H., 2013: Geoengineering cities to stabilise climate. *Proceedings of the Institution of Civil Engineers – Engineering Sustainability*, **166(5)**, 242–248, doi:10.1680/ensu.13.00002.

Finlayson, C., 2012: Forty years of wetland conservation and wise use. *Aquatic Conservation: Marine and Freshwater Ecosystems*, **22(2)**, 139–143, doi:10.1002/aqc.2233.

Finlayson, C.M. et al., 2017: Policy considerations for managing wetlands under a changing climate. *Marine and Freshwater Research*, **68(10)**, 1803–1815, doi:10.1071/mf16244.

Finley, R.J., 2014: An overview of the Illinois Basin – Decatur Project. *Greenhouse Gases: Science and Technology*, **4(5)**, 571–579, doi:10.1002/ghg.1433.

Finon, D., 2013: Towards a global governance of nuclear safety: an impossible quest? *Revue de l'Energie* (in French), **616**, 440–450.

Finon, D. and F. Roques, 2013: European Electricity Market Reforms: The "Visible Hand" of Public Coordination. *Economics of Energy & Environmental Policy*, **2(2)**, doi:10.5547/2160-5890.2.2.6.

Fischer, C., M. Greaker, and K.E. Rosendahl, 2017: Robust technology policy against emission leakage: The case of upstream subsidies. *Journal of Environmental Economics and Management*, **84**, 44–61, doi:10.1016/j.jeem.2017.02.001.

Fishman, E., S. Washington, and N. Haworth, 2015: Bikeshare's impact on active travel: Evidence from the United States, Great Britain, and Australia. *Journal of Transport & Health*, **2(2)**, 135–142, doi:10.1016/j.jth.2015.03.004.

Fishman, R., N. Devineni, and S. Raman, 2015: Can improved agricultural water use efficiency save India's groundwater? *Environmental Research Letters*, **10(8)**, 084022, doi:10.1088/1748-9326/10/8/084022.

Fleming, A., S.E. Park, and N.A. Marshall, 2015a: Enhancing adaptation outcomes for transformation: climate change in the Australian wine industry. *Journal of Wine Research*, **26(2)**, 99–114, doi:10.1080/09571264.2015.1031883.

Fleming, A., A.M. Dowd, E. Gaillard, S. Park, and M. Howden, 2015b: "Climate change is the least of my worries": Stress limitations on adaptive capacity. *Rural Society*, **24(1)**, 24–41, doi:10.1080/10371656.2014.1001481.

Fleurbaey, M. and P.J. Hammond, 2004: Interpersonally Comparable Utility. In: *Handbook of Utility Theory: Volume 2 Extensions* [Barbera, S., P.J. Hammond, and C. Seidl (eds.)]. Kluwer Academic Publishers, Dordrecht, The Netherlands, pp. 1179–1285, doi:10.1007/978-1-4020-7964-1_8.

Fleurbaey, M. et al., 2014: Sustainable Development and Equity. In: *Climate Change 2014: Mitigation of Climate Change. Contribution of Working Group III to the Fifth Assessment Report of the Intergovernmental Panel on Climate Change* [Edenhofer, O., R. Pichs-Madruga, Y. Sokona, E. Farahani, S. Kadner, K. Seyboth, A. Adler, I. Baum, S. Brunner, P. Eickemeier, B. Kriemann, J. Savolainen, S. Schlömer, C. von Stechow, T. Zwickel, and J.C. Minx (eds.)]. Cambridge University Press, Cambridge, United Kingdom and New York, NY, USA, pp. 283–350.

Floater, G. et al., 2014: *Cities and the New Climate Economy: the transformative role of global urban growth*. NCE Cities – Paper 01, LSE Cities. London School of Economics and Political Science, London, UK, 70 pp.

Flynn, M., J. Ford, T. Pearce, S. Harper, and IHACC Research Team, 2018: Participatory scenario planning and climate change impacts, adaptation, and vulnerability research in the Arctic. *Environmental Science & Policy*, **79**, 45–53, doi:10.1016/j.envsci.2017.10.012.

Foley, J.A. et al., 2011: Solutions for a cultivated planet. *Nature*, **478(7369)**, 337–342, doi:10.1038/nature10452.

Forbes, B.C. et al., 2009: High resilience in the Yamal-Nenets social–ecological system, West Siberian Arctic, Russia. *Proceedings of the National Academy of Sciences*, **106(52)**, 22041–22048, doi:10.1073/pnas.0908286106.

Forbis Jr, R. and K. Hayhoe, 2018: Does Arctic governance hold the key to achieving climate policy targets? *Environmental Research Letters*, **13(2)**, 020201, doi:10.1088/1748-9326/aaa359.

Ford, J.D., 2012: Indigenous health and climate change. *American Journal of Public Health*, **102(7)**, 1260–1266, doi:10.2105/ajph.2012.300752.

Ford, J.D. and D. King, 2015: Coverage and framing of climate change adaptation in the media: A review of influential North American newspapers during 1993-2013. *Environmental Science & Policy*, **48**, 137–146, doi:10.1016/j.envsci.2014.12.003.

Ford, J.D. and L. Berrang-Ford, 2016: The 4Cs of adaptation tracking: consistency, comparability, comprehensiveness, coherency. *Mitigation and Adaptation Strategies for Global Change*, **21(6)**, 839–859, doi:10.1007/s11027-014-9627-7.

Ford, J.D., G. McDowell, and J. Jones, 2014a: The state of climate change adaptation in the Arctic. *Environmental Research Letters*, **9(10)**, 104005, doi:10.1088/1748-9326/9/10/104005.

Ford, J.D., G. McDowell, and T. Pearce, 2015a: The adaptation challenge in the Arctic. *Nature Climate Change*, **5(12)**, 1046–1053, doi:10.1038/nclimate2723.

Ford, J.D., L. Berrang-Ford, A. Lesnikowski, M. Barrera, and S.J. Heymann, 2013: How to Track Adaptation to Climate Change: A Typology of Approaches for National-Level Application. *Ecology and Society*, **18(3)**, art40, doi:10.5751/es-05732-180340.

Ford, J.D. et al., 2014b: Adapting to the Effects of Climate Change on Inuit Health. *American Journal of Public Health*, **104(S3)**, e9–e17, doi:10.2105/ajph.2013.301724.

Ford, J.D. et al., 2015b: Adaptation tracking for a post-2015 climate agreement. *Nature Climate Change*, **5(11)**, 967–969, doi:10.1038/nclimate2744.

Ford, J.D. et al., 2015c: Evaluating climate change vulnerability assessments: a case study of research focusing on the built environment in northern Canada. *Mitigation and Adaptation Strategies for Global Change*, **20(8)**, 1267–1288, doi:10.1007/s11027-014-9543-x.

Ford, J.D. et al., 2016: Community-based adaptation research in the Canadian Arctic. *Wiley Interdisciplinary Reviews: Climate Change*, **7(2)**, 175–191, doi:10.1002/wcc.376.

Ford, J.D. et al., 2018: Preparing for the health impacts of climate change in Indigenous communities: The role of community-based adaptation. *Global Environmental Change*, **49**, 129–139, doi:10.1016/j.gloenvcha.2018.02.006.

Forman, C., I.K. Muritala, R. Pardemann, and B. Meyer, 2016: Estimating the global waste heat potential. *Renewable and Sustainable Energy Reviews*, **57**, 1568–1579, doi:10.1016/j.rser.2015.12.192.

Fornell, R., T. Berntsson, and A. Åsblad, 2013: Techno-economic analysis of a kraft pulp-mill-based biorefinery producing both ethanol and dimethyl ether. *Energy*, **50(1)**, 83–92, doi:10.1016/j.energy.2012.11.041.

Forsell, N. et al., 2016: Assessing the INDCs' land use, land use change, and forest emission projections. *Carbon Balance and Management*, **11(1)**, 26, doi:10.1186/s13021-016-0068-3.

Forster, J., I.R. Lake, A.R. Watkinson, and J.A. Gill, 2011: Marine biodiversity in the Caribbean UK overseas territories: Perceived threats and constraints to environmental management. *Marine Policy*, **35(5)**, 647–657, doi:10.1016/j.marpol.2011.02.005.

Fouquet, R., 2016: Lessons from energy history for climate policy: Technological change, demand and economic development. *Energy Research & Social Science*, **22**, 79–93, doi:10.1016/j.erss.2016.09.001.

Francesch-Huidobro, M., M. Dabrowski, Y. Tai, F. Chan, and D. Stead, 2017: Governance challenges of flood-prone delta cities: Integrating flood risk management and climate change in spatial planning. *Progress in Planning*, **114**, 1–27, doi:10.1016/j.progress.2015.11.001.

Francis, R. and B. Bekera, 2014: A metric and frameworks for resilience analysis of engineered and infrastructure systems. *Reliability Engineering & System Safety*, **121**, 90–103, doi:10.1016/j.ress.2013.07.004.

Frank, S. et al., 2013: How effective are the sustainability criteria accompanying the European Union 2020 biofuel targets? *GCB Bioenergy*, **5(3)**, 306–314, doi:10.1111/j.1757-1707.2012.01188.x.

Fraser, A. et al., 2017: Meeting the challenge of risk-sensitive and resilient urban development in sub-Saharan Africa: Directions for future research and practice. *International Journal of Disaster Risk Reduction*, **26**, 106–109, doi:10.1016/j.ijdrr.2017.10.001.

Frederiks, E.R., K. Stenner, and E. Hobman, 2015: Household energy use: Applying behavioural economics to understand consumer decision-making and behaviour. *Renewable and Sustainable Energy Reviews*, **41**, 1385–1394, doi:10.1016/j.rser.2014.09.026.

Freire, M.E., S. Lall, and D. Leipziger, 2014: *Africa's urbanization: challenges and opportunities*. Working Paper No. 7, The Growth Dialogue, Washington DC, USA, 44 pp.

Freire-González, J., 2017: Evidence of direct and indirect rebound effect in households in EU-27 countries. *Energy Policy*, **102**, 270–276, doi:10.1016/j.enpol.2016.12.002.

Fridahl, M., 2017: Socio-political prioritization of bioenergy with carbon capture and storage. *Energy Policy*, **104**, 89–99, doi:10.1016/j.enpol.2017.01.050.

Friedlingstein, P. et al., 2014: Uncertainties in CMIP5 climate projections due to carbon cycle feedbacks. *Journal of Climate*, **27(2)**, 511–526, doi:10.1175/jcli-d-12-00579.1.

Fritz, S. et al., 2011: Highlighting continued uncertainty in global land cover maps for the user community. *Environmental Research Letters*, **6(4)**, 044005, doi:10.1088/1748-9326/6/4/044005.

Froud, J., C. Haslam, S. Johal, and K. Williams, 2000: Shareholder value and Financialization: consultancy promises, management moves. *Economy and Society*, **29(1)**, 80–110, doi:10.1080/030851400360578.

Fryer, E., 2017: Digital infrastructure: And the impacts of climate change. *Journal of the Institute of Telecommunications Professionals*, **11(2)**, 8–13.

Fu, G. et al., 2017: Integrated Approach to Assess the Resilience of Future Electricity Infrastructure Networks to Climate Hazards. *IEEE Systems Journal*, 1–12, doi:10.1109/jsyst.2017.2700791.

Fudge, S., M. Peters, and B. Woodman, 2016: Local authorities as niche actors: the case of energy governance in the UK. *Environmental Innovation and Societal Transitions*, **18**, 1–17, doi:10.1016/j.eist.2015.06.004.

Fuhr, H., T. Hickmann, and K. Kern, 2018: The role of cities in multi-level climate governance: local climate policies and the 1.5°C target. *Current Opinion in Environmental Sustainability*, **30**, 1–6, doi:10.1016/j.cosust.2017.10.006.

Fujii, S. and R. Kitamura, 2003: What does a one-month free bus ticket do to habitual drivers? An experimental analysis of habit and attitude change. *Transportation*, **30(1)**, 81–95, doi:10.1023/a:1021234607980.

Fujimori, S. et al., 2016: Implication of Paris Agreement in the context of long-term climate mitigation goals. *SpringerPlus*, **5(1)**, 1620, doi:10.1186/s40064-016-3235-9.

Fulton, L., J. Mason, and D. Meroux, 2017: *Three Revolutions in Urban Transportation*. University of California – Davis (UC Davis) Sustainable Transportation Energy Pathways and Institute for Transporation & Development Policy (ITDP), 38 pp.

Fuss, S. et al., 2014: Betting on negative emissions. *Nature Climate Change*, **4(10)**, 850–853, doi:10.1038/nclimate2392.

Fuss, S. et al., 2018: Negative emissions – Part 2: Costs, potentials and side effects. *Environmental Research Letters*, **13(6)**, 063002, doi:10.1088/1748-9326/aabf9f.

Gajjar, S.P., C. Singh, and T. Deshpande, 2018: Tracing back to move ahead: a review of development pathways that constrain adaptation futures. *Climate and Development*, 1–15, doi:10.1080/17565529.2018.1442793.

Gambhir, A. et al., 2017: Assessing the Feasibility of Global Long-Term Mitigation Scenarios. *Energies*, **10(1)**, e89, doi:10.3390/en10010089.

Gandure, S., S. Walker, and J.J. Botha, 2013: Farmers' perceptions of adaptation to climate change and water stress in a South African rural community. *Environmental Development*, **5(1)**, 39–53, doi:10.1016/j.envdev.2012.11.004.

Gao, C., 2018: The future of CRISPR technologies in agriculture. *Nature Reviews Molecular Cell Biology*, **19(5)**, 275–276, doi:10.1038/nrm.2018.2.

Gao, Y. and J. Kenworthy, 2017: China. In: *The Urban Transport Crisis in Emerging Economies* [Pojani, D. and D. Stead (eds.)]. Springer, Cham, Switzerland, pp. 33–58, doi:10.1007/978-3-319-43851-1_3.

Gao, Y. and P. Newman, 2018: Beijing's Peak Car Transition: Hope for Emerging Cities in the 1.5°C Agenda. *Urban Planning*, **3(2)**, 82–93, doi:10.17645/up.v3i2.1246.

Gao, Y., J. Kenworthy, and P. Newman, 2015: Growth of a Giant: A Historical and Current Perspective on the Chinese Automobile Industry. *World Transport Policy and Practice*, **21.2**, 40–55, http://worldtransportjournal.com/wp-content/uploads/2015/05/30th-april-opt.pdf.

García Romero, H. and A. Molina, 2015: *Agriculture and Adaptation to Climate Change: The Role of Insurance in Risk Management: The Case of Colombia*. Inter-American Development Bank (IDB), Washington DC, USA, 49 pp., doi:10.18235/0000053.

García-Álvarez, M.T., L. Cabeza-García, and I. Soares, 2017: Analysis of the promotion of onshore wind energy in the EU: Feed-in tariff or renewable portfolio standard? *Renewable Energy*, **111**, 256–264, doi:10.1016/j.renene.2017.03.067.

Gardiner, S.M., 2013: Why geoengineering is not a 'global public good', and why it is ethically misleading to frame it as one. *Climatic Change*, **121(3)**, 513–525, doi:10.1007/s10584-013-0764-x.

Garg, A., J. Maheshwari, P.R. Shukla, and R. Rawal, 2017: Energy appliance transformation in commercial buildings in India under alternate policy

scenarios. *Energy*, **140**, 952–965, doi:10.1016/j.energy.2017.09.004.

Gasc, F., D. Guerrier, S. Barrett, and S. Anderson, 2014: *Assessing the effectiveness of investments in climate information services*. International Institute for Environment and Development (IIED), London, UK, 4 pp.

Gawith, D., A. Daigneault, and P. Brown, 2016: Does community resilience mitigate loss and damage from climate-related disasters? Evidence based on survey data. *Journal of Environmental Planning and Management*, **59(12)**, 2102–2123, doi:10.1080/09640568.2015.1126241.

GCEC, 2014: *Better growth, Better Climate: The New Climate Economy Report*. The Global Commission on the Economy and Climate, New Climate Economy, Washington DC, USA, 308 pp.

GEA, 2012: *Global Energy Assessment – Toward a sustainable future*. Global Energy Assessment (GEA). Cambridge University Press, Cambridge, United Kingdom and New York, NY, USA and the International Institute for Applied Systems Analysis, Laxenburg, Austria, 113 pp.

Gebrehiwot, T. and A. van der Veen, 2015: Farmers Prone to Drought Risk: Why Some Farmers Undertake Farm-Level Risk-Reduction Measures While Others Not? *Environmental Management*, **55(3)**, 588–602, doi:10.1007/s00267-014-0415-7.

Gebru, B., P. Kibaya, T. Ramahaleo, K. Kwena, and P. Mapfumo, 2015: *Improving access to climate-related information for adaptation*. International Development Research Centre (IDRC), Ottawa, ON, Canada, 4 pp.

Geden, O., V. Scott, and J. Palmer, 2018: Integrating carbon dioxide removal into EU climate policy: Prospects for a paradigm shift. *Wiley Interdisciplinary Reviews: Climate Change*, **9(4)**, e521, doi:10.1002/wcc.521.

Geels, F.W., 2014: Regime Resistance against Low-Carbon Transitions: Introducing Politics and Power into the Multi-Level Perspective. *Theory, Culture & Society*, **31(5)**, 21–40, doi:10.1177/0263276414531627.

Geels, F.W. and J. Schot, 2007: Typology of sociotechnical transition pathways. *Research Policy*, **36(3)**, 399–417, doi:10.1016/j.respol.2007.01.003.

Geels, F.W., F. Berkhout, and D.P. van Vuuren, 2016a: Bridging analytical approaches for low-carbon transitions. *Nature Climate Change*, **6(6)**, 576–583, doi:10.1038/nclimate2980.

Geels, F.W., B.K. Sovacool, T. Schwanen, and S. Sorrell, 2017: Sociotechnical transitions for deep decarbonization. *Science*, **357(6357)**, 1242–1244, doi:10.1126/science.aao3760.

Geels, F.W. et al., 2016b: The enactment of socio-technical transition pathways: A reformulated typology and a comparative multi-level analysis of the German and UK low-carbon electricity transitions (1990-2014). *Research Policy*, **45(4)**, 896–913, doi:10.1016/j.respol.2016.01.015.

Gemenne, F. and J. Blocher, 2017: How can migration serve adaptation to climate change? Challenges to fleshing out a policy ideal. *Geographical Journal*, **183**, 336–347, doi:10.1111/geoj.12205.

Georgeson, L., M. Maslin, M. Poessinouw, and S. Howard, 2016: Adaptation responses to climate change differ between global megacities. *Nature Climate Change*, **6(6)**, 584–588, doi:10.1038/nclimate2944.

Gerbens-Leenes, W., A.Y. Hoekstra, and T.H. van der Meer, 2009: The water footprint of bioenergy. *Proceedings of the National Academy of Sciences*, **106(25)**, 10219–10223, doi:10.1073/pnas.0812619106.

Gerber, P.J. et al., 2013: *Tackling climate change through livestock – A global assessment of emissions and mitigation opportunities*. Food and Agriculture Organization of the United Nations (FAO), Rome, Italy, 133 pp.

GeSI, 2015: *SMARTEr2030: ICT Solutions for 21st Century Challenges*. Global e-Sustainability Initiative (GeSI), Brussels, Belgium, 134 pp.

Gheewala, S.H., G. Berndes, and G. Jewitt, 2011: The bioenergy and water nexus. *Biofuels, Bioproducts and Biorefining*, **5(4)**, 353–360, doi:10.1002/bbb.295.

Ghorbani, N., A. Aghahosseini, and C. Breyer, 2017: Transition towards a 100% Renewable Energy System and the Role of Storage Technologies: A Case Study of Iran. *Energy Procedia*, **135**, 23–36, doi:10.1016/j.egypro.2017.09.484.

Gibbs, D. and K. O'Neill, 2014: Rethinking Sociotechnical Transitions and Green Entrepreneurship: The Potential for Transformative Change in the Green Building Sector. *Environment and Planning A: Economy and Space*, **46(5)**, 1088–1107, doi:10.1068/a46259.

Gillard, R., 2016: Questioning the Diffusion of Resilience Discourses in Pursuit of Transformational Change. *Global Environmental Politics*, **16(1)**, 13–20, doi:10.1162/glep_a_00334.

Gillard, R., A. Gouldson, J. Paavola, and J. Van Alstine, 2016: Transformational responses to climate change: beyond a systems perspective of social change in mitigation and adaptation. *Wiley Interdisciplinary Reviews: Climate Change*, **7(2)**, 251–265, doi:10.1002/wcc.384.

Gillingham, K. and K. Palmer, 2017: Bridging the Energy Efficiency Gap: Policy Insights from Economic Theory and Empirical Evidence. *Review of Environmental Economics and Policy*, **8(1)**, 18–38, doi:10.1093/reep/ret021.

Gillingham, K., M.J. Kotchen, D.S. Rapson, and G. Wagner, 2013: Energy policy: The rebound effect is overplayed. *Nature*, **493(7433)**, 475–476, doi:10.1038/493475a.

Giménez-Gómez, J.-M., J. Teixidó-Figueras, and C. Vilella, 2016: The global carbon budget: a conflicting claims problem. *Climatic Change*, **136(3–4)**, 693–703, doi:10.1007/s10584-016-1633-1.

Girod, B., T. Stucki, and M. Woerter, 2017: How do policies for efficient energy use in the household sector induce energy-efficiency innovation? An evaluation of European countries. *Energy Policy*, **103**, 223–237, doi:10.1016/j.enpol.2016.12.054.

Glaas, E., E.C.H. Keskitalo, and M. Hjerpe, 2017: Insurance sector management of climate change adaptation in three Nordic countries: the influence of policy and market factors. *Journal of Environmental Planning and Management*, **60(9)**, 1601–1621, doi:10.1080/09640568.2016.1245654.

Glachant, M. and A. Dechezleprêtre, 2016: What role for climate negotiations on technology transfer? *Climate Policy*, 1–15, doi:10.1080/14693062.2016.1222257.

Glazebrook, G. and P. Newman, 2018: The City of the Future. *Urban Planning*, **3(2)**, 1–20, doi:10.17645/up.v3i2.1247.

Global CCS Institute, 2017: *The Global Status of CCS 2016 Summary Report*. Global CCS Institute, Canberra, Australia, 28 pp.

Goeppert, A., M. Czaun, G.K. Surya Prakash, and G.A. Olah, 2012: Air as the renewable carbon source of the future: an overview of CO_2 capture from the atmosphere. *Energy & Environmental Science*, **5(7)**, 7833–7853, doi:10.1039/c2ee21586a.

Goldemberg, J., 2011: The Role of Biomass in the World's Energy System. In: *Routes to Cellulosic Ethanol* [Buckeridge, M.S. and G.H. Goldman (eds.)]. Springer, New York, NY, USA, pp. 3–14, doi:10.1007/978-0-387-92740-4_1.

Gölz, S. and U.J.J. Hahnel, 2016: What motivates people to use energy feedback systems? A multiple goal approach to predict long-term usage behaviour in daily life. *Energy Research & Social Science*, **21**, 155–166, doi:10.1016/j.erss.2016.07.006.

Gonzales, M.H., E. Aronson, and M.A. Costanzo, 1988: Using Social Cognition and Persuasion to Promote Energy Conservation: A Quasi-Experiment. *Journal of Applied Social Psychology*, **18(12)**, 1049–1066, doi:10.1111/j.1559-1816.1988.tb01192.x.

González, M.F. and T. Ilyina, 2016: Impacts of artificial ocean alkalinization on the carbon cycle and climate in Earth system simulations. *Geophysical Research Letters*, **43(12)**, 6493–6502, doi:10.1002/2016gl068576.

Goodwin, P. and K. Van Dender, 2013: 'Peak Car' – Themes and Issues. *Transport Reviews*, **33(3)**, 243–254, doi:10.1080/01441647.2013.804133.

Göpfert, C., C. Wamsler, and W. Lang, 2018: A framework for the joint institutionalization of climate change mitigation and adaptation in city administrations. *Mitigation and Adaptation Strategies for Global Change*, 1–21, doi:10.1007/s11027-018-9789-9.

Gorayeb, A., C. Brannstrom, A.J. de Andrade Meireles, and J. de Sousa Mendes, 2018: Wind power gone bad: Critiquing wind power planning processes in northeastern Brazil. *Energy Research & Social Science*, **40**, 82–88, doi:10.1016/j.erss.2017.11.027.

Gordon, D.J. and C.A. Johnson, 2017: The orchestration of global urban climate governance: conducting power in the post-Paris climate regime. *Environmental Politics*, **26(4)**, 694–714, doi:10.1080/09644016.2017.1320829.

Gota, S., C. Huizenga, K. Peet, N. Medimorec, and S. Bakker, 2018: Decarbonising transport to achieve Paris Agreement targets. *Energy Efficiency*, 1–24, doi:10.1007/s12053-018-9671-3.

Goulder, L.H., 1995: Effects of Carbon Taxes in an Economy with Prior Tax Distortions: An Intertemporal General Equilibrium Analysis. *Journal of Environmental Economics and Management*, **29(3)**, 271–297, doi:10.1006/jeem.1995.1047.

Goulder, L.H., 2013: Climate change policy's interactions with the tax system. *Energy Economics*, **40**, S3–S11, doi:10.1016/j.eneco.2013.09.017.

Goytia, S., M. Pettersson, T. Schellenberger, W.J. van Doorn-Hoekveld, and S. Priest, 2016: Dealing with change and uncertainty within the regulatory frameworks for flood defense infrastructure in selected European countries. *Ecology and Society*, **21(4)**, 23, doi:10.5751/es-08908-210423.

Grace, J., E. Mitchard, and E. Gloor, 2014: Perturbations in the carbon budget of the tropics. *Global Change Biology*, **20(10)**, 3238–3255, doi:10.1111/gcb.12600.

Graham, S., J. Barnett, R. Fincher, C. Mortreux, and A. Hurlimann, 2015: Towards fair local outcomes in adaptation to sea-level rise. *Climatic Change*, **130(3)**, 411–424, doi:10.1007/s10584-014-1171-7.

Granderson, A.A., 2017: The Role of Traditional Knowledge in Building Adaptive Capacity for Climate Change: Perspectives from Vanuatu. *Weather, Climate, and Society*, **9(3)**, 545–561, doi:10.1175/wcas-d-16-0094.1.

Grassi, G. et al., 2017: The key role of forests in meeting climate targets requires science for credible mitigation. *Nature Climate Change*, **7(3)**, 220–226, doi:10.1038/nclimate3227.

Grecequet, M., J. DeWaard, J.J. Hellmann, and G.J. Abel, 2017: Climate Vulnerability and Human Migration in Global Perspective. *Sustainability*, **9(5)**, 720, doi:10.3390/su9050720.

Green, D. and L. Minchin, 2014: Living on climate-changed country: Indigenous health, well-being and climate change in remote Australian communities. *EcoHealth*, **11(2)**, 263–272, doi:10.1007/s10393-013-0892-9.

Green, J. and P. Newman, 2017a: Citizen utilities: The emerging power paradigm. *Energy Policy*, **105**, 283–293, doi:10.1016/j.enpol.2017.02.004.

Green, J. and P. Newman, 2017b: Disruptive innovation, stranded assets and forecasting: the rise and rise of renewable energy. *Journal of Sustainable Finance & Investment*, **7(2)**, 169–187, doi:10.1080/20430795.2016.1265410.

Green, K.E., 2016: A political ecology of scaling: Struggles over power, land and authority. *Geoforum*, **74**, 88–97, doi:10.1016/j.geoforum.2016.05.007.

Green, O.O. et al., 2016: Adaptive governance to promote ecosystem services in urban green spaces. *Urban Ecosystems*, **19(1)**, 77–93, doi:10.1007/s11252-015-0476-2.

Greenblatt, J.B. and S. Saxena, 2015: Autonomous taxis could greatly reduce greenhouse-gas emissions of US light-duty vehicles. *Nature Climate Change*, **5(9)**, 860–863, doi:10.1038/nclimate2685.

Greene, C.H. et al., 2017: Geoengineering, marine microalgae, and climate stabilization in the 21st century. *Earth's Future*, **5(3)**, 278–284, doi:10.1002/2016ef000486.

Greene, D.L., 2011: Uncertainty, loss aversion, and markets for energy efficiency. *Energy Economics*, **33(4)**, 608–616, doi:10.1016/j.eneco.2010.08.009.

Greening, L.A., D.L. Greene, and C. Difiglio, 2000: Energy efficiency and consumption – the rebound effect – a survey. *Energy Policy*, **28(6–7)**, 389–401, doi:10.1016/s0301-4215(00)00021-5.

Gregorio, M. et al., 2016: *Integrating mitigation and adaptation in climate and land use policies in Brazil: a policy document analysis*. CCEP Working Papers no. 257, Centre for Climate Change Economics and Policy (CCCEP), Leeds, UK, 54 pp.

Gregory, R., S. Lichtenstein, and P. Slovic, 1993: Valuing environmental resources: A constructive approach. *Journal of Risk and Uncertainty*, **7(2)**, 177–197, doi:10.1007/bf01065813.

Grewe, V., E. Tsati, M. Mertens, C. Frömming, and P. Jöckel, 2017: Contribution of emissions to concentrations: the TAGGING 1.0 submodel based on the Modular Earth Submodel System (MESSy 2.52). *Geoscientific Model Development*, **10(7)**, 2615–2633, doi:10.5194/gmd-10-2615-2017.

Griscom, B.W. et al., 2017: Natural climate solutions. *Proceedings of the National Academy of Sciences*, **114(44)**, 11645–11650, doi:10.1073/pnas.1710465114.

Gross, R., W. Blyth, and P. Heptonstall, 2010: Risks, revenues and investment in electricity generation: Why policy needs to look beyond costs. *Energy Economics*, **32(4)**, 796–804, doi:10.1016/j.eneco.2009.09.017.

Grosse, G., S. Goetz, A.D. McGuire, V.E. Romanovsky, and E.A.G. Schuur, 2016: Changing permafrost in a warming world and feedbacks to the Earth system. *Environmental Research Letters*, **11(4)**, 040201, doi:10.1088/1748-9326/11/4/040201.

Grothmann, T. and F. Reusswig, 2006: People at Risk of Flooding: Why Some Residents Take Precautionary Action While Others Do Not. *Natural Hazards*, **38(1–2)**, 101–120, doi:10.1007/s11069-005-8604-6.

Grottera, C., W. Wills, and E.L. La Rovere, 2016: The Transition to a Low Carbon Economy and Its Effects on Jobs and Welfare – A Long-Term Scenario for Brazil. In: *The Fourth Green Growth Knowledge Platform Annual Conference, 6-7 September 2016, Jeju, Republic of Korea*. The Fourth Green Growth Knowledge Platform Annual Conference, 6-7 September 2016. Jeju, Republic of Korea, Geneva, Switzerland, pp. 1–7.

Grove, K.J., 2013: From Emergency Management to Managing Emergence: A Genealogy of Disaster Management in Jamaica. *Annals of the Association of American Geographers*, **103(3)**, 570–588, doi:10.1080/00045608.2012.740357.

Grubb, M., 1990: The Greenhouse Effect: Negotiating Targets. *International Affairs*, **66(1)**, 67–89, www.jstor.org/stable/2622190.

Grubb, M., J.C. Hourcade, and K. Neuhoff, 2014: *Planetary Economics: Energy, climate change and the three domains of sustainable development*. Routledge, Abingdon, UK, 548 pp.

Grubler, A., 2010: The costs of the French nuclear scale-up: A case of negative learning by doing. *Energy Policy*, **38(9)**, 5174–5188, doi:10.1016/j.enpol.2010.05.003.

Grubler, A. et al., 2012: Policies for the Energy Technology Innovation System (ETIS). In: *Global Energy Assessment – Toward a Sustainable Future*. Cambridge University Press, Cambridge, UK and New York, NY, USA and the International Institute for Applied Systems Analysis, Laxenburg, Austria, pp. 1665–1744.

Grubler, A. et al., 2018: A low energy demand scenario for meeting the 1.5°C target and sustainable development goals without negative emission technologies. *Nature Energy*, **3(6)**, 515–527, doi:10.1038/s41560-018-0172-6.

Guerra, A., 2017: La Crisis como Oportunidad, Análisis de la sequía en la costa sur de Guatemala en 2016 (in Spanish). *Red Nacional de Formación e Investigación Ambiental Guatemala C.A.*, **17**, 21–27.

Guivarch, C. and S. Hallegatte, 2011: Existing infrastructure and the 2°C target. *Climatic Change*, **109(3–4)**, 801–805, doi:10.1007/s10584-011-0268-5.

Gupta, J., 2014: *The History of Global Climate Governance*. Cambridge University Press, Cambridge, United Kingdom and New York, NY, USA, 262 pp., doi:10.1017/cbo9781139629072.

Gupta, S. et al., 2007: Policies, Instruments and Co-operative Arrangements. In: *Climate Change 2007: Mitigation of Climate Change. Contribution of Working Group III to the Fourth Assessment Report of the Intergovernmental Panel on Climate Change* [Metz, B., O.R. Davidson, P.R. Bosch, R. Dave, and L.A. Meyer (eds.)]. Cambridge University Press, Cambridge, United Kingdom and New York, NY, USA, pp. 746–807.

Gurara, D. et al., 2017: *Trends and Challenges in Infrastructure Investment in Low-Income Developing Countries*. WP/17/233, International Monetary Fund (IMF), Washington DC, USA, 31 pp.

GWA, 2016: *Statistics Digest 2015-16*. Government of Western Australia (GWA), Department of Mines and Petroleum, Perth, Australia, 74 pp.

Gwedla, N. and C.M. Shackleton, 2015: The development visions and attitudes towards urban forestry of officials responsible for greening in South African towns. *Land Use Policy*, **42**, 17–26, doi:10.1016/j.landusepol.2014.07.004.

Haas, W., F. Krausmann, D. Wiedenhofer, and M. Heinz, 2015: How Circular is the Global Economy?: An Assessment of Material Flows, Waste Production, and Recycling in the European Union and the World in 2005. *Journal of Industrial Ecology*, **19(5)**, 765–777, doi:10.1111/jiec.12244.

Haasnoot, M., J.H. Kwakkel, W.E. Walker, and J. ter Maat, 2013: Dynamic adaptive policy pathways: A method for crafting robust decisions for a deeply uncertain world. *Global Environmental Change*, **23(2)**, 485–498, doi:10.1016/j.gloenvcha.2012.12.006.

Hackmann, H., S.C. Moser, and A.L. St. Clair, 2014: The social heart of global environmental change. *Nature Climate Change*, **4(8)**, 653–655, doi:10.1038/nclimate2320.

Hadarits, M. et al., 2017: The interplay between incremental, transitional, and transformational adaptation: a case study of Canadian agriculture. *Regional Environmental Change*, **17(5)**, 1515–1525, doi:10.1007/s10113-017-1111-y.

Hagen, A., L. Kähler, and K. Eisenack, 2017: Transnational environmental agreements with heterogeneous actors. In: *Economics of International Environmental Agreements: A Critical Approach* [Kayalıca, M., S. Çağatay, and H. Mıhçı (eds.)]. Routledge, London, UK.

Hagen, B., A. Middel, and D. Pijawka, 2016: European Climate Change Perceptions: Public support for mitigation and adaptation policies. *Environmental Policy and Governance*, **26(3)**, 170–183, doi:10.1002/eet.1701.

Hahnel, U.J.J. et al., 2015: The power of putting a label on it: green labels weigh heavier than contradicting product information for consumers' purchase decisions and post-purchase behavior. *Frontiers in Psychology*, **6**, 1392, doi:10.3389/fpsyg.2015.01392.

Haines, A. et al., 2017: Short-lived climate pollutant mitigation and the Sustainable Development Goals. *Nature Climate Change*, **7(12)**, 863–869, doi:10.1038/s41558-017-0012-x.

Hallegatte, S. and K.J. Mach, 2016: Make climate-change assessments more relevant. *Nature*, **534(7609)**, 613–615, doi:10.1038/534613a.

Hallegatte, S. and J. Rozenberg, 2017: Climate change through a poverty lens. *Nature Climate Change*, **7**, 250–256, doi:10.1038/nclimate3253.

Hallegatte, S., J.-C. Hourcade, and P. Dumas, 2007: Why economic dynamics matter in assessing climate change damages: Illustration on extreme events. *Ecological Economics*, **62(2)**, 330–340, doi:10.1016/j.ecolecon.2006.06.006.

Hallegatte, S., C. Green, R.J. Nicholls, and J. Corfee-Morlot, 2013: Future flood losses in major coastal cities. *Nature Climate Change*, **3(9)**, 802–806, doi:10.1038/nclimate1979.

Hallegatte, S., A. Vogt-Schilb, M. Bangalore, and J. Rozenberg, 2017: *Unbreakable: Building the Resilience of the Poor in the Face of Natural Disasters*. The World Bank, Washington DC, USA, 201 pp., doi:10.1596/978-1-4648-1003-9.

Hallegatte, S. et al., 2016: *Shock waves: Managing the Impacts of Climate Change on Poverty*. World Bank Group, Washington DC, USA, 227 pp., doi:10.1596/978-1-4648-0673-5.

Halpern, B.S. et al., 2015: Spatial and temporal changes in cumulative human impacts on the world's ocean. *Nature Communications*, **6**, 7615, doi:10.1038/ncomms8615.

Hammill, A., R. Matthew, and E. McCarter, 2008: Microfinance and climate change adaptation. *IDS bulletin*, **39(4)**, 113–122.

Handgraaf, M.J.J., M.A. Lidth, D. Jeude, and K.C. Appelt, 2013: Public praise vs. private pay: Effects of rewards on energy conservation in the workplace. *Ecological Economics*, **86**, 86–92, doi:10.1016/j.ecolecon.2012.11.008.

Hangx, S.J.T. and C.J. Spiers, 2009: Coastal spreading of olivine to control atmospheric CO_2 concentrations: A critical analysis of viability. *International Journal of Greenhouse Gas Control*, **3(6)**, 757–767, doi:10.1016/j.ijggc.2009.07.001.

Haoqi, Q., W. Libo, and T. Weiqi, 2017: "Lock-in" effect of emission standard and its impact on the choice of market based instruments. *Energy Economics*, **63**, 41–50, doi:10.1016/j.eneco.2017.01.005.

Harberger, A.C., 1984: Basic Needs versus Distributional Weights in Social Cost-Benefit Analysis. *Economic Development and Cultural Change*, **32(3)**, 455–474, doi:10.1086/451400.

Harding, A. and J.B. Moreno-Cruz, 2016: Solar geoengineering economics: From incredible to inevitable and half-way back. *Earth's Future*, **4(12)**, 569–577, doi:10.1002/2016ef000462.

Hardoy, J. and L.S. Velásquez Barrero, 2014: Re-thinking "Biomanizales": addressing climate change adaptation in Manizales, Colombia. *Environment and Urbanization*, **26(1)**, 53–68.

Hardoy, J. and L.S. Velásquez Barrero, 2016: Manizales, Colombia. In: *Cities on a finite planet: Towards transformative responses to climate change* [Bartlett, S. and D. Satterthwaite (eds.)]. Routledge, Abingdon, UK and New York, NY, USA, pp. 274.

Harjanne, A., 2017: Servitizing climate science-Institutional analysis of climate services discourse and its implications. *Global Environmental Change*, **46**, 1–16, doi:10.1016/j.gloenvcha.2017.06.008.

Harris, Z.M., R. Spake, and G. Taylor, 2015: Land use change to bioenergy: A meta-analysis of soil carbon and GHG emissions. *Biomass and Bioenergy*, **82**, 27–39, doi:10.1016/j.biombioe.2015.05.008.

Harrison, D.P., 2013: A method for estimating the cost to sequester carbon dioxide by delivering iron to the ocean. *International Journal of Global Warming*, **5(3)**, 231–254, doi:10.1504/ijgw.2013.055360.

Harrison, D.P., 2017: Global negative emissions capacity of ocean macronutrient fertilization. *Environmental Research Letters*, **12(3)**, doi:10.1088/1748-9326/aa5ef5.

Hartley, P.R. and K.B. Medlock, 2013: The Valley of Death for New Energy Technologies. *The Energy Journal*, **38(3)**, 1–61, www.iaee.org/en/publications/ejarticle.aspx?id=2926.

Hartmann, J. and S. Kempe, 2008: What is the maximum potential for CO_2 sequestration by "stimulated" weathering on the global scale? *Naturwissenschaften*, **95(12)**, 1159–1164, doi:10.1007/s00114-008-0434-4.

Hartmann, J. et al., 2013: Enhanced chemical weathering as a geoengineering strategy to reduce atmospheric carbon dioxide, supply nutrients, and mitigate ocean acidification: Enhanced weathering. *Reviews of Geophysics*, **51(2)**, 113–149, doi:10.1002/rog.20004.

Harvey, C.A. et al., 2014: Climate-Smart Landscapes: Opportunities and Challenges for Integrating Adaptation and Mitigation in Tropical Agriculture. *Conservation Letters*, **7(2)**, 77–90, doi:10.1111/conl.12066.

Harvey, L.D.D., 2008: Mitigating the atmospheric CO_2 increase and ocean acidification by adding limestone powder to upwelling regions. *Journal of Geophysical Research: Oceans*, **113(C4)**, C04028, doi:10.1029/2007jc004373.

Hasanbeigi, A., M. Arens, and L. Price, 2014: Alternative emerging ironmaking technologies for energy-efficiency and carbon dioxide emissions reduction: A technical review. *Renewable and Sustainable Energy Reviews*, **33**, 645–658, doi:10.1016/j.rser.2014.02.031.

Haselip, J., U.E. Hansen, D. Puig, S. Trærup, and S. Dhar, 2015: Governance, enabling frameworks and policies for the transfer and diffusion of low carbon and climate adaptation technologies in developing countries. *Climatic Change*, **131(3)**, 363–370, doi:10.1007/s10584-015-1440-0.

Hauck, J., P. Köhler, D. Wolf-Gladrow, and C. Völker, 2016: Iron fertilisation and century-scale effects of open ocean dissolution of olivine in a simulated CO_2 removal experiment. *Environmental Research Letters*, **11(2)**, 024007, doi:10.1088/1748-9326/11/2/024007.

Havet, A. et al., 2014: Review of livestock farmer adaptations to increase forages in crop rotations in western France. *Agriculture, Ecosystems & Environment*, **190**, 120–127, doi:10.1016/j.agee.2014.01.009.

Havlik, P. et al., 2014: Climate change mitigation through livestock system transitions. *Proceedings of the National Academy of Sciences*, **111(10)**, 3709–3714, doi:10.1073/pnas.1308044111.

Hayashi, A., F. Sano, Y. Nakagami, and K. Akimoto, 2018: Changes in terrestrial water stress and contributions of major factors under temperature rise constraint scenarios. *Mitigation and Adaptation Strategies for Global Change*, 1–27, doi:10.1007/s11027-018-9780-5.

Heard, B.P., B.W. Brook, T.M.L. Wigley, and C.J.A. Bradshaw, 2017: Burden of proof: A comprehensive review of the feasibility of 100% renewable-electricity systems. *Renewable and Sustainable Energy Reviews*, **76**, 1122–1133, doi:10.1016/j.rser.2017.03.114.

Heck, V., D. Gerten, W. Lucht, and A. Popp, 2018: Biomass-based negative emissions difficult to reconcile with planetary boundaries. *Nature Climate Change*, **8(2)**, 151–155, doi:10.1038/s41558-017-0064-y.

Heckendorn, P. et al., 2009: The impact of geoengineering aerosols on stratospheric temperature and ozone. *Environmental Research Letters*, **4(4)**, 045108, doi:10.1088/1748-9326/4/4/045108.

Heede, R., 2014: Tracing anthropogenic carbon dioxide and methane emissions to fossil fuel and cement producers, 1854–2010. *Climatic Change*, **122(1)**, 229–241, doi:10.1007/s10584-013-0986-y.

Heidrich, O. et al., 2016: National climate policies across Europe and their impacts on cities strategies. *Journal of Environmental Management*, **168**, 36–45, doi:10.1016/j.jenvman.2015.11.043.

Hekkert, M.P., R.A.A. Suurs, S.O. Negro, S. Kuhlmann, and R.E.H.M. Smits, 2007: Functions of innovation systems: A new approach for analysing technological change. *Technological Forecasting and Social Change*, **74(4)**, 413–432, doi:10.1016/j.techfore.2006.03.002.

Henly-Shepard, S., S.A. Gray, and L.J. Cox, 2015: The use of participatory modeling to promote social learning and facilitate community disaster planning. *Environmental Science & Policy*, **45**, 109–122, doi:10.1016/j.envsci.2014.10.004.

Henry, R.C. et al., 2018: Food supply and bioenergy production within the global cropland planetary boundary. *PLOS ONE*, **13(3)**, e0194695, doi:10.1371/journal.pone.0194695.

Henry, R.K., Z. Yongsheng, and D. Jun, 2006: Municipal solid waste management challenges in developing countries – Kenyan case study. *Waste Management*, **26(1)**, 92–100, doi:10.1016/j.wasman.2005.03.007.

Henstra, D., 2016: The tools of climate adaptation policy: analysing instruments and instrument selection. *Climate Policy*, **16(4)**, 496–521, doi:10.1080/14693062.2015.1015946.

Hering, J.G., D.A. Dzombak, S.A. Green, R.G. Luthy, and D. Swackhamer, 2014: Engagement at the Science–Policy Interface. *Environmental Science & Technology*, **48(19)**, 1031–11033, doi:10.1021/es504225t.

Hermwille, L., 2016: Climate Change as a Transformation Challenge. A New Climate Policy Paradigm? *GAIA – Ecological Perspectives for Science and Society*, **25(1)**, 19–22, doi:10.14512/gaia.25.1.6.

Hermwille, L., W. Obergassel, H.E. Ott, and C. Beuermann, 2017: UNFCCC before and after Paris – what's necessary for an effective climate regime? *Climate Policy*, **17(2)**, 150–170, doi:10.1080/14693062.2015.1115231.

Hernández-Morcillo, M., P. Burgess, J. Mirck, A. Pantera, and T. Plieninger, 2018: Scanning agroforestry-based solutions for climate change mitigation and adaptation in Europe. *Environmental Science & Policy*, **80**, 44–52, doi:10.1016/j.envsci.2017.11.013.

Herrero, M. et al., 2015: Livestock and the Environment: What Have We Learned in the Past Decade? *Annual Review of Environment and Resources*, **40(1)**, 177–202, doi:10.1146/annurev-environ-031113-093503.

Herwehe, L. and C.A. Scott, 2018: Drought adaptation and development: small-scale irrigated agriculture in northeast Brazil. *Climate and Development*, **10(4)**, 337–346, doi:10.1080/17565529.2017.1301862.

Hess, J.S. and I. Kelman, 2017: Tourism Industry Financing of Climate Change Adaptation: Exploring the Potential in Small Island Developing States. *Climate, Disaster and Development Journal*, **2(2)**, 34–45, doi:10.18783/cddj.v002.i02.a04.

Hesse, C., 2016: *Decentralising climate finance to reach the most vulnerable*. International Institute for Environment and Development (IIED), London, UK, 4 pp.

Hetz, K., 2016: Contesting adaptation synergies: political realities in reconciling climate change adaptation with urban development in Johannesburg, South Africa. *Regional Environmental Change*, **16(4)**, 1171–1182, doi:10.1007/s10113-015-0840-z.

Heutel, G., J. Moreno-Cruz, and S. Shayegh, 2018: Solar geoengineering, uncertainty, and the price of carbon. *Journal of Environmental Economics and Management*, **87**, 24–41, doi:10.1016/j.jeem.2017.11.002.

Hewitt, C., S. Mason, and D. Walland, 2012: The Global Framework for Climate Services. *Nature Climate Change*, **2(12)**, 831–832, doi:10.1038/nclimate1745.

Heyen, D., T. Wiertz, and P.J. Irvine, 2015: Regional disparities in SRM impacts: the challenge of diverging preferences. *Climatic Change*, **133(4)**, 557–563, doi:10.1007/s10584-015-1526-8.

Hiç, C., P. Pradhan, D. Rybski, and J.P. Kropp, 2016: Food Surplus and Its Climate Burdens. *Environmental Science & Technology*, **50(8)**, 4269–4277, doi:10.1021/acs.est.5b05088.

Hill, J., E. Nelson, D. Tilman, S. Polasky, and D. Tiffany, 2006: Environmental, economic, and energetic costs and benefits of biodiesel and ethanol biofuels. *Proceedings of the National Academy of Sciences*, **103(30)**, 11206–11210, doi:10.1073/pnas.0604600103.

Hill Clarvis, M. and N.L. Engle, 2015: Adaptive capacity of water governance arrangements: a comparative study of barriers and opportunities in Swiss and US states. *Regional Environmental Change*, **15(3)**, 517–527, doi:10.1007/s10113-013-0547-y.

Hirschberg, S. et al., 2016: Health effects of technologies for power generation: Contributions from normal operation, severe accidents and terrorist threat. *Reliability Engineering & System Safety*, **145**, 373–387, doi:10.1016/j.ress.2015.09.013.

Hirth, L. and J.C. Steckel, 2016: The role of capital costs in decarbonizing the electricity sector. *Environmental Research Letters*, **11(11)**, 114010, doi:10.1088/1748-9326/11/11/114010.

Hiwasaki, L., E. Luna, Syamsidik, and J.A. Marçal, 2015: Local and indigenous knowledge on climate-related hazards of coastal and small island communities in Southeast Asia. *Climatic Change*, **128(1–2)**, 35–56, doi:10.1007/s10584-014-1288-8.

Hof, A.F. et al., 2017: Global and regional abatement costs of Nationally Determined Contributions (NDCs) and of enhanced action to levels well below 2°C and 1.5°C. *Environmental Science & Policy*, **71**, 30–40, doi:10.1016/j.envsci.2017.02.008.

Hogarth, J.R., 2017: Evolutionary models of sustainable economic change in Brazil: No-till agriculture, reduced deforestation and ethanol biofuels. *Environmental Innovation and Societal Transitions*, **24**, 130–141, doi:10.1016/j.eist.2016.08.001.

Höglund-Isaksson, L., 2012: Global anthropogenic methane emissions 2005-2030: technical mitigation potentials and costs. *Atmospheric Chemistry and Physics*, **12(19)**, 9079–9096, doi:10.5194/acp-12-9079-2012.

Höglund-Isaksson, L. et al., 2017: Cost estimates of the Kigali Amendment to phase-down hydrofluorocarbons. *Environmental Science & Policy*, **75**, 138–147, doi:10.1016/j.envsci.2017.05.006.

Högy, P. et al., 2009: Effects of elevated CO_2 on grain yield and quality of wheat: results from a 3-year free-air CO_2 enrichment experiment. *Plant Biology*, **11(s1)**, 60–69, doi:10.1111/j.1438-8677.2009.00230.x.

Höhne, N., H. Fekete, M.G.J. den Elzen, A.F. Hof, and T. Kuramochi, 2017: Assessing the ambition of post-2020 climate targets: a comprehensive framework. *Climate Policy*, 1–16, doi:10.1080/14693062.2017.1294046.

Holland, B., 2017: Procedural justice in local climate adaptation: political capabilities and transformational change. *Environmental Politics*, **26(3)**, 391–412, doi:10.1080/09644016.2017.1287625.

Holland, R.A. et al., 2015: A synthesis of the ecosystem services impact of second generation bioenergy crop production. *Renewable and Sustainable Energy Reviews*, **46**, 30–40, doi:10.1016/j.rser.2015.02.003.

Holmes, G. and D.W. Keith, 2012: An air-liquid contactor for large-scale capture of CO_2 from air. *Philosophical Transactions of the Royal Society A: Mathematical,*

4

Physical and Engineering Sciences, **370(1974)**, 4380–403, doi:10.1098/rsta.2012.0137.

Holmes, G. et al., 2013: Outdoor prototype results for direct atmospheric capture of carbon dioxide. *Energy Procedia*, **37**, 6079–6095, doi:10.1016/j.egypro.2013.06.537.

Hölscher, K., N. Frantzeskaki, and D. Loorbach, 2018: Steering transformations under climate change: capacities for transformative climate governance and the case of Rotterdam, the Netherlands. *Regional Environmental Change*, 1–15, doi:10.1007/s10113-018-1329-3.

Holz, C., S. Kartha, and T. Athanasiou, 2018: Fairly sharing 1.5: national fair shares of a 1.5°C-compliant global mitigation effort. *International Environmental Agreements: Politics, Law and Economics*, **18(1)**, 117–134, doi:10.1007/s10784-017-9371-z.

Honegger, M., K. Sugathapala, and A. Michaelowa, 2013: Tackling climate change: where can the generic framework be located? *Carbon & Climate Law Review*, **7(2)**, 125–135, doi:10.5167/uzh-86551.

Hong, N.B. and M. Yabe, 2017: Improvement in irrigation water use efficiency: a strategy for climate change adaptation and sustainable development of Vietnamese tea production. *Environment, Development and Sustainability*, **19(4)**, 1247–1263, doi:10.1007/s10668-016-9793-8.

Höök, M., J. Li, N. Oba, and S. Snowden, 2011: Descriptive and Predictive Growth Curves in Energy System Analysis. *Natural Resources Research*, **20(2)**, 103–116, doi:10.1007/s11053-011-9139-z.

Hornsey, M.J., E.A. Harris, P.G. Bain, and K.S. Fielding, 2016: Meta-analyses of the determinants and outcomes of belief in climate change. *Nature Climate Change*, **6(6)**, 622–626, doi:10.1038/nclimate2943.

Horton, J.B., 2011: Geoengineering and the Myth of Unilateralism: Pressures and Prospects for International Cooperation. *Stanford Journal of Law, Science & Policy*, **IV**, 56–69, doi:10.1017/cbo9781139161824.010.

Horton, J.B., D.W. Keith, and M. Honegger, 2016: *Implications of the Paris Agreement for Carbon Dioxide Removal and Solar Geoengineering*. Harvard Project on Climate Agreements, The Belfer Center for Science and International Affairs, Harvard University, Cambridge, MA, USA, 10 pp.

Horton, J.B. et al., 2018: Solar Geoengineering and Democracy. *Global Environmental Politics*, **18(3)**, 5–24, doi:10.1162/glep_a_00466.

Houghton, R.A., 2013: The emissions of carbon from deforestation and degradation in the tropics: past trends and future potential. *Carbon Management*, **4(5)**, 539–546, doi:10.4155/cmt.13.41.

Houghton, R.A., B. Byers, and A.A. Nassikas, 2015: A role for tropical forests in stabilizing atmospheric CO_2. *Nature Climate Change*, **5(12)**, 1022–1023, doi:10.1038/nclimate2869.

Hourcade, J.-C., P.-R. Shukla, and C. Cassen, 2015: Climate policy architecture for the Cancun paradigm shift: building on the lessons from history. *International Environmental Agreements: Politics, Law and Economics*, **15(4)**, 353–367, doi:10.1007/s10784-015-9301-x.

Hourcade, J.-C. et al., 2001: Global, Regional, and National Costs and Ancillary Benefits of Mitigation. In: *Climate Change 2001: Mitigation. Contribution of Working Group III to the Third Assessment Report of the Intergovernmental Panel on Climate Change* [Metz, B., O.R. Davidson, P.R. Bosch, R. Dave, and L.A. Meyer (eds.)]. Cambridge University Press, Cambridge, United Kingdom and New York, NY, USA, pp. 409–559.

House, K.Z., C.H. House, D.P. Schrag, and M.J. Aziz, 2007: Electrochemical acceleration of chemical weathering as an energetically feasible approach to mitigating anthropogenic climate change. *Environmental Science & Technology*, **41(24)**, 8464–8470, doi:10.1021/es0701816.

House, K.Z. et al., 2011: Economic and energetic analysis of capturing CO_2 from ambient air. *Proceedings of the National Academy of Sciences*, **108(51)**, 20428–20433, doi:10.1073/pnas.1012253108.

Hovi, J., D.F. Sprinz, H. Sælen, and A. Underdal, 2016: Climate change mitigation: a role for climate clubs? *Palgrave Communications*, **2**, 16020, doi:10.1057/palcomms.2016.20.

Howes, M. et al., 2015: Towards networked governance: improving interagency communication and collaboration for disaster risk management and climate change adaptation in Australia. *Journal of Environmental Planning and Management*, **58(5)**, 757–776, doi:10.1080/09640568.2014.891974.

Hoy, M.B., 2016: Smart Buildings: An Introduction to the Library of the Future. *Medical Reference Services Quarterly*, **35(3)**, 326–331, doi:10.1080/02763869.2016.1189787.

Hristov, A.N. et al., 2015: An inhibitor persistently decreased enteric methane emission from dairy cows with no negative effect on milk production. *Proceedings of the National Academy of Sciences*, **112(34)**, 10663–10668, doi:10.1073/pnas.1504124112.

HSRTF, 2013: Rebuild by Design Competition. Hurricane Sandy Rebuilding Task Force (HSRTF). Retrieved from: www.hud.gov/sandyrebuilding/rebuildbydesign.

Hsu, A., A.J. Weinfurter, and K. Xu, 2017: Aligning subnational climate actions for the new post-Paris climate regime. *Climatic Change*, **142(3–4)**, 419–432, doi:10.1007/s10584-017-1957-5.

Hu, A., Y. Xu, C. Tebaldi, W.M. Washington, and V. Ramanathan, 2013: Mitigation of short-lived climate pollutants slows sea-level rise. *Nature Climate Change*, **3**, 730–734, doi:10.1038/nclimate1869.

Huang, W., W. Chen, and G. Anandarajah, 2017a: The role of technology diffusion in a decarbonizing world to limit global warming to well below 2°C: An assessment with application of Global TIMES model. *Applied Energy*, **208**, 291–301, doi:10.1016/j.apenergy.2017.10.040.

Huang, W., D. Ma, and W. Chen, 2017b: Connecting water and energy: Assessing the impacts of carbon and water constraints on China's power sector. *Applied Energy*, **185**, 1497–1505, doi:10.1016/j.apenergy.2015.12.048.

Huang-Lachmann, J.-T. and J.C. Lovett, 2016: How cities prepare for climate change: Comparing Hamburg and Rotterdam. *Cities*, **54**, 36–44, doi:10.1016/j.cities.2015.11.001.

Huebner, G.M. and D. Shipworth, 2017: All about size? – The potential of downsizing in reducing energy demand. *Applied Energy*, **186**, 226–233, doi:10.1016/j.apenergy.2016.02.066.

Hufty, M., 2011: Investigating policy processes: The Governance Analytical Framework (GAF). In: *Research for Sustainable Development: Foundations, Experiences, and Perspectives* [Wiesmann, U. and H. Hurni (eds.)]. Perspectives of the Swiss National Centre of Competence in Research (NCCR) North-South Vol. 6, Geographica Bernensia, Bern, Switzerland, pp. 403–424.

Hughes, A.D. et al., 2012: Does seaweed offer a solution for bioenergy with biological carbon capture and storage? *Greenhouse Gases: Science and Technology*, **2(6)**, 402–407, doi:10.1002/ghg.1319.

IAEA, 2017: *Nuclear Technology Review 2017*. GC(61)/INF/4, International Atomic Energy Agency (IAEA), Vienna, Austria, 45 pp.

IAEA, 2018: Power Reactor Information System – Country Statistics: France. Retrieved from: www.iaea.org/pris/countrystatistics/countrydetails.aspx?current=fr.

Ickowitz, A., E. Sills, and C. de Sassi, 2017: Estimating Smallholder Opportunity Costs of REDD+: A Pantropical Analysis from Households to Carbon and Back. *World Development*, **95**, 15–26, doi:10.1016/j.worlddev.2017.02.022.

Idem, R. et al., 2015: Practical experience in post-combustion CO_2 capture using reactive solvents in large pilot and demonstration plants. *International Journal of Greenhouse Gas Control*, **40**, 6–25, doi:10.1016/j.ijggc.2015.06.005.

IEA, 2014a: *World Energy Investment Outlook Special Report*. International Energy Agency (IEA), Paris, France, 190 pp.

IEA, 2014b: *World Energy Outlook 2014*. International Energy Agency (IEA), Paris, France, 748 pp.

IEA, 2016a: *20 Years of Carbon Capture and Storage – Accelerating Future Deployment*. International Energy Agency (IEA), Paris, France, 115 pp.

IEA, 2016b: *Solid State Lighting Annex. Task 7: Smart Lighting – New Features Impacting Energy Consumption. First Status Report*. Energy Efficient End-Use Equipment (4E). International Energy Agency (IEA), Paris, France, 41 pp.

IEA, 2016c: *World Energy Outlook 2016*. International Energy Agency (IEA), Paris, France, 684 pp.

IEA, 2017a: *Digitalization & Energy*. International Energy Agency (IEA), Paris, France, 188 pp.

IEA, 2017b: *Energy Access Outlook 2017*. International Energy Agency (IEA), Paris, France, 144 pp.

IEA, 2017c: *Energy Technology Perspectives 2017: Catalysing Energy Technology Transformations*. International Energy Agency (IEA), Paris, France, 443 pp.

IEA, 2017d: *Getting Wind and Sun onto the Grid A Manual for Policy Makers*. International Energy Agency (IEA), Paris, France, 64 pp.

IEA, 2017e: *The Future of Trucks: Implications for energy and the environment*. International Energy Agency (IEA), Paris, France, 161 pp.

IEA, 2017f: *Tracking Clean Energy Progress 2017*. Energy Technology Perspectives 2017 Excerpt. International Energy Agency (IEA), Paris, France, 112 pp.

IEA, 2017g: *World Energy Outlook 2017*. International Energy Agency (IEA), Paris, France, 748 pp.

IEA, 2018: *Global EV Outlook 2018: Towards cross-modal electrification*. International Energy Agency (IEA), Paris, France, 141 pp.,

4

doi:10.1787/9789264302365-en.

Iguma, H. and H. Kidoshi, 2015: *Why Toyota can sell Mirai at 7 million Yen?* (in Japanese). Nikkan-Kogyo Press, Tokyo, Japan, 176 pp.

IIF, 2017: Global Debt Monitor. Institute of International Finance (IIF). Retrieved from: www.iif.com/publication/global-debt-monitor/global-debt-monitor-april-2018.

Iiyama, M. et al., 2017: Understanding patterns of tree adoption on farms in semi-arid and sub-humid Ethiopia. *Agroforestry Systems*, **91(2)**, 271–293, doi:10.1007/s10457-016-9926-y.

IMF, 2014: *World Economic Outlook October 2014: Legacies, Clouds, Uncertainties.* International Monetary Fund (IMF), Washington DC, USA, 222 pp.

IMO, 1972: *Convention on the Prevention of Marine Pollution by Dumping of Wastes and Other Matter.* International Maritime Organization (IMO), London, UK, 16 pp.

IMO, 1996: *1996 Protocol to the Convention on the Prevention of Marine Pollution by Dumping of Wastes and Other Matter, 1972.* International Maritime Organization (IMO), London, United Kingdom, 25 pp.

Inamara, A. and V. Thomas, 2017: Pacific climate change adaptation: The use of participatory media to promote indigenous knowledge. *Pacific Journalism Review*, **23(1)**, 113–132, doi:10.24135/pjr.v23i1.210.

Ingold, K. and M. Fischer, 2014: Drivers of collaboration to mitigate climate change: An illustration of Swiss climate policy over 15 years. *Global Environmental Change*, **24**, 88–98, doi:10.1016/j.gloenvcha.2013.11.021.

Ingty, T., 2017: High mountain communities and climate change: adaptation, traditional ecological knowledge, and institutions. *Climatic Change*, **145(1–2)**, 41–55, doi:10.1007/s10584-017-2080-3.

IPCC, 1995: Climate Change 1995: Economic and Social Dimensions of Climate Change Contribution of Working Group III to the IPCC Second Assessment Report. [Bruce, J.P., H. Lee, and E.F. Haites (eds.)]. Cambridge University Press, Cambridge, United Kingdom and New York, NY, USA, 339 pp.

IPCC, 2001: Climate Change 2001: Mitigation. Contribution of Working Group III to the IPCC Third Assessment Report. [Metz, B., O. Davidson, R. Swart, and J. Pan (eds.)]. Cambridge University Press, Cambridge, United Kingdom and New York, NY, USA, 753 pp.

IPCC, 2005a: IPCC Special Report on Carbon Dioxide Capture and Storage. Prepared by Working Group III of the Intergovernmental Panel on Climate Change. [Metz, B., O. Davidson, H.C. de Coninck, M. Loos, and L.A. Meyer (eds.)]. 442 pp.

IPCC, 2005b: Special Report on Carbon Dioxide Capture and Storage. [Metz, B., O. Davidson, H.C. de Coninck, M. Loos, and L.A. Meyer (eds.)]. Prepared by Working Group III of the Intergovernmental Panel on Climate Change. Cambridge University Press, Cambridge, United Kingdom and New York, NY, USA, 442 pp.

IPCC, 2012: Managing the Risks of Extreme Events and Disasters to Advance Climate Change Adaptation. A Special Report of Working Groups I and II of IPCC Intergovernmental Panel on Climate Change. [Field, C.B., V. Barros, T.F. Stocker, Q. Dahe, D.J. Dokken, K.L. Ebi, M.D. Mastrandrea, K.J. Mach, G.-K. Plattner, S.K. Allen, and Others (eds.)]. Cambridge University Press, Cambridge, United Kingdom and New York, NY, USA, 594 pp.

IPCC, 2014a: Climate Change 2014: Impacts, Adaptation, and Vulnerability. Contribution of Working Group II to the Fifth Assessment Report of the Intergovernmental Panel on Climate Change. [Barros, V.R., C.B. Field, D.J. Dokken, M.D. Mastrandrea, K.J. Mach, T.E. Bilir, M. Chatterjee, K.L. Ebi, Y.O. Estrada, R.C. Genova, B. Girma, E.S. Kissel, A.N. Levy, S. MacCracken, P.R. Mastrandrea, and L.L. White (eds.)]. Cambridge University Press, Cambridge, United Kingdom and New York, NY, USA, 1820 pp.

IPCC, 2014b: Climate Change 2014: Mitigation of Climate Change. Contribution of Working Group III to the Fifth Assessment Report of the Intergovernmental Panel on Climate Change. [Edenhofer, O., R. Pichs-Madruga, Y. Sokona, E. Farahani, S. Kadner, K. Seyboth, A. Adler, I. Baum, S. Brunner, P. Eickemeier, B. Kriemann, J. Savolainen, S. Schlömer, C. von Stechow, T. Zwickel, and J.C. Minx (eds.)]. Cambridge University Press, Cambridge, United Kingdom and New York, NY, USA, 1454 pp.

IPCC, 2014c: Climate Change 2014: Synthesis Report. Contribution of Working Groups I, II and III to the Fifth Assessment Report of the Intergovernmental Panel on Climate Change. [Core Writing Team, R.K. Pachauri, and L.A. Meyer (eds.)]. Intergovernmental Panel on Climate Change (IPCC), Geneva, Switzerland, 169 pp.

IRENA, 2013: *Smart Grids and Renewables: A Guide for Effective Deployment.* International Renewable Energy Agency (IRENA), Abu Dhabi, UAE, 47 pp.

IRENA, 2015: *Renewable energy options for shipping: Technology Brief.* International Renewable Energy Agency (IRENA), Bonn, Germany, 58 pp.

IRENA, 2017a: *Electricity storage and renewables: Costs and markets to 2030.* International Renewable Energy Agency (IRENA), Abu Dhabi, UAE, 131 pp.

IRENA, 2017b: *Renewable Energy and Jobs: Annual Review 2017.* International Renewable Energy Agency (IRENA), Abu Dhabi, United Arab Emirates, 24 pp.

IRENA, 2018: *Renewable Power Generation Costs in 2017.* International Renewable Energy Agency (IRENA), Abu Dhabi, UAE, 158 pp.

Irvine, P.J., A. Ridgwell, and D.J. Lunt, 2011: Climatic effects of surface albedo geoengineering. *Journal of Geophysical Research: Atmospheres*, **116(D24)**, D24112, doi:10.1029/2011jd016281.

Irvine, P.J., B. Kravitz, M.G. Lawrence, and H. Muri, 2016: An overview of the Earth system science of solar geoengineering. *Wiley Interdisciplinary Reviews: Climate Change*, **7(6)**, 815–833, doi:10.1002/wcc.423.

Isayama, K. and N. Ono, 2015: Steps towards sustainable and resilient disaster management in Japan: Lessons from Cuba. *International Journal of Health System and Disaster Management*, **3(2)**, 54–60, www.ijhsdm.org/article.asp?issn=2347-9019;year=2015;volume=3;issue=2;spage=54;epage=60;aulast=isayama;t=6.

International Social Science Council (ISSC) and United Nations Educational, Scientic and Cultural Organization (UNESCO), 2013: *World Social Science Report 2013: Changing Global Environments.* ISSC and UNESCO, Paris, France, 612 pp.

ITF, 2015: *Urban Mobility System Upgrade: How shared self-driving cars could change city traffic.* International Transport Forum. OECD Publishing, Paris, France, 34 pp.

Izrael, Y.A., E.M. Volodin, S. Kostrykin, A.P. Revokatova, and A.G. Ryaboshapko, 2014: The ability of stratospheric climate engineering in stabilizing global mean temperatures and an assessment of possible side effects. *Atmospheric Science Letters*, **15(2)**, 140–148, doi:10.1002/asl2.481.

Jabeen, H., 2014: Adapting the built environment: the role of gender in shaping vulnerability and resilience to climate extremes in Dhaka. *Environment & Urbanization*, **26(1)**, 147–165, doi:10.1177/0956247813517851.

Jaboyedoff, P., C.A. Roulet, V. Dorer, A. Weber, and A. Pfeiffer, 2004: Energy in air-handling units – results of the AIRLESS European Project. *Energy and Buildings*, **36(4)**, 391–399, doi:10.1016/j.enbuild.2004.01.047.

Jackson, L.S., J.A. Crook, and P.M. Forster, 2016: An intensified hydrological cycle in the simulation of geoengineering by cirrus cloud thinning using ice crystal fall speed changes. *Journal of Geophysical Research: Atmospheres*, **121(12)**, 6822–6840, doi:10.1002/2015jd024304.

Jackson, L.S. et al., 2015: Assessing the controllability of Arctic sea ice extent by sulfate aerosol geoengineering. *Geophysical Research Letters*, **42(4)**, doi:10.1002/2014gl062240.

Jackson, T. and P. Senker, 2011: Prosperity without growth: Economics for a finite planet. *Energy & Environment*, **22(7)**, 1013–1016, doi:10.1260/0958-305x.22.7.1013.

Jacobi, J., S. Rist, and M.A. Altieri, 2017: Incentives and disincentives for diversified agroforestry systems from different actors' perspectives in Bolivia. *International Journal of Agricultural Sustainability*, **15(4)**, 365–379, doi:10.1080/14735903.2017.1332140.

Jacobson, M.Z. and J.E. Ten Hoeve, 2012: Effects of urban surfaces and white roofs on global and regional climate. *Journal of Climate*, **25(3)**, 1028–1044, doi:10.1175/jcli-d-11-00032.1.

Jacobson, M.Z., M.A. Delucchi, M.A. Cameron, B.A. Frew, and S. Polasky, 2015: Low-cost solution to the grid reliability problem with 100% penetration of intermittent wind, water, and solar for all purposes. *Proceedings of the National Academy of Sciences*, 15060–15065, doi:10.1073/pnas.1510028112.

Jacobson, M.Z. et al., 2017: 100% Clean and Renewable Wind, Water, and Sunlight All-Sector Energy Roadmaps for 139 Countries of the World. *Joule*, **1(1)**, 108–121, doi:10.1016/j.joule.2017.07.005.

Jaeger, C.C., A. Haas, and K. Töpfer, 2013: *Sustainability, Finance, and a Proposal from China.* Institute for Advanced Sustainability Studies, Potsdam, Germany, 31 pp., doi:10.2312/iass.2013.004.

Jägermeyr, J. et al., 2015: Water savings potentials of irrigation systems: global simulation of processes and linkages. *Hydrology and Earth System Sciences*, **19(7)**, 3073–3091, doi:10.5194/hess-19-3073-2015.

Jaglin, S., 2014: Regulating Service Delivery in Southern Cities: Rethinking urban heterogeneity. In: *The Routledge Handbook on Cities of the Global South* [Parnell, S. and S. Oldfield (eds.)]. Routledge, Abingdon, UK, pp. 434–447, doi:10.4324/9780203387832.ch37.

Jahandideh-Tehrani, M., O. Bozorg Haddad, and H.A. Loáiciga, 2014: Hydropower Reservoir Management Under Climate Change: The Karoon Reservoir System. *Water Resources Management*, **29(3)**, 749–770, doi:10.1007/s11269-014-0840-7.

Jaiswal, D. et al., 2017: Brazilian sugarcane ethanol as an expandable green

4

alternative to crude oil use. *Nature Climate Change*, **7(11)**, 788–792, doi:10.1038/nclimate3410.

Jakob, M. and J. Hilaire, 2015: Climate science: Unburnable fossil-fuel reserves. *Nature*, **517(7533)**, 150–152, doi:10.1038/517150a.

Jakob, M., C. Chen, S. Fuss, A. Marxen, and O. Edenhofer, 2015: Development incentives for fossil fuel subsidy reform. *Nature Climate Change*, **5(8)**, 709–712, doi:10.1038/nclimate2679.

Jakob, M. et al., 2016: Carbon Pricing Revenues Could Close Infrastructure Access Gaps. *World Development*, **84**, 254–265, doi:10.1016/j.worlddev.2016.03.001.

Jakovcevic, A. and L. Steg, 2013: Sustainable transportation in Argentina: Values, beliefs, norms and car use reduction. *Transportation Research Part F: Traffic Psychology and Behaviour*, **20**, 70–79, doi:10.1016/j.trf.2013.05.005.

James, R., R. Washington, C.-F. Schleussner, J. Rogelj, and D. Conway, 2017: Characterizing half-a-degree difference: a review of methods for identifying regional climate responses to global warming targets. *Wiley Interdisciplinary Reviews: Climate Change*, **8(2)**, e457, doi:10.1002/wcc.457.

Jänicke, M. and R. Quitzow, 2017: Multi-level Reinforcement in European Climate and Energy Governance: Mobilizing economic interests at the sub-national levels. *Environmental Policy and Governance*, **27(2)**, 122–136, doi:10.1002/eet.1748.

Janif, S.Z. et al., 2016: Value of traditional oral narratives in building climate-change resilience: insights from rural communities in Fiji. *Ecology and Society*, **21(2)**, 7, doi:10.5751/es-08100-210207.

Jansson, J., 2011: Consumer eco-innovation adoption: Assessing attitudinal factors and perceived product characteristics. *Business Strategy and the Environment*, **20(3)**, 192–210, doi:10.1002/bse.690.

Jantke, K., J. Müller, N. Trapp, and B. Blanz, 2016: Is climate-smart conservation feasible in Europe? Spatial relations of protected areas, soil carbon, and land values. *Environmental Science & Policy*, **57**, 40–49, doi:10.1016/j.envsci.2015.11.013.

Jasanoff, S., 2007: Technologies of humility. *Nature*, **450(7166)**, 33–33, doi:10.1038/450033a.

Jayachandran, S. et al., 2017: Cash for carbon: A randomized trial of payments for ecosystem services to reduce deforestation. *Science*, **357(6348)**, 267–273, doi:10.1126/science.aan0568.

Jenkins, K., B.K. Sovacool, and D. McCauley, 2018: Humanizing sociotechnical transitions through energy justice: An ethical framework for global transformative change. *Energy Policy*, **117**, 66–74, doi:10.1016/j.enpol.2018.02.036.

Jenkins, K., S. Surminski, J. Hall, and F. Crick, 2017: Assessing surface water flood risk and management strategies under future climate change: Insights from an Agent-Based Model. *Science of The Total Environment*, **595**, 159–168, doi:10.1016/j.scitotenv.2017.03.242.

Jennings, V., L. Larson, and J. Yun, 2016: Advancing Sustainability through Urban Green Space: Cultural Ecosystem Services, Equity, and Social Determinants of Health. *International Journal of Environmental Research and Public Health*, **13(2)**, 196, doi:10.3390/ijerph13020196.

Jensen, C.B. and J.J. Spoon, 2011: Testing the 'Party Matters' Thesis: Explaining Progress towards Kyoto Protocol Targets. *Political Studies*, **59(1)**, 99–115, doi:10.1111/j.1467-9248.2010.00852.x.

Jensen, N. and C. Barrett, 2017: Agricultural index insurance for development. *Applied Economic Perspectives and Policy*, **39(2)**, 199–219, doi:10.1093/aepp/ppw022.

Jeong, S. and Y.-Y. An, 2016: Climate change risk assessment method for electrical facility. In: *2016 International Conference on Information and Communication Technology Convergence, ICTC 2016*, doi:10.1109/ictc.2016.7763464.

Jeuland, M., S.K. Pattanayak, and R. Bluffstone, 2015: The Economics of Household Air Pollution. *Annual Review of Resource Economics*, **7(1)**, 81–108, doi:10.1146/annurev-resource-100814-125048.

Ji, Z. and F. Sha, 2015: The challenges of the post-COP21 regime: interpreting CBDR in the INDC context. *International Environmental Agreements: Politics, Law and Economics*, **15**, 421–430, doi:10.1007/s10784-015-9303-8.

Jiang, K., K. Tamura, and T. Hanaoka, 2017: Can we go beyond INDCs: Analysis of a future mitigation possibility in China, Japan, EU and the US. *Advances in Climate Change Research*, **8(2)**, 117–122, doi:10.1016/j.accre.2017.05.005.

Jiang, Y. et al., 2016: Importance of soil thermal regime in terrestrial ecosystem carbon dynamics in the circumpolar north. *Global and Planetary Change*, **142**, 28–40, doi:10.1016/j.gloplacha.2016.04.011.

Jiang, Z., T. Xiao, V.L. Kuznetsov, and P.P. Edwards, 2010: Turning carbon dioxide into fuel. *Philosophical Transactions of the Royal Society A: Mathematical, Physical and Engineering Sciences*, **368(1923)**, 3343–3364, doi:10.1098/rsta.2010.0119.

Jimenez, R., 2017: *Development Effects of Rural Electrification*. Policy Brief No IDB-PB-261, Infrastructure and Energy Sector – Energy Division, Inter-American Development Bank (IDB), Washington DC, USA, 20 pp.

Jin, J., W. Wang, and X. Wang, 2016: Farmers' Risk Preferences and Agricultural Weather Index Insurance Uptake in Rural China. *International Journal of Disaster Risk Science*, **7(4)**, 366–373, doi:10.1007/s13753-016-0108-3.

Johannessen, S.C. and R.W. Macdonald, 2016: Geoengineering with seagrasses: is credit due where credit is given? *Environmental Research Letters*, **11(11)**, 113001, doi:10.1088/1748-9326/11/11/113001.

Johnson, C. and J. Walker, 2016: *Peak car ownership report: The Market Opportunity of Electric Automated Mobility Services*. Rocky Mountain Institute (RMI), Colorado, USA, 30 pp.

Johnson, N. et al., 2015: The contributions of Community-Based monitoring and traditional knowledge to Arctic observing networks: Reflections on the state of the field. *Arctic*, **68(5)**, 1–13, doi:10.14430/arctic4447.

Jones, A. and J.M. Haywood, 2012: Sea-spray geoengineering in the HadGEM2-ES earth-system model: Radiative impact and climate response. *Atmospheric Chemistry and Physics*, **12(22)**, 10887–10898, doi:10.5194/acp-12-10887-2012.

Jones, A., J. Haywood, and O. Boucher, 2011: A comparison of the climate impacts of geoengineering by stratospheric SO$_2$ injection and by brightening of marine stratocumulus cloud. *Atmospheric Science Letters*, **12(2)**, 176–183, doi:10.1002/asl.291.

Jones, A. et al., 2013: The impact of abrupt suspension of solar radiation management (termination effect) in experiment G2 of the Geoengineering Model Intercomparison Project (GeoMIP). *Journal of Geophysical Research: Atmospheres*, **118(17)**, 9743–9752, doi:10.1002/jgrd.50762.

Jones, A.C. et al., 2018: Regional climate impacts of stabilizing global warming at 1.5K using solar geoengineering. *Earth's Future*, 1–22, doi:10.1002/2017ef000720.

Jones, A.D., K. Calvin, W.D. Collins, and J. Edmonds, 2015: Accounting for radiative forcing from albedo change in future global land-use scenarios. *Climatic Change*, **131(4)**, 691–703, doi:10.1007/s10584-015-1411-5.

Jones, C.D. et al., 2016: Simulating the Earth system response to negative emissions. *Environmental Research Letters*, **11(9)**, 095012, doi:10.1088/1748-9326/11/9/095012.

Jones, I.S.F., 2014: The cost of carbon management using ocean nourishment. *International Journal of Climate Change Strategies and Management*, **6(4)**, 391–400, doi:10.1108/ijccsm-11-2012-0063.

Jones, L., B. Harvey, and R. Godfrey-Wood, 2016: *The changing role of NGOs in supporting climate services*. Building Resilience and Adaptation to Climate Extremes and Disasters (BRACED), London, UK, 24 pp.

Jordan, A.J. et al., 2015: Emergence of polycentric climate governance and its future prospects. *Nature Climate Change*, **5(11)**, 977–982, doi:10.1038/nclimate2725.

Joyette, A.R.T., L.A. Nurse, and R.S. Pulwarty, 2015: Disaster risk insurance and catastrophe models in risk-prone small Caribbean islands. *Disasters*, **39(3)**, 467–492, doi:10.1111/disa.12118.

Juhl, H.J., M.H.J. Fenger, and J. Thøgersen, 2017: Will the Consistent Organic Food Consumer Step Forward? An Empirical Analysis. *Journal of Consumer Research*, 1–17, doi:10.1093/jcr/ucx052.

Juhola, S., E. Glaas, B.O. Linnér, and T.S. Neset, 2016: Redefining maladaptation. *Environmental Science & Policy*, **55**, 135–140, doi:10.1016/j.envsci.2015.09.014.

Kabir, E., K.-H. Kim, and J. Szulejko, 2017: Social Impacts of Solar Home Systems in Rural Areas: A Case Study in Bangladesh. *Energies*, **10(10)**, 1615, doi:10.3390/en10101615.

Kahan, D., 2010: Fixing the communications failure. *Nature*, **463(7279)**, 296–297, doi:10.1038/463296a.

Kahneman, D., 2003: A perspective on judgment and choice: Mapping bounded rationality. *American Psychologist*, **58(9)**, 697–720, doi:10.1037/0003-066x.58.9.697.

Kaika, M., 2017: 'Don't call me resilient again!' The New Urban Agenda as immunology … or … what happens when communities refuse to be vaccinated with 'smart cities' and indicators. *Environment & Urbanization*, **29(1)**, 89–102, doi:10.1177/0956247816684763.

Kale, E., 2015: Problematic Uses and Practices of Farm Ponds in Maharashtra.

Economic and Political Weekly, **52(3)**, 7–8, www.epw.in/journal/2017/3/ commentary/problematic-uses-and-practices-farm-ponds-maharashtra.html.

Kammann, C. et al., 2017: Biochar as a tool to reduce the agricultural greenhouse-gas burden – knowns, unknowns and future research needs. *Journal of Environmental Engineering and Landscape Management*, **25(2)**, 114–139, doi:10.3846/16486897.2017.1319375.

Kärcher, B., 2017: Cirrus Clouds and Their Response to Anthropogenic Activities. *Current Climate Change Reports*, **3(1)**, 45–57, doi:10.1007/s40641-017-0060-3.

Kardooni, R., S.B. Yusoff, and F.B. Kari, 2016: Renewable energy technology acceptance in Peninsular Malaysia. *Energy Policy*, **88**, 1–10, doi:10.1016/j.enpol.2015.10.005.

Kärki, J., E. Tsupari, and A. Arasto, 2013: CCS feasibility improvement in industrial and municipal applications by heat utilization. *Energy Procedia*, **37**, 2611–2621, doi:10.1016/j.egypro.2013.06.145.

Karlin, B., J.F. Zinger, and R. Ford, 2015: The effects of feedback on energy conservation: A meta-analysis.. *Psychological Bulletin*, **141(6)**, 1205–1227, doi:10.1037/a0039650.

Kashima, Y., A. Paladino, and E.A. Margetts, 2014: Environmentalist identity and environmental striving. *Journal of Environmental Psychology*, **38**, 64–75, doi:10.1016/j.jenvp.2013.12.014.

Kashimura, H. et al., 2017: Shortwave radiative forcing, rapid adjustment, and feedback to the surface by sulfate geoengineering: Analysis of the Geoengineering Model Intercomparison Project G4 scenario. *Atmospheric Chemistry and Physics*, **17(5)**, 3339–3356, doi:10.5194/acp-17-3339-2017.

Kasser, T. and K.M. Sheldon, 2002: What Makes for a Merry Christmas? *Journal of Happiness Studies*, **3(4)**, 313–329, doi:10.1023/a:1021516410457.

Kasser, T., S. Cohn, A.D. Kanner, and R.M. Ryan, 2007: Some Costs of American Corporate Capitalism: A Psychological Exploration of Value and Goal Conflicts. *Psychological Inquiry*, **18(1)**, 1–22, doi:10.1080/10478400701386579.

Kastner, I. and P.C. Stern, 2015: Examining the decision-making processes behind household energy investments: A review. *Energy Research & Social Science*, **10**, 72–89, doi:10.1016/j.erss.2015.07.008.

Kates, R.W., W.R. Travis, and T.J. Wilbanks, 2012: Transformational adaptation when incremental adaptations to climate change are insufficient. *Proceedings of the National Academy of Sciences*, **109(19)**, 7156–61, doi:10.1073/pnas.1115521109.

Kato, T. and J. Ellis, 2016: *Communicating Progress in National and Global Adaptation to Climate Change*. OECD/IEA Climate Change Expert Group Papers No. 47, OECD Publishing, Paris, France, 47 pp., doi:10.1787/5jlww009v1hj-en.

Kauffman, S.A., 2002: *Investigations*. Oxford University Press, Oxford, UK, 308 pp.

KCCA, 2016: *Kampala Climate Change Action Plan*. Kampala Capital City Authority, Kampala, Uganda, 50 pp.

Keen, M., I. Parry, and J. Strand, 2013: Planes, ships and taxes: charging for international aviation and maritime emissions. *Economic Policy*, **28(76)**, 701–749, doi:10.1111/1468-0327.12019.

Keim, M.E., 2008: Adaptation to Climate Change. *American Journal of Preventive Medicine*, **35(5)**, 508–516, doi:10.1016/j.amepre.2008.08.022.

Keith, D.W. and P.J. Irvine, 2016: Solar geoengineering could substantially reduce climate risks – a research hypothesis for the next decade. *Earth's Future*, **4(11)**, 549–559, doi:10.1002/2016ef000465.

Keith, D.W., M. Ha-Duong, and J.K. Stolaroff, 2006: Climate Strategy with CO_2 Capture from the Air. *Climatic Change*, **74(1–3)**, 17–45, doi:10.1007/s10584-005-9026-x.

Keith, D.W., G. Wagner, and C.L. Zabel, 2017: Solar geoengineering reduces atmospheric carbon burden. *Nature Climate Change*, **7(9)**, 617–619, doi:10.1038/nclimate3376.

Keller, D.P., E.Y. Feng, and A. Oschlies, 2014: Potential climate engineering effectiveness and side effects during a high carbon dioxide-emission scenario. *Nature Communications*, **5**, 3304, doi:10.1038/ncomms4304.

Keller, D.P. et al., 2018: The Carbon Dioxide Removal Model Intercomparison Project (CDRMIP): rationale and experimental protocol for CMIP6. *Geoscientific Model Development*, **11(3)**, 1133–1160, doi:10.5194/gmd-11-1133-2018.

Kelleway, J.J. et al., 2017: Review of the ecosystem service implications of mangrove encroachment into salt marshes. *Global Change Biology*, **23(10)**, 3967–3983, doi:10.1111/gcb.13727.

Kelman, I., 2015: Difficult decisions: Migration from Small Island Developing States under climate change. *Earth's Future*, **3(4)**, 133–142, doi:10.1002/2014ef000278.

Kelman, I., 2017: Linking disaster risk reduction, climate change, and the sustainable development goals. *Disaster Prevention and Management: An International Journal*, **26(3)**, 254–258, doi:10.1108/dpm-02-2017-0043.

Kelman, I., J.C. Gaillard, and J. Mercer, 2015: Climate Change's Role in Disaster Risk Reduction's Future: Beyond Vulnerability and Resilience. *International Journal of Disaster Risk Science*, **6(1)**, 21–27, doi:10.1007/s13753-015-0038-5.

Kemper, J., 2015: Biomass and carbon dioxide capture and storage: A review. *International Journal of Greenhouse Gas Control*, **40**, 401–430, doi:10.1016/j.ijggc.2015.06.012.

Kennedy, C., I.D. Stewart, M.I. Westphal, A. Facchini, and R. Mele, 2018: Keeping global climate change within 1.5°C through net negative electric cities. *Current Opinion in Environmental Sustainability*, **30**, 18–25, doi:10.1016/j.cosust.2018.02.009.

Kennedy, C.A. et al., 2015: Energy and material flows of megacities. *Proceedings of the National Academy of Sciences*, **112(19)**, 5985–5990, doi:10.1073/pnas.1504315112.

Keohane, N., A. Petsonk, and A. Hanafi, 2017: Toward a club of carbon markets. *Climatic Change*, **144(1)**, 81–95, doi:10.1007/s10584-015-1506-z.

Keohane, R.O. and D.G. Victor, 2011: The Regime Complex for Climate Change. *Perspectives on Politics*, **9(1)**, 7–23, doi:10.1017/s1537592710004068.

Kern, F. and K.S. Rogge, 2016: The pace of governed energy transitions: Agency, international dynamics and the global Paris agreement accelerating decarbonisation processes? *Energy Research & Social Science*, **22**, 13–17, doi:10.1016/j.erss.2016.08.016.

Kern, F., J. Gaede, J. Meadowcroft, and J. Watson, 2016: The political economy of carbon capture and storage: An analysis of two demonstration projects. *Technological Forecasting and Social Change*, **102**, 250–260, doi:10.1016/j.techfore.2015.09.010.

Kern, K. and G. Alber, 2009: Governing Climate Change in Cities: Modes of Urban Climate Governance in Multi-level Systems. In: *The international conference on Competitive Cities and Climate Change, Milan, Italy*. Organisation for Economic Co-operation and Development (OECD), Paris, France, pp. 171–196.

Kerstholt, J., H. Duijnhoven, and D. Paton, 2017: Flooding in The Netherlands: How people's interpretation of personal, social and institutional resources influence flooding preparedness. *International Journal of Disaster Risk Reduction*, **24**, 52–57, doi:10.1016/j.ijdrr.2017.05.013.

Keskitalo, E.C.H., H. Dannevig, G.K. Hovelsrud, J.J. West, and G. Swartling, 2011: Adaptive capacity determinants in developed states: Examples from the Nordic countries and Russia. *Regional Environmental Change*, **11(3)**, 579–592, doi:10.1007/s10113-010-0182-9.

Kettle, N.P. et al., 2014: Integrating scientific and local knowledge to inform risk-based management approaches for climate adaptation. *Climate Risk Management*, **4**, 17–31, doi:10.1016/j.crm.2014.07.001.

Khan, M.R., 2013: *Toward a binding climate change adaptation regime: a proposed framework*. Routledge, Abingdon, UK and New York, NY, USA, 280 pp.

Kholod, N. and M. Evans, 2016: Reducing black carbon emissions from diesel vehicles in Russia: An assessment and policy recommendations. *Environmental Science & Policy*, **56**, 1–8, doi:10.1016/j.envsci.2015.10.017.

Kilpeläinen, A. et al., 2017: Effects of Initial Age Structure of Managed Norway Spruce Forest Area on Net Climate Impact of Using Forest Biomass for Energy. *BioEnergy Research*, **10(2)**, 499–508, doi:10.1007/s12155-017-9821-z.

Kim, S. and W. Shin, 2017: Understanding American and Korean Students' Support for Pro-environmental Tax Policy: The Application of the Value-Belief-Norm Theory of Environmentalism. *Environmental Communication*, **11**, 311–331, doi:10.1080/17524032.2015.1088458.

Kimball, S. et al., 2015: Cost-effective ecological restoration. *Restoration Ecology*, **23(6)**, 800–810, doi:10.1111/rec.12261.

Kimura, S., 2016: When a Seawall Is Visible: Infrastructure and Obstruction in Post-tsunami Reconstruction in Japan. *Science as Culture*, **25(1)**, 23–43, doi:10.1080/09505431.2015.1081501.

King, S.D., 2011: *Losing control: the emerging threats to Western prosperity*. Yale University Press, New Haven, CT, USA and London, UK, 304 pp.

Kinley, R., 2017: Climate change after Paris: from turning point to transformation. *Climate Policy*, **17(1)**, 9–15, doi:10.1080/14693062.2016.1191009.

Kirk, E.J., 2017: Alternatives–Dealing with the perfect storm: Cuban disaster management. *Studies in Political Economy*, **98(1)**, 93–103, doi:10.1080/07078552.2017.1297047.

Kita, S.M., 2017: "Government Doesn't Have the Muscle": State, NGOs, Local Politics, and Disaster Risk Governance in Malawi. *Risk, Hazards & Crisis in Public Policy*, **8(3)**, 244–267, doi:10.1002/rhc3.12118.

4

Kiunsi, R., 2013: The constraints on climate change adaptation in a city with a large development deficit: the case of Dar es Salaam. *Environment and Urbanization*, 25(2), 321–337, doi:10.1177/0956247813489617.

Kivimaa, P. and F. Kern, 2016: Creative destruction or mere niche support? Innovation policy mixes for sustainability transitions. *Research Policy*, 45(1), 205–217, doi:10.1016/j.respol.2015.09.008.

Klepper, G. and W. Rickels, 2014: Climate Engineering: Economic Considerations and Research Challenges. *Review of Environmental Economics and Policy*, 8(2), 270–289, doi:10.1093/reep/reu010.

Klimont, Z. et al., 2017: Global anthropogenic emissions of particulate matter including black carbon. *Atmospheric Chemistry and Physics*, 17(14), 8681–8723, doi:10.5194/acp-17-8681-2017.

Kline, K.L. et al., 2017: Reconciling food security and bioenergy: priorities for action. *GCB Bioenergy*, 9(3), 557–576, doi:10.1111/gcbb.12366.

Kloeckner, C.A., E. Matthies, and M. Hunecke, 2003: Operationalizing Habits and Integrating Habits in Normative Decision-Making Models. *Journal of Applied Social Psychology*, 33(2), 396–417.

Klotz, L., 2011: Cognitive biases in energy decisions during the planning, design, and construction of commercial buildings in the United States: An analytical framework and research needs. *Energy Efficiency*, 4(2), 271–284, doi:10.1007/s12053-010-9089-z.

Knight, P.J., T. Prime, J.M. Brown, K. Morrissey, and A.J. Plater, 2015: Application of flood risk modelling in a web-based geospatial decision support tool for coastal adaptation to climate change. *Natural Hazards and Earth System Sciences*, 15(7), 1457–1471, doi:10.5194/nhess-15-1457-2015.

Knoop, K. and S. Lechtenböhmer, 2017: The potential for energy efficiency in the EU Member States – A comparison of studies. *Renewable and Sustainable Energy Reviews*, 68, 1097–1105, doi:10.1016/j.rser.2016.05.090.

Knowlton, K. et al., 2014: Development and Implementation of South Asia's First Heat-Health Action Plan in Ahmedabad (Gujarat, India). *International Journal of Environmental Research and Public Health*, 11(4), 3473–3492, doi:10.3390/ijerph110403473.

Koch, N., S. Fuss, G. Grosjean, and O. Edenhofer, 2014: Causes of the EU ETS price drop: Recession, CDM, renewable policies or a bit of everything? – New evidence. *Energy Policy*, 73, 676–685, doi:10.1016/j.enpol.2014.06.024.

Koch, N., G. Grosjean, S. Fuss, and O. Edenhofer, 2016: Politics matters: Regulatory events as catalysts for price formation under cap-and-trade. *Journal of Environmental Economics and Management*, 78, 121–139, doi:10.1016/j.jeem.2016.03.004.

Koerth, J., A.T. Vafeidis, and J. Hinkel, 2017: Household-Level Coastal Adaptation and Its Drivers: A Systematic Case Study Review. *Risk Analysis*, 37(4), 629–646, doi:10.1111/risa.12663.

Koezjakov, A., D. Urge-Vorsatz, W. Crijns-Graus, and M. van den Broek, 2018: The relationship between operational energy demand and embodied energy in Dutch residential buildings. *Energy and Buildings*, 165, 233–245, doi:10.1016/j.enbuild.2018.01.036.

Köhler, P., J. Hartmann, and D.A. Wolf-Gladrow, 2010: Geoengineering potential of artificially enhanced silicate weathering of olivine. *Proceedings of the National Academy of Sciences*, 107(47), 20228–20233, doi:10.1073/pnas.1000545107.

Köhler, P., J.F. Abrams, C. Volker, J. Hauck, and D.A. Wolf-Gladrow, 2013: Geoengineering impact of open ocean dissolution of olivine on atmospheric CO_2, surface ocean pH and marine biology. *Environmental Research Letters*, 8(1), 014009, doi:10.1088/1748-9326/8/1/014009.

Kolstad, C. et al., 2014: Social, Economic and Ethical Concepts and Methods. In: *Climate Change 2014: Mitigation of Climate Change. Contribution of Working Group III to the Fifth Assessment Report of the Intergovernmental Panel on Climate Change* [Edenhofer, O., R. Pichs-Madruga, Y. Sokona, E. Farahani, S. Kadner, K. Seyboth, A. Adler, I. Baum, S. Brunner, P. Eickemeier, B. Kriemann, J. Savolainen, S. Schlömer, C. von Stechow, T. Zwickel, and J.C. Minx (eds.)]. Cambridge University Press, Cambridge, United Kingdom and New York, NY, USA, pp. 207–282.

Komarnicki, P., 2016: Energy storage systems: power grid and energy market use cases. *Archives of Electrical Engineering*, 65, 495–511, doi:10.1515/aee-2016-0036.

Kona, A., P. Bertoldi, F. Monforti-Ferrario, S. Rivas, and J.F. Dallemand, 2018: Covenant of mayors signatories leading the way towards 1.5 degree global warming pathway. *Sustainable Cities and Society*, 41, 568–575, doi:10.1016/j.scs.2018.05.017.

Kooiman, J., 2003: *Governing as governance*. Sage, London, UK, 264 pp., doi:10.4135/9781446215012.

Koomey, J.G., H.S. Matthews, and E. Williams, 2013: Smart Everything: Will Intelligent Systems Reduce Resource Use? *Annual Review of Environment and Resources*, 38(1), 311–343, doi:10.1146/annurev-environ-021512-110549.

Koomey, J.G., N.E. Hultman, and A. Grubler, 2017: A reply to "Historical construction costs of global nuclear power reactors". *Energy Policy*, 102, 640–643, doi:10.1016/j.enpol.2016.03.052.

Koornneef, J., A. Ramírez, W. Turkenburg, and A. Faaij, 2012: The environmental impact and risk assessment of CO_2 capture, transport and storage – An evaluation of the knowledge base. *Progress in Energy and Combustion Science*, 38(1), 62–86, doi:10.1016/j.pecs.2011.05.002.

Kossoy, A. et al., 2015: *State and Trends of Carbon Pricing 2015 (September)*. The World Bank, Washington DC, USA, 92 pp., doi:10.1596/978-1-4648-0725-1.

Kosugi, T., 2013: Fail-safe solar radiation management geoengineering. *Mitigation and Adaptation Strategies for Global Change*, 18(8), 1141–1166, doi:10.1007/s11027-012-9414-2.

Kothari, U., 2014: Political discourses of climate change and migration: resettlement policies in the Maldives. *The Geographical Journal*, 180(2), 130–140, doi:10.1111/geoj.12032.

Kotilainen, K. and U.A. Saari, 2018: Policy Influence on Consumers' Evolution into Prosumers – Empirical Findings from an Exploratory Survey in Europe. *Sustainability*, 2018(10), 186, doi:10.3390/su10010186.

Kraucunas, I. et al., 2015: Investigating the nexus of climate, energy, water, and land at decision-relevant scales: the Platform for Regional Integrated Modeling and Analysis (PRIMA). *Climatic Change*, 129(3–4), 573–588, doi:10.1007/s10584-014-1064-9.

Kravitz, B. et al., 2011: The Geoengineering Model Intercomparison Project (GeoMIP). *Atmospheric Science Letters*, 12(2), 162–167, doi:10.1002/asl.316.

Kravitz, B. et al., 2013: Sea spray geoengineering experiments in the geoengineering model intercomparison project (GeoMIP): Experimental design and preliminary results. *Journal of Geophysical Research: Atmospheres*, 118(19), 11175–11186, doi:10.1002/jgrd.50856.

Kraxner, F. et al., 2014: BECCS in South Korea – Analyzing the negative emissions potential of bioenergy as a mitigation tool. *Renewable Energy*, 61, 102–108, doi:10.1016/j.renene.2012.09.064.

Kreidenweis, U. et al., 2016: Afforestation to mitigate climate change: impacts on food prices under consideration of albedo effects. *Environmental Research Letters*, 11(8), 085001, doi:10.1088/1748-9326/11/8/085001.

Kreuz, S. and F. Müsgens, 2017: The German Energiewende and its roll-out of renewable energies: An economic perspective. *Frontiers in Energy*, 11(2), 126–134, doi:10.1007/s11708-017-0467-5.

Kriegler, E., O. Edenhofer, L. Reuster, G. Luderer, and D. Klein, 2013: Is atmospheric carbon dioxide removal a game changer for climate change mitigation? *Climatic Change*, 118(1), 45–57, doi:10.1007/s10584-012-0681-4.

Kristjánsson, J.E., H. Muri, and H. Schmidt, 2015: The hydrological cycle response to cirrus cloud thinning. *Geophysical Research Letters*, 42(24), 10807–10815, doi:10.1002/2015gl066795.

Krugman, P., 2009: *The Return of Depression Economics and the Crisis of 2008*. W.W. Norton & Company Inc, New York, NY, USA, 207 pp.

Kryvasheyeu, Y. et al., 2016: Rapid assessment of disaster damage using social media activity. *Science Advances*, 2(3), e1500779–e1500779, doi:10.1126/sciadv.1500779.

Kull, D. et al., 2016: Building Resilience: World Bank Group Experience in Climate and Disaster Resilient Development. In: *Climate Change Adaptation Strategies – An Upstream-downstream Perspective* [Salzmann, N., C. Huggel, S.U. Nussbaumer, and G. Ziervogel (eds.)]. Springer International Publishing, Cham, Switzerland, pp. 255–270, doi:10.1007/978-3-319-40773-9_14.

Kumari Rigaud, K. et al., 2018: *Groundswell: Preparing for Internal Climate Migration*. The World Bank, Washington DC, USA, 222 pp.

Kuramochi, T. et al., 2018: Ten key short-term sectoral benchmarks to limit warming to 1.5°C. *Climate Policy*, 18(3), 287–305, doi:10.1080/14693062.2017.1397495.

Kurdgelashvili, L., J. Li, C.-H. Shih, and B. Attia, 2016: Estimating technical potential for rooftop photovoltaics in California, Arizona and New Jersey. *Renewable Energy*, 95, 286–302, doi:10.1016/j.renene.2016.03.105.

Kuruppu, N. and D. Liverman, 2011: Mental preparation for climate adaptation: The role of cognition and culture in enhancing adaptive capacity of water management in Kiribati. *Global Environmental Change*, 21(2), 657–669, doi:10.1016/j.gloenvcha.2010.12.002.

Kuruppu, N. and R. Willie, 2015: Barriers to reducing climate enhanced disaster risks in Least Developed Country-Small Islands through anticipatory adaptation. *Weather and Climate Extremes*, 7, 72–83, doi:10.1016/j.wace.2014.06.001.

Kverndokk, S. and K.E. Rosendahl, 2007: Climate policies and learning by doing:

Impacts and timing of technology subsidies. *Resource and Energy Economics*, **29(1)**, 58–82, doi:10.1016/j.reseneeco.2006.02.007.

La Rovere, E.L., C. Gesteira, C. Grottera, and W. William, 2017a: Pathways to a low carbon economy in Brazil. In: *Brazil in the Anthropocene: Conflicts between predatory development and environmental policies* [Issberner, L.-R. and P. Léna (eds.)]. Routledge, Abingdon, UK and New York, NY, USA, pp. 243–266.

La Rovere, E.L., J.-C. Hourcade, S. Priyadarshi, E. Espagne, and B. Perrissin-Fabert, 2017b: *Social Value of Mitigation Activities and forms of Carbon Pricing*. CIRED Working Paper No. 2017-60, CIRED, Paris, France, 8 pp.

Labanca, N. (ed.), 2017: *Complex Systems and Social Practices in Energy Transitions: Framing Energy Sustainability in the Time of Renewables*. Springer Nature, Cham, Switzerland, 337 pp., doi:10.1007/978-3-319-33753-1.

Labbé, J., J.D. Ford, M. Araos, and M. Flynn, 2017: The government-led climate change adaptation landscape in Nunavut, Canada. *Environmental Reviews*, **25(1)**, 12–25, doi:10.1139/er-2016-0032.

Lacasse, K., 2015: The Importance of Being Green. *Environment and Behavior*, **47(1)**, 754–781, doi:10.1177/0013916513520491.

Lacasse, K., 2016: Don't be satisfied, identify! Strengthening positive spillover by connecting pro-environmental behaviors to an "environmentalist" label. *Journal of Environmental Psychology*, **48**, 149–158, doi:10.1016/j.jenvp.2016.09.006.

Lachapelle, E., R. MacNeil, and M. Paterson, 2017: The political economy of decarbonisation: from green energy 'race' to green 'division of labour'. *New Political Economy*, **22(3)**, 311–327, doi:10.1080/13563467.2017.1240669.

Lackner, K.S. et al., 2012: The urgency of the development of CO_2 capture from ambient air. *Proceedings of the National Academy of Sciences*, **109(33)**, 13156–13162, doi:10.1073/pnas.1108765109.

Läderach, P. et al., 2017: Climate change adaptation of coffee production in space and time. *Climatic Change*, **141(1)**, 47–62, doi:10.1007/s10584-016-1788-9.

Lahn, B., 2018: In the light of equity and science: scientific expertise and climate justice after Paris. *International Environmental Agreements: Politics, Law and Economics*, **18(1)**, 29–43, doi:10.1007/s10784-017-9375-8.

Lahn, B. and G. Sundqvist, 2017: Science as a "fixed point"? Quantification and boundary objects in international climate politics. *Environmental Science & Policy*, **67**, 8–15, doi:10.1016/j.envsci.2016.11.001.

Lam, N.S.-N., Y. Qiang, H. Arenas, P. Brito, and K.-B. Liu, 2015: Mapping and assessing coastal resilience in the Caribbean region. *Cartography and Geographic Information Science*, **42(4)**, 315–322, doi:10.1080/15230406.2015.1040999.

Lamb, W.F. et al., 2014: Transitions in pathways of human development and carbon emissions. *Environmental Research Letters*, **9(1)**, 014011, doi:10.1088/1748-9326/9/1/014011.

Lamond, J.E., C.B. Rose, and C.A. Booth, 2015: Evidence for improved urban flood resilience by sustainable drainage retrofit. *Proceedings of the Institution of Civil Engineers – Urban Design and Planning*, **168(2)**, 101–111, doi:10.1680/udap.13.00022.

Lampert, D.J., H. Cai, and A. Elgowainy, 2016: Wells to wheels: water consumption for transportation fuels in the United States. *Energy & Environmental Science*, **9(3)**, 787–802, doi:10.1039/c5ee03254g.

Lampin, L.B.A., F. Nadaud, F. Grazi, and J.-C. Hourcade, 2013: Long-term fuel demand: Not only a matter of fuel price. *Energy Policy*, **62**, 780–787, doi:10.1016/j.enpol.2013.05.021.

Larsen, J.N. et al., 2014: Polar Regions. In: *Climate Change 2014: Impacts, Adaptation, and Vulnerability. Part B: Regional Aspects. Contribution of Working Group II to the Fifth Assessment Report of the Intergovernmental Panel on Climate Change* [Barros, V.R., C.B. Field, D.J. Dokken, M.D. Mastrandrea, and K.J. Mach (eds.)]. Cambridge University Press, Cambridge, United Kingdom and New York, NY, USA, pp. 1567–1612.

Larsen, K. and U. Gunnarsson-Östling, 2009: Climate change scenarios and citizen-participation: Mitigation and adaptation perspectives in constructing sustainable futures. *Habitat International*, **33(3)**, 260–266, doi:10.1016/j.habitatint.2008.10.007.

Larson, W. and W. Zhao, 2017: Telework: Urban Form, Energy Consumption, and Greenhouse Gas Implications. *Economic Inquiry*, **55(2)**, 714–735, doi:10.1111/ecin.12399.

Larsson, L., E.C.H. Keskitalo, and J. Åkermark, 2016: Climate Change Adaptation and Vulnerability Planning within the Municipal and Regional System: Examples from Northern Sweden. *Journal of Northern Studies*, **10(1)**, 67–90.

Lasco, R.D., R.J.P. Delfino, and M.L.O. Espaldon, 2014: Agroforestry systems: helping smallholders adapt to climate risks while mitigating climate change. *Wiley Interdisciplinary Reviews: Climate Change*, **5(6)**, 825–833, doi:10.1002/wcc.301.

Lashley, J.G. and K. Warner, 2015: Evidence of demand for microinsurance for coping and adaptation to weather extremes in the Caribbean. *Climatic Change*, **133(1)**, 101–112, doi:10.1007/s10584-013-0922-1.

Latham, J., B. Parkes, A. Gadian, and S. Salter, 2012: Weakening of hurricanes via marine cloud brightening (MCB). *Atmospheric Science Letters*, **13(4)**, 231–237, doi:10.1002/asl.402.

Latham, J., J. Kleypas, R. Hauser, B. Parkes, and A. Gadian, 2013: Can marine cloud brightening reduce coral bleaching? *Atmospheric Science Letters*, **14(4)**, 214–219, doi:10.1002/asl2.442.

Laurance, W.F. and G.B. Williamson, 2001: Positive Feedbacks among Forest Fragmentation, Drought, and Climate Change in the Amazon. *Conservation Biology*, **15(6)**, 1529–1535, doi:10.1046/j.1523-1739.2001.01093.x.

Lauren, N., K.S. Fielding, L. Smith, and W.R. Louis, 2016: You did, so you can and you will: Self-efficacy as a mediator of spillover from easy to more difficult pro-environmental behaviour. *Journal of Environmental Psychology*, **48**, 191–199, doi:10.1016/j.jenvp.2016.10.004.

Laurice Jamero, M. et al., 2017: Small-island communities in the Philippines prefer local measures to relocation in response to sea-level rise. *Nature Climate Change*, **7**, 581–586, doi:10.1038/nclimate3344.

Lauvset, S.K., J. Tjiputra, and H. Muri, 2017: Climate engineering and the ocean: effects on biogeochemistry and primary production. *Biogeosciences*, **14(24)**, 5675–5691, doi:10.5194/bg-14-5675-2017.

Lawrence, M.G. and P.J. Crutzen, 2017: Was breaking the taboo on research on climate engineering via albedo modification a moral hazard, or a moral imperative? *Earth's Future*, **5(2)**, 136–143, doi:10.1002/2016ef000463.

Lazurko, A. and H.D. Venema, 2017: Financing High Performance Climate Adaptation in Agriculture: Climate Bonds for Multi-Functional Water Harvesting Infrastructure on the Canadian Prairies. *Sustainability*, **9(7)**, 1237, doi:10.3390/su9071237.

Le Dang, H., E. Li, J. Bruwer, and I. Nuberg, 2014: Farmers' perceptions of climate variability and barriers to adaptation: lessons learned from an exploratory study in Vietnam. *Mitigation and Adaptation Strategies for Global Change*, **19(5)**, 531–548, doi:10.1007/s11027-012-9447-6.

Le Page, Y. et al., 2013: Sensitivity of climate mitigation strategies to natural disturbances. *Environmental Research Letters*, **8(1)**, 015018, doi:10.1088/1748-9326/8/1/015018.

Leal Filho, W. et al., 2018: Implementing climate change research at universities: Barriers, potential and actions. *Journal of Cleaner Production*, **170(1)**, 269–277, doi:10.1016/j.jclepro.2017.09.105.

Leape, J., 2006: The London Congestion Charge. *Journal of Economic Perspectives*, **20(4)**, 157–176, doi:10.1257/jep.20.4.157.

Leck, H. and D. Roberts, 2015: What lies beneath: understanding the invisible aspects of municipal climate change governance. *Current Opinion in Environmental Sustainability*, **13**, 61–67, doi:10.1016/j.cosust.2015.02.004.

Leck, H., D. Conway, M. Bradshaw, and J. Rees, 2015: Tracing the Water–Energy–Food Nexus: Description, Theory and Practice. *Geography Compass*, **9(8)**, 445–460, doi:10.1111/gec3.12222.

Lecuyer, O. and P. Quirion, 2013: Can uncertainty justify overlapping policy instruments to mitigate emissions? *Ecological Economics*, **93**, 177–191, doi:10.1016/j.ecolecon.2013.05.009.

Lee, T. and M. Painter, 2015: Comprehensive local climate policy: The role of urban governance. *Urban Climate*, **14**, 566–577, doi:10.1016/j.uclim.2015.09.003.

Leeson, D., N. Mac Dowell, N. Shah, C. Petit, and P.S. Fennell, 2017: A Techno-economic analysis and systematic review of carbon capture and storage (CCS) applied to the iron and steel, cement, oil refining and pulp and paper industries, as well as other high purity sources. *International Journal of Greenhouse Gas Control*, **61**, 71–84, doi:10.1016/j.ijggc.2017.03.020.

Lefevre, C.E. et al., 2015: Heat protection behaviors and positive affect about heat during the 2013 heat wave in the United Kingdom. *Social Science & Medicine*, **128**, 282–289, doi:10.1016/j.socscimed.2015.01.029.

Lefèvre, J., W. Wills, and J.-C. Hourcade, 2018: Combining low-carbon economic development and oil exploration in Brazil? An energy–economy assessment. *Climate Policy*, 1–10, doi:10.1080/14693062.2018.1431198.

Leijten, F.R.M. et al., 2014: Factors that influence consumers' acceptance of future energy systems: the effects of adjustment type, production level, and price. *Energy Efficiency*, **7(6)**, 973–985e, doi:10.1007/s12053-014-9271-9.

Lemos, M.C., 2015: Usable climate knowledge for adaptive and co-managed water governance. *Current Opinion in Environmental Sustainability*, **12**, 48–52, doi:10.1016/j.cosust.2014.09.005.

Lemos, M.C., Y.J. Lo, D.R. Nelson, H. Eakin, and A.M. Bedran-Martins, 2016: Linking development to climate adaptation: Leveraging generic and specific capacities to reduce vulnerability to drought in NE Brazil. *Global Environmental Change*, **39**, 170–179, doi:10.1016/j.gloenvcha.2016.05.001.

Lempert, R.J. and D. Prosnitz, 2011: *Governing geoengineering research: a political and technical vulnerability analysis of potential near-term options*. RAND Corporation, Santa Monica, CA, USA, 93 pp.

Lenferna, G.A., R.D. Russotto, A. Tan, S.M. Gardiner, and T.P. Ackerman, 2017: Relevant climate response tests for stratospheric aerosol injection: A combined ethical and scientific analysis. *Earth's Future*, 577–591, doi:10.1002/2016ef000504.

Lenton, T.M., 2010: The potential for land-based biological CO_2 removal to lower future atmospheric CO_2 concentration. *Carbon Management*, **1(1)**, 145–160, doi:10.4155/cmt.10.12.

Lenton, T.M., 2014: The Global Potential for Carbon Dioxide Removal. In: *Geoengineering of the Climate System* [Harrison, R.M. and R.E. Hester (eds.)]. The Royal Society of Chemistry (RSC), Cambridge, UK, pp. 52–79, doi:10.1039/9781782621225-00052.

Lesnikowski, A.C., J.D. Ford, L. Berrang-Ford, M. Barrera, and J. Heymann, 2015: How are we adapting to climate change? A global assessment. *Mitigation and Adaptation Strategies for Global Change*, **20(2)**, 277–293, doi:10.1007/s11027-013-9491-x.

Lesnikowski, A.C., J.D. Ford, R. Biesbroek, L. Berrang-Ford, and S.J. Heymann, 2016: National-level progress on adaptation. *Nature Climate Change*, **6**, 261–266, doi:10.1038/nclimate2863.

Lesnikowski, A.C. et al., 2017: What does the Paris Agreement mean for adaptation? *Climate Policy*, **17(7)**, 825–831, doi:10.1080/14693062.2016.1248889.

Levy, D.L. and D. Egan, 2003: A Neo-Gramscian Approach to Corporate Political Strategy: Conflict and Accommodation in the Climate Change Negotiations. *Journal of Management Studies*, **40(4)**, 803–829, doi:10.1111/1467-6486.00361.

Lewandowski, M., 2016: Designing the Business Models for Circular Economy – Towards the Conceptual Framework. *Sustainability*, **8(1)**, 43, doi:10.3390/su8010043.

Li, F. et al., 2017: Urban ecological infrastructure: an integrated network for ecosystem services and sustainable urban systems. *Journal of Cleaner Production*, **163(S1)**, S12–S18, doi:10.1016/j.jclepro.2016.02.079.

Li, H., J. He, Z.P. Bharucha, R. Lal, and J. Pretty, 2016: Improving China's food and environmental security with conservation agriculture. *International Journal of Agricultural Sustainability*, **14(4)**, 377–391, doi:10.1080/14735903.2016.1170330.

Li, J. and X. Wang, 2012: Energy and climate policy in China's twelfth five-year plan: A paradigm shift. *Energy Policy*, **41**, 519–528, doi:10.1016/j.enpol.2011.11.012.

Li, L. and B.P.Y. Loo, 2017: Railway Development and Air Patronage in China, 1993-2012: Implications for Low-Carbon Transport. *Journal of Regional Science*, **57(3)**, 507–522, doi:10.1111/jors.12276.

Li, N., D. Ma, and W. Chen, 2017: Quantifying the impacts of decarbonisation in China's cement sector: A perspective from an integrated assessment approach. *Applied Energy*, **185**, 1840–1848, doi:10.1016/j.apenergy.2015.12.112.

Lilford, R.J. et al., 2017: Improving the health and welfare of people who live in slums. *Lancet*, **389(10068)**, 559–570, doi:10.1016/s0140-6736(16)31848-7.

Lillemo, S., 2014: Measuring the effect of procrastination and environmental awareness on households' energy-saving behaviours: An empirical approach. *Energy Policy*, **66**, 249–256, doi:10.1016/j.enpol.2013.10.077.

Lim-Camacho, L. et al., 2015: Facing the wave of change: stakeholder perspectives on climate adaptation for Australian seafood supply chains. *Regional Environmental Change*, **15(4)**, 595–606, doi:10.1007/s10113-014-0670-4.

Lin, A.C., 2013: Does Geoengineering Present a Moral Hazard? *Ecology Law Quarterly*, **40**, 673–712, doi:10.2307/24113611.

Lin, B.B. et al., 2017: How green is your garden?: Urban form and socio-demographic factors influence yard vegetation, visitation, and ecosystem service benefits. *Landscape and Urban Planning*, **157**, 239–246, doi:10.1016/j.landurbplan.2016.07.007.

Lin, C.S.K. et al., 2013: Food waste as a valuable resource for the production of chemicals, materials and fuels. Current situation and global perspective. *Energy & Environmental Science*, **6(2)**, 426–464, doi:10.1039/c2ee23440h.

Lindenberg, S. and L. Steg, 2013: What makes organizations in market democracies

adopt environmentally-friendly policies? In: *Green Organizations: Driving Change with I-O Psychology* [Huffmann, A.H. and S.R. Klein (eds.)]. Routledge, New York, NY, USA and Hove, UK, pp. 93–114.

Linder, M. and M. Williander, 2017: Circular Business Model Innovation: Inherent Uncertainties. *Business Strategy and the Environment*, **26(2)**, 182–196, doi:10.1002/bse.1906.

Lindoso, D.P. et al., 2014: Integrated assessment of smallholder farming's vulnerability to drought in the Brazilian Semi-arid: a case study in Ceará. *Climatic Change*, **127(1)**, 93–105, doi:10.1007/s10584-014-1116-1.

Linnenluecke, M.K., A. Griffiths, and P.J. Mumby, 2015: Executives' engagement with climate science and perceived need for business adaptation to climate change. *Climatic Change*, **131(2)**, 321–333, doi:10.1007/s10584-015-1387-1.

Linnér, B.-O. and V. Wibeck, 2015: Dual high-stake emerging technologies: a review of the climate engineering research literature. *Wiley Interdisciplinary Reviews: Climate Change*, **6(2)**, 255–268, doi:10.1002/wcc.333.

Linnerooth-Bayer, J. and S. Hochrainer-Stigler, 2015: Financial instruments for disaster risk management and climate change adaptation. *Climatic Change*, **133(1)**, 85–100, doi:10.1007/s10584-013-1035-6.

Liu, W., W. Chen, and C. Peng, 2014: Assessing the effectiveness of green infrastructures on urban flooding reduction: A community scale study. *Ecological Modelling*, **291**, 6–14, doi:10.1016/j.ecolmodel.2014.07.012.

Liu, W., Z. Yu, X. Xie, K. von Gadow, and C. Peng, 2018: A critical analysis of the carbon neutrality assumption in life cycle assessment of forest bioenergy systems. *Environmental Reviews*, **26(1)**, 93–101, doi:10.1139/er-2017-0060.

Liu, X. et al., 2017: Microgrids for Enhancing the Power Grid Resilience in Extreme Conditions. *IEEE Transactions on Smart Grid*, **8(2)**, 589–597, doi:10.1109/tsg.2016.2579999.

Lizarralde, G. et al., 2015: A systems approach to resilience in the built environment: the case of Cuba. *Disasters*, **39(s1)**, s76–s95, doi:10.1111/disa.12109.

Lloyd, I.D. and M. Oppenheimer, 2014: On the Design of an International Governance Framework for Geoengineering. *Global Environmental Politics*, **14(2)**, 45–63, doi:10.1162/glep_a_00228.

Loboda, T., 2014: Adaptation strategies to climate change in the Arctic: a global patchwork of reactive community-scale initiatives. *Environmental Research Letters*, **9(11)**, 7–10, doi:10.1088/1748-9326/9/11/111006.

Locatelli, B., V. Evans, A. Wardell, A. Andrade, and R. Vignola, 2011: Forests and Climate Change in Latin America: Linking Adaptation and Mitigation. *Forests*, **2(4)**, 431–450, doi:10.3390/f2010431.

Locatelli, B. et al., 2015: Tropical reforestation and climate change: beyond carbon. *Restoration Ecology*, **23(4)**, 337–343, doi:10.1111/rec.12209.

Lockley, A., 2012: Comment on "Review of Methane Mitigation Technologies with Application to Rapid Release of Methane from the Arctic". *Environmental Science & Technology*, **46(24)**, 13552–13553, doi:10.1021/es303074j.

Lockley, A. and D.M. Coffman, 2016: Distinguishing morale hazard from moral hazard in geoengineering. *Environmental Law Review*, **18(3)**, 194–204, doi:10.1177/1461452916659830.

Löffler, K. et al., 2017: Designing a model for the global energy system-GENeSYS-MOD: An application of the Open-Source Energy Modeling System (OSeMOSYS). *Energies*, **10(10)**, 1–29, doi:10.3390/en10101468.

Loftus, P., A. Cohen, J.C. Long, and J.D. Jenkins, 2014: A critical review of global decarbonization scenarios: what do they tell us about feasibility? *Wiley Interdisciplinary Reviews: Climate Change*, **6(1)**, 93–112, doi:10.1002/wcc.324.

Lohmann, U. and B. Gasparini, 2017: A cirrus cloud climate dial? *Science*, **357(6348)**, 248–249, doi:10.1126/science.aan3325.

Lokhorst, A.M., C. Werner, H. Staats, E. van Dijk, and J.L. Gale, 2013: Commitment and Behavior Change: A Meta-Analysis and Critical Review of Commitment-Making Strategies in Environmental Research. *Environment and Behavior*, **45(1)**, 3–34, doi:10.1177/0013916511411477.

Loladze, I., 2014: Hidden shift of the ionome of plants exposed to elevated CO_2 depletes minerals at the base of human nutrition. *eLife*, **2014(3)**, 1–29, doi:10.7554/elife.02245.

Lomax, G., M. Workman, T. Lenton, and N. Shah, 2015: Reframing the policy approach to greenhouse gas removal technologies. *Energy Policy*, **78**, 125–136, doi:10.1016/j.enpol.2014.10.002.

Lombardi, M., P. Pazienza, and R. Rana, 2016: The EU environmental-energy policy for urban areas: The Covenant of Mayors, the ELENA program and the role of ESCos. *Energy Policy*, **93**, 33–40, doi:10.1016/j.enpol.2016.02.040.

Long, J. and J. Shepherd, 2014: The Strategic Value of Geoengineering Research. In: *Global Environmental Change* [Freedman, B. (ed.)]. Springer Netherlands, Dordrecht, The Netherlands, pp. 757–770, doi:10.1007/978-94-007-5784-4_24.

Loock, C.-, T. Staake, and F. Thiesse, 2013: Motivating Energy-Efficient Behavior with Green IS: An Investigation of Goal Setting and the Role of Defaults. *MIS Quarterly*, **37(4)**, 1313–1332, https://misq.org/motivating-energy-efficient-behavior-with-green-is-an-investigation-of-goal-setting-and-the-role-of-defaults.html.

López, P.Y. and S. Álvarez, 2016: Bioindicadores y conocimiento ancestral/tradicional para el pronóstico meteorológico en comunicades indígenas Maya – K'iche' de Nahualá, Sololá (in Spanish). Retrieved from: https://icc.org.gt/wp-content/uploads/2017/08/poster-ancestral-2-ilovepdf-compressed.pdf.

Loring, P.A., S.C. Gerlach, and H.J. Penn, 2016: "Community work" in a climate of adaptation: Responding to change in rural Alaska. *Human Ecology*, **44(1)**, 119–128, doi:10.1007/s10745-015-9800-y.

Lourenço, T.C., R. Swart, H. Goosen, and R. Street, 2016: The rise of demand-driven climate services. *Nature Climate Change*, **6(1)**, 13–14, doi:10.1038/nclimate2836.

Lövbrand, E., M. Hjerpe, and B.-O. Linnér, 2017: Making climate governance global: how UN climate summitry comes to matter in a complex climate regime. *Environmental Politics*, **26(4)**, 1–20, doi:10.1080/09644016.2017.1319019.

Lovering, J.R., A. Yip, and T. Nordhaus, 2016: Historical construction costs of global nuclear power reactors. *Energy Policy*, **91**, 371–382, doi:10.1016/j.enpol.2016.01.011.

Lovins, A.B., T. Palazzi, R. Laemel, and E. Goldfield, 2018: Relative deployment rates of renewable and nuclear power: A cautionary tale of two metrics. *Energy Research & Social Science*, **38**, 188–192, doi:10.1016/j.erss.2018.01.005.

LTA, 2013: *Land Transport Master Plan 2013*. Land Transport Authority of Singapore (LTA), Singapore, 55 pp.

LTA, 2015: *Singapore Land Transport: Statistics In Brief 2014*. Land Transport Authority of Singapore (LTA), Singapore, 2 pp.

LTA, 2017: *Annual Vehicle Statistics 2016: Motor vehicle population by vehicle type*. Land Transport Authority of Singapore (LTA), Singapore, 1 pp.

Lu, L., Z. Huang, G.H. Rau, and Z.J. Ren, 2015: Microbial Electrolytic Carbon Capture for Carbon Negative and Energy Positive Wastewater Treatment. *Environmental Science & Technology*, **49(13)**, 8193–8201, doi:10.1021/acs.est.5b00875.

Luderer, G. et al., 2012: The economics of decarbonizing the energy system – results and insights from the RECIPE model intercomparison. *Climatic Change*, **114(1)**, 9–37, doi:10.1007/s10584-011-0105-x.

Luderer, G. et al., 2016: *Deep decarbonisation towards 1.5°C–2°C stabilisation. Policy findings from the ADVANCE project (first edition)*. The ADVANCE Consortium, 42 pp.

Luderer, G. et al., 2018: Residual fossil CO_2 emissions in 1.5–2°C pathways. *Nature Climate Change*, **8(7)**, 626–633, doi:10.1038/s41558-018-0198-6.

Lupíšek, A., M. Vaculíková, ManĽík, J. Hodková, and J. RůžiĽka, 2015: Design Strategies for Low Embodied Carbon and Low Embodied Energy Buildings: Principles and Examples. *Energy Procedia*, **83**, 147–156, doi:10.1016/j.egypro.2015.12.205.

Lusiana, B., M. van Noordwijk, and G. Cadisch, 2012: Land sparing or sharing? Exploring livestock fodder options in combination with land use zoning and consequences for livelihoods and net carbon stocks using the FALLOW model. *Agriculture, Ecosystems & Environment*, **159**, 145–160, doi:10.1016/j.agee.2012.07.006.

Lustick, I.S., D. Nettle, D.S. Wilson, H. Kokko, and B.A. Thayer, 2011: Institutional Rigidity and Evolutionary Theory: Trapped on a Local Maximum. *Cliodynamics: The Journal of Theoretical and Mathematical History*, **2(2)**, 1–20, doi:10.21237/c7clio2211722.

Lutz, W. and R. Muttarak, 2017: Forecasting societies' adaptive capacities through a demographic metabolism model. *Nature Climate Change*, **7(3)**, 177–184, doi:10.1038/nclimate3222.

Lwasa, S., 2010: Adapting urban areas in Africa to climate change: the case of Kampala. *Current Opinion in Environmental Sustainability*, **2(3)**, 166–171, doi:10.1016/j.cosust.2010.06.009.

Lwasa, S., 2017: Options for reduction of greenhouse gas emissions in the low-emitting city and metropolitan region of Kampala. *Carbon Management*, **8(3)**, 263–276, doi:10.1080/17583004.2017.1330592.

Lwasa, S. et al., 2014: Urban and peri-urban agriculture and forestry: Transcending poverty alleviation to climate change mitigation and adaptation. *Urban Climate*, **7**, 92–106, doi:10.1016/j.uclim.2013.10.007.

Lwasa, S. et al., 2015: A meta-analysis of urban and peri-urban agriculture and forestry in mediating climate change. *Current Opinion in Environmental Sustainability*, **13**, 68–73, doi:10.1016/j.cosust.2015.02.003.

Ma, D., A. Hasanbeigi, L. Price, and W. Chen, 2015: Assessment of energy-saving and emission reduction potentials in China's ammonia industry. *Clean Technologies and Environmental Policy*, **17(6)**, 1633–1644, doi:10.1007/s10098-014-0896-3.

Ma, Y., 2014: A Study on Carbon Financing Innovation of Financial Institutions in China. *International Journal of Business Administration*, **5(103)**, 1923–4007, doi:10.5430/ijba.v5n4p103.

Mac Dowell, N., P.S. Fennell, N. Shah, and G.C. Maitland, 2017: The role of CO_2 capture and utilization in mitigating climate change. *Nature Climate Change*, **7(4)**, 243–249, doi:10.1038/nclimate3231.

MacDonald, A.E. et al., 2016: Future cost-competitive electricity systems and their impact on US CO_2 emissions. *Nature Climate Change*, **6(5)**, 526–531, doi:10.1038/nclimate2921.

MacDonald, J.P. et al., 2015: Protective factors for mental health and well-being in a changing climate: Perspectives from Inuit youth in Nunatsiavut, Labrador. *Social Science & Medicine*, **141**, 133–141, doi:10.1016/j.socscimed.2015.07.017.

Macedo, I.C., J.E.A. Seabra, and J.E.A.R. Silva, 2008: Green house gases emissions in the production and use of ethanol from sugarcane in Brazil: The 2005/2006 averages and a prediction for 2020. *Biomass and Bioenergy*, **32(7)**, 582–595, doi:10.1016/j.biombioe.2007.12.006.

MacKay, D.J.C., 2013: Could energy-intensive industries be powered by carbon-free electricity? *Philosophical Transactions of the Royal Society A: Mathematical, Physical and Engineering Sciences*, **371(1986)**, doi:10.1098/rsta.2011.0560.

Mackerron, G., 2014: *Costs and economics of geoengineering*. Climate Geoengineering Governance Working Paper Series: 013, Climate Geoengineering Governance (CCG), Oxford, UK, 28 pp.

Maclean, J., J. Tan, D. Tirpak, V. Sonntag-O'Brien, and E. Usher, 2008: *Public Finance Mechanisms to Mobilise Investment in Climate Change Mitigation*. United Nations Environment Programme (UNEP), 40 pp.

MacMartin, D.G., K. Caldeira, and D.W. Keith, 2014: Solar geoengineering to limit the rate of temperature change. *Philosophical Transactions of the Royal Society A: Mathematical, Physical and Engineering Sciences*, **372(2031)**, 20140134–20140134, doi:10.1098/rsta.2014.0134.

MacMartin, D.G., K.L. Ricke, and D.W. Keith, 2018: Solar geoengineering as part of an overall strategy for meeting the 1.5°C Paris target. *Philosophical Transactions of the Royal Society A: Mathematical, Physical and Engineering Sciences*, **376(2119)**, 20160454, doi:10.1098/rsta.2016.0454.

Maggioni, V. et al., 2016: A Review of Merged High-Resolution Satellite Precipitation Product Accuracy during the Tropical Rainfall Measuring Mission (TRMM) Era. *Journal of Hydrometeorology*, **17(4)**, 1101–1117, doi:10.1175/jhm-d-15-0190.1.

Maghari, B.M. and A.M. Ardekani, 2011: Genetically Modified Foods and Social Concerns. *Avicenna Journal of Medical Biotechnology*, **3(3)**, 109–117, www.ajmb.org/article?id=64.

Magnan, A., T. Ribera, and S. Treyer, 2015: *National adaptation is also a global concern*. Working Papers N°04/15, Institut du Développement Durable et des Relations Internationales (IDDRI), Paris, France, 16 pp.

Magnan, A.K., 2016: Metrics needed to track adaptation. *Nature*, **530(7589)**, 160–160, doi:10.1038/530160d.

Magnan, A.K. and T. Ribera, 2016: Global adaptation after Paris. *Science*, **352(6291)**, 1280–1282, doi:10.1126/science.aaf5002.

Magnan, A.K. et al., 2016: Addressing the risk of maladaptation to climate change. *Wiley Interdisciplinary Reviews: Climate Change*, **7(5)**, 646–665, doi:10.1002/wcc.409.

Magni, G., 2017: Indigenous knowledge and implications for the sustainable development agenda. *European Journal of Education*, **52(4)**, 437–447, doi:10.1111/ejed.12238.

Mahlkow, N. and J. Donner, 2017: From Planning to Implementation? The Role of Climate Change Adaptation Plans to Tackle Heat Stress: A Case Study of Berlin, Germany. *Journal of Planning Education and Research*, **37(4)**, 385–396, doi:10.1177/0739456x16664787.

Mahoo, H. et al., 2013: Seasonal weather forecasting: integration of indigenous and scientific knowledge. In: *Innovation in smallholder farming in Africa:*

4

recent advances and recommendations. Proceedings of the International Workshop on Agricultural Innovation Systems in Africa (AISA), Nairobi, Kenya, 29-31 May 2013 [Triomphe, B., A. Waters-Bayer, L. Klerkx, M. Schut, B. Cullen, G. Kamau, and E. Borgne (eds.)]. French Agricultural Research Centre for International Development (CIRAD), Montpellier, France, 137–142 pp.

Mahoo, H. et al., 2015: *Integrating indigenous knowledge with scientific seasonal forecasts for climate risk management in Lushoto district in Tanzania*. CCAFS Working Paper no. 103, CGIAR Research Program on Climate Change, Agriculture and Food Security (CCAFS), Copenhagen, Denmark, 32 pp.

Majzoobi, A. and A. Khodaei, 2017: Application of microgrids in providing ancillary services to the utility grid. *Energy*, **123**, 555–563, doi:10.1016/j.energy.2017.01.113.

Maki, A., R.J. Burns, L. Ha, and A.J. Rothman, 2016: Paying people to protect the environment: A meta-analysis of financial incentive interventions to promote proenvironmental behaviors. *Journal of Environmental Psychology*, **47**, 242–255, doi:10.1016/j.jenvp.2016.07.006.

Malmodin, J., D. Lunden, A. Moberg, G. Andersson, and M. Nilsson, 2014: Life Cycle Assessment of ICT Carbon Footprint and Operational Electricity Use from the Operator, National, and Subscriber Perspective in Sweden. *Journal of Industrial Ecology*, **18(6)**, 829–845, doi:10.1111/jiec.12145.

Manning, D.A. and P. Renforth, 2013: Passive sequestration of atmospheric CO_2 through coupled plant-mineral reactions in urban soils. *Environmental Science & Technology*, **47(1)**, 135–141, doi:10.1021/es301250j.

Mannke, F., 2011: Key themes of local adaptation to climate change: results from mapping community-based initiatives in Africa. In: *Experiences of Climate Change Adaptation in Africa* [Walter Leal Filho (ed.)]. Springer, Berlin and Heidelberg, Germany, pp. 17–32, doi:10.1007/978-3-642-22315-0_2.

Manville, M., D.A. King, and M.J. Smart, 2017: The Driving Downturn: A Preliminary Assessment. *Journal of the American Planning Association*, **83(1)**, 42–55, doi:10.1080/01944363.2016.1247653.

Maor, M., J. Tosun, and A. Jordan, 2017: Proportionate and disproportionate policy responses to climate change: core concepts and empirical applications. *Journal of Environmental Policy & Planning*, 1–13, doi:10.1080/1523908x.2017.1281730.

Mapfumo, P., F. Mtambanengwe, and R. Chikowo, 2016: Building on indigenous knowledge to strengthen the capacity of smallholder farming communities to adapt to climate change and variability in southern Africa. *Climate and Development*, **8(1)**, 72–82, doi:10.1080/17565529.2014.998604.

Maréchal, K., 2010: Not irrational but habitual: The importance of "behavioural lock-in" in energy consumption. *Ecological Economics*, **69(5)**, 1104–1114, doi:10.1016/j.ecolecon.2009.12.004.

Marengo, J.A. and J.C. Espinoza, 2016: Extreme seasonal droughts and floods in Amazonia: causes, trends and impacts. *International Journal of Climatology*, **36(3)**, 1033–1050, doi:10.1002/joc.4420.

Marengo, J.A. et al., 2013: Recent Extremes of Drought and Flooding in Amazonia: Vulnerabilities and Human Adaptation. *American Journal of Climate Change*, **02(02)**, 87–96, doi:10.4236/ajcc.2013.22009.

Margerum, R.D. and C.J. Robinson, 2015: Collaborative partnerships and the challenges for sustainable water management. *Current Opinion in Environmental Sustainability*, **12**, 53–58, doi:10.1016/j.cosust.2014.09.003.

Maru, Y.T., M. Stafford Smith, A. Sparrow, P.F. Pinho, and O.P. Dube, 2014: A linked vulnerability and resilience framework for adaptation pathways in remote disadvantaged communities. *Global Environmental Change*, **28**, 337–350, doi:10.1016/j.gloenvcha.2013.12.007.

Massey, E. and D. Huitema, 2016: The emergence of climate change adaptation as a new field of public policy in Europe. *Regional Environmental Change*, **16(2)**, 553–564, doi:10.1007/s10113-015-0771-8.

Massey, E., R. Biesbroek, D. Huitema, and A. Jordan, 2014: Climate policy innovation: The adoption and diffusion of adaptation policies across Europe. *Global Environmental Change*, **29**, 434–443, doi:10.1016/j.gloenvcha.2014.09.002.

Matear, R.J., 2004: Enhancement of oceanic uptake of anthropogenic CO_2 by macronutrient fertilization. *Journal of Geophysical Research: Oceans*, **109(C4)**, C04001, doi:10.1029/2000jc000321.

Mathioudakis, V., P.W. Gerbens-Leenes, T.H. Van der Meer, and A.Y. Hoekstra, 2017: The water footprint of second-generation bioenergy: A comparison of biomass feedstocks and conversion techniques. *Journal of Cleaner Production*, **148**, 571–582, doi:10.1016/j.jclepro.2017.02.032.

Mathy, S., P. Criqui, K. Knoop, M. Fischedick, and S. Samadi, 2016: Uncertainty management and the dynamic adjustment of deep decarbonization pathways. *Climate Policy*, **16(sup1)**, S47–S62, doi:10.1080/14693062.2016.1179618.

Mavhura, E., A. Collins, and P.P. Bongo, 2017: Flood vulnerability and relocation readiness in Zimbabwe. *Disaster Prevention and Management: An International Journal*, **26(1)**, 41–54, doi:10.1108/dpm-05-2016-0101.

Mazzucato, M., 2013: *The Entrepreneurial State: Debunking Public vs. Private Sector Myths*. Anthem Press, London, UK and New York, NY, USA, 237 pp.

Mazzucato, M., 2018: *Mission-Oriented Research & Innovation in the European Union: A problem-solving approach to fuel innovation-led growth*. European Commission, Brussels, Belgium, 36 pp., doi:10.2777/36546.

Mazzucato, M. and G. Semieniuk, 2017: Public financing of innovation: new questions. *Oxford Review of Economic Policy*, **33(1)**, 24–48, doi:10.1093/oxrep/grw036.

Mbow, C., P. Smith, D. Skole, L. Duguma, and M. Bustamante, 2014: Achieving mitigation and adaptation to climate change through sustainable agroforestry practices in Africa. *Current Opinion in Environmental Sustainability*, **6(1)**, 8–14, doi:10.1016/j.cosust.2013.09.002.

McCarl, B.A., C. Peacocke, R. Chrisman, C.-C. Kung, and R.D. Sands, 2009: Economics of Biochar Production, Utilization and Greenhouse Gas Offsets. In: *Biochar for Environmental Management: Science and Technology*. Routledge, London, UK, pp. 341–358.

McClellan, J., D.W. Keith, and J. Apt, 2012: Cost analysis of stratospheric albedo modification delivery systems. *Environmental Research Letters*, **7(3)**, 034019, doi:10.1088/1748-9326/7/3/034019.

McCormick, K. and T. Kåberger, 2007: Key barriers for bioenergy in Europe: Economic conditions, know-how and institutional capacity, and supply chain co-ordination. *Biomass and Bioenergy*, **31(7)**, 443–452, doi:10.1016/j.biombioe.2007.01.008.

McCright, A.M. and R.E. Dunlap, 2011: Cool dudes: The denial of climate change among conservative white males in the United States. *Global Environmental Change*, **21(4)**, 1163–1172, doi:10.1016/j.gloenvcha.2011.06.003.

McCright, A.M., R.E. Dunlap, and C. Xiao, 2013: Perceived scientific agreement and support for government action on climate change in the USA. *Climatic Change*, **119**, 511–518, doi:10.1007/s10584-013-0704-9.

Mccubbin, S.G., T. Pearce, J.D. Ford, and B. Smit, 2017: Social-ecological change and implications for food security in Funafuti, Tuvalu. *Ecology and Society*, **22(1)**, 53–65, doi:10.5751/es-09129-220153.

McCusker, K.E., K.C. Armour, C.M. Bitz, and D.S. Battisti, 2014: Rapid and extensive warming following cessation of solar radiation management. *Environmental Research Letters*, **9(2)**, 024005, doi:10.1088/1748-9326/9/2/024005.

McGlade, C. and P. Ekins, 2015: The geographical distribution of fossil fuels unused when limiting global warming to 2°C. *Nature*, **517(7533)**, 187–190, doi:10.1038/nature14016.

McGlashan, N., N. Shah, B. Caldecott, and M. Workman, 2012: High-level techno-economic assessment of negative emissions technologies. *Process Safety and Environmental Protection*, **90(6)**, 501–510, doi:10.1016/j.psep.2012.10.004.

McGranahan, G., D. Schensul, and G. Singh, 2016: Inclusive urbanization: Can the 2030 Agenda be delivered without it? *Environment and Urbanization*, **28(1)**, 13–34, doi:10.1177/0956247815627522.

McKay, B., S. Sauer, B. Richardson, and R. Herre, 2016: The political economy of sugarcane flexing: initial insights from Brazil, Southern Africa and Cambodia. *The Journal of Peasant Studies*, **43(1)**, 195–223, doi:10.1080/03066150.2014.992016.

McKinnon, C., 2018: Sleepwalking into lock-in? Avoiding wrongs to future people in the governance of solar radiation management research. *Environmental Politics*, 1–19, doi:10.1080/09644016.2018.1450344.

McLaren, D., 2012: A comparative global assessment of potential negative emissions technologies. *Process Safety and Environmental Protection*, **90(6)**, 489–500, doi:10.1016/j.psep.2012.10.005.

McLaren, D., 2016: Mitigation deterrence and the "moral hazard" of solar radiation management. *Earth's Future*, **4(12)**, 596–602, doi:10.1002/2016ef000445.

McMillen, H.L. et al., 2014: Small islands, valuable insights: Systems of customary resource use and resilience to climate change in the Pacific. *Ecology and Society*, **19(4)**, 44, doi:10.5751/es-06937-190444.

McNamara, K.E. and S.S. Prasad, 2014: Coping with extreme weather: Communities in Fiji and Vanuatu share their experiences and knowledge. *Climatic Change*,

4

123(2), 121–132, doi:10.1007/s10584-013-1047-2.

McPhearson, T. et al., 2016: Scientists must have a say in the future of cities. *Nature*, **538(7624)**, 165–166, doi:10.1038/538165a.

Group of Multilateral Development Banks (MDBs): 2016: *Joint Report on Multilateral Development Banks' Climate Finance*. Inter-American Development Bank, Inter-American Investment Corporation, African Development Bank, Asian Development Bank, European Bank for Reconstruction and Development, European Investment Bank, The World Bank, 45 pp., doi:10.18235/0000806.

Measham, T.G. et al., 2011: Adapting to climate change through local municipal planning: barriers and challenges. *Mitigation and Adaptation Strategies for Global Change*, **16(8)**, 889–909, doi:10.1007/s11027-011-9301-2.

Medek, D.E., J. Schwartz, and S.S. Myers, 2017: Estimated Effects of Future Atmospheric CO_2 Concentrations on Protein Intake and the Risk of Protein Deficiency by Country and Region. *Environmental Health Perspectives*, **125(8)**, 1–8, doi:10.1289/ehp41.

Mehta, L. et al., 2014: The politics of IWRM in Southern Africa. *International Journal of Water Resources Development*, **30(3)**, 528–542, doi:10.1080/07900627.2014.916200.

Meijer, S.S., D. Catacutan, O.C. Ajayi, G.W. Sileshi, and M. Nieuwenhuis, 2015: The role of knowledge, attitudes and perceptions in the uptake of agricultural and agroforestry innovations among smallholder farmers in sub-Saharan Africa. *International Journal of Agricultural Sustainability*, **13(1)**, 40–54, doi:10.1080/14735903.2014.912493.

Méjean, A., F. Lecocq, and Y. Mulugetta, 2015: Equity, burden sharing and development pathways: reframing international climate negotiations. *International Environmental Agreements: Politics, Law and Economics*, **15(4)**, 387–402, doi:10.1007/s10784-015-9302-9.

Melde, S., F. Laczko, and F. Gemenne, 2017: *Making mobility work for adaptation to environmental changes: Results from the MECLEP global research*. International Organization for Migration (IOM), Geneva, Switzerland, 122 pp.

Melica, G. et al., 2018: Multilevel governance of sustainable energy policies: The role of regions and provinces to support the participation of small local authorities in the Covenant of Mayors. *Sustainable Cities and Society*, **39**, 729–739, doi:10.1016/j.scs.2018.01.013.

Melvin, A.M. et al., 2017: Climate change damages to Alaska public infrastructure and the economics of proactive adaptation. *Proceedings of the National Academy of Sciences*, **114(2)**, E122–E131, doi:10.1073/pnas.1611056113.

Mercado, L.M. et al., 2009: Impact of changes in diffuse radiation on the global land carbon sink. *Nature*, **458(7241)**, 1014–1017, doi:10.1038/nature07949.

Mercer, D.E., 2004: Adoption of agroforestry innovations in the tropics: A review. *Agroforestry Systems*, **61–62(1–3)**, 311–328, doi:10.1023/b:agfo.0000029007.85754.70.

Mercure, J.-F. et al., 2018: Macroeconomic impact of stranded fossil fuel assets. *Nature Climate Change*, **8(7)**, 588–593, doi:10.1038/s41558-018-0182-1.

Merk, C., G. Pönitzsch, and K. Rehdanz, 2016: Knowledge about aerosol injection does not reduce individual mitigation efforts. *Environmental Research Letters*, **11(5)**, 054009, doi:10.1088/1748-9326/11/5/054009.

Merk, C., G. Pönitzsch, C. Kniebes, K. Rehdanz, and U. Schmidt, 2015: Exploring public perceptions of stratospheric sulfate injection. *Climatic Change*, **130(2)**, 299–312, doi:10.1007/s10584-014-1317-7.

Merrey, D.J., 2009: African Models for Transnational River Basin Organisations in Africa: An Unexplored Dimension. *Water Alternatives*, **2(2)**, 183–204, www.water-alternatives.org/index.php/alldoc/volume2/v2issue2/50-a2-2-2/file.

Methmann, C. and A. Oels, 2015: From 'fearing' to 'empowering' climate refugees: Governing climate-induced migration in the name of resilience. *Security Dialogue*, **46(1)**, 51–68, doi:10.1177/0967010614552548.

Meze-Hausken, E., 2004: Contrasting climate variability and meteorological drought with perceived drought and climate change in northern Ethiopia. *Climate Research*, **27**, 19–31, www.jstor.org/stable/24868730.

Mguni, P., L. Herslund, and M.B. Jensen, 2016: Sustainable urban drainage systems: examining the potential for green infrastructure-based stormwater management for Sub-Saharan cities. *Natural Hazards*, **82(S2)**, 241–257, doi:10.1007/s11069-016-2309-x.

Michaelowa, A., 2013: The politics of climate change in Germany: ambition versus lobby power. *Wiley Interdisciplinary Reviews: Climate Change*, **4(4)**, 315–320, doi:10.1002/wcc.224.

Michaelowa, A., M. Allen, and F. Sha, 2018: Policy instruments for limiting global temperature rise to 1.5°C – can humanity rise to the challenge? *Climate Policy*, **18(3)**, 275–286, doi:10.1080/14693062.2018.1426977.

Michaelowa, K. and A. Michaelowa, 2017: Transnational Climate Governance Initiatives: Designed for Effective Climate Change Mitigation? *International Interactions*, **43(1)**, 129–155, doi:10.1080/03050629.2017.1256110.

Midden, C. and J. Ham, 2012: Persuasive technology to promote pro-environmental behaviour. In: *Environmental Psychology: An Introduction* [Steg, L., A.E. Berg, and J.I.M.E. de Groot (eds.)]. John Wiley & Sons, Oxford, UK, pp. 243–254.

Middlemiss, L., 2011: The effects of community-based action for sustainability on participants' lifestyles. *Local Environment*, **16(3)**, 265–280, doi:10.1080/13549839.2011.566850.

Mikunda, T. et al., 2014: Designing policy for deployment of CCS in industry. *Climate Policy*, **14(5)**, 665–676, doi:10.1080/14693062.2014.905441.

Miles, D., 1993: Testing for Short Termism in the UK Stock Market. *The Economic Journal*, **103(421)**, 1379–1396, doi:10.2307/2234472.

Milinski, M., D. Semmann, H.-J. Krambeck, and J. Marotzke, 2006: Stabilizing the Earth's climate is not a losing game: Supporting evidence from public goods experiments. *Proceedings of the National Academy of Sciences*, **103(11)**, 3994–3998, doi:10.1073/pnas.0504902103.

Millard-Ball, A. and L. Schipper, 2011: Are We Reaching Peak Travel? Trends in Passenger Transport in Eight Industrialized Countries. *Transport Reviews*, **31(3)**, 357–378, doi:10.1080/01441647.2010.518291.

Mills, M.P., 2015: Shale 2.0: Technology and the Coming Big-Data Revolution in America's Shale Oil Fields. Energy Policy & The Environment Report No. 16 May 2015, Center for Energy Policy and the Environment (CEPE), Manhattan Institute, New York, NY, USA, 17 pp.

Milner, A.M. et al., 2017: Glacier shrinkage driving global changes in downstream systems. *Proceedings of the National Academy of Sciences*, **114(37)**, 9770–9778, doi:10.1073/pnas.1619807114.

Mimura, N. et al., 2014: Adaptation Planning and Implementation. In: *Climate Change 2014: Impacts, Adaptation, and Vulnerability. Part A: Global and Sectoral Aspects. Contribution of Working Group II to the Fifth Assessment Report of the Intergovernmental Panel on Climate Change* [Field, C.B., V.R. Barros, D.J. Dokken, K.J. Mach, and M.D. Mastrandrea (eds.)]. Cambridge University Press, Cambridge, United Kingdom and New York, NY, USA, pp. 869–898.

Minville, M., F. Brissette, S. Krau, and R. Leconte, 2009: Adaptation to climate change in the management of a Canadian water-resources system exploited for hydropower. *Water Resources Management*, **23(14)**, 2965–2986, doi:10.1007/s11269-009-9418-1.

Minx, J., W.F. Lamb, M.W. Callaghan, L. Bornmann, and S. Fuss, 2017: Fast growing research on negative emissions. *Environmental Research Letters*, **12(3)**, 035007, doi:10.1088/1748-9326/aa5ee5.

Minx, J.C. et al., 2018: Negative emissions – Part 1: Research landscape and synthesis. *Environmental Research Letters*, **13(6)**, 063001, doi:10.1088/1748-9326/aabf9b.

Mir-Artigues, P. and P. del Río, 2014: Combining tariffs, investment subsidies and soft loans in a renewable electricity deployment policy. *Energy Policy*, **69**, 430–442, doi:10.1016/j.enpol.2014.01.040.

Mistry, J. and A. Berardi, 2016: Bridging indigenous and scientific knowledge. *Science*, **352(6291)**, 1274–1275, doi:10.1126/science.aaf1160.

Mitchell, D. et al., 2017: Half a degree additional warming, prognosis and projected impacts (HAPPI): background and experimental design. *Geoscientific Model Development*, **10(2)**, 571–583, doi:10.5194/gmd-10-571-2017.

Mitchell, R., 2013: *Agreement on the Establishment of the Limpopo Watercourse Commission*. South African Department of Foreign Affairs, Office of the Chief State Law Adviser (IL), Treaty and Information Management Section.

Mitlin, D., 2005: Understanding chronic poverty in urban areas. *International Planning Studies*, **10(1)**, 3–19, doi:10.1080/13563470500159220.

Mitlin, D. and D. Satterthwaite, 2013: *Urban Poverty in the Global South: Scale and Nature*. Routledge, Abingdon, UK and New York, NY, USA, 368 pp.

Mittal, S., H. Dai, and P.R. Shukla, 2016: Low carbon urban transport scenarios for China and India: A comparative assessment. *Transportation Research Part D: Transport and Environment*, **44**, 266–276, doi:10.1016/j.trd.2015.04.002.

Moffatt, S., 2014: Resilience and competing temporalities in cities. *Building Research & Information*, **42(2)**, 202–220, doi:10.1080/09613218.2014.869894.

Möllersten, K., J. Yan, and J. R. Moreira, 2003: Potential market niches for biomass energy with CO_2 capture and storage-Opportunities for energy supply with negative CO_2 emissions. *Biomass and Bioenergy*, **25(3)**, 273–285, doi:10.1016/s0961-9534(03)00013-8.

4

Monstadt, J. and A. Wolff, 2015: Energy transition or incremental change? Green policy agendas and the adaptability of the urban energy regime in Los Angeles. *Energy Policy*, **78**, 213–224, doi:10.1016/j.enpol.2014.10.022.

Montserrat, F. et al., 2017: Olivine Dissolution in Seawater: Implications for CO_2 Sequestration through Enhanced Weathering in Coastal Environments. *Environmental Science & Technology*, **51**(7), 3960–3972, doi:10.1021/acs.est.6b05942.

Montzka, S.A., E.J. Dlugokencky, and J.H. Butler, 2011: Non-CO_2 Greenhouse Gases and Climate Change. *Nature*, **476**, 43–50, doi:10.1038/nature10322.

Mooij, R.A., 2000: *Environmental Taxation and the Double Dividend*. Emerald Group Publishing Ltd, Bingley, UK, 292 pp.

Moore, J.C., S. Jevrejeva, and A. Grinsted, 2010: Efficacy of geoengineering to limit 21st century sea-level rise. *Proceedings of the National Academy of Sciences*, **107**(36), 15699–15703, doi:10.1073/pnas.1008153107.

Moreno-Cruz, J.B. and D.W. Keith, 2013: Climate policy under uncertainty: a case for solar geoengineering. *Climatic Change*, **121**(3), 431–444, doi:10.1007/s10584-012-0487-4.

Morita, K. and K. Matsumoto, 2015: Financing Adaptation to Climate Change in Developing Countries. In: *Handbook of Climate Change Adaptation* [Leal Filho, W. (ed.)]. Springer-Verlag Berlin Heidelberg, Germany, pp. 983–1005, doi:10.1007/978-3-642-38670-1_22.

Moriyama, R. et al., 2016: The cost of stratospheric climate engineering revisited. *Mitigation and Adaptation Strategies for Global Change*, 1–22, doi:10.1007/s11027-016-9723-y.

Morrow, D.R., 2014: Starting a flood to stop a fire? Some moral constraints on solar radiation management. *Ethics, Policy & Environment*, **17**(2), 123–138, doi:10.1080/21550085.2014.926056.

Mortreux, C. and J. Barnett, 2009: Climate change, migration and adaptation in Funafuti, Tuvalu. *Global Environmental Change*, **19**(1), 105–112, doi:10.1016/j.gloenvcha.2008.09.006.

Moser, S.C., 2014: Communicating adaptation to climate change: The art and science of public engagement when climate change comes home. *Wiley Interdisciplinary Reviews: Climate Change*, **5**(3), 337–358, doi:10.1002/wcc.276.

Moss, R.H. et al., 2013: Hell and High Water: Practice-Relevant Adaptation Science. *Science*, **342**(6159), 696–698, doi:10.1126/science.1239569.

Mota-Babiloni, A., P. Makhnatch, and R. Khodabandeh, 2017: Recent investigations in HFCs substitution with lower GWP synthetic alternatives: Focus on energetic performance and environmental impact. *International Journal of Refrigeration*, **82**, 288–301, doi:10.1016/j.ijrefrig.2017.06.026.

Mouratiadou, I. et al., 2016: The impact of climate change mitigation on water demand for energy and food: An integrated analysis based on the Shared Socioeconomic Pathways. *Environmental Science & Policy*, **64**, 48–58, doi:10.1016/j.envsci.2016.06.007.

Mousavi, B. and M. Blesl, 2018: Analysis of the Relative Roles of Supply-Side and Demand-Side Measures in Tackling the Global 1.5°C Target. In: *Limiting Global Warming to Well Below 2°C: Energy System Modelling and Policy Development* [Giannakidis, G., K. Karlsson, M. Labriet, and B. Gallachóir (eds.)]. Springer International Publishing, Cham, Switzerland, pp. 67–83, doi:10.1007/978-3-319-74424-7_5.

Muench, S., S. Thuss, and E. Guenther, 2014: What hampers energy system transformations? The case of smart grids. *Energy Policy*, **73**, 80–92, doi:10.1016/j.enpol.2014.05.051.

Mullaney, J., T. Lucke, and S.J. Trueman, 2015: A review of benefits and challenges in growing street trees in paved urban environments. *Landscape and Urban Planning*, **134**, 157–166, doi:10.1016/j.landurbplan.2014.10.013.

Mulville, M., K. Jones, G. Huebner, and J. Powell-Greig, 2017: Energy-saving occupant behaviours in offices: change strategies. *Building Research & Information*, **45**(8), 861–874, doi:10.1080/09613218.2016.1212299.

Munck, J., J.G. Rozema, and L.A. Frye-levine, 2014: Institutional inertia and climate change: a review of the new institutionalist literature. *Wiley Interdisciplinary Reviews: Climate Change*, **5**, 639–648, doi:10.1002/wcc.292.

Mundaca, L., 2007: Transaction costs of Tradable White Certificate schemes: The Energy Efficiency Commitment as case study. *Energy Policy*, **35**, 4340–4354, doi:10.1016/j.enpol.2007.02.029.

Mundaca, L. and A. Markandya, 2016: Assessing regional progress towards a 'Green Energy Economy'. *Applied Energy*, **179**, 1372–1394, doi:10.1016/j.apenergy.2015.10.098.

Murphy, A.G., J. Hartell, V. Cárdenas, and J.R. Skees, 2012: *Risk Management Instruments for Food Price Volatility and Weather Risk in Latin America and the Caribbean: The Use of Risk Management Instruments*. Discussion Paper, Inter-American Development Bank (IDB), Washington DC, USA, 110 pp.

Murrant, D., A. Quinn, and L. Chapman, 2015: The water-energy nexus: Future water resource availability and its implications on UK thermal power generation. *Water and Environment Journal*, **29**(3), 307–319, doi:10.1111/wej.12126.

Murthy, I.K., 2013: Carbon Sequestration Potential of Agroforestry Systems in India. *Journal of Earth Science & Climatic Change*, **04**(01), 1–7, doi:10.4172/2157-7617.1000131.

Musall, F.D. and O. Kuik, 2011: Local acceptance of renewable energy-A case study from southeast Germany. *Energy Policy*, **39**(6), 3252–3260, doi:10.1016/j.enpol.2011.03.017.

Mushtaq, S., 2016: Economic and policy implications of relocation of agricultural production systems under changing climate: Example of Australian rice industry. *Land Use Policy*, **52**, 277–286, doi:10.1016/j.landusepol.2015.12.029.

Mushtaq, S., 2018: Managing climate risks through transformational adaptation: Economic and policy implications for key production regions in Australia. *Climate Risk Management*, **19**, 48–60, doi:10.1016/j.crm.2017.12.001.

Mycoo, M. and M.G. Donovan, 2017: *A Blue Urban Agenda: Adapting to Climate Change in the Coastal Cities of Caribbean and Pacific Small Island Developing States*. Inter-American Development Bank (IDB), Washington DC, USA, 215 pp., doi:10.18235/0000690.

Myers, C.D., T. Ritter, and A. Rockway, 2017: Community Deliberation to Build Local Capacity for Climate Change Adaptation: The Rural Climate Dialogues Program. In: *Climate Change Adaptation in North America: Fostering Resilience and the Regional Capacity to Adapt* [Leal Filho, W. and J.M. Keenan (eds.)]. Springer International Publishing, Cham, Switzerland, pp. 9–26, doi:10.1007/978-3-319-53742-9_2.

Myers, T.A., E.W. Maibach, C. Roser-Renouf, K. Akerlof, and A.A. Leiserowitz, 2012: The relationship between personal experience and belief in the reality of global warming. *Nature Climate Change*, **3**, 343–347, doi:10.1038/nclimate1754.

Myhre, G. et al., 2013: Anthropogenic and Natural Radiative Forcing. In: *Climate Change 2013: The Physical Science Basis. Contribution of Working Group I to the Fifth Assessment Report of the Intergovernmental Panel on Climate Change* [Stocker, T.F., D. Qin, G.-K. Plattner, M. Tignor, S.K. Allen, J. Boschung, A. Nauels, Y. Xia, V. Bex, and P.M. Midgley (eds.)]. Cambridge University Press, Cambridge, United Kingdom and New York, NY, USA, pp. 659–740.

Nabernegg, S. et al., 2017: The Deployment of Low Carbon Technologies in Energy Intensive Industries: A Macroeconomic Analysis for Europe, China and India. *Energies*, **10**(3), 360, doi:10.3390/en10030360.

Nadaï, A. and O. Labussière, 2017: Landscape commons, following wind power fault lines. The case of Seine-et-Marne (France). *Energy Policy*, **109**, 807–816, doi:10.1016/j.enpol.2017.06.049.

Nägele, F., T. Kasper, and B. Girod, 2017: Turning up the heat on obsolete thermostats: A simulation-based comparison of intelligent control approaches for residential heating systems. *Renewable and Sustainable Energy Reviews*, **75**, 1254–1268, doi:10.1016/j.rser.2016.11.112.

Nair, S., B. George, H.M. Malano, M. Arora, and B. Nawarathna, 2014: Water–energy–greenhouse gas nexus of urban water systems: Review of concepts, state-of-art and methods. *Resources, Conservation and Recycling*, **89**, 1–10, doi:10.1016/j.resconrec.2014.05.007.

Nakashima, D.J., K. Galloway McLean, H.D. Thulstrup, A. Ramos Castillo, and J.T. Rubis, 2012: *Weathering Uncertainty: Traditional Knowledge for Climate Change Assessment and Adaptation*. UNESCO, Paris, France and UNU, Darwin, Australia, 120 pp.

Nakhooda, S. and C. Watson, 2016: *Adaptation finance and the infrastructure agenda*. Overseas Development Institute (ODI), London, UK, 40 pp.

Nalau, J., B.L. Preston, and M.C. Maloney, 2015: Is adaptation a local responsibility? *Environmental Science & Policy*, **48**, 89–98, doi:10.1016/j.envsci.2014.12.011.

Napp, T.A., A. Gambhir, T.P. Hills, N. Florin, and P.S. Fennell, 2014: A review of the technologies, economics and policy instruments for decarbonising energy-intensive manufacturing industries. *Renewable and Sustainable Energy Reviews*, **30**, 616–640, doi:10.1016/j.rser.2013.10.036.

Napp, T.A. et al., 2017: Exploring the Feasibility of Low-Carbon Scenarios Using Historical Energy Transitions Analysis. *Energies*, **10**(1), 116, doi:10.3390/en10010116.

Natcher, D.C. et al., 2007: Notions of time and sentience: Methodological considerations for Arctic climate change research. *Arctic Anthropology*, **44**(2), 113–126, doi:10.1353/arc.2011.0099.

NCE, 2016: *The Sustainable Infrastructure Imperative: Financing for Better Grown and Development*. The 2016 New Climate Economy Report, New Climate

Economy (NCE), Washington DC and London, UK, 152 pp.

NEC, 2011: *Second National Communication to the UNFCCC*. National Environment Commission, Royal Government of Bhutan, Thimphu, Bhutan, 160 pp.

NEC, 2015: *Communication of INDC of the Kingdom of Bhutan*. National Environment Commission, Royal Government of Bhutan, 8 pp.

Nemet, G.F., 2014: Solar Photovoltaics: Multiple Drivers of Technological Improvement. In: *Energy Technology Innovation* [Grubler, A. and C. Wilson (eds.)]. Cambridge University Press, Cambridge, United Kingdom and New York, NY, USA, pp. 206–218, doi:10.1017/cbo9781139150880.020.

Nemet, G.F., M. Jakob, J.C. Steckel, and O. Edenhofer, 2017: Addressing policy credibility problems for low-carbon investment. *Global Environmental Change*, **42**, 47–57, doi:10.1016/j.gloenvcha.2016.12.004.

Nemet, G.F. et al., 2018: Negative emissions – Part 3: Innovation and upscaling. *Environmental Research Letters*, **13(6)**, 063003, doi:10.1088/1748-9326/aabff4.

Newell, P. and M. Paterson, 1998: A Climate For Business: Global Warming, the State and Capital. *Review of International Political Economy*, **5**, 679–703, www.jstor.org/stable/4177292.

Newell, R.G. and W.A. Pizer, 2003: Regulating stock externalities under uncertainty. *Journal of Environmental Economics and Management*, **45(2)**, 416–432, doi:10.1016/s0095-0696(02)00016-5.

Newham, M. and B. Conradie, 2013: *A Critical Review of South Africa's Carbon Tax Policy Paper: Recommendations for the Implementation of an Offset Mechanism*. CSSR Working Paper, Centre for Social Science Research, University of Cape Town, Cape Town, South Africa, 23 pp.

Newman, P., 2017: Decoupling Economic Growth from Fossil Fuels. *Modern Economy*, **8(6)**, 791–805, doi:10.4236/me.2017.86055.

Newman, P. and J. Kenworthy, 2011: 'Peak Car Use': Understanding the Demise of Automobile Dependence. *World Transport Policy and Practice*, **17.2**, 31–42, http://worldtransportjournal.com/wp-content/uploads/2015/02/wtpp17.2.pdf.

Newman, P. and J.R. Kenworthy, 2015: *The End of Automobile Dependence: How Cities are Moving Beyond Car-based Planning*. Island Press, Washington DC, USA, 320 pp., doi:10.5822/978-1-61091-613-4.

Newman, P., L. Kosonen, and J. Kenworthy, 2016: Theory of urban fabrics: planning the walking, transit/public transport and automobile/motor car cities for reduced car dependency. *Town Planning Review*, **87(4)**, 429–458, doi:10.3828/tpr.2016.28.

Newman, P., T. Beatley, and H. Boyer, 2017: *Resilient Cities: Overcoming Fossil Fuel Dependence (Second edition)*. Island Press, Washington DC, USA, 264 pp.

Ng, A.K.Y., J. Andrews, D. Babb, Y. Lin, and A. Becker, 2018: Implications of climate change for shipping: Opening the Arctic seas. *Wiley Interdisciplinary Reviews: Climate Change*, **9(2)**, e507, doi:10.1002/wcc.507.

Ngendakumana, S. et al., 2017: Implementing REDD+: learning from forest conservation policy and social safeguards frameworks in Cameroon. *International Forestry Review*, **19(2)**, 209–223, doi:10.1505/146554817821255187.

Nguyen, T.T.T., P.J. Bowman, M. Haile-Mariam, J.E. Pryce, and B.J. Hayes, 2016: Genomic selection for tolerance to heat stress in Australian dairy cattle. *Journal of Dairy Science*, **99(4)**, 2849–2862, doi:10.3168/jds.2015-9685.

Nicholson, S., 2013: The Promises and Perils of Geoengineering. In: *State of the World 2013: Is Sustainability Still Possible?* [Worldwatch Institute (ed.)]. Island Press/Center for Resource Economics, Washington DC, USA, pp. 317–331.

Nicholson, S., S. Jinnah, and A. Gillespie, 2018: Solar radiation management: a proposal for immediate polycentric governance. *Climate Policy*, **18(3)**, 322–334, doi:10.1080/14693062.2017.1400944.

Nicolson, M., G. Huebner, and D. Shipworth, 2017: Are consumers willing to switch to smart time of use electricity tariffs? The importance of loss-aversion and electric vehicle ownership. *Energy Research & Social Science*, **23**, 82–96, doi:10.1016/j.erss.2016.12.001.

Niemeier, U. and C. Timmreck, 2015: What is the limit of climate engineering by stratospheric injection of SO_2? *Atmospheric Chemistry and Physics*, **15(16)**, 9129–9141, doi:10.5194/acp-15-9129-2015.

Nierop, S.C.A., 2014: Envisioning resilient electrical infrastructure: A policy framework for incorporating future climate change into electricity sector planning. *Environmental Science & Policy*, **40**, 78–84, doi:10.1016/j.envsci.2014.04.011.

Nieto, J., Carpintero, and L.J. Miguel, 2018: Less than 2°C? An Economic-Environmental Evaluation of the Paris Agreement. *Ecological Economics*, **146**, 69–84, doi:10.1016/j.ecolecon.2017.10.007.

Nightingale, A.J., 2017: Power and politics in climate change adaptation efforts: Struggles over authority and recognition in the context of political instability. *Geoforum*, **84**, 11–20, doi:10.1016/j.geoforum.2017.05.011.

Nijland, H. and J. Meerkerk, 2017: Environmental Innovation and Societal Transitions Mobility and environmental impacts of car sharing in the Netherlands. *Environmental Innovation and Societal Transitions*, **23**, 84–91, doi:10.1016/j.eist.2017.02.001.

Nikiforos, M. and G. Zezza, 2017: *Stock-flow Consistent Macroeconomic Models: A Survey*. LEI Working Paper No. 891, Levy Economics Institute, Annandale-on-Hudson, NY, USA, 71 pp.

Nissinen, A. et al., 2015: Combinations of policy instruments to decrease the climate impacts of housing, passenger transport and food in Finland. *Journal of Cleaner Production*, **107**, 455–466, doi:10.1016/j.jclepro.2014.08.095.

NITI Aayog and RMI, 2017: *India Leaps Ahead: Transformative Mobility Solutions for All*. National Institution for Transforming India (NITI Aayog) and Rocky Mountain Institute (RMI), New Delhi and Colorado, 134 pp.

Nitschke, M., A. Krackowizer, L.A. Hansen, P. Bi, and R.G. Tucker, 2017: Heat Health Messages: A Randomized Controlled Trial of a Preventative Messages Tool in the Older Population of South Australia. *International Journal of Environmental Research and Public Health*, **14(9)**, 992, doi:10.3390/ijerph14090992.

Nitschke, M. et al., 2016: Evaluation of a heat warning system in Adelaide, South Australia, using case-series analysis. *BMJ open*, **6(7)**, e012125, doi:10.1136/bmjopen-2016-012125.

Noble, I. et al., 2014: Adaptation needs and options. In: *Climate Change 2014: Impacts, Adaptation, and Vulnerability. Part A: Global and Sectoral Aspects. Contribution of Working Group II to the Fifth Assessment Report of the Intergovernmental Panel on Climate Change* [Field, C.B., V.R. Barros, D.J. Dokken, K.J. Mach, M.D. Mastrandrea, T.E. Bilir, M. Chatterjee, K.L. Ebi, Y.O. Estrada, R.C. Genova, B. Girma, E.S. Kissel, A.N. Levy, S. MacCracken, P.R. Mastrandrea, and L.L. White (eds.)]. Cambridge University Press, Cambridge, United Kingdom and New York, NY, USA, pp. 659–708.

Nolan, J.M., P.W. Schultz, R.B. Cialdini, N.J. Goldstein, and V. Griskevicius, 2008: Normative Social Influence is Underdetected. *Personality and Social Psychology Bulletin*, **34(7)**, 913–923, doi:10.1177/0146167208316691.

Noppers, E.H., K. Keizer, J.W. Bolderdijk, and L. Steg, 2014: The adoption of sustainable innovations: Driven by symbolic and environmental motives. *Global Environmental Change*, **25**, 52–62, doi:10.1016/j.gloenvcha.2014.01.012.

Nordhaus, W., 2015: Climate clubs: Overcoming free-riding in international climate policy. *American Economic Review*, **105(4)**, 1339–1370, doi:10.1257/aer.15000001.

North, D.C., 1990: *Institutions, institutional change and economic performance*. Cambridge University Press, Cambridge, United Kingdom and New York, NY, USA, 152 pp.

Nowak, D.J., D.E. Crane, and J.C. Stevens, 2006: Air pollution removal by urban trees and shrubs in the United States. *Urban Forestry & Urban Greening*, **4(3–4)**, 115–123, doi:10.1016/j.ufug.2006.01.007.

NPCC, 2015: Building the Knowledge Base for Climate Resiliency: New York City Panel on Climate Change 2015 Report. *Annals of the New York Academy of Sciences*, **1336(1)**, 1–150, doi:10.1111/nyas.12625.

NRC, 2015a: *Climate Intervention: Carbon Dioxide Removal and Reliable Sequestration*. National Research Council (NRC). The National Academies Press, Washington DC, USA, 154 pp., doi:10.17226/18805.

NRC, 2015b: *Climate Intervention: Reflecting Sunlight to Cool Earth*. National Research Council (NRC). The National Academies Press, Washington DC, USA, 260 pp., doi:10.17226/18988.

NRC, 2015c: Governance of Research and Other Sociopolitical Considerations. In: *Climate Intervention: Reflecting Sunlight to Cool Earth*. National Research Council (NRC). The National Academies Press, Washington DC, USA, pp. 149–175, doi:10.17226/18988.

Nunn, P.D., J. Runman, M. Falanruw, and R. Kumar, 2017: Culturally grounded responses to coastal change on islands in the Federated States of Micronesia, northwest Pacific Ocean. *Regional Environmental Change*, **17(4)**, 959–971, doi:10.1007/s10113-016-0950-2.

Nurse, L.A. et al., 2014: Small islands. In: *Climate Change 2014: Impacts, Adaptation, and Vulnerability. Part B: Regional Aspects. Contribution of Working Group II to the Fifth Assessment Report of the Intergovernmental Panel on Climate Change* [Barros, V.R., C.B. Field, D.J. Dokken, M.D. Mastrandrea, K.J. Mach, T.E. Bilir, M. Chatterjee, K.L. Ebi, Y.O. Estrada, R.C. Genova, B. Girma, E.S. Kissel, A.N. Levy, S. MacCracken, P.R. Mastrandrea, and L.L. White (eds.)]. Cambridge

4

University Press, Cambridge, United Kingdom and New York, NY, USA, pp. 1613–1654.

Nyagwambo, N.L., E. Chonguiça, D. Cox, and F. Monggae, 2008: *Local Governments and IWRM in the SADC Region*. Institute of Water and Sanitation Development (IWSD), Harare, Zimbabwe, 58 pp.

Nyantakyi-Frimpong, H. and R. Bezner-Kerr, 2015: The relative importance of climate change in the context of multiple stressors in semi-arid Ghana. *Global Environmental Change*, **32**, 40–56, doi:10.1016/j.gloenvcha.2015.03.003.

NYC Parks, 2010: *Designing the Edge: Creating a Living Urban Shore at Harlem River Park*. NYC Department of Parks & Recreation, Metropolitan Waterfront Alliance, Harlem River Park Task Force, NY Department of State Division of Coastal Resources, New York, NY, USA, 52 pp.

Nyholm, E., M. Odenberger, and F. Johnsson, 2017: An economic assessment of distributed solar PV generation in Sweden from a consumer perspective – The impact of demand response. *Renewable Energy*, **108**, 169–178, doi:10.1016/j.renene.2017.02.050.

Nykvist, B. and M. Nilsson, 2015: Rapidly falling costs of battery packs for electric vehicles. *Nature Climate Change*, **5(4)**, 329–332.

Nyong, A., F. Adesina, and B. Osman Elasha, 2007: The value of indigenous knowledge in climate change mitigation and adaptation strategies in the African Sahel. *Mitigation and Adaptation Strategies for Global Change*, **12(5)**, 787–797, doi:10.1007/s11027-007-9099-0.

O'Hare, P., I. White, and A. Connelly, 2016: Insurance as maladaptation: Resilience and the 'business as usual' paradox. *Environment and Planning C: Government and Policy*, **34(6)**, 1175–1193, doi:10.1177/0263774x15602022.

O'Leary, J.K. et al., 2017: The Resilience of Marine Ecosystems to Climatic Disturbances. *BioScience*, **67(3)**, 208–220, doi:10.1093/biosci/biw161.

O'Neill, B.C. et al., 2015: The roads ahead: Narratives for shared socioeconomic pathways describing world futures in the 21st century. *Global Environmental Change*, **42**, 169–180, doi:10.1016/j.gloenvcha.2015.01.004.

O'Neill, B.C. et al., 2017: IPCC reasons for concern regarding climate change risks. *Nature Climate Change*, **7(1)**, 28–37, doi:10.1038/nclimate3179.

O'Neill, S. and S. Nicholson-Cole, 2009: "Fear Won't Do It" Visual and Iconic Representations. *Science Communication*, **30(3)**, 355–379, doi:10.1177/1075547008329201.

O'Neill, S.J., M. Boykoff, S. Niemeyer, and S.A. Day, 2013: On the use of imagery for climate change engagement. *Global Environmental Change*, **23**, 413–421, doi:10.1016/j.gloenvcha.2012.11.006.

Obergassel, W. et al., 2016: *Phoenix from the Ashes: An Analysis of the Paris Agreement to the United Nations Framework Convention on Climate Change – Part II*. Wuppertal Institute for Climate, Environment and Energy, Wuppertal, Germany, 10 pp.

Ockwell, D. and R. Byrne, 2016: Improving technology transfer through national systems of innovation: climate relevant innovation-system builders (CRIBs). *Climate Policy*, **16(7)**, 836–854, doi:10.1080/14693062.2015.1052958.

Ockwell, D., A. Sagar, and H. de Coninck, 2015: Collaborative research and development (R&D) for climate technology transfer and uptake in developing countries: towards a needs driven approach. *Climatic Change*, **131(3)**, 401–415, doi:10.1007/s10584-014-1123-2.

OECD, 2015a: *Aligning Policies for a Low-carbon Economy*. OECD Publishing, Paris, France, 240 pp., doi:10.1787/9789264233294-en.

OECD, 2015b: *Climate Finance in 2013-14 and the USD 100 billion goal: A Report by the OECD in Collaboration with Climate Policy Initiative*. Organisation for Economic Co-operation and Development (OECD), Paris, France, 64 pp.

OECD, 2016: *2020 Projections of Climate Finance Towards the USD 100 Billion Goal: Technical Note*. OECD Publishing, Paris, France, 40 pp.

OECD, 2017a: *Investing in Climate, Investing in Growth*. Organisation for Economic Co-operation and Development (OECD). OECD Publishing, Paris, France, 314 pp., doi:10.1787/9789264273528-1-en.

OECD, 2017b: *Mobilising Bond Markets for a Low-Carbon Transition*. Organisation for Economic Co-operation and Development (OECD) Publishing, Paris, 132 pp., doi:10.1787/9789264272323-en.

OECD, 2017c: *The Next Production Revolution: Implications for Governments and Business*. Organisation for Economic Co-operation and Development (OECD) Publishing, Paris, France, 444 pp., doi:10.1787/9789264271036-en.

OECD/FAO, 2017: *OECD-FAO Agricultural Outlook 2017-2026*. Organisation for Economic Co-operation and Development (OECD) Publishing, Paris, France, 150 pp., doi:10.1787/agr_outlook-2017-en.

Ogunbode, C.A., Y. Liu, and N. Tausch, 2017: The moderating role of political affiliation in the link between flooding experience and preparedness to

reduce energy use. *Climatic Change*, **145(3–4)**, 445–458, doi:10.1007/s10584-017-2089-7.

Ohndorf, M., J. Blasch, and R. Schubert, 2015: Emission budget approaches for burden sharing: some thoughts from an environmental economics point of view. *Climatic Change*, **133(3)**, 385–395, doi:10.1007/s10584-015-1442-y.

Ölander, F. and J. Thøgersen, 2014: Informing Versus Nudging in Environmental Policy. *Journal of Consumer Policy*, **37(3)**, 341–356, doi:10.1007/s10603-014-9256-2.

Oldenbroek, V., L.A. Verhoef, and A.J.M. van Wijk, 2017: Fuel cell electric vehicle as a power plant: Fully renewable integrated transport and energy system design and analysis for smart city areas. *International Journal of Hydrogen Energy*, **42(12)**, 8166–8196, doi:10.1016/j.ijhydene.2017.01.155.

Oliver, T.H. and M.D. Morecroft, 2014: Interactions between climate change and land use change on biodiversity: attribution problems, risks, and opportunities. *Wiley Interdisciplinary Reviews: Climate Change*, **5(3)**, 317–335, doi:10.1002/wcc.271.

OneNYC Team, 2015: *One New York: The Plan for a Strong and Just City*. Office of the Mayor of New York City, New York, NY, USA, 354 pp.

Opio, C. et al., 2013: *Greenhouse gas emissions from ruminant supply chains – A global life cycle assessment*. Food and Agriculture Organization of the United Nations (FAO), Rome, Italy, 191 pp.

ORR, 2018: *Climate Resiliency Design Guidelines*. NYC Mayor's Office of Recovery & Resiliency (ORR), New York, NY, USA, 56 pp.

Osbaldiston, R. and J.P. Schott, 2012: Environmental Sustainability and Behavioral Science: Meta-Analysis of Proenvironmental Behavior Experiments. *Environment and Behavior*, **44(2)**, 257–299, doi:10.1177/0013916511402673.

Osei, P.D., 2007: Policy responses, institutional networks management and post-Hurricane Ivan reconstruction in Jamaica. *Disaster Prevention and Management: An International Journal*, **16(2)**, 217–234, doi:10.1108/09653560710739540.

Ostrom, E., 2009: A general framework for analyzing sustainability of social-ecological systems. *Science*, **325(5939)**, 419–422, doi:10.1126/science.1172133.

Ostrom, E. and J. Walker (eds.), 2005: *Trust and Reciprocity: Interdisciplinary Lessons for Experimental Research*. Russell Sage Foundation, New York, NY, USA, 424 pp.

Ostrom, E., J. Burger, C.B. Field, R.B. Norgaard, and D. Policansky, 1999: Revisiting the Commons: Local Lessons, Global Challenges. *Science*, **284(5412)**, 278–282, doi:10.1126/science.284.5412.278.

Oteros-Rozas, E. et al., 2015: Participatory scenario planning in place-based social-ecological research: insights and experiences from 23 case studies. *Ecology and Society*, **20(4)**, 32, doi:10.5751/es-07985-200432.

Ouyang, M. and L. Dueñas-Osorio, 2014: Multi-dimensional hurricane resilience assessment of electric power systems. *Structural Safety*, **48**, 15–24, doi:10.1016/j.strusafe.2014.01.001.

Overmars, K.P. et al., 2014: Estimating the opportunity costs of reducing carbon dioxide emissions via avoided deforestation, using integrated assessment modelling. *Land Use Policy*, **41**, 45–60, doi:10.1016/j.landusepol.2014.04.015.

Owen, R., 2014: Solar Radiation Management and the Governance of Hubris. In: *Geoengineering of the Climate System* [Harrison, R.M. and R.E. Hester (eds.)]. The Royal Society of Chemistry, pp. 212–248, doi:10.1039/9781782621225-00212.

Pablo-Romero, M.P., A. Sánchez-Braza, J. Salvador-Ponce, and N. Sánchez-Labrador, 2017: An overview of feed-in tariffs, premiums and tenders to promote electricity from biogas in the EU-28. *Renewable and Sustainable Energy Reviews*, **73**, 1366–1379, doi:10.1016/j.rser.2017.01.132.

Pahl, S., J. Goodhew, C. Boomsma, and S.R.J. Sheppard, 2016: The Role of Energy Visualization in Addressing Energy Use : Insights from the eViz Project. *Frontiers in Psychology*, **7**, 92, doi:10.3389/fpsyg.2016.00092.

PAHO/WHO, 2016: *Strategy for Technical Cooperation with the United Kingdom Overseas Territories (UKOTs) in the Caribbean 2016-2022*. Pan American Health Organization (PAHO) and World Health Organization (WHO) Regional office for the Americas, Washington DC, USA, 57 pp.

Pan, X., M. Elzen, N. Höhne, F. Teng, and L. Wang, 2017: Exploring fair and ambitious mitigation contributions under the Paris Agreement goals. *Environmental Science & Policy*, **74**, 49–56, doi:10.1016/j.envsci.2017.04.020.

Panagopoulos, T., J.A. González Duque, and M. Bostenaru Dan, 2016: Urban planning with respect to environmental quality and human well-being. *Environmental Pollution*, **208**, 137–144, doi:10.1016/j.envpol.2015.07.038.

Pant, L.P., B. Adhikari, and K.K. Bhattarai, 2015: Adaptive transition for transformations to sustainability in developing countries. *Current Opinion in*

Environmental Sustainability, **14**, 206–212, doi:10.1016/j.cosust.2015.07.006.

Panteli, M. and P. Mancarella, 2015: Influence of extreme weather and climate change on the resilience of power systems: Impacts and possible mitigation strategies. *Electric Power Systems Research*, **127**, 259–270, doi:10.1016/j.epsr.2015.06.012.

Panteli, M., D.N. Trakas, P. Mancarella, and N.D. Hatziargyriou, 2016: Boosting the Power Grid Resilience to Extreme Weather Events Using Defensive Islanding. *IEEE Transactions on Smart Grid*, **7(6)**, 2913–2922, doi:10.1109/tsg.2016.2535228.

Papargyropoulou, E., R. Lozano, J. K. Steinberger, N. Wright, and Z. Ujang, 2014: The food waste hierarchy as a framework for the management of food surplus and food waste. *Journal of Cleaner Production*, **76**, 106–115, doi:10.1016/j.jclepro.2014.04.020.

Parag, Y. and B.K. Sovacool, 2016: Electricity market design for the prosumer era. *Nature Energy*, **1(4)**, 16032, doi:10.1038/nenergy.2016.32.

Paraiso, M.L.S. and N. Gouveia, 2015: Health risks due to pre-harvesting sugarcane burning in São Paulo State, Brazil. *Revista Brasileira de Epidemiologia*, **18(3)**, 691–701, doi:10.1590/1980-5497201500030014.

Pardo, P. et al., 2014: A review on high temperature thermochemical heat energy storage. *Renewable and Sustainable Energy Reviews*, **32**, 591–610, doi:10.1016/j.rser.2013.12.014.

Parikh, K.S. and J.K. Parikh, 2016: Realizing potential savings of energy and emissions from efficient household appliances in India. *Energy Policy*, **97**, 102–111, doi:10.1016/j.enpol.2016.07.005.

Parker, A., 2014: Governing solar geoengineering research as it leaves the laboratory. *Philosophical Transactions of the Royal Society A: Mathematical, Physical and Engineering Sciences*, **372(2031)**, 20140173–20140173, doi:10.1098/rsta.2014.0173.

Parker, A. and P.J. Irvine, 2018: The Risk of Termination Shock From Solar Geoengineering. *Earth's Future*, **6(2)**, 1–12, doi:10.1002/2017ef000735.

Parkes, B., A. Challinor, and K. Nicklin, 2015: Crop failure rates in a geoengineered climate: impact of climate change and marine cloud brightening. *Environmental Research Letters*, **10(8)**, 084003, doi:10.1088/1748-9326/10/8/084003.

Parkhill, K., N. Pidgeon, A. Corner, and N. Vaughan, 2013: Deliberation and Responsible Innovation: A Geoengineering Case Study. In: *Responsible Innovation* [Owen, R., J. Bessant, and M. Heintz (eds.)]. John Wiley & Sons Ltd, Chichester, UK, pp. 219–239, doi:10.1002/9781118551424.ch12.

Parkinson, S.C. and N. Djilali, 2015: Robust response to hydro-climatic change in electricity generation planning. *Climatic Change*, **130(4)**, 475–489, doi:10.1007/s10584-015-1359-5.

Parnell, S., 2015: Fostering Transformative Climate Adaptation and Mitigation in the African City: Opportunities and Constraints of Urban Planning. In: *Urban Vulnerability and Climate Change in Africa: A Multidisciplinary Approach*. Springer, Cham, Switzerland, pp. 349–367, doi:10.1007/978-3-319-03982-4_11.

Parry, I.W.H., D. Evans, and W.E. Oates, 2014: Are energy efficiency standards justified? *Journal of Environmental Economics and Management*, **67(2)**, 104–125, doi:10.1016/j.jeem.2013.11.003.

Parson, E.A., 2017: *Starting the Dialogue on Climate Engineering Governance: A World Commission*. Fixing Climate Governance Series Policy Brief No. 8, Centre for International Governance Innovation, Waterloo, ON, Canada, 8 pp.

Pasimeni, M.R. et al., 2014: Scales, strategies and actions for effective energy planning: A review. *Energy Policy*, **65**, 165–174, doi:10.1016/j.enpol.2013.10.027.

Pasquini, L., G. Ziervogel, R.M. Cowling, and C. Shearing, 2015: What enables local governments to mainstream climate change adaptation? Lessons learned from two municipal case studies in the Western Cape, South Africa. *Climate and Development*, **7(1)**, 60–70, doi:10.1080/17565529.2014.886994.

Patel, R., G. Walker, M. Bhatt, and V. Pathak, 2017: The Demand for Disaster Microinsurance for Small Businesses in Urban Slums: The Results of Surveys in Three Indian Cities. *PLOS Currents Disasters*.

Patt, A., 2017: Beyond the tragedy of the commons: Reframing effective climate change governance. *Energy Research & Social Science*, **34**, 1–3, doi:10.1016/j.erss.2017.05.023.

Patt, A.G. and D. Schröter, 2008: Perceptions of climate risk in Mozambique: Implications for the success of adaptation strategies. *Global Environmental Change*, **18(3)**, 458–467, doi:10.1016/j.gloenvcha.2008.04.002.

Pattanayak, S.K., D.E. Mercer, E. Sills, and J.C. Yang, 2003: Taking stock of agroforestry adoption studies. *Agroforestry Systems*, **57(3)**, 173–186, doi:10.1023/a:1024809108210.

Patterson, J.J. et al., 2018: Political feasibility of 1.5 C societal transformations: the role of social justice. *Current Opinion in Environmental Sustainability*, **31**, 1–9, doi:10.1016/j.cosust.2017.11.002.

Paul, K.I. et al., 2016: Managing reforestation to sequester carbon, increase biodiversity potential and minimize loss of agricultural land. *Land Use Policy*, **51**, 135–149, doi:10.1016/j.landusepol.2015.10.027.

Pauw, W.P., R.J.T. Klein, P. Vellinga, and F. Biermann, 2016a: Private finance for adaptation: do private realities meet public ambitions? *Climatic Change*, **134(4)**, 489–503, doi:10.1007/s10584-015-1539-3.

Pauw, W.P. et al., 2016b: NDC Explorer. German Development Institute/Deutsches Institut für Entwicklungspolitik (DIE), African Centre for Technology Studies (ACTS), Stockholm Environment Institute (SEI). Retrieved from: https://klimalog.die-gdi.de/ndc.

Pauw, W.P. et al., 2018: Beyond headline mitigation numbers: we need more transparent and comparable NDCs to achieve the Paris Agreement on climate change. *Climatic Change*, **147(1–2)**, 23–29, doi:10.1007/s10584-017-2122-x.

Peake, S. and P. Ekins, 2017: Exploring the financial and investment implications of the Paris Agreement. *Climate Policy*, **17(7)**, 832–852, doi:10.1080/14693062.2016.1258633.

Pearce, T.C.L., 2018: Incorporating Indigenous Knowledge in Research. In: *Routledge Handbook of Environmental Migration and Displacement* [McLeman, R. and F. Gemenne (eds.)]. Taylor & Francis Group, New York and London, pp. 125–134.

Pearce, T.C.L., J.D. Ford, A.C. Willox, and B. Smit, 2015: Inuit Traditional Ecological Knowledge (TEK), Subsistence Hunting and Adaptation to Climate Change in the Canadian Arctic. *Arctic*, **68(2)**, 233–245, www.jstor.org/stable/43871322.

Pearce, T.C.L., R. Currenti, A. Mateiwai, and B. Doran, 2017: Adaptation to climate change and freshwater resources in Vusama village, Viti Levu, Fiji. *Regional Environmental Change*, 1–10, doi:10.1007/s10113-017-1222-5.

Pellegrino, M., M. Simonetti, and G. Chiesa, 2016: Reducing thermal discomfort and energy consumption of Indian residential buildings: Model validation by in-field measurements and simulation of low-cost interventions. *Energy and Buildings*, **113**, 145–158, doi:10.1016/j.enbuild.2015.12.015.

Pelletier, L.G., K.M. Tuson, I. Green-Demers, K. Noels, and A.M. Beaton, 1998: Why Are You Doing Things for the Environment? The Motivation Toward the Environment Scale (MTES). *Journal of Applied Social Psychology*, **28(5)**, 437–468, doi:10.1111/j.1559-1816.1998.tb01714.x.

Pelling, M., K. O'Brien, and D. Matyas, 2015: Adaptation and transformation. *Climatic Change*, **133(1)**, 113–127, doi:10.1007/s10584-014-1303-0.

Pelling, M. et al., 2018: Africa's urban adaptation transition under a 1.5° climate. *Current Opinion in Environmental Sustainability*, **31**, 10–15, doi:10.1016/j.cosust.2017.11.005.

Peng, J. et al., 2016: Markedly enhanced absorption and direct radiative forcing of black carbon under polluted urban environments. *Proceedings of the National Academy of Sciences*, **113(16)**, 4266–4271, doi:10.1073/pnas.1602310113.

Perlaviciute, G. and L. Steg, 2014: Contextual and psychological factors shaping evaluations and acceptability of energy alternatives: Integrated review and research agenda. *Renewable and Sustainable Energy Reviews*, **35**, 361–381, doi:10.1016/j.rser.2014.04.003.

Perrels, A., T. Frei, F. Espejo, L. Jamin, and A. Thomalla, 2013: Socio-economic benefits of weather and climate services in Europe. *Advances in Science and Research*, **10(1)**, 65–70, doi:10.5194/asr-10-65-2013.

Perrier, Q., 2016: *A robust nuclear strategy for France*. Association des Economistes de l'Energie (FAEE), 29 pp.

Persson, U.M., 2015: The impact of biofuel demand on agricultural commodity prices: a systematic review. *Wiley Interdisciplinary Reviews: Energy and Environment*, **4(5)**, 410–428, doi:10.1002/wene.155.

Perugini, L. et al., 2017: Biophysical effects on temperature and precipitation due to land cover change. *Environmental Research Letters*, **12(5)**, 053002, doi:10.1088/1748-9326/aa6b3f.

Peters, A.M., E. van der Werff, and L. Steg, 2018: Beyond purchasing: Electric vehicle adoption motivation and consistent sustainable energy behaviour in the Netherlands. *Energy Research & Social Science*, **39**, 234–247, doi:10.1016/j.erss.2017.10.008.

Peters, G.P. and O. Geden, 2017: Catalysing a political shift from low to negative carbon. *Nature Climate Change*, **7(9)**, 619–621, doi:10.1038/nclimate3369.

Peters, G.P. et al., 2017: Key indicators to track current progress and future ambition of the Paris Agreement. *Nature Climate Change*, **7(2)**, 118–122, doi:10.1038/nclimate3202.

Pfeiffer, A., R. Millar, C. Hepburn, and E. Beinhocker, 2016: The '2°C capital stock'

4

for electricity generation: Committed cumulative carbon emissions from the electricity generation sector and the transition to a green economy. *Applied Energy*, **179**, 1395–1408, doi:10.1016/j.apenergy.2016.02.093.

Pfeiffer, L. and C.-Y.C. Lin, 2014: Does efficient irrigation technology lead to reduced groundwater extraction? Empirical evidence. *Journal of Environmental Economics and Management*, **67(2)**, 189–208, doi:10.1016/j.jeem.2013.12.002.

Philibert, C., 2017: *Renewable Energy for Industry – From green energy to green materials and fuels*. International Energy Agency (IEA), Paris, France, 72 pp.

Pichert, D. and K. Katsikopoulos, 2008: Green defaults: Information presentation and pro-environmental behaviour. *Journal of Environmental Psychology*, **28(1)**, 63–73, doi:10.1016/j.jenvp.2007.09.004.

Pichler, A. and E. Striessnig, 2013: Differential Vulnerability to Hurricanes in Cuba, Haiti, and the Dominican Republic: The Contribution of Education. *Ecology and Society*, **18(3)**, 31, doi:10.5751/es-05774-180331.

Pichler, M., A. Schaffartzik, H. Haberl, and C. Görg, 2017: Drivers of society-nature relations in the Anthropocene and their implications for sustainability transformations. *Current Opinion in Environmental Sustainability*, **26–27**, 32–36, doi:10.1016/j.cosust.2017.01.017.

Pickering, N.K. et al., 2015: Animal board invited review: genetic possibilities to reduce enteric methane emissions from ruminants. *Animal*, **9(09)**, 1431–1440, doi:10.1017/s1751731115000968.

Pielke, R.A., 2009: An idealized assessment of the economics of air capture of carbon dioxide in mitigation policy. *Environmental Science & Policy*, **12(3)**, 216–225, doi:10.1016/j.envsci.2009.01.002.

Pierpaoli, E., G. Carli, E. Pignatti, and M. Canavari, 2013: Drivers of Precision Agriculture Technologies Adoption: A Literature Review. *Procedia Technology*, **8**, 61–69, doi:10.1016/j.protcy.2013.11.010.

Pierrehumbert, R.T., 2014: Short-Lived Climate Pollution. *Annual Review of Earth and Planetary Sciences*, **42**, 341–79, doi:10.1146/annurev-earth-060313-054843.

Piketty, T., 2014: *Capital in the Twenty-first Century*. The Belknap Press of Harvard University Press, Cambridge, MA, USA, 696 pp.

Pindyck, R.S., 2017: The Use and Misuse of Models for Climate Policy. *Review of Environmental Economics and Policy*, **11(1)**, 100–114, doi:10.1093/reep/rew012.

Pinho, P.F., J.A. Marengo, and M.S. Smith, 2015: Complex socio-ecological dynamics driven by extreme events in the Amazon. *Regional Environmental Change*, **15(4)**, 643–655, doi:10.1007/s10113-014-0659-z.

Pinho, P.F. et al., 2014: Ecosystem protection and poverty alleviation in the tropics: Perspective from a historical evolution of policy-making in the Brazilian Amazon. *Ecosystem Services*, **8**, 97–109, doi:10.1016/j.ecoser.2014.03.002.

Pisano, I. and M. Lubell, 2017: Environmental Behavior in Cross-National Perspective: A Multilevel Analysis of 30 Countries. *Environment and Behavior*, **49(1)**, 31–58, doi:10.1177/0013916515600494.

Pitari, G. et al., 2014: Stratospheric ozone response to sulfate geoengineering: Results from the Geoengineering Model Intercomparison Project (GeoMIP). *Journal of Geophysical Research: Atmospheres*, **119(5)**, 2629–2653, doi:10.1002/2013jd020566.

Pittelkow, C.M. et al., 2014: Productivity limits and potentials of the principles of conservation agriculture. *Nature*, **517(7534)**, 365–368, doi:10.1038/nature13809.

Pittman, J., D. Armitage, S. Alexander, D. Campbell, and M. Alleyne, 2015: Governance fit for climate change in a Caribbean coastal-marine context. *Marine Policy*, **51**, 486–498, doi:10.1016/j.marpol.2014.08.009.

Pittock, J., H. Bjornlund, R. Stirzaker, and A. van Rooyen, 2017: Communal irrigation systems in South-Eastern Africa: findings on productivity and profitability. *International Journal of Water Resources Development*, **33(5)**, 839–847, doi:10.1080/07900627.2017.1324768.

Pizer, W.A., 2002: Combining price and quantity controls to mitigate global climate change. *Journal of Public Economics*, **85(3)**, 409–434, doi:10.1016/s0047-2727(01)00118-9.

Place, F. et al., 2012: Improved Policies for Facilitating the Adoption of Agroforestry. In: *Agroforestry for Biodiversity and Ecosystem Services – Science and Practice* [Kaonga, M. (ed.)]. IntechOpen, London, UK, pp. 113–128, doi:10.5772/34524.

PlaNYC, 2013: *A Stronger, More Resilient New York*. PlaNYC, New York, NY, USA, 438 pp.

Platinga, A. and B. Scholtens, 2016: *The Financial Impact of Divestment from Fossil Fuels*. SOM Research Reports; no. 16005-EEF, SOM research school, University of Groningen, Groningen, Netherlands, 47 pp.

Plazzotta, M., R. Séférian, H. Douville, B. Kravitz, and J. Tjiputra, 2018: Land Surface Cooling Induced by Sulfate Geoengineering Constrained by Major

Volcanic Eruptions. *Geophysical Research Letters*, **45(11)**, 5663–5671, doi:10.1029/2018gl077583.

Plevin, R.J., M. O'Hare, A.D. Jones, M.S. Torn, and H.K. Gibbs, 2010: Greenhouse Gas Emissions from Biofuels' Indirect Land Use Change Are Uncertain but May Be Much Greater than Previously Estimated. *Environmental Science & Technology*, **44(21)**, 8015–8021, doi:10.1021/es101946t.

Poeplau, C. and A. Don, 2015: Carbon sequestration in agricultural soils via cultivation of cover crops – A meta-analysis. *Agriculture, Ecosystems & Environment*, **200**, 33–41, doi:10.1016/j.agee.2014.10.024.

Poff, N.L.R. et al., 2016: Sustainable water management under future uncertainty with eco-engineering decision scaling. *Nature Climate Change*, **6(1)**, 25–34, doi:10.1038/nclimate2765.

Pollak, M., B. Meyer, and E. Wilson, 2011: Reducing greenhouse gas emissions: Lessons from state climate action plans. *Energy Policy*, **39(9)**, 5429–5439, doi:10.1016/j.enpol.2011.05.020.

Pollitt, H. and J.-F. Mercure, 2017: The role of money and the financial sector in energy-economy models used for assessing climate and energy policy. *Climate Policy*, 1–14, doi:10.1080/14693062.2016.1277685.

Pollitt, H., 2018: Policies for limiting climate change to well below 2°C. (in press).

Poortinga, W., L. Steg, and C. Vlek, 2002: Environmental Risk Concern and Preferences for Energy-Saving Measures. *Environment and Behavior*, **34(4)**, 455–478, doi:10.1177/00116502034004003.

Poortinga, W., L. Steg, C. Vlek, and G. Wiersma, 2003: Household preferences for energy-saving measures: A conjoint analysis. *Journal of Economic Psychology*, **24(1)**, 49–64, doi:10.1016/s0167-4870(02)00154-x.

Popp, A. et al., 2014a: Land-use protection for climate change mitigation. *Nature Climate Change*, **4(12)**, 1095–1098, doi:10.1038/nclimate2444.

Popp, A. et al., 2014b: Land-use transition for bioenergy and climate stabilization: model comparison of drivers, impacts and interactions with other land use based mitigation options. *Climatic Change*, **123(3–4)**, 495–509, doi:10.1007/s10584-013-0926-x.

Pradhan, A., C. Chan, P.K. Roul, J. Halbrendt, and B. Sipes, 2018: Potential of conservation agriculture (CA) for climate change adaptation and food security under rainfed uplands of India: A transdisciplinary approach. *Agricultural Systems*, **163**, 27–35, doi:10.1016/j.agsy.2017.01.002.

Preston, B.L., J. Mustelin, and M.C. Maloney, 2013: Climate adaptation heuristics and the science/policy divide. *Mitigation and Adaptation Strategies for Global Change*, **20(3)**, 467–497, doi:10.1007/s11027-013-9503-x.

Preston, B.L., L. Rickards, H. Fünfgeld, and R.J. Keenan, 2015: Toward reflexive climate adaptation research. *Current Opinion in Environmental Sustainability*, **14**, 127–135, doi:10.1016/j.cosust.2015.05.002.

Preston, C.J., 2013: Ethics and geoengineering: reviewing the moral issues raised by solar radiation management and carbon dioxide removal. *Wiley Interdisciplinary Reviews: Climate Change*, **4(1)**, 23–37, doi:10.1002/wcc.198.

Preston, C.J. (ed.), 2016: *Climate Justice and Geoengineering: Ethics and Policy in the Atmospheric Anthropocene*. Rowman & Littlefield International, Lanham, MD, USA, 234 pp.

Pretty, J. and Z.P. Bharucha, 2014: Sustainable intensification in agricultural systems. *Annals of Botany*, **114(8)**, 1571–1596, doi:10.1093/aob/mcu205.

Pretzsch, H. et al., 2017: Climate change accelerates growth of urban trees in metropolises worldwide. *Scientific Reports*, **7(1)**, 15403, doi:10.1038/s41598-017-14831-w.

Pritchard, C., A. Yang, P. Holmes, and M. Wilkinson, 2015: Thermodynamics, economics and systems thinking: What role for air capture of CO_2? *Process Safety and Environmental Protection*, **94**, 188–195, doi:10.1016/j.psep.2014.06.011.

Pritoni, M., A.K. Meier, C. Aragon, D. Perry, and T. Peffer, 2015: Energy efficiency and the misuse of programmable thermostats: The effectiveness of crowdsourcing for understanding household behavior. *Energy Research & Social Science*, **8**, 190–197, doi:10.1016/j.erss.2015.06.002.

Proost, S. and D. Van Regemorter, 1995: The double dividend and the role of inequality aversion and macroeconomic regimes. *International Tax and Public Finance*, **2(2)**, 207–219, doi:10.1007/bf00877497.

Pueyo, A. and R. Hanna, 2015: *What level of electricity access is required to enable and sustain poverty reduction? Annex 1 Literature review*. Practical Action Consulting, Bourton-on-Dunsmore, UK, 65 pp.

Puppim de Oliveira, J.A. et al., 2013: Promoting win–win situations in climate change mitigation, local environmental quality and development in Asian cities through co-benefits. *Journal of Cleaner Production*, **58**, 1–6, doi:10.1016/j.jclepro.2013.08.011.

Pyörälä, P. et al., 2014: Effects of Management on Economic Profitability of Forest

4

Biomass Production and Carbon Neutrality of Bioenergy Use in Norway Spruce Stands Under the Changing Climate. *Bioenergy Research*, **7(1)**, 279–294, doi:10.1007/s12155-013-9372-x.

Qazi, S. and W. Young Jr., 2014: Disaster relief management and resilience using photovoltaic energy. In: *2014 International Conference on Collaboration Technologies and Systems (CTS)*. IEEE, pp. 628–632, doi:10.1109/cts.2014.6867637.

Qin, Z., J.B. Dunn, H. Kwon, S. Mueller, and M.M. Wander, 2016: Soil carbon sequestration and land use change associated with biofuel production: empirical evidence. *GCB Bioenergy*, **8(1)**, 66–80, doi:10.1111/gcbb.12237.

Quaas, M.F., J. Quaas, W. Rickels, and O. Boucher, 2017: Are there reasons against open-ended research into solar radiation management? A model of intergenerational decision-making under uncertainty. *Journal of Environmental Economics and Management*, **84**, 1–17, doi:10.1016/j.jeem.2017.02.002.

Quandt, A., H. Neufeldt, and J.T. McCabe, 2017: The role of agroforestry in building livelihood resilience to floods and drought in semiarid Kenya. *Ecology and Society*, **22(3)**, 10, doi:10.5751/es-09461-220310.

Quitzow, R., 2015: Assessing policy strategies for the promotion of environmental technologies: A review of India's National Solar Mission. *Research Policy*, **44(1)**, 233–243, doi:10.1016/j.respol.2014.09.003.

Rabitz, F., 2016: Going rogue? Scenarios for unilateral geoengineering. *Futures*, **84**, 98–107, doi:10.1016/j.futures.2016.11.001.

Radhakrishnan, M., A. Pathirana, R.M. Ashley, B. Gersonius, and C. Zevenbergen, 2018: Flexible adaptation planning for water sensitive cities. *Cities*, **78**, 87–95, doi:10.1016/j.cities.2018.01.022.

Rai, V., D.C. Reeves, and R. Margolis, 2016: Overcoming barriers and uncertainties in the adoption of residential solar PV. *Renewable Energy*, **89**, 498–505, doi:10.1016/j.renene.2015.11.080.

Rajamani, L., 2016: Ambition and Differentiation in the 2015 Paris Agreement: Interpretative Possibilities and Underlying Politics. *International and Comparative Law Quarterly*, **65(02)**, 493–514, doi:10.1017/s0020589316000130.

Rajan, R.G., 2011: *Fault Lines: How Hidden Fractures Still Threaten the World Economy*. Princeton University Press, Princeton, NJ, USA, 280 pp.

Rakatama, A., R. Pandit, C. Ma, and S. Iftekhar, 2017: The costs and benefits of REDD+: A review of the literature. *Forest Policy and Economics*, **75**, 103–111, doi:10.1016/j.forpol.2016.08.006.

Ramachandran, S., V. Ramaswamy, G.L. Stenchikov, and A. Robock, 2000: Radiative impact of the Mount Pinatubo volcanic eruption: Lower stratospheric response. *Journal of Geophysical Research: Atmospheres*, **105(D19)**, 24409–24429, doi:10.1029/2000jd900355.

Ramage, M.H. et al., 2017: The wood from the trees: The use of timber in construction. *Renewable and Sustainable Energy Reviews*, **68**, 333–359, doi:10.1016/j.rser.2016.09.107.

Ramankutty, N. et al., 2018: Trends in Global Agricultural Land Use: Implications for Environmental Health and Food Security. *Annual Review of Plant Biology*, **69(1)**, 789–815, doi:10.1146/annurev-arplant-042817-040256.

Ran, F.A. et al., 2013: Genome engineering using the CRISPR-Cas9 system. *Nature Protocols*, **8(11)**, 2281–2308, doi:10.1038/nprot.2013.143.

Rand, J. and B. Hoen, 2017: Thirty years of North American wind energy acceptance research: What have we learned? *Energy Research & Social Science*, **29**, 135–148, doi:10.1016/j.erss.2017.05.019.

Ranganathan, J. et al., 2016: *Shifting Diets for a Sustainable Food Future*. Installment 11 of Creating a Sustainable Food Future, World Resources Institute (WRI), Washington DC, USA, 90 pp.

Ranjan, M. and H.J. Herzog, 2011: Feasibility of air capture. *Energy Procedia*, **4**, 2869–2876, doi:10.1016/j.egypro.2011.02.193.

Rasul, G. and B. Sharma, 2016: The nexus approach to water–energy–food security: an option for adaptation to climate change. *Climate Policy*, **16(6)**, 682–702, doi:10.1080/14693062.2015.1029865.

Rathore, D., A.-S. Nizami, A. Singh, and D. Pant, 2016: Key issues in estimating energy and greenhouse gas savings of biofuels: challenges and perspectives. *Biofuel Research Journal*, **3(2)**, 380–393, doi:10.18331/brj2016.3.2.3.

Rau, G.H., 2008: Electrochemical splitting of calcium carbonate to increase solution alkalinity: implications for mitigation of carbon dioxide and ocean acidity. *Environmental Science & Technology*, **42(23)**, 8935–8940, doi:10.1021/es800366q.

Rau, G.H., 2011: CO_2 Mitigation via Capture and Chemical Conversion in Seawater. *Environmental Science & Technology*, **45(3)**, 1088–1092, doi:10.1021/es102671x.

Rau, G.H. and K. Caldeira, 1999: Enhanced carbonate dissolution: a means of sequestering waste CO_2 as ocean bicarbonate. *Energy Conversion and Management*, **40(17)**, 1803–1813, doi:10.1016/s0196-8904(99)00071-0.

Rau, G.H. et al., 2013: Direct electrolytic dissolution of silicate minerals for air CO_2 mitigation and carbon-negative H_2 production. *Proceedings of the National Academy of Sciences*, **110(25)**, 10095–100, doi:10.1073/pnas.1222358110.

Ray, A., L. Hughes, D.M. Konisky, and C. Kaylor, 2017: Extreme weather exposure and support for climate change adaptation. *Global Environmental Change*, **46**, 104–113, doi:10.1016/j.gloenvcha.2017.07.002.

RCI, 2008: *Rotterdam Climate Proof: The Rotterdam Challenge on Water and Climate Adaptation*. Rotterdam Climate Initiative (RCI), Rotterdam, The Netherlands, 11 pp.

RCI, 2013: *Rotterdam Climate Change Adaptation Strategy*. Rotterdam Climate Initiative (RCI), Rotterdam, The Netherlands, 70 pp.

RCI, 2017: Rotterdam Climate Intitiative. Rotterdam Climate Initiative (RCI). Retrieved from: www.rotterdamclimateinitiative.nl.

Reckien, D., J. Flacke, M. Olazabal, and O. Heidrich, 2015: The Influence of Drivers and Barriers on Urban Adaptation and Mitigation Plans – An Empirical Analysis of European Cities. *PLOS ONE*, **10(8)**, e0135597, doi:10.1371/journal.pone.0135597.

Reckien, D. et al., 2014: Climate change response in Europe: what's the reality? Analysis of adaptation and mitigation plans from 200 urban areas in 11 countries. *Climatic Change*, **122(1)**, 331–340, doi:10.1007/s10584-013-0989-8.

Reckien, D. et al., 2017: Climate change, equity and the Sustainable Development Goals: an urban perspective. *Environment & Urbanization*, **29(1)**, 159–182, doi:10.1177/0956247816677778.

Refsgaard, K. and K. Magnussen, 2009: Household behaviour and attitudes with respect to recycling food waste – experiences from focus groups. *Journal of Environmental Management*, **90(2)**, 760–771, doi:10.1016/j.jenvman.2008.01.018.

Rehmatulla, N. and T. Smith, 2015: Barriers to energy efficiency in shipping: A triangulated approach to investigate the principal agent problem. *Energy Policy*, **84**, 44–57, doi:10.1016/j.enpol.2015.04.019.

Reid, H., 2016: Ecosystem- and community-based adaptation: learning from community-based natural resource management. *Climate and Development*, **8(1)**, 4–9, doi:10.1080/17565529.2015.1034233.

Reid, H. et al., 2009: Community-based adaptation to climate change: An overview. *Participatory Learning and Action*, **60**, 11–38, http://pubs.iied.org/pdfs/g02608.pdf.

REN21, 2017: *Renewables 2017 Global Status Report*. Renewable Energy Policy Network for the 21st Century, Paris, France, 302 pp.

Renforth, P., 2012: The potential of enhanced weathering in the UK. *International Journal of Greenhouse Gas Control*, **10**, 229–243, doi:10.1016/j.ijggc.2012.06.011.

Renforth, P. and G. Henderson, 2017: Assessing ocean alkalinity for carbon sequestration. *Reviews of Geophysics*, **55(3)**, 636–674, doi:10.1002/2016rg000533.

Repo, A., J.-P. Tuovinen, and J. Liski, 2015: Can we produce carbon and climate neutral forest bioenergy? *GCB Bioenergy*, **7(2)**, 253–262, doi:10.1111/gcbb.12134.

Revi, A., 2016: Afterwards: Habitat III and the Sustainable Development Goals. *Urbanisation*, **1(2)**, x–xiv, doi:10.1177/2455747116682899.

Revi, A., 2017: Re-imagining the United Nations' Response to a Twenty-first-century Urban World. *Urbanisation*, **2(2)**, 1–7, doi:10.1177/2455747117740438.

Revi, A. et al., 2014a: Towards transformative adaptation in cities: the IPCC's Fifth Assessment. *Environment and Urbanization*, **26(1)**, 11–28, doi:10.1177/0956247814523539.

Revi, A. et al., 2014b: Urban Areas. In: *Climate Change 2014: Impacts, Adaptation, and Vulnerability. Part A: Global and Sectoral Aspects. Contribution of Working Group II to the Fifth Assessment Report of the Intergovernmental Panel on Climate Change* [Field, C.B., V.R. Barros, D.J. Dokken, K.J. Mach, M.D. Mastrandrea, T.E. Bilir, M. Chatterjee, K.L. Ebi, Y.O. Estrada, R.C. Genova, B. Girma, E.S. Kissel, A.N. Levy, S. MacCracken, P.R. Mastrandrea, and L.L. White (eds.)]. Cambridge University Press, Cambridge, United Kingdom and New York, NY, USA, pp. 535–612.

Rey Benayas, J.M. et al., 2009: Enhancement of biodiversity and ecosystem services by ecological restoration: a meta-analysis. *Science*, **325(5944)**, 1121–4, doi:10.1126/science.1172460.

Reynolds, J.L., 2015: An Economic Analysis of Liability and Compensation for Harm from Large-Scale Field Research in Solar Climate Engineering. *Climate Law*, **5(2–4)**, 182–209, doi:10.1163/18786561-00504004.

Reynolds, J.L., J.L. Contreras, and J.D. Sarnoff, 2018: Intellectual property policies for solar geoengineering. *Wiley Interdisciplinary Reviews: Climate Change*, **9(2)**, 1–7, doi:10.1002/wcc.512.

Reynolds, T.W., A. Bostrom, D. Read, and M.G. Morgan, 2010: Now What Do People Know About Global Climate Change? Survey Studies of Educated Laypeople. *Risk Analysis*, **30(10)**, 1520–1538, doi:10.1111/j.1539-6924.2010.01448.x.

RGoB, 2012: *The Report of the High-Level Meeting on Wellbeing and Happiness: Defining a New Economic Paradigm*. Royal Government of Bhutan (RGoB). The Permanent Mission of the Kingdom of Bhutan to the United Nations, New York, NY, USA and Office of the Prime Minister, Thimphu, Bhutan, 166 pp.

RGoB, 2013: *Eleventh Five Year Plan Volume I: Main Document*. Gross National Happiness Commission. Royal Government of Bhutan (RGoB), Thimphu, Bhutan, 399 pp.

RGoB, 2016: *Economic Development Policy*. Royal Government of Bhutan (RGoB), Thimphu, Bhutan, 49 pp.

Riahi, K. et al., 2015: Locked into Copenhagen pledges – Implications of short-term emission targets for the cost and feasibility of long-term climate goals. *Technological Forecasting and Social Change*, **90(Part A)**, 8–23, doi:10.1016/j.techfore.2013.09.016.

Riahi, K. et al., 2017: The Shared Socioeconomic Pathways and their energy, land use, and greenhouse gas emissions implications: An overview. *Global Environmental Change*, **42**, 153–168, doi:10.1016/j.gloenvcha.2016.05.009.

Ribeiro, H., 2008: Sugar cane burning in Brazil: respiratory health effects (in Spanish). *Revista de Saúde Pública*, **42(2)**, 1–6, doi:10.1590/s0034-89102008005000009.

Rickards, L. and S.M. Howden, 2012: Transformational adaptation: agriculture and climate change. *Crop and Pasture Science*, **63(3)**, 240–250, doi:10.1071/cp11172.

Ricke, K.L., J.B. Moreno-Cruz, and K. Caldeira, 2013: Strategic incentives for climate geoengineering coalitions to exclude broad participation. *Environmental Research Letters*, **8(1)**, 014021, doi:10.1088/1748-9326/8/1/014021.

Ricke, K.L., J.B. Moreno-Cruz, J. Schewe, A. Levermann, and K. Caldeira, 2015: Policy thresholds in mitigation. *Nature Geoscience*, **9(1)**, 5–6, doi:10.1038/ngeo2607.

Ringel, M., 2017: Energy efficiency policy governance in a multi-level administration structure – evidence from Germany. *Energy Efficiency*, **10(3)**, 753–776, doi:10.1007/s12053-016-9484-1.

Rinkevich, B., 2014: Rebuilding coral reefs: does active reef restoration lead to sustainable reefs? *Current Opinion in Environmental Sustainability*, **7**, 28–36, doi:10.1016/j.cosust.2013.11.018.

Rippke, U. et al., 2016: Timescales of transformational climate change adaptation in sub-Saharan African agriculture. *Nature Climate Change*, **6(6)**, 605–609, doi:10.1038/nclimate2947.

Ritzenhofen, I. and S. Spinler, 2016: Optimal design of feed-in-tariffs to stimulate renewable energy investments under regulatory uncertainty – A real options analysis. *Energy Economics*, **53**, 76–89, doi:10.1016/j.eneco.2014.12.008.

Rivas, S. et al., 2015: *The Covenant of Mayors: In-depth Analysis of Sustainable Energy Action Plans*. EUR 27526 EN, Joint Research Centre of the European Commission, 170 pp., doi:10.2790/182945.

Roberts, D., 2016: The New Climate Calculus: 1.5°C = Paris Agreement, Cities, Local Government, Science and Champions (PLSC²). *Urbanisation*, **1(2)**, 71–78, doi:10.1177/2455747116672474.

Roberts, J.T. and B. Parks, 2006: *A climate of injustice: Global inequality, north-south politics, and climate policy*. MIT Press, 424 pp.

Roberts, J.T. and R. Weikmans, 2017: Postface: fragmentation, failing trust and enduring tensions over what counts as climate finance. *International Environmental Agreements: Politics, Law and Economics*, **17(1)**, 129–137, doi:10.1007/s10784-016-9347-4.

Roberts, J.T. et al., 2017: How Will We Pay for Loss and Damage? *Ethics, Policy & Environment*, **20(2)**, 208–226, doi:10.1080/21550085.2017.1342963.

Robertson, J.L. and J. Barling (eds.), 2015: *The Psychology of Green Organizations*. Oxford University Press, Oxford, UK, 408 pp.

Robinson et al., 2014: How deep is deep enough? Ocean iron fertilization and carbon sequestration in the Southern Ocean. *Geophysical Research Letters*, **41(7)**, 2489–2495, doi:10.1002/2013gl058799.

Robinson, S.-A., 2017: Climate change adaptation trends in small island developing states. *Mitigation and Adaptation Strategies for Global Change*, **22(4)**, 669–691, doi:10.1007/s11027-015-9693-5.

Robinson, S.-A. and M. Dornan, 2017: International financing for climate change adaptation in small island developing states. *Regional Environmental Change*, **17(4)**, 1103–1115, doi:10.1007/s10113-016-1085-1.

Robiou du Pont, Y. et al., 2017: Equitable mitigation to achieve the Paris Agreement goals. *Nature Climate Change*, **7(1)**, 38–43, doi:10.1038/nclimate3186.

Robledo-Abad, C. et al., 2017: Bioenergy production and sustainable development: science base for policymaking remains limited. *GCB Bioenergy*, **9(3)**, 541–556, doi:10.1111/gcbb.12338.

Robock, A., 2012: Is Geoengineering Research Ethical? *Sicherheit und Frieden (S+F)/Security and Peace*, **30(4)**, 226–229, www.jstor.org/stable/24233207.

Robock, A., L. Oman, and G.L. Stenchikov, 2008: Regional climate responses to geoengineering with tropical and Arctic SO₂ injections. *Journal of Geophysical Research: Atmospheres*, **113(D16)**, D16101, doi:10.1029/2008jd010050.

Robock, A., A. Marquardt, B. Kravitz, and G. Stenchikov, 2009: Benefits, risks, and costs of stratospheric geoengineering. *Geophysical Research Letters*, **36(19)**, L19703, doi:10.1029/2009gl039209.

Rochedo, P.R.R. et al., 2016: Carbon capture potential and costs in Brazil. *Journal of Cleaner Production*, **131**, 280–295, doi:10.1016/j.jclepro.2016.05.033.

Rockström, J. et al., 2017: A roadmap for rapid decarbonization. *Science*, **355(6331)**, 1269–1271, doi:10.1126/science.aah3443.

Rode, P. et al., 2014: *Accessibility in Cities: Transport and Urban Form*. NCE Cities – Paper 03, LSE Cities, London School of Economics and Political Science (LSE), London, UK, 61 pp.

Rodríguez, H., 2015: Risk and Trust in Institutions That Regulate Strategic Technological Innovations: Challenges for a Socially Legitimate Risk Analysis. In: *New Perspectives on Technology, Values, and Ethics* [Gonzalez, W. (ed.)]. Springer, Cham, Switzerland, pp. 147–166, doi:10.1007/978-3-319-21870-0_8.

Roe, M.J., 1996: *Strong Managers, Weak Owners: The Political Roots of American Corporate Finance*. Princeton University Press, Princeton, NJ, USA, 342 pp.

Roelfsema, M., M. Harmsen, J.J.G. Olivier, A.F. Hof, and D.P. van Vuuren, 2018: Integrated assessment of international climate mitigation commitments outside the UNFCCC. *Global Environmental Change*, **48**, 67–75, doi:10.1016/j.gloenvcha.2017.11.001.

Rogelj, J. et al., 2014: Disentangling the effects of CO₂ and short-lived climate forcer mitigation. *Proceedings of the National Academy of Sciences*, **111(46)**, 16325–16330, doi:10.1073/pnas.1415631111.

Rogelj, J. et al., 2016: Paris Agreement climate proposals need a boost to keep warming well below 2°C. *Nature*, **534(7609)**, 631–639, doi:10.1038/nature18307.

Rogelj, J. et al., 2017: Understanding the origin of Paris Agreement emission uncertainties. *Nature Communications*, **8**, 15748, doi:10.1038/ncomms15748.

Rogers, D. and V. Tsirkunov, 2010: *Costs and Benefits of Early Warning Systems*. The World Bank and The United Nations Office for Disaster Risk Reduction (UNISDR), 16 pp.

Rogge, K.S., F. Kern, and M. Howlett, 2017: Conceptual and empirical advances in analysing policy mixes for energy transitions. *Energy Research & Social Science*, **33**, 1–10, doi:10.1016/j.erss.2017.09.025.

Roh, S., 2017: Big Data Analysis of Public Acceptance of Nuclear Power in Korea. *Nuclear Engineering and Technology*, **49(4)**, 850–854, doi:10.1016/j.net.2016.12.015.

Romero-Lankao, P., S. Hughes, A. Rosas-Huerta, R. Borquez, and D.M. Gnatz, 2013: Institutional capacity for climate change responses: An examination of construction and pathways in Mexico City and Santiago. *Environment and Planning C: Government and Policy*, **31(5)**, 785–805, doi:10.1068/c12173.

Romero-Lankao, P. et al., 2018: Governance and policy. In: *Climate Change and Cities: Second Assessment Report of the Urban Climate Change Research Network* [Rosenzweig, C., W. Solecki, P. Romero-Lankao, S. Mehrotra, S. Dhakal, and S. Ali Ibrahim (eds.)]. Cambridge University Press, Cambridge, United Kingdom and New York, NY, USA, pp. 585–606, doi:10.1017/9781316563878.023.

Röös, E. et al., 2017: Protein futures for Western Europe: potential land use and climate impacts in 2050. *Regional Environmental Change*, **17(2)**, 367–377, doi:10.1007/s10113-016-1013-4.

Roques, F.A., D.M. Newbery, and W.J. Nuttall, 2008: Fuel mix diversification incentives in liberalized electricity markets: A Mean--Variance Portfolio theory approach. *Energy Economics*, **30(4)**, 1831–1849, doi:10.1016/j.eneco.2007.11.008.

Rosa, E.A. and T. Dietz, 2012: Human drivers of national greenhouse-gas emissions.

4

Nature Climate Change, **2**, 581–586, doi:10.1038/nclimate1506.

Rose, A., 2016: *Capturing the co-benefits of disaster risk management on the private sector side*. Policy Research Working Paper No. 7634, World Bank, Washington DC, USA, 33 pp.

Rose, S.K., R. Richels, G. Blanford, and T. Rutherford, 2017: The Paris Agreement and next steps in limiting global warming. *Climatic Change*, **142(1–2)**, 255–270, doi:10.1007/s10584-017-1935-y.

Rosenbloom, D., 2017: Pathways: An emerging concept for the theory and governance of low-carbon transitions. *Global Environmental Change*, **43**, 37–50, doi:10.1016/j.gloenvcha.2016.12.011.

Rosenow, J., F. Kern, and K. Rogge, 2017: The need for comprehensive and well targeted instrument mixes to stimulate energy transitions: The case of energy efficiency policy. *Energy Research & Social Science*, **33**, 95–104, doi:10.1016/j.erss.2017.09.013.

Rosenzweig, C. and W. Solecki, 2014: Hurricane Sandy and adaptation pathways in New York: Lessons from a first-responder city. *Global Environmental Change*, **28**, 395–408, doi:10.1016/j.gloenvcha.2014.05.003.

Rosenzweig, C. et al., 2015: *ARC3.2 Summary for city leaders*. Urban Climate Change Research Network. Columbia University, New York, NY, USA, 28 pp.

Roser, D., C. Huggel, M. Ohndorf, and I. Wallimann-Helmer, 2015: Advancing the interdisciplinary dialogue on climate justice. *Climatic Change*, **133(3)**, 349–359, doi:10.1007/s10584-015-1556-2.

Ross, M.L., 2015: What Have We Learned about the Resource Curse? *Annual Review of Political Science*, **18(1)**, 239–259, doi:10.1146/annurev-polisci-052213-040359.

Rothman, D.S., P. Romero-Lankao, V.J. Schweizer, and B.A. Bee, 2014: Challenges to adaptation: a fundamental concept for the shared socio-economic pathways and beyond. *Climatic Change*, **122(3)**, 495–507, doi:10.1007/s10584-013-0907-0.

Roudier, P., A. Alhassane, C. Baron, S. Louvet, and B. Sultan, 2016: Assessing the benefits of weather and seasonal forecasts to millet growers in Niger. *Agricultural and Forest Meteorology*, **223**, 168–180, doi:10.1016/j.agrformet.2016.04.010.

Rozenberg, J., S. Hallegatte, B. Perrissin-Fabert, and J.-C. Hourcade, 2013: Funding low-carbon investments in the absence of a carbon tax. *Climate Policy*, **13(1)**, 134–141, doi:10.1080/14693062.2012.691222.

Rubin, E.S., J.E. Davison, and H.J. Herzog, 2015: The cost of CO_2 capture and storage. *International Journal of Greenhouse Gas Control*, **40**, 378–400, doi:10.1016/j.ijggc.2015.05.018.

Ruepert, A. et al., 2016: Environmental considerations in the organizational context: A pathway to pro-environmental behaviour at work. *Energy Research & Social Science*, **17**, 59–70, doi:10.1016/j.erss.2016.04.004.

Ruparathna, R., K. Hewage, and R. Sadiq, 2016: Improving the energy efficiency of the existing building stock: A critical review of commercial and institutional buildings. *Renewable and Sustainable Energy Reviews*, **53**, 1032–1045, doi:10.1016/j.rser.2015.09.084.

Russell, L.M. et al., 2012: Ecosystem impacts of geoengineering: A review for developing a science plan. *Ambio*, **41(4)**, 350–369, doi:10.1007/s13280-012-0258-5.

Ryaboshapko, A.G. and A.P. Revokatova, 2015: Technical Capabilities for Creating an Aerosol Layer In the Stratosphere for Climate Stabilization Purpose (in Russian). *Problems of Environmental monitoring and Ecosystem modeling*, **T 26(2)**, 115–127.

Sælen, H. and S. Kallbekken, 2011: A choice experiment on fuel taxation and earmarking in Norway. *Ecological Economics*, **70(11)**, 2181–2190, doi:10.1016/j.ecolecon.2011.06.024.

Safarzyńska, K. and J.C.J.M. van den Bergh, 2017: Financial stability at risk due to investing rapidly in renewable energy. *Energy Policy*, **108**, 12–20, doi:10.1016/j.enpol.2017.05.042.

Salon, D., M.G. Boarnet, S. Handy, S. Spears, and G. Tal, 2012: How do local actions affect VMT? A critical review of the empirical evidence. *Transportation Research Part D: Transport and Environment*, **17(7)**, 495–508, doi:10.1016/j.trd.2012.05.006.

Salter, S., G. Sortino, and J. Latham, 2008: Sea-going hardware for the cloud albedo method of reversing global warming. *Philosophical Transactions of the Royal Society A: Mathematical, Physical and Engineering Sciences*, **366(1882)**, 3989–4006, doi:10.1098/rsta.2008.0136.

Salvo, A., J. Brito, P. Artaxo, and F.M. Geiger, 2017: Reduced ultrafine particle levels in São Paulo's atmosphere during shifts from gasoline to ethanol use. *Nature Communications*, **8(1)**, 77, doi:10.1038/s41467-017-00041-5.

Samset, B.H. et al., 2018: Climate Impacts From a Removal of Anthropogenic Aerosol Emissions. *Geophysical Research Letters*, **45(2)**, 1020–1029, doi:10.1002/2017gl076079.

Sanchez, D.L. and V. Sivaram, 2017: Saving innovative climate and energy research: Four recommendations for Mission Innovation. *Energy Research & Social Science*, **29**, 123–126, doi:10.1016/j.erss.2017.05.022.

Sand, M. et al., 2015: Response of Arctic temperature to changes in emissions of short-lived climate forcers. *Nature Climate Change*, **6(3)**, 286–289, doi:10.1038/nclimate2880.

Sanderson, B.M., B.C. O'Neill, and C. Tebaldi, 2016: What would it take to achieve the Paris temperature targets? *Geophysical Research Letters*, **43(13)**, 7133–7142, doi:10.1002/2016gl069563.

Sandler, T., 2017: Collective action and geoengineering. *The Review of International Organizations*, 1–21, doi:10.1007/s11558-017-9282-3.

Sands, R., 2018: U.S. Carbon Tax Scenarios and Bioenergy. *Climate Change Economics (CCE)*, **9(1)**, 1–12, doi:10.1142/s2010007818400109.

Sanesi, G., G. Colangelo, R. Lafortezza, E. Calvo, and C. Davies, 2017: Urban green infrastructure and urban forests: a case study of the Metropolitan Area of Milan. *Landscape Research*, **42(2)**, 164–175, doi:10.1080/01426397.2016.1173658.

Sanna, A., M. Dri, M.R. Hall, and M. Maroto-Valer, 2012: Waste materials for carbon capture and storage by mineralisation (CCSM) – A UK perspective. *Applied Energy*, **99**, 545–554, doi:10.1016/j.apenergy.2012.06.049.

Santangeli, A. et al., 2016: Global change synergies and trade-offs between renewable energy and biodiversity. *GCB Bioenergy*, **8(5)**, 941–951, doi:10.1111/gcbb.12299.

Santos, G., 2008: The London experience. In: *Pricing in Road Transport: A Multi-Disciplinary Perspective* [Verhoef, E., M. Bliemer, L. Steg, and B. van Wee (eds.)]. Edward Elgar Publishing, Cheltenham, UK, pp. 273–292.

Sanz-Pérez, E.S., C.R. Murdock, S.A. Didas, and C.W. Jones, 2016: Direct Capture of CO_2 from Ambient Air. *Chemical Reviews*, **116(19)**, 11840–11876, doi:10.1021/acs.chemrev.6b00173.

Sarmiento, H. and C. Tilly, 2018: Governance Lessons from Urban Informality. *Politics and Governance*, **6(1)**, 199–202, doi:10.17645/pag.v6i1.1169.

Sarmiento, J.L. and J.C. Orr, 1991: Three-dimensional simulations of the impact of Southern Ocean nutrient depletion on atmospheric CO_2 and ocean chemistry. *Limnology and Oceanography*, **36(8)**, 1928–1950, doi:10.4319/lo.1991.36.8.1928.

Sartor, O., 2013: *Carbon Leakage in the Primary Aluminium Sector: What Evidence after 6.5 Years of the EU ETS?* Working Paper No. 2012-12, CDC Climat Research, 24 pp.

Satterthwaite, D. and S. Bartlett, 2017: Editorial: The full spectrum of risk in urban centres: changing perceptions, changing priorities. *Environment and Urbanization*, **29(1)**, 3–14, doi:10.1177/0956247817691921.

Satterthwaite, D., D. Archer, S. Colenbrander, D. Dodman, and J. Hardoy, 2018: *Responding to climate change in cities and in their informal settlements and economies*. International Institute for Environment and Development (IIED) and IIED-América Latina, 61 pp.

Saunders, M.I. et al., 2014: Interdependency of tropical marine ecosystems in response to climate change. *Nature Climate Change*, **4(8)**, 724–729, doi:10.1038/nclimate2274.

Savo, V. et al., 2016: Observations of climate change among subsistence-oriented communities around the world. *Nature Climate Change*, **6(5)**, 462–473, doi:10.1038/nclimate2958.

Schaeffer, R. et al., 2012: Energy sector vulnerability to climate change: A review. *Energy*, **38(1)**, 1–12, doi:10.1016/j.energy.2011.11.056.

Schaeffer, S.M. and P.A. Nakata, 2015: CRISPR/Cas9-mediated genome editing and gene replacement in plants: Transitioning from lab to field. *Plant Science*, **240**, 130–142, doi:10.1016/j.plantsci.2015.09.011.

Schäfer, S. et al., 2013: Field tests of solar climate engineering. *Nature Climate Change*, **3**, 766, doi:10.1038/nclimate1987.

Scheben, A., Y. Yuan, and D. Edwards, 2016: Advances in genomics for adapting crops to climate change. *Current Plant Biology*, **6**, 2–10, doi:10.1016/j.cpb.2016.09.001.

Scheer, D. and O. Renn, 2014: Public Perception of geoengineering and its consequences for public debate. *Climatic Change*, **125(3–4)**, 305–318, doi:10.1007/s10584-014-1177-1.

Schelling, T.C., 1991: Economic responses to global warming: prospects for cooperative approaches. In: *Global Warming, Economic Policy Responses* [Dornbusch, R. and J.M. Poterba (eds.)]. MIT Press, Cambridge, MA, USA.

Schiller, P.L. and J.R. Kenworthy, 2018: An Introduction to Sustainable

4

Transportation: Policy, Planning and Implementation (Second edition). Routledge, London, UK, 442 pp., doi:10.4324/9781315644486.

Schimmelpfennig, D. and R. Ebel, 2016: Sequential Adoption and Cost Savings from Precision Agriculture. *Journal of Agricultural and Resource Economics*, **41(1)**, 97–115, www.waeaonline.org/userfiles/file/jarejanuary20166schimmelpfennigpp97-115.pdf.

Schlenker, W. and M.J. Roberts, 2009: Nonlinear temperature effects indicate severe damages to US crop yields under climate change. *Proceedings of the National Academy of Sciences*, **106(37)**, 15594–15598, doi:10.1073/pnas.0906865106.

Schleussner, C.-F. et al., 2016: Differential climate impacts for policy-relevant limits to global warming: the case of 1.5°C and 2°C. *Earth System Dynamics*, **7(2)**, 327–351, doi:10.5194/esd-7-327-2016.

Schmale, J., D. Shindell, E. von Schneidemesser, I. Chabay, and M. Lawrence, 2014: Clean up our skies. *Nature*, **515**, 335–337, doi:10.1038/515335a.

Schmeier, S., 2014: International River Basin Organizations Lost in Translation? Transboundary River Basin Governance Between Science and Policy. In: *The Global Water System in the Anthropocene*. Springer International Publishing, Cham, pp. 369–383, doi:10.1007/978-3-319-07548-8_24.

Schmitt, M.T., L.B. Aknin, J. Axsen, and R.L. Shwom, 2018: Unpacking the Relationships Between Pro-environmental Behavior, Life Satisfaction, and Perceived Ecological Threat. *Ecological Economics*, **143**, 130–140, doi:10.1016/j.ecolecon.2017.07.007.

Schneider, M. et al., 2017: *The World Nuclear Industry Status Report 2017*. Mycle Schneider Consulting, Paris, France, 267 pp.

Schoenfeld, J.J., M. Hildén, and A.J. Jordan, 2016: The challenges of monitoring national climate policy: learning lessons from the EU. *Climate Policy*, 1–11, doi:10.1080/14693062.2016.1248887.

Schoenmaker, D. and R. van Tilburg, 2016: *Financial risks and opportunities in the time of climate change*. Bruegel Policy Brief 2016/02, Bruegel, Brussels, Belgium, 8 pp.

Schroeder, R. and K. Schroeder, 2014: Happy Environments: Bhutan, Interdependence and the West. *Sustainability*, **6(6)**, 3521–3533, doi:10.3390/su6063521.

Schuiling, R.D. and P. Krijgsman, 2006: Enhanced Weathering: An Effective and Cheap Tool to Sequester CO_2. *Climatic Change*, **74(1)**, 349–354, doi:10.1007/s10584-005-3485-y.

Schuiling, R.D. and P.L. de Boer, 2010: Coastal spreading of olivine to control atmospheric CO_2 concentrations: A critical analysis of viability. Comment: Nature and laboratory models are different. *International Journal of Greenhouse Gas Control*, **4(5)**, 855–856, doi:10.1016/j.ijggc.2010.04.012.

Schuitema, G. and L. Steg, 2008: The role of revenue use in the acceptability of transport pricing policies. *Transportation Research Part F: Traffic Psychology and Behaviour*, **11(3)**, 221–231, doi:10.1016/j.trf.2007.11.003.

Schuitema, G., L. Steg, and S. Forward, 2010a: Explaining differences in acceptability before and acceptance after the implementation of a congestion charge in Stockholm. *Transportation Research Part A: Policy and Practice*, **44(2)**, 99–109, doi:10.1016/j.tra.2009.11.005.

Schuitema, G., L. Steg, and J.A. Rothengatter, 2010b: The acceptability, personal outcome expectations, and expected effects of transport pricing policies. *Journal of Environmental Psychology*, **30(4)**, 587–593, doi:10.1016/j.jenvp.2010.05.002.

Schuitema, G., L. Steg, and M. van Kruining, 2011: When Are Transport Pricing Policies Fair and Acceptable? *Social Justice Research*, **24(1)**, 66–84, doi:10.1007/s11211-011-0124-9.

Schuler, M.L., O. Mantegazza, and A.P.M. Weber, 2016: Engineering C4 photosynthesis into C3 chassis in the synthetic biology age. *The Plant Journal*, **87(1)**, 51–65, doi:10.1111/tpj.13155.

Schulze, E.-D., C. Körner, B.E. Law, H. Haberl, and S. Luyssaert, 2012: Large-scale bioenergy from additional harvest of forest biomass is neither sustainable nor greenhouse gas neutral. *GCB Bioenergy*, **4(6)**, 611–616, doi:10.1111/j.1757-1707.2012.01169.x.

Schwan, S. and X. Yu, 2017: Social protection as a strategy to address climate-induced migration. *International Journal of Climate Change Strategies and Management*, IJCCSM–01–2017–0019, doi:10.1108/ijccsm-01-2017-0019.

Schwartz, D., W. Bruine de Bruin, B. Fischhoff, and L. Lave, 2015: Advertising energy saving programs: The potential environmental cost of emphasizing monetary savings. *Journal of Experimental Psychology: Applied*, **21(2)**, 158–166, doi:10.1037/xap0000042.

Scott, C.A., M. Kurian, and J.L. Wescoat, 2015: The Water–Energy–Food Nexus: Enhancing Adaptive Capacity to Complex Global Challenges. In: *Governing the Nexus* [Kurian, M. and R. Ardakanian (eds.)]. Springer International Publishing, Cham, Switzerland, pp. 15–38, doi:10.1007/978-3-319-05747-7_2.

Scott, M.J. et al., 2015: Calculating impacts of energy standards on energy demand in U.S. buildings with uncertainty in an integrated assessment model. *Energy*, **90**, 1682–1694, doi:10.1016/j.energy.2015.06.127.

Scott, V. and O. Geden, 2018: The challenge of carbon dioxide removal for EU policy-making. *Nature Energy*, 1–3, doi:10.1038/s41560-018-0124-1.

Scott, V., R.S. Haszeldine, S.F.B. Tett, and A. Oschlies, 2015: Fossil fuels in a trillion tonne world. *Nature Climate Change*, **5(5)**, 419–423, doi:10.1038/nclimate2578.

Scovronick, N., C. Dora, E. Fletcher, A. Haines, and D. Shindell, 2015: Reduce short-lived climate pollutants for multiple benefits. *The Lancet*, **386(10006)**, e28–e31, doi:10.1016/s0140-6736(15)61043-1.

SDSN, 2013: *The Urban Opportunity: Enabling Transformative and Sustainable Development*. Background Paper for the High-Level Panel of Eminent Persons on the Post-2015 Development Agenda. Sustainable Development Solutions Network (SDSN) Thematic Group on Sustainable Cities, 47 pp.

Searchinger, T.D., T. Beringer, and A. Strong, 2017: Does the world have low-carbon bioenergy potential from the dedicated use of land? *Energy Policy*, **110**, 434–446, doi:10.1016/j.enpol.2017.08.016.

Searchinger, T.D. et al., 2008: Use of U.S. Croplands for Biofuels Increases Greenhouse Gases Through Emissions from Land-Use Change. *Science*, **319(5867)**, 1238–1240, doi:10.1126/science.1151861.

Seba, T., 2014: *Clean Disruption of Energy and Transportation: How Silicon Valley Will Make Oil, Nuclear, Natural Gas, Coal, Electric Utilities and Conventional Cars Obsolete by 2030*. Clean Planet Ventures, Silicon Valley, CA, USA, 290 pp.

Seidl, R. et al., 2017: Forest disturbances under climate change. *Nature Climate Change*, **7**, 395–402, doi:10.1038/nclimate3303.

Seigo, S.L.O., S. Dohle, and M. Siegrist, 2014: Public perception of carbon capture and storage (CCS): A review. *Renewable and Sustainable Energy Reviews*, **38**, 848–863, doi:10.1016/j.rser.2014.07.017.

Sejersen, F., 2015: *Rethinking Greenland and the Arctic in the Era of Climate Change: New Northern Horizons*. Routledge, London, UK, 248 pp., doi:10.4324/9781315728308.

Semprini, G., R. Gulli, and A. Ferrante, 2017: Deep regeneration vs shallow renovation to achieve nearly Zero Energy in existing buildings. *Energy and Buildings*, **156**, 327–342, doi:10.1016/j.enbuild.2017.09.044.

Sen, B., M. Noori, and O. Tatari, 2017: Will Corporate Average Fuel Economy (CAFE) Standard help? Modeling CAFE's impact on market share of electric vehicles. *Energy Policy*, **109**, 279–287, doi:10.1016/j.enpol.2017.07.008.

Sendzimir, J., C.P. Reija, and P. Magnuszewski, 2011: Rebuilding Resilience in the Sahel. *Ecology and Society*, **16(3)**, 1–29, doi:10.5751/es-04198-160301.

Seneviratne, S.I. et al., 2018: Land radiative management as contributor to regional-scale climate adaptation and mitigation. *Nature Geoscience*, **11(2)**, 88–96, doi:10.1038/s41561-017-0057-5.

Serrao-Neumann, S., F. Crick, B. Harman, G. Schuch, and D.L. Choy, 2015: Maximising synergies between disaster risk reduction and climate change adaptation: Potential enablers for improved planning outcomes. *Environmental Science & Policy*, **50**, 46–61, doi:10.1016/j.envsci.2015.01.017.

Sewando, P.T., K.D. Mutabazi, and N.Y.S. Mdoe, 2016: Vulnerability of agro-pastoral farmers to climate risks in northern and central Tanzania. *Development Studies Research*, **3(1)**, 11–24, doi:10.1080/21665095.2016.1238311.

Seyfang, G. and A. Haxeltine, 2012: Growing grassroots innovations: Exploring the role of community-based initiatives in governing sustainable energy transitions. *Environment and Planning C: Politics and Space*, **30(3)**, 381–400, doi:10.1068/c10222.

Shaffer, L.J., 2014: Making Sense of Local Climate Change in Rural Tanzania Through Knowledge Co-Production. *Journal of Ethnobiology*, **34(3)**, 315–334, doi:10.2993/0278-0771-34.3.315.

Shah, N., M. Wei, V. Letschert, and A. Phadke, 2015: *Benefits of Leapfrogging to Superefficiency and Low Global Warming Potential Refrigerants in Room Air Conditioning*. LBNL-1003671, Lawrence Berkeley National Laboratory (LBL), Berkeley, CA, USA, 39 pp.

Shang, Y. et al., 2018: China's energy-water nexus: Assessing water conservation synergies of the total coal consumption cap strategy until 2050. *Applied Energy*, **210**, 643–660, doi:10.1016/j.apenergy.2016.11.008.

4

Shapiro, S., 2016: The realpolitik of building codes: overcoming practical limitations to climate resilience. *Building Research & Information*, **44(5–6)**, 490–506, doi:10.1080/09613218.2016.1156957.

Sharadin, N., 2018: Rational Coherence in Environmental Policy: Paris, Montreal, and Kigali. *Ethics, Policy & Environment*, **21(1)**, 4–8, doi:10.1080/21550085.2018.1447885.

Sharma, R., 2018: Financing Indian Urban Rail through Land Development: Case Studies and Implications for the Accelerated Reduction in Oil Associated with 1.5°C. *Urban Planning*, **3(2)**, 21–34, doi:10.17645/up.v3i2.1158.

Shaw, C., S. Hales, P. Howden-Chapman, and R. Edwards, 2014: Health co-benefits of climate change mitigation policies in the transport sector. *Nature Climate Change*, **4(6)**, 427–433, doi:10.1038/nclimate2247.

Shayegh, S., D.L. Sanchez, and K. Caldeira, 2017: Evaluating relative benefits of different types of R&D for clean energy technologies. *Energy Policy*, **107**, 532–538, doi:10.1016/j.enpol.2017.05.029.

Sheeran, K., 2006: Who Should Abate Carbon Emissions? A Note. *Environmental and Resource Economics*, **35(2)**, 89–98, doi:10.1007/s10640-006-9007-1.

Sherman, M., J. Ford, A. Llanos-Cuentas, and M.J. Valdivia, 2016: Food system vulnerability amidst the extreme 2010--2011 flooding in the Peruvian Amazon: a case study from the Ucayali region. *Food Security*, **8(3)**, 551–570, doi:10.1007/s12571-016-0583-9.

Shi, L. et al., 2016: Roadmap towards justice in urban climate adaptation research. *Nature Climate Change*, **6(2)**, 131–137, doi:10.1038/nclimate2841.

Shindell, D., J.S. Fuglestvedt, and W.J. Collins, 2017a: The Social Cost of Methane: Theory and Applications. *Faraday Discussions*, doi:10.1039/c7fd00009j.

Shindell, D. et al., 2012: Simultaneously Mitigating Near-Term Climate Change and Improving Human Health and Food Security. *Science*, **335(6065)**, 183–189, doi:10.1126/science.1210026.

Shindell, D. et al., 2017b: A climate policy pathway for near- and long-term benefits. *Science*, **356(6337)**, 493–494, doi:10.1126/science.aak9521.

Shine, T. and G. Campillo, 2016: *The Role of Development Finance in Climate Action Post-2015*. OECD Development Co-operation Working Papers No. 31, OECD Publishing, Paris, France, 38 pp.

Shleifer, A. and R.W. Vishny, 1990: Equilibrium Short Horizons of Investors and Firms. *The American Economic Review*, **80(2)**, 148–153, www.jstor.org/stable/2006560.

Shrivastava, P. and S. Persson, 2018: Silent transformation to 1.5°C-with China's encumbered leading. *Current Opinion in Environmental Sustainability*, **31**, 130–136, doi:10.1016/j.cosust.2018.02.014.

Shue, H., 2018: Mitigation gambles: uncertainty, urgency and the last gamble possible. *Philosophical Transactions of the Royal Society A: Mathematical, Physical and Engineering Sciences*, **376(2119)**, 20170105, doi:10.1098/rsta.2017.0105.

Shukla, P.R., 2005: Aligning Justice and Efficiency in the Global Climate Change Regime: A Developing Country Perspective. *Advances in the Economics of Environmental Resources*, **5**, 121–144, doi:10.1016/s1569-3740(05)05006-6.

Shukla, P.R., J.-C. Hourcade, E. La Rovere, E. Espagne, and B. Perrissin-Fabert, 2017: *Revisiting the Carbon Pricing Challenge after COP21 and COP22*. CIRED Working Papers No. 2017-59, Centre International de Recherche sur l'Environnement et le Développement (CIRED), Paris, France, 8 pp.

Sida, T.S., F. Baudron, H. Kim, and K.E. Giller, 2018: Climate-smart agroforestry: *Faidherbia albida* trees buffer wheat against climatic extremes in the Central Rift Valley of Ethiopia. *Agricultural and Forest Meteorology*, **248**, 339–347, doi:10.1016/j.agrformet.2017.10.013.

Siders, A.R., 2017: A role for strategies in urban climate change adaptation planning: Lessons from London. *Regional Environmental Change*, **17(6)**, 1801–1810, doi:10.1007/s10113-017-1153-1.

Siegmeier, J. et al., 2018: The fiscal benefits of stringent climate change mitigation: an overview. *Climate Policy*, **18(3)**, 352–367, doi:10.1080/14693062.2017.1400943.

SIFMA, 2017: *2017 Fact book*. Securities Industry and Financial Markets Association (SIFMA), New York, NY, USA, 96 pp.

Sikka, A.K., A. Islam, and K.V. Rao, 2018: Climate-Smart Land and Water Management for Sustainable Agriculture. *Irrigation and Drainage*, **67(1)**, 72–81, doi:10.1002/ird.2162.

Simon, A.J., N.B. Kaahaaina, S. Julio Friedmann, and R.D. Aines, 2011: Systems analysis and cost estimates for large scale capture of carbon dioxide from air. *Energy Procedia*, **4**, 2893–2900, doi:10.1016/j.egypro.2011.02.196.

Simon, D. and H. Leck, 2015: Understanding climate adaptation and transformation challenges in African cities. *Current Opinion in Environmental Sustainability*, **13**, 109–116, doi:10.1016/j.cosust.2015.03.003.

Simone, A.M. and E.A. Pieterse, 2017: *New urban worlds: inhabiting dissonant times*. Polity Press, Cambridge, UK and Malden, MA, USA, 192 pp.

Sims, R. et al., 2014: Transport. In: *Climate Change 2014: Mitigation of Climate Change. Contribution of Working Group III to the Fifth Assessment Report of the Intergovernmental Panel on Climate Change* [Edenhofer, O., R. Pichs-Madruga, Y. Sokona, E. Farahani, S. Kadne, K. Seyboth, A. Adler, I. Baum, S. Brunner, P. Eickemeier, B. Kriemann, J. Savolainen, S. Schlömer, C. Stechow, T. Zwickel, and J.C. Minx (eds.)]. Cambridge University Press, Cambridge, United Kingdom and New York, NY, USA, pp. 599–670.

Singh, C., P. Dorward, and H. Osbahr, 2016a: Developing a holistic approach to the analysis of farmer decision-making: Implications for adaptation policy and practice in developing countries. *Land Use Policy*, **59**, 329–343, doi:10.1016/j.landusepol.2016.06.041.

Singh, C., P. Urquhart, and E. Kituyi, 2016b: *From pilots to systems: Barriers and enablers to scaling up the use of climate information services in smallholder farming communities*. CARIAA Working Paper no. 3, Collaborative Adaptation Research Initiative in Africa and Asia, International Development Research Centre, Ottawa, ON, Canada, 56 pp.

Singh, C. et al., 2017: The utility of weather and climate information for adaptation decision-making: current uses and future prospects in Africa and India. *Climate and Development*, 1–17, doi:10.1080/17565529.2017.1318744.

Sinha, A., L.A. Darunte, C.W. Jones, M.J. Realff, and Y. Kawajiri, 2017: Systems Design and Economic Analysis of Direct Air Capture of CO_2 through Temperature Vacuum Swing Adsorption Using MIL-101(Cr)-PEI-800 and mmen-Mg2 (dobpdc) MOF Adsorbents. *Industrial & Engineering Chemistry Research*, **56(3)**, 750–764, doi:10.1021/acs.iecr.6b03887.

Sirkis, A. et al., 2015: *Moving the trillions: a debate on positive pricing of mitigation actions*. Brasil no Clima, Rio de Janeiro, Brazil, 157 pp.

Sivakumar, M.V.K., C. Collins, A. Jay, and J. Hansen, 2014: *Regional priorities for strengthening climate services for farmers in Africa and South Asia*. CCAFS Working Paper no. 71, CGIAR Research Program on Climate Change, Agriculture and Food Security (CCAFS), Copenhagen, Denmark, 36 pp.

Sjöberg, L. and B.-M. Drottz-Sjöberg, 2001: Fairness, risk and risk tolerance in the siting of a nuclear waste repository. *Journal of Risk Research*, **4(1)**, 75–101, doi:10.1080/136698701456040.

Skougaard Kaspersen, P., N. Høegh Ravn, K. Arnbjerg-Nielsen, H. Madsen, and M. Drews, 2015: Influence of urban land cover changes and climate change for the exposure of European cities to flooding during high-intensity precipitation. *Proceedings of the International Association of Hydrological Sciences*, **370**, 21–27, doi:10.5194/piahs-370-21-2015.

Skuce, P.J., E.R. Morgan, J. van Dijk, and M. Mitchell, 2013: Animal health aspects of adaptation to climate change: beating the heat and parasites in a warming Europe. *Animal*, **7(s2)**, 333–345, doi:10.1017/s175173111300075x.

Slade, R., A. Bauen, and R. Gross, 2014: Global bioenergy resources. *Nature Climate Change*, **4(2)**, 99–105, doi:10.1038/nclimate2097.

Slee, B., 2015: Is there a case for community-based equity participation in Scottish on-shore wind energy production? Gaps in evidence and research needs. *Renewable and Sustainable Energy Reviews*, **41**, 540–549, doi:10.1016/j.rser.2014.08.064.

Sleenhoff, S., E. Cuppen, and P. Osseweijer, 2015: Unravelling emotional viewpoints on a bio-based economy using Q methodology. *Public Understanding of Science*, **24(7)**, 858–877, doi:10.1177/0963662513517071.

Smith, C.J. et al., 2017: Impacts of stratospheric sulfate geoengineering on global solar photovoltaic and concentrating solar power resource. *Journal of Applied Meteorology and Climatology*, JAMC–D–16–0298.1, doi:10.1175/jamc-d-16-0298.1.

Smith, H. and P. Jenkins, 2015: Trans-disciplinary research and strategic urban expansion planning in a context of weak institutional capacity: Case study of Huambo, Angola. *Habitat International*, **46**, 244–251, doi:10.1016/j.habitatint.2014.10.006.

Smith, H., E. Kruger, J. Knot, and J. Blignaut, 2017: Conservation Agriculture in South Africa: Lessons from Case Studies. In: *Conservation Agriculture for Africa: Building Resilient Farming Systems in a Changing Climate* [Kassam, A.H., S. Mkomwa, and T. Friedrich (eds.)]. Centre for Agriculture and Biosciences International (CABI), Wallingford, UK, pp. 214–245.

Smith, H.A. and K. Sharp, 2012: Indigenous climate knowledges. *Wiley*

Interdisciplinary Reviews: Climate Change, **3(5)**, 467–476, doi:10.1002/wcc.185.

Smith, J.B., J.M. Vogel, and J.E. Cromwell III, 2009: An architecture for government action on adaptation to climate change. An editorial comment. *Climatic Change*, **95(1–2)**, 53–61, doi:10.1007/s10584-009-9623-1.

Smith, K.R. et al., 2014: Human health: impacts, adaptation, and co-benefits. In: *Climate Change 2014: Impacts, Adaptation, and Vulnerability. Part A: Global and Sectoral Aspects. Contribution of Working Group II to the Fifth Assessment Report of the Intergovernmental Panel on Climate Change* [Field, C.B., V.R. Barros, D.J. Dokken, K.J. Mach, M.D. Mastrandrea, and T.E. Bilir (eds.)]. Cambridge University Press, Cambridge, United Kingdom and New York, NY, USA, pp. 709–754.

Smith, L.J. and M.S. Torn, 2013: Ecological limits to terrestrial biological carbon dioxide removal. *Climatic Change*, **118(1)**, 89–103, doi:10.1007/s10584-012-0682-3.

Smith, P., 2016: Soil carbon sequestration and biochar as negative emission technologies. *Global Change Biology*, **22(3)**, 1315–1324, doi:10.1111/gcb.13178.

Smith, P., R.S. Haszeldine, and S.M. Smith, 2016a: Preliminary assessment of the potential for, and limitations to, terrestrial negative emission technologies in the UK. *Environmental Science: Processes & Impacts*, **18(11)**, 1400–1405, doi:10.1039/c6em00386a.

Smith, P. et al., 2014: Agriculture, Forestry and Other Land Use (AFOLU). In: *Climate Change 2014: Mitigation of Climate Change. Contribution of Working Group III to the Fifth Assessment Report of the Intergovernmental Panel on Climate Change* [Edenhofer, O., R. Pichs-Madruga, Y. Sokona, E. Farahani, S. Kadner, K. Seyboth, A. Adler, I. Baum, S. Brunner, P. Eickemeier, B. Kriemann, J. Savolainen, S. Schlömer, C. von Stechow, T. Zwickel, and J.C. Minx (eds.)]. Cambridge University Press, Cambridge, United Kingdom and New York, NY, USA, pp. 811–922.

Smith, P. et al., 2016b: Biophysical and economic limits to negative CO_2 emissions. *Nature Climate Change*, **6(1)**, 42–50, doi:10.1038/nclimate2870.

Smith, S.J. and A. Mizrahi, 2013: Near-term climate mitigation by short-lived forcers. *Proceedings of the National Academy of Sciences*, **110(35)**, 14202–14206, doi:10.1073/pnas.1308470110.

Smith, S.J. and P.J. Rasch, 2013: The long-term policy context for solar radiation management. *Climatic Change*, **121(3)**, 487–497, doi:10.1007/s10584-012-0577-3.

Smucker, T.A. et al., 2015: Differentiated livelihoods, local institutions, and the adaptation imperative: Assessing climate change adaptation policy in Tanzania. *Geoforum*, **59**, 39–50, doi:10.1016/j.geoforum.2014.11.018.

Snow, J.T. et al., 2016: *A New Vision for Weather and Climate Services in Africa*. United Nations Development Programme (UNDP), New York, NY, USA, 137 pp.

Soares-Filho, B. et al., 2010: Role of Brazilian Amazon protected areas in climate change mitigation. *Proceedings of the National Academy of Sciences*, **107(24)**, 10821–10826, doi:10.1073/pnas.0913048107.

Soccol, C.R. et al., 2010: Bioethanol from lignocelluloses: Status and perspectives in Brazil. *Bioresource Technology*, **101(13)**, 4820–4825, doi:10.1016/j.biortech.2009.11.067.

Socolow, R. et al., 2011: *Direct air capture of CO_2 with chemicals: A technology assessment for the APS Panel on Public Affairs*. American Physical Society (APS), 91 pp.

Soderlund, J. and P. Newman, 2015: Biophilic architecture: a review of the rationale and outcomes. *AIMS Environmental Science*, **2(4)**, 950–969, doi:10.3934/environsci.2015.4.950.

Sohlberg, J., 2017: The Effect of Elite Polarization: A Comparative Perspective on How Party Elites Influence Attitudes and Behavior on Climate Change in the European Union. *Sustainability*, **9(1)**, 1–13, doi:10.3390/su9010039.

Solecki, W., K.C. Seto, and P.J. Marcotullio, 2013: It's Time for an Urbanization Science. *Environment: Science and Policy for Sustainable Development*, **55(1)**, 12–17, doi:10.1080/00139157.2013.748387.

Solecki, W., M. Pelling, and M. Garschagen, 2017: Transitions between risk management regimes in cities. *Ecology And Society*, **22(2)**, 38, doi:10.5751/es-09102-220238.

Solecki, W. et al., 2018: City transformations in a 1.5°C warmer world. *Nature Climate Change*, **8(3)**, 177–181, doi:10.1038/s41558-018-0101-5.

Somanathan, E. et al., 2014: National and Sub-national Policies and Institutions. In: *Climate Change 2014: Mitigation of Climate Change. Contribution of Working Group III to the Fifth Assessment Report of the Intergovernmental Panel on Climate Change* [Edenhofer, O., R. Pichs-Madruga, Y. Sokona, E. Farahani, S.

Kadner, K. Seyboth, A. Adler, I. Baum, S. Brunner, P. Eickemeier, B. Kriemann, J. Savolainen, S. Schlömer, C. von Stechow, T. Zwickel, and J.C. Minx (eds.)]. Cambridge University Press, Cambridge, United Kingdom and New York, NY, USA, pp. 1141–1205.

Sorrell, S., J. Dimitropoulos, and M. Sommerville, 2009: Empirical estimates of the direct rebound effect: A review. *Energy Policy*, **37(4)**, 1356–1371, doi:10.1016/j.enpol.2008.11.026.

Sovacool, B.K., 2016: How long will it take? Conceptualizing the temporal dynamics of energy transitions. *Energy Research & Social Science*, **13**, 202–215, doi:10.1016/j.erss.2015.12.020.

Sovacool, B.K., B.-O. Linnér, and M.E. Goodsite, 2015: The political economy of climate adaptation. *Nature Climate Change*, **5(7)**, 616–618, doi:10.1038/nclimate2665.

Sovacool, B.K., B.-O. Linnér, and R.J.T. Klein, 2017: Climate change adaptation and the Least Developed Countries Fund (LDCF): Qualitative insights from policy implementation in the Asia-Pacific. *Climatic Change*, **140(2)**, 209–226, doi:10.1007/s10584-016-1839-2.

Spence, A., W. Poortinga, C. Butler, and N.F. Pidgeon, 2011: Perceptions of climate change and willingness to save energy related to flood experience. *Nature Climate Change*, **1(1)**, 46–49, doi:10.1038/nclimate1059.

Spencer, B. et al., 2017: Case studies in co-benefits approaches to climate change mitigation and adaptation. *Journal of Environmental Planning and Management*, **60(4)**, 647–667, doi:10.1080/09640568.2016.1168287.

Spencer, T., R. Pierfederici, and others, 2015: *Beyond the numbers: understanding the transformation induced by INDCs*. Study N°05/15, Institut du Développement Durable et des Relations Internationales (IDDRI) – MILES Project Consortium, Paris, France, 80 pp.

Springmann, M., H.C.J. Godfray, M. Rayner, and P. Scarborough, 2016: Analysis and valuation of the health and climate change cobenefits of dietary change. *Proceedings of the National Academy of Sciences*, **113(15)**, 4146–4151, doi:10.1073/pnas.1523119113.

Stadelmann, M., 2017: Mind the gap? Critically reviewing the energy efficiency gap with empirical evidence. *Energy Research & Social Science*, **27**, 117–128, doi:10.1016/j.erss.2017.03.006.

Stavins, R.N., 1988: *Project 88 – Harnessing Market Forces to Protect Our Environment*. A Public Policy Study sponsored by Senator Timothy E. Wirth, Colorado, and Senator John Heinz, Pennsylvania, Washington DC, USA, 86 pp.

Stavins, R.N. et al., 2014: International Cooperation: Agreements & Instruments. In: *Climate Change 2014: Mitigation of Climate Change. Contribution of Working Group III to the Fifth Assessment Report of the Intergovernmental Panel on Climate Change* [Edenhofer, O., R. Pichs-Madruga, Y. Sokona, E. Farahani, S. Kadner, K. Seyboth, A. Adler, I. Baum, S. Brunner, P. Eickemeier, B. Kriemann, J. Savolainen, S. Schlömer, C. von Stechow, T. Zwickel, and J.C. Minx (eds.)]. Cambridge University Press, Cambridge, United Kingdom and New York, NY, USA, pp. 1001 – 1081.

Steenhof, P. and E. Sparling, 2011: The Role of Codes, Standards, and Related Instruments in Facilitating Adaptation to Climate Change. In: *Climate Change Adaptation in Developed Nations: From Theory to Practice* [Ford, J.D. and L. Berrang-Ford (eds.)]. Advances in Global Change Research, Springer, Dordrecht, The Netherlands, pp. 243–254, doi:10.1007/978-94-007-0567-8_17.

Steg, L., 2016: Values, Norms, and Intrinsic Motivation to Act Proenvironmentally. *Annual Review of Environment and Resources*, **41(1)**, 277–292, doi:10.1146/annurev-environ-110615-085947.

Steg, L. and C. Vlek, 2009: Encouraging pro-environmental behaviour: An integrative review and research agenda. *Journal of Environmental Psychology*, **29(3)**, 309–317, doi:10.1016/j.jenvp.2008.10.004.

Steg, L. and J. de Groot, 2010: Explaining prosocial intentions: Testing causal relationships in the norm activation model. *British Journal of Social Psychology*, **49(4)**, 725–743, doi:10.1348/014466609x477745.

Steg, L., L. Dreijerink, and W. Abrahamse, 2005: Factors influencing the acceptability of energy policies: A test of VBN theory. *Journal of Environmental Psychology*, **25(4)**, 415–425, doi:10.1016/j.jenvp.2005.08.003.

Steg, L., L. Dreijerink, and W. Abrahamse, 2006: Why are Energy Policies Acceptable and Effective? *Environment and Behavior*, **38(1)**, 92–111, doi:10.1177/0013916505278519.

Steg, L., G. Perlaviciute, and E. van der Werff, 2015: Understanding the human dimensions of a sustainable energy transition. *Frontiers in Psychology*, **6**, 1–17, doi:10.3389/fpsyg.2015.00805.

4

Steg, L., R. Shwom, and T. Dietz, 2018: What Drives Energy Consumers?: Engaging People in a Sustainable Energy Transition. *IEEE Power and Energy Magazine*, **16(1)**, 20–28, doi:10.1109/mpe.2017.2762379.

Steinhoff, D.F., A.J. Monaghan, and M.P. Clark, 2014: Projected impact of twenty-first century ENSO changes on rainfall over Central America and northwest South America from CMIP5 AOGCMs. *Climate Dynamics*, **44(5–6)**, 1329–1349, doi:10.1007/s00382-014-2196-3.

Sterling, E.J. et al., 2017: Biocultural approaches to well-being and sustainability indicators across scales. *Nature Ecology & Evolution*, **1(12)**, 1798–1806, doi:10.1038/s41559-017-0349-6.

Sterman, J.D., L. Siegel, and J.N. Rooney-Varga, 2018: Does replacing coal with wood lower CO_2 emissions? Dynamic lifecycle analysis of wood bioenergy. *Environmental Research Letters*, **13(1)**, 015007, doi:10.1088/1748-9326/aaa512.

Stern, N., 2013: The Structure of Economic Modeling of the Potential Impacts of Climate Change: Grafting Gross Underestimation of Risk onto Already Narrow Science Models. *Journal of Economic Literature*, **51(3)**, 838–859, doi:10.1257/jel.51.3.838.

Stern, N., 2015: Economic development, climate and values: making policy. *Proceedings of the Royal Society of London B: Biological Sciences*, **282(1812)**, doi:10.1098/rspb.2015.0820.

Stern, P.C., 2011: Design principles for global commons: Natural resources and emerging technologies. *International Journal of the Commons*, **5(2)**, 213–232, doi:10.18352/ijc.305.

Stern, P.C., 2014: Individual and household interactions with energy systems: Toward integrated understanding. *Energy Research & Social Science*, **1**, 41–48, doi:10.1016/j.erss.2014.03.003.

Stern, P.C. and G.T. Gardner, 1981: Psychological research and energy policy. *American Psychologist*, **36(4)**, 329–342, doi:10.1037/0003-066x.36.4.329.

Stern, P.C. et al., 2016a: consumption by households and organizations. *Nature Energy*, **1**, 16043, doi:10.1038/nenergy.2016.43.

Stern, P.C. et al., 2016b: Opportunities and insights for reducing fossil fuel consumption by households and organizations. *Nature Energy*, **1(5)**, 16043, doi:10.1038/nenergy.2016.43.

Stevanović, M. et al., 2017: Mitigation Strategies for Greenhouse Gas Emissions from Agriculture and Land-Use Change: Consequences for Food Prices. *Environmental Science & Technology*, **51(1)**, 365–374, doi:10.1021/acs.est.6b04291.

Stevenson, H. and J.S. Dryzek, 2014: *Democratizing Global Climate Governance*. Cambridge University Press, Cambridge, United Kingdom and New York, NY, USA, 256 pp., doi:10.1017/cbo9781139208628.

Stevenson, M. et al., 2016: Land use, transport, and population health: estimating the health benefits of compact cities. *The Lancet*, **388(10062)**, 2925–2935, doi:10.1016/s0140-6736(16)30067-8.

Stiglitz, J.E., 2002: *Globalization and its Discontents*. W.W. Norton & Company, New York, NY, USA, 304 pp.

Stiglitz, J.E. et al., 2017: *Report of the High-Level Commission on Carbon Prices*. Carbon Pricing Leadership Coalition, 68 pp.

Stjern, C.W. et al., 2017: Rapid Adjustments Cause Weak Surface Temperature Response to Increased Black Carbon Concentrations. *Journal of Geophysical Research: Atmospheres*, **122(21)**, 11,411–462,481, doi:10.1002/2017jd027326.

Stohl, A. et al., 2015: Evaluating the climate and air quality impacts of short-lived pollutants. *Atmospheric Chemistry and Physics*, **15(18)**, 10529–10566, doi:10.5194/acp-15-10529-2015.

Stolaroff, J.K. et al., 2012: Review of methane mitigation technologies with application to rapid release of methane from the Arctic. *Environmental Science & Technology*, **46**, 6455–6469, doi:10.1021/es204686w.

Storelvmo, T., W.R. Boos, and N. Herger, 2014: Cirrus cloud seeding: a climate engineering mechanism with reduced side effects? *Philosophical Transactions of the Royal Society A: Mathematical, Physical and Engineering Sciences*, **372(2031)**, 20140116–20140116, doi:10.1098/rsta.2014.0116.

Strassburg, B.B.N. et al., 2014: Biophysical suitability, economic pressure and land-cover change: a global probabilistic approach and insights for REDD+. *Sustainability Science*, **9(2)**, 129–141, doi:10.1007/s11625-013-0209-5.

Strefler, J., T. Amann, N. Bauer, E. Kriegler, and J. Hartmann, 2018a: Potential and costs of carbon dioxide removal by enhanced weathering of rocks.

Environmental Research Letters, **13(3)**, 034010, doi:10.1088/1748-9326/aaa9c4.

Strefler, J. et al., 2018b: Between Scylla and Charybdis: Delayed mitigation narrows the passage between large-scale CDR and high costs. *Environmental Research Letters*, **13(4)**, 044015, doi:10.1088/1748-9326/aab2ba.

Strong, A., J.J. Cullen, and S.W. Chisholm, 2009: Ocean Fertilization. *Oceanography*, **22(3)**, 236–261, doi:10.5670/oceanog.2009.83.

Stua, M., 2017: *From the Paris Agreement to a Low-Carbon Bretton Woods*. Springer International Publishing, Cham, Switzerland, 239 pp., doi:10.1007/978-3-319-54699-5.

Studart, R. and K. Gallagher, 2015: Guaranteeing Finance for Sustainable Infrastructure: A Proposal. In: *Moving the Trillions – a debate on positive pricing of mitigation actions*. Centro Brasil No Clima, Rio de Janeiro, Brazil, pp. 92–113.

Stults, M. and S.C. Woodruff, 2017: Looking under the hood of local adaptation plans: shedding light on the actions prioritized to build local resilience to climate change. *Mitigation and Adaptation Strategies for Global Change*, **22(8)**, 1249–1279, doi:10.1007/s11027-016-9725-9.

Su, Y., P. Zhang, and Y. Su, 2015: An overview of biofuels policies and industrialization in the major biofuel producing countries. *Renewable and Sustainable Energy Reviews*, **50**, 991–1003, doi:10.1016/j.rser.2015.04.032.

Suarez, P. and M. van Aalst, 2017: Geoengineering: a humanitarian concern. *Earth's Future*, **5**, 183–195, doi:10.1002/eft2.181.

Sugiyama, M., Y. Arino, T. Kosugi, A. Kurosawa, and S. Watanabe, 2017a: Next steps in geoengineering scenario research: limited deployment scenarios and beyond. *Climate Policy*, 1–9, doi:10.1080/14693062.2017.1323721.

Sugiyama, M. et al., 2017b: Transdisciplinary co-design of scientific research agendas: 40 research questions for socially relevant climate engineering research. *Sustainability Science*, **12(1)**, 31–44, doi:10.1007/s11625-016-0376-2.

Sullivan-Wiley, K.A. and A.G. Short Gianotti, 2017: Risk Perception in a Multi-Hazard Environment. *World Development*, **97**, 138–152, doi:10.1016/j.worlddev.2017.04.002.

Summers, L.H., 2016: The Age of Secular Stagnation: What It Is and What to Do About It. *Foreign Affairs*, **95**, 2, www.foreignaffairs.com/articles/united-states/2016-02-15/age-secular-stagnation.

Sun, Y., G. Huang, X. Xu, and A.C.-K. Lai, 2018: Building-group-level performance evaluations of net zero energy buildings with non-collaborative controls. *Applied Energy*, **212**, 565–576, doi:10.1016/j.apenergy.2017.11.076.

Sunderlin, W.D. et al., 2014: How are REDD+ Proponents Addressing Tenure Problems? Evidence from Brazil, Cameroon, Tanzania, Indonesia, and Vietnam. *World Development*, **55**, 37–52, doi:10.1016/j.worlddev.2013.01.013.

Sundqvist, E., P. Crill, M. Mölder, P. Vestin, and A. Lindroth, 2012: Atmospheric methane removal by boreal plants. *Geophysical Research Letters*, **39(21)**, L21806, doi:10.1029/2012gl053592.

Sunikka-Blank, M. and Y. Iwafune, 2011: Sustainable Building in Japan – Observations on a Market Transformation Policy. *Environmental Policy and Governance*, **21(5)**, 351–363, doi:10.1002/eet.580.

Surminski, S., 2013: Private-sector adaptation to climate risk. *Nature Climate Change*, **3(11)**, 943–945, doi:10.1038/nclimate2040.

Sutherland, R.J., 1991: Market Barriers to Energy-Efficiency Investments. *Energy Journal*, **12(3)**, 15–34, www.jstor.org/stable/41322426.

Svoboda, T., 2017: *The Ethics of Climate Engineering: Solar Radiation Management and Non-Ideal Justice*. Routledge, Abingdon, UK and New York, NY, USA, 176 pp.

Swatuk, L.A., 2015: Water conflict and cooperation in Southern Africa. *Wiley Interdisciplinary Reviews: Water*, **2(3)**, 215–230, doi:10.1002/wat2.1070.

Szabó, S., M. Moner-Girona, I. Kougias, R. Bailis, and K. Bódis, 2016: Identification of advantageous electricity generation options in sub-Saharan Africa integrating existing resources. *Nature Energy*, **1**, 16140, doi:10.1038/nenergy.2016.140.

Szerszynski, B., M. Kearnes, P. Macnaghten, R. Owen, and J. Stilgoe, 2013: Why solar radiation management geoengineering and democracy won't mix. *Environment and Planning A: Economy and Space*, **45(12)**, 2809–2816, doi:10.1068/a45649.

Tacoli, C., B. Bukhari, and S. Fisher, 2013: *Urban poverty, food security and climate change*. International Institute for Environment and Development (IIED) Human Settlements Group, London, UK, 29 pp.

Tait, L. and M. Euston-Brown, 2017: What role can African cities play in low-carbon development? A multilevel governance perspective of Ghana,

Uganda and South Africa. *Journal of Energy in Southern Africa*, **28**(3), 43–53, doi:10.17159/2413-3051/2017/v28i3a1959.

Takahashi, B., M. Burnham, C. Terracina-Hartman, A.R. Sopchak, and T. Selfa, 2016: Climate Change Perceptions of NY State Farmers: The Role of Risk Perceptions and Adaptive Capacity. *Environmental Management*, **58**(6), 946–957, doi:10.1007/s00267-016-0742-y.

Takahashi, N. et al., 2015: Community Trial on Heat Related-Illness Prevention Behaviors and Knowledge for the Elderly. *International Journal of Environmental Research and Public Health*, **12**(3), 3188–3214, doi:10.3390/ijerph120303188.

Talberg, A., P. Christoff, S. Thomas, and D. Karoly, 2018: Geoengineering governance-by-default: an earth system governance perspective. *International Environmental Agreements: Politics, Law and Economics*, **18**(2), 229–253, doi:10.1007/s10784-017-9374-9.

Tall, A. et al., 2014: *Scaling up climate services for farmers: Mission possible. Learning from good practice in Africa and South Asia*. CCAFS Report No. 13, CGIAR Research Program on Climate Change, Agriculture and Food Security (CCAFS), Copenhagen, Denmark, 44 pp.

Tallis, M., G. Taylor, D. Sinnett, and P. Freer-Smith, 2011: Estimating the removal of atmospheric particulate pollution by the urban tree canopy of London, under current and future environments. *Landscape and Urban Planning*, **103**(2), 129–138, doi:10.1016/j.landurbplan.2011.07.003.

Tassey, G., 2014: Competing in Advanced Manufacturing: The Need for Improved Growth Models and Policies. *Journal of Economic Perspectives*, **28**(1), 27–48, doi:10.1257/jep.28.1.27.

Taub, D.R., B. Miller, and H. Allen, 2008: Effects of elevated CO_2 on the protein concentration of food crops: a meta-analysis. *Global Change Biology*, **14**(3), 565–575, doi:10.1111/j.1365-2486.2007.01511.x.

Taufik, D., J.W. Bolderdijk, and L. Steg, 2015: Acting green elicits a literal warm glow. *Nature Climate Change*, **5**(1), 37–40, doi:10.1038/nclimate2449.

Taufik, D., J.W. Bolderdijk, and L. Steg, 2016: Going green? The relative importance of feelings over calculation in driving environmental intent in the Netherlands and the United States. *Energy Research & Social Science*, **22**, 52–62, doi:10.1016/j.erss.2016.08.012.

Taylor, A.L., S. Dessai, and W. Bruine de Bruin, 2014: Public perception of climate risk and adaptation in the UK: A review of the literature. *Climate Risk Management*, **4–5**, 1–16, doi:10.1016/j.crm.2014.09.001.

Taylor, L.L. et al., 2016: Enhanced weathering strategies for stabilizing climate and averting ocean acidification. *Nature Climate Change*, **6**(4), 402–406, doi:10.1038/nclimate2882.

TEC, 2016: *Updated guidance on technology action plans*. TEC/2016/12/7, Technology Executive Committee (TEC), United Nations Framework Convention on Climate Change (UNFCCC), 21 pp.

Teferi, Z.A. and P. Newman, 2017: Slum Regeneration and Sustainability: Applying the Extended Metabolism Model and the SDGs. *Sustainability*, **9**(12), 2273, doi:10.3390/su9122273.

Teferi, Z.A. and P. Newman, 2018: Slum Upgrading: Can the 1.5°C Carbon Reduction Work with SDGs in these Settlements? *Urban Planning*, **3**(2), 52–63, doi:10.17645/up.v3i2.1239.

Tehranian, H. and J.F. Waegelein, 1985: Market reaction to short-term executive compensation plan adoption. *Journal of Accounting and Economics*, **7**(1–3), 131–144, doi:10.1016/0165-4101(85)90032-1.

ten Berge, H.F.M. et al., 2012: Olivine weathering in soil, and its effects on growth and nutrient uptake in Ryegrass (*Lolium perenne* L.): a pot experiment. *PLOS ONE*, **7**(8), e42098, doi:10.1371/journal.pone.0042098.

Termeer, C.J.A.M., A. Dewulf, and G.R. Biesbroek, 2017: Transformational change: governance interventions for climate change adaptation from a continuous change perspective. *Journal of Environmental Planning and Management*, **60**(4), 558–576, doi:10.1080/09640568.2016.1168288.

Tessler, Z.D. et al., 2015: Profiling risk and sustainability in coastal deltas of the world. *Science*, **349**(6248), 638–643, doi:10.1126/science.aab3574.

Teulings, C. and R. Baldwin (eds.), 2014: *Secular Stagnation: Facts, Causes and Cures*. Centre for Economic Policy Research Press (CEPR), London, UK, 154 pp.

Thi Hong Phuong, L., G.R. Biesbroek, and A.E.J. Wals, 2017: The interplay between social learning and adaptive capacity in climate change adaptation: A systematic review. *NJAS – Wageningen Journal of Life Sciences*, **82**, 1–9, doi:10.1016/j.njas.2017.05.001.

Thieme, T.A., 2018: The hustle economy: Informality, uncertainty and the geographies of getting by. *Progress in Human Geography*, **42**(4), 529–548, doi:10.1177/0309132517690039.

Thøgersen, J. and F. Ölander, 2003: Spillover of environment-friendly consumer

behaviour. *Journal of Environmental Psychology*, **23**, 225–236, doi:10.1016/s0272-4944(03)00018-5.

Thomas, S., L.-A. Brischke, J. Thema, L. Leuser, and M. Kopatz, 2017: Energy sufficiency policy: how to limit energy consumption and per capita dwelling size in a decent way. In: *Proceedings of the ECEEE 2017 Summer Study on Consumption, Efficiency & Limits*. European Council for an Energy Efficient Economy (ECEEE) Secretariat, Stockholm, Sweden, pp. 103–112.

Thomas, S.D., 1988: *The realities of nuclear power: international economic and regulatory experience*. Cambridge University Press, Cambridge, United Kingdom and New York, NY , USA, 289 pp.

Thomson, G. and P. Newman, 2016: Geoengineering in the Anthropocene through Regenerative Urbanism. *Geosciences*, **6**(4), 46, doi:10.3390/geosciences6040046.

Thornton, P.K. and M. Herrero, 2014: Climate change adaptation in mixed crop-livestock systems in developing countries. *Global Food Security*, **3**(2), 99–107, doi:10.1016/j.gfs.2014.02.002.

Thornton, P.K. and M. Herrero, 2015: Adapting to climate change in the mixed crop and livestock farming systems in sub-Saharan Africa. *Nature Climate Change*, **5**(9), 830–836, doi:10.1038/nclimate2754.

Thornton, P.K. et al., 2018: A Qualitative Evaluation of CSA Options in Mixed Crop-Livestock Systems in Developing Countries. In: *Climate Smart Agriculture: Building Resilience to Climate Change* [Lipper, L., N. McCarthy, D. Zilberman, S. Asfaw, and G. Branca (eds.)]. Springer International Publishing, Cham, pp. 385–423, doi:10.1007/978-3-319-61194-5_17.

Thornton, T.F. and N. Manasfi, 2010: Adaptation – Genuine and Spurious: Demystifying Adaptation Processes in Relation to Climate Change. *Environment and Society*, **1**(1), 132–155, doi:10.3167/ares.2010.010107.

Thornton, T.F. and C. Comberti, 2017: Synergies and trade-offs between adaptation, mitigation and development. *Climatic Change*, **140**(1), 5–18, doi:10.1007/s10584-013-0884-3.

Tiefenbeck, V. et al., 2016: Overcoming Salience Bias: How Real-Time Feedback Fosters Resource Conservation. *Management Science*, **64**(3), 1458–1476, doi:10.1287/mnsc.2016.2646.

Tilleard, S. and J. Ford, 2016: Adaptation readiness and adaptive capacity of transboundary river basins. *Climatic Change*, **137**(3–4), 575–591, doi:10.1007/s10584-016-1699-9.

Tilman, D. and M. Clark, 2014: Global diets link environmental sustainability and human health. *Nature*, **515**(7528), 518–522, doi:10.1038/nature13959.

Tilmes, S., B.M. Sanderson, and B.C. O'Neill, 2016: Climate impacts of geoengineering in a delayed mitigation scenario. *Geophysical Research Letters*, **43**(15), 8222–8229, doi:10.1002/2016gl070122.

Tilmes, S. et al., 2012: Impact of very short-lived halogens on stratospheric ozone abundance and UV radiation in a geo-engineered atmosphere. *Atmospheric Chemistry and Physics*, **12**(22), 10945–10955, doi:10.5194/acp-12-10945-2012.

Tjiputra, J.F., A. Grini, and H. Lee, 2016: Impact of idealized future stratospheric aerosol injection on the large-scale ocean and land carbon cycles. *Journal of Geophysical Research: Biogeosciences*, **121**(1), 2–27, doi:10.1002/2015jg003045.

Tobler, C., V.H.M. Visschers, and M. Siegrist, 2011: Organic Tomatoes Versus Canned Beans: How Do Consumers Assess the Environmental Friendliness of Vegetables? *Environment and Behavior*, **43**(5), 591–611, doi:10.1177/0013916510372865.

Tobler, C., V.H.M. Visschers, and M. Siegrist, 2012: Consumers' knowledge about climate change. *Climatic Change*, **114**(2), 189–209, doi:10.1007/s10584-011-0393-1.

Tokarska, K.B. and K. Zickfeld, 2015: The effectiveness of net negative carbon dioxide emissions in reversing anthropogenic climate change. *Environmental Research Letters*, **10**(9), 094013, doi:10.1088/1748-9326/10/9/094013.

Toleikyte, A., L. Kranzl, and A. Müller, 2018: Cost curves of energy efficiency investments in buildings – Methodologies and a case study of Lithuania. *Energy Policy*, **115**, 148–157, doi:10.1016/j.enpol.2017.12.043.

Tollefson, J., 2017: World's carbon emissions set to spike by 2% in 2017. *Nature*, **551**(7680), 283–283, doi:10.1038/nature.2017.22995.

Torssonen, P. et al., 2016: Effects of climate change and management on net climate impacts of production and utilization of energy biomass in Norway spruce with stable age-class distribution. *GCB Bioenergy*, **8**(2), 419–427, doi:10.1111/gcbb.12258.

Trabacchi, C. and B.K. Buchner, 2017: Adaptation Finance: Setting the Ground for

4

Post-Paris Action. In: *Climate Finance* [Markandya, A., I. Galarraga, and D. Rübbelke (eds.)]. World Scientific Series on the Economics of Climate Change Volume 2, World Scientific, Singapore, pp. 35–54, doi:10.1142/9789814641814_0003.

Trenberth, K.E., M. Marquis, and S. Zebiak, 2016: The vital need for a climate information system. *Nature Climate Change*, **6(12)**, 1057–1059, doi:10.1038/nclimate3170.

Trisos, C.H. et al., 2018: Potentially dangerous consequences for biodiversity of solar geoengineering implementation and termination. *Nature Ecology & Evolution*, **2(3)**, 475–482, doi:10.1038/s41559-017-0431-0.

Truelove, H., A.R. Carrico, and L. Thabrew, 2015: A socio-psychological model for analyzing climate change adaptation: A case study of Sri Lankan paddy farmers. *Global Environmental Change*, **31**, 85–97, doi:10.1016/j.gloenvcha.2014.12.010.

Truelove, H.B., A.R. Carrico, E.U. Weber, K.T. Raimi, and M.P. Vandenbergh, 2014: Positive and negative spillover of pro-environmental behavior: An integrative review and theoretical framework. *Global Environmental Change*, **29**, 127–138, doi:10.1016/j.gloenvcha.2014.09.004.

Tschakert, P. et al., 2017: Climate change and loss, as if people mattered: values, places, and experiences. *Wiley Interdisciplinary Reviews: Climate Change*, **8(5)**, e476, doi:10.1002/wcc.476.

Tukker, A. et al., 2011: Environmental impacts of changes to healthier diets in Europe. *Ecological Economics*, **70(10)**, 1776–1788, doi:10.1016/j.ecolecon.2011.05.001.

Tuomisto, H.L. and M.J. Teixeira de Mattos, 2011: Environmental Impacts of Cultured Meat Production. *Environmental Science & Technology*, **45(14)**, 6117–6123, doi:10.1021/es200130u.

Turner, P.A. et al., 2018: The global overlap of bioenergy and carbon sequestration potential. *Climatic Change*, **148(1–2)**, 1–10, doi:10.1007/s10584-018-2189-z.

Turnhout, E. et al., 2017: Envisioning REDD+ in a post-Paris era: between evolving expectations and current practice. *Wiley Interdisciplinary Reviews: Climate Change*, **8(1)**, e425, doi:10.1002/wcc.425.

UN, 2016: *Transforming our World: The 2030 Agenda for Sustainable Development*. A/RES/70/1, United Nations (UN), New York, NY, USA, 41 pp.

UN DESA, 2014: *World Urbanization Prospects: The 2014 Revision*. ST/ESA/SER.A/366, United Nations Department of Economic and Social Affairs (UN DESA) Population Division, New York, NY, USA, 493 pp.

UN DESA, 2017: *World Population Prospects – 2017 Revision*. Working Paper No. ESA/P/WP/248, United Nations Department of Economic and Social Affairs (UN DESA) Population Division, New York, NY, USA, 46 pp.

Ünal, A.B., L. Steg, and M. Gorsira, 2017: Values Versus Environmental Knowledge as Triggers of a Process of Activation of Personal Norms for Eco-Driving. *Environment and Behavior*, **50(10)**, 1092–1118, doi:10.1177/0013916517728991.

Underwood, B.S., Z. Guido, P. Gudipudi, and Y. Feinberg, 2017: Increased costs to US pavement infrastructure from future temperature rise. *Nature Climate Change*, **7(10)**, 704–707, doi:10.1038/nclimate3390.

UNEP, 2005: *Enhancing Capacity Building for Integrated Policy Design and Implementation for Sustainable Development*. Economics and Trade Branch, Division of Technology, Industry and Economics, United Nations Environment Programme, Geneva, Switzerland, 65 pp.

UNEP, 2011: *Near-term Climate Protection and Clean Air Benefits: Actions for Controlling Short-Lived Climate Forcers*. United Nations Environment Programme (UNEP), Nairobi, Kenya, 78 pp.

UNEP, 2014: *Decoupling 2: Technologies, Opportunities and Policy Options*. United Nations Environment Programme, Nairobi, Kenya, 174 pp.

UNEP, 2015: *The financial system we need; aligning the financial system with sustainable development*. United Nations Environment Programme, Geneva, Switzerland, 112 pp.

UNEP, 2016: *The Adaptation Finance Gap Report 2016*. United Nations Environment Programme (UNEP), Nairobi, Kenya, 50 pp.

UNEP, 2017a: *The Adaptation Gap Report 2017*. United Nations Environment Programme (UNEP), Nairobi, Kenya, 62 pp.

UNEP, 2017b: *The Emissions Gap Report 2017*. United Nations Environment Programme (UNEP), Nairobi, Kenya, 116 pp., doi:978-92-807-3673-1.

UNEP/Climate Analytics, 2015: *Africa's Adaptation Gap 2: Bridging the gap – mobilising sources*. United Nations Environment Programme (UNEP) and Climate Analytics, 67 pp.

UNESCO, 2017: What is Local and Indigenous Knowledge? United Nations Educational, Scientific and Cultural Organization (UNESCO). Retrieved from: www.unesco.org/new/en/natural-sciences/priority-areas/links/related-information/what-is-local-and-indigenous-knowledge.

UNFCCC, 1992: *United Nations Framework Convention on Climate Change*. FCCC/INFORMAL/84, United Nations Framework Convention on Climate Change (UNFCCC), 24 pp.

UNFCCC, 2016: Decision 1/CP.21: Adoption of the Paris Agreement. In: *Report of the Conference of the Parties on its twenty-first session, held in Paris from 30 November to 13 December 2015. Addendum: Part two: Action taken by the Conference of the Parties at its twenty-first session*. FCCC/CP/2015/10/Add.1, United Nations Framework Convention on Climate Change (UNFCCC), pp. 1–36.

UN-Habitat, 2011: *Cities and Climate Change: Global Report on Human Settlements 2011*. Earthscan, London, UK and Washington, DC, USA, 300 pp.

UN-Habitat, 2017: *Sustainable Urbanisation in the Paris Agreement*. United Nations Human Settlements Programme (UN-Habitat), Nairobi, Kenya, 46 pp.

UNISDR, 2009: *Global Assessment Report on Disaster Risk Reduction 2009 – Risk and Poverty in a Changing Climate: Invest Today for a Safer Tomorrow*. United Nations International Strategy for Disaster Reduction (UNISDR), Geneva, Switzerland, 14 pp.

Upadhyay, A.P. and A. Bijalwan, 2015: Climate Change Adaptation: Services and Role of Information Communication Technology (ICT) in India. *American Journal of Environmental Protection*, **4(1)**, 70–74, doi:10.11648/j.ajep.20150401.20.

Upton, G.B. and B.F. Snyder, 2017: Funding Renewable Energy: An Analysis of Renewable Portfolio Standards. *Energy Economics*, **66(0)**, 205–216, doi:10.1016/j.eneco.2017.06.003.

Ura, K., 2015: *The Experience of Gross National Happiness as Development Framework*. ADB South Asia Working Paper Series No. 42, Asian Development Bank (ADB), Manila, Philippines, 38 pp.

Ürge-Vorsatz, D., S.T. Herrero, N.K. Dubash, and F. Lecocq, 2014: Measuring the Co-Benefits of Climate Change Mitigation. *Annual Review of Environment and Resources*, **39(1)**, 549–582, doi:10.1146/annurev-environ-031312-125456.

Valdivia, C., C. Barbieri, and M.A. Gold, 2012: Between Forestry and Farming: Policy and Environmental Implications of the Barriers to Agroforestry Adoption. *Canadian Journal of Agricultural Economics*, **60(2)**, 155–175, doi:10.1111/j.1744-7976.2012.01248.x.

van Asselt, H., P. Pauw, and H. Sælen, 2015: *Assessment and Review under a 2015 Climate Change Agreement*. Nordic Council of Ministers, Copenhagen, Denmark, 141 pp., doi:10.6027/tn2015-530.

van der Brugge, R. and R. Roosjen, 2015: An institutional and sociocultural perspective on the adaptation pathways approach. *Journal of Water and Climate Change*, **6(4)**, 743–758, doi:10.2166/wcc.2015.001.

van der Giesen, C. et al., 2017: A Life Cycle Assessment Case Study of Coal-Fired Electricity Generation with Humidity Swing Direct Air Capture of CO_2 versus MEA-Based Postcombustion Capture. *Environmental Science & Technology*, **51(2)**, 1024–1034, doi:10.1021/acs.est.6b05028.

van der Keur, P. et al., 2016: Identification and analysis of uncertainty in disaster risk reduction and climate change adaptation in South and Southeast Asia. *International Journal of Disaster Risk Reduction*, **16**, 208–214, doi:10.1016/j.ijdrr.2016.03.002.

van Der Werff, E. and L. Steg, 2015: One model to predict them all: Predicting energy behaviours with the norm activation model. *Energy Research & Social Science*, **6**, 8–14, doi:10.1016/j.erss.2014.11.002.

van der Werff, E., L. Steg, and K. Keizer, 2013a: It is a moral issue: The relationship between environmental self-identity, obligation-based intrinsic motivation and pro-environmental behaviour. *Global Environmental Change*, **23(5)**, 1258–1265, doi:10.1016/j.gloenvcha.2013.07.018.

van der Werff, E., L. Steg, and K. Keizer, 2013b: The value of environmental self-identity: The relationship between biospheric values, environmental self-identity and environmental preferences, intentions and behaviour. *Journal of Environmental Psychology*, **34**, 55–63, doi:10.1016/j.jenvp.2012.12.006.

van der Werff, E., L. Steg, and K. Keizer, 2014a: Follow the signal: When past pro-environmental actions signal who you are. *Journal of Environmental Psychology*, **40**, 273–282, doi:10.1016/j.jenvp.2014.07.004.

van der Werff, E., L. Steg, and K. Keizer, 2014b: I Am What I Am, by Looking Past the Present: The Influence of Biospheric Values and Past Behavior on Environmental. *Environment and Behavior*, **46(5)**, 626–657, doi:10.1177/0013916512475209.

van Dooren, C., M. Marinussen, H. Blonk, H. Aiking, and P. Vellinga, 2014: Exploring dietary guidelines based on ecological and nutritional values: A comparison of six dietary patterns. *Food Policy*, **44**, 36–46, doi:10.1016/j.foodpol.2013.11.002.

van Kasteren, Y., 2014: How are householders talking about climate change adaptation? *Journal of Environmental Psychology*, **40**, 339–350, doi:10.1016/j.jenvp.2014.09.001.

van Oppen, M.J.H., J.K. Oliver, H.M. Putnam, and R.D. Gates, 2015: Building coral reef resilience through assisted evolution. *Proceedings of the National Academy of Sciences*, **112(8)**, 2307–2313, doi:10.1073/pnas.1422301112.

van Sluisveld, M.A.E. et al., 2015: Comparing future patterns of energy system change in 2°C scenarios with historically observed rates of change. *Global Environmental Change*, **35**, 436–449, doi:10.1016/j.gloenvcha.2015.09.019.

van Soest, D.P. and E.H. Bulte, 2001: Does the Energy-Efficiency Paradox Exist? Technological Progress and Uncertainty. *Environmental and Resource Economics*, **18(1)**, 101–112, doi:10.1023/a:1011112406964.

van Soest, H.L. et al., 2017: Early action on Paris Agreement allows for more time to change energy systems. *Climatic Change*, **144(2)**, 165–179, doi:10.1007/s10584-017-2027-8.

van Vliet, M.T.H., D. Wiberg, S. Leduc, and K. Riahi, 2016: Power-generation system vulnerability and adaptation to changes in climate and water resources. *Nature Climate Change*, **6(4)**, 375–380, doi:10.1038/nclimate2903.

van Vuuren, D.P., O.Y. Edelenbosch, D.L. McCollum, and K. Riahi, 2017: A special issue on model-based long-term transport scenarios: Model comparison and new methodological developments to improve energy and climate policy analysis. *Transportation Research Part D: Transport and Environment*, **55**, 277–280, doi:10.1016/j.trd.2017.05.003.

van Vuuren, D.P. et al., 2018: Alternative pathways to the 1.5°C target reduce the need for negative emission technologies. *Nature Climate Change*, **8(5)**, 391–397, doi:10.1038/s41558-018-0119-8.

Vandentorren, S. et al., 2006: Heat-related mortality August 2003 Heat Wave in France: Risk Factors for Death of Elderly People Living at Home. *European Journal of Public Health*, **16(6)**, 583–591, doi:10.1093/eurpub/ckl063.

Vandyck, T., K. Keramidas, B. Saveyn, A. Kitous, and Z. Vrontisi, 2016: A global stocktake of the Paris pledges: Implications for energy systems and economy. *Global Environmental Change*, **41**, 46–63, doi:10.1016/j.gloenvcha.2016.08.006.

Varela-Ortega, C. et al., 2016: How can irrigated agriculture adapt to climate change? Insights from the Guadiana Basin in Spain. *Regional Environmental Change*, **16(1)**, 59–70, doi:10.1007/s10113-014-0720-y.

Vaughan, C. and S. Dessai, 2014: Climate services for society: Origins, institutional arrangements, and design elements for an evaluation framework. *Wiley Interdisciplinary Reviews: Climate Change*, **5(5)**, 587–603, doi:10.1002/wcc.290.

Vaughan, C., S. Dessai, and C. Hewitt, 2018: Surveying Climate Services: What Can We Learn from a Bird's-Eye View? *Weather, Climate, and Society*, **10(2)**, 373–395, doi:10.1175/wcas-d-17-0030.1.

Vaughan, N.E. and C. Gough, 2016: Expert assessment concludes negative emissions scenarios may not deliver. *Environmental Research Letters*, **11(9)**, 095003, doi:10.1088/1748-9326/11/9/095003.

Velásquez Barrero, L.S., 1998: Agenda 21: a form of joint environmental management in Manizales, Colombia. *Environment and Urbanization*, **10(2)**, 9–36, doi:10.1177/095624789801000218.

Velders, G.J.M., D.W. Fahey, J.S. Daniel, S.O. Andersen, and M. McFarland, 2015: Future atmospheric abundances and climate forcings from scenarios of global and regional hydrofluorocarbon (HFC) emissions. *Atmospheric Environment*, **123**, 200–209, doi:10.1016/j.atmosenv.2015.10.071.

Veldman, J.W. et al., 2015: Where Tree Planting and Forest Expansion are Bad for Biodiversity and Ecosystem Services. *BioScience*, **65(10)**, 1011–1018, doi:10.1093/biosci/biv118.

Venables, A.J., 2016: Using Natural Resources for Development: Why Has It Proven So Difficult? *Journal of Economic Perspectives*, **30(1)**, 161–184, doi:10.1257/jep.30.1.161.

Venhoeven, L.A., J.W. Bolderdijk, and L. Steg, 2013: Explaining the paradox: How pro-environmental behaviour can both thwart and foster well-being. *Sustainability*, **5(4)**, 1372–1386, doi:10.3390/su5041372.

Venhoeven, L.A., J.W. Bolderdijk, and L. Steg, 2016: Why acting environmentally-friendly feels good: Exploring the role of self-image. *Frontiers in Psychology*, **7(NOV)**, 1990–1991, doi:10.3389/fpsyg.2016.01846.

Venkataraman, C., S. Ghosh, and M. Kandlikar, 2016: Breaking out of the Box: India and Climate Action on Short-Lived Climate Pollutants. *Environmental Science & Technology*, **50(23)**, 12527–12529,

doi:10.1021/acs.est.6b05246.

Vergara, W., A.R. Rios, L.M. Galindo, and J. Samaniego, 2015: Physical Damages Associated with Climate Change Impacts and the Need for Adaptation Actions in Latin America and the Caribbean. In: *Handbook of Climate Change Adaptation* [Leal Filho, W. (ed.)]. Springer Berlin Heidelberg, Germany, pp. 479–491, doi:10.1007/978-3-642-38670-1_101.

Verhoef, L.A., B.W. Budde, C. Chockalingam, B. García Nodar, and A.J.M. van Wijk, 2018: The effect of additive manufacturing on global energy demand: An assessment using a bottom-up approach. *Energy Policy*, **112**, 349–360, doi:10.1016/j.enpol.2017.10.034.

Verplanken, B. and D. Roy, 2013: "My Worries Are Rational, Climate Change Is Not": Habitual Ecological Worrying Is an Adaptive Response. *PLOS ONE*, **8(9)**, e74708, doi:10.1371/journal.pone.0074708.

Verplanken, B. and D. Roy, 2016: Empowering interventions to promote sustainable lifestyles: Testing the habit discontinuity hypothesis in a field experiment. *Journal of Environmental Psychology*, **45**, 127–134, doi:10.1016/j.jenvp.2015.11.008.

Verplanken, B., H. Aarts, and A. Van Knippenberg, 1997: Habit, information acquisition, and the process of making travel mode choices. *European Journal of Social Psychology*, **27(5)**, 539–560, doi:10.1002/(sici)1099-0992(199709/10)27:5<539::aid-ejsp831>3.0.co;2-a.

Verstegen, J. et al., 2015: What can and can't we say about indirect land-use change in Brazil using an integrated economic – land-use change model? *GCB Bioenergy*, **8(3)**, 561–578, doi:10.1111/gcbb.12270.

Veugelers, R., 2012: Which policy instruments to induce clean innovating? *Research Policy*, **41(10)**, 1770–1778, doi:10.1016/j.respol.2012.06.012.

Vicente-Vicente, J.L., R. García-Ruiz, R. Francaviglia, E. Aguilera, and P. Smith, 2016: Soil carbon sequestration rates under Mediterranean woody crops using recommended management practices: A meta-analysis. *Agriculture, Ecosystems & Environment*, **235**, 204–214, doi:10.1016/j.agee.2016.10.024.

Viegas, J., L. Martinez, P. Crist, and S. Masterson, 2016: *Shared Mobility: Inovation for Liveable Cities*. International Transport Forum, Paris, France, 54 pp., doi:10.1787/5jlwvz8bd4mx-en.

Villarroel Walker, R., M.B. Beck, J.W. Hall, R.J. Dawson, and O. Heidrich, 2014: The energy-water-food nexus: Strategic analysis of technologies for transforming the urban metabolism. *Journal of Environmental Management*, **141**, 104–115, doi:10.1016/j.jenvman.2014.01.054.

Vinke-de Kruijf, J. and C. Pahl-Wostl, 2016: A multi-level perspective on learning about climate change adaptation through international cooperation. *Environmental Science & Policy*, **66**, 242–249, doi:10.1016/j.envsci.2016.07.004.

Vision 2020 Project Team, 2011: *Vision 2020: New York City Comprehensive Waterfront Plan*. Department of City Planning (DCP), City of New York, New York, NY, USA, 192 pp.

Visioni, D., G. Pitari, and V. Aquila, 2017a: Sulfate geoengineering: a review of the factors controlling the needed injection of sulfur dioxide. *Atmospheric Chemistry and Physics*, **17(6)**, 3879–3889, doi:10.5194/acp-17-3879-2017.

Visioni, D. et al., 2017b: Sulfate geoengineering impact on methane transport and lifetime: results from the Geoengineering Model Intercomparison Project (GeoMIP). *Atmospheric Chemistry and Physics*, **17(18)**, 11209–11226, doi:10.5194/acp-17-11209-2017.

Visschers, V.H.M., J. Shi, M. Siegrist, and J. Arvai, 2017: Beliefs and values explain international differences in perception of solar radiation management: insights from a cross-country survey. *Climatic Change*, **142(3–4)**, 531–544, doi:10.1007/s10584-017-1970-8.

Vlek, C.A.J. and L. Steg, 2007: Human behavior and environmental sustainability: Problems, driving forces, and research topics. *Journal of Social Issues*, **63(1)**, 1–19, doi:10.1111/j.1540-4560.2007.00493.x.

Vogel, C., S.C. Moser, R.E. Kasperson, and G.D. Dabelko, 2007: Linking vulnerability, adaptation, and resilience science to practice: Pathways, players, and partnerships. *Global Environmental Change*, **17(3–4)**, 349–364, doi:10.1016/j.gloenvcha.2007.05.002.

von der Assen, N., J. Jung, and A. Bardow, 2013: Life-cycle assessment of carbon dioxide capture and utilization: avoiding the pitfalls. *Energy & Environmental Science*, **6(9)**, 2721–2734, doi:10.1039/c3ee41151f.

Voskamp, I.M. and F.H.M. Van de Ven, 2015: Planning support system for climate adaptation: Composing effective sets of blue-green measures to reduce urban vulnerability to extreme weather events. *Building and Environment*, **83**, 159–167, doi:10.1016/j.buildenv.2014.07.018.

4

Vrontisi, Z. et al., 2018: Enhancing global climate policy ambition towards a 1.5°C stabilization: a short-term multi-model assessment. *Environmental Research Letters*, **13(4)**, 044039, doi:10.1088/1748-9326/aab53e.

Wachsmuth, D., D.A. Cohen, and H. Angelo, 2016: Expand the frontiers of urban sustainability. *Nature*, **536(7617)**, 391–393, doi:10.1038/536391a.

Wadud, Z., D. MacKenzie, and P. Leiby, 2016: Help or hindrance? The travel, energy and carbon impacts of highly automated vehicles. *Transportation Research Part A: Policy and Practice*, **86**, 1–18, doi:10.1016/j.tra.2015.12.001.

Waisman, H., J. Rozenberg, and J.C. Hourcade, 2013: Monetary compensations in climate policy through the lens of a general equilibrium assessment: The case of oil-exporting countries. *Energy Policy*, **63**, 951–961, doi:10.1016/j.enpol.2013.08.055.

Waisman, H., C. Guivarch, F. Grazi, and J.C. Hourcade, 2012: The Imaclim-R model: infrastructures, technical inertia and the costs of low carbon futures under imperfect foresight. *Climatic Change*, **114(1)**, 101–120, doi:10.1007/s10584-011-0387-z.

Walker, W. and M. Lönnroth, 1983: *Nuclear power struggles: industrial competition and proliferation control*. Allen & Unwin, London, UK and Boston, USA, 204 pp.

Wallace, B., 2017: A framework for adapting to climate change risk in coastal cities. *Environmental Hazards*, **16(2)**, 149–164, doi:10.1080/17477891.2017.1298511.

Walsh, J.E., 2014: Intensified warming of the Arctic: Causes and impacts on middle latitudes. *Global and Planetary Change*, **117**, 52–63, doi:10.1016/j.gloplacha.2014.03.003.

Wamsler, C., 2007: Bridging the gaps: stakeholder-based strategies for risk reduction and financing for the urban poor. *Environment & Urbanization*, **19(1)**, 115–152, doi:10.1177/0956247807077029.

Wamsler, C., 2017: Stakeholder involvement in strategic adaptation planning: Transdisciplinarity and co-production at stake? *Environmental Science & Policy*, **75**, 148–157, doi:10.1016/j.envsci.2017.03.016.

Wamsler, C. and E. Brink, 2014a: Interfacing citizens' and institutions' practice and responsibilities for climate change adaptation. *Urban Climate*, **7**, 64–91, doi:10.1016/j.uclim.2013.10.009.

Wamsler, C. and E. Brink, 2014b: Moving beyond short-term coping and adaptation. *Environment & Urbanization*, **26(6)**, 86–111, doi:10.1177/0956247813516061.

Wang, G. et al., 2017: Continued increase of extreme El Nino frequency long after 1.5°C warming stabilization. *Nature Climate Change*, **7(8)**, 568–572, doi:10.1038/nclimate3351.

Wang, Q., F. Xiao, F. Zhang, and S. Wang, 2013: Labile soil organic carbon and microbial activity in three subtropical plantations. *Forestry*, **86(5)**, 569–574, doi:10.1093/forestry/cpt024.

Wang, Q., X. Chen, A.N. Jha, and H. Rogers, 2014: Natural gas from shale formation – The evolution, evidences and challenges of shale gas revolution in United States. *Renewable and Sustainable Energy Reviews*, **30**, 1–28, doi:10.1016/j.rser.2013.08.065.

Wang, Q. et al., 2015: Structural Evolution of Household Energy Consumption: A China Study. *Sustainability*, **7**, 3919–3932, doi:10.3390/su7043919.

Wang, X. et al., 2016: Taking account of governance: Implications for land-use dynamics, food prices, and trade patterns. *Ecological Economics*, **122**, 12–24, doi:10.1016/j.ecolecon.2015.11.018.

Wangui, E.E. and T.A. Smucker, 2017: Gendered opportunities and constraints to scaling up: a case study of spontaneous adaptation in a pastoralist community in Mwanga District, Tanzania. *Climate and Development*, 1–8, doi:10.1080/17565529.2017.1301867.

Ward, J., S. Fankhauser, C. Hepburn, H. Jackson, and R. Rajan, 2009: *Catalysing low-carbon growth in developing economies: Public Finance Mechanisms to scale up private sector investment in the climate solution*. United Nations Environment Programme (UNEP), Paris, France, 28 pp.

Ward, P.J., W.P. Pauw, M.W. van Buuren, and M.A. Marfai, 2013: Governance of flood risk management in a time of climate change: the cases of Jakarta and Rotterdam. *Environmental Politics*, **22(3)**, 518–536, doi:10.1080/09644016.2012.683155.

Warren, C.R. et al., 2005: 'Green On Green': Public perceptions of wind power in Scotland and Ireland. *Journal of Environmental Planning and Management*, **48(6)**, 853–875, doi:10.1080/09640560500294376.

Waters, J. and W.N. Adger, 2017: Spatial, network and temporal dimensions of the determinants of adaptive capacity in poor urban areas. *Global Environmental Change*, **46**, 42–49, doi:10.1016/j.gloenvcha.2017.06.011.

Watkins, K., 2015: *Power, People, Planet: Seizing Africa's Energy and Climate Opportunities*. Africa Progress Report 2015. Africa Progress Panel, Geneva, Switzerland, 179 pp.

WBCSD, 2017: *LCTPi Progress Report 2017*. World Business Council for Sustainable Development (WBCSD), Geneva, Switzerland, 47 pp.

Weber, E.U., 2015: Climate Change Demands Behavioral Change: What Are the Challenges? *Social Research: An International Quarterly*, **82(3)**, 561–580, http://muse.jhu.edu/article/603150.

Wedlock, D.N., P.H. Janssen, S.C. Leahy, D. Shu, and B.M. Buddle, 2013: Progress in the development of vaccines against rumen methanogens. *Animal*, **7(S2)**, 244–252, doi:10.1017/s1751731113000682.

Wee, B., 2015: Peak car: The first signs of a shift towards ICT-based activities replacing travel? A discussion paper. *Transport Policy*, **42**, 1–3, doi:10.1016/j.tranpol.2015.04.002.

Weenig, M.W.H. and C.J.H. Midden, 1991: Communication Network Influences on Information Diffusion and Persuasion. *Journal of Personality and Social Psychology*, **61(5)**, 734–742, doi:10.1037/0022-3514.61.5.734.

WEF, 2013: *The Green Investment Report: ways and means to unlock private finance for green growth*. World Economic Forum (WEF), Geneva, Switzerland, 40 pp.

WEF, 2015: *Industrial Internet of Things*. World Economic Forum (WEF), Geneva, Switzerland, 39 pp.

Wehkamp, J., N. Koch, S. Lübbers, and S. Fuss, 2018a: Governance and deforestation – a meta-analysis in economics. *Ecological Economics*, **144**, 214–227, doi:10.1016/j.ecolecon.2017.07.030.

Wehkamp, J. et al., 2018b: Accounting for institutional quality in global forest modeling. *Environmental Modelling & Software*, **102**, 250–259, doi:10.1016/j.envsoft.2018.01.020.

Wehner, M.F., K.A. Reed, B. Loring, D. Stone, and H. Krishnan, 2018: Changes in tropical cyclones under stabilized 1.5 and 2.0°C global warming scenarios as simulated by the Community Atmospheric Model under the HAPPI protocols. *Earth System Dynamics*, **9(1)**, 187–195, doi:10.5194/esd-9-187-2018.

Wei, M., S.J. Smith, and M.D. Sohn, 2017: Experience curve development and cost reduction disaggregation for fuel cell markets in Japan and the US. *Applied Energy*, **191**, 346–357, doi:10.1016/j.apenergy.2017.01.056.

Wei, Y., D. Tang, Y. Ding, and G. Agoramoorthy, 2016: Incorporating water consumption into crop water footprint: A case study of China's South–North Water Diversion Project. *Science of The Total Environment*, **545–546**, 601–608, doi:10.1016/j.scitotenv.2015.12.062.

Weindl, I. et al., 2015: Livestock in a changing climate: production system transitions as an adaptation strategy for agriculture. *Environmental Research Letters*, **10(9)**, 094021, doi:10.1088/1748-9326/10/9/094021.

Weischer, L., J. Morgan, and M. Patel, 2012: Climate Clubs: Can Small Groups of Countries make a Big Difference in Addressing Climate Change? *Review of European Community & International Environmental Law*, **21(3)**, 177–192, doi:10.1111/reel.12007.

Weitzman, M.L., 2015: A Voting Architecture for the Governance of Free-Driver Externalities, with Application to Geoengineering. *Scandinavian Journal of Economics*, **117(4)**, 1049–1068, doi:10.1111/sjoe.12120.

Wejs, A., K. Harvold, S.V. Larsen, and I.-L. Saglie, 2014: Legitimacy building in weak institutional settings: climate change adaptation at local level in Denmark and Norway. *Environmental Politics*, **23(3)**, 490–508, doi:10.1080/09644016.2013.854967.

Weldegebriel, Z.B. and M. Prowse, 2013: Climate-Change Adaptation in Ethiopia: To What Extent Does Social Protection Influence Livelihood Diversification? *Development Policy Review*, **31**, o35–o56, doi:10.1111/dpr.12038.

Well, M. and A. Carrapatoso, 2017: REDD+ finance: policy making in the context of fragmented institutions. *Climate Policy*, **17(6)**, 687–707, doi:10.1080/14693062.2016.1202096.

Wells, L., B. Rismanchi, and L. Aye, 2018: A review of Net Zero Energy Buildings with reflections on the Australian context. *Energy and Buildings*, **158**, 616–628, doi:10.1016/j.enbuild.2017.10.055.

Wenzel, G.W., 2009: Canadian Inuit subsistence and ecological instability – If the climate changes, must the Inuit? *Polar Research*, **28(1)**, 89–99, doi:10.1111/j.1751-8369.2009.00098.x.

Werner, C., H.-P. Schmidt, D. Gerten, W. Lucht, and C. Kammann, 2018: Biogeochemical potential of biomass pyrolysis systems for limiting global warming to 1.5°C. *Environmental Research Letters*, **13(4)**, 44036, doi:10.1088/1478-3975/aa9768.

Wesseling, J.H. et al., 2017: The transition of energy intensive processing industries

4

towards deep decarbonization: Characteristics and implications for future research. *Renewable and Sustainable Energy Reviews*, **79**, 1303–1313, doi:10.1016/j.rser.2017.05.156.

West, S.E. and R.C. Williams, 2004: Estimates from a consumer demand system: implications for the incidence of environmental taxes. *Journal of Environmental Economics and Management*, **47(3)**, 535–558, doi:10.1016/j.jeem.2003.11.004.

Westphal, M.I., S. Martin, L. Zhou, and D. Satterthwaite, 2017: *Powering Cities in the Global South: How Energy Access for All Benefits the Economy and the Environment*. World Resources Institute (WRI), Washington DC, USA, 55 pp.

White, C.J. et al., 2017: Potential applications of subseasonal-to-seasonal (S2S) predictions. *Meteorological Applications*, **24(3)**, 315–325, doi:10.1002/met.1654.

White, R., J. Turpie, and G. Letley, 2017: *Greening Africa's Cities: Enhancing the Relationship between Urbanization, Environmental Assets, and Ecosystem Services*. The World Bank, Washington DC, USA, 56 pp.

Whitmarsh, L., G. Seyfang, and S.O.N. Workspace., 2011: Public engagement with carbon and climate change: To what extent is the public 'carbon capable'? *Global Environmental Change*, **21(1)**, 56–65, doi:10.1016/j.gloenvcha.2010.07.011.

WHO, 2015: *Lessons learned on health adaptation to climate variability and change: experiences across low- and middle-income countries*. World Health Organization (WHO), Geneva, Switzerland, 72 pp.

Whyte, K.P., 2012: Now This! Indigenous Sovereignty, Political Obliviousness and Governance Models for SRM Research. *Ethics, Policy & Environment*, **15(2)**, 172–187, doi:10.1080/21550085.2012.685570.

Wibeck, V. et al., 2017: Making sense of climate engineering: a focus group study of lay publics in four countries. *Climatic Change*, **145(1–2)**, 1–14, doi:10.1007/s10584-017-2067-0.

Wigand, C. et al., 2017: A Climate Change Adaptation Strategy for Management of Coastal Marsh Systems. *Estuaries and Coasts*, **40(3)**, 682–693, doi:10.1007/s12237-015-0003-y.

Wilcox, J., P.C. Psarras, and S. Liguori, 2017: Assessment of reasonable opportunities for direct air capture. *Environmental Research Letters*, **12(6)**, 065001, doi:10.1088/1748-9326/aa6de5.

Williams, M. et al., 2011: Options for policy responses and their impacts. In: *Integrated Assessment of Black Carbon and Tropospheric Ozone*. United Nations Environment Programme (UNEP), Nairobi, Kenya, pp. 171–250.

Williams, P. et al., 2017: Community-based observing networks and systems in the Arctic: Human perceptions of environmental change and instrument-derived data. *Regional Environmental Change*, **18(2)**, 547–559, doi:10.1007/s10113-017-1220-7.

Williams, P.M. and E.R.M. Druffel, 1987: Radiocarbon in dissolved organic matter in the central North Pacific Ocean. *Nature*, **330(6145)**, 246–248, doi:10.1038/330246a0.

Williamson, P. and R. Bodle, 2016: *Update on Climate Geoengineering in Relation to the Convention on Biological Diversity: Potential Impacts and Regulatory Framework*. CBD Technical Series No. 84, Secretariat of the Convention on Biological Diversity, Montreal, QC, Canada, 158 pp.

Williamson, P. et al., 2012: Ocean fertilization for geoengineering: A review of effectiveness, environmental impacts and emerging governance. *Process Safety and Environmental Protection*, **90(6)**, 475–488, doi:10.1016/j.psep.2012.10.007.

Willis, R., 2017: How Members of Parliament understand and respond to climate change. *The Sociological Review*, 1–17, doi:10.1177/0038026117731658.

Wilson, C., 2012: Up-scaling, formative phases, and learning in the historical diffusion of energy technologies. *Energy Policy*, **50**, 81–94, doi:10.1016/j.enpol.2012.04.077.

Wilson, C. and H. Dowlatabadi, 2007: Models of Decision Making and Residential Energy Use. *Annual Review of Environment and Resources*, **32(1)**, 169–203, doi:10.1146/annurev.energy.32.053006.141137.

Wilson, C., A. Grubler, N. Bauer, V. Krey, and K. Riahi, 2013: Future capacity growth of energy technologies: Are scenarios consistent with historical evidence? *Climatic Change*, **118(2)**, 381–395, doi:10.1007/s10584-012-0618-y.

Wilson, S.A. et al., 2009: Carbon Dioxide Fixation within Mine Wastes of Ultramafic-Hosted Ore Deposits: Examples from the Clinton Creek and Cassiar Chrysotile Deposits, Canada. *Economic Geology*, **104(1)**, 95–112, doi:10.2113/gsecongeo.104.1.95.

Windeatt, J.H. et al., 2014: Characteristics of biochars from crop residues: Potential for carbon sequestration and soil amendment. *Journal of Environmental*

Management, **146**, 189–197, doi:10.1016/j.jenvman.2014.08.003.

Winkler, H., 2017: Reducing energy poverty through carbon tax revenues in South Africa. *Journal of Energy in Southern Africa*, **28(3)**, 12–26, doi:10.17159/2413-3051/2017/v28i3a2332.

Winkler, H. and N.K. Dubash, 2015: Who determines transformational change in development and climate finance? *Climate Policy*, 783–791, doi:10.1080/14693062.2015.1033674.

Winkler, H., T. Letete, and A. Marquard, 2013: Equitable access to sustainable development: operationalizing key criteria. *Climate Policy*, **13(4)**, 411–432, doi:10.1080/14693062.2013.777610.

Winkler, H. et al., 2017: Countries start to explain how their climate contributions are fair: more rigour needed. *International Environmental Agreements: Politics, Law and Economics*, **18(1)**, 99–115, doi:10.1007/s10784-017-9381-x.

Wise, M., M. Muratori, and P. Kyle, 2017: Biojet fuels and emissions mitigation in aviation: An integrated assessment modeling analysis. *Transportation Research Part D: Transport and Environment*, **52**, 244–253, doi:10.1016/j.trd.2017.03.006.

WMO, 2015: *Valuing Weather and Climate: Economic Assessment of Meteorological and Hydrological Services*. WMO-No. 1153, World Meteorological Organization (WMO), Geneva, Switzerland, 308 pp.

Wolak, F.A., 2011: Do residential customers respond to hourly prices? Evidence from a dynamic pricing experiment. *American Economic Review*, **101(3)**, 83–87, doi:10.1257/aer.101.3.83.

Wolfrom, L. and M. Yokoi-Arai, 2015: Financial instruments for managing disaster risks related to climate change. *OECD Journal: Financial Market Trends*, **2015(1)**, 25–47, doi:10.1787/fmt-2015-5jrqdkpxk5d5.

Wollenberg, E. et al., 2016: Reducing emissions from agriculture to meet the 2°C target. *Global Change Biology*, **22(12)**, 3859–3864, doi:10.1111/gcb.13340.

Wolske, K.S., P.C. Stern, and T. Dietz, 2017: Explaining interest in adopting residential solar photovoltaic systems in the United States: Toward an integration of behavioral theories. *Energy Research & Social Science*, **25**, 134–151, doi:10.1016/j.erss.2016.12.023.

Wong, P.-H., 2014: Maintenance Required: The Ethics of Geoengineering and Post-Implementation Scenarios. *Ethics, Policy & Environment*, **17(2)**, 186–191, doi:10.1080/21550085.2014.926090.

Wong, S., 2012: Overcoming obstacles against effective solar lighting interventions in South Asia. *Energy Policy*, **40**, 110–120, doi:10.1016/j.enpol.2010.09.030.

Wood, B.T., C.H. Quinn, L.C. Stringer, and A.J. Dougill, 2017: Investigating Climate Compatible Development Outcomes and their Implications for Distributive Justice: Evidence from Malawi. *Environmental Management*, **1**, 1–18, doi:10.1007/s00267-017-0890-8.

Wood, P. and F. Jotzo, 2011: Price floors for emissions trading. *Energy Policy*, **39(3)**, 1746–1753, doi:10.1016/j.enpol.2011.01.004.

Wood, S.A., A.S. Jina, M. Jain, P. Kristjanson, and R.S. DeFries, 2014: Smallholder farmer cropping decisions related to climate variability across multiple regions. *Global Environmental Change*, **25**, 163–172, doi:10.1016/j.gloenvcha.2013.12.011.

Woodcock, J. et al., 2009: Public health benefits of strategies to reduce greenhouse-gas emissions: urban land transport. *The Lancet*, **374(9705)**, 1930–1943, doi:10.1016/s0140-6736(09)61714-1.

Woodruff, S.C. and M. Stults, 2016: Numerous strategies but limited implementation guidance in US local adaptation plans. *Nature Climate Change*, **6(8)**, 796–802, doi:10.1038/nclimate3012.

Woods, B.A., H. Nielsen, A.B. Pedersen, and D. Kristofersson, 2017: Farmers' perceptions of climate change and their likely responses in Danish agriculture. *Land Use Policy*, **65**, 109–120, doi:10.1016/j.landusepol.2017.04.007.

Woolf, D., J.E. Amonette, A. Street-Perrott, J. Lehmann, and S. Joseph, 2010: Sustainable bio-char to mitigate global climate change. *Nature Communications*, **1**, 56, doi:10.1038/ncomms1053.

World Bank, 2016: *World Bank Group Climate Action Plan*. The World Bank, Washington DC, USA, 59 pp.

World Bank, 2017a: *Pacific Possible: Long-term Economic Opportunities and Challenges for Pacific Island Countries*. World Bank Group, Washington DC, USA, 158 pp.

World Bank, 2017b: Understanding Poverty: Safety Nets. The World Bank, Washington DC, USA. Retrieved from: www.worldbank.org/en/topic/safetynets.

World Bank, 2018a: *Global Financial Development Report 2017/2018: Bankers without Borders*. The World Bank, Washington DC, USA, 159 pp., doi:10.1596/978-1-4648-1148-7.

World Bank, 2018b: Market capitalization of listed domestic companies (current US$). World Federation of Exchanges database. Retrieved from: https://data.worldbank.org/indicator/cm.mkt.lcap.cd.

Wright, H. et al., 2014: Farmers, food and climate change: ensuring community-based adaptation is mainstreamed into agricultural programmes. *Climate and Development*, **6(4)**, 318–328, doi:10.1080/17565529.2014.965654.

Wu, P., J. Ridley, A. Pardaens, R. Levine, and J. Lowe, 2015: The reversibility of CO_2 induced climate change. *Climate Dynamics*, **45(3)**, 745–754, doi:10.1007/s00382-014-2302-0.

Xia, L., A. Robock, S. Tilmes, and R.R. Neely, 2016: Stratospheric sulfate geoengineering could enhance the terrestrial photosynthesis rate. *Atmospheric Chemistry and Physics*, **16(3)**, 1479–1489, doi:10.5194/acp-16-1479-2016.

Xiao, J.J., H. Li, J. Jian, and X. Haifeng, 2011: Sustainable Consumption and Life Satisfaction. *Social Indicators Research*, **104(2)**, 323–329, doi:10.1007/s11205-010-9746-9.

Xie, J. et al., 2017: An integrated assessment of urban flooding mitigation strategies for robust decision making. *Environmental Modelling & Software*, **95**, 143–155, doi:10.1016/j.envsoft.2017.06.027.

Xu, Y. and V. Ramanathan, 2017: Well below 2°C: Mitigation strategies for avoiding dangerous to catastrophic climate changes. *Proceedings of the National Academy of Sciences*, **114(39)**, 10315–10323, doi:10.1073/pnas.1618481114.

Xue, X. et al., 2015: Critical insights for a sustainability framework to address integrated community water services: Technical metrics and approaches. *Water Research*, **77**, 155–169, doi:10.1016/j.watres.2015.03.017.

Yamamoto, L., D.A. Serraglio, and F.S. Cavedon-Capdeville, 2017: Human mobility in the context of climate change and disasters: a South American approach. *International Journal of Climate Change Strategies and Management*, **10(1)**, 65–85, doi:10.1108/ijccsm-03-2017-0069.

Yang, Y.C.E., S. Wi, P.A. Ray, C.M. Brown, and A.F. Khalil, 2016: The future nexus of the Brahmaputra River Basin: Climate, water, energy and food trajectories. *Global Environmental Change*, **37**, 16–30, doi:10.1016/j.gloenvcha.2016.01.002.

Yangka, D. and M. Diesendorf, 2016: Modeling the benefits of electric cooking in Bhutan: A long term perspective. *Renewable and Sustainable Energy Reviews*, **59**, 494–503, doi:10.1016/j.rser.2015.12.265.

Yangka, D. and P. Newman, 2018: Bhutan: Can the 1.5°C Agenda Be Integrated with Growth in Wealth and Happiness? *Urban Planning*, **3(2)**, 94–112, doi:10.17645/up.v3i2.1250.

Yenneti, K. and R. Day, 2016: Distributional justice in solar energy implementation in India: The case of Charanka solar park. *Journal of Rural Studies*, **46**, 35–46, doi:10.1016/j.jrurstud.2016.05.009.

Young, O.R., 2016: *Governing Complex Systems: Social Capital for the Anthropocene*. The MIT Press, Cambridge, MA, USA and London, UK, 296 pp.

Young, W. et al., 2015: Changing Behaviour: Successful Environmental Programmes in the Workplace. *Business Strategy and the Environment*, **24(8)**, 689–703, doi:10.1002/bse.1836.

Yuan, X., J. Zuo, R. Ma, and Y. Wang, 2017: How would social acceptance affect nuclear power development? A study from China. *Journal of Cleaner Production*, **163**, 179–186, doi:10.1016/j.jclepro.2015.04.049.

Zahariev, K., J.R. Christian, and K.L. Denman, 2008: Preindustrial, historical, and fertilization simulations using a global ocean carbon model with new parameterizations of iron limitation, calcification, and N2 fixation. *Progress in Oceanography*, **77(1)**, 56–82, doi:10.1016/j.pocean.2008.01.007.

Zanchi, G., N. Pena, and N. Bird, 2012: Is woody bioenergy carbon neutral? A comparative assessment of emissions from consumption of woody bioenergy and fossil fuel. *GCB Bioenergy*, **4(6)**, 761–772, doi:10.1111/j.1757-1707.2011.01149.x.

Zangheri, P., R. Armani, M. Pietrobon, and L. Pagliano, 2018: Identification of cost-optimal and NZEB refurbishment levels for representative climates and building typologies across Europe. *Energy Efficiency*, **11(2)**, 337–369, doi:10.1007/s12053-017-9566-8.

Zaval, L., E.M. Markowitz, and E.U. Weber, 2015: How Will I Be Remembered? Conserving the Environment for the Sake of One's Legacy. *Psychological Science*, **26(2)**, 231–236, doi:10.1177/0956797614561266.

Zeebe, R.E., 2005: Feasibility of ocean fertilization and its impact on future atmospheric CO_2 levels. *Geophysical Research Letters*, **32(9)**, L09703, doi:10.1029/2005gl022449.

Zelli, F., 2011: The fragmentation of the global climate governance architecture. *Wiley Interdisciplinary Reviews: Climate Change*, **2(1)**, 255–270, doi:10.1002/wcc.104.

Zeman, F.S., 2014: Reducing the Cost of Ca-Based Direct Air Capture of CO_2. *Environmental Science & Technology*, **48(19)**, 11730–11735, doi:10.1021/es502887y.

Zeng, S. and Z. Chen, 2016: Impact of fossil fuel subsidy reform in China: Estimations of household welfare effects based on 2007–2012 data. *Economic and Political Studies*, **4(3)**, 299–318, doi:10.1080/20954816.2016.1218669.

Zhang, H., 2016: Towards global green shipping: the development of international regulations on reduction of GHG emissions from ships. *International Environmental Agreements: Politics, Law and Economics*, **16(4)**, 561–577, doi:10.1007/s10784-014-9270-5.

Zhang, H., W. Chen, and W. Huang, 2016: TIMES modelling of transport sector in China and USA: Comparisons from a decarbonization perspective. *Applied Energy*, **162**, 1505–1514, doi:10.1016/j.apenergy.2015.08.124.

Zhang, W. et al., 2017: Perception, knowledge and behaviors related to typhoon: A cross sectional study among rural residents in Zhejiang, China. *International Journal of Environmental Research and Public Health*, **14(5)**, 1–12, doi:10.3390/ijerph14050492.

Ziervogel, G., A. Cowen, and J. Ziniades, 2016a: Moving from Adaptive to Transformative Capacity: Building Foundations for Inclusive, Thriving, and Regenerative Urban Settlements. *Sustainability*, **8(9)**, 955, doi:10.3390/su8090955.

Ziervogel, G., J. Waddell, W. Smit, and A. Taylor, 2016b: Flooding in Cape Town's informal settlements: barriers to collaborative urban risk governance. *South African Geographical Journal*, **98(1)**, 1–20, doi:10.1080/03736245.2014.924867.

Ziervogel, G. et al., 2017: Inserting rights and justice into urban resilience: a focus on everyday risk. *Environment and Urbanization*, **29(1)**, 123–138, doi:10.1177/0956247816686905.

Zilberman, D., 2017: Indirect land use change: much ado about (almost) nothing. *GCB Bioenergy*, **9(3)**, 485–488, doi:10.1111/gcbb.12368.

Zomer, R.J. et al., 2016: Global Tree Cover and Biomass Carbon on Agricultural Land: The contribution of agroforestry to global and national carbon budgets. *Scientific Reports*, **6**, 29987, doi:10.1038/srep29987.

Zusman, E., A. Miyatsuka, J. Romero, and M. Arif, 2015: *Aligning Interests around Mitigating Short Lived Climate Pollutants (SLCP) in Asia: A Stepwise Approach*. IGES Discussion Paper, Institute for Global Environmental Strategies (IGES), Hayama, Japan, 14 pp.

4

Sustainable Development, Poverty Eradication and Reducing Inequalities

5

Coordinating Lead Authors:

Joyashree Roy (India), Petra Tschakert (Australia/Austria), Henri Waisman (France)

Lead Authors:

Sharina Abdul Halim (Malaysia), Philip Antwi-Agyei (Ghana), Purnamita Dasgupta (India), Bronwyn Hayward (New Zealand), Markku Kanninen (Finland), Diana Liverman (USA), Chukwumerije Okereke (UK/Nigeria), Patricia Fernanda Pinho (Brazil), Keywan Riahi (Austria), Avelino G. Suarez Rodriguez (Cuba)

Contributing Authors:

Fernando Aragón-Durand (Mexico), Mustapha Babiker (Sudan), Mook Bangalore (USA), Paolo Bertoldi (Italy), Bishwa Bhaskar Choudhary (India), Edward Byres (Austria/Brazil), Anton Cartwright (South Africa), Riyanti Djalante (Japan/Indonesia), Kristie L. Ebi (USA), Neville Ellis (Australia), Francois Engelbrecht (South Africa), Maria Figueroa (Denmark/Venezuela), Mukesh Gupta (India), Diana Hinge Salili (Vanuatu), Daniel Huppmann (Austria), Saleemul Huq (Bangladesh/UK), Daniela Jacob (Germany), Rachel James (UK), Debora Ley (Guatemala/Mexico), Peter Marcotullio (USA), Omar Massera (Mexico), Reinhard Mechler (Germany), Haileselassie Amaha Medhin (Ethiopia), Shagun Mehrotra (USA/India), Peter Newman (Australia), Karen Paiva Henrique (Brazil), Simon Parkinson (Canada), Aromar Revi (India), Wilfried Rickels (Germany), Lisa Schipper (UK/Sweden), Jörn Schmidt (Germany), Seth Schultz (USA), Pete Smith (UK), William Solecki (USA), Shreya Some (India), Nenenteiti Teariki-Ruatu (Kiribati), Adelle Thomas (Bahamas), Penny Urquhart (South Africa), Margaretha Wewerinke-Singh (Netherlands)

Review Editors:

Svitlana Krakovska (Ukraine), Ramon Pichs Madruga (Cuba), Roberto Sanchez (Mexico)

Chapter Scientist:

Neville Ellis (Australia)

This chapter should be cited as:

Roy, J., P. Tschakert, H. Waisman, S. Abdul Halim, P. Antwi-Agyei, P. Dasgupta, B. Hayward, M. Kanninen, D. Liverman, C. Okereke, P.F. Pinho, K. Riahi, and A.G. Suarez Rodriguez, 2018: Sustainable Development, Poverty Eradication and Reducing Inequalities. In: *Global Warming of 1.5°C. An IPCC Special Report on the impacts of global warming of 1.5°C above pre-industrial levels and related global greenhouse gas emission pathways, in the context of strengthening the global response to the threat of climate change, sustainable development, and efforts to eradicate poverty* [Masson-Delmotte, V., P. Zhai, H.-O. Pörtner, D. Roberts, J. Skea, P.R. Shukla, A. Pirani, W. Moufouma-Okia, C. Péan, R. Pidcock, S. Connors, J.B.R. Matthews, Y. Chen, X. Zhou, M.I. Gomis, E. Lonnoy, T. Maycock, M. Tignor, and T. Waterfield (eds.)]. Cambridge University Press, Cambridge, UK and New York, NY, USA, pp. 445-538. https://doi.org/10.1017/9781009157940.007.

Table of Contents

Executive Summary

This chapter takes sustainable development as the starting point and focus for analysis. It considers the broad and multifaceted bi-directional interplay between sustainable development, including its focus on eradicating poverty and reducing inequality in their multidimensional aspects, and climate actions in a 1.5°C warmer world. These fundamental connections are embedded in the Sustainable Development Goals (SDGs). The chapter also examines synergies and trade-offs of adaptation and mitigation options with sustainable development and the SDGs and offers insights into possible pathways, especially climate-resilient development pathways towards a 1.5°C warmer world.

Sustainable Development, Poverty and Inequality in a 1.5°C Warmer World

Limiting global warming to 1.5°C rather than 2°C above pre-industrial levels would make it markedly easier to achieve many aspects of sustainable development, with greater potential to eradicate poverty and reduce inequalities (*medium evidence, high agreement*). Impacts avoided with the lower temperature limit could reduce the number of people exposed to climate risks and vulnerable to poverty by 62 to 457 million, and lessen the risks of poor people to experience food and water insecurity, adverse health impacts, and economic losses, particularly in regions that already face development challenges (*medium evidence, medium agreement*). {5.2.2, 5.2.3} Avoided impacts expected to occur between 1.5°C and 2°C warming would also make it easier to achieve certain SDGs, such as those that relate to poverty, hunger, health, water and sanitation, cities and ecosystems (SDGs 1, 2, 3, 6, 11, 14 and 15) (*medium evidence, high agreement*). {5.2.3, Table 5.2 available at the end of the chapter}

Compared to current conditions, 1.5°C of global warming would nonetheless pose heightened risks to eradicating poverty, reducing inequalities and ensuring human and ecosystem well-being (*medium evidence, high agreement*). Warming of 1.5°C is not considered 'safe' for most nations, communities, ecosystems and sectors and poses significant risks to natural and human systems as compared to the current warming of 1°C (*high confidence*). {Cross-Chapter Box 12 in Chapter 5} The impacts of 1.5°C of warming would disproportionately affect disadvantaged and vulnerable populations through food insecurity, higher food prices, income losses, lost livelihood opportunities, adverse health impacts and population displacements (*medium evidence, high agreement*). {5.2.1} Some of the worst impacts on sustainable development are expected to be felt among agricultural and coastal dependent livelihoods, indigenous people, children and the elderly, poor labourers, poor urban dwellers in African cities, and people and ecosystems in the Arctic and Small Island Developing States (SIDS) (*medium evidence, high agreement*). {5.2.1, Box 5.3, Chapter 3, Box 3.5, Cross-Chapter Box 9 in Chapter 4}

Climate Adaptation and Sustainable Development

Prioritization of sustainable development and meeting the SDGs is consistent with efforts to adapt to climate change (*high

confidence). Many strategies for sustainable development enable transformational adaptation for a 1.5°C warmer world, provided attention is paid to reducing poverty in all its forms and to promoting equity and participation in decision-making (*medium evidence, high agreement*). As such, sustainable development has the potential to significantly reduce systemic vulnerability, enhance adaptive capacity, and promote livelihood security for poor and disadvantaged populations (*high confidence*). {5.3.1}

Synergies between adaptation strategies and the SDGs are expected to hold true in a 1.5°C warmer world, across sectors and contexts (*medium evidence, medium agreement*). Synergies between adaptation and sustainable development are significant for agriculture and health, advancing SDGs 1 (extreme poverty), 2 (hunger), 3 (healthy lives and well-being) and 6 (clean water) (*robust evidence, medium agreement*). {5.3.2} Ecosystem- and community-based adaptation, along with the incorporation of indigenous and local knowledge, advances synergies with SDGs 5 (gender equality), 10 (reducing inequalities) and 16 (inclusive societies), as exemplified in drylands and the Arctic (*high evidence, medium agreement*). {5.3.2, Box 5.1, Cross-Chapter Box 10 in Chapter 4}

Adaptation strategies can result in trade-offs with and among the SDGs (*medium evidence, high agreement*). Strategies that advance one SDG may create negative consequences for other SDGs, for instance SDGs 3 (health) versus 7 (energy consumption) and agricultural adaptation and SDG 2 (food security) versus SDGs 3 (health), 5 (gender equality), 6 (clean water), 10 (reducing inequalities), 14 (life below water) and 15 (life on the land) (*medium evidence, medium agreement*). {5.3.2}

Pursuing place-specific adaptation pathways towards a 1.5°C warmer world has the potential for significant positive outcomes for well-being in countries at all levels of development (*medium evidence, high agreement*). Positive outcomes emerge when adaptation pathways (i) ensure a diversity of adaptation options based on people's values and the trade-offs they consider acceptable, (ii) maximize synergies with sustainable development through inclusive, participatory and deliberative processes, and (iii) facilitate equitable transformation. Yet such pathways would be difficult to achieve without redistributive measures to overcome path dependencies, uneven power structures, and entrenched social inequalities (*medium evidence, high agreement*). {5.3.3}

Mitigation and Sustainable Development

The deployment of mitigation options consistent with 1.5°C pathways leads to multiple synergies across a range of sustainable development dimensions. At the same time, the rapid pace and magnitude of change that would be required to limit warming to 1.5°C, if not carefully managed, would lead to trade-offs with some sustainable development dimensions (*high confidence*). The number of synergies between mitigation response options and sustainable development exceeds the number of trade-offs in energy demand and supply sectors; agriculture, forestry and other land use (AFOLU); and for oceans (*very high confidence*). {Figure 5.2, Table 5.2 available at the end of the chapter} The 1.5°C pathways indicate robust synergies, particularly for the SDGs 3 (health), 7 (energy), 12 (responsible consumption and production) and 14 (oceans) (*very high confidence*). {5.4.2, Figure 5.3} For SDGs 1 (poverty), 2 (hunger), 6 (water) and 7 (energy), there is a risk of trade-offs or negative side effects from stringent mitigation actions compatible with 1.5°C of warming (*medium evidence, high agreement*). {5.4.2}

Appropriately designed mitigation actions to reduce energy demand can advance multiple SDGs simultaneously. Pathways compatible with 1.5°C that feature low energy demand show the most pronounced synergies and the lowest number of trade-offs with respect to sustainable development and the SDGs (*very high confidence*). Accelerating energy efficiency in all sectors has synergies with SDGs 7 (energy), 9 (industry, innovation and infrastructure), 11 (sustainable cities and communities), 12 (responsible consumption and production), 16 (peace, justice and strong institutions), and 17 (partnerships for the goals) (*robust evidence, high agreement*). {5.4.1, Figure 5.2, Table 5.2} Low-demand pathways, which would reduce or completely avoid the reliance on bioenergy with carbon capture and storage (BECCS) in 1.5°C pathways, would result in significantly reduced pressure on food security, lower food prices and fewer people at risk of hunger (*medium evidence, high agreement*). {5.4.2, Figure 5.3}

The impacts of carbon dioxide removal options on SDGs depend on the type of options and the scale of deployment (*high confidence*). If poorly implemented, carbon dioxide removal (CDR) options such as bioenergy, BECCS and AFOLU would lead to trade-offs. Appropriate design and implementation requires considering local people's needs, biodiversity and other sustainable development dimensions (*very high confidence*). {5.4.1.3, Cross-Chapter Box 7 in Chapter 3}

The design of the mitigation portfolios and policy instruments to limit warming to 1.5°C will largely determine the overall synergies and trade-offs between mitigation and sustainable development (*very high confidence*). Redistributive policies that shield the poor and vulnerable can resolve trade-offs for a range of SDGs (*medium evidence, high agreement*). Individual mitigation options are associated with both positive and negative interactions with the SDGs (*very high confidence*). {5.4.1} However, appropriate choices across the mitigation portfolio can help to maximize positive side effects while minimizing negative side effects (*high confidence*). {5.4.2, 5.5.2} Investment needs for complementary policies resolving trade-offs with a range of SDGs are only a small fraction of the overall mitigation investments in 1.5°C pathways (*medium evidence, high agreement*). {5.4.2, Figure 5.4} Integration of mitigation with adaptation and sustainable development compatible with 1.5°C warming requires a systems perspective (*high confidence*). {5.4.2, 5.5.2}

Mitigation consistent with 1.5°C of warming create high risks for sustainable development in countries with high dependency on fossil fuels for revenue and employment generation (*high confidence*). These risks are caused by the reduction of global demand affecting mining activity and export revenues and challenges to rapidly decrease high carbon intensity of the domestic economy (*robust

evidence, high agreement). {5.4.1.2, Box 5.2} Targeted policies that promote diversification of the economy and the energy sector could ease this transition (*medium evidence, high agreement*). {5.4.1.2, Box 5.2}

Sustainable Development Pathways to 1.5°C

Sustainable development broadly supports and often enables the fundamental societal and systems transformations that would be required for limiting warming to 1.5°C above pre-industrial levels (*high confidence*). Simulated pathways that feature the most sustainable worlds (e.g., Shared Socio-Economic Pathways (SSP) 1) are associated with relatively lower mitigation and adaptation challenges and limit warming to 1.5°C at comparatively lower mitigation costs. In contrast, development pathways with high fragmentation, inequality and poverty (e.g., SSP3) are associated with comparatively higher mitigation and adaptation challenges. In such pathways, it is not possible to limit warming to 1.5°C for the vast majority of the integrated assessment models (*medium evidence, high agreement*). {5.5.2} In all SSPs, mitigation costs substantially increase in 1.5°C pathways compared to 2°C pathways. No pathway in the literature integrates or achieves all 17 SDGs (*high confidence*). {5.5.2} Real-world experiences at the project level show that the actual integration between adaptation, mitigation and sustainable development is challenging as it requires reconciling trade-offs across sectors and spatial scales (*very high confidence*). {5.5.1}

Without societal transformation and rapid implementation of ambitious greenhouse gas reduction measures, pathways to limiting warming to 1.5°C and achieving sustainable development will be exceedingly difficult, if not impossible, to achieve (*high confidence*). The potential for pursuing such pathways differs between and within nations and regions, due to different development trajectories, opportunities and challenges (*very high confidence*). {5.5.3.2, Figure 5.1} Limiting warming to 1.5°C would require all countries and non-state actors to strengthen their contributions without delay. This could be achieved through sharing efforts based on bolder and more committed cooperation, with support for those with the least capacity to adapt, mitigate and transform (*medium evidence, high agreement*). {5.5.3.1, 5.5.3.2} Current efforts towards reconciling low-carbon trajectories and reducing inequalities, including those that avoid difficult trade-offs associated with transformation, are partially successful yet demonstrate notable obstacles (*medium evidence, medium agreement*). {5.5.3.3, Box 5.3, Cross-Chapter Box 13 in this chapter}

Social justice and equity are core aspects of climate-resilient development pathways for transformational social change. Addressing challenges and widening opportunities between and within countries and communities would be necessary to achieve sustainable development and limit warming to 1.5°C, without making the poor and disadvantaged worse off (*high confidence*). Identifying and navigating inclusive and socially acceptable pathways towards low-carbon, climate-resilient futures is a challenging yet important endeavour, fraught with moral, practical and political difficulties and inevitable trade-offs (*very high confidence*). {5.5.2, 5.5.3.3, Box 5.3} It entails deliberation and problem-solving processes to negotiate societal values, well-being, risks and resilience and to determine what is desirable and fair, and to whom (*medium evidence, high agreement*). Pathways that encompass joint, iterative planning and transformative visions, for instance in Pacific SIDS like Vanuatu and in urban contexts, show potential for liveable and sustainable futures (*high confidence*). {5.5.3.1, 5.5.3.3, Figure 5.5, Box 5.3, Cross-Chapter Box 13 in this chapter}

The fundamental societal and systemic changes to achieve sustainable development, eradicate poverty and reduce inequalities while limiting warming to 1.5°C would require meeting a set of institutional, social, cultural, economic and technological conditions (*high confidence*). The coordination and monitoring of policy actions across sectors and spatial scales is essential to support sustainable development in 1.5°C warmer conditions (*very high confidence*). {5.6.2, Box 5.3} External funding and technology transfer better support these efforts when they consider recipients' context-specific needs (*medium evidence, high agreement*). {5.6.1} Inclusive processes can facilitate transformations by ensuring participation, transparency, capacity building and iterative social learning (*high confidence*). {5.5.3.3, Cross-Chapter Box 13, 5.6.3} Attention to power asymmetries and unequal opportunities for development, among and within countries, is key to adopting 1.5°C-compatible development pathways that benefit all populations (*high confidence*). {5.5.3, 5.6.4, Box 5.3} Re-examining individual and collective values could help spur urgent, ambitious and cooperative change (*medium evidence, high agreement*). {5.5.3, 5.6.5}

5

5.1   Scope and Delineations

This chapter takes sustainable development as the starting point and focus for analysis, considering the broader bi-directional interplay and multifaceted interactions between development patterns and climate actions in a 1.5°C warmer world and in the context of eradicating poverty and reducing inequality. It assesses the impacts of keeping temperatures at or below 1.5°C of global warming above pre-industrial levels on sustainable development and compares the impacts avoided at 1.5°C compared to 2°C (Section 5.2). It then examines the interactions, synergies and trade-offs of adaptation (Section 5.3) and mitigation (Section 5.4) measures with sustainable development and the Sustainable Development Goals (SDGs). The chapter offers insights into possible pathways towards a 1.5°C warmer world, especially through climate-resilient development pathways providing a comprehensive vision across different contexts (Section 5.5). The chapter also identifies the conditions that would be needed to simultaneously achieve sustainable development, poverty eradication, the reduction of inequalities, and the 1.5°C climate objective (Section 5.6).

5.1.1   Sustainable Development, SDGs, Poverty Eradication and Reducing Inequalities

Chapter 1 (see Cross-Chapter Box 4 in Chapter 1) defines sustainable development as 'development that meets the needs of the present and future generations' through balancing economic, social and environmental considerations, and then introduces the United Nations (UN) 2030 Agenda for Sustainable Development, which sets out 17 ambitious goals for sustainable development for all countries by 2030. These SDGs are: no poverty (SDG 1), zero hunger (SDG 2), good health and well-being (SDG 3), quality education (SDG 4), gender equality (SDG 5), clean water and sanitation (SDG 6), affordable and clean energy (SDG 7), decent work and economic growth (SDG 8), industry, innovation and infrastructure (SDG 9), reduced inequalities (SDG 10), sustainable cities and communities (SDG 11), responsible consumption and production (SDG 12), climate action (SDG 13), life below water (SDG 14), life on land (SDG 15), peace, justice and strong institutions (SDG 16) and partnerships for the goals (SDG 17).

The IPCC Fifth Assessment Report (AR5) included extensive discussion of links between climate and sustainable development, especially in Chapter 13 (Olsson et al., 2014) and Chapter 20 (Denton et al., 2014) in Working Group II and Chapter 4 (Fleurbaey et al., 2014) in Working Group III. However, the AR5 preceded the 2015 adoption of the SDGs and the literature that argues for their fundamental links to climate (Wright et al., 2015; Salleh, 2016; von Stechow et al., 2016; Hammill and Price-Kelly, 2017; ICSU, 2017; Maupin, 2017; Gomez-Echeverri, 2018).

The SDGs build on efforts under the UN Millennium Development Goals to reduce poverty, hunger, and other deprivations. According to the UN, the Millennium Development Goals were successful in reducing poverty and hunger and improving water security (UN, 2015a). However, critics argued that they failed to address within-country disparities, human rights and key environmental concerns, focused only on developing countries, and had numerous measurement and attribution problems

(Langford et al., 2013; Fukuda-Parr et al., 2014). While improvements in water security, slums and health may have reduced some aspects of climate vulnerability, increases in incomes were linked to rising greenhouse gas (GHG) emissions and thus to a trade-off between development and climate change (Janetos et al., 2012; UN, 2015a; Hubacek et al., 2017).

While the SDGs capture many important aspects of sustainable development, including the explicit goals of poverty eradication and reducing inequality, there are direct connections from climate to other measures of sustainable development including multidimensional poverty, equity, ethics, human security, well-being and climate-resilient development (Bebbington and Larrinaga, 2014; Robertson, 2014; Redclift and Springett, 2015; Barrington-Leigh, 2016; Helliwell et al., 2018; Kirby and O'Mahony, 2018) (see Glossary). The UN proposes sustainable development as 'eradicating poverty in all its forms and dimensions, combating inequality within and among countries, preserving the planet, creating sustained, inclusive and sustainable economic growth and fostering social inclusion' (UN, 2015b). There is *robust evidence* of the links between climate change and poverty (see Chapter 1, Cross-Chapter Box 4). The AR5 concluded with *high confidence* that disruptive levels of climate change would preclude reducing poverty (Denton et al., 2014; Fleurbaey et al., 2014). International organizations have since stated that climate changes 'undermine the ability of all countries to achieve sustainable development' (UN, 2015b) and can reverse or erase improvements in living conditions and decades of development (Hallegatte et al., 2016).

Climate warming has unequal impacts on different people and places as a result of differences in regional climate changes, vulnerabilities and impacts, and these differences then result in unequal impacts on sustainable development and poverty (Section 5.2). Responses to climate change also interact in complex ways with goals of poverty reduction. The benefits of adaptation and mitigation projects and funding may accrue to some and not others, responses may be costly and unaffordable to some people and countries, and projects may disadvantage some individuals, groups and development initiatives (Sections 5.3 and 5.4, Cross-Chapter Box 11 in Chapter 4).

5.1.2   Pathways to 1.5°C

Pathways to 1.5°C (see Chapter 1, Cross-Chapter Box 1 in Chapter 1, Glossary) include ambitious reductions in emissions and strategies for adaptation that are transformational, as well as complex interactions with sustainable development, poverty eradication and reducing inequalities. The AR5 WGII introduced the concept of climate-resilient development pathways (CRDPs) (see Glossary) which combine adaptation and mitigation to reduce climate change and its impacts, and emphasize the importance of addressing structural and intersecting inequalities, marginalization and multidimensional poverty to 'transform […] the development pathways themselves towards greater social and environmental sustainability, equity, resilience, and justice' (Olsson et al., 2014). This chapter assesses literature on CRDPs relevant to 1.5°C global warming (Section 5.5.3), to understand better the possible societal and systems transformations (see Glossary) that reduce inequality and increase well-being

(Figure 5.1). It also summarizes the knowledge on conditions to achieve such transformations, including changes in technologies, culture, values, financing and institutions that support low-carbon and resilient pathways and sustainable development (Section 5.6).

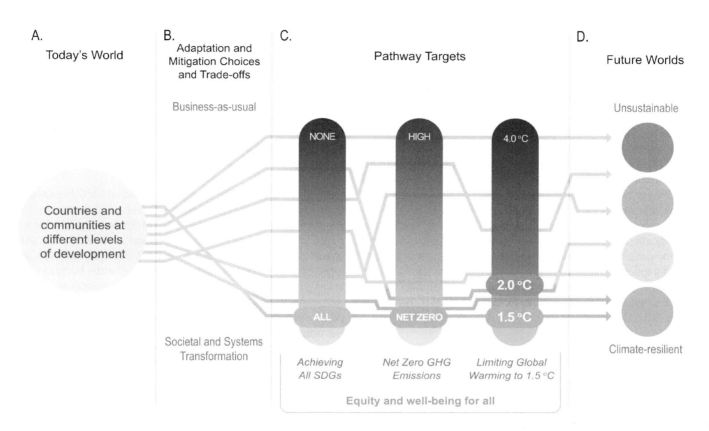

Figure 5.1 | Climate-resilient development pathways (CRDPs) (green arrows) between a current world in which countries and communities exist at different levels of development (A) and future worlds that range from climate-resilient (bottom) to unsustainable (top) (D). CRDPs involve societal transformation rather than business-as-usual approaches, and all pathways involve adaptation and mitigation choices and trade-offs (B). Pathways that achieve the Sustainable Development Goals by 2030 and beyond, strive for net zero emissions around mid-21st century, and stay within the global 1.5°C warming target by the end of the 21st century, while ensuring equity and well-being for all, are best positioned to achieve climate-resilient futures (C). Overshooting on the path to 1.5°C will make achieving CRDPs and other sustainable trajectories more difficult; yet, the limited literature does not allow meaningful estimates.

5.1.3   Types of Evidence

A variety of sources of evidence are used to assess the interactions of sustainable development and the SDGs with the causes, impacts and responses to climate change of 1.5°C warming. This chapter builds on Chapter 3 to assess the sustainable development implications of impacts at 1.5°C and 2°C, and on Chapter 4 to examine the implications of response measures. Scientific and grey literature, with a post-AR5 focus, and data that evaluate, measure and model sustainable development–climate links from various perspectives, quantitatively and qualitatively, across scales, and through well-documented case studies are assessed.

Literature that explicitly links 1.5°C global warming to sustainable development across scales remains scarce; yet we find relevant insights in many recent publications on climate and development that assess impacts across warming levels, the effects of adaptation and mitigation response measures, and interactions with the SDGs. Relevant evidence also stems from emerging literature on possible pathways, overshoot

and enabling conditions (see Glossary) for integrating sustainable development, poverty eradication and reducing inequalities in the context of 1.5°C.

5.2   Poverty, Equality and Equity Implications of a 1.5°C Warmer World

Climate change could lead to significant impacts on extreme poverty by 2030 (Hallegatte et al., 2016; Hallegatte and Rozenberg, 2017). The AR5 concluded, with *very high confidence*, that climate change and climate variability worsen existing poverty and exacerbate inequalities, especially for those disadvantaged by gender, age, race, class, caste, indigeneity and (dis)ability (Olsson et al., 2014). New literature on these links is substantial, showing that the poor will continue to experience climate change severely, and climate change will exacerbate poverty (*very high confidence*) (Fankhauser and Stern, 2016; Hallegatte et al., 2016; O'Neill et al., 2017a; Winsemius et al., 2018). The understanding of regional impacts and risks of 1.5°C global warming and interactions with patterns of societal vulnerability and poverty remains limited. Yet identifying and addressing poverty and inequality is at the core of staying within a safe and just space for humanity (Raworth, 2017; Bathiany et al., 2018). Building on relevant findings from Chapter 3 (see Section 3.4), this section examines anticipated impacts and risks of 1.5°C and higher warming on sustainable development, poverty, inequality and equity (see Glossary).

5.2.1 Impacts and Risks of a 1.5°C Warmer World: Implications for Poverty and Livelihoods

Global warming of 1.5°C will have consequences for sustainable development, poverty and inequalities. This includes residual risks, limits to adaptation, and losses and damages (Cross-Chapter Box 12 in this chapter; see Glossary). Some regions have already experienced a 1.5°C warming, with impacts on food and water security, health and other components of sustainable development (*medium evidence, medium agreement*) (see Chapter 3, Section 3.4). Climate change is also already affecting poorer subsistence communities through decreases in crop production and quality, increases in crop pests and diseases, and disruption to culture (Savo et al., 2016). It disproportionally affects children and the elderly and can increase gender inequality (Kaijser and Kronsell, 2014; Vinyeta et al., 2015; Carter et al., 2016; Hanna and Oliva, 2016; Li et al., 2016).

At 1.5°C warming, compared to current conditions, further negative consequences are expected for poor people, and inequality and vulnerability (*medium evidence, high agreement*). Hallegatte and Rozenberg (2017) report that by 2030 (roughly approximating a 1.5°C warming), 122 million additional people could experience extreme poverty, based on a 'poverty scenario' of limited socio-economic progress, comparable to the Shared Socio-Economic Pathway (SSP) 4 (inequality), mainly due to higher food prices and declining health, with substantial income losses for the poorest 20% across 92 countries. Pretis et al. (2018) estimate negative impacts on economic growth in lower-income countries at 1.5°C warming, despite uncertainties. Impacts are likely to occur simultaneously across livelihood, food, human, water and ecosystem security (*limited evidence, high agreement*) (Byers et al., 2018), but the literature on interacting and cascading effects remains scarce (Hallegatte et al., 2014; O'Neill et al., 2017b; Reyer et al., 2017a, b).

Chapter 3 outlines future impacts and risks for ecosystems and human systems, many of which could also undermine sustainable development and efforts to eradicate poverty and hunger, and to protect health and ecosystems. Chapter 3 findings (see Section 3.5.2.1) suggest increasing Reasons for Concern from moderate to high at a warming of 1.1° to 1.6°C, including for indigenous people and their livelihoods, and ecosystems in the Arctic (O'Neill et al., 2017b). In 2050, based on the Hadley Centre Climate Prediction Model 3 (HadCM3) and the Special Report on Emission Scenarios A1b scenario (roughly comparable to 1.5°C warming), 450 million more flood-prone people would be exposed to doubling in flood frequency, and global flood risk would increase substantially (Arnell and Gosling, 2016). For droughts, poor people are expected to be more exposed (85% in population terms) in a warming scenario greater than 1.5°C for several countries in Asia and southern and western

Africa (Winsemius et al., 2018). In urban Africa, a 1.5°C warming could expose many households to water poverty and increased flooding (Pelling et al., 2018). At 1.5°C warming, fisheries-dependent and coastal livelihoods, of often disadvantaged populations, would suffer from the loss of coral reefs (see Chapter 3, Box 3.4).

Global heat stress is projected to increase in a 1.5°C warmer world, and by 2030, compared to 1961–1990, climate change could be responsible for additional annual deaths of 38,000 people from heat stress, particularly among the elderly, and 48,000 from diarrhoea, 60,000 from malaria, and 95,000 from childhood undernutrition (WHO, 2014). Each 1°C increase could reduce work productivity by 1 to 3% for people working outdoors or without air conditioning, typically the poorer segments of the workforce (Park et al., 2015).

The regional variation in the 'warming experience at 1.5°C' (see Chapter 1, Section 1.3.1) is large (see Chapter 3, Section 3.3.2). Declines in crop yields are widely reported for Africa (60% of observations), with serious consequences for subsistence and rain-fed agriculture and food security (Savo et al., 2016). In Bangladesh, by 2050, damages and losses are expected for poor households dependent on freshwater fish stocks due to lack of mobility, limited access to land and strong reliance on local ecosystems (Dasgupta et al., 2017). Small Island Developing States (SIDS) are expected to experience challenging conditions at 1.5°C warming due to increased risk of internal migration and displacement and limits to adaptation (see Chapter 3, Box 3.5, Cross-Chapter Box 12 in this chapter). An anticipated decline of marine fisheries of 3 million metric tonnes per degree warming would have serious regional impacts for the Indo-Pacific region and the Arctic (Cheung et al., 2016).

5.2.2 Avoided Impacts of 1.5°C versus 2°C Warming for Poverty and Inequality

Avoided impacts between 1.5°C and 2°C warming are expected to have significant positive implications for sustainable development, and reducing poverty and inequality. Using the SSPs (see Chapter 1, Cross-Chapter Box 1 in Chapter 1, Section 5.5.2), Byers et al. (2018) model the number of people exposed to multi-sector climate risks and vulnerable to poverty (income < $10/day), comparing 2°C and 1.5°C; the respective declines are from 86 million to 24 million for SSP1 (sustainability), from 498 million to 286 million for SSP2 (middle of the road), and from 1220 million to 763 million for SSP3 (regional rivalry), which suggests overall 62–457 million fewer people exposed and vulnerable at 1.5°C warming. Across the SSPs, the largest populations exposed and vulnerable are in South Asia (Byers et al., 2018). The avoided impacts on poverty at 1.5°C relative to 2°C are projected to depend at least as much or more on development scenarios than on warming (Wiebe et al., 2015; Hallegatte and Rozenberg, 2017).

Limiting warming to 1.5°C is expected to reduce the number of people exposed to hunger, water stress and disease in Africa (Clements, 2009). It is also expected to limit the number of poor people exposed to floods and droughts at higher degrees of warming, especially in African and Asian countries (Winsemius et al., 2018). Challenges for poor populations – relating to food and water security, clean energy access and environmental well-being – are projected to be less at 1.5°C, particularly for vulnerable people in Africa and Asia (Byers et al., 2018). The overall projected socio-economic losses compared to the present day are less at 1.5°C (8% loss of gross domestic product per capita) compared to 2°C (13%), with lower-income countries projected to experience greater losses, which may increase economic inequality between countries (Pretis et al., 2018).

5.2.3   Risks from 1.5°C versus 2°C Global Warming and the Sustainable Development Goals

The risks that can be avoided by limiting global warming to 1.5°C rather than 2°C have many complex implications for sustainable development (ICSU, 2017; Gomez-Echeverri, 2018). There is *high confidence* that constraining warming to 1.5°C rather than 2°C would reduce risks for unique and threatened ecosystems, safeguarding the services they provide for livelihoods and sustainable development and making adaptation much easier (O'Neill et al., 2017b), particularly in Central America, the Amazon, South Africa and Australia (Schleussner et al., 2016; O'Neill et al., 2017b; Reyer et al., 2017b; Bathiany et al., 2018).

In places that already bear disproportionate economic and social challenges to their sustainable development, people will face lower risks at 1.5°C compared to 2°C. These include North Africa and the Levant (less water scarcity), West Africa (less crop loss), South America and Southeast Asia (less intense heat), and many other coastal nations and island states (lower sea level rise, less coral reef loss) (Schleussner et al., 2016; Betts et al., 2018). The risks for food, water and ecosystems, particularly in subtropical regions such as Central America and countries such as South Africa and Australia, are expected to be lower at 1.5°C than at 2°C warming (Schleussner et al., 2016). Fewer people would be exposed to droughts and

heat waves and the associated health impacts in countries such as Australia and India (King et al., 2017; Mishra et al., 2017).

Limiting warming to 1.5°C would make it markedly easier to achieve the SDGs for poverty eradication, water access, safe cities, food security, healthy lives and inclusive economic growth, and would help to protect terrestrial ecosystems and biodiversity (*medium evidence, high agreement*) (Table 5.2 available at the end of the chapter). For example, limiting species loss and expanding climate refugia will make it easier to achieve SDG 15 (see Chapter 3, Section 3.4.3). One indication of how lower temperatures benefit the SDGs is to compare the impacts of Representative Concentration Pathway (RCP) 4.5 (lower emissions) and RCP8.5 (higher emissions) on the SDGs (Ansuategi et al., 2015). A low emissions pathway allows for greater success in achieving SDGs for reducing poverty and hunger, providing access to clean energy, reducing inequality, ensuring education for all and making cities more sustainable. Even at lower emissions, a medium risk of failure exists to meet goals for water and sanitation, and marine and terrestrial ecosystems.

Action on climate change (SDG 13), including slowing the rate of warming, would help reach the goals for water, energy, food and land (SDGs 6, 7, 2 and 15) (Obersteiner et al., 2016; ICSU, 2017) and contribute to poverty eradication (SDG 1) (Byers et al., 2018). Although the literature that connects 1.5°C to the SDGs is limited, a pathway that stabilizes warming at 1.5°C by the end of the century is expected to increase the chances of achieving the SDGs by 2030, with greater potential to eradicate poverty, reduce inequality and foster equity (*limited evidence, medium agreement*). There are no studies on overshoot and dimensions of sustainable development, although literature on 4°C of warming suggests the impacts would be severe (Reyer et al., 2017b).

Table 5.1 | Sustainable development implications of avoided impacts between 1.5°C and 2°C global warming.

Impacts	Chapter 3 Section	1.5°C	2°C	Sustainable Development Goals (SDGs) More Easily Achieved when Limiting Warming to 1.5°C
Water scarcity	3.4.2.1	4% more people exposed to water stress	8% more people exposed to water stress, with 184–270 million people more exposed	SDG 6 water availability for all
	Table 3.4	496 (range 103–1159) million people exposed and vulnerable to water stress	586 (range 115–1347) million people exposed and vulnerable to water stress	
Ecosystems	3.4.3, Table 3.4	Around 7% of land area experiences biome shifts	Around 13% (range 8–20%) of land area experiences biome shifts	SDG 15 to protect terrestrial ecosystems and halt biodiversity loss
	Box 3.5	70–90% of coral reefs at risk from bleaching	99% of coral reefs at risk from bleaching	
Coastal cities	3.4.5.1	31–69 million people exposed to coastal flooding	32–79 million exposed to coastal flooding	SDG 11 to make cities and human settlements safe and resilient
	3.4.5.2	Fewer cities and coasts exposed to sea level rise and extreme events	More people and cities exposed to flooding	
Food systems	3.4.6, Box 3.1	Significant declines in crop yields avoided, some yields may increase	Average crop yields decline	SDG 2 to end hunger and achieve food security
	Table 3.4	32–36 million people exposed to lower yields	330–396 million people exposed to lower yields	
Health	3.4.5.1	Lower risk of temperature-related morbidity and smaller mosquito range	Higher risks of temperature-related morbidity and mortality and larger geographic range of mosquitoes	SDG 3 to ensure healthy lives for all
	3.4.5.2	3546–4508 million people exposed to heat waves	5417–6710 million people exposed to heat waves	

Cross-Chapter Box 12 | Residual Risks, Limits to Adaptation and Loss and Damage

Lead Authors:

Riyanti Djalante (Japan/Indonesia), Kristie L. Ebi (USA), Debora Ley (Guatemala/Mexico), Reinhard Mechler (Germany), Patricia Fernanda Pinho (Brazil), Aromar Revi (India), Petra Tschakert (Australia/Austria)

Contributing Authors:

Karen Paiva Henrique (Brazil), Saleemul Huq (Bangladesh/UK), Rachel James (UK), Adelle Thomas (Bahamas), Margaretha Wewerinke-Singh (Netherlands)

Introduction

Residual climate-related risks, limits to adaptation, and loss and damage (see Glossary) are increasingly assessed in the scientific literature (van der Geest and Warner, 2015; Boyd et al., 2017; Mechler et al., 2019). The AR5 (IPCC, 2013; Oppenheimer et al., 2014) documented impacts that have been detected and attributed to climate change, projected increasing climate-related risks with continued global warming, and recognized barriers and limits to adaptation. It recognized that adaptation is constrained by biophysical, institutional, financial, social and cultural factors, and that the interaction of these factors with climate change can lead to soft adaptation limits (adaptive actions currently not available) and hard adaptation limits (adaptive actions appear infeasible leading to unavoidable impacts) (Klein et al., 2014).

Loss and damage: concepts and perspectives

'Loss and Damage' (L&D) has been discussed in international climate negotiations for three decades (INC, 1991; Calliari, 2016; Vanhala and Hestbaek, 2016). A work programme on L&D was established as part of the Cancun Adaptation Framework in 2010 supporting developing countries particularly vulnerable to climate change impacts (UNFCCC, 2011a). In 2013, the Conference of the Parties (COP) 19 established the Warsaw International Mechanism for Loss and Damage (WIM) as a formal part of the United Nations Framework Convention on Climate Change (UNFCCC) architecture (UNFCCC, 2014). It acknowledges that L&D 'includes, and in some cases involves more than, that which can be reduced by adaptation' (UNFCCC, 2014). The Paris Agreement recognized 'the importance of averting, minimizing and addressing loss and damage associated with the adverse effects of climate change' through Article 8 (UNFCCC, 2015).

There is no one definition of L&D in climate policy, and analysis of policy documents and stakeholder views has demonstrated ambiguity (Vanhala and Hestbaek, 2016; Boyd et al., 2017). UNFCCC documents suggest that L&D is associated with adverse impacts of climate change on human and natural systems, including impacts from extreme events and slow-onset processes (UNFCCC, 2011b, 2014, 2015). Some documents focus on impacts in developing or particularly vulnerable countries (UNFCCC, 2011b, 2014). They refer to economic (loss of assets and crops) and non-economic (biodiversity, culture, health) impacts, the latter also being an action area under the WIM workplan, and irreversible and permanent loss and damage. Lack of clarity of what the term addresses (avoidance through adaptation and mitigation, unavoidable losses, climate risk management, existential risk) was expressed among stakeholders, with further disagreement ensuing about what constitutes anthropogenic climate change versus natural climate variability (Boyd et al., 2017).

Limits to adaptation and residual risks

The AR5 described adaptation limits as points beyond which actors' objectives are compromised by intolerable risks threatening key objectives such as good health or broad levels of well-being, thus requiring transformative adaptation for overcoming soft limits (see Chapter 4, Sections 4.2.2.3, 4.5.3 and Cross-Chapter Box 9, Section 5.3.1) (Dow et al., 2013; Klein et al., 2014). The AR5 WGII risk tables, based on expert judgment, depicted the potential for, and the limits of, additional adaptation to reduce risk. Near-term (2030–2040) risks can be used as a proxy for 1.5°C warming by the end of the century and compared to longer-term (2080–2100) risks associated with an approximate 2°C warming. Building on the AR5 risk approach, Cross-Chapter Box 12, Figure 1 provides a stylised application example to poverty and inequality.

Cross-Chapter Box 12 (continued)

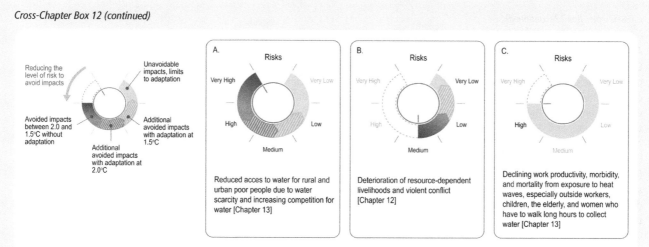

Cross-Chapter Box 12, Figure 1 | Stylized reduced risk levels due to avoided impacts between 2°C and 1.5°C warming (in solid red-orange), additional avoided impacts with adaptation under 2°C (striped orange) and under 1.5°C (striped yellow), and unavoidable impacts (losses) with no or very limited potential for adaptation (grey), extracted from the AR5 WGII risk tables (Field et al., 2014), and underlying chapters by Adger et al. (2014) and Olsson et al. (2014). For some systems and sectors (A), achieving 1.5°C could reduce risks to low (with adaptation) from very high (without adaptation) and high (with adaptation) under 2°C. For other areas (C), no or very limited adaptation potential is anticipated, suggesting limits, with the same risks for 1.5°C and 2°C. Other risks are projected to be medium under 2°C with further potential for reduction, especially with adaptation, to very low levels (B).

Limits to adaptation, residual risks, and losses in a 1.5°C warmer world

The literature on risks at 1.5°C (versus 2°C and more) and potentials for adaptation remains limited, particularly for specific regions, sectors, and vulnerable and disadvantaged populations. Adaptation potential at 1.5°C and 2°C is rarely assessed explicitly, making an assessment of residual risk challenging. Substantial progress has been made since the AR5 to assess which climate change impacts on natural and human systems can be attributed to anthropogenic emissions (Hansen and Stone, 2016) and to examine the influence of anthropogenic emissions on extreme weather events (NASEM, 2016), and on consequent impacts on human life (Mitchell et al., 2016), but less so on monetary losses and risks (Schaller et al., 2016). There has also been some limited research to examine local-level limits to adaptation (Warner and Geest, 2013; Filho and Nalau, 2018). What constitutes losses and damages is context-dependent and often requires place-based research into what people value and consider worth protecting (Barnett et al., 2016; Tschakert et al., 2017). Yet assessments of non-material and intangible losses are particularly challenging, such as loss of sense of place, belonging, identity, and damage to emotional and mental well-being (Serdeczny et al., 2017; Wewerinke-Singh, 2018a). Warming of 1.5°C is not considered 'safe' for most nations, communities, ecosystems and sectors, and poses significant risks to natural and human systems as compared to the current warming of 1°C (high confidence) (see Chapter 3, Section 3.4, Box 3.4, Box 3.5, Table 3.5, Cross-Chapter Box 6 in Chapter 3). Table 5.2, drawing on findings from Chapters 3, 4 and 5, presents examples of soft and hard limits in natural and human systems in the context of 1.5°C and 2°C of warming.

Cross-Chapter Box 12, Table 1 | Soft and hard adaptation limits in the context of 1.5°C and 2°C of global warming.

System/Region	Example	Soft Limit	Hard Limit
Coral reefs	Loss of 70–90% of tropical coral reefs by mid-century under 1.5°C scenario (total loss under 2°C scenario) (see Chapter 3, Sections 3.4.4 and 3.5.2.1, Box 3.4)		✓
Biodiversity	6% of insects, 8% of plants and 4% of vertebrates lose over 50% of the climatically determined geographic range at 1.5°C (18% of insects, 16% of plants and 8% of vertebrates at 2°C) (see Chapter 3, Section 3.4.3.3)		✓
Poverty	24–357 million people exposed to multi-sector climate risks and vulnerable to poverty at 1.5°C (86–1220 million at 2°C) (see Section 5.2.2)	✓	
Human health	Twice as many megacities exposed to heat stress at 1.5°C compared to present, potentially exposing 350 million additional people to deadly heat wave conditions by 2050 (see Chapter 3, Section 3.4.8)	✓	✓
Coastal livelihoods	Large-scale changes in oceanic systems (temperature and acidification) inflict damage and losses to livelihoods, income, cultural identity and health for coastal-dependent communities at 1.5°C (potential higher losses at 2°C) (see Chapter 3, Sections 3.4.4, 3.4.5, 3.4.6.3, Box 3.4, Box 3.5, Cross-Chapter Box 6, Chapter 4, Section 4.3.5; Section 5.2.3)	✓	✓
Small Island Developing States	Sea level rise and increased wave run up combined with increased aridity and decreased freshwater availability at 1.5°C warming potentially leaving several atoll islands uninhabitable (see Chapter 3, Sections 3.4.3, 3.4.5, Box 3.5, Chapter 4, Cross-Chapter Box 9)		✓

5

Cross-Chapter Box 12 (continued)

Approaches and policy options to address residual risk and loss and damage
Conceptual and applied work since the AR5 has highlighted the synergies and differences with adaptation and disaster risk reduction policies (van der Geest and Warner, 2015; Thomas and Benjamin, 2017), suggesting more integration of existing mechanisms, yet careful consideration is advised for slow-onset and potentially irreversible impacts and risk (Mechler and Schinko, 2016). Scholarship on justice and equity has provided insight on compensatory, distributive and procedural equity considerations for policy and practice to address loss and damage (Roser et al., 2015; Wallimann-Helmer, 2015; Huggel et al., 2016). A growing body of legal literature considers the role of litigation in preventing and addressing loss and damage and finds that litigation risks for governments and business are bound to increase with improved understanding of impacts and risks as climate science evolves (high confidence) (Mayer, 2016; Banda and Fulton, 2017; Marjanac and Patton, 2018; Wewerinke-Singh, 2018b). Policy proposals include international support for experienced losses and damages (Crosland et al., 2016; Page and Heyward, 2017), addressing climate displacement, donor-supported implementation of regional public insurance systems (Surminski et al., 2016) and new global governance systems under the UNFCCC (Biermann and Boas, 2017).

5.3 Climate Adaptation and Sustainable Development

Adaptation will be extremely important in a 1.5°C warmer world since substantial impacts will be felt in every region (*high confidence*) (Chapter 3, Section 3.3), even if adaptation needs will be lower than in a 2°C warmer world (see Chapter 4, Sections 4.3.1 to 4.3.5, 4.5.3, Cross-Chapter Box 10 in Chapter 4). Climate adaptation options comprise structural, physical, institutional and social responses, with their effectiveness depending largely on governance (see Glossary), political will, adaptive capacities and availability of finance (see Chapter 4, Sections 4.4.1 to 4.4.5) (Betzold and Weiler, 2017; Sonwa et al., 2017; Sovacool et al., 2017). Even though the literature is scarce on the expected impacts of future adaptation measures on sustainable development specific to warming experiences of 1.5°C, this section assesses available literature on how (i) prioritising sustainable development enhances or impedes climate adaptation efforts (Section 5.3.1); (ii) climate adaptation measures impact sustainable development and the SDGs in positive (synergies) or negative (trade-offs) ways (Section 5.3.2); and (iii) adaptation pathways towards a 1.5°C warmer world affect sustainable development, poverty and inequalities (Section 5.3.3). The section builds on Chapter 4 (see Section 4.3.5) regarding available adaptation options to reduce climate vulnerability and build resilience (see Glossary) in the context of 1.5°C-compatible trajectories, with emphasis on sustainable development implications.

5.3.1 Sustainable Development in Support of Climate Adaptation

Making sustainable development a priority, and meeting the SDGs, is consistent with efforts to adapt to climate change (*very high confidence*). Sustainable development is effective in building adaptive capacity if it addresses poverty and inequalities, social and economic exclusion, and inadequate institutional capacities (Noble et al., 2014; Abel et al., 2016; Colloff et al., 2017). Four ways in which sustainable development leads to effective adaptation are described below.

First, sustainable development enables transformational adaptation (see Chapter 4, Section 4.2.2.2) when an integrated approach is adopted, with inclusive, transparent decision-making, rather than addressing current vulnerabilities as stand-alone climate problems (Mathur et al., 2014; Arthurson and Baum, 2015; Shackleton et al., 2015; Lemos et al., 2016; Antwi-Agyei et al., 2017b). Ending poverty in its multiple dimensions (SDG 1) is often a highly effective form of climate adaptation (Fankhauser and McDermott, 2014; Leichenko and Silva, 2014; Hallegatte and Rozenberg, 2017). However, ending poverty is not sufficient, and the positive outcome as an adaptation strategy depends on whether increased household wealth is actually directed towards risk reduction and management strategies (Nelson et al., 2016), as shown in urban municipalities (Colenbrander et al., 2017; Rasch, 2017) and agrarian communities (Hashemi et al., 2017), and whether finance for adaptation is made available (Section 5.6.1).

Second, local participation is effective when wider socio-economic barriers are addressed via multiscale planning (McCubbin et al., 2015; Nyantakyi-Frimpong and Bezner-Kerr, 2015; Toole et al., 2016). This is the case, for instance, when national education efforts (SDG 4) (Muttarak and Lutz, 2014; Striessnig and Loichinger, 2015) and indigenous knowledge (Nkomwa et al., 2014; Pandey and Kumar, 2018) enhance information sharing, which also builds resilience (Santos et al., 2016; Martinez-Baron et al., 2018) and reduces risks for maladaptation (Antwi-Agyei et al., 2018; Gajjar et al., 2018).

Third, development promotes transformational adaptation when addressing social inequalities (Section 5.5.3, 5.6.4), as in SDGs 4, 5, 16 and 17 (O'Brien, 2016; O'Brien, 2017). For example, SDG 5 supports measures that reduce women's vulnerabilities and allow women to benefit from adaptation (Antwi-Agyei et al., 2015; Van Aelst and Holvoet, 2016; Cohen, 2017). Mobilization of climate finance, carbon taxation and environmentally motivated subsidies can reduce inequalities (SDG 10), advance climate mitigation and adaptation (Chancel and Picketty, 2015), and be conducive to strengthening and enabling environments for resilience building (Nhamo, 2016; Halonen et al., 2017).

Fourth, when sustainable development promotes livelihood security, it enhances the adaptive capacities of vulnerable communities and households. Examples include SDG 11 supporting adaptation in cities to reduce harm from disasters (Kelman, 2017; Parnell, 2017); access to water and sanitation (SDG 6) with strong institutions (SDG 16) (Rasul and Sharma, 2016); SDG 2 and its targets that promote adaptation in agricultural and food systems (Lipper et al., 2014); and targets for SDG 3 such as reducing infectious diseases and providing health cover are consistent with health-related adaptation (ICSU, 2017; Gomez-Echeverri, 2018).

Sustainable development has the potential to significantly reduce systemic vulnerability, enhance adaptive capacity and promote livelihood security for poor and disadvantaged populations (*high confidence*). Transformational adaptation (see Chapter 4, Sections 4.2.2.2 and 4.5.3) would require development that takes into consideration multidimensional poverty and entrenched inequalities, local cultural specificities and local knowledge in decision-making, thereby making it easier to achieve the SDGs in a 1.5°C warmer world (*medium evidence, high agreement*).

5.3.2 Synergies and Trade-Offs between Adaptation Options and Sustainable Development

There are short-, medium-, and long-term positive impacts (synergies) and negative impacts (trade-offs) between the dual goals of keeping temperatures below 1.5°C global warming and achieving sustainable development. The extent of synergies between development and adaptation goals will vary by the development process adopted for a particular SDG and underlying vulnerability contexts (*medium evidence, high agreement*). Overall, the impacts of adaptation on sustainable development, poverty eradication and reducing inequalities in general, and the SDGs specifically, are expected to be largely positive, given that the inherent purpose of adaptation is to lower risks. Building on Chapter 4 (see Section 4.3.5), this section examines synergies and trade-offs between adaptation and sustainable development for some key sectors and approaches.

Agricultural adaptation: The most direct synergy is between SDG 2 (zero hunger) and adaptation in cropping, livestock and food systems, designed to maintain or increase production (Lipper et al., 2014; Rockström et al., 2017). Farmers with effective adaptation strategies tend to enjoy higher food security and experience lower levels of poverty (FAO, 2015; Douxchamps et al., 2016; Ali and Erenstein, 2017). Vermeulen et al. (2016) report strong positive returns on investment across the world from agricultural adaptation with side benefits for environment and economic well-being. Well-adapted agricultural systems contribute to safe drinking water, health, biodiversity and equity goals (DeClerck et al., 2016; Myers et al., 2017). Climate-smart agriculture has synergies with food security, though it can be biased towards technological solutions, may not be gender sensitive, and can create specific challenges for institutional and distributional aspects (Lipper et al., 2014; Arakelyan et al., 2017; Taylor, 2017).

At the same time, adaptation options increase risks for human health, oceans and access to water if fertiliser and pesticides are used without regulation or when irrigation reduces water availability for other purposes (Shackleton et al., 2015; Campbell et al., 2016). When agricultural insurance and climate services overlook the poor, inequality may rise (Dinku et al., 2014; Carr and Owusu-Daaku, 2015; Georgeson

et al., 2017a; Carr and Onzere, 2018). Agricultural adaptation measures may increase workloads, especially for women, while changes in crop mix can result in loss of income or culturally inappropriate food (Carr and Thompson, 2014; Thompson-Hall et al., 2016; Bryan et al., 2017), and they may benefit farmers with more land to the detriment of land-poor farmers, as seen in the Mekong River Basin (see Chapter 3, Cross-Chapter Box 6 in Chapter 3).

Adaptation to protect human health: Adaptation options in the health sector are expected to reduce morbidity and mortality (Arbuthnott et al., 2016; Ebi and Otmani del Barrio, 2017). Heat-early-warning systems help lower injuries, illnesses and deaths (Hess and Ebi, 2016), with positive impacts for SDG 3. Institutions better equipped to share information, indicators for detecting climate-sensitive diseases, improved provision of basic health care services and coordination with other sectors also improve risk management, thus reducing adverse health outcomes (Dasgupta et al., 2016; Dovie et al., 2017). Effective adaptation creates synergies via basic public health measures (K.R. Smith et al., 2014; Dasgupta, 2016) and health infrastructure protected from extreme weather events (Watts et al., 2015). Yet trade-offs can occur when adaptation in one sector leads to negative impacts in another sector. Examples include the creation of urban wetlands through flood control measures which can breed mosquitoes, and migration eroding physical and mental well-being, hence adversely affecting SDG 3 (K.R. Smith et al., 2014; Watts et al., 2015). Similarly, increased use of air conditioning enhances resilience to heat stress (Petkova et al., 2017), yet it can result in higher energy consumption, undermining SDG 13.

Coastal adaptation: Adaptation to sea level rise remains essential in coastal areas even under a climate stabilization scenario of 1.5°C (Nicholls et al., 2018). Coastal adaptation to restore ecosystems (for instance by planting mangrove forests) supports SDGs for enhancing life and livelihoods on land and oceans (see Chapter 4, Sections 4.3.2.3). Synergistic outcomes between development and relocation of coastal communities are enhanced by participatory decision-making and settlement designs that promote equity and sustainability (van der Voorn et al., 2017). Limits to coastal adaptation may rise, for instance in low-lying islands in the Pacific, Caribbean and Indian Ocean, with attendant implications for loss and damage (see Chapter 3 Box 3.5, Chapter 4, Cross-Chapter Box 9 in Chapter 4, Cross-Chapter Box 12 in Chapter 5, Box 5.3).

Migration as adaptation: Migration has been used in various contexts to protect livelihoods from challenges related to climate change (Marsh, 2015; Jha et al., 2017), including through remittances (Betzold and Weiler, 2017). Synergies between migration and the achievement of sustainable development depend on adaptive measures and conditions in both sending and receiving regions (Fatima et al., 2014; McNamara, 2015; Entzinger and Scholten, 2016; Ober and Sakdapolrak, 2017; Schwan and Yu, 2017). Adverse developmental impacts arise when vulnerable women or the elderly are left behind or if migration is culturally disruptive (Wilkinson et al., 2016; Albert et al., 2017; Islam and Shamsuddoha, 2017).

Ecosystem-based adaptation: Ecosystem-based adaptation (EBA) can offer synergies with sustainable development (Morita and Matsumoto,

2015; Ojea, 2015; Szabo et al., 2015; Brink et al., 2016; Butt et al., 2016; Conservation International, 2016; Huq et al., 2017), although assessments remain difficult (see Chapter 4, Section 4.3.2.2) (Doswald et al., 2014). Examples include mangrove restoration reducing coastal vulnerability, protecting marine and terrestrial ecosystems, and increasing local food security, as well as watershed management reducing flood risks and improving water quality (Chong, 2014). In drylands, EBA practices, combined with community-based adaptation, have shown how to link adaptation with mitigation to improve livelihood conditions of poor farmers (Box 5.1). Synergistic developmental outcomes arise where EBA is cost effective, inclusive of indigenous and local knowledge and easily accessible by the poor (Ojea, 2015; Daigneault et al., 2016; Estrella et al., 2016). Payment for ecosystem services can provide incentives to land owners and natural resource managers to preserve environmental services with synergies with SDGs 1 and 13 (Arriagada et al., 2015), when implementation challenges are overcome (Calvet-Mir et al., 2015; Wegner, 2016; Chan et al., 2017). Trade-offs include loss of other economic land use types, tension between biodiversity and adaptation priorities, and conflicts over governance (Wamsler et al., 2014; Ojea, 2015).

Community-based adaptation: Community-based adaptation (CBA) (see Chapter 4, Sections 4.3.3.2) enhances resilience and sustainability of adaptation plans (Ford et al., 2016; Fernandes-Jesus et al., 2017; Grantham and Rudd, 2017; Gustafson et al., 2017). Yet negative impacts occur if it fails to fairly represent vulnerable populations and to foster long-term social resilience (Ensor, 2016; Taylor Aiken et al., 2017). Mainstreaming CBA into planning and decision-making enables the attainment of SDGs 5, 10 and 16 (Archer et al., 2014; Reid and Huq, 2014; Vardakoulias and Nicholles, 2014; Cutter, 2016; Kim et al., 2017). Incorporating multiple forms of indigenous and local knowledge is an important element of CBA, as shown for instance in the Arctic region (see Chapter 4, Section 4.3.5.5, Box 4.3, Cross-Chapter Box 9) (Apgar et al., 2015; Armitage, 2015; Pearce et al., 2015; Chief et al., 2016; Cobbinah and Anane, 2016; Ford et al., 2016). Indigenous and local knowledge can be synergistic with achieving SDGs 2, 6 and 10 (Ayers et al., 2014; Lasage et al., 2015; Regmi and Star, 2015; Berner et al., 2016; Chief et al., 2016; Murtinho, 2016; Reid, 2016).

There are clear synergies between adaptation options and several SDGs, such as poverty eradication, elimination of hunger, clean water and health (*robust evidence, high agreement*), as well-integrated adaptation supports sustainable development (Eakin et al., 2014; Weisser et al., 2014; Adam, 2015; Smucker et al., 2015). Substantial synergies are observed in the agricultural and health sectors, and in ecosystem-based adaptations. However, particular adaptation strategies can lead to adverse consequences for developmental outcomes (*medium evidence, high agreement*). Adaptation strategies that advance one SDG can result in trade-offs with other SDGs; for instance, agricultural adaptation to enhance food security (SDG 2) causing negative impacts for health, equality and healthy ecosystems (SDGs 3, 5, 6, 10, 14 and 15), and resilience to heat stress increasing energy consumption (SDGs 3 and 7) and high-cost adaptation in resource-constrained contexts (*medium evidence, medium agreement*).

5.3.3 Adaptation Pathways towards a 1.5°C Warmer World and Implications for Inequalities

In a 1.5°C warmer world, adaptation measures and options would need to be intensified, accelerated and scaled up. This entails not only the right 'mix' of options (asking 'right for whom and for what?') but also a forward-looking understanding of dynamic trajectories, that is adaptation pathways (see Chapter 1, Cross-Chapter Box 1 in Chapter 1), best understood as decision-making processes over sets of potential action sequenced over time (Câmpeanu and Fazey, 2014; Wise et al., 2014). Given the scarcity of literature on adaptation pathways that navigate place-specific warming experiences at 1.5°C, this section presents insights into current local decision-making for adaptation futures. This grounded evidence shows that choices between possible pathways, at different scales and for different groups of people, are shaped by uneven power structures and historical legacies that create their own, often unforeseen change (Fazey et al., 2016; Bosomworth et al., 2017; Lin et al., 2017; Murphy et al., 2017; Pelling et al., 2018).

Pursuing a place-specific adaptation pathway approach towards a 1.5°C warmer world harbours the potential for significant positive outcomes, with synergies for well-being possibilities to 'leap-frog the SDGs' (J.R.A. Butler et al., 2016), in countries at all levels of development (*medium evidence, high agreement*). It allows for identifying local, socially salient tipping points before they are crossed, based on what people value and trade-offs that are acceptable to them (Barnett et al., 2014, 2016; Gorddard et al., 2016; Tschakert et al., 2017). Yet evidence also reveals adverse impacts that reinforce rather than reduce existing social inequalities and hence may lead to poverty traps (*medium evidence, high agreement*) (Nagoda, 2015; Warner et al., 2015; Barnett et al., 2016; J.R.A. Butler et al., 2016; Godfrey-Wood and Naess, 2016; Pelling et al., 2016; Albert et al., 2017; Murphy et al., 2017).

Past development trajectories as well as transformational adaptation plans can constrain adaptation futures by reinforcing dominant political-economic structures and processes, and narrowing option spaces; this leads to maladaptive pathways that preclude alternative, locally relevant and sustainable development initiatives and increase vulnerabilities (Warner and Kuzdas, 2017; Gajjar et al., 2018). Such dominant pathways tend to validate the practices, visions and values of existing governance regimes and powerful members of a community while devaluing those of less privileged stakeholders. Examples from Romania, the Solomon Islands and Australia illustrate such pathway dynamics in which individual economic gains and prosperity matter more than community cohesion and solidarity; this discourages innovation, exacerbates inequalities and further erodes adaptive capacities of the most vulnerable (Davies et al., 2014; Fazey et al., 2016; Bosomworth et al., 2017). In the city of London, United Kingdom, the dominant adaptation and disaster risk management pathway promotes resilience that emphasizes self-reliance; yet it intensifies the burden on low-income citizens, the elderly, migrants and others unable to afford flood insurance or protect themselves against heat waves (Pelling et al., 2016). Adaptation pathways in the Bolivian Altiplano have transformed subsistence farmers into world-leading quinoa producers, but loss of social cohesion and traditional values, dispossession and loss of ecosystem services now constitute undesirable trade-offs (Chelleri et al., 2016).

A narrow view of adaptation decision-making, for example focused on technical solutions, tends to crowd out more participatory processes (Lawrence and Haasnoot, 2017; Lin et al., 2017), obscures contested values and reinforces power asymmetries (Bosomworth et al., 2017; Singh, 2018). A situated and context-specific understanding of adaptation pathways that galvanizes diverse knowledge, values and joint initiatives helps to overcome dominant path dependencies, avoid trade-offs that intensify inequities and challenge policies detached from place (Fincher et al., 2014; Wyborn et al., 2015; Murphy et al., 2017; Gajjar et al., 2018). These insights suggest that adaptation pathway approaches to prepare for 1.5°C warmer futures would be difficult to achieve without considerations for inclusiveness, place-specific trade-off deliberations, redistributive measures and procedural justice mechanisms to facilitate equitable transformation (*medium evidence, high agreement*).

Box 5.1 | Ecosystem- and Community-Based Practices in Drylands

Drylands face severe challenges in building climate resilience (Fuller and Lain, 2017), yet small-scale farmers can play a crucial role as agents of change through ecosystem- and community-based practices that combine adaptation, mitigation and sustainable development.

Farmer managed natural regeneration (FMNR) of trees in cropland is practised in 18 countries across sub-Saharan Africa, Southeast Asia, Timor-Leste, India and Haiti and has, for example, permitted the restoration of over five million hectares of land in the Sahel (Niang et al., 2014; Bado et al., 2016). In Ethiopia, the Managing Environmental Resources to Enable Transitions programme, which entails community-based watershed rehabilitation in rural landscapes, supported around 648,000 people, resulting in the rehabilitation of 25,400,000 hectares of land in 72 severely food-insecure districts across Ethiopia between 2012 and 2015 (Gebrehaweria et al., 2016). In India, local farmers have benefitted from watershed programmes across different agro-ecological regions (Singh et al., 2014; Datta, 2015).

These low-cost, flexible community-based practices represent low-regrets adaptation and mitigation strategies. These strategies often contribute to strengthened ecosystem resilience and biodiversity, increased agricultural productivity and food security, reduced household poverty and drudgery for women, and enhanced agency and social capital (Niang et al., 2014; Francis et al., 2015; Kassie et al., 2015; Mbow et al., 2015; Reij and Winterbottom, 2015; Weston et al., 2015; Bado et al., 2016; Dumont et al., 2017). Small check dams in dryland areas and conservation agriculture can significantly increase agricultural output (Kumar et al., 2014; Agoramoorthy and Hsu, 2016; Pradhan et al., 2018). Mitigation benefits have also been quantified (Weston et al., 2015); for example, FMNR of more than five million hectares in Niger has sequestered 25–30 Mtonnes of carbon over 30 years (Stevens et al., 2014).

However, several constraints hinder scaling-up efforts: inadequate attention to the socio-technical processes of innovation (Grist et al., 2017; Scoones et al., 2017), difficulties in measuring the benefits of an innovation (Coe et al., 2017), farmers' inability to deal with long-term climate risk (Singh et al., 2017), and difficulties for matching practices with agro-ecological conditions and complementary modern inputs (Kassie et al., 2015). Key conditions to overcome these challenges include: developing agroforestry value chains and markets (Reij and Winterbottom, 2015) and adaptive planning and management (Gray et al., 2016). Others include inclusive processes giving greater voice to women and marginalized groups (MRFCJ, 2015a; UN Women and MRFCJ, 2016; Dumont et al., 2017), strengthening community land and forest rights (Stevens et al., 2014; Vermeulen et al., 2016), and co-learning among communities of practice at different scales (Coe et al., 2014; Reij and Winterbottom, 2015; Sinclair, 2016; Binam et al., 2017; Dumont et al., 2017; Epule et al., 2017).

5.4   Mitigation and Sustainable Development

The AR5 WGIII examined the potential of various mitigation options for specific sectors (energy supply, industry, buildings, transport, and agriculture, forestry, and other land use; AFOLU); it provided a narrative of dimensions of sustainable development and equity as a framing for evaluating climate responses and policies, respectively, in Chapters 4, 7, 8, 9, 10 and 11 (IPCC, 2014a). This section builds on the analyses of Chapters 2 and 4 of this report to re-assess mitigation and sustainable development in the context of 1.5°C global warming as well as the SDGs.

5.4.1   Synergies and Trade-Offs between Mitigation Options and Sustainable Development

Adopting stringent climate mitigation options can generate multiple positive non-climate benefits that have the potential to reduce the costs of achieving sustainable development (IPCC, 2014b; Ürge-Vorsatz et al., 2014, 2016; Schaeffer et al., 2015; von Stechow et al., 2015). Understanding the positive impacts (synergies) but also the negative impacts (trade-offs) is key for selecting mitigation options and policy choices that maximize the synergies between mitigation and developmental actions (Hildingsson and Johansson, 2015; Nilsson et al., 2016; Delponte et al., 2017; van Vuuren et al., 2017b; McCollum et al., 2018b). Aligning mitigation response options to sustainable development objectives can ensure public acceptance (IPCC, 2014a), encourage faster action (Lechtenboehmer and Knoop, 2017) and support the design of equitable mitigation (Holz et al., 2018; Winkler et al., 2018) that protect human rights (MRFCJ, 2015b) (Section 5.5.3).

This sub-section assesses available literature on the interactions of individual mitigation options (see Chapter 2, Section 2.3.1.2, Chapter 4, Sections 4.2 and 4.3) with sustainable development and the SDGs and underlying targets. Table 5.2 presents an assessment of these synergies and trade-offs and the strength of the interaction using an SDG-interaction score (see Glossary) (McCollum et al., 2018b), with evidence and agreements levels. Figure 5.2 presents the information of Table 5.2, showing gross (not net) interactions with the SDGs. This detailed assessment of synergies and trade-offs of individual mitigation options with the SDGs (Table 5.2 a–d and Figure 5.2) reveals that the number of synergies exceeds that of trade-offs. Mitigation response options in the energy demand sector, AFOLU and oceans have more positive interactions with a larger number of SDGs compared to those on the energy supply side (*robust evidence, high agreement*).

5.4.1.1 Energy Demand: Mitigation Options to Accelerate Reduction in Energy Use and Fuel Switch

For mitigation options in the energy demand sectors, the number of synergies with all sixteen SDGs exceeds the number of trade-offs (Figure 5.2 and Table 5.2) (*robust evidence, high agreement*). Most of the interactions are of a reinforcing nature, hence facilitating the achievement of the goals.

Accelerating energy efficiency in all sectors, which is a necessary condition for a 1.5°C warmer world (see Chapters 2 and 4), has synergies with a large number of SDGs (*robust evidence, high agreement*) (Figure 5.2 and Table 5.2). The diffusion of efficient equipment and appliances across end use sectors has synergies with international partnership (SDG 17) and participatory and transparent institutions (SDG 16) because innovations and deployment of new technologies require transnational capacity building and knowledge sharing. Resource and energy savings support sustainable production and consumption (SDG 12), energy access (SDG 7), innovation and infrastructure development (SDG 9) and sustainable city development (SDG 11). Energy efficiency supports the creation of decent jobs by new service companies providing services for energy efficiency, but the net employment effect of efficiency improvement remains uncertain due to macro-economic feedback (SDG 8) (McCollum et al., 2018b).

In the buildings sector, accelerating energy efficiency by way of, for example, enhancing the use of efficient appliances, refrigerant transition, insulation, retrofitting and low- or zero-energy buildings generates benefits across multiple SDG targets. For example, improved cook stoves make fuel endowments last longer and hence reduce deforestation (SDG 15), support equal opportunity by reducing school absences due to asthma among children (SDGs 3 and 4) and empower rural and indigenous women by reducing drudgery (SDG 5) (*robust evidence, high agreement*) (Derbez et al., 2014; Lucon et al., 2014; Maidment et al., 2014; Scott et al., 2014; Cameron et al.,

2015; Fay et al., 2015; Liddell and Guiney, 2015; Shah et al., 2015; Sharpe et al., 2015; Wells et al., 2015; Willand et al., 2015; Hallegatte et al., 2016; Kusumaningtyas and Aldrian, 2016; Berrueta et al., 2017; McCollum et al., 2018a).

In energy-intensive processing industries, 1.5°C-compatible trajectories require radical technology innovation through maximum electrification, shift to other low emissions energy carriers such as hydrogen or biomass, integration of carbon capture and storage (CCS) and innovations for carbon capture and utilization (CCU) (see Chapter 4, Section 4.3.4.5). These transformations have strong synergies with innovation and sustainable industrialization (SDG 9), supranational partnerships (SDGs 16 and 17) and sustainable production (SDG 12). However, possible trade-offs due to risks of CCS-based carbon leakage, increased electricity demands, and associated price impacts affecting energy access and poverty (SDGs 7 and 1) would need careful regulatory attention (Wesseling et al., 2017). In the mining industry, energy efficiency can be synergetic or face trade-offs with sustainable management (SDG 6), depending on the option retained for water management (Nguyen et al., 2014). Substitution and recycling are also an important driver of 1.5°C-compatible trajectories in industrial systems (see Chapter 4, Section 4.3.4.2). Structural changes and reorganization of economic activities in industrial park/clusters following the principles of industrial symbiosis (circular economy) improves the overall sustainability by reducing energy and waste (Fan et al., 2017; Preston and Lehne, 2017) and reinforces responsible production and consumption (SDG 12) through recycling, water use efficiency (SDG 6), energy access (SDG 7) and ecosystem protection and restoration (SDG 15) (Karner et al., 2015; Zeng et al., 2017).

In the transport sector, deep electrification may trigger increases of electricity prices and adversely affect poor populations (SDG 1), unless pro-poor redistributive policies are in place (Klausbruckner et al., 2016). In cities, governments can lay the foundations for compact, connected low-carbon cities, which are an important component of 1.5°C-compatible transformations (see Chapter 4, Section 4.3.3) and show synergies with sustainable cities (SDG 11) (Colenbrander et al., 2016).

Behavioural responses are important determinants of the ultimate outcome of energy efficiency on emission reductions and energy access (SDG 7) and their management requires a detailed understanding of the drivers of consumption and the potential for and barriers to absolute reductions (Fuchs et al., 2016). Notably, the rebound effect tends to offset the benefits of efficiency for emissions reductions through growing demand for energy services (Sorrell, 2015; Suffolk and Poortinga, 2016). However, high rebound can help in providing faster access to affordable energy (SDG 7.1) where the goal is to reduce energy poverty and unmet energy demand (see Chapter 2, Section 2.4.3) (Chakravarty et al., 2013). Comprehensive policy design – including rebound supressing policies, such as carbon pricing and policies that encourage awareness building and promotional material design – is needed to tap the full potential of energy savings, as applicable to a 1.5°C warming context (Chakravarty and Tavoni, 2013; IPCC, 2014b; Karner et al., 2015; Zhang et al., 2015; Altieri et al., 2016; Santarius et al., 2016) and to address policy-related trade-offs and welfare-enhancing benefits (*robust evidence, high agreement*) (Chakravarty et al., 2013; Chakravarty and Roy, 2016; Gillingham et al., 2016).

Other behavioural responses will affect the interplay between energy efficiency and sustainable development. Building occupants reluctant to change their habits may miss out on welfare-enhancing energy efficiency opportunities (Zhao et al., 2017). Preferences for new products and premature obsolescence for appliances is expected to adversely affect sustainable consumption and production (SDG 12) with ramifications for resource use efficiency (Echegaray, 2016). Changes in user behaviour towards increased physical activity, less reliance on motorized travel over short distances and the use of public transport would help to decarbonize the transport sector in a synergetic manner with SDGs 3, 11 and 12 (Shaw et al., 2014; Ajanovic, 2015; Chakrabarti and Shin, 2017), while reducing inequality in access to basic facilities (SDG 10) (Lucas and Pangbourne, 2014; Kagawa et al., 2015). However, infrastructure design and regulations would need to ensure road safety and address risks of road accidents for pedestrians (Hwang et al., 2017; Khreis et al., 2017) to ensure sustainable infrastructure growth in human settlements (SDGs 9 and 11) (Lin et al., 2015; SLoCaT, 2017).

5.4.1.2 Energy Supply: Accelerated Decarbonization

Decreasing the share of coal in energy supply in line with 1.5°C-compatible scenarios (see Chapter 2, Section 2.4.2) reduces adverse impacts of upstream supply-chain activities, in particular air and water pollution and coal mining accidents, and enhances health by reducing air pollution, notably in cities, showing synergies with SDGs 3, 11 and 12 (Yang et al., 2016; UNEP, 2017).

Fast deployment of renewables such as solar, wind, hydro and modern biomass, together with the decrease of fossil fuels in energy supply (see Chapter 2, Section 2.4.2.1), is aligned with the doubling of renewables in the global energy mix (SDG 7.2). Renewables could also support progress on SDGs 1, 10, 11 and 12 and supplement new technology (*robust evidence, high agreement*) (Chaturvedi and Shukla, 2014; Rose et al., 2014; Smith and Sagar, 2014; Riahi et al., 2015; IEA, 2016; van Vuuren et al., 2017a; McCollum et al., 2018a). However, some trade-offs with the SDGs can emerge from offshore installations, particularly SDG 14 in local contexts (McCollum et al., 2018a). Moreover, trade-offs between renewable energy production and affordability (SDG 7) (Labordena et al., 2017) and other environmental objectives would need to be scrutinised for potential negative social outcomes. Policy interventions through regional cooperation-building (SDG 17) and institutional capacity (SDG 16) can enhance affordability (SDG 7) (Labordena et al., 2017). The deployment of small-scale renewables, or off-grid solutions for people in remote areas (Sánchez and Izzo, 2017), has strong potential for synergies with access to energy (SDG 7), but the actualization of these potentials requires measures to overcome technology and reliability risks associated with large-scale deployment of renewables (Giwa et al., 2017; Heard et al., 2017). Bundling energy-efficient appliances and lighting with off-grid renewables can lead to substantial cost reduction while increasing reliability (IEA, 2017). Low-income populations in industrialized countries are often left out of renewable energy generation schemes, either because of high start-up costs or lack of home ownership (UNRISD, 2016).

Nuclear energy, the share of which increases in most of the 1.5°C-compatible pathways (see Chapter 2, Section 2.4.2.1), can increase the risks of proliferation (SDG 16), have negative environmental effects (e.g., for water use; SDG 6) and have mixed effects for human health when replacing fossil fuels (SDGs 7 and 3) (see Table 5.2). The use of fossil CCS, which plays an important role in deep mitigation pathways (see Chapter 2, Section 2.4.2.3), implies continued adverse impacts of upstream supply-chain activities in the coal sector, and because of lower efficiency of CCS coal power plants (SDG 12), upstream impacts and local air pollution are likely to be exacerbated (SDG 3). Furthermore, there is a non-negligible risk of carbon dioxide leakage from geological storage and the carbon dioxide transport infrastructure (SDG 3) (Table 5.2).

Economies dependent upon fossil fuel-based energy generation and/or export revenue are expected to be disproportionally affected by future restrictions on the use of fossil fuels under stringent climate goals and higher carbon prices; this includes impacts on employment, stranded assets, resources left underground, lower capacity use and early phasing out of large infrastructure already under construction (*robust evidence, high agreement*) (Box 5.2) (Johnson et al., 2015; McGlade and Ekins, 2015; UNEP, 2017; Spencer et al., 2018). Investment in coal continues to be attractive in many countries as it is a mature technology and provides cheap energy supplies, large-scale employment and energy security (Jakob and Steckel, 2016; Vogt-Schilb and Hallegatte, 2017; Spencer et al., 2018). Hence, accompanying policies and measures would be required to ease job losses and correct for relatively higher prices of alternative energy (Oosterhuis and Ten Brink, 2014; Oei and Mendelevitch, 2016; Garg et al., 2017; HLCCP, 2017; Jordaan et al., 2017; OECD, 2017; UNEP, 2017; Blondeel and van de Graaf, 2018; Green, 2018). Research on historical transitions shows that managing the impacts on workers through retraining programmes is essential in order to align the phase-down of mining industries with meeting ambitious climate targets, and the objectives of a 'just transition' (Galgóczi, 2014; Caldecott et al., 2017; Healy and Barry, 2017). This aspect is even more important in developing countries where the mining workforce is largely semi- or unskilled (Altieri et al., 2016; Tung, 2016). Ambitious emissions reduction targets can unlock very strong decoupling potentials in industrialized fossil exporting economies (Hatfield-Dodds et al., 2015).

5

Box 5.2 | Challenges and Opportunities of Low-Carbon Pathways in Gulf Cooperative Council Countries

The Gulf Cooperative Council (GCC) region (Bahrain, Kuwait, Oman, Qatar, Saudi Arabia and United Arab Emirates) is characterized by high dependency on hydrocarbon resources (natural oil and gas), with high risks of socio-economic impacts of policies and response measures to address climate change. The region is also vulnerable to the decrease of the global demand and price of hydrocarbons as a result of climate change response measures. The projected declining use of oil and gas under low emissions pathways creates risks of significant economic losses for the GCC region (e.g., Waisman et al., 2013; Van de Graaf and Verbruggen, 2015; Al-Maamary et al., 2016; Bauer et al., 2016), given that natural gas and oil revenues contributed to about 70% of government budgets and > 35% of the gross domestic product in 2010 (Callen et al., 2014).

The current high energy intensity of the domestic economies (Al-Maamary et al., 2017), triggered mainly by low domestic energy prices (Alshehry and Belloumi, 2015), suggests specific challenges for aligning mitigation towards 1.5°C-consistent trajectories, which would require strong energy efficiency and economic development for the region.

The region's economies are highly reliant on fossil fuel for their domestic activities. Yet the renewables deployment potentials are large, deployment is already happening (Cugurullo, 2013; IRENA, 2016) and positive economic benefits can be envisaged (Sgouridis et al., 2016). Nonetheless, the use of renewables is currently limited by economics and structural challenges (Lilliestam and Patt, 2015; Griffiths, 2017a). Carbon capture and storage (CCS) is also envisaged with concrete steps towards implementation (Alsheyab, 2017; Ustadi et al., 2017); yet the real potential of this technology in terms of scale and economic dimensions is still uncertain.

Beyond the above mitigation-related challenges, the region's human societies and fragile ecosystems are highly vulnerable to the impacts of climate change, such as water stress (Evans et al., 2004; Shaffrey et al., 2009), desertification (Bayram and Öztürk, 2014), sea level rise affecting vast low coastal lands, and high temperature and humidity with future levels potentially beyond adaptive capacities (Pal and Eltahir, 2016). A low-carbon pathway that manages climate-related risks within the context of sustainable development requires an approach that jointly addresses both types of vulnerabilities (Al Ansari, 2013; Lilliestam and Patt, 2015; Babiker, 2016; Griffiths, 2017b).

The Nationally Determined Contributions (NDCs) for GCC countries identified energy efficiency, deployment of renewables and technology transfer to enhance agriculture, food security, protection of marine resources, and management of water and costal zones (Babiker, 2016). Strategic vision documents, such as Saudi Arabia's 'Vision 2030', identify emergent opportunities for energy price reforms, energy efficiency, turning emissions into valuable products, and deployment of renewables and other clean technologies, if accompanied with appropriate policies to manage the transition and in the context of economic diversification (Luomi, 2014; Atalay et al., 2016; Griffiths, 2017b; Howarth et al., 2017).

5.4.1.3 Land-based agriculture, forestry and ocean: mitigation response options and carbon dioxide removal

In the AFOLU sector, dietary change towards global healthy diets, that is, a shift from over-consumption of animal-related to plant-related diets, and food waste reduction (see Chapter 4, Section 4.3.2.1) are in synergy with SDGs 2 and 6, and SDG 3 through lower consumption of animal products and reduced losses and waste throughout the food system, contributing to achieving SDGs 12 and 15 (Bajželj et al., 2014; Bustamante et al., 2014; Tilman and Clark, 2014; Hiç et al., 2016).

Power dynamics play an important role in achieving behavioural change and sustainable consumption (Fuchs et al., 2016). In forest management (see Chapter 4, Section 4.3.2.2), encouraging responsible sourcing of forest products and securing indigenous land tenure has the potential to increase economic benefits by creating decent jobs (SDG 8), maintaining biodiversity (SDG 15), facilitating innovation and upgrading technology (SDG 9), and encouraging responsible and just decision-making (SDG 16) (*medium evidence, high agreement*) (Ding et al., 2016; WWF, 2017).

Emerging evidence indicates that future mitigation efforts that would be required to reach stringent climate targets, particularly those associated with carbon dioxide removal (CDR) (e.g., afforestation and reforestation and bioenergy with carbon capture and storage; BECCS), may also impose significant constraints upon poor and vulnerable communities (SDG 1) via increased food prices and competition for arable land, land appropriation and dispossession (Cavanagh and Benjaminsen, 2014; Hunsberger et al., 2014; Work, 2015; Muratori et al., 2016; Smith et al., 2016; Burns and Nicholson, 2017; Corbera et al., 2017) with disproportionate negative impacts upon rural poor and indigenous populations (SDG 1) (*robust evidence, high agreement*) (Section 5.4.2.2, Table 5.2, Figure 5.2) (Grubert et al., 2014; Grill et al., 2015; Zhang and Chen, 2015; Fricko et al., 2016; Johansson et al., 2016; Aha and Ayitey, 2017; De Stefano et al., 2017; Shi et al., 2017). Crops for bioenergy may increase irrigation needs and exacerbate water stress with negative associated impacts on SDGs 6 and 10 (Boysen et al., 2017).

Ocean iron fertilization and enhanced weathering have two-way interactions with life under water and on land and food security (SDGs

2, 14 and 15) (Table 5.2). Development of blue carbon resources through coastal (mangrove) and marine (seaweed) vegetative ecosystems encourages: integrated water resource management (SDG 6) (Vierros, 2017); promotes life on land (SDG 15) (Potouroglou et al., 2017); poverty

reduction (SDG 1) (Schirmer and Bull, 2014; Lamb et al., 2016); and food security (SDG 2) (Ahmed et al., 2017a, b; Duarte et al., 2017; Sondak et al., 2017; Vierros, 2017; Zhang et al., 2017).

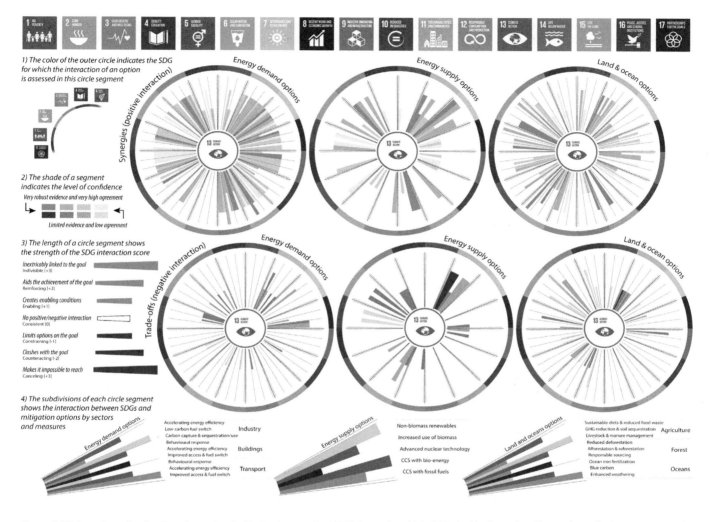

Figure 5.2 | Synergies and trade-offs and gross Sustainable Development Goal (SDG)-interaction with individual mitigation options. The top three wheels represent synergies and the bottom three wheels show trade-offs. The colours on the border of the wheels correspond to the SDGs listed above, starting at the 9 o'clock position, with reading guidance in the top-left corner with the quarter circle (Note 1). Mitigation (climate action, SDG 13) is at the centre of the circle. The coloured segments inside the circles can be counted to arrive at the number of synergies (green) and trade-offs (red). The length of the coloured segments shows the strength of the synergies or trade-offs (Note 3) and the shading indicates confidence (Note 2). Various mitigation options within the energy demand sector, energy supply sector, and land and ocean sector, and how to read them within a segment are shown in grey (Note 4). See also Table 5.2.

5.4.2 Sustainable Development Implications of 1.5°C and 2°C Mitigation Pathways

While previous sections have focused on individual mitigation options and their interaction with sustainable development and the SDGs, this section takes a systems perspective. Emphasis is on quantitative pathways depicting path-dependent evolutions of human and natural systems over time. Specifically, the focus is on fundamental transformations and thus stringent mitigation policies consistent with 1.5°C or 2°C, and the differential synergies and trade-offs with respect to the various sustainable development dimensions.

Both 1.5°C and 2°C pathways would require deep cuts in greenhouse gas (GHG) emissions and large-scale changes of energy supply and demand, as well as in agriculture and forestry systems (see Chapter 2, Section 2.4). For the assessment of the sustainable development implications of these pathways, this chapter draws upon studies that show the aggregated impact of mitigation for multiple sustainable development dimensions (Grubler et al., 2018; McCollum et al., 2018b; Rogelj et al., 2018) and across multiple integrated assessment modelling (IAM) frameworks. Often these tools are linked to disciplinary models covering specific SDGs in more detail (Cameron et al., 2016; Rao et al., 2017; Grubler et al., 2018; McCollum et al.,

2018b). Using multiple IAMs and disciplinary models is important for a robust assessment of the sustainable development implications of different pathways. Emphasis is on multi-regional studies, which can be aggregated to the global scale. The recent literature on 1.5°C mitigation pathways has begun to provide quantifications for a range of sustainable development dimensions, including air pollution and health, food security and hunger, energy access, water security, and multidimensional poverty and equity.

5.4.2.1   Air pollution and health

GHGs and air pollutants are typically emitted by the same sources. Hence, mitigation strategies that reduce GHGs or the use of fossil fuels typically also reduce emissions of pollutants, such as particulate matter (e.g., PM2.5 and PM10), black carbon (BC), sulphur dioxide (SO_2), nitrogen oxides (NO_x) and other harmful species (Clarke et al., 2014) (Figure 5.3), causing adverse health and ecosystem effects at various scales (Kusumaningtyas and Aldrian, 2016).

Mitigation pathways typically show that there are significant synergies for air pollution, and that the synergies increase with the stringency of the mitigation policies (Amann et al., 2011; Rao et al., 2016; Klimont et al., 2017; Shindell et al., 2017; Markandya et al., 2018). Recent multimodel comparisons indicate that mitigation pathways consistent with 1.5°C would result in higher synergies with air pollution compared to pathways that are consistent with 2°C (Figures 5.4 and 5.5). Shindell et al. (2018) indicate that health benefits worldwide over the century of 1.5°C pathways could be in the range of 110 to 190 million fewer premature deaths compared to 2°C pathways. The synergies for air pollution are highest in the developing world, particularly in Asia. In addition to significant health benefits, there are also economic benefits from mitigation, reducing the investment needs in air pollution control technologies by about 35% globally (or about 100 billion USD2010 per year to 2030 in 1.5°C pathways; McCollum et al., 2018b) (Figure 5.4).

5.4.2.2   Food security and hunger

Stringent climate mitigation pathways in line with 'well below 2°C' or '1.5°C' goals often rely on the deployment of large-scale land-related measures, like afforestation and/or bioenergy supply (Popp et al., 2014; Rose et al., 2014; Creutzig et al., 2015). These land-related measures can compete with food production and hence raise food security concerns (Section 5.4.1.3) (P. Smith et al., 2014). Mitigation studies indicate that so-called 'single-minded' climate policy, aiming solely at limiting warming to 1.5°C or 2°C without concurrent measures in the food sector, can have negative impacts for global food security (Hasegawa et al., 2015; McCollum et al., 2018b). Impacts of 1.5°C mitigation pathways can be significantly higher than those of 2°C pathways (Figures 5.4 and 5.5). An important driver of the food security impacts in these scenarios is the increase of food prices and the effect of mitigation on disposable income and wealth due to GHG pricing. A recent study indicates that, on aggregate, the price and income effects on food may be bigger than the effect due to competition over land between food and bioenergy (Hasegawa et al., 2015).

In order to address the issue of trade-offs with food security, mitigation policies would need to be designed in a way that shields the population

at risk of hunger, including through the adoption of different complementary measures, such as food price support. The investment needs of complementary food price policies are found to be globally relatively much smaller than the associated mitigation investments of 1.5°C pathways (Figure 5.3) (McCollum et al., 2018b). Besides food support price, other measures include improving productivity and efficiency of agricultural production systems (FAO and NZAGRC, 2017a, b; Frank et al., 2017) and programmes focusing on forest land-use change (Havlík et al., 2014). All these lead to additional benefits of mitigation, improving resilience and livelihoods.

Van Vuuren et al. (2018) and Grubler et al. (2018) show that 1.5°C pathways without reliance on BECCS can be achieved through a fundamental transformation of the service sectors which would significantly reduce energy and food demand (see Chapter 2, Sections 2.1.1, 2.3.1 and 2.4.3). Such low energy demand (LED) pathways would result in significantly reduced pressure on food security, lower food prices and fewer people at risk of hunger. Importantly, the trade-offs with food security would be reduced by the avoided impacts in the agricultural sector due to the reduced warming associated with the 1.5°C pathways (see Chapter 3, Section 3.5). However, such feedbacks are not comprehensively captured in the studies on mitigation.

5.4.2.3   Lack of energy access/energy poverty

A lack of access to clean and affordable energy (especially for cooking) is a major policy concern in many countries, especially in those in South Asia and Africa where major parts of the population still rely primarily on solid fuels for cooking (IEA and World Bank, 2017). Scenario studies which quantify the interactions between climate mitigation and energy access indicate that stringent climate policy which would affect energy prices could significantly slow down the transition to clean cooking fuels, such as liquefied petroleum gas or electricity (Cameron et al., 2016).

Estimates across six different IAMs (McCollum et al., 2018b) indicate that, in the absence of compensatory measures, the number of people without access to clean cooking fuels may increase. Redistributional measures, such as subsidies on cleaner fuels and stoves, could compensate for the negative effects of mitigation on energy access. Investment costs of the redistributional measures in 1.5°C pathways (on average around 120 billion USD2010 per year to 2030; Figure 5.4) are much smaller than the mitigation investments of 1.5°C pathways (McCollum et al., 2018b). The recycling of revenues from climate policy might act as a means to help finance the costs of providing energy access to the poor (Cameron et al., 2016).

5.4.2.4   Water security

Transformations towards low emissions energy and agricultural systems can have major implications for freshwater demand as well as water pollution. The scaling up of renewables and energy efficiency as depicted by low emissions pathways would, in most instances, lower water demands for thermal energy supply facilities ('water-for-energy') compared to fossil energy technologies, and thus reinforce targets related to water access and scarcity (see Chapter 4, Section 4.2.1). However, some low-carbon options such as bioenergy, centralized solar

a) Scenario ranges for selected sustainable development dimensions (2050)

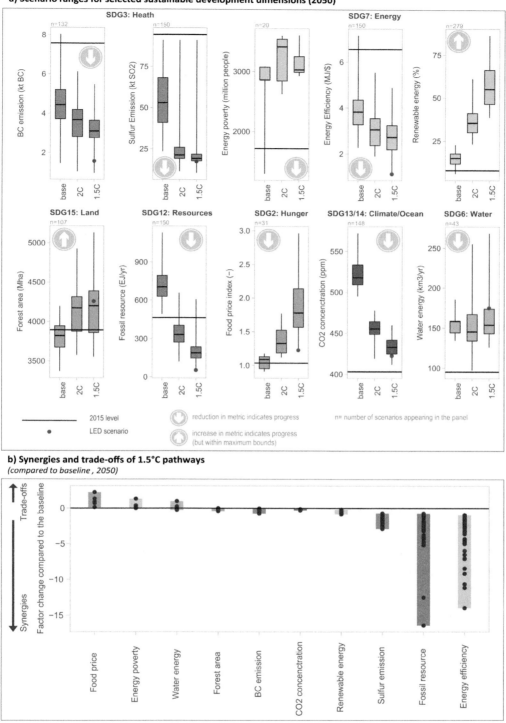

b) Synergies and trade-offs of 1.5°C pathways
(compared to baseline , 2050)

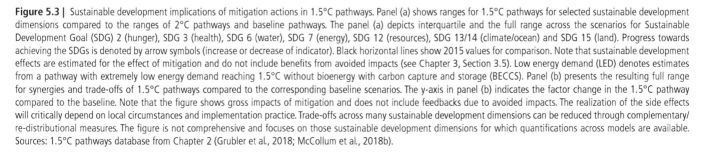

Figure 5.3 | Sustainable development implications of mitigation actions in 1.5°C pathways. Panel (a) shows ranges for 1.5°C pathways for selected sustainable development dimensions compared to the ranges of 2°C pathways and baseline pathways. The panel (a) depicts interquartile and the full range across the scenarios for Sustainable Development Goal (SDG) 2 (hunger), SDG 3 (health), SDG 6 (water), SDG 7 (energy), SDG 12 (resources), SDG 13/14 (climate/ocean) and SDG 15 (land). Progress towards achieving the SDGs is denoted by arrow symbols (increase or decrease of indicator). Black horizontal lines show 2015 values for comparison. Note that sustainable development effects are estimated for the effect of mitigation and do not include benefits from avoided impacts (see Chapter 3, Section 3.5). Low energy demand (LED) denotes estimates from a pathway with extremely low energy demand reaching 1.5°C without bioenergy with carbon capture and storage (BECCS). Panel (b) presents the resulting full range for synergies and trade-offs of 1.5°C pathways compared to the corresponding baseline scenarios. The y-axis in panel (b) indicates the factor change in the 1.5°C pathway compared to the baseline. Note that the figure shows gross impacts of mitigation and does not include feedbacks due to avoided impacts. The realization of the side effects will critically depend on local circumstances and implementation practice. Trade-offs across many sustainable development dimensions can be reduced through complementary/re-distributional measures. The figure is not comprehensive and focuses on those sustainable development dimensions for which quantifications across models are available. Sources: 1.5°C pathways database from Chapter 2 (Grubler et al., 2018; McCollum et al., 2018b).

power, nuclear and hydropower technologies could, if not managed properly, have counteracting effects that compound existing water-related problems in a given locale (Byers et al., 2014; Fricko et al., 2016; IEA, 2016; Fujimori et al., 2017a; Wang, 2017; McCollum et al., 2018a).

Under stringent mitigation efforts, the demand for bioenergy can result in a substantial increase of water demand for irrigation, thereby potentially contributing to water scarcity in water-stressed regions (Berger et al., 2015; Bonsch et al., 2016; Jägermeyr et al., 2017). However, this risk can be reduced by prioritizing rain-fed production of bioenergy (Hayashi et al., 2015, 2018; Bonsch et al., 2016), but might have adverse effects for food security (Boysen et al., 2017).

Reducing food and energy demand without compromising the needs of the poor emerges as a robust strategy for both water conservation and GHG emissions reductions (von Stechow et al., 2015; IEA, 2016; Parkinson et al., 2016; Grubler et al., 2018). The results underscore the importance of an integrated approach when developing water, energy and climate policy (IEA, 2016).

Estimates across different models for the impacts of stringent mitigation pathways on energy-related water uses seem ambiguous. Some pathways show synergies (Mouratiadou et al., 2018) while others indicate trade-offs and thus increases of water use due to mitigation (Fricko et al., 2016). The synergies depend on the adopted policy implementation or mitigation strategies and technology portfolio. A number of adaptation options exist (e.g., dry cooling), which can effectively reduce electricity-related water trade-offs (Fricko et al., 2016; IEA, 2016). Similarly, irrigation water use will depend on the regions where crops are produced, the sources of bioenergy (e.g., agriculture vs. forestry) and dietary change induced by climate policy. Overall, and also considering other water-related SDGs, including access to safe drinking water and sanitation as well as waste-water treatment, investments into the water sector seem to be only modestly affected by stringent climate policy compatible with 1.5°C (Figure 5.4) (McCollum et al., 2018b).

In summary, the assessment of mitigation pathways shows that to meet the 1.5°C target, a wide range of mitigation options would need to be deployed (see Chapter 2, Sections 2.3 and 2.4). While pathways aiming at 1.5°C are associated with high synergies for some sustainable development dimensions (such as human health and air pollution, forest preservation), the rapid pace and magnitude of the required changes would also lead to increased risks for trade-offs for other sustainable development dimensions (particularly food security) (Figures 5.4 and 5.5). Synergies and trade-offs are expected to be unevenly distributed between regions and nations (Box 5.2), though little literature has formally examined such distributions under 1.5°C-consistent mitigation scenarios. Reducing these risks requires smart policy designs and mechanisms that shield the poor and redistribute the burden so that the most vulnerable are not disproportionately affected. Recent scenario analyses show that associated investments for reducing the trade-offs for, for example, food, water and energy access to be significantly lower than the required mitigation investments (McCollum et al., 2018b). Fundamental transformation of demand, including efficiency and behavioural changes, can help to significantly reduce the reliance on risky technologies, such as BECCS, and thus reduce the risk of potential

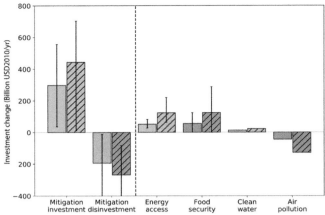

Figure 5.4 | Investment into mitigation up until 2030 and implications for investments for four sustainable development dimensions. Cross-hatched bars show the median investment in 1.5°C pathways across results from different models, and solid bars for 2°C pathways, respectively. Whiskers on bars represent minima and maxima across estimates from six models. Clean water and air pollution investments are available only from one model. Mitigation investments show the change in investments across mitigation options compared to the baseline. Negative mitigation investments (grey bars) denote disinvestment (reduced investment needs) into fossil fuel sectors compared to the baseline. Investments for different sustainable development dimensions denote the investment needs for complementary measures in order to avoid trade-offs (negative impacts) of mitigation. Negative sustainable development investments for air pollution indicate cost savings, and thus synergies of mitigation for air pollution control costs. The values compare to about 2 trillion USD2010 (range of 1.4 to 3 trillion) of total energy-related investments in the 1.5°C pathways. Source: Estimates from CD-LINKS scenarios summarised by McCollum et al., 2018b.

trade-offs between mitigation and other sustainable development dimensions (von Stechow et al., 2015; Grubler et al., 2018; van Vuuren et al., 2018). Reliance on demand-side measures only, however, would not be sufficient for meeting stringent targets, such as 1.5°C and 2°C (Clarke et al., 2014).

5.5 Sustainable Development Pathways to 1.5°C

This section assesses what is known in the literature on development pathways that are sustainable and climate-resilient and relevant to a 1.5°C warmer world. Pathways, transitions from today's world to achieving a set of future goals (see Chapter 1, Section 1.2.3, Cross-Chapter Box 1), follow broadly two main traditions: first, as integrated pathways describing the required societal and systems transformations, combining quantitative modelling and qualitative narratives at multiple spatial scales (global to sub-national); and second, as country- and community-level, solution-oriented trajectories and decision-making processes about context- and place-specific opportunities, challenges and trade-offs. These two notions of pathways offer different, though complementary, insights into the nature of 1.5°C-relevant trajectories and the short-term actions that enable long-term goals. Both highlight to varying degrees the urgency, ethics and equity dimensions of possible trajectories and society- and system-wide transformations, yet at different scales, building on Chapter 2 (see Section 2.4) and Chapter 4 (see Section 4.5).

5.5.1   Integration of Adaptation, Mitigation and Sustainable Development

Insights into climate-compatible development (see Glossary) illustrate how integration between adaptation, mitigation and sustainable development works in context-specific projects, how synergies are achieved and what challenges are encountered during implementation (Stringer et al., 2014; Suckall et al., 2014; Antwi-Agyei et al., 2017a; Bickersteth et al., 2017; Kalafatis, 2017; Nunan, 2017). The operationalization of climate-compatible development, including climate-smart agriculture and carbon-forestry projects (Lipper et al., 2014; Campbell et al., 2016; Quan et al., 2017), shows multilevel and multisector trade-offs involving 'winners' and 'losers' across governance levels (*high confidence*) (Kongsager and Corbera, 2015; Naess et al., 2015; Karlsson et al., 2017; Tanner et al., 2017; Taylor, 2017; Wood, 2017; Ficklin et al., 2018). Issues of power, participation, values, equity, inequality and justice transcend case study examples of attempted integrated approaches (Nunan, 2017; Phillips et al., 2017; Stringer et al., 2017; Wood, 2017), also reflected in policy frameworks for integrated outcomes (Stringer et al., 2014; Di Gregorio et al., 2017; Few et al., 2017; Tanner et al., 2017).

Ultimately, reconciling trade-offs between development needs and emissions reductions towards a 1.5°C warmer world requires a dynamic view of the interlinkages between adaptation, mitigation and sustainable development (Nunan, 2017). This entails recognition of the ways in which development contexts shape the choice and effectiveness of interventions, limit the range of responses afforded to communities and governments, and potentially impose injustices upon vulnerable groups (UNRISD, 2016; Thornton and Comberti, 2017). A variety of approaches, both quantitative and qualitative, exist to examine possible sustainable development pathways under which climate and sustainable development goals can be achieved, and synergies and trade-offs for transformation identified (Sections 5.3 and 5.4).

5.5.2   Pathways for Adaptation, Mitigation and Sustainable Development

This section focuses on the growing body of pathways literature describing the dynamic and systemic integration of mitigation and adaptation with sustainable development in the context of a 1.5°C warmer world. These studies are critically important for the identification of 'enabling' conditions under which climate and the SDGs can be achieved, and thus help the design of transformation strategies that maximize synergies and avoid potential trade-offs (Sections 5.3 and 5.4). Full integration of sustainable development dimensions is, however, challenging, given their diversity and the need for high temporal, spatial and social resolution to address local effects, including heterogeneity related to poverty and equity (von Stechow et al., 2015). Research on long-term climate change mitigation and adaptation pathways has covered individual SDGs to different degrees. Interactions between climate and other SDGs have been explored for SDGs 2, 3, 4, 6, 7, 8, 12, 14 and 15 (Clarke et al., 2014; Abel et al., 2016; von Stechow et al., 2016; Rao et al., 2017), while interactions with SDGs 1, 5, 11 and 16 remain largely underexplored in integrated long-term scenarios (Zimm et al., 2018).

Quantitative pathways studies now better represent 'nexus' approaches to assess sustainable development dimensions. In such approaches (see Chapter 4, Section 4.3.3.8), a subset of sustainable development dimensions are investigated together because of their close relationships (Welsch et al., 2014; Conway et al., 2015; Keairns et al., 2016; Parkinson et al., 2016; Rasul and Sharma, 2016; Howarth and Monasterolo, 2017). Compared to single-objective climate–SDG assessments (Section 5.4.2), nexus solutions attempt to integrate complex interdependencies across diverse sectors in a systems approach for consistent analysis. Recent pathways studies show how water, energy and climate (SDGs 6, 7 and 13) interact (Parkinson et al., 2016; McCollum et al., 2018b) and call for integrated water–energy investment decisions to manage systemic risks. For instance, the provision of bioenergy, important in many 1.5°C-consistent pathways, can help resolve 'nexus challenges' by alleviating energy security concerns, but can also have adverse 'nexus impacts' on food security, water use and biodiversity (Lotze-Campen et al., 2014; Bonsch et al., 2016). Policies that improve resource use efficiency across sectors can maximize synergies for sustainable development (Bartos and Chester, 2014; McCollum et al., 2018b; van Vuuren et al., 2018). Mitigation compatible with 1.5°C can significantly reduce impacts and adaptation needs in the nexus sectors compared to 2°C (Byers et al., 2018). In order to avoid trade-offs due to high carbon pricing of 1.5°C pathways, regulation in specific areas may complement price-based instruments. Such combined policies generally lead also to more early action maximizing synergies and avoiding some of the adverse climate effects for sustainable development (Bertram et al., 2018).

The comprehensive analysis of climate change in the context of sustainable development requires suitable reference scenarios that lend themselves to broader sustainable development analyses. The Shared Socio-Economic Pathways (SSPs) (Chapter 1, Cross-Chapter Box 1 in Chapter 1) (O'Neill et al., 2017a; Riahi et al., 2017) constitute an important first step in providing a framework for the integrated assessment of adaptation and mitigation and their climate–development linkages (Ebi et al., 2014). The five underlying SSP narratives (O'Neill et al., 2017a) map well into some of the key SDG dimensions, with one of the pathways (SSP1) explicitly depicting sustainability as the main theme (van Vuuren et al., 2017b).

To date, no pathway in the literature proves to achieve all 17 SDGs because several targets are not met or not sufficiently covered in the analysis, hence resulting in a sustainability gap (Zimm et al., 2018). The SSPs facilitate the systematic exploration of different sustainable dimensions under ambitious climate objectives. SSP1 proves to be in line with eight SDGs (3, 7, 8, 9, 10, 11, 13 and 15) and several of their targets in a 2°C warmer world (van Vuuren et al., 2017b; Zimm et al., 2018). However, important targets for SDGs 1, 2 and 4 (i.e., people living in extreme poverty, people living at the risk of hunger and gender gap in years of schooling) are not met in this scenario.

The SSPs show that sustainable socio-economic conditions will play a key role in reaching stringent climate targets (Riahi et al., 2017; Rogelj et al., 2018). Recent modelling work has examined 1.5°C-consistent, stringent mitigation scenarios for 2100 applied to the SSPs, using six different IAMs. Despite the limitations of these models, which are coarse approximations of reality, robust trends can be identified

5

(Rogelj et al., 2018). SSP1 – which depicts broader 'sustainability' as well as enhancing equity and poverty reductions – is the only pathway where all models could reach 1.5°C and is associated with the lowest mitigation costs across all SSPs. A decreasing number of models was successful for SSP2, SSP4 and SSP5, respectively, indicating distinctly higher risks of failure due to high growth and energy intensity as well as geographical and social inequalities and uneven regional development. And reaching 1.5°C has even been found infeasible in the less sustainable SSP3 – 'regional rivalry' (Fujimori et al., 2017b; Riahi et al., 2017). All these conclusions hold true if a 2°C objective is considered (Calvin et al., 2017; Fujimori et al., 2017b; Popp et al., 2017; Riahi et al., 2017). Rogelj et al. (2018) also show that fewer scenarios are, however, feasible across different SSPs in case of 1.5°C, and mitigation costs substantially increase in 1.5°C pathways compared to 2°C pathways.

There is a wide range of SSP-based studies focusing on the connections between adaptation/impacts and different sustainable development dimensions (Hasegawa et al., 2014; Ishida et al., 2014; Arnell et al., 2015; Bowyer et al., 2015; Burke et al., 2015; Lemoine and Kapnick, 2016; Rozenberg and Hallegatte, 2016; Blanco et al., 2017; Hallegatte and Rozenberg, 2017; O'Neill et al., 2017a; Rutledge et al., 2017; Byers et al., 2018). New methods for projecting inequality and poverty (downscaled to sub-national rural and urban levels as well as spatially explicit levels) have enabled advanced SSP-based assessments of locally sustainable development implications of avoided impacts and related adaptation needs. For instance, Byers et al. (2018) find that, in a 1.5°C warmer world, a focus on sustainable development can reduce the climate risk exposure of populations vulnerable to poverty by more than an order of magnitude (Section 5.2.2). Moreover, aggressive reductions in between-country inequality may decrease the emissions intensity of global economic growth (Rao and Min, 2018). This is due to the higher potential for decoupling of energy from income growth in lower-income countries, due to high potential for technological advancements that reduce the energy intensity of growth of poor countries – critical also for reaching 1.5°C in a socially and economically equitable way. Participatory downscaling of SSPs in several European Union countries and in Central Asia shows numerous possible pathways of solutions to the 2°C–1.5°C goal, depending on differential visions (Tàbara et al., 2018). Other participatory applications of the SSPs, for example in West Africa (Palazzo et al., 2017) and the southeastern United States (Absar and Preston, 2015), illustrate the potentially large differences in adaptive capacity within regions and between sectors.

Harnessing the full potential of the SSP framework to inform sustainable development requires: (i) further elaboration and extension of the current SSPs to cover sustainable development objectives explicitly; (ii) the development of new or variants of current narratives that would facilitate more SDG-focused analyses with climate as one objective (among other SDGs) (Riahi et al., 2017); (iii) scenarios with high regional resolution (Fujimori et al., 2017b); (iv) a more explicit representation of institutional and governance change associated with the SSPs (Zimm et al., 2018); and (v) a scale-up of localized and spatially explicit vulnerability, poverty and inequality estimates, which have emerged in recent publications based on the SSPs (Byers et al., 2018) and are essential to investigate equity dimensions (Klinsky and Winkler, 2018).

5.5.3　Climate-Resilient Development Pathways

This section assesses the literature on pathways as solution-oriented trajectories and decision-making processes for attaining transformative visions for a 1.5°C warmer world. It builds on climate-resilient development pathways (CRDPs) introduced in the AR5 (Section 5.1.2) (Olsson et al., 2014) as well as growing literature (e.g., Eriksen et al., 2017; Johnson, 2017; Orindi et al., 2017; Kirby and O'Mahony, 2018; Solecki et al., 2018) that uses CRDPs as a conceptual and aspirational idea for steering societies towards low-carbon, prosperous and ecologically safe futures. Such a notion of pathways foregrounds decision-making processes at local to national levels to situate transformation, resilience, equity and well-being in the complex reality of specific places, nations and communities (Harris et al., 2017; Ziervogel et al., 2017; Fazey et al., 2018; Gajjar et al., 2018; Klinsky and Winkler, 2018; Patterson et al., 2018; Tàbara et al., 2018).

Pathways compatible with 1.5°C warming are not merely scenarios to envision possible futures but processes of deliberation and implementation that address societal values, local priorities and inevitable trade-offs. This includes attention to politics and power that perpetuate business-as-usual trajectories (O'Brien, 2016; Harris et al., 2017), the politics that shape sustainability and capabilities of everyday life (Agyeman et al., 2016; Schlosberg et al., 2017), and ingredients for community resilience and transformative change (Fazey et al., 2018). Chartering CRDPs encourages locally situated and problem-solving processes to negotiate and operationalize resilience 'on the ground' (Beilin and Wilkinson, 2015; Harris et al., 2017; Ziervogel et al., 2017). This entails contestation, inclusive governance and iterative engagement of diverse populations with varied needs, aspirations, agency and rights claims, including those most affected, to deliberate trade-offs in a multiplicity of possible pathways (*high confidence*) (see Figure 5.5) (Stirling, 2014; Vale, 2014; Walsh-Dilley and Wolford, 2015; Biermann et al., 2016; J.R.A. Butler et al., 2016; O'Brien, 2016, 2018; Harris et al., 2017; Jones and Tanner, 2017; Mapfumo et al., 2017; Rosenbloom, 2017; Gajjar et al., 2018; Klinsky and Winkler, 2018; Lyon, 2018; Tàbara et al., 2018).

5.5.3.1　Transformations, equity and well-being

Most literature related to CRDPs invokes the concept of transformation, underscoring the need for urgent and far-reaching changes in practices, institutions and social relations in society. Transformations towards a 1.5°C warmer world would need to address considerations for equity and well-being, including in trade-off decisions (see Figure 5.1).

To attain the anticipated *transformations*, all countries as well as non-state actors would need to strengthen their contributions, through bolder and more committed cooperation and equitable effort-sharing (*medium evidence, high agreement*) (Rao, 2014; Frumhoff et al., 2015; Ekwurzel et al., 2017; Millar et al., 2017; Shue, 2017; Holz et al., 2018; Robinson and Shine, 2018). Sustaining decarbonization rates at a 1.5°C-compatible level would be unprecedented and not possible without rapid transformations to a net-zero-emissions global economy by mid-century or the later half of the century (see Chapters 2 and 4). Such efforts would entail overcoming technical, infrastructural, institutional and behavioural barriers across all sectors and levels

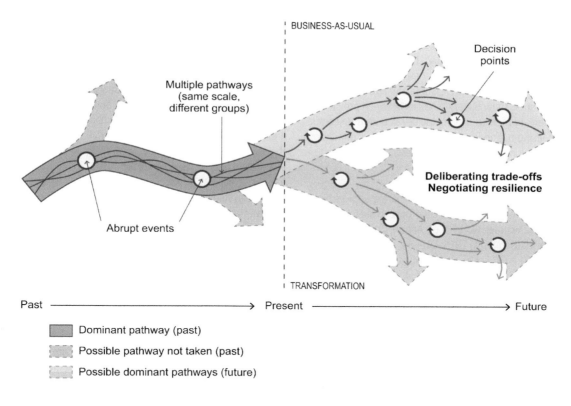

BUSINESS-AS-USUAL

Decision points

Multiple pathways
(same scale,
different groups)

Deliberating trade-offs
Negotiating resilience

Abrupt events

TRANSFORMATION

Past ⟶ Present ⟶ Future

▨ Dominant pathway (past)

▨ Possible pathway not taken (past)

▨ Possible dominant pathways (future)

Figure 5.5 | Pathways into the future, with path dependencies and iterative problem-solving and decision-making (after Fazey et al., 2016).

of society (Pfeiffer et al., 2016; Seto et al., 2016) and defeating path dependencies, including poverty traps (Boonstra et al., 2016; Enqvist et al., 2016; Lade et al., 2017; Haider et al., 2018). Transformation also entails ensuring that 1.5°C-compatible pathways are inclusive and desirable, build solidarity and alliances, and protect vulnerable groups, including against disruptions of transformation (Patterson et al., 2018).

There is growing emphasis on the role of *equity, fairness* and *justice* (see Glossary) regarding context-specific transformations and pathways to a 1.5°C warmer world (*medium evidence, high agreement*) (Shue, 2014; Thorp, 2014; Dennig et al., 2015; Moellendorf, 2015; Klinsky et al., 2017b; Roser and Seidel, 2017; Sealey-Huggins, 2017; Klinsky and Winkler, 2018; Robinson and Shine, 2018). Consideration for what is equitable and fair suggests the need for stringent decarbonization and up-scaled adaptation that do not exacerbate social injustices, locally and at national levels (Okereke and Coventry, 2016), uphold human rights (Robinson and Shine, 2018), are socially desirable and acceptable (von Stechow et al., 2016; Rosenbloom, 2017), address values and beliefs (O'Brien, 2018), and overcome vested interests (Normann, 2015; Patterson et al., 2016). Attention is often drawn to huge disparities in the cost, benefits, opportunities and challenges involved in transformation within and between countries, and the fact that the suffering of already poor, vulnerable and disadvantaged populations may be worsened, if care to protect them is not taken (Holden et al., 2017; Klinsky and Winkler, 2018; Patterson et al., 2018).

Well-being for all (Dearing et al., 2014; Raworth, 2017) is at the core of an ecologically safe and socially just space for humanity, including health and housing, peace and justice, social equity, gender

equality and political voices (Raworth, 2017). It is in alignment with transformative social development (UNRISD, 2016) and the 2030 Agenda of 'leaving no one behind'. The social conditions to enable well-being for all are to reduce entrenched inequalities within and between countries (Klinsky and Winkler, 2018); rethink prevailing values, ethics and behaviours (Holden et al., 2017); allow people to live a life in dignity while avoiding actions that undermine capabilities (Klinsky and Golub, 2016); transform economies (Popescu and Ciurlau, 2016; Tàbara et al., 2018); overcome uneven consumption and production patterns (Dearing et al., 2014; Häyhä et al., 2016; Raworth, 2017) and conceptualize development as well-being rather than mere economic growth (*medium evidence, high agreement*) (Gupta and Pouw, 2017).

5.5.3.2 Development trajectories, sharing of efforts and cooperation

The potential for pursuing sustainable and climate-resilient development pathways towards a 1.5°C warmer world differs between and within nations, due to differential development achievements and trajectories, and opportunities and challenges (*very high confidence*) (Figure 5.1). There are clear differences between high-income countries where social achievements are high, albeit often with negative effects on the environment, and most developing nations where vulnerabilities to climate change are high and social support and life satisfaction are low, especially in the Least Developed Countries (LDCs) (Sachs et al., 2017; O'Neill et al., 2018). Differential starting points for CRDPs between and within countries, including path dependencies (Figure 5.5), call for sensitivity to context (Klinsky and Winkler, 2018). For the developing world, limiting warming to 1.5°C also means potentially

5

severely curtailed development prospects (Okereke and Coventry, 2016) and risks to human rights from both climate action and inaction to achieve this goal (Robinson and Shine, 2018) (Section 5.2). Within-country development differences remain, despite efforts to ensure inclusive societies (Gupta and Arts, 2017; Gupta and Pouw, 2017). Cole et al. (2017), for instance, show how differences between provinces in South Africa constitute barriers to sustainable development trajectories and for operationalising nation-level SDGs, across various dimensions of social deprivation and environmental stress, reflecting historic disadvantages.

Moreover, various equity and effort- or burden-sharing approaches to climate stabilization in the literature describe how to sketch national potentials for a 1.5°C warmer world (e.g., Anand, 2004; CSO Equity Review, 2015; Meinshausen et al., 2015; Okereke and Coventry, 2016; Bexell and Jönsson, 2017; Otto et al., 2017; Pan et al., 2017; Robiou du Pont et al., 2017; Holz et al., 2018; Kartha et al., 2018; Winkler et al., 2018;). Many approaches build on the AR5 'responsibility – capacity – need' assessment (Clarke et al., 2014), complement other proposed national-level metrics for capabilities, equity and fairness (Heyward and Roser, 2016; Klinsky et al., 2017a), or fall under the wider umbrella of fair share debates on responsibility, capability and the right to development in climate policy (Fuglestvedt and Kallbekken, 2016). Importantly, different principles and methodologies generate different calculated contributions, responsibilities and capacities (Skeie et al., 2017).

The notion of nation-level fair shares is now also discussed in the context of limiting global warming to 1.5°C and the Nationally Determined Contributions (NDCs) (see Chapter 4, Cross-Chapter Box 11 in Chapter 4) (CSO Equity Review, 2015; Mace, 2016; Pan et al., 2017; Robiou du Pont et al., 2017; Holz et al., 2018; Kartha et al., 2018; Winkler et al., 2018). A study by Pan et al. (2017) concluded that all countries would need to contribute to ambitious emissions reductions and that current pledges for 2030 by seven out of eight high-emitting countries would be insufficient to meet 1.5°C. Emerging literature on justice-centred pathways to 1.5°C points towards ambitious emissions reductions domestically and committed cooperation internationally whereby wealthier countries support poorer ones, technologically, financially and otherwise to enhance capacities (Okereke and Coventry, 2016; Holz et al., 2018; Robinson and Shine, 2018; Shue, 2018). These findings suggest that equitable and 1.5°C-compatible pathways would require fast action across all countries at all levels of development rather than late accession of developing countries (as assumed under SSP3, see Chapter 2), with external support for prompt mitigation and resilience-building efforts in the latter (*medium evidence, medium agreement*).

Scientific advances since the AR5 now also make it possible to determine contributions to climate change for non-state actors (see Chapter 4, Section 4.4.1) and their potential to contribute to CRDPs (*medium evidence, medium agreement*). These non-state actors includes cities (Bulkeley et al., 2013, 2014; Byrne et al., 2016), businesses (Heede, 2014; Frumhoff et al., 2015; Shue, 2017), transnational initiatives (Castro, 2016; Andonova et al., 2017) and industries. Recent work demonstrates the contributions of 90 industrial carbon producers to global temperature and sea level rise, and their responsibilities to

contribute to investments in and support for mitigation and adaptation (Heede, 2014; Ekwurzel et al., 2017; Shue, 2017) (Sections 5.6.1 and 5.6.2).

At the level of groups and individuals, equity in pursuing climate resilience for a 1.5°C warmer world means addressing disadvantage, inequities and empowerment that shape transformative processes and pathways (Fazey et al., 2018), and deliberate efforts to strengthen the capabilities, capacities and well-being of poor, marginalized and vulnerable people (Byrnes, 2014; Tokar, 2014; Harris et al., 2017; Klinsky et al., 2017a; Klinsky and Winkler, 2018). Community-driven CRDPs can flag potential negative impacts of national trajectories on disadvantaged groups, such as low-income families and communities of colour (Rao, 2014). They emphasize social equity, participatory governance, social inclusion and human rights, as well as innovation, experimentation and social learning (see Glossary) (*medium evidence, high agreement*) (Sections 5.5.3.3 and 5.6).

5.5.3.3 Country and community strategies and experiences

There are many possible pathways towards climate-resilient futures (O'Brien, 2018; Tàbara et al., 2018). Literature depicting different sustainable development trajectories in line with CRDPs is growing, with some of it being specific to 1.5°C global warming. Most experiences to date are at local and sub-national levels (Cross-Chapter Box 13 in this chapter), while state-level efforts align largely with green economy trajectories or planning for climate resilience (Box 5.3). Due to the fact that these strategies are context-specific, the literature is scarce on comparisons, efforts to scale up and systematic monitoring.

States can play an enabling or hindering role in a transition to a 1.5°C warmer world (Patterson et al., 2018). The literature on strategies to reconcile low-carbon trajectories with sustainable development and ecological sustainability through green growth, inclusive growth, de-growth, post-growth and development as well-being *shows low agreement* (see Chapter 4, Section 4.5). Efforts that align best with CRDPs are described as 'transformational' and 'strong' (Ferguson, 2015). Some view 'thick green' perspectives as enabling equity, democracy and agency building (Lorek and Spangenberg, 2014; Stirling, 2014; Ehresman and Okereke, 2015; Buch-Hansen, 2018), others show how green economy and sustainable development pathways can align (Brown et al., 2014; Georgeson et al., 2017b), and how a green economy can help link the SDGs with NDCs, for instance in Mongolia, Kenya and Sweden (Shine, 2017). Others still critique the continuous reliance on market mechanisms (Wanner, 2014; Brockington and Ponte, 2015) and disregard for equity and distributional and procedural justice (Stirling, 2014; Bell, 2015).

Country-level pathways and achievements vary significantly (*robust evidence, medium agreement*). For instance, the Scandinavian countries rank at the top of the Global Green Economy Index (Dual Citizen LLC, 2016), although they also tend to show high spill-over effects (Holz et al., 2018) and transgress their biophysical boundaries (O'Neill et al., 2018). State-driven efforts in non-member countries of the Organisation for Economic Co-operation and Development include Ethiopia's 'Climate-resilient Green Economy Strategy', Mozambique's 'Green Economy Action Plan' and Costa Rica's ecosystem- and conservation-driven green transition paths. China and India have adopted technology and renewables pathways (Brown et al., 2014; Death, 2014, 2015, 2016; Khanna et al., 2014; Chen et al., 2015; Kim and Thurbon, 2015; Wang et al., 2015; Weng et al., 2015). Brazil promotes low per capita GHG emissions, clean energy sources, green jobs, renewables and sustainable transportation, while slowing rates of deforestation (see Chapter 4, Box 4.7) (Brown et al., 2014; La Rovere, 2017). Yet concerns remain regarding persistent inequalities, ecosystem monetization, lack of participation in green-style projects (Brown et al., 2014) and labour conditions and risk of displacement in the sugarcane ethanol sector (McKay et al., 2016). Experiences with low-carbon development pathways in LDCs highlight the crucial role of identifying synergies across scale, removing institutional barriers and ensuring equity and fairness in distributing benefits as part of the right to development (Rai and Fisher, 2017).

In small islands states, for many of which climate change hazards and impacts at 1.5°C pose significant risks to sustainable development (see Chapter 3 Box 3.5, Chapter 4 Box 4.3, Box 5.3), examples of CRDPs have emerged since the AR5. This includes the SAMOA Pathway: SIDS Accelerated Modalities of Action (see Chapter 4, Box 4.3) (UNGA, 2014; Government of Kiribati, 2016; Steering Committee on Partnerships for SIDS and UN DESA, 2016; Lefale et al., 2017) and the Framework for Resilient Development in the Pacific, a leading example of integrated regional climate change adaptation planning for mitigation and sustainable development, disaster risk management and low-carbon economies (SPC, 2016). Small islands of the Pacific vary significantly in their capacity and resources to support effective integrated planning (McCubbin et al., 2015; Barnett and Walters, 2016; Cvitanovic et al., 2016; Hemstock et al., 2017; Robinson and Dornan, 2017). Vanuatu (Box 5.3) has developed a significant coordinated national adaptation plan to advance the 2030 Agenda for Sustainable Development, respond to the Paris Agreement and reduce the risk of disasters in line with the Sendai targets (UNDP, 2016; Republic of Vanuatu, 2017).

Box 5.3 | Republic of Vanuatu – National Planning for Development and Climate Resilience

The Republic of Vanuatu is leading Pacific Small Island Developing States (SIDS) to develop a nationally coordinated plan for climate-resilient development in the context of high exposure to hazard risk (MoCC, 2016; UNU-EHS, 2016). The majority of the population depends on subsistence, rain-fed agriculture and coastal fisheries for food security (Sovacool et al., 2017). Sea level rise, increased prolonged drought, water shortages, intense storms, cyclone events and degraded coral reef environments threaten human security in a 1.5°C warmer world (see Chapter 3, Box 3.5) (SPC, 2015; Aipira et al., 2017). Given Vanuatu's long history of climate hazards and disasters, local adaptive capacity is relatively high, despite barriers to the use of local knowledge and technology, and low rates of literacy and women's participation (McNamara and Prasad, 2014; Aipira et al., 2017; Granderson, 2017). However, the adaptive capacity of Vanuatu and other SIDS is increasingly constrained due to more frequent severe weather events (see Chapter 3, Box 3.5, Chapter 4, Cross-Chapter Box 9 in Chapter 4) (Gero et al., 2013; Kuruppu and Willie, 2015; SPC, 2015; Sovacool et al., 2017).

Vanuatu has developed a national sustainable development plan for 2016–2030: the People's Plan (Republic of Vanuatu, 2016). This coordinated, inclusive plan of action on economy, environment and society aims to strengthen adaptive capacity and resilience to climate change and disasters. It emphasizes rights of all Ni-Vanuatu, including women, youth, the elderly and vulnerable groups (Nalau et al., 2016). Vanuatu has also developed a Coastal Adaptation Plan (Republic of Vanuatu, 2016), an integrated Climate Change and Disaster Risk Reduction Policy (2016–2030) (SPC, 2015) and the first South Pacific National Advisory Board on Climate Change & Disaster Risk Reduction (SPC, 2015; UNDP, 2016).

Vanuatu aims to integrate planning at multiple scales, and increase climate resilience by supporting local coping capacities and iterative processes of planning for sustainable development and integrated risk assessment (Aipira et al., 2017; Eriksson et al., 2017; Granderson, 2017). Climate-resilient development is also supported by non-state partnerships, for example, the 'Yumi stap redi long climate change'–the Vanuatu non-governmental organization Climate Change Adaptation Program (Maclellan, 2015). This programme focuses on equitable governance, with particular attention to supporting women's voices in decision-making through allied programmes addressing domestic violence, and rights-based education to reduce social marginalization; alongside institutional reforms for greater transparency, accountability and community participation in decision-making (Davies, 2015; Maclellan, 2015; Sterrett, 2015; Ensor, 2016; UN Women, 2016).

Power imbalances embedded in the political economy of development (Nunn et al., 2014), gender discrimination (Aipira et al., 2017) and the priorities of climate finance (Cabezon et al., 2016) may marginalize the priorities of local communities and influence how local risks are understood, prioritised and managed (Kuruppu and Willie, 2015; Baldacchino, 2017; Sovacool et al., 2017). However, the experience of the low death toll after Cyclone Pam suggests effective use of local knowledge in planning and early warning may support resilience at least in the absence of storm surge flooding (Handmer and Iveson, 2017; Nalau et al., 2017). Nevertheless, the very severe infrastructure damage of Cyclone Pam 2015 highlights the limits of individual Pacific SIDS efforts and the need for global and regional responses to a 1.5°C warmer world (see Chapter 3, Box 3.5, Chapter 4, Box 4.3) (Dilling et al., 2015; Ensor, 2016; Shultz et al., 2016; Rey et al., 2017).

5

Communities, towns and cities also contribute to low-carbon pathways, sustainable development and fair and equitable climate resilience, often focused on processes of power, learning and contestation as entry points to more localised CRDPs (*medium evidence, high agreement*) (Cross-Chapter Box 13 in this chapter, Box 5.2). In the Scottish Borders Climate Resilient Communities Project (United Kingdom), local flood management is linked with national policies to foster cross-scalar and inclusive governance, with attention to systemic disadvantages, shocks and stressors, capacity building, learning for change and climate narratives to inspire hope and action, all of which are essential for community resilience in a 1.5°C warmer world (Fazey et al., 2018). Narratives and storytelling are vital for realizing place-based 1.5°C futures as they create space for agency, deliberation, co-constructing meaning, imagination and desirable and dignified pathways (Veland et al., 2018). Engagement with possible futures, identity and self-reliance is also documented for Alaska, where warming has already exceeded 1.5°C and indigenous communities invest in renewable energy, greenhouses for food security and new fishing practices to overcome loss of sea ice, flooding and erosion (Chapin et al., 2016; Fazey et al., 2018). The Asian Cities Climate Change Resilience Network facilitates shared learning dialogues, risk-to-resilience workshops, and

iterative, consultative planning in flood-prone cities in India; vulnerable communities, municipal governmental agents, entrepreneurs and technical experts negotiate different visions, trade-offs and local politics to identify desirable pathways (Harris et al., 2017).

Transforming our societies and systems to limit global warming to 1.5°C and ensuring equity and well-being for human populations and ecosystems in a 1.5°C warmer world would require ambitious and well-integrated adaptation–mitigation–development pathways that deviate fundamentally from high-carbon, business-as-usual futures (Okereke and Coventry, 2016; Arts, 2017; Gupta and Arts, 2017; Sealey-Huggins, 2017). Identifying and negotiating socially acceptable, inclusive and equitable pathways towards climate-resilient futures is a challenging, yet important, endeavour, fraught with complex moral, practical and political difficulties and inevitable trade-offs (*very high confidence*). The ultimate questions are: what futures do we want (Bai et al., 2016; Tàbara et al., 2017; Klinsky and Winkler, 2018; O'Brien, 2018; Veland et al., 2018), whose resilience matters, for what, where, when and why (Meerow and Newell, 2016), and 'whose vision … is being pursued and along which pathways' (Gillard et al., 2016).

Cross-Chapter Box 13 | Cities and Urban Transformation

Lead Authors:
Fernando Aragon-Durand (Mexico), Paolo Bertoldi (Italy), Anton Cartwright (South Africa), François Engelbrecht (South Africa), Bronwyn Hayward (New Zealand), Daniela Jacob (Germany), Debora Ley (Guatemala/Mexico), Shagun Mehrotra (USA/India), Peter Newman (Australia), Aromar Revi (India), Seth Schultz (USA), William Solecki (USA), Petra Tschakert (Australia/Austria)

Contributor:
Peter Marcotullio (USA)

Global Urbanization in a 1.5°C Warmer World

The concentration of economic activity, dense social networks, human resource capacity, investment in infrastructure and buildings, relatively nimble local governments, close connection to surrounding rural and natural environments, and a tradition of innovation provide urban areas with transformational potential (see Chapter 4, Section 4.3.3) (Castán Broto, 2017). In this sense, the urbanization megatrend that will take place over the next three decades, and add approximately 2 billion people to the global urban population (UN, 2014), offers opportunities for efforts to limit warming to 1.5°C.

Cities can also, however, concentrate the risks of flooding, landslides, fire and infectious and parasitic disease that are expected to heighten in a 1.5°C warmer world (Chapter 3). In African and Asian countries where urbanization rates are highest, these risks could expose and amplify pre-existing stresses related to poverty, exclusion, and governance (Gore, 2015; Dodman et al., 2017; Jiang and O'Neill, 2017; Pelling et al., 2018; Solecki et al., 2018). Through its impact on economic development and investment, urbanization often leads to increased consumption and environmental degradation and enhanced vulnerability and risk (Rosenzweig et al., 2018). In the absence of innovation, the combination of urbanization and urban economic development could contribute 226 GtCO2 in emissions by 2050 (Bai et al., 2018). At the same time, some new urban developments are demonstrating combined carbon and Sustainable Development Goals (SDG) benefits (Wiktorowicz et al., 2018), and it is in towns and cities that building renovation rates can be most easily accelerated to support the transition to 1.5°C pathways (Kuramochi et al., 2018), including through voluntary programmes (Van der Heijden, 2018).

Urban transformations and emerging climate-resilient development pathways

The 1.5°C pathways require action in all cities and urban contexts. Recent literature emphasizes the need to deliberate and negotiate how resilience and climate-resilient pathways can be fostered in the context of people's daily lives, including the failings of everyday development such as unemployment, inadequate housing and a growing informal sector and settlements (informality), in order to acknowledge local priorities and foster transformative learning (Vale, 2014; Shi et al., 2016; Harris et al., 2017; Ziervogel et al., 2017; Fazey et al., 2018; Macintyre et al., 2018). Enhancing deliberate transformative capacities in urban contexts also entails new and relational forms of envisioning agency, equity, resilience, social cohesion and well-being (Section 5.5.3) (Gillard et al., 2016; Ziervogel et al., 2016). Two examples of urban transformation are explored here.

The built environment, spatial planning, infrastructure, energy services, mobility and urban–rural linkages necessary in rapidly growing cities in South Asia and Africa in the next three decades present mitigation, adaptation and development opportunities that are crucial for a 1.5°C world (Newman et al., 2017; Lwasa et al., 2018; Teferi and Newman, 2018). Realizing these opportunities would require the structural challenges of poverty, weak and contested local governance, and low levels of local government investment to be addressed on an unprecedented scale (Wachsmuth et al., 2016; Chu et al., 2017; van Noorloos and Kloosterboer, 2017; Pelling et al., 2018).

Urban governance is critical to ensuring that the necessary urban transitions deliver economic growth and equity (Hughes et al., 2018). The proximity of local governments to citizens and their needs can make them powerful agents of climate action (Melica et al., 2018), but urban governance is enhanced when it involves multiple actors (Ziervogel et al., 2016; Pelling et al., 2018), supportive national governments (Tait and Euston-Brown, 2017), and sub-national climate networks (see Chapter 4, Section 4.4.1). Governance is complicated for the urban population currently living in informality. This population is expected to triple, to three billion, by 2050 (Satterthwaite et al., 2018), placing a significant portion of the world's population beyond the direct reach of formal climate mitigation and adaptation policies (Revi et al., 2014). How to address the co-evolved and structural conditions that lead to urban informality and associated vulnerability to 1.5°C of warming is a central question for this report. Brown and McGranahan (2016) cite evidence that the informal urban 'green economy' that has emerged out of necessity in the absence of formal service provisions is frequently low-carbon and resource-efficient.

Realising the potential for low carbon transitions in informal urban settlements would require an express recognition of the unpaid-for contributions of women in the informal economy, and new partnerships between the state and communities (Ziervogel et al., 2017; Pelling et al., 2018; Satterthwaite et al., 2018). There is no guarantee that these partnerships will evolve or cohere into the type of service delivery and climate governance system that could steer the change on a scale required to limit to warming to 1.5°C (Jaglin, 2014). However, work by transnational networks, such as Shack/Slum Dwellers International, C40, the Global Covenant of Mayors, and the International Council for Local Environmental Initiatives, as well as efforts to combine in-country planning for Nationally Determined Contributions (NDCs) (Andonova et al., 2017; Fuhr et al., 2018) with those taking place to support the New Urban Agenda and National Urban Policies, represent one step towards realizing the potential (Tait and Euston-Brown, 2017). So too do 'old urban agendas', such as slum upgrading and universal water and sanitation provision (McGranahan et al., 2016; Satterthwaite, 2016; Satterthwaite et al., 2018).

Transition Towns (TTs) are a type of urban transformation that have emerged mainly in high-income countries. The grassroots TT movement (origin in the United Kingdom) combines adaptation, mitigation and just transitions, mainly at the level of communities and small towns. It now has more than 1,300 registered local initiatives in more than 40 countries (Grossmann and Creamer, 2017), many of them in the United Kingdom, the United States, and other high-income countries. TTs are described as 'progressive localism' (Cretney et al., 2016), aiming to foster a 'communitarian ecological citizenship' that goes beyond changes in consumption and lifestyle (Kenis, 2016). They aspire to promote equitable communities resilient to the impacts of climate change, peak oil and unstable global markets; re-localization of production and consumption; and transition pathways to a post-carbon future (Feola and Nunes, 2014; Evans and Phelan, 2016; Grossmann and Creamer, 2017).

TT initiatives typically pursue lifestyle-related low-carbon living and economies, food self-sufficiency, energy efficiency through renewables, construction with locally sourced material and cottage industries (Barnes, 2015; Staggenborg and Ogrodnik, 2015; Taylor Aiken, 2016). Social and iterative learning through the collective involves dialogue, deliberation, capacity building, citizen science engagements, technical re-skilling to increase self-reliance, for example canning and preserving food and permaculture, future visioning and emotional training to share difficulties and loss (Feola and Nunes, 2014; Barnes, 2015; Boke, 2015; Taylor Aiken, 2015; Kenis, 2016; Mehmood, 2016; Grossmann and Creamer, 2017).

Important conditions for successful transition groups include flexibility, participatory democracy, care ethics, inclusiveness and consensus-building, assuming bridging or brokering roles, and community alliances and partnerships (Feola and Nunes, 2014; Mehmood, 2016; Taylor Aiken, 2016; Grossmann and Creamer, 2017). Smaller scale rural initiatives allow for more experimentation

5

Cross-Chapter Box 13 (continued)

(Cretney et al., 2016), while those in urban centres benefit from stronger networks and proximity to power structures (North and Longhurst, 2013; Nicolosi and Feola, 2016). Increasingly, TTs recognize the need to participate in policymaking (Kenis and Mathijs, 2014; Barnes, 2015).

Despite high self-ratings of success, some TT initiatives are too inwardly focused and geographically isolated (Feola and Nunes, 2014), while others have difficulties in engaging marginalized, non-white, non-middle-class community members (Evans and Phelan, 2016; Nicolosi and Feola, 2016; Grossmann and Creamer, 2017). In the United Kingdom, expectations of innovations growing in scale (Taylor Aiken, 2015) and carbon accounting methods required by funding bodies (Taylor Aiken, 2016) undermine local resilience building. Tension between explicit engagements with climate change action and efforts to appeal to more people have resulted in difficult trade-offs and strained member relations (Grossmann and Creamer, 2017) though the contribution to changing an urban culture that prioritizes climate change is sometimes underestimated (Wiktorowicz et al., 2018).

Urban actions that can highlight the 1.5°C agenda include individual actions within homes (Werfel, 2017; Buntaine and Prather, 2018); demonstration zero carbon developments (Wiktorowicz et al., 2018); new partnerships between communities, government and business to build mass transit and electrify transport (Glazebrook and Newman, 2018); city plans to include climate outcomes (Millard-Ball, 2013); and support for transformative change across political, professional and sectoral divides (Bai et al., 2018).

5.6   Conditions for Achieving Sustainable Development, Eradicating Poverty and Reducing Inequalities in 1.5°C Warmer Worlds

This chapter has described the fundamental, urgent and systemic transformations that would be needed to achieve sustainable development, eradicate poverty and reduce inequalities in a 1.5°C warmer world, in various contexts and across scales. In particular, it has highlighted the societal dimensions, putting at the centre people's needs and aspirations in their specific contexts. Here we synthesize some of the most pertinent enabling conditions (see Glossary) to support these profound transformations. These conditions are closely interlinked and connected by the overarching concept of governance, which broadly includes institutional, socio-economic, cultural and technological elements (see Chapter 1, Cross-Chapter Box 4 in Chapter 1).

5.6.1   Finance and Technology Aligned with Local Needs

Significant gaps in green investment constrain transitions to a low-carbon economy aligned with development objectives (Volz et al., 2015; Campiglio, 2016). Hence, unlocking new forms of public, private and public–private financing is essential to support environmental sustainability of the economic system (Croce et al., 2011; Blyth et al., 2015; Falcone et al., 2018) (see Chapter 4, Section 4.4.5). To avoid risks of undesirable trade-offs with the SDGs caused by national budget constraints, improved access to international climate finance is essential for supporting adaptation, mitigation and sustainable development, especially for LDCs and SIDS (*medium evidence, high agreement*) (Shine and Campillo, 2016; Wood, 2017). Care needs to be taken when international donors or partnership arrangements influence project financing structures (Kongsager and Corbera, 2015; Purdon, 2015; Phillips et al., 2017; Ficklin et al., 2018). Conventional climate funding schemes, especially the Clean Development Mechanism (CDM), have shown positive effects on sustainable development but also adverse consequences, for example, on adaptive capacities of rural households and uneven distribution of costs and benefits, often exacerbating inequalities (*robust evidence, high agreement*) (Aggarwal, 2014; Brohé, 2014; He et al., 2014; Schade and Obergassel, 2014; Smits and Middleton, 2014; Wood et al., 2016a; Horstmann and Hein, 2017; Kreibich et al., 2017). Close consideration of recipients' context-specific needs when designing financial support helps to overcome these limitations as it better aligns community needs, national policy objectives and donors' priorities; puts the emphasis on the increase of transparency and predictability of support; and fosters local capacity building (*medium evidence, high agreement*) (Barrett, 2013; Boyle et al., 2013; Shine and Campillo, 2016; Ley, 2017; Sánchez and Izzo, 2017).

The development and transfer of technologies is another enabler for developing countries to contribute to the requirements of the 1.5°C objective while achieving climate resilience and their socio-economic development goals (see Chapter 4, Section 4.4.4). International-level governance would be needed to boost domestic innovation and the deployment of new technologies, such as negative emission technologies, towards the 1.5°C objective (see Chapter 4, Section 4.3.7), but the alignment with local needs depends on close consideration of the specificities of the domestic context in countries at all levels of development (de Coninck and Sagar, 2015; IEA, 2015; Parikh et al., 2018). Technology transfer supporting development in developing countries would require an understanding of local and national actors and institutions (de Coninck and Puig, 2015; de Coninck and Sagar, 2017; Michaelowa et al., 2018), careful attention to the capacities in the entire innovation chain (Khosla et al., 2017; Olawuyi, 2017) and transfer of not only equipment but also knowledge (*medium evidence, high agreement*) (Murphy et al., 2015).

5.6.2   Integration of Institutions

Multilevel governance in climate change has emerged as a key enabler for systemic transformation and effective governance (see Chapter 4,

Section 4.4.1). On the one hand, low-carbon and climate-resilient development actions are often well aligned at the lowest scale possible (Suckall et al., 2015; Sánchez and Izzo, 2017), and informal, local institutions are critical in enhancing the adaptive capacity of countries and marginalized communities (Yaro et al., 2015). On the other hand, international and national institutions can provide incentives for projects to harness synergies and avoid trade-offs (Kongsager et al., 2016).

Governance approaches that coordinate and monitor multiscale policy actions and trade-offs across sectoral, local, national, regional and international levels are therefore best suited to implement goals towards 1.5°C warmer conditions and sustainable development (Ayers et al., 2014; Stringer et al., 2014; von Stechow et al., 2016; Gwimbi, 2017; Hayward, 2017; Maor et al., 2017; Roger et al., 2017; Michaelowa et al., 2018). Vertical and horizontal policy integration and coordination is essential to take into account the interplay and trade-offs between sectors and spatial scales (Duguma et al., 2014; Naess et al., 2015; von Stechow et al., 2015; Antwi-Agyei et al., 2017a; Di Gregorio et al., 2017; Runhaar et al., 2018), enable the dialogue between local communities and institutional bodies (Colenbrander et al., 2016), and involve non-state actors such as business, local governments and civil society operating across different scales (*robust evidence, high agreement*) (Hajer et al., 2015; Labriet et al., 2015; Hale, 2016; Pelling et al., 2016; Kalafatis, 2017; Lyon, 2018).

5.6.3   Inclusive Processes

Inclusive governance processes are critical for preparing for a 1.5°C warmer world (Fazey et al., 2018; O'Brien, 2018; Patterson et al., 2018). These processes have been shown to serve the interests of diverse groups of people and enhance empowerment of often excluded stakeholders, notably women and youth (MRFCJ, 2015a; Dumont et al., 2017). They also enhance social- and co-learning which, in turn, facilitates accelerated and adaptive management and the scaling up of capacities for resilience building (Ensor and Harvey, 2015; Reij and Winterbottom, 2015; Tschakert et al., 2016; Binam et al., 2017; Dumont et al., 2017; Fazey et al., 2018; Lyon, 2018; O'Brien, 2018), and provides opportunities to blend indigenous, local and scientific knowledge (*robust evidence, high agreement*) (see Chapter 4, Section 4.3.5.5, Box 4.3, Section 5.3) (Antwi-Agyei et al., 2017a; Coe et al., 2017; Thornton and Comberti, 2017) . Such co-learning has been effective in improving deliberative decision-making processes that incorporate different values and world views (Cundill et al., 2014; C. Butler et al., 2016; Ensor, 2016; Fazey et al., 2016; Gorddard et al., 2016; Aipira et al., 2017; Chung Tiam Fook, 2017; Maor et al., 2017), and create space for negotiating diverse interests and preferences (*robust evidence, high agreement*) (O'Brien et al., 2015; Gillard et al., 2016; DeCaro et al., 2017; Harris et al., 2017; Lahn, 2018).

5.6.4   Attention to Issues of Power and Inequality

Societal transformations to limit global warming to 1.5°C and strive for equity and well-being for all are not power neutral (Section 5.5.3). Development preferences are often shaped by powerful interests that determine the direction and pace of change, anticipated benefits and beneficiaries, and acceptable and unacceptable trade-offs (Newell et

al., 2014; Fazey et al., 2016; Tschakert et al., 2016; Winkler and Dubash, 2016; Wood et al., 2016b; Karlsson et al., 2017; Quan et al., 2017; Tanner et al., 2017). Each development pathway, including legacies and path dependencies, creates its own set of opportunities and challenges and winners and losers, both within and across countries (Figure 5.5) (*robust evidence, high agreement*) (Mathur et al., 2014; Phillips et al., 2017; Stringer et al., 2017; Wood, 2017; Ficklin et al., 2018; Gajjar et al., 2018).

Addressing the uneven distribution of power is critical to ensure that societal transformation towards a 1.5°C warmer world does not exacerbate poverty and vulnerability or create new injustices but rather encourages equitable transformational change (Patterson et al., 2018). Equitable outcomes are enhanced when they pay attention to just outcomes for those negatively affected by change (Newell et al., 2014; Dilling et al., 2015; Naess et al., 2015; Sovacool et al., 2015; Cervigni and Morris, 2016; Keohane and Victor, 2016) and promote human rights, increase equality and reduce power asymmetries within societies (*robust evidence, high agreement*) (UNRISD, 2016; Robinson and Shine, 2018).

5.6.5   Reconsidering Values

The profound transformations that would be needed to integrate sustainable development and 1.5°C-compatible pathways call for examining the values, ethics, attitudes and behaviours that underpin societies (Hartzell-Nichols, 2017; O'Brien, 2018; Patterson et al., 2018). Infusing values that promote sustainable development (Holden et al., 2017), overcome individual economic interests and go beyond economic growth (Hackmann, 2016), encourage desirable and transformative visions (Tàbara et al., 2018), and care for the less fortunate (Howell and Allen, 2017) is part and parcel of climate-resilient and sustainable development pathways. This entails helping societies and individuals to strive for sufficiency in resource consumption within planetary boundaries alongside sustainable and equitable well-being (O'Neill et al., 2018). Navigating 1.5°C societal transformations, characterized by action from local to global, stresses the core commitment to social justice, solidarity and cooperation, particularly regarding the distribution of responsibilities, rights and mutual obligations between nations (*medium evidence, high agreement*) (Patterson et al., 2018; Robinson and Shine, 2018).

5.7   Synthesis and Research Gaps

The assessment in Chapter 5 illustrates that limiting global warming to 1.5°C above pre-industrial levels is fundamentally connected with achieving sustainable development, poverty eradication and reducing inequalities. It shows that avoided impacts between 1.5°C and 2°C temperature stabilization would make it easier to achieve many aspects of sustainable development, although important risks would remain at 1.5°C (Section 5.2). Synergies between adaptation and mitigation response measures with sustainable development and the SDGs can often be enhanced when attention is paid to well-being and equity while, when unaddressed, poverty and inequalities may be exacerbated (Section 5.3 and 5.4). Climate-resilient development pathways (CRDPs) open up routes towards socially desirable futures that are sustainable and liveable, but concrete evidence reveals complex trade-offs along a continuum of different pathways, highlighting the role of societal values, internal contestations and political dynamics (Section 5.5). The transformations towards sustainable development in a 1.5°C warmer world, in all contexts, involve fundamental societal and systemic changes over time and across scale, and a set of enabling conditions without which the dual goal is difficult if not impossible to achieve (Sections 5.5 and 5.6).

This assessment is supported by growing knowledge on the linkages between a 1.5°C warmer world and different dimensions of sustainable development. However, several gaps in the literature remain:

Limited evidence exists that explicitly examines the real-world implications of a 1.5°C warmer world (and overshoots) as well as avoided impacts between 1.5°C versus 2°C for the SDGs and sustainable development more broadly. Few projections are available for households, livelihoods and communities. And literature on differential localized impacts and their cross-sector interacting and cascading effects with multidimensional patterns of societal vulnerability, poverty and inequalities remains scarce. Hence, caution is needed when global-level conclusions about adaptation and mitigation measures in a 1.5°C warmer world are applied to sustainable development in local, national and regional settings.

Limited literature has systematically evaluated context-specific synergies and trade-offs between and across adaptation and mitigation response measures in 1.5°C-compatible pathways and the SDGs. This hampers the ability to inform decision-making and fair and robust policy packages adapted to different local, regional or national circumstances. More research is required to understand how trade-offs and synergies will intensify or decrease, differentially across geographic regions and time, in a 1.5°C warmer world and as compared to higher temperatures.

Limited availability of interdisciplinary studies also poses a challenge for connecting the socio-economic transformations and the governance aspects of low emissions, climate-resilient transformations. For example, it remains unclear how governance structures enable or hinder different groups of people and countries to negotiate pathway options, values and priorities.

The literature does not demonstrate the existence of 1.5°C-compatible pathways achieving the 'universal and indivisible' agenda of the 17 SDGs, and hence does not show whether and how the nature and pace of changes that would be required to meet 1.5°C climate stabilization could be fully synergetic with all the SDGs.

The literature on low emissions and CRDPs in local, regional and national contexts is growing. Yet the lack of standard indicators to monitor such pathways makes it difficult to compare evidence grounded in specific contexts with differential circumstances, and therefore to derive generic lessons on the outcome of decisions on specific indicators. This knowledge gap poses a challenge for connecting local-level visions with global-level trajectories to better understand key conditions for societal and systems transformations that reconcile urgent climate action with well-being for all.

Frequently Asked Questions

FAQ 5.1 | What are the Connections between Sustainable Development and Limiting Global Warming to 1.5°C above Pre-Industrial Levels?

Summary: Sustainable development seeks to meet the needs of people living today without compromising the needs of future generations, while balancing social, economic and environmental considerations. The 17 UN Sustainable Development Goals (SDGs) include targets for eradicating poverty; ensuring health, energy and food security; reducing inequality; protecting ecosystems; pursuing sustainable cities and economies; and a goal for climate action (SDG 13). Climate change affects the ability to achieve sustainable development goals, and limiting warming to 1.5°C will help meet some sustainable development targets. Pursuing sustainable development will influence emissions, impacts and vulnerabilities. Responses to climate change in the form of adaptation and mitigation will also interact with sustainable development with positive effects, known as synergies, or negative effects, known as trade-offs. Responses to climate change can be planned to maximize synergies and limit trade-offs with sustainable development.*

For more than 25 years, the United Nations (UN) and other international organizations have embraced the concept of sustainable development to promote well-being and meet the needs of today's population without compromising the needs of future generations. This concept spans economic, social and environmental objectives including poverty and hunger alleviation, equitable economic growth, access to resources, and the protection of water, air and ecosystems. Between 1990 and 2015, the UN monitored a set of eight Millennium Development Goals (MDGs). They reported progress in reducing poverty, easing hunger and child mortality, and improving access to clean water and sanitation. But with millions remaining in poor health, living in poverty and facing serious problems associated with climate change, pollution and land-use change, the UN decided that more needed to be done. In 2015, the UN Sustainable Development Goals (SDGs) were endorsed as part of the 2030 Agenda for Sustainable Development. The 17 SDGs (Figure FAQ 5.1) apply to all countries and have a timeline for success by 2030. The SDGs seek to eliminate extreme poverty and hunger; ensure health, education, peace, safe water and clean energy for all; promote inclusive and sustainable consumption, cities, infrastructure and economic growth; reduce inequality including gender inequality; combat climate change and protect oceans and terrestrial ecosystems.

Climate change and sustainable development are fundamentally connected. Previous IPCC reports found that climate change can undermine sustainable development, and that well-designed mitigation and adaptation responses can support poverty alleviation, food security, healthy ecosystems, equality and other dimensions of sustainable development. Limiting global warming to 1.5°C would require mitigation actions and adaptation measures to be taken at all levels. These adaptation and mitigation actions would include reducing emissions and increasing resilience through technology and infrastructure choices, as well as changing behaviour and policy.

These actions can interact with sustainable development objectives in positive ways that strengthen sustainable development, known as synergies. Or they can interact in negative ways, where sustainable development is hindered or reversed, known as trade-offs.

An example of a synergy is sustainable forest management, which can prevent emissions from deforestation and take up carbon to reduce warming at reasonable cost. It can work synergistically with other dimensions of sustainable development by providing food (SDG 2) and clean water (SDG 6) and protecting ecosystems (SDG 15). Other examples of synergies are when climate adaptation measures, such as coastal or agricultural projects, empower women and benefit local incomes, health and ecosystems.

An example of a trade-off can occur if ambitious climate change mitigation compatible with 1.5°C changes land use in ways that have negative impacts on sustainable development. An example could be turning natural forests, agricultural areas, or land under indigenous or local ownership to plantations for bioenergy production. If not managed carefully, such changes could undermine dimensions of sustainable development by threatening food and water security, creating conflict over land rights and causing biodiversity loss. Another trade-off could occur for some countries, assets, workers and infrastructure already in place if a switch is made from fossil fuels to other energy sources without adequate planning for such a transition. Trade-offs can be minimized if effectively managed, as when care is taken to improve bioenergy crop yields to reduce harmful land-use change or where workers are retrained for employment in lower carbon sectors.

(continued on next page)

5

FAQ 5.1 (continued)

Limiting temperature increase to 1.5°C can make it much easier to achieve the SDGs, but it is also possible that pursuing the SDGs could result in trade-offs with efforts to limit climate change. There are trade-offs when people escaping from poverty and hunger consume more energy or land and thus increase emissions, or if goals for economic growth and industrialization increase fossil fuel consumption and greenhouse gas emissions. Conversely, efforts to reduce poverty and gender inequalities and to enhance food, health and water security can reduce vulnerability to climate change. Other synergies can occur when coastal and ocean ecosystem protection reduces the impacts of climate change on these systems. The sustainable development goal of affordable and clean energy (SDG 7) specifically targets access to renewable energy and energy efficiency, which are important to ambitious mitigation and limiting warming to 1.5°C.

The link between sustainable development and limiting global warming to 1.5°C is recognized by the SDG for climate action (SDG 13), which seeks to combat climate change and its impacts while acknowledging that the United Nations Framework Convention on Climate Change (UNFCCC) is the primary international, intergovernmental forum for negotiating the global response to climate change.

The challenge is to put in place sustainable development policies and actions that reduce deprivation, alleviate poverty and ease ecosystem degradation while also lowering emissions, reducing climate change impacts and facilitating adaptation. It is important to strengthen synergies and minimize trade-offs when planning climate change adaptation and mitigation actions. Unfortunately, not all trade-offs can be avoided or minimized, but careful planning and implementation can build the enabling conditions for long-term sustainable development.

FAQ5.1: The United Nations Sustainable Development Goals (SDGs)

The link between sustainable development and limiting global warming to 1.5°C is recognised by the Sustainable Development Goal for climate action (SDG 13)

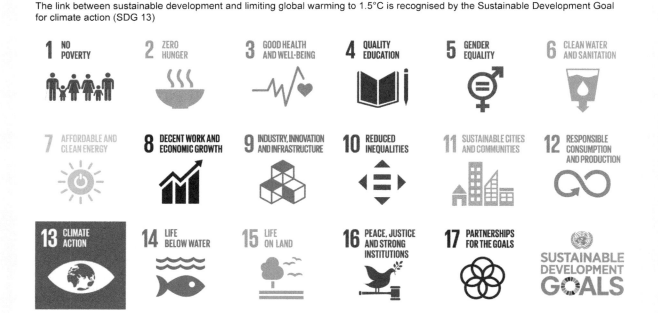

FAQ 5.1, Figure 1 | Climate change action is one of the United Nations Sustainable Development Goals (SDGs) and is connected to sustainable development more broadly. Actions to reduce climate risk can interact with other sustainable development objectives in positive ways (synergies) and negative ways (trade-offs).

Frequently Asked Questions

FAQ 5.2 | What are the Pathways to Achieving Poverty Reduction and Reducing Inequalities while Reaching a 1.5°C World?

Summary: There are ways to limit global warming to 1.5°C above pre-industrial levels. Of the pathways that exist, some simultaneously achieve sustainable development. They entail a mix of measures that lower emissions and reduce the impacts of climate change, while contributing to poverty eradication and reducing inequalities. Which pathways are possible and desirable will differ between and within regions and nations. This is due to the fact that development progress to date has been uneven and climate-related risks are unevenly distributed. Flexible governance would be needed to ensure that such pathways are inclusive, fair and equitable to avoid poor and disadvantaged populations becoming worse off. Climate-resilient development pathways (CRDPs) offer possibilities to achieve both equitable and low-carbon futures.

Issues of equity and fairness have long been central to climate change and sustainable development. Equity, like equality, aims to promote justness and fairness for all. This is not necessarily the same as treating everyone equally, since not everyone comes from the same starting point. Often used interchangeably with fairness and justice, equity implies implementing different actions in different places, all with a view to creating an equal world that is fair for all and where no one is left behind.

The Paris Agreement states that it 'will be implemented to reflect equity... in the light of different national circumstances' and calls for 'rapid reductions' of greenhouse gases to be achieved 'on the basis of equity, and in the context of sustainable development and efforts to eradicate poverty'. Similarly, the UN SDGs include targets to reduce poverty and inequalities, and to ensure equitable and affordable access to health, water and energy for all.

Equity and fairness are important for considering pathways that limit warming to 1.5°C in a way that is liveable for every person and species. They recognize the uneven development status between richer and poorer nations, the uneven distribution of climate impacts (including on future generations) and the uneven capacity of different nations and people to respond to climate risks. This is particularly true for those who are highly vulnerable to climate change, such as indigenous communities in the Arctic, people whose livelihoods depend on agriculture or coastal and marine ecosystems, and inhabitants of small island developing states. The poorest people will continue to experience climate change through the loss of income and livelihood opportunities, hunger, adverse health effects and displacement.

Well-planned adaptation and mitigation measures are essential to avoid exacerbating inequalities or creating new injustices. Pathways that are compatible with limiting warming to 1.5°C and aligned with the SDGs consider mitigation and adaptation options that reduce inequalities in terms of who benefits, who pays the costs and who is affected by possible negative consequences. Attention to equity ensures that disadvantaged people can secure their livelihoods and live in dignity, and that those who experience mitigation or adaptation costs have financial and technical support to enable fair transitions.

CRDPs describe trajectories that pursue the dual goal of limiting warming to 1.5°C while strengthening sustainable development. This includes eradicating poverty as well as reducing vulnerabilities and inequalities for regions, countries, communities, businesses and cities. These trajectories entail a mix of adaptation and mitigation measures consistent with profound societal and systems transformations. The goals are to meet the short-term SDGs, achieve longer-term sustainable development, reduce emissions towards net zero around the middle of the century, build resilience and enhance human capacities to adapt, all while paying close attention to equity and well-being for all.

The characteristics of CRDPs will differ across communities and nations, and will be based on deliberations with a diverse range of people, including those most affected by climate change and by possible routes towards transformation. For this reason, there are no standard methods for designing CRDPs or for monitoring their progress towards climate-resilient futures. However, examples from around the world demonstrate that flexible and inclusive governance structures and broad participation often help support iterative decision-making, continuous learning and experimentation. Such inclusive processes can also help to overcome weak institutional arrangements and power structures that may further exacerbate inequalities.

(continued on next page)

5

FAQ 5.2 (continued)

FAQ5.2: **Climate-resilient development pathways**

Decision-making that achieves the United Nation Sustainable Development Goals (SDGs), lowers greenhouse gas emissions, limits global warming and enables adaptation could help lead to a climate-resilient world.

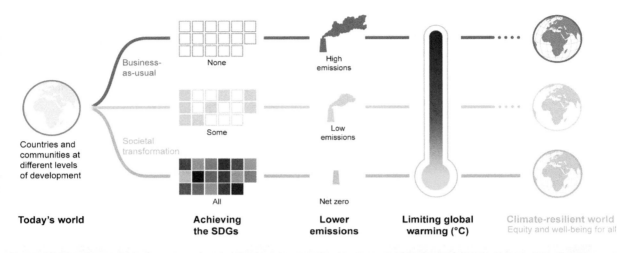

FAQ 5.2, Figure 1 | Climate-resilient development pathways (CRDPs) describe trajectories that pursue the dual goals of limiting warming to 1.5°C while strengthening sustainable development. Decision-making that achieves the SDGs, lowers greenhouse gas emissions and limits global warming could help lead to a climate-resilient world, within the context of enhancing adaptation.

Ambitious actions already underway around the world can offer insight into CRDPs for limiting warming to 1.5°C. For example, some countries have adopted clean energy and sustainable transport while creating environmentally friendly jobs and supporting social welfare programmes to reduce domestic poverty. Other examples teach us about different ways to promote development through practices inspired by community values. For instance, *Buen Vivir*, a Latin American concept based on indigenous ideas of communities living in harmony with nature, is aligned with peace; diversity; solidarity; rights to education, health, and safe food, water, and energy; and well-being and justice for all. The Transition Movement, with origins in Europe, promotes equitable and resilient communities through low-carbon living, food self-sufficiency and citizen science. Such examples indicate that pathways that reduce poverty and inequalities while limiting warming to 1.5°C are possible and that they can provide guidance on pathways towards socially desirable, equitable and low-carbon futures.

5

Table 5.2 | Mitigation – SDG table
Social-Demand

Industry	SDG 1 NO POVERTY					SDG 2 ZERO HUNGER					SDG 3 GOOD HEALTH AND WELL-BEING					SDG 4 QUALITY EDUCATION				
	Interaction	Score	Evidence	Agreement	Confidence	Interaction	Score	Evidence	Agreement	Confidence	Interaction	Score	Evidence	Agreement	Confidence	Interaction	Score	Evidence	Agreement	Confidence
Accelerating Energy Efficiency Improvement	Reduces Poverty ↑	[+2]	▣	☺	★		[0]				Air, Water Pollution Reduction and Better Health (3.9) ↑	[+2]	▣▣▣	☺☺	★★	Technical Education, Vocational Training, Education for Sustainability (4.3/4.4/4.5/4.7) ←	[+1]	▣▣▣	☺☺	★★
	% of people living below poverty line declines from 49% to 18% in South African context.						No direct interaction				People living in deprived communities feel positive and predict considerable financial savings. Efficiency changes in the industrial sector that lead to reduced energy demand can lead to reduced requirements on energy supply. As water is used to convert energy into useful forms, the reduction in industrial demand is anticipated to reduce water consumption and wastewater, resulting in more clean water for other sectors and the environment. In extractive industries there are trade-off unless strategically managed. Behavioural changes in the industrial sector that lead to reduced energy demand can lead to reduced requirements on energy supply. As water is used to convert energy into useful forms, the reduction in industrial demand is anticipated to reduce water consumption and wastewater, resulting in more clean water for other sectors and the environment.					Awareness, knowledge, technical and managerial capability are closely linked, energy audit, information for trade unions, product/appliance labeling help in sustainability education.				
	Altieri et al., 2016										Vassolo and Döll, 2005; Xi et al., 2013; Nguyen et al., 2014; Holland et al., 2015; Zhang et al., 2015; Fricko et al., 2016					Apeaning and Thollander, 2013; Fernando and Evans, 2015; Roy et al., 2018				
Low-carbon Fuel Switch		[0]					[0]				Water and Air Pollution Reduction and Better Health (3.9) ↑	[+2]	▣▣▣	☺☺	★★	Technical Education, Vocational Training, Education for Sustainability (4.b/4.7) ←	[+1]	▣▣▣	☺☺	★★
	No direct interaction					No direct interaction					Industries are becoming suppliers of energy, waste heat, water and roof tops for solar energy generation, and hence helping to improve air and water quality.					New technology deployment creates demand for awareness and knowledge with technical and managerial capability; otherwise acts as barrier for rapid expansion.				
											Vassolo and Döll, 2005; Nguyen et al., 2014; Holland et al., 2015; Karner et al., 2015; Fricko et al., 2016					Apeaning and Thollander, 2013; Fernando and Evans, 2015; Roy et al., 2018				
Decarbonisation/CCS/CCU		[0]					[0]				Disease and Mortality (3.1/3.2/3.3/3.4) ↓	[-1]	▣▣	☺☺☺	★★★		[0]			
	No direct interaction					No direct interaction					There is a risk of CO_2 leakage both from geological formations as well as from the transportation infrastructure from source to sequestration locations.					No direct interaction				
											Wang and Jaffe, 2004; Hertwich et al., 2008; Apps et al., 2010; Veltman et al., 2010; Koornneef et al., 2011; Singh et al., 2011; Sirila et al., 2012; Atchley et al., 2013; Corsten et al., 2013; IPCC, 2014									

5

Social-Demand (continued)

Sector rows (left axis): Buildings → Behavioural Response / Accelerating Energy Efficiency Improvement / Improved Access and Fuel Switch to Modern Low-carbon Energy

Each SDG column has sub-columns: Interaction | Score | Evidence | Agreement | Confidence

SDG 1 — NO POVERTY

Behavioural Response — Poverty Reduction via Financial Savings (1.1)
Interaction: ← | Score: [+2] | Agreement: ● | Confidence: ★

People living in deprived communities feel positive and predict considerable financial savings.

Scott et al., 2014

Accelerating Energy Efficiency Improvement — Poverty and Development (1.1/1.2/1.3/1.4)
Interaction: ←/↑ | Score: [+2,-1] | Agreement: ●●● | Confidence: ★★★

Energy efficiency interventions lead to cost savings which are realized due to reduced energy bills that further lead to poverty reduction. Participants with low incomes experience greater benefits. 'Energy efficiency and biomass strategies benefitted the poor more than wind and solar, whose benefits are captured by industry. Carbon mitigation can increase or decrease inequalities. The distributional costs of new energy policies (e.g., supporting renewables and energy efficiency) are dependent on instrument design. If costs fall disproportionately on the poor, then this could impair progress towards universal energy access and, by extension, counteract the fight to eliminate poverty. (Quote from McCollum et al., 2018).

Casillas and Kammen, 2012; Hirth and Ueckerdt, 2013; Jakob and Steckel, 2014; Maidment et al., 2014; Scott et al., 2014; Fay et al., 2015; Cameron et al., 2016; Hallegatte et al., 2016; Berrueta et al., 2017; McCollum et al., 2018

Improved Access and Fuel Switch to Modern Low-carbon Energy — Poverty and Development (1.1/1.2/1.3/1.4)
Interaction: ← | Score: [+2] | Agreement: ●●● | Confidence: ★★★★

Access to modern energy forms (electricity, clean stoves, high-quality lighting) is fundamental to human development since the energy services made possible by them help alleviate chronic and persistent poverty. Strength of the impact varies in the literature. (Quote from McCollum et al., 2018)

Kirubi et al., 2009; Casillas and Kammen, 2010; Cook, 2011; Pachauri et al., 2012; Pode, 2013; Pueyo et al., 2013; Zulu and Richardson, 2013; Bonan et al., 2014; Rao et al., 2014; Burlig and Preonas, 2016; McCollum et al., 2018

SDG 2 — ZERO HUNGER

Behavioural Response
Score: [0] — No direct interaction

Accelerating Energy Efficiency Improvement — Food Security (2.1)
Interaction: ← | Score: [+2] | Agreement: ● | Confidence: ★

Using the improved stoves supports local food security and has significantly impacted on food security. By making fuel last longer, the improved stoves help improve food security and also provide a better buffer against fuel shortages induced by climate change-related events such as droughts, floods or hurricanes (Berrueta et al. 2017).

Berrueta et al., 2017

Improved Access and Fuel Switch to Modern Low-carbon Energy — Food Security and Agricultural Productivity (2.1/2.4)
Interaction: ~/↓ | Score: [0,-1] | Agreement: ●●● | Confidence: ★★

Modern energy access is critical to enhance agricultural yields/productivity, decrease post-harvest losses and mechanize agri-processing – all of which can aid food security. However, large-scale bioenergy and food production may compete for scarce land and other inputs (e.g., water, fertilizers), depending on how and where biomass supplies are grown and the indirect land use change impacts that result. If not implemented thoughtfully, this could lead to higher food prices globally, and thus reduce access to affordable food for the poor. Enhanced agricultural productivities can ameliorate the situation by allowing as much bioenergy to be produced on as little land as possible.

Cabraal et al., 2005; Tilman et al., 2009; van Vuuren et al., 2009; Asaduzzaman et al., 2010; Finco and Doppler, 2010; Msangi et al., 2010; Smith et al., 2013, 2014; Lotze-Campen et al., 2014; Hasegawa et al., 2015; Sola et al., 2016; McCollum et al., 2018

SDG 3 — GOOD HEALTH AND WELL-BEING

Behavioural Response — Improved Warmth and Comforts
Interaction: ← | Score: [+2] | Agreement: ●●● | Confidence: ★★★

Home occupants reported warmth as the most important aspect of comfort which was largely temperature-related and low in energy costs. Residents living in deprived areas expect improved warmth in their properties after energy efficiency measures are employed.

Huebner et al., 2013; Yue et al., 2013; Scott et al., 2014; Zhao et al., 2017

Accelerating Energy Efficiency Improvement — Healthy Lives and Well-being for All at All Ages (3.2/3.9)
Interaction: ← | Score: [+2] | Agreement: ●●● | Confidence: ★★★★

Efficient stoves improve health, especially for indigenous and poor rural communities. Household energy efficiency has positive health impacts on children's respiratory health, weight and susceptibility to illness, and the mental health of adults. Household energy efficiency improves winter warmth, lowers relative humidity with benefits for cardiovascular and respiratory health. Further improved indoor air quality by thermal regulation and occupant comfort are realised. However, in one instance, negative health impacts (asthma) of increased household energy efficiency were also noted when housing upgrades took place without changes in occupant behaviours. Home occupants reported warmth as the most important aspect of comfort which was largely temperature-related and low in energy costs. Residents living in the deprived areas expect improved warmth in their properties after energy efficiency measures are employed.

Djamila et al., 2013; Huebner et al., 2013; Yue et al., 2013; Bhojvaid et al., 2014; Derbez et al., 2014; Maidment et al., 2014; Scott et al., 2014; Cameron et al., 2015; Liddell and Guiney, 2015; Sharpe et al., 2015; Wells et al., 2015; Willand et al., 2015; Berrueta et al., 2017; Zhao et al., 2017

Improved Access and Fuel Switch to Modern Low-carbon Energy — Disease and Mortality (3.1/3.2/3.3/3.4)
Interaction: ← | Score: [+2] | Agreement: ●●● | Confidence: ★★★★

Access to modern energy services can contribute to fewer injuries and diseases related to traditional solid fuel collection and burning, as well as utilization of kerosene lanterns. Access to modern energy services can facilitate improved health care provision, medicine and vaccine storage, utilization of powered medical equipment, and dissemination of health-related information and education. Such services can also enable thermal comfort in homes and contribute to food preservation and safety. (Quote from McCollum et al., 2018)

Lam et al., 2012; Lim et al., 2012; Smith et al., 2013; Aranda et al., 2014; McCollum et al., 2018

SDG 4 — QUALITY EDUCATION

Behavioural Response
Score: [0] — No direct interaction

Accelerating Energy Efficiency Improvement — Equal Access to Educational Institutions (4.1/4.2/4.3/4.5)
Interaction: ← | Score: [+2] | Agreement: ● | Confidence: ★

Household energy efficiency measures reduce school absences for children with asthma due to indoor pollution.

Maidment et al., 2014

Improved Access and Fuel Switch to Modern Low-carbon Energy — Equal Access to Educational Institutions (4.1/4.2/4.3/4.5)
Interaction: ← | Score: [+1] | Agreement: ●●● | Confidence: ★★

Access to modern energy is necessary for schools to have quality lighting and thermal comfort, as well as modern information and communication technologies. Access to modern lighting and energy allows for studying after sundown and frees constraints on time management that allow for higher school enrolment rates and better literacy outcomes. (Quote from McCollum et al., 2018)

Lipscomb et al., 2013; van de Walle et al., 2013; McCollum et al., 2018

Social-Demand (continued)

Transport

1 NO POVERTY

Behavioural Response — Equal Right to Economic Resources Access Basic Services (1.1/1.4/1.a/1.b)

Interaction: ↑/↓ | Score [+2,-1] | Evidence ▣▣▣▣ | Agreement ●●● | Confidence ★★★

The costs of daily mobility can have important economic stress impacts, not only impacting carless families with low-mobility, but in countries with high levels of car dependence, the costs of motoring can be burdensome, raising questions of affordability for households with limited economic resources. During economic crisis, public transport authorities may react by reducing levels of service and increasing fares, likely exacerbating the situation for low-income households.

Dodson et al., 2004; Cascajo et al., 2017

Accelerating Energy Efficiency Improvement — End Poverty in all its Forms Everywhere (1.1/1.4/1.a/1.b)

Interaction: ↑/↓ | Score [+2,-1] | Evidence ▣▣▣▣ | Agreement ●●● | Confidence ★★★

Decarbonization of public buses in Sweden is receiving attention more than efficiency improvement. With more electrification, electricity prices go up and affordability can worsen for the poor unless redistributive policies are in place.

Xylia and Silveira, 2017

Improved Access and Fuel Switch to Modern Low-carbon Energy — End Poverty in all its Forms Everywhere (1.1/1.4/1.a/1.b)

Interaction: ↑/↓ | Score [+2,-1] | Evidence ▣▣▣▣ | Agreement ●●● | Confidence ★★★

Increasingly volatile global oil prices have raised concerns for the vulnerability of households to fuel price increases. Pricing measures as a key component of sustainable transport policy need to consider equity. Pro-poor mitigation policies are needed to reduce climate impact and reduce threat; for example, investing more and better in infrastructure by leveraging private resources and using designs that account for future climate change and the related uncertainty. Communities in poor areas cope with and adapt to multiple-stressors including climate change. Coping strategies provide short-term relief but in the long-term may negatively affect development goals. And responses generate a trade-off between adaptation, mitigation and development. For African cities with slums, due to high commuting costs, many walk to work places which limit access. In Latin America triple informality leading to low productivity and living standards.

Dodson and Sipe, 2008; Suckall et al., 2014; Suckall et al., 2014; Hallegatte et al., 2016a; Klausbruckner et al., 2016; CAF, 2017; Lall et al., 2017

2 ZERO HUNGER

Behavioural Response — Ensure Access to Safe Nutritious Food (2.1/2.2)

Interaction: ↑ | Score [+2] | Evidence ▣ | Agreement ⊕ | Confidence ★

Low-income community residents (non-white) who lack local access to affordable, quality sources of nutrition have to travel outside their immediate neighbourhood to find better sources of food to feed themselves and their families. Lack of locally available healthy food often exacerbates the rates of obesity in many of these communities since it is often difficult or expensive to travel long distances on a regular basis to shop for food.

Clifton, 2004; Hillier, 2011; Krukowski et al., 2013; LeDoux and Vojnovic, 2013; Ghosh-Dastidar et al., 2014; Zenk et al., 2015; Lowery et al., 2016

Accelerating Energy Efficiency Improvement

Score [0] — No direct interaction

Improved Access and Fuel Switch to Modern Low-carbon Energy — Ensure Access to Food Security (2.1/2.2.3/2.a/2.b/2.c)

Interaction: ~ | Score [0] | Evidence ▣ | Agreement ⊕

21 projects aiming at resilient transport infrastructure development to improve access (e.g., C40 Cities Clean Bus Declaration, UITP Declaration on Climate Leadership, Cycling Delivers on the Global Goals, Global Sidewalk Challenge) do not substantially contribute to realizing the (indirect) transport targets with mostly a rural focus: agricultural productivity (SDG 2) and access to safe drinking water (SDG 6).

SloCaT, 2017

3 GOOD HEALTH AND WELL-BEING

Behavioural Response — Road Traffic Accidents (3.4/3.6)

Interaction: ↑/↓ | Score [+2,-1] | Evidence ▣▣▣▣ | Agreement ●●● | Confidence ★★

Active travel modes, such as walking and cycling, represent strategies not only for boosting energy efficiency but also, potentially, for improving health and well-being (e.g., lowering rates of diabetes, obesity, heart disease, dementia and some cancers). However, a risk associated with these measures is that they could increase rates of road traffic accidents, if the existing infrastructure is unsatisfactory. Overall health effects will depend on the severity of the injuries sustained from these potential accidents relative to the health benefits accruing from increased exercise (McCollum et al., 2018).

Woodcock et al., 2009; Creutzig et al., 2012; Haines and Dora, 2012; Saunders et al., 2013; Shaw et al., 2014, 2017; Chakrabarti and Shin, 2017; Hwang et al., 2017; McCollum et al., 2018

Accelerating Energy Efficiency Improvement — Reduce Illnesses from Hazardous Air, Water and Soil Pollution (3.9)

Interaction: ↑ | Score [+2] | Evidence ▣▣▣▣ | Agreement ●●● | Confidence ★★★

Locally relevant policies targeting traffic reductions and ambitious diffusion of electric vehicles results in measured changes in non-climatic exposure for population, including ambient air pollution, physical activity and noise. The transition to low-carbon equitable and sustainable transport can be fostered by numerous short- and medium-term strategies that would benefit energy security, health, productivity and sustainability. An evidence-based approach that takes into account GHG emissions, ambient air pollutants, economic factors (affordability, cost optimization), social factors (poverty alleviations, public health benefits) and political acceptability is needed to tackle these challenges.

Figueroa et al., 2014; Schucht et al., 2015; Klausbruckner et al., 2016; Peng et al., 2017

Improved Access and Fuel Switch to Modern Low-carbon Energy — Reduce Illnesses from Hazardous Air Pollution (3.9)

Interaction: ↑ | Score [+2] | Evidence ▣ | Agreement ⊕

Projects aiming at resilient transport infrastructure development (e.g., C40 Cities Clean Bus Declaration, UITP Declaration on Climate Leadership, Cycling Delivers on the Global Goals, Global Sidewalk Challenge) are targeted at reducing air pollution; electric vehicles using electricity from renewables or low carbon sources combined with e-mobility options such as trolley buses, metros, trams and electro buses, as well as promoting walking and biking, especially for short distances, need consideration.

Ajanovic, 2015; SloCaT, 2017

4 QUALITY EDUCATION

Behavioural Response — Equal Safe Access to Educational Institutions (4.1/4.2/4.3/4.5)

Interaction: ↑ | Score [+1] | Evidence ▣ | Agreement ●●● | Confidence ★★

Poor road quality affects school travel safety, so collaborative efforts need to address safety issues from a dual perspective, first by working to change the existing infrastructure and use of roads to better address the traffic problems that children currently face walking to school, and then to better situate schools and control the roadways and land uses around them in the future.

Yu, 2015

Accelerating Energy Efficiency Improvement

Score [0] — No direct interaction

Improved Access and Fuel Switch to Modern Low-carbon Energy

Score [0] — No direct interaction

Social-Supply

Replacing Coal

SDG 1 — NO POVERTY

Non-biomass Renewables – solar, wind, hydro

Interaction: Poverty and Development (1.1/1.2/1.3/1.4) | Score: [+2] | Evidence: ●●● | Agreement: ☺☺ | Confidence: ★★★

Deployment of renewable energy and improvements in energy efficiency globally will aid climate change mitigation efforts, and this, in turn, can help to reduce the exposure of the world's poor to climate-related extreme events, negative health impacts and other environmental shocks (McCollum et al., 2018).

Riahi et al., 2012; IPCC, 2014; Hallegatte et al., 2016b; McCollum et al., 2018

Increased Use of Biomass

Interaction: Poverty and Development (1.1/1.2/1.3/1.4) | Score: [+2,-2] | Evidence: ●●● | Agreement: ☺☺☺ | Confidence: ★★★

Large-scale bioenergy production could lead to the creation of agricultural jobs, as well as higher farm wages and more diversified income streams for farmers. Modern energy access can make marginal lands more cultivable, thus potentially generating on-farm jobs and incomes; on the other hand, greater farm mechanization can also displace labour. However, large-scale bioenergy production could alter the structure of global agricultural markets in a way that is, potentially, unfavourable to small-scale food producers. See SDG2 (McCollum et al., 2018).

Balishter and Singh, 1991; Gohin, 2008; de Moraes et al., 2010; van der Horst and Vermeylen, 2011; Corbera and Pascual, 2012; Rud, 2012; Creutzig et al., 2013; Davis et al., 2013; Satolo and Bacchi, 2013; Muys et al., 2014; Ertem et al., 2017; McCollum et al., 2018

SDG 2 — ZERO HUNGER

Non-biomass Renewables – solar, wind, hydro

Score: [0] — No direct interaction

Increased Use of Biomass

Interaction: Farm Employment and Incomes (2.3) | Score: [+2,-2] | Evidence: ●●● | Agreement: ☺☺☺ | Confidence: ★★★

Large-scale bioenergy production could lead to the creation of agricultural jobs, as well as higher farm wages and more diversified income streams for farmers. Modern energy access can make marginal lands more cultivable, thus potentially generating on-farm jobs and incomes; on the other hand, greater farm mechanization can also displace labour. However, large-scale bioenergy production could alter the structure of global agricultural markets in a way that is, potentially, unfavourable to small-scale food producers. The distributional effects of bioenergy production are underexplored in the literature (McCollum et al., 2018).

Balishter and Singh, 1991; Gohin, 2008; de Moraes et al., 2010; van der Horst and Vermeylen, 2011; Corbera and Pascual, 2012; Rud, 2012; Creutzig et al., 2013; Davis et al., 2013; Satolo and Bacchi, 2013; Muys et al., 2014; Ertem et al., 2017; McCollum et al., 2018

SDG 3 — GOOD HEALTH AND WELL-BEING

Non-biomass Renewables – solar, wind, hydro

Interaction: Air Pollution (3.9) | Score: [+2] | Evidence: ●●●● | Agreement: ☺☺☺ | Confidence: ★★★★

Promoting most types of renewables and boosting efficiency greatly aids the achievement of targets to reduce local air pollution and improve air quality; however, the order of magnitude of the effects, both in terms of avoided emissions and monetary valuation, varies significantly between different parts of the world. Benefits would especially accrue to those living in the dense urban centres of rapidly developing countries. Utilization of biomass and biofuels might not lead to any air pollution benefits, however, depending on the control measures applied. In addition, household air quality can be significantly improved through lowered particulate emissions from access to modern energy services (McCollum et al., 2018).

Haines et al., 2007; Nemet et al., 2010; Kaygusuz, 2011; Riahi et al., 2012; van Vliet et al., 2012; Anenberg et al., 2013; Rafaj et al., 2013; Rao et al., 2013, 2016; West et al., 2013; Chaturvedi and Shukla, 2014; Rose et al., 2014; Smith and Sagar, 2014; IEA, 2016; McCollum et al., 2018

Increased Use of Biomass

Interaction: Disease and Mortality (3.1/3.2/3.3/3.4), Air Pollution (3.9) | Score: [+2] | Evidence: ●●●● | Agreement: ☺☺☺ | Confidence: ★★★

Replacing coal by biomass can reduce adverse impacts of upstream supply-chain activities, in particular local air and water pollution, and prevent coal mining accidents. Improvements to local air pollution in power generation compared to coal-fired power plants depend on the technology and fuel of biomass power plants, but could be significant when switching from outdated coal combustion technologies to state-of-the-art biogas power generation.

IPCC, 2005, 2014; Miller et al., 2007; Hertwich et al., 2008; de Best-Waldhober et al., 2009; Shackley et al., 2009; Wallquist et al., 2009, 2010; Wong-Parodi and Ray, 2009; Chan and Griffiths, 2010; Veltman et al., 2010; Epstein et al., 2011; Koornneef et al., 2011; Reiner and Nuttall, 2011; Singh et al., 2011; Ashworth et al., 2012; Burgher et al., 2012; Chen et al., 2012; Aslaw et al., 2013; Corsten et al., 2013; Einsiedel et al., 2013

SDG 4 — QUALITY EDUCATION

Non-biomass Renewables – solar, wind, hydro

Interaction: Vocational Training, Education for Sustainability (4.b/4.7) | Score: [+1] | Evidence: ● | Agreement: ☺ | Confidence: ★

Decentralized renewable energy systems (e.g., home- or village-scale solar power) can support education and vocational training.

Anderson et al., 2017

Increased Use of Biomass

Score: [0] — No direct interaction

Social-Supply (continued)

		1 NO POVERTY					2 ZERO HUNGER					3 GOOD HEALTH AND WELL-BEING					4 QUALITY EDUCATION				
		Interaction	Score	Evidence	Agreement	Confidence	Interaction	Score	Evidence	Agreement	Confidence	Interaction	Score	Evidence	Agreement	Confidence	Interaction	Score	Evidence	Agreement	Confidence
Replacing Coal	**Nuclear/Advanced Nuclear**	**Poverty and Development (1.1/1.2/1.3/1.4)** ↑/↓ [+2,-2] ◫◫ •• ★ See effects of increased bioenergy use.	[+2,-2]				No direct interaction	[0]				**Disease and Mortality (3.1/3.2/3.3/3.4)** → [-1] ◫◫◫◫ ••• ★★★ In spite of the industry's overall safety track record, a non-negligible risk for accidents in nuclear power plants and waste treatment facilities remains. The long-term storage of nuclear waste is a politically fraught subject, with no large-scale long-term storage operational worldwide. Negative impacts from upstream uranium mining and milling are comparable to those of coal, hence replacing fossil fuel combustion by nuclear power would be neutral in that aspect. Increased occurrence of childhood leukaemia in populations living within 5 km of nuclear power plants was identified by some studies, even though a direct causal relation to ionizing radiation could not be established and other studies could not confirm any correlation (*low evidence/agreement* on this issue). Abdelouas, 2006; Cardis et al., 2006; Kaatsch et al., 2008; Al-Zoughool and Krewski, 2009; Heinävaara et al., 2010; Schneizer et al., 2010; Brugge and Buchner, 2011; Møller and Mousseau, 2011; Møller et al., 2011, 2012; Moomaw et al., 2011; UNSCEAR, 2011; Sermage-Faure et al., 2012; Ten Hoeve and Jacobson, 2012; Tirmarche et al., 2012; Hiyama et al., 2013; Mousseau and Møller, 2013; Smith et al., 2013; WHO, 2013; IPCC, 2014; von Stechow et al., 2016	[-1]				No direct interaction	[0]			
	CCS: Bioenergy		[0]				**Farm Employment and Incomes (2.3)** ↑/↓ [+1,-2] ◫◫◫◫ ••• ★★★ See increased use of biomass effects. In addition, the concern that more bioenergy (for BECCS) necessarily leads to unacceptably high food prices is not founded on large agreement in the literature. AR5, for example, finds a significantly lower effect of large-scale bioenergy deployment on food prices by mid-century than the effect of climate change on crop yields. Also, Muratori et al. (2016) show that BECCS reduces the upward pressure on food crop prices by lowering carbon prices and lowering the total biomass demand in climate change mitigation scenarios. On the other hand, competition for land use may increase food prices and thereby increase risk of hunger. Use of agricultural residue for bioenergy can reduce soil carbon, thereby threatening agricultural productivity. See literature on increased biomass use: IPCC, 2014; Muratori et al., 2016; Dooley and Kartha, 2018	[+1,-2]				**Disease and Mortality (3.1/3.2/3.3/3.4)** ↑/↓ [+2,-1] ◫◫ ••• ★★★ See positive impacts of increased biomass use. At the same time, there is a non-negligible risk of CO_2 leakage both from geological formations as well as from the transportation infrastructure from source to sequestration locations. Wang and Jaffe, 2004; Hertwich et al., 2008; Apps et al., 2010; Veltman et al., 2010; Koornneef et al., 2011; Singh et al., 2011; Siirila et al., 2012; Atchley et al., 2013; Corsten et al., 2013; IPCC, 2014	[+2,-1]				No direct interaction	[0]			
Advanced Coal	**CCS: Fossil**	No direct interaction	[0]				No direct interaction	[0]				**Disease and Mortality (3.1/3.2/3.3/3.4)** → [-1] ◫◫ ••• ★★★ The use of fossil CCS implies continued adverse impacts of upstream supply-chain activities in the coal sector, and because of lower efficiency of CCS coal power plants, upstream impacts and local air pollution are likely to be exacerbated. Furthermore, there is a non-negligible risk of CO_2 leakage from geological storage or the CO_2 transport infrastructure from source to sequestration location. Wang and Jaffe, 2004; Hertwich et al., 2008; Apps et al., 2010; Veltman et al., 2010; Koornneef et al., 2011; Singh et al., 2011; Siirila et al., 2012; Atchley et al., 2013; Corsten et al., 2013; IPCC, 2014	[-1]				No direct interaction	[0]			

Social-Other

Agriculture and Livestock

SDG 1 — NO POVERTY

Behavioural Response: Sustainable Healthy Diets and Reduced Food Waste

Poverty and Development (1.1/1.2/1.3/1.4) — Interaction: ~ / ↓ ; Score [0,-1]; Agreement ⊕⊕⊕; Confidence ★★

Cutting livestock consumption can increase food security for some if land grows food not feed, but can also undermine livelihoods and culture where livestock has long been the best use of land, such as in parts of Sub-Saharan Africa.

IPCC, 2014

Land-based GHG Reduction and Soil Carbon Sequestration

Poverty and Development (1.1/1.2/1.3/1.4) — Interaction: ↑ ; Score [+2]; Confidence ★★★★

Many CSA interventions aim to improve rural livelihoods, thereby contributing to poverty alleviation. Agroforestry or integrated crop–livestock–biogas systems can substitute costly, external inputs, saving on household expenditures – or even lead to the selling of some of the products, providing the farmer with extra income, leading to increased adaptive capacity (Bogdanski, 2012).

Branca et al., 2011; Bogdanski, 2012; Scher et al., 2012; Vermeulen et al., 2012; Campbell et al., 2014; Lipper et al., 2014; Mbow et al., 2014; Steenwerth et al., 2014; Hammond et al., 2017

Greenhouse Gas Reduction from Improved Livestock Production and Manure Management Systems

Poverty Reduction and Minimize Exposure to Risk (1.5) — Interaction: ↑ ; Score [+2]; Confidence «

With mixed-farming systems farmers can not only mitigate risks by producing a multitude of commodities, but they can also increase the productivity of both crops and animals in a more profitable and sustainable way.

Sansoucy, 1995

SDG 2 — ZERO HUNGER

Food Security, Promoting Sustainable Agriculture (2.1/2.4/2a) — Interaction: ↑ ; Score [+2]; Agreement ⊕⊕⊕⊕; Confidence ★★★★

Curbing consumer waste of major food crops (i.e., wheat, rice and vegetables) and meats (i.e., beef, pork and poultry) in China, USA and India alone could feed ~413 million people per year (West et al., 2014). One billion extra people could be fed if food crop losses could be halved (Kummu et al., 2012). Reducing waste, especially from meat and dairy, could play a role in delivering food security and reduce the need for sustainable intensification (Smith, 2013). Dietary change toward global healthy diets could improve nutritional health, food security and reduce emissions.

Garnet, 2011; Beddington et al., 2012; Kummu et al., 2012; Smith, 2013; Bajželj et al., 2014; Tilman and Clark, 2014; West et al., 2014; Lamb et al., 2016

Food Security, Promoting Sustainable Agriculture (2.1/2.4/2a) — Interaction: ↑ ; Score [+2]; Agreement ⊕⊕⊕⊕; Confidence ★★★★

Safe application of biotechnology, both conventional and modern methods, can help to improve agricultural productivity, improving crop adaptability and thereby catering to food security. Reducing tillage, eliminating fallow and keeping the soil covered with residue, cover crops or perennial vegetation helps prevent soil erosion and has the potential to increase soil organic matter. Efficient land-management techniques can help in increasing crop yields, and so food security issues can be addressed. Yield projections are actually higher for developing countries than for developed countries, reflecting the fact that they have more 'catch-up' potential (Evenson, 1999). Action is needed throughout the food system on moderating demand, reducing waste, improving governance and producing more food (Godfray and Garnett, 2014). Improving cropland management is the key to increase crop productivity without further degrading soil and water resources (Branca et al., 2011). CSA practices increase productivity and prioritize food security.

Evenson, 1999; West and Post, 2002; Johnson et al., 2007; Branca et al., 2011; McCarthy et al., 2011; Beinassi et al., 2014; Campbell et al., 2014; Godfray and Garnett, 2014; Harvey et al., 2014; Lipper et al., 2014

Food Security, Promoting Sustainable Agriculture (2.1/2.4/2a) — Interaction: ↑ ; Score [+2]; Confidence «««

Fostering transitions towards more productive livestock production systems targeting land-use change appears to be the most efficient lever to deliver food availability outcomes. Genomic selection should be able to at least double the rate of genetic gain in the dairy industry. Given the prevalence of mixed crop–livestock systems in many parts of the world, closer integration of crops and livestock in such systems can give rise to increased productivity and increased soil fertility (Thornton, 2010). Managing the indirect effects of livestock systems intensification is critical for the sustainability of the global food system: such as improving productivity and the close link to land sparing (Herrero and Thornton, 2013). In East Africa pastoralists have shifted from cows to camels, which are better adapted to survive periods of water scarcity and able to consistently provide more milk (Steenwerth et al., 2014). Scenarios where zero human-edible concentrate feed is used for livestock, soil erosion potential reduces by 12%.

Thornton, 2010, 2013; Herrero and Thornton, 2013; Havlik et al., 2014; Steenwerth et al., 2014; Schader et al., 2015

SDG 3 — GOOD HEALTH AND WELL-BEING

Tobacco Control (3.a/3.a.1) — Interaction: ↑ ; Score [+1]; Agreement ⊕; Confidence ★

Consume fewer foods with low nutritional value, e.g., alcohol (Garnett, 2011). Demand-side measures aimed at reducing the proportion of animal products in human diets, where the consumption of animal products is higher than recommended, are associated with multiple health benefits, especially in industrialized countries (Bustamante et al., 2014).

Garnett, 2011; Bustamante et al., 2014

Ensure Healthy Lives (3.c) — Interaction: ↑ / ↓ ; Score [+2,-2]; Agreement ⊕⊕; Confidence ★★

Growing crops such as cassava, sorghum and millet, even in harsh conditions, is important to the diets of very poor people. Policy scenarios show that reduced research support, delayed industrialization, delayed biotechnology and climate change will delay progress in reducing childhood malnutrition. The global effects are small, but local effects for some countries, e.g., Bangladesh and Nigeria, are significant (Evenson, 1999).

Evenson, 1999; Godfray and Garnett, 2014

Ensure Healthy Lives (3.c) — Interaction: ↑ ↓ ; Score [+2,-2]; Agreement &&; Confidence ««

Biodigestion, which has positive public health aspects, particularly where toilets are coupled with the biodigester; anaerobic conditions kill pathogenic organisms as well as digestive toxins. Separation processes can improve or worsen health risks related to food crops or to livestock.

Sansoucy, 1995; Burton, 2007

SDG 4 — QUALITY EDUCATION

Ensure Inclusive and Quality Education (4.4/4.7) — Score [0]

No direct interaction

Ensure Inclusive and Quality Education (4.4/4.7) — Interaction: ↑ / ↓ ; Score [+2,-2]; Agreement ☺; Confidence ★

Science-based action within CSA is required to integrate data sets and sound metrics for testing hypotheses about feedback regarding climate, weather data products and agricultural productivity, such as the nonlinearity of temperature effects on crop yield and the assessment of trade-offs and synergies that arise from different agricultural intensification strategies (Steenwerth et al., 2014). Low commodity prices have led to declining investment in research and development, farmer education, etc. (Lamb et al., 2016).

Steenwerth et al., 2014; Lamb et al., 2016

Score [0] — No direct interaction

Social-Other (continued)

		SDG 1 NO POVERTY					SDG 2 ZERO HUNGER					SDG 3 GOOD HEALTH AND WELL-BEING					SDG 4 QUALITY EDUCATION				
		Interaction	Score	Evidence	Agreement	Confidence	Interaction	Score	Evidence	Agreement	Confidence	Interaction	Score	Evidence	Agreement	Confidence	Interaction	Score	Evidence	Agreement	Confidence

Forest

Reduced Deforestation, REDD+

SDG 1 — Poverty Reduction (1.5): ← [+2] 🔲 ☺ ★ — Partnerships between local forest managers, community enterprises and private sector companies can support local economies and livelihoods, and boost regional and national economic growth. Katila et al., 2017

SDG 2 — Food Security, Promoting Sustainable Agriculture (2.1/2.4/2a): ←/↓ [+1,-2] 🔲🔲 ☺ ★ — Food security may lead to the conversion of productive land under forest, including community forests, into agricultural production. In a similar fashion, the production of biomass for energy purposes (SDG 7) may reduce land available for food production and/or for community forest activities. Efforts by the Government of Zambia to reduce emissions by REDD+ have contributed erosion control, ecotourism and pollination valued at 2.5% of the country's GDP. Turpie et al., 2015; Epstein and Theuer, 2017; Katila et al., 2017; Dooley and Kartha, 2018

SDG 3: [0] — No direct interaction

SDG 4 — Ensure Inclusive and Quality Education (4.4/4.7): ← [+1] 🔲 ☺ ★ — Local forest users learn to understand laws, regulations and policies which facilitate their participation in society. Education and capacity building provide technical skill and knowledge (Katila et al., 2017). Katila et al., 2017

Afforestation and Reforestation

SDG 1 — Poverty and Development (1.1/1.2/1.3/1.4): ←/↓ [+2,-2] 🔲🔲🔲 ☺☺ ★★★ — Clean Development Mechanism (CDM) can have different implications on local community livelihoods. For example, willingness to adopt afforestation is influenced in particular by Australian landholder's perceptions of its potential to provide a diversified income stream, and its impacts on flexibility of land management ; land sparing would have far reaching implications for the UK countryside and would affect landowners and rural communities ; and livelihoods could be threatened if subsistence agriculture is targeted. Zomer et al., 2008; Schirmer and Bull, 2014; Lamb et al., 2016; Dooley and Kartha, 2018

SDG 2 — Food Security (2.1): ←/↓ [+1,-1] 🔲 ☺ ★ — CDM can have different implications on local to regional food security and local community livelihoods. Zomer et al., 2008; Dooley and Kartha, 2018

SDG 3: [0] — No direct interaction

SDG 4 — Promote Knowledge and Skill to Promote SD (4.7): → [-1] 🔲 ☺ ★ — Most landholders reported having low levels of knowledge about tree planting for carbon sequestration – particularly available programmes, prices and markets, and government rules and regulations . Schirmer and Bull, 2014

Behavioural Response (Responsible)

SDG 1: [0] — No direct interaction
SDG 2: [0] — No direct interaction
SDG 3: [0] — No direct interaction
SDG 4: [0] — No direct interaction

Oceans

Ocean Iron Fertilization

SDG 1: [0] — No direct interaction

SDG 2 — Food Security (2.2/2.3): ←/↓ [+1,-1] 🔲🔲 ☺ ★ — OIF can have different implications on fish stocks and aquaculture, and it might actually increase food availability for fish stocks (increasing yields); but potentially at the cost of reducing the yields of fisheries outside the enhancement region by depleting other nutrients. Lampitt et al., 2008; Smetacek and Naqvi, 2008; Williamson et al., 2012

SDG 3 — Ensure Healthy Lives (3.c): ← [+1] 🔲 ☺ — Urban trees are increasingly seen as a way to reduce harmful air pollutants and therefore improve cardio-respiratory health. Jones and McDermott, 2018

SDG 4: [0] — No direct interaction

Blue Carbon

SDG 1 — Poverty and Development (1.1/1.2/1.5): ← [+3] 🔲🔲🔲 ☺☺☺ ★★★ — Avoiding loss of mangroves and maintaining the 2000 stock could save a value of ecosystem services from mangroves in South East Asia of approximately 2.16 billion USD until 2050 (2007 prices), with a 95% prediction interval of 1.58–2.76 billion USD (case study area South East Asia); seaweed aquaculture will enhance carbon uptake and provide employment; traditional management systems provide benefits for blue carbon and support livelihoods for local communities; greening of aquaculture can significantly enhance carbon storage; PES schemes could help capture the benefits derived from multiple ecosystem services beyond carbon sequestration. Zomer et al., 2008; Schirmer and Bull, 2014; Lamb et al., 2016

SDG 2 — Food Production (2.3/2.4): ← [+3] 🔲🔲🔲🔲 ☺☺☺ ★★★ — Avoiding loss of mangroves and maintaining the 2000 stock could save a value of ecosystem services from mangroves in South East Asia including fisheries; seaweed aquaculture will provide employment; traditional management systems provide livelihoods for local communities; greening of aquaculture can increase income and well-being; and mariculture is a promising approach for China. Brander et al., 2012; Ahmed et al., 2017a, 2017b; Duarte et al., 2017; Sondak et al., 2017; Vierros, 2017; Zhang et al., 2017

SDG 3: [0] — No direct interaction

SDG 4: [0] — No direct interaction

Enhanced Weathering

SDG 1: [0] — No direct interaction
SDG 2: [0] — No direct interaction
SDG 3: [0] — No direct interaction
SDG 4: [0] — No direct interaction

Social 2-Demand

		5 GENDER EQUALITY					10 REDUCED INEQUALITIES					16 PEACE, JUSTICE AND STRONG INSTITUTIONS					17 PARTNERSHIPS FOR THE GOALS				
		Interaction	Score	Evidence	Agreement	Confidence	Interaction	Score	Evidence	Agreement	Confidence	Interaction	Score	Evidence	Agreement	Confidence	Interaction	Score	Evidence	Agreement	Confidence
Industry	**Accelerating Energy Efficiency Improvement**		[0]				**Knowledge and Skills Needed to Promote SD (4.7)** There is need for skill in managing in-house energy efficiency. Sometimes ESCOs also help. Energy audits, but many times absence of skill acts as barrier for energy efficiency improvement. In many countries, especially developing countries, these act as barriers. Apeaning and Thollander, 2013; Johansson and Thollander, 2018	[+1]	▢	●●●	★★★		[0]				**Global Partnership (17.6/17.7)** A driving force for energy efficiency is collaboration among companies, networks, experience sharing and management tools. Sharing among countries can help accelerate managerial action. Absence of information, budgetary funding, lack of access to capital, etc. are. Apeaning and Thollander, 2013; Griffin et al., 2018; Johansson and Thollander, 2018; Lawrence et al., 2018	[+2]	▢▢	●●●	★★★
				No direct interaction										No direct interaction							
	Low-carbon Fuel Switch		[0]					[0]					[0]				**Global Partnership (17.6/17.7)** Ultra-low carbon steel making and breakthrough technologies are under trial across many countries and helping in enhancing the learning. Abdul Quader et al., 2016	[+2]	▢	●●	★★
				No direct interaction					No direct interaction					No direct interaction							
	Decarbonization/CCS/CCU		[0]					[0]					[0]				**Global Partnership (17.6/17.7)** EPI plants are capital intensive and are mostly operated by multinationals with long investment cycles. In developed countries new innovation investments are happening in brown fields. Such large innovation investments need strong collaboration among partners/competitors which can be facilitated by public funds. They happen at national and supranational scales and across sectors, needs fresh revisit at Intellectual Property Rights issues. Global production of bio-based polymers increasingly need public support and incentives to push forward. Wesseling et al., 2017; Griffin et al., 2018	[+2]	▢	●●●	★★★
				No direct interaction					No direct interaction					No direct interaction							

5

Social 2-Demand (continued)

Buildings

SDG 5 — GENDER EQUALITY

Row	Interaction	Score	Evidence	Agreement	Confidence
Behavioural Response	No direct interaction	[0]			
Accelerating Energy Efficiency Improvement	Gender Equality and Women's Empowerment (5.1/5.4) — Efficient stoves lead to empowerment of rural and indigenous women. *Bhojvaid et al., 2014; Berrueta et al., 2017*	[+1]	▫▫	◔◔	★★
Improved Access and Fuel Switch to Modern Low-carbon Energy	Women's Safety and Worth (5.1/5.2/5.4)/Opportunities for Women (5.1/5.5) — Improved access to electric lighting can improve women's safety and girls' school enrolment. Cleaner cooking fuel and lighting access can reduce health risks and drudgery, which women disproportionately face. Access to modern energy services has the potential to empower women by improving their income-earning and entrepreneurial opportunities and reducing drudgery. Participating in energy supply chains can increase women's opportunities and agency and improve business outcomes. *Chowdhury, 2010; Dinkelman, 2011; Kayousuz, 2011; Köhlin et al., 2011; Clancy et al., 2012; Haves, 2012; Matinga, 2012; Anenberg et al., 2013; Pachauri and Rao, 2013; Burney et al., 2017; McCollum et al.,*	[+1]	▫▫	◔◔	★★

SDG 10 — REDUCED INEQUALITIES

Row	Interaction	Score	Evidence	Agreement	Confidence
Behavioural Response	No direct interaction	[0]			
Accelerating Energy Efficiency Improvement	Empowerment and Inclusion (10.1/10.2/10.3/10.4) — Energy efficiency measures and the provision of energy access can free up resources that can then be put towards other productive uses (e.g., educational and employment opportunities), especially for women and children in poor, rural areas. The distributional costs of new energy policies are dependent on instrument design. If costs fall disproportionately on the poor, then this could work against the promotion of social, economic and political equality for all. The impacts of energy efficiency measures and policies on inequality can be both positive, if they reduce energy costs, or negative, if mandatory standards increase the need for purchasing more expensive equipment and appliances. *Dinkelman, 2011; Casillas and Kammen, 2012; Pachauri et al., 2012; Cayla and Osso, 2013; Hirth and Ueckerdt, 2013; Pueyo et al., 2013; Jakob and Steckel, 2014; Fay et al., 2015; Cameron et al., 2016; Hallegatte et al., 2016b; McCollum et al., 2018*	[1,-1]	▫▫▫	◔◔◔	★★★
Improved Access and Fuel Switch to Modern Low-carbon Energy	No direct interaction	[0]			

SDG 16 — PEACE, JUSTICE AND STRONG INSTITUTIONS

Row	Interaction	Score	Evidence	Agreement	Confidence
Behavioural Response	Environmental Justice (16.7) — Consumption perspectives strengthen environmental justice discourse (as it claims to be a more just way of calculating global and local environmental effects) while possibly also increasing the participatory environmental discourse. *Hult and Larsson, 2016*	[+2]	▫	◔	★
Accelerating Energy Efficiency Improvement	Capacity and Accountability (16.1/16.3/16.5/16.6/16.7/16.8) — Institutions that are effective, accountable and transparent are needed at all levels of government (local to national to international) for providing energy access, promoting modern renewables and boosting efficiency. Strengthening the participation of developing countries in international institutions (e.g., international energy agencies, UN organizations, WTO, regional development banks and beyond) will be important for issues related to energy trade, foreign direct investment, labour migration and knowledge and technology transfer. Reducing corruption, where it exists, will help these bodies and related domestic institutions maximize their societal impacts. Limiting armed conflict and violence will aid most efforts related to sustainable development, including progress in the energy dimension. *Acemoglu, 2009; Tabellini, 2010; Acemoglu et al., 2014; ICSU and ISSC, 2015; McCollum et al., 2018*	[+2]	▫▫▫▫	◔◔◔	★★★★
Improved Access and Fuel Switch to Modern Low-carbon Energy	Capacity and Accountability (16.1/16.3/16.5/16.6/16.7/16.8) — Institutions that are effective, accountable and transparent are needed at all levels of government (local to national to international) for providing energy access, promoting modern renewables and boosting efficiency. Strengthening the participation of developing countries in international institutions (e.g., international energy agencies, UN organizations, WTO, regional development banks and beyond) will be important for issues related to energy trade, foreign direct investment, labour migration, and knowledge and technology transfer. Reducing corruption, where it exists, will help these bodies and related domestic institutions maximize their societal impacts. Limiting armed conflict and violence will aid most efforts related to sustainable development, including progress in the energy dimension. *Acemoglu, 2009; Tabellini, 2010; Acemoglu et al., 2014; ICSU and ISSC, 2015; McCollum et al., 2018*	[+2]	▫▫▫▫	◔◔◔	★★★★

SDG 17 — PARTNERSHIPS FOR THE GOALS

Row	Interaction	Score	Evidence	Agreement	Confidence
Behavioural Response	No direct interaction	[0]			
Accelerating Energy Efficiency Improvement	Enhance Policy Coherence for Sustainable Development (17.4) — Implementing refrigerant transition and energy efficiency improvement policies in parallel for room ACs, roughly doubles the benefit of either policy implemented in isolation. *Shah et al., 2015*	[+2]	▫	◔	★
Improved Access and Fuel Switch to Modern Low-carbon Energy	Promote Transfer and Diffusion of Technology (17.6/17.7) — Green building technology in Kazakhstan was based on transfer of knowledge among various parties. *Kim and Sun, 2017*	[+2]	▫	◔	★

5

Social 2-Demand (continued)

Transport

	Interaction	Score	Evidence	Agreement	Confidence
SDG 5 GENDER EQUALITY					
Behavioural response	Recognize Women's Unpaid Work (5.1/5.4)/Opportunities for Women (5.1/5.5)	[+1]	▢▢▢	◉◉	★★

The woman's average trip to work differs markedly from the man's average trip. Working-poor women rely on extensive social networks creating communities of spatial necessity, bartering for basic needs to overcome transportation constraints. Women earn lower wages and so are less likely to justify longer commutes. Many women need to manage dual roles as workers and mothers. Women tend to perform multi-purpose commuting, combining both work and household needs.

Crane, 2007; Rogalsky, 2010

	Interaction	Score	Evidence	Agreement	Confidence
Accelerating Energy Efficiency Improvement		[0]			

No direct interaction

	Interaction	Score	Evidence	Agreement	Confidence
Improved Access and Fuel Switch to Modern Low-carbon Energy		[0]			

No direct interaction

SDG 10 REDUCED INEQUALITIES

	Interaction	Score	Evidence	Agreement	Confidence
Behavioural response	Reduce Inequality (10.2)	[+2]	▢▢	◉◉	★★

The equity impacts of climate change mitigation measures for transport, and indeed of transport policy intervention overall, are poorly understood by policymakers. This is in large part because standard assessment of these impacts is not a statutory requirement of current policymaking. Managing transport energy demand growth will have to be advanced alongside efforts in passenger travel towards reducing the deep inequalities in access to transport services that currently affect the poor worldwide. Free provision of roads and parking spaces converts vast amounts of public land and capital into under-priced space for cars, in extreme cases like Los Angeles, USA, roads and streets free for parking and driving are 20% of land area; as governments give drivers free land, people drive more than they would otherwise. High levels of car dependence and the costs of motoring can be burdensome, and lead to increasing debt, raising questions of affordability for households with limited resources, particularly low-income houses located in suburban areas.

Figueroa et al., 2014; Lucas and Pangbourne, 2014; Walks, 2015; Manville, 2017; Belton Chevallier et al., 2018

	Interaction	Score	Evidence	Agreement	Confidence
Accelerating Energy Efficiency Improvement		[0]			

No direct interaction

	Interaction	Score	Evidence	Agreement	Confidence
Improved Access and Fuel Switch to Modern Low-carbon Energy	Reduce Inequality (10.2)	[+2]	▢▢	◉◉	★★

The equity impacts of climate change mitigation measures for transport, and indeed of transport policy intervention overall, are poorly understood by policymakers. This is in large part because standard assessment of these impacts is not a statutory requirement of current policymaking. Managing transport energy demand growth will have to be advanced alongside efforts in passenger travel towards reducing the deep inequalities in access to transport services that currently affect the poor worldwide.

Figueroa et al., 2014; Lucas and Pangbourne, 2014

SDG 16 PEACE, JUSTICE AND STRONG INSTITUTIONS

	Interaction	Score	Evidence	Agreement	Confidence
Behavioural response	Accountable and Transparent Institutions at All Levels (16.6/16.8)	[+1, -1]	▢▢	◉	★

With behavioural change towards walking for short distances, pedestrian safety on the road might reduce, unless public policy is appropriately formulated. Prevalence of high levels of triple forms of informality, in jobs, housing and transportation, are responsible for low productivity and low standards of living, and are a major challenge for policies targeting urban growth in Latin America.

CAF, 2017; SLoCaT, 2017

	Interaction	Score	Evidence	Agreement	Confidence
Accelerating Energy Efficiency Improvement	Responsive, Inclusive, Participatory Decision-making (16.7)	[+2]	▢▢	◉◉	★★

In transport mitigation it is necessary to conduct needs assessments and stakeholder consultation to determine plausible challenges, prior to introducing desired planning reforms. Further, the involved personnel should actively engage transport-based stakeholders during policy identification and its implementation to achieve the desired results. User behaviour and stakeholder integration are key for successful transport policy implementation.

Aggarwal, 2017; AlSabbagh et al., 2017

	Interaction	Score	Evidence	Agreement	Confidence
Improved Access and Fuel Switch to Modern Low-carbon Energy	Responsive, Inclusive, Participatory Decision-making (16.7)	[+1, -1]	▢	◉	★

Formal transport infrastructure improvement in many cities in developing countries leads to eviction from informal settlements; need for appropriate redistributive policies and cooperation and partnerships with all stakeholders.

Colenbrander et al., 2016

SDG 17 PARTNERSHIPS FOR THE GOALS

	Interaction	Score	Evidence	Agreement	Confidence
Behavioural response	Help Promote Global Partnership (17.1/17.3/17.5/17.6/17.7)	[+2]	▢	◉	★

Projects aiming at resilient transport infrastructure development (e.g., C40 Cities Clean Bus Declaration, UITP Declaration on Climate Leadership, Cycling Delivers on the Global Goals, Global Sidewalk Challenge) are happening through multi-stakeholder coalitions.

SLoCaT, 2017

	Interaction	Score	Evidence	Agreement	Confidence
Accelerating Energy Efficiency Improvement	Help Promote Global Partnership (17.1/17.3/17.5/17.6/17.7)	[+2]	▢	◉	★

Projects aiming at resilient transport infrastructure development and technology adoption (e.g. C40 Cities Clean Bus Declaration, UITP Declaration on Climate Leadership, Cycling Delivers on the Global Goals, Global Sidewalk Challenge) are happening through multi-stakeholder coalitions.

SLoCaT, 2017

	Interaction	Score	Evidence	Agreement	Confidence
Improved Access and Fuel Switch to Modern Low-carbon Energy	Help Promote Global Partnership (17.1/17.3/17.5/17.6/17.7)	[+2]	▢	◉	★

Projects aiming at resilient transport infrastructure development (e.g. C40 Cities Clean Bus Declaration, UITP Declaration on Climate Leadership, Cycling Delivers on the Global Goals, Global Sidewalk Challenge) are happening through multi-stakeholder coalitions.

SLoCaT, 2017

Social 2-Supply

	SDG 5 – GENDER EQUALITY					**SDG 10 – REDUCED INEQUALITIES**					**SDG 16 – PEACE, JUSTICE AND STRONG INSTITUTIONS**					**SDG 17 – PARTNERSHIPS FOR THE GOALS**				
	Gender Equality and Women's Empowerment (5.1/5.4)					**Empowerment and Inclusion (10.1/10.2/10.3/10.4)**					**Energy Justice** / **Reduce Illicit Arms Trade (16.4)**					**International Cooperation (All Goals)**				
	Interaction	Score	Evidence	Agreement	Confidence	Interaction	Score	Evidence	Agreement	Confidence	Interaction	Score	Evidence	Agreement	Confidence	Interaction	Score	Evidence	Agreement	Confidence
Replacing Coal — Non-biomass Renewables – solar, wind, hydro	←	[+1]	▭	☺	★ Decentralized renewable energy systems (e.g., home- or village-scale solar power) can reduce the burden on girls and women of procuring traditional biomass. Schwerhoff and Sy, 2017	←	[+1]	▭▭▭	☺☺	★★ Decentralized renewable energy systems (e.g., home- or village-scale solar power) can enable a more participatory, democratic process for managing energy-related decisions within communities. Walker and Devine-Wright, 2008; Cass et al., 2010; Cumbers, 2012; Kunze and Becker, 2015; McCollum et al., 2018	←	[+2]	▭	☺	★ *Energy Justice.* The energy justice framework serves as an important decision-making tool in order to understand how different principles of justice can inform energy systems and policies. Islar et al. (2017) state that off-grid and micro-scale energy development offers an alternative path to fossil-fuel use and top-down resource management as they democratize the grid and increase marginalized communities' access to renewable energy, education and health care. Islar et al., 2017	←/~	[+2,0]	▭▭	☺☺	★★ International cooperation (in policy) and collaboration (in science) is required for the protection of shared resources. Fragmented approaches have been shown to be more costly. Specific to SDG7, to achieve the targets for energy access, renewables and efficiency, it will be critical that all countries: (i) are able to mobilize the necessary financial resources (e.g., via taxes on fossil energy, sustainable financing, foreign direct investment, financial transfers from industrialized to developing countries); (ii) are willing to disseminate knowledge and share innovative technologies between each other; (iii) follow recognized international trade rules while at the same time ensuring that the least developed countries are able to take part in that trade; (iv) respect each other's policy space and decisions; (v) forge new partnerships between their public and private entities and within civil society; and (vi) support the collection of high-quality, timely and reliable data relevant to furthering their missions. There is some disagreement in the literature on the effect of some of the above strategies, such as free trade. Regarding international agreements, 'no-regrets options', where all sides gain through cooperation, are seen as particularly beneficial (e.g., nuclear test ban treaties) (McCollum et al., 2018). UN, 1989; Ramaker et al., 2003; Clarke et al., 2009; NCE, 2015; Riahi et al., 2015, 2017; Eis et al., 2016; O'Neill et al., 2017; McCollum et al., 2018
Replacing Coal — Increased use of Biomass		[0]			No direct interaction		[0]			No direct interaction		[0]			No direct interaction		[0]			No direct interaction
Replacing Coal — Nuclear/Advanced Nuclear		[0]			No direct interaction		[0]			No direct interaction	↓	[-1]	▭▭	☺☺	★★ *Reduce Illicit Arms Trade (16.4).* Continued use of nuclear power poses a constant risk of proliferation. Adamantiades and Kessides, 2009; Rogner, 2010; Sagan, 2011; von Hippel et al., 2012; Yim and Li, 2013; IPCC, 2014		[0]			No direct interaction
Replacing Coal — CCS: Bioenergy		[0]			No direct interaction		[0]			No direct interaction		[0]			No direct interaction		[0]			No direct interaction
Advanced Coal — CCS: Fossil		[0]			No direct interaction		[0]			No direct interaction		[0]			No direct interaction		[0]			No direct interaction

5

491

Social 2-Other

5

Agriculture and Livestock

Row categories (left side):
- **Behavioural Response: Sustainable Healthy Diets and Reduced Food Waste**
- **Land-based Greenhouse Gas Reduction and Soil Carbon Sequestration**
- **Greenhouse Gas Reduction from Improved Livestock Production and Manure Management Systems**

5 GENDER EQUALITY

	Interaction	Score	Evidence	Agreement	Confidence
Behavioural Response		[0]			
Land-based GHG Reduction	↑/~	[+2,0]	🔲🔲🔲	⊕⊕	★★★
GHG Reduction from Livestock	↑/~	[+2,0]	🔲🔲🔲	⊕	★

Equal Access, Empowerment of Women (5.5)

Many programmes for CSA have been used to empower women and to improve gender equality. Women often have an especially important role to play in adaptation, because of their gendered indigenous knowledge on matters such as agriculture (Terry, 2009). Without access to land, credit and agricultural technologies, women farmers face major constraints in their capacity to diversify into alternative livelihoods (Demetriades and Esplen, 2008).

Denton, 2002; Nelson et al., 2002; Morton, 2007; Demetriades and Esplen, 2009; Terry, 2009; Bernier et al., 2013; Jost et al, 2016

Equal Access to Economic Resources, Promote Empowerment of Women (5.5/5.a/5.b)

Most of the animal farming activities such as fodder collection and feeding are performed by women. Alongside the considerable involvement and contribution of women, gender inequalities are pervasive in Indian villages in terms of accessing natural resources, extension services, marketing opportunities and financial services as well as in exercising their decision-making powers. Therefore, there is a need to correct gender bias in the farming sector. Efforts are needed to increase the capacity of women to negotiate with confidence and meet their strategic needs. Access to and control and management of small ruminants, grazing areas and feed resources empower women and lead to an overall positive impact on the welfare of the household.

Patel et al., 2016

10 REDUCED INEQUALITIES

	Interaction	Score	Evidence	Agreement	Confidence
Behavioural Response		No direct interaction			
Land-based GHG Reduction	↑/~	[+1,0]	🔲🔲	⊕⊕	★★
GHG Reduction from Livestock	↑/~	[+1,0]	🔲	⊕	★

Empower Economic and Political Inclusion of All, Irrespective of Sex (10.2)

In many rural societies women are side-lined from decisions regarding agriculture even when male household heads are absent, and they often lack access to important inputs such as irrigation water, credit, tools and fertilizer. To be effective, agricultural mitigation strategies need to take these and other aspects of local gender relations into account (Terry, 2009). Women's key role in maintaining biodiversity, through conserving and domesticating wild edible plant seed, and in food crop breeding, is not sufficiently recognized for agricultural and economic policymaking; nor is the importance of biodiversity to sustainable rural livelihoods in the face of predicted climate changes (Nelson et al., 2002).

Nelson et al., 2002; Demetriades and Esplen, 2009; Terry, 2009

Empower Economic and Political Inclusion of All, Irrespective of Sex (10.2)

Livestock ownership is increasing women's decision-making and economic power within both the household and the community. Access to and control and management of small ruminants, grazing areas and feed resources empower women and lead to an overall positive impact on the welfare of the household.

Patel et al., 2016

16 PEACE, JUSTICE AND STRONG INSTITUTIONS

	Interaction	Score	Evidence	Agreement	Confidence
Behavioural Response	↑/↓	[+1,-1]	🔲🔲🔲	⊕⊕	★★
Land-based GHG Reduction	~/↓	[0,-1]	🔲🔲🔲🔲	⊕⊕	★★
GHG Reduction from Livestock	↑	[+1]	🔲	⊕	★

making (16.6/16.7/16.a)

Appropriate incentives to reduce food waste may require some policy innovation and experimentation, but a strong commitment for devising and monitoring them seems essential.
A financial incentive to minimize waste could be created through effective taxation (e.g., by taxing foods with the highest wastage rates, or by increasing taxes on waste disposal). Decision makers should try to integrate agricultural, environmental and nutritional objectives through appropriate policy measures to achieve sustainable healthy diets coupled with reduction in food waste. It is surprising that politicians and policymakers demonstrate little regarding the need to have strategies to reduce meat consumption and to encourage more sustainable eating practices.

Garnett, 2011; Dagevos and Voordouw, 2013; Bajželj et al., 2014; Lamb et al., 2016

Build Effective, Accountable and Inclusive Institutions (16.6/16.7/16.8)

Action is needed throughout the food system for improving governance and producing more food (Godfray and Garnett, 2014). CSA requires policy intervention for careful adjustment of agricultural practices to natural conditions, a knowledge-intensive approach, huge financial investment, etc., so having strong institutional frameworks is very important. The main source of climate finance for CSA in developing countries is the public sector. Lack of institutional capacity (as a means for securing creation of equal institutions among social groups and individuals) can reduce feasibility of AFOLU mitigation measures in the near future, especially in areas where small-scale farmers or forest users are the main stakeholders (Bustamante et al., 2014).

Behnassi et al., 2014; Bustamante et al., 2014; Godfray and Garnett, 2014; Lipper et al., 2014; Steenwerth et al., 2014

Responsible Decision-making (16.7)

To minimize the economic and social cost, policies should target emissions at their source—on the supply side—rather than on the demand side as supply-side policies have lower calorie cost than demand-side policies. The role of livestock system transitions in emission reductions depends on the level of the carbon price and which emissions sector is targeted by the policies.

Havlik et al., 2014

17 PARTNERSHIPS FOR THE GOALS

	Interaction	Score	Evidence	Agreement	Confidence
Behavioural Response	↑/↓	[+1,-1]	🔲	⊕	★
Land-based GHG Reduction	↑/↓	[+2]	🔲🔲🔲🔲	⊕⊕⊕	★★★
GHG Reduction from Livestock	↑	[+2]	🔲🔲	⊕⊕	★★

Resource Mobilization and Strengthen Partnership (17.14/17.17)

Decision makers should try to integrate agricultural, environmental and nutritional objectives through appropriate policy measures to achieve sustainable healthy diets coupled with reduction in food waste. It is surprising that politicians and policymakers demonstrate little regarding the need to have strategies to reduce meat consumption and to encourage more sustainable eating practices.

Garnett, 2011; Dagevos and Voordouw, 2013

Resource Mobilization and Strengthen Multi-stakeholder Partnership (17.1/17.3/17.5/17.17)

CSA requires more careful adjustment of agricultural practices to natural conditions, a knowledge-intensive approach, huge financial investment and policy and institutional innovation, etc. Besides private investment, quality of public investment is also important (Behnass et al., 2014). Sources of climate finance for CSA in developing countries include bilateral donors and multilateral financial institutions, besides public sector finance. CSA is committed to new ways of engaging in participatory research and partnerships with producers (Steenwerth et al., 2014).

Behnassi et al., 2014; Lipper et al., 2014; Steenwerth et al., 2014

Improve Domestic Capacity for Tax Collection (17.1)

The role of livestock system transitions in emission reductions depends on the level of the carbon price and which emissions sector is targeted by the policies (Havlik et al., 2014). Mechanisms for affecting behavioural change in livestock systems need to be better understood by implementing combinations of incentives and taxes simultaneously in different parts of the world (Herrero and Thornton, 2013).

Herrero and Thornton, 2013; Havlik et al., 2014

Social 2-Other (continued)

5 GENDER EQUALITY | 10 REDUCED INEQUALITIES | 16 PEACE, JUSTICE AND STRONG INSTITUTIONS | 17 PARTNERSHIPS FOR THE GOALS

Forest

Reduced Deforestation, REDD+

SDG 5 — Opportunities for Women (5.1/5.5) — Interaction: ↑/↓ — Score [+1,-1] — Confidence ★

Women have been less involved in REDD+ initiative (pilot project) design decisions and processes than men. Girls and women have an important role in forestry activities, related to fuel-wood, forest-food and pharmaceutical. Their empowerment contributes to sustainable forestry as well as reducing inequality.

Brown, 2011; Larson et al., 2014; Katila et al., 2017

SDG 10 — Reduced Inequality, Empowerment and Inclusion (10.1/10.2/10.3/10.4) — Interaction: ← — Score [+2] — Confidence ★

Urges developed countries to support, through multilateral and bilateral channels, the development of REDD+ national strategies or action plans and implementation. Girls and women have an important role in forestry activities, related to fuel-wood, forest-food and medicine. Their empowerment contributes to sustainable forestry as well as reducing inequality.

Bastos Lima et al., 2017; Katila et al., 2017

SDG 16 — Build Effective, Accountable and Inclusive Institutions, Responsible Decision-making (16.6/16.7/16.8) — Interaction: ← — Score [+2] — Confidence ★★★

Institutional building (National Forest Monitoring Systems, Safeguard Information Systems, etc.), with full and effective participation of all relevant countries. REDD+ actions also deliver non-carbon benefits (e.g. local socioeconomic benefits, governance improvements). Forest governance is another central aspect in recent studies, including the debate on decentralization of forest management, logging concessions in public-owned commercially valuable forests and timber certification, primarily in temperate forests.

Bustamante et al., 2014; Bastos Lima et al., 2015, 2017

SDG 17 — Resource Mobilization and Strengthen Multi-stakeholder Partnership (17.1/17.3/17.5/17.17) — Interaction: ↑/↓ — Score [+1,-1] — Confidence ★

To provide finance and technology to developing countries to support emissions reductions. Be supported by adequate and predictable financial and technology support, including support for capacity building. Partnerships in the form of significant aid money from, e.g., Norway, other bilateral donors and the World Bank's Forest Carbon Partnership Facility (FCPF) are forthcoming. Estimates of opportunity cost for REDD+ are very low. Lower costs and/or higher carbon prices could combine to protect more forests, including those with lower carbon content. Conversely, where the cost of action is high, a large amount of additional funding would be required for the forest to be protected (Miles and Kapos, 2008). Forest governance is another central aspect in recent studies, including debate on decentralization of forest management, logging concessions in public-owned commercially valuable forests and timber certification, primarily in temperate forests. Partnerships between local forest managers, community enterprises and private sector companies can support local economies and livelihoods and boost regional and national economic growth.

Miles and Kapos, 2008; Bustamante et al., 2014; Andrew, 2017; Bastos Lima et al., 2017; Katila et al., 2017

Afforestation and Reforestation

SDG 5 — Opportunities for Women (5.1/5.5) — Interaction: ← — Score [+1] — Confidence ★

Many women in developing countries are already prominently engaged in economic sectors related to climate adaptation and mitigation efforts such as agriculture, renewable energy and forest management and are important drivers and leaders in climate responses that are innovative and effective, benefitting not only their families but also their wider communities. Women's participation in the decision-making process of forest management, for example, has been shown to increase rates of reforestation while decreasing the illegal extraction of forest products.

UN-Women et al., 2015

SDG 10 — Empower Economic and Political Inclusion of All, Irrespective of Sex (10.2) — Interaction: ← — Score [+1] — Confidence ★

Women's participation in the decision-making process of forest management, for example, has been shown to increase rates of reforestation while decreasing the illegal extraction of forest products.

UN-Women et al., 2015

SDG 16 — Responsible Decision-making (16.7) — Interaction: ← — Score [+1] — Confidence ★

Land-related mitigation, such as biofuel production, as well as conservation and reforestation action can increase competition for land and natural resources, so these measures should be accompanied by complementary policies. (Quoted from Epstein and Theur, 2017)

Epstein and Theur, 2017

SDG 17 — Resource Mobilization and Strengthen Partnership (17.1/17.14) — Interaction: ← — Score [+2] — Confidence ★★

Financing at the national and international level is required to grow more seedlings/sapling, restore land, create awareness and education factsheets, provide training to local communities regarding the benefits of afforestation and reforestation. Article 12 of the Kyoto Protocol further sets a Clean Development Mechanism through which countries in Annex I earn 'certified emissions reductions' through projects implemented in developing countries (Montanarella and Alva, 2015). Afforestation and reforestation in India are being carried out under various programmes, namely social forestry initiated in the early 1980s, the Joint Forest Management Programme initiated in 1990, afforestation under National Afforestation and Eco-development Board programmes since 1992, and private farmer and industry initiated plantation forestry. If the current rate of afforestation and reforestation is assumed to continue, the carbon stock could increase by 11% by 2030 (Ravindranath et al., 2008).

Ravindranath et al., 2008; Kibria, 2015; Montanarella and Alva, 2015

Behavioural Response (Responsible Sourcing)

SDG 5 — Score [0] — No direct interaction

SDG 10 — Score [0] — No direct interaction

SDG 16 — Responsible Decision-making (16.7) — Interaction: ← — Score [+1] — Confidence ★★

Indonesian factories may seek advantages through non-price competition—perhaps by highlighting decent working conditions or the existence of a union—or to see trade associations or government agencies promoting the country as a responsible sourcing location (Bartley, 2010). In the absence of domestic legal instruments providing incentives to improve sustainability of sourcing, it appears that initiatives to engage the major importing enterprises in developing responsible sourcing practices and policies is a practical approach. Unless initiatives involve all the major importers, they are unlikely to be successful since the high costs associated with accreditation would increase production costs for these firms relative to their competitors (Huang et al., 2013).

Bartley, 2010; Huang et al., 2013

SDG 17 — Finance and Trade (17.1/17.10) — Interaction: ← — Score [+1] — Confidence ★★

Private certification initiatives for wood product and biomass sourcing may extend their schemes with criteria for 'leakage' (external GHG effects). Also recycling of waste wood in pellets is not yet practiced, due to unclear rules in the EU Waste Directive about overseas shipping (Sikkema et al., 2014). Engagement of Chinese government and private sector stakeholders in supply-country sustainability initiatives may be the best way to support this gradual process of improvement. Although carrying out due diligence in timber sourcing can require considerable internal resources, it may be substantially less of a financial burden than the potential fines and reputational damage resulting from sourcing unknown or controversial timber (Huang et al., 2013).

Huang et al., 2013; Sikkema et al., 2014

5

Social 2-Other (continued)

		5 GENDER EQUALITY				10 REDUCED INEQUALITIES				16 PEACE, JUSTICE AND STRONG INSTITUTIONS				17 PARTNERSHIPS FOR THE GOALS			
		Interaction	Score	Evidence	Agreement	Confidence	Interaction	Score	Evidence	Agreement	Confidence	Interaction	Score	Evidence	Agreement	Confidence	Interaction	Score	Evidence	Agreement	Confidence
Oceans	Ocean Iron Fertilization		[0]	No direct interaction				[0]	No direct interaction				[0]	No direct interaction				[0]	No direct interaction		
	Blue Carbon		[0]	No direct interaction				[0]	No direct interaction				[0]	No direct interaction				[0]	No direct interaction		
	Enhanced Weathering		[0]	No direct interaction				[0]	No direct interaction				[0]	No direct interaction				[0]	No direct interaction		

Environment-Demand

Industry

SDG 6 — Clean Water and Sanitation

Industry option	Interaction	Score	Evidence	Agreement	Confidence
Accelerating Energy Efficiency Improvement — Water Efficiency and Pollution Prevention (6.3/6.4/6.6)	↗/↘	[+2,-1]		☺☺	★★
Low-carbon Fuel Switch — Water Efficiency and Pollution Prevention (6.3/6.4/6.6)	↗/↘	[+2,-2]		☺☺	★★★
Decarbonisation/CCS/CCU — Water Efficiency and Pollution Prevention (6.3/6.4/6.6)	↗/↘	[+1,-1]		☺	★★

Accelerating Energy Efficiency Improvement: Efficiency and behavioural changes in the industrial sector that lead to reduced energy demand can lead to reduced requirements on energy supply. As water is used to convert energy into useful forms, the reduction in industrial demand is anticipated to reduce water consumption and waste water, resulting in more clean water for other sectors and the environment. Likewise, reducing material inputs for industrial processes through efficiency and behavioural changes will reduce water inputs in the material supply chains. In extractive industries there can be a trade-off with production unless strategically managed.
Vassolo and Doll, 2005; Nguyen et al., 2014; Holland et al., 2015; Fricko et al., 2016

Low-carbon Fuel Switch: A switch to low-carbon fuels can lead to a reduction in water demand and waste water if the existing higher-carbon fuel is associated with a higher water intensity than the lower-carbon fuel. However, in some situations the switch to a low-carbon fuel such as, for example, biofuel could increase water use compared to existing conditions if the biofuel comes from a water-intensive feedstock.
Hejazi et al., 2015; Fricko et al., 2016; Song et al., 2016

Decarbonisation/CCS/CCU: CCUS requires access to water for cooling and processing which could contribute to localized water stress. CCS/U processes can potentially be configured for increased water efficiency compared to a system without carbon capture via process integration.
Meldrum et al., 2013; Byers et al., 2016; Fricko et al., 2016; Brandl et al., 2017

SDG 12 — Responsible Consumption and Production

Industry option	Interaction	Score	Evidence	Agreement	Confidence
Accelerating Energy Efficiency Improvement — Sustainable and Efficient Resource (12.2/12.5/12.6/12.7/12.a)	↗	[+1]		☺☺☺	★★★
Low-carbon Fuel Switch — Sustainable Production (12.2/12.3/12.a)	↗	[+2]		☺☺☺☺	★★★★
Decarbonisation/CCS/CCU — Sustainable Production and Consumption (12.1/12.6/12.a)	↗	[+2]		☺☺☺☺	★★★★

Accelerating Energy Efficiency Improvement: Once started leads to chain of actions within the sector and policy space to sustain the effort. Helps in expansion of sustainable industrial production (Ghana).
Apeaning and Thollander, 2013; Fernando et al., 2017

Low-carbon Fuel Switch: A circular economy instead of a linear global economy can achieve climate goals and can help in economic growth through industrialization which saves on resources, the environment and supports small, medium and even large industries, and can lead to employment generation. So new regulations, incentives and a tax regime can help in achieving the goal, especially in newly emerging developing countries - although also applicable for large industrialized countries.
Liu and Bai, 2014; Lieder and Rashid, 2016; Stahel, 2016; Supino et al., 2016; Fan et al., 2017; Shi et al., 2017; Zeng et al., 2017

Decarbonisation/CCS/CCU: EPI plants are capital intensive and are mostly operated by multinationals with long investment cycles. In developed countries new investments are happening in brown fields, while in developing countries these are in green fields. Collaboration among partners and user demand change, policy change is essential for encouraging these large risky investments.
Wesseling et al., 2017

SDG 14 — Life Below Water

Industry option	Interaction	Score	Evidence	Agreement	Confidence
Accelerating Energy Efficiency Improvement		[0]			
Low-carbon Fuel Switch		[0]			
Decarbonisation/CCS/CCU — Conserve and Sustainably Use Ocean (14.1/14.5)	↘	[-1]		☺	★

Accelerating Energy Efficiency Improvement: No direct interaction

Low-carbon Fuel Switch: No direct interaction

Decarbonisation/CCS/CCU: CCUS in the chemical industry faces challenges for transport costs and storage. In the UK cluster region have been identified for storage under sea.
Griffin et al., 2018

SDG 15 — Life on Land

Industry option	Interaction	Score	Evidence	Agreement	Confidence
Accelerating Energy Efficiency Improvement		[0]			
Low-carbon Fuel Switch — Sustainable Production (15.1/15.5/15.9/15.10)	↗	[+1,-1]		☺	★
Decarbonisation/CCS/CCU		[0]			

Accelerating Energy Efficiency Improvement: No direct interaction

Low-carbon Fuel Switch: A circular economy help in managing local biodiversity better by having less resource use footprint
Shi et al., 2017

Decarbonisation/CCS/CCU: No direct interaction

Environment-Demand (continued)

Buildings

Behavioural Response

6 CLEAN WATER AND SANITATION

Interaction	Score	Evidence	Agreement	Confidence

Water Efficiency and Pollution Prevention (6.3/6.4/6.6)

↑ [+2] ▢▢ ⊕⊕⊕ ★★★

Behavioural changes in the residential sector that lead to reduced energy demand can lead to reduced requirements on energy supply. As water is used to convert energy into useful forms, the reduction in residential demand is anticipated to reduce water consumption and waste water, resulting in more clean water for other sectors and the environment.

Bartos and Chester (2014); Fricko et al. (2016); Holland et al. (2016)

12 RESPONSIBLE CONSUMPTION AND PRODUCTION

Interaction	Score	Evidence	Agreement	Confidence

Responsible and Sustainable Consumption

↑ [+2] ▢▢▢▢ ⊕⊕⊕ ★★★

Technological improvements alone are not sufficient to increase energy savings. Zhao et al. (2017) found that building technology and occupant behaviours interact with each other and finally affect energy consumption from home. They found that occupant habits could not take advantage of more than 50% of energy efficiency potential allowed by an efficient building. In the electronic segment, product obsolescence represents a key challenge for sustainability. Echegaray (2016) discusses the dissonance between consumers' product durability experience, orientations to replace devices before terminal technical failure, and perceptions of industry responsibility and performance. The results from their urban sample survey indicate that technical failure is far surpassed by subjective obsolescence as a cause for fast product replacement. At the same time Liu et al. (2017) suggest that we need to go beyond individualist and structuralist perspectives to analyse sustainable consumption (i.e., combines both human agency paradigm and social structural perspective).

Sweeney et al., 2013; Webb et al., 2013; Allen et al., 2015; Echegaray (2015); He et al., 2016; Hult and Larsson, 2016; Isenhour and Feng, 2016; van Sluisveld et al., 2016; Zhao et al., 2017; Liu et al., 2017; Sommerfeld et al., 2017

14 LIFE BELOW WATER

Interaction	Score	Evidence	Agreement	Confidence
[0]				

No direct interaction

15 LIFE ON LAND

Interaction	Score	Evidence	Agreement	Confidence
[0]				

No direct interaction

Accelerating Energy Efficiency Improvement

6 CLEAN WATER AND SANITATION

Water Efficiency and Pollution Prevention (6.3/6.4/6.6)

↑ [+2] ▢▢▢▢ ⊕⊕⊕ ★★★

Efficiency changes in the residential sector that lead to reduced energy demand can lead to reduced requirements on energy supply. As water is used to convert energy into useful forms, the reduction in residential demand is anticipated to reduce water consumption and waste water, resulting in more clean water for other sectors and the environment. A switch to low-carbon fuels in the residential sector can lead to a reduction in water demand and waste water if the existing higher-carbon fuel is associated with a higher water intensity than the lower-carbon fuel. However, in some situations the switch to a low-carbon fuel such as, for example, biofuel could increase water use compared to existing conditions if the biofuel comes from a water-intensive feedstock. As water is used to convert energy into useful forms, energy efficiency is anticipated to reduce water consumption and waste water, resulting in more clean water for other sectors and the environment. Subsidies for renewables are anticipated to lead to the benefits and trade-offs outlined when deploying renewables. Subsidies for renewables could lead to improved water access and treatment if subsidies support projects that provide both water and energy services (e.g., solar desalination).

Bilton et al., 2011; Scott, 2011; Kumar et al., 2012; Meldrum et al., 2013; Bartos and Chester, 2014; Hendrickson and Horvath, 2014; Kern et al., 2014; Holland et al., 2015; Fricko et al., 2016; Kim et al., 2017

12 RESPONSIBLE CONSUMPTION AND PRODUCTION

Sustainable Practices and Lifestyles (12.6/12.7/12.8)

↑ [+1] ▢▢▢▢ ⊕⊕⊕ ★★★

Sustainable practices adopted by public and private bodies in their operations (e.g., for goods procurement, supply chain management and accounting) create an enabling environment in which renewable energy and energy efficiency measures may gain greater traction (McCollum et al., 2018).

Stefan and Paul, 2008; ECF, 2014; CDP, 2015; Khan et al., 2015; NCE, 2015; McCollum et al., 2018

14 LIFE BELOW WATER

Interaction	Score	Evidence	Agreement	Confidence
[0]				

No direct interaction

15 LIFE ON LAND

Reduced Deforestation (15.2)

↑ [+2] ⊕⊕⊕ ★★★

Improved stoves has helped halt deforestation in rural India.

Bhojvaid et al., 2014

Environment-Demand (continued)

Sector	Intervention	SDG	Interaction	Score	Evidence	Agreement	Confidence	Description	References
Buildings	Improved Access and Fuel Switch to Modern Low-carbon Energy	6 CLEAN WATER AND SANITATION — Access to Improved Water and Sanitation (6.1/6.2), Water Efficiency and Pollution Prevention (6.3/6.4/6.6)	↑/↓	[+2,-1]	▦▦	●●	★★★	A switch to low-carbon fuels in the residential sector can lead to a reduction in water demand and waste water if the existing higher-carbon fuel is associated with a higher water intensity than the lower-carbon fuel. However, in some situations the switch to a low-carbon fuel such as, for example, biofuel could increase water use compared to existing conditions if the biofuel comes from a water-intensive feedstock. Improved access to energy can support clean water and sanitation technologies. If energy access is supported with water-intensive energy sources, there could be trade-offs with water efficiency targets.	Hejazi et al., 2015; Cibin et al., 2016; Fricko et al., 2016; Song et al., 2016; Rao and Pachauri, 2017
Buildings	Improved Access and Fuel Switch to Modern Low-carbon Energy	12 RESPONSIBLE CONSUMPTION AND PRODUCTION — Sustainable Use and Management of Natural Resource (12.2)	↑/↓	[+2,-1]	▦▦	●●	★★★	A switch to low-carbon fuels in the residential sector can lead to a reduction in water demand and waste water if the existing higher-carbon fuel is associated with a higher water intensity than the lower-carbon fuel. However, in some situations the switch to a low-carbon fuel such as, for example, biofuel could increase water use compared to existing conditions if the biofuel comes from a water-intensive feedstock. Improved access to energy can support clean water and sanitation technologies. If energy access is supported with water-intensive energy sources, there could be trade-offs with water efficiency targets.	Hejazi et al., 2015; Cibin et al., 2016; Fricko et al., 2016; Song et al., 2016; Rao and Pachauri, 2017
Buildings	Improved Access and Fuel Switch to Modern Low-carbon Energy	14 LIFE BELOW WATER	No direct interaction	[0]					
Buildings	Improved Access and Fuel Switch to Modern Low-carbon Energy	15 LIFE ON LAND — Healthy Terrestrial Ecosystems (15.1/15.2/15.4/15.5/15.8)	↑	[+2]	▦▦	●●●	★★★	Ensuring that the world's poor have access to modern energy services would reinforce the objective of halting deforestation, since firewood taken from forests is a commonly used energy resource among the poor (McCollum et al., 2018).	Bazilian et al., 2011; Karekezi et al., 2012; Bailis et al., 2015; Winter et al., 2015; McCollum et al., 2018
Transport	Behavioural Response	6 CLEAN WATER AND SANITATION — Water Efficiency and Pollution Prevention (6.3/6.4/6.6)	↑	[+2]	▦▦	●●	★★	Behavioural changes in the transport sector that lead to reduced transport demand can lead to reduced transport energy supply. As water is used to produce a number of important transport fuels, the reduction in transport demand is anticipated to reduce water consumption and waste water, resulting in more clean water for other sectors and the environment.	Vidic et al., 2013; Holland et al., 2015; Fricko et al., 2016; Tiedeman et al., 2016
Transport	Behavioural Response	12 RESPONSIBLE CONSUMPTION AND PRODUCTION — Ensure Sustainable Consumption and Production Patterns (12.3)	↑	[+2]	▦▦	●●	★★	Urban carbon mitigation must consider the supply chain management of imported goods, the production efficiency within the city, the consumption patterns of urban consumers, and the responsibility of the ultimate consumers outside the city. Important for climate policy of monitoring the CO_2, clusters that dominate CO_2, emissions in global supply chains, because they offer insights on where climate policy can be effectively directed.	Kagawa et al., 2015; Lin et al., 2015; Creutzig et al., 2016
Transport	Behavioural Response	14 LIFE BELOW WATER	No direct interaction	[0]					
Transport	Behavioural Response	15 LIFE ON LAND	No direct interaction	[0]					
Transport	Accelerating Energy Efficiency Improvement	6 CLEAN WATER AND SANITATION — Water Efficiency and Pollution Prevention (6.3/6.4/6.6)	↑	[+2]	▦▦▦	●●●	★★★	Similar to behavioural changes, efficiency measures in the transport sector that lead to reduced transport demand can lead to reduced transport energy supply. As water is used to produce a number of important transport fuels, the reduction in transport demand is anticipated to reduce water consumption and waste water, resulting in more clean water for other sectors and the environment.	Vidic et al., 2013; Holland et al., 2015; Fricko et al., 2016; Tiedeman et al., 2016
Transport	Accelerating Energy Efficiency Improvement	12 RESPONSIBLE CONSUMPTION AND PRODUCTION — Sustainable Consumption (12.2/12.8)	↑	[+2]	▦▦▦	●●●	★★★	Relational complex transport behaviour resulting in significant growth in energy-inefficient car choices, as well as differences in mobility patterns (distances driven, driving styles) and actual fuel consumption between different car segments all affect non-progress on transport decarbonization. Consumption choices and individual lifestyles are situated and tied to the form of the surrounding urbanization. Major behavioural changes and emissions reductions require understanding of this relational complexity, consideration of potential interactions with other policies, and the local context and implementation of both command-and-control as well as market-based measures.	Stanley et al., 2011; Gallego et al., 2013; Heinonen et al., 2013; Aamaas and Peters, 2017; Azevedo and Leal, 2017; Gössling and Metzler, 2017
Transport	Accelerating Energy Efficiency Improvement	14 LIFE BELOW WATER	No direct interaction	[0]					
Transport	Accelerating Energy Efficiency Improvement	15 LIFE ON LAND	No direct interaction	[0]					

5

5

Environment-Demand (continued)

		6 CLEAN WATER AND SANITATION				12 RESPONSIBLE CONSUMPTION AND PRODUCTION					14 LIFE BELOW WATER					15 LIFE ON LAND				
		Interaction	Score	Evidence	Agreement	Confidence	Interaction	Score	Evidence	Agreement	Confidence	Interaction	Score	Evidence	Agreement	Confidence	Interaction	Score	Evidence	Agreement	Confidence
Transport	**Improved Access and Fuel Switch to Modern Low-carbon Energy**	**Water Efficiency and Pollution Prevention (6.3/6.4/6.6)** ↑/↓ [+2,-1] ▨▨ ●● ★★★ A switch to low-carbon fuels in the transport sector can lead to a reduction in water demand and waste water if the existing higher-carbon fuel is associated with a higher water intensity than the lower-carbon fuel. However, in some situations the switch to a low-carbon fuel such as, for example, biofuel could increase water use compared to existing conditions if the biofuel comes from a water-intensive feedstock. Transport electrification could lead to trade-offs with water use if the electricity is provided with water intensive power generation. Hejazi et al., 2015; Fricko et al., 2016; Song et al., 2016					**Ensure Sustainable Consumption and Production Patterns (12.3)** ↑ [+2] ▨▨▨▨ ●●● ★★★ Due to persistent reliance on fossil fuels, it is posited that transport is more difficult to decarbonize than other sectors. This study partially confirms that transport is less reactive to a given carbon tax than the non-transport sectors: in the first half of the century, transport mitigation is delayed by 10–30 years compared to non-transport mitigation. The extent to which earlier mitigation is possible strongly depends on implemented technologies and model structures. Figueroa et al., 2014; IPCC, 2014; Pietzcker et al., 2014; Creutzig et al., 2015						[0]	No direct interaction				[0]	No direct interaction		

Environnement-Supply

Replacing Coal

SDG 6 — Clean Water and Sanitation

Water Efficiency and Pollution Prevention (6.3/6.4/6.6)/ Access to Improved Water and Sanitation (6.1/6.2)

Category	Interaction	Score	Evidence	Agreement	Confidence
Non-biomass Renewables - solar, wind hydro	↑/↓	[+2,2]	(4)	●●	★★★★
Increased Use of Biomass	↑/↓	[+1,-2]	(1)	●●	★★★★

Non-biomass Renewables: Wind/solar renewable energy technologies are associated with very low water requirements compared to existing thermal power plant technologies. Widespread deployment is therefore anticipated to lead to improved water efficiency and avoided thermal pollution. However, managing wind and solar variability can increase water use at thermal power plants and can cause poor water quality downstream from hydropower plants. Access to distributed renewables can provide power to improve water access, but could also lead to increased groundwater pumping and stress if mismanaged. Developing dams to support reliable hydropower production can result in disputes for water in basins with up- and down-stream users. Storing water in reservoirs increases evaporation, which could offset water conservation targets and reduce availability of water downstream. However, hydropower plays an important role in energy access for water supply in developing regions, can support water security, and has the potential to reduce water demands if used without reservoir storage to displace other water intensive energy processes.
Bilton et al., 2011; Scott et al., 2011; Kumar et al., 2012; Ziv et al., 2012; Meldrum et al., 2013; Kern et al., 2014; Grill et al., 2015; Fricko et al., 2016; Gruber, 2016; De Stefano et al., 2017

Increased Use of Biomass: Biomass expansion could lead to increased water stress when irrigated feedstocks and water-intensive processing steps are used. Bioenergy crops can alter flow over land and through soils as well as require fertilizer, and this can reduce water availability and quality. Planting bioenergy crops on marginal lands or in some situations to replace existing crops can lead to reductions in soil erosion and fertilizer inputs, improving water quality.
Hejazi et al., 2015; Bonsch et al., 2016; Cibin et al., 2016; Song et al., 2016; Gao and Bryan, 2017; Griffiths et al., 2017; Ha and Wu, 2017; Taniwaki et al., 2017; Woodbury et al., 2018

SDG 12 — Responsible Consumption and Production

Natural Resource Protection (12.2/12.3/12.4/12.5)

Category	Interaction	Score	Evidence	Agreement	Confidence
Non-biomass Renewables	←	[+2]	(4)	●●●	★★★
Increased Use of Biomass	←	[+2]	(3)	●●●	★★★★

Non-biomass Renewables: Renewable energy and energy efficiency slow the depletion of several types of natural resources, namely coal, oil, natural gas and uranium. In addition, the phasing-out of fossil fuel subsidies encourages less wasteful energy consumption; but if that is done, then the policies implemented must take care to minimize any counteracting adverse side effects on the poor (e.g., fuel price rises). (Quote from McCollum et al., 2018)
Banerjee et al., 2012; Riahi et al., 2012; Schwanitz et al., 2014; Bhattacharyya et al., 2016; Cameron et al., 2016; McCollum et al., 2018

Increased Use of Biomass: Switching to renewable energy reduces the depletion of finite natural resources.
Banerjee et al., 2012; Riahi et al., 2012; Schwanitz et al., 2014; Bhattacharyya et al., 2016; Cameron et al., 2016; McCollum et al., 2018

SDG 14 — Life Below Water

Marine Economies (14.7)/ Marine Protection (14.1/14.2/14.4/14.5)

Category	Interaction	Score	Evidence	Agreement	Confidence
Non-biomass Renewables	↑/↓	[2,-1]	(2)	●●●	★★★
Increased Use of Biomass		[0]			

Non-biomass Renewables: Ocean-based energy from renewable sources (e.g., offshore wind farms, wave and tidal power) are potentially significant energy resource bases for island countries and countries situated along coastlines. Multi-use platforms combining renewable energy generation, aqua-culture, transport services and leisure activities can lay the groundwork for more diversified marine economies. Depending on the local context and prevailing regulations, ocean-based energy installations could either induce spatial competition with other marine activities, such as tourism, shipping, resources exploitation, and marine and coastal habitats and protected areas, or provide further grounds for protecting those exact habitats, therefore enabling marine protection. (Quote from McCollum et al., 2018) Hydropower disrupts the integrity and connectivity of aquatic habitats and impacts the productivity of inland waters and their fisheries.
Inger et al., 2009; Michler-Cieluch et al., 2009; Buck and Krause, 2012; WBGU, 2013; Cooke et al., 2016; Matthews and McCartney, 2018; McCollum et al., 2018

Increased Use of Biomass: No direct interaction

SDG 15 — Life on Land

Healthy Terrestrial Ecosystems (15.1/15.2/15.4/15.5/15.8)

Category	Interaction	Score	Evidence	Agreement	Confidence
Non-biomass Renewables	↓	[-1]	(3)	●●●	★★★
Increased Use of Biomass	↑/↓	[+1,-2]	(1)	●●●	★★

Non-biomass Renewables: Landscape and wildlife impact for wind; habitat impact for hydropower.
Alho, 2011; Garvin et al., 2011; Grodsky et al., 2011; Jain et al., 2011; Kumar et al., 2011; Kunz et al., 2011; Wiser et al., 2011; Dahl et al., 2012; de Lucas et al., 2012; Ziv et al., 2012; Lovich and Ennen, 2013; Smith et al., 2013; Matthews and McCartney, 2018

Increased Use of Biomass: Protecting terrestrial ecosystems, sustainably managing forests, halting deforestation, preventing biodiversity loss and controlling invasive alien species could potentially clash with renewable energy expansion, if that would mean constraining large-scale utilization of bioenergy or hydropower. Good governance, cross-jurisdictional coordination and sound implementation practices are critical for minimizing trade-offs (McCollum et al., 2018).
Smith et al., 2010, 2014; Acheampong et al., 2017; McCollum et al., 2018

Environement-Supply (continued)

		6 CLEAN WATER AND SANITATION					**12 RESPONSIBLE CONSUMPTION AND PRODUCTION**					**14 LIFE BELOW WATER**					**15 LIFE ON LAND**				
		Interaction	Score	Evidence	Agreement	Confidence	Interaction	Score	Evidence	Agreement	Confidence	Interaction	Score	Evidence	Agreement	Confidence	Interaction	Score	Evidence	Agreement	Confidence
Replacing Coal	**Nuclear/Advanced Nuclear**	**Water Efficiency and Pollution Prevention (6.3/6.4/6.6)** ↑/↓ [+2,-1] &&& JJJ ««« Nuclear power generation requires water for cooling which can lead to localized water stress and the resulting cooling effluents can cause thermal pollution in rivers and oceans. Webster et al., 2013; Holland et al., 2015; Fricko et al., 2016; Raptis et al., 2016						[0] No direct interaction					[0] No direct interaction				**Healthy Terrestrial Ecosystems (15.1/15.2/15.4/15.5/15.8)** ↓ [-1] && JJ ««« Safety and waste concerns from uranium mining and milling. Bickerstaff et al., 2008; Sjoberg and Sjoberg, 2009; Ahearne, 2011; Corner et al., 2011; Visschers and Siegrist, 2012; IPCC, 2014				
	CCS: Bioenergy	**Water Efficiency and Pollution Prevention (6.3/6.4/6.6)** ↑/↓ [+1,-2] ▢▢ ☺ ★★ CCUS requires access to water for cooling and processing which could contribute to localized water stress. However, CCS/U processes can potentially be configured for increased water efficiency compared to a system without carbon capture via process integration. The bioenergy component adds the additional trade-offs associated with bioenergy use. Large-scale bioenergy increases input demand, resulting in environmental degradation and water stress. Meldrum et al., 2013; Byers et al., 2016; Fricko et al., 2016; Brandl et al., 2017; Dooley and Kartha, 2018					**Natural Resource Protection (12.2/12.3/12.4/12.5)** ↑ [+1] ▢▢▢ ☺☺ ★★ Switching to renewable energy reduces the depletion of finite natural resources. On the other hand, the availability of underground storage is limited and therefore reduces the benefits of switching from finite resources to bioenergy. Banerjee et al., 2012; Riahi et al., 2012; Schwanitz et al., 2014; Bhattacharyya et al., 2016; Cameron et al., 2016; McCollum et al., 2018					[0] No direct interaction					**Healthy Terrestrial Ecosystems (15.1/15.2/15.4/15.5/15.8)** ↑/↓ [+1,-2] ▢ ☺☺☺ ★★ Protecting terrestrial ecosystems, sustainably managing forests, halting deforestation, preventing biodiversity loss and controlling invasive alien species could potentially clash with renewable energy expansion, if that would mean constraining large-scale utilization of bioenergy or hydropower. Good governance, cross-jurisdictional coordination and sound implementation practices are critical for minimizing trade-offs (McCollum et al., 2018). Large-scale bioenergy increases input demand, resulting in environmental degradation and water stress. Smith et al., 2010, 2014; Acheampong et al., 2017; Dooley and Kartha, 2018, McCollum et al., 2018				
Advanced Coal	**CCS: Fossil**	**Water Efficiency and Pollution Prevention (6.3/6.4/6.6)** ↑/↓ [+1,-2] ▢▢▢▢ ☺ ★★ CCUS requires access to water for cooling and processing which could contribute to localized water stress. However, CCS/U processes can potentially be configured for increased water efficiency compared to a system without carbon capture via process integration. Coal mining to support clean coal CCS will negatively impact water resources due to the associated water demands, waste water and land-use requirements. Meldrum et al., 2013; Byers et al., 2016; Fricko et al., 2016; Brandl et al., 2017					[0] No direct interaction					[0] No direct interaction					[0] No direct interaction				

Environment-Other

Agriculture and Livestock

Behavioural Response: Sustainable Healthy Diets and Reduced Food Waste

	Interaction	Score	Evidence	Agreement	Confidence
6 CLEAN WATER AND SANITATION — Water Efficiency and Pollution Prevention (6.3/6.4/6.6)	↑/↓	[+2,-1]	robust	high	★★★★

Reduced food waste avoids direct water demand and waste water for crops and food processing, and avoids water used for energy supply by reducing agricultural, food processing and waste management energy inputs. Healthy diets will support water efficiency targets if the shift towards healthy foods results in food supply chains that are less water intensive than the historical dietary pattern.

Khan et al., 2009; Ingram, 2011; Kummu et al., 2012; Haileselassie et al., 2013; Bajželj et al., 2014; Tilman and Clark, 2014; Walker et al., 2014; Ran et al., 2016

	Interaction	Score	Evidence	Agreement	Confidence
12 RESPONSIBLE CONSUMPTION AND PRODUCTION — Ensure Sustainable Consumption and Production Patterns, Sustainable Practices and Lifestyle (12.3/12.4/12.6/12.7/12.8)	↑	[+2]	robust	high	★★★★

Reduce loss and waste in food systems, processing, distribution and by changing household habits. To reduce environmental impact of livestock both production and consumption trends in this sector should be traced. Livestock production needs to be intensified in a responsible way (i.e., be made more efficient in the way that it uses natural resources). Wasted food represents a waste of all the emissions generated during the course of producing and distributing that food. Mitigation measures include: eat no more than needed to maintain a healthy body weight; eat seasonal, robust, field-grown vegetables rather than protected, fragile foods prone to spoilage and requiring heating and lighting in their cultivation, refrigeration stage; consume fewer foods with low nutritional value e.g., alcohol, tea, coffee, chocolate and bottled water (these foods are not needed in our diet and need not be produced); shop on foot or over the Internet (reduced energy use). Reduction in food waste will not only pave the path for sustainable production but will also help in achieving sustainable consumption (Garnett, 2011). Reduce meat consumption to encourage more sustainable eating practices.

Stehfest et al., 2009; Steinfeld and Gerber, 2010; Garnett, 2011; Ingram, 2011; Beddington et al., 2012; Kummu et al., 2012; Bellarby et al., 2013; Dagevos and Voordouw, 2013; Smith, 2013; Bajželj et al., 2014; Hedenus et al., 2014; Tilman and Clark, 2014; West et al., 2014; Hiç et al., 2016, Lamb et al., 2016

	Interaction	Score	Evidence	Agreement	Confidence
14 LIFE BELOW WATER		[0]			

No direct interaction

	Interaction	Score	Evidence	Agreement	Confidence
15 LIFE ON LAND — Conservation of Biodiversity and Restoration of Land (15.1/15.5/15.9)	↑	[+1]	medium	medium	★

Reducing food waste has secondary benefits like protecting soil from degradation, and decreasing pressure for land conversion into agriculture and thereby protecting biodiversity. The agricultural area that becomes redundant through the dietary transitions can be used for other agricultural purposes such as energy crop production, or will revert to natural vegetation. A global food transition to less meat, or even a complete switch to plant-based protein food, could have a dramatic effect on land use. Up to 2,700 Mha of pasture and 100 Mha of crop land could be abandoned (Quoted from Stehfest et al., 2009)

Stehfest et al., 2009; Kummu et al., 2012

Land-based Greenhouse Gas Reduction and Soil Carbon Sequestration

	Interaction	Score	Evidence	Agreement	Confidence
6 CLEAN WATER AND SANITATION — Water Efficiency and Pollution Prevention (6.3/6.4/6.6)	↑/↓	[+1,-1]	medium	medium	★★★

Soil carbon sequestration can alter the capacity of soils to store water, which impacts the hydrological cycle and could be positive or negative from a water perspective, dependent on existing conditions. CSA enrich linkages across sectors including management of water resources. Minimum tillage systems have been reported to reduce water erosion and thus sedimentation of water courses (Bustamante et al., 2014).

Behnassi et al., 2014; Bustamante et al., 2014; P. Smith et al., 2016b

	Interaction	Score	Evidence	Agreement	Confidence
12 RESPONSIBLE CONSUMPTION AND PRODUCTION — Ensure Sustainable Production Patterns (12.3)	↑	[+1]	medium	medium	★

Millet or sorghum yield can double as compared with unimproved land by more than 1 tonne per hectare due to sustainable intensification. An integrated approach to safe applications of both conventional and modern agricultural biotechnologies will contribute to increased yield (Lakshmi et al., 2015).

Campbell et al., 2014; Lakshmi et al., 2015

	Interaction	Score	Evidence	Agreement	Confidence
14 LIFE BELOW WATER		[0]			

No direct interaction

	Interaction	Score	Evidence	Agreement	Confidence
15 LIFE ON LAND — Conservation of Biodiversity and Restoration of Land (15.1/15.5/15.9)	↑/↓	[+1,-1]	robust	high	★★★

Agricultural intensification can promote conservation of biological diversity by reducing deforestation, and by rehabilitation and restoration of biodiverse communities on previously developed farm or pasture land. However, planting monocultures on biodiversity hot spots can have adverse side-effects, reducing biodiversity. Genetically modified crops reduce demand for cultivated land. Adaptation of integrated landscape approaches can provide various ecosystem services. CSA enrich linkages across sectors including management of land and bio-resources. Land sparing has the potential to be beneficial for biodiversity, including for many species of conservation concern, but benefits will depend strongly on the use of spared land. In addition, high yield farming involves trade-offs and is likely to be detrimental for wild species associated with farm land (Lamb et al., 2016).

Lybbert and Sumner, 2010; Behnassi et al., 2014; Harvey et al., 2014; IPCC, 2014; Lamb et al., 2016

Environement-Other (continued)

Agriculture and Livestock

Greenhouse Gas Reduction from Improved Livestock Production and Manure Management Systems

SDG 6 — Clean Water and Sanitation: Water Efficiency and Pollution Prevention (6.3/6.4/6.6)

Interaction	Score	Evidence	Agreement	Confidence
↑/↓	[+2,-1]	▪▪▪▪	●●●	★★★

Livestock efficiency measures are expected to reduce water required for livestock systems as well as associated livestock waste water flows. However, efficiency measures that include agricultural intensification could increase water demands locally, leading to increased water stress if the intensification is mismanaged. In scenarios where zero human-edible concentrate feed is used for livestock, freshwater use reduces by 21%.

Haileselassie et al., 2013; Schader et al., 2015; Kong et al., 2016; Ran et al., 2016

SDG 12 — Responsible Consumption and Production: Ensure Sustainable Production Patterns and Restructing Taxation (12.3/12c)

Interaction	Score	Evidence	Agreement	Confidence
↑	[+1]	▪▪▪	●●	★★

In the future, many developed countries will see a continuing trend in which livestock breeding focuses on other attributes in addition to production and productivity, such as product quality, increasing animal welfare, disease resistance (Thornton, 2010). Diet composition and quality are key determinants of the productivity and feed-use efficiency of farm animals (Herrero, et al., 2013). Mechanisms for effecting behavioural change in livestock systems need to be better understood by implementing combinations of incentives and taxes simultaneously in different parts of the world (Herrero and Thornton, 2013). Reducing the amount of human-edible crops that are fed to livestock represents a reversal of the current trend of steep increases in livestock production, and especially of monogastrics, so would require drastic changes in production and consumption (Schader et al., 2015).

Thornton, 2010; Herrero and Thornton, 2013; Herrero et al., 2013; Schader et al., 2015

SDG 14 — Life Below Water

Interaction	Score	Evidence	Agreement	Confidence
	[0]	No direct interaction		

SDG 15 — Life on Land: Restoration of Land (15.1)

Interaction	Score	Evidence	Agreement	Confidence
↑	[+1]	▪▪	●	★

Grasslands are valuable, but improved management is required as grass accounts for close to 50% of feed use in livestock systems . The scenario with 100% reduction of food-competing-feedstuffs resulted in a 335 Mha decrease in arable land area, which corresponds to a decrease of 22% in arable and 7% in the total agricultural area.

Herrero et al., 2013; Schader et al., 2015

Forest

Reduced Deforestation, REDD+

SDG 6 — Clean Water and Sanitation: Water Efficiency and Pollution Prevention (6.3/6.4/6.6)

Interaction	Score	Evidence	Agreement	Confidence
↑/↓	[+1,-1]	▪▪	●●	★★

Forest management alters the hydrological cycle which could be positive or negative from a water perspective and is dependent on existing conditions. Conservation of ecosystem services indirectly could help countries maintain watershed integrity. Forests provide sustainable and regulated provision and help in water purification.

Zomer et al., 2008; Kibria, 2015; Bonsch et al., 2016; Gao and Bryan, 2017; Griffiths et al., 2017; Katila et al., 2017

SDG 12 — Responsible Consumption and Production: Ensure Sustainable Consumption (12.3)

Interaction	Score	Evidence	Agreement	Confidence
↑	[+1]	▪	●	★

Reduce the human pressure on forests, including actions to address drivers of deforestation.

Bastos Lima et al., 2017

SDG 14 — Life Below Water

Interaction	Score	Evidence	Agreement	Confidence
	[0]	No direct interaction		

SDG 15 — Life on Land: Conservation of Biodiversity, Sustainability of Terrestrial Ecosystems (15.2/15.3/15.4/15.5/15.9)

Interaction	Score	Evidence	Agreement	Confidence
↑	[+1]	▪▪▪	●●●	★★★

Policies and programmes for reducing deforestation and forest degradation for rehabilitation and restoration of degraded lands can promote conservation of biological diversity. Reduce the human pressure on forests, including actions to address drivers of deforestation. Efforts by the Government of Zambia to reduce emissions by REDD+ have contributed erosion control, ecotourism and pollination valued at 2.5% of the country's GDP.

Miles and Kapos, 2008; IPCC, 2014; Bastos Lima et al., 2015; Turpie et al., 2015; Epstein and Theuer, 2017; Katila et al., 2017

Afforestation and Reforestation

SDG 6 — Clean Water and Sanitation: Enhance Water Quality (6.3)

Interaction	Score	Evidence	Agreement	Confidence
↑/↓	[+2,-1]	▪▪▪▪	●●●	★★★

Similar to REDD+, forest management alters the hydrological cycle which could be positive or negative from a water perspective and is dependent on existing conditions. Forest landscape restoration can have a large impact on water cycles. Strategic placement of tree belts in lands affected by dryland salinity can remediate the affected zero by modifying landscape water balances. Watershed scale reforestation can result in the restoration of water quality. Fast-growing species can increase nutrient input and water inputs that can cause ecological damage and alter local hydrological patterns. Reforestation of mixed native species and in carefully chosen sites could increase biodiversity and restore waterways, reducing run-off and erosion (Dooley and Kartha, 2018).

Zomer et al., 2008; Bustamante et al., 2014; Kibria, 2015; Lamb et al., 2016; Dooley and Kartha, 2018

SDG 12 — Responsible Consumption and Production

No direct interaction

SDG 14 — Life Below Water: Marine Economies (14.7)/Marine Protection and Income Generation (14.1/14.2/14.4/14.5)

Interaction	Score	Evidence	Agreement	Confidence
↑	[+2]	▪	●	★

Mangroves would help to enhance fisheries and tourism businesses.

Kibria, 2015

SDG 15 — Life on Land: Conservation of Biodiversity and Restoration of Land (15.1/15.5/15.9)

Interaction	Score	Evidence	Agreement	Confidence
↑	[+2]	▪▪▪▪	●●●●	★★★★

Identified large amounts of land (749 Mha) globally as biophysically suitable and meeting the CDM eligibility criteria . Forest landscape restoration can conserve biodiversity and reduce land degradation. Mangroves reduce impacts of disasters (cyclones/storms/floods) acting as live seawalls and enhance forest resources/biodiversity. Forest goal can conserve/restore 3.9–8.8 m ha/year average, 77.2–176.9 m ha in total and 7.7–17.7 m ha /year in 2030 of forest area by 2030 (Wolosin, 2014). Forest and biodiversity conservation, protected area formation and forestry-based afforestation are practices that enhance resilience of forest ecosystems to climate change (IPCC, 2014). Strategic placement of tree belts in lands affected by dryland salinity can remediate the affected lands by modifying landscape water balances and protect livestock. It can restore biologically diverse communities on previously developed farmland . Large-scale restoration is likely to benefit ecosystem service provision, including recreation, biodiversity, conservation and flood mitigation. Reforestation of mixed native species and in carefully chosen sites could increase biodiversity, reducing run-off and erosion .

Zomer et al., 2008; Bustamante et al., 2014; IPCC, 2014; Kibria, 2015; Lamb et al., 2016; Epstein and Theuer, 2017; Dooley and Kartha, 2018

Environment-Other (continued)

		6 CLEAN WATER AND SANITATION	12 RESPONSIBLE CONSUMPTION AND PRODUCTION	14 LIFE BELOW WATER	15 LIFE ON LAND
Forest	**Behavioural Response (Responsible Sourcing)**	**Water Efficiency and Pollution Prevention (6.3/6.4/6.6)** Interaction ↑/↓ | Score [+2,-1] | Evidence [robust] | Agreement ⊕⊕ | Confidence ★★ Responsible sourcing will have co-benefits for water efficiency and pollution prevention if the sourcing strategies incorporate water metrics. There is a risk that shifting supply sources could lead to increased water use in another part of the economy. At local levels, forest certification programmes and practicing sustainable forest management provide freshwater supplies. van Oel and Hoekstra, 2012; Launiainen et al., 2014; Hontelez, 2016	**Ensure Sustainable Production Patterns (12.3)** Interaction ↑ | Score [+1] | Evidence [robust] | Agreement ⊕ | Confidence ★ At local levels, forest certification programmes and practicing sustainable forest management provide the provision of raw materials for a 'low ecological footprint' economy. Hontelez, 2016	Interaction | Score [0] | Evidence | Agreement | Confidence No direct interaction	**Sustainability and Conservation (15.1/15.2/15.3)** Interaction ↑/↓ | Score [+1,-1] | Evidence [medium] | Agreement ⊕ | Confidence ★ At the macro level, forest certification has done little to stem the tide of forest degradation, conversion of forest land to agriculture, and illegal logging—all of which remain serious threats to Indonesian forests (Bartley, 2010). At local levels, forest certification programmes and practicing sustainable forest management help in biodiversity protection. Bartley, 2010; Hontelez, 2016
Oceans	**Ocean Iron Fertilization**	Score [0] No direct interaction	Score [0] No direct interaction	**Nutrient Pollution, Ocean Acidification, Fish Stocks, MPAs, SISD (14.1/14.3/14.4/14.5/14.7)** Interaction ↑/↓ | Score [+1,-2] | Evidence [robust] | Agreement ⊕⊕ | Confidence ★ OIF could exacerbate or reduce nutrient pollution, increase the likelihood of mid-water deoxygenation, increase ocean acidification, might contribute to the rebuilding of fish stocks in producing plankton, therefore generating benefits for SISD, but might also be in conflict with designing MPAs. Gnanadesikan et al., 2003; Jin and Gruber, 2003; Denman, 2008; Lampitt et al., 2008; Smetacek and Naqvi, 2008; Gussow et al., 2010; Oschlies et al., 2010; Trick et al., 2010; Williamson et al., 2012	Score [0] No direct interaction
	Blue Carbon	**Integrated Water Resources Management (6.3/6.5)** Interaction ↑ | Score [+2] | Evidence [medium] | Agreement ⊕ | Confidence ★ Development of blue carbon resources (coastal and marine vegetated ecosystems) can lead to coordinated management of water in coastal areas. Vierros et al., 2015	Score [0] No direct interaction	**Ocean Acidification, Nutrient Pollution (14.3/14.1)** Interaction ↑/~ | Score [+2,0] | Evidence [medium] | Agreement ⊕⊕⊕ | Confidence ★★★ Mangroves could buffer acidification in their immediate vicinity; seaweeds have not been able to mitigate the effect on ocean foraminifera. Pettit et al., 2015; Sippo et al., 2016	**Conservation of Biodiversity and Restoration of Land (15.1/15.2/15.3/15.4/15.9)** Interaction ↑ | Score [+3] | Evidence [robust] | Agreement ⊕⊕⊕⊕ | Confidence ★★★★ Average difference of 31 mm per year in elevation rates between areas with seagrass and unvegetated areas (case study areas: Scotland, Kenya, Tanzania and Saudi Arabia); mangroves fostering sediment accretion of about 5mm a year. Alongi, 2012; Potouroglou et al., 2017
	Enhanced Weathering	Score [0] No direct interaction	Score [0] No direct interaction	**Ocean Acidification, Nutrient Pollution (14.3/14.1)** Interaction ↑/↓ | Score [+2,-1] | Evidence [robust] | Agreement ⊕⊕⊕ | Confidence ★★★ Enhanced weathering (either by spreading lime or quicklime, in combination with CCS, over the ocean or olivine at beaches or the catchment area of rivers) opposes ocean acidification. "End-of-century ocean acidification is reversed under RCP4.5 and reduced by about two-thirds under RCP8.5; additionally, surface ocean aragonite saturation state, a key control on coral calcification rates, is maintained above 3.5 throughout the low latitudes, thereby helping maintain the viability of tropical coral reef ecosystems." However, marine biology would also be affected, in particular if spreading olivine is used, which works like ocean (iron) fertilization. Köhler et al., 2010, 2013; Hartmann et al., 2013; Paquay and Zeebe, 2013; P. Smith et al., 2016a; Taylor et al., 2016	**Protect Inland Freshwater Systems (14.1)** Interaction ↓ | Score [-2] | Evidence [robust] | Agreement ⊕ | Confidence ★ Olivine can contain toxic metals such as nickel which could accumulate in the environment or disrupt the local ecosystem by changing the pH of the water (in case of spreading in the catchment area of rivers). Hartmann et al., 2013

Economic-Demand

Industry

SDG 7 — Affordable and Clean Energy

Interaction	Score	Evidence	Agreement	Confidence

Accelerating Energy Efficiency Improvement — Energy Savings (7.1/7.3/7.a/7.b)
↑ | [+2] | (evidence) | ⊕⊕⊕ | ★★★★

Energy efficiency leads to reduced energy demand and hence energy supply and energy security, reduces import. Positive rebound effect can raise demand but to a very less extent due to low rebound effect in industry sector in many countries and by appropriate mix of industries (China) can maintain energy savings gain. Supplying surplus energy to cities is also happening, proving maintenance culture, switching off idle equipment helps in saving energy (e.g Ghana).

Apeaning and Thollander, 2013; Chakravarty et al., 2013; IPCC, 2014; Karner et al., 2015; Zhang et al., 2015; Li et al., 2016; Fernando et al., 2017; Wesseling et al., 2017

Low-Carbon Fuel Switch — Sustainable and Modern (7.2/7.a)
↑ | [+2] | | ⊕ | ★

Industries are becoming suppliers of energy, waste heat, water and roof tops used for solar energy generation, and therefore helping to reduce primary energy demand. CHP in chemical industries can help in providing surplus power in the grid.

Karner et al., 2015; Griffin et al., 2018

Decarbonization/CCS/CCU — Affordable and Sustainable Energy Sources
↑ / ↓ | [+2,-2] | (evidence) | ⊕⊕ | ★★

CCS for EPIs can be incremental, but need additional space and can need additional energy, sometimes compensating for higher efficiency. For example, recirculating blast R furnace and CCS for iron steel means high energy demand; electric melting in glass can mean higher electricity prices; in the paper industry, new separation and drying technologies are key to reducing the energy intensity, allowing for carbon neutral operation in the future; bio-refineries can reduce petro-refineries; DRI in iron and steel with H2 encourages innovation in hydrogen infrastructure; and the chemicals industry also encourage renewable electricity and hydrogen as bio-based polymers can increase biomass price.

Griffin et al., 2017; Wesseling et al., 2017

SDG 8 — Decent Work and Economic Growth

Interaction	Score	Evidence	Agreement	Confidence

Accelerating Energy Efficiency Improvement — Reduces Unemployment (8.2/8.3/8.4/8.5/8.6)
↑ | [+1] | (evidence) | ⊕⊕⊕ | ★★★

Unemployment rate reduction from 25% to 12% in South Africa. Enhances firm productivity and technical and managerial capacity of the employees. New jobs for managing energy efficiency opens up opportunities in energy service delivery sector.

Altieri et al., 2016; Fernando et al., 2017; Johansson and Thollander, 2018

Low-Carbon Fuel Switch — Economic Growth with Decent Employment (8.1/8.2/8.3/8.4)
↑ | [+2] | (evidence) | ⊕⊕⊕⊕ | ★★★★

The circular economy instead of linear global economy can achieve climate goals and can help in economic growth through industrialization, which saves on resources and the environment and supports small, medium and even large industries, which can lead to employment generation. So new regulations, incentives and a revised tax regime can help in achieving the goal.

Stahel, 2013, 2017; Liu et al., 2014; Leider et al., 2015; Supino et al., 2015; Zheng et al., 2016; Fan et al., 2017; Shi et al., 2017

Decarbonization/CCS/CCU — Decouple Growth from Environmental Degradation (8.1/8.2/8.4)
↑ | [+2] | (evidence) | ⊕⊕⊕ | ★★★

EPI s are important players for economic growth. Deep decarbonization of EPIs through radical innovation is consistent with well-below 2°C scenarios.

Denis-Ryan et al., 2016; Åhman et al., 2017; Wesseling et al., 2017

SDG 9 — Industry, Innovation and Infrastructure

Interaction	Score	Evidence	Agreement	Confidence

Accelerating Energy Efficiency Improvement — Infrastructure Renewal (9.1/9.3/9.5/9.a)
↑ | [+1] | (evidence) | ⊕⊕ | ★★

Transitioning to a more renewables-based energy system that is highly energy efficient is well-aligned with the goal of upgrading energy infrastructure and making the energy industry more sustainable. At the same time, infrastructure upgrades in other parts of the economy, such as modernized telecommunications networks, can create the conditions for a successful expansion of renewable energy and energy efficiency measures (e.g., smart metering and demand-side management; McCollum et al., 2018).

Riahi et al., 2012; Apeaning and Thollander, 2013; Goldthau, 2014; Bhattacharyya et al., 2016; Meltzer, 2016; McCollum et al., 2018

Low-Carbon Fuel Switch — Innovation and New Infrastructure (9.2/9.3/9.4/9.5/9.a)
↑ | [+2] | (evidence) | ⊕⊕⊕⊕ | ★★★★

A circular economy instead of linear global economy is helping new innovation, and infrastructure can achieve climate goals and can help in economic growth through industrialization which saves on resources and the environment and supports small, medium and even large industries, which can lead to employment generation. So new regulations, incentives and revised tax regime can help in achieving the goal.

Stahel, 2013, 2017; Liu et al., 2014; Leider et al., 2015; Supino et al., 2015; Zheng et al., 2016; Fan et al., 2017; Shi et al., 2017

Decarbonization/CCS/CCU — Innovation and New Infrastructure (9.2/9.4/9.5)
↑ | [+2] | (evidence) | ⊕⊕⊕⊕ | ★★★★

Deep decarbonization through radical technological change in EPI will lead to radical innovations, for example, in completely changing industries' innovation strategies, skills, production techniques, design, etc. Radical CCS will need new infrastructure to transport CO_2.

Denis-Ryan et al., 2016; Åhman et al., 2017; Wesseling et al., 2017; Griffin et al., 2018

SDG 11 — Sustainable Cities and Communities

Interaction	Score	Evidence	Agreement	Confidence

Accelerating Energy Efficiency Improvement — Sustainable Cities (15.6/15.8/15.9)
↑ | [+2] | (evidence) | ⊕ | ★

Industries are becoming suppliers of energy, waste heat and water to neighbourial human settlements, and therefore there is a reduced primary energy demand, which also makes towns and cities grow sustainably.

Karner et al., 2015

Low-Carbon Fuel Switch — Sustainable Cities (15.6/15.8/15.9)
↑ | [+2] | (evidence) | ⊕ | ★

Industries are becoming suppliers of energy, waste heat, water and roof tops used for solar energy generation, and supply to neighbourial human settlements, therefore reducing primary energy demand, which also makes towns and cities grow sustainably.

Karner et al., 2015

Decarbonization/CCS/CCU
[0] — No direct interaction

5

Economic-Demand (continued)

Buildings

---

7 — AFFORDABLE AND CLEAN ENERGY

Behavioural Response

Saving Energy, Improvement in Energy Efficiency (7.3/7.a/7.b)

Interaction	Score	Evidence	Agreement	Confidence
←	[+2]	▦▦▦▦	●●●●	★★★

Lifestyle change measures and adoption behaviour affect residential energy use and implementation of efficient technologies as residential HVAC systems. Also, social influence can drive energy savings in users exposed to energy consumption feedback. Effect of autonomous motivation on energy savings is greater than that of other more established predictors, such as intentions, subjective norms, perceived behavioural control and past behaviour. Use of a hybrid engineering approach using social psychology and economic behaviour models are suggested for residential peak electricity demand response. However, some take-back in energy savings can happen due to rebound effects unless managed appropriately or accounted for welfare improvement. Adjusting thermostats helps in saving energy. Uptake of energy efficient appliances by households with an introduction to appliance standards, training, promotional material dissemination and the desire to save on energy bills are helping to change acquisition behaviour.

Chakravarty et al., 2013; Gyamfi et al., 2013; Hori et al., 2013; Huebner et al., 2013; Jain et al., 2013; Sweeney et al., 2013; Webb et al., 2013; Yue et al., 2013; Anda and Temmen, 2014; Allen et al., 2015; Noonan et al., 2015; de Koning et al., 2016; Isenhour and Feng, 2016; Santarius et al., 2016; Song et al., 2016; van Sluisveld et al., 2016; Sommerfeld et al., 2017; Zhao et al., 2018

Accelerating Energy Efficiency Improvement

Increase in Energy Savings (7.3)

Interaction	Score	Evidence	Agreement	Confidence
←	[+2]	▦▦▦▦	●●●	★★★★

There is high agreement among researchers based on a great deal of evidence across various countries that energy efficiency improvement reduces energy consumption and therefore leads to energy savings (e.g., efficient stoves save bioenergy). Countries with higher hours of use due to higher ambient temperatures or more carbon intensive electricity grids benefit more from available improvements in energy efficiency and use of refrigerant transition.

McLeod et al., 2013; Noris et al., 2013; Bhojvaid et al., 2014; Holopainen et al., 2014; Kwong et al., 2014; Yang et al., 2014; Cameron et al., 2015; Liddell and Guiney, 2015; Shah et al., 2015; Berrueta et al., 2017; Kim et al., 2017; Salvalai et al., 2017

---

8 — DECENT WORK AND ECONOMIC GROWTH

Behavioural Response

Progressively Improve Resource Efficiency (8.4), Employment Opportunities (8.2/8.3/8.5/8.6)

Interaction	Score	Evidence	Agreement	Confidence
←	[+2]	▦	◉	★

Behavioural change programmes help in sustaining energy savings through new infrastructure developments.

Anda and Temmen, 2014

Accelerating Energy Efficiency Improvement

Employment Opportunities (8.2/8.3/8.5/8.6)/Strong Financial Institutions (8.10)

Interaction	Score	Evidence	Agreement	Confidence
↑ / ↓	[+2, -1]	▦▦	◉	★★

Deploying renewables and energy efficient technologies, when combined with other targeted monetary and fiscal policies, can help spur innovation and reinforce local, regional and national industrial and employment objectives. Gross employment effects seem likely to be positive; however, uncertainty remains regarding the net employment effects due to several uncertainties surrounding macro-economic feedback loops playing out at the global level. Moreover, the distributional effects experienced by individual actors may vary significantly. Strategic measures may need to be taken to ensure that a large-scale switch to renewable energy minimizes any negative impacts on those currently engaged in the business of fossil fuels (e.g., government support could help businesses re-tool and workers re-train). To support clean energy and energy efficiency efforts, strengthened financial institutions in developing country communities are necessary for providing capital, credit and insurance to local entrepreneurs attempting to enact change (McCollum et al., 2018).

Babiker and Eckaus, 2007; Fankhauser and Tepic, 2007; Gohin, 2008; Frondel et al., 2010; Dinkelman, 2011; Guivarch et al., 2011; Jackson and Senker, 2011; Borenstein, 2012; Creutzig et al., 2013; Blyth et al., 2014; Clarke et al., 2014; Dechezleprêtre and Sato, 2014; Bertram et al., 2015; Johnson et al., 2015; IRENA, 2016; A. Smith et al., 2016; Berrueta et al., 2017; McCollum et al., 2018

---

9 — INDUSTRY, INNOVATION AND INFRASTRUCTURE

Behavioural Response

Innovation and New Infrastructure (9.2/9.4/9.5)

Interaction	Score	Evidence	Agreement	Confidence
←	[+2]	▦▦	●●	★★

Adoption of smart meters and smart grids following community-based social marketing help with infrastructure expansion. People are adopting solar rooftops, white roof/vertical garden/green roofs at much faster rates due to new innovations and regulations.

Anda and Temmen, 2014; Roy et al., 2018

Accelerating Energy Efficiency Improvement

Innovation and New Infrastructure (9.2/9.4/9.5)

Interaction	Score	Evidence	Agreement	Confidence
←	[+2]	▦▦	●●	★★

Adoption of smart meters and smart grids following community-based social marketing help in infrastructure expansion. Statutory norms to enhance energy and resource efficiency in buildings is encouraging green building projects.

Anda and Temmen, 2014; Roy et al., 2018

---

11 — SUSTAINABLE CITIES AND COMMUNITIES

Behavioural Response

Sustainable Cities (15.6/15.8/15.9)

Interaction	Score	Evidence	Agreement	Confidence
←	[+2]	▦▦	●●	★★

Behavioural change programmes help in making cities more sustainable.

Anda and Temmen, 2014; Roy et al., 2018

Accelerating Energy Efficiency Improvement

Urban Environmental Sustainability (11.3/11.6/11.b/11.c)

Interaction	Score	Evidence	Agreement	Confidence
←	[+2]	▦▦▦	●●●	★★★★

Renewable energy technologies and energy efficient urban infrastructure solutions (e.g., public transit) can also promote urban environmental sustainability by improving air quality and reducing noise. Efficient transportation technologies powered by renewably based energy carriers will be a key building block of any sustainable transport system (McCollum et al., 2018). Green buildings help in sustainable construction.

Creutzig et al., 2012; Kahn Ribeiro et al., 2012; Riahi et al., 2012; Bongardt et al., 2013; Grubler and Fisk, 2013; Raji et al., 2015; Kim et al., 2017; McCollum et al., 2018

Economic-Demand (continued)

5

SDG 7 — AFFORDABLE AND CLEAN ENERGY

	Interaction	Score	Evidence	Agreement	Confidence
Buildings — Improved Access and Fuel Switch to Modern Low-carbon Energy	**Meeting Energy Demand** — Renewable energies could potentially serve as the main source to meet energy demand in rapidly growing developing country cities. Ali et al. (2015) estimated the potential of solar, wind and biomass renewable energy options to meet part of the electricity demand in Karachi, Pakistan. *Li et al., 2013; Peng and Lu, 2013; Pietzcker, 2013; Pode, 2013; Yanine and Sauma, 2013; Zulu and Richardson, 2013; Connolly et al., 2014; Creutzig et al., 2014; Pietzcker et al., 2014; Ali et al., 2015; O'Mahony and Dufour, 2015; Abanda et al., 2016; Mittelehldt, 2016; Bligli et al., 2017; Byravan et al., 2017; Islar et al., 2017; Ozturk et al., 2017*	[+2]	▢▢▢▢	●●●	★★★
Transport — Behavioural Response	**Energy Savings (7.3/7.a/7.b)** — Behavioural responses will reduce the volume of transport needs and, by extension, energy demand. *Figueroa and Ribeiro, 2013; Ahmad and Puppim de Oliveira, 2016*	[+2]	▢▢	●●	★★
Transport — Accelerating Energy Efficiency Improvement	**Energy Savings (7.3/7.a/7.b)** — Accelerating efficiency in tourism transport reduces energy demand (China). *Shukxin et al., 2016*	[+2]	▢	●	★
Transport — Improved Access and Fuel Switch to Modern Low-carbon Energy	**Increase Share of Renewable (7.2)** — Biofuel increases share of the renewables but can perform poorly if too many countries increase their use of biofuel, whereas electrification performs best when many other countries implement this technology. The strategies are not mutually exclusive and simultaneous implementation of some provides synergies for national energy security. Therefore, it is important to consider the results of material and contextual factors that co-evolve. Electric vehicles using electricity from renewables or low carbon sources combined with e-mobility options such as trolley buses, metros, trams and electro buses, as well as promote walking and biking, especially for short distances, need consideration. *Ajanovic, 2015; Mânsson, 2016; Alahakoon, 2017; Wolfram et al., 2017*	[+2,-2]	▢▢▢	●●	★★

SDG 8 — DECENT WORK AND ECONOMIC GROWTH

	Interaction	Score	Evidence	Agreement	Confidence
Buildings	**Sustainable Economic Growth and Employment** — Creutzig et al. (2014) assessed the potential for renewable energies in the European region. They found that a European energy transition with a high-level of renewable energy installations in the periphery could act as an economic stimulus, decrease trade deficits and possibly have positive employment effects. Provision of energy access can play a critical enabling role for new productive activities, livelihoods and employment. Reliable access to modern energy services can have an important influence on productivity and earnings (McCollum et al., 2018). *Grogan and Sadanand, 2013; Pueyo et al., 2013; Rao et al., 2013; Chakravorty et al., 2014; Creutzig et al., 2014; Ali et al., 2015; Bernard and Torero, 2015; Byravan et al., 2017; McCollum et al., 2018*	[+2]	▢▢	●●●	★★
Transport — Behavioural Response	**Promote Sustained, Inclusive Economic Growth (8.3)** — Policy contradictions (e.g., standards, efficient technologies leading to increased electricity prices leading the poor to switch away from clean(er) fuels) and unintended outcomes (e.g., redistribution of income generated by carbon taxes) results in contradictions of the primary aims of (productive) job creation and poverty alleviation, and in trade-offs between mitigation, adaptation and development policies. Detailed assessments of mitigation policies consequences requires developing methods and reliable evidence to enable policymakers to more systematically identify how different social groups may be affected by the different available policy options. *Lucas and Pangbourne, 2014; Suckall et al., 2014; Klausbruckner et al., 2016*	[-2]	▢▢▢▢	●●●	★★★
Transport — Accelerating Energy Efficiency Improvement	**Promote Sustained, Inclusive Economic Growth (8.3)** — Significant opportunities to slow travel growth and improve efficiency exist and, similarly, alternatives to petroleum exist but have different characteristics in terms of availability, cost, distribution, infrastructure, storage and public acceptability. Production of new technologies, fuels and infrastructure can favour economic growth; however, efficient financing of increased capital spending and infrastructure is critical. *Gouldson et al., 2015; Karkatsoulis et al., 2016*	[+2,-2]	▢▢▢	●●	★★
Transport — Improved Access and Fuel Switch to Modern Low-carbon Energy	**Promote Sustained, Inclusive Economic Growth (8.3)** — The decarbonization of the freight sector tends to occur in the second part of the century, and the sector decarbonizes by a lower extent than the rest of the economy. Decarbonizing road freight on a global scale remains a challenge even when notable progress in biofuels and electric vehicles has been accounted for. *IPCC, 2014; Creutzig et al., 2015; Carrara and Longden, 2017*	[+2,-2]			

SDG 9 — INDUSTRY, INNOVATION AND INFRASTRUCTURE

	Interaction	Score	Evidence	Agreement	Confidence
Buildings	**Innovation and New Infrastructure (9.2/9.4/9.5)** — Adoption of smart meters and smart grids following community-based social marketing help in infrastructure expansion. Statutory norms to enhance energy and resource efficiency in buildings is encouraging green building projects. Introduction of incentives and norms for solar rooftops/white/green roofs in cities are helping to accelerate innovation and the expansion of infrastructure. *Roy et al., 2018; Anda and Temmen, 2014*	[+2]	▢▢▢	●●	★★
Transport — Behavioural Response	**Build Resilient Infrastructure (9.1)** — As people prefer more mass transportation – train lines, tram lines, BRTs, gondola lift systems, bicycle-sharing systems and hybrid buses – and telecommuting, the need for new infrastructure increases. *Dulac, 2013; Aamaas and Peters, 2017; Martinez-Jaramillo et al., 2017; Xylia and Silveira, 2017*	[+2]	▢▢▢	●●	★★
Transport — Accelerating Energy Efficiency Improvement	**Build Resilient Infrastructure (9.1)** — Combining promotion of mass transportation – train lines, tram lines, BRTs, gondola lift systems, bicycle-sharing systems and hybrid buses – and telecommuting reduces traffic and significantly contributes to meeting climate targets. A comprehensive package of complementary mitigation options is necessary for deep and sustained emissions reductions. In Sweden, a public bus fleet is aiming more towards decarbonization than efficiency. *Dulac, 2013; Aamaas and Peters, 2017; Martinez-Jaramillo et al., 2017; Xylia and Silveira, 2017*	[+2]	▢▢▢	●●	★★
Transport — Improved Access and Fuel Switch to Modern Low-carbon Energy	**Help Building Inclusive Infrastructure (9.1/9.a)** — Lack of appropriate infrastructure leads to limited access to jobs for the urban poor (Africa, Latin America, India). *Figueroa et al., 2013; Gouldson et al., 2015; Vasconcellos and Mendonça, 2016; Lali et al., 2017*	[+2]	▢▢▢	●●●	★★★

SDG 11 — SUSTAINABLE CITIES AND COMMUNITIES

	Interaction	Score	Evidence	Agreement	Confidence
Buildings	**Housing (11.1)** — Ensuring access to basic housing services implies that households have access to modern energy forms. (Quote from McCollum et al., 2018) Solar roof tops in Macau make cities sustainable. Introduction of incentives and norms for solar/white/green rooftops in cities are helping to accelerate the expansion of the infrastructure. *Bhattacharyya et al., 2016; Song et al., 2016; UN, 2016; McCollum et al., 2018; Roy et al., 2018*	[+3]	▢▢▢	●●●	★★★★
Transport — Behavioural Response	**Make Cities and Human Settlements Inclusive, Safe, Resilient** — Climate change threatens to worsen poverty, therefore pro-poor mitigation policies are needed to reduce this threat; for example, investing more and better in infrastructure by leveraging private resources and using designs that account for future climate change and the related uncertainty. *Ahmad and Puppim de Oliveira, 2016; Hallegatte et al., 2016a*	[+2]		●●	★★
Transport — Accelerating Energy Efficiency Improvement	**Make Cities Sustainable (11.2/11.3)** — The two most important elements of making cities sustainable are efficient buildings and transport (e.g. Macau). *Song et al., 2016*	[+2]		●●	★
Transport — Improved Access and Fuel Switch to Modern Low-carbon Energy	**Make Cities and Human Settlements Inclusive, Safe, Resilient** — In rapidly growing cities, the carbon savings from investments at scale, in cost-effective low-carbon measures, could be quickly overwhelmed – in as little as 7 years – by the impacts of sustained population and economic growth, highlighting the need to build capacities that enable the exploitation not only of the economically attractive options in the short term but also of those deeper and more structural changes that are likely to be needed in the longer term. With hybrid electric vehicles and plug-in electric vehicles, there is the emergence of new concepts in transportation, such as electric highways. *Figueroa et al., 2013; Gouldson et al., 2015; Vasconcellos and Mendonça, 2016; Alahakoon, 2017*	[+2]		●●	★★

Economic-Supply

SDG 7 — Affordable and Clean Energy

Row	Interaction	Score	Evidence	Agreement	Confidence	Text	References
Non-biomass Renewables – solar, wind, hydro (Replacing Coal)	←	[+3]	●●●●	●●●	★★★★	**Sustainable and Modern Energy (7.2/7.a)** Decarbonization of the energy system through an upscaling of renewables will greatly facilitate access to clean, affordable and reliable energy. Hydropower plays an increasingly important role for the global electricity supply. This mitigation option is in line with the targets of SDG7 under the caveat of a transition to modern biomass.	Rogelj et al., 2013; Cherian, 2015; Jingura and Kamusoko, 2016
Increased Use of Biomass (Replacing Coal)	←	[+3]	●●●●	●●●	★★★	**Sustainable and Modern Energy (7.2/7.a)** Increased use of modern biomass will facilitate access to clean, affordable and reliable energy. This mitigation option is in line with the targets of SDG7.	Rogelj et al., 2013; Cherian, 2015; Jingura and Kamusoko, 2016
Nuclear/Advanced Nuclear (Replacing Coal)	←	[1]	●●	●	★★	**Sustainable and Modern Energy (7.2/7.a)** Increased use of nuclear power can provide stable baseload power supply and reduce price volatility.	IPCC, 2014
CCS: Bioenergy (Replacing Coal)	←	[+2]	●●●●	●●●	★★★	**Sustainable and Modern Energy (7.2/7.a)** Increased use of modern biomass will facilitate access to clean, affordable and reliable energy.	IPCC, 2014
CCS: Fossil (Advanced Coal)	←	[+2]	●●●●	●●●	★★★	**Ensure energy access and promote investment in new technologies (7.1/7.b)** Advanced and cleaner fossil fuel technology is in line with the targets of SDG7.	IPCC, 2014

SDG 8 — Decent Work and Economic Growth

Row	Interaction	Score	Evidence	Agreement	Confidence	Text	References
Non-biomass Renewables – solar, wind, hydro (Replacing Coal)	~	[0]	●●●	●●	★★	**Innovation and Growth (8.1/8.2/8.4)** Decarbonization of the energy system through an upscaling of renewables and energy efficiency is consistent with sustained economic growth and resource decoupling. Long-term scenarios point towards slight consumption losses caused by a rapid and pervasive expansion of such energy solutions. Whether sustainable growth, as an overarching concept, is attainable or not is more disputed in the literature. Existing literature is also undecided as to whether or not access to modern energy services causes economic growth (McCollum et al., 2018).	Jackson and Senker, 2011; Bonan et al., 2014; Clarke et al., 2014; NCE, 2014; OECD, 2017; York and McGee, 2017; McCollum et al., 2018
Increased Use of Biomass (Replacing Coal)	←	[+1]	●	●	★	**Innovation and Growth (8.1/8.2/8.4)** Decarbonization of the energy system through an upscaling of renewables will greatly facilitate access to clean, affordable and reliable energy.	Jingura and Kamusoko, 2016
Nuclear/Advanced Nuclear (Replacing Coal)	←	[1]	●	●	★	**Innovation and Growth (8.1/8.2/8.4)** Local employment impact and reduced price volatility.	IPCC, 2014
CCS: Bioenergy (Replacing Coal)	←	[+1]	●	●	★	**Innovation and Growth (8.1/8.2/8.4)** See positive impacts of bioenergy use.	
CCS: Fossil (Advanced Coal)	→	[-1]	●●	●●●	★★★	**Innovation and Growth (8.1/8.2/8.4)** Lock-in of human and physical capital in the fossil resources industry.	IPCC, 2005, 2014; Benson and Cole, 2008; Fankhaeser et al., 2008; Vergragt et al., 2011; Markusson et al., 2012; Shackley and Thompson, 2012; Bertram et al., 2015; Johnson et al., 2015

SDG 9 — Industry, Innovation and Infrastructure

Row	Interaction	Score	Evidence	Agreement	Confidence	Text	References
Non-biomass Renewables – solar, wind, hydro (Replacing Coal)	~ / ↓	[0,-1]	●●	●●●	★★	**Inclusive and Sustainable Industrialization (9.2/9.4)** A rapid upscaling of renewable energies could necessitate the early retirement of fossil energy infrastructure (e.g., power plants, refineries, pipelines) on a large scale. The implications of this could in some cases be negative, unless targeted policies can help alleviate the burden on industry (McCollum et al., 2018).	Fankhaeser et al., 2008; McCollum et al., 2008; Guivarch et al., 2011; Bertram et al., 2015; Johnson et al., 2015
Increased Use of Biomass (Replacing Coal)	←	[+1]	●●	●●●	★★	**Innovation and New Infrastructure (9.2/9.4/9.5)** Access to modern and sustainable energy will be critical to sustain economic growth.	Jingura and Kamusoko, 2016; Shahbaz et al., 2016
Nuclear/Advanced Nuclear (Replacing Coal)	→	[-1]	●●	●●●	★★★	**Innovation and New Infrastructure (9.2/9.4/9.5)** Legacy cost of waste and abandoned reactors.	Marra and Palmer, 2011; Greenberg et al., 2013; Schwenk-Ferrero, 2013; Skipperud et al., 2013; Tyler et al., 2013; IPCC, 2014
CCS: Bioenergy (Replacing Coal)	←	[+1]	●	●	★	**Innovation and New Infrastructure (9.2/9.4/9.5)** See positive impacts of bioenergy use and CCS/CCU in industrial demand.	
CCS: Fossil (Advanced Coal)	←	[+1]	●	●	★	**Innovation and New Infrastructure (9.2/9.4/9.5)** See positive impacts of CCS/CCU in industrial demand.	

SDG 11 — Sustainable Cities and Communities

Row	Interaction	Score	Evidence	Agreement	Confidence	Text	References
Non-biomass Renewables – solar, wind, hydro (Replacing Coal)	←	[+2]	●●●●	●●	★★★	**Disaster Preparedness and Prevention (11.5)** Deployment of renewable energy and improvements in energy efficiency globally will aid climate change mitigation efforts, and this, in turn, can help to reduce the exposure of people to certain types of disasters and extreme events (McCollum et al., 2018).	Tully, 2006; Riahi et al., 2012; Daut et al., 2013; IPCC, 2014; Hallegatte et al., 2016b; McCollum et al., 2018
Increased Use of Biomass (Replacing Coal)		[0]				No direct interaction	
Nuclear/Advanced Nuclear (Replacing Coal)		[0]				No direct interaction	
CCS: Bioenergy (Replacing Coal)		[0]				No direct interaction	
CCS: Fossil (Advanced Coal)		[0]				No direct interaction	

Economic-Other

Agriculture and Livestock

SDG 7 — Affordable and Clean Energy

Response row	Interaction	Score	Evidence	Agreement	Confidence
Behavioural Response: Sustainable Healthy Diets and Reduced Food Waste	Energy Efficiency, Universal Access (7.1/7.3) — Reducing global food supply chain losses have several important secondary benefits like conserving energy. *Kummu et al., 2012*	[+1]	▪	⊙	★
Land-based Greenhouse Gas Reduction and Soil Carbon Sequestration	Sustainable and Modern Energy (7.b) — Conventional agricultural biotechnology methods such as energy efficient farming can help in sequestration of soil carbon. Modern biotechnologies such as green energy and N-efficient GM crops can also help in C-sequestration. Biotech crops allow farmers to use less – and environmentally friendly – energy and practice soil carbon sequestration. Biofuels, both from traditional and GMO crops, such as sugar cane, oilseed, rapeseed and jatropha, can be produced. Green energy programmes through plantations of perennial nonedible oilseed producing plants and production of biodiesel for direct use in the energy sector or blending biofuels with fossil fuels in certain proportions can thereby minimize fossil fuel use. (Quoted from Lakshmi et al., 2015) GM crops reduce demand for fossil fuel-based inputs. *Johnson et al., 2007; Sarin et al., 2007; Treasury, 2009; Jain and Sharma, 2010; Lybbert and Sumner, 2010; Mtui, 2011; Lakshmi et al., 2015*	[+1]	▪▪▪▪	●●●	★★★
Greenhouse Gas Reduction from Improved Livestock Production and Manure	Energy Efficiency (7.3) — Scenarios where zero human-edible concentrate feed is used for livestock, non-renewable energy use is reduced by 36%. *Schader et al., 2015*	[+1]	&	J	«

SDG 8 — Decent Work and Economic Growth

Response row	Interaction	Score	Evidence	Agreement	Confidence
Behavioural Response: Sustainable Healthy Diets and Reduced Food Waste	Sustained and Inclusive Economic Growth (8.2) — 23–24% of total cropland and fertilizers are used to produce losses. So reduction in food losses will help to diversify these valuable resources into other productive activities. *Kummu et al., 2012; Hiç, 2016*	[+1]	▪▪▪▪	●●●	★★★
Land-based Greenhouse Gas Reduction and Soil Carbon Sequestration	Sustainable Growth (8.2) — Many developing countries including Gulf States will benefit from CSA given the central role of agriculture in their economic and social development. (Quoted from Behnassi et al. 2014). Low commodity prices have reduced the incentive to invest in yield growth and have led to declining farm labour and farm capital investment. (Quoted from Lamb et al., 2016) *Behnassi et al., 2014; Lamb et al., 2016*	[+2,-1]	▪▪	●●	★★
Greenhouse Gas Reduction from Improved Livestock Production and Manure	Sustainable Economic Growth (8.4) — Exploiting the increasingly decoupled interactions between crops and livestock could be beneficial for promoting structural changes in the livestock sector and is a prerequisite for the sustainable growth of the sector. (Quoted from Herrero et al., 2013) *Herrero and Thornton, 2013; Herrero et al., 2013*	[+1]	&	J	«

SDG 9 — Industry, Innovation and Infrastructure

Response row	Interaction	Score	Evidence	Agreement	Confidence
Behavioural Response: Sustainable Healthy Diets and Reduced Food Waste	Infrastructure Building and Promotion of Inclusive Industrialization (9.1/9.2) — By targeting infrastructure, processing and distribution losses, wastage in food systems can be minimized. 23–24% of total cropland and fertilizers are used to produce losses. So reduction in food losses will help to diversify these valuable resources into other productive activities. *Ingram, 2011; Beddington et al., 2012; Kummu et al., 2012; Hiç et al., 2016; Lamb et al., 2016*	[+1]	▪▪▪▪	●●●	★★★
Land-based Greenhouse Gas Reduction and Soil Carbon Sequestration	Infrastructure Building, Promotion of Inclusive Industrialization and Innovation (9.1/9.2/9.5/9.b) — Reduced research support and delayed industrialization will have an adverse effect on food security and nourishment of children. Organic farming technologies utilizing bio-based fertilizers (composted human and animal manure) are some of the conventional biotechnological options for reducing artificial fertilizer use (Lakshmi et al., 2015). CSA requires huge financial investment and institutional innovation. CSA is committed to new ways of engaging in participatory research and partnerships with producers (Steenwerth et al., 2014). Technologies used on-farm and during food processing to increase productivity which also helps in adaptation and/or mitigation are new, so convincing potential customers is difficult. Also, low-awareness of CSA, inaccessible language, high costs, lack of verified impact of technologies, hard to reach and train farmers, low consumer demand and unequal distribution of costs/benefits across supply chains are barriers to CSA technology adoption (Long et al., 2016). Low commodity prices have reduced the incentive to invest in yield growth and have led to declining investment in research and development (Lamb et al., 2016). *Evenson, 1999; Behnassi et al., 2014; Steenwerth et al., 2014; Lakshmi et al., 2015; Lamb et al., 2016; Long et al., 2016*	[+2,-2]	▪▪▪▪	●●●	★★★
Greenhouse Gas Reduction from Improved Livestock Production and Manure	Technological Upgradation and Innovation (9.2) — Complete genome maps for poultry and cattle now exist, and these open up the way to possible advances in evolutionary biology, animal breeding and animal models for human diseases. Genomic selection should be able to at least double the rate of genetic gain in the dairy industry. (Quoted from Thornton, 2010) Nanotechnology, biogas technology and separation technologies are disruptive technologies that enhance biogas production from anaerobic digesters or to reduce odours. *Sansoucy, 1995; Burton, 2007; Thornton, 2010*	[+2]	▪▪▪▪	●●●	★★★

SDG 11 — Sustainable Cities and Communities

Response row	Interaction	Score	Evidence	Agreement	Confidence
Behavioural Response: Sustainable Healthy Diets and Reduced Food Waste	No interaction	[0]			
Land-based Greenhouse Gas Reduction and Soil Carbon Sequestration	no direct interaction	[0]			
Greenhouse Gas Reduction from Improved Livestock Production and Manure	No direct interaction	[0]			

Economic-Other (continued)

		7 AFFORDABLE AND CLEAN ENERGY					8 DECENT WORK AND ECONOMIC GROWTH					9 INDUSTRY, INNOVATION AND INFRASTRUCTURE					11 SUSTAINABLE CITIES AND COMMUNITIES				
		Interaction	Score	Evidence	Agreement	Confidence	Interaction	Score	Evidence	Agreement	Confidence	Interaction	Score	Evidence	Agreement	Confidence	Interaction	Score	Evidence	Agreement	Confidence
Forest	**Reduced Deforestation, REDD+**	**Energy Efficiency (7.3)** ↑/↓	[+1,-1] ★				**Sustainable Economic Growth (8.4)** ↑	[+1] ★				**Infrastructure, Promotion of Inclusive Industrialization (9.1/9.2/9.5)** ↑/↓	[+1,-1] ★				No direct interaction	[0]			
		Consider the entire sinks and reservoirs of GHG while developing the nationally appropriate mitigation actions. For countries with a significant contribution of forest degradation (and GHG emissions) from wood fuels, this should be considered. (Quoted from Bastos Lima et al., 2017). Biomass for energy is recognized as often being inefficient, and is often harvested in an unsustainable manner, but is a renewable energy source. Bastos Lima et al., 2017; Katila et al., 2017					Efforts by the Government of Zambia to reduce emissions by REDD+ have contributed to erosion control, ecotourism and pollination valued at 2.5% of the country's GDP. Partnerships between local forest managers, community enterprises and private sector companies can support local economies and livelihoods, and boost regional and national economic growth. Turpie et al., 2015; Epstein and Theuer, 2017; Katila et al., 2017					Expanding road networks are recognized as one of the main drivers of deforesting and forest degradation, diminishing forest benefits to communities. On the other hand, roads can enhance market access, thereby boosting local benefits (SDG 1) from the commercialization of forest products. (Quoted from Katila et al., 2017). Efforts by the Government of Zambia to reduce emissions by REDD+ have contributed to erosion control, ecotourism and pollination valued at 2.5% of the country's GDP. Turpie et al., 2015; Epstein and Theuer, 2017; Katila et al., 2017									
	Afforestation and Reforestation	**Energy Conservation (7.3/7.b)** ↑	[+1] ★				**Decent Job Creation and Sustainable Economic Growth (8.3/8.4)** ↑	[+2] ★★				No direct interaction	[0]				**Improving Air Quality, Green and Public Spaces (11.6/11.7/11.a/11.b)** ↑	[+2] ★★★★			
		The US Forest Service estimates that an average NYC street tree (urban afforestation) produces 209 USD in annual benefits, which is primarily driven by aesthetic (90 USD per tree) and energy savings (from shade) benefits (47.63 USD per tree). Jones and McDermott, 2018					Many tree plantations worldwide have higher growth rates which can provide higher rates of returns for investors. Agroforestry initiatives that offer significant opportunities for projects to provide benefits to smallholder farmers can also help address land degradation through community-based efforts in more marginal areas. Mangroves reduce impacts of disasters (cyclones/storms/floods) and enhance water quality, fisheries, tourism businesses and livelihoods. Zomer et al., 2008; Kibria, 2015										Many urban tree plantations worldwide are created with a focus on multiple benefits, like air quality improvement, cultural preference for green nature, healthy community interaction as well as temperature control and biodiversity enhancement goals. Chen and Qi, 2018; Fu et al., 2018; Kowarik, 2018; McKinney and Ingo, 2018; McPherson et al., 2018; Pei et al., 2018				
	Behavioural Response (Responsible Sourcing)	**Universal Access (7.3)** ↑	[+1] ★				**Decent Job Creation and Sustainable Economic Growth (8.3/8.4)** ↑	[+2] ★				**Technological Upgradation and Innovation, Promotion of Inclusive Industrialization (9.1/9.2/9.5)** ↑	[+2] ★				**Improving Air Quality, Green and Public Spaces, Peri-urban Spaces (11.6/11.7/11.a/11.b)** ↑	[+2] ★★★★			
		The trade of wood pellets from clean wood waste should be facilitated with less administrative import barriers by the EU, in order to have this new option seriously accounted for as a future resource for energy. (Quoted from Sikkema et al., 2014) Recommends further harmonization of legal harvesting, sustainable sourcing and cascaded use requirements for woody biomass for energy with the current requirements of voluntary SFM certification schemes. Sikkema et al., 2014					Some standards seek primarily to coordinate global trade, many purport to promote ecological sustainability and social justice or to institutionalize CSR, for example, labour standards developed in the wake of sweatshops and child labour scandals. Environmental standards for pollution control, etc. Indonesian factories may seek advantages through non-price competition—perhaps by highlighting decent working conditions or the existence of a union—or to see trade associations or government promoting the country as a responsible sourcing location. Bartley, 2010					Capacity for processing certified timber is often underutilized, due to the limited supply available. As a result, manufacturing firms that are seeking to tap into green markets often turn to other sources of timber. (Quoted from Bartley, 2010) Responsible sourcing, when integrated into business practices, can enable retailers to better manage brand value and reputation by avoiding negative public relations, as well as maintaining and enhancing brand integrity (Huang et al., 2013). Bartley, 2010; Huang et al., 2013					Many urban tree plantations worldwide are created with a focus on multiple benefits, like air quality improvement, cultural preference for green nature, healthy community interaction as well as temperature control and biodiversity enhancement goals. People's preference for urban forest gardens are encouraging new urban green spaces, and tree selection helps in building resilience to disaster. Chen and Qi, 2018; Fu et al., 2018; Kowarik, 2018; McKinney and Ingo, 2018; McPherson et al., 2018; Pei et al., 2018				
Oceans	**Ocean Iron Fertilization**	No direct interaction	[0]				No direct interaction	[0]				No direct interaction	[0]				No direct interaction	[0]			
	Blue Carbon	No direct interaction	[0]				No direct interaction	[0]				No direct interaction	[0]				No direct interaction	[0]			
	Enhanced Weathering	No direct interaction	[0]				No direct interaction	[0]				No direct interaction	[0]				No direct interaction	[0]			

5

References

Note that this reference list does not account for the references in Table 5.2, for which a separate reference list is provided.

Abel, G.J., B. Barakat, S. KC, and W. Lutz, 2016: Meeting the Sustainable Development Goals leads to lower world population growth. *Proceedings of the National Academy of Sciences*, **113(50)**, 14294–14299, doi:10.1073/pnas.1611386113.

Absar, S.M. and B.L. Preston, 2015: Extending the Shared Socioeconomic Pathways for sub-national impacts, adaptation, and vulnerability studies. *Global Environmental Change*, **33**, 83–96, doi:10.1016/j.gloenvcha.2015.04.004.

Adam, H.N., 2015: Mainstreaming adaptation in India – the Mahatma Gandhi National Rural Employment Guarantee Act and climate change. *Climate and Development*, **7(2)**, 142–152, doi:10.1080/17565529.2014.934772.

Adger, W.N. et al., 2014: Human security. In: *Climate Change 2014: Impacts, Adaptation, and Vulnerability. Part A: Global and Sectoral Aspects. Contribution of Working Group II to the Fifth Assessment Report of the Intergovernmental Panel on Climate Change* [Field, C.B., V.R. Barros, D.J. Dokken, K.J. Mach, M.D. Mastrandrea, T.E. Bilir, M. Chatterjee, K.L. Ebi, Y.O. Estrada, R.C. Genova, B. Girma, E.S. Kissel, A.N. Levy, S. MacCracken, P.R. Mastrandrea, and L.L. White (eds.)]. Cambridge University Press, Cambridge, United Kingdom and New York, NY, USA, pp. 755–791.

Aggarwal, A., 2014: How sustainable are forestry clean development mechanism projects? A review of the selected projects from India. *Mitigation and Adaptation Strategies for Global Change*, **19(1)**, 73–91, doi:10.1007/s11027-012-9427-x.

Agoramoorthy, G. and M.J. Hsu, 2016: Small dams revive dry rivers and mitigate local climate change in India's drylands. *International Journal of Climate Change Strategies and Management*, **8(2)**, 271–285, doi:10.1108/ijccsm-12-2014-0141.

Agyeman, J., D. Schlosberg, L. Craven, and C. Matthews, 2016: Trends and directions in environmental justice: from inequity to everyday life, community, and just sustainabilities. *Annual Review of Environment and Resources*, **41**, 321–340, doi:10.1146/annurev-environ-110615-090052.

Aha, B. and J.Z. Ayitey, 2017: Biofuels and the hazards of land grabbing: tenure (in)security and indigenous farmers' investment decisions in Ghana. *Land Use Policy*, **60**, 48–59, doi:10.1016/j.landusepol.2016.10.012.

Ahmed, N., W.W.L. Cheung, S. Thompson, and M. Glaser, 2017a: Solutions to blue carbon emissions: Shrimp cultivation, mangrove deforestation and climate change in coastal Bangladesh. *Marine Policy*, **82**, 68–75, doi:10.1016/j.marpol.2017.05.007.

Ahmed, N., S.W. Bunting, M. Glaser, M.S. Flaherty, and J.S. Diana, 2017b: Can greening of aquaculture sequester blue carbon? *Ambio*, **46(4)**, 468–477, doi:10.1007/s13280-016-0849-7.

Aipira, C., A. Kidd, and K. Morioka, 2017: Climate Change Adaptation in Pacific Countries: Fostering Resilience Through Gender Equality. In: *Climate Change Adaptation in Pacific Countries: Fostering Resilience and Improving the Quality of Life* [Leal Filho, W. (ed.)]. Springer International Publishing AG, Cham, Switzerland, pp. 225–239, doi:10.1007/978-3-319-50094-2_13.

Ajanovic, A., 2015: The future of electric vehicles: prospects and impediments. *Wiley Interdisciplinary Reviews: Energy and Environment*, **4(6)**, 521–536, doi:10.1002/wene.160.

Al Ansari, M.S., 2013: Climate change policies and the potential for energy efficiency in the Gulf Cooperation Council (GCC) Economy. *Environment and Natural Resources Research*, **3(4)**, 106–117, doi:10.5539/enrr.v3n4p106.

Albert, S. et al., 2017: Heading for the hills: climate-driven community relocations in the Solomon Islands and Alaska provide insight for a 1.5°C future. *Regional Environmental Change*, 1–12, doi:10.1007/s10113-017-1256-8.

Ali, A. and O. Erenstein, 2017: Assessing farmer use of climate change adaptation practices and impacts on food security and poverty in Pakistan. *Climate Risk Management*, **16**, 183–194, doi:10.1016/j.crm.2016.12.001.

Al-Maamary, H.M.S., H.A. Kazem, and M.T. Chaichan, 2016: Changing the Energy Profile of the GCC States: A Review. *International Journal of Applied Engineering Research*, **11(3)**, 1980–1988, www.ripublication.com/volume/ijaerv11n3.htm.

Al-Maamary, H.M.S., H.A. Kazem, and M.T. Chaichan, 2017: The impact of oil price fluctuations on common renewable energies in GCC countries. *Renewable and Sustainable Energy Reviews*, **75**, 989–1007, doi:10.1016/j.rser.2016.11.079.

Alshehry, A.S. and M. Belloumi, 2015: Energy consumption, carbon dioxide emissions and economic growth: the case of Saudi Arabia. *Renewable and Sustainable Energy Reviews*, **41**, 237–247, doi:10.1016/j.rser.2014.08.004.

Alsheyab, M.A.T., 2017: Qatar's effort for the deployment of Carbon Capture and Storage. *Global Nest Journal*, **19(3)**, 453–457, https://journal.gnest.org/sites/default/files/submissions/gnest_02269/gnest_02269_published.pdf.

Altieri, K.E. et al., 2016: Achieving development and mitigation objectives through a decarbonization development pathway in South Africa. *Climate Policy*, **16(sup1)**, S78–S91, doi:10.1080/14693062.2016.1150250.

Amann, M. et al., 2011: Cost-effective control of air quality and greenhouse gases in Europe: Modeling and policy applications. *Environmental Modelling & Software*, **26(12)**, 1489–1501, doi:10.1016/j.envsoft.2011.07.012.

Anand, R., 2004: *International environmental justice: A North-South Dimension*. Routledge, London, UK, 161 pp.

Andonova, L.B., T.N. Hale, and C.B. Roger, 2017: National policy and transnational governance of climate change: substitutes or complements? *International Studies Quarterly*, **61(2)**, 253–268, doi:10.1093/isq/sqx014.

Ansuategi, A. et al., 2015: *The impact of climate change on the achievement of the post-2015 sustainable development goals*. Metroeconomica, HR Wallingford and CDKN, 84 pp.

Antwi-Agyei, P., A.J. Dougill, and L.C. Stringer, 2015: Impacts of land tenure arrangements on the adaptive capacity of marginalized groups: The case of Ghana's Ejura Sekyedumase and Bongo districts. *Land Use Policy*, **49**, 203–212, doi:10.1016/j.landusepol.2015.08.007.

Antwi-Agyei, P., A. Dougill, and L. Stringer, 2017a: Assessing Coherence between Sector Policies and Climate Compatible Development: Opportunities for Triple Wins. *Sustainability*, **9(11)**, 2130, doi:10.3390/su9112130.

Antwi-Agyei, P., A.J. Dougill, L.C. Stringer, and S.N.A. Codjoe, 2018: Adaptation opportunities and maladaptive outcomes in climate vulnerability hotspots of northern Ghana. *Climate Risk Management*, **19**, 83–93, doi:10.1016/j.crm.2017.11.003.

Antwi-Agyei, P. et al., 2017b: Perceived stressors of climate vulnerability across scales in the Savannah zone of Ghana: a participatory approach. *Regional Environmental Change*, **17(1)**, 213–227, doi:10.1007/s10113-016-0993-4.

Apgar, M.J., W. Allen, K. Moore, and J. Ataria, 2015: Understanding adaptation and transformation through indigenous practice: the case of the Guna of Panama. *Ecology and Society*, **20(1)**, 45, doi:10.5751/es-07314-200145.

Arakelyan, I., D. Moran, and A. Wreford, 2017: Climate smart agriculture: a critical review. In: *Making Climate Compatible Development Happen* [Nunan, F. (ed.)]. Routledge, Abingdon, UK and New York, NY, USA, pp. 66–86.

Arbuthnott, K., S. Hajat, C. Heaviside, and S. Vardoulakis, 2016: Changes in population susceptibility to heat and cold over time: assessing adaptation to climate change. *Environmental Health*, **15(S1)**, S33, doi:10.1186/s12940-016-0102-7.

Archer, D. et al., 2014: Moving towards inclusive urban adaptation: approaches to integrating community-based adaptation to climate change at city and national scale. *Climate and Development*, **6(4)**, 345–356, doi:10.1080/17565529.2014.918868.

Armitage, D.R., 2015: Social-ecological change in Canada's Arctic: coping, adapting, learning for an uncertain future. In: *Climate Change and the Coast: Building Resilient Communities* [Glavovic, B., M. Kelly, R. Kay, and A. Travers (eds.)]. CRC Press, Boca Raton, FL, USA, pp. 103–124.

Arnell, N.W. and S.N. Gosling, 2016: The impacts of climate change on river flood risk at the global scale. *Climatic Change*, **134(3)**, 387–401, doi:10.1007/s10584-014-1084-5.

Arnell, N.W. et al., 2015: *The global impacts of climate change under pathways that reach 2°, 3° and 4°C above pre-industrial levels*. Report from AVOID2 project to the Committee on Climate Change, 34 pp.

Arriagada, R.A., E.O. Sills, P.J. Ferraro, and S.K. Pattanayak, 2015: Do payments pay off? Evidence from participation in Costa Rica's PES program. *PLOS ONE*, **10(7)**, 1–17, doi:10.1371/journal.pone.0131544.

Arthurson, K. and S. Baum, 2015: Making space for social inclusion in conceptualising climate change vulnerability. *Local Environment*, **20(1)**, 1–17, doi:10.1080/13549839.2013.818951.

Arts, K., 2017: Inclusive sustainable development: a human rights perspective. *Current Opinion in Environmental Sustainability*, **24**, 58–62, doi:10.1016/j.cosust.2017.02.001.

5

Atalay, Y., F. Biermann, and A. Kalfagianni, 2016: Adoption of renewable energy technologies in oil-rich countries: explaining policy variation in the Gulf Cooperation Council states. *Renewable Energy*, **85**, 206–214, doi:10.1016/j.renene.2015.06.045.

Ayers, J.M., S. Huq, H. Wright, A.M. Faisal, and S.T. Hussain, 2014: Mainstreaming climate change adaptation into development in Bangladesh. *Climate and Development*, **6(4)**, 293–305, doi:10.1002/wcc.226.

Babiker, M.H., 2016: Options for climate change policy in MENA countries after Paris. Policy Perspective No. 18, The Economic Research Forum (ERF), Giza, Egypt, 7 pp.

Bado, B.V., P. Savadogo, and M.L.S. Manzo, 2016: *Restoration of Degraded Lands in West Africa Sahel: Review of experiences in Burkina Faso and Niger*. CGIAR, 16 pp.

Bai, X. et al., 2016: Plausible and desirable futures in the Anthropocene: A new research agenda. *Global Environmental Change*, **39**, 351–362, doi:10.1016/j.gloenvcha.2015.09.017.

Bai, X. et al., 2018: Six research priorities for cities and climate change. *Nature*, **555**, 23–25, doi:10.1038/d41586-018-02409-z.

Bajželj, B. et al., 2014: Importance of food-demand management for climate mitigation. *Nature Climate Change*, **4(10)**, 924–929, doi:10.1038/nclimate2353.

Baldacchino, G., 2017: Seizing history: development and non-climate change in Small Island Developing States. *International Journal of Climate Change Strategies and Management*, 10(2), 217–228, doi:10.1108/ijccsm-02-2017-0037.

Banda, M.L. and S. Fulton, 2017: Litigating Climate Change in National Courts: Recent Trends and Developments in Global Climate Law. *Environmental Law Reporter*, **47(10121)**, 10121–10134, www.eli.org/sites/default/files/elr.featuredarticles/47.10121.pdf.

Barnes, P., 2015: The political economy of localization in the transition movement. *Community Development Journal*, **50(2)**, 312–326, doi:10.1093/cdj/bsu042.

Barnett, J. and E. Walters, 2016: Rethinking the Vulnerability of Small Island States: Climate Change and Development in the Pacific Islands. In: *The Palgrave Handbook of International Development* [Grugel, J. and D. Hammett (eds.)]. Palgrave, London, UK, pp. 731–748, doi:10.1057/978-1-137-42724-3_40.

Barnett, J., P. Tschakert, L. Head, and W.N. Adger, 2016: A science of loss. *Nature Climate Change*, **6**, 976–978, doi:10.1038/nclimate3140.

Barnett, J. et al., 2014: A local coastal adaptation pathway. *Nature Climate Change*, **4(12)**, 1103–1108, doi:10.1038/nclimate2383.

Barrett, S., 2013: Local level climate justice? Adaptation finance and vulnerability reduction. *Global Environmental Change*, **23(6)**, 1819–1829, doi:10.1016/j.gloenvcha.2013.07.015.

Barrington-Leigh, C., 2016: Sustainability and Well-Being: A Happy Synergy. *Development*, **59**, 292–298, doi:10.1057/s41301-017-0113-x.

Bartos, M.D. and M. Chester, 2014: The Conservation Nexus: Valuing Interdependent Water and Energy Savings in Arizona. *Environmental Science & Technology*, **48(4)**, 2139–2149, doi:10.1021/es4033343.

Bathiany, S., V. Dakos, M. Scheffer, and T.M. Lenton, 2018: Climate models predict increasing temperature variability in poor countries. *Science Advances*, **4(5)**, eaar5809, doi:10.1126/sciadv.aar5809.

Bauer, N. et al., 2016: Global fossil energy markets and climate change mitigation – an analysis with REMIND. *Climatic Change*, **136(1)**, 69–82, doi:10.1007/s10584-013-0901-6.

Bayram, H. and A.B. Öztürk, 2014: Global climate change, desertification, and its consequences in Turkey and the Middle East. In: *Global Climate Change and Public Health* [Pinkerton, K.E. and W.N. Rom (eds.)]. Springer, New York, NY, USA, pp. 293–305, doi:10.1007/978-1-4614-8417-2_17.

Bebbington, J. and C. Larrinaga, 2014: Accounting and sustainable development: an exploration. *Accounting, Organizations and Society*, **39(6)**, 395–413, doi:10.1016/j.aos.2014.01.003.

Beilin, R. and C. Wilkinson, 2015: Introduction: Governing for urban resilience. *Urban Studies*, **52(7)**, 1205–1217, doi:10.1177/0042098015574955.

Bell, K., 2015: Can the capitalist economic system deliver environmental justice? *Environmental Research Letters*, **10(12)**, 125017, doi:10.1088/1748-9326/10/12/125017.

Berger, M., S. Pfister, V. Bach, and M. Finkbeiner, 2015: Saving the planet's climate or water resources? The trade-off between carbon and water footprints of European biofuels. *Sustainability*, **7(6)**, 6665–6683, doi:10.3390/su7066665.

Berner, J. et al., 2016: Adaptation in Arctic circumpolar communities: food and water security in a changing climate. *International Journal of Circumpolar Health*, **75(1)**, 33820, doi:10.3402/ijch.v75.33820.

Berrueta, V.M., M. Serrano-Medrano, C. Garcia-Bustamante, M. Astier, and O.R. Masera, 2017: Promoting sustainable local development of rural communities and mitigating climate change: the case of Mexico's Patsari improved cookstove project. *Climatic Change*, **140(1)**, 63–77, doi:10.1007/s10584-015-1523-y.

Bertram, C. et al., 2018: Targeted policies can compensate most of the increased sustainability risks in 1.5°C mitigation scenarios. *Environmental Research Letters*, **13(6)**, 064038, doi:10.1088/1748-9326/aac3ec.

Betts, R.A. et al., 2018: Changes in climate extremes, fresh water availability and vulnerability to food insecurity projected at 1.5°C and 2°C global warming with a higher-resolution global climate model. *Philosophical Transactions of the Royal Society A: Mathematical, Physical and Engineering Sciences*, **376(2119)**, 20160452, doi:10.1098/rsta.2016.0452.

Betzold, C. and F. Weiler, 2017: Allocation of aid for adaptation to climate change: do vulnerable countries receive more support? *International Environmental Agreements: Politics, Law and Economics*, **17**, 17–36, doi:10.1007/s10784-016-9343-8.

Bexell, M. and K. Jönsson, 2017: Responsibility and the United Nations' Sustainable Development Goals. *Forum for Development Studies*, **44(1)**, 13–29, doi:10.1080/08039410.2016.1252424.

Bickersteth, S. et al., 2017: *Mainstreaming climate compatible development: Insights from CDKN's first seven years*. Climate and Development Knowledge Network (CDKN), London, 199 pp.

Biermann, F. and I. Boas, 2017: Towards a global governance system to protect climate migrants: taking stock. In: *Research Handbook on Climate Change, Migration and the Law* [Mayer, B. and F. Crepeau (eds.)]. Edward Elgar Publishing, Cheltenham, UK and Northampton, MA, USA, pp. 405–419, doi:10.4337/9781785366598.00026.

Biermann, M., K. Hillmer-Pegram, C.N. Knapp, and R.E. Hum, 2016: Approaching a critical turn? A content analysis of the politics of resilience in key bodies of resilience literature. *Resilience*, **4(2)**, 59–78, doi:10.1080/21693293.2015.1094170.

Binam, J.N., F. Place, A.A. Djalal, and A. Kalinganire, 2017: Effects of local institutions on the adoption of agroforestry innovations: evidence of farmer managed natural regeneration and its implications for rural livelihoods in the Sahel. *Agricultural and Food Economics*, **5(1)**, 2, doi:10.1186/s40100-017-0072-2.

Blanco, V., C. Brown, S. Holzhauer, G. Vulturius, and M.D.A. Rounsevell, 2017: The importance of socio-ecological system dynamics in understanding adaptation to global change in the forestry sector. *Journal of Environmental Management*, **196**, 36–47, doi:10.1016/j.jenvman.2017.02.066.

Blondeel, M. and T. van de Graaf, 2018: Toward a global coal mining moratorium? A comparative analysis of coal mining policies in the USA, China, India and Australia. *Climatic Change*, 1–13, doi:10.1007/s10584-017-2135-5.

Blyth, W., R. McCarthy, and R. Gross, 2015: Financing the UK power sector: is the money available? *Energy Policy*, **87**, 607–622, doi:10.1016/j.enpol.2015.08.028.

Boke, C., 2015: Resilience's problem of the present: reconciling social justice and future-oriented resilience planning in the Transition Town movement. *Resilience*, **3(3)**, 207–220, doi:10.1080/21693293.2015.1072313.

Bonsch, M. et al., 2016: Trade-offs between land and water requirements for large-scale bioenergy production. *GCB Bioenergy*, **8(1)**, 11–24, doi:10.1111/gcbb.12226.

Boonstra, W.J., E. Björkvik, L.J. Haider, and V. Masterson, 2016: Human responses to social-ecological traps. *Sustainability Science*, **11(6)**, 877–889, doi:10.1007/s11625-016-0397-x.

Bosomworth, K., P. Leith, A. Harwood, and P.J. Wallis, 2017: What's the problem in adaptation pathways planning? The potential of a diagnostic problem-structuring approach. *Environmental Science & Policy*, **76**, 23–28, doi:10.1016/j.envsci.2017.06.007.

Bowyer, P., M. Schaller, S. Bender, and D. Jacob, 2015: Adaptation as ClimateRisk Management: Methods and Approaches. In: *Handbook of Climate Change Adaptation* [Filho, L. (ed.)]. Springer, Berlin and Heidelberg, Germany, pp. 71–92, doi:10.1007/978-3-642-38670-1_28.

Boyd, E., R.A. James, R.G. Jones, H.R. Young, and F.E.L. Otto, 2017: A typology of loss and damage perspectives. *Nature Climate Change*, **7(10)**, 723–729, doi:10.1038/nclimate3389.

5

Boyle, J. et al., 2013: *Exploring Trends in Low-Carbon, Climate-Resilient Development.* International Institute for Sustainable Development (IISD), 37 pp.

Boysen, L.R., W. Lucht, and D. Gerten, 2017: Trade-offs for food production, nature conservation and climate limit the terrestrial carbon dioxide removal potential. *Global Change Biology*, **23(10)**, 4303–4317, doi:10.1111/gcb.13745.

Brink, E. et al., 2016: Cascades of green: A review of ecosystem-based adaptation in urban areas. *Global Environmental Change*, **36**, 111–123, doi:10.1016/j.gloenvcha.2015.11.003.

Brockington, D. and S. Ponte, 2015: The Green Economy in the global South: experiences, redistributions and resistance. *Third World Quarterly*, **36(12)**, 2197–2206, doi:10.1080/01436597.2015.1086639.

Brohé, A., 2014: Whither the CDM? Investment outcomes and future prospects. *Environment, Development and Sustainability*, **16(2)**, 305–322, doi:10.1007/s10668-013-9478-5.

Brown, D. and G. McGranahan, 2016: The urban informal economy, local inclusion and achieving a global green transformation. *Habitat International*, **53**, 97–105, doi:10.1016/j.habitatint.2015.11.002.

Brown, E., J. Cloke, D. Gent, P.H. Johnson, and C. Hill, 2014: Green growth or ecological commodification: debating the green economy in the global south. *Geografiska Annaler, Series B: Human Geography*, **96(3)**, 245–259, doi:10.1111/geob.12049.

Bryan, E., Q. Bernier, M. Espinal, and C. Ringler, 2017: Making climate change adaptation programmes in sub-Saharan Africa more gender responsive: insights from implementing organizations on the barriers and opportunities. *Climate and Development*, 1–15, doi:10.1080/17565529.2017.1301870.

Buch-Hansen, H., 2018: The Prerequisites for a Degrowth Paradigm Shift: Insights from Critical Political Economy. *Ecological Economics*, **146**, 157–163, doi:10.1016/j.ecolecon.2017.10.021.

Bulkeley, H., G.A.S. Edwards, and S. Fuller, 2014: Contesting climate justice in the city: examining politics and practice in urban climate change experiments. *Global Environmental Change*, **25(1)**, 31–40, doi:10.1016/j.gloenvcha.2014.01.009.

Bulkeley, H., J.A. Carmin, V. Castán Broto, G.A.S. Edwards, and S. Fuller, 2013: Climate justice and global cities: mapping the emerging discourses. *Global Environmental Change*, **23(5)**, 914–925, doi:10.1016/j.gloenvcha.2013.05.010.

Buntaine, M.T. and L. Prather, 2018: Preferences for Domestic Action Over International Transfers in Global Climate Policy. *Journal of Experimental Political Science*, 1–15, doi:10.1017/xps.2017.34.

Burke, M., S.M. Hsiang, and E. Miguel, 2015: Global non-linear effect of temperature on economic production. *Nature*, **527(7577)**, 235–239, doi:10.1038/nature15725.

Burns, W. and S. Nicholson, 2017: Bioenergy and carbon capture with storage (BECCS): the prospects and challenges of an emerging climate policy response. *Journal of Environmental Studies and Sciences*, **7(4)**, 527–534, doi:10.1007/s13412-017-0445-6.

Bustamante, M. et al., 2014: Co-benefits, trade-offs, barriers and policies for greenhouse gas mitigation in the agriculture, forestry and other land use (AFOLU) sector. *Global Change Biology*, **20(10)**, 3270–3290, doi:10.1111/gcb.12591.

Butler, C., K.A. Parkhill, and N.F. Pidgeon, 2016: Energy consumption and everyday life: choice, values and agency through a practice theoretical lens. *Journal of Consumer Culture*, **16(3)**, 887–907, doi:10.1177/1469540514553691.

Butler, J.R.A. et al., 2016: Scenario planning to leap-frog the Sustainable Development Goals: an adaptation pathways approach. *Climate Risk Management*, **12**, 83–99, doi:10.1016/j.crm.2015.11.003.

Butt, N. et al., 2016: Challenges in assessing the vulnerability of species to climate change to inform conservation actions. *Biological Conservation*, **199**, 10–15, doi:10.1016/j.biocon.2016.04.020.

Byers, E.A., J.W. Hall, and J.M. Amezaga, 2014: Electricity generation and cooling water use: UK pathways to 2050. *Global Environmental Change*, **25**, 16–30, doi:10.1016/j.gloenvcha.2014.01.005.

Byers, E.A. et al., 2018: Global exposure and vulnerability to multi-sector development and climate change hotspots. *Environmental Research Letters*, **13(5)**, 055012, doi:10.1088/1748-9326/aabf45.

Byrne, J. et al., 2016: Could urban greening mitigate suburban thermal inequity?: the role of residents' dispositions and household practices. *Environmental Research Letters*, **11(9)**, 095014, doi:10.1088/1748-9326/11/9/095014.

Byrnes, W.M., 2014: Climate justice, Hurricane Katrina, and African American environmentalism. *Journal of African American Studies*, **18(3)**, 305–314, doi:10.1007/s12111-013-9270-5.

Cabezon, E., L. Hunter, P. Tumbarello, K. Washimi, and Yiqun Wu, 2016: Strengthening Macro-Fiscal Resilience to Natural Disasters and Climate Change in the Small States of the Pacific. In: *Resilience and Growth in the Small States of the Pacific* [Khor, H.E., R.P. Kronenberg, and P. Tumbarello (eds.)]. International Monetary Fund (IMF), Washington DC, USA, pp. 71–94.

Caldecott, B., O. Sartor, and T. Spencer, 2017: *Lessons from previous 'coal transitions': High-level summary for decision-makers*. IDDRI and Climate Strategies, 24 pp.

Callen, T., R. Cherif, F. Hasanov, A. Hegazy, and P. Khandelwal, 2014: Economic Diversification in the GCC: Past, Present, and Future. IMF Staff Discussion Note SDN/14 /12, International Monetary Fund (IMF), 32 pp.

Calliari, E., 2016: Loss and damage: a critical discourse analysis of Parties' positions in climate change negotiations. *Journal of Risk Research*, **9877**, 1–23, doi:10.1080/13669877.2016.1240706.

Calvet-Mir, L., E. Corbera, A. Martin, J. Fisher, and N. Gross-Camp, 2015: Payments for ecosystem services in the tropics: a closer look at effectiveness and equity. *Current Opinion in Environmental Sustainability*, **14**, 150–162, doi:10.1016/j.cosust.2015.06.001.

Calvin, K. et al., 2017: The SSP4: A world of deepening inequality. *Global Environmental Change*, **42**, 284–296, doi:10.1016/j.gloenvcha.2016.06.010.

Cameron, C. et al., 2016: Policy trade-offs between climate mitigation and clean cook-stove access in South Asia. *Nature Energy*, **1**, 1–5, doi:10.1038/nenergy.2015.10.

Cameron, R.W.F., J. Taylor, and M. Emmett, 2015: A Hedera green facade – energy performance and saving under different maritime-temperate, winter weather conditions. *Building and Environment*, **92**, 111–121, doi:10.1016/j.buildenv.2015.04.011.

Campbell, B.M. et al., 2016: Reducing risks to food security from climate change. *Global Food Security*, **11**, 34–43, doi:10.1016/j.gfs.2016.06.002.

Câmpeanu, C.N. and I. Fazey, 2014: Adaptation and pathways of change and response: a case study from Eastern Europe. *Global Environmental Change*, **28**, 351–367, doi:10.1016/j.gloenvcha.2014.04.010.

Campiglio, E., 2016: Beyond carbon pricing: The role of banking and monetary policy in financing the transition to a low-carbon economy. *Ecological Economics*, **121**, 220–230, doi:10.1016/j.ecolecon.2015.03.020.

Carr, E.R. and M.C. Thompson, 2014: Gender and climate change adaptation in agrarian settings. *Geography Compass*, **8(3)**, 182–197, doi:10.1111/gec3.12121.

Carr, E.R. and K.N. Owusu-Daaku, 2015: The shifting epistemologies of vulnerability in climate services for development: The case of Mali's agrometeorological advisory programme. *Area*, 7–17, doi:10.1111/area.12179.

Carr, E.R. and S.N. Onzere, 2018: Really effective (for 15% of the men): Lessons in understanding and addressing user needs in climate services from Mali. *Climate Risk Management*, **22**, 82–95, doi:10.1016/j.crm.2017.03.002.

Carter, T.R. et al., 2016: Characterising vulnerability of the elderly to climate change in the Nordic region. *Regional Environmental Change*, **16(1)**, 43–58, doi:10.1007/s10113-014-0688-7.

Castán Broto, V., 2017: Urban Governance and the Politics of Climate change. *World Development*, **93**, 1–15, doi:10.1016/j.worlddev.2016.12.031.

Castro, P., 2016: Common but differentiated responsibilities beyond the nation state: how is differential treatment addressed in transnational Climate governance initiatives? *Transnational Environmental Law*, **5(02)**, 379–400, doi:10.1017/s2047102516000224.

Cavanagh, C. and T.A. Benjaminsen, 2014: Virtual nature, violent accumulation: The 'spectacular failure' of carbon offsetting at a Ugandan National Park. *Geoforum*, **56**, 55–65, doi:10.1016/j.geoforum.2014.06.013.

Cervigni, R. and M. Morris (eds.), 2016: *Confronting Drought in Africa's Drylands: Opportunities for Enhancing Resilience*. World Bank, Washington DC, USA, 299 pp.

Chakrabarti, S. and E.J. Shin, 2017: Automobile dependence and physical inactivity: insights from the California Household Travel Survey. *Journal of Transport and Health*, **6**, 262–271, doi:10.1016/j.jth.2017.05.002.

Chakravarty, D. and M. Tavoni, 2013: Energy poverty alleviation and climate change mitigation: Is there a trade off? *Energy Economics*, **40**, S67–S73, doi:10.1016/j.eneco.2013.09.022.

Chakravarty, D. and J. Roy, 2016: The Global South: New Estimates and Insights from Urban India. In: *Rethinking Climate and Energy Policies: New Perspectives on the Rebound Phenomenon* [Santarius, T., H.J. Walnum, and A. Carlo (eds.)]. Springer, Cham, Switzerland, pp. 55–72, doi:10.1007/978-3-319-38807-6_4.

Chakravarty, D., S. Dasgupta, and J. Roy, 2013: Rebound effect: how much to worry? *Current Opinion in Environmental Sustainability*, **5(2)**, 216–228, doi:10.1016/j.cosust.2013.03.001.

5

Chan, K.M.A., E. Anderson, M. Chapman, K. Jespersen, and P. Olmsted, 2017: Payments for ecosystem services: rife with problems and potential-for transformation towards sustainability. *Ecological Economics*, **140**, 110–122, doi:10.1016/j.ecolecon.2017.04.029.

Chancel, L. and T. Picketty, 2015: *Carbon and inequality: from Kyoto to Paris. Trends in the global inequality of carbon emissions (1998–2013) & prospects for an equitable adaptation fund*. Paris School of Economics, Paris, France, 50 pp.

Chapin, F.S., C.N. Knapp, T.J. Brinkman, R. Bronen, and P. Cochran, 2016: Community-empowered adaptation for self-reliance. *Current Opinion in Environmental Sustainability*, **19**, 67–75, doi:10.1016/j.cosust.2015.12.008.

Chaturvedi, V. and P.R. Shukla, 2014: Role of energy efficiency in climate change mitigation policy for India: assessment of co-benefits and opportunities within an integrated assessment modeling framework. *Climatic Change*, **123(3)**, 597–609, doi:10.1007/s10584-013-0898-x.

Chelleri, L., G. Minucci, and E. Skrimizea, 2016: Does community resilience decrease social-ecological vulnerability? Adaptation pathways trade-off in the Bolivian Altiplano. *Regional Environmental Change*, **16(8)**, 2229–2241, doi:10.1007/s10113-016-1046-8.

Chen, X., X. Liu, and D. Hu, 2015: Assessment of sustainable development: A case study of Wuhan as a pilot city in China. *Ecological Indicators*, **50**, 206–214, doi:10.1016/j.ecolind.2014.11.002.

Cheung, W.W.L., G. Reygondeau, and T.L. Frölicher, 2016: Large benefits to marine fisheries of meeting the 1.5°C global warming target. *Science*, **354(6319)**, 1591–1594, doi:10.1126/science.aag2331.

Chief, K., A. Meadow, and K. Whyte, 2016: Engaging southwestern tribes in sustainable water resources topics and management. *Water*, **8(8)**, 1–21, doi:10.3390/w8080350.

Chong, J., 2014: Ecosystem-based approaches to climate change adaptation: progress and challenges. *International Environmental Agreements: Politics, Law and Economics*, **14(4)**, 391–405, doi:10.1007/s10784-014-9242-9.

Chu, E., I. Anguelovski, and D. Roberts, 2017: Climate adaptation as strategic urbanism: assessing opportunities and uncertainties for equity and inclusive development in cities. *Cities*, **60**, 378–387, doi:10.1016/j.cities.2016.10.016.

Chung Tiam Fook, T., 2017: Transformational processes for community-focused adaptation and social change: a synthesis. *Climate and Development*, **9(1)**, 5–21, doi:10.1080/17565529.2015.1086294.

Clarke, L.E. et al., 2014: Assessing transformation pathways. In: *Climate Change 2014: Mitigation of Climate Change. Contribution of Working Group III to the Fifth Assessment Report of the Intergovernmental Panel on Climate Change* [Edenhofer, O., R. Pichs-Madruga, Y. Sokona, E. Farahani, S. Kadner, K. Seyboth, A. Adler, I. Baum, S. Brunner, P. Eickemeier, B. Kriemann, J. Savolainen, S. Schlömer, C. von Stechow, T. Zwickel, and J.C. Minx (eds.)]. Cambridge University Press, Cambridge, United Kingdom and New York, NY, USA, pp. 413–510.

Clements, R., 2009: *The Economic Cost of Climate Change in Africa*. Pan African Climate Justice Alliance (PACJA), 52 pp.

Cobbinah, P.B. and G.K. Anane, 2016: Climate change adaptation in rural Ghana: indigenous perceptions and strategies. *Climate and Development*, **8(2)**, 169–178, doi:10.1080/17565529.2015.1034228.

Coe, R., F. Sinclair, and E. Barrios, 2014: Scaling up agroforestry requires research 'in' rather than 'for' development. *Current Opinion in Environmental Sustainability*, **6(1)**, 73–77, doi:10.1016/j.cosust.2013.10.013.

Coe, R., J. Njoloma, and F. Sinclair, 2017: To control or not to control: how do we learn more about how agronomic innovations perform on farms? *Experimental Agriculture*, 1–7, doi:10.1017/s0014479717000102.

Cohen, M.G. (ed.), 2017: *Climate change and gender in rich countries: work, public policy and action*. Routledge, Abingdon, UK and New York, NY, USA, 322 pp.

Cole, M.J., R.M. Bailey, and M.G. New, 2017: Spatial variability in sustainable development trajectories in South Africa: provincial level safe and just operating spaces. *Sustainability Science*, **12(5)**, 829–848, doi:10.1007/s11625-016-0418-9.

Colenbrander, S., D. Dodman, and D. Mitlin, 2017: Using climate finance to advance climate justice: the politics and practice of channelling resources to the local level. *Climate Policy*, 1–14, doi:10.1080/14693062.2017.1388212.

Colenbrander, S. et al., 2016: Can low-carbon urban development be pro-poor? The case of Kolkata, India. *Environment and Urbanization*, **29(1)**, 139–158, doi:10.1177/0956247816677775.

Colloff, M.J. et al., 2017: An integrative research framework for enabling transformative adaptation. *Environmental Science & Policy*, **68**, 87–96, doi:10.1016/j.envsci.2016.11.007.

Conservation International, 2016: *Ecosystem Based Adaptation: Essential for Achieving the Sustainable Development Goals*. Conservation International, Arlington, VA, USA, 4 pp.

Conway, D. et al., 2015: Climate and southern Africa's water-energy-food nexus. *Nature Climate Change*, **5(9)**, 837–846, doi:10.1038/nclimate2735.

Corbera, E., C. Hunsberger, and C. Vaddhanaphuti, 2017: Climate change policies, land grabbing and conflict: perspectives from Southeast Asia. *Canadian Journal of Development Studies*, **38(3)**, 297–304, doi:10.1080/02255189.2017.1343413.

Cretney, R.M., A.C. Thomas, and S. Bond, 2016: Maintaining grassroots activism: Transition Towns in Aotearoa New Zealand. *New Zealand Geographer*, **72(2)**, 81–91, doi:10.1111/nzg.12114.

Creutzig, F. et al., 2015: Bioenergy and climate change mitigation: an assessment. *GCB Bioenergy*, **7(5)**, 916–944, doi:10.1111/gcbb.12205.

Croce, D., C. Kaminker, and F. Stewart, 2011: he Role of Pension Funds in Financing Green Growth Initiatives. OECD Working Papers on Finance, Insurance and Private Pensions No. 10, Organisation for Economic Co-operation and Development (OECD) Publishing, Paris, France, doi:10.1787/5kg58j1lwdjd-en.

Crosland, T., A. Meyer, and M. Wewerinke-singh, 2016: The Paris Agreement Implementation Blueprint: a practical guide to bridging the gap between actions and goal and closing the accountability deficit (Part 1). *Environmental liability*, **25(2)**, 114–125, https://ssrn.com/abstract=2952215.

CSO Equity Review, 2015: *Fair Shares: A Civil Society Equity Review of INDCs*. CSO Equity Review Coalition, Manila, London, Cape Town, Washington, et al. 36 pp.

Cugurullo, F., 2013: How to build a sandcastle: an analysis of the genesis and development of Masdar City. *Journal of Urban Technology*, **20(1)**, 23–37, doi:10.1080/10630732.2012.735105.

Cundill, G. et al., 2014: *Social learning for adaptation: a descriptive handbook for practitioners and action researchers*. IRDC, Rhodes University, Ruliv, 118 pp.

Cutter, S.L., 2016: Resilience to What? Resilience for Whom? *Geographical Journal*, **182(2)**, 110–113, doi:10.1111/geoj.12174.

Cvitanovic, C. et al., 2016: Linking adaptation science to action to build food secure Pacific Island communities. *Climate Risk Management*, **11**, 53–62, doi:10.1016/j.crm.2016.01.003.

Daigneault, A., P. Brown, and D. Gawith, 2016: Dredging versus hedging: Comparing hard infrastructure to ecosystem-based adaptation to flooding. *Ecological Economics*, **122**, 25–35, doi:10.1016/j.ecolecon.2015.11.023.

Dasgupta, P., 2016: *Climate Sensitive Adaptation in Health: Imperatives for India in a Developing Economy Context*. Springer India, 194 pp.

Dasgupta, P., K. Ebi, and I. Sachdeva, 2016: Health sector preparedness for adaptation planning in India. *Climatic Change*, **138(3–4)**, 551–566, doi:10.1007/s10584-016-1745-7.

Dasgupta, S., M. Huq, M.G. Mustafa, M.I. Sobhan, and D. Wheeler, 2017: The impact of aquatic salinization on fish habitats and poor communities in a changing climate: evidence from southwest coastal Bangladesh. *Ecological Economics*, **139**, 128–139, doi:10.1016/j.ecolecon.2017.04.009.

Datta, N., 2015: Evaluating impacts of watershed development program on agricultural productivity, income, and livelihood in bhalki watershed of Bardhaman District, West Bengal. *World Development*, **66**, 443–456, doi:10.1016/j.worlddev.2014.08.024.

Davies, K., 2015: Kastom, climate change and intergenerational democracy: experiences from Vanuatu. In: *Climate change in the Asia-Pacific region* [Leal Filho, W. (ed.)]. Springer, Cham, Switzerland, pp. 49–66, doi:10.1007/978-3-319-14938-7_4.

Davies, T.E., N. Pettorelli, W. Cresswill, and I.R. Fazey, 2014: Who are the poor? Measuring wealth inequality to aid understanding of socioeconomic contexts for conservation: a case-study from the Solomon Islands. *Environmental Conservation*, **41(04)**, 357–366, doi:10.1017/s0376892914000058.

de Coninck, H.C. and D. Puig, 2015: Assessing climate change mitigation technology interventions by international institutions. *Climatic Change*, **131(3)**, 417–433, doi:10.1007/s10584-015-1344-z.

de Coninck, H.C. and A. Sagar, 2015: Making sense of policy for climate technology development and transfer. *Climate Policy*, **15(1)**, 1–11, doi:10.1080/14693062.2014.953909.

de Coninck, H.C. and A. Sagar, 2017: Technology development and transfer (Article 10). In: *The Paris Agreement on Climate Change* [Klein, D., M.P. Carazo, M. Doelle, J. Bulmer, and A. Higham (eds.)]. Oxford University Press, Oxford, UK, pp. 258–276.

De Stefano, L., J.D. Petersen-Perlman, E.A. Sproles, J. Eynard, and A.T. Wolf, 2017: Assessment of transboundary river basins for potential hydro-political tensions. *Global Environmental Change*, **45**, 35–46, doi:10.1016/j.gloenvcha.2017.04.008.

5

Dearing, J.A. et al., 2014: Safe and just operating spaces for regional social-ecological systems. *Global Environmental Change*, **28(1)**, 227–238, doi:10.1016/j.gloenvcha.2014.06.012.

Death, C., 2014: The Green Economy in South Africa: Global Discourses and Local Politics. *Politikon*, **41(1)**, 1–22, doi:10.1080/02589346.2014.885668.

Death, C., 2015: Four discourses of the green economy in the global South. *Third World Quarterly*, **36(12)**, 2207–2224, doi:10.1080/01436597.2015.1068110.

Death, C., 2016: Green states in Africa: beyond the usual suspects. *Environmental Politics*, **25:1**, 116–135, doi:10.1080/09644016.2015.1074380.

DeCaro, D.A., C. Anthony, T. Arnold, E.F. Boamah, and A.S. Garmestani, 2017: Understanding and applying principles of social cognition and decision making in adaptive environmental governance. *Ecology and Society*, **22(1)**, 33, doi:10.5751/es-09154-220133.

DeClerck, F.A.J. et al., 2016: Agricultural ecosystems and their services: the vanguard of sustainability? *Current Opinion in Environmental Sustainability*, **23**, 92–99, doi:10.1016/j.cosust.2016.11.016.

Delponte, I., I. Pittaluga, and C. Schenone, 2017: Monitoring and evaluation of Sustainable Energy Action Plan: practice and perspective. *Energy Policy*, **100**, 9–17, doi:10.1016/j.enpol.2016.10.003.

Dennig, F., M.B. Budolfson, M. Fleurbaey, A. Siebert, and R.H. Socolow, 2015: Inequality, climate impacts on the future poor, and carbon prices. *Proceedings of the National Academy of Sciences*, **112(52)**, 15827–15832, doi:10.1073/pnas.1513967112.

Denton, F. et al., 2014: Climate-resilient pathways: adaptation, mitigation, and sustainable development. In: *Climate Change 2014: Impacts, Adaptation, and Vulnerability. Part A: Global and Sectoral Aspects. Contribution of Working Group II to the Fifth Assessment Report of the Intergovernmental Panel on Climate Change* [Field, C.B., V.R. Barros, D.J. Dokken, K.J. Mach, M.D. Mastrandrea, T.E. Bilir, M. Chatterjee, K.L. Ebi, Y.O. Estrada, R.C. Genova, B. Girma, E.S. Kissel, A.N. Levy, S. MacCracken, P.R. Mastrandrea, and L.L. White (eds.)]. Cambridge University Press, Cambridge, United Kingdom and New York, NY, USA, pp. 1101–1131.

Derbez, M. et al., 2014: Indoor air quality and comfort in seven newly built, energy-efficient houses in France. *Building and Environment*, **72**, 173–187, doi:10.1016/j.buildenv.2013.10.017.

Di Gregorio, M. et al., 2017: Climate policy integration in the land use sector: Mitigation, adaptation and sustainable development linkages. *Environmental Science & Policy*, **67**, 35–43, doi:10.1016/j.envsci.2016.11.004.

Dilling, L., M.E. Daly, W.R. Travis, O. Wilhelmi, and R.A. Klein, 2015: The dynamics of vulnerability: why adapting to climate variability will not always prepare us for climate change. *Wiley Interdisciplinary Reviews: Climate Change*, **6(4)**, 413–425, doi:10.1002/wcc.341.

Ding, H. et al., 2016: *Climate Benefits, Tenure Costs: The Economic Case For Securing Indigenous Land Rights in the Amazon*. World Resources Institute, Washington DC, USA, 98 pp.

Dinku, T. et al., 2014: Bridging critical gaps in climate services and applications in Africa. *Earth Perspectives*, **1(1)**, 15, doi:10.1186/2194-6434-1-15.

Dodman, D., H. Leck, M. Rusca, and S. Colenbrander, 2017: African urbanisation and urbanism: implications for risk accumulation and reduction. *International Journal of Disaster Risk Reduction*, **26**, 7–15, doi:10.1016/j.ijdrr.2017.06.029.

Doswald, N. et al., 2014: Effectiveness of ecosystem-based approaches for adaptation: review of the evidence-base. *Climate and Development*, **6(2)**, 185–201, doi:10.1080/17565529.2013.867247.

Douxchamps, S. et al., 2016: Linking agricultural adaptation strategies, food security and vulnerability: evidence from West Africa. *Regional Environmental Change*, **16(5)**, 1305–1317, doi:10.1007/s10113-015-0838-6.

Dovie, D.B.K., M. Dzodzomenyo, and O.A. Ogunseitan, 2017: Sensitivity of health sector indicators' response to climate change in Ghana. *Science of the Total Environment*, **574**, 837–846, doi:10.1016/j.scitotenv.2016.09.066.

Dow, K. et al., 2013: Limits to adaptation. *Nature Climate Change*, **3(4)**, 305–307, doi:10.1038/nclimate1847.

Dual Citizen LLC, 2016: *The Global Green Economy Index – GGEI 2016: Measuring National Performance in the Green Economy*. Dual Citizen LLC, Washington DC, USA and New York, NY, USA, 58 pp.

Duarte, C.M., J. Wu, X. Xiao, A. Bruhn, and D. Krause-Jensen, 2017: Can Seaweed Farming Play a Role in Climate Change Mitigation and Adaptation? *Frontiers in Marine Science*, **4**, 100, doi:10.3389/fmars.2017.00100.

Duguma, L.A., P.A. Minang, and M. Van Noordwijk, 2014: Climate change mitigation and adaptation in the land use sector: from complementarity to synergy. *Environmental Management*, **54(3)**, 420–432, doi:10.1007/s00267-014-0331-x.

Dumont, E.S., S. Bonhomme, T.F. Pagella, and F.L. Sinclair, 2017: Structured stakeholder engagement leads to development of more diverse and inclusive agroforestry options. *Experimental Agriculture*, 1–23, doi:10.1017/s0014479716000788.

Eakin, H.C., M.C. Lemos, and D.R. Nelson, 2014: Differentiating capacities as a means to sustainable climate change adaptation. *Global Environmental Change*, **27(27)**, 1–8, doi:10.1016/j.gloenvcha.2014.04.013.

Ebi, K.L. and M. Otmani del Barrio, 2017: Lessons Learned on Health Adaptation to Climate Variability and Change: Experiences Across Low- and Middle-Income Countries. *Environmental Health Perspectives*, **125(6)**, 065001, doi:10.1289/ehp405.

Ebi, K.L. et al., 2014: A new scenario framework for climate change research: background, process, and future directions. *Climatic Change*, **122(3)**, 363–372, doi:10.1007/s10584-013-0912-3.

Echegaray, F., 2016: Consumers' reactions to product obsolescence in emerging markets: the case of Brazil. *Journal of Cleaner Production*, **134**, 191–203, doi:10.1016/j.jclepro.2015.08.119.

Ehresman, T.G. and C. Okereke, 2015: Environmental justice and conceptions of the green economy. *International Environmental Agreements: Politics, Law and Economics*, **15(1)**, 13–27, doi:10.1007/s10784-014-9265-2.

Ekwurzel, B. et al., 2017: The rise in global atmospheric CO_2, surface temperature, and sea level from emissions traced to major carbon producers. *Climatic Change*, 1–12, doi:10.1007/s10584-017-1978-0.

Enqvist, J., M. Tengö, and W.J. Boonstra, 2016: Against the current: rewiring rigidity trap dynamics in urban water governance through civic engagement. *Sustainability Science*, **11(6)**, 919–933, doi:10.1007/s11625-016-0377-1.

Ensor, J., 2016: Adaptation and resilience in Vanuatu: Interpreting community perceptions of vulnerability, knowledge and power for community-based adaptation programming. Report by SEI for Oxfam Australia, Carlton, 32 pp.

Ensor, J. and B. Harvey, 2015: Social learning and climate change adaptation: evidence for international development practice. *Wiley Interdisciplinary Reviews: Climate Change*, **6(5)**, 509–522, doi:10.1002/wcc.348.

Entzinger, H. and P. Scholten, 2016: *Adapting to Climate Change through Migration. A case study of the Vietnamese Mekong River Delta*. International Organization for Migration (IOM), Geneva, Switzerland, 62 pp.

Epule, T.E., J.D. Ford, S. Lwasa, and L. Lepage, 2017: Climate change adaptation in the Sahel. *Environmental Science & Policy*, **75**, 121–137, doi:10.1016/j.envsci.2017.05.018.

Eriksen, S., L.O. Naess, R. Haug, L. Lenaerts, and A. Bhonagiri, 2017: Courting catastrophe? Can humanitarian actions contribute to climate change adaptation? *IDS Bulletin*, **48(4)**, doi:10.19088/1968-2017.149.

Eriksson, H. et al., 2017: The role of fish and fisheries in recovering from natural hazards: lessons learned from Vanuatu. *Environmental Science & Policy*, **76**, 50–58, doi:10.1016/j.envsci.2017.06.012.

Estrella, M., F.G. Renaud, K. Sudmeier-Rieux, and U. Nehren, 2016: Defining New Pathways for Ecosystem-Based Disaster Risk Reduction and Adaptation in the Post-2015 Sustainable Development Agenda. In: *Ecosystem-Based Disaster Risk Reduction and Adaptation in Practice* [Renaud, F.G., K. Sudmeier-Rieux, M. Estrella, and U. Nehren (eds.)]. Springer, Cham, Switzerland, pp. 553–591, doi:10.1007/978-3-319-43633-3.

Evans, G. and L. Phelan, 2016: Transition to a post-carbon society: linking environmental justice and just transition discourses. *Energy Policy*, **99**, 329–339, doi:10.1016/j.enpol.2016.05.003.

Evans, J.P., R.B. Smith, and R.J. Oglesby, 2004: Middle East climate simulation and dominant precipitation processes. *International Journal of Climatology*, **24(13)**, 1671–1694, doi:10.1002/joc.1084.

Falcone, P.M., P. Morone, and E. Sica, 2018: Greening of the financial system and fuelling a sustainability transition: a discursive approach to assess landscape pressures on the Italian financial system. *Technological Forecasting and Social Change*, **127**, 23–37, doi:10.1016/j.techfore.2017.05.020.

Fan, Y., Q. Qiao, L. Fang, and Y. Yao, 2017: Energy analysis on industrial symbiosis of an industrial park: a case study of Hefei economic and technological development area. *Journal of Cleaner Production*, **141**, 791–798, doi:10.1016/j.jclepro.2016.09.159.

Fankhauser, S. and T.K.J. McDermott, 2014: Understanding the adaptation deficit: Why are poor countries more vulnerable to climate events than rich countries? *Global Environmental Change*, **27**, 9–18, doi:10.1016/j.gloenvcha.2014.04.014.

Fankhauser, S. and N. Stern, 2016: Climate change, development, poverty and economics. Grantham Research Institute on Climate Change and the Environment Working Paper No. 253, Grantham Research Institute, London, UK, 25 pp.

FAO, 2015: *Adaptation to climate risk and food security: Evidence from smallholder farmers in Ethiopia*. Food and Agriculture Organization (FAO), Rome, Italy, 50 pp.

FAO and NZAGRC, 2017a: *Low emissions development of the beef cattle sector in Uruguay – reducing enteric methane for food security and livelihoods*. Food and

Agriculture Organization of the United Nations and New Zealand Agricultural Greenhouse Gas Research Centre (NZAGRC), Rome, Italy, 34 pp.

FAO and NZAGRC, 2017b: *Supporting low emissions development in the Ethiopian dairy cattle sector – reducing enteric methane for food security and livelihoods*. Food and Agriculture Organization of the United Nations and New Zealand Agricultural Greenhouse Gas Research Centre (NZAGRC), Rome, Italy, 34 pp.

Fatima, R., A.J. Wadud, and S. Coelho, 2014: *Human Rights, Climate Change, Environmental Degradation and Migration: A New Paradigm*. International Organization for Migration and Migration Policy Institute, Bangkok, Thailand and Washington DC, USA, 12 pp.

Fay, M. et al., 2015: *Decarbonizing Development: Three Steps to a Zero-Carbon Future*. World Bank, Washington DC, USA, 185 pp, doi:10.1596/978-1-4648-0479-3.

Fazey, I. et al., 2016: Past and future adaptation pathways. *Climate and Development*, **8(1)**, 26–44, doi:10.1080/17565529.2014.989192.

Fazey, I. et al., 2018: Community resilience for a 1.5°C world. *Current Opinion in Environmental Sustainability*, **31**, 30–40, doi:10.1016/j.cosust.2017.12.006.

Feola, G. and R. Nunes, 2014: Success and failure of grassroots innovations for addressing climate change: The case of the Transition Movement. *Global Environmental Change*, **24**, 232–250, doi:10.1016/j.gloenvcha.2013.11.011.

Ferguson, P., 2015: The green economy agenda: business as usual or transformational discourse? *Environmental Politics*, **24(1)**, 17–37, doi:10.1080/09644016.2014.919748.

Fernandes-Jesus, M., A. Carvalho, L. Fernandes, and S. Bento, 2017: Community engagement in the Transition movement: views and practices in Portuguese initiatives. *Local Environment*, **22(12)**, 1546–1562, doi:10.1080/13549839.2017.1379477.

Few, R., A. Martin, and N. Gross-Camp, 2017: Trade-offs in linking adaptation and mitigation in the forests of the Congo Basin. *Regional Environmental Change*, **17(3)**, 851–863, doi:10.1007/s10113-016-1080-6.

Ficklin, L., L.C. Stringer, A.J. Dougill, and S.M. Sallu, 2018: Climate compatible development reconsidered: calling for a critical perspective. *Climate and Development*, **10(3)**, 193–196, doi:10.1080/17565529.2017.1372260.

Field, C.B. et al., 2014: Technical Summary. In: *Climate Change 2014: Impacts, Adaptation, and Vulnerability. Part A: Global and Sectoral Aspects. Contribution of Working Group II to the Fifth Assessment Report of the Intergovernmental Panel on Climate Change* [Field, C.B., V.R. Barros, D.J. Dokken, K.J. Mach, M.D. Mastrandrea, T.E. Bilir, M. Chatterjee, K.L. Ebi, Y.O. Estrada, R.C. Genova, B. Girma, E.S. Kissel, A.N. Levy, S. MacCracken, P.R. Mastrandrea, and L.L. White (eds.)]. Cambridge University Press, Cambridge, United Kingdom and New York, NY, USA, pp. 35–94.

Filho, L. and J. Nalau (eds.), 2018: *Limits to climate change adaptation*. Springer International Publishing, Cham, Switzerland, 410 pp., doi:10.1007/978-3-319-64599-5.

Fincher, R., J. Barnett, S. Graham, and A. Hurlimann, 2014: Time stories: making sense of futures in anticipation of sea-level rise. *Geoforum*, **56**, 201–210, doi:10.1016/j.geoforum.2014.07.010.

Fleurbaey, M. et al., 2014: Sustainable development and equity. In: *Climate Change 2014: Mitigation of Climate Change. Contribution of Working Group III to the Fifth Assessment Report of the Intergovernmental Panel on Climate Change* [Edenhofer, O., R. Pichs-Madruga, Y. Sokona, E. Farahani, S. Kadner, K. Seyboth, A. Adler, I. Baum, S. Brunner, P. Eickemeier, B. Kriemann, J. Savolainen, S. Schlömer, C. Stechow, T. Zwickel, and J.C. Minx (eds.)]. Cambridge University Press, Cambridge, United Kingdom and New York, NY, USA, pp. 283–350.

Ford, J.D. et al., 2016: Community-based adaptation research in the Canadian Arctic. *Wiley Interdisciplinary Reviews: Climate Change*, **7(2)**, 175–191, doi:10.1002/wcc.376.

Francis, R., P. Weston, and J. Birch, 2015: *The social, environmental and economics benefits of Farmer Managed Natural Regeneration (FMNR)*. World Vision Australia, Melbourne, Australia, 44 pp.

Frank, S. et al., 2017: Reducing greenhouse gas emissions in agriculture without compromising food security? *Environmental Research Letters*, **12(10)**, 105004, doi:10.1088/1748-9326/aa8c83.

Fricko, O. et al., 2016: Energy sector water use implications of a 2°C degree climate policy. *Environmental Research Letters*, **11(3)**, 034011, doi:10.1088/1748-9326/11/3/034011.

Frumhoff, P.C., R. Heede, and N. Oreskes, 2015: The climate responsibilities of industrial carbon producers. *Climatic Change*, **132(2)**, 157–171, doi:10.1007/s10584-015-1472-5.

Fuchs, D. et al., 2016: Power: the missing element in sustainable consumption and absolute reductions research and action. *Journal of Cleaner Production*, **132**, 298–307, doi:10.1016/j.jclepro.2015.02.006.

Fuglestvedt, J.S. and S. Kallbekken, 2016: Climate responsibility: fair shares? *Nature Climate Change*, **6(1)**, 19–20, doi:10.1038/nclimate2791.

Fuhr, H., T. Hickmann, and K. Kern, 2018: The role of cities in multi-level climate governance: local climate policies and the 1.5°C target. *Current Opinion in Environmental Sustainability*, **30**, 1–6, doi:10.1016/j.cosust.2017.10.006.

Fujimori, S., N. Hanasaki, and T. Masui, 2017a: Projections of industrial water withdrawal under shared socioeconomic pathways and climate mitigation scenarios. *Sustainability Science*, **12**, 275–292, doi:10.1007/s11625-016-0392-2.

Fujimori, S. et al., 2017b: SSP3: AIM implementation of Shared Socioeconomic Pathways. *Global Environmental Change*, **42**, 268–283, doi:10.1016/j.gloenvcha.2016.06.009.

Fukuda-Parr, S., A.E. Yamin, and J. Greenstein, 2014: The power of numbers: a critical review of Millennium Development Goal targets for human development and human rights. *Journal of Human Development and Capabilities*, **15(2–3)**, 105–117, doi:10.1080/19452829.2013.864622.

Fuller, R. and J. Lain, 2017: *Building Resilience: A meta-analysis of Oxfam's resilience*. Oxfam Effectiveness Reviews, Oxfam, 35 pp.

Gajjar, S.P., C. Singh, and T. Deshpande, 2018: Tracing back to move ahead: a review of development pathways that constrain adaptation futures. *Climate and Development*, 1–15, doi:10.1080/17565529.2018.1442793.

Galgóczi, B., 2014: The Long and Winding Road from Black to Green: Decades of Structural Change in the Ruhr Region. *International Journal of Labour Research*, **6(2)**, 217–240, www.ilo.org/wcmsp5/groups/public/---ed_dialogue/---actrav/documents/publication/wcms_375223.pdf.

Garg, A., P. Mohan, S. Shukla, B. Kankal, and S.S. Vishwanathan, 2017: *High impact opportunities for energy efficiency in India*. UNEP DTU Partnerhsip, Copenhagen, Denmark, 49 pp.

Gebrehaweria, G., A. Dereje Assefa, G. Girmay, M. Giordano, and L. Simon, 2016: An assessment of integrated watershed management in Ethiopia. IWMI Working Paper 170, International Water Management Institute (IWMI), Colombo, Sri Lanka, 28 pp., doi:10.5337/2016.214.

Georgeson, L., M. Maslin, and M. Poessinouw, 2017a: Global disparity in the supply of commercial weather and climate information services. *Science Advances*, **3**, e1602632, doi:10.1126/sciadv.1602632.

Georgeson, L., M. Maslin, and M. Poessinouw, 2017b: The global green economy: a review of concepts, definitions, measurement methodologies and their interactions. *Geo: Geography and Environment*, **4(1)**, e00036, doi:10.1002/geo2.36.

Gero, A. et al., 2013: *Understanding the Pacific's adaptive capacity to emergencies in the context of climate change: Country Report – Vanuatu*. Report prepared for NCCARF by the Institute for Sustainable Futures, and WHO Collaborating Centre, University of Technology, Sydney, Australia, 36 pp.

Gillard, R., A. Gouldson, J. Paavola, and J. Van Alstine, 2016: Transformational responses to climate change: beyond a systems perspective of social change in mitigation and adaptation. *Wiley Interdisciplinary Reviews: Climate Change*, **7(2)**, 251–265, doi:10.1002/wcc.384.

Gillingham, K., D. Rapson, and G. Wagner, 2016: The rebound effect and energy efficiency policy. *Review of Environmental Economics and Policy*, **10(1)**, 68–88, doi:10.1093/reep/rev017.

Giwa, A., A. Alabi, A. Yusuf, and T. Olukan, 2017: A comprehensive review on biomass and solar energy for sustainable energy generation in Nigeria. *Renewable and Sustainable Energy Reviews*, **69**, 620–641, doi:10.1016/j.rser.2016.11.160.

Glazebrook, G. and P. Newman, 2018: The City of the Future. *Urban Planning*, **3(2)**, 1–20, doi:10.17645/up.v3i2.1247.

Godfrey-Wood, R. and L.O. Naess, 2016: Adapting to Climate Change: Transforming Development? *IDS Bulletin*, **47(2)**, 49–62, doi:10.19088/1968-2016.131.

Gomez-Echeverri, L., 2018: Climate and development: enhancing impact through stronger linkages in the implementation of the Paris Agreement and the Sustainable Development Goals (SDGs). *Philosophical Transactions of the Royal Society A: Mathematical, Physical and Engineering Sciences*, **376(2119)**, 20160444, doi:10.1098/rsta.2016.0444.

Gorddard, R., M.J. Colloff, R.M. Wise, D. Ware, and M. Dunlop, 2016: Values, rules and knowledge: Adaptation as change in the decision context. *Environmental Science & Policy*, **57**, 60–69, doi:10.1016/j.envsci.2015.12.004.

Gore, C., 2015: Climate Change Adaptation and African Cities: Understanding the Impact of Government and Governance on Future Action. In: *The Urban Climate Challenge: Rethinking the Role of Cities in the Global Climate Regime* [Johnson, C., N. Toly, and H. Schroeder (eds.)]. Routledge, New York, NY, USA and London, UK, pp. 205–226.

Government of Kiribati, 2016: *Kiribati development plan 2016–19*. Government of Kiribati, 75 pp.

5

Granderson, A.A., 2017: Value conflicts and the politics of risk: challenges in assessing climate change impacts and risk priorities in rural Vanuatu. *Climate and Development*, 1–14, doi:10.1080/17565529.2017.1318743.

Grantham, R.W. and M.A. Rudd, 2017: Household susceptibility to hydrological change in the Lower Mekong Basin. *Natural Resources Forum*, **41(1)**, 3–17, doi:10.1111/1477-8947.12113.

Gray, E., N. Henninger, C. Reij, R. Winterbottom, and P. Agostini, 2016: *Integrated landscape approaches for Africa's drylands*. World Bank, Washington DC, USA, 184 pp., doi:10.1596/978-1-4648-0826-5.

Green, F., 2018: Anti-fossil fuel norms. *Climatic Change*, 1–14, doi:10.1007/s10584-017-2134-6.

Griffiths, S., 2017a: A review and assessment of energy policy in the Middle East and North Africa region. *Energy Policy*, **102**, 249–269, doi:10.1016/j.enpol.2016.12.023.

Griffiths, S., 2017b: Renewable energy policy trends and recommendations for GCC countries. *Energy Transitions*, **1(1)**, 3, doi:10.1007/s41825-017-0003-6.

Grill, G. et al., 2015: An index-based framework for assessing patterns and trends in river fragmentation and flow regulation by global dams at multiple scales. *Environmental Research Letters*, **10(1)**, 015001, doi:10.1088/1748-9326/10/1/015001.

Grist, N. et al., 2017: Framing innovations for climate resilience for farmers in Sahel. *Resilience Intel*, **9**, 20, www.odi.org/sites/odi.org.uk/files/resource-documents/11647.pdf.

Grossmann, M. and E. Creamer, 2017: Assessing diversity and inclusivity within the Transition Movement: an urban case study. *Environmental Politics*, **26(1)**, 161–182, doi:10.1080/09644016.2016.1232522.

Grubert, E.A., A.S. Stillwell, and M.E. Webber, 2014: Where does solar-aided seawater desalination make sense? A method for identifying sustainable sites. *Desalination*, **339**, 10–17, doi:10.1016/j.desal.2014.02.004.

Grubler, A. et al., 2018: A low energy demand scenario for meeting the 1.5°C target and sustainable development goals without negative emission technologies. *Nature Energy*, **3(6)**, 515–527, doi:10.1038/s41560-018-0172-6.

Gupta, J. and K. Arts, 2017: Achieving the 1.5°C objective: just implementation through a right to (sustainable) development approach. *International Environmental Agreements: Politics, Law and Economics*, doi:10.1007/s10784-017-9376-7.

Gupta, J. and N. Pouw, 2017: Towards a trans-disciplinary conceptualization of inclusive development. *Current Opinion in Environmental Sustainability*, **24**, 96–103, doi:10.1016/j.cosust.2017.03.004.

Gustafson, S. et al., 2017: Merging science into community adaptation planning processes: a cross-site comparison of four distinct areas of the Lower Mekong Basin. *Climatic Change*, 1–16, doi:10.1007/s10584-016-1887-7.

Gwimbi, P., 2017: Mainstreaming national adaptation programmes of action into national development plans in Lesotho: lessons and needs. *International Journal of Climate Change Strategies and Management*, **9(3)**, 299–315, doi:10.1108/ijccsm-11-2015-0164.

Hackmann, B., 2016: Regime learning in global environmental governance. *Environmental Values*, **25(6)**, 663–686, doi:10.3197/096327116x14736981715625.

Haider, L.J., W.J. Boonstra, G.D. Peterson, and M. Schlüter, 2018: Traps and Sustainable Development in Rural Areas: A Review. *World Development*, **101(2013)**, 311–321, doi:10.1016/j.worlddev.2017.05.038.

Hajer, M. et al., 2015: Beyond cockpit-ism: Four insights to enhance the transformative potential of the sustainable development goals. *Sustainability*, **7(2)**, doi:10.3390/su7021651.

Hale, T., 2016: "All Hands on Deck": The Paris Agreement and Nonstate Climate Action. *Global Environmental Politics*, **16(3)**, 12–22, doi:10.1162/glep_a_00362.

Hallegatte, S. and J. Rozenberg, 2017: Climate change through a poverty lens. *Nature Climate Change*, **7(4)**, 250–256, doi:10.1038/nclimate3253.

Hallegatte, S. et al., 2014: Climate Change and Poverty – An Analytical Framework. WPS7126, World Bank Group, Washington DC, USA, 47 pp.

Hallegatte, S. et al., 2016: *Shock Waves: Managing the Impacts of Climate Change on Poverty*. The World Bank, Washington, DC, USA, 227 pp., doi:10.1596/978-1-4648-0673-5.

Halonen, M. et al., 2017: *Mobilizing climate finance flows: Nordic approaches and opportunities*. TemaNord 2017:519, Nordic Council of Ministers, 151 pp., doi:10.6027/TN2017-519.

Hammill, B.A. and H. Price-Kelly, 2017: *Using NDCs, NAPs and the SDGs to Advance Climate-Resilient Development*. NDC Expert perspectives, NDC Partnership, Washington DC, USA and Bonn, Germany, 10 pp.

Handmer, J. and H. Iveson, 2017: Cyclone Pam in Vanuatu: Learning from the low death toll. *Australian jouranl of Emergency Management*, **22(2)**, 60–65, https://ajem.infoservices.com.au/items/ajem-32-02-22.

Hanna, R. and P. Oliva, 2016: Implications of Climate Change for Children in Developing Countries. *The Future of Children*, **26(1)**, 115–132, doi:10.1353/foc.2016.0006.

Hansen, G. and D. Stone, 2016: Assessing the observed impact of anthropogenic climate change. *Nature Climate Change*, **6(5)**, 532–537, doi:10.1038/nclimate2896.

Harris, L.M., E.K. Chu, and G. Ziervogel, 2017: Negotiated resilience. *Resilience*, **3293**, 1–19, doi:10.1080/21693293.2017.1353196.

Hartzell-Nichols, L., 2017: *A Climate of Risk: Precautionary Principles, Catastrophes, and Climate Change*. Routledge, New York, NY, USA, 168 pp.

Hasegawa, T. et al., 2014: Climate change impact and adaptation assessment on food consumption utilizing a new scenario framework. *Environmental Science & Technology*, **48(1)**, 438–445, doi:10.1021/es4034149.

Hasegawa, T. et al., 2015: Consequence of climate mitigation on the risk of hunger. *Environmental Science & Technology*, **49(12)**, 7245–7253, doi:10.1021/es5051748.

Hashemi, S., A. Bagheri, and N. Marshall, 2017: Toward sustainable adaptation to future climate change: insights from vulnerability and resilience approaches analyzing agrarian system of Iran. *Environment, Development and Sustainability*, **19(1)**, 1–25, doi:10.1007/s10668-015-9721-3.

Hatfield-Dodds, S. et al., 2015: Australia is 'free to choose' economic growth and falling environmental pressures. *Nature*, **527(7576)**, 49–53, doi:10.1038/nature16065.

Havlík, P. et al., 2014: Climate change mitigation through livestock system transitions. *Proceedings of the National Academy of Sciences*, **111(10)**, 3709–14, doi:10.1073/pnas.1308044111.

Hayashi, A., F. Akimoto, F. Sano, and T. Tomoda, 2015: Evaluation of global energy crop production potential up to 2100 under socioeconomic development and climate change scenarios. *Journal of the Japan Institute of Energy*, **94(6)**, 548–554, doi:10.3775/jie.94.548.

Hayashi, A., F. Sano, Y. Nakagami, and K. Akimoto, 2018: Changes in terrestrial water stress and contributions of major factors under temperature rise constraint scenarios. *Mitigation and Adaptation Strategies for Global Change*, 1–27, doi:10.1007/s11027-018-9780-5.

Häyhä, T., P.L. Lucas, D.P. van Vuuren, S.E. Cornell, and H. Hoff, 2016: From Planetary Boundaries to national fair shares of the global safe operating space – How can the scales be bridged? *Global Environmental Change*, **40**, 60–72, doi:10.1016/j.gloenvcha.2016.06.008.

Hayward, B., 2017: *Sea change: climate politics and New Zealand*. Bridget Williams Books, Wellington, NZ, 120 pp.

He, J., Y. Huang, and F. Tarp, 2014: Has the clean development mechanism assisted sustainable development? *Natural Resources Forum*, **38(4)**, 248–260, doi:10.1111/1477-8947.12055.

Healy, N. and J. Barry, 2017: Politicizing energy justice and energy system transitions: Fossil fuel divestment and a "just transition". *Energy Policy*, **108**, 451–459, doi:10.1016/j.enpol.2017.06.014.

Heard, B.P., B.W. Brook, T.M.L. Wigley, and C.J.A. Bradshaw, 2017: Burden of proof: A comprehensive review of the feasibility of 100% renewable-electricity systems. *Renewable and Sustainable Energy Reviews*, **76**, 1122–1133, doi:10.1016/j.rser.2017.03.114.

Heede, R., 2014: Tracing anthropogenic carbon dioxide and methane emissions to fossil fuel and cement producers, 1854–2010. *Climatic Change*, **122(1–2)**, 229–241, doi:10.1007/s10584-013-0986-y.

Helliwell, J., R. Layard, and J. Sachs (eds.), 2018: *World Happiness Report*. Sustainable Development Solutions Network, New York, NY, USA, 167 pp.

Hemstock, S.L. et al., 2017: A Case for Formal Education in the Technical, Vocational Education and Training (TVET) Sector for Climate Change Adaptation and Disaster Risk Reduction in the Pacific Islands Region. In: *Climate Change Adaptation in Pacific Countries: Fostering Resilience and Improving the Quality of Life* [Filho, W. (ed.)]. Springer Nature, Cham, Switzerland, pp. 309–324, doi:10.1007/978-3-319-50094-2_19.

Hess, J.J. and K.L. Ebi, 2016: Iterative management of heat early warning systems in a changing climate. *Annals of the New York Academy of Sciences*, **1382(1)**, 21–30, doi:10.1111/nyas.13258.

Heyward, C. and D. Roser (eds.), 2016: *Climate justice in a non-ideal world*. Oxford University Press, Oxford UK, 352 pp.

Hiç, C., P. Pradhan, D. Rybski, and J.P. Kropp, 2016: Food Surplus and Its Climate Burdens. *Environmental Science & Technology*, **50(8)**, 4269–4277, doi:10.1021/acs.est.5b05088.

Hildingsson, R. and B. Johansson, 2015: Governing low-carbon energy transitions in sustainable ways: potential synergies and conflicts between climate and environmental policy objectives. *Energy Policy*, **88**, 245–252, doi:10.1016/j.enpol.2015.10.029.

HLCCP, 2017: *Report of the High-Level Commission on Carbon Prices*. High-Level Commission on Carbon Prices (HLCCP). World Bank, Washington DC, USA, 61 pp.

Holden, E., K. Linnerud, and D. Banister, 2017: The imperatives of sustainable development. *Sustainable Development*, **25(3)**, 213–226, doi:10.1002/sd.1647.

Holz, C., S. Kartha, and T. Athanasiou, 2018: Fairly sharing 1.5: national fair shares of a 1.5°C-compliant global mitigation effort. *International Environmental Agreements: Politics, Law and Economics*, **18(1)**, 117–134, doi:10.1007/s10784-017-9371-z.

Horstmann, B. and J. Hein, 2017: *Aligning climate change mitigation and sustainable development under the UNFCCC: A critical assessment of the Clean Development Mechanism, the Green Climate Fund and REDD+*. German Development Institute, Bonn, 154 pp.

Howarth, C. and I. Monasterolo, 2017: Opportunities for knowledge co-production across the energy-food-water nexus: making interdisciplinary approaches work for better climate decision making. *Environmental Science & Policy*, **75**, 103–110, doi:10.1016/j.envsci.2017.05.019.

Howarth, N., M. Galeotti, A. Lanza, and K. Dubey, 2017: Economic development and energy consumption in the GCC: an international sectoral analysis. *Energy Transitions*, **1(2)**, 6, doi:10.1007/s41825-017-0006-3.

Howell, R. and S. Allen, 2017: People and Planet: values, motivations and formative influences of individuals acting to mitigate climate change. *Environmental Values*, **26(2)**, 131–155, doi:10.3197/096327117x14847335385436.

Hubacek, K., G. Baiocchi, K. Feng, and A. Patwardhan, 2017: Poverty eradication in a carbon constrained world. *Nature Communications*, **8(1)**, 1–8, doi:10.1038/s41467-017-00919-4.

Huggel, C., I. Wallimann-Helmer, D. Stone, and W. Cramer, 2016: Reconciling justice and attribution research to advance climate policy. *Nature Climate Change*, **6(10)**, 901–908, doi:10.1038/nclimate3104.

Hughes, S., E.K. Chu, and S.G. Mason (eds.), 2018: *Climate Change in Cities: Innovations in Multi-Level Governance*. Springer International Publishing, Cham, Switzerland, 378 pp.

Hunsberger, C., S. Bolwig, E. Corbera, and F. Creutzig, 2014: Livelihood impacts of biofuel crop production: implications for governance. *Geoforum*, **54**, 248–260, doi:10.1016/j.geoforum.2013.09.022.

Huq, N., A. Bruns, L. Ribbe, and S. Huq, 2017: Mainstreaming ecosystem services based climate change adaptation (EbA) in bangladesh: status, challenges and opportunities. *Sustainability*, **9(6)**, 926, doi:10.3390/su9060926.

Hwang, J., K. Joh, and A. Woo, 2017: Social inequalities in child pedestrian traffic injuries: Differences in neighborhood built environments near schools in Austin, TX, USA. *Journal of Transport and Health*, **6**, 40–49, doi:10.1016/j.jth.2017.05.003.

Griggs, D.J., M. Nilsson, A. Stevance, and D. McCollum (eds.), 2017: *A Guide to SDG interactions: from Science to Implementation*. International Council for Science (ICSU), Paris, France, 239 pp., doi:10.24948/2017.01.

IEA, 2015: *India Energy Outlook*. International Energy Agency (IEA), Paris, France, 187 pp.

IEA, 2016: *Energy and Air Pollution: World Energy Outlook Special Report*. International Energy Agency (IEA), Paris, France, 266 pp.

IEA, 2017: *Energy Access Outlook 2017: From Poverty to Prosperity*. International Energy Agency (IEA), Paris, France, 144 pp., doi:10.1787/9789264285569-en.

IEA and World Bank, 2017: *Sustainable Energy for All 2017 – Progress towards Sustainable Energy*. International Energy Agency (IEA) and International Bank for Reconstruction and Development / The World Bank, Washington DC, USA, 208 pp., doi:10.1596/ 978-1-4648-1084-8.

INC, 1991: *Vanuatu: Draft annex relating to Article 23 (Insurance) for inclusion in the revised single text on elements relating to mechanisms (A/AC.237/WG.II/ Misc.13) submitted by the Co-Chairmen of Working Group II*. A/AC.237/WG.II/ CRP.8, Intergovernmental Negotiating Committee for a Framework Convention on Climate Change: Working Group II.

IPCC, 2013: Climate Change 2013: The Physical Science Basis. Contribution of Working Group I to the Fifth Assessment Report of the Intergovernmental Panel on Climate Change. [Stocker, T.F., D. Qin, G.-K. Plattner, M. Tignor, S.K. Allen, J. Boschung, A. Nauels, Y. Xia, V. Bex, and P.M. Midgley (eds.)]. Cambridge University Press, Cambridge, United Kingdom and New York, NY, USA, 1535 pp.

IPCC, 2014a: Climate Change 2014: Mitigation of Climate Change. Contribution of Working Group III to the Fifth Assessment Report of the Intergovernmental Panel on Climate Change. [Edenhofer, O., R. Pichs-Madruga, Y. Sokona, E. Farahani, S. Kadner, K. Seyboth, A. Adler, I. Baum, S. Brunner, P. Eickemeier, B. Kriemann, J. Savolainen, S. Schlömer, C. von Stechow, T. Zwickel, and J.C. Minx (eds.)]. Cambridge University Press, Cambridge, United Kingdom and New York, NY, USA, 1454 pp.

IPCC, 2014b: Summary for Policymakers. In: *Climate Change 2014: Mitigation of Climate Change. Contribution of Working Group III to the Fifth Assessment Report of the Intergovernmental Panel on Climate Change* [Edenhofer, O., R. Pichs-Madruga, Y. Sokona, E. Farahani, S. Kadner, K. Seyboth, A. Adler, I. Baum, S. Brunner, P. Eickemeier, B. Kriemann, J. Savolainen, S. Schlömer, C. Stechow, T. Zwickel, and J.C. Minx (eds.)]. Cambridge University Press, Cambridge, United Kingdom and New York, NY, USA, pp. 1–30.

IRENA, 2016: *Renewable Energy Market Analysis: The GCC Region*. International Renewable Energy Agency (IRENA), Abu Dhabi, United Arab Emirates, 96 pp.

Ishida, H. et al., 2014: Global-scale projection and its sensitivity analysis of the health burden attributable to childhood undernutrition under the latest scenario framework for climate change research. *Environmental Research Letters*, **9(6)**, 064014, doi:10.1088/1748-9326/9/6/064014.

Islam, M.R. and M. Shamsuddoha, 2017: Socioeconomic consequences of climate induced human displacement and migration in Bangladesh. *International Sociology*, **32(3)**, 277–298, doi:10.1177/0268580917693173.

Jägermeyr, J., A. Pastor, H. Biemans, and D. Gerten, 2017: Reconciling irrigated food production with environmental flows for Sustainable Development Goals implementation. *Nature Communications*, **8**, 1–9, doi:10.1038/ncomms15900.

Jaglin, S., 2014: Regulating service delivery in southern cities: rethinking urban heterogeneity. In: *The Routledge Handbook on Cities of the Global South* [Parnell, S. and S. Oldfield (eds.)]. pp. 434–447, doi:10.4324/9780203387832.ch37.

Jakob, M. and J. C. Steckel, 2016: Implications of climate change mitigation for sustainable development. *Environmental Research Letters*, **11(10)**, 104010, doi:10.1088/1748-9326/11/10/104010.

Janetos, A.C., E. Malone, E. Mastrangelo, K. Hardee, and A. de Bremond, 2012: Linking climate change and development goals: framing, integrating, and measuring. *Climate and Development*, **4(2)**, 141–156, doi:10.1080/17565529.2012.726195.

Jha, C.K., V. Gupta, U. Chattopadhyay, and B. Amarayil Sreeraman, 2017: Migration as adaptation strategy to cope with climate change: A study of farmers' migration in rural India. *International Journal of Climate Change Strategies and Management*, 10(1), 121–141, doi:10.1108/ijccsm-03-2017-0059.

Jiang, L. and B.C. O'Neill, 2017: Global urbanization projections for the Shared Socioeconomic Pathways. *Global Environmental Change*, **42**, 193–199, doi:10.1016/j.gloenvcha.2015.03.008.

Johansson, E.L., M. Fader, J.W. Seaquist, and K.A. Nicholas, 2016: Green and blue water demand from large-scale land acquisitions in Africa. *Proceedings of the National Academy of Sciences*, **113(41)**, 11471–11476, doi:10.1073/pnas.1524741113.

Johnson, C.A., 2017: Resilient cities? The global politics of urban climate adaptation. In: *The Power of Cities in Global Climate Politics* [Johnson, C.A. (ed.)]. Palgrave Macmillan, London, UK, pp. 91–146, doi:10.1057/978-1-137-59469-3_4.

Johnson, N. et al., 2015: Stranded on a low-carbon planet: implications of climate policy for the phase-out of coal-based power plants. *Technological Forecasting and Social Change*, **90(Part A)**, 89–102, doi:10.1016/j.techfore.2014.02.028.

Jones, L. and T. Tanner, 2017: 'Subjective resilience': using perceptions to quantify household resilience to climate extremes and disasters. *Regional Environmental Change*, **17(1)**, 229–243, doi:10.1007/s10113-016-0995-2.

Jordaan, S.M. et al., 2017: The role of energy technology innovation in reducing greenhouse gas emissions: a case study of Canada. *Renewable and Sustainable Energy Reviews*, **78**, 1397–1409, doi:10.1016/j.rser.2017.05.162.

Kagawa, S. et al., 2015: CO_2 emission clusters within global supply chain networks: Implications for climate change mitigation. *Global Environmental Change*, **35**, 486–496, doi:10.1016/j.gloenvcha.2015.04.003.

Kaijser, A. and A. Kronsell, 2014: Climate change through the lens of intersectionality. *Environmental Politics*, **23(3)**, 417–433, doi:10.1080/09644016.2013.835203.

Kalafatis, S., 2017: Identifying the Potential for Climate Compatible Development Efforts and the Missing Links. *Sustainability*, **9(9)**, 1642, doi:10.3390/su9091642.

Karlsson, L., A. Nightingale, L.O. Naess, and J. Thompson, 2017: 'Triple wins' or 'triple faults'? Analysing policy discourse on climate-smart agriculture (CSA).

Working Paper no.197, CGIAR Research Program on Climate Change, Agriculture and Food Security (CCAFS), Copenhagen, Denmark, 43 pp.

Karner, K., M. Theissing, and T. Kienberger, 2015: Energy efficiency for industries through synergies with urban areas. *Journal of Cleaner Production*, **2020**, 1–11, doi:10.1016/j.jclepro.2016.02.010.

Kartha, S. et al., 2018: Inequitable mitigation: cascading biases against poorer countries. *Nature Climate Change*, **8**, 348–349, doi:10.1038/s41558-018-0152-7.

Kassie, M., H. Teklewold, M. Jaleta, P. Marenya, and O. Erenstein, 2015: Understanding the adoption of a portfolio of sustainable intensification practices in eastern and southern Africa. *Land Use Policy*, **42**, 400–411, doi:10.1016/j.landusepol.2014.08.016.

Keairns, D.L., R.C. Darton, and A. Irabien, 2016: The energy-water-food nexus. *Annual Review of Chemical and Biomolecular Engineering*, **7(1)**, 239–262, doi:10.1146/annurev-chembioeng-080615-033539.

Kelman, I., 2017: Linking disaster risk reduction, climate change, and the sustainable development goals. *Disaster Prevention and Management: An International Journal*, **26(3)**, 254–258, doi:10.1108/dpm-02-2017-0043.

Kenis, A., 2016: Ecological citizenship and democracy: Communitarian versus agonistic perspectives. *Environmental Politics*, **4016**, 1–22, doi:10.1080/09644016.2016.1203524.

Kenis, A. and E. Mathijs, 2014: (De)politicising the local: The case of the Transition Towns movement in Flanders (Belgium). *Journal of Rural Studies*, **34**, 172–183, doi:10.1016/j.jrurstud.2014.01.013.

Keohane, R.O. and D.G. Victor, 2016: Cooperation and discord in global climate policy. *Nature Climate Change*, **6(6)**, 570–575, doi:10.1038/nclimate2937.

Khanna, N., D. Fridley, and L. Hong, 2014: China's pilot low-carbon city initiative: A comparative assessment of national goals and local plans. *Sustainable Cities and Society*, **12**, 110–121, doi:10.1016/j.scs.2014.03.005.

Khosla, R., A. Sagar, and A. Mathur, 2017: Deploying Low-carbon Technologies in Developing Countries: A view from India's buildings sector. *Environmental Policy and Governance*, **27(2)**, 149–162, doi:10.1002/eet.1750.

Khreis, H., A.D. May, and M.J. Nieuwenhuijsen, 2017: Health impacts of urban transport policy measures: a guidance note for practice. *Journal of Transport & Health*, **6**, 209–227, doi:10.1016/j.jth.2017.06.003.

Kim, S.-Y. and E. Thurbon, 2015: Developmental Environmentalism: Explaining South Korea's Ambitious Pursuit of Green Growth. *Politics & Society*, **43(2)**, 213–240, doi:10.1177/0032329215571287.

Kim, Y. et al., 2017: A perspective on climate-resilient development and national adaptation planning based on USAID's experience. *Climate and Development*, **9(2)**, 141–151, doi:10.1080/17565529.2015.1124037.

King, A.D., D.J. Karoly, and B.J. Henley, 2017: Australian climate extremes at 1.5°C and 2°C of global warming. *Nature Climate Change*, **7**, 412–416, doi:10.1038/nclimate3296.

Kirby, P. and T. O'Mahony, 2018: Development Models: Lessons from International Development. In: *The Political Economy of the Low-Carbon Transition: Pathways Beyond Techno-Optimism*. Springer International Publishing, Cham, Switzerland, pp. 89–114, doi:10.1007/978-3-319-62554-6_4.

Klausbruckner, C., H. Annegarn, L.R.F. Henneman, and P. Rafaj, 2016: A policy review of synergies and trade-offs in South African climate change mitigation and air pollution control strategies. *Environmental Science & Policy*, **57**, 70–78, doi:10.1016/j.envsci.2015.12.001.

Klein, R.J.T. et al., 2014: Adaptation opportunities, constraints, and limits. In: *Climate Change 2014: Impacts, Adaptation, and Vulnerability. Part A: Global and Sectoral Aspects. Contribution of Working Group II to the Fifth Assessment Report of the Intergovernmental Panel on Climate Change* [Field, C.B., V.R. Barros, D.J. Dokken, K.J. Mach, M.D. Mastrandrea, T.E. Bilir, M. Chatterjee, K.L. Ebi, Y.O. Estrada, R.C. Genova, B. Girma, E.S. Kissel, A.N. Levy, S. MacCracken, P.R. Mastrandrea, and L.L. White (eds.)]. Cambridge University Press, Cambridge, United Kingdom and New York, NY, USA, pp. 899–943.

Klimont, Z. et al., 2017: Bridging the gap – the role of short-lived climate pollutants. In: *The Emissions Gap Report 2017: A UN Environmental Synethesis Report*. United Nations Environment Programme (UNEP), Nairobi, Kenya, pp. 48–57.

Klinsky, S. and A. Golub, 2016: Justice and Sustainability. In: *Sustainability Science: An introduction* [Heinrichs, H., P. Martens, G. Michelsen, and A. Wiek (eds.)]. Springer Netherlands, Dordrecht, Netherlands, pp. 161–173, doi:10.1007/978-94-017-7242-6.

Klinsky, S. and H. Winkler, 2018: Building equity in: strategies for integrating equity into modelling for a 1.5°C world. *Philosophical Transactions of the Royal Society A: Mathematical, Physical and Engineering Sciences*, **376(2119)**, 20160461, doi:10.1098/rsta.2016.0461.

Klinsky, S., D. Waskow, E. Northrop, and W. Bevins, 2017a: Operationalizing equity and supporting ambition: identifying a more robust approach to 'respective capabilities'. *Climate and Development*, **9(4)**, 1–11, doi:10.1080/17565529.2016.1146121.

Klinsky, S. et al., 2017b: Why equity is fundamental in climate change policy research. *Global Environmental Change*, **44**, 170–173, doi:10.1016/j.gloenvcha.2016.08.002.

Kongsager, R. and E. Corbera, 2015: Linking mitigation and adaptation in carbon forestry projects: Evidence from Belize. *World Development*, **76**, 132–146, doi:10.1016/j.worlddev.2015.07.003.

Kongsager, R., B. Locatelli, and F. Chazarin, 2016: Addressing climate change mitigation and adaptation together: a global assessment of agriculture and forestry projects. *Environmental Management*, **57(2)**, 271–282, doi:10.1007/s00267-015-0605-y.

Kreibich, N., L. Hermwille, C. Warnecke, and C. Arens, 2017: An update on the Clean Development Mechanism in Africa in times of market crisis. *Climate and Development*, **9(2)**, 178–190, doi:10.1080/17565529.2016.1145102.

Kumar, N.S. et al., 2014: *Climatic Risks and Strategizing Agricultural Adaptation in Climatically Challenged Regions*. TB-ICN: 136/2014, Indian Agriculture Research Institute, New Delhi, India, 106 pp.

Kuramochi, T. et al., 2018: Ten key short-term sectoral benchmarks to limit warming to 1.5°C. *Climate Policy*, **18(3)**, 287–305, doi:10.1080/14693062.2017.1397495.

Kuruppu, N. and R. Willie, 2015: Barriers to reducing climate enhanced disaster risks in least developed country-small islands through anticipatory adaptation. *Weather and Climate Extremes*, **7**, 72–83, doi:10.1016/j.wace.2014.06.001.

Kusumaningtyas, S.D.A. and E. Aldrian, 2016: Impact of the June 2013 Riau province Sumatera smoke haze event on regional air pollution. *Environmental Research Letters*, **11(7)**, 075007, doi:10.1088/1748-9326/11/7/075007.

La Rovere, E.L., 2017: Low-carbon development pathways in Brazil and 'Climate Clubs'. *Wiley Interdisciplinary Reviews: Climate Change*, **8(1)**, 1–7, doi:10.1002/wcc.439.

Labordena, M., A. Patt, M. Bazilian, M. Howells, and J. Lilliestam, 2017: Impact of political and economical barriers for concentrating solar power in Sub-Saharan Africa. *Energy Policy*, **102**, 52–72, doi:10.1016/j.enpol.2016.12.008.

Labriet, M., C. Fiebig, and M. Labrousse, 2015: *Working towards a Smart Energy Path: Experience from Benin, Mali and Togo*. Inside Stories on climate compatible development, Climate and Development Knowlegdge Network (CDKN), 6 pp.

Lade, S.J., L.J. Haider, G. Engström, and M. Schlüter, 2017: Resilience offers escape from trapped thinking on poverty alleviation. *Science Advances*, **3(5)**, e1603043, doi:10.1126/sciadv.1603043.

Lahn, B., 2018: In the light of equity and science: scientific expertise and climate justice after Paris. *International Environmental Agreements: Politics, Law and Economics*, **18(1)**, 29–43, doi:10.1007/s10784-017-9375-8.

Lamb, A. et al., 2016: The potential for land sparing to offset greenhouse gas emissions from agriculture. *Nature Climate Change*, **6**, 488–492, doi:10.1038/nclimate2910.

Langford, M., A. Sumner, and A.E. Yamin (eds.), 2013: *The Millennium Development Goals and Human Rights: Past, Present and Future*. Cambridge University Press, New York, NY, USA, 571 pp.

Lasage, R. et al., 2015: A Stepwise, participatory approach to design and implement community based adaptation to drought in the Peruvian Andes. *Sustainability*, **7(2)**, 1742–1773, doi:10.3390/su7021742.

Lawrence, J. and M. Haasnoot, 2017: What it took to catalyse uptake of dynamic adaptive pathways planning to address climate change uncertainty. *Environmental Science & Policy*, **68**, 47–57, doi:10.1016/j.envsci.2016.12.003.

Lechtenboehmer, S. and K. Knoop, 2017: *Realising long-term transitions towards low carbon societies. Impulses from the 8th Annual Meeting of the International Research Network for Low Carbon Societies*. Wuppertal Spezial no. 53, Wuppertal Institut für Klima, Umwelt, Energie, Wuppertal, Germany, 100 pp.

Lefale, P., P. Faiva, and A. C, 2017: *Living with Change (LivC): An Integrated National Strategy for Enhancing the Resilience of Tokelau to Climate Change and Related Hazards, 2017–2030*. Government of Tokelau, Apia, Soamoa, 16 pp.

Leichenko, R. and J.A. Silva, 2014: Climate change and poverty: vulnerability, impacts, and alleviation strategies. *Wiley Interdisciplinary Reviews: Climate Change*, **5(4)**, 539–556, doi:10.1002/wcc.287.

Lemoine, D. and S. Kapnick, 2016: A top-down approach to projecting market impacts of climate change. *Nature Climate Change*, **6(1)**, 51–55, doi:10.1038/nclimate2759.

Lemos, C.M., Y. Lo, D.R. Nelson, H. Eakin, and A.M. Bedran-Martins, 2016: Linking development to climate adaptation: Leveraging generic and specific capacities to reduce vulnerability to drought in NE Brazil. *Global Environmental Change*, **39**, 170–179, doi:10.1016/j.gloenvcha.2016.05.001.

Ley, D., 2017: Sustainable Development, Climate Change, and Renewable Energy in Rural Central America. In: *Evaluating Climate Change Action for Sustainable Development* [Uitto, J. I., J. Puri, and R. D. van den Berg (eds.)]. Springer International Publishing, Cham, Switzerland, pp. 187–212, doi:10.1007/978-3-319-43702-6.

Li, T. et al., 2016: Aging will amplify the heat-related mortality risk under a changing climate: projection for the elderly in Beijing, China. *Scientific Reports*, **6(1)**, 28161, doi:10.1038/srep28161.

Liddell, C. and C. Guiney, 2015: Living in a cold and damp home: Frameworks for understanding impacts on mental well-being. *Public Health*, **129(3)**, 191–199, doi:10.1016/j.puhe.2014.11.007.

Lilliestam, J. and A. Patt, 2015: Barriers, risks and policies for renewables in the Gulf states. *Energies*, **8(8)**, 8263–8285, doi:10.3390/en8088263.

Lin, B.B. et al., 2017: Adaptation Pathways in Coastal Case Studies: Lessons Learned and Future Directions. *Coastal Management*, **45(5)**, 384–405, doi:10.1080/08920753.2017.1349564.

Lin, J., Y. Hu, S. Cui, J. Kang, and A. Ramaswami, 2015: Tracking urban carbon footprints from production and consumption perspectives. *Environmental Research Letters*, **10(5)**, 054001, doi:10.1088/1748-9326/10/5/054001.

Lipper, L. et al., 2014: Climate-smart agriculture for food security. *Nature Climate Change*, **4(12)**, 1068–1072, doi:10.1038/nclimate2437.

Lorek, S. and J.H. Spangenberg, 2014: Sustainable consumption within a sustainable economy – Beyond green growth and green economies. *Journal of Cleaner Production*, **63**, 33–44, doi:10.1016/j.jclepro.2013.08.045.

Lotze-Campen, H. et al., 2014: Impacts of increased bioenergy demand on global food markets: an AgMIP economic model intercomparison. *Agricultural Economics*, **45(1)**, 103–116, doi:10.1111/agec.12092.

Lucas, K. and K. Pangbourne, 2014: Assessing the equity of carbon mitigation policies for transport in Scotland. *Case Studies on Transport Policy*, **2(2)**, 70–80, doi:10.1016/j.cstp.2014.05.003.

Lucon, O. et al., 2014: Buildings. In: *Climate Change 2014: Mitigation of Climate Change. Contribution of Working Group III to the Fifth Assessment Report of the Intergovernmental Panel on Climate Change* [Edenhofer, O., R. Pichs-Madruga, Y. Sokona, E. Farahani, S. Kadner, K. Seyboth, A. Adler, S. I. Baum, P. Brunner, B. Eickemeier, J. Kriemann, J. Savolainen, C. Schlömer, V. Stechow, T. Zwickel, and J.C. Minx (eds.)]. Cambridge University Press, Cambridge, United Kingdom and New York, NY, USA, pp. 671–738.

Luomi, M., 2014: Mainstreaming climate policy in the Gulf Cooperation Council States. OIES Paper: MEP 7, The Oxford Institute for Energy Studies, Oxford, UK, 73 pp.

Lwasa, S., K. Buyana, P. Kasaija, and J. Mutyaba, 2018: Scenarios for adaptation and mitigation in urban Africa under 1.5°C global warming. *Current Opinion in Environmental Sustainability*, **30**, 52–58, doi:10.1016/j.cosust.2018.02.012.

Lyon, C., 2018: Complexity ethics and UNFCCC practices for 1.5°C climate change. *Current Opinion in Environmental Sustainability*, **31**, 48–55, doi:10.1016/j.cosust.2017.12.008.

Mace, M.J., 2016: Mitigation commitments under the Paris Agreement and the way forward. *Climate Law*, **6(1–2)**, 21–39, doi:10.1163/18786561-00601002.

Macintyre, T., H. Lotz-Sisitka, A. Wals, C. Vogel, and V. Tassone, 2018: Towards transformative social learning on the path to 1.5°C degrees. *Current Opinion in Environmental Sustainability*, **31**, 80–87, doi:10.1016/j.cosust.2017.12.003.

Maclellan, N., 2015: *Yumi stap redi long klaemet jenis: Lessons from the Vanuatu NGO Climate Change Adaptation Program*. Oxfam Australia, 48 pp.

Maidment, C.D., C.R. Jones, T.L. Webb, E.A. Hathway, and J.M. Gilbertson, 2014: The impact of household energy efficiency measures on health: A meta-analysis. *Energy Policy*, **65**, 583–593, doi:10.1016/j.enpol.2013.10.054.

Maor, M., J. Tosun, and A. Jordan, 2017: Proportionate and disproportionate policy responses to climate change: core concepts and empirical applications. *Journal of Environmental Policy and Planning*, 1–13, doi:10.1080/1523908x.2017.1281730.

Mapfumo, P. et al., 2017: Pathways to transformational change in the face of climate impacts: an analytical framework. *Climate and Development*, **9(5)**, 439–451, doi:10.1080/17565529.2015.1040365.

Marjanac, S. and L. Patton, 2018: Extreme weather event attribution science and climate change litigation: an essential step in the causal chain? *Journal of Energy & Natural Resources Law*, 1–34, doi:10.1080/02646811.2018.1451020.

Markandya, A. et al., 2018: Health co-benefits from air pollution and mitigation costs of the Paris Agreement: a modelling study. *The Lancet Planetary Health*, **2(3)**, e126–e133, doi:10.1016/s2542-5196(18)30029-9.

Marsh, J., 2015: Mixed motivations and complex causality in the Mekong. *Forced Migration Review*, 68–69, www.fmreview.org/climatechange-disasters/marsh.

Martinez-Baron, D., G. Orjuela, G. Renzoni, A.M. Loboguerrero Rodríguez, and S.D. Prager, 2018: Small-scale farmers in a 1.5°C future: the importance of local social dynamics as an enabling factor for implementation and scaling of climate-smart agriculture. *Current Opinion in Environmental Sustainability*, **31**, 112–119, doi:10.1016/j.cosust.2018.02.013.

Mathur, V.N., S. Afionis, J. Paavola, A.J. Dougill, and L.C. Stringer, 2014: Experiences of host communities with carbon market projects: towards multi-level climate justice. *Climate Policy*, **14(1)**, 42–62, doi:10.1080/14693062.2013.861728.

Maupin, A., 2017: The SDG13 to combat climate change: an opportunity for Africa to become a trailblazer? *African Geographical Review*, **36(2)**, 131–145, doi:10.1080/19376812.2016.1171156.

Mayer, B., 2016: The relevance of the no-harm principle to climate change law and politics. *Asia Pacific Journal of Environmental Law*, **19(1)**, 79–104, doi:10.4337/apjel.2016.01.04.

Mbow, C., C. Neely, and P. Dobie, 2015: How can an integrated landscape approach contribute to the implementation of the Sustainable Development Goals (SDGs) and advance climate-smart objectives? In: *Climate-Smart Landscapes: Multifunctionality in Practice* [Minang, P.A., M. van Noordwijk, C. Mbow, J. de Leeuw, and D. Catacutan (eds.)]. World Agroforestry Centre (ICRAF), Nairobi, Kenya, pp. 103–117.

McCollum, D.L. et al., 2018a: Connecting the Sustainable Development Goals by their energy inter-linkages. *Environmental Research Letters*, **13(3)**, 033006, doi:10.1088/1748-9326/aaafe3.

McCollum, D.L. et al., 2018b: Energy investment needs for fulfilling the Paris Agreement and achieving the Sustainable Development Goals. *Nature Energy*, **3(7)**, 589–599, doi:10.1038/s41560-018-0179-z.

McCubbin, S., B. Smit, and T. Pearce, 2015: Where does climate fit? Vulnerability to climate change in the context of multiple stressors in Funafuti, Tuvalu. *Global Environmental Change*, **30**, 43–55, doi:10.1016/j.gloenvcha.2014.10.007.

McGlade, C. and P. Ekins, 2015: The geographical distribution of fossil fuels unused when limiting global warming to 2°C. *Nature*, **517(7533)**, 187–190, doi:10.1038/nature14016.

McGranahan, G., D. Schensul, and G. Singh, 2016: Inclusive urbanization: can the 2030 Agenda be delivered without it? *Environment and Urbanization*, **28(1)**, 13–34, doi:10.1177/0956247815627522.

McKay, B., S. Sauer, B. Richardson, and R. Herre, 2016: The political economy of sugarcane flexing: initial insights from Brazil, Southern Africa and Cambodia. *The Journal of Peasant Studies*, **43(1)**, 195–223, doi:10.1080/03066150.2014.992016.

McNamara, K.E., 2015: Cross-border migration with dignity in Kiribati. *Forced Migration Review*, **49**, 62, www.fmreview.org/climatechange-disasters/mcnamara.

McNamara, K.E. and S.S. Prasad, 2014: Coping with extreme weather: Communities in Fiji and Vanuatu share their experiences and knowledge. *Climatic Change*, **123(2)**, 121–132, doi:10.1007/s10584-013-1047-2.

Mechler, R. and T. Schinko, 2016: Identifying the policy space for climate loss and damage. *Science*, **354(6310)**, 290–292, doi:10.1126/science.aag2514.

Mechler, R., L.M. Bouwer, T. Schinko, S. Surminski, and J. Linnerooth-Bayer (eds.), 2019: *Loss and Damage from Climate Change: Concepts, Methods and Policy Options*. Springer International Publishing, 561 pp, doi:10.1007/978-3-319-72026-5.

Meerow, S. and J.P. Newell, 2016: Urban resilience for whom, what, when, where, and why? *Urban Geography*, 1–21, doi:10.1080/02723638.2016.1206395.

Mehmood, A., 2016: Of resilient places: planning for urban resilience. *European Planning Studies*, **24(2)**, 407–419, doi:10.1080/09654313.2015.1082980.

Meinshausen, M. et al., 2015: National post-2020 greenhouse gas targets and diversity-aware leadership. *Nature Climate Change*, **5(12)**, 1098–1106, doi:10.1038/nclimate2826.

Melica, G. et al., 2018: Multilevel governance of sustainable energy policies: The role of regions and provinces to support the participation of small local authorities in the Covenant of Mayors. *Sustainable Cities and Society*, **39**, 729–739, doi:10.1016/j.scs.2018.01.013.

Michaelowa, A., M. Allen, and F. Sha, 2018: Policy instruments for limiting global temperature rise to 1.5°C – can humanity rise to the challenge? *Climate Policy*, **18(3)**, 275–286, doi:10.1080/14693062.2018.1426977.

Millar, R.J. et al., 2017: Emission budgets and pathways consistent with limiting warming to 1.5°C. *Nature Geoscience*, 1–8, doi:10.1038/ngeo3031.

5

Millard-Ball, A., 2013: The limits to planning: causal impacts of city climate action plans. *Journal of Planning Education and Research*, **33(1)**, 5–19, doi:10.1177/0739456x12449742.

Mishra, V., S. Mukherjee, R. Kumar, and D. Stone, 2017: Heat wave exposure in India in current, 1.5°C, and 2.0°C worlds. *Environmental Research Letters*, **12**, 124012, doi:10.1088/1748-9326/aa9388.

Mitchell, D. et al., 2016: Attributing human mortality during extreme heat waves to anthropogenic climate change. *Environmental Research Letters*, **11(7)**, 074006, doi:10.1088/1748-9326/11/7/074006.

MoCC, 2016: *Corporate Plan 2016–2018*. Ministry of Climate Chanage and Adaptation (MoCC), Government of Vanuatu, Vanuatu, 98 pp.

Moellendorf, D., 2015: Climate change justice. *Philosophy Compass*, **10**, 173–186, doi:10.3197/096327111x12997574391887.

Morita, K. and K. Matsumoto, 2015: Enhancing Biodiversity Co-benefits of Adaptation to Climate Change. In: *Handbook of Climate Change Adaptation* [Filho, W.L. (ed.)]. Springer Berlin Heidelberg, Berlin, Heidelberg, pp. 953–972, doi:10.1007/978-3-642-38670-1_21.

Mouratiadou, I. et al., 2018: Water demand for electricity in deep decarbonisation scenarios: a multi-model assessment. *Climatic Change*, **147(1)**, 91–106, doi:10.1007/s10584-017-2117-7.

MRFCJ, 2015a: *Women's Participation – An Enabler of Climate Justice*. Mary Robinson Foundation Climate Justice (MRFCJ), Dublin, Ireland, 24 pp.

MRFCJ, 2015b: *Zero Carbon, Zero Poverty The Climate Justice Way: Achieving an equitable phase-out of carbon emissions by 2050 while protecting human rights*. Mary Robinson Foundation Climate Justice (MRFCJ), Dublin, Ireland, 70 pp.

Muratori, M., K. Calvin, M. Wise, P. Kyle, and J. Edmonds, 2016: Global economic consequences of deploying bioenergy with carbon capture and storage (BECCS). *Environmental Research Letters*, **11(9)**, 1–9, doi:10.1088/1748-9326/11/9/095004.

Murphy, D.J., L. Yung, C. Wyborn, and D.R. Williams, 2017: Rethinking climate change adaptation and place through a situated pathways framework: a case study from the Big Hole Valley, USA. *Landscape and Urban Planning*, **167**, 441–450, doi:10.1016/j.landurbplan.2017.07.016.

Murphy, K., G.A. Kirkman, S. Seres, and E. Haites, 2015: Technology transfer in the CDM: an updated analysis. *Climate Policy*, **15(1)**, 127–145, doi:10.1080/14693062.2013.812719.

Murtinho, F., 2016: What facilitates adaptation? An analysis of community-based adaptation to environmental change in the Andes. *International Journal of the Commons*, **10(1)**, 119–141, doi:10.18352/ijc.585.

Muttarak, R. and W. Lutz, 2014: Is education a key to reducing vulnerability to natural disasters and hence unavoidable climate change? *Ecology and Society*, **19(1)**, 42, doi:10.5751/es-06476-190142.

Myers, S.S. et al., 2017: Climate Change and Global Food Systems: Potential Impacts on Food Security and Undernutrition. *Annual Review of Public Health*, 1–19, doi:10.1146/annurev-publhealth-031816-044356.

Naess, L.O. et al., 2015: Climate policy meets national development contexts: insights from Kenya and Mozambique. *Global Environmental Change*, **35**, 534–544, doi:10.1016/j.gloenvcha.2015.08.015.

Nagoda, S., 2015: New discourses but same old development approaches? Climate change adaptation policies, chronic food insecurity and development interventions in northwestern Nepal. *Global Environmental Change*, **35**, 570–579, doi:10.1016/j.gloenvcha.2015.08.014.

Nalau, J., J. Handmer, and M. Dalesa, 2017: The role and capacity of government in a climate crisis: Cyclone Pam in Vanuatu. In: *Climate Change Adaptation in Pacific Countries: Fostering Resilience and Improving the Quality of Life* [Leal Filho, W. (ed.)]. Springer International Publishing, Cham, Switzerland, pp. 151–161, doi:10.1007/978-3-319-50094-2_9.

Nalau, J. et al., 2016: The practice of integrating adaptation and disaster risk reduction in the south-west Pacific. *Climate and Development*, **8(4)**, 365–375, doi:10.1080/17565529.2015.1064809.

NASEM, 2016: *Attribution of Extreme Weather Events in the Context of Climate Change*. National Academies of Sciences, Engineering, and Medicine (NASEM). The National Academies Press, Washington DC, USA, 186 pp., doi:10.17226/21852.

Nelson, D.R., M.C. Lemos, H. Eakin, and Y.-J. Lo, 2016: The limits of poverty reduction in support of climate change adaptation. *Environmental Research Letters*, **11(9)**, 094011, doi:10.1088/1748-9326/11/9/094011.

Newell, P. et al., 2014: The Political Economy of Low Carbon Energy in Kenya. IDS Working Papers Vol 2014 No 445, Institute of Development Studies (IDS), Brighton, UK, 38 pp., doi:10.1111/j.2040-0209.2014.00445.x.

Newman, P., T. Beatley, and H. Boyer, 2017: *Resilient Cities: Overcoming Fossil Fuel Dependence (2nd edition)*. Island Press, Washington DC, USA, 264 pp.

Nguyen, M.T., S. Vink, M. Ziemski, and D.J. Barrett, 2014: Water and energy synergy and trade-off potentials in mine water management. *Journal of Cleaner Production*, **84(1)**, 629–638, doi:10.1016/j.jclepro.2014.01.063.

Nhamo, G., 2016: New Global Sustainable Development Agenda: A Focus on Africa. *Sustainable Development*, **25**, 227–241, doi:10.1002/sd.1648.

Niang, I. et al., 2014: Africa. In: *Climate Change 2014: Impacts, Adaptation, and Vulnerability. Part B: Regional Aspects. Contribution of Working Group II to the Fifth Assessment Report of the Intergovernmental Panel on Climate Change* [Barros, V.R., C.B. Field, D.J. Dokken, M.D. Mastrandrea, K.J. Mach, T.E. Bilir, M. Chatterjee, K.L. Ebi, Y.O. Estrada, R.C. Genova, B. Girma, E.S. Kissel, A.N. Levy, S. MacCracken, P.R. Mastrandrea, and L.L. White (eds.)]. Cambridge University Press, Cambridge, United Kingdom and New York, NY, USA, pp. 1199–1265.

Nicholls, R.J. et al., 2018: Stabilization of global temperature at 1.5°C and 2.0°C: implications for coastal areas. *Philosophical Transactions of the Royal Society A: Mathematical, Physical and Engineering Sciences*, **376(2119)**, 20160448, doi:10.1098/rsta.2016.0448.

Nicolosi, E. and G. Feola, 2016: Transition in place: Dynamics, possibilities, and constraints. *Geoforum*, **76**, 153–163, doi:10.1016/j.geoforum.2016.09.017.

Nilsson, M., D. Griggs, and M. Visback, 2016: Map the interactions between Sustainable Development Goals. *Nature*, **534(7607)**, 320–322, doi:10.1038/534320a.

Nkomwa, E.C., M.K. Joshua, C. Ngongondo, M. Monjerezi, and F. Chipungu, 2014: Assessing indigenous knowledge systems and climate change adaptation strategies in agriculture: a case study of Chagaka Village, Chikhwawa, Southern Malawi. *Physics and Chemistry of the Earth, Parts A/B/C*, **67–69**, 164–172, doi:10.1016/j.pce.2013.10.002.

Noble, I. et al., 2014: Adaptation needs and options. In: *Climate Change 2014: Impacts, Adaptation, and Vulnerability. Part A: Global and Sectoral Aspects. Contribution of Working Group II to the Fifth Assessment Report of the Intergovernmental Panel on Climate Change* [Field, C.B., V.R. Barros, D.J. Dokken, K.J. Mach, M.D. Mastrandrea, T.E. Bilir, M. Chatterjee, K.L. Ebi, Y.O. Estrada, R.C. Genova, B. Girma, E.S. Kissel, A.N. Levy, S. MacCracken, P.R. Mastrandrea, and L.L. White (eds.)]. Cambridge University Press, Cambridge, United Kingdom and New York, NY, USA, pp. 659–708.

Normann, H.E., 2015: The role of politics in sustainable transitions: The rise and decline of offshore wind in Norway. *Environmental Innovation and Societal Transitions*, **15(2015)**, 180–193, doi:10.1016/j.eist.2014.11.002.

North, P. and N. Longhurst, 2013: Grassroots localisation? The scalar potential of and limits of the 'transition' approach to climate change and resource constraint. *Urban Studies*, **50(7)**, 1423–1438, doi:10.1177/0042098013480966.

Nunan, F. (ed.), 2017: *Making Climate Compatible Development Happen*. Routledge, Abingdon, UK and New York, NY, USA, 262 pp.

Nunn, P.D., W. Aalbersberg, S. Lata, and M. Gwilliam, 2014: Beyond the core: Community governance for climate-change adaptation in peripheral parts of Pacific Island Countries. *Regional Environmental Change*, **14(1)**, 221–235, doi:10.1007/s10113-013-0486-7.

Nyantakyi-Frimpong, H. and R. Bezner-Kerr, 2015: The relative importance of climate change in the context of multiple stressors in semi-arid Ghana. *Global Environmental Change*, **32**, 40–56, doi:10.1016/j.gloenvcha.2015.03.003.

O'Brien, K.L., 2016: Climate change and social transformations: is it time for a quantum leap? *Wiley Interdisciplinary Reviews: Climate Change*, **7(5)**, 618–626, doi:10.1002/wcc.413.

O'Brien, K.L., 2017: Climate Change Adaptation and Social Transformation. In: *International Encyclopedia of Geography: People, the Earth, Environment and Technology*. John Wiley & Sons, Ltd, Oxford, UK, pp. 1–8, doi:10.1002/9781118786352.wbieg0987.

O'Brien, K.L., 2018: Is the 1.5°C target possible? Exploring the three spheres of transformation. *Current Opinion in Environmental Sustainability*, **31**, 153–160, doi:10.1016/j.cosust.2018.04.010.

O'Brien, K.L., S. Eriksen, T.H. Inderberg, and L. Sygna, 2015: Climate change and development: adaptation through transformation. In: *Climate Change Adaptation and Development: Changing Paradigms and Practices* [Inderberg, T.H., S. Eriksen, K. O'Brien, and L. Sygna (eds.)]. Routledge, Abingdon, UK and New York, NY, USA, pp. 273–289.

O'Neill, B.C. et al., 2017a: The roads ahead: narratives for shared socioeconomic pathways describing world futures in the 21st century. *Global Environmental Change*, **42**, 169–180, doi:10.1016/j.gloenvcha.2015.01.004.

O'Neill, B.C. et al., 2017b: IPCC reasons for concern regarding climate change risks. *Nature Climate Change*, **7(1)**, 28–37, doi:10.1038/nclimate3179.

O'Neill, D.W., A.L. Fanning, W.F. Lamb, and J.K. Steinberger, 2018: A good life for all within planetary boundaries. *Nature Sustainability*, **1(2)**, 88–95, doi:10.1038/s41893-018-0021-4.

Ober, K. and P. Sakdapolrak, 2017: How do social practices shape policy? Analysing the field of 'migration as adaptation' with Bourdieu's 'Theory of Practice'. *The Geographical Journal*, **183(4)**, 359–369, doi:10.1111/geoj.12225.

Obersteiner, M. et al., 2016: Assessing the land resource-food price nexus of the Sustainable Development Goals. *Science Advances*, **2(9)**, e1501499–e1501499, doi:10.1126/sciadv.1501499.

OECD, 2017: The Government Role in Mobilising Investment and Innovation in Renewable Energy. *OECD Investment Insights,* August 2017, Organisation for Economic Co-operation and Development (OECD), Paris, France, 4 pp.

Oei, P.-Y. and R. Mendelevitch, 2016: European Scenarios of CO_2 Infrastructure Investment until 2050. *The Energy Journal*, **37(01)**, doi:10.5547/01956574.37.si3.poei.

Ojea, E., 2015: Challenges for mainstreaming Ecosystem-based Adaptation into the international climate agenda. *Current Opinion in Environmental Sustainability*, **14**, 41–48, doi:10.1016/j.cosust.2015.03.006.

Okereke, C. and P. Coventry, 2016: Climate justice and the international regime: before, during, and after Paris. *Wiley Interdisciplinary Reviews: Climate Change*, **7(6)**, 834–851, doi:10.1002/wcc.419.

Olawuyi, D.S., 2017: From technology transfer to technology absorption: addressing climate technology gaps in Africa. Fixing Climate Governance Series Paper No. 5, Centre for International Governance Innovation, Waterloo, Canada, 16 pp.

Olsson, L. et al., 2014: Livelihoods and Poverty. In: *Climate Change 2014: Impacts, Adaptation, and Vulnerability. Part A: Global and Sectoral Aspects. Contribution of working Group II to the Fifth Assessment Report of the Intergovernmental Panel on Climate Change* [Field, C.B., V.R. Barros, D.J. Dokken, K.J. Mach, M.D. Mastrandrea, T.E. Bilir, M. Chatterjee, K.L. Ebi, Y.O. Estrada, R.C. Genova, B. Girma, E.S. Kissel, A.N. Levy, S. MacCracken, P.R. Mastrandrea, and L.L. White (eds.)]. Cambridge University Press, Cambridge, United Kingdom and New York, NY, USA, pp. 793–832.

Oosterhuis, F.H. and P. Ten Brink (eds.), 2014: *Paying the Polluter: Environmentally Harmful Subsidies and their Reform*. Edward Elgar Publishing, Cheltenham, UK and Northampton, MA, USA, 368 pp.

Oppenheimer, M., M. Campos, and R. Warren, 2014: Emergent risks and key vulnerabilities. In: *Climate Change 2014: Impacts, Adaptation, and Vulnerability. Part A: Global and Sectoral Aspects. Contribution of Working Group II to the Fifth Assessment Report of the Intergovernmental Panel on Climate Change* [Field, C.B., V.R. Barros, D.J. Dokken, K.J. Mach, M.D. Mastrandrea, T.E. Bilir, M. Chatterjee, K.L. Ebi, Y.O. Estrada, R.C. Genova, B. Girma, E.S. Kissel, A.N. Levy, S. MacCracken, P.R. Mastrandrea, and L.L. White (eds.)]. Cambridge University Press, Cambridge, United Kingdom and New York, NY, USA, pp. 659–708.

Orindi, V., Y. Elhadi, and C. Hesse, 2017: Democratising climate finance at local levels. In: *Building a Climate Resilient Economy and Society: Challenges and Opportunities* [Ninan, K.N. and M. Inoue (eds.)]. Edward Elgar Publishing, Cheltenham, UK and Northampton, MA, USA, pp. 250–264.

Otto, F.E.L., R.B. Skeie, J.S. Fuglestvedt, T. Berntsen, and M.R. Allen, 2017: Assigning historic responsibility for extreme weather events. *Nature Climate Change*, **7(11)**, 757–759, doi:10.1038/nclimate3419.

Page, E.A. and C. Heyward, 2017: Compensating for climate change Loss and Damage. *Political Studies*, **65(2)**, 356–372, doi:10.1177/0032321716647401.

Pal, J.S. and E.A.B. Eltahir, 2016: Future temperature in southwest Asia projected to exceed a threshold for human adaptability. *Nature Climate Change*, **18203**, 1–4, doi:10.1038/nclimate2833.

Palazzo, A. et al., 2017: Linking regional stakeholder scenarios and shared socioeconomic pathways: Quantified West African food and climate futures in a global context. *Global Environmental Change*, **45**, 227–242, doi:10.1016/j.gloenvcha.2016.12.002.

Pan, X., M. Elzen, N. Höhne, F. Teng, and L. Wang, 2017: Exploring fair and ambitious mitigation contributions under the Paris Agreement goals. *Environmental Science & Policy*, **74**, 49–56, doi:10.1016/j.envsci.2017.04.020.

Pandey, U.C. and C. Kumar, 2018: Emerging Paradigms of Capacity Building in the Context of Climate Change. In: *Climate Literacy and Innovations in Climate Change Education: Distance Learning for Sustainable Development* [Azeiteiro, U.M., W. Leal Filho, and L. Aires (eds.)]. Springer International Publishing, Cham, pp. 193–214, doi:10.1007/978-3-319-70199-8_11.

Parikh, K.S., J.K. Parikh, and P.P. Ghosh, 2018: Can India grow and live within a 1.5 degree CO_2 emissions budget? *Energy Policy*, **120**, 24–37, doi:10.1016/j.enpol.2018.05.014.

Park, J., S. Hallegatte, M. Bangalore, and E. Sandhoefner, 2015: Households and heat stress estimating the distributional consequences of climate change. Policy Research Working Paper no. WPS7479, World Bank Group, Washington, DC, USA, 58 pp.

Parkinson, S.C. et al., 2016: Impacts of Groundwater Constraints on Saudi Arabia's Low-Carbon Electricity Supply Strategy. *Environmental Science & Technology*, **50(4)**, 1653–1662, doi:10.1021/acs.est.5b05852.

Parnell, S., 2017: Africa's urban risk and resilience. *International Journal of Disaster Risk Reduction*, **26**, 1–6, doi:10.1016/j.ijdrr.2017.09.050.

Patterson, J.J. et al., 2016: Exploring the governance and politics of transformations towards sustainability. *Environmental Innovation and Societal Transitions*, 1–16, doi:10.1016/j.eist.2016.09.001.

Patterson, J.J. et al., 2018: Political feasibility of 1.5°C societal transformations: the role of social justice. *Current Opinion in Environmental Sustainability*, **31**, 1–9, doi:10.1016/j.cosust.2017.11.002.

Pearce, T., J. Ford, A.C. Willox, and B. Smit, 2015: Inuit Traditional Ecological Knowledge (TEK), subsistence hunting and adaptation to climate change in the Canadian Arctic. *Arctic*, **68(2)**, 233–245, doi:10.2307/43871322.

Pelling, M., T. Abeling, and M. Garschagen, 2016: Emergence and Transition in London's Climate Change Adaptation Pathways. *Journal of Extreme Events*, **3(3)**, 1650012, doi:10.1142/s2345737616500123.

Pelling, M. et al., 2018: Africa's urban adaptation transition under a 1.5° climate. *Current Opinion in Environmental Sustainability*, **31**, 10–15, doi:10.1016/j.cosust.2017.11.005.

Petkova, E.P. et al., 2017: Towards more comprehensive projections of urban heat-related mortality: estimates for New York city under multiple population, adaptation, and climate scenarios. *Environmental Health Perspectives*, **125(1)**, 47–55, doi:10.1289/ehp166.

Pfeiffer, A., R. Millar, C. Hepburn, and E. Beinhocker, 2016: The '2°C carbon stock' for electricity generation: cumulative committed carbon emissions from the electricity generation sector and the transition to a green economy. *Applied Energy*, **179**, 1395–1408, doi:10.1016/j.apenergy.2016.02.093.

Phillips, J., P. Newell, and A. Pueyo, 2017: Triple wins? Prospects for pro-poor, low carbon, climate resilient energy services in Kenya. In: *Making Climate Compatible Development Happen* [Nunan, F. (ed.)]. Routledge, Abingdon, UK and New York, NY, USA, pp. 114–129.

Popescu, G.H. and F.C. Ciurlau, 2016: Can environmental sustainability be attained by incorporating nature within the capitalist economy? *Economics, Management, and Financial Markets*, **11(4)**, 75–81.

Popp, A. et al., 2014: Land-use transition for bioenergy and climate stabilization: Model comparison of drivers, impacts and interactions with other land use based mitigation options. *Climatic Change*, **123(3–4)**, 495–509, doi:10.1007/s10584-013-0926-x.

Popp, A. et al., 2017: Land-use futures in the shared socio-economic pathways. *Global Environmental Change*, **42**, 331–345, doi:10.1016/j.gloenvcha.2016.10.002.

Potouroglou, M. et al., 2017: Measuring the role of seagrasses in regulating sediment surface elevation. *Scientific Reports*, **7(1)**, 1–11, doi:10.1038/s41598-017-12354-y.

Pradhan, A., C. Chan, P.K. Roul, J. Halbrendt, and B. Sipes, 2018: Potential of conservation agriculture (CA) for climate change adaptation and food security under rainfed uplands of India: a transdisciplinary approach. *Agricultural Systems*, **163**, 27–35, doi:10.1016/j.agsy.2017.01.002.

Preston, F. and J. Lehne, 2017: *A wider circle? The circular economy in developing countries*. Chatham House: The Royal Institute of International Affairs, London, 24 pp.

Pretis, F., M. Schwarz, K. Tang, K. Haustein, and M.R. Allen, 2018: Uncertain impacts on economic growth when stabilizing global temperatures at 1.5°C or 2°C warming. *Philosophical Transactions of the Royal Society A: Mathematical, Physical and Engineering Sciences*, **376(2119)**, 20160460, doi:10.1098/rsta.2016.0460.

Purdon, M., 2015: Opening the black box of carbon finance "additionality": the political economy of carbon finance effectiveness across Tanzania, Uganda, and Moldova. *World Development*, **74**, 462–478, doi:10.1016/j.worlddev.2015.05.024.

Quan, J., L.O. Naess, A. Newsham, A. Sitoe, and M.C. Fernandez, 2017: The Political Economy of REDD+ in Mozambique: Implications for Climate Compatible Development. In: *Making Climate Compatible Development Happen* [Nunan, F. (ed.)]. Routledge, Abingdon, UK and New York, NY, USA, pp. 151–181.

Rai, N. and S. Fisher (eds.), 2017: *The Political Economy of Low Carbon Resilient Development: Planning and implementation*. Routledge, Abingdon, UK and New York, NY, USA, 172 pp.

Rao, N.D., 2014: International and intranational equity in sharing climate change mitigation burdens. *International Environmental Agreements: Politics, Law and Economics*, **14(2)**, 129–146, doi:10.1007/s10784-013-9212-7.

5

Rao, N.D. and J. Min, 2018: Less global inequality can improve climate outcomes. *Wiley Interdisciplinary Reviews: Climate Change*, **9**, e513, doi:10.1002/wcc.513.

Rao, N.D., B.J. van Ruijven, V. Bosetti, and K. Riahi, 2017: Improving poverty and inequality modeling in climate research. *Nature Climate Change*, **7**, 857–862, doi:10.1038/s41558-017-0004-x.

Rao, S. et al., 2016: Future Air Pollution in the Shared Socio-Economic Pathways. *Global Environmental Change*, **42**, 346–358, doi:10.1016/j.gloenvcha.2016.05.012.

Rasch, R., 2017: Income Inequality and Urban Vulnerability to Flood Hazard in Brazil. *Social Science Quarterly*, **98(1)**, 299–325, doi:10.1111/ssqu.12274.

Rasul, G. and B. Sharma, 2016: The nexus approach to water-energy-food security: an option for adaptation to climate change. *Climate Policy*, **16(6)**, 682–702, doi:10.1080/14693062.2015.1029865.

Raworth, K., 2017: A Doughnut for the Anthropocene: humanity's compass in the 21st century. *The Lancet Planetary Health*, **1(2)**, e48–e49, doi:10.1016/s2542-5196(17)30028-1.

Redclift, M. and D. Springett (eds.), 2015: *Routledge International Handbook of Sustainable Development*. Routledge, Abingdon, UK and New York, NY, USA, 448 pp.

Regmi, B.R. and C. Star, 2015: Exploring the policy environment for mainstreaming community-based adaptation (CBA) in Nepal. *International Journal of Climate Change Strategies and Management*, **7(4)**, 423–441, doi:10.1108/ijccsm-04-2014-0050.

Reid, H., 2016: Ecosystem- and community-based adaptation: learning from community-based natural resource management. *Climate and Development*, **8(1)**, 4–9, doi:10.1080/17565529.2015.1034233.

Reid, H. and S. Huq, 2014: Mainstreaming community-based adaptation into national and local planning. *Climate and Development*, **6(4)**, 291–292, doi:10.1080/17565529.2014.973720.

Reij, C. and R. Winterbottom, 2015: *Scaling up Regreening: Six Steps to Success. A Practical Approach to Forest Landscape Restoration*. World Resource Institute (WRI), Washington DC, USA, 72 pp.

Republic of Vanuatu, 2016: *Vanuatu 2030: The People's Plan*. Government of the Republic of Vanuatu, Port Villa, Vanuatu, 28 pp.

Republic of Vanuatu, 2017: *Vanuatu 2030: The People's Plan. National Sustainable Development Plan 2016–2030. Monitoring and Evaluation Framework*. The Government of the Republic of Vanuatu, Port Vila, Vanuatu, 48 pp.

Revi, A. et al., 2014: Urban areas. In: *Climate Change 2014: Impacts, Adaptation, and Vulnerability. Part A: Global and Sectoral Aspects. Contribution of Working Group II to the Fifth Assessment Report of the Intergovernmental Panel on Climate Change* [Field, C.B., V.R. Barros, D.J. Dokken, K.J. Mach, M.D. Mastrandrea, T.E. Bilir, M. Chatterjee, K.L. Ebi, Y.O. Estrada, R.C. Genova, B. Girma, E.S. Kissel, A.N. Levy, S. MacCracken, P.R. Mastrandrea, and L.L. White (eds.)]. Cambridge University Press, Cambridge, UK, and New York, NY, USA, pp. 535–612.

Rey, T., L. Le De, F. Leone, and D. Gilbert, 2017: An integrative approach to understand vulnerability and resilience post-disaster: the 2015 cyclone Pam in urban Vanuatu as case study. *Disaster Prevention and Management: An International Journal*, **26(3)**, 259–275, doi:10.1108/dpm-07-2016-0137.

Reyer, C.P.O. et al., 2017a: Climate change impacts in Latin America and the Caribbean and their implications for development. *Regional Environmental Change*, **17(6)**, 1601–1621, doi:10.1007/s10113-015-0854-6.

Reyer, C.P.O. et al., 2017b: Turn down the heat: regional climate change impacts on development. *Regional Environmental Change*, **17(6)**, 1563–1568, doi:10.1007/s10113-017-1187-4.

Riahi, K. et al., 2015: Locked into Copenhagen pledges – Implications of short-term emission targets for the cost and feasibility of long-term climate goals. *Technological Forecasting and Social Change*, **90**, 8–23, doi:10.1016/j.techfore.2013.09.016.

Riahi, K. et al., 2017: The Shared Socioeconomic Pathways and their energy, land use, and greenhouse gas emissions implications: an overview. *Global Environmental Change*, **42**, 153–168, doi:10.1016/j.gloenvcha.2016.05.009.

Robertson, M., 2014: *Sustainability Principles and Practice*. Routledge, London, UK, 392 pp., doi:10.4324/9780203768747.

Robinson, M. and T. Shine, 2018: Achieving a climate justice pathway to 1.5°C. *Nature Climate Change*, **8(7)**, 564–569, doi:10.1038/s41558-018-0189-7.

Robinson, S. and M. Dornan, 2017: International financing for climate change adaptation in small island developing states. *Regional Environmental Change*, **17(4)**, 1103–1115, doi:10.1007/s10113-016-1085-1.

Robiou du Pont, Y. et al., 2017: Equitable mitigation to achieve the Paris Agreement goals. *Nature Climate Change*, **7(1)**, 38–43, doi:10.1038/nclimate3186.

Rockström, J. et al., 2017: Sustainable intensification of agriculture for human prosperity and global sustainability. *Ambio*, **46(1)**, 4–17, doi:10.1007/s13280-016-0793-6.

Rogelj, J. et al., 2018: Scenarios towards limiting global mean temperature increase below 1.5°C. *Nature Climate Change*, 1–8, doi:10.1038/s41558-018-0091-3.

Roger, C., T. Hale, and L. Andonova, 2017: The comparative politics of transnational climate governance. *International Interactions*, **43(1)**, 1–25, doi:10.1080/03050629.2017.1252248.

Rose, S.K. et al., 2014: Bioenergy in energy transformation and climate management. *Climatic Change*, **123(3–4)**, 477–493, doi:10.1007/s10584-013-0965-3.

Rosenbloom, D., 2017: Pathways: An emerging concept for the theory and governance of low-carbon transitions. *Global Environmental Change*, **43**, 37–50, doi:10.1016/j.gloenvcha.2016.12.011.

Rosenzweig, C. et al., 2018: *Climate Change and Cities: Second Assessment Report of the Urban Climate Change Research Network*. Cambridge University Press, Cambridge, United Kingdom and New York, NY, USA, 811 pp., doi:10.1017/9781316563878.

Roser, D. and C. Seidel, 2017: *Climate justice*. Routledge, Abingdon, UK, 230 pp.

Roser, D., C. Huggel, M. Ohndorf, and I. Wallimann-Helmer, 2015: Advancing the interdisciplinary dialogue on climate justice. *Climatic Change*, **133(3)**, 349–359, doi:10.1007/s10584-015-1556-2.

Rozenberg, J. and S. Hallegatte, 2016: Modeling the Impacts of Climate Change on Future Vietnamese Households: A Micro-Simulation Approach. World Bank Policy Research Working Paper No. 7766, World Bank, 20 pp.

Runhaar, H., B. Wilk, Persson, C. Uittenbroek, and C. Wamsler, 2018: Mainstreaming climate adaptation: taking stock about "what works" from empirical research worldwide. *Regional Environmental Change*, **18(4)**, 1201–1210, doi:10.1007/s10113-017-1259-5.

Rutledge, D. et al., 2017: *Identifying Feedbacks, Understanding Cumulative Impacts and Recognising Limits: A National Integrated Assessment. Synthesis Report RA3. Climate Changes, Impacts and Implications for New Zealand to 2100*. CCII (Climate Changes, Impacts & Implications), 84 pp.

Sachs, J., G. Schmidt-Traub, C. Kroll, D. Durand-Delacre, and K. Teksoz, 2017: *SDG Index and Dashboards Report 2017*. Bertelsmann Stiftung and Sustainable Development Solutions Network (SDSN), New York, NY, USA, 479 pp.

Salleh, A., 2016: Climate, Water, and Livelihood Skills: A Post-Development Reading of the SDGs. *Globalizations*, **13(6)**, 952–959, doi:10.1080/14747731.2016.1173375.

Sánchez, A. and M. Izzo, 2017: Micro hydropower: an alternative for climate change mitigation, adaptation, and development of marginalized local communities in Hispaniola Island. *Climatic Change*, **140(1)**, 79–87, doi:10.1007/s10584-016-1865-0.

Santarius, T., H.J. Walnum, and C. Aall (eds.), 2016: *Rethinking Climate and Energy Policies: New Perspectives on the Rebound Phenomenon*. Springer, Cham, Switzerland, 294 pp., doi:10.1007/978-3-319-38807-6.

Santos, P., P. Bacelar-Nicolau, M.A. Pardal, L. Bacelar-Nicolau, and U.M. Azeiteiro, 2016: Assessing Student Perceptions and Comprehension of Climate Change in Portuguese Higher Education Institutions. In: *Implementing Climate Change Adaptation in Cities and Communities: Integrating Strategies and Educational Approaches* [Filho, W.L., K. Adamson, R.M. Dunk, U.M. Azeiteiro, S. Illingworth, and F. Alves (eds.)]. Springer, Cham, Switzerland, pp. 221–236, doi:10.1007/978-3-319-28591-7_12.

Satterthwaite, D., 2016: Missing the Millennium Development Goal targets for water and sanitation in urban areas. *Environment and Urbanization*, **28(1)**, 99–118, doi:10.1177/0956247816628435.

Satterthwaite, D. et al., 2018: *Responding to climate change in cities and in their informal settlements and economies*. International Institute for Environment and Development, Edmonton, Canada, 61 pp.

Savo, V. et al., 2016: Observations of climate change among subsistence-oriented communities around the world. *Nature Climate Change*, **6(5)**, 462–473, doi:10.1038/nclimate2958.

Schade, J. and W. Obergassel, 2014: Human rights and the Clean Development Mechanism. *Cambridge Review of International Affairs*, **27(4)**, 717–735, doi:10.1080/09557571.2014.961407.

Schaeffer, M. et al., 2015: *Feasibility of limiting warming to 1.5 and 2°C*. Climate Analytics, Berlin, Germany, 20 pp.

Schaller, N. et al., 2016: Human influence on climate in the 2014 southern England winter floods and their impacts. *Nature Climate Change*, **6(6)**, 627–634, doi:10.1038/nclimate2927.

5

Schirmer, J. and L. Bull, 2014: Assessing the likelihood of widespread landholder adoption of afforestation and reforestation projects. *Global Environmental Change*, **24**, 306–320, doi:10.1016/j.gloenvcha.2013.11.009.

Schleussner, C.-F. et al., 2016: Differential climate impacts for policy-relevant limits to global warming: the case of 1.5°C and 2°C. *Earth System Dynamics*, **7(2)**, 327–351, doi:10.5194/esd-7-327-2016.

Schlosberg, D., L.B. Collins, and S. Niemeyer, 2017: Adaptation policy and community discourse: risk, vulnerability, and just transformation. *Environmental Politics*, **26(3)**, 1–25, doi:10.1080/09644016.2017.1287628.

Schwan, S. and X. Yu, 2017: Social protection as a strategy to address climate-induced migration. *International Journal of Climate Change Strategies and Management*, 10(1), 43–64, doi:10.1108/ijccsm-01-2017-0019.

Scoones, I. et al., 2017: *Pathways to Sustainable Agriculture*. Routledge, Abingdon, UK and New York, NY, USA, 132 pp.

Scott, F.L., C.R. Jones, and T.L. Webb, 2014: What do people living in deprived communities in the UK think about household energy efficiency interventions? *Energy Policy*, **66**, 335–349, doi:10.1016/j.enpol.2013.10.084.

Sealey-Huggins, L., 2017: '1.5°C to stay alive': climate change, imperialism and justice for the Caribbean. *Third World Quarterly*, **6597**, 1–20, doi:10.1080/01436597.2017.1368013.

Serdeczny, O.M., S. Bauer, and S. Huq, 2017: Non-economic losses from climate change: opportunities for policy-oriented research. *Climate and Development*, 1–5, doi:10.1080/17565529.2017.1372268.

Seto, K.C. et al., 2016: Carbon Lock-In: Types, Causes, and Policy Implications. *Annual Review of Environment and Resources*, **41(1)**, 425–452, doi:10.1146/annurev-environ-110615-085934.

Sgouridis, S. et al., 2016: RE-mapping the UAE's energy transition: an economy-wide assessment of renewable energy options and their policy implications. *Renewable and Sustainable Energy Reviews*, **55**, 1166–1180, doi:10.1016/j.rser.2015.05.039.

Shackleton, S., G. Ziervogel, S. Sallu, T. Gill, and P. Tschakert, 2015: Why is socially-just climate change adaptation in sub-Saharan Africa so challenging? A review of barriers identified from empirical cases. *Wiley Interdisciplinary Reviews: Climate Change*, **6(3)**, 321–344, doi:10.1002/wcc.335.

Shaffrey, L.C. et al., 2009: U.K. HiGEM: the new U.K. high-resolution global environment model – model description and basic evaluation. *Journal of Climate*, **22(8)**, 1861–1896, doi:10.1175/2008jcli2508.1.

Shah, N., M. Wei, V. Letschert, and A. Phadke, 2015: *Benefits of leapfrogging to superefficiency and low global warming potential refrigerants in room air conditioning*. LBNL-1003671, Ernest Orlando Lawrence Berkeley National Laboratory(LBNL), Berkeley, CA, USA, 58 pp.

Sharpe, R.A., C.R. Thornton, V. Nikolaou, and N.J. Osborne, 2015: Higher energy efficient homes are associated with increased risk of doctor diagnosed asthma in a UK subpopulation. *Environment International*, **75**, 234–244, doi:10.1016/j.envint.2014.11.017.

Shaw, C., S. Hales, P. Howden-Chapman, and R. Edwards, 2014: Health co-benefits of climate change mitigation policies in the transport sector. *Nature Climate Change*, **4(6)**, 427–433, doi:10.1038/nclimate2247.

Shi, L. et al., 2016: Roadmap towards justice in urban climate adaptation research. *Nature Climate Change*, **6(2)**, 131–137, doi:10.1038/nclimate2841.

Shi, Y., J. Liu, H. Shi, H. Li, and Q. Li, 2017: The ecosystem service value as a new eco-efficiency indicator for industrial parks. *Journal of Cleaner Production*, **164**, 597–605, doi:10.1016/j.jclepro.2017.06.187.

Shindell, D.T., G. Faluvegi, K. Seltzer, and C. Shindell, 2018: Quantified, localized health benefits of accelerated carbon dioxide emissions reductions. *Nature Climate Change*, **8(4)**, 291–295, doi:10.1038/s41558-018-0108-y.

Shindell, D.T. et al., 2017: A climate policy pathway for near- and long-term benefits. *Science*, **356(6337)**, 493–494, doi:10.1126/science.aak9521.

Shine, T., 2017: *Integrating climate action into national development planning – Coherent Implementation of the Paris Agreement and Agenda 2030*. Swedish International Development Cooperation Agency (SIDA), Stockholm, Sweden, 26 pp.

Shine, T. and G. Campillo, 2016: The Role of Development Finance in Climate Action Post-2015. OECD Development Co-operation Working Papers, No. 31, OECD Publishing, Paris, France, 38 pp.

Shue, H., 2014: *Climate Justice: Vulnerability and Protection*. Oxford University Press, Oxford, UK, 368 pp.

Shue, H., 2017: Responsible for what? Carbon producer CO_2 contributions and the energy transition. *Climatic Change*, 1–6, doi:10.1007/s10584-017-2042-9.

Shue, H., 2018: Mitigation gambles: uncertainty, urgency and the last gamble possible. *Philosophical Transactions of the Royal Society A: Mathematical, Physical and Engineering Sciences*, **376(2119)**, 20170105, doi:10.1098/rsta.2017.0105.

Shultz, J.M., M.A. Cohen, S. Hermosilla, Z. Espinel, and Andrew McLean, 2016: Disaster risk reduction and sustainable development for small island developing states. *Disaster Health*, **3(1)**, 32–44, doi:10.1080/21665044.2016.1173443.

Sinclair, F.L., 2016: Systems science at the scale of impact: Reconciling bottom-up participation with the production of widely applicable research outputs. In: *Sustainable Intensification in Smallholder Agriculture: An Integrated Systems Research Approach* [Oborn, I., B. Vanlauwe, M. Phillips, R. Thomas, W. Brooijmans, and K. Atta-Krah (eds.)]. Earthscan, London, UK, pp. 43–57.

Singh, C., 2018: Is participatory watershed development building local adaptive capacity? Findings from a case study in Rajasthan, India. *Environmental Development*, **25**, 43–58, doi:10.1016/j.envdev.2017.11.004.

Singh, C. et al., 2017: The utility of weather and climate information for adaptation decision-making: current uses and future prospects in Africa and India. *Climate and Development*, 1–17, doi:10.1080/17565529.2017.1318744.

Singh, R., K.K. Garg, S.P. Wani, R.K. Tewari, and S.K. Dhyani, 2014: Impact of water management interventions on hydrology and ecosystem services in Garhkundar-Dabar watershed of Bundelkhand region, Central India. *Journal of Hydrology*, **509**, 132–149, doi:10.1016/j.jhydrol.2013.11.030.

Skeie, R.B. et al., 2017: Perspective has a strong effect on the calculation of historical contributions to global warming. *Environmental Research Letters*, **12(2)**, 024022, doi:10.1088/1748-9326/aa5b0a.

SLoCaT, 2017: *Marrakech Partnership for Global Climate Action (MPGCA) Transport Initiatives: Stock-take on action toward implementation of the Paris Agreement and the 2030 Agenda on Sustainable Development. Second Progress Report*. Partnership on Sustainable Low Carbon Transport (SLoCaT), Bonn, Germany, 72 pp.

Smith, K.R. and A. Sagar, 2014: Making the clean available: escaping India's Chulha Trap. *Energy Policy*, **75**, 410–414, doi:10.1016/j.enpol.2014.09.024.

Smith, K.R. et al., 2014: Human health: impacts, adaptation, and co-benefits. In: *Climate Change 2014: Impacts, Adaptation, and Vulnerability. Part A: Global and Sectoral Aspects. Contribution of Working Group II to the Fifth Assessment Report of the Intergovernmental Panel on Climate Change* [Field, C.B., V.R. Barros, D.J. Dokken, K.J. Mach, M.D. Mastrandrea, T.E. Bilir, M. Chatterjee, K.L. Ebi, Y.O. Estrada, R.C. Genova, B. Girma, E.S. Kissel, A.N. Levy, S. MacCracken, P.R. Mastrandrea, and L.L. White (eds.)]. Cambridge University Press, Cambridge, United Kingdom and New York, NY, USA, pp. 709–754.

Smith, P. et al., 2014: Agriculture, Forestry and Other Land Use (AFOLU). In: *Climate Change 2014: Mitigation of Climate Change. Contribution of Working Group III to the Fifth Assessment Report of the Intergovernmental Panel on Climate Change* [Edenhofer, O., R. Pichs-Madruga, Y. Sokona, E. Farahani, S. Kadner, K. Seyboth, A. Adler, I. Baum, S. Brunner, P. Eickemeier, B. Kriemann, J. Savolainen, S. Schlömer, C. Stechow, T. Zwickel, and J.C. Minx (eds.)]. Cambridge University Press, Cambridge, United Kingdom and New York, NY, USA, pp. 811–922.

Smith, P. et al., 2016: Biophysical and economic limits to negative CO_2 emissions. *Nature Climate Change*, **6(1)**, 42–50, doi:10.1038/nclimate2870.

Smits, M. and C. Middleton, 2014: New Arenas of Engagement at the Water Governance-Climate Finance Nexus? An Analysis of the Boom and Bust of Hydropower CDM Projects in Vietnam. *Water Alternatives*, **7(3)**, 561–583, www.water-alternatives.org/index.php/alldoc/articles/vol7/v8issue3/264-a7-3-7/file.

Smucker, T.A. et al., 2015: Differentiated livelihoods, local institutions, and the adaptation imperative: Assessing climate change adaptation policy in Tanzania. *Geoforum*, **59**, 39–50, doi:10.1016/j.geoforum.2014.11.018.

Solecki, W. et al., 2018: City transformations in a 1.5°C warmer world. *Nature Climate Change*, **8(3)**, 177–181, doi:10.1038/s41558-018-0101-5.

Sondak, C.F.A. et al., 2017: Carbon dioxide mitigation potential of seaweed aquaculture beds (SABs). *Journal of Applied Phycology*, **29(5)**, 2363–2373, doi:10.1007/s10811-016-1022-1.

Sonwa, D.J. et al., 2017: Drivers of climate risk in African agriculture. *Climate and Development*, **9(5)**, 383–398, doi:10.1080/17565529.2016.1167659.

Sorrell, S., 2015: Reducing energy demand: A review of issues, challenges and approaches. *Renewable and Sustainable Energy Reviews*, **47**, 74–82, doi:10.1016/j.rser.2015.03.002.

Sovacool, B.K., B.-O. Linnér, and M.E. Goodsite, 2015: The political economy of climate adaptation. *Nature Climate Change*, **5(7)**, 616–618, doi:10.1038/nclimate2665.

Sovacool, B.K., B.-O. Linnér, and R.J.T. Klein, 2017: Climate change adaptation and the Least Developed Countries Fund (LDCF): qualitative insights from policy implementation in the Asia-Pacific. *Climatic Change*, **140(2)**, 209–226, doi:10.1007/s10584-016-1839-2.

5

SPC, 2015: *Vanuatu climate change and disaster risk reduction policy 2016–2030*. Secretariat of the Pacific Community (SPC), Suva, Fiji, 48 pp.

SPC, 2016: *Framework for Resilient Development in the Pacific: An Integrated Approach to Address Climate Change and Disaster Risk Management (FRDP) 2017–2030*. Pacific Community (SPC), Suva, Fiji, 40 pp.

Spencer, T. et al., 2018: The 1.5°C target and coal sector transition: at the limits of societal feasibility. *Climate Policy*, **18(3)**, 335–331, doi:10.1080/14693062.2017.1386540.

Staggenborg, S. and C. Ogrodnik, 2015: New environmentalism and Transition Pittsburgh. *Environmental Politics*, **24(5)**, 723–741, doi:10.1080/09644016.2015.1027059.

Steering Committee on Partnerships for SIDS and UN DESA, 2016: *Partnerships for Small Island Developing States 2016*. The Steering Committee on Partnerships for Small Island Developing States and the United Nations Department of Economic and Social Affairs, 49 pp.

Sterrett, C.L., 2015: *Final evaluation of the Vanuatu NGO Climate Change Adaptation Program*. Climate Concern, 96 pp.

Stevens, C., R. Winterbottom, J. Springer, and K. Reytar, 2014: *Securing Rights, Combating Climate Change: How Strengthening Community Forest Rights Mitigates Climate Change*. World Resources Institute, Washington DC, USA, 64 pp.

Stirling, A., 2014: *Emancipating Transformations: From controlling 'the transition' to culturing plural radical progress*. STEPS Centre (Social, Technological and Environmental Pathways to Sustainability), Brighton, UK, 48 pp.

Striessnig, E. and E. Loichinger, 2015: Future differential vulnerability to natural disasters by level of education. *Vienna Yearbook of Population Research*, **13**, 221–240, www.jstor.org/stable/24770031.

Stringer, L.C., S.M. Sallu, A.J. Dougill, B.T. Wood, and L. Ficklin, 2017: Reconsidering climate compatible development as a new development landscape in southern Africa. In: *Making Climate Compatible Development Happen* [Nunan, F. (ed.)]. Routledge, Abingdon, UK and New York, NY, USA, pp. 22–43.

Stringer, L.C. et al., 2014: Advancing climate compatible development: Lessons from southern Africa. *Regional Environmental Change*, **14(2)**, 713–725, doi:10.1007/s10113-013-0533-4.

Suckall, N., E. Tompkins, and L. Stringer, 2014: Identifying trade-offs between adaptation, mitigation and development in community responses to climate and socio-economic stresses: Evidence from Zanzibar, Tanzania. *Applied Geography*, **46**, 111–121, doi:10.1016/j.apgeog.2013.11.005.

Suckall, N., L.C. Stringer, and E.L. Tompkins, 2015: Presenting Triple-Wins? Assessing Projects That Deliver Adaptation, Mitigation and Development Co-benefits in Rural Sub-Saharan Africa. *Ambio*, **44(1)**, 34–41, doi:10.1007/s13280-014-0520-0.

Suffolk, C. and W. Poortinga, 2016: Behavioural changes after energy efficiency improvements in residential properties. In: *Rethinking Climate and Energy Policies: New Perspectives on the Rebound Phenomenon* [Santarius, T., H.J. Walnum, and C. Aall (eds.)]. Springer International Publishing, Cham, Switzerland, pp. 121–142, doi:10.1007/978-3-319-38807-6_8.

Surminski, S., L.M. Bouwer, and J. Linnerooth-Bayer, 2016: How insurance can support climate resilience. *Nature Climate Change*, **6(4)**, 333–334, doi:10.1038/nclimate2979.

Szabo, S. et al., 2015: Sustainable Development Goals Offer New Opportunities for Tropical Delta Regions. *Environment: Science and Policy for Sustainable Development*, **57(4)**, 16–23, doi:10.1080/00139157.2015.1048142.

Tàbara, J.D., A.L. St. Clair, and E.A.T. Hermansen, 2017: Transforming communication and knowledge production processes to address high-end climate change. *Environmental Science & Policy*, **70**, 31–37, doi:10.1016/j.envsci.2017.01.004.

Tàbara, J.D. et al., 2018: Positive tipping points in a rapidly warming world. *Current Opinion in Environmental Sustainability*, **31**, 120–129, doi:10.1016/j.cosust.2018.01.012.

Tait, L. and M. Euston-Brown, 2017: What role can African cities play in low-carbon development? A multilevel governance perspective of Ghana, Uganda and South Africa. *Journal of Energy in Southern Africa*, **28(3)**, 43–53, doi:10.17159/2413-3051/2017/v28i3a1959.

Tanner, T. et al., 2017: Political economy of climate compatible development: artisanal fisheries and climate change in Ghana. In: *Making Climate Compatible Development Happen* [Nunan, F. (ed.)]. Routledge, Abingdon, UK and New York, NY, USA, pp. 223–241, doi:10.1111/j.2040-0209.2014.00446.x.

Taylor, M., 2017: Climate-smart agriculture: what is it good for? *The Journal of Peasant Studies*, **45(1)**, 89–107, doi:10.1080/03066150.2017.1312355.

Taylor Aiken, G., 2015: (Local-) community for global challenges: carbon conversations, transition towns and governmental elisions. *Local Environment*, **20(7)**, 764–781, doi:10.1080/13549839.2013.870142.

Taylor Aiken, G., 2016: Prosaic state governance of community low carbon transitions. *Political Geography*, **55**, 20–29, doi:10.1016/j.polgeo.2016.04.002.

Taylor Aiken, G., L. Middlemiss, S. Sallu, and R. Hauxwell-Baldwin, 2017: Researching climate change and community in neoliberal contexts: an emerging critical approach. *Wiley Interdisciplinary Reviews: Climate Change*, **8(4)**, e463, doi:10.1002/wcc.463.

Teferi, Z.A. and P. Newman, 2018: Slum upgrading: can the 1.5°C carbon reduction work with SDGs in these settlements? *Urban Planning*, **3(2)**, 52–63, doi:10.17645/up.v3i2.1239.

Thomas, A. and L. Benjamin, 2017: Management of loss and damage in small island developing states: implications for a 1.5°C or warmer world. *Regional Environmental Change*, **17(81)**, 1–10, doi:10.1007/s10113-017-1184-7.

Thompson-Hall, M., E.R. Carr, and U. Pascual, 2016: Enhancing and expanding intersectional research for climate change adaptation in agrarian settings. *Ambio*, **45(s3)**, 373–382, doi:10.1007/s13280-016-0827-0.

Thornton, T.F. and C. Comberti, 2017: Synergies and trade-offs between adaptation, mitigation and development. *Climatic Change*, 1–14, doi:10.1007/s10584-013-0884-3.

Thorp, T.M., 2014: *Climate Justice: A Voice for the Future*. Palgrave Macmillan UK, New York, 439 pp., doi:10.1057/9781137394644.

Tilman, D. and M. Clark, 2014: Global diets link environmental sustainability and human health. *Nature*, **515(7528)**, 518–522, doi:10.1038/nature13959.

Tokar, B., 2014: *Toward climate justice: perspectives on the climate crisis and social change (2nd edition)*. New Compass Press, Porsgrunn, Norway, 182 pp.

Toole, S., N. Klocker, and L. Head, 2016: Re-thinking climate change adaptation and capacities at the household scale. *Climatic Change*, **135(2)**, 203–209, doi:10.1007/s10584-015-1577-x.

Tschakert, P. et al., 2016: Micropolitics in collective learning spaces for adaptive decision making. *Global Environmental Change*, **40**, 182–194, doi:10.1016/j.gloenvcha.2016.07.004.

Tschakert, P. et al., 2017: Climate change and loss, as if people mattered: Values, places, and experiences. *Wiley Interdisciplinary Reviews: Climate Change*, **8(5)**, e476, doi:10.1002/wcc.476.

Tung, R.L., 2016: Opportunities and Challenges Ahead of China's "New Normal". *Long Range Planning*, **49(5)**, 632–640, doi:10.1016/j.lrp.2016.05.001.

UN, 2014: *World urbanisation prospects, 2014 revisions*. Department of Economic and Social Affairs, New York, NY, USA.

UN, 2015a: *The Millennium Development Goals Report 2015*. United Nations (UN), New York, NY, USA, 75 pp.

UN, 2015b: *Transforming Our World: The 2030 Agenda for Sustainable Development*. A/RES/70/1, United Nations General Assembly (UNGA), New York, 35 pp.

UN Women, 2016: *Time to Act on Gender, Climate Change and Disaster Risk Reduction: An overview of progress in the Pacific region with evidence from The Republic of Marshall Islands, Vanuatu and Samoa*. United Nations Entity for Gender Equality and the Empowerment of Women (UN Women) Regional Office for Asia and the Pacific, Bangkok, Thailand, 92 pp.

UN Women and MRFCJ, 2016: *The Full View: Ensuring a comprehensive approach to achieve the goal of gender balance in the UNFCCC process*. UN Women and Mary Robinson Foundation Climate Justice (MRFCJ), 80 pp.

UNDP, 2016: *Risk Governance: Building Blocks for Resilient Development in the Pacific*. Policy Brief: October 2016, United Nations Development Programme (UNDP) and Pacific Risk Resilience Programme (PRRP), Suva, Fiji, 20 pp.

UNEP, 2017: *The Emissions Gap Report 2017*. United Nations Environment Programme (UNEP), Nairobi, Kenya, 89 pp.

UNFCCC, 2011a: Decision 1/CP.16: *The Cancun Agreements: Outcome of the work of the Ad Hoc Working Group on Long-term Cooperative Action under the Convention*. United Nations Framework Convention on Climate Change (UNFCCC), 31 pp.

UNFCCC, 2011b: Decision 7/CP.17: Work programme on loss and damage. In: *Report of the Conference of the Parties on its seventeenth session, held in Durban from 28 November to 11 December 2011. Addendum. Part Two: Action taken by the Conference of the Parties at its seventeenth session*. FCCC/CP/2011/9/Add.2, United Nations Framework Convention on Climate Change (UNFCCC), pp. 5–8.

UNFCCC, 2014: Decision 2/CP.19: Warsaw international mechanism for loss and damage associated with climate change impact. In: *Report of the Conference of the Parties on its nineteenth session, held in Warsaw from 11 to 23 November 2013. Addendum. Part two: Action taken by the Conference of the Parties at its nineteenth session*. FCCC/CP/2013/10/Add.1, United Nations Framework Convention on Climate Change (UNFCCC), pp. 6–8.

UNFCCC, 2015: *Adoption of the Paris Agreement*. FCCC/CP/2015/L.9, United Nations Framework Convention on Climate Change (UNFCCC), 32 pp.

UNGA, 2014: *Resolution adopted by the General Assembly on 14 November 2014: SIDS accelerated modalities of action (SAMOA) pathway*. A/RES/69/15, United Nations General Assembly (UNGA), 30 pp.

UNRISD, 2016: *Policy Innovations for Transformative Change: Implementing the 2030 Agenda for Sustainable Development*. United Nations Research Institute for Social Development (UNRISD), Geneva, Switzerland, 248 pp.

UNU-EHS, 2016: *World Risk Report 2016. Focus: Logistics and infrastructure*. United Nations University, Institute for Environment and Human Security (UNU-EHS), Bonn, Germany, 74 pp.

Ürge-Vorsatz, D., S.T. Herrero, N.K. Dubash, and F. Lecocq, 2014: Measuring the co-benefits of climate change mitigation. *Annual Review of Environment and Resources*, **39**, 549–582, doi:10.1146/annurev-environ-031312-125456.

Ürge-Vorsatz, D. et al., 2016: Measuring multiple impacts of low-carbon energy options in a green economy context. *Applied Energy*, **179**, 1409–1426, doi:10.1016/j.apenergy.2016.07.027.

Ustadi, I., T. Mezher, and M.R.M. Abu-Zahra, 2017: The effect of the carbon capture and storage (CCS) technology deployment on the natural gas market in the United Arab Emirates. *Energy Procedia*, **114**, 6366–6376, doi:10.1016/j.egypro.2017.03.1773.

Vale, L.J., 2014: The politics of resilient cities: whose resilience and whose city? *Building Research and Information*, **42(2)**, 191–201, doi:10.1080/09613218.2014.850602.

Van Aelst, K. and N. Holvoet, 2016: Intersections of gender and marital status in accessing climate change adaptation: evidence from rural Tanzania. *World Development*, **79**, 40–50, doi:10.1016/j.worlddev.2015.11.003.

Van de Graaf, T. and A. Verbruggen, 2015: The oil endgame: strategies of oil exporters in a carbon-constrained world. *Environmental Science & Policy*, **54**, 456–462, doi:10.1016/j.envsci.2015.08.004.

van der Geest, K. and K. Warner, 2015: Editorial: Loss and damage from climate change: emerging perspectives. *International Journal of Global Warming*, **8(2)**, 133–140, doi:10.1504/ijgw.2015.071964.

Van der Heijden, J., 2018: The limits of voluntary programs for low-carbon buildings for staying under 1.5°C. *Current Opinion in Environmental Sustainability*, **30**, 59–66, doi:10.1016/j.cosust.2018.03.006.

van der Voorn, T., J. Quist, C. Pahl-Wostl, and M. Haasnoot, 2017: Envisioning robust climate change adaptation futures for coastal regions: a comparative evaluation of cases in three continents. *Mitigation and Adaptation Strategies for Global Change*, **22(3)**, 519–546, doi:10.1007/s11027-015-9686-4.

van Noorloos, F. and M. Kloosterboer, 2017: Africa's new cities: The contested future of urbanisation. *Urban Studies*, 1–19, doi:10.1177/0042098017700574.

van Vuuren, D.P. et al., 2017a: The Shared Socio-economic Pathways: Trajectories for human development and global environmental change. *Global Environmental Change*, **42**, 148–152, doi:10.1016/j.gloenvcha.2016.10.009.

van Vuuren, D.P. et al., 2017b: Energy, land-use and greenhouse gas emissions trajectories under a green growth paradigm. *Global Environmental Change*, **42**, 237–250, doi:10.1016/j.gloenvcha.2016.05.008.

van Vuuren, D.P. et al., 2018: Alternative pathways to the 1.5°C target reduce the need for negative emission technologies. *Nature Climate Change*, **8**, 1–7, doi:10.1038/s41558-018-0119-8.

Vanhala, L. and C. Hestbaek, 2016: Framing climate change Loss and Damage in UNFCCC negotiations. *Global Environmental Politics*, **16(4)**, 111–129, doi:10.1162/glep_a_00379.

Vardakoulias, O. and N. Nicholles, 2014: *Managing uncertainty: An economic evaluation of community-based adaptation in Dakoro, Niger*. nef consulting, London, UK, 53 pp.

Veland, S. et al., 2018: Narrative matters for sustainability: the transformative role of storytelling in realizing 1.5°C futures. *Current Opinion in Environmental Sustainability*, **31**, 41–47, doi:10.1016/j.cosust.2017.12.005.

Vermeulen, S. et al., 2016: *The Economic Advantage: Assessing the value of climate change actions in agriculture*. International Fund for Agricultural Development (IFAD), Rome, Italy, 77 pp.

Vierros, M., 2017: Communities and blue carbon: the role of traditional management systems in providing benefits for carbon storage, biodiversity conservation and livelihoods. *Climatic Change*, **140(1)**, 89–100, doi:10.1007/s10584-013-0920-3.

Vinyeta, K., K.P. Whyte, and K. Lynn, 2015: *Climate Change Through an Intersectional Lens: Gendered Vulnerability and Resilience in Indigenous Communities in the United States*. General Technical Report PNW-GTR-923, US Department of Agriculture Forest Service, Pacific Northwest Research Station, Corvallis, Oregon, USA, 72 pp.

Vogt-Schilb, A. and S. Hallegatte, 2017: Climate Policies and Nationally Determined Contributions: Reconciling the Needed Ambition with the Political Economy. IDB Working Paper Series No. IDB-WP-818, Inter-American Development Bank, Washington DC, USA, 35 pp., doi:10.18235/0000714.

Volz, U. et al., 2015: *Financing the Green Transformation: How to Make Green Finance Work in Indonesia*. Palgrave Macmillan, Basingstoke, Hampshire, UK, 174 pp., doi:10.1057/9781137486127.

von Stechow, C. et al., 2015: Integrating Global Climate Change Mitigation Goals with Other Sustainability Objectives: A Synthesis. *Annual Review of Environment and Resources*, **40(1)**, 363–394, doi:10.1146/annurev-environ-021113-095626.

von Stechow, C. et al., 2016: 2°C and SDGs: United they stand, divided they fall? *Environmental Research Letters*, **11(3)**, 034022, doi:10.1088/1748-9326/11/3/034022.

Wachsmuth, D., D. Cohen, and H. Angelo, 2016: Expand the frontiers of urban sustainability. *Nature*, **536**, 391–393, doi:10.1038/536391a.

Waisman, H.-D., C. Guivarch, and F. Lecocq, 2013: The transportation sector and low-carbon growth pathways: modelling urban, infrastructure, and spatial determinants of mobility. *Climate Policy*, **13(sup01)**, 106–129, doi:10.1080/14693062.2012.735916.

Wallimann-Helmer, I., 2015: Justice for climate loss and damage. *Climatic Change*, **133(3)**, 469–480, doi:10.1007/s10584-015-1483-2.

Walsh-Dilley, M. and W. Wolford, 2015: (Un)Defining resilience: subjective understandings of 'resilience' from the field. *Resilience*, **3(3)**, 173–182, doi:10.1080/21693293.2015.1072310.

Wamsler, C., C. Luederitz, E. Brink, C. Wamsler, and C. Luederitz, 2014: Local levers for change: mainstreaming ecosystem-based adaptation into municipal planning to foster sustainability transitions. *Global Environmental Change*, **29**, 189–201, doi:10.1016/j.gloenvcha.2014.09.008.

Wang, X., 2017: The role of attitudinal motivations and collective efficacy on Chinese consumers' intentions to engage in personal behaviors to mitigate climate change. *The Journal of Social Psychology*, 1–13, doi:10.1080/00224545.2017.1302401.

Wang, Y., Q. Song, J. He, and Y. Qi, 2015: Developing low-carbon cities through pilots. *Climate Policy*, **15**, 81–103, doi:10.1080/14693062.2015.1050347.

Wanner, T., 2014: The new 'Passive Revolution' of the green economy and growth discourse: Maintaining the 'Sustainable Development' of Neoliberal capitalism. *New Political Economy*, **20(1)**, 1–21, doi:10.1080/13563467.2013.866081.

Warner, B.P. and C.P. Kuzdas, 2017: The role of political economy in framing and producing transformative adaptation. *Current Opinion in Environmental Sustainability*, **29**, 69–74, doi:10.1016/j.cosust.2017.12.012.

Warner, B.P., C. Kuzdas, M.G. Yglesias, and D.L. Childers, 2015: Limits to adaptation to interacting global change risks among smallholder rice farmers in Northwest Costa Rica. *Global Environmental Change*, **30**, 101–112, doi:10.1016/j.gloenvcha.2014.11.002.

Warner, K. and K. Geest, 2013: Loss and damage from climate change: local-level evidence from nine vulnerable countries. *International Journal of Global Warming*, **5(4)**, 367–386, doi:10.1504/ijgw.2013.057289.

Watts, N. et al., 2015: Health and climate change: policy responses to protect public health. *The Lancet*, **386(10006)**, 1861–1914, doi:10.1016/s0140-6736(15)60854-6.

Wegner, G.I., 2016: Payments for ecosystem services (PES): a flexible, participatory, and integrated approach for improved conservation and equity outcomes. *Environment, Development and Sustainability*, **18(3)**, 617–644, doi:10.1007/s10668-015-9673-7.

Weisser, F., M. Bollig, M. Doevenspeck, and D. Müller-Mahn, 2014: Translating the 'adaptation to climate change' paradigm: the politics of a travelling idea in Africa. *The Geographical Journal*, **180(2)**, 111–119, doi:10.1111/geoj.12037.

Wells, E.M. et al., 2015: Indoor air quality and occupant comfort in homes with deep versus conventional energy efficiency renovations. *Building and Environment*, **93(Part 2)**, 331–338, doi:10.1016/j.buildenv.2015.06.021.

Welsch, M. et al., 2014: Adding value with CLEWS – Modelling the energy system and its interdependencies for Mauritius. *Applied Energy*, **113**, 1434–1445, doi:10.1016/j.apenergy.2013.08.083.

Weng, X., Z. Dong, Q. Wu, and Y. Qin, 2015: *China's path to a green economy: decoding China's green economy concepts and policies*. IIED Country Report, International Institute for Environment and Development (IIED), London, UK, 40 pp.

Werfel, S.H., 2017: Household behaviour crowds out support for climate change policy when sufficient progress is perceived. *Nature Climate Change*, **7(7)**, 512–515, doi:10.1038/nclimate3316.

Wesseling, J.H. et al., 2017: The transition of energy intensive processing industries towards deep decarbonization: Characteristics and implications for future research. *Renewable and Sustainable Energy Reviews*, **79**, 1303–1313, doi:10.1016/j.rser.2017.05.156.

Weston, P., R. Hong, C. Kaboré, and C.A. Kull, 2015: Farmer-managed natural regeneration enhances rural livelihoods in dryland West Africa. *Environmental Management*, **55(6)**, 1402–1417, doi:10.1007/s00267-015-0469-1.

Wewerinke-Singh, M., 2018a: Climate migrants' right to enjoy their culture. In: *Climate Refugees: Beyond the Legal Impasse?* [Behrman, S. and A. Kent (eds.)]. Earthscan/Routledge, Abingdon, UK and New York, NY, USA, pp. 194–213.

Wewerinke-Singh, M., 2018b: State Responsibility for Human Rights Violations Associated with Climate Change. In: *Routledge Handbook of Human Rights and Climate Governance* [Duyck, S., S. Jodoin, and A. Johl (eds.)]. Routledge, Abingdon, UK and New York, NY, USA, pp. 75–89.

WHO, 2014: Quantitative risk assessment of the effects of climate change on selected causes of death, 2030s and 2050s. [Hales, S., S. Kovats, S. Lloyd, and D. Campbell-Lendrum (eds.)]. World Health Organization (WHO), Geneva, Switzerland, 128 pp.

Wiebe, K. et al., 2015: Climate change impacts on agriculture in 2050 under a range of plausible socioeconomic and emissions scenarios. *Environmental Research Letters*, **10(8)**, 085010, doi:10.1088/1748-9326/10/8/085010.

Wiktorowicz, J., T. Babaeff, J. Eggleston, and P. Newman, 2018: WGV: an Australian urban precinct case study to demonstrate the 1.5C agenda including multiple SDGs. *Urban Planning*, **3(2)**, 64–81, doi:10.17645/up.v3i2.1245.

Wilkinson, E., A. Kirbyshire, L. Mayhew, P. Batra, and A. Milan, 2016: *Climate-induced migration and displacement: closing the policy gap*. Overseas Development Institute (ODI), London, UK, 12 pp.

Willand, N., I. Ridley, and C. Maller, 2015: Towards explaining the health impacts of residential energy efficiency interventions – A realist review. Part 1: Pathways. *Social Science and Medicine*, **133**, 191–201, doi:10.1016/j.socscimed.2015.02.005.

Winkler, H. and N.K. Dubash, 2016: Who determines transformational change in development and climate finance? *Climate Policy*, **16(6)**, 783–791, doi:10.1080/14693062.2015.1033674.

Winkler, H. et al., 2018: Countries start to explain how their climate contributions are fair: more rigour needed. *International Environmental Agreements: Politics, Law and Economics*, **18(1)**, 99–115, doi:10.1007/s10784-017-9381-x.

Winsemius, H.C. et al., 2018: Disaster risk, climate change, and poverty: assessing the global exposure of poor people to floods and droughts. *Environment and Development Economics*, **17**, 1–21, doi:10.1017/s1355770X17000444.

Wise, R.M. et al., 2014: Reconceptualising adaptation to climate change as part of pathways of change and response. *Global Environmental Change*, **28**, 325–336, doi:10.1016/j.gloenvcha.2013.12.002.

Wood, B.T., 2017: Socially just triple-wins? An evaluation of projects that pursue climate compatible development goals in Malawi. PhD Thesis, School of Earth and Environment, University of Leeds, Leeds, UK, 278 pp.

Wood, B.T., S.M. Sallu, and J. Paavola, 2016a: Can CDM finance energy access in Least Developed Countries? Evidence from Tanzania. *Climate Policy*, **16(4)**, 456–473, doi:10.1080/14693062.2015.1027166.

Wood, B.T., A.J. Dougill, C.H. Quinn, and L.C. Stringer, 2016b: Exploring Power and Procedural Justice Within Climate Compatible Development Project Design: Whose Priorities Are Being Considered? *The Journal of Environment & Development*, **25(4)**, 363–395, doi:10.1177/1070496516664179.

Work, C., 2015: Intersections of Climate Change Mitigation Policies, Land Grabbing and Conflict in a Fragile State: Insights from Cambodia. MOSAIC Working Paper Series No. 2, MOSAIC Research project, International Institute of Social Studies (IISS), RCSD Chiang Mai University, 34 pp.

Wright, H., S. Huq, and J. Reeves, 2015: Impact of climate change on least developed countries: are the SDGs possible? IIED Briefing May 2015, International Institute for Environment and Development (IIED), London, UK, 4 pp.

WWF, 2017: *Responsible sourcing of forest products: The business case for retailers*. World Wide Fund for Nature (WWF), Gland, Switzerland, 47 pp.

Wyborn, C., L. Yung, D. Murphy, and D.R. Williams, 2015: Situating adaptation: how governance challenges and perceptions of uncertainty influence adaptation in the Rocky Mountains. *Regional Environmental Change*, **15(4)**, 669–682, doi:10.1007/s10113-014-0663-3.

Yang, S., B. Chen, and S. Ulgiati, 2016: Co-benefits of CO_2 and PM2.5 Emission Reduction. *Energy Procedia*, **104**, 92–97, doi:10.1016/j.egypro.2016.12.017.

Yaro, J.A., J. Teye, and S. Bawakyillenuo, 2015: Local institutions and adaptive capacity to climate change/variability in the northern savannah of Ghana. *Climate and Development*, **7(3)**, 235–245, doi:10.1080/17565529.2014.951018.

Zeng, H., X. Chen, X. Xiao, and Z. Zhou, 2017: Institutional pressures, sustainable supply chain management, and circular economy capability: Empirical evidence from Chinese eco-industrial park firms. *Journal of Cleaner Production*, **155**, 54–65, doi:10.1016/j.jclepro.2016.10.093.

Zhang, H. and W. Chen, 2015: The role of biofuels in China's transport sector in carbon mitigation scenarios. *Energy Procedia*, **75**, 2700–2705, doi:10.1016/j.egypro.2015.07.682.

Zhang, S., E. Worrell, and W. Crijns-Graus, 2015: Cutting air pollution by improving energy efficiency of China's cement industry. *Energy Procedia*, **83**, 10–20, doi:10.1016/j.egypro.2015.12.191.

Zhang, Y. et al., 2017: Processes of coastal ecosystem carbon sequestration and approaches for increasing carbon sink. *Science China Earth Sciences*, **60(5)**, 809–820, doi:10.1007/s11430-016-9010-9.

Zhao, D., A.P. McCoy, J. Du, P. Agee, and Y. Lu, 2017: Interaction effects of building technology and resident behavior on energy consumption in residential buildings. *Energy and Buildings*, **134**, 223–233, doi:10.1016/j.enbuild.2016.10.049.

Ziervogel, G., A. Cowen, and J. Ziniades, 2016: Moving from adaptive to transformative capacity: Building foundations for inclusive, thriving, and regenerative urban settlements. *Sustainability*, **8(9)**, 955, doi:10.3390/su8090955.

Ziervogel, G. et al., 2017: Inserting rights and justice into urban resilience: a focus on everyday risk. *Environment and Urbanization*, **29(1)**, 123–138, doi:10.1177/0956247816686905.

Zimm, C., F. Sperling, and S. Busch, 2018: Identifying sustainability and knowledge gaps in socio-economic pathways vis-à-vis the Sustainable Development Goals. *Economies*, **6(2)**, 20, doi:10.3390/economies6020020.

REF

5

Reference list for Table 5.2.

Aamaas, B. and G.P. Peters, 2017: The climate impact of Norwegians' travel behavior. *Travel Behaviour and Society*, **6**, 10–18, doi:10.1016/j.tbs.2016.04.001.

Abanda, F.H., M.B. Manjia, K.E. Enongene, J.H.M. Tah, and C. Pettang, 2016: A feasibility study of a residential photovoltaic system in Cameroon. *Sustainable Energy Technologies and Assessments*, **17**, 38–49, doi:10.1016/j.seta.2016.08.002.

Abdelouas, A., 2006: Uranium mill tailings: geochemistry, mineralogy, and environmental impact. *Elements*, **2(6)**, 335–341, doi:10.2113/gselements.2.6.335.

Abdul Quader, M., S. Ahmed, S.Z. Dawal, and Y. Nukman, 2016: Present needs, recent progress and future trends of energy-efficient Ultra-Low Carbon Dioxide (CO_2) Steelmaking (ULCOS) program. *Renewable and Sustainable Energy Reviews*, **55**, 537–549, doi:10.1016/j.rser.2015.10.101.

Acemoglu, D., 2009: *Introduction to modern economic growth*. Princeton University Press, Princeton, NJ, USA, 1008 pp.

Acemoglu, D., F.A. Gallego, and J.A. Robinson, 2014: Institutions, human Capital, and development. *Annual Review of Economics*, **6(1)**, 875–912, doi:10.1146/annurev-economics-080213-041119.

Acheampong, M., F.C. Ertem, B. Kappler, and P. Neubauer, 2017: In pursuit of Sustainable Development Goal (SDG) number 7: Will biofuels be reliable? *Renewable and Sustainable Energy Reviews*, **75(7)**, 927–937, doi:10.1016/j.rser.2016.11.074.

Adamantiades, A. and I. Kessides, 2009: Nuclear power for sustainable development: Current status and future prospects. *Energy Policy*, **37(12)**, 5149–5166, doi:10.1016/j.enpol.2009.07.052.

Aggarwal, P., 2017: 2°C target, India's climate action plan and urban transport sector. *Travel Behaviour and Society*, **6**, 110–116, doi:10.1016/j.tbs.2016.11.001.

Ahearne, J.F., 2011: Prospects for nuclear energy. *Energy Economics*, **33(4)**, 572–580, doi:10.1016/j.eneco.2010.11.014.

Ahmad, S. and J.A. Puppim de Oliveira, 2016: Determinants of urban mobility in India: lessons for promoting sustainable and inclusive urban transportation in developing countries. *Transport Policy*, **50**, 106–114, doi:10.1016/j.tranpol.2016.04.014.

Åhman, M., L.J. Nilsson, and B. Johansson, 2017: Global climate policy and deep decarbonization of energy-intensive industries. *Climate Policy*, **17(5)**, 634–649, doi:10.1080/14693062.2016.1167009.

Ahmed, N., W.W.L. Cheung, S. Thompson, and M. Glaser, 2017a: Solutions to blue carbon emissions: Shrimp cultivation, mangrove deforestation and climate change in coastal Bangladesh. *Marine Policy*, **82**, 68–75, doi:10.1016/j.marpol.2017.05.007.

Ahmed, N., S.W. Bunting, M. Glaser, M.S. Flaherty, and J.S. Diana, 2017b: Can greening of aquaculture sequester blue carbon? *Ambio*, **46(4)**, 468–477, doi:10.1007/s13280-016-0849-7.

Ajanovic, A., 2015: The future of electric vehicles: prospects and impediments. *Wiley Interdisciplinary Reviews: Energy and Environment*, **4(6)**, 521–536, doi:10.1002/wene.160.

Alahakoon, S., 2017: Significance of energy storages in future power networks. *Energy Procedia*, **110**, 14–19, doi:10.1016/j.egypro.2017.03.098.

Alho, C.J.R., 2011: Environmental Effects of Hydropower Reservoirs on Wild Mammals and Freshwater Turtles in Amazonia: A Review. *Oecologia Australis*, **15(3)**, 593–604, doi:10.4257/oeco.2011.1503.11.

Ali, S.M.H., M.J.S. Zuberi, M.A. Tariq, D. Baker, and A. Mohiuddin, 2015: A study to incorporate renewable energy technologies into the power portfolio of Karachi, Pakistan. *Renewable and Sustainable Energy Reviews*, **47**, 14–22, doi:10.1016/j.rser.2015.03.009.

Allen, S., T. Dietz, and A.M. McCright, 2015: Measuring household energy efficiency behaviors with attention to behavioral plasticity in the United States. *Energy Research & Social Science*, **10**, 133–140, doi:10.1016/j.erss.2015.07.014.

Alongi, D.M., 2012: Carbon sequestration in mangrove forests. *Carbon Management*, **3(3)**, 313–322, doi:10.4155/cmt.12.20.

AlSabbagh, M., Y.L. Siu, A. Guehnemann, and J. Barrett, 2017: Integrated approach to the assessment of CO_2-mitigation measures for the road passenger transport sector in Bahrain. *Renewable and Sustainable Energy Reviews*, **71**, 203–215, doi:10.1016/j.rser.2016.12.052.

Altieri, K.E. et al., 2016: Achieving development and mitigation objectives through a decarbonization development pathway in South Africa. *Climate Policy*, **16(sup1)**, S78–S91, doi:10.1080/14693062.2016.1150250.

Al-Zoughool, M. and D. Krewski, 2009: Health effects of radon: A review of the literature. *International Journal of Radiation Biology*, **85(1)**, 57–69, doi:10.1080/09553000802635054.

Anda, M. and J. Temmen, 2014: Smart metering for residential energy efficiency: the use of community based social marketing for behavioural change and smart grid introduction. *Renewable Energy*, **67**, 119–127, doi:10.1016/j.renene.2013.11.020.

Anderson, A. et al., 2017: Empowering smart communities: electrification, education, and sustainable entrepreneurship in IEEE Smart Village Initiatives. *IEEE Electrification Magazine*, **5(2)**, 6–16, doi:10.1109/mele.2017.2685738.

Andrew, D., 2017: *Trade and Sustainable Development Goal (SDG) 15: Promoting "Life on Land" through Mandatory and Voluntary Approaches*. ADBI Working Paper Series, No. 700, Asian Development Bank Institute (ADBI), Tokyo, Japan, 33 pp.

Anenberg, S.C. et al., 2013: Cleaner cooking solutions to achieve health, climate, and economic cobenefits. *Environmental Science & Technology*, **47(9)**, 3944–3952, doi:10.1021/es304942e.

Apeaning, R.W. and P. Thollander, 2013: Barriers to and driving forces for industrial energy efficiency improvements in African industries: a case study of Ghana's largest industrial area. *Journal of Cleaner Production*, **53**, 204–213, doi:10.1016/j.jclepro.2013.04.003.

Apps, J.A., L. Zheng, Y. Zhang, T. Xu, and J.T. Birkholzer, 2010: Evaluation of Potential Changes in Groundwater Quality in Response to CO_2 Leakage from Deep Geologic Storage. *Transport in Porous Media*, **82(1)**, 215–246, doi:10.1007/s11242-009-9509-8.

Aranda, C., A.C. Kuesel, and E.R. Fletcher, 2014: *A systematic review of linkages between access to electricity in healthcare facilities, health services delivery, and health outcomes: findings for emergency referrals, maternal and child services*. UBS Optimus Foundation & Liberian Institute for Biomedical Research, Zurich, Switzerland.

Asaduzzaman, M., D.F. Barnes, and S.R. Khandker, 2010: *Restoring Balance: Bangladesh's Rural Energy Realities*. World Bank Working Paper: No. 181, World Bank, Washington DC, USA, 170 pp., doi:10.1596/978-0-8213-7897-7.

Asfaw, A., C. Mark, and R. Pana-Cryan, 2013: Profitability and occupational injuries in US underground coal mines. *Accident Analysis & Prevention*, **50**, 778–786, doi:10.1016/j.aap.2012.07.002.

Ashworth, P. et al., 2012: What's in store: Lessons from implementing CCS. *International Journal of Greenhouse Gas Control*, **9**, 402–409, doi:10.1016/j.ijggc.2012.04.012.

Atchley, A.L., R.M. Maxwell, and A.K. Navarre-Sitchler, 2013: Human health risk assessment of CO_2 leakage into overlying aquifers using a stochastic, geochemical reactive transport approach. *Environmental Science & Technology*, **47(11)**, 5954–5962.

Azevedo, I. and V.M.S. Leal, 2017: Methodologies for the evaluation of local climate change mitigation actions: A review. *Renewable and Sustainable Energy Reviews*, **79**, 681–690, doi:10.1016/j.rser.2017.05.100.

Babiker, M.H. and R.S. Eckaus, 2007: Unemployment effects of climate policy. *Environmental Science & Policy*, **10(7–8)**, 600–609, doi:10.1016/j.envsci.2007.05.002.

Bailis, R., R. Drigo, A. Ghilardi, and O. Masera, 2015: The carbon footprint of traditional woodfuels. *Nature Climate Change*, **5(3)**, 266–272, doi:10.1038/nclimate2491.

Bajželj, B. et al., 2014: Importance of food-demand management for climate mitigation. *Nature Climate Change*, **4(10)**, 924–929, doi:10.1038/nclimate2353.

Balishter, G.V.K. and R. Singh, 1991: Impact of mechanization on employment and farm productivity. *Productivity*, **32(3)**, 484–489.

Banerjee, R. et al., 2012: Energy End-Use: Industry. In: *Global Energy Assessment – Toward a Sustainable Future* [Johansson, T.B., N. Nakicenovic, A. Patwardhan, and L. Gomez-Echeverri (eds.)]. Cambridge University Press, Cambridge, United Kingdom and New York, NY, USA and the International Institute for Applied Systems Analysis, pp. 513–574.

Bartley, T., 2010: Transnational Private Regulation in Practice: The Limits of Forest and Labor Standards Certification in Indonesia. *Business and Politics*, **12(03)**, 1–34, doi:10.2202/1469-3569.1321.

Bartos, M.D. and M. Chester, 2014: The Conservation Nexus: Valuing Interdependent Water and Energy Savings in Arizona. *Environmental Science & Technology*, **48(4)**, 2139–2149, doi:10.1021/es4033343.

Bastos Lima, M.G., W. Ashely-Cantello, I. Visseren-Hamakers, A. Gupta, and J. Braña-Varela, 2015: *Forests Post-2015: Maximizing Synergies between the Sustainable Development Goals and REDD+.* WWF-WUR Policy Brief No. 3, World Wildlife Fund (WWF) and Wageningen University & Research (WUR), 5 pp.

Bastos Lima, M.G., G. Kissinger, I.J. Visseren-Hamakers, J. Braña-Varela, and A. Gupta, 2017: The Sustainable Development Goals and REDD+: assessing institutional interactions and the pursuit of synergies. *International Environmental Agreements: Politics, Law and Economics*, **17(4)**, 589–606, doi:10.1007/s10784-017-9366-9.

Bazilian, M. et al., 2011: Considering the energy, water and food nexus: towards an integrated modelling approach. *Energy Policy*, **39(12)**, 7896–7906, doi:10.1016/j.enpol.2011.09.039.

Beddington, J. et al., 2012: *Achieving food security in the face of climate change: Final report from the Commission on Sustainable Agriculture and Climate Change.* CGIAR Research Program on Climate Change, Agriculture and Food Security (CCAFS), Copenhagen, Denmark, 59 pp.

Behnassi, M., M. Boussaid, and R. Gopichandran, 2014: Achieving Food Security in a Changing Climate: The Potential of Climate-Smart Agriculture. In: *Environmental Cost and Face of Agriculture in the Gulf Cooperation Council Countries* [Shahid, S.A. and M. Ahmed (eds.)]. Springer International Publishing, Cham, Switzerland, pp. 27–42, doi:10.1007/978-3-319-05768-2_2.

Bellarby, J. et al., 2013: Livestock greenhouse gas emissions and mitigation potential in Europe. *Global Change Biology*, **19(1)**, 3–18, doi:10.1111/j.1365-2486.2012.02786.x.

Belton Chevallier, L., B. Motte-Baumvol, S. Fol, and Y. Jouffe, 2018: Coping with the costs of car dependency: A system of expedients used by low-income households on the outskirts of Dijon and Paris. *Transport Policy*, **65**, 79–88, doi:10.1016/j.tranpol.2017.06.006.

Benson, S.M. and D.R. Cole, 2008: CO_2 Sequestration in Deep Sedimentary Formations. *Elements*, **4(5)**, 325–331, doi:10.2113/gselements.4.5.325.

Bernard, T. and M. Torero, 2015: Social interaction effects and connection to electricity: experimental evidence from rural Ethiopia. *Economic Development and Cultural Change*, **63(3)**, 459–484, doi:10.1086/679746.

Bernier, Q. et al., 2013: *Addressing Gender in Climate-Smart Smallholder Agriculture.* ICRAF Policy Brief 14, World Agroforestry Centre (ICRAF), Nairobi, Kenya, 4 pp.

Berrueta, V.M., M. Serrano-Medrano, C. Garcia-Bustamante, M. Astier, and O.R. Masera, 2017: Promoting sustainable local development of rural communities and mitigating climate change: the case of Mexico's Patsari improved cookstove project. *Climatic Change*, **140(1)**, 63–77, doi:10.1007/s10584-015-1523-y.

Bertram, C. et al., 2015: Carbon lock-in through capital stock inertia associated with weak near-term climate policies. *Technological Forecasting and Social Change*, **90(PA)**, 62–72, doi:10.1016/j.techfore.2013.10.001.

Bhattacharyya, A., J.P. Meltzer, J. Oppenheim, Z. Qureshi, and N. Stern, 2016: *Delivering on better infrastructure for better development and better climate.* The Brookings Institution, The New Climate Economy, and Grantham Research Institute on Climate Change and the Environment, 160 pp.

Bhojvaid, V. et al., 2014: How do People in Rural India Perceive Improved Stoves and Clean Fuel? Evidence from Uttar Pradesh and Uttarakhand. *International Journal of Environmental Research and Public Health*, **11(2)**, 1341–1358, doi:10.3390/ijerph110201341.

Bickerstaff, K., P. Simmons, and N. Pidgeon, 2008: Constructing Responsibilities for Risk: Negotiating Citizen – State Relationships. *Environment and Planning A: Economy and Space*, **40(6)**, 1312–1330, doi:10.1068/a39150.

Bilton, A.M., R. Wiesman, A.F.M. Arif, S.M. Zubair, and S. Dubowsky, 2011: On the feasibility of community-scale photovoltaic-powered reverse osmosis desalination systems for remote locations. *Renewable Energy*, **36(12)**, 3246–3256, doi:10.1016/j.renene.2011.03.040.

Blyth, W. et al., 2014: *Low carbon jobs: the evidence for net job creation from policy support for energy efficiency and renewable energy.* Technical Report, UK Energy Research Centre, London, UK, 66 pp.

Bogdanski, A., 2012: Integrated food-energy systems for climate-smart agriculture. *Agriculture & Food Security*, **1(1)**, 9, doi:10.1186/2048-7010-1-9.

Bonan, J., S. Pareglio, and M. Tavoni, 2014: *Access to Modern Energy: A Review of Impact Evaluations.* Fondazione Eni Enrico Mattei, Milan, Italy, 30 pp.

Bongardt, D. et al., 2013: *Low-Carbon Land Transport: Policy Handbook.* Routledge, Abingdon, UK and New York, NY, USA, 264 pp.

Bonsch, M. et al., 2016: Trade-offs between land and water requirements for large-scale bioenergy production. *GCB Bioenergy*, **8(1)**, 11–24, doi:10.1111/gcbb.12226.

Borenstein, S., 2012: The private and public economics of renewable electricity generation. *Journal of Economic Perspectives*, **26**, 67–92, doi:10.1257/jep.26.1.67.

Branca, G., N. McCarthy, L. Lipper, and M.C. Jolejole, 2011: *Climate-smart agriculture: a synthesis of empirical evidence of food security and mitigation benefits from improved cropland management.* Mitigation of Climate Change in Agriculture Series 3, Food and Agriculture Organization of the United Nations (FAO), Rome, Italy, 35 pp.

Brandl, P., S.M. Soltani, P.S. Fennell, and N. Mac Dowell, 2017: Evaluation of cooling requirements of post-combustion CO_2 capture applied to coal-fired power plants. *Chemical Engineering Research and Design*, **122**, 1–10, doi:10.1016/j.cherd.2017.04.001.

Brown, H.C.P., 2011: Gender, climate change and REDD+ in the Congo Basin forests of Central Africa. *International Forestry Review*, **13(2)**, 163–176, doi:10.1505/146554811797406651.

Brugge, D. and V. Buchner, 2011: Health effects of uranium: new research findings. *Reviews on Environmental Health*, **26(4)**, 231–249, doi:10.1515/reveh.2011.032.

Buck, B.H. and G. Krause, 2012: Integration of Aquaculture and Renewable Energy Systems. In: *Encyclopedia of Sustainability Science and Technology Vol. 1.* Springer Science+ Business Media, New York, NY, USA, pp. 511–533, doi:10.1007/978-1-4419-0851-3_180.

Burgherr, P., P. Eckle, and S. Hirschberg, 2012: Comparative assessment of severe accident risks in the coal, oil and natural gas chains. *Reliability Engineering & System Safety*, **105**, 97–103, doi:10.1016/j.ress.2012.03.020.

Burlig, F. and L. Preonas, 2016: *Out of the darkness and into the light? Development effects of rural electrification.* EI @ Haas WP 268, Energy Institute at Haas, University of California, Berkeley, CA, USA, 52 pp.

Burney, J., H. Alaofè, R. Naylor, and D. Taren, 2017: Impact of a rural solar electrification project on the level and structure of women's empowerment. *Environmental Research Letters*, **12(9)**, 095007, doi:10.1088/1748-9326/aa7f38.

Burton, C.H., 2007: The potential contribution of separation technologies to the management of livestock manure. *Livestock Science*, **112(3)**, 208–216, doi:10.1016/j.livsci.2007.09.004.

Bustamante, M. et al., 2014: Co-benefits, trade-offs, barriers and policies for greenhouse gas mitigation in the agriculture, forestry and other land use (AFOLU) sector. *Global Change Biology*, **20(10)**, 3270–3290, doi:10.1111/gcb.12591.

Byers, E.A., J.W. Hall, J.M. Amezaga, G.M. O'Donnell, and A. Leathard, 2016: Water and climate risks to power generation with carbon capture and storage. *Environmental Research Letters*, **11(2)**, 024011, doi:10.1088/1748-9326/11/2/024011.

Byravan, S. et al., 2017: Quality of life for all: A sustainable development framework for India's climate policy reduces greenhouse gas emissions. *Energy for Sustainable Development*, **39**, 48–58, doi:10.1016/j.esd.2017.04.003.

Cabraal, A.R., D.F. Barnes, and S.G. Agarwal, 2005: Productive uses of energy for rural development. *Annual Review of Environment and Resources*, **30(1)**, 117–144, doi:10.1146/annurev.energy.30.050504.144228.

CAF, 2017: *Crecimiento urbano y accesso a Oportunidades: un desafío para América Latina (in Spanish).* Corporacion Andina de Fomento (CAF), Bogotá, Colombia, 287 pp.

Cameron, C. et al., 2016: Policy trade-offs between climate mitigation and clean cook-stove access in South Asia. *Nature Energy*, **1**, 1–5, doi:10.1038/nenergy.2015.10.

Cameron, R.W.F., J. Taylor, and M. Emmett, 2015: A Hedera green facade – energy performance and saving under different maritime-temperate, winter weather conditions. *Building and Environment*, **92**, 111–121, doi:10.1016/j.buildenv.2015.04.011.

Campbell, B.M., P. Thornton, R. Zougmoré, P. van Asten, and L. Lipper, 2014: Sustainable intensification: What is its role in climate smart agriculture? *Current Opinion in Environmental Sustainability*, **8**, 39–43, doi:10.1016/j.cosust.2014.07.002.

Cardis, E. et al., 2006: Estimates of the cancer burden in Europe from radioactive fallout from the Chernobyl accident. *International Journal of Cancer*, **119(6)**, 1224–1235, doi:10.1002/ijc.22037.

Carrara, S. and T. Longden, 2017: Freight futures: The potential impact of road freight on climate policy. *Transportation Research Part D: Transport and Environment*, **55**, 359–372, doi:10.1016/j.trd.2016.10.007.

5

Cascajo, R., A. Garcia-Martinez, and A. Monzon, 2017: Stated preference survey for estimating passenger transfer penalties: design and application to Madrid. *European Transport Research Review*, **9(3)**, 42, doi:10.1007/s12544-017-0260-x.

Casillas, C.E. and D.M. Kammen, 2010: The Energy-Poverty-Climate Nexus. *Science*, **330(6008)**, 1181–1182, doi:10.1126/science.1197412.

Casillas, C.E. and D.M. Kammen, 2012: Quantifying the social equity of carbon mitigation strategies. *Climate Policy*, **12(6)**, 690–703, doi:10.1080/14693062.2012.669097.

Cass, N., G. Walker, and P. Devine-Wright, 2010: Good neighbours, public relations and bribes: the politics and perceptions of community benefit provision in renewable energy development in the UK. *Journal of Environmental Policy & Planning*, **12(3)**, 255–275, doi:10.1080/1523908x.2010.509558.

Cayla, J.-M. and D. Osso, 2013: Does energy efficiency reduce inequalities? Impact of policies in Residential sector on household budget. In: *ECEEE Summer Study Proceedings*. European Council for an Energy Efficient Economy (ECEEE), Toulon/ Hyeres, France, pp. 1247–1257.

CDP, 2015: *CDP Global Climate Change Report 2015 – At the tipping point?* Carbon Disclosure Project (CDP) Worldwide, 91 pp.

Chakrabarti, S. and E.J. Shin, 2017: Automobile dependence and physical inactivity: insights from the California Household Travel Survey. *Journal of Transport and Health*, **6**, 262–271, doi:10.1016/j.jth.2017.05.002.

Chakravarty, D., S. Dasgupta, and J. Roy, 2013: Rebound effect: how much to worry? *Current Opinion in Environmental Sustainability*, **5(2)**, 216–228, doi:10.1016/j.cosust.2013.03.001.

Chakravorty, U., M. Pelli, and B. Ural Marchand, 2014: Does the quality of electricity matter? Evidence from rural India. *Journal of Economic Behavior & Organization*, **107**, 228–247, doi:10.1016/j.jebo.2014.04.011.

Chan, E.Y.Y. and S.M. Griffiths, 2010: The epidemiology of mine accidents in China. *The Lancet*, **376(9741)**, 575–577, doi:10.1016/s0140-6736(10)60660-5.

Chaturvedi, V. and P.R. Shukla, 2014: Role of energy efficiency in climate change mitigation policy for India: assessment of co-benefits and opportunities within an integrated assessment modeling framework. *Climatic Change*, **123(3)**, 597–609, doi:10.1007/s10584-013-0898-x.

Chen, B. and X. Qi, 2018: Protest response and contingent valuation of an urban forest park in Fuzhou City, China. *Urban Forestry & Urban Greening*, **29**, 68–76, doi:10.1016/j.ufug.2017.11.005.

Chen, H., H. Qi, R. Long, and M. Zhang, 2012: Research on 10-year tendency of China coal mine accidents and the characteristics of human factors. *Safety science*, **50(4)**, 745–750, doi:10.1016/j.ssci.2011.08.040.

Cherian, A., 2015: *Energy and Global Climate Change: Bridging the Sustainable Development Divide*. John Wiley & Sons Ltd, Chichester, UK, 304 pp., doi:10.1002/9781118846070.

Chowdhury, S.K., 2010: Impact of infrastructures on paid work opportunities and unpaid work burdens on rural women in Bangladesh. *Journal of International Development*, **22(7)**, 997–1017, doi:10.1002/jid.1607.

Cibin, R., E. Trybula, I. Chaubey, S.M. Brouder, and J.J. Volenec, 2016: Watershed-scale impacts of bioenergy crops on hydrology and water quality using improved SWAT model. *GCB Bioenergy*, **8(4)**, 837–848, doi:10.1111/gcbb.12307.

Clancy, J.S., T. Winther, M. Matinga, and S. Oparaocha, 2012: *Gender equity in access to and benefits from modern energy and improved energy technologies: world development report background paper*. Gender and Energy WDR Background Paper 44, ETC/ENERGIA in association Nord/Sør-konsulentene, 44 pp.

Clarke, L.E. et al., 2009: International climate policy architectures: Overview of the EMF 22 International Scenarios. *Energy Economics*, **31**, S64–S81, doi:10.1016/j.eneco.2009.10.013.

Clarke, L.E. et al., 2014: Assessing transformation pathways. In: *Climate Change 2014: Mitigation of Climate Change. Contribution of Working Group III to the Fifth Assessment Report of the Intergovernmental Panel on Climate Change* [Edenhofer, O., R. Pichs-Madruga, Y. Sokona, E. Farahani, S. Kadner, K. Seyboth, A. Adler, I. Baum, S. Brunner, P. Eickemeier, B. Kriemann, J. Savolainen, S. Schlömer, C. von Stechow, T. Zwickel, and J.C. Minx (eds.)]. Cambridge University Press, Cambridge, United Kingdom and New York, NY, USA, pp. 413–510.

Clifton, K.J., 2004: Mobility Strategies and Food Shopping for Low-Income Families: A Case Study. *Journal of Planning Education and Research*, **23(4)**, 402–413, doi:10.1177/0739456x04264919.

Colenbrander, S. et al., 2016: Can low-carbon urban development be pro-poor? The case of Kolkata, India. *Environment and Urbanization*, **29(1)**, 139–158, doi:10.1177/0956247816677775.

Connolly, D. et al., 2014: Heat Roadmap Europe: Combining district heating with heat savings to decarbonise the EU energy system. *Energy Policy*, **65**, 475–489, doi:10.1016/j.enpol.2013.10.035.

Cook, P., 2011: Infrastructure, rural electrification and development. *Energy for Sustainable Development*, **15(3)**, 304–313, doi:10.1016/j.esd.2011.07.008.

Cooke, S.J. et al., 2016: On the sustainability of inland fisheries: finding a future for the forgotten. *Ambio*, **45(7)**, 753–764, doi:10.1007/s13280-016-0787-4.

Corbera, E. and U. Pascual, 2012: Ecosystem Services: Heed Social Goals. *Science*, **335(6069)**, 655–656, doi:10.1126/science.335.6069.655-c.

Corner, A. et al., 2011: Nuclear power, climate change and energy security: Exploring British public attitudes. *Energy Policy*, **39(9)**, 4823–4833, doi:10.1016/j.enpol.2011.06.037.

Corsten, M., A. Ramirez, L. Shen, J. Koornneef, and A. Faaij, 2013: Environmental impact assessment of CCS chains – lessons learned and limitations from LCA literature. *International Journal of Greenhouse Gas Control*, **13**, 59–71.

Crane, R., 2007: Is There a Quiet Revolution in Women's Travel? Revisiting the Gender Gap in Commuting. *Journal of the American Planning Association*, **73(3)**, 298–316, doi:10.1080/01944360708977979.

Creutzig, F., R. Mühlhoff, and J. Römer, 2012: Decarbonizing urban transport in European cities: four cases show possibly high co-benefits. *Environmental Research Letters*, **7(4)**, 044042, doi:10.1088/1748-9326/7/4/044042.

Creutzig, F., C. Esteve, B. Simon, and H. Carol, 2013: Integrating place-specific livelihood and equity outcomes into global assessments of bioenergy deployment. *Environmental Research Letters*, **8(3)**, 035047, doi:10.1088/1748-9326/8/3/035047.

Creutzig, F. et al., 2014: Catching two European birds with one renewable stone: mitigating climate change and Eurozone crisis by an energy transition. *Renewable and Sustainable Energy Reviews*, **38**, 1015–1028, doi:10.1016/j.rser.2014.07.028.

Creutzig, F. et al., 2015: Transport: A roadblock to climate change mitigation? *Science*, **350(6263)**, 911–912, doi:10.1126/science.aac8033.

Creutzig, F. et al., 2016: Beyond technology: demand-side solutions for climate change mitigation. *Annual Review of Environment and Resources*, **41(1)**, 173–198, doi:10.1146/annurev-environ-110615-085428.

Cumbers, A., 2012: *Reclaiming Public Ownership: Making Space for Economic Democracy*. Zed Books Ltd, London, UK and New York, NY, USA, 192 pp.

Dagevos, H. and J. Voordouw, 2013: Sustainability and meat consumption: is reduction realistic? *Sustainability: Science, Practice, & Policy*, **9(2)**, 1–10, doi:10.1080/15487733.2013.11908115.

Dahl, E.L., K. Bevanger, T. Nygård, E. Røskaft, and B.G. Stokke, 2012: Reduced breeding success in white-tailed eagles at Smøla windfarm, western Norway, is caused by mortality and displacement. *Biological Conservation*, **145(1)**, 79–85, doi:10.1016/j.biocon.2011.10.012.

Daut, I., M. Adzrie, M. Irwanto, P. Ibrahim, and M. Fitra, 2013: Solar Powered Air Conditioning System. *Energy Procedia*, **36**, 444–453, doi:10.1016/j.egypro.2013.07.050.

Davis, S.C. et al., 2013: Management swing potential for bioenergy crops. *GCB Bioenergy*, **5(6)**, 623–638, doi:10.1111/gcbb.12042.

de Best-Waldhober, M., D. Daamen, and A. Faaij, 2009: Informed and uninformed public opinions on CO_2 capture and storage technologies in the Netherlands. *International Journal of Greenhouse Gas Control*, **3(3)**, 322–332, doi:10.1016/j.ijggc.2008.09.001.

de Koning, J.I.J.C., T.H. Ta, M.R.M. Crul, R. Wever, and J.C. Brezet, 2016: GetGreen Vietnam: towards more sustainable behaviour among the urban middle class. *Journal of Cleaner Production*, **134(Part A)**, 178–190, doi:10.1016/j.jclepro.2016.01.063.

de Lucas, M., M. Ferrer, M.J. Bechard, and A.R. Muñoz, 2012: Griffon vulture mortality at wind farms in southern Spain: distribution of fatalities and active mitigation measures. *Biological Conservation*, **147(1)**, 184–189, doi:10.1016/j.biocon.2011.12.029.

de Moraes, M.A.F.D., C.C. Costa, J.J.M. Guilhoto, L.G.A. Souza, and F.C.R. Oliveira, 2010: Social Externalities of Fuels. In: *Ethanol and Bioelectricity: Sugarcane in the Future of the Energy Matrix* [Leão de Sousa, E.L. and I. de Carvalho Macedo (eds.)]. UNICA Brazilian Sugarcane Industry Association, São Paulo, Brazil, pp. 44–75.

De Stefano, L., J.D. Petersen-Perlman, E.A. Sproles, J. Eynard, and A.T. Wolf, 2017: Assessment of transboundary river basins for potential hydro-political tensions. *Global Environmental Change*, **45**, 35–46, doi:10.1016/j.gloenvcha.2017.04.008.

5

Dechezleprêtre, A. and M. Sato, 2014: *The impacts of environmental regulations on competitiveness*. London School of Economics (LSE) and Global Green Growth Institute (GGGI), 28 pp.

Demetriades, J. and E. Esplen, 2009: The Gender Dimensions of Poverty and Climate Change Adaptation. *IDS Bulletin*, **39(4)**, 24–31, doi:10.1111/j.1759-5436.2008.tb00473.x.

Denis-Ryan, A., C. Bataille, and F. Jotzo, 2016: Managing carbon-intensive materials in a decarbonizing world without a global price on carbon. *Climate Policy*, **16(sup1)**, S110–S128, doi:10.1080/14693062.2016.1176008.

Denman, K.L., 2008: Climate change, ocean processes and ocean iron fertilization. *Marine Ecology Progress Series*, **364**, 219–225, doi:10.3354/meps07542.

Denton, F., 2002: Climate Change Vulnerability, Impacts, and Adaptation: Why Does Gender Matter? *Gender and Development*, **10(2)**, 10–20, www.jstor.org/stable/4030569.

Derbez, M. et al., 2014: Indoor air quality and comfort in seven newly built, energy-efficient houses in France. *Building and Environment*, **72**, 173–187, doi:10.1016/j.buildenv.2013.10.017.

Dinkelman, T., 2011: The Effects of Rural Electrification on Employment: New Evidence from South Africa. *The American Economic Review*, **101(7)**, 3078–3108, doi:10.1257/aer.101.7.3078.

Djamila, H., C.-M. Chu, and S. Kumaresan, 2013: Field study of thermal comfort in residential buildings in the equatorial hot-humid climate of Malaysia. *Building and Environment*, **62**, 133–142, doi:10.1016/j.buildenv.2013.01.017.

Dodson, J. and N. Sipe, 2008: Shocking the Suburbs: Urban Location, Homeownership and Oil Vulnerability in the Australian City. *Housing Studies*, **23(3)**, 377–401, doi:10.1080/02673030802015619.

Dodson, J., B. Gleeson, and N. Sipe, 2004: *Transport Disadvantage and Social Status: A review of literature and methods*. Urban Policy Program Research Monograph 5, Griffith University, Brisbane, Australia, 55 pp.

Dooley, K. and S. Kartha, 2018: Land-based negative emissions: risks for climate mitigation and impacts on sustainable development. *International Environmental Agreements: Politics, Law and Economics*, **18(1)**, 79–98, doi:10.1007/s10784-017-9382-9.

Duarte, C.M., J. Wu, X. Xiao, A. Bruhn, and D. Krause-Jensen, 2017: Can Seaweed Farming Play a Role in Climate Change Mitigation and Adaptation? *Frontiers in Marine Science*, **4**, 100, doi:10.3389/fmars.2017.00100.

Dulac, J., 2013: *Global land transport infrastructure requirements: Estimating road and railway infrastructure capacity and costs to 2050*. 50 pp.

ECF, 2014: *Europe's low-carbon transition: Understanding the challenges and opportunities for the chemical sector*. European Climate Foundation (ECF), 60 pp.

Echegaray, F., 2016: Consumers' reactions to product obsolescence in emerging markets: the case of Brazil. *Journal of Cleaner Production*, **134**, 191–203, doi:10.1016/j.jclepro.2015.08.119.

Einsiedel, E.F., A.D. Boyd, J. Medlock, and P. Ashworth, 2013: Assessing socio-technical mindsets: Public deliberations on carbon capture and storage in the context of energy sources and climate change. *Energy Policy*, **53**, 149–158, doi:10.1016/j.enpol.2012.10.042.

Eis, J., R. Bishop, and P. Gradwell, 2016: *Galvanising Low-Carbon Innovation. A New Climate Economy working paper for Seizing the Global Opportunity: Partnerships for Better Growth and a Better Climate*. New Climate Economy (NCE), Washington DC, USA and London, UK, 28 pp.

Epstein, A.H. and S.L.H. Theuer, 2017: Sustainable development and climate action: thoughts on an integrated approach to SDG and climate policy implementation. In: *Papers from Interconnections 2017*. Interconnections 2017.

Epstein, P.R. et al., 2011: Full cost accounting for the life cycle of coal. *Annals of the New York Academy of Sciences*, **1219(1)**, 73–98, doi:10.1111/j.1749-6632.2010.05890.x.

Ertem, F.C., P. Neubauer, and S. Junne, 2017: Environmental life cycle assessment of biogas production from marine macroalgal feedstock for the substitution of energy crops. *Journal of Cleaner Production*, **140**, 977–985, doi:10.1016/j.jclepro.2016.08.041.

Evenson, R.E., 1999: Global and local implications of biotechnology and climate change for future food supplies. *Proceedings of the National Academy of Sciences*, **96(11)**, 5921–8, doi:10.1073/pnas.96.11.5921.

Fan, Y., Q. Qiao, L. Fang, and Y. Yao, 2017: Energy analysis on industrial symbiosis of an industrial park: a case study of Hefei economic and technological development area. *Journal of Cleaner Production*, **141**, 791–798, doi:10.1016/j.jclepro.2016.09.159.

Fankhaeser, S., F. Sehlleier, and N. Stern, 2008: Climate change, innovation and jobs. *Climate Policy*, **8(4)**, 421–429, doi:10.3763/cpol.2008.0513.

Fankhauser, S. and S. Tepic, 2007: Can poor consumers pay for energy and water? An affordability analysis for transition countries. *Energy Policy*, **35(2)**, 1038–1049, doi:10.1016/j.enpol.2006.02.003.

Fay, M. et al., 2015: *Decarbonizing Development: Three Steps to a Zero-Carbon Future*. World Bank, Washington DC, USA, 185 pp., doi:10.1596/978-1-4648-0479-3.

Fernando, L. and S. Evans, 2015: Case Studies in Transformation towards Industrial Sustainability. *International Journal of Knowledge and Systems Science*, **6(3)**, 1–17, doi:10.4018/ijkss.2015070101.

Figueroa, M.J. and S.K. Ribeiro, 2013: Energy for road passenger transport and sustainable development: assessing policies and goals interactions. *Current Opinion in Environmental Sustainability*, **5(2)**, 152–162, doi:10.1016/j.cosust.2013.04.004.

Figueroa, M.J., L. Fulton, and G. Tiwari, 2013: Avoiding, transforming, transitioning: Pathways to sustainable low carbon passenger transport in developing countries. *Current Opinion in Environmental Sustainability*, **5(2)**, 184–190, doi:10.1016/j.cosust.2013.02.006.

Figueroa, M.J., O. Lah, L.M. Fulton, A. McKinnon, and G. Tiwari, 2014: Energy for transport. *Annual Review of Environment and Resources*, **39**, 295–325, doi:10.1146/annurev-environ-031913-100450.

Finco, M.V.A. and W. Doppler, 2010: Bioenergy and sustainable development: The dilemma of food security and climate change in the Brazilian savannah. *Energy for Sustainable Development*, **14**, 194–199, doi:10.1016/j.esd.2010.04.006.

Fricko, O. et al., 2016: Energy sector water use implications of a 2°C degree climate policy. *Environmental Research Letters*, **11(3)**, 034011, doi:10.1088/1748-9326/11/3/034011.

Frondel, M., N. Ritter, C.M. Schmidt, and C. Vance, 2010: Economic impacts from the promotion of renewable energy technologies: The German experience. *Energy Policy*, **38(8)**, 4048–4056, doi:10.1016/j.enpol.2010.03.029.

Fu, W., Y. Lü, P. Harris, A. Comber, and L. Wu, 2018: Peri-urbanization may vary with vegetation restoration: A large scale regional analysis. *Urban Forestry & Urban Greening*, **29**, 77–87, doi:10.1016/j.ufug.2017.11.006.

Gallego, F., J.P. Montero, and C. Salas, 2013: The effect of transport policies on car use: A bundling model with applications. *Energy Economics*, **40**, S85–S97, doi:10.1016/j.eneco.2013.09.018.

Gao, L. and B.A. Bryan, 2017: Finding pathways to national-scale land-sector sustainability. *Nature*, **544(7649)**, 217–222, doi:10.1038/nature21694.

Garnett, T., 2011: Where are the best opportunities for reducing greenhouse gas emissions in the food system (including the food chain)? *Food Policy*, **36**, S23–S32, doi:10.1016/j.foodpol.2010.10.010.

Garvin, J.C., C.S. Jennelle, D. Drake, and S.M. Grodsky, 2011: Response of raptors to a windfarm. *Journal of Applied Ecology*, **48(1)**, 199–209, doi:10.1111/j.1365-2664.2010.01912.x.

Ghosh-Dastidar, B. et al., 2014: Distance to Store, Food Prices, and Obesity in Urban Food Deserts. *American Journal of Preventive Medicine*, **47(5)**, 587–595, doi:10.1016/j.amepre.2014.07.005.

Gnanadesikan, A., J.L. Sarmiento, and R.D. Slater, 2003: Effects of patchy ocean fertilization on atmospheric carbon dioxide and biological production. *Global Biogeochemical Cycles*, **17(2)**, 1050, doi:10.1029/2002gb001940.

Godfray, H.C.J. and T. Garnett, 2014: Food security and sustainable intensification. *Philosophical Transactions of the Royal Society B: Biological Sciences*, **369(1639)**, doi:10.1098/rstb.2012.0273.

Gohin, A., 2008: Impacts of the European Biofuel Policy on the Farm Sector: A General Equilibrium Assessment. *Review of Agricultural Economics*, **30(4)**, 623–641, www.jstor.org/stable/30225908.

Goldthau, A., 2014: Rethinking the governance of energy infrastructure: Scale, decentralization and polycentrism. *Energy Research & Social Science*, **1**, 134–140, doi:10.1016/j.erss.2014.02.009.

Gössling, S. and D. Metzler, 2017: Germany's climate policy: Facing an automobile dilemma. *Energy Policy*, **105**, 418–428, doi:10.1016/j.enpol.2017.03.019.

Gouldson, A. et al., 2015: Exploring the economic case for climate action in cities. *Global Environmental Change*, **35**, 93–105, doi:10.1016/j.gloenvcha.2015.07.009.

Greenberg, H.R., J.A. Blink, and T.A. Buscheck, 2013: *Repository Layout and Required Ventilation Trade Studies in Clay/Shale using the DSEF Thermal Analytical Model*. LLNL-TR-638880, Lawrence Livermore National Laboratory (LLNL), Livermore, CA, USA, 33 pp.

Griffin, P.W., G.P. Hammond, and J.B. Norman, 2018: Industrial energy use and carbon emissions reduction in the chemicals sector: A UK perspective. *Applied Energy*, **227**, 587–602, doi:10.1016/j.apenergy.2017.08.010.

Griffiths, N.A. et al., 2017: Water quality effects of short-rotation pine management for bioenergy feedstocks in the southeastern United States. *Forest Ecology and Management*, **400**, 181–198, doi:10.1016/j.foreco.2017.06.011.

Grill, G. et al., 2015: An index-based framework for assessing patterns and trends in river fragmentation and flow regulation by global dams at multiple scales. *Environmental Research Letters*, **10(1)**, 015001, doi:10.1088/1748-9326/10/1/015001.

Grodsky, S.M. et al., 2011: Investigating the causes of death for wind turbine-associated bat fatalities. *Journal of Mammalogy*, **92(5)**, 917–925, doi:10.1644/10-mamm-a-404.1.

Grogan, L. and A. Sadanand, 2013: Rural Electrification and Employment in Poor Countries: Evidence from Nicaragua. *World Development*, **43**, 252–265, doi:10.1016/j.worlddev.2012.09.002.

Grubert, E.A., 2016: Water consumption from hydroelectricity in the United States. *Advances in Water Resources*, **96**, 88–94, doi:10.1016/j.advwatres.2016.07.004.

Grubler, A. and D. Fisk (eds.), 2013: *Energizing Sustainable Cities: Assessing Urban Energy*. Routledge Earthscan, Abingdon, UK and New York, NY, USA, 222 pp.

Guivarch, C., R. Crassous, O. Sassi, and S. Hallegatte, 2011: The costs of climate policies in a second-best world with labour market imperfections. *Climate Policy*, **11(1)**, 768–788, doi:10.3763/cpol.2009.0012.

Güssow, K., A. Proelss, A. Oschlies, K. Rehdanz, and W. Rickels, 2010: Ocean iron fertilization: why further research is needed. *Marine Policy*, **34(5)**, 911–918, doi:10.1016/j.marpol.2010.01.015.

Gyamfi, S., S. Krumdieck, and T. Urmee, 2013: Residential peak electricity demand response – Highlights of some behavioural issues. *Renewable and Sustainable Energy Reviews*, **25**, 71–77, doi:10.1016/j.rser.2013.04.006.

Ha, M. and M. Wu, 2017: Land management strategies for improving water quality in biomass production under changing climate. *Environmental Research Letters*, **12(3)**, 034015, doi:10.1088/1748-9326/aa5f32.

Haileselassie, M., H. Taddele, K. Adhana, and S. Kalayou, 2013: Food safety knowledge and practices of abattoir and butchery shops and the microbial profile of meat in Mekelle City, Ethiopia. *Asian Pacific Journal of Tropical Biomedicine*, **3(5)**, 407–412, doi:10.1016/s2221-1691(13)60085-4.

Haines, A. and C. Dora, 2012: How the low carbon economy can improve health. *BMJ*, **344**, 1–6, doi:10.1136/bmj.e1018.

Haines, A. et al., 2007: Policies for accelerating access to clean energy, improving health, advancing development, and mitigating climate change. *The Lancet*, **370(9594)**, 1264–1281, doi:10.1016/s0140-6736(07)61257-4.

Hallegatte, S. et al., 2016a: *Shock Waves: Managing the Impacts of Climate Change on Poverty*. The World Bank, Washington, DC, USA, 227 pp., doi:10.1596/978-1-4648-0673-5.

Hallegatte, S. et al., 2016b: Mapping the climate change challenge. *Nature Climate Change*, **6(7)**, 663–668, doi:10.1038/nclimate3057.

Hammond, J. et al., 2017: The Rural Household Multi-Indicator Survey (RHoMIS) for rapid characterisation of households to inform climate smart agriculture interventions: Description and applications in East Africa and Central America. *Agricultural Systems*, **151**, 225–233, doi:10.1016/j.agsy.2016.05.003.

Hartmann, J. et al., 2013: Enhanced chemical weathering as a geoengineering strategy to reduce atmospheric carbon dioxide, supply nutrients, and mitigate ocean acidification. *Reviews of Geophysics*, **51(2)**, 113–149, doi:10.1002/rog.20004.

Harvey, C.A. et al., 2014: Climate-Smart Landscapes: Opportunities and Challenges for Integrating Adaptation and Mitigation in Tropical Agriculture. *Conservation Letters*, **7(2)**, 77–90, doi:10.1111/conl.12066.

Hasegawa, T. et al., 2015: Consequence of climate mitigation on the risk of hunger. *Environmental Science & Technology*, **49(12)**, 7245–7253, doi:10.1021/es5051748.

Haves, E., 2012: *Does energy access help women? Beyond anecdotes: a review of the evidence*. Ashden, London, UK, 9 pp.

Havlík, P. et al., 2014: Climate change mitigation through livestock system transitions. *Proceedings of the National Academy of Sciences*, **111(10)**, 3709–14, doi:10.1073/pnas.1308044111.

He, R., Y. Xiong, and Z. Lin, 2016: Carbon emissions in a dual channel closed loop supply chain: the impact of consumer free riding behavior. *Journal of Cleaner Production*, **134(Part A)**, 384–394, doi:10.1016/j.jclepro.2016.02.142.

Hedenus, F., S. Wirsenius, and D.J.A. Johansson, 2014: The importance of reduced meat and dairy consumption for meeting stringent climate change targets. *Climatic Change*, **124(1–2)**, 79–91, doi:10.1007/s10584-014-1104-5.

Heinävaara, S. et al., 2010: Cancer incidence in the vicinity of Finnish nuclear power plants: an emphasis on childhood leukemia. *Cancer Causes & Control*, **21(4)**, 587–595, doi:10.1007/s10552-009-9488-7.

Heinonen, J., M. Jalas, J.K. Juntunen, S. Ala-Mantila, and S. Junnila, 2013: Situated lifestyles: II. The impacts of urban density, housing type and motorization on the greenhouse gas emissions of the middle-income consumers in Finland. *Environmental Research Letters*, **8(3)**, 035050, doi:10.1088/1748-9326/8/3/035050.

Hejazi, M.I. et al., 2015: 21st century United States emissions mitigation could increase water stress more than the climate change it is mitigating. *Proceedings of the National Academy of Sciences*, **112(34)**, 10635–40, doi:10.1073/pnas.1421675112.

Hendrickson, T.P. and A. Horvath, 2014: A perspective on cost-effectiveness of greenhouse gas reduction solutions in water distribution systems. *Environmental Research Letters*, **9(2)**, 024017, doi:10.1088/1748-9326/9/2/024017.

Herrero, M. and P.K. Thornton, 2013: Livestock and global change: Emerging issues for sustainable food systems. *Proceedings of the National Academy of Sciences*, **110(52)**, 20878–20881, doi:10.1073/pnas.1321844111.

Herrero, M. et al., 2013: Biomass use, production, feed efficiencies, and greenhouse gas emissions from global livestock systems. *Proceedings of the National Academy of Sciences*, **110(52)**, 20888–20893, doi:10.1073/pnas.1308149110.

Hertwich, E.G., M. Aaberg, B. Singh, and A.H. Strømman, 2008: Life-cycle Assessment of Carbon Dioxide Capture for Enhanced Oil Recovery. *Chinese Journal of Chemical Engineering*, **16(3)**, 343–353, doi:10.1016/s1004-9541(08)60085-3.

Hiç, C., P. Pradhan, D. Rybski, and J.P. Kropp, 2016: Food Surplus and Its Climate Burdens. *Environmental Science & Technology*, **50(8)**, 4269–4277, doi:10.1021/acs.est.5b05088.

Hillier, J., 2011: Strategic navigation across multiple planes: Towards a Deleuzean-inspired methodology for strategic spatial planning. *Town Planning Review*, **82(5)**, 503–527, doi:10.3828/tpr.2011.30.

Hirth, L. and F. Ueckerdt, 2013: Redistribution effects of energy and climate policy: The electricity market. *Energy Policy*, **62**, 934–947, doi:10.1016/j.enpol.2013.07.055.

Hiyama, A. et al., 2013: The Fukushima nuclear accident and the pale grass blue butterfly: evaluating biological effects of long-term low-dose exposures. *BMC evolutionary biology*, **13(1)**, 168, doi:10.1186/1471-2148-13-168.

Holland, R.A. et al., 2015: Global impacts of energy demand on the freshwater resources of nations. *Proceedings of the National Academy of Sciences*, **112(48)**, E6707–E6716, doi:10.1073/pnas.1507701112.

Holopainen, R. et al., 2014: Comfort assessment in the context of sustainable buildings: Comparison of simplified and detailed human thermal sensation methods. *Building and Environment*, **71**, 60–70, doi:10.1016/j.buildenv.2013.09.009.

Hontelez, J., 2016: Advancing SDG Implementation through Forest Certification. International Institute for Sustainable Development (IISD), Winnipeg, MN, Canada. Retrieved from: http://sdg.iisd.org/commentary/guest-articles/advancing-sdg-implementation-through-forest-certification.

Hori, S., K. Kondo, D. Nogata, and H. Ben, 2013: The determinants of household energy-saving behavior: Survey and comparison in five major Asian cities. *Energy Policy*, **52**, 354–362, doi:10.1016/j.enpol.2012.09.043.

Huang, W., A. Wilkes, X. Sun, and A. Terheggen, 2013: Who is importing forest products from Africa to China? An analysis of implications for initiatives to enhance legality and sustainability. *Environment, Development and Sustainability*, **15(2)**, 339–354, doi:10.1007/s10668-012-9413-1.

Huebner, G.M., J. Cooper, and K. Jones, 2013: Domestic energy consumption – What role do comfort, habit, and knowledge about the heating system play? *Energy and Buildings*, **66**, 626–636, doi:10.1016/j.enbuild.2013.07.043.

Hult, A. and J. Larsson, 2016: Possibilities and problems with applying a consumption perspective in local climate strategies – the case of Gothenburg, Sweden. *Journal of Cleaner Production*, **134**, 434–442, doi:10.1016/j.jclepro.2015.10.033.

Hwang, J., K. Joh, and A. Woo, 2017: Social inequalities in child pedestrian traffic injuries: Differences in neighborhood built environments near schools in Austin, TX, USA. *Journal of Transport and Health*, **6**, 40–49, doi:10.1016/j.jth.2017.05.003.

ICSU and ISSC, 2015: *Review of targets for the Sustainable Development Goals: The Science Perspective*. International Council for Science (ICSU), Paris, France, 92 pp., doi:978-0-930357-97-9.

5

IEA, 2016: *Energy and Air Pollution: World Energy Outlook Special Report.* International Energy Agency (IEA), Paris, France, 266 pp.

Inger, R. et al., 2009: Marine renewable energy: potential benefits to biodiversity? An urgent call for research. *Journal of Applied Ecology*, **46(6)**, 1145–1153, doi:10.1111/j.1365-2664.2009.01697.x.

Ingram, J., 2011: A food systems approach to researching food security and its interactions with global environmental change. *Food Security*, **3(4)**, 417–431, doi:10.1007/s12571-011-0149-9.

IPCC, 2005: IPCC Special Report on Carbon Dioxide Capture and Storage. [Metz, B., O. Davidson, H.C. de Coninck, M. Loos, and L.A. Meyer (eds.)]. Prepared by Working Group III of the Intergovernmental Panel on Climate Change. Cambridge University Press, Cambridge, United Kingdom and New York, NY, USA, 442 pp.

IPCC, 2014: Climate Change 2014: Mitigation of Climate Change. Contribution of Working Group III to the Fifth Assessment Report of the Intergovernmental Panel on Climate Change. [Edenhofer, O., R. Pichs-Madruga, Y. Sokona, E. Farahani, S. Kadner, K. Seyboth, A. Adler, I. Baum, S. Brunner, P. Eickemeier, B. Kriemann, J. Savolainen, S. Schlömer, C. von Stechow, T. Zwickel, and J.C. Minx (eds.)]. Cambridge University Press, Cambridge, United Kingdom and New York, NY, USA, 1454 pp.

IRENA, 2016: *Renewable Energy and Jobs – Annual Review 2016.* International Renewable Energy Agency (IRENA), Abu Dhabi, UAE, 19 pp.

Isenhour, C. and K. Feng, 2016: Decoupling and displaced emissions: on Swedish consumers, Chinese producers and policy to address the climate impact of consumption. *Journal of Cleaner Production*, **134(Part A)**, 320–329, doi:10.1016/j.jclepro.2014.12.037.

Islar, M., S. Brogaard, and M. Lemberg-Pedersen, 2017: Feasibility of energy justice: Exploring national and local efforts for energy development in Nepal. *Energy Policy*, **105**, 668–676, doi:10.1016/j.enpol.2017.03.004.

Jackson, T. and P. Senker, 2011: Prosperity without Growth: Economics for a Finite Planet. *Energy & Environment*, **22(7)**, 1013–1016, doi:10.1260/0958-305x.22.7.1013.

Jain, A.A., R.R. Koford, A.W. Handcock, and G.G. Zenner, 2011: Bat mortality and activity at a northern Iowa wind resource area. *The American Midland Naturalist*, **165(1)**, 185–200, doi:10.1674/0003-0031-165.1.185.

Jain, R.K., R. Gulbinas, J.E. Taylor, and P.J. Culligan, 2013: Can social influence drive energy savings? Detecting the impact of social influence on the energy consumption behavior of networked users exposed to normative eco-feedback. *Energy and Buildings*, **66**, 119–127, doi:10.1016/j.enbuild.2013.06.029.

Jain, S. and M.P. Sharma, 2010: Prospects of biodiesel from Jatropha in India: A review. *Renewable and Sustainable Energy Reviews*, **14(2)**, 763–771, doi:10.1016/j.rser.2009.10.005.

Jakob, M. and J.C. Steckel, 2014: How climate change mitigation could harm development in poor countries. *Wiley Interdisciplinary Reviews: Climate Change*, **5(2)**, 161–168, doi:10.1002/wcc.260.

Jin, X. and N. Gruber, 2003: Offsetting the radiative benefit of ocean iron fertilization by enhancing N_2O emissions. *Geophysical Research Letters*, **30(24)**, 1–4, doi:10.1029/2003gl018458.

Jingura, R. and R. Kamusoko, 2016: The energy-development nexus in Sub-Saharan Africa. In: *Handbook on Africa: Challenges and Issues for the 21st Century* [Sherman, W. (ed.)]. Nova Science Publishers, Hauppauge, NY, USA, pp. 25–46.

Johansson, M.T. and P. Thollander, 2018: A review of barriers to and driving forces for improved energy efficiency in Swedish industry – Recommendations for successful in-house energy management. *Renewable and Sustainable Energy Reviews*, **82**, 618–628, doi:10.1016/j.rser.2017.09.052.

Johnson, J.M.-F., A.J. Franzluebbers, S.L. Weyers, and D.C. Reicosky, 2007: Agricultural opportunities to mitigate greenhouse gas emissions. *Environmental Pollution*, **150(1)**, 107–124, doi:10.1016/j.envpol.2007.06.030.

Johnson, N. et al., 2015: Stranded on a low-carbon planet: implications of climate policy for the phase-out of coal-based power plants. *Technological Forecasting and Social Change*, **90(Part A)**, 89–102, doi:10.1016/j.techfore.2014.02.028.

Jones, B.A. and S.M. McDermott, 2018: The economics of urban afforestation: Insights from an integrated bioeconomic-health model. *Journal of Environmental Economics and Management*, **89**, 116–135, doi:10.1016/j.jeem.2018.03.007.

Jost, C. et al., 2016: Understanding gender dimensions of agriculture and climate change in smallholder farming communities. *Climate and Development*, **8(2)**, 133–144, doi:10.1080/17565529.2015.1050978.

Kaatsch, P., C. Spix, R. Schulze-Rath, S. Schmiedel, and M. Blettner, 2008: Leukaemia in young children living in the vicinity of German nuclear power plants. *International Journal of Cancer*, **122(4)**, 721–726, doi:10.1002/ijc.23330.

Kagawa, S. et al., 2015: CO_2 emission clusters within global supply chain networks: Implications for climate change mitigation. *Global Environmental Change*, **35**, 486–496, doi:10.1016/j.gloenvcha.2015.04.003.

Kahn Ribeiro, S. et al., 2012: Energy End-Use: Transport. In: *Global Energy Assessment – Toward a Sustainable Future* [Johansson, T.B., N. Nakicenovic, A. Patwardhan, and L. Gomez-Echeverri (eds.)]. Cambridge University Press, Cambridge, United Kingdom and New York, NY, USA and the International Institute for Applied Systems Analysis, Laxenburg, Austria, pp. 575–648.

Karekezi, S., S. McDade, B. Boardman, and J. Kimani, 2012: Energy, Poverty and Development. In: *Global Energy Assessment – Toward a Sustainable Future* [Johansson, T.B., N. Nakicenovic, A. Patwardhan, and L. Gomez-Echeverri (eds.)]. Cambridge University Press, Cambridge, United Kingdom and New York, NY, USA and the International Institute for Applied Systems Analysis, Laxenburg, Austria, pp. 151–190.

Karner, K., M. Theissing, and T. Kienberger, 2015: Energy efficiency for industries through synergies with urban areas. *Journal of Cleaner Production*, **2020**, 1–11, doi:10.1016/j.jclepro.2016.02.010.

Katila, P., W. de Jong, G. Galloway, B. Pokorny, and P. Pacheco, 2017: *Building on synergies: Harnessing community and smallholder forestry for Sustainable Development Goals.* International Union of Forest Research Organizations (IUFRO), 23 pp.

Kaygusuz, K., 2011: Energy services and energy poverty for sustainable rural development. *Renewable and Sustainable Energy Reviews*, **15**, 936–947, doi:10.1016/j.rser.2010.11.003.

Kern, J.D., D. Patino-Echeverri, and G.W. Characklis, 2014: The Impacts of Wind Power Integration on Sub-Daily Variation in River Flows Downstream of Hydroelectric Dams. *Environmental Science & Technology*, **48(16)**, 9844–9851, doi:10.1021/es405437h.

Khan, M., G. Srafeim, and A. Yoon, 2015: *Corporate Sustainability: First Evidence on Materiality.* HBS Working Paper Number: 15-073, Harvard Business School, Cambridge, MA, USA, 55 pp.

Khan, S., M.A. Hanjra, and J. Mu, 2009: Water management and crop production for food security in China: A review. *Agricultural Water Management*, **96(3)**, 349–360, doi:10.1016/j.agwat.2008.09.022.

Kibria, G., 2015: *Climate Resilient Development (CRD), Sustainable Development Goals (SDGs) & Climate Finance (CF) – A Case Study.* ResearchGate, 4 pp., doi:10.13140/rg.2.1.4393.2240.

Kim, Y. and C. Sun, 2017: The Energy-Efficient Adaptation Scheme for Residential Buildings in Kazakhstan. *Energy Procedia*, **118**, 28–34, doi:10.1016/j.egypro.2017.07.005.

Kim, Y. et al., 2017: A perspective on climate-resilient development and national adaptation planning based on USAID's experience. *Climate and Development*, **9(2)**, 141–151, doi:10.1080/17565529.2015.1124037.

Kirubi, C., A. Jacobson, D.M. Kammen, and A. Mills, 2009: Community-Based Electric Micro-Grids Can Contribute to Rural Development: Evidence from Kenya. *World Development*, **37(7)**, 1208–1221, doi:10.1016/j.worlddev.2008.11.005.

Klausbruckner, C., H. Annegarn, L.R.F. Henneman, and P. Rafaj, 2016: A policy review of synergies and trade-offs in South African climate change mitigation and air pollution control strategies. *Environmental Science & Policy*, **57**, 70–78, doi:10.1016/j.envsci.2015.12.001.

Köhler, P., J. Hartmann, and D.A. Wolf-Gladrow, 2010: Geoengineering potential of artificially enhanced silicate weathering of olivine. *Proceedings of the National Academy of Sciences*, **107(47)**, 20228–20233, doi:10.1073/pnas.1000545107.

Köhler, P., J.F. Abrams, C. Völker, J. Hauck, and D.A. Wolf-Gladrow, 2013: Geoengineering impact of open ocean dissolution of olivine on atmospheric CO_2, surface ocean pH and marine biology. *Environmental Research Letters*, **8(1)**, 014009, doi:10.1088/1748-9326/8/1/014009.

Köhlin, G., E.O. Sills, S.K. Pattanayak, and C. Wilfong, 2011: *Energy, Gender and Development: What are the Linkages? Where is the Evidence?* A background paper for the World Development Report 2012 on Gender Equality and Development. Paper No. 125, World Bank, Washington DC, USA, 75 pp.

Kong, X. et al., 2016: Groundwater Depletion by Agricultural Intensification in China's HHH Plains, Since 1980s. *Advances in Agronomy*, **135**, 59–106, doi:10.1016/bs.agron.2015.09.003.

Koornneef, J., A. Ramírez, W. Turkenburg, and A. Faaij, 2011: The environmental impact and risk assessment of CO_2 capture, transport and storage-an evaluation of the knowledge base using the DPSIR framework. *Energy Procedia*, **4**, 2293–2300, doi:10.1016/j.egypro.2011.02.119.

Kowarik, I., 2018: Urban wilderness: Supply, demand, and access. *Urban Forestry & Urban Greening*, **29**, 336–347, doi:10.1016/j.ufug.2017.05.017.

5

Krukowski, R.A., C. Sparks, M. Dicarlo, J. McSweeney, and D.S. West, 2013: There's more to food store choice than proximity: A questionnaire development study. *BMC Public Health*, **13(1)**, 586, doi:10.1186/1471-2458-13-586.

Kumar, A. et al., 2011: Hydropower. In: *IPCC Special Report on Renewable Energy Sources and Climate Change Mitigation* [Edenhofer, O., R. Pichs-Madruga, Y. Sokona, K. Seyboth, P. Matschoss, S. Kadner, T. Zwickel, P. Eickemeier, G. Hansen, S. Schlömer, and C. von Stechow (eds.)]. Prepared by Working Group III of the Intergovernmental Panel on Climate Change. Cambridge University Press, Cambridge, United Kingdom and New York, NY, USA, pp. 437–496.

Kumar, N., P. Besuner, S. Lefton, D. Agan, and D. Hilleman, 2012: *Power Plant Cycling Costs*. NREL/SR-5500-55433, National Renewable Energy Laboratory (NREL), Golden, CO, USA, 80 pp.

Kummu, M. et al., 2012: Lost food, wasted resources: Global food supply chain losses and their impacts on freshwater, cropland, and fertiliser use. *Science of The Total Environment*, **438**, 477–489, doi:10.1016/j.scitotenv.2012.08.092.

Kunz, M.J., A. Wüest, B. Wehrli, J. Landert, and D.B. Senn, 2011: Impact of a large tropical reservoir on riverine transport of sediment, carbon, and nutrients to downstream wetlands. *Water Resources Research*, **47(12)**, doi:10.1029/2011wr010996.

Kunze, C. and S. Becker, 2015: Collective ownership in renewable energy and opportunities for sustainable degrowth. *Sustainability Science*, **10**, 425–437, doi:10.1007/s11625-015-0301-0.

Kwong, Q.J., N.M. Adam, and B.B. Sahari, 2014: Thermal comfort assessment and potential for energy efficiency enhancement in modern tropical buildings: A review. *Energy and Buildings*, **68(Part A)**, 547–557, doi:10.1016/j.enbuild.2013.09.034.

Lakshmi, K., C. Anuradha, K. Boomiraj, and A. Kalaivani, 2015: Applications of Biotechnological Tools to Overcome Climate Change and its Effects on Agriculture. *Research News For U (RNFU)*, **20**, 218–222.

Lall, S.V. et al., 2017: *Africa's Cities: Opening Doors to the World*. World Bank, Washington DC, USA, 162 pp., doi:10.1596/978-1-4648-1044-2.

Lam, N.L., K.R. Smith, A. Gauthier, and M.N. Bates, 2012: Kerosene: A Review of Household Uses and their Hazards in Low- and Middle-Income Countries. *Journal of Toxicology and Environmental Health, Part B*, **15(6)**, 396–432, doi:10.1080/10937404.2012.710134.

Lamb, A. et al., 2016: The potential for land sparing to offset greenhouse gas emissions from agriculture. *Nature Climate Change*, **6**, 488–492, doi:10.1038/nclimate2910.

Lampitt, R.S. et al., 2008: Ocean fertilization: a potential means of geoengineering? *Philosophical Transactions of the Royal Society A: Mathematical, Physical and Engineering Sciences*, **366(1882)**, 3919–3945, doi:10.1098/rsta.2008.0139.

Larson, A.M., T. Dokken, and A.E. Duchelle, 2014: The role of women in early REDD+ implementation. *REDD+ on the ground: A case book of subnational initiatives across the globe*, **17(1)**, 440–441, doi:10.1505/146554815814725031.

Launiainen, S. et al., 2014: Is the Water Footprint an Appropriate Tool for Forestry and Forest Products: The Fennoscandian Case. *Ambio*, **43(2)**, 244–256, doi:10.1007/s13280-013-0380-z.

Lawrence, A., M. Karlsson, and P. Thollander, 2018: Effects of firm characteristics and energy management for improving energy efficiency in the pulp and paper industry. *Energy*, **153**, 825–835, doi:10.1016/j.energy.2018.04.092.

LeDoux, T.F. and I. Vojnovic, 2013: Going outside the neighborhood: The shopping patterns and adaptations of disadvantaged consumers living in the lower eastside neighborhoods of Detroit, Michigan. *Health & Place*, **19**, 1–14, doi:10.1016/j.healthplace.2012.09.010.

Li, T. et al., 2016: Aging will amplify the heat-related mortality risk under a changing climate: projection for the elderly in Beijing, China. *Scientific Reports*, **6(1)**, 28161, doi:10.1038/srep28161.

Liddell, C. and C. Guiney, 2015: Living in a cold and damp home: Frameworks for understanding impacts on mental well-being. *Public Health*, **129(3)**, 191–199, doi:10.1016/j.puhe.2014.11.007.

Lieder, M. and A. Rashid, 2016: Towards circular economy implementation: A comprehensive review in context of manufacturing industry. *Journal of Cleaner Production*, **115**, 36–51, doi:10.1016/j.jclepro.2015.12.042.

Lim, S.S. et al., 2012: A comparative risk assessment of burden of disease and injury attributable to 67 risk factors and risk factor clusters in 21 regions, 1990-2010: a systematic analysis for the Global Burden of Disease Study 2010. *The Lancet*, **380(9859)**, 2224–2260, doi:10.1016/s0140-6736(12)61766-8.

Lin, J., Y. Hu, S. Cui, J. Kang, and A. Ramaswami, 2015: Tracking urban carbon footprints from production and consumption perspectives. *Environmental Research Letters*, **10(5)**, 054001, doi:10.1088/1748-9326/10/5/054001.

Lipper, L. et al., 2014: Climate-smart agriculture for food security. *Nature Climate Change*, **4(12)**, 1068–1072, doi:10.1038/nclimate2437.

Lipscomb, M., A.M. Mobarak, and T. Barham, 2013: Development Effects of Electrification: Evidence from the Topographic Placement of Hydropower Plants in Brazil. *American Economic Journal: Applied Economics*, **5(2)**, 200–231, doi:10.1257/app.5.2.200.

Liu, Y. and Y. Bai, 2014: An exploration of firms' awareness and behavior of developing circular economy: An empirical research in China. *Resources, Conservation and Recycling*, **87**, 145–152, doi:10.1016/j.resconrec.2014.04.002.

Long, T.B., V. Blok, and I. Coninx, 2016: Barriers to the adoption and diffusion of technological innovations for climate-smart agriculture in Europe: evidence from the Netherlands, France, Switzerland and Italy. *Journal of Cleaner Production*, **112**, 9–21, doi:10.1016/j.jclepro.2015.06.044.

Lotze-Campen, H. et al., 2014: Impacts of increased bioenergy demand on global food markets: an AgMIP economic model intercomparison. *Agricultural Economics*, **45(1)**, 103–116, doi:10.1111/agec.12092.

Lovich, J.E. and J.R. Ennen, 2013: Assessing the state of knowledge of utility-scale wind energy development and operation on non-volant terrestrial and marine wildlife. *Applied Energy*, **103**, 52–60, doi:10.1016/j.apenergy.2012.10.001.

Lowery, B., D. Sloane, D. Payán, J. Illum, and L. Lewis, 2016: Do Farmers' Markets Increase Access to Healthy Foods for All Communities? Comparing Markets in 24 Neighborhoods in Los Angeles. *Journal of the American Planning Association*, **82(3)**, 252–266, doi:10.1080/01944363.2016.1181000.

Lucas, K. and K. Pangbourne, 2014: Assessing the equity of carbon mitigation policies for transport in Scotland. *Case Studies on Transport Policy*, **2(2)**, 70–80, doi:10.1016/j.cstp.2014.05.003.

Lybbert, T. and D. Sumner, 2010: *Agricultural Technologies for Climate Change Mitigation and Adaptation in Developing Countries: Policy Options for Innovation and Technology Diffusion*. ICTSD–IPC Platform on Climate Change, Agriculture and Trade, Issue Brief No.6, International Centre for Trade and Sustainable Development, Geneva, Switzerland and International Food & Agricultural Trade Policy Council, Washington DC, USA, 32 pp.

Brander, L.M. et al., 2012: Ecosystem service values for mangroves in Southeast Asia: a meta-analysis and value transfer application. *Ecosystem Services*, **1(1)**, 62–69, doi:10.1016/j.ecoser.2012.06.003.

Maidment, C.D., C.R. Jones, T.L. Webb, E.A. Hathway, and J.M. Gilbertson, 2014: The impact of household energy efficiency measures on health: A meta-analysis. *Energy Policy*, **65**, 583–593, doi:10.1016/j.enpol.2013.10.054.

Månsson, A., 2016: Energy security in a decarbonised transport sector: A scenario based analysis of Sweden's transport strategies. *Energy Strategy Reviews*, **13–14**, 236–247, doi:10.1016/j.esr.2016.06.004.

Manville, M., 2017: Travel and the Built Environment: Time for Change. *Journal of the American Planning Association*, **83(1)**, 29–32, doi:10.1080/01944363.2016.1249508.

Markusson, N. et al., 2012: A socio-technical framework for assessing the viability of carbon capture and storage technology. *Technological Forecasting and Social Change*, **79(5)**, 903–918, doi:10.1016/j.techfore.2011.12.001.

Marra, J.E. and R.A. Palmer, 2011: Radioactive Waste Management. In: *Waste* [Letcher, T.M. and D.A. Vallero (eds.)]. Academic Press, Boston, MA, USA, pp. 101–108, doi:10.1016/b978-0-12-381475-3.10007-5.

Martínez-Jaramillo, J.E., S. Arango-Aramburo, K.C. Álvarez-Uribe, and P. Jaramillo-Álvarez, 2017: Assessing the impacts of transport policies through energy system simulation: The case of the Medellin Metropolitan Area, Colombia. *Energy Policy*, **101**, 101–108, doi:10.1016/j.enpol.2016.11.026.

Matinga, M.N., 2012: *A socio-cultural perspective on transformation of gender roles and relations, and non-change in energy-health perceptions following electrification in rural South Africa (Case Study for World Development Report 2012)*. ETC/ENERGIA in association Nord/Sør-konsulentene, 17 pp.

Matthews, N. and M. McCartney, 2018: Opportunities for building resilience and lessons for navigating risks: Dams and the water energy food nexus. *Environmental Progress & Sustainable Energy*, **37(1)**, 56–61, doi:10.1002/ep.12568.

Mbow, C. et al., 2014: Agroforestry solutions to address food security and climate change challenges in Africa. *Current Opinion in Environmental Sustainability*, **6**, 61–67, doi:10.1016/j.cosust.2013.10.014.

McCarthy, N., L. Lipper, and G. Branca, 2011: *Climate-Smart Agriculture: Smallholder Adoption and Implications for Climate Change Adaptation and Mitigation*. Mitigation of Climate Change Series 4, Food and Agricultural Organization of the United Nations (FAO), 25 pp.

5

McCollum, D.L. et al., 2018: Connecting the sustainable development goals by their energy inter-linkages. *Environmental Research Letters*, **13(3)**, 033006, doi:10.1088/1748-9326/aaafe3.

McKinney, M.L. and K. Ingo, 2018: The contribution of wild urban ecosystems to liveable cities. *Urban Forestry & Urban Greening*, **29**, 334–335, doi:10.1016/j.ufug.2017.09.004.

McLeod, R.S., C.J. Hopfe, and A. Kwan, 2013: An investigation into future performance and overheating risks in Passivhaus dwellings. *Building and Environment*, **70**, 189–209, doi:10.1016/j.buildenv.2013.08.024.

McPherson, E.G., A.M. Berry, and N.S. van Doorn, 2018: Performance testing to identify climate-ready trees. *Urban Forestry & Urban Greening*, **29**, 28–39, doi:10.1016/j.ufug.2017.09.003.

Meldrum, J., S. Nettles-Anderson, G. Heath, and J. Macknick, 2013: Life cycle water use for electricity generation: a review and harmonization of literature estimates. *Environmental Research Letters*, **8(1)**, 015031.

Meltzer, J.P., 2016: *Financing Low Carbon, Climate Resilient Sustainable Infrastructure: The Role of Climate Finance and Green Financial Systems*. Global Economy & Development Working Paper 96, Brookings Institution, Washington DC, USA, 52 pp.

Michler-Cieluch, T., G. Krause, and B.H. Buck, 2009: Reflections on integrating operation and maintenance activities of offshore wind farms and mariculture. *Ocean & Coastal Management*, **52(1)**, 57–68, doi:10.1016/j.ocecoaman.2008.09.008.

Miles, L. and V. Kapos, 2008: Reducing Greenhouse Gas Emissions from Deforestation and Forest Degradation: Global Land-Use Implications. *Science*, **320(5882)**, 1454–1455, doi:10.1126/science.1155358.

Miller, E., L.M. Bell, and L. Buys, 2007: Public understanding of carbon sequestration in Australia: socio-demographic predictors of knowledge, engagement and trust. *International Journal of Emerging Technologies and Society*, **5(1)**, 15–33.

Mittlefehldt, S., 2016: Seeing forests as fuel: How conflicting narratives have shaped woody biomass energy development in the United States since the 1970s. *Energy Research & Social Science*, **14**, 13–21, doi:10.1016/j.erss.2015.12.023.

Møller, A.P. and T.A. Mousseau, 2011: Conservation consequences of Chernobyl and other nuclear accidents. *Biological Conservation*, **144(12)**, 2787–2798, doi:10.1016/j.biocon.2011.08.009.

Møller, A.P., F. Barnier, and T.A. Mousseau, 2012: Ecosystems effects 25 years after Chernobyl: pollinators, fruit set and recruitment. *Oecologia*, **170(4)**, 1155–1165, doi:10.1007/s00442-012-2374-0.

Møller, A.P., A. Bonisoli-Alquati, G. Rudolfsen, and T.A. Mousseau, 2011: Chernobyl Birds Have Smaller Brains. *PLOS ONE*, **6(2)**, e16862, doi:10.1371/journal.pone.0016862.

Montanarella, L. and I.L. Alva, 2015: Putting soils on the agenda: the three Rio Conventions and the post-2015 development agenda. *Current Opinion in Environmental Sustainability*, **15**, 41–48, doi:10.1016/j.cosust.2015.07.008.

Moomaw, W. et al., 2011: Annex II: Methodology. In: *IPCC Special Report on Renewable Energy Sources and Climate Change Mitigation* [Edenhofer, O., R. Pichs-Madruga, Y. Sokona, K. Seyboth, P. Matschoss, S. Kadner, T. Zwickel, P. Eickemeier, G. Hansen, S. Schlömer, and C. von Stechow (eds.)]. Prepared by Working Group III of the Intergovernmental Panel on Climate Change. Cambridge University Press, Cambridge, United Kingdom and New York, NY, USA, pp. 973–1000.

Morton, J.F., 2007: The impact of climate change on smallholder and subsistence agriculture. *Proceedings of the National Academy of Sciences*, **104(50)**, 19680–19685, doi:10.1073/pnas.0701855104.

Mousseau, T.A. and A.P. Møller, 2013: Elevated Frequency of Cataracts in Birds from Chernobyl. *PLOS ONE*, **8(7)**, e66939, doi:10.1371/journal.pone.0066939.

Msangi, S., M. Ewing, M.W. Rosegrant, and T. Zhu, 2010: Biofuels, Food Security, and the Environment: A 2020/2050 Perspective. In: *Global Change: Impacts on Water and food Security* [Ringler, C., A.K. Biswas, and S. Cline (eds.)]. Springer Berlin Heidelberg, Berlin and Heidelberg, Germany, pp. 65–94, doi:10.1007/978-3-642-04615-5_4.

Mtui, G.Y.S., 2011: Involvement of biotechnology in climate change adaptation and mitigation: Improving agricultural yield and food security. *International Journal for Biotechnology and Molecular Biology Research*, **2(13)**, 222–231, www.academicjournals.org/ijbmbr/abstracts/abstracts/abstracts2011/30dec/mtui.htm.

Muratori, M., K. Calvin, M. Wise, P. Kyle, and J. Edmonds, 2016: Global economic consequences of deploying bioenergy with carbon capture and storage (BECCS). *Environmental Research Letters*, **11(9)**, 1–9, doi:10.1088/1748-9326/11/9/095004.

Muys, B. et al., 2014: Integrating mitigation and adaptation into development: the case of Jatropha curcas in sub-Saharan Africa. *GCB Bioenergy*, **6**, 169–171, doi:10.1111/gcbb.12070.

NCE, 2014: *Better Growth, Better Climate: Global Report*. New Climate Economy (NCE), Washington DC, USA, 308 pp.

NCE, 2015: *Seizing the Global Opportunity: Partnerships for Better Growth and a Better Climate*. New Climate Economy (NCE), Washington DC, USA and London, UK, 76 pp.

Nelson, V., K. Meadows, T. Cannon, J. Morton, and A. Martin, 2002: Uncertain Predictions, Invisible Impacts, and the Need to Mainstream Gender in Climate Change Adaptations. *Gender and Development*, **10(2)**, 51–59, www.jstor.org/stable/4030574.

Nemet, G.F., T. Holloway, and P. Meier, 2010: Implications of incorporating air-quality co-benefits into climate change policymaking. *Environmental Research Letters*, **5(1)**, 014007, doi:10.1088/1748-9326/5/1/014007.

Nguyen, M.T., S. Vink, M. Ziemski, and D.J. Barrett, 2014: Water and energy synergy and trade-off potentials in mine water management. *Journal of Cleaner Production*, **84(1)**, 629–638, doi:10.1016/j.jclepro.2014.01.063.

Noonan, D.S., L.-H.C. Hsieh, and D. Matisoff, 2015: Economic, sociological, and neighbor dimensions of energy efficiency adoption behaviors: Evidence from the U.S residential heating and air conditioning market. *Energy Research & Social Science*, **10**, 102–113, doi:10.1016/j.erss.2015.07.009.

Noris, F. et al., 2013: Indoor environmental quality benefits of apartment energy retrofits. *Building and Environment*, **68**, 170–178, doi:10.1016/j.buildenv.2013.07.003.

O'Mahony, T. and J. Dufour, 2015: Tracking development paths: Monitoring driving forces and the impact of carbon-free energy sources in Spain. *Environmental Science & Policy*, **50(2007)**, 62–73, doi:10.1016/j.envsci.2015.02.005.

O'Neill, B.C. et al., 2017: The roads ahead: narratives for shared socioeconomic pathways describing world futures in the 21st century. *Global Environmental Change*, **42**, 169–180, doi:10.1016/j.gloenvcha.2015.01.004.

OECD, 2017: *Economic Outlook for Southeast Asia, China and India 2017: Addressing Energy Challenges*. Organisation for Economic Co-operation and Development (OECD) Publishing, Paris, France, 261 pp.

Oschlies, A., W. Koeve, W. Rickels, and K. Rehdanz, 2010: Side effects and accounting aspects of hypothetical large-scale Southern Ocean iron fertilization. *Biogeosciences*, **7(12)**, 4014–4035, doi:10.5194/bg-7-4017-2010.

Ozturk, M. et al., 2017: Biomass and bioenergy: An overview of the development potential in Turkey and Malaysia. *Renewable and Sustainable Energy Reviews*, **79**, 1285–1302, doi:10.1016/j.rser.2017.05.111.

Pachauri, S. and N.D. Rao, 2013: Gender impacts and determinants of energy poverty: are we asking the right questions? *Current Opinion in Environmental Sustainability*, **5(2)**, 205–215, doi:10.1016/j.cosust.2013.04.006.

Pachauri, S. et al., 2012: Energy Access for Development. In: *Global Energy Assessment – Toward a Sustainable Future* [Johansson, T.B., N. Nakicenovic, A. Patwardhan, and L. Gomez-Echeverri (eds.)]. Cambridge University Press, Cambridge, United Kingdom and New York, NY, USA and the International Institute for Applied Systems Analysis (IIASA), Laxenburg, Austria, pp. 1401–1458.

Paquay, F.S. and R.E. Zeebe, 2013: Assessing possible consequences of ocean liming on ocean pH, atmospheric CO_2 concentration and associated costs. *International Journal of Greenhouse Gas Control*, **17**, 183–188, doi:10.1016/j.ijggc.2013.05.005.

Patel, S.J., J.H. Patel, A. Patel, and R.N. Gelani, 2016: Role of women gender in livestock sector: A review. *Journal of Livestock Science*, **7**, 92–96.

Pei, N. et al., 2018: Long-term afforestation efforts increase bird species diversity in Beijing, China. *Urban Forestry & Urban Greening*, **29**, 88–95, doi:10.1016/j.ufug.2017.11.007.

Peng, J. and L. Lu, 2013: Investigation on the development potential of rooftop PV system in Hong Kong and its environmental benefits. *Renewable and Sustainable Energy Reviews*, **27**, 149–162, doi:10.1016/j.rser.2013.06.030.

Peng, W., J. Yang, F. Wagner, and D.L. Mauzerall, 2017: Substantial air quality and climate co-benefits achievable now with sectoral mitigation strategies in China. *Science of The Total Environment*, **598**, 1076–1084, doi:10.1016/j.scitotenv.2017.03.287.

Pettit, L.R., C.W. Smart, M.B. Hart, M. Milazzo, and J.M. Hall-Spencer, 2015: Seaweed fails to prevent ocean acidification impact on foraminifera along a shallow-water CO_2 gradient. *Ecology and Evolution*, **5(9)**, 1784–1793, doi:10.1002/ece3.1475.

Pietzcker, R.C. et al., 2014: Long-term transport energy demand and climate policy: Alternative visions on transport decarbonization in energy-economy models. *Energy*, **64**, 95–108, doi:10.1016/j.energy.2013.08.059.

Pode, R., 2013: Financing LED solar home systems in developing countries. *Renewable and Sustainable Energy Reviews*, **25**, 596–629, doi:10.1016/j.rser.2013.04.004.

Potouroglou, M. et al., 2017: Measuring the role of seagrasses in regulating sediment surface elevation. *Scientific Reports*, **7(1)**, 1–11, doi:10.1038/s41598-017-12354-y.

Pueyo, A., F. Gonzalez, C. Dent, and S. DeMartino, 2013: *The Evidence of Benefits for Poor People of Increased Renewable Electricity Capacity: Literature Review.* Brief supporting Evidence Report 31, Institute of Development Studies (IDS), Brighton, UK, 8 pp.

Rafaj, P., W. Schöpp, P. Russ, C. Heyes, and M. Amann, 2013: Co-benefits of post-2012 global climate mitigation policies. *Mitigation and Adaptation Strategies for Global Change*, **18(6)**, 801–824, doi:10.1007/s11027-012-9390-6.

Raji, B., M.J. Tenpierik, and A. van den Dobbelsteen, 2015: The impact of greening systems on building energy performance: A literature review. *Renewable and Sustainable Energy Reviews*, **45**, 610–623, doi:10.1016/j.rser.2015.02.011.

Ramaker, J., J. Mackby, P.D. Marshall, and R. Geil, 2003: *The final test: a history of the comprehensive nuclear-test-ban treaty negotiations.* Provisional Technical Secretariat, Preparatory Commission for the Comprehensive Nuclear-Test-Ban Treaty Organization, 291 pp.

Ran, Y., M. Lannerstad, M. Herrero, C.E. Van Middelaar, and I.J.M. De Boer, 2016: Assessing water resource use in livestock production: A review of methods. *Livestock Science*, **187**, 68–79, doi:10.1016/j.livsci.2016.02.012.

Rao, N.D. and S. Pachauri, 2017: Energy access and living standards: some observations on recent trends. *Environmental Research Letters*, **12(2)**, 025011, doi:10.1088/1748-9326/aa5b0d.

Rao, N.D., K. Riahi, and A. Grubler, 2014: Climate impacts of poverty eradication. *Nature Climate Change*, **4(9)**, 749–751, doi:10.1038/nclimate2340.

Rao, S. et al., 2013: Better air for better health: forging synergies in policies for energy access, climate change and air pollution. *Global Environmental Change*, **23**, 1122–1130, doi:10.1016/j.gloenvcha.2013.05.003.

Rao, S. et al., 2016: A multi-model assessment of the co-benefits of climate mitigation for global air quality. *Environmental Research Letters*, **11(12)**, 124013, doi:10.1088/1748-9326/11/12/124013.

Raptis, C.E., M.T.H. van Vliet, and S. Pfister, 2016: Global thermal pollution of rivers from thermoelectric power plants. *Environmental Research Letters*, **11(10)**, 104011, doi:10.1088/1748-9326/11/10/104011.

Ravindranath, N.H., R.K. Chaturvedi, and I.K. Murthy, 2008: Forest conservation, afforestation and reforestation in India: Implications for forest carbon stocks. *Current Science*, **95(2)**, 216–222.

Reiner, D.M. and W.J. Nuttall, 2011: Public Acceptance of Geological Disposal of Carbon Dioxide and Radioactive Waste: Similarities and Differences. In: *Geological Disposal of Carbon Dioxide and Radioactive Waste: A Comparative Assessment* [Toth, F.L. (ed.)]. Springer Netherlands, Dordrecht, The Netherlands, pp. 295–315, doi:10.1007/978-90-481-8712-6_10.

Riahi, K. et al., 2012: Energy Pathways for Sustainable Development. In: *Global Energy Assessment – Toward a Sustainable Future* [Johansson, T.B., N. Nakicenovic, A. Patwardhan, and L. Gomez-Echeverri (eds.)]. Cambridge University Press, Cambridge, United Kingdom and New York, NY, USA and the International Institute for Applied Systems Analysis (IIASA), Laxenburg, Austria, pp. 1203–1306.

Riahi, K. et al., 2015: Locked into Copenhagen pledges – Implications of short-term emission targets for the cost and feasibility of long-term climate goals. *Technological Forecasting and Social Change*, **90**, 8–23, doi:10.1016/j.techfore.2013.09.016.

Riahi, K. et al., 2017: The Shared Socioeconomic Pathways and their energy, land use, and greenhouse gas emissions implications: an overview. *Global Environmental Change*, **42**, 153–168, doi:10.1016/j.gloenvcha.2016.05.009.

Rogalsky, J., 2010: The working poor and what GIS reveals about the possibilities of public transit. *Journal of Transport Geography*, **18(2)**, 226–237, doi:10.1016/j.jtrangeo.2009.06.008.

Rogelj, J., D.L. McCollum, and K. Riahi, 2013: The UN's 'Sustainable Energy for All' initiative is compatible with a warming limit of 2°C. *Nature Climate Change*, **3(6)**, 545–551, doi:10.1038/nclimate1806.

Rogner, H.-H., 2010: Nuclear Power and Sustainable Development. *Journal of International Affairs*, **64(1)**, 137–163, www.jstor.org/stable/24385190.

Rose, S.K. et al., 2014: Bioenergy in energy transformation and climate management. *Climatic Change*, **123(3–4)**, 477–493, doi:10.1007/s10584-013-0965-3.

Roy, J. et al., 2018: Where is the hope? Blending modern urban lifestyle with cultural practices in India. *Current Opinion in Environmental Sustainability*, **31**, 96–103, doi:10.1016/j.cosust.2018.01.010.

Rud, J.P., 2012: Electricity provision and industrial development: Evidence from India. *Journal of Development Economics*, **97(2)**, 352–367, doi:10.1016/j.jdeveco.2011.06.010.

Sagan, S.D., 2011: The Causes of Nuclear Weapons Proliferation. *Annual Review of Political Science*, **14(1)**, 225–244, doi:10.1146/annurev-polisci-052209-131042.

Salvalai, G., M.M. Sesana, and G. Iannaccone, 2017: Deep renovation of multi-storey multi-owner existing residential buildings: a pilot case study in Italy. *Energy and Buildings*, **148**, 23–36, doi:10.1016/j.enbuild.2017.05.011.

Sansoucy, R., 1995: Livestock – a driving force for food security and sustainable development. *World Animal Review*, **84/85(2)**.

Santarius, T., H.J. Walnum, and C. Aall (eds.), 2016: *Rethinking Climate and Energy Policies: New Perspectives on the Rebound Phenomenon.* Springer, Cham, Switzerland, 294 pp., doi:10.1007/978-3-319-38807-6.

Sarin, R., M. Sharma, S. Sinharay, and R.K. Malhotra, 2007: Jatropha-Palm biodiesel blends: An optimum mix for Asia. *Fuel*, **86(10–11)**, 1365–1371, doi:10.1016/j.fuel.2006.11.040.

Satolo, L. and M. Bacchi, 2013: Impacts of the Recent Expansion of the Sugarcane Sector on Municipal per Capita Income in São Paulo State. *ISRN Economics*, **2013**, 828169, doi:10.1155/2013/828169.

Saunders, L.E., J.M. Green, M.P. Petticrew, R. Steinbach, and H. Roberts, 2013: What Are the Health Benefits of Active Travel? A Systematic Review of Trials and Cohort Studies. *PLOS ONE*, **8**, e69912, doi:10.1371/journal.pone.0069912.

Schader, C. et al., 2015: Impacts of feeding less food-competing feedstuffs to livestock on global food system sustainability. *Journal of The Royal Society Interface*, **12(113)**, doi:10.1098/rsif.2015.0891.

Scherr, S.J., S. Shames, and R. Friedman, 2012: From climate-smart agriculture to climate-smart landscapes. *Agriculture & Food Security*, **1(1)**, 12, doi:10.1186/2048-7010-1-12.

Schirmer, J. and L. Bull, 2014: Assessing the likelihood of widespread landholder adoption of afforestation and reforestation projects. *Global Environmental Change*, **24**, 306–320, doi:10.1016/j.gloenvcha.2013.11.009.

Schnelzer, M., G.P. Hammer, M. Kreuzer, A. Tschense, and B. Grosche, 2010: Accounting for smoking in the radon-related lung cancer risk among German uranium miners: results of a nested case-control study. *Health Physics*, **98(1)**, 20–28, doi:10.1097/hp.0b013e3181b8ce81.

Schucht, S. et al., 2015: Moving towards ambitious climate policies: Monetised health benefits from improved air quality could offset mitigation costs in Europe. *Environmental Science & Policy*, **50**, 252–269, doi:10.1016/j.envsci.2015.03.001.

Schwanitz, V.J., F. Piontek, C. Bertram, and G. Luderer, 2014: Long-term climate policy implications of phasing out fossil fuel subsidies. *Energy Policy*, **67**, 882–894, doi:10.1016/j.enpol.2013.12.015.

Schwenk-Ferrero, A., 2013: German Spent Nuclear Fuel Legacy: Characteristics and High-Level Waste Management Issues. *Science and Technology of Nuclear Installations*, 293792, doi:10.1155/2013/293792.

Schwerhoff, G. and M. Sy, 2017: Financing renewable energy in Africa – Key challenge of the sustainable development goals. *Renewable and Sustainable Energy Reviews*, **75**, 393–401, doi:10.1016/j.rser.2016.11.004.

Scott, C.A., 2011: The water-energy-climate nexus: Resources and policy outlook for aquifers in Mexico. *Water Resources Research*, **47(6)**, doi:10.1029/2011wr010805.

Scott, C.A. et al., 2011: Policy and institutional dimensions of the water–energy nexus. *Energy Policy*, **39(10)**, 6622–6630, doi:10.1016/j.enpol.2011.08.013.

Scott, F.L., C.R. Jones, and T.L. Webb, 2014: What do people living in deprived communities in the UK think about household energy efficiency interventions? *Energy Policy*, **66**, 335–349, doi:10.1016/j.enpol.2013.10.084.

Sermage-Faure, C. et al., 2012: Childhood leukemia around French nuclear power plants – The geocap study, 2002-2007. *International Journal of Cancer*, **131(5)**, E769–E780, doi:10.1002/ijc.27425.

Shackley, S. and M. Thompson, 2012: Lost in the mix: will the technologies of carbon dioxide capture and storage provide us with a breathing space as we strive to make the transition from fossil fuels to renewables? *Climatic Change*, **110(1)**, 101–121, doi:10.1007/s10584-011-0071-3.

5

Shackley, S. et al., 2009: The acceptability of CO_2 capture and storage (CCS) in Europe: An assessment of the key determining factors: Part 2. The social acceptability of CCS and the wider impacts and repercussions of its implementation. *International Journal of Greenhouse Gas Control*, **3(3)**, 344–356, doi:10.1016/j.ijggc.2008.09.004.

Shah, N., M. Wei, V. Letschert, and A. Phadke, 2015: *Benefits of leapfrogging to superefficiency and low global warming potential refrigerants in room air conditioning*. LBNL-1003671, Ernest Orlando Lawrence Berkeley National Laboratory (LBNL), Berkeley, CA, USA, 58 pp.

Shahbaz, M., G. Rasool, K. Ahmed, and M.K. Mahalik, 2016: Considering the effect of biomass energy consumption on economic growth: fresh evidence from BRICS region. *Renewable and Sustainable Energy Reviews*, **60**, 1442–1450, doi:10.1016/j.rser.2016.03.037.

Sharpe, R.A., C.R. Thornton, V. Nikolaou, and N.J. Osborne, 2015: Higher energy efficient homes are associated with increased risk of doctor diagnosed asthma in a UK subpopulation. *Environment International*, **75**, 234–244, doi:10.1016/j.envint.2014.11.017.

Shaw, C., S. Hales, P. Howden-Chapman, and R. Edwards, 2014: Health co-benefits of climate change mitigation policies in the transport sector. *Nature Climate Change*, **4(6)**, 427–433, doi:10.1038/nclimate2247.

Shaw, C., S. Hales, R. Edwards, and P. Howden-Chapman, 2017: Health Co-Benefits of Policies to Mitigate Climate Change in the Transport Sector: Systematic Review. *Journal of Transport & Health*, **5**, S107–S108, doi:10.1016/j.jth.2017.05.268.

Shi, Y., J. Liu, H. Shi, H. Li, and Q. Li, 2017: The ecosystem service value as a new eco-efficiency indicator for industrial parks. *Journal of Cleaner Production*, **164**, 597–605, doi:10.1016/j.jclepro.2017.06.187.

Siirila, E.R., A.K. Navarre-Sitchler, R.M. Maxwell, and J.E. McCray, 2012: A quantitative methodology to assess the risks to human health from CO_2 leakage into groundwater. *Advances in Water Resources*, **36**, 146–164, doi:10.1016/j.advwatres.2010.11.005.

Sikkema, R. et al., 2014: Legal Harvesting, Sustainable Sourcing and Cascaded Use of Wood for Bioenergy: Their Coverage through Existing Certification Frameworks for Sustainable Forest Management. *Forests*, **5(9)**, 2163–2211, doi:10.3390/f5092163.

Singh, B., A.H. Strømman, and E.G. Hertwich, 2011: Comparative life cycle environmental assessment of CCS technologies. *International Journal of Greenhouse Gas Control*, **5(4)**, 911–921, doi:10.1016/j.ijggc.2011.03.012.

Sippo, J.Z., D.T. Maher, D.R. Tait, C. Holloway, and I.R. Santos, 2016: Are mangroves drivers or buffers of coastal acidification? Insights from alkalinity and dissolved inorganic carbon export estimates across a latitudinal transect. *Global Biogeochemical Cycles*, **30(5)**, 753–766, doi:10.1002/2015gb005324.

Sjoberg, L. and B.M.D. Sjoberg, 2009: Public risk perception of nuclear waste. *International Journal of Risk Assessment and Management*, **11(3/4)**, doi:10.1504/ijram.2009.023156.

Skipperud, L. et al., 2013: Environmental impact assessment of radionuclide and metal contamination at the former U sites Taboshar and Digmai, Tajikistan. *Journal of Environmental Radioactivity*, **123**, 50–62, doi:10.1016/j.jenvrad.2012.05.007.

SLoCaT, 2017: *Marrakech Partnership for Global Climate Action (MPGCA) Transport Initiatives: Stock-take on action toward implementation of the Paris Agreement and the 2030 Agenda on Sustainable Development. Second Progress Report.* Partnership on Sustainable Low Carbon Transport (SLoCaT), Bonn, Germany, 72 pp.

Smetacek, V. and S.W.A. Naqvi, 2008: The next generation of iron fertilization experiments in the Southern Ocean. *Philosophical Transactions of the Royal Society A: Mathematical, Physical and Engineering Sciences*, **366(1882)**, 3947–3967, doi:10.1098/rsta.2008.0144.

Smith, A., A. Pridmore, K. Hampshire, C. Ahlgren, and J. Goodwin, 2016: *Scoping study on the co-benefits and possible adverse side effects of climate change mitigation: Final report*. Report to the UK Department of Energy and Climate Change (DECC). Aether Ltd, Oxford, UK, 91 pp.

Smith, J. et al., 2014: What is the potential for biogas digesters to improve soil fertility and crop production in Sub-Saharan Africa? *Biomass and Bioenergy*, **70**, 58–72, doi:10.1016/j.biombioe.2014.02.030.

Smith, K.R. and A. Sagar, 2014: Making the clean available: escaping India's Chulha Trap. *Energy Policy*, **75**, 410–414, doi:10.1016/j.enpol.2014.09.024.

Smith, P., 2013: Delivering food security without increasing pressure on land. *Global Food Security*, **2(1)**, 18–23, doi:10.1016/j.gfs.2012.11.008.

Smith, P. et al., 2010: Competition for land. *Philosophical Transactions of the Royal Society B: Biological Sciences*, **365**, 2941–2957, doi:10.1098/rstb.2010.0127.

Smith, P. et al., 2013: How much land-based greenhouse gas mitigation can be achieved without compromising food security and environmental goals? *Global Change Biology*, **19(8)**, 2285–2302, doi:10.1111/gcb.12160.

Smith, P. et al., 2016a: Biophysical and economic limits to negative CO_2 emissions. *Nature Climate Change*, **6(1)**, 42–50, doi:10.1038/nclimate2870.

Smith, P. et al., 2016b: Global change pressures on soils from land use and management. *Global Change Biology*, **22(3)**, 1008–1028, doi:10.1111/gcb.13068.

Sola, P., C. Ochieng, J. Yila, and M. Iiyama, 2016: Links between energy access and food security in sub Saharan Africa: an exploratory review. *Food Security*, **8(3)**, 635–642, doi:10.1007/s12571-016-0570-1.

Sommerfeld, J., L. Buys, and D. Vine, 2017: Residential consumers' experiences in the adoption and use of solar PV. *Energy Policy*, **105**, 10–16, doi:10.1016/j.enpol.2017.02.021.

Sondak, C.F.A. et al., 2017: Carbon dioxide mitigation potential of seaweed aquaculture beds (SABs). *Journal of Applied Phycology*, **29(5)**, 2363–2373, doi:10.1007/s10811-016-1022-1.

Song, Y. et al., 2016: The Interplay Between Bioenergy Grass Production and Water Resources in the United States of America. *Environmental Science & Technology*, **50(6)**, 3010–3019, doi:10.1021/acs.est.5b05239.

Stahel, W.R., 2013: Policy for material efficiency – sustainable taxation as a departure from the throwaway society. *Philosophical Transactions of the Royal Society A: Mathematical, Physical and Engineering Sciences*, **371(1986)**, doi:10.1098/rsta.2011.0567.

Stahel, W.R., 2016: The circular economy. *Nature*, **531(7595)**, 435–438, doi:10.1038/531435a.

Stanley, J.K., D.A. Hensher, and C. Loader, 2011: Road transport and climate change: Stepping off the greenhouse gas. *Transportation Research Part A: Policy and Practice*, **45(10)**, 1020–1030, doi:10.1016/j.tra.2009.04.005.

Steenwerth, K.L. et al., 2014: Climate-smart agriculture global research agenda: scientific basis for action. *Agriculture & Food Security*, **3(1)**, 11, doi:10.1186/2048-7010-3-11.

Stefan, A. and L. Paul, 2008: Does It Pay to Be Green? A Systematic Overview. *Academy of Management Perspectives*, **22(4)**, 45–62, doi:10.5465/amp.2008.35590353.

Stehfest, E. et al., 2009: Climate benefits of changing diet. *Climatic Change*, **95(1–2)**, 83–102, doi:10.1007/s10584-008-9534-6.

Steinfeld, H. and P. Gerber, 2010: Livestock production and the global environment: Consume less or produce better? *Proceedings of the National Academy of Sciences*, **107(43)**, 18237–18238, doi:10.1073/pnas.1012541107.

Suckall, N., E. Tompkins, and L. Stringer, 2014: Identifying trade-offs between adaptation, mitigation and development in community responses to climate and socio-economic stresses: Evidence from Zanzibar, Tanzania. *Applied Geography*, **46**, 111–121, doi:10.1016/j.apgeog.2013.11.005.

Supino, S., O. Malandrino, M. Testa, and D. Sica, 2016: Sustainability in the EU cement industry: The Italian and German experiences. *Journal of Cleaner Production*, **112**, 430–442, doi:10.1016/j.jclepro.2015.09.022.

Sweeney, J.C., J. Kresling, D. Webb, G.N. Soutar, and T. Mazzarol, 2013: Energy saving behaviours: Development of a practice-based model. *Energy Policy*, **61**, 371–381, doi:10.1016/j.enpol.2013.06.121.

Tabellini, G., 2010: Culture and Institutions: Economic Development in the Regions of Europe. *Journal of the European Economic Association*, **8(4)**, 677–716, doi:10.1111/j.1542-4774.2010.tb00537.x.

Taniwaki, R.H. et al., 2017: Impacts of converting low-intensity pastureland to high-intensity bioenergy cropland on the water quality of tropical streams in Brazil. *Science of The Total Environment*, **584**, 339–347, doi:10.1016/j.scitotenv.2016.12.150.

Taylor, L.L. et al., 2016: Enhanced weathering strategies for stabilizing climate and averting ocean acidification. *Nature Climate Change*, **6(4)**, 402–406, doi:10.1038/nclimate2882.

Ten Hoeve, J.E. and M.Z. Jacobson, 2012: Worldwide health effects of the Fukushima Daiichi nuclear accident. *Energy & Environmental Science*, **5(9)**, 8743–8757, doi:10.1039/c2ee22019a.

Terry, G., 2009: No climate justice without gender justice: an overview of the issues. *Gender and Development*, **17(1)**, 5–18, www.jstor.org/stable/27809203.

Thornton, P.K., 2010: Livestock production: recent trends, future prospects. *Philosophical Transactions of the Royal Society B: Biological Sciences*, **365(1554)**, 2853–2867, doi:10.1098/rstb.2010.0134.

Tiedeman, K., S. Yeh, B.R. Scanlon, J. Teter, and G.S. Mishra, 2016: Recent Trends in Water Use and Production for California Oil Production. *Environmental Science & Technology*, **50(14)**, 7904–7912, doi:10.1021/acs.est.6b01240.

5

Tilman, D. and M. Clark, 2014: Global diets link environmental sustainability and human health. *Nature*, **515(7528)**, 518–522, doi:10.1038/nature13959.

Tilman, D. et al., 2009: Beneficial Biofuels – The Food, Energy, and Environment Trilemma. *Science*, **325(5938)**, 270–271, doi:10.1126/science.1177970.

Tirmarche, M. et al., 2012: Risk of lung cancer from radon exposure: contribution of recently published studies of uranium miners. *Annals of the ICRP*, **41(3)**, 368–377, doi:10.1016/j.icrp.2012.06.033.

Treasury, H.M., 2009: *Green biotechnology and climate change*. Euro Bio, 12 pp.

Trick, C.G. et al., 2010: Iron enrichment stimulates toxic diatom production in high-nitrate, low-chlorophyll areas. *Proceedings of the National Academy of Sciences*, **107(13)**, 5887–5892, doi:10.1073/pnas.0910579107.

Tully, S., 2006: The Human Right to Access Electricity. *The Electricity Journal*, **19**, 30–39, doi:10.1016/j.tej.2006.02.003.

Turpie, J., B. Warr, and J.C. Ingram, 2015: *Benefits of Forest Ecosystems in Zambia and the Role of REDD+ in a Green Economy Transformation*. United Nations Environment Programme (UNEP), 98 pp.

Tyler, A.N. et al., 2013: The radium legacy: Contaminated land and the committed effective dose from the ingestion of radium contaminated materials. *Environment International*, **59**, 449–455, doi:10.1016/j.envint.2013.06.016.

UN, 1989: The Montreal Protocol on Substances that Deplete the Ozone Layer. In: *United Nations – Treaty Series No. 26369*. United Nations (UN), pp. 29–111.

UNSCEAR, 2011: Health effects due to radiation from the Chernobyl accident (Annex D). In: *UNSCEAR 2008 Report Vol. II: Effects. Scientific Annexes C, D and E*. United Nations Scientific Committee on the Effects of Atomic Radiation (UNSCEAR), New York, NY, USA, pp. 45–220.

UN-Women, UNDESA, and UNFCCC, 2015: *Implementation of gender-responsive climate action in the context of sustainable development*. EGM/GR-CR/Report, UN-Women, UNDESA and UNFCCC, 47 pp.

van de Walle, D., M. Ravallion, V. Mendiratta, and G. Koolwal, 2013: *Long-Term Impacts of Household Electrification in Rural India*. Policy Research working paper no. WPS 6527, World Bank, Washington DC, USA, 54 pp.

van der Horst, D. and S. Vermeylen, 2011: Spatial scale and social impacts of biofuel production. *Biomass and Bioenergy*, **35(6)**, 2435–2443, doi:10.1016/j.biombioe.2010.11.029.

van Oel, P.R. and A.Y. Hoekstra, 2012: Towards Quantification of the Water Footprint of Paper: A First Estimate of its Consumptive Component. *Water Resources Management*, **26(3)**, 733–749, doi:10.1007/s11269-011-9942-7.

van Sluisveld, M.A.E., S.H. Martínez, V. Daioglou, and D.P. van Vuuren, 2016: Exploring the implications of lifestyle change in 2°C mitigation scenarios using the IMAGE integrated assessment model. *Technological Forecasting and Social Change*, **102**, 309–319, doi:10.1016/j.techfore.2015.08.013.

van Vliet, O. et al., 2012: Synergies in the Asian energy system: Climate change, energy security, energy access and air pollution. *Energy Economics*, **34, Supple**, S470–S480, doi:10.1016/j.eneco.2012.02.001.

van Vuuren, D.P., J. van Vliet, and E. Stehfest, 2009: Future bio-energy potential under various natural constraints. *Energy Policy*, **37(11)**, 4220–4230, doi:10.1016/j.enpol.2009.05.029.

Vasconcellos, E.A. and A. Mendonça, 2016: *Observatorio de Movilidad Urbana: Informe final 2015-2016 (in Spanish)*. Corporacion Andina de Fomento (CAF) Banco de Desarrollo de América Latina, 25 pp.

Vassolo, S. and P. Döll, 2005: Global-scale gridded estimates of thermoelectric power and manufacturing water use. *Water Resources Research*, **41(4)**, W04010, doi:10.1029/2004wr003360.

Veltman, K., B. Singh, and E.G. Hertwich, 2010: Human and Environmental Impact Assessment of Postcombustion CO_2 Capture Focusing on Emissions from Amine-Based Scrubbing Solvents to Air. *Environmental Science & Technology*, **44(4)**, 1496–1502, doi:10.1021/es902116r.

Vergragt, P.J., N. Markusson, and H. Karlsson, 2011: Carbon capture and storage, bio-energy with carbon capture and storage, and the escape from the fossil-fuel lock-in. *Global Environmental Change*, **21(2)**, 282–292, doi:10.1016/j.gloenvcha.2011.01.020.

Vermeulen, S. et al., 2012: Climate change, agriculture and food security: a global partnership to link research and action for low-income agricultural producers and consumers. *Current Opinion in Environmental Sustainability*, **4(1)**, 128–133, doi:10.1016/j.cosust.2011.12.004.

Vidic, R.D., S.L. Brantley, J.M. Vandenbossche, D. Yoxtheimer, and J.D. Abad, 2013: Impact of Shale Gas Development on Regional Water Quality. *Science*, **340(6134)**, 1235009, doi:10.1126/science.1235009.

Vierros, M., 2017: Communities and blue carbon: the role of traditional management systems in providing benefits for carbon storage, biodiversity conservation and livelihoods. *Climatic Change*, **140(1)**, 89–100, doi:10.1007/s10584-013-0920-3.

Vierros, M., C. Salpin, C. Chiarolla, and S. Aricò, 2015: Emerging and unresolved issues: the example of marine genetic resources of areas beyond national jurisdiction. In: *Ocean Sustainability in the 21st Century* [Aricò, S. (ed.)]. Cambridge University Press, Cambridge, United Kingdom and New York, NY, USA, pp. 198–231, doi:10.1017/cbo9781316164624.012.

Visschers, V.H.M. and M. Siegrist, 2012: Fair play in energy policy decisions: Procedural fairness, outcome fairness and acceptance of the decision to rebuild nuclear power plants. *Energy Policy*, **46**, 292–300, doi:10.1016/j.enpol.2012.03.062.

von Hippel, D., T. Suzuki, J.H. Williams, T. Savage, and P. Hayes, 2011: Energy security and sustainability in Northeast Asia. *Energy Policy*, **39(11)**, 6719–6730, doi:10.1016/j.enpol.2009.07.001.

von Hippel, F., R. Ewing, R. Garwin, and A. Macfarlane, 2012: Time to bury plutonium. *Nature*, **485**, 167–168, doi:10.1038/485167a.

von Stechow, C. et al., 2016: 2°C and SDGs: United they stand, divided they fall? *Environmental Research Letters*, **11(3)**, 034022, doi:10.1088/1748-9326/11/3/034022.

Walker, G. and P. Devine-Wright, 2008: Community renewable energy: What should it mean? *Energy Policy*, **36**, 497–500, doi:10.1016/j.enpol.2007.10.019.

Walker, R.V., M.B. Beck, J.W. Hall, R.J. Dawson, and O. Heidrich, 2014: The energy-water-food nexus: Strategic analysis of technologies for transforming the urban metabolism. *Journal of Environmental Management*, **141**, 104–115, doi:10.1016/j.jenvman.2014.01.054.

Walks, A., 2015: *The Urban Political Economy and Ecology of Automobility: Driving Cities, Driving Inequality, Driving Politics*. Routledge, London, UK, 348 pp.

Wallquist, L., V.H.M. Visschers, and M. Siegrist, 2009: Lay concepts on CCS deployment in Switzerland based on qualitative interviews. *International Journal of Greenhouse Gas Control*, **3(5)**, 652–657, doi:10.1016/j.ijggc.2009.03.005.

Wallquist, L., V.H.M. Visschers, and M. Siegrist, 2010: Impact of Knowledge and Misconceptions on Benefit and Risk Perception of CCS. *Environmental Science & Technology*, **44(17)**, 6557–6562, doi:10.1021/es1005412.

Wang, S. and P.R. Jaffe, 2004: Dissolution of a mineral phase in potable aquifers due to CO_2 releases from deep formations; effect of dissolution kinetics. *Energy Conversion and Management*, **45(18–19)**, 2833–2848, doi:10.1016/j.enconman.2004.01.002.

WBGU, 2013: *World in Transition: Governing the Marine Heritage*. German Advisory Council on Global Change (WBGU), Berlin, Germany, 390 pp.

Webb, D., G.N. Soutar, T. Mazzarol, and P. Saldaris, 2013: Self-determination theory and consumer behavioural change: Evidence fromahousehold energy-saving behaviour study. *Journal of Environmental Psychology*, **35**, 59–66, doi:10.1016/j.jenvp.2013.04.003.

Webster, M., P. Donohoo, and B. Palmintier, 2013: Water-CO_2 trade-offs in electricity generation planning. *Nature Climate Change*, **3(12)**, 1029–1032, doi:10.1038/nclimate2032.

Wells, E.M. et al., 2015: Indoor air quality and occupant comfort in homes with deep versus conventional energy efficiency renovations. *Building and Environment*, **93(Part 2)**, 331–338, doi:10.1016/j.buildenv.2015.06.021.

Wesseling, J.H. et al., 2017: The transition of energy intensive processing industries towards deep decarbonization: Characteristics and implications for future research. *Renewable and Sustainable Energy Reviews*, **79**, 1303–1313, doi:10.1016/j.rser.2017.05.156.

West, J.J. et al., 2013: Co-benefits of Global Greenhouse Gas Mitigation for Future Air Quality and Human Health. *Nature Climate Change*, **3(10)**, 885–889, doi:10.1038/nclimate2009.

West, P.C. et al., 2014: Leverage points for improving global food security and the environment. *Science*, **345(6194)**, 325–328, doi:10.1126/science.1246067.

West, T.O. and W.M. Post, 2002: Soil Organic Carbon Sequestration Rates by Tillage and Crop Rotation. *Soil Science Society of America Journal*, **66**, 1930–1946, doi:10.2136/sssaj2002.1930.

WHO, 2013: *Health risk assessment from the nuclear accident after the 2011 Great East Japan earthquake and tsunami, based on a preliminary dose estimation*. World Health Organization (WHO), Geneva, Switzerland, 172 pp.

Willand, N., I. Ridley, and C. Maller, 2015: Towards explaining the health impacts of residential energy efficiency interventions – A realist review. Part 1: Pathways. *Social Science and Medicine*, **133**, 191–201, doi:10.1016/j.socscimed.2015.02.005.

5

Williamson, P. et al., 2012: Ocean fertilization for geoengineering: A review of effectiveness, environmental impacts and emerging governance. *Process Safety and Environmental Protection*, **90(6)**, 475–488, doi:10.1016/j.psep.2012.10.007.

Winter, E., A. Faße, and K. Frohberg, 2015: Food security, energy equity, and the global commons: a computable village model applied to sub-Saharan Africa. *Regional Environmental Change*, **15(7)**, 1215–1227, doi:10.1007/s10113-014-0674-0.

Wiser, R. et al., 2011: Wind Energy. In: *IPCC Special Report on Renewable Energy Sources and Climate Change Mitigation* [Edenhofer, O., R. Pichs-Madruga, Y. Sokona, K. Seyboth, P. Matschoss, S. Kadner, T. Zwickel, P. Eickemeier, G. Hansen, S. Schlömer, and C. von Stechow (eds.)]. Cambridge University Press, Cambridge, United Kingdom and New York, NY, USA, pp. 535–608.

Wolosin, M., 2014: *Quantifying Benefits of the New York Declaration on Forests*. Climate Advisers, 17 pp.

Wong-Parodi, G. and I. Ray, 2009: Community perceptions of carbon sequestration: insights from California. *Environmental Research Letters*, **4(3)**, 034002, doi:10.1088/1748-9326/4/3/034002.

Woodbury, P.B., A.R. Kemanian, M. Jacobson, and M. Langholtz, 2018: Improving water quality in the Chesapeake Bay using payments for ecosystem services for perennial biomass for bioenergy and biofuel production. *Biomass and Bioenergy*, **114**, 132–142, doi:10.1016/j.biombioe.2017.01.024.

Woodcock, J. et al., 2009: Public health benefits of strategies to reduce greenhouse-gas emissions: urban land transport. *The Lancet*, **374(9705)**, 1930–1943, doi:10.1016/s0140-6736(09)61714-1.

Xi, Y., T. Fei, and W. Gehua, 2013: Quantifying co-benefit potentials in the Chinese cement sector during 12th Five Year Plan: An analysis based on marginal abatement cost with monetized environmental effect. *Journal of Cleaner Production*, **58**, 102–111, doi:10.1016/j.jclepro.2013.07.020.

Xylia, M. and S. Silveira, 2017: On the road to fossil-free public transport: the case of Swedish bus fleets. *Energy Policy*, **100**, 397–412, doi:10.1016/j.enpol.2016.02.024.

Yang, L., H. Yan, and J.C. Lam, 2014: Thermal comfort and building energy consumption implications – A review. *Applied Energy*, **115**, 164–173, doi:10.1016/j.apenergy.2013.10.062.

Yanine, F.F. and E.E. Sauma, 2013: Review of grid-tie micro-generation systems without energy storage: Towards a new approach to sustainable hybrid energy systems linked to energy efficiency. *Renewable and Sustainable Energy Reviews*, **26**, 60–95, doi:10.1016/j.rser.2013.05.002.

Yim, M.-S. and J. Li, 2013: Examining relationship between nuclear proliferation and civilian nuclear power development. *Progress in Nuclear Energy*, **66**, 108–114, doi:10.1016/j.pnucene.2013.03.005.

York, R. and J.A. McGee, 2017: Does Renewable Energy Development Decouple Economic Growth from CO_2 Emissions? *Socius*, **3**, doi:10.1177/2378023116689098.

Yu, C.-Y., 2015: How Differences in Roadways Affect School Travel Safety. *Journal of the American Planning Association*, **81(3)**, 203–220, doi:10.1080/01944363.2015.1080599.

Yue, T., R. Long, and H. Chen, 2013: Factors influencing energy-saving behavior of urban households in Jiangsu Province. *Energy Policy*, **62**, 665–675, doi:10.1016/j.enpol.2013.07.051.

Zeng, H., X. Chen, X. Xiao, and Z. Zhou, 2017: Institutional pressures, sustainable supply chain management, and circular economy capability: Empirical evidence from Chinese eco-industrial park firms. *Journal of Cleaner Production*, **155**, 54–65, doi:10.1016/j.jclepro.2016.10.093.

Zenk, S.N. et al., 2015: Prepared Food Availability in U.S. Food Stores. *American Journal of Preventive Medicine*, **49(4)**, 553–562, doi:10.1016/j.amepre.2015.02.025.

Zhang, S., E. Worrell, and W. Crijns-Graus, 2015: Cutting air pollution by improving energy efficiency of China's cement industry. *Energy Procedia*, **83**, 10–20, doi:10.1016/j.egypro.2015.12.191.

Zhang, Y. et al., 2017: Processes of coastal ecosystem carbon sequestration and approaches for increasing carbon sink. *Science China Earth Sciences*, **60(5)**, 809–820, doi:10.1007/s11430-016-9010-9.

Zhao, D., A.P. McCoy, J. Du, P. Agee, and Y. Lu, 2017: Interaction effects of building technology and resident behavior on energy consumption in residential buildings. *Energy and Buildings*, **134**, 223–233, doi:10.1016/j.enbuild.2016.10.049.

Ziv, G., E. Baran, S. Nam, I. Rodríguez-Iturbe, and S.A. Levin, 2012: Trading-off fish biodiversity, food security, and hydropower in the Mekong River Basin. *Proceedings of the National Academy of Sciences*, **109(15)**, 5609–5614, doi:10.1073/pnas.1201423109.

Zomer, R.J., A. Trabucco, D.A. Bossio, and L. Verchot, 2008: Climate change mitigation: A spatial analysis of global land suitability for clean development mechanism afforestation and reforestation. *Agriculture, Ecosystems & Environment*, **126(1–2)**, 67–80, doi:10.1016/j.agee.2008.01.014.

Zulu, L.C. and R.B. Richardson, 2013: Charcoal, livelihoods, and poverty reduction: Evidence from sub-Saharan Africa. *Energy for Sustainable Development*, **17(2)**, 127–137, doi:10.1016/j.esd.2012.07.007.

5

Annexes

AI

Annex I: Glossary

Coordinating Editor:
J.B. Robin Matthews (France/UK)

Editorial Team:
Mustafa Babiker (Sudan), Heleen de Coninck (Netherlands/EU), Sarah Connors (France/UK), Renée van Diemen (UK/Netherlands), Riyanti Djalante (Japan/Indonesia), Kristie L. Ebi (USA), Neville Ellis (Australia), Andreas Fischlin (Switzerland), Tania Guillén Bolaños (Germany/Nicaragua), Kiane de Kleijne (Netherlands/EU), Valérie Masson-Delmotte (France), Richard Millar (UK), Elvira S. Poloczanska (Germany/UK), Hans-Otto Pörtner (Germany), Andy Reisinger (New Zealand), Joeri Rogelj (Austria/Belgium), Sonia I. Seneviratne (Switzerland), Chandni Singh (India), Petra Tschakert (Australia/Austria), Nora M. Weyer (Germany)

Notes:
Note that subterms are in italics beneath main terms.

This glossary defines some specific terms as the Lead Authors intend them to be interpreted in the context of this report. Blue, italicized words indicate that the term is defined in the Glossary.

This annex should be cited as:
IPCC, 2018: Annex I: Glossary [Matthews, J.B.R. (ed.)]. In: *Global Warming of 1.5°C. An IPCC Special Report on the impacts of global warming of 1.5°C above pre-industrial levels and related global greenhouse gas emission pathways, in the context of strengthening the global response to the threat of climate change, sustainable development, and efforts to eradicate poverty* [Masson-Delmotte, V., P. Zhai, H.-O. Pörtner, D. Roberts, J. Skea, P.R. Shukla, A. Pirani, W. Moufouma-Okia, C. Péan, R. Pidcock, S. Connors, J.B.R. Matthews, Y. Chen, X. Zhou, M.I. Gomis, E. Lonnoy, T. Maycock, M. Tignor, and T. Waterfield (eds.)]. Cambridge University Press, Cambridge, UK and New York, NY, USA, pp. 541-562. https://doi.org/10.1017/9781009157940.008.

1.5°C pathway   See *Pathways*.

1.5°C warmer worlds   *Projected* worlds in which *global warming* has reached and, unless otherwise indicated, been limited to 1.5°C above *pre-industrial* levels. There is no single 1.5°C warmer world, and *projections* of 1.5°C warmer worlds look different depending on whether it is considered on a near-term transient trajectory or at climate equilibrium after several millennia, and, in both cases, if it occurs with or without *overshoot*. Within the 21st century, several aspects play a role for the assessment of risk and potential *impacts* in 1.5°C warmer worlds: the possible occurrence, magnitude and duration of an overshoot; the way in which emissions reductions are achieved; the ways in which policies might be able to influence the *resilience* of human and natural systems; and the nature of the regional and sub-regional risks. Beyond the 21st century, several elements of the *climate system* would continue to change even if the global mean temperatures remain stable, including further increases of sea level.

2030 Agenda for Sustainable Development   A UN resolution in September 2015 adopting a plan of action for people, planet and prosperity in a new global development framework anchored in 17 *Sustainable Development Goals* (UN, 2015). See also *Sustainable Development Goals (SDGs)*.

Acceptability of policy or system change   The extent to which a policy or system change is evaluated unfavourably or favourably, or rejected or supported, by members of the general public (public acceptability) or politicians or governments (political acceptability). Acceptability may vary from totally unacceptable/fully rejected to totally acceptable/fully supported; individuals may differ in how acceptable policies or system changes are believed to be.

Adaptability   See *Adaptive capacity*.

Adaptation   In *human systems*, the process of adjustment to actual or expected *climate* and its effects, in order to moderate harm or exploit beneficial opportunities. In natural systems, the process of adjustment to actual climate and its effects; human intervention may facilitate adjustment to expected climate and its effects.

> *Incremental adaptation*
> Adaptation that maintains the essence and integrity of a system or process at a given scale. In some cases, incremental adaptation can accrue to result in *transformational adaptation* (Termeer et al., 2017; Tàbara et al., 2018).
>
> *Transformational adaptation*
> Adaptation that changes the fundamental attributes of a *socio-ecological system* in anticipation of *climate change* and its *impacts*.
>
> *Adaptation limits*
> The point at which an actor's objectives (or system needs) cannot be secured from intolerable risks through adaptive actions.
> - Hard adaptation limit: No adaptive actions are possible to avoid intolerable risks.
> - Soft adaptation limit: Options are currently not available to avoid intolerable risks through adaptive action.
>
> See also *Adaptation options*, *Adaptive capacity* and *Maladaptive actions (Maladaptation)*.

Adaptation behaviour   See *Human behaviour*.

Adaptation limits   See *Adaptation*.

Adaptation options   The array of strategies and measures that are available and appropriate for addressing *adaptation*. They include a wide range of actions that can be categorized as structural, institutional, ecological or behavioural. See also *Adaptation*, *Adaptive capacity* and *Maladaptive actions (Maladaptation)*.

Adaptation pathways   See *Pathways*.

Adaptive capacity   The ability of systems, *institutions*, humans and other organisms to adjust to potential damage, to take advantage of opportunities, or to respond to consequences. This glossary entry builds from definitions used in previous IPCC reports and the Millennium Ecosystem Assessment (MEA, 2005). See also *Adaptation*, *Adaptation options* and *Maladaptive actions (Maladaptation)*.

Adaptive governance   See *Governance*.

Aerosol   A suspension of airborne solid or liquid particles, with a typical size between a few nanometres and 10 μm that reside in the *atmosphere* for at least several hours. The term aerosol, which includes both the particles and the suspending gas, is often used in this report in its plural form to mean aerosol particles. Aerosols may be of either natural or *anthropogenic* origin. Aerosols may influence *climate* in several ways: through both interactions that scatter and/or absorb radiation and through interactions with cloud microphysics and other cloud properties, or upon deposition on snow- or ice-covered surfaces thereby altering their *albedo* and contributing to *climate feedback*. Atmospheric aerosols, whether natural or anthropogenic, originate from two different pathways: emissions of primary particulate matter (PM), and formation of secondary PM from gaseous *precursors*. The bulk of aerosols are of natural origin. Some scientists use group labels that refer to the chemical composition, namely: sea salt, organic carbon, *black carbon (BC)*, mineral species (mainly desert dust), sulphate, nitrate, and ammonium. These labels are, however, imperfect as aerosols combine particles to create complex mixtures. See also *Short-lived climate forcers (SLCF)* and *Black carbon (BC)*.

Afforestation   Planting of new *forests* on lands that historically have not contained forests. For a discussion of the term forest and related terms such as afforestation, *reforestation* and *deforestation*, see the IPCC Special Report on Land Use, Land-Use Change, and Forestry (IPCC, 2000), information provided by the United Nations Framework Convention on Climate Change (UNFCCC, 2013) and the report on Definitions and Methodological Options to Inventory Emissions from Direct Human-induced Degradation of Forests and Devegetation of Other Vegetation Types (IPCC, 2003). See also *Reforestation*, *Deforestation*, and *Reducing Emissions from Deforestation and Forest Degradation (REDD+)*.

Agreement   In this report, the degree of agreement within the scientific body of knowledge on a particular finding is assessed based on multiple lines of *evidence* (e.g., mechanistic understanding, theory, data, models, expert judgement) and expressed qualitatively (Mastrandrea et al., 2010). See also *Evidence*, *Confidence*, *Likelihood* and *Uncertainty*.

Air pollution   Degradation of air quality with negative effects on human health or the natural or built environment due to the introduction, by natural processes or human activity, into the *atmosphere* of substances (gases, *aerosols*) which have a direct (primary pollutants) or indirect (secondary pollutants) harmful effect. See also *Aerosol* and *Short-lived climate forcers (SLCF)*.

Albedo   The fraction of solar radiation reflected by a surface or object, often expressed as a percentage. Snow-covered surfaces have a high albedo, the surface albedo of soils ranges from high to low, and vegetation-covered surfaces and the oceans have a low albedo. The Earth's planetary albedo changes mainly through varying cloudiness and changes in snow, ice, leaf area and land cover.

Ambient persuasive technology   Technological systems and environments that are designed to change human cognitive processing,

AI

attitudes and behaviours without the need for the user's conscious attention.

Anomaly   The deviation of a variable from its value averaged over a *reference period*.

Anthropocene   The 'Anthropocene' is a proposed new geological epoch resulting from significant human-driven changes to the structure and functioning of the Earth System, including the *climate system*. Originally proposed in the Earth System science community in 2000, the proposed new epoch is undergoing a formalization process within the geological community based on the stratigraphic *evidence* that human activities have changed the Earth System to the extent of forming geological deposits with a signature that is distinct from those of the *Holocene*, and which will remain in the geological record. Both the stratigraphic and Earth System approaches to defining the Anthropocene consider the mid-20th Century to be the most appropriate starting date, although others have been proposed and continue to be discussed. The Anthropocene concept has been taken up by a diversity of disciplines and the public to denote the substantive influence humans have had on the state, dynamics and future of the Earth System. See also *Holocene*.

Anthropogenic   Resulting from or produced by human activities. See also *Anthropogenic emissions* and *Anthropogenic removals*.

Anthropogenic emissions   Emissions of *greenhouse gases (GHGs)*, *precursors* of GHGs and aerosols caused by human activities. These activities include the burning of *fossil fuels*, *deforestation*, *land use* and *land-use changes (LULUC)*, livestock production, fertilisation, waste management and industrial processes. See also *Anthropogenic* and *Anthropogenic removals*.

Anthropogenic removals   Anthropogenic removals refer to the withdrawal of *GHGs* from the *atmosphere* as a result of deliberate human activities. These include enhancing biological *sinks of CO$_2$* and using chemical engineering to achieve long-term removal and storage. *Carbon capture and storage (CCS)* from industrial and energy-related sources, which alone does not remove CO$_2$ in the atmosphere, can reduce atmospheric CO$_2$ if it is combined with *bioenergy* production *(BECCS)*. See also *Anthropogenic emissions*, *Bioenergy with carbon dioxide capture and storage (BECCS)* and *Carbon dioxide capture and storage (CCS)*.

Artificial intelligence (AI)   Computer systems able to perform tasks normally requiring human intelligence, such as visual perception and speech recognition.

Atmosphere   The gaseous envelope surrounding the earth, divided into five layers – the *troposphere* which contains half of the Earth's atmosphere, the *stratosphere*, the mesosphere, the thermosphere, and the exosphere, which is the outer limit of the atmosphere. The dry atmosphere consists almost entirely of nitrogen (78.1% volume mixing ratio) and oxygen (20.9% volume mixing ratio), together with a number of trace gases, such as argon (0.93 % volume mixing ratio), helium and radiatively active *greenhouse gases (GHGs)* such as *carbon dioxide (CO$_2$)* (0.04% volume mixing ratio) and *ozone (O$_3$)*. In addition, the atmosphere contains the GHG water vapour (H$_2$O), whose amounts are highly variable but typically around 1% volume mixing ratio. The atmosphere also contains clouds and aerosols. See also *Troposphere*, *Stratosphere*, *Greenhouse gas (GHG)* and *Hydrological cycle*.

Atmosphere–ocean general circulation model (AOGCM)   See *Climate model*.

Attribution   See *Detection and attribution*.

Baseline scenario   In much of the literature the term is also synonymous with the term business-as-usual (BAU) *scenario*, although

the term BAU has fallen out of favour because the idea of business as usual in century-long socio-economic *projections* is hard to fathom. In the context of *transformation pathways*, the term baseline scenarios refers to scenarios that are based on the assumption that no mitigation *policies* or measures will be implemented beyond those that are already in force and/or are legislated or planned to be adopted. Baseline scenarios are not intended to be predictions of the future, but rather counterfactual constructions that can serve to highlight the level of emissions that would occur without further policy effort. Typically, baseline scenarios are then compared to *mitigation scenarios* that are constructed to meet different goals for *greenhouse gas (GHG)* emissions, atmospheric concentrations or temperature change. The term baseline scenario is often used interchangeably with reference scenario and no policy scenario. See also *Emission scenario* and *Mitigation scenario*.

Battery electric vehicle (BEV)   See *Electric vehicle (EV)*.

Biochar   Stable, carbon-rich material produced by heating *biomass* in an oxygen-limited environment. Biochar may be added to soils to improve soil functions and to reduce *greenhouse gas* emissions from biomass and soils, and for *carbon sequestration*. This definition builds from IBI (2018).

Biodiversity   Biological diversity means the variability among living organisms from all sources, including, inter alia, terrestrial, marine and other aquatic *ecosystems* and the ecological complexes of which they are part; this includes diversity within species, between species and of ecosystems (UN, 1992).

Bioenergy   Energy derived from any form of *biomass* or its metabolic by-products. See also *Biomass* and *Biofuel*.

Bioenergy with carbon dioxide capture and storage (BECCS)   *Carbon dioxide capture and storage (CCS)* technology applied to a *bioenergy* facility. Note that depending on the total emissions of the BECCS supply chain, *carbon dioxide (CO$_2$)* can be removed from the *atmosphere*. See also *Bioenergy* and *Carbon dioxide capture and storage (CCS)*.

Biofuel   A fuel, generally in liquid form, produced from *biomass*. Biofuels currently include bioethanol from sugarcane or maize, biodiesel from canola or soybeans, and black liquor from the paper-manufacturing process. See also *Biomass* and *Bioenergy*.

Biomass   Living or recently dead organic material. See also *Bioenergy* and *Biofuel*.

Biophilic urbanism   Designing cities with green roofs, green walls and green balconies to bring nature into the densest parts of cities in order to provide *green infrastructure* and human health benefits. See also *Green infrastructure*.

Black carbon (BC)   Operationally defined *aerosol* species based on measurement of light absorption and chemical reactivity and/or thermal stability. It is sometimes referred to as soot. BC is mostly formed by the incomplete combustion of *fossil fuels*, *biofuels* and *biomass* but it also occurs naturally. It stays in the *atmosphere* only for days or weeks. It is the most strongly light-absorbing component of particulate matter (PM) and has a warming effect by absorbing heat into the atmosphere and reducing the *albedo* when deposited on snow or ice. See also *Aerosol*.

Blue carbon   Blue carbon is the carbon captured by living organisms in coastal (e.g., mangroves, salt marshes, seagrasses) and marine *ecosystems*, and stored in *biomass* and sediments.

Burden sharing (also referred to as Effort sharing)   In the context of *mitigation*, burden sharing refers to sharing the effort of reducing the sources or enhancing the sinks of *greenhouse gases (GHGs)* from historical or *projected* levels, usually allocated by some criteria, as well as sharing the cost burden across countries.

Business as usual (BAU)   See *Baseline scenario*.

Carbon budget   This term refers to three concepts in the literature: (1) an assessment of *carbon cycle* sources and *sinks* on a global level, through the synthesis of *evidence* for *fossil fuel* and cement emissions, *land-use change* emissions, ocean and land CO_2 sinks, and the resulting atmospheric CO_2 growth rate. This is referred to as the global carbon budget; (2) the estimated cumulative amount of global carbon dioxide emissions that that is estimated to limit global surface temperature to a given level above a *reference period*, taking into account global surface temperature contributions of other *GHGs* and climate forcers; (3) the distribution of the carbon budget defined under (2) to the regional, national, or sub-national level based on considerations of *equity*, costs or efficiency. See also *Remaining carbon budget*.

Carbon cycle   The term used to describe the flow of carbon (in various forms, e.g., as *carbon dioxide (CO_2)*, carbon in *biomass*, and carbon dissolved in the ocean as carbonate and bicarbonate) through the *atmosphere*, hydrosphere, terrestrial and marine biosphere and lithosphere. In this report, the reference unit for the global carbon cycle is $GtCO_2$ or GtC (Gigatonne of carbon = 1 GtC = 10^{15} grams of carbon. This corresponds to 3.667 $GtCO_2$).

Carbon dioxide (CO_2)   A naturally occurring gas, CO_2 is also a by-product of burning *fossil fuels* (such as oil, gas and coal), of burning *biomass*, of *land-use changes (LUC)* and of industrial processes (e.g., cement production). It is the principal *anthropogenic* greenhouse gas *(GHG)* that affects the Earth's radiative balance. It is the reference gas against which other GHGs are measured and therefore has a global warming potential (GWP) of 1. See also *Greenhouse gas (GHG)*.

Carbon dioxide capture and storage (CCS)   A process in which a relatively pure stream of *carbon dioxide (CO_2)* from industrial and energy-related sources is separated (captured), conditioned, compressed and transported to a storage location for long-term isolation from the *atmosphere*. Sometimes referred to as Carbon capture and storage. See also *Carbon dioxide capture and utilisation (CCU)*, *Bioenergy with carbon dioxide capture and storage (BECCS)* and *Uptake*.

Carbon dioxide capture and utilisation (CCU)   A process in which CO_2 is captured and then used to produce a new product. If the CO_2 is stored in a product for a *climate*-relevant time horizon, this is referred to as carbon dioxide capture, utilisation and storage (CCUS). Only then, and only combined with CO_2 recently removed from the *atmosphere*, can CCUS lead to *carbon dioxide removal*. CCU is sometimes referred to as carbon dioxide capture and use. See also *Carbon dioxide capture and storage (CCS)*.

Carbon dioxide capture, utilisation and storage (CCUS)   See *Carbon dioxide capture and utilisation (CCU)*.

Carbon dioxide removal (CDR)   *Anthropogenic* activities removing CO_2 from the *atmosphere* and durably storing it in geological, terrestrial, or ocean reservoirs, or in products. It includes existing and potential anthropogenic enhancement of biological or geochemical sinks and direct air capture and storage, but excludes natural CO_2 *uptake* not directly caused by human activities. See also *Mitigation (of climate change)*, *Greenhouse gas removal (GGR)*, *Negative emissions*, *Direct air carbon dioxide capture and storage (DACCS)* and *Sink*.

Carbon intensity   The amount of emissions of *carbon dioxide (CO_2)* released per unit of another variable such as *gross domestic product (GDP)*, output energy use or transport.

Carbon neutrality   See *Net zero CO_2 emissions*.

Carbon price   The price for avoided or released *carbon dioxide (CO_2)* or CO_2*-equivalent emissions*. This may refer to the rate of a carbon tax, or the price of emission permits. In many models that are used to assess the economic costs of *mitigation*, carbon prices are used as a proxy to represent the level of effort in mitigation *policies*.

Carbon sequestration   The process of storing carbon in a carbon pool. See also *Blue carbon*, *Carbon dioxide capture and storage (CCS)*, *Uptake* and *Sink*.

Carbon sink   See *Sink*.

Clean Development Mechanism (CDM)   A mechanism defined under Article 12 of the *Kyoto Protocol* through which investors (governments or companies) from developed (Annex B) countries may finance *greenhouse gas (GHG)* emission reduction or removal projects in developing countries (Non-Annex B), and receive Certified Emission Reduction Units (CERs) for doing so. The CERs can be credited towards the commitments of the respective developed countries. The CDM is intended to facilitate the two objectives of promoting *sustainable development (SD)* in developing countries and of helping *industrialised countries* to reach their emissions commitments in a cost-effective way.

Climate   Climate in a narrow sense is usually defined as the average weather, or more rigorously, as the statistical description in terms of the mean and variability of relevant quantities over a period of time ranging from months to thousands or millions of years. The classical period for averaging these variables is 30 years, as defined by the World Meteorological Organization. The relevant quantities are most often surface variables such as temperature, precipitation and wind. Climate in a wider sense is the state, including a statistical description, of the *climate system*.

Climate change   Climate change refers to a change in the state of the *climate* that can be identified (e.g., by using statistical tests) by changes in the mean and/or the variability of its properties and that persists for an extended period, typically decades or longer. Climate change may be due to natural internal processes or external *forcings* such as modulations of the solar cycles, volcanic eruptions and persistent *anthropogenic* changes in the composition of the *atmosphere* or in *land use*. Note that the *Framework Convention on Climate Change (UNFCCC)*, in its Article 1, defines climate change as: 'a change of climate which is attributed directly or indirectly to human activity that alters the composition of the global atmosphere and which is in addition to natural climate variability observed over comparable time periods.' The UNFCCC thus makes a distinction between climate change attributable to human activities altering the atmospheric composition and climate variability attributable to natural causes. See also *Climate variability*, *Global warming*, *Ocean acidification (OA)* and *Detection and attribution*.

Climate change commitment   Climate change commitment is defined as the unavoidable future *climate change* resulting from inertia in the geophysical and socio-economic systems. Different types of climate change commitment are discussed in the literature (see subterms). Climate change commitment is usually quantified in terms of the further change in temperature, but it includes other future changes, for example in the *hydrological cycle*, in *extreme weather events*, in extreme climate events, and in sea level.

Constant composition commitment
The constant composition commitment is the remaining *climate change* that would result if atmospheric composition, and hence *radiative forcing*, were held fixed at a given value. It results from the thermal inertia of the ocean and slow processes in the cryosphere and land surface.

Constant emissions commitment

The constant emissions commitment is the committed *climate change* that would result from keeping *anthropogenic emissions* constant.

Zero emissions commitment

The zero emissions commitment is the climate change commitment that would result from setting *anthropogenic emissions* to zero. It is determined by both inertia in physical *climate system* components (ocean, cryosphere, land surface) and *carbon cycle* inertia.

Feasible scenario commitment

The feasible scenario commitment is the *climate change* that corresponds to the lowest *emission scenario* judged feasible.

Infrastructure commitment

The infrastructure commitment is the *climate change* that would result if existing *greenhouse gas* and *aerosol* emitting infrastructure were used until the end of its expected lifetime.

Climate-compatible development (CCD)   A form of development building on climate strategies that embrace development goals and development strategies that integrate climate *risk management*, *adaptation* and *mitigation*. This definition builds from Mitchell and Maxwell (2010).

Climate extreme (extreme weather or climate event)   The occurrence of a value of a weather or *climate* variable above (or below) a threshold value near the upper (or lower) ends of the range of observed values of the variable. For simplicity, both *extreme weather events* and extreme climate events are referred to collectively as 'climate extremes'. See also *Extreme weather event*.

Climate feedback   An interaction in which a perturbation in one *climate* quantity causes a change in a second and the change in the second quantity ultimately leads to an additional change in the first. A negative feedback is one in which the initial perturbation is weakened by the changes it causes; a positive feedback is one in which the initial perturbation is enhanced. The initial perturbation can either be externally forced or arise as part of internal variability.

Climate governance   See *Governance*.

Climate justice   See *Justice*.

Climate model   A numerical representation of the *climate system* based on the physical, chemical and biological properties of its components, their interactions and *feedback* processes, and accounting for some of its known properties. The climate system can be represented by models of varying complexity; that is, for any one component or combination of components a spectrum or hierarchy of models can be identified, differing in such aspects as the number of spatial dimensions, the extent to which physical, chemical or biological processes are explicitly represented, or the level at which empirical parametrizations are involved. There is an evolution towards more complex models with interactive chemistry and biology. *Climate models* are applied as a research tool to study and simulate the *climate* and for operational purposes, including monthly, seasonal and interannual climate predictions. See also *Earth system model (ESM)*.

Climate neutrality   Concept of a state in which human activities result in no net effect on the *climate system*. Achieving such a state would require balancing of residual emissions with emission (*carbon dioxide*) removal as well as accounting for regional or local biogeophysical effects of human activities that, for example, affect surface *albedo* or local *climate*. See also *Net zero CO₂ emissions*.

Climate projection   A climate *projection* is the simulated response of the *climate system* to a *scenario* of future emission or concentration of

greenhouse gases (GHGs) and *aerosols*, generally derived using *climate models*. Climate projections are distinguished from climate predictions by their dependence on the emission/concentration/*radiative forcing* scenario used, which is in turn based on assumptions concerning, for example, future socioeconomic and technological developments that may or may not be realized.

Climate-resilient development pathways (CRDPs)   Trajectories that strengthen *sustainable development* and efforts to eradicate *poverty* and reduce inequalities while promoting *fair* and cross-scalar *adaptation* to and *resilience* in a changing *climate*. They raise the *ethics*, equity and *feasibility* aspects of the deep *societal transformation* needed to drastically reduce emissions to limit *global warming* (e.g., to 1.5°C) and achieve desirable and liveable futures and *well-being* for all.

Climate-resilient pathways   Iterative processes for managing change within complex systems in order to reduce disruptions and enhance opportunities associated with *climate change*. See also *Development pathways* (under *Pathways*), *Transformation pathways* (under *Pathways*), and *Climate-resilient development pathways (CRDPs)*.

Climate sensitivity   Climate sensitivity refers to the change in the annual *global mean surface temperature* in response to a change in the atmospheric CO_2 concentration or other radiative forcing.

Equilibrium climate sensitivity

Refers to the equilibrium (steady state) change in the annual *global mean surface temperature* following a doubling of the atmospheric *carbon dioxide (CO₂)* concentration. As a true equilibrium is challenging to define in *climate models* with dynamic oceans, the equilibrium climate sensitivity is often estimated through experiments in AOGCMs where CO_2 levels are either quadrupled or doubled from *pre-industrial* levels and which are integrated for 100-200 years. The climate sensitivity parameter (units: °C (W m⁻²)⁻¹) refers to the equilibrium change in the annual global mean surface temperature following a unit change in *radiative forcing*.

Effective climate sensitivity

An estimate of the *global mean surface temperature* response to a doubling of the atmospheric *carbon dioxide (CO₂)* concentration that is evaluated from model output or observations for evolving non-equilibrium conditions. It is a measure of the strengths of the *climate feedbacks* at a particular time and may vary with *forcing* history and *climate* state, and therefore may differ from *equilibrium climate sensitivity*.

Transient climate response

The change in the *global mean surface temperature*, averaged over a 20-year period, centered at the time of atmospheric CO_2 doubling, in a *climate model* simulation in which CO_2 increases at 1% yr-1 from *pre-industrial*. It is a measure of the strength of *climate feedbacks* and the timescale of ocean heat uptake.

Climate services   Climate services refers to information and products that enhance users' knowledge and understanding about the *impacts* of *climate change* and/or *climate variability* so as to aid decision-making of individuals and organizations and enable preparedness and early climate change action. Products can include climate data products.

Climate-smart agriculture (CSA)   Climate-smart agriculture (CSA) is an approach that helps to guide actions needed to transform and reorient agricultural systems to effectively support development and ensure *food security* in a changing *climate*. CSA aims to tackle three main objectives: sustainably increasing agricultural productivity and incomes, *adapting* and building *resilience* to *climate change*, and reducing and/or removing *greenhouse gas* emissions, where possible (FAO, 2018).

Climate system   The climate system is the highly complex system consisting of five major components: the *atmosphere*, the hydrosphere,

AI

the cryosphere, the lithosphere and the biosphere and the interactions between them. The climate system evolves in time under the influence of its own internal dynamics and because of external *forcings* such as volcanic eruptions, solar variations and *anthropogenic* forcings such as the changing composition of the atmosphere and *land-use change*.

Climate target Climate target refers to a temperature limit, concentration level, or emissions reduction goal used towards the aim of avoiding dangerous *anthropogenic* interference with the *climate system*. For example, national climate targets may aim to reduce *greenhouse gas* emissions by a certain amount over a given time horizon, for example those under the *Kyoto Protocol*.

Climate variability Climate variability refers to variations in the mean state and other statistics (such as standard deviations, the occurrence of extremes, etc.) of the *climate* on all spatial and temporal scales beyond that of individual weather events. Variability may be due to natural internal processes within the *climate system* (internal variability), or to variations in natural or *anthropogenic* external *forcing* (external variability). See also *Climate change*.

CO₂ equivalent (CO₂-eq) emission The amount of *carbon dioxide (CO₂)* emission that would cause the same integrated *radiative forcing* or temperature change, over a given time horizon, as an emitted amount of a *greenhouse gas (GHG)* or a mixture of GHGs. There are a number of ways to compute such equivalent emissions and choose appropriate time horizons. Most typically, the CO₂-equivalent emission is obtained by multiplying the emission of a GHG by its global warming potential (GWP) for a 100-year time horizon. For a mix of GHGs it is obtained by summing the CO₂-equivalent emissions of each gas. CO₂-equivalent emission is a common scale for comparing emissions of different GHGs but does not imply equivalence of the corresponding *climate change* responses. There is generally no connection between CO₂-equivalent emissions and resulting CO₂-equivalent concentrations.

Co-benefits The positive effects that a policy or measure aimed at one objective might have on other objectives, thereby increasing the total benefits for society or the environment. Co-benefits are often subject to *uncertainty* and depend on local circumstances and implementation practices, among other factors. Co-benefits are also referred to as ancillary benefits.

Common but Differentiated Responsibilities and Respective Capabilities (CBDR-RC) Common but Differentiated Responsibilities and Respective Capabilities (CBDR–RC) is a key principle in the *United Nations Framework Convention on Climate Change (UNFCCC)* that recognises the different capabilities and differing responsibilities of individual countries in tacking *climate change*. The principle of CBDR–RC is embedded in the 1992 UNFCCC treaty. The convention states: "… the global nature of climate change calls for the widest possible cooperation by all countries and their participation in an effective and appropriate international response, in accordance with their common but differentiated responsibilities and respective capabilities and their social and economic conditions." Since then the CBDR-RC principle has guided the UN climate negotiations.

Conference of the Parties (COP) The supreme body of UN conventions, such as the *United Nations Framework Convention on Climate Change (UNFCCC)*, comprising parties with a right to vote that have ratified or acceded to the convention. See also *United Nations Framework Convention on Climate Change (UNFCCC)*.

Confidence The robustness of a finding based on the type, amount, quality and consistency of *evidence* (e.g., mechanistic understanding, theory, data, models, expert judgment) and on the degree of *agreement* across multiple lines of evidence. In this report, confidence is expressed qualitatively (Mastrandrea et al., 2010). See Section 1.6 for the list of confidence levels used. See also *Agreement*, *Evidence*, *Likelihood* and *Uncertainty*.

Conservation agriculture A coherent group of agronomic and soil management practices that reduce the disruption of soil structure and biota.

Constant composition commitment See *Climate change commitment*.

Constant emissions commitment See *Climate change commitment*.

Coping capacity The ability of people, *institutions*, organizations, and systems, using available skills, values, beliefs, resources, and opportunities, to address, manage, and overcome adverse conditions in the short to medium term. This glossary entry builds from the definition used in UNISDR (2009) and IPCC (2012a). See also *Resilience*.

Cost–benefit analysis Monetary assessment of all negative and positive impacts associated with a given action. Cost–benefit analysis enables comparison of different interventions, investments or strategies and reveals how a given investment or policy effort pays off for a particular person, company or country. Cost–benefit analyses representing society's point of view are important for *climate change* decision-making, but there are difficulties in aggregating costs and benefits across different actors and across timescales. See also *Discounting*.

Cost-effectiveness A measure of the cost at which policy goal or outcome is achieved. The lower the cost the greater the cost-effectiveness.

Coupled Model Intercomparison Project (CMIP) The Coupled Model Intercomparison Project (CMIP) is a climate modelling activity from the World Climate Research Programme (WCRP) which coordinates and archives *climate model* simulations based on shared model inputs by modelling groups from around the world. The CMIP3 multimodel data set includes *projections* using SRES *scenarios*. The CMIP5 data set includes projections using the *Representative Concentration Pathways (RCPs)*. The CMIP6 phase involves a suite of common model experiments as well as an ensemble of CMIP-endorsed model intercomparison projects (MIPs).

Cumulative emissions The total amount of emissions released over a specified period of time. See also *Carbon budget*, and *Transient climate response to cumulative CO₂ emissions (TCRE)*.

Decarbonization The process by which countries, individuals or other entities aim to achieve zero fossil carbon existence. Typically refers to a reduction of the carbon emissions associated with electricity, industry and transport.

Decoupling Decoupling (in relation to *climate change*) is where economic growth is no longer strongly associated with consumption of *fossil fuels*. Relative decoupling is where both grow but at different rates. Absolute decoupling is where economic growth happens but fossil fuels decline.

Deforestation Conversion of *forest* to non-forest. For a discussion of the term forest and related terms such as *afforestation*, *reforestation* and *deforestation*, see the IPCC Special Report on Land Use, Land-Use Change, and Forestry (IPCC, 2000). See also information provided by the United Nations Framework Convention on Climate Change (UNFCCC, 2013) and the report on Definitions and Methodological Options to Inventory Emissions from Direct Human-induced Degradation of Forests and Devegetation of Other Vegetation Types (IPCC, 2003). See also *Afforestation*, *Reforestation* and *Reducing Emissions from Deforestation and Forest Degradation (REDD+)*.

Deliberative governance See *Governance*.

Demand- and supply-side measures

Demand-side measures
Policies and programmes for influencing the demand for goods and/or services. In the energy sector, demand-side management aims at reducing the demand for electricity and other forms of energy required to deliver energy services.

Supply-side measures
Policies and programmes for influencing how a certain demand for goods and/or services is met. In the energy sector, for example, supply-side *mitigation measures* aim at reducing the amount of *greenhouse gas* emissions emitted per unit of energy produced.

See also *Mitigation measures*.

Demand-side measures   See *Demand- and supply-side measures*.

Detection   See *Detection and attribution*.

Detection and attribution   Detection of change is defined as the process of demonstrating that *climate* or a system affected by climate has changed in some defined statistical sense, without providing a reason for that change. An identified change is detected in observations if its *likelihood* of occurrence by chance due to internal variability alone is determined to be small, for example, <10%. Attribution is defined as the process of evaluating the relative contributions of multiple causal factors to a change or event with a formal assessment of *confidence*.

Development pathways   See *Pathways*.

Direct air carbon dioxide capture and storage (DACCS)   Chemical process by which CO_2 is captured directly from the ambient air, with subsequent storage. Also known as direct air capture and storage (DACS).

Disaster   Severe alterations in the normal functioning of a community or a society due to hazardous physical events interacting with vulnerable social conditions, leading to widespread adverse human, material, economic or environmental effects that require immediate emergency response to satisfy critical human needs and that may require external support for recovery. See also *Hazard* and *Vulnerability*.

Disaster risk management (DRM)   Processes for designing, implementing, and evaluating strategies, policies, and measures to improve the understanding of *disaster* risk, foster disaster risk reduction and transfer, and promote continuous improvement in disaster preparedness, response, and recovery practices, with the explicit purpose of increasing *human security*, *well-being*, *quality of life*, and *sustainable development*.

Discount rate   See *Discounting*.

Discounting   A mathematical operation that aims to make monetary (or other) amounts received or expended at different times (years) comparable across time. The discounter uses a fixed or possibly time-varying discount rate from year to year that makes future value worth less today (if the discount rate is positive). The choice of discount rate(s) is debated as it is a judgement based on hidden and/or explicit values.

(Internal) Displacement   Internal displacement refers to the forced movement of people within the country they live in. Internally displaced persons (IDPs) are 'Persons or groups of persons who have been forced or obliged to flee or to leave their homes or places of habitual residence, in particular as a result of or in order to avoid the effects of armed conflict, situations of generalized violence, violations of human rights or natural or human-made disasters, and who have not crossed an internationally recognized State border.' (UN, 1998). See also *Migration*.

Disruptive innovation   Disruptive innovation is demand-led technological change that leads to significant system change and is characterized by strong exponential growth.

Distributive equity   See *Equity*.

Distributive justice   See *Justice*.

Double dividend   The extent to which revenues generated by *policy* instruments, such as carbon taxes or auctioned (tradeable) emission permits can (1) contribute to *mitigation* and (2) offset part of the potential welfare losses of climate policies through recycling the revenue in the economy by reducing other distortionary taxes.

Downscaling   Downscaling is a method that derives local- to regional-scale (up to 100 km) information from larger-scale models or data analyses. Two main methods exist: dynamical downscaling and empirical/statistical downscaling. The dynamical method uses the output of regional *climate models*, global models with variable spatial resolution, or high-resolution global models. The empirical/statistical methods are based on observations and develop statistical relationships that link the large-scale atmospheric variables with local/regional *climate* variables. In all cases, the quality of the driving model remains an important limitation on quality of the downscaled information. The two methods can be combined, e.g., applying empirical/statistical downscaling to the output of a regional climate model, consisting of a dynamical downscaling of a global climate model.

Drought   A period of abnormally dry weather long enough to cause a serious hydrological imbalance. Drought is a relative term, therefore any discussion in terms of precipitation deficit must refer to the particular precipitation-related activity that is under discussion. For example, shortage of precipitation during the growing season impinges on crop production or *ecosystem* function in general (due to *soil moisture* drought, also termed agricultural drought), and during the *runoff* and percolation season primarily affects water supplies (hydrological drought). Storage changes in soil moisture and groundwater are also affected by increases in actual evapotranspiration in addition to reductions in precipitation. A period with an abnormal precipitation deficit is defined as a meteorological drought. See also *Soil moisture*.

Megadrought
A megadrought is a very lengthy and pervasive drought, lasting much longer than normal, usually a decade or more.

Early warning systems (EWS)   The set of technical, financial and *institutional capacities* needed to generate and disseminate timely and meaningful warning information to enable individuals, communities and organizations threatened by a *hazard* to prepare to act promptly and appropriately to reduce the possibility of harm or loss. Dependent upon context, EWS may draw upon scientific and/or *Indigenous knowledge*. EWS are also considered for ecological applications e.g., conservation, where the organization itself is not threatened by hazard but the *ecosystem* under conservation is (an example is coral bleaching alerts), in agriculture (for example, warnings of ground frost, hailstorms) and in fisheries (storm and tsunami warnings). This glossary entry builds from the definitions used in UNISDR (2009) and IPCC (2012a).

Earth system feedbacks   See *Climate feedback*.

Earth system model (ESM)   A coupled atmosphere–ocean general circulation model in which a representation of the *carbon cycle* is included, allowing for interactive calculation of atmospheric CO_2 or compatible emissions. Additional components (e.g., atmospheric chemistry, *ice sheets*, dynamic vegetation, nitrogen cycle, but also urban or crop models) may be included. See also *Climate model*.

AI

Ecosystem An ecosystem is a functional unit consisting of living organisms, their non-living environment and the interactions within and between them. The components included in a given ecosystem and its spatial boundaries depend on the purpose for which the ecosystem is defined: in some cases they are relatively sharp, while in others they are diffuse. Ecosystem boundaries can change over time. Ecosystems are nested within other ecosystems and their scale can range from very small to the entire biosphere. In the current era, most ecosystems either contain people as key organisms, or are influenced by the effects of human activities in their environment. See also *Ecosystem services*.

Ecosystem services Ecological processes or functions having monetary or non-monetary value to individuals or society at large. These are frequently classified as (1) supporting services such as productivity or *biodiversity* maintenance, (2) provisioning services such as food or fibre, (3) regulating services such as climate regulation or *carbon sequestration*, and (4) cultural services such as tourism or spiritual and aesthetic appreciation.

Effective climate sensitivity See *Climate sensitivity*.

Effective radiative forcing See *Radiative forcing*.

El Niño-Southern Oscillation (ENSO) The term El Niño was initially used to describe a warm-water current that periodically flows along the coast of Ecuador and Peru, disrupting the local fishery. It has since become identified with warming of the tropical Pacific Ocean east of the dateline. This oceanic event is associated with a fluctuation of a global-scale tropical and subtropical surface pressure pattern called the Southern Oscillation. This coupled atmosphere–ocean phenomenon, with preferred time scales of two to about seven years, is known as the El Niño-Southern Oscillation (ENSO). It is often measured by the surface pressure anomaly difference between Tahiti and Darwin and/or the *sea surface temperatures* in the central and eastern equatorial Pacific. During an ENSO event, the prevailing trade winds weaken, reducing upwelling and altering ocean currents such that the sea surface temperatures warm, further weakening the trade winds. This phenomenon has a great impact on the wind, sea surface temperature and precipitation patterns in the tropical Pacific. It has climatic effects throughout the Pacific region and in many other parts of the world, through global teleconnections. The cold phase of ENSO is called La Niña.

Electric vehicle (EV) A vehicle whose propulsion is powered fully or mostly by electricity.

Battery electric vehicle (BEV)
A vehicle whose propulsion is entirely electric without any internal combustion engine.

Plug-in hybrid electric vehicle (PHEV)
A vehicle whose propulsion is mostly electric with batteries re-charged from an electric source but extra power and distance are provided by a hybrid internal combustion engine.

Emission pathways See *Pathways*.

Emission scenario A plausible representation of the future development of emissions of substances that are radiatively active (e.g., *greenhouse gases (GHGs)*, *aerosols*) based on a coherent and internally consistent set of assumptions about driving forces (such as demographic and socio-economic development, technological change, energy and *land use*) and their key relationships. Concentration *scenarios*, derived from emission scenarios, are often used as input to a *climate model* to compute *climate projections*. See also *Baseline scenario*, *Mitigation scenario*, *Socio-economic scenario*, *Scenario*, *Representative Concentration Pathways (RCPs)* (under *Pathways*), *Shared Socio-economic Pathways (SSPs)* (under *Pathways*) and *Transformation pathways* (under *Pathways*).

Emission trajectories A *projected* development in time of the emission of a *greenhouse gas (GHG)* or group of GHGs, *aerosols*, and GHG *precursors*. See also *Emission pathways* (under *Pathways*).

Emissions trading A market-based instrument aiming at meeting a *mitigation* objective in an efficient way. A cap on *GHG* emissions is divided in tradeable emission permits that are allocated by a combination of auctioning and handing out free allowances to entities within the jurisdiction of the trading scheme. Entities need to surrender emission permits equal to the amount of their emissions (e.g., tonnes of CO_2). An entity may sell excess permits to entities that can avoid the same amount of emissions in a cheaper way. Trading schemes may occur at the intra-company, domestic, or international level (e.g., the flexibility mechanisms under the *Kyoto Protocol* and the EU-ETS) and may apply to carbon dioxide (CO_2), other greenhouse gases (GHGs), or other substances.

Enabling conditions Conditions that affect the *feasibility* of *adaptation* and *mitigation* options, and can accelerate and scale-up systemic transitions that would limit temperature increase to 1.5°C and enhance capacities of systems and societies to adapt to the associated *climate change*, while achieving *sustainable development*, eradicating *poverty* and reducing *inequalities*. Enabling conditions include finance, technological innovation, strengthening *policy* instruments, *institutional capacity*, *multilevel governance*, and changes in *human behaviour* and lifestyles. They also include inclusive processes, attention to power asymmetries and unequal opportunities for development and reconsideration of values. See also *Feasibility*.

Energy efficiency The ratio of output or useful energy or energy services or other useful physical outputs obtained from a system, conversion process, transmission or storage activity to the input of energy (measured as kWh kWh^{-1}, tonnes kWh^{-1} or any other physical measure of useful output like tonne-km transported). Energy efficiency is often described by energy intensity. In economics, energy intensity describes the ratio of economic output to energy input. Most commonly energy efficiency is measured as input energy over a physical or economic unit, i.e., kWh USD^{-1} (energy intensity), kWh tonne^{-1}. For buildings, it is often measured as kWh m^{-2}, and for vehicles as km liter^{-1} or liter km^{-1}. Very often in policy 'energy efficiency' is intended as the measures to reduce energy demand through technological options such as insulating buildings, more efficient appliances, efficient lighting, efficient vehicles, etc.

Energy security The goal of a given country, or the global community as a whole, to maintain an adequate, stable and predictable energy supply. Measures encompass safeguarding the sufficiency of energy resources to meet national energy demand at competitive and stable prices and the *resilience* of the energy supply; enabling development and deployment of technologies; building sufficient infrastructure to generate, store and transmit energy supplies; and ensuring enforceable contracts of delivery.

Enhanced weathering Enhancing the removal of *carbon dioxide (CO_2)* from the *atmosphere* through dissolution of silicate and carbonate rocks by grinding these minerals to small particles and actively applying them to soils, coasts or oceans.

(Model) Ensemble A group of parallel model simulations characterising historical *climate* conditions, *climate predictions*, or climate projections. Variation of the results across the ensemble members may give an estimate of modelling-based *uncertainty*. Ensembles made with the same model but different initial conditions only characterize the uncertainty associated with internal *climate variability*, whereas multimodel ensembles including simulations by several models also include the impact of model differences. Perturbed parameter ensembles, in which model parameters are varied in a systematic manner, aim to assess the uncertainty resulting from internal model specifications within a single model. Remaining sources of uncertainty unaddressed with model ensembles are related to systematic model errors or biases, which may be assessed from systematic comparisons of model simulations with observations wherever available. See also *Climate projection*.

Equality   A principle that ascribes equal worth to all human beings, including equal opportunities, rights, and obligations, irrespective of origins.

Inequality
Uneven opportunities and social positions, and processes of discrimination within a group or society, based on gender, class, ethnicity, age, and (dis) ability, often produced by uneven development. Income inequality refers to gaps between highest and lowest income earners within a country and between countries. See also *Equity*, *Ethics* and *Fairness*.

Equilibrium climate sensitivity   See *Climate sensitivity*.

Equity   Equity is the principle of *fairness* in burden sharing and is a basis for understanding how the *impacts* and responses to *climate change*, including costs and benefits, are distributed in and by society in more or less equal ways. It is often aligned with ideas of *equality*, fairness and *justice* and applied with respect to equity in the responsibility for, and distribution of, climate *impacts* and *policies* across society, generations, and gender, and in the sense of who participates and controls the processes of decision-making.

Distributive equity
Equity in the consequences, outcomes, costs and benefits of actions or policies. In the case of *climate change* or climate *policies* for different people, places and countries, including equity aspects of sharing burdens and benefits for *mitigation* and *adaptation*.

Gender equity
Ensuring equity in that women and men have the same rights, resources and opportunities. In the case of *climate change* gender equity recognizes that women are often more vulnerable to the *impacts* of climate change and may be disadvantaged in the process and outcomes of climate *policy*.

Inter-generational equity
Equity between generations that acknowledges that the effects of past and present emissions, *vulnerabilities* and policies impose costs and benefits for people in the future and of different age groups.

Procedural equity
Equity in the process of decision-making, including recognition and inclusiveness in participation, equal representation, bargaining power, voice and equitable access to knowledge and resources to participate.

See also *Equality*, *Ethics* and *Fairness*.

Ethics   Ethics involves questions of *justice* and value. Justice is concerned with right and wrong, *equity* and *fairness*, and, in general, with the rights to which people and living beings are entitled. Value is a matter of worth, benefit, or good. See also *Equality*, *Equity* and *Fairness*.

Evidence   Data and information used in the scientific process to establish findings. In this report, the degree of evidence reflects the amount, quality and consistency of scientific/technical information on which the Lead Authors are basing their findings. See also *Agreement*, *Confidence*, *Likelihood* and *Uncertainty*.

Exposure   The presence of people; *livelihoods*; species or *ecosystems*; environmental functions, services, and resources; infrastructure; or economic, social, or cultural assets in places and settings that could be adversely affected. See also *Hazard*, *Risk* and *Vulnerability*.

Extratropical cyclone   Any cyclonic-scale storm that is not a *tropical cyclone*. Usually refers to a middle- or high-latitude migratory storm system formed in regions of large horizontal temperature variations. Sometimes called extratropical storm or extratropical low. See also *Tropical cyclone*.

Extreme weather event   An extreme weather event is an event that is rare at a particular place and time of year. Definitions of rare vary, but an extreme weather event would normally be as rare as or rarer than the 10th or 90th percentile of a probability density function estimated from observations. By definition, the characteristics of what is called extreme weather may vary from place to place in an absolute sense. When a pattern of extreme weather persists for some time, such as a season, it may be classed as an extreme climate event, especially if it yields an average or total that is itself extreme (e.g., *drought* or heavy rainfall over a season). See also *Heatwave* and *Climate extreme (extreme weather or climate event)*.

Extreme weather or climate event   See *Climate extreme (extreme weather or climate event)*.

Fairness   Impartial and just treatment without favouritism or discrimination in which each person is considered of equal worth with equal opportunity. See also *Equity*, *Equality* and *Ethics*.

Feasibility   The degree to which climate goals and response options are considered possible and/or desirable. Feasibility depends on geophysical, ecological, technological, economic, social and *institutional* conditions for change. Conditions underpinning feasibility are dynamic, spatially variable, and may vary between different groups. See also *Enabling conditions*.

Feasible scenario commitment   See *Climate change commitment*.

Feedback   See *Climate feedback*.

Flexible governance   See *Governance*.

Flood   The overflowing of the normal confines of a stream or other body of water, or the accumulation of water over areas that are not normally submerged. Floods include river (fluvial) floods, flash floods, urban floods, pluvial floods, sewer floods, coastal floods, and glacial lake outburst floods.

Food security   A situation that exists when all people, at all times, have physical, social and economic access to sufficient, safe and nutritious food that meets their dietary needs and food preferences for an active and healthy life (FAO, 2001).

Food wastage   Food wastage encompasses food loss (the loss of food during production and transportation) and food waste (the waste of food by the consumer) (FAO, 2013).

Forcing   See *Radiative forcing*.

Forest   A vegetation type dominated by trees. Many definitions of the term forest are in use throughout the world, reflecting wide differences in biogeophysical conditions, social structure and economics. For a discussion of the term forest and related terms such as *afforestation*, *reforestation* and *deforestation*, see the IPCC Special Report on Land Use, Land-Use Change, and Forestry (IPCC, 2000). See also information provided by the United Nations Framework Convention on Climate Change (UNFCCC, 2013) and the Report on Definitions and Methodological Options to Inventory Emissions from Direct Human-induced Degradation of Forests and Devegetation of Other Vegetation Types (IPCC, 2003). See also *Afforestation*, *Deforestation* and *Reforestation*.

Fossil fuels   Carbon-based fuels from fossil hydrocarbon deposits, including coal, oil, and natural gas.

AI

Framework Convention on Climate Change   See *United Nations Framework Convention on Climate Change (UNFCCC)*.

Gender equity   See *Equity*.

General purpose technologies (GPT)   General purpose technologies can be or are used pervasively in a wide range of sectors in ways that fundamentally change the modes of operation of those sectors (Helpman, 1998). Examples include the steam engine, power generator and motor, *ICT*, and biotechnology.

Geoengineering   In this report, separate consideration is given to the two main approaches considered as 'geoengineering' in some of the literature: *solar radiation modification (SRM)* and *carbon dioxide removal (CDR)*. Because of this separation, the term 'geoengineering' is not used in this report. See also *Carbon dioxide removal (CDR)* and *Solar radiation modification (SRM)*.

Glacier   A perennial mass of ice, and possibly firn and snow, originating on the land surface by the recrystallisation of snow and showing *evidence* of past or present flow. A glacier typically gains mass by accumulation of snow, and loses mass by melting and ice discharge into the sea or a lake if the glacier terminates in a body of water. Land ice masses of continental size (>50,000 km²) are referred to as *ice sheets*. See also *Ice sheet*.

Global climate model (also referred to as general circulation model, both abbreviated as GCM)   See *Climate model*.

Global mean surface temperature (GMST)   Estimated global average of near-surface air temperatures over land and sea-ice, and *sea surface temperatures* over ice-free ocean regions, with changes normally expressed as departures from a value over a specified *reference period*. When estimating changes in GMST, near-surface air temperature over both land and oceans are also used.[1] See also *Land surface air temperature*, *Sea surface temperature (SST)* and *Global mean surface air temperature (GSAT)*.

Global mean surface air temperature (GSAT)   Global average of near-surface air temperatures over land and oceans. Changes in GSAT are often used as a measure of global temperature change in *climate models* but are not observed directly. See also *Global mean surface temperature (GMST)* and *Land surface air temperature*.

Global warming   The estimated increase in *global mean surface temperature (GMST)* averaged over a 30-year period, or the 30-year period centered on a particular year or decade, expressed relative to *pre-industrial* levels unless otherwise specified. For 30-year periods that span past and future years, the current multi-decadal warming trend is assumed to continue. See also *Climate change* and *Climate variability*.

Governance   A comprehensive and inclusive concept of the full range of means for deciding, managing, implementing and monitoring policies and measures. Whereas government is defined strictly in terms of the nation-state, the more inclusive concept of governance recognizes the contributions of various levels of government (global, international, regional, sub-national and local) and the contributing roles of the private sector, of nongovernmental actors, and of civil society to addressing the many types of issues facing the global community.

Adaptive governance
An emerging term in the literature for the evolution of formal and informal *institutions* of governance that prioritize *social learning* in planning, implementation and evaluation of policy through iterative social learning to steer the use and protection of natural resources, *ecosystem services* and common pool natural resources, particularly in situations of complexity and *uncertainty*.

Climate governance
Purposeful mechanisms and measures aimed at steering social systems towards preventing, mitigating, or adapting to the risks posed by *climate change* (Jagers and Stripple, 2003).

Deliberative governance
Deliberative governance involves decision-making through inclusive public conversation, which allows opportunity for developing policy options through public discussion rather than collating individual preferences through voting or referenda (although the latter governance mechanisms can also be proceeded and legitimated by public deliberation processes).

Flexible governance
Strategies of governance at various levels, which prioritize the use of *social learning* and rapid feedback mechanisms in planning and policy making, often through incremental, experimental and iterative management processes.

Governance capacity
The ability of governance *institutions*, leaders, and non-state and civil society to plan, co-ordinate, fund, implement, evaluate and adjust policies and measures over the short, medium and long term, adjusting for *uncertainty*, rapid change and wide-ranging impacts and multiple actors and demands.

Multilevel governance
Multilevel governance refers to negotiated, non-hierarchical exchanges between *institutions* at the transnational, national, regional and local levels. Multilevel governance identifies relationships among governance processes at these different levels. Multilevel governance does include negotiated relationships among institutions at different institutional levels and also a vertical 'layering' of governance processes at different levels. Institutional relationships take place directly between transnational, regional and local levels, thus bypassing the state level (Peters and Pierre, 2001)

Participatory governance
A governance system that enables direct public engagement in decision-making using a variety of techniques for example, referenda, community deliberation, citizen juries or participatory budgeting. The approach can be applied in formal and informal *institutional* contexts from national to local, but is usually associated with devolved decision-making. This definition builds from Fung and Wright (2003) and Sarmiento and Tilly (2018).

Governance capacity   See *Governance*.

Green infrastructure   The interconnected set of natural and constructed ecological systems, green spaces and other landscape features. It includes planted and indigenous trees, wetlands, parks, green open spaces and original grassland and woodlands, as well as possible building and street-level design interventions that incorporate vegetation. Green infrastructure provides services and functions in the same way as conventional infrastructure. This definition builds from Culwick and Bobbins (2016).

Greenhouse gas (GHG)   Greenhouse gases are those gaseous constituents of the *atmosphere*, both natural and *anthropogenic*, that absorb and emit radiation at specific wavelengths within the spectrum of terrestrial radiation emitted by the Earth's surface, the atmosphere itself and by clouds. This property causes the greenhouse effect. Water vapour (H_2O), *carbon dioxide (CO_2)*, *nitrous oxide (N_2O)*, *methane (CH_4)* and

---

[1]   Past IPCC reports, reflecting the literature, have used a variety of approximately equivalent metrics of GMST change.

ozone (O₃) are the primary GHGs in the Earth's atmosphere. Moreover, there are a number of entirely human-made GHGs in the atmosphere, such as the *halocarbons* and other chlorine- and bromine-containing substances, dealt with under the Montreal Protocol. Beside CO_2, N_2O and CH_4, the *Kyoto Protocol* deals with the GHGs sulphur hexafluoride (SF_6), hydrofluorocarbons (HFCs) and perfluorocarbons (PFCs). See also *Carbon dioxide (CO₂)*, *Methane (CH₄)*, *Nitrous oxide (N₂O)* and *Ozone (O₃)*.

Greenhouse gas removal (GGR) Withdrawal of a *GHG* and/or a *precursor* from the *atmosphere* by a *sink*. See also *Carbon dioxide removal (CDR)* and *Negative emissions*.

Gross domestic product (GDP) The sum of gross value added, at purchasers' prices, by all resident and non-resident producers in the economy, plus any taxes and minus any subsidies not included in the value of the products in a country or a geographic region for a given period, normally one year. GDP is calculated without deducting for depreciation of fabricated assets or depletion and degradation of natural resources.

Gross fixed capital formation (GFCF) One component of the *GDP* that corresponds to the total value of acquisitions, minus disposals of fixed assets during one year by the business sector, governments and households, plus certain additions to the value of non-produced assets (such as subsoil assets or major improvements in the quantity, quality or productivity of land).

Halocarbons A collective term for the group of partially halogenated organic species, which includes the chlorofluorocarbons (CFCs), hydrochlorofluorocarbons (HCFCs), hydrofluorocarbons (HFCs), halons, methyl chloride and methyl bromide. Many of the halocarbons have large global warming potentials. The chlorine and bromine-containing halocarbons are also involved in the depletion of the ozone layer.

Hazard The potential occurrence of a natural or human-induced physical event or trend that may cause loss of life, injury, or other health impacts, as well as damage and loss to property, infrastructure, *livelihoods*, service provision, *ecosystems* and environmental resources. See also *Disaster*, *Exposure*, *Risk*, and *Vulnerability*.

Heatwave A period of abnormally hot weather. Heatwaves and warm spells have various and in some cases overlapping definitions. See also *Extreme weather event*.

Heating, ventilation, and air conditioning (HVAC) Heating, ventilation and air conditioning technology is used to control temperature and humidity in an indoor environment, be it in buildings or in vehicles, providing thermal comfort and healthy air quality to the occupants. HVAC systems can be designed for an isolated space, an individual building or a distributed heating and cooling network within a building structure or a district heating system. The latter provides economies of scale and also scope for integration with solar heat, natural seasonal cooling/heating etc.

Holocene The Holocene is the current interglacial geological epoch, the second of two epochs within the Quaternary period, the preceding being the Pleistocene. The International Commission on Stratigraphy defines the start of the Holocene at 11,650 years before 1950. See also *Anthropocene*.

Human behaviour The way in which a person acts in response to a particular situation or stimulus. Human actions are relevant at different levels, from international, national, and *sub-national actors*, to NGO, firm-level actors, and communities, households, and individual actions.

Adaptation behaviour
Human actions that directly or indirectly affect the risks of climate change *impacts*.

Mitigation behaviour
Human actions that directly or indirectly influence *mitigation*.

Human behavioural change A transformation or modification of human actions. Behaviour change efforts can be planned in ways that mitigate *climate change* and/or reduce negative consequences of climate change *impacts*.

Human rights Rights that are inherent to all human beings, universal, inalienable, and indivisible, typically expressed and guaranteed by law. They include the right to life; economic, social, and cultural rights; and the right to development and self-determination. Based upon the definition by the UN Office of the High Commissioner for Human Rights (UNOHCHR, 2018).

Procedural rights
Rights to a legal procedure to enforce *substantive rights*.

Substantive rights
Basic human rights, including the right to the substance of being human such as life itself, liberty and happiness.

Human security A condition that is met when the vital core of human lives is protected, and when people have the freedom and capacity to live with dignity. In the context of *climate change*, the vital core of human lives includes the universal and culturally specific, material and non-material elements necessary for people to act on behalf of their interests and to live with dignity.

Human system Any system in which human organizations and *institutions* play a major role. Often, but not always, the term is synonymous with society or social system. Systems such as agricultural systems, urban systems, political systems, technological systems and economic systems are all human systems in the sense applied in this report.

Hydrological cycle The cycle in which water evaporates from the oceans and the land surface, is carried over the earth in atmospheric circulation as water vapour, condenses to form clouds, precipitates as rain or snow, which on land can be intercepted by trees and vegetation, potentially accumulates as snow or ice, provides *runoff* on the land surface, infiltrates into soils, recharges groundwater, discharges into streams, flows out into the oceans, and ultimately evaporates again from the ocean or land surface. The various systems involved in the hydrological cycle are usually referred to as hydrological systems.

Ice sheet A mass of land ice of continental size that is sufficiently thick to cover most of the underlying bed, so that its shape is mainly determined by its dynamics (the flow of the ice as it deforms internally and/or slides at its base). An ice sheet flows outward from a high central ice plateau with a small average surface slope. The margins usually slope more steeply, and most ice is discharged through fast flowing ice streams or outlet *glaciers*, in some cases into the sea or into ice shelves floating on the sea. There are only two ice sheets in the modern world, one on Greenland and one on Antarctica. During glacial periods there were others. See also *Glacier*.

(climate change) Impact assessment The practice of identifying and evaluating, in monetary and/or non-monetary terms, the effects of *climate change* on natural and *human systems*.

Impacts (consequences, outcomes) The consequences of realized risks on natural and *human systems*, where risks result from the interactions of climate-related *hazards* (including *extreme weather and climate events*), *exposure*, and *vulnerability*. Impacts generally refer to effects on lives; *livelihoods*; health and *well-being*; *ecosystems* and species; economic, social and cultural assets; services (including *ecosystem services*); and infrastructure. Impacts may be referred to as consequences or outcomes, and can be adverse or beneficial. See also *Adaptation*, *Exposure*, *Hazard*, *Loss and Damage, and losses and damages*, and *Vulnerability*.

Incremental adaptation   See *Adaptation*.

Indigenous knowledge   Indigenous knowledge refers to the understandings, skills and philosophies developed by societies with long histories of interaction with their natural surroundings. For many Indigenous peoples, Indigenous knowledge informs decision-making about fundamental aspects of life, from day-to-day activities to longer term actions. This knowledge is integral to cultural complexes, which also encompass language, systems of classification, resource use practices, social interactions, values, ritual and spirituality. These distinctive ways of knowing are important facets of the world's cultural diversity. This definition builds on UNESCO (2018).

Indirect land-use change (iLUC)   See *Land-use change (LUC)*.

Industrial revolution   A period of rapid industrial growth with far-reaching social and economic consequences, beginning in Britain during the second half of the 18th century and spreading to Europe and later to other countries, including the United States. The invention of the steam engine was an important trigger of this development. The industrial revolution marks the beginning of a strong increase in the use of *fossil fuels*, initially coal, and hence emission of *carbon dioxide (CO_2)*. See also *Pre-industrial*.

Industrialized/developed/developing countries   There are a diversity of approaches for categorizing countries on the basis of their level of development, and for defining terms such as industrialized, developed, or developing. Several categorizations are used in this report. (1) In the United Nations system, there is no established convention for designation of developed and developing countries or areas. (2) The United Nations Statistics Division specifies developed and developing regions based on common practice. In addition, specific countries are designated as Least Developed Countries (LDC), landlocked developing countries, *small island developing states*, and transition economies. Many countries appear in more than one of these categories. (3) The World Bank uses income as the main criterion for classifying countries as low, lower middle, upper middle and high income. (4) The UNDP aggregates indicators for life expectancy, educational attainment, and income into a single composite Human Development Index (HDI) to classify countries as low, medium, high or very high human development.

Inequality   See *Equality*.

Information and communication technology (ICT)   An umbrella term that includes any information and communication device or application, encompassing: computer systems, network hardware and software, cell phones, etc.

Infrastructure commitment   See *Climate change commitment*.

Institution   Institutions are rules and norms held in common by social actors that guide, constrain and shape human interaction. Institutions can be formal, such as laws and policies, or informal, such as norms and conventions. Organizations – such as parliaments, regulatory agencies, private firms and community bodies – develop and act in response to institutional frameworks and the incentives they frame. Institutions can guide, constrain and shape human interaction through direct control, through incentives, and through processes of socialization. See also *Institutional capacity*.

Institutional capacity   *Institutional* capacity comprises building and strengthening individual organizations and providing technical and management training to support integrated planning and decision-making processes between organizations and people, as well as empowerment, social capital, and an enabling environment, including the culture, values and power relations (Willems and Baumert, 2003).

Integrated assessment   A method of analysis that combines results and models from the physical, biological, economic and social sciences and the interactions among these components in a consistent framework to evaluate the status and the consequences of environmental change and the policy responses to it. See also *Integrated assessment model (IAM)*.

Integrated assessment model (IAM)   Integrated assessment models (IAMs) integrate knowledge from two or more domains into a single framework. They are one of the main tools for undertaking *integrated assessments*.

One class of IAM used in respect of climate change *mitigation* may include representations of: multiple sectors of the economy, such as energy, *land use* and *land-use change*; interactions between sectors; the economy as a whole; associated *GHG* emissions and *sinks*; and reduced representations of the *climate system*. This class of model is used to assess linkages between economic, social and technological development and the evolution of the climate system.

Another class of IAM additionally includes representations of the costs associated with climate change *impacts*, but includes less detailed representations of economic systems. These can be used to assess impacts and mitigation in a *cost–benefit* framework and have been used to estimate the *social cost of carbon*.

Integrated water resources management (IWRM)   A process which promotes the coordinated development and management of water, land and related resources in order to maximize economic and social welfare in an equitable manner without compromising the sustainability of vital *ecosystems*.

Inter-generational equity   See *Equity*.

Inter-generational justice   See *Justice*.

Internal variability   See *Climate variability*.

Internet of Things (IoT)   The network of computing devices embedded in everyday objects such as cars, phones and computers, connected via the internet, enabling them to send and receive data.

Iron fertilization   See *Ocean fertilization*.

Irreversibility   A perturbed state of a dynamical system is defined as irreversible on a given timescale, if the recovery time scale from this state due to natural processes is substantially longer than the time it takes for the system to reach this perturbed state. See also *Tipping point*.

Justice   Justice is concerned with ensuring that people get what is due to them, setting out the moral or legal principles of *fairness* and *equity* in the way people are treated, often based on the *ethics* and values of society.

Climate justice

Justice that links development and *human rights* to achieve a human-centred approach to addressing *climate change*, safeguarding the rights of the most vulnerable people and sharing the burdens and benefits of climate change and its impacts *equitably* and *fairly*. This definition builds upon the one used by the Mary Robinson Foundation – Climate Justice (MRFCJ, 2018).

Distributive justice

Justice in the allocation of economic and non-economic costs and benefits across society.

Inter-generational justice
Justice in the distribution of economic and non-economic costs and benefits across generations.

Procedural justice
Justice in the way outcomes are brought about including who participates and is heard in the processes of decision-making.

Social justice
Just or fair relations within society that seek to address the distribution of wealth, access to resources, opportunity, and support according to principles of justice and fairness.

See also *Equity*, *Ethics*, *Fairness*, and *Human rights*.

Kyoto Protocol   The Kyoto Protocol to the *United Nations Framework Convention on Climate Change (UNFCCC)* is an international treaty adopted in December 1997 in Kyoto, Japan, at the Third Session of the *Conference of the Parties* (COP3) to the UNFCCC. It contains legally binding commitments, in addition to those included in the UNFCCC. Countries included in Annex B of the Protocol (mostly OECD countries and countries with economies in transition) agreed to reduce their anthropogenic *greenhouse gas (GHG)* emissions *(carbon dioxide (CO₂)*, *methane (CH₄)*, *nitrous oxide (N₂O)*, hydrofluorocarbons (HFCs), perfluorocarbons (PFCs), and sulphur hexafluoride (SF₆)) by at least 5% below 1990 levels in the first commitment period (2008–2012). The Kyoto Protocol entered into force on 16 February 2005 and as of May 2018 had 192 Parties (191 States and the European Union). A second commitment period was agreed in December 2012 at COP18, known as the Doha Amendment to the Kyoto Protocol, in which a new set of Parties committed to reduce GHG emissions by at least 18% below 1990 levels in the period from 2013 to 2020. However, as of May 2018, the Doha Amendment had not received sufficient ratifications to enter into force. See also *United Nations Framework Convention on Climate Change (UNFCCC)* and *Paris Agreement*.

Land surface air temperature   The near-surface air temperature over land, typically measured at 1.25–2 m above the ground using standard meteorological equipment.

Land use   Land use refers to the total of arrangements, activities and inputs undertaken in a certain land cover type (a set of human actions). The term land use is also used in the sense of the social and economic purposes for which land is managed (e.g., grazing, timber extraction, conservation and city dwelling). In national *greenhouse gas* inventories, land use is classified according to the IPCC land use categories of forest land, cropland, grassland, wetland, settlements, other. See also *Land-use change (LUC)*.

Land-use change (LUC)   Land-use change involves a change from one *land use* category to another.

Indirect land-use change (iLUC)
Refers to market-mediated or policy-driven shifts in *land use* that cannot be directly attributed to land-use management decisions of individuals or groups. For example, if agricultural land is diverted to fuel production, *forest* clearance may occur elsewhere to replace the former agricultural production.

Land use, land-use change and forestry (LULUCF)
In the context of national *greenhouse gas (GHG)* inventories under the *UNFCCC*, LULUCF is a GHG inventory sector that covers *anthropogenic emissions* and removals of GHG from carbon pools in managed lands, excluding non-CO₂ agricultural emissions. Following the 2006 IPCC Guidelines for National GHG Inventories, 'anthropogenic' land-related GHG fluxes are defined as all those occurring on 'managed land', i.e., 'where human interventions and practices have been applied to perform

production, ecological or social functions'. Since managed land may include *CO₂* removals not considered as 'anthropogenic' in some of the scientific literature assessed in this report (e.g., removals associated with CO₂ fertilization and N deposition), the land-related net GHG emission estimates included in this report are not necessarily directly comparable with LULUCF estimates in National GHG Inventories.

See also *Afforestation*, *Deforestation*, *Reforestation*, and the IPCC Special Report on Land Use, Land-Use Change, and Forestry (IPCC, 2000).

Land use, land-use change and forestry (LULUCF)   See *Land-use change (LUC)*.

Life cycle assessment (LCA)   Compilation and evaluation of the inputs, outputs and the potential environmental impacts of a product or service throughout its life cycle. This definition builds from ISO (2018).

Likelihood   The chance of a specific outcome occurring, where this might be estimated probabilistically. Likelihood is expressed in this report using a standard terminology (Mastrandrea et al., 2010). See Section 1.6 for the list of likelihood qualifiers used. See also *Agreement*, *Evidence*, *Confidence* and *Uncertainty*.

Livelihood   The resources used and the activities undertaken in order to live. Livelihoods are usually determined by the entitlements and assets to which people have access. Such assets can be categorised as human, social, natural, physical or financial.

Local knowledge   Local knowledge refers to the understandings and skills developed by individuals and populations, specific to the places where they live. Local knowledge informs decision-making about fundamental aspects of life, from day-to-day activities to longer-term actions. This knowledge is a key element of the social and cultural systems which influence observations of, and responses to *climate change*; it also informs *governance* decisions. This definition builds on UNESCO (2018).

Lock-in   A situation in which the future development of a system, including infrastructure, technologies, investments, *institutions*, and behavioural norms, is determined or constrained ('locked in') by historic developments.

Long-lived climate forcers (LLCF)   Long-lived climate forcers refer to a set of well-mixed *greenhouse gases* with long atmospheric lifetimes. This set of compounds includes *carbon dioxide (CO₂)* and *nitrous oxide (N₂O)*, together with some fluorinated gases. They have a warming effect on *climate*. These compounds accumulate in the *atmosphere* at decadal to centennial time scales, and their effect on climate hence persists for decades to centuries after their emission. On time scales of decades to a century, already emitted emissions of long-lived climate forcers can only be abated by *greenhouse gas removal (GGR)*. See also *Short-lived climate forcers (SLCF)*.

Loss and Damage, and losses and damages   Research has taken Loss and Damage (capitalized letters) to refer to political debate under the *UNFCCC* following the establishment of the Warsaw Mechanism on Loss and Damage in 2013, which is to 'address loss and damage associated with impacts of climate change, including extreme events and slow onset events, in developing countries that are particularly vulnerable to the adverse effects of climate change.' Lowercase letters (losses and damages) have been taken to refer broadly to harm from (observed) *impacts* and (projected) *risks* (see Mechler et al., in press).

Maladaptive actions (Maladaptation)   Actions that may lead to increased *risk* of adverse climate-related outcomes, including via increased *GHG* emissions, increased *vulnerability* to *climate change*, or diminished welfare, now or in the future. Maladaptation is usually an unintended consequence.

Market exchange rate (MER)   The rate at which a currency of one country can be exchanged with the currency of another country. In most economies such rates evolve daily while in others there are official conversion rates that are adjusted periodically. See also *Purchasing power parity (PPP)*.

Market failure   When private decisions are based on market prices that do not reflect the real scarcity of goods and services but rather reflect market distortions, they do not generate an efficient allocation of resources but cause welfare losses. A market distortion is any event in which a market reaches a market clearing price that is substantially different from the price that a market would achieve while operating under conditions of perfect competition and state enforcement of legal contracts and the ownership of private property. Examples of factors causing market prices to deviate from real economic scarcity are environmental externalities, public goods, monopoly power, information asymmetry, transaction costs and non-rational behaviour.

Measurement, Reporting and Verification (MRV)

Measurement
'Processes of data collection over time, providing basic datasets, including associated accuracy and precision, for the range of relevant variables. Possible data sources are field measurements, field observations, detection through remote sensing and interviews.' (UN-REDD, 2009).

Reporting
'The process of formal reporting of assessment results to the UNFCCC, according to predetermined formats and according to established standards, especially the IPCC [Intergovernmental Panel on Climate Change] Guidelines and GPG [Good Practice Guidance].' (UN-REDD, 2009)

Verification
'The process of formal verification of reports, for example the established approach to verify national communications and national inventory reports to the UNFCCC.' (UN-REDD, 2009)

Megadrought   See *Drought*.

Methane (CH$_4$)   One of the six *greenhouse gases (GHGs)* to be mitigated under the *Kyoto Protocol* and is the major component of natural gas and associated with all hydrocarbon fuels. Significant emissions occur as a result of animal husbandry and agriculture, and their management represents a major *mitigation* option.

Migrant   See *Migration*.

Migration   The International Organization for Migration (IOM) defines migration as 'The movement of a person or a group of persons, either across an international border, or within a State. It is a population movement, encompassing any kind of movement of people, whatever its length, composition and causes; it includes migration of refugees, displaced persons, economic migrants, and persons moving for other purposes, including family reunification.' (IOM, 2018).

Migrant
The International Organization for Migration (IOM) defines a migrant as 'any person who is moving or has moved across an international border or within a State away from his/her habitual place of residence, regardless of (1) the person's legal status; (2) whether the movement is voluntary or involuntary; (3) what the causes for the movement are; or (4) what the length of the stay is.' (IOM, 2018).

See also *(Internal) Displacement*.

Millennium Development Goals (MDGs)   A set of eight time-bound and measurable goals for combating *poverty*, hunger, disease,

illiteracy, discrimination against women and environmental degradation. These goals were agreed at the UN Millennium Summit in 2000 together with an action plan to reach the goals by 2015.

Mitigation (of climate change)   A human intervention to reduce emissions or enhance the *sinks* of *greenhouse gases*.

Mitigation behaviour   See *Human behaviour*.

Mitigation measures   In climate *policy*, mitigation measures are technologies, processes or practices that contribute to *mitigation*, for example, renewable energy (RE) technologies, waste minimization processes and public transport commuting practices. See also *Mitigation option*, and *Policies (for climate change mitigation and adaptation)*.

Mitigation option   A technology or practice that reduces *GHG* emissions or enhances *sinks*.

Mitigation pathways   See *Pathways*.

Mitigation scenario   A plausible description of the future that describes how the (studied) system responds to the implementation of *mitigation* policies and measures. See also *Emission scenario*, *Pathways*, *Socio-economic scenario* and *Stabilization (of GHG or CO$_2$-equivalent concentration)*.

Monitoring and evaluation (M&E)   Monitoring and evaluation refers to mechanisms put in place at national to local scales to respectively monitor and evaluate efforts to reduce *greenhouse gas* emissions and/or adapt to the *impacts* of *climate change* with the aim of systematically identifying, characterizing and assessing progress over time.

Motivation (of an individual)   An individual's reason or reasons for acting in a particular way; individuals may consider various consequences of actions, including financial, social, affective and environmental consequences. Motivation can come from outside (extrinsic) or from inside (intrinsic) the individual.

Multilevel governance   See *Governance*.

Narratives   Qualitative descriptions of plausible future world evolutions, describing the characteristics, general logic and developments underlying a particular quantitative set of *scenarios*. Narratives are also referred to in the literature as 'storylines'. See also *Scenario*, *Scenario storyline* and *Pathways*.

Nationally Determined Contributions (NDCs)   A term used under the *United Nations Framework Convention on Climate Change (UNFCCC)* whereby a country that has joined the *Paris Agreement* outlines its plans for reducing its emissions. Some countries' NDCs also address how they will adapt to climate change impacts, and what support they need from, or will provide to, other countries to adopt low-carbon pathways and to build climate resilience. According to Article 4 paragraph 2 of the Paris Agreement, each Party shall prepare, communicate and maintain successive NDCs that it intends to achieve. In the lead up to 21st *Conference of the Parties* in Paris in 2015, countries submitted Intended Nationally Determined Contributions (INDCs). As countries join the Paris Agreement, unless they decide otherwise, this INDC becomes their first Nationally Determined Contribution (NDC). See also *United Nations Framework Convention on Climate Change (UNFCCC)* and *Paris Agreement*.

Negative emissions   Removal of *greenhouse gases (GHGs)* from the atmosphere by deliberate human activities, i.e., in addition to the removal that would occur via natural *carbon cycle* processes. See also *Net negative emissions*, *Net zero emissions*, *Carbon dioxide removal (CDR)* and *Greenhouse gas removal (GGR)*.

Net negative emissions A situation of net negative emissions is achieved when, as result of human activities, more *greenhouse gases* are removed from the *atmosphere* than are emitted into it. Where multiple greenhouse gases are involved, the quantification of *negative emissions* depends on the climate metric chosen to compare emissions of different gases (such as global warming potential, global temperature change potential, and others, as well as the chosen time horizon). See also *Negative emissions*, *Net zero emissions* and *Net zero CO₂ emissions*.

Net zero CO₂ emissions Net zero *carbon dioxide (CO₂)* emissions are achieved when *anthropogenic* CO_2 emissions are balanced globally by anthropogenic CO_2 removals over a specified period. Net zero CO_2 emissions are also referred to as carbon neutrality. See also *Net zero emissions* and *Net negative emissions*.

Net zero emissions Net zero emissions are achieved when *anthropogenic emissions* of *greenhouse gases* to the *atmosphere* are balanced by *anthropogenic removals* over a specified period. Where multiple greenhouse gases are involved, the quantification of net zero emissions depends on the climate metric chosen to compare emissions of different gases (such as global warming potential, global temperature change potential, and others, as well as the chosen time horizon). See also *Net zero CO₂ emissions*, *Negative emissions* and *Net negative emissions*.

Nitrous oxide (N₂O) One of the six *greenhouse gases (GHGs)* to be mitigated under the *Kyoto Protocol*. The main *anthropogenic* source of N_2O is agriculture (soil and animal manure management), but important contributions also come from sewage treatment, *fossil fuel* combustion, and chemical industrial processes. N_2O is also produced naturally from a wide variety of biological sources in soil and water, particularly microbial action in wet tropical *forests*.

Non-CO₂ emissions and radiative forcing Non-CO_2 emissions included in this report are all *anthropogenic emissions* other than CO_2 that result in *radiative forcing*. These include *short-lived climate forcers*, such as *methane (CH₄)*, some fluorinated gases, *ozone (O₃)* precursors, *aerosols* or aerosol *precursors*, such as *black carbon* and sulphur dioxide, respectively, as well as long-lived *greenhouse gases*, such as *nitrous oxide (N₂O)* or other fluorinated gases. The radiative forcing associated with non-CO_2 emissions and changes in surface *albedo* is referred to as non-CO_2 radiative forcing.

Non-overshoot pathways See *Pathways*.

Ocean acidification (OA) Ocean acidification refers to a reduction in the *pH* of the ocean over an extended period, typically decades or longer, which is caused primarily by uptake of *carbon dioxide (CO₂)* from the *atmosphere*, but can also be caused by other chemical additions or subtractions from the ocean. *Anthropogenic* ocean acidification refers to the component of pH reduction that is caused by human activity (IPCC, 2011, p. 37).

Ocean fertilization Deliberate increase of nutrient supply to the near-surface ocean in order to enhance biological production through which additional *carbon dioxide (CO₂)* from the *atmosphere* is sequestered. This can be achieved by the addition of micro-nutrients or macro-nutrients. Ocean fertilization is regulated by the London Protocol.

Overshoot See *Temperature overshoot*.

Overshoot pathways See *Pathways*.

Ozone (O₃) Ozone, the triatomic form of oxygen (O_3), is a gaseous atmospheric constituent. In the *troposphere*, it is created both naturally and by photochemical reactions involving gases resulting from human activities (smog). Tropospheric ozone acts as a *greenhouse gas*. In the *stratosphere*, it is created by the interaction between solar ultraviolet radiation and molecular oxygen (O_2). Stratospheric ozone plays a dominant role in the stratospheric radiative balance. Its concentration is highest in the ozone layer.

Paris Agreement The Paris Agreement under the *United Nations Framework Convention on Climate Change (UNFCCC)* was adopted on December 2015 in Paris, France, at the 21st session of the *Conference of the Parties (COP)* to the UNFCCC. The agreement, adopted by 196 Parties to the UNFCCC, entered into force on 4 November 2016 and as of May 2018 had 195 Signatories and was ratified by 177 Parties. One of the goals of the Paris Agreement is 'Holding the increase in the global average temperature to well below 2°C above pre-industrial levels and pursuing efforts to limit the temperature increase to 1.5°C above pre-industrial levels', recognising that this would significantly reduce the risks and impacts of climate change. Additionally, the Agreement aims to strengthen the ability of countries to deal with the impacts of climate change. The Paris Agreement is intended to become fully effective in 2020. See also *United Nations Framework Convention on Climate Change (UNFCCC)*, *Kyoto Protocol* and *Nationally Determined Contributions (NDCs)*.

Participatory governance See *Governance*.

Pathways The temporal evolution of natural and/or *human systems* towards a future state. Pathway concepts range from sets of quantitative and qualitative *scenarios* or *narratives* of potential futures to solution-oriented decision-making processes to achieve desirable societal goals. Pathway approaches typically focus on biophysical, techno-economic, and/or socio-behavioural trajectories and involve various dynamics, goals and actors across different scales.

1.5°C pathway
A pathway of emissions of *greenhouse gases* and other climate forcers that provides an approximately one-in-two to two-in-three chance, given current knowledge of the climate response, of *global warming* either remaining below 1.5°C or returning to 1.5°C by around 2100 following an *overshoot*. See also *Temperature overshoot*.

Adaptation pathways
A series of *adaptation* choices involving trade-offs between short-term and long-term goals and values. These are processes of deliberation to identify solutions that are meaningful to people in the context of their daily lives and to avoid potential *maladaptation*.

Development pathways
Development pathways are trajectories based on an array of social, economic, cultural, technological, *institutional* and biophysical features that characterise the interactions between human and natural systems and outline visions for the future, at a particular scale.

Emission pathways
Modelled trajectories of global *anthropogenic emissions* over the 21st century are termed emission pathways.

Mitigation pathways
A mitigation pathway is a temporal evolution of a set of *mitigation scenario* features, such as *greenhouse gas* emissions and socio-economic development.

Overshoot pathways
Pathways that exceed the stabilization level (concentration, *forcing*, or temperature) before the end of a time horizon of interest (e.g., before 2100) and then decline towards that level by that time. Once the target level is exceeded, removal by sinks of *greenhouse gases* is required. See also *Temperature overshoot*.

AI

Non-overshoot pathways
Pathways that stay below the stabilization level (concentration, *forcing*, or temperature) during the time horizon of interest (e.g., until 2100).

Representative Concentration Pathways (RCPs)
Scenarios that include time series of emissions and concentrations of the full suite of *greenhouse gases (GHGs)* and *aerosols* and chemically active gases, as well as *land use*/land cover (Moss et al., 2008). The word representative signifies that each RCP provides only one of many possible scenarios that would lead to the specific *radiative forcing* characteristics. The term pathway emphasizes the fact that not only the long-term concentration levels but also the trajectory taken over time to reach that outcome are of interest (Moss et al., 2010). RCPs were used to develop *climate projections* in CMIP5.

- RCP2.6: One pathway where radiative forcing peaks at approximately 3 W m^{-2} and then declines to be limited at 2.6 W m^{-2} in 2100 (the corresponding Extended Concentration Pathway, or ECP, has constant emissions after 2100).
- RCP4.5 and RCP6.0: Two intermediate stabilization pathways in which radiative forcing is limited at approximately 4.5 W m^{-2} and 6.0 W m^{-2} in 2100 (the corresponding ECPs have constant concentrations after 2150).
- RCP8.5: One high pathway which leads to >8.5 W m^{-2} in 2100 (the corresponding ECP has constant emissions after 2100 until 2150 and constant concentrations after 2250).

See also *Coupled Model Intercomparison Project (CMIP)* and *Shared Socio-economic Pathways (SSPs)*.

Shared Socio-economic Pathways (SSPs)
Shared Socio-economic Pathways (SSPs) were developed to complement the RCPs with varying socio-economic challenges to *adaptation* and *mitigation* (O'Neill et al., 2014). Based on five *narratives*, the SSPs describe alternative socio-economic futures in the absence of climate *policy* intervention, comprising sustainable development (SSP1), regional rivalry (SSP3), inequality (SSP4), fossil–fuelled development (SSP5) and middle-of-the-road development (SSP2) (O'Neill, 2000; O'Neill et al., 2017; Riahi et al., 2017). The combination of SSP-based socio-economic scenarios and Representative Concentration Pathway (RCP)-based *climate projections* provides an integrative frame for climate *impact* and policy analysis.

Transformation pathways
Trajectories describing consistent sets of possible futures of *greenhouse gas (GHG)* emissions, atmospheric concentrations, or *global mean surface temperatures* implied from *mitigation* and *adaptation* actions associated with a set of broad and irreversible economic, technological, societal and behavioural changes. This can encompass changes in the way energy and infrastructure are used and produced, natural resources are managed and *institutions* are set up and in the pace and direction of technological change.

See also *Scenario*, *Scenario storyline*, *Emission scenario*, *Mitigation scenario*, *Baseline scenario*, *Stabilization (of GHG or CO$_2$-equivalent concentration)* and *Narratives*.

Peri-urban areas   Peri-urban areas are those parts of a city that appear to be quite rural but are in reality strongly linked functionally to the city in its daily activities.

Permafrost   Ground (soil or rock and included ice and organic material) that remains at or below 0°C for at least two consecutive years.

pH   pH is a dimensionless measure of the acidity of a solution given by its concentration of hydrogen ions ([H$^+$]). pH is measured on a logarithmic scale where pH = -log$_{10}$[H$^+$]. Thus, a pH decrease of 1 unit corresponds to a 10-fold increase in the concentration of H$^+$, or acidity.

Plug-in hybrid electric vehicle (PHEV)   See *Electric vehicle (EV)*.

Policies (for climate change mitigation and adaptation)
Policies are taken and/or mandated by a government – often in conjunction with business and industry within a single country, or collectively with other countries – to accelerate *mitigation* and *adaptation* measures. Examples of policies are support mechanisms for renewable energy supplies, carbon or energy taxes, fuel efficiency standards for automobiles, etc.

Political economy   The set of interlinked relationships between people, the state, society and markets as defined by law, politics, economics, customs and power that determine the outcome of trade and transactions and the distribution of wealth in a country or economy.

Poverty   Poverty is a complex concept with several definitions stemming from different schools of thought. It can refer to material circumstances (such as need, pattern of deprivation or limited resources), economic conditions (such as standard of living, *inequality* or economic position) and/or social relationships (such as social class, dependency, exclusion, lack of basic security or lack of entitlement). See also *Poverty eradication*.

Poverty eradication   A set of measures to end *poverty* in all its forms everywhere. See also *Sustainable Development Goals (SDGs)*.

Precursors   Atmospheric compounds that are not *greenhouse gases (GHGs)* or *aerosols*, but that have an effect on GHG or aerosol concentrations by taking part in physical or chemical processes regulating their production or destruction rates. See also *Aerosol* and *Greenhouse gas (GHG)*.

Pre-industrial   The multi-century period prior to the onset of large-scale industrial activity around 1750. The *reference period* 1850–1900 is used to approximate pre-industrial *global mean surface temperature (GMST)*. See also *Industrial revolution*.

Procedural equity   See *Equity*.

Procedural justice   See *Justice*.

Procedural rights   See *Human rights*.

Projection   A projection is a potential future evolution of a quantity or set of quantities, often computed with the aid of a model. Unlike predictions, projections are conditional on assumptions concerning, for example, future socio-economic and technological developments that may or may not be realized. See also *Climate projection*, *Scenario* and *Pathways*.

Purchasing power parity (PPP)   The purchasing power of a currency is expressed using a basket of goods and services that can be bought with a given amount in the home country. International comparison of, for example, *gross domestic products (GDPs)* of countries can be based on the purchasing power of currencies rather than on current exchange rates. PPP estimates tend to lower the gap between the per capita GDP in *industrialized* and developing countries. See also *Market exchange rate (MER)*.

Radiative forcing   Radiative forcing is the change in the net, downward minus upward, radiative flux (expressed in W m^{-2}) at the tropopause or top of *atmosphere* due to a change in a driver of *climate change*, such as a change in the concentration of *carbon dioxide (CO$_2$)* or the output of the Sun. The traditional radiative forcing is computed with all tropospheric properties held fixed at their unperturbed values, and after allowing for stratospheric temperatures, if perturbed, to readjust to radiative-dynamical equilibrium. Radiative forcing is called instantaneous if no change in stratospheric temperature is accounted for. The radiative forcing once rapid adjustments are accounted for is termed

AI

the effective radiative forcing. Radiative forcing is not to be confused with cloud radiative forcing, which describes an unrelated measure of the impact of clouds on the radiative flux at the top of the atmosphere.

Reasons for Concern (RFCs) Elements of a classification framework, first developed in the IPCC Third Assessment Report, which aims to facilitate judgments about what level of *climate change* may be dangerous (in the language of Article 2 of the *UNFCCC*) by aggregating risks from various sectors, considering *hazards*, *exposures*, *vulnerabilities*, capacities to adapt, and the resulting impacts.

Reducing Emissions from Deforestation and Forest Degradation (REDD+) An effort to create financial value for the carbon stored in *forests*, offering incentives for developing countries to reduce emissions from forested lands and invest in low-carbon paths to *sustainable development (SD)*. It is therefore a mechanism for *mitigation* that results from avoiding *deforestation*. REDD+ goes beyond deforestation and forest degradation, and includes the role of conservation, sustainable management of forests and enhancement of forest carbon stocks. The concept was first introduced in 2005 in the 11th Session of the *Conference of the Parties (COP)* in Montreal and later given greater recognition in the 13th Session of the COP in 2007 at Bali and inclusion in the Bali Action Plan, which called for 'policy approaches and positive incentives on issues relating to reducing emissions from deforestation and forest degradation in developing countries (REDD) and the role of conservation, sustainable management of forests and enhancement of forest carbon stock in developing countries.' Since then, support for REDD has increased and has slowly become a framework for action supported by a number of countries.

Reference period The period relative to which *anomalies* are computed. See also *Anomaly*.

Reference scenario See *Baseline scenario*.

Reforestation Planting of *forests* on lands that have previously contained forests but that have been converted to some other use. For a discussion of the term forest and related terms such as *afforestation*, reforestation and *deforestation*, see the IPCC Special Report on Land Use, Land-Use Change, and Forestry (IPCC, 2000), information provided by the *United Nations Framework Convention on Climate Change* (UNFCCC, 2013), the report on Definitions and Methodological Options to Inventory Emissions from Direct Human-induced Degradation of Forests and Devegetation of Other Vegetation Types (IPCC, 2003). See also *Deforestation*, *Afforestation* and *Reducing Emissions from Deforestation and Forest Degradation (REDD+)*.

Region A region is a relatively large-scale land or ocean area characterized by specific geographical and climatological features. The *climate* of a land-based region is affected by regional and local scale features like topography, *land use* characteristics and large water bodies, as well as remote influences from other regions, in addition to global climate conditions. The IPCC defines a set of standard regions for analyses of observed climate trends and climate model *projections* (see Figure 3.2; AR5, SREX).

Remaining carbon budget Estimated cumulative net global *anthropogenic* CO_2 emissions from the start of 2018 to the time that anthropogenic CO_2 emissions reach net zero that would result, at some probability, in limiting *global warming* to a given level, accounting for the impact of other *anthropogenic emissions*.

Representative Concentration Pathways (RCPs) See *Pathways*.

Resilience The capacity of social, economic and environmental systems to cope with a hazardous event or trend or disturbance, responding or reorganizing in ways that maintain their essential function,

identity and structure while also maintaining the capacity for *adaptation*, learning and *transformation*. This definition builds from the definition used by Arctic Council (2013). See also *Hazard*, *Risk* and *Vulnerability*.

Risk The potential for adverse consequences where something of value is at stake and where the occurrence and degree of an outcome is uncertain. In the context of the assessment of climate *impacts*, the term risk is often used to refer to the potential for adverse consequences of a climate-related *hazard*, or of *adaptation* or *mitigation* responses to such a hazard, on lives, *livelihoods*, health and *well-being*, *ecosystems* and species, economic, social and cultural assets, services (including *ecosystem services*), and infrastructure. Risk results from the interaction of *vulnerability* (of the affected system), its *exposure* over time (to the hazard), as well as the (climate-related) hazard and the *likelihood* of its occurrence.

Risk assessment The qualitative and/or quantitative scientific estimation of *risks*. See also *Risk*, *Risk management* and *Risk perception*.

Risk management Plans, actions, strategies or policies to reduce the *likelihood* and/or consequences of risks or to respond to consequences. See also *Risk*, *Risk assessment* and *Risk perception*.

Risk perception The subjective judgment that people make about the characteristics and severity of a *risk*. See also *Risk*, *Risk assessment* and *Risk management*.

Runoff The flow of water over the surface or through the subsurface, which typically originates from the part of liquid precipitation and/or snow/ice melt that does not evaporate or refreeze, and is not transpired. See also *Hydrological cycle*.

Scenario A plausible description of how the future may develop based on a coherent and internally consistent set of assumptions about key driving forces (e.g., rate of technological change, prices) and relationships. Note that scenarios are neither predictions nor forecasts, but are used to provide a view of the implications of developments and actions. See also *Baseline scenario*, *Emission scenario*, *Mitigation scenario* and *Pathways*.

Scenario storyline A *narrative* description of a *scenario* (or family of scenarios), highlighting the main scenario characteristics, relationships between key driving forces and the dynamics of their evolution. Also referred to as 'narratives' in the scenario literature. See also *Narratives*.

SDG-interaction score A seven-point scale (Nilsson et al., 2016) used to rate interactions between *mitigation options* and the *SDGs*. Scores range from +3 (indivisible) to −3 (cancelling), with a zero score indicating 'consistent' but with neither a positive or negative interaction. The scale, as applied in this report, also includes direction (whether the interaction is uni- or bi-directional) and *confidence* as assessed per IPCC guidelines.

Sea ice Ice found at the sea surface that has originated from the freezing of seawater. Sea ice may be discontinuous pieces (ice floes) moved on the ocean surface by wind and currents (pack ice), or a motionless sheet attached to the coast (land-fast ice). Sea ice concentration is the fraction of the ocean covered by ice. Sea ice less than one year old is called first-year ice. Perennial ice is sea ice that survives at least one summer. It may be subdivided into second-year ice and multi-year ice, where multi-year ice has survived at least two summers.

Sea level change (sea level rise/sea level fall) Sea level can change, both globally and locally (relative sea level change) due to (1) a change in ocean volume as a result of a change in the mass of water in the ocean, (2) changes in ocean volume as a result of changes in ocean water density, (3) changes in the shape of the ocean basins and changes

AI

in the Earth's gravitational and rotational fields, and (4) local subsidence or uplift of the land. Global mean sea level change resulting from change in the mass of the ocean is called barystatic. The amount of barystatic sea level change due to the addition or removal of a mass of water is called its sea level equivalent (SLE). Sea level changes, both globally and locally, resulting from changes in water density are called steric. Density changes induced by temperature changes only are called thermosteric, while density changes induced by salinity changes are called halosteric. Barystatic and steric sea level changes do not include the effect of changes in the shape of ocean basins induced by the change in the ocean mass and its distribution.

Sea surface temperature (SST)   The sea surface temperature is the subsurface bulk temperature in the top few meters of the ocean, measured by ships, buoys, and drifters. From ships, measurements of water samples in buckets were mostly switched in the 1940s to samples from engine intake water. Satellite measurements of skin temperature (uppermost layer; a fraction of a millimeter thick) in the infrared or the top centimeter or so in the microwave are also used, but must be adjusted to be compatible with the bulk temperature.

Sendai Framework for Disaster Risk Reduction   The Sendai Framework for Disaster Risk Reduction 2015–2030 outlines seven clear targets and four priorities for action to prevent new, and to reduce existing, disaster risks. The voluntary, non-binding agreement recognizes that the State has the primary role to reduce disaster risk but that responsibility should be shared with other stakeholders, including local government and the private sector. Its aim is to achieve 'substantial reduction of disaster risk and losses in lives, *livelihoods* and health and in the economic, physical, social, cultural and environmental assets of persons, businesses, communities and countries.'

Sequestration   See *Uptake*.

Shared Socio-economic Pathways (SSPs)   See *Pathways*.

Short-lived climate forcers (SLCF)   Short-lived climate forcers refers to a set of compounds that are primarily composed of those with short lifetimes in the *atmosphere* compared to well-mixed *greenhouse gases*, and are also referred to as near-term climate forcers. This set of compounds includes *methane (CH_4)*, which is also a well-mixed greenhouse gas, as well as *ozone (O_3)* and *aerosols*, or their *precursors*, and some halogenated species that are not well-mixed greenhouse gases. These compounds do not accumulate in the atmosphere at decadal to centennial time scales, and so their effect on *climate* is predominantly in the first decade after their emission, although their changes can still induce long-term climate effects such as *sea level change*. Their effect can be cooling or warming. A subset of exclusively warming short-lived climate forcers is referred to as short-lived climate pollutants. See also *Long-lived climate forcers (LLCF)*.

Short-lived climate pollutants (SLCP)   See *Short-lived climate forcers (SLCF)*.

Sink   A reservoir (natural or human, in soil, ocean, and plants) where a *greenhouse gas*, an *aerosol* or a *precursor* of a greenhouse gas is stored. Note that *UNFCCC* Article 1.8 refers to a sink as any process, activity or mechanism which removes a greenhouse gas, an aerosol or a precursor of a greenhouse gas from the *atmosphere*. See also *Uptake*.

Small island developing states (SIDS)   Small island developing states (SIDS), as recognised by the United Nations OHRLLS (Office of the High Representative for the Least Developed Countries, Landlocked Developing Countries and Small Island Developing States), are a distinct group of developing countries facing specific social, economic and environmental vulnerabilities (UN-OHRLLS, 2011). They were recognized

as a special case both for their environment and development at the Rio Earth Summit in Brazil in 1992. Fifty-eight countries and territories are presently classified as SIDS by the UN OHRLLS, with 38 being UN member states and 20 being Non-UN Members or Associate Members of the Regional Commissions (UN-OHRLLS, 2018).

Social cost of carbon (SCC)   The net present value of aggregate climate damages (with overall harmful damages expressed as a number with positive sign) from one more tonne of carbon in the form of *carbon dioxide (CO_2)*, conditional on a global emissions trajectory over time.

Social costs   The full costs of an action in terms of social welfare losses, including external costs associated with the impacts of this action on the environment, the economy (*GDP*, employment) and on the society as a whole.

Social-ecological systems   An integrated system that includes human societies and *ecosystems*, in which humans are part of nature. The functions of such a system arise from the interactions and interdependence of the social and ecological subsystems. The system's structure is characterized by reciprocal feedbacks, emphasising that humans must be seen as a part of, not apart from, nature. This definition builds from Arctic Council (2016) and Berkes and Folke (1998).

Social inclusion   A process of improving the terms of participation in society, particularly for people who are disadvantaged, through enhancing opportunities, access to resources, and respect for rights (UN DESA, 2016).

Social justice   See *Justice*.

Social learning   A process of social interaction through which people learn new behaviours, capacities, values and attitudes.

Social value of mitigation activities (SVMA)   Social, economic and environmental value of *mitigation* activities that include, in addition to their climate benefits, their *co-benefits* to *adaptation* and *sustainable development* objectives.

Societal (social) transformation   See *Transformation*.

Socio-economic scenario   A *scenario* that describes a possible future in terms of population, *gross domestic product (GDP)*, and other socio-economic factors relevant to understanding the implications of *climate change*. See also *Baseline scenario*, *Emission scenario*, *Mitigation scenario* and *Pathways*.

Socio-technical transitions   Socio-technical transitions are where technological change is associated with social systems and the two are inextricably linked.

Soil carbon sequestration (SCS)   Land management changes which increase the soil organic carbon content, resulting in a net removal of CO_2 from the *atmosphere*.

Soil moisture   Water stored in the soil in liquid or frozen form. Root-zone soil moisture is of most relevance for plant activity.

Solar radiation management   See *Solar radiation modification (SRM)*.

Solar radiation modification (SRM)   Solar radiation modification refers to the intentional modification of the Earth's shortwave radiative budget with the aim of reducing warming. Artificial injection of stratospheric *aerosols*, marine cloud brightening and land surface *albedo* modification are examples of proposed SRM methods. SRM does not fall within the definitions of *mitigation* and *adaptation* (IPCC, 2012b, p. 2). Note that in the literature SRM is also referred to as solar radiation management or albedo enhancement.

Stabilization (of GHG or CO₂-equivalent concentration) A state in which the atmospheric concentrations of one *greenhouse gas (GHG)* (e.g., *carbon dioxide*) or of a CO₂-equivalent basket of GHGs (or a combination of GHGs and *aerosols*) remains constant over time.

Stranded assets Assets exposed to devaluations or conversion to 'liabilities' because of unanticipated changes in their initially expected revenues due to innovations and/or evolutions of the business context, including changes in public regulations at the domestic and international levels.

Stratosphere The highly stratified region of the *atmosphere* above the *troposphere* extending from about 10 km (ranging from 9 km at high latitudes to 16 km in the tropics on average) to about 50 km altitude. See also *Atmosphere*, and *Troposphere*.

Sub-national actor Sub-national actors include state/provincial, regional, metropolitan and local/municipal governments as well as non-party stakeholders, such as civil society, the private sector, cities and other sub-national authorities, local communities and indigenous peoples.

Substantive rights See *Human rights*.

Supply-side measures See *Demand- and supply-side measures*.

Surface temperature See *Global mean surface temperature (GMST)*, *Land surface air temperature*, *Global mean surface air temperature (GSAT)* and *Sea surface temperature (SST)*.

Sustainability A dynamic process that guarantees the persistence of natural and *human systems* in an equitable manner.

Sustainable development (SD) Development that meets the needs of the present without compromising the ability of future generations to meet their own needs (WCED, 1987) and balances social, economic and environmental concerns. See also *Sustainable Development Goals (SDGs)* and *Development pathways* (under *Pathways*).

Sustainable Development Goals (SDGs) The 17 global goals for development for all countries established by the United Nations through a participatory process and elaborated in the *2030 Agenda for Sustainable Development*, including ending *poverty* and hunger; ensuring health and *well-being*, education, gender *equality*, clean water and energy, and decent work; building and ensuring *resilient* and sustainable infrastructure, cities and consumption; reducing *inequalities*; protecting land and water *ecosystems*; promoting peace, *justice* and partnerships; and taking urgent action on *climate change*. See also *Sustainable development (SD)*.

Technology transfer The exchange of knowledge, hardware and associated software, money and goods among stakeholders, which leads to the spread of technology for *adaptation* or *mitigation*. The term encompasses both diffusion of technologies and technological cooperation across and within countries.

Temperature overshoot The temporary exceedance of a specified level of *global warming*, such as 1.5°C. Overshoot implies a peak followed by a decline in global warming, achieved through *anthropogenic removal* of *CO₂* exceeding remaining CO₂ emissions globally. See also *Overshoot pathways* and *Non-overshoot pathways* (both under *Pathways*).

Tipping point A level of change in system properties beyond which a system reorganizes, often abruptly, and does not return to the initial state even if the drivers of the change are abated. For the *climate system*, it refers to a critical threshold when global or regional *climate* changes from one stable state to another stable state. See also *Irreversibility*.

Transformation A change in the fundamental attributes of natural and human systems.

Societal (social) transformation
A profound and often deliberate shift initiated by communities toward sustainability, facilitated by changes in individual and collective values and behaviours, and a fairer balance of political, cultural, and *institutional* power in society.

Transformation pathways See *Pathways*.

Transformational adaptation See *Adaptation*.

Transformative change A system-wide change that requires more than technological change through consideration of social and economic factors that, with technology, can bring about rapid change at scale.

Transient climate response See *Climate sensitivity*.

Transient climate response to cumulative CO₂ emissions (TCRE) The transient global average surface temperature change per unit cumulative *CO₂* emissions, usually 1000 GtC. TCRE combines both information on the airborne fraction of cumulative CO₂ emissions (the fraction of the total CO₂ emitted that remains in the *atmosphere*, which is determined by *carbon cycle* processes) and on the *transient climate response (TCR)*. See also *Transient climate response* (under *Climate sensitivity*).

Transit-oriented development (TOD) An approach to urban development that maximizes the amount of residential, business and leisure space within walking distance of efficient public transport, so as to enhance mobility of citizens, the viability of public transport and the value of urban land in mutually supporting ways.

Transition The process of changing from one state or condition to another in a given period of time. Transition can be in individuals, firms, cities, regions and nations, and can be based on incremental or transformative change.

Tropical cyclone The general term for a strong, cyclonic-scale disturbance that originates over tropical oceans. Distinguished from weaker systems (often named tropical disturbances or depressions) by exceeding a threshold wind speed. A tropical storm is a tropical cyclone with one-minute average surface winds between 18 and 32 m s⁻¹. Beyond 32 m s⁻¹, a tropical cyclone is called a hurricane, typhoon, or cyclone, depending on geographic location. See also *Extratropical cyclone*.

Troposphere The lowest part of the *atmosphere*, from the surface to about 10 km in altitude at mid-latitudes (ranging from 9 km at high latitudes to 16 km in the tropics on average), where clouds and weather phenomena occur. In the troposphere, temperatures generally decrease with height. See also *Atmosphere* and *Stratosphere*.

Uncertainty A state of incomplete knowledge that can result from a lack of information or from disagreement about what is known or even knowable. It may have many types of sources, from imprecision in the data to ambiguously defined concepts or terminology, incomplete understanding of critical processes, or uncertain *projections* of *human behaviour*. Uncertainty can therefore be represented by quantitative measures (e.g., a probability density function) or by qualitative statements (e.g., reflecting the judgment of a team of experts) (see Moss and Schneider, 2000; IPCC, 2004; Mastrandrea et al., 2010). See also *Confidence* and *Likelihood*.

United Nations Framework Convention on Climate Change (UNFCCC) The UNFCCC was adopted in May 1992 and opened for signature at the 1992 Earth Summit in Rio de Janeiro. It entered into force in March 1994 and as of May 2018 had 197 Parties (196 States and the European Union). The Convention's ultimate objective is the 'stabilisation of greenhouse gas concentrations in the atmosphere at a level that would prevent dangerous anthropogenic interference with the climate

AI

system.' The provisions of the Convention are pursued and implemented by two treaties: the *Kyoto Protocol* and the *Paris Agreement*. See also *Kyoto Protocol* and *Paris Agreement*.

Uptake   The addition of a substance of concern to a reservoir. See also *Carbon sequestration* and *Sink*.

Vulnerability   The propensity or predisposition to be adversely affected. Vulnerability encompasses a variety of concepts and elements including sensitivity or susceptibility to harm and lack of capacity to cope and adapt. See also *Exposure*, *Hazard* and *Risk*.

Water cycle   See *Hydrological cycle*.

Well-being   A state of existence that fulfils various human needs, including material living conditions and quality of life, as well as the ability to pursue one's goals, to thrive, and feel satisfied with one's life. Ecosystem well-being refers to the ability of *ecosystems* to maintain their diversity and quality.

Zero emissions commitment   See *Climate change commitment*.

References

Arctic Council, 2013: Glossary of terms. In: *Arctic Resilience Interim Report 2013*. Stockholm Environment Institute and Stockholm Resilience Centre, Stockholm, Sweden, pp. viii.

Carson, M. and G. Peterson (eds.), 2016: *Arctic Resilience Report 2016*. Stockholm Environment Institute and Stockholm Resilience Centre, Stockholm, Sweden, 218 pp.

Berkes, F. and C. Folke, 1998: *Linking Social and Ecological Systems: Management Practices and Social Mechanisms for Building Resilience*. Cambridge University Press, Cambridge, United Kingdom and New York, NY, USA, 459 pp.

Culwick, C. and K. Bobbins, 2016: *A Framework for a Green Infrastructure Planning Approach in the Gauteng City–Region*. GCRO Research Report No. 04, Gauteng City–Region Observatory (GRCO), Johannesburg, South Africa, 127 pp.

FAO, 2001: Glossary. In: *The State of Food Insecurity in the World 2001*. Food and Agriculture Organisation of the United Nations (FAO), Rome, Italy, pp. 49–50.

FAO, 2013: *Food wastage footprint: Impacts on natural resources. Summary report*. Food and Agriculture Organization of the United Nations (FAO), Rome, Italy, 63 pp.

FAO, 2018: Climate–Smart Agriculture. Food and Agriculture Organization of the United Nations (FAO). Retrieved from: www.fao.org/climate–smart–agriculture.

Fung, A. and E.O. Wright (eds.), 2003: *Deepening Democracy: Institutional Innovations in Empowered Participatory Governance*. Verso, London, UK, 312 pp.

Helpman, E. (ed.), 1998: *General Purpose Technologies and Economic Growth*. MIT Press, Cambridge, MA, USA, 315 pp.

IBI, 2018: Frequently Asked Questions About Biochar: What is biochar? International Biochar Initiative (IBI). Retrieved from: https://biochar–international.org/faqs.

IOM, 2018: Key Migration Terms. International Organization for Migration (IOM). Retrieved from: www.iom.int/key–migration–terms.

IPCC, 2000: Land Use, Land–Use Change, and Forestry: A Special Report of the IPCC. [Watson, R.T., I.R. Noble, B. Bolin, N.H. Ravindranath, D.J. Verardo, and D.J. Dokken (eds.)]. Cambridge University Press, Cambridge, UK, 375 pp.

IPCC, 2003: Definitions and Methodological Options to Inventory Emissions from Direct Human–induced Degradation of Forests and Devegetation of Other Vegetation Types. [Penman, J., M. Gytarsky, T. Hiraishi, T. Krug, D. Kruger, R. Pipatti, L. Buendia, K. Miwa, T. Ngara, K. Tanabe, and F. Wagner (eds.)]. Institute for Global Environmental Strategies (IGES), Hayama, Kanagawa, Japan, 32 pp.

IPCC, 2004: *IPCC Workshop on Describing Scientific Uncertainties in Climate Change to Support Analysis of Risk of Options. Workshop Report*. Intergovernmental Panel on Climate Change (IPCC), Geneva, Switzerland, 138 pp.

IPCC, 2011: Workshop Report of the Intergovernmental Panel on Climate Change Workshop on Impacts of Ocean Acidification on Marine Biology and Ecosystems. [Field, C.B., V. Barros, T.F. Stocker, D. Qin, K.J. Mach, G.–K. Plattner, M.D. Mastrandrea, M. Tignor, and K.L. Ebi (eds.)]. IPCC Working Group II Technical Support Unit, Carnegie Institution, Stanford, California, United States of America, 164 pp.

IPCC, 2012a: Managing the Risks of Extreme Events and Disasters to Advance Climate Change Adaptation. A Special Report of Working Groups I and II of the Intergovernmental Panel on Climate Change (IPCC). [Field, C.B., V. Barros, T.F. Stocker, D. Qin, D.J. Dokken, K.L. Ebi, M.D. Mastrandrea, K.J. Mach, G.–K. Plattner, S.K. Allen, M. Tignor, and P.M. Midgley (eds.)]. Cambridge University Press, Cambridge, UK and New York, NY, USA, 582 pp.

IPCC, 2012b: *Meeting Report of the Intergovernmental Panel on Climate Change Expert Meeting on Geoengineering*. IPCC Working Group III Technical Support Unit, Potsdam Institute for Climate Impact Research, Potsdam, Germany, 99 pp.

ISO, 2018: ISO 14044:2006. Environmental management – Life cycle assessment – Requirements and guidelines. International Standards Organisation (ISO). Retrieved from: www.iso.org/standard/38498.html.

Jagers, S.C. and J. Stripple, 2003: Climate Governance Beyond the State. *Global Governance*, **9(3)**, 385–399, www.jstor.org/stable/27800489.

Mastrandrea, M.D. et al., 2010: *Guidance Note for Lead Authors of the IPCC Fifth Assessment Report on Consistent Treatment of Uncertainties*. Intergovernmental Panel on Climate Change (IPCC), Geneva, Switzerland, 6 pp.

MEA, 2005: Appendix D: Glossary. In: *Ecosystems and Human Well–being: Current States and Trends. Findings of the Condition and Trends Working Group* [Hassan, R., R. Scholes, and N. Ash (eds.)]. Millennium Ecosystem Assessment (MEA). Island Press, Washington DC, USA, pp. 893–900.

Mechler, R., L.M. Bouwer, T. Schinko, S. Surminski, and J. Linnerooth–Bayer (eds.), in press: *Loss and Damage from Climate Change: Concepts, Methods and Policy Options*. Springer International Publishing, 561 pp.

Mitchell, T. and S. Maxwell, 2010: Defining climate compatible development. CDKN ODI Policy Brief November 2010/A, Climate & Development Knowledge Network (CDKN), 6 pp.

Moss, R.H. and S.H. Schneider, 2000: Uncertainties in the IPCC TAR: Recommendations to Lead Authors for More Consistent Assessment and Reporting. In: *Guidance Papers on the Cross Cutting Issues of the Third Assessment Report of the IPCC* [Pachauri, R., T. Taniguchi, and K. Tanaka (eds.)]. Intergovernmental Panel on Climate Change (IPCC), Geneva, Switzerland, pp. 33–51.

Moss, R.H. et al., 2008: *Towards New Scenarios for Analysis of Emissions, Climate Change, Impacts, and Response Strategies*. Technical Summary. Intergovernmental Panel on Climate Change (IPCC), Geneva, Switzerland, 25 pp.

Moss, R.H. et al., 2010: The next generation of scenarios for climate change research and assessment. *Nature*, **463(7282)**, 747–756, doi:10.1038/nature08823.

MRFCJ, 2018: Principles of Climate Justice. Mary Robinson Foundation For Climate Justice (MRFCJ). Retrieved from: www.mrfcj.org/principles–of–climate–justice.

Nilsson, M., D. Griggs, and M. Visbeck, 2016: Policy: Map the interactions between Sustainable Development Goals. *Nature*, **534(7607)**, 320–322, doi:10.1038/534320a.

O'Neill, B.C., 2000: The Jury is Still Out on Global Warming Potentials. *Climatic Change*, **44(4)**, 427–443, doi:10.1023/A:1005582929198.

O'Neill, B.C. et al., 2014: A new scenario framework for climate change research: the concept of shared socioeconomic pathways. *Climatic Change*, **122(3)**, 387–400, doi:10.1007/s10584–013–0905–2.

O'Neill, B.C. et al., 2017: The roads ahead: Narratives for shared socioeconomic pathways describing world futures in the 21st century. *Global Environmental Change*, **42**, 169–180, doi:10.1016j.gloenvcha.2015.01.004.

Peters, B.G. and J. Pierre, 2001: Developments in intergovernmental relations: towards multi–level governance. *Policy & Politics*, **29(2)**, 131–135, doi:10.1332/0305573012501251.

Riahi, K. et al., 2017: The Shared Socioeconomic Pathways and their energy, land use, and greenhouse gas emissions implications: An overview. *Global Environmental Change*, **42**, 153–168, doi:10.1016/j.gloenvcha.2016.05.009.

Sarmiento, H. and C. Tilly, 2018: Governance Lessons from Urban Informality. *Politics and Governance*, **6(1)**, 199–202, doi:10.17645/pag.v6i1.1169.

Tàbara, J.D., J. Jäger, D. Mangalagiu, and M. Grasso, 2018: Defining transformative climate science to address high–end climate change. *Regional Environmental Change*, 1–12, doi:10.1007/s10113–018–1288–8.

Termeer, C.J.A.M., A. Dewulf, and G.R. Biesbroek, 2017: Transformational change: governance interventions for climate change adaptation from a continuous change perspective. *Journal of Environmental Planning and Management*, **60(4)**, 558–576, doi:10.1080/09640568.2016.1168288.

UN, 1992: Article 2: Use of Terms. In: *Convention on Biological Diversity*. United Nations (UN), pp. 3–4.

UN, 1998: *Guiding Principles on Internal Displacement*. E/CN.4/1998/53/Add.2, United Nations (UN) Economic and Social Council, 14 pp.

UN, 2015: *Transforming Our World: The 2030 Agenda for Sustainable Development*. A/RES/70/1, United Nations General Assembly (UNGA), New York, NY, USA, 35 pp.

UN DESA, 2016: Identifying social inclusion and exclusion. In: *Leaving no one behind: the imperative of inclusive development. Report on the World Social Situation 2016*. ST/ESA/362, United Nations Department of Economic and Social Affairs (UN DESA), New York, NY, USA, pp. 17–31.

UNESCO, 2018: Local and Indigenous Knowledge Systems. United Nations Educational, Scientific and Cultural Organization (UNESCO). Retrieved from: www.unesco.org/new/en/natural–sciences/priority–areas/links/related–information/what–is–local–and–indigenous–knowledge.

UNFCCC, 2013: Reporting and accounting of LULUCF activities under the Kyoto Protocol. United Nations Framework Convention on Climatic Change (UNFCCC), Bonn, Germany. Retrieved from: http://unfccc.int/methods/lulucf/items/4129.php.

UNISDR, 2009: *2009 UNISDR Terminology on Disaster Risk Reduction*. United Nations International Strategy for Disaster Reduction (UNISDR), Geneva, Switzerland, 30 pp.

UNOHCHR, 2018: What are Human rights? UN Office of the High Commissioner for Human Rights (UNOHCHR). Retrieved from: www.ohchr.org/EN/Issues/Pages/whatarehumanrights.aspx.

UN–OHRLLS, 2011: *Small Island Developing States: Small Islands Big(ger) Stakes*. Office for the High Representative for the Least Developed Countries, Landlocked Developing Countries and Small Island Developing States (UN–OHRLLS), New York, NY, USA, 32 pp.

UN–OHRLLS, 2018: Small Island Developing States: Country profiles. Office for the High Representative for the Least Developed Countries, Landlocked Developing Countries and Small Island Developing States (UN–OHRLLS). Retrieved from: http://unohrlls.org/about–sids/country–profiles.

UN–REDD, 2009: *Measurement, Assessment, Reporting and Verification (MARV): Issues and Options for REDD*. Draft Discussion Paper, United Nations Collaborative Programme on Reducing Emissions from Deforestation and Forest Degradation in Developing Countries (UN–REDD), Geneva, Switzerland, 12 pp.

WCED, 1987: *Our Common Future*. World Commission on Environment and Development (WCED), Geneva, Switzerland, 400 pp., doi:10.2307/2621529.

Willems, S. and K. Baumert, 2003: *Institutional Capacity and Climate Actions*. COM/ENV/EPOC/IEA/SLT(2003)5, Organisation for Economic Co–operation and Development (OECD) International Energy Agency (IEA), Paris, France, 50 pp.

AII

Annex II: Acronyms

This annex should be cited as:
IPCC, 2018: Annex II: Acronyms. In: *Global Warming of 1.5°C. An IPCC Special Report on the impacts of global warming of 1.5°C above pre-industrial levels and related global greenhouse gas emission pathways, in the context of strengthening the global response to the threat of climate change, sustainable development, and efforts to eradicate poverty* [Masson-Delmotte, V., P. Zhai, H.-O. Pörtner, D. Roberts, J. Skea, P.R. Shukla, A. Pirani, W. Moufouma-Okia, C. Péan, R. Pidcock, S. Connors, J.B.R. Matthews, Y. Chen, X. Zhou, M.I. Gomis, E. Lonnoy, T. Maycock, M. Tignor, and T. Waterfield (eds.)]. Cambridge University Press, Cambridge, UK and New York, NY, USA, pp. 563-572.

µatm	Microatmospheres	BET	Basic Energy systems, Economy, Environment, and End-use Technology Model
1.5DS	1.5 Degree Scenario	BEV	Battery Electric Vehicle
2DS	2 Degree Scenario	BNEF	Bloomberg New Energy Finance
ACCESS	Australian Community Climate and Earth-System Simulator	BNU	Beijing Normal University
ACCMIP	Atmospheric Chemistry and Climate Model Intercomparison Project	BRT	Bus Rapid Transit
		cm	Centimetres
ACCRN	The Asian Cities Climate Change Resilience Network	C	Carbon
ACOLA	Australian Council of Learned Academies	CA	Conservation Agriculture
ACs	Air Conditioners	CAF	Corporacion Andina de Fomento (Development Bank of Latin America)
ADB	Asian Development Bank	CAM	Central America/Mexico or Community Atmosphere Model
ADVANCE	Advanced Model Development and Validation for the Improved Analysis of Costs and Impacts of Mitigation Policies	CAMx	Comprehensive Air Quality Model with Extensions
AEZ	Agro-Ecological Zone	CanESM	Canadian Earth System Model
AfDB	African Development Bank	CanRCM	Canadian Regional Climate Model
AFOLU	Agriculture, Forestry and Other Land-Use	CAR	Small Islands Regions Caribbean
AGCM	Atmospheric General Circulation Model	CAS	Central Asia
AI	Artificial Intelligence	Cat-HM	Catchment-scale Hydrological Models
AIM	Asia-Pacific Integrated Model	CbA	Community-based Adaptation
ALA	Alaska/Northwest Canada	CBA	Cost-Benefit Analysis
AMAP	Arctic Monitoring and Assessment Programme	CBD	Convention on Biological Diversity
AMOC	Atlantic Meridional Overturning Circulation	CBDR-RC	Common But Differentiated Responsibilities and Respective Capabilities
AMP	Adjusting Mitigation Pathway	CBS & GNH	Centre for Bhutan Studies and Gross National Happiness Research
AMZ	Amazon		
ANT	Antarctica	CC	Carbon Capture
APEX	Air Pollutants Exposure Model	CCAM	Conformal Cubic Atmospheric Model
AR	Afforestation and Reforestation	CCC	Constant Composition Commitment
AR4	IPCC Fourth Assessment Report	CCCma	Canadian Centre for Climate Modelling and Analysis
AR5	IPCC Fifth Assessment Report		
AR6	IPCC Sixth Assessment Report	CCRIF	Caribbean Catastrophe Risk Insurance Facility
ARC	Arctic	CCS	Carbon dioxide Capture and Storage
ASEAN	Association of Southeast Asian Nations	CCSM	Community Climate System Model
ASIA	Non-OECD Asia	CCT	Cirrus Cloud Thinning
AUD	Australian Dollar	CCU	Carbon dioxide Capture and Utilisation
B2DS	Beyond 2 Degrees Scenario	CCUS	Carbon dioxide Capture, Utilisation and Storage
BASIC	Brazil, South Africa, India, China	CDD	Consecutive Dry Days
BC	Black Carbon	CD-LINKS	Linking Climate and Development Policies – Leveraging International Networks and Knowledge Sharing
BCC-CSM	Beijing Climate Center Climate System Model		
BCM	Bergen Climate Model	CDM	Clean Development Mechanism
BECCS	Bioenergy with Carbon dioxide Capture and Storage	CDP	Carbon Disclosure Project
		CDR	Carbon Dioxide Removal

AII

CEA	Cost-Effectiveness Analysis
CEC	Clean Energy Council
CEDS	Community Emissions Data System
CEMICS	Contextualizing Climate Engineering and Mitigation: Illusion, Complement or Substitute?
CES	Constant Elasticity of Substitution
CESM	Community Earth System Model
CEU	Central Europe
CFCs	Chlorofluorocarbons
CGCM	Coupled Global Climate Model
CGE	Computable General Equilibrium
CGI	Canada/Greenland/Iceland
CGIAR	Consultative Group on International Agricultural Research
CH_4	Methane
CHP	Combined Heat and Power
CI	Confidence Interval
CIRED	Centre International de Recherche sur l'Environnement et le Développement
CISL	Cambridge Institute for Sustainability Leadership
CLM	Climate Limited-area Modelling
CMAQ	Community Multiscale Air Quality Modeling System
CMIP3	Coupled Model Intercomparison Project Phase 3
CMIP5	Coupled Model Intercomparison Project Phase 5
CMIP6	Coupled Model Intercomparison Project Phase 6
CNA	Central North America
CNRM	Centre National de Recherches Météorologiques
CO	Carbon monoxide
CO_2	Carbon dioxide
CO_2e	Carbon dioxide equivalent
CO_2eq	Carbon dioxide equivalent
CoM	Covenant of Mayors
COP	Conference of the Parties
COPPE-COFFEE	Programa de Planejamento Energético – COmputable Framework For Energy and the Environment
CORDEX	Coordinated Regional Climate Downscaling Experiment
COSMO	Consortium for Small-scale Modeling
CRCM	Canadian Regional Climate Model
CRDPs	Climate-Resilient Development Pathways
CRIEPI	Institut Central de Recherche des Industries Électriques

CRISPR	Clustered Regularly Interspaced Short Palindromic Repeats
C-ROADS	Climate Rapid Overview And Decision-support Simulator
CRU	Climatic Research Unit
CSA	Climate-Smart Agriculture
CSC	Climate Service Center Germany
CSDI	Cold Spell Duration Index
CSIRO	Commonwealth Scientific and Industrial Research Organisation
CSP	Concentrated Solar Power
CSR	Corporate Social Responsibility
CTC	Covenant Territorial Coordinator
CWD	Consecutive Wet Days
DACCS	Direct Air Carbon dioxide Capture and Storage
DACS	Direct Air Capture and Storage
DALY	Disability Adjusted Life Year
DICE	Dynamic Integrated Climate-Economy model
DJF	December, January, February
DM8H	Daily Maximum 8-Hour exposure
DNE21+	Dynamic New Earth 21 model
DOE	Department of Energy (USA)
DRI	Direct Reduced Iron
DRM	Disaster Risk Management
DTU	Technical University of Denmark
E	Equilibrium, Evaporation or Evapotranspiration
EAF	East Africa
EAIS	East Antarctic Ice Sheet
EAS	East Asia
EbA	Ecosystems-based Adaptation
EC	European Commission
ECF	European Climate Foundation
ECMWF	European Centre for Medium-Range Weather Forecasts
ECS	Equilibrium Climate Sensitivity
EDGAR	Emission Database for Global Atmospheric Research
EEA	European Environment Agency
EGMAM	ECHO-G Middle Atmosphere Model
E-HYPE	European Hydrological Predictions for the Environment
EJ	Exajoules
EMEP	European Monitoring and Evaluation Programme

AII

EMF	Energy Modeling Forum	F-gas	Fluorinated gases
EMIC	Earth-system Model of Intermediate Complexity	FGOALS	Flexible Global Ocean-Atmosphere-Land System model
ENA	East North America	FIO	First Institute of Oceanography
ENSO	El Niño-Southern Oscillation	FMNR	Farmer Managed Natural Regeneration
EOR	Enhanced Oil Recovery	FUND	Climate Framework for Uncertainty, Negotiation, and Distribution model
EPA	Environmental Protection Agency (USA)		
EPIs	Energy-Intensive Processing Industries	FUSSR	Former Union of Soviet Socialist Republics
ERA	ECMWF ReAnalysis	g	Grams
ERF	Effective Radiative Forcing	GAMS	General Algebraic Modeling System
ERFaci	Effective Radiative Forcing from aerosol-cloud interactions	GBAM	Ground-Based Albedo Modification
		GCAM	Global Change Assessment Model
ERFari	Effective Radiative Forcing from aerosol-radiation interactions	GCC	Gulf Cooperative Council
		GCEC	Global Commission on the Economy and Climate
ESCOs	Energy Service Companies	GCM	General Circulation Model or Global Climate Model
ESL	Extreme Sea Level		
ESM	Earth System Model	GCP	Global Carbon Project
ESR	Empirical Scaling Relationship	GDP	Gross Domestic Product
ESRB ASC	European Systemic Risk Board Advisory Scientific Committee	GE	General Equilibrium
		GEA	Global Energy Assessment
ESRL	NOAA Earth System Research Laboratory	GEM-E3	General Equilibrium Model for Economy - Energy - Environment
Eta-CPTEC	Eta Centro de Previsão do Tempo e Estudos Climáticos		
		GENeSYS-MOD	Global Energy System Model
ETP	Pacific Islands region [3] or Energy Technology Perspectives model	GeSI	Global e-Sustainability Initiative
		GFCF	Gross Fixed Capital Formation
ETS	Emission Trading Scheme	GFDL	Geophysical Fluid Dynamic Laboratory
EU	European Union	GFR	Grid Flexibility Resources
EU-FP6	European Union Sixth Framework Programme	Gha	Gigahectares
EUG4	France, Germany, Italy, United Kingdom	GHCNDEX	Global Historical Climatology Network – Daily climate Extremes
EURO-CORDEX	European branch of the Coordinated Regional Climate Downscaling Experiment		
		GHGs	Greenhouse Gases
EV	Electric Vehicle	GHM	Global Hydrological Models
EW	Enhanced Weathering	GIS	Greenland Ice Sheet
FAIR	Finite Amplitude Impulse Response model	GISS	Goddard Institute for Space Studies
FAO	Food and Agriculture Organization of the United Nations	GISTEMP	Goddard Institute for Space Studies Surface Temperature Analysis
		GJ	Gigajoules
FAOSTAT	Database Collection of the Food and Agriculture Organization of the United Nations	GLEAM	Global Livestock Environmental Assessment Model
		Glob-HM	Global Hydrological Model
FAQ	Frequently Asked Questions	GLOBIOM	GLObal BIOsphere Management model
FARM	Future Agricultural Resources Model	GLOFs	Glacial Lake Outburst Floods
Fe	Iron	GM	Genetically Modified
FE	Final Energy	GMO	Genetically Modified Organism
FEMA	Federal Emergency Management Agency (USA)	GMSL	Global Mean Sea Level
FF	Fossil Fuel		
FF&I	Fossil-Fuel combustion and Industrial processes		

All

GMST	Global Mean Surface Temperature		ICSU	International Council for Science
GMT	Global Mean Temperature		ICT	Information and Communication Technology
GNHI	Gross National Happiness Index		IEA	International Energy Agency
GPP	Gross Primary Productivity		IEAGHG	IEA Greenhouse Gas R&D Programme
GPT	General Purpose Technologies		IFAD	International Fund for Agricultural Development
GRAPE	Global Relationship Assessment to Protect the Environment model		IFPRI	International Food Policy Research Institute
GSAT	Global mean Surface Air Temperature		IGCC	Integrated Gasification Combined Cycle
Gt	Gigatonne		IIASA	International Institute for Applied Systems Analysis
GTP	Global Temperature-change Potential		IIF	Institute of International Finance
GWA	Government of Western Australia		iLUC	Indirect Land-Use Change
GWP	Global Warming Potential or Gross World Product		IMACLIM-NLU	IMpact Assessment of CLIMate policies model – Nexus Land-Use model
ha	Hectares		IMAGE	Integrated Model to Assess the Global Environment
H_2	Hydrogen		IMF	International Monetary Fund
HadCM	Hadley Centre Coupled Model		IMO	International Maritime Organization
HadCRUT	Hadley Centre Climatic Research Unit Gridded Surface Temperature Data Set		IMPACT2C	Quantifying Projected Impacts under 2°C Warming
HadEX	Hadley Centre Global Climate Extremes index		INDCs	Intended Nationally Determined Contributions
HadGEM	Hadley Centre Global Environmental Model		INM	Russian Institute for Numerical Mathematics
HadRM	Hadley Centre Regional Model		IOM	International Organization for Migration
HAPPI	Half a degree Additional warming, Prognosis and Projected Impacts		IoT	Internet of Things
HDV	Heavy-Duty Vehicle		IPCC	Intergovernmental Panel on Climate Change
HEV	Hybrid Electric Vehicle		IPSL	Institute Pierre Simon Laplace
HFCs	Hydrofluorocarbons		IRENA	International Renewable Energy Agency
HLCCP	High-Level Commission on Carbon Prices		ISIMIP	Inter-Sectoral Impact Model Intercomparison Project
HLPE	High Level Panel of Experts on Food Security and Nutrition		ISO	International Standards Organisation
HSRTF	Hurricane Sandy Rebuilding Task Force		ISSC	International Social Science Council
HTM	Holocene Thermal Maximum		ITF	International Transport Forum
HYMOD	HYdrological MODel		IUCN	International Union for Conservation of Nature
IAEA	International Atomic Energy Agency		IWG	Interagency Working Group on Social Cost of Greenhouse Gases
IAMC	Integrated Assessment Modelling Consortium		IWRM	Integrated Water Resources Management
IAMs	Integrated Assessment Models		JeDi	Jena Diversity-Dynamic Global Vegetation Model
IBA	International Bar Association		JJA	June, July, August
IBI	International Biochar Initiative		JMA	Japan Meteorological Agency
ICAMS	Integrated Climate and Air Quality Modeling System		JRA-55	Japanese 55-year Reanalysis
ICEM	International Centre for Environmental Management		JRC	European Commission – Joint Research Centre
ICLEI	International Council for Local Environmental Initiatives		JULES	Joint United Kingdom Land Environment Simulator
			kcal cap^{-1} day^{-1}	Kilocalories per capita per day
ICPDR	International Commission for the Protection of the Danube River		km	Kilometres
			kt	Kilotonnes
			kWh	Kilowatt hours

All

KNMI	Koninklijk Nederlands Meteorologisch Instituut (Royal Netherlands Meteorological Institute)		**MER**	Market Exchange Rates
L	Litres		**MERET**	Managing Environmental Resources to Enable Transitions
L&D	Loss and Damage		**MERGE-ETL**	Model for Evaluating Regional and Global Effects of greenhouse gas reduction policies – Endogenous Technology Learning
LAM	Latin America and Caribbean			
LDCs	Least Developed Countries			
LDMz-INCA	Laboratoire de Météorologie Dynamique – INteractions between Chemistry and Aerosols		**MESSAGE**	Model for Energy Supply Systems And their General Environmental impact
LDV	Light-Duty Vehicle		**Mha**	Megahectare
LE	Limited Evidence		**MIROC**	Model for Interdisciplinary Research on Climate
LED	Low Energy Demand or Light Emitting Diode		**MISI**	Marine Ice Sheet Instability
LGM	Last Glacial Maximum		**MIT IGSM**	Massachusetts Institute of Technology Integrated Global System Model
LIG	Last Interglacial			
LLCFs	Long-Lived Climate Forcers		**MJ**	Megajoules
LNG	Liquefied Natural Gas		**MoCC**	Ministry of Climate Change and Adaptation (Government of Vanuatu)
LPG	Liquefied Petroleum Gas			
LPJmL	Lund-Potsdam-Jena managed Land model		**MOHC**	Met Office Hadley Centre
LTA	Land Transport Authority of Singapore		**MOPEX**	Model Parameter Estimation Experiment
LTGG	Long-Term Global Goal		**MPAs**	Marine Protected Areas
LUC	Land-Use Change		**MPI**	Max-Planck-Institut für Meteorologie (Max Planck Institute for Meteorology)
LULUCF	Land Use, Land-Use Change, and Forestry			
m	Metres		**MPWP**	Mid Pliocene Warm Period
m³ cap⁻¹ yr⁻¹	Cubic metres per capita per year		**MRFCJ**	Mary Robinson Foundation – Climate Justice
mg	Milligrams		**MRI**	Meteorological Research Institute of Japan Meteorological Agency
mL	Millilitres			
mm	Millimetres		**MRV**	Measurement, Reporting and Verification
M&E	Monitoring and Evaluation		**MSR**	Multi-Sector Risk score
Ma	Million years ago		**Mt**	Megatonnes
MAC	Marginal Abatement Cost		**N**	Nitrogen
MacPDM	Macro-scale – Probability-Distributed Moisture Model		**N₂O**	Nitrous oxide
			NAP	National Adaptation Plan
MAGICC	Model for the Assessment of Greenhouse Gas Induced Climate Change		**NAS**	North Asia
			NASA	National Aeronautics and Space Administration
MAgPIE	Model of Agricultural Production and its Impact on the Environment		**NASEM**	National Academies of Sciences, Engineering, and Medicine
MAM	March, April, May		**NAU**	North Australia
MCB	Marine Cloud Brightening		**NCAR**	National Center for Atmospheric Research
MCCA	Mercado Común Centroamericano		**NCCARF**	National Climate Change Adaptation Research Facility
MDB	Group of Multilateral Development Banks			
MDGs	Millennium Development Goals		**NCE**	New Climate Economy
MEA	Millennium Ecosystem Assessment		**NDCs**	Nationally Determined Contributions
MED	South Europe/Mediterranean		**NEA**	Nuclear Energy Agency
MEPS	Minimum Energy Performance Standards		**NEB**	North-East Brazil
			NEC	National Environment Commission (Royal Government of Bhutan)

$m^3\ cap^{-1}\ yr^{-1}$

AII

NETL	National Energy Technology Laboratory (US Department of Energy)		**PDSI**	Palmer Drought Severity Index
NEU	North Europe		**PE**	Primary Energy or Partial Equilibrium
NF_3	Nitrogen trifluoride		**PET**	Physiologically Equivalent Temperature or Potential Evapo-Transpiration
NGO	Non-Governmental Organization		**PFCs**	Perfluorocarbons
NH3	Ammonia		**Pg**	Petagrams
NHD	Number of Hot Days		**PHEV**	Plug-in Hybrid Electric Vehicle
NITI Aayog	National Institution for Transforming India		**PIK**	Potsdam-Institut für Klimafolgenforschung (Potsdam Institute for Climate Impact Research)
NMVOC	Non-Methane Volatile Organic Compounds		PM_{10}	Particulate Matter with Aerodynamic Diameter <10 μm
NOAA	National Oceanic and Atmospheric Administration		$PM_{2.5}$	Particulate Matter with Aerodynamic Diameter <2.5 μm
NorESM	Norwegian Earth System Model			
NO_x	Nitrogen oxides		**POLES**	Prospective Outlook on Long-term Energy Systems model
NPCC	New York City Panel on Climate Change		**PPP**	Purchasing Power Parity
NPP	Net Primary Productivity		**PR**	Probability Ratio
NPV	Net Present Value		**PV**	Photovoltaics
NRC	National Research Council		**R&D**	Research and Development
NSR	Northern Sea Route		**RCA**	Rossby Centre Regional Atmospheric Model
NTP	Pacific Islands region [2]		**RCI**	Rotterdam Climate Initiative
NYC	New York City		**RCM**	Regional Climate Model
NZAGRC	New Zealand Agricultural Greenhouse Gas Research Center		**RCPs**	Representative Concentration Pathways
O_2	Oxygen		**REDD+**	Reducing Emissions from Deforestation and forest Degradation; and the role of conservation, sustainable management of forests and enhancement of forest carbon stocks in developing countries
O_3	Ozone			
OA	Ocean Acidification or Ocean Alkalinization			
OC	Organic Carbon			
OECD	Organisation for Economic Co-operation and Development		**ReEDS-IPM**	Regional Electricity Deployment System model – Integrated Planning Model
OGCC	Optimal Gasification Combined Cycle		**RegCM**	Regional Climate Model system
OHCHR	Office of the United Nations High Commissioner for Human Rights		**REMIND**	REgional Model of INvestments and Development
OIF	Ocean Iron Fertilisation		**REN21**	Renewable Energy Policy Network for the 21st Century
ORCHIDEE	ORganising Carbon and Hydrology In Dynamic EcosystEms model		**RF**	Radiative Forcing
ORR	NYC Mayor's Office of Recovery & Resiliency		**RFC**	Reason for Concern
OS	Overshoot		**RGoB**	Royal Government of Bhutan
pp	People		**RMI**	Rocky Mountain Institute
ppb	Parts per billion		**RNCFC**	Reference Non-CO_2 Forcing Contribution
ppm	Parts per million		**RNCTC**	Reference Non-CO_2 Temperature Contribution
ppt	Parts per thousand		**Rx1day**	Annual maximum 1-day precipitation
P	Precipitation or Phosphorous		**Rx5day**	Annual maximum 5-day precipitation
PAGE	Policy Analysis of the Greenhouse Effect model		**SAF**	Southern Africa
PAHO	Pan American Health Organization		**SAH**	Sahara
PCM	Parallel Climate Model		**SAI**	Stratospheric Aerosol Injection

AII

SAMS	South American Monsoon System	SREX	IPCC Special Report on Managing the Risks of Extreme Events and Disasters to Advance Climate Change Adaptation
SAR	IPCC Second Assessment Report		
SAS	South Asia		
SAT	Surface Air Temperature	SRM	Solar Radiation Modification
SAU	South Australia/New Zealand	SROCC	IPCC Special Report on the Ocean and Cryosphere in a Changing Climate
SBSTA	Subsidiary Body for Scientific and Technological Advice (UNFCCC)		
		SSA	Southeastern South America
SCC	Social Cost of Carbon	SSPs	Shared Socioeconomic Pathways
SCS	Soil Carbon Sequestration	SST	Sea Surface Temperature
SD	Sustainable Development	STP	Southern Tropical Pacific
SDGs	Sustainable Development Goals	SWAT	Soil & Water Assessment Tool
SDGVM	Sheffield Dynamic Global Vegetation Model	SWF	Social Welfare Function
SDII	Simple Daily Intensity Index	SYR	IPCC Synthesis Report
SDSN	Sustainable Development Solutions Network	t	Tonnes
SEA	Southeast Asia	tDM	Tonnes Dry Matter
SEAPs	Sustainable Energy Action Plans	tril$	Trillion dollars
SED	Structured Expert Dialogue	T	Temperature or Transient
SEM	Semi-Empirical Model	T&D	Transmission and Distribution
SF_6	Sulphur hexafluoride	TCR	Transient Climate Response
SFM	Sustainable Forest Management	TCRE	Transient Climate Response to cumulative CO_2 Emissions
SIDS	Small Island Developing States		
SIFMA	Securities Industry and Financial Markets Association	TEAP	Technology and Economic Assessment Panel
		TFE	Thematic Focus Element
SLCFs	Short-Lived Climate Forcers	TFP	Total Factor Productivity
SLCPs	Short-Lived Climate Pollutants	Tg	Teragrams
SLR	Sea Level Rise	TIB	Tibetan Plateau
SM	Supplementary Material	TNn	Coldest night-time temperature of the year
SMA	Soil Moisture Anomalies	TOD	Transit Oriented Development
SMHI	Swedish Meteorological and Hydrological Institute	TS	Technical Summary
SO_2	Sulphur dioxide	Tt	Teratonnes
SOLARIS HEPPA	SOLARIS High Energy Particle Precipitation in the Atmosphere	TTs	Transition Towns
		TXx	Hottest daytime temperature of the year
SON	September, October, November	UCCRN	Urban Climate Change Research Network
SO_x	Sulphur oxides	UHI	Urban Heat Islands
SPAs	Shared climate Policy Assumptions	UITP	Union Internationale des Transports Publics (International Association of Public Transport)
SPC	Secretariat of the Pacific Community		
SPEI	Standardized Precipitation Evapotranspiration Index	UKCP	United Kingdom Climate Projections
		UN	United Nations
SPI	Standardised Precipitation Index	UN DESA	United Nations Department of Economic and Social Affairs
SPM	Summary for Policymakers		
SR1.5	IPCC Special Report on Global Warming of 1.5°C	UNCBD	United Nations Convention on Biological Diversity
SRCCL	IPCC Special Report on Climate Change and Land	UNDP	United Nations Development Programme
SRES	IPCC Special Report on Emissions Scenarios	UNEP	UN Environment

All

UNEP-WCMC	UNEP World Conservation Monitoring Centre	WGII	IPCC Working Group II
UNESCO	United Nations Educational, Scientific and Cultura Organization	WGIII	IPCC Working Group III
		WHO	World Health Organization
UNFCCC	United Nations Framework Convention on Climate Change	WIM	Warsaw International Mechanism for Loss and Damage
UNGA	United Nations General Assembly	WIO	West Indian Ocean
UNICEF	United Nations International Children's Emergency Fund	WITCH	World Induced Technical Change Hybrid Model
		WMGHGs	Well-Mixed Greenhouse Gases
UNISDR	United Nations Office for Disaster Risk Reduction	WMO	World Meteorological Organization
UN-OHRLLS	Office of the High Representative for the Least Developed Countries, Landlocked Developing Countries and Small Island Developing States	WNA	West North America
		WRF	Weather Research and Forecasting
UNRISD	United Nations Research Institute for Social Development	WSA	West Coast South America
		WSDI	Warm Spell Duration Index
UNSCEAR	United Nations Scientific Committee on the Effects of Atomic Radiation	WTO	World Trade Organization
		yr	Year
UNU	United Nations University	ZEC	Zero Emissions Commitment
UNU-EHS	United Nations University – Institute for Environment and Human Security		
USD	United States Dollars		
UV	Ultraviolet		
VISIT	Vegetation Integrative Simulator for Trace Gases		
VKT	Vehicle Kilometres of Travel		
VOCs	Volatile Organic Compounds		
w/	With		
w/o	Without		
w.r.t.	With respect to		
W	Watts		
WAF	West Africa		
WAIS	West Antarctic Ice Sheet		
WAS	West Asia		
WBCSD	World Business Council for Sustainable Development		
WBGU	Wissenschaftlicher Beirat der Bundesregierung Globale Umweltveränderungen (German Advisory Council on Global Change)		
WBM	Water Balance Model		
WCED	World Commission on Environment and Development		
WCRP	World Climate Research Programme		
WEC	World Energy Council		
WEF	Water-Energy-Food or World Economic Forum		
WEM	World Energy Model		
WEO	World Energy Outlook		
WFP	World Food Programme		
WGI	IPCC Working Group I		

All

AIII

Annex III: Contributors to the IPCC Special Report on Global Warming of 1.5°C

ABDUL HALIM, Sharina
Institute for Environment and
Development-LESTARI
Malaysia

ABDULLA, Amjad
Climate Change Department
Ministry of Environment and Energy
Maldives

ACHLATIS, Michelle
University of Queensland
Australia/ Greece

ALEXANDER, Lisa
University of New South Wales
Australia

ALLEN, Myles R.
University of Oxford
UK

ANTWI-AGYEI, Philip
Kwame Nkrumah University of
Science and Technology
Ghana

ARAGÓN-DURAND, Fernando
Institute for Environmental Studies
Program on Sustainable Development
El Colegio de México
Mexico

ARAOS, Malcolm
Ministry of Environment and Energy
Canada

BABIKER, Mustafa
Saudi Aramco
Sudan

BAKKER, Stefan
Netherlands

BANGALORE, Mook
London School of Economics
USA

BAZAZ, Amir
Indian Institute for Human Settlements
India

BELFER, Ella
McGill University
USA

BENTON, Tim
University of Leeds
UK

BERRY, Peter
Climate Change and Innovation Bureau
Health Canada
Canada

BERTOLDI, Paolo
European Commission
Italy

BHASKAR CHOUDHARY, Bishwa
National Institute of Agricultural
Economics and Policy Research
India

BINDI, Marco
University of Florence
Italy

BOER, Rizaldi
Center for Climate Risk and
Opportunity Management
Indonesia

BOYER, Christopher
University of Washington
USA

BRILLI, Lorenzo
University of Florence
IBIMET-CNR
Italy

BROWN, Sally
University of Southampton
UK

BUCKERIDGE, Marcos
University of São Paulo
Biosciences Institute
Brazil

BYERS, Edward
International Institute for
Applied Systems Analysis
Austria/Brazil

CALVIN, Katherine
Pacific Northwest National Laboratory
USA

CAMILLONI, Ines
Centro de Investigaciones del
Mar y la Atmósfera
University of Buenos Aires
Argentina

CARTWRIGHT, Anton
African Centre for Cities
South Africa

CHEN, Yang
IPCC WGI Technical Support Unit
Chinese Academy of Meteorological Sciences
China

CHEUNG, William
University of British Columbia
Canada

CONNORS, Sarah
IPCC WGI Technical Support Unit
Université Paris-Saclay
France/UK

**CORREIA DE OLIVEIRA DE
PORTUGAL PEREIRA, Joana**
IPCC WGIII Technical Support Unit
Imperial College London's Centre
for Environmental Policy
UK/Portugal

CRAIG, Marlies
IPCC WGII Technical Support Unit
School of Life Sciences
University of KwaZulu-Natal
South Africa

CRAMER, Wolfgang
CNRS
Mediterranean Institute for
Biodiversity and Ecology
Aix-Marseille University
France/Germany

DASGUPTA, Dipak
The Energy and Resources Institute (TERI)
India

DASGUPTA, Purnamita
Institute of Economic Growth
India

DE CONINCK, Heleen
Radboud University
Netherlands/EU

DE KLEIJNE, Kiane
Radboud University
Netherlands/EU

DEL MAR ZAMORA DOMINGUEZ, Maria
Mercator Research Institute on Global
Commons and Climate Change
Mexico

DEN ELZEN, Michel
PBL Netherlands Environmental
Assessment Agency
Netherlands

DIEDHIOU, Arona
Institut de Recherche pour le Développement
Ivory Coast/Senegal

DJALANTE, Riyanti
United Nations University
Institute for the Advanced Study
of Sustainability (UNU-IAS)
Japan/Indonesia

DONG, Wenjie
Sun Yat-sen University
China

DUBE, Opha Pauline
University of Botswana
Botswana

EAKIN, Hallie
Julie Ann Wrigley Global
Institute of Sustainability
USA

EBI, Kristie L.
University of Washington
USA

EDELENBOSCH, Oreane
Politecnico di Milano
Italy/Netherlands

ELGIZOULI IDRIS, Ismail
Private Consultancy
Sudan

ELLIS, Neville
University of Western Australia
Australia

EMMERLING, Johannes
Fondazione Eni Enrico Mattei
Italy/Germany

ENGELBRECHT, Francois
Council for Scientific and Industrial Research
South Africa

EVANS, Jason
University of New South Wales
Australia

FERRAT, Marion
IPCC WGIII Technical Support Unit
Imperial College London's Centre
for Environmental Policy
UK/France

FIFITA, Solomone
Pacific Community
Fiji/Tonga

FIGUEROA, Maria
Copenhagen Business School
Denmark/Venezuela

FINON, Dominique
Centre International de Recherches sur
l'Environnement et le Développement
France

FISCHER, Hubertus
Universität Bern
Switzerland

FISCHLIN, Andreas
Swiss Federal Institute of Technology
Switzerland

FLATO, Greg
Environment and Climate Change Canada
Canada

FORD, James
University of Leeds
UK/ Canada

FORSTER, Piers
University of Leeds
UK

FRAEDRICH, Klaus
Universität Hamburg
Germany

FUGLESTVEDT, Jan
CICERO Center for International
Climate Research
Norway

FUSS, Sabine
Mercator Research Institute on Global
Commons and Climate Change
Germany

GANASE, Anjani
University of Queensland
Australia/Trinidad and Tobago

GAO, Xuejie
Institute of Atmospheric Physics
China

GASSER, Thomas
International Institute for
Applied Systems Analysis
Austria/France

GATTUSO, Jean-Pierre
Institut du développement durable et
des relations internationales (IDDRI)
CNRS
Sorbonne Université
France

GHERSI, Frédéric
Centre International de Recherches sur
l'Environnement et le Développement (CIRED)
CNRS
France

GILLETT, Nathan
Environment and Climate Change Canada
Canada

GINZBURG, Veronika
Institute of Global Climate and
Ecology Roshydromet
Russian Federation

GRANDIS, Adriana
University of São Paulo
Biosciences Institute
Brazil

GREVE, Peter
International Institute for
Applied Systems Analysis
Austria/Germany

GUILLEN BOLAÑOS, Tania
Climate Service Center (GERICS)
Helmholtz-Zentrum Geesthacht
Germany/Nicaragua

GUIOT, Joel
CNRS
Aix-Marseille University
France

GUPTA, Mukesh
Asian University for Women of Bangladesh
India

HANASAKI, Naota
National Institute for Environmental Studies
Japan

HANDA, Collins
Technical University of Kenya
Kenya

HAROLD, Jordan
University of East Anglia
UK

HASEGAWA, Tomoko
National Institute for Environmental Studies
Japan

HAUGHEY, Eamon
School of Natural Sciences
Trinity College Dublin
Ireland

HAYES, Katie
University of Toronto
Canada

AIII

HAYWARD, Bronwyn
University of Canterbury
New Zealand

HE, Chenmin
Peking University
China

HERTWICH, Edgar
Yale University
USA/Austria

HIJIOKA, Yasuaki
National Institute for Environmental Studies
Japan

HINGE SALILI, Diana
University of the South Pacific Laucala Campus
Vanuatu

HIRSCH, Annette
ETH Zurich
University of New South Wales
Switzerland/Australia

HOEGH-GULDBERG, Ove
Global Change Institute
University of Queensland
Australia

HÖGLUND-ISAKSSON, Lena
International Institute for
Applied Systems Analysis
Austria/Sweden

HOURCADE, Jean-Charles
Centre International de Recherches sur
l'Environnement et le Développement (CIRED)
CNRS
France

HOWDEN, Mark
Climate Change Institute
Australian National University
Australia

HUMPHREYS, Stephen
London School of Economics
and Political Science
UK/Ireland

HUPPMANN, Daniel
International Institute for
Applied Systems Analysis
Austria

HUQ, Saleemul
International Centre for Climate
Change & Development
UK/Bengladesh

JACOB, Daniela
Climate Service Center Germany (GERICS)
Helmholtz-Zentrum Geesthacht (HZG)
Germany

JAMES, Rachel
Environmental Change Institute
University of Oxford
UK

JIANG, Kejun
Energy Research Institute
China

JOHANSEN, Tom Gabriel
Infodesign Lab
Norway

JONES, Chris
Met Office Hadley Centre
UK

JUNG, Thomas
Alfred Wegener Institute
Helmholtz Centre for Polar
and Marine Research
Germany

KAINUMA, Mikiko
Institute for Global Environmental Strategies
Japan

KALA, Jatin
Murdoch University
Australia

KANNINEN, Markku
University of Helsinki
Finland

KHESHGI, Haroon
ExxonMobil Research and
Engineering Company
USA

KLEIN, Richard
Stockholm Environment Institute
Netherlands/Germany

KOBAYASHI, Shigeki
Transport Institute of Central Japan
Japan

KRAKOVSKA, Svitlana
Ukrainian Hydrometeorological Institute
Ukraine

KRIEGLER, Elmar
Potsdam Institute for Climate Impact Research
Germany

KRINNER, Gerhard
Institut de Géosciences de l'Environnement
France

LAWRENCE, David
National Center for Athmospheric Research
USA

LENTON, Tim
University of Exeter
UK

LEY, Debora
Latinoamérica Renovable
Guatemala/Mexico

LIVERMAN, Diana
University of Arizona
USA

LUDERER, Gunnar
Potsdam Institute for Climate Impact Research
Germany

MAHOWALD, Natalie
Cornell University
USA

MARCOTULLIO, Peter
CUNY Institute for Sustainable Cities
USA

MARENGO, Jose Antonio
National Centre for Monitoring and
Warning of Natural Disasters
Brazil/Peru

MARKANDYA, Anil
Basque Centre for Climate Change
Spain/UK

MASSERA, Omar
Instituto de Investigaciones en
Ecosistemas y Sustentabilidad
Universidad Nacional Autonoma de México
Mexico

MASSON-DELMOTTE, Valérie
Co-Chair IPCC WGI
France

MATTHEWS, J. B. Robin
IPCC WGI Technical Support Unit
Université Paris-Saclay
France/UK

MCCOLLUM, David
International Institute for
Applied Systems Analysis
Austria/USA

MCINNES, Kathleen
Commonwealth Scientific and
Industrial Research Organisation
Australia

MECHLER, Reinhard
International Institute for
Applied Systems Analysis
Germany

MEDHIN, Haileselassie Amaha
Ethiopian Development Research Institute
Ethiopia

MEHROTRA, Shagun
The World Bank
The New School
USA, India

MEINSHAUSEN, Malte
The University of Melbourne
Australia/Germany

MEISSNER, Katrin J.
University of New South Wales Sydney
Australia

MILLAR, Richard
University of Oxford
UK

MINTENBECK, Katja
IPCC WGII Technical Support Unit
Alfred-Wegener-Institut Bremen
Germany

MITCHELL, Dann
University of Bristol
UK

MIX, Alan C
Oregon State University
USA

MORELLI, Angela
Infodesign Lab
Norway

MOUFOUMA-OKIA, Wilfran
IPCC WGI Technical Support Unit
Université Paris-Saclay
France/Congo

MRABET, Rachid
National Institute of Agricultural
Research (INRA)
Morocco

MULUGETTA, Yacob
University College London
UK/Ethiopia

MUNDACA, Luis
Lund University
Sweden/Chile

NEWMAN, Peter
Curtin University
Australia

NICOLAI, Maike
IPCC WGII Technical Support Unit
Alfred-Wegener-Institut Bremen
Germany

NOTZ, Dirk
Max-Planck-Institut für Meteorologie
Germany

NURSE, Leonard
The University of West Indies
Barbados

OKEM, Andrew
IPCC WGII Technical Support Unit
School of Life Sciences
University of KwaZulu Natal
South Africa/Nigeria

OKEREKE Chukwumerije
University of Reading
UK, Nigeria

OLSSON, Lennart
Uppsala University
Sweden

OPIO, Carolyn
Food and Agriculture Organization
Uganda

OPPENHEIMER, Michael
Princeton University
USA

PAIVA HENRIQUE, Karen
University of Western Australia
Brazil

PARKINSON, Simon
International Institute for
Applied Systems Analysis
Canada

PATHAK, Minal
IPCC WGIII Technical Support Unit
Ahmedabad University
India

PAYNE, Antony
University of Bristol
UK

PAZ, Shlomit
University of Haifa
Israel

PEREIRA, Joy
Institute for Environment and Development
Malaysia

PEREZ, Rosa
Manila Observatory
Philippines

PETERSEN, Juliane
Climate Service Center Germany (GERICS)
Helmholtz-Zentrum Geesthacht
Germany

PETZOLD, Jan
IPCC WGII Technical Support Unit
Alfred-Wegener-Institut Bremen
Germany

PICHS-MADRUGA, Ramon
Centre for World Economy Studies
Cuba

PIDCOCK, Roz
IPCC WGI Technical Support Unit
Université Paris-Saclay
France/UK

PINHO, Patricia Fernanda
University of Sao Paulo
Brazil

PIRANI, Anna
IPCC WGI Technical Support Unit
Université Paris-Saclay
The Abdus Salam International
Centre for Theoretical Physics
Italy/UK

PLAZZOTTA, Maxime
Météo-France
Centre Naional de Recherches
Météorologiques
France

POLOCZANSKA, Elvira
IPCC WGII Technical Support Unit
Alfred-Wegener-Institut Bremen
Germany, UK

POPP, Alexander
Potsdam Institute for Climate Impact Research
Germany

PÖRTNER, Hans-Otto
Co-Chair IPCC WGII
Germany

AIII

PREUSCHMANN, Swantje
Climate Service Center Germany (GERICS)
Helmholtz-Zentrum Geesthacht
Germany

PUROHIT, Pallav
International Institute for
Applied Systems Analysis
Austria/India

RAGA, Graciela
Universidad Nacional Autonoma de Mexico
Mexico/Argentina

RAHMAN, Mohammad Feisal
International Centre for Climate
Change and Development
Bengladesh

REISINGER, Andy
New Zealand Agricultural Greenhouse
Gas Research Centre
New Zealand

REVI, Aromar
Indian Institute for Human Settlements
India

REVOKATOVA, Anastasia
Hydrometeorological Research
Centre of Russian Federation
Russian Federation

RHINEY, Kevon
Rutgers University
Jamaica

RIAHI, Keywan
International Institute for
Applied Systems Analysis
Austria

RIBES, Aurélien
Centre National de Recherches
Météorologiques
France

RICHARDSON, Mark
USA/UK

RICKELS, Wilfried
Kiel Institute for the World Economy
Germany

ROBERTS, Debra
Co-Chair IPCC WGII
South Africa

ROBERTS, Timmons
Brown University
USA

ROGELJ, Joeri
International Institute for
Applied Systems Analysis
Belgium/Austria

ROJAS, Maisa
University of Chile
Chile

ROY, Joyashree
Jadavpur University
Asian Institute of Technology
Thailand/India

SANCHEZ, Roberto
El Colegio de la Frontera Norte
Mexico

SAUNDERS, Harry
Independent
Canada/USA

SCHÄDEL, Christina
Northern Arizona University
USA/Switzerland

SCHAEFFER, Roberto
Universidade Federal do Rio de Janeiro
Brazil

SCHEUFFELE, Hanna
IPCC WGII Technical Support Unit
Alfred-Wegener-Institut Bremen
Germany

SCHIPPER, Lisa
Environmental Change Institute
University of Oxford
UK/Sweden

SCHLEUSSNER, Carl-Friedrich
Potsdam Institute for Climate Impact Research
Humboldt University
Germany

SCHMIDT, Jörn
Christian-Albrechts-Universität zu Kiel
Germany

SCHULTZ, Seth
C40 Cities Climate Leadership Group
USA

SCOTT, Daniel
University of Waterloo
Canada

SÉFÉRIAN, Roland
Centre National de Recherches
Météorologiques
France

SENEVIRATNE, Sonia I.
Swiss Federal Institute of Technology
Switzerland

SHERSTYUKOV, Boris
Russian Research Institute of
Hydrometeorological Information
World Data Centre
Russian Federation

SHINDELL, Drew
Duke University
USA

SHUKLA, Priyadarshi R.
Co-Chair IPCC WGIII
India

SILLMANN, Jana
Center for International Climate Research
Germany/Norway

SINGH, Chandni
Indian Institute for Human Settlements
India

SKEA, Jim
Co-Chair IPCC WGIII
UK

SLADE, Raphael
IPCC WGIII Technical Support Unit
Imperial College London's Centre
for Environmental Policy
UK

SMITH, Chris
University of Leeds
UK

SMITH, Christopher
UK

SMITH, Pete
University of Aberdeen
UK

SOLECKI, William
City University of New York
USA

SOME, Shreya
Jadavpur University
India

SPAROVEK, Gerd
Universidade de São Paulo
Escola Superior de Agricultura Luiz de Queiroz
Brazil

AIII

STEFFEN, Will
The Australian National University
Australia

STEG, Linda
University of Groningen
Netherlands

STEPHENSON, Kimberly
The University of the West Indies
Jamaica

STEPHENSON, Tannecia
The University of the West Indies
Jamaica

SUAREZ RODRIGUEZ, Avelino G.
Research Center for the World Economy
Cuba

SUAREZ, Pablo
Red Cross Climate Centre
Argentina

SUGIYAMA, Taishi
The Canon Institute for Global Studies
Japan

SYLLA, Mouhamadou B
West African Science Service Center on
Climate Change and Adapted Land Use
Senegal

TAYLOR, Michael
The University of the West Indies
Jamaica

TEARIKI-RUATU, Nenenteiti
University of the South Pacific Laucala Campus
Kiribati

TEBBOTH, Mark
University of East Anglia
UK

THOMAS, Adelle
University of The Bahamas
Bahamas

THORNE, Peter
Maynooth University
Ireland/UK

TRUTNEVYTE, Evelina
University of Geneva
Switzerland/Lithuania

TSCHAKERT, Petra
University of Western Australia
Australia/Austria

ÜRGE-VORSATZ, Diana
Department of Environmental
Sciences and Policy
Central European University
Hungary

URQUHART, Penny
Freelance climate resilient
development specialist
South Africa

VAN DIEMEN, Renee
IPCC WGIII Technical Support Unit
Imperial College London's Centre
for Environmental Policy
UK/Netherlands

VAN ROOIJ, Arjan
Netherlands

VAN VALKENGOED, Anne
University of Groningen
Netherlands

VAUTARD, Robert
Laboratory for Sciences of
Climate and Environment
France

VILARIÑO, Maria Virginia
Argentinean Business Council for
Sustainable Development
Argentina

WAIRIU, Morgan
University of the South Pacific
Solomon Islands

WAISMAN, Henri
Institut du développement durable et
des relations internationales (IDDRI)
France

WARREN, Rachel
Tyndall Centre and School of
Environmental Sciences
UK

WARTENBURGER, Richard
ETH Zürich
Switzerland/Germany

WEHNER, Michael
Lawrence Berkeley National Laboratory
USA

WEWERINKE-SINGH, Margaretha
University of Leiden
Netherlands

WEYER, Nora M
IPCC WGII Technical Support Unit
Alfred-Wegener-Institut Bremen
Germany

WHYTE, Felicia
The University of the West Indies
Jamaica

WOLLENBERG, Eva
University of Vermont and the CGIAR
Research Program on Climate Change
Agriculture and Food Security
USA

XIU, Yang
National Center for Climate Change
Strategy and International Cooperation
China

YOHE, Gary
Wesleyan University
USA

ZHAI, Panmao
Co-Chair IPCC WGI
China

ZHANG, Xuebin
Environment and Climate Change Canada
Canada

ZHOU, Guangsheng
Chinese Academy of Meteorological Sciences
China

ZHOU, Wenji
International Institute for
Applied Systems Analysis
Austria/China

ZICKFELD, Kirsten
Simon Fraser University
Canada/Germany

ZOUGMORÉ, Robert B
Consultative Group for International
Agricultural Research
International Crops Research Institute
for the Semi-Arid Tropics
Burkina Faso/Mali

AIII

AIV

Annex IV: Expert Reviewers of the IPCC Special Report on Global Warming of 1.5°C

This annex should be cited as:

IPCC, 2018: Annex IV: Expert Reviewers of the IPCC Special Report on Global Warming of 1.5°C. In: *Global Warming of 1.5°C. An IPCC Special Report on the impacts of global warming of 1.5°C above pre-industrial levels and related global greenhouse gas emission pathways, in the context of strengthening the global response to the threat of climate change, sustainable development, and efforts to eradicate poverty* [Masson-Delmotte, V., P. Zhai, H.-O. Pörtner, D. Roberts, J. Skea, P.R. Shukla, A. Pirani, W. Moufouma-Okia, C. Péan, R. Pidcock, S. Connors, J.B.R. Matthews, Y. Chen, X. Zhou, M.I. Gomis, E. Lonnoy, T. Maycock, M. Tignor, and T. Waterfield (eds.)]. Cambridge University Press, Cambridge, UK and New York, NY, USA, pp. 581-600.

AAMAAS, Borgar
Center for International Climate Research
Norway

ABABNEH, Linah
Swedish University of Agriculture
and Cornell University
USA

ABANADES Carlos,
Consejo Superior de Investigaciones
Científicas
Spain

ACOSTA NAVARRO, Juan Camilo
Barcelona Supercomputing Center
Spain

ADU-BOATENG, Afua
Training and Research Network
UK

ADVANI, Nikhil
World Wildlife Fund
USA

AGUILAR-AMUCHASTEGUI, Naikoa
World Wildlife Fund
USA

AHMAD, Ijaz
Applied Systems Analysis Division
Pakistan Atomic Energy Commission
Pakistan

AHMADI, Mohammad
Regional Meteorological MetOffice
Iran

AHN, Young-Hwan
Korea Energy Economics Institute
Republic of Korea

AKHTAR, Farhan
U.S. Department of State
Office of Global Change
USA

AKIMOTO, Keigo
Research Institute of Innovative
Technology for the Earth
Japan

AKPAN, Archibong
University of Ibadan
Nigeria

ALAM, Lubna
Institute for Environment and
Development (LESTARI)
Bangladesh

ALBEROLA, Emilie
Eco-Act
France

ALCARAZ, Olga
Universitat Politècnica de Catalunya
Spain

ALDRICH, Elizabeth
USA

ALEXEEVA, Victoria
International Atomic Energy Agency
Austria

ALFIERI, Lorenzo
European Commission
Italy

ALI, Shaukat
Global Change Impact Studies Centre
Ministry of climate change
Pakistan

ALLEN, Myles
University of Oxford
Environmental Change Institute
UK

ALLWRIGHT, Gavin
International Windship Association
UK

ALPERT, Alice
United States Department of State
USA

AN, Nazan
Bogazici University Center for Climate
Change and Policy Studies
Turkey

ANDERSON, Cheryl
LeA International Consultants Ltd
New Zealand

ANDERSSON, Peter
Uppsala University
Sweden

ANDREWS, Nadine
IPCC WGII Technical Support Unit
Alfred-Wegener-Institut Bremen
Germany/UK

ANORUO, Chukwuma
Imo State University Owerri
Nigeria

AQUINO, Sergio
University of British Columbia
Canada

ARIKAN, Yunus
ICLEI - Local Governments for Sustainability
Germany

ARIMA, Jun
University of Tokyo
Japan

ARTINANO, Begoña
Centro de Investigaciones Energéticas
Medioambientales y Tecnológicas
Spain

AYEB-KARLSSON, Sonja
University of Sussex
UNU-EHS
UK

AZEITEIRO, Ulisses
Departamento de Biologia & CESAM
University of Aveiro
Portugal

BABAEIAN, Iman
Climatological Research Institute
Iran

BABIKER, Mustafa
Saudi Aramco
Sudan

BAHAMONDES DOMINGUEZ,
Angela Andrea
University of Southampton
UK/Chile

BAKHTIARI, Fatemeh
Researcher at UNEP DTU partnership
Denmark

BALA, Govindasamy
Indian Institute of Science
India

BARAU, Aliyu
Faculty of Earth and Environmental Sciences
Bayero University Kano
Nigeria

BARBOSA ARAUJO SOARES
SNIEHOTTA, Vera
Newcastle University
UK

BARDHAN, Suchandra
Jadavpur University
India

BARKER, Timothy
DIYNGO.org
UK

AIV

BARRETT, Ko
National Oceanographic and
Atmospheric Administration
USA

BASTOS, Ana
Laboratoire des sciences du climat
et de l'environnement
France

BATAILLE, Christopher
Institute for Sustainable Development
and International Relations
Simon Fraser University
Canada

BAUER, Nico
Potsdam Institut for Climate Impact Research
Germany

BAYAS, Dilipsing
International Consultant (Roster)
UN Global Pulse Lab
India

BELLAMY, Rob
University of Oxford
UK

BENNACEUR, Kamel
Abu Dhabi National Oil Company
United Arab Emirates

BENNETT, Simon
International Energy Agency
France

BENTO, Nuno
ISCTE-Instituto Universitário de Lisboa
Portugal

BENVENISTE, Hélène
Woodrow Wilson School of Public
and International Affairs
Princeton University
USA

BERDALET, Elisa
Consejo Superior de Investigaciones Científicas
Spain

BERTOLDI, Paolo
European Commission
Italy

BISHOP, Justin
University of Cambridge
UK

BLOK, Kornelis
Delft University of Technology
The Netherlands

BLUM, Mareike
University of Freiburg
Germany

BOBIN, Jean Louis
Université Pierre et Marie Curie
France

BODMAN, Roger
The University of Melbourne
Australia

BONDUELLE, Antoine
E&E Consultant
France

BOONEEADY, Prithiviraj
Mauritius Meteorological Services
Mauritius

BOOTH, Mary
Partnership for Policy Integrity
USA

BORGES LANDAEZ, Pedro Alfredo
Venezuelan Institute for Scientific Research
UNFCCC Technology Executive Committee
Venezuela

BORGFORD-PARNELL, Nathan
Climate and Clean Air Coalition
Switzerland

BOSETTI, Valentina
Bocconi University
Fondazione Eni Enrico Mattei
Italy

BOUCHER, Olivier
Institut Pierre Simon Laplace
France

BOYER-VILLEMAIRE, Ursule
Université du Québec à Montréal
Canada

BRANDÃO, Miguel
Royal Institute of Technology (KTH)
Sweden

BRANDER, Keith
DTU Aqua
Denmark

BREGMAN, Bram
Radboud University
The Netherlands

BREON, Francois-Marie
Laboratoire des sciences du climat
et de l'environnement
France

BREYER, Christian
Lappeenranta University of Technology
Finland

BRICEÑO-ELIZONDO, Elemer
Instituto Tecnológico de Costa Rica
Costa Rica

BROWN, Louis
Manchester Business School
UK

BROWN, Sally
University of Southampton
UK

BRUCKNER, Thomas
University of Leipzig
Germany

BRUNNER, Beat
Lightning MultiCom SA
Switzerland

BUDINIS, Sara
Imperial College London
Sustainable Gas Institute
UK

BULLOCK, Simon
Friends of the Earth, England,
Wales and Northern Ireland
UK

BUZÁSI, Attila
Budapest University of
Technology and Economics
Hungary

BYERS, Edward
International Institute for
Applied Systems Analysis
Austria

BUTT, Nathalie
The University of Queensland
Australia

CAESAR, John
Met Office Hadley Centre
UK

CAI, Rongshuo
Third Institute of Oceanography
State Oceanic Administration of China
China

CALLEN, Jessica
International Institute for
Applied Systems Analysis
Austria

AIV

CAMES, Martin
Öko-Institut
Germany

CAMPBELL, Donovan
Jamaica

CAMPBELL, Kristin
Institute for Governance and
Sustainable Development
USA

CAMPBELL-DURUFLÉ, Christopher
Center for International Sustainable
Development Law
Canada

CANEILL, Jean-Yves
IETA non-profit business organisation
France

CANEY, Simon
University of Warwick
UK

CAPSTICK, Stuart
Cardiff University
UK

CARAZO ORTIZ, Maria Pia
University for Peace
Germany

CARNICER, Jofre
University of Barcelona
Spain

CARPENTER, Mike
Gassnova
Norway

CARRASCO, Jorge
Universidad de Magallanes
Chile

CARTER, Peter
Climate Emergency Institute
Canada

CARTER, Timothy
Finnish Environment Institute (SYKE)
Finland

CARTWRIGHT, Anton
African Centre for Cities
University of Cape Town
South Africa

CASSOTTA, Sandra
Denmark

CASTANEDA, Fátima
Konrad Adenauer S.
Guatemala

CAZZOLA, Pierpaolo
International Energy Agency
France

CEARRETA, Alejandro
Universidad del Pais Vasco/EHU
Spain

CENTELLA, Abel
Instituto de Meteorologia
Cuba

CERDÁ, Emilio
University Complutense of Madrid
Spain

CHACÓN, Noemi
Instituto Venezolano de
Investigaciones Científicas
Venezuela

CHADBURN, Sarah
University of Leeds
UK

CHAN, Yi-Chieh
Delta Electronics Foundation
China

CHARPENTIER LJUNGQVIST, Fredrik
Stockholm University
Sweden

CHEN, Wenting
Norwegian Institute for Water Research
Norway

CHEN, Wenying
Tsinghua University
China

CHEN, Ying
Chinese Academy of Social Sciences
China

CHERKAOUI, Ayman Bel Hassan
Center for International Sustainable
Development Law
Morocco

CHERNOKULSKY, Alexander
A.M. Obukhov Institute of Atmospheric Physics
Russian Academy of Sciences
Russian Federation

CHINWEZE, Chizoba
Chemtek Associates Limited
Nigeria

CHOI, Woonsup
University of Wisconsin-Milwaukee
USA

CHOW, Winston
National University of Singapore
Singapore

CHRISTOPHER, Barrington-Leigh
McGill University
Canada

CHUST, Guillem
AZTI Marine Research Division
Spain

CIOT, Marco
Italian Peace Civil Corps FOCSIV
Italy

CLARK, Christopher
US Government
USA

CLARKE, David
Canada

CLARKE, Jamie
Climate Outreach
UK

CLAYTON, Susan
The College of Wooster
USA

CLEMMER, Steve
Union of Concerned Scientists
USA

COLLINS, Mat
University of Exeter
UK

COLLINS, William
University of Reading
UK

CONNORS, Sarah
IPCC WGI Technical Support Unit
Université Paris-Saclay
France/UK

CONVERSI, Alessandra
Consiglio Nazionale delle Ricerche
Italy

COOK, Lindsey
Friends World Committee for Consultation
Germany

COOPER, David
Convention on Biological Diversity
Canada

CORFEE-MORLOT, Jan
New Climate Economy
France

CORNELIUS, Stephen
World Wildlife Fund
UK

COROBOV, Roman
Eco-Tiras International Association
of River Keepers
Republic of Moldova

COULTER, Liese
Griffith University
Australia

COURAULT, Romain
Sorbonne-Universités Paris IV
France

CREMADES, Roger
Climate Service Center Germany (GERICS)
Germany

CREUTZIG, Felix
Mercator Research Institute on Global
Commons and Climate Change
Technical University Berlin
Germany

CURRIE-ALDER, Bruce
International Development Research Centre
Canada

CUSACK, Geraldine Ann
Siemens Ltd
Ireland

CUTTING, Hunter
Cimate Nexus/Climate Signals
USA

CZERNICHOWSKI-LAURIOL, Isabelle
Bureau de recherches géologiques et minières
France

DAALDER, Henk
Pak de Wind
Netherlands

DAGNET, Yamide
World Resources Institute
USA

DAI, Zhen
Harvard University
USA/China

DAIOGLOU, Vassilis
PBL Netherlands Environmental
Assessment Agency
Netherlands

DALIAKOPOULOS, Ioannis
Technical University of Crete
Greece

DAMASSA, Thomas
Oxfam America
USA

DAÑO, Elenita
Action Group on Erosion
Technology and Concentration
Philippines

DAVIES, Elizabeth Penelope
Ford Foundation
USA

DE BEAUVILLE-SCOTT, Susanna
Sustainable Development Department
Government of Saint Lucia
Saint Lucia

DE CONINCK, Heleen
Radboud University
Netherlands

DE FRENNE, Pieter
Ghent University
Belgium

DE OLIVEIRA, Gabriel
University of Kansas
Brazil

DEISSENBERG, Christophe
Aix-Marseille University
Groupement de recherche en
économie quantitative
Luxembourg

DELUSCA, Kenel
Institut des sciences, des technologies
et des études avancées d'Haïti
University of Montreal
Haiti

DEMKINE, Volodymyr
Kenya

DEN ELZEN, Michel
PBL Netherlands Environmental
Assessment Agency
Netherlands

DENG, Xiangzheng
Institute of Geographic Sciences
and Natural Resources Research
Chinese Academy of Sciences
China

DENIS-RYAN, Amandine
Monash University
Australia

DESCONSI, Cristiano
Federal University of Goiás
Brazil

DI BELLA, Jose
ParlAmericas
Canada

DIAZ, Julio
National School of Public Health
Carlos III Institute of Health
Spain

DIOP, Cherif
Agence nationale de l'aviation
civile et de la météorologie
Senegal

DIOSEY RAMON, Lugo-Morin
Universidad Intercultural del Estado de Puebla
Mexico

DIXON, Tim
IEA Greenhouse Gas
UK

DOCQUIER, David
Université catholique de Louvain
Belgium

DOELLE, Meinhard
Dalhousie University
Canada

DONEV, Jason
University of Calgary
Canada

DONNELLY, Chantal
Bureau of Meteorology
Australia

DOOLEY, Kate
University of Melbourne
Australia

DOYLE, Paul
BC Rivers Consulting Ltd
Canada

DRIOUECH, Fatima
Direction de la météorologie nationale
Morocco

DROEGE, Susanne
German Institute for International
and Security Affairs (SWP)
Germany

DUBE, Lokesh Chandra
Ministry of Environment, Forest
and Climate Change
India

DUNPHY, Brendon
The University of Auckland
New Zealand

DURAND, Frédéric
Toulouse II University
France

DYKEMA, John
Paulson School
Harvard University
USA

EASTHAM, Sebastian
Massachusetts Institute of Technology
USA/UK

EGBENDEWE, Aklesso
University of Lome
Togo

EHARA, Makoto
Forestry and Forest Products Research Institute
Japan

EHSANI, Nima
Saint Louis University
USA

EISEN, Olaf
Alfred-Wegener-Institut
Helmholtz-Zentrum für Polar-
und Meeresforschung
Germany

EKHOLM, Tommi
Aalto University
Finland

EL ZEREY, Wael
Djillali Liabes University
Algeria

ELBASIOUNY, Heba
Al-Azhar University
Egypt

ELBEHIRY, Fathy
Central laboratory for Environmental Studies
Kafrelsheikh University
Egypt

ELBEHRI, Aziz
Italy

ELSHAROUNY, Mohamed
Cairo University
Egypt

EMORI, Seita
National Institute for Environmental Studies
Japan

ENOMOTO, Hiroyuki
Japan

ERB, Karlheinz
Austria

ERIKSSON, Flintull Annica
Statistics Denmark
Sweden

ERLANIA, Erlania
Center for Fisheries Research
Indonesia

ESPARTA, Adelino Ricardo Jacintho
Universidade de São Paulo
Brazil

FÆHN, Taran
Statistics Norway Research Department
Norway

FANG, Kai
Zhejiang University
China

FARAGO, Tibor
Eötvös L. University (ELTE)
Hungary

FARIA, Sergio Henrique
Basque Centre for Climate Change (BC3)
Spain

FAST, Stewart
Institute for Science, Society and Policy
University of Ottawa
Canada

FAUSET, Sophie
University of Leeds
UK

FICHERA, Alberto
University of Catania
Italy

FINNVEDEN, Göran
Royal Institute of Technology (KTH)
Fortum Värme
Sweden

FISCHLIN, Andreas
ETH Zurich
Switzerland

FLATO, Greg
Canadian Centre for Climate
Modelling and Analysis
Environment Canada
Canada

FLEMING, John
Center for Biological Diversity
USA

FLEMING, Sean
University of British Columbia
USA

FODA, Rabiz
Hydro One Networks Inc.
Canada

FOLTESCU, Valentin
United Nations Environment
France

FORD, James
Department of Geography
McGill University
Canada

FORSTER, Piers
University of Leeds
UK

FREI, Thomas
Research and Consulting
Switzerland

FUGLESTVEDT, Jan
Center for International Climate Research
Norway

FUHR, Lili
Heinrich Böll Foundation
Germany

FUJIMORI, Shinichiro
National Institute for Environmental Studies
Japan

FUSS, Sabine
Mercator Research Institute on Global
Commons and Climate Change
Germany

GADIAN, Alan
National Centre for Atmospheric Sciences
University of Leeds
UK

GAILL, Françoise
Plateforme océan climat
France

GAJJAR, Sumetee Pahwa
Indian Institute for Human Settlements
India

GALOS, Borbala
University of Sopron
Hungary

GALYNA, Trypolska
Institute for Economics and Forecasting, UNAS
Ukraine

GAN, Thian
University of Alberta
Canada

GARCI SOTO, Carlos
Spanish Institute of Ocenography
Spain

GECK, Angela
Institute of Political Science
University of Freiburg
Germany

GEDEN, Oliver
German Institute for International and Security
Affairs (Stiftung Wissenschaft und Politik)
Germany

GEORGIADIS, Teodoro
Italian national Research Council
Instituto of Biometeorology (CNR-IBIMET)
Italy

GIARDINO, Alessio
Deltares
Netherlands

GIAROLA, Sara
Imperial College London
UK

GILLE, Sarah
Scripps Institution of Oceanography
University of California San Diego
USA

GILLETT, Nathan
Environment and Climate Change Canada
Canada

GLENN, Aaron
Agriculture and Agri-Food Canada
Canada

GOLSTON, Levi
Princeton University
USA

GÓMEZ CANTERO, Jonathan
Universidad de Alicante
Castilla-La Mancha Media
Spain

GONZALEZ, Miguel
Global Education and
Infrastructure Services Ltd
Nigeria

GONZALEZ, Patrick
University of California
USA

GONZÁLEZ-EGUINO, Mikel
Basque Centre for Climate Change
Spain

GOODWIN, Philip
University of Southampton
UK

GORNER, Marine
Organisation for Economic Co-operation and
Development, International Energy Agency
France

GRANT, Jonathan
PwC
UK

GRASSI, Giacomo
Italy

GRAU, Marion
MF Norwegian School of Theology
Norway

GRILLAKIS, Manolis
Technical University of Crete
Greece

GROVER, Samantha
La Trobe University
Australia

GUIVARCH, Céline
Centre international de recherche sur
l'environnement et le développement,
Ecole des Ponts ParisTech
France

GUPTA, Himangana
NATCOM Cell
Ministry of Environment, Forest
and Climate Change
India

HABERL, Helmut
Institute of Social Ecology
University of Natural Resources
and Life Sciences (BOKU)
Austria

HAFEZ, Marwa
Institute for Graduate Studies & Research
Alexandria Governor office
Egypt

HAGEN, Achim
Resource Economics Group
Humboldt-Universität
Germany

HAITES, Erik
Margaree Consultants Inc.
Canada

HALLAM, Samantha
University of Southampton
UK

HALSNAES, Kirsten
The Danish Technical University
Denmark

HAMDI, Rafiq
Royal Meteorological Institute of Belgium
Belgium

HANN, Veryan
University of Tasmania
Australia

HARA, Masayuki
Center for Environmental Science in Saitama
Japan

HARE, Bill
Climate Analytics
Germany

HARMELING, Sven
CARE International
Germany

HAROLD, Jordan
University of East Anglia
UK

HARPER, Anna
University of Exeter
UK

AIV

HARRISON, Jonathan
University of Southampton
UK

HASEGAWA, Tomoko
National Institute for Environmental Studies
Japan

HASHIMOTO, Shoji
Forestry and Forest Products Research Institute
Japan

HAYMAN, Garry
NERC
Centre for Ecology & Hydrology
UK

HAYWOOD, Jim
University of Exeter
UK

HEAD, Erica
Fisheries and Oceans Canada
Canada

HEBBINGHAUS, Heike
Landesamt für Natur, Umwelt und
Verbraucherschutz Nordrhein-
Westfalen (LANUV)
Germany

HENSON, Stephanie
National Oceanography Centre
UK

HERBERT, Annika
University of Sydney
Australia

HERRALA, Risto
International Monetary Fund
USA

HERTWICH, Edgar
Yale University
USA

HEYD, Thomas
University of Victoria
Canada

HIDALGO, Julia
National Center of Scientific Research
France

HILDÉN, Mikael
Finnish Environment Institute
Finland

HILMI, Nathalie
France

HIRVONEN, Janne
Aalto University
Finland

HITE, Kristen
American University
School of International Service
USA

HOF, Andries
PBL Netherlands Environmental
Assessment Agency
Netherlands

HOLZ, Christian
Climate Equity Reference Project
Canada

HONDA, Yasushi
Faculty of Health and Sport Sciences
University of Tsukuba
Japan

HONEGGER, Matthias
Research Scientist at Institute for Advanced
Sustainability Studies Potsdam
Germany

HONG, Jinkyu
Department of Atmospheric Sciences
Yonsei University
Republic of Korea

HONGO, Takashi
Mitsui Global Strategic Studies Institute
Japan

HOREN GREENFORD, Daniel
Concordia University
Canada

HORTON, Joshua
Harvard University
Havard Kennedy School
USA

HOSSEN, Mohammad Anwar
University of Dhaka
Bangladesh

HOWDEN, Mark
Australian National University
Australia

HUANG, Yuanyuan
Laboratoire des sciences du climat
et de l'environnement
France

HUEBENER, Heike
Hessian Agency for Nature Conservation,
Environment and Geology
Germany

HURTADO ALBIR, Francisco Javier
European Patent Office
Germany

HUSSEIN, Amal
National Research Centre
Egypt

IGLESIAS BRIONES, Maria Jesus
Universidad de Vigo
Spain

IIZUMI, Toshichika
Institute for Agro-Environmental Sciences
National Agriculture and Food
Research Organization
Japan

INFIELD, David
University of Strathclyde
UK

INSAROV, Gregory
Institute of Geography
Russian Academy of Sciences
Russian Federation

INYANG, Hilary
Global Education and
Infrastructure Services LLC
Nigeria

IQBAL, Muhammad Mohsin
Global Change Impact Studies Centre
Pakistan

IRVINE, Peter
School of Engineering and Applied Sciences
Harvard University
USA

ISHIMOTO, Yuki
The Institute of Applied Energy
Norway

ISLAM, Akm Saiful
Bangladesh University of
Engineering and Technology
Bangladesh

ISLAM, Md. Sirajul
North South University Bangladesh

ITO, Akihiko
National Institute for Environmental Studies
Japan

AIV

IZZET, Ari
Head of Department of Sustainable
Development and Environment
Turkey

JACOBSON, Mark
Stanford University
USA

JAEGER-WALDAU, Arnulf
European Commission
Italy

JAKOB, Grandin
University of Bergen
Norway

JAMES, Rachel
University of Oxford
UK

JEREZ, Sonia
University of Murcia
Spain

JERSTAD, Heid
University of Edinburgh
UK

JIA, Gensuo
China

JIM ILHAM, Jasmin Irisha
Jeffrey Sachs Center on
Sustainable Development
Malaysia

JINNAH, Sikina
UC Santa Cruz
USA

JOHANSEN, Tom Gabriel
InfoDesignLab
Norway

JOHNSTON, Eleanor
Climate Interactive
USA

JONES, Ian
University of Sydney
Australia

JONES, Lindsey
London School of Economics
Overseas Development Institute
UK

JOSEY, Simon
National Oceanography Centre
UK

JOUZEL, Jean
Centre d'énergie atomique de Saclay
France

KABIDI, Khadija
Direction de la météorologie nationale
Morocco

KACHI, Aki
Carbon Market Watch
Germany

KADITI, Eleni
Organization of the Petroleum
Exporting Countries
Austria

KAINUMA, Mikiko
Institute for Global Environmental Strategies
Japan

KALLBEKKEN, Steffen
Center for International Climate Research
Norway

KALLIOKOSKI, Tuomo
University of Helsinki
Finland

KALUGIN, Andrey
Water Problems Institute of
Russian Academy of Sciences
Russian Federation

KANAKO, Morita
Forestry and Forest Products Research Institute
Japan

KARIMIAN, Maryam
Climatology Research Institute
National Center of Climatology
Iran

KARMALKAR, Ambarish
University of Massachusetts Amherst
USA

KARTADIKARIA, Aditya
Bandung Institute of Technology
Indonesia

KATBEHBADER, Nedal
Environment Quality Authority
State of Palestine
Switzerland

KAWAMIYA, Michio
Japan Agency for Marine-Earth
Science and Technology
Japan

KAY, Robert
ICF
USA

KEENAN, Jesse
Harvard University
USA

KEITH, David
Harvard University
USA/Canada/UK

KEMPER, Jasmin
IEA Greenhouse Gas R&D Programme
UK

KERKHOVEN, John
Partner Quintel Intelligence
Netherlands

KERSTING, Diego Kurt
Freie Universität Berlin
Germany

KHAZAEI, Mahnaz
Iran Meteorological Organization
Iran

KHAZANEDARI, Leili
Climate Research Institute
Iran

KHENNAS, Smail
Independent energy and climate change expert
UK

KHESHGI, Haroon
ExxonMobil Research and
Engineering Company
USA

KIENDLER-SCHARR, Astrid
IEK-8: Troposphere, Forschungszentrum
Jülich GmbH
Germany

KILKIS, Siir
The Scientific and Technological Research
Council of Turkey (TUBITAK)
International Centre for Sustainable
Development of Energy, Water and
Environment Systems (SDEWES Centre)
Turkey

KIM, Hyung Ju
Green Technology Center Korea
Republic of Korea

KINN, Moshe
The University of Salford
UK

AIV

KJELLSTRÖM, Erik
Swedish Meteorological and
Hydrological Institute
Sweden

KNOPF, Brigitte
Mercator Research Institute on Global
Commons and Climate Change
Germany

KOBAYAHI, Shigeki
Transport Institute of Central Japan
Japan

KOCHTITZKY, William
University of Maine
USA

KOLL, Roxy Mathew
India

KONDO, Hiroaki
National Institute of Advanced
Industrial Science and Technology
Japan

KOPPU, Robert
Rutgers University
USA

KOUTROULIS, Aristeidis
Technical University of Crete
School of Environmental Engineering
Greece

KRAVITZ, Ben
Pacific Northwest National Laboratory
USA

KREUTER, Judith
Technische Universität
Germany

KREY, Volker
International Institute for
Applied Systems Analysis
Austria

KRIEGLER, Elmar
Potsdam Institute for Climate Impact Research
Germany

KRINNER, Gerhard
Centre national de la recherche scientifique
France

KRISHNASWAMY, Jagdish
India

KÜHNE, Kjell
Leave it in the Ground Initiative
Mexico

KUHNHENN, Kai
Konzeptwerk Neue Ökonomie
Germany

KUSCH, Sigrid
University of Padua
Germany

KUYLENSTIERNA, Johan Carl Ivar
Stockholm Environment Institute
UK

LABRIET, Maryse
Eneris Environment Energy Consultants
Spain

LAHA, Priyanka
Indian Institute of the Technology Kharagur
India

LAHIRI, Souparna
Global Forest Coalition
India

LAHN, Bård
Center for International Climate Research
Norway

LATIF, Muhammad
Applied Systems Analysis Division
Pakistan Atomic Energy Commission
Pakistan

LAW, Matt
Bath Spa University
UK

LAWRENCE, Mark
Institute for Advanced Sustainability Studies
Germany

LE BRIS, Nadine
Sorbonne University
France

LE QUÉRÉ, Corinne
Tyndall Centre
University of East Anglia
UK

LEAHY, Paul
University College Cork
Ireland

LEAL, Walter
Hamburg University of Applied Sciences
Germany

LECOCQ, Noé
Inter-Environnement Wallonie
Belgium

LEE, Arthur
Chevron Energy Technology Company
USA

LEE, Sai Ming
Hong Kong Observatory
China

LEFALE, Penehuro Fatu
LeA International
Joint Centre for Disaster Research
Massey University
New Zealand

LEFFERTSTRA, Harold
Climate Consultant
Norway

LEHOCZKY, Annamaria
Centre for Climate Change
Universitat Rovira i Virgili
Spain

LESLIE, Michelle
Canadian Nuclear Association
Canada

LEVIHN, Fabian
Royal Institute of Technology (KTH)
Fortum Värme
Sweden

LEVINA, Ellina
International Energy Agency
France

LEY, Debora
Latinoamérica Renovable
Guatemala/Mexico

LICKER, Rachel
Union of Concerned Scientists
USA

LIJUAN, Ma
National Climate Center
China

LINARES, Cristina
Carlos III Institute of Health
National School of Public Health
Spain

LITTLE, Conor
University of Limerick
Ireland

LLASAT, Maria-Carmen
University of Barcelona
Spain

LLOYD, Philip
Cape Peninsula University of Technology
Beijing Agricultural University
South Africa

LOBELLE, Delphine
University of Southampton
UK

LOCKIE, Stewart
James Cook University
Australia

LOCKLEY, Andrew
UK

LOMBROSO, Luca
University of Modena and Reggio Emilia
Italy

LONGDEN, Thomas
Centre for Health Economics
Research and Evaluation
University of Technology Sydney
Australia

LOPEZ-BUSTINS, Joan A.
University of Barcelona
Spain

LOUGHMAN, Joshua
Arizona State University
USA

LOUREIRO, Carlos
Ulster University
University of KwaZulu-Natal
UK/South Africa

LOVERA-BILDERBEEK, Simone
Global Forest Coalition
University of Amsterdam
Paraguay

LOVINS, Amory
Rocky Mountain Institute
USA

LUCERO, Lisa
University of Illinois at Urbana-Champaign
USA

LUDERER, Gunnar
Potsdam Institute for Climate Impact Research
Germany

LUENING, Sebastian
Institute for Hydrography, Geoecology
and Climate Sciences
Portugal

LUPO, Anthony
University of Missouri
USA

LYNN, Jonathan
IPCC Secretariat
World Meteorological Organization
Switzerland

LYONS, Lorcan
France

MAAS, Wilfried
Royal Dutch Shell
Netherlands

MACCRACKEN, Michael
Climate Institute
USA

MACDOUGALL, Andrew
St. Francis Xavier University
Canada

MACMARTIN, Douglas
Cornell University
USA

MAHAL, Snaliah
Department of Sustainable Development
Saint Lucia

MAHOWALD, Natalie
Cornell University
USA

MANTYKA-PRINGLE, Chrystal
School of Environment and Sustainability
University of Saskatchewan
Canada

MARBAIX, Philippe
Université catholique de Louvain
Belgium

MARCOTULLIO, Peter
Hunter College
City University of New York
USA

MARTIN, Eric
Irstea
France

MARTINI, Catherine
Yale Program on Climate
Change Communication
USA

MARTYR-KOLLER, Rosanne
University of California Berkeley
Germany

MARX, Andreas
Helmholtz Centre for Environmental
Research GmbH (UFZ)
Germany

MASOOD, Amjad
Global Change Impact Studies Centre
Pakistan

MASSON-DELMOTTE, Valérie
IPCC Co-Chair WGI
France

MATA, Érika
IVL Swedish Environmental Research Institute
Sweden

MATSUMOTO, Katsumi
University of Minnesota
USA

MATSUMOTO, Ken'ichi
Nagasaki University
Japan

MATTHEWS, J. B. Robin
IPCC WGI Technical Support Unit
Université Paris-Saclay
France/UK

MAY, Wilhelm
Lund University
Denmark

MAZAUD, Alain
Laboratoire des sciences du climat
et de l'environnement
France

MAZZOTTI, Marco
ETH Zurich
Switzerland

MBEVA, Kennedy
African Centre for Technology Studies
Australia

MCCAFFREY, Mark
National University for Public Service
Hungary

MCKINNON, Catriona
University of Reading
UK

MECHLER, Reinhard
International Institute for
Applied Systems Analysis
Austria/Germany

AIV

MEFTAH, Mustapha
Centre national de la recherche scientifique
France

MELAMED, Megan
International Global Atmospheric Chemistry
University of Colorado
Cooperative Institute for Research
in Environmental Sciences
USA

MELIA, Nathanael
Scion
New Zealand

METZ, Bert
European Climate Foundation
Netherlands

MEYA, Jasper
Humboldt-Universtiät
Germany

MICHAELIS, Laurence
Living Witness (Quakers)
UK

MICHAELOWA, Axel
University of Zurich
Switzerland

MIDGLEY, Pauline
Independent Consultant
Germany

MIN, Seung-Ki
Pohang University of Science and Technology
Republic of Korea

MINDENBECK, Katja
IPCC WGII Technical Support Unit
Alfred-Wegener-Institut Bremen
Germany

MITCHELL, Dann
University of Bristol
UK

MIZUNO, Yuji
Institute for Global Environmental Strategies
Japan

MKWAMBISI, David
Lilongwe University of Agriculture
and Natural Resources
Malawi

MODIRIAN, Rahele
Climatological Research Institute
Iran

MOHAMED ABULEIF, Khalid
Sustainability Advisor to the Minister Ministry
of Petroleum and Mineral Resources
Saudi Arabia

MOLERO, Francisco
Centro de Investigaciones Energéticas,
Medioambientales y Tecnológicas
Spain

MÖLLER, Ina
Lund University
Sweden

MÖLLERSTEN, Kenneth
Swedish Energy Agency
Sweden

MONFORTI-FERRARIO, Fabio
Joint Research Centre
European Commission
Italy

MONTT, Guillermo
International Labour Organization
Switzerland

MOORE, Robert Daniel
University of British Columbia
Canada

MORALES, Manuel
Université Clermont Auvergne
France

MORECROFT, Mike
Natural England
UK

MORELLI, Angela
InfoDesignLab
Norway

MORENO, Meimalin
Venezuelan Institute for Scientific Research
Venezuela

MORGAN, Jennifer
Greenpeace
The Netherlands

MORIN, Samuel
France

MORROW, David
American University
George Mason University
USA

MORTON, John
Natural Resources Institute
University of Greenwich
UK

MORTON, Oliver
University College London
UK

MOUFOUMA OKIA, Wilfran
IPCC WGI Technical Support Unit
Université Paris-Saclay
France/Congo

MOUSSADEK, Rachid
National Agricultural Research Institute
Morocco

MOVAGHARI, Alireza
Urmia University
Iran

MUÑOZ SOBRINO, Castor
Universidade de Vigo
Spain

MURI, Helene
University of Oslo
Norway

MUSOLIN, Dmitry L.
Saint Petersburg State Forest
Technical University
Russian Federation

MYCOO, Michelle
The University of the West Indies
Trinidad and Tobago

NALAU, Johanna
Griffith University
Australia

NANGOMBE, Shingirai Shepard
University of Chinese Academy of Science
Zimbabwe

NATALINI, Davide
Global Sustainability Institute
Anglia Ruskin University
UK

NAUELS, Alexander
University of Melbourne
Australia

NDIONE, Jacques-André
Centre de Suivi Ecologique
Senegal

NEDJRAOUI, Dalila
Algeria

NERILIE, Abram
Australian National University
Australia

NEU, Urs
ProClim
Swiss Academy of Sciences
Switzerland

NICHOLLS, Neville
Monash University
Australia

NICOLAU, Mariana
Collaborating Centre on Sustainable
Consumption and Production
Germany

NIEVES, Barros
University of Santiago de Compostela
Spain

NIFENECKER, Herve
Global Initiative to Save Our Climate (GISOC)
France

NISHIOKA, Shuzo
Institute for Global Environmental Strategies
Japan

NOGUEIRA DA SILVA, Milton
Climate Change & Technology Consultant
Brazil

NOH, Dong-Woon
Korea Energy Economics Institute
Republic of Korea

NORMAN, Barbara
University of Canberra
Australia

NUGRAHA, Adi
Pacific Northwest National Laboratory
USA

NUNES, Ana Raquel
University of Warwick
UK

NUNEZ-RIBONI, Ismael
Thünen Institute of Sea Fisheries
Germany

OGDEN, Nicholas
Public Health Agency of Canada
Canada

OKPALA, Denise
The Institution of Environmental Sciences
Nigeria

OLA, Kalen
Swedish Meteorological and
Hydrological Institute
Sweden

OLHOFF, Anne
UNEP DTU Partnership
Denmark

OLSEN, Karen
UNEP DTU Partnership
Denmark

O'MAHONY, Tadhg
Finland Futures Research Centre
Finland

ONGOMA, Victor
South Eastern Kenya University
Kenya

ONUOHA, Mgbeodichinma Eucharia
TU Bergakademie Freiberg Saxony
Germany

OOGJES, Justin
University of Melbourne
Australia

OPPENHEIMER, Michael
Princeton University
USA

OSCHLIES, Andreas
GEOMAR
Germany

OTTO, Friederike
University of Oxford
UK

OURBAK, Timothée
Agence française de développement
France

PAGNIEZ, Capucine
Plateforme océan et climat
France

PAJARES, Erick
The Biosphere Group
Peru

PALTER, Jaime
University of Rhode Island
Graduate School of Oceanography
USA

PÁNTANO, Vanesa
Department of Atmosphere
and Ocean Sciences
University of Buenos Aires
Argentina

PAREDES, Franklin
Institute of Atmospheric Sciences (ICAT)
Federal University of Alagoas
Brazil

PARK, Go Eun
National Institute of Forest Science
Republic of Korea

PARKER, Andrew
University of Bristol
UK

PASTOR, Amandine
Institut de recherche pour le développement
France

PATERSON, Matthew
University of Manchester
UK

PATT, Anthony
ETH Zurich
Switzerland

PATWARDHAN, Anand
University of Maryland
USA

PAUL ANTONY, Anish
Massachusetts Institute of Technology
USA

PAUW, Willem Pieter
German Development Institute
Germany

PEARSON, Pamela
International Cryosphere Climate Initiative
USA

PEBAYLE, Antoine
Plateforme océan et climat
France

PEDACE, Alberto
Climate Action Network LatinoAmerica
Argentina

PERDINAN
Bogor Agricultural University
Indonesia

PERLMAN, Kelsey
Carbon Market Watch
France

PERRIER, Quentin
Centre international de recherche sur
l'environnement et le développement
France

AIV

PETERS, Glen
Center for International Climate Research
Norway

PETRASEK MACDONALD, Joanna
Inuit Circumpolar Council
Canada

PETZOLD, Jan
IPCC WGII Technical Support Unit
Alfred-Wegener-Institut Bremen
Germany

PHILIBERT, Cedric
International Energy Agency
France

PIACENTINI, Rubén
Institute of Physics Rosario
Consejo Nacional de Investigaciones
Científicas y Técnicas
National University of Rosario
Argentina

PIANA, Valentino
Economics Web Institute
Italy

PIGUET, Etienne
University of Neuchâtel
Switzerland

PIRLO, Giacomo
Consiglio per la ricerca in agricotura
e l'analisi dell'economia agraria
Italy

PISANI, Bruno
Civil Engineering School
University of A Coruña
Spain

PISKOZUB, Jacek
Institute of Oceanology Polish
Academy of Sciences
Poland

PITARI, Giovanni
Department of Physical and Chemical Sciences
Università L'Aquila
Italy

PLANTON, Serge
Météo-France
France

POITOU, Jean
Sauvons Le Climat
France

POLOCZANSKA, Elvira
IPCC WGII Technical Support Unit
Alfred-Wegener-Institut Bremen
Germany

POMPEU PAVANELLI, João Arthur
Instituto Nacional de Pesquisas Espaciais
Brazil

POOT-DELGADO, Carlos
Instituto Tecnologico Superior de Champotón
Mexico

POPKOSTOVA, Yana
European Centre for Energy
and Geopolitical Analysis
France

PÖRTNER, Hans-Otto
IPCC Co-Chair WGII
Germany

PRAG, Andrew
International Energy Agency
France

PRAJAL, Pradhan
Potsdam Institute for Climate Impact Research
Germany

PRICE, Lynn
Lawrence Berkeley National Laboratory
USA

PUIG, Daniel
Technical University of Denmark
Denmark

PUIG ARNAVAT, Maria
Technical University of Denmark
Denmark

PULIDO-VELAZQUEZ, David
Instituto Geológico y Minero
Spanish Geological Survey
Spain

PUPPIM DE OLIVEIRA, Jose Antonio
Fondation Getúlio Vargas
Brazil

QIAN, Budong
Agriculture and Agri-Food Canada
Canada

RABITZ, Florian
Kaunas University of Technology
Lithuania

RADUNSKY, Klaus
Umweltbundesamt
Austria

RAHIMI, Mohammad
Faculty of Desert Studies
Semnan University
Iran

RASUL, Golam
International Centre for Integrated
Mountain Development
Nepal

RAU, Greg
University of California, Santa Cruz
USA

RAWE, Tonya
CARE
USA

RAYMOND, Colin
Columbia University
USA

REAY, David
University of Edinburgh
UK

REES, Morien
The International Council of Museums
Norway

REINECKE, Sabine
University of Freiburg
Forest and Environmental Policy
Germany

REISINGER, Andy
New Zealand Agricultural
GHG Research Centre
New Zealand

RETUERTO, Rubén
Universidade de Santiago de Compostela
Spain

REYER, Christopher
Potsdam Institute for Climate Impact Reserach
Germany

REYNOLDS, Jesse
Utrecht University
The Netherlands

RINKEVITCH, Baruch
Israel

RIXEN, Tim
Leibniz Zentrum für Marine Tropenforschung
Germany

ROBERTS, Debra
IPCC Co-Chair WGII
South Africa

AIV

ROBERTS, Erin
King's College London
UK

ROBIOU DU PONT, Yann
University of Melbourne
France

ROBLEDO ABAD, Carmenza
USYS-TdLab
ETH Zurich
Switzerland

ROBOCK, Alan
Rutgers University
USA

ROCKMAN, Marcy
National Park Service
USA

RODRÍGUEZ AÑÓN, José Antonio
University of Santiago de Compostela
Spain

ROEHM, Charlotte
Terralimno LLC
USA

ROGELJ, Joeri
International Institute for
Applied Systems Analysis
Austria/Belgium

ROMERI, Mario Valentino
Italy

ROSE, Steven
Electric Power Research Institute
USA

ROSEN, Richard
Germany

ROSENZWEIG, Cynthia
National Aeronautics and
Space Administration
Goddard Institute for Space Studies
USA

ROTLLANT, Guiomar
Instituto de Ciencias del mar
Consell Superior d'Investigacions Científiques
Spain

ROUDIER, Philippe
French Development Agency
France

ROY, Joyashree
Jadavpur University
Institute of Technology, Bangkok
Thailand/India

ROY, Shouraseni
University of Miami
USA

ROYER, Marie-Jeanne S.
Universite de Montreal
Canada

RUTH, Urs
Robert Bosch GmbH
Germany

SAGNI, Regasa
Malole consults
Ethiopia

SAHEB, Yamina
OpenExp
Ecole des Mines of Paris
France

SALA, Hernan Edgardo
Argentine Antarctic Institute
National Antarctic Directorate
Argentina

SALANAVE, Jean-Luc
Ecole CENTRALE SUPELEC
France

SALAT, Jordi
Instituto de Ciencias del mar
Consell Superior d'Investigacions Científiques
Spain

SALAWITCH, Ross
University of Maryland
USA

SALTER, Stephen
University of Edinburgh
UK

SALVADOR, Pedro
Centro de Investigaciones Energéticas,
Medioambientales y Tecnológicas
Spain

SAMSET, Bjørn
Center for International Climate Research
Norway

SANCHEZ, Jose Luis
University of Leon
Spain

SÁNCHEZ-MOREIRAS, Adela M
University of Vigo
Spain

SANDER, Sylvia
Section of Marine Environmental
Studies Laboratory
Monaco

SANTOSH KUMAR, Mishra
S. N. D. T. Women's University
India

SANZ SANCHEZ, Maria Jose
Basque Centre for Climate Change
Spain

SARGENT, Philip
Cambridge Energy Forum
UK

SAUNDERS, Harry
Decision Processes Inc
USA

SAVARESE, Stephan
ForCES SAS
France

SAVÉ, Robert
Institut de Recerca i Tecnologia
Agroalimentaries
Spain

SAVOLAINEN, Ilkka
Technical Research Centre of Finland
Finland

SAYGIN, Deger
Turkey

SCHAEFFER, Michiel
Climate Analytics
University of Wageningen
The Netherlands

SCHEWE, Jacob
Potsdam Institute for Climate Impact Research
Germany

SCHIPPER, Lisa
Stockholm Environment Institute
Overseas Development Institute
Vietnam

SCHISMENOS, Spyros
National Yunlin University of
Science and Technology
Eastern Macedonia and Thrace Institute
China

SCHLEUSSNER, Carl-Friedrich
Climate Analytics
Germany

AIV

SCHNEIDER, Linda
Heinrich Boell Foundation
Germany

SCHOEMAN, David
University of the Sunshine Coast
Australia

SCHULZ, Astrid
Wissenschaftliche Beirat der Bundesregierung
Globale Umweltveränderungen
Germany

SCOWCROFT, John
Global Carbon Capture and Storage Institute
Belgium

SEHATKASHANI, Saviz
Academic member of Atmospheric Science
and Meteorological Research Center
Iran

SEILER, Jean Marie
Retired from Commissariat à l'énergie
atomique et aux énergies alternatives)
France

SEITZINGER, Sybil
University Victoria
Pacific Institute for Climate Solutions
Canada

SEMENOV, Sergey
Institute of Global Climate and Ecology
Russian Federation

SEMENOVA, Inna
Odessa State Environmental University
Ukraine

SENEVIRATNE, Sonia I.
ETH Zurich
Switzerland

SENSOY, Serhat
Turkish State Meteorological Service
Turkey

SERRAO-NEUMANN, Silvia
Cities Research Institute
Griffith University
Australia

SETTELE, Josef
Helmholtz Centre for Environmental
Research (UFZ)
Germany

SHAH, Shipra
Fiji National University
Fiji

SHAPIRO, Robert
Climate Mobilization Outer Cape
USA

SHAWOO, Zoha
University of Oxford
UK

SHEPARD, Isaac
University of Maine
USA

SHINDELL, Drew
Duke University
USA

SHINE, Keith
Department of Meteorology
University of Reading
UK

SHINE, Tara
Mary Robinson Foundation
Climate Justice
Ireland

SHIOGAMA, Hideo
National Institute for Environmental Studies
Japan

SHOAI-TEHRANI, Bianka
Research Institute of Innovative
Technology for the Earth
Japan

SHUE, Henry
University of Oxford
UK

SIETZ, Diana
Wageningen University
Netherlands

SIHI, Debjani
University of Maryland Center for
Environmental Science Appalachian Laboratory
USA

SIKAND, Monika
Bronx Community College
City University of New York
USA

SIMMONS, Adrian
European Centre for Medium-
Range Weather Forecasts
UK

SIMS, Ralph
Massey University
New Zealand

SINGER, Stephan
Climate Action Network International
Belgium

SINGH, Chandni
Indian Institute for Human Settlements
Myanmar/India

SINGH, Neelam
World Resources Institute
USA

SKEA Jim
IPCC Co-Chair WGIII
UK

SKEIE, Ragnhild
Center for International Climate Research
Norway

SMEDLEY, Andrew
University of Manchester
UK

SMITH, Alison
University of Oxford
UK

SMITH, Sharon
Geological Survey of Canada
Natural Resources Canada
Canada

SMITHERS, Richard J.
Ricardo Energy & Environment
UK

SMOLKER, Rachel
Biofuelwatch
USA

SOLAYMANI OSBOOEI, Hamidreza
Forest, Range and Watershed
Management Org.
Iran

SOLERA UREÑA, Miriam
Universidad Nacional de Educación
a Distancia (Spain)
Germany

SOORA, Naresh Kumar
Indian Agricultural Research Institute
India

SÖRENAAON, Anna
Centro de Investigaciones del
Mar y la Atmósfera
Argentina

AIV

SREENIVAS, Ashok
Prayas (Energy Group)
India

STABINSKY, Doreen
College of the Atlantic
USA

STANGELAND, Aage
The Research Council of Norway
Norway

STANLEY, Janet
University of Melbourne
Australia

STEFANO, Caserini
Politecnico di Milano, Dipartimento di
Ingegneria Civile ed Ambientale
Italy

STENMARK, Aurora
Norwegian Environment Agency
Norway

STOCKER, Thomas
University of Bern
Switzerland

STONE, Kelly
ActionAid USA
USA

STOTT, Peter
University of Exeter and Met Office
UK

STRANDBERG, Gustav
Swedish Meteorological and
Hydrological Institute
Sweden

STRAPASSON, Alexandre
Harvard University
Brazil

SU, Mingshah
National Center for Climate Change
Strategy and International Cooperation
China

SUGIYAMA, Masahiro
The University of Tokyo
Japan

SULISTYAWATI, Linda Yanti
Universitas Ahmad Dahlan
Indonesia

SUN, Junying
Chinese Academy of Meteorological Sciences
China

SUN, Yongping
Center of Hubei Cooperative Innovation
for Emissions Trading System
China

SUSANTO, Raden Dwi
USA

SUSATYA, Agus
UNIB
Indonesia

SUTHERLAND, Michael
Trinidad and Tobago

SUTTER, Daniel
ETH Zurich
Switzerland

SWART, Rob
Wageningen Environmental Research
The Netherlands

SWEENEY, John
Maynooth University
Ireland

SYRI, Sanna
Aalto University
Finland

TABARA, J. David
Autonomous University of Barcelona
Spain

TABATABAEI, Seyed Muhammadreza
University of Tehran
Iran

TACHIIRI, Kaoru
Japan Agency for Marine
Earth Science and Technology
Japan

TAKAGI, Hiroshi
Tokyo Institute of Technology
Japan

TAKAGI, Masato
Research Institute of Innovative
Technology for the Earth
Japan

TAKAHASHI, Kiyoshi
National Institute for Environmental Studies
Japan

TAKANO, Kohei
Nagano Environmental Conservation
Research Institute
Japan

TAKAYABU, Izuru
Japan Meteorological Agency
Meteorological Research Institute
Japan

TAKEMURA, Toshihiko
Kyushu University
Japan

TAM, Chi Keung
Newcastle University
Singapore

TAMAKI, Tetsuya
Kyushu University
Japan

TAMURA, Makoot
Ibaraki University
Japan

TANAKA, Katsumasa
National Institute for Environmental Studies
Japan

TESKE, Sven
University of Technology Sydney
Australia

TEXTOR, Christiane
German Aerospace Centre
Germany

THALER, Thomas
Institute of Mountain Risk Engineering
University of Natural Resources
and Life Sciences
Austria

THIERY, Wim
ETH Zurich
Switzerland

THOBER, Stephan
Helmholtz Centre for Environmental
Research (UFZ)
Germany

THOMPSON, Michael
Forum for Climate Engineering Assessment
USA

THORNE, Peter
Maynooth University
Ireland

AIV

THORNTON, Thomas
Environmental Change Institute
University of Oxford
UK

THWAITES, Joe
World Resources Institute
USA

TIBIG, Lourdes
Climate Change Commission
Philippines

TILCHE, Andrea
European Union
Belgium

TINDALL, David
Department of Sociology
University of British Columbia
Canada

TOKARSKA, Katarzyna B
University of Victoria
UK

TORVANGER, Asbjørn
Center for International Climate Research
Norway

TREBER, Manfred
Germanwatch
Germany

TREGUER, Paul
Université de Bretagne Occidentale
France

TSCHAKERT, Petra
University of Western Australia
Australia/Austria

TSUTSUI, Junichi
Central Research Institute of
Electric Power Industry
Japan

TUITT, Cate
Honourable society of Inner Temple
UK

TULKKI, Ville
VTT Technical Research Centre of Finland Ltd
Finland

TURCO, Marco
University of Barcelona
Spain

TURP, Mustafa Tufan
Bogazici University Center for Climate
Change and Policy Studies
Turkey

TYLER, Emily
University of Cape Town
South Africa

UDDIN, Noim
CPMA International
Australia

UDO, Keiko
Tohoku University
Japan

URQUHART, Penny
Independent climate resilient
development specialist
South Africa

VAILLES, Charlotte
Institute for Climate Economics
France

VALDES, Luis
Instituto Español de Oceanografía
Spain

VAN DE WAL, Roderik
Netherlands

VAN DEN HURK, Bart
Koninklijk Nederlands Meteorologisch
Instituut
Netherlands

VAN MUNSTER, Birgit
Homo Sapiens Foundation
UK

VAN RUIJVEN, Bastiaan
International Institute for
Applied Systems Analysis
Austria

VAN VELTHOVEN, Peter
Koninklijk Nederlands Meteorologisch
Instituut
Netherlands

VAN YPERSELE, Jean-Pascal
Université catholique de Louvain
Earth and Life Institute
Belgium

VAUTARD, Robert
Institut Pierre Simon Laplace
France

VELDORE, Vidyunmala
DNV-GL
Norway

VENEMA, Henry David
International Institute for
Sustainable Development
Prairie Climate Centre
Canada

VERA, Carolina
Centro de Investigaciones del
Mar y la Atmosfera
University of Buenos Aires
Comision Nacional de Investigaciones
Cientifico Tecnologicas
Argentina

VERHOEF, Leendert
University of Technology Delft
Netherlands

VICTOR, David
University of California San Diego
USA

VIDALENC, Eric
Agence de l'environnement et
de la maîtrise de l'énergie
France

VINCENT, Ceri
British Geological Survey
UK

VINER, David
Mott MacDonald
UK

VIVIAN, Scott
University of Edinburgh
UK

VLADU, Iulain Florin
United Nations Framework
Convention on Climate Change
Germany

VON SCHUCKMANN, Karina
France

WACHSMUTH, Jakob
Fraunhofer Institute for Systems
and Innovation Research
Germany

WACKERNAGEL, Mathis
Global Footprint Network
USA

WADLEIGH, Michael
Closed Mass
USA

AIV

WAGNER, Gernot
Harvard John A. Paulson School of
Engineering and Applied Sciences
Harvard Kennedy School
USA

WANG, Bin
University of Virginia
USA

WANG, Junye
Athabasca University
Canada

WANG, Xiaojun
Research Center for Climate Change,
Ministry of Water Resources
China

WANG, Zhen-Yi
Delta Electronics Foundation
China

WARNER, Koko
United Nations Framework
Convention on Climate Change
Germany

WARRILOW, David
Royal Meteorological Society
UK

WASHBOURNE, Carla-Leanne
University College London
UK

WASKOW, David
World Resources Institute
USA

WEBB, Jeremy
Department of Science, Technology
Engineering and Public Policy
University College London
UK

WEBER, Christopher
1982
USA

WEHNER, Michael
Lawrence Berkeley National Laboratory
USA

WEI, Taoyuan
Center for International Climate Research
Norway

WEISENSTEIN, Debra
School of Engineering and Applied
Science, Harvard University
USA

WEST, Thales A. P.
University of Florida
Brazil

WESTPHAL, Michael
World Resources Institute
USA

WESTRA, Seth
University of Adelaide
Australia

WHITEFORD, Ross
University of Southampton
UK

WHITLEY, Shelagh
Overseas Development Institute
UK

WICHMANN, Janine
University of Pretoria
South Africa

WIEL, Stephen
CLASP
USA

WILDENBORG, Ton
Netherlands Organisation for Applied
Scientific Research (TNO)
Netherlands

WILLIAMS, Jonny
The National Institute of Water
and Atmospheric Research
New Zealand

WILLIAMS, Richard
Liverpool University
UK

WINIGER, Patrik
Netherlands

WINKLER, Harald
University of Cape Town
Energy Research Centre
South Africa

WINROTH, Mats
Chalmers University of Technology
Sweden

WISSENBURG, Marcel
Radboud University
Netherlands

WITHANACHCHI, Sisira S.
University of Kassel
Germany

WOLF, Shaye
Center for Biological Diversity
USA

WOOLLACOTT, Jared
RTI International
USA

WRATT, David
National Institute of Water &
Atmospheric Research
New Zealand

WRIGHT, Helena
E3G
UK

WU, Jianguo
Chinese Research Academy of
Environmental Sciences
China

WURZLER, Sabine
North Rhine Westphalian State Agency for
Nature, Environment, and Consumer Protection
Germany

XENIAS, Dimitrios
Cardiff University
UK

XU, Yangyang
Texas A&M University
USA

XU, Yinlong
China

YAMAGUCHI, Mitsutsune
Research Institute of Innovative
Technology for the Earth
Japan

YANG, Hong
Swiss Federal Institute of Aquatic
Science and Technology
Switzerland

YANG, Tao
Jiangxi Normal University
China

YANG, Xiu
National Center for Climate Change
Strategy and International Cooperation
China

YOON, Soonuk
Green Technology Center
Republic of Korea

AIV

YOSEPH-PAULUS, Rahayu
Local Government of Buton Regency
Indonesia

YU, Rita Man Sze
CSR Asia
China

YU, Yau Hing
Sovran Environment & Energy Corp.
China

ZABOL ABBASI, Fatemeh
Climatological Research Institute
Iran

ZAELKE, Durwood
Institute for Governance and
Sustainable Development
USA

ZAFAR, Qudsia
Global Change Impact Studies Centre
Pakistan

ZARIN, Daniel
Climate and Land Use Alliance
USA

ZAVIALOV, Petr
Shirshov Institute of Oceanology
Russian Federation

ZEREFOS, Christos
Academy of Athens
Greece

ZHANG, Jingyong
Institute of Atmospheric Physics
Chinese Academy of Sciences
China

ZHANG, Wei
IIHR Hydroscience and Engineering
University of Iowa
USA

ZHANG, Xiaolin
Florida State University
China

ZHAO, Zong-Ci
National Climate Center
China Meteorological Administration
China

ZHOU, Tianjun
Institute of Atmospheric Physics
Chinese Academy of Sciences
China

ZICKFELD, Kirsten
Simon Fraser University
Canada/Germany

ZIELINSKI, Tymon
Institute of Oceanology
Polish Academy of Sciences
Poland

ZINKE, Jens
Freie Universitaet Berlin
Germany

ZOBAA, Ahmed
Brunel University London
UK

ZRINKA, Mendas
University of Bolton
UK

AIV

Index

This index should be cited as:
IPCC, 2018: Index. In: *Global Warming of 1.5°C. An IPCC Special Report on the impacts of global warming of 1.5°C above pre-industrial levels and related global greenhouse gas emission pathways, in the context of strengthening the global response to the threat of climate change, sustainable development, and efforts to eradicate poverty* [Masson-Delmotte, V., P. Zhai, H.-O. Pörtner, D. Roberts, J. Skea, P.R. Shukla, A. Pirani, W. Moufouma-Okia, C. Péan, R. Pidcock, S. Connors, J.B.R. Matthews, Y. Chen, X. Zhou, M.I. Gomis, E. Lonnoy, T. Maycock, M. Tignor, and T. Waterfield (eds.)]. Cambridge University Press, Cambridge, UK and New York, NY, USA, pp. 601-616.

Index

Index

Index